Relative Atomic Masses of the Elements

The atomic masses of many elements are not invariant but depend on the origin and treatmen[illegible] re apply to elements as they exist naturally on earth. Values in parentheses are used for radioacti[illegible] not be quoted precisely without knowledge of the origins of the elements; the value given is the [illegible] that element having the longest half-life. Elements 112, 113, 114, 115, 116, and 118 have not yet b[illegible] International Union of Pure and Applied Chemistry (IUPAC). In these six cases, either only prelim[illegible] data exist, so the numbers in parentheses contain either only the latest available mass number or the mass number of the main isotope that was studied.

Name	Symbol	Atomic Number	Relative Atomic Mass
Actinium	Ac	89	(227.0278)
Aluminum	Al	13	26.9815386
Americium	Am	95	(243.0614)
Antimony	Sb	51	121.760
Argon	Ar	18	39.948
Arsenic	As	33	74.92160
Astatine	At	85	(209.9871)
Barium	Ba	56	137.327
Berkelium	Bk	97	(247.0703)
Beryllium	Be	4	9.012182
Bismuth	Bi	83	208.98040
Bohrium	Bh	107	(272.1380)
Boron	B	5	10.811
Bromine	Br	35	79.904
Cadmium	Cd	48	112.411
Calcium	Ca	20	40.078
Californium	Cf	98	(251.0796)
Carbon	C	6	12.0107
Cerium	Ce	58	140.116
Cesium	Cs	55	132.9054519
Chlorine	Cl	17	35.453
Chromium	Cr	24	51.9961
Cobalt	Co	27	58.933195
Copper	Cu	29	63.546
Curium	Cm	96	(247.0704)
Darmstadtium	Ds	110	(281.162)
Dubnium	Db	105	(268.1255)
Dysprosium	Dy	66	162.500
Einsteinium	Es	99	(252.0830)
Erbium	Er	68	167.259
Europium	Eu	63	151.964
Fermium	Fm	100	(257.0951)
Fluorine	F	9	18.9984032
Francium	Fr	87	(223.0197)
Gadolinium	Gd	64	157.25
Gallium	Ga	31	69.723
Germanium	Ge	32	72.64
Gold	Au	79	196.966569
Hafnium	Hf	72	178.49
Hassium	Hs	108	(277.150)
Helium	He	2	4.002602
Holmium	Ho	67	164.93032
Hydrogen	H	1	1.00794
Indium	In	49	114.818
Iodine	I	53	126.90447
Iridium	Ir	77	192.217
Iron	Fe	26	55.845
Krypton	Kr	36	83.798
Lanthanum	La	57	138.90547
Lawrencium	Lr	103	(262.1096)
Lead	Pb	82	207.2
Lithium	Li	3	6.941
Lutetium	Lu	71	174.967
Magnesium	Mg	12	24.3050
Manganese	Mn	25	54.938045
Meitnerium	Mt	109	(276.1512)
Mendelevium	Md	101	(258.0984)
Mercury	Hg	80	200.59
Molybdenum	Mo	42	95.94
Neodymium	Nd	60	144.242
Neon	Ne	10	20.1797
Neptunium	Np	93	(237.0482)
Nickel	Ni	28	58.6934
Niobium	Nb	41	92.90638
Nitrogen	N	7	14.00674
Nobelium	No	102	(259.1010)
Osmium	Os	76	190.23
Oxygen	O	8	15.9994
Palladium	Pd	46	106.42
Phosphorus	P	15	30.973762
Platinum	Pt	78	195.084
Plutonium	Pu	94	(244.0642)
Polonium	Po	84	(208.9824)
Potassium	K	19	39.0983
Praseodymium	Pr	59	140.90765
Promethium	Pm	61	(144.9127)
Protactinium	Pa	91	231.03588
Radium	Ra	88	(226.0254)
Radon	Rn	86	(222.0176)
Rhenium	Re	75	186.207
Rhodium	Rh	45	102.90550
Roentgenium	Rg	111	(280.1645)
Rubidium	Rb	37	85.4678
Ruthenium	Ru	44	101.07
Rutherfordium	Rf	104	(267.1215)
Samarium	Sm	62	150.36
Scandium	Sc	21	44.955912
Seaborgium	Sg	106	(271.1335)
Selenium	Se	34	78.96
Silicon	Si	14	28.0855
Silver	Ag	47	107.8682
Sodium	Na	11	22.98976928
Strontium	Sr	38	87.62
Sulfur	S	16	32.065
Tantalum	Ta	73	180.94788
Technetium	Tc	43	(97.9064)
Tellurium	Te	52	127.60
Terbium	Tb	65	158.92535
Thallium	Tl	81	204.3833
Thorium	Th	90	232.03806
Thulium	Tm	69	168.93421
Tin	Sn	50	118.710
Titanium	Ti	22	47.867
Tungsten	W	74	183.84
Ununbium	Uub	112	(285)
Ununhexium	Uuh	116	(293)
Ununoctium	Uuo	118	(294)
Ununpentium	Uup	115	(288.192)
Ununquadium	Uuq	114	(289.189)
Ununtrium	Uut	113	(285.174)
Uranium	U	92	238.02891
Vanadium	V	23	50.9415
Xenon	Xe	54	131.293
Ytterbium	Yb	70	173.04
Yttrium	Y	39	88.90585
Zinc	Zn	30	65.409
Zirconium	Zr	40	91.224

Source: Commission on Atomic Weights and Isotopic Abundances, International Union of Pure and Applied Chemistry, *Pure and Applied Chemistry,* Vol. 78, 2051–2066 (2006).

STREAMLINE YOUR COURSE

AND STRENGTHEN YOUR STUDENTS' UNDERSTANDING

The Most Popular Chemistry Mastery System In Use Today!

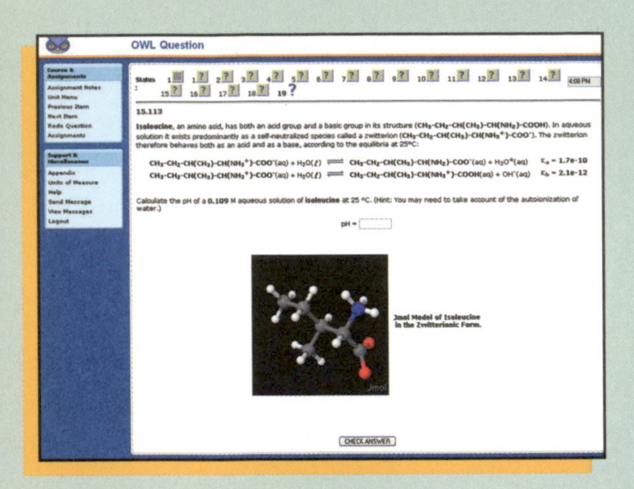

OWL for General Chemistry

Developed at the University of Massachusetts, Amherst.

Used by more than 300 institutions and proven reliable for tens of thousands of students, **OWL** offers unsurpassed ease of use, reliability, and dedicated training and service. **OWL** makes homework management a breeze and helps students improve their problem-solving skills and visualize concepts, providing instant analysis and feedback on a variety of homework problems such as tutors, simulations, and chemically or numerically parameterized short-answer questions. **OWL** is the only system specifically designed to support mastery learning, where students work as long as they need to master each chemical concept and skill. A fee-based access code is required for **OWL**.

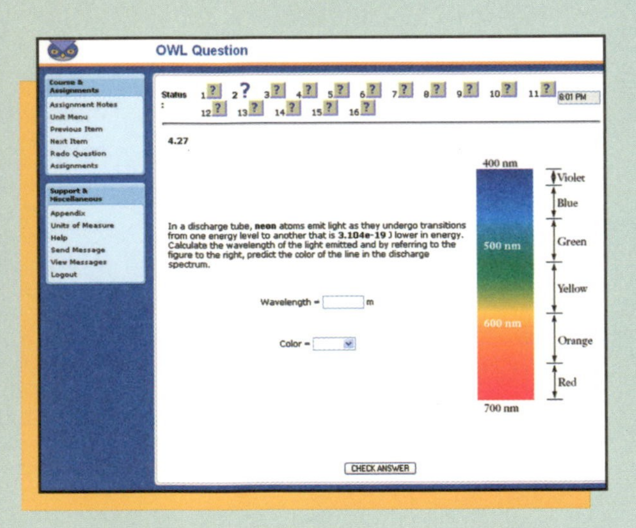

NEW!

Fully parameterized end-of-chapter questions, written specifically for this text, are available in **OWL,** ensuring that instructors will always find corresponding online exercises at the same level of rigor as the text.

A Complete e-Book!

The *Principles of Modern Chemistry* **e-Book in OWL** includes the complete textbook as an assignable resource that is fully linked to **OWL** homework content. This new **e-Book in OWL** is an exclusive option that will be available to all your students if you choose it. It can be bundled with the text and/or ordered as a text replacement. Please consult your Thomson Brooks/Cole representative for pricing details.

To learn more about **OWL,**
visit ***http://owl.thomsonlearning.com***
or contact your Thomson Brooks/Cole representative.

ORDERING OPTIONS

Student Edition of Oxtoby's
Principles of Modern Chemistry, 6e

Packaged with access to OWL
order 0-495-41599-5

Packaged with access to e-Book in OWL
order 0-495-41906-0

SIXTH EDITION

Principles of Modern Chemistry

SIXTH EDITION

Principles of Modern Chemistry

DAVID W. OXTOBY
Pomona College

H.P. GILLIS
University of California–Los Angeles

ALAN CAMPION
The University of Texas at Austin

Images of orbitals in Chapters 4, 5, 6 and 8 contributed by

HATEM H. HELAL
California Institute of Technology

KELLY P. GAITHER
The University of Texas at Austin

Australia • Brazil • Canada • Mexico • Singapore • Spain
United Kingdom • United States

Principles of Modern Chemistry, Sixth Edition
David W. Oxtoby, H.P. Gillis, Alan Campion

Acquisitions Editor: *Lisa Lockwood*
Development Editor: *Jay Campbell*
Assistant Editor: *Sylvia Krick*
Editorial Assistant: *Toriana Holmes*
Technology Project Manager: *Lisa Weber*
Marketing Manager: *Amee Mosley*
Marketing Communications Manager: *Brian Vann*
Content Project Manager, Editorial Production: *Teresa Trego*
Creative Director: *Rob Hugel*
Art Director: *John Walker*
Print Buyer: *Doreen Suruki*
Permissions Editor: *Roberta Broyer*
Photo Researcher: *Dena Digilio Betz*
Production Service: *Graphic World Inc.*
Text Designer: *Carolyn Deacy*
Illustrator: *Greg Gambino, 2064design*
OWL Producers: *Stephen Battisti, Cindy Stein, David Hart (Center for Educational Software Development, University of Massachusetts, Amherst)*
Cover Designer: *Andrew Ogus*
Cover Image: *Eric Heller*
Cover Printer: *Courier-Kendallville*
Compositor: *Graphic World Inc.*
Printer: *Courier-Kendallville*

Printed in the United States of America
1 2 3 4 5 6 7 11 10 09 08 07

Library of Congress Control Number 2006941020
Student Edition:
ISBN-13: 978-0-534-49366-0
ISBN-10: 0-534-49366-1

Thomson Higher Education
10 Davis Drive
Belmont, CA 94002-3098
USA

For more information about our products, contact us at:
Thomson Learning Academic Resource Center
1-800-423-0563

For permission to use material from this text or product, submit a request online at **http://www.thomsonrights.com.** Any additional questions about permissions can be submitted by e-mail to **thomsonrights@thomson.com.**

Asia
Thomson Learning
5 Shenton Way
#01-01 UIC Building
Singapore 068808

Australia/New Zealand
Thomson Learning Australia
102 Dodds Street
Southbank, Victoria 3006
Australia

Canada
Thomson Nelson
1120 Birchmount Road
Toronto, Ontario M1K 5G4
Canada

UK/Europe/Middle East/Africa
Thomson Learning
High Holborn House
50/51 Bedford Row
London WC1R 4LR
United Kingdom

Latin America
Thomson Learning
Seneca, 53
Colonia Polanco
11560 Mexico D.F.
Mexico

Spain/Portugal
Thomson Paraninfo
Calle Magallanes, 25
28015 Madrid, Spain

IN APPRECIATION OF

Harry B. Gray, Bruce H. Mahan[†]*, and George C. Pimentel*[†]
for the legacy of their textbooks and lectures

The search for truth is in one way hard and in another easy,
for it is evident that no one can master it fully or miss it completely.
But each adds a little to our knowledge of nature, and from all
the facts assembled there arises a certain grandeur.

(Greek inscription, taken from Aristotle, on the facade of the National Academy of Sciences building in Washington, D.C.)

Brief Contents

Preface

When the first edition of *Principles of Modern Chemistry* appeared in 1986, the standard sequence of topics in honors and high-level mainstream general chemistry courses began with macroscopic descriptions of chemical phenomena and proceeded to interpret these in terms of molecular structure. This traditional "macro-to-micro" approach has shifted in recent years, and today the central topics in these courses are chemical bonding and molecular structure. The relation of molecular structure to function and properties requires the introduction of molecular structure early in the course and the use of structural arguments in presenting the remaining topics.

In preparing the sixth edition, we have revised the textbook extensively to meet these present-day needs. In particular, we believe that the most logical sequence of topics begins with the physical properties and structure of atoms; is followed by structure, bonding, and properties of molecules; proceeds to describe macroscopic collections of atoms and molecules; continues with a discussion of chemical properties and reactions under equilibrium conditions; and finishes with dynamics and kinetics.

Significant Changes in This Edition

- *New Treatment and Placement of Structure and Bonding* Chemical bonding and molecular structure are now at the beginning of the book. We describe the classical elements of bonding theory—ionic, covalent, and polar bonds; dipole moments; Lewis electron diagrams; and Valence Shell Electron Pair Repulsion (VSEPR) theory. We present a unified and thorough treatment of quantum bonding theory, presenting the molecular orbital (MO) and valence bond (VB) models on equal footing and at the same intellectual and conceptual level. We provide detailed comparisons of these two models and show how either one can be the starting point for the development of computational quantum chemistry and molecular simulation programs that our students will encounter soon in subsequent chemistry courses.
- *New Molecular Art* Molecular shapes are rendered with quantitative accuracy and in modern graphical style. All illustrations of atomic and molecular orbitals, charge density, and electrostatic potential energy maps were generated from accurate quantum chemistry calculations carried out at the California Institute of Technology. For this edition, the orbitals were plotted using state-of-the-art software at the Texas Advanced Computing Center at the University of Texas at Austin. The colors, lighting effects, and viewing angles were chosen to display three-dimensional objects with maximum clarity and to provide chemical insight.

- *Revised Writing Style without Loss of Rigor* The language is more modern and less formal. We have introduced a more conversational writing style, designed to engage our students as active participants in developing the presentation. We have examined every sentence in the book to simplify and lighten the language without compromising intellectual integrity.
- *Greater Flexibility in Topic Coverage* In response to user and reviewer comments, greater modularity and flexibility have been built into the text to make it compatible with alternative sequences of topics. While moving the discussion of bonding and structure to the beginning of the book, we have been careful to maintain the option to follow the "macro-to-micro" approach used in previous editions. Selecting alternative approaches is facilitated by the Unit structure of the book; we offer several suggestions in the **Teaching Options** section.
- *New End-of-Chapter Student Aids* In response to suggestions by users and reviewers, we provide a *Chapter Summary*, *Chapter Review*, and list of *Key Equations* (with citations to the sections in which they appear) at the end of each chapter. These are integrated with the *Cumulative Exercises* and *Concepts & Skills* from previous editions to provide a comprehensive set of tools for reviewing and studying the contents of each chapter.
- *New Problems* Approximately 85 new problems have been added, mostly in Unit II on bonding and structure. These follow the tradition established in previous editions that problems are based on real data for real chemical systems. We intend the problems to guide our students to develop intuition for chemical results and the magnitudes of chemical quantities, as well as facility in manipulating the equations in the problems.
- *OWL Online Homework System* Homework management is now included in the text's instructional package. Approximately 15 problems from each chapter are available for assignment in the OWL program. See the section on Supporting Materials for a description of OWL.

Major Changes in Content and Organization

Chapter 1: The Atom in Modern Chemistry

This chapter has been reorganized to place greater emphasis on the physical structure of the atom, as determined from the classic experiments of Thomson, Millikan, and Rutherford. The chapter ends with direct scanning tunneling microscopy images of individual atoms in chemical reactions. Section 1.6 in *Principles of Modern Chemistry,* fifth edition (mole, density, molecular volume), has been moved to Chapter 2, which now gives a comprehensive treatment of formulas, stoichiometry, and chemical equations.

Chapter 3: Chemical Bonding: The Classical Description

This chapter provides a substantial introduction to molecular structure by coupling experimental observation with interpretation through simple classical models. Today, the tools of classical bonding theory—covalent bonds, ionic bonds, polar covalent bonds, electronegativity, Lewis electron dot diagrams, and VSEPR Theory—have all been explained by quantum mechanics. It is a matter of taste whether to present the classical theory first and then gain deeper insight from the quantum explanations, or to cover the quantum theory first and then see the classical theory as a limiting case. We have found that presenting the classical description first enables our students to bring considerably greater sophistication to their first encounter with quantum mechanics and therefore to develop a deeper appreciation for that subject. In our classroom experience, we have seen that this approach offers definitive pedagogical advantages by enabling students to

- learn the language and vocabulary of the chemical bond starting from familiar physical concepts
- become familiar with the properties of a broad array of real molecules *before* attempting to explain these results using quantum mechanics
- develop experience in using physical concepts and equations to describe the behavior of atoms and molecules

We have revised this chapter to more effectively meet these goals. Changes include the following:

- Section 3.2 is completely new. It illustrates the Coulomb potential with several quantitative applications and introduces the screened potential in many-electron atoms.
- In Section 3.4 the description of electron affinity has been extended and clarified, and the Pauling and Mulliken descriptions of electronegativity are discussed together.
- Section 3.5 describing forces and potential energy in molecules is completely new. We identify the driving force for the formation of chemical bonds between atoms as a reduction of the total energy of the system. We introduce the virial theorem to analyze the separate contributions of potential and kinetic energy to this total energy reduction in various bonding models.
- The role of Coulomb stabilization in ionic bonding has been substantially simplified and clarified.

Chapter 4: Introduction to Quantum Mechanics

This chapter is the revision of Sections 15.1–15.5 in *Principles of Modern Chemistry,* fifth edition. It presents a significant introduction to the concepts and vocabulary of quantum mechanics through very careful choice of language, illustrations with experimental data, interpretation with aid of simple models, and extensive use of graphical presentations. We highlight five features of this new chapter:

- We present quantitative, computer-generated plots of the solutions to the particle-in-a-box models in two and three dimensions and use these examples to introduce contour plots and three-dimensional isosurfaces as tools for visual representation of wave functions. We show our students how to obtain physical insight into quantum behavior from these plots without relying on equations. In the succeeding chapters we expect them to use this skill repeatedly to interpret quantitative plots for more complex cases.
- The discussion of Planck's analysis of blackbody radiation has been greatly simplified and clarified.
- The description of the wavelike behavior of electrons has been extended and clarified, based on a careful description of an electron diffraction experiment.
- The explanation of uncertainty and indeterminacy has been extended and clarified.
- Section 4.7 introduces the quantum harmonic oscillator and provides the groundwork for subsequent discussions of vibrational spectroscopy. This section is completely new.

Chapter 5: Quantum Mechanics and Atomic Structure

This chapter is the revision of Sections 15.7–15.9 in *Principles of Modern Chemistry,* fifth edition. Three features are new:

- The very long Section 15.8 has been broken into three parts (Sec. 5.2–5.4) for greater ease of presentation.
- Photoelectron spectroscopy has been moved into Section 5.4, where it fits logically with the discussion of the shell structure of the atom and the periodic table.
- All atomic orbitals have been re-calculated and rendered in modern style.

Chapter 6: Quantum Mechanics and Molecular Structure

This revision has the same logical structure as Chapter 16 of *Principles of Modern Chemistry,* fifth edition. It achieves more uniform coverage and proper depth, and it adds several important new features. The mathematical level is uniform throughout the chapter. Notable features of the present version include:

- An expanded treatment of H_2^+ as the starting point for describing molecular quantum mechanics and for motivating the development of the linear combination of atomic orbitals (LCAO) approximation. This expanded treatment includes:
  - A simplified and more thorough description of the Born–Oppenheimer approximation.
  - New graphical representations of the exact molecular orbitals for H_2^+ that make it easier to visualize these orbitals and interpret their meanings. These images provide a foundation for developing MO theory for the first- and second-period diatomic molecules.
  - Careful attention has been paid to the description and analysis of potential energy curves, zero-point energy, and total energy.
  - **"A Deeper Look"** section shows how the H_2^+ molecular orbitals correlate with sums and differences of hydrogen atomic orbitals at large internuclear separations. This correlation provides a very clear motivation for developing the LCAO approximation to exact molecular orbitals.
- Application of the virial theorem to reveal the interplay between kinetic and potential energy in the mechanism of bond formation described by MO theory.
- Application of the LCAO approximation to small polyatomic molecules to demonstrate the generality of the method.
- Molecular photoelectron spectroscopy to connect molecular orbital energy levels to experimental measurements.
- Much more thorough treatment of the VB method than in the fifth edition; it is now on equal footing with that for LCAO.
- Detailed comparison between LCAO and VB methods, including ways to improve both, showing how each is the starting point for high-level computational quantum chemistry.

Throughout this revision we have simplified notation to the maximum extent possible without sacrificing clarity, and we have devoted considerable attention to graphical explanations of the concepts.

Chapter 7: Bonding in Organic Molecules

This chapter represents a significant reorganization and extension of Chapter 20 and Section 16.4 in the fifth edition. We describe the bonding and nomenclature in alkanes, alkenes, alkynes, aromatics, and conjugated hydrocarbons and in the major functional groups. Our main goal is to illustrate the bonding theories from Chapter 6 with examples from organic chemistry that can be used in conjunction with Chapter 6. Our secondary goal is to provide sufficient material for a brief introduction to systematic organic chemistry. These goals are developed in these contexts:

- Petroleum refining as the source of starting materials for organic processes organizes a survey of organic structures and reactions.
- Organic synthesis organizes a survey of organic functional groups and reaction types.
- Pesticides and pharmaceuticals introduce a variety of organic compounds and structures in familiar contexts.

Chapter 8: Bonding in Transition Metal Compounds and Coordination Complexes

We have presented a comprehensive introduction to bonding in transition metal compounds and coordination complexes using MO and VB theory as developed in Chapter 6. Our goal was to demonstrate that MO theory is not limited to the first- and second-period diatomic molecules and that it provides the most satisfactory method for describing bonding in coordination complexes. The material covered in this chapter now provides a self-contained introduction to structure and bonding in inorganic chemistry that should provide sound preparation for an advanced inorganic chemistry course.

Chapter 9: The Gaseous State

A new **"Deeper Look"** section introduces the Boltzmann energy distribution and applies it to determine the relative populations of molecular energy states.

Chapter 10: Solids, Liquids, and Phase Transitions

Section 10.2 on Intermolecular Forces includes an introduction to electrostatic potential energy maps. We define these surfaces very carefully to provide a solid foundation for our students when they encounter these representations in their organic chemistry courses.

Chapter 14: Chemical Equilibrium

The language of this chapter has been completely revised, but the contents are essentially the same as in Chapter 9 of the fifth edition. To provide flexibility for instructors, this chapter was written to allow thermodynamics to be taught either before or after equilibrium. Each topic is introduced first from the empirical point of view then followed immediately with the thermodynamic treatment of the same topic. Instructors who prefer to treat thermodynamics first can use the chapter as written, whereas those who prefer the empirical approach can skip appropriate sections, then come back and pick up the thermo-based equilibrium sections after they cover basic thermodynamics. "Signposts" are provided in each section to guide these two groups of readers; the options are clearly marked. Specific examples of this flexible approach are:

- Section 14.2 provides a thorough discussion of procedures for writing the empirical law of mass action for gas-phase, solution, and heterogeneous reactions, with specific examples for each.

- Section 14.3 follows with the thermodynamic prescription for calculating the equilibrium constant from tabulated Gibbs free energy values for gas-phase, solution, and heterogeneous reactions, with specific examples for each.
- Sections 14.4 and 14.5 present a variety of equilibrium calculations based on the empirical law of mass action.
- Section 14.6 discusses direction of change in terms of the empirical reaction quotient *Q*, with illustrations in gas-phase, solution, and heterogeneous reactions.
- Section 14.7 discusses direction of change from the point of view of thermodynamics, relating *Q* to the Gibbs free energy change and the equilibrium constant.

Chapter 22: Inorganic Materials

Chapters 22 and 23 in the fifth edition have been combined, reorganized, and revised to provide a systematic introduction to ceramics, electronic materials, and optical materials. Structural, electrical, and optical properties are related to the nature of the chemical bonds in each class of materials.

Teaching Options

The text is structured and written to give instructors significant flexibility in choosing the order in which topics are presented. We suggest several such possibilities here. In all cases we recommend starting with Chapter 1 to provide a contemporary introduction to the structure and properties of the atom, followed by Chapter 2 to establish a secure foundation in "chemical accounting methods" that is necessary for studying all the remaining chapters. Particularly well-prepared students can skip Chapter 2, especially if diagnostics are available to ascertain satisfactory background.

Classical Bonding before Introduction to Quantum Theory

Chapters 1, 2, 3, 4, 5, 6; selections from Chapter 7 and Chapter 8; Chapters 9–23

This is the sequence we have found most effective because it enables our students to bring substantially greater maturity to their first exposure to quantum theory. This leads to deeper and quicker mastery of quantum theory and its applications to atomic and molecular structure. Instructors who wish to introduce molecular spectroscopy earlier can easily cover Sections 20.1–20.4 immediately after Chapter 6.

Introduction to Quantum Theory before Bonding

Chapters 1, 2, 4, 5, 3, 6; selections from Chapter 7 and Chapter 8; Chapters 9–23

Chapters 1, 2, 4, 5, 6, 3; selections from Chapter 7 and Chapter 8; Chapters 9–23

These sequences are appropriate for instructors who prefer to establish background in quantum theory before discussing ionic and covalent bonding, Lewis diagrams, and VSEPR theory. Instructors who prefer to cover these classical bonding topics after quantum mechanics but before MO and VB theory would cover

Chapter 3 before Chapter 6. Those who want to present the full quantum story first and then present the classical description as the limiting case would cover Chapter 3 after Chapter 6. We recommend that both of these sequences cover Section 3.2 (force and potential energy in atoms) before Chapter 4 to give a good physical feeling for Rutherford's planetary model of the atom in preparation for the quantum theory. Instructors who wish to introduce molecular spectroscopy earlier can easily cover Sections 20.1–20.4 immediately after Chapter 6.

Traditional "Macro-to-Micro" Approach

Chapters 1, 2, 9–19, 3–8, 20–23

This sequence covers fully the macroscopic descriptions of chemical phenomena and then begins to interpret them in terms of molecular structure. Instructors could choose either of the two bonding approaches suggested earlier for the specific order of Chapters 3–6 late in this course. This sequence represents a rather pure form of the "macro-to-micro" approach that was followed in the first three editions. Alternatively, they could cover Chapter 3 between Chapter 2 and Chapter 9, as was done in the fourth and fifth editions. This approach has the advantage of building a substantial foundation in structure—and a complete discussion of chemical nomenclature—as the basis for the macroscopic descriptions, while leaving the quantum theory of bonding to come later in the course.

Thermodynamics before Chemical Equilibrium

Chapters 12, 13, 14, 15, 16, 17

This is the sequence we have found to be the most effective. If our students first have a good understanding for the physical basis of equilibrium, then the facts and trends of chemical equilibrium quickly begin to form patterns around molecular structure. Changes in entropy and bond energy immediately organize chemical equilibrium.

Empirical Chemical Equilibrium before Thermodynamics

Chapter 14 (omit Sections 14.3, 14.7); Chapters 15, 16, 12, 13; Sections 14.3, 14.7; Chapter 17

Perhaps to provide background for quantitative laboratory work, others may wish to present chemical equilibrium earlier in the course in a more empirical fashion, before the presentation of thermodynamics. Chapter 14 is clearly marked with "signposts" to facilitate this sequence.

General Aspects of Flexibility

Certain topics may be omitted without loss of continuity. For example, a principles-oriented course might cover the first 20 chapters thoroughly and then select one or two specific topics in the last chapters for close attention. A course with a more descriptive orientation might omit the sections entitled **"A Deeper Look,"** which are more advanced conceptually and mathematically than the sections in the main part of the book, and cover the last three chapters more systematically. Additional suggestions are given in the *Instructor's Manual* that accompanies the book.

Features

Mathematical Level

This book presupposes a solid high school background in algebra and coordinate geometry. The concepts of slope and area are introduced in the physical and chemical contexts in which they arise, and differential and integral notation is used only when necessary. The book is designed to be fully self-contained in its use of mathematical methods. In this context, Appendix C should prove particularly useful to the student and the instructor.

Key equations in the text are highlighted in color and numbered on the right side of the text column. Students should practice using them for chemical calculations. These key equations appear again in a special section at the end of each chapter. Other equations, such as intermediate steps in mathematical derivations, are less central to the overall line of reasoning in the book.

Updated Design and New Illustrations and Photographs

This sixth edition features a modern design, whose elements have been carefully arranged for maximum clarity and whose aesthetics should engage today's visually oriented students. We have selected photographs and illustrations to amplify and illuminate concepts in the narrative text. All illustrations of atomic and molecular orbitals, charge density, and electrostatic potential energy maps were generated expressly for this textbook. The orbitals and charge densities were calculated by Mr. Hatem Helal in the Materials Simulation Center at the California Institute of Technology, directed by Professor William A. Goddard III. Dr. Kelly Gaither plotted the images using state-of-the-art software at the Scientific Visualization Laboratory at The University of Texas at Austin. The colors, lighting effects, and viewing angles were chosen to display three-dimensional objects with maximum clarity and chemical insight. In many cases quantitative contour plots accompany the three-dimensional isosurfaces representing orbitals to help our students understand how the appearances of isosurfaces depend on choices made by scientists and that these isosurfaces are neither unique nor definitive.

Worked Examples

This textbook includes worked examples that demonstrate the methods of reasoning applied in solving chemical problems. The examples are inserted immediately after the presentation of the corresponding principles, and cross-references are made to related problems appearing at the end of the chapter.

A Deeper Look

Sections entitled **"A Deeper Look"** provide students with a discussion of the physical origins of chemical behavior. The material that they present is sometimes more advanced mathematically than that in the main parts of the book. The material provided in these sections allows instructors to more easily tailor the breadth and depth of their courses to meet their specific objectives.

Key Terms

Key terms appear in boldface where they are first introduced. Definitions for most key terms are also included in the Index/Glossary for ready reference.

NEW Chapter Summary

Immediately at the end of each chapter is a summary that ties together the main themes of the chapter in a retrospective manner. This complements the introductory passage at the beginning of the chapter in a manner that conveys the importance of the chapter. The summary is the first in a set of six end-of-chapter features that constitute a comprehensive set of tools for organizing, studying, and evaluating mastery of the chapter.

Cumulative Exercise

At the end of each of Chapters 2 through 21 is a cumulative exercise that focuses on a problem of chemical interest and draws on material from the entire chapter for its solution. Working through a chapter's cumulative exercise provides a useful review of material in the chapter, helps students put principles into practice, and prepares them to solve the problems that follow.

NEW Chapter Review

The chapter review is a concise summary of the main ideas of the chapter. It provides a checklist for students to review their mastery of these topics and return to specific points that need further study.

Concepts & Skills

Each chapter concludes with a list of concepts and skills for review by our students. Included in this list are cross-references to the section in which the topic was covered and to problems that help test mastery of the particular skill involved. This feature is helpful for self-testing and review of material.

NEW Key Equations

All the equations that are highlighted in color in the chapter text are collected here, with references to the section in which they appeared. This list guides our students to greater familiarity with these important equations and enables quick location of additional information related to each equation.

Problems

Problems are grouped into three categories. Answers to odd-numbered "paired problems" are provided in Appendix G; they enable students to check the answer to the first problem in a pair before tackling the second problem. The *Additional Problems*, which are unpaired, illustrate further applications of the principles developed in the chapter. The *Cumulative Problems* integrate material from the chapter with topics presented earlier in the book. We integrate more challenging problems throughout the problems sets and identify them with asterisks.

Appendices

Appendices A, B, and C are important pedagogically. Appendix A discusses experimental error and scientific notation. Appendix B introduces the SI system of units used throughout the book and describes the methods used for converting units. Appendix B also provides a brief review of some fundamental principles in physics, which may be particularly helpful to students in understanding topics covered in Chapters 3, 4, 5, 6, 9, 10, 12, and 13. Appendix C provides a review of mathematics for general chemistry. Appendices D, E, and F are compilations of thermodynamic, electrochemical, and physical data, respectively.

Index/Glossary

The Index/Glossary at the back of the book provides brief definitions of key terms, as well as cross-references to the pages on which the terms appear.

Supporting Materials

Student Resources

Student Solutions Manual (ISBN: 0-495-11226-7)
The *Student Solutions Manual,* written by Wade A. Freeman of the University of Illinois at Chicago, presents detailed solutions to all of the odd-numbered problems in this book.

OWL: Online Web-based Learning
Written by Roberta Day and Beatrice Botch of the University of Massachusetts, Amherst, and William Vining of the State University of New York at Oneonta. Used by more than 300 institutions and proven reliable for tens of thousands of students, **OWL** offers unsurpassed ease of use, reliability, and dedicated training and service. **OWL** makes homework management a breeze and helps students improve their problem-solving skills and visualize concepts, providing instant analysis and feedback on a variety of homework problems, including tutors, simulations, and chemically and/or numerically parameterized short-answer questions and questions that employ graphic rendering programs such as Jmol. **OWL** is the only system specifically designed to support mastery learning, where students work as long as they need to master each chemical concept and skill. *Approximately 15 problems from each chapter are available for assignment in* **OWL**.

NEW A Complete e-Book!
The **Oxtoby e-Book in OWL** includes the complete textbook as an assignable resource that is fully linked to **OWL** homework content, including the Oxtoby problems mentioned earlier. This new **e-Book in OWL** is an exclusive option that will be available to all your students if you choose it. The e-Book in OWL can be packaged with the text and/or ordered as a text replacement. Please consult your Thomson Brooks/Cole representative for pricing details.

To learn more about **OWL**, *visit* ***http://owl.thomsonlearning.com*** *or contact your Thomson Brooks/Cole representative. OWL is only available for use by adopters in North America.*

Instructor Resources

eBank Instructor's Manual
The *Instructor's Manual,* written by Wade A. Freeman of the University of Illinois at Chicago, contains solutions to the even-numbered problems, as well as suggestions for ways to use this textbook in courses with different sequences of topics. *Contact your Thomson Brooks/Cole representative to download your copy.*

NEW eBank Testbank
The test bank includes problems and questions representing every chapter in the text. *Contact your Thomson Brooks/Cole representative to download your copy.*

NEW ExamView (Windows/Macintosh)
This easy-to-use software allows professors to create, print, and customize exams. The test bank includes problems and questions representing every chapter in the text. Answers are provided on a separate grading key, making it easy to use the questions for tests, quizzes, or homework assignments. *ExamView* is packaged as a hybrid CD for both Windows' and Macintosh formats.

OWL: Online Web-based Learning and e-Book in OWL
See description under Student Resources.

Presentation Tools

Online Images for Oxtoby/Gillis/Campion's
Principles of Modern Chemistry, Sixth Edition
This digital library of artwork from the text is available in PowerPoint® format on the book's website. Download at www.thomsonedu.com/chemistry/oxtoby.

Laboratory Resources

Customize your own lab manual for your general chemistry course using the wide range of high-quality experiments available from CER, Outernet, and Brooks/Cole. Work with your Brooks/Cole or Thomson Custom Publishing representative and/or visit www.textchoice.com/chemistry to select the specific experiments you want, collate in any order, and combine them with materials of your own for a lab manual that is perfectly suited to your particular course needs. The full-service website allows you to search by course and experiment and view each lab in its entirety—including new labs as they are developed—before selecting. Add your own material—course notes, lecture outlines, articles, and more at the beginning and end of any experiment—online in minutes. It's easy! Your custom lab manual can be packaged with any Brooks/Cole text for even greater value and convenience. Visit www.textchoice.com/chemistry.

Acknowledgments

In preparing the sixth edition, we have benefited greatly from the comments of students who used the first five editions over the years. We would also like to acknowledge the many helpful suggestions of colleagues at Pomona College, The University of Chicago, the University of California–Los Angeles, the University of Texas at Austin and other colleges and universities who have taught from this book. We are particularly grateful to Professor Samir Anz of California Polytechnic State University, Pomona, and to Professor Andrew Pounds of Mercer University for their comments and advice. Professors Eric Anslyn, Ray Davis, Brad Holliday, Brent Iverson, Richard Jones, Peter Rossky, Jason Shear, John Stanton, David Vanden Bout, Grant Willson, and Robert Wyatt of The University of Texas at Austin were unfailingly generous with their time and advice.

We extend special thanks to the following professors who offered comments on the fifth edition or reviewed manuscript for the sixth edition:

Joseph J. Belbruno, Dartmouth College

Laurie J. Butler, The University of Chicago

Patricia D. Christie, Massachusetts Institute of Technology

Regina F. Frey, Washington University, Saint Louis
Roberto A. Garza, Pomona College
Graeme C. Gerrans, University of Virginia
Henry C. Griffin, University of Michigan, Ann Arbor
Jeffrey Krause, University of Florida
Adam List, Vanderbilt University
Andrew J. Pounds, Mercer University
George C. Schatz, Northwestern University
John E. Straub, Boston University
Michael R. Topp, University of Pennsylvania
John Weare, University of California–San Diego
Peter M. Weber, Brown University
John S. Winn, Dartmouth College

We are grateful to Dr. Justin Fermann for his very careful attention to detail as accuracy reviewer of the sixth edition.

We are much indebted to our longtime friend Professor Eric J. Heller of Harvard University for the beautiful and striking image that graces the cover of our book. Professor Heller's work demonstrates that images of great beauty can arise from scientific research and that artistic renderings effectively convey the meaning of scientific results. We are certain this image will entice readers to peek between the covers of our book, and we hope they find scientific beauty on the inside as well as on the cover!

We are particularly grateful to friends and colleagues who provided original scientific illustrations for the book. They are Professor Wilson Ho (University of California–Irvine), Dr. Gilberto Medeiros-Ribeiro and Dr. R. Stanley Williams (Hewlett-Packard Research Laboratories), Professor Leonard Fine (Columbia University), Professor Andrew J. Pounds (Mercer University) and Dr. Mark Iken (Scientific Visualization Laboratory, Georgia Institute of Technology), Dr. Stuart Watson and Professor Emily Carter (Princeton University), Professor Nathan Lewis (California Institute of Technology), Dr. Don Eigler (IBM Almaden Research Center), Dr. Gerard Parkinsen and Mr. William Gerace (OMICRON Vakuumphysik), Dr. Richard P. Muller and Professor W.A. Goddard III (California Institute of Technology), Professor Moungi Bawendi and Ms. Felice Frankel (Massachusetts Institute of Technology), Professor Graham Fleming (University of California–Berkeley), Professor Donald Levy (The University of Chicago), Professor W.E. Moerner (Stanford University), Dr. Jane Strouse (University of California–Los Angeles), Professor James Speck and Professor Stephen Den Baars (University of California–Santa Barbara), and Professor John Baldeschwieler (California Institute of Technology).

We are especially grateful to Mr. Hatem H. Helal (California Institute of Technology) who carried out all the quantum chemistry calculations for the orbital illustrations in Chapters 4, 5, 6, and 8 and to Dr. Kelly P. Gaither (Texas Advanced Computing Center, The University of Texas at Austin) who generated these illustrations from the results of the calculations. Our longtime friend and colleague Professor William A. Goddard III (California Institute of Technology) very generously made his computational facilities available for these calculations and provided much good advice as we selected and prepared these illustrations. Sarah Chandler (The University of Texas–Austin) was very helpful in generating a number of graphs and two-dimensional surfaces.

We are also indebted to Professor Charles M. Knobler of the University of California–Los Angeles, Professor Jurg Waser formerly of the California Institute of Technology, and Mrs. Jean T. Trueblood (widow of the late Professor Kenneth N.

Trueblood of the University of California–Los Angeles) for permission to incorporate selected problems from their distinguished textbook *ChemOne,* Second Edition, McGraw-Hill, New York (1980).

On a personal note, it gives us genuine pleasure to dedicate this sixth edition of our textbook to Professor Harry Gray and to the memory of Professors Bruce Mahan and George Pimentel. We have enjoyed and been inspired by their textbooks, lectures, research papers, and seminars since our student days in the 1970s. Our own education owes much to their pioneering explanations of the role of quantum mechanics in chemical bonding.

The staff members at Brooks/Cole have been most helpful in preparing this sixth edition. In particular, we acknowledge the key role of our Acquisitions Editor Lisa Lockwood and our Developmental Editor Jay Campbell for guiding us toward revisions in this edition. Assistant Editor Sylvia Krick and Editorial Assistant Toriana Holmes coordinated production of the ancillary materials. Technology Project Manager Lisa Weber handled the media products. Senior Content Project Manager Teresa L. Trego of Brooks/Cole and Production Editor Alison Trulock of Graphic World Publishing Services kept the schedule moving smoothly. We acknowledge the contributions of Art Director Rob Hugel, who shepherded the new illustrations through the production process in this edition, and of Photo Researcher Dena Digilio Betz, who assisted in obtaining key photographs. Jim Smith, color consultant, made important contributions to the development of the color palette for the book. We are grateful to Marketing Manager Amee Mosley for helping us obtain valuable comments from users and reviewers. We gratefully acknowledge the continuing support of Publisher David Harris.

Finally, Alan Campion would like to acknowledge his parents, Alice and Harold Campion, for their support and encouragement during the course of his education and career. And special thanks go to his wife, Ellen, and daughters, Blair and Ali, for putting up with him for the past 18 months with more patience and grace than he deserves.

David W. Oxtoby
Pomona College

H.P. Gillis
University of California–Los Angeles

Alan Campion
The University of Texas at Austin

November 2006

Contents

UNIT II

Chemical Bonding and Molecular Structure 52

CHAPTER 3

Chemical Bonding: The Classical Description 54

CHAPTER 4

Introduction to Quantum Mechanics 114

CHAPTER 5

Quantum Mechanics and Atomic Structure 169

CHAPTER 6

Quantum Mechanics and Molecular Structure 211

CHAPTER 7

Bonding in Organic Molecules 275

CHAPTER 8

Bonding in Transition Metal Compounds and Coordination Complexes 313

UNIT III

Kinetic Molecular Description of the States of Matter 362

UNIT IV

Equilibrium in Chemical Reactions 484

CHAPTER 12

Thermodynamic Processes and Thermochemistry 486

CHAPTER 13

Spontaneous Processes and Thermodynamic Equilibrium 529

CHAPTER 14

Chemical Equilibrium 569

UNIT V

Rates of Chemical and Physical Processes 748

About the Authors

David W. Oxtoby

David W. Oxtoby is a physical chemist who studies the statistical mechanics of liquids, including nucleation, phase transitions, and liquid-state reaction and relaxation. He received his B.A. (Chemistry and Physics) from Harvard University and his Ph.D. (Chemistry) from the University of California at Berkeley. After a postdoctoral position at the University of Paris, he joined the faculty at The University of Chicago, where he taught general chemistry, thermodynamics, and statistical mechanics and served as Dean of Physical Sciences. Since 2003 he has been President and Professor of Chemistry at Pomona College in Claremont, California.

H.P. Gillis

H.P. Gillis is an experimental physical chemist who studies the surface chemistry of electronic and optical materials, including fabrication and characterization of nanostructures. He received his B.S. (Chemistry and Physics) at Louisiana State University and his Ph.D. (Chemical Physics) at The University of Chicago. After postdoctoral research at the University of California–Los Angeles and 10 years on the technical staff at Hughes Research Laboratories in Malibu, California, he joined the faculty of Georgia Institute of Technology, and now serves as Adjunct Professor of Materials Science and Engineering at the University of California–Los Angeles. He has taught general chemistry, physical chemistry, quantum mechanics, surface science, and materials science at Georgia Tech and at UCLA.

Alan Campion

Alan Campion is an experimental physical chemist who develops and applies novel methods of molecular spectroscopy to study the physical and chemical properties of solid surfaces, thin films, adsorbed molecules, and nanostructured materials. He received his B.A. (Chemistry) from New College of Florida and his Ph.D. (Chemical Physics) from the University of California–Los Angeles. After a postdoctoral position at the University of California at Berkeley, he joined the faculty of The University of Texas at Austin where he is Dow Chemical Company Professor of Chemistry and University Distinguished Teaching Professor. He teaches the honors general chemistry course, chemistry in context for students not majoring in science or engineering, physical chemistry, and advanced topics graduate courses in molecular spectroscopy.

UNIT I

Introduction to the Study of Modern Chemistry

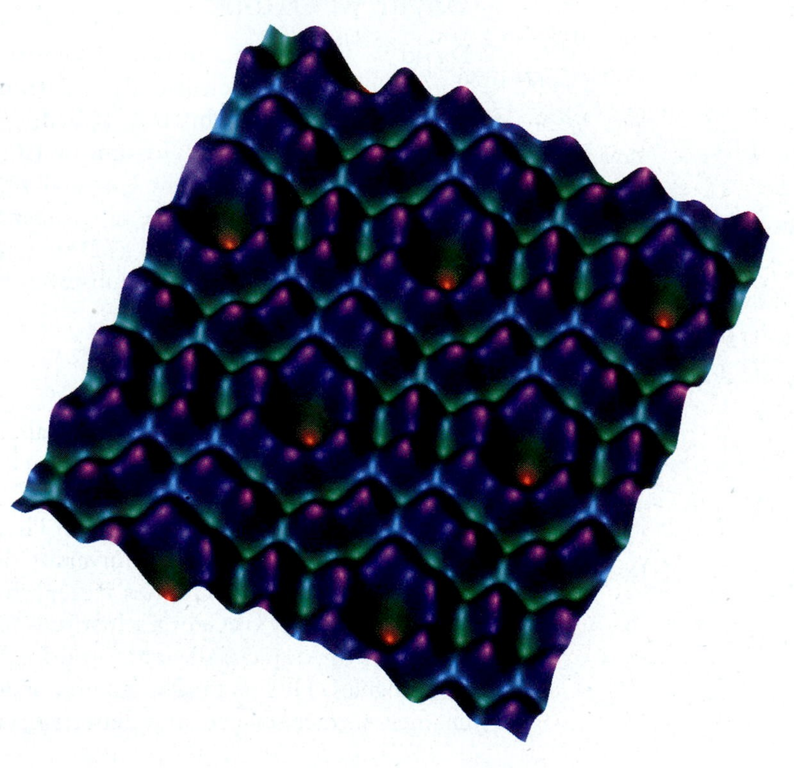

Photo courtesy of Wilson Ho, University of California, Irvine. Reprinted by permission of *Physical Review Letters*. 79, 4397–4400 (1997).

The surface of a silicon crystal imaged using a scanning tunneling microscope. Individual silicon atoms appear as purple protrusions above the background. The surface was cleaned in ultrahigh vacuum to remove all impurity atoms and the image was taken at very low temperatures (−220°C) to obtain the high resolution shown here. There are two kinds of surface silicon atoms shown in this image: "corner" silicon atoms that form hexagonal rings around a hole in the surface layer and "center" silicon atoms that appear as pairs arranged around the hexagonal rings.

Modern chemistry explores the world of atoms and molecules, seeking to explain not only their bonding, structures, and properties but also how these very structures are transformed in chemical reactions. The search for atoms and molecules began with the speculations of ancient philosophers and—stimulated by the classic experiments of the 18th- and 19th-centuries—led to John Dalton's famed atomic hypothesis in 1808. The quest continues. Thanks to the invention of the scanning tunneling microscope (STM) in the 1980s, today's scientists can detect and manipulate individual atoms and molecules.

UNIT CHAPTERS

UNIT GOALS

- To describe the key experiments, and the underlying physical models, that justify the central role of the atom in modern chemistry
  - Indirect (chemical) evidence for the existence of atoms and molecules
  - Direct (physical) evidence for the existence of atoms and molecules
  - The modern, planetary model of the atom
- To convey the established quantitative procedures for describing chemical reactions as rearrangements of atoms from reactants to products
  - The mole concept that connects weighing and counting molecules and atoms
  - Balanced chemical equations that connect moles of reactants to moles of products

CHAPTER 1

The Atom in Modern Chemistry

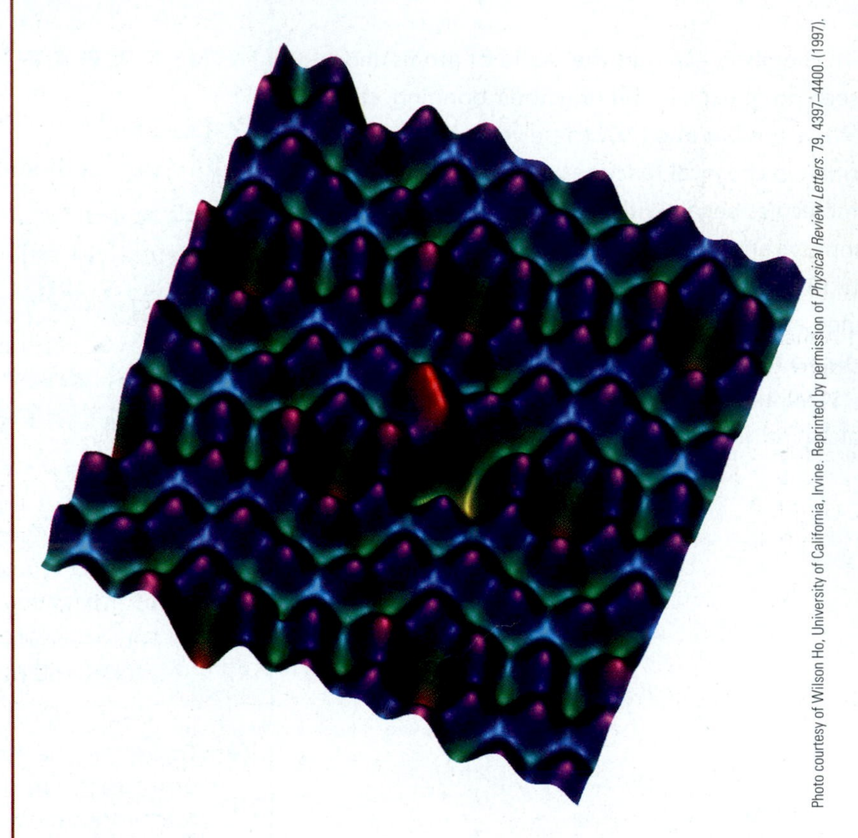

Reversible single atom transfer using the scanning tunneling microscope. This image was taken under the same conditions as the one shown opposite page 1. One of the "center" silicon atoms (imaged in red) has been transferred halfway to another center atom site by the scanning tunneling microscope tip. The atom is stable in this position at low temperatures but returns to its home site as the temperature is raised above −100°C.

1.1 The Nature of Modern Chemistry

Chemists study the properties of substances, their aim being to understand that reactions can transform substances into still other substances. Chemistry thus provides ways to tailor the properties of existing substances to meet a particular need—and even to create entirely new materials designed to have specific properties. This is how chemistry has improved agricultural production, helped prevent and cure many diseases, increased the efficiency of energy production, and reduced environmental pollution, to cite just a few advances. A particularly exciting challenge for modern chemical research is to understand the molecular dynamics of

Painting *The Alchemist* by Hendrick Heerschop, 1671; Courtesy of Dr. Alfred Bader

FIGURE 1.1 Alchemists searched in vain for procedures that would turn base metals into gold. Their apparatus foreshadowed equipment in modern chemical laboratories.

these chemical transformations, for they govern phenomena as diverse as the evolution of small carbon-containing molecules in interstellar space, changes in terrestrial atmospheric and climatic patterns caused by pollutants, and the unfolding of life processes in living organisms. Perhaps no other science covers as broad a range of topics as does chemistry; it influences disciplines from solid-state physics to molecular biology. Within a single modern chemistry department, you're apt to find chemists studying high-temperature superconductors, detecting and identifying single molecules, tailoring the properties of catalytic antibodies, and developing highly selective integrated sensors for a variety of applications in science and technology. Despite the diversity of these areas of scientific inquiry, they are all unified by a single set of fundamental scientific principles, which we will introduce to you in this textbook.

Chemistry is a relatively young science and its foundations weren't established until the last quarter of the 18th century. Before that, most chemists were known as *alchemists*—early entrepreneurs who sought to transform the properties of materials for economic gain (Fig. 1.1). For many centuries their obsession was to transform "base" metals, such as lead, into gold. They boldly assumed that the properties of one material could somehow be extracted from that material and transferred to another. If the essential properties—such as yellow color, softness, and ductility—could be assembled from various inexpensive sources, then gold could be created at great profit.

The alchemists persisted in their efforts for more than a thousand years. Although they collected many useful, empirical results that have since been incorporated into modern chemistry, they never transformed base metals into gold. Toward the middle of the 17th century, a number of individuals began to challenge the validity of the basic assumptions of the alchemists. These doubts culminated in publication of *The Sceptical Chymist* by Robert Boyle in England in the 1660s, which is one of the pivotal events that began the evolution of modern chemistry. Another century was required to complete the conceptual foundations of modern chemistry, which then flourished throughout the 19th and 20th centuries.

To observers in the early 21st century, the mistake of the alchemists is immediately clear: They did not follow the scientific method. In the scientific method, a new idea is accepted only temporarily, in the form of a **hypothesis.** It is then subjected to rigorous testing, in carefully controlled experiments. Only by surviving many such tests is a hypothesis elevated to become a **scientific law.** In addition to having explained the results of numerous experiments, a scientific law must be predictive; failure to accurately predict the results of a new experiment is sufficient to invalidate a scientific law. Concepts or ideas that have earned the status of scientific laws by direct and repeated testing then can be applied with confidence in new environments. Had a proper set of tests been made in separate, independent experiments, the alchemists would have recognized that the properties of a material are, in fact, intrinsic, inherent characteristics of that material and cannot be separated from it.

The history of the alchemists shows the origin of a certain duality in the nature of modern chemistry, which persists to the present. Because chemistry contributes to the foundations of numerous professions and industries, we see the urge to apply established chemical knowledge for profit. But we also see the urge to create new chemical knowledge, driven by both intellectual curiosity and by the desire to have reliable information for applications. Both aspects involve numerous scientists and engineers in addition to professional chemists. No matter what the specific context, the second aspect requires scrupulous adherence to the scientific method, in which new knowledge is subjected to rigorous scrutiny before it earns the confidence of the scientific community.

During their professional careers, most students who learn chemistry will be more concerned with applying chemistry than with generating new chemical knowledge. Still, a useful strategy for learning to think like an experienced chemist is to assume that you are personally responsible for establishing the scientific foundations of chemistry for the very first time. Upon encountering a new topic, try this: imagine that you are the first person ever to see the laboratory results on which it is based. Imagine that you must construct the new concepts and explanations to interpret these results, and that you will present and defend your conclusions before the scientific community. Be suspicious. Cross check everything. Demand independent confirmations. Always remain, with Boyle, the "skeptical chemist." Follow the scientific method in your acquisition of knowledge, even from textbooks. In this way, you will make the science of chemistry your own, and you will experience the intellectual joys of discovery and interpretation. Most important, you will recognize that chemistry is hardly a closed set of facts and formulas. Quite the contrary, it is a living, growing method for investigating all aspects of human experience that depend on the changes in the composition of substances.

Conservation of Matter and Energy

The science of chemistry rests on two well-established principles: the conservation of matter and the conservation of energy. What this means with respect to matter is absolute: The total amount of matter involved in any chemical reaction is *conserved*—that is, it remains constant throughout the reaction. Matter is neither created nor destroyed in chemical reactions; its components are simply *rearranged* to transform one substance into another.

These rearrangements are inevitably accompanied by changes in energy, which brings us to the second principle. The amounts of chemical energy stored in the molecules of two different substances are intrinsically different, and chemical energy may be converted into thermal, electrical, or mechanical energy during reactions. Energy may also flow in the opposite direction. But energy is neither created nor destroyed during chemical reactions. The total amount of energy involved in a chemical reaction has always been found to be conserved.

These two core principles must be modified slightly for nuclear reactions, which occur at energies so high that matter and energy can be converted into one another through Einstein's relation, $E = mc^2$. The *sum* of mass and energy is conserved in nuclear reactions.

Macroscopic Methods and Nanoscopic Models

Chemical reasoning, both in applications and in basic research, resembles a detective story in which tangible clues lead to a mental picture of events never directly witnessed by the detective. Chemical experiments are conducted in laboratories equipped with beakers, flasks, analytical balances, pipettes, optical spectrophotometers, lasers, vacuum pumps, pressure gauges, mass spectrometers, centrifuges, and other apparatus. Each of these devices exists on the *macroscopic* scale—that is, it is perceptible to ordinary human senses. Macroscopic sizes reach from 1 meter (m) down to 1 millimeter (mm), which is 1×10^{-3} m. But the actual chemical transformation events occur in the *nanoscopic* world of atoms and molecules—objects far too small to be detected by the naked eye, even with the aid of a first-class microscope. One nanometer (nm) is 1×10^{-9} m. So our modern laboratory instruments are the bridge between these worlds, giving us the means not only to influence the actions of the atoms and molecules but also to measure their response. Figure 1.2 shows both worlds simultaneously. In illustrating the chemical decomposition of water into gaseous hydrogen and oxygen by electrolysis, it shows the relation between events on the macroscale and the nanoscale. Chemists *think* in the highly visual nanoscopic world of atoms and molecules, but they *work* in the tangible world of macroscopic laboratory apparatus. These two aspects of chemical science cannot be divorced, and we will emphasize their interplay throughout this textbook. Students of chemistry must master not only the fascinating concepts of chemistry, which describe the nanoscopic world of atoms and molecules, but also the macroscopic procedures of chemistry on which those concepts are founded.

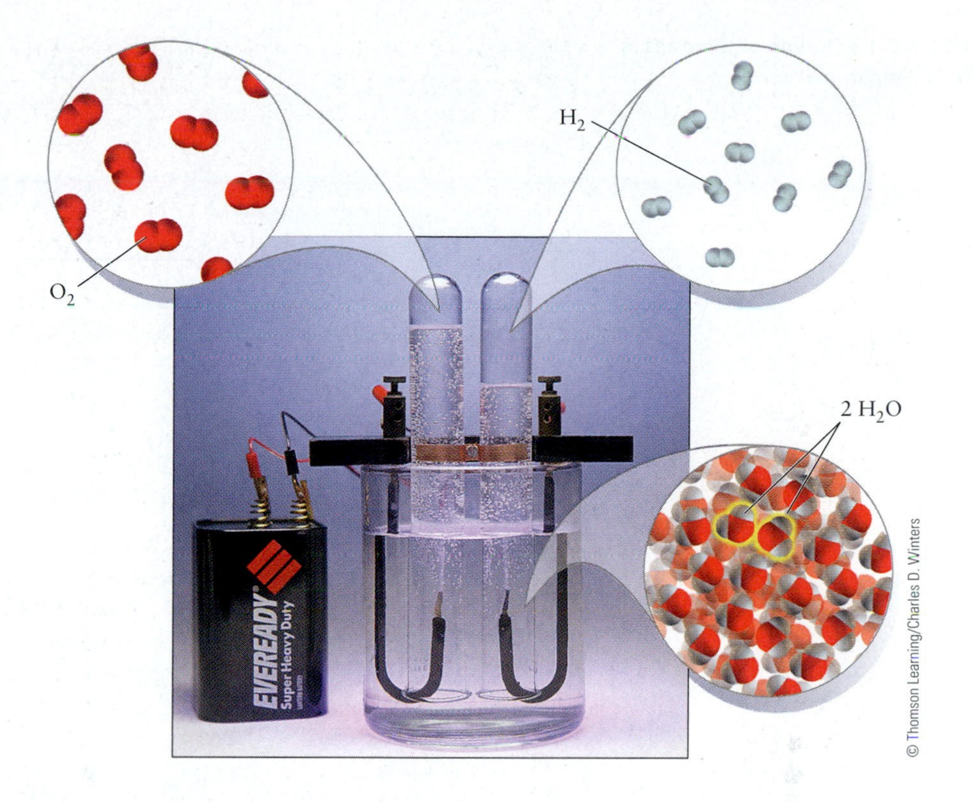

FIGURE 1.2 As electric current passes through water containing dissolved sulfuric acid, gaseous hydrogen and oxygen form as bubbles at the electrodes, producing the two gases in the 2 : 1 ratio by volume. The chemical transformation, induced by the macroscopic apparatus, proceeds by rearrangement of atoms at the nanoscale.

1.2 Macroscopic Methods for Classifying Matter

Chemists study how one set of pure substances will transform into another set of pure substances in a chemical reaction. This study involves two traditions—**analysis** (taking things apart) and **synthesis** (putting things together)—that go back to early Greek philosophers, who sought to analyze the constituents of all matter for four elements: air, earth, fire, and water. Contemporary chemists classify matter using a very different set of fundamental building blocks, but the analysis and synthesis steps are basically unchanged.

Substances and Mixtures

Investigating chemical reactions can be greatly complicated and often obscured by the presence of extraneous materials. So, the first step, therefore, is to learn how to analyze and classify materials to be sure you are working with *pure* substances before commencing with reactions (Fig. 1.3). Suppose you take a sample of a material—some gas, liquid, or solid—and examine its various properties or distinguishing characteristics, such as its color, odor, or density. How uniform are those properties? Different regions of a piece of wood, for example, have different properties, such as variations in color. Wood, then, is said to be **heterogeneous.** Other materials, such as air or a mixture of salt and water, are classified as **homogeneous** because their properties do not vary throughout the sample. We cannot call them pure substances, however. We still have to call them **mixtures,** because it is possible to separate them into components by ordinary physical means such as melting, freezing, boiling, or dissolving in solvents (Fig. 1.4). These operations provide ways of separating materials from one another by their properties, such as freezing point, boiling point, and solubility. For example, air is a mixture of several components—oxygen, nitrogen, argon, and various other gases. If air is liquefied and then warmed slowly, the gases with the lowest boiling points will evaporate first, leaving behind in the liquid those with higher boiling points. Such a separation would not be perfect, but the processes of liquefaction and evaporation could

FIGURE 1.3 Outline of the steps in the analysis of matter.

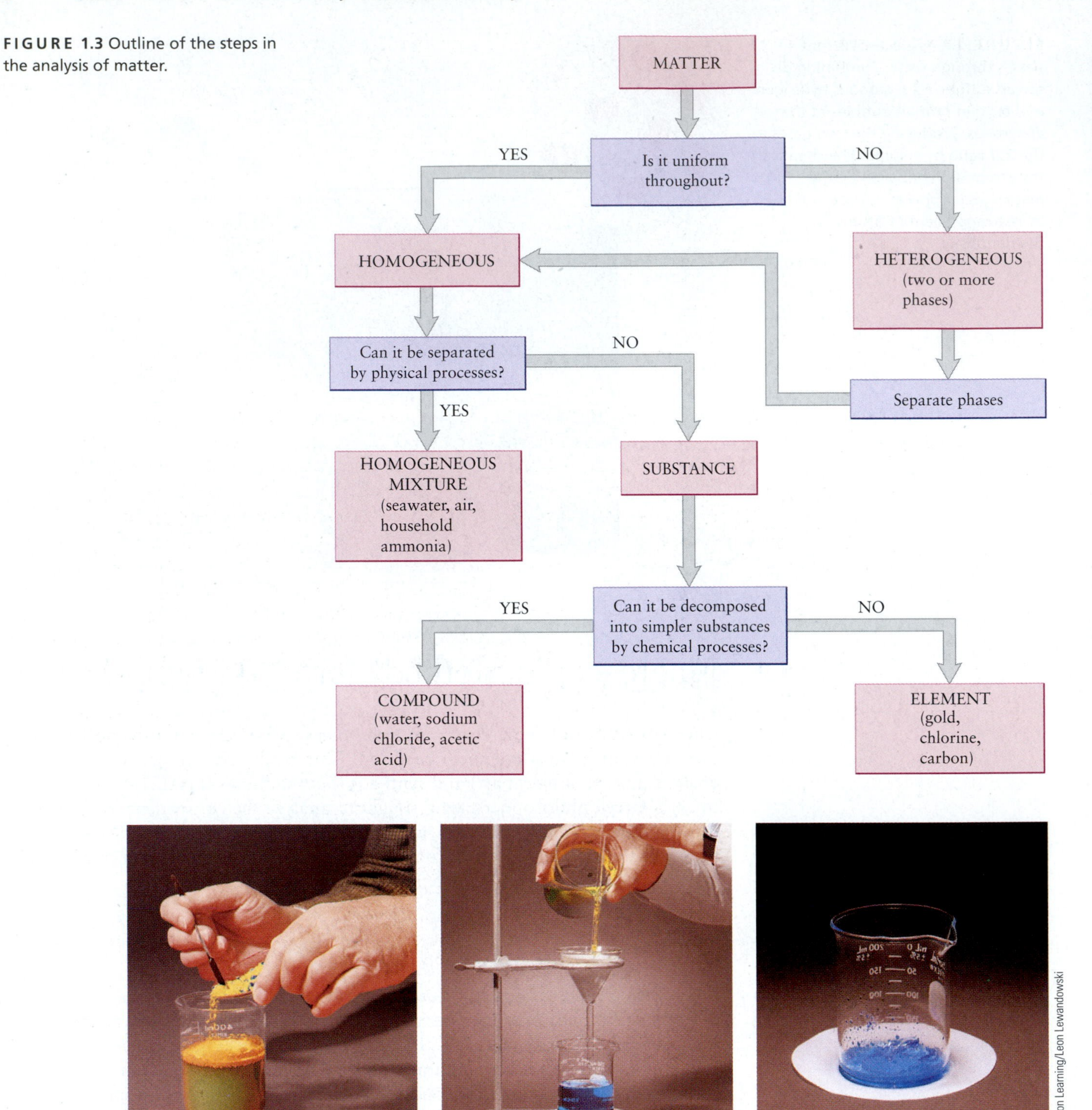

FIGURE 1.4 (a) A solid mixture of blue $Cu(NO_3)_2 \cdot 6H_2O$ and yellow CdS is added to water. (b) Although the $Cu(NO_3)_2 \cdot 6H_2O$ dissolves readily and passes through the filter, the CdS remains largely undissolved and is held on the filter. (c) Evaporation of the solution leaves nearly pure crystals of $Cu(NO_3)_2 \cdot 6H_2O$.

be repeated to improve the resolution of air into its component gases to any required degree of purity.

If all these physical procedures (and many more) fail to separate matter into portions that have different properties, the material is said to be a **substance.** What about the common material sodium chloride, which we call table salt? Is it a substance? The answer is yes if we use the term *sodium chloride,* but no if we use the

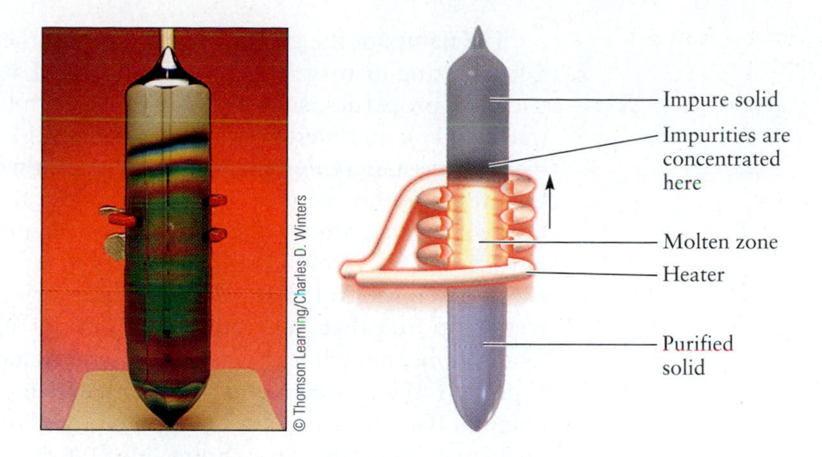

FIGURE 1.5 Nearly pure elemental silicon is produced by pulling a 10-inch-long solid cylinder (called a boule) out of the melt, leaving most of the impurities behind.

term *table salt.* Table salt is a mixture of sodium chloride with small additives of sodium iodide (needed by the thyroid gland) and magnesium carbonate (needed to prevent the salt from caking). Even if these two components were not added, table salt would contain other impurities that had not been removed in its preparation, so to that extent, table salt is a mixture. In contrast, when we refer to sodium chloride, we imply that all other materials are absent, so it qualifies as a substance.

In practice, nothing is absolutely pure, so the word *substance* is an idealization. Among the purest materials ever prepared are silicon (Fig. 1.5) and germanium. These elements are used in electronic devices and solar cells, and their electronic properties require either high purity or else precisely controlled concentrations of deliberately added impurities. Meticulous chemical and physical methods have enabled scientists to prepare germanium and silicon with concentrations less than one part per billion of impurities. Anything more would alter their electrical properties.

Elements

Literally millions of substances have so far been either discovered or synthesized and formally identified. Are these the fundamental building blocks of matter? Happily not, for their classification alone would pose an insurmountable task. In fact, all these substances are merely combinations of much smaller numbers of building blocks called **elements.** Elements are substances that cannot be decomposed into two or more simpler substances by ordinary physical or chemical means. The word *ordinary* excludes the processes of radioactive decay, whether natural or artificial, and high-energy nuclear reactions that *do* transform one element into another. When a substance contains two or more chemical elements, we call it a ***compound.*** For example, hydrogen and oxygen are elements because no further chemical separation is possible, whereas water is a compound because it can be separated into hydrogen and oxygen by passing an electric current through it (see Fig. 1.2). *Binary* compounds are substances, such as water, that contain two elements, *ternary* compounds contain three elements, *quaternary* compounds contain four elements, and so on.

At present, scientists have identified some 112 chemical elements. A few have been known since before recorded history, principally because they occur in nature as elements rather than in combination with one another in compounds. Gold, silver, lead, copper, and sulfur are chief among them. Gold is found in streams in the form of little granules (placer gold) or nuggets in loosely consolidated rock. Sulfur is associated with volcanoes, and copper often can be found in its native state in shallow mines. Iron occurs in its elemental state only rarely (in meteorites); it usually is combined with oxygen or other elements. In the second millennium B.C., ancient metallurgists somehow learned to reduce iron oxide to iron with charcoal in forced-draft fires, and the Iron Age was born.

The names of the chemical elements and the symbols that designate them have a fascinating history. Many elements have Latin roots that describe physical or chemical properties, such as gold (*aurum,* symbol Au), copper (*cuprum,* Cu), iron (*ferrum,* Fe), and mercury (*hydrargyrum,* Hg). Hydrogen (H) means "water former." Potassium (*kalium,* K) takes its common name from potash (potassium carbonate), a useful chemical obtained in early times by leaching the ashes of wood fires with water. Many elements take their names from Greek and Roman mythology: cerium (Ce) from Ceres, goddess of plenty; tantalum (Ta) from Tantalus, who was condemned in the afterlife to an eternity of hunger and thirst while close to water and fruit that were always tantalizingly just out of reach; and niobium (Nb) from Niobe, daughter of Tantalus. Some elements are named for continents: europium (Eu) and americium (Am). Other elements are named after countries: germanium (Ge), francium (Fr), and polonium (Po). Cities provide the names of other elements: holmium (Stockholm, Ho), ytterbium (Ytterby, Yb), and berkelium (Berkeley, Bk). Still more elements are named for the planets: uranium (U), plutonium (Pu), and neptunium (Np). Other elements take their names from colors: praseodymium (green, Pr), rubidium (red, Rb), and cesium (sky blue, Cs). Still others honor great scientists: curium (Marie Curie, Cm), mendelevium (Dmitri Mendeleev, Md), fermium (Enrico Fermi, Fm), einsteinium (Albert Einstein, Es), and seaborgium (Glenn Seaborg, Sg).

1.3 Indirect Evidence for the Existence of Atoms: Laws of Chemical Combination

How did we acquire the chemical evidence for the existence of atoms and the scale of relative atomic masses? It is an instructive story, both in its own right and as an illustration of how science progresses.

We may know the elements to be the most fundamental substances, and we may know they can be combined chemically to form compound substances, but that knowledge provides us no information on the nanoscopic structure of matter or how that nanoscopic structure controls and is revealed by chemical reactions. Ancient philosophers dealt with these fascinating questions by proposing assumptions, or *postulates,* about the structure of matter. The Greek philosopher Democritus (c. 460–370 B.C.) postulated the existence of unchangeable *atoms* of the elements, which he imagined to undergo continuous random motion in the vacuum, a remarkably modern point of view. It follows from this postulate that matter is not divisible without limit; there is a lower limit to which a compound can be divided before it becomes separated into atoms of the elements from which it is made. Lacking both experimental capabilities and the essentially modern scientific view that theories must be tested and refined by experiment, the Greek philosophers were content to leave their views in the form of assertions.

More than 2000 years passed before a group of European chemists demonstrated experimentally that elements combine only in masses with definite ratios when forming compounds, and that compounds react with each other only in masses with definite ratios. These results could be interpreted only by inferring that smallest indivisible units of the elements (atoms) combined to form smallest indivisible units of the compounds (molecules). The definite mass ratios involved in reactions were interpreted as a convenient means for counting the number of atoms of each element participating in the reaction. These results, summarized as the **laws of chemical combination,** provided overwhelming, if indirect, evidence for the existence of atoms and molecules.

For more than a century, we have become so accustomed to speaking of atoms that we rarely stop to consider the experimental evidence for their existence collected in the 18th and 19th centuries. Twentieth-century science developed a

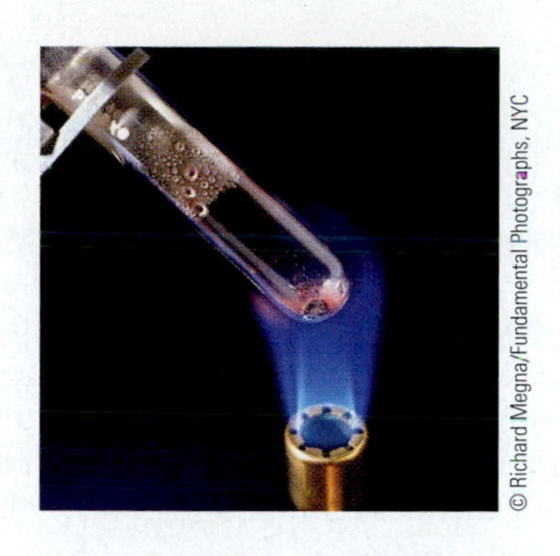

FIGURE 1.6 When the red solid mercury(II) oxide is heated, it decomposes to mercury and oxygen. Note the drops of liquid mercury condensing on the side of the test tube.

number of sophisticated techniques to measure the properties of single atoms, and powerful microscopes even allow us to observe them (see Section 1.5). But long before single atoms were detected, chemists could speak with confidence about their existence and the ways in which they combine to form molecules. Moreover, although the absolute masses of single atoms of oxygen and hydrogen were not measured until the early 20th century, chemists could assert (correctly) some 50 years earlier that the *ratio* of the two masses was close to 16:1.

Law of Conservation of Mass

The first key steps toward formulating the laws of chemical composition were taken during the 18th century in the course of studies of heat and combustion. It had been observed that an organic material, such as wood, left a solid residue of ash when burned; similarly, a metal heated in air was transformed into a "calx," which we now call an oxide. The popular explanation for these phenomena in the early 18th century was that a property called *phlogiston* was driven out of wood or metal by the heat of a fire. From the modern perspective, this seems absurd, because the ash weighs less than the original wood, whereas the calx weighs more than the metal. But at the time, the principle of conservation of mass had not yet been established, and people saw no reason why the mass of a material should not change on heating.

Further progress could be made only by carefully measuring the changes in mass[1] that occur in chemical reactions. The balance had been known since antiquity, but it had been used principally as an assayer's tool and for verifying the masses of coins or commodities in commerce. The analytical balance developed in the 18th century, however, was accurate to perhaps 1 part in 10,000, enabling much more accurate measurements of mass changes accompanying chemical reactions than had been possible previously. French chemist Antoine Lavoisier used the analytical balance (See the photo on page 29 to demonstrate that the sum of the masses of the products of a chemical reaction equals the sum of the masses of the reactants to the high degree of accuracy provided by the instrument. Lavoisier heated mercury in a sealed flask that contained air. After several days, a red substance, mercury(II) oxide, was produced. The gas remaining in the flask was reduced in mass and could no longer support life or combustion; a candle was extinguished by it, and animals suffocated when forced to breathe it. We now know that this residual gas was nitrogen, and that the oxygen in the air had reacted with the mercury. Lavoisier then took a carefully weighed amount of the red oxide of mercury and heated it strongly (Fig. 1.6). He weighed both the mercury and the gas that were produced and showed that their combined mass was the same as that of the mercury(II) oxide with which he had started. After further experiments, Lavoisier was able to state the **law of conservation of mass:**

> *In every chemical operation an equal quantity of matter exists before and after the operation.*

Lavoisier was the first to observe that a chemical reaction is analogous to an algebraic equation. We would write his second reaction as

$$2\ HgO \longrightarrow 2\ Hg + O_2$$

although during Lavoisier's lifetime, the identity of the gas (oxygen) was not known.

[1]Chemists sometimes use the term *weight* in place of *mass*. Strictly speaking, weight and mass are not the same. The mass of a body is an invariant quantity, but its weight is the force exerted on it by gravitational attraction (usually by the Earth). Newton's second law relates the two ($w = m \times g$, where g is the acceleration due to gravity). As g varies from place to place on the Earth's surface, so does the weight of a body. In chemistry, we deal mostly with ratios, which are the same for masses and weights. In this textbook we use the term *mass* exclusively, but *weight* is still in colloquial chemical use.

Law of Definite Proportions

Rapid progress ensued as chemists began to make accurate determinations of the masses of reactants and products. A controversy arose between two schools of thought, led by a pair of French chemists, Claude Berthollet and Joseph Proust. Berthollet believed that the proportions (by mass) of the elements in a particular compound were not fixed, but could actually vary over a certain range. Water, for example, rather than containing 11.1% by mass of hydrogen, might have somewhat less or more than this mass percentage. Proust disagreed, arguing that any apparent variation was due to impurities and experimental errors. He also stressed the difference between homogeneous mixtures and chemical compounds. In 1794, Proust published the fundamental **law of definite proportions:**

> *In a given chemical compound, the proportions by mass of the elements that compose it are fixed, independent of the origin of the compound or its mode of preparation.*

Pure sodium chloride contains 60.66% chlorine by mass, whether we obtain it from salt mines, crystallize it from waters of the oceans or inland salt seas, or synthesize it from its elements, sodium and chlorine.[2]

The law of definite proportions was a crucial step in the development of modern chemistry, and by 1808, Proust's conclusions had become widely accepted. We now recognize that this law is not strictly true in all cases. Although all gaseous compounds obey Proust's law, certain solids exist with a small range of compositions and are called **nonstoichiometric compounds.** An example is wüstite, which has the nominal chemical formula FeO (with 77.73% iron by mass), but the composition of which, in fact, ranges continuously from $Fe_{0.95}O$ (with 76.8% iron) down to $Fe_{0.85}O$ (74.8% iron), depending on the method of preparation. Such compounds are called **berthollides,** in honor of Berthollet. We now know, on the atomic level, why they are nonstoichiometric (see the discussion in Section 21.6).

This account illustrates a common pattern of scientific progress: Experimental observation of parallel behavior leads to the establishment of a law, or principle. More accurate studies may then demonstrate exceptions to the general principle. The following explanation of the exceptions leads to deeper understanding.

Dalton's Atomic Theory

English scientist John Dalton was by no means the first person to propose the existence of atoms; as we have seen, speculations about them date back to Greek times (the word *atom* is derived from Greek *a-* ["not"] plus *tomos* ["cut"], meaning "not divisible"). Dalton's major contribution to chemistry was to marshal the evidence for the existence of atoms. He showed that the mass relationships found by Lavoisier and Proust could be interpreted most simply by postulating the existence of atoms of the various elements.

In 1808, Dalton published *A New System of Chemical Philosophy,* in which the following five postulates comprise the **atomic theory of matter:**

1. Matter consists of indivisible atoms.
2. All the atoms of a given chemical element are identical in mass and in all other properties.
3. Different chemical elements have different kinds of atoms; in particular, their atoms have different masses.

[2]This statement needs some qualification. As explained in the next section, many elements have several *isotopes,* which are species whose atoms have almost identical chemical properties but different masses. Natural variation in isotope abundance leads to small variations in the mass proportions of elements in a compound, and larger variations can be induced by artificial isotopic enrichment.

4. Atoms are indestructible and retain their identities in chemical reactions.
5. A compound forms from its elements through the combination of atoms of unlike elements in small whole-number ratios.

Dalton's fourth postulate clearly is related to the law of conservation of mass. The fifth aims to explain the law of definite proportions. Perhaps Dalton's reasoning went something like this: Suppose you reject the atomic theory and believe instead that compounds are subdivisible without limit. What, then, ensures the constancy of composition of a substance such as sodium chloride? Nothing! But if each sodium atom in sodium chloride is matched by one chlorine atom, then the constancy of composition can be understood. So in this argument for the law of definite proportions, it does not matter how small the atoms of sodium and chlorine are. It is important merely that there be some lower bound to the subdivisibility of matter, because the moment we put in such a lower bound, arithmetic steps in. Matter becomes countable, and the units of counting are simply atoms. Believing in the law of definite proportions as an established experimental fact, Dalton *postulated* the existence of the atom.

Law of Multiple Proportions

The composition of a compound is shown by its **chemical formula.** The symbol H_2O for water indicates that the substance water contains two atoms of hydrogen for each atom of oxygen. It is now known that in water the atoms in each group of three (two H and one O) are linked by attractive forces strong enough to keep the group together for a reasonable period. Such a group is called a **molecule.** In the absence of knowledge about a compound's molecules, the numerical subscripts in the chemical formula simply give the relative proportions of the elements in the compound. How do we know that these are the true proportions? The determination of chemical formulas (and the accompanying determination of relative atomic masses), building on the atomic hypothesis of Dalton, was a major accomplishment of 19th-century chemistry.

In the simplest type of compound, two elements combine, contributing equal numbers of atoms to the union to form **diatomic molecules,** which consist of two atoms each. Eighteenth- and 19th-century chemists knew, however, that two elements will often combine in different proportions, thus forming more than one compound.

For example, carbon (C) and oxygen (O) combine under different conditions to form two different compounds, which we will call A and B. Analysis shows that A contains 1.333 grams (g) of oxygen per 1.000 g of carbon, and B contains 2.667 g of oxygen per 1.000 g of carbon. Although at this point we know nothing about the chemical formulas of the two oxides of carbon, we can say immediately that molecules of compound A contain half as many oxygen atoms per carbon atom as do molecules of compound B. The evidence for this is that the ratio of the masses of oxygen in A and B, for a fixed mass of carbon in each, is 1.333:2.667, or 1:2. If the formula of compound A were CO, then the formula of compound B would have to be CO_2, C_2O_4, C_3O_6, or some other multiple of CO_2. If compound A were CO_2, then compound B would be CO_4 or C_2O_8, and so on. From these data, we cannot say which of these (or an infinite number of other possibilities) are the true formulas of the molecules of compounds A and B, but we do know this: The number of oxygen atoms per carbon atom in the two compounds is the *quotient of integers*.

Consider another example. Arsenic (As) and sulfur (S) combine to form two sulfides, A and B, in which the masses of sulfur per 1.000 g of arsenic are 0.428 and 0.642 g, respectively. The ratio of these sulfur masses is 0.428:0.642 = 2:3. We conclude that *if* the formula of compound A is a multiple of AsS, then the formula of compound B must be a multiple of As_2S_3.

These two examples illustrate the **law of multiple proportions:**

When two elements form a series of compounds, the masses of one element that combine with a fixed mass of the other element are in the ratio of small integers to each other.

In the first example, the ratio of the masses of oxygen in the two compounds, for a given mass of carbon, was 1:2. In the second example, the ratio of the masses of sulfur in the two compounds, for a given mass of arsenic, was 2:3. Today, we know that the carbon oxides are CO (carbon monoxide) and CO_2 (carbon dioxide), and the arsenic sulfides are As_4S_4 and As_2S_3. Dalton could not have known this, however, because he had no information from which to decide how many atoms of carbon and oxygen are in one molecule of the carbon–oxygen compounds or how many atoms of arsenic and sulfur are in the arsenic–sulfur compounds.

EXAMPLE 1.1

Chlorine (Cl) and oxygen form four different binary compounds. Analysis gives the following results:

Compound	Mass of O Combined with 1.0000 g Cl
A	0.22564 g
B	0.90255 g
C	1.3539 g
D	1.5795 g

(a) Show that the law of multiple proportions holds for these compounds.

(b) If the formula of compound A is a multiple of Cl_2O, then determine the formulas of compounds B, C, and D.

SOLUTION

(a) Form ratios by dividing each mass of oxygen by the smallest, which is 0.22564 g:

$$0.22564 \text{ g} : 0.22564 \text{ g} = 1.0000 \text{ for compound A}$$

$$0.90255 \text{ g} : 0.22564 \text{ g} = 4.0000 \text{ for compound B}$$

$$1.3539 \text{ g} : 0.22564 \text{ g} = 6.0003 \text{ for compound C}$$

$$1.5795 \text{ g} : 0.22564 \text{ g} = 7.0001 \text{ for compound D}$$

The ratios are whole numbers to a high degree of precision, and the law of multiple proportions is satisfied. Ratios of whole numbers also would have satisfied that law.

(b) If compound A has a formula that is some multiple of Cl_2O, then compound B is Cl_2O_4 (or ClO_2, or Cl_3O_6, and so forth) because it is four times richer in oxygen than is compound A. Similarly, compound C, which is six times richer in oxygen than compound A, is Cl_2O_6 (or ClO_3, or Cl_3O_9, and so forth), and compound D, which is seven times richer in oxygen than compound A, is Cl_2O_7 (or a multiple thereof).

Related Problems: 7, 8, 9, 10

Dalton made a sixth assumption to resolve the dilemma of the absolute number of atoms present in a molecule; he called it the "rule of greatest simplicity." It states that if two elements form only a single compound, its molecules will have the simplest possible formula: AB. Thus, he assumed that when hydrogen and oxygen combine to form water, the reaction is

$$\text{H} + \text{O} \longrightarrow \text{HO}$$

However, Dalton was wrong, as we now know, and the correct reaction is

$$2\,\text{H}_2 + \text{O}_2 \longrightarrow 2\,\text{H}_2\text{O}$$

Law of Combining Volumes

At this time, French chemist Joseph Gay-Lussac conducted some important experiments on the volumes of gases that react with one another to form new gases. He discovered the **law of combining volumes:**

The volumes of two reacting gases (at the same temperature and pressure) are in the ratio of simple integers. Moreover, the ratio of the volume of each product gas to the volume of either reacting gas is the ratio of simple integers.

Here are three examples:

2 volumes of hydrogen + 1 volume of oxygen ⟶ 2 volumes of water vapor
1 volume of nitrogen + 1 volume of oxygen ⟶ 2 volumes of nitrogen oxide
3 volumes of hydrogen + 1 volume of nitrogen ⟶ 2 volumes of ammonia

Avogadro's Hypothesis

Gay-Lussac did not theorize on his experimental findings, but in 1811 shortly after their publication the Italian chemist Amedeo Avogadro used them to formulate an important postulate since known as **Avogadro's hypothesis:**

Equal volumes of different gases at the same temperature and pressure contain equal numbers of particles.

The question immediately arose; Are "particles" of the elements the same as Dalton's atoms? Avogadro believed that they were not; rather, he proposed that elements could exist as diatomic molecules. Avogadro's hypothesis could explain Gay-Lussac's law of combining volumes (Fig. 1.7). Thus, the reactions we wrote out in words become

$$2\,H_2 + O_2 \longrightarrow 2\,H_2O$$

$$N_2 + O_2 \longrightarrow 2\,NO$$

$$3\,H_2 + N_2 \longrightarrow 2\,NH_3$$

The coefficients of the above reactions are proportional to the volumes of the reactant and product gases in Gay-Lussac's experiments, and the chemical formulas

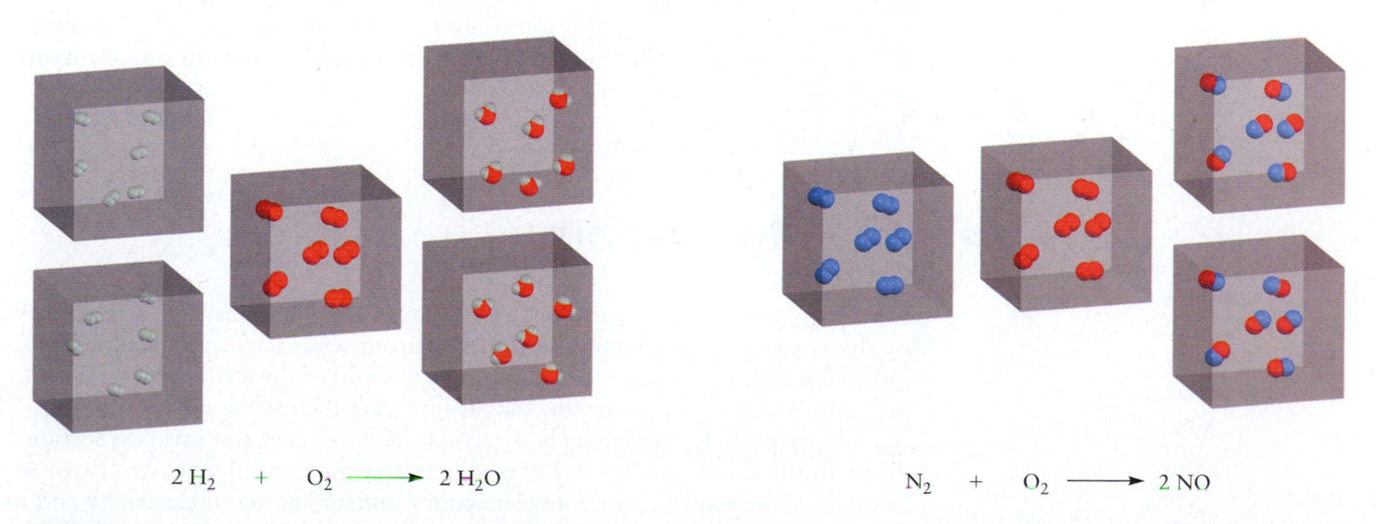

FIGURE 1.7 Each cube represents a container of equal volume under the same conditions. If each cube contains the same number of molecules (Avogadro's hypothesis), and if hydrogen, oxygen, and nitrogen exist as diatomic molecules, then the combining volumes that Gay-Lussac observed in the two reactions can be understood.

of the reactions agree with modern results. Dalton, on the other hand, would have written

$$H + O \longrightarrow OH$$

$$N + O \longrightarrow NO$$

$$H + N \longrightarrow NH$$

The coefficients of the previous reactions disagree with Gay-Lussac's observations of their relative volumes.

Besides predicting correct molecular formulas, Avogadro's hypothesis gives correct results for the relative atomic masses of the elements. Analysis by chemists during the 18th century had demonstrated that 1 g of hydrogen reacts completely with 8 g of oxygen to produce 9 g of water. If Dalton's formula for water, HO, were correct, then an atom of oxygen would have to weigh 8 times as much as an atom of hydrogen; that is, Dalton's assumption requires the **relative atomic mass** of oxygen to be 8 on a scale where the relative atomic mass of hydrogen is set at 1. Avogadro's hypothesis predicted, however, that each water molecule has twice as many atoms of hydrogen as oxygen; therefore, to explain the observed experimental mass relation, the relative mass for oxygen must be 16, a result consistent with modern measurements.

We might expect that Dalton would have welcomed Avogadro's brilliant hypothesis, but he did not. Dalton and others insisted that elements could not exist as diatomic molecules. One reason for their belief was the then-popular idea that a force called *affinity* held molecules together. Affinity expressed the attraction of opposites, just as we think of the attraction between positive and negative electric charges. If the affinity theory were true, why should two *like* atoms be held together in a molecule? Moreover, if like atoms somehow did hold together in pairs, why should they not aggregate further to form molecules with three, four, or six atoms, and so forth? With so many chemists accepting the affinity theory, Avogadro's reasoning did not attract the attention it deserved. Because different chemists adopted different chemical formulas for molecules, confusion reigned. A textbook published by the German chemist August Kekulé in 1861 gave 19 different chemical formulas for acetic acid!

In 1860, 50 years after Avogadro's work, Italian chemist Stanislao Cannizzaro presented a paper at the First International Chemical Congress in Karlsruhe, Germany, that convinced others to accept Avogadro's approach. Cannizzaro had analyzed many gaseous compounds and was able to show that their chemical formulas could be established with a consistent scheme that used Avogadro's hypothesis and avoided any extra assumptions about molecular formulas. Gaseous hydrogen, oxygen, and nitrogen (as well as fluorine, chlorine, bromine, and iodine), indeed, turn out to consist of diatomic molecules under ordinary conditions.

1.4 The Physical Structure of Atoms

In the original Greek conception, carried over into Dalton's time and reenforced by the laws of chemical combination, the atom was considered the ultimate and indivisible building block of matter. But by the end of the 19th century, this notion began to be replaced by the view that atoms were themselves composed of smaller, *elementary* particles. Scientists had carried the process of analysis (see Section 1.2) down to the subatomic level. Because many of the experiments described in the following paragraphs require an elementary understanding of electricity and magnetism, we suggest that you review the relevant sections of Appendix B before continuing.

Electrons

One key piece of evidence for the existence of subatomic particles came from studies of the effects of large electric fields on atoms and molecules. In these experiments, a gas was enclosed in a glass tube that had two conducting plates inside. When a large electrical potential difference (voltage) was established between the plates, current passed through the gas. The magnitude of this current could be measured in the external circuit connecting the two metal plates. This result suggested that the electric field had broken down the atoms into new species that carry charge. (The same effect occurs when an electrical discharge from a lightning bolt passes through air.) The magnitude of the current was proportional to the amount of gas in the tube. But when the gas was nearly all removed, the current did not go to zero; this suggested that the current originated from one of the metal plates. The mysterious invisible current carriers appeared to travel in straight lines from the cathode (the plate at negative potential) and produced a luminous spot where they impinged on the glass tube near the anode (the plate at positive potential). These current carriers were called **cathode rays,** or **beta rays.** After several decades of research, experimenters learned that cathode rays could be deflected by both magnetic and electric fields and could heat a piece of metal foil in the tube until it glowed. Influenced by Maxwell's electromagnetic theory of light, one school of physicists believed the cathode rays to be a strange form of invisible light, whereas another group considered them to be a stream of negatively charged particles.

In 1897, British physicist J. J. Thomson performed a series of experiments that resolved the controversy. He proved that cathode rays are negatively charged particles; these particles were subsequently named **electrons.** Thomson's key experiment is depicted schematically in Figure 1.8. A beam of cathode rays was produced by a cathode and anode in the usual way in a highly evacuated tube. A hole in the anode allowed some of the rays to pass between a second pair of plates that could be charged positively and negatively to establish an electric field oriented perpendicular to the cathode ray trajectory. For the arrangement shown in Figure 1.8, the cathode rays were deflected downward (indicating that they carried a negative charge), and the deviation could be measured accurately from the displacement of the luminous spot on a screen at the end of the tube. The only sensible explanation of these results was to view the cathode rays as material particles with negative charge and unknown mass. Thomson calculated e/m_e (the charge-to-mass ratio of the electron) by relating the net deflection to the forces applied to the particle, through Newton's second law of motion. Thomson's method is explained in the next few paragraphs.

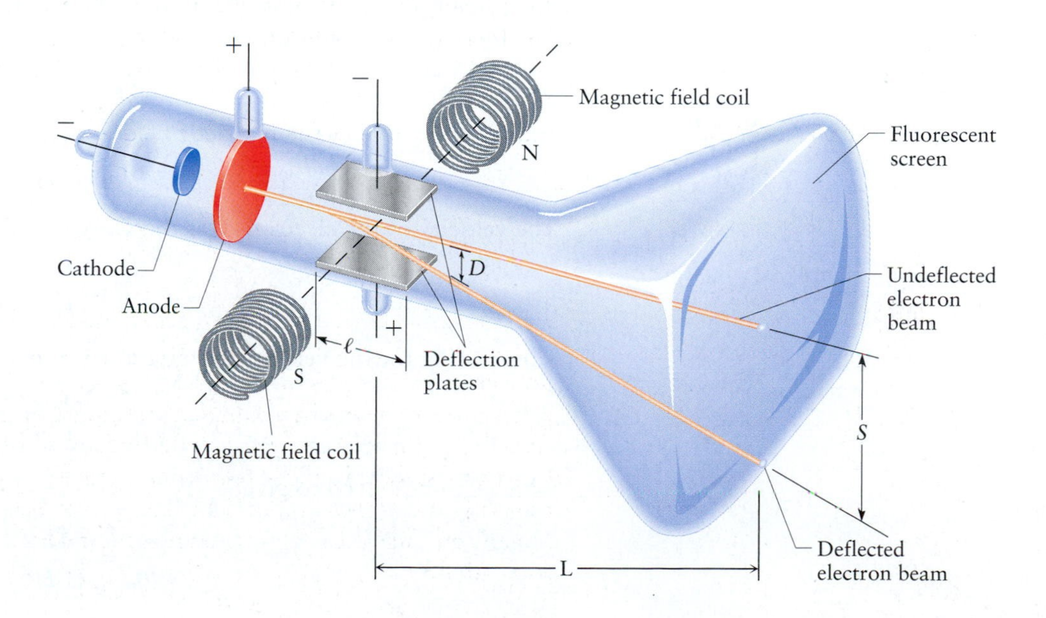

FIGURE 1.8 Thomson's apparatus to measure the electron charge-to-mass ratio, e/m_e. Electrons (cathode rays) stream across the tube from left to right. The electric field alone deflects the beam down, and the magnetic field alone deflects it up. By adjusting the two field strengths, Thomson could achieve a condition of zero net deflection. (ℓ indicates the length of the deflection plates.)

As soon as the electron flies into the space between the plates, it begins to experience a constant downward force, given by

$$F_E = eE \qquad [1.1]$$

where E is the electric field between the plates. By the time the electron flies out of the space between the plates, it has experienced a downward deflection, D, given by Newton's second law as

$$D = \frac{1}{2}at^2 \qquad [1.2]$$

where t is the time required to travel the distance ℓ, the length of the plates. The value of t can be determined from the velocity of the electron because $v = \ell/t$, and a can be determined from Newton's second law:

$$F_E = m_e a = eE \qquad [1.3]$$

The net downward deflection of the electron by the time it escapes from the plates is then

$$D = \frac{1}{2}at^2 = \frac{1}{2}\left(\frac{e}{m_e}\right)\left(\frac{\ell}{v}\right)^2 E \qquad [1.4]$$

After the electron escapes from the plates, it experiences no further forces, so it continues in straight-line motion toward the fluorescent screen. This motion carries the electron farther from the undeflected path and "magnifies" the displacement by the factor $2L/\ell$, where L is the distance from the center of the plates to the screen. When the electron arrives at the screen, the net displacement will be

$$S = 2\frac{L}{\ell}D = \left(\frac{e}{m_e}\right)\left(\frac{\ell}{v}\right)^2\left(\frac{L}{\ell}\right)E \qquad [1.5]$$

All of these quantities could be read off the apparatus except for the velocity of the electron, which was hard to measure directly.

So Thomson took one additional ingenious experimental step to determine e/m_e. He established a magnetic field in the same region as the electric deflection plates by passing an electric current through a pair of coils, located to make the magnetic field direction perpendicular to that of the electric field and to the flight path of the electrons. The magnetic field deflected the electrons in the direction opposite to that caused by the electric field. By varying the strengths of the two fields, Thomson could pass the electron beam through the tube without deflection. Under these conditions, the beam experienced two equal but opposing forces; from this force balance, Thomson determined the velocity of the electrons in the beam. The force due to the electric field E was

$$F_E = eE \qquad [1.6]$$

and the force due to the magnetic field H was

$$F_H = evH \qquad [1.7]$$

The velocity of the electrons was therefore

$$v = \frac{E}{H} \qquad [1.8]$$

Substituting for the velocity in Equation 1.5 gives the net deflection as

$$S = \left(\frac{e}{m_e}\right)\left(\frac{\ell H}{E}\right)\left(\frac{L}{\ell}\right)E \qquad [1.9]$$

which can be solved to give

$$\frac{e}{m_e} = \frac{SE}{\ell L H^2} \qquad [1.10]$$

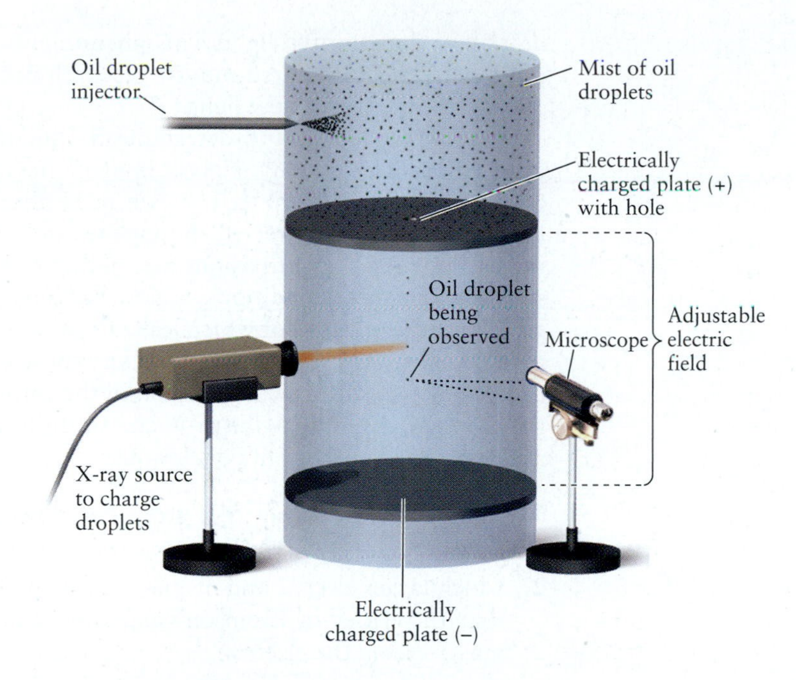

FIGURE 1.9 Millikan's apparatus to measure the charge on an electron, *e*. By adjusting the electric field strength between the charged plates, Millikan could halt the fall of negatively charged oil drops and determine their net charge.

From Equation 1.10 the charge-to-mass ratio for the electron could be determined from quantities read directly off Thomson's apparatus. The currently accepted value is $e/m_e = 1.7588202 \times 10^{11}$ C kg^{-1}, where charge is measured in coulombs and mass in kilograms. (See Appendix B for a full discussion of units of measure.)

Thomson's apparatus is the forerunner of the modern cathode ray tube (crt) display widely used as a video monitor. Electrons emitted from the cathode are steered by rapidly varying electric and magnetic fields to "write" the video image on light-emitting materials deposited on the inside wall of the tube. The image is viewed through the glass wall at the end of the tube.

Thomson's experiment determined only the *ratio* of the charge to the mass of the electron. The actual value of the electric charge was measured in 1906 by American physicist Robert Millikan with his student H. A. Fletcher. In Millikan and Fletcher's elegant experiment (Fig. 1.9), tiny drops of oil were charged by a source of ionizing radiation. A charged oil drop (with charge Q and mass M) situated in an electric field between two plates was subject to two forces: the force of gravity $-Mg$, causing it to fall, and a force QE from the electric field, causing it to rise. By adjusting the electric field to balance the two forces and independently determining the masses, M, of the drops from their falling speeds in the absence of an electric field, Millikan showed that the charge Q was always an integral multiple of the same basic charge, 1.59×10^{-19} C. He suggested that the different oil drops carried integral numbers of a fundamental charge, which he took to be the charge of a single electron. More accurate modern measurements led to the value $e = 1.60217646 \times 10^{-19}$ C. Combining this result with the e/m_e ratio found by Thomson gives $m_e = 9.1093819 \times 10^{-31}$ kg for the electron mass.

The Nucleus

It had been known as early as 1886 that light was emitted in gas discharge tubes, devices constructed much like the one Thomson used to study cathode rays. The chief difference between the cathode ray tube and gas discharge tubes was the pressure of the gases contained in the tube. Gas discharge tubes contained gases at moderate-to-low pressure, whereas the cathode ray tube operated under high vacuum (extremely low pressure). The electric field applied between the cathode and anode of a discharge tube caused "electrical breakdown" of the gas to form a *glow discharge* that emitted light. This phenomenon created great excitement among physicists and stimulated intense research projects to explain the nature of the glow and the origin of the light.

One approach was to determine the identity and properties of electrically charged particles within the glow. Imagine the following experiment, conducted in an apparatus similar to that shown in Figure 1.8, but with two modifications: (1) the voltages are reversed so that the anode (the plate with the hole) becomes the cathode, and (2) provisions are made to add different gases at different pressures. Under these conditions, a glow discharge was observed between the anode and the cathode, some of which leaked through the hole in the cathode and formed rays aimed toward the end of the tube. These came to be known as **canal rays** because they passed through the canal in the cathode. A contemporary of Thomson, the German physicist Wilhelm Wien, carefully studied the properties of these rays and drew three important conclusions:

1. The particles that emerged from the glow discharge to form canal rays were accelerated toward the cathode, so they must be positively charged.
2. Much larger electric and magnetic fields were required to deflect these particles than those used in Thomson's experiments, implying that they were much more massive than the electron.
3. If different gases were leaked into the apparatus, the magnitude of the fields required to displace canal rays of a different gas by the same amount differed, implying that the charged particles associated with each gas had different masses.

Wien's experiments suggested the existence of massive, positively charged particles in the glow discharge.

As an aside, studies of the glow discharge led to significant developments in physics and chemistry which continue to this day. Research on canal rays formed the basis for the mass spectrometer, our most accurate tool for measuring relative masses of atomic and molecular species. The physics and chemistry of **plasmas**—gases that contain charged particles—grew from understanding the structure of the glow itself. Plasma science is active in the 21st century, with fundamental and applied studies ranging from the nature of radiation in deep space to fabrication of nanometer-sized electronic and optical devices for computation and communication.

As exemplified in the results of the experiments of Thomson and Wien, physicists had discovered two quite different types of particles in matter: a light particle that was negatively charged, and a number of much heavier positively charged particles, particles whose relative masses depended on the element from which they were produced. Although it was generally agreed that these particles were the building blocks of atoms, it was not at all clear how they were assembled. That piece of the puzzle remained unsolved until Rutherford's pivotal discovery.

New Zealander Ernest Rutherford and his students at the University of Manchester made a startling discovery in 1911. They had been studying radioactive decay for a number of years and turned their attention to investigating the properties of alpha particles emitted from radium, an element that had been recently isolated by Marie and Pierre Curie. In the Rutherford experiment, a collimated beam of alpha particles irradiated an extremely thin (600 nm) piece of gold foil, and their deflections after colliding with the foil were measured by observing the scintillations they produced on a fluorescent ZnS screen (Fig. 1.10). Almost all of the alpha particles passed straight through the foil, but a few were deflected through large angles. Rarely, a particle was found to have been scattered backward! Rutherford was astounded, because the alpha particles were relatively massive and fast moving. In his words, "It was almost as incredible as if you fired a 15-inch shell at a piece of tissue paper and it came back and hit you." He and his students studied the frequency with which such large deflections occurred. They concluded

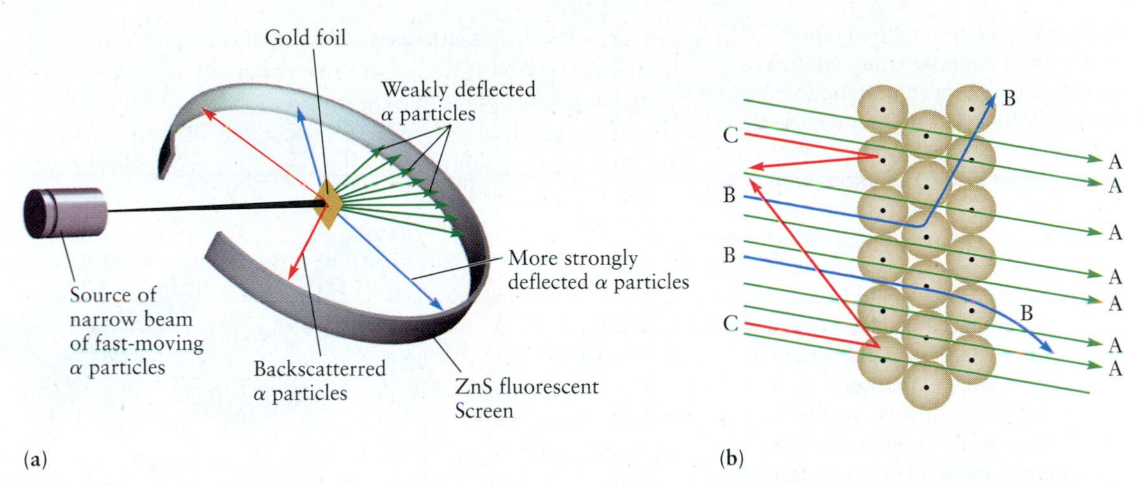

FIGURE 1.10 (a) Flashes of light mark the arrival of alpha particles at the detector screen. In the Rutherford experiment, the rate of hits on the screen varied from about 20 per minute at high angles to nearly 132,000 per minute at low angles. (b) Interpretation of the Rutherford experiment. Most of the alpha particles pass through the space between nuclei and undergo only small deflections (A). A few pass close to a nucleus and are more strongly deflected (B). Some are even scattered backward (C). The nucleus is far smaller proportionately than the dots suggest.

that most of the mass in the gold foil was concentrated in dense, extremely small, positively charged particles that they called **nuclei.** By analyzing the trajectories of the particles scattered by the foil, they estimated the radius of the gold nucleus to be less than 10^{-14} m and the positive charge on each nucleus to be approximately $+100e$ (the actual value is $+79e$).

Rutherford proposed a model of the atom in which the charge on the nucleus was $+Ze$, with Z electrons surrounding the nucleus out to a distance of about 10^{-10} m (0.1 nm). The Rutherford model for a gold atom has 79 electrons (each with a charge of $-1e$) arranged about a nucleus of charge $+79e$. The electrons occupy nearly the entire volume of the atom, whereas nearly all its mass is concentrated in the nucleus; this model is often called the "planetary model."

The Rutherford model has become the universally accepted picture of the structure of the atom. The properties of a given chemical element arise from the charge $+Ze$ on its nucleus and the presence of Z electrons around the nucleus. This integer Z is called the **atomic number** of the element. Atomic numbers are given on the inside back cover of this book.

Mass Spectrometry and the Measurement of Relative Masses

Mass spectrometry, besides being the chemist's most accurate method for determining relative atomic masses, also led to the discovery of other subatomic particles in addition to electrons and nuclei. In a mass spectrometer (Fig. 1.11), one or more electrons are removed from each such atom, usually by collision with a high-energy electron beam. The resulting positively charged species, called **ions,** are accelerated by an electric field and then passed through a magnetic field. The extent of curvature of the particle trajectories depends on the ratio of their charges to their masses, just as in Thomson's experiments on cathode rays (electrons) described earlier in this section. This technique allows species of different masses to be separated and detected. Early experiments in mass spectrometry demonstrated, for example, a mass ratio of 16:1 for oxygen relative to hydrogen, confirming by *physical* techniques a relationship deduced originally on *chemical* grounds. Although we focus here on the use of mass spectrometry to establish the relative

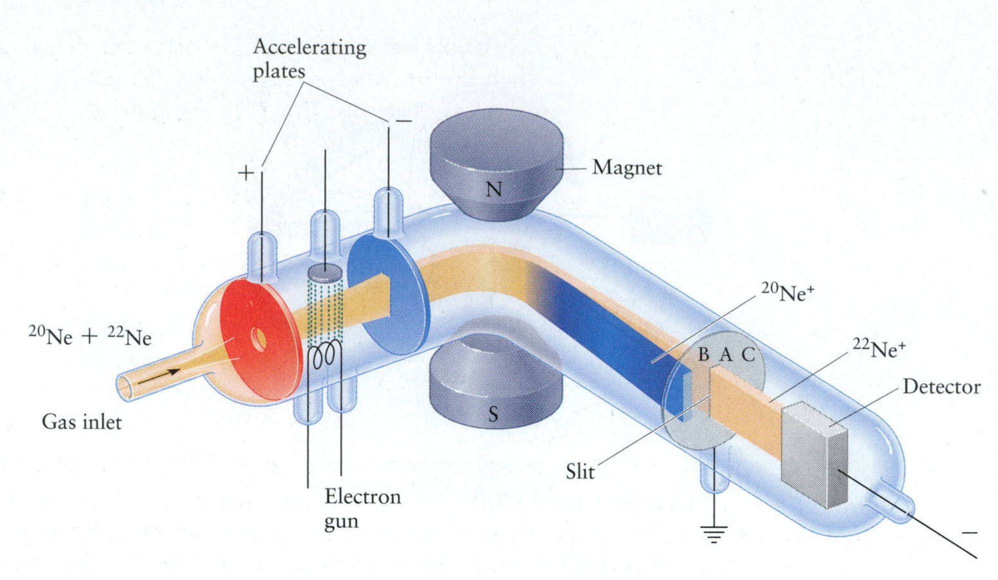

FIGURE 1.11 A simplified representation of a modern mass spectrometer. A gas mixture containing the isotopes ^{20}Ne and ^{22}Ne is introduced through the gas inlet. Some of these atoms are ionized by collisions with electrons as they pass through the electron gun. The resulting ions are accelerated to a particular kinetic energy by the electric field between the accelerating plates. The ion beam passes into a magnetic field, where it is separated into components, each containing ions with a characteristic charge-to-mass ratio. Here, the spectrometer has been adjusted to detect the less strongly deflected $^{22}Ne^+$ in the inlet mixture. By changing the magnitude of the electric or magnetic field, one can move the beam of $^{22}Ne^+$ from A to C, and the beam of $^{20}Ne^+$ from B to A, so that $^{20}Ne^+$ can be detected.

masses of atoms, it has become one of the chemist's most powerful tools for determining the masses and structures of molecules as well.

When a pure elemental gas, such as neon, was analyzed by a mass spectrometer, multiple peaks (two in the case of neon) were observed (see Fig. 1.11). Apparently, several kinds of atoms of the same element exist, differing only by their relative masses. Experiments on radioactive decay showed no differences in the *chemical* properties of these different forms of each element, so they all occupy the same place in the periodic table of the elements (see Chapter 3). Thus the different forms were named **isotopes.** Isotopes are identified by the chemical symbol for the element with a numerical superscript on the left side to specify the measured relative mass, for example ^{20}Ne and ^{22}Ne. Although the existence of isotopes of the elements had been inferred from studies of the radioactive decay paths of uranium and other heavy elements, mass spectrometry provided confirmation of their existence and their physical characterization. Later, we discuss the properties of the elementary particles that account for the mass differences of isotopes. Here, we discuss mass spectrometry as a tool for measuring atomic and molecular masses and the development of the modern atomic mass scale.

Because many elements have more than one naturally occurring isotope, the relationship between the chemist's and the physicist's relative atomic mass scales was not always simple. The atomic mass of chlorine determined by chemical means was 35.45 (relative to an oxygen atomic mass of 16), but instead of showing a single peak in the mass spectrum corresponding to a relative atomic mass of 35.45, chlorine showed 2 peaks, with relative masses near 35 and 37. Approximately three fourths of all chlorine atoms appear to have relative atomic mass 35 (the ^{35}Cl atoms), and one fourth have relative atomic mass 37 (the ^{37}Cl atoms). Naturally occurring chlorine is thus a mixture of two isotopes with different masses but nearly identical chemical properties. Elements in nature usually are mixtures, after all. Dalton's second assumption, that all atoms of a given element are identical in mass, thus is shown to be wrong in most cases.

Until 1900, chemists worked with a scale of relative atomic masses in which the average relative atomic mass of hydrogen was set at 1. At about that time, they changed to a scale in which the average relative atomic mass of naturally occurring oxygen (a mixture of ^{16}O, ^{17}O, and ^{18}O) was set at 16. In 1961, by international agreement, the atomic mass scale was revised further, with the adoption of exactly 12 as the relative atomic mass of ^{12}C. There are two stable isotopes of carbon: ^{12}C and ^{13}C (^{14}C and other isotopes of carbon are unstable and of very low terrestrial abundance). Natural carbon contains 98.892% ^{12}C and 1.108% ^{13}C by mass. The relative atomic masses of the elements as found in nature can be obtained as averages over the masses of the isotopes of each element, weighted by their observed fractional abundances. If an element consists of n isotopes, of which the ith isotope has a mass A_i and a fractional abundance p_i, then the average relative atomic mass of the element in nature (its chemical relative atomic mass) will be

$$A = A_1p_1 + A_2p_2 + \cdot\cdot\cdot + A_np_n \equiv \sum_{i=1}^{n} A_ip_i \qquad [1.11]$$

The relative atomic mass of a nuclide is close to (except for ^{12}C) but not exactly equal to its mass number.

EXAMPLE 1.2

Calculate the relative atomic mass of carbon, taking the relative atomic mass of ^{13}C to be 13.003354 on the ^{12}C scale.

SOLUTION

Set up the following table:

Isotope	Isotopic Mass × Abundance
^{12}C	12.000000 × 0.98892 = 11.867
^{13}C	13.003354 × 0.01108 = 00.144

Chemical relative atomic mass = 12.011

Related Problems: 15, 16, 17, 18

The number of significant figures in a table of chemical or natural relative atomic masses (see the inside back cover of this book) is limited not only by the accuracy of the mass spectrometric data but also by any variability in the natural abundances of the isotopes. If lead from one mine has a relative atomic mass of 207.18 and lead from another has a mass of 207.23, there is no way a result more precise than 207.2 can be obtained. In fact, geochemists are now able to use small variations in the $^{16}O:^{18}O$ isotopic abundance ratio as a "thermometer" to deduce the temperatures at which different oxygen-containing rocks were formed in the Earth's crust over geological time scales. They also find anomalies in the oxygen isotopic compositions of certain meteorites, implying that their origins may lie outside our solar system.

Relative atomic masses have no units because they are ratios of two masses measured in whatever units we choose (grams, kilograms, pounds, and so forth). The **relative molecular mass** of a compound is the sum of the relative atomic masses of the elements that constitute it, each one multiplied by the number of atoms of that element in a molecule. For example, the formula of water is H_2O, so its relative molecular mass is

$$2\ (\text{relative atomic mass of H}) + 1\ (\text{relative atomic mass of O}) = 2(1.0079) + 1(15.9994) = 18.0152$$

Protons, Neutrons, and Isotopes

The experiments described earlier led to the identification of the elementary particles that make up the atom. We discuss their properties in this section. The smallest and simplest nucleus is that of the hydrogen atom—the **proton.** It has a positive unit charge of exactly the same magnitude as the negative unit charge of the electron, but its mass is 1.67262×10^{-27} kg, which is 1836 times greater than the electron mass. Nuclei of other elements contain Z times the charge on the proton, but their masses are greater than Z times the proton mass. For example, the atomic number for helium is $Z = 2$, but its mass is approximately four times the mass of the proton. In 1920, Rutherford suggested the existence of an uncharged particle in nuclei, the **neutron,** with a mass close to that of the proton. In his model, the nucleus consists of Z protons and N neutrons. The **mass number** A is defined as $Z + N$.

A nuclear species **(nuclide)** is characterized by its atomic number Z (that is, the nuclear charge in units of e, or the number of protons in the nucleus) and its mass number A (the sum of the number of protons plus the number of neutrons in the nucleus). We denote an atom that contains such a nuclide with the symbol ${}^{A}_{Z}X$, where X is the chemical symbol for the element. The atomic number Z is sometimes omitted because it is implied by the chemical symbol for the element. Thus, ${}^{1}_{1}H$ (or ${}^{1}H$) is a hydrogen atom and ${}^{12}_{6}C$ (or ${}^{12}C$) is a carbon atom with a nucleus that contains six protons and six neutrons. Isotopes are nuclides of the same chemical species (that is, they have the same Z), but with different mass numbers A, and therefore different numbers of neutrons in the nucleus. The nuclear species of hydrogen, deuterium, and tritium, represented by ${}^{1}_{1}H$, ${}^{2}_{1}H$, and ${}^{3}_{1}H$, respectively, are all members of the family of isotopes that belong to the element hydrogen.

EXAMPLE 1.3

Radon-222 (${}^{222}Rn$) has recently received publicity because its presence in basements may increase the number of cancer cases in the general population, especially among smokers. State the number of electrons, protons, and neutrons that make up an atom of ${}^{222}Rn$.

SOLUTION

From the table on the inside back cover of the book, the atomic number of radon is 86; thus, the nucleus contains 86 protons and $222 - 86 = 136$ neutrons. The atom has 86 electrons to balance the positive charge on the nucleus.

Related Problems: 19, 20, 21, 22

1.5 Imaging Atoms, Molecules, and Chemical Reactions

The laws of chemical combination provided indirect evidence for the existence of atoms. The experiments of Thomson, Wien, and Rutherford provided direct physical evidence for the existence of the elementary particles that make up the atom. We conclude this chapter by describing an experimental method that allows us not only to image individual atoms and molecules but also to observe and control a chemical reaction at the single molecule level—a feat only dreamed of as recently as the mid-1980s.

Scanning Tunneling Microscopy Imaging of Atoms

Microscopy began with the fabrication of simple magnifying glasses and had evolved by the late 17th century to create the first optical microscopes through which single biological cells could be observed. By the 1930s, the electron microscope had been developed to detect objects too small to be seen in optical microscopes. The electron microscope showed single atoms, but at the cost of

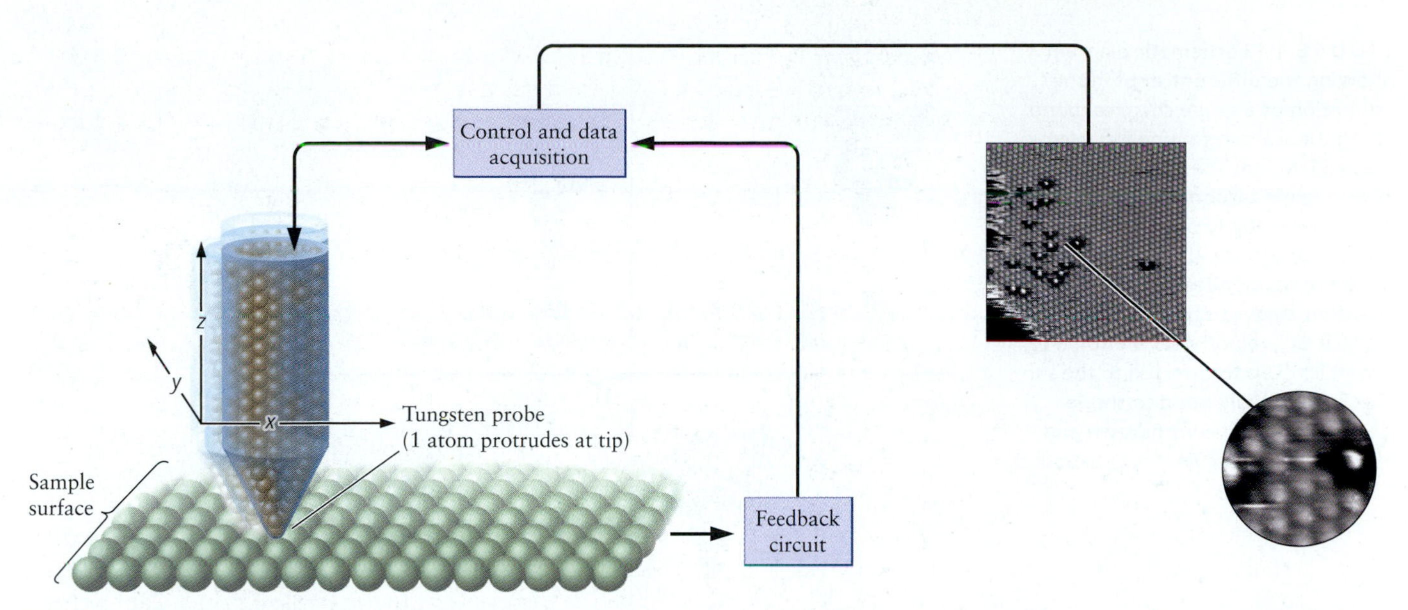

FIGURE 1.12 In a scanning tunneling microscope (STM), an electric current passes through a single atom or a small group of atoms in the probe tip, and then into the surface of the sample being examined. As the probe moves over the surface, its distance is adjusted to keep the current constant, allowing a tracing out of the shapes of the atoms or molecules on the surface.

Institut Für Allgemeine Physik, Technical University, Vienna, Austria; b: courtesy of Dr. Don Eigler/IBM Almaden Research Center, San Jose, CA

damage inflicted to the sample by the high-energy beam of electrons required to resolve such small objects. In the 1980s in Switzerland, Gerd Binnig and Heinrich Rohrer developed the scanning tunneling microscope (STM), which images atoms using low-energy electrons. For this accomplishment they received the 1986 Nobel Prize in Physics. Their device uses a sharp, electrically conducting tip that is passed over the surface of the sample being examined (Fig. 1.12a). When the tip approaches atoms of the sample, a small electrical current called the *tunneling current* can pass from the sample to the probe. The magnitude of this current is extremely sensitive to the distance of the probe from the surface, decreasing by a factor of 1000 as the probe moves away from the surface by a distance as small as 0.1 nm. Feedback circuitry holds the current constant while the probe is swept laterally across the surface, moving up and down as it passes over structural features in the surface. The vertical position of the tip is monitored, and that information is stored in a computer. By sweeping the probe tip along each of a series of closely spaced parallel tracks, one can construct and display a three-dimensional image of the surface (see the figure opposite page 1 and the figure on page 2).

Scanning tunneling microscope images visually confirm many features, such as size of atoms and the distances between them, which are already known from other techniques. But, much new information has been obtained as well. The STM images have shown the positions and shapes of molecules undergoing chemical reactions on surfaces, which helps guide the search for new ways of carrying out such reactions. They have also revealed the shape of the surface of the molecules of the nucleic acid DNA, which plays a central role in genetics.

Imaging and Controlling Reactions at the Single Molecule Level Using the Scanning Tunneling Microscope

The STM has been used to image the surfaces of materials since the mid-1980s, but only recently has it been used to image single molecules and initiate chemical reactions at the single molecule level, as we illustrate with the following

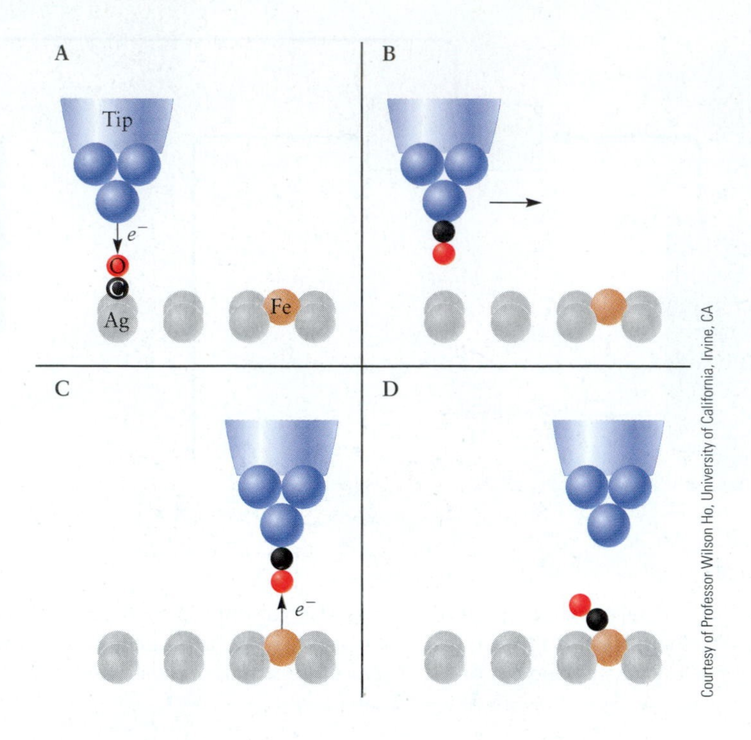

FIGURE 1.13 Schematic diagram showing the different steps in the formation of a single chemical bond using the scanning tunneling microscope (STM). (a) The tip is positioned over a single carbon monoxide (CO) molecule, ready to pluck it from the silver (Ag) surface. (b) CO is adsorbed onto the tip, bonded via the carbon (C) atom, and is translated across the surface to a region near an iron (Fe) atom. (c) CO is transferred to the surface where it will bond to the Fe atom. (d) The tip is withdrawn and the product molecule, FeCO, is bound to the Ag surface.

example. Although the STM can be used to image objects in air, the experiments described here were conducted in ultrahigh vacuum (ultralow pressure) to ensure that only the reactants of interest were present on the surface. Figure 1.13, a schematic of an STM probe hovering over a silver (Ag) surface on which an iron (Fe) atom and a carbon monoxide (CO) molecule have been chemically bonded (adsorbed), illustrates the steps leading to the formation of the product molecule Fe(CO). Figure 1.13a shows the tip approaching the CO molecule with the voltage and current adjusted so that the tip can pluck the molecule from the surface and attach it to the tip (see Fig. 1.13b). Note that the CO molecule has flipped and is bound to the tip via the carbon atom. In response to the voltage change (see Fig. 1.13c), the CO molecule inverts once again and is placed in position to bind to the Fe atom via the carbon atom, forming the molecule FeCO (see Fig. 1.13d).

The schematic in Figure 1.13 serves as a guide to the eye for interpreting the real STM images shown in Figure 1.14. Each image represents an area of the surface that is 6.3×6.3 nm. The false color scale reflects the height of the object above the plane of the silver surface atoms; the red end of the scale represents protrusions, whereas the purple end represents depressions. The identity of each chemical species was established by the nature of the image and also by the way in which the current varied with the applied voltage. That variation provides a chemical signature. In Figure 1.14a, five Fe atoms and five CO molecules are clearly seen; the red arrow identifies one Fe atom that is a bit difficult to see otherwise. The curved white arrow shows a CO molecule in close proximity to an Fe atom; Figure 1.14b shows the FeCO molecule formed as a result of the transfer of that CO molecule to the Fe atom by the tip, as well as another potentially reactive pair identified by the white curved arrow. From the shape of the resulting image in Figure 1.14c, we can see that another FeCO molecule has been formed. The white curved arrow suggests the possibility of adding an additional CO molecule to the first FeCO synthesized to form $Fe(CO)_2$, which, indeed, occurs as shown in Figure 1.14d. This remarkable sequence of images shows clearly the synthesis of a pair of distinct Fe(CO) molecules, as well as an $Fe(CO)_2$ molecule, from the reactants Fe and CO adsorbed onto a silver surface. These syntheses were accomplished by manipulating single CO molecules to place them sufficiently close to Fe atoms to initiate a chemical reaction, demonstrating our ability to observe and control chemical reactions at the single molecule level.

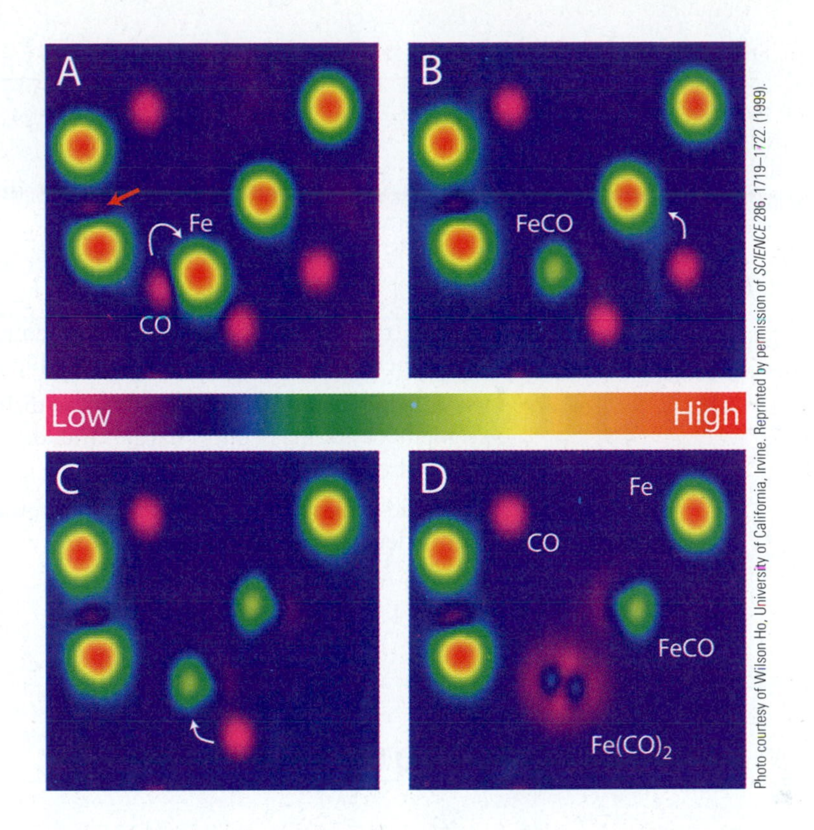

FIGURE 1.14 Scanning tunneling microscope (STM) images of chemical bond formation. Each image is 6.3 × 6.3 nm. The species are identified by their height above or below the surface silver (Ag) atoms and by the shapes of their current-voltage curves. (a) Five iron (Fe) atoms (concentric red-green circles) and five carbon monoxide (CO) molecules (purple) are observed in this region. The red arrow points to one CO molecule that is a little hard to see, and the white arrow shows a CO molecule about to react with an Fe atom. (b) Formation of the first FeCO molecule (green) and identification of another reactive Fe/CO pair (curved white arrow). (c) Formation of the second FeCO molecule (green) and identification of a second possible reaction to form $Fe(CO)_2$ (curved white arrow). (d) Image of individual $Fe(CO)_2$ and Fe(CO) molecules synthesized on a surface by the STM.

CHAPTER SUMMARY

We have come a long way since the attempts of the alchemists to turn base metals into gold, to transmute one element into another. Through the early chemical experiments of Dalton, Gay-Lussac, and Avogadro, we have learned that matter is ultimately indivisible, at least as far as its physical and chemical properties are concerned. The experiments of Thomson, Wien, and Rutherford confirmed, from the results of physical measurements, the existence of the atom. These experiments also identified and characterized the elementary particles from which the atom is made, and this led to the modern model of the atom as an object with a small, dense nucleus surrounded by a much larger volume occupied by the electrons. Physicists in the 21st century have developed tools of unprecedented power with which to analyze and synthesize single molecules, an achievement that has already led to exciting new applications in almost every area of modern science and engineering.

CHAPTER REVIEW

- Matter can be classified systematically by its uniformity and its response to a variety of separation processes; compounds and elements are the basic chemical building blocks.
- The existence of atoms was inferred from several lines of indirect evidence that all pointed to the conclusion that there was a fundamental limit below which no further division was possible.
- The existence of atoms was confirmed and the modern model of the atom developed from physical measurements that determined the properties of the electron and the nucleus and demonstrated the existence of isotopes.
- Individual atoms and molecules can be imaged and manipulated directly using the STM.

CONCEPTS & SKILLS

After studying this chapter and working the problems that follow, you should be able to:

1. Describe in operational terms how to distinguish among mixtures, compounds, and elements (Section 1.2).
2. Outline Dalton's atomic theory of matter and describe its experimental basis (Section 1.3).
3. Describe the reasoning that permits chemical formulas to be determined by purely chemical means (Section 1.3, Problems 7–10).
4. Describe the experiments that led to the discovery of the electron and the measurement of its mass and charge, and describe those that demonstrated the nature of the nucleus (Section 1.4).
5. Given the atomic masses and natural abundances of the isotopes of an element, calculate its chemical atomic mass (Section 1.4, Problems 15–18).
6. State the numbers of protons, neutrons, and electrons in particular atoms (Section 1.4, Problems 19–22).

KEY EQUATIONS

$$\frac{e}{m_e} = \frac{SE}{\ell LH^2}$$ Section 1.4

$$A = A_1p_1 + A_2p_2 + \cdots + A_np_n \equiv \sum_{i=1}^{n} A_ip_i$$ Section 1.4

PROBLEMS

Answers to problems whose numbers are boldface appear in Appendix G. Problems that are more challenging are indicated with asterisks.

Macroscopic Methods for Classifying Matter

1. Classify the following materials as substances or mixtures: table salt, wood, mercury, air, water, seawater, sodium chloride, and mayonnaise. If they are mixtures, subclassify them as homogeneous or heterogeneous; if they are substances, subclassify them as compounds or elements.

2. Classify the following materials as substances or mixtures: absolute (pure) alcohol, milk (as purchased in a store), copper wire, rust, barium bromide, concrete, baking soda, and baking powder. If they are mixtures, subclassify them as homogeneous or heterogeneous; if they are substances, subclassify them as compounds or elements.

3. A 17th-century chemist wrote of the "simple bodies which enter originally into the composition of mixtures and into which these mixtures resolve themselves or may be finally resolved." What is being discussed?

4. Since 1800, almost 200 sincere but erroneous reports of the discovery of new chemical elements have been made. Why have mistaken reports of new elements been so numerous? Why is it relatively easy to prove that a material is not a chemical element, but difficult to prove absolutely that a material is an element?

Indirect Evidence for the Existence of Atoms: Laws of Chemical Combination

5. A sample of ascorbic acid (vitamin C) is synthesized in the laboratory. It contains 30.0 g carbon and 40.0 g oxygen. Another sample of ascorbic acid, isolated from lemons (an excellent source of the vitamin), contains 12.7 g carbon. Compute the mass of oxygen (in grams) in the second sample.

6. A sample of a compound synthesized and purified in the laboratory contains 25.0 g hafnium and 31.5 g tellurium. The identical compound is discovered in a rock formation. A sample from the rock formation contains 0.125 g hafnium. Determine how much tellurium is in the sample from the rock formation.

7. Nitrogen (N) and silicon (Si) form two binary compounds with the following compositions:

Compound	Mass % N	Mass % Si
1	33.28	66.72
2	39.94	60.06

(a) Compute the mass of silicon that combines with 1.0000 g of nitrogen in each case.
(b) Show that these compounds satisfy the law of multiple proportions. If the second compound has the formula Si_3N_4, what is the formula of the first compound?

8. Iodine (I) and fluorine (F) form a series of binary compounds with the following compositions:

Compound	Mass % I	Mass % F
1	86.979	13.021
2	69.007	30.993
3	57.191	42.809
4	48.829	51.171

(a) Compute in each case the mass of fluorine that combines with 1.0000 g iodine.
(b) By figuring out small whole-number ratios among the four answers in part (a), show that these compounds satisfy the law of multiple proportions.

9. Vanadium (V) and oxygen (O) form a series of compounds with the following compositions:

Mass % V	Mass % O
76.10	23.90
67.98	32.02
61.42	38.58
56.02	43.98

What are the relative numbers of atoms of oxygen in the compounds for a given mass of vanadium?

10. Tungsten (W) and chlorine (Cl) form a series of compounds with the following compositions:

Mass % W	Mass % Cl
72.17	27.83
56.45	43.55
50.91	49.09
46.36	53.64

If a molecule of each compound contains only one tungsten atom, what are the formulas for the four compounds?

11. A liquid compound containing only hydrogen and oxygen is placed in a flask. Two electrodes are dipped into the liquid, and an electric current is passed between them. Gaseous hydrogen forms at one electrode and gaseous oxygen at the other. After a time, 14.4 mL hydrogen has evolved at the negative terminal, and 14.4 mL oxygen has evolved at the positive terminal.
(a) Assign a chemical formula to the compound in the cell.
(b) Explain why more than one formula is possible as the answer to part (a).

12. A sample of liquid N_2H_4 is decomposed to give gaseous N_2 and gaseous H_2. The two gases are separated, and the nitrogen occupies 13.7 mL at room conditions of pressure and temperature. Determine the volume of the hydrogen under the same conditions.

13. Pure nitrogen dioxide (NO_2) forms when dinitrogen oxide (N_2O) and oxygen (O_2) are mixed in the presence of a certain catalyst. What volumes of N_2O and oxygen are needed to produce 4.0 L NO_2 if all gases are held at the same conditions of temperature and pressure?

14. Gaseous methanol (CH_3OH) reacts with oxygen (O_2) to produce water vapor and carbon dioxide. What volumes of water vapor and carbon dioxide will be produced from 2.0 L methanol if all gases are held at the same temperature and pressure conditions?

Physical Structure of Atoms

15. The natural abundances and isotopic masses of the element silicon (Si) relative to $^{12}C = 12.00000$ are

Isotope	% Abundance	Isotopic Mass
^{28}Si	92.21	27.97693
^{29}Si	4.70	28.97649
^{30}Si	3.09	29.97376

Calculate the atomic mass of naturally occurring silicon.

16. The natural abundances and isotopic masses of the element neon (Ne) are

Isotope	% Abundance	Isotopic Mass
^{20}Ne	90.00	19.99212
^{21}Ne	0.27	20.99316
^{22}Ne	9.73	21.99132

Calculate the atomic mass of naturally occurring neon.

17. Only two isotopes of boron (B) occur in nature; their atomic masses and abundances are given in the following table. Complete the table by computing the relative atomic mass of ^{11}B to four significant figures, taking the tabulated relative atomic mass of natural boron as 10.811.

Isotope	% Abundance	Atomic Mass
^{10}B	19.61	10.013
^{11}B	80.39	?

18. More than half of all the atoms in naturally occurring zirconium are ^{90}Zr. The other four stable isotopes of zirconium have the following relative atomic masses and abundances:

Isotope	% Abundance	Atomic Mass
^{91}Zr	11.27	90.9056
^{92}Zr	17.17	91.9050
^{94}Zr	17.33	93.9063
^{96}Zr	2.78	95.9083

Compute the relative atomic mass of ^{90}Zr to four significant digits, using the tabulated relative atomic mass 91.224 for natural zirconium.

19. The isotope of plutonium used for nuclear fission is ^{239}Pu. Determine (a) the ratio of the number of neutrons in a ^{239}Pu nucleus to the number of protons, and (b) the number of electrons in a single plutonium atom.

20. The last "missing" element from the first six periods was promethium, which was finally discovered in 1947 among the fission products of uranium. Determine (a) the ratio of the number of neutrons in a ^{145}Pm nucleus to the number of protons, and (b) the number of electrons in a single promethium atom.

21. The americium isotope ^{241}Am is used in smoke detectors. Describe the composition of a neutral atom of this isotope for protons, neutrons, and electrons.

22. In 1982, the production of a single atom of $^{266}_{109}$Mt (meitnerium-266) was reported. Describe the composition of a neutral atom of this isotope for protons, neutrons, and electrons.

ADDITIONAL PROBLEMS

23. Soft wood chips weighing 17.2 kg are placed in an iron vessel and mixed with 150.1 kg water and 22.43 kg sodium hydroxide. A steel lid seals the vessel, which is then placed in an oven at 250°C for 6 hours. Much of the wood fiber decomposes under these conditions; the vessel and lid do not react.

(a) Classify each of the materials mentioned as a substance or mixture. Subclassify the substances as elements or compounds.

(b) Determine the mass of the contents of the iron vessel after the reaction.

* **24.** In a reproduction of the Millikan oil-drop experiment, a student obtains the following values for the charges on nine different oil droplets.

6.563×10^{-19} C	13.13×10^{-19} C	19.71×10^{-19} C
8.204×10^{-19} C	16.48×10^{-19} C	22.89×10^{-19} C
11.50×10^{-19} C	18.08×10^{-19} C	26.18×10^{-19} C

(a) Based on these data alone, what is your best estimate of the number of electrons on each of the above droplets? (*Hint:* Begin by considering differences in charges between adjacent data points, and see into what groups these are categorized.)

(b) Based on these data alone, what is your best estimate of the charge on the electron?

(c) Is it conceivable that the actual charge is half the charge you calculated in (b)? What evidence would help you decide one way or the other?

25. A rough estimate of the radius of a nucleus is provided by the formula $r = kA^{1/3}$, where k is approximately 1.3×10^{-13} cm and A is the mass number of the nucleus. Estimate the density of the nucleus of ^{127}I (which has a nuclear mass of 2.1×10^{-22} g) in grams per cubic centimeter. Compare with the density of solid iodine, 4.93 g cm^{-3}.

26. In a neutron star, gravity causes the electrons to combine with protons to form neutrons. A typical neutron star has a mass half that of the sun, compressed into a sphere of radius 20 km. If such a neutron star contains 6.0×10^{56} neutrons, calculate its density in grams per cubic centimeter. Compare this with the density inside a ^{232}Th nucleus, in which 142 neutrons and 90 protons occupy a sphere of radius 9.1×10^{-13} cm. Take the mass of a neutron to be 1.675×10^{-24} g and that of a proton to be 1.673×10^{-24} g.

27. Dalton's 1808 version of the atomic theory of matter included five general statements (see Section 1.3). According to modern understanding, four of those statements require amendment or extension. List the modifications that have been made to four of the five original postulates.

28. Naturally occurring rubidium (Rb) consists of two isotopes: ^{85}Rb (atomic mass 84.9117) and ^{87}Rb (atomic mass 86.9092). The atomic mass of the isotope mixture found in nature is 85.4678. Calculate the percentage abundances of the two isotopes in rubidium.

CHAPTER 2

Chemical Formulas, Chemical Equations, and Reaction Yields

An "assay balance of careful construction" of the type used by Lavoisier before 1788. This balance became the production model that served as a general, all-purpose balance for approximately 40 years. Users of this type of balance included Sir Humphrey Davy and his young assistant Michael Faraday.

Chapter 1 explained how chemical and physical methods are used to establish chemical formulas and relative atomic and molecular masses. This chapter begins our study of chemical reactions. We start by developing the concept of the mole, which allows us to count molecules by weighing macroscopic quantities of matter. We examine the balanced chemical equations that summarize these reactions and show how to relate the masses of substances consumed to the masses of substances produced. This is an immensely practical and important subject. The questions how much of a substance will react with a given amount of another substance and how much product will be generated are central to all chemical processes, whether industrial, geological, or biological.

2.1 The Mole: Weighing and Counting Molecules

The laws of chemical combination assert that chemical reactions occur in such a way that the number of atoms of a given type are conserved in every chemical reaction, except nuclear reactions. How do we weigh out a sample containing exactly the number of atoms or molecules needed for a particular chemical reaction? What is the mass of an atom or a molecule? These questions must be answered indirectly, because atoms and molecules are far too small to be weighed individually.

In the process of developing the laws of chemical combination, chemists had determined indirectly the relative masses of atoms of different elements; for example, 19th-century chemists concluded that oxygen atoms weigh 16 times as much as hydrogen atoms. In the early 20th century, relative atomic and molecular masses were determined much more accurately through the work of J. J. Thomson, F. W. Aston, and others using mass spectrometry. These relative masses on the atomic scale must be related to absolute masses on the gram scale by a conversion factor. Once this conversion factor is determined, we simply weigh out a sample with the mass required to provide the desired number of atoms or molecules of a substance for a particular reaction. These concepts and methods are developed in this section.

Relation between Atomic and Macroscopic Masses: Avogadro's Number

What are the actual masses (in grams) of individual atoms and molecules? To answer this question, we must establish a connection between the absolute macroscopic scale for mass used in the laboratory and the relative microscopic scale established for the masses of individual atoms and molecules. The link between the two is provided by **Avogadro's number** (N_A), defined as the number of atoms in exactly 12 g of ^{12}C. We consider below some of the many experimental methods devised to determine the numerical value of N_A. Its currently accepted value is

$$N_A = 6.0221420 \times 10^{23}$$

The mass of a single ^{12}C atom is then found by dividing exactly 12 g carbon (C) by N_A:

$$\text{Mass of a } ^{12}\text{C atom} = \frac{12.00000 \text{ g}}{6.0221420 \times 10^{23}} = 1.9926465 \times 10^{-23} \text{ g}$$

This is truly a small mass, reflecting the large number of atoms in a 12-g sample of carbon.

Avogadro's number is defined relative to the ^{12}C atom because that isotope has been chosen by international agreement to form the basis for the modern scale of relative atomic masses. We can apply it to other substances as well in a particularly simple fashion. Consider sodium, which has a relative atomic mass of 22.98977. A sodium atom is 22.98977/12 times as heavy as a ^{12}C atom. If the mass of N_A atoms of ^{12}C is 12 g, then the mass of N_A atoms of sodium must be

$$\frac{22.98977}{12}(12 \text{ g}) = 22.98977 \text{ g}$$

The mass (in grams) of N_A atoms of *any* element is numerically equal to the relative atomic mass of that element. The same conclusion applies to molecules. From the relative molecular mass of water calculated earlier, the mass of N_A molecules of water is 18.0152 g.

EXAMPLE 2.1

One of the heaviest atoms found in nature is ^{238}U. Its relative atomic mass is 238.0508 on a scale in which 12 is the atomic mass of ^{12}C. Calculate the mass (in grams) of one ^{238}U atom.

SOLUTION

Because the mass of N_A atoms of ^{238}U is 238.0508 g and N_A is 6.0221420×10^{23}, the mass of one ^{238}U atom must be

$$\frac{238.0508 \text{ g}}{6.0221420 \times 10^{23}} = 3.952926 \times 10^{-22} \text{ g}$$

Related Problems: 1, 2

The Mole

Because the masses of atoms and molecules are so small, laboratory scale chemical reactions must involve large numbers of atoms and molecules. It is convenient to group atoms or molecules in counting units of $N_A = 6.0221420 \times 10^{23}$ to measure the **number of moles** of a substance. One of these counting units is called a **mole** (abbreviated mol; derived from Latin *moles,* meaning "heap" or "pile"). One mole of a substance is the amount that contains Avogadro's number of atoms, molecules, or other entities. That is, 1 mol of ^{12}C contains N_A ^{12}C atoms, 1 mol of water contains N_A water molecules, and so forth. We must be careful in some cases, because a phrase such as "1 mol of oxygen" is ambiguous. We should refer instead to "1 mol of O_2" if there are N_A oxygen *molecules,* and "1 mol of O" if there are N_A oxygen *atoms.* Henceforth, for *any* species we use "number of moles of a particular species" to describe the number of moles in a sample of that species.

The mass of one mole of atoms of an element—the **molar mass,** with units of grams per mole—is numerically equal to the dimensionless relative atomic mass of that element, and the same relationship holds between the molar mass of a compound and its relative molecular mass. Thus, the relative molecular mass of water is 18.0152, and its molar mass is 18.0152 g mol^{-1}.

To determine the number of moles of a given substance, we use the chemist's most powerful tool, the laboratory balance. If a sample of iron weighs 8.232 g, then

$$\begin{aligned}\text{moles of iron} &= \frac{\text{number of grams of iron}}{\text{molar mass of iron}} \\ &= \frac{8.232 \text{ g Fe}}{55.847 \text{ g mol}^{-1}} \\ &= 0.1474 \text{ mol Fe}\end{aligned}$$

where the molar mass of iron was obtained from the periodic table of the elements or a table of relative atomic masses (see the inside front and back covers of this book). The calculation can be turned around as well. Suppose a certain amount, for example, 0.2000 mol, of water is needed in a chemical reaction. We have

$$(\text{moles of water}) \times (\text{molar mass of water}) = \text{mass of water}$$

$$(0.2000 \text{ mol } H_2O) \times (18.015 \text{ g mol}^{-1}) = 3.603 \text{ g } H_2O$$

We simply weigh 3.603 g water to get the 0.2000 mol needed for the reaction. In both cases, the molar mass is the conversion factor between the mass of the substance and the number of moles of the substance.

Although the number of moles in a sample is frequently determined by weighing, it is still preferable to think of a mole as a fixed number of particles (Avogadro's number) rather than as a fixed mass. The term *mole* is thus analogous to a term such as *dozen*: one dozen pennies weighs 26 g, which is substantially less than the mass of one dozen nickels, 60 g; but each group contains 12 coins. Figure 2.1 shows mole quantities of several substances.

EXAMPLE 2.2

Nitrogen dioxide (NO_2) is a major component of urban air pollution. For a sample containing 4.000 g NO_2, calculate (a) the number of moles of NO_2 and (b) the number of molecules of NO_2.

SOLUTION

(a) From the tabulated molar masses of nitrogen (14.007 g mol^{-1}) and oxygen (15.999 g mol^{-1}), the molar mass of NO_2 is

$$14.007 \text{ g mol}^{-1} + (2 \times 15.999 \text{ g mol}^{-1}) = 46.005 \text{ g mol}^{-1}$$

The number of moles of NO_2 is then

$$\text{mol of NO}_2 = \frac{4.000 \text{ g NO}_2}{46.005 \text{ g mol}^{-1}} = 0.08695 \text{ mol NO}_2$$

(b) To convert from moles to number of molecules, multiply by Avogadro's number:

$$\text{Molecules of NO}_2 = (0.08695 \text{ mol NO}_2) \times 6.0221 \times 10^{23} \text{ mol}^{-1}$$
$$= 5.236 \times 10^{22} \text{ molecules NO}_2$$

Related Problems: 7, 8

The fact that N_A is the ratio of the molar volume to the atomic volume of any element provides a route to measuring its value, and several methods have been used to determine this ratio. A new method to refine the value currently is under development. Nearly perfectly smooth spheres of highly crystalline silicon (Si) can be prepared and characterized. The surface roughness of these spheres (which affects the determination of their volume) is ± 1 silicon atom. The molar volume is determined by carefully measuring the mass and volume of the sphere, and the atomic volume is determined by measuring the interatomic distances directly using x-ray diffraction. (X-ray diffraction from solids is described in Chapter 21.) Avogadro's number is the ratio of these two quantities.

Density and Molecular Size

The **density** of a sample is the ratio of its mass to its volume:

$$\text{density} = \frac{\text{mass}}{\text{volume}} \qquad [2.1]$$

The base unit of mass in the International System of Units (SI; see discussion in Appendix B) is the kilogram (kg), but it is inconveniently large for most practical purposes in chemistry. The gram often is used instead; moreover, it is the standard unit for molar masses. Several units for volume are in frequent use. The base SI unit of the cubic meter (m^3) is also unwieldy for laboratory purposes (1 m^3 water weighs 1000 kg, or 1 metric ton). We will, therefore, use the liter (1 L = 10^{-3} m^3) and the cubic centimeter, which is identical to the milliliter

FIGURE 2.1 One-mole quantities of several substances. (Clockwise from top) Graphite (C), potassium permanganate ($KMnO_4$), copper sulfate pentahydrate ($CuSO_4 \cdot 5\ H_2O$), copper (Cu), sodium chloride (NaCl), and potassium dichromate ($K_2Cr_2O_7$). Antimony (Sb) is at the center.

($1\ cm^3 = 1\ mL = 10^{-3}\ L = 10^{-6}\ m^3$), for volume. Table 2.1 lists the densities of some substances in units of grams per cubic centimeter.

The density of a substance is not a fixed, invariant property of the substance; its value depends on the pressure and temperature at the time of measurement. For some substances (especially gases and liquids), the volume may be more convenient to measure than the mass, and when the density is known, it provides the conversion factor between volume and mass. For example, near room temperature, the density of liquid benzene (C_6H_6) is 0.8765 g cm^{-3}. Suppose that 0.2124 L benzene is measured into a container. The mass of benzene is then the product of the volume and the density:

$$m = \rho V$$

where m is the mass, ρ is the density, and V is the volume. Therefore, the value of the mass of benzene is

$$m = 0.2124\ \text{L} \times (1 \times 10^3\ \text{cm}^3\ \text{L}^{-1}) \times (0.8765\ \text{g cm}^{-3}) = 186.2\ \text{g}$$

TABLE 2.1 Densities of Some Substances

Substance	Density (g cm^{-3})
Hydrogen	0.000082
Oxygen	0.00130
Water	1.00
Magnesium	1.74
Sodium chloride	2.16
Quartz	2.65
Aluminum	2.70
Iron	7.86
Copper	8.96
Silver	10.5
Lead	11.4
Mercury	13.5
Gold	19.3
Platinum	21.4

These densities were measured at room temperature and at average atmospheric pressure near sea level.

Dividing the above amount by the molar mass of benzene (78.114 g mol^{-1}) gives the corresponding number of moles, 2.384 mol.

Knowing the density and molar mass of a substance, we can readily compute its **molar volume,** that is, the volume occupied by one mole of a substance:

$$V_m = \frac{\text{molar mass (g mol}^{-1})}{\text{density (g cm}^{-3})} = \text{molar volume (cm}^3\text{ mol}^{-1})$$

For example, near 0°C, ice has a density of 0.92 g cm^{-3}; thus, the molar volume of solid water under these conditions is

$$V_m = \frac{18.0 \text{ g mol}^{-1}}{0.92 \text{ g cm}^{-1}} = 20 \text{ cm}^3 \text{ mol}^{-1}$$

The molar volume of a gas is much larger than that of either a liquid or a solid. For O_2 under room conditions, the data in Table 2.1 give a molar volume of 24,600 cm^3 mol^{-1} = 24.6 L mol^{-1}, which is more than 1000 times larger than the molar volume just computed for ice under the same conditions of temperature and pressure. How can we interpret this fact on a microscopic level? We also note that the volumes of liquids and solids do not shift much with changes in temperature or pressure, but that the volumes of gases are quite sensitive to these changes. One hypothesis that would explain these observations is that the molecules in liquids and solids are close enough to touch one another, but that they are separated by large distances in gases. If this hypothesis is correct (as has been well established by further study), then the sizes of the molecules themselves can be estimated from the volume occupied per molecule in the liquid or solid state. The volume per molecule is the molar volume divided by Avogadro's number; for ice, this gives

$$\text{Volume per H}_2\text{O molecule} = \frac{20 \text{ cm}^3 \text{ mol}^{-1}}{6.02 \times 10^{23} \text{ mol}^{-1}} = 3.3 \times 10^{-23} \text{ cm}^3$$

This volume corresponds to that of a cube with edges about 3.2×10^{-8} cm (0.32 nm) on a side. We conclude from this and other density measurements that the characteristic size of atoms and small molecules is about 10^{-8} cm, or about 0.1 nm. Avogadro's number provides the link between the length and mass scales of laboratory measurements and the masses and volumes of single atoms and molecules.

2.2 Empirical and Molecular Formulas

According to the laws of chemical combination, each substance may be described by a chemical formula that specifies the relative numbers of atoms of the elements in that substance. We now distinguish between two types of formulas: the molecular formula and the empirical formula. The **molecular formula** of a substance specifies the number of atoms of each element in one molecule of that substance. Thus, the molecular formula of carbon dioxide is CO_2; each molecule of carbon dioxide contains 1 atom of carbon and 2 atoms of oxygen. The molecular formula of glucose is $C_6H_{12}O_6$; each glucose molecule contains 6 atoms of carbon, 6 of oxygen, and 12 of hydrogen. Molecular formulas can be defined for all gaseous substances and for those liquids or solids that, like glucose, possess well-defined molecular structures.

In contrast, the **empirical formula** of a compound is the simplest formula that gives the correct relative numbers of atoms of each kind in a compound. For example, the empirical formula of glucose is CH_2O, indicating that the numbers of atoms of carbon, hydrogen, and oxygen are in a ratio of 1:2:1. When a molecular formula is known, it is clearly preferable because it conveys more information. In some solids and liquids, however, distinct small molecules do not exist, and the

FIGURE 2.2 When cobalt(II) chloride crystallizes from solution, it brings with it six water molecules per formula unit, giving a red solid with the empirical formula $CoCl_2 \cdot 6\,H_2O$. This solid melts at 86°C; at greater than about 110°C, it loses some water and forms a lavender solid with the empirical formula $CoCl_2 \cdot 2\,H_2O$.

only meaningful chemical formula is an empirical one. Solid cobalt(II) chloride, which has the empirical formula $CoCl_2$, is an example. There are strong attractive forces between a cobalt atom and two adjoining chlorine (Cl) atoms in solid cobalt(II) chloride, but it is impossible to distinguish the forces *within* such a "molecule" of $CoCl_2$ from those operating *between* it and a neighbor; the latter are equally strong. The solid is, in effect, a single giant molecule. Consequently, cobalt(II) chloride is represented with an empirical formula and referred to by a **formula unit** of $CoCl_2$, rather than by "a molecule of $CoCl_2$." Many solids can be represented only by their formula units because it is not possible to identify a molecular unit in a unique way; other examples include sodium chloride (NaCl), the major component in table salt, and silicon dioxide (SiO_2), the major component of sand. In some cases, small molecules are incorporated into a solid structure, and the chemical formula is written to show this fact explicitly. Thus, cobalt and chlorine form not only the anhydrous salt $CoCl_2$ mentioned earlier but also the hexahydrate $CoCl_2 \cdot 6\,H_2O$, in which six water molecules are incorporated per $CoCl_2$ formula unit (Fig. 2.2). The dot in this formula is used to set off a well-defined molecular component of the solid, such as water.

2.3 Chemical Formula and Percentage Composition

The empirical formula H_2O specifies that for every atom of oxygen in water, there are two atoms of hydrogen. Equivalently, one mole of H_2O contains two moles of hydrogen atoms and one mole of oxygen atoms. The number of atoms and the number of moles of each element are present in the same ratio, namely, 2:1. The empirical formula for a substance is clearly related to the percentage composition by mass of that substance. This connection can be used in various ways.

Empirical Formula and Percentage Composition

The empirical formula of a compound can be simply related to the mass percentage of its constituent elements using the mole concept. For example, the empirical formula for ethylene (molecular formula C_2H_4) is CH_2. Its composition by mass is calculated from the masses of carbon and hydrogen in 1 mol of CH_2 formula units:

$$\text{Mass of C} = 1 \text{ mol C} \times (12.011 \text{ g mol}^{-1}) = 12.011 \text{ g}$$

$$\text{Mass of H} = 2 \text{ mol H} \times (1.00794 \text{ g mol}^{-1}) = 2.0159 \text{ g}$$

Adding these masses together gives a total mass of 14.027 g. The mass percentages of carbon and hydrogen in the compound are then found by dividing each of their masses by this total mass and multiplying by 100%, giving 85.628% C and 14.372% H by weight, respectively.

Determination of Empirical Formula from Measured Mass Composition

We can reverse the procedure just described and determine the empirical formula from the elemental analysis of a compound, as illustrated by Example 2.3.

EXAMPLE 2.3

A 60.00-g sample of a dry-cleaning fluid was analyzed and found to contain 10.80 g carbon, 1.36 g hydrogen, and 47.84 g chlorine. Determine the empirical formula of the compound using a table of atomic masses.

SOLUTION

The amounts of each element in the sample are

$$\text{carbon: } \frac{10.80 \text{ g C}}{12.011 \text{ g mol}^{-1}} = 0.8992 \text{ mol C}$$

$$\text{hydrogen: } \frac{1.36 \text{ g H}}{1.008 \text{ g mol}^{-1}} = 1.35 \text{ mol H}$$

$$\text{chlorine: } \frac{47.84 \text{ g Cl}}{35.453 \text{ g mol}^{-1}} = 1.349 \text{ mol Cl}$$

The ratio of the amount of carbon to that of chlorine (or hydrogen) is 0.8992:1.349 = 0.6666, which is close to 2:3. The numbers of moles form the ratio 2:3:3; therefore, the empirical formula is $C_2H_3Cl_3$. Additional measurements would be necessary to find the actual molecular mass and the correct *molecular* formula from among $C_2H_3Cl_3$, $C_4H_6Cl_6$, or any higher multiples $(C_2H_3Cl_3)_n$.

Related Problems: 19, 20, 21, 22, 23, 24

Empirical Formula Determined from Elemental Analysis by Combustion

A **hydrocarbon** is a compound that contains only carbon and hydrogen. Its empirical formula can be determined by using the combustion train shown in Figure 2.3. In this device, a known mass of the hydrocarbon is burned completely in oxygen,

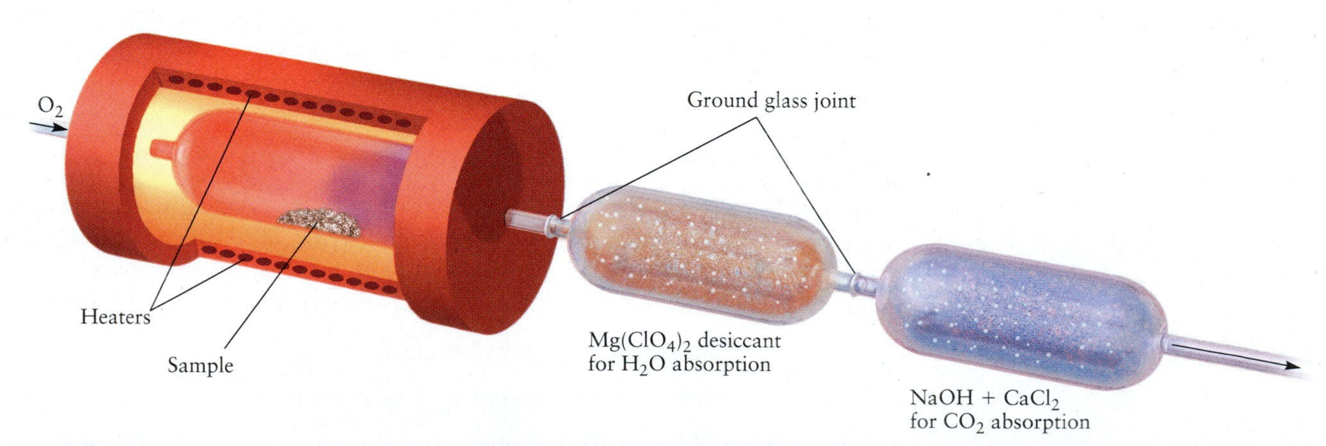

FIGURE 2.3 A combustion train to determine amounts of carbon and hydrogen in hydrocarbons. A weighed sample is burned in a flow of oxygen to form water and carbon dioxide. These products of combustion pass over a desiccant such as magnesium perchlorate, $Mg(ClO_4)_2$, which absorbs the water. In a second stage, the carbon dioxide passes over and is absorbed on finely divided particles of sodium hydroxide, NaOH, mixed with calcium chloride, $CaCl_2$. The changes in mass of the two absorbers give the amounts of water and carbon dioxide produced.

yielding carbon dioxide and water. The masses of water and carbon dioxide that result can then be determined, and from these data the empirical formula calculated, as illustrated in Example 2.4.

EXAMPLE 2.4

A certain compound, used as a welding fuel, contains only carbon and hydrogen. Burning a small sample of this fuel completely in oxygen produces 3.38 g CO_2, 0.692 g water, and no other products. What is the empirical formula of the compound?

SOLUTION

We first compute the amounts of CO_2 and H_2O. Because all the carbon has been converted to CO_2 and all the hydrogen to water, the amounts of C and H in the unburned gas can be determined:

$$\text{mol of C} = \text{mol of CO}_2 = \frac{3.38 \text{ g}}{44.01 \text{ g mol}^{-1}} = 0.0768 \text{ mol}$$

$$\text{mol of H} = 2(\text{mol of H}_2\text{O}) = \left(\frac{0.692 \text{ g}}{18.02 \text{ g mol}^{-1}}\right) = 0.0768 \text{ mol}$$

Because each water molecule contains two hydrogen atoms, it is necessary to multiply the number of moles of water by 2 to find the number of moles of hydrogen atoms. Having found that the compound contains equal numbers of moles of carbon and hydrogen, we have determined that its empirical formula is CH. Its molecular formula may be CH, C_2H_2, C_3H_3, and so on.

Related Problems: 25, 26

Connection between the Empirical Formula and the Molecular Formula

The molecular formula is some whole-number multiple of the empirical formula. To determine the molecular formula, you must know the approximate molar mass of the compound under study. From Avogadro's hypothesis, the ratio of molar masses of two gaseous compounds is the same as the ratio of their densities, provided that those densities are measured at the same temperature and pressure. (This is true because a given volume contains the same number of molecules of the two gases.) The density of the welding gas from Example 2.4 is 1.06 g L^{-1} at 25°C and atmospheric pressure. Under the same conditions, the density of gaseous oxygen (which exists as diatomic O_2 molecules with molar mass of 32.0 g mol^{-1}) is 1.31 g L^{-1}. The approximate molar mass of the welding gas is, therefore,

$$\text{Molar mass of welding gas} = \frac{1.06 \text{ g L}^{-1}}{1.31 \text{ g L}^{-1}}(32.0 \text{ g mol}^{-1}) = 25.9 \text{ g mol}^{-1}$$

The molar mass corresponding to the *empirical* formula CH is 13.0 g mol^{-1}. Because 25.9 g mol^{-1} is approximately twice this value, there must be two CH units per molecule; therefore, the molecular formula is C_2H_2. The gas is acetylene.

2.4 Writing Balanced Chemical Equations

Chemical reactions combine elements into compounds, decompose compounds back into elements, and transform existing compounds into new compounds. Because atoms are indestructible in chemical reactions, the same number of atoms (or moles of atoms) of each element must be present before and after any ordinary (as opposed to nuclear) chemical reaction. The conservation of matter in a chemical change is represented in a balanced chemical equation for that process. The study of the relationships between the numbers of reactant and product molecules is called **stoichiometry** (derived from the Greek *stoicheion,* meaning "element," and *metron,* meaning "measure"). Stoichiometry is fundamental to all aspects of chemistry.

An equation can be balanced using stepwise reasoning. Consider the decomposition of ammonium nitrate (NH_4NO_3) on gentle heating to produce dinitrogen oxide (N_2O) and water. The *unbalanced* equation for this process is

$$NH_4NO_3 \longrightarrow N_2O + H_2O$$

The formulas on the left side of the arrow denote **reactants,** and those on the right side denote **products.** This equation is unbalanced because there are 3 mol of oxygen atoms on the left side of the equation (and 4 of hydrogen), but only 2 mol of oxygen atoms and 2 mol of hydrogen atoms on the right side. To balance the equation, begin by assigning 1 as the coefficient of one species, usually the species that contains the most elements—in this case, NH_4NO_3. Next, seek out the elements that appear in only one other place in the equation and assign coefficients to balance the numbers of their atoms. Here, nitrogen appears in only one other place (N_2O), and a coefficient of 1 for the N_2O ensures that there are 2 mol of nitrogen atoms on each side of the equation. Hydrogen appears in H_2O; thus, its coefficient is 2 to balance the 4 mol of hydrogen atoms on the left side. This gives

$$NH_4NO_3 \longrightarrow N_2O + 2\ H_2O$$

Finally, verify that the last element, oxygen, is also balanced by noting that there are 3 mol of oxygen atoms on each side. The coefficients of 1 in front of the NH_4NO_3 and N_2O are omitted by convention.

As a second example, consider the reaction in which butane (C_4H_{10}) is burned in oxygen to form carbon dioxide and water:

$$\_\ C_4H_{10} + \_\ O_2 \longrightarrow \_\ CO_2 + \_\ H_2O$$

Spaces have been left for the coefficients that specify the number of moles of each reactant and product. Begin with 1 mol of butane, C_4H_{10}. It contains 4 mol of carbon atoms and must produce 4 mol of carbon dioxide molecules to conserve the number of carbon atoms in the reaction. Therefore, the coefficient for CO_2 is 4. In the same way, the 10 mol of hydrogen *atoms* must form 5 mol of water *molecules,* because each water molecule contains 2 hydrogen atoms; thus, the coefficient for the H_2O is 5:

$$C_4H_{10} + \_\ O_2 \longrightarrow 4\ CO_2 + 5\ H_2O$$

Four moles of CO_2 contain 8 mol of oxygen atoms, and 5 mol of H_2O contain 5 mol of oxygen atoms, resulting in a total of 13 mol of oxygen atoms. Thirteen moles of oxygen atoms are equivalent to $\frac{13}{2}$ moles of oxygen molecules; therefore, the coefficient for O_2 is $\frac{13}{2}$. The balanced equation is

$$C_4H_{10} + \tfrac{13}{2}\ O_2 \longrightarrow 4\ CO_2 + 5\ H_2O$$

There is nothing wrong with fractions such as $\frac{13}{2}$ in a balanced equation, because fractions of moles are perfectly meaningful. It is often customary, however, to eliminate such fractions. In this case, multiplying all coefficients in the equation by 2 gives

$$2\ C_4H_{10} + 13\ O_2 \longrightarrow 8\ CO_2 + 10\ H_2O$$

A summary of the steps in balancing a chemical equation is as follows:

1. Assign 1 as the coefficient of one species. The best choice is the most complicated species; that is, the species with the largest number of elements.
2. Identify, in sequence, elements that appear in only one chemical species, the coefficient of which has not yet been determined. Choose that coefficient to balance the number of moles of atoms of that element. Continue until all coefficients have been identified.
3. If desired, multiply the whole equation by the smallest integer that will eliminate any fractions.

This method of balancing equations "by inspection" works in many, but not all, cases. Section 11.4 presents techniques for balancing certain more complex chemical equations.

Once the reactants and products are known, balancing chemical equations is a routine, mechanical process of accounting. The difficult part (and the part where chemistry comes in) is to know which substances will react with each other and to determine which products are formed. We return to this question many times throughout this book.

EXAMPLE 2.5

Hargreaves process is an industrial procedure for making sodium sulfate (Na_2SO_4) for use in papermaking. The starting materials are sodium chloride (NaCl), sulfur dioxide (SO_2), water, and oxygen. Hydrogen chloride (HCl) is generated as a by-product. Write a balanced chemical equation for this process.

SOLUTION

The unbalanced equation is

$$\_\ NaCl + \_\ SO_2 + \_\ H_2O + \_\ O_2 \longrightarrow \_\ Na_2SO_4 + \_\ HCl$$

Begin by assigning a coefficient of 1 to Na_2SO_4 because it is the most complex species, composed of 3 different elements. There are 2 mol of sodium atoms on the right; therefore, the coefficient for NaCl must be 2. Following the same argument, the coefficient for SO_2 must be 1 to balance the 1 mol of sulfur on the right. This gives

$$2\ NaCl + SO_2 + \_\ H_2O + \_\ O_2 \longrightarrow Na_2SO_4 + \_\ HCl$$

Next, we note that there are 2 mol of Cl atoms on the left (reactant) side; therefore, the coefficient for HCl must be 2. Hydrogen is the next element to balance, with 2 mol on the right side, and therefore a coefficient of 1 for the H_2O:

$$2\ NaCl + SO_2 + H_2O + \_\ O_2 \longrightarrow Na_2SO_4 + 2\ HCl$$

Finally, the oxygen atoms must be balanced. There are 4 mol of oxygen atoms on the right side, but there are 2 mol from SO_2 and 1 mol from H_2O on the left side; therefore, 1 mol of oxygen *atoms* must come from O_2. Therefore, the coefficient for O_2 is $\frac{1}{2}$:

$$2\ NaCl + SO_2 + H_2O + \tfrac{1}{2}\ O_2 \longrightarrow Na_2SO_4 + 2\ HCl$$

Multiplying all coefficients in the equation by 2 gives

$$4\ NaCl + 2\ SO_2 + 2\ H_2O + O_2 \longrightarrow 2\ Na_2SO_4 + 4\ HCl$$

In balancing this equation, oxygen was considered last because it appears in several places on the left side of the equation.

Related Problems: 31, 32

2.5 Mass Relationships in Chemical Reactions

A balanced chemical equation makes a quantitative statement about the relative masses of the reacting substances. The chemical equation for the combustion of butane,

$$2\ C_4H_{10} + 13\ O_2 \longrightarrow 8\ CO_2 + 10\ H_2O$$

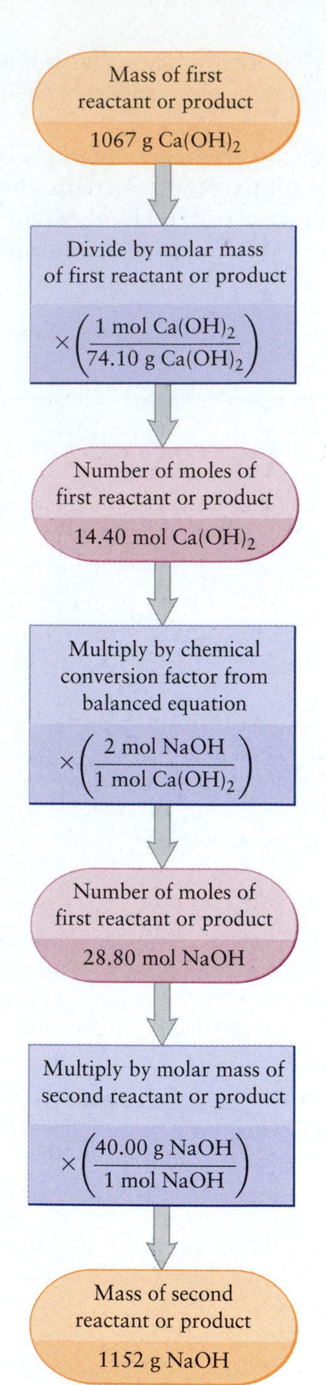

FIGURE 2.4 The steps in a stoichiometric calculation. In a typical calculation, the mass of one reactant or product is known and the masses of one or more other reactants or products are to be calculated using the balanced chemical equation and a table of relative atomic masses.

can be interpreted as either

$$2 \text{ molecules of } C_4H_{10} + 13 \text{ molecules of } O_2 \longrightarrow 8 \text{ molecules of } CO_2 + 10 \text{ molecules of } H_2O$$

or

$$2 \text{ mol of } C_4H_{10} + 13 \text{ mol of } O_2 \longrightarrow 8 \text{ mol of } CO_2 + 10 \text{ mol of } H_2O$$

Multiplying the molar mass of each substance in the reaction by the number of moles represented in the balanced equation gives

$$116.3 \text{ g } C_4H_{10} + 416.0 \text{ g } O_2 \longrightarrow 352.1 \text{ g } CO_2 + 180.2 \text{ g } H_2O$$

The coefficients in a balanced chemical equation give "chemical conversion factors" between the amounts of substances consumed in or produced by a chemical reaction. If 6.16 mol butane reacts according to the preceding equation, the amounts of O_2 consumed and CO_2 generated are

$$\text{mol } O_2 = 6.16 \text{ mol } C_4H_{10} \times \left(\frac{13 \text{ mol } O_2}{2 \text{ mol } C_4H_{10}}\right) = 40.0 \text{ mol } O_2$$

$$\text{mol } CO_2 = 6.16 \text{ mol } C_4H_{10} \times \left(\frac{8 \text{ mol } CO_2}{2 \text{ mol } C_4H_{10}}\right) = 24.6 \text{ } CO_2$$

For most practical purposes we are interested in the *masses* of reactants and products, because those are the quantities that are directly measured. In this case, the molar masses (calculated from a table of atomic masses) are used to convert the number of moles of a substance (in moles) to its mass (in grams), as illustrated by Example 2.6. Sometimes, however, we are also interested in knowing the number of molecules in a sample. The mole allows us to convert easily from mass to numbers of molecules as follows:

$$\textit{mass} \overset{\textit{molar mass}}{\longleftrightarrow} \textit{moles} \overset{N_A}{\longleftrightarrow} \textit{number of molecules}$$

Mass and moles are related by the molar mass; Avogadro's number N_A relates number and moles. You should practice using these relationships to calculate any desired quantity from any given quantity. You can use dimensional analysis to help figure out whether to divide or multiply in any given problem.

EXAMPLE 2.6

Calcium hypochlorite, $Ca(OCl)_2$, is used as a bleaching agent. It is produced from sodium hydroxide, calcium hydroxide, and chlorine according to the following overall equation:

$$2 \text{ NaOH} + Ca(OH)_2 + 2 \text{ } Cl_2 \longrightarrow Ca(OCl)_2 + 2 \text{ NaCl} + 2 \text{ } H_2O$$

How many grams of chlorine and sodium hydroxide react with 1067 g $Ca(OH)_2$, and how many grams of calcium hypochlorite are produced?

SOLUTION

The amount of $Ca(OH)_2$ consumed is

$$\frac{1067 \text{ g Ca } (OH)_2}{74.09 \text{ g mol}^{-1}} = 14.40 \text{ mol } Ca(OH)_2$$

where the molar mass of $Ca(OH)_2$ has been obtained from the molar masses of calcium, oxygen, and hydrogen as

$$40.08 + 2(15.999) + 2(1.0079) = 74.09 \text{ g mol}^{-1}$$

According to the balanced equation, 1 mol $Ca(OH)_2$ reacts with 2 mol NaOH and 2 mol Cl_2 to produce 1 mol $Ca(OCl)_2$. If 14.40 mol of $Ca(OH)_2$ reacts completely, then

$$\text{mol NaOH} = 14.40 \text{ mol Ca(OH)}_2\left(\frac{2 \text{ mol NaOH}}{1 \text{ mol Ca(OH)}_2}\right)$$

$$= 28.80 \text{ mol NaOH}$$

$$\text{mol Cl}_2 = 14.40 \text{ mol Ca(OH)}_2\left(\frac{2 \text{ mol Cl}_2}{1 \text{ mol Ca(OH)}_2}\right)$$

$$= 28.80 \text{ mol Cl}_2$$

$$\text{mol Ca(OCl)}_2 = 14.40 \text{ mol Ca(OH)}_2\left(\frac{1 \text{ mol Ca(OCl)}_2}{1 \text{ mol Ca(OH)}_2}\right)$$

$$= 14.40 \text{ mol Ca(OCl)}_2$$

From the number of moles and molar masses of reactants and products, the following desired masses are found:

$$\text{Mass NaOH reacting} = (28.80 \text{ mol})(40.00 \text{ g mol}^{-1}) = 1152 \text{ g}$$

$$\text{Mass Cl}_2 \text{ reacting} = (28.80 \text{ mol})(70.91 \text{ g mol}^{-1}) = 2042 \text{ g}$$

$$\text{Mass Ca(OCl)}_2 \text{ produced} = (14.40 \text{ mol})(142.98 \text{ g mol}^{-1}) = 2059 \text{ g}$$

Related Problems: 33, 34, 35, 36

In calculations such as the one illustrated in Example 2.6, we are given a known mass of one substance and are asked to calculate the masses of one or more of the other reactants or products. Figure 2.4 summarizes the three-step process used. With experience, it is possible to write down the answers in a shorthand form so that all three conversions are conducted at the same time. The amount of NaOH reacting in the preceding example can be written as

$$\left(\frac{1067 \text{ g Ca(OH)}_2}{74.10 \text{ g mol}^{-1}}\right) \times \left(\frac{2 \text{ mol NaOH}}{1 \text{ mol Ca(OH)}_2}\right) \times 40.00 \text{ g mol}^{-1} = 1152 \text{ g NaOH}$$

At first, however, it is better to follow a stepwise procedure for such calculations.

2.6 Limiting Reactant and Percentage Yield

In the cases we have considered so far, the reactants were present in the exact ratios necessary for them all to be completely consumed in forming products. This is not the usual case, however. It is necessary to have methods for describing cases in which one of the reactants may not be present in sufficient amount and in which conversion to products is less than complete.

Limiting Reactant

Suppose arbitrary amounts of reactants are mixed and allowed to react. The one that is used up first is called the **limiting reactant;** some quantity of the other reactants remains after the reaction has gone to completion. These other reactants are present **in excess.** An increase in the amount of the limiting reactant leads to an increase in the amount of product formed. This is not true of the other reactants. In an industrial process, the limiting reactant is often the most expensive one, to ensure that none of it is wasted. For instance, the silver nitrate used in preparing silver chloride for photographic film by the reaction

$$AgNO_3 + NaCl \longrightarrow AgCl + NaNO_3$$

is far more expensive than the sodium chloride (ordinary salt). Thus, it makes sense to perform the reaction with an excess of sodium chloride to ensure that as much of the silver nitrate as possible reacts to form products.

There is a systematic method to find the limiting reactant and determine the maximum possible amounts of products. Take each reactant in turn, assume that it is used up completely in the reaction, and calculate the mass of one of the products that will be formed. Whichever reactant gives the *smallest* mass of this product is the limiting reactant. Once it has reacted fully, no further product can be formed.

EXAMPLE 2.7

Sulfuric acid (H_2SO_4) forms in the chemical reaction

$$2\ SO_2 + O_2 + 2\ H_2O \longrightarrow 2\ H_2SO_4$$

Suppose 400 g SO_2, 175 g O_2, and 125 g H_2O are mixed and the reaction proceeds until one of the reactants is used up. Which is the limiting reactant? What mass of H_2SO_4 is produced, and what masses of the other reactants remain?

SOLUTION

The number of moles of each reactant originally present is calculated by dividing each mass by the corresponding molar mass:

$$\frac{400 \text{ g } SO_2}{64.06 \text{ g mol}^{-1}} = 6.24 \text{ mol } SO_2$$

$$\frac{175 \text{ g } O_2}{32.00 \text{ g mol}^{-1}} = 5.47 \text{ mol } O_2$$

$$\frac{125 \text{ g } H_2O}{18.02 \text{ g mol}^{-1}} = 6.94 \text{ mol } H_2O$$

If all the SO_2 reacted, it would give

$$6.24 \text{ mol } SO_2 \times \left(\frac{2 \text{ mol } H_3SO_4}{2 \text{ mol } SO_2}\right) = 6.24 \text{ mol } H_2SO_4$$

If all the O_2 reacted, it would give

$$5.47 \text{ mol } O_2 \times \left(\frac{2 \text{ mol } H_2SO_4}{1 \text{ mol } O_2}\right) = 10.94 \text{ mol } H_2SO_4$$

Finally, if all the water reacted, it would give

$$6.94 \text{ mol } H_2O \times \left(\frac{2 \text{ mol } H_2SO_4}{2 \text{ mol } H_2O}\right) = 6.94 \text{ mol } H_2SO_4$$

In this case, SO_2 is the limiting reactant because the computation based on its amount produces the smallest amount of product (6.24 mol H_2SO_4). Oxygen and water are present in excess. After reaction, the amount of each reactant that remains is the original amount minus the amount reacted:

$$\text{mol } O_2 = 5.47 \text{ mol } O_2 - \left(6.24 \text{ mol } SO_2 \times \frac{1 \text{ mol } O_2}{2 \text{ mol } SO_2}\right)$$

$$= 5.47 - 3.12 \text{ mol } O_2 = 2.35 \text{ mol } O_2$$

$$\text{mol } H_2O = 6.94 \text{ mol } H_2O - \left(6.24 \text{ mol } SO_2 \times \frac{2 \text{ mol } H_2O}{2 \text{ mol } SO_2}\right)$$

$$= 6.94 - 6.24 \text{ mol } H_2O = 0.70 \text{ mol } H_2O$$

The masses of reactants and products after the reaction are

$$\text{Mass } H_2SO_4 \text{ produced} = (6.24 \text{ mol})(98.07 \text{ g mol}^{-1}) = 612 \text{ g}$$

$$\text{Mass } O_2 \text{ remaining} = (2.35 \text{ mol})(32.00 \text{ g mol}^{-6}) = 75 \text{ g}$$

$$\text{Mass } H_2O \text{ remaining} = (0.70 \text{ mol})(18.02 \text{ g mol}^{-1}) = 13 \text{ g}$$

The total mass at the end is 612 g + 13 g + 75 g = 700 g, which is, of course, equal to the total mass originally present, 400 g + 175 g + 125 g = 700 g, as required by the law of conservation of mass.

Related Problems: 47, 48

Percentage Yield

The amounts of products calculated so far have been **theoretical yields,** determined by assuming that the reaction goes cleanly and completely. The **actual yield** of a product (that is, the amount present after separating it from other products and reactants and purifying it) is less than the theoretical yield. There are several possible reasons for this. The reaction may stop short of completion, so reactants remain unreacted. There may be competing reactions that give other products, and therefore reduce the yield of the desired one. Finally, in the process of separation and purification, some of the product is invariably lost, although that amount can be reduced by careful experimental techniques. The ratio of the actual yield to the theoretical yield (multiplied by 100%) gives the **percentage yield** for that product in the reaction.

EXAMPLE 2.8

The sulfide ore of zinc (ZnS) is reduced to elemental zinc by "roasting" it (heating it in air) to give ZnO, and then heating the ZnO with carbon monoxide. The two reactions can be written as

$$ZnS + \tfrac{1}{2}O_2 \longrightarrow ZnO + SO_2$$

$$ZnO + CO \longrightarrow Zn + CO_2$$

Suppose 5.32 kg ZnS is treated in this way and 3.30 kg pure Zn is obtained. Calculate the theoretical yield of zinc and its actual percentage yield.

SOLUTION

From the molar mass of ZnS (97.46 g mol^{-1}), the number of moles of ZnS initially present is

$$\frac{5320 \text{ g ZnS}}{97.46 \text{ g mol}^{-1}} = 54.6 \text{ mol ZnS}$$

Because each mole of ZnS gives 1 mol of ZnO in the first chemical equation, and each mole of ZnO then gives 1 mol of Zn, the theoretical yield of zinc is 54.6 mol. In grams, this is

$$54.6 \text{ mol Zn} \times 65.39 \text{ g mol}^{-1} = 3570 \text{ g Zn}$$

The ratio of actual yield to theoretical yield, multiplied by 100%, gives the percentage yield of zinc:

$$\% \text{ yield} = \left(\frac{3.30 \text{ kg}}{3.57 \text{ kg}}\right) \times 100\% = 92.4\%$$

Related Problems: 49, 50

It is clearly desirable to achieve the highest percentage yield of product possible to reduce the consumption of raw materials. In some synthetic reactions (especially in organic chemistry), the final product is the result of many successive reactions. In such processes, the yields in the individual steps must be quite high if the synthetic method is to be a practical success. Suppose, for example, that ten consecutive reactions must be performed to reach the product, and that each has a percentage yield of only 50% (a fractional yield of 0.5). The overall yield is the product of the fractional yields of the steps:

$$\underbrace{(0.5) \times (0.5) \times \ldots \times (0.5)}_{10\ \text{terms}} = (0.5) = (0.5)^{10} = 0.001$$

This overall percentage yield of 0.1% makes the process useless for synthetic purposes. If all the individual percentage yields could be increased to 90%, however, the overall yield would then be $(0.9)^{10} = 0.35$, or 35%. This is a much more reasonable result, and one that might make the process worth considering.

CHAPTER SUMMARY

We have shown you how chemists count molecules by weighing macroscopic quantities of substances. Avogadro's number connects the nanoscopic world of atoms and molecules to the macroscopic scale of the laboratory—1 mol = 6.02×10^{23} atoms or molecules. The relative number of atoms in a molecule or solid is given by its empirical or its molecular formula, and we have shown how these formulas are determined experimentally. The principle of conservation of mass has been sharpened a bit in our discussion of balancing chemical reactions. Not only is the total mass conserved in ordinary (as opposed to nuclear) chemical reactions but the total number of atoms of every element also is conserved. Balancing a chemical reaction requires nothing more than assuring that the same numbers of atoms (or moles of atoms) of each element appear on each side of the balanced equation. Because chemists weigh macroscopic quantities of reactants and products, it is important to understand how mass ratios relate to mole ratios in chemical reactions. Finally, we point out that not every reactant is completely consumed in a chemical reaction, and that the limiting reactant determines the maximum theoretical yield; the actual percentage yield may be somewhat less.

CUMULATIVE EXERCISE

A jet engine fan blade made of a single crystal titanium alloy.

Titanium

Metallic titanium and its alloys (especially those with aluminum and vanadium) combine the advantages of high strength and light weight and are therefore used widely in the aerospace industry for the bodies and engines of airplanes. The major natural source for titanium is the ore rutile, which contains titanium dioxide (TiO_2).

(a) An intermediate in the preparation of elemental titanium from TiO_2 is a volatile chloride of titanium (boiling point 136°C) that contains 25.24% titanium by mass. Determine the empirical formula of this compound.

(b) At 136°C and atmospheric pressure, the density of this gaseous chloride is 5.6 g L^{-1}. Under the same conditions, the density of gaseous nitrogen (N_2, molar mass 28.0 g mol^{-1}) is 0.83 g L^{-1}. Determine the molecular formula of this compound.

(c) The titanium chloride dealt with in parts (a) and (b) is produced by the reaction of chlorine with a hot mixture of titanium dioxide and coke (carbon), with carbon dioxide generated as a by-product. Write a balanced chemical equation for this reaction.

(d) What mass of chlorine is needed to produce 79.2 g of the titanium chloride?

(e) The titanium chloride then reacts with liquid magnesium at 900°C to give titanium and magnesium chloride ($MgCl_2$). Write a balanced chemical equation for this step in the refining of titanium.

(f) Suppose the reaction chamber for part (e) contains 351 g of the titanium chloride and 63.2 g liquid magnesium. Which is the limiting reactant? What maximum mass of titanium could result?

(g) Isotopic analysis of the titanium from a particular ore gave the following results:

Isotope	Relative Mass	Abundance (%)
^{46}Ti	45.952633	7.93
^{47}Ti	46.95176	7.28
^{48}Ti	47.947948	73.94
^{49}Ti	48.947867	5.51
^{50}Ti	49.944789	5.34

Calculate the mass of a single ^{48}Ti atom and the *average* mass of the titanium atoms in this ore sample.

Answers

(a) $TiCl_4$

(b) $TiCl_4$

(c) $TiO_2 + C + 2\ Cl_2 \longrightarrow TiCl_4 + CO_2$

(d) 59.2 g

(e) $TiCl_4 + 2\ Mg \longrightarrow Ti + 2\ MgCl_2$

(f) Mg; 62.3 g

(g) 7.961949×10^{-23} g; 7.950×10^{-23} g

CHAPTER REVIEW

- A mole is Avogadro's number of anything: $N_A = 6.022 \times 10^{23}$. It allows us to count molecules by weighing. N_A is the number of molecules in exactly 12 g ^{12}C, by international agreement.
- The molar mass is the mass of one mole of a substance.
- Mass, moles, and the number of molecules are related as follows:

$$\textit{mass} \overset{\textit{molar mass}}{\longleftrightarrow} \textit{moles} \overset{N_A}{\longleftrightarrow} \textit{number of molecules}$$

- The density of a substance is its mass divided by its volume; in chemistry, the units are generally grams per cubic centimeter (g cm^{-3}). The density of gases is typically 10^{-3} g cm^{-3}, whereas that of liquids and solids is generally in the range 1 to 20 g cm^{-3}.
- The molar volume, V_m, is the volume occupied by one mole of a substance; it is typically about a few tens of cubic centimeters for liquids and solids and about 25 L mol^{-1} for gases at room temperature and atmospheric pressure.

- The empirical formula for a substance is the simplest ratio of the number of atoms or moles of each element. The molecular formula gives the exact number of each atom or moles of atoms in a molecule, whereas the formula unit is the empirical formula for a solid for which no discrete molecules exist.
- Chemical equations are balanced by inspection, ensuring that the same number of atoms of each element appears on both sides of the equation.
- The limiting reagent is the one that is completely consumed first; it determines the degree to which the reaction goes to completion.
- The percentage yield determined by the limiting reagent is the theoretical yield; other losses may occur, resulting in a lower actual yield.

CONCEPTS & SKILLS

After studying this chapter and working the problems that follow, you should be able to:

1. Interconvert mass, number of moles, number of molecules, and (using density) the molar volume of a substance (Section 2.1, Problems 1–12).
2. Given the percentages by mass of the elements in a compound, determine its empirical formula and vice versa (Section 2.3, Problems 13–24).
3. Use ratios of gas densities to estimate molecular mass and determine molecular formulas (Section 2.3, Problems 27–30).
4. Balance simple chemical equations (Section 2.4, Problems 31 and 32).
5. Given the mass of a reactant or product in a chemical reaction, use a balanced chemical equation to calculate the masses of other reactants consumed and other products formed (Section 2.4, Problems 33–46).
6. Given a set of initial masses of reactants and a balanced chemical equation, determine the limiting reactant and calculate the masses of reactants and products after the reaction has gone to completion (Section 2.6, Problems 47 and 48).
7. Determine the percentage yield of a reaction from its calculated theoretical yield and its measured actual yield (Section 2.6, Problems 49 and 50).

PROBLEMS

Answers to problems whose numbers are boldface appear in Appendix G. Problems that are more challenging are indicated with asterisks.

The Mole: Weighing and Counting Molecules

1. Compute the mass (in grams) of a single iodine atom if the relative atomic mass of iodine is 126.90447 on the accepted scale of atomic masses (based on 12 as the relative atomic mass of ^{12}C).

2. Determine the mass (in grams) of exactly 100 million atoms of fluorine if the relative atomic mass of fluorine is 18.998403 on a scale on which exactly 12 is the relative atomic mass of ^{12}C.

3. Compute the relative molecular masses of the following compounds on the ^{12}C scale:
(a) P_4O_{10} (b) BrCl
(c) $Ca(NO_3)_2$ (d) $KMnO_4$
(e) $(NH_4)_2SO_4$

4. Compute the relative molecular masses of the following compounds on the ^{12}C scale:
(a) $[Ag(NH_3)_2]Cl$ (b) $Ca_3[Co(CO_3)_3]_2$
(c) OsO_4 (d) H_2SO_4
(e) $Ca_3Al_2(SiO_4)_3$

5. Suppose that a person counts out gold atoms at the rate of one each second for the entire span of an 80-year life. Has the person counted enough atoms to be detected with an ordinary balance? Explain.

6. A gold atom has a diameter of 2.88×10^{-10} m. Suppose the atoms in 1.00 mol of gold atoms are arranged just touching their neighbors in a single straight line. Determine the length of the line.

7. The vitamin A molecule has the formula $C_{20}H_{30}O$, and a molecule of vitamin A_2 has the formula $C_{20}H_{28}O$. Determine how many moles of vitamin A_2 contain the same number of atoms as 1.000 mol vitamin A.

8. Arrange the following in order of increasing mass: 1.06 mol SF_4; 117 g CH_4; 8.7×10^{23} molecules of Cl_2O_7; and 417×10^{23} atoms of argon (Ar).

9. Mercury is traded by the "flask," a unit that has a mass of 34.5 kg. Determine the volume of a flask of mercury if the density of mercury is 13.6 g cm^{-3}.

10. Gold costs $400 per troy ounce, and 1 troy ounce = 31.1035 g. Determine the cost of 10.0 cm^3 gold if the density of gold is 19.32 g cm^{-3} at room conditions.

11. Aluminum oxide (Al_2O_3) occurs in nature as a mineral called corundum, which is noted for its hardness and resistance to attack by acids. Its density is 3.97 g cm^{-3}. Calculate the number of atoms of aluminum in 15.0 cm^3 corundum.

12. Calculate the number of atoms of silicon (Si) in 415 cm^3 of the colorless gas disilane at 0°C and atmospheric pressure, where its density is 0.00278 g cm^{-3}. The molecular formula of disilane is Si_2H_6.

Chemical Formula and Percentage Composition

13. A newly synthesized compound has the molecular formula $ClF_2O_2PtF_6$. Compute, to four significant figures, the mass percentage of each of the four elements in this compound.

14. Acetaminophen is the generic name of the pain reliever in Tylenol and some other headache remedies. The compound has the molecular formula $C_8H_9NO_2$. Compute, to four significant figures, the mass percentage of each of the four elements in acetaminophen.

15. Arrange the following compounds from left to right in order of increasing percentage by mass of hydrogen: H_2O, $C_{12}H_{26}$, N_4H_6, LiH.

16. Arrange the following compounds from left to right in order of increasing percentage by mass of fluorine: HF, C_6HF_5, BrF, UF_6.

17. "Q-gas" is a mixture of 98.70% helium and 1.30% butane (C_4H_{10}) by mass. It is used as a filling for gas-flow Geiger counters. Compute the mass percentage of hydrogen in Q-gas.

18. A pharmacist prepares an antiulcer medicine by mixing 286 g Na_2CO_3 with water, adding 150 g glycine ($C_2H_5NO_2$), and stirring continuously at 40°C until a firm mass results. The pharmacist heats the mass gently until all the water has been driven away. No other chemical changes occur in this step. Compute the mass percentage of carbon in the resulting white crystalline medicine.

19. Zinc phosphate is used as a dental cement. A 50.00-mg sample is broken down into its constituent elements and gives 16.58 mg oxygen, 8.02 mg phosphorus, and 25.40 mg zinc. Determine the empirical formula of zinc phosphate.

20. Bromoform is 94.85% bromine, 0.40% hydrogen, and 4.75% carbon by mass. Determine its empirical formula.

21. Fulgurites are the products of the melting that occurs when lightning strikes the earth. Microscopic examination of a sand fulgurite shows that it is a globule with variable composition that contains some grains of the definite chemical composition Fe 46.01%, Si 53.99%. Determine the empirical formula of these grains.

22. A sample of a "suboxide" of cesium gives up 1.6907% of its mass as gaseous oxygen when gently heated, leaving pure cesium behind. Determine the empirical formula of this binary compound.

23. Barium and nitrogen form two binary compounds containing 90.745% and 93.634% barium, respectively. Determine the empirical formulas of these two compounds.

24. Carbon and oxygen form no fewer than five different binary compounds. The mass percentages of carbon in the five compounds are as follows: A, 27.29; B, 42.88; C, 50.02; D, 52.97; and E, 65.24. Determine the empirical formulas of the five compounds.

25. A sample of 1.000 g of a compound containing carbon and hydrogen reacts with oxygen at elevated temperature to yield 0.692 g H_2O and 3.381 g CO_2.
   (a) Calculate the masses of C and H in the sample.
   (b) Does the compound contain any other elements?
   (c) What are the mass percentages of C and H in the compound?
   (d) What is the empirical formula of the compound?

26. Burning a compound of calcium, carbon, and nitrogen in oxygen in a combustion train generates calcium oxide (CaO), carbon dioxide (CO_2), nitrogen dioxide (NO_2), and no other substances. A small sample gives 2.389 g CaO, 1.876 g CO_2, and 3.921 g NO_2. Determine the empirical formula of the compound.

27. The empirical formula of a gaseous fluorocarbon is CF_2. At a certain temperature and pressure, a 1-L volume holds 8.93 g of this fluorocarbon, whereas under the same conditions, the 1-L volume holds only 1.70 g gaseous fluorine (F_2). Determine the molecular formula of this compound.

28. At its boiling point (280°C) and at atmospheric pressure, phosphorus has a gas density of 2.7 g L^{-1}. Under the same conditions, nitrogen has a gas density of 0.62 g L^{-1}. How many atoms of phosphorus are there in one phosphorus molecule under these conditions?

29. A gaseous binary compound has a vapor density that is 1.94 times that of oxygen at the same temperature and pressure. When 1.39 g of the gas is burned in an excess of oxygen, 1.21 g water is formed, removing all the hydrogen originally present.
   (a) Estimate the molecular mass of the gaseous compound.
   (b) How many hydrogen atoms are there in a molecule of the compound?
   (c) What is the maximum possible value of the atomic mass of the second element in the compound?
   (d) Are other values possible for the atomic mass of the second element? Use a table of atomic masses to identify the element that best fits the data.
   (e) What is the molecular formula of the compound?

30. A gaseous binary compound has a vapor density that is 2.53 times that of nitrogen at 100°C and atmospheric pressure. When 8.21 g of the gas reacts with $AlCl_3$ at 100°C, 1.62 g gaseous nitrogen is produced, removing all of the nitrogen originally present.
    (a) Estimate the molecular mass of the gaseous compound.
    (b) How many nitrogen atoms are there in a molecule of the compound?
    (c) What is the maximum possible value of the atomic mass of the second element?
    (d) Are other values possible for the atomic mass of the second element? Use a table of atomic masses to identify the element that best fits the data.
    (e) What is the molecular formula of the compound?

Writing Balanced Chemical Equations

31. Balance the following chemical equations:
    (a) $H_2 + N_2 \longrightarrow NH_3$
    (b) $K + O_2 \longrightarrow K_2O_2$
    (c) $PbO_2 + Pb + H_2SO_4 \longrightarrow PbSO_4 + H_2O$
    (d) $BF_3 + H_2O \longrightarrow B_2O_3 + HF$
    (e) $KClO_3 \longrightarrow KCl + O_2$
    (f) $CH_3COOH + O_2 \longrightarrow CO_2 + H_2O$
    (g) $K_2O_2 + H_2O \longrightarrow KOH + O_2$
    (h) $PCl_5 + AsF_3 \longrightarrow PF_5 + AsCl_3$
32. Balance the following chemical equations:
    (a) $Al + HCl \longrightarrow AlCl_3 + H_2$
    (b) $NH_3 + O_2 \longrightarrow NO + H_2O$
    (c) $Fe + O_2 + H_2O \longrightarrow Fe(OH)_2$
    (d) $HSbCl_4 + H_2S \longrightarrow Sb_2S_3 + HCl$
    (e) $Al + Cr_2O_3 \longrightarrow Al_2O_3 + Cr$
    (f) $XeF_4 + H_2O \longrightarrow Xe + O_2 + HF$
    (g) $(NH_4)_2Cr_2O_7 \longrightarrow N_2 + Cr_2O_3 + H_2O$
    (h) $NaBH_4 + H_2O \longrightarrow NaBO_2 + H_2$

Mass Relationships in Chemical Reactions

33. For each of the following chemical reactions, calculate the mass of the underlined reactant that is required to produce 1.000 g of the underlined product.
    (a) $\underline{Mg} + 2\,HCl \longrightarrow \underline{H_2} + MgCl_2$
    (b) $2\,\underline{CuSO_4} + 4\,KI \longrightarrow 2\,CuI + \underline{I_2} + 2\,K_2SO_4$
    (c) $\underline{NaBH_4} + 2\,H_2O \longrightarrow NaBO_2 + \underline{4\,H_2}$
34. For each of the following chemical reactions, calculate the mass of the underlined product that is produced from 1.000 g of the underlined reactant.
    (a) $\underline{CaCO_3} + H_2O \longrightarrow \underline{Ca(OH)_2} + CO_2$
    (b) $\underline{C_3H_8} + 5\,O_2 \longrightarrow \underline{3\,CO_2} + 4\,H_2O$
    (c) $\underline{2\,MgNH_4PO_4} \longrightarrow \underline{Mg_2P_2O_7} + 2\,NH_3 + H_2O$
35. An 18.6-g sample of K_2CO_3 was treated in such a way that all of its carbon was captured in the compound $K_2Zn_3[Fe(CN)_6]_2$. Compute the mass (in grams) of this product.
36. A chemist dissolves 1.406 g pure platinum (Pt) in an excess of a mixture of hydrochloric and nitric acids and then, after a series of subsequent steps involving several other chemicals, isolates a compound of molecular formula $Pt_2C_{10}H_{18}N_2S_2O_6$. Determine the maximum possible yield of this compound.
37. Disilane (Si_2H_6) is a gas that reacts with oxygen to give silica (SiO_2) and water. Calculate the mass of silica that would form if 25.0 cm^3 disilane (with a density of 2.78×10^{-3} g cm^{-3}) reacted with excess oxygen.
38. Tetrasilane (Si_4H_{10}) is a liquid with a density of 0.825 g cm^{-3}. It reacts with oxygen to give silica (SiO_2) and water. Calculate the mass of silica that would form if 25.0 cm^3 tetrasilane reacted completely with excess oxygen.
39. Cryolite (Na_3AlF_6) is used in the production of aluminum from its ores. It is made by the reaction

$$6\,NaOH + Al_2O_3 + 12\,HF \longrightarrow 2\,Na_3AlF_6 + 9\,H_2O$$

    Calculate the mass of cryolite that can be prepared by the complete reaction of 287 g Al_2O_3.
40. Carbon disulfide (CS_2) is a liquid that is used in the production of rayon and cellophane. It is manufactured from methane and elemental sulfur via the reaction

$$CH_4 + 4\,S \longrightarrow CS_2 + 2\,H_2S$$

    Calculate the mass of CS_2 that can be prepared by the complete reaction of 67.2 g sulfur.
41. Potassium nitrate (KNO_3) is used as a fertilizer for certain crops. It is produced through the reaction

$$4\,KCl + 4\,HNO_3 + O_2 \longrightarrow 4\,KNO_3 + 2\,Cl_2 + 2\,H_2O$$

    Calculate the minimum mass of KCl required to produce 567 g KNO_3. What mass of Cl_2 will be generated as well?
42. Elemental phosphorus can be prepared from calcium phosphate via the overall reaction

$$2\,Ca_3(PO_4)_2 + 6\,SiO_2 + 10\,C \longrightarrow 6\,CaSiO_3 + P_4 + 10\,CO$$

    Calculate the minimum mass of $Ca_3(PO_4)_2$ required to produce 69.8 g P_4. What mass of $CaSiO_3$ is generated as a byproduct?
43. An element X has a dibromide with the empirical formula XBr_2 and a dichloride with the empirical formula XCl_2. The dibromide is completely converted to the dichloride when it is heated in a stream of chlorine according to the reaction

$$XBr_2 + Cl_2 \longrightarrow XCl_2 + Br_2$$

    When 1.500 g XBr_2 is treated, 0.890 g XCl_2 results.
    (a) Calculate the atomic mass of the element X.
    (b) By reference to a list of the atomic masses of the elements, identify the element X.

* 44. An element A has a triiodide with the formula AI_3 and a trichloride with the formula ACl_3. The triiodide is quantitatively converted to the trichloride when it is heated in a stream of chlorine, according to the reaction

$$AI_3 + \tfrac{3}{2}\,Cl_2 \longrightarrow ACl_3 + \tfrac{3}{2}\,I_2$$

    If 0.8000 g AI_3 is treated, 0.3776 g ACl_3 is obtained.
    (a) Calculate the atomic mass of the element A.
    (b) Identify the element A.

* 45. A mixture consisting of only sodium chloride (NaCl) and potassium chloride (KCl) weighs 1.0000 g. When the mixture is dissolved in water and an excess of silver nitrate is added, all the chloride ions associated with the original

mixture are precipitated as insoluble silver chloride (AgCl). The mass of the silver chloride is found to be 2.1476 g. Calculate the mass percentages of sodium chloride and potassium chloride in the original mixture.

* **46.** A mixture of aluminum and iron weighing 9.62 g reacts with hydrogen chloride in aqueous solution according to the parallel reactions

$$2\ Al + 6\ HCl \longrightarrow 2\ AlCl_3 + 3\ H_2$$

$$Fe + 2\ HCl \longrightarrow FeCl_2 + H_2$$

A 0.738-g quantity of hydrogen is evolved when the metals react completely. Calculate the mass of iron in the original mixture.

Limiting Reactant and Percentage Yield

47. When ammonia is mixed with hydrogen chloride (HCl), the white solid ammonium chloride (NH_4Cl) is produced. Suppose 10.0 g ammonia is mixed with the same mass of hydrogen chloride. What substances will be present after the reaction has gone to completion, and what will their masses be?

48. The poisonous gas hydrogen cyanide (HCN) is produced by the high-temperature reaction of ammonia with methane (CH_4). Hydrogen is also produced in this reaction.
(a) Write a balanced chemical equation for the reaction that occurs.
(b) Suppose 500.0 g methane is mixed with 200.0 g ammonia. Calculate the masses of the substances present after the reaction is allowed to proceed to completion.

49. The iron oxide Fe_2O_3 reacts with carbon monoxide (CO) to give iron and carbon dioxide:

$$Fe_2O_3 + 3\ CO \longrightarrow 2\ Fe + 3\ CO_2$$

The reaction of 433.2 g Fe_2O_3 with excess CO yields 254.3 g iron. Calculate the theoretical yield of iron (assuming complete reaction) and its percentage yield.

50. Titanium dioxide, TiO_2, reacts with carbon and chlorine to give gaseous $TiCl_4$:

$$TiO_2 + 2\ C + 2\ Cl_2 \longrightarrow TiCl_4 + 2\ CO$$

The reaction of 7.39 kg titanium dioxide with excess C and Cl_2 gives 14.24 kg titanium tetrachloride. Calculate the theoretical yield of $TiCl_4$ (assuming complete reaction) and its percentage yield.

ADDITIONAL PROBLEMS

51. Human parathormone has the impressive molecular formula $C_{691}H_{898}N_{125}O_{164}S_{11}$. Compute the mass percentages of all the elements in this compound.

52. A white oxide of tungsten is 79.2976% tungsten by mass. A blue tungsten oxide also contains exclusively tungsten and oxygen, but it is 80.8473% tungsten by mass. Determine the empirical formulas of white tungsten oxide and blue tungsten oxide.

53. A dark brown binary compound contains oxygen and a metal. It is 13.38% oxygen by mass. Heating it moderately drives off some of the oxygen and gives a red binary compound that is 9.334% oxygen by mass. Strong heating drives off more oxygen and gives still another binary compound, which is only 7.168% oxygen by mass.
(a) Compute the mass of oxygen that is combined with 1.000 g of the metal in each of these three oxides.
(b) Assume that the empirical formula of the first compound is MO_2 (where M represents the metal). Give the empirical formulas of the second and third compounds.
(c) Name the metal.

54. A binary compound of nickel and oxygen contains 78.06% nickel by mass. Is this a stoichiometric or a nonstoichiometric compound? Explain.

55. Two binary oxides of the element manganese contain, respectively, 30.40% and 36.81% oxygen by mass. Calculate the empirical formulas of the two oxides.

* **56.** A sample of a gaseous binary compound of boron and chlorine weighing 2.842 g occupies 0.153 L. This sample is decomposed to give 0.664 g solid boron and enough gaseous chlorine (Cl_2) to occupy 0.688 L at the same temperature and pressure. Determine the molecular formula of the compound.

57. A possible practical way to eliminate oxides of nitrogen (such as NO_2) from automobile exhaust gases uses cyanuric acid, $C_3N_3(OH)_3$. When heated to the relatively low temperature of 625°F, cyanuric acid converts to gaseous isocyanic acid (HNCO). Isocyanic acid reacts with NO_2 in the exhaust to form nitrogen, carbon dioxide, and water, all of which are normal constituents of the air.
(a) Write balanced equations for these two reactions.
(b) If the process described earlier became practical, how much cyanuric acid (in kilograms) would be required to absorb the 1.7×10^{10} kg NO_2 generated annually in auto exhaust in the United States?

58. Aspartame (molecular formula $C_{14}H_{18}N_2O_5$) is a sugar substitute in soft drinks. Under certain conditions, 1 mol of aspartame reacts with 2 mol of water to give 1 mol of aspartic acid (molecular formula $C_4H_7NO_4$), 1 mol of methanol (molecular formula CH_3OH), and 1 mol of phenylalanine. Determine the molecular formula of phenylalanine.

59. 3′-Methylphthalanilic acid is used commercially as a "fruit set" to prevent premature drop of apples, pears, cherries, and peaches from the tree. It is 70.58% carbon, 5.13% hydrogen, 5.49% nitrogen, and 18.80% oxygen. If eaten, the fruit set reacts with water in the body to produce an innocuous product, which contains carbon, hydrogen, and oxygen only, and *m*-toluidine ($NH_2C_6H_4CH_3$), which causes anemia and kidney damage. Compute the mass of the fruit set that would produce 5.23 g *m*-toluidine.

60. Aluminum carbide (Al_4C_3) reacts with water to produce gaseous methane (CH_4). Calculate the mass of methane formed from 63.2 g Al_4C_3.

61. Citric acid ($C_6H_8O_7$) is made by fermentation of sugars such as sucrose ($C_{12}H_{22}O_{11}$) in air. Oxygen is consumed and water generated as a by-product.
(a) Write a balanced equation for the overall reaction that occurs in the manufacture of citric acid from sucrose.
(b) What mass of citric acid is made from 15.0 kg sucrose?

62. A sample that contains only $SrCO_3$ and $BaCO_3$ weighs 0.800 g. When it is dissolved in excess acid, 0.211 g carbon dioxide is liberated. What percentage of $SrCO_3$ did the sample contain? Assume all the carbon originally present is converted to carbon dioxide.

63. A sample of a substance with the empirical formula XBr_2 weighs 0.5000 g. When it is dissolved in water and all its bromine is converted to insoluble AgBr by addition of an excess of silver nitrate, the mass of the resulting AgBr is found to be 1.0198 g. The chemical reaction is

$$XBr_2 + 2\ AgNO_3 \longrightarrow 2\ AgBr + X(NO_3)_2$$

(a) Calculate the molecular mass (that is, formula mass) of XBr_2.
(b) Calculate the atomic mass of X and give its name and symbol.

64. A newspaper article about the danger of global warming from the accumulation of greenhouse gases such as carbon dioxide states that "reducing driving your car by 20 miles a week would prevent release of over 1000 pounds of CO_2 per year into the atmosphere." Is this a reasonable statement? Assume that gasoline is octane (molecular formula C_8H_{18}) and that it is burned completely to CO_2 and H_2O in the engine of your car. Facts (or reasonable guesses) about your car's gas mileage, the density of octane, and other factors will also be needed.

65. In the Solvay process for producing sodium carbonate (Na_2CO_3), the following reactions occur in sequence:

$$NH_3 + CO_2 + H_2O \longrightarrow NH_4HCO_3$$

$$NH_4HCO_3 + NaCl \longrightarrow NaHCO_3 + NH_4Cl$$

$$2\ NaHCO_3 \xrightarrow{\text{heat}} Na_2CO_3 + H_2O + CO_2$$

How many metric tons of sodium carbonate would be produced per metric ton of NH_3 if the process were 100% efficient (1 metric ton = 1000 kg)?

66. A yield of 3.00 g $KClO_4$ is obtained from the (unbalanced) reaction

$$KClO_3 \longrightarrow KClO_4 + KCl$$

when 4.00 g of the reactant is used. What is the percentage yield of the reaction?

67. An industrial-scale process for making acetylene consists of the following sequence of operations:

$$\underset{\text{limestone}}{CaCO_3} \longrightarrow \underset{\text{lime}}{CaO} + \underset{\text{carbon dioxide}}{CO_2}$$

$$CaO + 3\ C \longrightarrow \underset{\text{calcium carbide}}{CaC_2} + \underset{\text{carbon monoxide}}{CO}$$

$$CaC_2 + 2\ H_2O \longrightarrow \underset{\text{calcium hydroxide}}{Ca(OH)_2} + \underset{\text{acetylene}}{C_2H_2}$$

What is the percentage yield of the overall process if 2.32 metric tons C_2H_2 is produced from 10.0 metric tons limestone (1 metric ton = 1000 kg)?

68. Silicon nitride (Si_3N_4), a valuable ceramic, is made by the direct combination of silicon and nitrogen at high temperature. How much silicon must react with excess nitrogen to prepare 125 g silicon nitride if the yield of the reaction is 95.0%?

UNIT II

Chemical Bonding and Molecular Structure

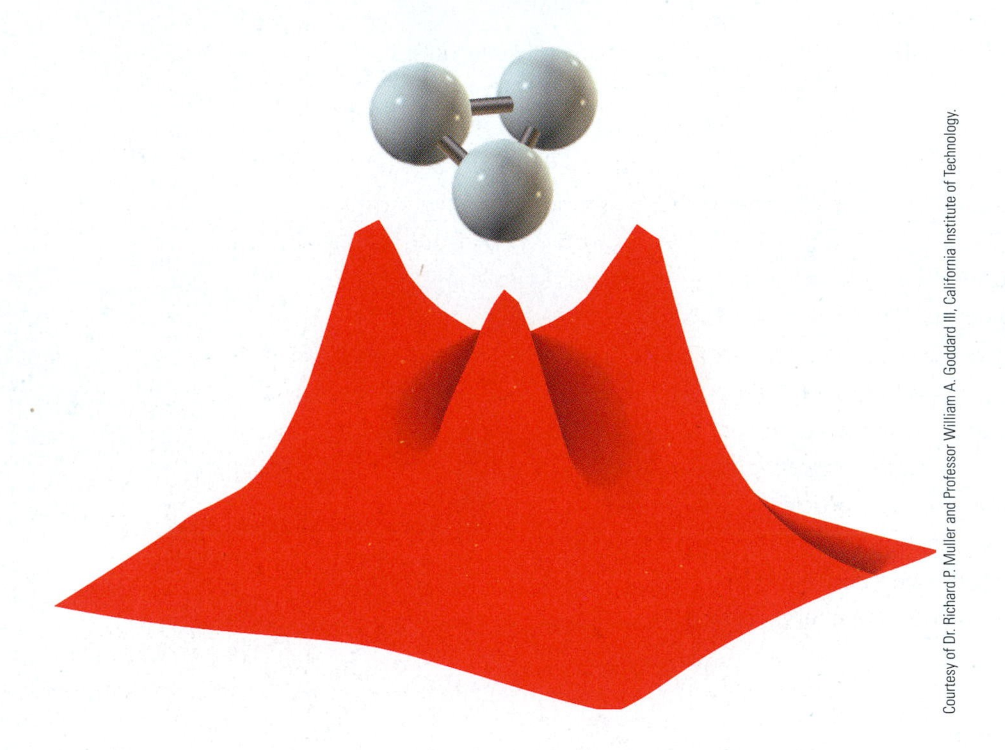

Courtesy of Dr. Richard P. Muller and Professor William A. Goddard III, California Institute of Technology.

The electron density in a delocalized three-center bond for H_3^+ calculated by quantum mechanics.

How do atoms bond together to form molecules, three-dimensional objects with unique structures? Rutherford's planetary model of atomic structure described in Section 1.4 (Z electrons moving around a dense nucleus of charge $+Ze$) suggests that the chemical bond involves the gain, loss, or sharing of electrons by atoms. This idea leads to the classical description of bonding, used daily by chemists worldwide. This classical description is neither quantitative nor complete. The existence of the chemical bond was fully explained only when a new theory called quantum mechanics replaced Newtonian mechanics for describing the nanoscopic world of molecules, atoms, and fundamental particles. Quantum mechanics explains the chemical bond as the distribution of electron density around the nuclei for which the energy of the molecule is lower than the energy of the separated atoms.

UNIT CHAPTERS

UNIT GOALS

- To introduce the classical description of chemical bonding as a tool for describing and understanding the structures and shapes of molecules
- To convey the basic concepts and methods of quantum mechanics that describe the discrete energy levels and the statistical behavior of microscopic systems
- To develop an intuition for the behavior of quantum systems and an appreciation for the magnitudes of the relevant physical quantities
- To use quantum mechanics to:
  - Describe the allowed energies and electron densities in atoms
  - Explain the structure of the periodic table and periodic trends in the properties of atoms
  - Describe covalent bond formation and the structures of diatomic and small polyatomic molecules
  - Describe covalent bond formation and the structures of organic molecules
  - Describe bonding in more complex structures that include transition metal complexes

CHAPTER 3

Chemical Bonding: The Classical Description

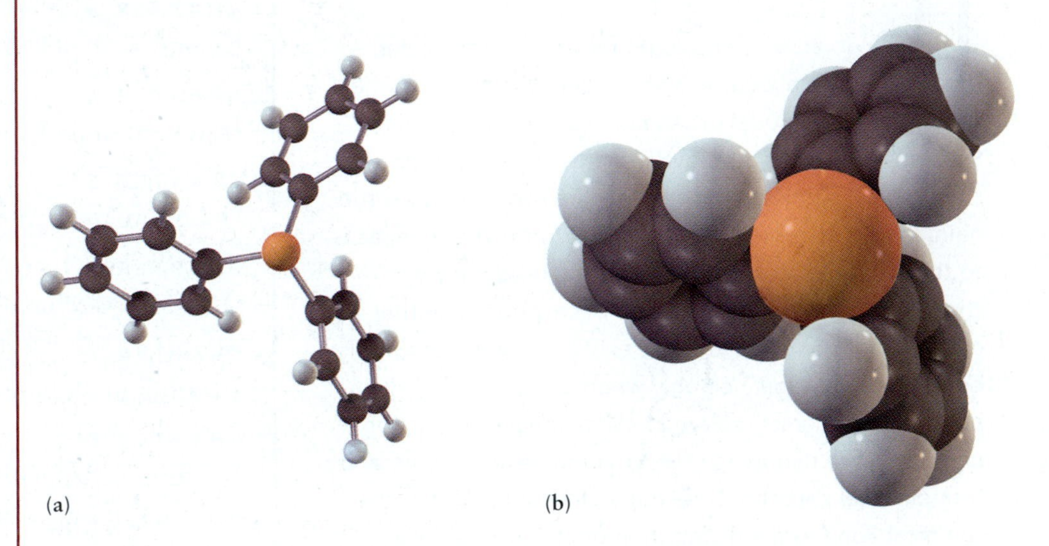

(a) The shape of the molecule triphenyl-phosphine, $(C_6H_5)_3P$, is determined by locating the valence shell electron pairs in those positions that minimize the overall energy of the molecule. (b) The space-filling representation aids the analysis and understanding of the steric environment responsible for the specific molecular geometry observed.

The previous chapters showed how the laws of conservation of mass and conservation of atomic identity, together with the concept of the mole, determine quantitative mass relationships in chemical reactions. That discussion assumed prior knowledge of the chemical formulas of the reactants and products in each equation. The far more open-ended questions of which compounds are found in nature (or which can be made in the laboratory) and what types of reactions they undergo now arise. Why are some elements and compounds violently reactive and others inert? Why are there compounds with chemical formulas H_2O and NaCl, but never H_3O or $NaCl_2$? Why are helium and the other noble gases monatomic, but molecules of hydrogen and chlorine diatomic? All of these questions can be answered by examining the formation of chemical bonds between atoms.

When two atoms come sufficiently close together, the electrons of each atom experience the additional attractive force of the other nucleus, the electrons repel each other, and the positively charged nuclei experience a mutually repulsive force. A stable chemical bond between two atoms in the gas phase is formed only when the total energy of the resulting molecule is lower than that of the two isolated atoms. A quantitative description of chemical bonding depends on the detailed arrangement of electrons in each atom and requires quantum mechanics for its explanation. **Quantum mechanics**—the fundamental branch of physics that describes the properties, interactions, and motions of atomic and subatomic particles—was established in 1926, and scientists immediately sought to explain chemical bond formation using its principles. More than 50 years of intense research were required to develop a comprehensive and useful quantum explanation of chemical bonding.

During that period, however, chemists developed a powerful suite of concepts and tools—covalent bonds, ionic bonds, polar covalent bonds, electronegativity, Lewis electron dot diagrams, and valence shell electron-pair repulsion (VSEPR) theory—that rationalized a great deal of information about the structure of molecules and patterns of chemical reactivity. This suite constitutes the *classical description of the chemical bond,* and it is part of the daily vocabulary of every working chemist, especially in organic and biological chemistry. These tools are the foundation of chemical intuition, by which we mean the ability to explain and even predict chemical phenomena. Intuition is judgment informed by experience. To guide the development of your own chemical intuition using these tools, we provide you with a comprehensive discussion of their conceptual basis and give you extensive practice in applying them to interpret factual information.

Today, all the tools of classical bonding theory have been explained by quantum mechanics. It is largely a matter of taste whether you first learn the classical theory and then gain deeper insight from the quantum explanations, or you first learn the quantum theory and then see the classical theory as a limiting case. We prefer to present the classical description first, and this chapter is devoted to that subject. That way, we establish the language and vocabulary of the chemical bond and allow you to become familiar with the properties of a broad array of real molecules *before* attempting to explain these results using quantum mechanics in Chapters 4, 5, and 6. Your instructor may prefer the opposite sequence, in which case you will read Chapters 4 and 5 before this chapter. This book was written to accommodate either approach.

The classical theory of chemical bonding and molecular shapes starts with conceptual models of the chemical bond. Much of structure and bonding can be understood on the basis of simple electrostatics. The distributions of electrons in atoms, molecules, and solids determine nearly all of their physical and chemical properties, with structure and reactivity being those of greatest interest to the chemist. Chemical bonds form by sharing or transferring electrons between atoms. The degree to which electrons are shared or transferred varies considerably in different chemical bonds, but chemists generally identify two extreme cases. In a **covalent** bond, the electrons are shared more or less equally between the two atoms comprising the bond. In an **ionic** bond, one or more electrons is essentially completely transferred from one atom to the other, and the dominant contribution to the strength of the bond is the electrostatic interaction between the resulting ions. Although many real chemical bonds are well described by these idealized models, most bonds are neither completely ionic nor completely covalent and are best described as having a mixture of ionic and covalent character. In **polar covalent** bonds, a partial transfer of charge from one atom to the other occurs. **Electronegativity,** the tendency of an atom in a molecule to draw electrons toward itself, explains whether a given pair of atoms form ionic, covalent, or polar covalent bonds.

Two simple tools, developed in the first half of the 20th century, are used to implement the classical theory of bonding and structure. The first tool is the

Lewis electron dot diagram, a schematic that shows the number of **valence** (outermost) electrons associated with each atom and whether they are bonding (shared) or nonbonding. These diagrams are useful in predicting connectivity—that is, which atoms are bonded to each other in polyatomic molecules. They do not, however, provide information about the three-dimensional shapes of molecules. The second tool, the **VSEPR theory,** is a simple, yet powerful, method to predict molecular shapes, based on the simple electrostatic argument that electron pairs in a molecule will arrange themselves to be as far apart as possible. These two physical tools allow for the organization of vast amounts of chemical information in a rational and systematic way.

We begin by describing the **periodic table,** a list of the elements arranged to display at a glance patterns of their physical properties and chemical reactivity. Relating bond formation to the positions of atoms in the periodic table reveals trends that build up chemical intuition. Next, we invoke Rutherford's planetary model of the atom and show how electrical forces control the gain or loss of electrons by the atom. We then examine the electrical forces within molecules and show how they lead to the ionic and covalent models of the chemical bond. The use of Lewis diagrams to describe bond formation and the VSEPR theory to describe molecular shapes completes the classical theory of bonding. We conclude with a brief survey of the procedures for assigning proper names to chemical compounds.

We urge you to keep Rutherford's planetary model of the atom (Section 1.4) in mind while reading this chapter. That model, with its consideration of the electrical forces within atoms and molecules, provides the foundation of the entire subject of chemical bonding and molecular structure.

3.1 The Periodic Table

The number of known chemical compounds is already huge and it continues to increase rapidly as the result of significant investments in chemical research. An unlimited number of chemical reactions is available among these compounds. The resultant body of chemical knowledge, viewed as a collection of facts, is overwhelming in its size, range, and complexity. It has been made manageable by the observation that the properties of the elements naturally display certain regularities. These regularities enable the classification of the elements into families whose members have similar chemical and physical properties. When the elements are arranged in order of increasing atomic number, Z, remarkable patterns emerge. Families of elements with similar chemical properties are easily identified by their locations in this arrangement. This discovery is summarized concisely by **the periodic law:**

> *The chemical properties of the elements are periodic functions of the atomic number Z.*

Consequently, the elements listed in order of increasing Z can be arranged in a chart called the periodic table, which displays, at a glance, the patterns of chemical similarity. The periodic table then permits systematic classification, interpretation, and prediction of all chemical information.

The modern periodic table (Fig. 3.1 and the inside front cover of this book) places elements in **groups** (arranged vertically) and **periods** (arranged horizontally).

There are eight groups of **representative elements,** or "main-group" elements. In addition to the representative elements, there are ten groups (and three periods) of **transition-metal elements,** a period of elements with atomic numbers 57 through 71 called the rare-earth or **lanthanide elements,** and a period of elements from atomic numbers 89 through 103 called the **actinides,** all of which are unstable and

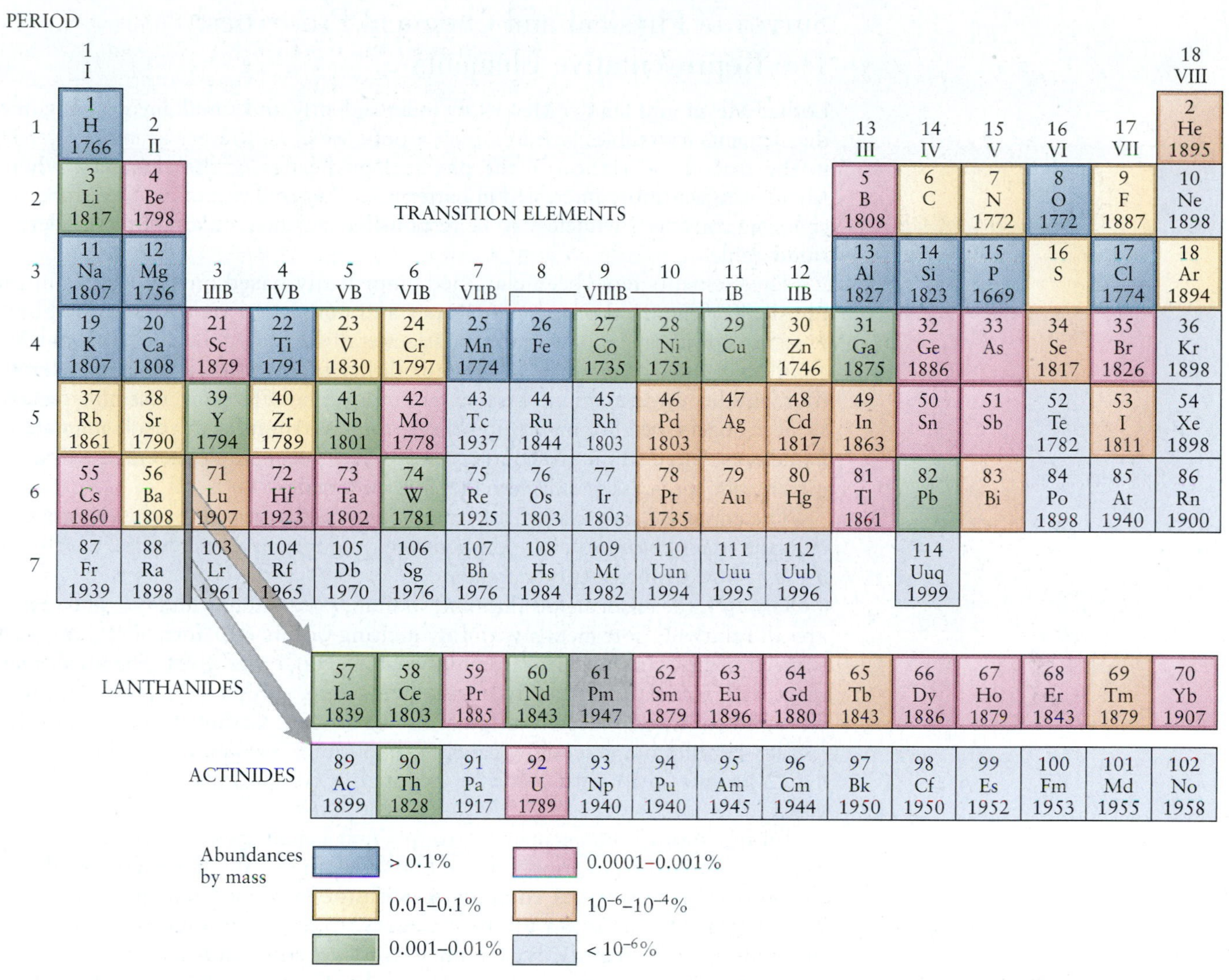

FIGURE 3.1 The modern periodic table of the elements. Below each symbol is the year in which that element was discovered; elements with no dates have been known since ancient times. Above each symbol is the atomic number. The color coding indicates the relative abundance by mass of the elements in the world (the atmosphere, oceans and fresh water bodies, and the Earth's crust to a depth of 40 km). Oxygen alone comprises almost 50% of the mass, and silicon comprises more than 25%.

most of which must be produced artificially. The lanthanide and actinide elements are usually placed below the rest of the table to conserve space. The groups of representative elements are numbered (using Roman numerals) from I to VIII, with the letter *A* sometimes added to differentiate them from the transition-metal groups, which are labeled from IB to VIIIB. This book uses group numbers exclusively for the representative elements (dropping the *A*) and refers to the transition-metal elements by the first element in the corresponding group. For example, the elements in the carbon group (C, Si, Ge, Sn, Pb) are designated as Group IV, and the elements chromium (Cr), molybdenum (Mo), and tungsten (W) as the chromium group.[1]

[1]Recently, several international organizations recommended a new system of group designation in which the main-group elements make up Groups 1, 2, and 13 through 18, and the transition-metal elements fill Groups 3 through 12.

Survey of Physical and Chemical Properties: The Representative Elements

Lothar Meyer and Dmitri Mendeleev independently and simultaneously organized the elements into tables based on their atomic weights. Meyer was more interested in the periodic variation in the physical properties of the elements, whereas Mendeleev was more interested in patterns of chemical reactivity. Therefore, most chemists consider Mendeleev to be responsible for the creation of the modern periodic table.

The elements have been classified empirically based on similarities in their physical or chemical properties. **Metals** and **nonmetals** are distinguished by the presence (or absence) of a characteristic metallic luster, good (or poor) ability to conduct electricity and heat, and malleability (or brittleness). Certain elements (boron, silicon, germanium, arsenic, antimony, and tellurium) resemble metals in some respects and nonmetals in others, and are therefore called **semimetals** or **metalloids.** Their ability to conduct electricity, for example, is much worse than metals, but is not essentially zero like the nonmetals.

The empirical formulas of the binary compounds formed by the elements with chlorine (their *chlorides*), with oxygen (their *oxides*), and with hydrogen (their *hydrides*) show distinct periodic trends.

Group I, the **alkali metals** (lithium, sodium, potassium, rubidium, and cesium), are all relatively soft metals with low melting points that form 1:1 compounds with chlorine, with chemical formulas such as NaCl and RbCl. The alkali metals react with water to liberate hydrogen; potassium, rubidium, and cesium liberate enough heat upon reaction to ignite the hydrogen. Group II, the **alkaline-earth metals** (beryllium, magnesium, calcium, strontium, barium, and radium), react in a 1:2 atomic ratio with chlorine, producing compounds such as $MgCl_2$ and $CaCl_2$.

Of the nonmetallic elements, Group VI, the **chalcogens** (oxygen, sulfur, selenium, and tellurium), forms 1:1 compounds with the alkaline-earth metals (such as CaO and BaS) but 2:1 compounds with the alkali metals (such as Li_2O and Na_2S). Members of Group VII, the **halogens** (fluorine, chlorine, bromine, and iodine), differ significantly in their physical properties (fluorine and chlorine are gases at room temperature, bromine is a liquid, and iodine a solid), but their chemical properties are similar. Any alkali metal will combine with any halogen in 1:1 proportion to form a compound such as LiF or RbI, which is called an **alkali halide.**

The remaining elements fall into three additional groups whose chemical and physical properties are somewhat less clearly delineated than those already discussed. Group III includes a semimetal (boron) and four metals (aluminum, gallium, indium, and thallium). All metals form 1:3 chlorides (such as $GaCl_3$) and 2:3 oxides (such as Al_2O_3). Group IV comprises the elements carbon, silicon, germanium, tin, and lead. All of these elements form 1:4 chlorides (such as $SiCl_4$), 1:4 hydrides (such as GeH_4), and 1:2 oxides (such as SnO_2). Tin and lead are metals with low melting points, and silicon and germanium are **semiconductors.** Although we classified silicon and germanium as semimetals earlier, their electrical properties can be finely tuned by incorporating small amounts of impurities. These two elements form the basis for the modern semiconductor industry, which manufactures computer chips and other solid-state devices. Several different allotropes of elemental carbon exist (for example, graphite, diamond, and the recently discovered fullerenes). **Allotropes** are modifications of an element with differing atomic arrangements that lead to different physical and chemical properties. For example, ozone (O_3) and ordinary diatomic oxygen (O_2) are also allotropes. Group V includes nitrogen, phosphorus, arsenic, antimony, and bismuth. These elements form binary compounds with hydrogen and oxygen that have empirical formulas such as PH_3 and N_2O_5. The hydrides become increasingly unstable as their molar masses increase, and BiH_3 is stable only below

−45°C. A similar trend exists for the oxides, and Bi_2O_5 has never been obtained in pure form. The lighter members of this group are clearly nonmetals (nitrogen and phosphorus), bismuth is clearly a metal, and arsenic and antimony are classified as semimetals.

Group VIII, the **noble gases** (helium, neon, argon, krypton, xenon, and radon), are sometimes called the **inert gases** because of their relative inertness toward chemical combination. They are all monatomic, in contrast with the other elements that exist as gases at room temperature and atmospheric pressure (hydrogen, oxygen, nitrogen, fluorine, chlorine), which are diatomic molecules.

Systematic trends in both the physical and chemical properties of the elements give important clues as to the structure of the atom. In addition to the properties that distinguish metals from nonmetals (electrical and thermal conductivity, malleability, luster, and ductility), there are a number of other physical properties that show clear periodic trends; these properties include melting and boiling points, densities, atomic sizes, and the energy changes that occur when an electron is added to or removed from a neutral atom. Numerical values for most of these properties are tabulated in Appendix F. In general, the elements on the left side of the table (especially in the later periods) are metallic solids and good conductors of electricity. On the right side (especially in earlier periods), they are poor electrical conductors of electricity and are generally gases at room temperature and atmospheric pressure. In between, the semimetals separate the metals from the nonmetals by a diagonal zigzag line (see the inside front cover of this book).

Patterns in chemical reactivity of the elements correlate with patterns in the physical structure of the atom; they are both periodic functions of Z. Reading across the periodic table (horizontally) shows that each main-group element (Groups I–VIII) in Period 3 has exactly 8 more electrons than the element immediately above it in Period 2. Similarly, each main-group element in Periods 4 and 5 has exactly 18 more electrons than the corresponding element in the period above. The sequence of numbers, 8, 8, 18, 18, and so forth, that organize the periodic table into groups (columns), whose elements have similar physical and chemical properties, arises from the quantum theory of atomic structure (see discussion in Chapter 5).

3.2 Forces and Potential Energy in Atoms

The atom arose in the domain of chemistry, its existence inferred indirectly from the laws of chemical combination. With the work of Thomson and Rutherford, the atom also became the province of physics, which sought to explain its structure and behavior as consequences of the electrical forces between the electrons and the nucleus. Modern chemistry combines these themes and uses the forces inside the atom to explain chemical behavior. The purpose of this section is to give you, a student of chemistry, a good appreciation for the nature of these forces, and the potential energy associated with them, in preparation for your studies of chemical bond formation. It is essential that you understand and learn to use potential energy diagrams for atoms. We suggest you review the background material on force, work, potential energy, potential energy diagrams, and electricity and magnetism in Appendix B2.

Rutherford's planetary model of the atom assumes that an atom of atomic number Z comprises a dense, central nucleus of positive charge $+Ze$ surrounded by a total of Z electrons moving around the nucleus. The attractive forces between each electron and the nucleus, and the repulsive forces between the electrons, are described by Coulomb's law. We first discuss Coulomb's law in general terms, and then apply it to the planetary atom.

According to Coulomb's law, the force of interaction between two charges, q_1 and q_2, separated by a distance, r, is

$$F(r) = \frac{q_1 q_2}{4\pi\epsilon_0 r^2} \qquad [3.1]$$

where ϵ_0, called the *permittivity of the vacuum,* is a proportionality constant with a numerical value of $8.854 \times 10^{-12}\ \mathrm{C^2\ J^{-1}\ m^{-1}}$.

In the International System of Units (SI), charge is expressed in coulombs (C), distance in meters (m), and force in newtons (N). In Equation 3.1 and related equations, the symbol q for each charge represents both the magnitude and the sign of the charge. The position of one of the particles is chosen as the origin of coordinates, and the displacement, r, runs outward to locate the second particle. Throughout physics, the sign convention for the direction of the force is chosen to be positive if the force is in the same direction as the displacement, and negative if the force is opposite to the displacement. If the particles have the same charge, their mutual repulsion pushes them apart in the same direction as r increases; thus, the repulsive force between them is assigned the positive sign. If they have opposite charges, their mutual attraction pulls them together in the direction *opposite* to the direction in which r increases. Thus, the attractive force between them is given the negative sign.

To determine how a particle responds to a force exerted on it, we normally solve Newton's second law, $F = ma$, to predict the new location of the particle after the force has been applied. Another approach, which is often easier, is to examine the *potential energy function* associated with the force. For example, if you compress a spring and hold it in position, you know it has the capability to push back on your hand as soon as you release it. The potential energy stored in the compressed spring measures how much force the spring can exert when it is released. We determine the amount of potential energy by measuring or calculating the amount of work done against the spring to compress it from its relaxed length to the particular length of interest.

The potential energy can be calculated from Coulomb's force law, and the result is

$$V(r) = \frac{q_1 q_2}{4\pi\epsilon_0 r} \qquad [3.2]$$

for the potential energy of two charges, q_1 and q_2, separated by a distance r. In SI units, energy is expressed in joules (J), charge in coulombs (C), and distance in meters (m), as noted earlier. By convention, $V(r) \longrightarrow 0$ as $r \longrightarrow \infty$. This is a logical choice for the zero of potential energy because there is no interaction between the particles at such large distances. It is also consistent with our definition that the change in potential energy is the work done on or by the two charges. The sign of the potential energy, like the force, depends on the signs of the charges. If the charges have the same sign, the potential energy, as expressed by Equation 3.2, is positive. This makes physical sense, because to increase the potential energy, we would have to push q_2 in from infinity against the repulsive force exerted by q_1. Work done *against* the force is positive, so the results agree with our physical intuition. If the charges have opposite signs, the expression in Equation 3.2 is negative. Again, the results agree with our physical intuition. When we start with q_2 at infinity, we have to rush madly after it and hold it back as the attractive force tries to pull it right up to q_1. This time, work is done *by* the force and is negative in sign.

Let's apply these insights to the planetary atom. Associated with each electron (of charge $-e$) and the nucleus (of charge $+Ze$) there is potential energy:

$$V(r) = -\frac{Ze^2}{4\pi\epsilon_0 r} \tag{3.3}$$

The minus sign indicates an attractive interaction; the potential energy becomes lower as the particles get closer together. We will see in Chapter 5 that the typical proton-electron distance in a hydrogen atom is about 10^{-10} m. This is an extremely small distance, and it appears throughout atomic and molecular physics. To avoid the inconvenience of always expressing powers of ten, this length has been given the special name angstrom (1 Å = 10^{-10} m). The potential energy of the hydrogen atom when the proton and electron are separated by a distance of 1 Å is

$$V(1\text{ Å}) = -\frac{(1.602 \times 10^{-19}\text{ C})^2}{4\pi(8.854 \times 10^{-12}\text{C}^2\text{ J}^{-1}\text{ m}^{-1})(1 \times 10^{-10}\text{ m})}$$

$$= -\frac{(8.988 \times 10^{9})(1.602 \times 10^{-19})^2}{(1 \times 10^{-10})}\text{ J}$$

$$V(1\text{ Å}) = -2.307 \times 10^{-18}\text{ J} \tag{3.4}$$

On the scale of ordinary human experience, this is an extremely small amount of energy. In comparison, one food calorie equals 4.184×10^3 J, so the amount of energy in one hydrogen atom is very small indeed. Energy values in this range appear regularly in atomic and molecular physics because of the small electrical charges involved, so it is appropriate to define a special energy unit for these applications. An electron accelerated through a potential difference of 1 V gains kinetic energy in the amount $\mathcal{T}$ = eV = $(1.60217646 \times 10^{-19}\text{ C})(1\text{ V}) = 1.60217646 \times 10^{-19}$ J. Therefore, it is convenient to define a unit of energy called the **electron volt (eV),** such that 1 eV = $1.60217646 \times 10^{-19}$ J.

Thus, the potential energy between the proton and electron separated by 1 Å is

$$V(1\text{ Å}) = -\frac{2.307 \times 10^{-18}\text{ J}}{1.602 \times 10^{-19}\text{ J(eV)}^{-1}} = -14.40\text{ eV} \tag{3.5}$$

Figure 3.2 plots the potential energy (in eV) versus distance (in Å) for proton–electron, proton–proton, electron–lithium nucleus, and helium nucleus–gold

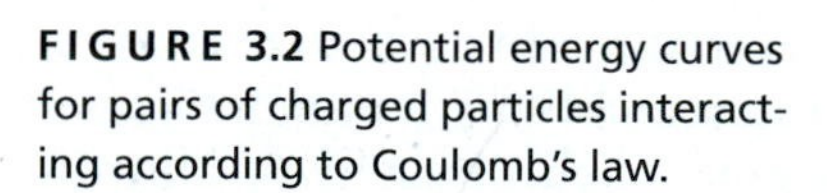

FIGURE 3.2 Potential energy curves for pairs of charged particles interacting according to Coulomb's law.

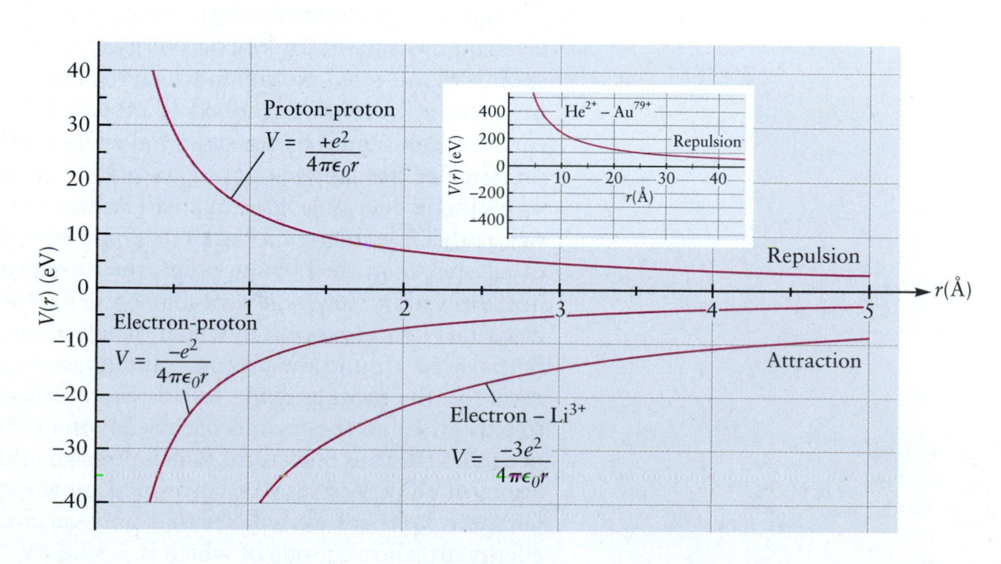

nucleus interactions. (The last pair was studied experimentally in Rutherford's experiment described in Section 1.4.) These plots summarize a great deal of physical information, and you should become skilled at interpreting them. The potential energy scale is defined to have the value zero when the particles are infinitely far apart and do not interact. At shorter separations, the sign of the potential energy depends on the signs of the charges, as explained earlier (see Fig. 3.2).

Once the potential energy curve is known, we can use it to predict the motions of the particles. The force between a pair of particles is the negative of the derivative of their potential energy (see Appendix B2). Therefore, regions in which the slope of the potential energy is *negative* are regions in which the force on the particle is *positive*. The particles will be pushed apart in these regions. Wherever the slope of the potential energy is *positive*, the force in that region is *negative*, and the particles will be attracted to one another. Let's illustrate this conclusion for the proton–electron interaction, for which the slope of the potential energy curve is positive everywhere:

$$F_{\text{coul}} = -\frac{d}{dr}\left(-\frac{Ze^2}{4\pi\epsilon_0 r}\right) = \frac{d}{dr}\left(\frac{Ze^2}{4\pi\epsilon_0 r}\right) = -\frac{Ze^2}{4\pi\epsilon_0 r^2} \qquad [3.6]$$

Equation 3.6 shows that the force is always attractive (as indicated by the negative sign) and decreases with increasing r. You should run through a similar analysis for each of the curves shown in Figure 3.2 to be sure that you understand how to interpret these diagrams. We make extensive use of potential energy curves to predict the motions of particles in many areas of chemistry discussed throughout this book.

Let's examine the motion of the electron in the hydrogen atom, which has only one proton and one electron. The total energy (kinetic and potential) of the electron in the atom is

$$E = \frac{1}{2}m_e v^2 - \frac{Ze^2}{4\pi\epsilon_0 r} \qquad [3.7]$$

Suppose the atom has a fixed total energy E. It is informative to represent this situation by a straight line on the potential energy curve and show the kinetic energy value $\mathcal{T}$ at each point as a vertical arrow connecting V to E. (See Appendix B2 for background.)

Figure 3.3a shows the total energy set to +10.0 eV. The graph shows that the electron has significant kinetic energy everywhere, so the case of positive E corresponds to *unbound motion* in which the electron approaches the proton but passes by without becoming trapped or attached. Imagine a helicopter flying above a canyon represented by the potential energy well; its motion is not affected by the presence of the canyon at all. Figure 3.3b shows the total energy set to −13.6 eV. The kinetic energy is large at small values of r and decreases to zero at the point where this line intersects V (the *turning point* of the electron's motion). For values of r larger than the turning point, the kinetic energy would be negative, which is not allowed in Newtonian mechanics. Therefore, the case of negative total energy describes *bound motion* in which the electron is said to be "trapped within a potential well around the proton," and its motions are limited to the range between zero and the turning point. In this case, we can imagine the helicopter being free to explore all of the regions of the canyon at the altitude of the red line until it approaches the canyon wall, at which point the pilot will turn around and head back! Some of these points will be refined by the quantum mechanical treatment of Chapter 5, where we will see that only certain specific values of the bound state energy are allowed, one of which is −13.6 eV.

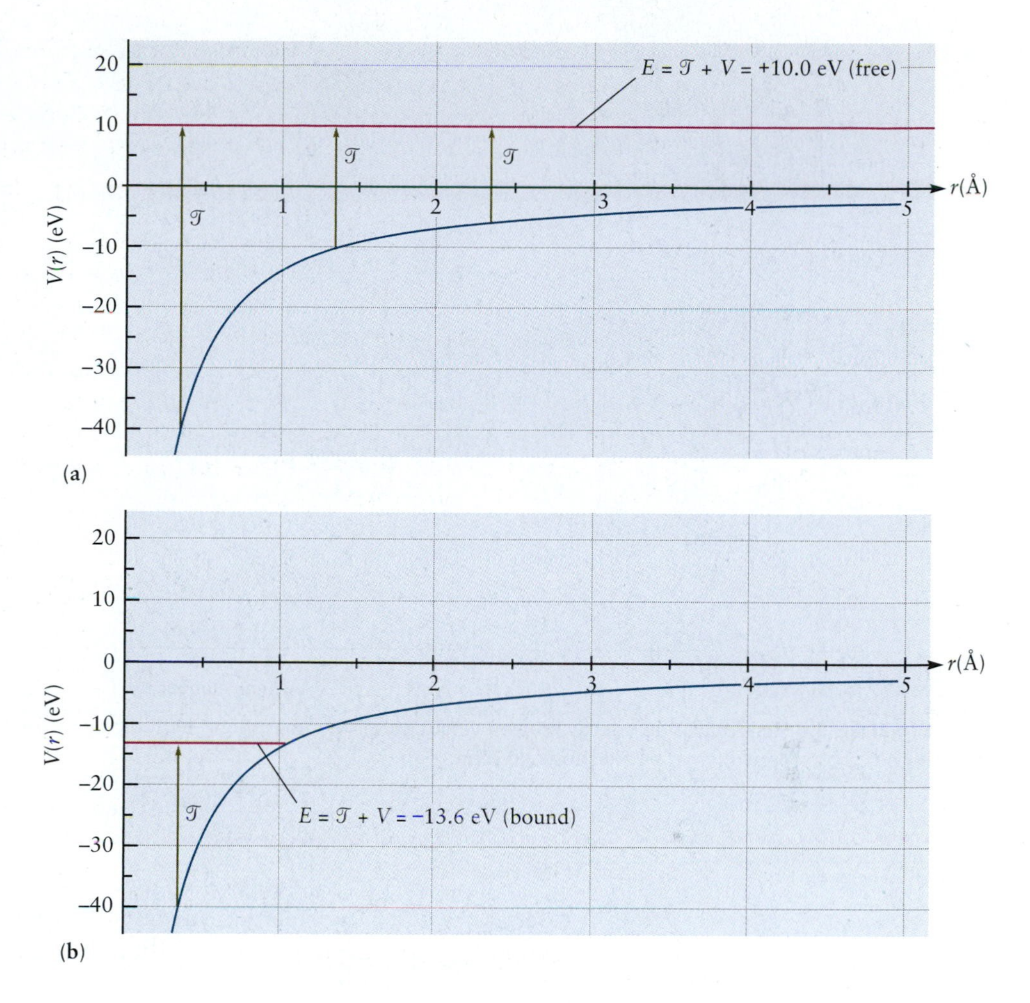

FIGURE 3.3 Potential energy, total energy, and kinetic energy for interaction of an electron with a proton. When the total energy is fixed, the kinetic energy at each point is represented by a vertical arrow from the potential energy curve to the value of the total energy. (a) Total energy $E > 0$ corresponds to unbound motion, characterized by significant kinetic energy at all positions. (b) Total energy $E < 0$ corresponds to bound motion where the electron is confined to distances smaller than the "turning point," at which the potential and total energy are equal, and the kinetic energy is 0.

The key point to keep in mind here is that the electron is bound within the atom whenever the total energy is less than zero. To remove the electron from the atom, it is necessary to add enough energy to make the total energy greater than zero. Thus, for the bound state of the hydrogen atom mentioned earlier, adding 13.6 eV of energy will set the electron free, but just barely. Adding 23.6 eV of energy to the atom would enable the electron to depart the atom with 10.0 eV of kinetic energy.

3.3 Ionization Energies and the Shell Model of the Atom

Electron distributions change during the course of all chemical reactions. The simplest possible chemical reactions are those in which an electron is either removed from or added to a neutral atom to form a positively charged **cation** or negatively charged **anion**, respectively. Although these might be considered to be physical processes, the reactants and products in both cases have different chemical properties, so these are clearly chemical changes. This section focuses on the process that creates positively charged ions, and Section 3.4 discusses the complementary process. The energy changes associated with each of these processes show clear periodic trends that correlate with the trends in chemical reactivity discussed in

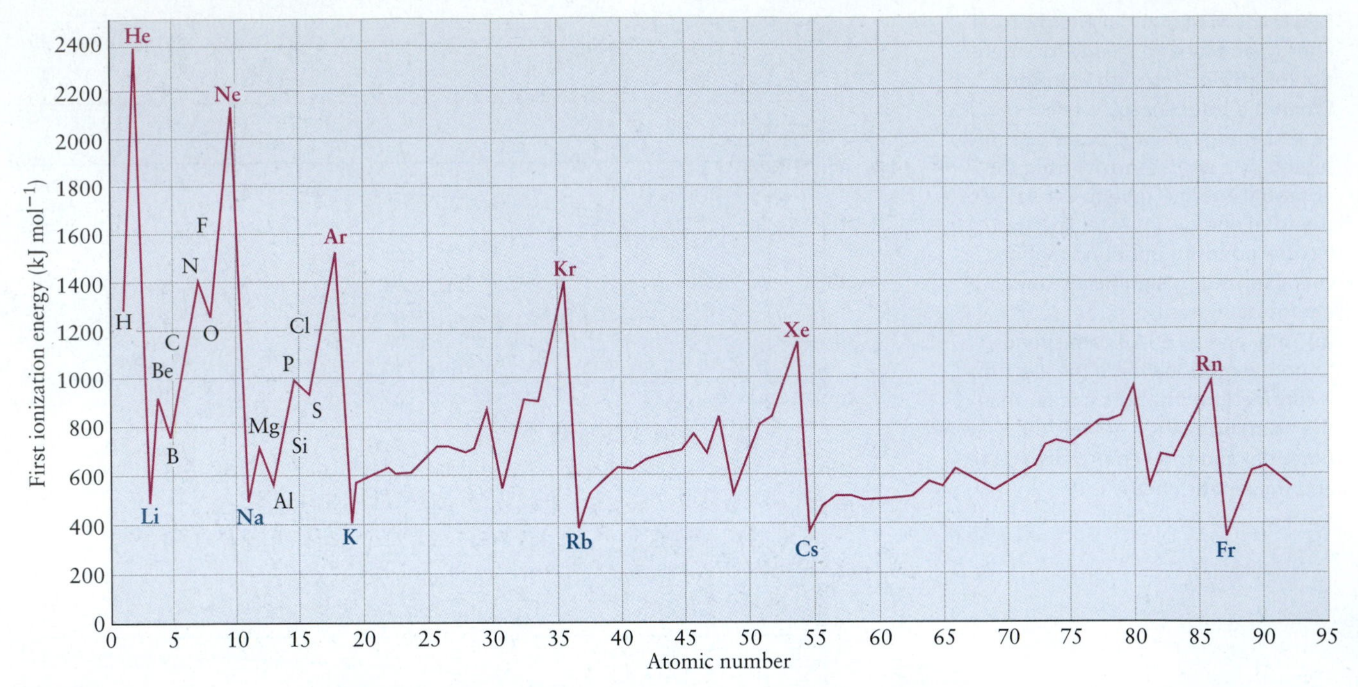

FIGURE 3.4 First ionization energy plotted versus atomic number shows periodic behavior. Symbols for the noble gases are shown in red; those for alkali metals are shown in blue.

Section 3.1. This correlation suggests that a qualitative explanation of chemical bonding may begin by understanding the factors that control the loss or gain of electrons by atoms.

The **ionization energy,** IE_1, of an atom (also referred to as the first ionization energy, or in some texts, the ionization potential) is the minimum energy necessary to detach an electron from the neutral gaseous atom and form a positively charged gaseous ion. It is the change in energy, ΔE, for the process

$$X(g) \longrightarrow X^+(g) + e^- \qquad \Delta E = IE_1$$

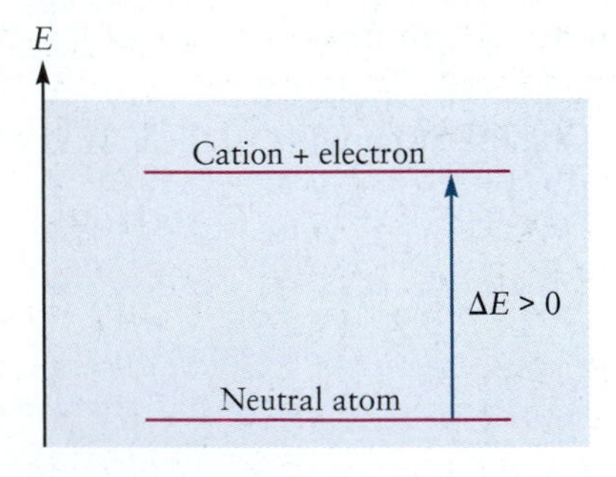

Ionization requires sufficient energy to enable the electron to escape from the potential energy well of the atom.

The Greek letter capital delta, Δ, is widely used to symbolize the difference in value of a property caused by a process or change carried out in the laboratory. Here, ΔE = [energy of reaction products] − [energy of reactants]. Thus, ΔE is positive when energy must be provided for the process to occur, and ΔE is negative if the process liberates energy. To achieve the ionization of X(g) to the products $X^+(g) + e^-$ in the previous reaction, it is necessary to add energy to the neutral atom X(g) to liberate the electron. This energy enables the electron to escape from the potential energy well that holds it in the atom. Therefore, the energy of the final state [free electron and ion $X^+(g)$] is greater than that of the initial state (neutral atom).

ΔE for ionization reactions is always positive. The ionization energy is a measure of the stability of the free atom. Those atoms with larger ionization energies are more stable than those with smaller ionization energies because their electrons must be removed from deeper potential energy wells.

Figure 3.4 shows the measured ionization energies of the elements plotted against their atomic numbers. Note that these ionization energies are reported in kJ per *mole* of atoms; following the discussion of Section 3.2, 1 eV per *atom* equals 96.48 kJ per mole.

The values generally increase moving across a period (from left to right), becoming large for each noble gas atom, and then fall abruptly for the alkali atom at the beginning of the next period. The large values for the noble gas atoms demonstrate that their electron configurations are extremely stable, and that considerable energy is required to liberate their electrons. Moreover, the electron configurations of the noble gas atoms are more stable than those of the atoms immediately before and after them in the periodic table.

Ionization energy is thus a periodic property of elemental atoms (see Section 3.1). The general trend of this periodicity in IE_1 correlates with the fact that each main-group element (Groups I–VIII) in Period 3 (through element 20, Ca) contains exactly 8 more electrons than the element immediately above it, and each main-group element in Periods 4 and 5 (from element 31, Ga, through element 56, Ba) contains exactly 18 more electrons than the element immediately above it. The small local increases and decreases observed across a period will be explained in detail by the quantum description of atomic structure in Chapter 5. Our primary objective here is to show that removing the first electron, as a simple model chemical reaction, is periodic in the atomic number Z. For this purpose, it is not necessary to consider the small local variations.

The *second* ionization energy, IE_2, is the minimum energy required to remove a second electron, or ΔE for the process

$$X^+(g) \longrightarrow X^{2+}(g) + e^- \qquad \Delta E = IE_2$$

The third, fourth, and higher ionization energies are defined in analogous fashion. Successive ionization energies always increase due to the greater electrostatic attraction of the electron to the product ions, which have increasingly greater positive charges.

Examination of successive ionization energies suggests that the electrons in an atom are organized in a very interesting structure. This pattern is revealed in Table 3.1, which shows the first ten ionization energies for the elements H through Ar. Note that in Table 3.1 the values of ionization energy are expressed in MJ mol^{-1}, rather than kJ mol^{-1}, as in Figure 3.4, to make it easier to display them in tabular form.

Let's first consider He. IE_1 for He is 2.37, which is much greater than that of H (1.31) or Li (0.52). The electronic structure of He is thus much more stable than that of either H or Li. Further disruption of the stable He structure by removing a second electron requires $IE_2 = 5.25$.

Next, let's consider Li. IE_1 for Li is 0.52, whereas IE_2 is 7.30, far greater than the difference between IE_1 and IE_2 for He. Note that the difference $IE_3 - IE_2$ for Li is comparable with the difference $IE_2 - IE_1$ for He. These results show that one electron is removed easily from Li to form Li^+, which has two electrons and is much more stable than the He atom.

As we proceed across Period 2, an interesting pattern develops. The ionization energies for Be show a large jump between IE_2 and IE_3, demonstrating that it easily loses two electrons to form Be^{2+}, which has two electrons and heliumlike stability. Boron ionization energies display a large jump between IE_3 and IE_4, showing that three electrons are easily removed, and C has a large jump between IE_4 and IE_5, showing that four electrons are easily removed. The pattern continues, showing that F has seven electrons that can be removed more easily than the last two, whereas Ne has eight. This pattern is shown in Table 3.1 through highlighting of the ionization energies for the more easily removed electrons in each atom. These results suggest that electrons in the atoms of Period 2 are arranged as a stable heliumlike *inner core,* surrounded by less tightly bound electrons. The number of less tightly bound electrons increases from one to eight as the atomic number increases from three to ten.

Examination of the atoms in Period 3, Na through Ar, reveals a similar pattern of relatively more easily removed electrons outside a stable core, which resembles the Ne atom. This is shown in Table 3.1 through highlighting of the ionization

TABLE 3.1 Successive Ionization Energies of the Elements Hydrogen through Argon

		Ionization Energy (MJ mol^{-1})									
Z	Element	IE_1	IE_2	IE_3	IE_4	IE_5	IE_6	IE_7	IE_8	IE_9	IE_{10}
1	H	1.31									
2	He	2.37	5.25								
3	Li	0.52	7.30	11.81							
4	Be	0.90	1.76	14.85	21.01						
5	B	0.80	2.42	3.66	25.02	32.82					
6	C	1.09	2.35	4.62	6.22	37.83	47.28				
7	N	1.40	2.86	4.58	7.48	9.44	53.27	64.36			
8	O	1.31	3.39	5.30	7.47	10.98	13.33	71.33	84.08		
9	F	1.68	3.37	6.05	8.41	11.02	15.16	17.87	92.04	106.43	
10	Ne	2.08	3.95	6.12	9.37	12.18	15.24	20.00	23.07	115.38	131.43
11	Na	0.50	4.56	6.91	9.54	13.35	16.61	20.11	25.49	28.93	141.37
12	Mg	0.74	1.45	7.73	10.54	13.62	17.99	21.70	25.66	31.64	35.46
13	Al	0.58	1.82	2.74	11.58	14.83	18.38	23.30	27.46	31.86	38.46
14	Si	0.79	1.58	3.23	4.36	16.09	19.78	23.79	29.25	33.87	38.73
15	P	1.06	1.90	2.91	4.96	6.27	21.27	25.40	29.85	35.87	40.96
16	S	1.00	2.25	3.36	4.56	7.01	8.49	27.11	31.67	36.58	43.14
17	Cl	1.26	2.30	3.82	5.16	6.54	9.36	11.02	33.60	38.60	43.96
18	Ar	1.52	2.67	3.93	5.77	7.24	8.78	11.99	13.84	40.76	46.19
19	K	0.42	3.05	4.40	5.87	7.96	9.63	11.32	...	...	...
20	Ca	0.59	1.14	4.90	6.46	8.13	10.48	12.30	...	...	...
21	Sc	0.63	1.23	2.38	7.08	8.82	10.70	13.29	...	...	...

energy values for the more easily removed electrons in each atom. (The ionization energy values highlighted show the beginning of this pattern for Period 4.) Na appears to have a single weakly bound electron outside a neonlike core. Further insight into this arrangement is obtained by plotting the successive ionization energies of Na versus *n*, the total number of electrons that have been removed at each step. It is convenient to plot the logarithm of ionization energy versus *n* to compress the vertical scale. The result for Na (Fig. 3.5) suggests that the electrons of the Na atom are arranged in three *shells*. The first electron is easily removed to produce Na^+, which has neonlike stability as indicated by the large jump between IE_1 and IE_2. Electrons 2 through 9 occupy the second shell, and all of them are more tightly bound than the first electron. A big jump between IE_9 and IE_{10} suggests that the last two electrons occupy a third shell, the electrons of which are the most tightly bound of all.

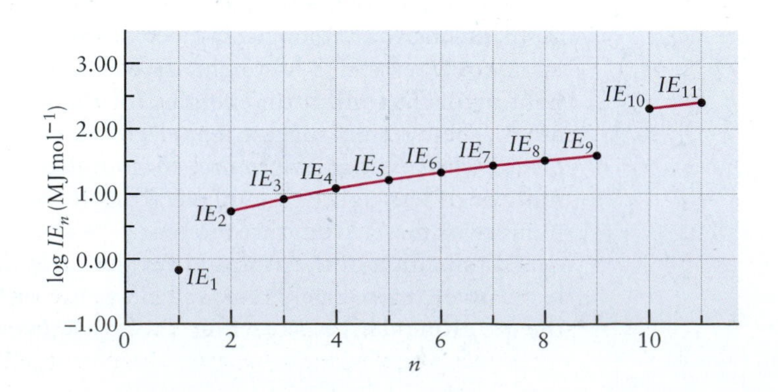

FIGURE 3.5 Logarithm of successive ionization energies for Na versus number of electrons removed suggests a three-shell electronic structure.

The Shell Model of the Atom

Examining the ionization process as a prototype simple chemical reaction leads us to conclude that electrons occupy a set of shells that surround the nucleus. This is a remarkable experimental result. All of the electrons in an atom are identical, and they all interact with the same nucleus. Why should they be arranged in shells, and what determines the number of electrons that can occupy a given shell?

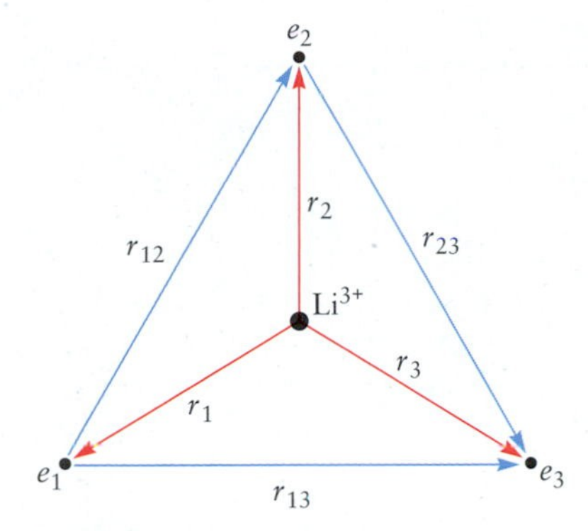

Li atom with three electrons.

We begin by considering the forces that act within and the potential energy functions for many-electron atoms. Consider Li, for which $Z = 3$.

Each of the electrons is attracted to the nucleus and repelled by the other electrons via the Coulomb interaction. The electrons are located relative to the nucleus by coordinates r_1, r_2, r_3, and the distances between the pairs of electrons are given by $r_{12} = r_1 - r_2$, $r_{13} = r_1 - r_3$ and $r_{23} = r_2 - r_3$. The potential energy is then given by

$$V = \frac{Ze^2}{4\pi\epsilon_0}\left(-\frac{1}{r_1} - \frac{1}{r_2} - \frac{1}{r_3} + \frac{1}{r_{12}} + \frac{1}{r_{13}} + \frac{1}{r_{23}}\right) \qquad [3.8]$$

We can calculate the potential energy for any configuration of the atom, for example, with the electrons at the vertices of an equilateral triangle with specific side lengths and the nucleus at the center. But unlike the case of the hydrogen atom, there is no simple way to relate this potential energy to the bound motion of the electrons in the atom and to describe how each electron is held inside the atom. There is no way to describe exactly the motions of more than two interacting particles, either in Newtonian mechanics or quantum mechanics. So we develop an approximate description in which we examine the behavior of one electron at the time. We imagine that (for example) electron 1 is fixed at r_1, whereas the other electrons move all around the atom. Their motion, in effect, averages all the terms in the potential energy except the first. The combination of this average and the first term gives an *effective potential energy* that governs the motion of electron 1. One way to generate this effective potential is to recognize that when the other electrons come between electron 1 and the nucleus, they *screen* or *shield* electron 1 from the full strength of the nuclear attraction. It is useful to think of the Coulomb interaction as strictly "line of sight," so intrusion by another electron will reduce its strength. In effect, the charge of the nucleus, as seen by electron 1, has been reduced by the presence of the other electrons. We can then take the effective potential energy to be the Coulomb potential energy with an *effective charge* Z_{eff} on the nucleus:

$$V_{eff}(r) = -\frac{Z_{eff}e^2}{4\pi\epsilon_0 r} \qquad [3.9]$$

Our approximate description of many-electron atoms thus relies on considering each electron to be bound within the atom by an effective potential well, where an appropriate value for Z_{eff} must be determined for each electron. In Chapter 5 we will see how to generate values for Z_{eff} systematically, and that they range from 1 to the full value of the nonscreened Z for the atom. Here, it is sufficient to get a sense of how much V_{eff} changes within a given atom. Consider Na, for which $Z = 11$. Figure 3.6 shows plots of V_{eff} curves for $Z_{eff} = 1, 5, 11$. Clearly, those electrons that experience the lower values of Z_{eff} are more weakly bound than those with higher values of Z_{eff}.

Our simple model for V_{eff} shows that the trend in successive ionization energies in a single atom arises naturally from screening and shielding. The key idea here is that the higher the value of Z_{eff}, the more energy is required to remove an electron from the atom. In Chapter 5 we combine this simple physical model with principles of quantum mechanics to understand why the electrons are organized in

FIGURE 3.6 Curves for the effective potential energy $V_{eff}(r)$ for electrons in Na ($Z = 11$) when $Z_{eff} = 1, 5, 11$. An electron at any location is more strongly bound in the atom as the value of Z_{eff} increases.

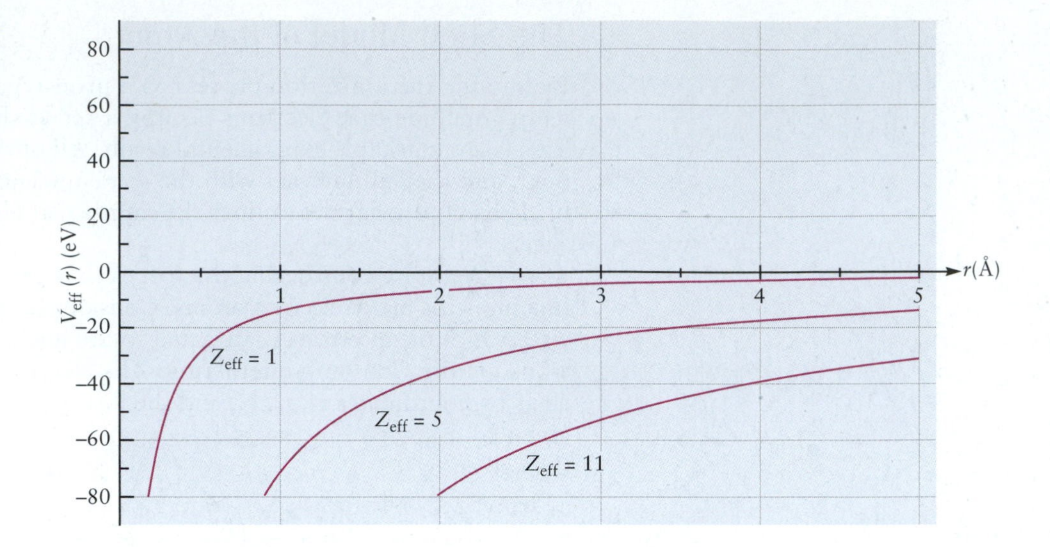

shells of 2, 8, and 18 electrons. Here, we consider the shell model to be justified by studies of successive ionization energies, and we use it to explain periodic trends in bond formation.

Electrons in the inner shells (called **core** electrons) do not participate significantly in chemical reactions. The outermost, partially filled shell (called the **valence shell**) contains the electrons involved in chemical bonding, the **valence electrons.** Progressing through the elements in order of increasing atomic number along a period, we see that stability increases from left to right, as indicated by increasing values of the ionization potential. The period ends in a filled shell, which is the stable configuration of the noble gases helium, neon, argon, and so on. Atoms with filled shells are extremely stable chemically, as shown by the large values of their ionization energies. The increase in stability observed when moving from left to right along a period (row) is easily explained classically. If the electrons in each shell are located at roughly the same distance from the nucleus, then the attractive electrostatic forces increase nearly monotonically as one unit of positive charge is added to the nucleus and one valence electron is added. The large decrease in ionization energy that occurs between a noble gas and the element whose valence electron occupies a new shell also is explained easily. If we envision these shells as a concentric series, each of which has a fixed radius that is larger than its predecessor, then it is clear why the electron-nuclear attraction decreases abruptly as an electron is added to the first empty shell outside a filled shell. This simple classical argument explains the major periodic trends in ionization energies; the small dips observed in going from left to right across a period require quantum mechanics for their explanation.

The number of valence electrons in a neutral atom of a main-group element (those in Groups I–VIII) of the second and third periods is equal to the group number of the element in the periodic table. However, the main-group elements that follow a series of transition-metal elements require some special attention. Atoms of bromine, for example, have 17 more electrons than atoms of argon, the preceding noble gas, but only 7 are considered to be valence electrons. This is true for two reasons. First, in the fourth, fifth, and sixth rows, the 10 electrons added to complete the transition metal series (although they are important for the bonding of those elements) have become *core* electrons by the time the end of the transition-metal series is reached. They are closer to the nucleus, on average, than the electrons that fill the rest of the shells of those periods, and it might be useful to visualize them as occupying a subshell. Second, and more importantly, the chemical properties of the main group elements in this part of the periodic table are characteristic of the group to which they belong. The bonding properties of an element such as bromine, for example, resemble those of the lighter elements in its group.

3.4 Electronegativity: The Tendency of Atoms to Attract Electrons

The type of bond formed between a pair of atoms is determined by the degree to which electrons are attracted to each nucleus. The ability of a free, isolated atom to lose an electron is measured by its ionization energy, whereas its ability to gain an electron is measured by its electron affinity, *EA*. The relative tendency of atoms to either gain or lose electrons is expressed succinctly by a new quantity called electronegativity. The most intuitively appealing definition of electronegativity for isolated atoms is simply the average of the ionization energy and the electron affinity. Atoms that are hard to ionize and have large electron affinities will tend to attract electrons; those that are easily ionized and have low electron affinities will tend to donate electrons. A comparison of the electronegativities of two atoms will suggest whether they will most likely form an ionic, covalent, or polar covalent bond.

This section defines electron affinity and two different electronegativity scales. Although the two electronegativity scales were developed using different physical models, they are essentially proportional to one another. Electronegativity, not surprisingly, is also a periodic property, and much can be learned about the nature of a particular bond simply by comparing the locations of its constituent elements in the periodic table. The remaining sections of this chapter apply this concept to describe systemically ionic, covalent, and polar covalent bonds.

Electron Affinity

Ionization energy, which measures the difficulty with which an atom gives up an electron to form a cation, is defined in Section 3.3. The energy change of the opposite reaction, in which an atom accepts an extra electron to form an anion, is the electron affinity of the atom.

An anion is formed by the electron attachment reaction,

$$X(g) + e^- \longrightarrow X^-(g),$$

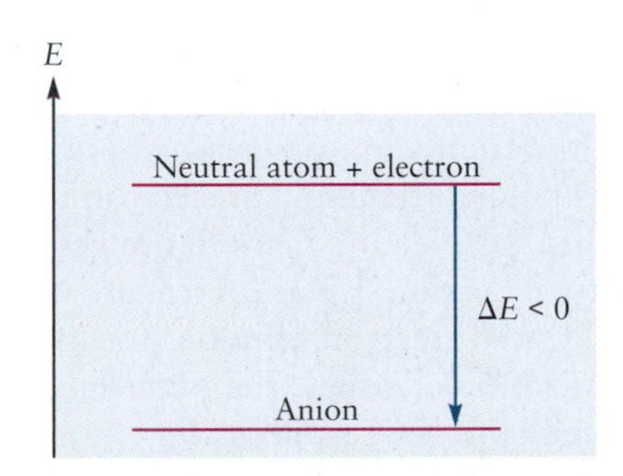

When an electron attaches to an atom to form a stable anion, the electron becomes trapped in the potential well of the atom, the energy of the products is lower than the energy of the reactants, and $\Delta E < 0$.

for which the energy change ΔE (see definition in Section 3.3) is called the **electron attachment energy.** This energy change is readily shown on a potential energy diagram where $X(g)$ and the electron are initially separated by large distances and do not interact, then approach to form the anion $X^-(g)$, whose energy is lower than that of the separated particles.

Because the energy of the products is lower than the energy of the reactants, ΔE is negative, energy is released in the reaction, and the anion is stable. This means that the neutral atom can accommodate an extra electron to form the anion in which the electron is strongly bound by the effective potential V_{eff}. The energy change, ΔE, for the reverse reaction, in which the electron is removed from $X^-(g)$ to give the neutral atom $X(g)$, is positive because energy must be supplied to overcome V_{eff} in the anion.

For historical reasons, **electron affinity** has been defined as the amount of energy *released* when an electron is attached to a neutral atom, but it is always expressed as a positive number. This is a time-honored, if frustrating, exception to the otherwise universal convention adopted by chemists and physicists that energy liberated in a process is assigned a negative number, an exception that must simply be remembered.

It is difficult to measure the electron attachment energy directly, and to obtain the electron affinity from $EA = -\Delta E$ (electron attachment). It is easier to obtain *EA* from another measurement that determines the stability of the gaseous anion in the same way that ionization energies are measured. The reaction

$$Cl^-(g) \longrightarrow Cl(g) + e^-$$

TABLE 3.2 Electron Affinity of Selected Atoms (in kJ mol^{-1})

H 73						
Li 60	Be *	B 27	C 122	N *	O 141	F 328
Na 53	Mg *	Al 42	Si 134	P 72	S 200	Cl 349
K 48	Ca 2	Ga 41	Ge 119	As 79	Se 195	Br 325
Rb 47	Sr 5	In 29	Sn 107	Sb 101	Te 190	I 295
Cs 46	Ba 14	Tl 19	Pb 35	Bi 91	Po 183	At 270

*No stable anion A^- exists for this element in the gas phase.

which is the reverse of the electron attachment reaction, has positive ΔE because energy must be supplied to overcome V_{eff} in the anion. By measuring the energy required to remove the electron from the ion, we can obtain the electron affinity for the atom as $EA = \Delta E$ (electron detachment). Removal of the electron from Cl^- requires $\Delta E = +349$ kJ mol^{-1} of energy, which shows that the Cl^- anion is stable. Thus, EA for Cl, defined as the energy released in the reaction

$$Cl(g) + e^- \longrightarrow Cl^-(g)$$

is 349 kJ mol^{-1}. We do not explicitly include the plus sign when using electron affinities to help you remember that they are always positive by convention. Values of electron affinity for selected elements are shown in Table 3.2. We choose not to consider "negative" electron affinities, which you might see tabulated in other texts or reference materials. An atom with a negative electron affinity would require energy be expended to hold the electron on the atom. Such an "anion" would be unstable with respect to dissociation; thus, the concept is not particularly useful in our view.

The periodic trends in electron affinity largely parallel those in ionization energy, increasing across a period to become large and positive for the halogens, then decreasing abruptly to essentially zero for the noble gases. A notable difference between the trends in ionization energy and electron affinity is that the dramatic changes in electron affinities occur between atoms whose atomic numbers are one lower than the corresponding breaks in the trends in ionization energy. The following examples illustrate and explain this point. Attaching an electron to F gives F^-, which has the same electron arrangement as Ne, and is therefore very stable. (Recall the discussion of ionization energy in Section 3.3 as a measure of stability.) Similarly, the noble gases have essentially zero electron affinities for the same reason that the alkali metals have small ionization energies; the outermost electron in Ne^- or Xe^- would reside in a new shell, and the resulting ion would be less stable than the neutral parent atom. Therefore, Ne^- is less stable than Ne for precisely the same reason that Na is less stable than Ne. Fluorine has the highest electron affinity, and that of Ne is nearly zero.

No gaseous atom has a positive electron affinity for a *second* electron, because a gaseous ion with a net charge of $-2e$ is always unstable with respect to ionization. Attaching a second electron means bringing it close to a species that is already negatively charged. The two repel each other, and the potential energy between them increases. In crystalline environments, however, doubly negative ions such as O^{2-} can be stabilized by electrostatic interactions with neighboring positive ions.

Mulliken's and Pauling's Electronegativity Scales

The values of IE_1 and EA determine how readily an atom might form a positive or a negative ion. To combine these propensities into a single quantity with predictive value, in 1934, the American physicist Robert Mulliken defined electronegativity as a measure of the relative tendency of atoms to attract electrons to one another in a chemical bond. Mulliken observed that elements located in the lower left corner of the periodic table have low ionization energies and small electron affinities. This means that they give up electrons readily (to form positive ions) but do not readily accept electrons (to form negative ions). They tend to act as electron *donors* in interactions with other elements. In contrast, elements in the upper right corner of the periodic table have large ionization energies and also (except for the noble gases) large electron affinities. As a result, these elements accept electrons easily but give them up only reluctantly; they act as electron *acceptors*.

Mulliken simply *defined* electronegativity as a quantity that is proportional to the average of the ionization energy and the electron affinity:

$$\text{EN (Mulliken)} \propto \tfrac{1}{2}(IE_1 + EA) \qquad [3.10]$$

Electron acceptors (such as the halogens) have both large ionization energies and large electron affinities; they are highly **electronegative.** Electron donors (such as the alkali metals) have small ionization energies and small electron affinities, and therefore low electronegativities; they are **electropositive.** The noble gases rarely participate in chemical bonding. Their large ionization energies and essentially zero electron affinities mean that they are reluctant either to give up or to accept electrons. Electronegativities, therefore, are not generally assigned to the noble gases.

Two years before Mulliken's publication, the American chemist Linus Pauling proposed a different electronegativity scale, based on a comparison of the bond energies of a large number of heteronuclear bond pairs with those of homonuclear diatomics comprising the same elements (for example, HF, HCl and HBr compared with H_2, F_2, Cl_2 and Br_2). Pauling observed that bonds formed between elements from opposite sides of the periodic table were stronger than those between identical elements or even those located in close proximity to one another. He suggested that this extra stability was provided by an *ionic* contribution to the bond strength and constructed an empirical formula for his electronegativity scale that explicitly took these contributions into account.

Pauling's argument goes as follows. Suppose the dissociation energy of an A—A bond is ΔE_{AA}, and that of a B—B bond is ΔE_{BB}; both bonds are covalent because the atoms in them are identical. Then an estimate of the *covalent* contribution to the dissociation energy of an A—B bond is the (geometric) mean of these two energies, $\sqrt{\Delta E_{AA}\Delta E_{BB}}$. (The postulate of the geometric mean was inspired by elementary quantum mechanical arguments but was retained only because it gave better fits than did the arithmetic mean.) The actual A—B bond, however, must include some ionic character because some charge transfer occurs between the atoms. The ionic character tends to strengthen the bond and to increase its value of ΔE_{AB}. Pauling suggested that the difference between the actual and covalent bond energies,

$$\Delta = \Delta E_{AB} - \sqrt{\Delta E_{AA}\Delta E_{BB}} \qquad [3.11]$$

called the **excess bond energy,** is a measure of the ionic contribution to the bond strength and should arise from the electronegativity difference between the two atoms A and B. He defined this electronegativity difference as

$$\chi_A - \chi_B = 0.102\Delta^{1/2} \qquad [3.12]$$

where χ_A and χ_B (Greek letter small chi) are the electronegativities of A and B, respectively. The coefficient 0.102 arises when Δ is measured in kilojoules per mole (kJ mol^{-1}). The atom that more readily accepts an electron (and thus tends to carry a partial negative charge) has the larger χ value. Note that Pauling's formula provides a recipe for calculating only differences in electronegativities, not their absolute values. Pauling himself, after proposing a number of absolute values, arbitrarily assigned the value 4.0 to fluorine. The modern Pauling table (Fig. 3.7) is the result of slight modifications made as better thermochemical data became available.

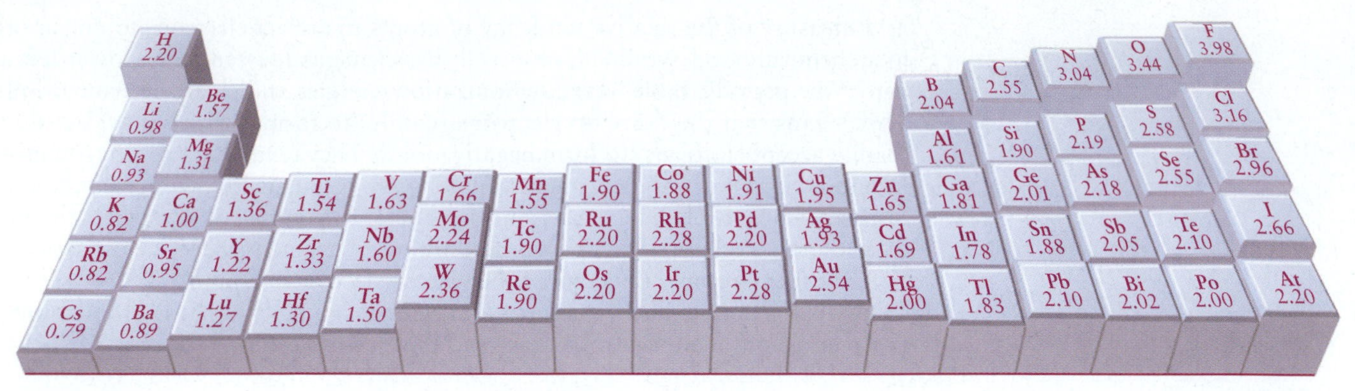

FIGURE 3.7 Average electronegativity of atoms, computed with the method that Linus Pauling developed. Electronegativity values have no units.

Let's compare Mulliken's and Pauling's approaches to the development of their respective electronegativity scales. Mulliken's approach was based on a simple physical model that focused on the properties of individual elements, which, although intuitively appealing, could not take into account any differences in the covalent contributions to the bond. In contrast, Pauling provided a model that explicitly separated the ionic and covalent contributions to the character of the bond, thus providing the framework for our contemporary understanding of the nature of the chemical bond. The other important contribution of his approach is that it averaged, in some sense, the electronegativity of a particular element when it is bonded to a variety of other elements. Despite the differences discussed earlier, Mulliken's and Pauling's scales produce values that are nearly proportional, and it has simply become customary to use the latter.

The periodic trends in electronegativity (Pauling; see Fig. 3.7) are quite interesting. Electronegativity increases across a period from left to right and decreases down a group from top to bottom. The latter trend is more pronounced for the representative elements. These trends can be rationalized semiclassically as before. Moving from left to right across a period, the increasing nuclear charge makes it energetically more difficult to remove an electron and also energetically more favorable to accommodate an additional electron. The trends observed moving down a group are much less dramatic (except for the differences between the second and third periods, in general, and the halogens) and less easy to rationalize. It is important to continue to emphasize, however, the predictive power of this remarkably simple concept, as we shall demonstrate in the sections that follow.

The essential points of this section can be summarized as follows: The tendency of an atom to donate or accept electrons in a chemical bond is expressed by its electronegativity. Highly electronegative atoms, on the right side of the periodic table, readily accept electrons to form negative ions. Highly electropositive atoms, on the left side of the table, readily donate electrons to form positive ions. Bonds formed by the complete transfer of an electron from one atom to another to form a pair of ions bound largely by electrostatic attraction are called ionic (see discussion in Section 3.5).

3.5 Forces and Potential Energy in Molecules: Formation of Chemical Bonds

What determines the stability of a chemical bond? Why is the H_2 molecule more stable than a pair of separated hydrogen atoms in the gas phase at normal temperatures and pressures? But what is the meaning of "more stable?" And how do bonds form spontaneously once the atoms are close enough together?

Experience shows that systems move spontaneously toward configurations that reduce their potential energy. A car rolls downhill, converting its gravitational potential energy into kinetic energy. We have already learned in Section 3.2 that microscopic charged particles move to reduce their electrostatic (Coulomb) potential energy. However, formation of a chemical bond is more subtle than ordinary motion of charged particles under Coulomb's force law, because it involves a special event. Two atoms flying toward each other have a certain total energy that includes contributions from their internal structure, their potential energy relative to each other, and their kinetic energy. To enter into what the distinguished chemist and author George C. Pimentel has characterized as "the blissful state of bondedness" in which the atoms fly together as a bonded pair forever after, they must give up some of their total energy. A diatomic molecule is more stable than the separated atoms from which it was formed because its *total energy* is less than that of the two atoms. You can reach the same conclusion by examining the reverse process; dissociation of a diatomic molecule requires energy. The formation of a chemical bond from a pair of atoms occurs spontaneously in the gas phase only if the reaction is **exothermic**—that is, one that releases heat into the surroundings. With the introduction of the second law of thermodynamics in Chapter 13, we show a deep connection between release of heat and spontaneity. But for this chapter, it is adequate to recognize that a reduction of the total energy of the system is the key to bond formation.

Let us interpret this fact in terms of the potential energy changes in formation of the molecule, illustrated with the specific example of H_2 shown in Figure 3.8. (We could equally well illustrate this with the atomic pairs Cl-Cl and Na-Cl, but the details would be more complicated.) The nuclei are labeled A and B, and the electrons are labeled 1 and 2. The distance between each electron and each proton (r_{1A}, r_{1B}, r_{2A}, r_{2B}) is shown in blue in Figure 3.8, while the distance between the two electrons (r_{12}) and the distance between the two protons (R_{AB}) are shown in red. The potential energy of the molecule is most conveniently expressed in terms of these distances.

$$V = -\frac{e^2}{4\pi\epsilon_0}\left(\frac{1}{r_{1A}} + \frac{1}{r_{2A}} + \frac{1}{r_{1B}} + \frac{1}{r_{2B}}\right) + \frac{e^2}{4\pi\epsilon_0}\left(\frac{1}{r_{12}}\right) + \frac{e^2}{4\pi\epsilon_0}\left(\frac{1}{R_{AB}}\right)$$

$$V = V_{en} + V_{ee} + V_{nn} \qquad [3.13]$$

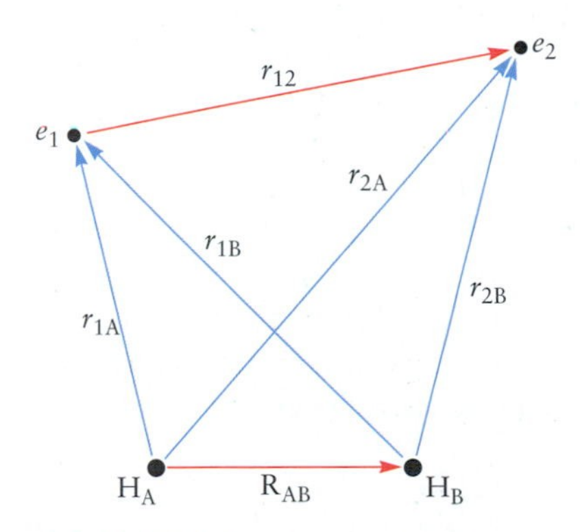

FIGURE 3.8 Coordinates for the hydrogen molecule. The nuclei are assumed stationary at fixed positions separated by the distance R_{AB}. The distance of electron 1 from nuclei A and B is given by r_{1A},r_{1B}; the distance of electron 2 from nuclei A and B is given by r_{2A},r_{2B}; the distance between the electrons is given by r_{12}.

The first four terms in Equation 3.13 represent the attractions between the electrons and the nuclei, and all are negative. The last two terms represent the repulsions between the pair of electrons and the pair of protons, and both are positive. The value of V can be calculated for any configuration of the molecule. But just as in the case of the lithium atom in Section 3.3, this potential energy function does not give a simple pictorial explanation of the stability of the molecule, because there is no exact solution for the motions of four interacting particles.

Out of these building blocks we must construct some new approximate potential energy function, V_{eff}, that holds the molecule together. That means that V_{eff} must depend on R_{AB}, which tracks the transition from two separated atoms to a diatomic molecule. At large distances, $V_{eff} \longrightarrow 0$ because the isolated atoms do not interact. As R_{AB} decreases, V_{eff} must become negative because the atoms begin to attract each other. At small distances, V_{eff} must become positive and large as $V_{eff} \longrightarrow \infty$ due to the repulsion between the protons. Therefore, at some intermediate value of R_{AB}, the potential function must reach a minimum negative value and change its slope as it heads toward positive values. Figure 3.9 is a sketch of a generic V_{eff} that shows all these features.

As explained in Section 3.2, the force between the protons is the negative of the slope of V_{eff}, that is, the negative of its derivative with respect to R_{AB}. For values of R_{AB} larger than the minimum, the attractive forces tend to reduce R_{AB}; for values of R_{AB} smaller than the minimum, the repulsive forces tend to increase R_{AB}. Thus, both the depth of the potential and the position of the minimum are

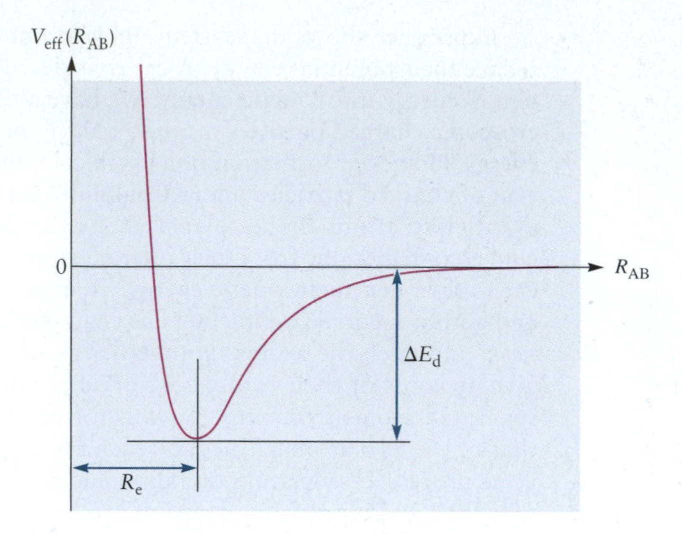

FIGURE 3.9 Dependence of the effective potential energy V_{eff} for a diatomic on the internuclear distance R_{AB}. The location of the minimum corresponds to the equilibrium bond length. The depth of the well relative to the separated atoms is the energy required to dissociate the molecule to form the atoms, and it measures the stability of the molecule.

determined by the competition between the attractive and repulsive forces along the internuclear direction. Therefore, we identify the value of R_{AB} at the minimum as the *equilibrium bond length* of the molecule. The depth of the minimum relative to the separated atoms is identified as the **bond dissociation energy,** ΔE_d, which is a measure of the strength of the bond and the extent to which the molecule is more stable than the separated atoms.

Formation of the chemical bond is associated with a reduction of the electrostatic potential energy and a release of this energy in the form of heat. The actual shape of V_{eff} must be determined by analyzing the Coulomb terms in Equation 3.13. Different classical models of bond formation achieve this goal in different ways. We see how this is done for the two main models for ionic and covalent bonding in Sections 3.6 and 3.7.

The Virial Theorem

We have asserted that bond formation from gas-phase atoms *reduces* the total energy of the system, and we have illustrated how bond formation reduces the electrostatic potential energy. To complete the story, we must see how this reduction in the total energy is partitioned between the kinetic and potential energies of the particles in the system and also gather what we can learn about the driving force for bond formation. For this, we invoke, without proof, the virial theorem, a powerful and quite generally applicable theorem of both classical and quantum mechanics that connects the kinetic, potential, and total energies of a system together, regardless of the details that characterize a particular system of interest. The **virial theorem** states that the average kinetic and the average potential energy of a system of particles interacting only through *electrostatic* forces are related as follows:

$$\overline{\mathcal{T}} = -\frac{1}{2}\overline{V} \qquad [3.14]$$

where $\overline{\mathcal{T}}$ and $\overline{V}$ are the average kinetic and potential energies, respectively. The bar above each symbol identifies it as an average quantity. Now, because

$$\overline{E} = \overline{\mathcal{T}} + \overline{V} \qquad [3.15]$$

we can state for any process that involves a change in the system that

$$\Delta\overline{E} = \Delta\overline{\mathcal{T}} + \Delta\overline{V} \qquad [3.16]$$

Therefore,

$$\Delta \overline{E} = \frac{1}{2} \Delta \overline{V} \tag{3.17}$$

and

$$\Delta \overline{\mathcal{T}} = -\frac{1}{2} \Delta \overline{V} \tag{3.18}$$

Several important conclusions can be drawn from Equations 3.17 and 3.18. First, Equation 3.17 shows clearly that the reduction in the total energy has the same sign as the reduction in the potential energy. In cases where potential energy makes the dominant contribution to the total bond energy (ionic bonds), the reduction in the potential energy can be thought of as the driving force for the formation of the bond. This is consistent with what you have been taught previously and with your own experience in the macroscopic world. Note, however, that Equation 3.18 requires the kinetic energy to increase, but only by half as much as the potential energy decreases. For bonds in which both the kinetic and potential energies play comparably important roles (covalent and polar covalent bonds), cause and effect become much more subtle, and their understanding requires some elementary notions from quantum mechanics. There is an interesting, delicate, and beautiful interplay between kinetic and potential energy during the formation of a covalent chemical bond (see Chapter 6 for a detailed discussion).

3.6 Ionic Bonding

Ionic bonds form between atoms with large differences in electronegativity, such as sodium and fluorine. A practical definition of an ionic bond is one in which the dominant contribution to the strength of the bond is the electrostatic attraction between the ions. Conceptually, the formation of an ionic bond from neutral gas-phase atoms can be thought of as the result of two sequential processes. The more electropositive ion transfers an electron to the more electronegative atom, forming an ion pair that is then drawn together by the attractive electrostatic force. Although we focus our discussion on ionic bonding in a gaseous diatomic molecule where we can clearly identify the forces responsible, most ionic compounds are solids under normal conditions. In an ionic solid, ions of one charge are surrounded by an orderly array of ions of the opposite charge, resulting in extremely large Coulomb stabilization energies. They generally have high melting and boiling points (for example, NaCl melts at 801°C and boils at 1413°C) and can form large crystals. Solid ionic compounds usually conduct electricity poorly, but their melts (molten salts) conduct well.

The charges of the most common ionic forms of the representative elements are determined easily by observing how many electrons must be added or removed to achieve a noble gas configuration, that is, a filled octet. The alkali metals, therefore, form cations of $+1$ charge, the alkaline-earth metals form cations with $+2$ charge, and the halogens form anions with -1 charge, for example. The total charge on the compound must be zero, thus the stoichiometry is determined by charge balance. Elemental cations retain the name of the parent element, whereas the suffix *-ide* is added to the root name of the element that forms the anion. For example, Cl^- is chloride and, the compound it forms with Na^+ is sodium chloride, the major ingredient in table salt. For this reason, ionic solids are often called salts. Simple, binary ionic compounds are easily named by inspection; if more than one ion is included in a compound, the Greek prefixes *mono-*, *di-*, *tri-*, and so forth are added for specificity. The preceding considerations allow us to write $CaBr_2$ as the molecular formula for calcium dibromide, for example. A more comprehensive discussion of inorganic nomenclature is presented in Section 3.11.

Let's consider the formation of an ionic bond from two neutral gas-phase atoms, potassium and fluorine to be specific. When the atoms are infinitely far apart, their interactions are negligible and we assign their potential energy of interaction as zero (see discussion in Section 3.5). Ionizing potassium requires energy, whereas attaching an electron to fluorine releases energy. The relevant reactions and their energy changes are

$$K \rightarrow K^{+} + e^{-} \qquad \Delta E = IE_1 = +419 \text{ kJ mol}^{-1}$$

and

$$F + e^{-} \rightarrow F^{-} \qquad \Delta E = -EA = -328 \text{ kJ mol}^{-1}$$

The total energy cost for the creation of this ion pair when the parent atoms are infinitely far apart, is

$$\Delta E_{\infty} = IE_1(K) - EA(F) = +91 \text{ kJ mol}^{-1}$$

Note that, even for this case, in which one element is highly electronegative and the other is highly electropositive, it still *costs* energy to transfer an electron from a potassium atom to a fluorine atom. This is always true. Because the smallest ionization energy of any element (Cs, 376 kJ mol^{-1}) is larger than the largest electron affinity of any element (Cl, 349 kJ mol^{-1}), creating an ion pair from neutral atoms always requires energy. Starting from an ion pair separated by a large distance, what interaction and which mechanism will lead to the reduction in the potential energy of the system required for the spontaneous formation of an ionic bond?

The ions are attracted to one another (because they have opposite charges) by the electrostatic force, and the potential energy of the system is described by Coulomb's law:

$$V(R_{12}) = \frac{q_1 q_2}{4\pi\epsilon_0 R_{12}} \text{ (J per ion pair)} \qquad [3.19]$$

where q_1 and q_2 are the charges on the ions, R_{12} is the separation between the ions, and ϵ_0 is defined in Equation 3.1. This energy, expressed in joules per ion pair, can be converted to kJ mol^{-1} by multiplying by Avogadro's number, N_A, and dividing by 10^3 to get

$$V(R_{12}) = \frac{q_1 q_2}{4\pi\epsilon_0 R_{12}} \cdot \frac{N_A}{10^3} \text{ (kJ mol}^{-1}\text{)} \qquad [3.20]$$

Consider the potential energy diagram shown in Figure 3.10. We have plotted the potential energy of the system as a function of the distance between the ions, choosing as our zero the potential energy of the neutral atoms when they are infinitely far apart, as before. We have plotted the function

$$V(R_{12}) = Ae^{-\alpha R_{12}} - B\left(\frac{(e)(-e)}{R_{12}}\right) + \Delta E_{\infty} \qquad [3.21]$$

where the first term represents the repulsion between the ions as they get very close together, the second term is the attractive Coulomb potential, and the third term is the energy required to make the ions from their respective neutral atoms (see earlier). We have written the potential in this way for simplicity; the constants A and B reflect the relative contributions made by the attractive and repulsive terms, and they are usually obtained by fitting to experiment, as is the value for α, which tells us at what distance repulsion becomes important.

Starting at the right side of the curve in Figure 3.10, notice that the potential energy of the system (the pair of ions) is greater than that of the neutral atoms by

$$\Delta E_{\infty} = IE_1(K) - EA(F) = +91 \text{ kJ mol}^{-1}$$

Moving toward the left of Figure 3.10, the potential energy of the system decreases rapidly due to the attractive **Coulomb stabilization energy** (the second term in Eq. 3.21), reaching a minimum at the equilibrium bond length, $R_e = 2.17$ Å. If

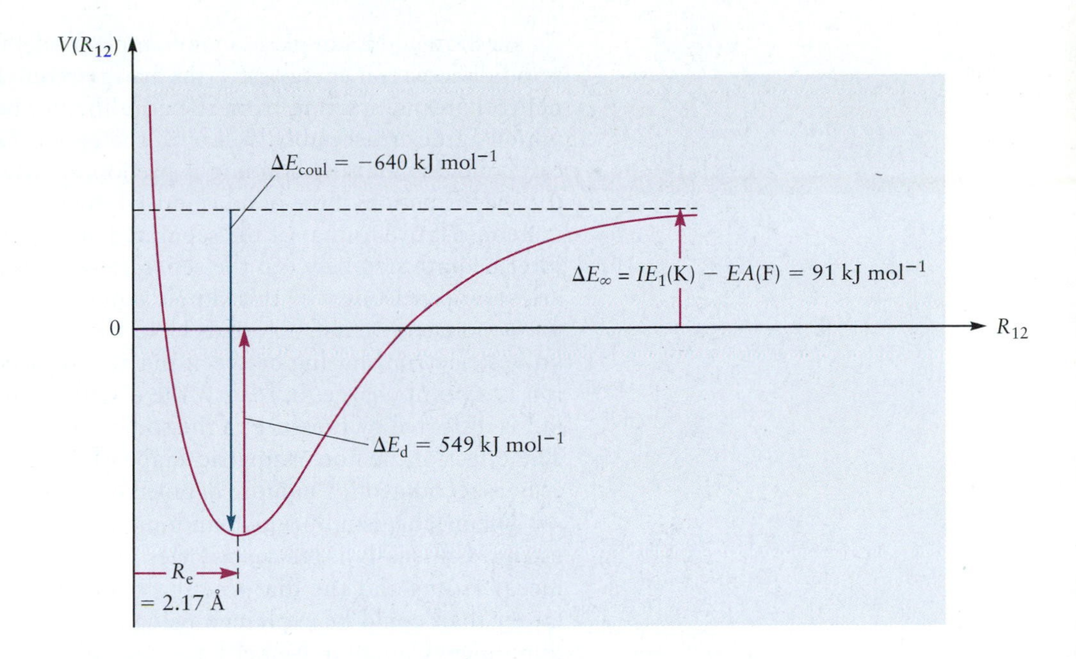

FIGURE 3.10 The potential energy of the system K^+ and F^- as a function of their internuclear separation R_{12}.

we were to try to force the ions to move closer together, we would encounter the resistance depicted by the steep repulsive wall on the left side of the curve, which arises from the repulsive interactions between the electrons of the two ions and is accounted for by the first term in Equation 3.3 The equilibrium bond length, R_e, is determined by the balance between the attractive and repulsive forces.

At last, we can estimate the stabilization of an ionic bond such as KF relative to the neutral atoms. From Figure 3.10, this energy difference is

$$\Delta E_d \approx \frac{q_1 q_2}{4\pi\epsilon_0 R_e} \cdot \frac{N_A}{10^3} - \Delta E_\infty \qquad [3.22]$$

where $\Delta E_\infty = IE_1(K) - EA(F)$. This stabilization energy measures the strength of the ionic bond and is approximately equal to the bond dissociation energy, which is the energy required to break the ionic bond and liberate neutral atoms.

EXAMPLE 3.1

Estimate the energy of dissociation to neutral atoms for KF, which has a bond length of 2.17×10^{-10} m. For KF, $\Delta E_\infty = IE_1(K) - EA(F) = 91$ kJ mol^{-1}.

SOLUTION

$$\Delta E_d \approx -\frac{q_1 q_2}{4\pi\epsilon_0 R_e} \cdot \frac{N_A}{10^3} - \Delta E_\infty$$

$$= -\frac{-(1.602 \times 10^{-19}\ \text{C})^2(6.022 \times 10^{23}\ \text{mol}^{-1})}{(4)(3.1416)(8.854 \times 10^{-12}\ \text{C}^2\ \text{J}^{-1}\ \text{m}^{-1})(2.17 \times 10^{-10}\text{m})(10^3\ \text{J kJ}^{-1})}$$
$$-91\ \text{kJ mol}^{-1}$$
$$= 640\ \text{kJ mol}^{-1} - 91\ \text{kJ mol}^{-1}$$
$$= 549\ \text{kJ mol}^{-1}$$

This estimate compares fairly well with the experimentally measured dissociation energy of 498 kJ mol^{-1}.

Related Problems: 25, 26

As shown in Example 3.1, our simple model for ionic bonding in KF predicts a bond dissociation energy ΔE_d (the energy required to dissociate the molecule into neutral atoms, starting from the equilibrium bond length R_e) of 549 kJ mol^{-1}, which agrees reasonably well with the experimental value of 498 kJ mol^{-1}. We can conclude that the bonding is predominantly ionic, and that the driving force for the formation of the bond is indeed the reduction of the potential energy of the system, relative to that of the separated atoms. The formation of the ions is a key intermediate step between the separated atoms and the stable ionic bond. There are several reasons why this simple model does not do a better job in calculating the bond energy. First, all bonds have some degree of covalent character, which involves electron sharing between the atoms. Second, we have assumed that each ion is a point charge. In reality, the distribution of electrons around the fluoride ion is distorted by presence of the sodium ion; this distortion is called **polarization.** The effect of the nonsymmetric shape of the charge distribution on the bond energy is accounted for in more detailed calculations.

The mechanism by which an ionic bond forms from gas-phase atoms is interesting. Unusually large reactivities are observed for collisions between alkali-metal atoms and the diatomic halogen molecules, reactivities that are much larger than could be explained using conventional theories of chemical reaction dynamics. Canadian Nobel Laureate John Polanyi proposed the following intriguing possibility. Electron transfer takes place at distances much greater than the distances at which most reactive molecular collisions occur. The strong coulombic attraction of the newly created ion pair then rapidly pulls the reactants together, where they form an ionic bond. The metal has sent its electron to "harpoon" the halogen, pulling it in with the "rope" of the Coulomb interaction. This **harpoon mechanism** has been studied extensively for a variety of systems and is generally agreed to provide a satisfactory semiquantitative description of the formation of gas-phase alkali halide molecules. You should keep in mind that although gas-phase molecules with predominantly ionic bonding can be prepared and are stable at high temperatures, most ionic bonds occur in ionic solids.

3.7 Covalent and Polar Covalent Bonding

We have discussed how ionic bonding results from electron transfer and Coulomb stabilization of the resulting ions and that the propensity of a pair of atoms to form an ionic bond is determined by the difference in their electronegativities. What kinds of bonds are formed between elements of identical or comparable electronegativities such as H_2 or CO, and what is the driving force for bond formation from separated atoms in the gas phase?

Section 3.5 provides a general argument for how a chemical bond forms from a pair of isolated atoms. We have used that argument to show that for ionic compounds the driving force is the Coulomb stabilization of the ion pair. We present below a plausibility argument to suggest why a covalent bond might form, by focusing on the *forces* acting on the nuclei due to the electrons.

Let's consider the simplest possible molecule, H_2^+ (the hydrogen molecule ion), which has only one electron. In Figure 3.11, H_A identifies the position of nucleus H_A, and H_B that of nucleus B. The distance between the two nuclei is R_{AB}. The distance between the electron and each nucleus is r_{Ae} and r_{Be}, respectively.

Consider the forces between the three particles. There is the internuclear repulsive force, $F_{AB} \propto (+Z_A e)(+Z_B e)/R_{AB}^2$, and two electron–nuclear attractive forces, $F_{Ae} \propto (-e)(+Z_A e)/r_{Ae}^2$ and $F_{Be} \propto (-e)(+Z_B e)/r_{Be}^2$. The internuclear repulsive

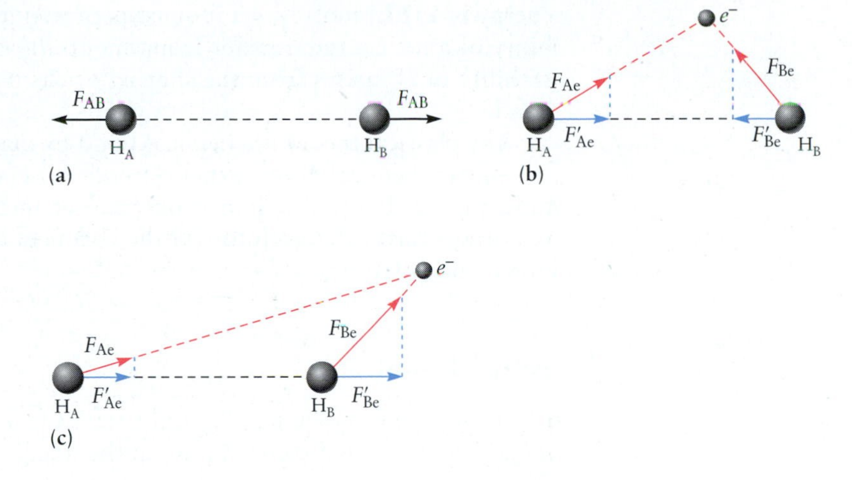

FIGURE 3.11 The forces between the particles in H_2^+. (a) The internuclear repulsion always opposes bonding the nuclei together. (b) An electron positioned in a region that will tend to bond the nuclei together. (c) An electron positioned in a region that will tend to pull the nuclei apart.
(Adapted from G.C. Pimentel and R.D. Spratley, Chemical Bonding Clarified through Quantum Mechanics, Holden-Day Inc., San Francisco. 1969. Page 74.)

force always opposes formation of a chemical bond (Figure 3.11a), so we must identify some force that overcomes this repulsion. We need only consider the attractive forces and ask, "Over what region in space does the electron exert forces on the nuclei that will tend to pull them together?" Only the component of the attractive force directed along the internuclear axis is important. Clearly, for all positions of the electron "between" the nuclei (for example, see Fig. 3.11b), the forces F'_{Ae} and F'_{Be} will tend to pull the nuclei together. In contrast, however, when the electron is outside the internuclear region (for example, see Fig. 3.11c), it exerts a greater force on the nearer nucleus than the farther, pulling the nuclei apart. It is straightforward (for some chemists!) to calculate the net forces using Coulomb's law and to identify a bonding and an antibonding region, the boundary of which is plotted in Figure 3.12. The curve that separates the bonding and antibonding regions approximates a hyperbola of revolution. Whenever the electron is found in the region between the two curves, the net force along the internuclear axis is attractive, encouraging bonding; when it is outside this region, the net force along the internuclear axis is repulsive, precluding bonding. This simple model is supported by experimental data for H_2^+. Its equilibrium bond length R_e is 1.06 Å, and its bond dissociation energy ΔE_d is 255.5 kJ mol^{-1}, which is characteristic of a stable covalent bond.

This picture of covalent bonding in the H_2^+ molecular ion can be applied to other molecules. For example, the H_2 molecule is quite stable (its bond dissociation

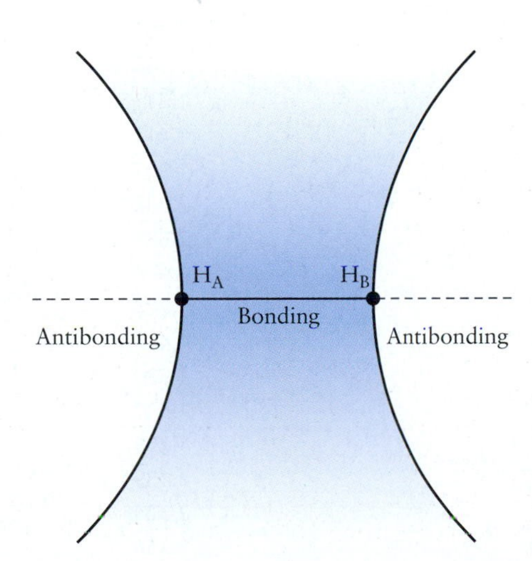

FIGURE 3.12 Bonding and antibonding regions in a homonuclear diatomic molecule. An electron located within the bonding region will tend to pull the nuclei together, whereas an electron in the antibonding regions will tend to pull the nuclei apart.
(Adapted from G.C. Pimentel and R.D. Spratley, Chemical Bonding Clarified through Quantum Mechanics, Holden-Day Inc., San Francisco. 1969. Page 75.)

energy is 432 kJ mol^{-1}), yet it consists of two identical atoms. There is no possibility of a net charge transfer from one to the other to form an ionic bond. The stability of H_2 arises from the sharing of electrons between atoms in a covalent bond.

Any classical theory of chemical bond formation must explain certain properties of the chemical bond, explain trends observed in bonding, and most important, predict likely bonding properties of molecules not yet made. The most important classical descriptors of the chemical bond are the bond length, energy, order, and polarity.

Bond Lengths

For a diatomic molecule, the only relevant structural parameter is the bond length, that is, the distance between the nuclei of the two atoms. Table 3.3 lists the bond lengths of a number of diatomic molecules, expressed in units of angstroms (1 Å = 10^{-10} m). Certain systematic trends are immediately obvious. Among the members of a group in the periodic table, bond lengths usually increase with increasing atomic number Z. The I_2 bond is longer than the F_2 bond, for example, and those of Cl_2 and Br_2 fall in line, as they should. A particularly significant result from experiment is that the length of a bond of the same type (see later) between a given pair of atoms changes little from one molecule to another. For example, C—H bond lengths in acetylsalicylic acid (aspirin, $C_9H_8O_4$) are about the same as they are in methane (CH_4), although the molecules have different structures and physical and chemical properties. Table 3.4 shows that the lengths of O—H, C—C, and C—H bonds in a number of molecules are constant to within a few percent.

Bond Energies

The stability of a molecule is determined by the energy required to dissociate the molecule into its constituent atoms. The greater the energy required, the more stable the molecule. The **bond energy,** also called the bond dissociation energy, is the

TABLE 3.3 Properties of Diatomic Molecules

Molecule	Bond Length (Å)	Bond Energy (kJ mol^{-1})
H_2	0.751	433
N_2	1.100	942
O_2	1.211	495
F_2	1.417	155
Cl_2	1.991	240
Br_2	2.286	190
I_2	2.669	148
HF	0.926	565
HCl	1.284	429
HBr	1.424	363
HI	1.620	295
ClF	1.632	252
BrF	1.759	282
BrCl	2.139	216
ICl	2.324	208
NO	1.154	629
CO	1.131	1073

TABLE 3.4 Reproducibility of Bond Lengths

Bond	Molecule	Bond Length (Å)
O—H	H_2O	0.958
	H_2O_2	0.960
	HCOOH	0.95
	CH_3OH	0.956
C—C	Diamond	1.5445
	C_2H_6	1.536
	CH_3CHF_2	1.540
	CH_3CHO	1.50
C—H	CH_4	1.091
	C_2H_6	1.107
	C_2H_4	1.087
	C_6H_6	1.084
	CH_3Cl	1.11
	CH_2O	1.06

energy required to break one mole of the particular bond under discussion (see Section 3.5). The bond energy is denoted by ΔE_d ("d" stands for *dissociation* here) and is measured directly in units of kJ mol^{-1}. Table 3.3 lists bond energies for selected diatomic molecules. Again, certain systematic trends with changes in atomic number are evident. Bonds generally grow weaker with increasing atomic number, as shown by the decrease in the bond energies of the hydrogen halides in the order HF > HCl > HBr > HI. Note, however, the unusual weakness of the bond in the fluorine molecule, F_2. Its bond energy is significantly *smaller* than that of Cl_2 and comparable with that of I_2. Bond strength decreases dramatically in the diatomic molecules from N_2 (942 kJ mol^{-1}) to O_2 (495 kJ mol^{-1}) to F_2 (155 kJ mol^{-1}). What accounts for this behavior? A successful theory of bonding must explain both the general trends and the reasons for particular exceptions.

Bond energies, like bond lengths, are fairly reproducible (within about 10%) from one compound to another. It is therefore possible to tabulate *average* bond energies from measurements on a series of compounds. The energy of any given bond in different compounds will deviate somewhat from those shown in Table 3.4, but in most cases, the deviations are small.

Bond Order

Sometimes, the length and energy of the bond between two specific kinds of atoms are *not* comparable among different compounds, but rather are sharply different. Table 3.5 shows the great differences in bond lengths and bond energies of carbon–carbon bonds in ethane (H_3CCH_3), ethylene (H_2CCH_2), and acetylene (HCCH). Carbon–carbon bonds from many other molecules all fit into one of the three classes given in the table (that is, some carbon–carbon bond lengths are close to 1.54 Å, others are close to 1.34 Å, and still others are close to 1.20 Å). This observation confirms the existence of not one, but three types of carbon–carbon bonds. We classify these as single, double, and triple bonds, respectively, based on their bond lengths and bond dissociation energies. The longest and weakest (as in ethane) is a single bond represented by C—C; that of intermediate strength (as in ethylene) is a double bond, C=C; and the shortest and strongest (as in acetylene) is a triple bond, C≡C. In Section 3.8, the **bond order** is shown to be the number of shared electron pairs for these bonds: 1, 2, and 3, respectively.

TABLE 3.5 Three Types of Carbon–Carbon Bonds

Bond	Molecule	Bond Length (Å)	Bond Energy (kJ mol^{-1})
C—C	C_2H_6 (or H_3CCH_3)	1.536	345
C=C	C_2H_4 (or H_2CCH_2)	1.337	612
C≡C	C_2H_2 (or HCCH)	1.204	809

Even these three types do not cover all the carbon–carbon bonds found in nature, however. In benzene (C_6H_6), the experimental carbon–carbon bond length is 1.397 Å, and its bond dissociation energy is 505 kJ mol^{-1}. This bond is intermediate between a single bond and a double bond (its bond order is $1\frac{1}{2}$). In fact, the bonding in compounds such as benzene differs from that in many other compounds (see Chapter 7). Although many bonds have properties that depend primarily on the two atoms that form the bond (and thus are similar from one compound to another), bonding in benzene and related molecules, and a few other classes of compounds, depends on the nature of the whole molecule.

Multiple bonds occur between atoms other than carbon and even between unlike atoms. Some representative bond lengths are listed in Table 3.6.

Polar Covalent Bonding: Electronegativity and Dipole Moments

Laboratory measurements show that most real bonds are neither fully ionic nor fully covalent, but instead possess a mixture of ionic and covalent character. Bonds in which there is a partial transfer of charge are called **polar covalent.** This section provides an approximate description of the polar covalent bond based on the relative abilities of each atom to attract the electron pair toward its nucleus. This ability is estimated by comparing the electronegativity values for the two atoms.

On the Pauling scale (see Fig. 3.7 and Appendix F), electronegativities range from 3.98 (for fluorine) to 0.79 (for cesium). These numerical values are useful for exploring periodic trends and for making semiquantitative comparisons. They represent the average tendency of an atom to attract electrons within a molecule, based on the properties of the bond it makes in a large range of compounds.

The absolute value of the difference in electronegativity of two bonded atoms tells the degree of *polarity* in their bond. A large difference (greater than about 2.0) means that the bond is ionic and that an electron has been transferred completely or nearly completely to the more electronegative atom. A small difference (less than about 0.4) means that the bond is largely covalent, with electrons in the bond shared fairly evenly. Intermediate values of the difference signify a polar covalent bond with intermediate character. These suggested dividing points between bond types are not sharply defined (see later) and your instructor may suggest alternatives.

TABLE 3.6 Average Bond Lengths (in Å)

C—C	1.54	N—N	1.45	C—H	1.10
C=C	1.34	N=N	1.25	N—H	1.01
C≡C	1.20	N≡N	1.10	O—H	0.96
C—O	1.43	N—O	1.43	C—N	1.47
C=O	1.20	N=O	1.18	C≡N	1.16

EXAMPLE 3.2

Using Figure 3.7, arrange the following bonds in order of decreasing polarity: H—C, O—O, H—F, I—Cl, Cs—Au.

SOLUTION

The differences in electronegativity among the five pairs of atoms (without regard to sign) are 0.35, 0.00, 1.78, 0.50, and 1.75, respectively. The order of decreasing polarity is the order of decrease in this difference: H—F, Cs—Au, I—Cl, H—C, and O—O. The last bond in this listing is nonpolar.

Related Problems: 33, 34

Dipole Moments and Percent Ionic Character

A bond that is almost purely ionic, such as that of KF, can be thought of as arising from the nearly complete transfer of one electron from the electropositive to the electronegative species. KF can be described fairly accurately as K^+F^-, with charges $+e$ and $-e$ on the two ions. However, characterizing the charge distribution for a molecule such as HF, which has significant covalent character, is more complex. If we wish to approximate the bond by its ionic character, it is best described as $H^{\delta+}F^{\delta-}$, where some fraction, δ, of the full charge, e, is on each nucleus. A useful measure of the ionic character of a bond, arising from electronegativity differences, especially for diatomic molecules, is the dipole moment of the molecule. If two charges of equal magnitude and opposite sign, $+q$ and $-q$, are separated by a distance R, the **dipole moment** μ (Greek letter lowercase mu) of that charge distribution is

$$\mu = qR \tag{3.23}$$

In SI units, μ is measured in coulomb meters, an inconveniently large unit for discussing molecules. The unit most often used is the debye (D), which is related to SI units by

$$1 \text{ D} = 3.336 \times 10^{-30} \text{ C m}$$

(This apparently peculiar definition arises from the transition from electrostatic units to SI units). The debye can also be defined as the dipole moment of two charges $\pm e$ separated by 0.2082 Å. If δ is the fraction of a unit charge on each atom in a diatomic molecule ($q = e\delta$) and R is the equilibrium bond length, then

$$\mu(\text{D}) = [R(\text{Å})/0.2082 \text{ Å}]\ \delta \tag{3.24}$$

This equation can, of course, be inverted to determine the fraction ionic character from the experimental value of the dipole moment. Dipole moments are measured experimentally by electrical and spectroscopic methods and provide useful information about the nature of bonding. In HF, for example, the value of δ calculated from the dipole moment ($\mu = 1.82$ D) and bond length ($R = 0.917$ Å) is 0.41, substantially less than the value of 1 for a purely ionic bond. We convert δ to a "percent ionic character" by multiplying by 100% and say that the bond in HF is 41% ionic. Deviations from 100% ionic bonding occur for two reasons: (1) covalent contributions lead to electron sharing between atoms, and (2) the electronic charge distribution around one ion may be distorted by the electric field of the other ion (polarization). When polarization is extreme, regarding the ions as point charges is no longer a good approximation, and a more accurate description of the distribution of electric charge is necessary.

Table 3.7 provides a scale of ionic character for diatomic molecules, based on the definition of δ. The degree of ionic character inferred from the dipole moment is reasonably well correlated with the Pauling electronegativity differences (Fig. 3.13). A great deal of ionic character usually corresponds to a large electronegativity difference, with the more electropositive atom carrying the charge $+\delta$. There are exceptions to this general trend, however. Carbon is less electronegative than oxygen, so one would predict a charge distribution of $C^{\delta+}O^{\delta-}$ in the CO molecule. In fact, the measured dipole moment is quite small in magnitude and is oriented in the opposite direction: $C^{\delta-}O^{\delta+}$, with $\delta = 0.02$. The discrepancy arises because of the lone-pair electron density on the carbon atom (which is reflected in the formal charge of -1 carried by that atom, as discussed in Section 3.8).

TABLE 3.7 Dipole Moments of Diatomic Molecules

Molecule	Bond Length (Å)	Dipole Moment (D)	Percent Ionic Character (100δ)
H_2	0.751	0	0
CO	1.131	0.112	2
NO	1.154	0.159	3
HI	1.620	0.448	6
ClF	1.632	0.888	11
HBr	1.424	0.828	12
HCl	1.284	1.109	18
HF	0.926	1.827	41
CsF	2.347	7.884	70
LiCl	2.027	7.129	73
LiH	1.604	5.882	76
KBr	2.824	10.628	78
NaCl	2.365	9.001	79
KCl	2.671	10.269	82
KF	2.176	8.593	82
LiF	1.570	6.327	84
NaF	1.931	8.156	88

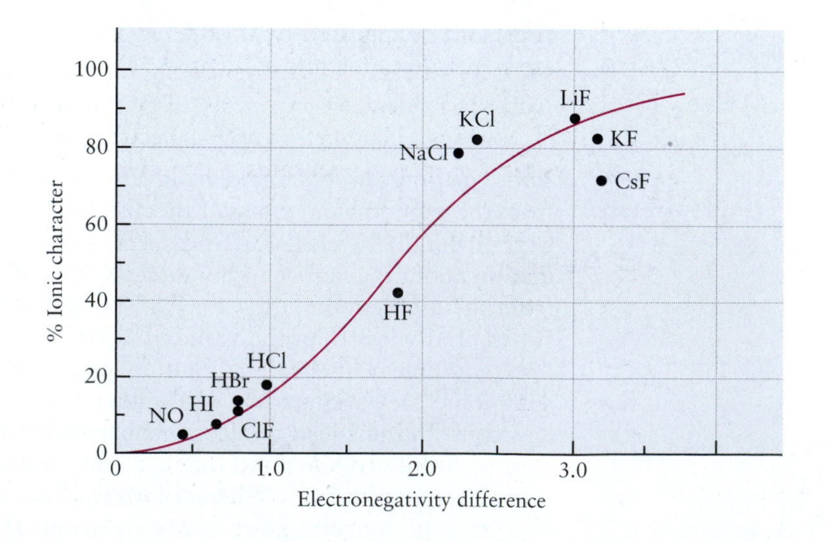

FIGURE 3.13 Two measures of ionic character for diatomic molecules are the electronegativity difference (from Fig. 3.7) and the percent ionic character 100δ, calculated from the observed dipole moment and bond length. The curve shows that the two correlate approximately but that there are many exceptions.

In summary, the properties of a chemical bond are often quite similar in a variety of compounds, but we must be alert for exceptions that may signal new types of chemical bonding.

The Lewis model (see Section 3.8) represents covalent bonds as shared valence-electron pairs positioned between two nuclei, where they presumably are involved in net attractive interactions that pull the nuclei together and contribute to the strengthening of the bond. The mechanism cannot be explained by classical physics, and is examined through quantum mechanics in Chapter 6.

3.8 Lewis Diagrams for Molecules

The American chemist G. N. Lewis introduced a useful model that describes the electronic structure of the atom and provides a starting point for describing chemical bonds. The Lewis model represents the valence electrons as dots arranged around the chemical symbol for the atom; the core electrons are not shown. The first four dots are displayed singly around the four sides of the elemental symbol. If the atom has more than four valence electrons, their dots are then paired with those already present. The result is a **Lewis dot symbol** for that atom. The Lewis notation for the elements of the first two periods is

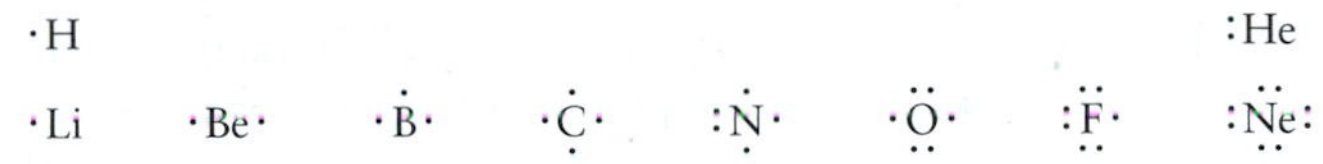

·H :He

·Li ·Be· ·Ḃ· ·Ċ· :Ṅ· ·Ö· :F̤· :N̤̈e:

The Lewis model for covalent bonding starts with the recognition that electrons are not transferred from one atom to another in a nonionic compound, but rather are *shared* between atoms to form covalent bonds. Hydrogen and chlorine combine, for example, to form the **covalent compound** hydrogen chloride. This result can be indicated with a **Lewis diagram** for the molecule of the product, in which the valence electrons from each atom are redistributed so that one electron from the hydrogen atom and one from the chlorine atom are now shared by the two atoms. The two dots that represent this electron pair are placed between the symbols for the two elements:

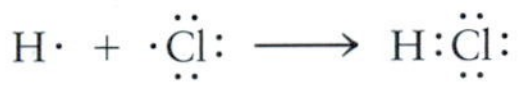

H· + ·C̤̈l: ⟶ H:C̤̈l:

The basic rule that governs Lewis diagrams is the **octet rule:** Whenever possible, the electrons in a covalent compound are distributed in such a way that each main-group element (except hydrogen) is surrounded by eight electrons (an *octet* of electrons). Hydrogen has two electrons in such a structure. When the octet rule is satisfied, the atom attains the special stability of a noble-gas shell. As a reminder, the special stability of the noble-gas configuration arises from the fact that electrons in a filled shell experience the maximum electron-nuclear attraction possible, because the number of protons (Z) is also the maximum allowed for a particular shell. In the structure for HCl shown earlier, the H nucleus has two valence electrons in its shell (like the noble gas, He), and Cl has eight (like Ar). Electrons that are shared between two atoms are counted as contributing to the filling of the valence shell of each atom.

A shared pair of electrons can also be represented by a short line (–):

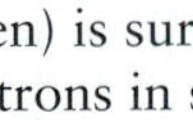

H—C̤̈l:

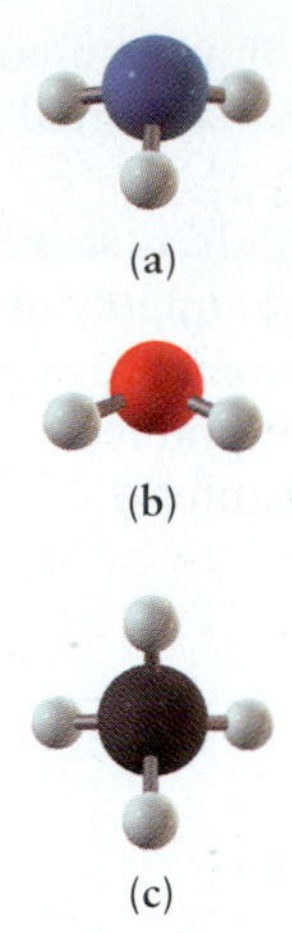

(a)

(b)

(c)

FIGURE 3.14 Molecules of three familiar substances, drawn in ball-and-stick fashion. The sizes of the balls have been reduced somewhat to show the bonds more clearly, but the relative sizes of the balls are correct. (a) Ammonia, NH_3. (b) Water, H_2O. (c) Methane, CH_4.

The unshared electron pairs around the chlorine atom in the Lewis diagram are called **lone pairs,** and they make no contribution to the bond between the atoms. Lewis diagrams of some simple covalent compounds are

```
  NH3          H2O          CH4

                              H
                             ..
  ..           ..
 H:N:H        H:O:H        H:C:H
  ..           ..           ..
  H                          H

                             H
  ..                         |
H—N—H       H—O—H        H—C—H
  |           ..             |
  H                          H
```

Lewis diagrams show how bonds connect the atoms in a molecule, but they do not show the spatial geometry of the molecule. The ammonia molecule is not planar, but pyramidal, for example, with the nitrogen atom at the apex. The water molecule is bent rather than straight. Three-dimensional geometries can be represented by ball-and-stick models (such as those shown in Fig. 3.14).

More than one pair of electrons may be shared by two atoms in a bond. For example, in the oxygen molecule, each atom has six valence electrons. Thus, for each to achieve an octet configuration, *two* pairs of electrons must be shared, making a **double bond** between the atoms:

```
 ..  ..        ..  ..
 O: :O    or   O=O
 ..  ..        ..  ..
```

Similarly, the N_2 molecule has a **triple bond,** involving three shared electron pairs:

```
:N:::N:    or    :N≡N:
```

In contrast, the F_2 molecule has only a single bond. The number of shared electron pairs in a bond determines the order of the bond, which has already been connected with bond energy and bond length in Section 3.7. The decrease in bond order from 3 to 2 to 1 explains the dramatic reduction in the bond energies of the sequence of diatomic molecules N_2, O_2, and F_2 pointed out in Section 3.7. A carbon–carbon bond can involve the sharing of one, two, or three electron pairs. A progression from single to triple bonding is found in the hydrocarbons ethane (C_2H_6), ethylene (C_2H_4), and acetylene (C_2H_2):

```
  H H       H  H
  .. ..     ..  ..
H:C:C:H     C::C       H:C:::C:H
  .. ..     ..  ..
  H H       H  H
```

Ethane, C_2H_6, can be burned in oxygen as a fuel, and if strongly heated, it reacts to form hydrogen and ethylene.

This progression corresponds to the three types of carbon–carbon bonds with properties that are related to bond order in Section 3.7 and are summarized in Tables 3.4 and 3.5.

Multiple bonding to attain an octet most frequently involves the elements carbon, nitrogen, oxygen, and to a lesser degree, sulfur. Double and triple bonds are shorter than a single bond between the same pair of atoms (see illustrative examples in Table 3.5).

Formal Charges

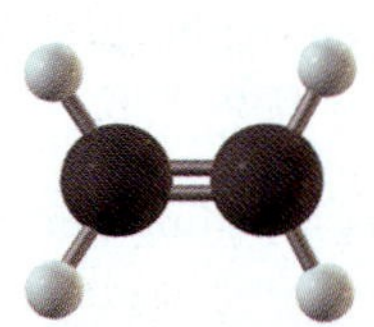

Ethylene, C_2H_4, is the largest volume organic (carbon-containing) chemical produced.

In **homonuclear** diatomic molecules (in which both atoms are the same, as in H_2 and Cl_2), the electrons are shared equally between the two atoms, and the covalency is nearly ideal for such molecules.

Consider, however, a molecule of carbon monoxide (CO). Its Lewis diagram has a triple bond

```
:C:::O:
```

Acetylene, C_2H_2, has a triple bond that makes it highly reactive.

Carbon monoxide, CO, is a colorless, odorless, and toxic gas produced by the incomplete burning of hydrocarbons in air. It is used in the production of elemental metals from their oxide ores.

that uses the ten valence electrons (four from the C and six from the O) and gives each atom an octet. If the six bonding electrons were shared equally, the carbon atom would own five valence electrons (one *more* than its group number) and the oxygen atom would own five electrons (one *less* than its group number). Equal sharing implies that, formally, the carbon atom must gain an electron and the oxygen atom must lose an electron. This situation is described by assigning a **formal charge** to each atom, defined as the charge an atom in a molecule would have if the electrons in its Lewis diagram were divided equally among the atoms that share them. Thus, in CO, C has a formal charge of −1 and O has a formal charge of +1:

$$\overset{-1}{:\text{C}}:::\overset{+1}{\text{O}:}$$

Carbon monoxide is a covalent compound, and the assignment of formal charges does not make it ionic.

We emphasize that the equal sharing of electrons in the bonds of such **heteronuclear** diatomic molecules is without experimental basis and is simply a postulate. It has a useful purpose in cases in which two or more Lewis diagrams are possible for the same molecule. When this happens, the diagram with the smallest formal charges is the preferred one, and it generally gives the best description of bonding in the molecule. This is yet another example where simple electrostatic arguments can provide great chemical insight.

The formal charge on an atom in a Lewis diagram is simple to calculate. If the valence electrons were removed from an atom, it would have a positive charge equal to its group number in the periodic table (elements in Group VI, the chalcogens, have six valence electrons, and therefore a charge of +6 when those electrons are removed). From this positive charge, subtract the number of lone-pair valence electrons possessed by the atom in the Lewis diagram, and then subtract half of the number of bonding electrons it shares with other atoms:

$$\text{formal charge} = \text{number of valence electrons} - \text{number of electrons in lone pairs} - \tfrac{1}{2}(\text{number of electrons in bonding pairs})$$

EXAMPLE 3.3

Compute the formal charges on the atoms in the following Lewis diagram, which represents the azide ion (N_3^-):

$$\left[\ddot{\underset{\cdot\cdot}{\text{N}}}\text{=N=}\ddot{\underset{\cdot\cdot}{\text{N}}}\right]^-$$

SOLUTION

Nitrogen is in Group V. Hence, each N atom contributes 5 valence electrons to the bonding, and the negative charge on the ion contributes one more electron. The Lewis diagram correctly represents 16 electrons.

Each of the terminal nitrogen atoms has four electrons in lone pairs and four bonding electrons (which comprise a double bond) associated with it. Therefore,

$$\text{formal charge}_{(\text{terminal N})} = 5 - 4 - \tfrac{1}{2}(4) = -1$$

The nitrogen atom in the center of the structure has no electrons in lone pairs. Its entire octet comprises the eight bonding electrons:

$$\text{formal charge}_{(\text{central N})} = 5 - 0 - \tfrac{1}{2}(8) = +1$$

The sum of the three formal charges is −1, which is the true overall charge on this polyatomic ion. Failure of this check indicates an error either in the Lewis diagram or in the arithmetic.

Related Problems: 39, 40

Drawing Lewis Diagrams

When drawing Lewis diagrams, we shall assume that the molecular "skeleton" (that is, a plan of the bonding of specific atoms to other atoms) is known. In this respect, it helps to know that hydrogen and fluorine only bond to one other atom and are always terminal atoms in Lewis diagrams. A systematic procedure for drawing Lewis diagrams can then be used, as expressed by the following rules:

1. Count the total number of valence electrons available by first using the group numbers to add the valence electrons from all the atoms present. If the species is a negative ion, *add* additional electrons to achieve the total charge. If it is a positive ion, *subtract* enough electrons to result in the total charge.
2. Calculate the total number of electrons that would be needed if each atom had its *own* noble-gas shell of electrons around it (two for hydrogen, eight for carbon and heavier elements).
3. Subtract the number in step 1 from the number in step 2. This is the number of shared (or bonding) electrons present.
4. Assign two bonding electrons (one pair) to each bond in the molecule or ion.
5. If bonding electrons remain, assign them in pairs by making some double or triple bonds. In some cases, there may be more than one way to do this. In general, double bonds form only between atoms of the elements C, N, O, and S. Triple bonds are usually restricted to C, N, or O.
6. Assign the remaining electrons as lone pairs to the atoms, giving octets to all atoms except hydrogen.
7. Determine the formal charge on each atom and write it next to that atom. Check that the formal charges add up to the correct total charge on the molecule or polyatomic ion. If more than one diagram is possible, choose the one with the smallest number of formal charges. This rule not only guides you to the best structure, it also provides a check for inadvertent errors (such as the wrong number of dots).

The use of these rules is illustrated by Example 3.4.

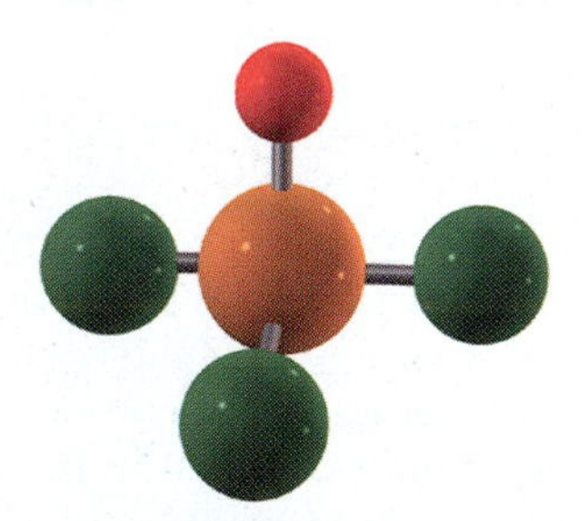

FIGURE 3.15 Phosphoryl chloride, $POCl_3$, is a reactive compound used to introduce phosphorus into organic molecules in synthesis reactions. Experimental studies show that the P—O bond is more like a double bond than a single bond. Lewis diagrams rationalizing the existence of the P=O double bond can be constructed as an example of valence shell expansion. See the discussion on p. 90.

EXAMPLE 3.4

Write a Lewis electron dot diagram for phosphoryl chloride, $POCl_3$ (Fig. 3.15). Assign formal charges to all the atoms.

SOLUTION

The first step is to calculate the total number of valence electrons available in the molecule. For $POCl_3$, it is

$$5 \text{ (from P)} + 6 \text{ (from O)} + [3 \times 7 \text{ (from Cl)}] = 32$$

Next, calculate how many electrons would be necessary if each atom were to have its own noble-gas shell of electrons around it. Because there are 5 atoms in the present case (none of them hydrogen), 40 electrons would be required. From the difference of these numbers ($40 - 32 = 8$), each atom can achieve an octet only if 8 electrons are shared between pairs of atoms. Eight electrons correspond to 4 electron pairs, so each of the four linkages in $POCl_3$ must be a single bond. (If the number of shared electron pairs were *larger* than the number of bonds, double or triple bonds would be present.)

The other 24 valence electrons are assigned as lone pairs to the atoms in such a way that each achieves an octet configuration. The resulting Lewis diagram is

```
         −1
         ..
        :O:
   ..    |+1  ..
  :Cl — P — Cl:
   ..    |    ..
        :Cl:
         ..
```

Formal charges are already indicated in this diagram. Phosphorus has the group number V, and it shares eight electrons with no lone-pair electrons, so

$$\text{formal charge on P} = 5 - 4 = +1$$

Oxygen has the group number VI with six lone-pair electrons and two shared electrons, so

$$\text{formal charge on O} = 6 - 6 - \tfrac{1}{2}(2) = -1$$

All three chlorine atoms have zero formal charge, computed by

$$\text{formal charge on Cl} = 7 - 6 - \tfrac{1}{2}(2) = 0$$

Related Problems: 45, 46, 47, 48

Resonance Forms

Ozone, O_3, is a pale blue gas with a pungent odor. It condenses to a dark blue liquid below −112°C.

For certain molecules or molecular ions, two or more equivalent Lewis diagrams can be written. An example is ozone (O_3), for which there are two possible Lewis diagrams:

$$\ddot{\text{O}}\!::\!\ddot{\text{O}}\!:\!\ddot{\text{O}}\!:^{-1}_{} \;(+1 \text{ on central O}) \qquad \text{and} \qquad :\!\ddot{\text{O}}^{-1}\!:\!\ddot{\text{O}}\!::\!\ddot{\text{O}} \;(+1 \text{ on central O})$$

These diagrams suggest that one O—O bond is a single bond and the other one is a double bond, so the molecule would be asymmetric. In fact, the two O—O bond lengths are found experimentally to be identical, with a bond length of 1.28 Å, which is intermediate between the O—O single bond length in H_2O_2 (1.49 Å) and the O=O double bond length in O_2 (1.21 Å).

The Lewis diagram picture fails, but it can be patched up by saying that the actual bonding in O_3 is represented as a **resonance hybrid** of the two Lewis diagrams in which each of the bonds is intermediate between a single bond and a double bond. This hybrid is represented by connecting the diagrams with a double-headed arrow:

$$\left\{ \ddot{\text{O}}\!::\!\ddot{\text{O}}\!:\!\ddot{\text{O}}\!:^{-1} \;\longleftrightarrow\; :\!\ddot{\text{O}}^{-1}\!:\!\ddot{\text{O}}\!::\!\ddot{\text{O}} \right\} \quad (+1 \text{ on central O in each})$$

The term *resonance* does not mean that the molecule physically oscillates back and forth from one of these bonding structures to the other. Rather, within the limitations of the Lewis dot model of bonding, the best representation of the actual bonding is a hybrid diagram that includes features of each of the acceptable individual diagrams. This awkwardness can be avoided by using molecular orbital theory to describe chemical bonding (see Chapter 6).

EXAMPLE 3.5

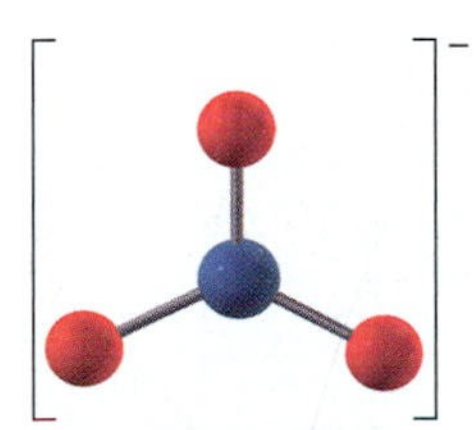

FIGURE 3.16 The nitrate ion, NO_3^-, has a symmetric planar structure.

Draw three resonance forms for the nitrate ion, NO_3^- (Fig. 3.16), and estimate the bond lengths.

SOLUTION

NO_3^- contains 24 valence electrons. For each atom to have its own octet, $4 \times 8 = 32$ electrons would be required. Therefore, $32 - 24 = 8$ electrons must be shared between atoms, implying a total of four bonding pairs. These can be distributed in one double and two single bonds, leading to the equivalent resonance diagrams:

$$\left\{ \ddot{\text{O}}\!::\!\text{N}\!:\!\ddot{\text{O}}\!: \;(\text{top }{:}\ddot{\text{O}}{:}^{-1};\ \text{N } +1,\ \text{right O } -1) \longleftrightarrow :\!\ddot{\text{O}}\!:\!\text{N}\!:\!\ddot{\text{O}}\!: \;(\text{top }{:}\text{O}{::};\ -1\ +1\ -1) \longleftrightarrow :\!\ddot{\text{O}}\!:\!\text{N}\!::\!\ddot{\text{O}} \;(\text{top }{:}\ddot{\text{O}}{:}^{-1};\ -1\ +1) \right\}$$

The two singly bonded oxygen atoms carry formal charges of −1, and the charge of the nitrogen atom is +1. The bond lengths should all be equal and should lie between the values given in Table 3.5 for N—O (1.43 Å) and N=O (1.18 Å). The experimentally measured value is 1.24 Å.

Related Problems: 51, 52, 53, 54, 55, 56

Breakdown of the Octet Rule

Lewis diagrams and the octet rule are useful tools for predicting the types of molecules that will be stable under ordinary conditions of temperature and pressure. For example, we can write a simple Lewis diagram for water (H_2O),

H:Ö:H

in which each atom has a noble-gas configuration. It is impossible to do this for OH or for H_3O, which suggests that these species are either unstable or highly reactive.

There are several situations in which the octet rule is *not* satisfied.

CASE 1: ODD-ELECTRON MOLECULES The electrons in a Lewis diagram that satisfies the octet rule must occur in pairs—bonding pairs or lone pairs. Any molecule that has an odd number of electrons cannot satisfy the octet rule. Most stable molecules have even numbers of electrons, but a few have odd numbers. An example is nitrogen oxide (NO), a stable (although reactive) molecule that is an important factor in air pollution. Nitrogen oxide has 11 electrons, and the best electron dot diagram for it is

:Ṅ: :Ö

in which only the oxygen atom has a noble-gas configuration. The stability of NO contradicts the octet rule.

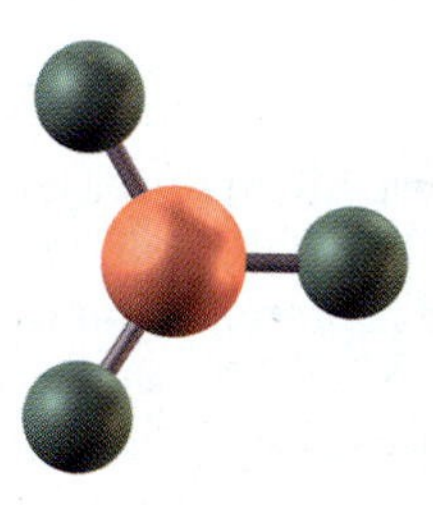

Boron trifluoride, BF_3, is a highly reactive gas that condenses to a liquid at −100°C. Its major use is in speeding up a large class of reactions that involve carbon compounds.

CASE 2: OCTET-DEFICIENT MOLECULES Some molecules are stable even though they have too few electrons to achieve an octet. For example, the standard rules for BF_3 would lead to the Lewis diagram

```
     :F̤̈:
      |   +1
:F̤̈—B=F̤̈
     −1
```

but experimental evidence strongly suggests there are no double bonds in BF_3 (fluorine never forms double bonds).

Moreover, the placement of a positive formal charge on fluorine is never correct. The following diagram avoids both problems:

```
     :F̤̈:
      |
:F̤̈—B—F̤̈:
```

Although this Lewis diagram denies an octet to the boron atom, it does at least assign zero formal charges to all atoms.

CASE 3: VALENCE SHELL EXPANSION Lewis diagrams become more complex in compounds formed from elements in the third and subsequent periods. For example, sulfur forms some compounds that are readily described by Lewis diagrams that give closed shells to all atoms. An example is hydrogen sulfide (H_2S),

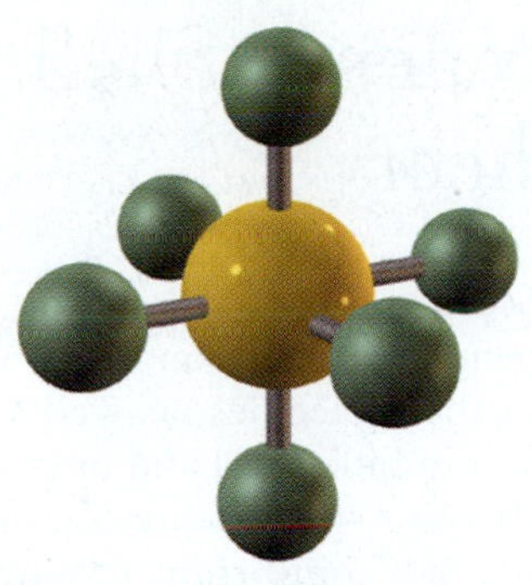

Sulfur hexafluoride, SF_6, is an extremely stable, dense, and unreactive gas. It is used as an insulator in high-voltage generators and switches.

which is analogous to water in its Lewis diagram. In sulfur hexafluoride (SF_6), however, the central sulfur is bonded to six fluorine atoms.

This molecule cannot be described by a Lewis diagram unless more than eight electrons are allowed around the sulfur atom, a process called **valence shell expansion.** The resulting Lewis diagram is

[Lewis diagram of SF_6: central S bonded to six F atoms, each F with three lone pairs]

The fluorine atoms have octets, but the central sulfur atom shares a total of 12 electrons.

In the standard procedure for writing Lewis diagrams, the need for valence shell expansion is signaled when the number of shared electrons is not sufficient to place a bonding pair between each pair of atoms that are supposed to be bonded. In SF_6, for example, 48 electrons are available, but 56 are needed to form separate octets on 7 atoms. This means that $56 - 48 = 8$ electrons would be shared. Four electron pairs are not sufficient to make even single bonds between the central S atom and the 6 terminal F atoms. In this case, we still follow rule 4 (assign one bonding pair to each bond in the molecule or ion; see earlier), even though it requires that we use more than 8 electrons. Rule 5 becomes irrelevant because there are no extra shared electrons. Rule 6 is now replaced with a new rule:

> ***Rule 6′:*** *Assign lone pairs to the terminal atoms to give them octets. If any electrons still remain, assign them to the central atoms as lone pairs.*

The effect of rule 6′ is to abandon the octet rule for the central atom but preserve it for the terminal atoms.

EXAMPLE 3.6

Write a Lewis diagram for the linear I_3^- (tri-iodide) ion.

SOLUTION

There are 7 valence electrons from each iodine atom plus 1 from the overall ion charge, giving a total of 22. Because $3 \times 8 = 24$ electrons would be needed in separate octets, $24 - 22 = 2$ are shared according to the original rules. Two electrons are not sufficient to make two different bonds, however, so valence expansion is necessary.

A pair of electrons is placed in each of the two bonds, and rule 6′ is used to complete the octets of the two terminal I atoms. This leaves

$$:\ddot{\underset{..}{I}}—I—\ddot{\underset{..}{I}}:$$

At this stage, 16 valence electrons have been used. The remaining 6 are placed as lone pairs on the central I atom. A formal charge of −1 then resides on this atom:

$$:\ddot{\underset{..}{I}}—\overset{-1}{\ddot{\underset{..}{I}}}—\ddot{\underset{..}{I}}:$$

Note the valence expansion on the central atom: It shares or owns a total of 10 electrons, rather than the 8 required by adherence to the octet rule.

Related Problems: 57, 58

3.9 The Shapes of Molecules: Valence Shell Electron-Pair Repulsion Theory

When two molecules approach one another to begin a chemical reaction, the probability of a successful encounter can depend critically on the three-dimensional shapes and the relative orientation of the molecules, as well as on their chemical identities. Shape is especially important in biological and biochemical reactions, in which molecules must fit precisely onto specific sites on membranes and templates; drug and enzyme activity are important examples. Characterization of molecular shape is therefore an essential part of the study of molecular structure.

The structure of a stable molecule is defined by the three-dimensional arrangement of its constituent atoms. Pictorial representations of molecular structure—for example, the familiar "ball-and-stick" models and sketches—show the positions of the nuclei of the constituent atoms, but not the positions of their electrons. The electrons are responsible for the chemical bonds that hold the atomic nuclei together as the molecule. Several properties characterize the three-dimensional structure of molecules (Fig. 3.17). The bond length measures the distance between the atomic nuclei in a particular bond; summing bond lengths projected along the three Cartesian axes provides a measure of the size and shape of the molecule. Bond angles, defined as the angle between the axes of adjacent bonds, provide a more detailed view of the three-dimensional structures of molecules. Finally, the relationships between planes defined by three atoms having one atom in common (the angle between these planes is the *dihedral angle*) provide additional insights into the topology of simple molecules. However, molecules are not rigid entities, with structures that are precisely defined by the coordinates of their nuclei. Their atoms vibrate about their equilibrium positions, albeit with relatively small displacements. Average bond lengths and angles are measured by spectroscopic techniques (see Chapter 21) and x-ray diffraction (see Section 22.1).

Molecular shape or geometry is governed by energetics; a molecule assumes the geometry that gives it the lowest potential energy. Sophisticated quantum mechanical calculations consider numerous possible geometrical arrangements for a molecule, calculate the total potential energy of the molecule for each arrangement, and identify the arrangement that gives the lowest potential energy. This procedure can be mimicked within the approximate classical model described in

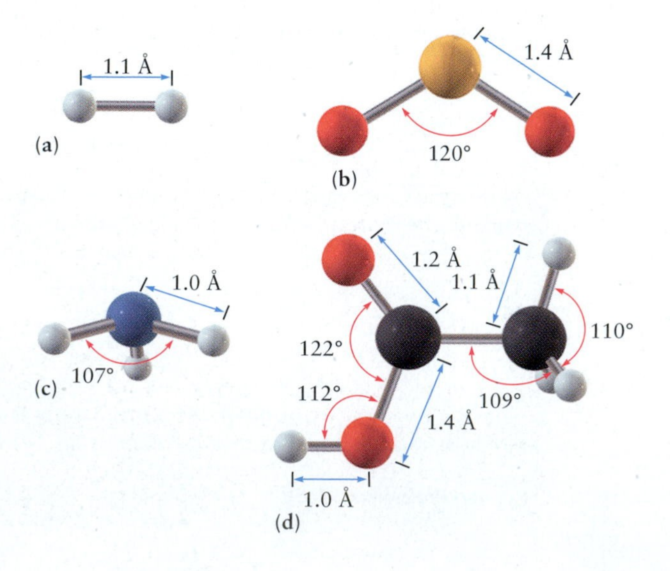

FIGURE 3.17 Three-dimensional molecular structures of (a) H_2, (b) SO_2, (c) NH_3, and (d) $C_2H_4O_2$, showing bond lengths and angles. *(Courtesy of Prof. Andrew J. Pounds, Mercer University, Macon, GA, and Dr. Mark A. Iken, Scientific Visualization Laboratory, Georgia Institute of Technology, Atlanta, GA.)*

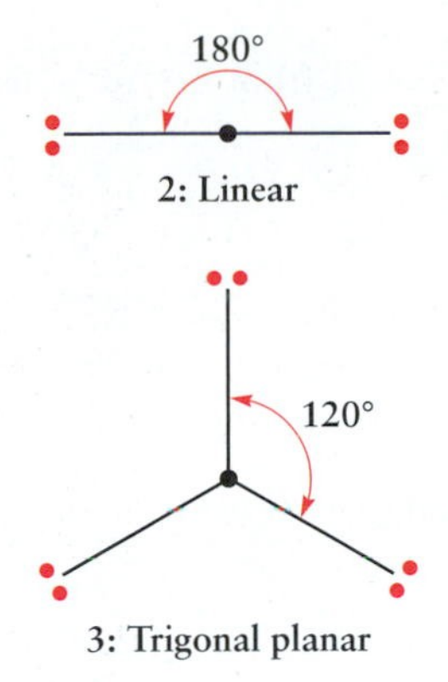

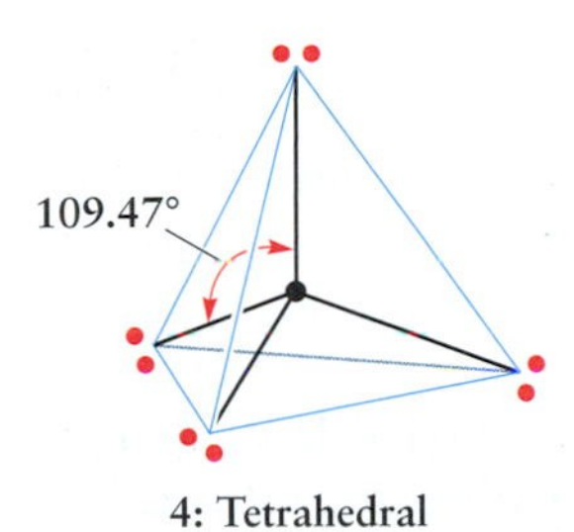

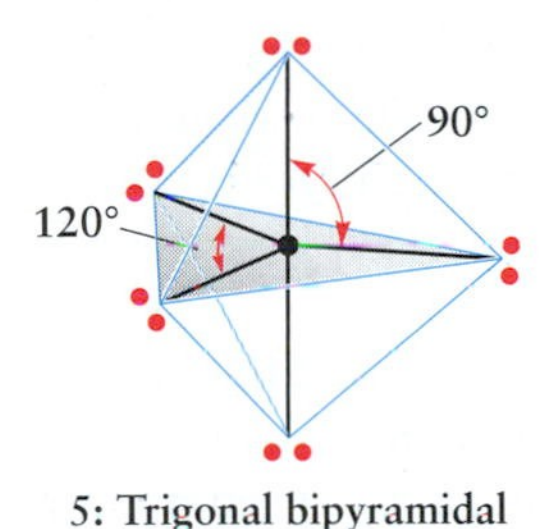

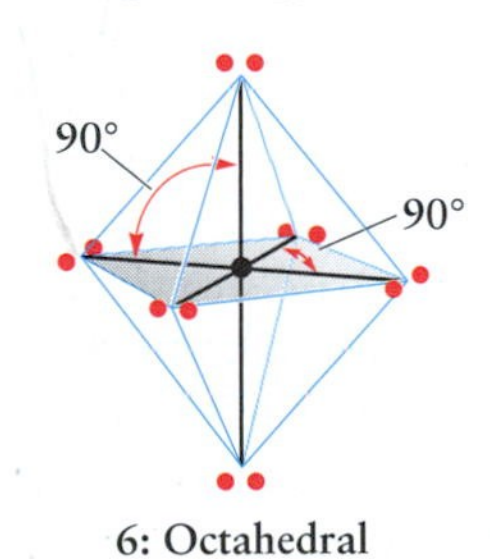

FIGURE 3.18 The positions of minimum energy for electron pairs on a sphere centered on the nucleus of an atom. The angles between the electron pairs are indicated. For two, three, and four electron pairs they are 180°, 120°, and 109.47°, respectively.

this chapter by considering numerous possible arrangements of bond angles and then identifying the one that corresponds to the lowest potential energy of the molecule. Because a covalent bond is formed by the sharing of a pair of electrons between two atoms (as described by the Lewis model in Section 3.8), changes in bond angles change the relative positions of the electron pairs around a given central atom. Electrons tend to repel each other through the electrostatic (Coulomb) repulsion between like charges and through quantum mechanical effects. Consequently, it is desirable in terms of energy for electrons to avoid each other. The VSEPR theory provides procedures for predicting molecular geometry by minimizing the potential energy due to electron-pair repulsions.

The Valence Shell Electron-Pair Repulsion Theory

The VSEPR theory starts with the fundamental idea that electron pairs in the valence shell of an atom repel each other. These include both lone pairs, which are localized on the atom and are not involved in bonding, and bonding pairs, which are shared covalently with other atoms. The electron pairs position themselves as far apart as possible to minimize their repulsions. The molecular geometry, which is defined by the positions of the *nuclei*, is then traced from the relative locations of the electron pairs.

The arrangement that minimizes repulsions naturally depends on the number of electron pairs. Figure 3.18 shows the minimum energy configuration for two to six electron pairs located around a central atom. Two electron pairs place themselves on opposite sides of the atom in a linear arrangement, three pairs form a trigonal planar structure, four arrange themselves at the corners of a tetrahedron, five define a trigonal bipyramid, and six define an octahedron. To find which geometry applies, we determine the **steric number,** *SN*, of the central atom, which is defined as

$$SN = \begin{pmatrix} \text{number of atoms} \\ \text{bonded to central atom} \end{pmatrix} + \begin{pmatrix} \text{number of lone pairs} \\ \text{on central atom} \end{pmatrix}$$

The steric number of an atom in a molecule can be determined by inspection from the Lewis diagram of the molecule.

EXAMPLE 3.7

Calculate steric numbers for iodine in IF_4^- and for bromine in BrO_4^-. These molecular ions have central I or Br surrounded by the other four atoms.

SOLUTION

The central I^- atom has eight valence electrons.

$$\left[\begin{array}{ccc} & :\ddot{F}: & \\ & | & \\ :\ddot{F}- & \ddot{\underset{..}{I}} & -\ddot{F}: \\ & | & \\ & :\underset{..}{F}: & \end{array} \right]^-$$

Each F atom has seven valence electrons of its own and needs to share one of the electrons from I^- to achieve a noble-gas configuration. Thus, four of the I^- valence electrons take part in covalent bonds, leaving the remaining four to form two lone pairs. The steric number is given by

$$SN = 4 \text{ (bonded atoms)} + 2 \text{ (lone pairs)} = 6$$

In BrO_4^-, each oxygen atom needs to share two of the electrons from Br^- to achieve a noble-gas configuration.

```
        ..
       :O:
        |
  ..    |     ..
[ :O—Br—O: ]⁻
  ..    |     ..
        |
       :O:
        ..
```

Because this assignment accounts for all eight of the Br^- valence electrons, there are no lone pairs on the central atom and

$$SN = 4 \text{ (bonded atoms)} + 0 \text{ (lone pairs)} = 4$$

Double-bonded or triple-bonded atoms count the same as single-bonded atoms in determining the steric number. In CO_2, for example, two double-bonded oxygen atoms are attached to the central carbon and there are no lone pairs on that atom, so $SN = 2$.

The steric number is used to predict molecular geometries. In molecules XY_n, in which there are no lone pairs on the central atom X (the simplest case),

$$SN = \text{number of bonded atoms} = n$$

The n bonding electron pairs (and therefore the outer atoms) position themselves (see Fig. 3.18) to minimize electron-pair repulsion. Thus, CO_2 is predicted (and found experimentally) to be linear, BF_3 is trigonal planar, CH_4 is tetrahedral, PF_5 is trigonal bipyramidal, and SF_6 is octahedral.

When lone pairs are present, the situation changes slightly. There can now be three different types of repulsions: (1) bonding pair against bonding pair, (2) bonding pair against lone pair, and (3) lone pair against lone pair. Consider the ammonia molecule (NH_3), which has three bonding electron pairs and one lone pair (Fig. 3.19a). The steric number is 4, and the electron pairs arrange themselves into an approximately tetrahedral structure. The lone pair is not identical to the three bonding pairs, however, so there is no reason for the electron-pair structure to be *exactly* tetrahedral. It is found that lone pairs tend to occupy more space than bonding pairs (because the bonding pairs are held closer to the central atom), so the angles of bonds opposite to them are reduced. The geometry of the *molecule,* as distinct from that of the electron pairs, is named for the sites occupied by actual atoms. The description of the molecular geometry makes no reference to lone pairs that may be present on the central atom, even though their presence affects that geometry. The structure of the ammonia molecule is thus predicted to be a trigonal pyramid in which the H—N—H bond angle is smaller than the tetrahedral angle of 109.5°. The observed structure has an H—N—H bond angle of 107.3°. The H—O—H bond angle in water, which has two lone pairs and two bonding pairs, is still smaller at 104.5° (see Fig. 3.19b).

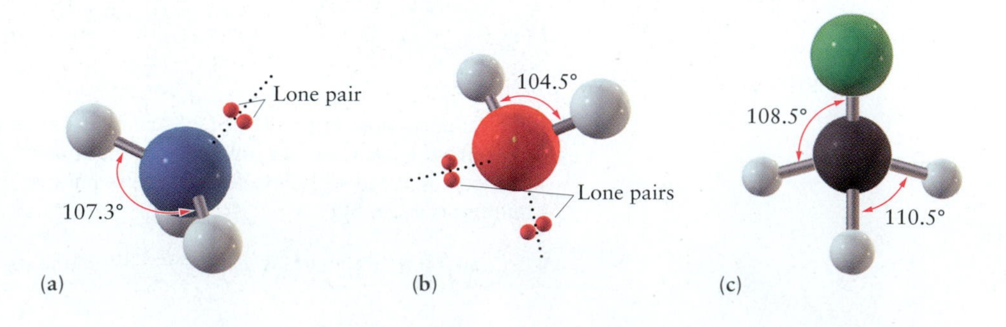

FIGURE 3.19 (a) Ammonia (NH_3) has a pyramidal structure in which the bond angles are less than 109.5°. (b) Water (H_2O) has a bent structure with a bond angle less than 109.5° and smaller than that of NH_3. (c) CH_3Cl has a distorted tetrahedral structure.

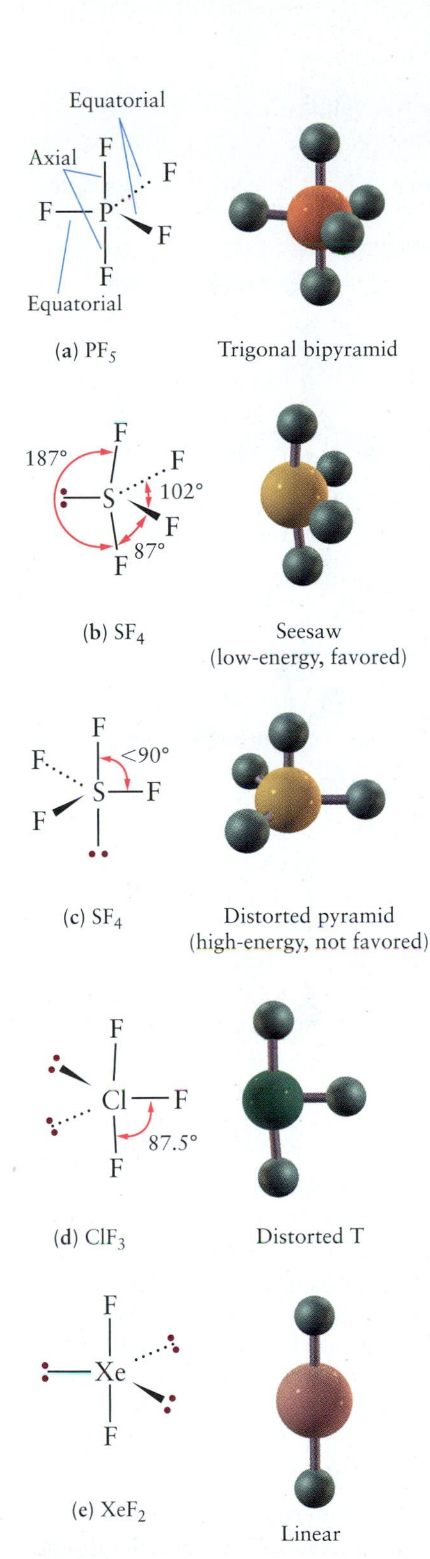

FIGURE 3.20 Molecules with steric number 5. The molecular geometry is named for the sites occupied by the *atoms*, not the underlying trigonal-bipyramidal structure of electron pairs.

A similar distortion takes place when two types of outer atoms are present. In CH_3Cl, the bonding electron pair in the C—Cl bond is not the same as those in the C—H bonds, so the structure is a distorted tetrahedron (see Fig. 3.19c). Because Cl is more electronegative than H, it tends to attract electrons away from the central atom, reducing the electron-pair repulsion. This allows the Cl—C—H bond angles to become 108.5°, which is smaller than tetrahedral, whereas the H—C—H angles become 110.5°, which is larger than tetrahedral. In effect, electropositive substituents repel other substituents more strongly than do electronegative substituents.

The fluorides PF_5, SF_4, ClF_3, and XeF_2 all have steric number 5 but have different numbers of lone pairs (0, 1, 2, and 3, respectively). What shapes do their molecules have? We have already mentioned that PF_5 is trigonal bipyramidal. Two of the fluorine atoms occupy **axial** sites (See Fig. 3.20a), and the other three occupy **equatorial** sites. Because the two kinds of sites are not equivalent, there is no reason for all of the P—F bond lengths to be equal. Experiment shows that the equatorial P—F bond length is 1.534 Å, which is shorter than the 1.577 Å axial P—F lengths.

SF_4 has four bonded atoms and one lone pair. Does the lone pair occupy an axial or an equatorial site? In the VSEPR theory, when electron pairs form a 90° angle to the central atom, they repel each other much more strongly than when the angle is larger. A single lone pair therefore finds a position that minimizes the number of 90° repulsions it has with bonding electron pairs. It occupies an equatorial position with two 90° repulsions (see Fig. 3.20b), rather than an axial position with three 90° repulsions (see Fig. 3.20c). The axial S—F bonds are bent slightly away from the lone pair; consequently, the molecular structure of SF_4 is a distorted seesaw. A second lone pair (in ClF_3, for example) also takes an equatorial position, leading to a distorted T-shaped molecular structure (see Fig. 3.20d). A third lone pair (in XeF_2 or I_3^-, for example) occupies the third equatorial position, and the molecular geometry is linear (see Fig. 3.20e). Table 3.8 summarizes the molecular shapes predicted by VSEPR and provides examples of each.

EXAMPLE 3.8

Predict the geometry of the following molecules and ions: (a) ClO_3^+, (b) ClO_2^+, (c) SiH_4, (d) IF_5.

SOLUTION

(a) The central Cl atom has all of its valence electrons (six, because the ion has a net positive charge) involved in bonds to the surrounding three oxygen atoms and has no lone pairs. Its steric number is 3. In the molecular ion, the central Cl should be surrounded by the three O atoms arranged in a trigonal planar structure.

(b) The central Cl atom in this ion also has a steric number of 3, comprising two bonded atoms and a single lone pair. The predicted molecular geometry is a bent molecule with an angle somewhat less than 120°.

(c) The central Si atom has a steric number of 4 and no lone pairs. The molecular geometry should consist of the Si atom surrounded by a regular tetrahedron of H atoms.

(d) Iodine has seven valence electrons, of which five are shared in bonding pairs with F atoms. This leaves two electrons to form a lone pair, so the steric number is 5 (bonded atoms) + 1 (lone pair) = 6. The structure will be based on the octahedron of electron pairs from Figure 3.18, with five F atoms and one lone pair. The lone pair can be placed on any one of the six equivalent sites and will cause the four F atoms to bend away from it toward the fifth F atom, giving the distorted structure shown in Figure 3.21.

Related Problems: 59, 60, 61, 62

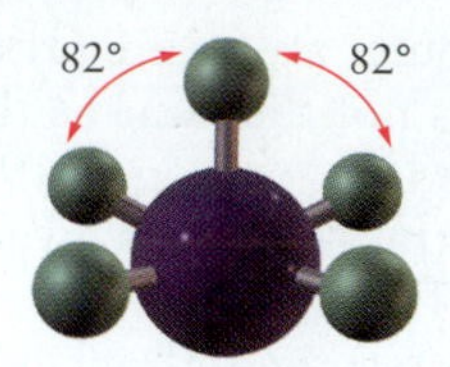

FIGURE 3.21 The structure of IF_5. Note the distortions of F—I—F bond angles from 90° because of the lone pair at the bottom (not shown).

TABLE 3.8 Molecular Shapes Predicted by the Valence Shell Electron-Pair Repulsion Theory

Molecule	Steric Number	Predicted Geometry		Example
AX_2	2	Linear	180°	CO_2
AX_3	3	Trigonal planar	120°	BF_3
AX_4	4	Tetrahedral	109.5°	CF_4
AX_5	5	Trigonal bipyramidal	120° 90°	PF_5
AX_6	6	Octahedral	90°	SF_6

The VSEPR theory is a simple but remarkably powerful model for predicting the geometries and approximate bond angles of molecules that have a central atom. Its success is even more remarkable when we realize that the VSEPR theory is based purely on empirical arguments, not theoretical calculations. However, it has its limitations. The VSEPR theory does not account for the fact that observed bond angles in the Group V and VI hydrides H_2S (92°), H_2Se (91°), PH_3 (93°), and AsH_3 (92°) are so far from tetrahedral (109.5°) and so close to right angles (90°).

Dipole Moments of Polyatomic Molecules

Polyatomic molecules, like the diatomic molecules considered in the previous section, may have dipole moments. Often, a good approximation is to assign a dipole moment (which is now a vector, shown as an arrow pointing from the positive charge to the negative charge) to each bond, and then obtain the total dipole moment by carrying out a vector sum of the bond dipoles. In CO_2, for example, each C—O has a bond dipole (Fig. 3.22a). However, because the dipoles are equal in magnitude and point in opposite directions, the total molecular dipole moment vanishes. In OCS, which is also linear with a central carbon atom, the C—O and C—S bond dipole moments have different magnitudes, leaving a net non-zero dipole moment (see Fig. 3.22b). In the water molecule (Fig. 3.22c), the two bond dipoles add vectorially to give a net molecular dipole moment. The more symmetric molecule CCl_4, on the other hand, has no net dipole moment (see Fig. 3.22d). Even though each of the four C—Cl bonds (pointing from the four corners of a tetrahedron) has

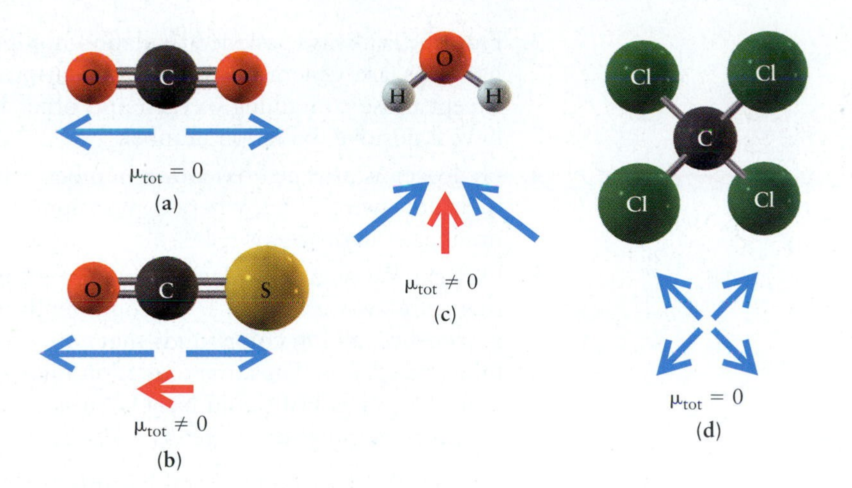

FIGURE 3.22 The total dipole moment of a molecule is obtained by vector addition of its bond dipoles. This operation is performed by adding the arrows when they lie pointing in the same direction, and subtracting the arrows if they lie pointing in different directions. (a) CO_2. (b) OCS. (c) H_2O. (d) CCl_4.

a dipole moment, their vector sum is zero. Molecules such as H_2O and OCS, with nonzero dipole moments, are **polar;** those such as CO_2 and CCl_4, with no net dipole moment, are **nonpolar,** even though they contain polar bonds. Intermolecular forces differ between polar and nonpolar molecules (see discussion in Chapter 10), a fact that significantly affects their physical and chemical properties.

EXAMPLE 3.9

Predict whether the molecules NH_3 and SF_6 will have dipole moments.

SOLUTION

The NH_3 molecule has a dipole moment because the three N—H bond dipoles add to give a net dipole pointing upward to the N atom from the base of the pyramid that defines the NH_3 structure. The SF_6 molecule has no dipole moment because each S—F bond dipole is balanced by one of equal magnitude pointing in the opposite direction on the other side of the molecule.

Related Problems: 65, 66

3.10 Oxidation Numbers

The terms *ionic* and *covalent* have been used in this chapter as though they represent the complete transfer of one or more electrons from one atom to another or the equal sharing of pairs of electrons between atoms in a molecule. In reality, both types of bonding are idealizations that rarely apply exactly. To account for the transfer of electrons from one molecule or ion to another in oxidation–reduction reactions and to name different binary compounds of the same elements, it is not necessary to have detailed knowledge of the exact electron distributions in molecules, whether they are chiefly ionic or covalent. Instead, we can assign convenient fictitious charges to the atoms in a molecule and call them **oxidation numbers,** making certain that the law of charge conservation is strictly observed. Oxidation numbers are chosen so that in ionic compounds the oxidation number coincides with the charge on the ion. The following simple conventions are useful:

1. The oxidation numbers of the atoms in a neutral molecule must add up to zero, and those in an ion must add up to the charge on the ion.
2. Alkali-metal atoms are assigned the oxidation number +1, and the alkaline-earth atoms +2, in their compounds.

3. Fluorine is always assigned oxidation number −1 in its compounds. The other halogens are generally assigned oxidation number −1 in their compounds, except those containing oxygen and other halogens in which the halogen can have a positive oxidation number.
4. Hydrogen is assigned oxidation number +1 in its compounds, except in metal hydrides such as LiH, where convention 2 takes precedence and the oxidation number of hydrogen is −1.
5. Oxygen is assigned oxidation number −2 in nearly all compounds. However, there are two exceptions: In compounds with fluorine, convention 3 takes precedence, and in compounds that contain O—O bonds, conventions 2 and 4 take precedence. Thus, the oxidation number for oxygen in OF_2 is +2; in peroxides (such as H_2O_2 and Na_2O_2), it is −1. In superoxides (such as KO_2), the oxidation number of oxygen is $-\frac{1}{2}$.

Convention 1 is fundamental because it guarantees charge conservation: The total number of electrons must remain constant in chemical reactions. This rule also makes the oxidation numbers of the neutral atoms of all elements zero. Conventions 2 to 5 are based on the principle that in ionic compounds the oxidation number should equal the charge on the ion. Note that fractional oxidation numbers, although uncommon, are allowed and, in fact, are necessary to be consistent with this set of conventions.

With the preceding conventions in hand, chemists can assign oxidation numbers to the atoms in most compounds. Apply conventions 2 through 5 as listed previously, noting the exceptions given, and then assign oxidation numbers to the other elements in such a way that convention 1 is always obeyed. Note that convention 1 applies not only to free ions, but also to the components that make up ionic solids. Chlorine has oxidation number −1 not only as a free Cl^- ion, but in the ionic solid AgCl and in covalent CH_3Cl. It is important to recognize common ionic species (especially molecular ions) and to know the total charges they carry. Table 3.9 lists the names and formulas of many common anions. Inspection of the table reveals that several elements exhibit different oxidation numbers in different compounds. In Ag_2S, sulfur appears as the sulfide ion and has oxidation number −2, but in Ag_2SO_4, it appears as part of a sulfate (SO_4^{2-}) ion. In this case,

$$(\text{oxidation number of S}) + [4 \times (\text{oxidation number of O})] = \text{total charge on ion}$$

$$x + [4(-2)] = -2$$

$$\text{oxidation number of S} = x = +6$$

TABLE 3.9 Formulas and Names of Some Common Anions

F^-	Fluoride	CO_3^{2-}	Carbonate
Cl^-	Chloride	HCO_3^-	Hydrogen carbonate
Br^-	Bromide	NO_2^-	Nitrite
I^-	Iodide	NO_3^-	Nitrate
H^-	Hydride	SiO_4^{4-}	Silicate
O^{2-}	Oxide	PO_4^{3-}	Phosphate
S^{2-}	Sulfide	HPO_4^{2-}	Hydrogen phosphate
O_2^{2-}	Peroxide	$H_2PO_4^-$	Dihydrogen phosphate
O_2^-	Superoxide	SO_3^{2-}	Sulfite
OH^-	Hydroxide	SO_4^{2-}	Sulfate
CN^-	Cyanide	HSO_4^-	Hydrogen sulfate
CNO^-	Cyanate	ClO^-	Hypochlorite
SCN^-	Thiocyanate	ClO_2^-	Chlorite
MnO_4^-	Permanganate	ClO_3^-	Chlorate
CrO_4^{2-}	Chromate	ClO_4^-	Perchlorate
$Cr_2O_7^{2-}$	Dichromate		

FIGURE 3.23 Several oxides of manganese. They are arranged in order of increasing oxidation number of the Mn, counterclockwise from the bottom left: MnO, Mn_3O_4, Mn_2O_3 and MnO_2. A compound of still higher oxidation state, Mn_2O_7, is a dark red liquid that explodes easily.

A convenient way to indicate the oxidation number of an atom is to write it directly above the corresponding symbol in the formula of the compound:

$$\overset{+1}{N_2}\overset{-2}{O} \qquad \overset{+1}{Li}\overset{-1}{H} \qquad \overset{0}{O_2} \qquad \overset{+6}{S}\overset{-2}{O_4^{2-}}$$

EXAMPLE 3.10

Assign oxidation numbers to the atoms in the following chemical compounds and ions: NaCl, ClO^-, $Fe_2(SO_4)_3$, SO_2, I_2, $KMnO_4$, CaH_2.

SOLUTION

$\overset{+1}{Na}\overset{-1}{Cl}$	From conventions 2 and 3.
$\overset{+1}{Cl}\overset{-2}{O^-}$	From conventions 1 and 5.
$\overset{+3}{Fe_2}(\overset{+6}{S}\overset{-2}{O_4})_3$	From conventions 1 and 5. This is solved by recognizing the presence of sulfate (SO_4^{2-}) groups.
$\overset{+4}{S}\overset{-2}{O_2}$	From conventions 1 and 5.
$\overset{0}{I_2}$	From convention 1. I_2 is an element.
$\overset{+1}{K}\overset{+7}{Mn}\overset{-2}{O_4}$	From conventions 1, 2, and 5.
$\overset{+2}{Ca}\overset{-1}{H_2}$	From conventions 1 and 2 (metal hydride case).

Related Problems: 71, 72

Oxidation numbers must not be confused with the formal charges on Lewis dot diagrams (see Section 3.8). They resemble formal charges to the extent that both are assigned, by arbitrary conventions, to symbols in formulas for specific purposes. The purposes differ, however. Formal charges are used solely to identify preferred Lewis diagrams. Oxidation numbers are used in nomenclature, in

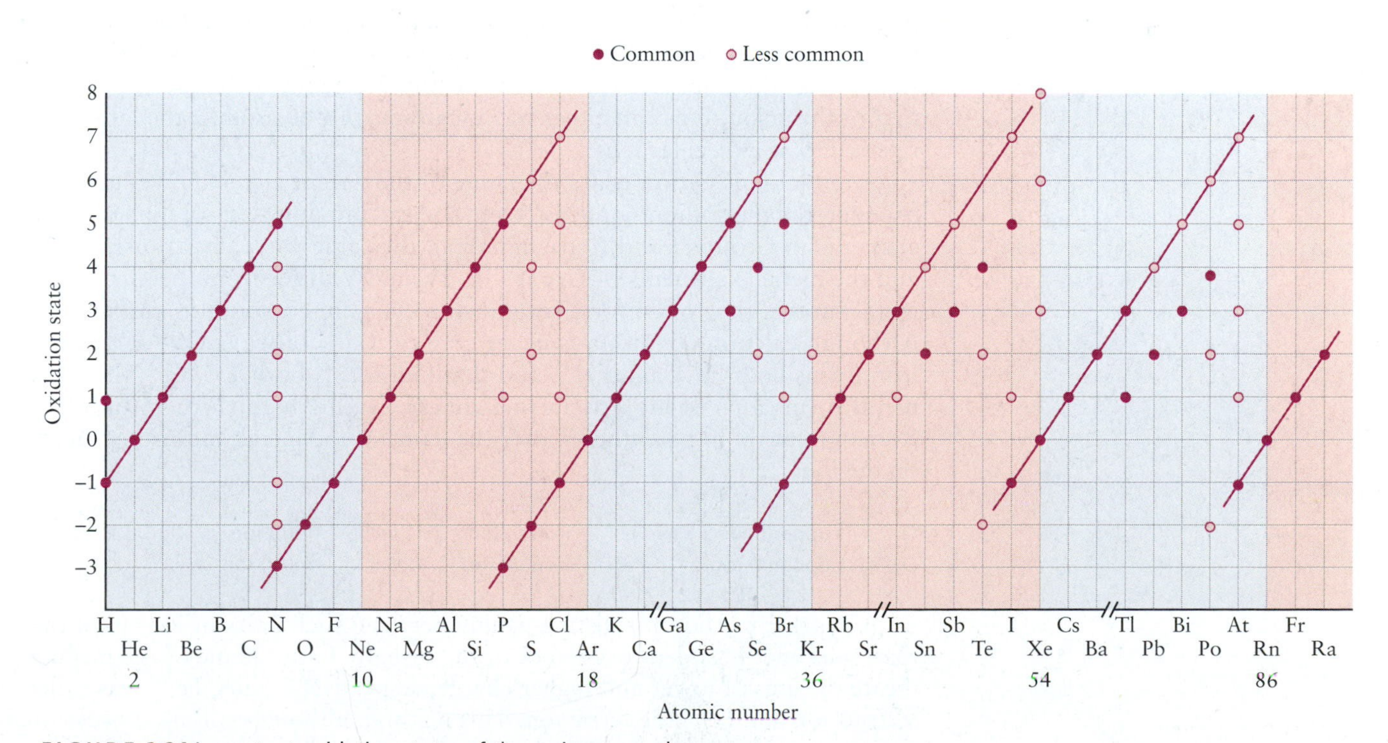

FIGURE 3.24 Important oxidation states of the main-group elements.

identifying oxidation–reduction reactions, and in exploring trends in chemical reactivity across the periodic table. Oxidation numbers are often not the same as the true charges on atoms. In $KMnO_4$ or Mn_2O_7, for example, Mn has the oxidation number +7, but the actual net charge on the atom is much less than +7. A high oxidation state usually indicates significant covalent character in the bonding of that compound. The oxides of manganese with lower oxidation states (Fig. 3.23) have more ionic character.

Figure 3.24 shows the most common oxidation states of the elements of the main groups. The strong diagonal lines for these elements reflect the stability of closed electron octets. For example, the oxidation states +6 and −2 for sulfur would correspond to losing or gaining enough electrons to attain a noble gas configuration, in this case that of Ar. The octet-based Lewis model is not useful for compounds of the transition elements (see Chapter 8 for further discussion).

3.11 Inorganic Nomenclature

Sections 3.6 and 3.7 used the periodic table and the Lewis model to describe the transfer and sharing of electrons in chemical compounds. We conclude this chapter by discussing the systematic nomenclature of inorganic compounds. The definitive source for the naming of inorganic compounds is *Nomenclature of Inorganic Chemistry-IUPAC Recommendations 2005* (N. G. Connelly and T. Damhus, Sr., Eds. Royal Society of Chemistry, 2005). Of the wide variety of inorganic compounds, we discuss only simple covalent and ionic compounds and emphasize current usage.

Names and Formulas of Ionic Compounds

Ionic compounds are made from ions, bound by electrostatic forces such that the total charge of the compound is zero. The stoichiometry of an ionic compound is determined by inspection, to satisfy charge neutrality. As mentioned earlier, gas-phase ionic compounds exist, but most ionic compounds are solids in which each ion is strongly bound to a number of the nearest neighbor ions of the opposite sign. An ionic compound is named by listing the name of the cation followed by that of the anion. Ions can either be monatomic or polyatomic; the latter are also referred to as molecular ions.

A monatomic cation bears the name of the parent element. We have already encountered the sodium ion (Na^+) and the calcium ion (Ca^{2+}); ions of the other elements in Groups I and II are named in the same way. The transition metals and the metallic elements of Groups III, IV, and V differ from the Group I and II metals in that they often form several stable ions in compounds and in solution. Although calcium compounds never contain Ca^{3+} (always Ca^{2+}), the element iron forms both Fe^{2+} and Fe^{3+}, and thallium forms both Tl^+ and Tl^{3+}. When a metal forms ions of more than one charge, we distinguish them by placing a Roman numeral in parentheses after the name of the metal, indicating the charge on the ion:

Cu^+	copper(I) ion	Fe^{2+}	iron(II) ion	Sn^{2+}	tin(II) ion
Cu^{2+}	copper(II) ion	Fe^{3+}	iron(III) ion	Sn^{4+}	tin(IV) ion

An earlier method for distinguishing between such pairs of ions used the suffixes *-ous* and *-ic* added to the root of the (usually Latin) name of the metal to indicate the ions of lower and higher charge, respectively. Thus, Fe^{2+} was called the ferrous ion and Fe^{3+} the ferric ion. This method, although still used occasionally, is not recommended for systematic nomenclature and is not used in this book.

A few polyatomic cations are important in inorganic chemistry. These include the ammonium ion, NH_4^+ (obtained by adding H^+ to ammonia); the hydronium ion, H_3O^+ (obtained by adding H^+ to water); and the particularly interesting molecular ion formed by mercury, Hg_2^{2+}, the mercury(I) ion. This species must be carefully distinguished from Hg^{2+}, the mercury(II) ion. The Roman numeral I in parentheses means, in this case, that the average charge on each of the two mercury atoms is +1. Compounds with the empirical formulas HgCl and HgBr have molecular formulas Hg_2Cl_2 and Hg_2Br_2, respectively.

A monatomic anion is named by adding the suffix *-ide* to the first portion of the name of the element. Thus, chlor*ine* becomes the chlor*ide* ion, and ox*ygen* becomes the ox*ide* ion. The other monatomic anions of Groups V, VI, and VII are named similarly. Many polyatomic anions exist, and the naming of these species is more complex. The names of the oxoanions (each contains oxygen in combination with a second element) are derived by adding the suffix *-ate* to the stem of the name of the second element. Some elements form two oxoanions. The *-ate* ending is then used for the oxoanion with the larger number of oxygen atoms (such as NO_3^-, nitr*ate*), and the ending *-ite* is added for the name of the anion with the smaller number (such as NO_2^-, nitr*ite*). For elements such as chlorine, which form more than two oxoanions, we use the additional prefixes *per-* (largest number of oxygen atoms) and *hypo-* (smallest number of oxygen atoms). An oxoanion that contains hydrogen as a third element includes that word in its name. For example, the HCO_3^- oxoanion is called the hydrogen carbonate ion in preference to its common (nonsystematic) name, "bicarbonate ion," and HSO_4^-, often called "bisulfate ion," is better designated as the hydrogen sulfate ion. Table 3.9 lists some of the most important anions. It is important to be able to recognize and name the ions from that table, bearing in mind that the electric charge is an essential part of the formula.

It is customary (and recommended) to name ionic compounds (salts) using the Roman numeral notation, based on group numbers or oxidation states. $CuSO_4$ is copper(I) sulfate, $Fe(NO_3)_2$ is iron(II) nitrate and $KMnO_4$ is potassium permanganate, the Roman numeral being omitted in the last example because potassium cations always carry only +1 charge.

The composition of an ionic compound is determined by overall charge neutrality. The total positive charge on the cations must exactly balance the total negative charge on the anions. The following names and formulas of ionic compounds illustrate this point:

Tin(II) bromide	One 2+ cation, two 1− anions	$SnBr_2$
Potassium permanganate	One 1+ cation, one 1− anion	$KMnO_4$
Ammonium sulfate	Two 1+ cations, one 2− anion	$(NH_4)_2SO_4$
Iron(II) dihydrogen phosphate	One 2+ cation, two 1− anions	$Fe(H_2PO_4)_2$

EXAMPLE 3.11

Give the chemical formulas of (a) calcium cyanide and (b) copper(II) phosphate.

SOLUTION

(a) Calcium cyanide is composed of Ca^{2+} and CN^-. For the overall charge to be 0, there must be two CN^- for each Ca^{2+}. Thus, the chemical formula of calcium cyanide is $Ca(CN)_2$.

(b) The ions present in copper(II) phosphate are Cu^{2+} and PO_4^{3-}. To ensure charge neutrality, there must be three Cu^{2+} (total charge +6) and two PO_4^{3-} (total charge −6) per formula unit. Thus, the chemical formula of copper(II) phosphate is $Cu_3(PO_4)_2$.

Related Problems: 77, 78

TABLE 3.10 Prefixes Used for Naming Binary Covalent Compounds

Number	Prefix
1	*mono-*
2	*di-*
3	*tri-*
4	*tetra-*
5	*penta-*
6	*hexa-*
7	*hepta-*
8	*octa-*
9	*nona-*
10	*deca-*

Naming Binary Covalent Compounds

How do we name nonionic (covalent) compounds? If a pair of elements forms only one compound, begin with the name of the element that appears first in the chemical formula, followed by the second element, with the suffix *-ide* added to its root. This is analogous to the naming of ionic compounds. Just as NaBr is sodium bromide, so the following names designate typical covalent compounds:

HBr	hydrogen bromide
$BeCl_2$	beryllium chloride
H_2S	hydrogen sulfide
BN	boron nitride

Many well-established nonsystematic names continue to be used (even the strictest chemist does not call water "hydrogen oxide"!) They include:

H_2O	water
NH_3	ammonia
N_2H_4	hydrazine
PH_3	phosphine
AsH_3	arsine
$COCl_2$	phosgene

If a pair of elements forms more than one compound, two methods can be used to distinguish between them:

1. Use Greek prefixes (Table 3.10) to specify the number of atoms of each element in the molecular formula of the compound (*di-* for two, *tri-* for three, and so forth). If the compound is a solid without well-defined molecules, name the empirical formula in this way. The prefix for one (*mono-*) is omitted, except in the case of carbon monoxide.
2. Write the oxidation number of the first-named element in Roman numerals and place it in parentheses after the name of that element.

Applying the two methods to the oxides of nitrogen gives

N_2O	dinitrogen oxide	nitrogen(I) oxide
NO	nitrogen oxide	nitrogen(II) oxide
N_2O_3	dinitrogen trioxide	nitrogen(III) oxide
NO_2	nitrogen dioxide	nitrogen(IV) oxide
N_2O_4	dinitrogen tetraoxide	nitrogen(IV) oxide
N_2O_5	dinitrogen pentaoxide	nitrogen(V) oxide

The first method has some advantages over the second and is recommended. It distinguishes between NO_2 (nitrogen dioxide) and N_2O_4 (dinitrogen tetraoxide), two distinct compounds that would both be called nitrogen(IV) oxide under the second system of nomenclature. Two of these oxides have common (nonsystematic) names that may be encountered elsewhere: N_2O is often called nitrous oxide, and NO is called nitric oxide.

CHAPTER SUMMARY

The classical description of chemical bonding provides the conceptual framework and language used by all chemists in their daily work. The classical description is based largely on simple electrostatics. We can understand a great deal about the nature of the chemical bond by examining the forces and potential energy of interaction between the electrons and the nuclei, which are governed by Coulomb's law. Potential energy diagrams will help you understand the interactions between and among electrons and nuclei that determine a wide variety of properties of atoms and molecules. We encourage you to learn to interpret and use these diagrams, because they will appear over and over again in this text and in subsequent chemistry courses.

The periodic table organizes the elements in a way that shows similar physical and chemical properties of groups and the variations of these properties across periods. Examination of periodic trends in ionization energies suggests that the electrons in an atom are organized in a series of concentric shells around the nucleus. These trends can be explained by the decreased electron–nuclear attraction in successive shells, as well as the screening or shielding of the outer electrons from the full nuclear charge by the inner electrons. Electron affinity is a measure of the tendency of an atom to form a stable anion. Periodic trends in electron affinities are explained in the same way as those observed for ionization energies.

Chemical bonds are generally classified according to the amount of charge separation in the bond. The character of a particular chemical bond—ionic, covalent, or polar covalent—can be predicted by comparing the electronegativities of the atoms involved. The electronegativity of an element is a measure of its relative propensity to attract electrons in a chemical bond. Elements with large ionization energies and large electron affinities tend to attract electrons, whereas those with small ionization energies and small electron affinities tend to donate electrons. Bonds formed between two atoms with large electronegativity differences tend to be ionic, whereas those formed between those atoms with nearly the same electronegativities tend to be covalent. Most bonds are somewhere in between—that is, they are polar covalent.

The arrangements of atoms in a molecule and the three-dimensional shape of a molecule can be rationalized or predicted using the Lewis dot model and the VSEPR theory, respectively. The Lewis model predicts the most likely arrangement of atoms in a molecule, the existence of multiple bonds, and charge distributions using the idea of formal charges. It is based on the idea that the representative elements (except hydrogen) are most stable with a filled octet of valence electrons (hydrogen only needs two). This stability can be understood from the empirical fact that eight electrons is the maximum each valence shell can hold, and that each electron added to a shell decreases the energy of the atom via the Coulomb interaction with the nucleus. Each element is represented by its chemical symbol, with its valence electrons arranged as dots. The atoms are combined in a way that creates the maximum number of filled octets around each representative element and a pair of electrons around each hydrogen atom. The VSEPR theory predicts the three-dimensional structures of molecules by arranging electron pairs to minimize their mutual repulsion.

Oxidation numbers are used to track the gain and loss of electrons in chemical reactions and are used in the systematic naming of inorganic compounds.

The concepts introduced in this chapter provide a sound basis for understanding a great deal of chemistry. A firm understanding of these concepts is essential for you to conduct and interpret experiments, read the chemical literature, and develop a deeper understanding based on quantum mechanics, which we develop in Chapters 4–6.

CUMULATIVE EXERCISE

Oxides and Peroxides

Consider the three compounds KO_2, BaO_2, and TiO_2. Each contains two oxygen atoms per metal atom, but the oxygen occurs in different forms in the three compounds.

(a) The oxygen in TiO_2 occurs as O^{2-} ions. What is the Lewis dot symbol for this ion? How many valence electrons does this ion have? What is the chemical name for TiO_2?

(b) Recall that Group II elements form stable 2+ ions. Using Table 3.9, identify the oxygen-containing ion in BaO_2 and give the name of the compound. Draw a Lewis diagram for the oxygen-containing ion, showing formal charge. Is the bond in this ion a single or a double bond?

(c) Recall that Group I elements form stable 1+ ions. Using Table 3.9, identify the oxygen-containing ion in KO_2 and give the name of the compound. Show that the oxygen-containing ion is an odd-electron species. Draw the best Lewis diagram you can for it.

Answers

(a) The ion $:\ddot{\underset{..}{O}}:^{2-}$ has eight valence electrons (an octet). TiO_2 is titanium(IV) oxide.

(b) The ion in BaO_2 must be the peroxide ion (O_2^{2-}), and the compound is barium peroxide. The Lewis diagram for the peroxide ion is

$$\overset{-1}{:\ddot{\underset{..}{O}}}\!:\!\overset{-1}{\ddot{\underset{..}{O}}}:$$

and the O—O bond is a single bond.

(c) The ion in KO_2 must be the superoxide ion (O_2^-), and the compound is potassium superoxide. The superoxide ion has 13 valence electrons. The best Lewis diagram is a pair of resonance diagrams

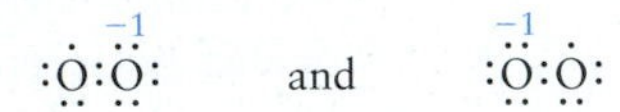

in which only one of the oxygen atoms attains an octet electron configuration.

CHAPTER REVIEW

- The periodic table organizes the elements into groups with similar physical and chemical properties.
- The physical structure of the atom, as determined by experiments, is summarized in the planetary model. In an atom with atomic number Z, there are Z electrons moving around a dense nucleus, which has positive charge $+Ze$.
- The physical structure of the atom can be understood using Coulomb's law, which describes the forces between each electron and the nucleus and between the electrons in many-electron atoms.
- Ionization energy is the minimum energy required to remove an electron from an atom and place it infinitely far away with zero kinetic energy.
- Periodic trends in ionization energies suggested the shell model of the atom in which electrons occupy concentric shells located at increasing distances from the nuclei.
- Electron affinity is the energy released when an electron attaches to an atom to form a stable anion; by convention, it is reported as a positive quantity.
- Ionization energies and electron affinities can be understood in terms of the effective potential energy of interaction between each electron and the nucleus, derived from Coulomb's law.
- Electronegativity is the relative tendency of atoms in a chemical bond to attract electrons. Mulliken's scale is based on a simple physical model for a gas-phase atom that averages the ionization energy and the electron affinity; Pauling's scale averages this tendency over a variety of bonding partners.
- Chemists generally distinguish among three types of chemical bonds based on the electronegativity differences between the atoms: covalent (nearly identical electronegativities), ionic (greatly different electronegativities), and polar

covalent (anything in between these extremes). The percent ionic character can be determined by measuring experimentally the dipole moments of molecules.

- The driving force for the formation of a chemical bond from a pair of atoms in the gas phase is the reduction in the total energy of the system. Bond formation is accompanied by a reduction in potential energy, relative to the free atoms. For ionic bonds, this is calculated easily using Coulomb's law.
- The formation of a covalent chemical bond can be understood classically by examining the forces between the electrons and the nuclei to discover bonding and antibonding regions in the locations of the electrons.
- The virial theorem relates the changes in average total energy in the formation of both ionic and covalent chemical bonds to the changes in the average potential energy and average kinetic energy.
- Important properties of the chemical bond include its length, dissociation energy, order, and dipole moment.
- Bond lengths and bond energies of the same type (for example, CH *single* bonds) are remarkably similar in all compounds in which they appear. Bond lengths decrease and bond dissociation energies increase for double and triple bonds formed between the same pairs of atoms.
- Lewis dot diagrams are a tool for predicting the most likely connectivity, or arrangement of bonds between atoms, in a molecule. They are also useful for predicting the existence of multiple bonds and for determining qualitatively the distribution of charges in a molecule.
- The VSEPR theory predicts the three-dimensional shapes of molecules. It is based on simple electrostatics—electron pairs in a molecule will arrange themselves in such a way as to minimize their mutual repulsion. The steric number determines the geometry of the electron pairs (linear, trigonal pyramidal, tetrahedral, and so forth), whereas the molecular geometry is determined by the arrangement of the nuclei and may be less symmetric than the geometry of the electron pairs.
- Oxidation numbers are assigned to elements to name inorganic compounds, to keep track of electrons in electron transfer (oxidation-reduction) reactions, and to explore trends in chemical reactivity across the periodic table.

CONCEPTS & SKILLS

After studying this chapter and working the problems that follow, you should be able to:

1. Describe the structure of the periodic table, and predict chemical and physical properties of an element based on those of others in its group and period (Section 3.1, Problems 1–4).
2. Calculate the force and potential energy between a pair of charged particles, and predict the direction of relative motion of the particles (Section 3.2, Problems 5–8).
3. Describe the trends in ionization energy across the periodic table (Section 3.3, Problems 9 and 10).
4. Describe the shell structure of the atom, and represent valence shell electrons of an atom by its Lewis electron dot symbol (Section 3.3, Problems 11 and 12).
5. Describe the trends in electron affinity and electronegativity across the periodic table (Section 3.4, Problems 13–16).
6. Correlate the shape of the effective potential energy function between the nuclei with trends in the bond length and bond energies of diatomic molecules (Section 3.5, Problems 17–20).

7. Use the principle of charge balance and Lewis diagrams to write chemical formulas for ionic compounds (Section 3.6, Problems 21–24).
8. Calculate the energy of dissociation of gaseous diatomic ionic compounds into neutral atoms and ions (Section 3.6, Problems 25 and 26).
9. Describe the relations among bond order, length, and energy (Section 3.7, Problems 27–32).
10. Estimate the percent ionic character of a bond from its dipole moment (Section 3.7, Problems 33–38).
11. Given a molecular formula, draw a Lewis diagram for the molecule and assign formal charge (Section 3.8, Problems 39–50).
12. Assign formal charges and identify resonance diagrams for a given Lewis diagram (Section 3.8, Problems 51–58).
13. Predict the geometries of molecules by use of the VSEPR model (Section 3.9, Problems 59–64).
14. Determine whether a polyatomic molecule is polar or nonpolar (Section 3.9, Problems 65–70).
15. Assign oxidation numbers to atoms in compounds (Section 3.10, Problems 71 and 72).
16. Name inorganic compounds, given their chemical formulas, and write chemical formulas for inorganic compounds (Section 3.11, Problems 73–84).

KEY EQUATIONS

Coulomb's law is the conceptual basis of this chapter. All the key equations are devoted to stating this law and using it to describe the physical structure of atoms (their ionization energies, electron affinities, and electronegativities) and the stabilization of chemical bonds.

$$F(r) = \frac{q_1 q_2}{4\pi\epsilon_0 r^2}$$ Section 3.2

$$V(r) = \frac{q_1 q_2}{4\pi\epsilon_0 r}$$ Section 3.2

$$V(r) = -\frac{Ze^2}{4\pi\epsilon_0 r}$$ Section 3.2

$$F_{\text{coul}} = -\frac{d}{dr}\left(-\frac{Ze^2}{4\pi\epsilon_0 r}\right) = \frac{d}{dr}\left(\frac{Ze^2}{4\pi\epsilon_0 r}\right) = -\frac{Ze^2}{4\pi\epsilon_0 r^2}$$ Section 3.2

$$E = \frac{1}{2}m_e v^2 - \frac{Ze^2}{4\pi\epsilon_0 r}$$ Section 3.2

$$V_{\text{eff}}(r) = -\frac{Z_{\text{eff}}e^2}{4\pi\epsilon_0 r}$$ Section 3.3

$$\text{EN (Mulliken)} \propto \tfrac{1}{2}(IE_1 + EA)$$ Section 3.4

$$\chi_{\text{A}} - \chi_{\text{B}} = 0.102\Delta^{1/2}$$ Section 3.4

$$V(R_{12}) = Ae^{-\alpha R_{12}} - B\left(\frac{(e)(-e)}{R_{12}}\right) + \Delta E_{\infty}$$ Section 3.6

$$\Delta E_{\text{d}} \approx \frac{q_1 q_2}{4\pi\epsilon_0 R_{\text{e}}} \cdot \frac{N_{\text{A}}}{10^3} - \Delta E_{\infty}$$ Section 3.6

PROBLEMS

Answers to problems whose numbers are boldface appear in Appendix G. Problems that are more challenging are indicated with asterisks.

The Periodic Table

1. Before the element scandium was discovered in 1879, it was known as "eka-boron." Predict the properties of scandium from averages of the corresponding properties of its neighboring elements in the periodic table. Compare your predictions with the observed values in Appendix F.

Element	Symbol	Melting Point (°C)	Boiling Point (°C)	Density (g cm^{-3})
Calcium	Ca	839	1484	1.55
Titanium	Ti	1660	3287	4.50
Scandium	Sc	?	?	?

2. The element technetium (Tc) is not found in nature but has been produced artificially through nuclear reactions. Use the data for several neighboring elements in the table below to estimate the melting point, boiling point, and density of technetium. Compare your predictions with the observed values in Appendix F.

Element	Symbol	Melting Point (°C)	Boiling Point (°C)	Density (g cm^{-3})
Manganese	Mn	1244	1962	7.2
Molybdenum	Mo	2610	5560	10.2
Rhenium	Re	3180	5627	20.5
Ruthenium	Ru	2310	3900	12.3

3. Use the group structure of the periodic table to predict the empirical formulas for the binary compounds that hydrogen forms with the elements antimony, bromine, tin, and selenium.

4. Use the group structure of the periodic table to predict the empirical formulas for the binary compounds that hydrogen forms with the elements germanium, fluorine, tellurium, and bismuth.

Forces and Potential Energy in Atoms

5. An electron is located at the origin of the coordinates, and a second electron is brought to a position 2 Å from the origin.
(a) Calculate the force between the two electrons.
(b) Calculate the potential energy of the two electrons.

6. A gold nucleus is located at the origin of coordinates, and an electron is brought to a position 2 Å from the origin.
(a) Calculate the force between the gold nucleus and the electron.
(b) Calculate the potential energy of the gold nucleus and the electron.

7. The electron in a hydrogen atom is initially at a distance 1 Å from the proton, and then moves to a distance 0.5 Å from the proton.
(a) Calculate the change in the force between the proton and the electron.
(b) Calculate the change in the potential energy between the proton and the electron.
(c) Calculate the change in the velocity of the electron.

8. A gold nucleus is located at the origin of coordinates, and a helium nucleus initially 2 Å from the origin is moved to a position 1 Å from the origin.
(a) Calculate the change in the force between the two nuclei.
(b) Calculate the change in the potential energy of the two nuclei.
(c) Calculate the change in the velocity of the helium nucleus.

Ionization Energies and the Shell Model of the Atom

9. For each of the following pairs of atoms, state which you expect to have the higher first ionization energy: (a) Rb or Sr; (b) Po or Rn; (c) Xe or Cs; (d) Ba or Sr.

10. For each of the following pairs of atoms, state which you expect to have the higher first ionization energy: (a) Bi or Xe; (b) Se or Te; (c) Rb or Y; (d) K or Ne.

11. Use the data in Table 3.1 to plot the logarithm of ionization energy versus the number of electrons removed for Be. Describe the electronic structure of the Be atom.

12. Use the data in Table 3.1 to plot the logarithm of ionization energy versus the number of electrons removed for Ne. Describe the electronic structure of the Ne atom.

Electronegativity: The Tendency of Atoms to Attract Electrons

13. For each of the following pairs of atoms, state which you expect to have the greater electron affinity: (a) Xe or Cs; (b) Pm or F; (c) Ca or K; (d) Po or At.

14. For each of the following pairs of atoms, state which you expect to have the higher electron affinity: (a) Rb or Sr; (b) I or Rn; (c) Ba or Te; (d) Bi or Cl.

15. Ignoring tables of electronegativity values and guided only by the periodic table, arrange these atoms in order of increasing electronegativity: O, F, S, Si, K. Briefly explain your reasoning.

16. Ignoring tables of electronegativity values and guided only by the periodic table, arrange these atoms in order of increasing electronegativity: S, Cl, Sb, Se, In. Briefly explain your reasoning.

Forces and Potential Energy in Molecules: Formation of Chemical Bonds

17. We will see later that H_2 has equilibrium bond length of 0.751 Å and bond dissociation energy of 433 kJ mol^{-1}, whereas F_2 has equilibrium bond length of 1.417 Å and bond dissociation energy of 155 kJ mol^{-1}. On the same graph show qualitative sketches of the effective potential energy curve V_{eff} for H_2 and F_2. (*Hint:* Convert the bond energy to electron volts (eV) before preparing your graphs.)

18. We will see later that N_2 has equilibrium bond length of 1.100 Å and bond dissociation energy of 942 kJ mol^{-1}, whereas O_2 has equilibrium bond length of 1.211 Å and bond dissociation energy of 495 kJ mol^{-1}. On the same graph show qualitative sketches of the effective potential energy curve V_{eff} for N_2 and O_2. (*Hint:* Convert the bond energy to electron volts (eV) before preparing your graphs.)

19. We will see later that HF has equilibrium bond length of 0.926 Å and bond dissociation energy of 565 kJ mol^{-1}. Compare the effective potential curve for HF with those for H_2 and F_2 in Problem 17.

20. We will see later that NO has equilibrium bond length of 1.154 Å and bond dissociation energy of 629 kJ mol^{-1}. Compare the effective potential curve for NO with those for N_2 and O_2 in Problem 18.

Ionic Bonding

21. For each of the following atoms or ions, state the total number of electrons, the number of valence electrons, and the number of core electrons.
(a) Rn (b) Sr^+ (c) Se^{2-} (d) Sb^-

22. For each of the following atoms or ions, state the total number of electrons, the number of valence electrons, and the number of core electrons.
(a) Ra^{2+} (b) Br (c) Bi^{2-} (d) Ga^+

23. Use the data in Figure 3.4 and Table 3.2 to calculate the energy changes (ΔE) for the following pairs of reactions:
(a) $K(g) + Cl(g) \longrightarrow K^+(g) + Cl^-(g)$
$K(g) + Cl(g) \longrightarrow K^-(g) + Cl^+(g)$
(b) $Na(g) + Cl(g) \longrightarrow Na^+(g) + Cl^-(g)$
$Na(g) + Cl(g) \longrightarrow Na^-(g) + Cl^+(g)$
Explain why K^+Cl^- and Na^+Cl^- form in preference to K^-Cl^+ and Na^-Cl^+.

24. Use the data in Figure 3.4 and Table 3.2 to calculate the energy changes (ΔE) for the following pairs of reactions:
(a) $Na(g) + I(g) \longrightarrow Na^+(g) + I^-(g)$
$Na(g) + I(g) \longrightarrow Na^-(g) + I^+(g)$
(b) $Rb(g) + Br(g) \longrightarrow Rb^+(g) + Br^-(g)$
$Rb(g) + Br(g) \longrightarrow Rb^-(g) + Br^+(g)$
Explain why Na^+I^- and Rb^+Br^- form in preference to Na^-I^+ and Rb^-Br^+.

25. In a gaseous KCl molecule, the internuclear distance is 2.67×10^{-10} m. Using data from Appendix F and neglecting the small, short-range repulsion between the ion cores of K^+ and Cl^-, estimate the dissociation energy of gaseous KCl into K and Cl atoms (in kJ mol^{-1}).

26. In a gaseous RbF molecule, the bond length is 2.274×10^{-10} m. Using data from Appendix F and neglecting the small, short-range repulsion between the ion cores of Rb^+ and F^-, estimate the dissociation energy of gaseous RbF into Rb and F atoms (in kJ mol^{-1}).

Covalent and Polar Covalent Bonding

27. The bond lengths of the X—H bonds in NH_3, PH_3, and SbH_3 are 1.02, 1.42, and 1.71 Å, respectively. Estimate the length of the As—H bond in AsH_3, the gaseous compound that decomposes on a heated glass surface in Marsh's test for arsenic. Which of these four hydrides has the weakest X—H bond?

28. Arrange the following covalent diatomic molecules in order of the lengths of the bonds: BrCl, ClF, IBr. Which of the three has the weakest bond (the smallest bond energy)?

29. The bond length in H—I (1.62 Å) is close to the sum of the atomic radii of H (0.37 Å) and I (1.33 Å). What does this fact indicate about the polarity of the bond?

30. The bond length in F_2 is 1.417 Å, instead of twice the atomic radius of F, which is 1.28 Å. What can account for the unexpected length of the F—F bond?

31. Use electronegativity values to arrange the following bonds in order of decreasing polarity: N—O, N—N, N—P, and C—N.

32. Use electronegativity values to rank the bonds in the following compounds from least ionic to most ionic in character: IF, ICl, ClF, BrCl, and Cl_2.

33. Ionic compounds tend to have higher melting and boiling points and to be less volatile (that is, have lower vapor pressures) than covalent compounds. For each of the following pairs, use electronegativity differences to predict which compound has the higher vapor pressure at room temperature.
(a) CI_4 or KI
(b) BaF_2 or OF_2
(c) SiH_4 or NaH

34. For each of the following pairs, use electronegativity differences to predict which compound has the higher boiling point.
(a) $MgBr_2$ or PBr_3
(b) OsO_4 or SrO
(c) Cl_2O or Al_2O_3

35. Estimate the percent ionic character of the bond in each of the following diatomic molecules, based on the dipole moment.

	Bond Length (Å)	Dipole Moment (D)
ClO	1.573	1.239
KI	3.051	10.82
TlCl	2.488	4.543
InCl	2.404	3.79

36. Estimate the percent ionic character of the bond in each of the following species. All the species are unstable or reactive under ordinary laboratory conditions, but they can be observed in interstellar space.

	Bond Length (Å)	Dipole Moment (D)
OH	0.980	1.66
CH	1.131	1.46
CN	1.175	1.45
C_2	1.246	0

37. The percent ionic character of a bond can be approximated by the formula $16\Delta + 3.5\Delta^2$, where Δ is the magnitude of the difference in the electronegativities of the atoms (see Fig. 3.7). Calculate the percent ionic character of HF, HCl, HBr, HI, and CsF, and compare the results with those in Table 3.7.

38. The percent ionic character of the bonds in several interhalogen molecules (as estimated from their measured dipole moments and bond lengths) are ClF (11%), BrF (15%), BrCl (5.6%), ICl (5.8%), and IBr (10%). Estimate the percent ionic characters for each of these molecules, using the equation in Problem 37, and compare them with the given values.

Lewis Diagrams for Molecules

39. Assign formal charges to all atoms in the following Lewis diagrams.

(a) SO_4^{2-}

```
[      :Ö:      ]2−
        |
[ :Ö—S—Ö:       ]
        |
[      :O:      ]
```

(b) $S_2O_3^{2-}$

```
[      :Ö:      ]2−
        |
[ :Ö—S—Ö:       ]
        |
[      :S:      ]
```

(c) SbF_3

```
:F—Sb—F:
    |
   :F:
```

(d) SCN^-

```
[:S—C≡N:]−
```

40. Assign formal charges to all atoms in the following Lewis diagrams.

(a) ClO_4^-

```
[      :Ö:      ]−
        |
[ :Ö—Cl—Ö:      ]
        |
[      :O:      ]
```

(b) SO_2

```
:Ö—S=Ö
```

(c) BrO_2^-

```
[:Ö—Br—Ö:]−
```

(d) NO_3^-

```
[ Ö=N—Ö: ]−
     |
[   :O:  ]
```

41. Determine the formal charges on all the atoms in the following Lewis diagrams.

```
H—N=O     and     H—O=N
```

Which one would best represent bonding in the molecule HNO?

42. Determine the formal charges on all the atoms in the following Lewis diagrams.

```
:Cl—Cl—O:     and     :Cl—O—Cl:
```

Which one would best represent bonding in the molecule Cl_2O?

43. In each of the following Lewis diagrams, Z represents a main-group element. Name the group to which Z belongs in each case and give an example of such a compound or ion that actually exists.

(a)

```
Ö=Z=Ö
```

(b)

```
    :Ö:     :Ö:
     |       |
:Ö—Z—Ö—Z—Ö:
     |       |
    :O:     :O:
```

(c)

```
[Ö=Z—Ö:]−
```

(d)

```
[         :Ö:   ]−
            |
[   H—Ö—Z—Ö:    ]
            |
[         :O:   ]
```

44. In each of the following Lewis diagrams, Z represents a main-group element. Name the group to which Z belongs in each case and give an example of such a compound or ion that actually exists.

(a)

```
[:C≡Z:]−
```

(b)

```
[      :Ö:      ]−
        |
[ :Ö—Z—Ö:       ]
        |
[      :O:      ]
```

(c) (d)

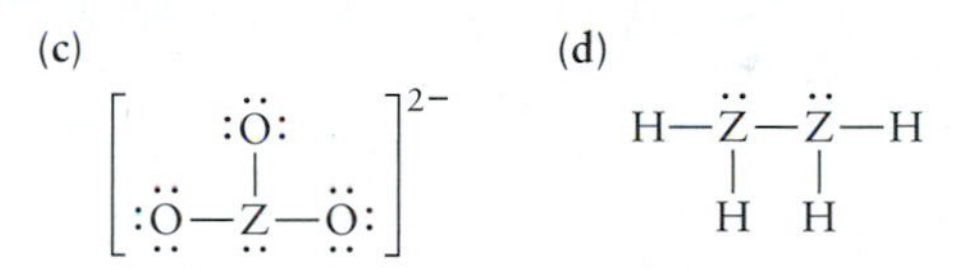

45. Draw Lewis electron dot diagrams for the following species: (a) AsH_3; (b) HOCl; (c) KrF^-; (d) $PO_2Cl_2^-$ (central P atom).

46. Draw Lewis electron dot diagrams for the following species: (a) methane; (b) carbon dioxide; (c) phosphorus trichloride; (d) perchlorate ion.

47. Urea is an important chemical fertilizer with the chemical formula $(H_2N)CO(NH_2)$. The carbon atom is bonded to both nitrogen atoms and the oxygen atom. Draw a Lewis diagram for urea and use Table 3.6 to estimate its bond lengths.

48. Acetic acid is the active ingredient of vinegar. Its chemical formula is CH_3COOH, and the second carbon atom is bonded to the first carbon atom and to both oxygen atoms. Draw a Lewis diagram for acetic acid and use Table 3.6 to estimate its bond lengths.

49. Under certain conditions, the stable form of sulfur consists of rings of eight sulfur atoms. Draw the Lewis diagram for such a ring.

50. White phosphorus (P_4) consists of four phosphorus atoms arranged at the corners of a tetrahedron. Draw the valence electrons on this structure to give a Lewis diagram that satisfies the octet rule.

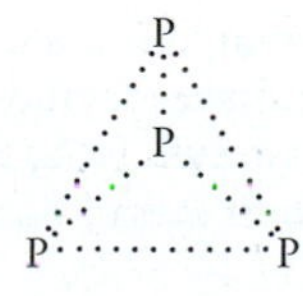

51. Draw Lewis electron dot diagrams for the following species, indicating formal charges and resonance diagrams where applicable.
(a) H_3NBF_3
(b) CH_3COO^- (acetate ion)
(c) HCO_3^- (hydrogen carbonate ion)

52. Draw Lewis electron dot diagrams for the following species, indicating formal charges and resonance diagrams where applicable.
(a) HNC (central N atom)
(b) SCN^- (thiocyanate ion)
(c) H_2CNN (the first N atom is bonded to the carbon and the second N)

53. Draw Lewis diagrams for the two resonance forms of the nitrite ion, NO_2^-. In what range do you expect the nitrogen–oxygen bond length to fall? (*Hint:* Use Table 3.6.)

54. Draw Lewis diagrams for the three resonance forms of the carbonate ion, CO_3^{2-}. In what range do you expect the carbon–oxygen bond length to fall? (*Hint:* Use Table 3.6.)

55. Methyl isocyanate, which was involved in the disaster in Bhopal, India, in 1984, has the chemical formula CH_3NCO. Draw its Lewis diagram, including resonance forms. (**Note:** The N atom is bonded to the two C atoms.)

56. Peroxyacetyl nitrate (PAN) is one of the prime irritants in photochemical smog. It has the formula $CH_3COOONO_2$, with the following structure:

$H_3C—C(=O)—O—O—NO_2$

Draw its Lewis diagram, including resonance forms.

57. Draw Lewis diagrams for the following compounds. In the formula the symbol of the central atom is given first. (*Hint:* The valence octet may be expanded for the central atom.)
(a) PF_5 (b) SF_4 (c) XeO_2F_2

58. Draw Lewis diagrams for the following ions. In the formula the symbol of the central atom is given first. (*Hint:* The valence octet may be expanded for the central atom.)
(a) BrO_4^- (b) PCl_6^- (c) XeF_6^+

The Shapes of Molecules: Valence Shell Electron-Pair Repulsion Theory

59. For each of the following molecules, give the steric number and sketch and name the approximate molecular geometry. In each case, the central atom is listed first and the other atoms are all bonded directly to it.
(a) CBr_4 (b) SO_3
(c) SeF_6 (d) $SOCl_2$
(e) ICl_3

60. For each of the following molecules or molecular ions, give the steric number and sketch and name the approximate molecular geometry. In each case, the central atom is listed first and the other atoms are all bonded directly to it.
(a) PF_3 (b) SO_2Cl_2
(c) PF_6^- (d) ClO_2^-
(e) GeH_4

61. For each of the following molecules or molecular ions, give the steric number, sketch and name the approximate molecular geometry, and describe the directions of any *distortions* from the approximate geometry due to lone pairs. In each case, the central atom is listed first and the other atoms are all bonded directly to it.
(a) ICl_4^- (b) OF_2
(c) BrO_3^- (d) CS_2

62. For each of the following molecules or molecular ions, give the steric number, sketch and name the approximate molecular geometry, and describe the direction of any *distortions* from the approximate geometry due to lone pairs. In each case, the central atom is listed first and the other atoms are all bonded directly to it.
(a) TeH_2 (b) AsF_3
(c) PCl_4^+ (d) XeF_5^+

63. Give an example of a molecule or ion having a formula of each of the following types and structures.
(a) B_3 (planar) (b) AB_3 (pyramidal)
(c) AB_2^- (bent) (d) AB_3^{2-} (planar)

64. Give an example of a molecule or ion having a formula of each of the following types and structures.
(a) AB_4^- (tetrahedral) (b) AB_2 (linear)
(c) AB_6^- (octahedral) (d) AB_3^- (pyramidal)

65. For each of the answers in Problem 59, state whether the species is polar or nonpolar.

66. For each of the answers in Problem 60, state whether the species is polar or nonpolar.

67. The molecules of a certain compound contain one atom each of nitrogen, fluorine, and oxygen. Two possible structures are NOF (O as central atom) and ONF (N as central atom). Does the information that the molecule is bent limit the choice to one of these two possibilities? Explain.

68. Mixing $SbCl_3$ and $GaCl_3$ in a 1:1 molar ratio (using liquid sulfur dioxide as a solvent) gives a solid ionic compound of empirical formula $GaSbCl_6$. A controversy arises over whether this compound is $(SbCl_2^+)(GaCl_4^-)$ or $(GaCl_2^+)(SbCl_4^-)$.
(a) Predict the molecular structures of the two anions.
(b) It is learned that the cation in the compound has a bent structure. Based on this fact, which formulation is more likely to be correct?

69. (a) Use the VSEPR theory to predict the structure of the NNO molecule.
(b) The substance NNO has a small dipole moment. Which end of the molecule is more likely to be the positive end, based only on electronegativity?

70. Ozone (O_3) has a nonzero dipole moment. In the molecule of O_3, one of the oxygen atoms is directly bonded to the other two, which are not bonded to each other.
(a) Based on this information, state which of the following structures are possible for the ozone molecule: symmetric linear, nonsymmetric linear (for example, different O—O bond lengths), and bent. (**Note:** Even an O—O bond can have a bond dipole if the two oxygen atoms are bonded to different atoms or if only one of the oxygen atoms is bonded to a third atom.)
(b) Use the VSEPR theory to predict which of the structures of part (a) is observed.

Oxidation Numbers

71. Assign oxidation numbers to the atoms in each of the following species: $SrBr_2$, $Zn(OH)_4^{2-}$, SiH_4, $CaSiO_3$, $Cr_2O_7^{2-}$, $Ca_5(PO_4)_3F$, KO_2, CsH.

72. Assign oxidation numbers to the atoms in each of the following species: NH_4NO_3, $CaMgSiO_4$, $Fe(CN)_6^{4-}$, B_2H_6, BaH_2, $PbCl_2$, $Cu_2O(SO_4)$, $S_4O_6^{2-}$.

Inorganic Nomenclature

73. Give the name and formula of an ionic compound involving only the elements in each pair that follows. Write Lewis symbols for the elements both before and after chemical combination.
(a) Chlorine and cesium (b) Calcium and astatine
(c) Aluminum and sulfur (d) Potassium and tellurium

74. Give the name and formula of an ionic compound involving only the elements in each pair that follows. Write Lewis symbols for the elements both before and after chemical combination.
(a) Gallium and bromine (b) Strontium and polonium
(c) Magnesium and iodine (d) Lithium and selenium

75. Give systematic names to the following compounds:
(a) Al_2O_3 (b) Rb_2Se
(c) $(NH_4)_2S$ (d) $Ca(NO_3)_2$
(e) Cs_2SO_4 (f) $KHCO_3$

76. Give systematic names to the following compounds:
(a) KNO_2 (b) $Sr(MnO_4)_2$
(c) $MgCr_2O_7$ (d) NaH_2PO_4
(e) $BaCl_2$ (f) $NaClO_3$

77. Write the chemical formulas for the following compounds:
(a) Silver cyanide
(b) Calcium hypochlorite
(c) Potassium chromate
(d) Gallium oxide
(e) Potassium superoxide
(f) Barium hydrogen carbonate

78. Write the chemical formulas for the following compounds:
(a) Cesium sulfite
(b) Strontium thiocyanate
(c) Lithium hydride
(d) Sodium peroxide
(e) Ammonium dichromate
(f) Rubidium hydrogen sulfate

79. Trisodium phosphate (TSP) is a heavy-duty cleaning agent. Write its chemical formula. What would be the systematic name for this ionic compound?

80. Monoammonium phosphate is the common name for a compound made up of NH_4^+ and $H_2PO_4^-$; it is used as a flame retardant. (Its use for this purpose was first suggested by Gay-Lussac in 1821.) Write its chemical formula. What is the systematic chemical name of this compound?

81. Write the chemical formula for each of the following compounds:
(a) Silicon dioxide
(b) Ammonium carbonate
(c) Lead(IV) oxide
(d) Diphosphorus pentaoxide
(e) Calcium iodide
(f) Iron(III) nitrate

82. Write the chemical formula for each of the following compounds:
(a) Lanthanum(III) sulfide
(b) Cesium sulfate
(c) Dinitrogen trioxide
(d) Iodine pentafluoride
(e) Chromium(III) sulfate
(f) Potassium permanganate

83. Give the systematic name for each of the following compounds:
(a) Cu_2S and CuS (b) Na_2SO_4
(c) As_4O_6 (d) $ZrCl_4$
(e) Cl_2O_7 (f) Ga_2O

84. Give the systematic name for each of the following compounds:
(a) Mg_2SiO_4 (b) $Fe(OH)_2$ and $Fe(OH)_3$
(c) As_2O_5 (d) $(NH_4)_2HPO_4$
(e) SeF_6 (f) Hg_2SO_4

Additional Problems

85. Refer to Figure 3.7 and compute the difference in electronegativity between the atoms in LiCl and those in HF. Based on their physical properties (see below), are the two similar or different in terms of bonding?

	LiCl	HF
Melting point	605°C	83.1°C
Boiling point	1350°C	19.5°C

86. Ordinarily, two metals, when mixed, form alloys that show metallic character. If the two metals differ sufficiently in electronegativity, they can form compounds with significant ionic character. Consider the solid produced by mixing equal chemical amounts of Cs and Rb, compared with that produced by mixing Cs and Au. Compute the electronegativity difference in each case, and determine whether either mixture has significant ionic character. If either compound is ionic or partially ionic, which atom carries the net negative charge? Are there alkali halides with similar or smaller electronegativity differences?

* 87. At large interatomic separations, an alkali halide molecule MX has a lower energy as two neutral atoms, M + X; at short separations, the ionic form (M+)(X−) has a lower energy. At a certain distance, R_c, the energies of the two forms become equal, and it is near this distance that the electron will jump from the metal to the halogen atom during a collision. Because the forces between neutral atoms are weak at large distances, a reasonably good approximation can be made by ignoring any variation in potential $V(R)$ for the neutral atoms between R_c and $R = \infty$. For the ions in this distance range, $V(R)$ is dominated by their Coulomb attraction.
(a) Express R_c for the first ionization energy of the metal M and the electron affinity of the halogen X.
(b) Calculate R_c for LiF, KBr, and NaCl using data from Appendix F.

88. Use the data in Appendix F to compute the energy changes (ΔE) of the following pairs of reactions:
(a) $Na(g) + I(g) \longrightarrow Na^+(g) + I^-(g)$ and $Na(g) + I(g) \longrightarrow Na^-(g) + I^+(g)$

(b) $K(g) + Cl(g) \longrightarrow K^+(g) + Cl^-(g)$ and $K(g) + Cl(g) \longrightarrow K^-(g) + Cl^+(g)$
Explain why Na^+I^- and K^+Cl^- form in preference to Na^-I^+ and K^-Cl^+.

89. The carbon–carbon bond length in C_2H_2 is 1.20 Å, that in C_2H_4 is 1.34 Å, and that in C_2H_6 is 1.53 Å. Near which of these values would you predict the bond length of C_2 to lie? Is the experimentally observed value, 1.31 Å, consistent with your prediction?

90. Two possible Lewis diagrams for sulfine (H_2CSO) are

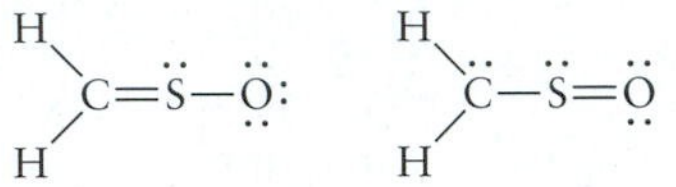

$H_2C{=}\ddot{S}{-}\ddot{O}{:}$ $\quad$ $H_2\ddot{C}{-}\ddot{S}{=}\ddot{O}$

(a) Compute the formal charges on all atoms.
(b) Draw a Lewis diagram for which all the atoms in sulfine have formal charges of zero.

91. There is persuasive evidence for the brief existence of the unstable molecule OPCl.
(a) Draw a Lewis diagram for this molecule in which the octet rule is satisfied on all atoms and the formal charges on all atoms are zero.
(b) The compound OPCl reacts with oxygen to give O2PCl. Draw a Lewis diagram of O2PCl for which all formal charges are equal to zero. Draw a Lewis diagram in which the octet rule is satisfied on all atoms.

92. The compound SF3N has been synthesized.
(a) Draw the Lewis diagram of this molecule, supposing that the three fluoride atoms and the nitrogen atom surround the sulfur atom. Indicate the formal charges. Repeat, but assume that the three fluorine atoms and the sulfur atom surround the nitrogen atom.
(b) From the results in part (a), speculate about which arrangement is more likely to correspond to the actual molecular structure.

93. In nitryl chloride (NO_2Cl), the chlorine atom and the two oxygen atoms are bonded to a central nitrogen atom, and all the atoms lie in a plane. Draw the two electron dot resonance forms that satisfy the octet rule and that together are consistent with the fact that the two nitrogen–oxygen bonds are equivalent.

94. The molecular ion $S_3N_3^-$ has the cyclic structure

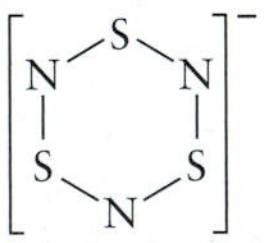

[ring: S–N–S–N–S–N, each S bonded to two N]$^-$

All S—N bonds are equivalent.
(a) Give six equivalent resonance hybrid Lewis diagrams for this molecular ion.
(b) Compute the formal charges on all atoms in the molecular ion in each of the six Lewis diagrams.
(c) Determine the charge on each atom in the polyatomic ion, assuming that the true distribution of electrons is the *average* of the six Lewis diagrams arrived at in parts (a) and (b).
(d) An advanced calculation suggests that the actual charge resident on each N atom is −0.375 and on each S atom is +0.041. Show that this result is consistent with the overall +1 charge on the molecular ion.

* 95. The two compounds nitrogen dioxide and dinitrogen tetraoxide are introduced in Section 3.11.
(a) NO_2 is an odd-electron compound. Draw the best Lewis diagrams possible for it, recognizing that one atom cannot achieve an octet configuration. Use formal charges to decide whether that should be the (central) nitrogen atom or one of the oxygen atoms.
(b) Draw resonance forms for N_2O_4 that obey the octet rule. The two N atoms are bonded in this molecule.

96. Although magnesium and the alkaline-earth metals situated below it in the periodic table form ionic chlorides, beryllium chloride ($BeCl_2$) is a covalent compound.
(a) Follow the usual rules to write a Lewis diagram for $BeCl_2$ in which each atom attains an octet configuration. Indicate formal charges.
(b) The Lewis diagram that results from part (a) is an extremely unlikely one because of the double bonds and formal charges it shows. By relaxing the requirement of placing an octet on the beryllium atom, show how a Lewis diagram without formal charges can be written.

97. (a) The first noble-gas compound, prepared by Neil Bartlett in 1962, was an orange-yellow ionic solid that consisted of XeF^+ and PtF_{11}^-. Draw a Lewis diagram for XeF^+.
(b) Shortly after the preparation of the ionic compound discussed in part (a), it was found that the irradiation of mixtures of xenon and fluorine with sunlight produced white crystalline XeF_2. Draw a Lewis diagram for this molecule, allowing valence expansion on the central xenon atom.

* 98. Represent the bonding in SF_2 (F—S—F) with Lewis diagrams. Include the formal charges on all atoms. The dimer of this compound has the formula S_2F_4. It was isolated in 1980 and shown to have the structure $F_3S{-}SF$. Draw a possible Lewis diagram to represent the bonding in the dimer, indicating the formal charges on all atoms. Is it possible to draw a Lewis diagram for S_2F_4 in which all atoms have valence octets? Explain why or why not.

* 99. A stable triatomic molecule can be formed that contains one atom each of nitrogen, sulfur, and fluorine. Three bonding structures are possible, depending on which is the central atom: NSF, SNF, and SFN.
(a) Write a Lewis diagram for each of these molecules, indicating the formal charge on each atom.
(b) Often, the structure with the least separation of formal charge is the most stable. Is this statement consistent with the observed structure for this molecule—namely, NSF, which has a central sulfur atom?
(c) Does consideration of the electronegativities of N, S, and F from Figure 3.7 help rationalize this observed structure? Explain.

100. The gaseous potassium chloride molecule has a measured dipole moment of 10.3 D, which indicates that it is a very polar molecule. The separation between the nuclei in this molecule is 2.67 Å. What would the dipole moment of a KCl molecule be if there were opposite charges of one fundamental unit (1.60×10^{-19} C) at the nuclei?

101. (a) Predict the geometry of the $SbCl_5^{2-}$ ion, using the VSEPR method.
(b) The ion $SbCl_6^{3-}$ is prepared from $SbCl_5^{2-}$ by treatment with Cl^-. Determine the steric number of the central

antimony atom in this ion, and discuss the extension of the VSEPR theory that would be needed for the prediction of its molecular geometry.

102. The element xenon (Xe) is by no means chemically inert; it forms a number of chemical compounds with electronegative elements such as fluorine and oxygen. The reaction of xenon with varying amounts of fluorine produces XeF_2 and XeF_4. Subsequent reaction of one or the other of these compounds with water produces (depending on conditions) XeO_3, XeO_4, and H_4XeO_6, as well as mixed compounds such as $XeOF_4$. Predict the structures of these six xenon compounds, using the VSEPR theory.

103. Predict the arrangement of the atoms about the sulfur atom in F_4SPO, assuming that double-bonded atoms require more space than single-bonded atoms.

104. Draw Lewis diagrams and predict the geometries of the following molecules. State which are polar and which are nonpolar.
(a) ONCl (b) O_2NCl
(c) XeF_2 (d) SCl_4
(e) CHF_3

*105. Suppose that any given kind of bond, such as O—H, has a characteristic electric dipole. That is, suppose that electric dipole moments can be assigned to bonds just as bond energies can be. Both are usefully accurate approximations. Consider the water molecule

$$H \overset{O}{\angle_{\theta}} H$$

Show that if μ_{OH} is the dipole moment of the OH bond, then the dipole moment of water is $\mu(H_2O) = 2\mu_{OH}\cos(\theta/2)$. What is the dipole moment μ_{OH} if $\mu(H_2O)$ is 1.86 D?

106. A good method of preparing pure oxygen on a small scale is the decomposition of $KMnO_4$ in a vacuum above 215°C:

$$2\,KMnO_4(s) \longrightarrow K_2MnO_4(s) + MnO_2(s) + O_2(g)$$

Assign an oxidation number to each atom and verify that the total number of electrons lost is equal to the total number gained.

107. Bismuth forms an ion with the formula Bi_5^{3+}. Arsenic and fluorine form a complex ion $[AsF_6]^-$, with fluorine atoms arranged around a central arsenic atom. Assign oxidation numbers to each of the atoms in the bright yellow crystalline solid with the formula $Bi_5(AsF_6)_3 \cdot 2SO_2$.

108. In some forms of the periodic table, hydrogen is placed in Group I; in others, it is placed in Group VII. Give arguments in favor of each location.

*109. (a) Determine the oxidation number of lead in each of the following oxides: PbO, PbO_2, Pb_2O_3, Pb_3O_4.
(b) The only known lead ions are Pb^{2+} and Pb^{4+}. How can you reconcile this statement with your answer to part (A)?

110. There have been some predictions that element 114 will be relatively stable in comparison with many other elements beyond uranium in the periodic table. Predict the maximum oxidation state of this element. Based on the trends in the oxidation states of other members of its group, is it likely that this oxidation state will be the dominant one?

CUMULATIVE PROBLEMS

111. A certain element, M, is a main-group metal that reacts with chlorine to give a compound with the chemical formula MCl_2 and with oxygen to give the compound MO.
(a) To which group in the periodic table does element M belong?
(b) The chloride contains 44.7% chlorine by mass. Name the element M.

*112. An ionic compound used as a chemical fertilizer has the composition (by mass) 48.46% O, 23.45% P, 21.21% N, 6.87% H. Give the name and chemical formula of the compound and draw Lewis diagrams for the two types of ions that make it up.

113. A compound is being tested for use as a rocket propellant. Analysis shows that it contains 18.54% F, 34.61% Cl, and 46.85% O.
(a) Determine the empirical formula for this compound.
(b) Assuming that the molecular formula is the same as the empirical formula, draw a Lewis diagram for this molecule. Review examples elsewhere in this chapter to decide which atom is most likely to lie at the center.
(c) Use the VSEPR theory to predict the structure of the molecule from part (b).

114. Many important fertilizers are ionic compounds that contain the elements nitrogen, phosphorus, and potassium because these are frequently the limiting plant-growth nutrients in soil.
(a) Write the chemical formulas for the following chemical fertilizers: ammonium phosphate, potassium nitrate, ammonium sulfate.
(b) Calculate the mass percentage of nitrogen, phosphorus, and potassium for each of the compounds in part (a).

CHAPTER 4

Introduction to Quantum Mechanics

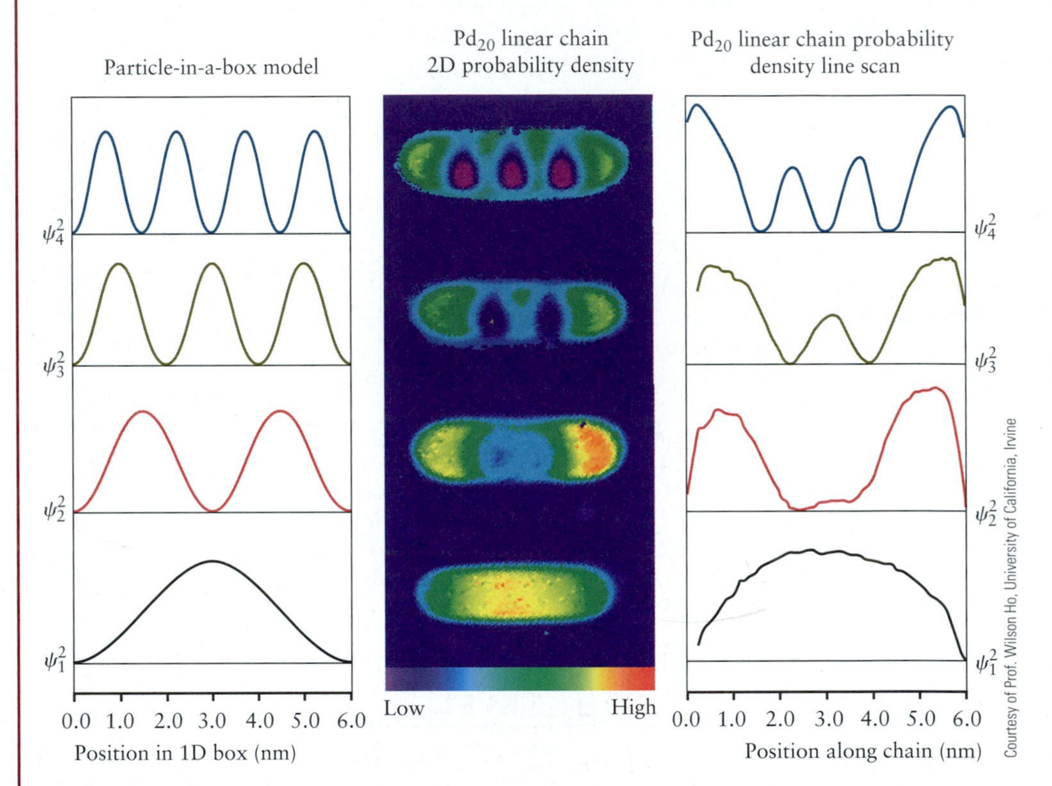

Particle-in-a-box states for an electron in a 20 Pd atom linear chain assembled on a single crystal NiAl surface. The left set of curves shows the predictions of the one-dimensional particle-in-a-box model, the center set of images is the 2D probability density distribution, and the right of curves is a line scan of the probability density distribution taken along the center of the chain. The chain was assembled and the probability densities measured using a scanning tunneling microscope.

Science can advance in different ways. Usually, the slow and steady accumulation of experimental results supports and refines existing models, which leads to a more satisfactory description of natural phenomena. Occasionally, however, the results of new experiments directly contradict previously accepted theories. In this case, a period of uncertainty ensues; it is resolved only through the eventual emergence of a new and more complete theory that explains both the previously understood results and the new experiments. This process is called a *scientific revolution*. In the first 25 years of the 20th century, a revolution in physics led to the development of the quantum theory, which also profoundly affected the science of chemistry.

One of the fundamental assumptions of early science was that nature is continuous; that is, nature does not make "jumps." On a macroscopic scale, this appears to be true enough. We can measure out an amount of graphite (carbon) of mass 9, or 8.23, or 6.4257 kg, and it appears that the mass can have any value provided that our balance is sufficiently accurate. On an atomic scale, however, this apparently continuous behavior breaks down. An analogy may be useful here. A sand beach from a distance appears smooth and continuous, but a close look reveals that it is made up of individual grains of sand. This same "graininess" is found in matter observed on the atomic scale. The mass of carbon (^{12}C) comes in "packets," each of which weighs 1.99265×10^{-26} kg. Although, in principle, two, three, or any integral number of such packets can be "weighed out," we cannot obtain $1\frac{1}{2}$ packets. Carbon is not a continuous material, but comes in chunks, each containing the minimum measurable mass of carbon—that of an atom. Similarly, electric charge comes in packets of size e, as shown in Section 1.4, and fractional charges are never observed in chemical reactions.

The central idea of quantum theory is that energy, like matter, is not continuous but it exists only in discrete packets. Whereas discreteness of matter and charge on the microscopic scale seems entirely reasonable and familiar to us, based on the modern picture of the physical structure of the atom, the idea that energy also exists only in discrete chunks is contrary to our experience of the macroscopic world. The motions of a soccer ball rolling up and down the sides of a gully involve arbitrary amounts of kinetic and potential energy; nothing in ordinary human experience suggests that the energy of a system should change abruptly by "jumps." Understanding quantum mechanics requires that we develop a new kind of physical intuition, based on the results of experiments that are impossible to understand using classical mechanics. These results are completely divorced from ordinary human experience in the macroscopic world around us, and our physical intuition from the macroscopic world cannot be transferred to the quantum domain. We must remain vigilant against the urge to interpret these quantum results in terms of ordinary experience.

To understand the far-reaching nature of the quantum revolution, you should consider the state of physics at the end of the 19th century. The 200 years that followed the seminal work of Isaac Newton were the classical period in the study of mechanics, the branch of physics that predicts the motions of particles and the collections of particles that make up working mechanisms. By the end of that period, about 1900, physicists had achieved a deep understanding that successfully dealt with problems ranging from the motions of the planets in their orbits to the design of a bicycle. These achievements make up the field now called **classical mechanics.**

Classical mechanics can predict the future positions of a group of particles from their present positions and velocities if the forces among them are known. At the end of the 19th century, it was naturally thought that the motion of elementary particles—such as the recently discovered electron—could be described by classical mechanics. Once the correct force laws had been discovered, the properties of atoms and molecules could be predicted to any desired accuracy by solving Newton's equations of motion. It was believed that all the fundamental laws of physics had been discovered. At the dedication of the Ryerson Physics Laboratory at the University of Chicago in 1894, the American physicist A. A. Michelson said, "Our future discoveries must be looked for in the sixth decimal place." Little did he imagine the revolutionary changes that would shake physics and chemistry during the following 30 years.

Central to those changes was not only the recognition that energy is quantized but also the discovery that all particles display wavelike properties in their motions. The effects of wavelike properties are most pronounced for small, low-mass particles such as electrons in atoms. **Quantum mechanics** incorporated both the ideas of "wave–particle duality" and energy quantization into a single comprehensive theory that superseded classical mechanics to describe the properties of matter on the nanometer length scale. Quantum mechanics is one of the greatest intellectual achievements of the 20th century.

This chapter describes the origins of the quantum theory, summarizes its techniques, and demonstrates their application to simple model systems. Our goals are to help you become skilled and confident in using the language, concepts, and tools of quantum theory. With these skills, we will guide you to develop an intuitive understanding of the behavior of quantum systems—so foreign to our ordinary human experience—and the magnitudes of the observable quantities (energy, momentum, length) in the quantum domain. Chapter 5 shows how quantum mechanics explains the structure of atoms and the periodic table and Chapter 6 shows how the quantum theory explains the formation of chemical bonds.

4.1 Preliminaries: Wave Motion and Light

Many kinds of waves are studied in physics and chemistry. Familiar examples include water waves stirred up by the winds over the oceans, set off by a stone dropped into a quiet pool, or created for teaching demonstrations by a laboratory water-wave machine. Sound waves are periodic compressions of the air that move from a source to a detector such as the human ear. Light waves, as discussed later in this chapter, consist of oscillating electric and magnetic fields moving through space. Even some chemical reactions occur in such a way that waves of color pass through the sample as the reaction proceeds. Common to all these wave phenomena is the oscillatory variation of some property with time at a given fixed location in space (Table 4.1). All of these waves are described by the same equations.

TABLE 4.1 Kinds of Waves

Wave	Oscillating Quantity
Water	Height of water surface
Sound	Density of air
Light	Electric and magnetic fields
Chemical	Concentrations of chemical species

A snapshot of a water wave (Fig. 4.1) records the crests and troughs present at some instant in time. The **amplitude** of the wave is the height or the displacement of the water surface compared with the undisturbed level of the water; this undisturbed height is customarily chosen as the reference height and assigned the value zero. Positive amplitudes describe displacements that increase the level of the water, whereas negative amplitudes describe those that decrease the level of the water. We define the **maximum amplitude** as either the height of a crest or the depth of a trough, and it is always given as an absolute value.[1] The distance between two successive crests (or troughs) is called the **wavelength,** λ (Greek lambda), of the wave, provided that this distance is reproducible from peak to peak. The **frequency** of a water wave can be measured by counting the number of peaks or troughs observed moving past a fixed point in space per second. The frequency, ν (Greek nu), is measured in units of waves (or cycles) per second, or simply s^{-1}. The fundamental frequency unit one cycle per second has been named the **hertz (Hz)** in honor of the German physicist Heinrich Hertz. For example, if 12 water-wave peaks are observed to pass a certain point in 30 seconds, the frequency is

$$\text{frequency} = \nu = \frac{12}{30\ \text{s}} = 0.40\ \text{s}^{-1} = 0.40\ \text{Hz}$$

The wavelength and frequency of a wave are related through its speed—the rate at which a particular wave crest moves through the medium. In Figure 4.1, the crest at the left end of the horizontal black arrow will move forward exactly one wavelength in one cycle of the wave. By definition, the time required for the crest to

[1]Most physics texts, especially older ones, define the amplitude as the quantity we call the maximum amplitude here. We have chosen the present definition to facilitate later discussions of the wave functions that describe atomic structure.

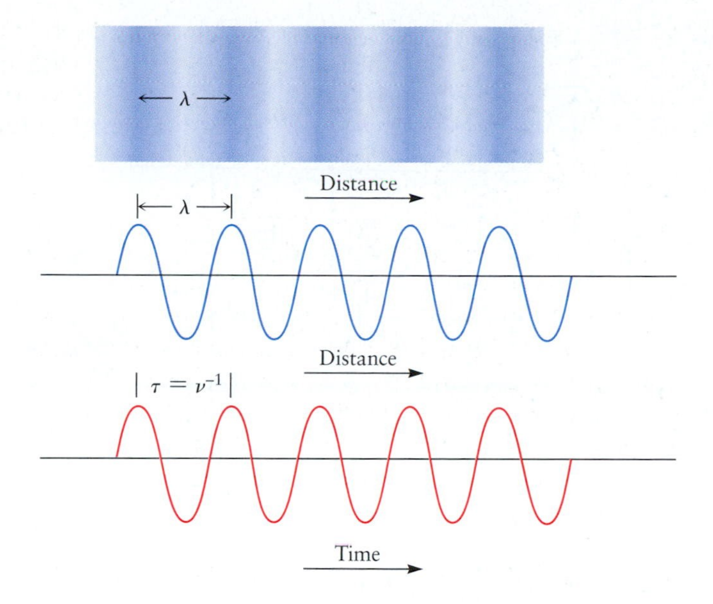

FIGURE 4.1 As a water wave moves across an otherwise calm tank, its maximum amplitude and wavelength can be determined. Its speed is found by dividing the travel distance of a particular wave crest by the time elapsed.

travel this distance is the reciprocal of the frequency, $\tau = \nu^{-1}$, so the speed (the distance traveled divided by the time elapsed) is given by

$$\text{speed} = \frac{\text{distance traveled}}{\text{time elapsed}} = \frac{\lambda}{\nu^{-1}} = \lambda\nu$$

The speed of a wave is the product of its wavelength and its frequency.

Electromagnetic Radiation

By the end of the 18th century, the behavior of light was well described by a wave model. The signature properties of light—diffraction, interference, and polarization—were understood as consequences of wave propagation. In 1865, the Scottish physicist James Clerk Maxwell proposed a theory that described visible light as a propagating wave of **electromagnetic radiation** that carries both energy and momentum. Unlike water and sound waves, electromagnetic waves are not sustained by some "propagating medium" such as water or air. Rather, a beam of light consists of oscillating *electric* and *magnetic fields* oriented perpendicular to one another and to the direction in which the light is propagating (Fig. 4.2). These

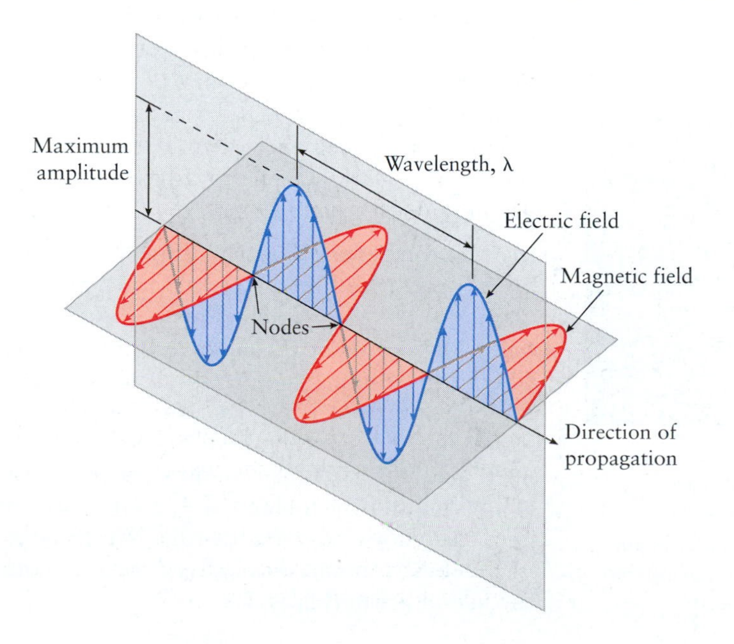

FIGURE 4.2 Light consists of waves of oscillating electric and magnetic fields that are perpendicular to each other and to the direction of propagation of the light.

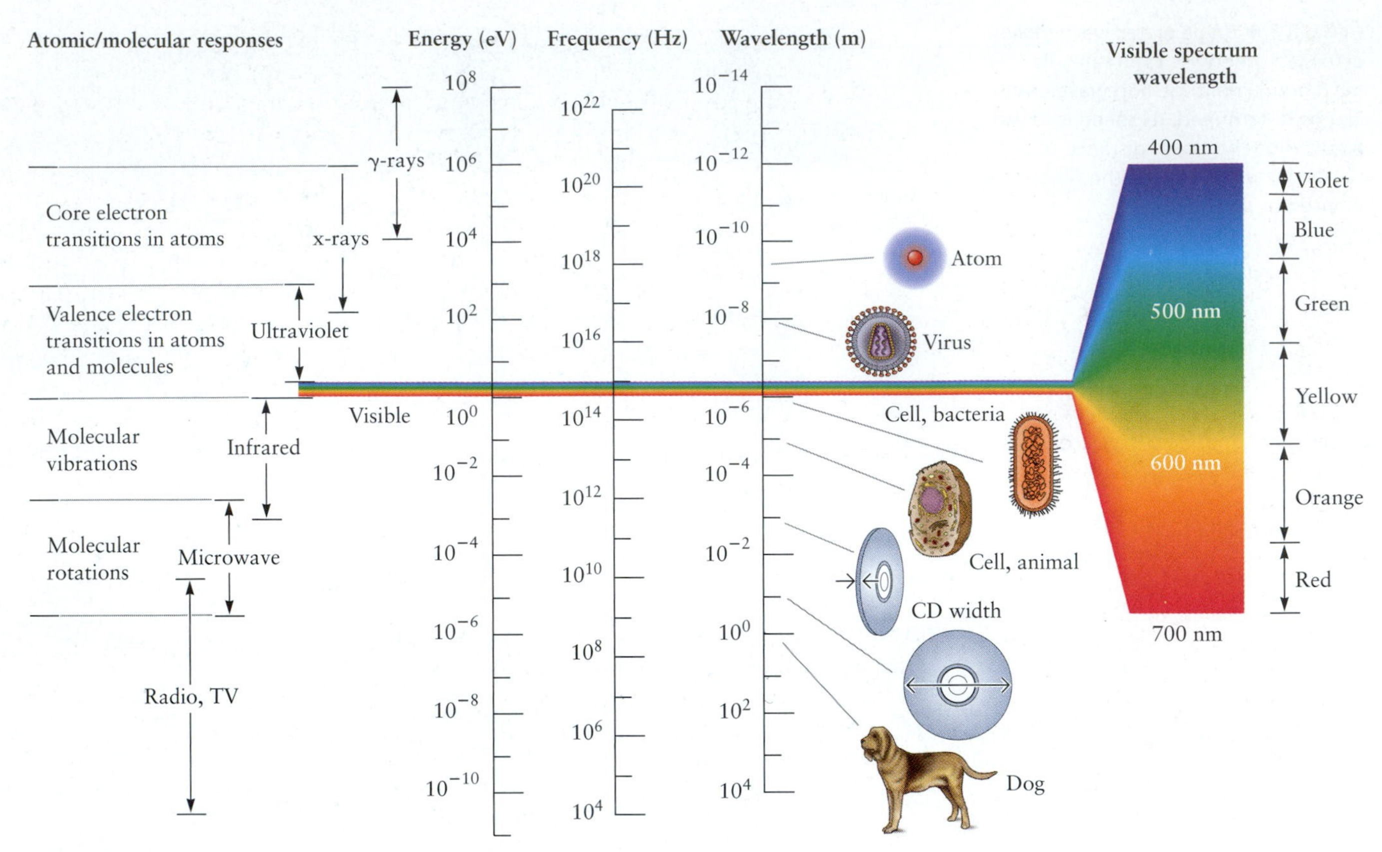

FIGURE 4.3 The electromagnetic spectrum. Note the small fraction that is visible to the human eye.

fields are produced by the motion of charged particles in the source of the light. These oscillating fields can transfer energy and momentum to other charged particles that intercept the beam in some location that is remote from the source. Electromagnetic waves carry information from a broadcast source to a remote receiver in wireless communication. Indeed, one of the early triumphs of Maxwell's theory was the development of radio, based largely on the experimental work of Heinrich Hertz. We will see that electromagnetic radiation is both emitted and absorbed by atoms and molecules. It is, therefore, one of our most effective tools for probing the nature of atoms and molecules.

Electromagnetic waves are described by the equations introduced earlier. The speed, c, of light passing through a vacuum is equal to the product $\lambda\nu$, and its value *by definition* is

$$c = \lambda\nu = 2.99792458 \times 10^8 \text{ m s}^{-1} \quad [4.1]$$

The speed, c, is a universal constant; it is the same for all types of radiation in the electromagnetic spectrum (Fig. 4.3). Regions of the electromagnetic spectrum are characterized by different values of wavelength and frequency. The region visible to the eye, which is a small fraction of the entire spectrum, comprises bands of colored light that cover particular ranges of wavelength and frequency. The band of light we perceive as green is centered about 5.7×10^{14} Hz with wavelengths near 5.3×10^{-7} m (530 nm). Red light is characterized by a lower frequency and a longer wavelength than green light, and violet light is characterized by a higher frequency and shorter wavelength than green light. A laser, such as the one shown in Figure 4.4, emits nearly monochromatic light (light of a single frequency and wavelength). White light contains the full range of visible wavelengths; it can be resolved into its component wavelengths by passing it through a prism (Fig. 4.5).

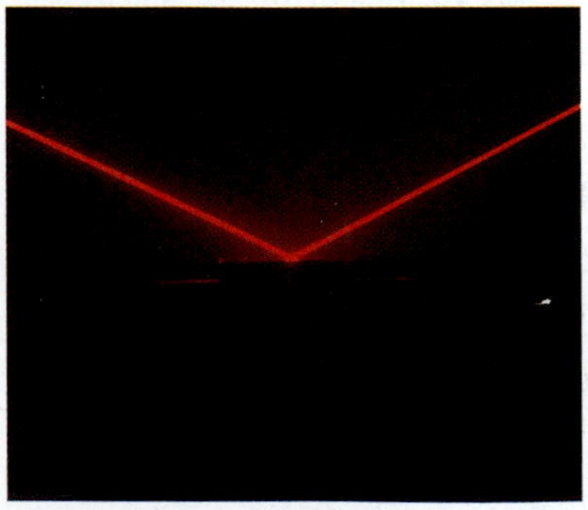

Thomson Learning/Henry Leap and Jim Lehman

FIGURE 4.4 A laser emits a well-collimated beam of light with a narrow range of wavelengths. The direction of motion of a laser beam can be manipulated by inserting mirrors in the path of the beam.

FIGURE 4.5 When white light is passed through slits to produce a narrow beam and then refracted in a glass prism, the various colors blend smoothly into one another.

Electromagnetic radiation that lies outside the visible region is also familiar to us (see Fig. 4.3). The warmth radiated from a stone pulled from a fire is largely due to infrared radiation, whose wavelength is longer than that of visible light. Microwave ovens use radiation whose wavelength is longer than infrared wavelengths, and radio communication uses still longer wavelengths. Radio stations are identified by their broadcast frequencies. FM stations typically broadcast at frequencies of tens to hundreds of megahertz ($1\ \text{MHz} = 10^6\ \text{s}^{-1}$), whereas AM stations broadcast at lower frequencies, from hundreds to thousands of kilohertz ($1\ \text{kHz} = 10^3\ \text{s}^{-1}$). You might check the frequencies of some of your favorite radio stations; ours include a classical music station broadcasting at 90.5 MHz (FM) and a sports station broadcasting at 1300 kHz (AM). Radiation with wavelengths shorter than that of visible light includes ultraviolet light, x-rays, and gamma rays; radiation in these regions of the electromagnetic spectrum (with wavelengths shorter than about 340 nm) can cause ionization and damage in biological tissue and are often collectively called **ionizing radiation.**

EXAMPLE 4.1

Almost all commercially available microwave ovens use radiation with a frequency of 2.45×10^9 Hz. Calculate the wavelength of this radiation.

SOLUTION

The wavelength is related to the frequency as follows:

$$\lambda = \frac{c}{\nu} = \frac{3.00 \times 10^8\ \text{s}^{-1}}{2.45 \times 10^9\ \text{s}^{-1}} = 0.122\ \text{m}$$

Thus, the wavelength is 12.2 cm.

Related Problems: 3, 4

4.2 Evidence for Energy Quantization in Atoms

Rutherford's planetary model of the atom was completely inconsistent with the laws of classical physics (see discussion in Section 3.2). According to Maxwell's electromagnetic theory, accelerated charges must emit electromagnetic radiation. An electron in orbit around the nucleus is accelerating because its direction is constantly changing. It must, therefore, emit electromagnetic radiation, lose energy, and eventually spiral into the nucleus. The very existence of stable atoms was perhaps the most fundamental conceptual challenge facing physicists in the early 1900s. The recognition that energy is quantized in atoms provided a path toward resolving the conceptual conflicts.

This section begins with a discussion of blackbody radiation, the experiment that introduced energy quantization into science. Two sets of experiments that demonstrated quantization of energy in free atoms in the gas phase then are described. We interpret these experiments using energy-level diagrams, which represent the discrete energy states of the atom. Our goal here is to introduce you to the relationship between the experimental evidence for energy quantization and the energy-level diagrams used to interpret these experiments. Later in the chapter, we use the quantum theory to explain how energy is quantized and to predict the allowed values of the energy for several model problems. We have organized our discussion to group key concepts together, for better coherence and to provide physical insight; it does not strictly follow the historical development of the field.

Blackbody Radiation and Planck's Hypothesis

We are about to discuss a monumental achievement in the development of modern science, which changed forever the way we look at the world. This was the recognition that objects cannot gain or lose energy in arbitrary or continuous amounts, but instead transfer energy only in discrete, discontinuous amounts that are multiples of some fundamental quantity of energy. The German physicist Max Planck achieved this insight in 1901 while trying to explain some puzzling new experimental measurements on the interaction of solid objects with radiant energy, which was known as **blackbody radiation.** You will shortly see that you are already familiar with blackbody radiation in various guises, and we relate the discussion closely to the experimental results so that you can always see the problem exactly as Planck saw it. We invite you to read and think along with Planck and to witness an important demonstration of how science advances. When experimental results do not agree with established scientific theories, the theories must be either modified or discarded and replaced with new ones, to account for both the new and the old experimental results. This process leads to the development of theories that provide a more fundamental understanding of a wider range of phenomena than their predecessors.

Every object emits energy from its surface in the form of thermal radiation. This energy is carried by electromagnetic waves; the distribution of the wavelengths of electromagnetic waves depends on the temperature of the object. At ordinary temperatures, thermal radiation falls within the infrared portion of the electromagnetic spectrum; imaging this radiation is used to map the surface of Earth from satellites in space and for tracking the movement of people in darkness using "night vision" detectors. As objects are heated to higher temperatures, the total intensity of radiation emitted over all frequencies increases, and the frequency distribution of the intensity also changes. The solid curves in Figure 4.6 show how the measured radiation intensity depends on frequency and temperature. There are two important features of these curves. First, the maximum in the radiation intensity distribution moves to higher frequency (shorter wavelength) as the temperature increases. This phenomenon is observed in familiar objects such as the heating element on an electric kitchen range or the filament in an incandescent lightbulb. As these objects are heated, they first glow red, then orange, then yellow, and finally, white. It also explains the differences in color among stars; the hottest stars appear to be nearly white, whereas the colors of cooler stars can range from red to yellow. Second—and this is a key result—the radiation intensity falls to zero at extremely high frequencies for objects heated to any temperature.

The sources of blackbody radiation, according to classical physics, are oscillating electrical charges in the surfaces of these objects that have been accelerated by ordinary thermal motion. Each motion persists for a certain period, producing radiation whose frequency is inversely related to that period. A number of scientists used different methods to calculate the radiation intensity curves using this simplified model and arrived at the following result:

$$\rho_T(\nu) = \frac{8\pi k_B T \nu^2}{c^3} \qquad [4.2]$$

where $\rho_T(\nu)$ is the intensity of the radiation at the frequency ν; k_B is a fundamental constant called the Boltzmann constant, which is discussed in Sections 9.5 and 9.6; T is the temperature in kelvins (K) and c is the speed of light.

These calculated results, shown for 5000 and 7000 K by the dashed curves in Figure 4.6, agree well with experiment at lower frequency. But the theory does not predict a maximum in the intensity distribution, and even worse, it disagrees badly with the experimental results at high frequencies. This feature of the result was called the "ultraviolet catastrophe" because it predicts an infinite intensity at very

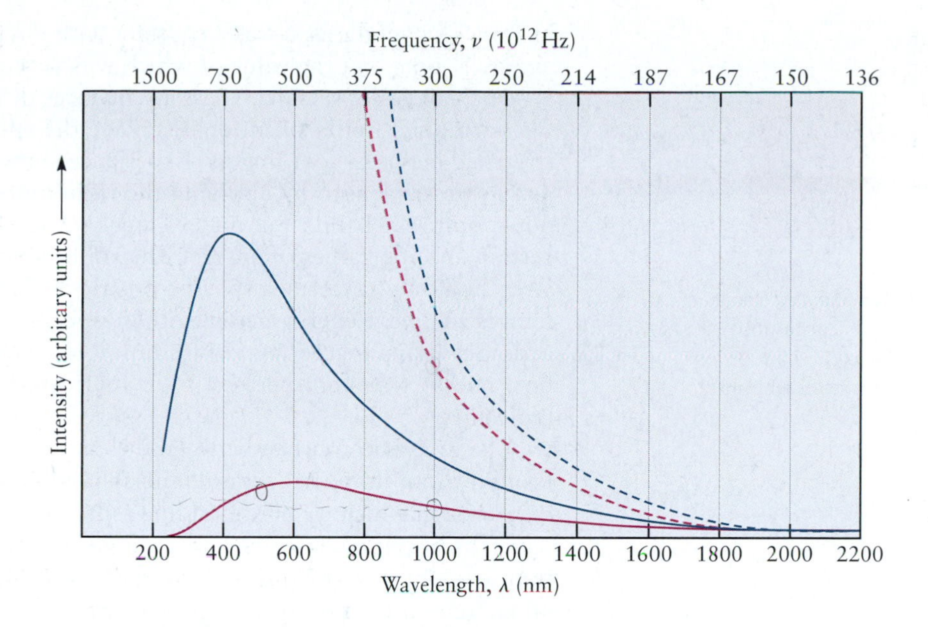

FIGURE 4.6 The dependence of the intensity of blackbody radiation on wavelength for two temperatures: 5000 K (red curve) and 7000 K (blue curve). The sun has a blackbody temperature near 5780 K, and its light-intensity curve lies between the two shown. The classical theory (dashed curves) disagrees with observation at shorter wavelengths.

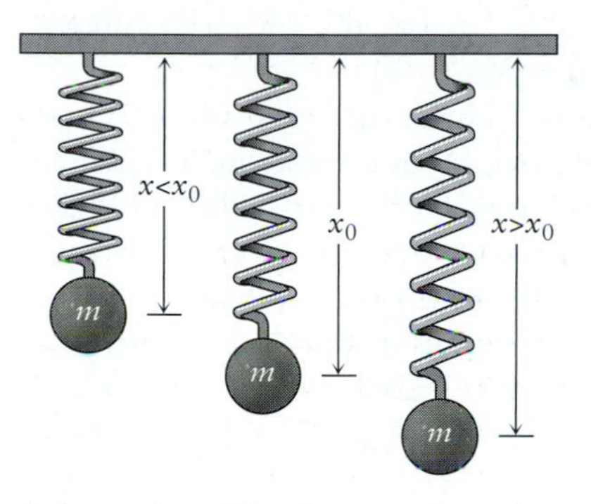

A charged particle of mass m bound to a solid surface by a spring is a model for the oscillatory motions of the surface atoms of a black body. The particle is shown at its rest position x_0, at a position closer to the surface ($x < x_0$), and at one further from the surface ($x > x_0$).

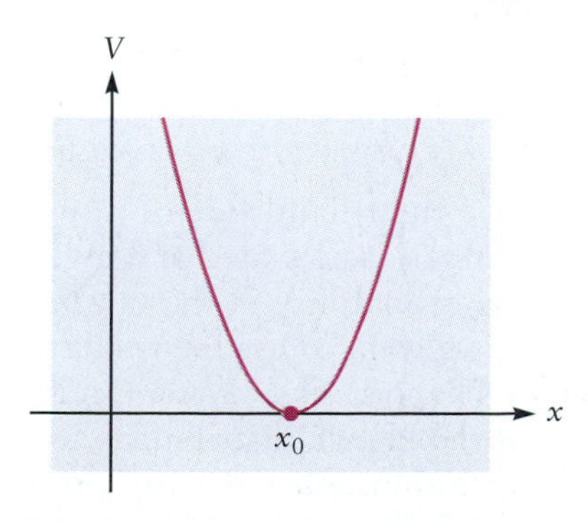

The potential energy curve for an oscillator is a consequence of the "restoring force" that always drives the oscillator toward its equilibrium position.

short wavelengths, whereas the experimental intensities always remain finite and actually fall to zero at very short wavelengths (very high frequencies). The calculated result failed completely to explain the frequency distribution in blackbody radiation; yet, it is a direct consequence of the laws of classical physics. How could this conflict be resolved?

Blackbody radiation was explained by Max Planck in 1901, but only by overthrowing the very foundations of classical mechanics. Planck reasoned that the very high-frequency oscillators must not be excited by the thermal energy of the hot body to the same degree as the lower frequency oscillators. This was a challenge to explain because classical mechanics allows an oscillator to have any energy. Planck's argument involved two steps, which are explained as follows.

For simplicity in following Planck's hypothesis, let us focus on just one of the oscillating charged particles and visualize it as a ball of mass, m, held in place by a spring. As the particle moves in response to the thermal motion of the atoms in the hot body, the spring exerts a "restoring force," F, which returns the particle to its equilibrium position, which we will call x_0. As discussed in Appendix B for this same model problem, the restoring force is directly proportional to the displacement, and the force law is $F = -k(x - x_0)$, where the constant k measures the "stiffness" of the spring. The displacement of the particle oscillates about x_0 in a periodic motion of frequency $\nu = (1/2\pi)\sqrt{k/m}$, and the associated potential energy of the particle is $V(x) = \frac{1}{2}k(x - x_0)^2$. This model is the simple harmonic oscillator described in Appendix B. Classical mechanics puts no restrictions on the value of the total energy, E. The total energy can be large or small, and it can be changed smoothly and continuously from one arbitrary value to another.

Planck's first step was to pose a daring hypothesis: It is *not* possible to put an arbitrary amount of energy into an oscillator of frequency ν. Instead, he postulated that the oscillator must gain and lose energy in "packets," or *quanta,* of magnitude $h\nu$, and that the total energy of an oscillator, ε_{osc}, can take only discrete values that are integral multiples of $h\nu$:

$$\varepsilon_{osc} = nh\nu \qquad n = 1, 2, 3, 4, \ldots \qquad [4.3]$$

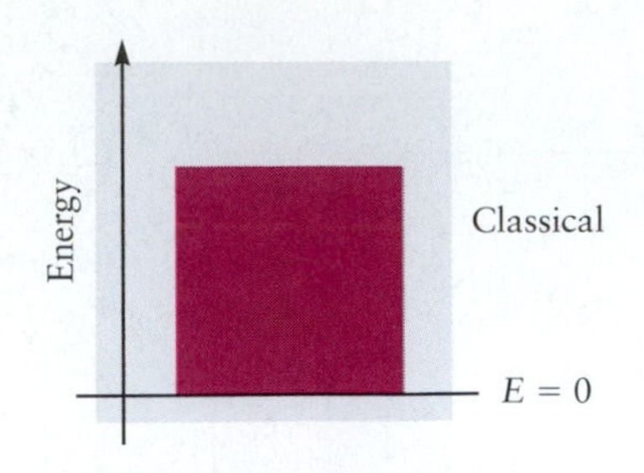

An oscillator obeying classical mechanics has continuous values of energy and can gain or lose energy in arbitrary amounts.

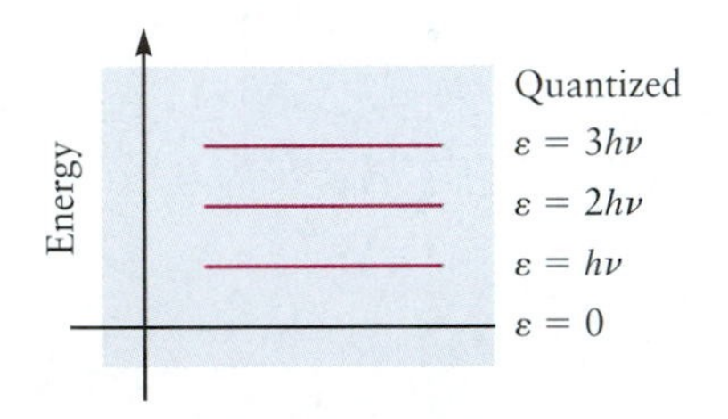

An oscillator described by Planck's postulate has discrete energy levels. It can gain or lose energy only in amounts that correspond to the difference between two energy levels.

In Planck's hypothesis, h was a constant with physical units *energy* × *frequency*$^{-1}$ = *energy* × *time,* but the value of which was yet to be determined.

You can easily visualize the consequences of Planck's hypothesis using the simple harmonic oscillator model. Replace the spring and ball with a rubber band stretched between your fingers. Experience shows that you can stretch the band to any arbitrary length by applying the right amount of energy (so long as you do not rupture the band). But under Planck's hypothesis, the band would accept only certain specific values of energy. The rubber band would behave as if it could be stretched only to certain specific positions. It simply would not respond to attempts to give it energy between these specific values. This fact is contrary to all ordinary human experience with tangible, macroscopic objects. And yet, this is how energy transfer operates in the microscopic world of atoms, electrons, and molecules.[2]

The dramatic contrast between the energy values allowed by classical mechanics and those that arise from Planck's postulate is illustrated using **energy-level diagrams,** in which a horizontal line represents an allowed energy value for a system. The height of each line above the zero of energy represents the total energy in that level. In macroscopic systems that are well described by classical mechanics, all energies are allowed; the upper energy level diagram in the margin represents the continuum of energies that the rubber band can accept up to the point where it breaks. For the quantum oscillators that Planck proposed, only those levels shown on the lower energy level diagram are allowed.

Planck's second step was to predict the radiation intensity curves by calculating the average energy in these quantized oscillators at each frequency as a function of temperature. The key idea is that the excitation of a particular oscillator is an all-or-nothing event; there is either enough thermal energy to cause it to oscillate or there is not. According to Planck, the falloff in the intensity with frequency at a given temperature of the blackbody radiation is due to a diminishing probability of exciting the high-frequency oscillators. Planck's intensity distribution is

$$\rho_T(\nu) = \frac{8\pi h\nu^3}{c^3} \frac{1}{e^{h\nu/k_B T} - 1} \qquad [4.4]$$

All of the symbols in Equation 4.4 have been identified earlier in this chapter. The value of h was determined by finding the best fit between this theoretical expression and the experimental results. Figure 4.7 shows the fit for $T = 1646$ K, resulting in the value $h = 6.63 \times 10^{-34}$ J s. The value of h, a fundamental constant of nature, has been measured to very high precision over the years by a number of other techniques. It is referred to as **Planck's constant,** and the currently accepted value is

$$h = 6.62608 \times 10^{-34} \text{ J s}$$

We ask you to accept that the second fraction on the right-hand side of Equation 4.4 is the probability that an oscillator of frequency ν is activated at a given temperature T. Chapter 9 presents the origin of this probability in the famous Maxwell–Boltzmann distribution, but in this chapter we want to use the result to demonstrate some additional consequences of Planck's hypothesis.

Before proceeding to explore the implications of the Planck distribution, we need to check whether it reduces to the classical expression under the appropriate conditions. It is always important to check whether new concepts can be

[2]We will see later that energy transfer into macroscopic objects is also quantized. However, the discrete values are so closely spaced that they appear continuous.

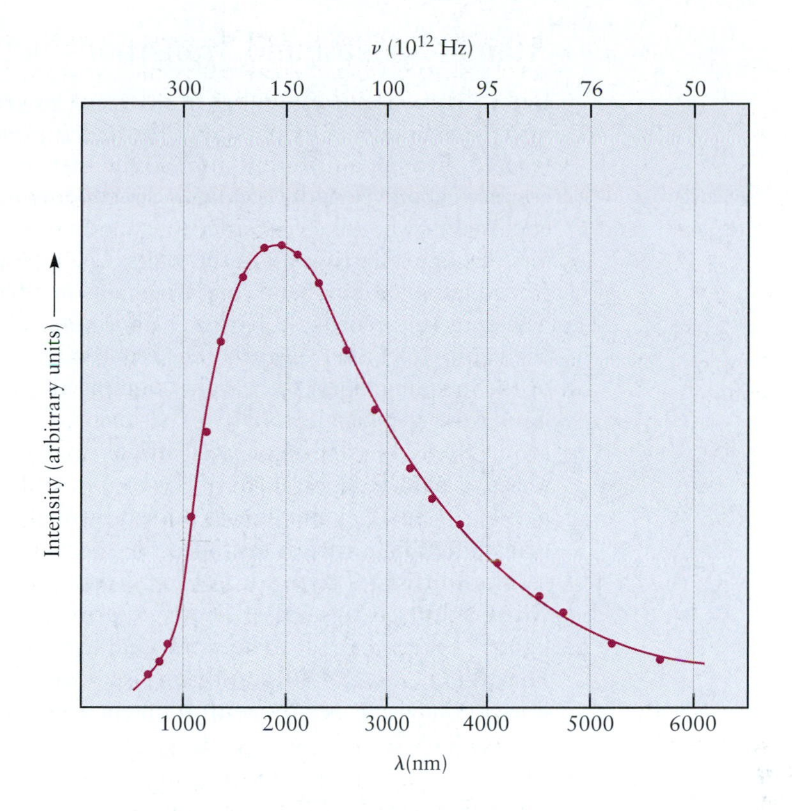

FIGURE 4.7 Experimental test of Planck's distribution for blackbody radiation. The dots represent experimental data acquired at $T = 1646$ K. The continuous curve represents Planck's predicted distribution, with the parameter $h = 6.63 \times 10^{-34}$ J s. Agreement between experiment and theory is spectacular, demonstrating the validity of Planck's theory and also determining the value of the previously unknown parameter h.

matched with old concepts under appropriate conditions; this demonstrates that the new concepts represent an orderly advance in knowledge. We can imagine that all of the oscillators would be excited at sufficiently high temperatures, in which case, the system should behave according to the laws of classical physics. We express this condition mathematically by setting $h\nu/k_BT \ll 1$, a ratio that is nearly zero. You will soon learn in calculus that most functions can be represented by simpler forms as the argument of the function approaches zero. For the exponential function, $\exp(x) \approx 1 + x$ when x is nearly zero. Using this approximation, we obtain the high temperature limit of Planck's distribution

$$\rho_T(\nu) = \frac{8\pi h\nu^3}{c^3}\frac{1}{e^{h\nu/k_BT} - 1} \approx \frac{8\pi h\nu^3}{c^3}\frac{1}{([1 + h\nu/k_BT] - 1)} = \frac{8\pi k_BT\nu^2}{c^3} \quad [4.5]$$

which is valid as $T \longrightarrow \infty$. This is indeed the classical result quoted in Equation 4.2.

Behind all the mathematics, Planck's dramatic explanation of blackbody radiation includes three fundamentally new ideas:

1. The energy of a system can take only discrete values, which are represented on its energy-level diagram.
2. A quantized oscillator can gain or lose energy only in discrete amounts $\Delta\varepsilon$, which are related to its frequency by $\Delta\varepsilon = h\nu$.
3. To emit energy from higher energy states, the temperature of a quantized system must be sufficiently high to excite those states.

These three ideas have permeated all areas of science and technology. They are the basis for our understanding that energy (like matter) is discrete, not continuous, and that it can be transferred only in discrete chunks and not by arbitrary amounts. Every system has its own energy-level diagram that describes the allowed energy values and the possible values of energy transfers.

Atomic Spectra and Transitions between Energy States

Light that contains a number of different wavelengths (Fig. 4.8) can be resolved into its components by passing it through a prism, because each wavelength is refracted through a different angle. One instrument used to separate light into its component wavelengths is called a **spectrograph** (Figs. 4.9a, b). The spectrograph is enclosed in a boxlike container to exclude stray light. The light to be analyzed enters through a narrow slit in the walls. The light passes to the prism, where it is dispersed into its components, and then falls on a photographic plate or other detector. The detector records the position and intensity of an image of the slit formed by each component wavelength. The recorded array of images is called the *spectrum* of the incoming light. If the light contains all frequencies, the spectrum is a continuous band (see Fig. 4.3). Early experiments showed that light emitted from gaseous atoms excited in flames or in electrical discharges gave discrete spectra, that is, a series of parallel lines. Each line was an image of the slit at a specific wavelength (see Fig. 4.9a). If white light is passed through a sample of gaseous atoms and the transmitted light then is sent into the spectrograph, the absorption spectrum that results consists of dark slit images superimposed on the continuous spectrum of white light (see Fig. 4.9b). These experiments show that atoms emit and absorb light at a discrete set of frequencies characteristic of a particular element (Fig. 4.10). For example, in 1885, J. J. Balmer discovered that hydrogen atoms emit a series of lines in the visible region, with frequencies given by the following simple formula:

$$\nu = \left[\frac{1}{4} - \frac{1}{n^2}\right] \times 3.29 \times 10^{15}\ \mathrm{s^{-1}} \qquad n = 3,\ 4,\ 5, \ldots \qquad [4.6]$$

The hydrogen atom lines shown in Figure 4.10 fit this equation, with $n = 3, 4, 5$, and 6 going from red to blue. Trying to understand the existence of discrete line spectra and the various empirical equations that relate the frequencies of the lines challenged physicists for more than three decades.

The first explanation for these surprising experimental results was provided in 1913 by the Danish physicist Niels Bohr. He proposed a model of the hydrogen atom that allowed only discrete energy states to exist. He also proposed that light absorption resulted from a *transition* of the atoms between two of these states. The frequency of the light absorbed is connected to the energy of the initial and final states by the expression

$$\nu = \frac{E_f - E_i}{h} \qquad \text{or} \qquad \Delta E = h\nu \qquad [4.7]$$

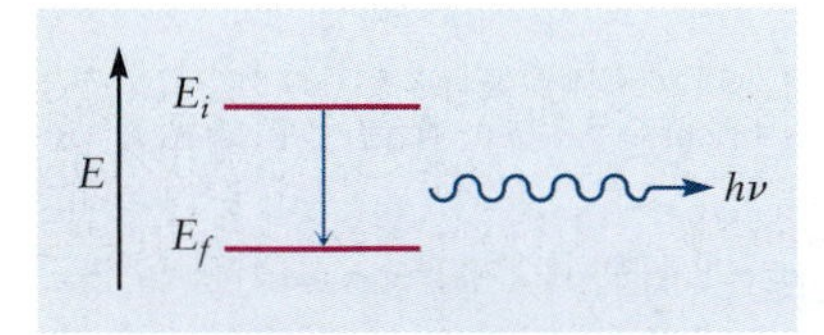

An atom makes a transition from state E_i to E_f and emits a photon of frequency $\nu = [E_i - E_f]/h$.

where h is Planck's constant. In absorption, the energy of the final state, E_f, is greater than that of the initial state so the signs work out correctly; ν is a positive number as it must be. For emission, however, $E_f < E_i$, and Equation 4.7 would predict a negative frequency, which is, of course, impossible. To account for both absorption and emission processes using the convention universally adopted by chemists that $\Delta E = E_f - E_i$, we use the more general expression that $|\Delta E| = h\nu$ and that $\Delta E > 0$ for absorption, whereas $\Delta E < 0$ for emission. The Bohr model also accounts for the values of the discrete energy levels in the hydrogen atom (see Section 4.3).

The atoms of every element can be represented by an energy-level diagram in which the energy difference between two levels is related by Equation 4.7 to the frequency of a specific line in the experimental spectrum of the atom. Except for the simplest case of hydrogen, however, constructing the energy-level diagram from the experimental spectrum is difficult because numerous transitions are involved. Nonetheless, spectroscopists have assigned the atomic spectra of most of the elements in the periodic table, and extensive tabulations of the results are readily available.

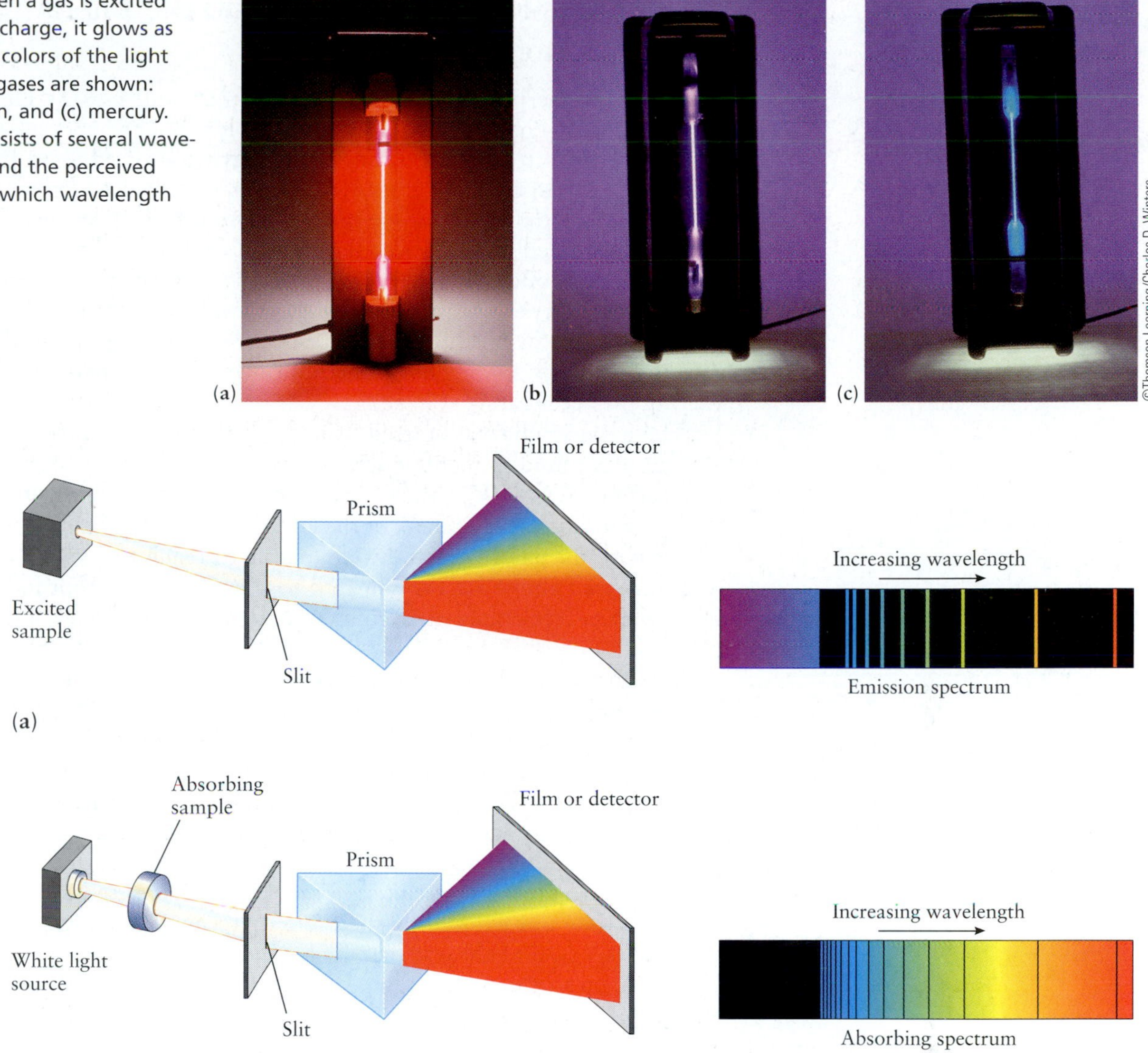

FIGURE 4.8 When a gas is excited in an electrical discharge, it glows as it emits light. The colors of the light emitted by three gases are shown: (a) neon, (b) argon, and (c) mercury. Each emission consists of several wavelengths of light, and the perceived color depends on which wavelength predominates.

FIGURE 4.9 (a) The emission spectrum of atoms or molecules is measured by passing the light emitted from an excited sample through a prism to separate it according to wavelength, then recording the image on photographic film or with another detector. (b) In absorption spectroscopy, white light from a source passes through the unexcited sample, which absorbs certain discrete wavelengths of light. Dark lines appear on a bright background.

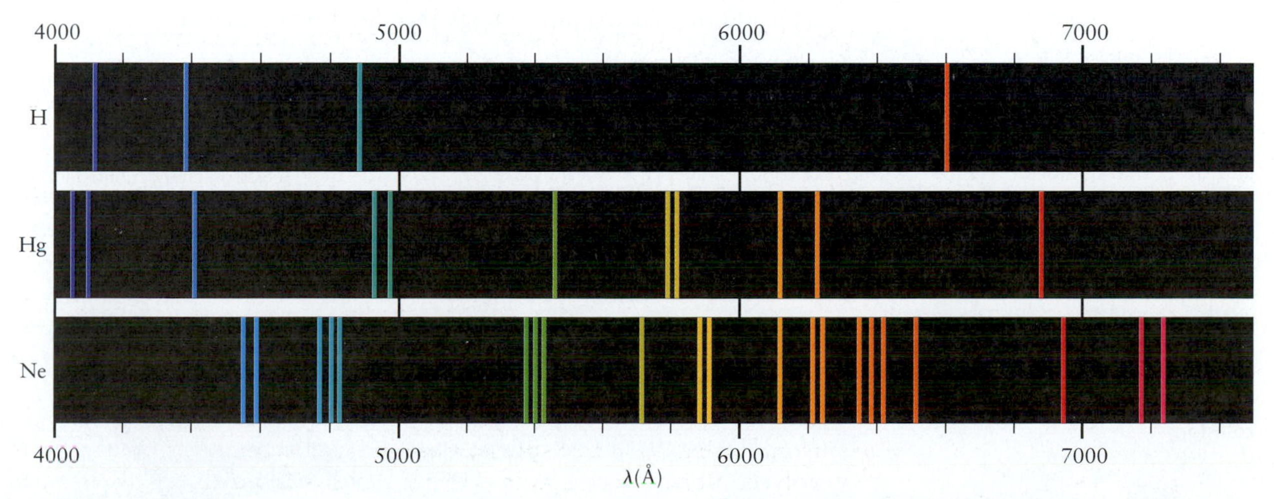

FIGURE 4.10 Atoms of hydrogen, mercury, and neon emit light at discrete wavelengths. The pattern seen is characteristic of the element under study. 1 Å = 10^{-10} m.

The Franck–Hertz Experiment and the Energy Levels of Atoms

In 1914, the German physicists James Franck and Gustav Hertz (nephew of Heinrich Hertz) conducted an experiment to test Bohr's hypothesis that the energy of atoms is quantized by measuring the energy transferred to an atom in collisions with electrons. In their apparatus (Fig. 4.11), electrons of known energy collided with gaseous atoms, and the energy lost from the electrons was measured. Electrons were emitted from the heated cathode C and accelerated toward the anode A. Holes in the anode allowed electrons to pass toward the collector plate P with known kinetic energy controlled by the accelerating voltage between C and A. The apparatus was filled to a low pressure with the gas to be studied. The current arriving at P was studied as a function of the kinetic energy of the electrons by varying the accelerating voltage.

The experiment was started using a very low accelerating voltage, and the current was found to increase steadily as the accelerating voltage was increased. At a certain voltage, V_{thr}, the current dropped sharply, going nearly to zero. This observation implied that most of the electrons had lost their kinetic energy in collisions with the gas atoms and were unable to reach the collector. As the voltage was increased above V_{thr}, the current rose again. This result indicated that electrons were reaccelerated after collisions and gained sufficient energy to reach the collector.

The current in the Franck–Hertz experiment shows a sharp change at a particular value of the accelerating voltage, corresponding to the threshold energy transfer from the electron to a gaseous atom.

The abrupt fall in the plot of current versus voltage at V_{thr} suggested that the kinetic energy of the electrons must reach a threshold eV_{thr} to transfer energy to the gas atoms, suggesting that the energy of the atoms must be quantized in discrete states. The **first excited state** must lie above the **ground state** (the state with lowest energy) by the amount eV_{thr}. Continuing the experiment with higher values of accelerating voltage revealed additional energy thresholds corresponding to excited states with higher energies.

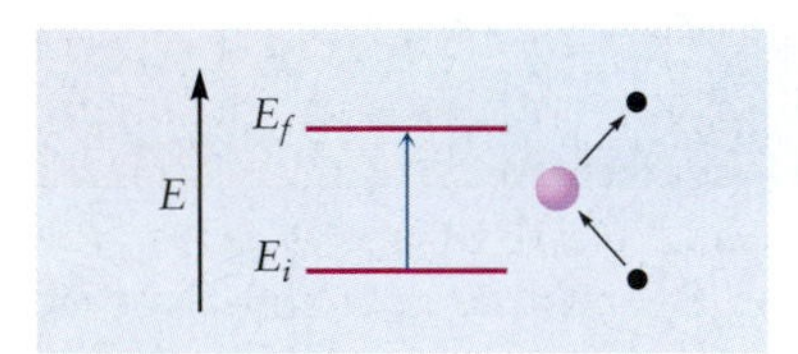

The Franck–Hertz experiment measures directly the separation between energy levels of the atom by measuring the energy lost by an electron colliding with the atom.

To confirm their interpretation, Franck and Hertz used a spectrograph to analyze light that was emitted by the excited atoms. When the accelerating voltage was below V_{thr}, no light was observed. When the accelerating voltage was slightly above V_{thr}, a single emission line was observed whose frequency was very nearly equal to

$$\nu = \frac{\Delta E}{h} = \frac{eV_{\text{thr}}}{h} \quad [4.8]$$

At higher accelerating voltages, additional spectral emission lines appeared as each additional excitation energy threshold was reached.

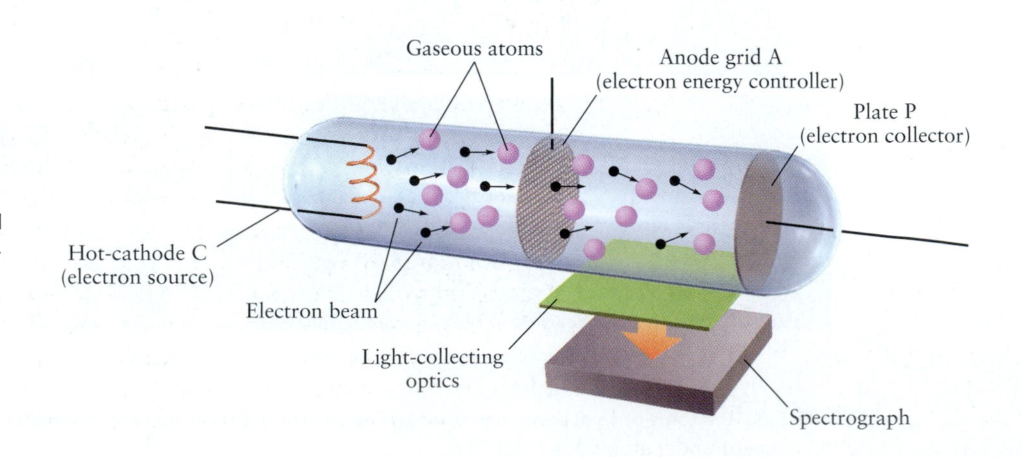

FIGURE 4.11 Apparatus of Franck and Hertz that demonstrates the quantization of energy in atoms. Gaseous atoms collide with electrons and gain energy by collisions only when the energy of the electron exceeds a certain threshold. The excited atom then emits a photon whose frequency is determined by the energy transferred to the atom during the collision.

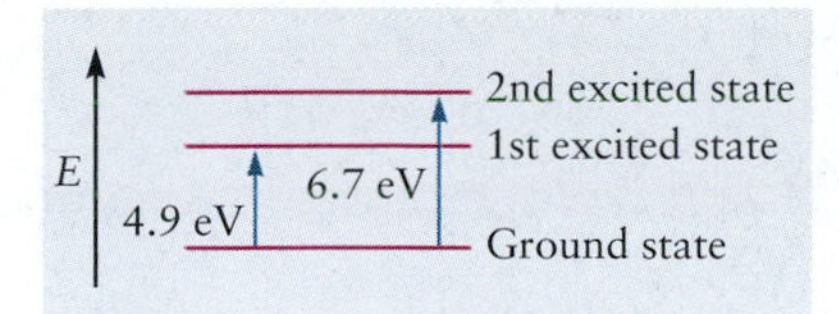

Simplified energy-level diagram for mercury.

EXAMPLE 4.2

The first two excitation voltage thresholds in the Franck–Hertz study of mercury vapor were found at 4.9 and 6.7 V. Calculate the wavelength of light emitted by mercury atoms after excitation past each of these thresholds.

SOLUTION

The emitted wavelength at a particular value of V_{thr} is given by the following equation:

$$\lambda = \frac{hc}{\Delta E} = \frac{hc}{eV_{thr}} = \frac{(6.6261 \times 10^{-34}\text{ J s})(2.9979 \times 10^{8}\text{ ms}^{-1})}{(1.6022 \times 10^{-19}\text{ C})(V_{thr}[\text{V}])}$$

$$= \frac{1239.8\text{ nm}}{V_{thr}[\text{V}]}$$

The value of each emission wavelength is calculated by substituting in a particular value of V_{thr} expressed in units of volts (V).

At $V_{thr} = 4.9$ V, the calculated wavelength is $\lambda = 250$ nm. The wavelength actually observed above this threshold was 253.7 nm.

At $V_{thr} = 6.7$ V, the calculated wavelength is $\lambda = 180$ nm. The wavelength actually observed above this threshold was 184.9 nm.

Energy differences measured by the Franck–Hertz method and by optical emission spectroscopy agree quite closely. The optical measurements are more precise. These results enable us to begin to construct the energy diagram for mercury, showing the location of the first two excited states relative to the ground state.

Related Problems: 17, 18

The significance of the Franck–Hertz experiment in the development of atomic physics cannot be exaggerated. It demonstrated that atoms absorb energy in collisions with electrons only in discrete, quantized amounts. The energy is then released only in discrete, quantized amounts by light emission. The Franck–Hertz experiment provided dramatic confirmation of Bohr's hypothesis that the energy of atoms is quantized in discrete states. It also provided a direct mechanical method for measuring the energy differences between these states and for constructing the energy-level diagram starting with the ground state. The technique continues to be used today to construct energy-level diagrams for molecules in the methods called "electron impact spectroscopy" or "electron energy loss spectroscopy."

4.3 The Bohr Model: Predicting Discrete Energy Levels

Atomic spectra and Franck–Hertz experiments measure the differences between energy levels, which can be calculated from the experimental data using Equations 4.7 and 4.8. In 1913, Niels Bohr developed the first theoretical model to predict the energy levels of the hydrogen atom and one-electron ions such as He^{+}, Li^{2+}, and Be^{3+}. The Bohr theory started from Rutherford's *planetary model* of the atom. (You should review Section 3.2 before continuing further in this chapter.) Pay careful attention to the definition of an absolute energy scale for atoms by choosing a reference state whose energy we set as zero. The logical choice, as discussed in Section 3.2, is the electron at rest located infinitely far from the nucleus.

Bohr supplemented Rutherford's planetary model of the atom with the assumption that an electron of mass m_e moves in a circular orbit of radius r about a fixed nucleus. The total energy of the hydrogen atom, kinetic plus potential, is given by Equation 3.7, which we reproduce and renumber here as Equation 4.9 for convenience:

$$E = \frac{1}{2} m_e v^2 - \frac{Ze^2}{4\pi\epsilon_0 r} \qquad [4.9]$$

The Coulomb force that attracts the electron to the nucleus, F_{coul}, is the negative derivative of the potential energy with respect to the separation r: $F_{coul} = -Ze^2/4\pi\epsilon_0 r^2$. Newton's second law relating force and acceleration is $F = m_e a$, and for uniform circular motion, the acceleration, a, of the electron is v^2/r. Combining these results gives the following relation for the *magnitude* of the force:

$$|F_{\text{Coulomb}}| = |m_e a| \qquad [4.10a]$$

$$\frac{Ze^2}{4\pi\epsilon_0 r^2} = m_e \frac{v^2}{r} \qquad [4.10b]$$

As mentioned earlier, classical physics requires that an accelerated electron emit electromagnetic radiation, thereby losing energy and eventually spiraling into the nucleus. Bohr avoided this conflict by simply *postulating* that only certain discrete orbits (characterized by radius r_n and energy E_n) are allowed, and that light is emitted or absorbed only when the electron "jumps" from one stable orbit to another. This bold assertion was Bohr's attempt to explain the existence of stable atoms, a well-established experimental fact. Faced with the contradiction between the experimental results and the requirements of classical electrodynamics, he simply discarded the latter in the formulation of his model.

The next step in the development of the Bohr model was his assertion that the angular momentum of the electron is quantized. This was an ad hoc assumption designed to produce stable orbits for the electron; it had no basis in either classical theory or experimental evidence. The **linear momentum** of an electron is the product of its mass and its velocity, $m_e v$. The **angular momentum,** L, is a different quantity that describes rotational motion about an axis. An introduction to angular momentum is provided in Appendix B. For the circular paths of the Bohr model, the angular momentum of the electron is the product of its mass, its velocity, and the radius of the orbit ($L = m_e vr$). Bohr postulated that the angular momentum is quantized in integral multiples of $h/2\pi$, where h is Planck's constant:

$$L = m_e vr = n\frac{h}{2\pi} \qquad n = 1, 2, 3, \ldots \qquad [4.11]$$

The existence of discrete orbits and quantized energies follows directly as a consequence of the quantization of angular momentum.

We can determine the properties of these discrete orbits as follows. Equations 4.9 and 4.10 contain two unknowns, v and r. Solving Equation 4.11 for v $(= nh/2\pi m_e r)$, inserting it into Equation 4.10, and solving for r gives the allowed values for radius of the orbits:

$$r_n = \frac{\epsilon_0 n^2 h^2}{\pi Z e^2 m_e} = \frac{n^2}{Z} a_0 \qquad [4.12]$$

where a_0, the **Bohr radius,** has the numerical value 5.29×10^{-11} m = 0.529 Å. (The Bohr radius ($a_0 = \epsilon_0 h^2/\pi e^2 m_e$) is a convenient, fundamental unit of length in

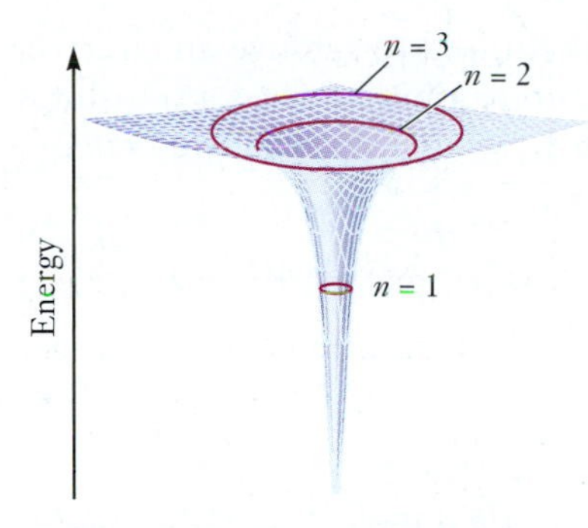

FIGURE 4.12 The potential energy of the electron and nucleus in the hydrogen atom has its lowest (most negative) value when the electron is closest to the nucleus. (Compare with the one-dimensional plot in Figure 3.2.) The electron moving away from the nucleus can be seen as moving up the sides of a steep potential energy well. In the Bohr theory, it can "catch" and stick on the sides only at certain allowed values of r, the radius, and E, the energy. The first three of these are shown by rings.

atomic physics that relieves us from the burden of carrying along all of the constants in Eq. 4.12.) This first prediction of the Bohr model is the existence of a series of orbits whose distances from the nucleus increase dramatically with increasing n. Substituting r_n from Equation 4.12 into Equation 4.11 allows us to calculate the velocity v_n corresponding to the orbit with radius r_n.

$$v_n = \frac{nh}{2\pi m_e r_n} = \frac{Ze^2}{2\epsilon_0 nh} \qquad [4.13]$$

The results obtained for r_n and v_n can now be substituted into Equation 4.9 to give us the allowed values of the energy:

$$E_n = \frac{-Z^2e^4m_e}{8\epsilon_0^2n^2h^2} = -(2.18 \times 10^{-18}\text{ J})\frac{Z^2}{n^2} \qquad n = 1, 2, 3, \ldots \qquad [4.14a]$$

That these energies are negative is a consequence of our choice for the zero of energy, as discussed in Section 3.2 and shown in Figure 3.2. For the same reason that we introduced the Bohr radius, it is convenient to express atomic energy levels in units of **rydbergs,** where 1 rydberg = 2.18×10^{-18} J. The energy level expression then becomes

$$E_n = -\frac{Z^2}{n^2} \quad \text{(rydberg)} \qquad n = 1, 2, 3, \ldots \qquad [4.14b]$$

The Bohr model thus predicts a discrete energy-level diagram for the one-electron atom (Figs. 4.12 and 4.13). The ground state is identified by $n = 1$, and the excited states have higher values of n (see Fig. 4.12).

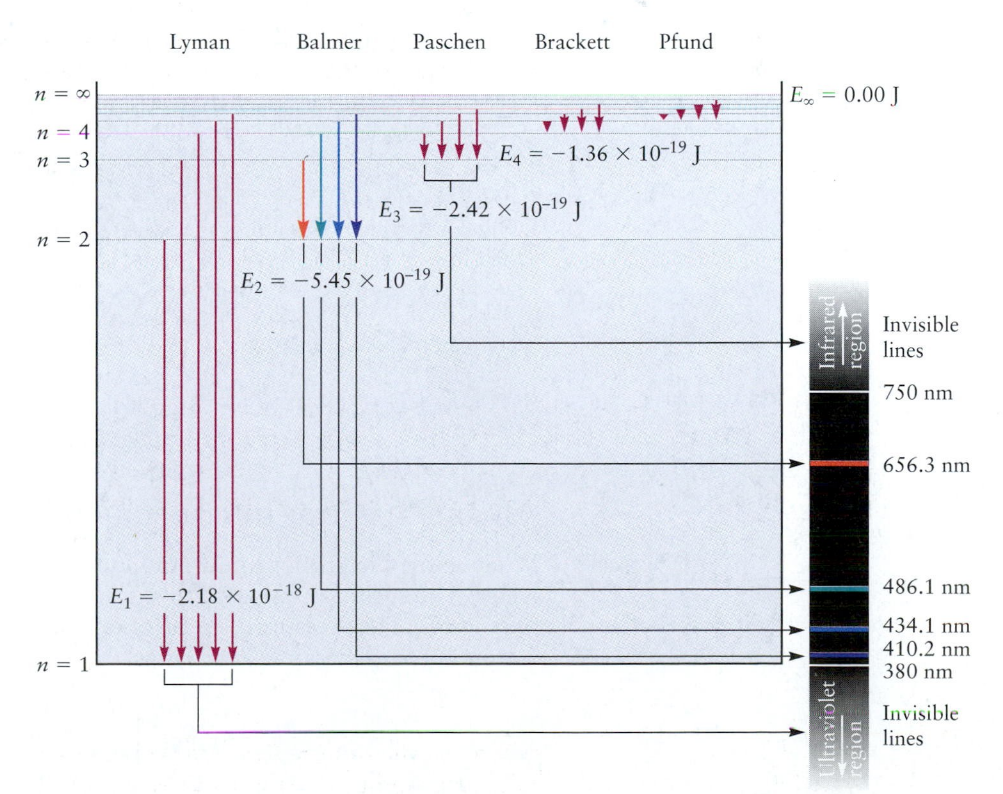

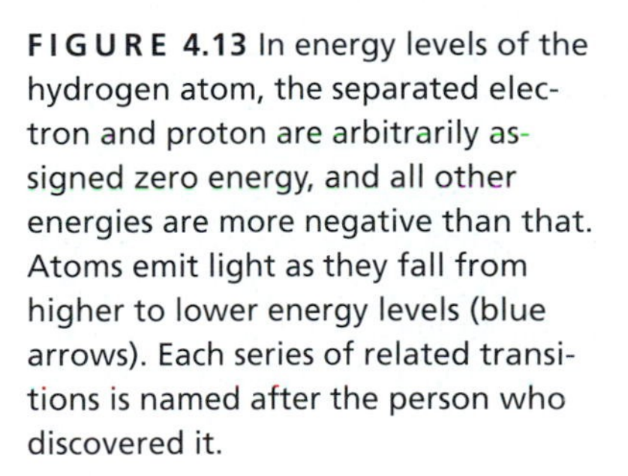

FIGURE 4.13 In energy levels of the hydrogen atom, the separated electron and proton are arbitrarily assigned zero energy, and all other energies are more negative than that. Atoms emit light as they fall from higher to lower energy levels (blue arrows). Each series of related transitions is named after the person who discovered it.

The **ionization energy** is the minimum energy required to remove an electron from an atom (see Section 3.3). In the Bohr model, ionization involves a transition from the $n = 1$ state to the $n = \infty$ state, in which $E_n = 0$. The associated energy change is

$$\Delta E = E_{\text{final}} - E_{\text{initial}} = 0 - (-2.18 \times 10^{-18}\ \text{J}) = 2.18 \times 10^{-18}\ \text{J}$$

Multiplying this result by Avogadro's number gives the ionization energy, *IE*, per mole of atoms:

$$IE = (6.022 \times 10^{23}\ \text{atoms mol}^{-1})(2.18 \times 10^{-18}\ \text{J atom}^{-1}) = 1.31 \times 10^{6}\ \text{J mol}^{-1} = 1310\ \text{kJ mol}^{-1}$$

This prediction agrees with the experimentally observed ionization energy of hydrogen atoms and provides confidence in the validity of the Bohr model. The discussion in Section 3.3 related measured ionization energies qualitatively to the effective potential energy binding electrons inside atoms. The Bohr model was the first physical theory that could predict ionization energies with remarkable accuracy.

EXAMPLE 4.3

Consider the $n = 2$ state of Li^{2+}. Using the Bohr model, calculate the radius of the electron orbit, the electron velocity, and the energy of the ion relative to that of the nucleus and electron separated by an infinite distance.

SOLUTION

Because $Z = 3$ for Li^{2+} (the nuclear charge is $+3e$) and $n = 2$, the radius is

$$r_2 = \frac{n^2}{Z} a_0 = \frac{4}{3} a_0 = \frac{4}{3}(0.529\ \text{Å}) = 0.705\ \text{Å}$$

The velocity is

$$v_2 = \frac{nh}{2\pi m_e r_2} = \frac{2(6.626 \times 10^{-34}\ \text{J s})}{2\pi(9.11 \times 10^{-31}\ \text{kg})(0.705 \times 10^{-10}\ \text{m})} = 3.28 \times 10^{6}\ \text{m s}^{-1}$$

The energy is

$$E_2 = -\frac{(3)^2}{(2)^2}(2.18 \times 10^{-18}\ \text{J}) = -4.90 \times 10^{-18}\ \text{J}$$

Typically, atomic sizes fall in the range of angstroms, and atomic excitation energies in the range of 10^{-18} J. This is consistent with the calculations of coulomb potential energies in electron volts and dimensions in angstroms in Section 3.2.

Related Problems: 19, 20

Atomic Spectra: Interpretation by the Bohr Model

When a one-electron atom or ion undergoes a transition from a state characterized by quantum number n_i to a state lower in energy with quantum number n_f ($n_i > n_f$), light is emitted to carry off the energy $h\nu$ lost by the atom. By conservation of energy, $E_i = E_f + h\nu$; thus,

$$h\nu = \frac{Z^2 e^4 m_e}{8\epsilon_0^2 h^2}\left[\frac{1}{n_f^2} - \frac{1}{n_i^2}\right] \qquad \text{(emission)} \qquad [4.15]$$

As n_i and n_f take on a succession of integral values, lines are seen in the emission spectrum (see Fig. 4.13) with frequencies

$$\nu = \frac{Z^2 e^4 m_e}{8\epsilon_0^2 h^3}\left(\frac{1}{n_f^2} - \frac{1}{n_i^2}\right) = (3.29 \times 10^{15}\ \text{s}^{-1})\, Z^2\left(\frac{1}{n_f^2} - \frac{1}{n_i^2}\right) \quad [4.16]$$

$$n_i > n_f = 1, 2, 3, \ldots \text{ (emission)}$$

Conversely, an atom can *absorb* energy $h\nu$ from a photon as it undergoes a transition to a *higher* energy state ($n_f > n_i$). In this case, conservation of energy requires $E_i + h\nu = E_f$; thus, the absorption spectrum shows a series of lines at frequencies

$$\nu = (3.29 \times 10^{15}\ \text{s}^{-1})\, Z^2\left(\frac{1}{n_i^2} - \frac{1}{n_f^2}\right) \quad [4.17]$$

$$n_f > n_i = 1, 2, 3, \ldots \text{ (absorption)}$$

For hydrogen, which has an atomic number of $Z = 1$, the predicted emission spectrum with $n_f = 2$ corresponds to the series of lines in the visible region measured by Balmer and shown in Figure 4.13. A series of lines at higher frequencies (in the ultraviolet region) is predicted for $n_f = 1$ (the Lyman series), and other series are predicted at lower frequencies (in the infrared region) for $n_f = 3, 4, \ldots$ In fact, the predicted and observed spectra of hydrogen and one-electron ions are in excellent agreement—a major triumph of the Bohr theory.

Despite these successes, the Bohr theory has a number of shortcomings. Most important, it cannot predict the energy levels and spectra of atoms and ions with more than one electron. Also, more fundamentally, it was an uncomfortable hybrid of classical and nonclassical concepts. The postulate of quantized angular momentum—which led to the circular orbits—had no fundamental basis and was simply grafted onto classical physics to force the predictions of classical physics to agree with the experimental results. In 1926, the Bohr theory was replaced by modern quantum mechanics in which the quantization of energy and angular momentum arise as natural consequences of the basic postulates and require no additional assumptions. The circular orbits of the Bohr theory do not appear in quantum mechanics. The Bohr theory provided the conceptual bridge from classical theoretical physics to the new quantum mechanics. Its historical and intellectual importance cannot be exaggerated.

4.4 Evidence for Wave–Particle Duality

The Bohr theory provided a prescription for calculating the discrete energy levels of a one-electron atom or ion, but it did not explain the origin of energy quantization. A key step toward the development of modern quantum mechanics was the concept of **wave–particle duality**—the idea that particles sometimes behave as waves, and vice versa. Experiments were forcing physicists to recognize that physical systems could display either particle or wave characteristics, depending on the experimental conditions to which they were subjected. German physicist Albert Einstein introduced wave–particle duality to explain the photoelectric effect, in which light acted as a particle. French physicist Louis de Broglie suggested that particles could exhibit wavelike properties, and the stage was set for the new quantum mechanics to synthesize wave–particle duality and energy quantization into a comprehensive new theory.

The Photoelectric Effect

In addition to the conceptual problems with the planetary model of the atom and the difficulties with blackbody radiation, another conflict between experiment and classical theory arose from the observation of the **photoelectric effect.** A beam of light shining onto a metal surface (called the *photocathode*) can eject electrons (called *photoelectrons*) and cause an electric current (called a *photocurrent*) to flow (Fig. 4.14). The photocurrent shows an extremely interesting dependence on the frequency and intensity of the incident light (Fig. 4.15). Regardless of the light intensity, no photocurrent flows until the frequency exceeds a particular threshold value ν_0, which is unique for each metal. Low-frequency (long wavelength; for example, red) light apparently cannot provide enough energy to eject the electrons, no matter how intense it is. When the frequency of the light is increased through the threshold value (corresponding, perhaps, to green or blue light), electrons are emitted and the photocurrent is directly proportional to the light intensity. The frequency of the light apparently is the key to delivering enough energy to eject the electrons; no electrons are emitted when the surface is excited by light whose frequency is below the threshold frequency, but electrons are readily emitted for

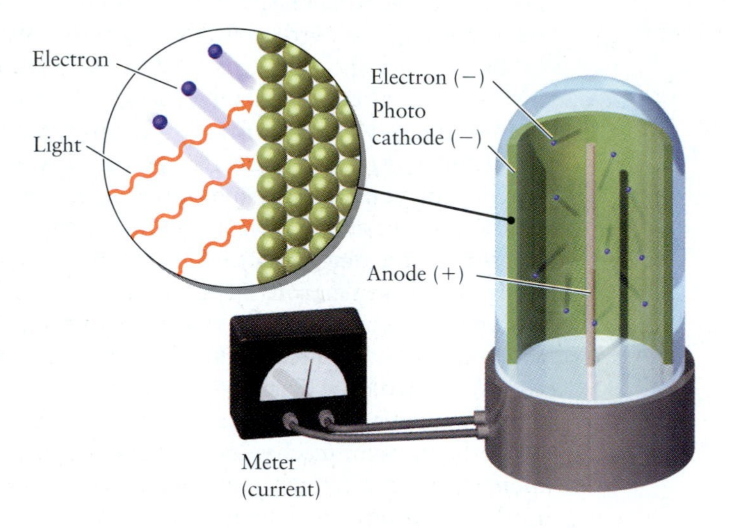

FIGURE 4.14 In a photoelectric cell (photocell), light strikes a metal surface in an evacuated space and ejects electrons. The electrons are attracted to a positively charged collector, and a current flows through the cell.

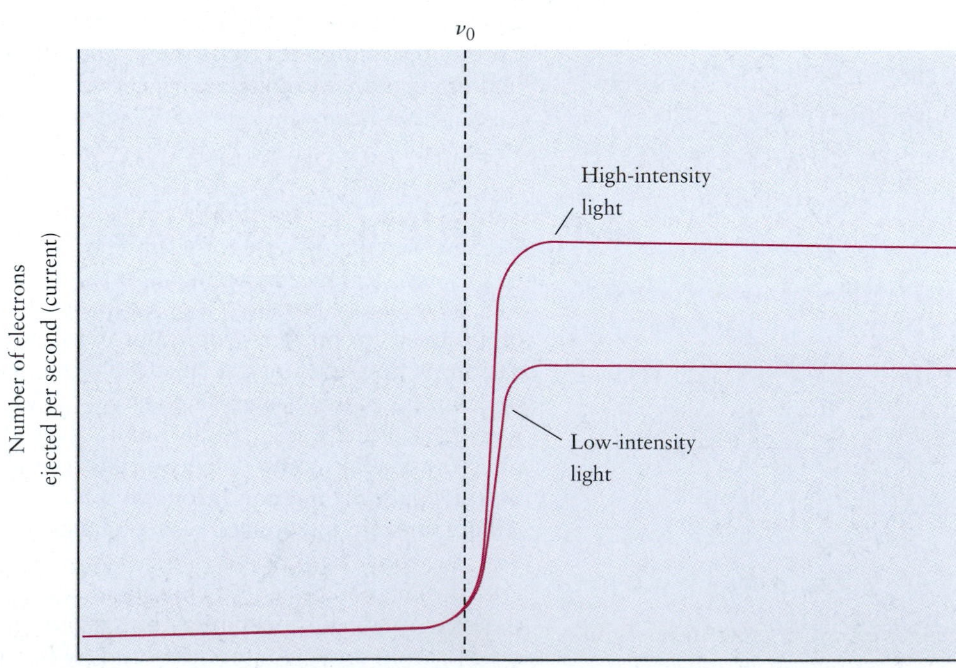

FIGURE 4.15 Frequency and intensity dependence of the photoelectric effect. Only light above the threshold frequency can eject photoelectrons from the surface. Once the frequency threshold has been passed, the total current of photoelectrons emitted depends on the intensity of the light, not on its frequency.

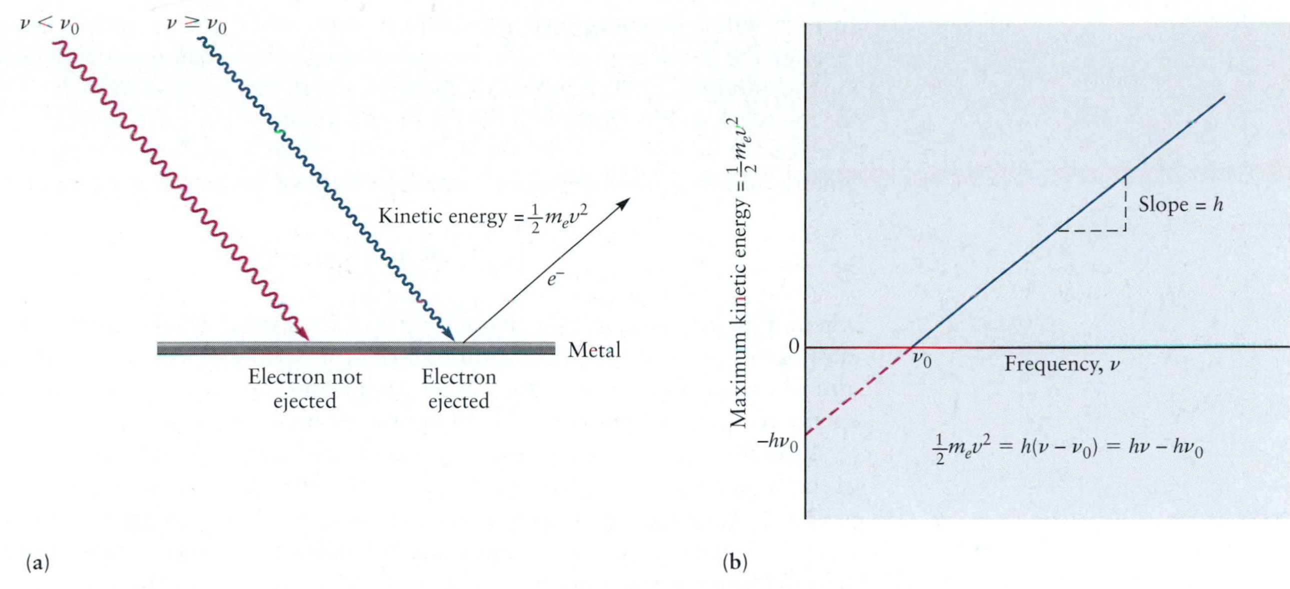

FIGURE 4.16 (a) Two key aspects of the photoelectric effect. Blue light is effective in ejecting electrons from the surface of this metal, but red light is not. (b) The maximum kinetic energy of the ejected electrons varies linearly with the frequency of light used.

all frequencies above the threshold frequency. These results could not be explained by classical physics. According to classical electromagnetic theory, the energy associated with electromagnetic radiation depends on only the intensity of the radiation, not on its frequency. Why, then, could a low-intensity (dim) beam of blue light (high frequency) eject electrons from sodium when a high-intensity (bright) beam of red light (low frequency) had no effect at all (Fig. 4.16a)? The key to the explanation is to relate the energy imparted by the light at ν_0 to the energy with which the photoelectrons are emitted from atoms in the metal.

The photoelectrons leaving the metal surface and traveling toward the detector have a range of kinetic energies. Let's assume that those photoelectrons arriving at the collector with E_{max} were emitted from atoms at the surface of the metal. Assume those arriving with lower kinetic energy were emitted deeper in the metal but lost some kinetic energy through collisions with other metal atoms before escaping from the surface. Then, the value of E_{max} should be directly related to the energy acquired by the photoelectron during the ejection process. We determine this maximum kinetic energy as follows. When the frequency and intensity of the beam are held constant, the magnitude of the photocurrent depends on the electrical potential (voltage) of the collector relative to the photocathode. At sufficiently positive potentials, all of the photoelectrons are attracted to the collector and the current–voltage curve becomes flat, or saturated. As the potential of the collector is made more negative, photoelectrons arriving with kinetic energies less than the maximum are repelled and the photocurrent decreases. Only those photoelectrons with sufficient kinetic energy to overcome this repulsion reach the collector. As the collector is made still more negative, the photocurrent drops sharply to zero at $-V_{max}$, identifying the maximum in the kinetic energy of the photoelectrons: $E_{max} = eV_{max}$. The potential required to stop all of the electrons from arriving at the collector is thus a direct measure of their maximum kinetic energy, expressed in units of electron volts (eV) (see Section 3.2). But what was the connection between ν_0 and E_{max}?

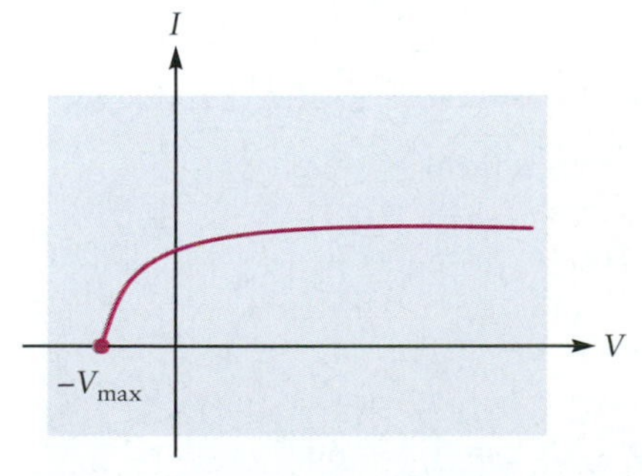

The current in a photocell depends on the potential between cathode and collector.

In 1905, Einstein used Planck's quantum hypothesis to explain the photoelectric effect. First, he suggested that a light wave of frequency ν consists of quanta of energy (later called **photons** by G. N. Lewis), each of which carries energy, $E_{photon} = h\nu$. Second, Einstein assumed that, in the photoelectric effect, an electron in the metal absorbs a photon of light and thereby gains the energy required to escape from the metal. A photoelectron emitted from beneath the surface will lose energy E' in collisions with other atoms and Φ in escaping through the surface, after which it travels through the vacuum to the detector with kinetic energy E. Conservation of energy leads to the relation $h\nu = E' + \Phi + E_k$ for the process. Electrons with the maximum kinetic energy are emitted at the surface, so for them $E' = 0$. Therefore, Einstein's theory predicts that the maximum kinetic energy of photoelectrons emitted by light of frequency ν is given by

$$E_{max} = \frac{1}{2}mv_e^2 = h\nu - \Phi \qquad [4.18]$$

where $\Phi = h\nu_0$ is a constant characteristic of the metal. The key idea of Einstein's explanation is that the interaction of a photon with an electron is a single event and the result is all or nothing; either the photon does or does not have enough energy to overcome the forces that bind the electron to the solid.

Einstein's theory predicts that the maximum kinetic energy is a linear function of the frequency, which provides a means for testing the validity of the theory. Experiments conducted at several frequencies demonstrated that the relation between E_{max} and frequency is indeed linear (see Fig. 4.16b). The slope of the experimental data determined the numerical value of h to be identical to the value that Planck found by fitting the experimental data to his theoretical blackbody radiation intensity distribution. Einstein's interpretation also provided a means to obtain the value of the quantity Φ from the experimental data as the "energy intercept" of the linear graph. Φ, called the **work function** of the metal, represents the binding energy, or energy barrier, that electrons must overcome to escape from the metal surface after they have absorbed a photon inside the metal. Φ governs the extraction of electrons from metal surfaces by heat and by electric fields, as well as by the photoelectric effect, and it is an essential parameter in the design of numerous electronic devices.

EXAMPLE 4.4

Light with a wavelength of 400 nm strikes the surface of cesium in a photocell, and the maximum kinetic energy of the electrons ejected is 1.54×10^{-19} J. Calculate the work function of cesium and the longest wavelength of light that is capable of ejecting electrons from that metal.

SOLUTION

The frequency of the light is

$$\nu = \frac{c}{\lambda} = \frac{3.00 \times 10^8 \text{ m s}^{-1}}{4.00 \times 10^{-7} \text{ m}} = 7.50 \times 10^{14} \text{ s}^{-1}$$

The binding energy $h\nu_0$ can be calculated from Einstein's formula:

$$E_{max} = h\nu - h\nu_0$$

$$1.54 \times 10^{-19} \text{ J} = (6.626 \times 10^{-34} \text{ J s})(7.50 \times 10^{14} \text{ s}^{-1}) - h\nu_0$$

$$= 4.97 \times 10^{-19} \text{ J} - h\nu_0$$

$$\Phi = h\nu_0 = (4.97 - 1.54) \times 10^{-19} \text{ J} = 3.43 \times 10^{-19} \text{ J}$$

The minimum frequency ν_0 for the light to eject electrons is then

$$\nu_0 = \frac{3.43 \times 10^{-19} \text{ J}}{6.626 \times 10^{-34}} = 5.18 \times 10^{14} \text{ s}^{-1}$$

From this, the maximum wavelength λ_0 is

$$\lambda_0 = \frac{c}{\nu_0} = \frac{3.00 \times 10^8 \text{ m s}^{-1}}{5.18 \times 10^{14} \text{ s}^{-1}} = 5.79 \times 10^{-7} \text{ m} = 579 \text{ nm}$$

Related Problems: 27, 28

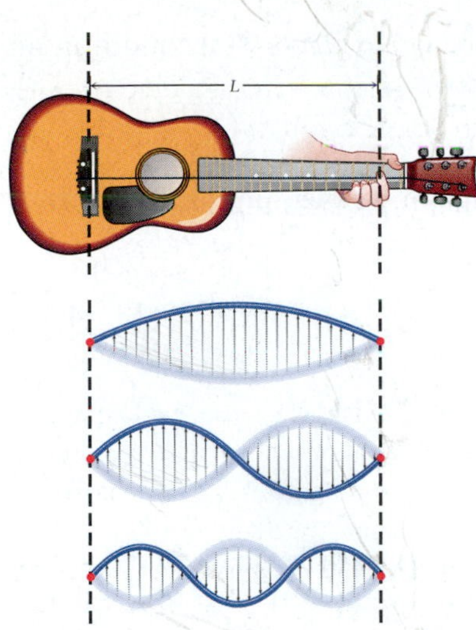

FIGURE 4.17 A guitar string of length L with fixed ends can vibrate in only a restricted set of ways. The positions of largest amplitude for the first three harmonics are shown here. In standing waves such as these, the whole string is in motion except at its end and at the nodes.

That two independent experiments involving two totally different phenomena gave the same value of h inspired great confidence in the validity of the quantum hypotheses that Planck and Einstein proposed, despite their unsettling implications. Einstein's bold assertion that light consisted of a stream of bundles of energy that appeared to transfer their energy through collisions like those of material particles was completely at odds with the classical wave representation of light, which had already been amply confirmed by experimental studies. How could light be both a wave and a particle?

By 1930, these paradoxes had been resolved by quantum mechanics, which superseded Newtonian mechanics. The classical wave description of light is adequate to explain phenomena such as interference and diffraction, but the emission of light from matter and the absorption of light by matter are described by the particlelike photon picture. A hallmark of quantum, as opposed to classical, thinking is not to ask "What *is* light?" but instead "How does light behave under particular experimental conditions?" Thus, wave–particle duality is not a contradiction, but rather part of the fundamental nature of light and also of matter.

De Broglie Waves

Thus far, this chapter has considered only one type of wave, a **traveling wave.** Electromagnetic radiation (light, x-rays, and gamma rays) is described by such a traveling wave moving through space at speed c. Another type of wave is a **standing wave,** of which a simple example is a guitar string with fixed ends (an example of a physical *boundary condition*). A plucked string vibrates, but only certain oscillations of the string are possible. Because the ends are fixed, the only oscillations that can persist are those in which an integral number of half-wavelengths fits into the length of string, L (Fig. 4.17). The condition on the allowed wavelengths is

$$n\frac{\lambda}{2} = L \qquad n = 1, 2, 3, \ldots \qquad [4.19]$$

It is impossible to create a wave with any other value of λ if the ends of the string are fixed. The oscillation with $n = 1$ is called the **fundamental** or first harmonic, and higher values of n correspond to higher harmonics. At certain points on the standing wave, the amplitude of oscillation is zero; these points are called **nodes.** (The fixed ends are not counted as nodes.) The higher the number of the harmonic n, the more numerous the nodes, the shorter the wavelength, the higher the frequency, and the higher the energy of the standing wave.

De Broglie realized that such standing waves are examples of quantization: Only certain discrete vibrational modes, characterized by the "quantum number" n, are allowed for the vibrating string. He suggested that the quantization of energy in a one-electron atom might have the same origin, and that the electron might be associated with a standing wave, in this case, a *circular* standing wave oscillating about the nucleus of the atom (Fig. 4.18). For the amplitude of the wave to be well defined (single valued and smooth), an integral number of

FIGURE 4.18 A circular standing wave on a closed loop. The state shown has $n = 7$, with seven full wavelengths around the circle.

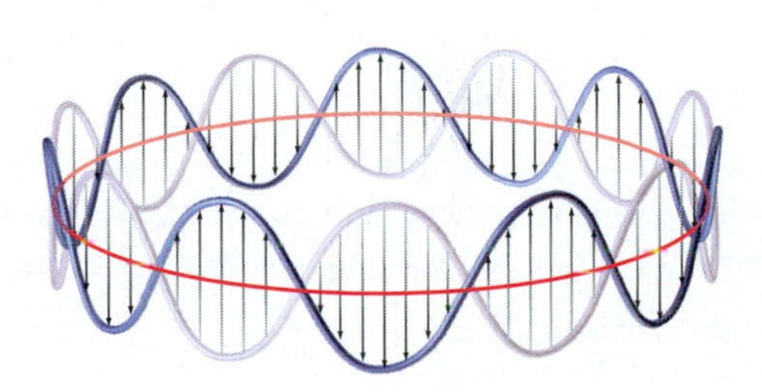

wavelengths must fit into the circumference of the circle ($2\pi r$). The condition on the allowed wavelengths for standing circular waves is

$$n\lambda = 2\pi r \qquad n = 1, 2, 3, \ldots \qquad [4.20]$$

Bohr's assumption about quantization of the angular momentum of the electron was

$$m_e vr = n\frac{h}{2\pi} \qquad [4.21]$$

which can be rewritten as

$$2\pi r = n\left[\frac{h}{m_e v}\right] \qquad [4.22]$$

Comparison of de Broglie's equation (see Eq. 4.20) with Bohr's equation (see Eq. 4.22) shows that the wavelength of the standing wave is related to the linear momentum, p, of the electron by the following simple formula:

$$\lambda = \frac{h}{m_e v} = \frac{h}{p} \qquad [4.23]$$

De Broglie used the theory of relativity to show that exactly the same relationship holds between the wavelength and momentum of a *photon*. De Broglie therefore proposed as a generalization that any particle—no matter how large or small—moving with linear momentum p has wavelike properties and a wavelength of $\lambda = h/p$ associated with its motion.

EXAMPLE 4.5

Calculate the de Broglie wavelengths of (**a**) an electron moving with velocity 1.0×10^6 m s^{-1} and (**b**) a baseball of mass 0.145 kg, thrown with a velocity of 30 m s^{-1}.

SOLUTION

(**a**) $\lambda = \dfrac{h}{p} = \dfrac{h}{m_e v} = \dfrac{6.626 \times 10^{-34}\text{ J s}}{(9.11 \times 10^{-31}\text{ kg})(1.0 \times 10^6\text{ m s}^{-1})}$

$= 7.3 \times 10^{-10}\text{ m} = 7.3\text{ Å}$

(**b**) $\lambda = \dfrac{h}{mv} = \dfrac{6.626 \times 10^{-34}\text{ J s}}{(0.145\text{ kg})(30\text{ ms}^{-1})}$

$= 1.5 \times 10^{-34}\text{ m} = 1.5 \times 10^{-24}\text{ Å}$

The latter wavelength is far too small to be observed. For this reason, we do not recognize the wavelike properties of baseballs or other macroscopic objects, even though they are always present. However, on a microscopic level, electrons moving in atoms show wavelike properties that are essential for explaining atomic structure.

Related Problems: 31, 32

Electron Diffraction

Under what circumstances does the wavelike nature of particles become apparent? When waves from two sources pass through the same region of space, they *interfere* with each other. Consider water waves as an example. When two crests meet,

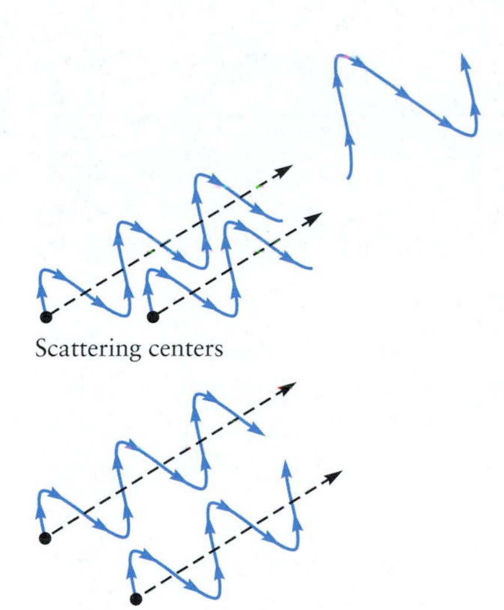

Scattering centers

FIGURE 4.19 A beam of x-rays (not shown) is striking two scattering centers that emit scattered radiation. The difference in the lengths of the paths followed by the scattered waves determines whether they interfere (a) constructively or (b) destructively.

constructive interference occurs and a higher crest (greater amplitude) appears; where a crest of one wave meets a trough of the other, *destructive* interference (smaller or even zero amplitude) occurs (Fig. 4.19).

As discussed in more detail in Chapter 21, it was known by 1914 that x-rays diffract from the lattice planes of a single crystal, because the spacing between the planes is comparable with the wavelength of the x-rays used in the experiment. Figure 4.20 shows the construction used to derive the scattering law. Clearly, waves scattered by planes located farther from the surface must travel a greater distance to reach the detector than those scattered from the surface plane. The condition for *constructive* interference is given by Bragg's law:

$$n\lambda = 2d \sin\theta \qquad [4.24]$$

which is derived in Chapter 21. The integer n tells which planes are responsible, λ is the x-ray wavelength, d is the spacing between the planes, and θ is the angle of incidence the impinging x-ray makes with the surface plane. If the x-ray source and detector are fixed in space and the sample is rotated about an axis that changes the angle θ, then a series of peaks appears whenever the diffraction condition is satisfied, allowing the lattice spacing d to be determined. X-ray diffraction was (and still is) among our most powerful tools for determining the three-dimensional structures of crystals.

If the de Broglie hypothesis was correct, then particles whose de Broglie wavelengths were comparable with lattice spacings should also diffract. This was indeed demonstrated to be the case in 1927 by the American physicists C. Davisson and L. H. Germer. In their experiment, a beam of low-energy electrons was directed toward a single crystal nickel sample in vacuum. The kinetic energy of the electrons could be varied continuously, and the sample could be rotated in space to change the angle θ. They conducted two experiments (the results of these experiments are shown in Fig. 4.21). They measured the angular dependence of the scattered electron current at fixed incident energy, and they measured the energy dependence of the scattered electron current at fixed θ. Let's see whether the de Broglie wavelength of the electrons used in this experiment is comparable with atomic lattice spacings, as required for the electrons to diffract from the planes of atoms in the solid.

The kinetic energy of an electron accelerated from rest to a final voltage V is $\mathcal{T} = eV$, where e is the charge on the electron. Recalling that $p = mv$ and

FIGURE 4.20 Constructive interference of x-rays scattered by atoms in lattice planes. Three beams of x-rays, scattered by atoms in three successive layers of a crystal are shown. Note that the phases of the waves are the same along the line CH, indicating constructive interference at this scattering angle 2θ.

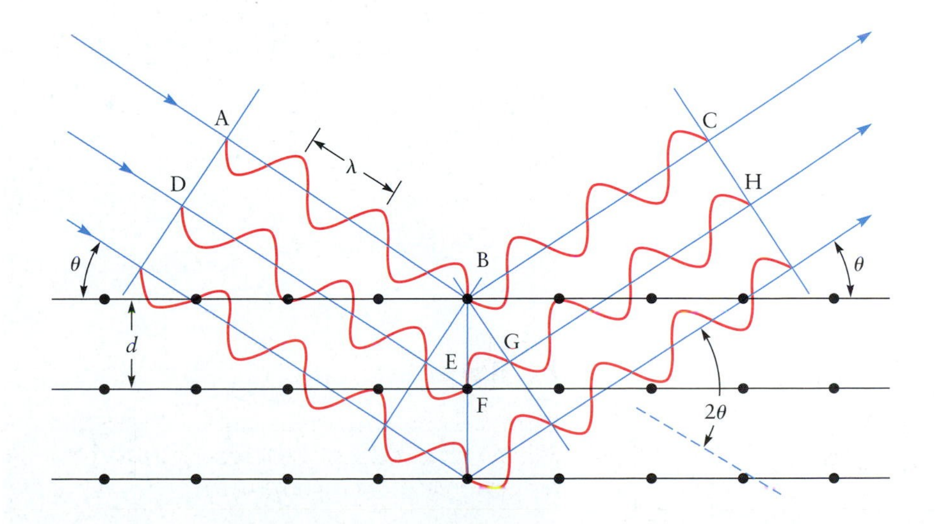

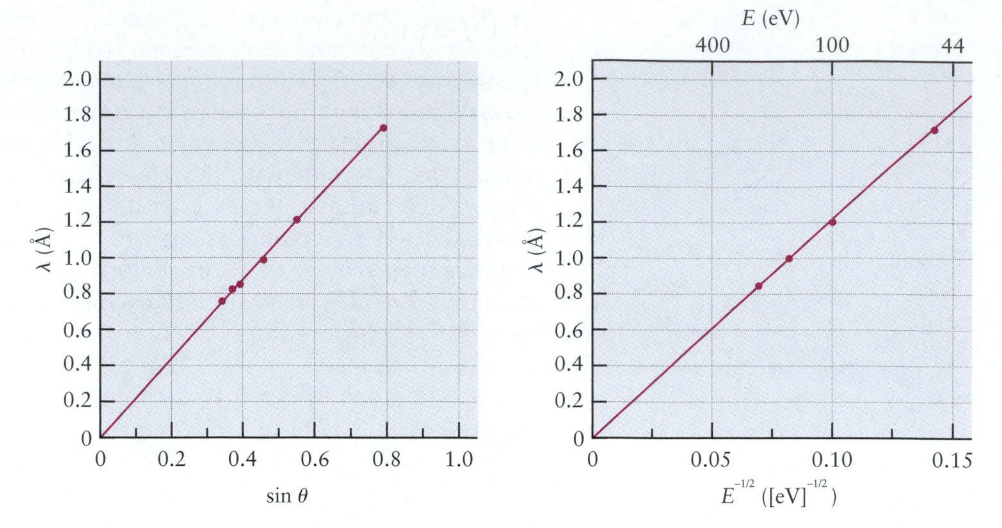

FIGURE 4.21 Results of the Davisson–Germer experiment showing (a) the sin θ dependence predicted by Bragg's law and (b) the dependence of the scattered intensity on the square root of the incident electron energy.

$\mathcal{T} = \frac{1}{2}mv^2$, we can set $\mathcal{T} = p^2/2m_e$ and solve for the momentum of the electron to get $p = \sqrt{2m_e eV}$. Calculations using these formulas must express the kinetic energy of the electron in joules (1 eV $= 1.6 \times 10^{-19}$ J) in order to obtain the proper units for momentum and wave length. The de Broglie wavelength of the electron is therefore $\lambda = h/\sqrt{2m_e eV}$. An electron accelerated through a voltage of 50 V (typical of the Davisson–Germer experiment) has a de Broglie wavelength of 1.73 Å, which is comparable to the spacing between atomic planes in metals. Davisson and Germer showed that, at a fixed angle, the scattered electron current depended on the square root of the incident energy and, at fixed energy, showed the expected sinθ dependence as the sample was rotated. These experiments demonstrated electron diffraction, and thereby provided a striking confirmation of de Broglie's hypothesis about the wavelike nature of matter. These experiments led to the development of a technique called low-energy electron diffraction (LEED) that now is used widely to study the atomic structure of solid surfaces (Fig. 4.22). It should be clear to you by now that "waves" and "particles" are idealized models that describe objects found in nature. Photons, electrons, and even helium atoms all have both wave and particle character; which aspect they display depends strongly on the conditions under which they are observed.

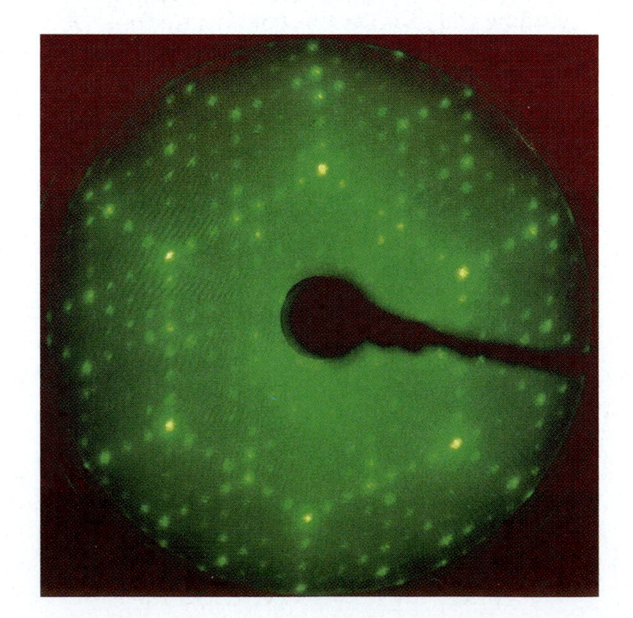

FIGURE 4.22 Low-energy electron diffraction pattern of the same silicon surface imaged by scanning tunneling microscopy in the figure opposite page 1 of this book.
(Courtesy of Dr. Gerard Parkinsen and Mr. William Gerace, OMICRON Vakuumphysik GMBH, Tanusstein, Germany)

Indeterminacy and Uncertainty: The Heisenberg Principle

An inevitable consequence of de Broglie's standing-wave description of an electron in an orbit around the nucleus is that the position and momentum of a particle cannot both be known precisely and simultaneously. The momentum of the circular standing wave shown in Figure 4.18 is given exactly by $p = h/\lambda$, but because the wave is spread uniformly around the circle, we cannot specify the angular position of the electron on the circle at all. We say the angular position is **indeterminate** because it has no definite value. This conclusion is in stark contrast with the principles of classical physics in which the positions and momenta are all known precisely and the trajectories of particles are well defined. How was this paradox resolved?

In 1927, the German physicist Werner Heisenberg proposed that **indeterminacy** is a general feature of quantum systems. Indeterminacy presents a fundamental limit to the "knowability" of the properties of these systems that is intrinsic and not just a limitation of our ability to make more precise measurements. In particular, it influences which combinations of properties can be measured together. Heisenberg identified pairs of properties that *cannot* be measured together with complete precision, and estimated the best precision we can hope to obtain when we do measure them. For example, we cannot measure position and momentum simultaneously and obtain sharp, definite values for each. The same is true for energy and time. Notice that the combination of dimensions *length* × *momentum* is the same as the combination *energy* × *time*. (You should verify this by simple dimensional analysis.) Either combination is called **action,** and it has the same dimensions as Planck's constant. The **Heisenberg indeterminacy principle** states that when we measure two properties, A and B, the product of which has dimensions of action, we will obtain a spread in results for each identified by ΔA and ΔB that will satisfy the following condition:

$$(\Delta A)(\Delta B) \geq h/4\pi \qquad [4.25]$$

If we try to measure A precisely and make ΔA nearly zero, then the spread in ΔB will have to increase to satisfy this condition. Trying to determine the angular position of the orbiting electron described by the de Broglie wave discussed earlier is a perfect illustration. Indeterminacy is intrinsic to the quantum description of matter and applies to all particles no matter how large or small.

The practical consequence of indeterminacy for the outcome of measurements is best seen by applying the Heisenberg principle in specific cases. How are the position and momentum of a macroscopic object such as a baseball in motion determined? The simplest way is to take a series of snapshots at different times, with each picture recording the light (photons) scattered by the baseball. This is true for imaging any object; we must scatter something (like a photon) from the object and then record the positions of the scattered waves or particles (Fig. 4.23). Scattering a photon from a baseball does not change the trajectory of the baseball appreciably (see Fig. 4.23a) because the momentum of the photon is negligible compared with that of the baseball, as shown in Example 4.6. Scattering photons from an electron, however, is another thing altogether. To locate the *position* of any object, we must use light with a wavelength that is comparable with or shorter than the size of the object. Thus, to measure the position of an electron in an atom to a precision of, say, 1% of the size of the atom, we would need a probe with a wavelength of order 10^{-12} m. The momentum of such a wave, given by the de Broglie relation, is 6.625×10^{-22} kg m s^{-1}. Using the virial theorem introduced in Chapter 3, we know that the kinetic energy, $\mathcal{T}$, of an electron in the ground state of the hydrogen atom is half the total energy or -1.14×10^{-18} J, corresponding to a momentum ($p = \sqrt{2m_e\mathcal{T}}$) of 1.44×10^{-24} kg m s^{-1}. Trying to measure the position of an electron to a precision of 10^{-12} m with a photon of sufficiently short wavelength turns out to be roughly equivalent to trying to

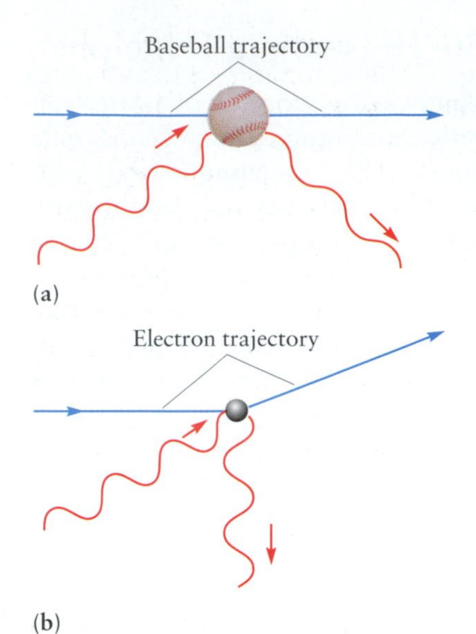

FIGURE 4.23 A photon, which has a negligible effect on the trajectory of a baseball (a), significantly perturbs the trajectory of the far less massive electron (b).

measure the position of a marble using a bowling ball as a probe! So on the length and mass scale of elementary particles, it is clear that we cannot measure simultaneously, to arbitrary precision, the values of the momentum and the position of a particle.

To make a rough estimate of the precision allowed by the indeterminacy principle, let us take as our spread of values in the position, Δx, the wavelength of our probe λ. This choice means that we can locate the particle somewhere between two crests of the wave. Let us take as our estimate of the spread in the momentum, Δp, the value of the momentum itself, p; that is, we know p to within $\pm p$. Their product is therefore $\Delta x \Delta p = h$, but because we have asserted that this is the *best* we can do, we write $\Delta x \Delta p \geq h$. A better choice for the spread in both variables is one standard deviation or the root mean square deviation from a series of measurements. For this choice, the result becomes

$$(\Delta x)(\Delta p) \geq h/4\pi \qquad [4.26]$$

which is in agreement with the Heisenberg principle.

At last, we can resolve the paradox between de Broglie waves and classical orbits, which started our discussion of indeterminacy. The indeterminacy principle places a *fundamental limit* on the precision with which the position and momentum of a particle can be known simultaneously. It has profound significance for how we think about the motion of particles. According to classical physics, the position and momentum are fully known simultaneously; indeed, we must know both to describe the classical trajectory of a particle. The indeterminacy principle forces us to abandon the classical concepts of trajectory and orbit. The most detailed information we can possibly know is the statistical spread in position and momentum allowed by the indeterminacy principle. In quantum mechanics, we think not about particle trajectories, but rather about the probability distribution for finding the particle at a specific location.

EXAMPLE 4.6

Suppose photons of green light (wavelength 5.3×10^{-7} m) are used to locate the position of the baseball from Example 4.5 with precision of one wavelength. Calculate the minimum spread in the *speed* of the baseball.

SOLUTION

The Heisenberg relation,

$$(\Delta x)(\Delta p) \geq h/4\pi$$

gives

$$\Delta p \geq \frac{h}{4\pi\Delta x} = \frac{6.626 \times 10^{-34}\text{ J s}}{4\pi(5.3 \times 10^{-7}\text{ m})} = 9.9 \times 10^{-29}\text{ kg m s}^{-1}$$

Because the momentum is just the mass (a constant) times the speed, the spread in the speed is

$$\Delta v = \frac{\Delta p}{m} \geq \frac{9.9 \times 10^{-29}\text{ kg ms}^{-1}}{(0.145\text{ kg})} = 6.8 \times 10^{-28}\text{ m s}^{-1}$$

This is such a tiny fraction of the speed of the baseball (30 m s^{-1}) that indeterminacy plays a negligible role in the measurement of the baseball's motion. Such is the case for all macroscopic objects.

Related Problems: 35, 36

Describing the indeterminacy principle presents certain challenges to language. Clearly, when a property of the system is indeterminate, measurements will produce a statistical spread of values for that property. In a colloquial sense, there will be *uncertainty* in the measurement, because its outcome is not precisely predictable, just as there is uncertainty in the outcome of playing a game of chance. In almost all English-language books on quantum mechanics, the spread in value of a property ΔA is called the *uncertainty in A*, and the relation in Equation 4.25 is called the *Heisenberg uncertainty principle.* A property is indeterminate if it has no definite value, whereas it is uncertain if it does have a definite value, but that value is not known to the experimenter. We prefer the phrase *indeterminacy principle* because it more accurately conveys that a fundamental limitation on measurements is being described, whereas *uncertainty principle* allows the suspicion to exist that more carefully designed experiments will make the problem disappear.

4.5 The Schrödinger Equation

de Broglie's work attributed wavelike properties to electrons in atoms, which inspired the Austrian physicist Erwin Schrödinger to think about how to describe electrons as waves. Schrödinger, a recognized authority on the theory of vibrations and the associated "quantization" of standing waves, reasoned that an electron (or any other particle with wavelike properties) might well be described by a wave function. A **wave function** maps out the amplitude of a wave in three dimensions; it may also be a function of time. Ocean waves have amplitudes that vary in both space and time, as do electromagnetic waves. Schrödinger's wave function, symbolized by the Greek letter psi (ψ), is the amplitude of the wave associated with the motion of a particle, at a position located by the coordinates x, y, z at time t. It is important to emphasize that the amplitude of a wave function (just like the amplitude of ordinary waves discussed at the beginning of this chapter) may be positive, negative, or zero. The sign of a wave function tells the direction of the displacement. If we assign zero as the amplitude of the undisturbed medium (or the value of the fields for electromagnetic radiation), then positive amplitude means that the wave is displaced "upward" (a crest), whereas negative amplitude means that the wave is displaced "downward" (a trough). Points or regions in space where the wave function goes through zero as it changes sign are called **nodes.** We cannot overemphasize the importance of both the magnitude and the sign of quantum mechanical wave functions, because they determine the extent to which two wave functions interfere. As discussed later, interference is an essential feature of the quantum description of atoms and molecules.

Schrödinger discovered the equation that bears his name in 1926, and it has provided the foundation for the wave-mechanical formulation of quantum mechanics. Heisenberg had independently, and somewhat earlier, proposed a matrix formulation of the problem, which Schrödinger later showed was an equivalent alternative to his approach. We choose to present Schrödinger's version because its physical interpretation is much easier to understand.

Origins of the Schrödinger Equation

Although it is beyond the scope of this text to explain the origins of the **Schrödinger equation,** it is nevertheless worthwhile to work through the logic that might have stimulated Schrödinger's thinking and, more importantly, to explore some of the properties of the mathematical form of the Schrödinger equation. Having been trained in the classical theory of waves and inspired by de Broglie's hypothesis, it was natural for Schrödinger to seek a wave equation that described the properties of matter on the atomic scale. Classical wave equations relate the second derivatives of the amplitude with respect to distance to the second derivatives with respect to time; for simplicity, we shall see if we can find a wave equation that relates the second derivative of a function with respect to displacement to the function itself, leaving the time dependence for more advanced work.

We begin by considering a particle moving freely in one dimension with classical momentum, p. Such a particle is associated with a wave of wavelength $\lambda = h/p$. Two "wave functions" that describe such a wave are

$$\psi(x) = A \sin \frac{2\pi x}{\lambda} \quad \text{and} \quad \psi(x) = B \cos \frac{2\pi x}{\lambda} \qquad [4.27]$$

where A and B are constants. Choosing the sine function, for example, let's see what its second derivative with respect to x looks like. From differential calculus, the derivative (or slope) of $\psi(x)$ is

$$\frac{d\psi(x)}{dx} = A\frac{2\pi}{\lambda} \cos \frac{2\pi x}{\lambda}$$

The slope of *this* function is given by the second derivative of ψ, written $\frac{d^2\psi(x)}{dx^2}$, which is equal to

$$\frac{d^2\psi(x)}{dx^2} = -A\left(\frac{2\pi}{\lambda}\right)^2 \sin \frac{2\pi x}{\lambda}$$

This is just a constant, $-(2\pi/\lambda)^2$, multiplied by the original wave function $\psi(x)$:

$$\frac{d^2\psi(x)}{dx^2} = -\left(\frac{2\pi}{\lambda}\right)^2 \psi(x)$$

This is an equation (called a *differential equation*) that is satisfied by the function $\psi(x) = A \sin (2\pi x/\lambda)$. It is easy to verify that this equation is also satisfied by the function $\psi(x) = B \cos (2\pi x/\lambda)$.

Let's now replace the wavelength λ with the momentum p from the de Broglie relation:

$$\frac{d^2\psi(x)}{dx^2} = -\left(\frac{2\pi}{h} p\right)^2 \psi(x) \qquad [4.28]$$

We can rearrange this equation into a suggestive form by multiplying both sides by $-h^2/8\pi^2 m$, giving

$$-\frac{h^2}{8\pi^2 m} \frac{d^2\psi(x)}{dx^2} = \frac{p^2}{2m} \psi(x) = \mathcal{T}\psi(x)$$

where $\mathcal{T} = p^2/2m$ is the kinetic energy of the particle. This form of the equation suggests that there is a fundamental relationship between the second derivative of the wave function (also called its *curvature*) and the kinetic energy, $\mathcal{T}$.

If external forces are present, a potential energy term $V(x)$ (due to the presence of walls enclosing the particle or to the presence of fixed charges, for example) must be included. Writing the total energy as $E = \mathcal{T} + V(x)$ and substituting the result in the previous equation gives

$$-\frac{h^2}{8\pi^2 m} \frac{d^2\psi(x)}{dx^2} + V(x)\psi(x) = E\psi(x) \qquad [4.29]$$

This is the Schrödinger equation for a particle moving in one dimension. The development provided here is not a derivation of this central equation of quantum mechanics; rather, it is a plausibility argument based on the idea that the motions of particles can be described by a wave function with the wavelength of the particle being given by the de Broglie relation.

The Validity of the Schrödinger Equation

The validity of any scientific theory must be tested by extensive comparisons of its predictions with a large body of experimental data. Although we have presented a plausibility argument that suggests how Schrödinger might have initially developed his equation, understanding the source of his inspiration is not nearly as important as evaluating the accuracy of the theory. It is the same for all great scientific discoveries; the story of Newton and the apple is not nearly as important as the fact that classical mechanics has been shown to describe the behavior of macroscopic systems to astonishingly high accuracy. Quantum mechanics superseded Newtonian mechanics because the latter failed to account for the properties of atoms and molecules. We believe that quantum mechanics is correct because its predictions agree with experiment to better than 10^{-10}%. It is generally considered to be among the most accurate theories of nature because of this astonishingly good agreement. But even quantum mechanics began to fail as scientists were able to make more accurate measurements than those made in the early part of the 20th century. Relativistic corrections to Schrödinger's equations improved the situation dramatically, but only with the development of **quantum electrodynamics**—in which matter and radiation are treated completely equivalently—did complete agreement between theory and experiment occur. Quantum electrodynamics is an extremely active field of research today, and it continues to ask questions such as, "How does the system know that it is being measured?" and "How can we use quantum mechanics to make computers of unprecedented power?" The fundamental ideas of quantum mechanics—energy quantization and wave–particle duality—appear to be universally true in science. These properties of nature are less evident in the macroscopic world, however, and the predictions of quantum mechanics agree well with those of classical mechanics on the relevant length and mass scales for macroscopic systems.

Interpretation of the Energy in the Schrödinger Equation

The Schrödinger equation can be solved exactly for any number of model problems and for a few real problems, notably the hydrogen atom. What do the solutions of this equation tell us about the energies and other properties of quantum systems? Or, to phrase the question slightly differently, how do we interpret ψ, and what information does it contain?

Let's focus initially on the energy. For all systems confined in space, solutions of the Schrödinger equation that are independent of time can be found only for certain discrete values of the energy; energy quantization is a natural consequence of the Schrödinger equation. States described by these time-independent wave functions are called **stationary states.** For a given system, there may be many states with different energies characterized by different wave functions. The solution that corresponds to the lowest energy is called the ground state (just as in the Bohr model), and higher energy solutions are called excited states.

Interpretation of the Wave Function in the Schrödinger Equation

What is the physical meaning of the wave function ψ? We have no way of measuring ψ directly, just as in classical wave optics we have no direct way of measuring the amplitudes of the electric and magnetic fields that constitute the light wave (see Fig. 4.2). What *can* be measured in the latter case is the intensity of the light wave, which, according to the classical theory of electromagnetism, is proportional to the square of the amplitude of the electric field:

$$\text{intensity} \propto (E_{\text{max}})^2$$

However, if we view electromagnetic radiation as a collection of *particles* (photons), then the intensity is simply proportional to the density of photons in a region of space. Connecting the wave and particle views of the electromagnetic field suggests that the *probability* of finding a photon is given by the square of the amplitude of the electric field.

By analogy, we interpret the square of the wave function ψ^2 for a particle as a probability density for that particle. That is, $[\psi(x, y, z)]^2\, dV$ is the probability that the particle will be found in a small volume $dV = dxdydz$ centered at the point (x, y, z). This probabilistic interpretation of the wave function, proposed by the German physicist Max Born, is now generally accepted because it provides a consistent picture of particle motion on a microscopic scale.

The probabilistic interpretation requires that any function must meet three mathematical conditions before it can be used as a wave function. The next section illustrates how these conditions are extremely helpful in solving the Schrödinger equation. To keep the equations simple, we will state these conditions for systems moving in only one dimension. All the conditions extend immediately to three dimensions when proper coordinates and notation are used. (You should read Appendix A6, which reviews probability concepts and language, before proceeding further with this chapter.)

When the possible outcomes of a probability experiment are continuous (for example, the position of a particle along the x-axis) as opposed to discrete (for example, flipping a coin), the distribution of results is given by the probability density function $P(x)$. The product $P(x)dx$ gives the probability that the result falls in the interval of width dx centered about the value x. The first condition, that the probability density must be **normalized,** ensures that probability density is properly defined (see Appendix A6), and that all possible outcomes are included. This condition is expressed mathematically as

$$\int_{-\infty}^{+\infty} P(x)dx = \int_{-\infty}^{+\infty} [\psi(x)]^2 dx = 1 \qquad [4.30]$$

The second and third conditions are subsidiary to the first, in that they must be satisfied to enable the first one to be satisfied. The second condition is that $P(x)$ must be continuous at each point x. At some specific point, call it x_a, the form of the probability density may change for physical reasons, but its value at x_a must be the same regardless of whether x_a is approached from the left or from the right. This translates into the condition that $\psi(x)$ and its first derivative $\psi'(x)$ are continuous at each point x. The third condition is that $\psi(x)$ must be bounded at large values of x. This is stated mathematically as

$$\psi \longrightarrow 0 \quad \text{as} \quad x \longrightarrow \pm\infty \qquad [4.31]$$

The second and third conditions are examples of **boundary conditions,** which are restrictions that must be satisfied by the solutions to differential equations such as the Schrödinger equation. A differential equation does not completely define a physical problem until the equation is supplemented with boundary conditions. These conditions invariably arise from physical analysis, and they help to select from the long list of possible solutions to the differential equation those that apply specifically to the problem being studied.

We must acknowledge that our information about the location of a particle is limited, and that it is statistical in nature. So not only are we restricted by the uncertainty principle as to what we can measure, but we must also come to grips with the fact that fundamental properties of quantum systems are unknowable, except in a statistical sense. If this notion troubles you, you are in good company. Many of the best minds of the 20th century, notably Einstein, never became comfortable with this central conclusion of the quantum theory.

Procedures for Solving the Schrödinger Equation

The application of quantum mechanics to solve for the properties of any particular system is straightforward in principle. You need only substitute the appropriate potential energy term for that system into the Schrödinger equation and solve the equation to obtain two principal results: the allowed energy values and the corresponding wave functions. You will find that solutions only exist for specific, discrete energy values. Energy quantization arises as a direct consequence of the boundary conditions imposed on the Schrödinger equation (see later discussion) with no need for extra assumptions to be grafted on. Each energy value corresponds to one or more wave functions; these wave functions describe the distribution of particles when the system has a specific energy value.

We illustrate this procedure in detail for a simplified model in the next section so that you will see how energy levels and wave functions are obtained. We also use the model problem to illustrate important general features of quantum mechanics including restrictions imposed on the form of the wave function by the Schrödinger equation and its physical interpretation.

4.6 Quantum Mechanics of Particle-in-a-Box Models

We are about to show you how to solve the Schrödinger equation for a simple but important model for which we can carry out every step of the complete solution using only simple mathematics. We will convert the equations into graphical form and use the graphs to provide physical interpretations of the solutions. The key point is for you to learn how to achieve a physical understanding from the graphical forms of the solution. Later in this textbook we present the solutions for more complex applications only in graphical form, and you will rely on the skills you develop here to see the physical interpretation for a host of important chemical applications of quantum mechanics. This section is, therefore, one of the most important sections in the entire textbook.

One-Dimensional Boxes

The simplest model problem for which the Schrödinger equation can be solved, and in which energy quantization appears, is the so-called particle in a box. It consists of a particle confined by potential energy barriers to a certain region of space (the "box"). In one dimension, the model is visualized easily as a bead sliding along a wire between barriers at the ends of the wire. The particle is located on the x-axis in the interval between 0 and L, where L is the length of the box (Fig. 4.24a). If the particle is to be completely confined in the box, the potential energy $V(x)$ must rise abruptly to an infinite value at the two end walls to prevent even fast-moving particles from escaping. Conversely, inside the box, the motion of the particle is free, so $V(x) = 0$ everywhere inside the box. This means that the total energy, $E = \mathcal{T} + V$, must be positive at each point inside the box. We will determine the possible values of E by solving the Schrödinger equation. The solution for this model illustrates the general methods used for other more difficult potential energy functions.

A quick inspection of the potential energy function tells us the general nature of the solution. Wherever the potential energy V is infinite, the probability of finding the particle must be zero. Hence, $\psi(x)$ and $\psi^2(x)$ must be zero in these regions:

$$\psi(x) = 0 \text{ for } x \leq 0 \text{ and } x \geq L \qquad [4.32]$$

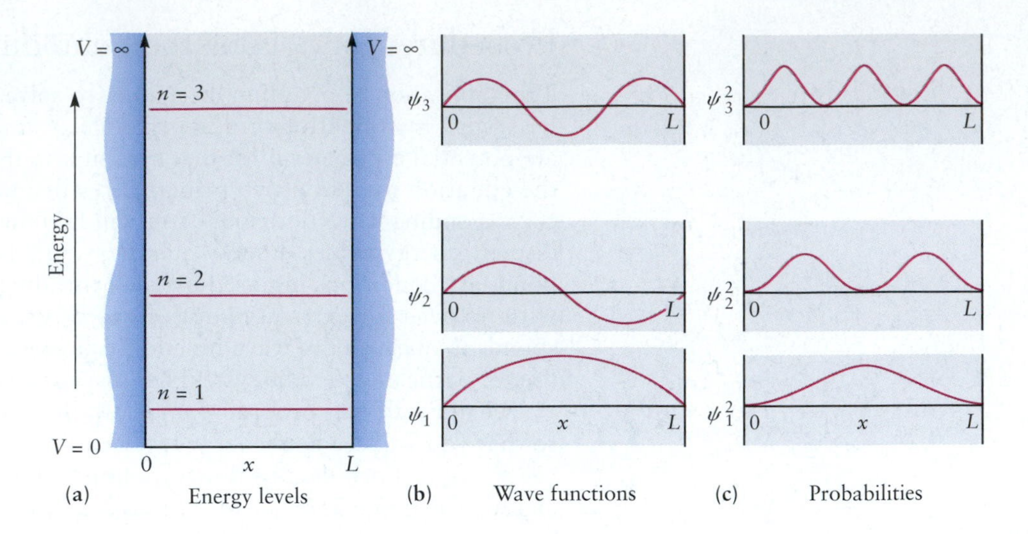

FIGURE 4.24 (a) The potential energy for a particle in a box of length L, with the first three energy levels marked. (b) Wave functions showing the ground state ψ_1 and the first two excited states. The more numerous the nodes, the higher the energy of the state. (c) The squares of the wave functions from (b), equal to the probability density for finding the particle at a particular point in the box.

Inside the box, where $V = 0$, the Schrödinger equation has the following form:

$$-\frac{h^2}{8\pi^2 m}\frac{d^2\psi(x)}{dx^2} = E\psi(x)$$

or equivalently,

$$\frac{d^2\psi(x)}{dx^2} = -\frac{8\pi^2 mE}{h^2}\psi(x)$$

As shown earlier, the sine and cosine functions are two possible solutions to this equation, because the second derivative of each function is the function itself multiplied by a (negative) constant.

Now let us apply the conditions defined in Section 4.5 to select the allowed solutions from these possibilities. The boundary conditions require that $\psi(x) = 0$ at $x = 0$ and $x = L$. An acceptable wave function must be continuous at both these points. The cosine function can be eliminated because it cannot satisfy the condition that $\psi(x)$ must be 0 at $x = 0$. The sine function does, however, satisfy this boundary condition since $\sin(0) = 0$ so

$$\psi(x) = A \sin kx \qquad [4.33]$$

is a potentially acceptable wave function.

If the wave function is also to be continuous at $x = L$, then we must have

$$\psi(L) = 0 \qquad [4.34]$$

or

$$\psi(L) = A \sin kL = 0$$

This can be true only if

$$kL = n\pi \qquad n = 1, 2, 3, \ldots$$

because $\sin(n\pi) = 0$. Thus, the combination of the boundary conditions and continuity requirement gives the allowed solutions as

$$\psi(x) = A \sin\left(\frac{n\pi x}{L}\right) \qquad n = 1, 2, 3, \ldots \qquad [4.35]$$

where the constant A is still to be determined. The restriction of the solutions to this form in which n is an integer quantizes the energy and the wave functions.

As explained in Section 4.5, the wave function must be normalized. This condition is not always satisfied by solutions to the Schrödinger equation; thus, we must see how we can enforce it. To normalize the wave function just obtained we set

$$A^2\int_0^L \sin^2\left(\frac{n\pi x}{L}\right)dx = 1$$

and solve for A. Evaluating the definite integral gives $L/2$, so

$$A^2\left(\frac{L}{2}\right) = 1$$

and

$$A = \sqrt{\frac{2}{L}}$$

The normalized wave function for the one-dimensional particle in a box is

$$\psi_n(x) = \sqrt{\frac{2}{L}} \sin\left(\frac{n\pi x}{L}\right) \quad n = 1, 2, 3, \ldots \qquad [4.36]$$

where n labels a particular allowed solution of the Schrödinger equation.

To find the energy E_n for a particle described by the wave function ψ_n, we calculate the second derivative:

$$\begin{aligned}\frac{d^2\psi_n(x)}{dx^2} &= \frac{d^2}{dx^2}\left[\sqrt{\frac{2}{L}} \sin\left(\frac{n\pi x}{L}\right)\right] \\ &= -\left(\frac{n\pi}{L}\right)^2\left[\sqrt{\frac{2}{L}} \sin\left(\frac{n\pi x}{L}\right)\right] \\ &= -\left(\frac{n\pi}{L}\right)^2 \psi_n(x)\end{aligned}$$

This must be equal to $-\dfrac{8\pi^2 mE_n}{h^2}\psi_n(x)$. Setting the coefficients equal to one another gives

$$\frac{8\pi^2 mE_n}{h^2} = \frac{n^2\pi^2}{L^2} \qquad \text{or}$$

$$E_n = \frac{n^2h^2}{8mL^2} \qquad n = 1, 2, 3, \ldots \qquad [4.37]$$

This solution of the Schrödinger equation demonstrates that the energy of a particle in the box is quantized. The energy, E_n, and wave function $\psi_n(x)$ are unique functions of the quantum number n, which must be a positive integer. These are the only allowed stationary states of the particle in a box. The allowed energy levels are plotted together with the potential energy function in Figure 4.24a. A system described by the particle-in-a-box model will have an emission or absorption spectrum that consists of a series of frequencies given by

$$h\nu = |E_n - E_{n'}| \qquad [4.38]$$

where n and n' are positive integers, and the E_n values are given by Equation 4.37.

The wave functions $\psi_n(x)$ plotted in Figure 4.24b for the quantum states n are the standing waves of Figure 4.17. The guitar string and the particle in the box are physically analogous. The boundary condition that the amplitude of the wave function ψ must be zero at each end of the guitar string or at each wall of the box is responsible for the quantization of energy and the restriction on the motions allowed.

Following Born's interpretation that the probability of finding the particle at a particular position is the square of its wave function evaluated at that position, we can study the probability distributions for the particle in a box in various quantum states. Figure 4.24c shows the probability distributions for the first three states of the particle in a box. For the ground state ($n = 1$), we see that the most likely place to find the particle is in the middle of the box, with a small chance of finding it near either wall. In the first excited state ($n = 2$), the probability is a maximum when it is near $L/4$ and $3L/4$ and zero near 0, $L/2$, and L. And for the $n = 3$ state, the maxima are located at $L/6$, $3L/6$, and $5L/6$, with nodes located at $L/3$ and $2L/3$. Wherever there is a node in the wave function, the probability is zero that the particle will be found at that location.

The number of nodes in the wave function is important not only for helping us understand probability distributions but also because it provides an important, and perfectly general, criterion for ordering the energy levels in any quantum system. The wave function ψ_n has $n - 1$ nodes, and it is clear from Figure 4.24c that the number of nodes increases with the energy of each state. This is a general feature in quantum mechanics: For a given system, the relative ordering of the energy levels can be determined simply by counting the number of nodes.

Can a particle in a box have zero energy? Setting $n = 0$ and solving for E using the equation $E_n = n^2h^2/8mL^2$ gives $E_0 = 0$. But this is not possible. Setting n equal to zero in $\psi_n(x) = A \sin(n\pi x/L)$ makes $\psi_0(x)$ zero everywhere. In this case, $\psi_0^2(x)$ would be zero everywhere in the box, and thus there would be no particle in the box. The same conclusion comes from the indeterminacy principle. If the energy of the lowest state could be zero, the momentum of the particle would also be zero. Moreover, the uncertainty or spread in the particle momentum, Δp_x, would also be zero, requiring that Δx be infinite, which contradicts our assertion that the particle is confined to a box of length L. Even at the absolute zero of temperature, where classical kinetic theory would suggest that all motion ceases, a finite quantity of energy remains for a bound system. It is required by the indeterminacy principle and is called the **zero-point energy.**

The wave functions for a particle in a box illustrate another important principle of quantum mechanics: the correspondence principle. We have already stated earlier (and will often repeat) that all successful physical theories must reproduce the explanations and predictions of the theories that preceded them on the length and mass scales for which they were developed. Figure 4.25 shows the probability density for the $n = 5$, 10, and 20 states of the particle in a box. Notice how the probability becomes essentially uniform across the box, and at $n = 20$ there is little evidence of quantization. The **correspondence principle** requires that the results of quantum mechanics reduce to those of classical mechanics for large values of the quantum numbers, in this case, n.

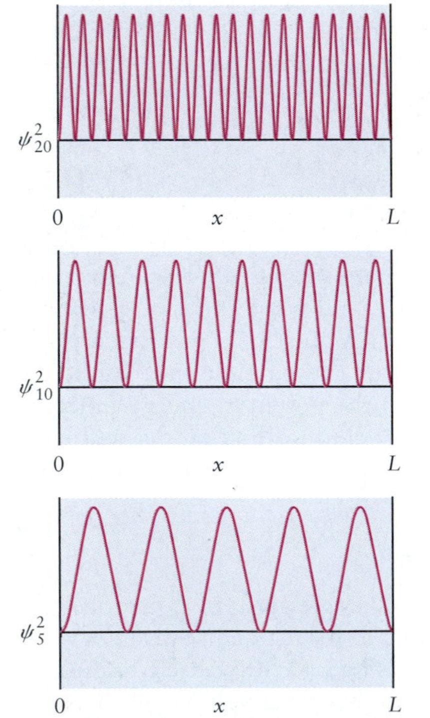

FIGURE 4.25 The probability distribution for a particle in a box of length L in the quantum states $n = 5$, 10, and 20. Compare these results with those shown in Figure 4.24c.

Energy Levels for Particles in Two- and Three-Dimensional Boxes

The Schrödinger equation is readily generalized to describe a particle in a box of two or three dimensions. A particle in a two-dimensional box can be visualized as a marble moving in the x-y plane at the bottom of a deep elevator shaft, with infinite potential walls confining its motion in the x and y directions. A particle in a three-dimensional rectangular box has infinite potential walls confining its motion in the x, y, and z directions. In both cases, the potential energy function is zero throughout the interior of the box. The wave function ψ and potential V now depend on as many as three coordinates (x, y, z), and derivatives with respect to each coordinate appear in the Schrödinger equation. Because the potential energy is constant in all directions inside the box, the motions in the x direction are independent of the motions in the y and z directions, and vice versa. For potential functions of this type, the Schrödinger equation can be solved by the method of *separation of variables,* and the results are quite interesting. The wave function is the product of the wave functions for independent motion in each direction, and the energy is the sum of the energies for independent motion in each direction. Therefore, we can immediately apply the results for the one-dimensional motions developed earlier to discuss, in turn, the energies and wave functions for multidimensional boxes.

The allowed energies for a particle in a three-dimensional rectangular box are

$$E_{n_x n_y n_z} = \frac{h^2}{8m}\left[\frac{n_x^2}{L_x^2} + \frac{n_y^2}{L_y^2} + \frac{n_z^2}{L_z^2}\right] \qquad \begin{cases} n_x = 1, 2, 3, \ldots \\ n_y = 1, 2, 3, \ldots \\ n_z = 1, 2, 3, \ldots \end{cases} \qquad [4.39]$$

where L_x, L_y, and L_z are the side lengths of the box. Here the state is designated by a set of three quantum numbers, (n_x, n_y, n_z). Each quantum number ranges independently over the positive integers. We can obtain the energy levels for a particle in a two-dimensional box in the x-y plane from Equation 4.39 by setting $n_z = 0$ and restrict the box to be a square by setting $L_x = L_y = L$. Similarly, we can specialize Equation 4.39 to a cubic box by setting $L_x = L_y = L_z = L$.

$$E_{n_x n_y n_z} = \frac{h^2}{8mL^2}\left[n_x^2 + n_y^2 + n_z^2\right] \qquad \begin{cases} n_x = 1, 2, 3, \ldots \\ n_y = 1, 2, 3, \ldots \\ n_z = 1, 2, 3, \ldots \end{cases} \qquad [4.40]$$

Figure 4.26 plots the first few energy levels from Equation 4.40. We see that certain energy values appear more than once because the squares of different sets of quantum numbers can add up to give the same total. Such energy levels, which correspond to more than one quantum state, are called **degenerate.** Degenerate energy levels appear only in systems with potential energy functions that have symmetric features. (You should convince yourself that none of the energy levels in Eq. 4.39 is degenerate.) In Chapters 5 and 6, we encounter many examples of degenerate energy levels in atomic and molecular systems, as consequences of symmetry. As an exercise, we suggest that you apply Equation 4.40 to determine the energy levels in a square box and examine their degeneracy. You should master the concept of degeneracy in these simple examples because it is used in all branches of science to describe the absorption and emission of electromagnetic radiation by atoms and molecules.

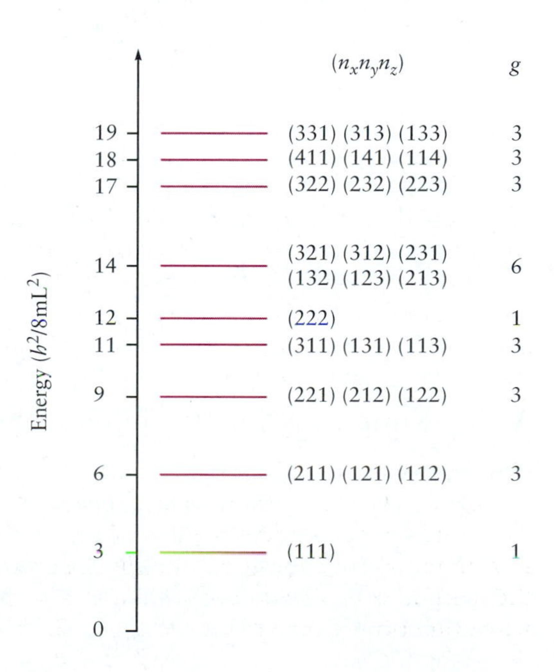

FIGURE 4.26 The energy levels for a particle in a cubic box. The quantum numbers identifying the quantum states and the degeneracy values are given for each energy level.

EXAMPLE 4.7

Consider the following two systems: (a) an electron in a one-dimensional box of length 1.0 Å and (b) a helium atom in a cube 30 cm on an edge. Calculate the energy difference between ground state and first excited state, expressing your answer in kJ mol^{-1}.

SOLUTION

(a) For a one-dimensional box,

$$E_{\text{ground state}} = \frac{h^2}{8mL^2}(1^2)$$

$$E_{\text{first excited state}} = \frac{h^2}{8mL^2}(2^2)$$

Then, for one electron in the box,

$$\Delta E = \frac{3h^2}{8mL^2}$$

$$= \frac{3(6.626 \times 10^{-34}\text{ J s})^2}{8(9.11 \times 10^{-31}\text{ kg})(1.0 \times 10^{-10}\text{ m})^2}$$

$$= 1.8 \times 10^{-17}\text{ J}$$

Multiplying this result by 10^{-3} kJ J^{-1} and by $N_A = 6.022 \times 10^{23}$ mol^{-1} gives

$$\Delta E = 11,000\text{ kJ mol}^{-1}$$

(b) For a three-dimensional cube, $L_x = L_y = L_z = L$, and

$$E_{\text{ground state}} = \frac{h^2}{8mL^2}(1^2 + 1^2 + 1^2)$$

$$E_{\text{first excited state}} = \frac{h^2}{8mL^2}(2^2 + 1^2 + 1^2)$$

In this case, the three states (2, 1, 1), (1, 2, 1), and (1, 1, 2) have the same energy.

$$\Delta E = \frac{3h^2}{8mL^2}$$

$$= \frac{3(6.626 \times 10^{-34}\text{ J s})^2}{8(6.64 \times 10^{-27}\text{ kg})(0.30\text{ m})^2}$$

$$= 2.8 \times 10^{-40}\text{ J}$$

$$= 1.7 \times 10^{-19}\text{ kJ mol}^{-1}$$

The energy levels are so close together in the latter case (due to the much larger dimensions of the box) that they appear continuous, and quantum effects play no role. The properties are almost those of a classical particle.

Related Problems: 37, 38

Wave Functions for Particles in Square Boxes

The wave function for a particle in a square box of length L on each side in the x-y plane, which we denote as Ψ, is given by

$$\Psi_{n_x n_y}(x, y) = \psi_{n_x}(x)\psi_{n_y}(y) = \frac{2}{L}\sin\left(\frac{n_x \pi x}{L}\right)\sin\left(\frac{n_y \pi y}{L}\right) \qquad [4.41]$$

as explained earlier. To generate a graphical representation, we calculate the value of Ψ at each point (x, y) in the plane and plot this value as a third dimension above the x-y plane. To make our graphs apply to square boxes of any size, we show them for dimensionless variables $\tilde{x} = x/L$ and $\tilde{y} = y/L$, which range from 0 to 1. We also plot the value of the wave function as a dimensionless variable $\tilde{\Psi}$, defined as the ratio of the value of Ψ to its maximum value, $\tilde{\Psi}(\tilde{x}, \tilde{y}) = \Psi(\tilde{x}, \tilde{y})/\Psi_{max}$. The value of $\tilde{\Psi}$ ranges from 0 to ± 1. We show three examples in Figure 4.27.

The wave function for the ground state $\tilde{\Psi}_{11}(\tilde{x}, \tilde{y})$ (see Fig. 4.27a) has no nodes and has its maximum at the center of the box, as you would expect from the one-dimensional results in Figure 4.24 from which $\tilde{\Psi}_{11}(\tilde{x}, \tilde{y})$ is constructed. Figure 4.27b shows $\tilde{\Psi}_{11}(\tilde{x}, \tilde{y})$ as a contour plot in the x-y plane, generated by choosing

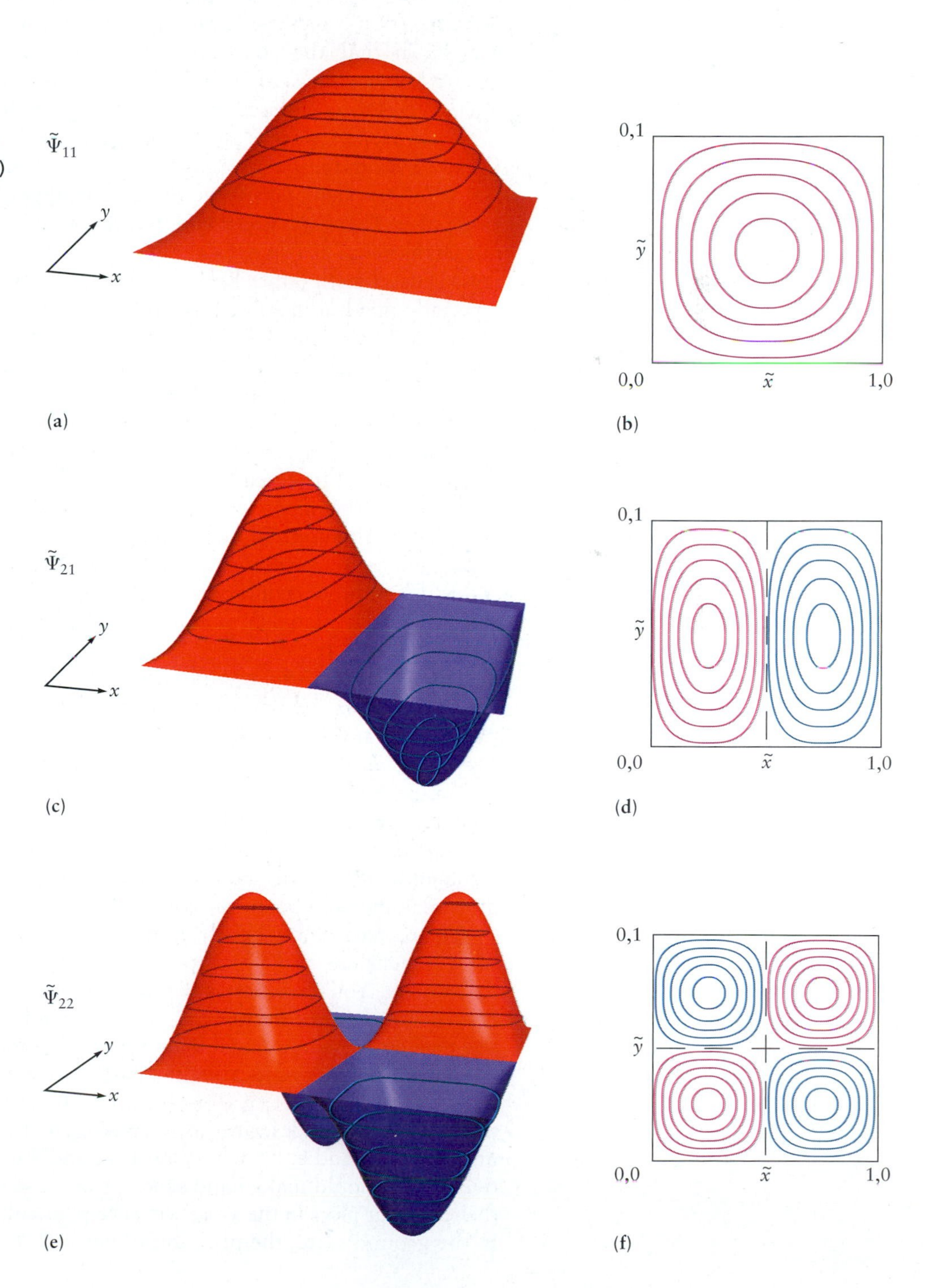

FIGURE 4.27 Wave function for a particle in a square box in selected quantum states. Dimensionless variables are used. (a) Three-dimensional plot for the ground state $\tilde{\Psi}_{11}(\tilde{x}, \tilde{y})$. (b) Contour plot for $\tilde{\Psi}_{11}(\tilde{x}, \tilde{y})$. (c) Three-dimensional plot for the first excited state $\tilde{\Psi}_{21}(\tilde{x}, \tilde{y})$. (d) Contour plot for $\tilde{\Psi}_{21}(\tilde{x}, \tilde{y})$. (e) Three-dimensional plot for the second excited state $\tilde{\Psi}_{22}(\tilde{x}, \tilde{y})$. (f) Contour plot for $\tilde{\Psi}_{22}(\tilde{x}, \tilde{y})$.

a particular value of the wave function in Figure 4.27a, "slicing" its three-dimensional image at that value, and projecting each point on the edge of that slice down to the x-y plane to form a closed contour in that plane. The contour then defines all points in the x-y plane for which $\tilde{\Psi}_{11}(\tilde{x}, \tilde{y})$ has the particular value selected. The process is continued by selecting other values of $\tilde{\Psi}_{11}(\tilde{x}, \tilde{y})$ until the entire three-dimensional image has been collapsed into a set of concentric contours in the two-dimensional x-y plane. Mountain climbers throughout the world use this method to generate contour maps of mountain ranges. The outermost contour identifies points at which the wave function has 10% of its maximum value. The second contour identifies points with 30% of the maximum value, and so on to the innermost contour that identifies points with 90% of the maximum value. Note that these contours, which correspond to uniform increases in amplitude, become much closer together as we approach the maximum. This indicates the value of the wave function is increasing rapidly as we approach the maximum. Notice in Figure 4.27b that the contours are circular at the large values of $\tilde{\Psi}_{11}(\tilde{x}, \tilde{y})$, but at lower values, they become squarish and approach perfect squares as $\tilde{\Psi}_{11}(\tilde{x}, \tilde{y}) \rightarrow 0$, as required by the boundary conditions imposed by the square box.

The wave function for the first excited state $\tilde{\Psi}_{21}(\tilde{x}, \tilde{y})$ is shown in Figure 4.27c. It has a maximum (positive) at $\tilde{x} = 0.25$, $\tilde{y} = 0.50$ and a minimum (negative) at $\tilde{x} = 0.75$, $\tilde{y} = 0.50$. The wave function changes sign as it moves along $\tilde{x}$ for any value of $\tilde{y}$. There is a **nodal line** that lies along $\tilde{x} = 0.5$. The wave function does not change sign as it moves along $\tilde{y}$ for any value of $\tilde{x}$. Make sure you see how these characteristics trace back to the one-dimensional solutions in Figure 4.24. Be especially mindful that nodal points in one dimension have become nodal lines in two dimensions. Figure 4.27d shows the contour plot for $\tilde{\Psi}_{21}(\tilde{x}, \tilde{y})$. Note that the contours are nearly circular near the maximum and minimum values, and they become ellipsoidal at smaller values of the wave function. This asymmetry in shape occurs because the motion in the x dimension occurs at a higher level of excitation than that in the y direction. At still lower values of the wave function, the contours begin to resemble rectangles. They approach perfect rectangles as $\tilde{\Psi}_{21}(\tilde{x}, \tilde{y}) \rightarrow 0$ to match the nodal line along $\tilde{x} = 0.5$ and the boundary conditions enforced by the box. To build up your expertise, we suggest that you construct and examine the wave function $\tilde{\Psi}_{12}(\tilde{x}, \tilde{y})$. Convince yourself it is degenerate with $\tilde{\Psi}_{21}(\tilde{x}, \tilde{y})$, and that its plots are the same as those of $\tilde{\Psi}_{21}(\tilde{x}, \tilde{y})$ rotated by 90 degrees in the x-y plane. Give a physical explanation why the two sets of plots are related in this way.

Finally, we plot the wave function for the second excited state $\tilde{\Psi}_{22}(\tilde{x}, \tilde{y})$ (see Fig. 4.27e). It has two maxima (positive) and two minima (negative) located at the values 0.25 and 0.75 for $\tilde{x}$ and $\tilde{y}$. There are two nodal lines, along $\tilde{x} = 0.5$ and $\tilde{y} = 0.5$. They divide the x-y plane into quadrants, each of which contains a single maximum (positive) or minimum (negative) value. Make sure that you see how these characteristics trace back to the one-dimensional solutions in Figure 4.24. Figure 4.27f shows the contour plots for $\tilde{\Psi}_{22}(\tilde{x}, \tilde{y})$. As the magnitude (absolute value) of $\tilde{\Psi}_{22}(\tilde{x}, \tilde{y})$ decreases, the contours distort from circles to squares to match the nodal lines and boundary conditions of the box.

The pattern is now clearly apparent. You can easily produce hand sketches, in three dimensions and as contour plots, for any wave function for a particle in a square box. You need only pay attention to the magnitude of the quantum numbers n_x and n_y, track the number of nodes that must appear along the x and y axes, and convert these into nodal lines in the x-y plane.

The probability of locating the particle in a small element of area of size $d\tilde{x}d\tilde{y}$ centered on the point $(\tilde{x}, \tilde{y})$ is given by $[\tilde{\Psi}_{n_x n_y}(\tilde{x}, \tilde{y})]^2 d\tilde{x}d\tilde{y}$ when the system is in the quantum state described by n_x and n_y. For the wave functions shown in Figures 4.27a, c, and e, $\tilde{\Psi}^2$ will show 1, 2, and 4 peaks above the x-y plane, respectively. You should make hand sketches of these probability functions and also of their contour plots in the x-y plane. The physical interpretation is straightforward. In the ground state, the probability has a global maximum at the center of the box. In progressively higher excited states, the probability spreads out from the center into a series of local maxima, just as the one-dimensional case in Figure 4.24. These local maxima are arranged in a pattern determined by the quantum numbers in the x and y directions. If $n_x = n_y$, the local maxima will form a highly symmetric arrangement with pairs separated by nodal lines. If $n_x \neq n_y$, the pattern will not be symmetric. As n_x and n_y take on larger values, the probability becomes more nearly uniform through the box, and the motion becomes more like that predicted by classical mechanics (see the one-dimensional case in Fig. 4.26). As an exercise, we suggest that you determine the number of nodal lines and the number of local probability maxima for the highly excited state $\tilde{\Psi}_{20,\,20}(\tilde{x}, \tilde{y})$ and predict the nature of the motion of the particle.

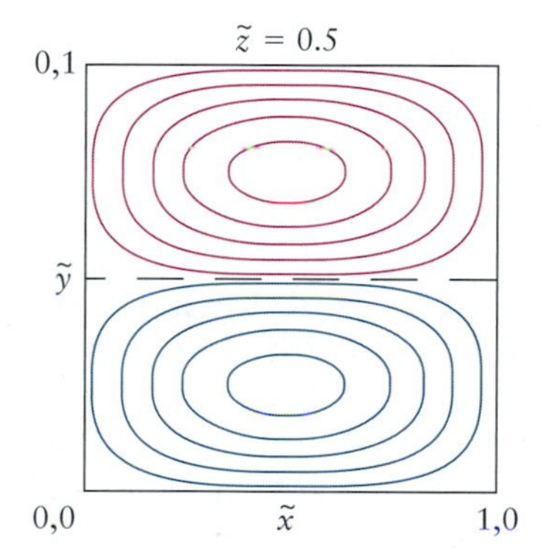

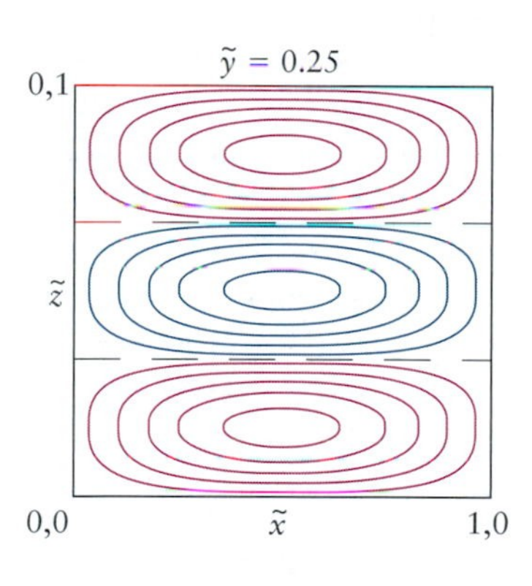

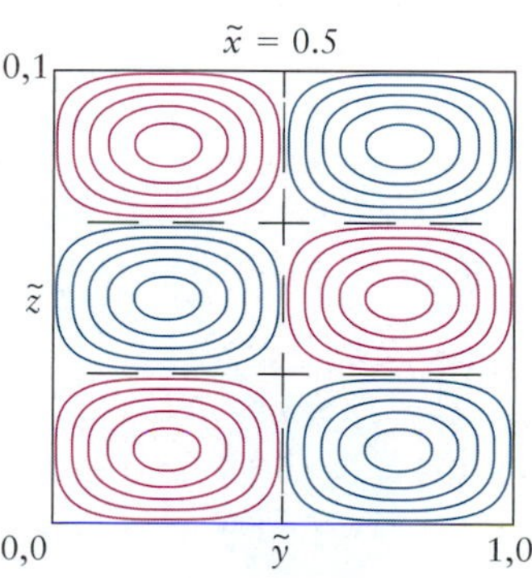

FIGURE 4.28 Contour plots for $\tilde{\Psi}_{123}(\tilde{x}, \tilde{y}, \tilde{z})$ for a particle in a cubic box. (a) Contours generated in a cut at $\tilde{z} = 0.5$. (b) Contours generated in a cut at $\tilde{y} = 0.25$. (c) Contours generated in a cut at $\tilde{x} = 0.5$. The location of nodal lines and shapes of the contours are explained in the text.

Wave Functions for Particles in Cubic Boxes

The wave function for a particle in a cubic box of length L on each side, with one corner located at the origin of coordinates, is given by

$$\Psi_{n_x n_y n_z}(x, y, z) = \left(\frac{2}{L}\right)^{3/2} \sin\left(\frac{n_x \pi x}{L}\right) \sin\left(\frac{n_y \pi y}{L}\right) \sin\left(\frac{n_z \pi z}{L}\right) \qquad [4.42]$$

where each of the quantum numbers n_x, n_y, and n_z can be any of the positive integers. Graphical representation of these wave functions requires some care. Equation 4.42 tells us simply to go to the point (x, y, z), evaluate the wave function there, and draw a graph showing the results of visiting many such points. However, all three spatial dimensions have already been used up to define the location; thus, we would need a fourth dimension to display the value of the wave function. Alternatively, we could set up a table of numbers giving the value of the wave function at each point (x, y, z), but it would be difficult to develop any intuition about shapes and structures from this table. We will get around these problems by slicing up three-dimensional space into various two- and one-dimensional regions, evaluating the wave function from Equation 4.42 at each point in these restricted regions, and generating graphical representations over these restricted regions. From the behavior of the wave function in these regions, we draw inferences about its overall behavior, even though we cannot graphically display its overall behavior in complete detail. For example, we could evaluate Equation 4.42 only at points in the x-y plane, and thereby generate contour maps of these wave functions in the x-y plane similar to those for the two-dimensional case shown in Figures 4.27b, d, and f. We could repeat this operation at several other "cut planes" through the box, and the resulting series of contour plots would provide considerable insight into the characteristics of the wave function.

In Figure 4.28, we examine the behavior of $\tilde{\Psi}_{123}(\tilde{x}, \tilde{y}, \tilde{z})$ using dimensionless variables defined earlier. Figure 4.28a shows a contour plot generated in a cut plane parallel to the x-y axis at $\tilde{z} = 0.5$. It demonstrates two sets of ellipses separated by one nodal line, arising because $n_x = 1$, $n_y = 2$. Figure 4.28b shows a contour plot generated in a cut plane at $\tilde{y} = 0.25$ or 0.75 parallel to the x-z plane. It shows three sets of ellipses separated by two nodal lines, as a consequence of $n_x = 1$, $n_z = 3$. Figure 4.28c shows a contour plot in the cut plane at $\tilde{x} = 0.5$. It shows six circles and three nodal lines, due to $n_y = 2$, $n_z = 3$. All of this suggests that $\tilde{\Psi}_{123}(\tilde{x}, \tilde{y}, \tilde{z})$ is an interesting object indeed!

How can we get some sense of the three-dimensional shape of a wave function? In Figure 4.28a, it is not necessary to have the cut plane $\tilde{z} = 0.5$ oriented parallel to the x-y axis. Let us imagine rotating this plane through a full 360-degree circle, always keeping the center of the plane anchored at $\tilde{z} = 0.5$, and let us generate contour plots at each of the angular orientations of the cut plane. The result will be that the ellipsoidal contours in Figure 4.28a generate a set of concentric

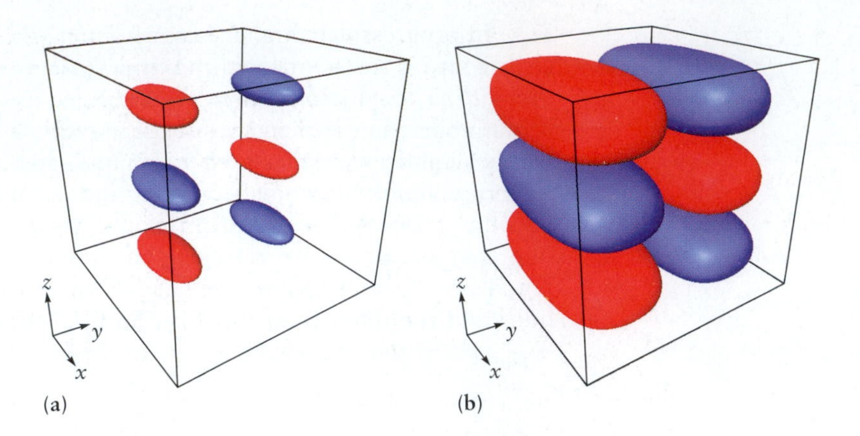

FIGURE 4.29 Isosurfaces for $\tilde{\Psi}_{123}(\tilde{x}, \tilde{y}, \tilde{z})$ for a particle in a cubic box. (a) Isosurfaces for wave function value $\tilde{\Psi}_{123} = \pm 0.8$. (b) Isosurfaces for wave function value $\tilde{\Psi}_{123} = \pm 0.2$. Each isosurface is shown in the same color as the corresponding contour in Figure 4.28.

"blimp-shaped" surfaces in three dimensions. Each of them identifies a surface of points (x, y, z) at every one of which $\tilde{\Psi}_{123}(\tilde{x}, \tilde{y}, \tilde{z})$ has the same value. These surfaces are called *isosurfaces* because the wave function has constant value at each point on them. In fact, we generate the isosurfaces in a more systematic way by evaluating $\tilde{\Psi}_{123}(\tilde{x}, \tilde{y}, \tilde{z})$ at every point in the cubic box and tracking in the computer all points that have, for example, the value $\tilde{\Psi} = \pm 0.9$. Then the computer plots the resulting isosurfaces in three dimensions. Figure 4.29 shows the isosurfaces for $\tilde{\Psi}_{123}(\tilde{x}, \tilde{y}, \tilde{z})$ at the values $\tilde{\Psi} = \pm 0.8, \pm 0.2$.

Figure 4.30 briefly summarizes key images for $\tilde{\Psi}_{222}(\tilde{x}, \tilde{y}, \tilde{z})$ for a particle in a cubic box. Figure 4.30a shows a contour plot in a cut plane at $\tilde{z} = 0.75$. Convince yourself that the contour plot in a cut at $\tilde{z} = 0.25$ would have the same pattern but each positive peak would become negative, and vice versa. Why should we not take a cut at $\tilde{z} = 0.5$? Be sure you understand the same concerns for cut planes perpendicular to $\tilde{x}$ and $\tilde{y}$. Figure 4.30bc shows the isosurfaces for the maxima and minima of $\tilde{\Psi}_{222}(\tilde{x}, \tilde{y}, \tilde{z})$ at the values $\tilde{\Psi} = \pm 0.9, \pm 0.3$.

Notice how in Figures 4.29 and 4.30 the shape depends on the value selected for the isosurface. This demonstrates an important point about plots of wave functions for a particle moving in three dimensions: It is not possible to show the shape of the wave function in three dimensions. You should be mindful that precisely because the wave function is a four-dimensional object, its appearance in three-dimensional representations depends strongly on choices made by the illustrator.

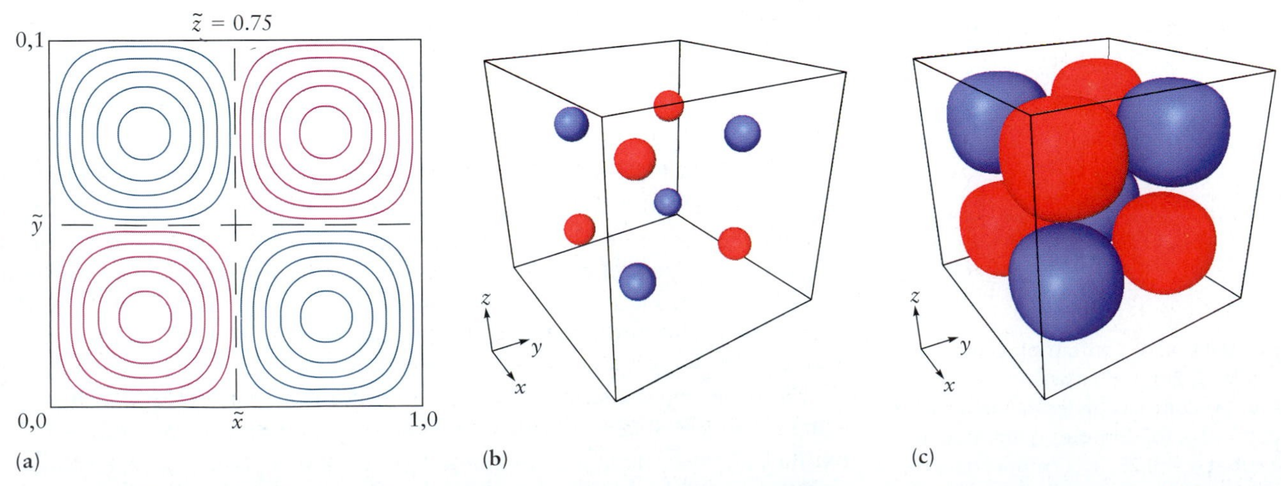

FIGURE 4.30 Representations of $\tilde{\Psi}_{222}(\tilde{x}, \tilde{y}, \tilde{z})$ for a particle in a cubic box. (a) Contour plots for a cut taken at $\tilde{z} = 0.75$. (b) Isosurfaces for wave function value $\tilde{\Psi}_{222} = \pm 0.9$. (c) Isosurfaces for wave function value $\tilde{\Psi}_{222} = \pm 0.3$. Each isosurface is shown in the same color as the corresponding contour in (a).

Be certain you understand these choices in each image you examine (or create!). These same issues appear in Chapter 5 when we discuss the wave functions for electrons in atoms, called *atomic orbitals*. Throughout this book, we have taken great care to generate accurate contour plots and isosurfaces for them from computer calculations to guide your thinking about the distribution of electrons in atoms and molecules.

4.7 Quantum Harmonic Oscillator

The quantum mechanical harmonic oscillator model system is used to describe the oscillators that are the sources of blackbody radiation previously discussed and also to describe the vibrational motions of molecules and solids. Moreover, it is the simplest model problem that illustrates the important but highly non-intuitive quantum process of tunneling. In classical mechanics, objects (cars, balls, particles, and so on) can make it up and over a hill only if they have enough kinetic energy to overcome the potential energy barrier imposed by the hill. But in the quantum world particles can **tunnel** through the hill; that is, they can go right through instead of over. This phenomenon is a consequence of the wave nature of matter, and it is one of the most difficult aspects of quantum mechanics to grasp. Because quantum mechanical tunneling is important in chemical kinetics, especially for reactions involving light particles such as electrons, protons, and hydrogen atoms, we want you to develop some familiarity with the concept here.

You have already encountered the classical version of the harmonic oscillator in our discussion of blackbody radiation in Section 4.2. The equations for the force and the potential energy of the harmonic oscillator are:

$$F(x) = -k(x - x_0)$$

and

$$V(x) = \frac{1}{2}k(x - x_0)^2$$

where k is the force constant that represents the "stiffness" of the spring, and $x - x_0$ is the displacement of the mass from its equilibrium position x_0, as defined in Section 4.2. If we choose the origin of our coordinate system by setting $x_0 = 0$, the potential energy of the harmonic oscillator (plotted on page 122) is a simple parabola centered at the origin. The vibrational frequency of a harmonic oscillator is given by

$$\nu = \frac{1}{2\pi}\sqrt{\frac{k}{m}}$$

where k is the force constant, and m is the mass.

To obtain the quantum version, we substitute the potential energy function for the harmonic oscillator into the Schrödinger equation to get

$$-(h^2/8\pi^2 m)d^2\psi(x)/dx^2 + \frac{1}{2}kx^2\psi(x) = E\psi(x)$$

We simply list the solutions, which you can verify by substituting them into the Schrödinger equation. The first four wave functions for the quantum harmonic oscillator are listed in Table 4.2 and plotted in Figure 4.31. The energy levels of the harmonic oscillator are given by

$$E_n = \left(n + \frac{1}{2}\right)h\nu \qquad n = 0, 1, 2, 3, \ldots \qquad [4.43]$$

TABLE 4.2 The First Four Harmonic Oscillator Wave Functions

$\alpha = \dfrac{2\pi}{h}(k\mu)^{1/2}$	$\psi_2(x) = \left(\dfrac{\alpha}{4\pi}\right)^{1/4}(2\alpha x^2 - 1)e^{-\alpha x^2/2}$
$\psi_0(x) = \left(\dfrac{\alpha}{\pi}\right)^{1/4} e^{-\alpha x^2/2}$	$\psi_3(x) = \left(\dfrac{\alpha^3}{9\pi}\right)^{1/4}(2\alpha x^3 - 3x)e^{-\alpha x^2/2}$
$\psi_1(x) = \left(\dfrac{4\alpha^3}{\pi}\right)^{1/4} x e^{-\alpha x^2/2}$	

Note that, unlike the particle in a box, $n = 0$ is perfectly acceptable because the term $\frac{1}{2}h\nu$ ensures that the energy of the oscillator never goes to zero, which would violate the indeterminacy principle. The frequency of the harmonic oscillator is given by

$$\nu = \frac{1}{2\pi}\sqrt{\frac{k}{m}} \qquad [4.44]$$

where m is the mass of the particle, and k (the force constant) represents the stiffness of the spring. The energy levels of a quantum harmonic oscillator are equally spaced; adjacent levels are separated by $\Delta E = h\nu$.

Potential energy diagrams for diatomic molecules were introduced in Section 3.5, and you can see that they are not parabolic over the entire region $0 < r < \infty$ (for example, see Fig. 3.9). Near the equilibrium internuclear separation the potential appears to be well approximated by a parabola. This similarity suggests that the harmonic oscillator should be a good model to describe the vibrations of diatomic molecules. The dependence of the vibrational frequency ν on the force constant k and the mass has the same form as Equation 4.44, but now the mass is the reduced mass μ of the two nuclei

$$\nu = \frac{1}{2\pi}\sqrt{\frac{k}{\mu}} \qquad [4.45a]$$

$$\mu = \frac{m_1 m_2}{m_1 + m_2} \qquad [4.45b]$$

(Vibrational frequencies are measured either by infrared or Raman spectroscopy, as discussed in Chapter 20.) The vibrational frequency is proportional to the square root of the force constant, a measure of the bond stiffness. Molecules with stiffer bonds are characterized by larger force constants (larger values of k) and greater vibrational frequencies. The vibrational frequencies of carbon–carbon single, double, and triple bonds are approximately 3×10^{13} Hz, 4.8×10^{13} Hz, and 6.6×10^{13} Hz respectively. This trend illustrates that stiffness of the bond correlates with bond order. For a given bond stiffness, the vibrational frequencies decrease as the masses of the atoms in the bond increase. To illustrate the mass dependence, compare the vibrational frequencies of two molecules with comparable bond stiffness, F_2 (2.7×10^{13} Hz) and I_2 (6.5×10^{12} Hz). Vibrational spectra of molecules, interpreted using the quantum harmonic oscillator model, are among our most powerful probes of molecular structure and dynamics.

Let's return to the characteristics of the harmonic oscillator wave functions (see Fig. 4.31). There are a number of interesting features to point out.

First, in contrast with the classical harmonic oscillator, the quantum harmonic oscillator in its ground state is most likely to be found at its equilibrium position. A classical harmonic oscillator spends most of its time at the classical turning points, the positions where it slows down, stops, and reverses directions. (It might

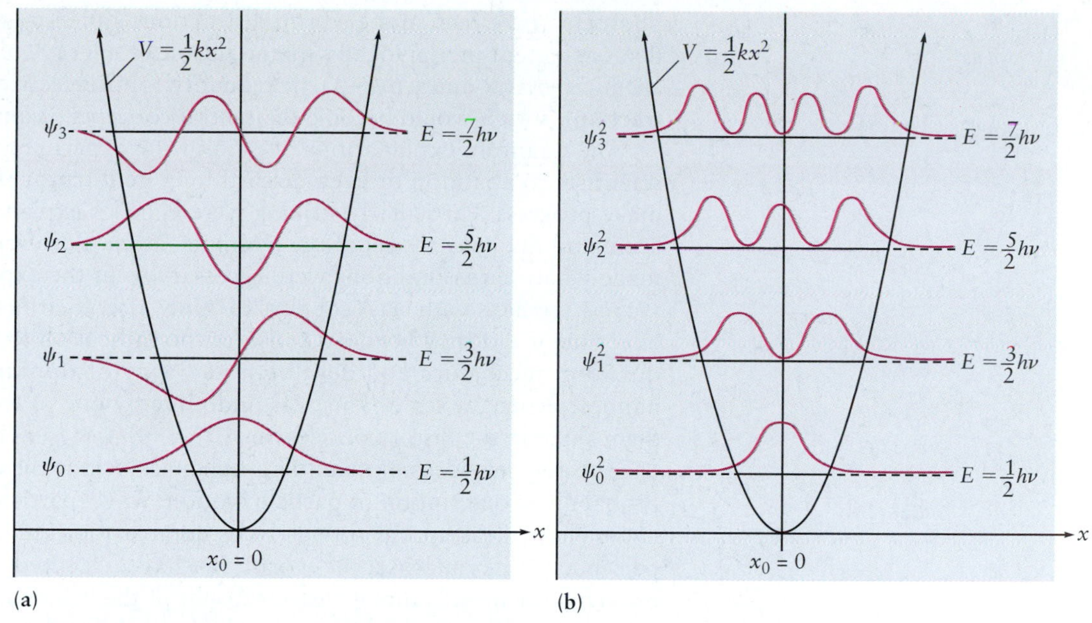

FIGURE 4.31 Solutions for the quantum harmonic oscillator. (a) The first four wave functions. (b) Probability densities corresponding to the first four wave functions.

be helpful to visualize the motions of a pendulum, which are described by the same equation.) The probability distribution function for the ground state quantum harmonic oscillator is different from its classical counterpart, just as that for the particle in a box differs dramatically from the classical prediction. Notice in Figure 4.31b how the probability density begins to accumulate near the classical turning points as the quantum number n increases. At higher energies, quantum systems begin to behave like their classical analogs.

Next, let's examine the energy of the quantum harmonic oscillator at the classical turning points. The horizontal line represents the total energy and the parabola represents the potential energy; at the classical turning point, the total energy equals the potential energy, which means that the kinetic energy is zero. This makes perfect sense classically. The oscillator must stop at some point while it changes direction, so its kinetic energy at that point goes to zero. The probability of the classical oscillator being outside that region is zero.

But it is clear from the figure that the amplitudes of the quantum harmonic oscillator wave functions do not go to zero past the classical turning point; thus, there is a finite probability of finding the particle in the classically forbidden regions. This "leaking out" of the probability is quantum mechanical tunneling, and it exists for every system that is not confined by an infinitely steep potential wall. If this concept is hard to grasp (which it is likely to be), focusing on the wavelike character of the oscillator might help. Just as light is partially reflected and partially transmitted through glass, quantum mechanical waves are partially transmitted through potential surfaces, like the parabolic well of the harmonic oscillator.

CHAPTER SUMMARY

We have introduced you to the concepts and methods of quantum mechanics; this branch of physics was developed to explain the behavior of matter on the nanometer length scale. The results of a number of key experiments demanded the creation of a new physical theory; classical mechanics and electrodynamics failed completely to account for these new observations. The pivotal experiments and observations included the spectrum and temperature dependence of blackbody radiation, the very existence of stable atoms and their discrete line spectra, the photoelectric effect, and electron diffraction. Taken together, these experiments demonstrated unequivocally quantization of energy (blackbody radiation and atomic spectra) and wave–particle duality (photoelectric effect and electron diffraction), which would become the central concepts of quantum mechanics.

The quantum explanations of each of the experiments listed earlier required scientists to abandon or even discard long held truths from classical physics to make progress. Particularly striking were Planck's explanation of blackbody radiation and the Bohr model of the hydrogen atom. Both scientists started fresh and made whatever assumptions were necessary to fit the experimental results, ignoring any conflicts with classical physics. Only after their models agreed so well with experiment did they begin to consider the radical philosophical implications of quantum mechanics and develop a new way of thinking about nature on the nanometer length scale. This was undoubtedly one of the most significant paradigm shifts in the history of science.

The key new ideas of quantum mechanics include the quantization of energy, a probabilistic description of particle motion, wave–particle duality, and indeterminacy. These ideas appear foreign to us because they are inconsistent with our experience of the macroscopic world. We have accepted them because they have provided the most comprehensive account of the behavior of matter and radiation and because the agreement between theory and the results of all experiments conducted to date has been astonishingly accurate.

Energy quantization arises for all systems that are confined by a potential. The one-dimensional particle-in-a-box model shows why quantization only becomes apparent on the atomic scale. Because the energy level spacing is inversely proportional to the mass and to the square of the length of the box, quantum effects become too small to observe for systems that contain more than a few hundred atoms or so.

Wave–particle duality accounts for the probabilistic nature of quantum mechanics and for indeterminacy. Once we accept that particles can behave as waves, then we can apply the results of classical electromagnetic theory to particles. By analogy, the probability is the square of the amplitude. Zero-point energy is a consequence of the Heisenberg uncertainty relation; all particles bound in potential wells have finite energy even at the absolute zero of temperature.

Particle-in-a-box models and the quantum harmonic oscillator illustrate a number of important features of quantum mechanics. The energy level structure depends on the nature of the potential; in the particle in a box, $E_n \propto n^2$, whereas for the harmonic oscillator, $E_n \propto n$. The probability distributions in both cases are different than for the classical analogs. The most probable location for the particle-in-a-box model in its ground state is the center of the box, rather than uniform over the box as predicted by classical mechanics. The most probable position for the quantum harmonic oscillator in the ground state is at its equilibrium position, whereas the classical harmonic oscillator is most likely to be found at the two classical turning points. Normalization ensures that the probabilities of finding the particle or the oscillator at all positions add up to one. Finally, for large values of n, the probability distribution looks much more classical, in accordance with the correspondence principle.

The concepts and principles we have discussed apply to any system of interest. In the next two chapters, we use quantum mechanics to explain atomic and molecular structure, respectively. In particular, it is important for you to have a firm grasp of these principles because they form the basis for our comprehensive discussion of chemical bonding in Chapter 6.

CUMULATIVE EXERCISE

An interesting class of carbon-containing molecules called *conjugated molecules* have structures that consist of a sequence of alternating single and double bonds. These chainlike molecules are represented as zigzag structures in which the angle between adjacent segments is determined by the geometry of the C—C double bond. Various properties to be explored in later chapters indicate that the electrons forming the double bonds are "de-localized" over the entire chain. Such molecules absorb light in the visible and ultraviolet regions of the electromagnetic spectrum. Many dyestuffs and molecules with biological significance have these structures. The properties of these molecules can be described approximately by the particle-in-a-box model in which we assume there is no interaction between the electrons, the potential energy is constant along the chain, and the potential energy is infinite at the ends of the chain. Assume the length of the potential well is Nd, where N is the number of carbon atoms in the chain, and d is half the sum of the lengths of a C—C single bond and a C—C double bond. In a molecule of N atoms, there will be N electrons involved in the double bonds.

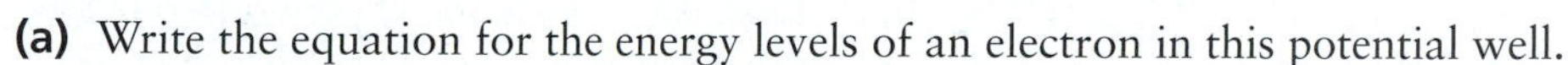

(a) Write the equation for the energy levels of an electron in this potential well.

(b) Write the equation for the wave function of an electron in this potential well. To describe the placement of N electrons in the energy levels, we anticipate a principle to be developed in Chapter 5 that requires that no more than two electrons can occupy a level. Therefore, levels will be occupied from the ground state up to level $n = N/2$. Absorption of light can cause one electron to move the next level $n = (N + 1)/2$.

(c) Write the equation for the frequency of light that will cause this transition.

(d) The molecule butadiene has four carbon atoms with conjugated structure; thus, $N = 4$. Calculate the wavelength of light in the first transition of butadiene.

(e) The molecular structure of vitamin A is conjugated with $N = 10$. Calculate the wavelength of light in the first transition of vitamin A.

(f) The molecule β-carotene has $N = 22$. Calculate the wavelength of light in the first transition of β-carotene.

We see in part (c) that the frequency of light absorbed should be inversely proportional to the length of the chain. Short-chain conjugated molecules absorb in the ultraviolet, whereas longer chain molecules absorb in the visible. This qualitative trend is predicted by the simple particle-in-a-box model. Later chapters detail how these results can be improved.

Answers

(a) $E_n = \dfrac{n^2h^2}{8mN^2d^2}$

(b) $\psi_n(x) = A \sin\left(\dfrac{n\pi x}{Nd}\right)$

(c) $h\nu = E_{N/2+1} - E_{N/2} = \dfrac{h^2(N+1)}{8md^2N^2} \simeq \dfrac{h^2}{8md^2N}$

(d) $\lambda = 2050$ Å

(e) $\lambda = 3150$ Å

(f) $\lambda = 4410$ Å

CHAPTER REVIEW

- Waves are disturbances in space and time. Electromagnetic waves are characterized by their amplitude, wavelength, and frequency; the latter are related by the equation $c = \lambda\nu$, where c is the speed of light (3×10^8 m s^{-1}) and λ and ν are the wavelength and frequency, respectively.
- Blackbody radiation emitted from hot sources has a characteristic frequency distribution that is temperature dependent. The spectrum of cooler objects has

a comparatively narrow band that peaks near the red end of the visible spectrum, whereas that of hotter objects has a much broader band that is shifted toward the blue.

- The peak observed in the frequency distribution of blackbody radiation is completely inconsistent with the predictions of classical electromagnetic theory. This failure of classical physics is called the ultraviolet catastrophe.
- The only way that Planck could fit the experimental spectrum was to postulate that the oscillating charges responsible for the radiation were restricted to discrete energies, that an oscillator was either excited or not, and that the probability of an oscillator being excited depends on the temperature.
- Line spectra of atoms: Atoms emit and absorb light in discrete amounts, photons. $\Delta E = h\nu$ is the energy difference between two quantum states, where ν is the frequency of the light absorbed or emitted.
- The Franck–Hertz experiment and atomic energy levels: Electrons can excite atoms from one quantum state to another by energy transferred during collisions. The threshold energy for excitation exactly matches the emission of light as the atom drops back down to the lower state, thus confirming the existence of quantized states and showing that they may be excited by either mechanical impact of electrons or absorption of photons.
- The Bohr model of one-electron atoms: Bohr *postulated* quantization of the angular momentum, $L = m_e vr = nh/2\pi$, substituted the result in the classical equations of motion, and correctly accounted for the spectrum of all one-electron atoms. $E = -Z^2/n^2$ (rydbergs). The model could not, however, account for the spectra of many-electron atoms.
- Wave–particle duality: de Broglie postulated that the motion of an electron in an atom could be described as a circular standing wave, which required an integral number of wavelengths to fit the circumference, $n\lambda = 2\pi r$. Combining this result with Bohr's quantization of angular momentum led to the de Broglie relation $\lambda = h/p$, which relates the wavelength and the momentum of a wave or a particle.
- Photoelectric effect: The photoelectric effect demonstrated the particle nature of electromagnetic radiation (formerly described only as waves).
- In the photoelectric effect, light shines on a metal surface in vacuum. The kinetic energy and photocurrent (number of electrons per second) emitted is measured as a function of frequency and intensity of the light.
- Experimental results of the photoelectric effect
  - No electrons are emitted below a threshold frequency ν_0, regardless of intensity.
  - Above threshold, the electron kinetic energy is $\frac{1}{2}mv^2 = h\nu - h\nu_0$, where h is a new constant, Planck's constant 6.625×10^{-34} J sec^{-1}. $h\nu_0 = \Phi$, the work function of the metal, which is the energy required to remove the electron from the metal. Φ is different for different metals.
  - Above threshold, the photocurrent is proportional to the light intensity.
- Interpretation of the photoelectric effect: Light behaves like a stream of particles called photons, each with an energy $E = h\nu$. A photon of a given energy either does or does not provide enough energy to overcome the work function; if it does, the excess energy goes into the kinetic energy of the photoelectron. Intensity is the number of photons passing a point per unit time, just like a current.
- Electron diffraction: Electron diffraction demonstrated the wave nature of particles.
- Heisenberg uncertainty relation: The Heisenberg uncertainty relation $(\Delta x)(\Delta p) \geq h/4\pi$ is a quantitative expression of indeterminacy—a fundamental limit on

our ability to know simultaneously the values of two properties of a particle (for example, position and momentum) with an arbitrarily high precision.

- The Schrödinger equation: This equation can be solved, in principle, for the energies and the wave functions for any quantum system of interest. The energy is quantized whenever a particle is confined by a potential. The square of the wave function gives the probability of finding the particle at a particular position in space. Normalization of the wave function by requiring that $\int_{all\ space} \psi^2 dV = 1$ ensures that the particle will be found somewhere.
- Particle-in-a-box problems: The energies of a particle in a one-dimensional box are given by $E = n^2h^2/8mL^2$, and the normalized wave functions by $\psi(x) = (2/L)^{1/2} \sin(n\pi x/L)$.
- The quantum harmonic oscillator is a model system that describes chemical and physical systems in which the restoring force is proportional to the displacement from the equilibrium position x_0 and is opposite to the direction of the displacement. The restoring force is given by $F(x) = -k(x - x_0)$ and the resulting potential energy is given by $V(x) = \frac{1}{2}k(x - x_0)^2$ where k is the force constant.
- The Schrödinger equation for the quantum harmonic oscillator can be solved exactly. The energy levels are given by $E_n = \left(n + \frac{1}{2}\right)h\nu$ with $n = 0, 1, 2, 3, \ldots$ The harmonic oscillator has zero point energy $E_0 = \frac{1}{2}h\nu$, which cannot be extracted from the oscillator, even at absolute zero temperature.
- The vibrational frequency of the quantum oscillator is given by $\nu = \left(\frac{1}{2\pi}\right)\sqrt{k/m}$.
- The quantum oscillator is a good model to describe the vibrations of a diatomic molecule. The frequency is given by the familiar equation but using the reduced mass of the two nuclei $\mu = (m_1m_2)/(m_1 + m_2)$ in place of m.
- The harmonic oscillator model enables measurement of the bond force constant through vibrational spectroscopy as follows. The measured frequency of the radiation absorbed is the same as the vibrational frequency of the molecule. The reduced mass is calculated for the molecule, and the force constant is evaluated as $k = (2\pi)^2\mu\nu^2$.
- The harmonic oscillator wave functions explain how a particle can tunnel through a potential barrier to arrive at regions forbidden by classical mechanics.

CONCEPTS & SKILLS

After studying this chapter and working the problems that follow, you should be able to:

1. Relate the frequency, wavelength, and speed of light waves. Do the same for other kinds of waves (Section 4.1, Problems 1–8).
2. Describe blackbody radiation, and discuss how related paradoxes of classical physics were resolved by quantum mechanics (Section 4.2, Problems 9 and 10).
3. Use experimental emission and absorption spectra to determine spacings between energy levels in atoms (Section 4.2, Problems 11–16).
4. Use the Franck–Hertz method to determine spacings between adjacent energy levels in atoms (Section 4.2, Problems 17 and 18).

5. Use the Bohr model to calculate the energy levels of one-electron atoms and to find the frequencies and wavelengths of light emitted in transitions between energy levels (Section 4.3, Problems 19–22).
6. Describe the photoelectric effect, and discuss how related paradoxes of classical physics were resolved by quantum mechanics (Section 4.4, Problems 23 and 24).
7. Using the law of conservation of energy, relate the work function of a metal to the wavelength of light used to eject electrons in the photoelectric effect and the kinetic energy of those electrons (Section 4.4, Problems 25–28).
8. Discuss the de Broglie relation and use it to calculate the wavelengths associated with particles in motion (Section 4.4, Problems 29–32).
9. Describe interference between wave functions for an electron. Explain how constructive and destructive interference influence the probability for finding the electron at a particular location (Section 4.5, Problems 33 and 34).
10. State the Heisenberg indeterminacy principle and use it to establish bounds within which the position and momentum of a particle can be known (Section 4.4, Problems 35 and 36).
11. State the conditions that a function must satisfy in order to be a solution of the Schrödinger equation. Explain how these conditions provide the probability interpretation of the wave function (Section 4.5).
12. Determine the energy levels for particles in rigid rectangular boxes (Section 4.6, Problems 37 and 38).
13. Prepare hand-drawn sketches for contour diagrams and isosurfaces of the wave functions for particles in square and cubic boxes. Relate the number and locations of the nodes, maxima, and minima to the quantum numbers for the wave function (Section 4.6, Problems 39–42).
14. Use the energy levels and the wave functions of the quantum harmonic oscillator model to describe the vibrational motions of diatomic molecules (Section 4.7, Problems 43 and 44).
15. Describe how particles tunnel through potential barriers (Section 4.7, Problems 45 and 46).

KEY EQUATIONS

$$c = \lambda\nu = 2.99792458 \times 10^8 \text{ m s}^{-1}$$ (Section 4.1)

$$E_{\text{osc}} = nh\nu \quad (n = 1, 2, 3, 4, \ldots)$$ (Section 4.2)

$$\nu = \left[\frac{1}{4} - \frac{1}{n^2}\right] \times 3.29 \times 10^{15} \text{ s}^{-1} \qquad n = 3, 4, 5, \ldots$$ (Section 4.2)

$$\nu = \frac{E_f - E_i}{h} \quad \text{or} \quad \Delta E = h\nu$$ (Section 4.2)

$$\nu = \frac{\Delta E}{h} = \frac{eV_{\text{thr}}}{h}$$ (Section 4.2)

$$E = \frac{1}{2}m_e v^2 - \frac{Ze^2}{4\pi\epsilon_0 r}$$ (Section 4.3)

$$|F_{\text{Coulomb}}| = |m_e a|$$ (Section 4.3)

$$\frac{Ze^2}{4\pi\epsilon_0 r^2} = m_e \frac{v^2}{r}$$ (Section 4.3)

$$L = m_e v r = n\frac{h}{2\pi} \qquad n = 1, 2, 3, \ldots \qquad \text{(Section 4.3)}$$

$$r_n = \frac{\epsilon_0 n^2 h^2}{\pi Z e^2 m_e} = \frac{n^2}{Z} a_0 \qquad \text{(Section 4.3)}$$

$$v_n = \frac{nh}{2\pi m_e r_n} = \frac{Ze^2}{2\epsilon_0 nh} \qquad \text{(Section 4.3)}$$

$$E_n = \frac{-Z^2 e^4 m_e}{8\epsilon_0^2 n^2 h^2} = -(2.18 \times 10^{-18}\ \text{J})\frac{Z^2}{n^2} \qquad n = 1, 2, 3, \ldots \qquad \text{(Section 4.3)}$$

$$E_n = -\frac{Z^2}{n^2} \quad \text{(rydberg)} \qquad n = 1, 2, 3, \ldots \qquad \text{(Section 4.3)}$$

$$\nu = \frac{Z^2 e^4 m_e}{8\epsilon_0^2 h^3}\left(\frac{1}{n_f^2} - \frac{1}{n_i^2}\right) = (3.29 \times 10^{15}\ \text{s}^{-1})Z^2\left(\frac{1}{n_f^2} - \frac{1}{n_i^2}\right)$$

$$n_i > n_f = 1, 2, 3, \ldots \ \text{(emission)} \qquad \text{(Section 4.3)}$$

$$\nu = (3.29 \times 10^{15}\ \text{s}^{-1})Z^2\left(\frac{1}{n_i^2} - \frac{1}{n_f^2}\right)$$

$$n_f > n_i = 1, 2, 3, \ldots \ \text{(absorption)} \qquad \text{(Section 4.3)}$$

$$E_{max} = \frac{1}{2} m v_e^2 = h\nu - \Phi \qquad \text{(Section 4.4)}$$

$$\lambda = \frac{h}{m_e v} = \frac{h}{p} \qquad \text{(Section 4.4)}$$

$$n\lambda = 2d \sin\theta \qquad \text{(Section 4.4)}$$

$$(\Delta x)(\Delta p) \geq h/4\pi \qquad \text{(Section 4.4)}$$

$$-\frac{h^2}{8\pi^2 m}\frac{d^2\psi(x)}{dx^2} + V(x)\psi(x) = E\psi(x) \qquad \text{(Section 4.5)}$$

$$\int_{-\infty}^{+\infty} P(x)dx = \int_{-\infty}^{+\infty} [\psi(x)]^2 dx = 1 \qquad \text{(Section 4.5)}$$

$$\psi \longrightarrow 0 \text{ as } x \longrightarrow \pm\infty \qquad \text{(Section 4.5)}$$

$$\psi_n(x) = \sqrt{\frac{2}{L}}\sin\left(\frac{n\pi x}{L}\right) \qquad (n = 1, 2, 3, \ldots) \qquad \text{(Section 4.6)}$$

$$E_n = \frac{h^2 n^2}{8mL^2} \qquad (n = 1, 2, 3, \ldots) \qquad \text{(Section 4.6)}$$

$$E_{n_x n_y n_z} = \frac{h^2}{8mL^2}[n_x^2 + n_y^2 + n_z^2] \qquad \begin{cases} n_x = 1, 2, 3, \ldots \\ n_y = 1, 2, 3, \ldots \\ n_z = 1, 2, 3, \ldots \end{cases} \qquad \text{(Section 4.6)}$$

$$\Psi_{n_x n_y}(x,y) = \psi_{n_x}(x)\psi_{n_y}(y) = \frac{2}{L}\sin\left(\frac{n_x \pi x}{L}\right)\sin\left(\frac{n_y \pi y}{L}\right) \qquad \text{(Section 4.6)}$$

$$\Psi_{n_x n_y n_z}(x,y,z) = \left(\frac{2}{L}\right)^{3/2}\sin\left(\frac{n_x \pi x}{L}\right)\sin\left(\frac{n_y \pi y}{L}\right)\sin\left(\frac{n_z \pi z}{L}\right) \qquad \text{(Section 4.6)}$$

$$E_n = \left(n + \frac{1}{2}\right) h\nu \qquad n = 0, 1, 2, 3, \ldots \qquad \text{(Section 4.7)}$$

$$\nu = \frac{1}{2\pi}\sqrt{\frac{k}{m}} \qquad \text{(Section 4.7)}$$

$$\nu = \frac{1}{2\pi}\sqrt{\frac{k}{\mu}} \qquad \text{(Section 4.7)}$$

$$\mu = \frac{m_1 m_2}{m_1 + m_2} \qquad \text{(Section 4.7)}$$

PROBLEMS

Answers to problems whose numbers are boldface appear in Appendix G. Problems that are more challenging are indicated with asterisks.

Preliminaries: Wave Motion and Light

1. Some water waves reach the beach at a rate of one every 3.2 s, and the distance between their crests is 2.1 m. Calculate the speed of these waves.

2. The spacing between bands of color in a chemical wave from an oscillating reaction is measured to be 1.2 cm, and a new wave appears every 42 s. Calculate the speed of propagation of the chemical waves.

3. An FM radio station broadcasts at a frequency of 9.86×10^7 s^{-1} (98.6 MHz). Calculate the wavelength of the radio waves.

4. The gamma rays emitted by ^{60}Co are used in radiation treatment of cancer. They have a frequency of 2.83×10^{20} s^{-1}. Calculate their wavelength, expressing your answer in meters and in angstroms.

5. Radio waves of wavelength 6.00×10^2 m can be used to communicate with spacecraft over large distances.
(a) Calculate the frequency of these radio waves.
(b) Suppose a radio message is sent home by astronauts in a spaceship approaching Mars at a distance of 8.0×10^{10} m from Earth. How long (in minutes) will it take for the message to travel from the spaceship to Earth?

6. An argon ion laser emits light of wavelength of 488 nm.
(a) Calculate the frequency of the light.
(b) Suppose a pulse of light from this laser is sent from Earth, is reflected from a mirror on the moon, and returns to its starting point. Calculate the time elapsed for the round trip, taking the distance from Earth to the moon to be 3.8×10^5 km.

7. The speed of sound in dry air at 20°C is 343.5 m s^{-1}, and the frequency of the sound from the middle C note on a piano is 261.6 s^{-1} (according to the American standard pitch scale). Calculate the wavelength of this sound and the time it will take to travel 30.0 m across a concert hall.

8. Ultrasonic waves have frequencies too high to be detected by the human ear, but they can be produced and detected by vibrating crystals. Calculate the wavelength of an ultrasonic wave of frequency 5.0×10^4 s^{-1} that is propagating through a sample of water at a speed of 1.5×10^3 m s^{-1}. Explain why ultrasound can be used to probe the size and position of the fetus inside the mother's abdomen. Could audible sound with a frequency of 8000 s^{-1} be used for this purpose?

Evidence for Energy Quantization in Atoms

9. The maximum in the blackbody radiation intensity curve moves to shorter wavelength as temperature increases. The German physicist Wilhelm Wien demonstrated the relation to be $\lambda_{max} \propto 1/T$. Later, Planck's equation showed the maximum to be $\lambda_{max} = 0.20\, hc/kT$. In 1965 scientists researching problems in telecommunication discovered "background radiation" with maximum wavelength 1.05 mm (microwave region of the EM spectrum) throughout space. Estimate the temperature of space.

10. Use the data in Figure 4.6 to estimate the ratio of radiation intensity at 10,000 Å (infrared) to that at 5000 Å (visible) from a blackbody at 5000 K. How will this ratio change with increasing temperature? Explain how this change occurs.

11. Excited lithium atoms emit light strongly at a wavelength of 671 nm. This emission predominates when lithium atoms are excited in a flame. Predict the color of the flame.

12. Excited mercury atoms emit light strongly at a wavelength of 454 nm. This emission predominates when mercury atoms are excited in a flame. Predict the color of the flame.

13. Barium atoms in a flame emit light as they undergo transitions from one energy level to another that is 3.6×10^{-19} J lower in energy. Calculate the wavelength of light emitted and, by referring to Figure 4.3, predict the color visible in the flame.

14. Potassium atoms in a flame emit light as they undergo transitions from one energy level to another that is 4.9×10^{-19} J lower in energy. Calculate the wavelength of light emitted and, by referring to Figure 4.3, predict the color visible in the flame.

15. The sodium D-line is actually a pair of closely spaced spectroscopic lines seen in the emission spectrum of sodium atoms. The wavelengths are centered at 589.3 nm. The intensity of this emission makes it the major source of light (and causes the yellow color) in the sodium arc light.
(a) Calculate the energy change per sodium atom emitting a photon at the D-line wavelength.
(b) Calculate the energy change per mole of sodium atoms emitting photons at the D-line wavelength.

(c) If a sodium arc light is to produce 1.000 kilowatt (1000 J s^{-1}) of radiant energy at this wavelength, how many moles of sodium atoms must emit photons per second?

16. The power output of a laser is measured by its wattage, that is, the number of joules of energy it radiates per second (1 W = 1 J s^{-1}). A 10-W laser produces a beam of green light with a wavelength of 520 nm (5.2×10^{-7} m).
(a) Calculate the energy carried by each photon.
(b) Calculate the number of photons emitted by the laser per second.

17. In a Franck–Hertz experiment on sodium atoms, the first excitation threshold occurs at 2.103 eV. Calculate the wavelength of emitted light expected just above this threshold. (**Note:** Sodium vapor lamps used in street lighting emit spectral lines with wavelengths 5891.8 and 5889.9 Å.)

18. In a Franck–Hertz experiment on hydrogen atoms, the first two excitation thresholds occur at 10.1 and 11.9 eV. Three optical emission lines are associated with these levels. Sketch an energy-level diagram for hydrogen atoms based on this information. Identify the three transitions associated with these emission lines. Calculate the wavelength of each emitted line.

The Bohr Model: Predicting Discrete Energy Levels

19. Use the Bohr model to calculate the radius and the energy of the B^{4+} ion in the $n = 3$ state. How much energy would be required to remove the electrons from 1 mol of B^{4+} in this state? What frequency and wavelength of light would be emitted in a transition from the $n = 3$ to the $n = 2$ state of this ion? Express all results in SI units.

20. He^+ ions are observed in stellar atmospheres. Use the Bohr model to calculate the radius and the energy of He^+ in the $n = 5$ state. How much energy would be required to remove the electrons from 1 mol of He^+ in this state? What frequency and wavelength of light would be emitted in a transition from the $n = 5$ to the $n = 3$ state of this ion? Express all results in SI units.

21. The radiation emitted in the transition from $n = 3$ to $n = 2$ in a neutral hydrogen atom has a wavelength of 656.1 nm. What would be the wavelength of radiation emitted from a doubly ionized lithium atom (Li^{2+}) if a transition occurred from $n = 3$ to $n = 2$? In what region of the spectrum does this radiation lie?

22. Be^{3+} has a single electron. Calculate the frequencies and wavelengths of light in the emission spectrum of the ion for the first three lines of each of the series that are analogous to the Lyman and the Balmer series of neutral hydrogen. In what region of the spectrum does this radiation lie?

Evidence for Wave–Particle Duality

23. Both blue and green light eject electrons from the surface of potassium. In which case do the ejected electrons have the higher average kinetic energy?

24. When an intense beam of green light is directed onto a copper surface, no electrons are ejected. What will happen if the green light is replaced with red light?

25. Cesium frequently is used in photocells because its work function (3.43×10^{-19} J) is the lowest of all the elements. Such photocells are efficient because the broadest range of wavelengths of light can eject electrons. What colors of light will eject electrons from cesium? What colors of light will eject electrons from selenium, which has a work function of 9.5×10^{-19} J?

26. Alarm systems use the photoelectric effect. A beam of light strikes a piece of metal in the photocell, ejecting electrons continuously and causing a small electric current to flow. When someone steps into the light beam, the current is interrupted and the alarm is triggered. What is the maximum wavelength of light that can be used in such an alarm system if the photocell metal is sodium, with a work function of 4.41×10^{-19} J?

27. Light with a wavelength of 2.50×10^{-7} m falls on the surface of a piece of chromium in an evacuated glass tube. If the work function of chromium is 7.21×10^{-19} J, determine (a) the maximum kinetic energy of the emitted photoelectrons and (b) the speed of photoelectrons that have this maximum kinetic energy.

28. Calculate the maximum wavelength of electromagnetic radiation if it is to cause detachment of electrons from the surface of metallic tungsten, which has a work function of 7.29×10^{-19} J. If the maximum speed of the emitted photoelectrons is to be 2.00×10^6 m s^{-1}, what should the wavelength of the radiation be?

29. A guitar string with fixed ends has a length of 50 cm.
(a) Calculate the wavelengths of its fundamental mode of vibration (that is, its first harmonic) and its third harmonic.
(b) How many nodes does the third harmonic have?

30. Suppose we picture an electron in a chemical bond as being a wave with fixed ends. Take the length of the bond to be 1.0 Å.
(a) Calculate the wavelength of the electron wave in its ground state and in its first excited state.
(b) How many nodes does the first excited state have?

31. Calculate the de Broglie wavelength of the following:
(a) an electron moving at a speed of 1.00×10^3 m s^{-1}
(b) a proton moving at a speed of 1.00×10^3 m s^{-1}
(c) a baseball with a mass of 145 g, moving at a speed of 75 km hr^{-1}

32. Calculate the de Broglie wavelength of the following:
(a) electrons that have been accelerated to a kinetic energy of 1.20×10^7 J mol^{-1}
(b) a helium atom moving at a speed of 353 m s^{-1} (the root-mean-square speed of helium atoms at 20 K)
(c) a krypton atom moving at a speed of 299 m s^{-1} (the root-mean-square speed of krypton atoms at 300 K)

33. In a particular Low Energy Electron Diffraction (LEED) study of a solid surface, electrons at 45 eV were diffracted at $\phi = 53°$.
(a) Calculate the crystal spacing d.
(b) Calculate the diffraction angle for 90 eV electrons on this same surface.

34. What electron energy is required to obtain the diffraction pattern for a surface with crystal spacing of 4.0 Å?

35. (a) The position of an electron is known to be within 10 Å (1.0×10^{-9} m). What is the minimum uncertainty in its velocity?
(b) Repeat the calculation of part (a) for a helium atom.

36. No object can travel faster than the speed of light, so it would appear evident that the uncertainty in the speed of any object is at most 3×10^8 m s^{-1}.
    (a) What is the minimum uncertainty in the position of an electron, given that we know nothing about its speed except that it is slower than the speed of light?
    (b) Repeat the calculation of part (a) for the position of a helium atom.

Quantum Mechanics of Particle-in-a-Box Models

37. Chapter 3 introduced the concept of a double bond between carbon atoms, represented by C=C, with a length near 1.34 Å. The motion of an electron in such a bond can be treated crudely as motion in a one-dimensional box. Calculate the energy of an electron in each of its three lowest allowed states if it is confined to move in a one-dimensional box of length 1.34 Å. Calculate the wavelength of light necessary to excite the electron from its ground state to the first excited state.

38. When metallic sodium is dissolved in liquid sodium chloride, electrons are released into the liquid. These dissolved electrons absorb light with a wavelength near 800 nm. Suppose we treat the positive ions surrounding an electron crudely as defining a three-dimensional cubic box of edge L, and we assume that the absorbed light excites the electron from its ground state to the first excited state. Calculate the edge length L in this simple model.

39. Write the wave function $\tilde{\Psi}_{12}(\tilde{x}, \tilde{y})$ for a particle in a square box.
    (a) Convince yourself it is degenerate with $\tilde{\Psi}_{21}(\tilde{x}, \tilde{y})$.
    (b) Convince yourself that its plots are the same as those of $\tilde{\Psi}_{21}(\tilde{x}, \tilde{y})$ rotated by 90 degrees in the x-y plane.
    (c) Give a physical explanation why the two sets of plots are related in this way.

40. Write the wave function for the highly excited state $\tilde{\Psi}_{100,100}(\tilde{x}, \tilde{y})$ for a particle in a square box.
    (a) Determine the number of nodal lines and the number of local probability maxima for this state.
    (b) Describe the motion of the particle in this state.

41. Consider the wave function$\tilde{\Psi}_{222}(\tilde{x}, \tilde{y}, \tilde{z})$ for a particle in a cubic box. Figure 4.30a shows a contour plot in a cut plane at $\tilde{z} = 0.75$.
    (a) Convince yourself that the contour plot in a cut at $\tilde{z} = 0.25$ would have the same pattern, but each positive peak would become negative, and vice versa.
    (b) Describe the shape of this wave function in a plane cut at $\tilde{z} = 0.5$

42. Consider the wave function $\tilde{\Psi}_{222}(\tilde{x}, \tilde{y}, \tilde{z})$ for a particle in a cubic box. Figure 4.30a shows a contour plot in a cut plane at $\tilde{z} = 0.75$.
    (a) Describe the shape of this wave function in a cut plane at $\tilde{x} = 0.5$.
    (b) Describe the shape of this wave function in a cut plane at $\tilde{y} = 0.5$.

Quantum Harmonic Oscillator

43. Calculate the natural frequency of vibration for a ball of mass 10 g attached to a spring with a force constant of 1 N m^{-1}.

44. Calculate the natural frequency of vibration for a system with two balls, one with mass 1 g and the other with mass 2 g, held together by a spring of length 10 cm and force constant 0.1 N m^{-1}.

45. Vibrational spectroscopic studies of HCl show that the radiation absorbed in a transition has frequency 8.63×10^{13} Hz.
    (a) Calculate the energy absorbed in the transition.
    (b) Calculate the vibrational frequency of the molecule in this transition.
    (c) Calculate the reduced mass for HCl.
    (d) Calculate the value of the force constant for HCl.

46. Assume the force constant for DCl is the same as that for HCl. Calculate the frequency of the infrared radiation that will be absorbed by DCl. Compare with the observed absorption frequency, 6.28×10^{13} Hz. (This result illustrates a general rule that replacing H by D lowers the vibrational frequency of a bond by the factor $\sqrt{2}$.)

ADDITIONAL PROBLEMS

47. A piano tuner uses a tuning fork that emits sound with a frequency of 440 s^{-1}. Calculate the wavelength of the sound from this tuning fork and the time the sound takes to travel 10.0 m across a large room. Take the speed of sound in air to be 343 m s^{-1}.

48. The distant galaxy called Cygnus A is one of the strongest sources of radio waves reaching Earth. The distance of this galaxy from Earth is 3×10^{24} m. How long (in years) does it take a radio wave of wavelength 10 m to reach Earth? What is the frequency of this radio wave?

49. Hot objects can emit blackbody radiation that appears red, orange, white, or bluish white, but never green. Explain.

50. Compare the energy (in joules) carried by an x-ray photon (wavelength $\lambda = 0.20$ nm) with that carried by an AM radio wave photon ($\lambda = 200$ m). Calculate the energy of 1.00 mol of each type of photon. What effect do you expect each type of radiation to have for inducing chemical reactions in the substances through which it passes?

51. The maximum in Planck's formula for the emission of blackbody radiation can be shown to occur at a wavelength $\lambda_{max} = 0.20\ hc/kT$. The radiation from the surface of the sun approximates that of a blackbody with $\lambda_{max} = 465$ nm. What is the approximate surface temperature of the sun?

52. Photons of wavelength 315 nm or less are needed to eject electrons from a surface of electrically neutral cadmium.
    (a) What is the energy barrier that electrons must overcome to leave an uncharged piece of cadmium?
    (b) What is the maximum kinetic energy of electrons ejected from a piece of cadmium by photons of wavelength 200 nm?
    (c) Suppose the electrons described in (b) were used in a diffraction experiment. What would be their wavelength?

53. When ultraviolet light of wavelength of 131 nm strikes a polished nickel surface, the maximum kinetic energy of ejected electrons is measured to be 7.04×10^{-19} J. Calculate the work function of nickel.

54. Express the velocity of the electron in the Bohr model for fundamental constants (m_e, e, h, ϵ_0), the nuclear charge Z,

and the quantum number n. Evaluate the velocity of an electron in the ground states of He^+ ion and U^{91+}. Compare these velocities with the speed of light c. As the velocity of an object approaches the speed of light, relativistic effects become important. In which kinds of atoms do you expect relativistic effects to be greatest?

55. Photons are emitted in the Lyman series as hydrogen atoms undergo transitions from various excited states to the ground state. If ground-state He^+ are present in the same gas (near stars, for example), can they absorb these photons? Explain.

* 56. Name a transition in C^{5+} that will lead to the absorption of green light.

57. The energies of macroscopic objects, as well as those of microscopic objects, are quantized, but the effects of the quantization are not seen because the difference in energy between adjacent states is so small. Apply Bohr's quantization of angular momentum to the revolution of Earth (mass 6.0×10^{24} kg), which moves with a velocity of 3.0×10^4 m s^{-1} in a circular orbit (radius 1.5×10^{11} m) about the sun. The sun can be treated as fixed. Calculate the value of the quantum number n for the present state of the Earth–sun system. What would be the effect of an increase in n by 1?

58. Sound waves, like light waves, can interfere with each other, giving maximum and minimum levels of sound. Suppose a listener standing directly between two loudspeakers hears the same tone being emitted from both. This listener observes that when one of the speakers is moved 0.16 m farther away, the perceived intensity of the tone decreases from a maximum to a minimum.
    (a) Calculate the wavelength of the sound.
    (b) Calculate its frequency, using 343 m s^{-1} as the speed of sound.

59. (a) If the kinetic energy of an electron is known to lie between 1.59×10^{-19} J and 1.61×10^{-19} J, what is the smallest distance within which it can be known to lie?
    (b) Repeat the calculation of part (a) for a helium atom instead of an electron.

60. By analyzing how the energy of a system is measured, Heisenberg and Bohr discovered that the uncertainty in the energy, ΔE, is related to the time, Δt, required to make the measurement by the relation $(\Delta E)(\Delta t) \geq h/4\pi$. The excited state of an atom responsible for the emission of a photon typically has an average life of 10^{-10} s. What energy uncertainty corresponds to this value? What is the corresponding uncertainty in the frequency associated with the photon?

61. It has been suggested that spacecraft could be powered by the pressure exerted by sunlight striking a sail. The force exerted on a surface is the momentum p transferred to the surface per second. Assume that photons of 6000 Å light strike the sail perpendicularly. How many must be reflected per second by 1 cm^2 of surface to produce a pressure of 10^{-6} atm?

62. A single particle of mass 2.30×10^{-26} kg held by a spring with force constant 150 N m^{-1} undergoes harmonic oscillations. Calculate the zero point energy of this oscillator.

63. A single particle of mass 1.30×10^{-26} kg undergoes harmonic oscillations, and the energy difference between adjacent levels is 4.82×10^{-21} J. Calculate the value of the force constant.

64. It is interesting to speculate on the properties of a universe with different values for the fundamental constants.
    (a) In a universe in which Planck's constant had the value h = 1 J s, what would be the de Broglie wavelength of a 145-g baseball moving at a speed of 20 m s^{-1}?
    (b) Suppose the velocity of the ball from part (a) is known to lie between 19 and 21 m s^{-1}. What is the smallest distance within which it can be known to lie?
    (c) Suppose that in this universe the mass of the electron is 1 g and the charge on the electron is 1 C. Calculate the Bohr radius of the hydrogen atom in this universe.

65. The normalized wave function for a particle in a one-dimensional box, in which the potential energy is zero, is $\psi(x) = \sqrt{2/L} \sin(n\pi x/L)$, where L is the length of the box. What is the probability that the particle will lie between $x = 0$ and $x = L/4$ if the particle is in its $n = 2$ state?

66. A particle of mass m is placed in a three-dimensional rectangular box with edge lengths $2L$, L, and L. Inside the box the potential energy is zero, and outside it is infinite; therefore, the wave function goes smoothly to zero at the sides of the box. Calculate the energies and give the quantum numbers of the ground state and the first five excited states (or sets of states of equal energy) for the particle in the box.

CHAPTER 5

Quantum Mechanics and Atomic Structure

Fireworks above Paris; La Grande Arche is in the foreground. Many of the colors in fireworks are produced from atomic emission: red from strontium, orange from calcium, yellow from sodium, green from barium, and blue from copper. The sharp lines observed in the emission spectra of atoms can only be explained using the quantum theory of atomic structure.

The atom is the most fundamental concept in the science of chemistry. A chemical reaction occurs by regrouping a set of atoms initially found in those molecules called reactants to form those molecules called products. Atoms are neither created nor destroyed in chemical reactions. Chemical bonds between atoms in the reactants are broken, and new bonds are formed between atoms in the products. We have traced the concept of the atom from the suppositions of the Greek philosophers to the physics experiments of Thomson and Rutherford and we have arrived at the planetary model of the atom. We have used the Coulomb force and potential energy laws describing the interactions among the nucleus and the electrons in the planetary atom to account for the gain and loss of electrons by atoms, and the formation of chemical bonds between atoms. These descriptions, based on the planetary model, accurately account for large amounts of experimental data.

Now we have to confront an inconvenient truth lurking quietly but ominously in the background of all these successful discussions. According to the laws of physics under which it was discovered, *the planetary atom cannot exist.* Newtonian mechanics says that an electron orbiting around a nucleus will be constantly accelerated. Maxwell's electromagnetic theory requires an accelerated charged particle to emit radiation. Thus, the electron should spiral into the nucleus, and the planetary atom should collapse in a fraction of a second. Clearly, real atoms are stable and do not behave as these theories predict. Real experimental data show that the internal physical structure of the atom is well described by the planetary model. The problem comes with attempts to analyze the planetary model using the classical physics of Newton and Maxwell. The physical picture is correct, but the equations are wrong for atoms.

The incompatibility of Rutherford's planetary model, based soundly on experimental data, with the principles of classical physics was the most fundamental of the conceptual challenges facing physicists in the early 1900s. The Bohr model was a temporary fix, sufficient for the interpretation of hydrogen (H) atomic spectra as arising from transitions between stationary states of the atom. The stability of atoms and molecules finally could be explained only after quantum mechanics had been developed.

The goal of this chapter is to describe the structure and properties of atoms using quantum mechanics. We couple the physical insight into the atom developed in Sections 3.2, 3.3, and 3.4 with the quantum methods of Chapter 4 to develop a quantitative description of atomic structure.

We begin with the hydrogen atom, for which the Schrödinger equation can be solved exactly because it has only one electron. We obtain exact expressions for the energy levels and the wave functions. The exact wave functions are called **hydrogen atomic orbitals.** The square of a wave function gives the probability of locating the electron at a specific position in space, determined by the properties of that orbital. The sizes and shapes of hydrogen atomic orbitals hold special interest, because they are the starting points for approximate solutions to more complex problems.

There is no exact solution for any other atom, so we must develop approximate solutions. We treat each electron in a many-electron atom as if it were moving in an *effective force field* that results from averaging its interactions with all the other electrons and the nucleus. The effective field was introduced by purely physical arguments in Section 3.2. Here, we develop this concept systematically and from it obtain approximate one-electron wave functions called **Hartree atomic orbitals** (to honor the English physicist Douglas Hartree who pioneered the method), which account for the effect of all the other electrons in the atom. The shapes of the Hartree orbitals are similar to those of the hydrogen atomic orbitals, but their sizes and their energy level patterns are quite different. We use the Hartree orbitals to explain the shell model of the atom, the structure of the periodic table, and the periodic behavior of atomic properties. The result is a comprehensive, approximate quantum description of atomic structure, which serves as the starting point for the quantum description of the chemical bond in Chapter 6.

The method of solving the Schrödinger equation for the hydrogen atom is the same as that used for the particle-in-a-box models in Section 4.6. Because the mathematics is more complicated, we do not show the details here, and we present the solutions only in graphical form. Your primary objective in this chapter should be to understand the shapes and structures of the hydrogen atomic orbitals and the Hartree orbitals from these graphical representations. You should be able to predict how the probability distribution for the electrons depends on the properties of the orbitals, as specified by their quantum numbers. Always keep in mind the distinction between the hydrogen atomic orbitals and the Hartree orbitals. The former apply only to the hydrogen atom, and the latter only to many-electron atoms. Be aware of their differences, as well as their similarities.

5.1 The Hydrogen Atom

The hydrogen atom is the simplest example of a one-electron atom or ion; other examples are He^{+}, Li^{2+}, and other ions in which all but one electron have been stripped off. They differ only in the charge $+Ze$ on the nucleus, and therefore in the attractive force experienced by the electron.

The potential energy for the one-electron atom, discussed in the context of the planetary model in Section 3.2 and of the Bohr model in Section 4.3, depends only on the distance of the electron from the nucleus; it does not depend on angular orientation (see Fig. 4.12). Solution of the Schrödinger equation is most easily carried out in coordinates that reflect the natural symmetry of the potential energy function. For an isolated one-electron atom or ion, spherical coordinates are more appropriate than the more familiar Cartesian coordinates. Spherical coordinates are defined in Figure 5.1: r is the distance of the electron at P from the nucleus at O, and the angles θ and ϕ are similar to those used to locate points on the surface of the globe; θ is related to the latitude, and ϕ is related to the longitude.

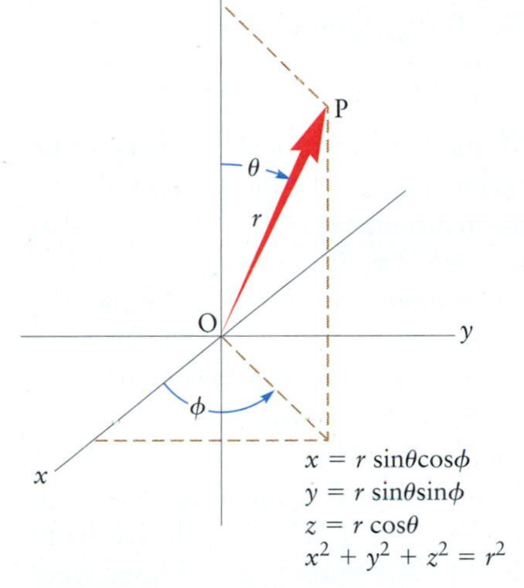

FIGURE 5.1 The relationship between spherical coordinates (r, θ, ϕ) and Cartesian coordinates (x, y, z). Here, θ is the angle with respect to the Cartesian z-axis, which ranges from 0 to π, and ϕ is the azimuthal angle (the angle between the x-axis and the projection onto the x-y plane of the arrow from the origin to P), which ranges from 0 to 2π. Here, r is the distance of the electron from the origin, and ranges from 0 to ∞.

The Schrödinger equation can be written out just as in Section 4.5, except in spherical polar coordinates, and the potential energy can be written as in Equation 3.2. The resulting equation is impressively complicated, but it nonetheless can be solved as in Section 4.6. The solution must be continuous in all three coordinates, and the radial portion must satisfy the boundary condition: $\psi \longrightarrow 0$ as $r \longrightarrow \infty$. This procedure leads naturally to quantization of the energy and the associated wave functions. We describe these parts of the solution in turn in the remainder of this section.

Energy Levels

Solutions of the Schrödinger equation for the one-electron atom exist only for particular values of the energy[1]:

$$E = E_n = -\frac{Z^2 e^4 m_e}{8\epsilon_0^2 n^2 h^2} \qquad n = 1, 2, 3, \ldots \qquad [5.1a]$$

In energy units of rydbergs (1 rydberg $= 2.18 \times 10^{-18}$ J), this equation becomes (see Eq. 4.9b):

$$E_n = -\frac{Z^2}{n^2} \quad \text{(rydberg)} \qquad n = 1, 2, 3, \ldots \qquad [5.1b]$$

The integer n, called the **principal quantum number,** indexes the individual energy levels. These are identical to the energy levels predicted by the Bohr theory. Here, however, quantization comes about naturally from the solution of the Schrödinger equation, rather than through an arbitrary assumption about the angular momentum.

The energy of a one-electron atom depends only on the principal quantum number n, because the potential energy depends only on the radial distance. The Schrödinger equation also quantizes L^2, the square magnitude of the angular momentum, as well as L_z, the projection of the angular momentum along the z-axis. (A review of elementary aspects of angular momentum in Appendix B provides useful background for this discussion.) Quantization of the square of the angular momentum as well as its projection along the z-axis requires two additional

[1]Strictly speaking, the electron mass m_e in the expressions for the energy levels of one-electron atoms should be replaced by the *reduced mass* μ, equal to $m_e m_N/(m_e + m_N)$, where m_N is the nuclear mass; μ differs from m_e by less than 0.1%.

quantum numbers. The **angular momentum quantum number** ℓ may take on any integral value from 0 to $n - 1$, and the angular momentum projection quantum number m may take on any integral value from $-\ell$ to ℓ. The quantum number m is referred to as the **magnetic quantum number** because its value governs the behavior of the atom in an external magnetic field. The allowed values of angular momentum L and its z projection are given by

$$L^2 = \ell(\ell + 1)\frac{h^2}{4\pi^2} \qquad \ell = 0, 1, \ldots, n - 1 \qquad [5.2a]$$

$$L_z = m\frac{h}{2\pi} \qquad m = -\ell, -\ell + 1, \ldots, 0, \ldots, \ell - 1, \ell \qquad [5.2b]$$

For $n = 1$ (the ground state), the only allowed quantum numbers are ($\ell = 0$, $m = 0$). For $n = 2$, there are $n^2 = 4$ allowed sets of quantum numbers:

$$(\ell = 0, m = 0), (\ell = 1, m = 1), (\ell = 1, m = 0), (\ell = 1, m = -1)$$

The restrictions on ℓ and m give rise to n^2 sets of quantum numbers for every value of n. Each set (n, ℓ, m) identifies a specific **quantum state** of the atom in which the electron has energy equal to E_n, angular momentum equal to $\sqrt{\ell(\ell + 1)}\, h/2\pi$, and z-projection of angular momentum equal to $mh/2\pi$. When $n > 1$, a total of n^2 specific quantum states correspond to the single energy level E_n; consequently, this set of states is said to be **degenerate.**

It is conventional to label specific states by replacing the angular momentum quantum number with a letter; we signify $\ell = 0$ with s, $\ell = 1$ with p, $\ell = 2$ with d, $\ell = 3$ with f, $\ell = 4$ with g, and on through the alphabet. Thus, a state with $n = 1$ and $\ell = 0$ is called a 1s state, one with $n = 3$ and $\ell = 1$ is a 3p state, one with $n = 4$ and $\ell = 3$ is a 4f state, and so forth. The letters s, p, d, and f derive from early (pre–quantum mechanics) spectroscopy, in which certain spectral lines were referred to as sharp, principal, diffuse, and fundamental. These terms are not used in modern spectroscopy, but the historical labels for the values of the quantum number ℓ are still followed. Table 5.1 summarizes the allowed combinations of quantum numbers.

The energy levels, including the degeneracy due to m and with states labeled by the spectroscopic notation, are conventionally displayed on a diagram as shown in Figure 5.2.

Wave Functions

For each quantum state (n, ℓ, m), solution of the Schrödinger equation provides a wave function of the form

$$\psi_{n\ell m}(r, \theta, \phi) = R_{n\ell}(r)Y_{\ell m}(\theta, \phi) \qquad [5.3]$$

TABLE 5.1 Allowed Values of Quantum Numbers for One-Electron Atoms

n	1	2		3		
ℓ	0	0	1	0	1	2
m	0	0	−1, 0, +1	0	−1, 0, +1	−2, −1, 0, +1, +2
Number of degenerate states for each ℓ	1	1	3	1	3	5
Number of degenerate states for each n	1	4		9		

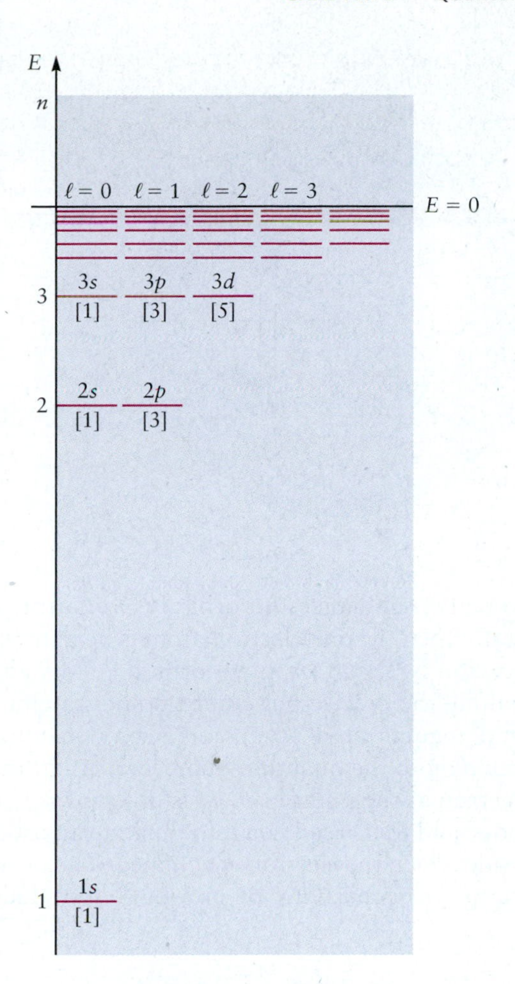

FIGURE 5.2 The energy-level diagram of the H atom predicted by quantum mechanics is arranged to show the degeneracy ($2\ell + 1$) for each value of ℓ.

in which the total wave function is the product of a radial part, $R_{n\ell}(r)$, and an angular part, $Y_{\ell m}(\theta, \phi)$. This product form is a consequence of the spherically symmetric potential energy function, and it enables separate examination of the angular and radial contributions to the wave function. The functions $Y_{\ell m}(\theta, \phi)$ are called **spherical harmonics.** They appear in many physical problems with spherical symmetry and were already well known before the advent of the Schrödinger equation. The angular motions of the electron described by θ and ϕ influence the *shape* of the wave function through the angular factor $Y_{\ell m}$, even though they do not influence the energy.

The wave function itself is not measured. It is to be viewed as an intermediate step toward calculating the physically significant quantity ψ^2, which is the probability density for locating the electron at a particular point in the atom. More precisely,

$$(\psi_{n\ell m})^2 dV = [R_{n\ell}(r)]^2[Y_{\ell m}(\theta, \phi)]^2 dV \qquad [5.4]$$

gives the probability of locating the electron within a small three-dimensional volume, dV, located at the position (r, θ, ϕ) when it is known that the atom is in the state n, ℓ, m. Specific examples are presented in succeeding paragraphs. The spherical volume element dV (Fig. 5.3) is defined as

$$dV = r^2 \sin\theta dr d\theta d\phi \qquad [5.5]$$

A wave function $\psi_{n\ell m}(r, \theta, \phi)$ for a one-electron atom in the state (n, ℓ, m) is called an **orbital.** This term recalls the circular orbits of the Bohr atom, but there is

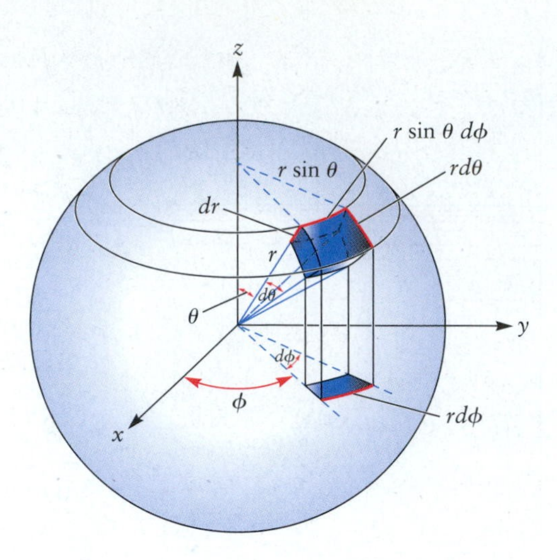

FIGURE 5.3 The differential volume element in spherical polar coordinates.

no real resemblance. An orbital is *not* a trajectory traced by an individual electron. When the one-electron atom is in state (n, ℓ, m), it is conventional to say the electron is "in an (n, ℓ, m) orbital." This phrase is merely a shorthand way of making the precise but cumbersome statement: "When an electron has energy, total angular momentum, and z component of angular momentum values corresponding to the quantum numbers n, ℓ, m, the probability density of finding the electron at the point (r, θ, ϕ) is given by $\psi^2_{n\ell m}(r, \theta, \phi)$." Do not allow this verbal shorthand to mislead you into thinking an orbital is some sort of "region in space" inside which the electron is confined. The orbitals are labeled 1*s*, 2*s*, 2*p*, 3*s*, ... by the spectroscopic notation previously introduced.

EXAMPLE 5.1

Give the names of all the orbitals with $n = 4$, and state how many m values correspond to each type of orbital.

SOLUTION

The quantum number ℓ may range from 0 to $n - 1$; thus, its allowed values in this case are 0, 1, 2, and 3. The labels for the groups of orbitals are then:

$\ell = 0$	4*s*	$\ell = 2$	4*d*
$\ell = 1$	4*p*	$\ell = 3$	4*f*

The quantum number m ranges from $-\ell$ to $+\ell$; thus, the number of m values is $2\ell + 1$. This gives one 4*s* orbital, three 4*p* orbitals, five 4*d* orbitals, and seven 4*f* orbitals for a total of $16 = 4^2 = n^2$ orbitals with $n = 4$. They all have the same energy, but they differ in shape.

Related Problems: 3, 4

Sizes and Shapes of Orbitals

The sizes and shapes of the hydrogen atom orbitals are important in chemistry because they provide the foundations for the quantum description of chemical bonding and the molecular shapes to which it leads. Sizes and shapes of the orbitals are revealed by graphical analysis of the wave functions, of which the first few are given in Table 5.2. Note that the radial functions are written in terms of the dimensionless variable σ, which is the ratio of Zr to a_0. For $Z = 1$, $\sigma = 1$ at the radius of the first Bohr orbit of the hydrogen atom.

TABLE 5.2 Angular and Radial Parts of Wave Functions for One-Electron Atoms

Angular Part $Y(\theta, \phi)$	Radial Part $R_{n\ell}(r)$
$\ell = 0$: $Y_s = \left(\frac{1}{4\pi}\right)^{1/2}$	$R_{1s} = 2\left(\frac{Z}{a_0}\right)^{3/2} \exp(-\sigma)$
	$R_{2s} = \frac{1}{2\sqrt{2}}\left(\frac{Z}{a_0}\right)^{3/2}(2 - \sigma)\exp(-\sigma/2)$
	$R_{3s} = \frac{2}{81\sqrt{3}}\left(\frac{Z}{a_0}\right)^{3/2}(27 - 18\sigma + 2\sigma^2)\exp(-\sigma/3)$
$\ell = 1$: $Y_{p_x} = \left(\frac{3}{4\pi}\right)^{1/2} \sin\theta\cos\phi$	$R_{2p} = \frac{1}{2\sqrt{6}}\left(\frac{Z}{a_0}\right)^{3/2}\sigma\exp(-\sigma/2)$
$Y_{p_y} = \left(\frac{3}{4\pi}\right)^{1/2} \sin\theta\sin\phi$	$R_{3p} = \frac{4}{81\sqrt{6}}\left(\frac{Z}{a_0}\right)^{3/2}(6\sigma - \sigma^2)\exp(-\sigma/3)$
$Y_{p_z} = \left(\frac{3}{4\pi}\right)^{1/2} \cos\theta$	
$\ell = 2$: $Y_{d_{z^2}} = \left(\frac{5}{16\pi}\right)^{1/2}(3\cos^2\theta - 1)$	
$Y_{d_{xz}} = \left(\frac{15}{4\pi}\right)^{1/2} \sin\theta\cos\theta\cos\phi$	
$Y_{d_{yz}} = \left(\frac{15}{4\pi}\right)^{1/2} \sin\theta\cos\theta\sin\phi$	$R_{3d} = \frac{4}{81\sqrt{30}}\left(\frac{Z}{a_0}\right)^{3/2}\sigma^2\exp(-\sigma/3)$
$Y_{d_{xy}} = \left(\frac{15}{16\pi}\right)^{1/2} \sin^2\theta\sin 2\phi$	
$Y_{d_{x^2-y^2}} = \left(\frac{15}{16\pi}\right)^{1/2} \sin^2\theta\cos 2\phi$	

$$\sigma = \frac{Zr}{a_0} \qquad a_0 = \frac{\epsilon_0 h^2}{\pi e^2 m_e} = 0.529 \times 10^{-10}\ \text{m}$$

Graphical representation of the orbitals requires some care. Equation 5.3 tells us simply to go to the point (r, θ, ϕ), evaluate the wave function there, and draw a graph showing the results of visiting many such points. But all three spatial dimensions have already been used to define the location; thus we would need a fourth dimension to display the value of the wave function at that point. Alternatively, we could create a table of numbers giving the value of the wave function at each point (r, θ, ϕ), but it would be difficult to develop intuition about shapes and structures from this table. We get around these problems by slicing up three-dimensional space into various two- and one-dimensional regions and examining the wave function at each point in these regions. For example, suppose in Figure 5.1 we look only at points in the x-y plane and evaluate the wave function at each point. Then, just as we did for the three-dimensional particle in a box in Figure 4.28, we can generate a contour map that represents the shape of the hydrogen wave function at locations in the x-y plane. We can generate contour maps in any other "cut planes" in the same way. A second approach is to look only at the radial behavior. We start at the origin in Figure 5.1 and move out along the direction (r, θ, ϕ), holding (θ, ϕ) constant, and plot the wave function at each value of r. We rely on contour plots to display the angular shapes of the wave function, and on two-dimensional graphs of the wave function versus r to display radial behavior. We use images that combine these angular and radial effects to display size and shape in three-dimensional space. We always state the conditions and limitations of such three-dimensional images. Because the wave function is a four-dimensional object, its appearance in three-dimensional representations depends strongly on choices made by the illustrator. Be certain you understand these choices in each image you examine (or create!).

s ORBITALS Let's begin with **s orbitals,** corresponding to $\psi_{n\ell m}$ with $\ell = 0$ (therefore, $m = 0$ as well). For all s orbitals, the angular part Y is a constant (see Table 5.2). Because ψ does not depend on either θ or ϕ, all s orbitals are spherically symmetric about the nucleus. This means that the amplitude of an s orbital (and therefore also the probability of finding the electron near some point in space) depends only on its distance, r, from the nucleus and not on its direction in space. There are several ways to visualize the radial variation of the ns orbitals with $n = 1, 2, 3, \ldots$, and the probability density they describe.

One way (Fig. 5.4a) is to prepare a contour plot in the x-y plane. The contours are circles because the wave function does not depend on the angles. The outermost contour identifies points at which the amplitude of ψ is 5% of its maximum value. The second circle identifies points with ψ at 10% of the maximum, and so on in steps of 20% to the innermost contour, which identifies points with ψ at 90% of the maximum. Contours with positive phase (positive sign for the amplitude) are shown in red, whereas negative phase (negative sign for the amplitude) is represented in blue. Note that these contours, which identify uniform increases in magnitude, become much closer together as we approach the origin. This indicates the amplitude of the wave function increases rapidly as we approach the nucleus. The x-y plane corresponds to $\theta = \pi/2$ in Figure 5.1. Because the wave function does not depend on angles, this same contour plot could be obtained by tilting the x-y plane to any value of θ. It could also be obtained by starting with the x-z plane and the y-z plane and tilting either of them to any value of θ. Therefore, we can rotate the contour plot in Figure 5.4a to generate a set of concentric spheres in three dimensions. Each sphere identifies a surface of points in (r, θ, ϕ) at each of which the wave function has the same value. These spheres are called *isosurfaces* because the amplitude of the wave function has the same value at each point on them.

A second way to visualize wave functions (see Fig. 5.4b) is to plot their radial portions directly: $\psi_{n00} \propto R_{n0}(r)$. These plots give the amplitude of the wave function at a certain distance from the nucleus. For 2s and 3s orbitals, the wave function has lobes with positive and negative phases. The transition between positive and negative lobes is a node, at which the value of the wave function is zero.

A third way (see Fig. 5.4c) is to plot the **radial probability density** $r^2[R_{n0}(r)]^2$. This is the probability density of finding the electron at any point in space at a distance r from the nucleus for all angles θ and ϕ. More precisely, the product $r^2[R_{n0}(r)]^2dr$ gives the probability of finding the electron anywhere within a thin spherical shell of thickness dr, located at distance r from the nucleus. This spherical shell is easily visualized with the aid of Figure 5.3. As the angle ϕ runs through its entire range from 0 to 2π, a circular annulus of width dr located between r and $r + dr$ is traced out in the x-y plane. This annulus will become a spherical shell as the angle θ runs through its range from 0 to π. The factor r^2 in front accounts for the increasing volume of spherical shells at greater distances from the nucleus. The radial probability distribution is small near the nucleus, where the shell volume (proportional to r^2) is small, and reaches its maximum value at the distance where the electron is most likely to be found.

Finally, we want to plot the size of the orbital. What is meant by the *size* of an orbital? Strictly speaking, the wave function of an electron in an atom stretches out to infinity, so an atom has no clear boundary. We might define the size of an atom as the contour at which ψ^2 has fallen off to some particular numerical value, or as the extent of a "balloon skin" inside which some definite fraction (for example, 90% or 99%) of the probability density of the electron is contained. We have adopted the former practice in representing the size of orbitals. Figure 5.4d shows spheres generated from the contours at which the wave functions for the 1s, 2s, and 3s orbitals of hydrogen have fallen to 0.05 of their maximum amplitude. Figure 5.4d shows that the size of an orbital increases with increasing quantum number n. A 3s orbital is larger than a 2s orbital, which, in turn, is larger than a 1s orbital. This is the quantum analog of the increase in radius of the Bohr orbits with increasing n.

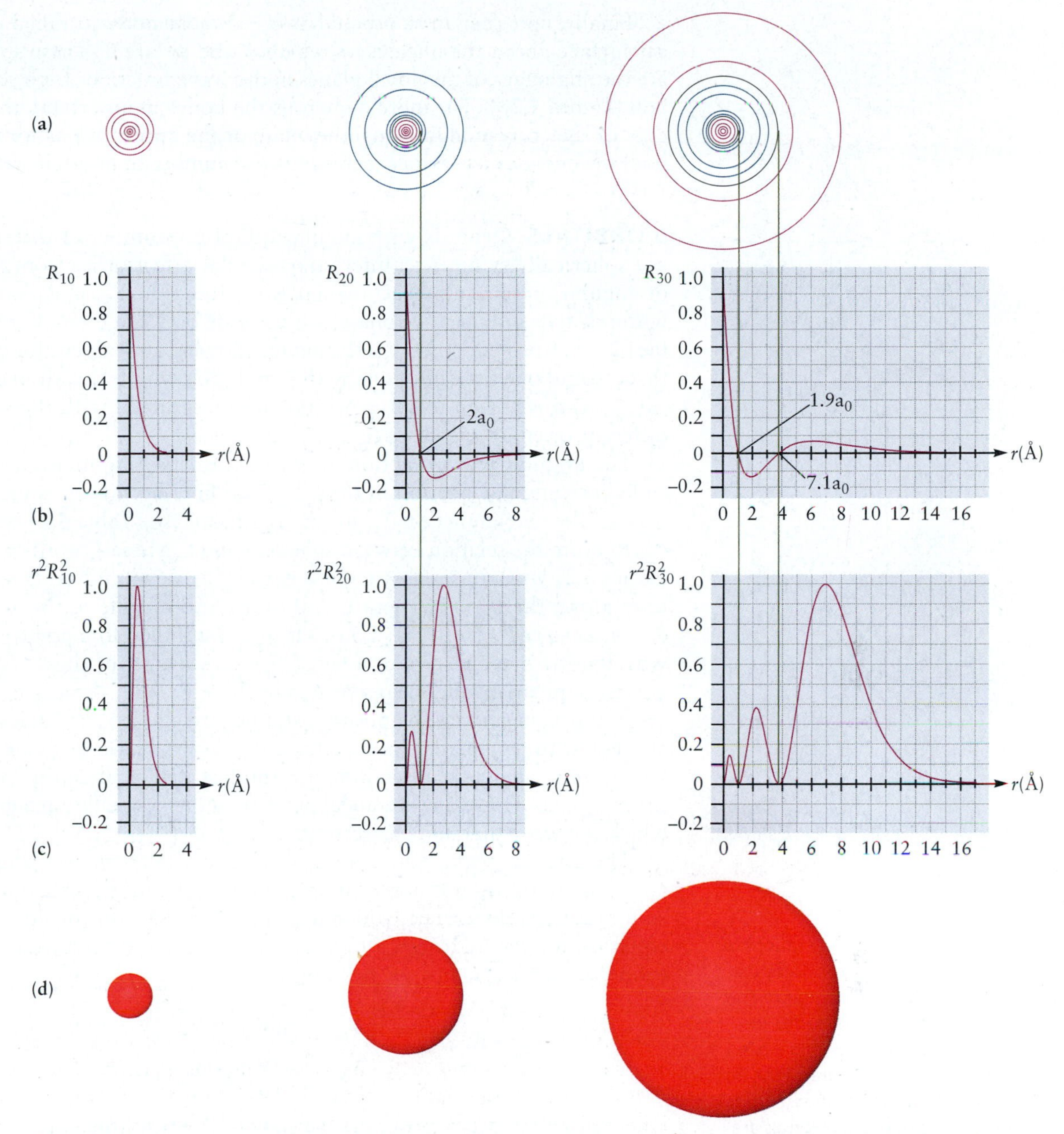

FIGURE 5.4 Four representations of hydrogen *s* orbitals. (a) A contour plot of the wave function amplitude for a hydrogen atom in its 1*s*, 2*s*, and 3*s* states. The contours identify points at which ψ takes on ± 0.05, ± 0.1, ± 0.3, ± 0.5, ± 0.7, and ± 0.9 of its maximum value. Contours with positive phase are shown in red; those with negative phase are shown in blue. Nodal contours, where the amplitude of the wave function is zero, are shown in black. They are connected to the nodes in the lower plots by the vertical green lines. (b) The radial wave functions plotted against distance from the nucleus, *r*. (c) The radial probability density, equal to the square of the radial wave function multiplied by r^2. (d) The "size" of the orbitals, as represented by spheres whose radius is the distance at which the probability falls to 0.05 of its maximum value.

Another measure of the size of an orbital is the most probable distance of the electron from the nucleus in that orbital. Figure 5.4c shows that the most probable location of the electron is progressively farther from the nucleus in *ns* orbitals for larger *n*. Nonetheless, there is a finite probability for finding the electron at the nucleus in both 2*s* and 3*s* orbitals. This happens because electrons in *s* orbitals have no angular momentum ($\ell = 0$), and thus can approach the nucleus along the radial direction. The ability of electrons in *s* orbitals to "penetrate" close to the nucleus has important consequences in the structure of many-electron atoms and molecules (see later).

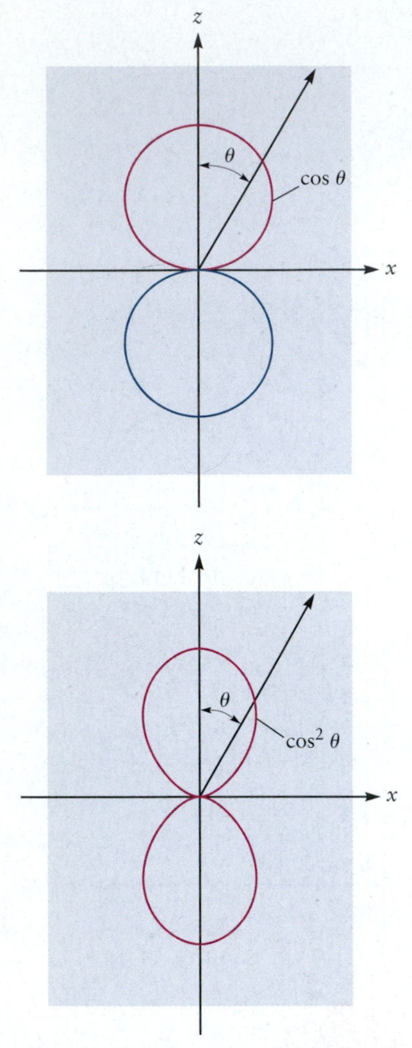

FIGURE 5.5 Two representations of hydrogen p orbitals. (a) The angular wave function for the p_z orbital. The p_x and p_y orbitals are the same, but are oriented along the x- and y-axis, respectively. (b) The square of the angular wave function for the p_z orbital. Results for the p_x and p_y orbitals are the same, but are oriented along the x- and y-axis, respectively.

Finally, note that an ns orbital has $n - 1$ **radial nodes;** a radial node is a spherical surface about the nucleus on which ψ and ψ^2 are 0. These spherical surfaces are the analogues of the nodal planes in the wave functions for a particle in a cubic box (Figure 4.28). The more numerous the nodes in an orbital, the higher the energy of the corresponding quantum state of the atom. Just as for the particle in a box, the energies of orbitals increase as the number of nodes increases.

***p* ORBITALS** Orbitals with angular quantum numbers ℓ different from 0 are not spherically symmetric. Interesting angular effects arise from the quantization of angular momentum. The angular wave function $Y_{\ell m}(\theta, \phi)$ has separate lobes with positive and negative phase, with a node between them. Equation 5.2b specifies $2\ell + 1$ projections of these angular momentum values along the z-axis. The three angular wave functions with $\ell = 1$, for which the allowed m values are -1, 0, and $+1$, lead to three orbitals (the ***p* orbitals**) with the same shapes but different orientations in space.

The angular wave function $Y_{10}(\theta, \phi)$ with the combination $\ell = 1$, $m = 0$ is called the angular portion of the p_z orbital here because it is oriented along the z-axis. The wave function Y_{p_z} for the p_z orbital (see Table 5.2) is proportional to $\cos\theta$. From the relation between spherical and Cartesian coordinates illustrated in Figure 5.1, you can see that $\cos\theta \propto z$; thus, this orbital has its maximum amplitude along the z-axis (where $\theta = 0$ or π) and a node in the x-y plane (where $\theta = \pi/2$, so $\cos\theta = 0$) (Fig. 5.5a). The p_z orbital therefore points along the z-axis, with its positive phase (red in Fig. 5.5a) on the side of the x-y plane where the z-axis is positive and negative phase (blue in Fig. 5.5a) on the side where the z-axis is negative. The positive and negative lobes are circles tangent to one another at the x-y plane. An electron in a p_z orbital has the greatest probability of being found at significant values of z and has zero probability of being found in the x-y plane. This plane is a nodal plane or, more generally, an **angular node** across which the wave function changes sign.

The angular wave functions $Y_{11}(\theta, \phi)$ for $\ell = 1$, $m = 1$ and $Y_{1,-1}(\theta, \phi)$ for $\ell = 1$, $m = -1$ do not have a simple geometrical interpretation. However, their sum and their difference, which are also allowed solutions of the Schrödinger equation for the hydrogen atom, do have simple interpretations. Therefore, we form two new angular wave functions:

$$
\begin{aligned}
Y_{p_x} &= c_1(Y_{11} + Y_{1,-1}) \\
Y_{p_y} &= c_2(Y_{11} - Y_{1,-1})
\end{aligned}
\qquad [5.6]
$$

where c_1 and c_2 are appropriate constants. The resulting expressions for Y_{p_x} and Y_{p_y} are given in Table 5.2. A comparison of these expressions with Figure 5.1 shows that Y_{p_x} lies along the x-axis and Y_{p_y} lies along the y-axis. The angular wave functions Y_{p_x} for the p_x orbital and Y_{p_y} for the p_y orbital thus have the same shape as Y_{p_z}, but point along the x- and y-axis, respectively. They have nodes at the y-z and x-z planes, respectively.

It is informative to examine the angular dependence of the probability density in the p orbitals, starting with p_z. The probability density for finding the electron at the position (θ, ϕ) with r constant is given by $Y_{p_x}^2$, the square of the angular wave function (see Fig. 5.5b). Notice that general shape is the same as Y_{p_z}, but the lobes are no longer circular. This happens because the values of $\cos\theta$, which are less than 1 except where $\theta = 0, \pi, -\pi$, become even smaller when squared and shrink the envelope away from the circular shape. The behavior of $Y_{p_x}^2$ and $Y_{p_y}^2$ are the same as $Y_{p_x}^2$. The radial parts of the np wave functions (represented by $R_{n\ell}$) are illustrated in Figure 5.6.

The p orbitals, like the s orbitals, may have radial nodes, at which the probability density vanishes at certain distances from the nucleus regardless of direction.

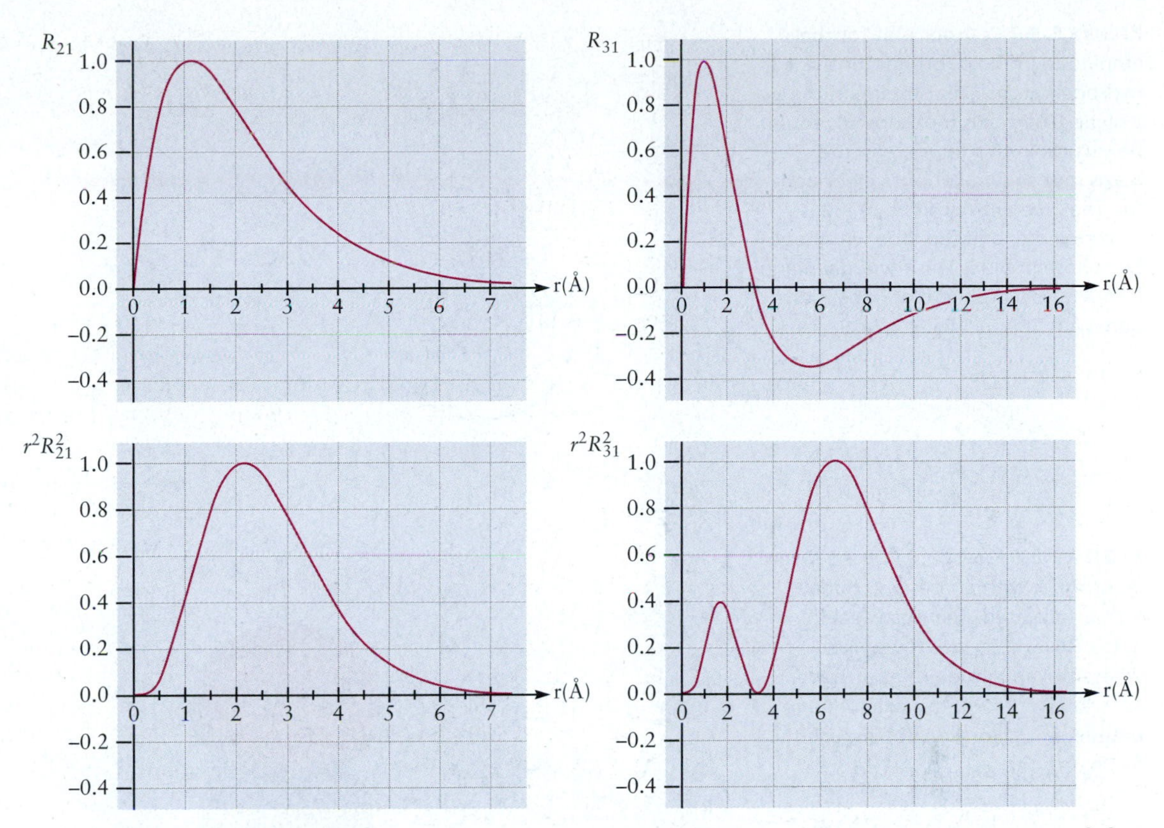

FIGURE 5.6 Radial wave functions $R_{n\ell}$ for *np* orbitals and the corresponding radial probability densities $r^2R_{n\ell}^2$.

From Figure 5.6, the R_{21} wave function has no radial nodes, and the R_{31} function has one radial node; the R_{41} (not shown) function has two radial nodes. The $R_{n\ell}$ wave functions have n-ℓ-1 radial nodes. Because the angular part of the *np* wave function always has a nodal plane, the total wave function has $n - 1$ nodes ($n - 2$ radial and 1 angular), which is the same number as an *s* orbital with the same principal quantum number. The $R_{21}(r)$ function (that is, R_{2p}) in Table 5.2 contains the factor σ, which is proportional to r ($\sigma = Zr/a_0$) and causes it to vanish at the nucleus. This is true of all the radial wave functions except the *ns* functions, and it means that the probability is zero for the electron to be at the nucleus for all wave functions with $\ell > 0$ (*p*, *d*, *f*, ...). Physically, electrons with angular momentum are moving around the nucleus, not toward it, and cannot "penetrate" toward the nucleus.

Finally, we combine the angular and radial dependence to get a sense of the shape of the complete orbital, $\psi_{n\ell m} = R_{n\ell}Y_{\ell m}$. Let's examine the $2p_z$ orbital at points (r, θ, ϕ) confined to the *x-z* plane. At each point, we calculate the value of R_{21} (as in Fig. 5.6) and the value of Y_{2p_z} (as in Fig. 5.5a). Then we multiply these values together to obtain the value of ψ_{2p_z} at that point. We continue this process and generate a contour plot for ψ_{2p_z} in the *x-z* plane. The results are shown in Figure 5.7. Contours identify points at which ψ_{2p_z} takes on ±0.1, ±0.3, ±0.5, ±0.7, and ±0.9 of its maximum value. Contours with positive phase are shown in red; blue contours represent negative phase. The radial wave function from Figure 5.6 has dramatically changed the circular angular wave function from Figure 5.5a. The circles have been flattened, especially on the sides nearest the *x-y* plane. The contours are not concentric, but rather bunch together on the sides nearest the *x-y* plane. This reflects the rapid decrease in amplitude near the nucleus (see Fig. 5.6) and the much slower decrease in amplitude at longer distances beyond the maximum in the radial function.

Finally, we can represent the $2p_z$ orbital as a three-dimensional object by rotating Figure 5.7 about the *z*-axis. Each of the closed contours in Figure 5.7 will then trace out a three-dimensional isosurface on which all the points (r, θ, ϕ) have the

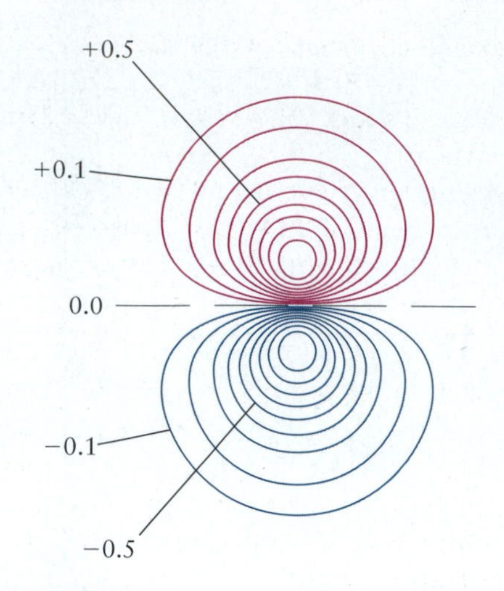

FIGURE 5.7 Contour plot for the amplitude in the p_z orbital for the hydrogen atom. This plot lies in the *x*-*z* plane. The *z*-axis (not shown) would be vertical in this figure, and the *x*-axis (not shown) would be horizontal. The lobe with positive phase is shown in red, and the lobe with negative phase in blue. The *x*-*y* nodal plane is shown as a dashed black line. Compare with Figure 5.5a.

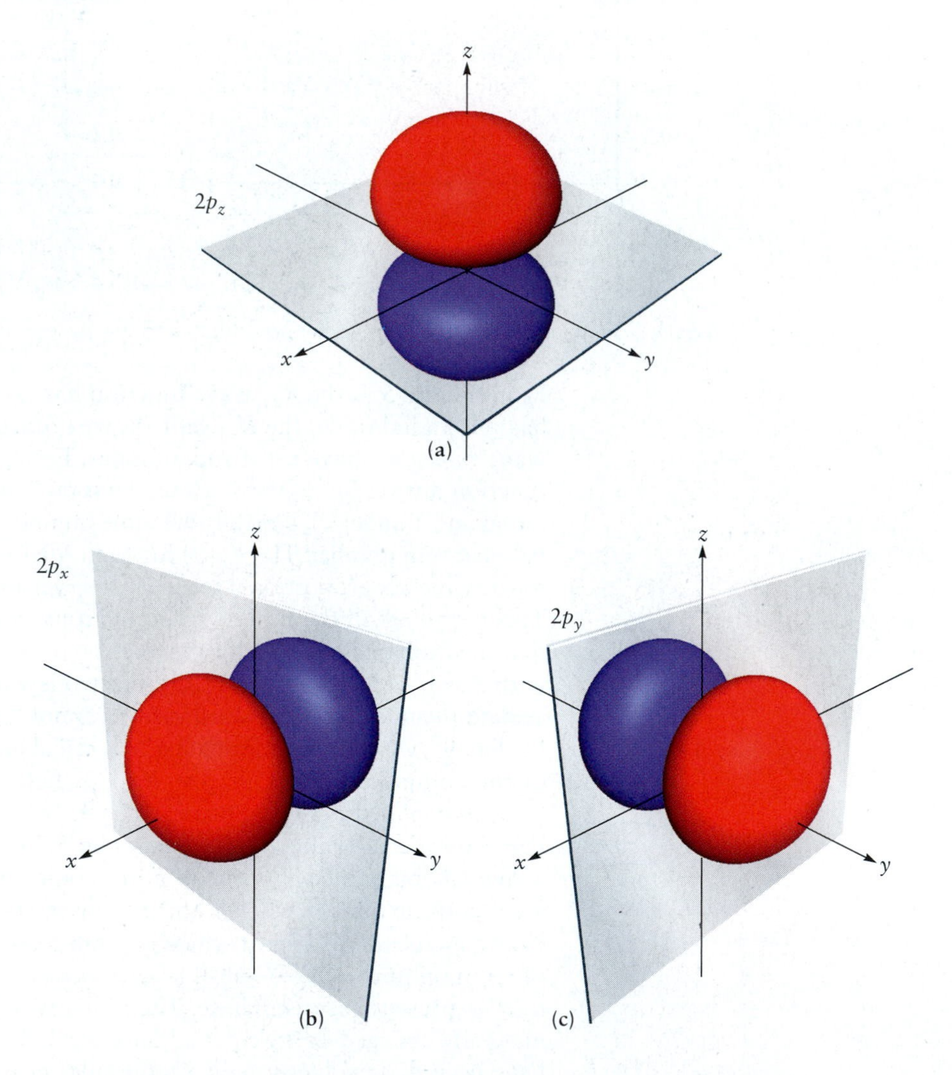

FIGURE 5.8 The shapes of the three 2*p* orbitals, with phases and nodal planes indicated. The isosurfaces in (a), (b), and (c) identify points where the amplitude of each wave function is ±0.2 of its maximum amplitude. (a) $2p_z$ orbital. (b) $2p_x$ orbital. (c) $2p_y$ orbital.

same amplitude and phase of the wave function. The same analysis generates isosurfaces for $2p_x$ and $2p_y$. Figure 5.8abc shows plots of all three, with the phases and nodal planes indicated, as isosurfaces at ±0.2 times the maximum amplitude.

Each 2*p* orbital appears, loosely speaking, as a pair of flattened and distorted hemispheres, with opposite phase, facing each other across their nodal plane. You can see how this shape would change dramatically if we selected for display isosurfaces with other values of amplitude. Sometimes you see the 2*p* orbitals represented (in a manner similar to Fig. 5.8) with no explanation of how the images were obtained. Although such pictures reliably indicate the gross shape and extent of the orbitals, they obscure completely all the rich structure *inside* this skin, in particular the location—even the existence—of the maximum. We encourage you to develop your intuition for orbitals by looking at detailed contour plots for angular information, and then visualizing three-dimensional isosurfaces at whatever value of amplitude is most appropriate for the problem you are investigating at the moment.

***d* ORBITALS** When $\ell = 2$ Equation 5.2b specifies five projections of the angular momentum along the z-axis. As with the p orbitals, we take linear combinations of the angular wave functions to obtain orbitals with specific orientations relative to the Cartesian axes. The conventionally chosen linear combinations of the solutions with $m = -2, -1, +1, +2$ give four orbitals with the same shape but different orientations with respect to the Cartesian axes: d_{xy}, d_{yz}, d_{xz}, and $d_{x^2-y^2}$ (Fig. 5.9).

For example, a d_{xy} orbital has four lobes, two with positive phase and two with negative phase; the maximum amplitude is at 45° to the x- and y-axes. The $d_{x^2-y^2}$ orbital has maximum amplitude along the x- and y-axes. The "down-axis" view of these four d orbitals, illustrated for one of them (d_{xy}) in Figure 5.9, shows that they all have the same shape when viewed down the appropriate axis. The fifth orbital, d_{z^2}, which corresponds to $m = 0$, has a different shape from the rest, with maximum amplitude along the z-axis and a little "doughnut" in the x-y plane. Each d orbital has two angular nodes (for example, the d_{xy} orbital has the x-z and y-z planes as its nodal surfaces). The radial functions, $R_{n2}(r)$, have $n - 3$ radial nodes, giving once again $n - 1$ total nodes.

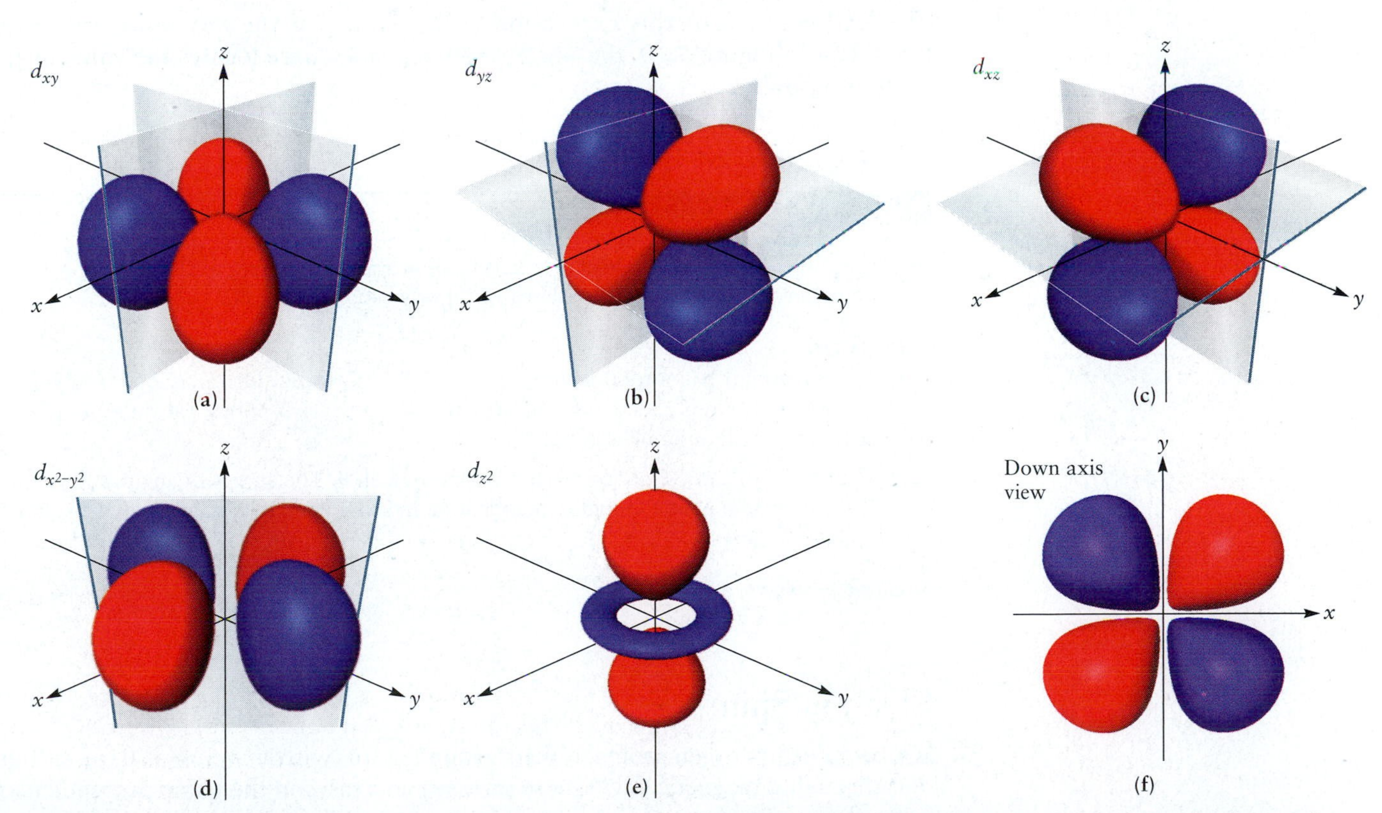

FIGURE 5.9 The shapes of the five 3d orbitals, with phases and nodal surfaces indicated. The "down-axis" view shows the shapes of the first four orbitals (a)–(d) when viewed down the appropriate axis; the specific example shown here is the d_{xy} orbital viewed down the z-axis.

The wave functions for f orbitals and orbitals of higher ℓ can be calculated, but they play a smaller role in chemistry than do the s, p, and d orbitals.

We summarize the important features of orbital shapes and sizes as follows:

1. For a given value of ℓ, an increase in n leads to an increase in the average distance of the electron from the nucleus, and therefore in the size of the orbital (see Figs. 5.4 and 5.6).
2. An orbital with quantum numbers n and ℓ has ℓ angular nodes and $n - \ell - 1$ radial nodes, giving a total of $n - 1$ nodes. An angular node is defined by a plane. A radial node is defined by a spherical surface. For a one-electron atom or ion, the energy depends only on the number of nodes—that is, on n but not on ℓ or m. The energy increases as the number of nodes increases.
3. As r approaches 0, $\psi(r, \theta, \phi)$ vanishes for all orbitals except s orbitals; thus, only an electron in an s orbital can "penetrate to the nucleus," that is, have a finite probability of being found right at the nucleus.

The next section shows that these general statements are important for determining the electronic structure of many-electron atoms even though they are deduced from the one-electron case.

The characteristics of the orbitals of a one-electron atom (or ion) are especially well displayed by a quantitative plot showing s, p, and d orbitals all on the same scale (Fig. 5.10). The best quantitative measure of the size of an orbital is $\bar{r}_{n\ell}$, the average value of the distance of the electron from the nucleus in that orbital. Quantum mechanics calculates $\bar{r}_{n\ell}$ as

$$\bar{r}_{n\ell} = \frac{n^2 a_0}{Z}\left\{1 + \frac{1}{2}\left[1 - \frac{\ell(\ell + 1)}{n^2}\right]\right\} \qquad [5.7]$$

The leading term of this expression is the radius of the nth Bohr orbit (see Eq. 4.12). In Figure 5.10, the small arrow on each curve locates the value of $\bar{r}_{n\ell}$ for that orbital.

EXAMPLE 5.2

Compare the $3p$ and $4d$ orbitals of a hydrogen atom with respect to the (a) number of radial and angular nodes and (b) energy of the corresponding atom.

SOLUTION

(a) The $3p$ orbital has a total of $n - 1 = 3 - 1 = 2$ nodes. Of these, one is angular ($\ell = 1$) and one is radial. The $4d$ orbital has $4 - 1 = 3$ nodes. Of these, two are angular ($\ell = 2$) and one is radial.

(b) The energy of a one-electron atom depends only on n. The energy of an atom with an electron in a $4d$ orbital is higher than that of an atom with an electron in a $3p$ orbital, because $4 > 3$.

Related Problems: 5, 6

Electron Spin

If a beam of hydrogen atoms in their ground state (with $n = 1$, $\ell = 0$, $m = 0$) is sent through a magnetic field whose intensity increases in the plane perpendicular to the flight of the beam, it splits into two beams, each containing half of the atoms (Fig. 5.11). The pioneering experiment of this type is called the Stern–Gerlach experiment after the German physicists who performed it, Otto Stern and Walther Gerlach.

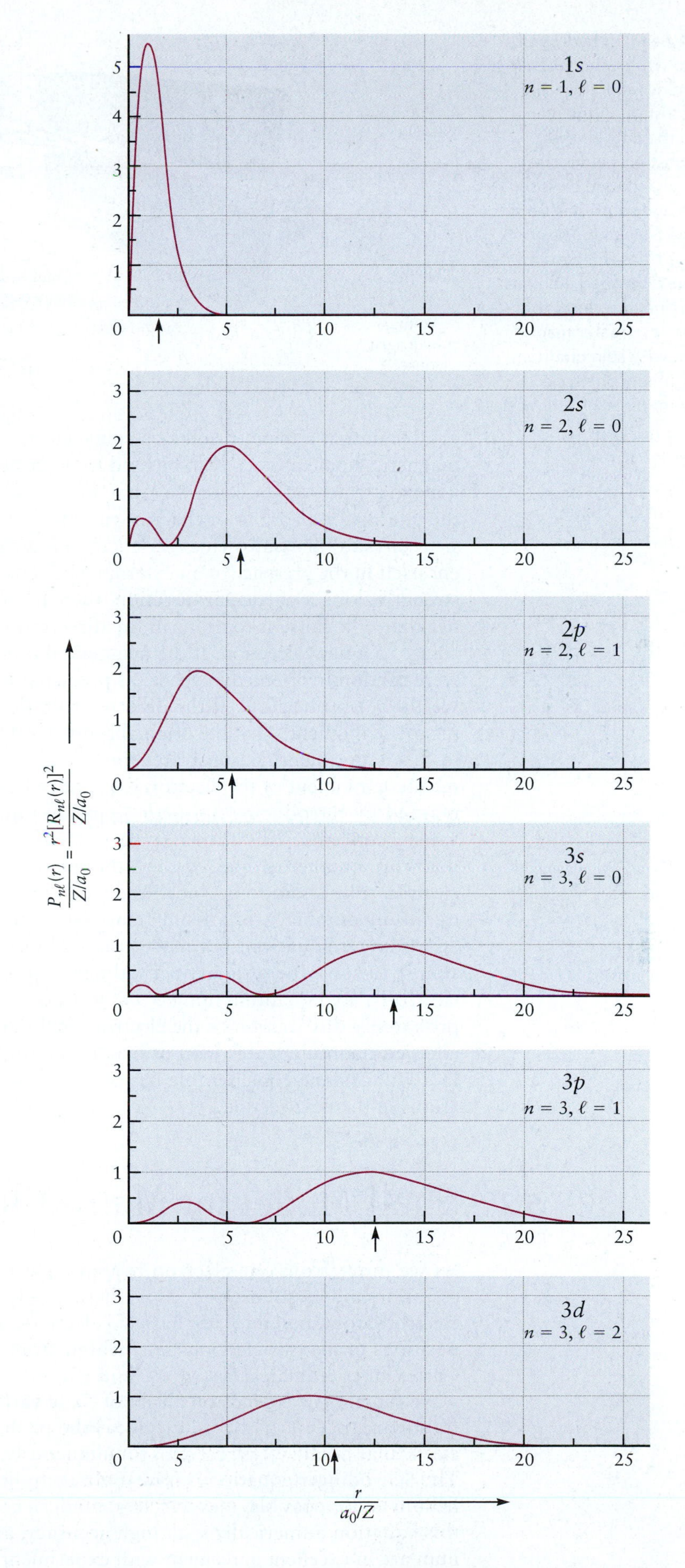

FIGURE 5.10 Dependence of radial probability densities on distance from the nucleus for one-electron orbitals with $n = 1, 2, 3$. The small arrow below each curve locates the value of $\bar{r}_{n\ell}$ for that orbital. The distance axis is expressed in the same dimensionless variable introduced in Table 5.1. The value 1 on this axis is the first Bohr radius for the hydrogen atom. Because the radial probability density has dimensions $(\text{length})^{-1}$, the calculated values of $r^2[R_{n\ell}(r)]^2$ are divided by $(a_0)^{-1}$ to give a dimensionless variable for the probability density axis.

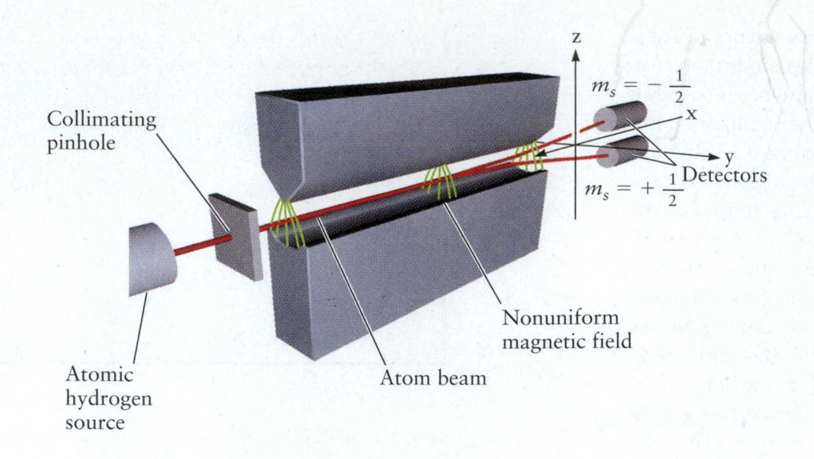

FIGURE 5.11 A beam of hydrogen atoms is split into two beams when it traverses a magnetic field, the value of which is not constant in the plane perpendicular to the path of the beam. The nonconstant field is created by the specially shaped cross section of the north and south poles of the magnet in the *z-y* plane. The green curved lines trace the pattern over which the field is varied. Regions where the green lines are closer together are regions of greater magnetic field. Atoms with spin quantum number $m_s = +\frac{1}{2}$ follow one trajectory, and those with $m_s = -\frac{1}{2}$ follow another.

Recall that a magnet, unlike a single electric charge, has two poles, and that a magnetic dipole moment can be used to describe the interactions of a magnet with a magnetic field in the same way that an electric dipole moment is used to describe the interaction of a pair of charges with an electric field. You may recall from classical physics (or your own experience) that a small bar magnet will rotate to orient itself in the presence of an external magnetic field. If the magnetic field changes strength along a particular direction, then a force will be exerted on the bar magnet that will cause it to *move* in the direction of the changing field, and not just rotate to a new direction. If the magnetic dipole moments of the hydrogen atoms were randomly oriented in space (as predicted by classical physics), then the beam would be smeared out at the detector to reflect all possible orientations of the magnetic moment. That the original beam is split into only two well-defined beams in this experiment demonstrates the unexpected fact that the *orientation* of the magnetic moment of the electron is quantized. The result of this experiment is explained by introducing a *fourth* quantum number, m_s, which can take on two values, conventionally chosen to be $+\frac{1}{2}$ and $-\frac{1}{2}$. For historical reasons, the fourth quantum number is referred to as the **spin quantum number.** When $m_s = +\frac{1}{2}$, the electron spin is said to be "up," and when $m_s = -\frac{1}{2}$, the spin is "down." The spin quantum number arises from relativistic effects that are not included in the Schrödinger equation. For most practical purposes in chemistry, it is sufficient simply to solve the ordinary Schrödinger equation, and then associate with each electron a spin quantum number $m_s = +\frac{1}{2}$ or $-\frac{1}{2}$ which does not affect the spatial probability distribution of the electron. Including the spin doubles the number of allowed quantum states with principal quantum number *n*, from n^2 to $2n^2$. This fact will assume considerable importance when considering the many-electron atoms in the next section.

5.2 Shell Model for Many-Electron Atoms

As we move from one-electron to many-electron atoms, both the Schrödinger equation and its solutions become increasingly complicated. The simplest many-electron atom, helium (He), has two electrons and a nuclear charge of $+2e$. The positions of the two electrons in a helium atom can be described using two sets of Cartesian coordinates, (x_1, y_1, z_1) and (x_2, y_2, z_2), relative to the same origin. The wave function ψ depends on all six of these variables: $\psi = \psi(x_1, y_1, z_1, x_2, y_2, z_2)$. Its square, $\psi^2(x_1, y_1, z_1, x_2, y_2, z_2)$, is the probability density of finding the first electron at point (x_1, y_1, z_1) and, simultaneously, the second electron at (x_2, y_2, z_2). The Schrödinger equation is now more complicated, and an explicit solution for helium is not possible. Nevertheless, modern computers have enabled us to solve this equation numerically with high accuracy, and the predicted properties of helium are in excellent agreement with experiment.

Although these numerical calculations demonstrate conclusively the usefulness of the Schrödinger equation for predicting atomic properties, they suffer from two defects. First, they are somewhat difficult to interpret physically, and second, they become increasingly difficult to solve, even numerically, as the number of electrons increases. As a result, approximate approaches to the many-electron Schrödinger equation have been developed.

Hartree Orbitals

The **self-consistent field (SCF) orbital approximation method** developed by Hartree is especially well suited for applications in chemistry. Hartree's method generates a set of approximate one-electron orbitals, φ_α, and associated energy levels, ε_α, reminiscent of those for the H atom. The subscript α represents the appropriate set of quantum numbers (see later in this chapter for a definition). The electronic structure of an atom with atomic number Z is then "built up" by placing Z electrons into these orbitals in accordance with certain rules (see later in this chapter for descriptions of these rules).

In this section, we introduce Hartree's method and use it to describe the electron arrangements and energy levels in many-electron atoms. Later sections detail how this approximate description rationalizes periodic trends in atomic properties and serves as a starting point for descriptions of chemical bond formation.

For any atom, Hartree's method begins with the exact Schrödinger equation in which each electron is attracted to the nucleus and repelled by all the other electrons in accordance with the Coulomb potential. The following three simplifying assumptions are made immediately:

1. Each electron moves in an *effective field* created by the nucleus and all the other electrons, and the effective field for electron i depends only on its position r_i.
2. The effective field for electron i is obtained by averaging its Coulomb potential interactions with each of the other electrons over all the positions of the other electrons so that r_i is the only coordinate in the description.
3. The effective field is spherically symmetric; that is, it has no angular dependence.

Under the first assumption, each electron moves as an independent particle and is described by a one-electron orbital similar to those of the hydrogen atom. The wave function for the atom then becomes a product of these one-electron orbitals, which we denote $\varphi_\alpha(r_i)$. For example, the wave function for lithium (Li) has the form $\psi_{\text{atom}} = \varphi_\alpha(r_1)\varphi_\beta(r_2)\varphi_\gamma(r_3)$. This product form is called **the orbital approximation for atoms.** The second and third assumptions in effect convert the exact Schrödinger equation for the atom into a set of simultaneous equations for the unknown effective field and the unknown one-electron orbitals. These equations must be solved by iteration until a self-consistent solution is obtained. (In spirit, this approach is identical to the solution of complicated algebraic equations by the method of iteration described in Appendix C.) Like any other method for solving the Schrödinger equation, Hartree's method produces two principal results: energy levels and orbitals.

These Hartree orbitals resemble the atomic orbitals of hydrogen in many ways. Their angular dependence is identical to that of the hydrogen orbitals, so quantum numbers ℓ and m are associated with each atomic orbital. The radial dependence of the orbitals in many-electron atoms differs from that of one-electron orbitals because the effective field differs from the Coulomb potential, but a principal quantum number n can still be defined. The lowest energy orbital is a 1*s* orbital and has no radial nodes, the next lowest *s* orbital is a 2*s* orbital and has one radial node, and so forth. Each electron in an atom has associated with it a set of four quantum numbers (n, ℓ, m, m_s). The first three quantum numbers describe its spatial distribution and the fourth specifies its spin state. The allowed quantum numbers follow the same pattern as those for the hydrogen atom. However, the number of states associated with each combination of (n, ℓ, m) is twice as large because of the two values for m_s.

Sizes and Shapes of Hartree Orbitals

The spatial properties of Hartree orbitals are best conveyed through a specific example. We present the results for argon (Ar), taken from Hartree's original work. The ground state of the argon atom has 18 electrons in the 1*s*, 2*s*, 2*p*, 3*s*, and 3*p* Hartree orbitals (see later). Figure 5.12 shows the radial probability density distributions for these five occupied orbitals as calculated by Hartree's method. The probability density distribution shown for the 2*p* level is the sum of the distributions for the $2p_x$, $2p_y$, and $2p_z$ orbitals; similarly, the 3*p* probability density distribution includes the $3p_x$, $3p_y$, and $3p_z$ orbitals. Comparing Figure 5.12 with Figure 5.10 shows that each Hartree orbital for argon is "smaller" than the corresponding orbital for hydrogen in the sense that the region of maximum probability density is closer to the nucleus. This difference occurs because the argon nucleus ($Z = 18$) exerts a much stronger attractive force on electrons than does the hydrogen nucleus ($Z = 1$). We develop a semiquantitative relation between orbital size and Z in the next subsection.

The fact that Hartree orbitals with the same value of n are large in the same narrow regions of space, despite their different values of ℓ, has interesting consequences. The total radial probability density function for a many-electron atom gives the probability of finding an electron at position r regardless of which orbital it occupies. We obtain this function by summing up the radial probability density functions of all the occupied orbitals. The resulting probability function is proportional to the *radial charge density distribution function* $\rho(r)$ for the atom. If the radial probability density functions in Figure 5.12 are all added together, the result reflects the contributions of electrons to the charge density $\rho(r)$ in a thin spherical shell of radius r, regardless of the orbital to which the electron belongs. A plot of $\rho(r)$ on the same scale as Figure 5.12 shows three peaks at r values of approximately 0.1, 0.3, and 1.2 in units of a_0 (Fig. 5.13) The total electron density

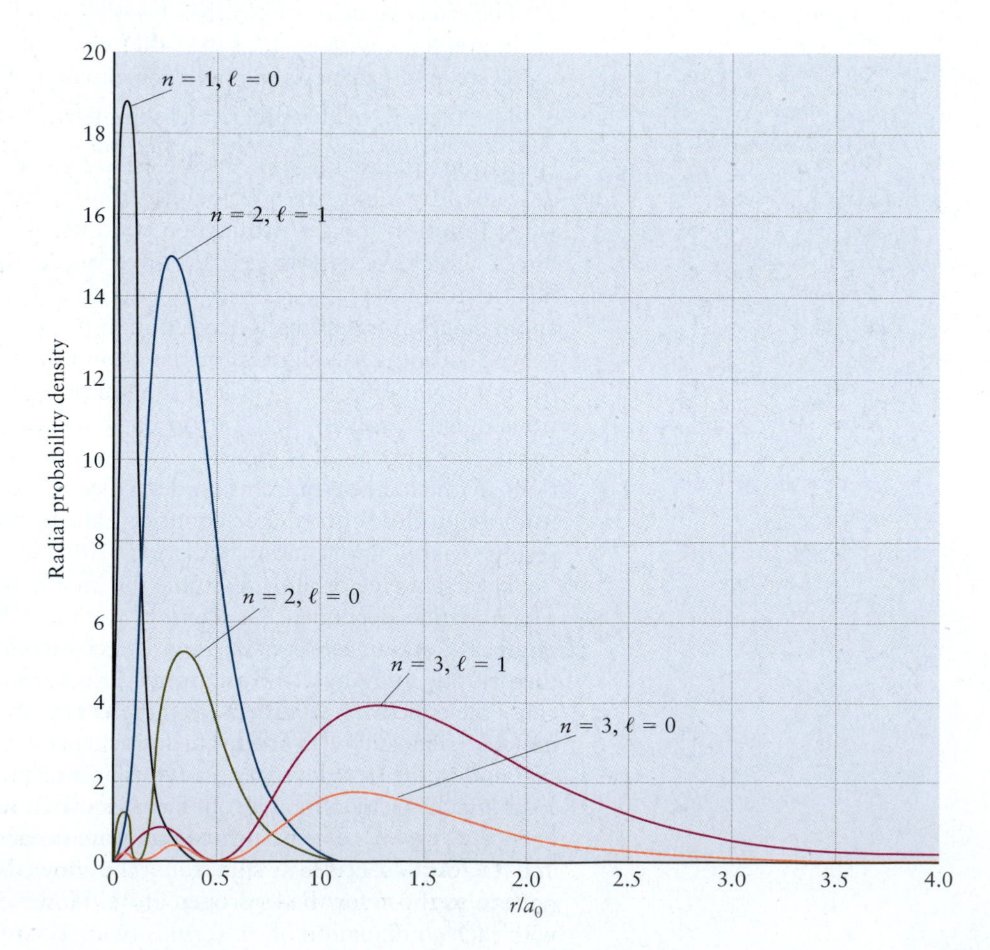

FIGURE 5.12 Dependence of radial probability densities on distance from the nucleus for Hartree orbitals in argon with $n = 1, 2, 3$. The results were obtained from self-consistent calculations using Hartree's method. Distance is plotted in the same dimensionless variable used in Figure 5.10 to facilitate comparison with the results for hydrogen. The fact that the radial probability density for all orbitals with the same value of n have maxima very near one another suggests that the electrons are arranged in "shells" described by these orbitals.

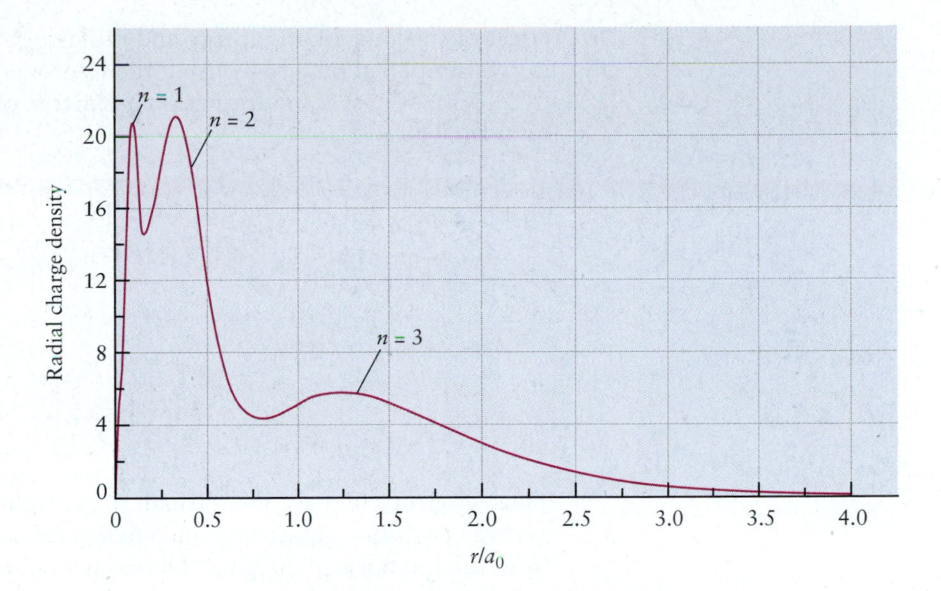

FIGURE 5.13 The radial charge density in the argon atom as calculated by Hartree's method. The charge is arrayed into three shells corresponding to the values 1, 2, and 3 for the principal quantum number *n*.

of the argon atom thus is concentrated in three concentric shells, where a **shell** is defined as all electrons with the same value of n. Each shell has a radius determined by the principal quantum number n. The shell model summarizes the coarse features of the electron density of an atom by averaging over all those local details not described by the principal quantum number n. Within each shell, a more detailed picture is provided by the **subshells,** defined as the set of orbitals with the same values of both n and ℓ.

The subshells (see Fig. 5.12) determine the structure of the periodic table and the formation of chemical bonds. In preparation for a discussion of these connections, it is necessary to describe the energy values for Hartree orbitals.

Shielding Effects: Energy Sequence of Hartree Orbitals

The energy-level diagrams calculated for many-electron atoms by Hartree's method resemble the diagram for the hydrogen atom (see Fig. 5.2), but differ in two important respects. First, the degeneracy of the p, d, and f orbitals is removed. Because the effective field in Hartree's method is different from the Coulomb field in the hydrogen atom, the energy levels of Hartree orbitals depend on both n and ℓ. Second, the energy values are distinctly shifted from the values of corresponding hydrogen orbitals because of the stronger attractive force exerted by nuclei with $Z > 1$.

These two effects can be explained qualitatively by a highly simplified one-electron model. Assume each of the electrons in shell n is moving in a Coulomb potential given approximately by

$$V_n^{\text{eff}}(r) \approx -\frac{Z_{\text{eff}}(n)e^2}{r} \qquad [5.8]$$

where $Z_{\text{eff}}(n)$ is the **effective nuclear charge** in that shell. To understand the origin and magnitude of $Z_{\text{eff}}(n)$, consider a particular electron e_1 in an atom. Inner electrons near the nucleus *shield* e_1 from the full charge Z of the nucleus by effectively canceling some of the positive nuclear charge. $Z_{\text{eff}}(n)$ is thus the net reduced nuclear charge experienced by a particular electron, due to the presence of the other electrons. (See Section 3.2.) For a neutral atom, $Z_{\text{eff}}(n)$ can range from a maximum value of Z near the nucleus (no screening) to a minimum value of 1 far from the nucleus (complete screening by the other $Z - 1$ electrons). Detailed Hartree calculations for argon show that $Z_{\text{eff}}(1) \sim 16$, $Z_{\text{eff}}(2) \sim 8$, and $Z_{\text{eff}}(3) \sim 2.5$. The effect of shielding on the energy and radius of a Hartree orbital is easily estimated in this simplified picture by using the hydrogen atom equations with Z replaced by $Z_{\text{eff}}(n)$. We use ε_n to distinguish a Hartree orbital energy from the H atom orbital E_n. Thus,

$$\varepsilon_n \approx -\frac{[Z_{\text{eff}}(n)]^2}{n^2} \quad \text{(rydbergs)} \tag{5.9}$$

and

$$\bar{r}_{n\ell} \approx \frac{n^2 a_0}{Z_{\text{eff}}(n)}\left\{1 + \frac{1}{2}\left[1 - \frac{\ell(\ell+1)}{n^2}\right]\right\} \tag{5.10}$$

Thus, electrons in inner shells (small n) are tightly bound to the nucleus, and their average position is quite near the nucleus because they are only slightly shielded from the full nuclear charge Z. Electrons in outer shells are only weakly attracted to the nucleus, and their average position is far from the nucleus because they are almost fully shielded from the nuclear charge Z.

EXAMPLE 5.3

Estimate the energy and the average value of r in the $1s$ orbital of argon. Compare the results with the corresponding values for hydrogen.

SOLUTION

Using Equation 5.9 and the value $Z_{\text{eff}}(1) \sim 16$ leads to $\varepsilon_{1s} \sim -256$ rydbergs for argon. The Ar($1s$) electron is more strongly bound than the H($1s$) electron by a factor 256. (Compare Equation 5.1b for the hydrogen atom.)

Using Equation 5.10 and the value $Z_{\text{eff}}(1) \sim 16$ leads to $\bar{r}_{1s} = \dfrac{3a_0}{2 \cdot 16}$ for argon. This is smaller by a factor of 16 than $\bar{r}_{1s}$ for hydrogen.

A comparison of Figure 5.12 with Figure 5.10 demonstrates that each Hartree orbital for argon is "smaller" than the corresponding orbital for hydrogen in the sense that the region of maximum probability is closer to the nucleus.

Related Problems: 9, 10, 11, 12, 13, 14

The dependence of the energy on ℓ in addition to n can be explained by comparing the extent of shielding in different subshells. Figures 5.4 through 5.10 show that only the s orbitals penetrate to the nucleus; both p and d orbitals have nodes at the nucleus. Consequently, the shielding will be smallest, and the electron most tightly bound, in s orbitals. Calculations show that

$$\varepsilon_{ns} < \varepsilon_{np} < \varepsilon_{nd}$$

The approximate energy-level diagram for Hartree orbitals showing dependence on both n and ℓ is presented in Figure 5.14. Values of Z_{eff} are determined in advanced computer calculations of atomic structure, using a SCF approach based on Hartree's method. Table 5.3 shows a representative sampling, and includes the actual Z value in parentheses after the symbol for the atom. Using these values and the method in Example 5.3, you can quickly generate an approximate energy-level diagram and estimate the size of the orbitals for any atom. How do we use these energy levels and orbitals to describe the electrons? The answer is the subject of the next section.

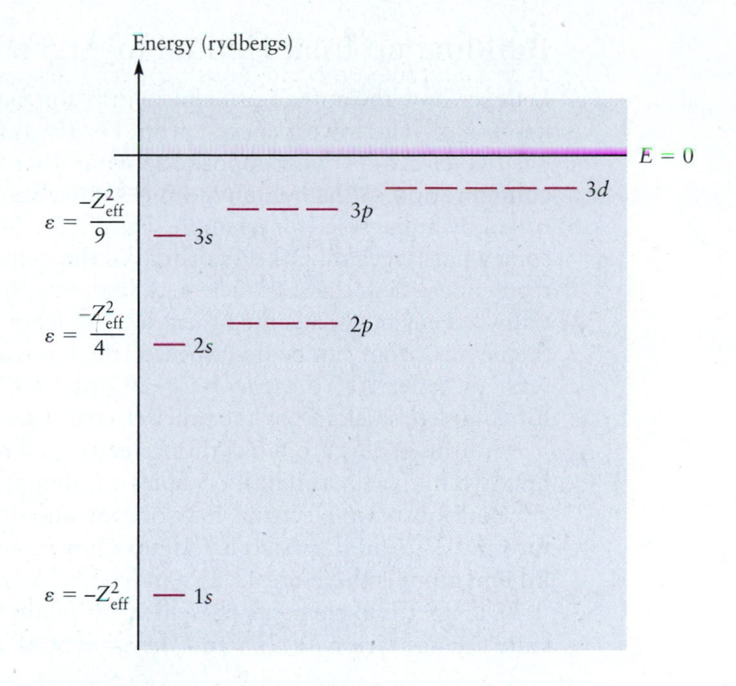

FIGURE 5.14 Approximate energy-level diagram for Hartree orbitals, estimated by incorporating values of Z_{eff}. Energy values are in units of rydbergs. The result of electron–electron repulsion is to remove the degeneracy of the hydrogen atom states with different ℓ values.

TABLE 5.3 Z_{eff} for Selected Atoms

	H(1)							He(2)
1s	1.00							1.69
	Li(3)	**Be(4)**	**B(5)**	**C(6)**	**N(7)**	**O(8)**	**F(9)**	**Ne(10)**
1s	2.69	3.68	4.68	5.67	6.66	7.66	8.65	9.64
2s	1.28	1.91	2.58	3.22	3.85	4.49	5.13	5.76
2p			2.42	3.14	3.83	4.45	5.10	5.76

5.3 Aufbau Principle and Electron Configurations

The ground-state electronic configuration of an atom with atomic number Z is built up by arranging the Hartree atomic orbitals in order of increasing energy and adding one electron at a time, starting with the lowest energy orbital, until all Z electrons are in place. The following additional restrictions are imposed at each step:

1. The **Pauli exclusion principle** states that no two electrons in an atom can have the same set of four quantum numbers (n, ℓ, m, m_s). Another way of stating this principle is that each Hartree atomic orbital (characterized by a set of three quantum numbers, n, ℓ, and m) holds at most two electrons, one with spin up and the other with spin down.
2. **Hund's rules** state that when electrons are added to Hartree orbitals of equal energy, a single electron enters each orbital before a second one enters any orbital. In addition, the lowest energy configuration is the one with parallel spins (see later discussion).

The energy level diagram for Hartree orbitals is shown qualitatively in Figure 5.14. Energy levels for *ns* orbitals, and for 2*p* orbitals up through F, can be estimated by using values for Z_{eff} in Equation 5.9. Values of Z_{eff} for each element have been obtained in advanced computer calculations based on Hartree's method. Table 5.3 shows the results for atoms in the first two periods. Starting with Ne, the simplified picture in Equations 5.8 and 5.9 is no longer adequate to estimate the *np* and *nd* energies, and full Hartree calculations are necessary.

Building up from Helium to Argon

Let's see how the aufbau principle works for the atoms from helium (He) through neon (Ne). The lowest energy orbital is always the 1*s* orbital; therefore, helium has two electrons (with opposite spins) in that orbital. The ground-state electron configuration of the helium atom is symbolized as $1s^2$ and is conveniently illustrated by a diagram (for example, Fig. 5.15). The 1*s* orbital in the helium atom is somewhat larger than the 1*s* orbital in the helium ion (He^+). In the ion, the electron experiences the full nuclear charge $+2e$, but in the atom, each electron partially screens or shields the other electron from the nuclear charge. The orbital in the helium atom can be described by the approximate equations given previously with an "effective" nuclear charge Z_{eff} of 1.69, which lies between +1 (the value for complete shielding by the other electron) and +2 (no shielding).

A lithium (Li) atom has three electrons. The third electron does not join the first two in the 1*s* orbital. It occupies a different orbital because, by the Pauli principle, at most two electrons may occupy any one orbital. The third electron goes into the 2*s* orbital, which is the next lowest in energy. The ground state of the lithium atom is therefore $1s^22s^1$, and $1s^22p^1$ is an excited state.

The next two elements present no difficulties. Beryllium (Be) has the ground-state configuration $1s^22s^2$, and the 2*s* orbital is now filled. Boron (B), with five

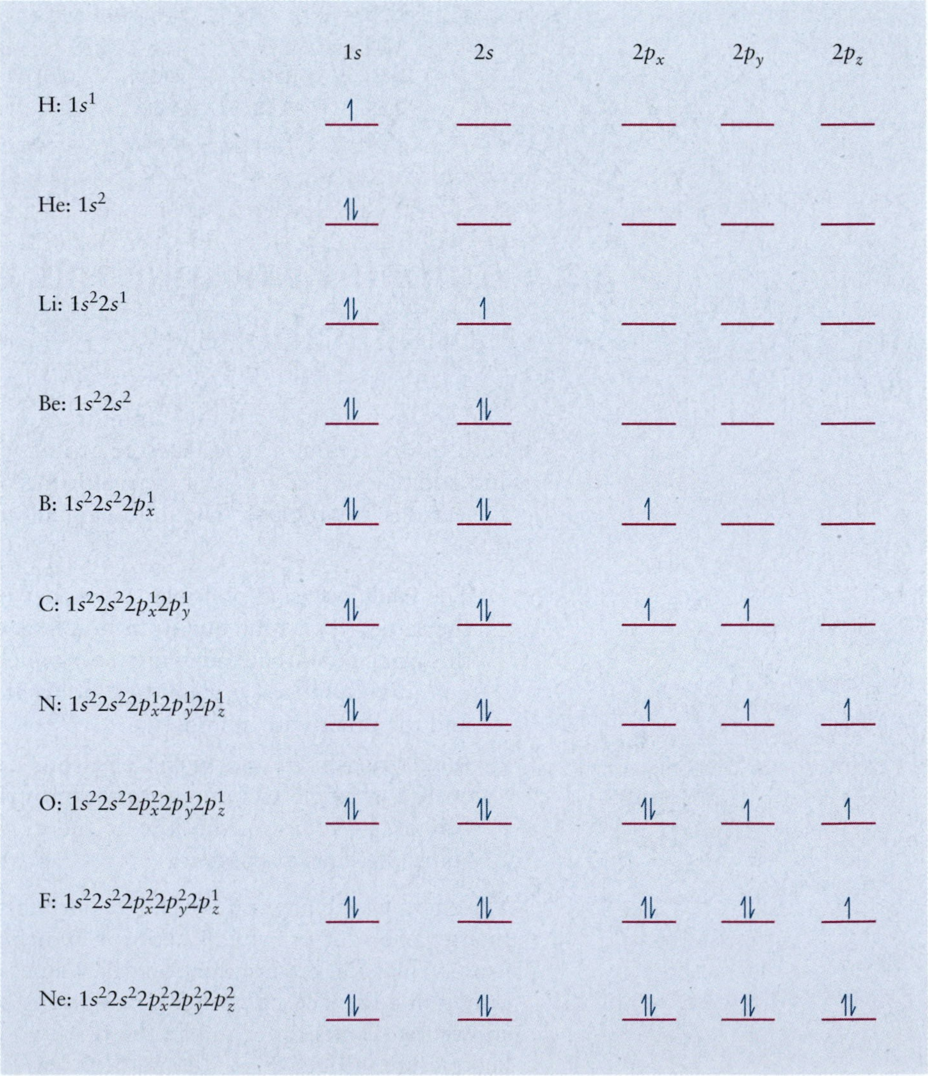

	$1s$	$2s$	$2p_x$	$2p_y$	$2p_z$
H: $1s^1$	↑				
He: $1s^2$	↑↓				
Li: $1s^22s^1$	↑↓	↑			
Be: $1s^22s^2$	↑↓	↑↓			
B: $1s^22s^22p_x^1$	↑↓	↑↓	↑		
C: $1s^22s^22p_x^12p_y^1$	↑↓	↑↓	↑	↑	
N: $1s^22s^22p_x^12p_y^12p_z^1$	↑↓	↑↓	↑	↑	↑
O: $1s^22s^22p_x^22p_y^12p_z^1$	↑↓	↑↓	↑↓	↑	↑
F: $1s^22s^22p_x^22p_y^22p_z^1$	↑↓	↑↓	↑↓	↑↓	↑
Ne: $1s^22s^22p_x^22p_y^22p_z^2$	↑↓	↑↓	↑↓	↑↓	↑↓

FIGURE 5.15 The ground-state electron configurations of first- and second-period atoms. Each horizontal line represents a specific atomic orbital. Arrows pointing up represent electrons with spin quantum number $m_s = +\frac{1}{2}$ and arrows pointing down represent electrons with spin quantum number $m_s = -\frac{1}{2}$.

electrons, has the ground-state configuration $1s^22s^22p^1$. Because the three $2p$ orbitals of boron have the same energy, there is an equal chance for the electron to be in each one.

With carbon (C), the sixth element, a new question arises. Will the sixth electron go into the same orbital as the fifth (for example, both into a $2p_x$ orbital with opposite spins), or will it go into the $2p_y$ orbital, which has equal energy? The answer is found in the observation that two electrons that occupy the same atomic orbital experience stronger electron–electron repulsion than they would if they occupied orbitals in different regions of space. Thus, putting the last two electrons of carbon into two different p orbitals, such as $2p_x$ and $2p_y$, in accordance with Hund's rule, leads to lower energy than putting them into the same p orbital. The electron configuration of carbon is then $1s^22s^22p_x^12p_y^1$, or more simply, $1s^22s^22p^2$. This configuration is shown in Figure 5.16.

The behavior of atoms in a magnetic field provides a test of their electron configuration. A substance is **paramagnetic** if it is attracted into a magnetic field. Any substance that has one or more unpaired electrons in the atoms, molecules, or ions that compose it is paramagnetic because a net magnetic moment arises from each of the unpaired electrons. (Recall the Stern–Gerlach experiment described in Section 5.1.) A substance in which all the electrons are paired is weakly **diamagnetic:** It is pushed *out* of a magnetic field, although the force it experiences is much smaller in magnitude than the force that pulls a typical paramagnetic substance into a magnetic field. Of the atoms discussed so far, hydrogen, lithium, boron, and carbon are known from experiments to be paramagnetic, whereas helium and beryllium are known to be diamagnetic. These results give us confidence in the validity of our description of atomic structure based on the orbital approximation and SCF calculations.

The electron configurations from nitrogen (N) through neon (Ne) follow from the stepwise filling of the $2p$ orbitals. The six elements from boron to neon are called ***p*-block** elements because their configurations involve filling of p orbitals in the building-up process. The four elements that precede them (hydrogen through beryllium) are called ***s*-block** elements.

The build-up of the third period, from sodium to argon, is an echo of what happened in the second; first the one $3s$ orbital is filled, and then the three $3p$ orbitals. As the number of electrons in an atom reaches 15 or 20, it is frequently the practice to explicitly include only those electrons added in the building up beyond the last preceding noble-gas element. The configuration of that noble gas is then represented by its chemical symbol enclosed in brackets. The ground-state configuration of silicon, for example, is written $[Ne]3s^23p^2$ using this system.

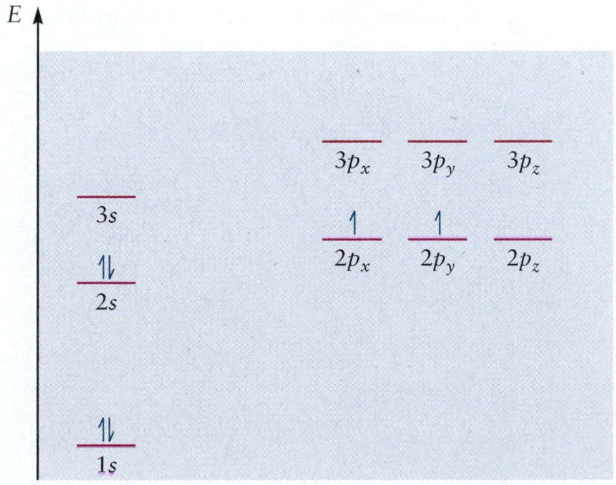

FIGURE 5.16 For a many-electron atom such as carbon, orbitals with different ℓ values (such as the 2*s* and 2*p* orbitals) have different energies (see Fig. 5.14). When two or more orbitals have the *same* energy (such as the three 2*p* orbitals here), electrons occupy different orbitals with parallel spins in the ground state.

EXAMPLE 5.4

Write the ground-state electron configurations for magnesium and sulfur. Are the gaseous atoms of these elements paramagnetic or diamagnetic?

SOLUTION

The noble-gas element preceding both elements is neon. Magnesium has two electrons beyond the neon core, which must be placed in the 3*s* orbital, the next higher in energy, to give the ground-state configuration [Ne]$3s^2$. A magnesium atom is diamagnetic because all of its electrons are paired in orbitals.

Sulfur has six electrons beyond the neon core; the first two of these are in the 3*s* orbital, and the next four are in the 3*p* orbitals. The ground-state configuration of sulfur is [Ne]$3s^2 3p^4$. When four electrons are put into three *p* orbitals, two electrons must occupy one of the orbitals, and the other two occupy different orbitals to reduce electron–electron repulsion. According to Hund's rules, the electrons' spins are parallel, and the sulfur atom is paramagnetic.

Related Problems: 15, 16, 17, 18

In summary, we remind you that the electron configuration for an atom is a concise, shorthand notation that represents a great deal of information about the structure and energy levels of the atom. Each configuration corresponds to an atomic wave function comprising a product of occupied Hartree orbitals. Each orbital has a well-defined energy (given by Equation 5.9 and shown in Figure 5.14) and average radius (given by Equation 5.10). The orbitals are grouped into subshells that are characterized by radial distribution functions (see Fig. 5.12). Chapter 6 describes the formation of chemical bonds by starting with the electron configurations of the participating atoms. We encourage you to become expert with atomic electronic configurations and all the information that they summarize.

EXAMPLE 5.5

The boron atom with $Z = 5$ has electron configuration $\mathrm{B}:(1s)^2(2s)^2(2p_x)^1$.

(a) Write the atomic wave function for a B atom.

(b) Estimate the energy level diagram for a B atom.

(c) Estimate the radius of the 2*s* and $2p_x$ orbitals.

SOLUTION

(a) The atomic wave function for a B atom is

$$\psi_{\mathrm{B}}(r_1, r_2, r_3, r_4, r_5) = [\varphi_{1s}(r_1)\varphi_{1s}(r_2)][\varphi_{2s}(r_3)\varphi_{2s}(r_4)][\varphi_{2p_x}(r_5)]$$

(b) Estimate the energy levels of a B atom.

$$\varepsilon_{1s} \approx -\frac{(4.68)^2}{1^2} = -21.90 \text{ Ry}$$

$$\varepsilon_{2s} \approx -\frac{(2.58)^2}{2^2} = -1.66 \text{ Ry}$$

$$\varepsilon_{2p} \approx -\frac{(2.42)^2}{2^2} = -1.46 \text{ Ry}$$

(c) Use Equation 5.10 and Z_{eff} values for boron from Table 5.3 to estimate orbital radii as follows:

$$\bar{r}_{2s} \approx \frac{2^2 a_0}{(2.58)}\left\{1 + \frac{1}{2}\left[1 - \frac{0(0+1)}{2^2}\right]\right\} = \frac{4a_0}{(2.58)}\left\{\frac{3}{2}\right\} = 2.33a_0$$

$$\bar{r}_{2p} \approx \frac{2^2 a_0}{(2.42)}\left\{1 + \frac{1}{2}\left[1 - \frac{1(1+1)}{2^2}\right]\right\} = \frac{4a_0}{(2.42)}\left\{\frac{5}{4}\right\} = 2.07a_0$$

Related Problems: 19, 20, 21, 22, 23, 24

Transition-Metal Elements and Beyond

After the $3p$ orbitals have been filled with six electrons, the natural next step is to continue the build-up process using the $3d$ subshell. Advanced calculations for elements 19 (K) through 30 (Zn) predict that ε_{3d} and ε_{4s} are very close, so the build-up process becomes rather subtle. For K and Ca, the calculations show that $\varepsilon_{4s} < \varepsilon_{3d}$. Optical spectroscopy confirms that the ground state of K is $[\text{Ar}]3d^0 4s^1$ and that of Ca is $[\text{Ar}]3d^0 4s^2$, as predicted by the sequence of calculated orbital energies. For Sc and the elements beyond, advanced calculations predict that $\varepsilon_{3d} < \varepsilon_{4s}$. Filling the $3d$ orbitals first would give the configurations $[\text{Ar}]3d^3 4s^0$ for Sc, $[\text{Ar}]3d^4 4s^0$ for Ti, and so on to $[\text{Ar}]3d^{10} 4s^0$ for Ni. These configurations are inconsistent with numerous optical, magnetic, and chemical properties of these elements, so some consideration besides the energies of the individual orbitals must also influence the build-up process. Let's consider the alternative configuration $[\text{Ar}]3d^1 4s^2$ for Sc and compare its total energy with that of the $[\text{Ar}]3d^3 4s^0$ configuration. This comparison must add the electrostatic repulsion energy between the electrons to the sum of the energies of the occupied one-electron orbitals to find the total energy of the atom. Because the $3d$ orbital is much more localized than the $4s$ orbital, the much greater repulsion energy of the two electrons in the $3d$ orbital outweighs the fact that $\varepsilon_{3d} < \varepsilon_{4s}$ and the configuration with two d electrons has higher energy. Thus, the configuration $[\text{Ar}]3d^1 4s^2$ has the lower energy and is the ground state for Sc. The same reasoning—minimizing the energy of the atom as a whole—predicts ground state electron configurations from $[\text{Ar}]3d^1 4s^2$ for Sc to $[\text{Ar}]3d^{10} 4s^2$ for Zn that agree with experimental results. The ten elements from scandium to zinc are called ***d*-block** elements because their configurations involve the filling of a d orbital in the building-up process.

Experimental evidence shows that chromium and copper do not fit this pattern. In its ground state, chromium has the configuration $[\text{Ar}]3d^5 4s^1$ rather than $[\text{Ar}]3d^4 4s^2$, and copper has the configuration $[\text{Ar}]3d^{10} 4s^1$ rather than $[\text{Ar}]3d^9 4s^2$. Similar anomalies occur in the fifth period, and others such as the ground-state configuration $[\text{Kr}]4d^7 5s^1$ that is observed for ruthenium in place of the expected $[\text{Kr}]4d^6 5s^2$.

In the sixth period, the filling of the $4f$ orbitals (and the generation of the ***f*-block** elements) begins as the rare-earth (lanthanide) elements from lanthanum to ytterbium are reached. The configurations determined from calculations and experiment can be recalled as needed by *assuming* that the orbitals are filled in the sequence $1s \rightarrow 2s \rightarrow 2p \rightarrow 3s \rightarrow 3p \rightarrow 4s \rightarrow 3d \rightarrow 4p \rightarrow 5s \rightarrow 4d \rightarrow 5p \rightarrow 6s \rightarrow 4f \rightarrow 5d \rightarrow 6p \rightarrow 7s \rightarrow 5f \rightarrow 6d$. The energies of the $4f$, $5d$, and $6s$ orbitals are comparable over much of the sixth period, and thus their order of filling is erratic.

The periodic table shown in Figure 5.17 classifies elements within periods according to the subshell that is being filled as the atomic number increases. Configurations are given explicitly for exceptions to this "standard" order of filling.

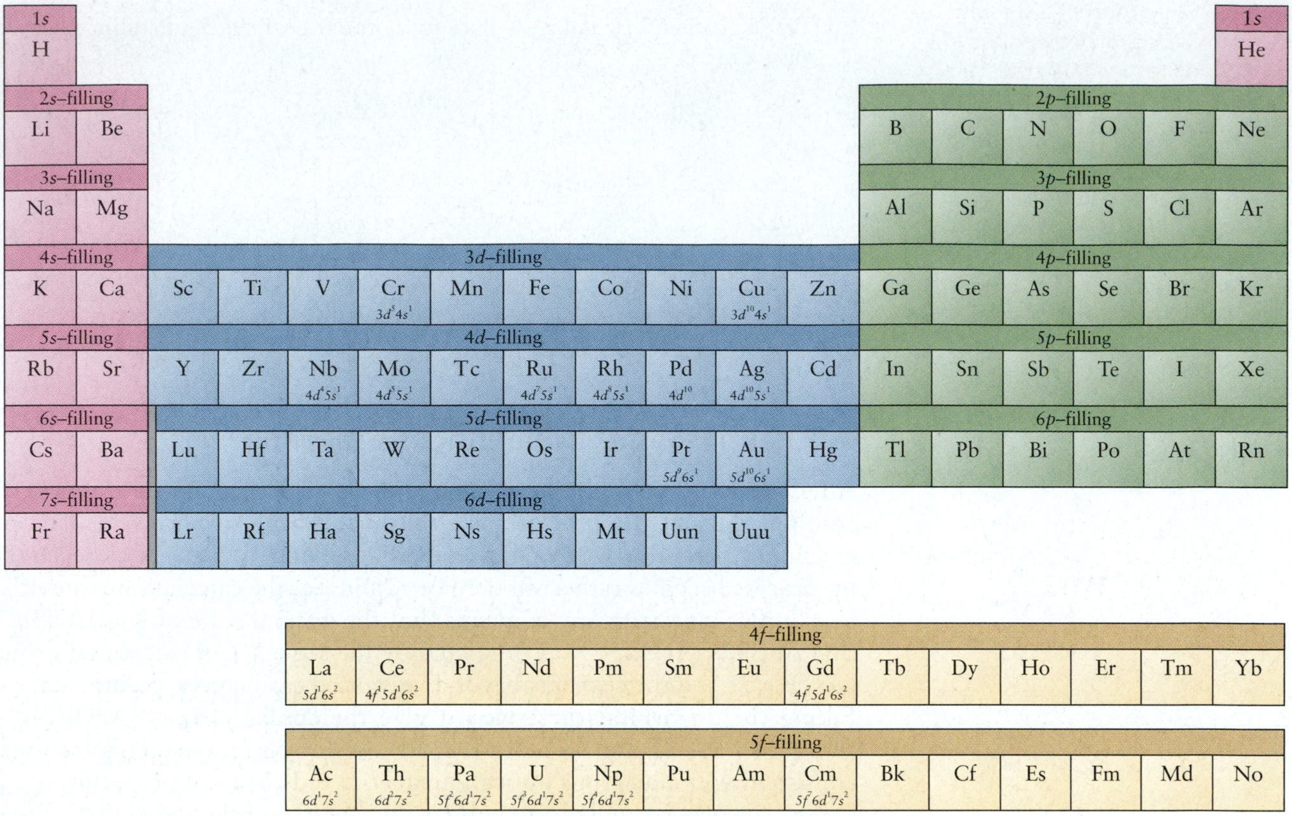

1s																	1s
H																	He
2s–filling												2p–filling					
Li	Be											B	C	N	O	F	Ne
3s–filling												3p–filling					
Na	Mg											Al	Si	P	S	Cl	Ar
4s–filling		3d–filling										4p–filling					
K	Ca	Sc	Ti	V	Cr $3d^5 4s^1$	Mn	Fe	Co	Ni	Cu $3d^{10} 4s^1$	Zn	Ga	Ge	As	Se	Br	Kr
5s–filling		4d–filling										5p–filling					
Rb	Sr	Y	Zr	Nb $4d^4 5s^1$	Mo $4d^5 5s^1$	Tc	Ru $4d^7 5s^1$	Rh $4d^8 5s^1$	Pd $4d^{10}$	Ag $4d^{10} 5s^1$	Cd	In	Sn	Sb	Te	I	Xe
6s–filling		5d–filling										6p–filling					
Cs	Ba	Lu	Hf	Ta	W	Re	Os	Ir	Pt $5d^9 6s^1$	Au $5d^{10} 6s^1$	Hg	Tl	Pb	Bi	Po	At	Rn
7s–filling		6d–filling															
Fr	Ra	Lr	Rf	Ha	Sg	Ns	Hs	Mt	Uun	Uuu							

4f–filling													
La $5d^1 6s^2$	Ce $4f^1 5d^1 6s^2$	Pr	Nd	Pm	Sm	Eu	Gd $4f^7 5d^1 6s^2$	Tb	Dy	Ho	Er	Tm	Yb
5f–filling													
Ac $6d^1 7s^2$	Th $6d^2 7s^2$	Pa $5f^2 6d^1 7s^2$	U $5f^3 6d^1 7s^2$	Np $5f^4 6d^1 7s^2$	Pu	Am	Cm $5f^7 6d^1 7s^2$	Bk	Cf	Es	Fm	Md	No

FIGURE 5.17 The filling of shells and the structure of the periodic table. Only the "anomalous" electron configurations are shown.

5.4 Shells and the Periodic Table: Photoelectron Spectroscopy

Our discussion of electronic structure began in Section 3.3 by analyzing patterns in the successive ionization energies of the atoms; these patterns suggested that the electrons are arranged in shells within the atom. In Section 5.2, we demonstrated that quantum theory *predicts* the shell structure of the atom. A shell is defined precisely as a set of orbitals that have the same principal quantum number, reflecting the fact that the average positions of the electrons in each of these shells are close to each other, but far from those of orbitals with different n values (see Fig. 5.12). Now, we can accurately interpret the results in Figure 3.5 as showing that Na has two electrons in the $n = 1$ shell, eight in the $n = 2$ shell, and one in the $n = 3$ shell.

The shell structure shows that two elements in the same group (column) of the periodic table have related valence electron configurations. For example, sodium (configuration $[Ne]3s^1$) and potassium (configuration $[Ar]4s^1$) each have a single valence electron in an s orbital outside a closed shell; consequently, the two elements closely resemble each other in their chemical properties. A major triumph of quantum mechanics is its ability to account for the periodic trends discovered by chemists many years earlier and organized empirically by Mendeleev and others in the periodic table (see Section 3.1). The ubiquitous octets in the Lewis electron dot diagrams of second- and third-period atoms and ions in Chapter 3 arise from the eight available sites for electrons in the one s orbital and three p orbitals of the valence shell. The special properties of the transition-metal elements are ascribed to the partial filling of their d orbitals (see Chapter 8 for further discussion of this feature).

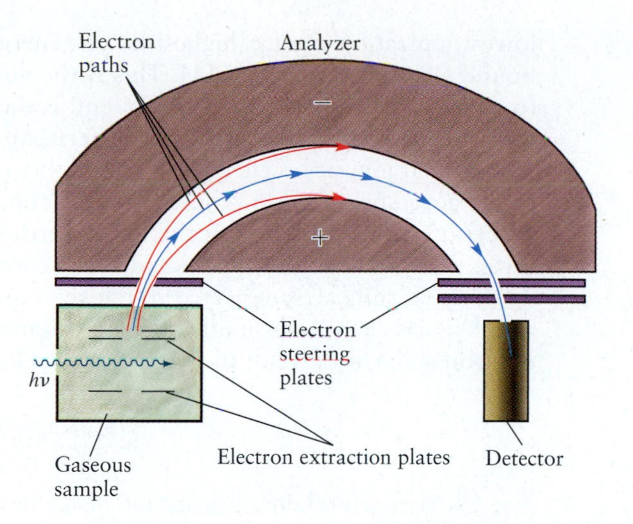

FIGURE 5.18 The energy of photoelectrons is determined by measuring the voltage required to deflect the electrons along a semicircular pathway between two charged metallic hemispherical plates in vacuum so they arrive at the detector.

The shell structure is confirmed directly by an important experimental technique called **photoelectron spectroscopy,** or PES. Photoelectron spectroscopy determines the energy level of each orbital by measuring the ionization energy required to remove each electron from the atom. Photoelectron spectroscopy is simply the photoelectric effect of Section 4.4 applied not to metals, but instead to free atoms. If radiation of sufficiently high frequency ν (in the ultraviolet or x-ray region of the spectrum) strikes an atom, an electron will be ejected with kinetic energy $\frac{1}{2}m_e v^2$. The kinetic energy of the ejected electrons is measured by an **energy analyzer,** which records the voltage required to deflect the electrons around a semicircular pathway in vacuum to reach the detector (Fig. 5.18). As the voltage between the hemispherical plates is changed, electrons with different values of kinetic energy will be deflected to the detector, and the spectrum of kinetic energy values can be recorded. Measuring the kinetic energy by deflection is analogous to measuring energy in the photoelectric effect experiments, and the results are conveniently expressed in units of electron volts (eV). Then the ionization energy spectrum, IE, is calculated by the principle of conservation of energy (see Section 4.4),

$$IE = h\nu_{\text{photon}} - \frac{1}{2}m_e v^2_{\text{electron}} \qquad [5.11]$$

Figure 5.19 shows the measured photoelectron spectrum for neon excited by x-rays with wavelength 9.890×10^{-10} m, and Example 5.6 shows how the spectrum is obtained and interpreted. Three peaks appear with kinetic energy values 383.4, 1205.2, and 1232.0 eV. The corresponding ionization energy is shown beneath each peak. (See Example 5.6 for details.) Note that ionization energy increases from right to left in Figure 5.19, opposite to kinetic energy. The peak at

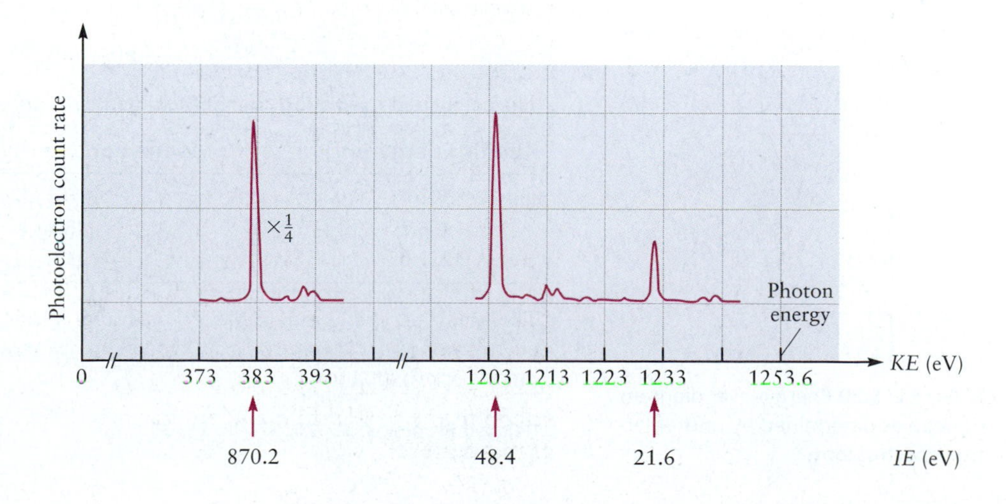

FIGURE 5.19 Photoelectron spectrum of neon. The spectrum shows three peaks, demonstrating that the electrons of neon are organized in three bonding states of distinct energy values. The peak at 383.4 eV has been reduced by a factor of 4 for display on the same scale as the other two.

lowest ionization energy (highest kinetic energy) is produced by the most weakly bound electrons (see Eq. 5.11). This is the minimum amount of energy required to detach an electron from an atom and is the same as the ionization energy IE_1 introduced in Section 3.3. Clearly, there can be no signal in the photoelectron spectrum at ionization energies less than this value. Peaks at higher ionization energies correspond to electrons removed from more strongly bound states. This spectrum demonstrates that the ten electrons of neon are arranged in bonding states that produce three distinct, discrete energy levels.

These results are connected to the shell model by **Koopmans's approximation,** which asserts in a form suitable for our discussion that the ionization energy of an electron is the negative of the energy of the Hartree orbital of the electron:

$$IE_\alpha = -\varepsilon_\alpha \qquad [5.12]$$

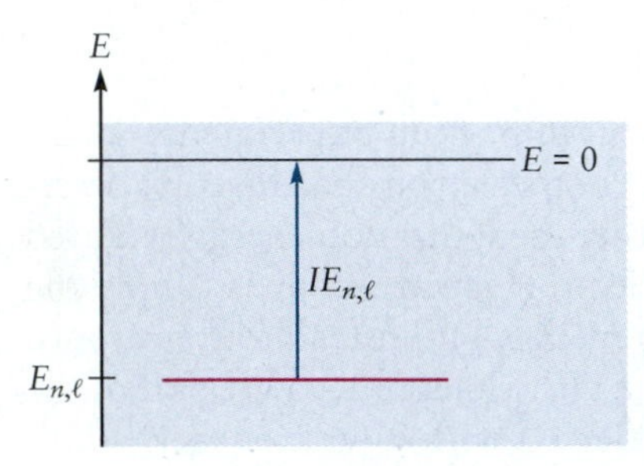

The Hartree orbital energies are intrinsically negative because they represent the energy stabilization of an electron bound in an atom relative to the free electron and a positive ion. The ionization energy is positive because it must be supplied to liberate the electron from the atom. Therefore, we should be able to read off the orbital energies directly from the measured spectrum of ionization energies. Koopmans's approximation is not strictly valid because it assumes the orbital energies are the same in the ion as in the parent atom, despite the loss of an electron. This is called the **frozen orbital approximation.** The theorem assumes no energy is lost to *relaxation* of the electronic structure during the ionization process. In fact, relaxation effects are usually no larger than 1 − 3 eV. They can be included with orbital energies calculated by the more advanced Hartree–Fock method. So Koopmans's approximation and PES provide a quantitative test for advanced theoretical models of electronic structure.

These experimental results for neon are consistent with the electron configuration Ne: $1s^2 2s^2 2p^6$ predicted by the aufbau principle. Ionization energies measured in this way are used to construct the energy-level diagram for atoms and to show explicitly the value of the ground-state energy.

EXAMPLE 5.6

Construct the energy-level diagram for neon from the data in Figure 5.19.

SOLUTION

The ionization energy of each level is calculated as $IE = E_{\text{photon}} - \mathcal{T}_{\text{electron}}$. Because the measured kinetic energy values for the photoelectrons are reported in units of electron volts, the most convenient approach is to calculate the photon energy in electron volts and then subtract the kinetic energy values. The energy of the photon is given by

$$E_{\text{photon}} = \frac{hc}{\lambda} = \frac{(6.6261 \times 10^{-34}\ \text{J s})(2.9979 \times 10^{8}\ \text{m s}^{-1})}{(9.890 \times 10^{-10}\ \text{m})(1.6022 \times 10^{-19}\ \text{J eV}^{-1})} = 1253.6\ \text{eV}$$

The calculated ionization energies for the peaks are summarized as follows:

Kinetic Energy (eV)	Ionization Energy (eV)
383.4	870.2
1205.2	48.4
1232.0	21.6

The energy-level diagram (Fig. 5.20) is drawn by showing the negative of each ionization energy value as the energy of an orbital, in accordance with Equation 5.12 and Koopmans's approximation.

Related Problems: 25, 26, 27, 28, 29, 30

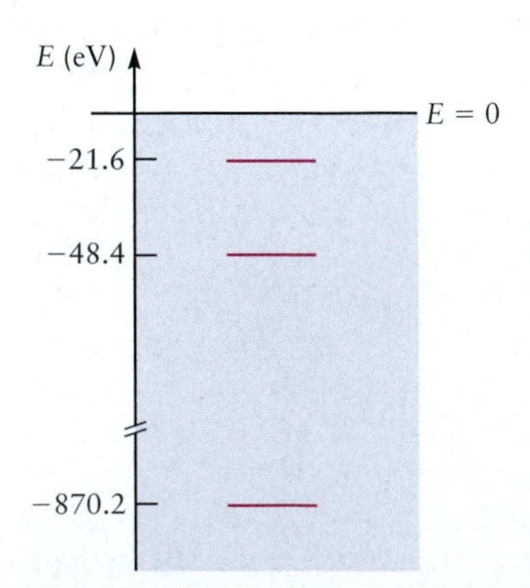

FIGURE 5.20 Energy-level diagram of neon as determined by photoelectron spectroscopy.

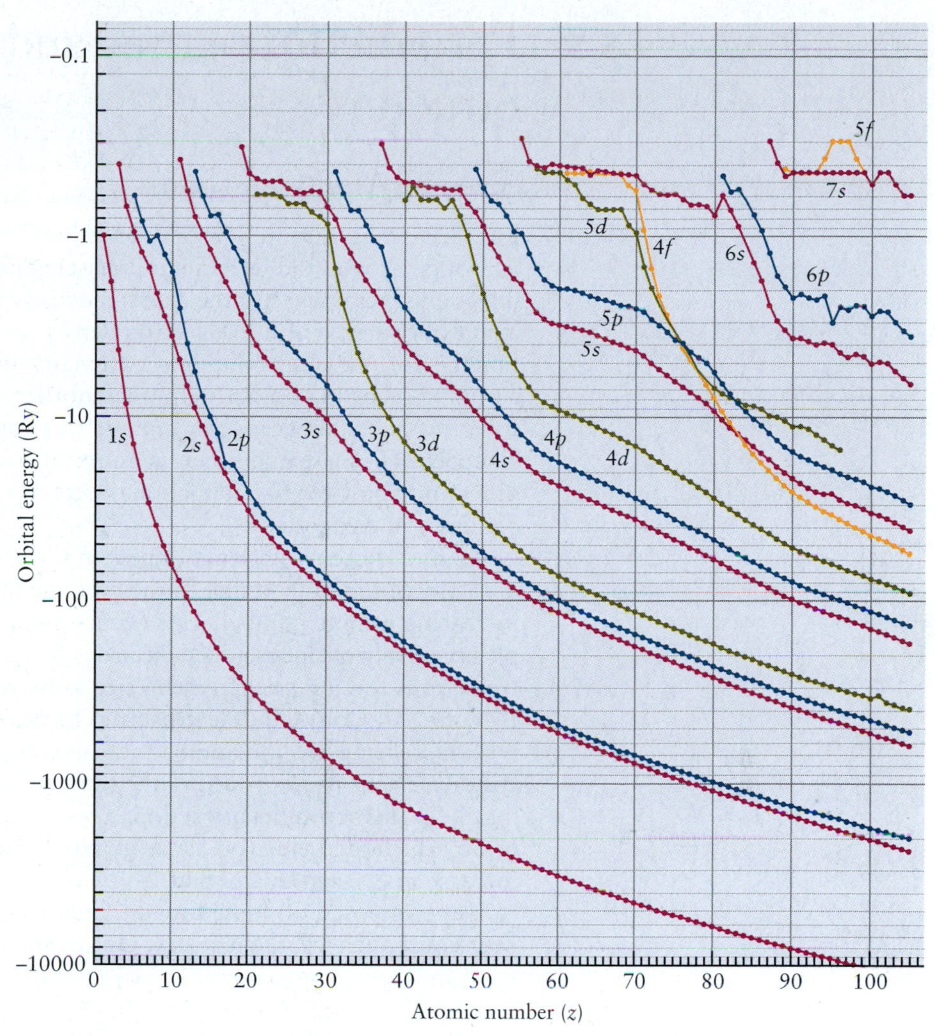

FIGURE 5.21 The energies of different subshells in the first 97 elements, as determined by photoelectron spectroscopy. Negative values on the vertical axis correspond to the bound state orbital energies. Subshells having the same principal quantum number n, such as 2*s* and 2*p*, have similar energies and are well separated from orbitals of different n. Significant exceptions do exist, as explained in the text. Note the logarithmic energy scale. One rydberg is 2.18×10^{-18} J.

This method has been used to determine the energy levels for orbitals in most neutral atoms (Fig. 5.21). The energies are reported in units of rydbergs and plotted on a logarithmic scale. These data confirm the existence of subshells, which are grouped into shells having similar energies. However, there are significant exceptions. The 3*d* subshell for elements 21 through 29 (scandium through copper) lies substantially higher than 3*s* and 3*p* and only slightly lower than 4*s*. This is consistent with the chemical observation that the 3*d* electrons are valence electrons in these transition metals. As Z goes above 30, the energy of the 3*d* subshell decreases rapidly, so the 3*d* electrons are not valence electrons for zinc and higher elements. The 4*d*, 5*d*, 4*f*, and 5*f* subshells all behave similarly, so electrons in filled *d* and *f* subshells are not valence electrons. We can develop an approximate criterion for distinguishing valence and core electrons by examining the noble gases (elements 2, 10, 18, 36, 54, and 86), which participate poorly or not at all in chemical bonding. The highest-energy subshell for each of them lies below −1 rydberg. Therefore, −1 rydberg is a reasonable approximate boundary for the difference between valence and core electrons.

5.5 Periodic Properties and Electronic Structure

Sizes of Atoms and Ions

The sizes of atoms and ions influence how they interact in chemical compounds. Although atomic radius is not a precisely defined concept, these sizes can be estimated in several ways. If the electron density is known from theory or experiment, a contour surface of fixed electron density can be drawn, as demonstrated in Section 5.1 for one-electron atoms. Alternatively, if the atoms or ions in a crystal are assumed to be in contact with one another, a size can be defined from the measured distances between their centers (this approach is explored in greater detail in Chapter 21). These and other measures of size are reasonably consistent with each other and allow for the tabulation of sets of atomic and ionic radii, many of which are listed in Appendix F.

Certain systematic trends appear in these radii. For a series of elements or ions in the same group (column) of the periodic table, the radius usually increases with increasing atomic number. This occurs mainly because the Pauli exclusion principle effectively excludes added electrons from the region occupied by the core electrons, thus forcing an increase in size as more distant electron shells are occupied. By contrast, Coulomb (electrostatic) forces cause the radii of atoms to *decrease* with increasing atomic number across a period. As the nuclear charge increases steadily, electrons are added to the same valence shell and are ineffective in shielding each other from its attraction. This "incomplete shielding" of the added proton by the added electron as we go from atomic number Z to $Z + 1$ leads to an increase in Z_{eff} across a period.

Superimposed on these broad trends are some subtler effects that have significant consequences in chemistry. One dramatic example is shown in Figure 5.22.

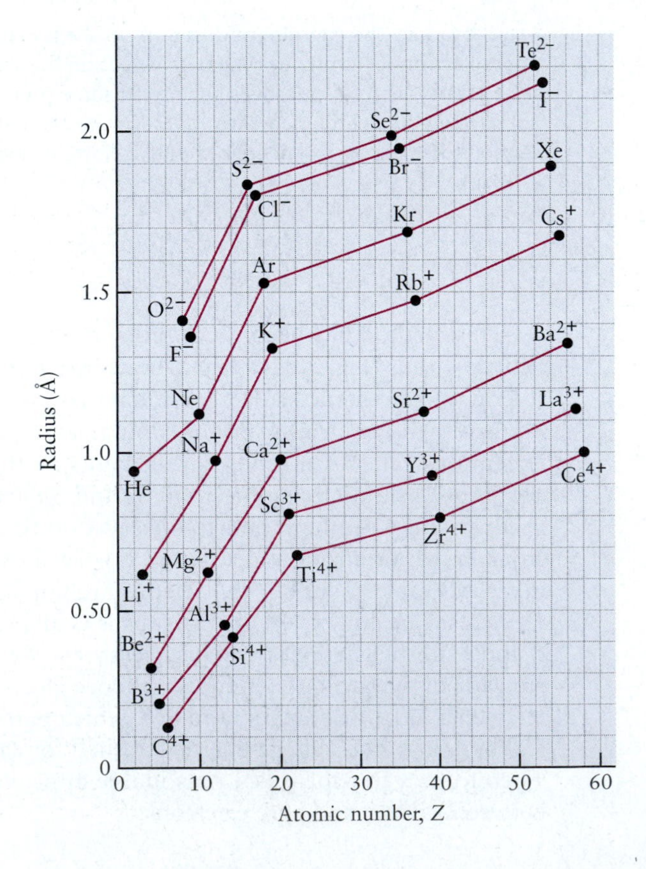

FIGURE 5.22 Ionic and atomic radii plotted versus atomic number. Each line connects a set of atoms or ions that have the same charge; all species have noble-gas configurations.

H 11.4																	He 21.0
Li 13.0	Be 4.85											B 4.39	C 3.42	N 13.5	O 17.4	F 11.2	Ne 13.2
Na 23.8	Mg 14.0											Al 10.0	Si 12.1	P 17.0	S 15.5	Cl 17.4	Ar 22.6
K 45.9	Ca 26.2	Sc 15.0	Ti 10.6	V 8.32	Cr 7.23	Mn 7.35	Fe 7.09	Co 6.67	Ni 6.59	Cu 7.11	Zn 9.16	Ga 11.8	Ge 13.6	As 13.0	Se 16.4	Br 19.8	Kr 28.0
Rb 55.8	Sr 33.9	Y 19.9	Zr 14.0	Nb 10.8	Mo 9.38	Tc 8.63	Ru 8.17	Rh 8.28	Pd 8.56	Ag 10.3	Cd 13.0	In 15.8	Sn 16.3	Sb 18.2	Te 20.5	I 25.7	Xe 35.9
Cs 70.9	Ba 38.2	Lu 17.8	Hf 13.4	Ta 10.9	W 9.47	Re 8.86	Os 8.42	Ir 8.52	Pt 9.09	Au 10.2	Hg 14.1	Tl 17.2	Pb 18.3	Bi 21.3	Po 23.0		Rn 50.5

FIGURE 5.23 The molar volumes (measured in $cm^3\ mol^{-1}$ of atoms) of some elements in their solid states. Note the large values for the alkali metals.

The radii of several sets of ions and atoms increase with atomic number in a given group (see earlier), but the *rate* of this increase changes considerably when the ions and atoms that contain the same number of electrons as argon are reached (S^{2-}, Cl^-, Ar, K^+, Ca^{2+}, Sc^{3+}, Ti^{4+}). For example, the change in size from Li^+ to Na^+ to K^+ is substantial, but the subsequent changes, to Rb^+ and Cs^+, are significantly smaller due to the filling of the *d* orbitals, which begins after K^+ is reached. Because atomic and ionic size decrease from left to right across a series of transition-metal elements (due to the increased effective nuclear charge), the radius of a main-group element is smaller than it would have been had the transition series not intervened. A similar phenomenon, called the **lanthanide contraction,** occurs during the filling of the 4*f* orbitals in the lanthanide series. Its effect on the sizes of transition-metal atoms is discussed in Section 8.1.

A different measure of atomic size is the volume occupied by a mole of atoms of the element in the solid phase. Figure 5.23 shows the pronounced periodicity of the molar volume, with maxima occurring for the alkali metals. Two factors affect the experimentally measured molar volume: the "size" of the atoms, and the geometry of the bonding that connects them. The large molar volumes of the alkali metals stem both from the large size of the atoms and the fact that they are organized in a rather open, loosely packed structure in the solid.

EXAMPLE 5.7

Predict which atom or ion in each of the following pairs should be larger: (a) Kr or Rb, (b) Y or Cd, (c) F^- or Br^-.

SOLUTION

(a) Rb should be larger because it has an extra electron in a 5*s* orbital beyond the Kr closed shell.

(b) Y should be larger because the effective nuclear charge increases through the transition series from Y to Cd.

(c) Br^- should be larger because the extra outer electrons are excluded from the core.

Related Problems: 31, 32, 33, 34

Periodic Trends in Ionization Energies

The ionization energy of an atom is defined as the minimum energy necessary to detach an electron from the neutral gaseous atom (see Section 3.3). It can be obtained directly from the photoelectron spectrum of an atomic gas. Appendix F lists measured ionization energies of the elements, and Figure 5.24 shows the periodic trends in first and second ionization energies with increasing atomic number.

Let's use insight from quantum mechanics to examine the periodic trends in the first ionization energy. We obtain deeper understanding of the stabilities of the various electron configurations using this approach than we did in our empirical discovery of shell structure in Section 3.3. There is a large reduction in IE_1 from helium to lithium for two reasons: (1) a $2s$ electron is much farther from the nucleus than a $1s$ electron, and (2) the $1s$ electrons screen the nucleus in lithium so effectively that the $2s$ electron "sees" a net positive charge close to $+1$, rather than the larger charge seen by the electrons in helium. Beryllium shows an increase in IE_1 compared with lithium because the effective nuclear charge has increased, but the electron being removed is still from a $2s$ orbital.

The IE_1 of boron is somewhat less than that of beryllium because the fifth electron is in a higher energy (and therefore less stable) $2p$ orbital. In carbon and nitrogen, the additional electrons go into $2p$ orbitals as the effective nuclear charge increases to hold the outer electrons more tightly; hence, IE_1 increases. The nuclear charge is higher in oxygen than in nitrogen, which would give it a higher ionization energy if this were the only consideration. However, oxygen must accommodate two electrons in the same $2p$ orbital, leading to greater electron–electron repulsion and diminished binding, thus more than compensating for the increased electron-nuclear interaction. Consequently, oxygen has a lower IE_1 than nitrogen. Fluorine and neon have successively higher first ionization energies because of increasing effective nuclear charge. The general trends of increasing ionization energy across a given period, as well as the dips that occur at certain points, can thus be understood through the orbital description of many-electron atoms.

The ionization energy tends to decrease down a group in the periodic table (for example, from lithium to sodium to potassium). As the principal quantum number increases, so does the distance of the outer electrons from the nucleus. There are some exceptions to this trend, however, especially for the heavier

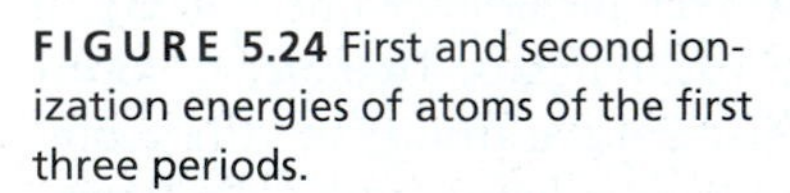
FIGURE 5.24 First and second ionization energies of atoms of the first three periods.

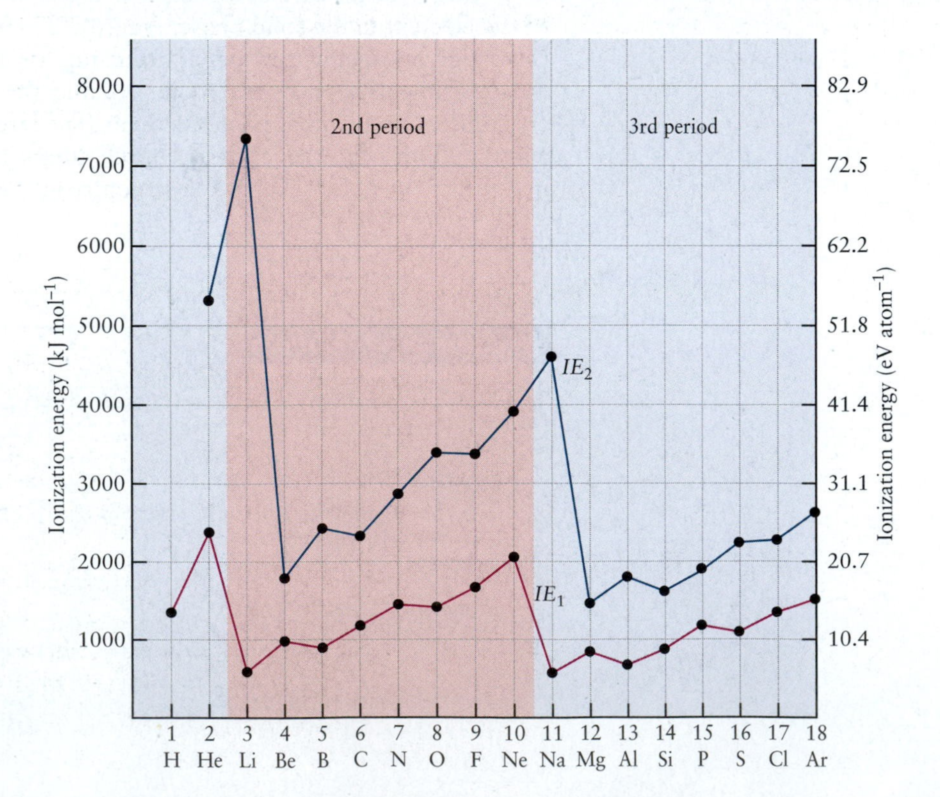

H 73																	He *
Li 60	Be *											B 27	C 122	N *	O 141	F 328	Ne *
Na 53	Mg *											Al 43	Si 134	P 72	S 200	Cl 349	Ar *
K 48	Ca 2	Sc 18	Ti 8	V 51	Cr 64	Mn *	Fe 16	Co 64	Ni 111	Cu 118	Zn *	Ga 29	Ge 116	As 78	Se 195	Br 325	Kr *
Rb 47	Sr 5	Y 30	Zr 41	Nb 86	Mo 72	Tc 53	Ru 99	Rh 110	Pd 52	Ag 126	Cd *	In 29	Sn 116	Sb 103	Te 190	I 295	Xe *
Cs 46	Ba 14	Lu 50	Hf *	Ta 31	W 79	Re 14	Os 106	Ir 151	Pt 214	Au 223	Hg *	Tl 19	Pb 35	Bi 91	Po 183	At 270	Rn *

FIGURE 5.25 Electron affinities (measured in kJ mol^{-1}) of gaseous atoms of the elements. An asterisk means that the element does not have a stable anion in the gas phase.

elements in the middle of the periodic table. For example, the first ionization energy of gold is higher than that of silver or copper. This fact is crucial in making gold a "noble metal"; that is, one that is resistant to attack by oxygen.

Similar trends are observed in the second ionization energies, but they are shifted higher in atomic number by one unit (see Fig. 5.24). Thus, IE_2 is large for lithium (because Li^+ has a filled $1s^2$ shell), but relatively small for beryllium (because Be^+ has a single electron in the outermost 2*s* orbital).

Electron Affinity

The **electron affinity,** *EA*, of an atom is the energy *released* when an electron is added to it (see Section 3.4). Appendix F lists the electron affinities of the elements.

The periodic trends in electron affinity (Fig. 5.25) parallel those in ionization energy for the most part, except that they are shifted one unit *lower* in atomic number. The reason is clear. Attaching an electron to F gives F^-, with the configuration $1s^22s^22p^6$, the same as that for neon. Fluorine has a large affinity for electrons because the resulting closed-shell configuration is stable. In contrast, the noble gases do not have well-defined electron affinities because the "extra" electron would reside in a new shell far from the nucleus and be almost totally screened from the nuclear charge.

EXAMPLE 5.8

Consider the elements selenium (Se) and bromine (Br). Which has the higher first ionization energy? Which has the higher electron affinity?

SOLUTION

These two atoms are adjacent to each other in the periodic table. Bromine has one more electron in the 4*p* subshell, and this electron should be more tightly bound than the 4*p* electrons in selenium because of the incomplete shielding and the extra unit of positive charge on the nucleus. Thus, IE_1 should be greater for bromine.

Bromine has a greater electron affinity than selenium. Gaining the extra electron changes the Br atom into Br^-, which has a particularly stable closed-shell electron configuration (the same as that of the noble gas atom krypton [Kr]). No such configuration is created when a Se atom gains an additional electron.

Related Problems: 35, 36, 37, 38

CHAPTER SUMMARY

The planetary model of the atom provides a reliable physical picture of the structure of the atom. This model is based securely on extensive experimental evidence obtained in the period roughly 1890 to 1915, part of an era of great excitement in the development of modern science. Although a great deal of physical insight into the structure and behavior of atoms can be obtained just by analyzing the consequences of the Coulomb force law between the nucleus and the electrons, theoretical explanation of the structure, properties, and behavior of atoms requires quantum mechanics. Quantum mechanics tells us that an atom can have only specific, discrete amounts of energy. Indeed, the very existence of quantum states explains the stability of the atom, which was predicted to collapse according to classical physics. Quantum mechanics also demonstrates that the concept of planetary orbits is simply not applicable on the atomic scale. We cannot know the detailed trajectory of an electron in the intuitive, classical sense familiar to ordinary human perception. Instead, we describe the probability for finding the electron at a particular location in the atom, based on knowing the quantum state of the atom.

For the hydrogen atom, we can solve the Schrödinger equation exactly to obtain the allowed energy levels and the hydrogen atomic orbitals. The sizes and shapes of these orbitals tell us the probability distribution for the electron in each quantum state of the atom. We are led to picture this distribution as a smeared cloud of electron density (probability density) with a shape that is determined by the quantum state.

For all other atoms, we have to generate approximations to solve the Schrödinger equation. The Hartree orbitals describe approximately the amplitude for each electron in the atom, moving under an effective force obtained by averaging over the interactions with all the other electrons. The Hartree orbitals have the same shapes as the hydrogen atomic orbitals—but very different sizes and energy values—and thus guide us to view the probability distribution for each electron as a smeared cloud of electron density.

The Hartree orbitals are the foundation of the quantum explanation of atomic structure. They justify the shell model of the atom, they explain the structure of the periodic table, and they provide the starting point for the quantum explanation of chemical bond formation in the following chapter.

CUMULATIVE EXERCISE

J. J. Hester (Arizona State University), and NASA

Clouds of gas surround hot stars in these galactic clusters. The red color arises from hydrogen radiation.

Interstellar Space

The vast stretches of space between the stars are by no means empty. They contain both gases and dust particles at very low concentrations. Interstellar space extends so far that these low-density species significantly affect the electromagnetic radiation arriving from distant stars and other sources, which is detected by telescopes. The gas in interstellar space consists primarily of hydrogen atoms (either neutral or ionized) at a concentration of about one atom per cubic centimeter. The dust (thought to be mostly solid water, methane, or ammonia) is even less concentrated, with typically only a few dust particles (each 10^{-4} to 10^{-5} cm in radius) per cubic kilometer.

(a) The hydrogen in interstellar space near a star is largely ionized by the high-energy photons from the star. Such regions are called H II regions. Suppose a ground-state hydrogen atom absorbs a photon with a wavelength of 65 nm. Calculate the kinetic energy of the electron ejected. (**Note:** This is the gas-phase analog of the photoelectric effect for solids.)

(b) What is the de Broglie wavelength of the electron from part (a)?

(c) Free electrons in H II regions can be recaptured by hydrogen nuclei. In such an event, the atom emits a series of photons of increasing energy as the electrons cascade down through the quantum states of the hydrogen atom. The particle densities are so low that extremely high quantum states can be detected in interstellar space. In particular, the transition from the state $n = 110$ to $n = 109$ for the hydrogen atom has been detected. What is the Bohr radius of an electron for hydrogen in the state $n = 110$?

(d) Calculate the wavelength of light emitted as an electron undergoes a transition from level $n = 110$ to $n = 109$. In what region of the electromagnetic spectrum does this lie?

(e) H II regions also contain ionized atoms that are heavier than hydrogen. Calculate the longest wavelength of light that will ionize a ground-state helium atom. Use data from Appendix F.

(f) The regions farther from stars are called H I regions. There, almost all of the hydrogen atoms are neutral rather than ionized and are in the ground state. Will such hydrogen atoms absorb light in the Balmer series emitted by atoms in H II regions?

(g) We stated in Section 5.1 that the energy of the hydrogen atom depends only on the quantum number n. In fact, this is not quite true. The electron spin (m_s quantum number) couples weakly with the spin of the nucleus, making the ground state split into two states of almost equal energy. The radiation emitted in a transition from the upper to the lower of these levels has a wavelength of 21.2 cm and is of great importance in astronomy because it allows the H I regions to be studied. What is the energy difference between these two levels, both for a single atom and for a mole of atoms?

(h) The gas and dust particles between a star and the earth scatter the star's light more strongly in the blue region of the spectrum than in the red. As a result, stars appear slightly redder than they actually are. Will an estimate of the temperature of a star based on its apparent color give too high or too low a number?

Answers

(a) 8.8×10^{-19} J

(b) 0.52 nm = 5.2 Å

(c) 6.40×10^{-7} m = 6400 Å

(d) 5.98×10^{-2} m = 5.98 cm, in the microwave region

(e) 50.4 nm

(f) No, because the lowest energy absorption for a ground-state hydrogen atom is in the ultraviolet region of the spectrum.

(g) 9.37×10^{-25} J; 0.564 J mol^{-1}

(h) Too low, because red corresponds to emitted light of lower energy. Experience with blackbody radiation curves would assign a lower temperature to a star emitting lower energy light.

CHAPTER REVIEW

- The physical structure of the atom, as determined by experiments, is summarized in the planetary model. In an atom with atomic number Z, there are Z electrons moving around a dense nucleus that has positive charge $+Ze$. Coulomb's law describes the forces and potential energy of interaction between each electron and the nucleus and between the electrons in many-electron atoms.

- Quantum mechanics explains the physical stability of the planetary atom, predicts its allowed energy levels, and defines the wave functions (also called atomic orbitals), which determine the probability density for finding the electrons at particular locations in the atom.
- The allowed energy levels for a one-electron atom or ion with atomic number Z are given by the expression $E_n \propto -Z/n^2$, where the quantum number $n = 1, 2, 3, \ldots$. These values are negative numbers because they measure the energy of the bound states of the stable atom relative to a separated electron and cation, which is defined to be the zero of the energy scale. The energy of the ground state of the hydrogen atom is -13.6 eV, a number worth remembering.
- The spacing between adjacent energy levels of a one-electron atom becomes narrower as n increases. As $n \longrightarrow \infty$, the value of $E_n \longrightarrow 0$, which corresponds to a separated electron and cation. Both of these important facts originate in the expression for the energy levels.
- Ionization energy is the amount of energy required to remove an electron from an atom and place it infinitely far away with zero kinetic energy. For any state n of a one-electron atom, the ionization energy IE_n is given by the transition $IE_n = E_\infty - E_n$. Ionization energy is intrinsically positive because the energy of the final state is higher than the energy of the initial state. The ionization energy of the ground state of the hydrogen atom is $+13.6$ eV.
- Because the potential energy in a one-electron atom depends only on r, the wave functions (atomic orbitals) have the product form $\psi_{n\ell m}(r, \theta, \phi) = R_{n\ell}(r)Y_{\ell m}(\theta, \varphi)$. The quantum number ℓ describes quantization of the total angular momentum of the electron, and the quantum number m describes quantization of the component of angular momentum along the z-axis. Whereas the quantum number n determines the allowed energy levels, the quantum numbers ℓ and m determine the shapes of the orbitals.
- The Stern–Gerlach experiment demonstrates that the electron has a property called spin, which leads to a magnetic dipole moment. Spin is quantized with only two allowed values described by the quantum number m_s. Complete determination of the quantum state of the electron required values for all four quantum numbers (n, ℓ, m_ℓ, m_s).
- Atoms with many electrons are described by Hartree's SCF method, in which each electron is assumed to move under the influence of an effective field $V_{\text{eff}}(r)$ due to the average positions of all the other electrons. This method generates a set of one-electron wave functions called the Hartree orbitals $\varphi_\alpha(r)$ with energy values ε_α, where α represents the proper set of quantum numbers. Hartree orbitals bear close relation to the hydrogen atomic orbitals but are not the same objects.
- Energies of the Hartree orbitals are different from those of the corresponding hydrogen atomic orbitals. For an atom with atomic number Z they can be estimated as $\varepsilon_n \propto -Z_{\text{eff}}/n^2$, where the effective nuclear charge experienced by each electron is determined by screening of that electron from the full nuclear charge by other electrons.
- The electron configuration for an atom with atomic number Z is determined by arranging the Hartree orbitals in order of increasing energy, then placing at most two electrons in each orbital in accordance with the Pauli exclusion principle and Hund's rule until all Z electrons have been placed. The configuration consists of specifying the set of four quantum numbers (n, ℓ, m_ℓ, m_s) for each electron in the atom.
- The Hartree orbitals have the same shapes as the corresponding hydrogen atomic orbitals, but their sizes are quite different.
- The Hartree orbitals and their electron configurations justify the shell model of the atom; that is, the electrons are grouped into shells of 2, 8, or 18 electrons

arranged concentrically around the nucleus at increasing distances from the nucleus. As we move outward from the nucleus, the electrons in each successive shell are bound progressively less strongly to the nucleus. Electrons in the outermost shell, called the valence electrons, are the least strongly bound, and they participate in the formation of chemical bonds.

- The shell model is verified experimentally by the technique of PES, in which ionization energy is measured for electrons in each shell. The results are connected to the shell model by Koopmans's theorem, which asserts that the orbital energy is the negative of the ionization energy.
- The Hartree orbitals and their electron configurations explain the structure of the periodic table.
- The Hartree orbitals and the shell model explain periodic trends in ionization energy, electron affinity, and the radii of atoms and ions. Small changes in these properties within a period are further explained by detailed changes in Z_{eff} within that period.

CONCEPTS & SKILLS

After studying this chapter and working the problems that follow, you should be able to:

1. Give the quantum numbers that characterize one-electron atoms, and discuss the shapes, sizes, and nodal properties of the corresponding orbitals (Section 5.1, Problems 1–6).
2. Prepare an approximate energy-level diagram for an atom using values for Z_{eff} (Section 5.2, Problems 9–14).
3. Use the aufbau principle to predict electron configurations of atoms and ions and to account for the structure of the periodic table (Section 5.3, Problems 15–24).
4. Construct the energy-level diagram for an atom using PES for orbital energies (Section 5.4, Problems 25–30).
5. Discuss the factors that lead to systematic variation of sizes of atoms and ions through the periodic table (Section 5.5, Problems 31–34).
6. Describe the trends in ionization energy and electron affinity across the periodic table and relate them to the electronic structure of atoms (Section 5.5, Problems 35–40).

KEY EQUATIONS

$$E = E_n = -\frac{Z^2e^4m_e}{8\epsilon_0^2n^2h^2} \qquad n = 1, 2, 3, \ldots \qquad \text{(Section 5.1)}$$

$$E_n = -\frac{Z^2}{n^2} \quad \text{(rydberg)} \qquad n = 1, 2, 3, \ldots \qquad \text{(Section 5.1)}$$

$$L^2 = \ell(\ell + 1)\frac{h^2}{4\pi^2} \qquad \ell = 0, 1, \ldots, n - 1 \qquad \text{(Section 5.1)}$$

$$L_z = m\frac{h}{2\pi} \qquad m = -\ell, -\ell + 1, \ldots, 0, \ldots, \ell - 1, \ell \qquad \text{(Section 5.1)}$$

$$\psi_{n\ell m}(r, \theta, \phi) = R_{n\ell}(r)Y_{\ell m}(\theta, \phi) \qquad \text{(Section 5.1)}$$

$$(\psi_{n\ell m})^2dV = [R_{n\ell}(r)]^2[Y_{\ell m}(\theta, \phi)]^2dV \qquad \text{(Section 5.1)}$$

$$dV = r^2 \sin\theta dr d\theta d\phi \qquad \text{(Section 5.1)}$$

$$\bar{r}_{n\ell} = \frac{n^2 a_0}{Z}\left\{1 + \frac{1}{2}\left[1 - \frac{\ell(\ell+1)}{n^2}\right]\right\} \qquad \text{(Section 5.1)}$$

$$V_n^{\text{eff}}(r) = -\frac{Z_{\text{eff}}(n)e^2}{r} \qquad \text{(Section 5.2)}$$

$$\varepsilon_n \approx -\frac{[Z_{\text{eff}}(n)]^2}{n^2} \quad \text{(rydbergs)} \qquad \text{(Section 5.2)}$$

$$\bar{r}_{n\ell} \approx \frac{n^2 a_0}{Z_{\text{eff}}(n)}\left\{1 + \frac{1}{2}\left[1 - \frac{\ell(\ell+1)}{n^2}\right]\right\} \qquad \text{(Section 5.2)}$$

$$IE = h\nu_{\text{photon}} - \tfrac{1}{2}m_e v_{\text{electron}}^2 \qquad \text{(Section 5.4)}$$

$$IE_\alpha = -\varepsilon_\alpha \qquad \text{(Section 5.4)}$$

PROBLEMS

Answers to problems whose numbers are boldface appear in Appendix G. Problems that are more challenging are indicated with asterisks.

The Hydrogen Atom

1. Which of the following combinations of quantum numbers are allowed for an electron in a one-electron atom? Which are not?
(a) $n = 2, \ell = 2, m = 1, m_s = \frac{1}{2}$
(b) $n = 3, \ell = 1, \text{m} = 0, m_s = -\frac{1}{2}$
(c) $n = 5, \ell = 1, m = 2, m_s = \frac{1}{2}$
(d) $n = 4, \ell = -1, m = 0, m_s = \frac{1}{2}$

2. Which of the following combinations of quantum numbers are allowed for an electron in a one-electron atom? Which are not?
(a) $n = 3, \ell = 2, m = 1, m_s = 0$
(b) $n = 2, \ell = 0, \text{m} = 0, m_s = -\frac{1}{2}$
(c) $n = 7, \ell = 2, m = -2, m_s = \frac{1}{2}$
(d) $n = 3, \ell = -3, m_s = -\frac{1}{2}$

3. Label the orbitals described by each of the following sets of quantum numbers:
(a) $n = 4, \ell = 1$
(b) $n = 2, \ell = 0$
(c) $n = 6, \ell = 3$

4. Label the orbitals described by each of the following sets of quantum numbers:
(a) $n = 3, \ell = 2$
(b) $n = 7, \ell = 4$
(c) $n = 5, \ell = 1$

5. How many radial nodes and how many angular nodes does each of the orbitals in Problem 3 have?

6. How many radial nodes and how many angular nodes does each of the orbitals in Problem 4 have?

7. Use the mathematical expression for the $2p_z$ wave function of a one-electron atom (see Table 5.2) to show that the probability of finding an electron in that orbital anywhere in the x-y plane is 0. What are the nodal planes for a d_{xz} orbital and for a $d_{x^2-y^2}$ orbital?

8. (a) Use the radial wave function for the $3p$ orbital of a hydrogen atom (see Table 5.2) to calculate the value of r for which a node exists.
(b) Find the values of r for which nodes exist for the $3s$ wave function of the hydrogen atom.

Shell Model for Many-Electron Atoms

9. Calculate the average distance of the electron from the nucleus in a hydrogen atom when the electron is in the $2s$ orbital. Repeat the calculation for an electron in the $2p$ orbital.

10. The helium ion He^+ is a one-electron system whose wave functions and energy levels are obtained from those for H by changing the atomic number to $Z = 2$. Calculate the average distance of the electron from the nucleus in the $2s$ orbital and in the $2p$ orbital. Compare your results with those in Problem 9 and explain the difference.

11. Spectroscopic studies show that Li can have electrons in its $1s$, $2s$, and $2p$ Hartree orbitals, and that $Z_{\text{eff}}(2s) = 1.26$. Estimate the energy of the $2s$ orbital of Li. Calculate the average distance of the electron from the nucleus in the $2s$ orbital of Li.

12. Spectroscopic studies of Li also show that $Z_{\text{eff}}(2p) = 1.02$. Estimate the energy of the $2p$ orbital of Li. Calculate the average distance of the electron from the nucleus in the $2p$ orbital of Li. Comparing your results with those in Problem 11 shows that the energy values differ by about 50%, whereas the average distances are nearly equal. Explain this observation.

13. Spectroscopic studies show that Na can have electrons in its $1s$, $2s$, $2p$, and $3s$ Hartree orbitals, and that $Z_{\text{eff}}(3s) = 1.84$. Using data from Problem 11, compare the energies of the Na $3s$ orbital, the Li $2s$ orbital, and the H $1s$ orbital.

14. Using data from Problems 11 and 13, calculate the average distance of the electron from the nucleus in the Na $3s$ orbital, the Li $2s$ orbital, and the H $1s$ orbital. Explain the trend in your results.

Aufbau Principle and Electron Configurations

15. Give the ground-state electron configurations of the following elements:
(a) C (b) Se (c) Fe

16. Give the ground-state electron configurations of the following elements:
(a) P (b) Tc (c) Ho

17. Write ground-state electron configurations for the ions Be^+, C^-, Ne^{2+}, Mg^+, P^{2+}, Cl^-, As^+, and I^-. Which do you expect will be paramagnetic due to the presence of unpaired electrons?

18. Write ground-state electron configurations for the ions Li^-, B^+, F^-, Al^{3+}, S^-, Ar^+, Br^+, and Te^-. Which do you expect to be paramagnetic due to the presence of unpaired electrons?

19. Identify the atom or ion corresponding to each of the following descriptions:
(a) an atom with ground-state electron configuration $[Kr]4d^{10}5s^25p^1$
(b) an ion with charge −2 and ground-state electron configuration $[Ne]3s^23p^6$
(c) an ion with charge +4 and ground-state electron configuration $[Ar]3d^3$

20. Identify the atom or ion corresponding to each of the following descriptions:
(a) an atom with ground-state electron configuration $[Xe]4f^{14}5d^66s^2$
(b) an ion with charge −1 and ground-state electron configuration $[He]2s^22p^6$
(c) an ion with charge +5 and ground-state electron configuration $[Kr]4d^6$

21. Predict the atomic number of the (as yet undiscovered) element in the seventh period that is a halogen.

22. (a) Predict the atomic number of the (as yet undiscovered) alkali-metal element in the eighth period.
(b) Suppose the eighth-period alkali-metal atom turned out to have atomic number 137. What explanation would you give for such a high atomic number (recall that the atomic number of francium is only 87)?

23. Suppose that the spin quantum number did not exist, and therefore only one electron could occupy each orbital of a many-electron atom. Give the atomic numbers of the first three noble-gas atoms in this case.

24. Suppose that the spin quantum number had three allowed values ($m_s = 0, +\frac{1}{2}, -\frac{1}{2}$). Give the atomic numbers of the first three noble-gas atoms in this case.

Shells and the Periodic Table: Photoelectron Spectroscopy

25. Photoelectron spectra of mercury (Hg) atoms acquired with radiation from a helium lamp at 584.4 Å show a peak in which the photoelectrons have kinetic energy of 11.7 eV. Calculate the ionization energy of electrons in that level.

26. Quantum mechanics predicts that the energy of the ground state of the H atom is −13.6 eV. Insight into the magnitude of this quantity is gained by considering several methods by which it can be measured.
(a) Calculate the longest wavelength of light that will ionize H atoms in their ground state.
(b) Assume the atom is ionized by collision with an electron that transfers all its kinetic energy to the atom in the ionization process. Calculate the speed of the electron before the collision. Express your answer in meters per second ($m\ s^{-1}$) and miles per hour ($miles\ h^{-1}$).
(c) Calculate the temperature required to ionize a H atom in its ground state by thermal excitation. (*Hint:* Recall the criterion for thermal excitation of an oscillator in Planck's theory of blackbody radiation is that $h\nu \approx k_BT$.)

27. Photoelectron spectroscopy studies of sodium atoms excited by x-rays with wavelength 9.890×10^{-10} m show four peaks in which the electrons have speeds $4.956 \times 10^6\ m\ s^{-1}$, $1.277 \times 10^7\ m\ s^{-1}$, $1.294 \times 10^7\ m\ s^{-1}$, and $1.310 \times 10^7\ m\ s^{-1}$. (Recall that $1\ J = 1\ kg\ m^2\ s^{-2}$.)
(a) Calculate the ionization energy of the electrons in each peak.
(b) Assign each peak to an orbital of the sodium atom.

28. Photoelectron spectroscopy studies of silicon atoms excited by X-rays with wavelength 9.890×10^{-10} m show four peaks in which the electrons have speeds $1.230 \times 10^7\ m\ s^{-1}$, $1.258 \times 10^7\ m\ s^{-1}$, $1.306 \times 10^7\ m\ s^{-1}$, and $1.308 \times 10^7\ m\ s^{-1}$. (Recall that $1\ J = 1\ kg\ m^2\ s^{-2}$.)
(a) Calculate the ionization energy of the electrons in each peak.
(b) Assign each peak to an orbital of the silicon atom.

29. Photoelectron spectroscopy studies have determined the orbital energies for fluorine atoms to be

1s	−689 eV
2s	−34 eV
2p	−12 eV

Estimate the value of Z_{eff} for F in each of these orbitals.

30. Photoelectron spectroscopy studies have determined the orbital energies for chlorine atoms to be

1s	−2,835 eV
2s	−273 eV
2p	−205 eV
s	−21 eV
3p	−10 eV

Estimate the value of Z_{eff} for Cl in each of these orbitals.

Periodic Properties and Electronic Structure

31. For each of the following pairs of atoms or ions, state which you expect to have the larger radius.
(a) Na or K (b) Cs or Cs^+
(c) Rb^+ or Kr (d) K or Ca
(e) Cl^- or Ar

32. For each of the following pairs of atoms or ions, state which you expect to have the larger radius.
(a) Sm or Sm^{3+} (b) Mg or Ca
(c) I^- or Xe (d) Ge or As
(e) Sr^+ or Rb

33. Predict the larger ion in each of the following pairs. Give reasons for your answers.
(a) O^-, S^{2-} (b) Co^{2+}, Ti^{2+}
(c) Mn^{2+}, Mn^{4+} (d) Ca^{2+}, Sr^{2+}

34. Predict the larger ion in each of the following pairs. Give reasons for your answers.
(a) S^{2-}, Cl^- (b) Tl^+, Tl^{3+}
(c) Ce^{3+}, Dy^{3+} (d) S^-, I^-

35. The first ionization energy of helium is 2370 kJ mol^{-1}, the highest for any element.
(a) Define *ionization energy* and discuss why for helium it should be so high.
(b) Which element would you expect to have the highest *second* ionization energy? Why?
(c) Suppose that you wished to ionize some helium by shining electromagnetic radiation on it. What is the maximum wavelength you could use?

36. The energy needed to remove one electron from a gaseous potassium atom is only about two-thirds as much as that needed to remove one electron from a gaseous calcium atom, yet nearly three times as much energy as that needed to remove one electron from K^+ as from Ca^+. What explanation can you give for this contrast? What do you expect to be the relation between the ionization energy of Ca^+ and that of neutral K?

37. Without consulting any tables, arrange the following substances in order and explain your choice of order:
(a) Mg^{2+}, Ar, Br^-, Ca^{2+} in order of increasing radius
(b) Na, Na^+, O, Ne in order of increasing ionization energy
(c) H, F, Al, O in order of increasing electronegativity

38. Both the electron affinity and the ionization energy of chlorine are higher than the corresponding quantities for sulfur. Explain why in terms of the electronic structure of the atoms.

39. The cesium atom has the lowest ionization energy, 375.7 kJ mol^{-1}, of all the neutral atoms in the periodic table. What is the longest wavelength of light that could ionize a cesium atom? In which region of the electromagnetic spectrum does this light fall?

40. Until recently, it was thought that Ca^- was unstable, and that the Ca atom therefore had a negative electron affinity. Some new experiments have now measured an electron affinity of +2.0 kJ mol^{-1} for calcium. What is the longest wavelength of light that could remove an electron from Ca^-? In which region of the electromagnetic spectrum does this light fall?

ADDITIONAL PROBLEMS

41. In the hydrogen atom, the transition from the 2*p* state to the 1*s* state emits a photon with energy 16.2×10^{-19} J. In an iron atom, the same transition emits x-rays with wavelength 0.193 nm. Calculate the energy difference between these two states in iron. Explain the difference in the 2*p*-1*s* energy level spacing in these two atoms.

42. The energy needed to ionize an atom of element X when it is in its most stable state is 500 kJ mol^{-1}. However, if an atom of X is in its lowest excited state, only 120 kJ mol^{-1} is needed to ionize it. What is the wavelength of the radiation emitted when an atom of X undergoes a transition from the lowest excited state to the ground state?

43. Suppose an atom in an excited state can return to the ground state in two steps. It first falls to an intermediate state, emitting radiation of wavelength λ_1, and then to the ground state, emitting radiation of wavelength λ_2. The same atom can also return to the ground state in one step, with the emission of radiation of wavelength λ. How are λ_1, λ_2, and λ related? How are the frequencies of the three radiations related?

44. For the Li atom, the energy difference between the ground state and the first excited state, in which the outermost electron is in a 2*p* orbital, is 2.96×10^{-19} J. In the Li^{2+} ion, the energy difference between the 2*s* and 2*p* levels is less than 0.00002 of this value. Explain this observation.

45. How does the $3d_{xy}$ orbital of an electron in O^{7+} resemble the $3d_{xy}$ orbital of an electron in a hydrogen atom? How does it differ?

* 46. The wave function of an electron in the lowest (that is, ground) state of the hydrogen atom is

$$\psi(r) = \left(\frac{1}{\pi a_0^3}\right)^{1/2} \exp\left(-\frac{r}{a_0}\right)$$

$$a_0 = 0.529 \times 10^{-10} \text{ m}$$

(a) What is the probability of finding the electron inside a sphere of volume 1.0 pm^3, centered at the nucleus (1 pm = 10^{-12} m)?
(b) What is the probability of finding the electron in a volume of 1.0 pm^3 at a distance of 52.9 pm from the nucleus, in a fixed but arbitrary direction?
(c) What is the probability of finding the electron in a spherical shell of 1.0 pm in thickness, at a distance of 52.9 pm from the nucleus?

47. An atom of sodium has the electron configuration $[Ne]6s^1$. Explain how this is possible.

48. (a) The nitrogen atom has one electron in each of the $2p_x$, $2p_y$, and $2p_z$ orbitals. By using the form of the angular wave functions, show that the total electron density, $\psi^2(2p_x) + \psi^2(2p_y) + \psi^2(2p_z)$, is spherically symmetric (that is, it is independent of the angles θ and ϕ). The neon atom, which has *two* electrons in each 2*p* orbital, is also spherically symmetric.
(b) The same result as in part (a) applies to *d* orbitals, thus a filled or half-filled subshell of *d* orbitals is spherically symmetric. Identify the spherically symmetric atoms or ions among the following: F^-, Na, Si, S^{2-}, Ar^+, Ni, Cu, Mo, Rh, Sb, W, Au.

49. Chromium(IV) oxide is used in making magnetic recording tapes because it is paramagnetic. It can be described as a solid made up of Cr^{4+} and O^{2-}. Give the electron configuration of Cr^{4+} in CrO_2, and determine the number of unpaired electrons on each chromium ion.

50. Use the data from Appendix F to graph the variation of atomic radius with atomic number for the rare-earth elements from lanthanum to lutetium.
(a) What is the general trend in these radii? How do you account for it?
(b) Which two elements in the series present exceptions to the trend?

51. Arrange the following seven atoms or ions in order of size, from smallest to largest: K, F^+, Rb, Co^{25+}, Br, F, Rb^-.

52. Which is higher, the third ionization energy of lithium or the energy required to eject a 1*s* electron from a Li atom in a PES experiment? Explain.

53. The outermost electron in an alkali-metal atom is sometimes described as resembling an electron in the corresponding state of a one-electron atom. Compare the first ionization energy of lithium with the binding energy of a 2*s* electron in a one-electron atom that has nuclear charge Z_{eff}, and determine the value of Z_{eff} that is necessary for the two energies to agree. Repeat the calculation for the 3*s* electron of sodium and the 4*s* electron of potassium.

* 54. In two-photon ionization spectroscopy, the combined energies carried by two different photons are used to remove an electron from an atom or molecule. In such an experiment, a K atom in the gas phase is to be ionized by two different light beams, one of which has a 650-nm wavelength. What is the maximum wavelength for the second beam that will cause two-photon ionization?

55. For the H atom, the transition from the 2*p* state to the 1*s* state is accompanied by the emission of a photon with an energy of 16.2×10^{-19} J. For an Fe atom, the same transition (2*p* to 1*s*) is accompanied by the emission of x-rays of 193-nm wavelengths. What is the energy difference between these states in iron? Comment on the reason for the variation (if any) in the 2*p*-1*s* energy-level spacing for these two atoms.

56. (a) Give the complete electron configuration ($1s^22s^22p$ ...) of aluminum in the ground state.
(b) The wavelength of the radiation emitted when the outermost electron of aluminum falls from the 4*s* state to the ground state is about 395 nm. Calculate the energy separation (in joules) between these two states in the Al atom.
(c) When the outermost electron in aluminum falls from the 3*d* state to the ground state, the radiation emitted has a wavelength of about 310 nm. Draw an energy-level diagram of the states and transitions discussed here and in (b). Calculate the separation (in joules) between the 3*d* and 4*s* states in aluminum. Indicate clearly which has higher energy.

57. What experimental evidence does the periodic table provide that an electron in a 5*s* orbital is slightly more stable than an electron in a 4*d* orbital for the elements with 37 and 38 electrons?

CHAPTER 6

Quantum Mechanics and Molecular Structure

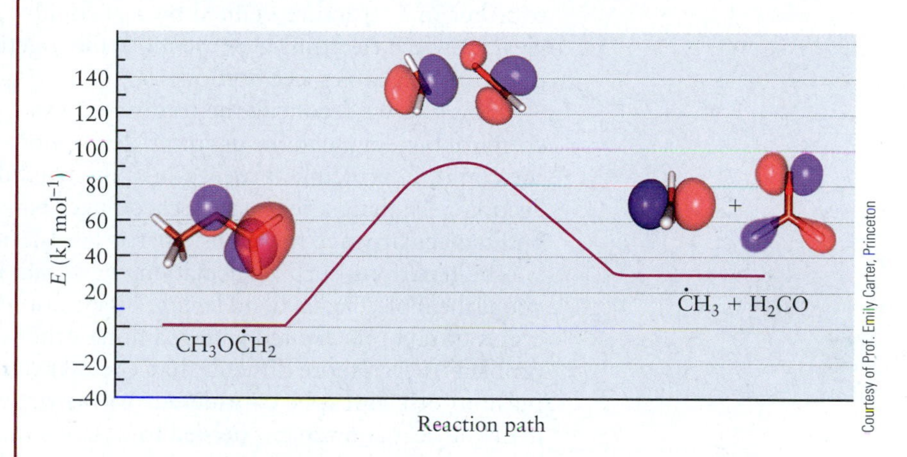

Potential energy diagram for the decomposition of the methylmethoxy radical, an important intermediate in the combustion of diethyl ether. Highly accurate computational quantum chemistry methods were used to calculate the configurations and the relative energies of the species shown.

Quantum mechanics embraces the experimental fact that matter and energy have a dual nature—part particle and part wave—and from this notion it correctly predicts the sizes, energy levels, and spectra of atoms and ions. This is a substantial achievement. But if quantum mechanics could do no more, the subject would hold little interest for chemists. Chemists seek to understand the atomic interactions that form molecules and extended solid structures, to understand intermolecular interactions, and most importantly, to understand chemical reactivity. Quantum mechanics provides a firm conceptual foundation on which an understanding of all of these phenomena can be built.

The concept of the chemical bond as an agent for holding atoms together in molecules—analogous to the connections between planets in our solar system—was formulated around the middle of the 19th century. In the 1860s, the chemists August Wilhelm Hofmann and Edward Frankland used three-dimensional arrays of colored wooden balls as models for molecular structure in their lectures in Berlin and London. In 1875, the Dutch chemist Jacobus van't Hoff popularized these developments in a book entitled *Chemistry in Space,* and by 1885, the three-dimensional representation of molecules was universally accepted.

Even today, chemists continue to use "ball-and-stick models," in which the balls represent atomic nuclei and the sticks represent chemical bonds, to help them think about the structures of molecules (see Fig. 3.17). After J. J. Thomson discovered the electron in 1897, and especially after Ernest Rutherford formulated the planetary model of the atom in 1912, physicists and chemists sought to explain the chemical bond as arrangements of the electrons around the nuclei. For example, G. N. Lewis considered the electrons to be "localized" in pairs to form covalent bonds between the nuclei. Nonetheless, electrons do not appear in the ball-and-stick figures, and they are not explicitly included in the description of molecular structure. The electrons function as the "glue" that holds the molecule together in a structure defined by a particular set of bond lengths and bond angles. How the electrons glue the molecule together can only be explained by using quantum mechanics (see Section 6.2).

How can we reconcile the traditional (and widely used to this day) picture of chemical bonding with the quantum description of atomic structure (see Chapter 5)? We could, in principle, proceed just like we did with the hydrogen (H) atom (see Section 5.1); that is, set up the Schrödinger equation for a molecule and solve it to find molecular wave functions, energy levels, and electron probability densities. If we could perform such a calculation, we would know all that it is possible to know about the molecule: its bond length, dissociation energy, dipole moment, and the energies of all of its excited states, among other characteristics. Unfortunately, this problem is even more difficult than the many-electron atom. Electron–electron repulsion forces must be considered. Worse yet, the solutions would depend on the many nuclear coordinates needed to describe the positions of even small molecules. Progress requires not only approximations but also new insights.

Fortunately, modern quantum chemistry provides good approximate solutions to the Schrödinger equation and also, perhaps more importantly, new qualitative concepts that we can use to represent and understand chemical bonds, molecular structure, and chemical reactivity. The quantum description of the chemical bond is a dramatic advance over the electron dot model, and it forms the basis for all modern studies in structural chemistry.

This chapter begins with a description of the quantum picture of the chemical bond for the simplest possible molecule, H_2^+, which contains only one electron. The Schrödinger equation for H_2^+ can be solved exactly, and we use its solutions to illustrate the general features of **molecular orbitals (MOs),** the one-electron wave functions that describe the electronic structure of molecules. Recall that we used the atomic orbitals (AOs) of the hydrogen atom to suggest approximate AOs for complex atoms. Similarly, we let the MOs for H_2^+ guide us to develop approximations for the MOs of more complex molecules.

Two powerful ways exist to construct approximate wave functions from AOs: the **linear combination of atomic orbitals (LCAO)** method and the **valence bond (VB)** method. The LCAO method generates MOs that are *de-localized* over the entire molecule and builds up the electronic configurations of molecules using an aufbau principle just like the one for atoms. In contrast, the VB method describes electron pairs that are *localized* between a pair of atoms, and it provides a quantum mechanical foundation for the valence shell electron-pair repulsion (VSEPR) theory. We apply both these methods to describe structure and bonding in a variety of molecules. We conclude the chapter by comparing the LCAO and VB methods and showing how each is the starting point for developing modern methods for computational quantum chemistry. These are now sufficiently accurate and so easy to use that they are becoming part of every chemist's set of tools for both research and education.

The central conceptual goal of this chapter is for you to understand how the wave functions of electrons initially localized on different atoms begin to interact with one another and form new wave functions that represent chemical bonds in molecules.

6.1 Quantum Picture of the Chemical Bond

The hydrogen molecular ion contains a single electron bound to two protons. It is a stable but highly reactive species produced by electrical discharge in H_2 gas. Its bond length is 1.060 Å, and its bond dissociation energy to produce H and H^+ is 2.791 eV. The similarity of these values to those for more familiar molecules (see Chapter 3) suggests that the exact quantum solutions for H_2^+ will provide insights into chemical bonding that can be transferred to more complex molecules. The solutions for H_2^+ introduce essential notation and terminology to guide our approximations for more complex molecules. Therefore, it is important to achieve a good understanding of H_2^+ as the foundation for the quantum explanation of chemical bonding.

6.1.1 The Simplest Molecule: H_2^+

The hydrogen molecular ion is sketched in Figure 6.1. The two nuclei, for convenience labeled A and B, are separated by the distance R_{AB} along the internuclear axis, chosen by convention to be the z-axis. The electron is located at distance r_A from nucleus A and at distance r_B from nucleus B. The angle ϕ describes rotation about the internuclear axis. For a fixed value of R_{AB}, the position of the electron is more conveniently specified by the values of (r_A, r_B, ϕ) than by (x, y, z) because the former set reflects the natural symmetry of the system. The internal potential energy is given by

$$V = -\frac{e^2}{4\pi\epsilon_0}\left(\frac{1}{r_A} + \frac{1}{r_B}\right) + \frac{e^2}{4\pi\epsilon_0}\left(\frac{1}{R_{AB}}\right) = V_{en} + V_{nn} \qquad [6.1]$$

The first two terms in Equation 6.1 represent the attractions between the electron and the two nuclei, and the last term represents the repulsion between the pair of protons. We can write the Schrödinger equation for this system, and we expect a solution of the form $\psi_{mol}(R_{AB}, r_A, r_B, \phi)$, because the potential energy is a function of all four of these coordinates.

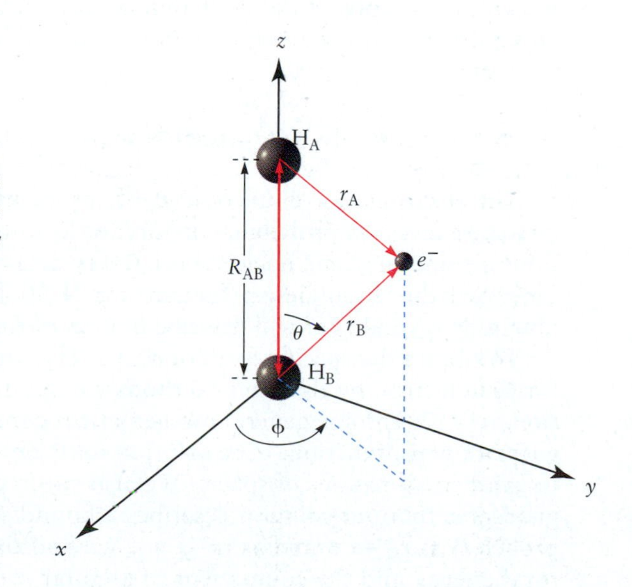

FIGURE 6.1 Coordinates for the H_2^+ molecular ion. The two nuclei are located along the z-axis, separated by the distance R_{AB}. The coordinates r_A and r_B are the distances of the electron from nuclei A and B, respectively; r_A and r_B range from 0 to ∞. The angle θ is determined from r_A, r_B, and R_{AB} by the law of cosines; it does not appear explicitly in the calculation. The angle ϕ varies from 0 to 2π.

The quantum treatment of this molecule appears to be a straightforward extension of that for the hydrogen atom (see Section 5.1 for the solutions). One more proton has been added, and symmetry of the system has changed from spherical to cylindrical. But this apparently small change has converted the system into the so-called three-body problem, for which there is no general solution, even in classical physics. Fortunately, there is an excellent approximation that treats the nuclear and electronic motions almost as if they were independent. This method allows us to solve the Schrödinger equation exactly for the electron in H_2^+ and provides an approximate solution for the protons as if they were almost independent.

6.1.2 Born–Oppenheimer Approximation

The **Born–Oppenheimer approximation** was developed in 1927 by the physicists Max Born (German) and J. Robert Oppenheimer (American), just one year after Schrödinger presented his quantum treatment of the hydrogen atom. This approximation method is the foundation for all of molecular quantum mechanics, so you should become familiar with it. The basic idea of the Born–Oppenheimer approximation is simple: because the nuclei are so much more massive than the electrons, they can be considered fixed for many periods of electronic motion. Let's see if this is a reasonable approximation. Using H_2 as a specific example, we estimate the velocity of the electrons to be roughly the same as that of an electron in the ground state ($n = 1$) of the hydrogen atom. From the Bohr formula, $v = e^2/2\epsilon_0 h$, we calculate an electron velocity of 2.2×10^6 m sec^{-1}. We can calculate a typical nuclear velocity from the vibrational spectrum of H_2 (see Section 4.7); the result is 1×10^4 m sec^{-1}. The electrons do indeed move much more rapidly than the nuclei, even for the lightest diatomic molecule, H_2. So, it is quite reasonable to think of the nuclei as fixed in space for many periods of electronic motion. We simply fix the nuclei in space, solve the Schrödinger equation for the electron with the nuclei fixed at that position, then move the nuclei a bit, and repeat the calculation until we have covered all reasonable values of the nuclear positions. This series of calculations gives $\psi_{el}(r_A, r_B, \phi; R_{AB})$, the *electronic wave function* for the electrons around the fixed nuclei. The semicolon inside the parentheses indicates that the nuclear coordinates are held fixed as a *parameter*, whereas the electronic coordinates range over all values as we seek the solution for ψ_{el}. Subsequently, we put in corrections to account for the sluggish motion of the nuclei by solving for the *nuclear wave function*, $\psi_{nuc}(R_{AB})$. By solving the electronic and nuclear motions separately, the Born–Oppenheimer approximation obtains the molecular wave function for H_2^+ in the form

$$\psi_{mol}(R_{AB}, r_A, r_B, \phi) \approx \psi_{el}(r_A, r_B, \phi; R_{AB}) \times \psi_{nuc}(R_{AB}) \qquad [6.2]$$

The electronic wave function ψ_{el} lies at the heart of chemical bonding because its square gives the probability density for locating the electrons at particular positions around the fixed nuclei. (Probability density is the probability per unit volume. See the discussion preceding Eq. 4.30.) Thus, we concentrate here on obtaining ψ_{el} and later will describe motion of the nuclei using ψ_{nuc}.

To obtain the specific form of ψ_{el} for H_2^+, we solve the Schrödinger equation for the electron by the same methods we applied to the hydrogen atom in Section 5.1. We enforce general mathematical conditions to ensure we obtain a legitimate wave function. (The solution must be smooth, single-valued, and finite in value in all regions of space.) We also enforce general boundary conditions to guarantee that our solution describes a bound state of the electron (ψ_{el} must approach 0 as $r_A \rightarrow \infty$ and as $r_B \rightarrow \infty$). We find that solutions exist only when the total energy and the component of angular momentum along the internuclear axis are quantized. The complete set of quantum numbers is more extensive than those for the hydrogen atom. In particular, the allowed energy values for the electron cannot be represented by simple, discrete diagrams as we obtained for atoms in Chapter 5. Rather, we generate an explicit mathematical relation between each allowed value $E_n^{(el)}$ of the electronic energy and the values of R_{AB}. This family of curves shows how the allowed energy values for the electron depend on the internuclear separation for each quantum state. We focus first on describing the electronic wave functions at a fixed internuclear separation, R_{AB}. We then discuss how the energy of the system changes as a function of the internuclear distance.

6.1.3 Electronic Wave Functions for H_2^+

We emphasize that the solutions for ψ_{el} are mathematically exact. Because of their ellipsoidal symmetry, they cannot be written as simple exponential and polynomial functions that are easily manipulated, as were the hydrogen atom solutions. Consequently, we present and interpret these exact solutions in graphical form. The first eight wave functions, starting with the ground state, are shown in Figure 6.2. These are plotted using the same coordinates as in Figure 6.1, where the two protons lie on the z-axis and the value of R_{AB} is the experimental bond length for H_2^+, which is 1.060 Å. Each wave function is shown in three different representations: (a) an isosurface comprising all those points in three-dimensional space where the wave function has a value equal to 0.1 of its maximum value; (b) a contour plot in a plane containing the internuclear axis with contours shown for ±0.1, ±0.3, ±0.5, ±0.7, and ±0.9 of the maximum amplitude; (c) a plot of the amplitude along the internuclear axis, which amounts to a "line scan" across the contour plot. Each wave function is identified by four labels: an integer, either σ or π, a subscript g or u, and some are labeled with a superscript asterisk. Each of these labels provides insight into the shape and symmetry of the wave function and its corresponding probability density for locating the electron. These labels are discussed in turn in the following paragraphs.

First, the Greek letter identifies the component of the angular momentum for the electron that is directed along the internuclear axis. This tells us how the electron probability density is distributed around the internuclear axis, as viewed in a plane perpendicular to the axis. The Greek letter identifies the value of the angular momentum as follows:

$$\sigma \longrightarrow \text{angular momentum component} = 0$$
$$\pi \longrightarrow \text{angular momentum component} = \pm h/2\pi$$
$$\delta \longrightarrow \text{angular momentum component} = \pm 2h/2\pi$$
$$\varphi \longrightarrow \text{angular momentum component} = \pm 3h/2\pi$$

Because the σ wave functions have no angular nodes (just like the s orbitals in an atom), they have finite amplitudes on the internuclear axis, and the probability of finding the electron on that axis is therefore finite. Although not shown in Figure 6.2, the σ wave functions are cylindrically symmetric about the internuclear axis. The π wave functions, in contrast, describe electron motion about the internuclear axis with angular momentum $+h/2\pi$ or $-h/2\pi$. This leads to two wave functions—π_{u2p_x} and π_{u2p_y}—with the same energy, one of which lies mainly along the x-axis and the other of which lies mainly along the y-axis (the internuclear axis is chosen to be the z-axis). Because the π wave functions have nodal planes that include the internuclear axis, there is zero probability of finding the electron anywhere along the z-axis, just as there is zero probability of finding the electron at the nucleus in an atomic p orbital. Viewed perpendicular to the z-axis, the π

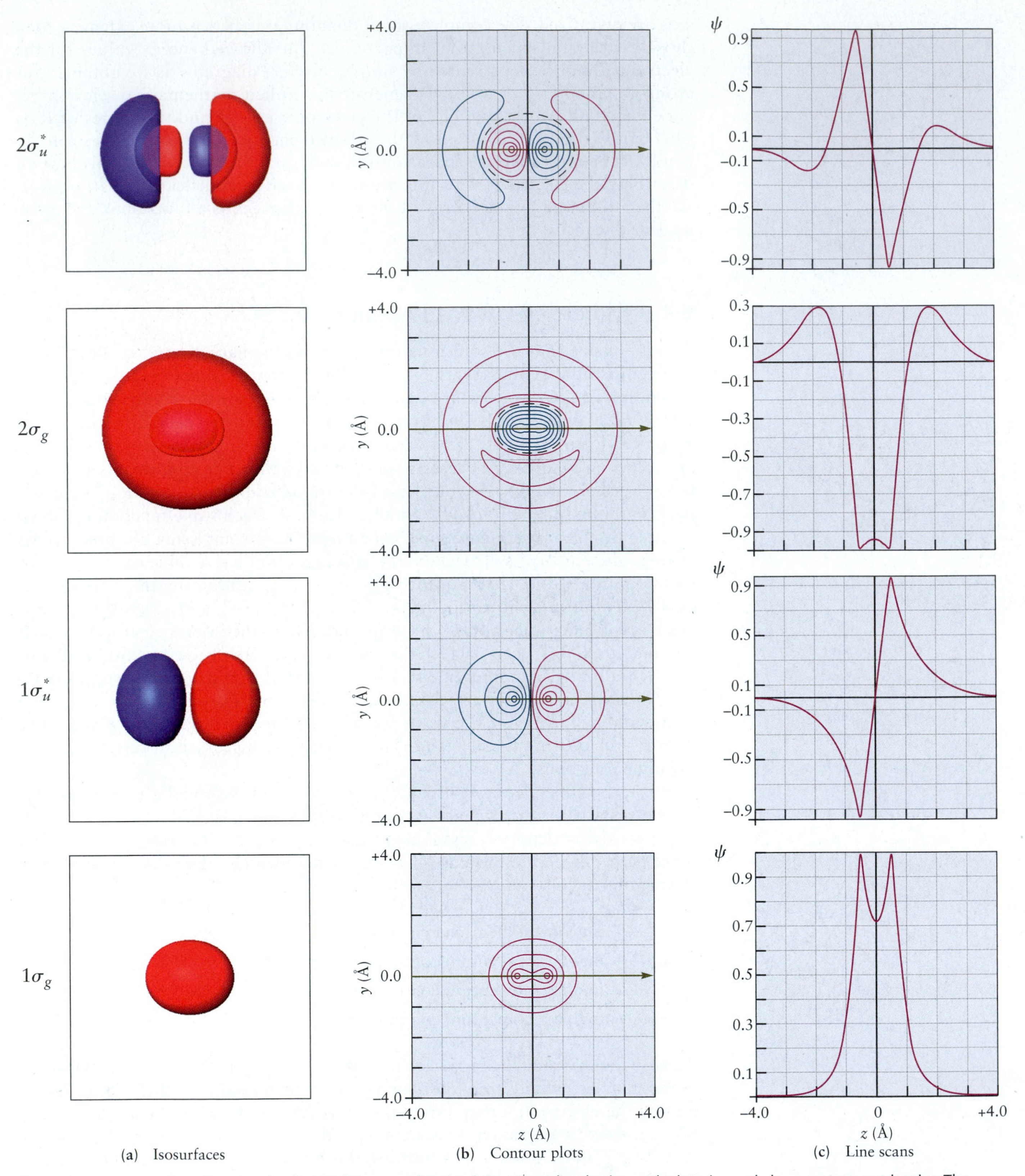

FIGURE 6.2 Wave functions for the first eight energy levels of the H_2^+ molecular ion, calculated exactly by quantum mechanics. The ground-state wave function is at the bottom of the figure; the others are arranged above it in order of increasing energy. The two nuclei lie along the *z*-axis, which is in the plane of the paper. Regions of positive and negative amplitude are shown in red and blue, respectively. The labels for each orbital are explained in the text. (a) Isosurfaces corresponding to contours at ± 0.1 of the maximum amplitude.

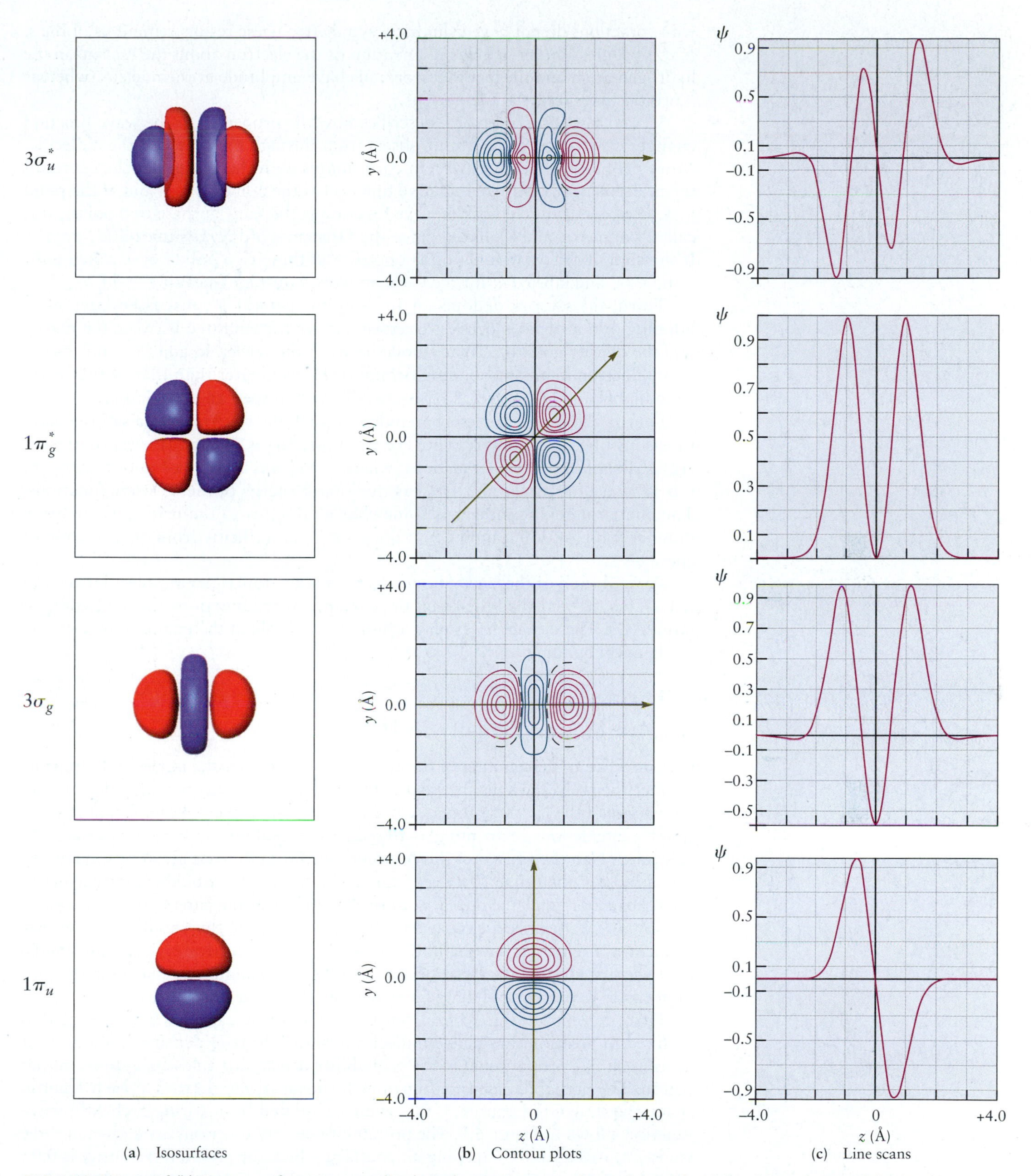

FIGURE 6.2 cont'd (b) Contours of constant amplitude in the *x-z* plane, at values ±0.1, ±0.3, ±0.5, ±0.7, and ±0.9 of the maximum amplitude. Radial nodes are represented by black dashed lines. (c) The amplitude along the *z*-axis, obtained as a "line scan" across the contour plot, along the direction indicated by the green arrow in (b).
(Courtesy of Mr. Hatem Helal and Professor William A. Goddard III, California Institute of Technology, and Dr. Kelly P. Gaither, University of Texas at Austin.)

wave functions do not have cylindrical symmetry. These results remind us of the *s*, *p*, *d*, *f* progression for angular momentum of the electron about the nucleus in the hydrogen atom and the fact that *s* orbitals have amplitude at the nucleus, whereas *p* orbitals have nodes at the nucleus.

Second, the subscript *g* or *u* describes how the properties of the wave function change as we invert our point of observation through the center of the molecule. More precisely, imagine Cartesian coordinates with their origin at the center of the molecule, and compare the wave function at the point (x, y, z) and at the point $(-x, -y, -z)$. If the sign of the wave function is the same at these two points, it is called symmetric and is labeled *g*, for the German word *gerade* (meaning "even"). If the sign of the wave function is opposite at these two points, it is called antisymmetric and labeled *u*, for the German word *ungerade* (meaning "odd").

Third, the asterisk denotes an antibonding orbital. As discussed later, antibonding orbitals have much less electron density concentrated between the nuclei, and the density goes to zero at a node between the nuclei. In addition, the energy of an electron in an antibonding orbital of H_2^+ is greater than that of the corresponding H and H^+ species, so it is unstable with respect to dissociation.

Fourth, the integer is merely an index to track the relative energies of the wave functions of each symmetry type. For example, $1\sigma_g$ is the first (lowest on the energy scale) of the σ_g wave functions, whereas $2\sigma_u^*$ has the second lowest energy of the σ_u^* wave functions, and $1\pi_u$ has the lowest energy of the π_u wave functions. The energy indexing integer is somewhat analogous to the principal quantum number *n* for AOs, but unlike *n*, it does not arise explicitly from quantization of energy.

We call each of these exact one-electron wave functions a molecular orbital (MO), just as we called the exact one-electron wave functions for the hydrogen atom AOs. These exact MOs play a fundamental role in the quantum description of chemical bonding.

6.1.4 Electron Density in H_2^+

Our intuitive understanding of the chemical bond is that the nuclei and electrons arrange themselves in a manner that reduces their total energy to a value lower than the energy of the isolated atoms. Achieving this arrangement requires that new attractive interactions come into play to reduce the total potential energy as the bond is formed. Naively, we expect this to occur when the electrons are arranged so they spend most of their time "between" the nuclei where they would experience maximum attraction to all the nuclei, not just the nucleus of the parent atom. We saw in Section 3.6 that classical electrostatics could not explain this stabilization mechanism, so we relied on the qualitative Lewis model of electron-shared pair bonds. Now we want to see how these ideas are handled by quantum mechanics in the simplest case. We recommend that you review Section 3.6 at this point.

First, let's explore the electron density around the nuclei, which are assumed to be fixed in position. In quantum mechanics, the electron density in a region of space is simply proportional to the probability density for an electron to be in that region. The probability density functions for locating the electron at each point in space are shown in Figure 6.3; they were calculated by squaring each MO wave function shown in Figure 6.2. The probability density functions are shown in three views: (a) isosurfaces comprising all points at which the probability density is 0.01 of its maximum value; (b) contour plots in a plane containing the internuclear axis with contours at 0.05, 0.01, 0.3, 0.5, 0.7, and 0.9 of the maximum value; (c) line scans across the contour plot show the variation in probability density along the internuclear axis. The ground-state wave function $1\sigma_g$ has much greater electron density in the region between the nuclei than at the extremes of the molecule, and so is consistent with our expectations about chemical bonding. But, the first excited state wave function $1\sigma_u^*$ has a node halfway between the nuclei, and thus appears inconsistent with a chemical bond.

It appears that $1\sigma_g$ supports formation of the bond by increasing electron density between the nuclei, whereas $1\sigma_u^*$ opposes bond formation by reducing electron density between the nuclei, relative to the density between noninteracting atoms. Yet, both functions are part of the exact quantum solution for H_2^+. To understand the role each of them plays in forming the bond, we need to determine how the energy of these states compares with the energy of the isolated atom and proton. Reducing the energy relative to the separated particles is the key to bond formation.

Solving the Schrödinger equation for H_2^+ within the Born–Oppenheimer approximation gives the electronic energy $E_n^{(el)}(R_{AB})$ as a function of the positions of the nuclei. The results, which we quote without verification, are as follows. At all distances, the energy in the $1\sigma_u^*$ MO is greater than that of the separated atoms and increases as the internuclear distance becomes shorter. This energy represents a repulsive interaction, under which the nuclei would fly apart rather than remain close together. In contrast, the energy in the $1\sigma_g$ MO is lower than that for noninteracting hydrogen atoms, and indeed reaches a minimum between the nuclei. In this orbital, the molecular ion is energetically stable with a bond length that corresponds to the minimum in the potential energy curve.

Considered together, the increased electron density and the lowered energy between the nuclei in the $1\sigma_g$ orbital indicate formation of a chemical bond. Consequently, $1\sigma_g$ is called a **bonding molecular orbital.** By contrast, the $1\sigma_u^*$ orbital shows increased energy and zero electron density between the nuclei and is called an **antibonding molecular orbital.** Similar analysis of the remaining six orbitals in Figure 6.2 shows that all those labeled with superscript asterisk (*) are antibonding, whereas those without the (*) labels are bonding, with increased electron density and decreased potential energy between the nuclei.

6.1.5 Summary: Key Features of the Quantum Picture of Chemical Bonding

In the remainder of this chapter, key features of the quantum picture of the chemical bond and the exact MOs for H_2^+ guide the development of approximate MO methods. These features are:

1. In representing molecular structures, we consider the nuclei to be fixed in specific positions while the electrons move rapidly around them (the Born–Oppenheimer approximation).
2. A molecular orbital is a one-electron wave function whose square describes the distribution of electron density around the nuclei.
3. Bonding orbitals describe arrangements with increased electron density in the region between the nuclei and decreased potential energy relative to that of the separated atoms.
4. Each bonding MO is related to an antibonding MO in which the amplitude is zero at some point corresponding to a node on the internuclear axis between the nuclei and the potential energy is increased relative to that of the separated atoms.
5. Bonding and antibonding MOs of type σ are those for which the electron has no component of angular momentum along the internuclear axis. A σ orbital is cylindrically symmetric about the internuclear axis.
6. Bonding and antibonding MOs of type π are those for which the electron has one unit of angular momentum along the internuclear axis. The internuclear axis lies in the nodal plane of a bonding π orbital. Consequently, the amplitude is concentrated "off the axis" and the orbital is not cylindrically symmetric.

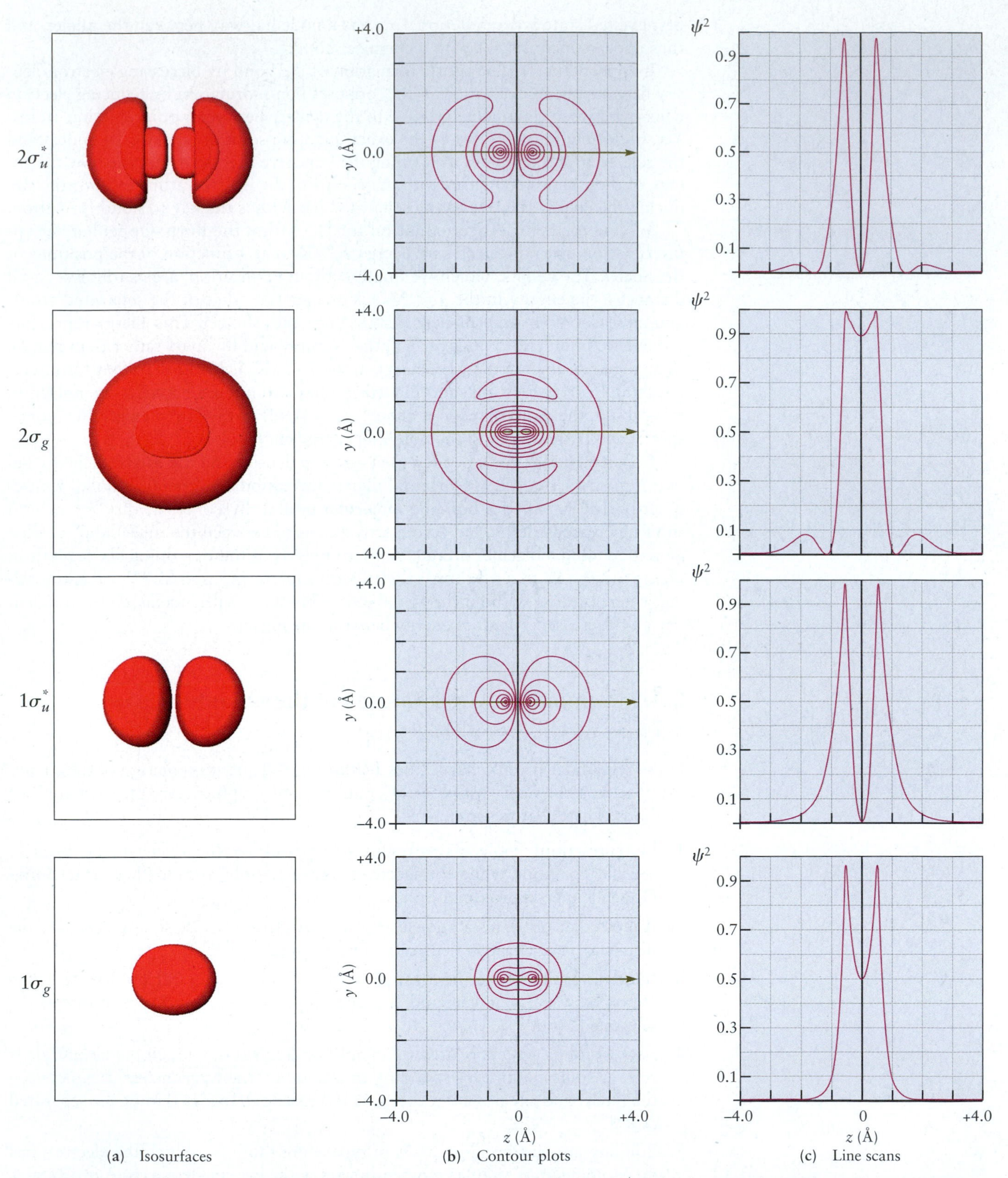

FIGURE 6.3 Probability density distributions for the first eight energy levels of the H_2^+ molecular ion, calculated exactly by quantum mechanics. (a) Isosurfaces comprising all points at which the probability density is 0.1 of its maximum value. (b) Contour plots in the *x-z* plane with contours at 0.01, 0.1, 0.3, 0.5, 0.7, and 0.9 of the maximum value. (c) Line scans across the contour plot, along the direction indicated by the green arrow in (b), showing the variation in probability density along the internuclear axis.
(Courtesy of Mr. Hatem Helal and Professor William A. Goddard III, California Institute of Technology, and Dr. Kelly P. Gaither, University of Texas at Austin.)

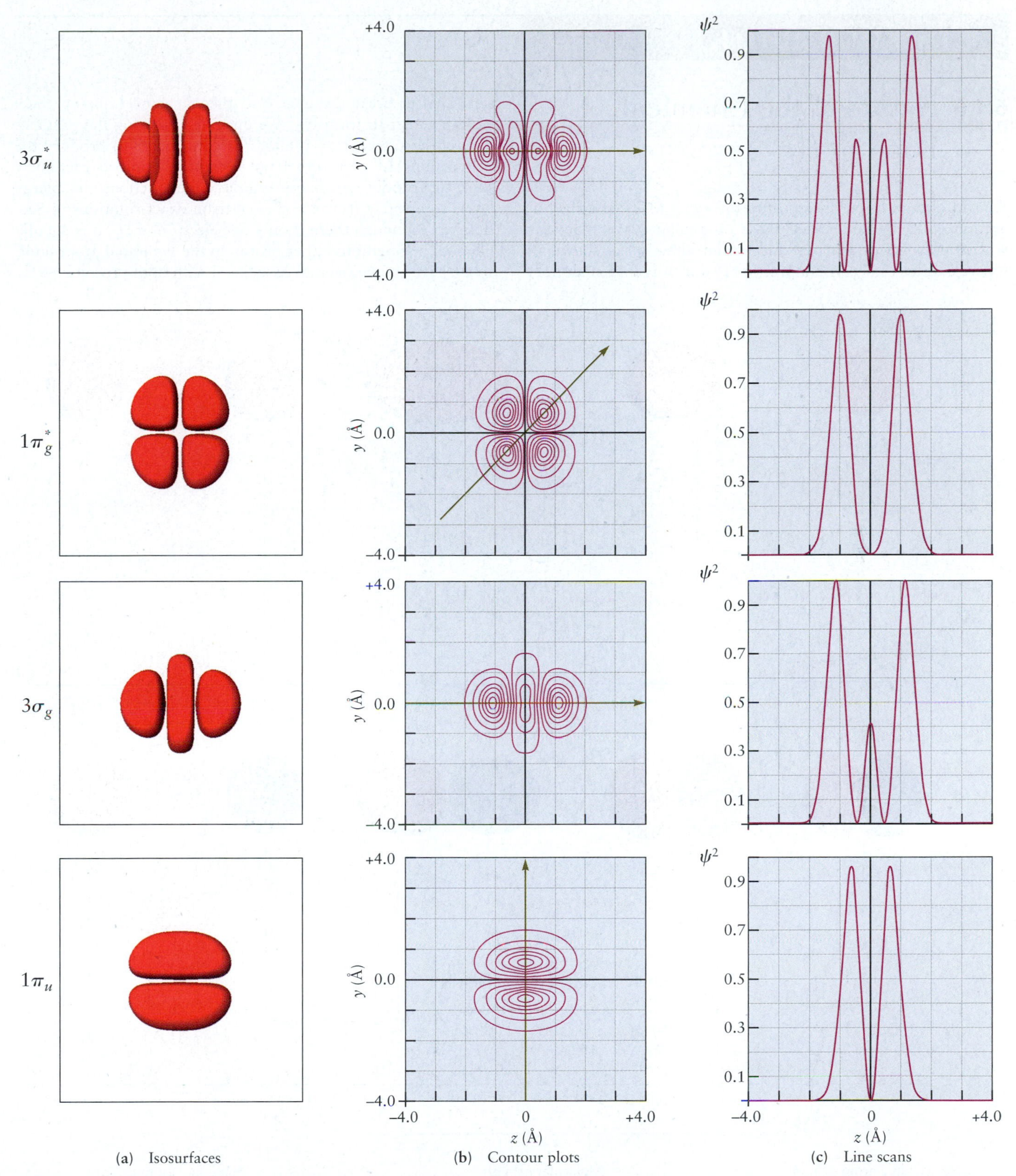

FIGURE 6.3 cont'd Probability density distributions for the first eight energy levels of the H_2^+ molecular ion, calculated exactly by quantum mechanics. (a) Isosurfaces comprising all points at which the probability density is 0.1 of its maximum value. (b) Contour plots in the *x-z* plane with contours at 0.01, 0.1, 0.3, 0.5, 0.7, and 0.9 of the maximum value. (c) Line scans across the contour plot, along the direction indicated by the green arrow in (b), showing the variation in probability density along the internuclear axis.
(Courtesy of Mr. Hatem Helal and Professor William A. Goddard III, California Institute of Technology, and Dr. Kelly P. Gaither, University of Texas at Austin.)

A DEEPER LOOK

6.1.6 Nature of the Chemical Bond in H_2^+

We can obtain deeper insight into the exact MOs described in Section 6.1.3 by comparing these unfamiliar wave functions with simpler cases that we understand already. As shown in Figure 6.4, let's start with a proton H^+ and a hydrogen atom H separated at great distance, and then imagine bringing them close together to form H_2^+. We imagine squeezing the particles even closer together, so the two protons merge to produce the helium ion He^+. At very large separations and at extremely short separations we understand the wave functions and energy levels completely from the one-electron exact solutions in Section 5.1. Although there is only one electron in H_2^+ it is equally likely to be bound to either proton in the separated atom limit, so we have shown orbitals associated with both protons.

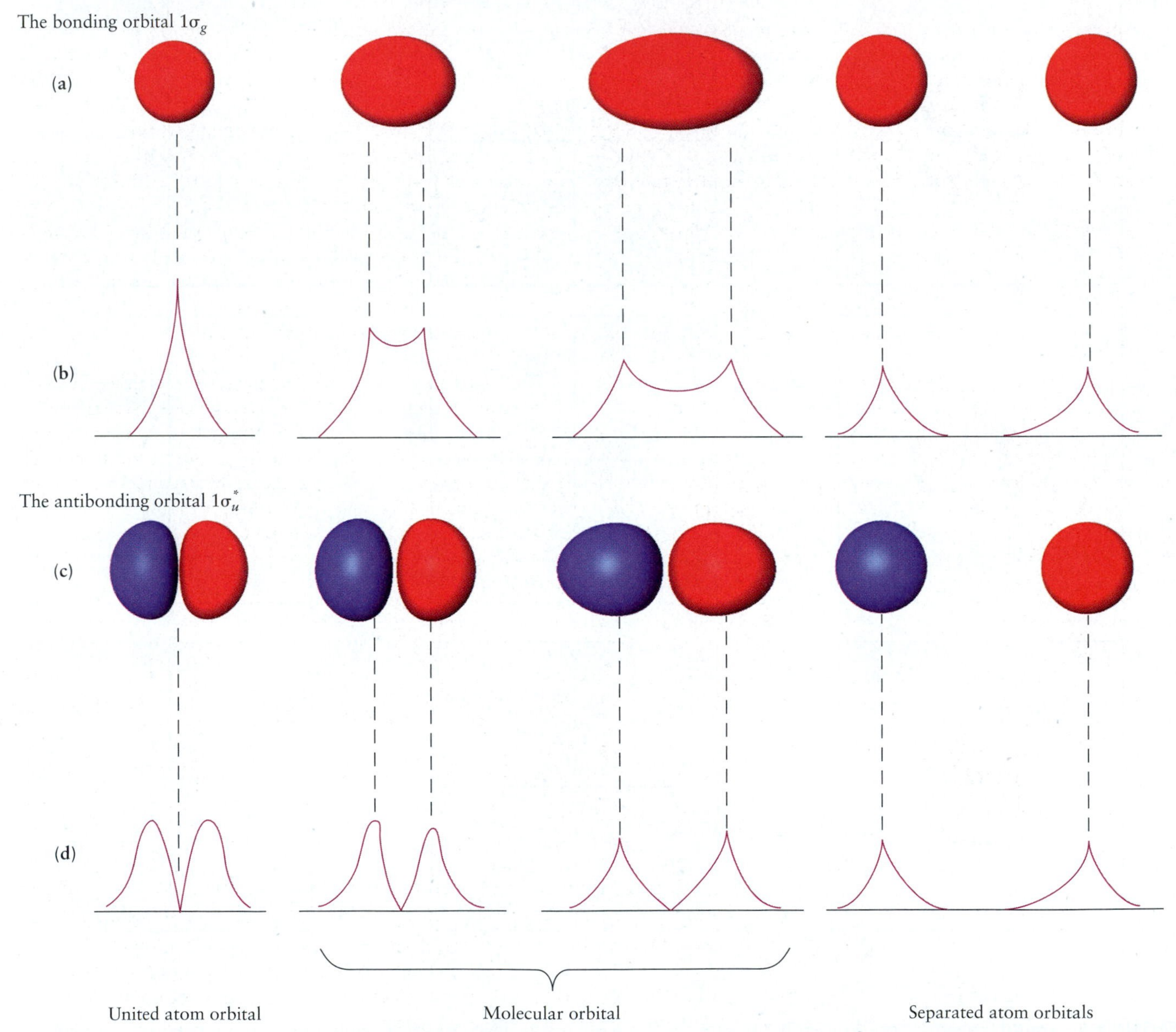

FIGURE 6.4 Evolution of the atomic orbitals on H and H^+ as the separation between them is decreased, eventually resulting in the atomic orbitals of He^+. The molecular ion H_2^+ appears at intermediate distances between the united atom and the separated atom limits. (a) Isosurfaces representing the changes in shape during the evolution of the bonding MO. (b) Line scans representing the changes in amplitude along the internuclear axis during the evolution of the bonding MO. (c) Isosurfaces representing the changes in shape during the evolution of the antibonding MO. (d) Line scans representing the changes in amplitude along their internuclear axis during the evolution of the antibonding MO.
(Adapted from G. C. Pimentel and R. D. Spratley, *Chemical Bonding Clarified through Quantum Mechanics*. San Francisco: Holden-Day, 1969, Figures 3.5 and 3.4, by permission.)

As the distance is initially decreased in Figure 6.4a, we expect the AOs of the H atom to distort smoothly into the wave functions of H_2^+, which, in turn, distort smoothly into the AOs of He^+. The reason is that this mental experiment is just another way of looking at the Born–Oppenheimer approximation. As we move the sluggish nuclei extremely slowly from one position to another, the electron density will adjust its shape to accommodate the changed internuclear distances, but will not be disrupted. In particular, the nodal structure of the wave function cannot change. As we know from solutions for the particle in box models (see Section 4.6) and the hydrogen atom (see Section 5.1), the number of nodes increases only when the system is moved into an excited state by absorbing energy. There is not sufficient energy in the experiment we are describing here to create excited states.

We can track exactly which AOs are involved in these changes by paying careful attention to the nodes in the wave functions. Suppose we start with the electron in the ground state of He^+ and pull that ion apart as the reverse of the process just described. The $He^+(1s)$ orbital is spherical and has no nodes. As the nuclei are separated, that orbital is distorted first into an elliptical shape, then at large internuclear distance it begins to resemble two separated atom wave functions with the same phase. These must be 1*s* orbitals; any others would require changes in the nodal structure. No new nodes have been created; the nodal pattern has remained constant through the mental experiment. The change in the wave function and the probability density in the experiment are shown in Figures 6.4a and b, respectively.

Now we have some insight into the structure of the ground-state wave function $1\sigma_g$. When the nuclei are infinitely far apart, it correlates with an in-phase or symmetric combination of the AOs, in which the orbitals have the same phase: $\varphi_{1s}^{A} + \varphi_{1s}^{B}$. In physical terms, when a proton and a hydrogen atom approach one another, at intermediate distances the electron begins to experience the attraction of both protons, and instead of residing in the orbitals of just one of them, it is now described by a new wave function characteristic of the entire molecule. We can visualize that this new wave function arose from constructive interference of the two AOs to produce amplitude spread over both nuclei. We have understood qualitatively how its electron density is greater between the nuclei than at the extremes of the molecule (see Fig. 6.3) and recognized that it is crudely approximated at large internuclear separations by the sum $\varphi_{1s}^{A} + \varphi_{1s}^{B}$.

The same reasoning provides insight into the first excited state $1\sigma_u^*$. Because it has a node between the two nuclei, at very small separations it must correlate with an orbital of He^+ that also has a nodal plane perpendicular to the *z*-axis, that is, in the *x*-*y* plane. The $2p_z$ orbital (see Fig. 5.8) is the proper choice. Now let's imagine pulling apart the protons in He^+ with its electron in the $He^+(2p_z)$ orbital. The orbital will distort into an elliptical shape along the *z*-axis, with a node in its center. At very large distances, this must resemble two separated atom wave functions. The only way to maintain the nodal structure is to correlate with the out-of-phase, antisymmetric combination in which the AOs have opposite phase: $\varphi_{1s}^{A} - \varphi_{1s}^{B}$. The change in the wave function and in the probability density are shown in Figures 6.4c and d, respectively. In physical terms, when a proton and a hydrogen atom approach each other at intermediate distances the electron still feels the attractions of both nuclei and is again described by a new MO characteristic of the molecule as a whole. We must also allow for the possibility that the AOs interfere destructively, giving reduced amplitude in the region between the nuclei, with a node at the midpoint of the internuclear separation. Thus, we have explained qualitatively how the MO $1\sigma_u^*$ has less electron density between the nuclei than in the extremes, and we have recognized that it is approximated crudely at large internuclear separation by the difference $\varphi_{1s}^{A} - \varphi_{1s}^{B}$.

A similar analysis can be applied to the remaining six exact MOs shown in Figure 6.2. Each of them is intermediate between an AO for He^+ (at very short separations) and the sum or difference of a pair of AOs for H (at very large separations). The results are summarized in Figure 6.5. We recommend that you work through each of these correlations in Figure 6.5 to develop experience and intuition for the exact MOs and their relation to AOs. We rely on Figure 6.5 to generate approximate MOs later in the chapter.

6.2 De-localized Bonds: Molecular Orbital Theory and the Linear Combination of Atomic Orbitals Approximation

The goal of this section is to introduce the first of the two main methods for generating approximate electronic wave functions for molecules. The methods we describe work well in qualitative descriptions of bonding and are used for that purpose in all branches of chemistry. They also serve as the starting point for sophisticated computer calculations of molecular electronic structure with readily available molecular modeling software packages.

The LCAO method extends to molecules the description developed for many-electron atoms in Section 5.2. Just as the wave function for a many-electron atom is written as a product of single-particle AOs, here the electronic wave function for a molecule is written as a product of single-particle MOs. This form is called the **orbital approximation for molecules.** We construct MOs, and we place electrons in them according to the Pauli exclusion principle to assign molecular electron configurations.

FIGURE 6.5 Correlation table showing how six of the exact H_2^+ MOs correlate at large separations to sums or differences of hydrogen atom orbitals and at short separations to the atomic orbitals of He^+.
(Adapted from R. S. Berry, S. A. Rice, and J. Ross, *Physical Chemistry* (2nd ed), New York: Oxford, 2000, Figure 6.2, by permission.)

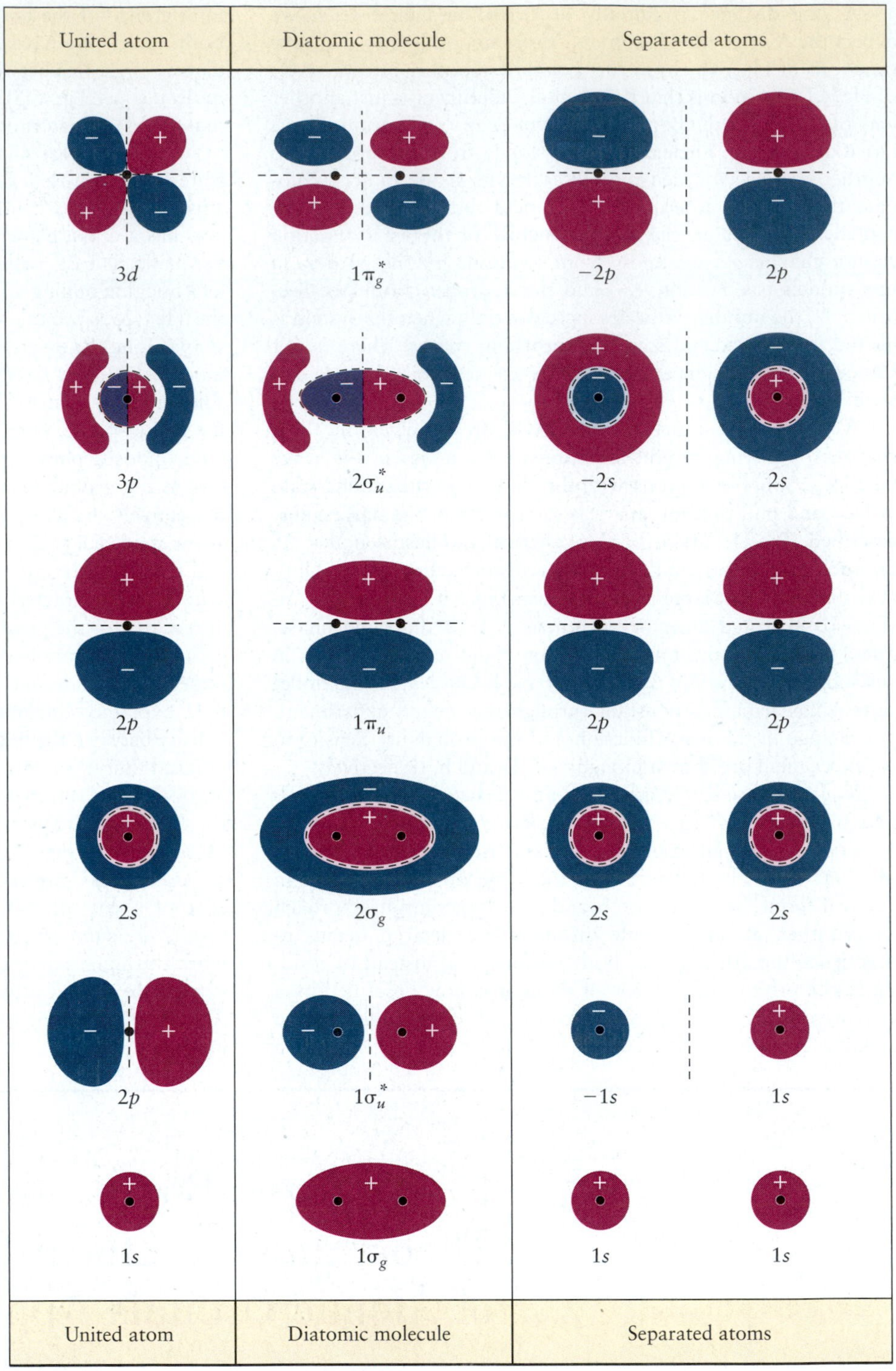

United atom	Diatomic molecule	Separated atoms	
$3d$	$1\pi_g^*$	$-2p$	$2p$
$3p$	$2\sigma_u^*$	$-2s$	$2s$
$2p$	$1\pi_u$	$2p$	$2p$
$2s$	$2\sigma_g$	$2s$	$2s$
$2p$	$1\sigma_u^*$	$-1s$	$1s$
$1s$	$1\sigma_g$	$1s$	$1s$
United atom	Diatomic molecule	Separated atoms	

What functions shall we use for the single-electron MOs? We could try the exact MOs for H_2^+ shown in Figure 6.2. However, these exact MOs are not described by simple equations, and they are inconvenient for applications. Therefore, we introduce the LCAO method to construct *approximate* MOs directly from the Hartree AOs for the atoms in the molecule, guided by molecular symmetry and chemical intuition. The essential new feature compared with the atomic case is that the (multicenter) approximate MOs are spread around all the nuclei in the molecule, so the electron density is *de-localized* over the entire molecule. The approximate MOs therefore differ considerably from the (single-center) AOs used in Section 5.2. Constructing the approximate MOs and using them in qualitative descriptions of bonding are the core objectives of this section.

Computer calculations of molecular electronic structure use the orbital approximation in exactly the same way. Approximate MOs are initially generated by starting with "trial functions" selected by symmetry and chemical intuition. The electronic wave function for the molecule is written in terms of trial functions, and then optimized through self-consistent field (SCF) calculations to produce the best values of the adjustable parameters in the trial functions. With these best values, the trial functions then become the optimized MOs and are ready for use in subsequent applications. Throughout this chapter, we provide glimpses of how the SCF calculations are carried out and how the optimized results are interpreted and applied.

6.2.1 Linear Combination of Atomic Orbitals Approximation for H_2^+

The LCAO method is best explained with the help of a specific example, so we start with the hydrogen molecular ion H_2^+. We can evaluate the success of the method by comparing its results with the exact solution described in Section 6.1, and gain confidence in applying the method to more complex molecules. The LCAO approximation is motivated by the "separated atom" limits of the exact MOs shown in Figure 6.2. Consider the $1\sigma_g$ MO. When the electron is close to nucleus A, it experiences a potential not very different from that in an isolated hydrogen atom. The ground-state wave function for the electron near A should therefore resemble a $1s$ atomic wave function φ_{1s}^{A}. Near B, the wave function should resemble φ_{1s}^{B}. The $1s$ orbitals at A and B are identical; the labels are attached to emphasize the presence of two nuclei. Note that A and B are labels, not exponents. A simple way to construct a MO with these properties is to approximate the $1\sigma_g$ MO as a sum of the H $1s$ orbitals with adjustable coefficients. Their values will change as the internuclear distance is reduced, to adjust the limiting form to values appropriate for each internuclear distance. The best value of the coefficients for each value of R_{AB} can be determined by self-consistent numerical calculations. For our purposes, it is adequate to ignore the dependence on R_{AB} when we evaluate the coefficients by normalizing the wave function. Similarly, we can approximate the other exact MOs in Figure 6.2 by forming linear combinations of the AOs to which they correlate in the separated atom limit in Figure 6.5.

The general form for an approximate MO is a linear combination of two AOs, obtained by adding or subtracting the two with coefficients whose values depend on R_{AB}:

$$\psi_{MO} = C_A\varphi_{1s}^{A} + C_B\varphi_{1s}^{B}$$

The coefficients C_A and C_B give the relative weights of the two AOs. If C_A were greater in magnitude than C_B, the φ_{1s}^{A} orbital would be more heavily weighted and the electron would be more likely to be found near nucleus A, and vice versa. But because the two nuclei in H_2^+ are identical, the electron is just as likely to be found near one nucleus as the other. Therefore, the magnitudes of C_A and C_B must be equal, and either $C_A = C_B$ or $C_A = -C_B$. For both these choices, $(\psi_{MO})^2$ is symmetric in the two nuclei.

To maintain the distinctions among the various orbitals, we use the following notation in describing the LCAO approximation. Atomic orbitals will be represented by φ and MOs by σ or π. Generic wave functions will be represented by ψ. Occasionally, ψ will represent some special wave function, in which case appropriate subscripts will be attached.

LCAO MOLECULAR ORBITALS FOR H_2^+ Proceeding as described in the preceding paragraphs, we construct approximate MOs for the exact $1\sigma_g$ and $1\sigma_u^*$ MOs in Figure 6.2:

$$1\sigma_g \approx \sigma_{g1s} = C_g(R_{AB})[\varphi_{1s}^A + \varphi_{1s}^B] \quad [6.3a]$$

$$1\sigma_u^* \approx \sigma_{u1s}^* = C_u(R_{AB})[\varphi_{1s}^A - \varphi_{1s}^B] \quad [6.3b]$$

where C_g and C_u are chosen to ensure that the total probability of finding the electron *somewhere* is one. Their values will depend on the choice of R_{AB} at which we have fixed the nuclei. Notice that we have introduced new symbols σ_{g1s} and σ_{u1s}^* for the approximate MOs, not only to distinguish them from the exact MOs but also to indicate explicitly the AOs from which they were constructed. The symbols for the exact and approximate MOs are summarized in Table 6.1. To simplify the notation, henceforth we omit the dependence of σ_{g1s} and σ_{u1s}^* on R_{AB}. The distribution of electron probability density is obtained by squaring each of the approximate MOs:

$$[\sigma_{g1s}]^2 = C_g^2[(\varphi_{1s}^A)^2 + (\varphi_{1s}^B)^2 + 2\varphi_{1s}^A\varphi_{1s}^B] \quad [6.4a]$$

$$[\sigma_{u1s}^*]^2 = C_u^2[(\varphi_{1s}^A)^2 + (\varphi_{1s}^B)^2 - 2\varphi_{1s}^A\varphi_{1s}^B] \quad [6.4b]$$

These probability distributions can be compared with the probability distribution for a noninteracting (n.i.) system (obtained by averaging the probabilities for $H_A + H_B^+$ and $H_A^+ + H_B$), which is

$$\psi_{n.i.}^2 = C_3^2[(\varphi_{1s}^A)^2 + (\varphi_{1s}^B)^2] \quad [6.5]$$

To describe the noninteracting system as one electron distributed over two possible sites, we set $C_3^2 = 0.5$. The interpretation of these approximate MOs in relation to the noninteracting system is best explained graphically. The plots of these various wave functions (left side) and their squares (right side) are shown in Figure 6.6. Compared with the noninteracting pair of atoms, the system described by the approximate MO σ_{g1s} shows increased electron density with a node between the nuclei. It is therefore a bonding orbital as defined in Section 6.1. By contrast, the approximate MO σ_{u1s}^* shows reduced probability for finding an electron between the nuclei and so is an antibonding orbital. Note in Figure 6.6 that σ_{u1s}^* has a node between the nuclei and is antisymmetric for inversion through the molecular center (see Section 6.1). Comparing Figure 6.6 with Figure 6.2 shows that the LCAO method has reproduced qualitatively the probability density in the first two exact wave functions for H_2^+.

ENERGY OF H_2^+ IN THE LCAO APPROXIMATION To complete the demonstration that σ_{g1s} and σ_{u1s}^* are bonding and antibonding MOs, respectively, we

TABLE 6.1 Molecular Orbitals for Homonuclear Diatomic Molecules

Exact MO Notation	LCAO MO Notation
$1\sigma_g$	σ_{g1s}
$1\sigma_u^*$	σ_{u1s}^*
$2\sigma_g$	σ_{g2s}
$2\sigma_u^*$	σ_{u2s}^*
$1\pi_u$	π_{u2p_x}, π_{u2p_y}
$3\sigma_g$	σ_{g2p_z}
$1\pi_g^*$	$\pi_{g2p_x}^*, \pi_{g2p_y}^*$
$3\sigma_u^*$	$\sigma_{u2p_z}^*$

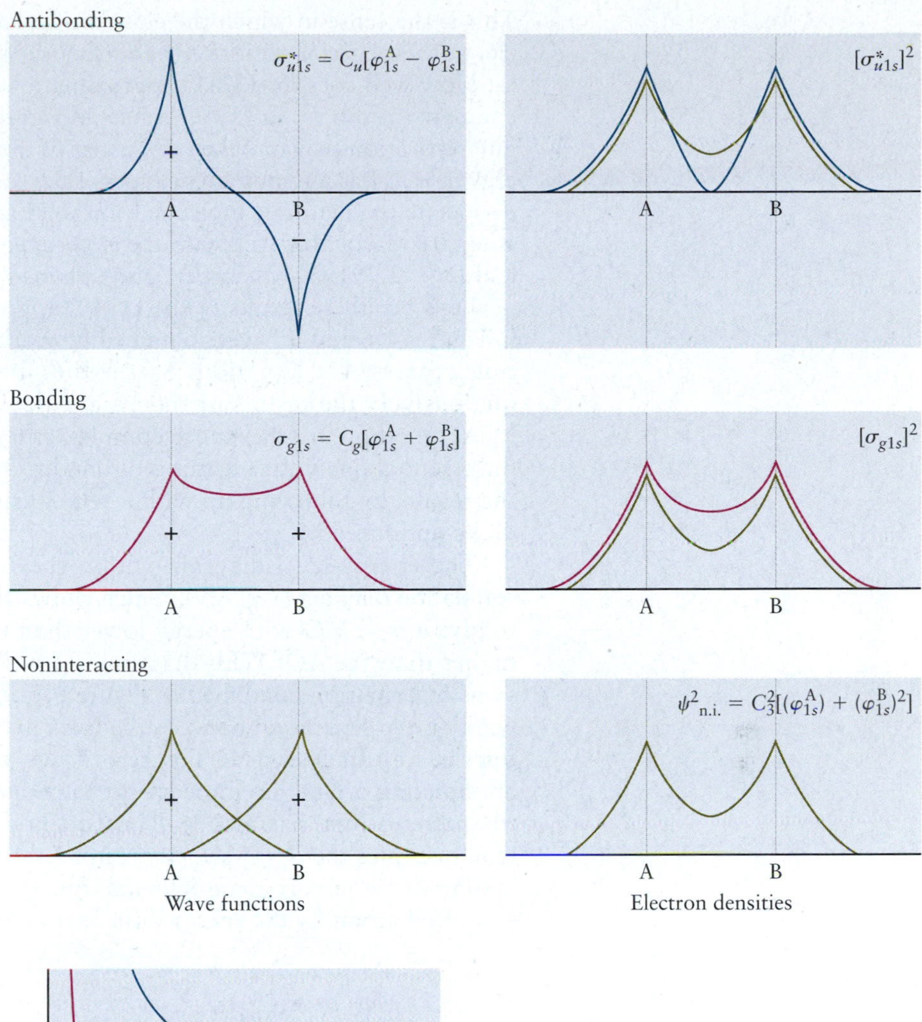

FIGURE 6.6 Antibonding and bonding molecular orbitals of H_2^+ along the internuclear axis in the linear combination of atomic orbitals (LCAO) approximation. For comparison, the green lines show the independent atomic orbitals and the electron probability distribution ψ^2(n.i.) for a noninteracting system. Compared with this reference system, the bonding orbital shows *increased* probability density between the nuclei, but the antibonding orbital shows *decreased* probability density in this region.

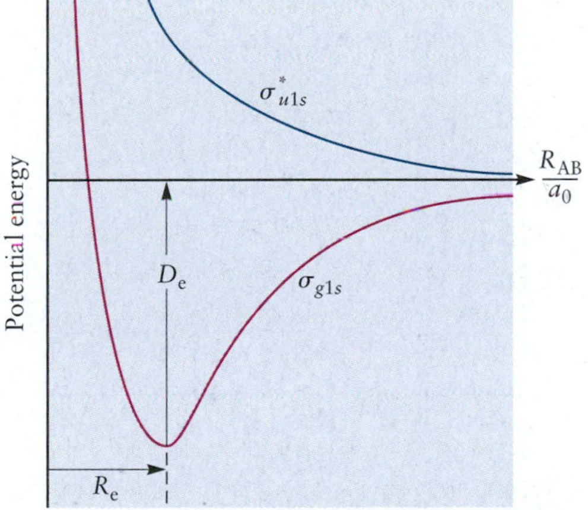

FIGURE 6.7 Potential energy of H_2^+ in a σ_{g1s} (bonding) and σ^*_{u1s} (antibonding) molecular orbital, shown as a function of internuclear separation R_{AB} in the LCAO approximation.

examine the potential energy in each of these approximate MOs. Figure 6.7 shows the potential energy of the H_2^+ ion in the LCAO approximation for the σ_{g1s} and σ^*_{u1s} MOs. The force between the nuclei in the antibonding state is everywhere repulsive, but in the bonding state the nuclei are attracted to each other and form a bound state at the distance corresponding to the lowest potential energy.

The energy minimum of the potential at R_e is called D_e, the **bond dissociation energy**—that is, the energy required to dissociate the molecule into separated atoms. At R_e, where the effective potential has its minimum value, the attractive and repulsive forces between the nuclei balance exactly and hold the internuclear distance at this value. The equilibrium bond length of the molecule is determined by the competition between attractive forces, which originate in electron–nuclear interactions, and repulsive forces, which originate in nuclear–nuclear interactions.

This is the sense in which the electrons function as the "glue" that holds the nuclei to their special positions that define the structure of a molecule.

How well does the LCAO approximation describe the energy values in H_2^+? We compare the exact and LCAO results in Figure 6.8, where the zero of energy at infinite separation is again taken to be that of the separated species H and H^+. The energy in σ_{g1s} has a minimum at R_{AB}= 1.32 Å, and the predicted energy required to dissociate the diatomic molecular ion to H and H^+ is D_e = 1.76 eV. These results compare reasonably well with the experimentally measured values R_{AB}= 1.060 Å and D_e = 2.791 eV, which were also obtained from the exact solution in Section 6.1.

Let's put these results of the LCAO approximation in perspective. The results in Figures 6.6 and 6.7 were obtained by working out the details of the approximation expressed in Equations 6.3a and 6.3b. These LCAO results have captured qualitatively the results of the exact calculation. Therefore, we can apply the LCAO method in other more complex cases and be confident we have included the essential qualitative features of bond formation. Also, we can always improve the results by following up with a self-consistent computer calculation that produces optimized MOs.

The energy-level diagram within the LCAO approximation is given by a **correlation diagram** (Fig. 6.9), which shows that two 1*s* AOs have been combined to give a σ_{g1s} MO with energy lower than the AOs and a σ^*_{u1s} MO with energy higher than the AOs. This diagram is a purely qualitative representation of the same information contained in Figure 6.7. The exact energy level values will depend on the separation between the fixed nuclei (as shown in Fig. 6.7). Even without the results shown in Figure 6.7, we would know that an electron in an antibonding orbital has more energy than one in a bonding orbital because the antibonding orbital has a node. Consequently, in the ground state of H_2^+, the electron occupies the σ_{g1s} MO. By forming the bond in the molecular ion, the total system of two hydrogen nuclei and one electron becomes more stable than the separated atoms by the energy difference $-\Delta E$ shown in Figure 6.9.

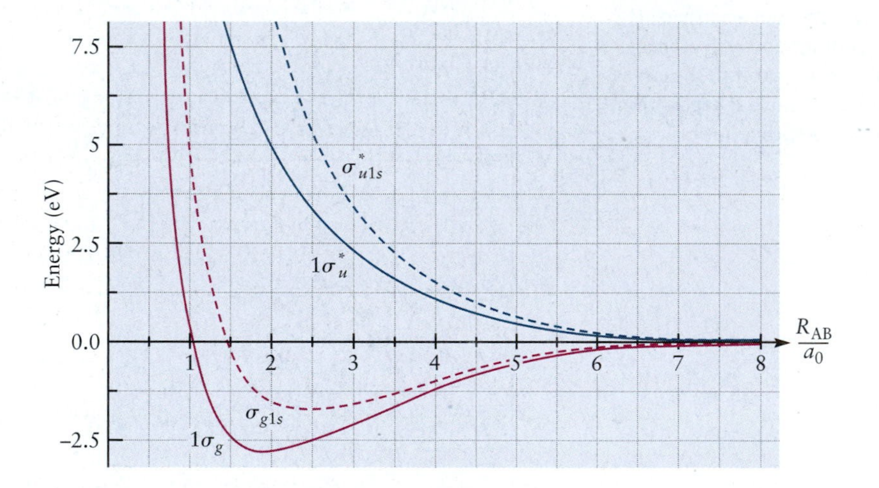

FIGURE 6.8 Comparison of potential energy for the σ_{g1s} and σ^*_{u1s} orbitals of H_2^+ in the LCAO approximation (dashed lines) with the exact results (solid lines). The internuclear separation is plotted in units of the Bohr radius.
(Adapted from R. S. Berry, S. A. Rice, and J. Ross, *Physical Chemistry* (2nd ed), New York: Oxford, 2000, Figure 6.6, by permission.)

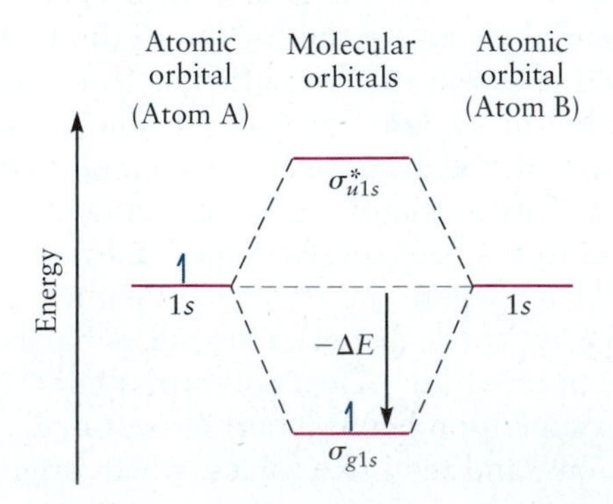

FIGURE 6.9 Correlation diagram for H_2^+ in the linear combination of atomic orbitals (LCAO) approximation. The bonding orbital is stabilized relative to the noninteracting system by the energy difference ΔE.

6.2.2 Homonuclear Diatomic Molecules: First-Period Atoms

We can combine the LCAO method for H_2^+ with an aufbau principle, analogous to that developed for atoms, to describe the electron configuration of more complex molecules. Electrons available from the two atoms are "fed" into the MOs, starting with the MO of lowest energy. At most, two electrons can occupy each MO. The ground-state H_2 molecule, therefore, accommodates two electrons with opposite spins in a σ_{g1s} bonding MO (Fig. 6.10). The diatomic molecule is more stable than the isolated atoms by the energy difference $-2\Delta E$.

The LCAO approximation can be applied in this same way to He_2^+ and He_2, with one change. The MOs must be generated as linear combinations of He(1s) AOs, not H(1s) orbitals. The reason is that when the electrons in He_2^+ and He_2 approach close to one of the nuclei, they experience a potential much closer to that in a helium atom than in a hydrogen atom. Therefore, the equations for the MOs are

$$\sigma_{g1s} = C_g[\varphi_{1s}^{A}(\text{He}) + \varphi_{1s}^{B}(\text{He})] \qquad [6.6a]$$

$$\sigma_{u1s}^{*} = C_u[\varphi_{1s}^{A}(\text{He}) - \varphi_{1s}^{B}(\text{He})] \qquad [6.6b]$$

We rewrite Equation 6.6ab using a simpler notation, which we adopt for the remainder of the text:

$$\sigma_{g1s} = C_g[1s^{A} + 1s^{B}] \qquad [6.7a]$$

$$\sigma_{u1s}^{*} = C_u[1s^{A} - 1s^{B}] \qquad [6.7b]$$

In Equations 6.7a and 6.7b, the symbol for the atomic wave function φ has been dropped and the AOs are identified by their hydrogenic labels 1s, 2s, 2p, and so on. The superscripts A and B are used to identify particular atoms of the same element in bonds formed from the same elements (**homonuclear** diatomics) and will be replaced by the symbols for the elements in bonds formed from different elements (**heteronuclear** diatomics). These helium MOs have the same general shapes and potential energy curves as shown in Figures 6.6 and 6.7 for the MOs constructed from H(1s), and a correlation diagram similar to Figure 6.10. Quantitative calculations of electron density and energy (these calculations are not performed in this book) would produce different values for the two sets of MOs in Equations 6.3a and 6.3b and Equations 6.7a and 6.7b. Keep in mind that we construct the MOs as combinations of all the AOs required to accommodate the electrons in the ground states of the atoms that form the molecule. This set of AOs is called the **minimum basis set** for that specific molecule. Therefore, quantitative calculations for each molecule are influenced by the detailed properties of the atoms in the molecule.

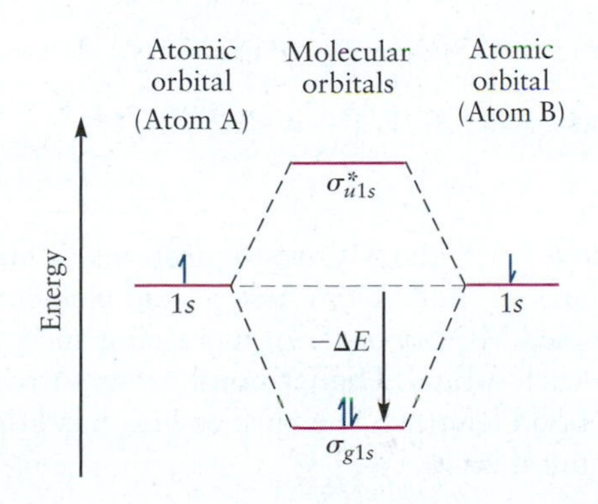

FIGURE 6.10 Correlation diagram for first-period diatomic molecules. Blue arrows indicate the electron filling for the H_2 molecule. All of the atomic electrons are pooled and used to fill the molecular orbitals using the aufbau principle. In the molecules, electrons are no longer connected to any particular atom.

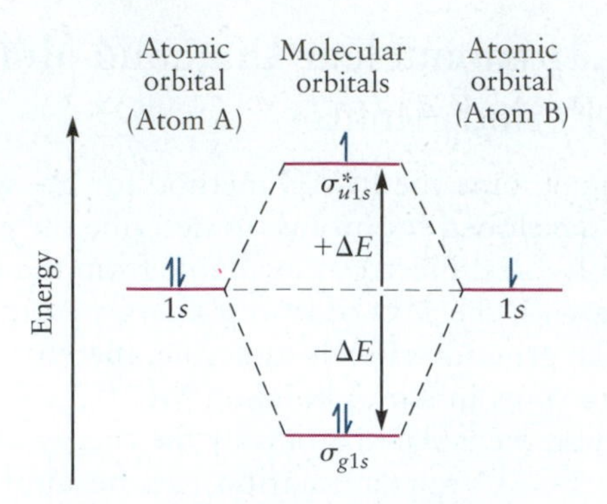

FIGURE 6.11 Correlation diagram for first-period diatomic molecules. Blue arrows indicate the electron filling for the He_2^+ molecule. The aufbau principle fills the bonding orbital with two electrons, so the third electron must go into the antibonding orbital, thus reducing the bond order compared with that in H_2.

Because He_2^+ and He_2 have more than two electrons, the aufbau principle requires them to have some electrons in the σ^*_{u1s} antibonding orbital. Each of these electrons contributes a destabilization energy in the amount $+\Delta E$ relative to the separated helium atoms (Fig. 6.11). This effect competes with the stabilization energy of $-\Delta E$ per electron in the σ_{g1s} bonding orbital, giving a weak bond in He_2^+ and no stable bond in He_2.

The general features of covalent bonding in the LCAO picture can be summarized as follows. In *bonding* MOs, bond formation arises from the sharing of electrons (most often electron pairs with opposite spins). The average electron density is greatest between the nuclei and tends to pull them together. Electrons that are shared in an *antibonding* MO tend to force the nuclei apart, reducing the bond strength. This competition is described by the **bond order,** which is defined as follows:

$$\text{Bond order} = \tfrac{1}{2}\,(\text{number of electrons in bonding molecular orbitals} - \text{number of electrons in antibonding molecular orbitals})$$

In the LCAO MO description, the H_2 molecule in its ground state has a pair of electrons in a bonding MO, and thus a single bond (that is, its bond order is 1). Later in this chapter, as we describe more complex diatomic molecules in the LCAO approximation, bond orders greater than 1 are discussed. This quantum mechanical definition of bond order generalizes the concept first developed in the Lewis theory of chemical bonding—a shared pair of electrons corresponds to a single bond, two shared pairs to a double bond, and so forth.

EXAMPLE 6.1

Give the ground-state electron configuration and the bond order of the He_2^+ molecular ion.

SOLUTION

He_2^+ has three electrons, which are placed in MOs to give the ground-state configuration $(\sigma_{g1s})^2(\sigma^*_{u1s})^1$, indicating that the ion has a doubly occupied σ_{g1s} orbital (bonding) and a singly occupied σ^*_{u1s} orbital (antibonding). The bond order is

$$\text{bond order} = \tfrac{1}{2}\,(2 \text{ electrons in } \sigma_{g1s} - 1 \text{ electron in } \sigma^*_{u1s}) = \tfrac{1}{2}$$

This should be a weaker bond than that in H_2.

Related Problems: 9, 10, 11, 12, 13, 14, 15, 16

Table 6.2 lists the MO configurations of homonuclear diatomic molecules and molecular ions made from first-period elements. These configurations are simply the occupied MOs in order of increasing energy, together with the number of electrons in each orbital. Higher bond order corresponds to higher bond energies and shorter bond lengths. The species He_2 has bond order zero and does not form a true chemical bond.

TABLE 6.2 Electron Configurations and Bond Orders for First-Row Diatomic Molecules

Species	Electron Configuration	Bond Order	Bond Energy (kJ mol^{-1})	Bond Length (Å)
H_2^+	$(\sigma_{g1s})^1$	$\frac{1}{2}$	255	1.06
H_2	$(\sigma_{g1s})^2$	1	431	0.74
He_2^+	$(\sigma_{g1s})^2(\sigma_{u1s}^*)^1$	$\frac{1}{2}$	251	1.08
He_2	$(\sigma_{g1s})^2(\sigma_{u1s}^*)^2$	0	~0	Large

The preceding paragraphs have illustrated the LCAO approximation with specific examples and shown how the character of the chemical bond is determined by the difference in the number of electrons in bonding and antibonding MOs. This subsection concludes with the following summary of the systematic procedure for applying the LCAO approximation to the MOs for any diatomic molecule:

1. Form linear combinations of the minimum basis set of AOs to generate MOs. The total number of MOs formed in this way must equal the number of AOs used.
2. Arrange the MOs in order from lowest to highest energy.
3. Put in electrons (at most two electrons per MO), starting from the orbital of lowest energy. Apply Hund's rules when appropriate.

6.2.3 Homonuclear Diatomic Molecules: Second-Period Atoms

How shall we generate approximate MOs to accommodate electrons from AOs higher than the 1*s* orbitals? The excited states of H_2^+ provide a clue, in their limiting behavior at large internuclear separation. Figure 6.5 correlated each of these excited states with sums or differences of the excited states of the hydrogen atom. Using these results, we can propose approximate MOs as linear combinations of the excited states of atoms. Because we must include in the minimum basis set all occupied AOs for the participating atoms, the question naturally arises whether these simple two-term approximate MOs will be sufficient, or whether each MO will require contributions from several AOs. For example, N_2 will require at least 7 approximate MOs to accommodate its 14 electrons. We could generate each of these MOs as a linear combination of the 1*s*, 2*s*, and 2*p* orbitals (the minimum basis set) with proper coefficients. The general form for these MOs is

$$\psi_{MO} = C_1[1s^A + 1s^B] + C_2[2s^A + 2s^B] + C_3[2p_x^A + 2p_x^B] + C_4[2p_y^A + 2p_y^B] + C_5[2p_z^A + 2p_z^B] \quad [6.8]$$

These MOs would be distinguished from one another by the values of their coefficients. The square of each of these MOs would give the electron density distribution for the electrons in that MO. This distribution function would include many cross terms, each of which represents the interaction between a pair of AOs on the two different atoms. It is difficult to visualize the result as the basis for qualitative arguments.

Two conclusions from more advanced quantum mechanics guide us to a simplified approach for constructing MOs for atoms with more than two electrons.

1. *Two AOs contribute significantly to bond formation only if their atomic energy levels are very close to one another.*

Consequently, we can ignore mixing between the core-shell 1*s* orbitals and the valence-shell 2*s* and 2*p* orbitals. Similarly, we can ignore mixing between the 2*s* and 2*p* orbitals, except in special cases to be described later.

2. *Two AOs on different atoms contribute significantly to bond formation only if they overlap significantly.*

Overlap must be defined somewhat precisely to understand the second statement. Two orbitals overlap significantly if they both have appreciable amplitudes over the same region of space. The *net* overlap may be positive or zero, depending on the relative phases of the orbitals involved. Bonding orbitals arise from positive overlap (constructive interference), whereas antibonding orbitals result from zero overlap (destructive interference).

For *s* orbitals, it is rather easy to guess the degree of overlap; the closer the nuclei, the greater the overlap. If the wave functions have the same phase, the overlap is positive; if they have opposite phases, the overlap is zero. For more complex cases, the overlap between participating AOs depends strongly on both the symmetry of the arrangement of the nuclei and on the phases of the orbitals. If the two orbitals are shaped so that neither has substantial amplitude in the region of interest, then their overlap is negligible. However, if they both have significant amplitude in the region of interest, it is important to know whether regions of positive overlap (where the two orbitals have the same phase) are canceled by regions of negative overlap (where the two orbitals have opposite phases). Such cancellation leads to negligible or zero overlap between the orbitals. Qualitative sketches that illustrate significant or negligible overlap in several common cases are shown in Figure 6.12. In particular, constructive interference and overlap between *s* and *p* orbitals is significant only in the case where an *s* orbital approaches a *p* orbital "end-on." The phase of the *p* orbital lobe pointing toward the *s* orbital must be the same as that of the *s* orbital. We recommend reviewing the "sizes and shapes" of hydrogenic orbitals discussed in Section 5.1 and depicted in Figure 5.4. A great deal of qualitative insight into the construction of MOs can be gleaned from these considerations.

The two conclusions stated earlier justify the following simplified LCAO MOs for the second-period homonuclear diatomic molecules. As with the first-period atoms, we use the new labels in Table 6.1 to distinguish the approximate MOs from the exact H_2^+ MOs in Figure 6.2 and to indicate their atomic parentage.

We combine the 2*s* AOs of the two atoms in the same fashion as 1*s* orbitals, giving a σ_{g2s} bonding orbital and a σ_{u2s}^* antibonding orbital:

$$\sigma_{g2s} = C_g[2s^A + 2s^B] \qquad [6.9a]$$

$$\sigma_{u2s}^* = C_u[2s^A - 2s^B] \qquad [6.9b]$$

The choice of appropriate combinations of the 2*p* orbitals is guided by the overlap arguments and by recalling that the bond axis is the *z*-axis. The 2*p* orbitals form different MOs depending on whether they are parallel or perpendicular to the internuclear (bond) axis. Consider first the $2p_z$ orbitals, which can be used to form two different kinds of σ orbitals. If the relative phases of the p_z orbitals are such that they interfere constructively in the internuclear region, then a bonding σ_{g2p_z} orbital is formed. If, conversely, lobes of opposite phases overlap, they form an antibonding MO labeled $\sigma_{u2p_z}^*$. These orbitals are shown in Figure 6.13.

$$\sigma_{g2p_z} = C_g[2p_z^A - 2p_z^B] \qquad [6.10a]$$

$$\sigma_{u2p_z}^* = C_u[2p_z^A + 2p_z^B] \qquad [6.10b]$$

The electron density in the bonding orbital has increased between the nuclei, whereas the antibonding orbital has a node.

$$\pi_{u2p_x} = C_u[2p_x^A + 2p_x^B] \qquad [6.11a]$$

$$\pi_{g2p_x}^* = C_g[2p_x^A - 2p_x^B] \qquad [6.11b]$$

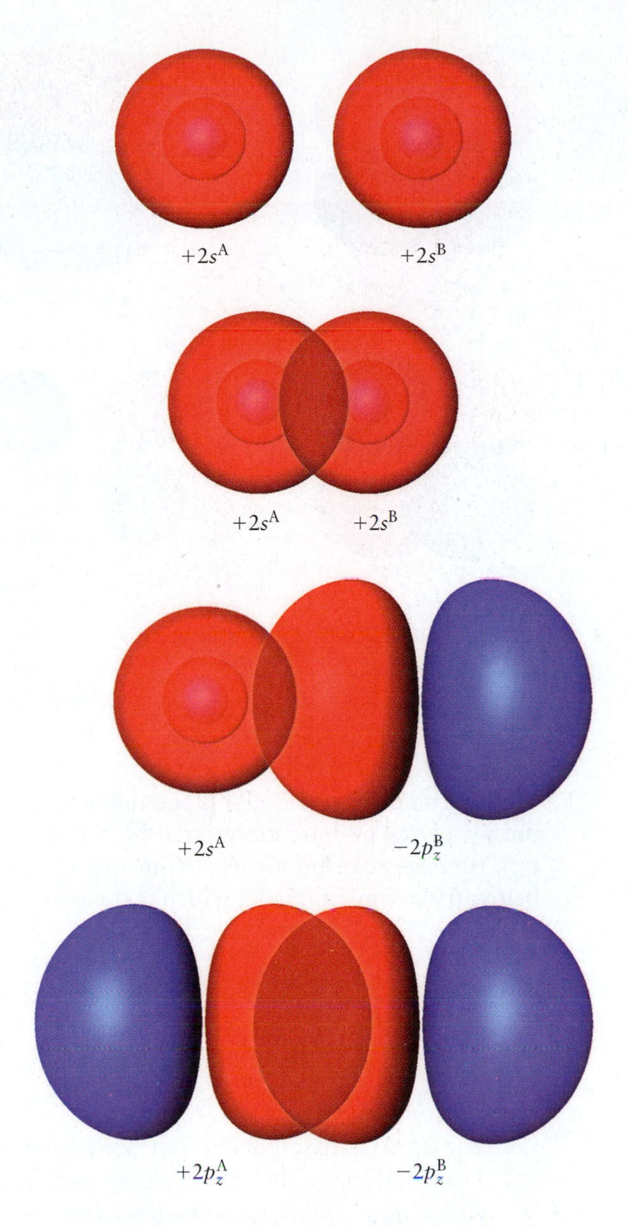

FIGURE 6.12 Overlap of orbitals in several common combinations. The magnitude of overlap can be estimated qualitatively from the relative size and symmetry of the two orbitals involved. (Note the radial nodes in the 2*s* orbitals, clearly visible in these images.)
(Courtesy of Mr. Hatem Helal and Professor William A. Goddard III, California Institute of Technology, and Dr. Kelly P. Gaither, University of Texas at Austin.)

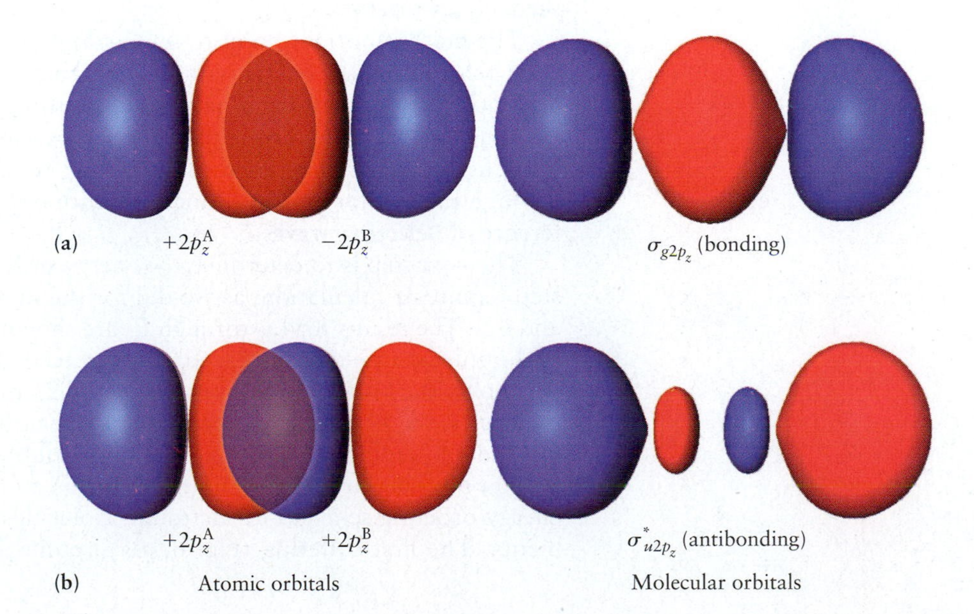

FIGURE 6.13 Formation of (a) σ_{g2p_z} bonding and (b) $\sigma^*_{u2p_z}$ antibonding molecular orbitals from $2p_z$ orbitals on atoms A and B. Regions with positive amplitude are shown in red, and those with negative amplitude are shown in blue.
(Courtesy of Mr. Hatem Helal and Professor William A. Goddard III, California Institute of Technology, and Dr. Kelly P. Gaither, University of Texas at Austin.)

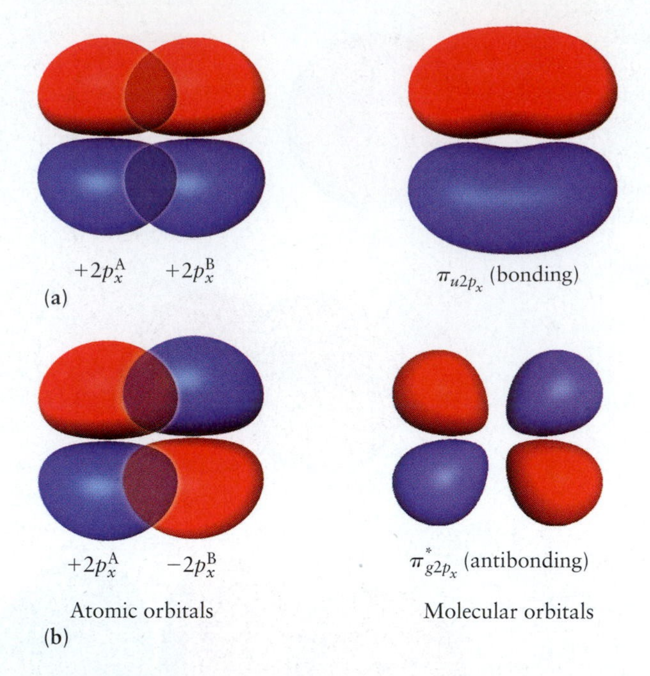

FIGURE 6.14 Formation of (a) π_{u2p_x} bonding and (b) $\pi^*_{g2p_x}$ antibonding molecular orbitals from $2p_x$ orbitals on atoms A and B.
(Courtesy of Mr. Hatem Helal and Professor William A. Goddard III, California Institute of Technology, and Dr. Kelly P. Gaither, University of Texas at Austin.)

These orbitals have a nodal plane that contains the internuclear axis (in this case, the y-z plane) and are designated by π rather than σ. In the same way, π_{u2p_y} and $\pi^*_{g2p_y}$ orbitals can be formed from the $2p_y$ AOs. Their lobes project above and below the x-z nodal plane, which is the plane of the page in Figure 6.14.

$$\pi_{u2p_y} = C_u[2p_y^A + 2p_y^B] \qquad [6.12a]$$

$$\pi^*_{g2p_y} = C_g[2p_y^A - 2p_y^B] \qquad [6.12b]$$

Like the $2p$ AOs from which they are constructed, the π_{2p} MOs are degenerate: π_{u2p_x} and π_{u2p_y} have the same energy, and $\pi^*_{g2p_x}$ and $\pi^*_{g2p_y}$ have the same energy. We expect π_{u2p_x} and π_{u2p_y} to be less effective than σ_{g2p_z} as bonding orbitals, because the overlap in the π orbitals occurs off the internuclear axis, and therefore has less tendency to increase the electron density between the nuclei and to pull them closer together.

The most important point to understand in constructing MOs and predicting their behavior by the overlap argument is that the relative phases of the two AOs are critical in determining whether the resulting MO is bonding or antibonding. Bonding orbitals form through the overlap of wave functions with the same phase, by constructive interference of "electron waves"; antibonding orbitals form through the overlap of wave functions with opposite phase, by destructive interference of "electron waves."

The next step is to determine the energy ordering of the MOs. In general, that step requires a calculation, as we did for the first-period diatomics in Figures 6.8 and 6.9. The results for Li_2 through F_2 are shown in Figure 6.15. The electrons for each molecule have been placed in MOs according to the aufbau principle. We show only the MOs formed from the $2s$ and $2p$ orbitals. In second-period diatomic molecules, the $1s$ orbitals of the two atoms barely overlap. Because the σ_{g1s} bonding and σ^*_{u1s} antibonding orbitals are both doubly occupied, they have little net effect on bonding properties and need not be considered. There are two different energy ordering schemes for diatomic molecules formed from second-period elements. The first ordering (Fig. 6.16a) applies to the molecules with atoms Li

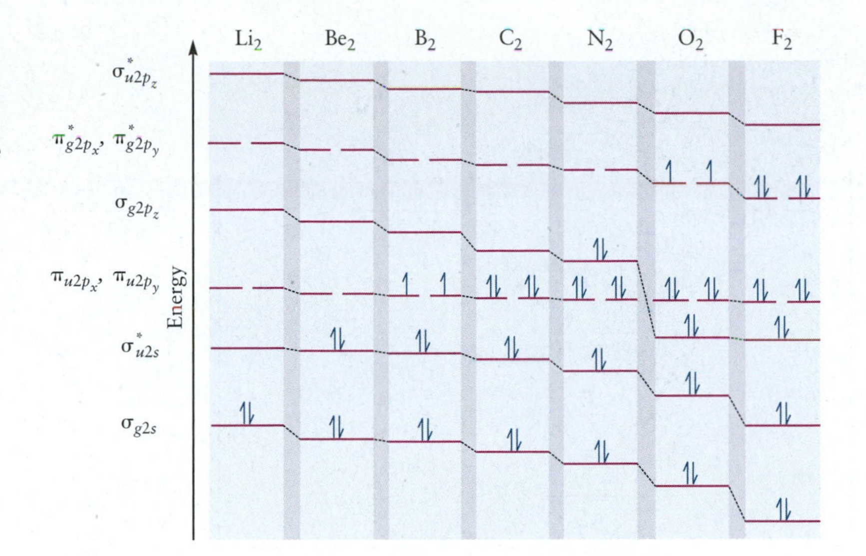

FIGURE 6.15 Energy levels for the homonuclear diatomics Li_2 through F_2. Notice how the highest occupied level changes with the number of valence electrons. Notice especially the change between N_2 and O_2.

through N (that is, the first part of the period) and their positive and negative ions. The second (see Fig. 6.16b) applies to the later elements, O, F, and Ne, and their positive and negative ions. The relative ordering of the σ_{g2p_z} and the π_{u2p_x}, π_{u2p_y} changes as we move across the periodic table. The energy of the π orbital remains essentially constant, whereas the energy of the σ orbital falls rapidly, dropping below that of the π orbital at O_2. This effect is largely caused by the increased nuclear charge felt by the electrons in the σ orbital in the heavier elements.

An important prediction comes from the correlation diagrams in Figure 6.15. Hund's rules require that, in the ground state, the electrons occupy different orbitals and have parallel spins; thus, B_2 and O_2 are predicted to be paramagnetic. This paramagnetism is exactly what is found experimentally (Fig. 6.17). In contrast, in the Lewis electron dot diagram for O_2,

:Ö: :O̤:

all the electrons appear to be paired. Moreover, the extremely reactive nature of molecular oxygen can be rationalized as resulting from the readiness of the two π^* electrons, unpaired and in different regions of space, to find additional bonding partners in other molecules.

The electron configurations in Figure 6.15 allow us to calculate the bond order for each molecule and correlate it with other properties of the molecules.

EXAMPLE 6.2

Determine the ground-state electron configuration and bond order of the F_2 molecule.

SOLUTION

Each atom of fluorine has 7 valence electrons, so 14 electrons are placed in the MOs to represent bonding in the F_2 molecule. The correlation diagram of Figure 6.15 gives the following electron configuration:

$$(\sigma_{g2s})^2(\sigma^*_{u2s})^2(\sigma_{g2p_z})^2(\pi_{u2p_x}, \pi_{u2p_y})^4(\pi^*_{g2p_x}, \pi^*_{g2p_y})^4$$

Because there are eight valence electrons in bonding orbitals and six in antibonding orbitals, the bond order is

$$\text{Bond order} = \tfrac{1}{2}(8 - 6) = 1$$

and the F_2 molecule has a single bond.

Related Problems: 17, 18, 19, 20, 21, 22, 23, 24

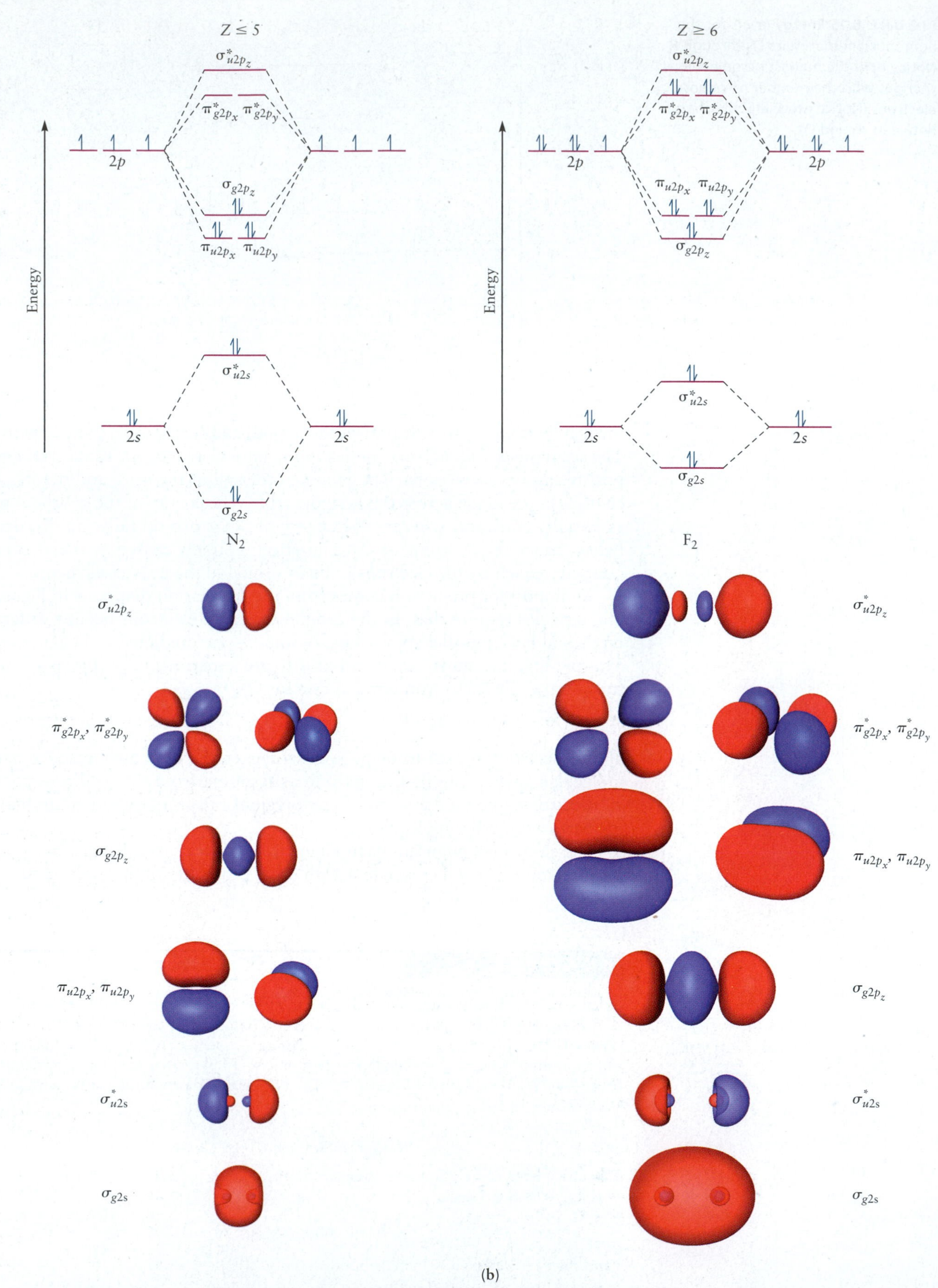

FIGURE 6.16 Correlation diagrams for second-period diatomic molecules. (a) Correlation diagram and molecular orbitals calculated for N_2. (b) Correlation diagram and molecular orbitals calculated for F_2.
(Courtesy of Mr. Hatem Helal and Professor William A. Goddard III, California Institute of Technology, and Dr. Kelly P. Gaither, University of Texas at Austin.)

(a)

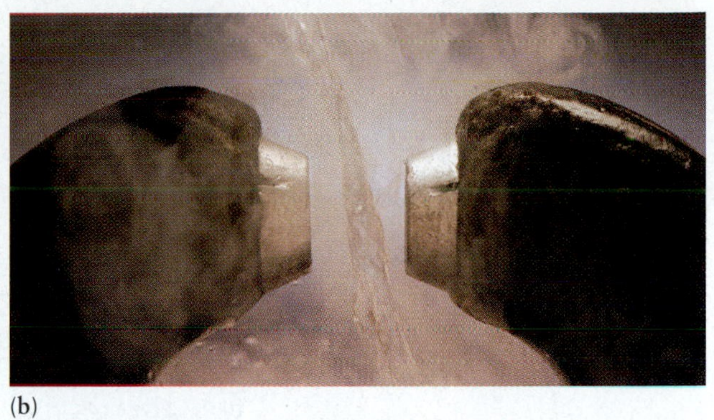
(b)

FIGURE 6.17 (a) Oxygen is paramagnetic; liquid oxygen (O_2) poured between the pole faces of a magnet is attracted and held there. (b) When the experiment is repeated with liquid nitrogen (N_2), which is diamagnetic, the liquid pours straight through. (Courtesy Larry Cameron.)

Table 6.3 summarizes the properties of second-period homonuclear diatomic molecules. Note the close relationship among bond order, bond length, and bond energy, and that the bond orders calculated from the MOs agree completely with the results of the Lewis electron dot model. How these properties depend on the number of electrons in the molecules is shown in Figure 6.18. The bond orders simply follow the filling of MOs in a given subshell, rising from 0 to 1, and then falling back to 0 for the first-period diatomics and also for Li_2, Be_2 and their ions. The bond orders move in half-integral steps if the molecular ions are included; they increase as the σ orbitals are filled and decrease as the σ^* orbitals begin to fill. These MOs are all constructed from *s* orbitals, so there is no possibility for multiple bonds. Moving from Be_2 through Ne_2, the bond orders move again in half-integer increments from 0 to 3 and back to 0 as the π orbitals and then the π^* orbitals are filled. Both bond energies and force constants are correlated directly with the bond order, whereas the bond length varies in the opposite direction. This makes perfect sense; multiple bonds between atoms should be stronger and shorter than single bonds. In summary, the simple LCAO method provides a great deal of insight into the nature of chemical bonding in homonuclear diatomic molecules and the trends in the properties that result. It is consistent with the predictions of simpler theories, such as that of G. N. Lewis, but clearly more powerful and more easily generalized to problems of greater complexity.

TABLE 6.3 Molecular Orbitals of Homonuclear Diatomic Molecules

Species	Number of Valence Electrons	Valence Electron Configuration	Bond Order	Bond Length (Å)	Bond Energy (kJ mol^{-1})
H_2	2	$(\sigma_{g1s})^2$	1	0.74	431
He_2	4	$(\sigma_{g1s})^2(\sigma^*_{u1s})^2$	0		
Li_2	2	$(\sigma_{g2s})^2$	1	2.67	105
Be_2	4	$(\sigma_{g2s})^2(\sigma^*_{u2s})^2$	0	2.45	9
B_2	6	$(\sigma_{g2s})^2(\sigma^*_{u2s})^2(\pi_{u2p})^2$	1	1.59	289
C_2	8	$(\sigma_{g2s})^2(\sigma^*_{u2s})^2(\pi_{u2p})^4$	2	1.24	599
N_2	10	$(\sigma_{g2s})^2(\sigma^*_{u2s})^2(\pi_{u2p})^4(\sigma_{g2p_z})^2$	3	1.10	942
O_2	12	$(\sigma_{g2s})^2(\sigma^*_{u2s})^2(\sigma_{g2p_z})^2(\pi_{u2p})^4(\pi^*_{g2p})^2$	2	1.21	494
F_2	14	$(\sigma_{g2s})^2(\sigma^*_{u2s})^2(\sigma_{g2p_z})^2(\pi_{u2p})^4(\pi^*_{g2p})^4$	1	1.41	154
Ne_2	16	$(\sigma_{g2s})^2(\sigma^*_{u2s})^2(\sigma_{g2p_z})^2(\pi_{u2p})^4(\pi^*_{g2p})^4(\sigma^*_{u2p_z})^2$	0		

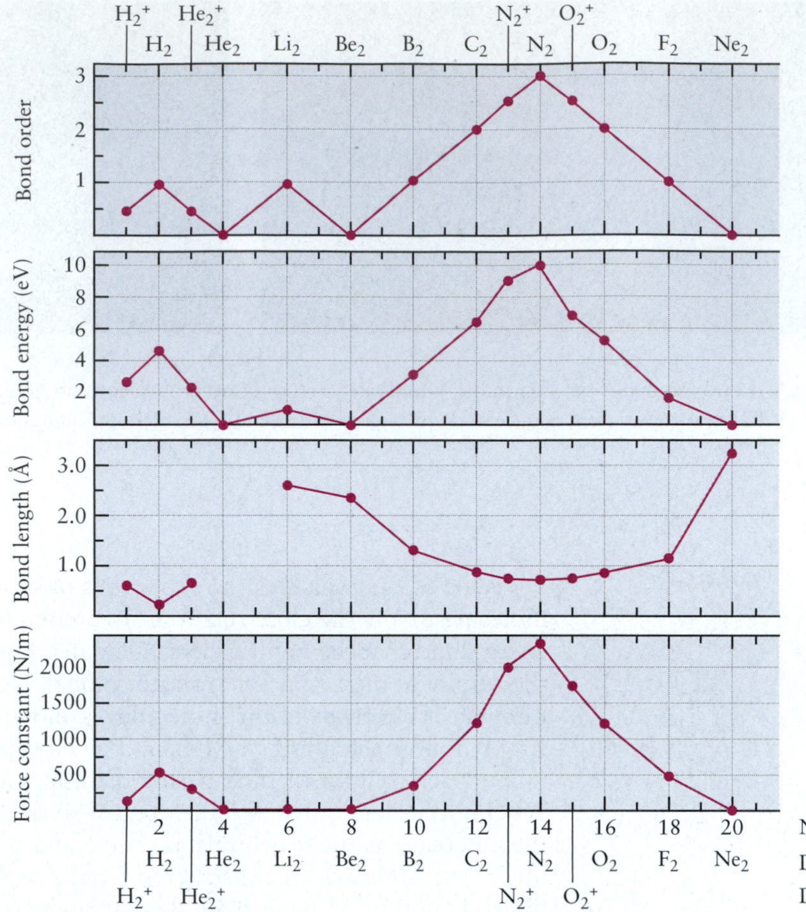

FIGURE 6.18 Trends in bond order, bond length, bond energy, and force constant with the number of valence electrons in the second-row diatomic molecules.

6.2.4 Heteronuclear Diatomic Molecules

Diatomic molecules such as CO and NO, formed from atoms of two different elements, are called heteronuclear. We construct MOs for such molecules by following the procedure described earlier, with two changes. First, we use a different set of labels because heteronuclear diatomic molecules lack the inversion symmetry of homonuclear diatomic molecules. We therefore drop the g and u subscripts on the MO labels. Second, we recognize that the AOs on the participating atoms now correspond to different energies. For example, we combine the 2s AO of carbon and the 2s AO of oxygen to produce a bonding MO (without a node),

$$\sigma_{2s} = C_A 2s^A + C_B 2s^B \quad [6.13a]$$

and an antibonding MO (with a node),

$$\sigma_{2s}^* = C'_A 2s^A - C'_B 2s^B \quad [6.13b]$$

where A and B refer to the two different atoms in the molecule. In the homonuclear case, we argued that $C_A = C_B$ and $C'_A = C'_B$ because the electron must have equal probability of being near each nucleus, as required by symmetry. When the two nuclei are different, this reasoning does not apply. If atom B is more electronegative than atom A, then $C_B > C_A$ for the bonding σ MO (and the electron spends more time on the electronegative atom); $C'_A > C'_B$ for the higher energy σ^* MO, and it will more closely resemble a $2s^A$ AO.

Molecular orbital correlation diagrams for heteronuclear diatomics start with the energy levels of the more electronegative atom displaced *downward*, because

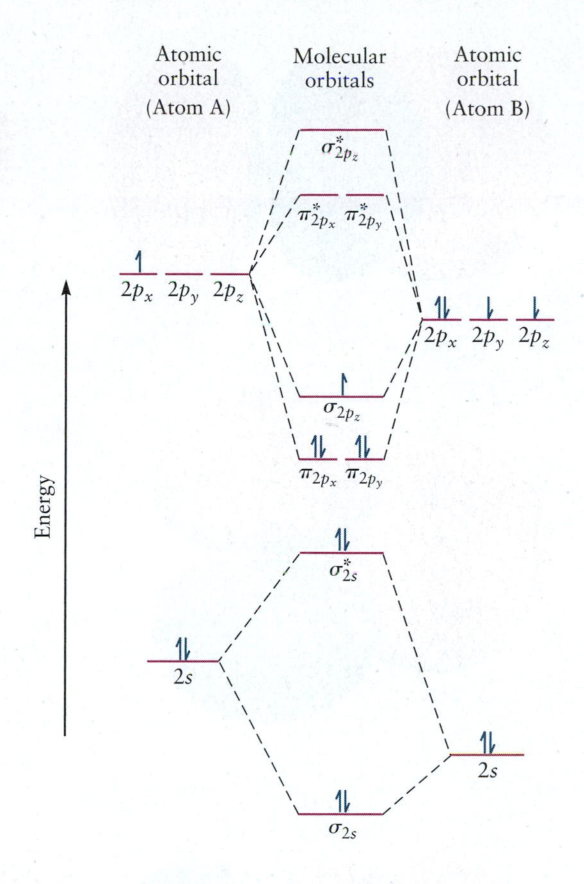

FIGURE 6.19 Correlation diagram for heteronuclear diatomic molecules, AB. The atomic orbitals for the more electronegative atom (B) are displaced downward because they have lower energies than those for A. The orbital filling shown is that for (boron monoxide) BO.

that atom attracts valence electrons more strongly than does the less electronegative atom. Figure 6.19 shows the diagram appropriate for many heteronuclear diatomic molecules of second-period elements (where the electronegativity difference is not too great). This diagram has been filled with the valence electrons for the ground state of the molecule BO. Another example, NO, with 11 valence electrons (5 from N, 6 from O), has the following ground-state configuration:

$$(\sigma_{2s})^2(\sigma^*_{2s})^2(\pi_{2p_x}, \pi_{2p_y})^4(\sigma_{2p_z})^2(\pi^*_{2p_x}, \pi^*_{2p_y})^1$$

With eight electrons in bonding orbitals and three in antibonding orbitals, the bond order of NO is $\frac{1}{2}(8 - 3) = 2\frac{1}{2}$ and it is paramagnetic. The bond energy of NO should be smaller than that of CO, which has one fewer electron but a bond order of 3; experiment agrees with this prediction.

We explained earlier that, in homonuclear diatomics, AOs mix significantly to form MOs only if they are fairly close in energy and have similar symmetries. The same reasoning is helpful in constructing MOs for heteronuclear diatomics. For example, in the HF molecule, both the 1s and 2s orbitals of the F atom are far too low in energy to mix with the H 1s orbital. Moreover, the overlap between the H 1s and F 2s is negligible (Fig. 6.20a). The net overlap of the H 1s orbital with the $2p_x$ or $2p_y$ F orbital is zero (see Fig. 6.20c) because the regions of positive and negative overlap sum to zero. This leaves only the $2p_z$ orbital of F to mix with the H 1s orbital to give both σ bonding and σ^* antibonding orbitals (see Figs. 6.20b, d). The correlation diagram for HF is shown in Figure 6.21. The 2s, $2p_x$, and $2p_y$ orbitals of fluorine do not mix with the 1s of hydrogen, and therefore remain as atomic (nonbonding) states denoted σ^{nb} and π^{nb}. Electrons in these orbitals do not contribute significantly to the chemical bonding. Because fluorine is more electronegative than hydrogen, its 2p orbitals lie below the 1s hydrogen orbital in energy. The σ orbital then contains more fluorine $2p_z$ character, and the σ^* orbital more closely resembles a hydrogen 1s AO. When the eight valence electrons are put in for HF, the result is the MO configuration:

$$(\sigma^{nb})^2(\sigma)^2(\pi_x^{nb}, \pi_y^{nb})^4$$

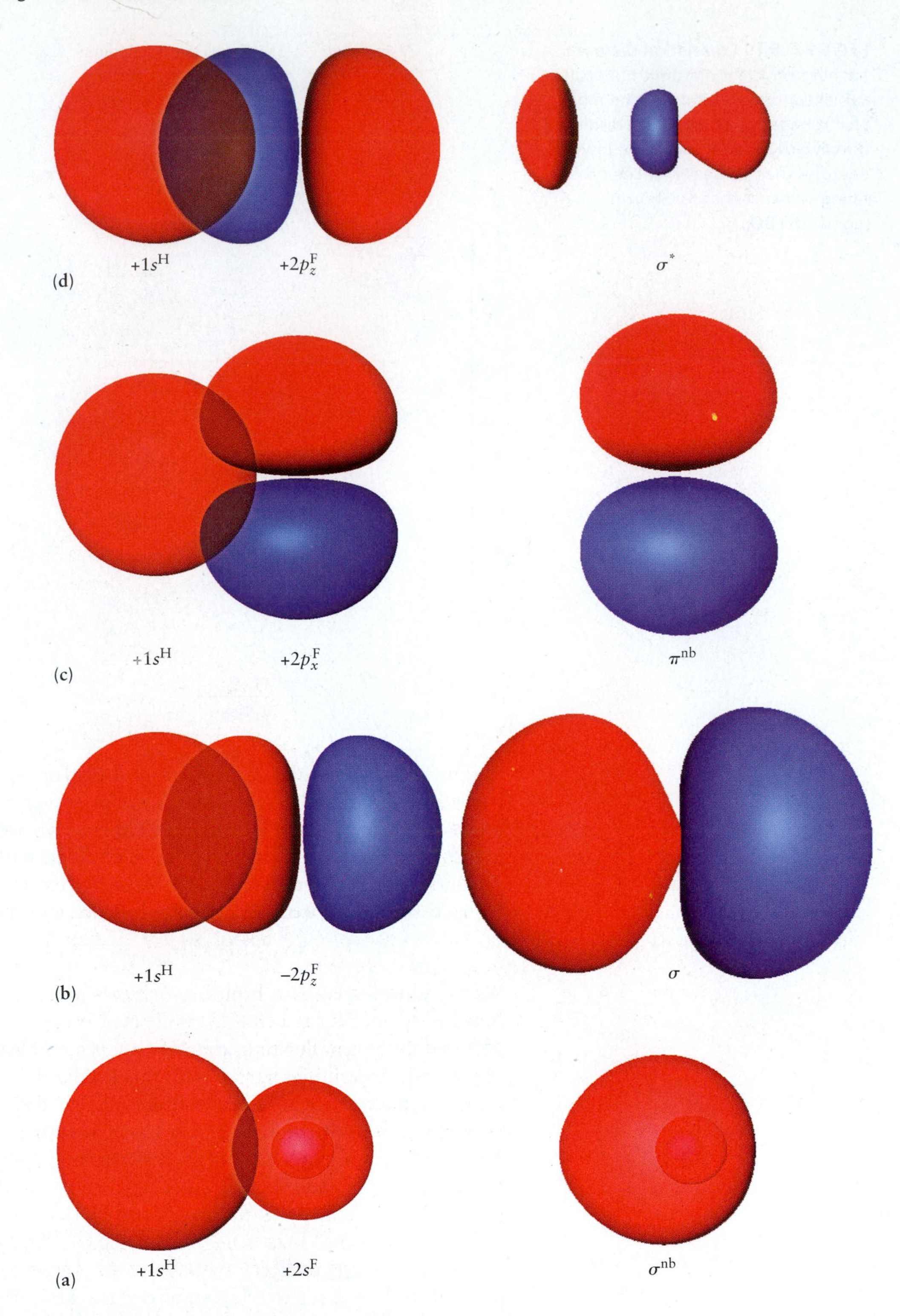

FIGURE 6.20 Overlap of atomic orbitals in HF.
(Courtesy of Mr. Hatem Helal and Professor William A. Goddard III, California Institute of Technology, and Dr. Kelly P. Gaither, University of Texas at Austin.)

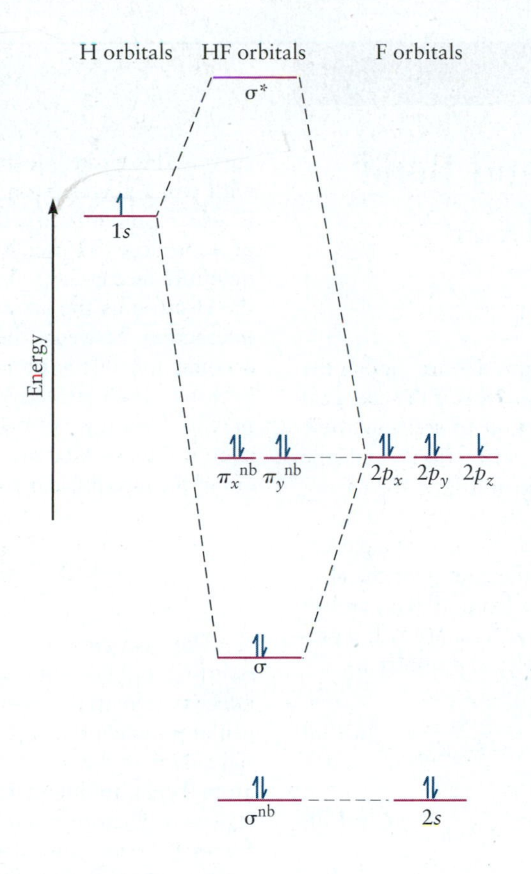

FIGURE 6.21 Correlation diagram for HF. The $2s$, $2p_x$, and $2p_y$ atomic orbitals of fluorine do not mix with the $1s$ atomic orbital of hydrogen, and therefore remain nonbonding.

The net bond order is 1, because electrons in **nonbonding** AOs do not affect bond order. The electrons in the 1σ orbital are more likely to be found near the fluorine atom than near the hydrogen atom; thus, HF has the dipole moment $H^{\delta+}F^{\delta-}$.

If a more electropositive atom (such as Na or K) is substituted for H, the energy of its outermost *s* orbital will be higher than that of the H atom, because its ionization energy is lower. In this case, the σ orbital will resemble a fluorine $2p_z$ orbital even more (that is, the coefficient C_F of the fluorine wave function will be close to 1, and C_A for the alkali atom will be very small). In this limit, the molecule can be described as having the valence electron configuration $(\sigma^{nb})^2(\pi_x^{nb}, \pi_y^{nb}, \pi_z^{nb})^6$, which corresponds to the ionic species Na^+F^- or K^+F^-. The magnitudes of the coefficients in the MO wave function are thus closely related to the ionic–covalent character of the bonding and to the dipole moment.

Summary Comments for the Linear Combination of Atomic Orbitals Method and Diatomic Molecules

The qualitative LCAO method can rationalize trends in bond length and bond energy in a group of molecules by relating both these properties to bond order, but it cannot predict bond energy or molecular geometry. Both of these properties require calculation of the electron energy as a function of the positions of the nuclei, as shown in Figure 6.7 for H_2^+. The equilibrium bond length is then identified as the minimum of this potential energy curve. The LCAO method can be extended to quantitative calculations of molecular properties using modern computer programs to calculate this potential energy curve, once the basis set of AOs has been chosen. Through a variety of software packages, such calculations are now routinely available to chemists engaged in both fundamental and applied work.

A DEEPER LOOK

6.2.5 Potential Energy and Bond Formation in the LCAO Approximation

The potential energy curves shown in Figure 6.7 are among the most important concepts in the quantum picture of the chemical bond and molecular structure. It is important to see how these curves arise from the electron–nucleus attractions and the nuclear–nuclear repulsions in the molecule and how the chemical bond is formed.

Although beyond the scope of this textbook, it is straightforward in quantum mechanics to calculate the energy of the electron in σ_{g1s} and σ^*_{u1s} when the nuclei are fixed at R_{AB}, and to include the value of the nuclear repulsion at R_{AB}. We will represent these values as E_{1g} and E_{1u}, respectively. The results are

$$E_{1g}(R_{AB}) = E_H(1s) + \frac{J + K}{1 + S_{AB}} + \frac{e^2}{4\pi\epsilon_0 R_{AB}} \quad [6.14a]$$

$$E_{1u}(R_{AB}) = E_H(1s) + \frac{J - K}{1 - S_{AB}} + \frac{e^2}{4\pi\epsilon_0 R_{AB}} \quad [6.14b]$$

where J, K, and S_{AB} represent collections of terms that depend on R_{AB} through the dimensionless variable $\rho = R_{AB}/a_0$:

$$S_{AB} = e^{-\rho}\left(1 + \rho - \frac{\rho^2}{3}\right)$$

$$J = -\frac{e^2}{4\pi a_0}\left[\frac{1}{\rho} - e^{-2\rho}\left(1 + \frac{1}{\rho}\right)\right]$$

$$K = -\frac{e^2}{4\pi a_0}e^{-\rho}(1 + \rho)$$

We will see that each term in Equations 6.14a and 6.14b has a straightforward interpretation and is easily understood in graphical form.

The first term in each equation is the energy of the H(1s) state, which is independent of R_{AB}. Therefore, it is just a constant added to the other terms.

As R_{AB} becomes large, approaching infinity, the second and third terms are negligible, so both E_{1g} and E_{1u} approach the value $E_H(1s)$. This is the correct energy for a hydrogen atom and a proton separated by a large distance. This result shows that both E_{1g} and E_{1u} have the simple limiting behavior that we expect.

The second term in each equation is the energy contribution arising from the interaction of the electron with both of the protons; note that this term differs in the two orbitals. We call this the *electronic bonding energy* of the electron in each orbital and denote it by $E^{(el)}_{1g}$ or $E^{(el)}_{1u}$.

The final term in each equation is the nuclear repulsion energy at R_{AB}. This term goes to zero at large values of R_{AB}, when the protons are very far apart. This term becomes large and positive at small values of R_{AB}, as the protons get very close together.

Now, let's put these pieces together to understand the overall behavior, first for the MO σ_{g1s}. In Figure 6.22a, we plot $E^{(el)}_{1g}$ as the lighter red curve and the nuclear repulsive term as the black curve. The electronic bonding energy represents the strength with which the electron is held by the two protons. Conversely, it also represents the effect of the electron in holding the two protons close together. Therefore, it is conventional in molecular quantum mechanics to interpret the electronic bonding energy of the electron as the *attractive* portion of the potential energy of interaction between the two protons. With this insight, we account for all the potential energy of interaction between the protons—both attractive and repulsive—and define the effective potential energy between the protons by adding the proton–proton repulsion to the electronic bonding energy at each point along the internuclear axis,

$$V^{(eff)}_{1g}(R_{AB}) = E^{(el)}_{1g}(R_{AB}) + V_{nn}(R_{AB}) \quad [6.15]$$

The heavy red curve in Figure 6.22a shows $V^{(eff)}_{1g}(R_{AB})$. By omitting $E_H(1s)$ from this plot, we are setting the zero of the effective potential energy as the energy of a proton and a hydrogen atom when they are infinitely far apart. The effective potential energy in the σ_{g1s} MO decreases as the proton and hydrogen atom begin to interact, reaching a minimum at the position $R_{AB} = R_e$ before rising again because of the strong repulsive forces between the nuclei. Considered together, the increased electron density and the lowered potential energy between the nuclei demonstrate that σ_{g1s} is a bonding MO.

Now let's see how the effective potential energy curve behaves in the σ^*_{u1s} MO. Proceeding exactly as before, we show the electron electronic bonding energy $E^{(el)}_{1u}$ as the light blue curve and the nuclear repulsive term as the black curve in Figure 6.22b. We combine these to obtain $V^{(eff)}_{1u}(R_{AB})$ (shown by the heavy blue curve in Fig. 6.22b). The effective potential energy in the σ^*_{u1s} state is positive at all distances, increasing as the internuclear distance becomes shorter. This potential generates a repulsive force that pushes the nuclei apart, so the system remains a separated proton and hydrogen atom. In contrast with σ_{g1s}, the σ^*_{u1s} MO shows potential energy higher than the energy of the separated proton and hydrogen atom and zero electron density at the midpoint between the nuclei. So, σ^*_{u1s} is an antibonding MO.

The effective potential energy $V(R_{AB})$ for a diatomic molecule has the general shape shown on the next page where D_e is the bond dissociation energy and R_e is the equilibrium bond length. How do we connect the effective potential energy curve to the total energy of the molecule, which must decrease during bond formation? Each point on this curve represents the potential energy stored in the molecule, by virtue of the interactions of the protons with the electron and with each other, as a function of the internuclear separation R_{AB}. We must now include the kinetic energy of the nuclei. The molecule displays translational kinetic energy as it flies through space, rotational kinetic energy as it rotates about its center of mass, and vibrational kinetic energy as the bond is stretched. Let's focus attention to the vibrational kinetic energy because it relates directly to the changes in the internuclear separation R_{AB}. Suppose the molecule contains a total amount of internal energy E_{tot}, which consists of the effective potential energy and the vibrational kinetic energy. At each value of R_{AB} the vibrational kinetic energy of the nuclei is given by $\mathcal{T} = E_{tot} - V_{eff}$. The kinetic energy goes to

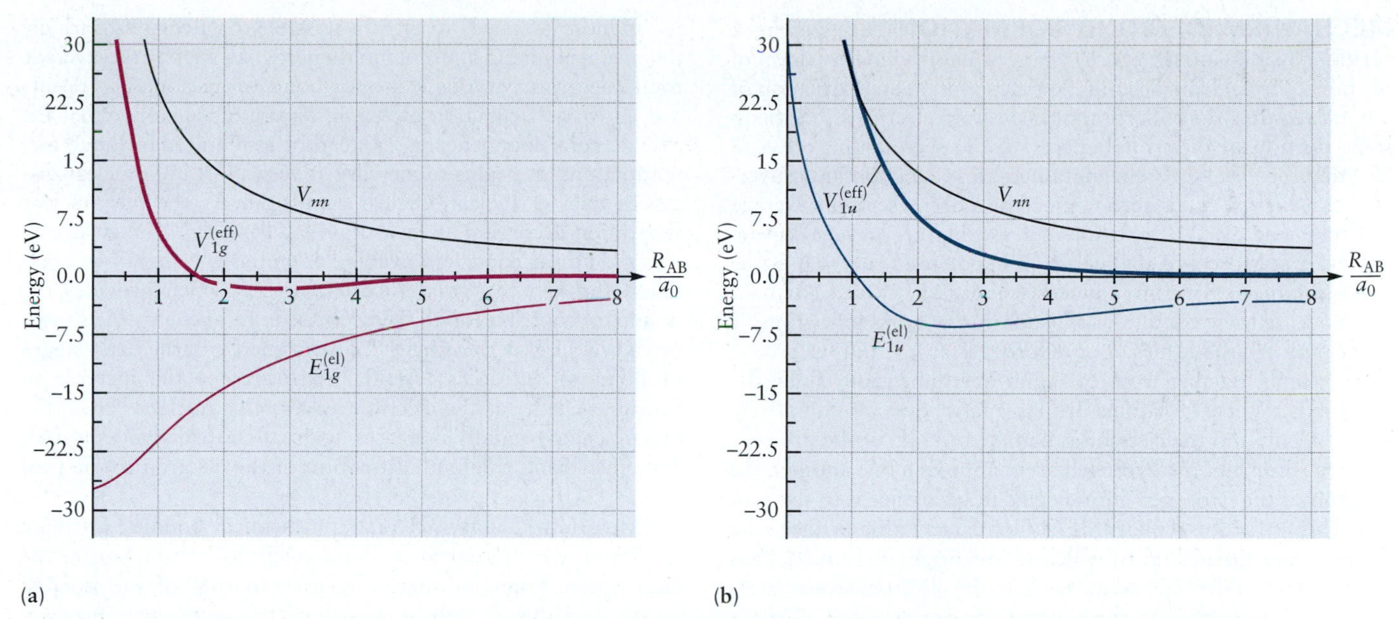

FIGURE 6.22 Dependence on internuclear distance of the contributions to the effective potential energy of the nuclei in H_2^+. (a) The effective potential energy in the σ_{g1s} MO. The electronic bonding energy for the electron is the lighter red curve, the internuclear repulsion between the protons is the black curve, and their sum is the heavier red curve. (b) The effective potential energy in the σ_{u1s}^* MO. The electronic bonding energy for the electron is the lighter blue curve, the internuclear repulsion between the protons is the black curve, and their sum is the heavier blue curve. The behavior of the effective potential energy shows that the σ_{g1s} MO is bonding and that the σ_{u1s}^* MO is antibonding.

(Adapted from R. S. Berry, S. A. Rice, and J. Ross, *Physical Chemistry* (2nd ed), New York: Oxford, 2000, Figure 6.5ab, by permission.).

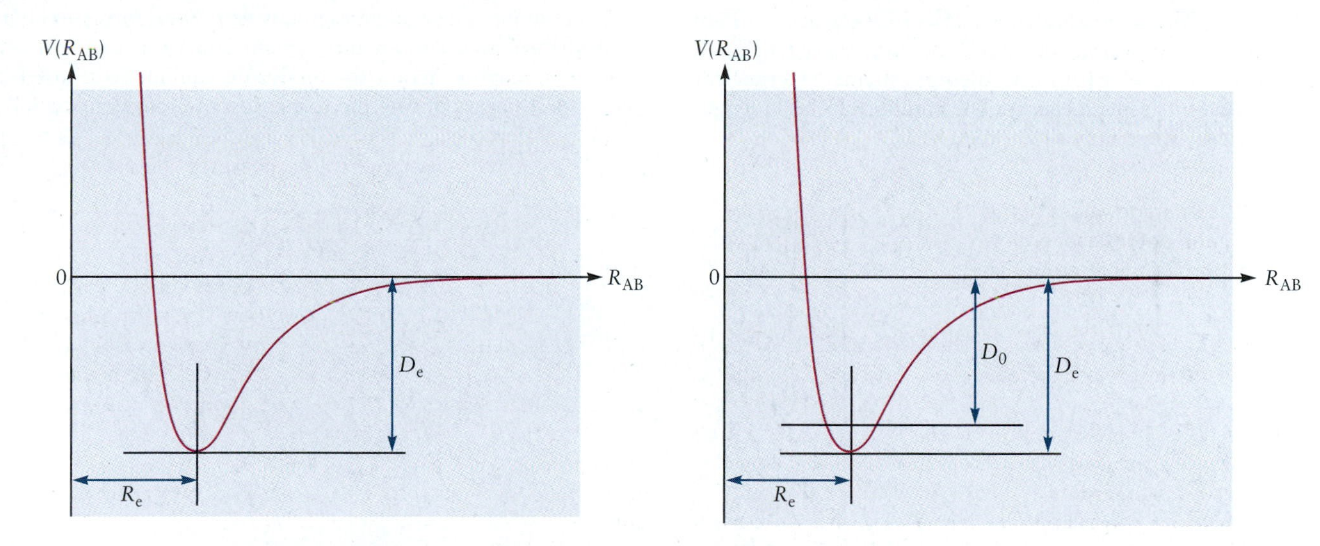

zero at the *classical turning points* where $E_{tot} = V_{eff}$, and the motion of the nuclei reverses direction. (See the discussion of vibrational motion in Appendix B2, especially Figure B.1.)

The total internal energy of the molecule is quantized, so we can visualize a set of energy levels superposed on the effective potential energy curve. Just as shown for the particle-in-a-box model (see Sec. 4.6), the uncertainty principle requires that there is a **zero-point energy** for the molecule (see Sec. 4.7). This is the lowest allowed value of the total internal energy, and is represented by the line D_0. Because the zero-point energy can never be removed from the molecule, it provides a reference point for the additional amount of energy required to dissociate the molecule. Relative to the zero-point energy, the dissociation energy is defined as D_0. Although both D_0 and D_e are called dissociation energies, only the former is measurable experimentally as the energy needed to dissociate the molecule. D_e is useful as a parameter to construct model potentials and optimize geometry in calculations.

In summary then, the structure and energetics of a diatomic molecule are defined by the parameters R_e, D_e, and D_0 on the effective potential energy curve for the molecule, all of which can be calculated within the Born–Oppenheimer approximation (see the figure above). In Chapter 20 we show how R_e and D_0 can be determined experimentally by molecular spectroscopy.

The idea of an effective potential function between the nuclei in a molecule can be formulated empirically (as we have done in Section 3.5 and Fig. 3.9), but it can be defined precisely and related to molecular parameters R_e, D_e, and D_0 only through quantum mechanics. The fundamental significance of the Born–Oppenheimer approximation is its separation of electronic and nuclear motions, which leads directly to the effective potential function.

MECHANISM OF BOND FORMATION Let's inquire a bit more deeply into the energy changes involved in formation of the bond. Recall from Section 3.6 that spontaneous formation of a bond requires the collection of nuclei and electrons to give up some portion of their total energy to the surroundings. Recall also that the virial theorem guarantees this reduction in the average total energy, $\Delta\bar{E}$, is accompanied by a decrease in the average potential energy, $\Delta\bar{V}$, and an increase in the average kinetic energy, $\Delta\bar{\mathcal{T}}$. Moreover, the reduction in potential energy must be twice as large as the gain in kinetic energy: $\Delta\bar{V} = -2\Delta\bar{\mathcal{T}}$.

How are these conditions satisfied during bond formation? It is helpful to think about a hydrogen atom and a proton approaching together to form H_2^+ in one dimension, along the bond axis. When the atoms are far apart, they are completely independent, and the electron is confined to one of the protons. As the proton and the hydrogen atom approach one another, the electron of the hydrogen atom begins to be attracted to the proton. This attraction reduces the Coulomb attraction between the electron and the proton to which it was originally bound, thus increasing its potential energy. Because the electron is now interacting with two protons, it can occupy a greater region of space than when it was part of an isolated hydrogen atom. So, its kinetic energy decreases as it becomes less confined (recall that the energies of the particle in a box decrease as the size of the box increases). The rapid decrease in kinetic energy more than offsets the initial increase in potential energy, so it is responsible for initiating bond formation.

As bond formation continues, the distance between the protons decreases. Then, the simultaneous electrostatic attraction of the electron to *two* protons decreases the potential energy, and the confinement of the electron to the now smaller internuclear region increases its kinetic energy. The equilibrium bond length is determined by these opposing forces.

In more advanced work, it is possible to calculate separately the average kinetic and potential energy, as well as the average total energy, at various stages of bond formation. The results are shown as a function of R_{AB} in Figure 6.23. Notice that the average total energy decreases steadily, as required for bond formation, until it begins to increase at very short internuclear distances due to nuclear–nuclear repulsion. As the internuclear separation decreases, at each step the change in average kinetic energy and the change in average potential energy can be compared through the virial theorem to see which provides the dominant contribution to the decrease in total energy. As shown in Figure 6.23, decreasing the kinetic energy in the early stages of bond formation is essential to overcome the increase in potential energy as the electron leaves "its" nucleus. Then, the kinetic energy rapidly increases again at short distances to balance the strong Coulomb attraction of the electron to the protons.

Most introductory accounts of chemical bonding attribute the stability of the covalent bond solely to a reduction in the electrostatic potential energy, relative to that of the isolated atoms. But that is only one part of the story. The interplay between kinetic and potential energy at each stage of bond formation determines the ultimate stability. The driving force for bond formation in an ionic bond is readily explained by the reduction in the potential energy alone, because the bonding in this case is well described by the *electrostatic* interaction between two charged ions. But in the covalent bond the charge distribution is *dynamic* and cannot be described by classical electrostatics alone. The energetics of covalent bond formation must be described by quantum mechanics. The virial theorem provides a conceptual guide for analyzing the subtle transfer of energy that occurs during the formation of a covalent chemical bond.

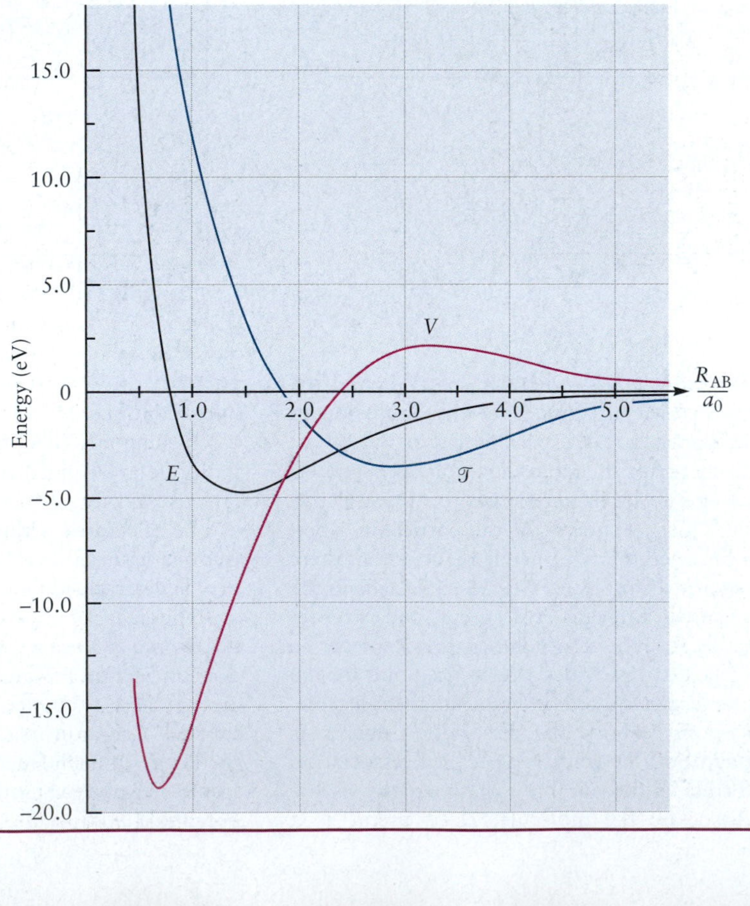

FIGURE 6.23 Average values of the total, kinetic, and potential energies of H_2 as functions of the internuclear distance.

A DEEPER LOOK

6.2.6 Small Polyatomic Molecules

In this section, we extend the LCAO treatment to the linear triatomics BeH_2 and CO_2 not only to illustrate its generality but also to compare the results with the predictions of VSEPR theory and (later) VB theory for these molecules.

BeH_2

Recall from the assertions made in Section 6.2.4 that AOs only make significant contributions to a MO if: (1) their atomic energy levels are close together, and (2) they overlap significantly. Consequently, we need to consider only the $2s$ and $2p_z$ orbitals on beryllium (Be) and the $1s$ orbitals on hydrogen. Formation of the bonding MOs is illustrated in Figure 6.24. Note the direction of the positive z-axis because it defines our sign convention. We form two bonding orbitals by taking linear combinations of either the beryllium $2s$ or $2p_z$ orbital with the hydrogen $1s$ orbitals, choosing the relative phases to concentrate the amplitude (and thus the electron density) between each pair of atoms. We use the simplified notation introduced in Equations 6.9a and 6.9b, but adopt the convention that the first AO belongs to beryllium and the second pair of orbitals belongs to hydrogen. Because the beryllium $2s$ orbital is spherically symmetric, it is clear that we form the bonding MO by adding the hydrogen orbitals in phase with the beryllium $2s$ orbital. Adding the hydrogen orbitals out of phase with the beryllium $2s$ orbital generates the antibonding MO. The results are

$$\sigma_s = C_1 2s + C_2(1s^A + 1s^B) \qquad [6.16a]$$

$$\sigma_s^* = C_3 2s - C_4(1s^A + 1s^B) \qquad [6.16b]$$

In these equations for small polyatomic molecules, the first AO belongs to the central atom, and the second set of equivalent starting orbitals from the surrounding atoms are identified by the labels A, B, C, D. In this case, A and B identify the left and right hydrogen atoms shown in Figure 6.24. Constructing σ orbitals from the beryllium $2p_z$ orbital, in contrast, requires a different choice of phases because it has a node at the nucleus. We must add the hydrogen orbitals with opposite phases to ensure constructive interference with each lobe of the beryllium $2p_z$ orbital to form a bonding MO; the opposite choice leads to the antibonding MO. Therefore, the resulting MOs are

$$\sigma_p = C_5 2p_z + C_6(1s^A - 1s^B) \qquad [6.17a]$$

$$\sigma_p^* = C_7 2p_z - C_8(1s^A - 1s^B) \qquad [6.17b]$$

The $2p_x$ and $2p_y$ beryllium orbitals do not participate in bonding because they would form π orbitals, but hydrogen does not have any p orbitals in its valence shell that can be used to form π bonds. So, they are labeled π_x^{nb} and π_y^{nb}.

Further progress in understanding these MOs requires values for the coefficients $C_1 - C_8$. We used an iterative, self-consistent computer calculation to identify the best values. The resulting MOs are shown in Figure 6.25.

The MO energy-level diagram for BeH_2 is shown in Figure 6.26. The relative displacements of the AOs are determined from the ionization energies of the respective atoms. The energy levels of the four MOs just constructed are shown in the center of the diagram and are linked to their parent AOs by the dashed lines. Note the energy levels of the two nonbonding orbitals arising from the beryllium p_x and p_y orbitals; they remain the same as in the beryllium atom. The ground-state electron configuration for BeH_2 is obtained in the usual way by applying the aufbau principle. Beryllium contributes two electrons, and each hydrogen atom contributes one electron; thus, the resulting electron configuration for the ground state is $(\sigma_s)^2(\sigma_p)^2$. Each bond is a single bond.

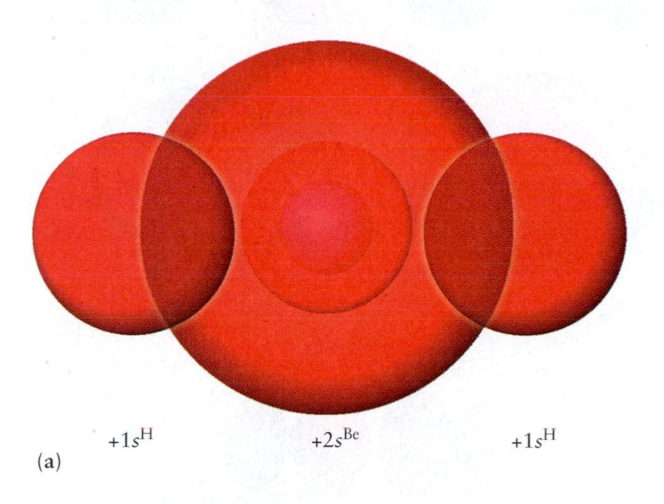

FIGURE 6.24 Overlap of atomic orbitals to form bonding molecular orbitals in BeH_2. (a) Overlap of H $1s$ orbitals with Be $2s$. (b) Overlap of H $1s$ orbitals with Be $2p_z$.
(Courtesy of Mr. Hatem Helal and Professor William A. Goddard III, California Institute of Technology, and Dr. Kelly P. Gaither, University of Texas at Austin.)

CO_2

CO_2 is a linear triatomic molecule whose MOs and energy-level structure are slightly more complex than those of BeH_2 due to the existence of π bonds. As for the diatomics, we separate the σ bonds from the π bonds based on their symmetry. The carbon $2s$ and $2p_z$ orbitals have the proper (cylindrical) symmetry to participate in the formation of σ bonds, as do the oxygen $2s$ and $2p_z$ orbitals. Including the oxygen $2s$ orbitals would certainly improve the accuracy of the approximation but would greatly complicate the development of the argument without adding any

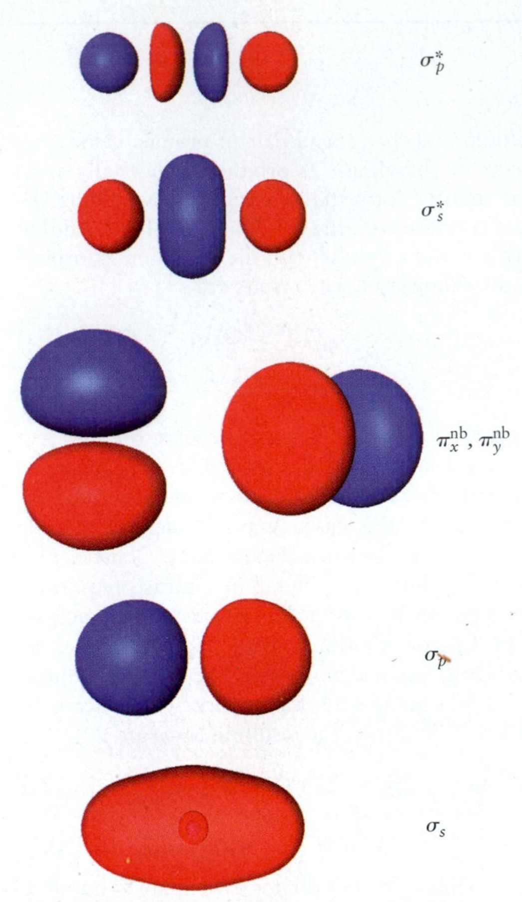

FIGURE 6.25 The isosurfaces shown are those for which the amplitude of the wave function is ±0.2 of the maximum amplitude.
(Courtesy of Mr. Hatem Helal and Professor William A. Goddard III, California Institute of Technology, and Dr. Kelly P. Gaither, University of Texas at Austin.)

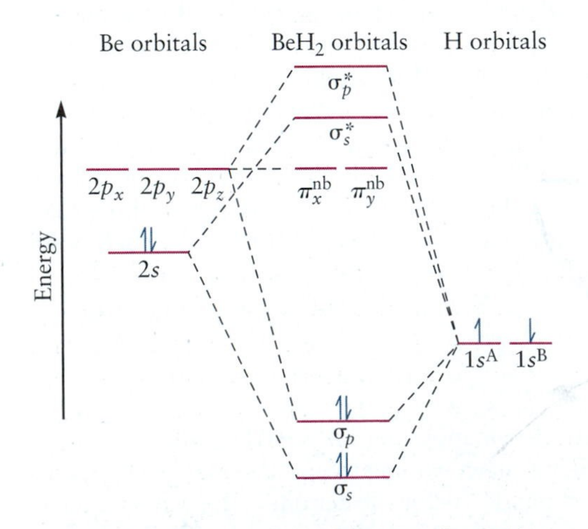

FIGURE 6.26 Energy-level diagram for BeH_2.

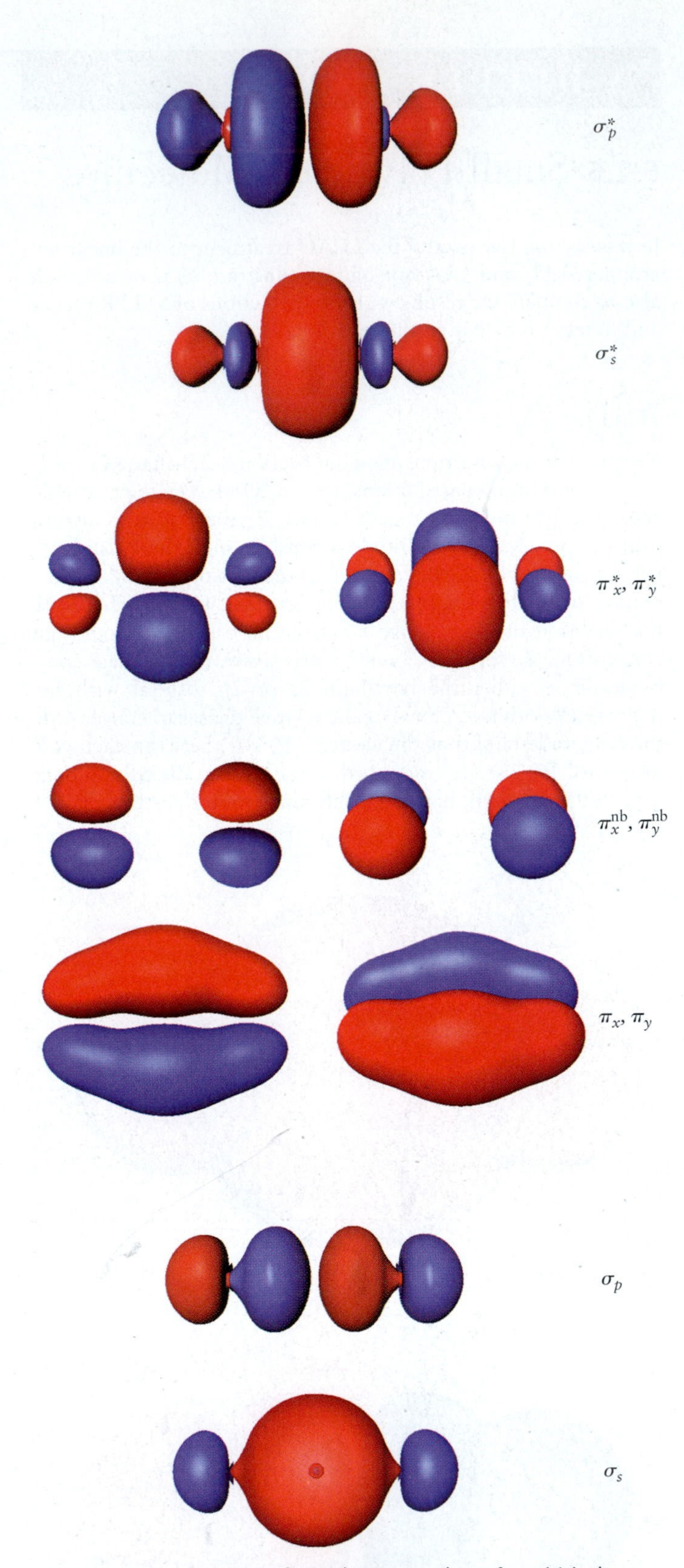

FIGURE 6.27 The isosurfaces shown are those for which the amplitude of the wave function is ±0.2 of the maximum amplitude.
(Courtesy of Mr. Hatem Helal and Professor William A. Goddard III, California Institute of Technology, and Dr. Kelly P. Gaither, University of Texas at Austin.)

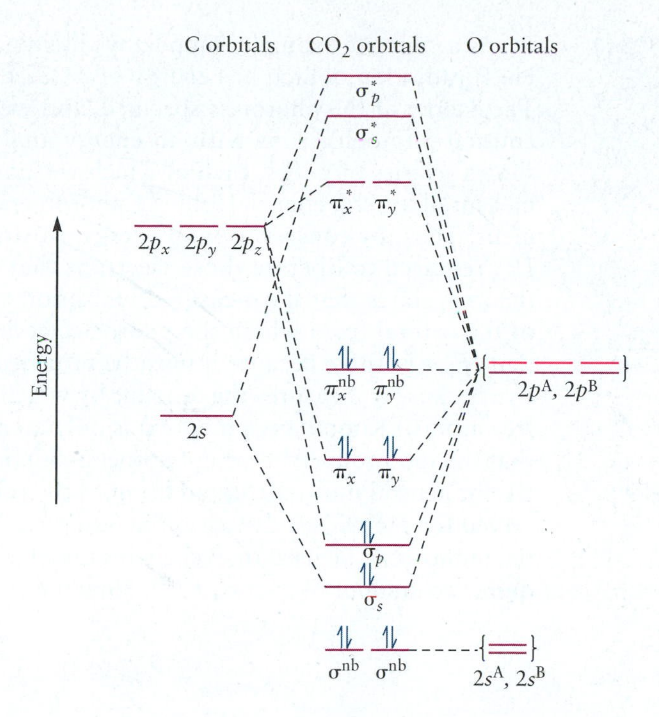

FIGURE 6.28 Energy-level diagram for CO_2.

new insights, so we omit them here for simplicity. The resulting bonding and antibonding MOs are

$$\sigma_s = C_1 2s + C_2(2p_z^A + 2p_z^B) \quad [6.18a]$$

$$\sigma_s^* = C_3 2s - C_4(2p_z^A + 2p_z^B) \quad [6.18b]$$

$$\sigma_p = C_5 2p_z + C_6(2p_z^A - 2p_z^B) \quad [6.19a]$$

$$\sigma_p^* = C_7 2p_z - C_8(2p_z^A - 2p_z^B) \quad [6.19b]$$

$$\pi_x = C_9 2p_x + C_{10}(2p_x^A + 2p_x^B) \quad [6.20a]$$

$$\pi_x^* = C_{11} 2p_x - C_{12}(2p_x^A + 2p_x^B) \quad [6.20b]$$

$$\pi_y = C_{13} 2p_y + C_{14}(2p_y^A + 2p_y^B) \quad [6.21a]$$

$$\pi_y^* = C_{15} 2p_y - C_{16}(2p_y^A + 2p_y^B) \quad [6.21b]$$

where, as for BeH_2, A and B are used to label the outer atoms, in this case, the oxygen atoms. In addition to the bonding and antibonding MOs, there is a pair of nonbonding MOs that result from no *net* overlap with the central carbon $2p_x$ or $2p_y$ orbitals.

$$\pi_x^{nb} = C_{17}(2p_x^A - 2p_x^B) \quad [6.22a]$$

$$\pi_y^{nb} = C_{18}(2p_y^A - 2p_y^B) \quad [6.22b]$$

We determined the optimum values for the coefficients by an iterative, self-consistent computer calculation. The resulting MOs for CO_2 are shown in Figure 6.27.

The energy-level structure of CO_2 is shown in Figure 6.28. With 16 valence electrons, the ground-state configuration of CO_2 is $(\sigma^{nb})^2(\sigma^{nb})^2(\sigma_s)^2(\sigma_p)^2(\pi_x, \pi_y)^4(\pi_x^{nb}, \pi_y^{nb})^4$. The bond order for each CO bond is 2 because there is a σ bond and a π bond between each pair of atoms.

6.3 Photoelectron Spectroscopy for Molecules

In photoelectron spectroscopy (PES), we illuminate a sample with high-frequency radiation (ultraviolet or x-ray) and measure the kinetic energy of the photoelectrons emitted from the sample (see Fig. 5.18.) We show the results as a graph or spectrum of the count rate (number of photoelectrons emitted per second) plotted against the kinetic energy (see Fig. 5.19). We used PES in Section 5.4 to confirm the shell structure of the atom predicted by quantum mechanics (see Fig. 5.21). For molecules, PES confirms the MO description of bonding and measures the energy, ε, for individual MOs. The bridge between PES results and MO theory is Koopmans's theorem (stated in Section 5.4). These three tools are used together to study the electronic structure of molecules in all branches of chemistry. You should master each of them and become expert in using them together.

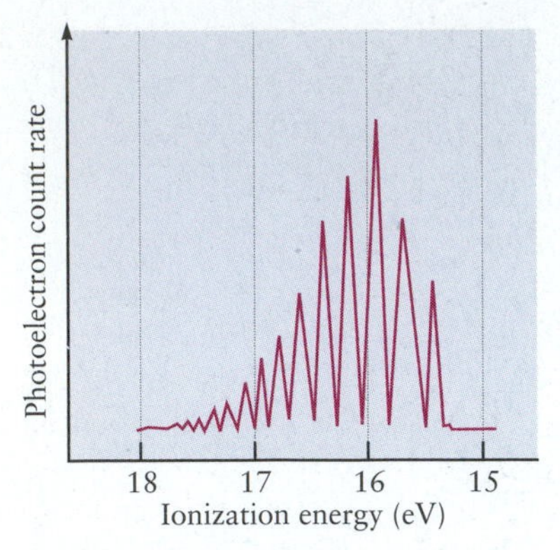

FIGURE 6.29 The photoelectron spectrum of H_2 shows a series of peaks corresponding to vibrational excitation of H_2^+.

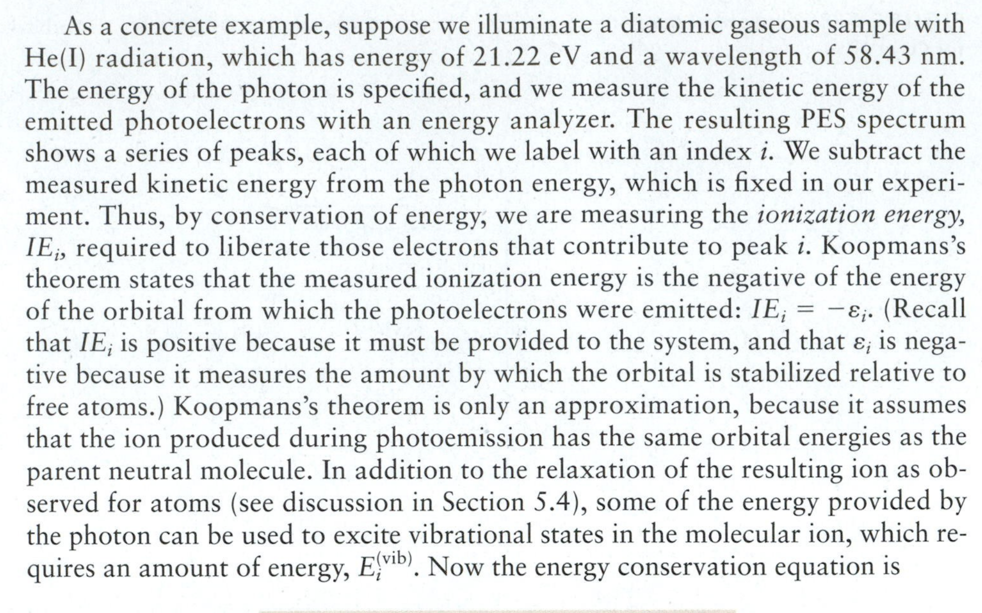

As a concrete example, suppose we illuminate a diatomic gaseous sample with He(I) radiation, which has energy of 21.22 eV and a wavelength of 58.43 nm. The energy of the photon is specified, and we measure the kinetic energy of the emitted photoelectrons with an energy analyzer. The resulting PES spectrum shows a series of peaks, each of which we label with an index i. We subtract the measured kinetic energy from the photon energy, which is fixed in our experiment. Thus, by conservation of energy, we are measuring the *ionization energy*, IE_i, required to liberate those electrons that contribute to peak i. Koopmans's theorem states that the measured ionization energy is the negative of the energy of the orbital from which the photoelectrons were emitted: $IE_i = -\varepsilon_i$. (Recall that IE_i is positive because it must be provided to the system, and that ε_i is negative because it measures the amount by which the orbital is stabilized relative to free atoms.) Koopmans's theorem is only an approximation, because it assumes that the ion produced during photoemission has the same orbital energies as the parent neutral molecule. In addition to the relaxation of the resulting ion as observed for atoms (see discussion in Section 5.4), some of the energy provided by the photon can be used to excite vibrational states in the molecular ion, which requires an amount of energy, $E_i^{(\mathrm{vib})}$. Now the energy conservation equation is

$$h\nu_{\mathrm{photon}} - \frac{1}{2}m_e v^2 = -\varepsilon_i + E_i^{(\mathrm{vib})} = IE_i \qquad [6.23]$$

We know from Section 4.7 that the vibrational energy is quantized, and we treat it as a simple harmonic oscillator: $E_i^{\mathrm{vib}} = nh\nu_{\mathrm{vib}}$, where $n = 0, 1, 2, 3, \ldots$ is the vibrational quantum number. As a result of the vibrational excitation, the peak i in the spectrum is actually a series of narrower peaks; the separation between adjacent peaks depends on the vibrational frequency of the diatomic ion:

$$h\nu_{\mathrm{photon}} - \frac{1}{2}m_e v^2 = -\varepsilon_i + nh\nu_{\mathrm{vib}} \qquad n = 0, 1, 2, 3, \ldots \qquad [6.24]$$

This pattern is illustrated in Figure 6.29, which shows the PES of hydrogen. Just as in Figure 5.19, the ionization energy of the electrons is plotted increasing to the left, whereas their kinetic energy (not shown in Fig. 6.29) increases to the right in the figure. The peak near 15.5 eV corresponds to $n = 0$ and represents the ionization energy for removing electrons with no vibrational excitation of the resulting molecular ion. As the energy increases along the axis toward 18 eV, the amount of vibrational excitation of the H_2^+ ion increases, and the spacing between vibrational levels becomes smaller. H_2^+ is approaching its dissociation limit.

We have just described the PES peaks for the σ_{g1s} MO of H_2; the correlation diagram is shown in Figure 6.10. The experimental results and the correlation diagram are displayed together in Figure 6.30 to show how PES measures the energies of MOs with the aid of Koopmans's theorem. The orbital energy becomes more negative (larger in magnitude) in the downward direction below zero along the axis. The ionization energy is positive and increases upward above the zero of energy. Purely for ease of interpreting the results, it is conventional in PES to flip the ionization energy axis to the opposite direction, so in Figure 6.30 ionization energy becomes larger and more positive in the *downward* direction along the black arrow. This procedure makes it easy to connect measured ionization energy values with the orbital energies. Note that the experimental peak with $n = 0$ is aligned with the energy of the MO, because this peak measures the energy of the orbital without vibrational excitation. In the other experimental peaks, some of the energy of the photon has been used to excite molecular vibrations and is not available to the outgoing photoelectron; therefore, these peaks appear to have orbital energies that are too large by the amount of their vibrational excitation.

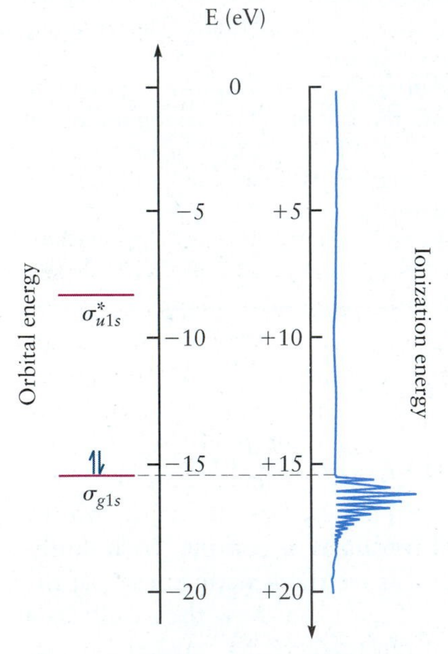

FIGURE 6.30 The photoelectron spectrum of H_2 measures the ionization energy of the σ_{g1s} molecular orbital. The procedures for obtaining orbital energies from measured ionization energies are described in the text. Photoelectron spectroscopy gives no information about the unoccupied σ^*_{u1s} orbital.

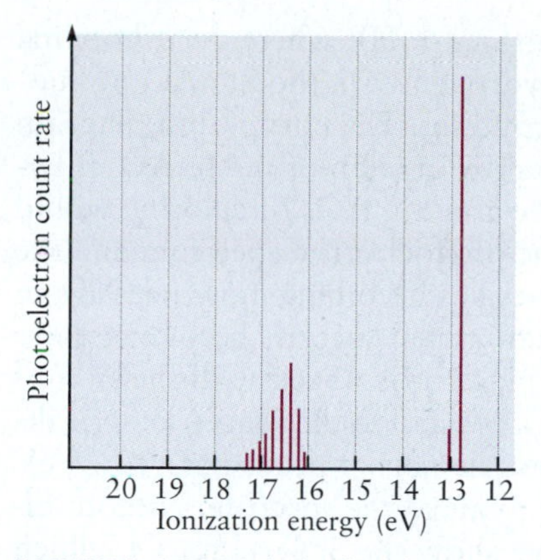

FIGURE 6.31 Peaks in the photoelectron spectrum of HCl measured with He(I) photons at 21.2 eV.

These vibrational "fine-structure" peaks on the PES data at first appear to be a nuisance, but in fact, they greatly aid in relating experimental data to particular MOs. The connection is made through the concept of bond order introduced in Section 6.2. We illustrate the procedure for three separate cases.

Case A

If the photoelectron is removed from a bonding orbital, the bond order of the positive ion will be smaller than the bond order of the parent molecule. Consequently, the bond in the molecular ion will be less stiff, and its vibrational frequency (determined directly from the PES fine structure) will be lower than that of the parent molecule (determined by vibrational spectroscopy). For example, in Figure 6.29, the vibrational frequency for H_2^+ is $6.78 \times 10^{13}\ s^{-1}$, compared with $12.84 \times 10^{13}\ s^{-1}$ for the parent molecule.

Case B

If the photoelectron is emitted from an antibonding orbital, the bond order of the positive ion will be larger than the bond order of the parent molecule. The bond in the diatomic molecular ion will be stiffer and will show a higher vibrational frequency.

Case C

If the photoelectron is emitted from a nonbonding orbital, there is no change in the bond order, and consequently little or no change in the vibrational frequency. The PES spectrum for the orbital will show few vibrational peaks, because the disturbance to the bond during photoemission is quite small. By contrast, the spectrum in Case A will show several vibrational fine structure peaks because removal of a bonding electron is a major disturbance that starts many vibrations of the bond. Case B is intermediate, with fewer vibrational subpeaks, because removing an antibonding electron disturbs the bond, but less so than in Case A.

We summarize these results in Table 6.4 and use them to interpret PES data for several additional molecules.

The PES data for gaseous hydrogen chloride (HCl) acquired with He(I) radiation are shown in Figure 6.31. The portion of the spectrum just below 13 eV shows few vibrational components, which suggests that it originates in a nonbonding orbital. The vibrational spacing in these components corresponds to a vibrational frequency of $7.98 \times 10^{13}\ s^{-1}$. The sequence of peaks starting at 16.25 eV shows numerous vibrational contributions, suggesting they originate in a bonding orbital; the vibrational separation is $4.83 \times 10^{13}\ s^{-1}$. The fundamental vibrational frequency for the parent HCl molecule is $8.66 \times 10^{13}\ s^{-1}$. The changes in vibrational energy are consistent with the peak assignments based on number of vibrational peaks. Now, which orbitals of this heteronuclear diatomic are actually

TABLE 6.4 Identification of Molecular Orbitals in Photoelectron Spectra of Diatomic Molecules

Vibrational Frequency of Ion Relative to Parent Molecule	Number of Vibrational Lines in Peak	Molecular Orbital from Which Electron Is Emitted
Much smaller	Many	Bonding
Similar	Few	Nonbonding
Similar to slightly larger	Intermediate	Antibonding

involved? Figure 6.23 shows the correlation diagram for HF, where the σ bonding orbital is formed by overlap of the $1s$ AO of hydrogen with the $2p_z$ AO of fluorine; the $2p_x$ and $2p_y$ AOs of fluorine are nonbonding. The correlation diagram for HCl would show a σ bonding orbital formed by overlap of the $1s$ AO of hydrogen with the $3p_z$ AO of chlorine (Cl); the $3p_x$ and $3p_y$ AOs of chlorine would be nonbonding. Figure 6.32 shows the schematic photoelectron spectrum and the correlation diagram for HCl, with the orbital energies becoming more negative in the downward direction along the axis and the ionization energies becoming more positive in the downward direction along the black axis. (See the discussion of Figure 6.30 for the reasons behind these choices.) We assign the sequence of peaks starting at 16.25 eV to the σ bonding orbital and the pair of peaks near 12.73 eV to the $3p_x$ and $3p_y$ AOs of Cl. So far, we have explained the spectrum without involving the $3s$ electrons of Cl. In Figure 6.32 we show the $3s$ level for Cl, which appears at a lower orbital energy characteristic of pure chlorine. This suggests that the $3s$ electrons are not involved in formation of the H—Cl bond. The corresponding peak does not appear in Figure 6.31 because this deeper energy level is not accessible to the ultraviolet lines from helium and requires higher energy photons for photoionization.

The photoelectron spectra for N_2 and O_2 are shown in Figures 6.33 and 6.34, respectively. The experimental peaks have been assigned to orbitals by slightly more complex versions of the arguments used previously. Note that for N_2, the energy for the σ_{g2p_z} MO is lower than that for the π_{u2p_x} and π_{u2p_y}, whereas the order is switched for O_2, as indicated in Figure 6.15, Table 6.3, and the related text. This switch is due to interaction between the $2s$ and $2p$ AOs.

The photoelectron spectrum for NO is shown in Figure 6.35. The orbital assignments are based on the arguments summarized in Table 6.4. Note that the $1s$ core levels for both N and O appear at the same orbital energies as they do in their respective elemental gases, N_2 in Figure 6.33 and O_2 in Figure 6.34. This experimental result demonstrates clearly that the core levels do not participate in chemical bond formation and can be neglected in the MO analysis of bond formation.

These examples show that photoelectron spectroscopy is useful in testing theoretical models for bonding because it directly measures ionization energies that can be correlated with theoretical orbital energies through Koopmans's theorem. These methods are readily extended to polyatomic molecules.

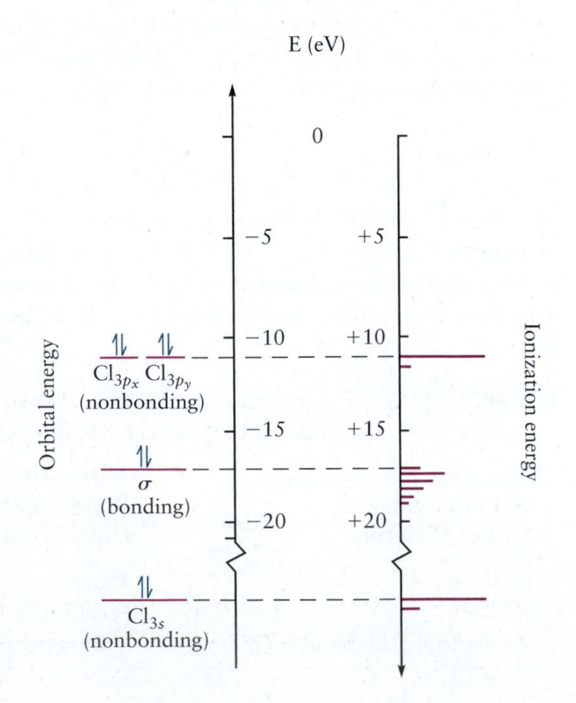

FIGURE 6.32 The photoelectron spectrum and the molecular orbital energy-level diagram for HCl. The peaks with ionization energies lower than 20 eV originate in the σ bonding molecular orbital and in the Cl_{3p} nonbonding orbitals. Photoionization of the Cl 3s electrons requires higher energy photons.

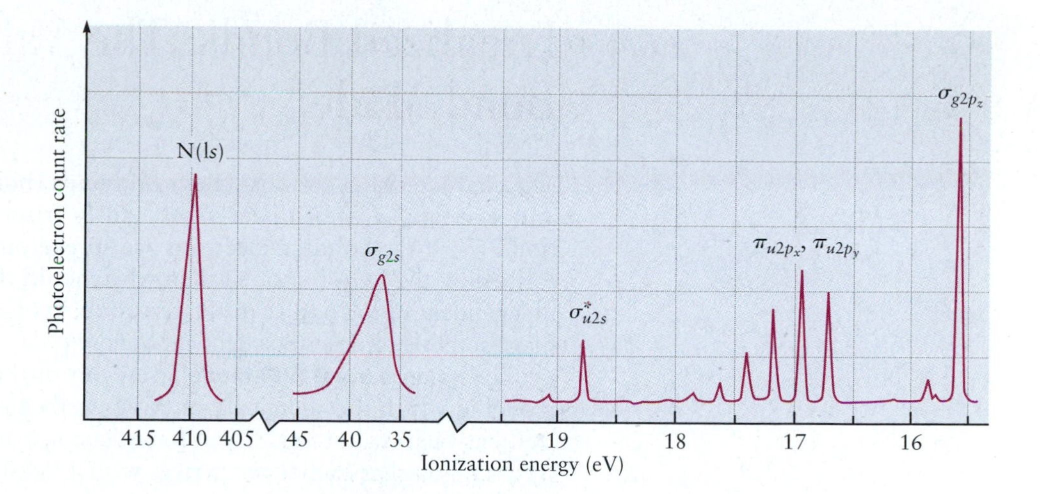

FIGURE 6.33 The photoelectron spectrum for N_2 shows the valence electrons in the occupied molecular orbitals and the N(1s) core electrons.

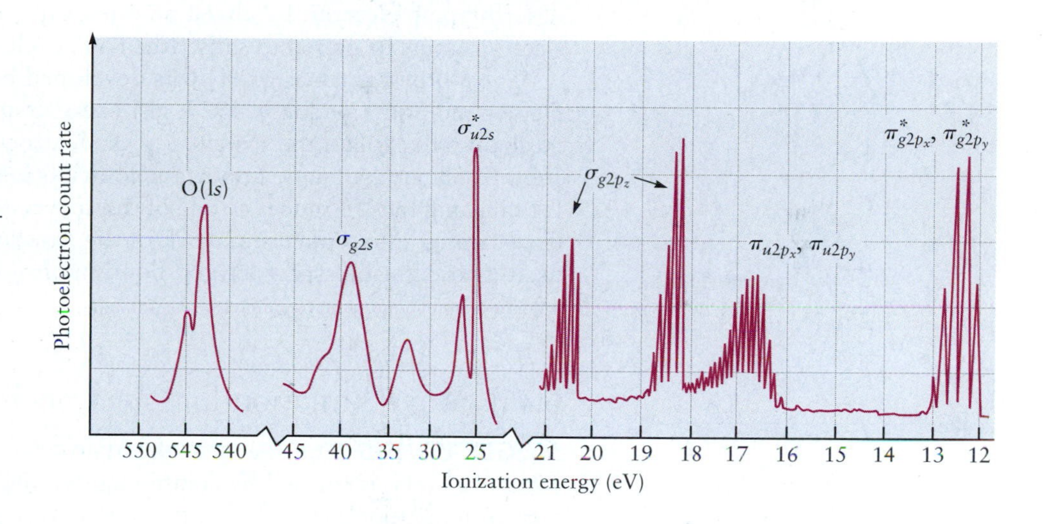

FIGURE 6.34 The photoelectron spectrum for O_2 shows valence electrons in the occupied molecular orbitals and the O(1s) core electrons. Note that the order of the σ_{g2p_z} and π_{u2p_x}, π_{u2p_y} orbitals has switched between N_2 and O_2. More advanced theory is required to explain why σ_{g2p_z} is split into two groups.

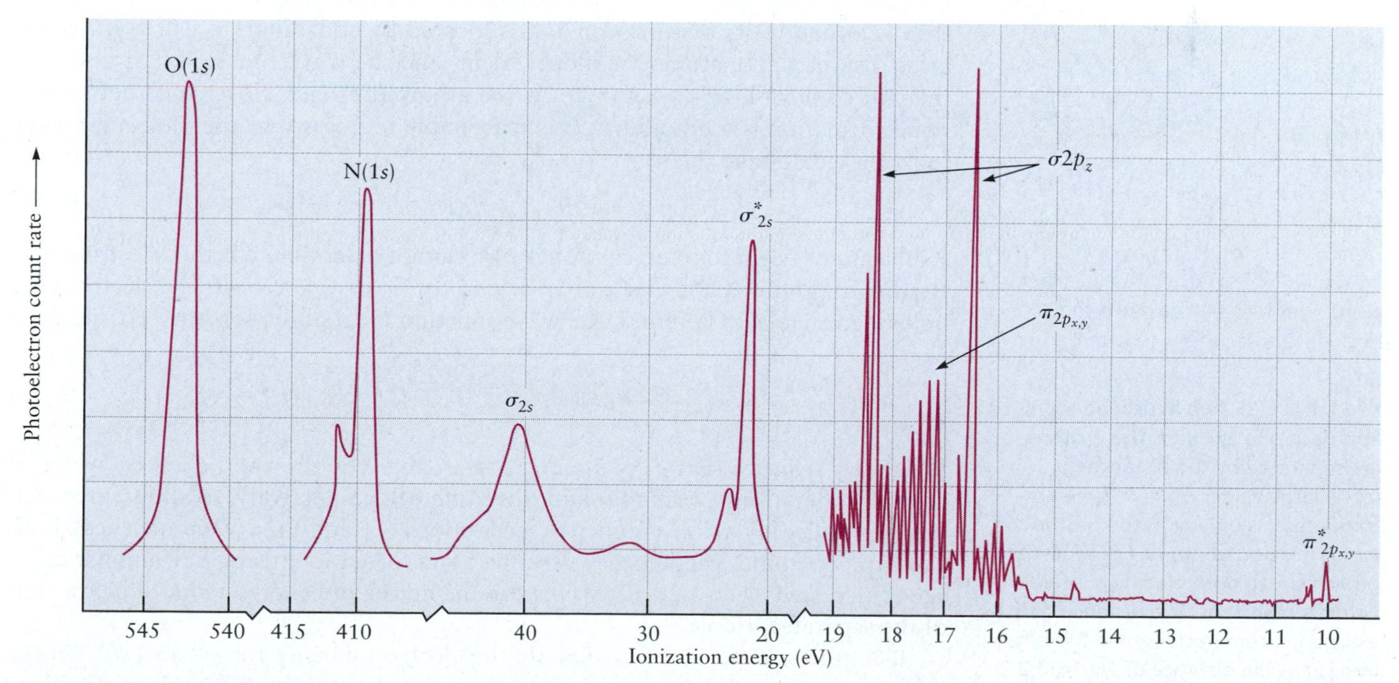

FIGURE 6.35 The photoelectron spectrum for NO. Note that the O(1s) and N(1s) orbitals remain at the same energies as in O_2 and N_2; these core electrons do not participate significantly in formation of the N—O bond. More advanced theory is required to explain why σ_{2p_z} is split into two groups.

6.4 Localized Bonds: The Valence Bond Model

The characteristics of most chemical bonds (bond length, bond energy, polarity, and so forth) do not differ significantly from molecule to molecule (see Section 3.7). If the bonding electrons are spread out over the entire molecule, as described by the LCAO model, then why should the properties of a bond be nearly independent of the nature of the rest of the molecule? Would some other model be more suitable to describe chemical bonds?

The **valence bond (VB) model** grew out of the qualitative Lewis electron-pair model in which the chemical bond is described as a pair of electrons that is localized between two atoms. The VB model constructs a wave function for the bond by assuming that each atom arrives with at least one unpaired electron in an AO. The resulting wave function is a *product* of two one-electron wave functions, each describing an electron localized on one of the arriving atoms. The spins of the electrons must be paired to satisfy the Pauli exclusion principle.

The VB description for H_2 was developed by the German physicists Walter Heitler and Fritz London in 1927, just one year after Schrödinger introduced wave mechanics to explain the structure of the hydrogen atom. The American physicist John C. Slater also made important contributions to developing the VB method. Establishing the VB model as one of the cornerstones of modern structural chemistry awaited the pioneering work of the American chemist Linus Pauling, who used it to describe structure and bonding in polyatomic molecules, starting in 1931.

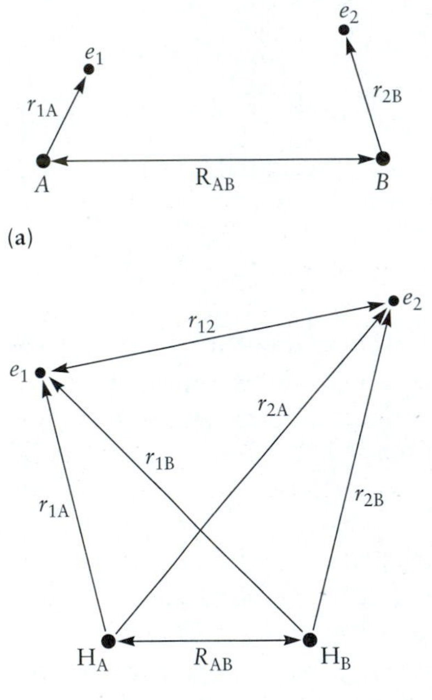

(a)

(b)

FIGURE 6.36 Two hydrogen atoms approach one another. The protons are separated by the distance R_{AB}. (a) At large values of R_{AB} each electron interacts only with the proton to which it is bound. (b) As the atoms approach closer, both electrons interact with both protons. The distance of electron 1 from nuclei A and B is given by r_{1A}, r_{1B}; the distance of electron 2 from nuclei A and B is given by r_{2A}, r_{2B}; the distance between the electrons is given by r_{12}.

6.4.1 Wave Function for Electron-Pair Bonds

SINGLE BONDS Consider that the hydrogen molecule, described by the Lewis structure H:H, is formed by combining two hydrogen atoms each with the electron configuration $H:(1s)^1$. The two atoms approach one another (Fig. 6.36), and the protons are separated by the distance R_{AB}. At very large separation, each electron is bound to its own proton and is located by coordinate r_{1A} or r_{2B}. At very large distances, the atoms are independent, and the wave function that describes the pair of them is $\varphi^A(r_{1A})\varphi^B(r_{2B})$. As the atoms approach closer together so that bond formation is a possibility, it is reasonable to guess that the molecular wave function would take the form

$$\psi^{el}(r_{1A}, r_{2B}; R_{AB}) = C(R_{AB})\varphi^A(r_{1A})\varphi^B(r_{2B}) \qquad [6.25]$$

As the atoms begin to interact strongly, we cannot determine whether electron 1 arrived with proton A and electron 2 with proton B, or vice versa. (The electrons are indistinguishable.) Therefore, the wave function must allow for both possibilities:

$$\psi^{el}(r_{1A}, r_{2B}; R_{AB}) = C_1(R_{AB})\varphi^A(r_{1A})\varphi^B(r_{2B}) + C_2(R_{AB})\varphi^A(r_{2A})\varphi^B(r_{1B}) \qquad [6.26]$$

Symmetry requires that $c_1 = c_2$ and $c_1 = -c_2$ are equally valid choices. We label these combinations *gerade* (g) and *ungerade* (u), respectively, to show how each behaves under inversion symmetry (see Sections 6.1 and 6.2). We must check both cases to determine whether they describe bond formation, using our familiar criteria of increased electron density between the nuclei and energy reduced below that of the separated atoms.

It requires some care to calculate the electron density for ψ_g^{el} and ψ_u^{el}. Unlike the wave functions we have seen earlier, these are *two-electron functions,* and their squares give the probability density for finding electron 1 at r_1 *and* electron 2 at r_2. To calculate the probability density of finding electron 1 at r_1, no matter where

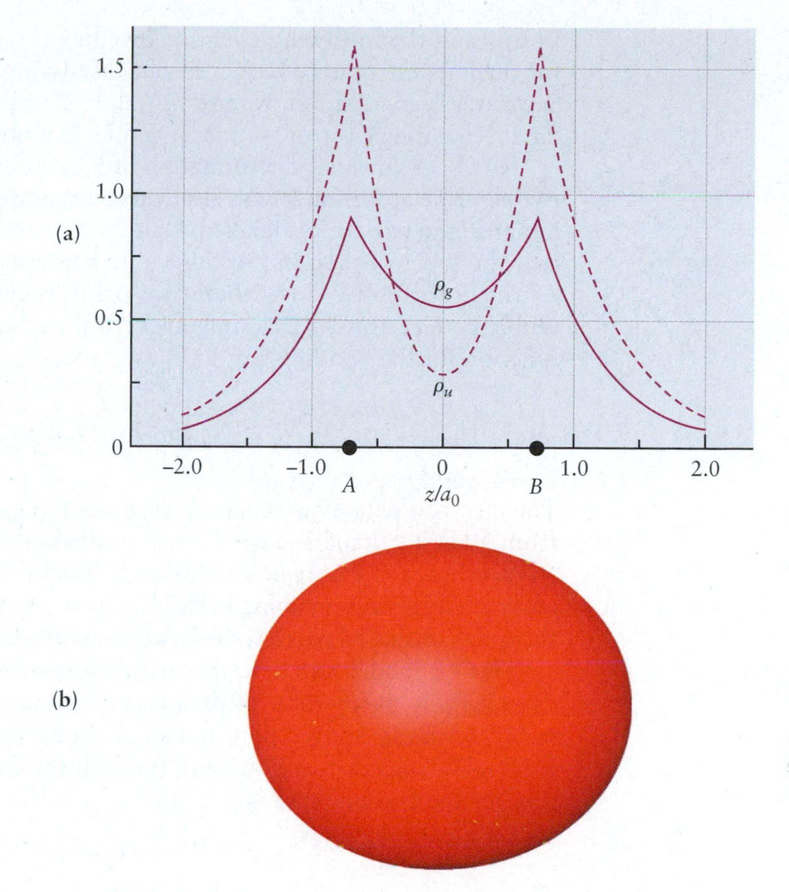

FIGURE 6.37 The electron density for the ψ_g^{el} and ψ_u^{el} wave functions in the simple valence bond model for H_2. (a) The electron density ρ_g for ψ_g^{el} and ρ_u for ψ_u^{el}, calculated analytically as described in the text. (b) Three-dimensional isosurface of the electron density for the ψ_g^{el} wave function, as calculated numerically by Generalized Valence Bond Theory (GVB).
(Courtesy of Mr. Hatem Helal and Professor William A. Goddard III, California Institute of Technology, and Dr. Kelly P. Gaither, University of Texas at Austin.)

electron 2 is located, we must square the function and then average over all possible locations for electron 2. Similarly, we calculate the probability density for finding electron 2 at r_2, regardless of the location of electron 1. Adding these results together gives the total electron density at each point in space, as a function of the internuclear distance R_{AB}. The results for ψ_g^{el} and ψ_u^{el} are shown in Figure. 6.37a. The combination ψ_g^{el} results in increased electron density between the nuclei, whereas the combination ψ_u^{el} results in reduced electron density between the nuclei. It is more convenient to obtain the electron density from computer calculations using a newer version of the VB method called Generalized Valence Bond Theory (GVB), which will be described later. Figure 6.37b shows the electron density for ψ_g^{el} obtained in a GVB calculation for H_2.

It is a straightforward exercise in quantum mechanics—although beyond the scope of this textbook—to calculate the total energy of the molecule as a function of R_{AB} for ψ_g^{el} and for ψ_u^{el}. The results give the potential energy of interaction between the protons for each value of R_{AB}. (Recall the discussion of potential energy in Section 6.2.1. See the discussion in Section 6.2.5 for a deeper analysis.) The two calculated potential energy curves (not shown) are qualitatively similar to those in Figure 6.7. They show that ψ_g^{el} describes a state with lower energy than that of the separated atoms, whereas ψ_u^{el} describes a state whose energy is higher than that of the separated atoms for all values of R_{AB}.

Taken together, the reduced potential energy and increased electron density between the nuclei demonstrate that ψ_g^{el} describes a stable chemical bond, whereas ψ_u^{el} describes a state that is strictly repulsive everywhere and does not lead to bond formation. Therefore, we take the proper VB electronic wave function for a pair bond to be

$$\psi_g^{el} = C_1[1s^A(1)1s^B(2) + 1s^A(2)1s^B(1)] \qquad [6.27]$$

In this and the following equations we have switched to the simplified notation for orbitals introduced earlier in our discussion of the LCAO approximation in Section 6.2. In addition, we use "1" and "2" as shorthand notation for the coordinates locating electrons 1 and 2, and C_1 is a normalization constant.

Now let's consider the formation of F_2, represented by its Lewis diagram from two fluorine atoms each with electron configuration F: $(1s)^2(2s)^2(2p_x)^2(2p_y)^2(2p_z)^1$. Suppose the two atoms labeled A and B approach each other along the z-axis so that the lobes of their $2p_z$ orbitals with the same phase point toward each other. As the atoms draw close, these two orbitals can overlap to form a single bond with two electrons. Reasoning as we did earlier for H_2, we write the VB wave function for the bonding pair in F_2 as

$$\psi_g^{\text{bond}} = C_1[2p_z^A(1)2p_z^B(2) + 2p_z^A(2)2p_z^B(1)] \qquad [6.28]$$

:F:F:

Lewis diagram for F_2.

The electron density associated with this two-electron wave function, obtained from a GVB calculation for F_2 at the experimentally measured bond length for R_{AB}, is shown in Figure 6.38. This wave function gives no information on the eight pairs of electrons remaining in their AOs on atoms A and B, six pairs of which are shown as unshared pairs in the Lewis diagram for F_2.

The VB model also describes bond formation in heteronuclear diatomics. We can combine the features of the two preceding examples to describe the bonding in HF, with one shared pair in a single bond produced by overlap of H(1s) and F(2p_z). We suggest that you work through the details to show that the wave function for the bonding pair is

$$\psi^{\text{bond}} = C_1[2p_z^F(1)1s^H(2)] + C_2[2p_z^F(2)1s^H(1)] \qquad [6.29]$$

The electron density in this bond, obtained from a GVB calculation for HF, is shown in Figure 6.39. Remember that the *gerade* and *ungerade* labels no longer apply, and $c_1 \neq c_2$, because HF is a heteronuclear diatomic molecule.

The bond-pair wave functions in Equations 6.27, 6.28, and 6.29 were specially constructed to describe two electrons localized between two atoms as a single chemical bond between the atoms. These wave functions should not be called MOs, because they are not single-electron functions and they are not de-localized over the entire molecule. The corresponding single bonds (see Figs. 6.37, 6.38, and 6.39) are called σ **bonds,** because their electron density is cylindrically symmetric about the bond axis. There is no simple correlation between this symmetry

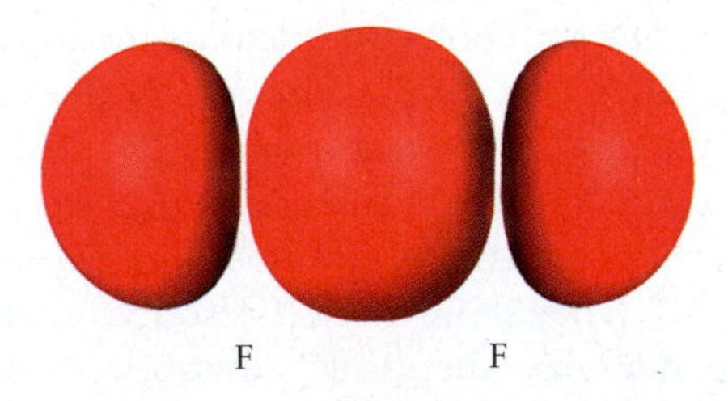

FIGURE 6.38 Isosurface representation of the electron density in the F_2 σ bond formed from a pair of electrons initially localized in a $2p_z$ orbital on each F atom.
(Courtesy of Mr. Hatem Helal and Professor William A. Goddard III, California Institute of Technology, and Dr. Kelly P. Gaither, University of Texas at Austin.)

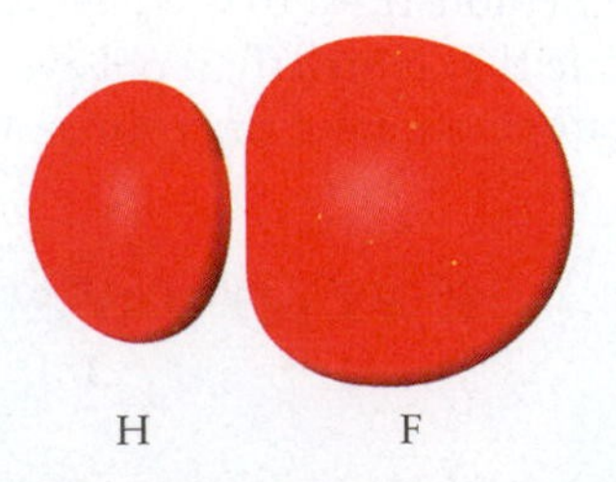

FIGURE 6.39 Isosurface representation of the electron density in the HF σ bond formed from a pair of electrons initially localized in a 2s orbital on H and in a $2p_z$ orbital on F.
(Courtesy of Mr. Hatem Helal and Professor William A. Goddard III, California Institute of Technology, and Dr. Kelly P. Gaither, University of Texas at Austin.)

and the angular momentum of electrons about the bond axis. Finally, electrons are not placed in these bonds by the aufbau principle. Rather, each bond is formed by overlap of two AOs, each of which is already half filled with one electron. The electrons in the two participating AOs must have opposite spin; thus, the bond corresponds to an electron pair with opposite, or "paired," spins.

:N:::N:

Lewis diagram for N_2.

MULTIPLE BONDS To see how the VB method describes multiple bonds, let's examine N_2. Suppose two nitrogen atoms with electron configuration $N:(1s)^2(2s)^2(2p_x)^1(2p_y)^1(2p_z)^1$ approach one another along the z-axis. The two $2p_z$ orbitals can overlap and form a σ bond, the wave function of which is

$$\psi_{\sigma}^{\text{bond}} = C_1[2p_z^A(1)2p_z^B(2) + 2p_z^A(2)2p_z^B(1)] \quad [6.30]$$

The $2p_x$ and $2p_y$ orbitals from the two atoms do not approach head-on in this configuration, but rather side by side. Therefore, the positive lobes of the $2p_x$ orbitals can overlap laterally, as can the negative lobes. Together, they form a π **bond,** which has a node through the plane containing the bond axis with electron density concentrated above and below the plane. The wave function for the bond pair is

$$\psi_{\pi x}^{\text{bond}}(1, 2; R_{AB}) = C_1(R_{AB})[2p_x^A(1)2p_x^B(2)] + C_1(R_{AB})[2p_x^A(2)2p_x^B(1)] \quad [6.31]$$

Similarly, the $2p_y$ orbitals on the two atoms can overlap to form a second π bond. Altogether, N_2 has a triple bond, for which the wave function is

$$\psi_{\sigma\pi\pi}^{\text{bond}}(1, 2, 3, 4, 5, 6; R_{AB}) = C_1(R_{AB})[2p_z^A(1)2p_z^B(2)][2p_x^A(3)2p_x^B(4)][2p_y^A(5)2p_y^B(6)] + C_1(R_{AB})[2p_z^A(2)2p_z^B(1)][2p_x^A(4)2p_x^B(3)][2p_y^A(6)2p_y^B(5)] \quad [6.32]$$

The electron densities in each bond contributing to the triple bond in N_2, as determined from a GVB calculation, are shown in Figure 6.40.

POLYATOMIC MOLECULES Specifying the three-dimensional structure of polyatomic molecules requires that we include bond angles and bond lengths. Any successful theory of bonding must explain and predict these structures. Let's test the VB approximation on the second-period hydrides, the structures of which we have already examined in Chapter 3 using VSEPR theory.

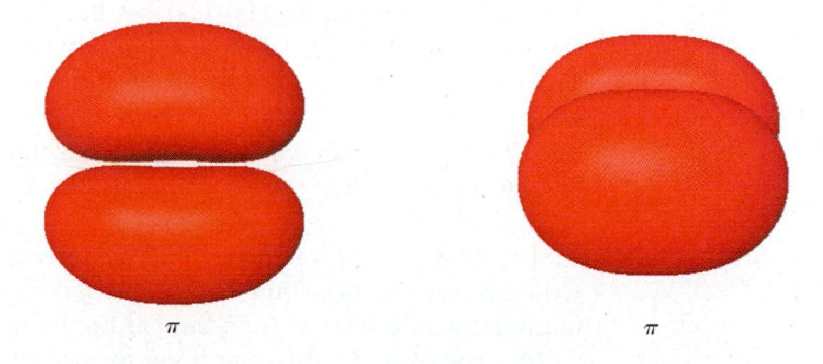

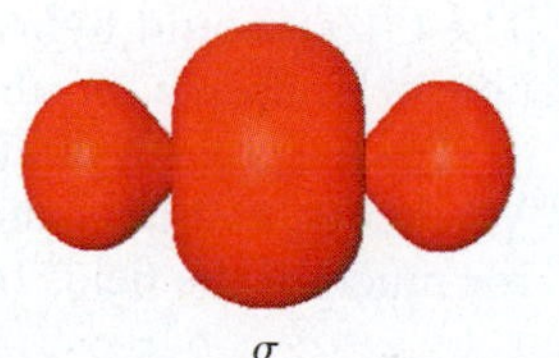

FIGURE 6.40 Isosurface representation of the electron densities in the σ bond and the two π bonds calculated by Generalized Valence Bond Theory (GVB).
(Courtesy of Mr. Hatem Helal and Professor William A. Goddard III, California Institute of Technology, and Dr. Kelly P. Gaither, University of Texas at Austin.)

H:Be:H

Lewis diagram for BeH_2.

Beryllium hydride, BeH_2, has four valence electrons, two from beryllium and one each from the two hydrogen atoms, all of which appear in its Lewis diagram. In VSEPR theory, the steric number is 2, so the molecule is predicted to be linear, and this prediction is verified by experiment. The electron configuration of the central atom is Be: $(1s)^2(2s)^2$. There are no unpaired electrons to overlap with H(1s) orbitals, so the VB model fails to predict the formation of BeH_2.

H:B:H
 ..
 H

Lewis diagram for BH_3.

Boron hydride, BH_3, has six valence electrons corresponding to steric number 3 and a trigonal planar structure. With the electron configuration B: $(1s)^2(2s)^2(2p)^1$ on the central atom, the VB model cannot account for the formation of BH_3 and, in fact, predicts that BH is the stable molecule, which does not agree with experimental results.

 H
 ..
H:C:H
 ..
 H

Lewis diagram for CH_4.

Methane, CH_4, has steric number 4, and VSEPR predicts a tetrahedral structure, which is confirmed by experimental results. Starting with the electron configuration C: $(1s)^2(2s)^2(2p)^2$, the VB model cannot account for the formation of CH_4 and predicts that CH_2 would be the stable hydride, which is again contrary to the experimental results.

H:N:H
 ..
 H

Lewis diagram for NH_3.

Ammonia, NH_3, has steric number 4 with three shared pairs and one unshared pair on the nitrogen atom. VSEPR predicts a trigonal pyramid structure, as a subcase of tetrahedral structure, with bond angles slightly less than 109° due to repulsion between the unshared pair and the three bonding pairs. Experiment verifies this structure with bond angles of 107°. The electron configuration N: $(1s)^2(2s)^2(2p_x)^1(2p_y)^1(2p_z)^1$ would permit the formation of three σ bonds by overlap of H(1s) orbitals with each of the 2p orbitals on nitrogen. Because these 2p orbitals are all mutually perpendicular, the VB model predicts a trigonal pyramidal structure, but one with bond angles of 90°.

Finally, water, H_2O, has steric number 4 with two shared pairs and two unshared pairs on the oxygen atom. VSEPR theory predicts a bent structure, as a subcase of tetrahedral structure, with angles significantly less than the tetrahedral value of 109° due to repulsion between the two unshared pairs and the bonding pairs. Experimentally determined bond angles of 104° verify this prediction.

The VB model does not accurately describe bonding in the second-period hydrides. It predicts the wrong valence for atoms in Groups IIA through IVA and the wrong structure for atoms in Groups VA and VIA. Clearly, the model had to be improved. Linus Pauling gave the answer in 1931 by introducing the concepts of promotion and hybridization.

Atoms such as Be, B, and C can have the correct valence for bonding by **promotion** of valence electrons from the ground state to excited states at higher energy. For example, Be: $(1s)^2(2s)^2 \rightarrow$ Be: $(1s)^2(2s)^1(2p)^1$ and C: $(1s)^2(2s)^2(2p)^2 \rightarrow$ C: $(1s)^2(2s)^1(2p_x)^1(2p_y)^1(2p_z)^1$ are ready to form BeH_2 and CH_4, respectively. These excited states are known from spectroscopy. The excited state of carbon lies about 8.26 eV (190 kJ mol^{-1}) above the ground state, and energy is clearly required for promotion. Pauling argued that this investment would be repaid by the energy released when the C—H bonds of methane form (about 100 kJ mol^{-1} for each bond).

Even though the valence would be correct after promotion, the structure still would be wrong. Beryllium hydride would have two different kinds of bonds, and methane would have three identical bonds formed by overlap of H(1s) with the C(2p) orbitals and a different bond formed by H(1s) and C(2s). Pauling proposed that new orbitals with the proper symmetry for bond formation could be formed by **hybridization** of 2s and 2p orbitals after promotion. The Be(2s) and Be(2p_z) orbitals would combine to form two equivalent hybrid orbitals oriented 180° apart. The C(2s) would hybridize with the three C(2p) orbitals to give four equivalent new orbitals in a tetrahedral arrangement around the carbon atom.

Pauling's achievements made it possible to describe polyatomic molecules by VB theory, and hybridization has provided the vocabulary and structural concepts for much of the fields of inorganic chemistry, organic chemistry, and biochemistry.

6.4.2 Orbital Hybridization for Polyatomic Molecules

Pauling developed the concept of hybrid orbitals to describe the bonding in molecules containing second-period atoms with steric numbers 2, 3, and 4. Let's discuss these hybridization schemes in sequence, starting with BeH_2. We will use the lowercase Greek letter *chi*, χ, to represent hybrid orbitals.

The BeH_2 molecule is known to linear. Let's define the z-axis of the coordinate system to lie along the H—Be—H bonds, and place the beryllium nucleus at the origin. We mix the $2s$ and $2p_z$ orbitals of beryllium to form two new orbitals on the beryllium atom:

$$\chi_1(r) = \frac{1}{\sqrt{2}}[2s + 2p_z] \qquad [6.33a]$$

$$\chi_2(r) = \frac{1}{\sqrt{2}}[2s - 2p_z] \qquad [6.33b]$$

The coefficient $1/\sqrt{2}$ is a normalization constant. We call these **sp hybrid atomic orbitals** because they are formed as the sum or difference of one s orbital and one p orbital. Like the familiar s and p orbitals, a hybrid AO is a one-electron wave function whose amplitude is defined at every point in space. Its amplitude at each point is the sum or difference of the other orbitals in the equation that defines the hybrid. Its square at each point gives the probability density for finding the electron at that point, when the electron is in the hybrid orbital.

The formation and the shapes of the sp hybrid orbitals and their participation in chemical bonds are shown in Figure 6.41. The first column shows the nonhybridized orbitals on the Be atom, and the second column shows the hybrid orbitals. The amplitude for each hybrid at any point r from the beryllium nucleus is easily visualized as the result of constructive and destructive interference of the $2s$ and $2p$ wave functions at that point. Because the sign of the $2s$ orbital is always positive, whereas that of the $2p_z$ orbital is different in the + and $-z$ directions, the amplitude of χ_1 is greatest along $+z$, and that of χ_2 is greatest along $-z$. Because the probabilities are the squares of the amplitudes, an electron in χ_2 is much more likely to be found on the left side of the nucleus than on the right; the opposite is

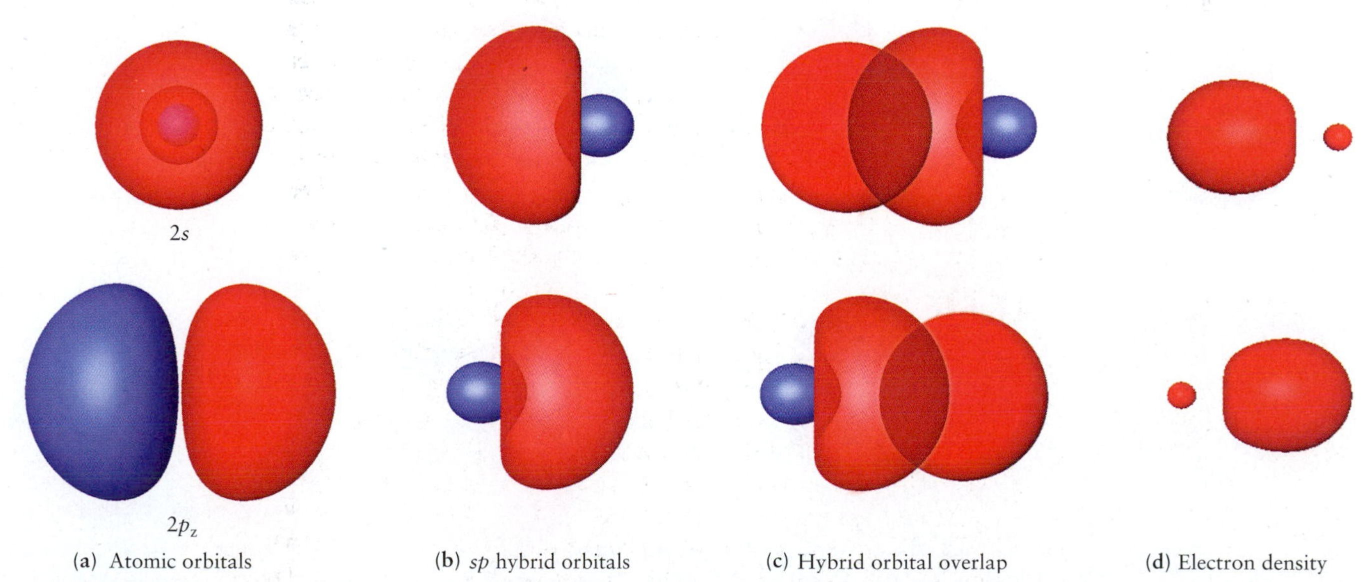

FIGURE 6.41 Formation, shapes, and bonding of the *sp* hybrid orbitals in the BeH_2 molecule. (a) The $2s$ and $2p$ orbitals of the Be atom. (b) The two *sp* hybrid orbitals formed from the $2s$ and $2p_z$ orbitals on the beryllium atom. (c) The two σ bonds that form from the overlap of the *sp* hybrid orbitals with the hydrogen $1s$ orbitals, making two single bonds in the BeH_2 molecule. (d) Electron density in the two σ bonds as calculated by Generalized Valence Bond (GVB) theory.
(Courtesy of Mr. Hatem Helal and Professor William A. Goddard III, California Institute of Technology, and Dr. Kelly P. Gaither, University of Texas at Austin.)

true for an electron in χ_1. Once the hybrid AOs form on the central atom, its electron configuration becomes Be: $(1s)^2(\chi_1)^1(\chi_2)^1$. As the two hydrogen atoms approach, each shares its electron with the corresponding hybrid orbital to form two localized single σ bonds (see Fig. 6.41). The wave functions for the two bonding pairs are

$$\psi_{\sigma 1}^{\text{bond}}(1,2) = c_+[\chi_1(1)1s^{\text{H}}(2) + \chi_1(2)1s^{\text{H}}(1)] \quad [6.34a]$$

$$\psi_{\sigma 2}^{\text{bond}}(3,4) = c_-[\chi_2(3)1s^{\text{H}}(4) + \chi_2(4)1s^{\text{H}}(3)] \quad [6.34b]$$

The third column in Figure 6.41 illustrates these σ bonds by locating Be and H atoms a distance apart equal to the experimental bond length of BeH_2, placing an *sp* hybrid on the Be atom and a 1*s* AO on the H atom and coloring the region where these orbitals overlap. Chemists have used such qualitative sketches since the advent of the VB method but have been hindered from more detailed representations by the mathematical complexity of obtaining the electron density from Equations 6.34a and 6.34b. This barrier has been overcome by GVB theory. The fourth column in Figure 6.41 shows the electron density in the σ bonds of BeH_2, as calculated by GVB.

The BH_3 molecule is known to have a trigonal planar structure with three equivalent bonds. Let's choose coordinates so that the structure lies in the

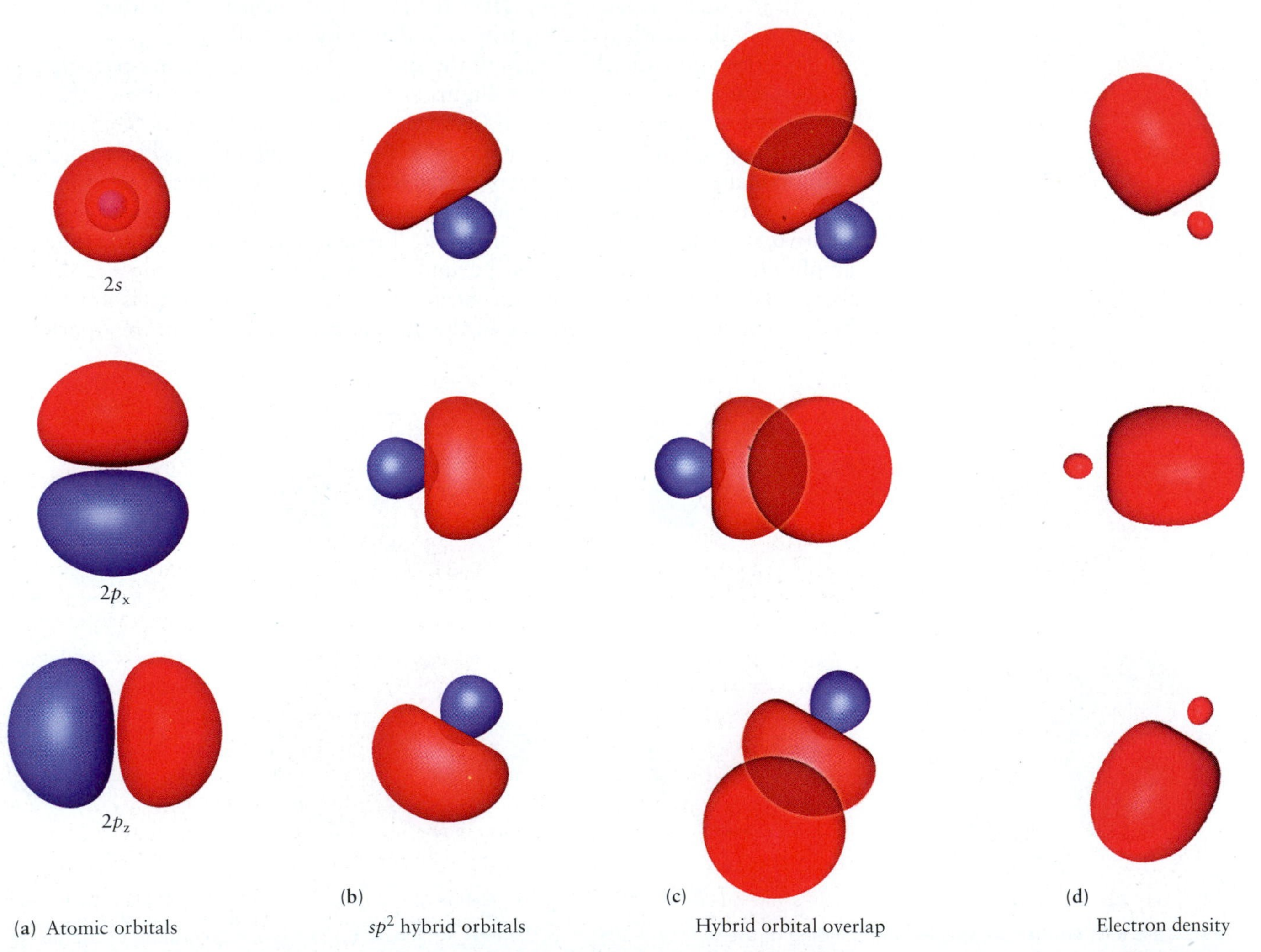

FIGURE 6.42 Formation, shapes, relative orientation, and bonding of the three sp^2 hybrid orbitals in the BH_3 molecule. (a) The 2*s*, $2p_x$, and $2p_y$ atomic orbitals on a boron atom. (b) The three sp^2 hybrid orbitals on a boron atom. (c) Overlap of the sp^2 hybrid orbitals with hydrogen 1*s* orbitals to form three σ bonds in BH_3. (d) Electron density in the three σ bonds as calculated by Generalized Valence Bond (GVB) theory.

(Courtesy of Mr. Hatem Helal and Professor William A. Goddard III, California Institute of Technology, and Dr. Kelly P. Gaither, University of Texas at Austin.)

x-y plane with the beryllium atom at the origin. Promotion of one of the $2s$ electrons creates the excited-state configuration Be: $2s^1 2p_x^1 2p_y^1$, and these three AOs can be mixed to form the three equivalent new orbitals:

$$\chi_1(r) = 2s + \left(\frac{1}{2}\right)^{1/2} 2p_y \quad [6.35a]$$

$$\chi_2(r) = 2s + \left(\frac{3}{2}\right)^{1/2} 2p_x - \left(\frac{1}{2}\right)^{1/2} 2p_y \quad [6.35b]$$

$$\chi_3(r) = 2s - \left(\frac{3}{2}\right)^{1/2} 2p_x - \left(\frac{1}{2}\right)^{1/2} 2p_y \quad [6.35c]$$

These are called **sp^2 hybrid atomic orbitals** because they are formed from one s and two p orbitals. The formation, shape, and orientation of the sp^2 hybrids are shown in Figure 6.42. They lie in the x-y plane with an angle of 120° between them. After hybridization, the electron configuration of the atom is B: $(1s)^2(\chi_1)^1(\chi_2)^1(\chi_3)^1$. Each of the sp^2 hybrids can overlap with a H($1s$) orbital to produce a σ bond. The wave functions for all bonding pairs would be the same, and they will have the same form as those in Equations 6.34a and 6.34b. The third column of Figure 6.42 shows the traditional qualitative sketches of the orbital overlap leading to σ bonds in BH_3, and the fourth column shows the electron density in these bonds as calculated by GVB. Experimentally, BH_3 molecules turn out to be unstable and react rapidly to form B_2H_6 or other higher compounds called "boranes." However, the closely related BF_3 molecule has the trigonal planar geometry characteristic of sp^2 hybridization. It forms three σ bonds by overlap of a boron sp^2 hybrid with a F($2p_z$).

To describe the known structure for CH_4, we combine the $2s$ and three $2p$ orbitals of the central carbon atom to form four equivalent **sp^3 hybrid atomic orbitals,** which point toward the vertices of a tetrahedron:

$$\chi_1(r) = \frac{1}{2}[2s + 2p_x + 2p_y + 2p_z] \quad [6.36a]$$

$$\chi_2(r) = \frac{1}{2}[2s - 2p_x - 2p_y + 2p_z] \quad [6.36b]$$

$$\chi_3(r) = \frac{1}{2}[2s + 2p_x - 2p_y - 2p_z] \quad [6.36c]$$

$$\chi_4(r) = \frac{1}{2}[2s - 2p_x + 2p_y - 2p_z] \quad [6.36d]$$

FIGURE 6.43 Shapes and relative orientations of the four sp^3 hybrid orbitals in CH_4 pointing at the corners of a tetrahedron with the carbon atom at its center. The "exploded view" at the bottom shows the tetrahedral geometry.
(Courtesy of Mr. Hatem Helal and Professor William A. Goddard III, California Institute of Technology, and Dr. Kelly P. Gaither, University of Texas at Austin.)

Figure 6.43 shows the shape and orientation of these four orbitals, pointing toward the vertices of a tetrahedron, which has the carbon atom at its center. The bottom image in Figure 6.43 shows an "exploded view" in which the orbitals have been displaced from one another to show the tetrahedral geometry. Each hybrid orbital can overlap a $1s$ orbital of one of the hydrogen atoms to give an overall tetrahedral structure for CH_4.

Because of their widespread use in chemistry, it is important to have a good sense of the sizes and shapes of the hybrid orbitals. The shapes of the sp hybrid orbitals in Figure 6.41 are quantitatively correct and properly scaled in size relative to the other orbitals shown. Chemists tend to sketch these orbitals by hand like those in Figure 6.44, which gives the misleading impression that the hybrids are thin, cigarlike shapes with highly directional electron density concentrated right along the direction of the bonds. A contour map of χ_1 (from Eq. 6.33a) shows that the orbital is rather diffuse and broadly spread out, despite its directional concentration (see Fig. 6.44). Because this plot is symmetric about the z-axis, each of these contours can be rotated about the z-axis to produce a three-dimensional isosurface at a specified fraction of the maximum amplitude; the isosurfaces in Figure 6.41 were generated in just that way. Visualizing these isosurfaces helps us see the real effect of sp hybridization: The amplitudes of the

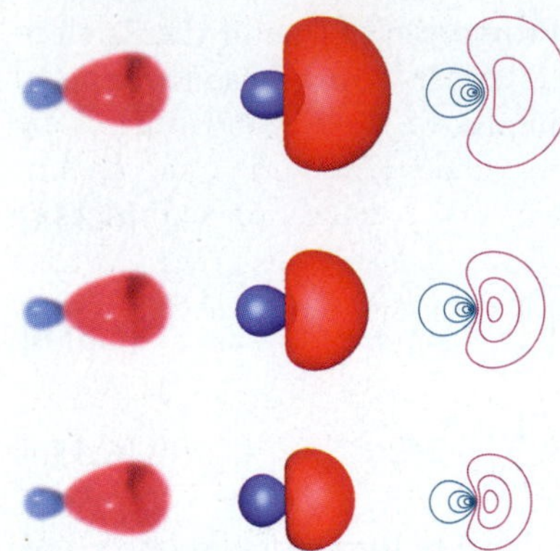

FIGURE 6.44 Exact and approximate representations of the hybrid orbital shapes. For each type of hybrid orbital shown the left column shows typical chemists' sketches, the center column shows isosurfaces, and the right column shows contour plots. The top row are the *sp* hybrid orbitals, the middle row are the sp^2 hybrid orbitals, and the bottom row are the sp^3 hybrid orbitals.
(Courtesy of Mr. Hatem Helal and Professor William A. Goddard III, California Institute of Technology, and Dr. Kelly P. Gaither, University of Texas at Austin.)

beryllium atom orbitals are now "pooched out" a bit in the $+z$ and $-z$ directions, but it has not been squeezed down into a thin tube. The $2p_x$ and $2p_y$ orbitals remain unchanged, oriented perpendicular to each other and to the *sp* hybrid orbitals

The chemist's sketches, which are typically drawn to emphasize directionality of the sp^2 hybrid orbitals, and a contour plot of the actual shape, are shown in Figure 6.44. Each of these contours can be rotated about the *x-y* plane to produce a three-dimensional isosurface whose amplitude is chosen to be a specific fraction of the maximum amplitude of the wave function. These isosurfaces demonstrate that sp^2 hybridization causes the amplitude of the boron atom to be "pooched out" at three equally spaced locations around the "equator" of the atom (see Fig. 6.42). The $2p_z$ orbital is not involved and remains perpendicular to the plane of the sp^2 hybrids. The standard chemist's sketches of the sp^3 hybrid orbitals and a contour plot that displays the exact shape and directionality of each orbital are shown in Figure 6.44. The isosurfaces shown in Figure 6.43 were generated from these contour plots.

Lone-pair electrons, as well as bonding pairs, can occupy hybrid orbitals. The nitrogen atom in NH_3 also has steric number 4, and its bonding can be described in terms of sp^3 hybridization. Of the eight valence electrons in NH_3, six are involved in σ bonds between nitrogen and hydrogen; the other two occupy the fourth sp^3 hybrid orbital as a lone pair (Fig. 6.45a). Oxygen in H_2O likewise has steric number 4 and can be described with sp^3 hybridization, with two lone pairs in sp^3 orbitals (see Fig. 6.45b). Placing the unshared pairs in sp^3 hybrid orbitals predicts bond angles of 109.5°, which are reasonably close to the measured values of 107° for NH_3 and 104° for H_2O.

There is a close relationship between the VSEPR theory discussed in Section 3.9 and the hybrid orbital approach, with steric numbers of 2, 3, and 4 corresponding to sp, sp^2, and sp^3 hybridization, respectively. The method can be extended to more complex structures; d^2sp^3 hybridization (see Sec. 8.7), which gives six equivalent hybrid orbitals pointing toward the vertices of a regular octahedron, is applicable to molecules with steric number 6. Both theories are based on minimizing the energy by reducing electron–electron repulsion.

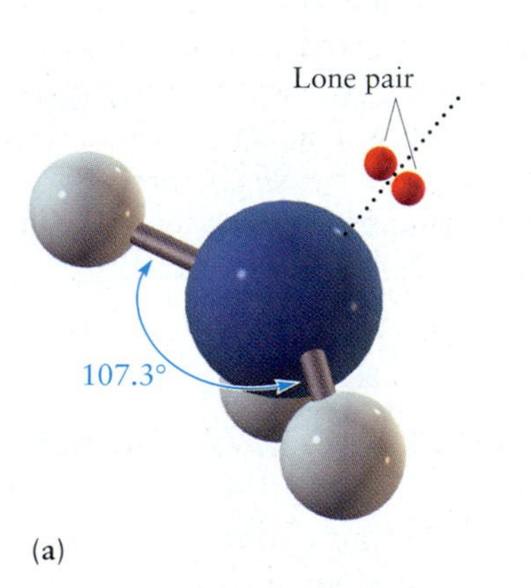

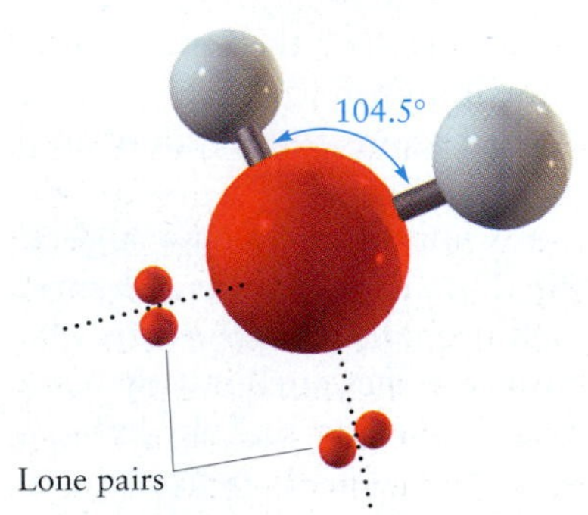

FIGURE 6.45 (a) Ammonia (NH_3) has a pyramidal structure in which the bond angles are less than 109.5°. (b) Water (H_2O) has a bent structure, with a bond angle less than 109.5° and smaller than that of NH_3.

EXAMPLE 6.3

Predict the structure of hydrazine (H_2NNH_2) by writing down its Lewis diagram and using the VSEPR theory. What is the hybridization of the two nitrogen atoms?

SOLUTION

The Lewis diagram is

```
       H
   ..  ..
 H:N:N:H
   ..  ..
   H
```

Lewis diagram for hydrazine, N_2H_4.

Both nitrogen atoms have steric number 4 and are sp^3 hybridized, with H—N—H and H—N—N angles of approximately 109.5°. The extent of rotation about the N—N bond cannot be predicted from the VSEPR theory or the hybrid orbital model. The full three-dimensional structure of hydrazine is shown in Figure 6.46.

Related Problems: 49, 50, 51, 52

The concept of orbital hybridization deserves a few summary comments. The method is used throughout basic and applied chemistry to give quick and convenient representations of molecular structure. The method provides a sound quantum mechanical basis for *organizing and correlating* vast amounts of experimental data for molecular structure. The simple examples discussed earlier all involved

FIGURE 6.46 The structure of hydrazine, N_2H_4.

symmetric molecules with identical ligands, and we simply "placed" lone pairs in hybrid orbitals when necessary. Constructing the hybrids for non-symmetrical molecules, bonding with different ligands, and giving explicit attention to lone pairs all involve considerable difficulty. The resulting models provide concrete images for visualizing and testing chemical reaction pathways by comparing the electron density at different possible "reactive sites" on a molecule. You will use hybridization extensively for these purposes in your subsequent chemistry classes, especially organic chemistry.

Hybridization is less successful as a tool for *predicting* molecular structure. The bond angle is usually known or assumed at the beginning. If the bond angle is not known in advance, various semiempirical schemes must be used to estimate the *s* and *p* character and search for the optimum value of bond angle. The calculations involved are less well suited to computer analysis than those done using the LCAO method; thus, extensive predictions of molecular geometry are quite difficult. In recent years, a newer version of the VB method called **Generalized Valence Bond Theory** has been developed by the American chemist William A. Goddard III as a powerful tool for large-scale practical calculations for localized orbitals.

Finally, orbital hybridization has inspired a great deal of discussion, some of it impassioned, on the meaning and significance of "promotion" and "return of the energy investment." Where does the energy input for promotion come from? How is the bond formed? How does the energy released on bond formation compensate for promotion? Do these concerns cast doubt on the validity and usefulness of the hybrid orbital representations of the chemical bond? These may be legitimate concerns if one is trying to describe the dynamical events by which the bond is actually formed. However, these concerns are largely side issues for our main question: Is hybridization a useful way to describe the *structure* of a chemical bond after it has been formed? Quantum mechanics provides a fundamental explanation of atomic structure for the allowed values of energy and angular momentum. One set of values is appropriate for describing free carbon atoms in the gas phase, and another set is appropriate for describing carbon atoms involved in tetravalent chemical bonds. Equations 6.36a–d provide the connections between these two sets. Pauling provided the following description: "If quantum theory had been developed by the chemist rather than the spectroscopist it is probable that the tetrahedral orbitals described above would play the fundamental role in the theory, in place of the *s* and *p* orbitals."[1]

6.5 Comparison of Linear Combination of Atomic Orbitals and Valence Bond Methods

The LCAO and VB methods start with different quantum mechanical approaches to the description of chemical bonding. The former constructs MOs that are delocalized over the entire molecule by taking linear combinations of the AOs. In contrast, the VB model provides a quantum mechanical description of the localized chemical bond between two atoms in the spirit of the Lewis dot model. The two methods look different at the beginning, and their results look quite different. Are the two methods equally valid? Are they equally applicable to a broad range of molecules? The accuracy of any particular quantum chemical method is ultimately judged by the degree to which its predictions agree with experimental results. The electronic wave function for the molecule is the key construct used for calculating the values of molecular properties that we can measure. Therefore, the best way to compare the LCAO and VB methods is to compare the electronic wave functions for the molecule generated by each.

[1]L. Pauling, *The Nature of the Chemical Bond,* 3rd ed. Ithaca, NY: Cornell University Press, 1960, p. 113.

Comparison for H_2

This comparison of LCAO with VB methods is most easily seen by explicitly writing out the electronic wave functions for the specific case of H_2 constructed using both methods.

In the LCAO method for H_2, a σ bonding orbital is constructed as a linear combination of H 1s orbitals centered on the two hydrogen atoms. This MO is de-localized over the entire molecule. The bonding orbital is given by the following equation:

$$\sigma_{g1s} = C_g(R_{AB})[\varphi_{1s}^{A} + \varphi_{1s}^{B}] \quad [6.37]$$

Neglecting the normalization constant and using the simplified notation from Section 6.2 give the following form:

$$\sigma_{g1s} = [1s^{A} + 1s^{B}] \quad [6.38]$$

Both electrons occupy this bonding orbital, satisfying the condition of indistinguishability and the Pauli principle. Recall from Section 6.2 that the electronic wave function for the entire molecule in the LCAO approximation is the *product* of all of the occupied MOs, just as an atomic wave function is the product of all occupied Hartree orbitals of an atom. Thus, we get

$$\psi_{MO}^{el} = \sigma_{g1s}(1)\sigma_{g1s}(2) = [1s^{A}(1) + 1s^{B}(1)][1s^{A}(2) + 1s^{B}(2)] \quad [6.39]$$

The VB model, in contrast, starts with the notion that a good approximation to the molecular electronic wave function for H_2 is the product of an H(1s) orbital centered on atom A, occupied by electron 1, and another H(1s) orbital centered on atom B, occupied by electron 2. As discussed in Section 6.4, the equation for this molecular electronic wave function is

$$\psi_{VB}^{el}(r_{1A}, r_{2B}; R_{AB}) = C_1(R_{AB})\varphi_A(r_{1A})\varphi_B(r_{2B}) + C_2(R_{AB})\varphi_A(r_{2A})\varphi_B(r_{1B}) \quad [6.40]$$

which on dropping the normalization factors and using the simplified notation introduced in Section 6.4 becomes

$$\psi_{VB}^{el} = 1s^{A}(1)1s^{B}(2) + 1s^{A}(2)1s^{B}(1) \quad [6.41]$$

Now we can compare the LCAO and VB versions of the electronic wave functions for the molecule directly by multiplying out ψ_{MO}^{el} and rearranging terms to obtain

$$\psi_{MO}^{el} = [1s^{A}(1)1s^{B}(2) + 1s^{A}(2)1s^{B}(1)] + [1s^{A}(1)1s^{A}(2) + 1s^{B}(1)1s^{B}(2)] \quad [6.42]$$

The first term in ψ_{MO}^{el} is identical to ψ_{VB}^{el}. The second term may be labeled ψ_{ionic} because it is a mixture of the ionic states $H_A^-H_B^+$ and $H_A^+H_B^-$, respectively. This can be seen by looking at the two terms in the second set of brackets; the first term puts both electrons on nucleus A (making it H^-), and the second term puts both electrons on nucleus B (making it H^-).

Our comparison shows that the LCAO method includes an ionic contribution to the bond, but the VB method does not. In fact, the simple MO approach suggests that the bond in H_2 is 50% covalent and 50% ionic, which is contrary both to experience and intuition. Because the electronegativities of the two atoms in a homonuclear diatomic molecule are the same, there is no reason to expect any ionic contribution to the bond, much less such a large one. The complete absence of ionic contributions in the VB wave function suggests this method is not well suited for polar molecules such as HF. Thus, the truth in describing the chemical bond and molecular structure appears to lie somewhere between the LCAO and

VB methods. It is also informative to compare these methods for describing chemical reactivity, which requires bonds to be broken. We already know that the VB wave function for H_2 correctly describes the long-distance limit as two separate hydrogen atoms. But, the LCAO wave function predicts that, in the long-distance limit H_2, dissociates into ionic species, as well as hydrogen atoms. Ionic products are not usually produced by dissociation under ordinary thermal conditions. Again, the best description must lie between the extremes provided by the simple LCAO and VB methods.

Improving the Linear Combination of Atomic Orbitals and Valence Bond Methods

The simple form of LCAO and VB methods, as presented in this chapter, must be refined to provide more accurate wave functions for molecules and solids from which measurable properties can be calculated. Both methods have been improved on significantly in a number of ways. We illustrate one approach, starting with ψ_{VB}, not only because it is easier to understand than methods for refining ψ_{MO}, but also because the method is generally applicable in many areas of quantum chemistry. The accuracy of the simple VB wave function can be improved by adding in (mixing) some ionic character. We write

$$\psi_{\text{improved}} = \psi_{VB} + \lambda\psi_{\text{ionic}} \qquad [6.43]$$

and then choose λ on the basis of some criterion. The most common way to do this is to adjust λ to minimize the energy of the orbital. One simply calculates the energy, using the methods developed in Chapter 5, with λ as a parameter, and then differentiates the result with respect to λ to find that value of λ that minimizes the energy, as is done in ordinary calculus. Using that special value of λ in Equation 6.43 gives a wave function that is a better approximation to the true wave function than the simple VB wave function. Moreover, the energy calculated for the ground state of the system using the "improved" wave function is guaranteed never to be lower than the true ground-state energy. These results are consequences of the **variational principle** in quantum mechanics, which gives an exceptionally powerful criterion for improving the accuracy of various approximations; lower energy is always better. Refinements of both approaches have led to highly accurate methods of modern computational quantum chemistry in which the distinction between the two starting points has completely disappeared.

Using the Linear Combination of Atomic Orbitals and Valence Bond Methods

The LCAO and VB approaches are both good starting points for describing bonding and reactivity. You can apply either one to set up a purely qualitative description of the problem of interest, confident that you can move on to a high-level quantitative calculation as your needs demand. Which method you choose at the beginning depends primarily on the area of chemistry in which you are working and the broad class of problems you are investigating. LCAO theory is most often used to describe the electronic states of molecules in contexts that require knowledge of energy levels. Examples include molecular spectroscopy, photochemistry, and phenomena that involve ionization (such as electron-induced reactions and PES). VB theory is more widely used to describe molecular structure, especially in pictorial ways.

Many chemists use a mixture of the two, where localized VB σ bonds describe the network holding the molecule together and de-localized LCAO π bonds describe the spread of electron density over the molecule. We illustrate this combination here for the case of bent triatomic molecules, and also much more extensively in Chapter 7 for organic molecules.

Nonlinear triatomic molecules can be described through sp^2 hybridization of the central atom. If the molecule lies in the x-y plane, then the s, p_x, and p_y orbitals of the central atom can be combined to form three sp^2 hybrid orbitals with an angle close to 120° between each pair. One of these orbitals holds a lone pair of electrons, and the other two take part in σ bonds with the outer atoms. The fourth orbital, a p_z orbital, takes part in de-localized π bonding (Fig. 6.47a). On the outer atoms, the p orbital pointing toward the central atom takes part in a localized σ bond, and the p_z orbital takes part in π bonding; the third p orbital and the s orbital are AOs that do not participate in bonding. The three p_z AOs can be combined into bonding, nonbonding, and antibonding π orbitals much as in the linear molecule case (see Fig. 6.47b). Here, there is only one of each type of orbital (π, π^{nb}, π^*) rather than two as for linear molecules.

Consider a specific example, NO_2^-, with 18 electrons. Two electrons are placed in each oxygen atom $2s$ orbital and 2 more in each nonbonding oxygen atom $2p$ orbital, so that a total of 8 electrons are localized on oxygen atoms. Two electrons also are placed as a lone pair in the third sp^2 orbital of the nitrogen atom. Of the remaining 8 electrons, 4 are involved in the 2 σ bonds between nitrogen and the oxygen atoms. The last 4 are placed into the π electron system: 2 into the bonding π orbital and 2 into the nonbonding π^{nb} orbital. Because a total of 6 electrons are in bonding orbitals and none are in antibonding orbitals, the net bond order for the molecule is 3, or $1\frac{1}{2}$ per bond. In the Lewis model, 2 resonance forms are needed to represent NO_2^-. The awkwardness of the resonance model is avoided by treating the electrons in the bonding π MO as de-localized over the 3 atoms in the molecule.

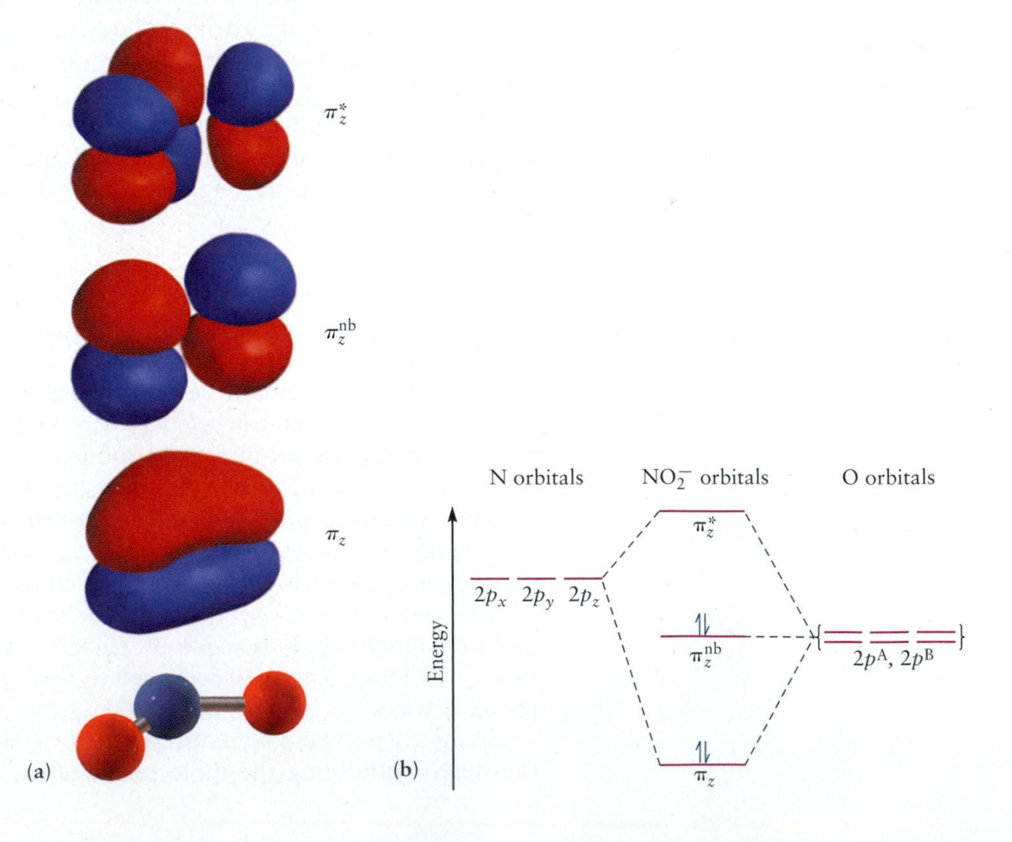

FIGURE 6.47 (a) Ball-and-stick model (bottom) and molecular orbitals for bent triatomic molecules. The central atom has three sp^2 hybrid orbitals (not shown) that would lie in the plane of the molecule. From the three p_z orbitals perpendicular to this plane, three π orbitals can be constructed. (b) Correlation diagram for the π orbitals.

Epilogue

At this point you might wonder why we presented such an extensive discussion of various approximate quantum mechanical approaches to describe chemical bonding when accurate, high-level calculational methods are widely available and relatively easy to use. The answer is that most of our understanding of chemical bonding, structure, and reactivity is based on the simple approaches to MO and VB theory presented in this chapter. They provide the foundation on which our chemical intuition has been built. The concepts we have introduced are central to all areas of modern chemistry. You will likely begin to use modern computational chemistry methods in your advanced chemistry courses. You will certainly use them in your career, if you pursue an advanced degree in the chemical sciences, molecular biology, or materials science and engineering. We believe it is essential for you to understand these fundamentals so you can maximize the benefits of the powerful tools of modern computational quantum chemistry.

CHAPTER SUMMARY

Modern methods of computational quantum chemistry are now sufficiently accurate and easy to use that they are indispensable tools in chemical research and education. Our goal has been to give you a comprehensive introduction to molecular quantum mechanics so you can easily read more advanced treatments and begin to use commercially available software with intelligence and confidence.

The Born–Oppenheimer approximation is the starting point for all of molecular quantum mechanics. The fact that electrons move so much faster than nuclei allows us to treat electronic motion independently of nuclear motion by solving an electronic Schrödinger equation for each value of the internuclear separation. The resulting MOs provide all of the information of interest, the probability density distributions and the electronic energies being the most important. The electronic bonding energy and the nuclear repulsion together define the effective potential energy function for the motion of the nuclei in the molecule.

The MOs for the simplest molecule, H_2^+, can be calculated exactly within the Born–Oppenheimer approximation. The results illustrate the general features of molecular quantum mechanics that form the basis for our understanding of structure and bonding in more complicated molecules. The orbitals are characterized by their symmetry, the number and nature of their nodes, and their energies. Orbitals in which the electron density increases between the nuclei lead to energies lower than that of separated H and H^+; these are called bonding MOs. Orbitals in which there is a node in the wave function and the electron density goes to zero midway between the nuclei are called antibonding orbitals. States of H_2^+ in which the electron resides in an antibonding orbital are unstable with respect to dissociation to H and H^+.

Additional approximations must be made to calculate the MOs for many-electron molecules. The most important approximation procedures are the LCAO method and the VB model. The LCAO method constructs de-localized one-electron MOs by taking linear combinations of AOs centered on different atoms and generates electron configurations by placing electrons in these MOs using an aufbau principle and invoking Hund's rules for the ground-state configuration. The VB model constructs a wave function for a localized pair bond starting with an "occupied" AO on each of the two atoms that form the bond. These two procedures provide the conceptual foundation and vocabulary for qualitative and even semiquantitative understanding of chemical bonding and molecular structure in contemporary chemistry. A variety of sophisticated computational methods have been developed using these procedures as starting points, and the results of these calculations are now sufficiently accurate to have both analytical and predictive value.

Iodine sublimes from the bottom of the beaker and condenses on the bottom of the chilled round-bottom flask.

Charles D. Winters

CUMULATIVE EXERCISE

Iodine

The shiny purple-black crystals of elemental iodine were first prepared in 1811 from the ashes of seaweed. Several species of seaweed concentrate the iodine that is present in low proportions in seawater, and for many years, seaweed was the major practical source of this element. Most iodine is now produced from natural brines via oxidation of iodide ion with chlorine.

(a) Iodine is an essential trace element in the human diet, and iodine deficiency causes goiter, the enlargement of the thyroid gland. Much of the salt intended for human consumption is "iodized" by the addition of small quantities of sodium iodide to prevent goiter. Calculate the electronegativity difference between sodium and iodine. Is sodium iodide an ionic or a covalent compound? What is its chemical formula?

(b) Iodine is an important reagent in synthetic organic chemistry because bonds between carbon and iodine form readily. Use electronegativities to determine whether the C—I bond is ionic, purely covalent, or polar covalent in character.

(c) Give the steric numbers for the iodine atom and identify the geometries of the following ions containing iodine and oxygen: IO_3^-, IO_6^{5-}, and IO_4^-.

(d) What is the ground-state electron configuration of the valence electrons of iodine molecules (I_2)? Is iodine paramagnetic or diamagnetic?

(e) What is the electron configuration of the I_2^+ molecular ion? Is its bond stronger or weaker than that in I_2? What is its bond order?

Answers

(a) The electronegativity difference is 1.73; thus, the compound is largely ionic, with formula NaI.

(b) The electronegativity difference is 0.11; thus, the C—I bond is largely covalent, with nearly equal sharing of electrons between the atoms.

(c) IO_3^-: $SN = 4$, structure is pyramidal; IO_6^{5-}: $SN = 6$, structure is octahedral; IO_4^-: $SN = 4$, structure is tetrahedral.

(d) $(\sigma_{g5s})^2(\sigma^*_{u5s})^2(\sigma_{g5p_z})^2(\pi_{u5p_x}\ \pi_{u5p_y})^4(\pi^*_{g5p_x}\ \pi^*_{g5p_y})^4$; iodine is diamagnetic.

(e) $(\sigma_{g5s})^2(\sigma^*_{u5s})^2(\sigma_{g5p_z})^2(\pi_{u5p_x}\ \pi_{u5p_y})^4(\pi^*_{g5p_x}\ \pi^*_{g5p_y})^3$; the bond is stronger; bond order is 3/2 versus 1.

CHAPTER REVIEW

The Born–Oppenheimer Approximation

- Nuclei are so much heavier than electrons that they may be considered fixed in space while the electrons move rapidly around them.
- The Born–Oppenheimer approximation allows us to solve the electronic Schrödinger equation for H_2^+ for a fixed internuclear separation R_{AB}. The result is a one-electron MO, which is analogous to the one-electron hydrogen AO.
- We calculate the electronic bonding energies for every value of R_{AB} and add to that the nuclear–nuclear repulsion energy to generate the effective potential energy function $V_{eff}(R_{AB})$ that governs the nuclear motion.
- We find the kinetic energy of the nuclei and add it to the potential energy described above to find the quantized ground-state energy of the molecule. The nuclei vibrate about the equilibrium bond length—this is the *zero point motion* required by the uncertainty principle. The energy required to dissociate the molecular ion from the ground state into a separated proton and hydrogen atom is the bond dissociation energy.

Exact Solutions for Molecular Orbitals: H_2^+

- We illustrate graphically the first eight MOs for H_2^+ to show their shapes and to characterize them by their energies and symmetry, just as we characterized the atomic orbitals for the hydrogen atom. The MOs are characterized by the component of the angular momentum along the internuclear axis: by analogy to the hydrogen atom, these are called σ for $L_Z = 0$, π for $L_Z = 1$, δ for $L_Z = 2$, and ϕ for $L_Z = 3$.

Linear Combination of Atomic Orbitals Approximation

- A good approximation to the one-electron MOs for a diatomic molecule is the sum or difference of AOs of the atoms of the molecule. The *sum* linear combination leads to increased electron density between the nuclei and bonding; the *difference* linear combination leads to a node between the nuclei and decreased electron density, and it is antibonding.
- Correlation diagrams show how pairs of AOs lead to bonding and antibonding pairs of MOs. An aufbau principle is used to build up electron configurations, just like for atoms. Hund's rules predict the lowest energy electron configurations and either paramagnetic or diamagnetic behavior.
- The bond order is found by counting the number of electrons in bonding orbitals, subtracting the number in antibonding orbitals, and dividing the result by two. Electrons in antibonding orbitals effectively cancel the bonding capacity of those in bonding orbitals. This scheme explains the trends in bond length, bond stiffness, and bond energy of the first- and second-row diatomic molecules.
- The energy sequence of the MOs is slightly more complicated in second-row homonuclear diatomic molecules, because the p orbitals can overlap in two different ways. Moving from left to right across the row, the energy-level ordering changes at N_2 because the energy of the π orbital remains nearly constant, whereas that of the σ orbital drops rapidly (see Fig. 6.15). Therefore two energy-level diagrams are required to explain the bonding in the second-period diatomic molecules.
- The MOs for heteronuclear diatomic molecules are obtained by the same approach, with the AO energies of the more electronegative element placed lower (more stable) than those of the other element. If the difference in AO energies is small, the MO energy sequence is given in Figure 6.19; for larger energy AO energy differences the MOs are described by Figure 6.21.
- The LCAO approximation can also be used to generate MOs for small polyatomic molecules, which are often also treated using the VB model. For small polyatomics the proper combinations of AOs for each MO can be identified by symmetry arguments, but iterative computer calculations are necessary to find the optimum value of the coefficients for the AOs.

Photoelectron Spectroscopy for Molecules

- Photoelectron spectroscopy confirms the validity of the orbital approximation by measuring the ionization energies of the MOs directly. The ionization energy of the orbital is obtained as the difference in the energy of the photon used to ionize the molecule and the measured kinetic energy of the emitted electrons. Koopmans's approximation states that the orbital energy ε in the LCAO method is the negative of the ionization energy.
- In addition to the orbital energies, PES provides a great deal of information about the nature of the orbital (bonding, nonbonding, or antibonding) from the vibrational fine structure observed in the spectra.

Valence Bond Model

- The VB model constructs wave functions to describe localized electron-pair bonds. The model describes bonding in diatomic molecules, including the formation of multiple (σ and π) bonds. It is most frequently applied to organize and correlate data on molecular structures, especially for molecules of the type AB_x, the geometries of which are described by VSEPR theory.

- The simple VB model is augmented with the concept of orbital hybridization to account for the valence of second-row atoms and the structures of their compounds. Hybrid orbitals are constructed by adding *s* and *p* orbitals with different coefficients (weights or percentage contributions) and phases. The number of hybrid orbitals produced equals the number of starting AOs; there are two *sp* hybrid orbitals, three sp^2 hybrid orbitals, and four sp^3 hybrid orbitals.

Comparison of Linear Combination of Atomic Orbitals and Valence Bond Methods

- Comparing the LCAO and VB treatments for the hydrogen molecule at the level of the *electronic wave function for the molecule* gives considerable insight into the differences between the methods and also suggests ways to improve each. The VB wave function predicts a purely covalent bond, whereas the LCAO wave function predicts a bond with an equal mixture of covalent and ionic character. Neither of these is the best representation of bonding in H_2, so refinements of both approaches are necessary to produce results that are in better agreement with experiment.
- Many methods have been developed to improve both the simple LCAO and VB models; it is easiest to illustrate one approach for improving the VB model. Let $\psi_{\text{improved}} = \psi_{\text{VB}} + \lambda\psi_{\text{MO}}$, where λ is chosen so that the energy of the orbital is in better agreement with experiment. The variational principle ensures that the true energy is always lower than the energy calculated using an approximate wave function. This provides a well-defined criterion to judge improvement—lower energy is always better.
- Many chemists combine the LCAO and VB methods to describe bonding in polyatomic molecules. They use the VB model to describe the localized σ bonds that provide "connectivity" for the molecule structure and use the LCAO method to describe the de-localized π bonds that distribute electrons over the entire structure.

CONCEPTS & SKILLS

After studying this chapter and working the problems that follow, you should be able to:

1. Give a general description of the quantum picture of the chemical bond and how it differs from the classical picture. Describe the key features of the quantum picture, including the nature of bonding and anti-bonding MOs, symmetry of MOs, and the energy sequence of MOs (Section 6.1, Problems 1–8).
2. Show how MOs can be constructed from the AOs of two atoms that form a chemical bond, and explain how the electron density between the atoms is related to the MO (Section 6.2, Problems 9–12).
3. Construct correlation diagrams for diatomic molecules formed from second- and third-period main-group elements. From these diagrams, give the electron configurations, work out the bond orders, and comment on their magnetic properties (Section 6.2, Problems 17–30).
4. Construct qualitative potential energy curves for diatomic molecules and relate trends in well depth (bond dissociation energies) and location of the

potential minimum (equilibrium bond length) with trends in bond order predicted by the LCAO method (Section 6.2, Problems 13–16).

5. Use symmetry arguments to find the proper combinations of AOs to construct MOs for small polyatomic molecules (Section 6.2, Problems 31–34).
6. Relate photoelectron spectra to correlation diagrams for MOs (Section 6.3, Problems 35–40).
7. Use the VB method to construct wave functions for localized electron pair bonds, including multiple bonds, and predict the molecular geometry from these bonds (Section 6.4, Problems 41–48).
8. Describe the hybrid/AO basis for representing the structures of molecules and the geometry predicted by each hybridization class (Section 6.4, Problems 49–56).
9. Compare the LCAO and VB approaches, and combine them to describe the molecular network and delocalized bonds in certain classes of molecules (Section 6.5, Problems 57–62).

KEY EQUATIONS

$1\sigma_g \approx \sigma_{g1s} = C_g(R_{AB})[\varphi_{1s}^{A} + \varphi_{1s}^{B}]$ (Section 6.2.1)

$1\sigma_u^* \approx \sigma_{u1s}^* = C_u(R_{AB})[\varphi_{1s}^{A} - \varphi_{1s}^{B}]$ (Section 6.2.1)

$[\sigma_{g1s}]^2 = C_g^2[(\varphi_{1s}^{A})^2 + (\varphi_{1s}^{B})^2 + 2\varphi_{1s}^{A}\varphi_{1s}^{B}]$ (Section 6.2.1)

$[\sigma_{u1s}^*]^2 = C_u^2[(\varphi_{1s}^{A})^2 + (\varphi_{1s}^{B})^2 - 2\varphi_{1s}^{A}\varphi_{1s}^{B}]$ (Section 6.2.1)

$\psi_{\text{n.i.}}^2 = C_3^2[(\varphi_{1s}^{A})^2 + (\varphi_{1s}^{B})^2]$ (Section 6.2.1)

$\sigma_{g1s} = C_g[1s^A + 1s^B]$ (Section 6.2.2)

$\sigma_{g1s}^* = C_g[1s^A - 1s^B]$ (Section 6.2.2)

$\sigma_{g2s} = C_g[2s^A + 2s^B]$ (Section 6.2.4)

$\sigma_{u2s}^* = C_u[2s^A - 2s^B]$ (Section 6.2.4)

$\sigma_{g2p_z} = C_g[2p_z^A - 2p_z^B]$ (Section 6.2.4)

$\sigma_{u2p_z}^* = C_u[2p_z^A + 2p_z^B]$ (Section 6.2.4)

$\pi_{u2p_x} = C_u[2p_x^A + 2p_x^B]$ (Section 6.2.4)

$\pi_{g2p_x}^* = C_g[2p_x^A - 2p_x^B]$ (Section 6.2.4)

$\pi_{u2p_y} = C_u[2p_y^A + 2p_y^B]$ (Section 6.2.4)

$\pi_{g2p_y}^* = C_g[2p_y^A - 2p_y^B]$ (Section 6.2.4)

$\sigma_{2s} = C_A 2s^A + C_B 2s^B$ (Section 6.2.4)

$\sigma_{2s}^* = C_A' 2s^A - C_B' 2s^B$ (Section 6.2.4)

$V_{1g}^{(\text{eff})}(R_{AB}) = E_{1g}^{(\text{el})}(R_{AB}) + V_{nn}(R_{AB})$ (Section 6.2.5)

$$h\nu_{\text{photon}} - \frac{1}{2}m_e v^2 = -\varepsilon_i + E_i^{(\text{vib})} = IE_i$$ (Section 6.3)

$$h\nu_{\text{photon}} - \frac{1}{2}m_e v^2 = -\varepsilon_i + nh\nu_{\text{vib}} \qquad n = 0, 1, 2, 3, \ldots$$ (Section 6.3)

$\psi^{\text{el}}(r_{1A}, r_{2B}; R_{AB}) = C_1(R_{AB})\varphi^A(r_{1A})\varphi^B(r_{2B}) + C_2(R_{AB})\varphi^A(r_{2A})\varphi^B(r_{1B})$ (Section 6.4.1)

$\psi_g^{\text{el}} = C_1[1s^A(1)1s^B(2) + 1s^A(2)1s^B(1)]$ (Section 6.4.1)

$\psi_g^{\text{bond}} = C_1[2p_z^A(1)2p_z^B(2) + 2p_z^A(2)2p_z^B(1)]$ (Section 6.4.1)

$$\psi_{\pi x}^{\text{bond}}(1, 2; R_{AB}) = C_1(R_{AB})[2p_x^A(1)2p_x^B(2)] + C_1(R_{AB})[2p_x^A(2)2p_x^B(1)] \quad \text{(Section 6.4.1)}$$

$$\psi_{MO}^{el} = \sigma_{g1s}(1)\sigma_{g1s}(2) = [1s^A(1) + 1s^B(1)][1s^A(2) + 1s^B(2)] \quad \text{(Section 6.5)}$$

$$\psi_{MO}^{el} = [1s^A(1)1s^B(2) + 1s^A(2)1s^B(1)] + [1s^A(1)1s^A(2) + 1s^B(1)1s^B(2)] \quad \text{(Section 6.5)}$$

$$\psi_{\text{improved}} = \psi_{VB} + \lambda\psi_{\text{ionic}} \quad \text{(Section 6.5)}$$

PROBLEMS

Answers to problems whose numbers are boldface appear in Appendix G. Problems that are more challenging are indicated with asterisks.

Quantum Picture of the Chemical Bond

1. Determine the number of nodes along the internuclear axis for each of the σ molecular orbitals for H_2^+ shown in Figure 6.2.

2. Determine the number of nodes along the internuclear axis and the number of nodal planes for each of the π molecular orbitals for H_2^+ shown in Figure 6.2

3. Sketch the shape of each of the σ molecular orbitals for H_2^+ shown in Figure 6.2 in a plane perpendicular to the internuclear axis located at the midpoint between the two nuclei. Repeat the sketches for a plane perpendicular to the internuclear axis located at a point one quarter of the distance between the two nuclei.

4. Sketch the shape of each of the π molecular orbitals for H_2^+ shown in Figure 6.2 in a plane perpendicular to the internuclear axis located at the midpoint between the two nuclei. Repeat the sketches for a plane perpendicular to the internuclear axis located at a point one quarter of the distance between the two nuclei.

5. Compare the electron density in the $1\sigma_g$ and $1\sigma_u^*$ molecular orbitals for H_2^+ shown in Figure 6.3 with the classical model for bonding for H_2^+ summarized in Figures 3.11 and 3.12. Which of these molecular orbitals describes the bond in H_2^+?

6. Explain why $1\sigma_g$ is the ground state for H_2^+. By combining your answer with the answer to Problem 5, what conclusions can you draw about the molecular orbital description of the bond in H_2^+?

7. The discussion summarized in Figure 6.4 explained how, at large internuclear separations, the $1\sigma_g$ molecular orbital for H_2^+ approaches the sum of two hydrogen atomic orbitals, and the $1\sigma_u^*$ molecular orbital approaches the difference of two hydrogen atomic orbitals. As illustrated in Figure 6.5, the 3σ molecular orbitals behave differently: $3\sigma_g$ approaches the *difference* of two hydrogen atomic orbitals, and $3\sigma_u^*$ approaches the *sum* of two hydrogen atomic orbitals. Explain this different behavior in the two sets of σ molecular orbitals.

8. The discussion summarized in Figure 6.4 explained how, at large internuclear separations, the $1\sigma_g$ molecular orbital for H_2^+ approaches the sum of two hydrogen atomic orbitals, and the $1\sigma_u^*$ molecular orbital approaches the difference of two hydrogen atomic orbitals. As illustrated in Figure 6.5, the 1π molecular orbitals behave differently: $1\pi_g^*$ approaches the *difference* of two hydrogen atomic orbitals, and $1\pi_u$ approaches the *sum* of two hydrogen atomic orbitals. Explain this different behavior in the two sets of molecular orbitals.

De-localized Bonds: Molecular Orbital Theory and the Linear Combination of Atomic Orbitals Approximation

9. Without consulting tables of data, predict which species has the higher bond energy, H_2 or He_2^+.

10. Without consulting tables of data, predict which species has the higher bond energy, H_2^+ or H_2.

11. Without consulting tables of data, predict which species has the greater bond length, H_2 or He_2^+.

12. Without consulting tables of data, predict which species has the greater bond length, H_2^+ or H_2.

13. Without consulting tables of data, on the same graph sketch the effective potential energy curves for H_2 and He_2^+.

14. Without consulting tables of data, on the same graph sketch the effective potential energy curves for H_2^+ and H_2.

15. Suppose we supply enough energy to H_2 to remove one of its electrons. Is the bond energy of the resulting ion larger or smaller than that of H_2? Is the bond length of the resulting ion larger or smaller than that of H_2?

16. Suppose we supply enough energy to He_2^+ to remove its most weakly bound electron. Is the bond energy of the resulting ion larger or smaller than that of He_2^+? Is the bond length of the resulting ion larger or smaller than that of He_2?

17. If an electron is removed from a fluorine molecule, an F_2^+ molecular ion forms.
(a) Give the molecular electron configurations for F_2 and F_2^+.
(b) Give the bond order of each species.
(c) Predict which species should be paramagnetic.
(d) Predict which species has the greater bond dissociation energy.

18. When one electron is added to an oxygen molecule, a superoxide ion (O_2^-) is formed. The addition of two electrons gives a peroxide ion (O_2^{2-}). Removal of an electron from O_2 leads to O_2^+.
(a) Construct the correlation diagram for O_2^-.
(b) Give the molecular electron configuration for each of the following species: O_2^+, O_2, O_2^-, O_2^{2-}.
(c) Give the bond order of each species.
(d) Predict which species are paramagnetic.
(e) Predict the order of increasing bond dissociation energy among the species.

19. Predict the valence electron configuration and the total bond order for the molecule S_2, which forms in the gas phase when sulfur is heated to a high temperature. Will S_2 be paramagnetic or diamagnetic?

20. Predict the valence electron configuration and the total bond order for the molecule I_2. Will I_2 be paramagnetic or diamagnetic?

21. For each of the following valence electron configurations of a homonuclear diatomic molecule or molecular ion, identify the element X, Q, or Z and determine the total bond order.
(a) $X_2: (\sigma_{g2s})^2(\sigma_{u2s}^*)^2(\sigma_{g2p_z})^2(\pi_{u2p})^4(\pi_{g2p}^*)^4$
(b) $Q_2^+: (\sigma_{g2s})^2(\sigma_{u2s}^*)^2(\pi_{u2p})^4(\sigma_{g2p_z}^*)^1$
(c) $Z_2^-: (\sigma_{g2s})^2(\sigma_{u2s}^*)^2(\sigma_{g2p_z})^2(\pi_{u2p})^4(\pi_{g2p}^*)^3$

22. For each of the following valence electron configurations of a homonuclear diatomic molecule or molecular ion, identify the element X, Q, or Z and determine the total bond order.
(a) $X_2: (\sigma_{g2s})^2(\sigma_{u2s}^*)^2(\sigma_{g2p_z})^2(\pi_{u2p})^4(\pi_{g2p}^*)^2$
(b) $Q_2^-: (\sigma_{g2s})^2(\sigma_{u2s}^*)^2(\pi_{u2p})^3$
(c) $Z_2^{2+}: (\sigma_{g2s})^2(\sigma_{u2s}^*)^2(\sigma_{g2p_z})^2(\pi_{u2p})^4(\pi_{g2p}^*)^2$

23. For each of the electron configurations in Problem 21, determine whether the molecule or molecular ion is paramagnetic or diamagnetic.

24. For each of the electron configurations in Problem 22, determine whether the molecule or molecular ion is paramagnetic or diamagnetic.

25. Following the pattern of Figure 6.19, work out the correlation diagram for the CN molecule, showing the relative energy levels of the atoms and the bonding and antibonding orbitals of the molecule. Indicate the occupation of the MOs with arrows. State the order of the bond and comment on the magnetic properties of CN.

26. Following the pattern of Figure 6.19, work out the correlation diagram for the BeN molecule, showing the relative energy levels of the atoms and the bonding and antibonding orbitals of the molecule. Indicate the occupation of the MOs with arrows. State the order of the bond and comment on the magnetic properties of BeN.

27. The bond length of the transient diatomic molecule CF is 1.291 Å; that of the molecular ion CF^+ is 1.173 Å. Explain why the CF bond shortens with the loss of an electron. Refer to the proper MO correlation diagram.

28. The compound nitrogen oxide (NO) forms when the nitrogen and oxygen in air are heated. Predict whether the nitrosyl ion (NO^+) will have a shorter or a longer bond than the NO molecule. Will NO^+ be paramagnetic like NO or diamagnetic?

29. What would be the electron configuration for a HeH^- molecular ion? What bond order would you predict? How stable should such a species be?

30. The molecular ion HeH^+ has an equilibrium bond length of 0.774 Å. Draw an electron correlation diagram for this ion, indicating the occupied MOs. Is HeH^+ paramagnetic? When HeH^+ dissociates, is a lower energy state reached by forming He + H^+ or He^+ + H?

31. Suppose we supply enough energy to BeH_2 to remove one of its electrons. Is the dissociation energy of the resulting ion larger or smaller than that of BeH_2? Explain your answer.

32. Suppose we supply enough energy to BeH_2 to remove one of its electrons. Is the length of the Be—H bonds in the resulting ion larger or smaller than those in BeH_2? Will both Be—H bonds change in the same way? Explain your answers.

33. Suppose we remove an electron from the highest energy occupied molecular orbital in CO_2. Is the dissociation energy of the resulting ion larger or smaller than that of CO_2? How will the C—O bonds change? Explain your answers.

34. Suppose we remove three electrons from CO_2 to create the ion CO_2^{3+}. Is the dissociation energy of the resulting ion larger or smaller than that of CO_2? How will the C—O bonds change? Explain your answers.

Photoelectron Spectroscopy for Molecules

35. Photoelectron spectra were acquired from a sample of gaseous N_2 using He(I) light with energy 21.22 eV as the ionization source. Photoelectrons were detected with kinetic energy values 5.63 eV and also with 4.53 eV. Calculate the ionization energy for each group of electrons. Identify the MOs that were most likely the sources of these two groups of electrons.

36. Photoelectron spectra were acquired from a sample of gaseous O_2 using x-ray radiation with wavelength 0.99 nm and energy 1253.6 eV. The spectrum contained a large peak for photoelectrons with speed of 1.57×10^7 m s^{-1}. Calculate the ionization energy of these electrons. Identify the orbital from which they were most likely emitted.

37. From the $n = 0$ peaks in the photoelectron spectrum for N_2 shown in Figure 6.33, prepare a quantitative energy level diagram for the molecular orbitals of N_2.

38. From the $n = 0$ peaks in the photoelectron spectrum for O_2 shown in Figure 6.34, prepare a quantitative energy level diagram for the molecular orbitals of O_2.

39. The photoelectron spectrum of HBr has two main groups of peaks. The first has ionization energy 11.88 eV. The next peak has ionization energy 15.2 eV, and it is followed by a long progression of peaks with higher ionization energies. Identify the molecular orbitals corresponding to these two groups of peaks.

40. The photoelectron spectrum of CO has four major peaks with ionization energies of 14.5, 17.2, 20.1, and 38.3 eV. Assign these peaks of molecular orbitals of CO, and prepare a quantitative energy level correlation diagram for CO. The ionization energy of carbon atoms is 11.26 eV, and the ionization energy of oxygen atoms is 13.62 eV.

Localized Bonds: The Valence Bond Model

41. Write simple valence bond wave functions for the diatomic molecules Li_2 and C_2. State the bond order predicted by the simple VB model and compare with the LCAO predictions in Table 6.3

42. Write simple valence bond wave functions for the diatomic molecules B_2 and O_2. State the bond order predicted by the simple VB model and compare with the LCAO predictions in Table 6.3

43. Both the simple VB model and the LCAO method predict that the bond order of Be_2 is 0. Explain how each arrives at that conclusion.

44. Both the simple VB model and the LCAO method predict that the bond order of Ne_2 is 0. Explain how each arrives at that conclusion.

45. Write simple valence bond wave functions for formation of bonds between B atoms and H atoms. What B—H compound does the VB model predict? What geometry does it predict for the molecules?

46. Write simple valence bond wave functions for formation of bonds between C and H atoms. What C—H compound does the VB model predict? What geometry does it predict for the molecules?

47. Write simple valence bond wave functions for the bonds in NH_3. What geometry does the VB model predict for NH_3?

48. Write simple valence bond wave functions for the bonds in H_2O. What geometry does the VB model predict for H_2O?

49. Formulate a localized bond picture for the amide ion (NH_2^-). What hybridization do you expect the central nitrogen atom to have, and what geometry do you predict for the molecular ion?

50. Formulate a localized bond picture for the hydronium ion (H_3O^+). What hybridization do you expect the central oxygen atom to have, and what geometry do you predict for the molecular ion?

51. Draw a Lewis electron dot diagram for each of the following molecules and ions. Formulate the hybridization for the central atom in each case and give the molecular geometry.
(a) CCl_4
(b) CO_2
(c) OF_2
(d) CH_3^-
(e) BeH_2

52. Draw a Lewis electron dot diagram for each of the following molecules and ions. Formulate the hybridization for the central atom in each case and give the molecular geometry.
(a) BF_3
(b) BH_4^-
(c) PH_3
(d) CS_2
(e) CH_3^+

53. Describe the hybrid orbitals on the chlorine atom in the ClO_3^+ and ClO_2^+ molecular ions. Sketch the expected geometries of these ions.

54. Describe the hybrid orbitals on the chlorine atom in the ClO_4^- and ClO_3^- molecular ions. Sketch the expected geometries of these ions.

55. The sodium salt of the unfamiliar orthonitrate ion (NO_4^{3-}) has been prepared. What hybridization is expected on the nitrogen atom at the center of this ion? Predict the geometry of the NO_4^{3-} ion.

56. Describe the hybrid orbitals used by the carbon atom in N≡C—Cl. Predict the geometry of the molecule.

Comparison of Linear Combination of Atomic Orbitals and Valence Bond Methods

57. Describe the bonding in the bent molecule NF_2. Predict its energy level diagram and electron configuration.

58. Describe the bonding in the bent molecule OF_2. Predict its energy level diagram and electron configuration.

59. The azide ion (N_3^-) is a weakly bound molecular ion. Formulate its MO structure for localized σ bonds and de-localized π bonds. Do you expect N_3 and N_3^+ to be bound as well? Which of the three species do you expect to be paramagnetic?

60. Formulate the MO structure of (NO_2^+) for localized π bonds and de-localized π bonds. Is it linear or nonlinear? Do you expect it to be paramagnetic? Repeat the analysis for NO_2 and for NO_2^-.

61. Discuss the nature of the bonding in the nitrite ion (NO_2^-). Draw the possible Lewis resonance diagrams for this ion. Use the VSEPR theory to determine the steric number, the hybridization of the central nitrogen atom, and the geometry of the ion. Show how the use of resonance structures can be avoided by introducing a de-localized π MO. What bond order does the MO model predict for the N—O bonds in the nitrite ion?

62. Discuss the nature of the bonding in the nitrate ion (NO_3^-). Draw the possible Lewis resonance diagrams for this ion. Use the VSEPR theory to determine the steric number, the hybridization of the central N atom, and the geometry of the ion. Show how the use of resonance structures can be avoided by introducing a de-localized π MO. What bond order is predicted by the MO model for the N—O bonds in the nitrate ion?

ADDITIONAL PROBLEMS

63. (a) Sketch the occupied MOs of the valence shell for the N_2 molecule. Label the orbitals as σ or π orbitals, and specify which are bonding and which are antibonding.
(b) If one electron is removed from the highest occupied orbital of N_2, will the equilibrium N—N distance become longer or shorter? Explain briefly.

64. Calcium carbide (CaC_2) is an intermediate in the manufacturing of acetylene (C_2H_2). It is the calcium salt of the carbide (also called acetylide) ion (C_2^{2-}). What is the electron configuration of this molecular ion? What is its bond order?

65. Show how that the B_2 molecule is paramagnetic indicates that the energy ordering of the orbitals in this molecule is given by Figure 6.16a rather than 6.16b.

66. The Be_2 molecule has been detected experimentally. It has a bond length of 2.45 Å and a bond dissociation energy of 9.46 kJ mol^{-1}. Write the ground-state electron configuration of Be_2 and predict its bond order using the theory developed in the text. Compare the experimental bonding data on Be_2 with those recorded for B_2, C_2, N_2, and O_2 in Table 6.3. Is the prediction that stems from the simple theory significantly incorrect?

* 67. (a) The ionization energy of molecular hydrogen (H_2) is *greater* than that of atomic hydrogen (H), but that of molecular oxygen (O_2) is *lower* than that of atomic oxygen (O). Explain. (*Hint:* Think about the stability of the molecular ion that forms in relation to bonding and antibonding electrons.)
(b) What prediction would you make for the relative ionization energies of atomic and molecular fluorine (F and F_2)?

68. The molecular ion HeH^+ has an equilibrium bond length of 0.774 Å. Draw an electron correlation diagram for this molecule, indicating the occupied MOs. If the lowest energy MO has the form $C_1\psi_{1s}^{H} + C_2\psi_{1s}^{He}$, do you expect C_2 to be larger or smaller than C_1?

* 69. The MO of the ground state of a heteronuclear diatomic molecule AB is

$$\psi_{mol} = C_A\varphi^A + C_B\varphi^B$$

If a bonding electron spends 90% of its time in an orbital φ^A on atom A and 10% of its time in φ^B on atom B, what are the values of C_A and C_B? (Neglect the overlap of the two orbitals.)

70. The stable molecular ion H_3^+ is triangular, with H—H distances of 0.87 Å. Sketch the molecule and indicate the region of greatest electron density of the lowest energy MO.

* 71. According to recent spectroscopic results, nitramide

```
H       O
 \     /
  N—N
 /     \
H       O
```

is a nonplanar molecule. It was previously thought to be planar.

(a) Predict the bond order of the N—N bond in the nonplanar structure.

(b) If the molecule really were planar after all, what would be the bond order of the N—N bond?

72. *trans*-Tetrazene (N_4H_4) consists of a chain of four nitrogen atoms with each of the two end atoms bonded to two hydrogen atoms. Use the concepts of steric number and hybridization to predict the overall geometry of the molecule. Give the expected structure of *cis*-tetrazene.

CHAPTER 7

Bonding in Organic Molecules

A petroleum refining tower.

Carbon (C) is unique among the elements in the large number of compounds it forms and in the variety of their structures. In combination with hydrogen (H), it forms molecules with single, double, and triple bonds; chains; rings; branched structures; and cages. There are thousands of stable hydrocarbons in sharp contrast to the mere two stable compounds between oxygen and hydrogen (water and hydrogen peroxide). Even the rather versatile elements nitrogen and oxygen form only six nitrogen oxides.

The unique properties of carbon relate to its position in the periodic table. As a second-period element, carbon atoms are relatively small. Therefore, it can easily form the double and triple bonds that are rare in the compounds of related elements, such as silicon. As a Group IV element, carbon can form four bonds, which is more than the other second-period elements; this characteristic gives it wide

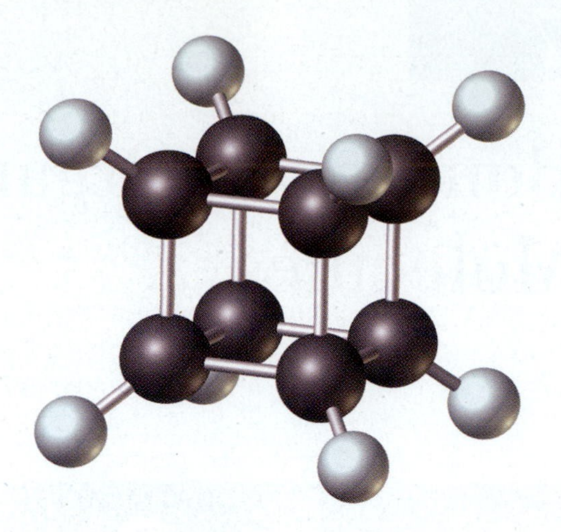

One simple and unusual hydrocarbon is cubane (C_8H_8), in which the eight carbon atoms are arranged at the corners of a cube. Recently, a derivative was made in which all eight hydrogen atoms were replaced by $-NO_2$ groups.

scope for structural elaboration. Finally, as an element of intermediate electronegativity, carbon forms covalent compounds both with relatively electronegative elements, such as oxygen, nitrogen, and the halogens, and with relatively electropositive elements, such as hydrogen and the heavy metals mercury and lead.

The study of the compounds of carbon is the discipline traditionally called **organic chemistry,** although the chemistry of carbon is intimately bound up with that of inorganic elements and with biochemistry. This chapter builds on the general principles of covalent bonding in carbon compounds presented in Sections 6.2, 6.4, and 6.5. The relation between molecular structure and properties of organic substances is illustrated by examining the composition, refining, and chemical processing of petroleum, the primary starting material for the production of hydrocarbons and their derivatives. This chapter continues with an introduction to the types of compounds that result when elements such as chlorine, oxygen, and nitrogen combine with carbon and hydrogen. The chapter concludes with a brief introduction to some organic molecules important to agriculture and to medicine.

7.1 Petroleum Refining and the Hydrocarbons

When the first oil well was drilled in 1859 near Titusville, Pennsylvania, the future effects of the exploitation of petroleum on everyday life could not have been anticipated. Today, the petroleum and petrochemical industries span the world and influence nearly every aspect of our daily lives. In the early years of the 20th century, the development of the automobile, fueled by low-cost gasoline derived from petroleum, dramatically changed many people's lifestyles. The subsequent use of gasoline and oil to power trains and planes, tractors and harvesters, and pumps and coolers transformed travel, agriculture, and industry. Natural gas and heating oil warm most homes in the United States. Finally, the spectacular growth of the petrochemical industry since 1945 has led to the introduction of innumerable new products, ranging from pharmaceuticals to plastics and synthetic fibers. More than half of the chemical compounds produced in greatest volumes are synthesized from petroleum feedstocks.

In the 21st century the prospects for the continued availability of cheap petroleum and petrochemicals are clouded. Many wells have been drained, and the remaining petroleum is relatively difficult and costly to extract. Petroleum is not easy to make. It originated from the deposition and decay of organic matter (of animal or vegetable origin) in oxygen-poor marine sediments. Subsequently, petroleum migrated to the porous sandstone rocks from which it is extracted today. Over the past 100 years, we have consumed a significant fraction of the petroleum accumulated in the earth over many millions of years. The imperative for the future is to save the remaining reserves for uses for which few substitutes are available (such as the manufacture of specialty chemicals) while finding other sources of heat and energy.

Although crude petroleum contains small amounts of oxygen, nitrogen, and sulfur, its major constituents are **hydrocarbons**—compounds of carbon and hydrogen. Isolating individual hydrocarbon substances from petroleum mixtures is an industrial process of central importance. Moreover, it provides a fascinating story that illustrates how the structures of molecules determines the properties of substances and the behavior of those substances in particular processes. The next three sections present a brief introduction to this story, emphasizing the structure–property correlations.

7.2 The Alkanes

Normal Alkanes

The most prevalent hydrocarbons in petroleum are the **straight-chain alkanes** (also called normal alkanes, or *n*-alkanes), which consist of chains of carbon atoms bonded to one another by single bonds, with enough hydrogen atoms on each carbon atom to bring it to the maximum bonding capacity of four. These alkanes have the generic formula C_nH_{2n+2}; Table 7.1 lists the names and formulas of the first few alkanes. The ends of the molecules are methyl ($-CH_3$) groups, with methylene ($-CH_2-$) groups between them. We could write pentane (C_5H_{12}) as $CH_3CH_2CH_2CH_2CH_3$ to indicate the structure more explicitly or, in abbreviated fashion, as $CH_3(CH_2)_3CH_3$.

TABLE 7.1 Straight-Chain Alkanes

Name	Formula
Methane	CH_4
Ethane	C_2H_6
Propane	C_3H_8
Butane	C_4H_{10}
Pentane	C_5H_{12}
Hexane	C_6H_{14}
Heptane	C_7H_{16}
Octane	C_8H_{18}
Nonane	C_9H_{20}
Decane	$C_{10}H_{22}$
Undecane	$C_{11}H_{24}$
Dodecane	$C_{12}H_{26}$
Tridecane	$C_{13}H_{28}$
Tetradecane	$C_{14}H_{30}$
Pentadecane	$C_{15}H_{32}$
.	.
.	.
.	.
Triacontane	$C_{30}H_{62}$

Bonding in the normal alkanes is explained by the valence bond (VB) model with orbital hybridization described in Section 6.4. The carbon atom in methane has four sp^3 hybridized orbitals, which overlap with hydrogen 1*s* orbitals to form four σ bonds pointing toward the vertices of a tetrahedron with the carbon atom at its center. These orbitals are represented in Figures 6.43 and 6.44; the methane molecule is shown in Figure 7.1a. The bonds in ethane are also described sp^3 hybridization. One hybrid orbital on each carbon atom overlaps another hybrid orbital to form the C—C σ bond. The remaining three hybrids on each carbon overlap with hydrogen 1*s* orbitals to form σ bonds. The ethane molecule is shown in Figure 7.1b.

The same bonding scheme applies to the larger straight-chain alkanes. Two of the sp^3 hybrid orbitals on each carbon atom overlap those of adjacent atoms to form the backbone of the chain, and the remaining two bond to hydrogen atoms. Although bond *lengths* change little through vibration, rotation about a C—C single bond occurs quite easily (Fig. 7.2). Thus, a hydrocarbon molecule in a gas or

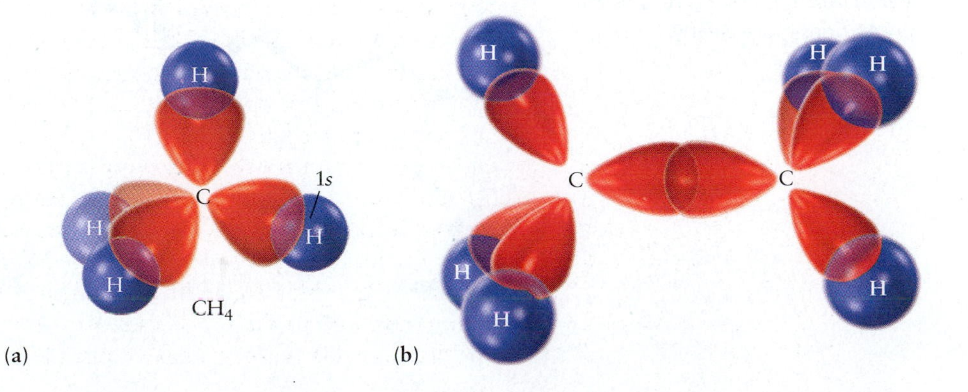

FIGURE 7.1 Bonding in the alkanes involves sp^3 hybridized orbitals on carbon. (a) Methane. (b) Ethane. The orbitals shown here are typical sketches used by organic chemists to describe bonding in organic molecules. Figure 6.44 compares these shapes to the actual shapes of the hybrid orbitals.

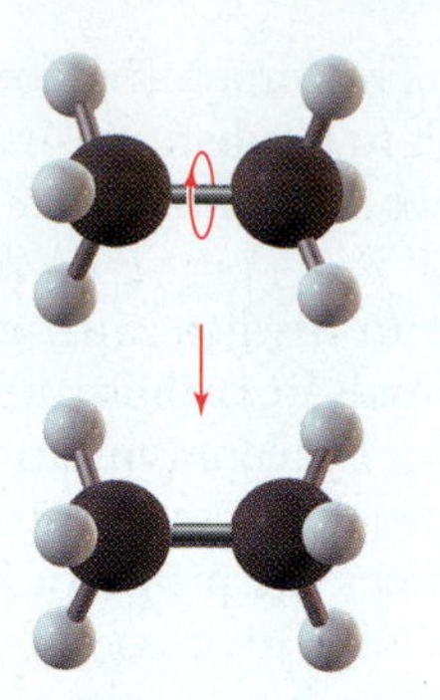

FIGURE 7.2 The two $-CH_3$ groups in ethane rotate easily about the bond that joins them.

liquid constantly changes its conformation, and through free rotation can become quite "balled up" in normal motions. The term *straight chain* refers only to the bonding pattern in which each carbon atom is bonded to the next one in a sequence; it does not mean that the carbon atoms are positioned along a straight line. An alkane molecule with 10 to 20 carbon atoms looks quite different when "balled up" than when its bonds are extended to give a "stretched" molecule (Fig. 7.3). These two extreme conformations and many others interconvert rapidly at room temperature.

Figure 7.4 shows the melting and boiling points of the straight-chain alkanes; both increase with the number of carbon atoms, and thus with molecular mass. This is a consequence of the increasing strength of *dispersion forces* between heavier molecules (see discussion in Section 10.2). Methane, ethane, propane, and butane are all gases at room temperature, but the hydrocarbons that follow them in the alkane series are liquids. Alkanes beyond about $C_{17}H_{36}$ are waxy solids at 20°C, whose melting points increase with the number of carbon atoms present. Paraffin wax, a low-melting solid, is a mixture of alkanes with 20 to 30 carbon atoms per molecule. Petrolatum (petroleum jelly, or Vaseline) is a different mixture that is semisolid at room temperature.

A mixture of hydrocarbons such as petroleum does not boil at a single, sharply defined temperature. Instead, as such a mixture is heated, the compounds with lower boiling points (the most volatile) boil off first, and as the temperature increases, more and more of the material vaporizes. The existence of a boiling-point range permits components of a mixture to be separated by distillation (see discussion in Section 11.6). The earliest petroleum distillation was a simple batch process: The crude oil was heated in a still, the volatile fractions were removed at the top and condensed to gasoline, and the still was cleaned for another batch. Modern petroleum refineries use much more sophisticated and efficient distillation methods, in which crude oil is added continuously and fractions of different volatility are tapped off at various points up and down the distillation column (Fig. 7.5). To save on energy costs, heat exchangers capture the heat liberated from condensation of the liquid products.

Distillation allows hydrocarbons to be separated by boiling point, and thus by molecular mass. A mixture of gases emerges from the top of the column, resembling the natural gas that collects in rock cavities above petroleum deposits. These gas mixtures contain ethane, propane, and butane, which can be separated further by redissolving them in a liquid solvent such as hexane. The methane-rich mixture of gases that remains is used for chemical synthesis or is shipped by pipeline to

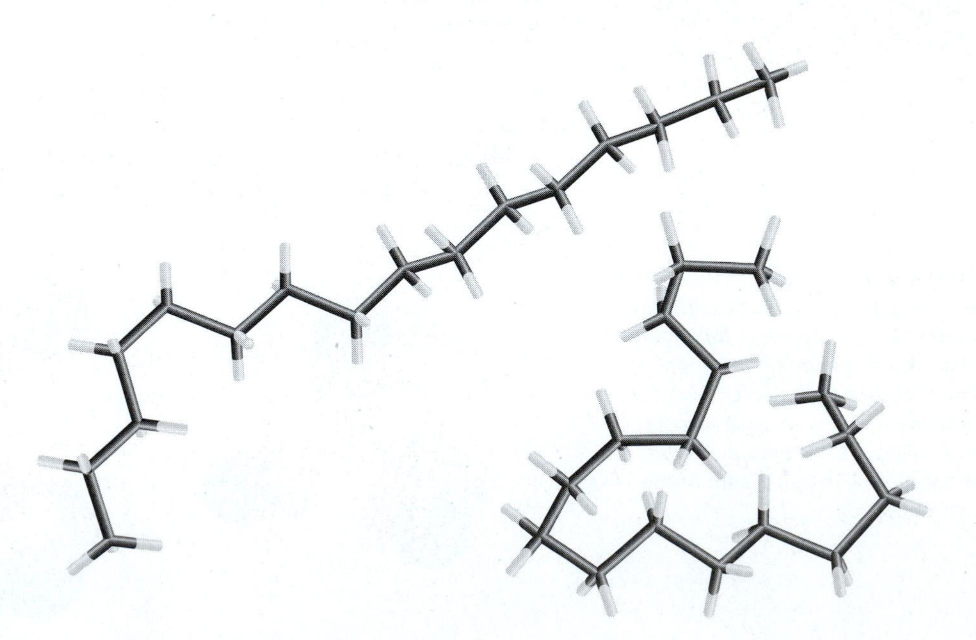

FIGURE 7.3 Two of the many possible conformations of the alkane $C_{16}H_{34}$. The carbon atoms are not shown explicitly, but they lie at the black intersections. Hydrogen atoms are at the white ends. Eliminating the spheres representing atoms in these tube (or Dreiding) models reveals the conformations more clearly.

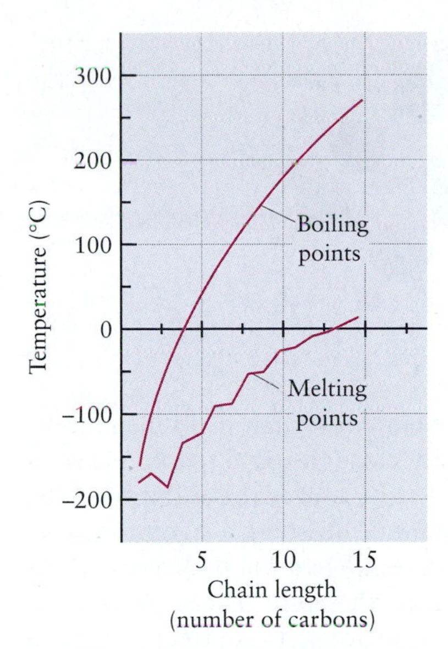

FIGURE 7.4 The melting and boiling points of the straight-chain alkanes increase with chain length *n*. Note the alternation in the melting points: Alkanes with *n* odd tend to have lower melting points because they are more difficult to pack into a crystal lattice.

heat homes. The gases dissolved in hexane can be separated by redistilling, after which they can be used as starting materials in chemical processes. Propane and butane are also bottled under pressure and sold as liquefied petroleum gas, which is used for fuel in areas where natural gas is not available from pipelines. After the gases, the next fraction to emerge from the petroleum distillation column is **naphtha,** which is used primarily in the manufacture of gasoline. Subsequent fractions of successively higher molecular mass are used for jet and diesel fuel, heating oil, and machine lubricating oil. The heavy, involatile sludge that remains at the bottom of the distillation unit is pitch or asphalt, which is used for roofing and paving.

Cyclic Alkanes

In addition to the straight-chain alkanes, the cyclic alkanes also appear in petroleum. A **cycloalkane** consists of at least one chain of carbon atoms attached at the ends to form a closed loop. In the formation of this additional C—C bond, two hydrogen atoms must be eliminated; thus, the general formula for cycloalkanes having one ring is C_nH_{2n} (Fig. 7.6). The cycloalkanes are named by adding the prefix *cyclo-* to the name of the straight-chain alkane that has the same number of carbon atoms as the ring compound. Bonding in the cycloalkanes involves sp^3 hybridization of the carbon atoms, just as in the straight-chain alkanes. But, coupling the tetrahedral angle of 109.5 degrees with the restriction of a cyclic structure leads to two new interesting structural features that introduce **strain energy** in the cycloalkanes and influence the stability of their conformations.

It is easy to see from inspection of molecular models that two distinct conformations of cyclohexane can be formed when the tetrahedral angle is maintained at each carbon atom. These are called the **boat** and **chair conformations** because of their resemblance to these objects (Fig. 7.7). The chair conformation has four carbon atoms in a plane with one above and one below that plane, located on opposite sides of the molecule. The boat conformation also has four carbon atoms in a plane, but both of the remaining atoms are located above this plane. Both conformations exist and appear to interconvert rapidly at room temperature through a sequence of rotations about single bonds (see Fig. 7.2). The chair conformation is significantly more stable than the boat, because the hydrogen atoms can become

FIGURE 7.5 In the distillation of petroleum, the lighter, more volatile hydrocarbon fractions are removed from higher up the column and the heavier fractions from lower down.

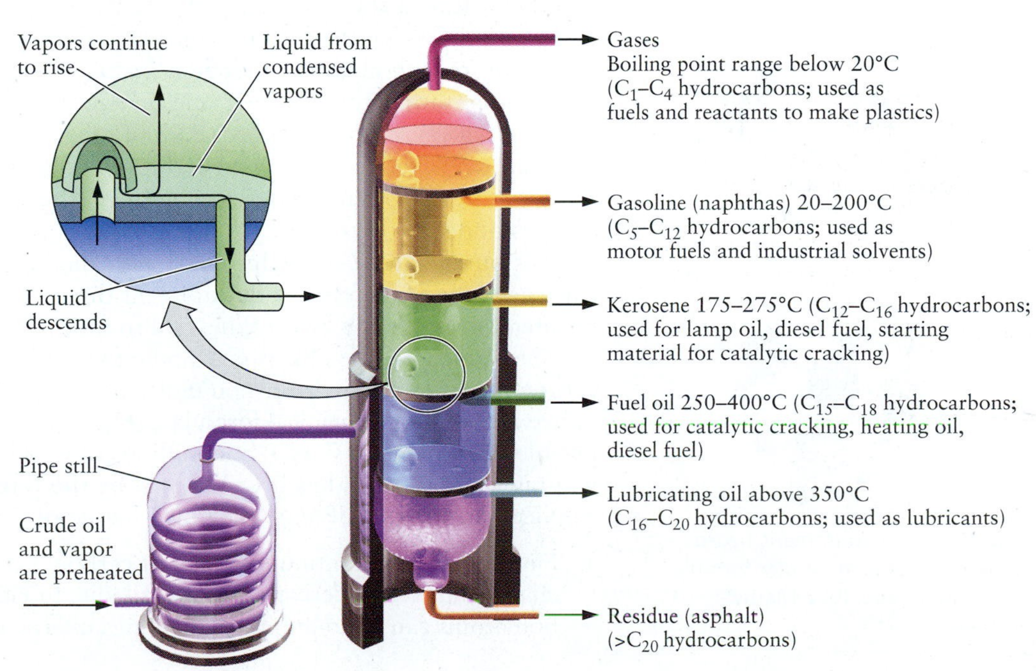

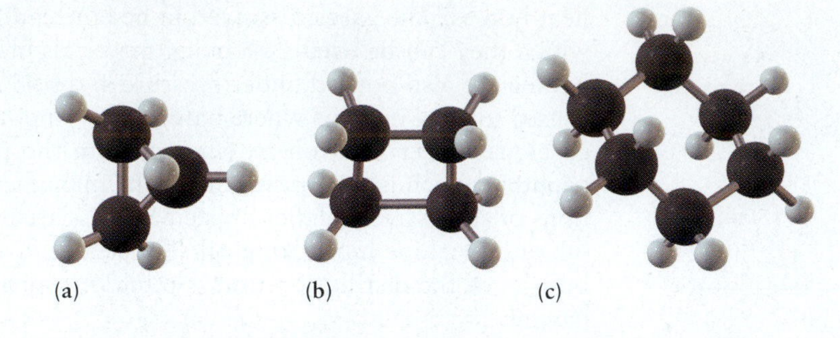

FIGURE 7.6 Three cyclic hydrocarbons. (a) Cyclopropane, C_3H_6. (b) Cyclobutane, C_4H_8. (c) Cyclohexane, C_6H_{12}.

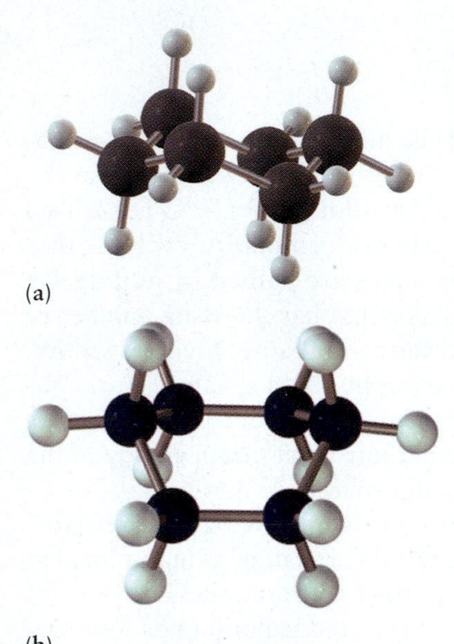

FIGURE 7.7 The conformations of cyclohexane. (a) Chair. (b) Boat.

quite close and interfere with one another in the boat conformation. When atoms that are not bonded to each other come sufficiently close in space to experience a repulsive interaction, this increase in potential energy reduces the stability of the molecule. Such interactions are called **steric strain,** and they play a significant role in determining the structure of polyatomic molecules. When the hydrogen atoms on cyclohexane are replaced with larger substituents, these effects can prevent interconversion between the boat and chair conformations. This effect is seen in many large molecules of biological significance, where the cyclohexane ring is an important structural unit, locked into one of its conformations.

Consider the possibility that cyclohexane could have a planar hexagonal structure. Then each C—C—C bond angle would be 120 degrees resulting in angle strain of 10.5 degrees. This distortion of the bond angle from the tetrahedral value increases the potential energy of the bond above its stable equilibrium value, and the resulting **angle strain** energy reduces the stability of the molecule. Cyclohexane minimizes this effect through rotation about single bonds. The smallest cycloalkanes, namely, cyclopropane and cyclobutane, have much less freedom to rotate about single bonds. Consequently, they are strained compounds because the C—C—C bond angle is 60 (in C_3H_6) or 90 degrees (in C_4H_8), which is far less than the normal tetrahedral angle of 109.5 degrees. As a result, these compounds are more reactive than the heavier cycloalkanes or their straight-chain analogs, propane and butane.

Branched-Chain Alkanes

Branched-chain alkanes are hydrocarbons that contain only C—C and C—H single bonds, but in which the carbon atoms are no longer arranged in a straight chain. One or more carbon atoms in each molecule is bonded to three or four other carbon atoms, rather than to only one or two as in the normal alkanes or cycloalkanes. The simplest branched-chain molecule (Fig. 7.8) is 2-methylpropane, sometimes referred to as isobutane. This molecule has the same molecular formula as butane (C_4H_{10}) but a different bonding structure in which the central carbon atom is bonded to three $-CH_3$ groups and only one hydrogen atom. The compounds butane and 2-methylpropane are called **geometrical isomers.** Their molecules have the same formula but different three-dimensional structures that could be interconverted only by breaking and re-forming chemical bonds.

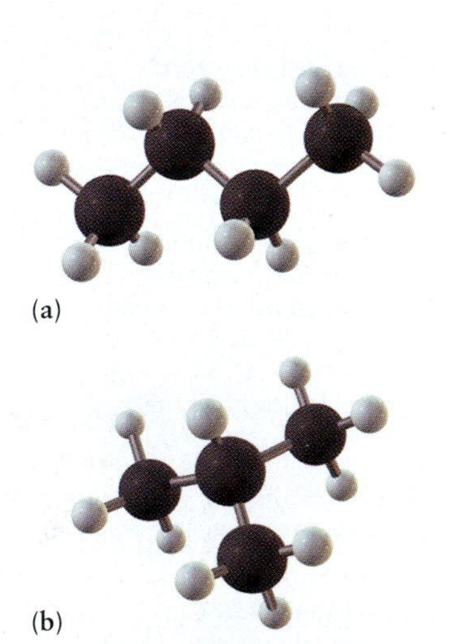

FIGURE 7.8 Two isomeric hydrocarbons with the molecular formula C_4H_{10}. (a) Butane. (b) 2-Methylpropane.

The number of possible isomers increases rapidly with increasing numbers of carbon atoms in the hydrocarbon molecule. Butane and 2-methylpropane are the only two isomers of chemical formula C_4H_{10}, but there are three isomers of C_5H_{12}, five of C_6H_{14}, nine of C_7H_{16}, and millions of $C_{30}H_{62}$. A systematic procedure for naming these isomers has been codified by the International Union of Pure and Applied Chemistry (IUPAC). The following set of rules is a part of that procedure:

1. Find the longest continuous chain of carbon atoms in the molecule. The molecule is named as a derivative of this alkane. In Figure 7.8b, a chain of three carbon atoms can be found; thus, the molecule is a derivative of propane.

TABLE 7.2 Alkyl Side Groups

Name	Formula
Methyl	$-CH_3$
Ethyl	$-CH_2CH_3$
Propyl	$-CH_2CH_2CH_3$
Isopropyl	$-CH(CH_3)_2$
Butyl	$-CH_2CH_2CH_2CH_3$

2. The hydrocarbon groups attached to the chain are called alkyl groups. Their names are obtained by dropping the ending *-ane* from the corresponding alkane and replacing it with *-yl* (Table 7.2). The methyl group, CH_3, is derived from methane (CH_4), for example. Note also the isopropyl group, which attaches by its middle carbon atom.
3. Number the carbon atoms along the chain identified in rule 1. Identify each alkyl group by the number of the carbon atom at which it is attached to the chain. The methyl group in the molecule in Figure 7.8b is attached to the second of the three carbon atoms in the propane chain; therefore, the molecule is called 2-methylpropane. The carbon chain is numbered from the end that gives the lowest number for the position of the first attached group.
4. If more than one alkyl group of the same type is attached to the chain, use the prefixes *di-* (two), *tri-* (three), *tetra-* (four), *penta-* (five), and so forth to specify the total number of such attached groups in the molecule. Thus, 2,2,3-trimethylbutane has two methyl groups attached to the second carbon atom and one to the third carbon atom of the four-atom butane chain. It is an isomer of heptane (C_7H_{16}).
5. If several types of alkyl groups appear, name them in alphabetical order. Ethyl is listed before methyl, which appears before propyl.

EXAMPLE 7.1

Name the following branched-chain alkane:

H_3C CH_2 CH_2-CH_3
C C
H_3C H CH_3 CH_2-CH_3

SOLUTION

The longest continuous chain of carbon atoms is six, so this is a derivative of hexane. Number the carbon atoms starting from the left.

H_3C 2 3 CH_2 4 5 6 CH_2-CH_3
C C
1 H_3C H H_3C CH_2-CH_3

Methyl groups are attached to carbon atoms 2 and 4, and an ethyl group is attached to atom 4. The name is thus 4-ethyl-2,4-dimethylhexane. Note that if we had started numbering from the right, the higher number 3 would have appeared for the position of the first methyl group; therefore, the numbering from the left is preferred.

Related Problems: 7, 8, 9, 10, 11, 12

A second type of isomerism characteristic of organic molecules is **optical isomerism,** or **chirality.** A carbon atom that forms single bonds to four different atoms or groups of atoms can exist in two forms that are mirror images of each other but that cannot be interconverted without breaking and re-forming bonds (Fig. 7.9). If a mixture of the two forms is resolved into its optical isomers, the two forms rotate the plane of polarized light in different directions; therefore, such molecules are said to be "optically active." Although paired optical isomers have identical physical properties, their chemical properties can differ when they interact with other optically active molecules. As discussed in Section 23.4, proteins and other biomolecules are optically active. One goal of pharmaceutical research is to prepare

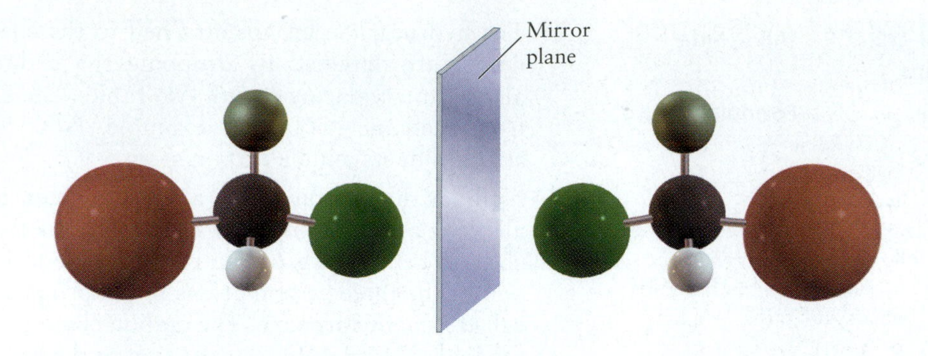

FIGURE 7.9 A molecule such as CHBrClF, which has four different atoms or groups of atoms bonded to a single carbon atom, exists in two mirror-image forms that cannot be superimposed by rotation. Such pairs of molecules are optical isomers; the carbon atom is called a chiral center.

particular optical isomers of carbon compounds for medicinal use. In many cases, one optical isomer is beneficial and the other is useless or even harmful.

The fraction of branched-chain alkanes in gasoline affects how it burns in an engine. Gasoline consisting entirely of straight-chain alkanes burns unevenly, causing "knocking" that can damage the engine. Blends that are richer in branched-chain and cyclic alkanes burn with much less knocking. Smoothness of combustion is rated quantitatively via the **octane number** of the gasoline, which was defined in 1927 by selecting as references one compound that causes large amounts of knocking and another that causes little to no knocking. Pure 2,2,4-trimethylpentane (commonly known as isooctane) burns smoothly and was assigned an octane number of 100. Of the compounds examined at the time, pure heptane caused the most knocking and was assigned octane number 0. Mixtures of heptane and isooctane cause intermediate amounts of knocking. Standard mixtures of these two compounds define a scale for evaluating the knocking caused by real gasolines, which are complex mixtures of branched- and straight-chain hydrocarbons. If a gasoline sample produces the same amount of knocking in a test engine as a mixture of 90% (by volume) 2,2,4-trimethylpentane and 10% heptane, it is assigned the octane number 90.

Certain additives increase the octane rating of gasoline. The least expensive of these is tetraethyllead, $Pb(C_2H_5)_4$, a compound that has weak bonds between the central lead atom and the ethyl carbon atoms. It readily releases ethyl radicals ($\cdot C_2H_5$) into the gasoline during combustion; these reactive species speed and smooth the combustion process, reducing knocking and giving better fuel performance. The lead released into the atmosphere is a long-term health hazard. Lead poisons catalytic converters (Section 18.7) rendering them ineffective. The use of lead in gasoline has been phased out and other low-cost additives have been developed to increase octane numbers. Chemical processing to make branched-chain compounds from straight-chain compounds is also used to control the octane number of gasoline.

7.3 The Alkenes and Alkynes

The hydrocarbons discussed so far in this chapter are referred to as **saturated,** because all the carbon–carbon bonds are single bonds. Hydrocarbons that have double and triple carbon–carbon bonds are referred to as **unsaturated** (Fig. 7.10). Ethylene (C_2H_4) has a double bond between its carbon atoms and is called an **alkene.** The simplest **alkyne** is acetylene (C_2H_2), which has a triple bond between its carbon atoms. In naming these compounds, the *-ane* ending of the corresponding alkane is replaced by *-ene* when a double bond is present and by *-yne* when a triple bond is present. Ethene is thus the systematic name for ethylene, and ethyne for acetylene, although we will continue to use their more common names. For any compound with a carbon backbone of four or more carbon atoms, it is necessary to specify the location of the double or triple bond. This is

FIGURE 7.10 One way to distinguish alkanes from alkenes is by their reactions with aqueous $KMnO_4$. This strong oxidizing agent undergoes no reaction with hexane and retains its purple color (left). But, when $KMnO_4$ is placed in contact with 1-hexene, a redox reaction occurs in which the brown solid MnO_2 forms (right) and $-OH$ groups are added to both sides of the double bond in the 1-hexene, giving a compound with the formula $CH_3(CH_2)_3CH(OH)CH_2OH$.

done by numbering the carbon–carbon bonds and putting the number of the lower numbered carbon involved in the multiple bond before the name of the alkene or alkyne. Thus, the two different isomeric alkynes with the formula C_4H_6 are

$$HC{\equiv}C{-}CH_2{-}CH_3 \qquad \text{1-butyne}$$

$$CH_3{-}C{\equiv}C{-}CH_3 \qquad \text{2-butyne}$$

Bonding in alkenes is described by the VB method with sp^2 hybrid orbitals on each carbon atom. (This method is described in Section 6.4 and shown in Figures 6.42 and 6.44. You should review that material before proceeding.) Figure 7.11a shows three sp^2 hybrid orbitals and Figure 7.11b shows the remaining nonhybridized $2p_z$ orbital with the sp^2 hybrid orbitals represented as shadows in the x-y plane. In ethylene, a σ orbital is formed between the two sp^2 hybrid orbitals on the carbon atoms (Fig. 7.12a) and the remaining four sp^2 hybrid orbitals are used to form bonds with four hydrogen atoms. The nonhybridized $2p_z$ orbitals on the two carbon atoms are parallel to each other and overlap to form a π bond (see example in Fig. 6.16). The result is a double bond between the two carbon atoms (see Fig. 7.12b).

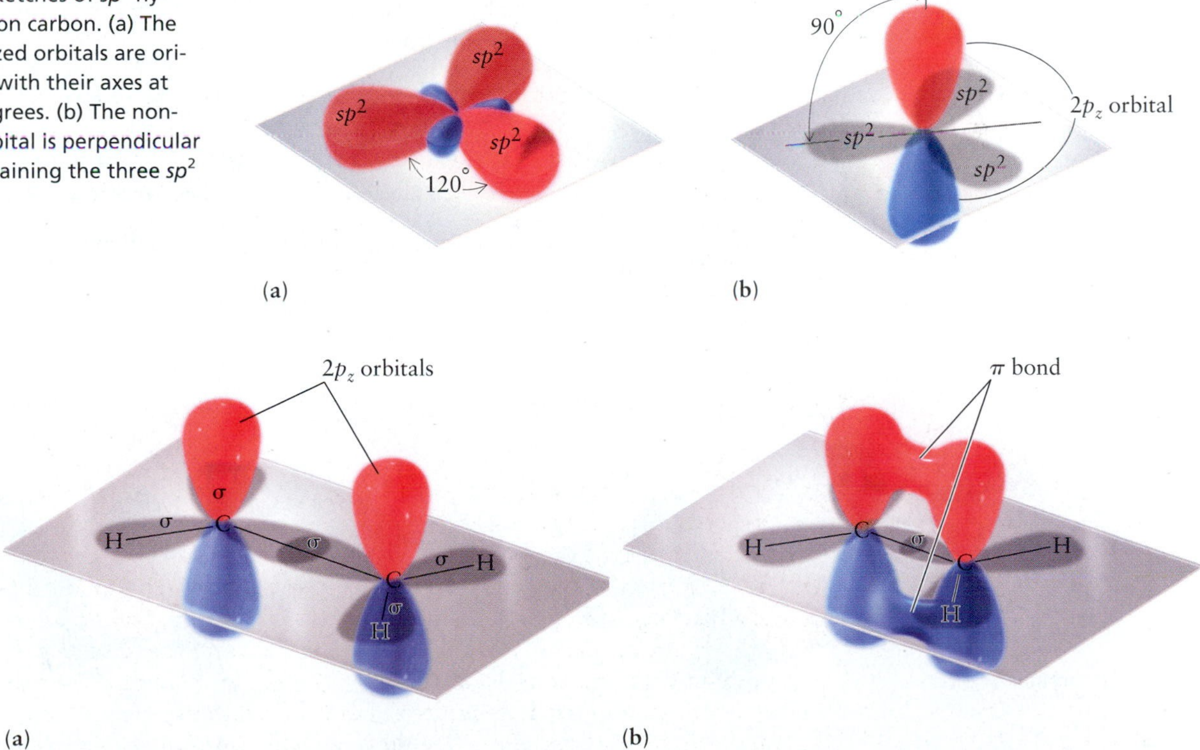

FIGURE 7.11 Sketches of sp^2 hybridized orbitals on carbon. (a) The three sp^2 hybridized orbitals are oriented in a plane with their axes at angles of 120 degrees. (b) The nonhybridized $2p$ orbital is perpendicular to the plane containing the three sp^2 hybrid orbitals.

FIGURE 7.12 Bond formation in ethylene. (a) Overlap of sp^2 hybrid orbitals forms a σ bond between the carbon atoms. (b) Overlap of parallel $2p$ orbitals forms a π bond.

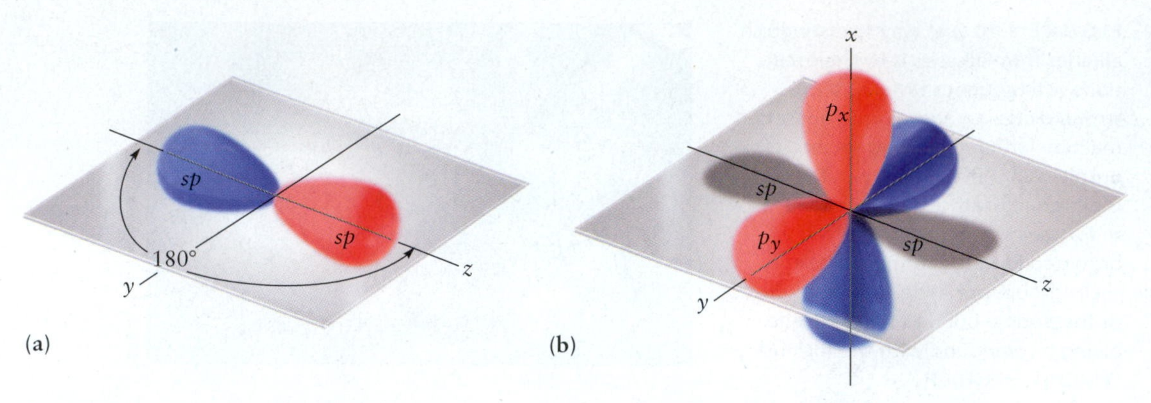

FIGURE 7.13 Sketches of *sp* hybridized orbitals on carbon. (a) The two *sp* hybridized orbitals are oriented in a plane with their axes at angles of 180 degrees. (b) Two nonhybridized *2p* orbitals are oriented perpendicular to the axes of the two *sp* hybrid orbitals.

Bonding in alkynes is explained by *sp* hybridization (see description in Section 6.4 and illustrations in Figs. 6.41 and 6.44). Figure 7.13a sketches the formation of the *sp* hybrid orbitals, and Figure 7.13b shows these together with the $2p_x$ and $2p_y$ nonhybridized atomic orbitals. Bond formation in acetylene is shown in Figures 7.14a and b. A σ bond between the two carbon atoms is formed by overlap of *sp* hybrids on each carbon, and each carbon forms a σ bond with one hydrogen atom using its other *sp* hybrid. The $2p_x$ and $2p_y$ nonhybridized atomic orbitals are parallel pairs on the two adjacent carbon atoms. Each pair overlaps to form a π bond (see Fig. 6.16). The result is a triple bond in acetylene, analogous to the triple bond in N_2 shown in Figure 6.40.

As explained later in this chapter, bond rotation does not occur readily about a carbon–carbon double bond. Many alkenes therefore exist in contrasting isomeric forms, depending on whether bonding groups are on the same (*cis*) or opposite sides (*trans*) of the double bond. There is only a single isomer of 1-butene but two of 2-butene, distinguished by the two possible placements for the outer CH_3 groups relative to the double bond:

H H
C=C
H_3C CH_3
cis-2-butene

H CH_3
C=C
H_3C H
trans-2-butene

These compounds differ in melting and boiling points, density, and other physical and chemical properties.

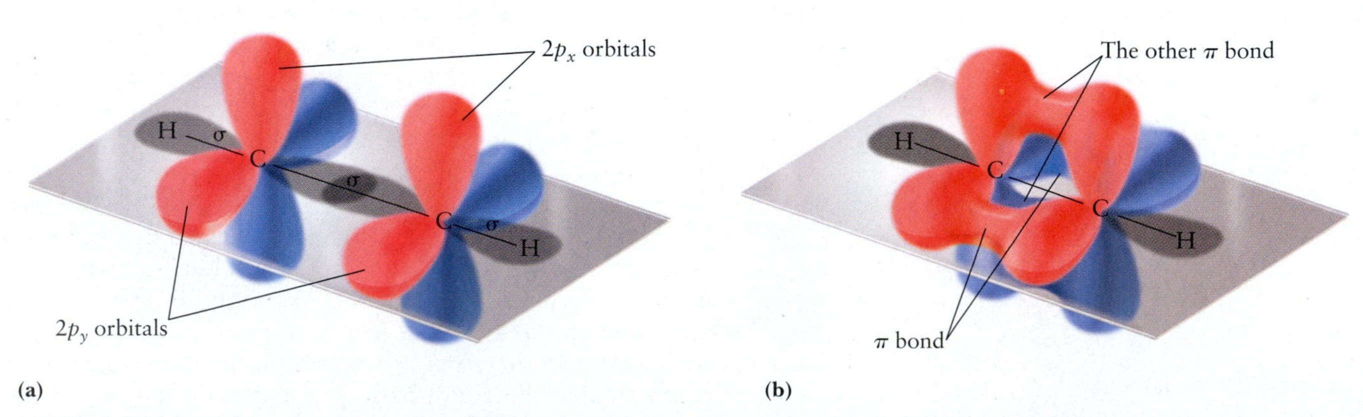

FIGURE 7.14 Bond formation in acetylene. (a) The σ bond framework and the two nonhybridized 2*p* orbitals on each carbon. (b) Overlap of two sets of parallel 2*p* orbitals forms two π bonds.

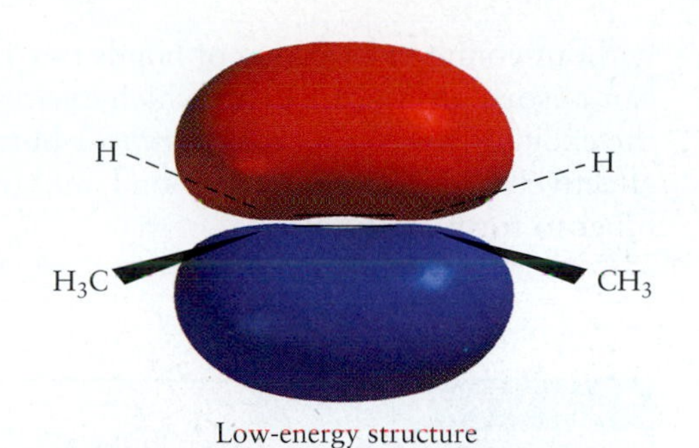

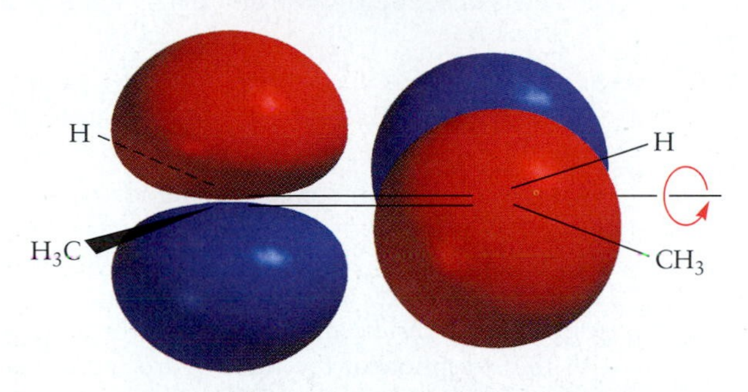

FIGURE 7.15 As the 2-butene molecule is twisted about the C═C bond, the overlap of the two *p* orbitals decreases, giving a higher energy.

The structures and bonding in substituted alkenes and alkynes can be described through the combined molecular orbital (MO) and VB picture presented in Section 6.5, which uses localized VB bonds to describe the molecular framework and delocalized MOs to describe the π electrons. Let's apply this method to 2-butene ($CH_3CHCHCH_3$) and gain deeper understanding of the isomers discussed earlier. The Lewis diagram for 2-butene is

$$\begin{array}{c} \text{H H}\;\;\text{H H} \\ \text{H:}\ddot{\text{C}}\text{:}\ddot{\text{C}}\text{::}\ddot{\text{C}}\text{:}\ddot{\text{C}}\text{:H} \\ \text{H}\qquad\;\;\text{H} \end{array}$$

From the valence shell electron-pair repulsion (VSEPR) theory, the steric number of the two outer carbon atoms is 4 (so they are sp^3 hybridized), and that of the two central carbon atoms is 3 (sp^2 hybridized). The bonding around the outer carbon atoms is tetrahedral, and that about the central ones is trigonal planar. Each localized σ bond uses two electrons, resulting in a single bond between each pair of bonding atoms. In the case of 2-butene, these placements use 22 of the 24 available valence electrons, forming a total of 11 single bonds.

Next, the remaining *p* orbitals that were not involved in hybridization are combined to form π MOs. The p_z orbitals from the two central carbon atoms can be mixed to form a π (bonding) MO and a π^* (antibonding) MO. The remaining two valence electrons are placed into the π orbital, resulting in a double bond between the central carbon atoms. If the p_z orbital of one of these atoms is rotated about the central C—C axis, its overlap with the p_z orbital of the other carbon atom changes (Fig. 7.15). The overlap is greatest and the energy lowest when the two p_z orbitals are parallel to each other. In the most stable molecular geometry, the hydrogen atoms on the central carbon atoms lie in the same plane as the C—C—C—C carbon skeleton. This prediction is verified by experiment.

Figure 7.16 shows the structures of the isomers *cis*-2-butene and *trans*-2-butene. Converting one form to the other requires breaking the central π bond (by rotating the two p_z orbitals 180 degrees with respect to each other as in Fig. 7.15), then re-forming it in the other configuration. Because breaking a π bond costs a significant amount of energy, both *cis* and *trans* forms are stable at room temperature, and interconversion between the two is slow. The change in molecular structure from *trans* to *cis* can be accomplished by photochemistry

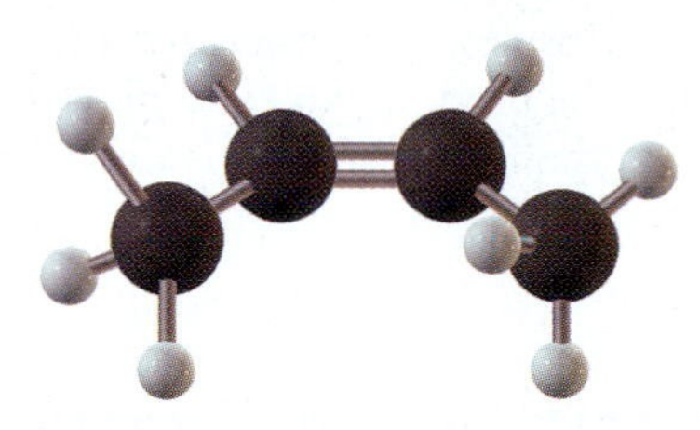

cis-2-butene

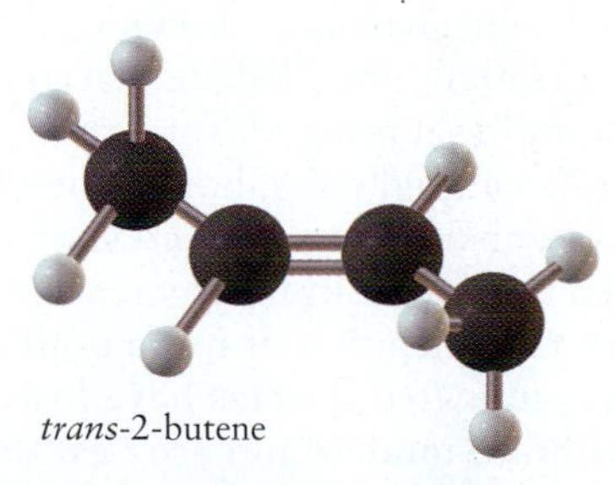

trans-2-butene

FIGURE 7.16 The two *cis–trans* isomers of 2-butene.

without complete breaking of bonds (see later). Molecules such as *trans*-2-butene can absorb ultraviolet light, which excites an electron from a π to a π^* MO. In the excited electronic state of *trans*-2-butene, the carbon–carbon double bond is effectively reduced to a single bond, and one CH_3 group can rotate relative to the other to form *cis*-2-butene.

EXAMPLE 7.2

The Lewis diagram for propyne (CH_3CCH) is

```
   H
   ..
H:C—C≡C:H
   ..
   H
```

Discuss its bonding and predict its geometry.

SOLUTION

The leftmost carbon atom in the structure is sp^3 hybridized, and the other two carbon atoms are sp hybridized. The atoms in the molecule are located on a single straight line, with the exception of the three hydrogen atoms on the leftmost carbon atom, which point outward toward three of the vertices of a tetrahedron. There is a σ bond between each pair of bonded atoms. The p_x and p_y orbitals on carbon atoms 2 and 3 combine to form two π orbitals and two π^* orbitals; only the former are occupied in the ground-state electron configuration.

Related Problems: 27, 28

Compounds with two double bonds are called *dienes,* those with three are called *trienes,* and so forth. The compound 1,3-pentadiene, for example, is a derivative of pentane with two double bonds:

$$CH_2{=}CH{-}CH{=}CH{-}CH_3$$

In such **polyenes,** each double bond may lead to *cis* and *trans* conformations, depending on its neighboring groups; therefore, several isomers may have the same bonding patterns but different molecular geometries and physical properties.

When two or more double or triple bonds occur close to each other in a molecule, a delocalized MO picture of the bonding should be used. As an example, let's examine 1,3-butadiene ($CH_2CHCHCH_2$), which has the following Lewis diagram

```
  H  H H  H
  ..  .. ..  ..
H:C::C:C::C:H
```

All four carbon atoms have steric number 3, so all are sp^2 hybridized. The remaining p_z orbitals have maximum overlap when the four carbon atoms lie in the same plane, so this molecule is predicted to be planar. From these four p_z atomic orbitals, four MOs can be constructed by combining their phases, as shown in Figure 7.17. The four electrons that remain after the σ orbitals are filled are placed in the two lowest π orbitals. The first of these is bonding among all four carbon atoms; the second is bonding between the outer carbon atom pairs, but antibonding between the central pair. Therefore, 1,3-butadiene has stronger and shorter bonds between the outer carbon pairs than between the two central carbon atoms. It is an example of a **conjugated π electron system,** in which two or more double or triple bonds alternate with single bonds. Such conjugated systems have lower energies than would be predicted from localized bond models and are best described with delocalized MOs extending over the entire π electron system.

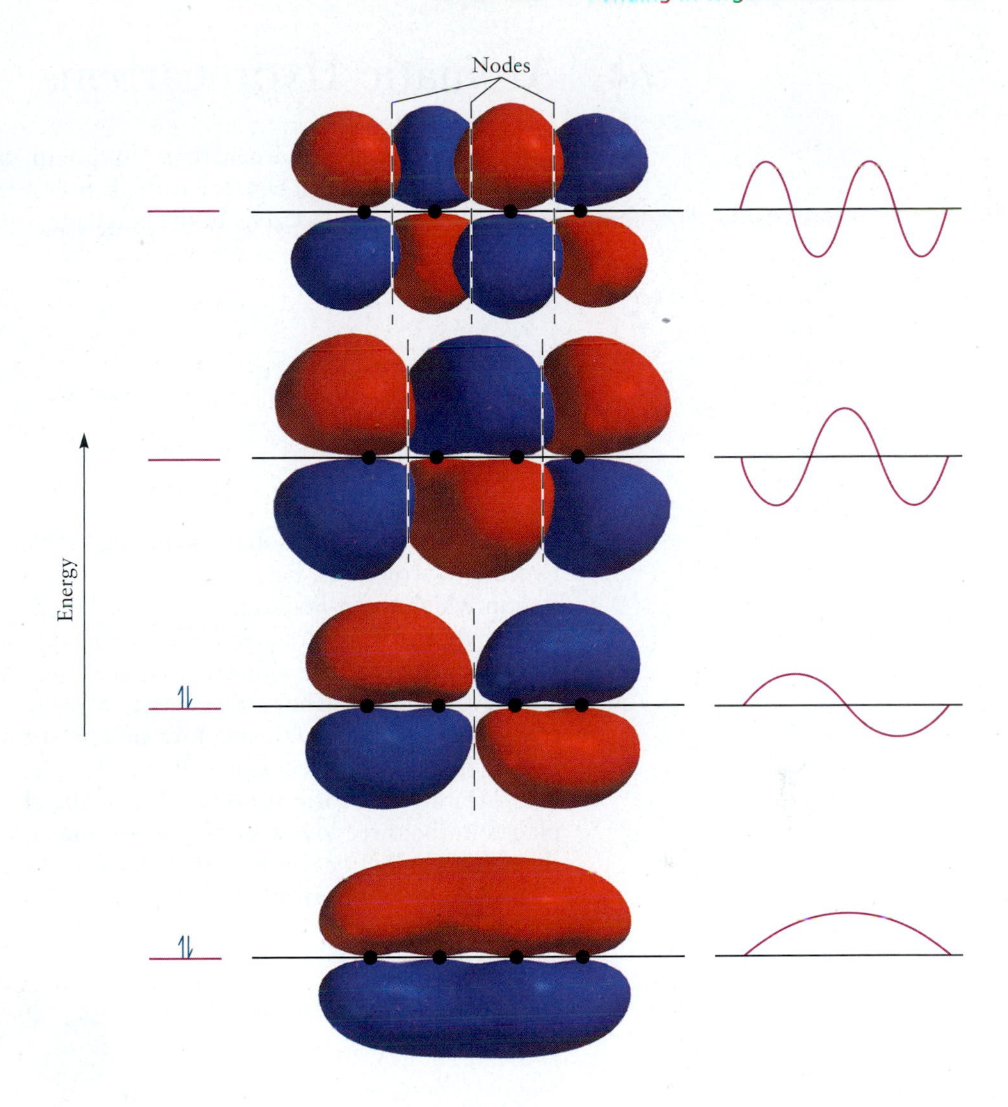

FIGURE 7.17 The four π molecular orbitals formed from four $2p_z$ atomic orbitals in 1,3-butadiene, viewed from the side. The dashed white lines represent nodes between the carbon atoms in the *y-z* plane. The horizontal black line represents the *x-y* nodal plane across which the *p* orbitals change sign. Note the similarity in the *y-z* nodal patterns to those of the first four modes of a vibrating string or the first four wave functions of the one-dimensional particle in a box (right). Only the lowest two orbitals are occupied in the ground state of 1,3-butadiene.

Alkenes are not present to a significant extent in crude petroleum. They are essential starting compounds for the synthesis of organic chemicals and polymers, so their production from alkanes is of great importance. One way to produce alkenes is by **cracking** the petroleum by heat or with catalysts. In **catalytic cracking,** the heavier fractions from the distillation column (compounds of C_{12} or higher) are passed over a silica–alumina catalyst at temperatures of 450°C to 550°C. Reactions such as

$$CH_3(CH_2)_{12}CH_3 \longrightarrow CH_3(CH_2)_4{-}CH{=}CH_2 + C_7H_{16}$$

occur to break the long chain into fragments. This type of reaction accomplishes two purposes. First, the shorter chain hydrocarbons have lower boiling points and can be added to gasoline. Second, the alkenes that result have higher octane numbers than the corresponding alkanes and perform better in the engine. Moreover, these alkenes can react with alkanes to give the more highly branched alkanes that are desirable in gasoline. **Thermal cracking** uses higher temperatures of 850°C to 900°C and no catalyst. It produces shorter chain alkenes, such as ethylene and propylene, through reactions such as

$$CH_3(CH_2)_{10}CH_3 \longrightarrow CH_3(CH_2)_8CH_3 + CH_2{=}CH_2$$

The short-chain alkenes are too volatile to be good components of gasoline, but they are among the most important starting materials for chemical synthesis.

7.4 Aromatic Hydrocarbons

One last group of hydrocarbons found in crude petroleum is the **aromatic hydrocarbons,** of which benzene is the simplest example. Benzene is a cyclic molecule with the formula C_6H_6. In the language of Chapter 3, benzene is represented as a resonance hybrid of two Lewis diagrams:

The modern view of resonant structures is that the molecule does not jump between two structures, but rather has a single, time-independent electron distribution in which the π bonding is described by delocalized MOs. Each carbon atom is sp^2 hybridized, and the remaining six p_z orbitals combine to give six MOs delocalized over the entire molecule. Figure 7.18 shows the π orbitals and their energy-level diagram. The lowest energy π orbital has no nodes, the next two have two nodes, the next two have four nodes, and the highest energy orbital has six nodes. The C_6H_6 molecule has 30 valence electrons, of which 24 occupy sp^2 hybrid orbitals and form σ bonds. When the six remaining valence electrons are placed in the three lowest energy π orbitals, the resulting electron distribution is the same in all six carbon–carbon bonds. As a result, benzene has six carbon–carbon bonds of equal length with properties intermediate between those of single and double bonds.

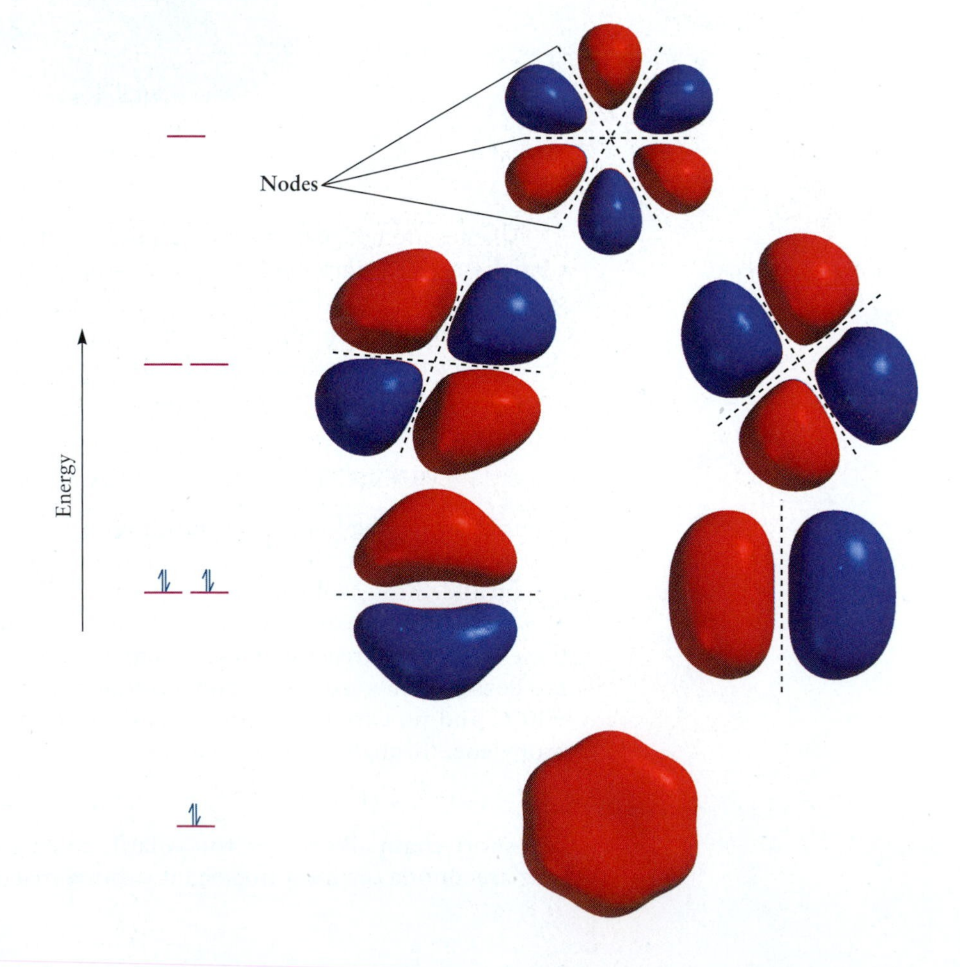

FIGURE 7.18 The six π molecular orbitals for benzene, viewed from the top, formed from the six $2p_z$ atomic orbitals lying perpendicular to the plane of the molecule. Note the similarity in nodal properties to the standing waves on a loop shown in Figure 4.18. Only the three lowest energy orbitals are occupied in molecules of benzene in the ground state.

Benzene is sometimes represented by its chemical formula C_6H_6 and sometimes (to show structure) by a hexagon with a circle inside it:

The six points of the hexagon represent the six carbon atoms, with the hydrogen atoms omitted for simplicity. The circle represents the de-localized π electrons, which are spread out evenly over the ring. The molecules of other aromatic compounds contain benzene rings with various side groups or two (or more) benzene rings linked by alkyl chains or fused side by side, as in naphthalene ($C_{10}H_8$):

Besides benzene, the most prevalent aromatic compounds in petroleum are toluene, in which one hydrogen atom on the benzene ring is replaced by a methyl group, and the xylenes, in which two such replacements are made:

CH_3 CH_3 CH_3 CH_3 CH_3 CH_3 CH_3

Toluene *o*-Xylene *m*-Xylene *p*-Xylene

This set of compounds is referred to as BTX (for *b*enzene-*t*oluene-*x*ylene). The BTX in petroleum is important to polymer synthesis (see Section 23.1). These components also significantly increase octane number and are used to make high-performance fuels with octane numbers above 100, as are required in modern aviation.

A major advance in petroleum refining has been the development of **reforming reactions,** which produce BTX aromatics from straight-chain alkanes that contain the same numbers of carbon atoms. A fairly narrow distillation fraction that contains only C_6 to C_8 alkanes is taken as the starting material. The reactions use high temperatures and transition-metal catalysts such as platinum or rhenium on alumina supports, and their detailed mechanisms are not fully understood. Apparently, a normal alkane such as hexane is cyclized and dehydrogenated to give benzene as the primary product (Fig. 7.19). Heptane yields mostly toluene, and octane yields a mixture of xylenes. Toluene is replacing benzene as a solvent in industrial applications because tests on laboratory animals show it to be far less carcinogenic (cancer-causing) than benzene. Benzene is more important for chemical synthesis; so a large fraction of the toluene produced is converted to benzene by **hydrodealkylation:**

CH_3

$+ H_2 \longrightarrow$ $+ CH_4$

Toluene Benzene

This reaction is conducted at high temperatures (550–650°C) and pressures of 40 to 80 atm.

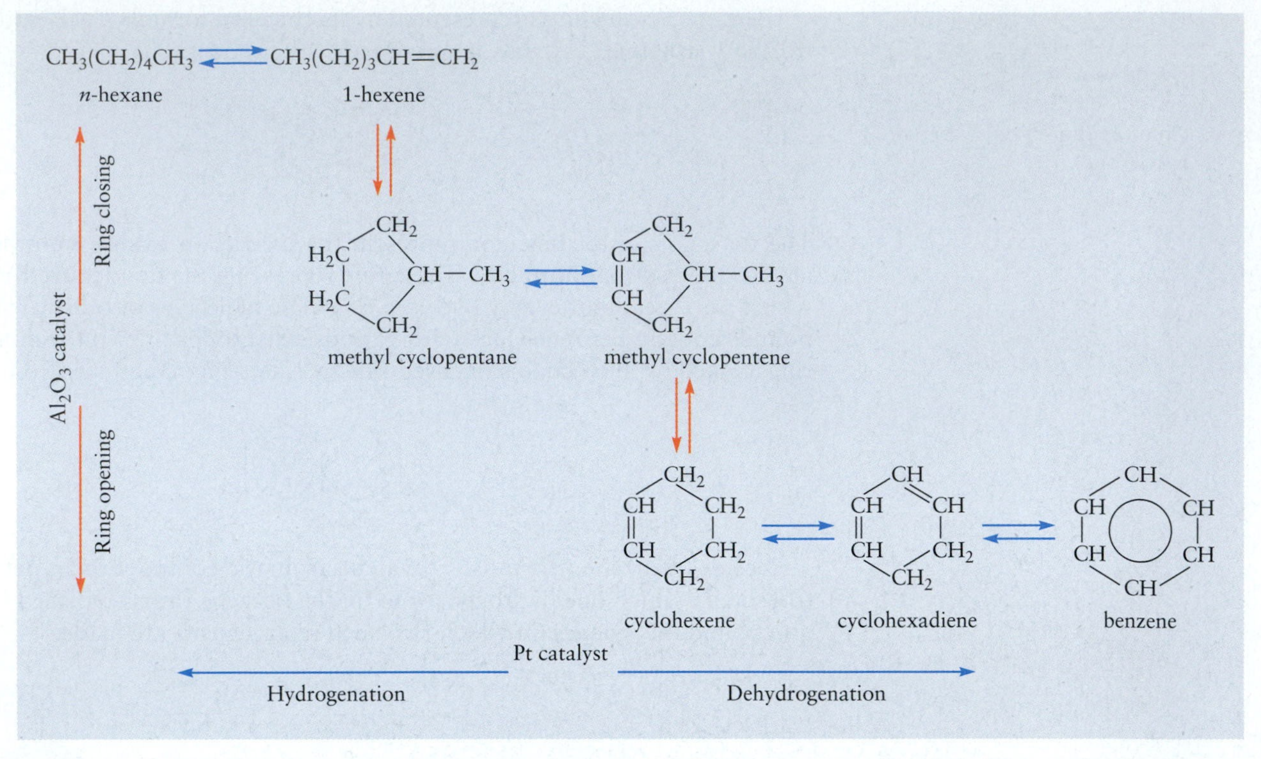

FIGURE 7.19 The reforming reaction that produces benzene from hexane uses a catalyst of platinum on alumina. The platinum facilitates the removal of hydrogen, and the alumina (Al_2O_3) facilitates the opening and closing of rings. One typical multistep sequence is shown; other intermediates can form as well.

7.5 Fullerenes

Among the most interesting conjugated π electron systems is a molecule discovered in 1985: buckminsterfullerene, C_{60}. Previously, only two forms of carbon (diamond and graphite) were known. In 1985, Harold Kroto, Robert Curl, and Richard Smalley were studying certain long-chain carbon molecules that had been discovered in the vicinity of red giant stars by radioastronomers using spectroscopy. They sought to duplicate the conditions near those stars by vaporizing a graphite target with a laser beam. Analysis of the products by mass spectrometry demonstrated not only the hoped-for molecules but a large proportion of molecules of molar mass 720 g mol^{-1}, which corresponds to the molecular formula C_{60}. Although the amounts of C_{60} present were far too small to isolate for direct determination of molecular structure, Kroto, Curl, and Smalley correctly suggested the cage structure shown in Figure 7.20a and named the molecule *buckminsterfullerene* after the architect Buckminster Fuller, the inventor of the geodesic dome, which the molecular structure of C_{60} resembles.

Molecules of C_{60} have a highly symmetric structure: 60 carbon atoms are arranged in a closed net with 20 hexagonal faces and 12 pentagonal faces. The pattern is exactly the design on the surface of a soccer ball (see Fig. 7.20b). Every carbon atom has a steric number of 3; all 60 atoms are sp^2 hybridized accordingly, although 1 of the 3 bond angles at each carbon atom must be distorted from the usual 120-degree sp^2 bond angle down to 108 degrees. The π electrons of the double bonds are de-localized: The 60 p orbitals (1 from each carbon atom) mix to give 60 MOs with amplitude spread over both the inner and outer surfaces of the molecule. The lowest 30 of these MOs are occupied by the 60 π electrons.

In 1990, scientists succeeded in synthesizing C_{60} in gram quantities by striking an electric arc between two carbon rods held under an inert atmosphere. The

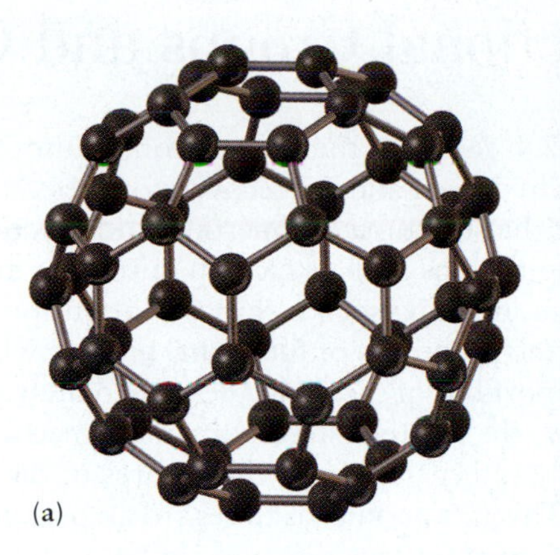
(a)

(b)

© Thomson Learning/George Semple

FIGURE 7.20 (a) The structure of C_{60}, buckminsterfullerene. Note the pattern of hexagons and pentagons. (b) The design on the surface of a soccer ball has the same pattern as the structure of C_{60}.

carbon vapors condensed to form soot, which was extracted with an organic solvent. Using chromatography, the scientists could separate the C_{60}—first as a solution of a delicate magenta hue (Fig. 7.21) and, finally, as a crystalline solid—from the various impurities in the growth batch. Since 1990, C_{60} has since been found in soot-forming flames when hydrocarbons are burned. Thus, the newest form of carbon has been (in the words of one of its discoverers) "under our noses since time immemorial." In 1994, the first buckminsterfullerene molecules were brought back from outer space in the form of the impact crater from a tiny meteorite colliding with an orbiting spacecraft.

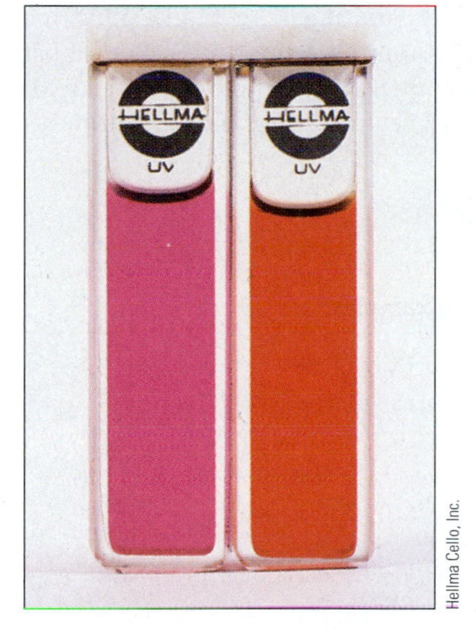

Hellma Cello, Inc.

FIGURE 7.21 (Left) A solution of C_{60} in an organic solvent has a delicate magenta color. (Right) The related fullerene C_{70} is distinctly orange in the same solvent.

Buckminsterfullerene is not the only new form of carbon to emerge from the chaos of carbon vapor condensing at high temperature. Synthesis of C_{60} simultaneously produces a whole family of closed-cage carbon molecules called **fullerenes.** All the fullerenes have even numbers of atoms, with formulas ranging up to C_{400} and higher. These materials offer exciting prospects for technical applications. For example, because C_{60} readily accepts and donates electrons to or from its π MOs, it has possible applications in batteries. It forms compounds (such as Rb_3C_{60}) that are superconducting (have zero resistance to the passage of an electric current) up to 30 K. Fullerenes also can encapsulate foreign atoms present during its synthesis. If a graphite disk is soaked with a solution of $LaCl_3$, dried, and used as a laser target, the substance La@C_{60} forms, where the symbol @ means that the lanthanum atom is trapped within the 60-atom carbon cage.

Condensation of the carbon vapor under certain conditions favors the formation of *nanotubes,* which consist of seamless, cylindrical shells of thousands of sp^2 hybridized carbon atoms arranged in hexagons. The ends of the tubes are capped by pentagons inserted into the hexagonal network. These structures all have a delocalized π-electron system that covers the inner and outer surfaces of the cage or cylinder. Nanotubes also offer exciting prospects for material science and technological applications. For example, the mechanical properties of nanotubes suggest applications as high-strength fibers (Fig. 7.22).

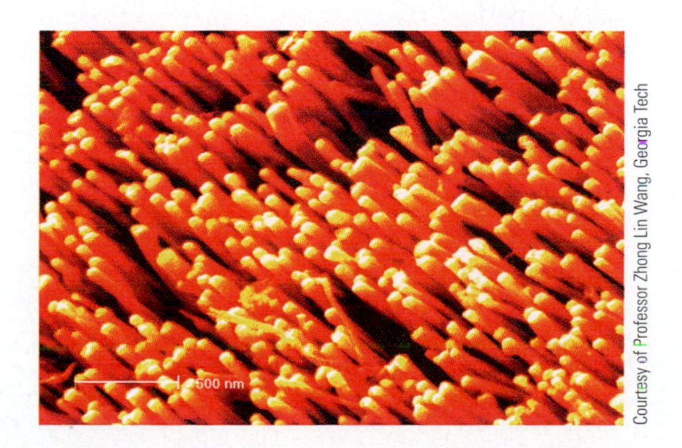

Courtesy of Professor Zhong Lin Wang, Georgia Tech

FIGURE 7.22 A bundle of carbon nanotubes, each about 1.4 nm in diameter. The bundle is 10 to 20 nm in thickness. Note that the tubes are packed in a triangular arrangement (or, alternatively, six tubes are arranged hexagonally around a central tube.)

7.6 Functional Groups and Organic Reactions

Section 7.1 describes the diverse compounds formed from the two elements carbon and hydrogen and the recovery of these compounds from petroleum. We now consider the structures, properties, and reactions of molecules formed by adding substituent atoms such as oxygen, nitrogen, and the halogens to the hydrocarbon backbone. In doing so, we shift our attention from the structures of entire molecules to the properties of **functional groups,** which consist of the noncarbon atoms plus the portions of the molecule immediately adjacent to them. Functional groups tend to be the reactive sites in organic molecules, and their chemical properties depend only rather weakly on the natures of the hydrocarbons to which they are attached. This fact permits us to regard an organic molecule as a hydrocarbon frame, which mainly governs size and shape, to which are attached functional groups that mainly determine the chemistry of the molecule. Table 7.3 shows some of the most important functional groups.

Industrial processes for the high-volume organic chemicals produced today require starting materials with appropriate functional groups. A small number of hydrocarbon building blocks from petroleum and natural gas (methane, ethylene, propylene, benzene, and xylene) are the starting points for the synthesis of most of these starting materials. This section introduces the common functional groups, describes their bonding, and illustrates typical reactions in their synthesis or applications.

TABLE 7.3 Common Functional Groups

Functional Group†	Type of Compound	Examples
R—F, —Cl, —Br, —I	Alkyl or aryl halide	CH_3CH_2Br (bromoethane)
R—OH	Alcohol	CH_3CH_2OH (ethanol)
	Phenol	C_6H_5—OH (phenol)
R—O—R′	Ether	$H_3C—O—CH_3$ (dimethyl ether)
R—C(=O)—H	Aldehyde	$CH_3CH_2CH_2—C(=O)—H$ (butyraldehyde, or butanal)
R(R′)C=O	Ketone	$H_3C—C(=O)—CH_3$ (propanone, or acetone)
R—C(=O)—OH	Carboxylic acid	CH_3COOH (acetic acid, or ethanoic acid)
R—C(=O)—O—R′	Ester	$H_3C—C(=O)—O—CH_3$ (methyl acetate)
R—NH_2	Amine	CH_3NH_2 (methylamine)
R—C(=O)—N(R′)(R″)	Amide	$H_3C—C(=O)—NH_2$ (acetamide)

†The symbols R, R′, and R″ stand for hydrocarbon radicals. In some cases, they may represent hydrogen.

Halides

One of the simplest functional groups consists of a single halogen atom, which we take to be chlorine for illustrative purposes. The chlorine atom forms a σ bond to a carbon atom by overlap of its $3p_z$ orbital with a hybridized orbital on the carbon. The hybridized orbital may be sp^3, sp^2, or sp depending on the bonding in the hydrocarbon frame. **Alkyl halides** form when mixtures of alkanes and halogens (except iodine) are heated or exposed to light.

$$\underset{\text{Methane}}{CH_4} + \underset{\text{Chlorine}}{Cl_2} \xrightarrow{250°C\text{–}400°C \text{ or light}} \underset{\text{Chloromethane}}{CH_3Cl} + \underset{\text{Hydrogen chloride}}{HCl}$$

The mechanism is a chain reaction (see description in Section 18.4). Ultraviolet light initiates the reaction by dissociating a small number of chlorine molecules into highly reactive atoms. These atoms take part in linked reactions of the following form:

$$Cl\cdot + CH_4 \longrightarrow HCl + \cdot CH_3 \qquad \text{(propagation)}$$

$$\cdot CH_3 + Cl_2 \longrightarrow CH_3Cl + Cl\cdot \qquad \text{(propagation)}$$

The chlorine atoms and the methyl species are called free radicals, and are denoted by the dots next to the chemical symbols. A **free radical** is a chemical species that contains an odd (unpaired) electron; it is usually formed by breaking a covalent bond to form a pair of such species. Radicals are often highly reactive and often appear as intermediates in reactions. Chloromethane (also called methyl chloride) is used in synthesis to add methyl groups to organic molecules. If sufficient chlorine is present, more highly chlorinated methanes form, providing a route for the industrial synthesis of dichloromethane (CH_2Cl_2, also called methylene chloride), trichloromethane ($CHCl_3$, chloroform), and tetrachloromethane (CCl_4, carbon tetrachloride). All three chloromethanes are used as solvents and have vapors with anesthetic or narcotic effects; environmental and health concerns about toxicity have reduced the use of these compounds in everyday life.

Adding chlorine to C=C bonds is a more important industrial route to alkyl halides than the free-radical reactions just described. Billions of kilograms of 1,2-dichloroethane (commonly called ethylene dichloride) are manufactured each year, making this compound the largest volume derived organic chemical. It is made by adding chlorine to ethylene over an iron(III) oxide catalyst at moderate temperatures (40–50°C), either in the vapor phase or in a solution of 1,2-dibromoethane:

$$CH_2{=}CH_2 + Cl_2 \longrightarrow ClCH_2CH_2Cl$$

Almost all of the 1,2-dichloroethane produced is used to make chloroethylene (vinyl chloride, $CH_2{=}CHCl$). This is accomplished by heating the 1,2-dichloroethane to 500°C over a charcoal catalyst to abstract HCl:

$$ClCH_2CH_2Cl \longrightarrow CH_2{=}CHCl + HCl$$

The HCl can be recovered and converted to Cl_2 for further production of 1,2-dichloroethane from ethylene. Vinyl chloride has a much lower boiling point than 1,2-dichloroethane (−13°C compared with 84°C), so the two are easily separated by fractional distillation. Vinyl chloride is used in the production of polyvinyl chloride plastic (see Section 23.1).

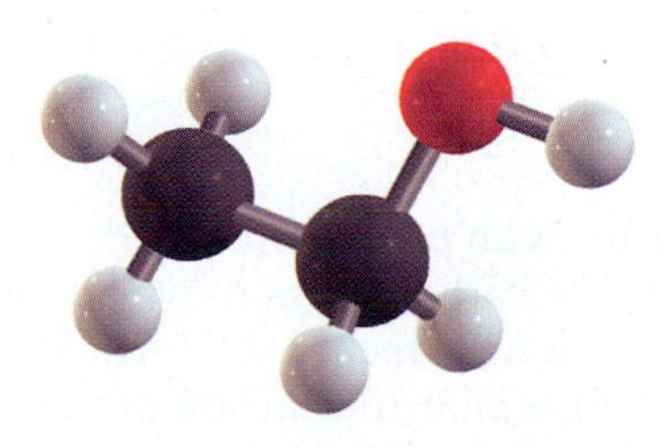

$$CH_3CH_2\ddot{\underset{\cdot\cdot}{O}}H$$

An alcohol (Ethanol)

Alcohols and Phenols

Alcohols have the —OH functional group attached to a tetrahedral carbon atom, that is, a carbon atom with single bonds to four other atoms. The carbon atom in the functional group is sp^3 hybridized. The oxygen atom is likewise sp^3 hybridized; two of the hybrid orbitals form σ bonds, whereas the other two hold lone pairs of electrons.

The simplest alcohol is methanol (CH_3OH), which is made from methane in a two-step process. The first is the **reforming reaction**

$$CH_4(g) + H_2O(g) \longrightarrow CO(g) + 3\,H_2(g)$$

conducted at high temperatures (750–1000°C) with a nickel catalyst. The gas mixture that results, called **synthesis gas,** reacts directly to form methanol at 300°C.

$$CO(g) + 2\,H_2(g) \longrightarrow CH_3OH(g)$$

The next higher alcohol, ethanol (CH_3CH_2OH), can be produced from the fermentation of sugars. Although fermentation is the major source of ethanol for alcoholic beverages and for "gasohol" (automobile fuel made up of 90% gasoline and 10% ethanol), it is not significant for industrial production, which relies on the direct hydration of ethylene:

$$CH_2{=}CH_2 + H_2O \longrightarrow CH_3CH_2OH$$

Temperatures of 300°C to 400°C and pressures of 60 to 70 atm are used with a phosphoric acid catalyst. Both methanol and ethanol are used widely as solvents and as intermediates for further chemical synthesis.

Two three-carbon alcohols exist, depending on whether the —OH group is attached to a terminal carbon atom or the central carbon atom. They are 1-propanol and 2-propanol:

$$\underset{\text{1-Propanol}}{CH_3CH_2CH_2OH} \qquad \underset{\text{2-Propanol}}{\underset{\displaystyle OH}{CH_3\overset{}{C}HCH_3}}$$

1-Propanol and 2-propanol are commonly referred to as *n*-propyl alcohol and isopropyl alcohol, respectively. The systematic names of alcohols are obtained by replacing the *-ane* ending of the corresponding alkane with *-anol* and using a numeric prefix, when necessary, to identify the carbon atom to which the —OH group is attached.[1] Isopropyl alcohol is made from propylene by means of an interesting hydration reaction that is catalyzed by sulfuric acid. The first step is addition of H^+ to the double bond,

$$H_3C{-}CH{=}CH_2 + H^+ \longrightarrow H_3C{-}\overset{+}{C}H{-}CH_3$$

producing a transient charged species in which the positive charge is centered on the central carbon atom. Attack by negative HSO_4^- occurs at this positive site and leads to a neutral intermediate that can be isolated:

$$H_3C{-}\underset{\displaystyle OSO_3H}{\underset{|}{C}H}{-}CH_3$$

Further reaction with water then causes the replacement of the $-OSO_3H$ group with an —OH group and the regeneration of the sulfuric acid:

$$H_2O + H_3C{-}\underset{\displaystyle OSO_3H}{\underset{|}{C}H}{-}CH_3 \longrightarrow H_3C{-}\underset{\displaystyle OH}{\underset{|}{C}H}{-}CH_3 + H_2SO_4$$

This mechanism explains why only 2-propanol forms, with no 1-propanol.

The compound 1-propanol is a **primary alcohol:** The carbon atom to which the —OH group is bonded has exactly one other carbon atom attached to it. The isomeric compound 2-propanol is a **secondary alcohol** because the carbon atom to which the —OH group is attached has two carbon atoms (in the two methyl

[1]Contrast the names of alcohols with the corresponding names of alkyl halides. If the —OH group were replaced by a chlorine atom, the names of these compounds would be 1-chloropropane and 2-chloropropane.

groups) attached to it. The simplest **tertiary alcohol** (in which the carbon atom attached to the —OH group is also bonded to three other carbon atoms) is 2-methyl-2-propanol:

$$\begin{array}{c} OH \\ | \\ H_3C{-}CH{-}CH_3 \\ | \\ CH_3 \end{array}$$

Primary, secondary, and tertiary alcohols differ in chemical properties.

Phenols are compounds in which an —OH group is attached directly to an aromatic ring. The simplest example is phenol itself (C_6H_5OH). As in the alcohols, the oxygen atom is sp^3 hybridized with two unshared pairs. The carbon atom, which is part of the aromatic ring, is sp^2 hybridized (see Section 7.4).

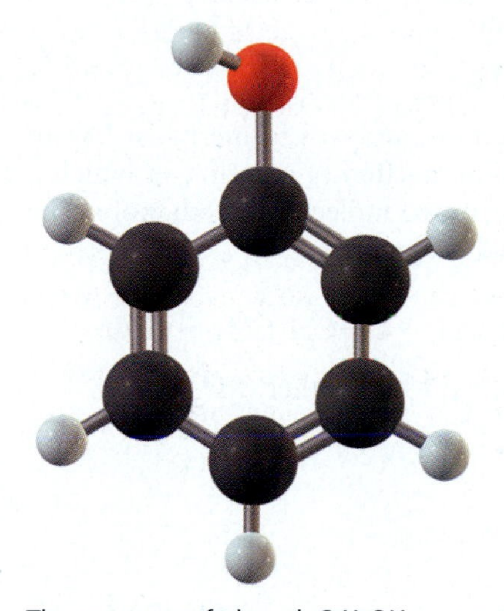

The structure of phenol, C_6H_5OH.

The manufacture of phenols uses quite different types of reactions from those used to make alcohols. One method, introduced in 1924 and still used to a small extent today, involves the chlorination of the benzene ring followed by reaction with sodium hydroxide:

$$C_6H_6 + Cl_2 \longrightarrow HCl + C_6H_5Cl$$

$$C_6H_5Cl + 2\ NaOH \longrightarrow C_6H_5O^-Na^+ + NaCl + H_2O$$

This approach illustrates a characteristic difference between the reactions of aromatics and alkenes. When chlorine reacts with an alkene, it *adds* across the double bond (as shown in the production of 1,2-dichloroethane). When an aromatic ring is involved, substitution of chlorine for hydrogen occurs instead and the aromatic π-bonding structure is preserved.

A different approach is used to make almost all phenol today. It involves, first, the acid-catalyzed reaction of benzene with propylene to give cumene, or isopropyl benzene:

$$C_6H_6 + \begin{array}{c} CH_2 \\ \| \\ C \\ H_3C \quad H \end{array} \xrightarrow{H^+} \begin{array}{c} CH_3 \\ | \\ C_6H_5{-}C{-}H \\ | \\ CH_3 \end{array}$$

Cumene

As in the production of 2-propanol, the first step is the addition of H^+ to propylene to give $CH_3{-}CH^+{-}CH_3$ (see page 294). This ion then attaches to the benzene ring through its central carbon atom to give the cumene and regenerate H^+. Subsequent reaction of cumene with oxygen (Fig. 7.23) gives phenol and acetone, an important compound that is discussed later in this section. The main use of phenol is in the manufacturing of polymers and aspirin.

$$O_2 + \text{Cumene} \longrightarrow \text{Cumene hydroperoxide} \xrightarrow{H_2SO_4} \text{Phenol} + \underset{\text{Acetone}}{H_3C{-}\overset{O}{\overset{\|}{C}}{-}CH_3}$$

FIGURE 7.23 The production of phenol and acetone from cumene is a two-step process that involves insertion of O_2 to make a peroxide, followed by acid-catalyzed migration of the —OH group to form the products.

(a)

```
    H     H
    |  ..  |
H—C—O—C—H
    |  ..  |
    H     H
```

(b)

FIGURE 7.24 Structure of dimethyl ether. (a) Lewis diagram. (b) Ball-and-stick model.

Ethers

Ethers are characterized by the —O— functional group, in which an oxygen atom provides a link between two separate alkyl or aromatic groups. Figure 7.24 shows the simplest ether, dimethyl ether. The oxygen atom is sp^3 hybridized. Two of these hybrid orbitals form σ bonds to the carbon atoms, whereas each of the other two holds an unshared pair. The C—O—C bond angle is 110.3 degrees, which is close to the tetrahedral value 109.5 degrees predicted by sp^3 hybridization.

One important ether is diethyl ether, often called simply ether, in which two ethyl groups are linked to the same oxygen atom:

$$C_2H_5{-}O{-}C_2H_5$$

Diethyl ether is a useful solvent for organic reactions, and was formerly used as an anesthetic. It can be produced by a **condensation reaction** (a reaction in which a small molecule such as water is split out) between two molecules of ethanol in the presence of concentrated sulfuric acid as a dehydrating agent:

$$CH_3CH_2OH + HOCH_2CH_3 \xrightarrow{H_2SO_4} CH_3CH_2{-}O{-}CH_2CH_3 + H_2O$$

Another ether of considerable importance is methyl *t*-butyl ether (MTBE):

```
          CH3
           |
H3C—O—C—CH3
           |
          CH3
```

This compound appeared to be a successful replacement for tetraethyllead as an additive to gasolines to increase their octane ratings. The bond between the oxygen and the *t*-butyl group is weak, and it breaks to form radicals that assist the smooth combustion of gasoline. MTBE is readily soluble in water, and has appeared in drinking water supplies through leaks from underground storage tanks for gasoline. Concern over possible health risks has caused MTBE to be phased out in various regions of the United States.

In a cyclic ether, oxygen forms part of a ring with carbon atoms, as in the common solvent tetrahydrofuran (Fig. 7.25). The smallest such ring has two carbon atoms bonded to each other and to the oxygen atom; it occurs in ethylene oxide,

```
      O
     / \
  H2C—CH2
```

which is made by direct oxidation of ethylene over a silver catalyst:

```
                                   O
                                  / \
H2C=CH2  +  1/2 O2  --Ag-->  H2C—CH2
```

Such ethers with three-membered rings are called **epoxides.** The major use of ethylene oxide is in the preparation of ethylene glycol:

```
    O
   / \
H2C—CH2  +  H2O  ---->  HO—H2C—CH2—OH
```

This reaction is conducted either at 195°C under pressure or at lower temperatures (50–70°C) with sulfuric acid as a catalyst. Ethylene glycol is a dialcohol, or **diol,** in which two —OH groups are attached to adjacent carbon atoms. Its primary use is as antifreeze to decrease the freezing point of water in automobile radiators.

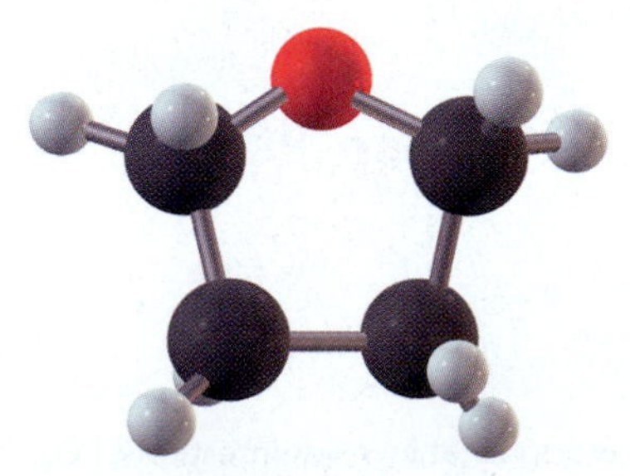

FIGURE 7.25 The structure of tetrahydrofuran, C_4H_8O.

Aldehydes and Ketones

An **aldehyde** contains the characteristic —C(=O)—H functional group in its molecules. Figure 7.26 shows the bonding in formaldehyde, which is the simplest organic

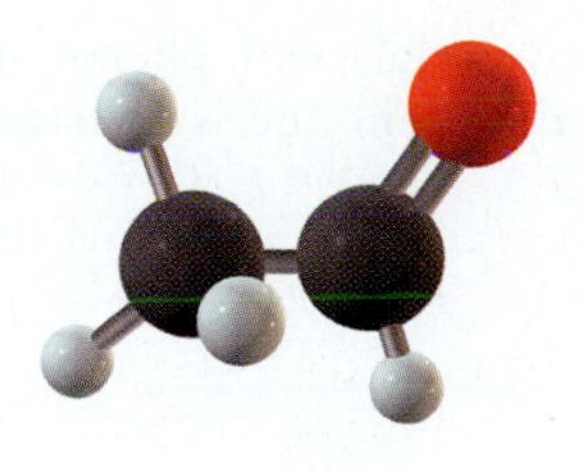

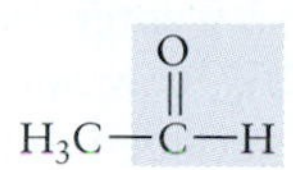

Acetaldehyde
(an aldehyde)

molecule with a double bond between carbon and oxygen. The carbon is sp^2 hybridized and forms σ bonds to two hydrogen atoms. The oxygen is also sp^2 hybridized. Carbon and oxygen form one σ bond by overlap of sp^2 orbitals and one π bond by overlap of parallel nonhybridized $2p$ orbitals. This is the same bonding model used to describe C=C double bonds in Figure 7.12.

Aldehydes can be prepared by the dehydrogenation of a primary alcohol. Formaldehyde results from the dehydrogenation of methanol at high temperatures with an iron oxide–molybdenum oxide catalyst:

$$\mathrm{CH_3OH} \longrightarrow \mathrm{H_2C{=}O} + \mathrm{H_2}$$

Another reaction that gives the same primary product is the oxidation reaction

$$\mathrm{CH_3OH} + \tfrac{1}{2}\mathrm{O_2} \longrightarrow \mathrm{H_2C{=}O} + \mathrm{H_2O}$$

Formaldehyde is readily soluble in water, and a 40% aqueous solution called *formalin* is used to preserve biological specimens. It is a component of wood smoke and helps to preserve smoked meat and fish, probably by reacting with nitrogen-containing groups in the proteins of attacking bacteria. Its major use is in making polymer adhesives and insulating foam.

The next aldehyde in the series is acetaldehyde, the structure of which is shown above. Industrially, acetaldehyde is produced not from ethanol but by the oxidation of ethylene, using a $PdCl_2$ catalyst.

Ketones have the $-\overset{\mathrm{O}}{\overset{\|}{\mathrm{C}}}-$ functional group in which a carbon atom forms a double bond to an oxygen atom and single bonds to two separate alkyl or aromatic groups. The simplest ketone is acetone, in which two methyl groups are bonded to the central carbon. The bonding scheme is the same as that in Figure 7.26, with alkyl or aromatic groups replacing hydrogen in the two single bonds. Such compounds can be prepared by dehydrogenation or oxidation of secondary alcohols, just as aldehydes come from primary alcohols. Acetone is made by the dehydrogenation of 2-propanol over a copper oxide or zinc oxide catalyst at 500°C:

$$\mathrm{H_3C{-}\underset{}{\overset{\mathrm{OH}}{\overset{|}{\mathrm{C}}}}H{-}CH_3} \longrightarrow \mathrm{H_3C{-}\overset{\mathrm{O}}{\overset{\|}{\mathrm{C}}}{-}CH_3} + \mathrm{H_2}$$

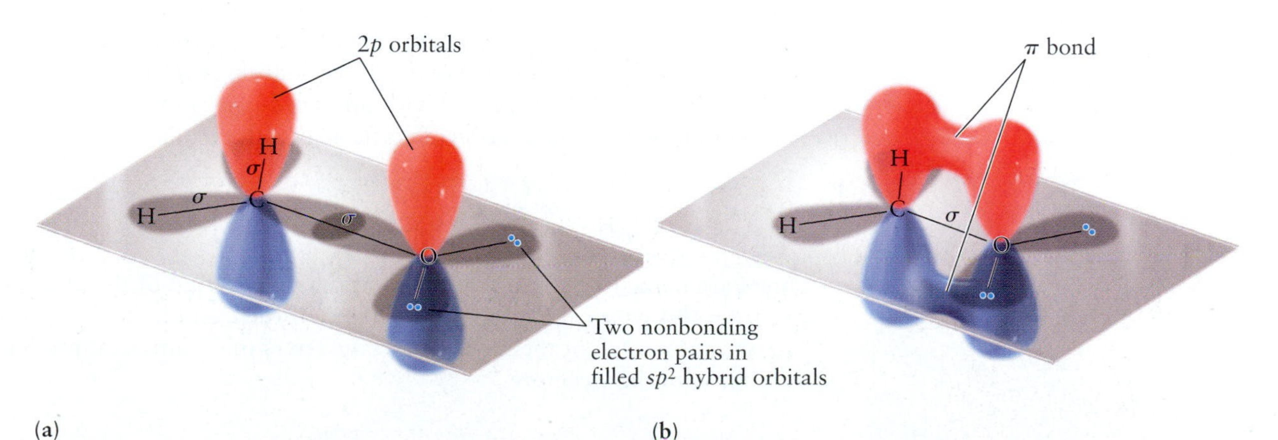

FIGURE 7.26 Bond formation in formaldehyde involves sp^2 hybridization of the carbon (C) and oxygen (O) atoms. (a) The σ bond framework and the parallel nonhybridized $2p$ orbitals on C and O. (b) Overlap of the parallel $2p$ orbitals to form a π bond.

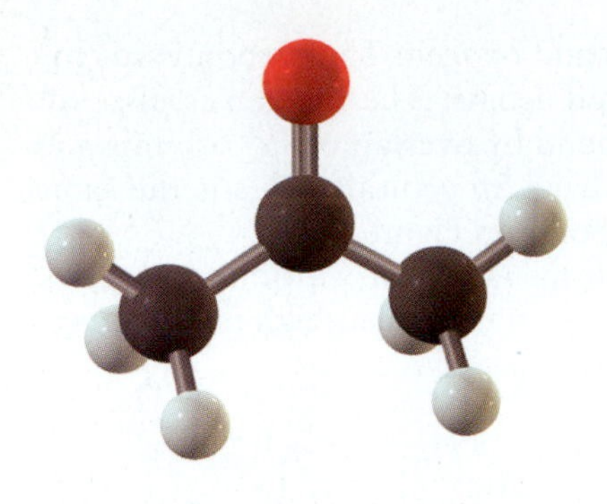

$H_3C-C(=O)-CH_3$

Acetone
(a ketone)

Acetone is also produced (in greater volume) as the coproduct with phenol of the oxidation of cumene (see earlier). It is a widely used solvent and is the starting material for the synthesis of a number of polymers.

Carboxylic Acids and Esters

Carboxylic acids contain the $-C(=O)-OH$ functional group (also written as $-COOH$). The bonding scheme is a variation of that in Figure 7.26, in which the doubly bonded oxygen atom is sp^2 hybridized and the singly bonded oxygen is sp^3 hybridized with unshared pairs in two of the hybrid orbitals. Carboxylic acids are the products of the oxidation of aldehydes, just as aldehydes are the products of the oxidation of primary alcohols. (The turning of wine to vinegar is a two-step oxidation leading from ethanol through acetaldehyde to acetic acid.) Industrially, acetic acid can be produced by the air oxidation of acetaldehyde over a manganese acetate catalyst at 55°C to 80°C:

$$H_3CC(=O)H + \tfrac{1}{2}O_2 \xrightarrow{Mn(CH_3COO)_2} H_3CC(=O)OH$$

The reaction now preferred on economic grounds for acetic acid production is the combination of methanol with carbon monoxide (both derived from natural gas) over a catalyst that contains rhodium and iodine. The overall reaction is

$$CH_3OH(g) + CO(g) \xrightarrow{Rh,I_2} CH_3COOH(g)$$

and can be described as a **carbonylation,** or the insertion of CO into the methanol C—O bond.

Acetic acid is a member of a series of carboxylic acids with formulas $H-(CH_2)_n-COOH$. Before acetic acid (with $n = 1$) comes the simplest of these carboxylic acids, formic acid (HCOOH), in which $n = 0$. This compound was first isolated from extracts of the crushed bodies of ants, and its name stems from the Latin word *formica,* meaning "ant." Formic acid is the strongest acid of the series, and acid strength decreases with increasing length of the hydrocarbon chain. (See Section 15.9.) The longer chain carboxylic acids are called fatty acids. Sodium stearate, the sodium salt of stearic acid, $CH_3(CH_2)_{16}COOH$, is a typical component of soap. It cuts grime by simultaneously interacting with grease particles at its hydrocarbon tail and with water at its carboxylate ion end group to make the grease soluble in water.

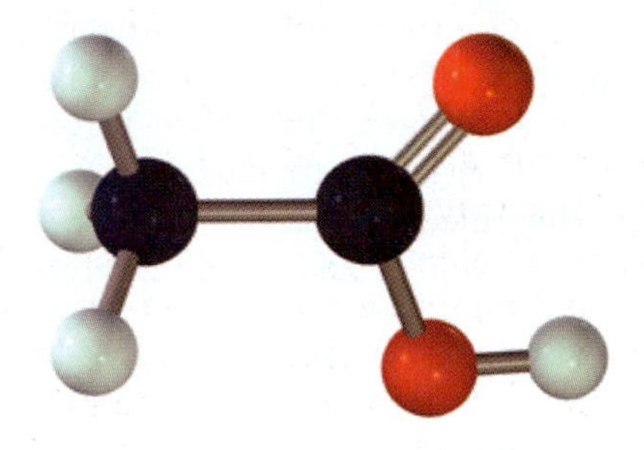

$H_3C-C(=O)-O-H$

Acetic acid
(a carboxylic acid)

Carboxylic acids react with alcohols or phenols to give **esters,** forming water as the coproduct. The bonding scheme in esters is a variation of that shown in Figure 7.26, in which the doubly bonded oxygen atom is sp^2 hybridized and the singly bonded oxygen is sp^3 hybridized with unshared pairs in two of the hybrid orbitals. An example is the condensation of acetic acid with methanol to give methyl acetate:

$$H_3CC(=O)-\boxed{OH + H}OCH_3 \longrightarrow H_3CC(=O)-OCH_3 + H_2O$$

Esters are named by stating the name of the alkyl group of the alcohol (the methyl group in this case), followed by the name of the carboxylic acid with the ending *-ate* (acetate). One of the most important esters in commercial production is vinyl acetate, with the structure

$$H_3C-C(=O)-O-CH{=}CH_2$$

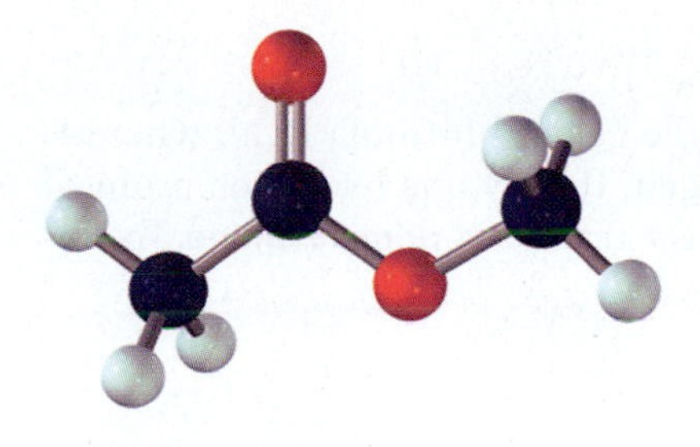

$$\mathrm{H_3C{-}\overset{\overset{\displaystyle O}{\|}}{C}{-}O{-}CH_3}$$

Methyl acetate
(an ester)

Despite its name, it is not prepared by the reaction of acetic acid with an alcohol but rather with ethylene and oxygen over a catalyst such as $CuCl_2$ and $PdCl_2$:

$$\mathrm{H_3C\overset{O}{\overset{/\!/}{C}}{-}OH + CH_2{=}CH_2 + \tfrac{1}{2}O_2 \xrightarrow{CuCl_2} H_3C\overset{O}{\overset{/\!/}{C}}{-}O{-}CH{=}CH_2 + H_2O}$$

Esters are colorless, volatile liquids that often have pleasant odors. Many occur naturally in flowers and fruits. Isoamyl acetate (Fig. 7.27a) is generated in apples as they ripen and contributes to the flavor and odor of the fruit. Benzyl acetate, the ester formed from acetic acid and benzyl alcohol (see Fig. 7.27b), is a major component of oil of jasmine and is used in the preparation of perfumes.

Animal fats and vegetable oils are triesters of long-chain carboxylic acids with glycerol, $HOCH_2CH(OH)CH_2OH$, a trialcohol; they are referred to as **triglycerides.** These are energy-storage molecules of biological origin. A large proportion of sunflower oil is an oily liquid composed of molecules with the structural formula

$$\begin{array}{l}
\mathrm{CH_3(CH_2)_4CH{=}CHCH_2CH{=}CH(CH_2)_7\overset{\overset{\displaystyle O}{\|}}{C}{-}OCH_2} \\
\mathrm{\;CH_3(CH_2)_4CH{=}CHCH_2CH{=}CH(CH_2)_7COOCH_2} \\
\mathrm{CH_3(CH_2)_4CH{=}CHCH_2CH{=}CH(CH_2)_7\underset{\underset{\displaystyle O}{\|}}{C}{-}OCH_2}
\end{array}$$

This is a **polyunsaturated oil** because there are six C=C bonds per molecule. Butter is a mixture of triglycerides, many of which are **saturated** because their hydrocarbon chains contain no double bonds. Hydrogen is used in food processing to convert unsaturated liquid vegetable oils to saturated solids. **Hydrogenation** of sunflower oil with 6 mol H_2 in the presence of a catalyst saturates it, and the product has a high enough melting point to make it a solid at room conditions. The use of solid fats (or solidified oils) has advantages in food processing and preservation; therefore, "hydrogenated vegetable oil" is an ingredient in many foodstuffs. Heavy consumption of saturated fats has been linked to diseases of the heart and circulatory system, and the presence of *un*hydrogenated (polyunsaturated) oils in foods is now extensively advertised. *Trans* fats, which have only one double bond with hydrocarbon chains in the *trans* configuration, have recently been shown to be as harmful as saturated fats, and a ban on their use in food preparation is under consideration. When a triglyceride is hydrolyzed through addition of sodium hydroxide, the ester bonds are broken and glycerol and sodium salts of long-chain carboxylic acids are produced. This reaction is the basis for traditional soap making through the addition of lye (sodium hydroxide) to animal fats.

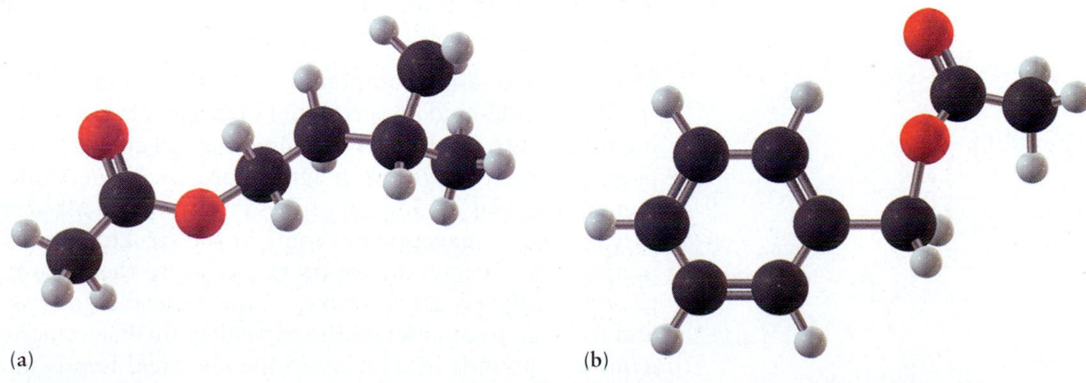

FIGURE 7.27 The structures of (a) isoamyl acetate, $CH_3COO(CH_2)_2CH(CH_3)_2$, and (b) benzyl acetate, $CH_3COOCH_2C_6H_5$.

Amines and Amides

The **amines** are derivatives of ammonia with the general formula R_3N, where R can represent a hydrocarbon group or hydrogen. If only one hydrogen atom of ammonia is replaced by a hydrocarbon group, the result is a **primary amine.** Examples are ethylamine and aniline:

Ethylamine: $(H)_2N{-}C_2H_5$ Aniline: $C_6H_5NH_2$ Dimethylamine: $H{-}N(CH_3)_2$ Trimethylamine: $H_3C{-}N(CH_3)_2$

If two hydrocarbon groups replace hydrogen atoms in the ammonia molecule, the compound is a **secondary amine** (such as dimethylamine), and three replacements make a **tertiary amine** (trimethylamine). You should draw Lewis dot diagrams for several amines and recognize that the nitrogen is sp^3 hybridized and forms three σ bonds with one unshared pair in a hybridized orbital. Amines are bases because the lone electron pair on the nitrogen can accept a hydrogen ion in the same way that the lone pair on the nitrogen in ammonia does (see Chapter 15).

A primary or secondary amine (or ammonia itself) can react with a carboxylic acid to form an **amide.** This is another example of a condensation reaction and is analogous to the formation of an ester from reaction of an alcohol with a carboxylic acid. An example of amide formation is

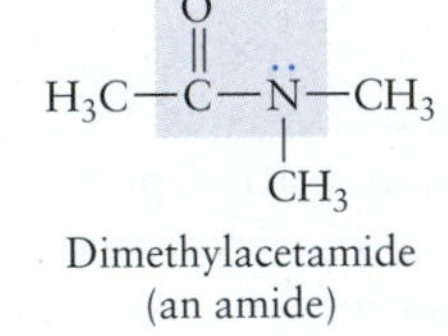

Dimethylacetamide (an amide)

$$H_3CC({=}O){-}\boxed{OH + H}{-}N(CH_3)_2 \longrightarrow H_3CC({=}O){-}N(CH_3)_2 + H_2O$$

If ammonia is the reactant, an $-NH_2$ group replaces the $-OH$ group in the carboxylic acid as the amide is formed:

$$H_3CC({=}O){-}OH + NH_3 \longrightarrow \underset{\text{Acetamide}}{H_3CC({=}O){-}NH_2} + H_2O$$

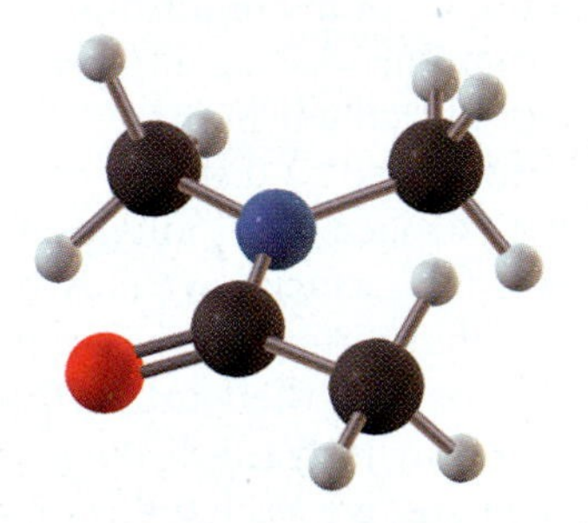

FIGURE 7.28 Bonding in dimethyl acetamide, an amide. The amide linkage is planar.

Amide linkages are present in the backbone of every protein molecule and are very important in biochemistry, where the structure of the molecule strongly influences its function (see Section 23.4). The bonding scheme in amides is a variation of that in Figure 7.26, in which the doubly bonded carbon and oxygen atoms are sp^2 hybridized and the singly bonded nitrogen atom is sp^3 hybridized with an unshared pair in one of the hybrid orbitals. As shown by the ball-and-stick model in Figure 7.28, the amide group is planar.

7.7 Pesticides and Pharmaceuticals

Most of the organic compounds discussed so far in this chapter have relatively small molecules and are produced in large volume. Molecules like these are starting materials for synthesis of numerous structurally more complex organic compounds with applications in agriculture, medicine, and consumer products. This section discusses a selection of these compounds, all of which are used in agriculture or as pharmaceuticals. Some of the structures and syntheses of these compounds are intricate; do not try to memorize them. Your goals instead, should be to note the hydrocarbon frames of the molecules, to recognize functional groups, and to begin to appreciate the extremely diverse structures and properties of organic compounds in relation to the chemical bonds in the frameworks and the functional groups.

Chemists have developed a shorthand notation to represent the structures of complex organic molecules. This notation, which focuses attention to the most

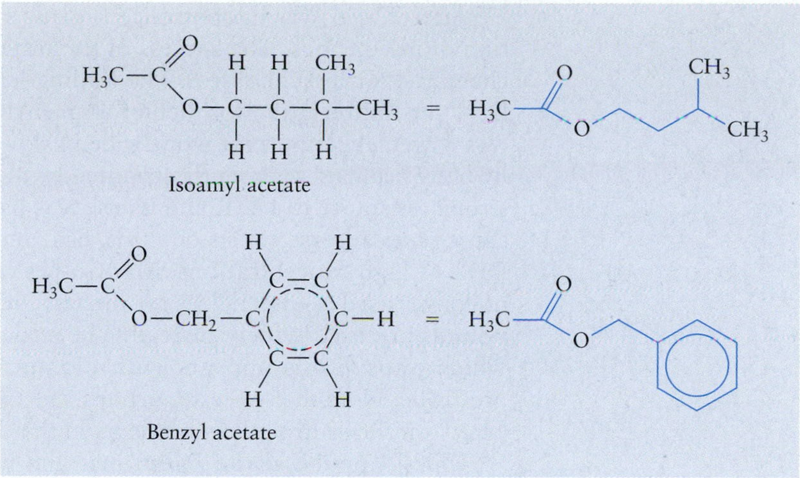

FIGURE 7.29 In the shorthand notation for the structures of organic compounds illustrated on the right side of this figure, carbon atoms are assumed to lie where the lines indicating bonds intersect. Furthermore, enough hydrogen atoms are assumed to be bonded to each carbon atom to give it a total valence of four. Terminal carbon atoms (those at the ends of chains) and their associated hydrogen atoms are shown explicitly, however.

Isoamyl acetate

Benzyl acetate

important aspects of structure, is illustrated in Figure 7.29. In this notation, the symbol "C" for a carbon atom is omitted, and only the C—C bonds are shown. A carbon atom is assumed to lie at each end of the line segments that represent bonds. In addition, symbols for the hydrogen atoms attached to carbon are omitted. Terminal carbon atoms (those at the end of chains) and their associated hydrogen atoms are shown explicitly. To generate the full structure (and the molecular formula) from such a shorthand formula, carbon atoms must be inserted at the end of each bond, and enough hydrogen atoms must be attached to each carbon atom to satisfy its valence of four.

Insecticides

The chemical control of insect pests dates back thousands of years. The earliest insecticides were inorganic compounds of copper, lead, and arsenic, as well as some naturally occurring organic compounds such as nicotine (Fig. 7.30a). Few of these "first-generation" insecticides are in use today because of their adverse side effects on plants, animals, and humans.

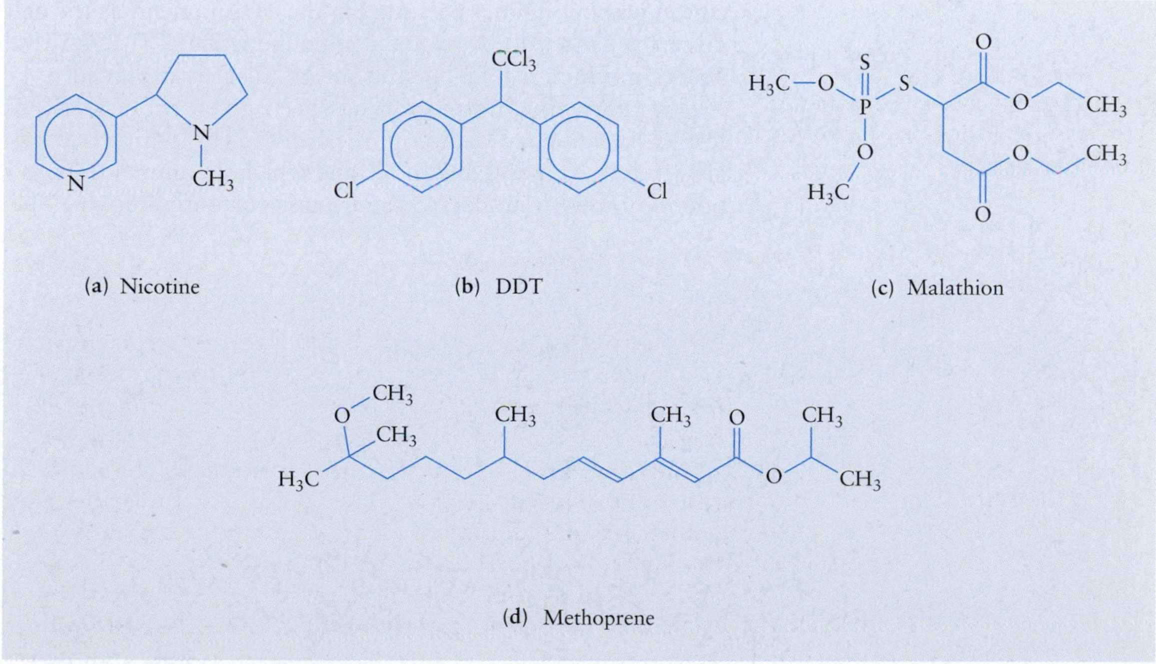

(a) Nicotine

(b) DDT

(c) Malathion

(d) Methoprene

FIGURE 7.30 Structures of several insecticides: (a) nicotine; (b) dichlorodiphenyltrichloroethane (DDT); (c) malathion; and (d) methoprene.

After World War II, controlled organic syntheses gave rise to a second generation of insecticides. The success of these agents led to rapid growth in the use of chemicals for insect control. The leading insecticide of the 1950s and 1960s was DDT (an abbreviation for **d**ichloro**d**iphenyl**t**richloroethane; see Fig. 7.30b). DDT was extremely important worldwide in slowing the spread of typhus (transmitted by body lice) and malaria (transmitted by mosquitoes). But, mosquitoes developed strong resistance to DDT, and its use was banned in the United States in 1972 because of its adverse effects on birds, fish, and other life-forms that can accumulate DDT to high concentrations. Many other chlorine-substituted hydrocarbons are no longer used as insecticides for the same reason. Today, organophosphorus compounds are used widely instead. The structure of the insecticide malathion, in which phosphorus appears with organic functional groups, is given in Figure 7.30c. Note the two ester groups, the two kinds of sulfur, and the "expanded octet" on the central phosphorus atom that lets it form five bonds.

Unless applied at the right times and in properly controlled doses, second-generation insecticides frequently kill beneficent insects together with the pests. Third-generation insecticides are more subtle. Many are based on sex attractants (to collect insects together in one place before exterminating them or to lead them to mate with sterile partners) or juvenile hormones (to prevent insects from maturing and reproducing). These compounds have the advantages of being specific against the pests and doing little or no harm to other organisms. Moreover, they can be used in small quantities, and they degrade rapidly in the environment. An example is the juvenile hormone methoprene (see Fig. 7.30d), which is used in controlling mosquitoes. It consists of a branched dialkene chain with a methyl ether (methoxy) and an isopropyl-ester functional group.

Herbicides

Chemical control of weeds, together with use of fertilizers, has contributed to the "green revolution" that began in the 1940s and during which agricultural productivity has increased dramatically throughout the world. The first herbicide of major importance, introduced in 1945 and still in use today, was 2,4-D (2,4-dichlorophenoxyacetic acid; Fig. 7.31a), a derivative of phenol with chlorine and carboxylic acid functional groups. 2,4-D kills broadleaf weeds in wheat, corn, and cotton without unduly persisting in the environment, as the chlorinated insecticides discussed earlier do. A related compound is 2,4,5-T (2,4,5-trichlorophenoxyacetic acid), in which a hydrogen atom in 2,4-D is replaced by a chlorine atom. In recent years, much attention has been given to TCDD ("dioxin," or 2,3,7,8-**t**etra**c**hloro**d**ibenzo-*p*-**d**ioxin; see Fig. 7.31b), which occurs as a trace impurity (10–20 ppb by mass) in 2,4,5-T and which, in animal tests, is the most toxic compound of low-to-moderate molar mass currently known. The use of 2,4,5-T as a

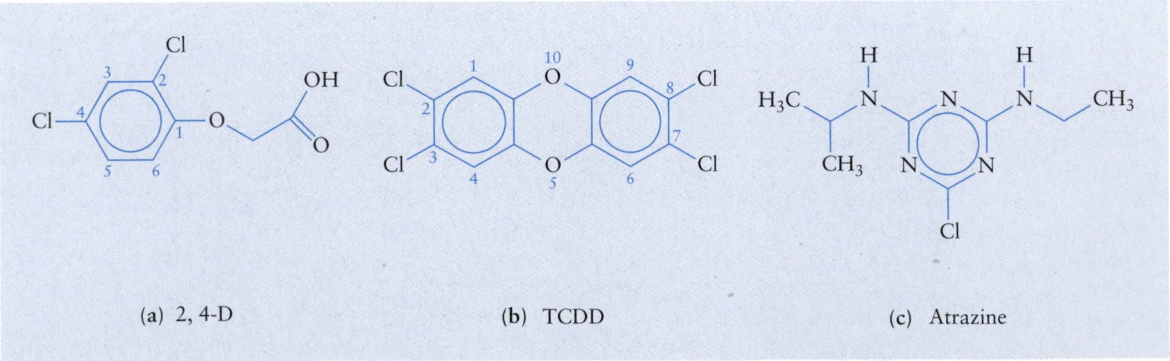

(a) 2, 4-D

(b) TCDD

(c) Atrazine

FIGURE 7.31 Structures of some herbicides. (a) 2,4-D (2,4-dichlorophenoxyacetic acid); (b) TCDD ("dioxin," or 2,3,7,8-tetrachlorodibenzo-*p*-dioxin); and (c) atrazine.

Crystals of 4-acetaminophenol (Tylenol) viewed under polarized light.

defoliant (Agent Orange) during the Vietnam War led to a lawsuit by veterans who claimed that health problems arose from contact with the traces of TCDD present in the 2,4,5-T. Although such a direct connection has never been proved, the use of chlorinated phenoxy herbicides has decreased, and that of other herbicides, such as atrazine (see Fig. 7.31c), has grown.

Analgesics

Drugs that relieve pain are called **analgesics.** The oldest and most widely used analgesic is aspirin, which has the chemical name acetylsalicylic acid (Fig. 7.32a). More than 15 million kg of aspirin is synthesized each year. Aspirin acts to reduce fevers and to relieve pain. Some recent studies have suggested that regular moderate consumption may reduce the chances of heart disease. As an acid, aspirin can irritate the stomach lining, a side effect that can be reduced by combining it in a buffer with a weak base such as sodium hydrogen carbonate. Another important pain reliever is 4-acetaminophenol, or acetaminophen (see Fig. 7.32b). This compound is sold under many trademarks, most prominently Tylenol. Both acetylsalicylic acid and 4-acetaminophenol are derivatives of phenol. During synthesis, the former is converted to an acetic acid ester with an additional carboxylic acid functional group, and the latter with an amide functional group.

A much more powerful pain reliever, which is available only by prescription because of its addictive properties, is morphine (see Fig. 7.32c). Morphine acts on

(a) Aspirin

(b) Acetaminophen (4-acetaminophenol)

(c) Morphine

FIGURE 7.32 The molecular structures of some analgesics: (a) aspirin; (b) acetaminophen; and (c) morphine. Note how slight the differences are among morphine, codeine, and heroin.

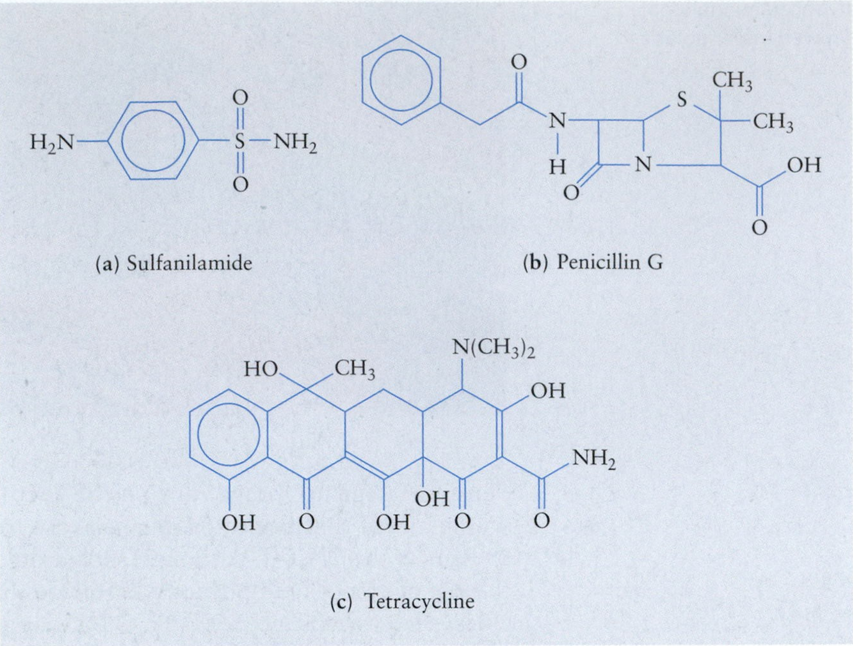

FIGURE 7.33 Molecular structures of some antibiotics: (a) sulfanilamide; (b) penicillin G; and (c) tetracycline.

(a) Sulfanilamide

(b) Penicillin G

(c) Tetracycline

the central nervous system, apparently because its shape fits a receptor site on the nerve cell, and blocks the transmission of pain signals to the brain. Its structure contains five interconnected rings. A small change (the replacement of one $-OH$ group by an $-OCH_3$ group, giving a methyl ether) converts morphine into codeine, a prescription drug used as a cough suppressant. Replacing *both* $-OH$ groups by acetyl groups ($-COCH_3$) generates the notoriously addictive substance heroin.

Antibacterial Agents

The advent of antibacterial agents changed the treatment of bacterial diseases such as tuberculosis and pneumonia dramatically beginning in the 1930s. The first "wonder drug" was sulfanilamide (Fig. 7.33a), a derivative of aniline. Other "sulfa drugs" are obtained by replacing one of the hydrogen atoms on the sulfonamide group by other functional groups. Bacteria mistake sulfanilamide for *p*-aminobenzoic acid, a molecule with a very similar shape but a carboxylic acid($-COOH$) group in place of the $-SO_2NH_2$ group. The drug then interferes with the bacterium's synthesis of folic acid, an essential biochemical, so the organism dies. Mammals do not synthesize folic acid (they obtain it from their diet), so they are not affected by sulfanilamide.

The penicillin molecule (see Fig. 7.33b) contains an amide linkage that connects a substituted double ring (including sulfur and nitrogen atoms) to a benzyl (phenylmethyl) group. It is a natural product formed by certain molds. Although the total synthesis of penicillin was achieved in 1957, that chemical route is not competitive economically with biosynthesis via fermentation. The mold grows for several days in tanks that may hold up to 100,000 L of a fermentation broth (Fig. 7.34). The penicillin is later separated by solvent extraction. Penicillin functions by deactivating enzymes responsible for building cell walls in the bacteria. Derivatives of natural penicillin have been developed and are commercially available.

FIGURE 7.34 Fermentation tanks used in modern penicillin production.

Finally we mention the tetracyclines, which are derivatives of the four-ring aromatic compound represented in Figure 7.33c. These drugs have the broadest spectrum of antibacterial activity found to date.

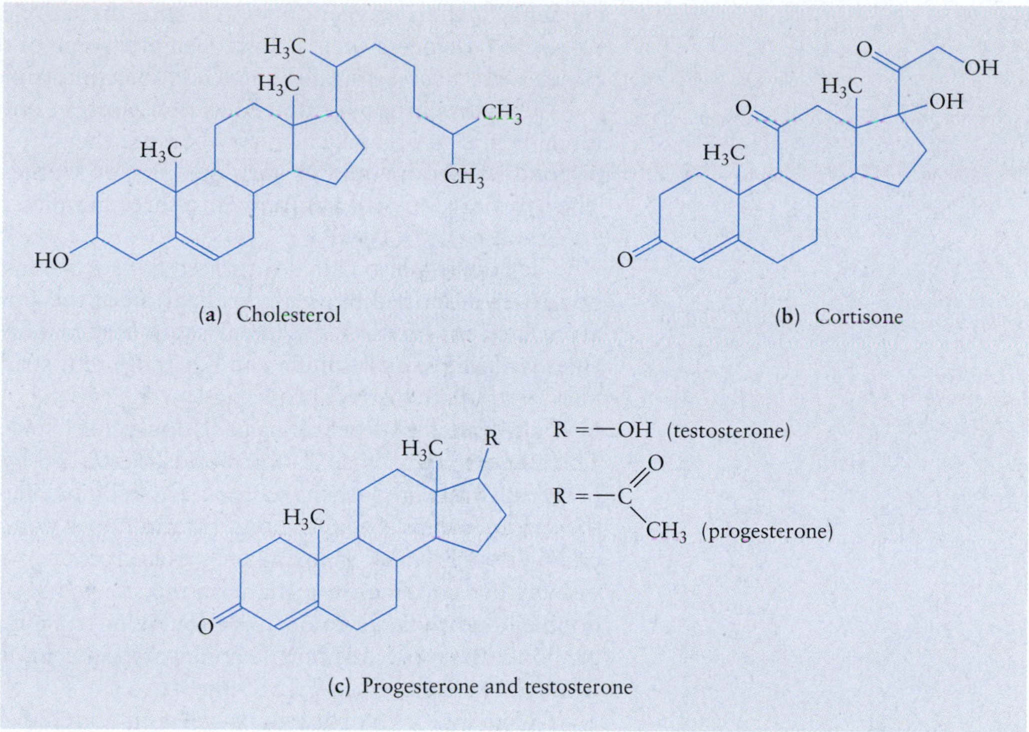

(a) Cholesterol (b) Cortisone

R = —OH (testosterone)

R = —C(=O)CH$_3$ (progesterone)

(c) Progesterone and testosterone

FIGURE 7.35 Molecular structures of some steroids: (a) cholesterol; (b) cortisone; and (c) progesterone and testosterone.

Steroids

The **steroids** are a family of naturally occurring compounds with a wide variety of functions. Most of them are synthesized from cholesterol. The structure of cholesterol (Fig. 7.35a) contains a group of four fused hydrocarbon rings (3 six-atom rings and 1 five-atom ring). All steroids possess this "steroid nucleus." Cholesterol itself is present in all tissues of the human body. When present in excess in the bloodstream, it can accumulate in the arteries, restricting the flow of blood and leading to heart attacks. Its derivatives have widely different functions. The hormone cortisone (see Fig. 7.35b), which is secreted by the adrenal glands, regulates the metabolism of sugars, fats, and proteins in all body cells. As a drug, cortisone reduces inflammation and moderates allergic responses. It often is prescribed to combat arthritic inflammation of the joints.

The human sex hormones are also derivatives of cholesterol. Here the resourcefulness of nature for building compounds with quite different functions from the same starting material is particularly evident. The female sex hormone progesterone (see Fig. 7.35c) differs from the male sex hormone testosterone only by replacing an acetyl (—$COCH_3$) group by a hydroxyl (—OH) group. Oral contraceptives are synthetic compounds with structures that are closely related to progesterone.

CHAPTER SUMMARY

The element carbon has a rich and varied chemistry because of its location in the periodic table. As a Group IV element, each carbon atom forms four covalent bonds, more than any other second-period element. In consequence of its intermediate value of electronegativity, carbon can bond with more electronegative elements such as oxygen, nitrogen, and the halogens, and also with more electropositive elements such as hydrogen and some of the heavy metals. Carbon also bonds with carbon, forming single, double, and triple bonds.

The hydrocarbons—molecules that contain only carbon and hydrogen—are fundamental to organic chemistry because they provide the archetypes for models of bond formation and the starting point for synthesis of other organic molecules. The hydrocarbons fall naturally into three families, based on their chemical properties and types of bonds.

The alkanes, also called saturated hydrocarbons, have only single bonds. The bonds are described by sp^3 hybridization of the carbon atoms. When the carbon atoms are not bonded in a linear sequence, cases can occur in which two molecules with the same formula can have different structures, called geometrical isomers, with quite distinct properties.

Unsaturated hydrocarbons have double or triple bonds between carbon atoms. The alkenes have C—C double bonds, described by sp^2 hybridization of the carbon atoms. The alkynes have C—C triple bonds, described by sp hybridization of the carbon atoms. Because bond rotation does not occur readily about a carbon–carbon double bond, many alkenes exist in contrasting isomeric forms, depending on whether bonding groups are on the same *(cis)* or opposite sides *(trans)* of the double bond. When two or more double or triple bonds are separated by one single bond, the p orbitals form a conjugated system, in which the de-localized π orbitals are best described by MO theory.

The aromatic hydrocarbons are conjugated cyclic structures in which the π bonding is described through de-localized MOs formed at carbon atoms that have sp^2 hybridization.

The fullerenes, which contain only carbon, are an allotropic form of carbon discovered in 1985. All the fullerenes have even numbers of atoms, with formulas ranging up to C_{400} and higher. Their π bonds are conjugated π electron systems.

Functional groups are sites of specific, heightened reactivity caused by insertion of noncarbon atoms into hydrocarbon structures or the attachment of noncarbon atoms to a hydrocarbon chain. Thus, organic molecules are conveniently viewed as carbon skeletal templates on which these highly reactive sites are located. Because of their reactivity, functional groups are key elements in strategies for synthesizing more complex organic structures.

The hydrocarbons recovered from petroleum, and their derivatives containing functional groups, are relatively small molecules with simple structures. Substances such as these provide starting materials for the synthesis of numerous structurally more complex organic compounds with applications in agriculture, medicine, and consumer products. These compounds include insecticides and herbicides for pest control and analgesics for controlling pain in the human body. Antibacterial agents fight disease. Steroids are naturally occurring compounds that derive from cholesterol. Hormones, including the human sex hormones, are derivatives of cholesterol. The bonding in these more elaborate structures is explained in the same way as the hydrocarbon skeletons and functional groups that comprise the structures.

CHAPTER REVIEW

- The alkanes, also called saturated hydrocarbons, have only single bonds, described by sp^3 hybridization of the carbon atoms.
- Carbon–carbon σ bonds form by overlap of sp^3 orbitals on adjacent atoms, and carbon–hydrogen σ bonds form by overlap of sp^3 orbitals with hydrogen 1s orbitals.
- In the straight-chain alkanes (general formula C_nH_{2n+2}), the carbon atoms are bonded in a linear sequence, and the molecules may take up numerous detailed

molecular conformations because of the ease of rotation about C—C single bonds.

- In cycloalkanes (general formula C_nH_{2n}), the carbon atoms at the end of a chain are attached together to form a closed loop. This connectivity reduces the ease with which the molecule changes conformations by rotation about single bonds and introduces both steric and angle strain into the total energy of the molecule. Consequently, cycloalkanes have particular preferred conformations such as boat and chair, and the molecule must absorb energy to move between them.
- In the branched-chain alkanes (general formula C_nH_{2n+2}), the carbon atoms are no longer arranged in a linear sequence, but instead can be bonded to three or four other carbon atoms. This possibility leads to a rich elaboration of structure in which two molecules with the same formula can have different structures, called geometrical isomers, and therefore quite distinct properties.
- Unsaturated hydrocarbons have double or triple bonds between carbon atoms.
- The alkenes have C—C double bonds, described by sp^2 hybridization of the carbon atoms. A σ bond forms by overlap of sp^2 orbitals on adjacent carbon atoms. The parallel nonhybridized $2p$ orbitals on adjacent carbon atoms overlap to form a C—C π bond.
- The alkynes have C—C triple bonds, described by sp hybridization of the carbon atoms. A σ bond forms by overlap of sp orbitals on adjacent carbon atoms. The two pairs of parallel nonhybridized $2p$ orbitals on adjacent carbon atoms overlap to form two C—C π bonds.
- Because bond rotation does not occur readily about a carbon–carbon double bond, many alkenes exist in contrasting isomeric forms, depending on whether bonding groups are on the same (*cis*) or opposite sides (*trans*) of the double bond.
- The structures and bonding in substituted alkenes and alkynes can be described by the combined MO-VB picture presented in Section 6.5, in which localized VB bonds describe the molecular framework and delocalized MOs describe the π electrons.
- When two or more double or triple bonds are separated by a single bond in a molecule, a conjugated system is formed, and a delocalized MO picture of the bonding should be used.
- The aromatic hydrocarbons are conjugated cyclic structures in which the π bonding is described by delocalized MOs. Each carbon atom has sp^2 hybrid orbitals that overlap to form C—C σ bonds that define the molecular framework. The remaining nonhybridized p_z orbitals combine to give π MOs delocalized over the entire molecule.
- The fullerenes, which contain only carbon, are an allotropic form of carbon discovered in 1985. All the fullerenes have even numbers of atoms, with formulas ranging up to C_{400} and higher. Their π bonds are conjugated π electron systems.
- Functional groups are sites of specific, heightened reactivity caused by insertion of noncarbon atoms into hydrocarbon structures or the attachment of noncarbon atoms to a hydrocarbon chain. The structures, shapes, and chemical reactivity of functional groups are largely independent of their location inorganic molecules. Thus, organic molecules are conveniently viewed as carbon skeletal templates on which the highly reactive sites are located. Because of their specific reactivity, functional groups are key elements in strategies for synthesizing more complex organic structures.
- Replacing a hydrogen atom by a halogen atom creates an alkyl halide. Replacing a hydrogen atom by an —OH group produces an alcohol. If the —OH group is attached to an aromatic hydrocarbon ring, the resulting structure is called a phenol.

- Ethers are characterized by the —O— functional group, where the oxygen atom links either alkyl or aromatic groups. Cyclic ethers, with various ring sizes, are called epoxides.
- An aldehyde has the functional group $-\overset{\text{O}}{\overset{\|}{\text{C}}}-\text{H}$ derived from dehydrogenation of an alcohol.
- A ketone has the functional group $-\overset{\text{O}}{\overset{\|}{\text{C}}}-$.
- Carboxylic acids contain the functional group $-\overset{\text{O}}{\overset{\|}{\text{C}}}-\text{OH}$ and react with alcohols to produce esters.
- The C—O double bonds in all these structures are described by sp^2 hybrid orbitals on both atoms for the σ bonds and nonhybridized $2p$ orbitals for the π bonds.
- The amines are organic derivatives of NH_3 in which one, two, or three hydrogens have been replaced by hydrocarbon groups to give primary, secondary, or tertiary amines, respectively. Primary or secondary amines react with carboxylic acids to give amides. The bonds in all these are described by sp^3 hybridization of the nitrogen atom.
- Controlled chemical syntheses have produced organic compounds that are used as pesticides and others to fight pain and disease in the human body.

CONCEPTS & SKILLS

After studying this chapter and working the problems that follow, you should be able to:

1. Identify important hydrocarbons in crude petroleum and describe their behavior in combustion (Sections 7.1–7.4, Problems 1–4).
2. Write names and structural formulas for hydrocarbons (Sections 7.2–7.4, Problems 5–12).
3. Describe the hybrid–atomic–orbital basis for representing the bonding and structure of organic molecules (Sections 7.2–7.4, Problems 13–14).
4. Discuss the delocalization of π electrons in organic molecules and fullerenes (Sections 7.3–7.5, Problems 12–16).
5. Identify important functional groups and outline chemical processes by which important chemical compounds are synthesized (Section 7.6, Problems 17–26).
6. Describe the bonding in important functional groups using the hybrid orbitals approach (Section 7.6, Problems 27–30).
7. Recognize and describe the shapes and functional groups for molecules used as pesticides and pharmaceuticals (Section 7.7, Problems 31–36).

PROBLEMS

Answers to problems whose numbers are boldface appear in Appendix G. Problems that are more challenging are indicated with asterisks.

Petroleum Refining and the Hydrocarbons

1. Is it possible for a gasoline to have an octane number exceeding 100? Explain.

2. Is it possible for a motor fuel to have a negative octane rating? Explain.

3. A gaseous alkane is burned completely in oxygen. The volume of the carbon dioxide that forms equals twice the volume of the alkane burned (the volumes are measured at the same temperature and pressure). Name the alkane and write a balanced equation for its combustion.

4. A gaseous alkyne is burned completely in oxygen. The volume of the water vapor that forms equals the volume of the alkyne burned (the volumes are measured at the same temperature and pressure). Name the alkyne and write a balanced equation for its combustion.

5. (a) Write a chemical equation involving structural formulas for the catalytic cracking of decane into an alkane and an alkene that contain equal numbers of carbon atoms. Assume that both products have straight chains of carbon atoms.
(b) Draw and name one other isomer of the alkene.

6. (a) Write an equation involving structural formulas for the catalytic cracking of 2,2,3,4,5,5-hexamethylhexane. Assume that the cracking occurs between carbon atoms 3 and 4.
(b) Draw and name one other isomer of the alkene.

7. Write structural formulas for the following:
(a) 2,3-Dimethylpentane
(b) 3-Ethyl-2-pentene
(c) Methylcyclopropane
(d) 2,2-Dimethylbutane
(e) 3-Propyl-2-hexene
(f) 3-Methyl-1-hexene
(g) 4-Ethyl-2-methylheptane
(h) 4-Ethyl-2-heptyne

8. Write structural formulas for the following:
(a) 2,3-Dimethyl-1-cyclobutene
(b) 2-Methyl-2-butene
(c) 2-Methyl-1,3-butadiene
(d) 2,3-Dimethyl-3-ethylhexane
(e) 4,5-Diethyloctane
(f) Cyclooctene
(g) Propadiene
(h) 2-pentyne

9. Write structural formulas for *trans*-3-heptene and *cis*-3-heptene.

10. Write structural formulas for *cis*-4-octene and *trans*-4-octene.

11. Name the following hydrocarbons:
(a) $H_2C{=}C{=}C(H){-}CH_2{-}CH_2{-}CH_3$
(b) $H_2C{=}C(H){-}C(H){=}C(H){-}C(H){=}CH_2$
(c) $H_2C{=}C(CH_3){-}CH_2{-}CH_2{-}CH_2{-}CH_3$
(d) $CH_3{-}CH_2{-}C{\equiv}C{-}CH_2{-}CH_3$

12. Name the following hydrocarbons:
(a) $H_2C{=}C(H_3C){-}C(CH_3){=}CH_2$
(b) $H_3C{-}C(H){=}C(H){-}C(H){=}C(H){-}CH_3$
(c) $H_3C{-}C(CH_3)_2{-}CH_2{-}CH_3$
(d) $H_3C{-}C({=}CH_2){-}CH_3$

13. State the hybridization of each of the carbon atoms in the hydrocarbon structures in Problem 11.

14. State the hybridization of each of the carbon atoms in the hydrocarbon structures in Problem 12.

Fullerenes

15. To satisfy the octet rule, fullerenes must have double bonds. How many? Give a simple rule for one way of placing them in the structure shown in Figure 7.20a.

16. It has been suggested that a compound of formula $C_{12}B_{24}N_{24}$ might exist and have a structure similar to that of C_{60} (buckminsterfullerene).
(a) Explain the logic of this suggestion by comparing the number of valence electrons in C_{60} and $C_{12}B_{24}N_{24}$.
(b) Propose the most symmetric pattern of carbon, boron, and nitrogen atoms in $C_{12}B_{24}N_{24}$ to occupy the 60 atom sites in the buckminsterfullerene structure. Where could the double bonds be placed in such a structure?

Functional Groups and Organic Reactions

17. In a recent year, the United States produced 6.26×10^9 kg ethylene dichloride (1,2-dichloroethane) and 15.87×10^9 kg ethylene. Assuming that all significant quantities of ethylene dichloride were produced from ethylene, what fraction of the ethylene production went into making ethylene dichloride? What mass of chlorine was required for this conversion?

18. In a recent year, the United States produced 6.26×10^9 kg ethylene dichloride (1,2-dichloroethane) and 3.73×10^9 kg vinyl chloride. Assuming that all significant quantities of vinyl chloride were produced from ethylene dichloride, what fraction of the ethylene dichloride production went into making vinyl chloride? What mass of hydrogen chloride was generated as a by-product?

19. Write balanced equations for the following reactions. Use structural formulas to represent the organic compounds.
(a) The production of butyl acetate from butanol and acetic acid
(b) The conversion of ammonium acetate to acetamide and water
(c) The dehydrogenation of 1-propanol
(d) The complete combustion (to CO_2 and H_2O) of heptane

20. Write balanced equations for the following reactions. Use structural formulas to represent the organic compounds.
(a) The complete combustion (to CO_2 and H_2O) of cyclopropanol
(b) The reaction of isopropyl acetate with water to give acetic acid and isopropanol
(c) The dehydration of ethanol to give ethylene
(d) The reaction of 1-iodobutane with water to give 1-butanol

21. Outline, using chemical equations, the synthesis of the following from easily available petrochemicals and inorganic starting materials.
(a) Vinyl bromide ($CH_2{=}CHBr$)
(b) 2-Butanol
(c) Acetone (CH_3COCH_3)

22. Outline, using chemical equations, the synthesis of the following from easily available petrochemicals and inorganic starting materials.
(a) Vinyl acetate ($CH_3COOCH{=}CH_2$)
(b) Formamide ($HCONH_2$)
(c) 1,2-Difluoroethane

23. Write a general equation (using R to represent a general alkyl group) for the formation of an ester by the condensation of a tertiary alcohol with a carboxylic acid.

24. Explain why it is impossible to form an amide by the condensation of a tertiary amine with a carboxylic acid.

25. Calculate the volume of hydrogen at 0°C and 1.00 atm that is required to convert 500.0 g linoleic acid ($C_{18}H_{32}O_2$) to stearic acid ($C_{18}H_{36}O_2$).

26. A chemist determines that 4.20 L hydrogen at 298 K and a pressure of 1.00 atm is required to completely hydrogenate 48.5 g of the unsaturated compound oleic acid to stearic acid ($C_{18}H_{36}O_2$). How many units of unsaturation (where a unit of unsaturation is one double bond) are in a molecule of oleic acid?

27. Acetic acid can be made by the oxidation of acetaldehyde (CH_3CHO). Molecules of acetaldehyde have a $-CH_3$ group, an oxygen atom, and a hydrogen atom attached to a carbon atom. Draw the Lewis diagram for this molecule, give the hybridization of each carbon atom, and describe the π orbitals and the number of electrons that occupy each one. Draw the three-dimensional structure of the molecule, showing all angles.

28. Acrylic fibers are polymers made from a starting material called acrylonitrile, $H_2C(CH)CN$. In acrylonitrile, a $-C{\equiv}N$ group replaces a hydrogen atom on ethylene. Draw the Lewis diagram for this molecule, give the hybridization of each carbon atom, and describe the π orbitals and the number of electrons that occupy each one. Draw the three-dimensional structure of the molecule, showing all angles.

29. Compare the bonding in formic acid (HCOOH) with that in its conjugate base formate ion ($HCOO^-$). Each molecule has a central carbon atom bonded to the two oxygen atoms and to a hydrogen atom. Draw Lewis diagrams, determine the steric numbers and hybridization of the central carbon atom, and give the molecular geometries. How do the π orbitals differ in formic acid and the formate molecular ion? The bond lengths of the C—O bonds in HCOOH are 1.23 (for the bond to the lone oxygen) and 1.36 Å (for the bond to the oxygen with a hydrogen atom attached). In what range of lengths do you predict the C—O bond length in the formate ion to lie?

30. Section 7.3 shows that the compound 2-butene exists in two isomeric forms, which can be interconverted only by breaking a bond (in that case, the central double bond). How many possible isomers correspond to each of the following chemical formulas? Remember that a simple rotation of an entire molecule does not give a different isomer. Each molecule contains a central C=C bond.
(a) $C_2H_2Br_2$
(b) C_2H_2BrCl
(c) $C_2HBrClF$

Pesticides and Pharmaceuticals

31. (a) The insecticide methoprene (see Fig. 7.30d) is an ester. Write the structural formulas for the alcohol and the carboxylic acid that react to form it. Name the alcohol.
(b) Suppose that the carboxylic acid from part (a) is changed chemically so that the OCH_3 group is replaced by a hydrogen atom and the COOH group is replaced by a CH_3 group. Name the hydrocarbon that would result.

32. (a) The herbicide 2,4-D (see Fig. 7.31a) is an ether. Write the structural formulas of the two alcohol or phenol compounds that, on condensation, would form this ether. (The usual method of synthesis does not follow this plan.)
(b) Suppose that hydrogen atoms replace the chlorine atoms and a $-CH_3$ group replaces the carboxylic acid group in the two compounds in part (a). Name the resulting compounds.

33. (a) Write the molecular formula of acetylsalicylic acid (see Fig. 7.32a).
(b) An aspirin tablet contains 325 mg acetylsalicylic acid. Calculate the number of moles of that compound in the tablet.

34. (a) Write the molecular formula of acetaminophen (see Fig. 7.32b).
(b) A tablet of Extra Strength Tylenol contains 500 mg acetaminophen. Calculate the chemical amount (in moles) of that compound in the tablet.

35. Describe the changes in hydrocarbon structure and functional groups that are needed to make cortisone from cholesterol (see Fig. 7.35).

36. Describe the changes in hydrocarbon structure and functional groups that are needed to make testosterone from cortisone (see Fig. 7.35).

ADDITIONAL PROBLEMS

37. *trans*-Cyclodecene boils at 193°C, but *cis*-cyclodecene boils at 195.6°C. Write structural formulas for these two compounds.

38. A compound $C_4H_{11}N$ is known from its reactivity and spectroscopic properties to have no hydrogen atoms attached directly to the nitrogen atom. Write all structural formulas consistent with this information.

39. A compound C_3H_6O has a hydroxyl group but no double bonds. Write a structural formula consistent with this information.

40. Consider the following proposed structures for benzene, each of which is consistent with the molecular formula C_6H_6.
(i)

H
H C H
C C
C C
H C H
H

(ii)  [benzene ring structure]

(iii)

```
        H
        |
       C
     //  \
H—C        C=CH2
    \      /
     C=C
    /    \
   H      H
```

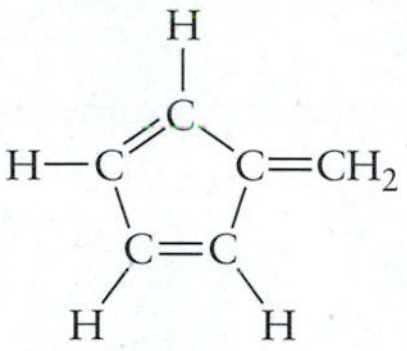

(iv) $CH_3-C\equiv C-C\equiv C-CH_3$

(v) $CH_2=CH-C\equiv C-CH=CH_2$

(a) When benzene reacts with chlorine to give C_6H_5Cl, only one isomer of that compound forms. Which of the five proposed structures for benzene are consistent with this observation?

(b) When C_6H_5Cl reacts further with chlorine to give $C_6H_4Cl_2$, exactly three isomers of the latter compound form. Which of the five proposed structures for benzene are consistent with this observation?

41. Acetyl chloride, CH_3COCl, reacts with the hydroxyl groups of alcohols to form ester groups with the elimination of HCl. When an unknown compound X with formula $C_4H_8O_3$ reacted with acetyl chloride, a new compound Y with formula $C_8H_{12}O_5$ was formed.
(a) How many hydroxyl groups were there in X?
(b) Assume that X is an aldehyde. Write a possible structure for X and a possible structure for Y consistent with your structure for X.

42. When an ester forms from an alcohol and a carboxylic acid, an oxygen atom links the two parts of each ester molecule. This atom could have come originally from the alcohol, from the carboxylic acid, or randomly from either. Propose an experiment using isotopes to determine which is the case.

43. Hydrogen can be added to a certain unsaturated hydrocarbon in the presence of a platinum catalyst to form hexane. When the same hydrocarbon is oxidized with $KMnO_4$, it yields acetic acid and butanoic acid. Identify the hydrocarbon and write balanced chemical equations for the reactions.

44. (a) It is reported that ethylene is released when pure ethanol is passed over alumina (Al_2O_3) that is heated to 400°C, but diethyl ether is obtained at a temperature of 230°C. Write balanced equations for both of these dehydration reactions.
(b) If the temperature is increased well above 400°C, an aldehyde forms. Write a chemical equation for this reaction.

* 45. The pyridine molecule (C_5H_5N) is obtained by replacing one C—H group in benzene with a nitrogen atom. Because nitrogen is more electronegative than the C—H group, orbitals with electron density on nitrogen are lower in energy. How do you expect the π MOs and energy levels of pyridine to differ from those of benzene?

* 46. For each of the following molecules, construct the π MOs from the $2p_z$ atomic orbitals perpendicular to the plane of the carbon atoms.
(a) Cyclobutadiene

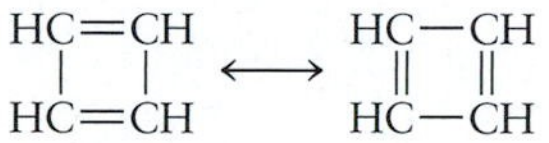

```
HC=CH          HC—CH
 |  |    ←→    ‖   ‖
HC=CH          HC—CH
```

(b) Allyl radical

```
   ·   ·   ·
H—C—C—C—H
   |   |   |
   H   H   H
```

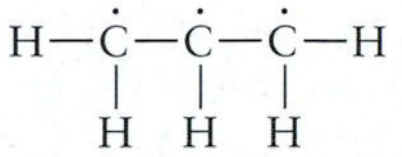

Indicate which, if any, of these orbitals have identical energies from symmetry considerations. Show the number of electrons occupying each π MO in the ground state, and indicate whether either or both of the molecules are paramagnetic. (*Hint:* Refer to Figs. 7.17 and 7.18.)

47. In what ways do the systematic developments of pesticides and of pharmaceuticals resemble each other, and in what ways do they differ? Consider such aspects as "deceptor" molecules, which are mistaken by living organisms for other molecules; side effects; and the relative advantages of a broad versus a narrow spectrum of activity.

48. The steroid stanolone is an androgenic steroid (a steroid that develops or maintains certain male sexual characteristics). It is derived from testosterone by adding a molecule of hydrogen across the C=C bond in testosterone.
(a) Using Figure 7.35c as a guide, draw the molecular structure of stanolone.
(b) What is the molecular formula of stanolone?

CUMULATIVE PROBLEMS

49. The structure of the molecule cyclohexene is

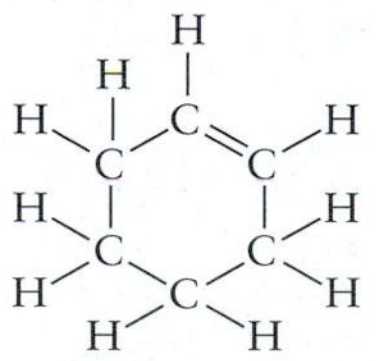

[cyclohexene structure with all H atoms]

Does the absorption of ultraviolet light by cyclohexene occur at longer or at shorter wavelengths than the absorption by benzene? Explain.

50. The naphthalene molecule has a structure that corresponds to two benzene molecules fused together:

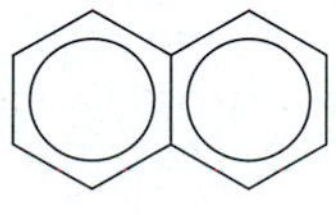

[naphthalene structure]

The π electrons in this molecule are delocalized over the entire molecule. The wavelength of maximum absorption in benzene is 255 nm. Will the corresponding wavelength in naphthalene be shorter or longer than 255 nm?

CHAPTER 8

Bonding in Transition Metal Compounds and Coordination Complexes

The colors of gemstones originate with transition-metal ions. Emerald is the mineral beryl (beryllium aluminum silicate $3BeO{\cdot}Al_2O_3{\cdot}6\ SiO_2$), in which some of the Al^{3+} ions have been replaced by Cr^{3+} ions. This structural environment splits the 3*d* orbitals to give a new energy separation that absorbs yellow and blue-violet light and transmits green light. The green color of emerald is the result.

The partially filled *d*-electron shells of the transition-metal elements are responsible for a range of physical properties and chemical reactions quite different from those of the main-group elements. The presence of unpaired electrons in the transition-metal elements and their compounds, the availability of low-energy unoccupied orbitals, and the facility with which transition-metal oxidation states change are important factors that determine their rich and fascinating chemistry. Transition-metal complexes are characterized by a wide variety of geometric structures, variable and striking colors, and magnetic properties that depend on subtle details of their structure and bonding. This chapter begins with a descriptive overview of the systematic trends in the properties of these metals, and then presents a comprehensive introduction to the classical and quantum mechanical models that describe bond formation in their compounds.

8.1 Chemistry of the Transition Metals

Let's begin by surveying some of the key physical and chemical properties of the transition-metal elements and interpreting trends in those properties using the quantum theory of atomic structure developed in Chapter 5. We focus initially on the fourth-period elements, also called the first transition series (those from scandium through zinc in which the 3*d* shell is progressively filled). Then we discuss the periodic trends in the melting points and atomic radii of the second and third transition series elements.

Physical Properties

Table 8.1 lists some key physical properties of the elements of the first transition series, taken mostly from Appendix F. The general trends in all of these properties can be understood by recalling that nuclear charge also increases across a period as electrons are being added to the same subshell, in this case, the *d* shell. The first and second ionization energies tend to increase across the period, but not smoothly. The energies of the 4*s* and 3*d* orbitals are so close to one another that the electron configurations of the neutral atoms and their ions are not easily predicted from the simplest model of atomic structure.

Electron-nuclear attraction increases as Z increases; thus, atomic and ionic radii generally decrease across each period as shown in Table 8.1 and Figure 8.1. But near the end of each period, repulsion between the electrons in the nearly filled *d* shell increases faster than the electron-nuclear attraction; thus, both the atomic and ionic radii begin to increase. Atoms of the second transition series (from yttrium to cadmium) are larger than those of the first transition series, as expected from the fact that the 4*d* orbitals are larger than the 3*d* orbitals. However, the atomic radii of the third transition series are not that much different from those of the second (see Fig. 8.1). This experimental fact is explained by the **lanthanide contraction.** The first and second transition series are separated by 18 elements, whereas the second and third are separated by 32 elements. The lanthanides have intervened, but their presence is not so obvious from the modern periodic table. Both the nuclear charge and electron count have increased between second- and third-period elements of the same group, but the *f* orbitals are much more diffuse than the *d* orbitals and much less effective in screening the nuclear charge. Elements in the third transition series experience a much greater effective nuclear charge than otherwise expected and thus are smaller. The atomic and ionic radii of hafnium, in the sixth period, are essentially the same as those of zirconium, in the fifth period. Because these elements are similar in both valence configuration *and* size, their properties are quite similar and they are difficult to separate from

TABLE 8.1 Properties of the Fourth-Period Transition Elements

Element	Sc	Ti	V	Cr	Mn	Fe	Co	Ni	Cu	Zn
IE_1 (kJ mol^{-1})	631	658	650	653	717	759	758	737	745	906
IE_2 (kJ mol^{-1})	1235	1310	1414	1592	1509	1562	1648	1753	1958	1733
Boiling point (°C)	2831	3287	3380	2672	1962	2750	2870	2732	2567	907
Melting point (°C)	1541	1660	1890	1857	1244	1535	1495	1453	1083	420
Atomic radius (Å)	1.61	1.45	1.31	1.25	1.37	1.24	1.25	1.25	1.28	1.34
M^{2+} Ionic radius (Å)	0.81	0.68	0.88	0.89	0.80	0.72	0.72	0.69	0.72	0.74
M^{2+} configuration	d^1	d^2	d^3	d^4	d^5	d^6	d^7	d^8	d^9	d^{10}
$\Delta H_{hyd}(M^{2+})^{\dagger}$ (kJ mol^{-1})				−2799	−2740	−2839	−2902	−2985	−2989	−2937

†Defined as $\Delta H_f^\circ(M^{2+}(aq)) - \Delta H_f^\circ(M^{2+}(g))$. Because aqueous species are defined relative to $\Delta H_f^\circ(H^+(aq)) = 0$, this is not an absolute enthalpy of hydration.

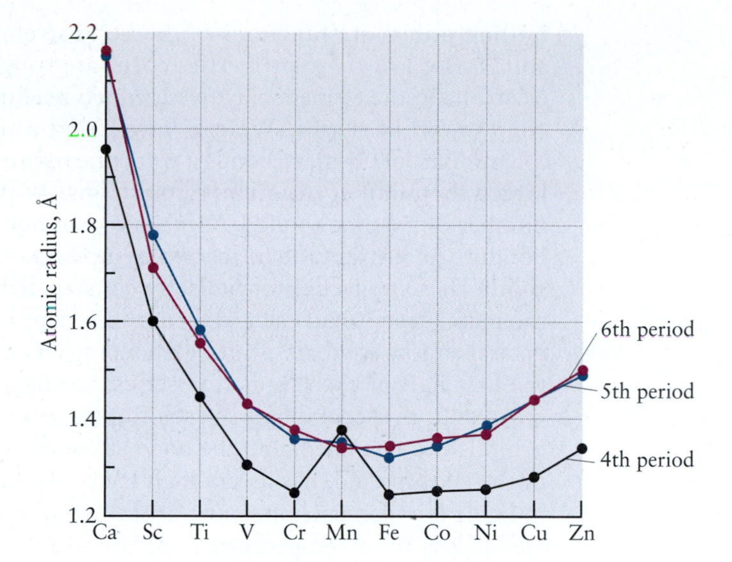

FIGURE 8.1 Variation of atomic radii through the fourth-, fifth-, and sixth-period transition-metal elements. Symbols shown are for the fourth-period elements.

one another. (A parallel effect, shown in Fig. 5.41 and discussed in Section 5.5 for main-group atoms and ions, arises from the filling of the 3*d* shell.)

The term *lanthanide contraction* is also used for another trend—the decrease in the atomic and ionic radii of the lanthanides from left to right along the sixth period. This trend has the same physical origin as that discussed previously in Chapter 5 for the second- and third-period elements and repeated earlier here: Nuclear charge is increasing while electrons are being added to the same subshell, in this case, the *f* shell. Using the same term for two different phenomena can cause confusion; therefore, we suggest you pay careful attention to the context when you see the term *lanthanide contraction*.

Metal-metal bond strengths first increase and then decrease going across each transition series, reaching a maximum in the middle. Evidence supporting this conclusion comes from the periodic variation in the melting and boiling points of the fourth-period elements shown in Table 8.1 and the corresponding trends in melting points for the three transition series shown in Figure 8.2. The melting and

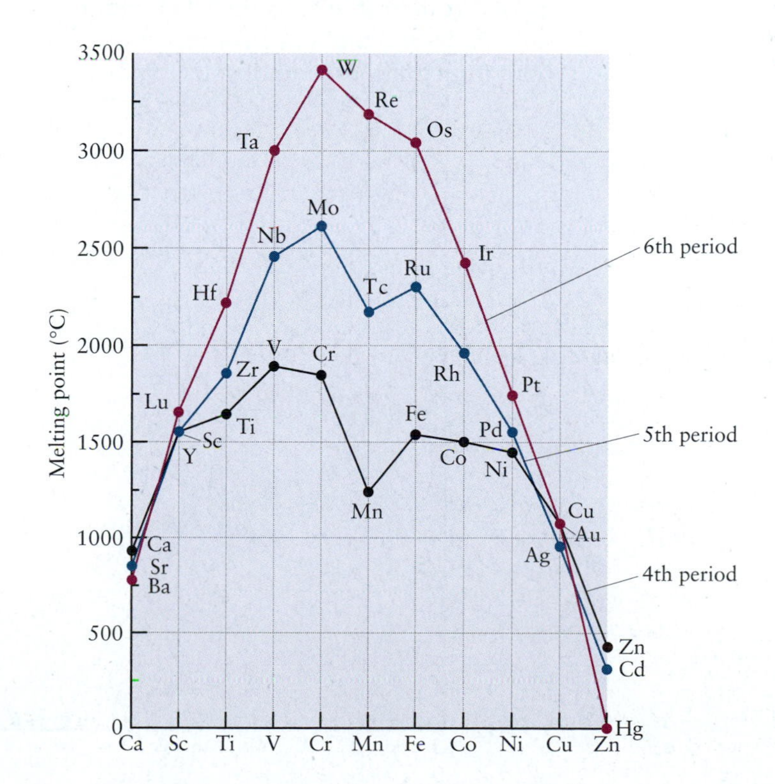

FIGURE 8.2 Variation of melting points through the three periods of transition-metal elements.

boiling points are functions of the bond strengths between the atoms or ions in solids. Both of these properties correlate roughly with the number of unpaired electrons in the elements involved, which reaches a maximum in the middle of the corresponding series. We can rationalize this correlation by thinking about the number of covalent bonds a given metal atom can form with its neighbors; the larger the number of unpaired electrons available for bonding, the greater the number of potential bonds. Another way to understand this trend is to think about bonding in terms of the formation of molecular orbitals delocalized over the entire solid. These molecular orbitals are constructed from the outermost *d* orbitals on the metal atoms, and each of them can accommodate two electrons. Lower energy orbitals in the solid are primarily bonding. As they are progressively filled through the first half of each transition series, the overall metal-metal bond strength increases. In the second half of each transition series, the higher energy antibonding orbitals become filled, and the bond strength decreases. Tungsten, near the middle of the sixth period, has a very high melting point (3410°C), which makes it useful in lightbulb filaments; mercury, at the end of the same period, has a melting point well below room temperature (−39°C).

Historically, most reactions of transition-metal ions have been carried out in aqueous solution; thus, it is important to understand the energy change that occurs when a metal ion is hydrated. These reactions are studied in solution under constant pressure. Under these conditions, the energy change is related to changes in the thermodynamic function *enthalpy*, denoted by *H*. Section 12.6 gives a proper discussion of enthalpies of reaction; for the present purpose, the enthalpy change can be treated simply as the energy change associated with the hydration of the gaseous ion

$$M^{2+}(g) \longrightarrow M^{2+}(aq)$$

Energies of hydration for the M2+ ions of the first transition series show an interesting trend (see Table 8.1 and Fig. 8.3). Although we have not yet discussed how ions interact with water in aqueous solutions (see Section 10.2), we might expect the strength of the interactions to increase as the ionic radii decrease, allowing the water molecules to approach the ions more closely. A linear trend that might be expected using this reasoning is shown as the red line in Figure 8.3. The experimental results shown as the black points connected by the blue line follow the same general trend, but clear deviations from linearity suggest that factors other than ionic radii are important. In particular, the experimental results show no

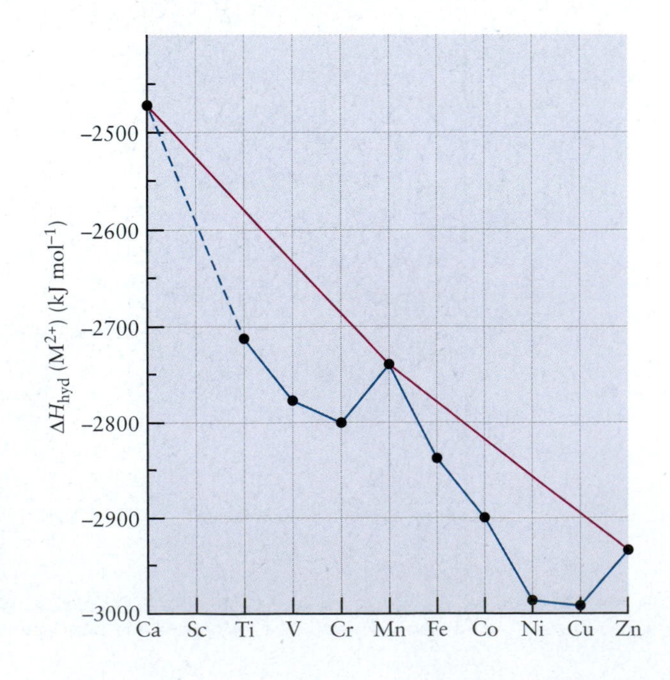

FIGURE 8.3 Enthalpies of hydration of M^{2+} ions, defined as $\Delta H_f^\circ(M^{2+}(aq)) - \Delta H_f^\circ(M^{2+}(g))$. The crystal field stabilization energy (discussed in Section 8.5) preferentially stabilizes certain ions, lowering ΔH_{hyd} from a line representing a linear change with increasing atomic number (red line) to the experimental (blue) line.

deviations for ions with filled shells (Ca^{2+}, Zn^{2+} with d^{10} configurations) or a shell that is exactly half filled (Mn^{2+} with a d^5 configuration). In Section 8.5, we connect this behavior to the way in which the shells are filled. Similar anomalies in the periodic trends in atomic radii and melting points occur for manganese (see Figs. 8.1 and 8.2); they are also related to the fact that manganese has a half-filled *d* shell.

Oxidation States of the Transition-Metal Elements

Figure 8.4 shows the characteristic oxidation states of the transition metals, which you should compare with those of the main-group elements shown in Figure 3.24. The maximum oxidation states of the early members of each period result from the participation of all the outer *s* and *d* electrons in ionic or covalent bonding. Thus, the maximum oxidation state for elements in the scandium group is +3, whereas manganese has a maximum oxidation state of +7, as do the other elements in its group. They form compounds such as $HReO_4$ and the dark red liquid Mn_2O_7. Higher oxidation states are more common in compounds of the heavier transition metals of a given group. The chemistry of iron is dominated by the +2 and +3 oxidation states, as in the common oxides FeO and Fe_2O_3, but the +8 state, which is nonexistent for iron, is important for the later members of the iron group, ruthenium and osmium. The oxide OsO_4, for example, is a volatile yellow solid that melts at 41°C and boils at 131°C. It selectively oxidizes C=C double bonds to *cis* diols, which makes it useful in organic synthesis (see Fig. 7.10) and as a biological stain, where it precipitates out easily seen black osmium metal. The chemistry of nickel is almost entirely that of the +2 oxidation state, but the higher oxidation states of the heavier elements in the nickel group, palladium and platinum, are commonly observed.

Compounds with transition metals in high oxidation states tend to be relatively covalent, whereas those with lower oxidation states are more ionic. Oxides are a good example: Mn_2O_7 is a covalent compound that is a liquid at room temperature (crystallizing only at 6°C), but Mn_3O_4 is an ionic compound, containing both Mn(II) and Mn(III), that melts at 1564°C. We can rationalize this important generalization by recalling that bonding configurations that have the smallest degree of charge separation tend to be the most stable. We used this principle (often called Pauling's principle of **electroneutrality**) as one way to choose among various possible Lewis dot diagrams (see Section 3.8).

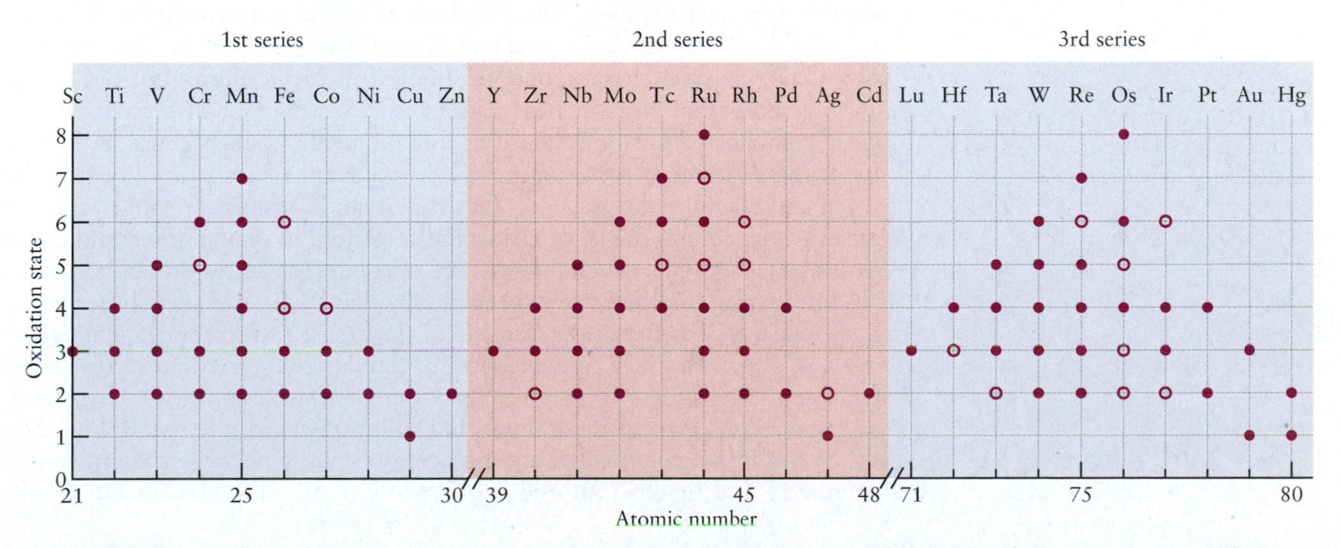

FIGURE 8.4 Some of the oxidation states found in compounds of the transition-metal elements. The more common oxidation states are represented by solid circles, and the less common ones are represented by open circles.

Transition metals exist in such a wide range of oxidations states because their partially filled *d* orbitals can either accept or donate electrons to form chemical bonds. For this same reason, many of their compounds are effective catalysts, substances that increase the rates of chemical reactions without undergoing permanent chemical changes themselves (see Chapter 18). Because an element such as iron can exist as Fe^{2+} or Fe^{3+} in solution, it can facilitate electron transfer reactions by shuttling back and forth between its oxidized and reduced forms without leaving solution. Elements like the alkali metal ions (K^+, for example) that have only one stable oxidation state in solution do not offer this possibility. Examples of catalysis by transition metal compounds include ammonia synthesis from elemental nitrogen and hydrogen catalyzed by iron oxides and the oxidation of SO_2 to SO_3 catalyzed by V_2O_5. The active sites of many enzymes (biological catalysts, see Section 18.7) are transition metal complexes that also facilitate electron transfer reactions by the mechanism just described.

Inorganic chemists use oxidation states to systematically organize patterns of structure and reactivity among the elements. Quantum chemical descriptions of the chemical bond (see Chapter 6) focus on the electron configurations of molecules. We can connect the two approaches by noting that metals in high oxidation states have relatively few *d* electrons, whereas those in low oxidation states are relatively rich in *d* electrons. Although this connection does not explain all of the patterns of bonding and reactivity in inorganic chemistry—it overlooks differences in the nuclear charge and energy levels of specific metals, for example—it does provide a useful way to organize our thinking about bonding in inorganic chemistry. Whether you focus on electron configurations or on oxidation states is a matter of choice for you (or your instructor). Either way, we suggest that you be mindful of the connection as we develop the molecular orbital theory of bonding in transition–metal compounds and coordination complexes

8.2 Bonding in Simple Molecules That Contain Transition Metals[1]

Homonuclear Diatomic Molecules

The elemental transition metals are all solids at normal temperatures and pressures, but many of them form stable homonuclear diatomic molecules at high temperatures in the gas phase. These include all of the elements in the first transition series and many from the second and third transition series. We can use molecular orbital theory to understand bonding in these molecules just as we did for the first- and second-period homonuclear diatomics in Section 6.2. The situation is slightly more complicated for molecules that contain metal atoms because of the availability of the *d* electrons for bonding. Two important issues should be kept in mind when constructing molecular orbitals that include atomic *d* orbitals. First, the energies of the *nd* and the $(n + 1)s$ orbitals are close to one another, and the relative ordering of these levels changes as we move across a row in the periodic table. We know from photoelectron spectroscopy that the energy level order is $\varepsilon_{4s} \cong \varepsilon_{3d} < \varepsilon_{4p}$ for the elements of the left side of the first transition series, changing to $\varepsilon_{3d} < \varepsilon_{4s} < \varepsilon_{4p}$ as we move to the right side. Similar changes occur in the second and third transition series. To simplify this discussion, we adopt the energy level order $\varepsilon_{3d} < \varepsilon_{4s} < \varepsilon_{4p}$ for most examples that follow. Second, the metal $3d_{z^2}$ and $4s$ orbitals can mix prior to forming molecular orbitals because they have the same symmetry and lie close to one another in energy. This mixing can result in an

[1]The discussion in this section follows closely the arguments developed in *Chemical Structure and Bonding*, Roger L. DeKock and Harry B. Gray, University Science Books, Sausalito, 1989, and was adapted with permission.

energy level structure that is quite different from one predicted without considering such mixing. We ignore *s*-*d* mixing as we develop our molecular orbital treatment of bonding in transition-metal compounds, but we do point out its consequences as they arise in examples. These approximations allow us to develop a simple conceptual framework that captures the essential features of bonding in these molecules; refinements can be added in more advanced work.

We construct molecular orbitals for the homonuclear diatomic metal molecules following exactly the same procedure developed in Chapter 6 for the main-group diatomic molecules. The relevant atomic orbitals for the first transition series are the valence orbitals 3*d*, 4*s*, and 4*p*. The 4*s* and $4p_z$ orbitals each generate a pair of bonding and antibonding σ and σ^* molecular orbitals, and the $4p_x$ and $4p_y$ orbitals each form a pair of bonding and antibonding π and π^* molecular orbitals, just as in the main-group diatomic molecules. The *d* orbitals can form six kinds of molecular orbitals (Fig. 8.5). The d_{z^2} orbitals are oriented along the

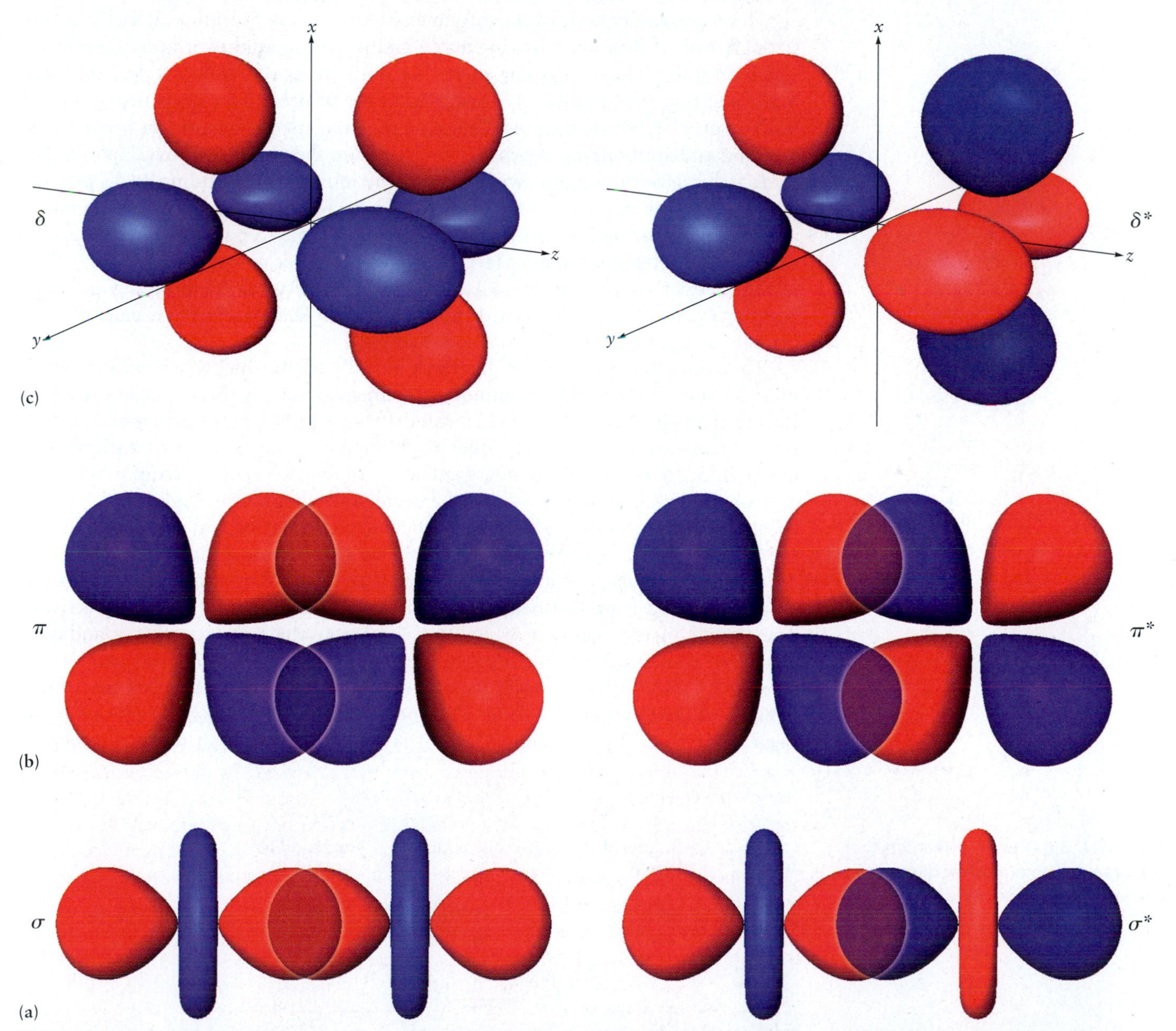

FIGURE 8.5 Bonding and antibonding molecular orbitals (MOs) formed from *d*-orbital overlap. (a) σ and σ^* orbitals formed by end-to-end overlap of d_{z^2} orbitals along the internuclear axis. (b) π and π^* MOs formed from side-by-side overlap of d_{xz} orbitals in the *x*-*z* plane (there is a corresponding pair of MOs formed from side-by-side overlap of d_{yz} orbitals in the *y*-*z* plane). (c) δ and δ^* orbitals from the face-to-face overlap of a pair of $d_{x^2-y^2}$ MOs (there is a corresponding pair of MOs constructed from the d_{xy} orbitals, which are rotated 45° from the set shown here).

internuclear axis (conventionally chosen to be the z-axis) and can overlap to form a pair of bonding and antibonding σ and σ^* molecular orbitals. Two pairs of bonding and antibonding π and π^* molecular orbitals can be formed from the side-by-side overlap of a pair of d_{xz} orbitals or a pair of d_{yz} orbitals. Finally, two molecular orbitals of a type we have not encountered earlier can be formed from the face-to-face overlap of a pair $d_{x^2-y^2}$ orbitals or a pair of d_{xy} orbitals. These **δ orbitals** have two units that contain the internuclear axis, and they have two units of angular momentum along that axis. The bonding δ orbital does not have a nodal plane perpendicular to the internuclear axis but the antibonding δ^* orbital does. Recall (see Section 6.1) that molecular orbitals are labeled according to the angular momentum component along the internuclear axis. They follow the Greek sequence σ, π, δ by analogy to s, p, d for atoms.

A typical energy-level diagram for homonuclear diatomic molecules of the first transition series is shown in Figure 8.6. The relative energy ordering for the molecular orbitals derived from the d orbitals is determined by precisely the same energy level proximity and overlap arguments presented in Section 6.2: Orbitals mix strongly only if they are close in energy and if their spatial overlap is significant. The d orbitals all have the same energy for both atoms of a homonuclear diatomic molecule; thus, the energy level ordering of the molecular orbitals is determined only by overlap considerations. Consequently, the energy separation between the bonding and antibonding σ orbitals derived from the end-to-end overlap of a pair of d_{z^2} orbitals is the largest, with σ being the most stable bonding molecular orbital (and σ^* the least stable), the separation between the bonding and antibonding π orbitals formed from the side-by-side overlap of a pair of d_{xz} orbitals or a pair of d_{yz} orbitals is the next largest (π lies at higher energy than σ, and π^* at lower energy than σ^*), and finally, the separation between the bonding and antibonding δ and δ^* orbitals formed from the face-to-face overlap of a pair of $d_{x^2-y^2}$ orbitals or a pair of d_{xy} orbitals is the smallest.

To make these energy-level diagrams broadly applicable, we introduce a simplified notation that will accommodate elements from all three transition series. Rather than specifying the atomic orbitals from which a particular molecular orbital is constructed, we simply label all of the molecular orbitals of a given symmetry in order of increasing energy. The set of atomic orbitals from which they are constructed is clearly shown in the correlation diagram in Figure 8.6, and the specific atomic orbitals leading to a particular molecular orbital are shown in Figure 8.5. The lowest energy σ orbital is labeled 1σ and it is derived from the metal $3d_{z^2}$ atomic orbitals. The lowest energy π and δ orbitals are labled 1π and 1δ and they are derived from the metal d_{xz}, d_{yz} and $d_{xy^2-y^2}$, d_{xy} orbital pairs, respectively. The second highest energy σ orbital is derived from the metal 4s orbitals and is labeled 2σ. The energy of the 2σ orbital relative to the set of molecular orbitals derived from the d orbitals varies depending on the relative energies of the atomic orbitals of the elements involved. The energy level order shown in Figure 8.6 has been determined by experiment and it is typical for the first transition period

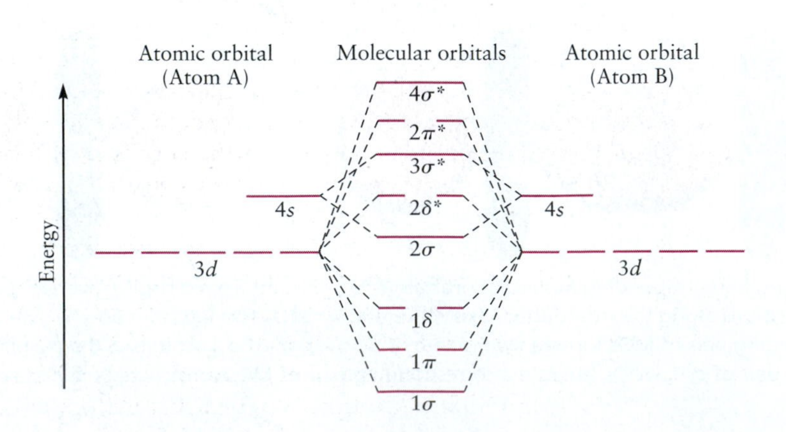

FIGURE 8.6 Orbital correlation diagram for homonuclear diatomic molecules of the transition metals
(Adapted from R. L. DeKock and H. B. Gray, *Chemical Bonding and Structure*, Sausalito: University Science Books, 1989, Figure 4-36, by permission).

Atomic orbitals interact strongly if they are close to one another in energy, have the appropriate symmetry, and overlap significantly in space. We have separated the energy levels of the degenerate set of $3d$ orbitals, as well as those of the degenerate set of $2p$ orbitals, to more clearly show which atomic orbitals mix to form specific MOs. Beginning at the bottom of the figure, we see no metal orbitals close in energy to the O $2s$ orbital. Thus, the lowest energy MO is essentially a nonbonding atomic $2s$ orbital localized on the oxygen (O) atom, just as the lowest energy MO in HF was essentially a nonbonding F $2s$ orbital (see Figure 6.20). Continuing upward, we see two kinds of bonds that can be formed from linear combinations of the metal d orbitals and the oxygen p orbitals. σ bonds result from overlap of the $3d_{z^2}$ and $2p_z$ orbitals, both of which point along the internuclear axis. π bonds are formed by overlap of the $3d_{xz}$ orbital with $2p_x$ and from overlap of $3d_{yz}$ with $2p_y$. Because there are no oxygen orbitals with the correct symmetry to mix with the metal $d_{x^2-y^2}$ or d_{xy} orbitals, they remain nonbonding atomic orbitals, which we label δ^{nb}. Recall that BeH_2 has a pair of nonbonding π orbitals that are essentially F $2p_x$ and $2p_y$ atomic orbitals (see Fig. 6.25). We labeled them π^{nb} orbitals to emphasize that they are still molecular orbitals with one unit of angular momentum quantized along the internuclear axis. Here, we choose the δ^{nb} notation for the same reasons. The nonbonding δ orbital has two units of angular momentum along the z-axis.

Let's use Figure 8.7 to obtain the ground-state electron configuration for a specific example, ScO, using the aufbau principle. ScO has nine valence electrons, three from $Sc(4s^2 3d^1)$ and six from $O(2s^2 2p^4)$. Using the aufbau principle, we expect the ground-state electronic configuration to be $(1\sigma^{nb})^2(2\sigma)^2(1\pi)^4(3\sigma^{nb})^1$. With two electrons occupying the bonding 2σ orbital and four electrons occupying the pair of bonding π orbitals, this simple MO picture predicts that the bond order is three and that ScO has a triple bond just like that in N_2. This prediction is confirmed by experiment, demonstrating the power of simple MO theory in describing bonding in molecules containing transition metal atoms.

We can examine the formation of the ScO bond from a different starting point in order to introduce specialized language used by inorganic chemists to describe bonding interactions in transition metal compounds and to connect that language to the language of molecular orbital theory developed earlier. Let's begin with a pair of widely separated Sc^{2+} and O^{2-} ions and see what happens as they come together to form an ScO molecule. At large distances the electron configurations of Sc^{2+} and O^{2-} are $3d^1$ and $2s^2 2p^6$, respectively, and the atomic orbitals do not overlap significantly. We can consider any bonding at large distances to be purely ionic and represent the molecule by the following Lewis dot diagram.

$$[\text{Sc}\cdot]^{2+}\ [:\ddot{\underset{..}{\text{O}}}:]^{2-}$$

As the ions approach one another their atomic orbitals begin to overlap, forming molecular orbitals. We imagine that the molecular orbitals are formed and filled sequentially purely for illustrative purposes. The O $2s$ orbital becomes the ScO $1\sigma^{nb}$ molecular orbital, which is nonbonding because it lies too low in energy to interact with any of the metal orbitals; it is filled by the O $2s$ electrons. Mixing of the O $2p_z$ orbital with the Sc $3d_{z^2}$ orbital produces the bonding 2σ and antibonding $4\sigma^*$ molecular orbitals. The 2σ orbital is mostly O $2p_z$ in character and lies at lower energy than the O $2p_z$ from which it was derived. Electrons will flow from the filled O $2p_z$ orbital to the empty, newly created 2σ molecular orbital in a process inorganic chemists call **ligand-to-metal (L→M) σ donation.** A bond that results from the transfer of a pair of electrons from one bonding partner to another is called a **dative bond.** This electron transfer introduces some covalent character into the ScO bond, which may now be represented by the following Lewis dot diagram:

$$\overset{+1}{\cdot\text{Sc}}-\overset{-1}{\ddot{\underset{..}{\text{O}}}}:$$

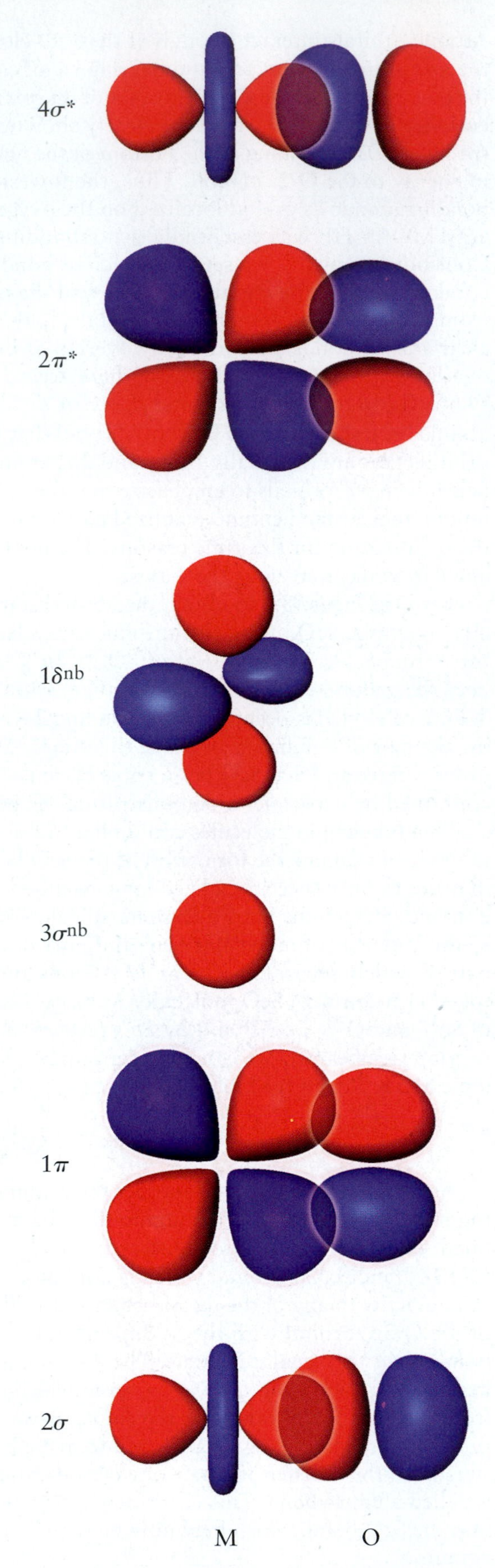

FIGURE 8.8 Orbital overlaps between metal and oxygen atom orbitals to form bonding, nonbonding, and antibonding molecular orbitals. (Adapted from R. L. DeKock and H. B. Gray, *Chemical Bonding and Structure*, Sausalito: University Science Books, 1989, Figure 4-44, by permission.)

The formal charge on Sc has been reduced from +2 to +1 and that on O from −2 to −1. (Recall from Section 3.8 that we assume the electrons in a bond are shared equally between the atoms when calculating formal charges.)

Continuing our thought experiment, we next imagine the formation of bonding and antibonding π molecular orbitals constructed from the O $2p$ and Sc $3d$ orbitals. Overlap between the O $2p_y$ and Sc $3d_{yz}$ orbitals leads to the formation of a bonding 1π and an antibonding $2\pi^*$ molecular orbital, which are shown in Figure 8.8. (A second set of orbitals constructed from the O $2p_x$ and Sc $3d_{xz}$ orbitals is oriented in the y-z plane and is not shown in the figure.) In our thought experiment, electrons will flow from the occupied O $2p$ orbitals to the newly created, initially empty, bonding 1π orbitals because they lie at lower energy than the O atomic orbitals. This interaction is called **ligand-to-metal (L→M) π donation.** O^{2-} can donate all six of its p electrons to Sc^{2+} to form a σ bond and a pair of π bonds resulting in a triple bond overall, in agreement with experiment. This bonding scheme is represented by the following Lewis dot diagram in which the formal charge on Sc is −1 and that on O is +1.

$$\overset{-1}{\cdot\text{Sc}}\equiv\overset{+1}{\text{O}}\text{:}$$

Bonding between a Transition-Metal Atom and an Unsaturated Ligand

A **ligand** is any atom or chemical fragment bonded to a transition-metal atom. An unsaturated ligand, like an unsaturated organic molecule, is not bonded to the maximum number of atoms allowed by its valence. Alternatively, an unsaturated ligand can be defined as one that can accept additional electrons without breaking any bonds in the ligand. The carbonyl (CO) ligand, which is extremely important in inorganic and coordination chemistry, can accept up to four additional electrons in its π^* orbitals, reducing the bond order from three to one, but not to zero. Many stable metal carbonyls exist ($Cr(CO)_6$, $Mn_2(CO)_{10}$, $Ni(CO)_4$, for example). For simplicity, we consider the bonding in a simple metal monocarbonyl, MCO, a number of which can be prepared and stabilized at very low temperatures.

The energy-level diagram for MCO is shown in Figure 8.9. The relative energies of the metal atomic orbitals and the CO molecular orbitals were determined by photoelectron spectroscopy. Recalling that significant interactions occur only

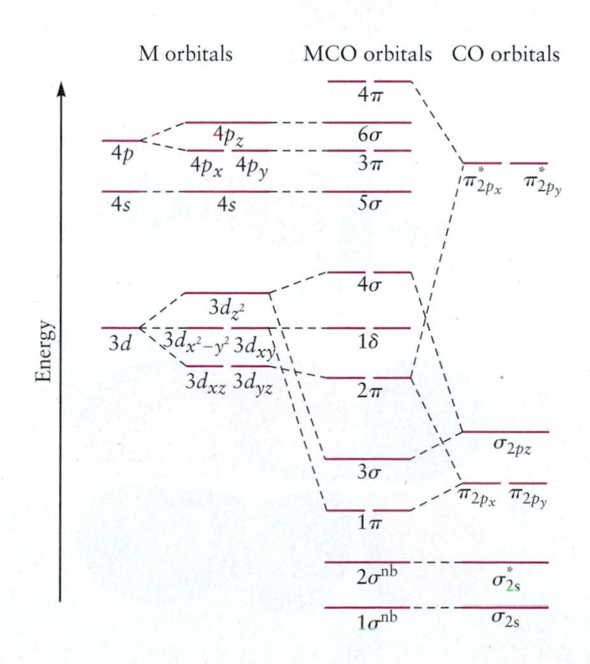

FIGURE 8.9 Orbital correlation diagram for a metal monocarbonyl.

between orbitals that are close in energy, we see that the dominant interactions between the metal atom and CO involve the metal d- electrons and the π_{2p_x} and π_{2p_y}, σ_{2p_z}, and $\pi^*_{2p_x}$ and $\pi^*_{2p_y}$ molecular orbitals of CO.

The orbital overlaps leading to the formation of the MCO molecular orbitals are shown in Figure 8.10. The 1σ and 2σ orbitals have been omitted because they are essentially unperturbed CO molecular orbitals at energies too low to interact with the atomic orbitals of the metal. Similar situations arise in heteronuclear diatomic molecules (ScO, earlier and HF in Fig. 6.20). Starting at the bottom of the diagram we see that a 1π orbital is formed by the overlap of a metal d_{yz}

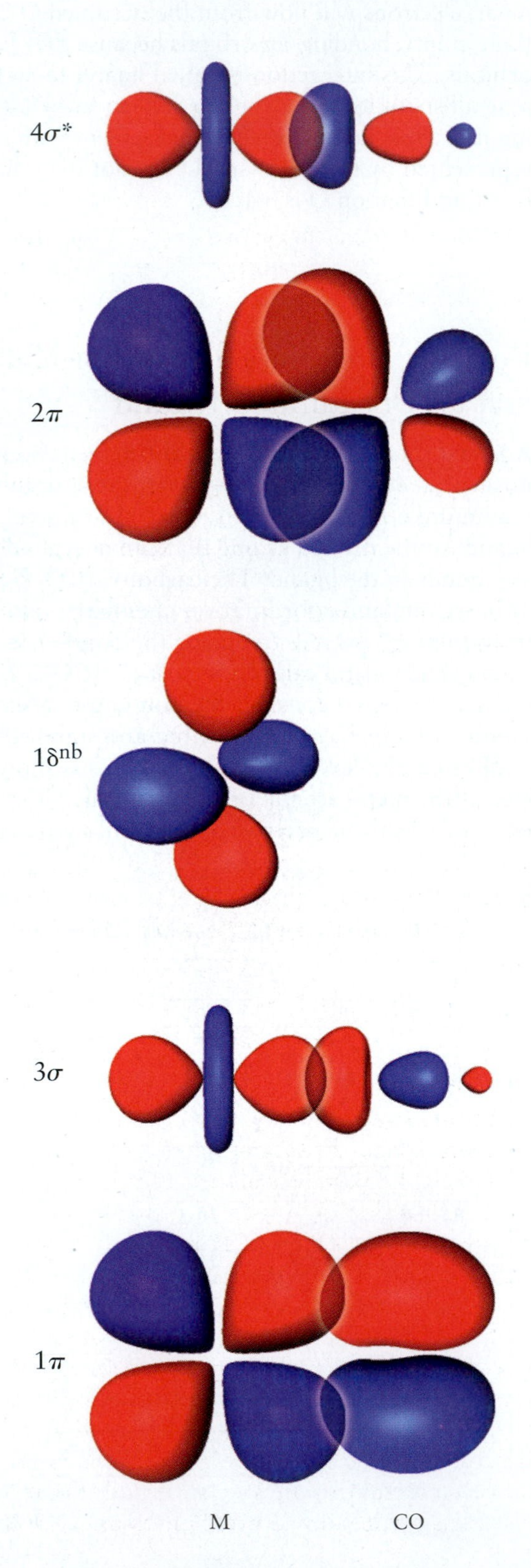

FIGURE 8.10 Overlaps between metal atomic orbitals and molecular orbitals (MOs) of CO leading to bonding, nonbonding and antibonding MOs.

orbital with the bonding CO π_{2p_y} orbital that is located in the y-z plane. A second 1π orbital, oriented in the plane perpendicular to the first, is formed from metal d_{xz} and the other CO π_{2p_y} orbital that is located in the x-z plane. For metal carbonyls in which the metal has few d electrons, electrons will flow from the occupied CO π orbitals to the empty metal d orbitals in the process characterized as ligand-to-metal (L→M) π donation, which was discussed earlier for ScO. The 3σ MCO molecular orbital is formed by the overlap of the $3\,\sigma_{2p_z}$ CO orbital with a metal orbital of the appropriate symmetry, which could be a $4s$, $4p_z$, or $3d_{z^2}$ orbital. Energy considerations favor the participation of the $3d_{z^2}$ orbital, which is illustrated in the figure. L→M σ donation transfers electron density from the filled $3\,\pi_{2p_z}$ CO orbital to the empty metal orbital. The following Lewis dot diagram represents the "ionic" structure M^+CO^- resulting from complete transfer of an electron pair from CO to the metal. The formal charge on the metal is -1 and that on the CO ligand is $+1$.

$$\overset{+1}{M}\leftarrow:\overset{-1}{C}\equiv O:$$

Continuing higher in energy, the next orbitals we encounter are a pair of δ^{nb} nonbinding orbitals derived from the metal $3d_{x^2-y^2}$ and $3d_{xy}$ orbitals. So far the bonding interactions are quite similar to those described earlier for metal oxides like ScO.

In addition to the lone pair of electrons on the carbon atom and the filled π orbitals, CO has empty π^* orbitals that can also participate in bonding. The energy of these orbitals is close to the energy of metal orbitals with the same symmetry so they can be expected to mix to form molecular orbitals. Figure 8.10 shows the formation of a molecular orbital that results from the overlap of the CO $\pi^*_{2p_y}$ orbital with the metal $3d_{yz}$ orbital. This molecular orbital is quite interesting in that it is bonding between M and O but antibonding between C and O. If the metal orbital is initially occupied and the CO $\pi^*_{2p_y}$ orbital is initially empty, electron density will flow from the metal to the ligand in a process called **metal-to-ligand (M→L) π donation.** M→L π donation transfers electron density from occupied metal atomic orbitals to unoccupied ligand molecular orbitals, in this case, from the metal $3d_{xz}$ and $3d_{yz}$ orbitals to the CO $\pi^*_{2p_x}$ and $\pi^*_{2p_y}$ orbitals, respectively, partially offsetting the effects of L→M σ donation described earlier. M→L π donation strengthens the M—C bond, weakens the C—O bond, and restores electron density on the O atom. L→M σ and M→L π bonding are synergistic; they strengthen the M—C bond and reduce excess negative charge on the metal (the electroneutrality principle again). Because M→L π bonding transfers charge density back to the ligand, it is generally referred to as **backbonding.** The Lewis dot diagram that results from both L→M σ donation and M→L π donation is shown below.

$$M{=}C{=}\ddot{O}:$$

Both the M—C and C—O bonds are now double bonds and the formal charges on both the metal and the ligand are now zero. These conclusions are supported by infrared spectroscopy (see Section 20.2), which shows that the CO bond does indeed weaken when coordinated to a metal atom. The experimental metal-carbon bond length observed is also consistent with a M—C double bond.

L→M π donation is certainly possible for complexes with ligands that have empty π^* orbitals because they also have filled π orbitals that can donate electron density to empty metal orbitals. The experimental evidence cited earlier suggests that L→M π donation is less important than M→L π donation for metal carbonyls. As a general rule, L→M π donation is more important for metals with few d electrons (high oxidation states), and M→L π backbonding is more important for metals with largely occupied d orbitals (low oxidation states).

Inorganic chemists have put their own special twist on the molecular orbital description of bonding developed in Chapter 6. Instead of first developing molecular orbitals from all the atomic orbitals in the molecule, and then putting in all the electrons by the aufbau principle to arrive at the ground-state configuration, they start by considering charge transfer between the metal atom and its ligands to form dative bonds. Because the energy of an electron pair initially localized in an atomic orbital decreases as the electrons become more delocalized on charge transfer, they reason that increased charge transfer by σ and π donation stabilizes the bonding molecular orbital and de-stabilizes the antibonding molecular orbital. (Recall our virial theorem arguments for the stabilizing effect of de-localization in Section 6.2.) Even so, when inorganic chemists refer to ligand-to-metal σ and π donation and metal-to-ligand π donation, they are implicitly considering two issues. First, formation of the molecular orbitals is governed by our familiar consideration of energy proximity and orbital overlap. Strong overlap of orbitals close in energy results in stable bonding molecular orbitals and unstable antibonding molecular orbitals. Weaker interactions lead to less stabilization of the bonding molecular orbitals and less destabilization of the antibonding molecular orbitals. The second issue is the source of the electrons: Instead of building up the ground state for the complete molecule, inorganic chemists obtain a great deal of qualitative insight into structure and bonding from assessing the relative energies and orbital occupancies of the atoms involved and the ability of each to donate or accept charge. The interactions in the molecule are always the same, but they can be described either in the language of basic molecular orbital theory or the language of inorganic chemistry. Choosing which one to use is a matter of convenience. In Section 8.6, we examine how the viewpoint of inorganic chemists effectively describes the bonding and energy levels in coordination complexes.

8.3 Introduction to Coordination Chemistry

Formation of Coordination Complexes

The Alsatian-Swiss chemist Alfred Werner pioneered the field of coordination chemistry in the late nineteenth century. At that time, a number of compounds of cobalt(III) chloride with ammonia were known. They had the following chemical formulas and colors:

Compound 1: $CoCl_3 \cdot 6NH_3$	Orange-yellow
Compound 2: $CoCl_3 \cdot 5NH_3$	Purple
Compound 3: $CoCl_3 \cdot 4NH_3$	Green
Compound 4: $CoCl_3 \cdot 3NH_3$	Green

Treatment of these compounds with aqueous hydrochloric acid did not remove the ammonia, which suggested that the ammonia was somehow closely bound with the cobalt ions. Treatment with aqueous silver nitrate at 0°C, however, gave interesting results. With compound 1, all of the chloride present precipitated as solid AgCl. With compound 2, only two thirds of the chloride precipitated, and with compound 3, only one third precipitated. Compound 4 did not react at all with the silver nitrate. These results suggested that there were two different kinds of species associated with the cobalt ions, which Werner called *valences* (recall that the electron, the key player in the formation of the chemical bond, was just being characterized by J.J. Thomson). The primary, or ionizable, valences were anions like Cl^- in simple salts such as $CoCl_3$, whereas the secondary valences could be either simple anions or neutral molecules such as NH_3. Werner assumed that the primary valences were nondirectional, whereas the secondary valences were oriented along well-defined directions in space. The picture that emerged was that of a metal atom coordinated to the secondary valences (ligands) in the **inner coordination sphere** surrounded by the primary valences and solvent in the **outer coordination sphere.** The primary valences neutralize the charge on the complex ion. Werner accounted for experimental results described earlier by positing the existence of **coordination**

complexes with six ligands (either chloride ions or ammonia molecules) attached to each Co^{3+} ion. Specifically, he wrote the formulas for compounds 1 through 4 as

$$\text{Compound 1: } [Co(NH_3)_6]^{3+}\ (Cl^-)_3$$

$$\text{Compound 2: } [Co(NH_3)_5Cl]^{2+}\ (Cl^-)_2$$

$$\text{Compound 3: } [Co(NH_3)_4Cl_2]^{+}\ (Cl^-)$$

$$\text{Compound 4: } [Co(NH_3)_3Cl_3]$$

with the charge on each complex ion being balanced by an equal number of Cl^- ions in the outer coordination sphere. Only these chloride ions, which were *not* bonded directly to cobalt, could react with the silver ions in cold aqueous silver nitrate to form the AgCl precipitate.

Werner realized that he could test his hypothesis by measuring the electrical conductivity of aqueous solutions of the salts of these complex ions. Ions are the electrical conductors in aqueous solutions, and the conductivity is proportional to the ion concentration. If Werner's proposal was correct, then an aqueous solution of Compound 1, for example, should have a molar conductivity close to that of an aqueous solution of $Al(NO_3)_3$, which also forms four ions per formula unit on complete dissociation in water (one 3+ ion and three 1− ions). His experiments confirmed that the conductivities of these two solutions were, indeed, similar. Furthermore the conductivity of aqueous solutions of compound 2 was close to those of $Mg(NO_3)_2$, and solutions of compound 3 conducted electricity about as well as those containing $NaNO_3$. Compound 4, in contrast, did not dissociate into ions when dissolved in water, producing a solution of very low electrical conductivity.

Werner and other chemists studied a variety of other coordination complexes, using both physical and chemical techniques. Their research has shown that the most common coordination number by far is 6, as in the cobalt complexes discussed earlier. Coordination numbers ranging from 1 to 16 are commonly observed, however. Examples include coordination numbers 2 (as in $[Ag(NH_2)_2]^+$), 4 (as in $[PtCl_4]^{2-}$), and 5 (as in $[Ni(CN)_5]^{3-}$).

A second example illustrates the ability of transition metals to form complexes with small molecules and ions. Copper metal and hot concentrated sulfuric acid ("oil of vitriol") react to form solid copper(II) sulfate, commonly called "blue vitriol" by virtue of its deep blue color. There is more to this compound than copper and sulfate, however; it contains water as well. When the water is driven away by heating, the blue color vanishes, leaving greenish white anhydrous copper(II) sulfate (Fig. 8.11). The blue color of blue vitriol comes from a

FIGURE 8.11 Hydrated copper(II) sulfate, $CuSO_4 \cdot 5H_2O$, is blue (left), but the anhydrous compound, $CuSO_4$, is greenish white (right). A structural study of the solid compound demonstrates that four of the water molecules are closely associated with the copper and the fifth is not. Thus, a better representation of the hydrated compound is $[Cu(H_2O)_4]SO_4 \cdot H_2O$.

TABLE 8.4 Common Ligands and Their Names

Ligand†	Name
$:NO_2^-$	Nitro
$:OCO_2^{2-}$	Carbonato
$:ONO^-$	Nitrito
$:CN^-$	Cyano
$:SCN^-$	Thiocyanato
$:NCS^-$	Isothiocyanato
$:OH^-$	Hydroxo
$:OH_2$	Aqua
$:NH_3$	Ammine
$:CO$	Carbonyl
$:NO^+$	Nitrosyl

†The ligating atom is indicated by a pair of dots (:) to show a lone pair of electrons. In the CO_3^{2-} ligand, either one or both of the O atoms can donate a lone pair to a metal.

coordination complex in which H_2O molecules bond directly to Cu^{2+} ions to form coordination complexes with the formula $[Cu(H_2O)_4]^{2+}$. Bonding in this complex ion can be described qualitatively using the Lewis theory of acids and bases (see Section 15.1). A Lewis acid is an electron-pair donor and a Lewis base is an electron-pair acceptor; thus, the transfer of a pair of electrons from water to Cu^{2+} to form a dative bond can be thought of as an acid–base reaction As a Lewis acid, the Cu^{2+} ion *coordinates* four water molecules into a group by accepting electron density from a lone pair on each. By acting as electron-pair donors and sharing electron density with the Cu^{2+} ion, the four water molecules are the ligands, occupying the inner coordination sphere of the ion. The chemical formula of blue vitriol is $[Cu(H_2O)_4]SO_4{\cdot}H_2O$; the fifth water molecule is not coordinated directly to copper.

The positive ions of every metal in the periodic table accept electron density to some degree and can therefore coordinate surrounding electron donors, even if only weakly. The solvation of Na^+ by H_2O molecules in aqueous solution (see Fig. 10.6) is an example of weak coordination. The ability to make fairly strong, *directional* bonds by accepting electron pairs from neighboring molecules or ions is characteristic of the transition-metal elements. Coordination occupies a middle place energetically between the weak intermolecular attractions in solids (see Chapter 10) and the stronger covalent and ionic bonding (see Chapters 3 and 6). Thus, heating blue vitriol disrupts the $Cu{-}H_2O$ bonds at temperatures well below those required to break the covalent bonds in the SO_4^{2-} group. The energy (more precisely, the enthalpy) required to break a $M^{2+}{-}H_2O$ bond in a transition metal coordination complex falls in the range between 170 and 210 kJ mol^{-1}. This bond dissociation energy is far less than the bond energies of the strongest chemical bonds (e.g., 942 kJ mol^{-1} for N_2), but it is by no means small. Trivalent metal cations (+3 charge) make still stronger coordinate bonds with water.

A wide variety of molecules and ions bond to metals as ligands; common ones include the halide ions (F^-, Cl^-, Br^-, I^-), ammonia (NH_3), carbon monoxide (CO), water, and a few other simple ligands listed in Table 8.4. Ligands that bond to a metal atom through a single point of attachment are called *monodentate* (derived from Latin *mono*, meaning "one," plus *dens*, meaning "tooth," indicating that they bind at only one point). More complex ligands can bond through two or more attachment points; they are referred to as *bidentate, tridentate,* and so forth. Ethylenediamine ($NH_2CH_2CH_2NH_2$), in which two NH_2 groups are held together by a carbon backbone, is a particularly important bidentate ligand. Both N atoms in ethylenediamine have lone electron pairs to share. If all the nitrogen donors of three ethylenediamine molecules bind to a single Co^{3+} ion, then that ion has a coordination number of 6 and the formula of the resulting complex is $[Co(en)_3]^{3+}$ (where "en" is the abbreviation for ethylenediamine). Complexes in which a ligand coordinates via two or more donors to the same central atom are called **chelates** (derived from Greek *chele*, meaning "claw," because the ligand grabs onto the central atom like a pincers). Figure 8.12 shows the structures of some important chelating ligands.

In the modern version of Werner's notation, brackets are used in writing cations or anions, but the net charges of the ions are not shown. In the formula $[Pt(NH_3)_6]Cl_4$, the portion in brackets represents a positively charged coordination complex in which Pt coordinates six NH_3 ligands. The brackets emphasize

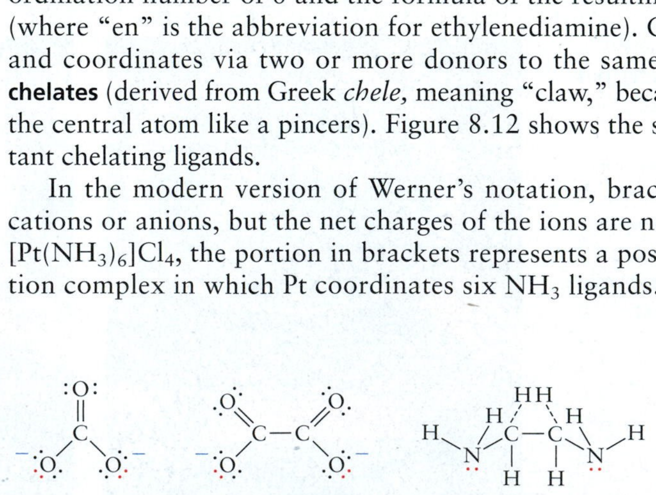

FIGURE 8.12 These three ligands are bidentate; each is capable of donating two pairs of electrons.

Carbonate ion, CO_3^{2-}

Oxalate ion, $C_2O_4^{2-}$

Ethylenediamine, $NH_2CH_2CH_2NH_2$

that a complex is a distinct chemical entity with its own properties. Within the brackets, the symbol of the central atom comes first. The electric charge on a coordination complex is the sum of the oxidation number of the metal ion and the charges of the ligands that surround it. Thus, the complex of copper(II) (Cu^{2+}) with four Br^- ions is an anion with a −2 charge, $[CuBr_4]^{2-}$.

EXAMPLE 8.1

Determine the oxidation state of the coordinated metal atom in each of the following compounds:

(a) $K[Co(NH_3)_2(CN)_4]$; (b) $Os(CO)_5$; (c) $Na[Co(H_2O)_3(OH)_3]$.

SOLUTION

(a) The oxidation state of K is known to be +1, so the complex in brackets is an anion with a −1 charge, $[Co(NH_3)_2(CN)_4]^-$. The charge on the two NH_3 ligands is 0, and the charge on each of the four CN^- ligands is −1. The oxidation state of the Co must then be +3, because 4×-1 (for the CN^-) + 2×0 (for the NH_3) + 3 (for Co) equals the required −1.

(b) The ligand CO has zero charge, and the complex has zero charge as well. Therefore, the oxidation state of the osmium is 0.

(c) There are three neutral ligands (the water molecules) and three ligands with −1 charges (the hydroxide ions). The Na^+ ion contributes only +1; thus, the oxidation state of the cobalt must be +2.

Related Problems: 9, 10

Coordination modifies the chemical and physical properties of both the central atom and the ligands. Consider the chemistry of aqueous cyanide (CN^-) and iron(II) (Fe^{2+}) ions. The CN^- ion reacts immediately with acid to generate gaseous HCN, a deadly poison, and Fe^{2+}, when mixed with aqueous base, instantly precipitates a gelatinous hydroxide. The reaction between Fe^{2+} and CN^- produces the complex ion $[Fe(CN)_6]^{4-}(aq)$, which undergoes neither of the two reactions just described nor any others considered characteristic of CN^- or Fe^{2+}. Ions or molecules may be present in multiple forms in the same compound. The two Cl^- ions in $[Pt(NH_3)_3Cl]Cl$ are chemically different, because one is coordinated and the other is not. Treatment of an aqueous solution of this substance with Ag^+ immediately precipitates the uncoordinated Cl^- as $AgCl(s)$, but *not* the coordinated Cl^- just as it did for Werner's complexes discussed earlier.

Ionic coordination complexes of opposite charges can combine with each other—just as any positive ion can combine with a negative ion—to form a salt. For example, the cation $[Pt(NH_3)_4]^{2+}$ and the anion $[PtCl_4]^{2-}$ form a doubly complex ionic compound of formula $[Pt(NH_3)_4][PtCl_4]$. This compound and the following four compounds

$$Pt(NH_3)_2Cl_2 \qquad [Pt(NH_3)_3Cl][Pt(NH_3)Cl_3]$$

$$[Pt(NH_3)_3Cl]_2[PtCl_4] \qquad [Pt(NH_3)_4][Pt(NH_3)Cl_3]_2$$

all contain Pt, NH_3, and Cl in the ratio of 1:2:2; that is, they have the same percentage composition. Two pairs even have the same molar mass. Yet, the five compounds differ in structure and in physical and chemical properties. The concept of coordination organizes an immense number of chemical compositions and patterns of reactivity by considering combinations of ligands linked in varied ratios with central metal atoms or ions.

Naming Coordination Compounds

So far, we have identified coordination compounds only by their chemical formulas, but names are also useful for many purposes. Some substances were named before their structures were known. Thus, $K_3[Fe(CN)_6]$ was called potassium ferricyanide, and $K_4[Fe(CN)_6]$ was potassium ferrocyanide [these are complexes of Fe^{3+} (ferric) and Fe^{2+} (ferrous) ions, respectively]. These older names are still used conversationally but systematic names are preferred to avoid ambiguity. The definitive source for the naming of inorganic compounds is *Nomenclature of Inorganic Chemistry-IUPAC Recommendations 2005* (N. G. Connelly and T. Damhus, Sr., Eds. Royal Society of Chemistry, 2005).

1. The name of a coordination complex is written as a single word built from the names of the ligands, a prefix before each ligand to indicate how many ligands of that kind are present in the complex, and a name for the central metal.
2. Compounds containing coordination complexes are named following the same rules as those for simple ionic compounds: The positive ion is named first, followed (after a space) by the name of the negative ion.
3. Anionic ligands are named by replacing the usual ending with the suffix *-o*. The names of neutral ligands are unchanged. Exceptions to the latter rule are aqua (for water), ammine (for NH_3), and carbonyl (for CO) (see Table 8.4).
4. Greek prefixes (*di-*, *tri-*, *tetra-*, *penta-*, *hexa-*) are used to specify the number of ligands of a given type attached to the central ion, if there is more than one. The prefix *mono-* (for one) is not used. If the name of the ligand itself contains a term such as *mono-* or *di-* (as in ethylenediamine), then the name of the ligand is placed in parentheses and the prefixes *bis-*, *tris-*, and *tetrakis-* are used instead of *di-*, *tri-*, and *tetra-*.
5. The ligands are listed in alphabetical order, without regard for the prefixes that tell how often each type of ligand occurs in the coordination sphere.
6. A Roman numeral, enclosed in parentheses placed immediately after the name of the metal, specifies the oxidation state of the central metal atom. If the complex ion has a net negative charge, the ending *-ate* is added to the stem of the name of the metal.

Examples of complexes and their systematic names follow.

Complex	Systematic Name
$K_3[Fe(CN)_6]$	Potassium hexacyanoferrate(III)
$K_4[Fe(CN)_6]$	Potassium hexacyanoferrate(II)
$Fe(CO)_5$	Pentacarbonyliron(0)
$[Co(NH_3)_5CO_3]Cl$	Penta-amminecarbonatocobalt(III) chloride
$K_3[Co(NO_2)_6]$	Potassium hexanitrocobaltate(III)
$[Cr(H_2O)_4Cl_2]Cl$	Tetra-aquadichlorochromium(III) chloride
$[Pt(NH_2CH_2CH_2NH_2)_3]Br_4$	Tris(ethylenediamine)platinum(IV) bromide
$K_2[CuCl_4]$	Potassium tetrachlorocuprate(II)

EXAMPLE 8.2

Interpret the names and write the formulas of these coordination compounds:

(a) sodium tricarbonatocobaltate(III)

(b) diamminediaquadichloroplatinum(IV) bromide

(c) sodium tetranitratoborate(III)

SOLUTION

(a) In the anion, three carbonate ligands (with −2 charges) are coordinated to a cobalt atom in the +3 oxidation state. Because the complex ion thus has an overall charge of −3, three sodium cations are required, and the correct formula is $Na_3[Co(CO_3)_3]$.

(b) The ligands coordinated to a Pt(IV) include two ammonia molecules, two water molecules, and two chloride ions. Ammonia and water are electrically neutral, but the two chloride ions contribute a total charge of $2 \times (-1) = -2$ that sums with the +4 of the platinum and gives the complex ion a +2 charge. Two bromide anions are required to balance this; thus, the formula is $[Pt(NH_3)_2(H_2O)_2Cl_2]Br_2$.

(c) The complex anion has four −1 nitrate ligands coordinated to a central boron(III). This gives a net charge of −1 on the complex ion and requires one sodium ion in the formula $Na[B(NO_3)_4]$.

Related Problems: 11, 12

Ligand Substitution Reactions

We discuss a few ligand substitution reactions to give you a feel for the properties and reactions of coordination complexes. These simple reactions are aptly named; one or more ligands are simply substituted for one another. We have already discussed one example of a series of ligand substitution reactions: the exchange of NH_3 and Cl^- in Werner's cobalt complexes.

If the yellow crystalline solid nickel(II) sulfate is exposed to moist air at room temperature, it takes up six water molecules per formula unit. These water molecules coordinate the nickel ions to form a bright green complex:

$$\underset{\text{Yellow}}{NiSO_4(s)} + \underset{\text{Colorless}}{6\,H_2O(g)} \rightleftharpoons \underset{\text{Green}}{[Ni(H_2O)_6]SO_4(s)}$$

Heating the green hexa-aquanickel(II) sulfate to temperatures well above the boiling point of water drives off the water and regenerates the yellow $NiSO_4$ in the reverse reaction. A different coordination complex forms when yellow $NiSO_4(s)$ is exposed to gaseous ammonia, $NH_3(g)$. This time, the product is a blue-violet complex:

$$\underset{\text{Yellow}}{NiSO_4(s)} + \underset{\text{Colorless}}{6\,NH_3(g)} \rightleftharpoons \underset{\text{Blue-violet}}{[Ni(NH_3)_6]SO_4(s)}$$

Heating the blue-violet product drives off ammonia, and the color of the solid returns to yellow. Given these facts, it is not difficult to explain the observation that a green $[Ni(H_2O)_6]^{2+}(aq)$ solution turns blue-violet when treated with $NH_3(aq)$ (Fig. 8.13). NH_3 must have displaced the H_2O ligands from the coordination sphere forming the blue-violet $[Ni(NH_3)_6]^{2+}$ complex.

$$\underset{\text{Green}}{[Ni(H_2O)_6]^{2+}(aq)} + \underset{\text{Colorless}}{6\,NH_3(aq)} \longrightarrow \underset{\text{Blue-violet}}{[Ni(NH_3)_6]^{2+}(aq)} + \underset{\text{Colorless}}{6\,H_2O}$$

FIGURE 8.13 When ammonia is added to the green solution of nickel(II) sulfate on the left (which contains $[Ni(H_2O)_6]^{2+}$ ions), ligand substitution occurs to give the blue–violet solution on the right (which contains $[Ni(NH_3)_6]^{2+}$ ions).

Labile complexes are those for which ligand substitution reactions proceed rapidly. Those in which substitution proceeds slowly or not at all are **inert.** These terms are used to describe the different kinetics (rates) of the reactions (see Chapter 18) that are thermodynamically allowed (see Chapter 13). Large activation energy barriers for ligand substitution reactions of inert complexes make those reactions slow even though there may be a thermodynamic tendency to proceed. In the substitution reaction

$$[Co(NH_3)_6]^{3+}(aq) + 6\ H_3O^+(aq) \rightleftharpoons [Co(H_2O)_6]^{3+}(aq) + 6\ NH_4^+(aq)$$

the products are favored thermodynamically by an enormous amount, yet the inert $[Co(NH_3)_6]^{3+}$ complex ion lasts for weeks in acidic solution because there is no low-energy pathway for the reaction. The cobalt(III) ion, $[Co(NH_3)_6]^{3+}$, is thermodynamically unstable relative to $[Co(H_2O)_6]^{3+}$, but kinetically stable (inert). The closely related cobalt(II) complex, $[Co(NH_3)_6]^{2+}$, reacts with water in a matter of seconds:

$$[Co(NH_3)_6]^{2+}(aq) + 6\ H_3O^+(aq) \longrightarrow [Co(H_2O)_6]^{2+}(aq) + 6\ NH_4^+(aq)$$

The hexa-aminecobalt(II) complex is thermodynamically unstable and also kinetically labile.

Ligand substitution reactions proceed in stages and they can usually be stopped at intermediate stages by controlling the reaction conditions. The following series of stable complexes represents all possible four-coordinate compositions of Pt(II) with the two ligands NH_3 and Cl^-.

$$[Pt(NH_3)_4]^{2+},\ [Pt(NH_3)_3Cl]^+,\ [Pt(NH_3)_2Cl_2],\ [Pt(NH_3)Cl_3]^-,\ \text{and}\ [PtCl_4]^{2-}$$

Such mixed-ligand complexes are wonderful examples of the variety and richness of coordination chemistry.

8.4 Structures of Coordination Complexes

Octahedral Geometries

What is the geometric structure of the complex $[Co(NH_3)_6]^{3+}$? This question naturally occurred to Werner, who suggested that the arrangement should be the simplest and most symmetric possible, with the ligands positioned at the six vertices of a regular octahedron (Fig. 8.14). Modern methods of x-ray diffraction (see Section 21.1) enable us to make precise determinations of atomic positions in crystals and have confirmed Werner's proposed octahedral structure for this complex. X-ray diffraction techniques were not available in the late 19th century, however, so Werner turned to a study of the properties of substituted complexes to test his hypothesis. Having a set of molecular ball-and-stick models at hand as you read this section will make it much easier for you to visualize the structures and transformations described.

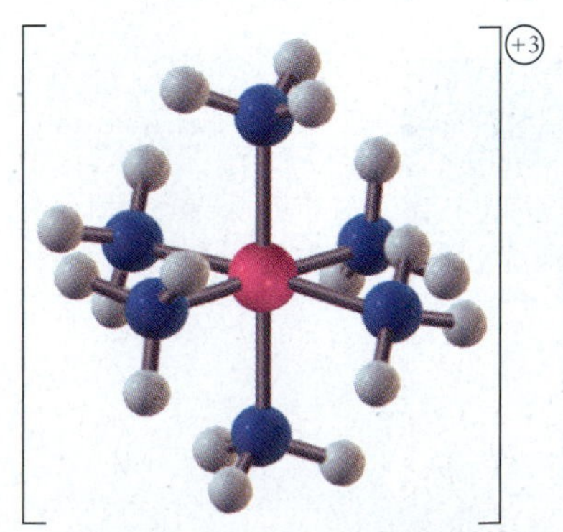

FIGURE 8.14 Octahedral structure of the $Co(NH_3)_6^{3+}$ ion. All six corners of the octahedron are equivalent.

Replacing one ammonia ligand by a chloride ion results in a complex with the formula $[Co(NH_3)_5Cl]^{2+}$, in which one vertex of the octahedron is occupied by Cl^- and the other five by NH_3. Only one structure of this type is possible, because all six vertices of a regular octahedron are equivalent and the various singly substituted complexes $[MA_5B]$ (where $A = NH_3$, $B = Cl^-$, $M = Co^{3+}$, for example) can be superimposed on one another. Now, suppose a second NH_3 ligand is replaced by Cl^-. The second Cl^- can lie in one of the four equivalent positions closest to the first Cl^- (in the horizontal plane; see Fig. 8.15a) or in position labeled 3, on the opposite side of the central metal atom (see Fig. 8.15b). The first of these **geometric isomers,** in which the two Cl^- ligands are closer to each other, is called *cis*-$[Co(NH_3)_4Cl_2]^+$, and the second, with the two Cl^- ligands farther apart, is called *trans*-$[Co(NH_3)_4Cl_2]^+$. The octahedral structure model predicts that there can be only two different ions with the chemical formula $[Co(NH_3)_4Cl_2]^+$. You

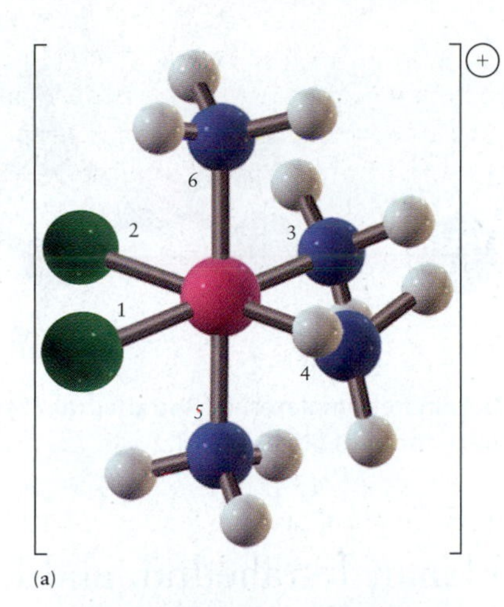

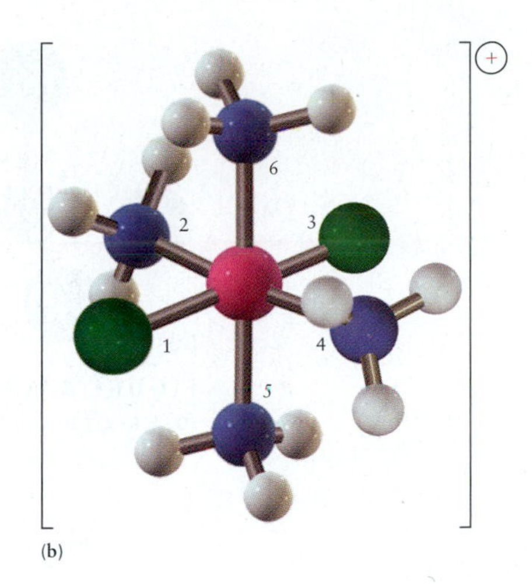

FIGURE 8.15 (a) The *cis*-$[Co(NH_3)_4Cl_2]^+$ and (b) *trans*-$[Co(NH_3)_4Cl_2]^+$ ions. The *cis* complex is purple in solution, but the *trans* complex is green.

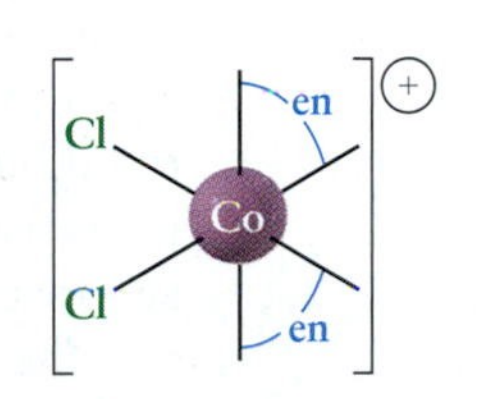

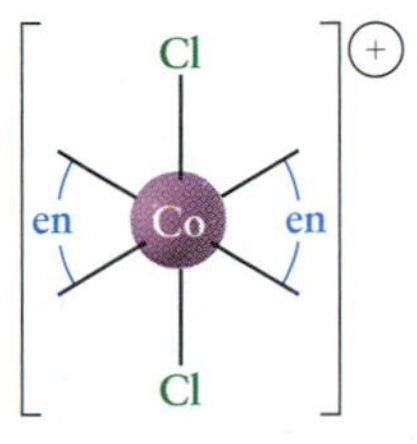

FIGURE 8.16 The complex ion $[CoCl_2(en)_2]^+$ is an octahedral complex that has *cis* and *trans* isomers, according to the relative positions of the two Cl^- ligands. Salts of the *cis* isomers are purple, and salts of the *trans* isomers are green.

may have already encountered geometric isomers in Chapter 7. When Werner began his work, only the green *trans* (across) form was known, but by 1907, he had prepared the *cis* (near) isomer and shown that it differed from the *trans* isomer in color (it was violet rather than green) and other physical properties. The isolation of two, and only two, geometric isomers of this ion was good (although not conclusive) evidence that the octahedral structure was correct. Similar isomerism is displayed by the complex ion $[CoCl_2(en)_2]^+$, which also exists in a purple *cis* form and a green *trans* form (Fig. 8.16).

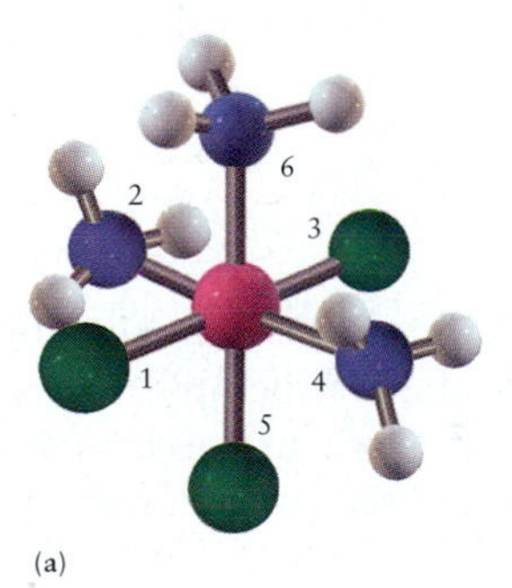

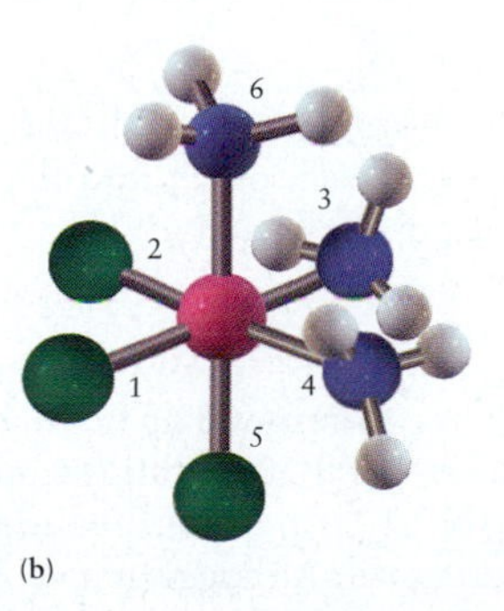

FIGURE 8.17 Two structural isomers of the coordination compound $Co(NH_3)_3Cl_3$.

EXAMPLE 8.3

How many geometric isomers does the octahedral coordination compound $[Co(NH_3)_3Cl_3]$ have?

SOLUTION

We begin with the two isomers of $[Co(NH_3)_4Cl_2]^+$ shown in Figure 8.15 and see how many different structures can be made by replacing one of the ammonia ligands with Cl^-.

Starting with the *trans* form (see Fig. 8.15b), it is clear that replacement of any of the four NH_3 ligands at site 2, 4, 5, or 6 gives a set of equivalent structures that can be superimposed on one another by rotation of the starting structure. Figure 8.17a shows the structure that results from substitution at the position 5. What isomers can be made from the *cis* form shown in Figure 8.17a? If either the ammonia ligand at site 3 or the one at site 4 (i.e., one of the two that are *trans* to existing Cl^- ligands) is replaced, the result is simply a rotated version of Figure 8.17a; therefore, these replacements do *not* give another isomer. Replacement of the ligand at site 5 or 6, however, gives a different structure (see Fig. 8.17b). (Note that Cl^- occupying sites 1, 2, 5 and 1, 2, 6 results in equivalent structures.)

We conclude that there are two, and only two, possible isomers of the octahedral complex structure MA_3B_3. In fact, only one form of $[Co(NH_3)_3Cl_3]$ has been prepared to date, presumably because the two isomers interconvert rapidly. However, two isomers are known for the closely related coordination complex $[Cr(NH_3)_3(NO_2)_3]$.

Related Problems: 17, 18

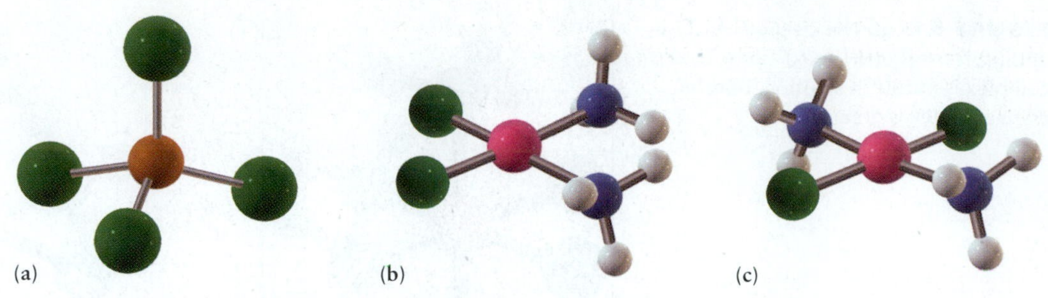

FIGURE 8.18 Four-coordinate complexes. (a) Tetrahedral, $[FeCl_4]^-$. (b, c) Square planar, illustrating (b) the *cis* and (c) the *trans* forms of $[Pt(NH_3)_2Cl_2]$.

Square-Planar, Tetrahedral, and Linear Geometries

Complexes with coordination numbers of 4 are typically either tetrahedral or square planar. The tetrahedral geometry (Fig. 8.18a) predominates for four-coordinate complexes of the early transition metals (those toward the left side of the *d* block of elements in the periodic table). Geometric isomerism is not possible for tetrahedral complexes of the general form MA_2B_2, because all four tetrahedral sites are completely equivalent.

The square-planar geometry (see Figs. 8.18b,c) is common for four-coordinate complexes of Au^{3+}, Ir^+, Rh^+, and especially common for ions with the d^8 valence electron configurations: Ni^{2+}, Pd^{2+}, and Pt^{2+}, for example. The Ni^{2+} ion forms a few tetrahedral complexes, but four-coordinate Pd^{2+} and Pt^{2+} are nearly always square planar. Square-planar complexes of the type MA_2B_2 can have isomers, as illustrated in Figures 8.18b and c for *cis*- and *trans*-$[Pt(NH_3)_2Cl_2]$. The *cis* form of this compound is a potent and widely used anticancer drug called cisplatin, but the *trans* form has no therapeutic properties.

Finally, linear complexes with coordination numbers of 2 are known, especially for ions with d^{10} configurations such as Cu^+, Ag^+, Au^+, and Hg^{2+}. The central Ag atom in a complex such as $[Ag(NH_3)_2]^+$ in aqueous solution strongly attracts several water molecules as well, however, so its actual coordination number under these circumstances may be greater than 2.

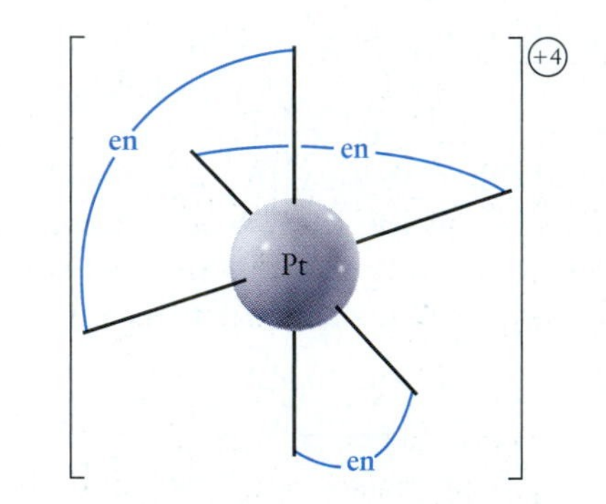

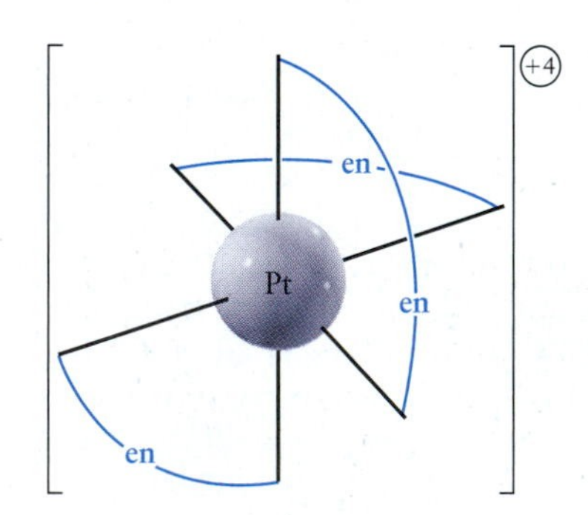

en = $H_2\ddot{N}–CH_2CH_2–\ddot{N}H_2$

FIGURE 8.19 Enantiomers of the $[Pt(en)_3]^{4+}$ ion. Reflection through the mirror plane transforms one enantiomer into the other. The two cannot be superimposed by simple rotation.

Chiral Structures

Molecules that rotate plane polarized light in opposite directions are called optical isomers (see Section 7.1, Fig. 7.9). They typically have chiral structures that cannot be superimposed on their mirror images by rotation. The two structures shown in Figure 8.19 for the complex ion $[Pt(en)_3]^{4+}$ are examples of such a mirror-image pair.

EXAMPLE 8.4

Suppose that the complex ion $[Co(NH_3)_2(H_2O)_2Cl_2]^+$ is synthesized with the two ammine ligands *cis* to each other, the two aqua ligands *cis* to each other, and the two chloro ligands *cis* to each other (Fig. 8.20a). Is this complex optically active?

SOLUTION

We represent a mirror by a shaded line and create the mirror image by making each point in the image lie at the same distance from the shaded line as the corresponding point in the original structure (see Fig. 8.20). Comparing the original (see Fig 8.20a) with the mirror image (see Fig. 8.20b) shows that *cis, cis*-$[Co(NH_3)_2(H_2O)_2Cl_2]^+$ is chiral, because the two structures cannot be superimposed even after they are turned. Although many chiral complexes contain chelating ligands, this example proves that nonchelates can be chiral.

Related Problems: 19, 20

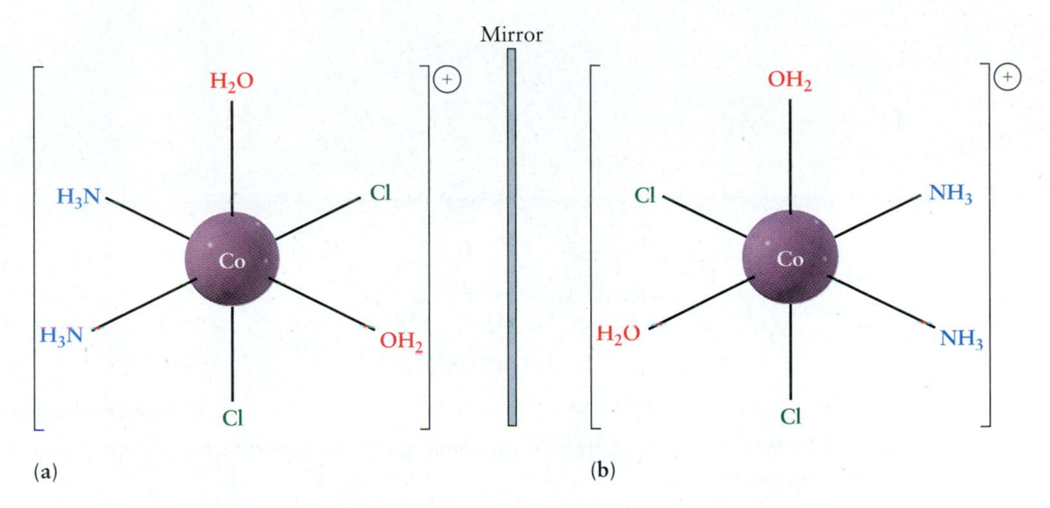

FIGURE 8.20 The structure of (a) the all-*cis* $[Co(NH_3)_2(H_2O)_2Cl_2]^+$ complex ion, together with (b) its mirror image.

The hexadentate ligand EDTA (ethylenediaminetetra-acetate ion) forms chiral complexes. Figure 8.21 shows the structure of this chelating ligand coordinated to a Co^{3+} ion. The central metal ion is literally "enveloped" by the ligand forming six coordinate covalent bonds with two nitrogen atoms and four oxygen anions. A chelating agent like EDTA has a strong affinity for certain metal ions and can **sequester** them effectively in solution. EDTA solubilizes the scummy precipitates that Ca^{2+} ion forms with anionic constituents of soap by forming a stable complex with Ca^{2+}. In so doing it breaks up the main contributor to bathtub rings and it is a "miracle ingredient" in some bathtub cleaners. EDTA is also used to recover trace contaminants from water (some metal ions, especially heavy ones, are toxic). It has been used as an antidote for lead poisoning because of its great affinity for Pb^{2+} ions. Iron complexes of EDTA in plant foods permit a slow release of iron to the plant. EDTA also sequesters copper and nickel ions in edible fats and oils. Because these metal ions catalyze the oxidation reactions that turn oils rancid, EDTA preserves freshness.

Coordination Complexes in Biology

Coordination complexes, particularly chelates, play fundamental roles in the biochemistry of both plants and animals. Trace amounts of at least nine transition elements are essential to life—vanadium, chromium, manganese, iron, cobalt, nickel, copper, zinc, and molybdenum.

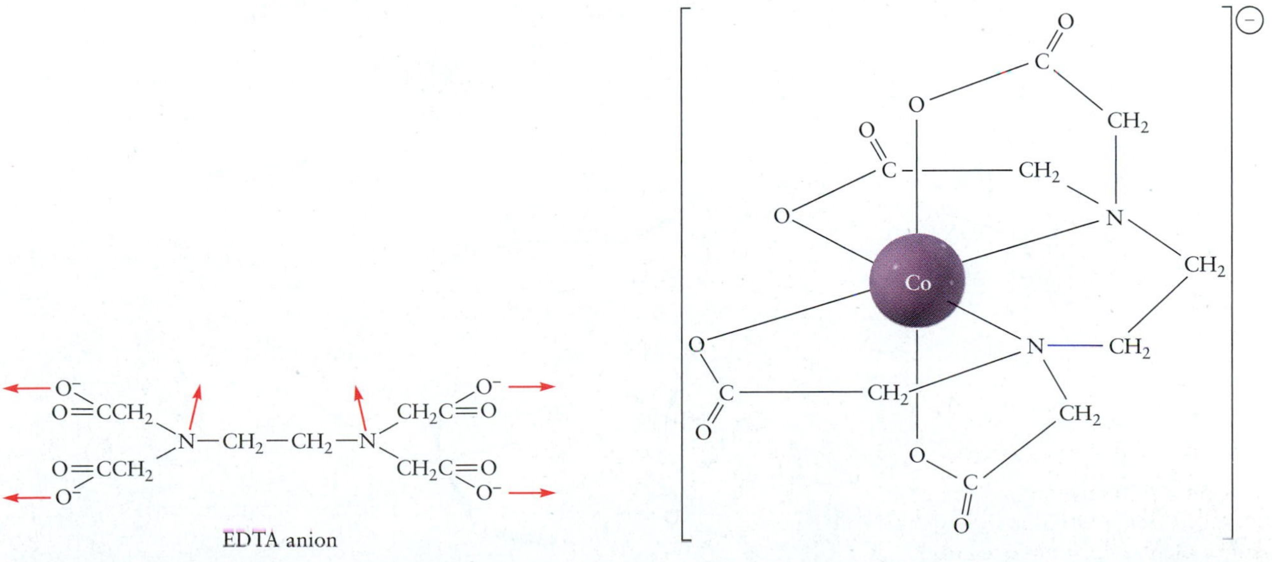

EDTA anion

FIGURE 8.21 The chelation complex of EDTA with cobalt(III). Each EDTA ion has six donor sites at which it can bind (by donating lone-pair electrons) to the central metal ion.

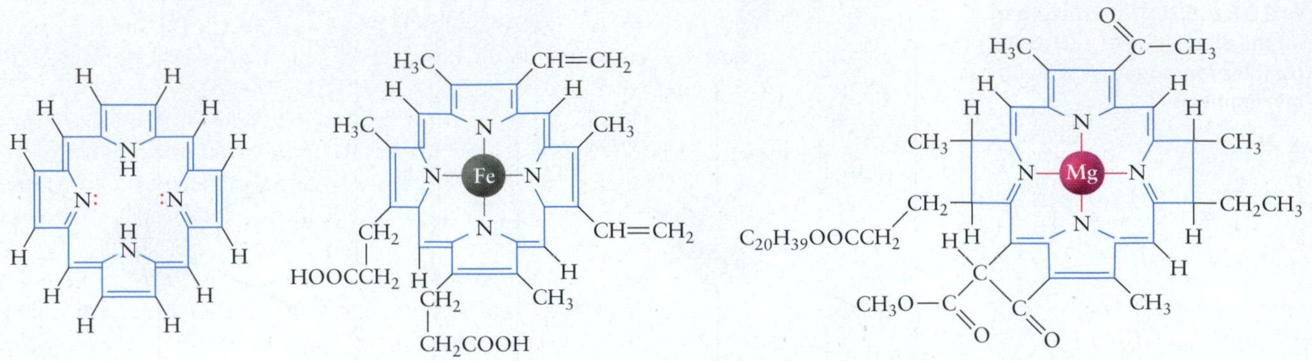

(a) Porphine (b) Heme (c) Bacteriochlorophyll

FIGURE 8.22 Structures of (a) porphine, (b) heme, and (c) bacteriochlorophyll. The porphine ring common to the three structures is highlighted in blue.

Several of the most important complexes are based on *porphine*, a compound of carbon, nitrogen, and hydrogen whose structure is shown in Figure 8.22a. Removal of two central H ions leaves four N atoms that can bind to a metal ion M^{2+} and form a tetradentate chelate structure. Such a structure, modified by the addition of several side groups, gives a complex with Fe^{2+} ions called *heme* (see Fig. 8.22b). *Hemoglobin,* the compound that transports oxygen in the blood, contains four such heme groups. The fifth coordination site of each iron(II) ion binds *globin* (a high-molar-mass protein) via N atoms on that protein, and the sixth site is occupied by water or molecular oxygen. Oxygen is a strong-field ligand that makes the d^6 Fe^{2+} diamagnetic and is responsible for the short-wavelength absorption (See Section 8.5) that leads to the red color of blood in the arteries. After the oxygen is delivered to the cells, it is replaced by water, a weaker field ligand, and a paramagnetic complex forms that absorbs at longer wavelengths, producing the bluish tint of blood in the veins. In cases of carbon monoxide (CO) poisoning,

FIGURE 8.23 Structure of vitamin B_{12}.

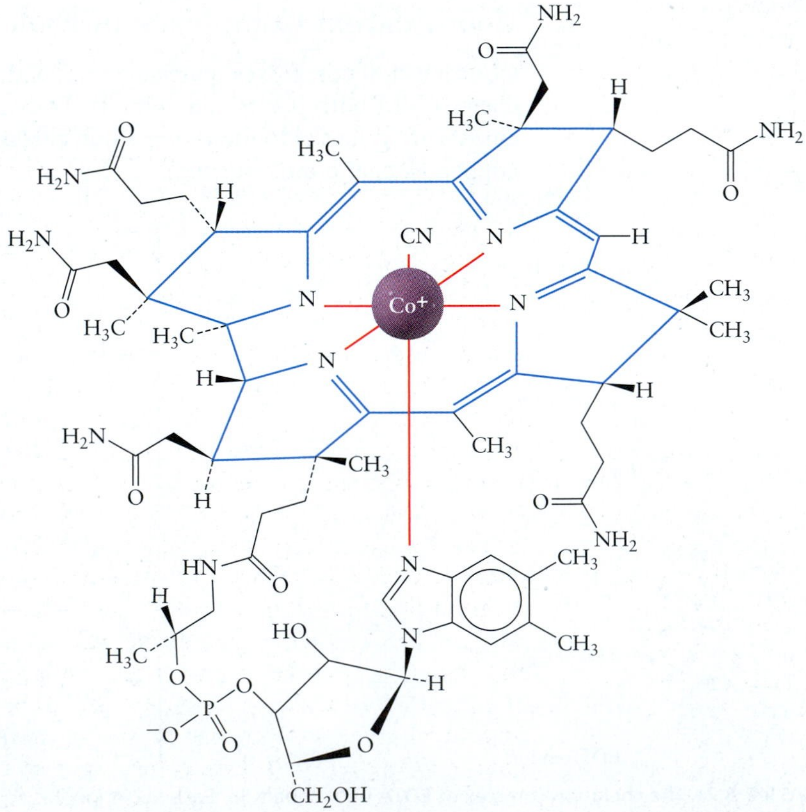

CO molecules occupy the sixth coordination site and block the binding and transport of oxygen; the equilibrium constant for CO binding is 200 times greater than that for O_2 binding (see Chapter 14).

Figure 8.22c shows the structure of a related compound, bacteriochlorophyll, which is part of the photosynthetic apparatus of photosynthetic bacteria. Section 20.6 describes the key role such structures play in photosynthesis, the process by which green plants, algae, and certain bacteria capture light and transform it into energy for use in chemical reactions.

Vitamin B_{12} is useful in the treatment of pernicious anemia and other diseases. Its structure (Fig. 8.23) has certain similarities to that of heme. Again, a metal ion is coordinated by a planar tetradentate ligand, with two other donors completing the coordination octahedron by occupying *trans* positions. In vitamin B_{12}, the metal ion is cobalt and the planar ring is *corrin,* which is somewhat similar to porphine. In the human body, enzymes derived from vitamin B_{12} accelerate a range of important reactions, including those involved in producing red blood cells.

8.5 Crystal Field Theory: Optical and Magnetic Properties

What is the nature of the bonding in coordination complexes of the transition metals that leads to their special properties? Why does Pt(IV) form only octahedral complexes, whereas Pt(II) forms square-planar ones, and under what circumstances does Ni(II) form octahedral, square-planar, and tetrahedral complexes? Can trends in the length and strength of metal–ligand bonds be understood? To answer these questions, we need a theoretical description of bonding in coordination complexes.

Crystal Field Theory

Crystal field theory, which is based on an ionic description of metal–ligand bonding, provides a simple and useful model for understanding the electronic structure, optical properties, and magnetic properties of coordination complexes. Crystal field theory was originally developed to explain these properties of ions in solids, for example, the red color of ruby, which arises from Cr^{3+} ions in an Al_2O_3 lattice. It was quickly applied to the related problem of understanding the bonding, structures, and other properties of coordination complexes. The theory treats the complex as a central metal ion perturbed by the approach of negatively charged ligands. In an octahedral complex, the six ligands are treated as negative point charges that are brought up along the $\pm x$, $\pm y$, and $\pm z$ coordinate axes toward a metal atom or ion at the origin. The energy of an electron in free space is the same in any of the five *d* orbitals in an atom or ion of a transition metal. When external charges are present, however (Fig. 8.24), the energies of electrons in the various *d* orbitals change by different amounts because of the Coulomb repulsion between the external charges and the electrons in the *d* orbitals. The magnitude of the Coulomb repulsion depends inversely on the separation between the charges, which differs for the different orbitals. An electron in a $d_{x^2-y^2}$ or d_{z^2} orbital on the central metal atom is most likely to be found along the coordinate axes, where it experiences a strong repulsive interaction with the electrons from the ligand, raising the energy of the orbital. In contrast, an electron in a d_{xy}, d_{yz}, or d_{xz} orbital of the metal is most likely to be found *between* the coordinate axes, and therefore experiences less repulsion from an octahedral array of approaching charges; the energy of these orbitals is also raised but not by as much as that of the $d_{x^2-y^2}$ and d_{z^2} orbitals. In the octahedral field of the ions, the *d* orbital energy

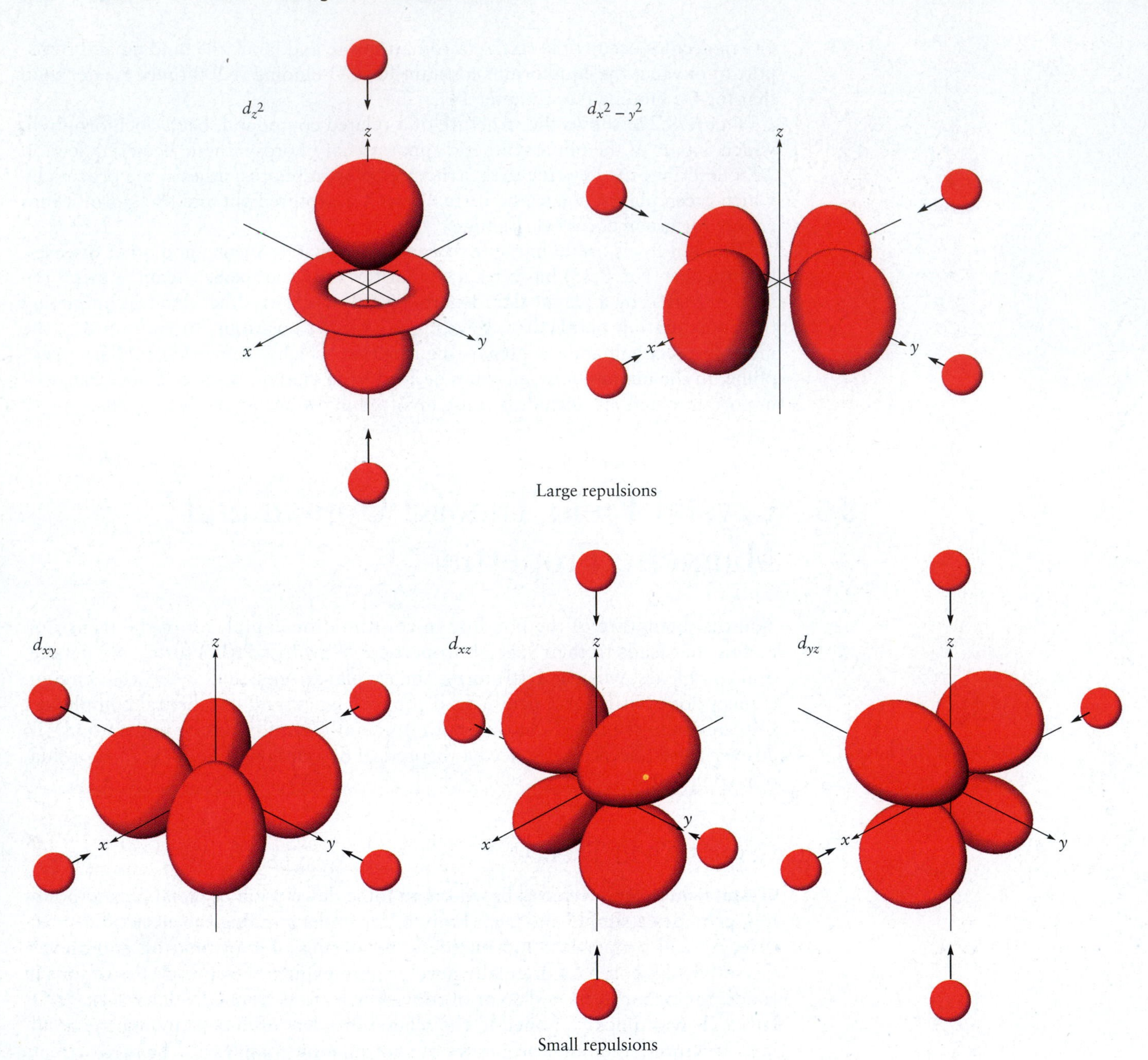

FIGURE 8.24 The basis for octahedral crystal field splitting of 3*d*-orbital energies by ligands. As the external charges approach the five 3*d* orbitals, the largest repulsions arise in the d_{z^2} and $d_{x^2-y^2}$ orbitals, which point directly at two or four of the approaching charges. External charges that have negligible interactions with the *d* electrons are not shown.

levels split into two groups (Fig. 8.25). The three lower energy orbitals, called t_{2g} orbitals, correspond to the d_{xy}, d_{yz}, and d_{xz} orbitals of the transition-metal atom; the two higher energy orbitals, called e_g orbitals, correspond to the $d_{x^2-y^2}$ and d_{z^2} orbitals. The labels t_{2g} and e_g specify the symmetry and degeneracy (number of orbitals with the same energy) of each set of orbitals. *t* orbitals are threefold (triply) degenerate, whereas *e* orbitals are twofold (doubly) degenerate. The subscript *g* has its usual meaning; *g* orbitals are symmetric with respect to inversion of the coordinates (see Section 6.1). The energy difference between the two sets of levels is Δ_o, the **crystal field splitting energy** for the octahedral complex that has formed.

FIGURE 8.25 An octahedral field increases the energies of all five d orbitals, but the increase is greater for the d_{z^2} and $d_{x^2-y^2}$ orbitals. As a result, the orbitals are split into two sets that differ by the energy Δ_o. The orbital occupancy shown is for (a) the high-spin (small Δ_o) and (b) the low-spin (large Δ_o) complexes of Mn(III).

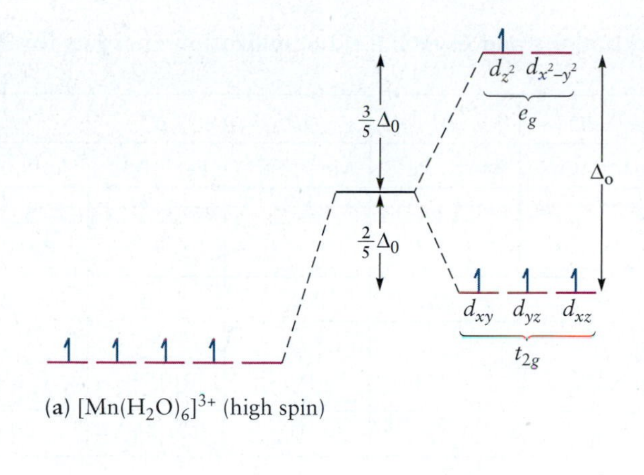

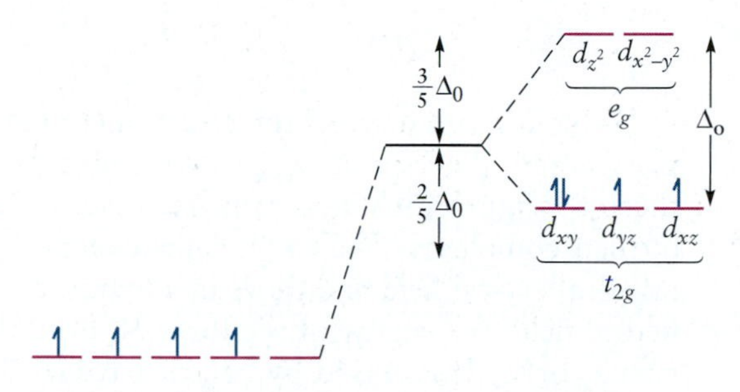

Let's consider the electron configuration expected for a transition-metal ion in the presence of such a crystal field. The Cr^{3+} ion, for example, has three d electrons. According to Hund's rules (see Section 5.3), the lowest energy configuration is one in which the three electrons occupy different t_{2g} levels with parallel spins, so complexes of Cr^{3+} are predicted to be paramagnetic. In an ion such as Mn^{3+}, which has a fourth d electron, two ground-state electron configurations are possible. The fourth electron can occupy either a t_{2g} level, with its spin opposite that of the electron already in that level (see Fig. 8.25a), or an e_g level, with its spin parallel to those of the three t_{2g} electrons (see Fig. 8.25b). The former configuration is the lower energy configuration when the splitting Δ_o is large, because it costs energy to promote the electron to the e_g level. More precisely, the cost of promoting the electron to the e_g orbital is greater than the electrostatic repulsion energy between two electrons occupying the same orbital. If the splitting is small, however, the e_g orbital will be preferentially occupied; the cost of promoting the electron to the higher energy orbital is less than the electrostatic repulsion energy.

In central metal atoms or ions with four to seven d electrons, two types of electron configurations are possible. When Δ_o is large, **low-spin complexes** are the most stable. Electrons are paired in the lower energy t_{2g} orbital, and the e_g orbitals are not occupied until the t_{2g} levels are filled. When Δ_o is small, **high-spin complexes** are the most stable. Electrons are placed one at a time, with parallel spins, in each of the three t_{2g} orbitals and then in each of the e_g orbitals, The terms *low spin* and *high spin* refer to the total number of unpaired electrons; configurations with a large number of unpaired electrons are called **high-spin configurations** and those with a small number of unpaired electrons are called **low-spin configurations.** Because low-spin configurations occur for large values of the crystal field Δ_o they are also called **strong field configurations,** and the ligands that produce such fields are called **strong field ligands. Weak field ligands** generate small crystal field splittings, resulting in **high-spin configurations.** These terms are used interchangeably (weak field↔high spin; strong field↔low spin); therefore, you should be alert when reading inorganic texts or the literature.

TABLE 8.5 Electron Configurations and Crystal Field Stabilization Energies for High- and Low-Spin Octahedral Complexes

	Configuration	d^1	d^2	d^3	d^4	d^5	d^6	d^7	d^8	d^9	d^{10}
	Examples	Ti^{3+}	Ti^{2+}, V^{3+}	V^{2+}, Cr^{3+}	Mn^{3+}, Re^{3+}	Mn^{2+}, Fe^{3+}	Fe^{2+}, Pd^{4+}	Co^{2+}, Rh^{2+}	Ni^{2+}, Pt^{2+}	Cu^{2+}	Zn^{2+}, Ag^{+}
HIGH SPIN	e_g	— —	— —	— —	↑ —	↑ ↑	↑ ↑	↑ ↑	↑ ↑	↑↓ ↑	↑↓ ↑↓
	t_{2g}	↑ — —	↑ ↑ —	↑ ↑ ↑	↑ ↑ ↑	↑ ↑ ↑	↑↓ ↑ ↑	↑↓ ↑↓ ↑	↑↓ ↑↓ ↑↓	↑↓ ↑↓ ↑↓	↑↓ ↑↓ ↑↓
	CFSE	$-\frac{2}{5}\Delta_o$	$-\frac{4}{5}\Delta_o$	$-\frac{6}{5}\Delta_o$	$-\frac{3}{5}\Delta_o$	0	$-\frac{2}{5}\Delta_o$	$-\frac{4}{5}\Delta_o$	$-\frac{6}{5}\Delta_o$	$-\frac{3}{5}\Delta_o$	0
LOW SPIN	e_g				— —	— —	— —	↑ —			
	t_{2g}				↑↓ ↑ ↑	↑↓ ↑↓ ↑	↑↓ ↑↓ ↑↓	↑↓ ↑↓ ↑↓			
	CFSE	Same as high spin			$-\frac{8}{5}\Delta_o$	$-\frac{10}{5}\Delta_o$	$-\frac{12}{5}\Delta_o$	$-\frac{9}{5}\Delta_o$	Same as high spin		

CFSE, Crystal field stabilization energies.

Table 8.5 summarizes the electron configurations possible for 10 electrons in an octahedral crystal field, provides specific examples of transition metals with these configurations, and tabulates the **crystal field stabilization energies (CFSE)** of their complexes. The CFSE is the energy difference between electrons in an octahedral crystal field and those in a hypothetical spherical crystal field. In an octahedral field, the energy of the three t_{2g} orbitals is lowered by $\frac{2}{5}\Delta_o$, and that of the two e_g orbitals is raised by $\frac{3}{5}\Delta_o$ relative to the spherical field (see Fig. 8.25). If these five orbitals are fully occupied or half occupied (as in d^{10} or high-spin d^5 complexes), then the energy of the ion is the same as in a spherical field: the CFSE is zero. If the lower energy levels are preferentially occupied, however, the configuration is stabilized. For example, in a low-spin d^4 complex, the energy of each of the four electrons in the t_{2g} orbitals is lowered by $\frac{2}{5}\Delta_o$, resulting in a total CFSE of $-\frac{8}{5}\Delta_o$.

The CFSE helps to explain the trends in the enthalpies of hydration of ions in the first transition series shown in Figure 8.3. If each measured value (blue curve) is adjusted by correcting for the CFSE of that complex ion, results quite close to the straight red lines are obtained. The relatively small magnitude of the enthalpy of hydration for Mn^{2+} arises from the high-spin d^5 configuration of Mn^{2+}, which results in a CFSE of zero. On either side of this ion, the negative CFSE lowers the enthalpy of hydration.

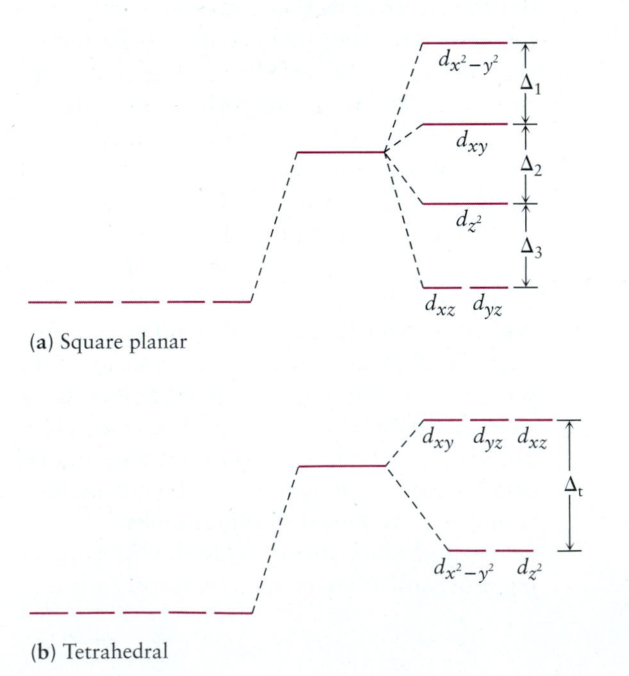

FIGURE 8.26 Energy-level structures of the 3*d* orbitals in square-planar and tetrahedral crystal fields.

Square-Planar and Tetrahedral Complexes

Crystal field theory applies to square-planar and tetrahedral complexes, as well as to octahedral complexes. Let's consider a square-planar complex in which four negative charges are brought toward the metal ion along the $\pm x$- and $\pm y$-axes. The relative *d*-orbital energy level ordering in a square-planar crystal field can be predicted using the same reasoning we applied to the octahedral case. The magnitude of the repulsive interaction between an electron in a given orbital and the electrons of the ligand depends on the degree to which the electron density is concentrated along the *x*- and *y*- axes. An electron in the $d_{x^2-y^2}$ orbital, which is oriented along these axes, experiences the greatest repulsion and lies at the highest energy. The energy of the d_{xy} orbital is lower than that of the $d_{x^2-y^2}$ orbital because the lobes of this orbital are oriented at 45 degrees to the axes. The d_{z^2} orbital energy level is lower still because its electron density is concentrated along the *z*-axis, with a small component in the *x*-*y* plane. Finally, the d_{xz} and d_{yz} orbitals experience the least repulsion. These orbitals are the most stable in a square-planar crystal field because they have nodes in the *x*-*y* plane. Figure 8.26a shows the resulting energy level diagram.

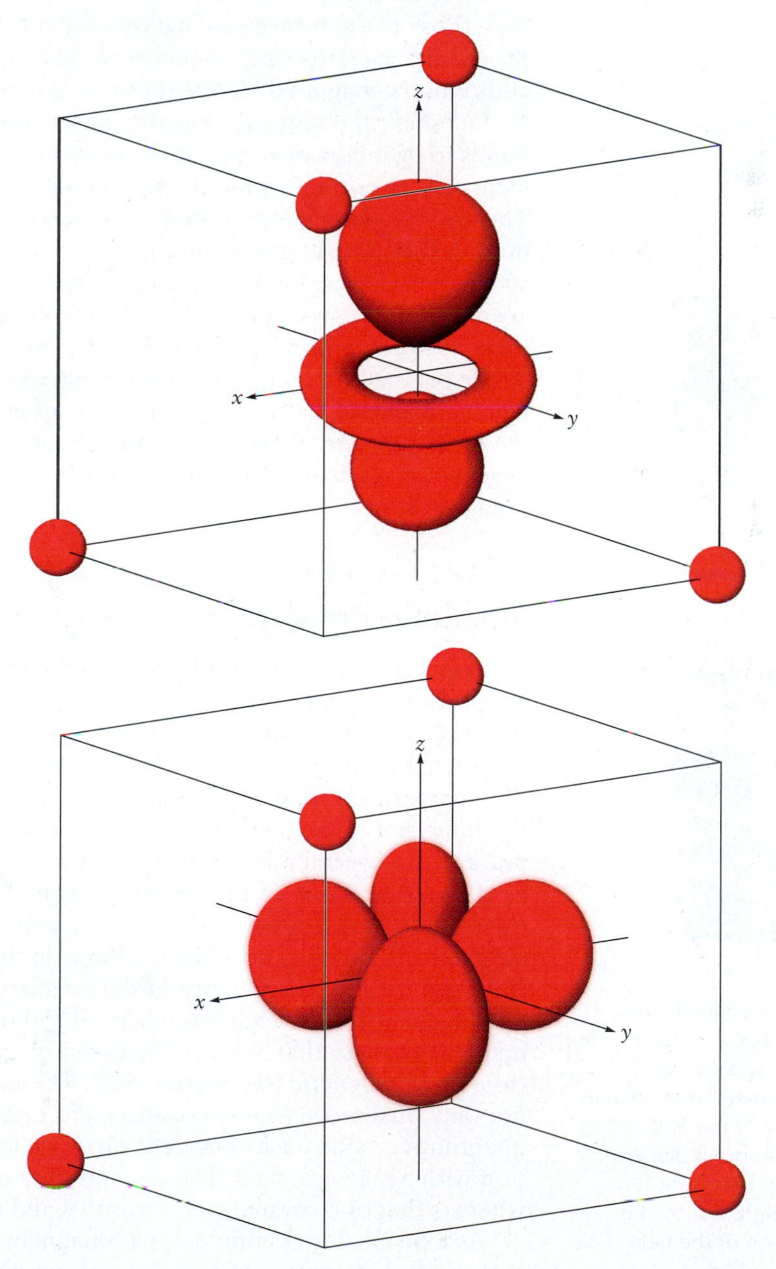

Another way to think of the square-planar crystal field splittings is to consider what happens to the octahedral energy levels shown in Figure 8.25 as the two ligands on the $\pm z$ axis move away from the metal. Let's consider each of the two degenerate levels separately. As the ligands retreat, the energy of the d_{z^2} orbital falls because of the decreased repulsion; the energy of the $d_{x^2-y^2}$ orbital must increase to conserve the total energy of the e_g level. Similarly, the energies of the d_{xz} and d_{yz} orbitals of the t_{2g} level are stabilized as the ligands are pulled away and the energy of the d_{xy} orbital increases. So the octahedral crystal field energy levels distort smoothly into the square planar levels as the z axis ligands are pulled away. Some coordination complexes are described as having distorted octahedral structures, with the two z-axis ligands moved outward but not removed completely. The level splittings observed in these cases are intermediate between the octahedral and the square-planar patterns.

Tetrahedral complexes result from bringing up ligands to four of the eight corners of an imaginary cube with the metal ion at its center. The $d_{x^2-y^2}$ and d_{z^2} orbitals point toward the centers of the cube faces, but the other three orbitals point toward the centers of the cube edges, which are closer to the corners occupied by the ligands. Electrons in the latter orbitals are therefore more strongly repelled than those in the former (see figures on page 343). Figure 8.25b shows the result; the energy level ordering is the reverse of that found for octahedral complexes. In addition, the magnitude of the splitting Δ_t is about half that of Δ_o.

Octahedral complexes are the most common because the formation of six bonds to ligands, rather than four, confers greater stability. Square-planar arrangements are important primarily for complexes of d^8 ions with strong field ligands. The low-spin configuration that leaves the high-energy $d_{x^2-y^2}$ orbital vacant is the most stable. The low-spin configuration of a square planar d^8 complex is more stable than the corresponding configuration of an octahedral complex because the energy of the highest occupied orbitals in the square planar complex is lower than those in the octahedral complex. Finally, tetrahedral complexes are less stable than either octahedral or square planar complexes for two reasons. First, the crystal field splitting is smaller so the lower set of energy levels is stabilized less than in the other geometries. Second, because the lower-energy set of levels is only doubly degenerate, electrons must be placed into the upper level at an earlier stage in the filling process.

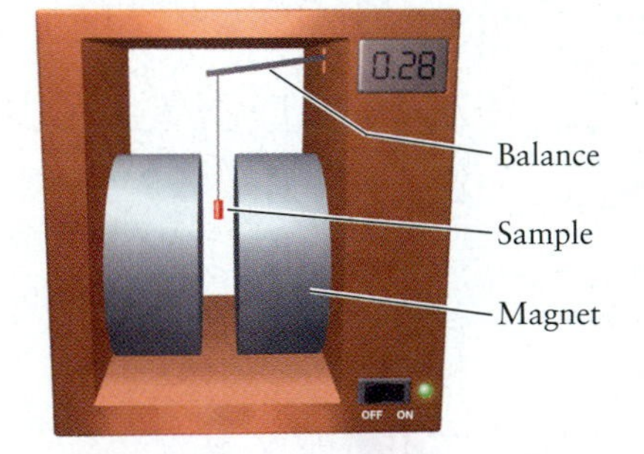

(a)

(b)

FIGURE 8.27 (a) If a sample of a paramagnetic substance is dipped into a magnetic field, it is drawn down into the field and weighs more than it does in the absence of the field. (b) If a diamagnetic substance is dipped into the field in the same way, it is buoyed up and weighs less than it would in the absence of the field.

Magnetic Properties

The existence of high- and low-spin configurations accounts for the magnetic properties of many different coordination compounds. As discussed in Section 6.2, substances are classified as paramagnetic or diamagnetic according to whether they are attracted into a magnetic field. Figure 8.27 shows a schematic of an experiment that demonstrates the universal *susceptibility* of substances to the influence of magnetic fields. A cylindrical sample is suspended between the poles of a powerful magnet. It is weighed accurately in the absence of a magnetic field and then again in the presence of the field. (There are several ways to do this; either the sample or the magnet may be moved, or the magnet may be switched off and then on.) The net force on the sample (apparent weight) is found to be different in the presence of the magnetic field. Substances that are repelled by a magnetic field weigh less when dipped into one and they are called *diamagnetic;* substances that are attracted by a magnetic field weigh more and they are called *paramagnetic* (see Section 6.2). The measurements just described provide not only qualitative characterization of a sample's magnetic properties, but also a quantitative value for its **magnetic susceptibility,** that is, the strength of its interaction with a magnetic field. The susceptibility of a diamagnet is negative and small, whereas that of a paramagnet is positive and can be quite large.

As explained in Section 6.2, paramagnetism is associated with atoms, ions, or molecules that contain one or more electrons with unpaired spins. Diamagnetic substances have the spins of all of their electrons paired. Thus, measurements of magnetic susceptibility show which substances have unpaired electron spins and which have completely paired electron spins. The number of unpaired electrons per molecule in a paramagnet can even be counted on the basis of the magnitude of the magnetic susceptibility of the sample. On a molar basis, a substance with two unpaired electrons per molecule is pulled into a magnetic field more strongly than a substance with only one unpaired electron per molecule.

These facts emerge in connection with coordination complexes, because paramagnetism is prevalent among transition-metal complexes, whereas most other chemical substances are diamagnetic. Among complexes of a given metal ion, the number of unpaired electrons, as observed by magnetic susceptibility, varies with the identities of the ligands. Both $[Co(NH_3)_6]^{3+}$ and $[CoF_6]^{3-}$ have six ligands surrounding a central Co^{3+} ion; yet, the former is diamagnetic (because it is a strong-field, low-spin complex), and the latter is paramagnetic to the extent of four unpaired electrons (because it is a weak-field, high-spin complex). Similarly, $[Fe(CN)_6]^{4-}$ is diamagnetic, but $[Fe(H_2O)_6]^{2+}$ has four unpaired electrons; these complexes also correspond to the two d^6 configurations shown in Table 8.4.

EXAMPLE 8.5

The octahedral complex ions $[FeCl_6]^{3-}$ and $[Fe(CN)_6]^{3-}$ are both paramagnetic, but the former is high spin and the latter is low spin. Identify the *d*-electron configurations in these two octahedral complex ions. In which is the octahedral field splitting greater? How does the CFSE differ between the complexes?

SOLUTION

The Fe^{3+} ion has five *d* electrons. A high-spin complex such as $[FeCl_6]^{3-}$ has five unpaired spins ($t_{2g}^3e_g^2$); a low-spin complex such as $[Fe(CN)_6]^{3-}$ has one unpaired spin (t_{2g}^5). The splitting Δ_o must be greater for cyanide than for chloride ion ligands. The CFSE for the $[FeCl_6]^{3-}$ complex is zero, whereas that for the $[Fe(CN)_6]^{3-}$ complex is $-2\Delta_o$.

Related Problems: 27, 28

8.6 Optical Properties and the Spectrochemical Series

FIGURE 8.28 Several colored coordination compounds. (clockwise from top left) They are $Cr(CO)_6$ (white), $K_3[Fe(C_2O_4)_3]$ (green), $[Co(en)_3]I_3$ (orange), $[Co(NH_3)_5(H_2O)]Cl_3$ (red), and $K_3[Fe(CN)_6]$ (red-orange).

Transition-metal complexes are characterized by their rich colors, which are often deep, vibrant, and saturated (Fig. 8.28). The colors depend on the oxidation state of the metal ion, the number and nature of the ligands, and the geometry of the complex. Earlier figures have shown the color changes that accompany dehydration and ligand substitution reactions and also the different colors of a pair of geometric isomers. The following series of Co(III) complexes shows how the colors of coordination complexes of the same ion can vary with different ligands:

$[Co(NH_3)_6]^{3+}$	Orange
$[Co(NH_3)_4Cl_2]^+$	A green form and a violet form
$[Co(NH_3)_5(H_2O)]^{3+}$	Purple

Coordination complexes appear colored when they absorb visible light. Recall that atoms absorb light when the energy of an incident photon exactly matches the energy difference between two atomic energy levels (see Fig. 4.9). The missing wavelengths appear as dark lines against the spectral rainbow. The transmitted light still appears quite white to our eyes, however, because atomic absorption lines are so narrow. Only a small fraction of the visible light has been absorbed. Coordination complexes, on the other hand, absorb light over significant regions of the visible spectrum. What we see is the color that is *complementary* to the color that is most strongly absorbed (see Section 20.3). The $[Co(NH_3)_5Cl]^{2+}$ ion, for example, absorbs greenish yellow light, with the strongest absorption occurring near 530 nm. Only the red and blue components of white light are transmitted through an aqueous solution of this ion, which appears purple to us. Materials that absorb all visible wavelengths appear gray or black, and those that absorb visible light weakly or not at all appear colorless.

Crystal field theory was developed, in part, to explain the colors of transition-metal complexes. It was not completely successful, however. Its failure to predict trends in the optical absorption of a series of related compounds stimulated the development of ligand field and molecular orbital theories and their application in coordination chemistry. The colors of coordination complexes are due to the excitation of the *d* electrons from filled to empty *d* orbitals (*d-d* transitions). In octahedral complexes, the electrons are excited from occupied t_{2g} levels to empty e_g levels. The crystal field splitting Δ_o is measured directly from the optical absorption spectrum of the complex. The wavelength of the strongest absorption is called λ_{max} and it is related to Δ_o as follows. $E = h\nu$, so $\Delta_o = h\nu = hc/\lambda_{max}$. Because energy is inversely proportional to wavelength, compounds with small crystal field splittings absorb light with longer wavelengths, toward the red end of the visible spectrum, and those with large crystal field splitting absorb light with shorter wavelengths, toward the blue end of the spectrum.

In $[Co(NH_3)_6]^{3+}$, an orange compound that absorbs most strongly in the violet region of the spectrum, the crystal field splitting Δ_o is larger than in $[Co(NH_3)_5Cl]^{2+}$, a violet compound that absorbs most strongly at lower frequencies (longer wavelengths) in the yellow–green region of the spectrum. d^{10} complexes (like those of Zn^{2+} or Cd^{2+}) are colorless because all of the *d* levels (both t_{2g} and e_g) are filled. Because there are no empty orbitals available to accept an excited electron, the transition is not allowed, which means that the absorption is weak or nonexistent. High-spin d^5 complexes such as $[Mn(H_2O)_6]^{2+}$ and $[Fe(H_2O)_6]^{3+}$ also show only weak absorption because promotion from a filled t_{2g} level to an empty e_g level would require a spin flip to satisfy the Pauli principle. (Recall that all of the spins are parallel to one another in high-spin complexes.) Light absorption rarely reverses the spin of an electron, so the optical absorption of these compounds is weak, as shown by the pale pink color of the hexa-aqua Mn^{2+} complex in Figure 8.29. Table 8.6 lists the crystal field splittings and absorption wavelengths for a number of coordination complexes to give you a feel for their diversity.

Experimental measurements of the optical absorption spectra and magnetic properties of transition-metal complexes provide a critical test of the validity of crystal field theory. The theory makes specific predictions about the strengths of crystal fields produced by different ligands. Charged ligands such as the halides

FIGURE 8.29 The colors of the hexa-aqua complexes of metal ions (from left) Mn^{2+}, Fe^{3+}, Co^{2+}, Ni^{2+}, Cu^{2+}, and Zn^{2+}, prepared from their nitrate salts. Note that the d^{10} Zn^{2+} complex is colorless. The green color of the Ni^{2+} is due to absorption of both red and blue light that passes through the solution. The yellow color of the solution containing $Fe(H_2O)_6^{3+}$ is caused by hydrolysis of that ion to form $Fe(OH)(H_2O)_5^{2+}$; if this reaction is suppressed, the solution is pale violet.

TABLE 8.6 Absorption Wavelengths for Selected Octahedral Transition-Metal Complexes

Octahedral Complexes	λ_{max} (nm)	Octahedral Complexes	λ_{max} (nm)
$[TiF_6]^{3-}$	588	$[Co(NH_3)_6]^{3+}$	437
$[Ti(H_2O)_6]^{3+}$	492	$[Co(CN)_6]^{3-}$	290
$[V(H_2O)_6]^{3+}$	560	$[Co(H_2O)_6]^{2+}$	1075
$[V(H_2O)_6]^{2+}$	806	$[Ni(H_2O)_6]^{2+}$	1176
$[Cr(H_2O)_6]^{3+}$	575	$[Ni(NH_3)_6]^{2+}$	926
$[Cr(NH_3)_6]^{3+}$	463	$[RhBr_6]^{3-}$	463
$[Cr(CN)_6]^{3-}$	376	$[RhCl_6]^{3-}$	439
$Cr(CO)_6$	311	$[Rh(NH_3)_6]^{3+}$	293
$[Fe(CN)_6]^{3-}$	310	$[Rh(CN)_6]^{3-}$	227
$[Fe(CN)_6]^{4-}$	296	$[IrCl_6]^{3-}$	362
$[Co(H_2O)_6]^{3+}$	549	$[Ir(NH_3)_6]^{3+}$	250

should produce much stronger crystal fields than neutral ligands. The interaction between the halides and a metal ion should increase with decreasing ionic radius. The smaller halide ions can approach the metal ion more closely, resulting in greater electrostatic repulsion. Crystal field theory predicts splittings that increase in the order $I^- < Br^- < Cl^- < F^-$. A systematic ranking of the strength of various ligands was obtained by comparing the optical absorption spectra of a series of complexes with the general formula $[Co(III)(NH_3)_5X]^{n+}$. The strength of the interaction between a single ligand X and the Co^{3+} ion could be measured directly because all other interactions and the geometry of the complex remained constant. Ligands were ranked from weakest to strongest in the **spectrochemical series** as follows:

$$I^- < Br^- < Cl^- < F^-, OH^- < H_2O < :NCS^- < N < H_3en < CO, CN^-$$

Weak-field ligands (high spin) Intermediate-field ligands Strong-field ligands (low spin)

Although this order is not followed for all metals ions, it is a useful generalization. More importantly, it illustrates the failure of crystal field theory to provide a satisfactory account of the factors that govern crystal field splitting. Neutral ligands with lone pairs, such as water and ammonia, produce larger splittings than any of the halides, and ligands with low-lying antibonding π molecular orbitals produce the largest splittings of all. A more comprehensive theory is clearly required to explain the spectrochemical series. Section 8.7 examines how molecular orbital theory correctly accounts for the trend observed in the spectrochemical series.

EXAMPLE 8.6

Predict which of the following octahedral complexes has the shortest λ_{max}: $[FeF_6]^{3-}$, $[Fe(CN)_6]^{3-}$, $[Fe(H_2O)_6]^{3+}$.

SOLUTION

$[Fe(CN)_6]^{3-}$ has the strongest field ligands of the three complexes; thus, its energy levels are split by the greatest amount. The frequency of the light absorbed should be greatest, and λ_{max} should be the shortest for this ion.

$[Fe(CN)_6]^{3-}$ solutions are red, which means that they absorb blue and violet light. Solutions of $[Fe(H_2O)_3]^{3+}$ are a pale violet due to the weak absorption of red light, and $[FeF_6]^{3-}$ solutions are colorless, indicating that the absorption lies beyond the long wavelength limit of the visible spectrum.

Related Problems: 37, 38, 39, 40

8.7 Bonding in Coordination Complexes

Valence Bond Theory

Valence bond (VB) theory is used widely in contemporary chemistry to describe structure and bonding in transition-metal compounds, especially coordination complexes. The VB model is intuitively appealing for this purpose for several reasons. Because transition-metal compounds, particularly coordination complexes, often comprise a central atom surrounded by ligands in a symmetric arrangement, forming hybrid orbitals on the central atom with the appropriate symmetry to bond to these ligands is a natural approach to the problem. Often, little interaction occurs among metal–ligand bonds, so the local description is reasonable. Participation of the *d* electrons enables a much more varied set of structures and hybrid orbitals than can be formed from only *s* and *p* orbitals. This section describes two sets of hybrid orbitals used to describe bonding in transition-metal compounds and complexes, and provides examples of each.

Hybridization is justified here for precisely the same reasons we laid out in Section 6.4. The lobes of the hybrid orbitals point toward the ligands and overlap the ligand orbitals more strongly than the standard atomic orbitals. The energy cost of promoting electrons from lower energy atomic orbitals (e.g., 3*d* and 4*s*) to the highest energy orbital (4*p*) to form a hybrid orbital is more than offset by the energy gained in forming a stronger bond with the ligand.

We construct the first set of hybrid orbitals from one *s* atomic orbital, the three *p* atomic orbitals and the d_{z^2} atomic orbital; they are called dsp^3 hybrid orbitals. The principal quantum numbers of the participating atomic orbitals depend on the particular metal atom under consideration; for Co, they would be the 3*d*, 4*s*, and 4*p* atomic orbitals. The dsp^3 hybrid orbitals in the most general case are written out as

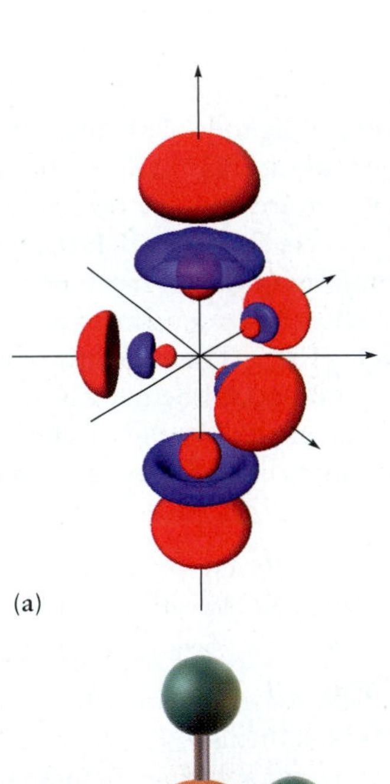

(a)

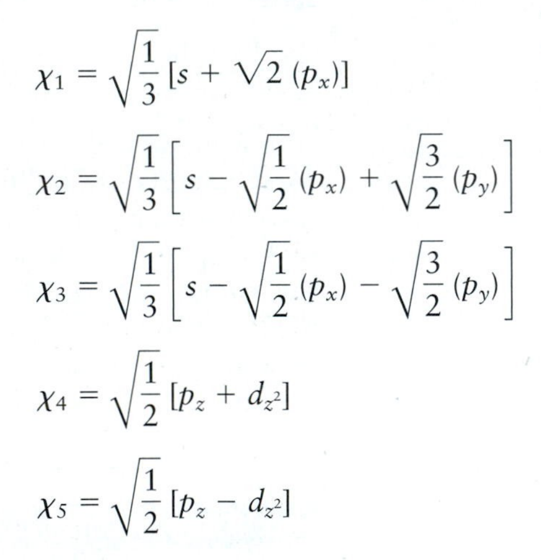

$$\chi_1 = \sqrt{\frac{1}{3}}\,[s + \sqrt{2}\,(p_x)]$$

$$\chi_2 = \sqrt{\frac{1}{3}}\left[s - \sqrt{\frac{1}{2}}\,(p_x) + \sqrt{\frac{3}{2}}\,(p_y)\right]$$

$$\chi_3 = \sqrt{\frac{1}{3}}\left[s - \sqrt{\frac{1}{2}}\,(p_x) - \sqrt{\frac{3}{2}}\,(p_y)\right]$$

$$\chi_4 = \sqrt{\frac{1}{2}}\,[p_z + d_{z^2}]$$

$$\chi_5 = \sqrt{\frac{1}{2}}\,[p_z - d_{z^2}]$$

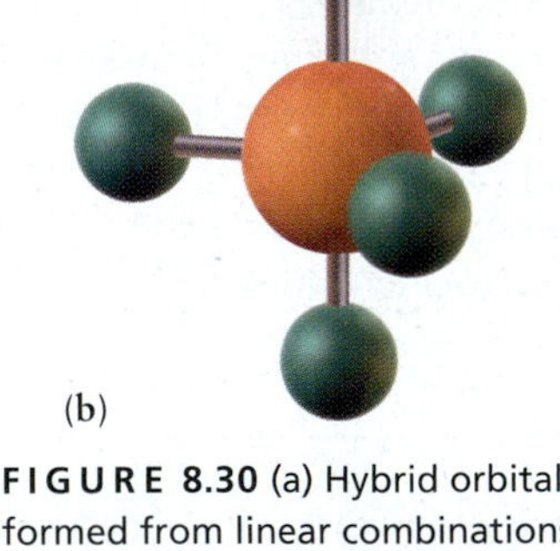

(b)

FIGURE 8.30 (a) Hybrid orbitals formed from linear combinations of d_{z^2}, *s*, p_x, p_y, and p_z orbitals. The pair of orbitals that point along the positive and negative *z*-axes are the same except for their orientation in space; they are called *axial* orbitals. The set of three orbitals in the *x*-*y* plane are equivalent to one another and are called *equatorial* orbitals. (b) This set of hybrid orbitals can be used to describe the bonding in PF_5, for example.

As shown in Figure 8.30a, these orbitals point to the vertices of a trigonal bipyramid; there are three equivalent equatorial hybrids and two equivalent axial hybrids. Examples of molecules whose shapes are described by dsp^3 hybridization include PF_5, which you have seen in Section 3.9, and $CuCl_5^{3-}$. PF_5 is shown in Figure 8.30b for comparison with the set of dsp^3 hybrid orbitals.

The second set of hybrid orbitals we construct are the d^2sp^3 hybrids; these are six equivalent orbitals directed toward the vertices of an octahedron (Fig. 8.31a). They describe the structures and bonding in all of the octahedral coordination complexes discussed in Section 8.6, as well as that in SF_6, which we show in Figure 8.31b.

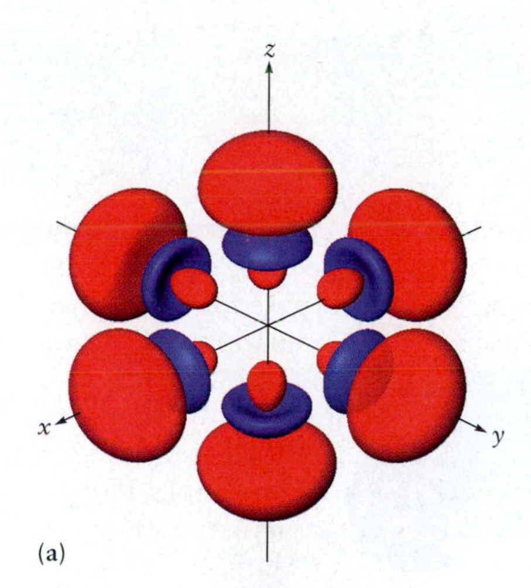

(a)

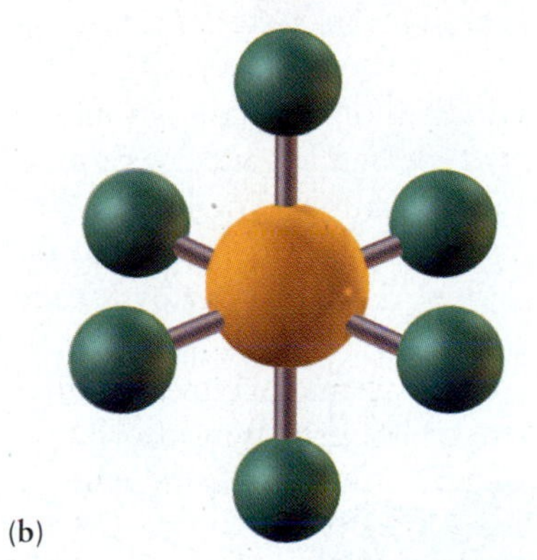

(b)

FIGURE 8.31 (a) Hybrid orbitals formed from linear combinations of d_{z^2}, $d_{x^2-y^2}$, s, p_x, p_y, and p_z orbitals. All six orbitals are equivalent except for their orientation in space. (b) This set of hybrid orbitals can be used to describe the bonding in SF_6, for example.

$$\chi_1 = \sqrt{\frac{1}{6}}\,[s + \sqrt{3}\,(p_z) + \sqrt{2}\,(d_{z^2})]$$

$$\chi_2 = \sqrt{\frac{1}{6}}\left[s + \sqrt{3}\,(p_z) - \sqrt{\frac{1}{2}}\,(d_{z^2}) + \sqrt{\frac{3}{2}}\,(d_{x^2-y^2})\right]$$

$$\chi_3 = \sqrt{\frac{1}{6}}\left[s + \sqrt{3}\,(p_z) - \sqrt{\frac{1}{2}}\,(d_{z^2}) - \sqrt{\frac{3}{2}}\,(d_{x^2-y^2})\right]$$

$$\chi_4 = \sqrt{\frac{1}{6}}\left[s - \sqrt{3}\,(p_z) - \sqrt{\frac{1}{2}}\,(d_{z^2}) + \sqrt{\frac{3}{2}}\,(d_{x^2-y^2})\right]$$

$$\chi_5 = \sqrt{\frac{1}{6}}\left[s - \sqrt{3}\,(p_z) - \sqrt{\frac{1}{2}}\,(d_{z^2}) - \sqrt{\frac{3}{2}}\,(d_{x^2-y^2})\right]$$

$$\chi_6 = \sqrt{\frac{1}{6}}\,[s - \sqrt{3}\,(p_z) + \sqrt{2}\,(d_{z^2})]$$

Table 8.7 shows the variety of hybrid orbitals that can be constructed from various combinations of *s*, *p*, and *d* orbitals, the shapes of the molecules that result, and selected examples.

VB theory with hybrid orbitals is widely used to rationalize the structures of coordination complexes. It complements classical valence shell electron-pair repulsion (VSEPR) theory by using methods of quantum mechanics to describe the geometry of coordination complexes. As with main-group elements, VB theory is better suited to rationalize structure and bonding after the fact than to predict structure. And by treating the bonds as local and equivalent, it fails completely to account for the colors and magnetic properties of coordination complexes. These shortcomings motivate the application of molecular orbital theory to describe structure and bonding in coordination chemistry.

Molecular Orbital Theory

The failure of crystal field theory and VB theory to explain the spectrochemical series stimulated the development of **ligand field theory,** which applies qualitative methods of molecular orbital theory to describe the bonding and structure of coordination complexes. The terms *ligand field theory* and *molecular orbital theory* are often used interchangeably in inorganic chemistry today.

We apply molecular orbital theory to octahedral coordination complexes just as we did for the simple metal carbonyl in Section 8.2. We begin by constructing the σ MOs from the valence *d*, *s*, and *p* orbitals of the central metal atom and the six ligand orbitals that point along the metal–ligand bond directions in an octahedral complex. In the case of the Cr^{3+} complexes we use as examples, the relevant

TABLE 8.7 Examples of Hybrid Orbitals and Bonding in Complexes

Coordination Number	Hybrid Orbital	Configuration	Examples
2	*sp*	Linear	$[Ag(NH_3)_2]^+$
3	sp^2	Trigonal planar	BF_3, NO_3^-, $[Ag(R_3P)_3]^+$
4	sp^3	Tetrahedral	$Ni(CO)_4$, $[MnO_4]^-$, $[Zn(NH_3)_4]^{2+}$
4	dsp^2	Planar	$[Ni(CN)_4]^{2-}$, $[Pt(NH_3)_4]^{2+}$
5	dsp^3	Trigonal bipyramid	TaF_5, $[CuCl_5]^{3-}$, $[Ni(PEt_3)_2Br_3]$
6	d^2sp^3	Octahedral	$[Co(NH_3)_6]^{3+}$, $[PtCl_6]^{2-}$

From G.E. Kimball, Directed valence. *J. Chem. Phys.* 1940, 8, 188.

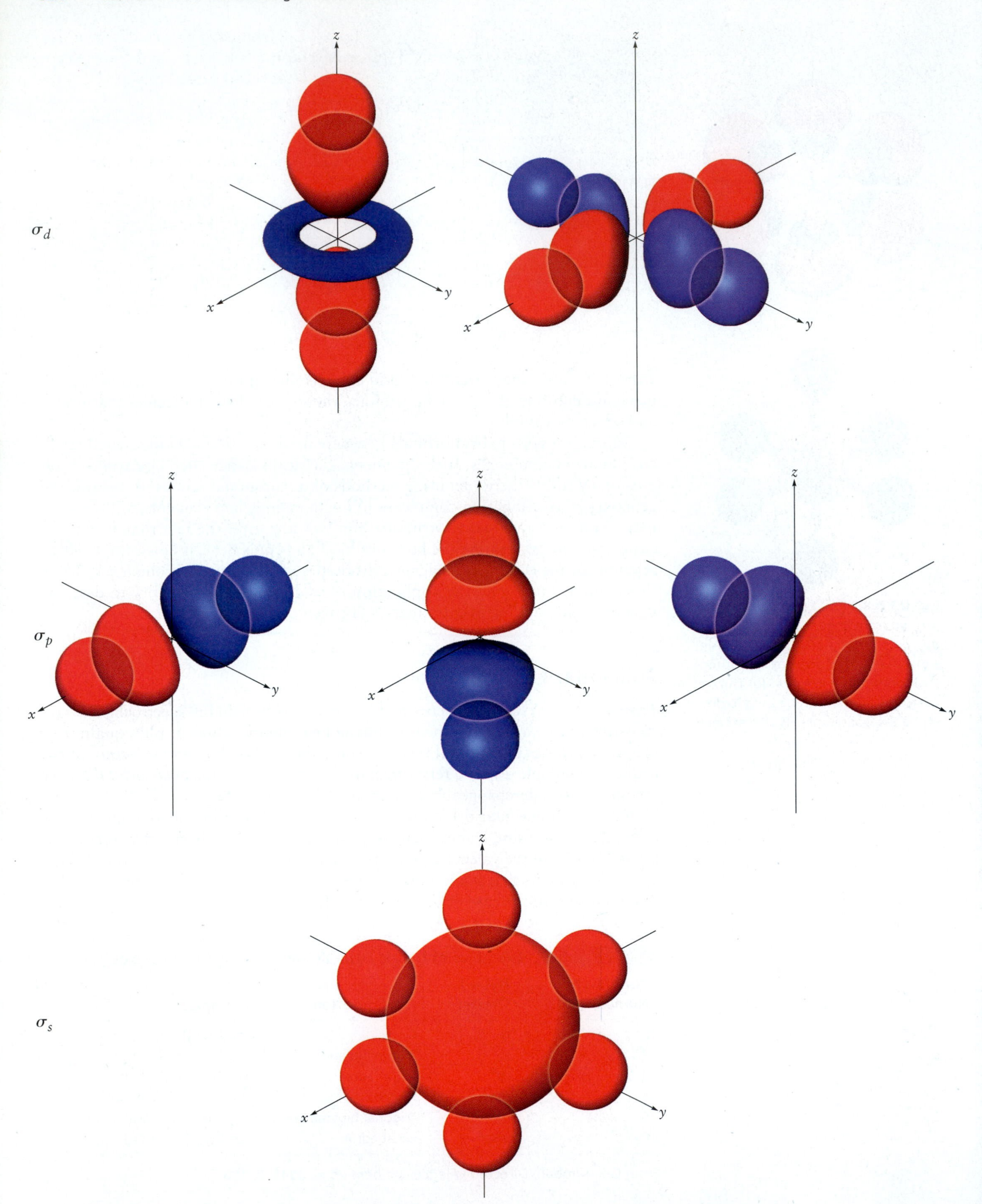

FIGURE 8.32 Overlap of metal orbitals with ligand orbitals to form σ bonds. The ligand orbitals can be either p or hybrid orbitals (e.g., sp^3 for water), and thus they are represented only schematically.

metal orbitals are the $3d$, $4s$, and $4p$ orbitals. The ligand orbitals could be sp^3 orbitals containing a lone pair in H_2O, or the p orbitals from F^- ions directed toward the metal ion. As shown in Figure 8.32, the $4s$ metal orbital can overlap with six ligand orbitals of the same phase to form a bonding σ molecular orbital. Overlap with six ligands orbitals of the opposite phase forms the corresponding antibonding σ^* molecular orbital (not shown). We label these orbitals σ_s and σ_s^*, respectively. In a similar fashion, we generate three bonding and three antibonding combinations using the metal p orbitals, which we label σ_p and σ_p^*, respectively. Only the d_{z^2} and $d_{x^2-y^2}$ orbitals are oriented to overlap with the ligand orbitals, and this overlap generates a pair of bonding and antibonding molecular orbitals, σ_d and σ_d^*. The d_{xy}, d_{yz}, and d_{xz} orbitals, whose lobes are oriented at 45 degrees to the bond axes, do not overlap with the ligand orbitals and are nonbonding. Thus, from our set of 9 metal orbitals and 6 ligand orbitals we have constructed 15 molecular orbitals, as required. Figure 8.33 is the resulting orbital correlation diagram.

Focusing on the set of three nonbonding δ_{xy}^{nb}, δ_{xz}^{nb}, and δ_{yz}^{nb} orbitals (t_{2g}) and the pair of antibonding σ^* (e_g) orbitals, we see that the energy level diagram is exactly the same as that predicted using crystal field theory (see Fig. 8.25). All of the results of that theory remain valid if we consider only σ bonding. The electron configuration of a coordination complex is built up by filling orbitals in order of increasing energy, just as for any other molecule that is characterized using molecular orbital theory. Using the complex ion $[CrCl_6]^{3-}$ as a specific example, there are 12 electrons available from the 6 ligands and 3 electrons from the metal, for a total of 15 electrons. 12 of these electrons fill the 6 bonding σ orbitals and the remaining 3 electrons are placed in the nonbonding δ_{xy}, δ_{xz}, and δ_{yz} orbitals (t_{2g}). In contrast to the VB treatment, which generates six localized bonding orbitals from

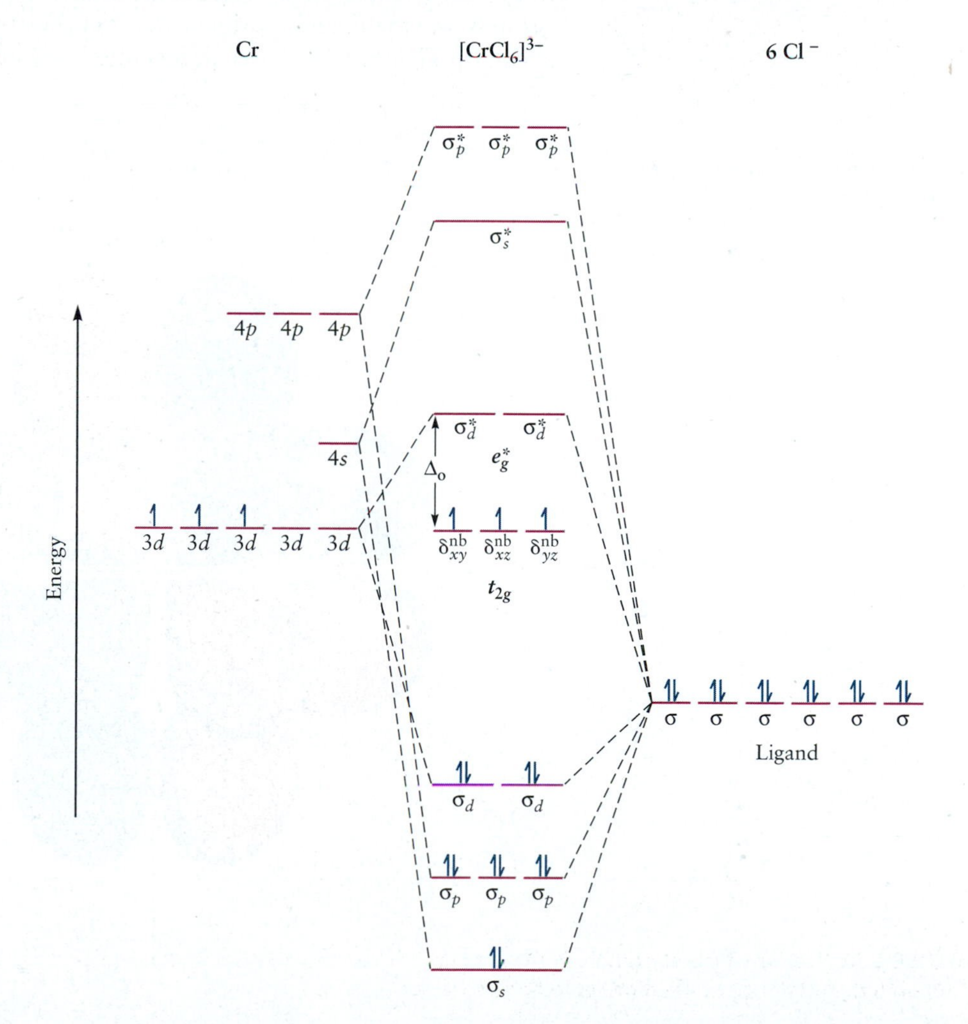

FIGURE 8.33 Orbital correlation diagram for an octahedral ligand field, showing the energy-level filling for a $[CrCl_6]^{3-}$ ion.

six hybrid orbitals on the metal and six ligand orbitals, the molecular orbital treatment produces six bonding molecular orbitals with different degrees of delocalization and with different spatial orientations. The σ_s orbital is delocalized over the entire complex. The three σ_p orbitals are each delocalized along one of the Cartesian axes. The σ_d orbital derived from the metal d_{z^2} orbital is delocalized along the z-axis whereas the σ_d orbital derived from the metal $d_{x^2-y^2}$ orbital is delocalized over the x-y plane.

There are important differences between crystal field theory and molecular orbital theory, even before we consider the important role of π bonding. The physical origin of the energy-level splitting is quite different in the two cases. Crystal field theory is essentially classical, and it attributes the splitting to the electrostatic repulsion between the electrons of the metal ion and the ionic charges of the ligands. Molecular orbital theory attributes the splitting to the mixing of the metal and ligand wave functions to form bonding, nonbonding, and antibonding molecular orbitals. The splitting Δ_o in molecular orbital theory is the energy difference between the nonbonding t_{2g} metal orbitals and the antibonding σ^* molecular orbitals that lie above them. We see immediately how this description explains the decrease in Δ_o as the strength of the metal–ligand interaction increases. Strong interactions, *whether ionic or covalent,* decrease the energy of (stabilize) the bonding σ_d orbital but increase the energy of (de-stabilize) the σ_d^* orbital.

The dramatically different strengths of ligands such as the halides and those containing unfilled antibonding π orbitals like CO and CN^- cannot be explained considering σ bonding alone. π bonding in octahedral coordination complexes may be treated by extending the arguments we presented in Section 8.2 for metal carbonyls. The only difference is that we must take into account the relative phases of the ligand orbitals as they overlap with the metal orbitals. The most important molecular orbitals for π bonding are constructed from the metal d_{xy}, d_{yz}, and d_{xz} orbitals and the ligand p orbitals that are oriented to overlap the metal orbitals as shown in Figure 8.34. Three degenerate bonding orbitals and a corresponding set

FIGURE 8.34 π bonding between a metal d_{xy} orbital and four ligand π orbitals with phases chosen for maximum constructive overlap.

of antibonding orbitals are formed in this way. They are located in the *x-y*, *y-z*, and *x-z* planes, respectively. Because each set of molecular orbitals is triply degenerate and each orbital has *g* symmetry, they are labeled t_{2g} and t_{2g}^*, respectively. The orbital overlap shown in Figure 8.34 explains both L→M π donation and M→L π backbonding. The former is important for metals in high oxidation states (with few *d* electrons and empty *d* orbitals) and ligands with filled orbitals whose energies lie close to those of the empty *d* orbitals (the halides, water, OH^- and ammonia). The latter interaction is important for metals in low oxidation states (many *d* electrons and few unfilled *d* orbitals) and for ligands with empty orbitals whose energies lie close to those of the filled *d* orbitals (*d* orbitals of P or S, or π^* orbitals in ligands such as CO and CN^-).

Figure 8.35 summarizes the molecular orbital picture of bonding in octahedral coordination complexes, identifies the interactions that determine the splitting energy Δ_o, and provides guidance for predicting the bonding and energy-level structure for any coordination compound of interest, based on the oxidation state of the metal (number of *d* electrons) and the nature and occupancy of the ligand orbitals. The center portion of the figure shows the orbital splitting predicted using the simple molecular orbital picture that includes only σ bonding; the three *d* electrons shown could be found in complexes of V^{2+}, Cr^{3+}, or Mn^{4+}, for example. Crystal field theory would make the same prediction for the energy-level structure of $[CrCl_6]^{3-}$, and it can be considered a limiting case of molecular orbital theory.

The left side of Figure 8.35 could represent the bonding interactions and energy-level structure for $[CrCl_6]^{3-}$ in which a set of filled ligand orbitals (in this case, Cl 2*p* orbitals) donate electrons to a set of empty (or only partially filled) metal *d* orbitals (L→M π donation). The ligand orbitals are assumed to lie lower in energy than the metal orbitals, so the resulting t_{2g} molecular orbital is mostly ligand in character and lies at a relatively low energy. There is a corresponding t_{2g}^* orbital, which is mostly metal-like, at higher energy. This antibonding orbital is

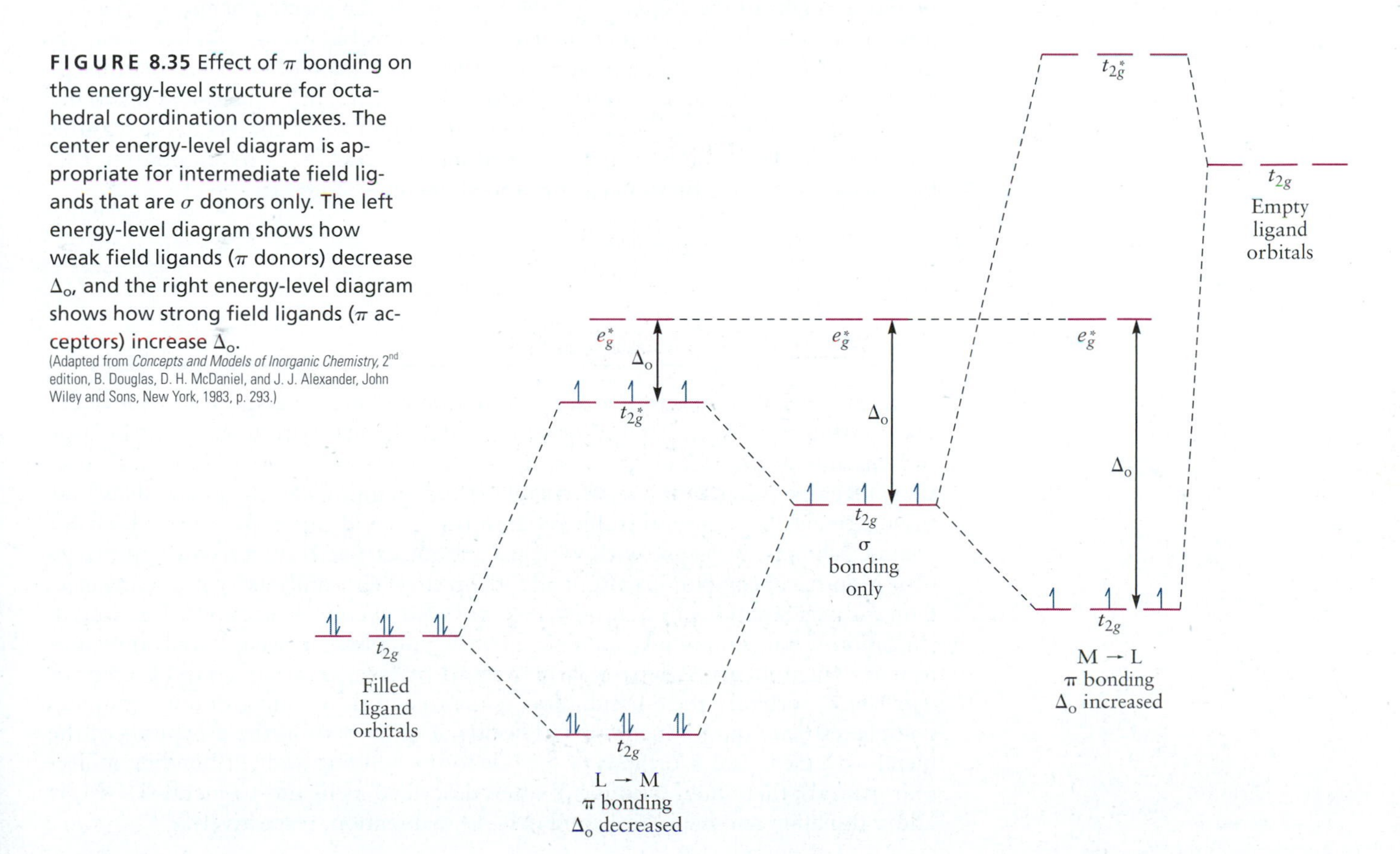

FIGURE 8.35 Effect of π bonding on the energy-level structure for octahedral coordination complexes. The center energy-level diagram is appropriate for intermediate field ligands that are σ donors only. The left energy-level diagram shows how weak field ligands (π donors) decrease Δ_o, and the right energy-level diagram shows how strong field ligands (π acceptors) increase Δ_o.
(Adapted from *Concepts and Models of Inorganic Chemistry*, 2nd edition, B. Douglas, D. H. McDaniel, and J. J. Alexander, John Wiley and Sons, New York, 1983, p. 293.)

the highest occupied molecular orbital in this example; six of the nine available electrons go into the bonding t_{2g} orbital, leaving three to occupy the t^*_{2g} orbital. There are two important differences between this picture and the one in the center of Figure 8.35. First, the t_{2g} orbitals that were nonbonding metal *d* orbitals in the absence of π bonding have now become antibonding MOs by virtue of L→M π donation. Second, the energy-level splitting, Δ_o, has been reduced from its value without π bonding interactions. So, we now have an explanation for the mechanism by which weak field ligands lead to the smallest field splittings and a criterion with which to classify them. Weak field ligands have filled low-energy *p* orbitals that can donate electron density to empty metal *d* orbitals to produce an occupied antibonding molecular orbital at higher energy than the nonbonding metal *d* orbitals. All of the weak field ligands identified empirically in the spectrochemical series (the halides) fall into this class.

The right side of Figure 8.35 illustrates the case in which the metal *d* orbitals are nearly filled and the ligand has empty π^* orbitals at higher energies. This situation leads to a bonding and antibonding pair of MOs as before, but in this case, the energy of the t^*_{2g} orbital is too high for it to be occupied. The highest occupied molecular orbital is the bonding t_{2g} orbital, and Δ_o is now the difference in energy between this orbital and the antibonding e^*_g orbital. In the language of inorganic chemistry, the t_{2g} orbital has been stabilized because of M→L π donation. These interactions are important for metals with filled or nearly filled t_{2g} orbitals (such as Fe, Co, and Ni) and ligands with empty, low-lying orbitals such as the *d* orbitals of P or S or the π^* orbitals of CO, CN^-, or NO^+. These ligands, as well as OH^-, en, and NH_3 are strong field ligands because they can accept electron density from the occupied *d* orbitals of the metal.

Molecular orbital theory provides a simple yet comprehensive way to understand and predict bonding patterns and energy levels in coordination complexes. Although we have worked out only the octahedral geometry in detail, the same considerations apply for all of the other geometries. Ligands classified empirically by the strengths of their interactions as revealed by the spectrochemical series can now be classified by the bonding interactions responsible. Weak field ligands are **π donors,** intermediate field ligands are **σ donors** with little to no π interactions at all, and strong field ligands are **π acceptors.** These classifications enable us to determine the nature of the ligand by simply looking for these characteristic features and to understand the bonding in coordination complexes by combining this knowledge with the oxidation state of the metal ion of interest.

CHAPTER SUMMARY

Transition-metal compounds and coordination complexes display a much wider variety of physical and chemical properties than the main-group elements in large part because of the participation of the *d* electrons in bonding and chemical reactions. Because the cost of transferring electrons to and from the *d* orbitals is low, these elements have several stable oxidation states and can make several kinds of chemical bonds. The ease with which electrons can be transferred to and from transition-metal compounds also makes them excellent catalysts by providing electron transfer reactions as alternate pathways for chemical reactions. Bonding in transition-metal compounds and coordination complexes is well described by molecular orbital theory. A type of bond formed by face-to-face overlap of a pair of $d_{x^2-y^2}$ or d_{xy} orbitals, the δ bond, plays a key role in the bonding of molecules and complexes that contain metal–metal bonds. The overlap of the *d* orbitals of the metal with the σ and π orbitals of ligands forms bonding and antibonding molecular orbitals; these interactions are often described as ligand-to-metal (L→M) σ and π donation and metal-to-ligand (M→L) π donation, respectively.

Coordination complexes are molecules or ions in which a central metal atom is bound to one or more ligands in a symmetric arrangement. Linear, tetrahedral, square-planar, trigonal bipyramidal, and octahedral arrangements are all known; the octahedral geometry is by far the most common. Crystal field theory accounts for the colors and magnetic properties of coordination complexes by considering the strengths of the repulsive interactions between the electrons in the various *d* orbitals and the ligands, represented as point charges located along the Cartesian axes. For octahedral geometries, the degenerate set of *d* orbitals splits into two levels, one set of three at lower energy and a set of two at higher energy. The colors of coordination complexes, as well as their magnetic properties, are rationalized using this model. Large energy differences between the two sets of levels result in the absorption of light in the blue or ultraviolet portion of the spectrum, and the complexes appear red in color. These strong fields also favor electron configurations in which the electrons preferentially occupy the lower levels in pairs, the low-spin configuration. Weak fields lead to optical absorption in the red or yellow regions of the spectrum, and the compounds appear blue. The electrons occupy both sets of orbitals with their spins parallel, the high-spin configuration.

Crystal field theory is only partially successful in explaining the optical and magnetic properties of the coordination complexes; it cannot explain the relative strengths of the ligands in the spectrochemical series. Molecular orbital theory (an earlier version of which is called ligand field theory) provides a more complete and quantitative description of bonding, optical, and magnetic properties by allowing for the formation of both σ and π bonds between the central metal ion and the ligands. Molecular orbital theory provides qualitative insight by classifying ligands by their bonding types and by considering the oxidation states of the metal ion.

CUMULATIVE EXERCISE

Platinum

ASTRA

Crystals of potassium hexachloroplatinate(IV) (K_2PtCl_6).

The precious metal platinum was first used by South American Indians, who found impure, native samples of it in the gold mines of what is now Ecuador and used the samples to make small items of jewelry. Platinum's high melting point (1772°C) makes it harder to work than gold (1064°C) and silver (962°C), but this same property and a high resistance to chemical attack make platinum suitable as a material for high-temperature crucibles. Although platinum is a noble metal, in the +4 and +2 oxidation states it forms a variety of compounds, many of which are coordination complexes. Its coordinating abilities make it an important catalyst for organic and inorganic reactions.

(a) The anticancer drug cisplatin, *cis*-$[Pt(NH_3)_2Cl_2]$ (see Fig. 8.18b), can be prepared from K_2PtCl_6 via reduction with N_2H_4 (hydrazine), giving K_2PtCl_4, followed by replacement of two chloride ion ligands with ammonia. Give systematic names to the three platinum complexes referred to in this statement.

(b) The coordination compound diamminetetracyanoplatinum(IV) has been prepared, but salts of the hexacyanoplatinate(IV) ion have not. Write the chemical formulas of these two species.

(c) Platinum forms organometallic complexes quite readily. In one of the simplest of these, Pt(II) is coordinated to two chloride ions and two molecules of ethylene (C_2H_4) to give an unstable yellow crystalline solid. Can this complex have more than one isomer? If so, describe the possible isomers.

(d) Platinum(IV) is readily complexed by ethylenediamine. Draw the structures of both enantiomers of the complex ion *cis*-$[Pt(Cl)_2(en)_2]^{2+}$. In this compound, the Cl^- ligands are *cis* to one another.

(e) In platinum(IV) complexes, the octahedral crystal field splitting Δ_o is relatively large. Is K_2PtCl_6 diamagnetic or paramagnetic? What is its *d* electron configuration?

(f) Is cisplatin diamagnetic or paramagnetic?

(g) The salt $K_2[PtCl_4]$ is red, but $[Pt(NH_3)_4]Cl_2{\cdot}H_2O$ is colorless. In what regions of the spectrum do the dominant absorptions for these compounds lie?

(h) When the two salts from part (g) are dissolved in water and the solutions mixed, a green precipitate called Magnus's green salt forms. Propose a chemical formula for this salt and assign the corresponding systematic name.

Answers

(a) *cis*-Diamminedichloroplatinum(II), potassium hexachloroplatinate(IV), and potassium tetrachloroplatinate(II)

(b) $[Pt(NH_3)_2(CN)_4]$ and $[Pt(CN)_6]^{2-}$

(c) Pt(II) forms square-planar complexes. There are two possible forms, arising from *cis* and *trans* placement of the ethylene molecules. These are analogous to the two isomers shown in Figures 8.18b and c.

(d) The structures are the *cis* form shown in Figure 8.16 and its mirror image.

(e) The six *d* electrons in Pt(IV) are all in the lower t_{2g} level in a low-spin, large-Δ_o complex. All are paired; thus, the compound is diamagnetic.

(f) Diamagnetic, with the bottom four levels in the square-planar configuration (all but $d_{x^2-y^2}$) occupied

(g) Red transmission corresponds to absorption of green light by $K_2[PtCl_4]$. A colorless solution has absorptions at either higher or lower frequency than visible. Because Cl^- is a weaker field ligand than NH_3, the absorption frequency should be higher for the $Pt(NH_3)_4^{2+}$ complex, putting it in the ultraviolet region of the spectrum.

(h) $[Pt(NH_3)_4][PtCl_4]$, tetra-ammineplatinum(II) tetrachloroplatinate(II)

CHAPTER REVIEW

- The properties of transition metal compounds and coordination complexes are determined, in large part, by the participation of *d* electrons in bonding and chemical reactions.
- Most transition metals have a number of stable oxidation states that lead to different kinds of chemical bonds and facilitate electron transfer reactions.
- Molecular orbital theory satisfactorily describes bonding in transition metal compounds and coordination complexes.

  δ bonds, formed by face-to-face overlap of metal $d_{x^2-y^2}$ or d_{xy} orbitals, are significantly more important in the chemistry of the transition metals than that of the main group elements.

  L→M σ donation, L→M π donation and M→L π donation are terms used by inorganic chemists to describe the formation and filling of molecular orbitals derived from the overlap of metal and ligand orbitals.
- Coordination complexes comprise a central metal atom or ion coordinated by dative bonds to a number of ligands in a symmetrical arrangement. Linear,

tetrahedral, square planar, trigonal bipyramidal, and octahedral geometries are all known, with octahedral being by far the most common.

- Crystal field theory accounts for the optical and magnetic properties of coordination complexes by considering the electrostatic repulsion between the metal d electrons and the charges on ionic ligands.

  For octahedral complexes, the degenerate d orbitals split into two levels, a set of three lower energy t_{2g} orbitals and a pair of higher energy e_g orbitals. The energy difference between the two levels is called the crystal field splitting energy Δ_o.

  Crystal field splitting is also observed for square-planar and tetrahedral geometries, but the energy level structure and the magnitudes of the splittings are different.

  The optical properties arise from electronic transitions between levels split by the crystal field. Electronic absorption spectra measure Δ_o directly.

  The magnetic properties depend on Δ_o. Large values of Δ_o (high field) lead to configurations in which electrons preferentially occupy the lower levels with paired spins (low spin). Small values of Δ_o lead to configurations in which electrons singly occupy the upper levels (with spins parallel to those in the lower levels) before doubly occupying any lower level.

- Crystal field theory does not account for the strengths of ligands revealed by the spectrochemical series.
- Molecular orbital theory that considers only σ bonding produces results similar to crystal field theory but with a different physical interpretation for the origin of the splitting.
- Including π bonding in molecular orbital theory is essential to understand the nature of bonding in coordination complexes and to account for the trends observed in the spectrochemical series.

CONCEPTS & SKILLS

After studying this chapter and working the problems that follow, you should be able to:

1. Discuss the systematic variation of physical and chemical properties of the transition elements through the periodic table (Section 8.1, Problems 1–6).
2. Define the terms *coordination compound, coordination number, ligand,* and *chelation* (Sections 8.3 and 8.5).
3. Name coordination compounds, given their molecular formulas (Section 8.3, Problems 11–14).
4. Draw geometric isomers of octahedral, tetrahedral, and square-planar complexes (Section 8.4, Problems 17–20).
5. Describe several coordination complexes that have roles in biology (Section 8.4).
6. Use crystal field theory to interpret the magnetic properties of coordination compounds in terms of the electron configurations of their central ions (Section 8.4, Problems 21–25).
7. Relate the colors of coordination compounds to their crystal field splitting energies and CFSE (Section 8.5, Problems 27-36).
8. Use ligand field theory to order the energy levels in coordination compounds and to account for the spectrochemical series (Section 8.7).

PROBLEMS

Answers to problems whose numbers are boldface appear in Appendix G. Problems that are more challenging are indicated with asterisks.

Chemistry of the Transition Metals

1. Of the compounds PtF_4 and PtF_6, predict (a) which is more soluble in water and (b) which is more volatile.

2. The melting point of $TiCl_4$ (−24°C) lies below those of TiF_4 (284°C) and $TiBr_4$ (38°C). Explain why by considering the covalent-ionic nature of these compounds and the intermolecular forces in each case.

3. The decavanadate ion is a complex species with chemical formula $V_{10}O_{28}^{6-}$. It reacts with an excess of acid to form the dioxovanadium ion, VO_2^+, and water. Write a balanced chemical equation for this reaction. In what oxidation state is vanadium before and after this reaction? What vanadium oxide has the same oxidation state?

4. What is the chemical formula of vanadium(III) oxide? Do you predict this compound to be more basic or more acidic than the vanadium oxide of Problem 3? Write a balanced chemical equation for the reaction of vanadium(III) oxide with a strong acid.

5. Titanium(III) oxide is prepared by reaction of titanium(IV) oxide with hydrogen at high temperature. Write a balanced chemical equation for this reaction. Which oxide do you expect to have stronger basic properties?

6. Treatment of cobalt(II) oxide with oxygen at high temperatures gives Co_3O_4. Write a balanced chemical equation for this reaction. What is the oxidation state of cobalt in Co_3O_4?

Introduction to Coordination Chemistry

7. Will methylamine (CH_3NH_2) be a monodentate or a bidentate ligand? With which of its atoms will it bind to a metal ion?

8. Show how the glycinate ion ($H_2N-CH_2-COO^-$) can act as a bidentate ligand. (Draw a Lewis diagram if necessary.) Which atoms in the glycinate ion will bind to a metal ion?

9. Determine the oxidation state of the metal in each of the following coordination complexes: $[V(NH_3)_4Cl_2]$, $[Mo_2Cl_8]^{4-}$, $[Co(H_2O)_2(NH_3)Cl_3]^-$, $[Ni(CO)_4]$.

10. Determine the oxidation state of the metal in each of the following coordination complexes: $Mn_2(CO)_{10}$, $[Re_3Br_{12}]^{3-}$, $[Fe(H_2O)_4(OH)_2]^+$, $[Co(NH_3)_4Cl_2]^+$.

11. Give the chemical formula that corresponds to each of the following compounds:
(a) Sodium tetrahydroxozincate(II)
(b) Dichlorobis(ethylenediamine)cobalt(III) nitrate
(c) Triaquabromoplatinum(II) chloride
(d) Tetra-amminedinitroplatinum(IV) bromide

12. Give the chemical formula of each of the following compounds:
(a) Silver hexacyanoferrate(II)
(b) Potassium tetraisothiocyanatocobaltate(II)
(c) Sodium hexafluorovanadate(III)
(d) Potassium trioxalatochromate(III)

13. Assign a systematic name to each of the following chemical compounds:
(a) $NH_4[Cr(NH_3)_2(NCS)_4]$
(b) $[Tc(CO)_5]I$
(c) $K[Mn(CN)_5]$
(d) $[Co(NH_3)_4(H_2O)Cl]Br_2$

14. Give the systematic name for each of the following chemical compounds:
(a) $[Ni(H_2O)_4(OH)_2]$
(b) $[HgClI]$
(c) $K_4[Os(CN)_6]$
(d) $[FeBrCl(en)_2]Cl$

Structures of Coordination Complexes

15. Suppose 0.010 mol of each of the following compounds is dissolved (separately) in 1.0 L water: KNO_3, $[Co(NH_3)_6]Cl_3$, $Na_2[PtCl_6]$, $[Cu(NH_3)_2Cl_2]$. Rank the resulting four solutions in order of conductivity, from lowest to highest.

16. Suppose 0.010 mol of each of the following compounds is dissolved (separately) in 1.0 L water: $BaCl_2$, $K_4[Fe(CN)_6]$, $[Cr(NH_3)_4Cl_2]Cl$, $[Fe(NH_3)_3Cl_3]$. Rank the resulting four solutions in order of conductivity, from lowest to highest.

17. Draw the structures of all possible isomers for the following complexes. Indicate which isomers are enantiomer pairs.
(a) Diamminebromochloroplatinum(II) (square-planar)
(b) Diaquachlorotricyanocobaltate(III) ion (octahedral)
(c) Trioxalatovanadate(III) ion (octahedral)

18. Draw the structures of all possible isomers for the following complexes. Indicate which isomers are enantiomer pairs.
(a) Bromochloro(ethylenediamine)platinum(II) (square-planar)
(b) Tetra-amminedichloroiron(III) ion (octahedral)
(c) Amminechlorobis(ethylenediamine)iron(III) ion (octahedral)

19. Iron(III) forms octahedral complexes. Sketch the structures of all the distinct isomers of $[Fe(en)_2Cl_2]^+$, indicating which pairs of structures are mirror images of each other.

20. Platinum(IV) forms octahedral complexes. Sketch the structures of all the distinct isomers of $[Pt(NH_3)_2Cl_2F_2]$, indicating which pairs of structures are mirror images of each other.

Crystal Field Theory: Optical and Magnetic Properties

21. For each of the following ions, draw diagrams like those in Figure 8.25 to show orbital occupancies in both weak and strong octahedral fields. Indicate the total number of unpaired electrons in each case.
(a) Mn^{2+} (b) Zn^{2+} (c) Cr^{3+} (d) Mn^{3+} (e) Fe^{2+}

22. Repeat the work of the preceding problem for the following ions:
(a) Cr^{2+} (b) V^{3+} (c) Ni^{2+} (d) Pt^{4+} (e) Co^{2+}

23. Experiments can measure not only whether a compound is paramagnetic, but also the number of unpaired electrons. It is found that the octahedral complex ion $[Fe(CN)_6]^{3-}$ has fewer unpaired electrons than the octahedral complex ion $[Fe(H_2O)_6]^{3+}$. How many unpaired electrons are present in each species? Explain. In each case, express the CFSE in terms of Δ_o.

24. The octahedral complex ion $[MnCl_6]^{3-}$ has more unpaired spins than the octahedral complex ion $[Mn(CN)_6]^{3-}$. How many unpaired electrons are present in each species? Explain. In each case, express the CFSE in terms of Δ_o.

25. Explain why octahedral coordination complexes with three and eight *d* electrons on the central metal atom are particularly stable. Under what circumstances would you expect complexes with five or six *d* electrons on the central metal atom to be particularly stable?

26. Mn, Fe, and Co in the +2 and +3 oxidation states all form hexaaquacomplexes in acidic aqueous solution. The reduction reactions of the three species are represented schematically below, where the water ligands are not shown for simplicity. It is an experimental fact from electrochemistry that Mn^{2+} and Co^{2+} are more easily reduced than Fe^{3+}; that is, they will more readily accept an electron. Based on the electron configurations of the ions involved, explain why Fe^{3+} is harder to reduce than Mn^{2+} and Co^{2+}.

$$Mn^{3+} + e \longrightarrow Mn^{2+}$$

$$Fe^{3+} + e \longrightarrow Fe^{2+}$$

$$Co^{3+} + e \longrightarrow Co^{2+}$$

Optical Properties and the Spectrochemical Series

27. An aqueous solution of zinc nitrate contains the $[Zn(H_2O)_6]^{2+}$ ion and is colorless. What conclusions can be drawn about the absorption spectrum of the $[Zn(H_2O)_6]^{2+}$ complex ion?

28. An aqueous solution of sodium hexaiodoplatinate(IV) is black. What conclusions can be drawn about the absorption spectrum of the $[PtI_6]^{2-}$ complex ion?

29. Estimate the wavelength of maximum absorption for the octahedral ion hexacyanoferrate(III) from the fact that light transmitted by a solution of it is red. Estimate the crystal field splitting energy Δ_o (in kJ mol^{-1}).

30. Estimate the wavelength of maximum absorption for the octahedral ion hexa-aquanickel(II) from the fact that its solutions are colored green by transmitted light. Estimate the crystal field splitting energy Δ_o (in kJ mol^{-1}).

31. Estimate the CFSE for the complex in Problem 29. (**Note:** This is a high-field (low-spin) complex.)

32. Estimate the CFSE for the complex in Problem 30.

33. The chromium(III) ion in aqueous solution is blue–violet.
- (a) What is the complementary color to blue–violet?
- (b) Estimate the wavelength of maximum absorption for a $Cr(NO_3)_3$ solution.
- (c) Will the wavelength of maximum absorption increase or decrease if cyano ligands are substituted for the coordinated water? Explain.

34. An aqueous solution containing the hexa-amminecobalt(III) ion is yellow.
- (a) What is the complementary color to yellow?
- (b) Estimate this solution's wavelength of maximum absorption in the visible spectrum.

35. (a) An aqueous solution of $Fe(NO_3)_3$ has only a pale color, but an aqueous solution of $K_3[Fe(CN)_6]$ is bright red. Do you expect a solution of $K_3[FeF_6]$ to be brightly colored or pale? Explain your reasoning.
- (b) Would you predict a solution of $K_2[HgI_4]$ to be colored or colorless? Explain.

36. (a) An aqueous solution of $Mn(NO_3)_2$ is very pale pink, but an aqueous solution of $K_4[Mn(CN)_6]$ is deep blue. Explain why the two differ so much in the intensities of their colors.
- (b) Predict which of the following compounds would be colorless in aqueous solution: $K_2[Co(NCS)_4]$, $Zn(NO_3)_2$, $[Cu(NH_3)_4]Cl_2$, $CdSO_4$, $AgClO_3$, $Cr(NO_3)_2$.

ADDITIONAL PROBLEMS

37. Of the ten fourth-period transition metal elements in Table 8.1, which one has particularly low melting and boiling points? How can you explain this in terms of the electronic configuration of this element?

* **38.** Although copper lies between nickel and zinc in the periodic table, the reduction potential of Cu^{2+} is above that for both Ni^{2+} and Zn^{2+} (see Table 8.1). Use other data from Table 8.1 to account for this observation. (*Hint:* Think of a multistep process to convert a metal atom in the solid to a metal ion in solution. For each step, compare the relevant energy or enthalpy changes for Cu with those for Ni or Zn.)

39. A reference book lists five different values for the electronegativity of molybdenum, a different value for each oxidation state from +2 through +6. Predict which electronegativity is highest and which is lowest.

40. If *trans*-$[Cr(en)_2(NCS)_2]SCN$ is heated, it forms gaseous ethylenediamine and solid $[Cr(en)_2(NCS)_2][Cr(en)(NCS)_4]$. Write a balanced chemical equation for this reaction. What are the oxidation states of the Cr ions in the reactant and in the two complex ions in the product?

41. A coordination complex has the molecular formula $[Ru_2(NH_3)_6Br_3](ClO_4)_2$. Determine the oxidation state of ruthenium in this complex.

42. Heating 2.0 mol of a coordination compound gives 1.0 mol NH_3, 2.0 mol H_2O, 1.0 mol HCl, and 1.0 mol $(NH_4)_3[Ir_2Cl_9]$. Write the formula of the original (six-coordinate) coordination compound and name it.

43. Explain why ligands are usually negative or neutral in charge and only rarely positive.

44. Match each compound in the group on the left with the compound on the right that is most likely to have the same electrical conductivity per mole in aqueous solution.

(a) $[Fe(H_2O)_5Cl]CO_3$	HCN
(b) $[Mn(H_2O)_6]Cl_3$	$Fe_2(SO_4)_3$
(c) $[Zn(H_2O)_3(OH)]Cl]$	NaCl
(d) $[Fe(NH_3)_6]_2(CO_3)_3$	$MgSO_4$
(e) $[Cr(NH_3)_3Br_3]$	Na_3PO_4
(f) $K_3[Fe(CN)_6]$	$GaCl_3$

45. Three different compounds are known to have the empirical formula $CrCl_3 \cdot 6H_2O$. When exposed to a dehydrating agent, compound 1 (which is dark green) loses 2 mol water per mole of compound, compound 2 (light green) loses 1 mol water, and compound 3 (violet) loses no water. What are the probable structures of these compounds? If an excess of silver nitrate solution is added to 100.0 g of each of these compounds, what mass of silver chloride will precipitate in each case?

46. The octahedral structure is not the only possible six-coordinate structure. Other possibilities include a planar hexagonal structure and a triangular prism structure. In the latter, the ligands are arranged in two parallel triangles, one lying above the metal atom and the other below the metal atom with its corners directly in line with the corners of the first triangle. Show that the existence of two and only two isomers of $[Co(NH_3)_4Cl_2]^+$ is evidence against both of these possible structures.

47. Cobalt(II) forms more tetrahedral complexes than any other ion except zinc(II). Draw the structure(s) of the tetrahedral complex $[CoCl_2(en)]$. Could this complex exhibit geometric or optical isomerism? If one of the Cl^- ligands is replaced by Br^-, what kinds of isomerism, if any, are possible in the resulting compound?

48. Is the coordination compound $[Co(NH_3)_6]Cl_2$ diamagnetic or paramagnetic?

49. A coordination compound has the empirical formula $PtBr(en)(SCN)_2$ and is diamagnetic.
(a) Examine the *d*-electron configurations on the metal atoms, and explain why the formulation $[Pt(en)_2(SCN)_2][PtBr_2(SCN)_2]$ is preferred for this substance.
(b) Name this compound.

50. We used crystal field theory to order the energy-level splittings induced in the five *d* orbitals. The same procedure could be applied to *p* orbitals. Predict the level splittings (if any) induced in the three *p* orbitals by octahedral and square-planar crystal fields.

51. The three complex ions $[Mn(CN)_6]^{5-}$, $[Mn(CN)_6]^{4-}$, and $[Mn(CN)_6]^{3-}$ have all been synthesized and all are low-spin octahedral complexes. For each complex, determine the oxidation number of Mn, the configuration of the *d* electrons (how many t_{2g} and how many e_g), and the number of unpaired electrons present.

52. On the basis of the examples presented in Problem 51, can you tell whether $Mn^{2+}(aq)$ is more easily oxidized or more easily reduced? What can you conclude about the stability of $Mn^{2+}(aq)$?

53. The following ionic radii (in angstroms) are estimated for the +2 ions of selected elements of the first transition-metal series, based on the structures of their oxides: Ca^{2+}(0.99), Ti^{2+}(0.71), V^{2+}(0.64), Mn^{2+}(0.80), Fe^{2+}(0.75), Co^{2+}(0.72), Ni^{2+}(0.69), Cu^{2+}(0.71), Zn^{2+}(0.74). Draw a graph of ionic radius versus atomic number in this series, and account for its shape. The oxides take the rock salt structure. Are these solids better described as high- or low-spin transition-metal complexes?

54. The coordination geometries of $[Mn(NCS)_4]^{2-}$ and $[Mn(NCS)_6]^{4-}$ are tetrahedral and octahedral, respectively. Explain why the two have the same room-temperature molar magnetic susceptibility.

55. The complex ion $CoCl_4^{2-}$ has a tetrahedral structure. How many *d* electrons are on the Co? What is its electronic configuration? Why is the tetrahedral structure stable in this case?

56. In the coordination compound $(NH_4)_2[Fe(H_2O)F_5]$, the Fe is octahedrally coordinated.
(a) Based on the fact that F^- is a weak-field ligand, predict whether this compound is diamagnetic or paramagnetic. If it is paramagnetic, tell how many unpaired electrons it has.
(b) By comparison with other complexes reviewed in this chapter, discuss the likely color of this compound.
(c) Determine the *d*-electron configuration of the iron in this compound.
(d) Name this compound.

57. The compound $Cs_2[CuF_6]$ is bright orange and paramagnetic. Determine the oxidation number of copper in this compound, the most likely geometry of the coordination around the copper, and the possible configurations of the *d* electrons of the copper.

58. In what directions do you expect the bond length and vibrational frequency of a free CO molecule to change when it becomes a CO ligand in a $Ni(CO)_4$ molecule? Explain your reasoning.

59. Give the number of valence electrons surrounding the central transition-metal ion in each of the following known organometallic compounds or complex ions: $[Co(C_5H_5)_2]^+$, $[Fe(C_5H_5)(CO)_2Cl]$, $[Mo(C_5H_5)_2Cl_2]$, $[Mn(C_5H_5)(C_6H_6)]$.

60. Molecular nitrogen (N_2) can act as a ligand in certain coordination complexes. Predict the structure of $[V(N_2)_6]$, which is isolated by condensing V with N_2 at 25 K. Is this compound diamagnetic or paramagnetic? What is the formula of the carbonyl compound of vanadium that has the same number of electrons?

61. What energy levels are occupied in a complex such as hexacarbonylchromium(0)? Are any electrons placed into antibonding orbitals that are derived from the chromium *d* orbitals?

* 62. The compound $WH_2(C_5H_5)_2$ acts as a base, but $TaH_3(C_5H_5)_2$ does not. Explain.

63. Discuss the role of transition-metal complexes in biology. Consider such aspects as their absorption of light, the existence of many different structures, and the possibility of multiple oxidation states.

CUMULATIVE PROBLEMS

64. Predict the volume of hydrogen generated at 1.00 atm and 25°C by the reaction of 4.53 g scandium with excess aqueous hydrochloric acid.

* 65. Mendeleev's early periodic table placed manganese and chlorine in the same group. Discuss the chemical evidence for these placements, focusing on the oxides of the two elements and their acid–base and redox properties. Is there a connection between the electronic structures of their atoms? In what ways are the elements different?

66. An orange–yellow osmium carbonyl compound is heated to release CO and leave elemental osmium behind. Treatment of 6.79 g of the compound releases 1.18 L CO(*g*) at 25°C and 2.00 atm pressure. What is the empirical formula of this compound? Propose a possible structure for it by comparing with the metal carbonyls shown on the right.

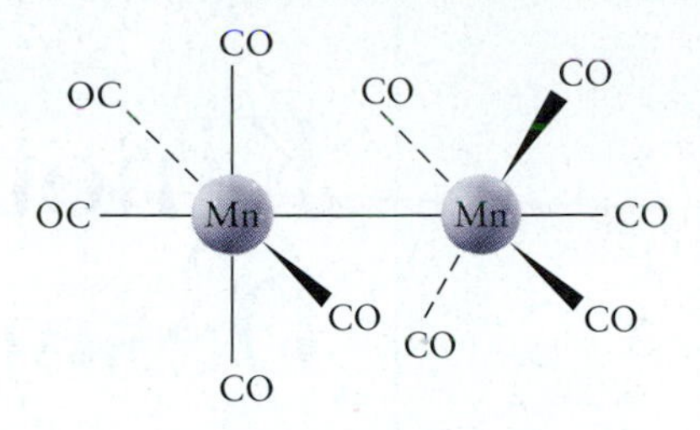

(a) $Mn_2(CO)_{10}$

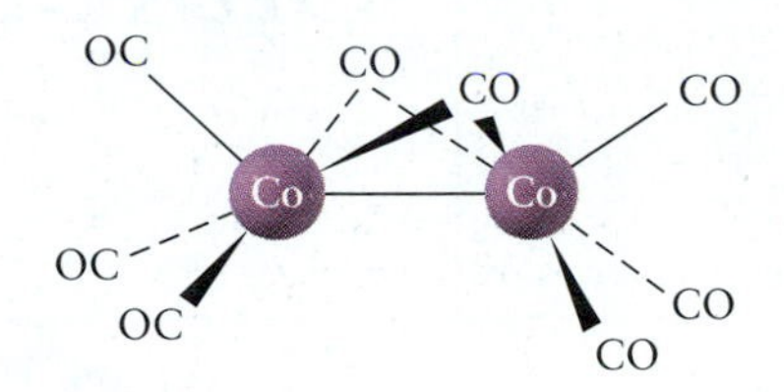

(b) $Co_2(CO)_8$

UNIT III

Kinetic Molecular Description of the States of Matter

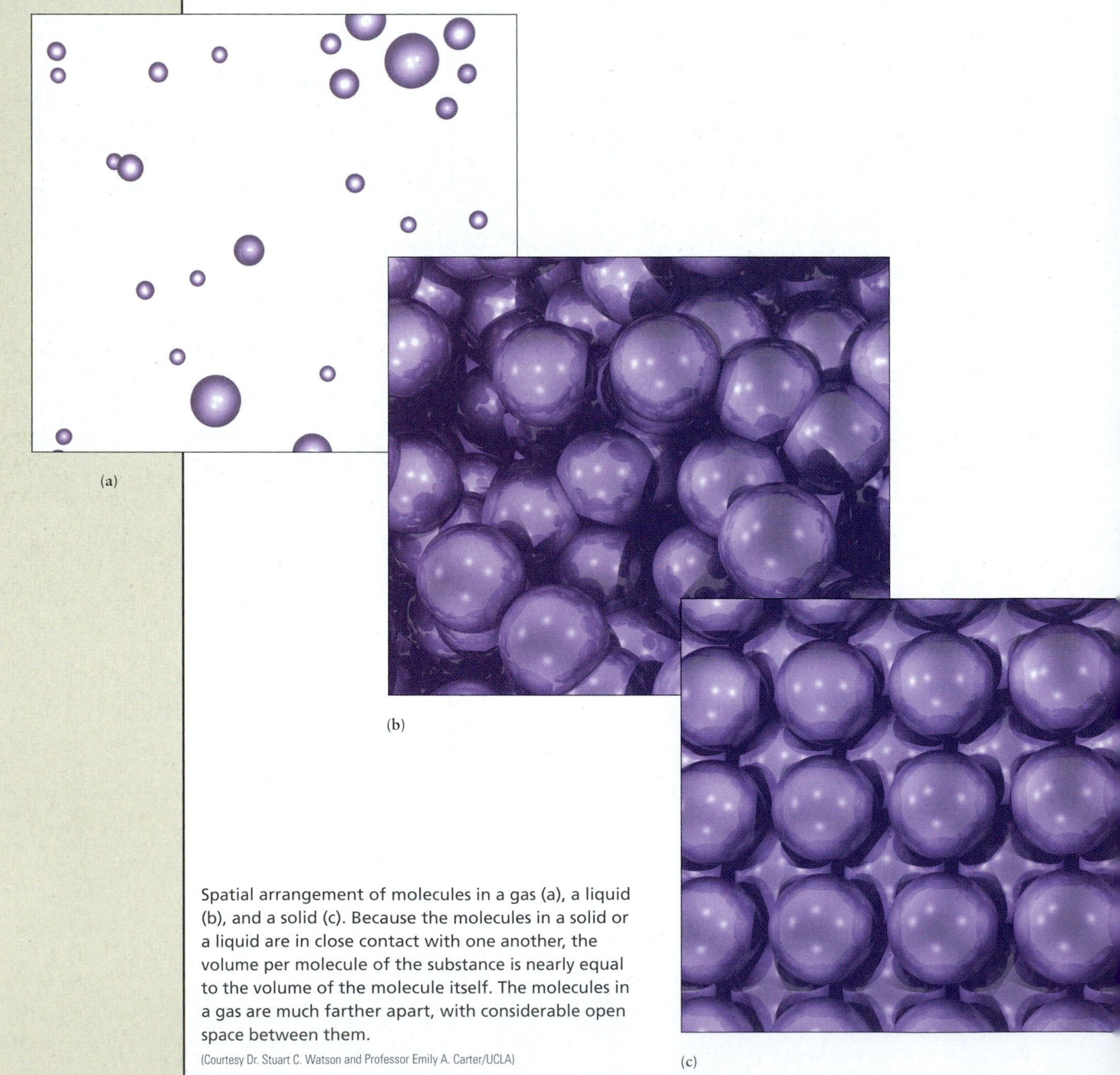

Spatial arrangement of molecules in a gas (a), a liquid (b), and a solid (c). Because the molecules in a solid or a liquid are in close contact with one another, the volume per molecule of the substance is nearly equal to the volume of the molecule itself. The molecules in a gas are much farther apart, with considerable open space between them.

(Courtesy Dr. Stuart C. Watson and Professor Emily A. Carter/UCLA)

We recognize gases, liquids, and solids quite easily because we are surrounded by important examples of each: air, water, and earth. Gases and liquids are fluid, but solids are rigid. A gas expands to fill any container it occupies. A liquid has a fixed volume but flows to conform to the shape of its container. A solid has a fixed volume *and* a fixed shape, both of which resist deformation. These characteristics originate in the arrangements and motions of molecules, which are determined by the forces between the molecules. In a solid or liquid, molecules (shown as spheres in the images on the previous page) are in very close contact. At these distances, molecules experience strong mutual forces of attraction. In a liquid, molecules can slide past one another, whereas in a solid, molecules remain attached to their home positions. In a gas, the molecules are much farther apart, and the forces between them are strong only during collisions. The molecules in a gas can roam freely throughout the container, unlike those in solids and liquids.

UNIT CHAPTERS

UNIT GOALS

- To define the essential properties of solids, liquids, and gases
- To relate the magnitudes of these properties to the motions of molecules and to the forces between molecules
- To define the properties of solutions and relate their magnitudes to composition of the solutions

CHAPTER 9

The Gaseous State

Gaseous nitrogen dioxide is generated by the reaction of copper with nitric acid.

The *kinetic molecular theory of matter* asserts that the macroscopic properties of matter are determined by the structures of its constituent molecules and the interactions between them. Starting in the latter half of the 19th century, scientists have developed a magnificent theoretical structure called **statistical mechanics** to explain the connection between microscopic structure and macroscopic properties. Statistical mechanics provides this bridge for all types of matter, ranging from biological materials to solid-state integrated circuits. Today, every student of science must be familiar with the concepts that make this connection. We give a qualitative introduction to these concepts, and apply them to numerous cases in the next four chapters.

We start the discussion with gases, because gases provide the simplest opportunity for relating macroscopic properties to the structures and interactions of molecules. This is possible because gases are much less dense than solids or liquids. On the macroscopic level, gases are distinguished from liquids and solids by their much smaller values of *mass density* (conveniently measured in grams per cubic centimeter). On the microscopic level, the *number density* (number of molecules in 1 cm^3 of the sample) is smaller—and the distances between molecules are much greater—than in liquids and solids. Molecules with no net electrical charge exert significant forces on each other only when they are close. Consequently, in the study of gases it is a legitimate simplification to ignore interactions between molecules until they collide and then to consider collisions between only two molecules at a time.

In this chapter we develop two themes that are key to relating macroscopic behavior to molecular structure. First, we show how to define and measure the macroscopic properties of gases as temperature, pressure, and volume are changed. This discussion leads us to develop and apply the ideal gas law, which—as verified by experiment—accurately represents the bulk properties of gases at low density. Second, we interpret and explain the ideal gas law in terms of the structures and motions of individual molecules. We obtain refinements to the ideal gas law by analyzing the consequences of collisions between pairs of molecules. The results and insights we obtain are essential background for later study of the rate and extent of chemical reactions since chemical reactions occur through molecular collisions.

9.1 The Chemistry of Gases

The ancient Greeks considered air to be one of the four fundamental elements in nature. European scientists began to study the properties of air (such as its resistance to compression) as early as the 17th century. The chemical composition of air (Table 9.1) was unknown until late in the 18th century, when Lavoisier, Priestley, and others showed that it consists primarily of two substances: oxygen and nitrogen. Oxygen was characterized by its ability to support life. Once the oxygen in a volume of air had been used up (by burning a candle in a closed container, for example), the nitrogen that remained could no longer keep animals alive. More

TABLE 9.1 Composition of Air

Constituent	Formula	Fraction by Volume
Nitrogen	N_2	0.78110
Oxygen	O_2	0.20953
Argon	Ar	0.00934
Carbon dioxide	CO_2	0.00038
Neon	Ne	1.82×10^{-5}
Helium	He	5.2×10^{-6}
Methane	CH_4	1.5×10^{-6}
Krypton	Kr	1.1×10^{-6}
Hydrogen	H_2	5×10^{-7}
Dinitrogen oxide	N_2O	3×10^{-7}
Xenon	Xe	8.7×10^{-8}

Air also contains other constituents, the abundances of which are quite variable in the atmosphere. Examples are water (H_2O), 0–0.07; ozone (O_3), 0–7 $\times 10^{-8}$; carbon monoxide (CO), 0–2 $\times 10^{-8}$; nitrogen dioxide (NO_2), 0–2 $\times 10^{-8}$; and sulfur dioxide (SO_2), 0–1 $\times 10^{-6}$.

Carbon dioxide, CO_2, is dissolved in aqueous solution to form carbonated beverages. Carbon dioxide also reacts with water to produce $H_2CO_3(aq)$, which provides some of the acidity in soft drinks. Solid carbon dioxide (dry ice) is used for refrigeration.

than 100 years elapsed before a careful reanalysis of the composition of air using multiple experimental techniques showed that oxygen and nitrogen account for only about 99% of the total volume, most of the remaining 1% being a new gas, given the name "argon." The other noble gases (helium, neon, krypton, and xenon) are present in air to lesser extents.

Other gases are found on the surface of the earth and in the atmosphere. Methane (CH_4), formerly known as "marsh gas," is produced by bacterial processes, especially in swampy areas. It is a major constituent of natural-gas deposits formed over many millennia by decay of plant matter beneath the surface of the earth. Recovery of methane from municipal landfills for use as a fuel is now a commercially feasible process. Gases also form when liquids evaporate. The most familiar example is water vapor in the air from the evaporation of liquid water; it provides the humidity of air.

Sulfur dioxide, SO_2, is produced by burning sulfur in air as the first step in the production of sulfuric acid.

Gases are also formed by chemical reactions. Some solids decompose upon heating to give gaseous products. One famous example is the decomposition of mercury(II) oxide to mercury and oxygen (Fig. 1.6):

$$2\ HgO(s) \xrightarrow{\text{Heat}} 2\ Hg(\ell) + O_2(g)$$

Joseph Priestly discovered the element oxygen while investigating this reaction. Even earlier, in 1756, Joseph Black showed that marble, which consists primarily of calcium carbonate ($CaCO_3$), decomposes upon heating to give quicklime (CaO) and carbon dioxide:

$$CaCO_3(s) \xrightarrow{\text{Heat}} CaO(s) + CO_2(g)$$

Sulfur trioxide, SO_3, is a corrosive gas produced by the further oxidation of sulfur dioxide.

Ammonium chloride (NH_4Cl) decomposes under heat to produce two gases: ammonia and hydrogen chloride:

$$NH_4Cl(s) \xrightarrow{\text{Heat}} NH_3(g) + HCl(g)$$

Some gas-forming reactions proceed explosively. The decomposition of nitroglycerin is a detonation in which all the products are gases:

$$4\ C_3H_5(NO_3)_3(\ell) \longrightarrow 6\ N_2(g) + 12\ CO_2(g) + O_2(g) + 10\ H_2O(g)$$

Several elements react with oxygen to form gaseous oxides. Carbon dioxide forms during animal respiration and is also produced by burning coal, oil, and other materials that contain carbon compounds. The role of carbon dioxide in global warming is the subject of intense research and policy debates. Oxides of sulfur are produced by burning elemental sulfur, and oxides of nitrogen arise from combustion of elemental nitrogen. Sulfur is a common impurity in fossil fuels, and nitrogen is burned in high-temperature environments like automobile engines:

$$S(s) + O_2(g) \longrightarrow SO_2(g)$$

$$2\ SO_2(g) + O_2(g) \longrightarrow 2\ SO_3(g)$$

$$N_2(g) + O_2(g) \longrightarrow 2\ NO(g)$$

$$2\ NO(g) + O_2(g) \longrightarrow 2\ NO_2(g)$$

The role of these compounds in air pollution is described in Section 20.5.

Gases can also be produced by the reactions of acids with ionic solids. Carbon dioxide is produced by the reaction of acids with carbonates (Fig. 9.1):

$$CaCO_3(s) + 2\ HCl(g) \longrightarrow CaCl_2(s) + CO_2(g) + H_2O(\ell)$$

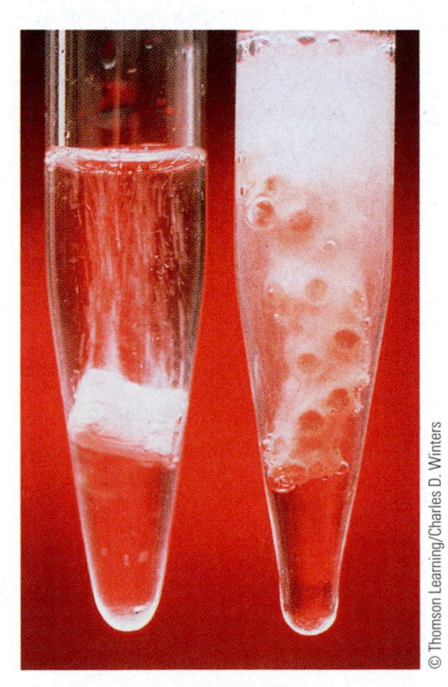

FIGURE 9.1 The calcium carbonate in a piece of chalk reacts with an aqueous solution of hydrochloric acid to produce bubbles of carbon dioxide.

Other examples of this type of reaction include:

$$Na_2S(s) + 2\ HCl(g) \longrightarrow 2\ NaCl(s) + H_2S(g)$$

$$K_2SO_3(s) + 2\ HCl(g) \longrightarrow 2\ KCl(s) + SO_2(g) + H_2O(\ell)$$

$$NaCl(s) + H_2SO_4(\ell) \longrightarrow NaHSO_4(s) + HCl(g)$$

In these reactions, the metal ions (sodium, calcium, and potassium) play no direct role; so, parallel reactions can be written for other metal ions.

The gases mentioned earlier vary greatly in their chemical properties. Some, such as HCl and SO_3, are reactive, acidic, and corrosive, whereas others, such as N_2O and N_2, are much less reactive. While the chemical properties of gases vary significantly, their physical properties are quite similar and much simpler to understand. At sufficiently low densities, all gases behave physically in the same way. Their properties are summarized and interpreted by a model system called the "ideal" gas, which is the subject of the following sections.

9.2 Pressure and Temperature of Gases

The macroscopic behavior of a fixed mass of a gas is completely characterized by three properties: volume (V), pressure (P), and temperature (T). Volume is self-evident and needs no comment; as stated in Section 2.1, we use the liter as a unit of volume. The definition of pressure and temperature require a little more care.

Pressure and Boyle's Law

The force exerted by a gas on a unit area of the walls of its container is called the **pressure** of that gas. We do not often stop to think that the air around us exerts a pressure on us and on everything else at the surface of the earth. Evangelista Torricelli (1608–1647), an Italian scientist who had been an assistant to Galileo, demonstrated this phenomenon in an ingenious experiment. He sealed a long glass tube at one end and filled it with mercury. He then covered the open end with his thumb, turned the tube upside down, and immersed the open end in a dish of liquid mercury (Fig. 9.2a), taking care that no air leaked in. The mercury in the tube fell, leaving a nearly perfect vacuum at the closed end, but it did not all run out of the tube. It stopped when its top was about 76 cm above the level of the mercury in the dish. Torricelli showed that the exact height varied somewhat from day to day and from place to place.

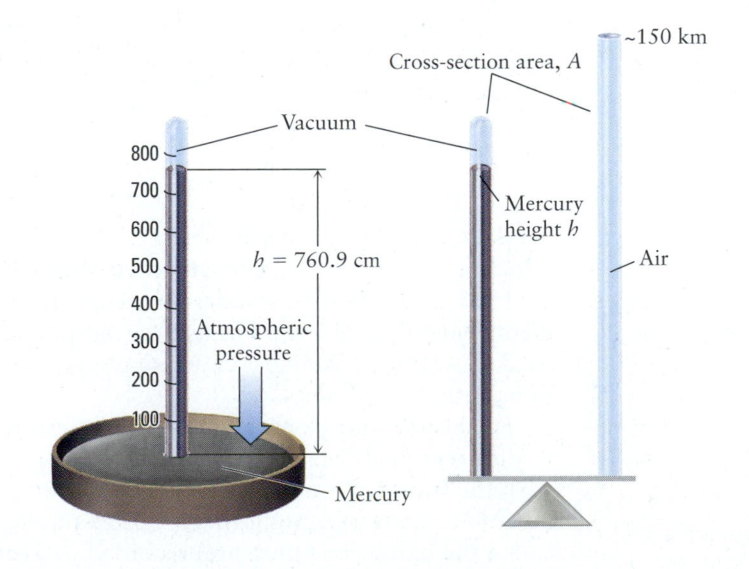

FIGURE 9.2 (a) In Torricelli's barometer, the top of the mercury in the tube is approximately 76 cm higher than that in the open beaker. (b) The mass of mercury in the column of height h exactly balances that of a column of air of the same diameter extending to the top of the atmosphere.

This simple device, called a **barometer,** works like a balance, one arm of which is loaded with the mass of mercury in the tube and the other with a column of air of the same cross-sectional area that extends to the top of the Earth's atmosphere, approximately 150 km above the surface of the earth. (see Fig. 9.2b). The height of the mercury column adjusts itself so that its mass and that of the air column become equal. This means that the two forces on the surface of the mercury in the dish are balanced. Day-to-day changes in the height of the column occur as the force exerted by the atmosphere varies with the weather. Atmospheric pressure varies strongly with altitude; it is lower at higher altitudes because the column of air pressing down is shorter and, therefore, has less mass.

How Torricelli's invention measures the pressure of the atmosphere is explained by Newton's second law of motion, which states:

$$\text{force} = \text{mass} \times \text{acceleration}$$

$$F = ma$$

in which the acceleration of a body (a) is the rate at which its velocity changes. The gravitational field of Earth exerts an attractive force that accelerates all bodies toward Earth. The standard acceleration due to the earth's gravitational field (usually denoted by g instead of a) is $g = 9.80665 \text{ m s}^{-2}$. Pressure is the force per unit area, or the total force, F, divided by the area, A:

$$P = \frac{F}{A} = \frac{mg}{A}$$

Because the volume of mercury in the barometer is $V = Ah$,

$$P = \frac{mg}{A} = \frac{mg}{V/h} = \frac{mgh}{V}$$

Writing the density as $\rho = m/V$ and substituting, we get

$$P = \rho gh \qquad [9.1]$$

We can use this equation to calculate the pressure exerted by the atmosphere. The density of mercury at 0°C, in SI units (see Appendix B for a description of SI units), is

$$\rho = 13.5951 \text{ g cm}^{-3} = 1.35951 \times 10^4 \text{ kg m}^{-3}$$

and the height of the mercury column under ordinary atmospheric conditions near sea level is close to 0.76 m (760 mm). Let's use exactly this value for the height in our computation:

$$P = \rho gh = (1.35951 \times 10^4 \text{ kg m}^{-3})(9.80665 \text{ m s}^{-2})(0.760000 \text{ m})$$

$$= 1.01325 \times 10^5 \text{ kg m}^{-1} \text{ s}^{-2}$$

Pressure is expressed in various units. The SI unit for pressure is the **pascal** (Pa), which is 1 kg m^{-1} s^{-2}. One **standard atmosphere** (1 atm) is defined as exactly 1.01325×10^5 Pa. The standard atmosphere is a useful unit because the pascal is inconveniently small and because "atmospheric pressure" is important as a standard of reference. We must express pressures in pascals when we perform calculations entirely in SI units.

For historical reasons, a number of different pressure units are commonly used in different fields of science and engineering. Although we will work primarily with the standard atmosphere, it is important that you recognize other units and be able to convert among them. For example, the atmospheric pressure (often called the barometric pressure) recorded in weather reports and forecasts is typically expressed as the height (in millimeters or inches) of the column of mercury it supports. One standard atmosphere supports a 760-mm column of mercury at

TABLE 9.2 Units of Pressure

Unit	Definition or Relationship
pascal (Pa)	1 kg m^{-1} s^{-2}
bar	1×10^5 Pa
atmosphere (atm)	101,325 Pa
torr	1/760 atm
760 mm Hg (at 0°C)	1 atm
14.6960 pounds per square inch (psi, lb in^{-2})	1 atm

0°C; thus, we often speak of 1 atm pressure as 760 mm or 30 inches (of mercury [Hg]). Because the density of mercury depends slightly on temperature, for accurate work it is necessary to specify the temperature and make the proper corrections to the density. A more precise term is the "torr," defined as 1 torr = 1/760 atm (or 760 torr = 1 atm) at *any* temperature. Only at 0°C do the torr and the millimeters of mercury (mm Hg) coincide. These and other units of pressure are summarized in Table 9.2.

Robert Boyle, an English natural philosopher and theologian, studied the properties of confined gases in the 17th century. He noted that a gas tends to spring back to its original volume after being compressed or expanded. Such behavior resembles that of metal springs, which were being investigated by his collaborator Robert Hooke. Boyle published his experiments on the compression and expansion of air in the 1662 monograph titled "The Spring of the Air and Its Effects."

Boyle studied how the volume of a confined gas responded to changes in pressure while the temperature was held constant. Boyle worked with a simple piece of apparatus: a J-tube in which air was trapped at the closed end by a column of mercury (Fig. 9.3). If the difference in height, *h*, between the two mercury levels in such a tube is 0, then the pressure of the air in the closed part exactly balances

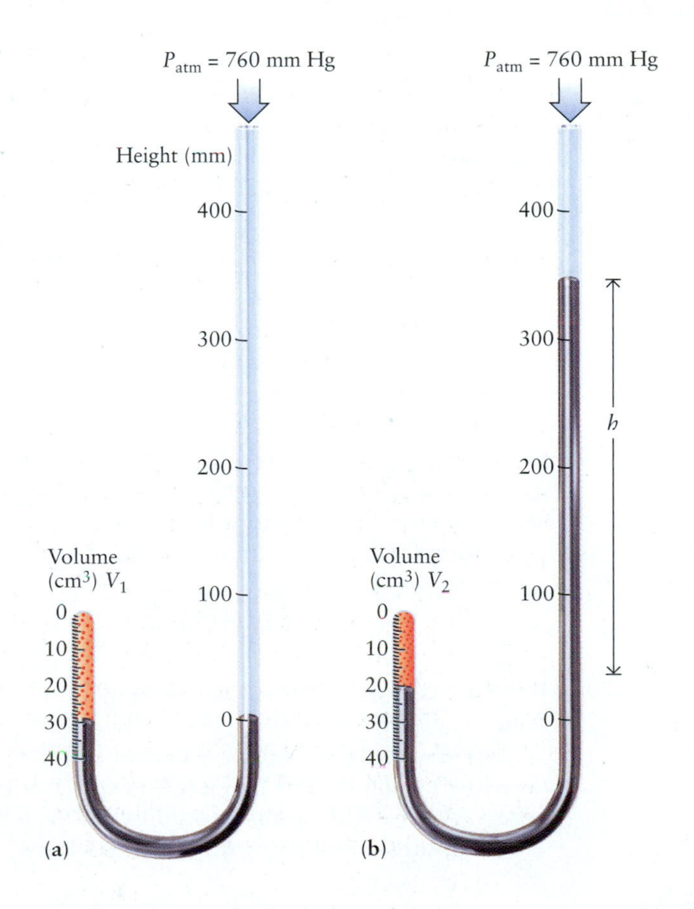

FIGURE 9.3 (a) Boyle's J-tube. When the heights of mercury on the two sides of the tube are the same, the pressure of the confined gas must equal that of the atmosphere, 1 atm or 760 mm Hg. (b) After mercury has been added, the pressure of the gas is increased by the number of millimeters of mercury in the height difference *h*. The compression of the gas causes it to occupy a smaller volume.

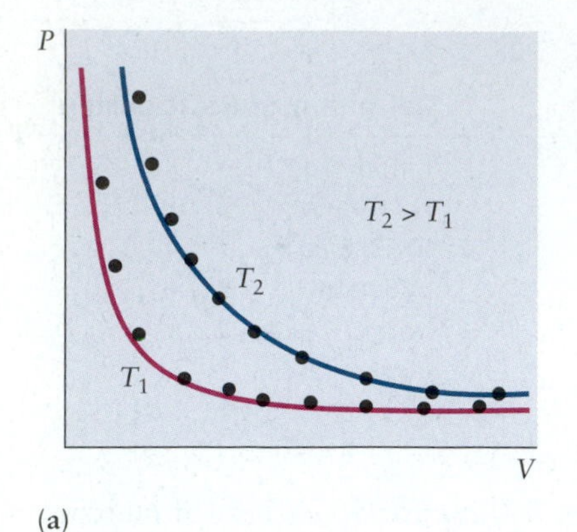

(a)

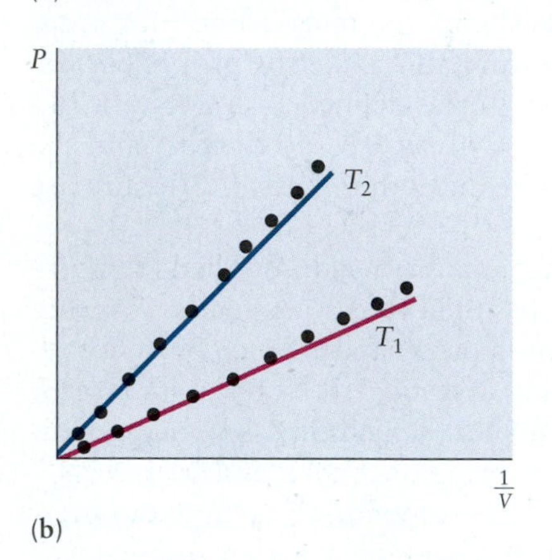

(b)

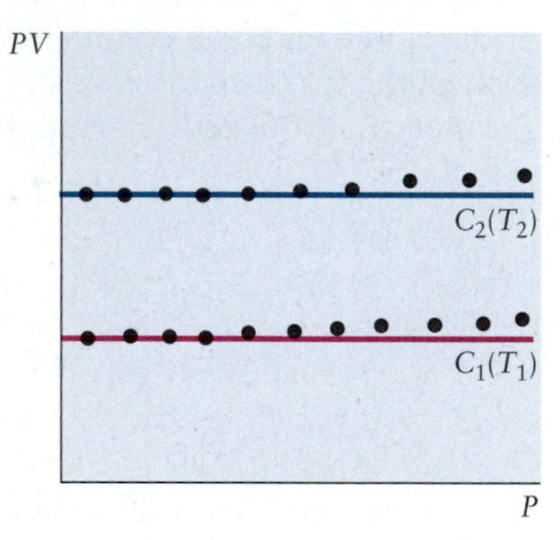

(c)

FIGURE 9.4 (a–c) Three ways of depicting Boyle's law. Small deviations from the law (especially apparent b and c) arise for real gases at higher pressures or smaller volumes.

that of the atmosphere; its pressure, P is 1 atm. Adding mercury to the open end of the tube increases the pressure applied to the confined air; its internal pressure increases by 1 mm Hg for every 1-mm difference between the levels of mercury on the open and closed sides of the tube. Expressed in atmospheres, the pressure in the closed end of the tube is

$$P = 1 \text{ atm} + \frac{h \text{ (mm)}}{760 \text{ mm atm}^{-1}}$$

The volume of the confined air can be read from the scale on the previously calibrated tube. The temperature is held constant by the ambient surroundings of the tube. Boyle discovered that the product of pressure and volume, PV, has the constant value C so long as temperature and the number of moles of confined gas remain fixed. The value of the constant C depends on the amount of gas and the temperature of the gas.

Boyle's data showed that P and V are inversely related, and *suggested* the relationship might be described by the equation $PV = C$, as shown in Figure 9.4a. Before the advent of graphing calculators and computers, it was not always easy to determine the mathematical function that best fit experimental data. With a limited amount of data, it is difficult to distinguish $P = C/V$ from $P = P^* \exp(-C/V)$, where P^* is some fixed reference value of pressure. How are we to know whether Boyle's experiments are best described by $PV = C$ or some more complicated equation? There are two ways of plotting the data to answer this question. The first is to rewrite Boyle's proposed equation in the form

$$P = \frac{C}{V} = C\left(\frac{1}{V}\right)$$

which makes P directly or linearly proportional to $1/V$. If Boyle's equation correctly fits his data, then replotting the data in the form P against $1/V$ (rather than V) should give a straight line passing through the origin with slope C (this is shown to be true in Figure 9.4b). Alternatively, if $PV = C$ is correct, replotting the data in the form PV against P should give a straight line independent of P (the results are shown in Figure 9.4c). Rewriting proposed equations to make them linear ($y = mx + b$) and plotting the data together with a proposed equation provides a good test of that equation's validity. The two plots studied here show convincingly that Boyle's equation accurately describes the relation between pressure and volume, at least over the range of temperatures and pressures measured.

$$PV = C \quad \text{(fixed temperature and fixed amount of gas)} \qquad [9.2]$$

This result is known as **Boyle's law.**

Keep in mind that the constant C depends on the temperature T and the amount (number of moles n) of gas in the closed container. For each combination of T and n, the limiting value of C at low P can be obtained by extrapolating plots (such as Fig. 9.4c) to zero pressure. For 1 mol gas (for example, 31.999 g O_2, 28.013 g N_2, or 2.0159 g H_2) at 0°C, extrapolation of PV to zero pressure (as shown by the red line in Fig. 9.4c) gives a limiting value of 22.414 L atm for C. Therefore, for these conditions, Boyle's law takes the special form:

$$PV = 22.414 \text{ L atm} \quad \text{(for 1 mol gas at 0°C)} \qquad [9.3]$$

If the pressure is 1.00 atm, the volume is 22.4 L; if P is 4.00 atm, V is 22.414/4.00 = 5.60 L.

Boyle's law is an idealized expression that is satisfied exactly by all gases at *very* low pressures. For real gases near 1 atm pressure, small corrections may be necessary for highly accurate studies of P-V-T behavior. At pressures beyond 50–100 atm, substantial corrections are necessary.

EXAMPLE 9.1

The long cylinder of a bicycle pump has a volume of 1131 cm^3 and is filled with air at a pressure of 1.02 atm. The outlet valve is sealed shut and the pump handle is pushed down until the volume of the air is 517 cm^3. Compute the pressure inside the pump. Express its value in atmospheres and pounds per square inch.

SOLUTION

Note that the temperature and amount of gas are not stated in this problem; thus, the value of 22.414 L atm cannot be used for the constant C. It is necessary only to assume that the temperature does not change as the pump handle is pushed down. If P_1 and P_2 are the initial and final pressures and V_1 and V_2 the initial and final volumes, then

$$P_1V_1 = P_2V_2$$

because the temperature and amount of air in the pump do not change. Substitution gives

$$(1.02 \text{ atm})(1131 \text{ cm}^3) = P_2(517 \text{ cm}^3)$$

which can be solved for P_2:

$$P_2 = 2.23 \text{ atm}$$

In pounds per square inch (see Table 9.2), this pressure is

$$P_2 = 2.23 \text{ atm} \times 14.696 \text{ psi atm}^{-1} = 32.8 \text{ psi}$$

Related Problems: 11, 12

Temperature and Charles's Law

Temperature is one of those elusive properties that we all think we understand but is, in fact, difficult to pin down in a quantitative fashion. We have an instinctive feeling (through the sense of touch) for *hot* and *cold*. Water at its freezing point is obviously colder than at its boiling point so we assign it a lower temperature. Both the Celsius and Fahrenheit temperature scales were defined using the freezing and boiling points of water. Water freezes at 0°C (32°F) and boils at 100°C (212°F). Although both scales use the same reference points, it is interesting to see how the size of the degree was defined and the reference temperatures were determined. Fahrenheit initially chose as reference points the freezing point for a saturated saltwater solution, which he assigned as 0° (then thought to be as cold as possible), and normal body temperature (that of his wife) assigned to be 96°. His choice of 96° for body temperature is thought to have been stimulated by the earlier work of Newton, who had devised a similar scale by dividing the interval between the boiling and freezing points of water into 12 units. As Fahrenheit's thermometers got better, he increased their resolution by several factors of 2 until the number 96 was reached. Further calibration led to the modern definition based on the freezing and boiling points of pure water cited earlier. The Celsius scale (originally called the centigrade scale) is a bit more logical. Celsius chose the same end points but divided the range into 100 units for ease of calculation.

Assigning two fixed points in this way does not show how to define a temperature scale. Ether boils at atmospheric pressure somewhere between 0°C and 100°C, but what temperature should be assigned to its boiling point? Further arbitrary choices are certainly not the answer. The problem is that temperature is not a mechanical quantity like pressure; therefore, it is more difficult to define.

One way around this problem is to find some mechanical property that depends on temperature and use it to define a temperature scale. If we measure the value of this property of an object immersed in boiling water and again when it is immersed in boiling ether, we can quantitatively compare the boiling points of these two liquids. A number of mechanical properties depend on temperature. For example, as liquid mercury is heated from 0°C to 100°C, its volume increases by 1.82%. This change in volume could be used to define a temperature scale. If we *assume* that the volume of mercury is a linear function of temperature, then we can simply measure temperature by measuring the volume of mercury in a tube (a mercury thermometer). The problem is that this definition is tied to the properties of a single substance, mercury, and rests on the assumption that the volume of a sample of mercury is directly proportional to the temperature. Can we define a temperature scale that is more universal, one that does not depend on the properties of a specific material or the assumption of linearity? To answer this question, let's examine the behavior of gases upon heating or cooling.

Boyle observed that the product of the pressure and volume of a confined gas changes on heating, but the first quantitative experiments on the temperature dependence of the properties of gases were performed by the French scientist Jacques Charles more than a century later. Charles observed that *all* gases expand by the same relative amount between the same initial and final temperatures, when studied at sufficiently low pressures. For example, heating a sample of N_2 from the freezing point of water to the boiling point causes the gas to expand to 1.366 times its original volume (Fig. 9.5). The same 36.6% increase in volume occurs for O_2, CO_2, and other gases. (In contrast, liquids and solids vary widely in their thermal expansion.) This universal behavior suggests that temperature is a linear function of gas volume. We write this function as

$$t = c\left(\frac{V}{V_0} - 1\right)$$

where V is the volume of the gas at temperature t, V_0 is its volume at the freezing point of water, and c is a constant that is the same for all gases. We have written the linear equation in this form to ensure that the freezing point of water (when $V = V_0$) will be at $t = 0$, corresponding to the zero of the Celsius scale of temperature. From the measured fractional increase in V to the boiling point (taken to be 100°C), the value of c can be determined. In 1802, Gay-Lussac reported a value for c of 267°C. Subsequent experiments have refined this result to give $c = 273.15°C$. The definition of temperature (in degrees Celsius) is then

$$t = 273.15°\text{C}\left(\frac{V}{V_0} - 1\right)$$

The temperature of a gas sample at low pressure can be measured by comparing its volume with the volume it occupies at the freezing point of water. For many

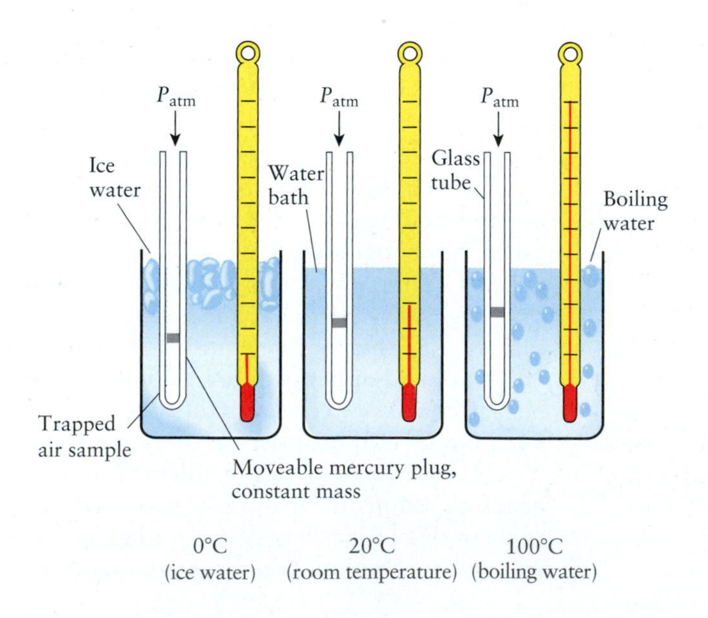

FIGURE 9.5 The volume of a gas confined at constant pressure increases as the temperature increases.

gases, atmospheric pressure is sufficiently low; but for highly accurate temperature determinations, it is necessary to use pressures below atmospheric or to apply small corrections.

With this definition of temperature in mind, let's return to mercury and measure its *actual* changes in volume with temperature. The result found is almost, but not quite, linear. If a mercury thermometer is calibrated to match the gas thermometer at 0°C and 100°C and if the scale in between is divided evenly into 100 parts to mark off degrees, a small error will result from using this thermometer. A temperature of 40.00°C on the gas thermometer will be read as 40.11°C on the mercury thermometer, because the volume of liquid mercury is not exactly a linear function of temperature.

We can rewrite the preceding equation to express the gas volume in terms of the temperature. The result is

$$V = V_0\left(1 + \frac{t}{273.15°\text{C}}\right) \qquad [9.4]$$

In words, the volume of a gas varies linearly with its temperature. This is the most common statement of **Charles's law,** but it is somewhat misleading because the linearity is built in through the definition of temperature. The key observation is the universal nature of the constant 273.15°C, which is the same for all gases at low pressures. Written in this form, Charles's law suggests that an interesting lower limit to the temperature exists. Negative temperatures on the Celsius scale correspond to temperatures below the freezing point of water and, of course, are meaningful. But what would happen as t approached −273.15°C? The volume would then approach zero (Fig. 9.6), and if t could go below this value, the volume would become negative, clearly an impossible result. We therefore surmise that $t = -273.15°\text{C}$ is a fundamental limit below which the temperature cannot be lowered. All real gases condense to liquid or solid form before they reach this **absolute zero** of temperature, so we cannot check for the existence of this limit simply by measuring the volume of gases. More rigorous arguments show that *no* substance (gas, liquid, or solid) can be cooled below −273.15°C. In fact, it becomes increasingly difficult to cool a substance as absolute zero is approached. The coldest temperatures reached to date are less than 1 nanodegree above absolute zero.

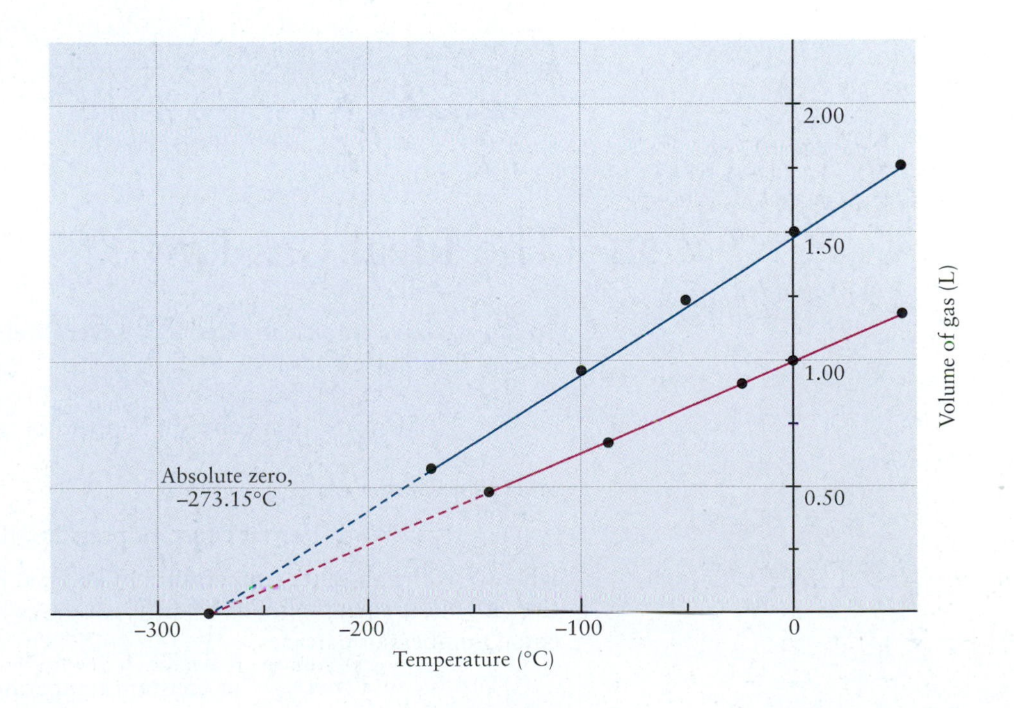

FIGURE 9.6 The volume of a sample of a gas is a function of temperature at constant pressure. The observed straight-line response of volume to temperature illustrates Charles's law. The volume of a particular sample of a gas (red line) is 1.0 L at a temperature of 0°C. Another sample of a gas (blue line) held at the same pressure takes up more volume at 0°C but shrinks faster as cooled. The *percentage* change in volume is the same as that of the first sample for every degree of temperature change. Extrapolation of the trends (dashed lines) predicts that the volumes of the samples go to zero at a temperature of −273.15°C. Similar observations are made regardless of the chemical identities of the gases.

The absolute zero of temperature is a compellingly logical choice as the zero point of a temperature scale. The easiest way to create such a new scale is to add 273.15 to the Celsius temperature, which leads to the **Kelvin temperature scale:**

$$T\ (\text{Kelvin}) = 273.15 + t\ (\text{Celsius}) \qquad [9.5]$$

The capital T signifies that this is an absolute scale, the unit of which is *kelvin* (K). Thus, a temperature of 25.00°C corresponds to 273.15 + 25.00 = 298.15 K. (**Note:** The unit is the kelvin, not the °K.) On this scale, Charles's law takes the following form:

$$V \propto T \qquad \text{(fixed temperature and fixed amount of gas)} \qquad [9.6]$$

where the proportionality constant is determined by the pressure and the amount (number of moles) of gas present. The ratios of volumes occupied at two different temperatures by a fixed amount of gas at fixed pressure are:

$$\frac{V_1}{V_2} = \frac{T_1}{T_2}$$

EXAMPLE 9.2

A scientist studying the properties of hydrogen at low temperature takes a volume of 2.50 L hydrogen at atmospheric pressure and a temperature of 25.00°C and cools the gas at constant pressure to −200.00°C. Predict the volume that the hydrogen occupies at the lower temperature.

SOLUTION

The first step is always to convert temperatures to kelvins:

$$t_1 = 25.00°\text{C} \Rightarrow T_1 = 273.15 + 25.00 = 298.15\ \text{K}$$

$$t_2 = -200.00°\text{C} \Rightarrow T_2 = 273.15 - 200.00 = 73.15\ \text{K}$$

The ratio in Charles's law is

$$\frac{V_1}{T_1} = \frac{2.50\ \text{L}}{298.15\ \text{K}} = \frac{V_2}{T_2} = \frac{V_2}{73.15\ \text{K}}$$

$$V_2 = \frac{(73.15\ \text{K})(2.50\ \text{L})}{298.15\ \text{K}} = 0.613\ \text{L}$$

Related Problems: 13, 14, 15, 16, 17, 18

9.3 The Ideal Gas Law

So far, we have empirically deduced several relationships between properties of gases. From Boyle's law,

$$V \propto \frac{1}{P} \qquad \text{(at constant temperature, fixed amount of gas)}$$

and from Charles's law,

$$V \propto T \qquad \text{(at constant pressure, fixed amount of gas)}$$

where T is the absolute temperature in kelvins. From Avogadro's hypothesis (Section 1.3) that equal volumes of different gases held at the same T and P contain equal numbers of particles,

$$V \propto n \qquad \text{(at constant temperature and pressure)}$$

where n is the number of moles of substance. These three statements may be combined in the form

$$V \propto \frac{nT}{P}$$

A proportionality constant called R converts this proportionality to an equation:

$$V = R\frac{nT}{P} \quad \text{or} \quad PV = nRT \tag{9.7}$$

Because Avogadro's hypothesis states that equal volumes of all gases contain the same number of molecules (or moles), R is a *universal* constant. Equation 9.7 states the **ideal gas law,** which holds approximately for all gases near atmospheric pressure and room temperature and becomes increasingly accurate at lower pressure and higher temperature. It is a limiting law that describes the behavior of gases at low *densities.* We obtained the ideal gas law empirically from experimental studies of the P-V-T behavior of gases, and we shall see in Section 9.5 that the same relation is predicted by a molecular model of gases. The ideal gas law is our first example of a direct connection between the experimentally observed macroscopic behavior of matter and the structure and interactions of its constituent molecules.

Situations frequently arise in which a gas undergoes a change that takes it from some initial condition (described by P_1, V_1, T_1, and n_1) to a final condition (described by P_2, V_2, T_2, and n_2). Because R is a constant,

$$\frac{P_1V_1}{n_1T_1} = \frac{P_2V_2}{n_2T_2} \tag{9.8}$$

This is a useful alternative form of the ideal gas law.

EXAMPLE 9.3

A weather balloon filled with helium (He) has a volume of 1.0×10^4 L at 1.00 atm and 30°C. It rises to an altitude at which the pressure is 0.60 atm and the temperature is −20°C. What is the volume of the balloon then? Assume that the balloon stretches in such a way that the pressure inside stays close to the pressure outside.

SOLUTION

Because the amount of helium does not change, we can set n_1 equal to n_2 and cancel it out of Equation 9.8, giving

$$\frac{P_1V_1}{T_1} = \frac{P_2V_2}{T_2}$$

Solving this for the only unknown quantity, V_2, gives

$$V_2 = V_1\left(\frac{P_1T_2}{P_2T_1}\right)$$

$$= 1.0 \times 10^4 \text{ L}\left(\frac{1.00 \text{ atm}}{0.60 \text{ atm}}\right)\left(\frac{253 \text{ K}}{303 \text{ K}}\right)$$

$$= 1.4 \times 10^4 \text{ L} = 14{,}000 \text{ L}$$

Remember that temperatures must always be converted to kelvins when using the ideal gas law.

Related Problems: 19, 20, 21, 22

FIGURE 9.7 In a hot-air balloon, the volume remains nearly constant (the balloon is rigid) and the pressure is nearly constant as well (unless the balloon rises very high). Thus, n is inversely proportional to T; as the air inside the balloon is heated, its amount decreases and its density falls. This reduced density gives the balloon its lift.

The approach outlined in Example 9.3 can be applied when other combinations of the four variables (P, V, T, and n) remain constant (Fig. 9.7). When n and T remain constant, they can be canceled from the equation to give Boyle's law, which was used in solving Example 9.1. When n and P remain constant, this relation reduces to Charles's law, used in Example 9.2.

The numerical value of R depends on the units chosen for P and V. At the freezing point of water ($T = 273.15$ K) the product PV for 1 mol of any gas approaches the value 22.414 L atm at low pressures; hence, R has the value

$$R = \frac{PV}{nT} = \frac{22.414 \text{ L atm}}{(1.000 \text{ mol})(273.15 \text{ K})} = 0.082058 \text{ L atm mol}^{-1} \text{ K}^{-1}$$

If P is measured in SI units of pascals (kg m^{-1} s^{-2}) and V in cubic meters, then R has the value

$$R = \frac{(1.01325 \times 10^5 \text{ kg m}^{-1} \text{ s}^{-2})(22.414 \times 10^{-3} \text{ m}^3)}{(1.0000 \text{ mol})(273.15 \text{ K})}$$

$$= 8.3145 \text{ kg m}^2 \, s^{-2} \text{ mol}^{-1} \text{ K}^{-1}$$

Because 1 kg m^2 s^{-2} is defined to be 1 *joule* (the SI unit of energy, abbreviated J), the gas constant may also be expressed as

$$R = 8.3145 \text{ J mol}^{-1} \text{ K}^{-1}$$

EXAMPLE 9.4

What mass of helium is needed to fill the weather balloon from Example 9.3?

SOLUTION

First, solve the ideal gas law for n:

$$n = \frac{PV}{RT}$$

If P is expressed in atmospheres and V is expressed in liters, then the value $R = 0.08206$ L atm mol^{-1} K^{-1} must be used.

$$n = \frac{(1.00 \text{ atm})(1.00 \times 10^4 \text{ L})}{(0.08206 \text{ L atm mol}^{-1} \text{ K}^{-1})(303.15 \text{ K})} = 402 \text{ mol}$$

Because the molar mass of helium is 4.00 g mol^{-1}, the mass of helium required is

$$(402 \text{ mol})(4.00 \text{ g mol}^{-1}) = 1610 \text{ g} = 1.61 \text{ kg}$$

Related Problems: 23, 24

Chemical Calculations for Gases

One of the most important applications of the gas laws in chemistry is to calculate the volumes of gases consumed or produced in chemical reactions. If the conditions of pressure and temperature are known, the ideal gas law can be used to convert between the number of moles and gas volume. Instead of working with the mass of each gas taking part in the reaction, we can then use its volume, which is easier to measure. This is illustrated by the following example.

EXAMPLE 9.5

Concentrated nitric acid acts on copper to give nitrogen dioxide and dissolved copper ions (Fig. 9.8) according to the balanced chemical equation

$$\text{Cu}(s) + 4\,\text{H}^+(aq) + 2\,\text{NO}_3^-(aq) \longrightarrow 2\,\text{NO}_2(g) + \text{Cu}^{2+}(aq) + 2\,\text{H}_2\text{O}(\ell)$$

FIGURE 9.8 When copper metal is immersed in concentrated nitric acid, the copper is oxidized and an aqueous solution of blue copper(II) nitrate forms. In addition, some of the nitrate ion is reduced to brown gaseous nitrogen dioxide, which bubbles off.

Suppose that 6.80 g copper is consumed in this reaction, and that the NO_2 is collected at a pressure of 0.970 atm and a temperature of 45°C. Calculate the volume of NO_2 produced.

SOLUTION

The first step (as in Fig. 2.4) is to convert from the mass of the known reactant or product (in this case, 6.80 g Cu) to number of moles by using the molar mass of copper, 63.55 g mol^{-1}:

$$\frac{6.80 \text{ g Cu}}{63.55 \text{ g mol}^{-1}} = 0.107 \text{ mol Cu}$$

Next, the number of moles of NO_2 generated in the reaction is calculated using the stoichiometric coefficients in the balanced equation:

$$0.107 \text{ mol Cu} \times \left(\frac{2 \text{ mol NO}_2}{1 \text{ mol Cu}}\right) = 0.214 \text{ mol NO}_2$$

Finally, the ideal gas law is used to find the volume from the number of moles (remember that the temperature must first be expressed in kelvins by adding 273.15):

$$V = \frac{nRT}{P} = \frac{(0.214 \text{ mol})(0.08206 \text{ L atm mol}^{-1} \text{ K}^{-1})(273.15 + 45)\text{K}}{0.970 \text{ atm}} = 5.76 \text{ L}$$

Therefore, 5.76 L NO_2 is produced under these conditions.

Related Problems: 25, 26, 27, 28, 29, 30, 31, 32

9.4 Mixtures of Gases

Suppose a mixture of gases occupies a container at a certain temperature. How does each gas contribute to the total pressure of the mixture? We define the **partial pressure** of each gas as the pressure that gas would exert if it alone were present in the container. John Dalton concluded, from experiment, that the total pressure measured, P_{tot}, is the sum of the partial pressures of the individual gases (Fig. 9.9). This should come as no surprise given the validity of Avogadro's hypothesis. Even so, it is an important result. **Dalton's Law** holds under the same conditions as the ideal gas law itself: It is approximate at moderate pressures and becomes increasingly more accurate as the pressure is lowered.

For a gas mixture at low pressure, the partial pressure of one component, A, is

$$P_A = n_A \frac{RT}{V}$$

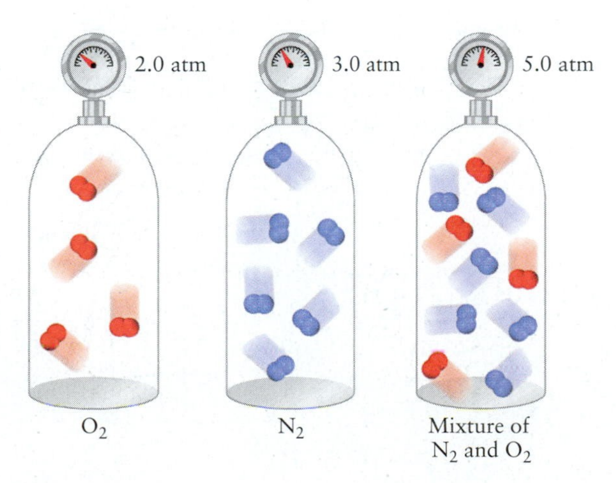

FIGURE 9.9 According to Dalton's law, the total pressure of a gas mixture is the sum of the pressures exerted by the individual gases. Note that the total volume is the same in all three containers.

The total pressure is the sum of the partial pressures:

$$P_{tot} = P_A + P_B + P_C + \cdots = (n_A + n_B + n_C + \cdots)\frac{RT}{V} = n_{tot}\frac{RT}{V} \quad [9.9]$$

where n_{tot} is the total number of moles in the gas mixture. Dividing the first equation by the second gives

$$\frac{P_A}{P_{tot}} = \frac{n_A}{n_{tot}} \quad \text{or} \quad P_A = \frac{n_A}{n_{tot}}P_{tot}$$

We define

$$X_A = \frac{n_A}{n_{tot}}$$

as the **mole fraction** of A in the mixture—that is, the number of moles of A divided by the total number of moles present. Then

$$P_A = X_A P_{tot} \quad [9.10]$$

The partial pressure of any component in a mixture of ideal gases is the total pressure multiplied by the mole fraction of that component. Note that the volume fractions in Table 9.1 are the same as mole fractions.

EXAMPLE 9.6

When NO_2 is cooled to room temperature, some of it reacts to form a *dimer,* N_2O_4, through the reaction

$$2\ NO_2(g) \longrightarrow N_2O_4(g)$$

Suppose 15.2 g of NO_2 is placed in a 10.0-L flask at high temperature and the flask is cooled to 25°C. The total pressure is measured to be 0.500 atm. What partial pressures and mole fractions of NO_2 and N_2O_4 are present?

SOLUTION

Initially, there is 15.2 g/46.01 g mol^{-1} = 0.330 mol NO_2, and therefore, the same number of moles of nitrogen atoms. If at 25°C there are n_{NO_2} moles NO_2 and $n_{N_2O_4}$ moles N_2O_4, then, because the total number of moles of nitrogen atoms is unchanged,

$$n_{NO_2} + 2n_{N_2O_4} = 0.330 \text{ mol} \quad \text{(a)}$$

To find a second relation between n_{NO_2} and $n_{N_2O_4}$, use Dalton's law:

$$P_{NO_2} + P_{N_2O_4} = 0.500 \text{ atm}$$

$$\frac{RT}{V}n_{NO_2} + \frac{RT}{V}n_{N_2O_4} = 0.500 \text{ atm}$$

$$n_{NO_2} + n_{N_2O_4} = 0.500 \text{ atm}\frac{V}{RT}$$

$$= \frac{(0.500 \text{ atm})(10.0 \text{ L})}{(0.08206 \text{ L atm mol}^{-1}\text{ K}^{-1})(298 \text{ K})}$$

$$n_{NO_2} + n_{N_2O_4} = 0.204 \text{ mol} \quad \text{(b)}$$

Subtracting Equation b from Equation a gives

$$n_{N_2O_4} = 0.126 \text{ mol}$$

$$n_{NO_2} = 0.078 \text{ mol}$$

From these results, NO_2 has a mole fraction of 0.38 and a partial pressure of (0.38)(0.500 atm) = 0.19 atm, and N_2O_4 has a mole fraction of 0.62 and a partial pressure of (0.62)(0.500 atm) = 0.31 atm.

Related Problems: 33, 34, 35, 36, 37, 38

9.5 The Kinetic Theory of Gases

The ideal gas law summarizes certain physical properties of gases at low pressures. It is an empirical law, the consequence of experimental observations, but its simplicity and generality prompt us to ask whether it has some underlying microscopic explanation that involves the properties of atoms and molecules in a gas. Such an explanation would allow other properties of gases at low pressures to be predicted and would clarify why real gases deviate from the ideal gas law to small but measurable extents. Such a theory was developed in the 19th century, notably by the physicists Rudolf Clausius, James Clerk Maxwell, and Ludwig Boltzmann. The **kinetic theory of gases** is one of the great milestones of science, and its success provides strong evidence for the atomic theory of matter (see discussion in Chapter 1).

This section introduces a type of reasoning different from that used so far. Instead of proceeding from experimental observations to empirical laws, we begin with a model and use the basic laws of physics with mathematical reasoning to show how this model helps explain the measured properties of gases. In this way, the kinetic theory of gases provides a microscopic understanding of Boyle's law and also a microscopic mechanical definition of temperature as a measure of the average kinetic energy of the molecules in a gas.

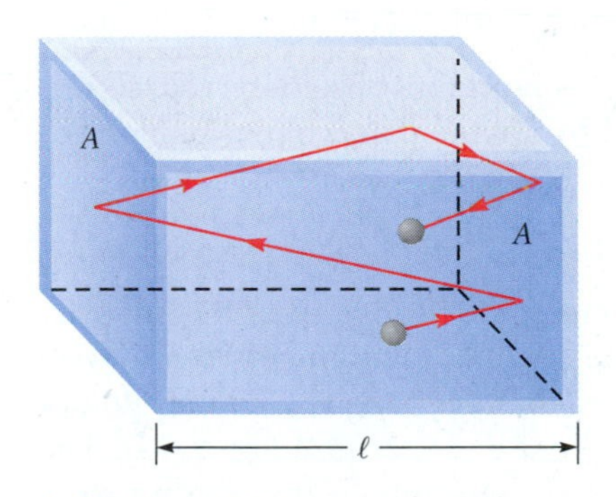

FIGURE 9.10 The path of a molecule in a box.

The underlying assumptions of the kinetic theory of gases are simple:

1. A pure gas consists of a large number of identical molecules separated by distances that are great compared with their size.
2. The gas molecules are constantly moving in random directions with a distribution of speeds.
3. The molecules exert no forces on one another between collisions, so between collisions they move in straight lines with constant velocities.
4. The collisions of molecules with the walls of the container are *elastic;* no energy is lost during a collision.

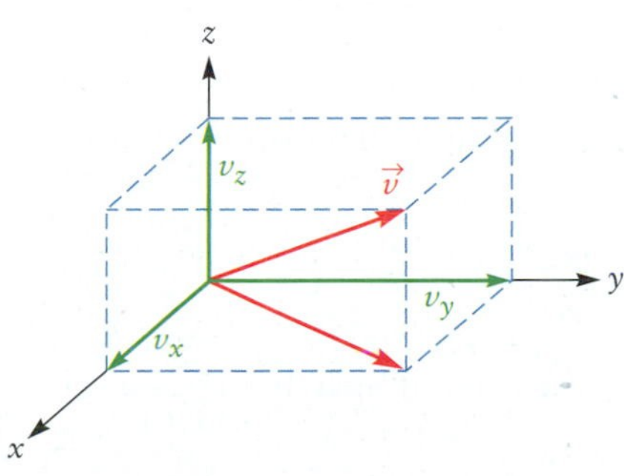

FIGURE 9.11 Velocity is shown by an arrow of length v. It can be separated into three components, v_x, v_y, and v_z, along the three Cartesian coordinate axes and projected into the x-y plane.

The Meaning of Temperature

We first use the kinetic theory of gases to find a relation among pressure, volume, and the motions of molecules in an ideal gas. Comparing the result obtained with the ideal gas law ($PV = nRT$) provides a deeper understanding of the meaning of temperature.

Suppose a container has the shape of a rectangular box of length ℓ, with end faces, each of which has area A (Fig. 9.10). A single molecule moving with speed u in some direction is placed in the box. It is important to distinguish between **speed** and **velocity.** The velocity of a molecule specifies both the rate at which it is moving (its speed, in meters per second) and the direction of motion. As shown in Figure 9.11, the velocity can be indicated by an arrow (vector) $\vec{v}$, which has a length

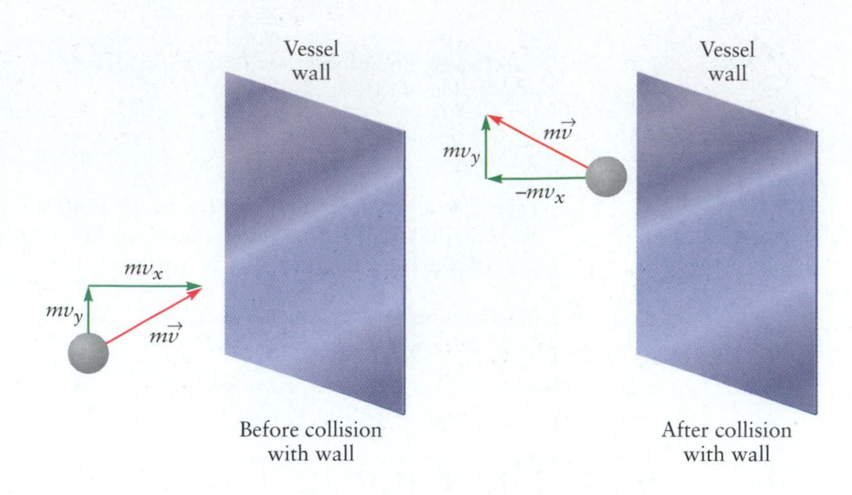

FIGURE 9.12 An elastic collision of a molecule with a wall. The component of the molecule's momentum perpendicular to the wall reverses sign, from mv_x to $-mv_x$. The component parallel to the wall, mv_y, is unchanged. The total momentum is shown by the red arrow. Although the direction of the red arrow is changed by the collision, its length, which represents the magnitude of the momentum, is not changed. Speeds of molecules are not affected by elastic collisions with the walls of the container.

equal to the speed u and which points in the direction of motion of the molecule. The velocity can also be represented by its components along three coordinate axes: v_x, v_y, and v_z. These are related to the speed, u, by the Pythagorean theorem:

$$u^2 = v_x^2 + v_y^2 + v_z^2$$

The **momentum** of a molecule, $\vec{p}$, is its velocity multiplied by its mass. When the molecule collides elastically with a wall of the box, such as one of the end faces of area A, the y and z components of the velocity, v_y and v_z, are unchanged, but the x component (perpendicular to A) reverses sign (Fig. 9.12). The change in the x component of the momentum of the molecule, $\Delta p_{x,\text{mol}}$, is

$$\Delta p_{x,\text{mol}} = \text{final momentum} - \text{initial momentum} = m(-v_x) - mv_x = -2\,mv_x$$

The total momentum of the system (molecule plus box) must be conserved, so this momentum change of the molecule is balanced by an equal and opposite momentum change given to the wall:

$$\Delta p_{x,\text{wall}} = 2mv_x$$

After colliding with the wall, the molecule reverses direction, strikes the opposite face of the box, and then again approaches the original face. In between, it may strike the top, bottom, and sides. These collisions do not change v_x, so they do not affect the time between collisions with the original end face (see Fig. 9.10). The distance traveled in the x direction is 2ℓ, and the magnitude of the velocity component in this direction is v_x, so the time elapsed between collisions with this end face is

$$\Delta t = \frac{2\ell}{v_x}$$

The momentum transferred to the wall per second is the momentum change per collision divided by Δt:

$$\frac{\Delta p_{x,\text{wall}}}{\Delta t} = \frac{2mv_x}{2\ell/v_x} = \frac{mv_x^2}{\ell}$$

From Newton's second law, the force exerted on the original face by repeated collisions of this molecule is:

$$f = ma = m\frac{\Delta v}{\Delta t} = \frac{\Delta p}{\Delta t} = \frac{mv_x^2}{\ell}$$

Suppose now that a large number, N, of molecules of mass m are moving independently in the box with x components of velocity, v_{x1}, v_{x2}, v_{x3}, and so forth. Then the total force exerted on the face by the N molecules is the sum of the forces exerted by the individual molecules:

$$F = \frac{mv_{x1}^2}{\ell} + \frac{mv_{x2}^2}{\ell} + \cdots + \frac{mv_{xN}^2}{\ell} = \frac{Nm}{\ell}\overline{v_x^2}$$

where

$$\overline{v_x^2} = \frac{1}{N}(v_{x1}^2 + v_{x2}^2 + \cdots + v_{xN}^2)$$

Here, $\overline{v_x^2}$ is the average of the square of the x component of the velocity of the N molecules, obtained by summing v_x^2 for the N molecules and dividing by N. The pressure is the total force on the wall divided by the area, A, so

$$P = \frac{F}{A} = \frac{Nm}{A\ell}\overline{v_x^2}$$

Because $A\ell$ is the volume, V, of the box, we conclude that

$$PV = Nm\overline{v_x^2}$$

There is no preferred direction of motion for the gas molecules; thus, $\overline{v_x^2}$, $\overline{v_y^2}$, and $\overline{v_z^2}$ should all be equal to one other. We therefore conclude that

$$\overline{u^2} = \overline{v_x^2} + \overline{v_y^2} + \overline{v_z^2} = 3\overline{v_x^2}$$

so

$$PV = \tfrac{1}{3}Nm\overline{u^2} \qquad [9.11]$$

where $\overline{u^2}$ is the **mean-square speed** of the gas molecules. From the ideal gas law,

$$PV = nRT$$

so, we conclude that

$$\tfrac{1}{3}Nm\overline{u^2} = nRT$$

We have achieved our major goal with the derivation of this equation: a relationship between the temperature and the speeds of molecules. It can be simplified to provide additional insights. The equation has the number of molecules, N, on the left and the number of moles, n, on the right. Because N is just n multiplied by Avogadro's number, N_A, we can divide both sides by n to find

$$\tfrac{1}{3}N_A m\overline{u^2} = RT \qquad [9.12]$$

Let's examine this equation in two ways. First, we note that the kinetic energy of a molecule of mass m moving at speed u is equal to $\frac{1}{2}mu^2$, so the *average* kinetic energy of N_A molecules (1 mol), which we denote by $\overline{E}$, is $\frac{1}{2}N_A m\overline{u^2}$. This quantity is exactly the same as that in the left side of Equation 9.12, with the factor $\frac{1}{2}$ replacing $\frac{1}{3}$:

$$\overline{E} = N_A m\overline{u^2} = \tfrac{3}{2} \times \left(\tfrac{1}{3}N_A m\overline{u^2}\right) = \tfrac{3}{2}RT \qquad [9.13]$$

We obtain the average kinetic energy per molecule, $\overline{\varepsilon}$, by dividing $\overline{E}$ by Avogadro's number:

$$\overline{\varepsilon} = \tfrac{3}{2}k_B T \qquad [9.14]$$

where k_B is **Boltzmann's constant** and is defined as R/N_A. The average kinetic energy of the molecules of a gas depends only on the temperature. It does not depend on the mass of the molecules or their number density in the gas. This relation is the most fundamental result of the kinetic theory of gases, and it is used in all branches of science.

A second way to look at the equation is to recall that if m is the mass of a single molecule, then $N_A m$ is the mass of 1 mol of molecules—the molar mass, abbreviated $\mathcal{M}$. Solving Equation 9.13 for the mean-square speed, we find that

$$\overline{u^2} = \frac{3RT}{\mathcal{M}} \qquad [9.15]$$

The mean-square speed of a gas molecule is proportional to temperature and inversely proportional to its mass. All molecules move faster at higher temperatures, and lighter molecules move faster than heavier ones at the same temperature.

Distribution of Molecular Speeds

One of the fundamental assumptions of the kinetic theory is that the molecules travel through the gas with a range of possible speeds. We would like to know how the molecules are distributed over the range of possible speeds.

As a first step, we can get some sense of the typical speeds in the gas by the following method. We define the **root-mean-square speed,** u_{rms}, as the square root of the mean-square speed $3RT/\mathcal{M}$:

$$u_{rms} = \sqrt{\overline{u^2}} = \sqrt{\frac{3RT}{\mathcal{M}}} \qquad [9.16]$$

This equation makes sense only when all of its terms are expressed in a self-consistent system of units such as the SI system. The appropriate value used for R is

$$R = 8.3145 \text{ J mol}^{-1} \text{ K}^{-1} = 8.3145 \text{ kg m}^2 \text{ s}^{-2} \text{ mol}^{-1} \text{ K}^{-1}$$

Note that molar masses $\mathcal{M}$ must be converted to *kilograms* per mole for use in the equation. The final result is expressed in the SI unit of speed, meters per second.

EXAMPLE 9.7

Calculate u_{rms} for (**a**) a He atom, (**b**) an oxygen molecule, and (**c**) a xenon atom at 298 K.

SOLUTION

Because the factor $3RT$ appears in all three expressions for u_{rms}, let's calculate it first:

$$3RT = (3)(8.3145 \text{ kg m}^2 \text{ s}^{-2} \text{ mol}^{-1} \text{ K}^{-1})(298 \text{ K}) = 7.43 \times 10^3 \text{ kg m}^2 \text{ s}^{-2} \text{ mol}^{-1}$$

The molar masses of He, O_2, and Xe are 4.00 g mol^{-1}, 32.00 g mol^{-1}, and 131.3 g mol^{-1}, respectively. Convert them to 4.00×10^{-3} kg mol^{-1}, 32.00×10^{-3} kg mol^{-1}, and 131.3×10^{-3} kg mol^{-1}, and insert them together with the value for $3RT$ into the equation for u_{rms}:

$$u_{rms}(\text{He}) = \sqrt{\frac{7.43 \times 10^3 \text{ kg m}^2 \text{ s}^{-2} \text{ mol}^{-1}}{4.00 \times 10^{-3} \text{ kg mol}^{-1}}} = 1360 \text{ m s}^{-1}$$

$$u_{rms}(\text{O}_2) = \sqrt{\frac{7.43 \times 10^3 \text{ kg m}^2 \text{ s}^{-2} \text{ mol}^{-1}}{32.00 \times 10^{-3} \text{ kg mol}^{-1}}} = 482 \text{ m s}^{-1}$$

$$u_{rms}(\text{Xe}) = \sqrt{\frac{7.43 \times 10^3 \text{ kg m}^2 \text{ s}^{-2} \text{ mol}^{-1}}{131.3 \times 10^{-3} \text{ kg mol}^{-1}}} = 238 \text{ m s}^{-1}$$

At the same temperature, the He, O_2, and Xe molecules all have the same average kinetic energy; lighter molecules move faster to compensate for their smaller masses. These rms speeds convert to 3050, 1080, and 532 mph, respectively. The average molecule moves along quite rapidly at room temperature!

Related Problems: 41, 42, 43, 44

It is useful to have a complete picture of the entire distribution of molecular speeds. This turns out to be important when we study chemical kinetics (see Chapter 18), where we will need to know what fraction of a sample of molecules has kinetic energy above the minimum necessary for a chemical reaction. In particular, we would like to know what fraction of molecules, $\Delta N/N$, have speeds between u and $u + \Delta u$. This fraction gives the speed distribution function $f(u)$:

$$\frac{\Delta N}{N} = f(u)\ \Delta u$$

The speed distribution of the molecules in a gas has been measured experimentally by an apparatus sketched in Figure 9.13. The entire apparatus is enclosed in a large vacuum chamber. The molecules leak out of their container to form a *molecular beam,* which passes into a speed analyzer. The analyzer consists of two rotating plates, each with a notch in its edge, separated by the fixed distance L. The plates are rotated so the notches align and permit molecules to pass through both to reach the detector only for a short time interval, $\Delta\tau$. Only those molecules with speeds in the range $\Delta u = L/\Delta\tau$ reach the detector and are counted. The entire speed distribution can be mapped out by progressively varying the duration of the measurement time interval, $\Delta\tau$.

The function $f(u)$ was predicted theoretically by Maxwell and Boltzmann about 60 years before it was first measured. It is called the **Maxwell–Boltzmann speed distribution** for a gas of molecules of mass m at temperature T and it has the following form:

$$f(u) = 4\pi\left(\frac{m}{2\pi k_B T}\right)^{3/2} u^2 \exp(-mu^2/2k_B T) \qquad [9.17]$$

where Boltzmann's constant k_B was defined in Equation 9.14. This distribution is plotted in Figure 9.14 for several temperatures. As the temperature is raised, the entire distribution of molecular speeds shifts toward higher values. Few molecules

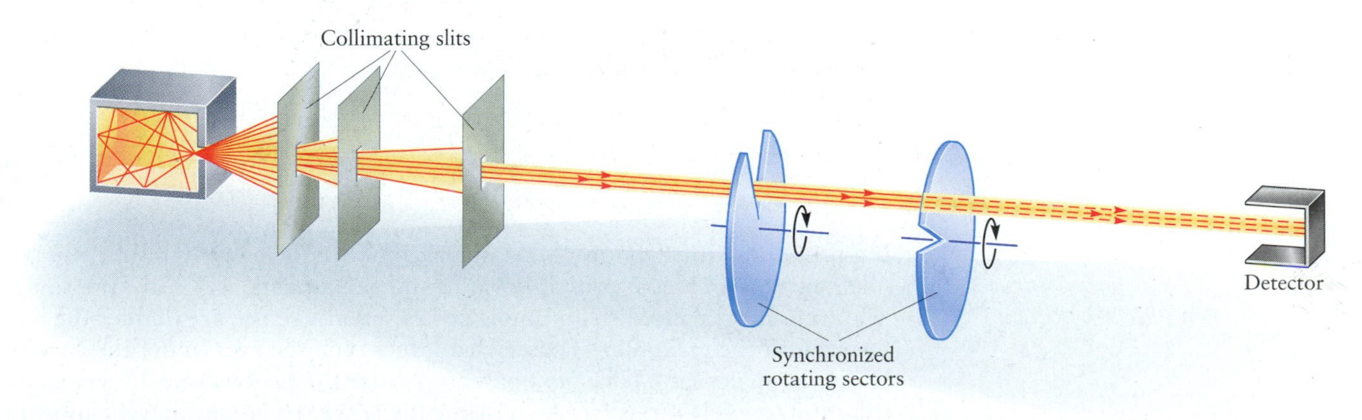

FIGURE 9.13 A device for measuring the distribution of molecular speeds. Only those molecules with the correct velocity to pass through *both* rotating sectors will reach the detector, where they will be counted. Changing the rate of rotation of the sectors allows the speed distribution to be determined.

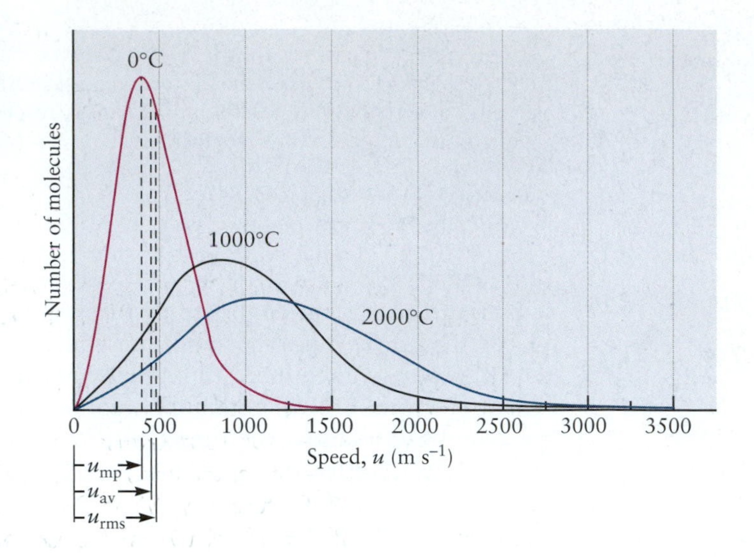

FIGURE 9.14 The Maxwell–Boltzmann distribution of molecular speeds in nitrogen at three temperatures. The peak in each curve gives the most probable speed, u_{mp}, which is slightly smaller than the root-mean-square speed, u_{rms}. The average speed u_{av} (obtained simply by adding the speeds and dividing by the number of molecules in the sample) lies in between. All three measures give comparable estimates of typical molecular speeds and show how these speeds increase with temperature.

have either very low or very high speeds; thus, $f(u)$ is small in these limits and has a maximum at some intermediate speed.

An alternative interpretation of the Maxwell-Boltzmann speed distribution is helpful in statistical analysis of the experiment. Experimentally, the probability that a molecule selected from the gas will have speed in the range Δu is defined as the fraction $\Delta N/N$ discussed earlier. Because $\Delta N/N$ is equal to $f(u)\,\Delta u$, we interpret this product as the probability predicted from theory that any molecule selected from the gas will have speed between u and $u + \Delta u$. In this way we think of the Maxwell-Boltzmann speed distribution $f(u)$ as a *probability distribution.* It is necessary to restrict Δu to very small ranges compared with u to make sure the probability distribution is a continuous function of u. An elementary introduction to probability distributions and their applications is given in Appendix C.6. We suggest you review that material now.

A probability distribution gives a quick visual indication of the likely outcome of the experiment it describes. The **most probable speed** u_{mp} is the speed at which $f(u)$ has its maximum. For the Maxwell–Boltzmann distribution function, this is

$$u_{mp} = \sqrt{\frac{2k_BT}{m}} = \sqrt{\frac{2RT}{\mathcal{M}}} \qquad [9.18]$$

A probability distribution enables us to calculate the average of the values obtained in several repetitions of the experiment it describes. The procedure is described in Appendix C.6. For the Maxwell–Boltzmann distribution, this calculation gives the **average speed** $\overline{u}$, which is

$$\overline{u} = \sqrt{\frac{8k_BT}{\pi m}} = \sqrt{\frac{8RT}{\pi\mathcal{M}}} \qquad [9.19]$$

If a probability distribution is symmetrical about its maximum, like the familiar "bell curve," the most probable value and the average value are the same. The Maxwell–Boltzmann distribution is not symmetrical; the area under the curve to the right of the maximum is somewhat larger than the area under the curve to the left of the maximum. (The next paragraphs use the mathematical form of the distribution to explain this fact.) Consequently, $\overline{u}$ will be larger than the most probable value of u.

The root-mean-square value can be calculated from the probability distribution, as shown in Appendix C.6. For a symmetrical distribution, this would be equal to the average value. For the Maxwell–Boltzmann distribution, we have already seen that

$$u_{rms} = \sqrt{\frac{3k_B T}{m}} = \sqrt{\frac{3RT}{\pi \mathcal{M}}}$$

which verifies that $\bar{u} < u_{rms}$.

There are several possible ways to characterize a non-symmetrical probability distribution by a single number. The three different speeds discussed above serve this purpose for the Maxwell–Boltzmann distribution. Because the distribution is non-symmetrical, they are close to each other but are not equal. They stand in the ratio:

$$u_{mp} : \bar{u} : u_{rms} = 1.000 : 1.128 : 1.1225$$

It is not important for you to memorize these ratios. But you should understand that each quantity is a measure of the "average" speed of the molecules described by the distribution. Different applications require different choices among these quantities. You will learn how to make these connections in more advanced work.

The Maxwell–Boltzmann distribution is not symmetrical because it has the following mathematical form

$$f(u) \propto u^2 \exp(-mu^2/2k_B T)$$

which describes a competition between the two factors that depend on u^2. The competition arises because these factors behave oppositely, for physical reasons, as the value of u changes. We can get a great deal of physical insight into the distribution by studying the behavior of these factors separately while T is held constant.

The exponential factor can be viewed graphically as the right half of a bell curve with its maximum at $u = 0$ (Fig. 9.15). At low values of u, this factor behaves as $\exp(-mu^2/2k_B T) \longrightarrow \exp(-0) = 1$. At very large values of u, this factor behaves as $\exp(-mu^2/k_B T) = 1/[\exp(mu^2/k_B T)] \longrightarrow 1/\infty = 0$. The role of this factor is to describe the statistical weight given to each value of u in relation to T. The limits we have just examined shows this factor gives large statistical weight to small values of u, and increasingly small weight to large values of u, eventually forcing the distribution to fall off to zero at extremely high values of u. This is exactly what we expect on physical grounds.

The factor u^2 can be viewed as the right half of a parabola with its minimum at $u = 0$ (see Fig. 9.15). The value of this factor approaches zero as u decreases towards 0, and it grows without bound as u becomes extremely large. Although we do not provide all the details, the role of this factor is to count the number of different ways molecules in the gas can achieve a particular value of the speed, u. With Avogadro's number of molecules moving around the vessel, it is physically sensible that many different combinations of velocity vectors correspond to a

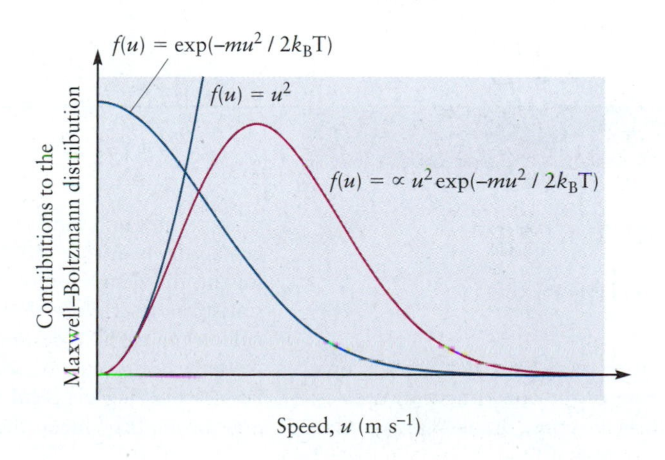

FIGURE 9.15 Mathematical form of the Maxwell–Boltzmann speed distribution. The factor u^2 cuts off the distribution at small values of u, whereas the exponential factor causes it to die off at large values of u. The competition between these effects causes the distribution to achieve its maximum value at intermediate values of u.

given value of the speed. And, we expect the number of such combinations to increase as the value of the speed increases. The shape of this factor strongly favors molecules with large values of u and it rapidly cuts off the distribution for small values of u.

The net result of these two competing factors is to keep the probability small for both extremely large and extremely small values of u. The probability will have a maximum at some intermediate value of u where the increasing effect of u^2 is just balanced by the decreasing effect of the exponential factor (see Fig. 9.15). This is the *most probable* value of u, denoted by u_{mp}, and it can be identified by setting to zero the derivative of the curve with respect to u. Because u^2 approaches zero for small values of u more rapidly than the exponential factor approaches zero for large values of u, the probability is larger to the right side of the maximum. The area under the curve to the right of the maximum is somewhat larger than the area under the curve to the left of the maximum. Consequently, the average value of u denoted by $\overline{u}$ will be larger than the most probable value of u. This is illustrated in Figure 9.14, which shows that $u_{mp} < \overline{u} < u_{rms}$.

The behavior of these competing factors also explains why the distribution becomes broader and its maximum moves to a higher value of u as the temperature increases (see Fig. 9.14). The maximum increases because the value of u at which the parabolic factor u^2 is cut off by the exponential factor increases as T increases. This happens because a particular value of u that would make $\exp(-mu^2/2k_BT) \ll 1$ at low T will now make $\exp(-mu^2/2k_BT) \longrightarrow \exp(-0) = 1$ at higher T. The distribution broadens because the falloff after the maximum is slower at high T than at low T. The reason is that as T increases, the value of u at which $\exp(-mu^2/2k_BT) \longrightarrow 0$ also increases. The net effect at higher T is that larger values of u become accessible, so the molecules are spread over a broader range of speeds.

The Maxwell–Boltzmann speed distribution defines temperature in the kinetic theory of gases as proportional to the average kinetic energy per molecule through Equation 9.14. Unless the molecular speed distribution for a given gas corresponds to the Maxwell–Boltzmann distribution, temperature has no meaning for the gas. Temperature describes a system of gaseous molecules only when their speed distribution is represented by the Maxwell–Boltzmann function. Consider a closed container filled with molecules whose speed distribution is not "Maxwellian." Such a situation is possible (for example, just after an explosion), but it cannot persist for long. Any distribution of molecular speeds other than a Maxwell–Boltzmann distribution quickly becomes Maxwellian through molecular collisions that exchange energy. Once attained, the Maxwell–Boltzmann distribution persists indefinitely (or at least until some new disturbance is applied). The gas molecules have come to **thermal equilibrium** with one another, and we can speak of a system as having a temperature only if the condition of thermal equilibrium exists.

A DEEPER LOOK

9.6 Distribution of Energy among Molecules

The kinetic molecular theory of gases relates the macroscopic properties of a gas to the structure of the constituent molecules, the forces between them, and their motions. Because the number of molecules in a sample of gas is so incredibly large—28 g nitrogen contains 6.02×10^{23} molecules—we give up the idea of following the detailed motions of any one molecule and rely on a statistical description that gives the *probability* of finding a molecule in the gas at a certain position, with a certain speed, with a certain value of energy, and so on. Treating the molecules as point masses obeying classical mechanics and using simple statistical arguments, the kinetic theory shows that the temperature of the gas is proportional to the average kinetic energy per molecule. This relation not only provides a microscopic interpretation of the concept of temperature, but it also indicates the typical values of molecular kinetic energy that occur in a gas at a particular temperature.

Now we want to determine the relation between temperature and the energy involved in other kinds of molecular motions that depend on molecular structure, not just the translation of the molecule. This relation is provided by the *Boltzmann energy distribution,* which relies on the quantum description of molecular motions. This section defines the Boltzmann distribution and uses it to describe the vibrational energy of diatomic molecules in a gas at temperature T.

The Boltzmann energy distribution is one of the most widely used relations in the natural sciences, because it provides a reliable way to interpret experimental results in terms of molecular behavior. You should become skilled in its applications.

The Boltzmann Energy Distribution

Just as in the previous section, we start with a model system in which gaseous molecules move around inside a container held at temperature T. The molecules collide with the container walls but not with one another. We can achieve this condition by setting up the experiment with sufficiently low pressure in the system. But this time we assume that the molecules have quantum states described by a quantum number n and represented on an energy level diagram where the energy of each state is labeled ε_n. After the system has settled down to equilibrium, how many of the molecules are in their ground state? To what extent are the excited states populated? The answers depend on the probability that a molecule in the gas is in the quantum state n, which is given by the **Boltzmann energy distribution:**

$$P(n) = C \exp(-\varepsilon_n/k_B T) \qquad [9.20]$$

where C is a normalization factor and k_B is Boltzmann's constant. This equation was derived for classical systems by Ludwig Boltzmann even before quantum mechanics had been invented. Max Planck used a version of the Boltzmann distribution in formulating his theory of blackbody radiation (see Section 4.2) to obtain the probability that his quantized oscillators would radiate energy when the blackbody was at temperature T. We do not derive the distribution, but illustrate its application and interpretation.

VIBRATIONAL ENERGY DISTRIBUTION

We apply the Boltzmann distribution to describe the probability of finding molecules in each of the vibrational states in a sample of CO held at temperature T. We describe the vibrational motions using the harmonic oscillator model, for which the allowed energy levels are

$$\varepsilon_n = \left(n + \frac{1}{2}\right)h\nu$$

where $n = 0, 1, 2, 3, \ldots$ and the vibrational frequency is related to the force constant by

$$\nu = \frac{1}{2\pi}\sqrt{\frac{k}{\mu}}$$

and μ is the reduced mass. These equations define the energy level diagram, which has uniformly spaced levels separated by

$$h\nu = \frac{h}{2\pi}\sqrt{\frac{k}{\mu}} \qquad [9.21]$$

We calculate the reduced mass of $^{12}C^{16}O$ using the isotopic masses in Table 19.1 to be

$$\mu = \frac{m_C m_O}{m_C + m_O}$$

$$= \frac{(12.00)(15.99)\text{amu}}{27.99}\left(\frac{1\text{ g}}{6.02 \times 10^{23}\text{ amu}}\right)\left(\frac{1\text{ kg}}{10^3\text{ g}}\right)$$

$$\mu = 1.14 \times 10^{-26}\text{ kg}$$

The value of the force constant for CO is 1902 N m^{-1}, as measured in vibrational spectroscopy. The value of the energy level separation is then

$$h\nu = \left(\frac{6.63 \times 10^{-34}\text{ J s}}{2\pi}\right)\left(\frac{1.902 \times 10^3\text{ N m}^{-1}}{1.14 \times 10^{-26}\text{ kg}}\right)^{1/2}$$

$$h\nu = 4.52 \times 10^{-20}\text{ J}$$

The relative probability of finding molecules in the excited state n and in the ground state $n = 0$ is given by

$$\frac{P(n)}{P(0)} = \frac{C\exp(-\varepsilon_n/k_B T)}{C\exp(-\varepsilon_0/k_B T)} = \exp(-[\varepsilon_n - \varepsilon_0]/k_B T) \qquad [9.22]$$

Inserting the energy level expression for the harmonic oscillator gives

$$\frac{P(n)}{P(0)} = \exp(-[(n + \frac{1}{2})h\nu - \frac{1}{2}h\nu]/k_B T) = \exp(-nh\nu/k_B T) \qquad [9.23]$$

The relative populations of the first excited state $n = 1$ and the ground state are determined by the ratio $h\nu/k_B T$. We know from Chapter 4 that $h\nu$ is the quantum of vibrational energy needed to put a CO molecule in its first excited state, and we have calculated that value to be $h\nu = 4.52 \times 10^{-20}$ J. From Section 9.5 we know that the average kinetic energy of a molecule in the gas is $(3/2)k_B T$, which is $(1/2)k_B T$ for each of the x, y, and z directions of motion. Therefore, we interpret $k_B T$ as a measure of the average energy available to each molecule in a gas at temperature T. So, the ratio $h\nu/k_B T$ determines whether there is sufficient energy in the gas to put the molecules into excited states. At 300 K, the value of $k_B T$ is 4.14×10^{-21} J, which is a factor of 10 smaller than the vibrational quantum of CO. Inserting these numbers into Equation 9.23 gives the relative probability as 3.03×10^{-5}. This means that only 3 molecules in a group of 100,000 are in the first excited state at 300 K. At 1000 K, the value of $k_B T$ is 1.38×10^{-20} J, which is closer to the value of the CO vibrational quantum and gives a relative population of 4.41×10^{-2}.

This case study shows that CO molecules do not have significant vibrational energy unless the temperature is quite high. This happens because CO has a triple bond and, therefore, a large force constant ($k = 1902$ N m^{-1}). The correlation between force constant and bond order in diatomic molecules is explained by molecular orbital theory, and is summarized in Figure 6.20. Other diatomic molecules will behave differently, as determined by their structure and the Boltzmann distribution.

EXAMPLE 9.8

Calculate the population of the first and second vibrational excited states, relative to the ground state, for Br_2 at $T = 300$ K and 1000 K. For Br_2, the measured vibrational frequency is $9.68 \times 10^{12}\ s^{-1}$. Interpret your results in relation to the chemical bond in Br_2.

SOLUTION

Use Equation 9.23 with $n = 1$ to obtain N_1/N_A and $n = 2$ to obtain N_2/N_A. It is convenient to evaluate the important quantities before substituting into the equation.

Evaluate the vibrational energy of Br_2:

$$h\nu = (6.63 \times 10^{-34}\ \text{J s})(9.68 \times 10^{12}\ \text{s}^{-1}) = 6.42 \times 10^{-21}\ \text{J}$$

Evaluate k_BT at 300 K and 1000 K:

$$k_BT = (300\ \text{K})(1.380 \times 10^{-23}\ \text{J K}^{-1}) = 4.14 \times 10^{-21}\ \text{J}$$

$$k_BT = (1000\ \text{K})(1.380 \times 10^{-23}\ \text{J K}^{-1}) = 1.38 \times 10^{-20}\ \text{J}$$

The population ratios at 300 K are:

$$\frac{P(1)}{P(0)} = \exp\left[-\frac{6.42 \times 10^{-21}\ \text{J}}{4.14 \times 10^{-21}\ \text{J}}\right] = 0.212$$

$$\frac{P(1)}{P(0)} = \exp\left[-\frac{2(6.42 \times 10^{-21}\ \text{J})}{4.14 \times 10^{-21}\ \text{J}}\right] = 0.045$$

The population ratios at 1000 K are:

$$\frac{P(2)}{P(0)} = \exp\left[-\frac{6.42 \times 10^{-21}\ \text{J}}{1.38 \times 10^{-20}\ \text{J}}\right] = 0.628$$

$$\frac{P(2)}{P(0)} = \exp\left[-\frac{2(6.42 \times 10^{-21}\ \text{J})}{1.38 \times 10^{-20}\ \text{J}}\right] = 0.395$$

The distribution for the first six states at 300 K and 1000 K is shown in Figure 9.16.

The quantized energy vibration is much less for Br_2 than for CO for two reasons: the single bond in Br_2 has a much smaller force constant than the triple bond in CO, and the Br atoms are much more massive than C and O atoms.

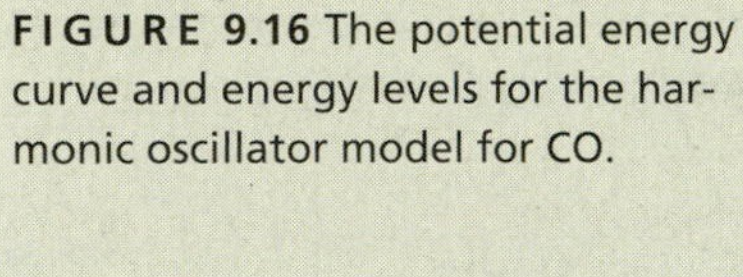

FIGURE 9.16 The potential energy curve and energy levels for the harmonic oscillator model for CO.

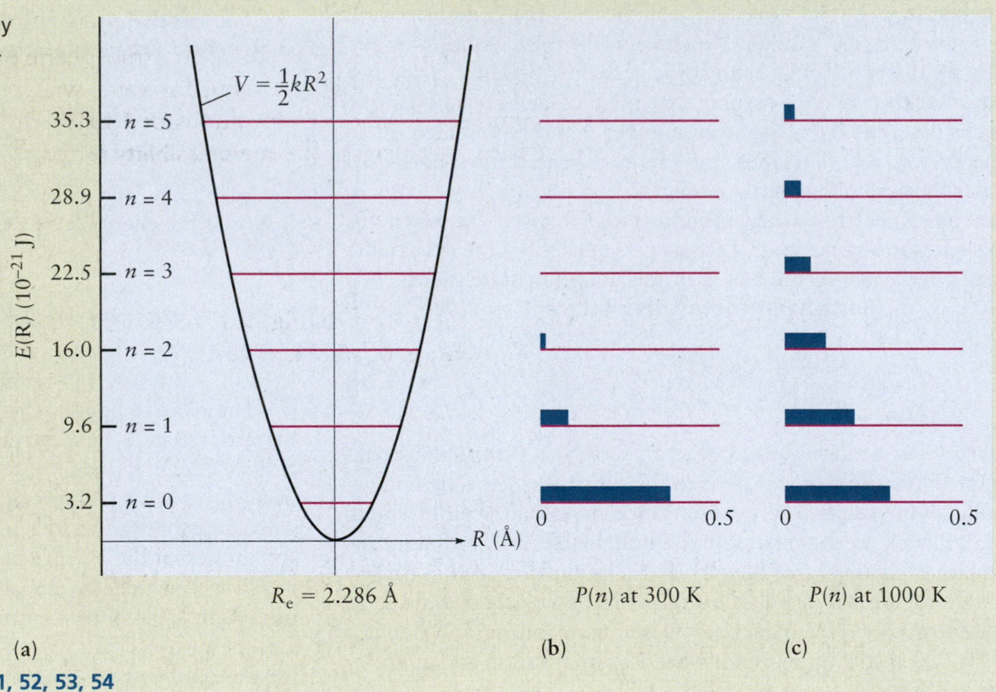

Related Problems: 47, 48, 49, 50, 51, 52, 53, 54

9.7 Real Gases: Intermolecular Forces

The ideal gas law, $PV = nRT$, is a particularly simple example of an **equation of state**—an equation relating the pressure, temperature, number of moles, and volume to one another. Equations of state can be obtained from either theory or experiment. They are useful not only for ideal gases but also for real gases, liquids, and solids.

Real gases follow the ideal gas equation of state only at sufficiently low densities. Deviations appear in a variety of forms. Boyle's law, $PV = C$, is no longer satisfied at high pressures, and Charles's law, $V \propto T$, begins to break down at low temperatures. Deviations from the predictions of Avogadro's hypothesis appear for

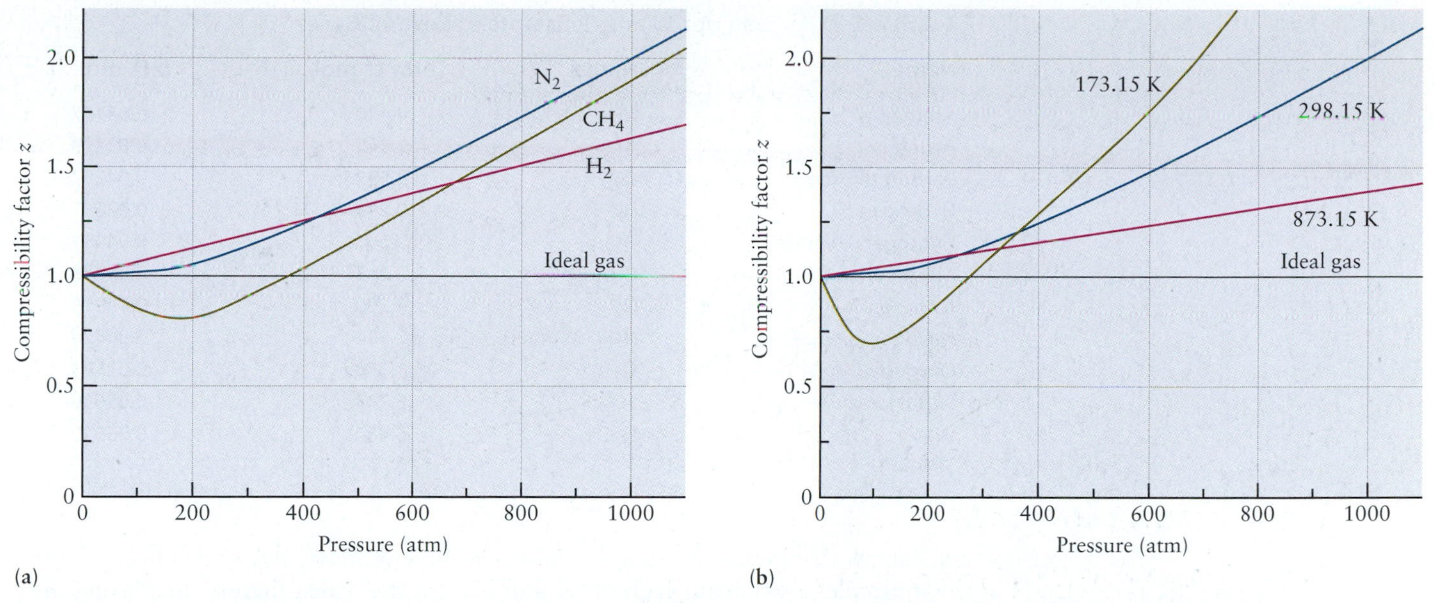

FIGURE 9.17 A plot of $z = PV/nRT$ against pressure shows deviations from the ideal gas law quite clearly, for an ideal gas, z is represented by the straight horizontal line. (a) Deviation of several real gases at 25°C. (b) Deviation of nitrogen at several temperatures.

real gases at moderate pressures. At atmospheric pressure, the ideal gas law is quite well satisfied for most gases, but for some with polar molecules (like water vapor and ammonia), there are deviations of 1 to 2%. The easiest way to detect these deviations is to calculate the **compressibility factor** z from experimental P-V-T data:

$$z = \frac{PV}{nRT} \qquad [9.24]$$

When z differs from 1 (Fig. 9.17), the ideal gas law is inadequate, and a more accurate equation of state is necessary.

The van der Waals Equation of State

One of the earliest and most important improvements on the ideal gas equation of state was proposed in 1873 by the Dutch physicist Johannes van der Waals. The **van der Waals equation of state** is:

$$\left(P + a\frac{n^2}{V^2}\right)(V - nb) = nRT \qquad [9.25a]$$

$$P = \frac{nRT}{V - nb} - a\frac{n^2}{V^2} \qquad [9.25b]$$

To obtain this equation, the ideal gas law—which ignores interactions between molecules—requires two modifications to describe the effects of the forces between molecules, which are repulsive at short distances and attractive at large distances. We know from Section 9.5 that pressure is determined by the product of the momentum transferred per collision with the walls of the container times the number of collisions per second. So, it is necessary to see how repulsive and attractive forces modify the collision rate away from the value it would have in the ideal gas. Because of repulsive forces, molecules cannot occupy the same space at the same time. They exclude other molecules from the volumes they occupy; in this way, the effective volume available to a given molecule is not V, but $V - nb$, where b is a

TABLE 9.3 van der Waals Constants of Several Gases

Name	Formula	a (atm L^2 mol^{-2})	b (L mol^{-1})
Ammonia	NH_3	4.170	0.03707
Argon	Ar	1.345	0.03219
Carbon dioxide	CO_2	3.592	0.04267
Hydrogen	H_2	0.2444	0.02661
Hydrogen chloride	HCl	3.667	0.04081
Methane	CH_4	2.253	0.04278
Nitrogen	N_2	1.390	0.03913
Nitrogen dioxide	NO_2	5.284	0.04424
Oxygen	O_2	1.360	0.03183
Sulfur dioxide	SO_2	6.714	0.05636
Water	H_2O	5.464	0.03049

constant describing the *excluded volume* per mole of molecules. This effect pushes the molecules away from each other and toward the walls, thereby increasing the rate of wall collisions. The result is a pressure higher than the ideal gas value, as shown in the first term of Equation 9.25b. Attractive forces hold pairs or groups of molecules together. Any tendency to cluster together reduces the effective number of independent molecules in the gas and, therefore, reduces the rate of collisions with the walls of the container. Having fewer wall collisions reduces the pressure below the ideal gas law prediction. Because this reduction depends on attractions between *pairs* of molecules, van der Waals argued that it should be proportional to the *square* of the number of molecules per unit volume (N^2/V^2) or, equivalently, proportional to n^2/V^2. Compared with the ideal gas, this intermolecular attraction reduces the pressure by an amount $a(n/V)^2$, where a is a positive constant that depends on the strength of the attractive forces. This effect gives the second term in Equation 9.25b. Rearranging Equation 9.25b gives the standard form of the van der Waals equation shown in Equation 9.25a.

The constants a and b are obtained by fitting experimental P-V-T data for real gases to Equation 9.25a or 9.25b (Table 9.3). The units for these constants are

$$a: \quad \text{atm L}^2\ \text{mol}^{-2}$$

$$b: \quad \text{L mol}^{-1}$$

when R has the units L atm mol^{-1} K^{-1}.

EXAMPLE 9.9

A sample of 8.00 kg gaseous nitrogen fills a 100-L flask at 300°C. What is the pressure of the gas, calculated from the van der Waals equation of state? What pressure would be predicted by the ideal gas equation?

SOLUTION

The molar mass of N_2 is 28 g mol^{-1}, so

$$n = \frac{8.00 \times 10^3\ \text{g}}{28.0\ \text{g mol}^{-1}} = 286\ \text{mol}$$

The temperature (in kelvins) is $T = 300 + 273 = 573$ K, and the volume, V, is 100 L. Using $R = 0.08206$ L atm mol^{-1} K^{-1} and the van der Waals constants for nitrogen given in Table 9.3, we calculate $P = 151 - 11 = 140$ atm. If the ideal gas law is used instead, a pressure of 134 atm is calculated. This illustrates the magnitude of deviations from the ideal gas law at higher pressures.

Related Problems: 55, 56, 57, 58

The effects of the two van der Waals parameters are clearly apparent in the compressibility factor for this equation of state:

$$z = \frac{PV}{nRT} = \frac{V}{V - nb} - \frac{a}{RT}\frac{n}{V} = \frac{1}{1 - bn/V} - \frac{a}{RT}\frac{n}{V} \qquad [9.26]$$

Repulsive forces (through b) increase z above 1, whereas attractive forces (through a) reduce z.

We illustrate the effects of a by comparing Equation 9.26 with the experimental data for the compressibility factor shown in Figure 9.17a. At lower pressures, for example 200 atm, the intermolecular forces reduce z for CH_4 to a value significantly below the ideal gas value. For N_2, the effect that decreases z is readily apparent but it is smaller than the effect that increases z. For H_2, the effect that decreases z is completely dominated by the forces that increase z. These results are consistent with the a-parameter value for CH_4 being about twice that for N_2 and about 10 times that for H_2 (see Table 9.3). The values of a originate in the structure of the molecules and vary significantly between highly polar molecules such as H_2O and nonpolar molecules such as H_2.

The constant b is the volume excluded by 1 mol of molecules and should therefore be close to V_m, the volume per mole in the liquid state, where molecules are essentially in contact with each other. For example, the density of liquid nitrogen is 0.808 g cm^{-3}. One mole of N_2 weighs 28.0 g, so

$$V_m \text{ of } N_2(\ell) = \frac{28.0 \text{ g mol}^{-1}}{0.808 \text{ g cm}^{-3}} = 34.7 \text{ cm}^3 \text{ mol}^{-1}$$

$$= 0.0347 \text{ L mol}^{-1}$$

This is reasonably close to the van der Waals b parameter of 0.03913 L mol^{-1}, obtained by fitting the equation of state to P-V-T data for nitrogen. In Table 9.3, the values for b are all quite similar. All the molecules in Table 9.3 are about the same size and have similar values of molar volume in the liquid state.

Intermolecular Forces

Deeper understanding of the attractive force parameter, a, and the excluded volume per mole, b, comes from examination of the forces acting between the atoms or molecules in a gas. As a pair of molecules approach one another, the forces between them generate potential energy, which competes with the kinetic energy associated with their speeds. This potential energy can increase the molar volume through intermolecular repulsions, or decrease the pressure by temporarily attracting molecules to form dimers and so reducing the rate of collisions with the walls. The potential energy is our means to describe systematically how intermolecular forces cause these two effects. The noble gases provide the simplest example. As two noble gas atoms approach one another, attractive forces dominate until the distance between their centers, R, becomes short enough for the repulsive forces to begin to become significant. If the atoms are forced still closer together, they repel each other with a strongly increasing force as the distance between them is reduced. These interactions can be described by a **potential energy curve** $V(R)$ (see Section 3.5, Section 6.2.5, and Appendix B) such as that shown for argon in Figure 9.18. If two molecules can lower their energy by moving closer together, then a net attractive force exists between them; if they can lower their energy only by moving apart, there is a net repulsive force. Graphically, this means that the force acting between the molecules is given by the negative slope (or derivative) of the potential energy curve $V(R)$. The force changes from attractive to repulsive at the minimum of $V(R)$, where the net force between atoms is zero. Potential energy curves for atoms are generated by fitting equations for $V(R)$ to measured properties of real gases. More accurate versions come from experiments in which beams of atoms collide with one another.

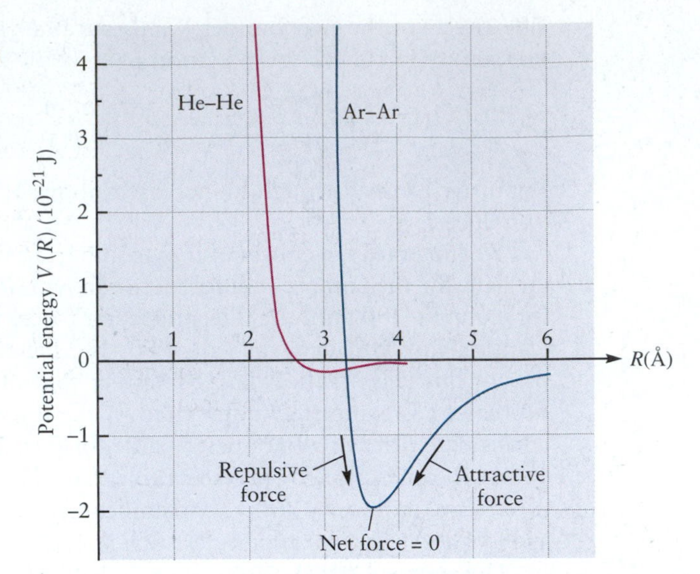

FIGURE 9.18 Potential energy curves $V(R)$ for pairs of helium atoms (red) and pairs of argon atoms (blue) obtained from atomic beam collision studies. At any point, the force between atoms is the negative of the slope of V (see Section 3.5, Section 6.2.5, and Appendix B.2). In regions where the slopes of the curves are negative, the atoms repel each other. In regions where the slopes are positive, the atoms attract one another. The greater well depth for Ar arises from stronger intermolecular attractions, and the location of the minimum for Ar at larger R correlates roughly with relative molecular size.

For many purposes, the detailed shape of the potential is less important than two characteristic parameters: the depth and location of the potential minimum. A simple expression frequently used to model these interactions between atoms is the **Lennard–Jones potential:**

$$V_{\mathrm{LJ}}(R) = 4\varepsilon\left[\left(\frac{\sigma}{R}\right)^{12} - \left(\frac{\sigma}{R}\right)^{6}\right] \qquad [9.27]$$

where ε is the depth and σ is the distance at which $V(R)$ passes through zero. This potential has an attractive part, proportional to R^{-6}, and a repulsive part, proportional to R^{-12}. The minimum is located at $2^{1/6}\sigma$, or 1.22σ. Table 9.4 lists Lennard–Jones parameters for a number of atoms. Note that the depth and range of the potential increase for the heavier noble-gas atoms. Molecules such as N_2 that are nearly spherical can also be described approximately with Lennard–Jones potentials. The two parameters ε and σ in the Lennard–Jones potential, like the van der Waals parameters a and b, are simple ways of characterizing the interactions between molecules in real gases.

The Lennard–Jones potential and the Boltzmann distribution (see Section 9.6) together explain how deviations from ideal gas behavior depend on temperature. Qualitatively, the most important effect is the ratio of the well depth ε to k_BT. When T is low enough to make $k_BT \ll \varepsilon$, a pair of molecules remain close together sufficiently long to reduce the rate of wall collisions, and thereby reduce the pressure below its ideal gas value. When T is such that $k_BT \gg \varepsilon$, the molecules experience only the repulsive part of the L–J potential, and the pressure is increased

TABLE 9.4 Lennard–Jones Parameters for Atoms and Molecules

Substance	σ(m)	ε(J)
He	2.56×10^{-10}	1.41×10^{-22}
Ne	2.75×10^{-10}	4.92×10^{-22}
Ar	3.40×10^{-10}	1.654×10^{-21}
Kr	3.60×10^{-10}	2.36×10^{-21}
Xe	4.10×10^{-10}	3.06×10^{-21}
H_2	2.93×10^{-10}	5.11×10^{-22}
O_2	3.58×10^{-10}	1.622×10^{-21}
CO	3.76×10^{-10}	1.383×10^{-21}
N_2	3.70×10^{-10}	1.312×10^{-21}
CH_4	3.82×10^{-10}	2.045×10^{-21}

above its ideal gas value. These effects are illustrated in Figure 9.17b, which shows that at low pressure the compressibility factor for N_2 is dominated by attractive forces at 173.15 K and by repulsive forces at 873.15 K.

The Lennard–Jones potential gives insight into the roles of the average kinetic energy of molecules (indicated by k_BT) and the potential energy between molecules (indicated by ε) in gases. At 300 K, the value of k_BT is 4.14×10^{-21} J, which is comparable to and slightly larger than the well depths for typical gases listed in Table 9.4. Consequently, the average kinetic energy of a molecule is larger than the greatest value of the potential energy that can occur between a pair of molecules. Because the molecules are only rarely close together in gases at ordinary pressures, the potential energy per pair averaged over all pairs in the gas is much smaller than the average kinetic energy. So, the behavior of gases is determined primarily by the average kinetic energy of the molecules.

A DEEPER LOOK

9.8 Molecular Collisions and Rate Processes

The kinetic theory we have developed can be applied to several important properties of gases. A study of the rates at which atoms and molecules collide with a wall and with one another helps to explain phenomena ranging from isotope separation based on gaseous diffusion to gas-phase chemical kinetics.

Molecule-Wall Collisions

Let's call Z_w the rate of collisions of gas molecules with a section of wall of area A. A full mathematical calculation of Z_w requires integral calculus and solid geometry. We present instead some simple physical arguments to show how this rate depends on the properties of the gas.

First of all, Z_w should be proportional to the area A, because doubling the area will double the number of collisions with the wall. Second, Z_w should be proportional to the average molecular speed, $\bar{u}$, because molecules moving twice as fast will collide twice as often with a given wall area. Finally, the wall collision rate should be proportional to the number density, N/V, because twice as many molecules in a given volume will have twice as many collisions with the wall. All of these arguments are consistent with the kinetic theory of gases and are confirmed by the full mathematical analysis. We conclude that

$$Z_w \propto \frac{N}{V}\bar{u}A$$

Note that the units of both sides is s^{-1}, as required for these expressions to represent a rate. The proportionality constant can be calculated from a complete analysis of the directions from which molecules impinge on the wall; it turns out to have the value $\frac{1}{4}$. So the wall collision rate is

$$Z_w \propto \frac{1}{4}\frac{N}{V}\bar{u}A = \frac{1}{4}\frac{N}{V}\sqrt{\frac{8RT}{\pi\mathcal{M}}}A \qquad [9.28]$$

We have used the result for $\bar{u}$ given earlier in Equation 9.19. This simple equation has many applications. It sets an upper limit on the rate at which a gas may react with a solid. It is also the basis for calculating the rate at which molecules effuse through a small hole in the wall of a vessel.

EXAMPLE 9.10

Calculate the number of collisions that oxygen molecules make per second on 1.00 cm^2 of the surface of the vessel containing them if the pressure is 1.00×10^{-6} atm and the temperature is 25°C (298 K).

SOLUTION

First compute the quantities that appear in the equation for Z_w:

$$\frac{N}{V} = \frac{N_A n}{V} = \frac{N_A P}{RT}$$

$$= \frac{(6.022 \times 10^{23}\ \text{mol}^{-1})(1.00 \times 10^{-6}\ \text{atm})}{(0.08206\ \text{L atm mol}^{-1}\ \text{K}^{-1})(298\ \text{K})}$$

$$= 2.46 \times 10^{16}\ \text{L}^{-1} = 2.46 \times 10^{19}\ \text{m}^{-3}$$

$$A = 1.00\ \text{cm}^2 = 1.00 \times 10^{-4}\ \text{m}^2$$

$$\bar{u} = \sqrt{\frac{8RT}{\pi\mathcal{M}}}$$

$$= \sqrt{\frac{8(8.3145\ \text{J mol}^{-1}\ \text{K}^{-1})(298\ \text{K})}{\pi(32.00 \times 10^{-3}\ \text{kg mol}^{-1})}}$$

$$= 444\ \text{m s}^{-1}$$

The collision rate is then

$$Z_w = \frac{1}{4}\frac{N}{V}\bar{u}A$$

$$= \tfrac{1}{4}(2.46 \times 10^{19}\ \text{m}^{-3})(444\ \text{m s}^{-1})(1.00 \times 10^{-4}\ \text{m}^2)$$

$$= 2.73 \times 10^{17}\ \text{s}^{-1}$$

Related Problems: 59, 60

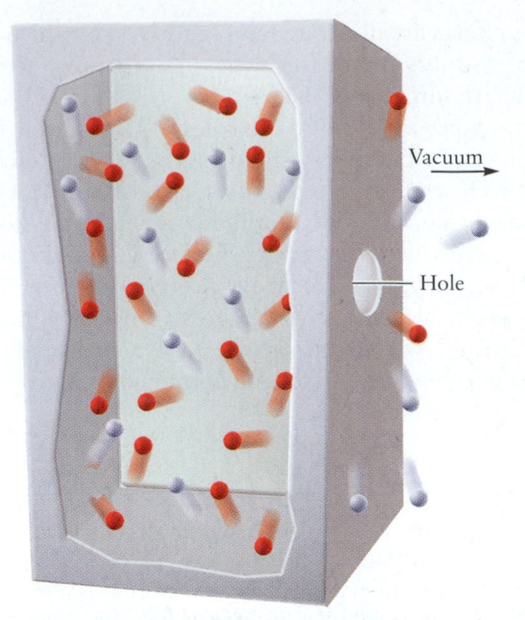

FIGURE 9.19 A small hole in the box permits molecules to effuse out into a vacuum. The less massive particles (here, helium atoms, red) effuse at greater rates than the more massive oxygen molecules (purple) because their speeds are greater on the average.

Equation 9.28 is the basis of explaining **Graham's law of effusion.** In 1846, Thomas Graham showed experimentally that the rate of effusion of a gas through a small hole into a vacuum (Fig. 9.19) is inversely proportional to the square root of its molar mass. Assuming that different gases are studied at the same temperature and pressure, their number density, N/V, is the same and the rate of effusion of each gas depends only on the factor $1/\sqrt{\mathcal{M}}$ in Equation 9.28, exactly as observed by Graham. Explaining this experimental result is yet another success for the kinetic theory of gases.

Graham's law also describes the effusion of a mixture of two gases through a small hole. The ratio of the rates of effusion of the two species, A and B, is

$$\frac{\text{rate of effusion of A}}{\text{rate of effusion of B}} = \frac{\frac{1}{4}\frac{N_A}{V}\sqrt{\frac{8RT}{\pi\mathcal{M}_A}}A}{\frac{1}{4}\frac{N_B}{V}\sqrt{\frac{8RT}{\pi\mathcal{M}_B}}A} = \frac{N_A}{N_B}\sqrt{\frac{\mathcal{M}_B}{\mathcal{M}_A}} \qquad [9.29]$$

This ratio is equal to the ratio of the numbers of molecules of the two species effusing through the hole in a short time interval. N_A and N_B are the number of molecules of species A and B, respectively. The emerging gas is enriched in the lighter component because lighter molecules effuse more rapidly than heavier ones. If B is heavier than A, the **enrichment factor** is $\sqrt{\mathcal{M}_B/\mathcal{M}_A}$ for the lighter species, A.

EXAMPLE 9.11

Calculate the enrichment factors from effusion for a mixture of $^{235}UF_6$ and $^{238}UF_6$, uranium hexafluoride with two different uranium isotopes. The atomic mass of ^{235}U is 235.04, and that of ^{238}U is 238.05. The atomic mass of fluorine is 19.00.

SOLUTION

The two molar masses are

$$\mathcal{M}(^{238}UF_6) = 238.05 + 6(19.00) = 352.05 \text{ g mol}^{-1}$$

$$\mathcal{M}(^{235}UF_6) = 235.04 + 6(19.00) = 349.04 \text{ g mol}^{-1}$$

$$\text{enrichment factor} = \sqrt{\frac{\mathcal{M}(^{238}UF_6)}{\mathcal{M}(^{235}UF_6)}} = \sqrt{\frac{352.05 \text{ g mol}^{-1}}{349.04 \text{ g mol}^{-1}}}$$

$$= 1.0043$$

Related Problems: 61, 62

Graham's law of effusion holds true only if the opening in the vessel is small enough and the pressure low enough that most molecules follow straight-line trajectories through the opening without colliding with one another. A related but more complex phenomenon is **gaseous diffusion through a porous barrier.** This differs from effusion in that molecules undergo many collisions with one another and with the barrier during their passage through it. Just as in effusion, the diffusion rate is inversely proportional to the square root of the molar mass of the gas. But the reasons for this dependence are not those outlined for the effusion process. If a mixture of gases is placed in contact with a porous barrier, the gas passing through is enriched in the lighter component, A, by a factor of $\sqrt{\mathcal{M}_B/\mathcal{M}_A}$, and the gas remaining behind is enriched in the heavier component.

In order to develop the atomic bomb, it was necessary to separate the more easily fissionable isotope ^{235}U from ^{238}U. Because the natural abundance of ^{235}U is only 0.7%, its isolation

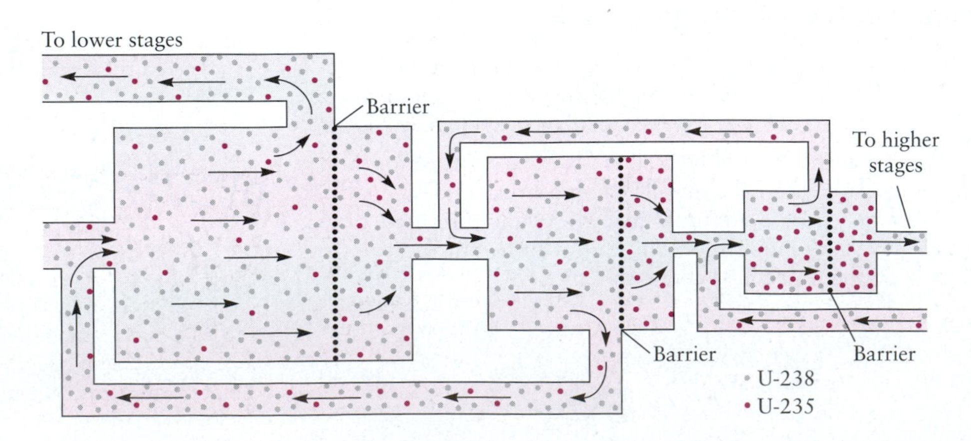

FIGURE 9.20 Schematic of one of 3122 diffusion stages in the gaseous diffusion plant at Oak Ridge National Laboratory. Gaseous UF_6 entering the left side of the stage is progressively enriched as it moves through the stage from left to right as indicated by an increase in the number of red dots representing $^{235}UF_6$ molecules. There are three diffusion barriers in each stage and gases are recycled through the barriers within a stage, as well as between stages, multiple times as shown by the arrows flowing back to the left.

in nearly pure form was a daunting task. The procedure adopted was to react uranium with F_2 gas to form the relatively volatile compound UF_6 (boiling point, 56°C), which can be enriched in ^{235}U by passing it through a porous barrier. As shown in Example 9.12, each passage gives an enrichment factor of only 1.0043. So, successive passage through many such barriers is necessary to provide sufficient enrichment of the ^{235}U component. A multi-state gaseous diffusion chamber was constructed in a short time at the Oak Ridge National Laboratories, and the enriched ^{235}U was used in the first atomic bomb. Similar methods are still used today to enrich uranium for use as fuel in nuclear power plants (Fig. 9.20).

EXAMPLE 9.12

How many diffusion stages are required if ^{235}U is to be enriched from 0.70 to 7.0% by means of the gaseous UF_6 diffusion process?

SOLUTION

From Example 9.11, the enrichment factor per stage is 1.0043; thus, the first stage is described by the relation

$$\left(\frac{N^{235}UF_6}{N^{238}UF_6}\right)_1 = \left(\frac{N^{235}UF_6}{N^{238}UF_6}\right)_0 (1.0043)$$

where the subscripts 0 and 1 denote the initial concentration and that after the first stage, respectively. The ratio of the numbers after n stages satisfies the equation

$$\left(\frac{N^{235}UF_6}{N^{238}UF_6}\right)_n = \left(\frac{N^{235}UF_6}{N^{238}UF_6}\right)_0 (1.0043)^n$$

$$\frac{7.0}{93.0} = \frac{0.70}{99.30}(1.0043)^n$$

$$10.677 = (1.0043)^n$$

This equation can be solved by taking logarithms of both sides (see Appendix C):

$$\log_{10}(10.677) = n \log_{10}(1.0043)$$

$$1.02845 = 0.0018635n$$

$$n = 552 \text{ stages}$$

Related Problems: 63, 64

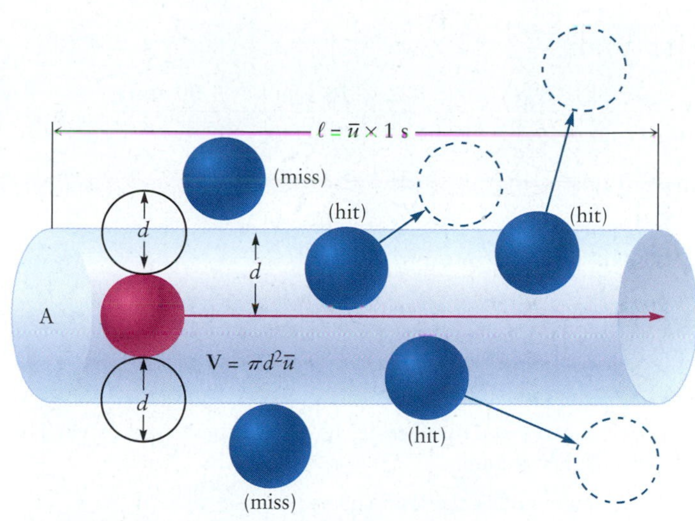

FIGURE 9.21 An average molecule (red) sweeps out a cylinder of volume $\pi d^2\bar{u}$ in 1 second. It will collide with any molecules whose centers lie within the cylinder. Using this construction, we can calculate the rate of collisions with other molecules.

Molecule–Molecule Collisions

Collisions lie at the very heart of chemistry, because chemical reactions can occur only when molecules collide with one another. The kinetic theory of gases provides methods for estimating the frequency of molecular collisions and the average distance traveled by a molecule between collisions, both important in understanding the rates of chemical reactions (See Chapter 18).

We assume molecules are approximately spherical, with diameters, d, on the order of 10^{-10} m (see Section 2.1). We initially suppose that a particular molecule moves through a gas of stationary molecules of the same diameter. Such a molecule "sweeps out" a cylinder with cross-sectional area, $A = \pi d^2$ (Fig. 9.21). This particular molecule collides with any molecule whose center lies inside the cylinder, within a distance, d, of the center of the moving molecule. In 1 second, the length of such a cylinder is $\bar{u}\,\Delta t = \bar{u} \times 1$ s, where $\bar{u}$ is the average molecular speed. It does not matter that the moving molecule is scattered in another direction when it collides with other molecules; these little cylinders are merely joined together, end to end, to define a cylinder whose length is $\bar{u} \times 1$ s.

In 1 second, the moving molecule therefore sweeps out the volume

$$V_{cyl} = \pi d^2 \bar{u}$$

If N/V is the number of molecules per unit volume in the gas (the number density of the gas), then the number of collisions per second experienced by the moving molecule is

$$\text{collision rate} = Z_1 = \frac{N}{V} V_{cyl} = \frac{N}{V}\pi d^2 \bar{u} \quad \text{(approximately)}$$

Actually, this equation is only approximate, because the other molecules are not standing still but are moving. A full calculation gives an extra factor of $\sqrt{2}$:

$$Z_1 = \sqrt{2}\frac{N}{V}\pi d^2\bar{u}$$

Inserting the result for $\bar{u}$ from Equation 9.19 gives

$$Z_1 = 4\frac{N}{V}d^2\sqrt{\frac{\pi RT}{\mathcal{M}}} \qquad [9.30]$$

EXAMPLE 9.13

Calculate the collision frequency for (a) a molecule in a sample of oxygen at 1.00 atm pressure and 25°C, and (b) a molecule of hydrogen in a region of interstellar space where the number density is 1.0×10^{10} molecules per cubic meter and the temperature is 30 K. Take the diameter of O_2 to be 2.92×10^{-10} m and that of H_2 to be 2.34×10^{-10} m.

SOLUTION

$$\frac{N}{V} = \frac{N_A P}{RT} = \frac{(6.022 \times 10^{23}\ \text{mol}^{-1})(1.00\ \text{atm})}{(0.08206\ \text{L atm mol}^{-1}\ \text{K}^{-1})(298\ \text{K})}$$

$$= 2.46 \times 10^{+22}\ \text{L}^{-1} = 2.46 \times 10^{25}\ \text{m}^{-3}$$

In Example 9.10, we found $\bar{u}$ to be 444 m s^{-1}. Thus, the collision frequency is

$$Z_1 = \sqrt{2}\pi(2.46 \times 10^{25}\ \text{m}^{-3}) \times (2.92 \times 10^{-10}\ \text{m})^2(444\ \text{m}^{-1})$$

$$= 4.14 \times 10^9\ \text{s}^{-1}$$

This is the average number of collisions experienced by each molecule per second.

(b) An analogous calculation gives

$$Z_1 = 1.4 \times 10^{-6}\ \text{s}^{-1}$$

In other words, the average molecule under these conditions waits 7.3×10^5 seconds, or 8.5 days, between collisions.

Mean Free Path and Diffusion

Z_1 is the rate at which a particular molecule collides with other molecules. Its inverse, Z_1^{-1}, therefore measures the average time between collisions. During this interval, a molecule travels an average distance $\bar{u}\, Z_1^{-1}$ which is called the **mean free path, λ.**

$$\lambda = \bar{u}Z_1^{-1} = \frac{\bar{u}}{\sqrt{2}(N/V)\pi d^2\bar{u}} = \frac{1}{\sqrt{2}\pi d^2 N/V} \quad [9.31]$$

The mean free path, unlike the collision frequency, does not depend on the molar mass. The mean free path must be much larger than the molecular diameter for a gas to show ideal gas behavior. For a molecule of diameter 3×10^{-10} m, the mean free path at 25°C and atmospheric pressure is 1×10^{-7} m, which is 300 times larger.

Molecules in a gas move in straight lines only for rather short distances before they are deflected by collisions and change direction (Fig. 9.22). Because each molecule follows a zigzag course, they take more time to travel a particular net distance from their starting points than they would if there were no collisions. This fact helps explain why diffusion is slow in gases. Recall that at room temperature the speed of a molecule is on the order of 1×10^3 m s^{-1}. If molecules traveled in straight-line trajectories, a perfume released in one part of a room would be noticed across the room almost instantaneously. Instead, there is a time lapse because the molecules follow irregular paths that we call "random walks."

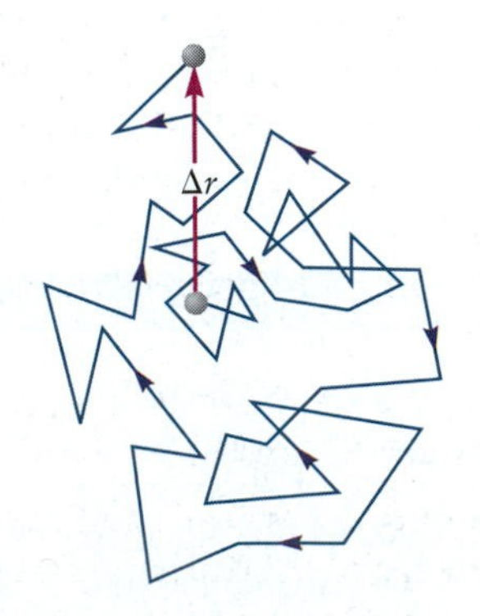

FIGURE 9.22 A gas molecule follows a straight-line path for only a short time before undergoing a collision, so its overall path is a zigzag one. The displacement Δr of a particular molecule in time, Δt, is shown. The path taken by the molecules is traced out in blue, and the red arrow represents the total displacement during the period Δt. The mean-square displacement $\overline{\Delta r^2}$, is equal to $6Dt$, where D is the diffusion constant of the molecules in the gas.

We can describe diffusion in a gas using averaged quantities such as the mean-square displacement, $\overline{\Delta r^2} = \overline{\Delta x^2} + \overline{\Delta y^2} + \overline{\Delta z^2}$, which is analogous to the mean-square speed considered earlier. If there are no gas currents to perturb the motion of the gas molecules—a rather strenuous condition requiring strict isolation of the experiment from the surroundings—then $\overline{\Delta r^2}$ is found to be proportional to the time elapsed, t:

$$\overline{\Delta r^2} = 6Dt \quad [9.32]$$

The proportionality constant is $6D$, where D is the **diffusion constant** of the molecules. So, the root-mean-square displacemen $\sqrt{\overline{\Delta r^2}}$ is equal to $\sqrt{6Dt}$.

The diffusion constant has units of m^2 s^{-1}. It is proportional to the mean free path, λ, and to the mean molecular speed $\bar{u}$, but the proportionality constant is difficult to calculate in general. For the simplest case, a single-component gas, the proportionality constant is $3\pi/16$, so

$$D = \frac{3\pi}{16}\lambda\bar{u} = \frac{3\pi}{16}\sqrt{\frac{8RT}{\pi\mathcal{M}}}\frac{1}{\sqrt{2}\pi d^2 N/V}$$

$$= \frac{3}{8}\sqrt{\frac{RT}{\pi\mathcal{M}}}\frac{1}{d^2 N/V} \quad [9.33]$$

EXAMPLE 9.14

Calculate the mean free path and the diffusion constant for the molecules in Example 9.13.

SOLUTION

(a) The mean free path is

$$\lambda = \frac{1}{\sqrt{2}\pi(2.92 \times 10^{-10}\ \text{m})^2(2.46 \times 10^{25}\ \text{m}^{-1})}$$

$$= 1.07 \times 10^{-7}\ \text{m}$$

The diffusion constant is

$$D = \frac{3\pi}{16}\lambda\bar{u} = \frac{3\pi}{16}(1.07 \times 10^{-7}\ \text{m})(444\ \text{m s}^{-1})$$

$$= 2.80 \times 10^{-5}\ \text{m}^2\ \text{s}^{-1}$$

(b) An analogous calculation for an average molecule of H_2 in interstellar space gives

$$\lambda = 4.1 \times 10^8\ \text{m}$$

For comparison, the distance from the Earth to the moon is 3.8×10^8 m. The diffusion constant is

$$D = 1.4 \times 10^{11}\ \text{m}^2\ \text{s}^{-1}$$

Related Problems: 65, 66

CHAPTER SUMMARY

Gases provide the simplest opportunity for relating macroscopic properties of matter to the structure, motions, and interactions of molecules. Because molecules in gases are quite far apart most of the time, we can neglect intermolecular forces and represent the molecules as point masses that have only kinetic energy and collide with the walls of the container but not with each other. The simplest treatment of this physical model predicts the ideal gas law, which was discovered empirically. More elaborate mathematical treatments of the same model produce the full probability distribution for molecular speeds. From this distribution, various average quantities can be calculated and used to interpret numerous experimental phenomena in gases at low density. At higher density, intermolecular forces can no longer be neglected. Their effect is described systematically by the intermolecular potential energy function, which includes both the attractive and repulsive effects. The well depth and location of the minimum in the potential energy curve are very useful parameters for summarizing these effects. They provide deeper insight into the attractive and repulsive constants that are obtained by fitting the van der Waals equation of state to empirical data. It is especially interesting to see that the value of k_BT at room temperature is larger than the well depths of the intermolecular potential. Because the molecules are far apart most of the time, the average kinetic energy per molecule exceeds the average potential energy per molecule. So, the properties of gases at room temperature are determined by the kinetic energy of the molecular motions.

CUMULATIVE EXERCISE

Ammonium Perchlorate

Ammonium perchlorate (NH_4ClO_4) is a solid rocket fuel used in space shuttles. When heated above 200°C, it decomposes to a variety of gaseous products, of which the most important are N_2, Cl_2, O_2, and water vapor.

(a) Write a balanced chemical equation for the decomposition of NH_4ClO_4, assuming the products just listed are the only ones generated.

(b) The sudden appearance of hot gaseous products in a small initial volume leads to rapid increases in pressure and temperature, which give the rocket its thrust. What total pressure of gas would be produced at 800°C by igniting 7.00×10^5 kg NH_4ClO_4 (a typical charge of the booster rockets in the space shuttle) and allowing it to expand to fill a volume of 6400 m^3 (6.40×10^6 L)? Use the ideal gas law.

(c) Calculate the mole fraction of chlorine and its partial pressure in the mixture of gases produced.

(d) The van der Waals equation applies strictly to pure real gases, not to mixtures. For a mixture like the one resulting from the reaction of part (a), it may still be possible to define effective a and b parameters to relate total pressure, volume, temperature, and total number of moles. Suppose the gas mixture has $a = 4.00$ atm L^2 mol^{-2} and $b = 0.0330$ L mol^{-1}. Recalculate the pressure of the gas mixture in part (b) using the van der Waals equation. Why is the result smaller than that in part (b)?

(e) Calculate and compare the root-mean-square speeds of water and chlorine molecules under the conditions of part (b).

(f) The gas mixture from part (b) cools and expands until it reaches a temperature of 200°C and a pressure of 3.20 atm. Calculate the volume occupied by the gas mixture after this expansion has occurred. Assume ideal gas behavior.

NASA

A space shuttle taking off.

Answers

(a) $2\ NH_4ClO_4(s) \longrightarrow N_2(g) + Cl_2(g) + 2\ O_2(g) + 4\ H_2O(g)$

(b) 328 atm

(c) $X_{Cl_2} = \frac{1}{8} = 0.125$ (exactly); $P_{Cl_2} = 41.0$ atm

(d) 318 atm. The real pressure is less than that calculated using the ideal gas law because half of the products are water molecules that have very strong intermolecular attractions, leading to the large value of the *a* parameters for the gas mixture.

(e) $u_{rms}(H_2O) = 1220\ m\ s^{-1}$; $u_{rms}(Cl_2) = 614\ m\ s^{-1}$

(f) $2.89 \times 10^5\ m^3$

CHAPTER REVIEW

- Composition of the Air

  Roughly 78% N_2, 21% O_2, 1% Ar.
  CO_2 is 380 ppm and increasing.

- Pressure and Temperature of Gases and the Ideal Gas Law

  Pressure is defined as force/area measured in atm, bar, pascal, and torr.
  The empirical gas laws of Boyle, Charles, and Avogadro combine to give the ideal gas law: $PV = nRT$.

- Gaseous Mixtures are described by Dalton's Law

  $P_A = n_A RT/V$
  $X_A = n_A/n_{tot}$
  $P_A = X_A P_{tot}$

- The Kinetic Theory of Gases

  Molecules are assumed to be widely separated point masses that interact only during collisions, move randomly with a distribution of speeds, and experience elastic collisions with walls
  The square of the speed is the sum of the squares of velocity components $u^2 = v_x^2 + v_y^2 + v_z^2$
  The rate of momentum transfer = (rate of collisions along any axis) × (momentum transfer per collision)
  The total number of wall collisions relates the pressure to the average kinetic energy per molecule: $PV = (1/3)Nm\overline{u^2}$
  The kinetic theory of gases predicts the ideal gas law from a simple molecular model
  Connecting the results of the kinetic theory of gases to the ideal gas law gives a microscopic, mechanical interpretation of temperature, the average kinetic energy per mole is equal to 3/2 *RT*.

- The Maxwell–Boltzmann (MB) distribution

  The MB distribution gives the probability of finding molecules with each value of speed from zero to infinity, as a function of the temperature of the gas. The MB distribution has the mathematical form $f(u) \propto u^2 \exp(-mu^2/2k_BT)$.

At a given temperature, lighter molecules have higher speeds than heavier molecules. The ratio of speeds is related to the ratio of masses as $u_2/u_1 = \sqrt{M_1/M_2}$.
The MB distribution gives low probability for very low and very high speeds, with a maximum called the most probable speed u_{mp}, which occurs at an intermediate value.
As temperature increases, the maximum shifts to higher speed values, and the distribution broadens.
The MB distribution is not symmetrical about its maximum value because it is the product of two competing factors: the statistical weight for each value of speed, and the number of ways of achieving that speed. The total probability is slightly greater for speed values higher than u_{mp} than for speed values lower than u_{mp}.
Like any nonsymmetrical probability distribution, the MB speed distribution can be characterized by several different single numbers: average speed $\overline{u}$, most probable speed u_{mp}, and root-mean-square speed u_{rms}. Because the MB has greater total probability to the high side of the most probable speed, these three numbers have this order: $u_{mp} < \overline{u} < u_{rms}$. We most often use $\overline{u}$, the last in applications of the kinetic theory.

- Rates of molecule-wall and molecule-molecule collisions from the kinetic theory of gases depend on the average molecular speed $\overline{u}$.

The rate of molecular collisions with a wall of area A given by $Z_w = \frac{1}{4}\frac{N}{V}\overline{u}A$ determines the rate of gaseous effusion through a small aperture and sets an upper bound for the rate of gas-surface chemical reactions
The rate of molecule-molecule collisions given by $Z_1 = \sqrt{2}\,(N/V)\pi d^2\overline{u}$ estimates the upper bound of gas phase chemical reaction rates and calculates the distance a molecule travels between collisions as well as the diffusion coefficient in gases

- The Boltzmann distribution

The probability is weighted by the energy of the state and the temperature of the gas $P(n) = C\exp(-\varepsilon_n/k_BT)$
The relative populations of two quantum states is given by
$$\frac{P(n)}{P(0)} = \frac{C\exp(-\varepsilon_n/k_BT)}{C\exp(-\varepsilon_0/k_BT)} = \exp(-[\varepsilon_n - \varepsilon_0]/k_BT)$$

- Deviations from ideal gas behavior and the van der Waals equation of state

Real gases do not obey the ideal gas law because attractive and repulsive forces between molecules introduce potential energy that competes with the purely kinetic energy of translation from which the ideal gas law arises.
The van der Waals equation of state is the simplest two-parameter equation of state for real gases. A simple physical model underlies the form of the equation and its two parameters.
Molecules have finite size and exclude volume to other molecules: the a parameter.
Molecules attract one another and reduce wall collision rate: the b parameter.
The equation is $P = \frac{nRT}{V - nb} - a\frac{n^2}{V^2}$
The Lennard–Jones potential energy function systematically describes the attractive and repulsive forces at all distances in terms of ε the well depth and σ the location of the potential minimum.

CONCEPTS & SKILLS

After studying this chapter and working the problems that follow, you should be able to:

1. Write chemical equations for several reactions that lead to gas formation (Section 9.1, Problems 1–4).
2. Describe how pressure and temperature are defined and measured (Section 9.2, Problems 5–10).
3. Use the ideal gas law to relate pressure, volume, temperature, and number of moles of an ideal gas and to do stoichiometric calculations involving gases (Section 9.3, Problems 19–32).
4. Use Dalton's law to calculate partial pressures in gas mixtures (Section 9.4, Problems 33–38).
5. Use the Maxwell–Boltzmann distribution of molecular speeds to calculate root-mean-square, most probable, and average speeds of molecules in a gas (Section 9.5, Problems 41–44).
6. Describe the connection between temperature and the speeds or kinetic energies of the molecules in a gas (Section 4.5, problems 45–46).
7. Use the Boltzmann distribution to determine the relative population of two quantum states in a gas at temperature T (Section 9.6, Problems 47–54).
8. Use the van der Waals equation to relate the pressure, volume, temperature, and number of moles of a nonideal gas (Section 9.7, Problems 55–58).
9. Discuss how forces between atoms and molecules vary with distance (Section 9.7).
10. Calculate the rate of collisions of molecules with a wall, and from that calculation determine the effusion rate of a gas through a small hole of known area (Section 9.8, Problems 59–60).
11. Calculate the enrichment factor for lighter molecules when a gas consisting of a mixture of light and heavy molecules effuses through a small aperture in a vessel wall (Section 9.8, Problems 61–64).
12. Calculate the collision frequency, mean free path, and diffusion constant for gases (Section 9.8, Problems 65–66).

KEY EQUATIONS

$P = \rho g h$ (Section 9.2)

$PV = C$ (fixed temperature and fixed amount of gas) (Section 9.2)

$PV = 22.414$ L atm (for 1 mol gas at 0°C) (Section 9.2)

$$V = V_0\left(1 + \frac{t}{273.15°\text{C}}\right)$$ (Section 9.2)

T (Kelvin) $= 273.15 + t$ (Celsius) (Section 9.2)

$V \propto T$ (fixed temperature and fixed amount of gas) (Section 9.2)

$$V = R\frac{nT}{P} \quad \text{or} \quad PV = nRT$$ (ideal gas) (Section 9.3)

$$\frac{P_1V_1}{n_1T_1} = \frac{P_2V_2}{n_2T_2}$$ (ideal gas) (Section 9.3)

$$P_{tot} = P_A + P_B + P_C + \cdots =$$ (Section 9.4)

$$(n_A + n_B + n_C + \cdots)\frac{RT}{V} = n_{tot}\frac{RT}{V}$$ (Section 9.4)

$$P_A = X_A P_{tot}$$ (Section 9.4)

$$PV = \tfrac{1}{3}Nm\overline{u^2}$$ (Section 9.5)

$$\tfrac{1}{3}N_A m\overline{u^2} = RT$$ (Section 9.5)

$$\overline{E} = N_A m\overline{u^2} = \tfrac{3}{2} \times (\tfrac{1}{3}N_A m\overline{u^2}) = \tfrac{3}{2}RT$$ (Section 9.5)

$$\overline{\varepsilon} = \frac{3}{2}k_B T$$ (Section 9.5)

$$\overline{u^2} = \frac{3RT}{\mathcal{M}}$$ (Section 9.5)

$$u_{rms} = \sqrt{\overline{u^2}} = \sqrt{\frac{3RT}{\mathcal{M}}}$$ (Section 9.5)

$$f(u) = 4\pi\left(\frac{m}{2\pi k_B T}\right)^{3/2} u^2 \exp(-mu^2/2k_B T)$$ (Section 9.5)

$$u_{mp} = \sqrt{\frac{2k_B T}{m}} = \sqrt{\frac{2RT}{\mathcal{M}}}$$ (Section 9.5)

$$\overline{u} = \sqrt{\frac{8k_B T}{\pi m}} = \sqrt{\frac{8RT}{\pi\mathcal{M}}}$$ (Section 9.5)

$$P(n) = C\exp(-E_n/k_B T)$$ (Section 9.6)

$$\frac{P(n)}{P(0)} = \frac{C\exp(-E_n/k_B T)}{C\exp(-E_0/k_B T)} = \exp(-[E_n - E_0]/k_B T)$$ (Section 9.6)

$$z = \frac{PV}{nRT}$$ (Section 9.7)

$$\left(P + a\frac{n^2}{V^2}\right)(V - nb) = nRT$$

$$P = \frac{nRT}{V - nb} - a\frac{n^2}{V^2}$$ (Section 9.7)

$$V_{LJ}(R) = 4\varepsilon\left[\left(\frac{\sigma}{R}\right)^{12} - \left(\frac{\sigma}{R}\right)^6\right]$$ (Section 9.7)

$$Z_w \propto \frac{1}{4}\frac{N}{V}\overline{u}A = \frac{1}{4}\frac{N}{V}\sqrt{\frac{8RT}{\pi\mathcal{M}}}A$$ (Section 9.8)

$$\frac{\text{rate of effusion of A}}{\text{rate of effusion of B}} = \frac{\frac{1}{4}\frac{N_A}{V}\sqrt{\frac{8RT}{\pi\mathcal{M}_A}}A}{\frac{1}{4}\frac{N_B}{V}\sqrt{\frac{8RT}{\pi\mathcal{M}_B}}A} = \frac{N_A}{N_B}\sqrt{\frac{\mathcal{M}_B}{\mathcal{M}_A}}$$ (Section 9.8)

$$Z_1 = 4\frac{N}{V}d^2\sqrt{\frac{\pi RT}{\mathcal{M}}}$$ (Section 9.8)

$$\lambda = \overline{u}Z_1^{-1} = \frac{\overline{u}}{\sqrt{2}(N/V)\pi d^2\overline{u}} = \frac{1}{\sqrt{2}\pi d^2 N/V}$$ (Section 9.8)

$$\overline{\Delta r^2} = 6Dt$$ (Section 9.8)

PROBLEMS

Answers to problems whose numbers are boldface appear in Appendix G. Problems that are more challenging are indicated with asterisks.

The Chemistry of Gases

1. Solid ammonium hydrosulfide (NH_4HS) decomposes entirely to gases when it is heated. Write a chemical equation representing this change.

2. Solid ammonium carbamate ($NH_4CO_2NH_2$) decomposes entirely to gases when it is heated. Write a chemical equation representing this change.

3. Ammonia (NH_3) is an important and useful gas. Suggest a way to generate it from ammonium bromide (NH_4Br). Include a balanced chemical equation.

4. Hydrogen cyanide (HCN) is a poisonous gas. Explain why solutions of potassium cyanide (KCN) should never be acidified. Include a balanced chemical equation.

Pressure and Temperature of Gases

5. Suppose a barometer were designed using water (with a density of 1.00 g cm^{-3}) rather than mercury as its fluid. What would be the height of the column of water balancing 1.00 atm pressure?

6. A vessel that contains a gas has two pressure gauges attached to it. One contains liquid mercury, and the other an oil such as dibutylphthalate. The difference in levels of mercury in the two arms of the mercury gauge is observed to be 9.50 cm. Given

$$\text{density of mercury} = 13.60 \text{ g cm}^{-3}$$

$$\text{density of oil} = 1.045 \text{ g cm}^{-3}$$

$$\text{acceleration due to gravity} = 9.806 \text{ m s}^{-2}$$

(a) What is the pressure of the gas?
(b) What is the difference in height of the oil in the two arms of the oil pressure gauge?

7. Calcium dissolved in the ocean is used by marine organisms to form $CaCO_3(s)$ in skeletons and shells. When the organisms die, their remains fall to the bottom. The amount of calcium carbonate that can be dissolved in seawater depends on the pressure. At great depths, where the pressure exceeds about 414 atm, the shells slowly redissolve. This reaction prevents all the Earth's calcium from being tied up as insoluble $CaCO_3(s)$ at the bottom of the sea. Estimate the depth (in feet) of water that exerts a pressure great enough to dissolve seashells.

8. Suppose that the atmosphere were perfectly uniform, with a density throughout equal to that of air at 0°C, 1.3 g L^{-1}. Calculate the thickness of such an atmosphere that would cause a pressure of exactly 1 standard atm at the Earth's surface.

9. The "critical pressure" of mercury is 172.00 MPa. Above this pressure mercury cannot be liquefied, no matter what the temperature. Express this pressure in atmospheres and in bars (1 bar = 10^5 Pa).

10. Experimental studies of solid surfaces and the chemical reactions that occur on them require very low gas pressures to avoid surface contamination. High-vacuum apparatus for such experiments can routinely reach pressures of 5×10^{-10} torr. Express this pressure in atmospheres and in pascals.

11. Some nitrogen is held in a 2.00-L tank at a pressure of 3.00 atm. The tank is connected to a 5.00-L tank that is completely empty (evacuated), and a valve is opened to connect the two tanks. No temperature change occurs in the process. Determine the total pressure in this two-tank system after the nitrogen stops flowing.

12. The Stirling engine, a heat engine invented by a Scottish minister, has been considered for use in automobile engines because of its efficiency. In such an engine, a gas goes through a four-step cycle: (1) expansion at constant T, (2) cooling at constant V, (3) compression at constant T to its original volume, and (4) heating at constant V to its original temperature. Suppose the gas starts at a pressure of 1.23 atm and the volume of the gas changes from 0.350 to 1.31 L during its expansion at constant T. Calculate the pressure of the gas at the end of this step in the cycle.

13. The absolute temperature of a 4.00-L sample of gas doubles at constant pressure. Determine the volume of the gas after this change.

14. The Celsius temperature of a 4.00-L sample of gas doubles from 20.0°C to 40.0°C at constant pressure. Determine the volume of the gas after this change.

15. The gill is an obscure unit of volume. If some $H_2(g)$ has a volume of 17.4 gills at 100°F, what volume would it have if the temperature were reduced to 0°F, assuming that its pressure stayed constant?

16. A gas originally at a temperature of 26.5°C is cooled at constant pressure. Its volume decreases from 5.40 L to 5.26 L. Determine its new temperature in degrees Celsius.

17. Calcium carbide (CaC_2) reacts with water to produce acetylene (C_2H_2), according to the following equation:

$$CaC_2(s) + 2\ H_2O(\ell) \longrightarrow Ca(OH)_2(s) + C_2H_2(g)$$

A certain mass of CaC_2 reacts completely with water to give 64.5 L C_2H_2 at 50°C and P = 1.00 atm. If the same mass of CaC_2 reacts completely at 400°C and P = 1.00 atm, what volume of C_2H_2 will be collected at the higher temperature?

18. A convenient laboratory source for high purity oxygen is the decomposition of potassium permanganate at 230°C:

$$2\ KMnO_4(s) \longrightarrow K_2MnO_4(s) + MnO_2(s) + O_2(g)$$

Suppose 3.41 L oxygen is needed at atmospheric pressure and a temperature of 20°C. What volume of oxygen should be collected at 230°C and the same pressure to give this volume when cooled?

The Ideal Gas Law

19. A bicycle tire is inflated to a gauge pressure of 30.0 psi at a temperature of t = 0°C. What will its gauge pressure be at 32°C if the tire is considered nonexpandable? (**Note:** The gauge pressure is the *difference* between the tire pressure and atmospheric pressure, 14.7 psi.)

20. The pressure of a poisonous gas inside a sealed container is 1.47 atm at 20°C. If the barometric pressure is 0.96 atm, to what temperature (in degrees Celsius) must the container

and its contents be cooled so that the container can be opened with no risk for gas spurting out?

21. A 20.6-L sample of "pure" air is collected in Greenland at a temperature of −20.0°C and a pressure of 1.01 atm and is forced into a 1.05-L bottle for shipment to Europe for analysis.
(a) Compute the pressure inside the bottle just after it is filled.
(b) Compute the pressure inside the bottle as it is opened in the 21.0°C comfort of the European laboratory.

22. Iodine heptafluoride (IF_7) can be made at elevated temperatures by the following reaction:

$$I_2(g) + 7\ F_2(g) \longrightarrow 2\ IF_7(g)$$

Suppose 63.6 L gaseous IF_7 is made by this reaction at 300°C and a pressure of 0.459 atm. Calculate the volume this gas will occupy if heated to 400°C at a pressure of 0.980 atm.

23. According to a reference handbook, "The weight of one liter of $H_2Te(g)$ is 6.234 g." Why is this information nearly valueless? Assume that $H_2Te(g)$ is an ideal gas, and calculate the temperature (in degrees Celsius) at which this statement is true if the pressure is 1.00 atm.

24. A scuba diver's tank contains 0.30 kg oxygen (O_2) compressed into a volume of 2.32 L.
(a) Use the ideal gas law to estimate the gas pressure inside the tank at 5°C, and express it in atmospheres and in pounds per square inch.
(b) What volume would this oxygen occupy at 30°C and a pressure of 0.98 atm?

25. Hydrogen is produced by the complete reaction of 6.24 g sodium with an excess of gaseous hydrogen chloride.
(a) Write a balanced chemical equation for the reaction that occurs.
(b) How many liters of hydrogen will be produced at a temperature of 50.0°C and a pressure of 0.850 atm?

26. Aluminum reacts with excess aqueous hydrochloric acid to produce hydrogen.
(a) Write a balanced chemical equation for the reaction. (*Hint:* Water-soluble $AlCl_3$ is the stable chloride of aluminum.)
(b) Calculate the mass of pure aluminum that will furnish 10.0 L hydrogen at a pressure of 0.750 atm and a temperature of 30.0°C.

27. The classic method for manufacturing hydrogen chloride, which is still in use today to a small extent, is the reaction of sodium chloride with excess sulfuric acid at elevated temperatures. The overall equation for this process is

$$NaCl(s) + H_2SO_4(\ell) \longrightarrow NaHSO_4(s) + HCl(g)$$

What volume of hydrogen chloride is produced from 2500 kg sodium chloride at 550°C and a pressure of 0.97 atm?

28. In 1783, the French physicist Jacques Charles supervised and took part in the first human flight in a hydrogen balloon. Such balloons rely on the low density of hydrogen relative to air for their buoyancy. In Charles's balloon ascent, the hydrogen was produced (together with iron(II) sulfate) from the action of aqueous sulfuric acid on iron filings.
(a) Write a balanced chemical equation for this reaction.
(b) What volume of hydrogen is produced at 300 K and a pressure of 1.0 atm when 300 kg sulfuric acid is consumed in this reaction?
(c) What would be the radius of a spherical balloon filled by the gas in part (b)?

29. Potassium chlorate decomposes when heated, giving oxygen and potassium chloride:

$$2\ KClO_3(s) \longrightarrow 2\ KCl(s) + 3\ O_2(g)$$

A test tube holding 87.6 g $KClO_3$ is heated, and the reaction goes to completion. What volume of O_2 will be evolved if it is collected at a pressure of 1.04 atm and a temperature of 13.2°C?

30. Elemental chlorine was first produced by Carl Wilhelm Scheele in 1774 using the reaction of pyrolusite (MnO_2) with sulfuric acid and sodium chloride:

$$4\ NaCl(s) + 2\ H_2SO_4(\ell) + MnO_2(s) \longrightarrow 2\ Na_2SO_4(s) + MnCl_2(s) + 2\ H_2O(\ell) + Cl_2(g)$$

Calculate the minimum mass of MnO_2 required to generate 5.32 L gaseous chlorine, measured at a pressure of 0.953 atm and a temperature of 33°C.

31. Elemental sulfur can be recovered from gaseous hydrogen sulfide (H_2S) through the following reaction:

$$2\ H_2S(g) + SO_2(g) \longrightarrow 3\ S(s) + 2\ H_2O(\ell)$$

(a) What volume of H_2S (in liters at 0°C and 1.00 atm) is required to produce 2.00 kg (2000 g) sulfur by this process?
(b) What minimum mass and volume (at 0°C and 1.00 atm) of SO_2 are required to produce 2.00 kg sulfur by this reaction?

32. When ozone (O_3) is placed in contact with dry, powdered KOH at −15°C, the red-brown solid potassium ozonide (KO_3) forms, according to the following balanced equation:

$$5\ O_3(g) + 2\ KOH(s) \longrightarrow 2\ KO_3(s) + 5\ O_2(g) + H_2O(s)$$

Calculate the volume of ozone needed (at a pressure of 0.134 atm and −15°C) to produce 4.69 g KO_3.

Mixtures of Gases

33. Sulfur dioxide reacts with oxygen in the presence of platinum to give sulfur trioxide:

$$2\ SO_2(g) + O_2(g) \longrightarrow 2\ SO_3(g)$$

Suppose that at one stage in the reaction, 26.0 mol SO_2, 83.0 mol O_2, and 17.0 mol SO_3 are present in the reaction vessel at a total pressure of 0.950 atm. Calculate the mole fraction of SO_3 and its partial pressure.

34. The synthesis of ammonia from the elements is conducted at high pressures and temperatures:

$$N_2(g) + 3\ H_2(g) \longrightarrow 2\ NH_3(g)$$

Suppose that at one stage in the reaction, 13 mol NH_3, 31 mol N_2, and 93 mol H_2 are present in the reaction vessel at a total pressure of 210 atm. Calculate the mole fraction of NH_3 and its partial pressure.

35. The atmospheric pressure at the surface of Mars is 5.92×10^{-3} atm. The Martian atmosphere is 95.3% CO_2 and 2.7% N_2 by volume, with small amounts of other gases also present. Compute the mole fraction and partial pressure of N_2 in the atmosphere of Mars.

36. The atmospheric pressure at the surface of Venus is 90.8 atm. The Venusian atmosphere is 96.5% CO_2 and 3.5% N_2 by volume, with small amounts of other gases also present. Compute the mole fraction and partial pressure of N_2 in the atmosphere of Venus.

37. A gas mixture at room temperature contains 10.0 mol CO and 12.5 mol O_2.
(a) Compute the mole fraction of CO in the mixture.
(b) The mixture is then heated, and the CO starts to react with the O_2 to give CO_2:

$$CO(g) + \frac{1}{2} O_2(g) \longrightarrow CO_2(g)$$

At a certain point in the heating, 3.0 mol CO_2 is present. Determine the mole fraction of CO in the new mixture.

38. A gas mixture contains 4.5 mol Br_2 and 33.1 mol F_2.
(a) Compute the mole fraction of Br_2 in the mixture.
(b) The mixture is heated above 150°C and starts to react to give BrF_5:

$$Br_2(g) + 5\ F_2(g) \longrightarrow 2\ BrF_5(g)$$

At a certain point in the reaction, 2.2 mol BrF_5 is present. Determine the mole fraction of Br_2 in the mixture at that point.

39. The partial pressure of water vapor in saturated air at 20°C is 0.0230 atm.
(a) How many molecules of water are in 1.00 cm^3 of saturated air at 20°C?
(b) What volume of saturated air at 20°C contains 0.500 mol water?

40. The partial pressure of oxygen in a mixture of oxygen and hydrogen is 0.200 atm, and that of hydrogen is 0.800 atm.
(a) How many molecules of oxygen are in a 1.500-L container of this mixture at 40°C?
(b) If a spark is introduced into the container, how many grams of water will be produced?

The Kinetic Theory of Gases

41. (a) Compute the root-mean-square speed of H_2 molecules in hydrogen at a temperature of 300 K.
(b) Repeat the calculation for SF_6 molecules in gaseous sulfur hexafluoride at 300 K.

42. Researchers recently reported the first optical atomic trap. In this device, beams of laser light replace the physical walls of conventional containers. The laser beams are tightly focused. They briefly (for 0.5 s) exert enough pressure to confine 500 sodium atoms in a volume of 1.0×10^{-15} m^3. The temperature of this gas is 0.00024 K, the lowest temperature ever reached for a gas. Compute the root-mean-square speed of the atoms in this confinement.

43. Compare the root-mean-square speed of helium atoms near the surface of the sun, where the temperature is approximately 6000 K, with that of helium atoms in an interstellar cloud, where the temperature is 100 K.

44. The "escape velocity" necessary for objects to leave the gravitational field of the Earth is 11.2 km s^{-1}. Calculate the ratio of the escape velocity to the root-mean-square speed of helium, argon, and xenon atoms at 2000 K. Does your result help explain the low abundance of the light gas helium in the atmosphere? Explain.

45. Chlorine dioxide (ClO_2) is used for bleaching wood pulp. In a gaseous sample held at thermal equilibrium at a particular temperature, 35.0% of the molecules have speeds exceeding 400 m s^{-1}. If the sample is heated slightly, will the percentage of molecules with speeds in excess of 400 m s^{-1} then be greater than or less than 35%? Explain.

46. The ClO_2 described in Problem 45 is heated further until it explodes, yielding Cl_2, O_2, and other gaseous products. The mixture is then cooled until the original temperature is reached. Is the percentage of *chlorine* molecules with speeds in excess of 400 m s^{-1} greater than or less than 35%? Explain.

Distribution of Energy among Molecules

47. Calculate the relative populations of two energy levels separated by 0.4×10^{-21} J in a gas at temperature 25°C.

48. Calculate the relative populations of two energy levels separated by 40×10^{-21} J in a gas at temperature 25°C.

49. Estimate the ratio of the number of molecules in the first excited vibrational state of the molecule N_2 to the number in the ground state, at a temperature of 450 K. The vibrational frequency of N_2 is 7.07×10^{13} s^{-1}.

50. The vibrational frequency of the ICl molecule is 1.15×10^{13} s^{-1}. For every million (1.00×10^6) molecules in the ground vibrational state, how many will be in the first excited vibrational state at a temperature of 300 K?

51. The force constant for HF is 966 N m^{-1}. Using the harmonic oscillator model, calculate the relative population of the first excited state and the ground state at 300 K.

52. The force constant for HBr is 412 N m^{-1}. Using the harmonic oscillator model, calculate the relative population of the first excited state and the ground state at 300 K.

53. The Boltzmann distribution describes the relative population of molecules at different altitudes in a column of air above the surface of the earth due to the differences in their gravitational potential energy at these altitudes. The most convenient form of this distribution—called the barometric equation—relates the pressure P_h in the column at altitude h to its value P_0 at the surface of the earth as $P_h = P_0 \exp[-\mathcal{M}gh/RT]$, where $\mathcal{M}$ is the molar mass of the gas and g is the acceleration of gravity. Calculate the pressure at an altitude of 1 km for a gas with average molar mass 29 g mol^{-1} when the temperature is 298 K and the pressure at the surface of the earth is 1 atm.

54. If we assume that the atmosphere of the earth is an ideal gas mixture, then each component is described by a separate barometric equation. This model predicts that the concentrations of heavier molecules, and therefore their partial pressures, decrease faster than for lighter molecules as the altitude increases. Assume the ratio of partial pressure of nitrogen to oxygen is 4 at the surface of the earth, corresponding to 80% nitrogen by moles, and assume the temperature throughout the column in 25 C. Calculate the value of P_{N_2}/P_{O_2} at the altitude of 10 km.

Real Gases: Intermolecular Forces

55. Oxygen is supplied to hospitals and chemical laboratories under pressure in large steel cylinders. Typically, such cylinders have an internal volume of 28.0 L and contain 6.80 kg oxygen. Use the van der Waals equation to estimate the pressure inside such cylinders at 20°C in atmospheres and in pounds per square inch.

56. Steam at high pressures and temperatures is used to generate electrical power in utility plants. A large utility boiler has a volume of 2500 m^3 and contains 140 metric tons (1 metric ton = 10^3 kg) of steam at a temperature of 540°C. Use the van der Waals equation to estimate the pressure of the steam under these conditions, in atmospheres and in pounds per square inch.

57. Using (a) the ideal gas law and (b) the van der Waals equation, calculate the pressure exerted by 50.0 g carbon dioxide in a 1.00-L vessel at 25°C. Do attractive or repulsive forces dominate?

58. When 60.0 g methane (CH_4) is placed in a 1.00-L vessel, the pressure is measured to be 130 atm. Calculate the temperature of the gas using (a) the ideal gas law and (b) the van der Waals equation. Do attractive or repulsive forces dominate?

Molecular Collisions and Rate Processes

59. A spherical bulb with a volume of 500 cm^3 is evacuated to a negligibly small residual gas pressure and then closed off. One hour later, the pressure in the vessel is found to be 1.00 $\times$ 10^{-7} atm because the bulb has a tiny hole in it. Assume that the surroundings are at atmospheric pressure, T = 300 K, and the average molar mass of molecules in the atmosphere is 28.8 g mol^{-1}. Calculate the radius of the hole in the vessel wall, assuming it to be circular.

60. A 200-cm^3 vessel contains hydrogen gas at a temperature of 25°C and a pressure of 0.990 atm. Unfortunately, the vessel has a tiny hole in its wall, and over a period of 1 hour, the pressure drops to 0.989 atm. What is the radius of the hole (assumed to be circular)?

61. Methane (CH_4) effuses through a small opening in the side of a container at the rate of 1.30 $\times$ 10^{-8} mol s^{-1}. An unknown gas effuses through the same opening at the rate of 5.42 $\times$ 10^{-9} mol s^{-1} when maintained at the same temperature and pressure as the methane. Determine the molar mass of the unknown gas.

62. Equal chemical amounts of two gases, fluorine and bromine pentafluoride, are mixed. Determine the ratio of the rates of effusion of the two gases through a small opening in their container.

63. Calculate the theoretical number of stages that would be needed to enrich ^{235}U to 95% purity by means of the barrier diffusion process, using $^{235}UF_6$ and $^{238}UF_6$ as the gaseous compounds. The natural abundance of ^{238}U is 99.27%, and that of ^{235}U is 0.72%. Take the relative atomic masses of ^{235}U and ^{238}U to be 235.04 and 238.05, respectively.

64. A mixture of H_2 and He at 300 K effuses from a tiny hole in the vessel that contains it. What is the mole fraction of H_2 in the original gas mixture if 3.00 times as many He atoms as H_2 molecules escape from the orifice in unit time? If the same mixture is to be separated by a barrier-diffusion process, how many stages are necessary to achieve H_2 of 99.9% purity?

65. At what pressure does the mean free path of krypton (Kr) atoms (d = 3.16 $\times$ 10^{-10} m) become comparable with the diameter of the 1-L spherical vessel that contains them at 300 K? Calculate the diffusion constant at this pressure.

66. At what pressure does the mean free path of Kr atoms (d = 3.16 $\times$ 10^{-10} m) become comparable with the diameter of a Kr atom if T = 300 K? Calculate the diffusion constant at this pressure. Assume that Kr obeys the ideal gas law even at these high pressures.

ADDITIONAL PROBLEMS

67. The Earth is approximately a sphere of radius 6370 km. Taking the average barometric pressure on the Earth's surface to be 730 mm Hg, estimate the total mass of the Earth's atmosphere.

* **68.** After a flood fills a basement to a depth of 9.0 feet and completely saturates the surrounding earth, the owner buys an electric pump and quickly pumps the water out of the basement. Suddenly, a basement wall collapses, the structure is severely damaged, and mud oozes in. Explain this event by estimating the difference between the outside pressure at the base of the basement walls and the pressure inside the drained basement. Assume that the density of the mud is 4.9 g cm^{-3}. Report the answer both in atmospheres and in pounds per square inch.

69. The density of mercury is 13.5955 g cm^{-3} at 0.0°C, but only 13.5094 g cm^{-3} at 35°C. Suppose that a mercury barometer is read on a hot summer day when the temperature is 35°C. The column of mercury is 760.0 mm long. Correct for the expansion of the mercury and compute the true pressure in atmospheres.

70. When a gas is cooled at constant pressure, the volume decreases according to the following equation:

$$V = 209.4 \text{ L} + \left(0.456 \frac{L}{°F}\right) \times t_F$$

where t_F is the temperature in degrees Fahrenheit. From this relationship, estimate the absolute zero of temperature in degrees Fahrenheit.

71. Amonton's law relates pressure to absolute temperature. Consider the ideal gas law and then write a statement of Amonton's law in a form that is analogous to the statements of Charles's law and Boyle's law in the text.

72. The density of a certain gas is 2.94 g L^{-1} at 50°C and P = 1.00 atm. What is its density at 150°C? Calculate the molar mass of the gas, assuming it obeys the ideal gas law.

73. A lighter-than-air balloon contains 1005 mol helium at 1.00 atm and 25.0°C.
(a) Compute the difference between the mass of the helium it contains and the mass of the air it displaces, assuming the molar mass of air to be 29.0 g mol^{-1}.
(b) The balloon now ascends to an altitude of 10 miles, where the temperature is −80.0°C. The walls of the balloon are elastic enough that the pressure inside it equals the pressure outside. Repeat the calculation of part (a).

74. Baseball reporters say that long fly balls that would have carried for home runs in July "die" in the cool air of October and are caught. The idea behind this observation is that a baseball carries better when the air is less dense. Dry air is a mixture of gases with an effective molar mass of 29.0 g mol^{-1}.
(a) Compute the density of dry air on a July day when the temperature is 95.0°F and the pressure is 1.00 atm.
(b) Compute the density of dry air on an October evening when the temperature is 50.0°F and the pressure is 1.00 atm.
(c) Suppose that the humidity on the July day is 100%; thus, the air is saturated with water vapor. Is the density of this hot, moist air less than, equal to, or greater than the density of the hot, dry air computed in part (a)? In other terms, does high humidity favor the home run?

75. Sulfuric acid reacts with sodium chloride to produce gaseous hydrogen chloride according to the following reaction:

$$NaCl(s) + H_2SO_4(\ell) \longrightarrow NaHSO_4(s) + HCl(g)$$

A 10.0-kg mass of NaCl reacts completely with sulfuric acid to give a certain volume of HCl(g) at 50°C and P = 1.00 atm. If the same volume of hydrogen chloride is collected at 500°C and P = 1.00 atm, what mass of NaCl has reacted?

76. Exactly 1.0 lb Hydrone, an alloy of sodium with lead, yields (at 0.0°C and 1.00 atm) 2.6 ft^3 of hydrogen when it is treated with water. All the sodium reacts according to the following reaction:

$$2\ Na_{\text{in alloy}} + 2\ H_2O(\ell) \longrightarrow 2\ NaOH(aq) + H_2(g)$$

and the lead does not react with water. Compute the percentage by mass of sodium in the alloy.

77. A sample of limestone (calcium carbonate, $CaCO_3$) is heated at 950 K until it is completely converted to calcium oxide (CaO) and CO_2. The CaO is then all converted to calcium hydroxide by addition of water, yielding 8.47 kg of solid $Ca(OH)_2$. Calculate the volume of CO_2 produced in the first step, assuming it to be an ideal gas at 950 K and a pressure of 0.976 atm.

78. A gas exerts a pressure of 0.740 atm in a certain container. Suddenly, a chemical change occurs that consumes half of the molecules originally present and forms two new molecules for every three consumed. Determine the new pressure in the container if the volume of the container and the temperature are unchanged.

79. The following arrangement of flasks is set up. Assuming no temperature change, determine the final pressure inside the system after all stopcocks are opened. The connecting tube has zero volume.

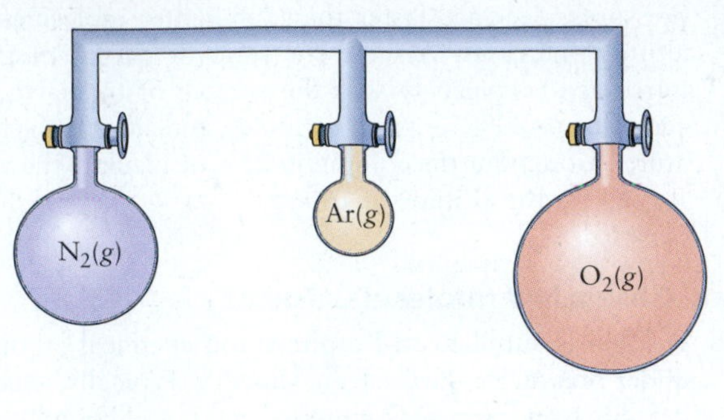

V = 4.00 L	V = 3.00 L	V = 5.00 L
P = 0.792 atm	P = 1.23 atm	P = 2.51 atm

* 80. A mixture of $CS_2(g)$ and excess $O_2(g)$ in a 10.0-L reaction vessel at 100.0°C is under a pressure of 3.00 atm. When the mixture is ignited by a spark, it explodes. The vessel successfully contains the explosion, in which all of the $CS_2(g)$ reacts to give $CO_2(g)$ and $SO_2(g)$. The vessel is cooled back to its original temperature of 100.0°C, and the pressure of the mixture of the two product gases and the unreacted $O_2(g)$ is found to be 2.40 atm. Calculate the mass (in grams) of $CS_2(g)$ originally present.

81. Acetylene reacts with hydrogen in the presence of a catalyst to form ethane according to the following reaction:

$$C_2H_2(g) + 2\ H_2(g) \longrightarrow C_2H_6(g)$$

The pressure of a mixture of acetylene and an excess of hydrogen decreases from 0.100 to 0.042 atm in a vessel of a given volume after the catalyst is introduced, and the temperature is restored to its initial value after the reaction reaches completion. What was the mole fraction of acetylene in the original mixture?

82. Refer to the atomic trap described in Problem 42.
(a) Assume ideal gas behavior to compute the pressure exerted on the "walls" of the optical bottle in this experiment.
(b) In this gas, the mean free path (the average distance traveled by the sodium atoms between collisions) is 3.9 m. Compare this with the mean free path of the atoms in gaseous sodium at room conditions.

83. Deuterium (^2H), when heated to sufficiently high temperature, undergoes a nuclear fusion reaction that results in the production of helium. The reaction proceeds rapidly at a temperature, T, at which the average kinetic energy of the deuterium atoms is 8×10^{-16} J. (At this temperature, deuterium molecules dissociate completely into deuterium atoms.)
(a) Calculate T in kelvins (atomic mass of ^2H = 2.015).
(b) For the fusion reaction to occur with ordinary H atoms, the average energy of the atoms must be about 32×10^{-16} J. By what factor does the average speed of the ^1H atoms differ from that of the ^2H atoms of part (a)?

84. Molecules of oxygen of the following isotopic composition are separated in an oxygen enrichment plant: $^{16}O^{16}O$, $^{16}O^{17}O$, $^{16}O^{18}O$, $^{17}O^{17}O$, $^{17}O^{18}O$, $^{18}O^{18}O$.
(a) Compare the average translational kinetic energy of the lightest and heaviest molecular oxygen species at 200°C and at 400°C.

(b) Compare the average speeds of the lightest and heaviest molecular oxygen species at the same two temperatures.

* 85. What is the probability that an O_2 molecule in a sample of oxygen at 300 K has a speed between 5.00×10^2 and 5.10×10^2 m s^{-1}? (*Hint:* Try approximating the area under the Maxwell–Boltzmann distribution by small rectangles.)

* 86. Molecules in a spherical container make wall collisions in a great circle plane of the sphere. All paths traveled between collisions are equal in length.

(a) Relate the path length $\Delta\ell$ to the angle θ and the sphere radius r.

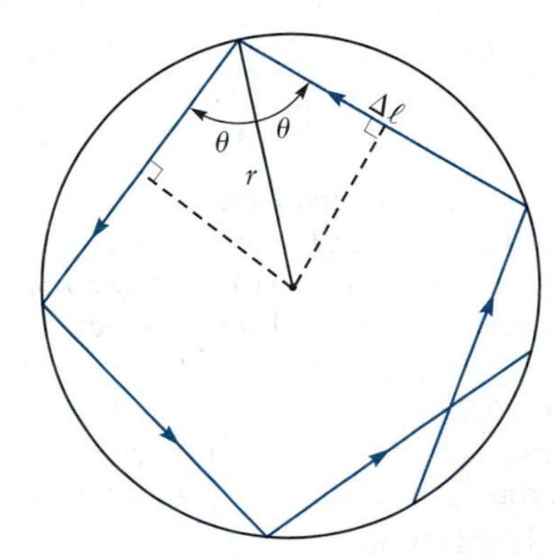

(b) Express the component of the molecular momentum transferred to the wall in a collision in terms of the molecular mass m, the speed u, and the angle θ.

(c) Using parts (a) and (b), calculate the average force exerted on the wall by one molecule.

(d) Derive the relation $PV = \overline{PV} = \frac{1}{3}Nm\overline{u^2}$ for a spherical container. This is the same relation found for a rectangular box in Section 9.5.

* 87. The van der Waals constant b is related to the volume excluded per mole of molecules, so it should be proportional to $N_A\sigma^3$, where σ is the distance parameter in the Lennard–Jones potential.

(a) Make a plot of b against $N_A\sigma^3$ for Ar, H_2, CH_4, N_2, and O_2, using data from Tables 9.3 and 9.4. Do you see an overall correlation between the two?

(b) The van der Waals constant a has dimensions of pressure times the square of the molar volume. Rewrite the units of a in terms of energy, length, and number of moles, and suggest a relation between a and some combination of the constants ε, σ, and N_A. Make a plot of a against this combination of constants for the gases of part (a). Do you see an overall correlation in this case?

* 88. Take the derivative of the Lennard–Jones potential to express the force exerted on one atom by another for the distance R between them. Calculate the forces (in joules per meter) on a pair of interacting argon atoms at distances of 3.0, 3.4, 3.8, and 4.2×10^{-10} m. Is the force attractive or repulsive at each of these distances?

89. A vessel with a small hole in its wall is filled with oxygen to a pressure of 1.00 atm at 25°C. In a 1.00-minute period, 3.25 g oxygen effuses out through the hole into a vacuum. The vessel is evacuated and filled with an unknown gas at the same pressure and volume. In this case, 5.39 g of the unknown gas effuses in 1.00 minute. Calculate the molar mass of the unknown gas.

90. A cylindrical storage tank for natural gas (mostly methane, CH_4) with a 20-ft radius and a 50-ft height is filled to a pressure of 2000 psi at 20°C. A small leak of 1.0-mm^2 area develops at one of the welds. Calculate the mass of CH_4 (in grams) that leaks out of the tank in one day. What fraction of the total gas escapes per day?

91. A thermos bottle (Dewar vessel) has an evacuated space between its inner and outer walls to diminish the rate of transfer of thermal energy to or from the bottle's contents. For good insulation, the mean free path of the residual gas (air; average molecular mass = 29) should be at least 10 times the distance between the inner and outer walls, which is about 1.0 cm. What should be the maximum residual gas pressure in the evacuated space if T = 300 K? Take an average diameter of $d = 3.1 \times 10^{-10}$ m for the molecules in the air.

92. A tanker truck carrying liquid ammonia overturns, releasing ammonia vapor into the air.

(a) Approximating ammonia, oxygen, and nitrogen as spheres of equal diameter (3×10^{-10} m), estimate the diffusion constant of ammonia in air at atmospheric pressure and 20°C.

(b) Calculate the time required for a 100-m root-mean-square displacement of ammonia from the truck, and express this time in everyday units (seconds, minutes, hours, days, or years). The actual time for the ammonia to travel this distance is far shorter because of the existence of air currents (even when there is no wind).

93. Molecules of UF_6 are approximately 175 times more massive than H_2 molecules; however, Avogadro's number of H_2 molecules confined at a set temperature exert the same pressure on the walls of the container as the same number of UF_6 molecules. Explain how this is possible.

94. The number density of atoms (chiefly hydrogen) in interstellar space is about 10 per cubic centimeter, and the temperature is about 100 K.

(a) Calculate the pressure of the gas in interstellar space, and express it in atmospheres.

(b) Under these conditions, an atom of hydrogen collides with another atom once every 1×10^9 seconds (that is, once every 30 years). By using the root-mean-square speed, estimate the distance traveled by a H atom between collisions. Compare this distance with the distance from the Earth to the Sun (150 million km).

95. A sample of 2.00 mol argon is confined at low pressure in a volume at a temperature of 50°C. Describe quantitatively the effects of each of the following changes on the pressure, the average energy per atom in the gas, the root-mean-square speed, the rate of collisions with a given area of wall, the frequency of Ar–Ar collisions, and the mean free path:

(a) The temperature is decreased to −50°C.

(b) The volume is doubled.

(c) The amount of argon is increased to 3.00 mol.

96. By assuming that the collision diameter of a CH_4 molecule is given by its Lennard–Jones σ parameter (see Table 9.4), estimate the rate at which methane molecules collide with one another at 25°C and a pressure of (a) 1.00 atm and (b) 1.0×10^{-7} atm.

CUMULATIVE PROBLEMS

97. A gaseous hydrocarbon, in a volume of 25.4 L at 400 K and a pressure of 3.40 atm, reacts in an excess of oxygen to give 47.4 g H_2O and 231.6 g CO_2. Determine the molecular formula of the hydrocarbon.

98. A sample of a gaseous binary compound of boron and chlorine weighing 2.842 g occupies 0.153 L at 0°C and 1.00 atm pressure. This sample is decomposed to give solid boron and gaseous chlorine (Cl_2). The chlorine occupies 0.688 L at the same temperature and pressure. Determine the molecular formula of the compound.

99. A mixture of calcium carbonate, $CaCO_3$, and barium carbonate, $BaCO_3$, weighing 5.40 g reacts fully with hydrochloric acid, HCl(*aq*), to generate 1.39 L $CO_2(g)$, measured at 50°C and 0.904 atm pressure. Calculate the percentages by mass of $CaCO_3$ and $BaCO_3$ in the original mixture.

100. A solid sample of Rb_2SO_3 weighing 6.24 g reacts with 1.38 L gaseous HBr, measured at 75°C and 0.953 atm pressure. The solid RbBr, extracted from the reaction mixture and purified, has a mass of 7.32 g.
(a) What is the limiting reactant?
(b) What is the theoretical yield of RbBr, assuming complete reaction?
(c) What is the actual percentage yield of product?

CHAPTER 10

Solids, Liquids, and Phase Transitions

Solid iodine is converted directly to a vapor (sublimes) when warmed. Here, purple iodine vapor is redeposited as a solid on the cooler upper surfaces of the vessel.

The bulk properties of gases, liquids, and solids—molar volume, density, compressibility, and thermal expansion, among others—differ widely, often by orders of magnitude. All of these properties depend on the temperature and pressure and describe the response of the system to changes in those variables. The average separation between molecules, and the nature of the intermolecular forces present, determine the properties of the bulk. The local structure—the arrangement of atoms or molecules on the nanometer length scale—is the key microscopic feature that distinguishes the three states of matter from one another and ultimately determines the differences in their bulk properties.

We begin with a brief survey of several important bulk properties and discuss how the number density of molecules and the strength of intermolecular forces rationalize the properties observed. Then, we describe the various kinds of intermolecular forces and show how these forces originate in the structures of molecules. Finally, we survey the transitions between the states of matter as consequences of intermolecular forces. In the previous chapter we derived the ideal gas law using kinetic molecular arguments by neglecting all intermolecular forces in the gas. In this chapter we apply similar kinetic molecular arguments to solids and liquids, but we cannot neglect the intermolecular forces. Consequently, the discussion of solids and liquids is more qualitative than that for gases.

10.1 Bulk Properties of Gases, Liquids, and Solids: Molecular Interpretation

Each of the following measurements provides clear distinctions among gases, liquids, and solids and also probes the strength of intermolecular forces, albeit indirectly. Each demonstrates that in gases at low densities, molecules are on average far apart and interact only weakly; in condensed phases, molecules are closely packed together and interact quite strongly. These general conclusions apply almost universally to substances consisting of small, nearly rigid molecules such as N_2, CO_2, CH_4, and acetic acid $C_2H_4O_2$. In contrast, many biological materials and synthetic polymers contain complex chainlike molecules that can become strongly entangled. Clear-cut classification of these materials as solids or liquids becomes difficult.

Interpreting bulk properties qualitatively on the basis of microscopic properties requires only consideration of the long-range attractive forces and short-range repulsive forces between molecules; it is not necessary to take into account the details of molecular shapes. We have already shown one kind of potential that describes these intermolecular forces, the Lennard–Jones 6–12 potential used in Section 9.7 to obtain corrections to the ideal gas law. In Section 10.2, we discuss a variety of intermolecular forces, most of which are derived from electrostatic (Coulomb) interactions, but which are expressed as a hierarchy of approximations to exact electrostatic calculations for these complex systems.

Molar Volume

One mole of a typical solid or liquid occupies a volume of 10 to 100 cm^3 at room conditions, but the **molar volume** of a gas under the same conditions is about 24,000 $cm^3\ mol^{-1}$. This large difference explains why solids and liquids are called the condensed states of matter. Because a mole of any substance contains Avogadro's number of molecules, the molar volume is inversely related to the **number density** (number of molecules per cubic centimeter) of the different phases. Liquids and solids have high number densities, and gases have very low number densities. On melting, most solids change volume by only 2% to 10%, showing that the solid and liquid states of a given substance are condensed, relative to the gaseous state, by roughly the same amount.

The similarity of the molar volumes of solid and liquid forms of the same substance suggests that the separation between neighboring molecules in the two states is approximately the same. Density measurements (see Section 2.1) show that the intermolecular contacts, the distances between the nuclei of atoms at the far edge of one molecule and the near edge of a neighbor, usually range from 3×10^{-10} m to 5×10^{-10} m in solids and liquids. At these distances, longer range

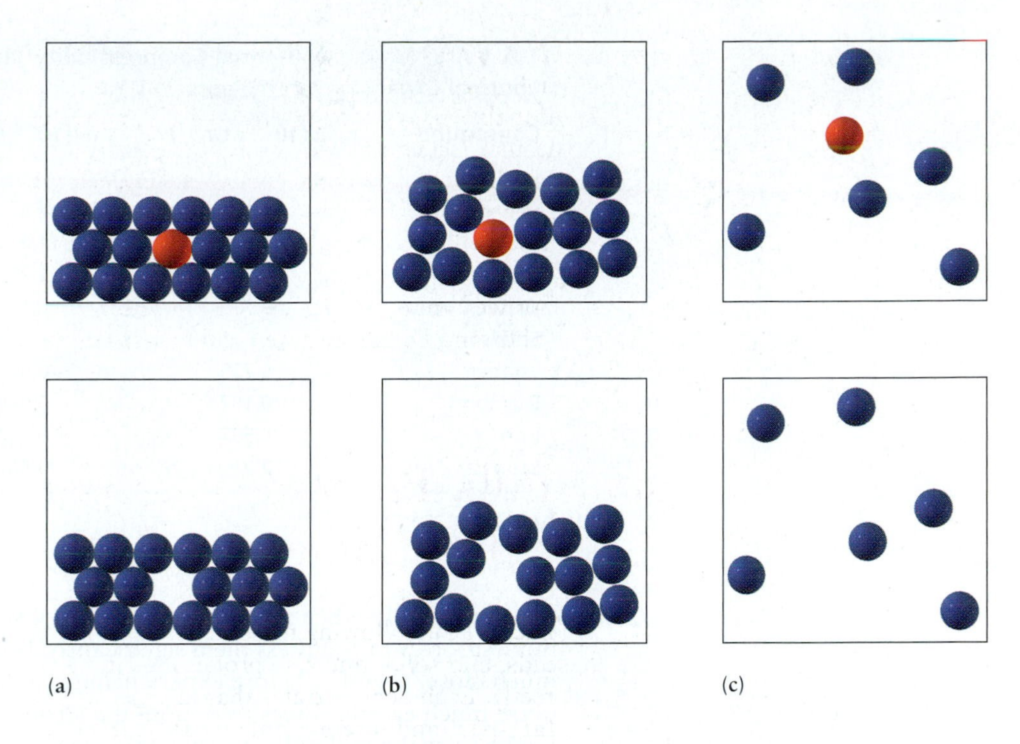

FIGURE 10.1 Intermolecular forces create structure in liquids and solids. If a single atom is removed from a snapshot of the atomic arrangement in a solid (a), it is easy to figure out exactly where to put it back in. For a liquid (b), the choices are limited. If a single atom is removed from a gas (c), no clue remains to tell where it came from.

attractive forces and shorter range repulsive forces just balance one another, resulting in a minimum in the potential energy (see Section 9.7). Although these intermolecular separations are significantly greater than chemical bond lengths (which range from $0.5\text{–}2.5 \times 10^{-10}$ m), they are much shorter than the intermolecular separations in gases, which average about 30×10^{-10} m under room conditions. The distinction is shown schematically in Figure 10.1.

Compressibility

The **compressibility** of a substance is defined as the fractional decrease in volume per unit increase in pressure. The compressibility is usually denoted by the Greek letter kappa, κ, and is defined operationally by its method of measurement: $\kappa = -(1/V)(\Delta V/\Delta P)$. A sample with volume V is subjected to a pressure increase, ΔP, and the resulting change in volume, ΔV, is measured. The result is divided by V; thus, the tabulated value depends only on the substance being measured and not on the geometry of the sample. Consequently, the unit of κ is P^{-1}. The minus sign is included in the definition to make κ positive because ΔV is negative. The measurements are always performed at fixed temperature to eliminate any thermal effects on the volume (see later). Consequently, κ is called the *isothermal compressibility*, and the temperature of measurement is quoted together with the tabulated values. Both solids and liquids are nearly incompressible, but gases are very compressible. According to Boyle's law, doubling the pressure exerted on an ideal gas from 1 to 2 atm reduces its volume by half (at constant temperature). The corresponding compressibility is 2 atm^{-1} or 20 MPa^{-1}. Doubling the pressure exerted on water or steel scarcely changes the volume at all; typical compressibilities for these materials are of the order 10^{-5} to 10^{-6} atm^{-1} (Table 10.1). Increasing the pressure on a liquid or solid by a factor of 2 changes its volume by 1% or less. The high compressibility of gases and the low compressibilities of solids and liquids suggest that in the gas phase there is substantial space between the molecules, but in the condensed states the particles of a substance are in contact or nearly in contact.

The much greater separation between molecules in the gas phase than in the condensed phases dramatically influences the effect of intermolecular forces. The repulsive force, although very strong, is very short-ranged and becomes

TABLE 10.1 Isothermal Compressibility† and Thermal Expansion Coefficients

Compound	$\kappa/(10^{-6}\ \text{atm}^{-1})$	$\alpha/(10^{-4}\ K^{-1})$
Liquids		
Benzene	92.1	12.4
Ethanol	76.8	11.2
Mercury	38.7	1.82
Water	49.7	2.1
Solids		
Copper	0.735	0.501
Diamond	0.187	0.030
Iron	0.597	0.354
Lead	2.21	0.861

†Values at 20°C.

significant only when molecules are very close together. So, gases require only modest forces to compress them significantly, because the molecules can be pushed much closer together before experiencing repulsive forces. Liquids and solids require much greater forces to oppose the strongly repulsive forces already operating because the molecules are in contact (see Section 9.7).

Thermal Expansion

The **coefficient of thermal expansion** α is defined as the fractional increase in the volume of a substance per degree increase in temperature. Like the compressibility, it is defined operationally by its method of measurement: $\alpha = (1/V)(\Delta V/\Delta T)$. The measurements are performed at constant pressure; thus, α is called the *isobaric coefficient of thermal expansion,* and the pressure at which the measurements were taken is quoted together with tabulated values. Charles's law shows that this coefficient is the same for all gases and takes the value $1/273.15(°C)^{-1}$ at 0°C. Increasing the temperature by 1°C thus causes a gas to expand by 1/273.15, or 0.366% of its original volume at 0°C, as long as the pressure is constant. The thermal expansion coefficients of liquids and solids are much smaller. Heating water from 20°C to 21°C increases its volume by only 0.0212%, and the volume of mercury goes up by only 0.0177% over the same temperature interval. The coefficients of thermal expansion of solids are mostly less than 0.02% per degree Celsius (see Table 10.1).

The difference in thermal expansion between condensed states and gases is explained by strong intermolecular forces (deep intermolecular potential wells) acting over short distances in the condensed states, but not in the gaseous state. Because of the much greater intermolecular separations, these forces are much weaker in gases. An increase in volume in a solid or liquid requires that attractive forces between each molecule and its neighbors be partially overcome. Because the intermolecular distances in a solid or liquid fall in the range where intermolecular attractive forces are strongest, relatively small expansion occurs when the temperature is increased. By contrast, molecules in a gas are so far apart that attractive forces are essentially negligible; the same temperature increase produces much greater expansion in a gas than in condensed phases.

Fluidity and Rigidity

The most characteristic property of gases and liquids is their **fluidity,** which contrasts with the **rigidity** of solids. Liquids possess definite volumes but keep no definite shapes of their own; they flow easily under stress (externally applied mechanical force). The resistance of a material to macroscopic flow is measured by its **shear viscosity.** On the microscopic level, shear viscosity arises from the resistance of one thin layer of molecules "dragged across" another thin layer. The shear viscosities of most liquids are about 16 orders of magnitude smaller than those of most solids, and those of gases are smaller yet. A *rigid* material retains its shape under stress; it manifests structural strength by resisting flow when stress is applied. The properties of **hardness** (resistance to indentation) and **elasticity** (capacity to recover shape when a deforming stress is removed) are closely related to rigidity, or high shear viscosity. Solids possess these properties in good measure; gases and liquids do not.

Diffusion

When two different substances are placed in contact—for example, a drop of red ink into a beaker of water—they start to mix. Molecules of one type migrate, or **diffuse,** into regions initially occupied only by the other type. Molecules dispersed in gases at room conditions diffuse at rates on the order of centimeters per second. If you have ever passed by a perfume counter, this rate should not surprise you. Molecules in liquids and solids diffuse far more slowly. The **diffusion constant** of a substance measures the rate of diffusive mixing. At room temperature and pressure, diffusion constants for the diffusion of liquids into liquids are about four orders of magnitude smaller than those for gases into gases; diffusion constants of solids into solids are many orders of magnitude smaller yet. Diffusion in solids is really quite slow. Values of the diffusion constant for selected materials are shown in Table 10.2.

Figure 10.1 shows a "snapshot" of a liquid and a solid, fixing the positions of the atoms at a particular instant in time. The paths followed by the molecules in these two states can also be examined over a short time interval (Fig. 10.2). In liquids, molecules are free to travel through the sample, changing neighbors constantly in the course of their diffusive motion. In a solid, the molecules constantly vibrate about their equilibrium positions, but remain quite close to those positions. The low shear viscosity of a liquid implies that its molecules can quickly change neighbors, finding new interactions as the liquid flows in response to an external stress. The rigidity of solids suggests, in contrast, a durable arrangement of neighbors about any given molecule. The durable arrangement of molecules in a solid, as opposed to the freedom of molecules to diffuse in a liquid at comparable packing density, is the critical difference between the solid and liquid states.

TABLE 10.2 Diffusion Constants

Diffusing Species	Host Material	Diffusion Constant (m^2/s^{-1})	Temperature (K)
Ar	Ar	2.3×10^{-6}	100
Ar	Ar	1.86×10^{-5}	300
N_2	N_2	2.05×10^{-5}	300
O_2	O_2	1.8×10^{-5}	273
CH_4	CH_4	2.06×10^{-5}	273
HCl	HCl	1.24×10^{-5}	295
Cu	Cu	4.2×10^{-19}	500
Al	Al	4.2×10^{-14}	500
Cu	Al	4.1×10^{-14}	500
Cu	Ni	1.3×10^{-22}	500
Fe	Fe	3.0×10^{-21}	500
Fe	Fe	1.8×10^{-15}	900
C	Fe	1.7×10^{-10}	900

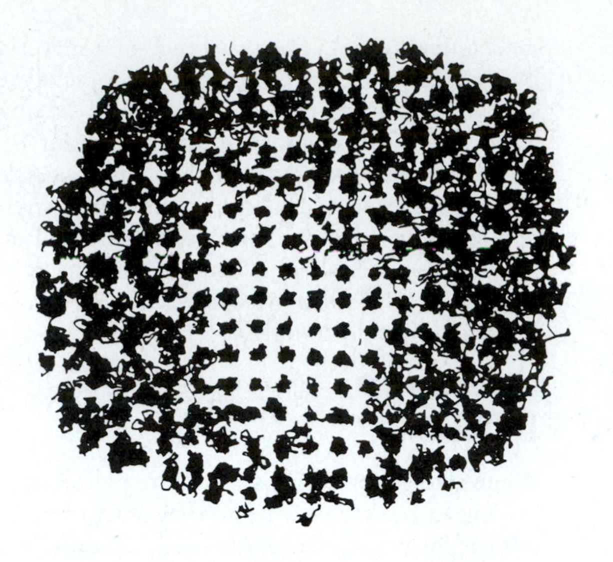

FIGURE 10.2 In this computer-simulated picture of the motions of atoms in a tiny melting crystal, the atoms at the center (in the solid) move erratically about particular sites. The atoms at the surface (in the liquid) move over much greater distances.

Short-range attractive intermolecular forces in liquids lead to well-defined local structures that persist for short periods. Individual molecules experience interactions with neighbors that lead, at any instant, to a local environment closely resembling that in a solid, but they quickly move on. Their trajectories consist of "rattling" motions in a temporary cage formed by neighbors and superimposed on erratic displacements over larger distances. In this respect, a liquid is intermediate between a gas and a solid. A gas (see Fig. 9.22) provides no temporary cages, so each molecule of a gas travels a longer distance before colliding with a second molecule. Consequently, the diffusion constant of a gas is larger than that of a liquid. In a solid, the cages are nearly permanent, thus diffusion is slow. Melting occurs as thermal energy increases the amplitude of vibration of the molecules around their equilibrium positions in a solid to such a degree that they are free to make major excursions. For these same reasons, liquids can dissolve substances much more rapidly than do solids. Individual molecules of a liquid quickly wander into contact with molecules of an added substance, and new attractions between the unlike molecules have an early chance to replace those existing originally in the pure liquid.

Liquids and gases may also mix through **convection,** as well as by diffusion. In convection, the net flow of a whole region of fluid with respect to another region leads to mixing at far greater rates than occurs through simple diffusion. Convection is the primary mechanism by which mixing occurs in the oceans and in the atmosphere. Convection is not observed in solids.

Surface Tension

Boundaries between phases have special importance in chemistry and biology. Each type of boundary has its own unique characteristics. The surface of water (or any liquid) in contact with air (or any gas) resists attempts to increase its area (Fig. 10.3a). This **surface tension** causes the surface to behave like a weak, elastic skin. Effects of surface tension are particularly apparent under zero gravity, where liquids float around as spherical drops because spheres contain the largest volume for the smallest surface area of any geometric shape. If two small drops encounter each other, they tend to coalesce into a larger drop because one large drop has a smaller surface area than two small drops. The surface tension of water is larger than that of most other liquids at room temperature, but it is about six times smaller than that for the liquid metal mercury, which has one of the highest values known for any liquid at room temperature and pressure (see Fig. 10.3b).

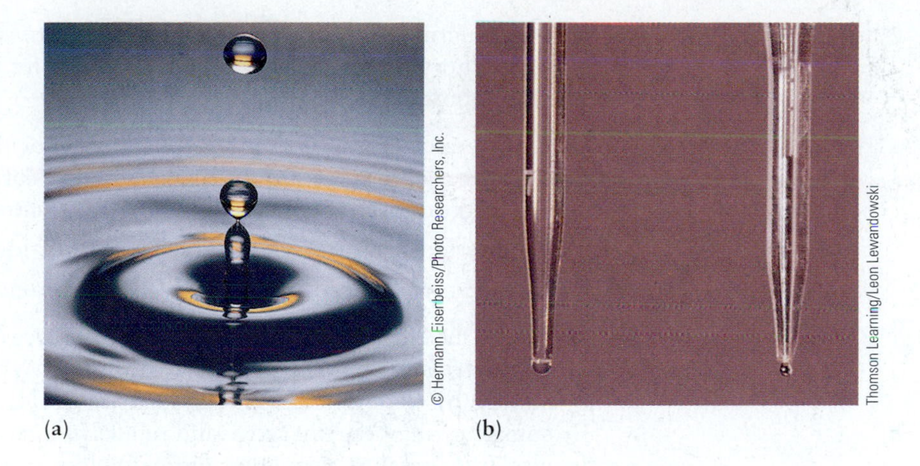

FIGURE 10.3 (a) Surface tension causes the spherical shape of the water droplet in this photograph, which was taken an instant after a drop of water hit the surface of a pool and bounced up, pulling with it a column of water. (b) The mercury drop at the dropper tip on the right is a nearly perfect sphere, whereas the water drop on the left sags slightly. This is evidence of the higher surface tension of the mercury, the drops of which resist the deforming pull of gravity more effectively than those of water.

Surface tension results from the intermolecular attractions among the molecules in a liquid (Fig. 10.4). Increasing the surface area of a liquid requires redistributing some of the molecules that were originally buried in the interior to positions at the enlarged boundaries. Molecules at the edges have no neighbors on one side and experience attractions only from molecules in the bulk of the liquid. Their potential energy is greater than it would be if they were in the interior. Thus, energy is required to increase the surface area of a liquid. Liquids such as water and mercury with high values of surface tension have particularly strong intermolecular attractions, as confirmed by measurement of other properties.

10.2 Intermolecular Forces: Origins in Molecular Structure

To provide a more quantitative explanation of the magnitudes of the properties of different materials, we must consider several types of intermolecular forces in greater detail than we gave to the Lennard–Jones model potential in Chapter 9. The Lennard–Jones potential describes net repulsive and attractive forces between molecules, but it does not show the origins of these forces. We discuss other intermolecular forces in the following paragraphs and show how they arise from molecular structure. Intermolecular forces are distinguished from *intramolecular* forces, which lead to the covalent chemical bonds discussed in Chapters 3 and 6. Intramolecular forces between atoms in the covalent bond establish and maintain

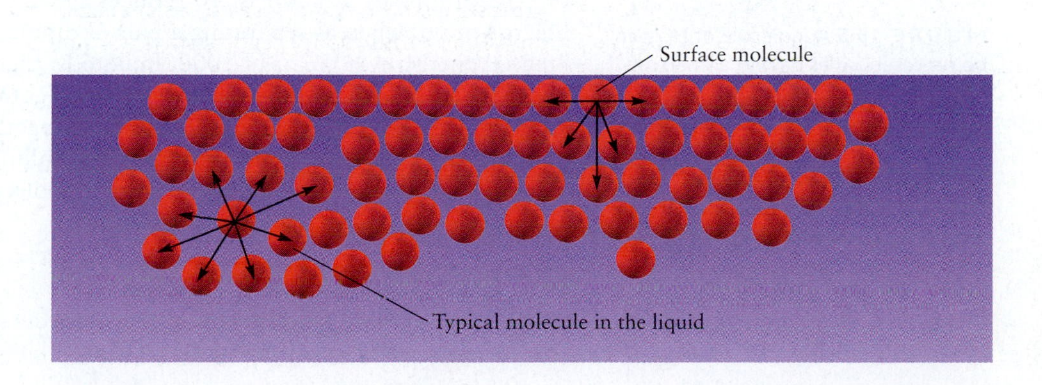

FIGURE 10.4 The intermolecular attractions acting on a molecule at the surface of a liquid pull it downward and to the sides but not upward. In the interior, a molecule is pulled more or less equally in all directions.

the structure of discrete molecules: They are strong, directional, and comparatively short ranged. Intermolecular forces differ from intramolecular forces in several important ways:

1. Intermolecular forces are generally weaker than covalent chemical bonds. For example, it takes 239 kJ to break 1 mol of Cl—Cl covalent bonds, but only 1.2 kJ to overcome 1 mol of Ar—Ar attractions.
2. Intermolecular forces are much less directional than covalent chemical bonds.
3. Intermolecular forces operate at longer range than covalent chemical bonds.

All intermolecular and intramolecular forces arise because matter is composed of electrically charged particles whose interactions with one another are all described by Coulomb's Law. Although we could, in principle, calculate the force by summing all of the attractive and repulsive interactions between the charged particles, it is useful to distinguish different classes of forces based on their strength, directionality, and range. The physical and chemical properties of liquids and solids can often be interpreted or even predicted by considering the types of intermolecular force that dominate in their internal structures.

Ion–Ion Forces

The units of organization in ionic solids and liquids are electrically charged entities, sometimes monatomic ions such as Na^+, Cl^-, and Ca^{2+}, and sometimes polyatomic ions such as NH_4^+ and SO_4^{2-}. The dominant interaction among these ions is the Coulomb force of electrostatic attraction or repulsion, which leads to the Coulomb potential described in Section 3.2. Ions of like charge repel one another, and ions of unlike charge attract one another. These **ion–ion forces** can be as strong as those in the covalent bond, and they are long ranged. The potential energy is proportional to R^{-1} and decreases much less rapidly with distance than do the strengths of other types of interactions. Ion–ion forces are not directional; each ion interacts equally strongly with neighboring ions on all sides. Ion–ion forces lead to the formation of ionic bonds through the Coulomb stabilization energy (see Section 3.6).

Dipole–Dipole Forces

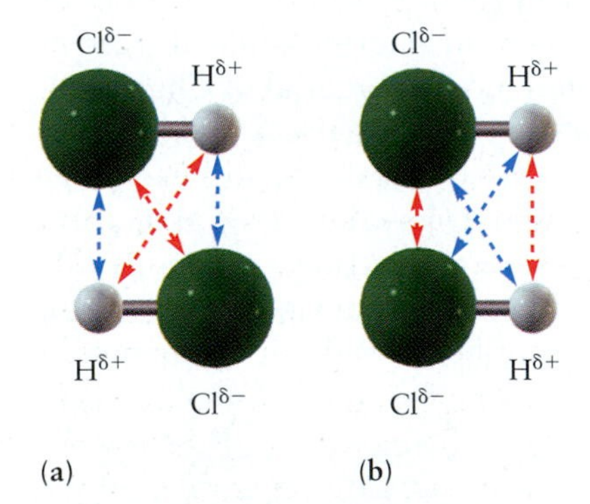

FIGURE 10.5 A molecule of HCl can be represented as having a small net negative charge on the Cl end, balanced by a small net positive charge on the H end. The forces between two HCl molecules depend on their orientations. (a) The oppositely charged ends (blue arrows) are closer than the ends with the same charge (red arrows). This gives a net attractive force. (b) Here, the opposite is true, and the net force is repulsive.

The dominant force between polar molecules is the **dipole–dipole force.** This is a second example of electrostatic forces that arise from interactions between fixed charges, in this case the magnitudes of the permanent dipole moments of polar molecules. As shown in Figure 10.5, these forces also depend on the orientations of the two molecules and can be either attractive or repulsive or zero. Random motions of polar molecules in gases and liquids lead to a variety of energetically favorable temporary dipole–dipole orientations. The potential energy between dipoles separated by the distance R falls off as R^{-3}. This potential decreases much more rapidly with separation than does the Coulomb potential between ions. Separating a pair of ions by a factor of 10 reduces the Coulomb potential energy by only a factor of 10, whereas separating a pair of dipoles by a factor of 10 reduces the potential energy by a factor of 1000. In liquids, thermal energy can overcome dipole–dipole attractions and disrupt favorable orientations; dipole–dipole interactions are too weak to hold molecules in a liquid together in a nearly rigid arrangement. Nonetheless, they are sufficiently strong to influence many physical properties, including boiling points, melting points, and molecular orientations in solids.

Ion–Dipole Forces

A third example of electrostatic forces occurs when a polar molecule is near an ion. The interaction between a polar solvent molecule, such as water, and a

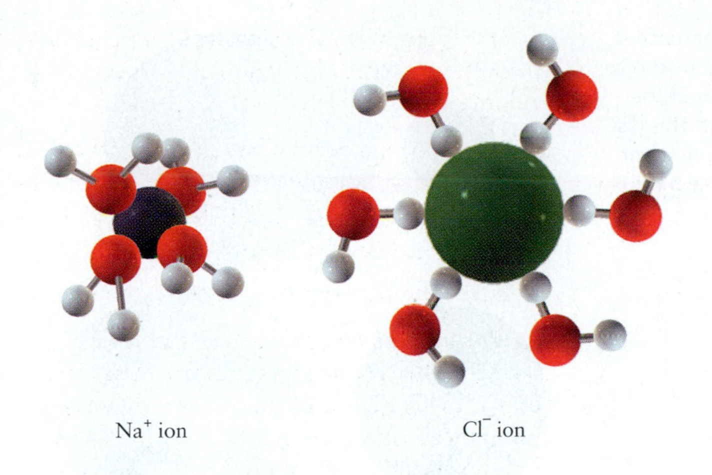

FIGURE 10.6 Solvation of ions in liquid water. The water molecules have dipole moments; thus, the oxygen (O) atoms bear small, negative charges, whereas the hydrogen (H) atoms bear small, positive charges. (a) Positive ions are attracted to neighboring water molecules in aqueous solution by ion–dipole forces. (b) Negative ions form hydrogen bonds with water, with a nearly linear bond from O to H to the anion.

dissolved ion is the most common case of ion–dipole interaction. Figure 10.6 shows dissolved Na^+ and Cl^- ions interacting with water dipoles. Positive ions are attracted by the negative end of the dipole and repelled by the positive end; thus, the ion is surrounded by a shell of water molecules whose oxygen (O) atoms are near the cation and whose hydrogen (H) atoms point outward into the solution. For many years it was believed that the opposite would be true for negative ions, that they would be surrounded by a shell of water dipoles whose H atoms were both near the anion. Since about 1980, neutron diffraction has been used to determine the distances between atoms in ionic aqueous solutions. A series of such studies has shown that the halide anion interacts with only one of the H atoms, and the atoms O—H—Cl lie nearly in a straight line. The other H atom points in a direction determined by the geometry of the water molecule. The solvation of the anion is not governed by ion–dipole forces. Rather, the O—H—Cl interaction is an example of the *hydrogen bond,* a special intermolecular force that occurs only in liquids. The hydrogen bond is discussed in the next section, and the solvation of ions is discussed more thoroughly in Chapter 11.

Induced Dipole Forces

The electrons in a nonpolar molecule or atom are distributed symmetrically, but the distribution can be distorted by an approaching electrical charge. An argon (Ar) atom has no dipole moment, but an approaching Na^+, with its positive charge, attracts the electrons on the side near it more strongly than those on the far side. By tugging on the nearby electrons harder, Na^+ induces a temporary dipole moment in the Ar atom (Fig. 10.7). The electron distribution of the nonpolar molecule is said to be *polarizable,* and the magnitude of the dipole moment induced measures the **polarizability** of the molecule. As long as the induced dipole is present, the interaction between molecules is similar to the ion–dipole case just described. **Induced dipole forces** also can be caused by a negative ion or by another dipole. These so-called induction forces differ from the electrostatic forces between permanent fixed charges such as ions or dipoles. Rather, they arise from interactions between the permanent charges or moments on one molecule and the induced moments, or the polarizability, of another molecule. These interactions are weak and are effective only at short range. The induced dipole moment closely tracks the motion of the charge or dipole moment of the inducing molecule. The induced dipole is dynamically correlated with the motion of the inducing molecule. A good way to study induced dipole forces quantitatively is to collide a beam of Na^+ ions and a beam of Ar atoms in vacuum, measure the energy and direction of their deflections, and deduce the correct potential function to explain these deflections.

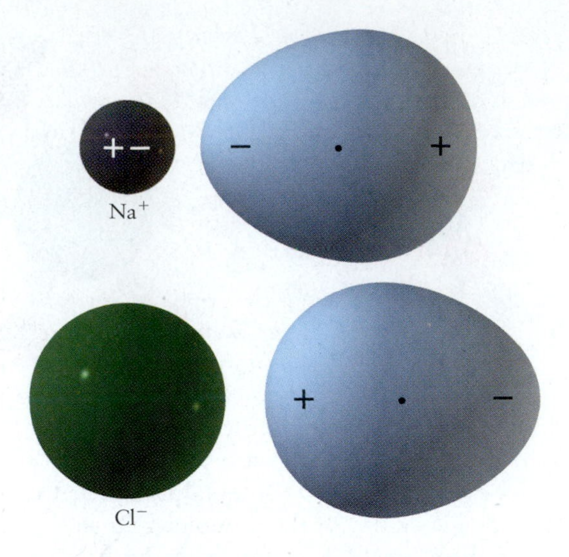

FIGURE 10.7 As an ion approaches an atom or molecule, its electrostatic field distorts the distribution of the outer electrons. The effect of this distortion is to create a dipole moment that exerts an attractive force back on the ion.

Induced Dipole–Induced Dipole Forces: London Dispersion Forces

Helium (He) atoms, like the atoms of the other noble gases, are electrically neutral and nonpolar, so none of the forces discussed so far explains the observed fact that there are attractions between He atoms. We know such attractions must exist, because helium becomes a liquid at 4.2 K and 1 atm. Attractions between neutral, nonpolar atoms or molecules arise from the **London dispersion forces** (often called van der Waals forces) that exist between all atoms and molecules. Dispersion forces are, in effect, a mutual interaction between the polarizable charge distributions on two separate molecules, and they are always attractive. Although the electron probability distribution around the molecule is described by the square of the wave function, dynamic motions of charge around the molecule can lead to an instantaneous, temporary dipole moment. Such a temporary dipole on one molecule will induce a temporary dipole in the other molecule. These transient, fluctuating dipoles attract one another in much the same way as do permanent dipoles. Figure 10.8 provides a simple view of the source of this interaction. The polarizability increases with the number of electrons in the atom or molecule. Heavier atoms or molecules interact more strongly by dispersion forces than do lighter ones because their electrons are located in shells farther from the nucleus. These electrons are less strongly bound than those of the lighter elements, because they are shielded from the full attraction of the nucleus by intervening

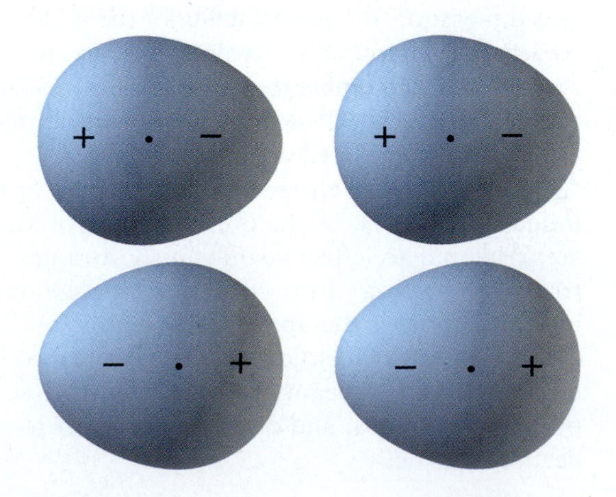

FIGURE 10.8 A fluctuation of the electron distribution on one atom induces a corresponding temporary dipole moment on a neighboring atom. The two dipole moments interact to give a net attractive force, called a "dispersion force."

electrons (see Sections 3.2 and 5.2). Consequently, they are more easily distorted by external fields of neighboring dipoles. Dispersion forces are always attractive and fall off as R^{-6}. These interactions are short ranged, much more so than dipole–dipole forces. Dispersion forces provide the attractive term in the Lennard–Jones potential (see description in Section 9.7).

Repulsive Forces

As atoms or molecules approach each other closely, **repulsive forces** come into play and can overcome the attractive forces considered so far. The source of these forces is the strong repulsion between the core (nonvalence) electron clouds when neighboring atoms are forced close to each other. This contribution is negligible until the distance between centers becomes small, at which point the repulsive energy increases rapidly as distance is reduced further. Two mathematical models are used to describe repulsive forces, although neither has a simple physical foundation to guide the choice of the parameters involved. The exponential form $Ae^{-R/\rho}$ is successful when A and ρ are chosen to fit experimental data such as compressibility measurements. The inverse power form R^{-n}, where n is quite large, is also successful. The choice $n = 12$ is widely used because, when combined with the $n = 6$ choice to describe attractive forces, the resulting Lennard–Jones potential reproduces the trends in experimental data over significant ranges. Regardless of mathematical form, this steep, repulsive interaction at extremely small distances justifies modeling atoms as hard, nearly incompressible spheres with characteristic dimensions called the **van der Waals radii.** This label honors the early contributions of Johannes van der Waals to the study of nonbonded interactions between molecules and their influence on the properties of materials. The minimum distance between molecules in a condensed phase is determined by the sum of the van der Waals radii of their atoms. Space-filling models and drawings are usually designed to approximate the van der Waals surface of molecules, which represents the distance of closest approach by neighboring molecules. Van der Waals radii for atoms are typically a few angstroms (Å).

EXAMPLE 10.1

State which attractive intermolecular forces are likely dominant in the following substances:

(a) $F_2(s)$

(b) $HBr(\ell)$

(c) $NH_4Cl(s)$

SOLUTION

(a) Molecules of F_2 are nonpolar, thus the predominant attractive forces between molecules in $F_2(s)$ come from dispersion.

(b) The HBr molecule has a permanent dipole moment. The predominant forces between molecules are dipole–dipole. Dispersion forces will also contribute to associations, especially because Br is a rather heavy atom.

(c) The ammonium ions are attracted to the chloride ions primarily by ion–ion forces.

Related Problems: 15, 16, 17, 18, 19, 20

Comparison of Potential Energy Curves

The relative strengths and effective ranges of several intermolecular forces are illustrated in Figure 10.9, which shows how the potential energy depends on the

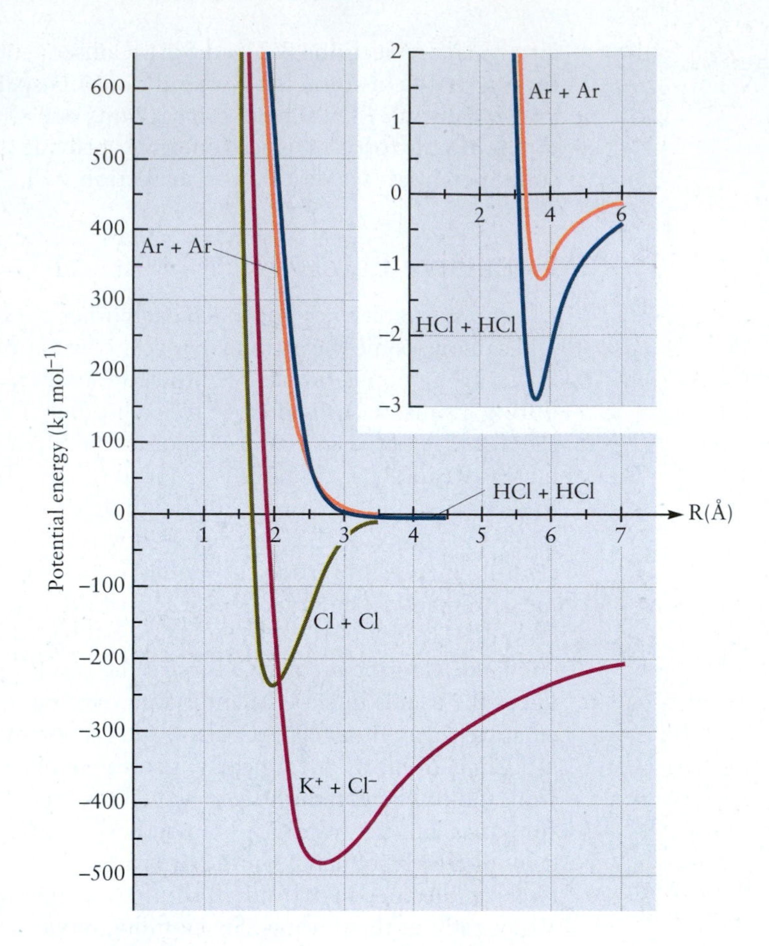

FIGURE 10.9 The potential energy of a pair of atoms, ions, or molecules depends on the distance between the members of the pair. Here, the potential energy at large separations (to the right side of the graph) is arbitrarily set to zero by convention (see Section 3.2). As pairs of particles approach each other, the potential energy becomes negative because attractive forces come into effect. The lowest point in each curve occurs at the distance where attractive and repulsive forces exactly balance. The relative potential energy values at these minima measure the relative strength of the attractive forces in the various cases illustrated. Note the shallow potential energy minimum for hydrogen chloride (HCl) and argon (Ar). (inset) The inset shows these same two curves with the vertical scale expanded by a factor of 100. (The HCl–HCl curve was computed for the relative orientations of Fig. 10.5a.)

intermolecular separation (center-to-center distance) for several pairs of ions, atoms, and molecules. The potentials illustrated here include Coulomb (R^{-1}), dipole–dipole (R^{-3}), dispersion (R^{-6}), and repulsive (R^{-12}) potentials. The species shown in Figure 10.9 were chosen so that the interacting atoms, ions, or molecules have the same number of electrons (Ar, Cl^-, K^+, HCl). For comparison, the covalent bond (*intra*molecular force) for Cl_2 is also shown. The ion–ion interaction of K^+ with Cl^- is the strongest (stronger even than the covalent interaction in Cl_2), followed by the interaction between two HCl molecules (dipole–dipole and dispersion) and the Ar–Ar interaction (dispersion only). The key points illustrated here (and detailed in the caption for Fig. 10.9) are the dramatically different depths of the wells (several orders of magnitude), as well as the distinctly different distances at which the minima occur.

The Shapes of Molecules and Electrostatic Forces

The potential energy diagrams in Figure 10.9 depend only on the distance between the two species. Interactions between complex molecules also depend strongly on their relative orientations, so we need a three-dimensional generalization of the potential energy diagram to describe these interactions more fully. This need could be met by constructing a potential energy *surface* where the interaction energy is plotted as a function of all three spatial coordinates that locate the center of one molecule relative to the center of the other. Because quantitative calculations of intermolecular forces are not yet available from quantum mechanics, we use approximate representations to describe the influence of shape and orientation as two molecules approach one another. One such approximate representation is the **electrostatic potential energy map,** which shows the shape and size of the molecule, as well as the sign and magnitude of the electrostatic potential at the "surface" of the molecule.

The electrostatic potential energy map for a molecule combines information from two different sources into one representation. The size and shape of the molecule come from the spatial distribution of the electron density represented as an isosurface. The electrostatic potential energy that a positive test charge would experience is indicated at each point on that isosurface. Let's construct each of these pieces in turn.

Isosurfaces of electron density are obtained from the probability density isosurfaces for molecules described in Chapter 6. These are surfaces in three-dimensional space that include all the points at which ψ^2 has a particular value. The value of electron density chosen to define the isosurface is selected by some definite, though arbitrary, criterion. There is broad acceptance of a standard density of 0.002 $e/(a_0)^3$, where a_0 is the Bohr radius. This value is thought to best represent the "sizes" and shapes of molecules because it corresponds to the van der Waals atomic radii discussed earlier in the context of repulsive forces. These are the same dimensions depicted in space-filling models of molecules.

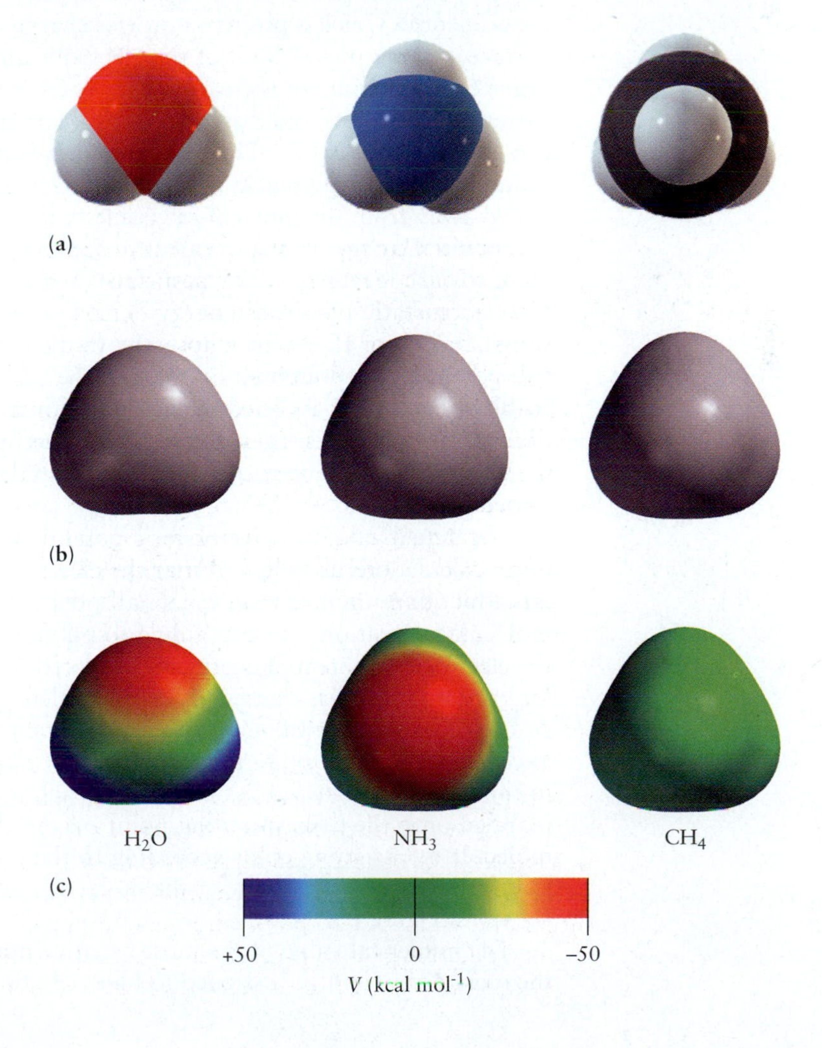

FIGURE 10.10 (a) Space-filling models, (b) 0.002 $e/(a_0)^3$ electron density isosurfaces, and (c) electrostatic potential energy surfaces for water, ammonia, and methane.

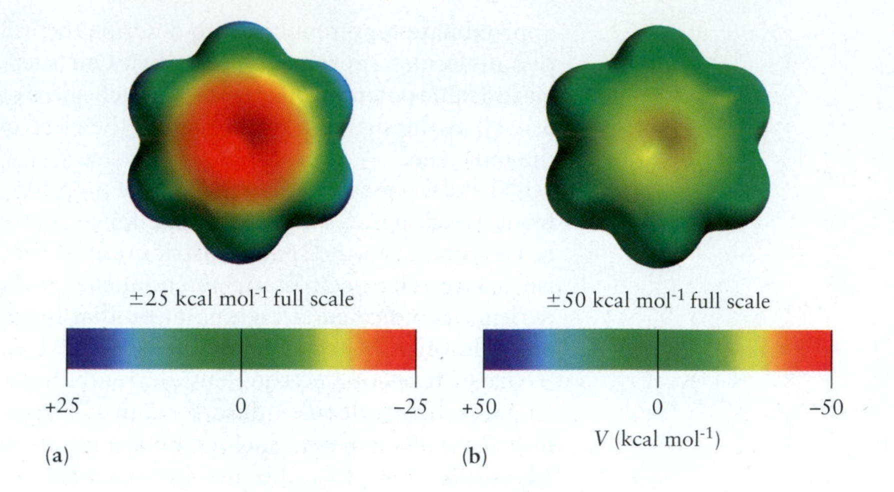

FIGURE 10.11 Electrostatic potential energy surfaces for benzene plotted at ±25 (a) and ±50 kcal mol^{-1} full scale (b), respectively.

Figures 10.10a and b show, respectively, space-filling models and electron density isosurfaces plotted at 0.002 $e/(a_0)^3$ for water, ammonia, and methane. The electron densities plotted here include all of the electrons in the molecule. They are calculated using state-of-the-art *ab initio* quantum chemical methods (see discussion in Chapter 6).

To understand how a value of the electrostatic potential energy can be associated with each point on the electron density isosurface, we imagine a thought experiment in which a positive unit test charge crawls like an insect over this isosurface, interacting with all of the electrons and the nuclei of the molecule as it visits every position on the isosurface. The magnitude of the test charge must be extremely small to avoid distorting and polarizing the electron density of the molecule (see Figure 10.7). The test particle is attracted to the molecule at those points where it experiences negative electrostatic potential energy, and it is repelled away from the molecule at points where it experiences positive electrostatic potential energy. It is convenient to summarize these explorations by assigning color to each location on the isosurface on the basis of the sign and magnitude of the electrostatic potential energy experienced by the test charge. The conventional choice for the color map is the visible spectrum, in which the most negative potential is represented by red and the most positive potential is represented by blue. The colors assigned to each point on the electron density isosurface vary continuously between these extremes on the basis of the sign and magnitude of the electrostatic potential energy. The potential energy values near zero are represented by green.

The actual values of electrostatic potential energy are obtained through computer calculations as follows. After the electron density isosurface has been calculated by quantum mechanics, a small positive test charge is moved around the molecule at locations corresponding to points on the isosurface. At each location the electrostatic potential energy is calculated using Coulomb's law to describe the interaction of the test charge with each nucleus and each electron in the molecule. At each location, the value of the electrostatic potential energy represents the balance between attraction of the test charge by electrons and repulsion by the nuclei in the molecule. The value obtained at each location of the test particle is then mapped onto the corresponding point on the electron density isosurface for the molecule by assigning color according to the color map. The electrostatic potential energy maps calculated for the molecules shown in Figures 10.10a and b are shown in Figure 10.10c.

The numerical value of the most negative potential value (assigned to red) and the most positive value (assigned to blue) can be adjusted to emphasize features of interest in a particular study. Figure 10.11 illustrates this fact by showing the electrostatic potential energy maps for benzene plotted for the ranges ± 25 and ± 50 kcal mol^{-1}. In the second case, where red is assigned to -50 kcal mol^{-1}, the values for the region in the center of the benzene ring appear in the yellow-green range and are difficult to distinguish from the surrounding regions. Maps for different compounds plotted using the same energy range can be compared immediately, and trends in the behavior of these molecules toward approaching charged particles will be readily apparent.

The electrostatic potential energy map for a given molecule, called the "target" molecule, shows the spatial shape of the electrostatic field around the molecule, and so it can be used to predict how the target molecule influences the motion of charged particles as they approach it. The images show at a glance which portions of the molecule are most likely to attract or repel a proton (see Figure 10.10c). To an approaching proton, those regions colored red act as if they were three-dimensional attractive wells (valleys), whereas those colored blue act as if they were three-dimensional repulsive walls (mountains). Positive ions will be attracted to the oxygen end of the water molecule and to the nitrogen end of the ammonia molecule; the opposite is true for negative ions. The interaction of a positive or a negative ion with methane is weak and shows no pronounced directionality. Comparing Figure 10.10c with Figure 10.11b shows that a positive ion experiences greater attraction to the oxygen end of a water molecule and to the nitrogen end of an ammonia molecule than to the center of a benzene molecule.

Electrostatic potential energy maps can be used to identify reactive sites on molecules. Locations with large negative values of the electrostatic potential are relatively rich in electron density, and those with large positive values are relatively depleted in electron density. These maps are now widely used in organic chemistry to predict patterns of reactivity for electrophilic (electron-loving) and nucleophilic (proton-loving) molecules and to explain how the presence of different functional groups in the molecule can affect these patterns. These methods are effective aids in identifying sites for chemical reactivity in more complicated molecules, including those of biological interest. They are widely used in molecular modeling simulations of drug design.

10.3 Intermolecular Forces in Liquids

The same intermolecular forces that make gases deviate from ideal behavior (see Sections 9.7 and 10.2) are responsible for the existence of solids and liquids. At very high temperatures, these forces are negligible because the high kinetic energy of the molecules disrupts all possible attractions; all materials are gaseous at sufficiently high temperatures. At lower temperatures, where materials are in the liquid state, molecules are close together and the details of the intermolecular potential energy determine their properties. Section 10.2 describes the influence of molecular structure on the intermolecular potential energy. This section surveys the correlation between the properties of liquids and the structure of their constituent molecules. Special attention is given to the unusual properties of water.

Substances with strong attractive intermolecular forces tend to remain liquids at higher temperatures than those with weaker intermolecular forces; they have higher normal boiling points, T_b. Ionic liquids generally have the strongest attractions, because of the Coulomb interaction among charged ions, and thus have high boiling points. Molten NaCl, for example, boils at 1686 K under atmospheric

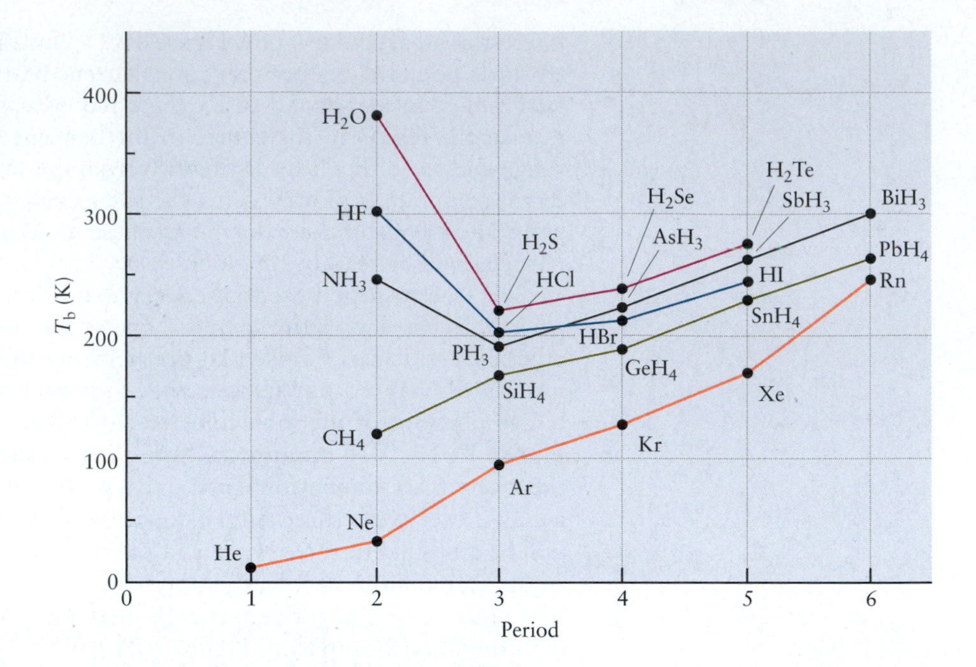

FIGURE 10.12 Trends in the boiling points of hydrides of some main-group elements and the noble gases.

pressure. At the opposite extreme, the normal boiling point of helium is only 4.2 K. Within a series of related compounds, those of higher molar mass tend to have higher normal boiling points. This trend arises from the increased polarizability of the heavier compounds, not from the increased mass per se. Progressing from helium to xenon, normal boiling points increase (Fig. 10.12), as do the strengths of the attractive dispersion forces among the noble gases. These forces, represented by the well depth ε in Table 9.4, arise from the polarizability of the atoms.

Between the noble-gas and ionic liquids falls a class of liquids called **polar liquids.** In liquid HCl, the molecules arrange themselves to the greatest extent possible with neighboring dipoles oriented to minimize the dipole–dipole potential energy. As described in Section 10.2, the dipole–dipole intermolecular forces in such polar liquids are weaker than the ion–ion Coulombic forces in ionic liquids but stronger than the dispersion forces in **nonpolar liquids** such as N_2. These three forces operate respectively between molecules in which the bonding is polar covalent, fully ionic, and fully covalent. As shown in Figure 10.12, HCl has a higher boiling point than argon (a nonpolar fluid of atoms with nearly the same molar mass) because of its polar nature. The magnitudes of the intermolecular forces in HCl and argon are compared explicitly in Figure 10.9.

Hydrogen Bonds

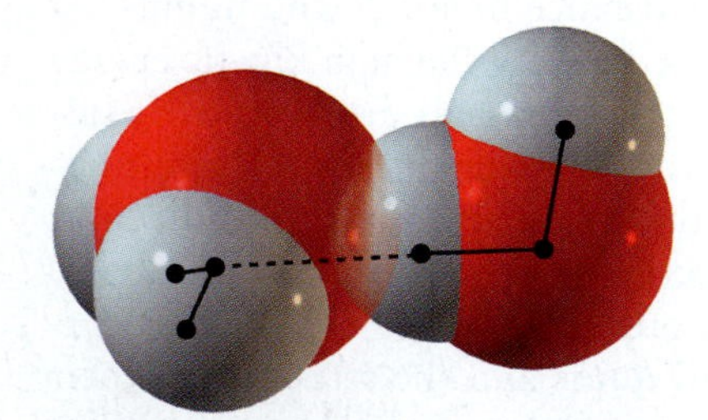

FIGURE 10.13 A single hydrogen bond between water molecules forms a dimer. This bond is far weaker than a covalent bond but still strong enough to resist dissociation at room temperature. The shared hydrogen (H) atom at the center approaches the neighboring oxygen (O) atom quite closely.

Figure 10.12 shows the normal boiling points of several series of hydrides, in which the boiling point increases with increasing molar mass in a series of related compounds. The dramatic deviations from these systematic trends shown by HF, NH_3, and especially H_2O indicate the strength and importance of the special type of bond that is common to these cases, a **hydrogen bond.** Such a bond forms when an H atom bonded to an O, N, or F atom (highly electronegative atoms) also interacts with the lone electron pair of another such atom nearby. Figure 10.13 shows the interaction of a pair of water molecules to form a dimer in the gas phase. The hydrogen bond that forms is weaker than an ordinary O—H covalent bond, but the interaction is significantly stronger than most other intermolecular interactions. Like most hydrogen bonds, that in water is nearly linear but asymmetric, with the H atom closer to and more strongly bound to one of the O atoms. It is indicated as O—H · · · O.

Water is a polar molecule, like HCl and H_2S. The water molecule is bent, as predicted by valence shell electron-pair repulsion theory and confirmed by experiment, and the orientation of its dipole moment (positive end toward the H atoms and negative end toward the O atom) has been related to its structure in Section 3.7. In the liquid, these molecules orient themselves in directions that minimize the potential energy between them; consequently, H atoms on one molecule are close to O atoms on neighboring molecules. The H atom in a bond such as O—H is surrounded by a relatively low density of negative charge because, unlike all other elements, it has no electrons other than valence electrons. As a result, it can approach close to the lone-pair electrons on a neighboring O atom, causing a strong electrostatic (Coulomb) interaction between the two. In addition, a small amount of covalent bonding arises from the sharing of electrons between the two O atoms and the intervening H atom. These effects combine to make the interaction unusually strong. For the same reason, hydrogen bonds form with anions in aqueous solution (see Fig. 10.6)

Special Properties of Water

Water makes up about 0.023% of the total mass of the earth. About 1.4×10^{21} kg of it is distributed above, on, and below the earth's surface. The volume of this vast amount of water is about 1.4 billion km^3. Most of the earth's water (97.7%) is contained in the oceans, with about 1.9% in the form of ice or snow and most of the remainder (a small fraction of the total) available as freshwater in lakes, rivers, underground sources, and atmospheric water vapor. A small but important fraction is bound to cations in certain minerals, such as clays and hydrated crystalline salts. More than 80% of the surface of the earth is covered with water—as ice and snow near the poles, as relatively pure water in lakes and rivers, and as a salt solution in the oceans.

The unusual properties of water, which come from its network of hydrogen bonds, have profound effects on life on earth. Figure 10.12 compares the boiling points of water and hydrides that lack hydrogen bonds. An extrapolation of the trends from the latter compounds would give a boiling point for "water without hydrogen bonds" near 150 K (−123°C). Life as we know it would not be possible under these circumstances.

If all possible hydrogen bonds form in a mole (N_A molecules) of pure water, then every oxygen atom is surrounded by four H atoms in a tetrahedral arrangement: its own two and two from neighboring molecules. This tetrahedral arrangement forms a three-dimensional network with a structure similar to that of diamond or SiO_2. The result is an array of interlocking six-membered rings of water molecules (Fig. 10.14) that manifests itself macroscopically in the characteristic sixfold symmetry of snowflakes.

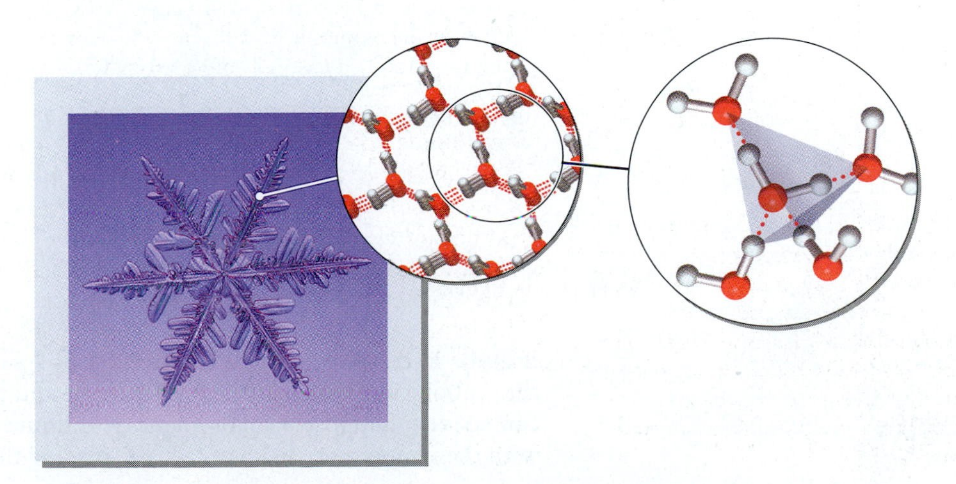

FIGURE 10.14 The structure of ice is quite open. Each water molecule has only four nearest neighbors with which it interacts by means of hydrogen bonds (red dashed lines).

(© Edward Kinsman/Photo Researchers, Inc.)

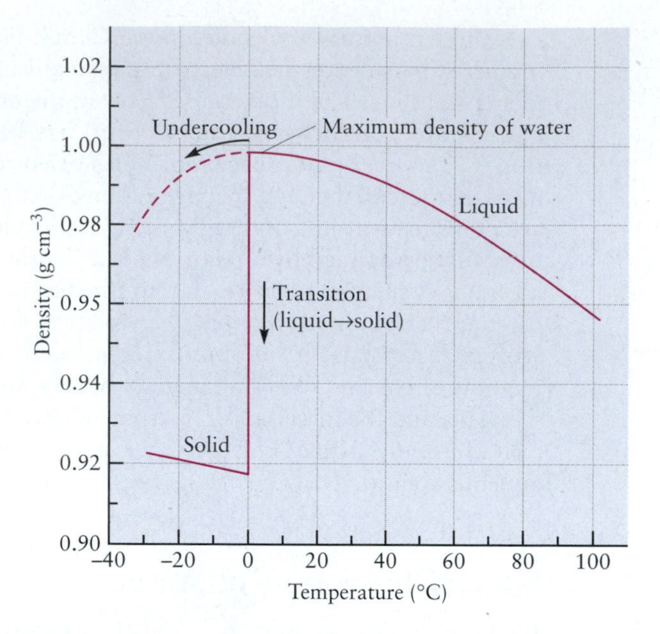

FIGURE 10.15 The density of water rises to a maximum as it is cooled to 3.98°C, then starts to decrease slowly. Undercooled water (water chilled below its freezing point but not yet converted to ice) continues the smooth decrease in density. When liquid water freezes, the density drops abruptly.

The density of water reaches its maximum at 4°C (Fig. 10.15), and it expands on freezing. This unusual behavior, which is seen in few other liquids, also is caused by hydrogen bonds. When ice melts, some of the hydrogen bonds that maintain the open structure shown in Figure 10.14 break and the structure partially collapses, producing a liquid with a smaller volume (higher density). The reverse process, a sudden expansion of water on freezing, can cause bursting of water pipes and freeze/thaw cracking of concrete. Such expansion also has many beneficial effects. If ice were denser than water, the winter ice that forms at the surface of a lake would sink to the bottom and the lake would freeze from the bottom up. Instead, the ice remains at the surface and the water near the bottom achieves a stable wintertime temperature near 4°C, which allows fish to survive.

EXAMPLE 10.2

Predict the order of increase in the normal boiling points of the following substances: F_2, HBr, NH_4Cl, and HF.

SOLUTION

As an ionic substance, NH_4Cl should have the highest boiling point of the four (measured value: 520°C). HF should have a higher boiling point than HBr, because its molecules form hydrogen bonds (see Fig. 10.10) that are stronger than the dipolar interactions in HBr (measured values: 20°C for HF, −67°C for HBr). Fluorine, F_2, is nonpolar and contains light atoms, and thus should have the lowest boiling point of the four substances (measured value: −188°C).

Related Problems: 23, 24

10.4 Phase Equilibrium

Liquids and solids, like gases, are **phases**—samples of matter that are uniform throughout in both chemical constitution and physical state. Two or more phases can coexist. Suppose a small quantity of liquid water is put in an evacuated flask, with the temperature held at 25°C by placing the system in a constant-temperature

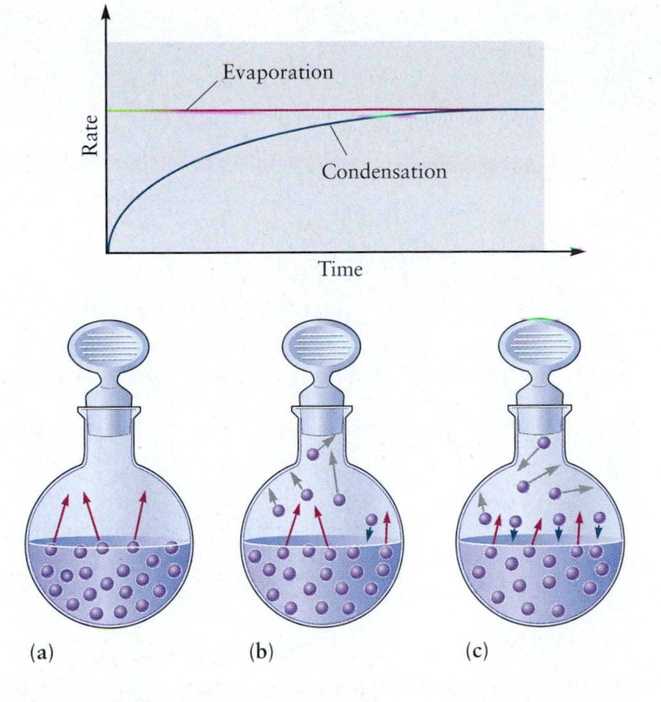

FIGURE 10.16 Approach to equilibrium in evaporation and condensation. Initially, the pressure above the liquid is very low, and many more molecules leave the liquid surface than return to it. As time passes, more molecules fill the gas phase until the equilibrium vapor pressure, P_{vap}, is approached; the rates of evaporation and condensation then become equal.

bath. A pressure gauge is used to monitor changes in the pressure of water vapor inside the flask. Immediately after the water enters the flask, the pressure of water vapor begins to rise from zero. It increases with time and gradually levels off at a value of 0.03126 atm, which is the **vapor pressure** of water at 25°C. The contents of the flask have reached **equilibrium,** a condition in which no further changes in macroscopic properties occur as long as the system remains isolated. This passage toward equilibrium is a spontaneous process, occurring in a closed system without any external influence. If some of the water vapor that has formed is removed, additional water evaporates from the liquid to reestablish the same vapor pressure, $P_{vap}(H_2O) = 0.03126$ atm.

What is happening on a microscopic scale to cause this spontaneous movement of the system toward equilibrium? According to the kinetic theory, the molecules of water in the liquid are in a constant state of thermal motion. Some of those near the surface are moving fast enough to escape the attractive forces holding them in the liquid; this process of **evaporation** causes the pressure of the water vapor to increase. As the number of molecules in the vapor phase increases, the reverse process begins to occur: Molecules in the vapor strike the surface of the liquid, and some are captured, leading to **condensation.** As the pressure of the gas increases, the rate of condensation increases until it balances the rate of evaporation from the surface (Fig. 10.16). Once this occurs, there is no further net flow of matter from one phase to the other; the system has reached **phase equilibrium,** characterized by a particular value of the water vapor pressure. Water molecules continue to evaporate from the surface of the liquid, but other water molecules return to the liquid from the vapor at an equal rate. A similar phase equilibrium is established between an ice cube and liquid water at the freezing point.

The vapor pressure of the water is independent of the size and shape of the container. If the experiment is duplicated in a larger flask, then a greater *amount* of water evaporates on the way to equilibrium, but the final pressure in the flask at 25°C is still 0.03126 atm as long as some liquid water is present. If the experiment is repeated at a temperature of 30.0°C, everything happens as just described, except that the pressure in the space above the water reaches 0.04187 atm. A higher temperature corresponds to a larger average kinetic energy for the water molecules. A new balance between the rates of evaporation and condensation is struck, but at a higher vapor pressure. The vapor pressure of water, and of all other substances, increases with rising temperature (Fig. 10.17; Table 10.3).

TABLE 10.3 Vapor Pressure of Water at Various Temperatures

Temperature (°C)	Vapor Pressure (atm)
15.0	0.01683
17.0	0.01912
19.0	0.02168
21.0	0.02454
23.0	0.02772
25.0	0.03126
30.0	0.04187
50.0	0.12170

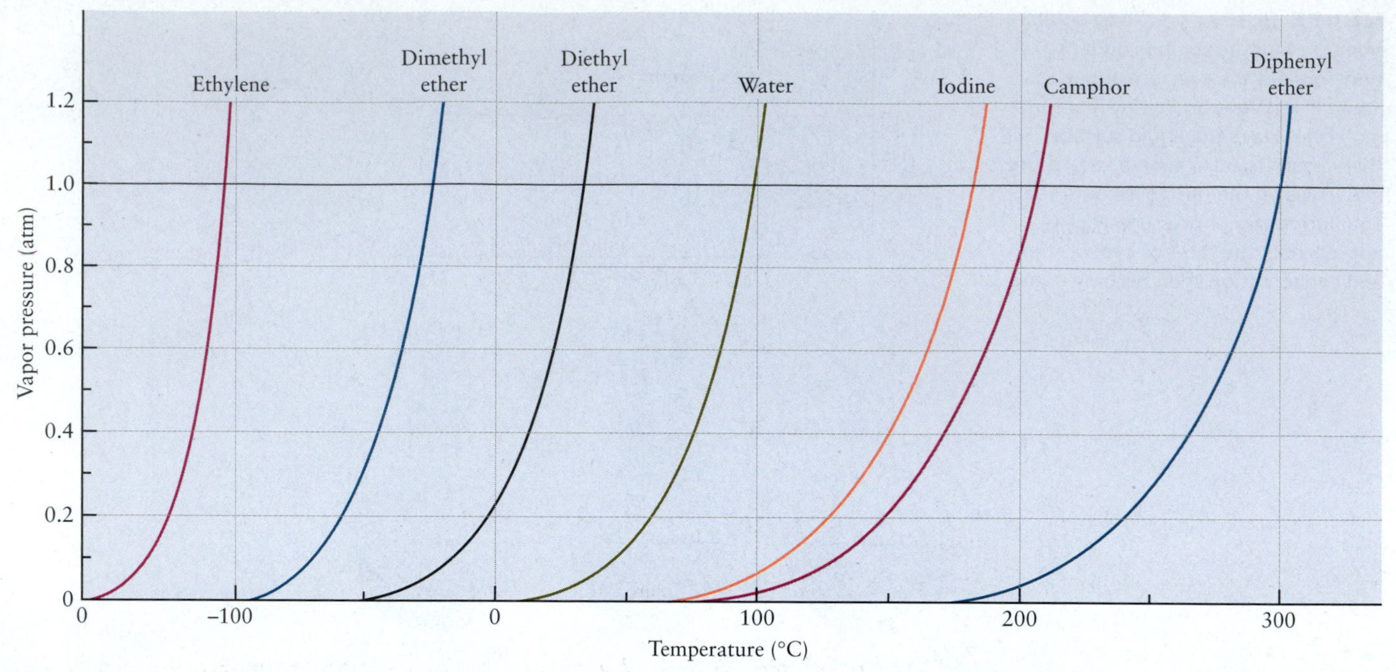

FIGURE 10.17 The vapor pressure of a solid or liquid depends strongly on temperature. The temperature at which the vapor pressure becomes 1 atm defines the normal boiling point of a liquid and the normal sublimation point of a solid.

Phase equilibrium is a *dynamic* process that is quite different from the static equilibrium achieved as a marble rolls to a stop after being spun into a bowl. In the equilibrium between liquid water and water vapor, the partial pressure levels off, not because evaporation and condensation stop, but because at equilibrium their rates become the same. The properties of a system at equilibrium are independent of the direction from which equilibrium is approached, a conclusion that can be drawn by observing the behavior of the liquid-vapor system. If we inject enough water vapor into the empty flask so that initially the pressure of the vapor is *above* the vapor pressure of liquid water, $P_{vap}(H_2O)$, then liquid water will condense until the same equilibrium vapor pressure is achieved (0.03126 atm at 25°C). Of course, if we do not use enough water vapor to exceed a pressure of 0.03126 atm, all the water will remain in the vapor phase and two-phase equilibrium will not be reached.

The presence of water vapor above an aqueous solution has an important practical consequence. If a reaction in aqueous solution generates gases, these gases are "wet," containing water vapor at a partial pressure given by the equilibrium vapor pressure of water at the temperature of the experiment. The amount of gas generated is determined not by the total pressure but by the partial pressure of the gas. Dalton's law (see Section 9.4) must be used to subtract the partial pressure of water as listed in Table 10.3. This correction is significant in quantitative work.

10.5 Phase Transitions

Suppose 1 mol of gaseous sulfur dioxide is compressed at a temperature fixed at 30.0°C. The volume is measured at each pressure, and a graph of volume against pressure is constructed (Fig. 10.18). At low pressures, the graph shows the inverse dependence ($V \propto 1/P$) predicted by the ideal gas law. As the pressure increases, deviations appear because the gas is not ideal. At this temperature, attractive forces dominate; therefore, the volume falls below its ideal gas value and approaches 4.74 L (rather than 5.50 L) as the pressure approaches 4.52 atm.

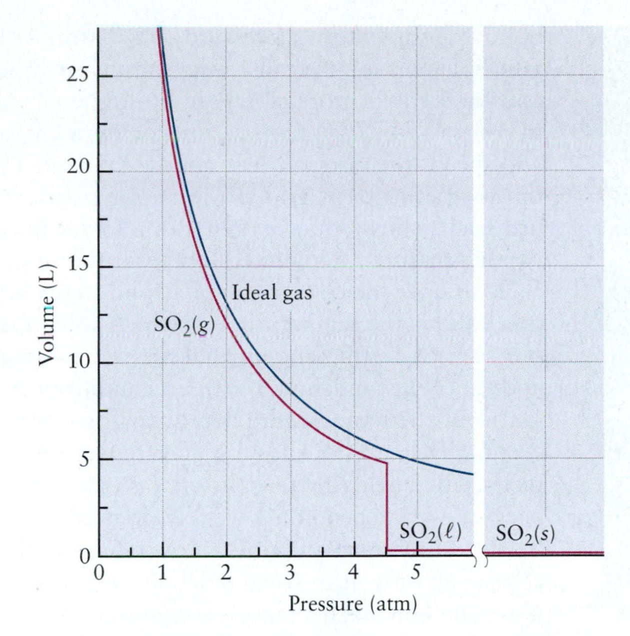

FIGURE 10.18 As 1 mol SO_2 is compressed at a constant temperature of 30°C, the volume at first falls somewhat below its ideal gas value. Then, at 4.52 atm, the volume decreases abruptly as the gas condenses to a liquid. At a much higher pressure, a further transition to the solid occurs.

For pressures up to 4.52 atm, this behavior is quite regular and can be described by the van der Waals equation. At 4.52 atm, something dramatic occurs: The volume decreases abruptly by a factor of 100 and remains small as the pressure is increased further. What has happened? The gas has been liquefied solely by the application of pressure. If the compression of SO_2 is continued, another abrupt (but small) change in volume will occur as the liquid freezes to form a solid.

Condensed phases also arise when the temperature of a gas is reduced at constant pressure. If steam (water vapor) is cooled at 1 atm pressure, it condenses to liquid water at 100°C and freezes to solid ice at 0°C. Liquids and solids form at low temperatures once the attractive forces between molecules become strong enough to overcome the kinetic energy of random thermal motion.

Six **phase transitions** occur among the three states of matter (Fig. 10.19). Solids typically melt to give liquids when they are heated, and liquids boil to give gases. **Boiling** is an extension of evaporation, in which the vapor escapes from the surface only. In boiling, gas bubbles form actively throughout the body of a liquid, and then rise to escape at the surface. Only when the vapor pressure of a liquid exceeds the external pressure can the liquid start to boil. The **boiling point** is the temperature at which the vapor pressure of a liquid equals the external pressure. The external pressure influences boiling points quite strongly; water boils at 25°C if the external pressure is reduced below 0.03126 atm (recall that the vapor pressure of water at 25°C is just this number) but requires a temperature of 121°C to

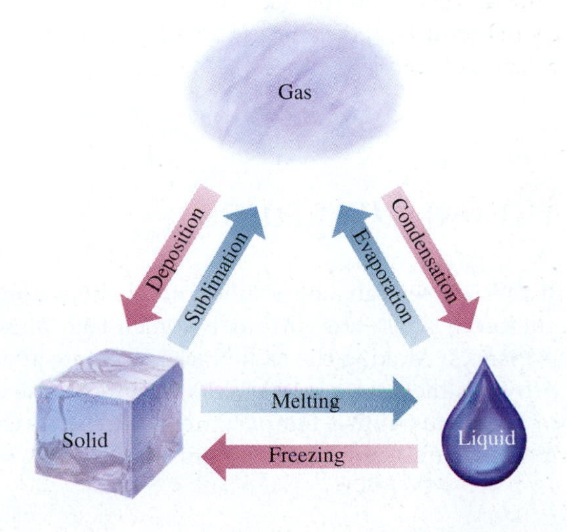

FIGURE 10.19 Direct transitions among all three states of matter not only are possible but are observed in everyday life.

(a)

(b)

FIGURE 10.20 When sugar (a) is heated, it melts and simultaneously decomposes to a dark-colored caramelized mixture (b).

boil under an external pressure of 2.0 atm. At high elevations, the pressure of the atmosphere is lower than 1 atm; thus, water boils at a temperature less than 100°C and food cooks more slowly in boiling water than it would at a lower elevation. In contrast, the use of a pressure cooker increases the boiling temperature of water and speeds the rate at which food is cooked. The **normal boiling point** is defined as the temperature at which the vapor pressure of the liquid equals 1 atm. Figure 10.12 shows that, in general, a lower normal boiling point implies a higher vapor pressure at any fixed temperature (and, therefore, a more volatile liquid).

Melting is the conversion of a solid to the liquid state. The **normal melting point** of a solid is the temperature at which solid and liquid are in equilibrium under a pressure of 1 atm. The normal melting point of ice is 0.00°C, thus liquid water and ice coexist indefinitely (are in equilibrium) at this temperature at a pressure of 1 atm. If the temperature is reduced by even a small amount, then all the water eventually freezes; if the temperature is raised infinitesimally, all the ice eventually melts. The qualifying term *normal* is often omitted in talking about melting points because they depend only weakly on pressure.

It is sometimes possible to overshoot a phase transition, with the new phase appearing only after some delay. An example is the **superheating** of a liquid. Liquid water can reach a temperature somewhat above 100°C if heated rapidly. When vaporization of a superheated liquid does occur, it can be quite violent, with liquid thrown out of the container. Boiling chips (pieces of porous fired clay) may be added to the liquid to avoid this superheating in the laboratory. They help initiate boiling as soon as the normal boiling point is reached, by providing sites where gas bubbles can form. Heating water in a microwave oven in a very clean container can also lead to superheating, and there have been reports of violent boiling resulting in injury as the container is removed from the oven. You should carefully monitor heating times when using a microwave oven to heat water for making coffee or tea. **Supercooling** of liquids below their freezing points is also possible. In careful experiments, supercooled liquid water has been studied at temperatures below −30°C (at atmospheric pressure).

Many materials react chemically when heated, before they have a chance to melt or boil. Substances whose chemical identities change before their physical state changes do not have normal melting or boiling points. For example, sucrose (table sugar) melts but quickly begins to darken and eventually chars (Fig. 10.20). Temperatures high enough to overcome the intermolecular attractions in sugar are also sufficient to break apart the sugar molecules themselves.

Intermolecular forces exert strong influences on phase transitions. Data presented in Section 10.3 illustrate the trend that the normal boiling point in a series of liquids increases as the strength of intermolecular forces in the liquids increases. The stronger the intermolecular attractions in a liquid, the lower its vapor pressure at any temperature and the higher its temperature must be raised to produce a vapor pressure equal to 1 atm. Melting points depend more strongly on molecular shapes and on the details of the molecular interactions than do boiling points. Consequently, their variation with the strength of the attractive forces is less systematic.

10.6 Phase Diagrams

If the temperature of a substance is held constant and the applied pressure is changed, phase transitions between two phases will be observed at particular pressures. Making the same measurements at a number of different temperatures provides the data necessary to draw the **phase diagram** for that substance—a plot of pressure against temperature that shows the stable state for every pressure–temperature combination. Figure 10.21 shows a sketch of the phase diagram for

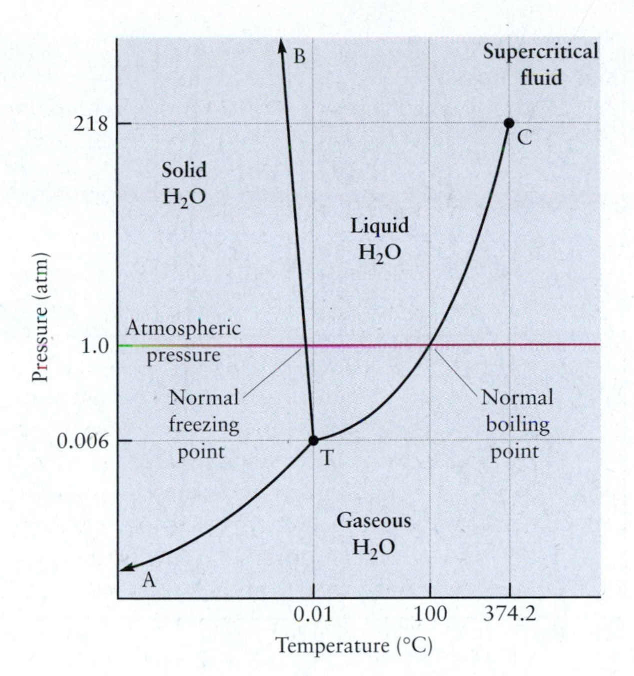

FIGURE 10.21 Phase diagram for water (the pressures and temperatures are not drawn to scale).

water. A great deal of information can be read from such diagrams. For each substance there is a unique combination of pressure and temperature, called the **triple point** (marked "T"), at which the gas, liquid, and solid phases coexist in equilibrium. Extending from the triple point are three lines, each denoting the conditions for the coexistence of two phases at equilibrium. Along the line TA, solid and gas are in equilibrium; along TB, solid and liquid; and along TC, liquid and gas. The regions bounded by these lines represent conditions where only one phase exists.

The gas–liquid coexistence curve extends upward in temperature and pressure from the triple point. This line, stretching from T to C in the phase diagrams, is the vapor pressure curve of the liquid substance, portions of which were shown in Figure 10.17. The gas–liquid coexistence curve does not continue indefinitely, but instead terminates at the **critical point** (point C in Fig. 10.21). Along this coexistence curve there is an abrupt, discontinuous change in the density and other properties from one side to the other. The differences between the properties of the liquid and the gas become smaller as the critical point is approached and disappear altogether at that point. If the substance is placed in a closed container and is gradually heated, a **meniscus** is observed at the boundary between liquid and gas (Fig. 10.22); at the critical point, this meniscus disappears. For pressures above the critical pressure (218 atm for water), it is no longer possible to identify a particular state as gas or liquid. A substance beyond its critical point is called a **supercritical fluid** because the term *fluid* includes both gases and liquids.

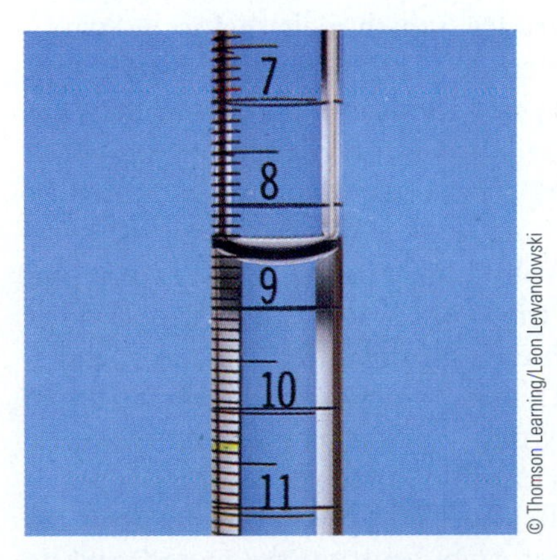

FIGURE 10.22 When a gas and liquid coexist, the interface between them is clearly visible as a meniscus. The meniscus is useful for reading the volume of a liquid in a buret. It disappears as the critical point is reached.

The liquid–solid coexistence curve does not terminate as the gas–liquid curve does at the critical point, but continues to indefinitely high pressures. In practice, such a curve is almost vertical because large changes in pressure are necessary to change the freezing temperature of a liquid. For most substances, this curve inclines slightly to the right (Figs. 10.23a, b): An increase in pressure increases the freezing point of the liquid. In other words, at constant temperature, an increase in pressure leads to the formation of a phase with higher density (smaller volume), and for most substances, the solid is denser than the liquid. Water and a few other substances are anomalous (see Fig. 10.23c); for them, the liquid–solid coexistence curve slopes up initially to the *left,* showing that an increase in pressure causes the solid to melt. This anomaly is related to the densities of the liquid and solid phases: Ice is less dense than water (which is why ice cubes float on water), so when ice is compressed at 0°C, it melts.

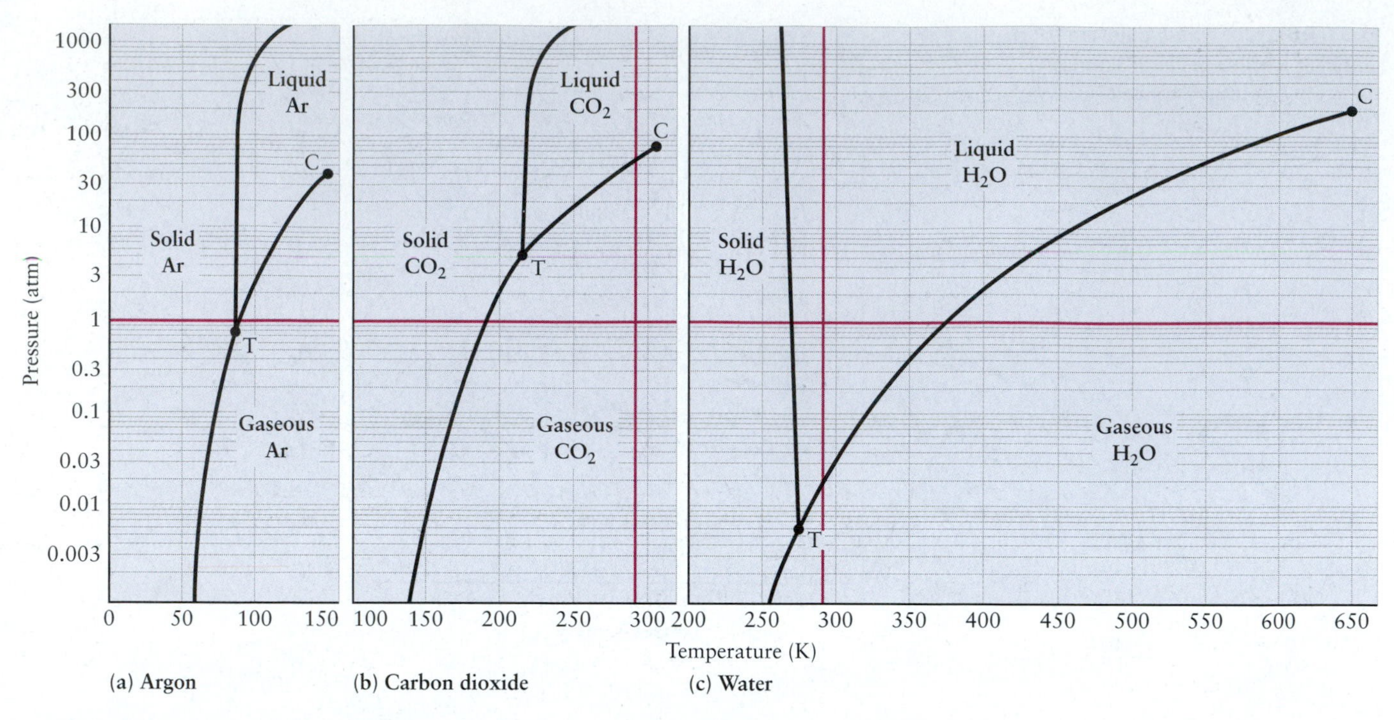

FIGURE 10.23 In these phase diagrams, the pressure increases by a factor of 10 at regular intervals along the vertical axis. This method of graphing allows large ranges of pressure to be plotted. The red horizontal and vertical lines mark a pressure of 1 atm and a temperature of 298.15 K, or 25°C. Their intersection identifies room conditions. Argon and carbon dioxide are gases at room conditions, but water is a liquid. The letter T marks the triple points of the substances, and the letter C marks their critical points. The region of stability of liquid water is larger than that of either carbon dioxide or argon.

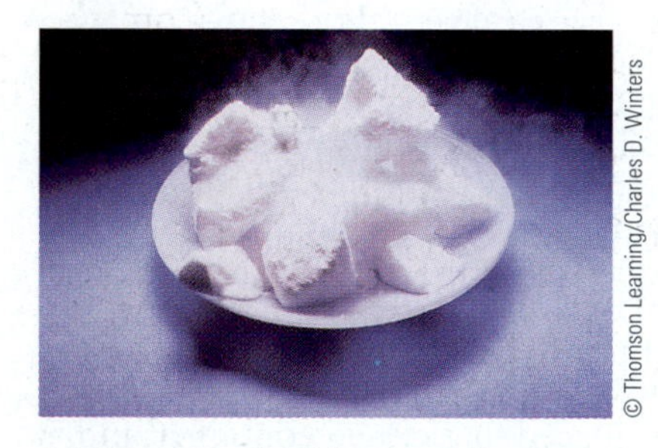

Sublimation of solid carbon dioxide (dry ice). The white clouds are drops of water vapor (moisture in the air) that condense at the low temperatures near the solid surface. Gaseous carbon dioxide itself is transparent.

For most substances, including water (see Fig. 10.23c), atmospheric pressure occurs somewhere between the triple-point pressure and the critical pressure, so in our ordinary experience, all three phases—gas, liquid, and solid—are observed. For a few substances, the triple-point pressure lies *above* $P = 1$ atm, and under atmospheric conditions, there is a direct transition called **sublimation** from solid to gas, without an intermediate liquid state. Carbon dioxide is such a substance (see Fig. 10.23b); its triple-point pressure is 5.117 atm (the triple-point temperature is −56.57°C). Solid CO_2 (dry ice) sublimes directly to gaseous CO_2 at atmospheric pressure. In this respect, it differs from ordinary ice, which melts before it evaporates and sublimes only at pressures below its triple-point pressure, 0.0060 atm. This fact is used in freeze-drying, a process in which foods are frozen and then put in a vacuum chamber at a pressure of less than 0.0060 atm. The ice crystals that formed on freezing then sublime, leaving a dried food that can be reconstituted by adding water.

Many substances exhibit more than one solid phase as the temperature and pressure are varied. At ordinary pressures, the most stable state of carbon is graphite, a rather soft, black solid; but at high enough pressures, the hard, transparent diamond form becomes more stable. That diamonds exist at all at atmospheric pressure is a consequence of how slowly they convert to graphite (we study the graphite-diamond case more thoroughly in our discussions of thermodynamics and spontaneous processes in Chapter 14). Below 13.2°C (and at atmospheric pressure), elemental tin undergoes a slow transformation from the metallic white form to a powdery gray form, a process referred to as "tin disease." This has caused the destruction of tin organ pipes in unheated buildings. No fewer than nine solid forms of ice are known, some of which exist only over a limited range of temperatures and pressures.

EXAMPLE 10.3

Consider a sample of argon held at $P = 600$ atm, $T = 100$ K in Figure 10.23a.

(a) What phase(s) is (are) present at equilibrium?

(b) Suppose the argon originally held at $P = 600$ atm, $T = 100$ K is heated at constant pressure. What happens?

(c) Describe a procedure to convert all the argon to gas without changing the temperature.

SOLUTION

(a) Because this point lies on the liquid–solid coexistence curve, both liquid and solid phases are present.

(b) An increase in temperature at constant pressure corresponds to a movement to the right in the figure, and the solid argon melts to a pure liquid.

(c) If the pressure is reduced sufficiently at constant temperature (below the triple-point pressure of 0.75 atm, for example) the argon will be converted completely to gas.

Related Problems: 43, 44, 45, 46, 47, 48, 49, 50, 51, 52

CHAPTER SUMMARY

The kinetic molecular theory of matter asserts that the macroscopic properties of a gas, liquid, or solid are determined by the number density of its molecules and the nature and strength of the forces between molecules. Intermolecular forces originate in the structures of the molecules and can be calculated from the Coulomb interactions among all the charged particles comprising the molecule. These forces give rise to potential energy between molecules, in magnitudes determined by the distance between the molecules. The repulsive and attractive forces are both included in potential energy functions, and their relative influence shown for each value of intermolecular separation. The same intermolecular forces that cause gas imperfection lead to the formation of liquids and solids. The three phases or states of matter can coexist in equilibrium. On the microscopic level, phase equilibrium is a dynamical balance in which each phase gains and loses molecules from the other phase at the same rate.

CUMULATIVE EXERCISE

Elemental bismuth.

Bismuth

Bismuth is a rather rare element in the earth's crust, but its oxides and sulfides appear at sufficient concentrations as impurities in lead and copper ores to make its recovery from these sources practical. Annual production of bismuth amounts to several million kilograms worldwide. Although elemental bismuth is a metal, its electrical conductivity is quite poor and it is relatively brittle. The major uses of bismuth arise from its low melting point (271.3°C) and the even lower melting points of its alloys, which range down to 47°C. These alloys are used as temperature sensors in fire detectors and automatic sprinkler systems because, in case of fire, they melt at a well-defined temperature, breaking an electrical connection and triggering an alarm or deluge.

(a) At its normal melting point, the density of solid bismuth is 9.73 g cm^{-3} and that of liquid bismuth is 10.05 g cm^{-3}. Does the volume of a sample of bismuth increase or decrease on melting? Does bismuth more closely resemble water or argon (see Fig. 10.23) in this regard?

(b) Since 1450 (10 years after Gutenberg), bismuth alloys have been used to cast metal type for printing. Explain why those alloys that share the melting behavior of bismuth discussed in part (a) would be especially useful for this application.

(c) A sample of solid bismuth is held at a temperature of 271.0°C and compressed. What will be observed?

(d) The vapor pressure of liquid bismuth has been measured to be 5.0 atm at a temperature of 1850°C. Does its normal boiling point lie above or below this temperature?

(e) At 1060°C, the vapor pressure of liquid bismuth is 0.013 atm. Calculate the number of bismuth atoms per cubic centimeter at equilibrium in the vapor above liquid bismuth at this temperature.

(f) The normal boiling point of liquid tin is 2270°C. Do you predict that liquid tin will be more volatile or less volatile than liquid bismuth at 1060°C?

(g) Bismuth forms two fluorides: BiF_3 and BiF_5. As is usually the case, the compound with the metal in the lower oxidation state has more ionic character, whereas that with the metal in the higher oxidation state has more covalent (molecular) character. Predict which bismuth fluoride will have the higher boiling point.

(h) Will AsF_5 have a higher or a lower normal boiling point than BiF_5?

Answers

(a) Bismuth resembles water in that its volume decreases on melting.

(b) The volume of bismuth increases on freezing; therefore, as the liquid alloy is cast, it fits tightly into its mold rather than shrinking away from the mold as most other metals do. This gives a more sharply defined metal type.

(c) The bismuth will melt.

(d) Its normal boiling point lies below this temperature.

(e) 7.2×10^{16} atoms cm^{-3}

(f) Liquid tin will be less volatile.

(g) BiF_3 will have the higher boiling point; in fact, BiF_3 boils at 900°C and BiF_5 boils at 230°C.

(h) It will have a lower normal boiling point.

CHAPTER REVIEW

The bulk properties of gases, liquids, and solids are defined by their methods of measurement.

- Their magnitudes depend on the structure of the molecules, the forces between the molecules, and the average distance between molecules.
- Molar volume: the volume per mole of a substance
  Ideal gases at standard temperature and pressure $V_m = 22.4$ L mol^{-1}
  Liquids and solids: V_m = molar mass/density, typically 10 – 100 cm^3 mol^{-1}
- Compressibility: fractional decrease in volume per unit increase in pressure
  Large for gases
  Small for liquids and solids

- Thermal expansion: fractional increase in volume per 1 K rise in temperature
  Same value for all gases at a given temperature
  Typical values for solids and liquids are of the order $\Delta V/V = 10^{-3}$/K
- Fluidity and rigidity: response to externally applied force (stress)
  Fluids flow, in amount determined by coefficient of viscosity
  Solids do not flow but show rigidity and mechanical strength
- Diffusion: the rate at which molecules of one substance move through the bulk of another substance.
  Quite slow in solids, faster in liquids, rapid in gases
- Surface tension: the energy required to increase the surface area of a liquid. Materials with high surface tensions readily form spherical drops, which have the highest possible surface area to volume ratio.

Intermolecular forces are determined by the structure of the molecules involved.

- Except for van der Waals forces, all intermolecular forces are *electrostatic*. They originate in the Coulomb interactions among all the charged particles in the molecules, and their magnitudes can be calculated from Coulomb's Law when we take proper account of the molecular structures.
- Ionic forces are the strongest and act over the longest distances (range). Their magnitude is easily calculated from Coulomb's Law. They are the dominant force in ionic compounds, which most commonly appear as solids but may be gases or molten salts.
- Ion–dipole forces act between ions and molecules with permanent dipole moments and are next strongest after ionic forces. They are relatively long ranged and are important in ionic solutions.
- Dipole–dipole forces act between neutral molecules that have permanent dipole moments. They are relatively weak and short ranged; the potential energy between two dipoles falls off as $1/R^3$.
- Induced dipole forces arise when an ion induces a dipole in a nonpolar atom or molecule and is then attracted to the opposite charge induced.
- Induced dipole–induced dipole forces arise between neutral nonpolar atoms or molecules and are the only source of attractive forces in substances such as Ar.
- Repulsive forces arise when two atoms or molecules are so close together that their respective electron clouds begin to interfere and overwhelm the attractive forces.

Intermolecular forces in liquids in general are described by the categories discussed above. In addition, hydrogen bonding occurs in substances where H is covalently bonded to N, O, or F.

- Trends in the boiling points of hydrides reveal the special nature of intermolecular forces when H is bonded to N, O, or F. The attractive forces are much stronger than normal dipole–dipole forces, and the orientation of the bond is linear, with the hydrogen located between the two heavy atoms. Hydrogen bonds are 2 to 5 times stronger than dipole–dipole interactions.
- Water has special properties that arise from hydrogen bonding.
  Boiling point is about 150 K higher than would occur without H-bonds.
  Solid is less dense than liquid due to a symmetrical, open, H-bonded network.

Matter is organized into phases, which may be in equilibrium with each other.

- Liquids, solids, and gases are the three normal phases of matter.
- Supercritical fluids and plasmas (ionized gases) are states of matter that appear under specialized conditions and have exotic properties.
- Two or more phases can coexist under given conditions of temperature and pressure.

- Transitions between phases include: evaporation and condensation (liquid to gas and reverse), melting and freezing (liquid to solid and reverse) and sublimation and condensation (solid to gas and reverse).
- Phase equilibria are dynamic events; molecules are constantly shuttling back and forth between phases but with equal rates so that no *macroscopic* changes are observed.

Phase diagrams describe the phases of a substance that exist under various combinations of temperature and pressure in a graphical representation.

- Pressure is plotted on the y-axis and temperature along the x-axis.
- Pressure and temperature conditions under which two phases coexist lie along the *coexistence curves* that divide the two phases.
- The one condition under which three phases can coexist is marked by a single point, the *triple point,* T.
- Above a certain temperature and pressure, marked by the *critical point,* a single *supercritical fluid* exists; its properties are markedly different from normal gases or liquids.

CONCEPTS & SKILLS

After studying this chapter and working the problems that follow, you should be able to:

1. Relate trends in values of bulk properties to the strength and range of intermolecular forces (Section 10.1, Problems 1–12).
2. Relate magnitudes and distance dependence of intermolecular forces to the structure of molecules (Section 10.2, Problems 13–20).
3. Describe the effects of different kinds of intermolecular forces on properties of liquids (Section 10.3, Problems 21–30).
4. Discuss the evidence that phase equilibrium is a dynamic process at the molecular level (Section 10.4, Problems 31–38).
5. Describe the effects of different kinds of intermolecular forces on phase transitions (Section 10.5, Problems 39–42).
6. Sketch the pressure–temperature phase diagram for a typical substance and identify the lines, areas, and singular points (Section 10.6, Problems 43–52).

PROBLEMS

Answers to problems whose numbers are boldface appear in Appendix G. Problems that are more challenging are indicated with asterisks.

Bulk Properties of Gases, Liquids, and Solids: Molecular Interpretation

1. A substance is nearly nonviscous and quite compressible, and it has a large coefficient of thermal expansion. Is it most likely to be a solid, a liquid, or a gas?

2. A substance is viscous, nearly incompressible, and not elastic. Is it most likely to be a solid, a liquid, or a gas?

3. A sample of volume 258 cm^3 has a mass of 2.71 kg.
(a) Is the material gaseous or condensed?
(b) If the molar mass of the material is 108 g mol^{-1}, calculate its molar volume.

4. A sample of volume 18.3 L has a mass of 57.9 g.
(a) Is the material gaseous or condensed?
(b) If the molar mass of the material is 123 g mol^{-1}, calculate its molar volume.

5. Heating a sample of matter from 20°C to 40°C at constant pressure causes its volume to increase from 546.0 to 547.6 cm^3. Classify the material as a nearly ideal gas, a nonideal gas, or condensed.

6. Cooling a sample of matter from 70°C to 10°C at constant pressure causes its volume to decrease from 873.6 to 712.6 cm^3. Classify the material as a nearly ideal gas, a nonideal gas, or condensed.

7. At 1.00 atm pressure and a temperature of 25°C, the volume of 1.0 g water is 1.0 mL. At the same pressure and a temperature of 101°C, the volume of 1.0 g water is nearly 1700 times larger. Give the reason for this large change in volume.

8. Doubling the absolute temperature of a gas essentially doubles its volume at constant pressure. Doubling the temperature of many metals, however, often increases their volumes by only a few percent. Explain.

9. Will solid sodium chloride be harder (that is, more resistant to indentation) or softer than solid carbon tetrachloride? Explain.

10. Will the surface tension of molten sodium chloride be higher than or lower than that of carbon tetrachloride? Explain.

11. Do you expect that the diffusion constant will increase or decrease as the density of a liquid is increased (by compressing it) at constant temperature? Explain. What will happen to the diffusion constant of a gas and a solid as the density increases?

12. Do you anticipate that the diffusion constant will increase as the temperature of a liquid increases at constant pressure? Why or why not? Will the diffusion constant increase with temperature for a gas and a solid? Explain.

Intermolecular Forces: Origins in Molecular Structure

13. Compare ion–dipole forces with induced dipole forces. In what ways are they similar and different? Give an example of each.

14. Compare dipole–dipole forces with dispersion forces. In what ways are they similar and different? Give an example of each.

15. Name the types of attractive forces that will contribute to the interactions among atoms, molecules, or ions in the following substances. Indicate the one(s) you expect to predominate.
(a) KF (b) HI (c) Rn (d) N_2

16. Name the types of attractive forces that will contribute to the interactions among atoms, molecules, or ions in the following substances. Indicate the one(s) you expect to predominate.
(a) Ne (b) ClF (c) F_2 (d) $BaCl_2$

17. Predict whether a sodium ion will be most strongly attracted to a bromide ion, a molecule of hydrogen bromide, or an atom of krypton.

18. Predict whether an atom of argon will be most strongly attracted to another atom of argon, an atom of neon, or an atom of krypton.

19. (a) Use Figure 10.9 to estimate the length of the covalent bond in Cl_2 and the length of the ionic bond in K^+Cl^-. **Note:** The latter corresponds to the distance between the atoms in an isolated single molecule of K^+Cl^-, not in KCl(*s*) (solid potassium chloride).
(b) A book states, "The shorter the bond, the stronger the bond." What features of Figure 10.9 show that this is not always true?

20. True or false: Any two atoms held together by nonbonded attractions must be farther apart than any two atoms held together by a chemical bond. Explain.

Intermolecular Forces in Liquids

21. Under room conditions, fluorine and chlorine are gases, bromine is a liquid, and iodine is a solid. Explain the origin of this trend in the physical state of the halogens.

22. The later halogens form pentafluorides: ClF_5, BrF_5, and IF_5. At 0°C, one of these is a solid, one a liquid, and one a gas. Specify which is which, and explain your reasoning.

23. List the following substances in order of increasing normal boiling points, T_b, and explain your reasoning: NO, NH_3, Ne, RbCl.

24. List the following substances in order of increasing normal boiling points, T_b, and explain your reasoning: SO_2, He, HF, CaF_2, Ar.

25. As a vapor, methanol exists to an extent as a tetramer, $(CH_3OH)_4$, in which four CH_3OH molecules are held together by hydrogen bonds. Propose a reasonable structure for this tetramer.

26. Hypofluorous acid (HOF) is the simplest possible compound that allows comparison between fluorine and oxygen in their abilities to form hydrogen bonds. Although F attracts electrons more strongly than O, solid HOF unexpectedly contains no H · · · F hydrogen bonds! Draw a proposed structure for chains of HOF molecules in the crystalline state. The bond angle in HOF is 101 degrees.

27. Hydrazine (N_2H_4) is used as a reducing agent and in the manufacture of rocket fuels. How do you expect its boiling point to compare with that of ethylene (C_2H_4)?

28. Hydrogen peroxide (H_2O_2) is a major industrial chemical that is produced on a scale approaching 10^9 kg per year. It is widely used as a bleach and in chemical manufacturing processes. How do you expect its boiling point to compare with those of fluorine (F_2) and hydrogen sulfide (H_2S), two substances with molar masses comparable with that of hydrogen peroxide?

29. A flask contains 1.0 L (1.0 kg) of room-temperature water. Calculate the number of possible hydrogen bonds among the water molecules present in this sample. Each water molecule can accept two hydrogen bonds and also furnish the H atoms for two hydrogen bonds.

30. What is the maximum number of hydrogen bonds that can be formed in a sample containing 1.0 mol (N_0 molecules) liquid HF? Compare with the maximum number that can form in 1.0 mol liquid water.

Phase Equilibrium

31. Hydrogen at a pressure of 1 atm condenses to a liquid at 20.3 K and solidifies at 14.0 K. The vapor pressure of liquid hydrogen is 0.213 atm at 16.0 K. Calculate the volume of 1.00 mol H_2 vapor under these conditions and compare it with the volume of 1.00 mol H_2 at standard temperature and pressure.

32. Helium condenses to a liquid at 4.224 K under atmospheric pressure and remains a liquid down to the absolute zero of temperature. (It is used as a coolant to reach very low temperatures.) The vapor pressure of liquid helium at 2.20 K is 0.05256 atm. Calculate the volume occupied by 1.000 mol helium vapor under these conditions and compare it with the volume of the same amount of helium at standard temperature and pressure.

33. The vapor pressure of liquid mercury at 27°C is 2.87×10^{-6} atm. Calculate the number of Hg atoms per cubic centimeter in the "empty" space above the top of the column of mercury in a barometer at 27°C.

34. The tungsten filament in an incandescent lightbulb ordinarily operates at a temperature of about 2500°C. At this temperature, the vapor pressure of solid tungsten is 7.0×10^{-9} atm. Estimate the number of gaseous tungsten atoms per cubic centimeter under these conditions.

35. Calcium carbide reacts with water to produce acetylene (C_2H_2) and calcium hydroxide. The acetylene is collected over water at 40.0°C under a total pressure of 0.9950 atm. The vapor pressure of water at this temperature is 0.0728 atm. Calculate the mass of acetylene per liter of "wet" acetylene collected in this way, assuming ideal gas behavior.

36. A metal reacts with aqueous hydrochloric acid to produce hydrogen. The hydrogen (H_2) is collected over water at 25°C under a total pressure of 0.9900 atm. The vapor pressure of water at this temperature is 0.0313 atm. Calculate the mass of hydrogen per liter of "wet" hydrogen above the water, assuming ideal gas behavior.

37. Carbon dioxide is liberated by the reaction of aqueous hydrochloric acid with calcium carbonate:

$$CaCO_3(s) + 2\,H^+(aq) \longrightarrow Ca^{2+}(aq) + CO_2(g) + H_2O(\ell)$$

A volume of 722 mL $CO_2(g)$ is collected over water at 20°C and a total pressure of 0.9963 atm. At this temperature, water has a vapor pressure of 0.0231 atm. Calculate the mass of calcium carbonate that has reacted, assuming no losses of carbon dioxide.

38. When an excess of sodium hydroxide is added to an aqueous solution of ammonium chloride, gaseous ammonia is produced:

$$NaOH(aq) + NH_4Cl(aq) \longrightarrow NaCl(aq) + NH_3(g) + H_2O(\ell)$$

Suppose 3.68 g ammonium chloride reacts in this way at 30°C and a total pressure of 0.9884 atm. At this temperature, the vapor pressure of water is 0.0419 atm. Calculate the volume of ammonia saturated with water vapor that will be produced under these conditions, assuming no leaks or other losses of gas.

Phase Transitions

39. High in the Andes, an explorer notes that the water for tea is boiling vigorously at a temperature of 90°C. Use Figure 10.17 to estimate the atmospheric pressure at the altitude of the camp. What fraction of the earth's atmosphere lies below the level of the explorer's camp?

40. The total pressure in a pressure cooker filled with water increases to 4.0 atm when it is heated, and this pressure is maintained by the periodic operation of a relief valve. Use Figure 10.23c to estimate the temperature of the water in the pressure cooker.

41. Iridium melts at a temperature of 2410°C and boils at 4130°C, whereas sodium melts at a temperature of 97.8°C and boils at 904°C. Predict which of the two molten metals has the larger surface tension at its melting point. Explain your prediction.

42. Aluminum melts at a temperature of 660°C and boils at 2470°C, whereas thallium melts at a temperature of 304°C and boils at 1460°C. Which metal will be more volatile at room temperature?

Phase Diagrams

43. At its melting point (624°C), the density of solid plutonium is 16.24 g cm^{-3}. The density of liquid plutonium is 16.66 g cm^{-3}. A small sample of liquid plutonium at 625°C is strongly compressed. Predict what phase changes, if any, will occur.

44. Phase changes occur between different solid forms, as well as from solid to liquid, liquid to gas, and solid to gas. When white tin at 1.00 atm is cooled below 13.2°C, it spontaneously changes (over a period of weeks) to gray tin. The density of gray tin is *less* than the density of white tin (5.75 g cm^{-3} vs 7.31 g cm^{-3}). Some white tin is compressed to a pressure of 2.00 atm. At this pressure, should the temperature be higher or lower than 13.2°C for the conversion to gray tin to occur? Explain your reasoning.

45. The following table gives several important points on the pressure-temperature diagram of ammonia:

	P (atm)	*T* (K)
Triple point	0.05997	195.42
Critical point	111.5	405.38
Normal boiling point	1.0	239.8
Normal melting point	1.0	195.45

Use this information to sketch the phase diagram of ammonia.

46. The following table gives several important points on the pressure–temperature diagram of nitrogen:

	P (atm)	*T* (K)
Triple point	0.123	63.15
Critical point	33.3978	126.19
Normal boiling point	1.0	77.35
Normal melting point	1.0	195.45

Use this information to sketch the phase diagram of nitrogen. The density of $N_2(s)$ is 1.03 g cm^{-3} and that of $N_2(\ell)$ is 0.808 g cm^{-3}.

47. Determine whether argon is a solid, a liquid, or a gas at each of the following combinations of temperature and pressure (use Fig. 10.23).

(a) 50 atm and 100 K (c) 1.5 atm and 25 K
(b) 8 atm and 150 K (d) 0.25 atm and 120 K

48. Some water starts out at a temperature of 298 K and a pressure of 1 atm. It is compressed to 500 atm at constant temperature, and then heated to 750 K at constant pressure. Next, it is decompressed at 750 K back to 1 atm and finally cooled to 400 K at constant pressure.
(a) What was the state (solid, liquid, or gas) of the water at the start of the experiment?
(b) What is the state (solid, liquid, or gas) of the water at the end of the experiment?
(c) Did any phase transitions occur during the four steps described? If so, at what temperature and pressure did they occur? (*Hint:* Trace out the various changes on the phase diagram of water [see Fig. 10.23].)

49. The vapor pressure of solid acetylene at −84.0°C is 760 torr.
(a) Does the triple-point temperature lie above or below −84.0°C? Explain.
(b) Suppose a sample of solid acetylene is held under an external pressure of 0.80 atm and heated from 10 to 300 K. What phase change(s), if any, will occur?

50. The triple point of hydrogen occurs at a temperature of 13.8 K and a pressure of 0.069 atm.
(a) What is the vapor pressure of solid hydrogen at 13.8 K?
(b) Suppose a sample of solid hydrogen is held under an external pressure of 0.030 atm and heated from 5 to 300 K. What phase change(s), if any, will occur?

51. The density of nitrogen at its critical point is 0.3131 g cm^{-3}. At a very low temperature, 0.3131 g solid nitrogen is sealed into a thick-walled glass tube with a volume of 1.000 cm^3. Describe what happens inside the tube as the tube is warmed past the critical temperature, 126.19 K.

52. At its critical point, ammonia has a density of 0.235 g cm^{-3}. You have a special thick-walled glass tube that has a 10.0-mm outside diameter, a wall thickness of 4.20 mm, and a length of 155 mm. How much ammonia must you seal into the tube if you wish to observe the disappearance of the meniscus as you heat the tube and its contents to a temperature higher than 132.23°C, the critical temperature?

ADDITIONAL PROBLEMS

53. Would you classify candle wax as a solid or a liquid? What about rubber? Discuss.

* **54.** When a particle diffuses, its *mean-square displacement* in a time interval Δt is $6D\Delta t$, where D is the diffusion constant. Its *root-mean-square displacement* is the square root of this (recall the analogous root-mean-square speed from Section 9.5). Calculate the root-mean-square displacement at 25°C after 1.00 hour of (a) an oxygen molecule in air ($D = 2.1 \times 10^{-5}$ m^2 s^{-1}), (b) a molecule in liquid water ($D = 2.26 \times 10^{-9}$ m^2 s^{-1}), and (c) an atom in solid sodium ($D = 5.8 \times 10^{-13}$ m^2 s^{-1}). Note that for solids with melting points higher than sodium, the diffusion constant can be many orders of magnitude smaller.

55. Liquid hydrogen chloride will dissolve many ionic compounds. Diagram how molecules of hydrogen chloride tend to distribute themselves about a negative ion and about a positive ion in such solutions.

* **56.** Section 9.7 explains that the van der Waals constant b (with units of L mol^{-1}) is related to the volume per molecule in the liquid, and thus to the sizes of the molecules. The combination of van der Waals constants a/b has units of L atm mol^{-1}. Because the liter atmosphere is a unit of energy (1 L atm = 101.325 J), a/b is proportional to the energy per mole for interacting molecules, and thus to the strength of the attractive forces between molecules, as shown in Figure 10.9. By using the van der Waals constants in Table 9.3, rank the following attractive forces from strongest to weakest: N_2, H_2, SO_2, and HCl.

57. Describe how the average kinetic and potential energies per mole change as a sample of water is heated from 10 to 1000 K at a constant pressure of 1 atm.

58. As a sample of water is heated from 0.0°C to 4.0°C, its density increases from 0.99987 to 1.00000 g cm^{-3}. What can you conclude about the coefficient of thermal expansion of water in this temperature range? Is water unusual in its behavior? Explain.

59. At 25°C, the equilibrium vapor pressure of water is 0.03126 atm. A humidifier is placed in a room of volume 110 m^3 and is operated until the air becomes saturated with water vapor. Assuming that no water vapor leaks out of the room and that initially there was no water vapor in the air, calculate the number of grams of water that have passed into the air.

60. The text states that at 1000°C, the vapor pressure of tungsten is 2×10^{-25} atm. Calculate the volume occupied per tungsten *atom* in the vapor under these conditions.

61. You boil a small quantity of water inside a 5.0-L metal can. When the can is filled with water vapor (and all air has been expelled), you quickly seal the can with a screw-on cap and remove it from the source of heat. It cools to 60°C and most of the steam condenses to liquid water. Determine the pressure inside the can. (*Hint:* Refer to Fig. 10.17.)

* **62.** The air over an unknown liquid is saturated with the vapor of that liquid at 25°C and a total pressure of 0.980 atm. Suppose that a sample of 6.00 L of the saturated air is collected and the vapor of the unknown liquid is removed from that sample by cooling and condensation. The pure air remaining occupies a volume of 3.75 L at −50°C and 1.000 atm. Calculate the vapor pressure of the unknown liquid at 25°C.

63. If it is true that all solids and liquids have vapor pressures, then at sufficiently low external pressures, every substance should start to boil. In space, there is effectively zero external pressure. Explain why spacecraft do not just boil away as vapors when placed in orbit.

64. Butane-fueled cigarette lighters, which give hundreds of lights each, typically contain 4 to 5 g butane (C_4H_{10}), which is confined in a 10-mL plastic container and exerts a pressure of 3.0 atm at room temperature (25°C). The butane boils at −0.5°C under normal pressure. Butane lighters have been known to explode during use, inflicting serious injury. A person hoping to end such accidents suggests that there be less butane placed in the lighters so that the pressure inside them does not exceed 1.0 atm. Estimate how many grams of butane would be contained in such a lighter.

65. A cooling bath is prepared in a laboratory by mixing chunks of solid CO_2 with ethanol. $CO_2(s)$ sublimes at $-78.5°C$ to $CO_2(g)$. Ethanol freezes at $-114.5°C$ and boils at $+78.4°C$. State the temperature of this cooling bath and describe what will be seen when it is prepared under ordinary laboratory conditions.

66. Oxygen melts at 54.8 K and boils at 90.2 K at atmospheric pressure. At the normal boiling point, the density of the liquid is 1.14 g cm^{-3} and the vapor can be approximated as an ideal gas. The critical point is defined by $T_c = 154.6$ K, $P_c = 49.8$ atm, and (density) $d_c = 0.436$ g cm^{-3}. The triple point is defined by $T_t = 54.4$ K, $P_t = 0.0015$ atm, a liquid density equal to 1.31 g cm^{-3}, and a solid density of 1.36 g cm^{-3}. At 130 K, the vapor pressure of the liquid is 17.25 atm. Use this information to construct a phase diagram showing P versus T for oxygen. You need not draw the diagram to scale, but you should give numeric labels to as many points as possible on both axes.

67. It can be shown that if a gas obeys the van der Waals equation, its critical temperature, its critical pressure, and its molar volume at the critical point are given by the equations

$$T_c = \frac{8a}{27Rb} \qquad P_c = \frac{a}{27b^2} \qquad \left(\frac{V}{n}\right)_c = 3b$$

where a and b are the van der Waals constants of the gas. Use the van der Waals constants for oxygen, carbon dioxide, and water (from Table 9.3) to estimate the critical-point properties of these substances. Compare with the observed values given in Figure 10.23 and in Problem 66.

* 68. Each increase in pressure of 100 atm decreases the melting point of ice by about 1.0°C.

(a) Estimate the temperature at which liquid water freezes under a pressure of 400 atm.

(b) One possible explanation of why a skate moves smoothly over ice is that the pressure exerted by the skater on the ice lowers its freezing point and causes it to melt. The pressure exerted by an object is the force (its mass × the acceleration of gravity, 9.8 m s^{-2}) divided by the area of contact. Calculate the change in freezing point of ice when a skater with a mass of 75 kg stands on a blade of area 8.0×10^{-5} m^2 in contact with the ice. Is this sufficient to explain the ease of skating at a temperature of, for example, $-5°C$ (23°F)?

* 69. (a) Sketch the phase diagram of *temperature* versus *molar volume* for carbon dioxide, indicating the region of each of the phases (gas, liquid, and solid) and the coexistence regions for two phases.

(b) Liquid water has a maximum at 4°C in its curve of density against temperature at $P = 1$ atm, and the solid is less dense than the liquid. What happens if you try to draw a phase diagram of T versus molar volume for water?

70. The critical temperature of HCl is 51°C, lower than that of HF, 188°C, and HBr, 90°C. Explain this by analyzing the nature of the intermolecular forces in each case.

71. The normal boiling points of the fluorides of the second-period elements are as follows: LiF, 1676°C; BeF_2, 1175°C; BF_3, $-100°C$; CF_4, $-128°C$; NF_3, $-129°C$; OF_2, $-145°C$; F_2, $-188°C$. Describe the nature of the intermolecular forces in this series of liquids, and account for the trends in boiling point.

CUMULATIVE PROBLEMS

72. At 20°C and a pressure of 1 atm, 1 mol argon gas occupies a volume of 24.0 L. Estimate the van de Waals radius for argon from the onset of the repulsive part of the argon intermolecular potential curve in Figure 9.18, and calculate the fraction of the gas volume that consists of argon atoms.

73. Other things being equal, ionic character in compounds of metals decreases with increasing oxidation number. Rank the following compounds from lowest to highest normal boiling points: AsF_5, SbF_3, SbF_5, F_2.

CHAPTER 11

Solutions

Dissolution of sugar in water.

Homogeneous systems that contain two or more substances are called **solutions.** Usually, we think of a solution as a liquid that contains some dissolved substance, such as a solid or gas, and we use the term in that sense in most of this chapter. But, solutions of one solid in another are also common, one example being an alloy of gold and silver. In fact, any homogeneous system of two or more substances (liquid, solid, or gas) is a solution. The major component is usually called the **solvent,** and minor components are called the **solutes.** The solvent is regarded as a "carrier" or medium for the solute, which can participate in chemical reactions in the solution or leave the solution through precipitation or evaporation. Throughout this chapter, it is helpful to keep in mind one guiding question: How are the reactions and the properties of the pure solute modified when it is dispersed in the solvent?

Description of these phenomena requires quantitative specifications of the amount of solute in the solution, or the *composition* of the solution. Solutions are formed by mixing two or more pure substances whose molecules interact directly in the mixed state. Molecules experience new intermolecular forces in moving from pure solute or solvent into the mixed state. The magnitude of these changes influences both the ease of formation and the stability of a solution.

Chemical reactions are frequently carried out in solution, and their description requires modifications to the rules of stoichiometry described in Chapter 2. We illustrate these modified rules by the important analytical techniques of titration in acid–base and oxidation–reduction reactions.

Just like pure substances, solutions can be in *phase equilibrium* with gases, solids, or other liquids. These equilibria frequently show interesting effects that depend on the molecular weight of the solute.

This chapter begins by explaining how the composition of solutions is defined and how solutions are prepared. It is important to master these concepts, because the properties and behavior of solutions are determined by their composition. With this background, we give quantitative descriptions of chemical reactions and phase equilibria in solutions, and relate these events to the nature of the species in the solution.

11.1 Composition of Solutions

Several measures are used to specify the composition of a solution. **Mass percentage** (colloquially called weight percentage), frequently used in everyday applications, is defined as the percentage by mass of a given substance in the solution. In quantitative chemistry, the most useful measures of composition are mole fraction, molarity, and molality.

The **mole fraction** of a substance in a mixture is the number of moles of that substance divided by the total number of moles present. This term was introduced in the discussion of gas mixtures and Dalton's law (see Section 9.4). In a binary mixture containing n_1 mol of species 1 and n_2 mol of species 2, the mole fractions X_1 and X_2 are

$$X_1 = \frac{n_1}{n_1 + n_2} \qquad [11.1a]$$

$$X_2 = \frac{n_2}{n_1 + n_2} = 1 - X_1 \qquad [11.1b]$$

The mole fractions of all the species present must add up to 1. When a clear distinction can be made between solvent and solutes, the label 1 denotes the solvent and higher numbers are given to the solutes. If comparable amounts of two liquids such as water and alcohol are mixed, the assignment of the labels 1 and 2 is arbitrary.

The **concentration** of a substance is the number of moles per unit volume. The SI units moles per cubic meter are inconveniently large for chemical work, so instead we use the **molarity,** defined as the number of moles of solute per liter of solution:

$$\text{molarity} = \frac{\text{moles solute}}{\text{liters solution}} = \text{mol L}^{-1} \qquad [11.2]$$

M is the abbreviation for "moles per liter." A 0.1 M (read "0.1 molar") solution of HCl has 0.1 mol of HCl (dissociated into ions, as explained later in this chapter) per liter of solution. Molarity is the most common way of specifying the compositions of dilute solutions. For accurate measurements it has the disadvantage of depending slightly on temperature. If a solution is heated or cooled, its volume changes, so the number of moles of solute per liter of solution also changes.

The **molality,** in contrast, involves the ratio of two masses, and so does not depend on temperature. Molality is defined as the number of moles of solute per kilogram of *solvent:*

$$\text{molality} = \frac{\text{moles solute}}{\text{kilograms solvent}} = \text{mol kg}^{-1} \qquad [11.3]$$

Because the density of water is 1.00 g cm^{-3} at 20°C, 1.00 L of water has mass of 1.00×10^3 g, or 1.00 kg. It follows that in a dilute aqueous solution, the number of moles of solute per liter is nearly the same as the number of moles per kilogram of water. Therefore, molarity and molality have nearly equal values. For nonaqueous solutions and concentrated aqueous solutions, this approximate equality is no longer valid.

EXAMPLE 11.1

A solution is prepared by dissolving 22.4 g of $MgCl_2$ in 0.200 L of water. Taking the density of pure water to be 1.00 g cm^{-3} and the density of the resulting solution to be 1.089 g cm^{-3}, calculate the mole fraction, molarity, and molality of $MgCl_2$ in this solution.

SOLUTION

We are given the *mass* of the $MgCl_2$ and the *volume* of the water. The number of moles for each are

$$\text{moles } MgCl_2 = \frac{22.4 \text{ g}}{95.22 \text{ g mol}^{-1}} = 0.235 \text{ mol}$$

$$\text{moles water} = \frac{(0.200 \text{ L})(1000 \text{ cm}^3 \text{ L}^{-1})(1.00 \text{ g cm}^{-3})}{18.02 \text{ g mol}^{-1}}$$

$$= 11.1 \text{ mol}$$

$$\text{mole fraction } MgCl_2 = \frac{0.235 \text{ mol}}{(11.1 + 0.235) \text{ mol}} = 0.0207$$

To calculate the molarity, we must first determine the volume of solution. Its mass is 200 g water + 22.4 g MgCl = 222.4 g, and its density is 1.089 g cm^{-3}, so the volume is

$$\text{volume solution} = \frac{222.4 \text{ g}}{1.089 \text{ g cm}^{-3}} = 204 \text{ cm}^3 = 0.204 \text{ L}$$

$$\text{molarity } MgCl_2 = \frac{0.235 \text{ mol } MgCl_2}{0.204 \text{ L}} = 1.15 \text{ M}$$

$$\text{molality } MgCl_2 = \frac{0.235 \text{ mol } MgCl_2}{0.200 \text{ kg } H_2O} = 1.18 \text{ mol kg}^{-1}$$

Related Problems: 3, 4

EXAMPLE 11.2

A 9.386 M aqueous solution of sulfuric acid has a density of 1.5091 g cm^{-3}. Calculate the molality, the percentage by mass, and the mole fraction of sulfuric acid in this solution.

SOLUTION

It is convenient to choose 1 L of the solution, whose mass is

$$(1000 \text{ cm}^3)(1.5091 \text{ g cm}^{-3}) = 1509.1 \text{ g} = 1.5091 \text{ kg}$$

One liter contains 9.386 mol H_2SO_4, or

$$9.386 \text{ mol } H_2SO_4 \times 98.08 \text{ g mol}^{-1} = 920.6 \text{ g } H_2SO_4$$

The mass of water in this liter of solution is then obtained by subtraction:

$$\text{Mass of water in 1 L of solution} = 1.5091 \text{ kg} - 0.9206 \text{ kg} = 0.5885 \text{ kg}$$

The molality is now directly obtained as

$$\text{Molality } H_2SO_4 = \frac{9.386 \text{ mol } H_2SO_4}{0.5885 \text{ kg } H_2O} = 15.95 \text{ mol kg}^{-1}$$

and the mass percentage is

$$\text{Mass percentage } H_2SO_4 = \frac{0.9206 \text{ kg}}{1.5091 \text{ kg}} \times 100\% = 61.00\%$$

The number of moles of water is

$$\text{moles } H_2O = \frac{588.5 \text{ g}}{18.02 \text{ g mol}^{-1}} = 32.66 \text{ mol}$$

so that the mole fraction of H_2SO_4 is

$$\text{Mole fraction } H_2SO_4 = X_2 = \frac{9.386 \text{ mol}}{9.386 + 32.66 \text{ mol}} = 0.2232$$

Related Problems: 5, 6

Preparation of Solutions

Examples 11.1 and 11.2 show that if a known mass of solute is added to a known volume of solvent, the molarity can be calculated only if the density of the resulting solution is known. If 1 L of solvent is used, the volume of the resulting solution is less than 1 L in some cases and more in others. If a solution is to have a given molarity, it is clearly inconvenient to need to know the solution density. We avoid this problem in practice by dissolving the measured amount of solute in a smaller amount of solvent, then adding solvent continuously until the desired total volume is reached. For accurate work, solutions are prepared in a **volumetric flask,** which has a distinct ring marked around its neck to indicate a carefully calibrated volume (Fig. 11.1). Filling the flask with solvent up to this mark controls the volume of the solution.

Sometimes it may be necessary to prepare a dilute solution of specified concentration from a more concentrated solution of known concentration by adding pure solvent to the concentrated solution. Suppose that the initial concentration (molarity) is c_i and the initial solution volume is V_i. The number of moles of solute is $(c_i \text{ mol L}^{-1})(V_i \text{ L}) = c_iV_i$ mol. This number does not change on dilution to a final

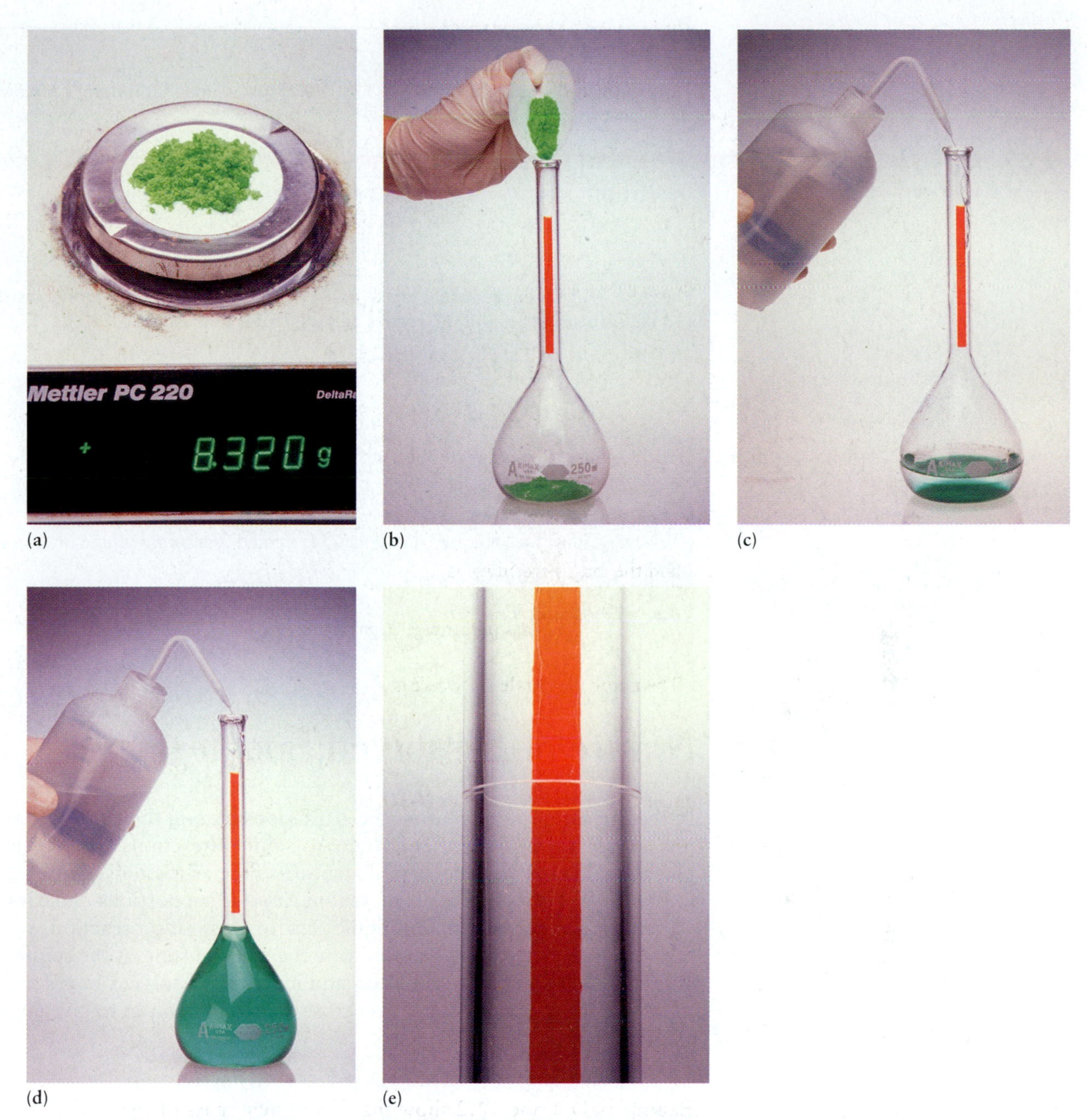

(a) (b) (c) (d) (e)

FIGURE 11.1 To prepare a solution of nickel chloride, $NiCl_2$, with accurately known concentration, weigh out an amount of the solid (a), transfer it to a volumetric flask (b), dissolve it in somewhat less than the required amount of water (c), and dilute to the total volume marked on the neck of the flask (d), (e).
(© Thomson Learning/Charles D. Winters)

solution volume, V_f, because only solvent, and not solute, is being added. Thus, $c_i V_i = c_f V_f$ and the final molarity is

$$c_f = \frac{\text{moles solute}}{\text{final solution volume}} = \frac{c_i V_i}{V_f} \qquad [11.4]$$

This equation can be used to calculate the final concentration after dilution to a given final volume and also to determine what final volume should be used to obtain a given concentration.

EXAMPLE 11.3

(a) Describe how to prepare 0.500 L of a 0.100 M aqueous solution of potassium hydrogen carbonate ($KHCO_3$).

(b) Describe how to dilute this solution to a final concentration of 0.0400 M $KHCO_3$.

SOLUTION

(a)

$$\text{moles solute} = (0.500 \text{ L})(0.100 \text{ mol L}^{-1}) = 0.0500 \text{ mol}$$

$$\text{grams solute} = (0.0500 \text{ mol})(100.12 \text{ g mol}^{-1}) = 5.01 \text{ g}$$

because 100.12 is the molar mass of $KHCO_3$. We would, therefore, dissolve 5.01 g $KHCO_3$ in a small amount of water and dilute the solution to 0.500 L.

(b) Rearranging Equation 11.4 gives

$$V_f = \frac{c_i}{c_f} V_i$$

$$= \left(\frac{0.100 \text{ mol L}^{-1}}{0.0400 \text{ mol L}^{-1}}\right)(0.500 \text{ L}) = 1.25 \text{ L}$$

To achieve this, the solution from part (a) is diluted to a total volume of 1.25 L by adding water.

Related Problems: 9, 10

11.2 Nature of Dissolved Species

In the formation of a solution the attractions among the particles in the original phases (solvent-to-solvent and solute-to-solute attractions) are broken up and replaced, at least in part, by new solvent-to-solute attractions. Unlike a compound, a solution has its components present in *variable* proportions and cannot be represented by a chemical formula. Equations for dissolution reactions do not include the solvent as a reactant. They indicate the original state of the solute in parentheses on the left side of the equation and identify the solvent in parentheses on the right side. For example, solid (*s*) sucrose dissolves in water to give an **aqueous** (*aq*) solution of sucrose:

$$C_{12}H_{22}O_{11}(s) \longrightarrow C_{12}H_{22}O_{11}(aq)$$

Although the solute and solvent can be any combination of solid, liquid, and gas phases, liquid water is indisputably the best known and most important solvent. Consequently, we emphasize aqueous solutions in this chapter, but you should always remember that dissolution also occurs in many other solvents. We describe formation of aqueous solutions by considering the intermolecular forces between the solute and water molecules. Because these forces can be quite different for molecular solutes and ionic solutes, we discuss these two cases separately.

Aqueous Solutions of Molecular Species

Molecular substances that have polar molecules are readily dissolved by water. Examples are the sugars, which have the general formula $C_m(H_2O)_n$. Specific cases are sucrose (table sugar), $C_{12}H_{22}O_{11}$; fructose (fruit sugar), $C_6H_{12}O_6$; and ribose, $C_5H_{10}O_5$, a subunit in the biomolecules known as ribonucleic acids. Despite their general formula, the sugars do not contain water molecules, but they do include polar OH (hydroxyl) groups bonded to carbon atoms, which provide sites for hydrogen-bonding interactions with water molecules. These attractions replace the

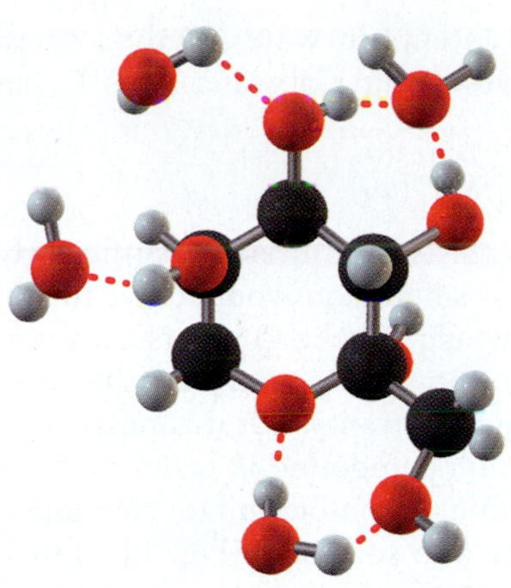

FIGURE 11.2 A molecule of fructose in aqueous solution. Note the attractions between the hydroxyl (O—H) groups of the fructose and molecules of water. The fructose molecule is aquated; the exact number and arrangement of the attached water molecules fluctuate. Also shown is one hydrogen bond between a water molecule and an oxygen atom in the fructose ring.

solute–solute interactions, and the individual aquated sugar molecules move off into the solution (Fig. 11.2). Many other molecular substances follow the same pattern, provided they are sufficiently polar. Nonpolar substances, such as carbon tetrachloride, octane, and the common oils and waxes, do not dissolve significantly in water.

Aqueous Solutions of Ionic Species (Electrolytes)

Potassium sulfate is an ionic solid that dissolves in water up to 120 g L^{-1} at 25°C; this maximum mass that can be dissolved in 1 L at 25°C is called the **solubility** in water. The chemical equation for this **dissolution reaction** is written as

$$K_2SO_4(s) \longrightarrow 2\,K^+(aq) + SO_4^{2-}(aq)$$

The dissolution of ionic species (Fig. 11.3) occurs through the ion–dipole forces described in Section 10.2. Each positive ion in solution is surrounded by water molecules oriented with the negative end of their dipole moments toward the positive ion. Each SO_4^{2-} anion in solution is surrounded by water molecules oriented with the positive end of their dipole moments toward the anion. When a halide such as KCl is dissolved, the anion forms a hydrogen bond with one of the H atoms in a water molecule that places the atoms O—H—Cl nearly in a straight line as described in Section 10.2.

Each ion dissolved in water and its surrounding **solvation shell** of water molecules constitute an entity held together by ion–dipole forces or by hydrogen bonds. These solvated ions can move as intact entities when an electric field is applied (Fig. 11.4). Because the resulting solution is a conductor of electricity, ionic species such as K_2SO_4 are called **electrolytes.**

In Example 11.1, it is important to note that, although the molarity of $MgCl_2$ is 1.15 M, the molarity of Cl^- ions in the solution is twice as large, or 2.30 M, because each formula unit of $MgCl_2$ dissociates to give two Cl^- ions.

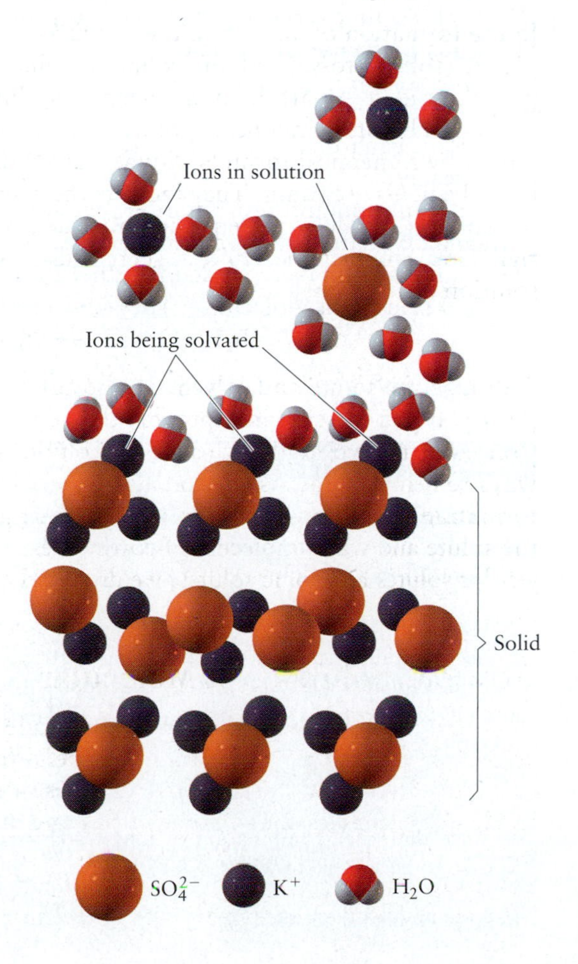

FIGURE 11.3 When an ionic solid (in this case, K_2SO_4) dissolves in water, the ions move away from their sites in the solid, where they were attracted strongly by ions of opposite electrical charge. New strong attractions replace those lost as each ion is surrounded by a group of water molecules. In a precipitation reaction, the process is reversed.

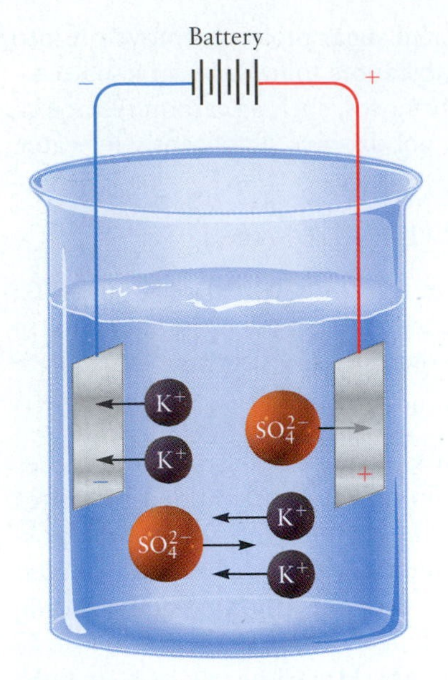

FIGURE 11.4 An aqueous solution of potassium sulfate conducts electricity. When metallic plates (electrodes) charged by a battery are put in the solution, positive ions (K^+) migrate toward the negative plate and negative ions (SO_4^{2-}) migrate toward the positive plate.

Different compounds dissolve to different extents in water. Only a small amount (0.0025 g) of solid barium sulfate dissolves per liter of water at 25°C, according to the following reaction:

$$BaSO_4(s) \longrightarrow Ba^{2+}(aq) + SO_4^{2-}(aq)$$

The near-total insolubility of barium sulfate suggests that mixing a sufficiently large amount of aqueous barium ion with aqueous sulfate ion would cause the reverse reaction to occur and solid barium sulfate would appear. Of course, it is impossible to prepare a solution containing ions of one charge only. Ions of both charges must be present to maintain overall charge neutrality. But it is possible to prepare one solution that contains a soluble barium compound in water (such as barium chloride) and a second solution that contains a soluble sulfate compound in water (such as potassium sulfate). Mixing the two solutions (Fig. 11.5) then produces solid barium sulfate through the following reaction:

$$Ba^{2+}(aq) + SO_4^{2-}(aq) \longrightarrow BaSO_4(s)$$

which is called a **precipitation reaction.**

Such precipitation reactions are sometimes written as

$$BaCl_2(aq) + K_2SO_4(aq) \longrightarrow BaSO_4(s) + 2\,KCl(aq)$$

which suggests ionic exchange—that is, the two anions exchange places. This is misleading, because $BaCl_2$, K_2SO_4, and KCl are all dissociated into ions in aqueous solution. It is more accurate to write

$$Ba^{2+}(aq) + 2\,Cl^-(aq) + 2\,K^+(aq) + SO_4^{2-}(aq) \longrightarrow BaSO_4(s) + 2\,K^+(aq) + 2\,Cl^-(aq)$$

The potassium and chloride ions appear on both sides of the equation. They are **spectator ions,** which ensure charge neutrality but do not take part directly in the chemical reaction. Omitting such spectator ions from the balanced chemical equation leads to the **net ionic equation:**

$$Ba^{2+}(aq) + SO_4^{2-}(aq) \longrightarrow BaSO_4(s)$$

A net ionic equation includes only the ions (and molecules) that actually take part in the reaction.

In dissolution and precipitation reactions, ions retain their identities, and in particular, oxidation states do not change. The ions simply exchange the positions they had in a solid (surrounded by other ions) for new positions in solution (surrounded by solvent molecules). They undergo the reverse process in precipitation.

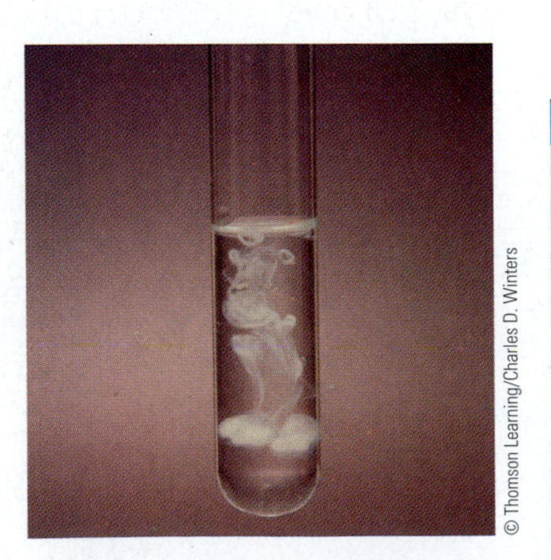

FIGURE 11.5 A solution of potassium sulfate is being added to one of barium chloride. A cloud of white solid barium sulfate is formed; the potassium chloride remains in solution.

EXAMPLE 11.4

An aqueous solution of sodium carbonate is mixed with an aqueous solution of calcium chloride, and a white precipitate immediately forms. Write a net ionic equation to account for this precipitate.

SOLUTION

Aqueous sodium carbonate contains $Na^+(aq)$ and $CO_3^{2-}(aq)$ ions, and aqueous calcium chloride contains $Ca^{2+}(aq)$ and $Cl^-(aq)$ ions. Mixing the two solutions places $Na^+(aq)$ and $Cl^-(aq)$ ions and also $Ca^{2+}(aq)$ and $CO_3^{2-}(aq)$ ions in contact for the first time. The precipitate forms by the following reaction:

$$Ca^{2+}(aq) + CO_3^{2-}(aq) \longrightarrow CaCO_3(s)$$

because the other combination of ions leads to sodium chloride, a compound that is known to be soluble in water.

Related Problems: 13, 14

11.3 Reaction Stoichiometry in Solutions: Acid–Base Titrations

Reactions in Solution

Most chemical reactions that occur on the earth's surface, whether in living organisms or among inorganic substances, take place in aqueous solution. Chemical reactions carried out between substances in solution obey the requirements of stoichiometry discussed in Chapter 2, in the sense that the conservation laws embodied in balanced chemical equations are always in force. But here we must apply these requirements in a slightly different way. Instead of a conversion between masses and number of moles, using the molar mass as a conversion factor, the conversion is now between *solution volumes* and number of moles, with the concentration as the conversion factor.

For instance, consider the reaction that is used commercially to prepare elemental bromine from its salts in solution:

$$2\ Br^-(aq) + Cl_2(aq) \longrightarrow 2\ Cl^-(aq) + Br_2(aq)$$

Suppose there is 50.0 mL of a 0.0600 M solution of NaBr. What volume of a 0.0500 M solution of Cl_2 is needed to react completely with the Br^-? To answer this, find the number of moles of bromide ion present:

$$0.0500\ L \times (0.0600\ mol\ L^{-1}) = 3.00 \times 10^{-3}\ mol\ Br^-$$

Next, use the chemical conversion factor 1 mol of Cl_2 per 2 mol of Br^- to find

$$\text{moles } Cl_2 \text{ reacting} = 3.00 \times 10^{-3}\ mol\ Br^- \left(\frac{1\ mol\ Cl_2}{2\ mol\ Br^-}\right) = 1.50 \times 10^{-3}\ mol\ Cl_2$$

Finally, find the necessary volume of aqueous chlorine:

$$\frac{1.50 \times 10^{-3}\ mol}{0.0500\ mol\ L^{-1}} = 3.00 \times 10^{-2}\ L\ \text{solution}$$

The reaction requires 3.00×10^{-2} L, or 30.0 mL, of the Cl_2 solution. (In practice, an excess of Cl_2 solution would be used to ensure more nearly complete conversion of the bromide ion to bromine.)

The chloride ion concentration after completion of the reaction might also be of interest. Because each mole of bromide ion that reacts gives 1 mol of chloride ion in the products, the number of moles of Cl^- produced is 3.00×10^{-3} mol. The final volume of the solution is 0.0800 L, so the final concentration of Cl^- is

$$[Cl^-] = \frac{3.00 \times 10^{-3}\ mol}{0.0800\ L} = 0.0375\ \text{M}$$

Square brackets around a chemical symbol signify the molarity of that species.

EXAMPLE 11.5

When potassium dichromate is added to concentrated hydrochloric acid, it reacts according to the following chemical equation

$$K_2Cr_2O_7(s) + 14\ HCl(aq) \longrightarrow 2\ K^+(aq) + 2\ Cr^{3+}(aq) + 8\ Cl^-(aq) + 7\ H_2O(\ell) + 3\ Cl_2(g)$$

producing a mixed solution of chromium(III) chloride and potassium chloride and evolving gaseous chlorine. Suppose that 6.20 g of $K_2Cr_2O_7$ reacts with concentrated HCl, and that the final volume of the solution is 100.0 mL. Calculate the final concentration of $Cr^{3+}(aq)$ and the number of moles of chlorine produced.

SOLUTION

The first step is to convert the mass of $K_2Cr_2O_7$ to moles:

$$\frac{6.20 \text{ g } K_2Cr_2O_7}{294.19 \text{ g mol}^{-1}} = 0.0211 \text{ mol } K_2Cr_2O_7$$

The balanced chemical equation states that 1 mol of $K_2Cr_2O_7$ reacts to give 2 mol of Cr^{3+} and 3 mol of Cl_2. Using these two chemical conversion factors gives

$$\text{moles } Cr^{3+} = 0.0211 \text{ mol } K_2Cr_2O_7\left(\frac{2 \text{ mol } Cr^{3+}}{1 \text{ mol } K_2Cr_2O_7}\right)$$

$$= 0.0422 \text{ mol } Cr^{3+}$$

$$\text{moles } Cl_2 = 0.0211 \text{ mol } K_2Cr_2O_7\left(\frac{3 \text{ mol } Cl_2}{1 \text{ mol } K_2Cr_2O_7}\right)$$

$$= 0.0633 \text{ mol } Cl_2$$

Because the final volume of the solution is 0.100 L, the concentration of $Cr^{3+}(aq)$ is

$$[Cr^{3+}] = \frac{0.0422 \text{ mol}}{0.100 \text{ L}} = 0.422 \text{ M}$$

Related Problems: 15, 16

Titration

One of the most important techniques in analytical chemistry is **titration**—the addition of a carefully measured volume of one solution, containing substance A in known concentration, to a second solution, containing substance B in unknown concentration. Solution A is added through a buret, an instrument that accurately measures the volume of solution transferred as a stopcock is opened and then closed. As the solutions are mixed, A and B react quantitatively. Completion of the reaction, the **end point,** is signaled by a change in some physical property, such as the color of the reacting mixture. End points can be detected in colorless reaction mixtures by adding a substance called an **indicator** that changes color at the end point. At the end point, the known number of moles of substance A that has been added is uniquely related to the unknown number of moles of substance B initially present by the balanced equation for the titration reaction. Titration enables chemists to determine the unknown amount of a substance present in a sample. The two most common applications of titrations involve acid–base neutralization reactions and oxidation–reduction (or redox) reactions. We describe both here briefly to illustrate the fundamental importance of solution stoichiometry calculations in titrations. Detailed discussion of acid–base and redox reactions, and the extent to which they go to completion, are presented in Chapter 15 and Chapter 17, respectively.

Background on Acid–Base Reactions

Table 11.1 lists the names and formulas of a number of important acids. Acids and bases have been known and characterized since ancient times. Chemical description and explanation of their properties and behavior has progressed through several stages of sophistication and generality. A broadly applicable modern treatment is presented in Chapter 15. Here, we introduce titrations using the treatment of the Swedish chemist Svante Arrhenius, who defined acids and bases by their behavior when dissolved in water.

TABLE 11.1 Names of Common Acids

Binary Acids	Oxoacids	Organic Acids
HF, hydrofluoric acid	H_2CO_3, carbonic acid	HCOOH, formic acid
HCl, hydrochloric acid	H_3PO_3, phosphorus acid	CH_3COOH, acetic acid
HCN, hydrocyanic acid†	H_3PO_4, phosphoric acid	C_6H_5COOH, benzoic acid
H_2S, hydrosulfuric acid	HNO_2, nitrous acid	HOOCCOOH, oxalic acid
	HNO_3, nitric acid	
	H_2SO_3, sulfurous acid	
	H_2SO_4, sulfuric acid	
	HClO, hypochlorous acid	
	$HClO_2$, chlorous acid	
	$HClO_3$, chloric acid	
	$HClO_4$, perchloric acid	

†Contains three elements but is named as a binary acid.

Sulfuric acid, H_2SO_4, is the industrial chemical produced on the largest scale in the world, in amounts exceeding 100 million tons per year.

Hydrochloric acid, HCl, is used in the pickling of steel and other metals to remove oxide layers on the surface.

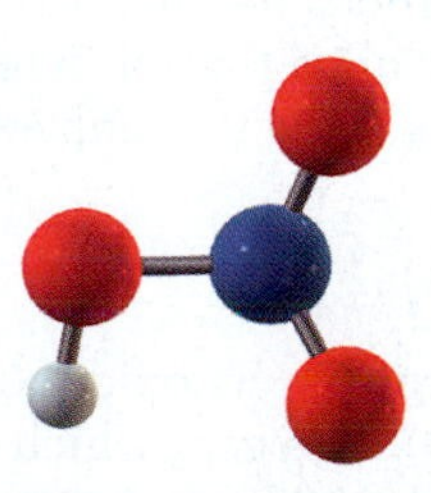

Nitric acid, HNO_3, is manufactured from ammonia.

In pure water, small but equal numbers of hydronium ions (H_3O^+) and hydroxide ions (OH^-) are present.[1] These arise from the partial ionization of water:

$$2\,H_2O(\ell) \longrightarrow H_3O^+(aq) + OH^-(aq)$$

Following Arrhenius, we define an **acid** as a substance that when dissolved in water increases the number of hydronium ions over the number present in pure water. Gaseous hydrogen chloride reacts with water to give hydrochloric acid:

$$H_2O(\ell) + HCl(g) \longrightarrow H_3O^+(aq) + Cl^-(aq)$$

A **base** is defined as a substance that when dissolved increases the number of hydroxide ions over the number present in pure water. Sodium hydroxide dissolves extensively in water according to the following reaction:

$$NaOH(s) \longrightarrow Na^+(aq) + OH^-(aq)$$

Ammonia is another base, as shown by the products of its reaction with water:

$$NH_3(aq) + H_2O(\ell) \longrightarrow NH_4^+(aq) + OH^-(aq)$$

When an acidic solution is mixed with a basic solution, a **neutralization reaction** occurs:

$$H_3O^+(aq) + OH^-(aq) \longrightarrow 2\,H_2O(\ell)$$

This is the reverse of the water ionization reaction shown earlier. If the spectator ions are put back into the equation, it reads

$$\underset{\text{Acid}}{HCl} + \underset{\text{Base}}{NaOH} \longrightarrow \underset{\text{Water}}{H_2O} + \underset{\text{Salt}}{NaCl}$$

showing that a salt can be defined as the product (other than water) of the reaction of an acid with a base. It is usually preferable to omit the spectator ions and to indicate explicitly only the reacting ions.

Acid–Base Titration

In most acid–base reactions, there is no sharp color change at the end point. In such cases it is necessary to add a small amount of an indicator, a dye that changes color when the reaction is complete (indicators are discussed in detail in

[1]As discussed in Chapter 15, a hydrogen ion in aqueous solution is closely held by a water molecule and is better represented as a hydronium ion, $H_3O^+(aq)$.

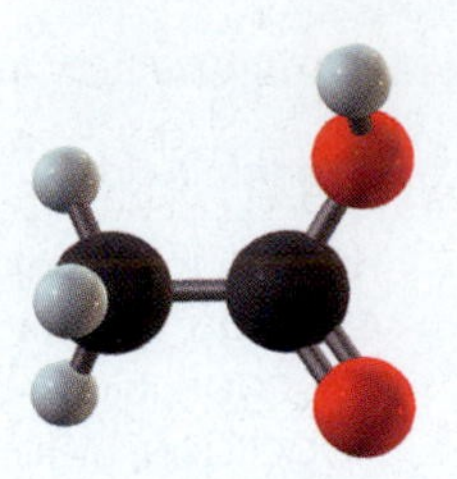

Acetic acid, CH_3COOH, is a common organic acid that is found in vinegar as a 3% to 5% solution by mass.

Section 15.3). Phenolphthalein is such an indicator, changing from colorless to pink when a solution changes from acidic to basic. The concentration of acetic acid in an aqueous solution can be determined by adding a few drops of a phenolphthalein solution and then titrating it with a solution of sodium hydroxide of accurately known concentration. At the first permanent appearance of a pink color, the stopcock of the buret is closed. At this point, the reaction

$$CH_3COOH(aq) + OH^-(aq) \longrightarrow CH_3COO^-(aq) + H_2O(\ell)$$

has gone stoichiometrically to completion.

EXAMPLE 11.6

A sample of vinegar is to be analyzed for its acetic acid content. A volume of 50.0 mL is measured out and titrated with a solution of 1.306 M NaOH; 31.66 mL of that titrant is required to reach the phenolphthalein end point. Calculate the concentration of acetic acid in the vinegar (in moles per liter).

SOLUTION

The number of moles of NaOH reacting is found by multiplying the volume of NaOH solution (31.66 mL = 0.03166 L) by its concentration (1.306 M):

$$0.03166 \text{ L} \times 1.306 \text{ mol L}^{-1} = 4.135 \times 10^{-2} \text{ mol NaOH}$$

Because 1 mol of acetic acid reacts with 1 mol of $OH^-(aq)$, the number of moles of acetic acid originally present must also have been 4.135×10^{-2} mol. Its concentration was then

$$[CH_3COOH] = \frac{4.135 \times 10^{-2} \text{ mol}}{0.0500 \text{ L}} = 0.827 \text{ M}$$

Related Problems: 25, 26

11.4 Reaction Stoichiometry in Solutions: Oxidation–Reduction Titrations

Background on Oxidation–Reduction (Redox) Reactions

In **oxidation–reduction** (or **redox**) **reactions,** electrons are transferred between reacting species as they combine to form products. This exchange is described as a change in the oxidation number of the reactants: The oxidation number of the species giving up electrons increases, whereas that for the species accepting electrons decreases. Oxidation numbers are defined and methods for their calculation are presented in Section 3.10. A prototype redox reaction is that of magnesium (Mg) with oxygen (O) (Fig. 11.6). When this reaction is carried to completion, the product is magnesium oxide:

$$2\ Mg(s) + O_2(g) \longrightarrow 2\ MgO(s)$$

Magnesium is **oxidized** in this process; it *gives up* electrons as its oxidation number *increases* from 0 (in elemental Mg) to +2 (in MgO). Oxygen, which accepts these electrons, is said to be reduced; its oxidation number decreases from 0 to −2. The transfer of electrons (e^-) can be indicated with arrows:

$$\underset{\substack{\downarrow \\ 2 \times 2e^-}}{2\ \overset{0}{Mg}} + \underset{\substack{\uparrow \\ = 2 \times 2e^-}}{\overset{0}{O_2}} \longrightarrow 2\ \overset{+2}{Mg}\overset{-2}{O}$$

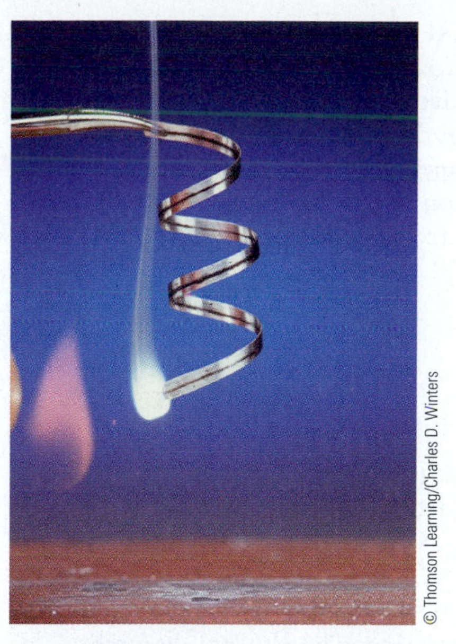

FIGURE 11.6 Magnesium burning in air gives off an extremely bright light. This characteristic led to the incorporation of magnesium into the flash powder used in early photography. Magnesium powder is still used in fireworks for the same reason.

The arrows point away from the species being oxidized (giving up electrons) and toward the species being reduced (accepting electrons). The electron "bookkeeping" beneath the equation ensures that the same number of electrons are taken up by oxygen as are given up by magnesium:

$$2 \text{ Mg atoms} \times 2 \text{ electrons per Mg atom} = 2 \text{ O atoms per formula unit} \times 2 \text{ electrons per O atom}$$

Originally, the term *oxidation* referred only to reactions with oxygen. It now is used to describe any process in which the oxidation number of a species increases, even if oxygen is not involved in the reaction. When calcium combines with chlorine to form calcium chloride,

$$\overset{0}{\mathrm{Ca}(s)} + \overset{0}{\mathrm{Cl_2}(g)} \longrightarrow \overset{+2\,-1}{\mathrm{CaCl_2}}$$

$$\downarrow \qquad \uparrow$$

$$2e^- = 2 \times 1e^-$$

the calcium has been oxidized and the chlorine has been reduced.

Oxidation–reduction reactions are among the most important in chemistry, biochemistry, and industry. Combustion of coal, natural gas, and gasoline for heat and power are redox reactions, as are the recovery of metals such as iron and aluminum from their oxide ores and the production of chemicals such as sulfuric acid from sulfur, air, and water. The human body metabolizes sugars through redox reactions to obtain energy; the reaction products are liquid water and gaseous carbon dioxide.

EXAMPLE 11.7

Determine whether the following equations represent oxidation–reduction reactions.

(a) $SnCl_2(s) + Cl_2(g) \longrightarrow SnCl_4(\ell)$

(b) $CaCO_3(s) \longrightarrow CaO(s) + CO_2(g)$

(c) $2\ H_2O_2(\ell) \longrightarrow 2\ H_2O(\ell) + O_2(g)$

SOLUTION

(a) Tin increases its oxidation number from +2 to +4, and the oxidation number of chlorine in Cl_2 is reduced from 0 to −1. Some other chlorine atoms are unchanged in oxidation number, but this is still a redox reaction.

(b) The oxidation state of Ca remains at +2, that of O at −2, and that of C at +4. Thus, this is not a redox reaction.

(c) The oxidation number of oxygen in H_2O_2 is −1. This changes to −2 in the product H_2O and 0 in O_2. This is a redox reaction in which the same element is both oxidized and reduced.

Related Problems: 27, 28, 29, 30

Balancing Oxidation–Reduction Equations in Aqueous Solution

Numerous redox reactions occur in aqueous solution. For example, consider the reaction between solid copper (Cu) and an aqueous solution of silver (Ag) nitrate:

$$Cu(s) + 2\ Ag^+(aq) \longrightarrow Cu^{2+}(aq) + 2\ Ag(s)$$

The nitrate ions are spectators that do not take part in the reaction, so they are omitted in the net equation. Two electrons are transferred from each reacting Cu atom to a pair of silver ions. Copper (the electron donor) is oxidized, and silver

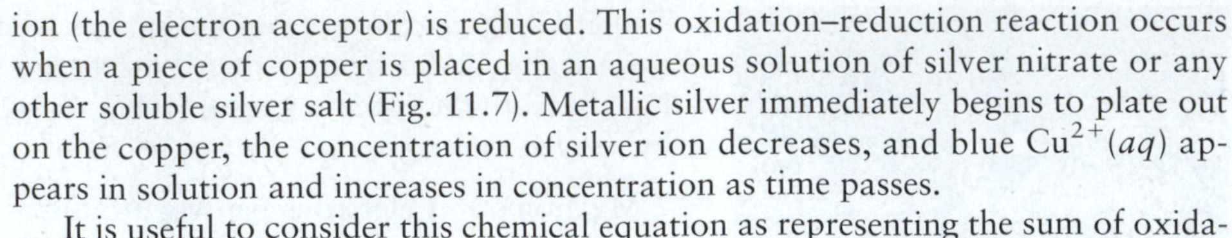

ion (the electron acceptor) is reduced. This oxidation–reduction reaction occurs when a piece of copper is placed in an aqueous solution of silver nitrate or any other soluble silver salt (Fig. 11.7). Metallic silver immediately begins to plate out on the copper, the concentration of silver ion decreases, and blue $Cu^{2+}(aq)$ appears in solution and increases in concentration as time passes.

FIGURE 11.7 When a piece of copper screen is inserted into a solution of silver nitrate, silver forms in a tree-like structure and the solution turns blue as Cu^{2+} ions form.

It is useful to consider this chemical equation as representing the sum of oxidation and reduction **half-reactions,** in which electrons (e^-) appear explicitly. The oxidation of copper is written as

$$Cu(s) \longrightarrow Cu^{2+}(aq) + 2\,e^-$$

and the reduction of silver ion as

$$Ag^+(aq) + e^- \longrightarrow Ag(s)$$

In the net equation, electrons must not appear explicitly. Thus, the second equation must be multiplied by 2 before being added to the first so that the electrons cancel out on both sides. As before, this gives

$$Cu(s) + 2\,Ag^+(aq) \longrightarrow Cu^{2+}(aq) + 2\,Ag(s)$$

Many redox reactions are too difficult to balance by the simple methods of logical reasoning described in Section 2.4. Here, we outline a systematic procedure based on half-reactions and apply it to reactions that occur in acidic or basic aqueous solution. In these reactions, water and H_3O^+ (acidic solution) or OH^- (basic solution) may take part either as reactants or as products; thus, it is necessary to *complete* the corresponding equations, as well as to balance them.

As an example, let's complete and balance the chemical equation for the dissolution of copper(II) sulfide in aqueous nitric acid (Fig. 11.8):

$$CuS(s) + NO_3^-(aq) \longrightarrow Cu^{2+}(aq) + SO_4^{2-}(aq) + NO(g)$$

Step 1 *Write two unbalanced half-equations, one for the species that is oxidized and its product and one for the species that is reduced and its product.*

Here, the unbalanced half-reaction involving CuS is

$$CuS \longrightarrow Cu^{2+} + SO_4^{2-}$$

The unbalanced half-reaction involving NO_3^- is

$$NO_3^- \longrightarrow NO$$

Step 2 *Insert coefficients to make the numbers of atoms of all elements except oxygen and hydrogen equal on the two sides of each equation.*

In this case, copper, sulfur, and nitrogen are already balanced in the two half-equations, so this step is already completed.

FIGURE 11.8 Copper(II) sulfide reacts with concentrated nitric acid to liberate nitrogen oxide and produce a solution of copper(II) sulfate, which displays the characteristic blue color of copper(II) ions in water.

Step 3 *Balance oxygen by adding H_2O to one side of each half-equation.*

$$CuS + 4\,H_2O \longrightarrow Cu^{2+} + SO_4^{2-}$$

$$NO_3^- \longrightarrow NO + 2\,H_2O$$

Step 4 *Balance hydrogen. For an acidic solution, add H_3O^+ to the side of each half-equation that is "deficient" in hydrogen and add an equal amount of H_2O to the other side. For a basic solution, add H_2O to the side of each half-equation that is "deficient" in hydrogen and add an equal amount of OH^- to the other side.*

Note that this step does not disrupt the oxygen balance achieved in step 3. In this case (acidic solution), the result is

$$CuS + 12\,H_2O \longrightarrow Cu^{2+} + SO_4^{2-} + 8\,H_3O^+$$

$$NO_3^- + 4\,H_3O^+ \longrightarrow NO + 6\,H_2O$$

Step 5 *Balance charge by inserting* e^- *(electrons) as a reactant or product in each half-equation.*

$$CuS + 12\ H_2O \longrightarrow Cu^{2+} + SO_4^{2-} + 8\ H_3O^+ + 8\ e^-$$
(oxidation)

$$NO_3^- + 4\ H_3O^+ + 3\ e^- \longrightarrow NO + 6\ H_2O$$
(reduction)

Step 6 *Multiply the two half-equations by numbers chosen to make the number of electrons given off by the oxidation equal the number taken up by the reduction. Then add the two half-equations, canceling electrons. If* H_3O^+, OH^-, *or* H_2O *appears on both sides of the final equation, cancel out the duplications.*

Here, the oxidation half-equation must be multiplied by 3 (so that 24 electrons are produced), and the reduction half-equation by 8 (so that the same 24 electrons are consumed):

$$3\ CuS + 36\ H_2O \longrightarrow 3\ Cu^{2+} + 3\ SO_4^{2-} + 24\ H_3O^+ + 24\ e^-$$
$$8\ NO_3^- + 32\ H_3O^+ + 24\ e^- \longrightarrow 8\ NO + 48\ H_2O$$
$$\overline{3\ CuS + 8\ NO_3^- + 8\ H_3O^+ \longrightarrow 3\ Cu^{2+} + 3\ SO_4^{2-} + 8\ NO + 12\ H_2O}$$

This procedure balances equations that are too difficult to balance by inspection. For basic solutions, remember to add H_2O and OH^-, rather than H_3O^+ and H_2O, at step 4.

EXAMPLE 11.8

Balance the following equation, which represents a reaction that occurs in basic aqueous solution:

$$Ag(s) + HS^-(aq) + CrO_4^{2-}(aq) \longrightarrow Ag_2S(s) + Cr(OH)_3(s)$$

SOLUTION

Step 1

$$Ag + HS^- \longrightarrow Ag_2S$$
$$CrO_4^{2-} \longrightarrow Cr(OH)_3$$

Step 2

$$2\ Ag + HS^- \longrightarrow Ag_2S$$

The other half-reaction is unchanged.

Step 3 H_2O is now added to the second half-reaction to balance oxygen:

$$CrO_4^{2-} \longrightarrow Cr(OH)_3 + H_2O$$

Step 4 The right side of the silver half-reaction is deficient by 1 H. Add 1 H_2O to the right and 1 OH^- to the left:

$$2\ Ag + HS^- + OH^- \longrightarrow Ag_2S + H_2O$$

In the chromium half-reaction, the left side is deficient by 5 H atoms, so 5 H_2O is added to that side and 5 OH^- to the right side:

$$CrO_4^{2-} + 4\ H_2O \longrightarrow Cr(OH)_3 + 5\ OH^-$$

(Notice that the H_2O on the right canceled out one of the five on the left, leaving four.)

Step 5 Electrons are added to the right side of the silver half-reaction and to the left side of the chromium half-reaction to balance charge:

$$2\ Ag + Hs^- + OH^- \longrightarrow Ag_2S + H_2O + 2\ e^-$$
(oxidation)

$$CrO_4^{2-} + 4\ H_2O + 3\ e^- \longrightarrow Cr(OH)_3 + 5\ OH^-$$
(reduction)

Step 6 The gain and loss of electrons are equalized. The first equation is multiplied by 3 so that it consumes six electrons, and the second is multiplied by 2 so that it produces six electrons.

$$6\ Ag + 3\ HS^- + 3\ OH^- \longrightarrow 3\ Ag_2S + 3\ H_2O + 6\ e^- \quad \text{(oxidation)}$$

$$2\ CrO_4^{2-} + 8\ H_2O + 6\ e^- \longrightarrow 2\ Cr(OH)_3 + 10\ OH^- \quad \text{(reduction)}$$

$$6\ Ag + 3\ HS^- + 2\ CrO_4^{2-} + 5\ H_2O \longrightarrow 3\ Ag_2S + 2\ Cr(OH)_3 + 7\ OH^-$$

Related Problems: 31, 32, 33, 34

EXAMPLE 11.9

Balance the following equation for the reaction of arsenic(III) sulfide with aqueous chloric acid:

$$As_2S_3(s) + ClO_3^-(aq) \longrightarrow H_3AsO_4(aq) + SO_4^{2-}(aq) + Cl^-(aq)$$

SOLUTION

Step 1

$$As_2S_3 \longrightarrow H_3AsO_4 + SO_4^{2-}$$

$$ClO_3^- \longrightarrow Cl^-$$

Step 2

$$As_2S_3 \longrightarrow 2\ H_3AsO_4 + 3\ SO_4^{2-}$$

Step 3

$$As_2S_3 + 20\ H_2O \longrightarrow 2\ H_3AsO_4 + 3\ SO_4^{2-}$$

$$ClO_3^- \longrightarrow Cl^- + 3\ H_2O$$

Step 4 The chloric acid makes this an acidic solution, so H_3O^+ and H_2O are used:

Step 5

$$As_2S_3 + 54\ H_2O \longrightarrow 2\ H_3AsO_4 + 3\ SO_4^{2-} + 34\ H_3O^+$$

$$ClO_3^- + 6\ H_3O^+ \longrightarrow Cl^- + 9\ H_2O$$

$$As_2S_3 + 54\ H_2O \longrightarrow 2\ H_3AsO_4 + 3\ SO_4^{2-} + 34\ H_3O^+ + 28\ e^-$$

$$ClO_3^- + 6\ H_3O^+ + 6\ e^- \longrightarrow Cl^- + 9\ H_2O$$

Step 6 We must find the least common multiple of 28 and 6. This is 84; therefore, we need to multiply the first equation by 84/28 = 3 and the second equation by 84/6 = 14:

$$3\ As_2S_3 + 162\ H_2O \longrightarrow 6\ H_3AsO_4 + 9\ SO_4^{2-} + 102\ H_3O^+ + 84\ e^-$$

$$14\ ClO_3^- + 84\ H_3O^+ + 84\ e^- \longrightarrow 14\ Cl^- + 126\ H_2O$$

$$3\ As_2S_3 + 14\ ClO_3^- + 36\ H_2O \longrightarrow 6\ H_3AsO_4 + 9\ SO_4^{2-} + 14\ Cl^- + 18\ H_3O^+$$

Disproportionation

An important type of redox reaction, **disproportionation,** occurs when a single substance is both oxidized and reduced. Example 11.7, part (c), provided such a reaction:

$$2\ \overset{-1}{H_2O_2}(\ell) \longrightarrow 2\ \overset{-2}{H_2O}(\ell) + \overset{0}{O_2}(g)$$

The oxygen in hydrogen peroxide (H_2O_2) is in an intermediate oxidation state of −1; some of it is oxidized to O_2 and some is reduced to H_2O. The balancing of equations for disproportionation reactions is described in the following example.

EXAMPLE 11.10

Balance the following equation for the reaction that occurs when chlorine is dissolved in basic solution:

$$Cl_2(g) \longrightarrow ClO_3^-(aq) + Cl^-(aq)$$

SOLUTION

Step 1 We solve this by writing the Cl_2 on the left sides of two half-equations:

$$Cl_2 \longrightarrow ClO_3^-$$

$$Cl_2 \longrightarrow Cl^-$$

Step 2

$$Cl_2 \longrightarrow 2\ ClO_3^-$$

$$Cl_2 \longrightarrow 2\ Cl^-$$

Step 3 The first half-equation becomes

$$Cl_2 + 6\ H_2O \longrightarrow 2\ ClO_3^-$$

Step 4 Now the first half-equation becomes

$$Cl_2 + 12\ OH^- \longrightarrow 2\ ClO_3^- + 6\ H_2O$$

Step 5

$$Cl_2 + 12\ OH^- \longrightarrow 2ClO_3^- + 6\ H_2O + 10\ e^- \quad \text{(oxidation)}$$

$$Cl_2 + 2\ e^- \longrightarrow 2\ Cl^- \quad \text{(reduction)}$$

Step 6 Multiply the second equation by 5 and add:

$$Cl_2 + 12\ OH^- \longrightarrow 2\ ClO_3^- + 6\ H_2O + 10\ e^-$$

$$5\ Cl_2 + 10\ e^- \longrightarrow 10\ Cl^-$$

$$6\ Cl_2 + 12\ OH^- \longrightarrow 2\ ClO_3^- + 10\ Cl^- + 6\ H_2O$$

Dividing this equation by 2 gives

$$3\ Cl_2 + 6\ OH^- \longrightarrow ClO_3^- + 5\ Cl^- + 3\ H_2O$$

Related Problems: 37, 38

Redox Titration

Redox titrations are illustrated by the reaction in which potassium permanganate oxidizes Fe^{2+} in acidic solution and the manganese (Mn) is reduced:

$$MnO_4^-(aq) + 5\ Fe^{2+}(aq) + 8\ H_3O^+(aq) \longrightarrow Mn^{2+}(aq) + 5\ Fe^{3+}(aq) + 12\ H_2O(\ell)$$

Redox titrations have the advantage that many involve intensely colored species that change color dramatically at the end point. For example, MnO_4^- is deep purple, whereas Mn^{2+} is colorless. Thus, when MnO_4^- has been added to Fe^{2+} in slight excess, the color of the solution changes permanently to purple.

The titration is begun by opening the stopcock of the buret and letting a small volume of permanganate solution run into the flask that contains the Fe^{2+} solution. A dash of purple colors the solution (Fig. 11.9a) but rapidly disappears as the permanganate ion reacts with Fe^{2+} to give the nearly colorless Mn^{2+} and Fe^{3+} products. The addition of incremental volumes of permanganate solution is continued until the Fe^{2+} is almost completely converted to Fe^{3+}. At this stage, the addition of just one drop of $KMnO_4$ imparts a pale purple color to the reaction mixture (see Fig. 11.9b) and signals the completion of the reaction. The volume of the titrant $KMnO_4$ solution is calculated by subtracting the initial reading of the solution meniscus in the buret from the final volume reading.

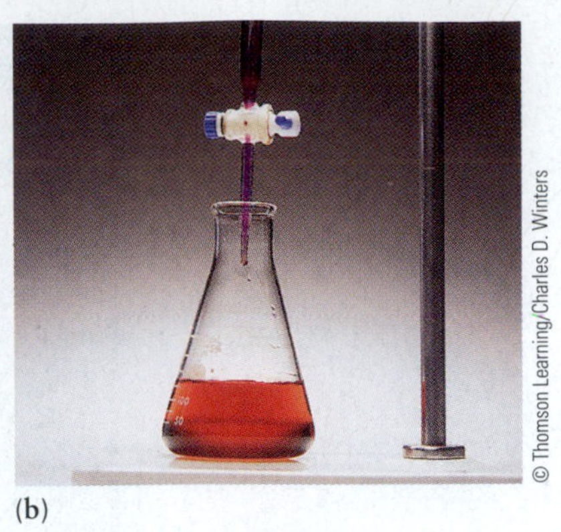

(a) (b)

FIGURE 11.9 (a) Addition of a small amount of potassium permanganate from a buret gives a dash of purple color to the Fe^{2+} solution, which disappears quickly as the permanganate ion reacts to give Fe^{3+} and Mn^{2+}. (b) When all the Fe^{2+} ions have been consumed, additional drops of permanganate solution give a pale purple color to the solution.

Suppose that the permanganate solution has a concentration of 0.09625 M, and 26.34 mL (0.02634 L) of it is added to reach the end point. The number of moles of Fe^{2+} in the original solution is calculated in two steps:

$$\text{Amount of } MnO_4^- \text{ reacting} = 0.02634 \text{ L} \times 0.09625 \text{ mol L}^{-1}$$
$$= 2.535 \times 10^{-3} \text{ mol } MnO_4^-$$

From the balanced chemical equation, each mole of MnO_4^- used causes the oxidation of 5 mol of Fe^{2+}, so

$$\text{Amount of } Fe^{2+} \text{ reacting} = 2.535 \times 10^{-3} \text{ mol } MnO_4^- \times \left(\frac{5 \text{ mol } Fe^{2+}}{1 \text{ mol } MnO_4^-}\right)$$
$$= 1.268 \times 10^{-2} \text{ mol } Fe^{2+}$$

These direct titrations form the basis of more complicated analytical procedures. Many analytical procedures are indirect and involve additional preliminary reactions of the sample before the titration can be carried out. For example, a soluble calcium salt will not take part in a redox reaction with potassium permanganate. But adding ammonium oxalate to the solution containing Ca^{2+} causes the quantitative precipitation of calcium oxalate:

$$Ca^{2+}(aq) + C_2O_4^{2-}(aq) \longrightarrow CaC_2O_4(s)$$

After the precipitate is filtered and washed, it is dissolved in sulfuric acid to form oxalic acid:

$$CaC_2O_4(s) + 2\ H_3O^+(aq) \longrightarrow Ca^{2+}(aq) + H_2C_2O_4(aq) + 2\ H_2O$$

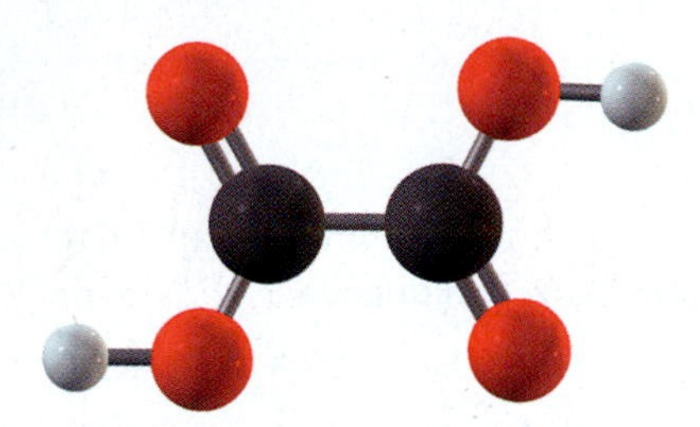

Oxalic acid, $H_2C_2O_4$, is an acid that can give up two H ions in aqueous solution to form the oxalate ion, $C_2O_4^{2-}$.

Finally, the oxalic acid is titrated with permanganate solution of accurately known concentration, based on the redox reaction

$$2\ MnO_4^-(aq) + 5\ H_2C_2O_4(aq) + 6\ H_3O^+(aq) \longrightarrow$$
$$2\ Mn^{2+}(aq) + 10\ CO_2(g) + 14\ H_2O(\ell)$$

In this way, the quantity of calcium can be determined indirectly by reactions that involve precipitation, acid–base, and redox steps.

11.5 Phase Equilibrium in Solutions: Nonvolatile Solutes

We turn now from the chemical reactions of solutions to such physical properties as their vapor pressures and phase diagrams. Consider first a solution made by dissolving a nonvolatile solute in a solvent. By "nonvolatile" we mean that the vapor pressure of the solute above the solution is negligible. An example is a

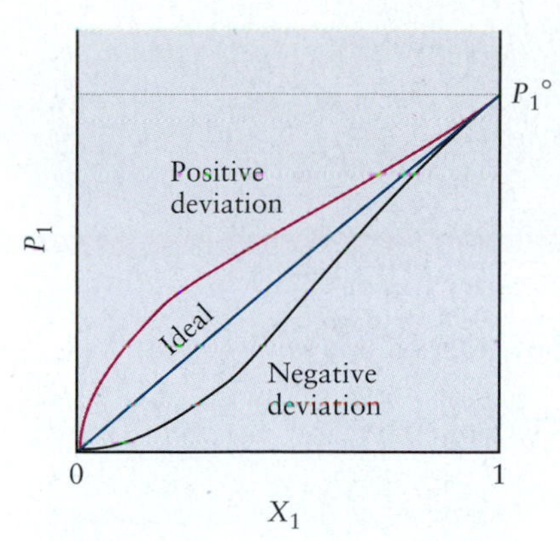

FIGURE 11.10 In an ideal solution, a graph of solvent vapor pressure P_1 versus mole fraction of solvent X_1 is a straight line. Nonideal solutions behave differently; examples of positive and negative deviations from the ideal solution are shown. The vapor pressure of pure solvent is P_1°.

solution of sucrose (cane sugar) in water, in which the vapor pressure of sucrose above the solution is zero. We study the case of a volatile solute in Section 11.6.

The *solvent* vapor pressure is not zero and changes with the composition of the solution at a fixed temperature. If the mole fraction of solvent (X_1) is 1, then the vapor pressure is P_1°, the vapor pressure of pure solvent at the temperature of the experiment. When X_1 approaches 0 (giving pure solute), the vapor pressure P_1 of the solvent must go to 0 also, because solvent is no longer present. As the mole fraction X_1 changes from 1 to 0, P_1 drops from P_1° to 0. What is the shape of the curve?

The French chemist François-Marie Raoult found that for some solutions a plot of solvent vapor pressure against solvent mole fraction can be fitted closely by a straight line (Fig. 11.10). Solutions that conform to this straight-line relationship obey the following simple equation:

$$P_1 = X_1 P_1^\circ \qquad [11.5]$$

which is known as **Raoult's law.** Such solutions are called **ideal solutions.** Other solutions deviate from straight-line behavior and are called **nonideal solutions.** They may show positive deviations (with vapor pressures higher than those predicted by Raoult's law) or negative deviations (with lower vapor pressures). On a molecular level, negative deviations arise when the solute attracts solvent molecules especially strongly, reducing their tendency to escape into the vapor phase. Positive deviations arise in the opposite case, when solvent and solute molecules are not strongly attracted to each other. Even nonideal solutions with nondissociating solutes approach Raoult's law as X_1 approaches 1, just as all real gases obey the ideal gas law at sufficiently low densities.

Raoult's law forms the basis for four properties of dilute solutions, which are called **colligative properties** (derived from Latin *colligare,* meaning "to collect together") because they depend on the collective effect of the *number* of dissolved particles rather than on the *nature* of the particular particles involved. These four properties are:

1. The lowering of the vapor pressure of a solution relative to pure solvent
2. The elevation of the boiling point
3. The depression of the freezing point
4. The phenomenon of osmotic pressure

Vapor-Pressure Lowering

Because $X_1 = 1 - X_2$ for a two-component solution, Raoult's law can be rewritten as

$$\Delta P_1 = P_1 - P_1^\circ = X_1 P_1^\circ - P_1^\circ = -X_2 P_1^\circ \qquad [11.6]$$

so that the *change* in vapor pressure of the solvent is proportional to the mole fraction of solute. The negative sign implies **vapor-pressure lowering;** the vapor pressure is always less above a dilute solution than it is above the pure solvent.

EXAMPLE 11.11

At 25°C, the vapor pressure of pure benzene is $P_1^\circ = 0.1252$ atm. Suppose 6.40 g of naphthalene, $C_{10}H_8$ (molar mass 128.17 g mol^{-1}), is dissolved in 78.0 g of benzene (molar mass 78.0 g mol^{-1}). Calculate the vapor pressure of benzene over the solution, assuming ideal behavior.

SOLUTION

The number of moles of solvent, n_1, is 1.00 mol in this case (because 78.0 g = 1.00 mol benzene has been used). The number of moles of solute is n_2 = 6.40 g/128.17 g mol^{-1} = 0.0499 mol $C_{10}H_8$; thus, the mole fraction X_1 is

$$X_1 = \frac{n_1}{n_1 + n_2} = \frac{1.00 \text{ mol}}{1.00 + 0.0499 \text{ mol}} = 0.9525$$

From Raoult's law, the vapor pressure of benzene above the solution is

$$P_1 = X_1 P_1^\circ = 0.9525 \times 0.1252 \text{ atm} = 0.119 \text{ atm}$$

Related Problems: 41, 42

Boiling-Point Elevation

The normal boiling point of a pure liquid T_b or a solution T_b' is the temperature at which the vapor pressure reaches 1 atm. Because a dissolved solute reduces the vapor pressure, the temperature of the solution must be increased to make it boil. That is, the boiling point of a solution is higher than that of the pure solvent. This phenomenon, referred to as **boiling-point elevation,** provides a method for determining molar masses.

The vapor-pressure curve of a dilute solution lies slightly below that for the pure solvent. In Figure 11.11, ΔP_1 is the decrease of vapor pressure at T_b and ΔT_b is the change in temperature necessary to hold the vapor pressure at 1 atm (that is, $\Delta T_b = T_b' - T_b$ is the increase in boiling point caused by addition of solute to the pure solvent). For small concentrations of nondissociating solutes, the two curves are parallel, so

$$-\frac{\Delta P_1}{\Delta T_b} = \text{slope of curve} = S$$

$$\Delta T_b = -\frac{\Delta P_1}{S} = \frac{X_2 P_1^\circ}{S}$$

$$= \frac{1}{S}\left(\frac{n_2}{n_1 + n_2}\right) \qquad \text{(from Raoult's law, with } P_1^\circ = 1 \text{ atm)}$$

The constant S is a property of the pure solvent only, because it is the slope of the vapor pressure curve $\Delta P_1/\Delta T_b$ near 1 atm pressure. That is, S is independent of the solute species involved.

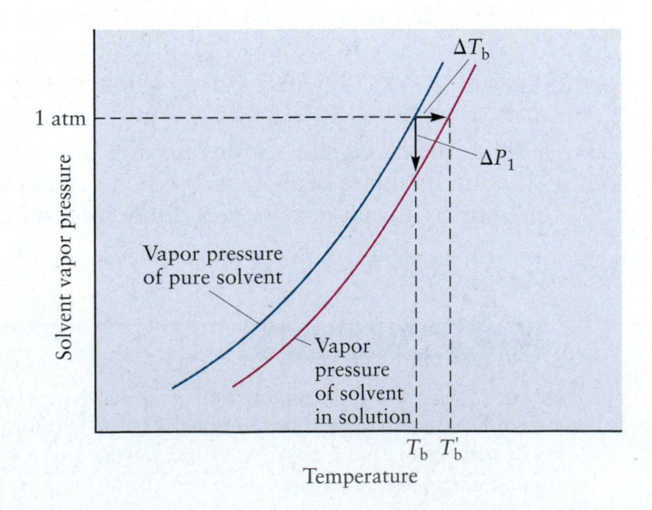

FIGURE 11.11 The vapor pressure of the solvent above a dilute solution is lower than that of the pure solvent at all temperatures. As a result, for the solution to boil (that is, for the vapor pressure to reach 1 atm), a higher temperature is required for the solution than for the pure solvent. This amounts to an elevation of the boiling point.

TABLE 11.2 Boiling-Point Elevation and Freezing-Point Depression Constants

Solvent	Formula	T_b (°C)	K_b (K kg mol^{-1})	T_f (°C)	K_f (K kg mol^{-1})
Acetic acid	CH_3COOH	118.1	3.07	17	3.9
Benzene	C_6H_6	80.1	2.53	5.5	4.9
Carbon tetrachloride	CCl_4	76.7	5.03	−22.9	32
Diethyl ether	$C_4H_{10}O$	34.7	2.02	−116.2	1.8
Ethanol	C_2H_5OH	78.4	1.22	−114.7	1.9
Naphthalene	$C_{10}H_8$	—	—	80.5	6.8
Water	H_2O	100.0	0.512	0.0	1.86

For very dilute solutions, $n_1 \gg n_2$; this may be simplified as follows:

$$\Delta T_b = \frac{1}{S}\frac{n_2}{n_1} = \frac{1}{S}\left(\frac{m_2/\mathcal{M}_2}{m_1/\mathcal{M}_1}\right)$$

where m_1 and m_2 are the masses of solvent and solute (in grams) and $\mathcal{M}_1$ and $\mathcal{M}_2$ are their molar masses in grams per mole. Because $\mathcal{M}_1$, like S, is a property of the solvent only, it is convenient to combine the two and define a new constant K_b through

$$K_b = \frac{\mathcal{M}_1}{(1000 \text{ g kg}^{-1})S}$$

Then

$$\Delta T_b = K_b\left(\frac{m_2/\mathcal{M}_2}{m_1/(1000 \text{ g kg}^{-1})}\right)$$

Because m_1 is measured in grams, $m_1/(1000 \text{ g kg}^{-1})$ is the number of kilograms of solvent. Also, $m_2/\mathcal{M}_2$ is the number of moles of solute. The expression in parentheses is, therefore, the molality (m) of the solution.

$$\Delta T_b = K_b m \qquad [11.7]$$

For a given solvent, K_b is obtained by measuring the boiling-point elevations for dilute solutions of known molality (that is, containing a known amount of solute with known molar mass). Table 11.2 gives values of K_b for a number of solvents. Once K_b has been found, it can be used either to predict boiling-point elevations for solutes of known molar mass or to determine molar masses from measured boiling-point elevations, as illustrated in part (b) of the following example.

EXAMPLE 11.12

(a) When 5.50 g of biphenyl ($C_{12}H_{10}$) is dissolved in 100.0 g of benzene, the boiling point increases by 0.903°C. Calculate K_b for benzene.

(b) When 6.30 g of an unknown hydrocarbon is dissolved in 150.0 g of benzene, the boiling point of the solution increases by 0.597°C. What is the molar mass of the unknown substance?

SOLUTION

(a) Because the molar mass of biphenyl is 154.2 g mol^{-1}, 5.50 g contains 5.50 g/154.2 g mol^{-1} = 0.0357 mol. The molality, m, is

$$m = \frac{\text{mol solute}}{\text{kg solvent}} = \frac{0.0357 \text{ mol}}{0.1000 \text{ kg}} = 0.357 \text{ mol kg}^{-1}$$

$$K_b = \frac{\Delta T_b}{m} = \frac{0.903 \text{ K}}{0.357 \text{ mol kg}^{-1}} = 2.53 \text{ K kg mol}^{-1} \text{ for benzene}$$

(b) Solving $\Delta T_b = K_b m$ for m gives

$$m = \frac{\Delta T_b}{K_b} = \frac{0.597 \text{ K}}{2.53 \text{ K kg mol}^{-1}} = 0.236 \text{ mol kg}^{-1}$$

The number of moles of solute is the product of the molality of the solution and the mass of the solvent, m_1:

$$n_2 = (0.236 \text{ mol kg}^{-1}) \times (0.1500 \text{ kg}) = 0.0354 \text{ mol}$$

Finally, the molar mass of the solute is its mass divided by its number of moles:

$$\text{molar mass of solute} = \mathcal{M}_2 = \frac{m_2}{n_2} = \frac{6.30 \text{ g}}{0.0354 \text{ mol}} = 178 \text{ g mol}^{-1}$$

The unknown hydrocarbon might be anthracene ($C_{14}H_{10}$), which has a molar mass of 178.24 g mol^{-1}.

Related Problems: 43, 44, 45, 46

So far, only *nondissociating* solutes have been considered. Colligative properties depend on the total number of moles per liter of dissolved species present. If a solute dissociates (as sodium chloride dissolves to furnish Na^+ and Cl^- ions in aqueous solution), then the molality, m, to be used is the *total* molality. One mole of NaCl dissolves to give 2 mol of ions; thus, the total molality and the boiling-point elevation are twice as large as they would be if NaCl molecules were present in solution. One mole of $Ca(NO_3)_2$ dissolves to give 3 mol of ions (1 mol of Ca^{2+} and 2 mol of NO_3^-), giving 3 times the boiling-point elevation. The corresponding vapor-pressure lowering is greater as well. Ions behave differently than neutral molecules in solution, however, and nonideal behavior appears at lower concentrations in solutions that contain ions.

EXAMPLE 11.13

Lanthanum(III) chloride ($LaCl_3$) is a salt that completely dissociates into ions in dilute aqueous solution,

$$LaCl_3(s) \longrightarrow La^{3+}(aq) + 3\ Cl^-(aq)$$

yielding 4 mol of ions per mole of $LaCl_3$. Suppose 0.2453 g of $LaCl_3$ is dissolved in 10.00 g of H_2O. What is the boiling point of the solution at atmospheric pressure, assuming ideal solution behavior?

SOLUTION

The molar mass of $LaCl_3$ is 245.3 g mol^{-1}.

$$\text{moles of LaCl}_3 = \frac{0.2453 \text{ g}}{245.3 \text{ g mol}^{-1}}$$

$$= 1.000 \times 10^{-3} \text{ mol}$$

$$\text{total molality} = m = \frac{(4)(1.000 \times 10^{-3}) \text{ mol of ions}}{0.0100 \text{ kg solvent}}$$

$$= 0.400 \text{ mol kg}^{-1}$$

This is inserted into the equation for the boiling-point elevation:

$$\Delta T_b = K_b m = (0.512 \text{ K kg mol}^{-1})(0.400 \text{ mol kg}^{-1}) = 0.205 \text{ K}$$

$$T_b = 100.205°\text{C}$$

The actual boiling point is slightly lower than this because the solution is nonideal.

Freezing-Point Depression

The phenomenon of **freezing-point depression** is analogous to that of boiling-point elevation. Here, we consider only cases in which the first solid that crystallizes from solution is the pure solvent. If solute crystallizes out with solvent, the situation is more complicated.

Pure solid solvent coexists at equilibrium with its characteristic vapor pressure, determined by the temperature (Section 10.4). Solvent in solution likewise coexists with a certain vapor pressure of solvent. If solid solvent and the solvent in solution are to coexist, they must have the *same* vapor pressure. This means that the freezing temperature of a solution can be identified as the temperature at which the vapor-pressure curve of the pure solid solvent intersects that of the solution (Fig. 11.12). As solute is added to the solution, the vapor pressure of the solvent falls and the freezing point, the temperature at which the first crystals of pure solvent begin to appear, drops. The difference $\Delta T_f = T_f' - T_f$ is therefore negative, and a freezing-point depression is observed.

The change in temperature, ΔT_f, is once again proportional to the change in vapor pressure, ΔP_1. For sufficiently small concentrations of solute, the freezing-point depression is related to the total molality, m (by analogy with the case of boiling-point elevation), through

$$\Delta T_f = T_f' - T_f = -K_f m \tag{11.8}$$

where K_f is a positive constant that depends only on the properties of the solvent (see Table 11.2). Freezing-point depression is responsible for the fact that seawater, containing dissolved salts, has a slightly lower freezing point than fresh water. Concentrated salt solutions have still lower freezing points. Salt spread on an icy road reduces the freezing point of the ice, so the ice melts.

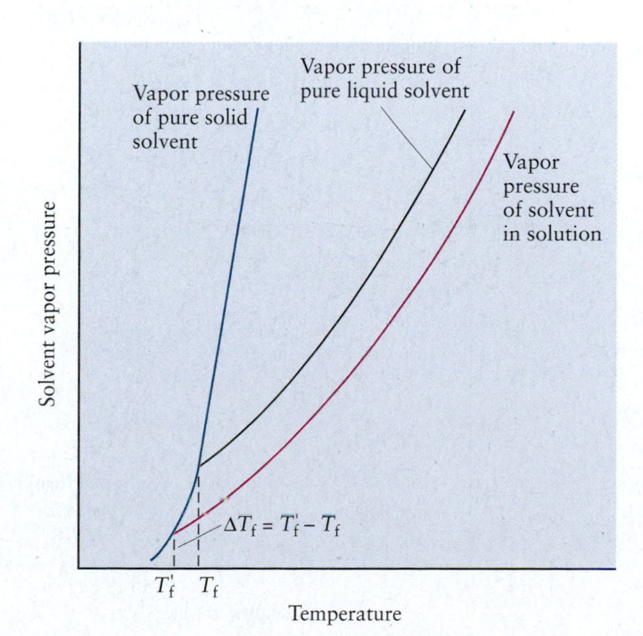

FIGURE 11.12 The vapor pressure of solvent above a dilute solution, compared with that above pure liquid and solid solvent. The depression of the freezing point from T_f to T_f' is shown.

Measurements of the drop in the freezing point, like those of elevation of the boiling point, can be used to determine molar masses of unknown substances. If a substance dissociates in solution, the *total* molality of all species present (ionic or neutral) must be used in the calculation.

EXAMPLE 11.14

The number of moles of the major dissolved species in a 1.000-L sample of seawater are as follows. Estimate the freezing point of the seawater, assuming $K_f = 1.86$ K kg mol^{-1} for water.

Na^+	0.458 mol	Cl^-	0.533 mol
Mg^{2+}	0.052 mol	SO_4^{2-}	0.028 mol
Ca^{2+}	0.010 mol	HCO_3^-	0.002 mol
K^+	0.010 mol	Br^-	0.001 mol
Neutral species	0.001 mol		

SOLUTION

Because water has a density of 1.00 g cm^{-3}, 1.00 L of water weighs 1.00 kg. For dilute *aqueous* solutions, the number of moles per kilogram of solvent (the molality, m) is therefore approximately equal to the number of moles per liter. The total molality, obtained by adding the individual species molalities just given, is $m = 1.095$ mol kg^{-1}. Then

$$\Delta T = -K_f m = -(1.86 \text{ K kg mol}^{-1})(1.095 \text{ mol kg}^{-1}) = -2.04 \text{ K}$$

The seawater should freeze at approximately −2°C. Nonideal solution effects make the actual freezing point slightly higher than this.

Related Problems: 47, 48

Both freezing-point depression and boiling-point elevation can be used to determine whether a species of known molar mass dissociates in solution (Fig. 11.13), as the following example shows.

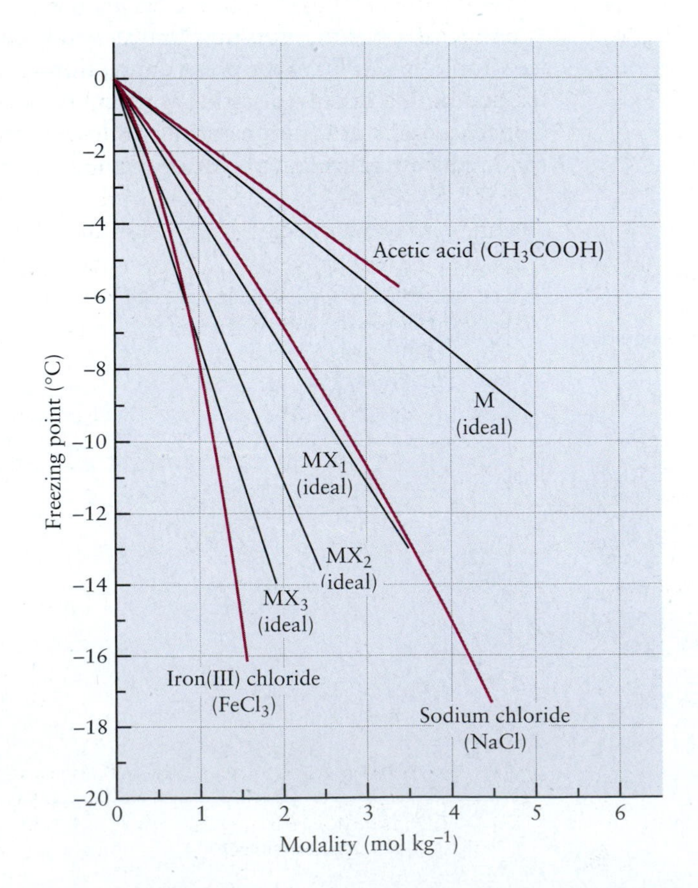

FIGURE 11.13 The heavy colored lines give the observed depression of the freezing point of water by acetic acid, NaCl, and $FeCl_3$ as the molality of the solutions increases. Straight black lines sketch the predicted ideal behavior for one through four moles of particles per mole in solution. The experimental curve for NaCl (which gives *two* moles of dissolved particles) stays close to the ideal straight line for MX; the experimental curve for $FeCl_3$ (which gives *four* moles of dissolved particles) stays fairly close to the ideal straight line for MX_3. The pattern suggests that acetic acid dissolves to give one mole of particles per mole of solute. As the molalities of the solutions increase, the observed freezing-point depressions deviate in varying ways from the straight lines.

EXAMPLE 11.15

When 0.494 g of $K_3Fe(CN)_6$ is dissolved in 100.0 g of water, the freezing point is found to be −0.093°C. How many ions are present for each formula unit of $K_3Fe(CN)_6$ dissolved?

SOLUTION

The total molality of all species in solution is

$$m = \frac{-\Delta T_f}{K_f} = \frac{0.093 \text{ K}}{1.86 \text{ K kg mol}^{-1}} = 0.050 \text{ mol kg}^{-1}$$

Because the molar mass of $K_3Fe(CN)_6$ is 329.25 g mol^{-1}, the total molality if *no* dissociation had taken place would be

$$\frac{\left(\dfrac{0.494 \text{ g}}{329.25 \text{ g mol}^{-1}}\right)}{0.100 \text{ kg}} = 0.0150 \text{ mol kg}^{-1}$$

This is between one fourth and one third of the measured total molality in solution, so each $K_3Fe(CN)_6$ must dissociate into three to four ions. In fact, the dissociation that occurs is

$$K_3Fe(CN)_6(s) \longrightarrow 3\,K^+(aq) + [Fe(CN)_6]^{3-}(aq)$$

Deviations from ideal solution behavior have reduced the effective total molality from 0.060 to 0.050 mol kg^{-1}.

Related Problems: 51, 52

Osmotic Pressure

The fourth colligative property is particularly important in cellular biology because it plays a vital role in the transport of molecules across cell membranes. Such membranes are **semipermeable,** allowing small molecules such as water to pass through but blocking the passage of large molecules such as proteins and carbohydrates. A semipermeable membrane (for example, common cellophane) can be used to separate small solvent molecules from large solute molecules.

Suppose a solution is contained in an inverted tube, the lower end of which is covered by a semipermeable membrane. This solution has a solute concentration of *c* moles per liter. When the end of the tube is inserted in a beaker of pure solvent (Fig. 11.14), solvent flows from the beaker into the tube. The volume of the solution increases, and the solvent rises in the tube until, at equilibrium, it reaches a height, *h*, above the solvent in the beaker. The pressure on the solution side of

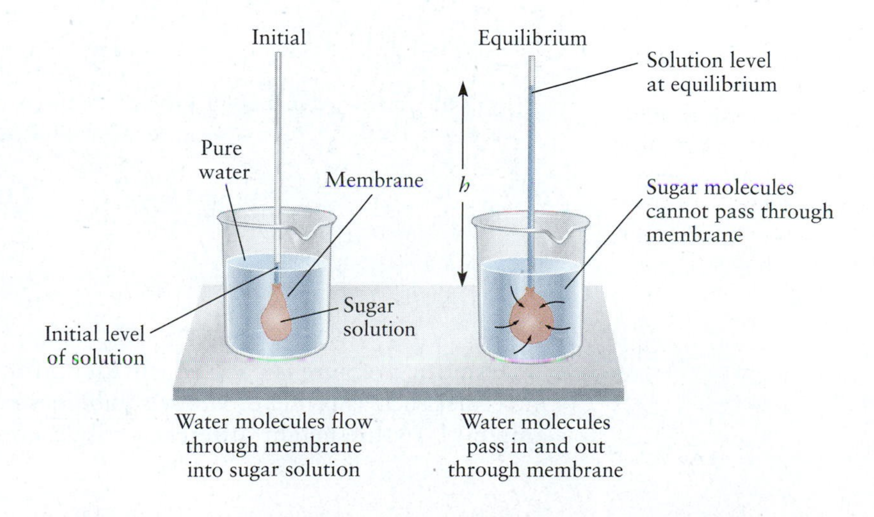

FIGURE 11.14 In this device to measure osmotic pressure, the semipermeable membrane allows solvent, but not solute, molecules to pass through. This results in a net flow of solvent into the tube until equilibrium is achieved, with the level of solution at a height, *h*, above the solvent in the beaker. Once this happens, the solvent molecules pass through the membrane at the same rate in both directions.

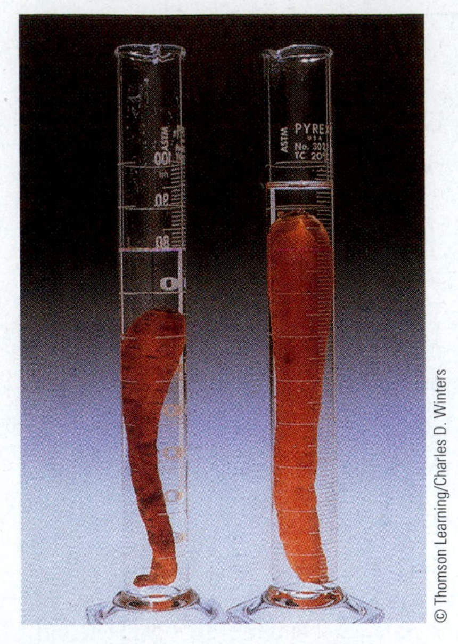

When a carrot is immersed in saltwater (left), water flows out into the solution, causing the carrot to shrink. The osmotic pressure outside the cells of the vegetable is greater than that inside. A carrot left in pure water (right) does not shrivel.

the membrane is greater than the atmospheric pressure on the surface of the pure solvent by an amount given by the **osmotic pressure,** π:

$$\pi = \rho g h \tag{11.9}$$

where ρ is the density of the solution (1.00 g cm^{-3} for a dilute aqueous solution) and g is the acceleration due to gravity (9.807 m s^{-2}).

For example, a height, h, of 0.17 m corresponds to an osmotic pressure for a dilute aqueous solution of

$$\begin{aligned}\pi &= [(1.00 \text{ g cm}^{-3})(10^{-3} \text{ kg g}^{-1})(10^{6} \text{ cm}^3 \text{ m}^{-3})](9.807 \text{ m s}^{-2})(0.17 \text{ m}) \\ &= 1.7 \times 10^3 \text{ kg m}^{-1} \text{ s}^{-2} = 1.7 \times 10^3 \text{ Pa}\end{aligned}$$

$$\pi(\text{atm}) = \frac{1.7 \times 10^3 \text{ Pa}}{1.013 \times 10^5 \text{ Pa atm}^{-1}} = 0.016 \text{ atm}$$

This example illustrates how accurately very small osmotic pressures can be measured.

In 1887, Jacobus van't Hoff discovered an important relation among osmotic pressure, π, concentration, c, and absolute temperature, T:

$$\pi = cRT \tag{11.10}$$

R is the gas constant, equal to 0.08206 L atm mol^{-1} K^{-1} if π is expressed in atmospheres and c in moles per liter. Because $c = n/V$, where n is the number of moles of solute and V is the volume of the solution, van't Hoff's equation can be rewritten as

$$\pi V = nRT$$

which bears a striking similarity to the ideal gas law. With this relation, the molar mass of a dissolved substance can be determined from the osmotic pressure of its solution.

EXAMPLE 11.16

A chemist dissolves 2.00 g of a protein in 0.100 L water. The osmotic pressure is 0.021 atm at 25°C. What is the approximate molar mass of the protein?

SOLUTION

The concentration in moles per liter is

$$\begin{aligned}c &= \frac{\pi}{RT} = \frac{0.021 \text{ atm}}{(0.08206 \text{ L atm mol}^{-1} \text{ K}^{-1})(298 \text{ K})} \\ &= 8.6 \times 10^{-4} \text{ mol L}^{-1}\end{aligned}$$

Now 2.00 g dissolved in 0.100 L gives the same concentration as 20.0 g in 1.00 L. Therefore, 8.6×10^{-4} mol of protein must weigh 20.0 g, and the molar mass is

$$\mathcal{M} = \frac{20.0 \text{ g}}{8.6 \times 10^{-4} \text{ mol}} = 23{,}000 \text{ g mol}^{-1}$$

Related Problems: 53, 54

Osmotic pressure is particularly useful for measuring molar masses of large molecules such as proteins, whose solubilities may be low. In the case given in Example 11.16, the height difference h is 22 cm, an easily measured quantity. By contrast, the other three colligative properties in this example would show small effects:

$$\text{Vapor-pressure lowering} = 4.8 \times 10^{-7} \text{ atm}$$
$$\text{Boiling-point elevation} = 0.00044 \text{ K}$$
$$\text{Freezing-point depression} = 0.0016 \text{ K}$$

All these changes are too small for accurate measurement. As with the other techniques, the *total* number of moles of solute species determines the osmotic pressure if dissociation occurs.

Osmosis has other important uses. In some parts of the world, potable water is a precious commodity. It can be obtained much more economically by desalinizing brackish waters, through a process called **reverse osmosis,** than by distillation. When an ionic solution in contact with a semipermeable membrane has a pressure applied to it that exceeds its osmotic pressure, water of quite high purity passes through. Reverse osmosis is also used to control water pollution.

11.6 Phase Equilibrium in Solutions: Volatile Solutes

The preceding section described the properties of solutions of nonvolatile solutes in liquid solvents. The concept of an ideal solution can be extended to mixtures of two or more components, each of which is volatile. In this case, an ideal solution is one in which the vapor pressure of *each* species present is proportional to its mole fraction in solution over the whole range of mole fraction:

$$P_i = X_i P_i^\circ$$

where P_i° is the vapor pressure (at a given temperature) of pure substance i, X_i is its mole fraction in solution, and P_i is its partial vapor pressure over the solution. This is a generalization of Raoult's law to each component of a solution.

For an ideal mixture of two volatile substances, the vapor pressure of component 1 is

$$P_1 = X_1 P_1^\circ$$

and that of component 2 is

$$P_2 = X_2 P_2^\circ = (1 - X_1) P_2^\circ$$

The vapor pressures for such an ideal solution are shown in Figure 11.15, together with typical vapor pressures for a solution that shows positive deviations from ideal behavior.

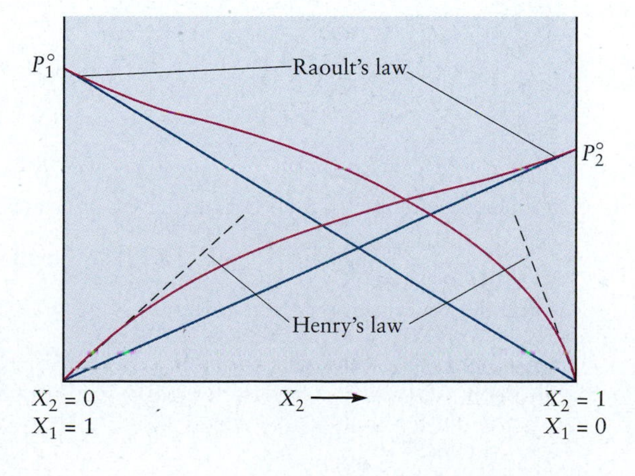

FIGURE 11.15 Vapor pressures above a mixture of two volatile liquids. Both ideal (blue lines) and nonideal behaviors (red curves) are shown. Positive deviations from ideal solution behavior are illustrated, although negative deviations are observed for other nonideal solutions. Raoult's and Henry's laws are shown as dilute solution limits for the nonideal mixture; the markers explicitly identify regions where Raoult's law and Henry's law represent actual behavior.

Henry's Law

At sufficiently low mole fraction X_2, the vapor pressure of component 2 (even in a nonideal solution) is proportional to X_2:

$$P_2 = k_2X_2 \qquad [11.11]$$

where k_2 is a constant. For X_1 small (X_2 near 1),

$$P_1 = k_1X_1 = k_1(1 - X_2)$$

This linear vapor pressure of what is called the solute (because it is present at small mole fraction) is known as **Henry's law:** The vapor pressure of a volatile dissolved substance is proportional to the mole fraction of that substance in solution. Whenever Raoult's law is valid for a solvent, Henry's law is valid for the solute (see Fig. 11.15).

One familiar application of Henry's law is in the carbonation of beverages. If the partial pressure of CO_2 above a solution is increased, the amount dissolved in the solution increases proportionately. When the beverage can is opened, dissolved gas bubbles out of solution in response to the lower CO_2 pressure outside. Henry's law is important in biology, where gases such as oxygen dissolve in blood and other bodily fluids, and in environmental chemistry, where volatile pollutants can move between bodies of water and the atmosphere.

EXAMPLE 11.17

The Henry's law constant for oxygen dissolved in water is 4.34×10^4 atm at 25°C. If the partial pressure of oxygen in air is 0.20 atm under ordinary atmospheric conditions, calculate the concentration (in moles per liter) of dissolved oxygen in water that is in equilibrium with air at 25°C.

SOLUTION

Henry's law is used to calculate the mole fraction of oxygen in water:

$$X_{O_2} = \frac{P_{O_2}}{k_{O_2}} = \frac{0.20 \text{ atm}}{4.34 \times 10^4 \text{ atm}} = 4.6 \times 10^{-6}$$

Next, the mole fraction is converted to molarity. One liter of water weighs 1000 g, so it contains

$$\frac{1000 \text{ g } H_2O}{18.02 \text{ g mol}^{-1}} = 55.5 \text{ mol water}$$

Because X_{O_2} is so small, $n_{H_2O} + n_{O_2}$ is close to n_{H_2O}, and it can be written as

$$X_{O_2} = \frac{n_{O_2}}{n_{H_2O} + n_{O_2}} \approx \frac{n_{O_2}}{n_{H_2O}}$$

$$4.6 \times 10^{-6} = \frac{n_{O_2}}{55.5 \text{ mol}}$$

Thus, the number of moles of oxygen in 1 L water is

$$n_{O_2} = (4.6 \times 10^{-6})(55.5 \text{ mol}) = 2.6 \times 10^{-4} \text{ mol}$$

and the concentration of dissolved O_2 is 2.6×10^{-4} M.

Related Problems: 57, 58

Distillation

The vapor pressures of the pure components of an ideal solution usually differ, and for this reason, such a solution has a composition different from that of the vapor phase with which it is in equilibrium. This can best be seen in an example.

Hexane (C_6H_{14}) and heptane (C_7H_{16}) form a nearly ideal solution over the whole range of mole fractions. At 25°C, the vapor pressure of pure hexane is $P_1^\circ = 0.198$ atm, and that of pure heptane is $P_2^\circ = 0.0600$ atm. Suppose a solution contains 4.00 mol of hexane and 6.00 mol of heptane, and that its mole fractions are, therefore, $X_1 = 0.400$ and $X_2 = 0.600$. The vapor in equilibrium with this ideal solution has partial pressures

$$P_{\text{hexane}} = P_1 = X_1 P_1^\circ = (0.400)(0.198 \text{ atm}) = 0.0792 \text{ atm}$$

$$P_{\text{heptane}} = P_2 = X_2 P_2^\circ = (0.600)(0.0600 \text{ atm}) = 0.0360 \text{ atm}$$

From Dalton's law, the total pressure is the sum of these partial pressures:

$$P_{\text{total}} = P_1 + P_2 = 0.1152 \text{ atm}$$

If X_1' and X_2' are the mole fractions in the vapor, then

$$X_1' = \frac{0.0792 \text{ atm}}{0.1152 \text{ atm}} = 0.688$$

$$X_2' = 1 - X_1' = \frac{0.0360 \text{ atm}}{0.1152 \text{ atm}} = 0.312$$

The liquid and the vapor with which it is in equilibrium have different compositions (Fig. 11.16), and the vapor is enriched in the more volatile component.

Suppose some of this vapor is removed and condensed to a liquid. The vapor in equilibrium with this new solution would be still richer in the more volatile component, and the process could be continued further (see Fig. 11.16). This progression underlies the technique of separating a mixture into its pure components by **fractional distillation,** a process in which the components are successively evaporated and recondensed. What we have described so far corresponds to a constant-temperature process, but actual distillation is conducted at constant total pressure. The vapor pressure–mole fraction plot is transformed into a boiling temperature–mole fraction plot (Fig. 11.17). Note that the component with the lower vapor pressure (component 2) has the higher boiling point, T_b^2. If the temperature of a solution of a certain composition is raised until it touches the liquid line in the plot, the vapor in equilibrium with the solution is richer in the more volatile component 1. Its composition lies at the intersection of the horizontal constant-temperature line and the equilibrium vapor curve.

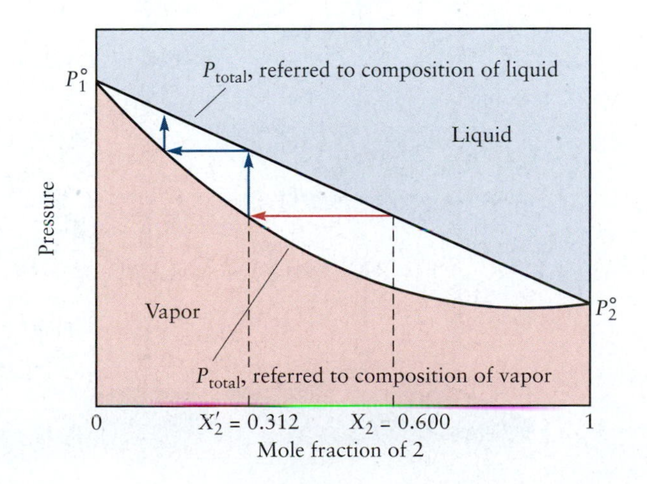

FIGURE 11.16 The composition of the vapor above a solution differs from the composition of the liquid with which it is in equilibrium. Here, the upper (straight) line is the total pressure of the vapor in equilibrium with an ideal solution having mole fraction X_2 of component 2. By moving horizontally from that line to a point of intersection with the lower curve, we can locate the mole fraction X_2' of component 2 in the vapor (red arrow). Subsequent condensations and vaporizations are shown by blue arrows.

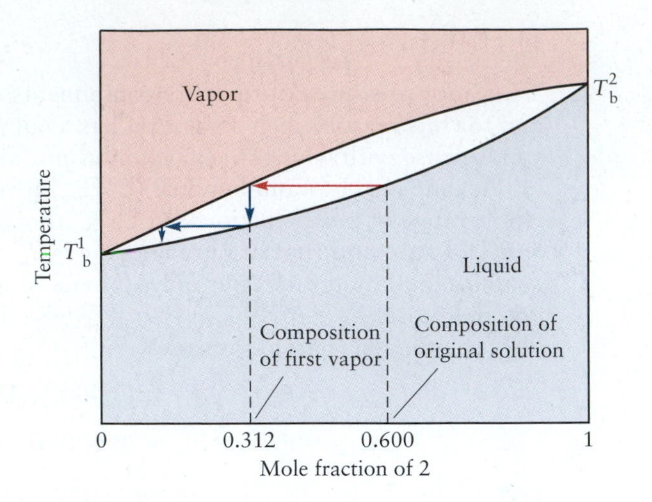

FIGURE 11.17 The boiling point of an ideal solution varies with the composition of the solution. The upper curve is the boiling temperature referred to the vapor composition, and the lower curve is the boiling temperature referred to the liquid composition. The vapors boiling off a solution that has a 0.600-mol fraction of component 2 are enriched in the more volatile component 1 to the extent that their mole fraction of component 2 is only 0.312 (red arrow). The subsequent blue arrows show the further steps used in obtaining nearly pure component 1 by fractional distillation.

A liquid can be vaporized in different ways. It can simply be boiled until it is entirely vaporized and the final composition of the vapor is the same as that of the original liquid. It is clear that such a mixture boils over a range of temperatures, rather than at a single T_b like a pure liquid. Alternatively, if the boiling is stopped midway, the vapor fraction that has boiled off can be collected and recondensed. The resulting liquid (the condensate) will be richer in component 1 than was the original solution. By repeating the process again and again, mixtures successively richer in component 1 will be obtained. This is the principle behind the **distillation column** (Fig. 11.18). Throughout the length of the tube, such evaporations and recondensations take place, and this allows mixtures to be separated into their constituent substances. Such a process is used to separate nitrogen and oxygen in air; the air is liquefied and then distilled, with the lower boiling nitrogen ($T_b = -196°C$) vaporizing before the oxygen ($T_b = -183°C$).

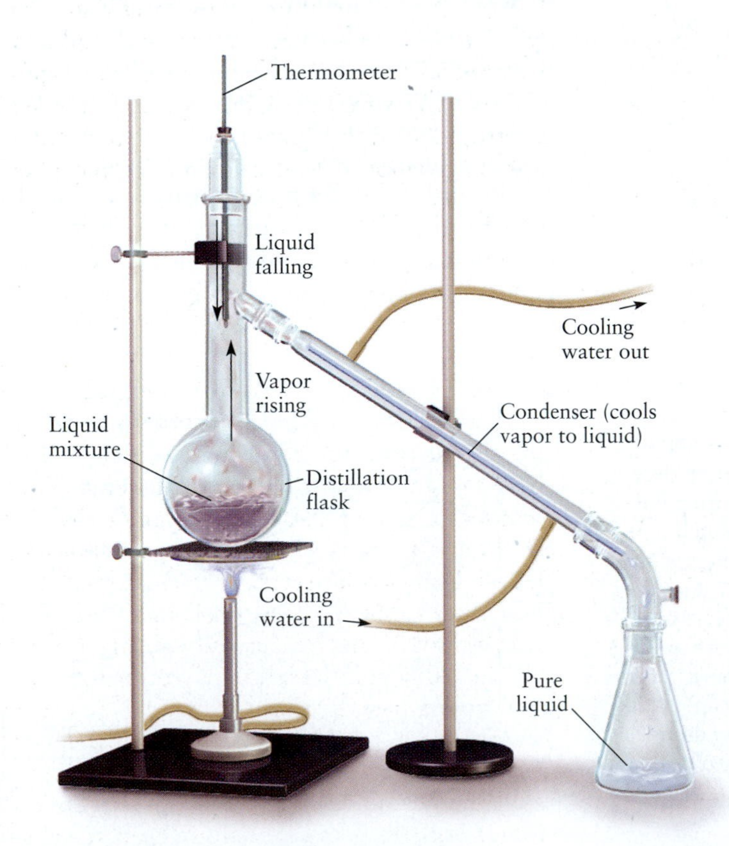

FIGURE 11.18 In a distillation column, temperature decreases with height in the column. The less volatile components condense and fall back to the flask, but the more volatile ones continue up the column into the water-cooled condenser, where they condense and are recovered in the receiver.

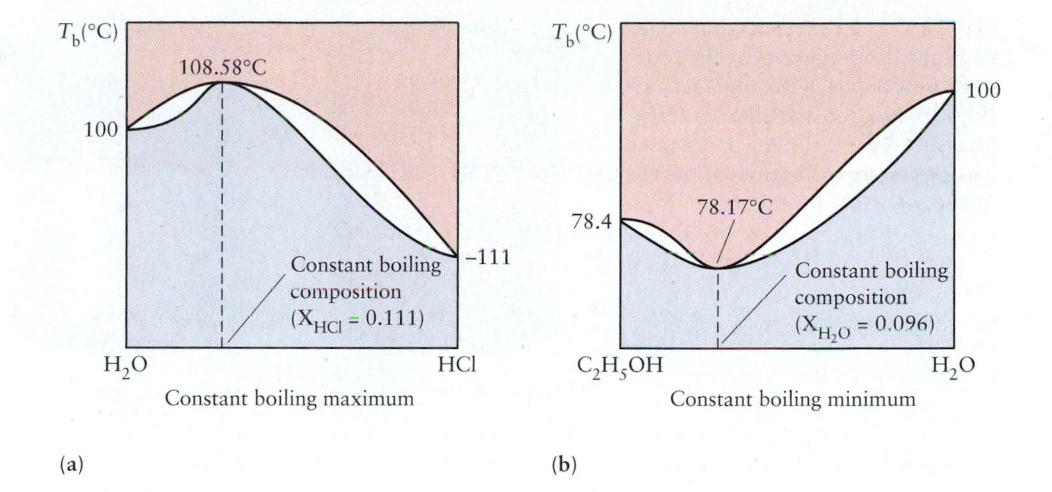

FIGURE 11.19 Dependence of boiling temperature on mole fraction for (a) maximum- and (b) minimum-boiling azeotropes. The coordinates are not to scale.

Nonideal solutions may have more complicated behavior. A mixture showing large *negative* deviations from Raoult's law (one in which solute–solvent forces are strongly attractive) will show a boiling-point *maximum* (Fig. 11.19a). A solution at the maximum is called a **maximum-boiling azeotrope;** an example is that formed by the H_2O/HCl system. The boiling-point maximum occurs in this case at 108.58°C and 1 atm pressure for a composition of 20.22% HCl by mass. A mixture showing large *positive* deviations from ideal behavior may show a boiling-point *minimum* (see Fig. 11.19b) and a corresponding **minimum-boiling azeotrope.** Ethanol and water form such an azeotrope with a normal boiling point of 78.17°C and a composition of 4% water by mass. In this case, attractive forces between ethanol molecules and between water molecules are stronger than those between ethanol and water, so the solution boils at a lower temperature than either pure component. An azeotrope behaves like a single-component fluid in that it boils at a well-defined temperature and the solution and vapor have the same composition. A mixture of two substances that form an azeotrope cannot be separated by fractional distillation into two pure substances, but rather into only one pure substance and a mixture with the azeotropic composition. A mixture of 50% ethanol and water, for example, can be distilled to obtain pure water and an azeotropic mixture containing 4% water and 96% ethanol. The last 4% of water cannot be removed by distillation at atmospheric pressure to obtain pure ethanol.

11.7 Colloidal Suspensions

A **colloid** is a mixture of two or more substances in which one phase is suspended as a large number of small particles in a second phase. The dispersed substance and the background medium may be any combination of gas, liquid, or solid. Examples of colloids include aerosol sprays (liquid suspended in gas), smoke (solid particles in air), milk (fat droplets and solids in water), mayonnaise (water droplets in oil), and paint (solid pigment particles in oil for oil-based paints, or pigment and oil dispersed in water for latex paints). Colloidal particles are larger than single molecules, but are too small to be seen by the eye; their dimensions typically range from 10^{-9} to 10^{-6} m in diameter. Their presence can be seen most dramatically in the way in which they scatter light; a familiar example is the passage of light from a movie projector through a suspension of small dust particles in air. The gemstone opal has remarkable optical properties that arise from colloidal water suspended in solid silicon dioxide (Fig. 11.20).

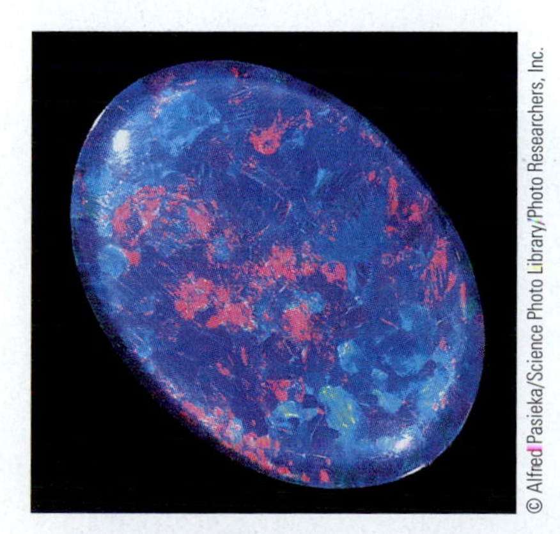

FIGURE 11.20 A natural opal.

Although some colloids settle out into two separate phases if left standing long enough, others persist indefinitely; a suspension of gold particles prepared by

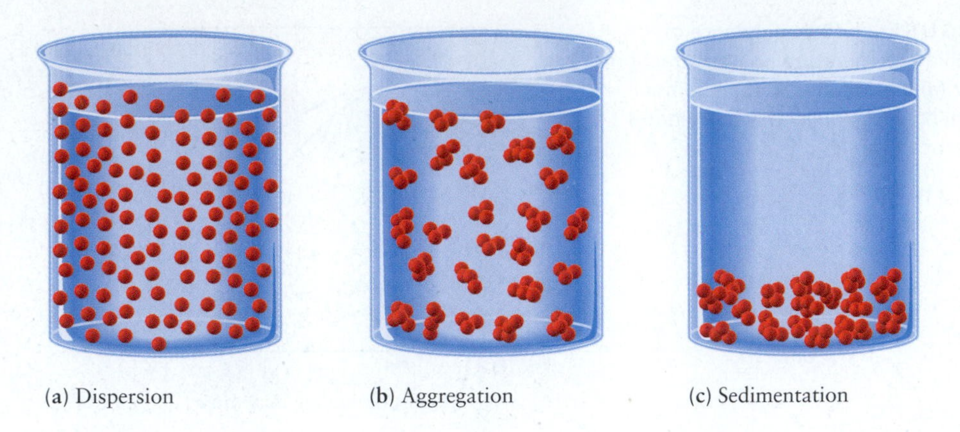

FIGURE 11.21 When a salt is added to a colloidal dispersion (a), the repulsive forces between the colloidal particles are reduced and aggregation occurs (b). Eventually, the aggregated particles fall to the bottom of the container as low-density sediment (c).

Michael Faraday in 1857 shows no apparent settling to date. In many colloids, the particles have net positive or negative charges on their surfaces, balanced by an opposite charge of ions in solution. The settling out of such colloids is speeded by dissolving salts in the solution, a process called **flocculation.** The salts reduce the repulsive electrostatic forces between the suspended particles, causing aggregation and sedimentation (Fig. 11.21). Flocculation occurs in river deltas; when river water containing suspended clay particles meets the salt water of the ocean, the clay settles out as open, low-density sediments. Flocculating agents are deliberately added to paints so that the pigment will settle in a loosely packed sediment. When the paint is stirred, the pigment is redispersed through the medium. In the absence of such agents, the suspended particles tend to settle in compact sediments that are difficult to resuspend.

In some cases, the formation of a colloid is not desirable, as in the precipitation of a solid from solution (see Section 11.2). Especially with metal sulfides, the solid precipitate may appear as a colloidal suspension with particles small enough to pass through ordinary filter paper (Fig. 11.22). If this happens, precipitated solid can be separated out only by flocculation, centrifugation, or forcing the suspension through a membrane, such as cellophane, that permits passage of only the small solvent molecules.

Suspended particles are in a constant state of motion, called **Brownian motion** after Robert Brown, a Scottish botanist who used a microscope to observe the motion of pollen particles in water. Brownian motion results from the constant random buffeting of the particles by solvent molecules. In 1905, Albert Einstein showed how the motion of Brownian particles could be described on a microscopic level; his work provided one of the most striking and convincing verifications of the molecular hypothesis and of the kinetic theory of matter and led to a fairly accurate determination of Avogadro's number.

(a)

(b)

FIGURE 11.22 (a) This colloidal suspension of $PbCrO_4$ appears cloudy. (b) After flocculation, the precipitate settles to the bottom.

CHAPTER SUMMARY

When pure substance A is mixed with pure substance B, the resulting solution has properties different from those of the pure substances because the intermolecular forces around each molecule are now quite different. In aqueous solutions, the dissolved species are described as solute ions or molecules surrounded by solvation shells of solvent molecules held in place by hydrogen bonding or ion–dipole forces. Solutions can be in equilibrium in solid, liquid, or vapor phases, but the conditions under which phases coexist are different from those for the pure solvent. Changes in vapor pressure, freezing point, boiling point, and osmotic pressure are explained quantitatively by the number of non-volatile solute species in the solution. When both species are volatile, the composition of the vapor phase is different from the composition of the solution, as described by Raoult's law and Henry's law, and the solution components can be separated by distillation. Many chemical reactions are carried out in liquid solutions, frequently by mixing solutions of the reactants. We write balanced chemical equations to describe the stoichiometry of reactions in solutions, and we count the number of moles of each reactant in a volume of solution through its concentration expressed in molarity. Solution stoichiometry describes many practical applications in acid–base and redox chemistry and is the basis for quantitative analytical techniques in these fields.

CUMULATIVE EXERCISE

Maple Syrup

The sap in a maple tree can be described as an approximately 3.0% (by mass) solution of sucrose ($C_{12}H_{22}O_{11}$) in water. Sucrose does not dissociate to any significant extent in aqueous solution.

(a) At 20°C, the density of sap is 1.010 g cm^{-3}. Calculate the molarity of sucrose in sap.

(b) A typical maple tree yields about 12 gallons of sap per year. Calculate how many grams of sucrose are contained in this volume of sap (1 gallon = 3.785 L).

(c) The rising of sap in trees is caused largely by osmosis; the concentration of dissolved sucrose in sap is higher than that of the groundwater outside the tree. Calculate the osmotic pressure of a sap solution and the height to which the sap should rise above the ground on a day when the temperature is 20°C. Approximate the groundwater as pure (although, in fact, it typically contains 0.01 to 0.03 M dissolved species). Express the answer in meters and in feet (1 m = 3.28 ft).

Tapping syrup from sugar maple trees.

(d) To produce maple syrup from sap, the sap is boiled to reduce its water content. Calculate the normal boiling point of a sap solution.

(e) Maple syrup is the concentrated sap solution that results when most of the water is boiled off. The syrup has a composition of approximately 64% (by mass) sucrose and 36% water, with flavoring components present in small concentrations. If the density of the maple syrup is 1.31 g cm^{-3}, calculate the mole fraction, molarity, and molality of sucrose in maple syrup.

(f) What volume (in gallons) of maple syrup can be obtained from the sap in one typical tree?

(g) In the presence of vanadium(V) oxide, dinitrogen tetroxide oxidizes sucrose to oxalic acid ($H_2C_2O_4$) according to the following equation:

$$C_{12}H_{22}O_{11}(aq) + 9\ N_2O_4(s) \longrightarrow 6\ H_2C_2O_4(aq) + 18\ NO(g) + 5\ H_2O(\ell)$$

Calculate the mass of N_2O_4 that will react completely with 7.00 L of the sap solution from part (a), and give the concentration of oxalic acid that results.

Answers

(a) 0.089 M

(b) 1.4×10^3 g sucrose

(c) $\pi = 2.1$ atm; height = 22 m = 72 ft

(d) 100.046°C

(e) Mole fraction = 0.086; molarity = 2.4 M; molality = 5.2 mol kg^{-1}

(f) 0.56 gallon

(g) 5.2×10^2 g N_2O_4; 0.53 M

CHAPTER REVIEW

Solutions are described by their composition and method of preparation.

- Mole fraction: $X_i = n_i/n_{tot}$ where n_i is the number of moles of component i and n_{tot} is the total number of moles of all components
- Molarity: M = moles of solute per liter of solution
- Molality: m = moles of solute per kg of solvent
- For *dilute, aqueous* solutions $m \approx$ M
- Special procedures have been defined to prepare solutions of desired X, M, or m.

Solutions are formed by breaking the bonds between molecules or ions in the solute and dispersing these species throughout the solvent.

- The solute species are surrounded by solvent molecules, and experience very different intermolecular forces than in their pure, undissolved state.
- When water is the solvent the solution is called *aqueous* and solutes are labeled (*aq*).
- Ionic solutes dissociate in solution, whereas molecular solutes remain intact when dispersed in solution.
- Energy is required to disrupt the structure of the pure solute and solvent, and energy is released due to the attractive interactions between the solute and solvent. Formation of a solution can be endothermic or exothermic, depending on the difference between these energies.
- In aqueous solutions, solute species are surrounded by solvation shells of solvent molecules held in place by intermolecular forces, primarily hydrogen bonding and ion–dipole forces.

In this chapter we use the Arrhenius definition of acids and bases.

- An *acid* donates one or more protons when dissolved in water.
- A *base* donates one or more hydroxide ions when dissolved in water.
- In the neutralization reaction $NaOH + HCl \longrightarrow NaCl + H_2O$, OH^- is the base and H^+ is the acid. The net reaction is $H^+ + OH^- \longrightarrow H_2O$.

Acid–base titration is an analytical technique for finding the concentration of acid or base in an unknown solution.

- Stoichiometry of reactions in solutions is described by balanced equations that relate the number of moles of each reactant and product. The number of moles of each reaction species in a volume of solution is given by $n_i = \text{M}_i V$.
- Both neutralization reactions and dilution operations are described by $\text{M}_1 V_1 = \text{M}_2 V_2$.
- In titrations a measured volume of a solution with known M reacts to neutralize a known volume of sample whose M is unknown. The end point, at which neutralization is achieved, is signified by an endpoint indicator, most often a color change. At the end point, we determine the molarity of the unknown sample using the earlier equation.

In oxidation–reduction (redox) reactions electrons are transferred between reacting species as they combine to form products.

- Reactants that lose electrons are oxidized.

- Reactants that gain electrons are reduced.
- These reactions are coupled; it is not possible to have oxidation without reduction and vice versa.
- Balancing the equations for redox reactions involves special procedures. The steps are:
  - Assign oxidation numbers as described in Section 3.10.
  - Write unbalanced half-equations for the species being oxidized and the species being reduced.
  - Balance all atoms except oxygen and hydrogen.
  - Determine whether the reaction takes place in acidic or basic solution.
  - In acidic solution, add one H_3O^+ to the side that needs an extra H atom and then H_2O to the other side.
  - In basic solution, add one H_2O to the side that needs an extra H atom and then OH^- to the other side.
  - Balance charge transfer in each half-equation by inserting electrons as reactants or products.
  - Multiply the two half-equations by numbers that make the number of electrons given off in oxidation the same as the number gained in reduction.
  - Add the two half-equations together.

Like pure substances, solutions can have solid, liquid, and gaseous phases in equilibrium with one another.

- The conditions where phases coexist—such as boiling points, freezing points, and vapor pressure—have different values in solution than in the pure solvent.
- The colligative properties of solutions describe how nonvolatile solvents cause these changes in properties of the solvent. These changes depend only on the number of solute particles, not on their nature.
  - Vapor pressure lowering, boiling point elevation, freezing point depression and osmotic pressure are all treated with *empirical* equations.
  - Vapor pressure lowering: $\Delta P_1 = -X_2 P_1^\circ$ where X_2 is the mole fraction of the solute and P_1° is the vapor pressure of the pure solvent.
  - Boiling point elevation: $\Delta T = mK_b$ (the constants have been tabulated)
  - Freezing point depression: $\Delta T = -mK_f$ (the constants have been tabulated)
  - A solution in contact with its pure solvent across a semi-permeable membrane experiences an increase in pressure as pure solvent flows through the membrane into the solution. This osmotic pressure can be measured quite accurately, and through the equation $\pi V = nRT$ permits determination of the molecular weight of the solute.

When the solute is volatile, the vapor pressure in equilibrium with the solution has contributions from both solute and solvent.

- Raoult's law: vapor pressure of the solvent $P_1 = X_1 P_1^\circ$
- Henry's law: vapor pressure of the solute $P_2 = X_2 k_2$
- The partial pressures of the solute and solvent in the vapor will be different from their concentrations in the solution. The vapor will be richer in the more volatile component than is the solution. The process of distillation exploits this fact to separate the components of the solution by heating the solution to the temperature where the more volatile component boils, removing the vapor, and condensing it to the liquid state.

CONCEPTS & SKILLS

After studying this chapter and working the problems that follow, you should be able to:

1. Express the concentration of a solute in solution in units of mass percentage, molarity, molality, and mole fraction (Section 11.1, Problems 3–8).

2. Describe how a solution of a given molarity is prepared and the effect on dilution of molarity (Section 11.1, Problems 9–12).
3. Describe the formation of a solution in molecular terms by comparing intermolecular forces in the pure phases and in the solution (Section 11.2).
4. Calculate the number of moles of substances reacting during a solution-phase reaction such as acid–base titration (Section 11.3, Problems 15–26).
5. Balance equations for redox reactions in aqueous solution, using the half-reaction method, and calculate the concentrations of substances during redox titrations (Section 11.4, Problems 27–40).
6. Calculate the molar mass of a nonvolatile solute from the changes it causes in the colligative properties (vapor-pressure lowering, boiling-point elevation, freezing-point lowering, or osmotic pressure) of its dilute solution (Section 11.5, Problems 41–56).
7. Discuss the meaning of Henry's law and use it to calculate the solubilities of gases in liquids (Section 11.6, Problems 57–60).
8. Relate the total pressure and composition of the vapor in equilibrium with an ideal two-component solution to the composition of the solution and the vapor pressures of its pure components (Section 11.6, Problems 61–64).
9. Explain how distillation is used to separate the volatile components of a binary liquid solution (Section 11.6).
10. Describe the physical properties of a colloidal suspension (Section 11.7).

KEY EQUATIONS

$$X_1 = \frac{n_1}{n_1 + n_2}$$

$$X_2 = \frac{n_2}{n_1 + n_2} = 1 - X_1$$

(Section 11.1)

$$\text{molarity} = \frac{\text{moles solute}}{\text{liters solution}} = \text{mol L}^{-1}$$

(Section 11.1)

$$\text{molality} = \frac{\text{moles solute}}{\text{kilograms solvent}} = \text{mol kg}^{-1}$$

(Section 11.1)

$$c_f = \frac{\text{moles solute}}{\text{final solution volume}} = \frac{c_i V_i}{V_f}$$

(Section 11.1)

$$P_1 = X_1 P_1^\circ$$

(Section 11.5)

$$P_1 = P_1 - P_1^\circ = X_1 P_1^\circ - P_1^\circ = -X_2 P_1^\circ$$

(Section 11.5)

$$\Delta T_b = K_b m$$

(Section 11.5)

$$\Delta T_f = T_f' - T_f = -K_f m$$

(Section 11.5)

$$\pi = \rho g h$$

(Section 11.5)

$$\pi = cRT$$

(Section 11.5)

$$P_2 = k_2 X_2$$

(Section 11.6)

PROBLEMS

Answers to problems whose numbers are boldface appear in Appendix G. Problems that are more challenging are indicated with asterisks.

Composition of Solutions

1. A patient has a "cholesterol count" of 214. Like many blood-chemistry measurements, this result is measured in units of milligrams per deciliter (mg dL^{-}).
(a) Determine the molar concentration of cholesterol in this patient's blood, taking the molar mass of cholesterol to be 386.64 g mol^{-1}.
(b) Estimate the molality of cholesterol in the patient's blood.
(c) If 214 is a typical cholesterol reading among men in the United States, determine the volume of such blood required to furnish 8.10 g of cholesterol.

2. In many states, a person is legally intoxicated if his or her blood has a concentration of 0.10 g (or more) of ethyl alcohol (C_2H_5OH) per deciliter. Express this "threshold concentration" in mol L^{-1}.

3. A solution of hydrochloric acid in water is 38.00% hydrochloric acid by mass. Its density is 1.1886 g cm^{-3} at 20°C. Compute its molarity, mole fraction, and molality at this temperature.

4. A solution of acetic acid and water contains 205.0 g L^{-1} of acetic acid and 820.0 g L^{-1} of water.
(a) Compute the density of the solution.
(b) Compute the molarity, molality, mole fraction, and mass percentage of acetic acid in this solution.
(c) Take the acetic acid as the solvent, and do the same for water as the solute.

5. A 6.0835 M aqueous solution of acetic acid ($C_2H_4O_2$) has a density of 1.0438 g cm^{-3}. Compute its molality.

6. A 1.241 M aqueous solution of $AgNO_3$ (used to prepare silver chloride photographic emulsions) has a density of 1.171 g cm^{-3}. Compute its molality.

7. Water is slightly soluble in liquid nitrogen. At −196°C (the boiling point of liquid nitrogen), the mole fraction of water in a saturated solution is 1.00×10^{-5}. Compute the mass of water that can dissolve in 1.00 kg of boiling liquid nitrogen.

8. Some water dissolves in liquid methane at −161°C to give a solution in which the mole fraction of water is 6.0×10^{-5}. Determine the mass of water dissolved in 1.00 L of this solution if the density of the solution is 0.78 g cm^{-3}.

9. Concentrated phosphoric acid as sold for use in the laboratory is usually 90% H_3PO_4 by mass (the rest is water). Such a solution contains 12.2 mol of H_3PO_4 per liter of solution at 25°C.
(a) Compute the density of this solution.
(b) What volume of this solution should be used in mixing 2.00 L of a 1.00 M phosphoric acid solution?

10. A perchloric acid solution is 60.0% $HClO_4$ by mass. It is simultaneously 9.20 M at 25°C.
(a) Compute the density of this solution.
(b) What volume of this solution should be used in mixing 1.00 L of a 1.00 M perchloric acid solution?

11. Suppose 25.0 g of solid NaOH is added to 1.50 L of an aqueous solution that is already 2.40 M in NaOH. Then water is added until the final volume is 4.00 L. Determine the concentration of the NaOH in the resulting solution.

12. Suppose 0.400 L of a solution of 0.0700 M nitric acid is added to 0.800 L of a solution of 0.0300 M nitric acid, giving a total volume of 1.200 L. Calculate the concentration (molarity) of nitric acid in the resulting solution.

Nature of Dissolved Species

13. Rewrite the following balanced equations as net ionic equations.
(a) $NaCl(aq) + AgNO_3(aq) \longrightarrow AgCl(s) + NaNO_3(aq)$
(b) $K_2CO_3(s) + 2\ HCl(aq) \longrightarrow$ $2\ KCl(aq) + CO_2(g) + H_2O(\ell)(s) + 2\ H_2O(\ell)$
(c) $2\ Cs(s) + 2\ H_2O(\ell) \longrightarrow 2\ CsOH(aq) + H_2(g)$
(d) $2\ KMnO_4(aq) + 16\ HCl(aq) \longrightarrow$ $5\ Cl_2(g) + 2\ MnCl_2(aq) + 2\ KCl(aq) + 8\ H_2O(\ell)$

14. Rewrite the following balanced equations as net ionic equations.
(a) $Na_2SO_4(aq) + BaCl_2(aq) \longrightarrow BaSO_4(s) + 2\ NaCl(aq)$
(b) $6\ NaOH(aq) + 3\ Cl_2(g) \longrightarrow$ $NaClO_3(aq) + 5\ NaCl(aq) + 3\ H_2O(\ell)$
(c) $Hg_2(NO_3)_2(aq) + 2\ KI(aq) \longrightarrow Hg_2I_2(s) + 2\ KNO_3(aq)$
(d) $3\ NaOCl(aq) + KI(aq) \longrightarrow$ $NaIO_3(aq) + 2\ NaCl(aq) + KCl(aq)$

Reaction Stoichiometry in Solutions: Acid–Base Titrations

15. When treated with acid, lead(IV) oxide is reduced to a lead(II) salt, with the liberation of oxygen:

$$2\ PbO_2(s) + 4\ HNO_3(aq) \longrightarrow 2\ Pb(NO_3)_2(aq) + 2\ H_2O(\ell) + O_2(g)$$

What volume of a 7.91 M solution of nitric acid is just sufficient to react with 15.9 g of lead(IV) oxide, according to this equation?

16. Phosphoric acid is made industrially by the reaction of fluorapatite, $Ca_5(PO_4)_3F$, in phosphate rock with sulfuric acid:

$$Ca_5(PO_4)_3F(s) + 5\ H_2SO_4(aq) + 10\ H_2O(\ell) \longrightarrow 3\ H_3PO_4(aq) + 5\ (CaSO_4{\cdot}2H_2O)(s) + HF(aq)$$

What volume of 6.3 M phosphoric acid is generated by the reaction of 2.2 metric tons (2200 kg) of fluorapatite?

17. The carbon dioxide produced (together with hydrogen) from the industrial-scale oxidation of methane in the presence of nickel is removed from the gas mixture in a scrubber containing an aqueous solution of potassium carbonate:

$$CO_2(g) + H_2O(\ell) + K_2CO_3(aq) \longrightarrow 2\ KHCO_3(aq)$$

Calculate the volume of carbon dioxide (at 50°C and 1.00 atm pressure) that will react with 187 L of a 1.36 M potassium carbonate solution.

18. Nitrogen oxide can be generated on a laboratory scale by the reaction of dilute sulfuric acid with aqueous sodium nitrite:

$$6\ NaNO_2(aq) + 3\ H_2SO_4(aq) \longrightarrow 4\ NO(g) + 2\ HNO_3(aq) + 2\ H_2O(\ell) + 3\ Na_2SO_4(aq)$$

What volume of 0.646 M aqueous $NaNO_2$ should be used in this reaction to generate 5.00 L of nitrogen oxide at a temperature of 20°C and a pressure of 0.970 atm?

19. Write a balanced equation for the acid–base reaction that leads to the production of each of the following salts. Name the acid, base, and salt.
(a) CaF_2
(b) Rb_2SO_4
(c) $Zn(NO_3)_2$
(d) KCH_3COO

20. Write a balanced equation for the acid–base reaction that leads to the production of each of the following salts. Name the acid, base, and salt.
(a) Na_2SO_3
(b) $Ca(C_6H_5COO)_2$
(c) $PbSO_4$
(d) $CuCl_2$

21. Hydrogen sulfide can be removed from natural gas by reaction with excess sodium hydroxide. Name the salt that is produced in this reaction. (**Note:** Hydrogen sulfide loses both its hydrogen atoms in the course of this reaction.)

22. During the preparation of viscose rayon, cellulose is dissolved in a bath containing sodium hydroxide and later reprecipitated as rayon using a solution of sulfuric acid. Name the salt that is a by-product of this process. Rayon production is, in fact, a significant commercial source for this salt.

23. Phosphorus trifluoride is a highly toxic gas that reacts slowly with water to give a mixture of phosphorous acid and hydrofluoric acid.
(a) Write a balanced chemical equation for this reaction.
(b) Determine the concentration (in moles per liter) of each of the acids that result from the reaction of 1.94 L of phosphorus trifluoride (measured at 25°C and 0.970 atm pressure) with water to give a solution volume of 872 mL.

24. Phosphorus pentachloride reacts violently with water to give a mixture of phosphoric acid and hydrochloric acid.
(a) Write a balanced chemical equation for this reaction.
(b) Determine the concentration (in moles per liter) of each of the acids that result from the complete reaction of 1.22 L of phosphorus pentachloride (measured at 215°C and 0.962 atm pressure) with enough water to give a solution volume of 697 mL.

25. To determine the concentration of a solution of nitric acid, a 100.0-mL sample is placed in a flask and titrated with a 0.1279 M solution of potassium hydroxide. A volume of 37.85 mL is required to reach the phenolphthalein end point. Calculate the concentration of nitric acid in the original sample.

26. The concentration of aqueous ammonia in a cleaning solution is determined by titration with hydrochloric acid. A volume of 23.18 mL of 0.8381 M HCl is needed to titrate a 50.0-mL sample of the ammonia solution to a methyl red end point. Calculate the concentration of ammonia in the cleaning solution.

Reaction Stoichiometry in Solutions: Oxidation–Reduction Titrations

27. For each of the following balanced equations, write the oxidation number above the symbol of each atom that changes oxidation state in the course of the reactions.
(a) $2\ PF_2I(\ell) + 2\ Hg(\ell) \longrightarrow P_2F_4(g) + Hg_2I_2(s)$
(b) $2\ KClO_3(s) \longrightarrow 2\ KCl(s) + 3\ O_2(g)$
(c) $4\ NH_3(g) + 5\ O_2(g) \longrightarrow 4\ NO(g) + 6\ H_2O(g)$
(d) $2\ As(s) + 6\ NaOH(\ell) \longrightarrow 2\ Na_3AsO_3(s) + 3\ H_2(g)$

28. For each of the following balanced equations, write the oxidation number above the symbol of each atom that changes oxidation state in the course of the reaction.
(a) $N_2O_4(g) + KCl(s) \longrightarrow NOCl(g) + KNO_3(s)$
(b) $H_2S(g) + 4\ O_2F_2(s) \longrightarrow SF_6(g) + 2\ HF(g) + 4\ O_2(g)$
(c) $2\ POBr_3(s) + 3\ Mg(s) \longrightarrow 2\ PO(s) + 3\ MgBr_2(s)$
(d) $4\ BCl_3(g) + 3\ SF_4(g) \longrightarrow 4\ BF_3(g) + 3\ SCl_2(\ell) + 3\ Cl_2(g)$

29. Selenic acid (H_2SeO_4) is a powerful oxidizing acid that dissolves not only silver (as does the related acid H_2SO_4) but gold, through the following reaction:

$$2\ Au(s) + 6\ H_2SeO_4(aq) \longrightarrow Au_2(SeO_4)_3(aq) + 3\ H_2SeO_3(aq) + 3\ H_2O(\ell)$$

Determine the oxidation numbers of the atoms in this equation. Which species is oxidized and which is reduced?

30. Diiodine pentaoxide oxidizes carbon monoxide to carbon dioxide under room conditions, yielding iodine as the second product:

$$I_2O_5(s) + 5\ CO(g) \longrightarrow I_2(s) + 5\ CO_2(g)$$

This can be used in an analytical method to measure the amount of carbon monoxide in a sample of air. Determine the oxidation numbers of the atoms in this equation. Which species is oxidized and which is reduced?

31. Complete and balance the following equations for reactions taking place in acidic solution.
(a) $VO_2^+(aq) + SO_2(g) \longrightarrow VO^{2+}(aq) + SO_4^{2-}(aq)$
(b) $Br_2(\ell) + SO_2(g) \longrightarrow Br^-(aq) + SO_4^{2-}(aq)$
(c) $Cr_2O_7^{2-}(aq) + Np^{4+}(aq) \longrightarrow Cr^{3+}(aq) + NpO_2^{2+}(aq)$
(d) $HCOOH(aq) + MnO_4^-(aq) \longrightarrow CO_2(g) + Mn^{2+}(aq)$
(e) $Hg_2HPO_4(s) + Au(s) + Cl^-(aq) \longrightarrow Hg(\ell) + H_2PO_4^-(aq) + AuCl_4^-(aq)$

32. Complete and balance the following equations for reactions taking place in acidic solution.
(a) $MnO_4^-(aq) + H_2S(aq) \longrightarrow Mn^{2+}(aq) + SO_4^{2-}(aq)$
(b) $Zn(s) + NO_3^-(aq) \longrightarrow Zn^{2+}(aq) + NH_4^+(aq)$
(c) $H_2O_2(aq) + MnO_4^-(aq) \longrightarrow O_2(g) + Mn^{2+}(aq)$
(d) $Sn(s) + NO_3^-(aq) \longrightarrow Sn^{4+}(aq) + N_2O(g)$
(e) $UO_2^{2+}(aq) + Te(s) \longrightarrow U^{4+}(aq) + TeO_4^{2-}(aq)$

33. Complete and balance the following equations for reactions taking place in basic solution.
(a) $Cr(OH)_3(s) + Br_2(aq) \longrightarrow CrO_4^{2-}(aq) + Br^-(aq)$
(b) $ZrO(OH)_2(s) + SO_3^{2-}(aq) \longrightarrow Zr(s) + SO_4^{2-}(aq)$
(c) $HPbO_2^-(aq) + Re(s) \longrightarrow Pb(s) + ReO_4^-(aq)$
(d) $HXeO_4^-(aq) \longrightarrow XeO_6^{4-}(aq) + Xe(g)$
(e) $N_2H_4(aq) + CO_3^{2-}(aq) \longrightarrow N_2(g) + CO(g)$

34. Complete and balance the following equations for reactions taking place in basic solution.
(a) $OCl^-(aq) + I^-(aq) \longrightarrow IO_3^-(aq) + Cl^-(aq)$
(b) $SO_3^{2-}(aq) + Be(s) \longrightarrow S_2O_3^{2-}(aq) + Be_2O_3^{2-}(aq)$
(c) $H_2BO_3^-(aq) + Al(s) \longrightarrow BH_4^-(aq) + H_2AlO_3^-(aq)$
(d) $O_2(g) + Sb(s) \longrightarrow H_2O_2(aq) + SbO_2^-(aq)$
(e) $Sn(OH)_6^{2-}(aq) + Si(s) \longrightarrow HSnO_2^-(aq) + SiO_3^{2-}(aq)$

35. The following balanced equations represent reactions that occur in aqueous acid. Break them down into balanced oxidation and reduction half-equations.
(a) $2\ H_3O^+(aq) + H_2O_2(aq) + 2\ Fe^{2+}(aq) \longrightarrow 2\ Fe^{3+}(aq) + 4\ H_2O(\ell)$
(b) $H_3O^+(aq) + H_2O(\ell) + 2\ MnO_4^-(aq) + 5\ SO_2(aq) \longrightarrow 2\ Mn^{2+}(aq) + 5\ HSO_4^-(aq)$
(c) $5\ ClO_2^-(aq) + 4\ H_3O^+(aq) \longrightarrow 4\ ClO_2(g) + Cl^-(aq) + 6\ H_2O(\ell)$

36. The following balanced equations represent reactions that occur in aqueous base. Break them down into balanced oxidation and reduction half-equations.
(a) $4\ PH_3(g) + 4\ H_2O(\ell) + 4\ CrO_4^{2-}(aq) \longrightarrow P_4(s) + 4\ Cr(OH)_4^-(aq) + 4\ OH^-(aq)$
(b) $NiO_2(s) + 2\ H_2O(\ell) + Fe(s) \longrightarrow Ni(OH)_2(s) + Fe(OH)_2(s)$
(c) $CO_2(g) + 2\ NH_2OH(aq) \longrightarrow CO(g) + N_2(g) + 3\ H_2O(g)$

37. Nitrous acid (HNO_2) disproportionates in acidic solution to nitrate ion (NO_3^-) and nitrogen oxide (NO). Write a balanced equation for this reaction.

38. Thiosulfate ion ($S_2O_3^{2-}$) disproportionates in acidic solution to give solid sulfur and aqueous hydrogen sulfite ion (HSO_3^-). Write a balanced equation for this reaction.

39. Potassium dichromate in acidic solution is used to titrate a solution of iron(II) ions, with which it reacts according to

$$Cr_2O_7^{2-}(aq) + 6\ Fe^{2+}(aq) + 14\ H_3O^+(aq) \longrightarrow 2\ Cr^{3+}(aq) + 6\ Fe^{3+}(aq) + 21\ H_2O(\ell)$$

A potassium dichromate solution is prepared by dissolving 5.134 g of $K_2Cr_2O_7$ in water and diluting to a total volume of 1.000 L. A total of 34.26 mL of this solution is required to reach the end point in a titration of a 500.0-mL sample containing $Fe^{2+}(aq)$. Determine the concentration of Fe^{2+} in the original solution.

40. Cerium(IV) ions are strong oxidizing agents in acidic solution, oxidizing arsenious acid to arsenic acid according to the following equation:

$$2\ Ce^{4+}(aq) + H_3AsO_3(aq) + 3\ H_2O(\ell) \longrightarrow 2\ Ce^{3+}(aq) + H_3AsO_4(aq) + 2\ H^+(aq)$$

A sample of As_2O_3 weighing 0.217 g is dissolved in basic solution and then acidified to make H_3AsO_3. Its titration with a solution of acidic cerium(IV) sulfate requires 21.47 mL. Determine the original concentration of $Ce^{4+}(aq)$ in the titrating solution.

Phase Equilibrium in Solutions: Nonvolatile Solutes

41. The vapor pressure of pure acetone (CH_3COCH_3) at 30°C is 0.3270 atm. Suppose 15.0 g of benzophenone, $C_{13}H_{10}O$, is dissolved in 50.0 g of acetone. Calculate the vapor pressure of acetone above the resulting solution.

42. The vapor pressure of diethyl ether (molar mass, 74.12 g mol^{-1}) at 30°C is 0.8517 atm. Suppose 1.800 g of maleic acid, $C_4H_4O_4$, is dissolved in 100.0 g of diethyl ether at 30°C. Calculate the vapor pressure of diethyl ether above the resulting solution.

43. Pure toluene (C_7H_8) has a normal boiling point of 110.60°C. A solution of 7.80 g of anthracene ($C_{14}H_{10}$) in 100.0 g of toluene has a boiling point of 112.06°C. Calculate K_b for toluene.

44. When 2.62 g of the nonvolatile solid anthracene, $C_{14}H_{10}$, is dissolved in 100.0 g of cyclohexane, C_6H_{12}, the boiling point of the cyclohexane is raised by 0.41°C. Calculate K_b for cyclohexane.

45. When 39.8 g of a nondissociating, nonvolatile sugar is dissolved in 200.0 g of water, the boiling point of the water is raised by 0.30°C. Estimate the molar mass of the sugar.

46. When 2.60 g of a substance that contains only indium and chlorine is dissolved in 50.0 g of tin(IV) chloride, the normal boiling point of the tin(IV) chloride is raised from 114.1°C to 116.3°C. If $K_b = 9.43$ K kg mol^{-1} for $SnCl_4$, what are the approximate molar mass and the probable molecular formula of the solute?

47. The Rast method for determining molar masses uses camphor as the solvent. Camphor melts at 178.4°C, and its large K_f (37.7 K kg mol^{-1}) makes it especially useful for accurate work. A sample of an unknown substance that weighs 0.840 g reduces the freezing point of 25.0 g of camphor to 170.8°C. What is its molar mass?

48. Barium chloride has a freezing point of 962°C and a freezing-point depression constant of 108 K kg mol^{-1}. If 12 g of an unknown substance dissolved in 562 g of barium chloride gives a solution with a freezing point of 937°C, compute the molar mass of the unknown, assuming no dissociation takes place.

49. Ice cream is made by freezing a liquid mixture that, as a first approximation, can be considered a solution of sucrose ($C_{12}H_{22}O_{11}$) in water. Estimate the temperature at which the first ice crystals begin to appear in a mix that consists of 34% (by mass) sucrose in water. As ice crystallizes out, the remaining solution becomes more concentrated. What happens to its freezing point?

50. The solution to Problem 49 shows that to make homemade ice cream, temperatures ranging downward from −3°C are needed. Ice cubes from a freezer have a temperature of about −12°C (+10°F), which is cold enough, but contact with the warmer ice cream mixture causes them to melt to liquid at 0°C, which is too warm. To obtain a liquid that is cold enough, salt (NaCl) is dissolved in water, and ice is added to the saltwater. The salt lowers the freezing point of the water enough so that it can freeze the liquid inside the ice cream maker. The instructions for an ice cream maker say to add one part salt to eight parts water (by mass). What is the freezing point of this solution (in degrees Celsius and degrees Fahrenheit)? Assume that the NaCl dissociates fully into ions, and that the solution is ideal.

51. An aqueous solution is 0.8402 molal in Na_2SO_4. It has a freezing point of $-4.218°C$. Determine the effective number of particles arising from each Na_2SO_4 formula unit in this solution.

52. The freezing-point depression constant of pure H_2SO_4 is 6.12 K kg mol^{-1}. When 2.3 g of ethanol (C_2H_5OH) is dissolved in 1.00 kg of pure sulfuric acid, the freezing point of the solution is 0.92 K lower than the freezing point of pure sulfuric acid. Determine how many particles are formed as 1 molecule of ethanol goes into solution in sulfuric acid.

53. A 200-mg sample of a purified compound of unknown molar mass is dissolved in benzene and diluted with that solvent to a volume of 25.0 cm^3. The resulting solution is found to have an osmotic pressure of 0.0105 atm at 300 K. What is the molar mass of the unknown compound?

54. Suppose 2.37 g of a protein is dissolved in water and diluted to a total volume of 100.0 mL. The osmotic pressure of the resulting solution is 0.0319 atm at 20°C. What is the molar mass of the protein?

55. A polymer of large molar mass is dissolved in water at 15°C, and the resulting solution rises to a final height of 15.2 cm above the level of the pure water, as water molecules pass through a semipermeable membrane into the solution. If the solution contains 4.64 g polymer per liter, calculate the molar mass of the polymer.

56. Suppose 0.125 g of a protein is dissolved in 10.0 cm^3 of ethyl alcohol (C_2H_5OH), whose density at 20°C is 0.789 g cm^{-3}. The solution rises to a height of 26.3 cm in an osmometer (an apparatus for measuring osmotic pressure). What is the approximate molar mass of the protein?

Phase Equilibrium in Solutions: Volatile Solutes

57. The Henry's law constant at 25°C for carbon dioxide dissolved in water is 1.65×10^3 atm. If a carbonated beverage is bottled under a CO_2 pressure of 5.0 atm:

(a) Calculate the number of moles of carbon dioxide dissolved per liter of water under these conditions, using 1.00 g cm^{-3} as the density of water.

(b) Explain what happens on a microscopic level after the bottle cap is removed.

58. The Henry's law constant at 25°C for nitrogen dissolved in water is 8.57×10^4 atm, that for oxygen is 4.34×10^4 atm, and that for helium is 1.7×10^5 atm.

(a) Calculate the number of moles of nitrogen and oxygen dissolved per liter of water in equilibrium with air at 25°C. Use Table 9.1.

(b) Air is dissolved in blood and other bodily fluids. As a deep-sea diver descends, the pressure increases and the concentration of dissolved air in the blood increases. If the diver returns to the surface too quickly, gas bubbles out of solution within the body so rapidly that it can cause a dangerous condition called "the bends." Use Henry's law to show why divers sometimes use a combination of helium and oxygen in their breathing tanks in place of compressed air.

59. At 25°C, some water is added to a sample of gaseous methane (CH_4) at 1.00 atm pressure in a closed vessel, and the vessel is shaken until as much methane as possible dissolves. Then 1.00 kg of the solution is removed and boiled to expel the methane, yielding a volume of 3.01 L of $CH_4(g)$ at 0°C and 1.00 atm. Determine the Henry's law constant for methane in water.

60. When exactly the procedure of Problem 59 is conducted using benzene (C_6H_6) in place of water, the volume of methane that results is 0.510 L at 0°C and 1.00 atm. Determine the Henry's law constant for methane in benzene.

61. At 20°C, the vapor pressure of toluene is 0.0289 atm and the vapor pressure of benzene is 0.0987 atm. Equal numbers of moles of toluene and benzene are mixed and form an ideal solution. Compute the mole fraction of benzene in the vapor in equilibrium with this solution.

62. At 90°C, the vapor pressure of toluene is 0.534 atm and the vapor pressure of benzene is 1.34 atm. Benzene (0.400 mol) is mixed with toluene (0.900 mol) to form an ideal solution. Compute the mole fraction of benzene in the vapor in equilibrium with this solution.

63. At 40°C, the vapor pressure of pure carbon tetrachloride (CCl_4) is 0.293 atm and the vapor pressure of pure dichloroethane ($C_2H_4Cl_2$) is 0.209 atm. A nearly ideal solution is prepared by mixing 30.0 g of carbon tetrachloride with 20.0 g of dichloroethane.

(a) Calculate the mole fraction of CCl_4 in the solution.

(b) Calculate the total vapor pressure of the solution at 40°C.

(c) Calculate the mole fraction of CCl_4 in the vapor in equilibrium with the solution.

64. At 300 K, the vapor pressure of pure benzene (C_6H_6) is 0.1355 atm and the vapor pressure of pure *n*-hexane (C_6H_{14}) is 0.2128 atm. Mixing 50.0 g of benzene with 50.0 g of *n*-hexane gives a solution that is nearly ideal.

(a) Calculate the mole fraction of benzene in the solution.

(b) Calculate the total vapor pressure of the solution at 300 K.

(c) Calculate the mole fraction of benzene in the vapor in equilibrium with the solution.

ADDITIONAL PROBLEMS

65. Veterinarians use Donovan's solution to treat skin diseases in animals. The solution is prepared by mixing 1.00 g of $AsI_3(s)$, 1.00 g of $HgI_2(s)$, and 0.900 g of $NaHCO_3(s)$ in enough water to make a total volume of 100.0 mL.

(a) Compute the total mass of iodine per liter of Donovan's solution, in grams per liter.

(b) You need a lot of Donovan's solution to treat an outbreak of rash in an elephant herd. You have plenty of mercury(II) iodide and sodium hydrogen carbonate, but the only arsenic(III) iodide you can find is 1.50 L of a 0.100 M aqueous solution. Explain how to prepare 3.50 L of Donovan's solution starting with these materials.

66. Relative solubilities of salts in liquid ammonia can differ significantly from those in water. Thus, silver bromide is soluble in ammonia, but barium bromide is not (the reverse of the situation in water).

(a) Write a balanced equation for the reaction of an ammonia solution of barium nitrate with an ammonia solution of silver bromide. Silver nitrate is soluble in liquid ammonia.

(b) What volume of a 0.50 M solution of silver bromide will react completely with 0.215 L of a 0.076 M solution of barium nitrate in ammonia?

(c) What mass of barium bromide will precipitate from the reaction in part (b)?

* 67. A 5.0-L flask contains a mixture of ammonia and nitrogen at 27°C and a total pressure of 3.00 atm. The sample of gas is allowed to flow from the flask until the pressure in the flask has fallen to 1.00 atm. The gas that escapes is passed through 1.50 L of 0.200 M acetic acid. All the ammonia in the gas that escapes is absorbed by the solution and turns out to be just sufficient to neutralize the acetic acid present. The volume of the solution does not change significantly.

(a) Will the electrical conductivity of the aqueous solution change significantly as the gas is absorbed? Give equations for any reactions, and calculate the final concentrations of the principal ions present (if any) at the end.

(b) Calculate the percentage by mass of ammonia in the flask initially.

* 68. It was desired to neutralize a certain solution X that had been prepared by mixing solutions of potassium chloride and hydrobromic acid. Titration of 10.0 mL X with 0.100 M silver nitrate required 50.0 mL of the latter. The resulting precipitate, containing a mixture of AgCl and AgBr, was dried and found to weigh 0.762 g. How much 0.100 M sodium hydroxide should be used to neutralize 10.0 mL solution X?

* 69. Vanadic ion, V^{3+}, forms green salts and is a good reducing agent, being itself changed in neutral solutions to the nearly colorless ion $V(OH)_4^+$. Suppose that 15.0 mL of a 0.200-M solution of vanadic sulfate, $V_2(SO_4)_3$, was needed to reduce completely a 0.540-g sample of an unknown substance X. If each molecule of X accepted just one electron, what is the molecular weight of X? Suppose that each molecule of X accepted three electrons; what would be the molecular weight of X then?

* 70. A new antibiotic, A, which is an acid, can readily be oxidized by hot aqueous permanganate; the latter is reduced to manganous ion, Mn^{2+}. The following experiments have been performed with A: (a) 0.293 g A consumes just 18.3 mL of 0.080 M $KMnO_4$; (b) 0.385 g A is just neutralized by 15.7 mL of 0.490 M NaOH. What can you conclude about the molecular weight of A from (a), from (b), and from both considered together?

71. Suppose 150 mL of a 10.00% by mass solution of sodium chloride (density = 1.0726 g cm^{-3}) is acidified with sulfuric acid and then treated with an excess of $MnO_2(s)$. Under these conditions, all the chlorine is liberated as $Cl_2(g)$. The chlorine is collected without loss and reacts with excess $H_2(g)$ to form HCl(g). The HCl(g) is dissolved in enough water to make 250 mL of solution. Compute the molarity of this solution.

* 72. The amount of ozone in a mixture of gases can be determined by passing the mixture through an acidic aqueous solution of potassium iodide, where the ozone reacts according to

$$O_3(g) + 3\,I^-(aq) + H_2O(\ell) \longrightarrow O_2(g) + I_3^-(aq) + 2\,OH^-(aq)$$

to form the triiodide ion I_3^-. The amount of triiodide produced is then determined by titrating with thiosulfate solution:

$$I_3^-(aq) + 2\,S_2O_3^{2-}(aq) \longrightarrow 3\,I^-(aq) + S_4O_6^{2-}(aq)$$

A small amount of starch solution is added as an indicator because it forms a deep-blue complex with the triiodide solution. Disappearance of the blue color thus signals the completion of the titration. Suppose 53.2 L of a gas mixture at a temperature of 18°C and a total pressure of 0.993 atm is passed through a solution of potassium iodide until the ozone in the mixture has reacted completely. The solution requires 26.2 mL of a 0.1359-M solution of thiosulfate ion to titrate to the endpoint. Calculate the mole fraction of ozone in the original gas sample.

73. The vapor pressure of pure liquid CS_2 is 0.3914 atm at 20°C. When 40.0 g of rhombic sulfur is dissolved in 1.00 kg of CS_2, the vapor pressure of CS_2 decreases to 0.3868 atm. Determine the molecular formula for the sulfur molecules dissolved in CS_2.

74. The expressions for boiling-point elevation and freezing-point depression apply accurately to *dilute* solutions only. A saturated aqueous solution of NaI (sodium iodide) in water has a boiling point of 144°C. The mole fraction of NaI in the solution is 0.390. Compute the molality of this solution. Compare the boiling-point elevation predicted by the expression in this chapter with the elevation actually observed.

75. You take a bottle of soft drink out of your refrigerator. The contents are liquid and stay liquid, even when you shake them. Presently, you remove the cap and the liquid freezes solid. Offer a possible explanation for this observation.

76. Mercury(II) chloride ($HgCl_2$) freezes at 276.1°C and has a freezing-point depression constant K_f of 34.3 K kg mol^{-1}. When 1.36 g of solid mercury(I) chloride (empirical formula HgCl) is dissolved in 100 g of $HgCl_2$, the freezing point is reduced by 0.99°C. Calculate the molar mass of the dissolved solute species and give its molecular formula.

* 77. The vapor pressure of an aqueous solution of $CaCl_2$ at 25°C is 0.02970 atm. The vapor pressure of pure water at the same temperature is 0.03126 atm. Estimate the freezing point of the solution.

78. Ethylene glycol (CH_2OHCH_2OH) is used in antifreeze because, when mixed with water, it lowers the freezing point below 0°C. What mass percentage of ethylene glycol in water must be used to reduce the freezing point of the mixture to −5.0°C, assuming ideal solution behavior?

79. A new compound has the empirical formula $GaCl_2$. This surprises some chemists who, based on the position of gallium in the periodic table, expect a chloride of gallium to have the formula $GaCl_3$ or possibly GaCl. They suggest that the "$GaCl_2$" is really Ga[$GaCl_4$], in which the bracketed group behaves as a unit with a −1 charge. Suggest experiments to test this hypothesis.

* 80. Suppose two beakers are placed in a small closed container at 25°C. One contains 400 mL of a 0.100-M aqueous solution of NaCl; the second contains 200 mL of a 0.250-M aqueous solution of NaCl. Small amounts of water evaporate from both solutions. As time passes, the volume of

solution in the second beaker gradually increases, and that in the first gradually decreases. Why? If we wait long enough, what will the final volumes and concentrations be?

* **81.** The walls of erythrocytes (red blood cells) are permeable to water. In a salt solution, they shrivel (lose water) when the outside salt concentration is high and swell (take up water) when the outside salt concentration is low. In an experiment at 25°C, an aqueous solution of NaCl that has a freezing point of 0.406°C causes erythrocytes neither to swell nor to shrink, indicating that the osmotic pressure of their contents is equal to that of the NaCl solution. Calculate the osmotic pressure of the solution inside the erythrocytes under these conditions, assuming that its molarity and molality are equal.

82. Silver dissolves in molten lead. Compute the osmotic pressure of a 0.010 M solution of silver in lead at 423°C. Compute the height of a column of molten lead (ρ = 11.4 g cm^{-3}) to which this pressure corresponds.

83. Henry's law is important in environmental chemistry, where it predicts the distribution of pollutants between water and the atmosphere. Benzene (C_6H_6) emitted in wastewater streams, for example, can pass into the air, where it is degraded by processes induced by light from the sun. The Henry's law constant for benzene in water at 25°C is 301 atm. Calculate the partial pressure of benzene vapor in equilibrium with a solution of 2.0 g of benzene per 1000 L of water. How many benzene molecules are present in each cubic centimeter?

* **84.** Refer to the data of Problem 62. Calculate the mole fraction of toluene in a mixture of benzene and toluene that boils at 90°C under atmospheric pressure.

85. What is the difference between a solution and a colloidal suspension? Give examples of each, and show how, in some cases, it may be difficult to classify a mixture as one or the other.

CUMULATIVE PROBLEMS

86. A student prepares a solution by dissolving 1.000 mol of Na_2SO_4 in water. She accidentally leaves the container uncovered and comes back the next week to find only a white, solid residue. The mass of the residue is 322.2 g. Determine the chemical formula of this residue.

87. Complete combustion of 2.40 g of a compound of carbon, hydrogen, and oxygen yielded 5.46 g CO_2 and 2.23 g H_2O. When 8.69 g of the compound was dissolved in 281 g of water, the freezing point of the solution was found to be −0.97°C. What is the molecular formula of the compound?

* **88.** Imagine that two 1-L beakers, A and B, each containing an aqueous solution of fructose (a nonvolatile sugar with molecular weight = 180) are placed together in a box, which is then sealed. (The concentrations of the solutions are not necessarily the same.) The temperature remains constant at 26°C. Initially, there is 600 mL of solution in A and 100 mL of solution in B. As the solutions stand in the sealed box, their volumes change slowly for a while. When they stop changing, beaker A contains 400 mL and beaker B contains 300 mL. It is then determined that the solution in A is 1.5 M in fructose and has a density of 1.10 g mL^{-1}.

(a) What is the molar concentration of fructose in the solution in beaker B at the end? Explain.

(b) Calculate the concentration of fructose in the solution in A at the start.

(c) Calculate the concentration of the fructose in the solution in B at the start.

(d) The vapor pressure of pure water at 26°C is 25.2 torr. What is the pressure of water vapor in the box at the end, after the volumes have stopped changing?

UNIT IV

Equilibrium in Chemical Reactions

© Theo Allofs/Zefa/Corbis

Stalactites (top) and stalagmites (bottom) consist of calcium carbonate. They form when a water solution containing Ca^{2+} and HCO_3^- ions enters a cave. Carbon dioxide is released, and calcium carbonate precipitates: $Ca^{2+}(aq) + 2\ HCO_3^-(aq) \longrightarrow CaCO_3(s) + H_2O + CO_2(g)$.

NOTE TO THE READER:
It is purely a matter of taste whether one should first study chemical equilibrium from the empirical point of view, and then study thermodynamics to provide the fundamental explanation of equilibrium, or learn thermodynamics first as essential background for the study of equilibrium. We have written this textbook to allow either approach. If your instructor prefers to cover thermodynamics before equilibrium, you should read Chapters 12, 13, and 14 straight through in the order written. If your instructor prefers to cover equilibrium from the empirical point of view before studying thermodynamics, you should skip now to Chapter 14 and omit those sections of Chapter 14 (clearly marked) that require background in thermodynamics. You should come back and read those sections later after you have studied Chapters 12 and 13.

How far do chemical reactions proceed toward completely consuming the reactants? What determines the extent of their progress? Experience shows that many reactions do not go to completion, but approach instead an *equilibrium state* in which products and unconsumed reactants are both present in specific, relative amounts. Once equilibrium has been achieved, the composition of the reaction mixture does not change any further. The equilibrium composition of the mixture is described quantitatively by the equilibrium constant for the reaction. If we know the equilibrium constant, we can calculate the equilibrium composition that will result from any initial composition. This is one of the most important tools available to chemists because it is used to predict and optimize the yield of reactions throughout fundamental and applied chemistry.

Heat influences the progress of chemical reactions—driving some forward while retarding others—and is quantitatively connected to chemical equilibrium by the science of thermodynamics. Thermodynamics predicts the equilibrium constant for a reaction from simple physical properties of the reactants and products and explains how the value of the equilibrium constant depends on the reaction temperature. In that way, thermodynamics shows how to increase the reaction yield by changing the reaction temperature.

UNIT CHAPTERS

UNIT GOALS

- To relate composition in the equilibrium state to the equilibrium constant
- To calculate composition in the equilibrium state from the equilibrium constant
- To describe the influence of temperature on the equilibrium constant
- To apply thermodynamics to explain these connections and optimize reaction yield

CHAPTER 12

Thermodynamic Processes and Thermochemistry

The steam locomotive operates by converting thermal energy into mechanical energy. The diesel locomotive converts chemical energy into electrical energy, then electrical energy into mechanical energy to generate motion. All these energy conversion processes are governed by thermodynamics. (a: ©DAJ/Getty Images; b: courtesy ©Kent Foster/Visuals Unlimited)

Experience shows that heat is the most important factor influencing the extent of chemical reactions. Heat drives some reactions toward completion, but retards the progress of others. Therefore, it is appropriate to launch our study of chemical equilibrium by learning how to measure the heat transfer in a chemical reaction. This objective leads us into the branch of physical science called thermodynamics, which describes the meaning of heat and gives procedures for measuring heat transfer quantitatively.

Thermodynamics is a broad and general subject with applications in all branches of the physical and biological sciences and engineering; thus, we limit our discussion to those aspects necessary for chemical equilibrium. In this chapter, we demonstrate that heat—which on first examination appears mysterious despite its familiarity—is just another form of energy, a form we call *thermal* energy. We describe how thermal energy is mutually convertible into mechanical and electrical energy, and how the total amount of energy is conserved during any such transfers. The previous sentence in more concise form is the first law of thermodynamics, which is the unifying theme of this chapter. We see how to measure quantitatively the heat transfer into a system through its *heat capacity*. We see how the transfer depends on experimental conditions such as constant pressure or constant volume. Finally, we apply these methods to measure the heat transfer in chemical reactions conducted at constant pressure. Then we are ready to connect heat transfer to chemical equilibrium at the beginning of Chapter 13.

We conclude this introduction with a general overview of thermodynamics, as context for the specific studies in this and the next chapter. Thermodynamics, in which a few apparently simple laws summarize a rich variety of observed behavior, is one of the surest and most powerful branches of science. The distinctive feature of thermodynamics is the universality of its basic laws, and the beauty of the subject is the many conclusions that can be deduced from those few laws. The laws of thermodynamics cannot themselves be derived or proved. Instead, they are generalizations drawn from a great many observations of the behavior of matter. The history of thermodynamics, like that of other fields of science, has been fraught with misconceptions. As we look back on the beginnings of the discipline in the 19th century, it appears to have developed with agonizing slowness. But it *has* developed, and its laws are the pillars on which much of modern science rests. The foundations of thermodynamics are completely understood today. It is being applied in research at the forefront of science, on systems ranging from black holes in distant parts of the universe to the growth and development of the biological cell. Many new results and insights are being acquired, but the foundations are not challenged.

Thermodynamics is an *operational* science, concerned with macroscopic, measurable properties and the relations among them. Its goal is to predict what types of chemical and physical processes are possible, and under what conditions, and to calculate quantitatively the properties of the equilibrium state that ensues when a process is conducted. For example, with thermodynamics we can answer the following types of chemical questions:

1. If hydrogen and nitrogen are mixed, is it possible for them to react? If so, what will be the percentage yield of ammonia?
2. How will a particular change in temperature or pressure affect the extent of the reaction?
3. How can the conditions for the reaction be optimized to maximize its yield?

Thermodynamics is an immensely practical subject. The knowledge from thermodynamics that a particular chemical process is impossible under certain proposed conditions can prevent great loss of time and resources spent vainly trying to conduct the reaction under those conditions. Thermodynamics can also suggest ways to change conditions so that a process becomes possible.

The power of thermodynamics lies in its generality: It rests on no particular model of the structure of matter. In fact, if the entire atomic theory of matter were to be found invalid and discarded (a *very* unlikely event!), the foundations of thermodynamics would remain unshaken. Nonetheless, thermodynamics has some important limitations. Thermodynamics asserts that substances have specific measurable macroscopic properties, but it cannot explain why a particular substance has particular numerical values for these properties. Thermodynamics can determine whether a process is possible, but it cannot say how rapidly the process will occur. For example, thermodynamics predicts that diamond is an unstable substance at atmospheric pressure and will eventually become graphite, but cannot predict how long this process will take.

12.1 Systems, States, and Processes

Thermodynamics uses abstract models to represent real-world systems and processes. These processes may appear in a rich variety of situations, including controlled laboratory conditions, industrial production facilities, living systems, the environment on Earth, and space. A key step in applying the methods of thermodynamics to such diverse processes is to formulate the thermodynamic model *for each process*. This step requires precise definitions of thermodynamic terms. Students (and professors!) of thermodynamics encounter—and sometimes create—apparent contradictions that arise from careless or inaccurate use of language. Part of the difficulty is that many thermodynamic terms also have everyday meanings different from their thermodynamic usage. This section provides a brief introduction to the language of thermodynamics.

A **system** is that part of the universe of immediate interest in a particular experiment or study. The system always contains a certain amount of matter and is described by specific parameters that are controlled in the experiment. For example, the gas confined in a closed box may constitute the system, characterized by the number of moles of the gas and the fixed volume of the box. But in other experiments, it would be more appropriate to consider the gas molecules in a particular cubic centimeter of space in the middle of a room to be the system. In the first case, the boundaries are physical walls, but in the second case, the boundaries are conceptual. We explain later that the two kinds of boundaries are treated the same way mathematically. In the second example, the system is characterized by its volume, which is definite, and by the number of moles of gas within it, which may fluctuate as the system exchanges molecules with the surrounding regions.

Systems are classified by the extent to which their boundaries permit exchange of matter and energy with the surrounding regions. In a **closed system,** the boundaries prevent the flow of matter into or out of it (the boundaries are **impermeable**), whereas the boundaries in an **open system** permit such flow. The amount of matter in an open system can change with time. An **isolated system** exchanges neither matter nor energy with the rest of the universe. **Rigid walls** prevent the system from gaining energy by mechanical processes such as compression and deformation; nonrigid walls permit mechanical energy transfer. **Adiabatic walls** prevent the system from gaining or losing thermal energy (described in detail later), whereas **diathermal walls** permit thermal energy transfer.

The definition of "the system" must be tailored to the specific process under consideration. Simple physical processes such as heating or cooling a metal object are modeled as closed systems (no matter is gained or lost) with diathermal (thermal energy is transferred) and nonrigid walls (the object may expand or contract). Most chemical reactions are modeled as open systems (matter is exchanged) with diathermal (thermal energy is transferred) and nonrigid walls (the density of the matter may change during the reaction). You will gain confidence in these classifications through experience as we examine many processes in this chapter.

The portion of the remainder of the universe that can exchange energy and matter with the system during the process of interest is called the **surroundings.** The surroundings provide the external forces that cause changes in the properties of the system during a process. The system and the surroundings together constitute the **thermodynamic universe** for that process. The thermodynamic universe for that process is isolated. Matter and energy are conserved in the thermodynamics universe while they are exchanged between the system and the surroundings during the process.

Thermodynamics is concerned with macroscopic properties of systems and changes in these properties during processes. Such properties are of two kinds: *extensive* and *intensive*. To distinguish between them, consider the following "thought experiment." Place a thin wall through the middle of a system, dividing it into two subsystems, each of which, like the system itself, is characterized by certain properties. An **extensive property** of the system can be written as the *sum* of the corresponding property in the two subsystems. Volume, mass, and energy are typical extensive properties; the volume of a system is the sum of the volumes of its subsystems. An **intensive property** of the system is the *same* as the corresponding property of each of the subsystems. Temperature and pressure are typical intensive properties; if a system at 298 K is divided in half, the temperature of each half will still be 298 K.

A **thermodynamic state** is a macroscopic condition of a system in which the properties of the system are held at selected fixed values independent of time. The properties of the system are held constant by its boundaries and the surroundings. For example, a system comprising 2 mol helium (He) gas can be held in a piston-cylinder apparatus that maintains the system pressure at 1.5 atm, and the apparatus may be immersed in a heat bath that maintains the system temperature at 298 K. The properties of pressure (P) and temperature (T) are then said to be **constrained** to the values 1 atm and 298 K, respectively. The piston-cylinder and the heat bath are the **constraints** that maintain the selected values of the properties P and T.

After the system has been prepared by establishing a set of constraints in the surroundings, after all disturbances caused by the preparation cease and none of its properties changes with time, the system is said to have reached **equilibrium.** The same equilibrium state can be reached from different directions. The thermodynamic state of a system comprising a given quantity of a liquid or gaseous pure substance is fixed when any two of its independent properties are given. Thus, the specification of P and T for 1 mol of a pure gas fixes not merely the volume, V, but all other properties of the material, such as the internal energy, U, which is defined in Section 12.2. These relations among properties can be displayed as three-dimensional plots of any property as a function of two other properties. For 1 mol of a gas, a plot of pressure against volume and temperature is shown in Figure 12.1. The resulting surface is represented by an equation giving P as a function of V and T; this is called the *equation of state* of the substance (see Section 9.7). If we avoid regions of small V and low T where gas nonideality becomes important, the experimentally determined surface is that shown in Figure 12.1, and the equation of state is given by the ideal gas law $PV = nRT$. The points on this surface (A, B, . . .) represent experimentally measured values of P in equilibrium thermodynamic states of the system fixed by particular values of V and T. Experience shows that the values of all other macroscopic properties take on definite values at each of these states. For example, we can visualize the internal energy, U, as a similar three-dimensional plot versus T and V.

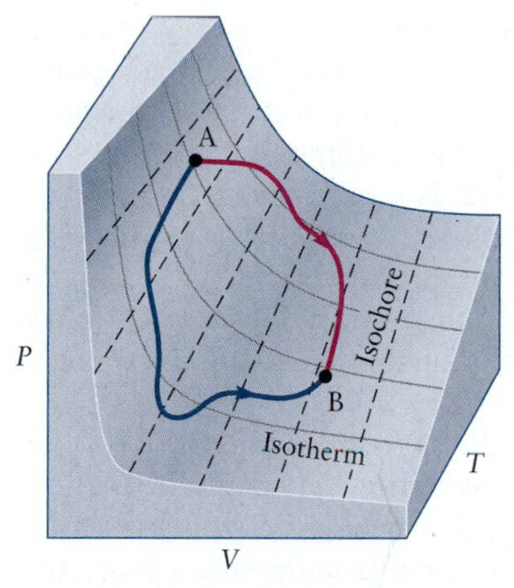

FIGURE 12.1 The *P-V-T* surface of 1 mol of ideal gas. Each point on the surface represents a combination of pressure (*P*), volume (*V*), and temperature (*T*) allowed by the equation of state of the gas. Along an isotherm (*T* constant), the pressure varies inversely with volume; along an isochore (*V* constant), it varies linearly with temperature. Two processes are shown connecting states A and B along paths that satisfy the equation of state at every point.

A **thermodynamic process** changes the thermodynamic state of a system. Such a process may be a *physical* (such as a change in the pressure of a gaseous system or the boiling of a liquid) or a *chemical* process, in which there is a change in the distribution of matter among different chemical species (for example, the reaction of solid $CaCO_3$, at 900 K and 1 atm pressure, to give solid CaO and gaseous CO_2 at the same temperature and pressure).

Because a process changes the state of a system, the process must start with the system in a particular equilibrium state and must also end with the system in a particular equilibrium state. Two such states A and B are indicated in Figure 12.1. You might wonder whether we can sketch a *path* on the surface of equilibrium thermodynamic states to summarize the progress of the system during a process. Only special processes of the type called *reversible* can be represented in this way (see discussion in the following paragraphs).

Many conditions of a system do not correspond to any equilibrium thermodynamic state. Suppose a gas is confined by a piston in a cylinder with volume V_1

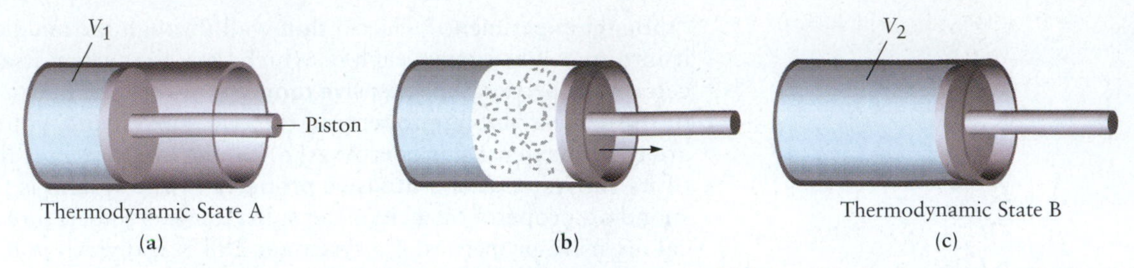

FIGURE 12.2 Stages in an irreversible expansion of a gas from an initial state (a) of volume V_1 to a final state (c) of volume V_2. In the intermediate stage shown (b), the gas is not in equilibrium; because of turbulence, pressure and temperature cannot be defined.

(thermodynamic state A). If the piston is abruptly pulled out to increase the volume to V_2 (Fig. 12.2), chaotic gas currents arise as the molecules begin to move into the larger volume. These intermediate stages are not thermodynamic states, because such properties as density and temperature are changing rapidly through space and in time. Eventually, the currents cease and the system approaches a new equilibrium thermodynamic state, B. States A and B are thermodynamic states, but the conditions in between cannot be described by only a few macroscopic variables, and therefore are not thermodynamic states. Such a process is called **irreversible.** An irreversible process cannot be represented as a path on a thermodynamic surface (as in Fig.12.1), because the intermediate stages are not thermodynamic equilibrium states and thus do not correspond to points on the equation-of-state surface.

In contrast, a **reversible process** proceeds through a continuous series of thermodynamic states, and thus can be shown as a path on the equation-of-state surface. Such a process is an idealization, because the final equilibrium state would be reached only after an infinite length of time; therefore, such a process could never occur in a finite time. If a real process is conducted slowly enough and in sufficiently small steps, the real (irreversible) process can be approximated by the idealized limiting reversible process. The term *reversible* is used because an infinitesimal change in external conditions suffices to reverse the direction of motion of the system. For example, if a gas is expanded by slowly pulling out a piston, only a tiny change in the forces exerted from the outside is required to change the direction of motion of the piston and begin to compress the gas. A gas confined inside a piston-cylinder arrangement will experience an irreversible compression when a kilogram of sand is suddenly dropped onto the piston. The same compression can be achieved (almost) reversibly by transferring the same kilogram of sand onto the piston one grain at the time.

An infinite number of reversible paths can be identified between any two thermodynamic states A and B. Two of them shown in Figure 12.1 could be realized by slowly changing the values of T and V in the proper sequence by manipulating the apparatus in the surroundings. We use such reversible paths throughout this textbook as a tool for calculating changes in properties caused by processes.

Certain properties of a system, called **state functions,** are uniquely determined by the thermodynamic state of the system. Volume, temperature, pressure, and the internal energy, U, are examples of state functions. The Greek letter delta (Δ) is used to indicate *changes* in state functions in a thermodynamic process. Thus, $\Delta V = V_{\text{final}} - V_{\text{initial}}$ (or $V_f - V_i$) is the change in volume between initial and final states, and $\Delta U = U_{\text{final}} - U_{\text{initial}}$ is the corresponding change in internal energy. Because ΔU (or ΔV or ΔT) depends only on the initial and final states, the same change, ΔU, will be measured no matter which path (reversible or irreversible) is followed between any given pair of thermodynamic states. *The change in any state function between two states is independent of path.* The converse statement is also true: If the change in a property of a system is independent of path, the property is a state function. Figure 12.3 illustrates two different paths that connect a given initial state and a common final state.

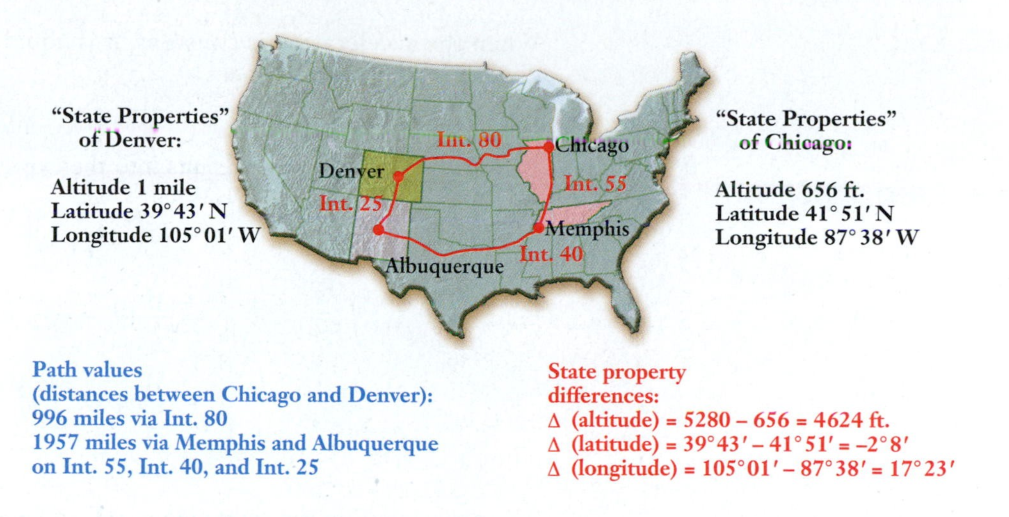

FIGURE 12.3 Differences in state properties (such as the difference in altitude between two points) are independent of the path followed. Other properties (such as the total distance traveled) depend on the particular path.

12.2 The First Law of Thermodynamics: Internal Energy, Work, and Heat

The first law of thermodynamics (which is stated at the end of this section) relates the energy change in a thermodynamic process to the amount of work done on the system and the amount of heat transferred to the system. It is first necessary to examine the ways in which amounts of heat and work are measured to understand the significance of this law. You will see that heat and work are simply different means by which energy is transferred into or out of a system.

Work

The mechanical definition of **work** is the product of the external force on a body times the distance through which the force acts. If a body moves in a straight line from point r_i to r_f with a constant force, F, applied along the direction of the path, the work done on the body is

$$w = F(r_f - r_i) \qquad \text{(force along direction of path)}$$

To illustrate how work can change the energy of a system, we will examine the relation between work and energy in an ordinary mechanical system.

As the first example, consider a block of mass, M, moving with initial velocity v_i along a frictionless surface. We know that a force acting on an object increases the velocity, and therefore the kinetic energy of the object. In the following derivation, we show how the kinetic energy of an object changes when work is done on it. If a constant force, F, is exerted on it in the direction of its motion, it will experience a constant acceleration, $a = F/M$. After a time, t, the velocity of the block will have increased from v_i to v_f, and its position will have changed from r_i to r_f. The work done on the block is

$$w = F(r_f - r_i) = Ma(r_f - r_i)$$

The distance traveled, $r_f - r_i$, is given by the average velocity, in this case $(v_i + v_f)/2$, multiplied by the elapsed time, t:

$$r_f - r_i = \left(\frac{v_i + v_f}{2}\right)t$$

When the acceleration is constant, it is equal to the change in velocity, $v_f - v_i$, divided by the elapsed time:

$$a = (v_f - v_i)/t$$

Substituting both of these results into the expression for the work done gives

$$\begin{aligned} w &= M\left(\frac{v_f - v_i}{t}\right)\left(\frac{v_i - v_f}{2}\right)t \\ &= \frac{M}{2}(v_f - v_i)(v_f + v_i) \\ &= \frac{M}{2}v_f^2 - \frac{M}{2}v_i^2 \\ &= \Delta E_{kin} \end{aligned}$$

The expression on the right side of the equations is the change in kinetic energy, $\frac{1}{2}Mv^2$, of the block. For this idealized example with a frictionless surface, the work done is equal to the change in energy (in this case, kinetic) of the block.

As a second mechanical example, consider the work done in lifting an object in a gravitational field. To raise a mass, M, from an initial height, h_i, to a final height, h_f, an upward force sufficient to counteract the downward force of gravity, Mg, must be exerted. The work done on the object in this case is

$$w = Mg(h_f - h_i) = Mg\Delta h = \Delta E_{pot}$$

This is the change in *potential* energy, Mgh, of the object, showing once again a connection between mechanical work done and a change in energy.

One important kind of mechanical work in chemistry is **pressure–volume work,** which results when a system is compressed or expanded under the influence of an outside pressure. Imagine that a gas has pressure P_i and is confined in a cylinder by a frictionless piston of cross-sectional area A and negligible mass (Fig. 12.4). The force exerted on the inside face of the piston by the gas is $F_i = P_iA$, because pressure is defined as force divided by area. If there is a gas on the outer side of the piston with pressure P_{ext} ("ext" for "external"), then if $P_{ext} = P_i$, the piston will experience no net force. If P_{ext} is increased, the gas will be compressed, and if it is decreased, the gas will expand. Consider first the case in which the external force is less than the initial force exerted by the gas, P_iA. Then the gas will expand and lift the piston from h_i to h_f. The work in this case is

$$w = -F_{ext}(h_f - h_i)$$

The negative sign is inserted because the force from the gas outside opposes the direction of displacement of the piston during expansion of the gas inside the cylinder. This is rewritten as

$$w = -P_{ext}A\Delta h$$

The product $A\Delta h$ is the volume change of the system, ΔV, so the work is

$$w = -P_{ext}\Delta V \qquad [12.1]$$

For an expansion, $\Delta V > 0$, thus $w < 0$ and the system does work; it pushes back the surroundings. For a compression (by making $P_{ext} > P_i$), work is done *on* the system; it is pushed back by the surroundings. Again, $w = -P_{ext}\Delta V$, but now $\Delta V < 0$, so $w > 0$. If there is no volume change, $\Delta V = 0$, and no pressure–volume work is done. Finally, if there is no mechanical link to the surroundings (that is, if $P_{ext} = 0$), then once again no pressure–volume work can be performed because the volume is not changed.

If the pressure P_{ext} is expressed in pascals and the volume in cubic meters, their product is in joules (J). These are the International System of Units (SI) units for

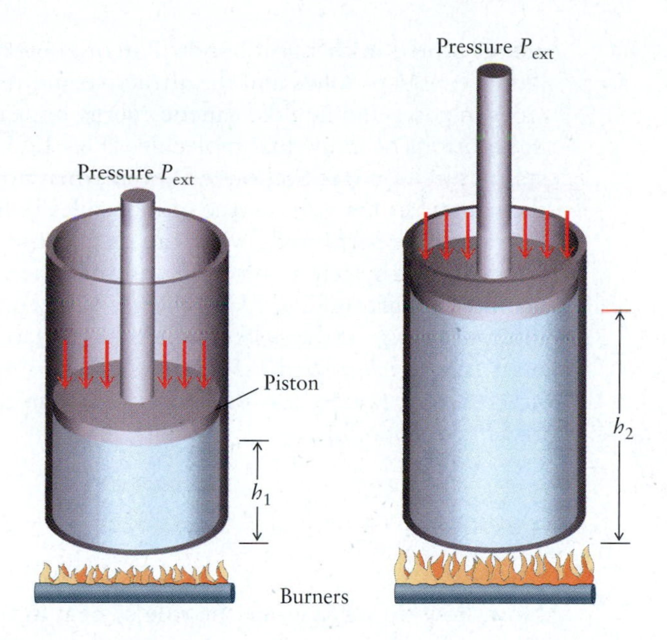

FIGURE 12.4 As the gas inside this cylinder is heated, it expands, pushing the piston against the pressure P_{ext} exerted by the gas outside. As the piston is displaced over a distance $h_f - \Delta h_i = \Delta h$, the volume of the cylinder increases by an amount $A\,\Delta h$, where A is the surface area of the piston.

these quantities. For many purposes, it is more convenient to express pressures in atmospheres and volumes in liters; therefore, work has the unit liter-atmospheres (L atm). The two work units are related by

$$1 \text{ L atm} = (10^{-3} \text{ m}^3)(1.01325 \times 10^5 \text{ kg m}^{-1} \text{ s}^{-2}) = 101.325 \text{ J}$$

EXAMPLE 12.1

A cylinder confines 2.00 L gas under a pressure of 1.00 atm. The external pressure is also 1.00 atm. The gas is heated slowly, with the piston sliding freely to maintain the pressure of the gas close to 1.00 atm. Suppose the heating continues until a final volume of 3.50 L is reached. Calculate the work done on the gas and express it in joules.

SOLUTION

This is an expansion of a system from 2.00 to 3.50 L against a constant external pressure of 1.00 atm. The work done on the system is then

$$w = -P_{ext}\,\Delta V = -\,(1.00 \text{ atm})(3.50 \text{ L} - 2.00 \text{ L}) = -1.50 \text{ L atm}$$

Conversion to joules gives

$$w = (-1.50 \text{ L atm})(101.325 \text{ J L}^{-1} \text{ atm}^{-1}) = -152 \text{ J}$$

Because w is negative, we see that -152 J of work was done *on* the gas. Put another way, $+152$ J of work was done *by* the gas as it expanded against atmospheric pressure.

Related Problems: 1, 2

Internal Energy

In the two simple mechanical cases discussed earlier, we saw how performing work changes the amount of two types of energy: the kinetic energy of a moving object and the potential energy of an object in a gravitational field. In the same way, performing work can change the amount of energy in more complex cases. A third type of energy, less apparent but equally important, is **internal energy,** defined as the total energy content of a system arising from the potential energy between molecules, from the kinetic energy of molecular motions, and from chemical

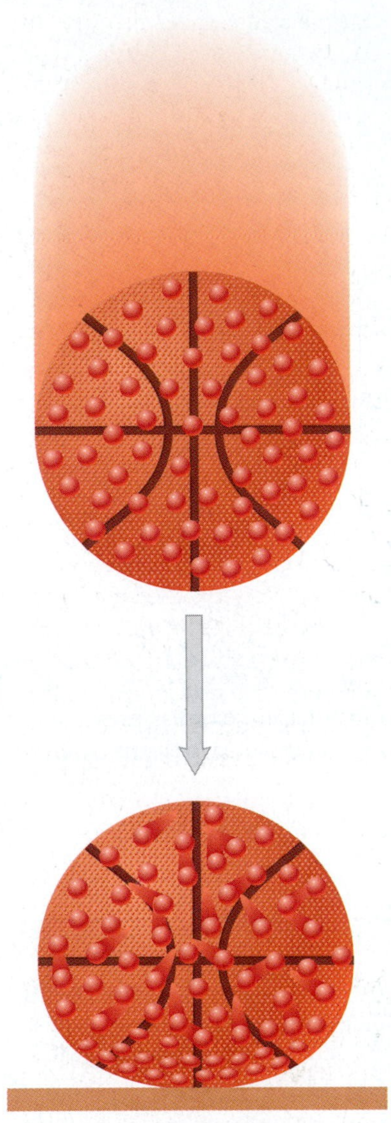

FIGURE 12.5 A ball dropped from a height increases its internal energy on impact with the ground. After impact, the molecules near the surface of the ball are pushed against one another, increasing the potential energy between the molecules. As the ball bounces, the molecules readjust their positions, after which they move a little faster. The kinetic energy of the molecules is higher than it was just before impact with the ground.

energy stored in chemical bonds. Potential energy between molecules includes the lattice energy of solids and the attractive and repulsive interactions between molecules in gases and liquids. Kinetic energy appears in the translation and the internal motions of individual molecules (Fig. 12.5). Gas molecules are in a constant state of motion (see Section 9.5) even when no overall gas flow is taking place in the container; the same is true of molecules in liquids and solids. Example 12.1 illustrates how performing work can change the internal energy of a system. In this example, the system reduced its internal energy by performing +152 J of work against the surroundings. Had the gas been compressed instead of expanding, the internal energy of the system would have increased by the amount of work done on it. *We conclude that P-V work is a means of changing the internal energy of a macroscopic system through purely mechanical interaction between the system and its surroundings.*

Heat

Now, how do we describe the role of heat in the process in Example 12.1? After all, heating the gas caused it to expand, which enabled it to move the piston and do work on the surroundings. The heater in Figure 12.4 has no mechanical "moving parts," yet it set in motion a train of events that led to a mechanical result. To explain this result, we interpret *heat as a means of increasing the internal energy of a system without mechanical interaction.* Justification for this procedure is provided later in this chapter.

The amount of energy transferred between two objects initially at different temperatures is called **heat,** or **thermal energy.** When a hot body is brought into contact with a colder body, the two temperatures change until they become equal. If a piece of hot metal is plunged into a container of water, the temperature of the water increases as its molecules begin to move faster, corresponding to an increase in the internal energy of the water. This process is sometimes described as the "flow" of heat from the hotter to the colder body. Although this picture is useful, it is based on the antiquated (and erroneous) notion that heat is a sort of fluid contained in matter.

The idea of heat flow has inspired methods for measuring the amount of energy transferred as heat; this branch of science is called **calorimetry.** One simple way is to use an **ice calorimeter** (Fig. 12.6), which consists of a bath containing ice and water, well insulated to prevent heat transfer to the surroundings. If heat is transferred to the bath from the system, some of the ice melts. Because a given mass of water has a smaller volume than the same mass of ice, the total volume of the ice–water mixture decreases as heat enters the bath. If twice as much heat is transferred, twice as much ice will melt and the volume change will be twice as great. The amount of heat transferred is determined from the change in volume of the contents of the calorimeter. Heat transferred from the ice bath *into* the system causes water to freeze and *increases* the total volume of the ice bath.

More contemporary versions of calorimetry use the fact that when heat is transferred to or removed from a substance in a single phase at constant pressure, the temperature changes in a reproducible way. The **specific heat capacity** of a material is the amount of heat required to increase the temperature of a 1-g mass by 1°C. If twice as much heat is transferred, the resulting temperature change will be twice as large (provided the specific heat capacity itself does not change appreciably with temperature). Thus, the temperature change of a fixed amount of a given substance is a measure of the thermal energy transferred to or from it. This is described by

$$q = Mc_s\Delta T \qquad [12.2]$$

where q is the heat transferred to a body of mass M with specific heat capacity c_s to cause a temperature change of ΔT.

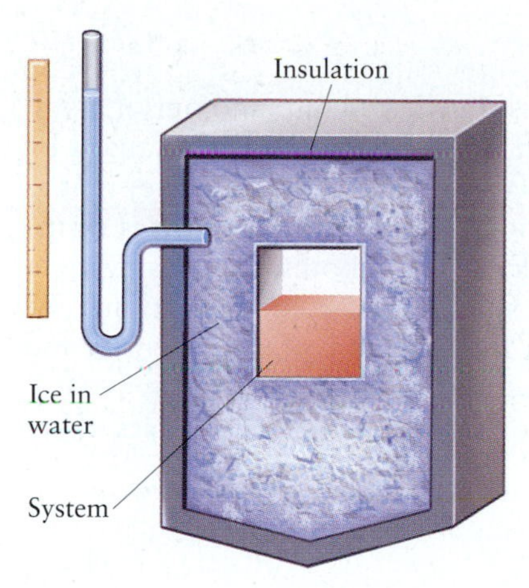

FIGURE 12.6 An ice calorimeter. As the ice melts, the volume of the ice–water mixture decreases, an effect that can be read off the scale on the left.

Because heat, like work, is a form in which energy is transferred, the appropriate unit for it is also the joule. Historically, however, the connections among work, heat, and energy were not appreciated until the middle of the 19th century, by which time a separate unit for heat, the calorie, was already well established. One calorie was defined as the amount of heat required to increase the temperature of 1 g water from 14.5°C to 15.5°C (or, in other words, the specific heat capacity of water, c_s, at 15°C was *defined* as 1.00 cal K^{-1} g^{-1}).

The equivalence of heat and work as means of energy transfer was suggested in 1798 by Benjamin Thompson, Count Rumford. In the course of his work as military advisor to the King of Bavaria, Thompson observed that the quantity of heat produced in boring cannons was proportional to the amount of work done in the process. Moreover, the operation could be continued indefinitely, demonstrating that heat was not a "fluid" contained in the metal of the cannon. More quantitative measurements were conducted by the German physician Julius Mayer and by the English physicist James Joule. In the 1840s, these scientists showed that the temperature of a substance could be increased by doing work on the substance, as well as by adding heat to it. Figure 12.7 shows an apparatus in which a paddle, driven by a falling weight, churns the water in a tank. Work is performed on the water, and the temperature increases. The work done is $-Mg\,\Delta h$, where Δh is the (negative) change in the height of the weight and M is its mass. The experiment is conducted in an insulated container, so no heat enters the container or leaks out to the surroundings. Because all this work goes to increase the water temperature, the specific heat capacity of the water in *joules* per gram per degree is equal to the quantity of work done divided by the product of the mass of water and its temperature increase.

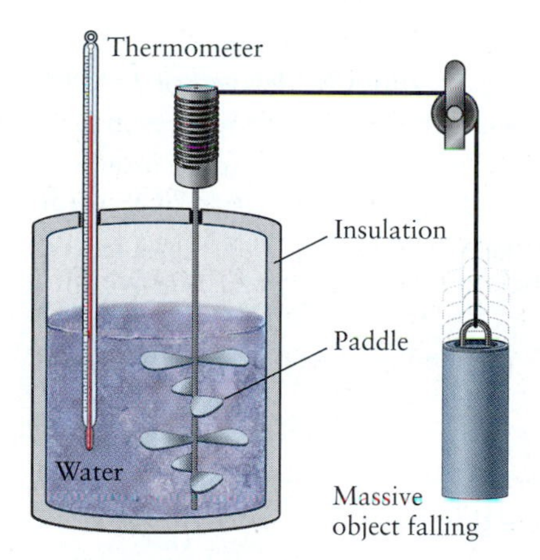

FIGURE 12.7 The falling weight turns a paddle that does work on the system (the water), causing an increase in its temperature.

EXAMPLE 12.2

Suppose a 10.00-kg mass drops through a height difference of 3.00 m, and the resulting work is used to turn a paddle in 200.0 g water, initially at 15.00°C. The final water temperature is found to be 15.35°C. Assuming that the work done is used entirely to increase the water temperature, calculate the conversion factor between joules and calories.

SOLUTION

The total work done is

$$w = -Mg\Delta h = -(10.00 \text{ kg})(9.807 \text{ m s}^{-2})(-3.00 \text{ m}) = 294 \text{ J}$$

The heat (in calories) required to increase the water temperature by the same amount is

$$q = Mc_s\Delta T = (200.0 \text{ g})(1.000 \text{ cal K}^{-1} \text{ g}^{-1})(0.35 \text{ K}) = 70 \text{ cal}$$

Because the work done has the same effect on the water as direct transfer of heat, these two expressions can be set equal to each other, giving

$$70 \text{ calories} = 294 \text{ joules}$$

$$1 \text{ calorie} \approx 4.2 \text{ joules}$$

Related Problems: 3, 4

These and other experiments eliminated the need for the calorie as an independent unit, and the calorie is now *defined* as

$$1 \text{ cal} = 4.184 \text{ J} \quad \text{(exactly)}$$

This book uses the joule as the primary unit for heat and energy. Because much of the chemical literature continues to use the calorie as the unit of heat, it is important to be familiar with both units.

The First Law of Thermodynamics

Both heat and work are forms in which energy is transferred into and out of a system; they can be thought of as energy in transit. If the energy change is caused by *mechanical* contact of the system with its surroundings, work is done; if it is caused by *thermal* contact (leading to equalization of temperatures), heat is transferred. In many processes, both heat and work cross the boundary of a system, and the change in the internal energy, U, is the sum of the two contributions. We denote the internal energy by U to distinguish it from the sum of the potential and kinetic energy in a simple mechanical process, for which we use the symbol E throughout this book. This statement, called the **first law of thermodynamics,** takes the mathematical form[1]

$$\Delta U = q + w \qquad [12.3]$$

A system cannot be said to "contain" work or heat, because both "work" and "heat" refer not to states of the system but to *processes* that transform one state into another. In the Joule experiment shown in Figure 12.7, the work done on the water (the system) by the falling weight increased the temperature of the water. Work was performed on the system, and no heat was transferred; thus, the first law for this process takes the form $\Delta U = w$. The same change in state of the system can be achieved by transferring heat to the system without work being done; for this process, $\Delta U = q$. Because q and w depend on the particular process (or path) connecting the states, they are not state functions. But their sum, $\Delta U = q + w$, is independent of path; therefore, internal energy is a function of state. The fundamental physical content of the first law of thermodynamics is the following observation:

> *Although q and w depend individually on the path followed between a given pair of states, their sum ΔU does not.*

We stated earlier that the laws of thermodynamics cannot be derived or proved; they are generalizations of the results of countless experiments on a tremendous variety of substances. It is not possible even to "check" the first law by independently measuring ΔU, w, and q, because no "energy gauges" exist to determine energy changes ΔU. But what we *can* do is to measure w and q for a series of different processes connecting the same initial and final states. Every time, we find that their sum, $q + w$, is always the same.

In any process, the heat *added to* the system is *removed from* the surroundings; thus,

$$q_{sys} = -q_{surr}$$

In the same way, the work done *on* the system is done *by* the surroundings; thus,

$$w_{sys} = -w_{surr}$$

Adding these two and invoking the first law give

$$\Delta U_{sys} = -\Delta U_{surr}$$

[1]Some books, especially older ones, define work as positive when it is done *by* the system. The reason is that many engineering applications focus on the work done by a particular heat engine; therefore, it is helpful to define that quantity as positive. Thus, the work is given by

$$w = P_{ext}\,\Delta V \text{ (older convention)}$$

and the first law reads

$$\Delta U = q - w \text{ (older convention)}$$

Although we do not use the older convention, you should check which convention is used when consulting other books.

Thus, the energy changes of system and surroundings have the same magnitude but opposite signs. The total energy change of the thermodynamic universe for a given process (system plus surroundings) is then

$$\Delta U_{\text{univ}} = \Delta U_{\text{sys}} + \Delta U_{\text{surr}} = 0 \qquad [12.4]$$

Our conclusion is that, in any process, the total energy of the thermodynamic universe remains unchanged; the total energy is conserved while it is exchanged between the system and the surroundings. This is another way to state the first law of thermodynamics.

12.3 Heat Capacity, Enthalpy, and Calorimetry

Section 12.2 defines specific heat capacity as the amount of heat required to increase the temperature of 1 g of material by 1 K. That definition is somewhat imprecise, because, in fact, the amount of heat required depends on whether the process is conducted at constant volume or at constant pressure. This section describes precise methods for measuring the amount of energy transferred as heat during a process and for relating this amount to the thermodynamic properties of the system under investigation.

Heat Capacity and Specific Heat Capacity

The **heat capacity, C,** is defined as the amount of energy that must be added to the system to increase its temperature by 1 K. The heat capacity is a property of the system as a whole and has units of J K^{-1}.

$$q = C\Delta T$$

Now consider a case in which two gaseous systems containing identical masses of the same substance are heated to produce identical changes in temperature. During the experiment, system 1 is held at constant volume, and system 2 at constant pressure. Which system absorbed more heat in these identical temperature changes? All the energy gained by system 1 contributed to increasing the temperature of the substance, and therefore the speed of the molecules, subject to the fixed volume. But in system 2, some of the energy gained was promptly lost as the system performed expansion work against the surroundings at constant pressure. Consequently, system 2 must absorb more thermal energy from the surroundings than does system 1 to achieve the identical temperature change. Two independent heat capacity functions must be defined: C_P, the heat capacity at constant pressure, and C_V, the heat capacity at constant volume. For any system, C_P is greater than C_V. This difference can be quite large for gases. It is usually negligible for solids and liquids, because their only volume change at constant pressure is the small expansion or contraction on heating and cooling.

In thermodynamics, the *molar* heat capacities c_V and c_P (the system heat capacities C_V and C_P divided by the number of moles of substance in the system) are particularly useful: c_V is the amount of heat required to increase the temperature of 1 mol of substance by 1 K at constant volume, and c_P is the corresponding amount required at constant pressure. If the total heat transferred to n moles at constant volume is q_V, then

$$q_V = nc_V(T_2 - T_1) = nc_V\Delta T \qquad [12.5]$$

TABLE 12.1 Specific Heat Capacities at Constant Pressure (at 25°C)

Substance	Specific Heat Capacity ($J\ K^{-1}\ g^{-1}$)
Hg(ℓ)	0.140
Cu(s)	0.385
Fe(s)	0.449
SiO_2(s)	0.739
$CaCO_3$(s)	0.818
O_2(g)	0.917
H_2O(ℓ)	4.18

If an amount q_P is transferred at constant pressure, then

$$q_P = nc_P\Delta T \quad [12.6]$$

provided that c_V and c_P do not change significantly between the initial and final temperatures. The *specific heat capacity* at constant V or constant P is the system heat capacity reported per gram of substance. Extensive tabulations of the molar and specific heat capacities are available. Representative values are listed in Table 12.1. We give a molecular interpretation of heat capacity for ideal gases in Section 12.4.

When two objects at different temperatures are brought into contact, energy in the form of heat is exchanged between them until they reach a common temperature. If the two objects together are insulated from their surroundings, the amount of heat q_2 taken up by the cooler object is equal to $-q_1$, the amount of heat given up by the hotter object. As always, the convention followed is that energy transferred to an object has a positive sign; thus, q_2 is positive when q_1 is negative. This analysis is broadly applicable; a typical example follows.

EXAMPLE 12.3

A piece of iron weighing 72.4 g is heated to 100.0°C and plunged into 100.0 g water that is initially at 10.0°C in a Styrofoam cup calorimeter. Assume no heat is lost to the Styrofoam cup or to the environment (Fig. 12.8). Calculate the final temperature that is reached.

SOLUTION

The "coffee cup calorimeter" operates at constant pressure determined by the atmosphere; therefore, we need specific heat data at constant pressure. Because the data involve masses, it is easier to work with specific heat capacities (see Table 12.1) than with molar heat capacities. If t_f is the final temperature (in degrees Celsius), then the equation for heat balance gives

$$M_1(c_{s1})\Delta T_1 = -M_2(c_{s2})\Delta T_2$$

$$(100.0\ g\ H_2O)(4.18\ J\ °C^{-1}\ g^{-1})(t_f - 10.0°C) =$$
$$-(72.4\ g\ Fe)(0.449\ J°C^{-1}\ g^{-1})(t_f - 100.0°C)$$

This is a linear equation for the unknown temperature t_f, and its solution is

$$418t_f - 4180 = -32.51t_f + 3251$$

$$t_f = 16.5°C$$

Note that specific heat capacities are numerically the same whether expressed in $J\ K^{-1}\ g^{-1}$ or $J\ (°C)^{-1}\ g^{-1}$ (because the degree Celsius and the kelvin have the same size). Converting 10.0°C and 100.0°C to kelvins and using specific heat capacities in units of $J\ K^{-1}\ g^{-1}$ gives t_f = 289.7 K, which is equivalent to the previous answer.

Related Problems: 11, 12

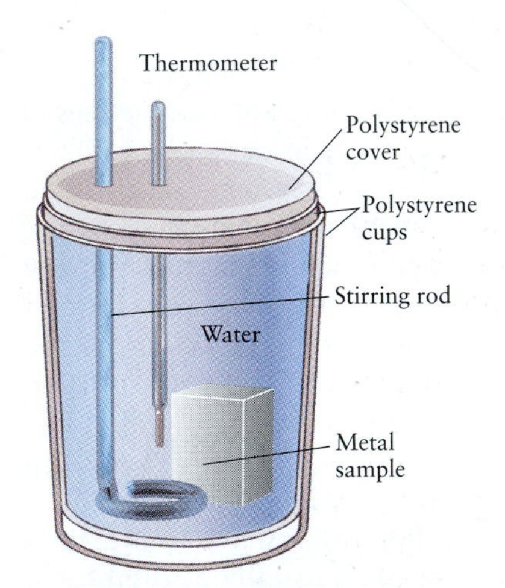

FIGURE 12.8 A Styrofoam cup calorimeter. As the piece of metal cools, it releases heat to the water. The amount of heat released can be determined from the temperature change of the water. The hot metal is the "system"; the water is the "surroundings." The Styrofoam cup wall prevents energy exchange with the remainder of the room and is the boundary of the "thermodynamic universe" for this problem.

Heat Transfer at Constant Volume: Bomb Calorimeters

Suppose some reacting species are sealed in a small closed container (called a *bomb*) and the container is placed in a calorimeter like the one in Figure 12.9. The reaction is initiated by a heated wire inside the bomb. As the molecules react chemically, heat is given off or taken up, and the change in temperature of the calorimetric fluid is measured. Because the container is sealed tightly, its volume is

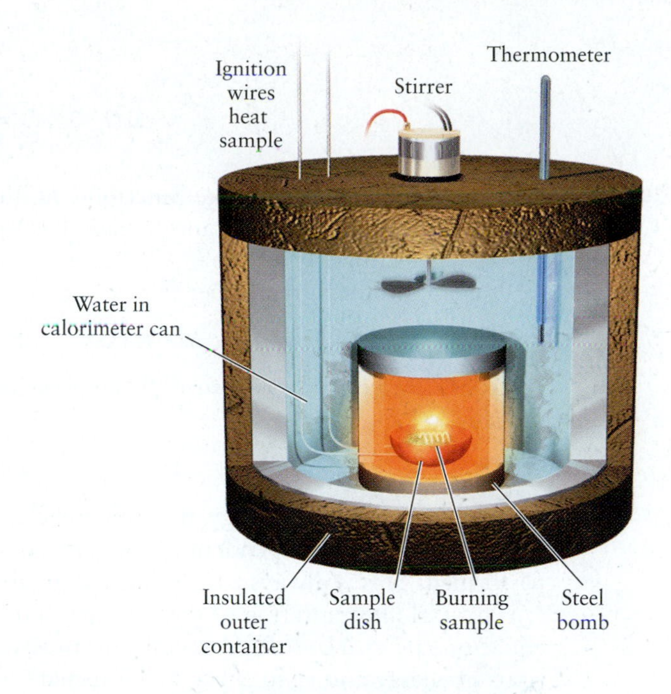

FIGURE 12.9 The combustion calorimeter is also called a "bomb calorimeter"; the combustion reaction in it is conducted at a fixed volume.

constant and no pressure–volume work is done. Therefore, the change in internal energy is equal to the measured heat absorbed from the chemical reaction at constant volume:

$$\Delta U = q_V$$

Such experiments at constant volume are often inconvenient or difficult to perform. They require the use of a well-constructed reaction vessel that can resist the large pressure changes that occur in many chemical reactions.

Heat Transfer at Constant Pressure: Enthalpy

Most chemical reactions are done under constant (atmospheric) *pressure* conditions rather than at constant volume, so it is desirable to relate the heat transferred at constant pressure, q_P, to some state property analogous to the internal energy, U.

If the work done is entirely pressure–volume work, and if the external pressure is held constant, then

$$\Delta U = q_P + w = q_P - P_{ext}\Delta V$$

If the external pressure is now assumed to be equal to the internal pressure of the system P, then

$$\Delta U = q_P - P\Delta V$$

$$q_P = \Delta U + P\Delta V$$

Because P is constant, $P\Delta V = \Delta(PV)$, and this equation becomes

$$q_P = \Delta(U + PV)$$

The combination $U + PV$ appearing on the right side is now defined as the **enthalpy** H:

$$H = U + PV \qquad [12.7a]$$

thus,

$$q_P = \Delta(U + PV) = \Delta H \qquad [12.7b]$$

Because U, P, and V are state functions, H must also be a state function. Heat transfer in a process at constant pressure has therefore been related to the change in a state function.

It is important to remember that

$$\Delta H = q_P = \Delta U + P\Delta V \qquad \text{(constant pressure)}$$

holds only at constant pressure. If the pressure changes, the more general relationship

$$\Delta H = \Delta U + \Delta(PV)$$

must be used. Like the energy, the enthalpy change is determined by the initial and final states and is independent of the particular path along which the process is performed. This is always true for a state function.

Physical interpretation of the enthalpy function follows immediately from the equation $\Delta H = \Delta U + P\Delta V$ at constant pressure. Clearly, H has physical dimensions of energy and is, in effect, a "corrected" internal energy that reflects the consequences of changing V while thermal energy is being absorbed at constant pressure. The "correction term" $P\Delta V$ accounts precisely for the energy used in expansion work, rather than for increasing the temperature of the system. Thus, the value of q in constant pressure processes where only pressure–volume work is done is equivalent to the change in the state function enthalpy, ΔH.

12.4 Illustrations of the First Law of Thermodynamics in Ideal Gas Processes

We stated the first law of thermodynamics in a general form, applicable to any process that begins and ends in equilibrium states. We analyzed the heat and work terms separately and presented methods for calculating, measuring, and interpreting each. All the concepts are now in place for applying thermodynamics to the discussion of specific processes. Applications require data on certain properties of the substance being studied, such as its equation of state and its heat capacities. Thermodynamic arguments alone cannot provide the actual values of such properties; instead, thermodynamics establishes universal relations among such properties. The actual values must be obtained by methods other than thermodynamics, such as experimental measurements or theoretical calculations in statistical thermodynamics. To illustrate these points, in the next few paragraphs we obtain data on the heat capacities of ideal gases by methods outside thermodynamics. Then we use these data to apply thermodynamics to analyze particular processes carried out on ideal gases.

Heat Capacities of Ideal Gases

The pair of molar heat capacities c_V and c_P for an ideal monatomic gas can be calculated from the results of the kinetic theory of gases and the ideal gas equation of state. From Section 9.5, the average translational kinetic energy of n moles of an ideal gas is

$$E_{\text{kin}} = \tfrac{3}{2}\, nRT$$

As shown in Chapter 17, the rotations and vibrations of diatomic or polyatomic molecules make additional contributions to the energy. In a *monatomic* gas, these other contributions are not present; thus, changes in the total internal energy ΔU measured in thermodynamics can be equated to changes in the translational kinetic energy of the atoms. If n moles of a monatomic gas is taken from a temperature T_1 to a temperature T_2, the internal energy change is

$$\Delta U = \tfrac{3}{2}\,nR(T_2 - T_1) = \tfrac{3}{2}\,nR\Delta T$$

Note that the pressure and volume do not affect U explicitly (except through temperature changes), so this result is independent of the change in pressure or volume of the gas.

Now, consider changing the temperature of an ideal gas at constant volume from the point of view of thermodynamics. Because the volume is constant (the gas is confined in a vessel with rigid, diathermal walls), the pressure–volume work, w, must be zero; therefore,

$$\Delta U = q_V = nc_V\Delta T \qquad \text{(ideal gas)}$$

where q_V is the heat transferred at constant volume. Equating this thermodynamic relation with the previous expression for ΔU (from kinetic theory) shows that

$$c_V = \tfrac{3}{2}\,R \qquad \text{(monatomic ideal gas)} \qquad [12.8]$$

Similarly, the molar heat capacity at constant pressure, c_P, is calculated by examining the heating of a monatomic ideal gas at constant pressure from temperature T_1 to T_2. Experimentally, such a process can be performed by placing the gas in a cylinder with a piston that moves out as the gas is heated, keeping the gas pressure equal to the outside pressure. In this case,

$$\Delta U = \tfrac{3}{2}\,nR\Delta T = nc_V\Delta T$$

still holds (because the energy change depends only on the temperatures for an ideal gas), but we now have

$$\Delta U = q_P + w$$

because the work is no longer zero. The work for a constant-pressure process is easily calculated from

$$w = -P\Delta V = -P(V_2 - V_1)$$

and the heat transferred is

$$q_P = nc_P\Delta T$$

Because w is negative, q_P is larger than q_V by the amount of work done by the gas as it expands. This gives

$$\Delta U = q + w$$
$$nc_V\Delta T = nc_P\Delta T - P(V_2 - V_1)$$

From the ideal gas law, $PV_1 = nRT_1$ and $PV_2 = nRT_2$; thus,

$$nc_V\Delta T = nc_P\Delta T - nR\Delta T$$
$$c_V = c_P - R$$
$$c_P = c_V + R$$

For a monatomic ideal gas, this shows that $c_P = \tfrac{5}{2}\,R$. It is important to use the proper units for R in these expressions for c_V and c_P. If heat is to be measured in joules, R must be expressed as

$$R = 8.315 \text{ J K}^{-1} \text{ mol}^{-1}$$

For a diatomic or polyatomic ideal gas, c_V is greater than $\frac{3}{2}R$, because energy can be stored in rotational and vibrational motions of the molecules; a greater amount of heat must be transferred to achieve a given temperature change. Even so, it is still true that

$$c_P = c_V + R \text{ (any ideal gas)} \qquad [12.9]$$

and that internal energy changes depend only on the temperature change; therefore, for a small temperature change ΔT,

$$\Delta U = nc_V\Delta T \text{ (any ideal gas)} \qquad [12.10]$$

For an ideal gas process,

$$\Delta H = \Delta U + \Delta(PV) = nc_V\Delta T + nR\Delta T$$
$$\Delta H = nc_P\Delta T \text{ (ideal gas)} \qquad [12.11]$$

because $c_P = c_V + R$. This result holds for any ideal gas process and shows that enthalpy changes, like internal energy changes, depend only on the temperature difference between initial and final states. These statements are not valid for systems other than ideal gases.

EXAMPLE 12.4

Suppose that 1.00 kJ of heat is transferred to 2.00 mol argon (at 298 K, 1 atm). What will the final temperature T_f be if the heat is transferred (**a**) at constant volume, or (**b**) at constant pressure? Calculate the energy change, ΔU, in each case.

SOLUTION

Because argon is a monatomic, approximately ideal gas,

$$c_V = \tfrac{3}{2}R = 12.47 \text{ J K}^{-1}\text{ mol}^{-1}$$
$$c_P = \tfrac{5}{2}R = 20.79 \text{ J K}^{-1}\text{ mol}^{-1}$$

At constant volume,

$$q_V = nc_V\Delta T$$
$$1000 \text{ J} = (2.00 \text{ mol})(12.47 \text{ J K}^{-1}\text{ mol}^{-1})\Delta T$$
$$\Delta T = 40.1 \text{ K};\ T_f = 298 + 40.1 = 338 \text{ K}$$
$$\Delta U = nc_V\Delta T = q_V = 1000 \text{ J}$$

At constant pressure,

$$q_P = nc_P\Delta T$$
$$1000 \text{ J} = (2.00 \text{ mol})(20.79 \text{ J K}^{-1}\text{ mol}^{-1})\Delta T$$
$$\Delta T = 24.0 \text{ K};\ T_f = 298 + 24.0 = 322 \text{ K}$$
$$\Delta U = nc_V\Delta T = (2.00 \text{ mol})(12.47 \text{ J K}^{-1}\text{ mol}^{-1})(24 \text{ K}) = 600 \text{ J}$$

Note that the expression for ΔU involves c_V even though the process is conducted at constant pressure. The difference of 400 J between the input q_P and ΔU is the work done by the gas as it expands.

Related Problems: 17, 18, 19, 20, 21, 22

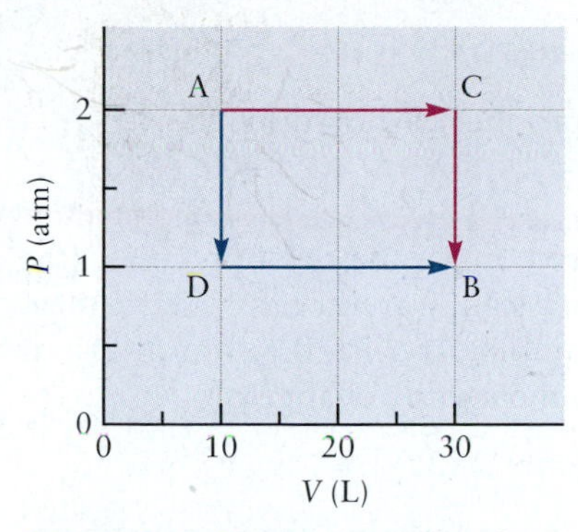

FIGURE 12.10 States A and B of a system are connected by two different ideal-gas processes, one passing through state C and the other through state D.

Heat and Work for Ideal Gases

Now we have all the data needed to calculate heat and work for a variety of processes involving an ideal gas. To illustrate that q and w individually depend on the path followed, but their sum does not, consider the expansion of 1.00 mol of an ideal monatomic gas following two different paths. The system begins at state A ($P_A = 2.00$ atm, $V_A = 10.0$ L) and reaches a final state, B ($P_B = 1.00$ atm, $V_B = 30.0$ L), via either of two paths shown in Figure 12.10. Along path ACB (red arrows in Fig. 12.10), the system is first heated at constant pressure ($P_{ext} = P_A = 2$ atm) until the volume has tripled; then it is cooled at constant volume until the pressure is halved. Along path ADB (blue arrows in Fig. 12.10), the system is cooled at constant volume until the pressure is halved, and then heated at constant pressure ($P_{ext} = P_B = 1$ atm) until the volume has tripled.

The calculations of heat and work for each step are of the type already performed and are straightforward. Thus,

$$w_{AC} = -P_{ext}\Delta V = -P_A(V_B - V_A)$$

$$w_{CB} = 0 \text{ because } V_C = V_B$$

$$q_{AC} = q_P = nc_P\Delta T = \tfrac{5}{2}\,nR(T_C - T_A)$$

$$q_{CB} = q_V = nc_V\Delta T = \tfrac{3}{2}\,nR(T_B - T_C)$$

From the ideal gas law, $nRT_A = P_AV_A$, $nRT_B = P_BV_B$, and $nRT_C = P_CV_C = P_AV_B$ (because $P_A = P_C$ and $V_B = V_C$). Using these relations and summing over the two steps give

$$w_{ACB} = w_{AC} + w_{CB} = -P_A(V_B - V_A) = -40.0 \text{ L atm} = -4050 \text{ J}$$

$$q_{ACB} = q_{AC} + q_{CB} = \tfrac{5}{2}\,nR(T_C - T_A) + \tfrac{3}{2}\,nR(T_B - T_C)$$

$$= \tfrac{5}{2}\,P_A(V_B - V_A) + \tfrac{3}{2}\,V_B(P_B - P_A) = (100.0 - 45.0) \text{ L atm} = 5570 \text{ J}$$

The sum of these is $\Delta U = w_{ACB} + q_{ACB} = 1520$ J. (This could also have been obtained by using the ideal gas law to calculate the initial and final temperatures T_A and T_B.)

The corresponding calculation for path ADB gives

$$w_{ADB} = -2030 \text{ J and } q_{ADB} = 3550 \text{ J}$$

Even though both the work and the heat have changed, their sum is still 1520 J, illustrating that U is a state function, whereas q and w are not.

12.5 Thermochemistry

Up to this point, all the energy changes we have considered are simple physical processes that involve direct mechanical work on a system (as in the paddle wheel driven by a falling weight) or direct thermal contact between two systems at different temperatures. The same methods apply to the heat given off or taken up in the course of a chemical reaction. The study of these heat transfers during chemical reactions is referred to as **thermochemistry.** Because chemical reactions are usually studied at constant pressure, heat transfers in reactions are measured at constant pressure. We obtain these numbers by conducting the chemical reaction in a constant-pressure calorimeter and measuring the heat transferred as q_P. This number is the enthalpy change of the reaction, viewed as a thermodynamic process:

$$q_P = \Delta H = H_f - H_i = H_{products} - H_{reactants} = \Delta H_{reaction}$$

The tabulated values are called **reaction enthalpies.**

Enthalpies of Reaction

When carbon monoxide is burned in oxygen to produce carbon dioxide,

$$CO(g) + \frac{1}{2} O_2(g) \longrightarrow CO_2(g)$$

heat is given off. Because this energy is transferred out of the reaction vessel (the system) and into the surroundings, it has a *negative* sign. Careful calorimetric measurements show that 1.000 mol CO reacted completely with 0.500 mol O_2, at 25°C and a constant pressure of 1 atm, leads to an enthalpy change of

$$\Delta H = q_P = -2.830 \times 10^5 \text{ J} = -283.0 \text{ kJ}$$

The kilojoule (kJ), equal to 10^3 J, is used because most enthalpy changes for chemical reactions lie in the range of thousands of joules per mole.

When heat is given off by a reaction (ΔH is negative), the reaction is said to be **exothermic** (Fig. 12.11). Reactions in which heat is taken up (ΔH positive) are called **endothermic** (Fig. 12.12). One example of an endothermic reaction is the preceding reaction written in the opposite direction:

$$CO_2(g) \longrightarrow CO(g) + \frac{1}{2} O_2(g) \qquad \Delta H = +283.0 \text{ kJ}$$

If the direction of a chemical reaction is reversed, the enthalpy change reverses sign. Heat is required to convert CO_2 to CO and O_2 at constant pressure. The decomposition of CO_2 into CO and O_2 is difficult to perform in the laboratory, whereas the reverse reaction is straightforward. Thermodynamics allows us to predict ΔH of the decomposition reaction with complete confidence, even if a calorimetric experiment is never actually performed for it.

Chemists have agreed on a convention for attaching reaction enthalpy values to balanced chemical equations. A reaction enthalpy written after a balanced chemical equation refers to the enthalpy change for the complete conversion of stoichiometric amounts of reactants to products; the numbers of moles of

(a)

(b)

(c)

FIGURE 12.11 The thermite reaction, $2\ Al(s) + Fe_2O_3(s) \longrightarrow 2\ Fe(s) + Al_2O_3(s)$, is among the most exothermic of all reactions, liberating 16 kJ of heat for every gram of aluminum that reacts. (a) A piece of burning magnesium acts as a source of ignition when inserted into a pot containing a finely divided mixture of aluminum powder and iron(III) oxide. (b) After the ignition, the reaction continues on its own. (c) Enough heat is generated to produce molten iron, which can be seen flowing out of the broken pot onto the protective mat and the bottom of the stand.

(a)

(b)

FIGURE 12.12 (a) When mixed in a flask, the two solids $Ba(OH)_2 \cdot 8H_2O(s)$ and $NH_4NO_3(s)$ undergo an acid–base reaction: $Ba(OH)_2 \cdot 8\,H_2O(s) + 2\,NH_4NO_3(s) \longrightarrow Ba(NO_3)_2\,(aq) + 2\,NH_3(aq) + 10\,H_2O(\ell)$. (b) The water produced dissolves excess ammonium nitrate in an endothermic reaction. The dissolution absorbs so much heat that the water on the surface of the wet wooden block freezes to the bottom of the flask, and the block can be lifted up with the flask.

reactants and products are given by the coefficients in the equation. The preceding equation shows the enthalpy change when 1 mol CO_2 is converted to 1 mol CO and $\frac{1}{2}$ mol O_2. If this equation is multiplied by a factor of 2, the enthalpy change must also be doubled because twice as many moles are then involved (enthalpy, like energy, is an *extensive* property).

$$2\,CO_2(g) \longrightarrow 2\,CO(g) + O_2(g) \qquad \Delta H = +566.0\text{ kJ}$$

The molar amounts need not be integers, as the following example illustrates.

EXAMPLE 12.5

Red phosphorus reacts with liquid bromine in an exothermic reaction (Fig. 12.13):

$$2\,P(s) + 3\,Br_2(\ell) \longrightarrow 2\,PBr_3(g) \qquad \Delta H = -243\text{ kJ}$$

Calculate the enthalpy change when 2.63 g phosphorus reacts with an excess of bromine in this way.

SOLUTION

First, convert from grams of phosphorus to moles, using the molar mass of phosphorus, 30.97 g mol^{-1}:

$$\text{moles P} = \frac{2.63\text{ g P}}{30.97\text{ g mol}^{-1}} = 0.0849\text{ mol}$$

Given that an enthalpy change of −243 kJ is associated with 2 mol P, it is readily seen that the enthalpy change associated with 0.0849 mol is

$$\Delta H = 0.0849\text{ mol P} \times \left(\frac{-243\text{ kJ}}{2\text{ mol P}}\right) = -10.3\text{ kJ}$$

Related Problems: 23, 24, 25, 26

FIGURE 12.13 Red phosphorus reacts exothermically in liquid bromine. The rising gases are a mixture of the product PBr_3 and unreacted bromine that has boiled off.

The enthalpy change for the reaction of 1 mol carbon monoxide with oxygen was stated to be −283.0 kJ. In a second experiment, the heat evolved when 1 mol carbon (graphite) is burned in oxygen to carbon dioxide at 25°C is readily measured to be

$$C(s,gr) + O_2(g) \longrightarrow CO_2(g) \qquad \Delta H = -393.5\text{ kJ}$$

Now, suppose we need to know the enthalpy change for the reaction

$$C(s,gr) + \tfrac{1}{2}\,O_2(g) \longrightarrow CO(g) \qquad \Delta H = ?$$

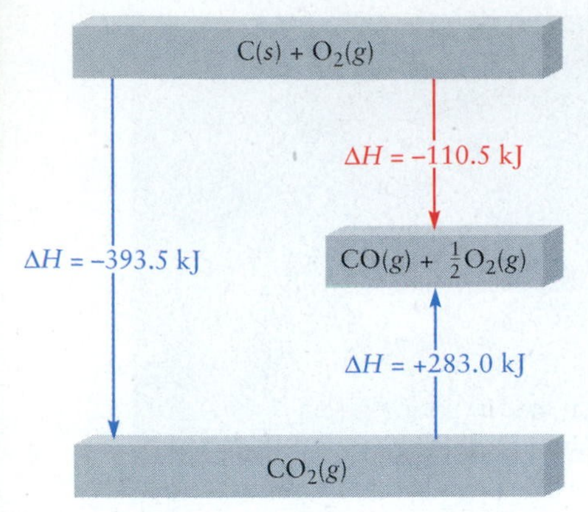

FIGURE 12.14 Because enthalpy is a state property, the enthalpy change for the reaction of carbon with oxygen to give carbon monoxide (red arrow) can be determined through measurements along a path that is less direct but easier to study (blue arrows). The enthalpy change sought is the sum of the enthalpy change to burn carbon to carbon dioxide and that to convert carbon dioxide to carbon monoxide and oxygen.

This reaction cannot be performed simply in the laboratory. If 1 mol graphite is heated with $\frac{1}{2}$ mol oxygen, almost half the carbon burns to $CO_2(g)$ and the remainder is left as unreacted carbon. Nevertheless, thermodynamics allows us to predict the heat that *would* evolve if we could perform the reaction as written. This is possible because H is a state function, and thus ΔH for the reaction is independent of the path followed from reactants to products. We are free to select any path for which we have all the data needed for the calculation. In Figure 12.14, we illustrate the path in which 1 mol C is burned with O_2 to CO_2 (with $\Delta H = -393.5$ kJ), and to this is added the calculated enthalpy change for the (hypothetical) process in which CO_2 is converted to CO and O_2 ($\Delta H = +283.0$ kJ). The total ΔH is the algebraic sum of the two known enthalpy changes, $-393.5 \text{ kJ} + 283.0 \text{ kJ} = -110.5$ kJ. To see this more clearly, the reactions are written out as follows:

$$\begin{array}{lr} C(s, gr) + O_2(g) \longrightarrow CO_2(g) & \Delta H_1 = -393.5 \text{ kJ} \\ CO_2(g) \longrightarrow CO(g) + \tfrac{1}{2} O_2(g) & \Delta H_2 = +283.0 \text{ kJ} \\ \hline C(s, gr) + \tfrac{1}{2} O_2(g) \longrightarrow CO(g) & \Delta H = \Delta H_1 + \Delta H_2 = -110.5 \text{ kJ} \end{array}$$

If two or more chemical equations are added to give another chemical equation, the corresponding enthalpies of reaction must be added.

This statement is known as **Hess's law** and derives from the fact that enthalpy is a state function. It is proper to include in the pathway any convenient reaction for which the enthalpy change is known, even if that reaction is difficult to study directly and its enthalpy change is known only from studies of the reverse reaction. The step labeled (hypothetical) above is a good example.

The corresponding internal energy change ΔU might be desired for this reaction. The quantity ΔU is simple to calculate because of the relation

$$\Delta H = \Delta U + \Delta(PV)$$

or

$$\Delta U = \Delta H - \Delta(PV)$$

The gases can be assumed to obey the ideal gas law, so

$$\Delta(PV) = \Delta(nRT) = RT\,\Delta n_g$$

because the temperature is constant at 25°C. Here, Δn_g is the change in the number of moles of *gas* in the reaction as written:

$$\begin{aligned} \Delta n_g &= \text{total moles of product gases} - \text{total moles of reactant gases} \\ &= 1 \text{ mol} - \tfrac{1}{2} \text{ mol} = \tfrac{1}{2} \text{ mol} \end{aligned}$$

(Graphite is a solid, and its volume is negligible compared with the volumes of the gases.) Hence,

$$\Delta(PV) = RT\,\Delta n_g = (8.315 \text{ J K}^{-1} \text{ mol}^{-1})(298 \text{ K})(\tfrac{1}{2} \text{ mol}) = 1.24 \times 10^3 \text{ J} = 1.24 \text{ kJ}$$

Note that R must be expressed in J K^{-1} mol^{-1} to obtain the result in joules. Therefore,

$$\Delta U = -110.5 \text{ kJ} - 1.24 \text{ kJ} = -111.7 \text{ kJ}$$

For reactions in which only liquids and solids are involved, or those in which the number of moles of gas does not change, the enthalpy and energy changes are almost equal and their difference can be neglected.

Phase changes are not chemical reactions, but their enthalpy changes can be analyzed in the same way. Heat must be absorbed by ice to transform to water, so the phase change is endothermic, with ΔH positive:

$$H_2O(s) \longrightarrow H_2O(\ell) \qquad \Delta H_{fus} = +6.007 \text{ kJ mol}^{-1}$$

TABLE 12.2 Enthalpy Changes of Fusion and Vaporization†

Substance	ΔH_{fus} (kJ mol^{-1})	ΔH_{vap} (kJ mol^{-1})
NH_3	5.650	23.35
HCl	1.992	16.15
CO	0.836	6.04
CCl_4	2.500	30.00
H_2O	6.007	40.66
NaCl	28.800	170.00

†The enthalpy changes are measured at the normal melting point and the normal boiling point, respectively.

Here, ΔH_{fus} is the **molar enthalpy of fusion,** the heat that must be transferred at constant pressure to melt 1 mole of substance. When a liquid freezes, the reaction is reversed and an equal amount of heat is given off to the surroundings; that is, $\Delta H_{freez} = -\Delta H_{fus}$. The vaporization of 1 mole of liquid at constant pressure and temperature requires an amount of heat called the **molar enthalpy of vaporization,** ΔH_{vap},

$$H_2O(\ell) \longrightarrow H_2O(g) \qquad \Delta H_{vap} = +40.7 \text{ kJ mol}^{-1}$$

whereas the condensation of a liquid from a vapor is an exothermic process, with $\Delta H_{cond} = -\Delta H_{vap}$. Table 12.2 lists enthalpies of fusion and vaporization.

EXAMPLE 12.6

To vaporize 100.0 g carbon tetrachloride at its normal boiling point, 349.9 K, and $P = 1$ atm, 19.5 kJ of heat is required. Calculate ΔH_{vap} for CCl_4 and compare it with ΔU for the same process.

SOLUTION

The molar mass of CCl_4 is 153.8 g mol^{-1}; thus, the number of moles in 100.0 g is

$$\frac{100.0 \text{ g } CCl_4}{153.8 \text{ g mol}^{-1}} = 0.6502 \text{ mol } CCl_4$$

The enthalpy change for 1 mol CCl_4 is then

$$\left(\frac{19.5 \text{ kJ}}{0.6502 \text{ mol } CCl_4}\right) \times 1.00 \text{ mol } CCl_4 = 30.0 \text{ kJ} = \Delta H_{vap}$$

The energy change is then

$$\Delta U = \Delta H_{vap} - \Delta(PV) = \Delta H_{vap} - RT\,\Delta n_g$$

Inserting $T = 349.9$ K and $\Delta n_g = 1$ (because there is an increase of 1 mol of gaseous products for each mole of liquid that is vaporized) gives

$$\Delta U = 30.0 \text{ kJ} - (8.315 \text{ J K}^{-1} \text{ mol}^{-1})(349.9 \text{ K})(1.00 \text{ mol})(10^{-3} \text{ kJ J}^{-1})$$

$$= (30.0 - 2.9) \text{ kJ} = +27.1 \text{ kJ mol}^{-1}$$

Thus, of the 30.0 kJ of energy transferred from the surroundings in the form of heat, 27.1 kJ is used to increase the internal energy of the molecules (ΔU) and 2.9 kJ is used to expand the resulting vapor, $\Delta(PV)$.

Related Problems: 27, 28, 29, 30

Standard-State Enthalpies

Absolute values of the enthalpy of a substance, like absolute values of the internal energy, cannot be measured or calculated. Only *changes* in enthalpy can be measured. Just as altitudes are measured relative to a standard altitude (sea level), it is necessary to adopt a reference state for the enthalpies of substances. To cope with this problem, chemists define **standard states** for chemical substances as follows:

> For solids and liquids, the standard state is the thermodynamically stable state at a pressure of 1 atm and at a specified temperature.
>
> For gases, the standard state is the gaseous phase at a pressure of 1 atm, at a specified temperature and exhibiting ideal gas behavior.
>
> For dissolved species, the standard state is a 1-M solution at a pressure of 1 atm, at a specified temperature and exhibiting ideal solution behavior.

Standard-state values of enthalpy and other quantities are designated by attaching a superscript ° (pronounced "naught") to the symbol for the quantity and writing the specified temperature as a subscript. Any temperature may be chosen as the "specified temperature." The most common choice is 298.15 K (25°C exactly); if the temperature of a standard state is not explicitly indicated, 298.15 K should be assumed to be the value.

Once standard states have been defined, the zero of the enthalpy scale is defined by arbitrarily setting the enthalpies of selected reference substances to zero in their standard states. This is completely analogous to assigning zero as the altitude at sea level. Chemists have agreed to the following: *The chemical elements in their standard states at 298.15 K have zero enthalpy.* A complication immediately arises because some elements exist in various allotropic forms that differ in structure and all physical properties, including enthalpy. For example, oxygen can be prepared as $O_2(g)$ or $O_3(g)$ (ozone), and carbon exists in numerous allotropic forms, including graphite, diamond, and the fullerenes (see Section 7.8). Chemists have agreed to assign zero enthalpy to the form that is most stable at 1 atm and 298.15 K. Thus, $O_2(g)$ is assigned zero enthalpy in its standard state at 298.15 K, whereas $O_3(g)$ has nonzero enthalpy in its standard state at 298.15 K. The most stable form of carbon at 1 atm and 298.15 K is graphite, which is assigned zero enthalpy; diamond and all the fullerenes are assigned nonzero enthalpy.[2]

The enthalpy change for a chemical reaction in which all reactants and products are in their standard states and at a specified temperature is called the **standard enthalpy** (written $\Delta H°$) for that reaction. The standard enthalpy is the central tool in thermochemistry because it provides a systematic means for comparing the energy changes due to bond rearrangements in different reactions. Standard enthalpies can be calculated from tables of reference data. For this purpose, we need one additional concept. The **standard enthalpy of formation** $\Delta H_f°$ of a compound is defined to be the enthalpy change for the reaction that produces 1 mol of the compound from its elements in their stable states, all at 25°C and 1 atm pressure. For example, the standard enthalpy of formation of liquid water is the enthalpy change for the reaction

$$H_2(g) + \tfrac{1}{2}\,O_2(g) \longrightarrow H_2O(\ell) \qquad \Delta H° = -285.83 \text{ kJ}$$

$$\Delta H_f°(H_2O(\ell)) = -285.83 \text{ kJ mol}^{-1}$$

Here, the superscript ° indicates standard-state conditions, and the subscript f stands for *formation*.

The $\Delta H_f°$ for an *element* that is already in its standard state is clearly zero, because no further change is needed to bring it to standard-state conditions. But the

[2]There is one exception to this choice of standard state. The standard state of phosphorus is taken to be white phosphorus, rather than the more stable red or black form, because the latter are less well characterized and less reproducible in their properties.

standard enthalpy of formation of a mole of *atoms* of an element is often a large positive quantity. That is, the reaction to generate them is endothermic:

$$\tfrac{1}{2}\,H_2(g) \longrightarrow H(g) \qquad \Delta H° = +217.96\ \text{kJ}$$

$$\Delta H_f°(H(g)) = 217.96\ \text{kJ mol}^{-1}$$

For dissolved species the standard state is defined as an ideal solution with a concentration of 1 M (this is obtained in practice by extrapolating the dilute solution behavior up to this concentration). A special comment is in order on the standard enthalpies of formation of ions. When a strong electrolyte dissolves in water, both positive and negative ions form; it is impossible to produce one without the other. It is therefore also impossible to measure the enthalpy change of formation of ions of only one charge. Only the sum of the enthalpies of formation of the positive and negative ions is accessible to calorimetric experiments. Therefore, chemists have agreed that $\Delta H_f°$ of $H^+(aq)$ is set to zero.

Tables of $\Delta H_f°$ for compounds are the most important data source for thermochemistry. From them it is easy to calculate $\Delta H°$ for reactions of the compounds, and thereby systematically compare the energy changes due to bond rearrangements in different reactions. Appendix D gives a short table of standard enthalpies of formation at 25°C. The following example shows how they can be used to determine enthalpy changes for reactions performed at 25°C and 1 atm pressure.

EXAMPLE 12.7

Using Appendix D, calculate $\Delta H°$ for the following reaction at 25°C and 1 atm pressure:

$$2\,NO(g) + O_2(g) \longrightarrow 2\,NO_2(g)$$

SOLUTION

Because enthalpy is a function of state, ΔH can be calculated along any convenient path. In particular, two steps can be chosen for which ΔH is found easily. In step 1, the reactants are decomposed into the elements in their standard states:

$$2\,NO(g) \longrightarrow N_2(g) + O_2(g) \qquad \Delta H_1 = -2\,\Delta H_f°(NO)$$

The minus sign appears because the process chosen is the reverse of the formation of NO; the factor of 2 is present because 2 mol NO is involved. Because oxygen is already an element in its standard state, it does not need to be changed [equivalently, $\Delta H_f°(O_2(g))$ is 0].

In step 2, the elements are combined to form products:

$$N_2(g) + 2\,O_2(g) \longrightarrow 2\,NO_2(g) \qquad \Delta H_1 = 2\,\Delta H_f°(NO_2)$$

The enthalpy change of the overall reaction is then the sum of these two enthalpies:

$$\Delta H° = \Delta H_1 + \Delta H_2 = -2\,\Delta H_f°(NO) + 2\,\Delta H_f°(NO_2)$$

$$= -(2\ \text{mol})(90.25\ \text{kJ mol}^{-1}) + (2\ \text{mol})(33.18\ \text{kJ mol}^{-1}) = -114.14\ \text{kJ}$$

Related Problems: 35, 36, 37, 38

The general pattern should be clear from this example. The change $\Delta H°$ for a reaction at atmospheric pressure and 25°C is the sum of the $\Delta H_f°$ for the *products* (multiplied by their coefficients in the balanced chemical equation) minus the sum of the $\Delta H_f°$ for the *reactants* (also multiplied by their coefficients). For a general reaction of the form

$$aA + bB \longrightarrow cC + dD$$

the standard enthalpy change is

$$\Delta H° = c\,\Delta H_f°(\text{C}) + d\,\Delta H_f°(\text{D}) - a\,\Delta H_f°(\text{A}) - b\,\Delta H_f°(\text{B})$$

This equation can be extended to calculate the standard-state enthalpy change for any chemical reaction by adding up the standard-state enthalpy of formation for all the products (each multiplied by its stoichiometric coefficient in the balanced chemical equation) and subtracting off the total for all the reactants (each multiplied by its stoichiometric coefficient in the balanced chemical equation). In mathematical form, this procedure is represented by the equation

$$\Delta H° = \sum_{i=1}^{prod} n_i \Delta H_i° - \sum_{j=1}^{react} n_j \Delta H_j° \qquad [12.12]$$

Bond Enthalpies

Chemical reactions between molecules require existing bonds to break and new ones to form in a new arrangement of the atoms. Chemists have developed methods to study highly reactive intermediates, species in which bonds have been broken and not yet re-formed. For example, a hydrogen atom can be removed from a methane molecule,

$$CH_4(g) \longrightarrow CH_3(g) + H(g)$$

leaving two fragments, neither of which has a stable valence electron structure in the Lewis electron dot picture. Both will go on to react rapidly with other molecules or fragments and eventually form the stable products of that reaction. Nonetheless, we can measure many of the properties of these reactive species during the short time they are present.

One such important measurable quantity is the enthalpy change when a bond is broken in the gas phase, called the **bond enthalpy**. This is invariably positive because heat must be added to a collection of stable molecules to break their bonds. For example, the bond enthalpy of a C—H bond in methane is 438 kJ mol^{-1}, measured as the standard enthalpy change for the reaction

$$CH_4(g) \longrightarrow CH_3(g) + H(g) \qquad \Delta H° = +438 \text{ kJ}$$

in which 1 mol of C—H bonds is broken, one for each molecule of methane. Bond enthalpies are fairly reproducible from one compound to another. Each of the following gas-phase reactions involves the breaking of a C—H bond:

$$C_2H_6(g) \longrightarrow C_2H_5(g) + H(g) \qquad \Delta H° = +410 \text{ kJ}$$

$$CHF_3(g) \longrightarrow CF_3(g) + H(g) \qquad \Delta H° = +429 \text{ kJ}$$

$$CHCl_3(g) \longrightarrow CCl_3(g) + H(g) \qquad \Delta H° = +380 \text{ kJ}$$

$$CHBr_3(g) \longrightarrow CBr_3(g) + H(g) \qquad \Delta H° = +377 \text{ kJ}$$

The approximate constancy of the measured enthalpy changes (all lie within 8% of their average value) suggests that the C—H bonds in all five molecules are similar. Because such bond enthalpies are reproducible from one molecule to another, it is useful to tabulate *average* bond enthalpies from measurements on a series of compounds (Table 12.3). Any given bond enthalpy will differ somewhat from those shown, but in most cases, the deviations are small. The reproducibility of bond energies in a series of molecules was introduced in Section 3.5, and representative values were listed in Table 3.4. Bond energy values are related to bond enthalpy values by the relation $\Delta H = \Delta U + \Delta(PV)$, described in the paragraphs following Example 12.5.

TABLE 12.3 Average Bond Enthalpies

	Molar Enthalpy of Atomization (kJ mol^{-1})‡	Bond Enthalpy (kJ mol^{-1})†							
		H—	C—	C=	C≡	N—	N=	N≡	O—	O=
H	218.0	436	413			391			463	
C	716.7	413	348	615	812	292	615	891	351	728
N	472.7	391	292	615	891	161	418	945		
O	249.2	463	351	728					139	498
S	278.8	339	259	477						
F	79.0	563	441			270			185	
Cl	121.7	432	328			200			203	
Br	111.9	366	276							
I	106.8	299	240							

†From Appendix D.
‡Data from L. Pauling. *The Nature of the Chemical Bond,* 3rd ed. Ithaca, NY: Cornell University Press, 1960.

FIGURE 12.15 Dichlorodifluoromethane, CCl_2F_2, also known as Freon-12.

The bond enthalpies in Table 12.3 can be used, together with enthalpies of atomization of the elements from the same table, to estimate standard enthalpies of formation ΔH_f° for molecules in the gas phase and enthalpy changes ΔH° for gas-phase reactions. This is illustrated by the following example.

EXAMPLE 12.8

Estimate the standard enthalpy of formation of dichlorodifluoromethane, $CCl_2F_2(g)$ (Fig. 12.15). This compound is also known as Freon-12 and has been used as a refrigerant because of its low reactivity and high volatility. It and other related chlorofluorocarbons (CFCs) are being phased out because of their role in depleting the ozone layer in the outer atmosphere, as discussed in Section 20.5.

SOLUTION

The standard enthalpy of formation of $CCl_2F_2(g)$ is the enthalpy change for the process in which it is formed from the elements in their standard states at 25°C:

$$C(s,gr) + Cl_2(g) + F_2(g) \longrightarrow CCl_2F_2(g)$$

This reaction can be replaced by a hypothetical two-step process: All the species appearing on the left are atomized, and then the atoms are combined to make CCl_2F_2.

$$C(s,gr) + Cl_2(g) + F_2(g) \longrightarrow C(g) + 2\ Cl(g) + 2\ F(g)$$

$$C(g) + 2\ Cl(g) + 2\ F(g) \longrightarrow CCl_2F_2(g)$$

The enthalpy change ΔH_1 for the first step is the sum of the atomization energies:

$$\Delta H_1 = \Delta H_f^\circ(C(g)) + 2\ \Delta H_f^\circ(Cl(g)) + 2\ \Delta H_f^\circ(F(g))$$

$$= 716.7 + 2(121.7) + 2(79.0) = 1118 \text{ kJ}$$

The enthalpy change ΔH_2 for the second step can be estimated from the bond enthalpies of Table 12.3. This step involves the formation (with release of heat and, therefore, negative enthalpy change) of two C—Cl and two C—F bonds per molecule. The net ΔH for this step is then

$$\Delta H_2 \approx -[2(328) + 2(441)] = -1538 \text{ kJ}$$

$$\Delta H_1 + \Delta H_2 = -1538 + 1118 = -420 \text{ kJ}$$

This ΔH_f°, -420 kJ mol^{-1}, compares fairly well with the experimental value, -477 kJ mol^{-1}. In general, much better agreement than this is not to be expected, because tabulated bond enthalpies are only average values.

Related Problems: 45, 46, 47, 48

12.6 Reversible Processes in Ideal Gases

Most thermodynamic processes conducted in laboratory work are irreversible, in the sense of Section 12.1. Except in the initial and final states, the system is not at equilibrium, and the equation of state relationship between observable properties does not exist. Consequently, changes in thermodynamic quantities during an irreversible process cannot in general be calculated. Nonetheless, the changes in those quantities that are state functions are well defined, as long as the initial and final equilibrium states are known. Because these changes are independent of the detailed path of the process, they can be evaluated for any known process that connects these initial and final states. Changes can be directly calculated along *reversible* paths, during which the system proceeds through a sequence of equilibrium states in which observable properties are related by the equation of state.

This section demonstrates calculations of changes in macroscopic properties caused during several specific reversible processes in ideal gases. These will serve as auxiliary calculation pathways for evaluating changes in state functions during irreversible processes. We use this procedure extensively in Chapter 13 on spontaneous processes and the second law of thermodynamics.

Recall from Section 12.1 that a true reversible process is an idealization; it is a process in which the system proceeds with infinitesimal speed through a series of equilibrium states. The external pressure P_{ext}, therefore, can never differ by more than an infinitesimal amount from the pressure, P, of the gas itself. The heat, work, energy, and enthalpy changes for ideal gases at constant volume (called **isochoric processes**) and at constant pressure **(isobaric processes)** have already been considered. This section examines *isothermal* (constant temperature) and *adiabatic* ($q = 0$) processes.

Isothermal Processes

An **isothermal process** is one conducted at constant temperature. This is accomplished by placing the system in a large reservoir (bath) at fixed temperature and allowing heat to be transferred as required between system and reservoir. The reservoir is large enough that its temperature is almost unchanged by this heat transfer. In Section 12.4, U for an ideal gas was shown to depend only on temperature; therefore, $\Delta U = 0$ for any isothermal ideal gas process. From the first law it follows that

$$w = -q \text{ (isothermal process, ideal gas)}$$

In a *reversible* process, $P_{ext} = P_{gas} \equiv P$. But the relation $w = -P_{ext}\,\Delta V$ from Section 12.2 cannot be used to calculate the work, because that expression applies only if the external pressure remains constant as the volume changes. In the reversible isothermal expansion of an ideal gas,

$$P_{ext} = P = \frac{nRT}{V} \qquad [12.13]$$

By Boyle's law, the pressure falls as the volume is increased from V_1 to V_2, as shown by the solid line in Figure 12.16. In this case, the work is calculated by approximating the process as a *series* of expansions by small amounts ΔV, during each of which P_{ext} is held constant at P_i, with $i = 1, 2, 3$, and so on labeling the step. The work done in step i is $-P_i\delta V$, and thus the total work done is the sum of the work in all steps:

$$w = -P_1\delta V - P_2\delta V - \cdots$$

The work in the complete process is the sum of the areas of the rectangles in the figure. As the step size δV is made smaller, it approaches the infinitesimal

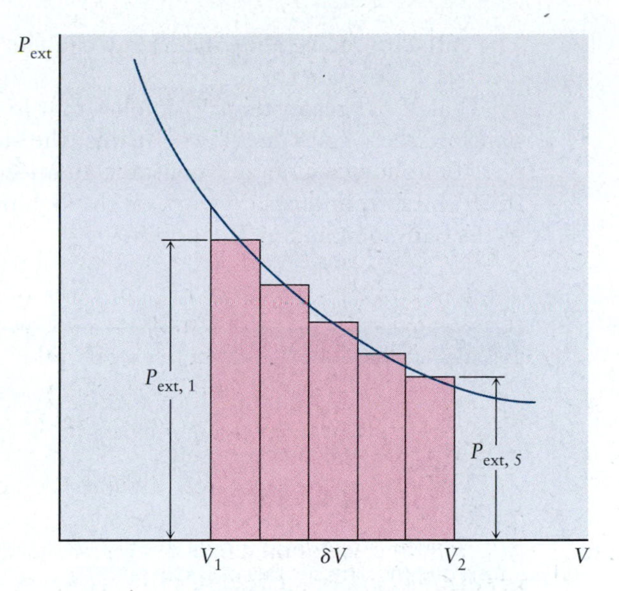

FIGURE 12.16 The area under the graph of external pressure against volume can be approximated as the sum of the areas of the rectangles shown.

volume change that is written dV, and the corresponding increment of work becomes

$$dw = -P_{ext}\, dV$$

In this limit the sum of the areas of the rectangles approaches the area under the graph of P versus V. This limiting sum is the *integral* of P_{ext} from V_1 to V_2 and is written symbolically using an integral sign:

$$w = -\int_{V_1}^{V_2} P\, dV \quad \text{(reversible process)} \qquad [12.14]$$

Note that we used Equation 12.13 to replace P_{ext} with the internal pressure P in this reversible process. This is a compact way of saying that the work, w, for a reversible expansion process is (with a minus sign) the area under the graph of P plotted against V from V_1 to V_2.

Mathematical methods of defining and calculating integrals are beyond the scope of this book, but Appendix C gives some supplementary information about several integrals that are important in chemistry, and it should be read by students unfamiliar with calculus. The main idea is the relation between a thermodynamic quantity (in this case, the work) and the area under a curve.

In this case, the ideal gas law is used to write this area as

$$w = -nRT\int_{V_1}^{V_2} \frac{1}{V}\, dV \qquad [12.15]$$

This integral is the area under a graph of $1/V$ against V (a hyperbola) from V_1 to V_2. It defines the *natural logarithm* function, symbolized "ln" (see Appendix C). In particular,

$$w = -nRT \ln \frac{V_2}{V_1}$$

$$q = -w = nRT \ln \frac{V_2}{V_1}$$

$$\Delta U = 0 \text{ because } \Delta T = 0$$

$$\Delta H = \Delta U + \Delta(PV) = \Delta U + \Delta(nRT) = 0 \qquad [12.16]$$

The enthalpy change, like the energy change, depends only on temperature for an ideal gas.

That V_2 is greater than V_1 implies that $w < 0$ and $q > 0$; in an isothermal expansion, the system does work against the surroundings and heat must be transferred into it to maintain T constant. In an isothermal compression, the reverse is true: The surroundings do work on the system, and the system must then lose heat to the bath to maintain T constant.

EXAMPLE 12.9

Calculate the heat and the work associated with a process in which 5.00 mol of gas expands reversibly at constant temperature $T = 298$ K from a pressure of 10.00 to 1.00 atm.

SOLUTION

At constant T and n, q and w are given by the equations just above Equation 12.16. Inserting the data for this example gives

$$\frac{V_2}{V_1} = \frac{P_1}{P_2} = \frac{10.0 \text{ atm}}{1.00 \text{ atm}} = 10.0$$

Thus,

$$w = -(5.00 \text{ mol})(8.315 \text{ J K}^{-1} \text{ mol}^{-1})(298 \text{ K}) \ln 10.0$$
$$= -2.85 \times 10^4 \text{ J} = -28.5 \text{ kJ}$$
$$q = -w = 28.5 \text{ kJ}$$

Related Problems: 51, 52

Adiabatic Processes

An **adiabatic process** is one in which there is no transfer of heat into or out of the system. This is accomplished by placing an adiabatic wall (thermal insulation) around the system to prevent heat flow.

$$q = 0$$
$$\Delta U = w$$

Consider a small adiabatic change. The volume changes by an amount dV and the temperature by an amount dT. Now U depends only on temperature for an ideal gas, so

$$dU = nc_V \, dT$$

As always, the work is given by $-P_{ext} \, dV$. Setting these equal gives

$$nc_V \, dT = -P_{ext} \, dV$$

In other words, the temperature change dT is related to the volume change dV in such a process.

If the process is *reversible,* as well as adiabatic, so that $P_{ext} \approx P$, the ideal gas law can be used to write

$$nc_V \, dT = -P \, dV = -\frac{nRT}{V} dV$$

The equation is simplified by dividing both sides through by nT, making the left side depend only on T and the right side only on V:

$$\frac{c_V}{T} dT = -\frac{R}{V} dV$$

Suppose now that the change is not infinitesimal but large. How are temperature and volume related in this case? If a series of such infinitesimal changes is added together, the result is an integral of both sides of the equation from the initial state (specified by T_1 and V_1) to the final state (specified by T_2 and V_2):

$$c_V \int_{T_1}^{T_2} \frac{1}{T}\, dT = -R \int_{V_1}^{V_2} \frac{1}{V}\, dV$$

Here, c_V has been assumed to be approximately independent of T over the range from T_1 to T_2. Evaluating the integrals gives

$$c_V \ln \frac{T_2}{T_1} = -R \ln \frac{V_2}{V_1} = R \ln \frac{V_1}{V_2}$$

A more useful form results from taking antilogarithms of both sides:

$$\left(\frac{T_2}{T_1}\right)^{c_V} = \left(\frac{V_1}{V_2}\right)^{R} = \left(\frac{V_1}{V_2}\right)^{c_P - c_V}$$

The last step used the fact that $R = c_P - c_V$. Thus,

$$\left(\frac{T_2}{T_1}\right) = \left(\frac{V_1}{V_2}\right)^{c_P/c_V - 1} = \left(\frac{V_1}{V_2}\right)^{\gamma - 1}$$

where $\gamma = c_P/c_V$ is the ratio of specific heats. This can be rearranged to give

$$T_1 V_1^{\gamma - 1} = T_2 V_2^{\gamma - 1} \qquad [12.17]$$

In many situations, the initial thermodynamic state and, therefore, T_1 and V_1 are known. If the final volume V_2 is known, T_2 can be calculated; if T_2 is known, V_2 can be calculated.

In some cases, only the final pressure, P_2, of an adiabatic process is known. In this case, the ideal gas law gives

$$\frac{P_1 V_1}{T_1} = \frac{P_2 V_2}{T_2}$$

Multiplying this by Equation 12.17 gives

$$P_1 V_1^{\gamma} = P_2 V_2^{\gamma} \qquad [12.18]$$

This can be used to calculate V_2 from a known P_2.

Once the pressure, temperature, and volume of the final state are known, the energy and enthalpy changes and the work done are straightforward to calculate:

$$\Delta U = n c_V (T_2 - T_1) = w \text{ (reversible adiabatic process for ideal gas)}$$

$$\Delta H = \Delta U + \Delta(PV) = \Delta U + (P_2 V_2 - P_1 V_1)$$

or more simply,

$$\Delta H = n c_P \Delta T$$

EXAMPLE 12.10

Suppose 5.00 mol of an ideal monatomic gas at an initial temperature of 298 K and pressure of 10.0 atm is expanded adiabatically and reversibly until the pressure has decreased to 1.00 atm. Calculate the final volume and temperature, the energy and enthalpy changes, and the work done.

SOLUTION

The initial volume is

$$V_1 = \frac{nRT_1}{P_1} = \frac{(5.00 \text{ mol})(0.08206 \text{ L atm K}^{-1} \text{ mol}^{-1})(298 \text{ K})}{(10.0 \text{ atm})} = 12.2 \text{ L}$$

and the heat capacity ratio for a monatomic gas is

$$\gamma = \frac{c_P}{c_V} = \frac{\frac{5}{2}R}{\frac{3}{2}R} = \frac{5}{3}$$

For a reversible adiabatic process,

$$\frac{P_1}{P_2} V_1^{\gamma} = V_2^{\gamma}$$

$$(10.0)(12.2 \text{ L})^{5/3} = V_2^{5/3}$$

$$V_2 = (12.2 \text{ L})(10.0)^{3/5} = 48.7 \text{ L}$$

The final temperature can now be calculated from the ideal gas law:

$$T_2 = \frac{P_2 V_2}{nR} = 119 \text{ K}$$

From this the work done and the energy change can be found,

$$w = \Delta U = nc_V \Delta T = (5.00 \text{ mol})(\tfrac{3}{2} \times 8.315 \text{ J K}^{-1} \text{ mol}^{-1})(119 \text{ K} - 298 \text{ K})$$
$$= -11{,}200 \text{ J}$$

as well as the enthalpy change:

$$\Delta H = nc_P \Delta T = (5.00 \text{ mol})(\tfrac{5}{2} \times 8.315 \text{ J K}^{-1} \text{ mol}^{-1})(119 \text{ K} - 298 \text{ K}) = -18{,}600 \text{ J}$$

Related Problems: 53, 54

Figure 12.17 compares the adiabatic expansion of this example to the isothermal expansion of Example 12.9. Note that the initial states were the same in the two

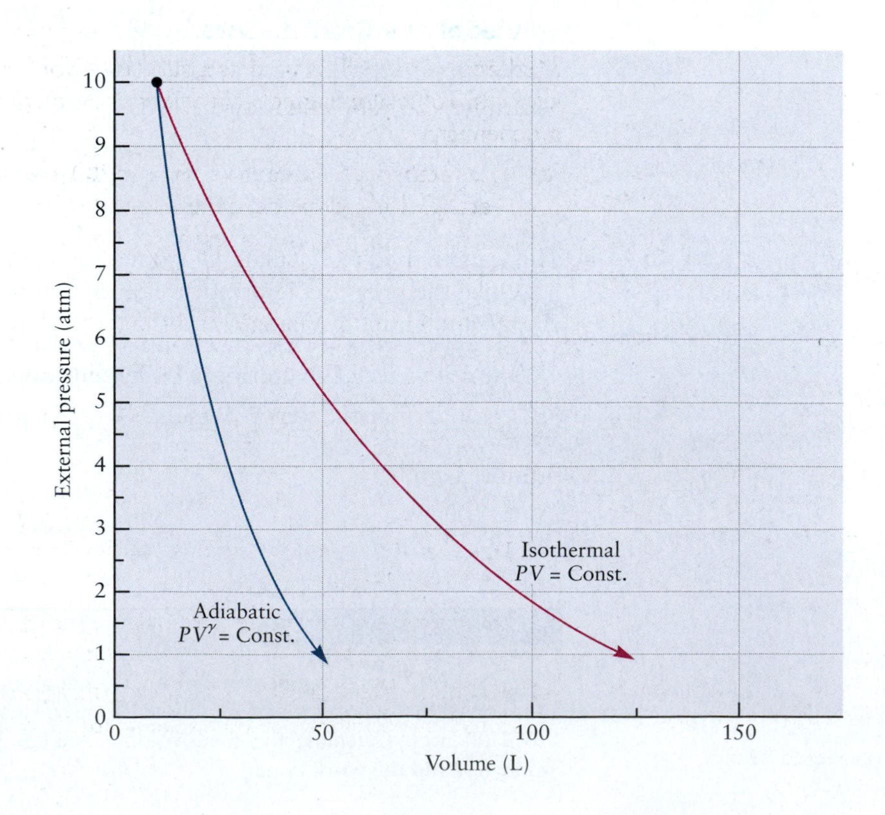

FIGURE 12.17 A comparison of reversible isothermal and adiabatic expansions. Using the technique shown in Figure 12.16, the adiabatic work is 40% of the isothermal work.

cases, as were the final pressures. However, the final volume is larger by more than a factor of 2 for the isothermal expansion, and the work output for the adiabatic case is only 40% of the output from the isothermal expansion. Because $\gamma > 1$, the adiabatic line falls off more rapidly with increasing volume than the isothermal line. Because no heat is transferred in an adiabatic expansion, the work comes from the internal energy of the gas; thus, the temperature declines.

CHAPTER SUMMARY

The central goal of this chapter is to define and measure the heat liberated or absorbed during a chemical reaction. We achieved that goal in Section 12.5 by introducing the enthalpy change of a reaction, which is measured by performing the reaction in a constant-pressure calorimeter. Tabulating data for reactions in which both products and reactants are in their standard states gives the standard enthalpy of reaction, which enables systematic comparison of the energy changes due to bond rearrangements in different reactions. The standard enthalpy change for a reaction is readily calculated from tables of the standard enthalpy of formation for the products and reactants. These basic tools of thermochemistry completely describe the energy transfers in chemical reactions.

The discussion of basic thermodynamics in Sections 12.1 through 12.4 can be viewed as background necessary for achieving the central goal. Even so, this material is important in its own right and will be used repeatedly throughout the book. Similarly, the discussion of reversible isothermal and adiabatic processes in Section 12.6 provides background needed later in the book to calculate changes in state functions for irreversible processes.

CUMULATIVE EXERCISE

Methanol as a Gasoline Substitute

Methanol (CH_3OH) is used as a substitute for gasoline in certain high-performance vehicles. To design engines that will run on methanol, we must understand its thermochemistry.

(a) The methanol in an automobile engine must be in the gas phase before it can react. Calculate the heat (in kilojoules) that must be added to 1.00 kg liquid methanol to increase its temperature from 25.0°C to its normal boiling point, 65.0°C. The molar heat capacity of liquid methanol is 81.6 J K^{-1} mol^{-1}.

(b) Once the methanol has reached its boiling point, it must be vaporized. The molar enthalpy of vaporization for methanol is 38 kJ mol^{-1}. How much heat must be added to vaporize 1.00 kg methanol?

(c) Once it is in the vapor phase, the methanol can react with oxygen in the air according to

$$CH_3OH(g) + \tfrac{3}{2}\,O_2(g) \longrightarrow CO_2(g) + 2\,H_2O(g)$$

Use average bond enthalpies to estimate the enthalpy change in this reaction, for 1 mol of methanol reacting.

(d) Use data from Appendix D to calculate the actual enthalpy change in this reaction, assuming it to be the same at 65°C as at 25°C.

(e) Calculate the heat released when 1.00 kg gaseous methanol is burned in air at constant pressure. Use the more accurate result of part (d), rather than that of part (c).

©Vanessa Vick/Photo Researchers, Inc.

A methanol-powered bus.

Methanol, CH_3OH.

(f) Calculate the *difference* between the change in enthalpy and the change in internal energy when 1.00 kg gaseous methanol is oxidized to gaseous CO_2 and H_2O at 65°C.

(g) Suppose now that the methanol is burned inside the cylinder of an automobile. Taking the radius of the cylinder to be 4.0 cm and the distance moved by the piston during one stroke to be 12 cm, calculate the work done on the gas per stroke as it expands against an external pressure of 1.00 atm. Express your answer in liter-atmospheres and in joules.

Answers

(a) 102 kJ

(b) 1.2×10^3 kJ

(c) -508 kJ

(d) -676.49 kJ

(e) 2.11×10^4 kJ

(f) $\Delta H - \Delta U = 43.9$ kJ

(g) -0.60 L atm $= -61$ J

CHAPTER REVIEW

Systems, States, and Processes

- The system is that part of the universe of interest, for example a chemical reaction, an engine, a human being.
- The surroundings are that part of the universe that exchange matter and energy with the system during a process.
- The thermodynamic universe is the combination of the system and the surroundings for a particular process of interest; it is assumed to be closed and isolated.
- A closed system is one in which no exchange of matter between system and surroundings is permitted.
- An adiabatic system is one in which no exchange of heat between system and surroundings is permitted.
- A thermodynamic state is a condition in which all macroscopic properties of a system are determined by the external conditions imposed on the system (for example, n, T, and P).
- A thermodynamic system is in thermodynamic equilibrium if none of its macroscopic properties is changing over time. (**Note:** Some states that appear not to be changing may not be true equilibrium states because the changes are too slow to be observed, for example, diamond turning into graphite.)
- Properties of a system may either be independent of the amount of material present (intensive, like T and P) or proportional to the quantity of material present (extensive, like internal energy and heat capacity).
- Processes can be either irreversible (a small change in the external condition will not reverse the course of the process) or reversible (a small change in the external condition will reverse the course of the process). The latter are idealizations, often called quasi-static, and would take an infinite amount of time to occur in a real sense.

- A state function is a property whose value depends only on the current state of the system and not on the path by which that state was reached (examples include T and P). A path-dependent function is one in which the value does depend on the details of the path taken, work and heat being the most common examples.

The First Law of Thermodynamics: Internal Energy, Work, and Heat

- Work (w) is force times displacement $w = Fd$. Perhaps the most important type of work in chemistry is pressure–volume work, in which a system either expands against or is compressed by the external pressure. If the external pressure is constant, then

$$w = -P_{ext}\,\Delta V$$

- Heat (q) is the transfer of thermal energy between the system and the surroundings.
- We chose the sign convention so that $+q$ and $+w$ indicate heat is added to the system and work is done on the system.
- Heat transfer is measured by calorimetry. $q = Mc_s\Delta T$, where M is the mass of the heat-absorbing substance in the calorimeter (usually water) and c_s is the specific heat of the substance. A commercial calorimeter will have calibration factors that take into account the heat capacity of materials other than water used in its construction (for example, sample holder, walls, and so forth).
- $\Delta U = q + w$. The internal energy of a system is a state function. Although q and w are functions of the path, their sum is a state function. Heat transferred and work done must leave the energy of the thermodynamic universe unchanged.

$$q_{sys} = -q_{surr}$$
$$w_{sys} = -w_{surr}$$
$$\Delta U_{sys} = -\Delta U_{surr}$$

Heat Capacity, Enthalpy, and Calorimetry

- Heat transfer at constant volume: $\Delta U = q + w = q_V$. Reactions to measure ΔU are performed in a constant volume calorimeter called a bomb calorimeter.
- Heat transfer at constant pressure: enthalpy.
  Define $H = U + PV$. Then at constant pressure

$$q_P = \Delta(U + PV) = \Delta H$$

  Unlike constant volume processes, work can be done in constant pressure processes, so the energy change may not all show up as heat.

Illustrations of the First Law of Thermodynamics in Ideal Gas Processes

- Heat capacities of ideal monatomic gases

$$c_V \text{ (ideal gas)} = 3/2 \text{ R mol}^{-1}$$
$$c_P = c_V + \text{R} = (5/2)\text{R}$$

  It takes 5/3 as much heat to increase the temperature of a gas by the same amount at constant pressure than at constant volume because some of the heat is used to pay for the work of expansion at constant pressure.
- Heat and work for ideal gases: The calculations on page 503 associated with Figure 12.10 demonstrate that the values of q and w depend on the path over which a process occurs.

Thermochemistry

- Because chemical reactions are most commonly studied at constant pressure, the heat absorbed or released in a reaction is measured as the enthalpy change of the reaction.
- Because enthalpy is a state function, enthalpy changes for any reaction of interest can be calculated by summing the enthalpy changes for a set of reactions that add up to give the reaction of interest.
- The most useful thermochemical data are tables of the standard enthalpy of formation ΔH_f° for compounds, defined as the enthalpy of formation of a compound in its standard state from the elements in their standard states at 1 atm and 298.15K.
- The change in standard state enthalpy for any reaction can be calculated from the standard state enthalpy of formation of its products and reactants as

$$\Delta H^\circ = \sum_{i=1}^{prod} n_i \Delta H_i^\circ - \sum_{j=1}^{react} n_j \Delta H_j^\circ$$

- Bond enthalpy is the enthalpy change associated with making or breaking a chemical bond. It is based on the idea that bond enthalpy, like bond energy (see Chapter 3), is approximately independent of the rest of the molecule. Average bond enthalpies are tabulated from measurements over a broad range of molecules in which the same bond appears.
- Heat transferred in phase transitions at constant pressure include ΔH_{fus}° and ΔH_{vap}° for melting and boiling, respectively.

Reversible Processes in Ideal Gases

- **Isothermal processes:** In this case, heat will flow (in either direction) to offset the cost of PV work, whereas T remains constant.

$$dw = -PdV$$

$$w = -\int_{V_1}^{V_2} PdV$$

- For an ideal gas, $P = nRT/V$. Because T is constant it comes outside the integral to give

$$w = -nRT\int_{V_1}^{V_2} (1/V)dV$$

$$w = -nRT \ln(V_2/V_1)$$

$$q = +nRT \ln(V_2/V_1)$$

$$\Delta U = 0$$

$$\Delta H = 0$$

- **Adiabatic processes:** No heat flows so all of the energy comes from or goes into the internal energy of the system. The key steps are:

$$q = 0, \text{ so } \Delta U = w$$

$nc_V dT = -PdV = nRTdV$ rearranging, integrating and "simplifying" gives

$$P_1V_1^\gamma = P_2V_2^\gamma, \text{ where } \gamma = c_p/c_V = 5/3 \text{ for ideal gases}$$

- The important result is that the pressure decreases faster in an adiabatic expansion than in an isothermal expansion because there is no heat to enable the expansion work and keep the pressure higher.

CONCEPTS & SKILLS

After studying this chapter and working the problems that follow, you should be able to:

1. Give precise definitions for the terms *thermodynamic system, open system, closed system, thermodynamic state,* and *reversible* and *irreversible process* (Section 12.1).
2. Define and give examples of properties that are state functions of a system (Section 12.1).
3. Calculate the work done on an ideal gas when it is compressed reversibly (Section 12.2, Problems 1–2).
4. Give a physical interpretation to the concept of heat, and calculate the change in temperature of a given quantity of a substance from its heat capacity and the amount of heat transferred to it (Section 12.2, Problems 5–8).
5. Calculate the final temperature reached when two substances of different mass, heat capacity, and temperature are placed in thermal contact (Section 12.3, Problems 11–16).
6. Calculate the amounts of heat and work and the change in energy of an ideal gas during expansions and compressions (Section 12.4, Problems 17–22).
7. Calculate the energy and enthalpy changes for chemical reactions from the standard molar enthalpies of formation of reactants and products (Section 12.5, Problems 35–40).
8. Use bond enthalpies to estimate enthalpies of formation of gaseous compounds (Section 12.5, Problems 45–48).
9. Calculate the heat absorbed and work done by an ideal gas when it expands reversibly and either isothermally or adiabatically (Section 12.6, Problems 51–54).

KEY EQUATIONS

$w = -P_{ext}\Delta V$	(Section 12.2)
$q = Mc_s\Delta T$	(Section 12.2)
$\Delta U = q + w$	(Section 12.2)
$\Delta U_{univ} = \Delta U_{sys} + \Delta U_{surr} = 0$	(Section 12.2)
$q_V = nc_V(T_2 - T_1) = nc_V\Delta T$	(Section 12.3)
$q_P = nc_P\Delta T$	(Section 12.3)
$H = U + PV$	(Section 12.3)
$q_P = \Delta(U + PV) = \Delta H$	(Section 12.3)
$c_V = \frac{3}{2}R$ (monatomic ideal gas)	(Section 12.4)
$c_P = c_V + R$ (any ideal gas)	(Section 12.4)
$\Delta U = nc_V\Delta T$ (any ideal gas)	(Section 12.4)
$\Delta H = \Delta U + \Delta(PV) = nc_V\Delta T + nR\Delta T$ $\Delta H = nc_P\Delta T$ (ideal gas)	(Section 12.4)

$$\Delta H^\circ = \sum_{i=1}^{prod} n_i \Delta H_i^\circ - \sum_{j=1}^{react} n_j \Delta H_j^\circ \qquad \text{(Section 12.5)}$$

$$w = -\int_{V_1}^{V_2} P\,dV \text{ (reversible process)} \qquad \text{(Section 12.6)}$$

$$w = -nRT \ln \frac{V_2}{V_1} \text{ (ideal gas, isothermal reversible process)}$$

$$q = -w = nRT \ln \frac{V_2}{V_1} \text{ (ideal gas, isothermal reversible process)} \qquad \text{(Section 12.6)}$$

$$\Delta U = 0 \text{ because } \Delta T = 0 \text{ (ideal gas, isothermal process)}$$

$$\Delta H = \Delta U + \Delta(PV) = \Delta U + \Delta(nRT) = 0 \text{ (ideal gas, isothermal process)}$$

$$T_1 V_1^{\gamma-1} = T_2 V_2^{\gamma-1} \text{ (ideal gas, reversible adiabatic process)} \qquad \text{(Section 12.6)}$$

$$P_1 V_1^{\gamma} = P_2 V_2^{\gamma} \text{ (ideal gas, reversible adiabatic process)} \qquad \text{(Section 12.6)}$$

PROBLEMS

Answers to problems whose numbers are boldface appear in Appendix G. Problems that are more challenging are indicated with asterisks.

The First Law of Thermodynamics: Internal Energy, Work, and Heat

1. Some nitrogen for use in synthesizing ammonia is heated slowly, maintaining the external pressure close to the internal pressure of 50.0 atm, until its volume has increased from 542 to 974 L. Calculate the work done on the nitrogen as it is heated, and express it in joules.

2. The gas mixture inside one of the cylinders of an automobile engine expands against a constant external pressure of 0.98 atm, from an initial volume of 150 mL (at the end of the compression stroke) to a final volume of 800 mL. Calculate the work done on the gas mixture during this process, and express it in joules.

3. When a ball of mass m is dropped through a height difference Δh, its potential energy changes by the amount $mg\,\Delta h$, where g is the acceleration of gravity, equal to 9.81 m s^{-2}. Suppose that when the ball hits the ground, all that energy is converted to heat, increasing the temperature of the ball. If the specific heat capacity of the material in the ball is 0.850 J K^{-1} g^{-1}, calculate the height from which the ball must be dropped to increase the temperature of the ball by 1.00°C.

4. During his honeymoon in Switzerland, James Joule is said to have used a thermometer to measure the temperature difference between the water at the top and at the bottom of a waterfall. Take the height of the waterfall to be Δh and the acceleration of gravity, g, to be 9.81 m s^{-2}. Assuming that all the potential energy change $mg\,\Delta h$ of a mass m of water is used to heat that water by the time it reaches the bottom, calculate the temperature difference between the top and the bottom of a waterfall 100 meters high. Take the specific heat capacity of water to be 4.18 J K^{-1} g^{-1}.

5. The specific heat capacities of Li(s), Na(s), K(s), Rb(s), and Cs(s) at 25°C are 3.57, 1.23, 0.756, 0.363, and 0.242 J K^{-1} g^{-1}, respectively. Compute the molar heat capacities of these elements and identify any periodic trend. If there is a trend, use it to predict the molar heat capacity of francium, Fr(s).

6. The specific heat capacities of $F_2(g)$, $Cl_2(g)$, $Br_2(g)$, and $I_2(g)$ are 0.824, 0.478, 0.225, and 0.145 J K^{-1} g^{-1}, respectively. Compute the molar heat capacities of these elements and identify any periodic trend. If there is a trend, use it to predict the molar heat capacity of astatine, $At_2(g)$.

7. The specific heat capacities of the metals nickel, zinc, rhodium, tungsten, gold, and uranium at 25°C are 0.444, 0.388, 0.243, 0.132, 0.129, and 0.116 J K^{-1} g^{-1}, respectively. Calculate the molar heat capacities of these six metals. Note how closely the molar heat capacities for these metals, which were selected at random, cluster about a value of 25 J K^{-1} mol^{-1}. The rule of Dulong and Petit states that the molar heat capacities of the metallic elements are approximately 25 J K^{-1} mol^{-1}.

8. Use the empirical rule of Dulong and Petit stated in Problem 7 to estimate the specific heat capacities of vanadium, gallium, and silver.

9. A chemical system is sealed in a strong, rigid container at room temperature, and then heated vigorously.
(a) State whether ΔU, q, and w of the system are positive, negative, or zero during the heating process.
(b) Next, the container is cooled to its original temperature. Determine the signs of ΔU, q, and w for the cooling process.
(c) Designate heating as step 1 and cooling as step 2. Determine the signs of $(\Delta U_1 + \Delta U_2)$, $(q_1 + q_2)$, and $(w_1 + w_2)$, if possible.

10. A battery harnesses a chemical reaction to extract energy in the form of useful electrical work.
(a) A certain battery runs a toy truck and becomes partially discharged. In the process, it performs a total of 117.0 J

of work on its immediate surroundings. It also gives off 3.0 J of heat, which the surroundings absorb. No other work or heat is exchanged with the surroundings. Compute q, w, and ΔU of the battery, making sure each quantity has the proper sign.

(b) The same battery is now recharged exactly to its original condition. This requires 210.0 J of electrical work from an outside generator. Determine q for the battery in this process. Explain why q has the sign that it does.

Heat Capacity, Enthalpy, and Calorimetry

11. Suppose 61.0 g hot metal, which is initially at 120.0°C, is plunged into 100.0 g water that is initially at 20.00°C. The metal cools down and the water heats up until they reach a common temperature of 26.39°C. Calculate the specific heat capacity of the metal, using 4.18 J K^{-1} g^{-1} as the specific heat capacity of the water.

12. A piece of zinc at 20.0°C that weighs 60.0 g is dropped into 200.0 g water at 100.0°C. The specific heat capacity of zinc is 0.389 J K^{-1} g^{-1}, and that of water near 100°C is 4.22 J K^{-1} g^{-1}. Calculate the final temperature reached by the zinc and the water.

13. Very early in the study of the nature of heat it was observed that if two bodies of equal mass but different temperatures are placed in thermal contact, their specific heat capacities depend inversely on the change in temperature each undergoes on reaching its final temperature. Write a mathematical equation in modern notation to express this fact.

14. Iron pellets with total mass 17.0 g at a temperature of 92.0°C are mixed in an insulated container with 17.0 g water at a temperature of 20.0°C. The specific heat capacity of water is 10 times greater than that of iron. What is the final temperature inside the container?

15. In their *Memoir on Heat,* published in 1783, Lavoisier and Laplace reported, "The heat necessary to melt ice is equal to three quarters of the heat that can raise the same mass of water from the temperature of the melting ice to that of boiling water" (English translation). Use this 18th-century observation to compute the amount of heat (in joules) needed to melt 1.00 g ice. Assume that heating 1.00 g water requires 4.18 J of heat for each 1.00°C throughout the range from 0°C to 100°C.

16. Galen, the great physician of antiquity, suggested scaling temperature from a reference point defined by mixing equal masses of ice and boiling water in an insulated container. Imagine that this is done with the ice at 0.00°C and the water at 100.0°C. Assume that the heat capacity of the container is negligible, and that it takes 333.4 J of heat to melt 1.000 g ice at 0.00°C to water at 0.00°C. Compute Galen's reference temperature in degrees Celsius.

Illustrations of the First Law of Thermodynamics in Ideal Gas Processes

17. If 0.500 mol neon at 1.00 atm and 273 K expands against a constant external pressure of 0.100 atm until the gas pressure reaches 0.200 atm and the temperature reaches 210 K, calculate the work done on the gas, the internal energy change, and the heat absorbed by the gas.

18. Hydrogen behaves as an ideal gas at temperatures greater than 200 K and at pressures less than 50 atm. Suppose 6.00 mol hydrogen is initially contained in a 100-L vessel at a pressure of 2.00 atm. The average molar heat capacity of hydrogen at constant pressure, c_P, is 29.3 J K^{-1} mol^{-1} in the temperature range of this problem. The gas is cooled reversibly at constant pressure from its initial state to a volume of 50.0 L. Calculate the following quantities for this process.

(a) Temperature of the gas in the final state, T_2
(b) Work done on the gas, w, in joules
(c) Internal energy change of the gas, ΔU, in joules
(d) Heat absorbed by the gas, q, in joules

19. Suppose 2.00 mol of an ideal, monatomic gas is initially at a pressure of 3.00 atm and a temperature T = 350 K. It is expanded *irreversibly* and *adiabatically* (q = 0) against a constant external pressure of 1.00 atm until the volume has doubled.

(a) Calculate the final volume.
(b) Calculate w, q, and ΔU for this process, in joules.
(c) Calculate the final temperature of the gas.

20. Consider the free, isothermal (constant T) expansion of an ideal gas. "Free" means that the external force is zero, perhaps because a stopcock has been opened and the gas is allowed to expand into a vacuum. Calculate ΔU for this irreversible process. Show that q = 0, so that the expansion is also adiabatic (q = 0) for an ideal gas. This is analogous to a classic experiment first performed by Joule.

21. If 6.00 mol argon in a 100-L vessel initially at 300 K is compressed adiabatically (q = 0) and irreversibly until a final temperature of 450 K is reached, calculate the energy change of the gas, the heat added to the gas, and the work done on the gas.

22. A gas expands against a constant external pressure of 2.00 atm until its volume has increased from 6.00 to 10.00 L. During this process, it absorbs 500 J of heat from the surroundings.

(a) Calculate the energy change of the gas, ΔU.
(b) Calculate the work, w, done on the gas in an irreversible adiabatic (q = 0) process connecting the same initial and final states.

Thermochemistry

23. For each of the following reactions, the enthalpy change written is that measured when the numbers of moles of reactants and products taking part in the reaction are as given by their coefficients in the equation. Calculate the enthalpy change when 1.00 *gram* of the underlined substance is consumed or produced.

(a) $4\ Na(s) + O_2(g) \longrightarrow \underline{2\ Na_2O(s)}$ $\quad \Delta H = -828$ kJ
(b) $CaMg(CO_3)_2(s) \longrightarrow CaO(s) + \underline{MgO(s)} + 2\ CO_2(g)$ $\quad \Delta H = +302$ kJ
(c) $H_2(g) + 2\ \underline{CO(g)} \longrightarrow H_2O_2(\ell) + 2\ C(s)$ $\quad \Delta H = +33.3$ kJ

24. For each of the following reactions, the enthalpy change given is that measured when the numbers of moles of reactants and products taking part in the reaction are as given by their coefficients in the equation. Calculate the enthalpy change when 1.00 *gram* of the underlined substance is consumed or produced.

(a) $Ca(s) + \underline{Br_2(\ell)} \longrightarrow CaBr_2(s)$ $\Delta H = -683$ kJ

(b) $6\ Fe_2O_3(s) \longrightarrow 4\ \underline{Fe_3O_4(s)} + O_2(g)$ $\Delta H = +472$ kJ

(c) $2\ \underline{NaHSO_4(s)} \longrightarrow 2\ NaOH(s) + 2\ SO_2(g) + O_2(g)$ $\Delta H = +806$ kJ

25. Liquid bromine dissolves readily in aqueous NaOH:

$$Br_2(\ell) + 2\ NaOH(aq) \longrightarrow NaBr(aq) + NaOBr(aq) + H_2O(\ell)$$

Suppose 2.88×10^{-6} mol of $Br_2(\ell)$ is sealed in a glass capsule that is then immersed in a solution containing excess $NaOH(aq)$. The capsule is broken, the mixture is stirred, and a measured 121.3 J of heat evolves. In a separate experiment, simply breaking an empty capsule and stirring the solution in the same way evolves 2.34 J of heat. Compute the heat evolved as 1.00 mol $Br_2(\ell)$ dissolves in excess $NaOH(aq)$.

26. A chemist mixes 1.00 g $CuCl_2$ with an excess of $(NH_4)_2HPO_4$ in dilute aqueous solution. He measures the evolution of 670 J of heat as the two substances react to give $Cu_3(PO_4)_2(s)$. Compute the ΔH that would result from the reaction of 1.00 mol $CuCl_2$ with an excess of $(NH_4)_2HPO_4$.

27. Calculate the enthalpy change when 2.38 g carbon monoxide (CO) vaporizes at its normal boiling point. Use data from Table 12.2.

28. Molten sodium chloride is used for making elemental sodium and chlorine. Suppose the electrical power to a vat containing 56.2 kg molten sodium chloride is cut off and the salt crystallizes (without changing its temperature). Calculate the enthalpy change, using data from Table 12.2.

29. Suppose an ice cube weighing 36.0 g at a temperature of −10°C is placed in 360 g water at a temperature of 20°C. Calculate the temperature after thermal equilibrium is reached, assuming no heat loss to the surroundings. The enthalpy of fusion of ice is $\Delta H_{fus} = 6.007$ kJ mol^{-1}, and the molar heat capacities c_P of ice and water are 38 and 75 J K^{-1} mol^{-1}, respectively.

30. You have a supply of ice at 0.0°C and a glass containing 150 g water at 25°C. The enthalpy of fusion for ice is $\Delta H_{fus} = 333$ J g^{-1}, and the specific heat capacity of water is 4.18 J K^{-1} g^{-1}. How many grams of ice must be added to the glass (and melted) to reduce the temperature of the water to 0°C?

31. The measured enthalpy change for burning ketene (CH_2CO)

$$CH_2CO(g) + 2\ O_2(g) \longrightarrow 2\ CO_2(g) + H_2O(g)$$

is $\Delta H_1 = -981.1$ kJ at 25°C. The enthalpy change for burning methane

$$CH_4(g) + 2\ O_2(g) \longrightarrow CO_2(g) + 2\ H_2O(g)$$

is $\Delta H_2 = -802.3$ kJ at 25°C. Calculate the enthalpy change at 25°C for the reaction

$$2\ CH_4(g) + 2\ O_2(g) \longrightarrow CH_2CO(g) + 3\ H_2O(g)$$

32. Given the following two reactions and corresponding enthalpy changes,

$$CO(g) + SiO_2(s) \longrightarrow SiO(g) + CO_2(g) \qquad \Delta H = +520.9 \text{ kJ}$$

$$8\ CO_2(g) + Si_3N_4(s) \longrightarrow 3\ SiO_2(s) + 2\ N_2O(g) + 8\ CO(g) \qquad \Delta H = +461.05 \text{ kJ}$$

compute the ΔH of the reaction

$$5\ CO_2(g) + Si_3N_4(s) \longrightarrow 3\ SiO(g) + 2\ N_2O(g) + 5\ CO(g)$$

33. The enthalpy change to make diamond from graphite is 1.88 kJ mol^{-1}. Which gives off more heat when burned—a pound of diamonds or a pound of graphite? Explain.

34. The enthalpy change of combustion of monoclinic sulfur to $SO_2(g)$ is −9.376 kJ g^{-1}. Under the same conditions, the rhombic form of sulfur has an enthalpy change of combustion to $SO_2(g)$ of −9.293 kJ g^{-1}. Compute the ΔH of the reaction

$$S(\text{monoclinic}) \longrightarrow S(\text{rhombic})$$

per gram of sulfur reacting.

35. Calculate the standard enthalpy change $\Delta H°$ at 25°C for the reaction

$$N_2H_4(\ell) + 3\ O_2(g) \longrightarrow 2\ NO_2(g) + 2\ H_2O(\ell)$$

using the standard enthalpies of formation ($\Delta H°_f$) of reactants and products at 25°C from Appendix D.

36. Using the data in Appendix D, calculate $\Delta H°$ for each of the following processes:

(a) $2\ NO(g) + O_2(g) \longrightarrow 2\ NO_2(g)$

(b) $C(s) + CO_2(g) \longrightarrow 2\ CO(g)$

(c) $2\ NH_3(g) + \frac{7}{2}\ O_2(g) \longrightarrow 2\ NO_2(g) + 3\ H_2O(g)$

(d) $C(s) + H_2O(g) \longrightarrow CO(g) + H_2(g)$

37. Zinc is commonly found in nature in the form of the mineral sphalerite (ZnS). A step in the smelting of zinc is the roasting of sphalerite with oxygen to produce zinc oxide:

$$2\ ZnS(s) + 3\ O_2(g) \longrightarrow 2\ ZnO(s) + 2\ SO_2(g)$$

(a) Calculate the standard enthalpy change $\Delta H°$ for this reaction, using data from Appendix D.

(b) Calculate the heat absorbed when 3.00 metric tons (1 metric ton = 10^3 kg) of sphalerite is roasted under constant-pressure conditions.

38. The thermite process (see Fig. 12.11) is used for welding railway track together. In this reaction, aluminum reduces iron(III) oxide to metallic iron:

$$2\ Al(s) + Fe_2O_3(s) \longrightarrow 2\ Fe(s) + Al_2O_3(s)$$

Igniting a small charge of barium peroxide mixed with aluminum triggers the reaction of a mixture of aluminum powder and iron(III) oxide; the molten iron produced flows into the space between the steel rails that are to be joined.

(a) Calculate the standard enthalpy change $\Delta H°$ for this reaction, using data from Appendix D.

(b) Calculate the heat given off when 3.21 g iron(III) oxide is reduced by aluminum at constant pressure.

39. The dissolution of calcium chloride in water

$$CaCl_2(s) \longrightarrow Ca^{2+}(aq) + 2\ Cl^-(aq)$$

is used in first-aid hot packs. In these packs, an inner pouch containing the salt is broken, allowing the salt to dissolve in the surrounding water.

(a) Calculate the standard enthalpy change $\Delta H°$ for this reaction, using data from Appendix D.

(b) Suppose 20.0 g $CaCl_2$ is dissolved in 0.100 L water at 20.0°C. Calculate the temperature reached by the solution, assuming it to be an ideal solution with a heat capacity close to that of 100 g pure water (418 J K^{-1}).

40. Ammonium nitrate dissolves in water according to the reaction

$$NH_4NO_3(s) \longrightarrow NH_4^+(aq) + NO_3^-(aq)$$

(a) Calculate the standard enthalpy change $\Delta H°$ for this reaction, using data from Appendix D.

(b) Suppose 15.0 g NH_4NO_3 is dissolved in 0.100 L water at 20.0°C. Calculate the temperature reached by the solution, assuming it to be an ideal solution with a heat capacity close to that of 100 g pure water (418 J K^{-1}).

(c) From a comparison with the results of Problem 39, can you suggest a practical application of this dissolution reaction?

41. The standard enthalpy change of combustion [to $CO_2(g)$ and $H_2O(\ell)$] at 25°C of the organic liquid cyclohexane, $C_6H_{12}(\ell)$, is -3923.7 kJ mol^{-1}. Determine the $\Delta H_f°$ of $C_6H_{12}(\ell)$. Use data from Appendix D.

42. The standard enthalpy change of combustion [to $CO_2(g)$ and $H_2O(\ell)$] at 25°C of the organic liquid cyclohexane, $C_6H_{10}(\ell)$, is -3731.7 kJ mol^{-1}. Determine the $\Delta H_f°$ of $C_6H_{10}(\ell)$.

43. A sample of pure solid naphthalene ($C_{10}H_8$) weighing 0.6410 g is burned completely with oxygen to $CO_2(g)$ and $H_2O(\ell)$ in a constant-volume calorimeter at 25°C. The amount of heat evolved is observed to be 25.79 kJ.

(a) Write and balance the chemical equation for the combustion reaction.

(b) Calculate the standard change in internal energy ($\Delta U°$) for the combustion of 1.000 mol naphthalene to $CO_2(g)$ and $H_2O(\ell)$.

(c) Calculate the standard enthalpy change ($\Delta H°$) for the same reaction as in part (b).

(d) Calculate the standard enthalpy of formation per mole of naphthalene, using data for the standard enthalpies of formation of $CO_2(g)$ and $H_2O(\ell)$ from Appendix D.

44. A sample of solid benzoic acid (C_6H_5COOH) that weighs 0.800 g is burned in an excess of oxygen to $CO_2(g)$ and $H_2O(\ell)$ in a constant-volume calorimeter at 25°C. The temperature increase is observed to be 2.15°C. The heat capacity of the calorimeter and its contents is known to be 9382 J K^{-1}.

(a) Write and balance the equation for the combustion of benzoic acid.

(b) Calculate the standard change in internal energy ($\Delta U°$) for the combustion of 1.000 mol benzoic acid to $CO_2(g)$ and $H_2O(\ell)$ at 25°C.

(c) Calculate the standard enthalpy change ($\Delta H°$) for the same reaction as in part (b).

(d) Calculate the standard enthalpy of formation per mole of benzoic acid, using data for the standard enthalpies of formation of $CO_2(g)$ and $H_2O(\ell)$ from Appendix D.

45. A second CFC used as a refrigerant and in aerosols (besides that discussed in Example 12.8) is CCl_3F. Use the atomization enthalpies and average bond enthalpies from Table 12.3 to estimate the standard enthalpy of formation ($\Delta H_f°$) of this compound in the gas phase.

46. The compound CF_3CHCl_2 (with a C—C bond) has been proposed as a substitute for CCl_3F and CCl_2F_2 because it decomposes more quickly in the atmosphere and is much less liable to reduce the concentration of ozone in the stratosphere. Use the atomization enthalpies and average bond enthalpies from Table 12.3 to estimate the standard enthalpy of formation ($\Delta H_f°$) of CF_3CHCl_2 in the gas phase.

47. Propane has the structure $H_3C—CH_2—CH_3$. Use average bond enthalpies from Table 12.3 to estimate the change in enthalpy $\Delta H°$ for the reaction

$$C_3H_8(g) + 5\ O_2(g) \longrightarrow 3\ CO_2(g) + 4\ H_2O(g)$$

48. Use average bond enthalpies from Table 12.3 to estimate the change in enthalpy $\Delta H°$ for the reaction

$$C_2H_4(g) + H_2(g) \longrightarrow C_2H_6(g)$$

Refer to the molecular structures on page 86.

49. The following reaction

$$BBr_3(g) + BCl_3(g) \longrightarrow BBr_2Cl(g) + BCl_2Br(g)$$

has a ΔH close to zero. Sketch the Lewis structures of the four compounds, and explain why ΔH is so small.

50. At 381 K, the following reaction takes place:

$$Hg_2Cl_4(g) + Al_2Cl_6(g) \longrightarrow 2\ HgAlCl_5(g) \qquad \Delta H = +10\ \text{kJ}$$

(a) Offer an explanation for the very small ΔH for this reaction for the known structures of the compounds

Cl—Hg(Cl)(Cl)Hg—Cl + Cl₂Al(Cl)(Cl)AlCl₂ ⟶ 2 Cl—Hg(Cl)(Cl)AlCl₂

(b) Explain why the small ΔH in this reaction is evidence against

Cl₂Hg—HgCl₂

as the structure of $Hg_2Cl_4(g)$.

Reversible Processes in Ideal Gases

51. If 2.00 mol of an ideal gas at 25°C expands isothermally and reversibly from 9.00 to 36.00 L, calculate the work done on the gas and the heat absorbed by the gas in the process. What are the changes in energy (ΔU) and in enthalpy (ΔH) of the gas in the process?

52. If 54.0 g argon at 400 K is compressed isothermally and reversibly from a pressure of 1.50 to 4.00 atm, calculate the work done on the gas and the heat absorbed by gas in the process. What are the changes in energy (ΔU) and in enthalpy (ΔH) of the gas?

53. Suppose 2.00 mol of a monatomic ideal gas ($c_V = \frac{3}{2} R$) is expanded adiabatically and reversibly from a temperature $T = 300$ K, where the volume of the system is 20.0 L, to a volume of 60.0 L. Calculate the final temperature of the gas, the work done on the gas, and the energy and enthalpy changes.

54. Suppose 2.00 mol of an ideal gas is contained in a heat-insulated cylinder with a moveable frictionless piston. Initially, the gas is at 1.00 atm and 0°C. The gas is compressed reversibly to 2.00 atm. The molar heat capacity at constant pressure, c_P, equals 29.3 J K^{-1} mol^{-1}. Calculate the final temperature of the gas, the change in its internal energy, ΔU, and the work done on the gas.

ADDITIONAL PROBLEMS

55. At one time it was thought that the molar mass of indium was near 76 g mol^{-1}. By referring to the law of Dulong and Petit (see Problem 7), show how the measured specific heat of metallic indium, 0.233 J K^{-1} g^{-1}, makes this value unlikely.

56. The following table shows how the specific heat at constant pressure of liquid helium changes with temperature. Note the sharp increase over this temperature range:

Temperature (K):

1.80	1.85	1.90	1.95	2.00	2.05	2.10	2.15

c_s (J K^{-1} g^{-1}):

2.81	3.26	3.79	4.42	5.18	6.16	7.51	9.35

Estimate how much heat it takes at constant pressure to increase the temperature of 1.00 g He(ℓ) from 1.8 to 2.15 K. (*Hint:* For each temperature interval of 0.05 K, take the average, c_s, as the sum of the values at the ends of the interval divided by 2.)

57. Imagine that 2.00 mol argon, confined by a moveable, frictionless piston in a cylinder at a pressure of 1.00 atm and a temperature of 398 K, is cooled to 298 K. Argon gas may be considered ideal, and its molar heat capacity at constant pressure is $c_P = (5/2)R$, where $R = 8.315$ J K^{-1} mol^{-1}. Calculate:

(a) The work done on the system, w
(b) The heat absorbed by the system, q
(c) The energy change of the system, ΔU
(d) The enthalpy change of the system, ΔH

58. Suppose 1.00 mol ice at −30°C is heated at atmospheric pressure until it is converted to steam at 140°C. Calculate q, w, ΔH, and ΔU for this process. For ice, water, and steam, c_P is 38, 75, and 36 J K^{-1} mol^{-1}, respectively, and can be taken to be approximately independent of temperature. ΔH_{fus} for ice is 6.007 kJ mol^{-1}, and ΔH_{vap} for water is 40.66 kJ mol^{-1}. Use the ideal gas law for steam, and assume that the volume of 1 mol ice or water is negligible relative to that of 1 mol steam.

59. The gas inside a cylinder expands against a constant external pressure of 1.00 atm from a volume of 5.00 L to a volume of 13.00 L. In doing so, it turns a paddle immersed in 1.00 L water. Calculate the temperature increase of the water, assuming no loss of heat to the surroundings or frictional losses in the mechanism. Take the density of water to be 1.00 g cm^{-3} and its specific heat to be 4.18 J K^{-1} g^{-1}.

60. Suppose 1.000 mol argon (assumed to be an ideal gas) is confined in a strong, rigid container of volume 22.41 L at 273.15 K. The system is heated until 3.000 kJ (3000 J) of heat has been added. The molar heat capacity of the gas does not change during the heating and equals 12.47 J K^{-1} mol^{-1}.

(a) Calculate the original pressure inside the vessel (in atmospheres).
(b) Determine q for the system during the heating process.
(c) Determine w for the system during the heating process.
(d) Compute the temperature of the gas after the heating, in degrees Celsius. Assume the container has zero heat capacity.
(e) Compute the pressure (in atmospheres) inside the vessel after the heating.
(f) Compute ΔU of the gas during the heating process.
(g) Compute ΔH of the gas during the heating process.
(h) The correct answer to part (g) exceeds 3.000 kJ. The increase in enthalpy (which at one time was misleadingly called the "heat content") in this system exceeds the amount of heat actually added. Why is this not a violation of the law of conservation of energy?

61. When glucose, a sugar, reacts fully with oxygen, carbon dioxide and water are produced:

$$C_6H_{12}O_6(s) + 6\,O_2(g) \longrightarrow 6\,CO_2(g) + 6\,H_2O(\ell) \qquad \Delta H° = -2820 \text{ kJ}$$

Suppose a person weighing 50 kg (mostly water, with specific heat capacity 4.18 J K^{-1} g^{-1}) eats a candy bar containing 14.3 g glucose. If all the glucose reacted with oxygen and the heat produced were used entirely to increase the person's body temperature, what temperature increase would result? (In fact, most of the heat produced is lost to the surroundings before such a temperature increase occurs.)

62. In walking 1 km, you use about 100 kJ of energy. This energy comes from the oxidation of foods, which is about 30% efficient. How much energy do you save by walking 1 km instead of driving a car that gets 8.0 km L^{-1} gasoline (19 miles/gal)? The density of gasoline is 0.68 g cm^{-3} and its enthalpy of combustion is −48 kJ g^{-1}.

63. Liquid helium and liquid nitrogen are both used as coolants; He(ℓ) boils at 4.21 K, and $N_2(\ell)$ boils at 77.35 K. The specific heat of liquid helium near its boiling point is 4.25 J K^{-1} g^{-1}, and the specific heat of liquid nitrogen near *its* boiling point is 1.95 J K^{-1} g^{-1}. The enthalpy of vaporization of He(ℓ) is 25.1 J g^{-1}, and the enthalpy of vaporization of $N_2(\ell)$ is 200.3 J g^{-1} (these data are calculated from the values in Appendix F). Discuss which liquid is the better coolant (on a per-gram basis) *near* its boiling point and which is better *at* its boiling point.

64. When 1.00 g potassium chlorate ($KClO_3$) is dissolved in 50.0 g water in a Styrofoam calorimeter of negligible heat capacity, the temperature decreases from 25.00°C to 23.36°C. Calculate q for the water and $\Delta H°$ for the process.

$$KClO_3(s) \longrightarrow K^+(aq) + ClO_3^-(aq)$$

The specific heat of water is 4.184 J K^{-1} g^{-1}.

65. The enthalpy of combustion and the standard enthalpy of formation of a fuel can be determined by measuring the temperature change in a calorimeter when a weighed amount of the fuel is burned in oxygen.
  (a) Write a balanced chemical equation for the combustion of isooctane, $C_8H_{18}(\ell)$, to $CO_2(g)$ and $H_2O(\ell)$. Isooctane is a component of gasoline and is used as a reference standard in determining the "octane rating" of a fuel mixture.
  (b) Suppose 0.542 g isooctane is placed in a fixed-volume (bomb) calorimeter, which contains 750 g water, initially at 20.450°C, surrounding the reaction compartment. The heat capacity of the calorimeter itself (excluding the water) has been measured to be 48 J K^{-1} in a separate calibration. After the combustion of the isooctane is complete, the water temperature is measured to be 28.670°C. Taking the specific heat of water to be 4.184 J K^{-1} g^{-1}, calculate ΔE for the combustion of 0.542 g isooctane.
  (c) Calculate ΔU for the combustion of 1 mol isooctane.
  (d) Calculate ΔH for the combustion of 1 mol isooctane.
  (e) Calculate $\Delta H°_f$ for the isooctane.

66. The enthalpy change to form 1 mol $Hg_2Br_2(s)$ from the elements at 25°C is −206.77 kJ mol^{-1}, and that of HgBr(g) is 96.23 kJ mol^{-1}. Compute the enthalpy change for the decomposition of 1 mol $Hg_2Br_2(s)$ to 2 mol HgBr(g):

$$Hg_2Br_2(s) \longrightarrow 2\ HgBr(g)$$

* 67. The gas most commonly used in welding is acetylene, $C_2H_2(g)$. When acetylene is burned in oxygen, the reaction is

$$C_2H_2(g) + \tfrac{5}{2}\ O_2(g) \longrightarrow 2\ CO_2(g) + H_2O(g)$$

  (a) Using data from Appendix D, calculate $\Delta H°$ for this reaction.
  (b) Calculate the total heat capacity of 2.00 mol $CO_2(g)$ and 1.00 mol $H_2O(g)$, using $c_P(CO_2(g))$ = 37 J K^{-1} mol^{-1} and $c_P(H_2O(g))$ = 36 J K^{-1} mol^{-1}.
  (c) When this reaction is performed in an open flame, almost all the heat produced in part (a) goes to increase the temperature of the products. Calculate the maximum flame temperature that is attainable in an open flame burning acetylene in oxygen. The actual flame temperature would be lower than this because heat is lost to the surroundings.

* 68. The enthalpy of reaction changes somewhat with temperature. Suppose we wish to calculate ΔH for a reaction at a temperature T that is different from 298 K. To do this, we can replace the direct reaction at T with a three-step process. In the first step, the temperature of the reactants is changed from T to 298 K. ΔH for this step can be calculated from the molar heat capacities of the reactants, which are assumed to be independent of temperature. In the second step, the reaction is conducted at 298 K with an enthalpy change $\Delta H°$. In the third step, the temperature of the products is changed from 298 K to T. The sum of these three enthalpy changes is ΔH for the reaction at temperature T.

An important process contributing to air pollution is the following chemical reaction

$$SO_2(g) + \tfrac{1}{2}\ O_2(g) \longrightarrow SO_3(g)$$

For $SO_2(g)$, the heat capacity c_P is 39.9, for $O_2(g)$ it is 29.4, and for $SO_3(g)$ it is 50.7 J K^{-1} mol^{-1}. Calculate ΔH for the preceding reaction at 500 K, using the enthalpies of formation at 298.15 K from Appendix D.

69. At the top of the compression stroke in one of the cylinders of an automobile engine (that is, at the minimum gas volume), the volume of the gas-air mixture is 150 mL, the temperature is 600 K, and the pressure is 12.0 atm. The ratio of the number of moles of octane vapor to the number of moles of air in the combustion mixture is 1.00:80.0. What is the maximum temperature attained in the gas if octane burns explosively before the power stroke of the piston (gas expansion) begins? The gases may be considered to be ideal, and their heat capacities at constant pressure (assumed to be temperature-independent) are

$$c_P(C_8H_{18}(g)) = 327\ \mathrm{J\ K^{-1}\ mol^{-1}}$$
$$c_P(O_2(g)) = 35.2\ \mathrm{J\ K^{-1}\ mol^{-1}}$$
$$c_P(N_2(g)) = 29.8\ \mathrm{J\ K^{-1}\ mol^{-1}}$$
$$c_P(CO_2(g)) = 45.5\ \mathrm{J\ K^{-1}\ mol^{-1}}$$
$$c_P(H_2O(g)) = 38.9\ \mathrm{J\ K^{-1}\ mol^{-1}}$$

The enthalpy of formation of $C_8H_{18}(g)$ at 600 K is −57.4 kJ mol^{-1}.

70. Initially, 46.0 g oxygen is at a pressure of 1.00 atm and a temperature of 400 K. It expands adiabatically and reversibly until the pressure is reduced to 0.60 atm, and it is then compressed isothermally and reversibly until the volume returns to its original value. Calculate the final pressure and temperature of the oxygen, the work done and heat added to the oxygen in this process, and the energy change ΔU. Take $c_P(O_2)$ = 29.4 J K^{-1} mol^{-1}.

71. A young chemist buys a "one-lung" motorcycle but, before learning how to drive it, wants to understand the processes that occur in its engine. The manual says the cylinder has a radius of 5.00 cm, a piston stroke of 12.00 cm, and a (volume) compression ratio of 8:1. If a mixture of gasoline vapor (taken to be C_8H_{18}) and air in mole ratio 1:62.5 is drawn into the cylinder at 80°C and 1.00 atm, calculate:
  (a) The temperature of the compressed gases just before the spark plug ignites them. (Assume the gases are ideal, the compression is adiabatic, and the average heat capacity of the mixture of gasoline vapor and air is c_P = 35 J K^{-1} mol^{-1}.)
  (b) The volume of the compressed gases just before ignition.
  (c) The pressure of the compressed gases just before ignition.
  (d) The maximum temperature of the combustion products, assuming combustion is completed before the piston begins its downstroke. Take $\Delta H°_f(C_8H_{18})$ = −57.4 kJ mol^{-1}.
  (e) The temperature of the exhaust gases, assuming the expansion stroke to be adiabatic.

72. Nitromethane, CH_3NO_2, is a good fuel. It is a liquid at ordinary temperatures. When the liquid is burned, the reaction involved is chiefly

$$2\ CH_3NO_2(\ell) + \tfrac{3}{2}\ O_2(g) \longrightarrow 2\ CO_2(g) + N_2(g) + 3\ H_2O(g)$$

The standard enthalpy of formation of liquid nitromethane at 25°C is -112 kJ mol^{-1}; other relevant values can be found in Appendix D.
(a) Calculate the enthalpy change in the burning of 1 mol liquid nitromethane to form gaseous products at 25°C. State explicitly whether the reaction is endothermic or exothermic.
(b) Would more or less heat be evolved if *gaseous* nitromethane were burned under the same conditions? Indicate what additional information (if any) you would need to calculate the exact amount of heat, and show just how you would use this information.

73. Dry air containing a small amount of CO was passed through a tube containing a catalyst for the oxidation of CO to CO_2. Because of the heat evolved in this oxidation, the temperature of the air increased by 3.2 K. Calculate the weight percentage of CO in the sample of air. Assume that the specific heat at constant pressure for air is 1.01 J $K^{-1}\ g^{-1}$.

74. When 1 mol isobutane, a gas with formula C_4H_8, is burned at 25°C and 1 atm to form CO_2 and gaseous water, the enthalpy change is -2528 kJ.
(a) Calculate, with the aid of any information needed from Table D-4 in Appendix D, the standard enthalpy of formation of isobutane.
(b) Suppose that 0.50 mol isobutane is burned adiabatically at constant pressure in the presence of an excess of oxygen, with 5.0 mol oxygen left at the end of the reaction. The heat capacity of the reaction vessel is 700 J K^{-1}, and pertinent molar heat capacities (in joules per kelvin per mole) are $CO_2(g)$, 37; $H_2O(g)$, 34; $O_2(g)$, 29. What is the approximate final temperature of this system (including the reaction vessel)?

75. Find the maximum possible temperature that may be reached when 0.050 mol $Ca(OH)_2(s)$ is allowed to react with 1.0 L of a 1.0-M HCl solution, both initially at 25°C. Assume that the final volume of the solution is 1.0 L, and that the specific heat at constant pressure of the solution is constant and equal to that of water, 4.18 J $K^{-1}\ g^{-1}$.

CUMULATIVE PROBLEMS

76. Suppose 32.1 g $ClF_3(g)$ and 17.3 g $Li(s)$ are mixed and allowed to react at atmospheric pressure and 25°C until one of the reactants is used up, producing $LiCl(s)$ and $LiF(s)$. Calculate the amount of heat evolved.

77. Calculate ΔH_f° and ΔU_f° for the formation of silane, $SiH_4(g)$, from its elements at 298 K, if 250 cm^3 of the gaseous compound at $T = 298$ K and $P = 0.658$ atm is burned in a constant-volume gas calorimeter in an excess of oxygen and causes the evolution of 9.757 kJ of heat. The combustion reaction is

$$SiH_4(g) + 2\ O_2(g) \longrightarrow SiO_2(s, \text{quartz}) + 2\ H_2O(\ell)$$

and the formation of silane from its elements is

$$Si(s) + 2\ H_2(g) \longrightarrow SiH_4(g)$$

78. (a) Draw Lewis diagrams for O_2, CO_2, H_2O, CH_4, C_8H_{18}, and C_2H_5OH. In C_8H_{18}, the eight carbon atoms form a chain with single bonds; in C_2H_5OH, the two carbon atoms are bonded to one another. Using average bond enthalpies from Table 12.3, compute the enthalpy change in each of the following reactions, if 1 mol of each carbon compound is burned, and all reactants and products are in the gas phase.
(b) $CH_4 + 2\ O_2 \longrightarrow CO_2 + 2\ H_2O$ (burning methane, or natural gas)
(c) $C_8H_{18} + \frac{25}{2}\ O_2 \longrightarrow 8\ CO_2 + 9\ H_2O$ (burning octane, in gasoline)
(d) $C_2H_5OH + 3\ O_2 \longrightarrow 2\ CO_2 + 3\ H_2O$ (burning ethanol, in gasohol)

79. By considering the nature of the intermolecular forces in each case, rank the following substances from smallest to largest enthalpy of vaporization: KBr, Ar, NH_3, and He. Explain your reasoning.

80. A supersonic nozzle is a cone-shaped object with a small hole in the end through which a gas is forced. As it moves through the nozzle opening, the gas expands in a manner that can be approximated as reversible and adiabatic. Such nozzles are used in molecular beams (see Section 17.4) and in supersonic aircraft engines to provide thrust, because as the gas cools, its random thermal energy is converted into directed motion of the molecules with average velocity v. Little thermodynamic work is done because the external pressure is low; thus, the net effect is to convert thermal energy to net translational motion of the gas molecules. Suppose the gas in the nozzle is helium; its pressure is 50 atm and its temperature is 400 K before it begins its expansion.
(a) What are the average speed and the average velocity of the molecules *before* the expansion?
(b) What will be the temperature of the gas after its pressure has decreased to 1.0 atm in the expansion?
(c) What is the average velocity of the molecules at this point in the expansion?

81. (a) Draw a Lewis diagram for carbonic acid, H_2CO_3, with a central carbon atom bonded to the three oxygen atoms.
(b) Carbonic acid is unstable in aqueous solution and converts to dissolved carbon dioxide. Use bond enthalpies to estimate the enthalpy change for the following reaction:

$$H_2CO_3 \longrightarrow H_2O + CO_2$$

CHAPTER 13

Spontaneous Processes and Thermodynamic Equilibrium

The reaction between solid sodium and gaseous chlorine proceeds imperceptibly, if at all, until the addition of a drop of water sets it off.

In both fundamental research and practical applications of chemistry, chemical reactions are performed by mixing the reactants and regulating external conditions such as temperature and pressure. Two questions arise immediately:

1. Is it possible for the reaction to occur at the selected conditions?
2. If the reaction is possible, what determines the ratio of products and reactants at equilibrium?

Predicting the equilibrium composition for a chemical reaction is the central goal of Unit 4, and in this chapter, we develop the conceptual basis for answering these two key questions.

We can find out whether a proposed reaction is possible by determining whether it is a *spontaneous* thermodynamic process. In this context, "spontaneous" has a precise technical meaning (see later for clarification) that should not be confused with its conversational meaning, such as describing the spontaneous behavior of people in social situations. Thermodynamics can tell us whether a proposed reaction is possible under particular conditions even before we attempt the reaction. If the reaction is spontaneous, thermodynamics can also predict the ratio of products and reactants at equilibrium. But, we cannot use thermodynamics to predict the rate of a spontaneous reaction or how long it will take to reach equilibrium. These questions are the subject of chemical kinetics. To obtain a large amount of product from a spontaneous reaction in a short time, we need a reaction that is spontaneous *and* fast.

This chapter develops the thermodynamic methods for predicting whether a reaction is spontaneous, and Chapter 14 uses these results to determine the equilibrium ratio of products and reactants. Chapter 18 discusses the rates of chemical reactions. Manipulating conditions to optimize the yield of chemical reactions in practical applications requires the concepts from all three chapters.

The criteria for predicting spontaneity of physical and chemical processes are provided by the second law of thermodynamics, which is a brilliant abstraction and generalization from the observed facts of directionality in processes involving heat transfer. This is accomplished by introducing a new state function called *entropy*, which is denoted by S. The entropy function is defined so that the algebraic sign of its change, when we consider the total value in the system and the surroundings, is positive in the direction of a spontaneous process. Sections 13.1 through 13.5 develop the second law and demonstrate methods for calculating ΔS_{tot} and predicting spontaneity for simple physical processes in which chemical reactions are not involved.

When processes are conducted at constant T and P, the criteria for spontaneity and for equilibrium are stated more conveniently in terms of another state function called the *Gibbs free energy* (denoted by G), which is derived from S. Because chemical reactions are usually conducted at constant T and constant P, their thermodynamic description is based on ΔG rather than ΔS. This chapter concludes by restating the criteria for spontaneity of chemical reactions in terms of ΔG. Chapter 14 shows how to identify the equilibrium state of a reaction, and calculate the equilibrium constant from ΔG.

For all these reasons, ΔG is the most important thermodynamic concept in the entire field of chemical equilibrium. Your goals in this chapter should be to understand the meaning of the state function G, to become skilled in calculating its changes, ΔG, and to interpret both the magnitude and the algebraic sign of these changes.

13.1 The Nature of Spontaneous Processes

In preparation for setting up the second law of thermodynamics and stating precisely the criteria for spontaneity, we will examine several familiar examples of spontaneous processes and describe their features in general terms. A spontaneous change is one that *can* occur by itself without outside intervention, once conditions have been established for its initiation. The change may be fast or slow, and we may have to wait a significant period to determine whether it *does* occur.

One of the most striking features of spontaneous change is that it follows a specific direction when starting from a particular initial condition. A quite

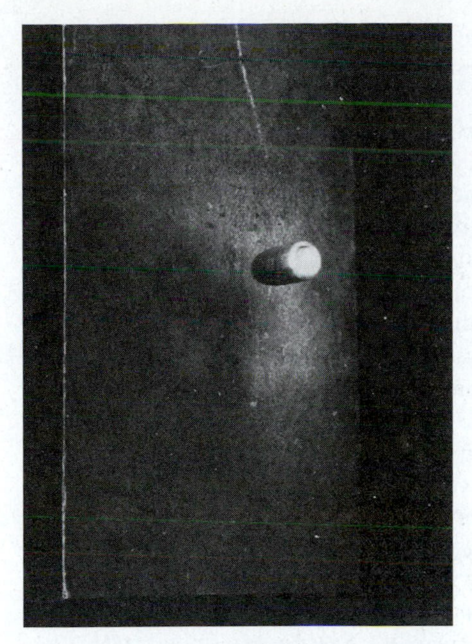

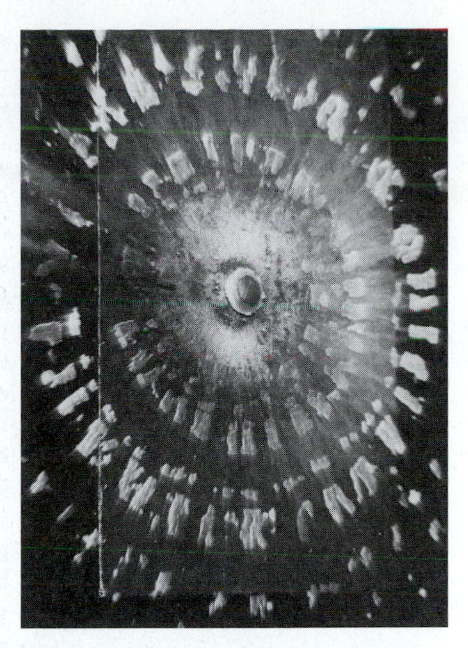

FIGURE 13.1 A bullet hitting a steel plate at a speed of 1600 ft/sec melts as its kinetic energy is converted to heat, and metal droplets spray in all directions. These three photographs make sense only in the order shown; the reverse process is unmistakably implausible.

(©Dr. Harold Edgerton/Palm Press, Inc.)

dramatic example is shown in the series of photographs in Figure 13.1, which show a speeding bullet that is fragmented on impact with a steel plate. We never observe a pile of metal fragments spontaneously assemble themselves into a speeding bullet. One of the goals of thermodynamics is to account for this *directionality* of spontaneous change. The first law of thermodynamics provides no guidance on this point. Energy is conserved both in a forward process and in its reverse; nothing in the first law indicates a preference for one direction of change over the other.

Spontaneous processes familiar in chemical laboratories also follow specific directions:

1. We measure *heat flowing* from a hot body to a cold one when they are brought into thermal contact, but we never detect heat flowing spontaneously in the opposite direction.
2. We observe a gas *expanding* into a region of lower pressure, but we never see the reverse process, a gas compressing itself spontaneously into a small part of its container.
3. We place a drop of red ink into a beaker of water and watch the color spread by *diffusion* of the ink particles until the water is uniformly pink, but we never see the ink spontaneously reorganize as a small red drop in a volume of otherwise colorless water.
4. We place 10 g sucrose (ordinary table sugar) in a beaker and add 100 mL water at 80°C. The sucrose *dissolves* to form a uniform solution. We never observe the spontaneous reappearance of a mound of sucrose at the bottom of a beaker of water.[1]
5. We open a container of acetone on the laboratory bench. We detect the aroma of acetone because some of the molecules have *evaporated* from the liquid and then diffused through the atmosphere to our position. We never observe the molecules to retreat spontaneously into the container.

[1]By intentional evaporation of the water, we can recover the sucrose. This constitutes a second process, which is beyond the spontaneous process of dissolution that is our immediate concern.

All the previous examples are physical processes that are both spontaneous and rapid. We initiate the process and see the result quickly thereafter, with no further intervention.

Like the simple physical processes described earlier, spontaneous chemical reactions are inherently directional. Ordinary experience shows that a mixture of hydrogen and oxygen gases exists indefinitely at room temperature. Yet, if the mixture is exposed to a small amount of powdered platinum metal, or an electric spark, the gases react explosively to produce water. The reverse reaction is not spontaneous; we never observe the spontaneous decomposition of water into gaseous hydrogen and oxygen. Figure 1.2 shows decomposition of water into hydrogen and oxygen gases by electrolysis in an electrochemical cell, where electrical energy is continually provided by the external circuit to drive this nonspontaneous reaction. Another example is shown in the photograph at the beginning of this chapter, where the reaction between sodium metal and chlorine gas occurs explosively after a drop of water is added. We never observe the spontaneous decomposition of sodium chloride into sodium metal and gaseous chlorine. A third example, less dramatic than the previous two, is the spontaneous reaction of copper metal with oxygen at room temperature. This is seen in many older municipal buildings, especially in Europe, where copper metal was used as roofing material. With time, these roofs develop the blue–green patina of copper oxides. We never observe the spontaneous reappearance of shiny metallic copper on these old roofs.

These three examples illustrate that the actual outcome of a spontaneous chemical reaction depends on the reaction rate. The possibility of reaction between hydrogen and oxygen was there all along, but the rate was too slow to be observed until the powdered metal or the electrical spark accelerated the reaction. The possibility of reaction between metallic copper and oxygen was there all along, and the rate was large enough to be observed, if not dramatically fast. Thermodynamics determines whether a reaction is possible, whereas chemical kinetics determines whether it is practical. At the end of this chapter, you will be able to predict whether a chemical reaction is spontaneous, and by the end of the next chapter, you will be able to predict its equilibrium state. But, you must wait until Chapter 18 to see whether a spontaneous reaction can be carried out at a useful rate.

The direction of each of these spontaneous processes is readily apparent by observing the initial and final states, regardless of their paths. This suggests the existence of a new *state function* that indicates the directionality of spontaneous processes. That state function will turn out to be **entropy,** and it will be defined so that the sign of its change indicates the direction in which a proposed process will be spontaneous. Entropy has the interesting property that we cannot predict spontaneity by considering the system alone; we must also consider the entropy changes in the surroundings during the process.

To develop the entropy function, we must first describe spontaneous processes in the language of thermodynamics summarized in Section 12.1. You should revisit each of the examples just discussed to see how it specifically fits this language. Initially, two objects in different thermodynamic states are brought into contact; one will be called "system" and the other "surroundings." Barriers (constraints) in place between them prevent their interaction. When the constraints are removed, a spontaneous process may occur in which the system and surroundings exchange energy and matter and in which the volume of both system and surroundings may change. Because the two together constitute a *thermodynamic universe,* the total amount of energy, volume, and matter shared between them is fixed. During the process, these quantities are redistributed between the system and surroundings.

Spontaneous processes are particular examples of irreversible processes defined in Section 12.1. In stark contrast with reversible processes, they do not proceed through a sequence of equilibrium states, and their direction cannot be reversed by an infinitesimal change in the direction of some externally applied force. Spontaneous processes cannot be represented as paths on the equation-of-state surface in Figure 12.1, but their initial and final equilibrium states can be represented as points on that surface.

What determines whether a process under consideration will be spontaneous? Where does a spontaneous process end? How are energy, volume, and matter partitioned between the system and surroundings at equilibrium? What is the nature of the final equilibrium state? These questions cannot be answered by the first law. Their answers require the second law and properties of the entropy, and a few developments are necessary before we can address these questions. We define entropy by molecular motions in Section 13.2 and by macroscopic process variables in Section 13.3. Finally, we present the methods for calculating entropy changes and for predicting spontaneity in Section 13.5.

13.2 Entropy and Spontaneity: A Molecular Statistical Interpretation

What is entropy and why should it be related to the spontaneity of processes in nature? These are deep questions that we can only begin to answer here. To do so, we step outside the confines of classical thermodynamics, which is concerned only with macroscopic properties, and examine the microscopic molecular basis for the second law. Such an approach, called **statistical thermodynamics,** shows that spontaneous change in nature can be understood by using probability theory to predict and explain the behavior of the many atoms and molecules that comprise a macroscopic sample of matter. Statistical thermodynamics also provides theoretical procedures for calculating the entropy of a system from molecular properties.

Spontaneity and Molecular Motions

Consider a particularly simple spontaneous process: the free *adiabatic* expansion of 1 mol of an ideal gas into a vacuum (Fig. 13.2). The gas is initially held in the left bulb in volume $V/2$, whereas the right bulb is evacuated. Before it is opened, the stopcock is a constraint that holds all the molecules in the left bulb. After the stopcock is opened, the gas expands to fill the entire volume, V.

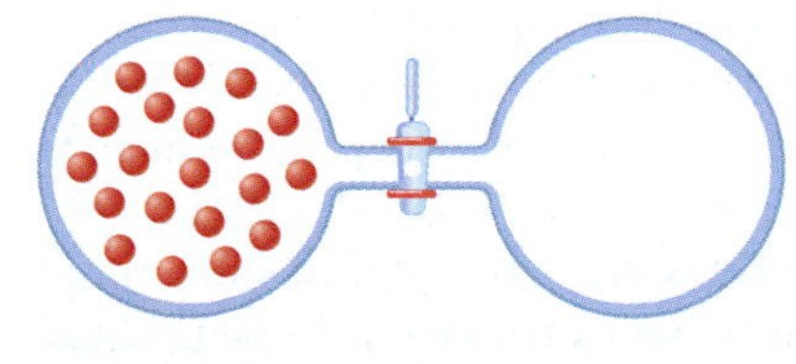

(a)

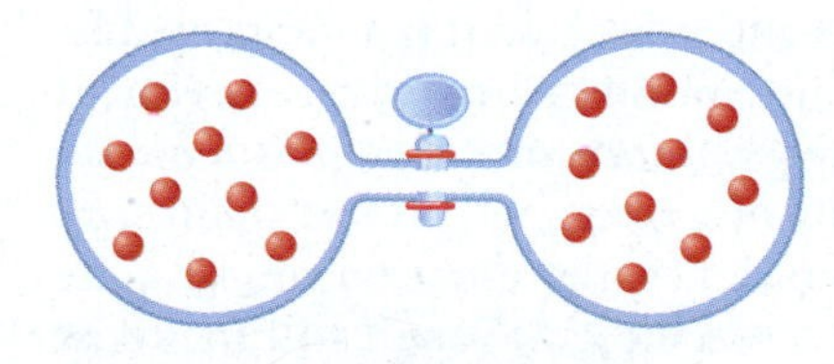

(b)

FIGURE 13.2 The free expansion of a gas into a vacuum. (a) The stopcock is closed with all gas in the left bulb. (b) With the stopcock open, half the gas is in each bulb.

Now examine the same free expansion from a microscopic point of view. Imagine that the path of one particular tagged molecule can be followed during the expansion and for some period of time after final equilibrium has been established, perhaps through a series of time-lapse snapshots showing the locations of all molecules in the gas. From its starting position on the left side at the beginning of the experiment, the tagged molecule will cross to the right, then back to the left, and so forth. If enough time elapses, the molecule will eventually spend equal amounts of time on the two sides, because there is no physical reason for it to prefer one side to the other. Recall from Section 9.5 that the molecules in an ideal gas do not collide with each other, and their kinetic energy does not change in collisions with the walls. The energy of the molecules remains the same at all locations, and the two sides of the container are identical. No experimental way exists to track the progress of one specific molecule in the gas; therefore, the best we can ask for is the *probability* that the molecule is on the left side at any given instant. (See Appendix C.6 for a review of probability concepts and methods.) The physical reasoning just summarized justifies that the probability that the molecule is on the left side will be $\frac{1}{2}$, the same as the probability that it is on the right side. Probability provides the key to understanding the direction of spontaneous change because it enables us to compare the likelihood of finding all the molecules on the left

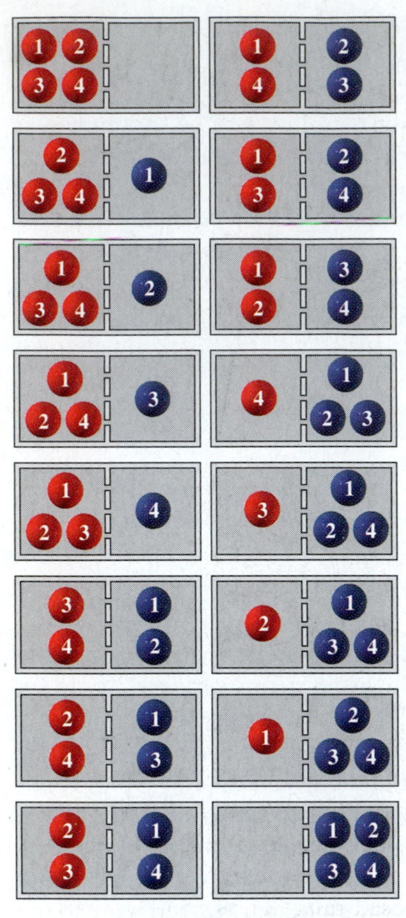

FIGURE 13.3 The 16 possible microstates of a system of 4 molecules that may occupy either side of a container. In only one of these are all four molecules on the left side.

side—after the constraint has been removed—with the likelihood of finding the molecules uniformly distributed over the combined volume of both sides.

Just how unlikely is a spontaneous compression of 1 mol gas from the combined volume back to the left side? The probability that one particular molecule is on the left side at a given time is $\frac{1}{2}$. A second specific molecule may be on either the left or the right, so the probability that both are on the left is $\frac{1}{2} \times \frac{1}{2} = \frac{1}{4}$. As shown in Figure 13.3, in a gas containing a total of four molecules, all four molecules will be on the left only $\frac{1}{2} \times \frac{1}{2} \times \frac{1}{2} \times \frac{1}{2} = \frac{1}{16}$ of the time. Continuing this argument for all $N_A = 6.0 \times 10^{23}$ molecules leads to the probability that all N_A molecules are on the left:

$$\frac{1}{2} \times \frac{1}{2} \times \cdots \times \frac{1}{2} = \left(\frac{1}{2}\right)^{6.0 \times 10^{23}}$$

To evaluate this number, it is helpful to rewrite it in scientific notation as 1 divided by 10 raised to the power a (or, equivalently, as 10^{-a}):

$$\frac{1}{10^a} = \left(\frac{1}{2}\right)^{6.0 \times 10^{23}} = \frac{1}{2^{6.0 \times 10^{23}}}$$

A calculator can handle numbers this large (or this small) only if logarithms are used. Taking the base-10 logarithms of both denominators gives

$$a = \log 2^{(6.0 \times 10^{23})} = 6.0 \times 10^{23} \log 2$$
$$= (6.0 \times 10^{23})(0.30)$$
$$= 1.8 \times 10^{23}$$

The probability that all the molecules will be on the left is 1 in $10^{1.8 \times 10^{23}}$.

This is a vanishingly small probability, because $10^{1.8 \times 10^{23}}$ is an unimaginably large number. It is vastly larger than the number 1.8×10^{23} (which is a large number already). To realize this, consider how such numbers would be written. The number 1.8×10^{23} is fairly straightforward to write out. It contains 22 zeros:

180,000,000,000,000,000,000,000

The number $10^{1.8 \times 10^{23}}$ is 1 followed by 1.8×10^{23} zeros. Written out, such a number would more than fill all the books in the world. Put in other terms, 1.8×10^{23} corresponds to the number of molecules in about 5 cm^3 water, but $10^{1.8 \times 10^{23}}$ is far larger than the number of molecules in the entire universe!

The statistical molecular picture explains that the gas expands to fill the whole available volume when the constraint is removed because this more uniform configuration of the molecules is overwhelmingly more probable than the initial configuration with all the molecules on the left side. The same explanation shows that the gas is never observed to compress spontaneously into a smaller volume because this nonuniform configuration of the molecules is overwhelmingly improbable in the absence of the constraint. Nothing in the laws of mechanics prevents a gas from compressing spontaneously. But this event is never seen because it is so improbable.

Thus, spontaneity in nature results from the random, statistical behavior of large numbers of molecules. *The directionality of spontaneous change is a consequence of the large numbers of molecules in the macroscopic systems treated by thermodynamics.* In systems containing fewer molecules, the situation can be quite different because the uniform configuration of molecules, although still the most probable, is no longer so overwhelmingly the most probable configuration. For example, if there were only 6 molecules, instead of 6×10^{23}, it would not be surprising to find them all on the left side at a given time. In fact, the probability that 6 molecules are on the left side is $\left(\frac{1}{2}\right)^6 = \frac{1}{64}$, so there is 1 chance in 64 that this will occur. Such small systems exhibit *statistical fluctuations* of the molecular configuration. Some of the fluctuations correspond to the initial nonuniform configuration maintained by the constraint, despite the absence of the constraint.

EXAMPLE 13.1

Calculate the probability of a spontaneous compression of 1.00 mol gas by 0.01%—that is, the probability that all the molecules will be found in a volume $V' = 0.9999V$ at a certain time.

SOLUTION

In this case, the probability that a given molecule is in V' is not $\frac{1}{2}$ but 0.9999 (the probability that it is in the remainder of V is 0.0001). The probability that all N_A molecules are in V' is

$$(0.9999)^{N_A} = 1/10^a$$

$$a = -N_A \log(0.9999) = (6.0 \times 10^{23})(4.3 \times 10^{-5})$$

$$= 2.6 \times 10^{19}$$

The chance of such a compression occurring spontaneously is 1 in $10^{2.6 \times 10^{19}}$, which is still vanishingly small. Thus, a spontaneous compression of even a fraction of a percent will not be seen.

Related Problems: 7, 8

Entropy and Molecular Motions

We want to define a new function called entropy that will increase when the system undergoes a spontaneous process. How can we relate entropy to the properties of the system and formulate a definition to achieve this goal? Spontaneous processes occur when constraints are removed from a system; the molecules respond by moving to explore the suddenly increased range of motions now available to them. In the free expansion of a gas, the molecules are initially confined in one part of the container. After the constraint is removed (the stopcock is opened), they are free to stay where they were, but they also are free to move throughout the larger combined volume of the two regions. Qualitatively, the numerical value of the entropy of a macroscopic system held in a particular thermodynamic state should depend on the *range of possible motions* (that is, the range of possible positions and momenta) available to the molecules while the system is held in that particular thermodynamic state. Any change in the macroscopic properties that enables the molecules to move out into larger regions of space or that increases the range of molecular speeds should increase the entropy of the system.

In preparation for defining entropy, we need a precise way to describe the "range of possible motions" of the molecules. This is accomplished by counting the number of microscopic, mechanical states, or **microstates**, available to molecules of the system. This number, denoted by Ω, counts all the possible combinations of positions and momenta available to the N molecules in the system when the system has internal energy U and volume V. For the simple model in Figure 13.3, $\Omega = 16$. In general, Ω is a function of U, V, and N, denoted as $\Omega(U, V, N)$. If the system is a monatomic ideal gas, the atoms do not interact. The position of each atom ranges freely throughout the entire volume, V. The internal energy consists of the total kinetic energy of the atoms, given as

$$U = \sum_{i=1}^{N} \varepsilon_i = \sum_{i=1}^{N} \frac{[p_{xi}^2 + p_{yi}^2 + p_{zi}^2]}{2m}$$

thus, the momenta of the atoms range through all values that satisfy the condition

$$2mU = \sum_{i=1}^{N} [p_{xi}^2 + p_{yi}^2 + p_{zi}^2]$$

FIGURE 13.4 The fundamental relation between entropy (S) and the number of microstates (W) was derived by Ludwig Boltzmann in 1868. On his tombstone in Vienna is carved the equation he obtained, $S = k \log W$. We would write "ln" instead of "log" for the natural logarithm.

For a monatomic ideal gas, the value of Ω is given by

$$\Omega(U, V, N) = g\, V^N U^{(3N/2)}$$

where g is a collection of constants. Now imagine the walls defining the system are manipulated to change the values of U and V. The range of positions and momenta available to the molecules will increase or decrease accordingly. Because N is a very large number—of order 10^{23}—the value of Ω will increase or decrease dramatically when the volume of the system is increased or decreased in an expansion or compression, respectively. It will increase or decrease dramatically when the internal energy of the system is increased or decreased by heating or cooling, respectively.

The equation connecting entropy S and the number of available microstates Ω is

$$S = k_B \ln \Omega \qquad [13.1]$$

which was originally discovered by the Austrian physicist Ludwig Boltzmann in the late 19th century (Fig. 13.4). **Boltzmann's constant** k_B is identified as R/N_A, the ratio of the universal gas constant R to Avogadro's number N_A. Thus, entropy has the physical dimensions J K^{-1}. It is impossible to overstate the importance of Boltzmann's relation, because it provides the link between the microscopic world of atoms and molecules and the macroscopic world of bulk matter and thermodynamics. Although this equation holds quite generally, it is difficult to apply because calculating Ω is a daunting theoretical task except for the simplest ideal systems. Other equations are used for practical applications in statistical thermodynamics. For our purposes here, the equation provides qualitative insight into the physical meaning of entropy and qualitative interpretation of the magnitude and sign of entropy changes caused by specific thermodynamic processes. The following example illustrates these insights in a simple case in which only the volume of the system changes in the process.

EXAMPLE 13.2

Consider the free expansion of 1 mol gas from $V/2$ to V (see Fig. 13.2). Use Boltzmann's relation to estimate the change in entropy for this process.

SOLUTION

Consider the entire apparatus consisting of the filled and the evacuated bulbs to be a thermodynamic universe, so any exchange of thermal energy occurs solely between them; the pair of bulbs is taken to be thermally insulated from their surroundings. Then examine the effects of doubling V on Ω. If the volume is doubled, the number of positions available to a given molecule is doubled also. Therefore, the number of states available to the molecule should be proportional to the volume, V:

$$\text{number of states available per molecule} = cV$$

where c is a proportionality constant. The state of a *two*-molecule system is given jointly by the states of the molecules in it, so the number of microscopic states available is just the product of the number of states for each molecule, $(cV) \times (cV) = (cV)^2$. For an N-molecule system,

$$\text{microscopic states available} = \Omega = (cV)^N$$

Inserting this expression into Boltzmann's relation gives the entropy change for the free expansion of 1 mol (N_0 atoms) gas from a volume $V/2$ to V:

$$\Delta S(\text{microscopic}) = N_A k_B \ln(cV) - N_A k_B \ln(cV/2)$$

$$= N_A k_B \ln\left(\frac{cV}{cV/2}\right) = N_A k_B \ln 2$$

Note that the constant c has dropped out. The calculated change in entropy on expansion of the gas is clearly positive, which is consistent with the increase in the number of available microstates on expansion.

In the next section, the entropy change for this process from the macroscopic view will be calculated to be

$$\Delta S \text{ (thermodynamic)} = R \ln 2$$

illustrating explicitly that Boltzmann's microscopic description accurately explains the measured macroscopic results.

Related Problems: 3, 4, 5, 6, 7, 8, 9, 10

This example illustrates why Boltzmann's relation must involve the logarithmic function. Because entropy is an extensive variable, its value is proportional to N. But Ω depends on N through the *power* to which cV is raised. Therefore, doubling N doubles S but leads to Ω being squared. The only mathematical function that can connect two such quantities is the logarithm.

Examples of other processes in which the entropy of the system increases include phase transitions such as the melting of a solid. In a solid, the atoms or molecules are constrained to stay near their equilibrium positions in the crystal lattice, whereas in the liquid they can move far away from these fixed positions. More microstates are available to the molecules in the liquid, corresponding to an increase in entropy; for melting of a solid, ΔS_{sys} is positive. In the same way, the entropy increases for evaporation of a liquid because the number of microstates available to the molecules increases enormously. Whereas the molecules in a liquid remain at the bottom of their container, those in a gas move throughout the container, so the number of microstates increases upon vaporization. In some solid-solid phase transformations it is difficult to predict qualitatively which phase has higher entropy. Nonetheless, calculating the relative number of microstates available correctly predicts the sign of the entropy change in a transformation from one such phase to the other.

So far, we have considered microstates that involve only the positions of the molecules in a system. The distribution of energies can also contribute to the number of microstates. For example, a gas in which all the molecules have the same speed has fewer microstates available to it than does a gas with a distribution of molecular speeds; a spontaneous process is observed experimentally in which a system initially constrained to a single molecular speed moves toward the Maxwell–Boltzmann speed distribution, because this distribution maximizes the number of microstates available for a given total energy. Similarly, changes in bonding structures by forming molecules through chemical reactions can change the number of microstates. A mixture of hydrogen and oxygen molecules can increase its available microstates (and thus its entropy) if chemical bonds break and re-form to make water molecules. This reaction may be slow (or it may need a spark to set if off), but once it starts, the entropy will rise with the number of accessible microstates. This textbook gives only qualitative microscopic interpretations of these more complex examples. In later courses you will see more thorough interpretations based on the more advanced equations that replace Equation 13.1.

13.3 Entropy and Heat: Experimental Basis of the Second Law of Thermodynamics

This section defines the entropy function in terms of measurable macroscopic quantities to provide a basis for calculating changes in entropy for specific processes. The definition is part of the second law of thermodynamics, which is stated as an abstraction and generalization of engineering observations on the efficiency of heat engines. We start the discussion by presenting a nonmathematical qualitative summary of the arguments on efficiency. Then we define entropy and state the second law. Section 13.5 applies the definition to calculate entropy changes and to predict spontaneity of processes.

Section 13.4 presents a mathematical version of the same arguments set forth here. Either Section 13.3 or Section 13.4 provides adequate background for the calculations in Section 13.5, which are the heart of the second law.

Background of the Second Law of Thermodynamics

The second law of thermodynamics originated in practical concerns over the efficiency of steam engines at the dawn of the Industrial Age, late in the 18th century, and required about a century for its complete development as an engineering tool. The central issues in that development are summarized as follows. In each stroke of a steam engine, a quantity of hot steam at high pressure is injected into a piston-cylinder assembly, where it immediately expands and pushes the piston outward against an external load, doing useful work by moving the load. The external mechanical arrangement to which the piston is attached includes a reciprocating mechanism that returns the piston to its original position at the end of the stroke so that a new quantity of steam can be injected to start the next stroke. The engine operates as a cyclic process, returning to the same state at the beginning of each stroke. In each stroke, the expansion process is highly irreversible because the steam is cooled and exhausted from the cylinder at the end of the stroke. In essence, the engine extracts thermal energy from a hot reservoir, uses some of this energy to accomplish useful work, then discards the remainder into a cooler reservoir (the environment). The energy lost to the environment cannot be recovered by the engine. The internal combustion engine operates in a similar manner, with injection of hot steam replaced by in situ ignition of combustible fuel that burns in highly exothermic reactions. Its expansion process is similarly irreversible, and some of the energy expended by the hot gases is unrecoverable.

In both engines the efficiency (that is, the ratio of work accomplished *by* the engine in a cycle to the heat invested to drive that cycle) can be improved by reducing the unrecoverable losses to the environment in each cycle. Seeking to maximize efficiency, Sadi Carnot, an officer in Napoleon's French Army Corps of Engineers, modeled operation of the engine with an idealized cyclic process now known as the *Carnot cycle.* He concluded that unrecoverable losses of energy to the environment cannot be completely eliminated, no matter how carefully the engine is designed. Even if the engine is operated as a reversible process (in which case, displacement of the external load is too slow to be of practical interest), its efficiency cannot exceed a fundamental limit known as the *thermodynamic efficiency.* Thus, an engine with 100% efficiency cannot be constructed.

Carnot's conclusion has been restated in more general terms by the German physicist Rudolf Clausius in the following form:

> *There is no device that can transfer heat from a colder to a warmer reservoir without net expenditure of work.*

and by the English physicist Lord Kelvin in the following form:

> *There is no device that can transform heat withdrawn from a reservoir completely into work with no other effect.*

These statements are consistent with ordinary experience that (1) heat always flows spontaneously from a hotter body to a colder body, and that (2) work is always required to refrigerate a body. With confidence based on experience, Clausius, Kelvin, and later scientists and engineers have assumed these statements to be valid for *all* heat transfer processes and labeled them as equivalent formulations of the second law of thermodynamics.

Definition of Entropy

How do we apply these general statements to chemical processes such as the examples described in Section 13.1, which at first glance bear no resemblance to heat engines? Following the same reasoning that led Clausius and Kelvin to these statements, we define entropy and obtain an equation for the entropy change during a process. Highlights of the argument are summarized here, and a more detailed development is presented in the following section.

Carnot's analysis of efficiency for a heat engine operating *reversibly* showed that in each cycle q/T at the high temperature reservoir and q/T at the low temperature environment summed to zero:

$$\frac{q_h}{T_h} + \frac{q_l}{T_l} = 0$$

This result suggests that q/T is a state function because its value does not change in a cyclic process. Clausius extended this result to show that the quantity $\int(1/T)\,dq_{rev}$ is independent of path in *any* reversible process and is, therefore, a state function. Clausius then *defined* the entropy change $\Delta S = S_f - S_i$ of a system in a process starting in state i and ending in state f by the equation

$$\Delta S = S_f - S_i = \int_i^f \frac{dq_{rev}}{T} \qquad [13.2]$$

Entropy is therefore a state function and has physical dimensions J K^{-1}. We calculate $\Delta S = S_f - S_i$ for a specific process by (1) identifying its initial and final equilibrium states i and f as points on the equation-of-state surface in Figure 12.1, (2) selecting *any convenient reversible path* between them along which dq_{rev} and T are known, and (3) evaluating this integral along the selected path. It does not matter that the actual process of interest may be irreversible. Because entropy is a state function, its change depends only on the initial and final states, not at all on the path. Therefore, we are free to choose any reversible path that connects the initial and final states, purely on grounds of convenience, for calculating ΔS.

This is a beautiful consequence of the fact that S is a state function. Section 13.5 provides detailed procedures for calculating ΔS for numerous types of systems and processes. The calculated values of ΔS are then used to predict whether a particular contemplated process will be spontaneous.

The preceding discussion emphasizes that the second law, like all other laws of science, is a bold extrapolation of the results of a great deal of direct experimental observation under controlled conditions. The second law is not proved to be true by a single definitive experiment, it is not derived as a consequence of some more general theory, and it is not handed down by some higher authority. Rather, it is invoked as one of the "starting points" of thermodynamics, and conclusions drawn from its application to a great variety of irreversible processes (not necessarily involving heat engines) are compared with the results obtained in experimental studies. To date, no disagreements have been found between predictions properly made from the second law and the results of properly designed experiments.

A DEEPER LOOK

13.4 Carnot Cycles, Efficiency, and Entropy

This section provides a mathematical development of the relation between entropy and heat already presented qualitatively in Section 13.3. No additional results are obtained, but considerably greater insight is provided.

The Carnot Cycle

In a Carnot cycle (Fig. 13.5), a system traverses two isothermal and two adiabatic paths to return to its original state. Each path is carried out reversibly (that is, in thermal equilibrium, with internal and external forces nearly balanced at every step). As the system proceeds from state A to C through state B, the system performs work (Fig. 13.5a):

$$w_{ABC} = -\int_{V_A}^{V_C} P\,dV$$

(Path ABC)

and it is clearly negative for this (expansion) process. If the system then returns from state C to A through state D,

$$w_{CDA} = -\int_{V_C}^{V_A} P\,dV$$

(Path CDA)

which is now a positive quantity for this (compression) process; work is performed on the system (see Fig. 13.5b). In the course of the cycle, the work performed by the system is the area under curve ABC, whereas that performed on the system is the (smaller) area under curve ADC. The net result of the whole cycle is that work is performed by the system, and the amount of this work is the difference between the two areas, which is the area enclosed by the cycle (see Fig. 13.5c). As in any cyclic process, the overall energy change ΔU is zero, thus the net work equals the negative of the total heat added to the system.

We have not yet specified the material contained in the system. Assume initially that it is an ideal gas, for which the results from Section 12.6 apply directly, with T_h defined to be the higher temperature in the cycle and T_l the lower temperature.

Path AB: Isothermal Expansion (temperature T_h)

$$w_{AB} = -q_{AB} = -nRT_h \ln\left(\frac{V_B}{V_A}\right)$$

Path BC: Adiabatic Expansion

$$q_{BC} = 0$$

$$w_{BC} = nc_V(T_l - T_h) = -nc_V - T_l)$$

Path CD: Isothermal Compression (temperature T_l)

$$w_{CD} = -q_{CD} = -nRT_l \ln\left(\frac{V_D}{V_C}\right)$$

$$= nRT_l \ln\left(\frac{V_C}{V_D}\right)$$

Path DA: Adiabatic Compression

$$q_{DA} = 0$$

$$w_{DA} = nc_V(T_h - T_l)$$

The net work done on the system is

$$w_{net} = w_{AB} + w_{BC} + w_{CD} + w_{DA}$$

$$= -nRT_h \ln\left(\frac{V_B}{V_A}\right) - nc_V(T_h - T_l)$$

$$+ nRT_l \ln\left(\frac{V_C}{V_D}\right) + nc_V(T_h - T_l)$$

$$= -nRT_h \ln\left(\frac{V_B}{V_A}\right) + nRT_l \ln\left(\frac{V_C}{V_D}\right)$$

This can be simplified by noting that V_B and V_C lie on one adiabatic path, and V_A and V_D lie on another. In Section 12.6, the relation for a reversible adiabatic process was found:

$$\frac{T_2}{T_1} = \left(\frac{V_1}{V_2}\right)^{\gamma-1}$$

Hence,

$$\frac{T_h}{T_l} = \left(\frac{V_C}{V_B}\right)^{\gamma-1} \quad \text{for path BC}$$

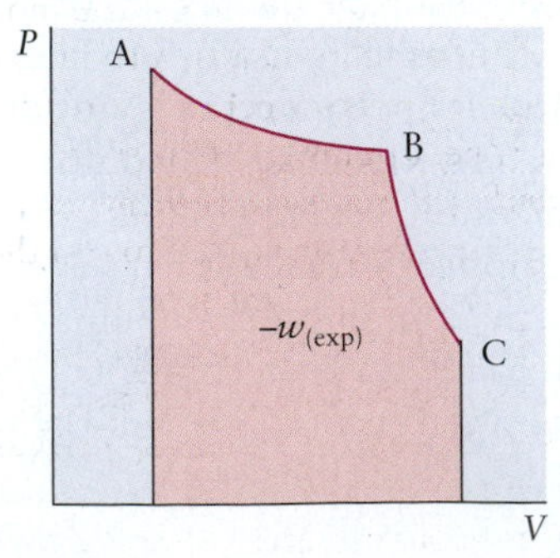

(a) Expansion

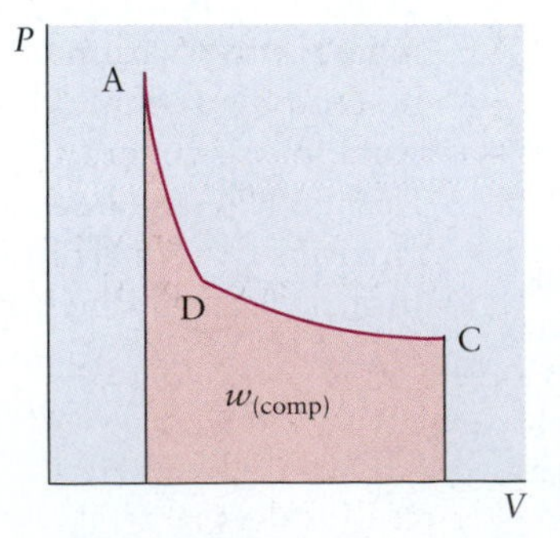

(b) Compression

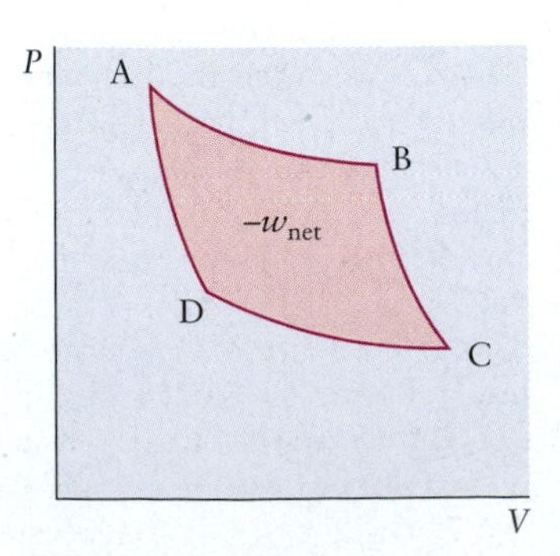

(c) Complete cycle

FIGURE 13.5 Stages of the Carnot cycle. The work done by the system in expansion (a) and on the system by compression (b) is shown by the shaded areas. (c) The net work done per cycle is the area enclosed by the curve ABCDA.

and

$$\frac{T_h}{T_l} = \left(\frac{V_D}{V_A}\right)^{\gamma-1} \quad \text{for path DA}$$

Equating these expressions gives

$$\left(\frac{V_C}{V_D}\right)^{\gamma-1} = \left(\frac{V_D}{V_A}\right)^{\gamma-1}$$

or

$$\frac{V_C}{V_B} = \frac{V_D}{V_A} \text{ and } \frac{V_B}{V_A} = \frac{V_C}{V_D}$$

Hence, the net work done in one passage around the Carnot cycle is

$$w_{net} = -nR(T_h - T_l) \ln \frac{V_B}{V_A} \qquad [13.3]$$

Heat Engines

The Carnot cycle is an idealized model for a heat engine. When a certain amount of heat, q_{AB}, is added to the system at the higher temperature, T_h, a net amount of work, $-w_{net}$, is obtained from the system. In addition, some heat q_{CD} is discharged at the lower temperature, but this energy is "degraded" and is no longer available for use in the engine. The **efficiency,** ϵ, of such an engine is the ratio of the negative of the net work done on the system, $-w_{net}$, to the heat added along the high-temperature isothermal path:

$$\epsilon = \frac{-w_{net}}{q_{AB}}$$

It is the net work that is available for the performance of useful mechanical tasks such as turning electrical generators or dynamos, but it is the heat, q_{AB}, absorbed at the higher temperature, T_h, that must be "paid for" in terms of coal or oil consumed to supply it. The efficiency, ϵ, must be maximized to get out the most work possible for the lowest cost.

For the ideal gas Carnot cycle, the efficiency is easily calculated:

$$\epsilon = \frac{-w_{net}}{q_{AB}} = \frac{nR(T_h - T_l)\ln(V_B/V_A)}{nRT_h \ln(V_B/V_A)}$$

$$\epsilon = \frac{T_h - T_l}{T_h} = 1 - \frac{T_l}{T_h} \qquad [13.4]$$

This result, called the Carnot efficiency or the **thermodynamic efficiency,** places a fundamental limit on the efficiency with which heat can be converted to mechanical work. Only if the high temperature, T_h, were infinite or the low temperature, T_l, were zero would it be possible to have a heat engine operate with 100% efficiency. To maximize efficiency, the greatest possible temperature difference should be used. Although we derived this result specifically for the ideal gas, we will show later in this section that it applies to *any* reversible engine operating between two temperatures. For a *real* engine, which must operate irreversibly, the actual efficiency must be lower than the thermodynamic efficiency.

EXAMPLE 13.3

Suppose a heat engine absorbs 10.0 kJ of heat from a high-temperature source at T_h = 450 K and discards heat to a low-temperature reservoir at T_l = 350 K. Calculate the thermodynamic efficiency, ϵ, of conversion of heat to work; the amount of work performed, $-w_{net}$; and the amount of heat discharged at T_l, q_{CD}.

SOLUTION

$$\epsilon = \frac{T_h - T_l}{T_h} = \frac{450 \text{ K} - 350 \text{ K}}{450 \text{ K}} = 0.222$$

Therefore, the engine can be, at most, 22.2% efficient. Because $\epsilon = -w_{net}/q_{AB}$,

$$w_{net} = -\epsilon q_{AB} = -(0.222)(10.0 \text{ kJ}) = -2.22 \text{ kJ}$$

Because ΔU for the whole cycle is 0,

$$\Delta U = 0 = q_{AB} + q_{CD} + w_{net}$$

$$q_{CD} = -q_{AB} - w_{net} = -10.0 + 2.22 = -7.8 \text{ kJ}$$

Therefore, 7.8 kJ is discharged at 350 K. This heat must be removed from the vicinity of the engine by a cooling system; otherwise, it will cause T_l to increase and reduce the efficiency of the engine.

Related Problems: 11, 12

Efficiency of General Carnot Engines

This subsection shows that all Carnot engines operating reversibly between two temperatures, T_h and T_l, have the same efficiency:

$$\varepsilon = \frac{-w_{net}}{q_h} = \frac{T_h - T_l}{T_h} \qquad [13.5]$$

That is, the efficiency calculated for an ideal gas applies equally to any other working fluid. To demonstrate this, we assume the *contrary* true and show that assumption leads to a contradiction with experience. The assumption is, therefore, deemed false.

Assume that there *are* two reversible machines operating between the same two temperatures, T_h and T_l, one of which has an efficiency, ϵ_1, that is greater than the efficiency, ϵ_2, of the other. The two machines are adjusted so that the total work output is the same for both. The more efficient machine is run as a heat engine so that it produces mechanical work $-w_1$. This work is used to operate the other machine in the opposite sense (as a heat pump) so that $w_2 = -w_1$. The net work input to the *combined* machines is then zero, because the work produced by the first machine is used to run the second.

Now, let's examine what happens to heat in this situation. Because engine 1 is more efficient than engine 2 and because the work is the same for both, engine 1 must withdraw less heat from the hot reservoir at T_h than is discharged into the same reservoir by engine 2; there is a net transfer of heat *into* the high-temperature reservoir. For the combined engines, $\Delta U_{tot} = w_{tot} = 0$; so q_{tot} must also be zero, and a net transfer of heat *out of* the low-temperature reservoir must therefore occur. By this reasoning we have devised an apparatus that can transfer heat from a low-temperature to a high-temperature reservoir with no net expenditure of work.

But that is impossible. All our experience shows we cannot make a device that transfers heat from a cold body to a hot body without doing work. In fact, exactly the opposite is seen: Heat flows spontaneously from hotter to colder bodies. Our experience is summarized and generalized in the following statement:

> *It is impossible to construct a device that will transfer heat from a cold reservoir to a hot reservoir in a continuous cycle with no net expenditure of work.*

This is one form of the second law of thermodynamics, as stated by Rudolf Clausius.

Further Discussion of Efficiency

Our assumption led to a conclusion that contradicts experience. Therefore, the original assumption must have been wrong, and there *cannot* be two reversible engines operating between the same two temperatures with different efficiencies. That is, all Carnot engines must have the same efficiency, which is

$$\epsilon = \frac{T_h - T_l}{T_h}$$

for the ideal gas. For any substance, ideal or not, undergoing a Carnot cycle,

$$\epsilon = \frac{-w_{net}}{q_h} = \frac{q_h + q_l}{q_h} = \frac{T_h - T_l}{T_h}$$

The Carnot cycle forms the basis for a *thermodynamic* scale of temperature. Because $\epsilon = 1 - (T_l/T_h)$, the Carnot efficiencies determine temperature ratios and thereby establish a temperature scale. The difficulty of operating real engines close to the reversible limit makes this procedure impractical. Instead, real gases at low pressures are used to define and determine temperatures (see Section 9.2).

The last two terms of the preceding equation can be rewritten as

$$1 + \frac{q_l}{q_h} = 1 - \frac{T_l}{T_h}$$

and simplified to

$$\frac{q_h}{T_h} + \frac{q_l}{T_l} = 0 \qquad [13.6]$$

This simple equation has profound importance because it contains the essence of the second law of thermodynamics, namely, that q/T is a state function. To see this, consider a *general* reversible cyclic process and draw a series of closely spaced adiabats, as shown in Figure 13.6. (An adiabat is a curve on the *PV* diagram showing those thermodynamic states connected by a particular reversible adiabatic process.) Now replace each segment along the given cycle (ABCDA) with a series of alternating isothermal and adiabatic segments. Clearly, we can construct a path that is arbitrarily close to the desired curve by taking more and more closely spaced adiabats.

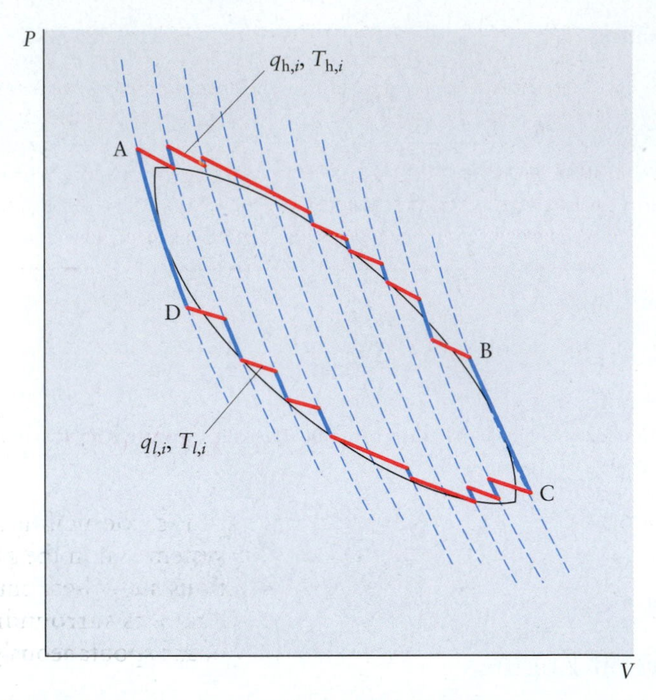

FIGURE 13.6 A general cyclic process (ABCDA) can be approximated to arbitrary accuracy by the sum of a series of Carnot cycles.

Now follow the evolution of $\Sigma_i(q_i/T_i)$ along this curve. The key observation is that the contributions q_i/T_i appear in pairs: For each $q_{h,i}/T_{h,i}$ from an isothermal segment along the ABC path, there is a $q_{l,i}/T_{l,i}$ along the CDA path. Any given pair forms two sides of a Carnot cycle, with the other two sides determined by adiabats; thus,

$$\frac{q_{h,i}}{T_{h,i}} + \frac{q_{l,i}}{T_{l,i}} = 0$$

Summing over i shows that

$$\sum_i \left(\frac{q_{h,i}}{T_{h,i}} + \frac{q_{l,i}}{T_{l,i}}\right) = 0$$

Because this summation follows the original reversible path to an arbitrary accuracy (if the lengths of the segments are made short enough), it follows that

$$\oint \frac{1}{T}\, dq_{rev} = 0$$

for *any* closed, reversible path. Although q is not a state function, q_{rev}/T is a state function because, like energy and enthalpy, its total change is zero for any process that begins and ends in the same state. From this result, Clausius *defined* the entropy change $\Delta S = S_f - S_i$ of a system in a process starting in state i and ending in state f by the equation

$$\Delta S = \int_i^f \frac{dq_{rev}}{T} \qquad [13.7]$$

As we have stated, the principles of thermodynamics are based on observations of nature and are not subject to mathematical proof. We have accomplished something quite significant, however. From the assumption—based on physical observation—that heat cannot be transferred from a low-temperature to a high-temperature body without expenditure of work, we have derived the result that $\int(1/T)\, dq_{rev}$ is independent of path and, therefore, is a state function. This result has been subjected to rigorous testing and no proper test has found it to be invalid. Therefore, we have great confidence in the generality of this result and are entirely comfortable in calling it a scientific law. Some critics might say, "If we have to make some assumption anyway, why don't we just assume that $\int(1/T)\, dq_{rev}$ is path independent to start with?" This is certainly possible and is practiced in many presentations of thermodynamics. Our approach is different; we prefer to base our assumptions directly on physical observation, not on abstract mathematical axioms.

13.5 Entropy Changes and Spontaneity

This section outlines procedures for calculating entropy changes that occur in the system and in the surroundings during several types of processes. The final subsections show how entropy changes calculated for the thermodynamic universe (system plus surroundings) predict whether a particular contemplated process can occur spontaneously when attempted in the laboratory or in a chemical plant.

ΔS_{sys} for Isothermal Processes

If the reversible process selected as the pathway connecting the initial and final states is isothermal, the calculation simplifies immediately. Because T is constant, it comes outside the integral:

$$\Delta S = \int_i^f \frac{dq_{rev}}{T} = \frac{1}{T}\int_i^f dq_{rev} = \frac{q_{rev}}{T} \tag{13.8}$$

Here, q_{rev} is the *finite* amount of heat absorbed by the system during the entire reversible isothermal process.

COMPRESSION/EXPANSION OF AN IDEAL GAS Consider an ideal gas enclosed in a piston-cylinder arrangement that is maintained at constant temperature in a heat bath. The gas can be compressed (or expanded) reversibly by changing the position of the piston to accomplish a specified change in volume. In Section 12.6, the heat transferred between system and bath when the gas is expanded (or compressed) isothermally and reversibly from volume V_1 to V_2 is shown to be

$$q_{rev} = nRT \ln\left(\frac{V_2}{V_1}\right)$$

The resulting change in entropy, therefore, is

$$\Delta S = nR \ln\left(\frac{V_2}{V_1}\right) \quad \text{(constant } T\text{)} \tag{13.9}$$

From Equation 13.9 we see that the entropy of a gas increases during an isothermal expansion ($V_2 > V_1$) and decreases during a compression ($V_2 < V_1$). Boltzmann's relation (see Eq. 13.1) provides the molecular interpretation of these results. The number of microstates available to the system, Ω, increases as the volume of the system increases and decreases as volume decreases, and the entropy of the system increases or decreases accordingly.

PHASE TRANSITIONS Another type of constant-temperature process is a phase transition such as the melting of a solid at constant pressure. This occurs reversibly at the fusion temperature, T_f, because an infinitesimal change in external conditions, such as reducing the temperature, can reverse the melting process. The reversible heat when 1 mol of substance melts is $q_{rev} = \Delta H_{fus}$, so

$$\Delta S_{fus} = \frac{q_{rev}}{T_f} = \frac{\Delta H_{fus}}{T_f} \qquad [13.10]$$

The entropy increases when a solid melts or a liquid vaporizes, and it decreases when the phase transition occurs in the opposite direction. Again, Boltzmann's relation provides the molecular interpretation. When a solid melts or a liquid vaporizes, the number of accessible microstates Ω increases, and thus the entropy increases.

EXAMPLE 13.4

Calculate the entropy change when 3.00 mol benzene vaporizes reversibly at its normal boiling point of 80.1°C. The molar enthalpy of vaporization of benzene at this temperature is 30.8 kJ mol^{-1}.

SOLUTION

The entropy change when 1 mol benzene is vaporized at 80.1°C (=353.25 K) is

$$\Delta S_{vap} = \frac{\Delta H_{vap}}{T_b} = \frac{30{,}800 \text{ J mol}^{-1}}{353.25 \text{ K}} = +87.2 \text{ J K}^{-1}$$

When 3.00 mol is vaporized, the entropy change is three times as great:

$$\Delta S = (3.00 \text{ mol})(+87.2 \text{ J K}^{-1} \text{ mol}^{-1}) = +262 \text{ J K}^{-1}$$

Related Problems: 13, 14

Remarkably, most liquids have similar values for the molar entropy of vaporization. **Trouton's rule** summarizes this observation:

$$\Delta S_{vap} = 88 \pm 5 \text{ J K}^{-1} \text{ mol}^{-1} \qquad [13.11]$$

Note that the ΔS_{vap} of benzene, calculated in Example 13.4, is within this range. The constancy of ΔS_{vap} means that ΔH_{vap} and T_b, which vary widely from substance to substance, must do so in the same proportion. Trouton's rule allows enthalpies of vaporization to be estimated from boiling temperatures. However, exceptions exist; the molar entropy of vaporization for water is 109 J K^{-1} mol^{-1}. The value for water is unusually high because hydrogen bonding in liquid water means there are many fewer allowed configurations (lower entropy) than in other liquids; thus, water shows a much greater increase in the number of microstates on vaporization.

ΔS_{sys} for Processes with Changing Temperature

Now consider a reversible process in which the temperature changes. In this case, Equation 13.2 must be used:

$$\Delta S = \int_A^B \frac{1}{T}\, dq_{rev}$$

In the integral, A and B represent, respectively, the initial and final equilibrium states for the process. The calculation must be conducted along a reversible path connecting A and B. For a reversible *adiabatic* process, $q = 0$ and, therefore, $\Delta S = 0$. Such a process is also called **isentropic** (that is, the entropy is constant).

In a reversible *isochoric* process, the volume is held constant and the system is heated or cooled by contact with a reservoir whose temperature differs from that of the system by an infinitesimal amount, dT. The heat transferred in this case is

$$dq_{\text{rev}} = nc_{\text{V}}\, dT$$

and the entropy change of the system as it is heated from T_1 to T_2 is

$$\Delta S = \int_{T_1}^{T_2} \frac{1}{T}\, dq_{\text{rev}} = \int_{T_1}^{T_2} \frac{nc_{\text{V}}}{T}\, dT$$

If c_{V} is independent of T over the temperature range of interest, it can be removed from the integral, giving the result

$$\Delta S = nc_{\text{V}} \int_{T_1}^{T_2} \frac{1}{T}\, dT = nc_{\text{V}} \ln\left(\frac{T_2}{T_1}\right) \quad \text{(constant } V\text{)} \qquad [13.12]$$

The analogous result for the entropy change of the system in a reversible *isobaric* process (constant *pressure*) is

$$\Delta S = \int_{T_1}^{T_2} \frac{nc_{\text{P}}}{T}\, dT = nc_{\text{P}} \ln\left(\frac{T_2}{T_1}\right) \quad \text{(constant } P\text{)} \qquad [13.13]$$

Entropy always increases with increasing temperature. From the kinetic theory of ideal gases in Chapter 9, it is clear that increasing the temperature of the gas increases the magnitude of the average kinetic energy per molecule and, therefore, the range of momenta available to molecules. This, in turn, increases Ω for the gas and, by Boltzmann's relation, the entropy of the gas.

Now consider an experiment in which identical samples of a gas are taken through identical temperature increases, one sample at constant V and the other at constant P. Let's compare the entropy changes in the two processes. From the previous discussion it follows that $\Delta S_{\text{P}} > \Delta S_{\text{V}}$ because $c_{\text{P}} > c_{\text{V}}$. The molecular interpretation is based on the discussion of c_{P} and c_{V} in Section 12.3. The gas heated at constant P increases in volume, as well as in temperature; its molecules therefore gain access to a greater range of positions, as well as a greater range of momenta. Consequently, the gas heated at constant P experiences a *greater* increase in Ω than does the gas heated at constant V and, therefore, a greater increase in S.

The following example illustrates that the entropy is a state function, for which changes are independent of the path followed.

EXAMPLE 13.5

(a) Calculate the entropy change for the process described in Example 12.9: 5.00 mol argon expands reversibly at a constant temperature of 298 K from a pressure of 10.0 to 1.00 atm.

(b) Calculate the entropy change for the same initial and final states as in part (a) but along a different path. First, the 5.00 mol argon expands reversibly and *adiabatically* between the same two pressures. This is the path followed in Example 12.10; it causes the temperature to decrease to 118.6 K. Then the gas is heated at constant pressure back to 298 K.

SOLUTION

(a) At constant temperature, the entropy change is

$$\Delta S = nR \ln\left(\frac{V_2}{V_1}\right) = nR \ln\left(\frac{P_1}{P_2}\right)$$

$$= (5.00 \text{ mol})(8.315 \text{ J K}^{-1} \text{ mol}^{-1}) \ln 10.0$$

$$= +95.7 \text{ J K}^{-1}$$

(b) For the adiabatic part of this path, the entropy change is zero. When the gas is then heated reversibly at constant pressure from 118.6 to 298 K, the entropy change is

$$\Delta S = nc_P \ln\left(\frac{T_2}{T_1}\right)$$

$$= (5.00 \text{ mol})\left(\frac{5}{2} \times 8.315 \text{ J K}^{-1} \text{ mol}^{-1}\right) \ln \frac{298 \text{ K}}{118.6 \text{ } K}$$

$$= +95.7 \text{ J K}^{-1}$$

This is the same as the result from part (a), an illustration of the fact that the entropy is a state function. By contrast, the amounts of heat for the two paths are different: 28.5 and 18.6 kJ.

Related Problems: 17, 18

ΔS for Surroundings

Usually, the surroundings can be treated as a large *heat bath* that transfers heat to or from the system at the fixed temperature of the bath. In such cases, the heat capacity of the surroundings (heat bath) must be so large that the heat transferred during the process does not change the temperature of the bath. The heat gained by the surroundings during a process is the heat lost from the system. If the process occurs at constant P, then

$$q_{\text{surr}} = -\Delta H_{\text{sys}}$$

and the entropy change of the surroundings is

$$\Delta S_{\text{surr}} = \frac{-\Delta H_{\text{sys}}}{T_{\text{surr}}} \qquad [13.14]$$

If the process occurring in the system is exothermic, the surroundings gain heat and the entropy change of the surroundings is positive. Similarly, an endothermic process in the system is accompanied by a negative entropy change in the surroundings, because the surroundings give up heat during the process to keep the system at the temperature of the heat bath.

If the surroundings lack sufficient heat capacity to maintain constant temperature during the process, then entropy changes for the surroundings must be calculated by the methods just demonstrated for the system, taking explicit account of the temperature change of the surroundings. Examples of both cases are included in the problems at the end of this chapter.

ΔS_{tot} for System Plus Surroundings

The tools are now in place for using the second law to predict whether specific processes will be spontaneous. We illustrate the procedure first for spontaneous cooling of a hot body, and then for irreversible expansion of an ideal gas.

SPONTANEOUS COOLING OF A HOT BODY Consider a spontaneous process in which a sample of hot metal is cooled by sudden immersion in a cold bath. Heat flows from the metal into the bath until they arrive at the same temperature. This spontaneous process is accompanied by an increase in the total entropy for the thermodynamic universe of the process, as illustrated by the following example.

EXAMPLE 13.6

A well-insulated ice-water bath at 0.0°C contains 20 g ice. Throughout this experiment, the bath is maintained at the constant pressure of 1 atm. When a piece of nickel at 100°C is dropped into the bath, 10.0 g of the ice melts. Calculate the total entropy change for the thermodynamic universe of this process. (Specific heats at constant *P:* nickel, 0.46 J K^{-1} g^{-1}; water, 4.18 J K^{-1} g^{-1}; ice, 2.09 J K^{-1} g^{-1}. Enthalpy of fusion of ice, 334 J K^{-1} g^{-1}.)

SOLUTION

Consider the nickel to be the *system* and the ice-water bath to be the *surroundings* in this experiment. Heat flows from the nickel into the bath and melts some of the ice. Consequently, the entropy of the nickel decreases and the entropy of the bath increases. The final equilibrium temperature of both system and bath is 0.0°C, as indicated by the presence of some ice in the bath at equilibrium.

Before calculating ΔS_{Ni} it is necessary to calculate the mass of the nickel from the calorimetry equation as follows:

$$\text{heat lost by Ni} = \text{heat gained by ice bath} = \text{heat used in melting ice}$$

$$-M\,(0.46\text{ J K}^{-1}\text{ g}^{-1})(273.15\text{ K} - 373.15\text{ K}) = (10.0\text{ g})(334\text{ J g}^{-1})$$

$$M = 73\text{ g}$$

Because the nickel is cooled at constant *P*, the entropy change for the nickel is calculated as

$$\Delta S_{Ni} = (73\text{ g})(0.46\text{ J g}^{-1}\text{ K}^{-1})\ln(0.73) = -10\text{ J K}^{-1}$$

Because the ice bath has remained at constant *T* throughout the experiment, it can be treated as a "large heat bath," and its entropy change is calculated as

$$\Delta S_{bath} = \frac{-\Delta H_{sys}}{T_{bath}} = \frac{-(-334\text{ J g}^{-1})(10.0\text{ g})}{273.15\text{ K}} = 12\text{ J K}^{-1}$$

Now,

$$\Delta S_{tot} = \Delta S_{Ni} + \Delta S_{bath} = -10 + 12 = +2\text{ J K}^{-1}$$

Thus, the process is spontaneous, driven by the fact that the entropy gain of the melting ice exceeds the entropy loss of the cooling metal.

Related Problems: 20, 21, 22

IRREVERSIBLE EXPANSION OF AN IDEAL GAS Consider a gas confined within a piston-cylinder arrangement and held at constant temperature in a heat bath. Suppose the external pressure is abruptly reduced and held constant at the new lower value. The gas immediately expands against the piston until its internal pressure declines to match the new external pressure. The total entropy of system plus surroundings will increase during this expansion. In preparation for a quantitative example, a general comparison of irreversible and reversible processes connecting the same initial and final states provides insight into why the total entropy increases in a spontaneous process.

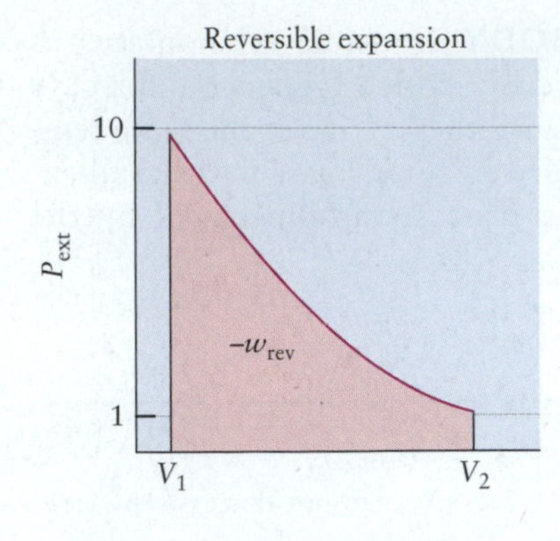

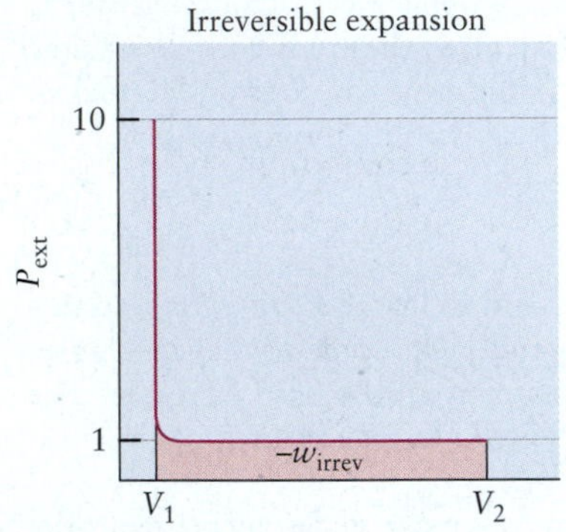

FIGURE 13.7 Work done by a system in reversible and irreversible expansions between the same initial and final states. The work performed is greater for the reversible process.

The work performed *by* a system ($-w$) as it undergoes an irreversible isothermal expansion is always less than when the expansion is conducted reversibly. To see this, return to the definition of work done *on* a system:

$$w = -\int P_{ext}\, dV$$

During an expansion, P_{ext} must be less than P, the pressure of the gas. For a reversible expansion, P_{ext} is only infinitesimally smaller (so the system is always close to equilibrium); but for an irreversible expansion, P_{ext} is measurably smaller. Therefore, the area under a graph of P_{ext} plotted against V is less than that of a graph of P against V (Fig. 13.7), so

$$-w_{irrev} = \int P_{ext}\, dV < \int P\, dV = -w_{rev}$$

and the work performed by the system, $-w_{irrev}$, is algebraically less than $-w_{rev}$. If the system is viewed as an "engine" for performing useful work on the surroundings, a reversible process is always more efficient than an irreversible one.

If the reversible and irreversible processes have the same initial and final states, then ΔU is the same for both.

$$\Delta U = w_{irrev} + q_{irrev} = w_{rev} + q_{rev}$$

But because

$$-w_{irrev} < -w_{rev}$$

we must have

$$w_{irrev} > w_{rev}$$

and

$$q_{irrev} < q_{rev}$$

The heat absorbed is a *maximum* when the process is conducted reversibly.

The following example illustrates these inequalities.

EXAMPLE 13.7

Calculate the heat absorbed and the work done on a system of 5.00 mol of an ideal gas as it expands irreversibly at constant temperature T = 298 K from a pressure of 10.0 to 1.00 atm. The external pressure is held constant at 1.00 atm.

SOLUTION

The initial volume V_1 is

$$V_1 = \frac{nRT}{P_1} = \frac{(5.00 \text{ mol})(0.08206 \text{ L atm K}^{-1} \text{ mol}^{-1})(298 \text{ K})}{10.0 \text{ atm}} = 12.2 \text{ L}$$

The final volume is 10 times this, or

$$V_2 = 122 \text{ L}$$

For a constant external pressure,

$$w_{irrev} = -P_{ext}\Delta V = -(1.00 \text{ atm})(122 \text{ L} - 12.2L) = -110 \text{ L atm}$$
$$= -11.1 \text{ kJ}$$

At constant T, $\Delta U = 0$, however, so

$$q_{irrev} = -w_{irrev} = 11.1 \text{ kJ}$$

In Example 12.9, a reversible expansion between the same two states was carried out, with the result that

$$w_{rev} = -28.5 \text{ kJ}$$

This demonstrates that

$$-w_{irrev} < -w_{rev}$$

and

$$q_{irrev} < q_{rev}$$

For reversible and irreversible processes connecting the same pair of initial and final states, it is always true that

$$q_{rev} > q_{irrev}$$

Dividing this expression by T, the temperature at which the heat is transferred, gives

$$\frac{q_{rev}}{T} > \frac{q_{irrev}}{T}$$

The left side is the entropy change ΔS,

$$\Delta S = \frac{q_{rev}}{T}$$

so

$$\Delta S > \frac{q_{irrev}}{T}$$

The last two equations can be combined as

$$\Delta S \geq \frac{q}{T}$$

where the equality applies only to a reversible process. This expression, called the **inequality of Clausius,** states that in any spontaneous process the heat absorbed by the system from surroundings at the same temperature is always less than $T\Delta S$. In a reversible process, the heat absorbed is equal to $T\Delta S$.

Now let's apply Clausius's inequality to processes occurring within an *isolated* system. In this case, there is no transfer of heat into or out of the system, and $q = 0$. Therefore, for spontaneous processes within an isolated system, $\Delta S > 0$.

The thermodynamic universe of a process (that is, a system plus its surroundings) is clearly an isolated system to which Clausius's inequality can be applied. It follows that

1. *In a* reversible *process the total entropy of a system plus its surroundings is unchanged.*
2. *In an* irreversible *process the total entropy of a system plus its surroundings must increase.*
3. *A process for which $\Delta S_{total} < 0$ is impossible.*

These statements constitute the heart of the second law, because they provide its predictive power.

EXAMPLE 13.8

Calculate $\Delta S_{tot} = \Delta S_{sys} + \Delta S_{surr}$ for the reversible and irreversible isothermal expansions of Examples 12.9 and 13.7.

SOLUTION

For the reversible expansion,

$$q_{rev} = 28.5 \text{ kJ}$$

$$\Delta S_{sys} = \frac{q_{rev}}{T} = \frac{28{,}500 \text{ J}}{298 \text{ K}} = +95.7 \text{ J K}^{-1}$$

The surroundings give up the same amount of heat at the same temperature. Hence,

$$\Delta S_{surr} = \frac{-28{,}500 \text{ J}}{298 \text{ K}} = -95.7 \text{ J K}^{-1}$$

or $\Delta S_{tot} = 95.7 - 95.7 = 0$ for the reversible process. For the irreversible expansion, it is still true that

$$\Delta S_{sys} = +95.7 \text{ J K}^{-1}$$

because S is a function of state, and the initial and final states are the same as for the reversible expansion. From Example 13.7, only 11.1 kJ of heat is given up by the surroundings in this case.[2] Hence,

$$\Delta S_{surr} = \frac{-11{,}100 \text{ J}}{298 \text{ K}} = -37.2 \text{ J K}^{-1}$$

and $\Delta S_{tot} = 95.7 - 37.2 = 58.5 \text{ J K}^{-1} > 0$ for the irreversible process.

13.6 The Third Law of Thermodynamics

In thermodynamic processes, only *changes* in entropy, ΔS, are measured, just as only changes in internal energy, ΔU, or enthalpy, ΔH, are measured. It is nevertheless useful to define absolute values of entropy relative to some reference state. An important experimental observation that simplifies the choice of reference state is:

> *In any thermodynamic process involving only pure phases in their equilibrium states, the entropy change ΔS approaches zero as T approaches 0 K.*

This observation is the **Nernst heat theorem,** named after its discoverer, the German physicist Walther Nernst. It immediately suggests a choice of reference state: The entropy of any pure element in its equilibrium state is defined to approach zero as T approaches 0 K. From the Nernst theorem, the entropy change for any chemical reaction, including one in which elements react to give a pure compound, approaches zero at 0 K. The most general form of this statement is the **third law of thermodynamics:**

> *The entropy of any pure substance (element or compound) in its equilibrium state approaches zero at the absolute zero of temperature.*

Absolute zero can never actually be reached; therefore, a small extrapolation is needed to make use of this result.

[2]How can the heat from Example 13.7, which is irreversible from the perspective of the system, be reversible from the perspective of the surroundings? This can be accomplished by enclosing the gas in a material (such as a metal) that can efficiently transfer heat to and from the surroundings, and thus remain close to equilibrium, at the same time that the gas itself is far from equilibrium due to the gas currents that occur during the irreversible expansion.

The third law, like the two laws that precede it, is a macroscopic law based on experimental measurements. It is consistent with the microscopic interpretation of the entropy presented in Section 13.2. From quantum mechanics and statistical thermodynamics, we know that the number of microstates available to a substance at equilibrium falls rapidly toward one as the temperature approaches absolute zero. Therefore, the absolute entropy defined as $k_B \ln \Omega$ should approach zero. The third law states that the entropy of a substance in its equilibrium state approaches zero at 0 K. In practice, equilibrium may be difficult to achieve at low temperatures, because particle motion becomes very slow. In solid CO, molecules remain randomly oriented (CO or OC) as the crystal is cooled, even though in the equilibrium state at low temperatures, each molecule would have a definite orientation. Because a molecule reorients slowly at low temperatures, such a crystal may not reach its equilibrium state in a measurable period. A nonzero entropy measured at low temperatures indicates that the system is not in equilibrium.

Standard-State Entropies

Because the entropy of any substance in its equilibrium state is zero at absolute zero, its entropy at any other temperature T is given by the entropy increase as it is heated from 0 K to T. If heat is added at constant pressure,

$$\Delta S = n \int_{T_1}^{T_2} \frac{c_P}{T}\, dT$$

Thus, S_T, the absolute entropy of 1 mol of substance at temperature T, is given by

$$S_T = \int_0^T \frac{c_P}{T}\, dT$$

It is necessary merely to measure c_P as a function of temperature and determine the area under a plot of c_P/T versus T from 0 K to any desired temperature. If a substance melts, boils, or undergoes some other phase change before reaching the temperature T, the entropy change for that process must be added to $\int (c_P/T)\, dT$.

To calculate entropy changes for chemical reactions, we find it convenient to use the same standard state already selected for enthalpy calculations in Section 12.3. For this purpose, we define the **standard molar entropy** to be the absolute molar entropy $S°$ at 298.15 K and 1 atm pressure (Fig. 13.8):

$$S° = \int_0^{298.15} \frac{c_P}{T}\, dT + \Delta S \text{ (phase changes between 0 and 298.15 K)} \qquad [13.15]$$

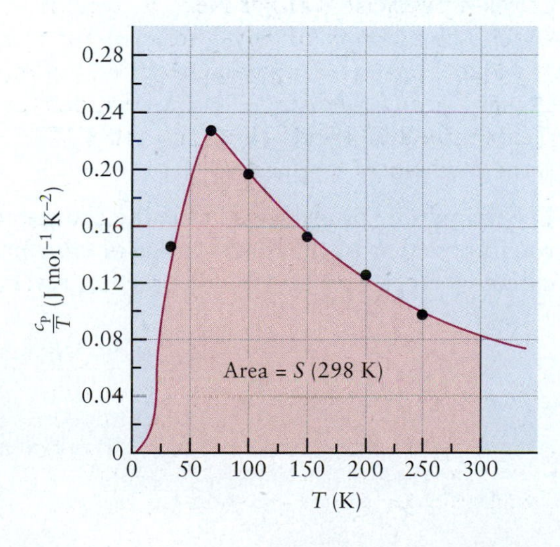

FIGURE 13.8 A graph of c_P/T versus T for platinum. The black dots represent experimental measurements. The area up to any temperature (here, 298 K) is the molar entropy at that temperature.

Standard molar entropies $S°$ are tabulated for a number of elements and compounds in Appendix D. If c_P is measured in J K^{-1} mol^{-1}, then the entropy $S°$ will have the same units. For dissolved ions, the arbitrary convention $S°\ (H^+(aq)) = 0$ is applied (just as for the standard enthalpy of formation of H^+ discussed in Section 12.3). For this reason, some $S°$ values are negative for aqueous ions—an impossibility for substances.

Tabulated standard molar entropies are used to calculate entropy changes in chemical reactions at 25°C and 1 atm, just as standard enthalpies of formation are combined to obtain enthalpies of reaction according to Hess's law (see Section 12.3).

EXAMPLE 13.9

Using the table of standard molar entropies in Appendix D, calculate $\Delta S°$ for the chemical reaction

$$N_2(g) + 2\ O_2(g) \longrightarrow 2\ NO_2(g)$$

with reactants and products at a temperature of 25°C and a pressure of 1 atm.

SOLUTION

From the table,

$$S°(N_2(g)) = 191.50 \text{ J K}^{-1}\text{ mol}^{-1}$$

$$S°(O_2(g)) = 205.03 \text{ J K}^{-1}\text{ mol}^{-1}$$

$$S°(NO_2(g)) = 239.95 \text{ J K}^{-1}\text{ mol}^{-1}$$

The entropy change for the reaction is the sum of the entropies of the products, minus the sum of entropies of the reactants, each multiplied by its coefficient in the balanced chemical equation:

$$\begin{aligned}\Delta S° &= 2S°(NO_2(g)) - S°(N_2(g)) - 2S°(O_2(g)) \\ &= (2\text{ mol})(239.95 \text{ J K}^{-1}\text{ mol}^{-1}) - (1\text{ mol})(191.50 \text{ J K}^{-1}\text{ mol}^{-1}) - \\ &\quad (2\text{ mol})(205.03 \text{ J K}^{-1}\text{ mol}^{-1}) \\ &= -121.66 \text{ J K}^{-1}\end{aligned}$$

The factors of 2 multiply $S°$ for NO_2 and O_2 because 2 mol of each appears in the chemical equation. Note that standard molar entropies, unlike standard molar enthalpies of formation $\Delta H_f°$, are not zero for elements at 25°C. The negative $\Delta S°$ results because this is the entropy change of the system only. The surroundings must undergo a positive entropy change in such a way that $\Delta S_{tot} \geq 0$.

Related Problems: 23, 24, 25, 26

13.7 The Gibbs Free Energy

In Section 13.5, we showed that the change in entropy of a system plus its surroundings (that is, the total change of entropy, ΔS_{tot}) provides a criterion for deciding whether a process is spontaneous, reversible, or impossible:

$$\Delta S_{tot} > 0 \quad \text{spontaneous} \qquad [13.16a]$$

$$\Delta S_{tot} = 0 \quad \text{reversible} \qquad [13.16b]$$

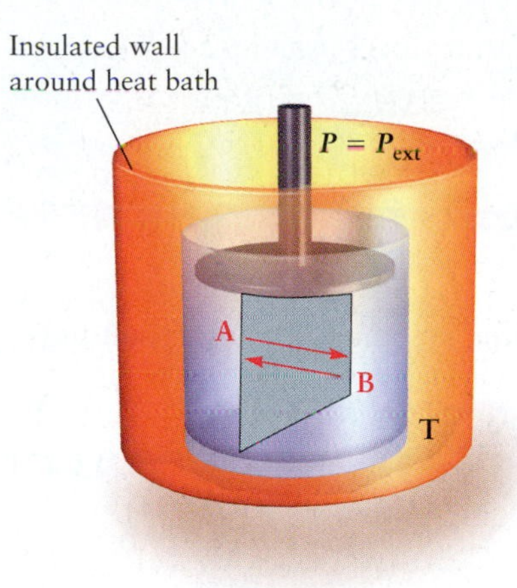

FIGURE 13.9 After the constraint between phases A and B is removed, matter can flow spontaneously between phases inside a system held at constant *T* (temperature) and *P* (pressure) by its surroundings.

$$\Delta S_{tot} < 0 \quad \text{nonspontaneous} \qquad [13.16c]$$

Although the algebraic sign of ΔS_{tot} is a completely general criterion for the spontaneity or impossibility of a process, it requires calculating the entropy change for the surroundings, as well as for the system. It would be much more convenient to have a state function that predicts the feasibility of a process in the system without explicit calculations for the surroundings.

For the special case of processes at constant temperature and pressure, the most important in chemistry, such a state function exists. It is called the *Gibbs free energy* and is denoted by G. We start the discussion of G with a qualitative examination of spontaneous laboratory processes at fixed T and P. Then we define G and develop its properties. Finally, we apply ΔG to determine spontaneity in phase transitions and chemical reactions.

Nature of Spontaneous Processes at Fixed T and P

Consider a system enclosed in a piston-cylinder assembly, which constrains pressure at the value P. The assembly is immersed in a heat bath, which constrains temperature at the value T. Experience shows that spontaneous processes under these conditions consist of spontaneous flow of molecules across a boundary completely internal to the system, separating different regions (called *phases*) of the system (Fig. 13.9).

Under these conditions, we visualize starting a spontaneous process by bringing phases A and B in Figure 13.9 into contact—both already prepared at T and P—but separated by an impermeable membrane (a constraint) that prevents exchange of matter between the phases. Removing the constraint allows spontaneous flow of molecules across the interface between phases. The system does not exchange matter with the surroundings, and the distribution of energy and volume between system and surroundings is not described explicitly. The only function of the surroundings is to maintain T and P constant throughout the experiment. Consequently, as we show in the next subsection, spontaneity of the process is determined by the change in Gibbs free energy of the *system only* while T and P remain constant.

These processes lead to changes in the structure or composition of the phases. Solutes are redistributed between immiscible solvents. Phase transitions occur between the solid, liquid, and gaseous states. Reactants become products. The system may gain or lose heat from the large heat reservoir while these rearrangements occur in its phases, but T remains constant. For example, the latent heat of fusion released during freezing of a liquid in the system is absorbed by the bath. Endothermic processes in the system will take heat from the bath, and exothermic processes will give heat to the bath. P–V work may be done on or by the system at constant P, depending on whether its density increases or decreases through the rearrangements of its phases.

Whether molecules flow spontaneously from phase A to B, or vice versa, is determined by the associated change in Gibbs free energy, as we show in the following subsection.

Gibbs Free Energy and Its Properties

During any process conducted at constant T and P as described earlier, the heat gained by the system is $\Delta H_{sys} = q_P$ and the heat transferred to the surroundings is $-q_P = -\Delta H_{sys}$. Because the surroundings remain at constant temperature during the process, the transfer of process heat must have the same effect on them as would a reversible transfer of the same amount of heat. Their entropy change is then

$$\Delta S_{surr} = \frac{-\Delta H_{sys}}{T_{surr}}$$

The total entropy change is

$$\Delta S_{tot} = \Delta S_{sys} + \Delta S_{surr} = \Delta S_{sys} - \frac{\Delta H_{sys}}{T_{surr}}$$

$$= \frac{-(\Delta H_{sys} - T_{surr}\,\Delta S_{sys})}{T_{surr}}$$

Because the temperature, T, is the same for both the system and the surroundings, we can rewrite this as

$$\Delta S_{tot} = \frac{-\Delta(H_{sys} - TS_{sys})}{T} \qquad [13.17]$$

We define the **Gibbs free energy** G as

$$G = H - TS \qquad [13.18]$$

therefore, Equation 13.17 becomes

$$\Delta S_{tot} = \frac{-\Delta G_{sys}}{T} \qquad [13.19]$$

Because the absolute temperature T is always positive, ΔS_{tot} and ΔG_{sys} must have the opposite sign for processes occurring at constant T and P. It follows that

$$\Delta G_{sys} < 0 \quad \text{spontaneous processes} \qquad [13.20a]$$

$$\Delta G_{sys} = 0 \quad \text{reversible processes} \qquad [13.20b]$$

$$\Delta G_{sys} > 0 \quad \text{nonspontaneous processes} \qquad [13.20c]$$

for processes conducted at constant temperature and pressure. If $\Delta G_{sys} > 0$ for a proposed process, then $\Delta G_{sys} < 0$ for the *reverse* of the proposed process, and that reverse process can occur spontaneously. Experimentally, one prepares the system initially at chosen values of T and P, and then releases the appropriate constraint to allow a process. The resulting "flow" of matter between phases A and B goes in the direction that reduces the value of G for the system, so that $\Delta G < 0$ at the selected values of T and P.

Gibbs Free Energy and Phase Transitions

As a simple application of the Gibbs free energy, consider the freezing of 1 mol liquid water to form ice:

$$H_2O(\ell) \longrightarrow H_2O(s)$$

Let us first examine what thermodynamics predicts when this process occurs at the ordinary freezing point of water under atmospheric pressure, 273.15 K. The measured enthalpy change (the heat absorbed at constant pressure) is

$$\Delta H_{273} = q_P = -6007 \text{ J}$$

At $T_f = 273.15$ K, water freezes reversibly—the system remains close to equilibrium as it freezes. Therefore, the entropy change is

$$\Delta S_{273} = \frac{q_{rev}}{T_f} = \frac{-6007 \text{ J}}{273.15 \text{ K}} = -21.99 \text{ J K}^{-1}$$

The Gibbs free energy change is

$$\Delta G_{273} = \Delta H_{273} - T\,\Delta S_{273} = -6007\text{ J} - (273.15\text{ K})(-21.99\text{ J K}^{-1}) = 0$$

At the normal freezing point, the Gibbs free energy change is zero because the freezing of water under these conditions is an equilibrium, reversible process.

Now, let's see what thermodynamics predicts as the water is cooled below 273.15 to 263.15 K (−10.00°C). Let's calculate the change in Gibbs free energy as water freezes at this lower temperature.

Assume that ΔH and ΔS for the freezing process do not depend on temperature. Then we can write

$$\Delta G_{263} = -6007\text{ J} - (263.15\text{ K})(-21.99\text{ J K}^{-1}) = -220\text{ J}$$

An exact calculation takes into account that ΔH and ΔS *do* depend slightly on temperature, and it leads to

$$\Delta G_{263} = -213\text{ J}$$

for the process. Because $\Delta G < 0$, thermodynamics predicts the undercooled water will freeze spontaneously at 263.15 K. At a temperature *higher* than T_f, ΔG is greater than 0, predicting that the liquid would not freeze. This agrees with our experience in nature. Water does not freeze at atmospheric pressure if the temperature is held greater than 273.15 K; instead, the reverse process occurs, and ice melts spontaneously.

Writing the Gibbs free energy change as $\Delta G = \Delta H - T\,\Delta S$ shows that a negative value of ΔG—and therefore a spontaneous process—is favored by a negative value of ΔH and a positive value of ΔS. For freezing a liquid, ΔH is negative, but ΔS for freezing is also negative. Whether a liquid freezes depends on the competition between two factors: an enthalpy change that favors freezing and an entropy change that disfavors freezing (Fig. 13.10). At temperatures less than T_f, the former dominates and the liquid freezes spontaneously, but at temperatures greater than T_f, the latter dominates and freezing does not occur. At T_f, the Gibbs free energies of the two phases are equal ($\Delta G = 0$), and the phases coexist at equilibrium. Similar types of analysis apply to other phase transitions, such as condensing a gas to a liquid.

Gibbs Free Energy and Chemical Reactions

The change in the Gibbs free energy provides a criterion for the spontaneity of any process occurring at constant temperature and pressure (Fig. 13.11). To predict whether a chemical reaction is spontaneous at given values of T and P, it is

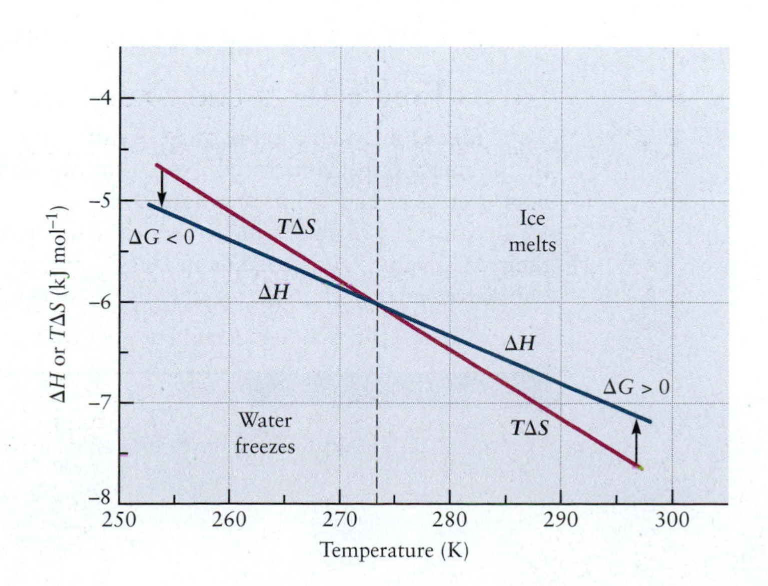

FIGURE 13.10 Plots of ΔH and $T\,\Delta S$ versus temperature for the freezing of water. At 273.15 K, the two curves cross, meaning that at this temperature, $\Delta G = 0$ and ice and water coexist. Below this temperature, the freezing of water to ice is spontaneous; above it, the reverse process, the melting of ice to water, is spontaneous.

FIGURE 13.11 The dissolution of hydrogen chloride in water, $HCl(g) \rightarrow HCl(aq)$, is a spontaneous process, with $\Delta G° = -35.9$ kJ. In this demonstration, the upper flask is filled with gaseous hydrogen chloride and a small amount of water is injected into it. As the hydrogen chloride dissolves spontaneously, its pressure declines. The resulting air pressure difference draws water up the tube from the lower flask, allowing more hydrogen chloride to dissolve. The change is so fast that a vigorous fountain of water plays into the upper flask. The free energy change of the process appears as work, raising the water.

necessary to determine only the sign of ΔG for the reaction at these same conditions. From experience with other state functions, we would expect to calculate ΔG for reactions by consulting appropriate tabulations of free energy data. Because we cannot know the absolute value of the Gibbs free energy of a substance (just as we cannot know the absolute value of its internal energy U), it is convenient to define a **standard molar Gibbs free energy of formation,** ΔG_f°, analogous to the standard molar enthalpy of formation ΔH_f° introduced in Section 12.5. From tables of ΔG_f° we can calculate $\Delta G°$ for a wide range of chemical reactions, just as we used Hess's law to calculate $\Delta H°$ for a reaction from tables of ΔH_f° for products and reactants. The next few paragraphs show how tables of ΔG_f° are generated and how the data are used to determine spontaneity of reactions.

STANDARD-STATE FREE ENERGIES The change in the Gibbs free energy for a chemical reaction performed at constant temperature is

$$\Delta G = \Delta H - T\,\Delta S \qquad [13.21]$$

where ΔH is the enthalpy change in the reaction (considered in Section 12.5) and ΔS is the entropy change in the reaction (see Section 13.5). The standard molar Gibbs free energy of formation ΔG_f° of a compound is the change in Gibbs free energy for the reaction in which 1 mol of the compound in its standard state is formed from its elements in their standard states. (You should review the definition of standard states in Section 12.5.) For example, ΔG_f° for $CO_2(g)$ is given by the Gibbs free energy change for the reaction

$$C(s) + O_2(g) \longrightarrow CO_2(g) \qquad \Delta G_f^\circ = ?$$

The value of ΔG_f° can be constructed from Equation 13.21 by using ΔH_f° and $\Delta S°$ values for this reaction and setting $T = 298.15$ K. For this reaction, $\Delta H°$ is simply ΔH_f° for $CO_2(g)$, because graphite and oxygen are elements in their standard states,

$$\Delta H° = \Delta H_f^\circ(CO_2) = -393.51 \text{ kJ}$$

The value of $\Delta S°$ can be obtained from the absolute entropies of the substances involved at 25°C and 1 atm pressure (both elements and compounds, because the absolute entropy $S°$ of an element is not zero in its standard state).

$$\begin{aligned}\Delta S° &= S°(CO_2) - S°(C) - S°(O_2) = 213.63 - 5.74 - 205.03 \text{ J K}^{-1}\\ &= +2.86 \text{ J K}^{-1}\end{aligned}$$

The ΔG_f° for CO_2 is then

$$\begin{aligned}\Delta G_f^\circ &= \Delta H_f^\circ - T\,\Delta S° = -393.51 \text{ kJ} - (298.15 \text{ K})(2.86 \text{ J K}^{-1})(10^{-3} \text{ kJ J}^{-1})\\ &= -394.36 \text{ kJ}\end{aligned}$$

Appendix D includes a table of ΔG_f° values for numerous substances, all obtained by the method just illustrated. Note that our definition makes $\Delta G_f^\circ = 0$ for an *element* that is already in its standard state.

Because G is a state function, chemical equations can be added together—with their ΔG_f° values combined as in Hess's law for changes in enthalpy—to calculate Gibbs free energy changes for chemical reactions under standard-state conditions.

EXAMPLE 13.10

Calculate $\Delta G°$ for the following reaction, using tabulated values for ΔG_f° from Appendix D.

$$3\,NO(g) \longrightarrow N_2O(g) + NO_2(g)$$

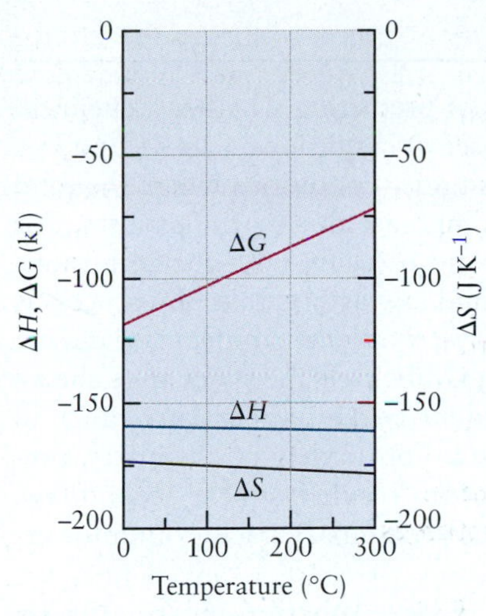

FIGURE 13.12 The entropy change of the reaction 3 NO(g) $\longrightarrow$ N_2O(g) + NO_2(g) varies less than 5% between 0°C and 300°C; the enthalpy of reaction is even closer to constancy. The free energy change in the reaction shifts greatly over the temperature range, however, as the magnitude of $T\,\Delta S$ increases. Note that the units are very different on the left- (energy) and right-hand vertical axes (entropy).

SOLUTION

$$\Delta G° = \Delta G_f°(N_2O) + \Delta G_f°(NO_2) - 3\,\Delta G_f°(NO)$$

$$= (1\text{ mol})(104.18\text{ kJ mol}^{-1}) + (1\text{ mol})(51.29\text{ kJ mol}^{-1}) - (3\text{ mol})(86.55\text{ kJ mol}^{-1})$$

$$= -104.18\text{ kJ}$$

EFFECTS OF TEMPERATURE ON ΔG Values of $\Delta G°$ calculated from the data in Appendix D are accurate only at T = 298.15 K. Values of $\Delta G°$ can be estimated for reactions at other temperatures and at P = 1 atm using the equation

$$\Delta G° = \Delta H° - T\,\Delta S° \qquad [13.22]$$

and tables of standard entropies and standard enthalpies of formation. The estimates will be close to the true value if $\Delta H°$ and $\Delta S°$ are not strongly dependent on T, which is usually the case (Fig. 13.12).

The value and the sign of $\Delta G°$ can depend strongly on T, even when the values of $\Delta H°$ and $\Delta S°$ do not, because of the competition between $\Delta H°$ and $T\,\Delta S°$ in Equation 13.22. If $\Delta H°$ is negative and $\Delta S°$ is positive, then the reaction is spontaneous at all temperatures when reactants and products are at atmospheric pressure. If $\Delta H°$ is positive and $\Delta S°$ is negative, the reaction is never spontaneous. For the other possible combinations, there exists a special temperature T^*, defined by

$$T^* = \frac{\Delta H°}{\Delta S°} \qquad [13.23]$$

at which $\Delta G°$ equals zero. If both $\Delta H°$ and $\Delta S°$ are positive, the reaction will be spontaneous at temperatures greater than T^*. If both are negative, the reaction will be spontaneous at temperatures less than T^* (Fig. 13.13). This discussion demonstrates an important result from chemical thermodynamics with enormous practical importance: With knowledge of $\Delta H°$ and $\Delta S°$, we can manipulate conditions to make a reaction spontaneous.

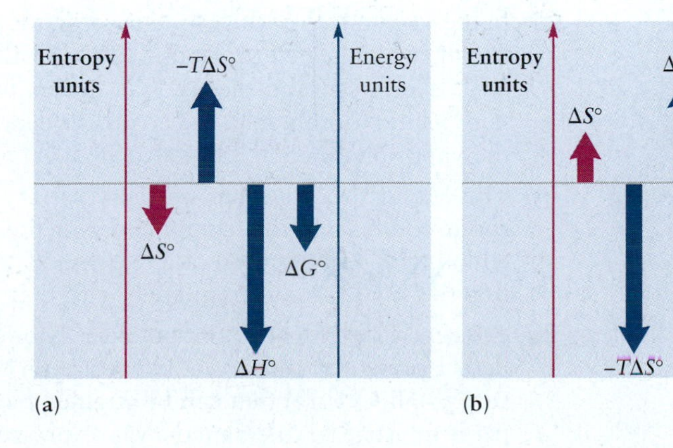

FIGURE 13.13 The competition between $\Delta H°$ and $\Delta S°$ determines the temperature range in which a reaction is spontaneous. (a) If both $\Delta H°$ and $\Delta S°$ are negative, the reaction is spontaneous at temperatures less than $T^* = \Delta H°/\Delta S°$. (b) If both $\Delta H°$ and $\Delta S°$ are positive, the reaction is spontaneous at temperatures greater than $T^* = \Delta H°/\Delta S°$.

CHAPTER SUMMARY

This chapter opened with the quest for methods of predicting whether a chemical reaction can occur spontaneously under a given set of conditions. The second law provides the answer: Any process is spontaneous under conditions where the total entropy of the *system and its surroundings* can increase during the process. For the particular case of constant temperature (T) and pressure (P)—the conditions most widely used for chemical reaction—the second law asserts that any process is spontaneous when the Gibbs free energy of the *system alone* can decrease during the process. The temperature dependence of the Gibbs free energy change shows that with knowledge of $\Delta H°$ and $\Delta S°$, we can identify the temperature range in which a given reaction is spontaneous. From the point of view of chemistry, predicting this temperature range is the most important result from the second law. All other material in this chapter can be viewed as preliminary background for arriving at this one crucial result.

The Gibbs free energy is the thermodynamic state function most naturally suited to describing the progress of chemical reactions at constant T and P. It provides the basis for predicting the equilibrium composition of the reaction mixture in Chapter 14.

CUMULATIVE EXERCISE

Purifying Nickel

Impure nickel, obtained from the smelting of its sulfide ores in a blast furnace, can be converted to metal of 99.90% to 99.99% purity by the Mond process, which relies on the equilibrium

$$Ni(s) + 4\ CO(g) \rightleftharpoons Ni(CO)_4(g)$$

The standard enthalpy of formation of nickel tetracarbonyl, $Ni(CO)_4(g)$, is -602.9 kJ mol^{-1}, and its absolute entropy $S°$ is 410.6 J K^{-1} mol^{-1}.

(a) Predict (without referring to a table) whether the entropy change of the system (the reacting atoms and molecules) is positive or negative in this process.

(b) At a temperature where this reaction is spontaneous, predict whether the entropy change of the surroundings is positive or negative.

(c) Use the data in Appendix D to calculate $\Delta H°$ and $\Delta S°$ for this reaction.

(d) At what temperature is $\Delta G° = 0$ for this reaction?

(e) The first step in the Mond process is the equilibration of impure nickel with CO and $Ni(CO)_4$ at about 50°C. In this step, the goal is to draw as much nickel as possible into the vapor-phase complex. Calculate $\Delta G°$ for the preceding reaction at 50°C.

(f) In the second step of the Mond process, the gases are removed from the reaction chamber and heated to about 230°C. At high enough temperatures, the sign of $\Delta G°$ is reversed and the reaction occurs in the opposite direction, depositing pure nickel. In this step, the goal is to deposit as much nickel as possible from the vapor-phase complex. Calculate $\Delta G°$ for the preceding reaction at 230°C.

(g) The Mond process relies on the volatility of $Ni(CO)_4$ for its success. Under room conditions, this compound is a liquid, but it boils at 42.2°C with an enthalpy of vaporization of 29.0 kJ mol^{-1}. Calculate the entropy of vaporization of $Ni(CO)_4$, and compare it with that predicted by Trouton's rule.

(h) A recently developed variation of the Mond process carries out the first step at higher pressures and at a temperature of 150°C. Estimate the maximum pressure of $Ni(CO)_4(g)$ that can be attained before the gas will liquefy at this temperature (that is, calculate the vapor pressure of $Ni(CO)_4(\ell)$ at 150°C).

Answers

(a) Negative
(b) Positive
(c) $\Delta H° = -160.8$ kJ; $\Delta S° = -409.5$ J K^{-1}
(d) 392.7 K = 119.5°C
(e) $\Delta G° = -28.4$ kJ
(f) $\Delta G° = +46.0$ kJ
(g) 92.0 J K^{-1} mol^{-1}, close to the Trouton's rule value of 88 J K^{-1} mol^{-1}
(h) 16.7 atm

CHAPTER REVIEW

- In general and qualitative terms, a spontaneous process is one that can occur by itself—given enough time—without outside intervention, once conditions have been established for its initiation.
- The outcome of a spontaneous process depends on the rate of the process. If it is slow, considerable time may be required before the results are seen.
- Spontaneous processes are described precisely in the language of thermodynamics. Initially, some barrier or constraint between the system and surroundings prevents their interaction. Once the constraint is removed, the spontaneous process is allowed to begin. During the process, the system may gain or lose energy, matter, and volume in exchange with the surroundings.
- A spontaneous process occurs in that direction which increases the total entropy of the combination system plus surroundings.
- A molecular statistical analysis shows that a spontaneous process occurs in the direction that gives the system access to more microstates. The extremely high probability that the system moves to the conditions with more microstates is a consequence of the large number of molecules in the system.
- In the molecular statistical analysis, Boltzmann defined the entropy S in any thermodynamic state as $S = k_B \ln \Omega$, where Ω is the number of microstates available to the system in that same thermodynamic state. This equation is used for qualitative interpretations of entropy changes. It shows that any process that increases Ω will increase S, and any process that decreases Ω will decrease S.
- By analyzing the Carnot cycle description of macroscopic energy transfer processes, Clausius demonstrated that the quantity $\int(1/T)dq_{rev}$ is a state function, because its value for any reversible process is independent of the path. Based on this result, Clausius defined the procedure for calculating the entropy change $\Delta S = S_f - S_i$ for a system between any thermodynamic states i and f as $\Delta S = \int_i^f (1/T)dq_{rev}$. The integral can be evaluated along any reversible path between i and f so long as T and dq_{rev} are known along the path. Because S is a state function, this procedure is valid for any process—even an irreversible one—that connects states i and f.
- It is easy to calculate entropy changes for isothermal processes, because T is constant and comes outside the integral to give $\Delta S = q_{rev}/T$. A specific example is the isothermal compression or expansion of an ideal gas, for which $\Delta S = nR \ln(V_f/V_i)$. A second example is any phase transition at constant pressure for which $q_{rev} = \Delta H_{trans}$. The entropy change is then $\Delta S_{trans} = \Delta H_{trans}/T_{trans}$.

- When the temperature changes during the process, we account for the variation of T along the process path by writing $dq_{\text{rev}} = nc_X dT$, where X represents V or P for a constant volume or a constant pressure process, respectively. When c_X is constant, the integral gives $\Delta S = nc_X \ln(T_f/T_i)$.
- Usually the surroundings is sufficiently large that it can be considered a constant temperature heat bath during the process. The heat lost by the system during the process is gained by the surroundings, so the entropy change for the surroundings is $\Delta S_{\text{surr}} = -\Delta H_{\text{sys}}/T_{\text{surr}}$. If the surroundings is not large enough to be treated in this way, ΔS_{surr} is calculated by the same procedures as ΔS_{sys}.
- To determine whether a process is spontaneous, we must calculate the total entropy change, $\Delta S_{\text{tot}} = \Delta S_{\text{sys}} + \Delta S_{\text{surr}}$. If $\Delta S_{\text{tot}} > 0$, the process is spontaneous. If $\Delta S_{\text{tot}} < 0$, the process cannot occur spontaneously. If $\Delta S_{\text{tot}} = 0$, the system is at equilibrium, and no process will occur.
- The third law of thermodynamics states that the entropy of any pure substance in equilibrium approaches zero at the absolute zero of temperature. Consequently, the entropy of every pure substance has a fixed value at each temperature and pressure, which can be calculated by starting with the low-temperature values and adding the results of all phase transitions that occur at intervening temperatures. This leads to tabulations of standard molar entropy $S°$ at 298.15 K and 1 atm pressure, which can be used to calculate entropy changes for chemical reactions in which the reactants and products are in these standard states.
- When processes are conducted at fixed temperature and pressure, spontaneity is determined by changes in the Gibbs free energy, $G = H - TS$, for the system with no consideration of changes in the surroundings. If $\Delta G < 0$, the process is spontaneous. If $\Delta G > 0$, the process cannot occur spontaneously. If $\Delta G = 0$, the system is at equilibrium, and no process will occur.
- Tabulations of the standard Gibbs free energy of formation for each substance in its standard state have been prepared by combining absolute entropy values with standard enthalpy of formation values. We can determine whether any chemical reaction is spontaneous by calculating its value of ΔG from the tabulated standard Gibbs free energy for its reactants and products. The algebraic sign of ΔG tells us whether the reaction is spontaneous.
- Whether a reaction is spontaneous depends on the relation between $\Delta H°$ and $\Delta S°$ at the temperature of the reaction. If $\Delta H° < 0$ and $\Delta S° > 0$, the reaction is spontaneous at all temperatures. If $\Delta H° > 0$ and $\Delta S° < 0$, the reaction is never spontaneous. For other combinations, there is a special temperature $T^* = \Delta H°/\Delta S°$ at which $\Delta G° = 0$. If both $\Delta H°$ and $\Delta S°$ are positive, the reaction is spontaneous for temperatures greater than T^*. If both $\Delta H°$ and $\Delta S°$ are negative, the reaction is spontaneous for temperatures less than T^*.

CONCEPTS & SKILLS

After studying this chapter and working the problems that follow, you should be able to:

1. Identify the system and surroundings involved in a spontaneous process and identify the constraint that was removed to enable the process to occur (Section 13.1, Problems 1–2).
2. Provide a statistical interpretation of the change in entropy that occurs when a gas undergoes a volume change (Section 13.2, Problems 3–10).
3. Summarize the justification that entropy is a state function (Sections 13.3 and 13.4).
4. Calculate the entropy change for the system and the surroundings for reversible and irreversible processes (Section 13.5, Problems 13–22).

5. Describe measurements of absolute entropy, and calculate standard-state entropy changes for chemical reactions (Section 13.6, Problems 23–30).
6. Define the Gibbs free energy function and state the criterion it provides for the spontaneity of a process (Section 13.7).
7. Calculate the change in Gibbs free energy for reversible and spontaneous phase transformations (Section 13.7, Problems 31–34).
8. Calculate the change in Gibbs free energy for chemical reactions and identify temperature ranges in which a particular reaction is spontaneous (Section 13.7, Problems 35–40).

KEY EQUATIONS

$S = k_B \ln \Omega$ (Section 13.2)

$w_{net} = -nR(T_h - T_l) \ln \frac{V_B}{V_A}$ (Section 13.4)

$\epsilon = \frac{-w_{net}}{q_h} = \frac{T_h - T_l}{T_h}$ (Section 13.4)

$\frac{q_h}{T_h} + \frac{q_l}{T_l} = 0$ (Section 13.4)

$\Delta S = \int_i^f \frac{dq_{rev}}{T}$ (Section 13.4)

$\Delta S = \int_i^f \frac{dq_{rev}}{T} = \frac{1}{T}\int_i^f dq_{rev} = \frac{q_{rev}}{T}$ (constant T) (Section 13.5)

$\Delta S = nR \ln\left(\frac{V_2}{V_1}\right)$ (constant T) (Section 13.5)

$\Delta S_{fus} = \frac{q_{rev}}{T_f} = \frac{\Delta H_{fus}}{T_f}$ (Section 13.5)

$\Delta S_{vap} = 88 \pm 5 \text{ J K}^{-1} \text{ mol}^{-1}$ (Section 13.5)

$\Delta S = \int_{T_1}^{T_2} \frac{nc_V}{T} dT = nc_V \ln\left(\frac{T_2}{T_1}\right)$ (constant V) (Section 13.5)

$\Delta S = \int_{T_1}^{T_2} \frac{nc_P}{T} dT = nc_P \ln\left(\frac{T_2}{T_1}\right)$ (constant P) (Section 13.5)

$\Delta S_{surr} = \frac{-\Delta H_{sys}}{T_{surr}}$ (Section 13.5)

$S° = \int_0^{298.15} \frac{c_P}{T} dT + \Delta S$ (phase changes between 0 and 298.15 K) (Section 13.6)

$\Delta S_{tot} > 0$ spontaneous (Section 13.7)

$\Delta S_{tot} = 0$ reversible (Section 13.7)

$\Delta S_{tot} < 0$ impossible (Section 13.7)

$G = H - TS$ (Section 13.7)

$\Delta G_{sys} < 0$ spontaneous processes (Section 13.7)

$\Delta G_{sys} = 0$ reversible processes (Section 13.7)

$\Delta G_{sys} > 0$ nonspontaneous processes (Section 13.7)

$\Delta G = \Delta H - T\,\Delta S$ (Section 13.7)

$\Delta G° = \Delta H° - T\,\Delta S°$ (Section 13.7)

PROBLEMS

Answers to problems whose numbers are boldface appear in Appendix G. Problems that are more challenging are indicated with asterisks.

The Nature of Spontaneous Processes

1. For each of the following processes, identify the system and the surroundings. Identify those processes that are spontaneous. For each spontaneous process, identify the constraint that has been removed to enable the process to occur:
(a) Ammonium nitrate dissolves in water.
(b) Hydrogen and oxygen explode in a closed bomb.
(c) A rubber band is rapidly extended by a hanging weight.
(d) The gas in a chamber is slowly compressed by a weighted piston.
(e) A glass shatters on the floor.

2. For each of the following processes, identify the system and the surroundings. Identify those processes that are spontaneous. For each spontaneous process, identify the constraint that has been removed to enable the process to occur:
(a) A solution of hydrochloric acid is titrated with a solution of sodium hydroxide.
(b) Zinc pellets dissolve in aqueous hydrochloric acid.
(c) A rubber band is slowly extended by a hanging weight.
(d) The gas in a chamber is rapidly compressed by a weighted piston.
(e) A tray of water freezes in the freezing compartment of an electric refrigerator.

Entropy and Spontaneity: A Molecular Statistical Interpretation

3. (a) How many "microstates" are there for the numbers that come up on a pair of dice?
(b) What is the probability that a roll of a pair of dice will show two sixes?

4. (a) Suppose a volume is divided into three equal parts. How many microstates can be written for all possible ways of distributing four molecules among the three parts?
(b) What is the probability that all four molecules are in the leftmost third of the volume at the same time?

5. When $H_2O(\ell)$ and $D_2O(\ell)$ are mixed, the following reaction occurs spontaneously:

$$H_2O(\ell) + D_2O(\ell) \rightarrow 2\,HOD(\ell)$$

There is little difference between the enthalpy of an O—H bond and that of an O—D bond. What is the main driving force for this reaction?

6. The two gases $BF_3(g)$ and $BCl_3(g)$ are mixed in equal molar amounts. All B—F bonds have about the same bond enthalpy, as do all B—Cl bonds. Explain why the mixture tends to react to form $BF_2Cl(g)$ and $BCl_2F(g)$.

7. Two large glass bulbs of identical volume are connected by means of a stopcock. One bulb initially contains 1.00 mol H_2; the other contains 1.00 mol helium (He). The stopcock is opened and the gases are allowed to mix and reach equilibrium. What is the probability that all the H_2 in the first bulb will diffuse into the second bulb and all the He gas in the second bulb will diffuse into the first bulb?

8. A mixture of 2.00 mol nitrogen and 1.00 mol oxygen is in thermal equilibrium in a 100-L container at 25°C. Calculate the probability that at a given time all the nitrogen will be found in the left half of the container and all the oxygen in the right half.

9. Predict the sign of the system's entropy change in each of the following processes.
(a) Sodium chloride melts
(b) A building is demolished
(c) A volume of air is divided into three separate volumes of nitrogen, oxygen, and argon, each at the same pressure and temperature as the original air

10. Predict the sign of the system's entropy change in each of the following processes.
(a) A computer is constructed from iron, copper, carbon, silicon, gallium, and arsenic
(b) A container holding a compressed gas develops a leak and the gas enters the atmosphere
(c) Solid carbon dioxide (dry ice) sublimes to gaseous carbon dioxide

A DEEPER LOOK . . . Carnot Cycles, Efficiency, and Entropy

11. A thermodynamic engine operates cyclically and reversibly between two temperature reservoirs, absorbing heat from the high-temperature bath at 450 K and discharging heat to the low-temperature bath at 300 K.
(a) What is the thermodynamic efficiency of the engine?
(b) How much heat is discarded to the low-temperature bath if 1500 J of heat is absorbed from the high-temperature bath during each cycle?
(c) How much work does the engine perform in one cycle of operation?

12. In each cycle of its operation, a thermal engine absorbs 1000 J of heat from a large heat reservoir at 400 K and discharges heat to another large heat sink at 300 K. Calculate:
    (a) The thermodynamic efficiency of the heat engine, operated reversibly
    (b) The quantity of heat discharged to the low-temperature sink each cycle
    (c) The maximum amount of work the engine can perform each cycle

Entropy Changes and Spontaneity

13. Tungsten melts at 3410°C and has an enthalpy change of fusion of 35.4 kJ mol^{-1}. Calculate the entropy of fusion of tungsten.
14. Tetraphenylgermane, $(C_6H_5)_4Ge$, has a melting point of 232.5°C, and its enthalpy increases by 106.7 J g^{-1} during fusion. Calculate the molar enthalpy of fusion and molar entropy of fusion of tetraphenylgermane.
15. The normal boiling point of acetone is 56.2°C. Use Trouton's rule to estimate its molar enthalpy of vaporization.
16. The molar enthalpy of vaporization of liquid hydrogen chloride is 16.15 kJ mol^{-1}. Use Trouton's rule to estimate its normal boiling point.
17. If 4.00 mol hydrogen (c_P = 28.8 J K^{-1} mol^{-1}) is expanded reversibly and isothermally at 400 K from an initial volume of 12.0 L to a final volume of 30.0 L, calculate ΔU, q, w, ΔH, and ΔS for the gas.
18. Suppose 60.0 g hydrogen bromide, HBr(*g*), is heated reversibly from 300 to 500 K at a constant volume of 50.0 L, and then allowed to expand isothermally and reversibly until the original pressure is reached. Using $c_P(HBr(g))$ = 29.1 J K^{-1} mol^{-1}, calculate ΔU, q, w, ΔH, and ΔS for this process. Assume that HBr is an ideal gas under these conditions.
19. Exactly 1 mol ice is heated reversibly at atmospheric pressure from −20°C to 0°C, melted reversibly at 0°C, and then heated reversibly at atmospheric pressure to 20°C. ΔH_{fus} = 6007 J mol^{-1}; c_P(ice) = 38 J K^{-1} mol^{-1}; and c_P(water) = 75 J K^{-1} mol^{-1}. Calculate ΔS for the system, the surroundings, and the thermodynamic universe for this process.
20. Suppose 1.00 mol water at 25°C is flash-evaporated by allowing it to fall into an iron crucible maintained at 150°C. Calculate ΔS for the water, ΔS for the iron crucible, and ΔS_{tot}, if $c_P(H_2O(\ell))$ = 75.4 J K^{-1} mol^{-1} and $c_P(H_2O(g))$ = 36.0 J K^{-1} mol^{-1}. Take ΔH_{vap} = 40.68 kJ mol^{-1} for water at its boiling point of 100°C.
21. In Example 12.3, a process was considered in which 72.4 g iron initially at 100.0°C was added to 100.0 g water initially at 10.0°C, and an equilibrium temperature of 16.5°C was reached. Take c_P(Fe) to be 25.1 J K^{-1} mol^{-1} and $c_P(H_2O)$ to be 75.3 J K^{-1} mol^{-1}, independent of temperature. Calculate ΔS for the iron, ΔS for the water, and ΔS_{tot} in this process.
22. Iron has a heat capacity of 25.1 J K^{-1} mol^{-1}, approximately independent of temperature between 0°C and 100°C.
    (a) Calculate the enthalpy and entropy change of 1.00 mol iron as it is cooled at atmospheric pressure from 100°C to 0°C.
    (b) A piece of iron weighing 55.85 g and at 100°C is placed in a large reservoir of water held at 0°C. It cools irreversibly until its temperature equals that of the water. Assuming the water reservoir is large enough that its temperature remains close to 0°C, calculate the entropy changes for the iron and the water and the total entropy change in this process.

The Third Law of Thermodynamics

23. (a) Use data from Appendix D to calculate the standard entropy change at 25°C for the reaction.

$$N_2H_4(\ell) + 3\,O_2(g) \longrightarrow NO_2(g) + 2\,H_2O(\ell)$$

    (b) Suppose the hydrazine (N_2H_4) is in the gaseous, rather than liquid, state. Will the entropy change for its reaction with oxygen be higher or lower than that calculated in part (a)? (*Hint:* Entropies of reaction can be added when chemical equations are added, in the same way that Hess's law allows enthalpies to be added.)
24. (a) Use data from Appendix D to calculate the standard entropy change at 25°C for the reaction.

$$CH_3COOH(g) + NH_3(g) \longrightarrow CH_3NH_2(g) + CO_2(g) + H_2(g)$$

    (b) Suppose that 1.00 mol each of solid acetamide, $CH_3CONH_2(s)$, and water, $H_2O(\ell)$, react to give the same products. Will the standard entropy change be larger or smaller than that calculated for the reaction in part (a)?
25. The alkali metals react with chlorine to give salts:

$$2\,Li(s) + Cl_2(g) \longrightarrow 2\,LiCl(s)$$
$$2\,Na(s) + Cl_2(g) \longrightarrow 2\,NaCl(s)$$
$$2\,K(s) + Cl_2(g) \longrightarrow 2\,KCl(s)$$
$$2\,Rb(s) + Cl_2(g) \longrightarrow 2\,RbCl(s)$$
$$2\,Cs(s) + Cl_2(g) \longrightarrow 2\,CsCl(s)$$

    Using the data in Appendix D, compute $\Delta S°$ of each reaction and identify a periodic trend, if any.
26. All of the halogens react directly with $H_2(g)$ to give binary compounds. The reactions are

$$F_2(g) + H_2(g) \longrightarrow 2\,HF(g)$$
$$Cl_2(g) + H_2(g) \longrightarrow 2\,HCl(g)$$
$$Br_2(g) + H_2(g) \longrightarrow 2\,HBr(g)$$
$$I_2(g) + H_2(g) \longrightarrow 2\,HI(g)$$

    Using the data in Appendix D, compute $\Delta S°$ of each reaction and identify a periodic trend, if any.
27. The dissolution of calcium chloride in water

$$CaCl_2(s) \longrightarrow Ca^{2+}(aq) + 2\,Cl^-(aq)$$

    is a spontaneous process at 25°C, even though the standard entropy change of the preceding reaction is negative ($\Delta S°$ = −44.7 J K^{-1}). What conclusion can you draw about the change in entropy of the surroundings in this process?

28. Quartz, $SiO_2(s)$, does not spontaneously decompose to silicon and oxygen at 25°C in the reaction

$$SiO_2(s) \longrightarrow Si(s) + O_2(g)$$

even though the standard entropy change of the reaction is large and positive ($\Delta S° = +182.02 \text{ J K}^{-1}$). Explain.

29. Use the microscopic interpretation of entropy from Section 13.2 to explain why the entropy change of the system in Problem 28 is positive.

30. (a) Why is the entropy change of the system negative for the reaction in Problem 27, when the ions become dispersed through a large volume of solution? (*Hint:* Think about the role of the solvent, water.)
(b) Use Appendix D to calculate $\Delta S°$ for the corresponding dissolution of $CaF_2(s)$. Explain why this value is even more negative than that given in Problem 27.

The Gibbs Free Energy

31. The molar enthalpy of fusion of solid ammonia is 5.65 kJ mol^{-1}, and the molar entropy of fusion is 28.9 J K^{-1} mol^{-1}.
(a) Calculate the Gibbs free energy change for the melting of 1.00 mol ammonia at 170 K.
(b) Calculate the Gibbs free energy change for the conversion of 3.60 mol solid ammonia to liquid ammonia at 170 K.
(c) Will ammonia melt spontaneously at 170 K?
(d) At what temperature are solid and liquid ammonia in equilibrium at a pressure of 1 atm?

32. Solid tin exists in two forms: white and gray. For the transformation

$$Sn(s, \text{white}) \longrightarrow Sn(s, \text{gray})$$

the enthalpy change is −2.1 kJ and the entropy change is -7.4 J K^{-1}.
(a) Calculate the Gibbs free energy change for the conversion of 1.00 mol white tin to gray tin at −30°C.
(b) Calculate the Gibbs free energy change for the conversion of 2.50 mol white tin to gray tin at −30°C.
(c) Will white tin convert spontaneously to gray tin at −30°C?
(d) At what temperature are white and gray tin in equilibrium at a pressure of 1 atm?

33. Ethanol's enthalpy of vaporization is 38.7 kJ mol^{-1} at its normal boiling point, 78°C. Calculate q, w, ΔU, ΔS_{sys}, and ΔG when 1.00 mol ethanol is vaporized reversibly at 78°C and 1 atm. Assume that the vapor is an ideal gas and neglect the volume of liquid ethanol relative to that of its vapor.

34. Suppose 1.00 mol superheated ice melts to liquid water at 25°C. Assume the specific heats of ice and liquid water have the same value and are independent of temperature. The enthalpy change for the melting of ice at 0°C is 6007 J mol^{-1}. Calculate ΔH, ΔS_{sys}, and ΔG for this process.

35. At 1200°C, the reduction of iron oxide to elemental iron and oxygen is not spontaneous:

$$2\ Fe_2O_3(s) \longrightarrow 4\ Fe(s) + 3\ O_2(g) \qquad \Delta G = +840 \text{ kJ}$$

Show how this process can be made to proceed if all the oxygen generated reacts with carbon:

$$C(s) + O_2(g) \longrightarrow CO_2(g) \quad \Delta G = -400 \text{ kJ}$$

This observation is the basis for the smelting of iron ore with coke to extract metallic iron.

36. The primary medium for free energy storage in living cells is adenosine triphosphate (ATP). Its formation from adenosine diphosphate (ADP) is not spontaneous:

$$ADP^{3-}(aq) + HPO_4^{2-}(aq) + H^+(aq) \longrightarrow ATP^{4-}(aq) + H_2O(\ell) \quad \Delta G = +34.5 \text{ kJ}$$

Cells couple ATP production with the metabolism of glucose (a sugar):

$$C_6H_{12}O_6(aq) + 6\ O_2(g) \longrightarrow 6\ CO_2(g) + 6\ H_2O(\ell) \quad \Delta G = -2872 \text{ kJ}$$

The reaction of 1 molecule of glucose leads to the formation of 38 molecules of ATP from ADP. Show how the coupling makes this reaction spontaneous. What fraction of the free energy released in the oxidation of glucose is stored in the ATP?

37. A process at constant T and P can be described as spontaneous if $\Delta G < 0$ and nonspontaneous if $\Delta G > 0$. Over what range of temperatures is each of the following processes spontaneous? Assume that all gases are at a pressure of 1 atm. (*Hint:* Use Appendix D to calculate ΔH and ΔS [assumed independent of temperature and equal to $\Delta H°$ and $\Delta S°$, respectively], and then use the definition of ΔG.)
(a) The rusting of iron, a complex reaction that can be approximated as

$$4\ Fe(s) + 3\ O_2(g) \longrightarrow 2\ Fe_2O_3(s)$$

(b) The preparation of $SO_3(g)$ from $SO_2(g)$, a step in the manufacture of sulfuric acid:

$$SO_2(g) + \tfrac{1}{2}\ O_2(g) \longrightarrow SO_3(g)$$

(c) The production of the anesthetic dinitrogen oxide through the decomposition of ammonium nitrate:

$$NH_4NO_3(s) \longrightarrow N_2O(g) + 2\ H_2O(g)$$

38. Follow the same procedure used in Problem 37 to determine the range of temperatures over which each of the following processes is spontaneous.
(a) The preparation of the poisonous gas phosgene:

$$CO(g) + Cl_2(g) \longrightarrow COCl_2(g)$$

(b) The laboratory-scale production of oxygen from the decomposition of potassium chlorate:

$$2\ KClO_3(s) \longrightarrow 2\ KCl(s) + 3\ O_2(g)$$

(c) The reduction of iron(II) oxide (wüstite) by coke (carbon), a step in the production of iron in a blast furnace:

$$FeO(s) + C(s, \text{gr}) \longrightarrow Fe(s) + CO(g)$$

39. Explain how it is possible to reduce tungsten(VI) oxide (WO_3) to metal with hydrogen at an elevated temperature. Over what temperature range is this reaction spontaneous? Use the data of Appendix D.

40. Tungsten(VI) oxide can also be reduced to tungsten by heating it with carbon in an electric furnace:

$$2\ WO_3(s) + 3\ C(s) \longrightarrow 2\ W(s) + 3\ CO_2(g)$$

(a) Calculate the standard free energy change ($\Delta G°$) for this reaction, and comment on the feasibility of the process at room conditions.
(b) What must be done to make the process thermodynamically feasible, assuming ΔH and ΔS are nearly independent of temperature?

ADDITIONAL PROBLEMS

41. Ethanol (CH_3CH_2OH) has a normal boiling point of 78.4°C and a molar enthalpy of vaporization of 38.74 kJ mol^{-1}. Calculate the molar entropy of vaporization of ethanol and compare it with the prediction of Trouton's rule.

42. A quantity of ice is mixed with a quantity of hot water in a sealed, rigid, insulated container. The insulation prevents heat exchange between the ice–water mixture and the surroundings. The contents of the container soon reach equilibrium. State whether the total *internal energy* of the contents decreases, remains the same, or increases in this process. Make a similar statement about the total *entropy* of the contents. Explain your answers.

43. (a) If 2.60 mol $O_2(g)$ (c_P = 29.4 J K^{-1} mol^{-1}) is compressed reversibly and adiabatically from an initial pressure of 1.00 atm and 300 K to a final pressure of 8.00 atm, calculate ΔS for the gas.
(b) Suppose a different path from that in part (a) is used. The gas is first heated at constant pressure to the same final temperature, and then compressed reversibly and isothermally to the same final pressure. Calculate ΔS for this path and show that it is equal to that found in part (a).

44. One mole of a monatomic ideal gas begins in a state with P = 1.00 atm and T = 300 K. It is expanded reversibly and adiabatically until the volume has doubled; then it is expanded irreversibly and isothermally into a vacuum until the volume has doubled again; and then it is heated reversibly at constant volume to 400 K. Finally, it is compressed reversibly and isothermally until a final state with P = 1.00 atm and T = 400 K is reached. Calculate ΔS_{sys} for this process. (*Hint:* There are two ways to solve this problem—an easy way and a hard way.)

* 45. The motion of air masses through the atmosphere can be approximated as adiabatic (because air is a poor conductor of heat) and reversible (because pressure differences in the atmosphere are small). To a good approximation, air can be treated as an ideal gas with average molar mass 29 g mol^{-1} and average heat capacity 29 J K^{-1} mol^{-1}.
(a) Show that the displacement of the air masses occurs at constant entropy ($\Delta S = 0$).
(b) Suppose the average atmospheric pressure near the earth's surface is P_0 and the temperature is T_0. The air is displaced upward until its temperature is T and its pressure is P. Determine the relation between P and T. (*Hint:* Consider the process as occurring in two steps: first a cooling from T_0 to T at constant pressure, and then an expansion from P_0 to P at constant temperature. Equate the sum of the two entropy changes to $\Delta S_{tot} = 0$.)
(c) In the lower atmosphere, the dependence of pressure on height, h, above the earth's surface can be approximated as

$$\ln (P/P_0) = -\mathcal{M}gh/RT$$

where $\mathcal{M}$ is the molar mass (kg mol^{-1}), g the acceleration due to gravity (9.8 m s^{-2}), and R the gas constant. If the air temperature at sea level near the equator is 38°C (~100°F), calculate the air temperature at the summit of Mount Kilimanjaro, 5.9 km above sea level. (For further discussion of this problem, see L. K. Nash, *J. Chem. Educ.* 61:23, 1984.)

46. Calculate the entropy change that results from mixing 54.0 g water at 273 K with 27.0 g water at 373 K in a vessel whose walls are perfectly insulated from the surroundings. Consider the specific heat of water to be constant over the temperature range from 273 to 373 K and to have the value 4.18 J K^{-1} g^{-1}.

47. Problem 22 asked for the entropy change when a piece of iron is cooled by immersion in a reservoir of water at 0°C.
(a) Repeat Problem 22(b), supposing that the iron is cooled to 50°C in a large water reservoir held at that temperature before being placed in the 0°C reservoir.
(b) Repeat the calculation supposing that four water reservoirs at 75°C, 50°C, 25°C, and 0°C are used.
(c) As more reservoirs are used, what happens to ΔS for the iron, for the water, and for the universe? How would you attempt to conduct a reversible cooling of the iron?

48. Problem 42 in Chapter 9 described an optical atomic trap. In one experiment, a gas of 500 sodium atoms is confined in a volume of 1000 μm^3. The temperature of the system is 0.00024 K. Compute the probability that, by chance, these 500 slowly moving sodium atoms will all congregate in the left half of the available volume. Express your answer in scientific notation.

* 49. Suppose we have several different ideal gases, i = 1, 2, 3, . . . , N, each occupying its own volume V_i, all at the same pressure and temperature. The boundaries between the volumes are removed so that the gases mix at constant temperature in the total volume $V = \Sigma_i V_i$.
(a) Using the microscopic interpretation of entropy, show that

$$\Delta S = -nR \sum_i X_i \ln X_i$$

for this process, where n is the total number of moles of gas and X_i is the mole fraction of gas i.
(b) Calculate the entropy change when 50 g each of $O_2(g)$, $N_2(g)$, and Ar(g) are mixed at 1 atm and 0°C.
(c) Using Table 9.1, calculate the entropy change when 100 L of air (assumed to be a mixture of ideal gases) at 1 atm and 25°C is separated into its component gases at the same pressure and temperature.

* 50. The N_2O molecule has the structure N—N—O. In an ordered crystal of N_2O, the molecules are lined up in a regular fashion, with the orientation of each determined by its position in the crystal. In a random crystal (formed on rapid freezing), each molecule has two equally likely orientations.
(a) Calculate the number of microstates available to a random crystal of N_A (Avogadro's number) of molecules.
(b) Calculate the entropy change when 1.00 mol of a random crystal is converted to an ordered crystal.

51. By examining the following graphs, predict which element—copper or gold—has the higher absolute entropy at a temperature of 200 K.

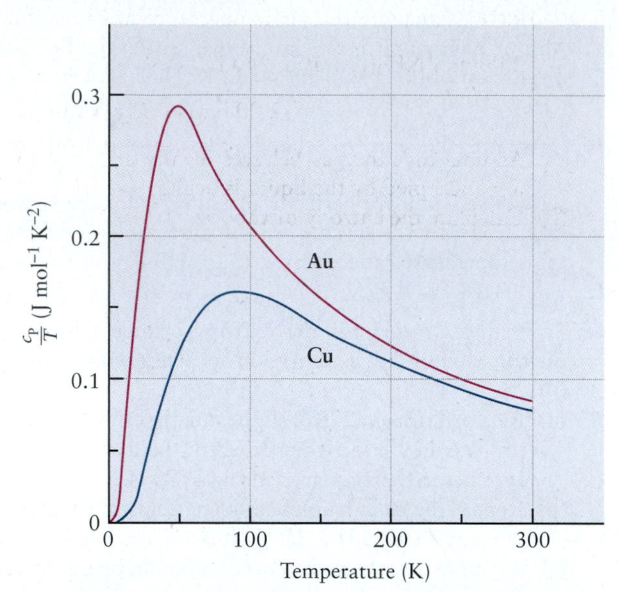

* 52. Consider the process described in Problem 19 in Chapter 12. Use the results from that problem to do the following.
(a) Calculate ΔS for the system, the surroundings, and the universe.
(b) If the absolute entropy per mole of the gas *before* the expansion is 158.2 J K^{-1} mol^{-1}, calculate ΔG_{sys} for the process.

53. Two different crystalline forms of sulfur are the rhombic form and the monoclinic form. At atmospheric pressure, rhombic sulfur undergoes a transition to monoclinic when it is heated above 368.5 K:

$$S(s, \text{rhombic}) \longrightarrow S(s, \text{monoclinic})$$

(a) What is the sign of the entropy change (ΔS) for this transition?
(b) $|\Delta H|$ for this transition is 400 J mol^{-1}. Calculate ΔS for the transition.

* 54. Use data from Appendix D to estimate the temperature at which $I_2(g)$ and $I_2(s)$ are in equilibrium at a pressure of 1 atm. Can this equilibrium actually be achieved? Refer to Appendix F for data on iodine.

55. The molar enthalpy of fusion of ice at 0°C is 6.02 kJ mol^{-1}; the molar heat capacity of undercooled water is 75.3 J mol^{-1} K^{-1}. (a) One mole of undercooled water at −10°C is induced to crystallize in a heat-insulated vessel. The result is a mixture of ice and water at 0°C. What fraction of this mixture is ice? (b) Calculate ΔS for the system.

56. A certain substance consists of two modifications A and B; $\Delta G°$ for the transition from A to B is positive. The two modifications produce the same vapor. Which has the higher vapor pressure? Which is the more soluble in a solvent common to both?

57. From the values in Appendix D, calculate the values of $\Delta G°$ and $\Delta H°$ for the reaction.

$$3\ Fe_2O_3(s) \longrightarrow 2\ Fe_3O_4(s) + \tfrac{1}{2}\ O_2(g)$$

at 25°C. Which of the two oxides is more stable at 25°C and P_{O_2} = 1 atm?

58. The strongest known chemical bond is that in carbon monoxide, CO, with bond enthalpy of 1.05×10^3 kJ mol^{-1}. Furthermore, the entropy increase in a gaseous dissociation of the kind AB $\rightleftarrows$ A + B is about 110 J mol^{-1} K^{-1}. These factors establish a temperature above which there is essentially no chemistry of molecules. Show why this is so, and find the temperature.

59. The $\Delta G_f°$ of $Si_3N_4(s)$ is −642.6 kJ mol^{-1}. Use this fact and the data in Appendix D to compute $\Delta G°$ of the reaction.

$$3\ CO_2(g) + Si_3N_4(s) \longrightarrow 3\ SiO_2(\text{quartz}) + 2\ N_2(g) + 3\ C(s, \text{gr})$$

60. The compound $Pt(NH_3)_2I_2$ comes in two forms, the cis and the trans, which differ in their molecular structure. The following data are available:

	$\Delta H_f°$ (kJ mol^{-1})	$\Delta G_f°$ (kJ mol^{-1})
cis	−286.56	−130.25
trans	−316.94	−161.50

Combine these data with data from Appendix D to compute the standard entropies ($S°$) of both of these compounds at 25°C.

61. (a) Use data from Appendix D to calculate $\Delta H°$ and $\Delta S°$ at 25°C for the reaction

$$2\ CuCl_2(s) \rightleftarrows 2\ CuCl(s) + Cl_2(g)$$

(b) Calculate ΔG at 590 K, assuming $\Delta H°$ and $\Delta S°$ are independent of temperature.
(c) Careful high-temperature measurements show that when this reaction is performed at 590 K, ΔH_{590} is 158.36 kJ and ΔS_{590} is 177.74 J K^{-1}. Use these facts to compute an improved value of ΔG_{590} for this reaction. Determine the percentage error in ΔG_{590} that comes from using the 298-K values in place of 590-K values in this case.

62. (a) The normal boiling point of carbon tetrachloride (CCl_4) is 76.5°C. A student looks up the standard enthalpies of formation of $CCl_4(\ell)$ and of $CCl_4(g)$ in Appendix D. They are listed as −135.44 and −102.9 kJ mol^{-1}, respectively. By subtracting the first from the second, she computes $\Delta H°$ for the vaporization of CCl_4 to be 32.5 kJ mol^{-1}. But Table 7.2 states that the ΔH_{vap} of CCl_4 is 30.0 kJ mol^{-1}. Explain the discrepancy.
(b) Calculate the molar entropy change of vaporization (ΔS_{vap}) of CCl_4 at 76.5°C.

63. The typical potassium ion concentration in the fluid outside a cell is 0.0050 M, whereas that inside a muscle cell is 0.15 M.
(a) What is the spontaneous direction of motion of ions through the cell wall?
(b) In *active transport*, cells use free energy stored in ATP (see Problem 36) to move ions in the direction opposite their spontaneous direction of flow. Calculate the cost in free energy to move 1.00 mol K^+ through the cell wall by active transport. Assume no change in K^+ concentrations during this process.

CUMULATIVE PROBLEMS

64. When a gas undergoes a reversible adiabatic expansion, its entropy remains constant even though the volume increases. Explain how this can be consistent with the microscopic interpretation of entropy developed in Section 13.2. (*Hint:* Consider what happens to the distribution of velocities in the gas.)

65. The normal boiling point of liquid ammonia is 240 K; the enthalpy of vaporization at that temperature is 23.4 kJ mol^{-1}. The heat capacity of gaseous ammonia at constant pressure is 38 J mol^{-1} K^{-1}.
   (a) Calculate q, w, ΔH, and ΔU for the following change in state:

$$2.00 \text{ mol } NH_3(\ell, 1 \text{ atm}, 240 \text{ K}) \longrightarrow 2.00 \text{ mol } NH_3(g, 1 \text{ atm}, 298 \text{ K})$$

   Assume that the gas behaves ideally and that the volume occupied by the liquid is negligible.
   (b) Calculate the entropy of vaporization of NH_3 at 240 K.

CHAPTER 14

Chemical Equilibrium

Gaseous ammonia, NH_3, and gaseous hydrogen chloride, HCl, react to form solid NH_4Cl, the white smoke. In the reverse reaction, solid NH_4Cl decomposes when heated to form gaseous NH_3 and HCl.

Every time we carry out a chemical reaction—from fundamental research studies to practical industrial applications—the *yield* of the reaction is extremely important. Did we obtain all the product we could expect? Chapter 2 shows how to calculate the amount of product expected when we start a reaction with particular amounts of the reactants. This calculation assumes that the reaction goes to completion—that is, all of the limiting reagent is consumed. The resulting number, called the theoretical yield, represents the maximum amount of product that could be obtained from that reaction.

In practice, many reactions do not go to completion but rather approach a state or position of **equilibrium.** This equilibrium position, at which the reaction apparently comes to an end, is a mixture of products and unconsumed reactants present in fixed relative amounts. Once equilibrium has been achieved, there is no further net conversion of reactants to products unless the experimental conditions of the reaction (temperature and pressure) are changed. The equilibrium state is characterized by the **equilibrium constant,** which has a unique value for each reaction. Knowing the equilibrium constant and the initial amounts of reactants and products, we can calculate the composition of the equilibrium reaction mixture. Knowing the equilibrium constant and its dependence on experimental conditions, we can manipulate conditions to maximize the practical yield of that reaction. Calculating the equilibrium composition for a particular reaction and its dependence on experimental conditions is therefore a practical skill of enormous importance in chemistry.

In this chapter we describe the equilibrium constant, its dependence on conditions, and its role in manipulating the yield of reactions, emphasizing those aspects broadly applicable to all chemical reactions. We illustrate these general principles with applications to reactions in the gas phase and to heterogeneous reactions. Detailed applications to reactions in aqueous solutions and electrochemical reactions are presented in the three following chapters.

The fact that reactions go to the equilibrium position was discovered empirically, and the equilibrium constant was first defined empirically. All the aforementioned applications can be accomplished with empirically determined equilibrium constants. Nonetheless, the empirical approach leaves unanswered several important fundamental questions: Why should the equilibrium state exist? Why does the equilibrium constant take its particular mathematical form? These and related questions are answered by recognizing that the chemical equilibrium position is the *thermodynamic equilibrium state* of the reaction mixture. Once we have made that connection, thermodynamics explains the existence and the mathematical form of the equilibrium constant. Thermodynamics also gives procedures for calculating the value of the equilibrium constant from the thermochemical properties of the pure reactants and products, as well as procedures for predicting its dependence on experimental conditions.

Some instructors prefer to introduce equilibrium from the empirical viewpoint and later use thermodynamics to explain the empirical developments. Others prefer to develop the background of thermodynamics first and then apply it to chemical equilibrium. We have organized this chapter to allow either approach. After an introductory section on the general nature of chemical equilibrium, we introduce the equilibrium constant empirically in Section 14.2, and then give a thermodynamic description in Section 14.3. Similarly, we discuss the direction of change in chemical reactions empirically in Section 14.6 and give a thermodynamic treatment in Section 14.7. Readers who have studied thermodynamics before starting this chapter should read the sections in the order presented. Readers who have not yet studied thermodynamics should skip over Sections 14.3 and 14.7 and return to them after studying Chapters 12 and 13 on thermodynamics. We provide signposts for both sets of readers at the end of each section.

14.1 The Nature of Chemical Equilibrium

Approach to Equilibrium

Most chemical reactions are carried out by mixing the selected reactants in a vessel and adjusting T and P until the desired products appear. Once reaction conditions have been identified, the greatest concern is the yield of the reaction. How

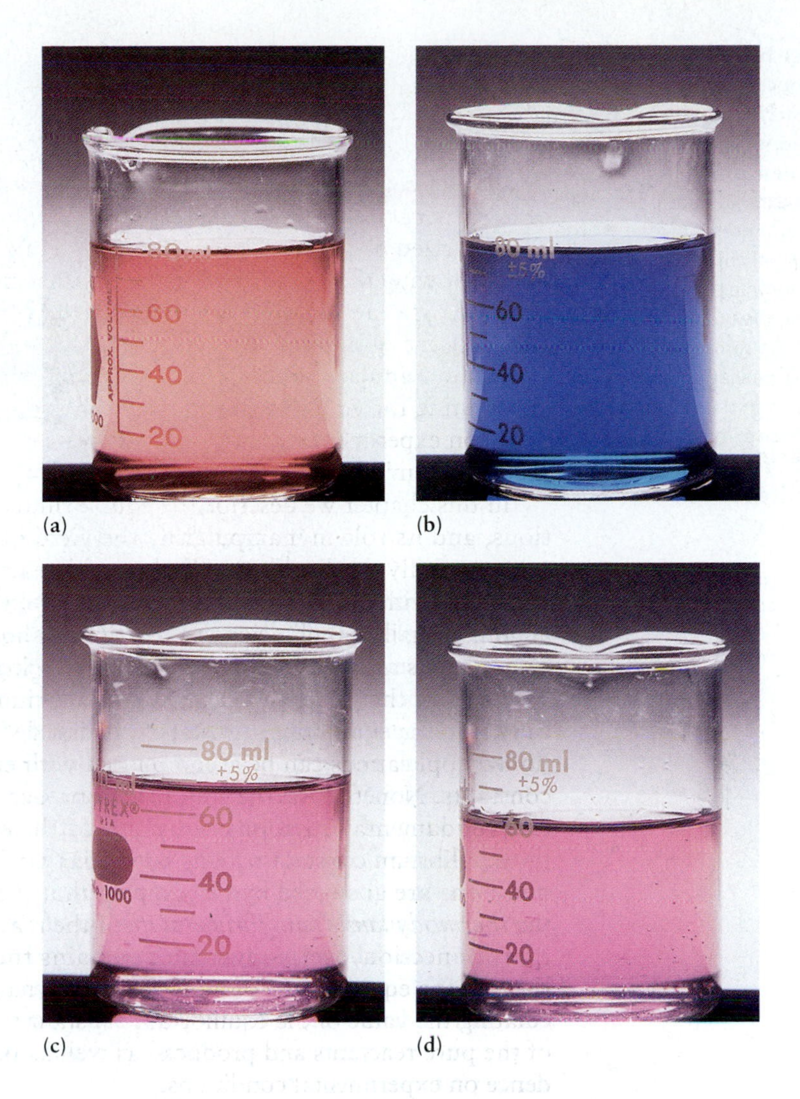

FIGURE 14.1 Chemical equilibrium in the cobalt chloride–HCl system. (a) The pink color is due to the hexaaqua complex ion $[Co(H_2O)_6]^{2+}$. (b) The blue color is due to the tetrachloro complex ion $[CoCl_4]^{2-}$. (c) Adding HCl to the pink solution in (a) converts some of the Co(II) to the tetrachloro complex. The lavender color is produced by the combination of pink hexaaqua species and blue tetrachloro species. (d) Adding water to the blue solution in (b) converts some of the Co(II) to the hexaaqua species. The combination of the two gives the lavender color. The same equilibrium state is reached by running the reaction from the left (c) and from the right (d).

do we obtain the greatest amount of the desired product? How close to completion does the reaction proceed?

Let's start by observing a particular reaction that illustrates all the aspects of equilibrium we need to understand. We have selected the reactions of cobalt(II) ions in aqueous solutions in which chloride ion is also present because the progress and outcome of the reactions are directly visible. The cobalt(II) ions can form various different complex ions, depending on the amount of chloride present. For example, if $CoCl_2{\cdot}6H_2O$ is dissolved in pure water to the concentration 0.08 M, the resulting solution is pale pink in color due to the hexaaquacobalt(II) complex ion $[Co(H_2O)_6]^{2+}$ (Fig. 14.1a). If $CoCl_2{\cdot}6H_2O$ is dissolved in 10 M HCl to the concentration 0.08 M, the solution is deep blue due to the tetrachlorocobalt(II) complex ion $[CoCl_4]^{2-}$ (see Fig. 14.1b). Solutions containing a mixture of both Co(II) species are purple.

The hexaaqua complex can be converted into the tetrachloro complex by addition of chloride ion through the reaction

$$[Co(H_2O)_6]^{2+} + 4\ Cl^- \rightleftharpoons [CoCl_4]^{2-} + 6\ H_2O$$

If we add concentrated HCl (one of the reactants, on the left side of the equation) to the pink solution in Figure 14.1a until the Co(II) concentration is 0.044 M and the HCl concentration is 5.5 M, the result is a lavender-colored solution, a sample of which is shown in Figure 14.1c. Optical absorption spectroscopy measurements to be described in Chapter 20 confirm the presence of both Co(II) species in the solution in Figure 14.1c. Ninety-eight percent of the Co is found in

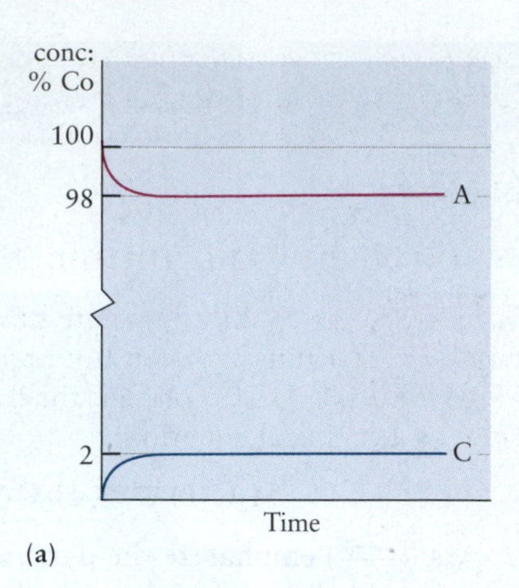

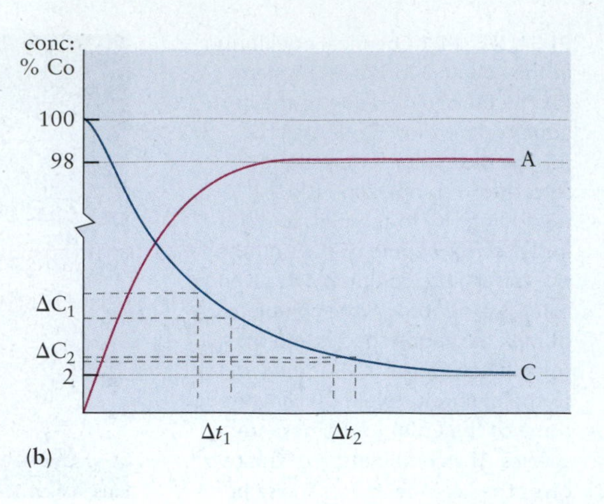

FIGURE 14.2 Sketch of the change with time of the concentrations of products and reactants in the spontaneous reactions illustrated in Figure 14.1. For ease of display, concentrations are expressed as percent of the total Co(II) present in each species. (a) Partial conversion of pink hexaaqua complex A into blue tetrachloro complex C. (b) Partial conversion of blue tetrachloro complex C into pink hexaaqua complex A. After changes in the slope of each species concentration become imperceptibly small, we say the reaction has arrived at chemical equilibrium.

the pink hexaaqua complex, and the remaining two percent is in the blue tetrachloro complex.[1] If the lavender solution is allowed to stand at constant temperature for several hours and the measurements repeated, the results are the same.

The reaction can also be carried out from "right to left" as written. If we add pure water (one of the products, on the right side of the equation) to the blue solution in Figure 14.1b until the cobalt concentration is 0.044 M and the HCl concentration is 5.5 M, the result is a lavender-colored solution, a sample of which is shown in Figure 14.1d. Just as we saw after running the reaction "left to right," optical absorption spectroscopy confirms that 98% of the Co(II) is present in the pink hexaaqua complex and 2% is present in the blue tetrachloro complex. Again, we find no further change in composition of the mixture after a long wait.

These data show that the reaction has not gone to completion but has apparently halted at an intermediate state containing products as well as unconsumed reactants. Moreover, the same final state can be achieved from either direction. This result is typical of most chemical reactions, even though it is not generally as readily apparent as here.

The questions raised in the first paragraph require quantitative investigations of the reaction mixture, which we carry out as follows. In the preceding reaction let A represent the pink hexaaqua complex, B the chloride ion, and C the blue tetrachloro complex. In the first experiment, we start the reaction by mixing initial concentrations of A and B, denoted as $[A]_0$ and $[B]_0$. As the reaction proceeds, we periodically sample the reaction mixture. For each sample, we measure the concentration of A, B, and C and plot concentration of each species versus time. The results of the first experiment are represented schematically in Figure 14.2a, which shows the consumption of A and the production of C. Similarly, we start the second experiment with the initial concentration $[C]_0$, and add water. The results are represented schematically in Figure 14.2b, which shows the consumption of C and the production of A.

As the reaction proceeds, we see that the concentration of each species changes progressively more slowly. This fact is indicated by the decreasing values of the slope $[m = \Delta(\text{conc.})/\Delta t]$ sketched on the concentration curves during progressively later time intervals. Eventually, the slopes become close enough to zero to show that the concentration of each species has become constant in time. When this condition has been achieved, we say the reaction is at *chemical equilibrium,* and

[1]It may appear surprising to find such imbalance in the concentration of the Co(II) species when the color intensities of the pink and lavender solutions in Figure 14.1a and Figure 14.1c appear quite similar. This difference is explained by the fact that the blue tetrachloro complex absorbs light much more efficiently than the pink hexaaqua complex.

the reaction mixture is at the *equilibrium composition.*[2] At later times, there is no further change in composition of the reaction mixture; the concentration of each of the species X = A, B, C remains at its equilibrium value denoted by $[X]_{eq}$.

Characteristics of the Equilibrium State

We can gain insight into the equilibrium state by comparing chemical reactions to the familiar phase equilibrium between liquid water and water vapor (see Section 10.4 and Figure 10.19). Let's represent the transfer of water molecules in this phase equilibrium as a chemical equation:

$$H_2O(\ell) \rightleftarrows H_2O(g)$$

The double arrows ($\rightleftarrows$) emphasize the dynamic nature of phase equilibrium: Liquid water evaporates to form water vapor and at the same time vapor condenses to give liquid. An analogous dynamic description applies to a chemical equilibrium, in which bonds are broken or formed as atoms move back and forth between reactant and product molecules. When the initial concentrations of the reactants are high, collisions between their molecules cause product molecules to form. Once the concentrations of the products have increased sufficiently, the reverse reaction (forming "reactants" from "products") begins to occur. As the equilibrium state is approached, the forward and backward rates of reaction become equal and there is no further net change in reactant or product concentrations. Just as the equilibrium between liquid water and water vapor is a dynamic process on the molecular scale, with evaporation and condensation taking place simultaneously, the chemical equilibrium between reactants and products also occurs through the continuous formation of molecules of product from reactant molecules and their reaction back into reactant molecules with equal rates. Chemical equilibrium is not a static condition, although macroscopic properties such as concentrations do stop changing when equilibrium is attained. Chemical equilibrium is the consequence of a dynamic balance between forward and backward reactions.

The experimental results shown in Figures 14.1 and 14.2 demonstrate that the same equilibrium state is reached whether one starts with the reactants or with the products. This fact can be used to test whether a system is truly in equilibrium or whether the reaction is just so slow that changes in concentration are unmeasurably small, even though the system is far from equilibrium. If the same state is reached from either reactants or products, that state is a true equilibrium state.

Equilibrium states have four fundamental characteristics:

1. They display no macroscopic evidence of change.
2. They are reached through spontaneous (in the sense of Chapter 13) processes.
3. They show a dynamic balance of forward and reverse processes.
4. They are the same regardless of direction of approach.

We frequently also encounter so-called *steady states* in which the macroscopic concentrations of species are not changing with time, even though the system is not at equilibrium. Steady states are maintained not by a dynamic balance between forward and reverse processes but rather by the competition between a process that supplies the species to the system and a process that removes the species from the system. Many chemical reactions occur in living systems in steady states and do not represent an equilibrium between reactants and products. You must be certain that a reaction is at equilibrium and not in steady state before applying the methods of this chapter to explain the relative concentrations of reactants and products.

[2]You must exercise judgment in deciding when the slope is "sufficiently close to zero" and the concentrations are "effectively constant." There is no one instant at which equilibrium is achieved.

14.2 The Empirical Law of Mass Action

Let us examine again the approach to equilibrium, as represented in Figure 14.2 and the related discussion. Extensive studies of this type for broad classes of reactions represented generally as

$$a\text{A} + b\text{B} \rightleftharpoons c\text{C} + d\text{D}$$

have demonstrated a most remarkable result. No matter what initial concentrations of reactants are selected at the beginning of the experiment, the value of the ratio

$$\frac{[\text{C}]_{\text{eq}}^{c}[\text{D}]_{\text{eq}}^{d}}{[\text{A}]_{\text{eq}}^{a}[\text{B}]_{\text{eq}}^{b}}$$

measured *at equilibrium* is always the same. Even if the experiment is started with an arbitrary initial mixture of reactants and products, the reaction will consume some species and produce others until it achieves this same value of this ratio at equilibrium. Consequently, this ratio is called the **empirical equilibrium constant** for the reaction and denoted as K_{C}. The results of these studies are summarized in the following equation called the **law of mass action,** first stated in approximate form in 1864 by two Norwegians, C. M. Guldberg (a mathematician) and his brother-in-law P. Waage (a chemist):

$$\frac{[\text{C}]_{\text{eq}}^{c}[\text{D}]_{\text{eq}}^{d}}{[\text{A}]_{\text{eq}}^{a}[\text{B}]_{\text{eq}}^{b}} = K_{\text{C}} \qquad [14.1\text{a}]$$

The subscript C denotes that the reaction is carried out in solution and that the empirical equilibrium constant K_{C} is evaluated by directly measuring the concentration of each species in the equilibrium state of the reaction. In general, K_{C} has dimensions $(\text{concentration})^{c+d-a-b}$; it will be dimensionless only for those reactions for which $a + b = c + d$.

Similar results have been obtained for reactions carried out in the gas phase, where the amount of each reactant and product in the reaction mixture is measured by its partial pressure P_{X}. For gas-phase reactions, the empirical law of mass action takes the form

$$\frac{(P_{\text{C}})_{\text{eq}}^{c}(P_{\text{D}})_{\text{eq}}^{d}}{(P_{\text{A}})_{\text{eq}}^{a}(P_{\text{B}})_{\text{eq}}^{b}} = K_{\text{P}} \qquad [14.1\text{b}]$$

In general, K_{P} has dimensions $(\text{pressure})^{c+d-a-b}$; it will be dimensionless only for those reactions for which $a + b = c + d$.

The significance of the empirical law of mass action is twofold. First, the numerical value of K_{C} or K_{P} is an inherent property of the chemical reaction itself and does not depend on the specific initial concentrations of reactants and products selected. Second, the magnitude of K_{P} or K_{C} gives direct information about the nature of the equilibrium state or position of the reaction. If the equilibrium constant is very large, then at equilibrium the concentration or partial pressures of products are large compared with those of the reactants. In this case, stoichiometry can be used to estimate the number of moles or the masses of product formed because the reaction is near completion. If the equilibrium constant is very small, the concentration or partial pressures of reactants are large compared with those for products, and the extent of reaction is very limited. If the equilibrium constant has a value close to 1, both reactants and products are present in significant proportions at equilibrium.

The law of mass action is the basis for equilibrium calculations, which pervade the science of chemistry. All follow the same general pattern: Suppose we know the numerical value of K_C or K_P for a reaction of interest. Furthermore, suppose we run an experiment by mixing selected initial concentrations of reactants. We can then use Equation 14.1 to calculate the concentration of the reaction products and reactants that will appear at equilibrium, without doing the laborious measurements described in Figure 14.2. In this and succeeding chapters in Unit 4, we demonstrate the methods of these calculations for reactions in the gas phase, for acid–base reactions involving ionic species in aqueous solutions, and for heterogeneous reactions occurring at the interface between different states of matter. In preparation for these calculations, we now show how to write the empirical law of mass action for these classes of reactions.

Law of Mass Action for Gas-Phase Reactions

Many reactions are conveniently carried out entirely in the gas phase. Molecules in the gas phase are highly mobile, and the collisions necessary for chemical reactions occur frequently. Two examples are the key steps in the production of hydrogen from the methane in natural gas. The first is the **reforming reaction:**

$$CH_4(g) + H_2O(g) \longrightarrow CO(g) + 3H_2(g)$$

The second step is called the **shift reaction:**

$$CO(g) + H_2O(g) \longrightarrow CO_2(g) + H_2(g)$$

This process is currently the main industrial method for the preparation of hydrogen. Its success relies critically on application of the equilibrium principles developed in this chapter.

To write the mass action law, a first glance at Equation 14.1b suggests we need only examine the balanced equation for the reaction and insert the partial pressure of each reactant and product into Equation 9.1b and raise it to a power equivalent to its stoichiometric coefficient in the balanced equation, to obtain a ratio with dimensions of $(\text{press.})^{c+d-a-b}$. A deeper study of equilibrium shows that instead of inserting just the partial pressure for each reactant or product, we must insert the value of the partial pressure *relative to* a specified reference pressure P_{ref}. The result is the following expression, denoted by K with no subscript

$$\frac{(P_C/P_{ref})^c(P_D/P_{ref})^d}{(P_A/P_{ref})^a(P_B/P_{ref})^b} = K$$

Note that K is dimensionless. Collecting the terms involving P_{ref} on the right side of the equation gives

$$\frac{P_C^c P_D^d}{P_A^a P_B^b} = K(P_{ref})^{(c+d-a-b)}$$

which must be equal to K_P. It is customary to choose P_{ref} as atmospheric pressure, which can be expressed as $P_{ref} = 1$ atm, $P_{ref} = 760$ torr, or $P_{ref} = 101{,}325$ Pa. If all pressures are expressed in atmospheres, then $P_{ref} = 1$ atm, and the right side shows that K_P has the same *numerical value* as the dimensionless quantity K, with P_{ref} factors serving only to make the equation dimensionally correct. If some other unit is chosen for pressures, the P_{ref} factors no longer have a numerical value of unity and must be inserted explicitly into the equilibrium expression.

The dimensionless quantity K is the **thermodynamic equilibrium constant,** which Section 14.3 shows can be calculated from tabulated data on the products and reactants, even if the empirical equilibrium constant defined in Equation 14.1b is not known. Therefore, K is the preferred tool for analyzing reaction equilibria in general. The informal argument by which we replaced K_P with K is made rigorous by the thermodynamic treatment of equilibrium in Section 14.3. Meantime, we can freely use the result in advance of formal justification.

The convention we follow in this book is to describe chemical equilibrium in terms of the thermodynamic equilibrium constant K, even when analyzing reactions empirically. Consequently, for gaseous reactions we will state values of K without dimensions, and we will express all pressures in atmospheres. The P_{ref} factors will not be explicitly included because their value is unity with these choices of pressure unit and reference pressure. Following this convention, we write the mass action law for a general reaction involving ideal gases as

$$\frac{(P_C)^c(P_D)^d}{(P_A)^a(P_B)^b} = K \qquad [14.2a]$$

with K dimensionless. The following example illustrates these practices.

EXAMPLE 14.1

Write equilibrium expressions for the following gas-phase chemical equilibria.

(a) $2\ NOCl(g) \rightleftharpoons 2\ NO(g) + Cl_2(g)$

(b) $CO(g) + \frac{1}{2}O_2(g) \rightleftharpoons CO_2(g)$

SOLUTION

(a)
$$\frac{(P_{NO})^2(P_{Cl_2})}{(P_{NOCl_2})^2} = K$$

The powers of 2 come from the factors of 2 in the balanced equation.

(b)
$$\frac{(P_{CO_2})}{(P_{CO})(P_{O_2})^{1/2}} = K$$

Fractional powers appear in the equilibrium expression whenever they are present in the balanced equation.

Related Problems: 1, 2, 3, 4

Law of Mass Action for Reactions in Solution

A great many reactions are carried out in a convenient solvent for reactants and products. Dissolved reactants can be rapidly mixed, and the reaction process is easily handled. Water is a specially favored solvent because its polar structure allows a broad range of polar and ionic species to be dissolved. Water itself is partially ionized in solution, liberating H^+ and OH^- ions that can participate in reactions with the dissolved species. This leads to the important subject of acid–base equilibria in aqueous solutions (see Chapter 15), which is based on the equilibrium principles developed in this chapter. We limit the discussion in this subsection to cases in which the solvent does not participate in the reaction.

The procedure for reactions in solution follows the same informal discussion just given for gas-phase reactions. A deeper study of equilibrium shows that instead of inserting just the concentration for each reactant or product into Equation 14.1a, we must insert the value of concentration *relative to* a specified reference concentration c_{ref}. The result is the following expression, denoted by K with no subscript

$$\frac{([C]/c_{ref})^c([D]/c_{ref})^d}{([A]/c_{ref})^a([B]/c_{ref})^b} = K$$

The square bracket [X] represents the concentration of species X in units of mol L^{-1}. If all factors containing c_{ref} are collected on the right side and the reference state for each reactant and product is defined to be an ideal solution with a concentration $c_{ref} = 1$ M, then the same arguments used before for gas-phase reactions show that the dimensionless thermodynamic equilibrium constant K is numerically equal to K_C.

Just as with gaseous reactions, the convention we follow in this book is to describe solution equilibria in terms of the thermodynamic equilibrium constant K rather than the empirical K_C. Thus, we express solution concentrations in units of mol L^{-1} with the reference state as $c_{ref} = 1$ M, and we state values of K as dimensionless quantities. For these conditions the mass action law for solution reactions becomes

$$\frac{[C]^c[D]^d}{[A]^a[B]^b} = K \qquad [14.2b]$$

The following example illustrates these practices.

EXAMPLE 14.2

Household laundry bleach is a solution of sodium hypochlorite (NaOCl) prepared by adding gaseous Cl_2 to a solution of sodium hydroxide:

$$Cl_2(aq) + 2\ OH^-(aq) \rightleftharpoons ClO^-(aq) + Cl^-(aq) + H_2O(\ell)$$

The active bleaching agent is the hypochlorite ion, which can decompose to chloride and chlorate ions in a side reaction that competes with bleaching:

$$3\ ClO^-(aq) \rightleftharpoons 2\ Cl^-(aq) + ClO_3^-(aq)$$

Write the equilibrium expression for the decomposition reaction.

SOLUTION

$$\frac{[Cl^-]^2[ClO_3^-]}{[ClO^-]^3} = K$$

The exponents come from the coefficients in the balanced chemical equation.

Related Problems: 7, 8

Law of Mass Action for Reactions Involving Pure Substances and Multiple Phases

A variety of equilibria occur in heterogeneous systems that involve solids and liquids as well as gases and dissolved species. Molecular species cross the interfaces between phases in order to participate in reactions. A whole class of examples (Chapter 16) is based on the dissolution of slightly soluble salts, where the dissolved ions are in equilibrium with the pure solid. Another class includes the reaction of pure metals with acids to produce hydrogen gas:

$$Zn(s) + 2\ H_3O^+(aq) \rightleftharpoons Zn^{2+}(aq) + H_2(g) + 2\ H_2O(\ell)$$

Yet another example shows how iron sulfide residues from mining operations introduce iron and sulfur as pollutants in water streams:

$$4\ FeS_2(s) + 15\ O_2(g) + 6\ H_2O(\ell) \rightleftharpoons 4\ Fe(OH)^{2+}(aq) + 8\ HSO_4^-(aq)$$

Reactions in aqueous solution sometimes involve water as a direct participant. In addition to familiar aqueous acid–base chemistry, many organic reactions fall into this class. One example is the hydrolysis of ethyl acetate to produce acetic acid and ethanol:

$$CH_3COOC_2H_5(aq) + H_2O(\ell) \rightleftharpoons CH_3COOH(aq) + C_2H_5OH(aq)$$

Several examples illustrate the procedures for writing the mass action law for heterogeneous reactions and for reactions involving pure solids or liquids.

1. Recall the phase equilibrium between liquid water and water vapor from Section 5.4:

$$H_2O(\ell) \rightleftharpoons H_2O(g) \qquad P_{H_2O} = K$$

Experiments show that as long as *some* liquid water is in the container, the pressure of water vapor at 25°C is 0.03126 atm. The position of this equilibrium is not affected by the amount of liquid water present, and therefore liquid water should not appear in the mass action law. Recall that for a gas or solute, a ratio of pressures or concentrations appears in the law of mass action. This ratio is equal to 1 when the gas or solute is in its reference state (1 atm or 1 M). For a pure liquid appearing in an equilibrium chemical equation, the convention is to take that pure liquid as the reference state, so the liquid water contributes only a factor of 1 to the equilibrium expression and can thus be entirely omitted. We postpone justification of this rule to Section 14.3.

2. An analogous situation occurs in the equilibrium between solid iodine and iodine dissolved in aqueous solution:

$$I_2(s) \rightleftharpoons I_2(aq) \qquad [I_2] = K$$

Experiment shows that the position of the equilibrium (given by the concentration of I_2 dissolved at a given temperature) is independent of the amount of solid present, as long as there is some. The pure solid iodine does not appear in the mass action law, for the same reason pure liquid water did not appear in the preceding case.

3. Another example involves a chemical reaction, the decomposition of calcium carbonate:

$$CaCO_3(s) \rightleftharpoons CaO(s) + CO_2(g) \qquad P_{CO_2} = K$$

If calcium carbonate is heated, it decomposes into calcium oxide (lime) and carbon dioxide; the reverse reaction is favored at sufficiently high pressures of carbon dioxide. The equilibrium can be studied experimentally, and it is found that at any given temperature the pressure of $CO_2(g)$ is constant, independent of the amounts of $CaCO_3(s)$ and $CaO(s)$, as long as some of each is present (Fig. 14.3). The two pure solids do not appear in the mass action law, which reduces to the partial pressure of CO_2.

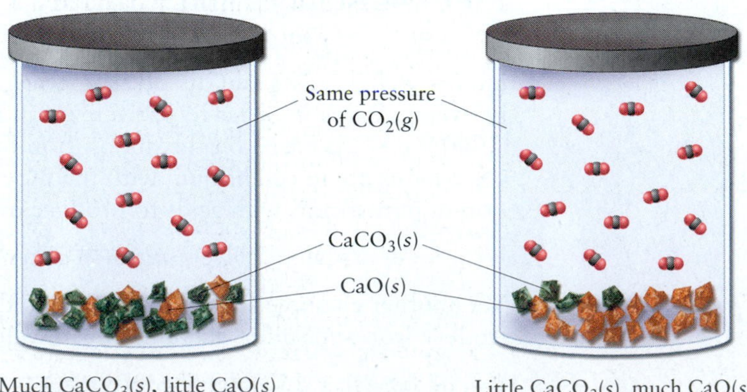

FIGURE 14.3 As long as both CaO(*s*) and $CaCO_3(s)$ are present at equilibrium in a closed container, the partial pressure of $CO_2(g)$ at a fixed temperature does not depend on the amounts of the two solids present.

The general procedure for writing the mass action law for these more complex reactions is the following:

1. Gases enter the equilibrium expression as partial pressures, measured in atmospheres.
2. Dissolved species enter as concentrations, in moles per liter.
3. Pure solids and pure liquids do not appear in equilibrium expressions; neither does a solvent taking part in a chemical reaction, provided the solution is dilute.
4. Partial pressures and concentrations of products appear in the numerator, and those of reactants in the denominator; each is raised to a power equal to its coefficient in the balanced chemical equation for the reaction.

This procedure gives the dimensionless thermodynamic equilibrium constant K because each species has entered the equilibrium expression relative to its standard reference state. This procedure will be justified by the thermodynamic treatment of equilibrium in Section 14.3, but it can be used with confidence in advance of that justification.

The following example illustrates these practices.

EXAMPLE 14.3

Hypochlorous acid (HOCl) is produced by bubbling chlorine through an agitated suspension of mercury(II) oxide in water. The chemical equation for this process is

$$2\ Cl_2(g) + 2\ HgO(s) + H_2O(\ell) \rightleftharpoons HgO{\cdot}HgCl_2(s) + 2\ HOCl(aq)$$

Write the equilibrium expression for this reaction.

SOLUTION

$$\frac{[HOCl]^2}{P_{Cl_2}^2} = K$$

The HgO and $HgO{\cdot}HgCl_2$ do not appear because they are solids, and water does not appear because it is the solvent. Chlorine, as a gas, enters as its partial pressure in atmospheres. The HOCl appears as its concentration, in moles per liter. Both the concentration of HOCl and the partial pressure of Cl_2 are raised to the second power because their coefficients in the balanced chemical equation are 2.

Related Problems: 9, 10, 11, 12

The preceding discussion has specified the procedures for setting up the mass action law for broad classes of chemical reactions. The resulting expressions are ready to be used in equilibrium calculations.

Despite their success in giving the correct expressions for the mass action law, the empirical procedures leave unanswered numerous fundamental questions about chemical equilibrium. Why should the law of mass action exist in the first place, and why should it take the particular mathematical form shown here? Why should the equilibrium constant take a unique value for each individual chemical reaction? What factors determine that value? Why does the value of the equilibrium constant change slightly when studied over broad ranges of concentration? Why should the equilibrium constant depend on temperature? Is there a quantitative explanation for the temperature dependence?

All these questions are answered by the thermodynamic description of the equilibrium constant, provided in the next section. Readers who have already studied thermodynamics should continue to Section 14.3.

Readers who have not yet studied thermodynamics should go to Sections 14.4 and 14.5, which give detailed analysis of equilibrium calculations based on the mass action law procedures just described. Sections 14.4 and 14.5 do not require background in thermodynamics.

14.3 Thermodynamic Description of the Equilibrium State

In this section we use thermodynamics to demonstrate why the mass action law takes its special mathematical form and why the thermodynamic equilibrium constant K is a dimensionless quantity. This demonstration justifies the procedures we presented in Section 14.2 for writing down the mass action law by inspection for any chemical reaction. In addition, thermodynamics gives a method for calculating the value of K from tabulated properties of the reactants and products. Consequently, the value of K can be obtained for a reaction, even if the empirical equilibrium constant K_C or K_P has not been measured. Thermodynamics also explains how K changes when the reaction is run under different experimental conditions. With this information, we can manipulate reaction conditions to obtain maximum yield from the reaction.

Thermodynamics views a chemical reaction as a process in which atoms "flow" from reactants to products. If the reaction is spontaneous and is carried out at constant T and P, thermodynamics requires that $\Delta G < 0$ for the process (see Section 13.7). Consequently, G always *decreases* during a spontaneous chemical reaction. When a chemical reaction has reached equilibrium, $\Delta G = 0$; that is, there is no further tendency for the reaction to occur in either the forward or the reverse direction. We will use the condition $\Delta G = 0$ in the following three subsections to develop the mass action law and the thermodynamic equilibrium constant for gaseous, solution, and heterogeneous reactions.

Reactions among Ideal Gases

Before we develop the mass action law, it is necessary to investigate how the Gibbs free energy changes with pressure at constant temperature, because in chemical equilibria, the partial pressures of gases can differ from 1 atm.

DEPENDENCE OF GIBBS FREE ENERGY OF A GAS ON PRESSURE If the pressure of an ideal gas is changed from P_1 to P_2 with the temperature held constant, the free energy change is

$$\Delta G = \Delta(H - TS) = \Delta H - T\Delta S = -T\Delta S$$

The last equality is true because $\Delta H = 0$ when the pressure of an ideal gas is changed at constant temperature. The entropy change for an ideal gas in an isothermal process was calculated in Section 13.5:

$$\Delta S = nR \ln\left(\frac{V_2}{V_1}\right) = nR \ln\left(\frac{P_1}{P_2}\right) = -nR \ln\left(\frac{P_2}{P_1}\right)$$

so

$$\Delta G = nRT \ln\left(\frac{P_2}{P_1}\right) \qquad [14.3a]$$

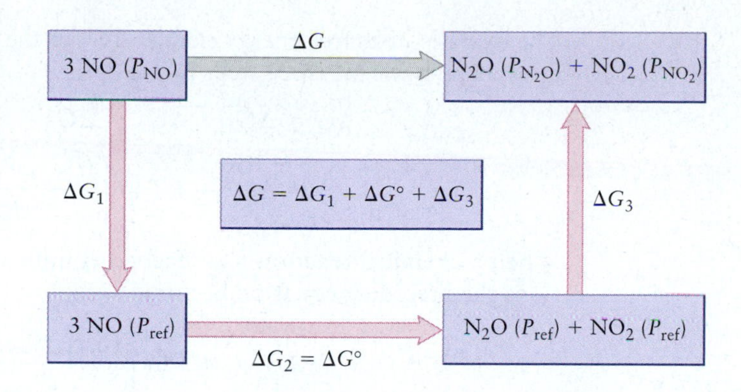

FIGURE 14.4 A three-step process (red arrows) to calculate ΔG of a reaction (blue arrow) for which reactants and products are not in their standard states of 1 atm.

Let us choose the value of P_1 to be 1 atm, which has already been defined as the standard state of gaseous substances for measurements both of enthalpy of formation (see Section 12.5) and Gibbs free energy of formation (see Section 13.7). Then, Equation 14.3a relates the Gibbs free energy of the gas at any pressure P_2 to its value at 1 atm. This result can be expressed compactly as follows. If we call 1 atm the *reference state* for the gas, then the change in Gibbs free energy in taking the gas from the reference state to any pressure P is given by

$$\Delta G = nRT \ln\left(\frac{P}{P_{ref}}\right) = nRT \ln P \qquad [14.3b]$$

Equation 14.3b is a shorthand version that can be used only when the pressure P is expressed in atm. The presence of P_{ref} in the denominator makes the argument of the natural logarithm function dimensionless. Choosing P_{ref} = 1 atm gives P_{ref} the numerical value 1, which for convenience we do not write explicitly. Nonetheless, you should always remember this (invisible) P_{ref} is required to make the equation dimensionally correct when the general pressure P in the equation is expressed in atm. If some unit of pressure other than atm is selected, P_{ref} no longer has value 1 and the P_{ref} selected must be carried explicitly in the equations.

THE EQUILIBRIUM EXPRESSION FOR REACTIONS IN THE GAS PHASE

Consider now a mixture of gases that react chemically, such as the NO, N_2O, and NO_2 given in Example 13.10:

$$3\ NO(g) \rightleftharpoons N_2O(g) + NO_2(g)$$

If all of the partial pressures are 1 atm, then ΔG for this reaction is just $\Delta G°$ at 25°C. If the pressures differ from 1 atm, ΔG must be calculated from a three-step process (Fig. 14.4). In step 1 the partial pressure of the reactant (in this case, 3 mol of NO) is changed from its initial value, P_{NO}, to the reference pressure P_{ref} = 1 atm:

$$\Delta G_1 = 3RT \ln\left(\frac{P_{ref}}{P_{NO}}\right) = RT \ln\left(\frac{P_{ref}}{P_{NO}}\right)^3$$

In step 2 the reaction is carried out with all reactants and products at partial pressures of P_{ref} = 1 atm:

$$\Delta G_2 = \Delta G°$$

In step 3 the partial pressures of the products (in this case, 1 mol of N_2O and 1 mol of NO_2) are changed from P_{ref} = 1 atm to P_{N_2O} and P_{NO_2}:

$$\Delta G_3 = RT \ln\left(\frac{P_{N_2O}}{P_{ref}}\right) + RT \ln\left(\frac{P_{NO_2}}{P_{ref}}\right) = RT \ln\left[\left(\frac{P_{N_2O}}{P_{ref}}\right)\left(\frac{P_{NO_2}}{P_{ref}}\right)\right]$$

The overall Gibbs free energy change ΔG for the reaction is the sum of the free energy changes for the three steps in the path:

$$\Delta G = \Delta G_1 + \Delta G_2 + \Delta G_3$$

$$= \Delta G^\circ + RT \ln\left[\frac{(P_{N_2O}/P_{ref})(P_{NO_2}/P_{ref})}{(P_{NO}/P_{ref})^3}\right]$$

When a chemical reaction has reached equilibrium, $\Delta G = 0$. Under these conditions the preceding equation becomes

$$-\Delta G^\circ = RT \ln\left[\frac{(P_{N_2O}/P_{ref})(P_{NO_2}/P_{ref})}{(P_{NO}/P_{ref})^3}\right]$$

Because ΔG° depends only on temperature, the quantity $\Delta G^\circ/RT$ must be a constant at each value of T. Therefore, in the last equation, the ratio of partial pressures inside the natural logarithm function must also be constant *at equilibrium* at each value of T. Consequently, this ratio of partial pressures is denoted by $K(T)$ and is called the *thermodynamic equilibrium constant* for the reaction. Finally, we have

$$-\Delta G^\circ = RT \ln K(T) \qquad [14.4]$$

For the general reaction

$$aA + bB \longrightarrow cC + dD$$

the result, obtained in the same way, is

$$-\Delta G^\circ = RT \ln\left[\frac{(P_C/P_{ref})^c(P_D/P_{ref})^d}{(P_A/P_{ref})^a(P_B/P_{ref})^b}\right] = RT \ln K(T) \qquad \text{(at equilibrium)}$$

The expression for the thermodynamic equilibrium constant for reactions involving ideal gases,

$$\left[\frac{(P_C/P_{ref})^c(P_D/P_{ref})^d}{(P_A/P_{ref})^a(P_B/P_{ref})^b}\right] = K$$

has the same form as the empirical law of mass action with K_P for gaseous reactions introduced in Section 14.2. The thermodynamic expression provides deeper understanding of that empirical result in three important ways. First, the law of mass action, previously introduced on purely empirical grounds to describe chemical equilibrium, is seen to be a consequence of the reaction system being in thermodynamic equilibrium. Second, the thermodynamic equilibrium constant can be calculated from ΔG° (that is, from ΔH° and ΔS°), so the *extent* of any equilibrium chemical reaction can be deduced from calorimetric data alone. Third, unlike K_P, the thermodynamic equilibrium constant K is always a dimensionless quantity because the pressure of a reactant or product always appears as a ratio to the reference pressure P_{ref}. The algebraic rearrangements leading to Equation 14.2a have already shown that K is numerically equal to K_P if we express all partial pressures in units of atm and select the reference state to be $P_{ref} = 1$ atm. Other choices of pressure units or reference states require the P_{ref} factors to be carried explicitly.

The convention we follow in this book, already stated in Section 14.2, is to describe chemical equilibrium in terms of the thermodynamic equilibrium constant K and to emphasize the fact that K can be obtained easily from tabulated thermodynamic data. Following this convention, the mass action law for a general reaction involving ideal gases is written as

$$\frac{(P_C)^c(P_D)^d}{(P_A)^a(P_B)^b} = K$$

with K dimensionless. The value of K can be calculated from Equation 14.4 using tables of data in Appendix D. The following example illustrates these practices.

EXAMPLE 14.4

The $\Delta G°$ of the chemical reaction

$$3\ NO(g) \longrightarrow N_2O(g) + NO_2(g)$$

was calculated in Example 13.10. Now calculate the equilibrium constant of this reaction at 25°C.

SOLUTION

The standard free energy change for the conversion of 3 mol of NO to 1 mol of N_2O and 1 mol of NO_2 was found to be −104.18 kJ. To keep the units of the calculation correct, the $\Delta G°$ is rewritten as $-104.18\ kJ\ mol^{-1}$, where "per mole" signifies "per mole of the reaction as it is written," that is, per 3 mol of NO, 1 mol of N_2O, and 1 mol of NO_2, the number of moles in the balanced equation. Substitution gives

$$\ln K = \frac{-\Delta G°}{RT} = \frac{-(-104{,}180\ J\ mol^{-1})}{(8.315\ J\ K^{-1}mol^{-1})(298.15\ K)} = 42.03$$

$$K = e^{42.03} = 1.8 \times 10^{18}$$

Related Problems: 13, 14

The conversion of $NO(g)$ to $N_2O(g)$ plus $NO_2(g)$ is spontaneous under standard conditions. The forward reaction under these conditions is scarcely observed because its rate is so slow. Nonetheless, its equilibrium constant can be calculated! Such calculations often have enormous impact in evaluating proposed solutions to practical problems. For example, the calculation shows that this reaction could be used to reduce the amount of NO in cooled exhaust gases from automobiles. The fundamental reaction tendency is there, but successful application requires finding a route to increasing the reaction rate at standard conditions. Had the equilibrium constant calculated from thermodynamics been small, this proposed application would be doomed at the outset and investment in it would not be justified.

Reactions in Ideal Solutions

The thermodynamic description of reactions in solution parallels the discussion just completed for ideal gas reactions. Although the result is not derived here, the Gibbs free energy change for n mol of a solute, as an ideal (dilute) solution changes in concentration from c_1 to c_2 mol L^{-1}, is

$$\Delta G = nRT \ln\left(\frac{c_2}{c_1}\right) \qquad [14.5a]$$

If the reference state for the solute is defined to be an ideal solution with a concentration $c_{ref} = 1$ M, the change in Gibbs free energy for taking the solution from the reference state to the concentration c is given by

$$\Delta G = nRT \ln\left(\frac{c}{c_{ref}}\right) = nRT \ln c \qquad [14.5b]$$

The same arguments used for gas-phase reactions show that the overall change in Gibbs free energy for a reaction in solution is

$$\Delta G = \Delta G° + RT \ln\left[\frac{([C]/c_{ref})^c([D]/c_{ref})^d}{([A]/c_{ref})^a([B]/c_{ref})^b}\right]$$

where the square bracket [X] represents the concentration of species X in units of mol L^{-1}. When the reaction arrives at equilibrium, $\Delta G = 0$, and the equilibrium constant is given by

$$-\Delta G° = RT \ln K \qquad [14.6]$$

for reactions involving dissolved species.

Following the convention already stated for gaseous reactions, in this book we describe solution equilibria in terms of the thermodynamic equilibrium constant K rather than the empirical K_C introduced in Section 14.2. Thus, we give solution concentrations in units of mol L^{-1} with the reference state as $c_{ref} = 1$ M, and state values of K as dimensionless quantities. For this convention, the mass action law for solution reactions becomes

$$\frac{[C]^c[D]^d}{[A]^a[B]^b} = K$$

and is numerically equivalent to K_C. When working with this expression, you must keep in mind the role of the (invisible) c_{ref} factors in making the equation dimensionally correct. This expression is the foundation for the discussions of acid–base equilibria in aqueous solutions in the next chapter.

Solution-phase equilibrium constants can be calculated from tables of standard free energies for solutes in aqueous solution at 25°C (see Appendix D). The procedure is demonstrated by the following example.

EXAMPLE 14.5

Calculate $\Delta G°$ and the equilibrium constant at 25°C for the chemical reaction

$$3\ ClO^-(aq) \rightleftharpoons 2\ Cl^-(aq) + ClO_3^-(aq)$$

whose equilibrium expression we developed in Example 14.2

SOLUTION

We calculate the change in standard Gibbs free energy using the ΔG_f° values for these aqueous ions in their standard states tabulated in Appendix D:

$$\begin{aligned}\Delta G° &= 2\ \Delta G_f^\circ(Cl^-(aq)) + \Delta G_f^\circ(ClO_3^-(aq)) - 3\ \Delta G_f^\circ(ClO^-(aq)) \\ &= (2\text{ mol})(-131.23\text{ kJ mol}^{-1}) + (1\text{ mol})(-7.95\text{ kJ mol}^{-1}) - \\ &\quad (3\text{ mol})(-36.8\text{ kJ mol}^{-1}) \\ &= -160.01\text{ kJ}\end{aligned}$$

As in Example 14.4, we calculate K by inserting the value for $\Delta G°$ into the equation

$$\ln K = \frac{-\Delta G°}{RT} = \frac{-(-160{,}010\text{ J mol}^{-1})}{(8.315\text{ J K}^{-1}\text{mol}^{-1})(298.15\text{ K})} = 64.54$$

$$K = e^{64.54} = 1.1 \times 10^{28}$$

The decomposition of hypochlorite ion in bleach is thus a spontaneous process, with a very large equilibrium constant at 25°C. But, this reaction is very slow at room temperature, so the laundry bleach solution remains stable. If the temperature is raised to about 75°C, the reaction occurs rapidly and the bleach solution decomposes.

Reactions Involving Pure Solids and Liquids and Multiple Phases: The Concept of Activity

The mass action law for homogeneous reactions in ideal gases and ideal solutions was written in Section 14.2 by straightforward inspection of the balanced equation for the reaction under study. If one or more of the reactants or products was a solid or liquid in its pure state, the procedure was less obvious, because "concentration" has no meaning for a pure species. This apparent difficulty is resolved by the concept of **activity,** which is a convenient means for comparing the properties of a substance in a general thermodynamic state with its properties in a specially selected reference state. A full treatment of equilibrium in terms of activity requires thermodynamic results beyond the scope of this book. Here we sketch the essential ideas leading to the more general form of the mass action law and merely state the range of validity of the idealized expressions presented.

The activity concept arises from the dependence of the Gibbs free energy on the pressure of a pure substance or on the composition of a solution, regardless of the phase of the system. The discussion just before Equation 14.3 shows that the change in Gibbs free energy when a gas is taken from a reference state P_{ref} to any pressure P is given by

$$\Delta G = nRT \ln\left(\frac{P}{P_{ref}}\right) = nRT \ln P$$

The last form is used only when pressure P is expressed in atmospheres and $P_{ref} = 1$ atm. A similar equation can be developed for more complex systems—and keep the same simple mathematical form—if we *define* the activity by the equation

$$\Delta G = nRT \ln a \qquad [14.7]$$

Of course if a system is already in its reference state, then $\Delta G = 0$ and the activity in this state is 1. The change in Gibbs free energy in taking a system from the reference state to any general thermodynamic state determines the activity a in the general state.

The activity is connected to pressure or to concentration by the **activity coefficient.** The *activity coefficient* γ_i of a nonideal gaseous species at pressure P_i is defined by the equation

$$a_i = \frac{\gamma_i P_i}{P_{ref}} \qquad [14.8a]$$

The reference state is chosen to be an ideal gas at one atmosphere. Comparing Equations 14.7 and 14.8a with Equation 14.3 shows that γ_i takes the value 1 for ideal gases. The activity of an ideal gas is the ratio of its pressure to the selected reference pressure. If pressures are given in units of atmospheres, then the activity of an ideal gas is numerically equal to its pressure. Similarly, the activity coefficient γ_i for solute i in a solution at concentration c_i is defined by the equation

$$a_i = \frac{\gamma_i c_i}{c_{ref}} \qquad [14.8b]$$

The activity coefficient γ_i equals 1 in the reference state, selected as an ideal solution with convenient concentration. For solute species in the dilute solutions considered in this book, concentration is most conveniently expressed in molarity, and the reference state is selected to be an ideal solution at concentration $c_{ref} = 1$ M.

The reference states for pure solids and liquids are chosen to be those forms stable at 1 atm, just as in the definition of standard states for enthalpy of formation (see Chapter 12) and Gibbs free energy of formation (see Chapter 13). Pure substances in their reference states are assigned activity of value 1.

Once reference states have been defined, the activity coefficients γ_i can be determined from experimental P-V-T and calorimetric data by procedures that are not described in this book. Then, Equation 14.7 can be written explicitly in terms of pressure, temperature, and concentrations when needed for specific calculations. In its present version, Equation 14.7 is especially well suited for general discussions, because it summarizes much complicated information about nonideal systems in a simple and compact form.

Starting from Equation 14.7, arguments similar to those preceding Equation 14.4 lead to the following expression for the thermodynamic equilibrium constant K regardless of the phase of each product or reactant:

$$\frac{a_C^c \cdot a_D^d}{a_A^a \cdot a_B^b} = K \qquad [14.9]$$

Substituting the appropriate ideal expression for the activity of gaseous or dissolved species from Equation 14.8a or 14.8b leads to the forms of the mass action law and the equilibrium constant K already derived earlier in Section 14.3 for reactions in ideal gases or in ideal solutions. We write the mass action law for reactions involving pure solids and liquids and multiple phases by substituting unity for the activity of pure liquids or solids and the appropriate ideal expression for the activity of each gaseous or dissolved species into Equation 14.9. Once a proper reference state and concentration units have been identified for *each* reactant and product, we use tabulated free energies based on these reference states to calculate the equilibrium constant.

The following example illustrates the essential points.

EXAMPLE 14.6

The compound urea, important in biochemistry, can be prepared in aqueous solution by the following reaction:

$$CO_2(g) + 2\ NH_3(g) \rightleftharpoons CO(NH_2)_2(aq) + H_2O(\ell)$$

(a) Write the mass action law for this reaction. (b) Calculate $\Delta G°$ for this reaction at 25°C. (c) Calculate K for this reaction at 25°C.

SOLUTION

(a) In terms of activities for each species, the equilibrium expression is

$$\frac{a_{\text{urea}} \cdot a_{H_2O}}{a_{CO_2} \cdot a_{NH_3}^2} = K$$

Choose the reference state for the gases to be $P_{\text{ref}} = 1$ atm, for the urea to be $c_{\text{ref}} = 1$ M, and for water to be the pure liquid with unit activity. Then at low pressures and concentrations, the limiting idealized mass action law is

$$\frac{[\text{urea}]}{P_{CO_2} \cdot P_{NH_3}^2} = K$$

with [urea] in units of mol L^{-1} and the partial pressures in atm. The (invisible) P_{ref} and c_{ref} factors must always be kept in mind, because they make K dimensionless in this expression. This expression can be extended to higher pressures and concentrations, where nonideality becomes important, by inserting the appropriate activity coefficients.

(b) The procedure is the same as in Examples 14.4 and 14.5. The ΔG_f° values for gaseous CO_2 and NH_3 and liquid H_2O with the reference states just specified are obtained directly from Appendix D. Because the data for urea in Appendix D are based on a *solid* reference state, we consult an alternate source[3] to obtain $\Delta G_f^\circ(\text{urea}(aq)) = -203.84$ kJ mol^{-1} with the reference state an ideal solution with $c = 1$ M. From these data obtain

$$\Delta G° = -203.84 - 237.18 + 394.36 + 2\,(16.48) = -13.70 \text{ kJ mol}^{-1}$$

(c) Following the procedure in Example 14.5 we find $K = 251.1$.

This example illustrates a very important point: Before using ΔG_f° values to estimate equilibrium constants, *be certain* you know the reference state for the tabulated values.

Related Problems: 15, 16

[3]Thermodynamic data for many biological molecules in solution are given in F. H. Carpenter, *J. Am. Chem. Soc.*, *82*, 1120 (1960).

The procedure for writing the mass action law for this class of reactions given at the end of Section 14.2 was based on informal discussion and included one rote rule. The procedure is now justified by the activity concept.

Equilibrium expressions written in this way are accurate to about 5% when the pressures of gases do not exceed several atmospheres and the concentrations of solutes do not exceed 0.1 M. These cases cover all the applications discussed in this book, so we do not need correction factors. For concentrated ionic solutions, the corrections can become very large. In accurate studies of solution equilibria, especially in biochemical applications, activities must be used in place of partial pressures or concentrations. You should consult more advanced books when you need these techniques.

14.4 The Law of Mass Action for Related and Simultaneous Equilibria

Each time we write the mass action law, it is based on a specific balanced chemical equation in which the reaction is carried out as written "left to right." Chemical reactions can be carried out in various alternative ways, including "in reverse" and in concert with other reactions. These variations lead to relationships among equilibrium expressions, which are best illustrated in a series of examples.

Relationships among Equilibrium Expressions

(1) Suppose a reaction is written in two opposing directions:

$$2\,H_2(g) + O_2(g) \rightleftharpoons 2\,H_2O(g) \qquad \frac{P^2_{H_2O}}{P^2_{H_2}P_{O_2}} = K_1$$

$$2\,H_2O(g) \rightleftharpoons 2\,H_2(g) + O_2(g) \qquad \frac{P^2_{H_2}P_{O_2}}{P^2_{H_2O}} = K_2$$

In the first reaction, hydrogen and oxygen combine to form water vapor, whereas in the second, water vapor dissociates into hydrogen and oxygen. The equilibrium constants K_1 and K_2 are clearly the inverse of each other, so their product $K_1K_2 = 1$. This is true quite generally: The equilibrium constant for a reverse reaction is the reciprocal of the equilibrium constant for the corresponding forward reaction.

(2) What happens if a balanced chemical equation is multiplied by a constant? In the preceding example, multiplying the first equation by $\frac{1}{2}$ gives

$$H_2(g) + \tfrac{1}{2}O_2(g) \rightleftharpoons H_2O(g)$$

This is a perfectly satisfactory way to write the equation for the chemical reaction; it says that 1 mol of hydrogen reacts with $\frac{1}{2}$ mol of oxygen to yield 1 mol of water vapor. The corresponding equilibrium expression is

$$\frac{P_{H_2O}}{P_{H_2}P^{1/2}_{O_2}} = K_3$$

Comparison with the expression for K_1 shows that

$$K_3 = K_1^{1/2}$$

When a balanced chemical equation is multiplied by a constant factor, the corresponding equilibrium constant is raised to a power equal to that factor.

(3) A further variation is to add two equations to give a third. In this case the equilibrium constant for the third equation is the product of the equilibrium constants for the first two. For example, at 25°C,

$$2\ BrCl(g) \rightleftharpoons Cl_2(g) + Br_2(g) \qquad \frac{P_{Cl_2}P_{Br_2}}{P^2_{BrCl}} = K_1 = 0.45$$

$$Br_2(g) + I_2(g) \rightleftharpoons 2\ IBr(g) \qquad \frac{P^2_{IBr}}{P_{Br_2}P_{I_2}} = K_2 = 0.051$$

Adding the two chemical equations gives

$$2\ BrCl(g) + Br_2(g) + I_2(g) \rightleftharpoons 2\ IBr(g) + Cl_2(g) + Br_2(g)$$

Removing $Br_2(g)$ from both sides leaves

$$2\ BrCl(g) + I_2(g) \rightleftharpoons 2\ IBr(g) + Cl_2(g) \qquad \frac{P^2_{IBr}P_{Cl_2}}{P^2_{BrCl}P_{I_2}} = K_3$$

and, by inspection, $K_3 = K_1K_2 = (0.45)(0.051) = 0.023$.

Bromine chloride, BrCl (top), and iodine bromide, IBr (bottom), are two examples of *interhalogens,* combinations of two or more halogen elements.

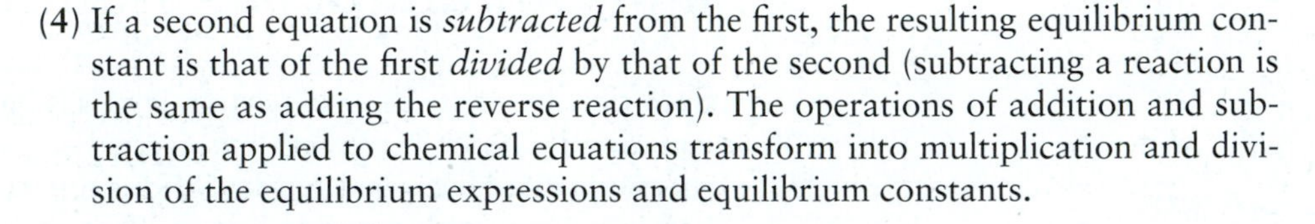

(4) If a second equation is *subtracted* from the first, the resulting equilibrium constant is that of the first *divided* by that of the second (subtracting a reaction is the same as adding the reverse reaction). The operations of addition and subtraction applied to chemical equations transform into multiplication and division of the equilibrium expressions and equilibrium constants.

EXAMPLE 14.7

The concentrations of the oxides of nitrogen are monitored in air-pollution reports. At 25°C, the equilibrium constant for the reaction

$$NO(g) + \tfrac{1}{2}\ O_2(g) \rightleftharpoons NO_2(g)$$

is

$$\frac{P_{NO_2}}{P_{NO}P^{1/2}_{O_2}} = K_1 = 1.3 \times 10^6$$

and that for

$$\tfrac{1}{2}\ N_2(g) + \tfrac{1}{2}\ O_2(g) \rightleftharpoons NO(g)$$

is

$$\frac{P_{NO_2}}{P^{1/2}_{N_2}P^{1/2}_{O_2}} = K_2 = 6.5 \times 10^{-16}$$

Find the equilibrium constant K_3 for the reaction

$$N_2(g) + 2\ O_2(g) \rightleftharpoons 2\ NO_2(g)$$

SOLUTION

Adding the chemical equations for the first two reactions gives

$$\tfrac{1}{2}\ N_2(g) + O_2(g) \rightleftharpoons NO_2(g)$$

The equilibrium constant for this reaction, K_3', is just the product of K_1 and K_2, or K_1K_2. The constant sought, K_3, is defined by a chemical equation that is twice this, so K_3' must be raised to the power 2 (that is, squared) to give K_3:

$$K_3 = (K_3')^2 = (K_1K_2)^2 = 7.1 \times 10^{-19}$$

Related Problems: 17, 18, 19, 20

Consecutive Equilibria

Many real-world applications of chemistry and biochemistry involve fairly complex sets of reactions occurring in sequence and/or in parallel. Each of these individual reactions is governed by its own equilibrium constant. How do we describe the overall progress of the entire coupled set of reactions? We write all the involved equilibrium expressions and treat them as a set of *simultaneous algebraic equations,* because the concentrations of various chemical species appear in several expressions in the set. Examination of relative values of equilibrium constants shows that some reactions dominate the overall coupled set of reactions, and this chemical insight enables mathematical simplifications in the simultaneous equations. We study coupled equilibria in considerable detail in Chapter 15 on acid–base equilibrium. Here, we provide a brief introduction to this topic in the context of an important biochemical reaction.

Hemoglobin and Oxygen Transport

The cells of the human body need oxygen for life, and the bloodstream satisfies this need by transporting oxygen from the lungs. Whole blood can carry as much as 0.01 mol of O_2 per liter because the compound hemoglobin (Hb) in the red blood cells binds O_2 chemically. In contrast, blood plasma, which contains no hemoglobin, dissolves only about 0.0001 mol of O_2 per liter, a value that is close to the solubility of O_2 in ordinary water at room conditions. As highly oxygenated blood from the lungs passes through the capillaries, the binding of oxygen to hemoglobin loosens and the freed O_2 is taken up by a different oxygen-binding compound, called myoglobin (Mb), which is found in nearby cells. Myoglobin then carries the oxygen to the cells, where it is converted to carbon dioxide.

Why is oxygen held so tightly by hemoglobin in the lungs, then released so readily to myoglobin in the capillaries? The answer starts with a comparison of the heterogeneous equilibria by which hemoglobin and myoglobin bind oxygen. Hemoglobin consists of a large protein (the globin part) in which four iron-containing **heme** groups are embedded (see Fig. 8.22b). Oxygen can bind at each heme group, so one molecule of hemoglobin binds up to four molecules of O_2 in a series of consecutive equilibria:

$$\mathrm{Hb}(aq) + \mathrm{O_2}(g) \rightleftharpoons \mathrm{Hb(O_2)}(aq) \qquad K_1 = \frac{[\mathrm{Hb(O_2)}]}{[\mathrm{Hb}]P_{\mathrm{O_2}}} = 4.88$$

$$\mathrm{Hb(O_2)}(aq) + \mathrm{O_2}(g) \rightleftharpoons \mathrm{Hb(O_2)_2}(aq) \qquad K_2 = \frac{[\mathrm{Hb(O_2)_2}]}{[\mathrm{HbO_2}]P_{\mathrm{O_2}}} = 15.4$$

$$\mathrm{Hb(O_2)_2}(aq) + \mathrm{O_2}(g) \rightleftharpoons \mathrm{Hb(O_2)_3}(aq) \qquad K_3 = \frac{[\mathrm{Hb(O_2)_3}]}{[\mathrm{Hb(O_2)_2}]P_{\mathrm{O_2}}} = 6.49$$

$$\mathrm{Hb(O_2)_3}(aq) + \mathrm{O_2}(g) \rightleftharpoons \mathrm{Hb(O_2)_4}(aq) \qquad K_4 = \frac{[\mathrm{Hb(O_2)_4}]}{[\mathrm{Hb(O_2)_3}]P_{\mathrm{O_2}}} = 1750$$

The four K's are not equal, and K_4 is far larger than the other three.

Myoglobin is different. The globin portion of myoglobin embeds only one heme group, and each myoglobin molecule can bind only one O_2.

$$\mathrm{Mb}(aq) + \mathrm{O_2}(g) \rightleftharpoons \mathrm{Mb(O_2)}(aq) \qquad K = \frac{[\mathrm{Mb(O_2)}]}{[\mathrm{Mb}]P_{\mathrm{O_2}}} = 271$$

Changing the partial pressure of O_2 should affect the myoglobin equilibrium and all four of the hemoglobin equilibria similarly. Increasing P_{O_2} should shift all of the equilibria to the right; decreasing P_{O_2} should shift them all to the left. One way to check this is to plot the fraction of the binding sites occupied by O_2 as a function of P_{O_2}. Figure 14.5 shows such **fractional saturation** plots for hemoglobin and myoglobin. Clearly, a reduction of P_{O_2} in fact causes less O_2 to be bound, but hemoglobin and myoglobin respond to the reduction in P_{O_2} to very different extents. At $P_{O_2} = 0.13$ atm, both hemoglobin and myoglobin are more than 95% saturated with O_2 at equilibrium. By contrast, at $P_{O_2} = 0.040$ atm, saturation of hemoglobin has been markedly reduced to 55%, while that of myoglobin stays above 90%. The first of these partial pressure values is typical of arterial blood from the lungs, and the second is typical of venous blood. Thus, as P_{O_2} decreases in the capillaries, hemoglobin releases a substantial amount of oxygen, but myoglobin in the nearby cells can take it up. The net result is transfer of O_2 from blood (hemoglobin) to cells (myoglobin).

The S-shaped curve for hemoglobin in Figure 14.5 can be derived mathematically from the different *K*'s for the four equilibria by which hemoglobin binds O_2. The large K_4 means that the affinity of hemoglobin for oxygen rises with uptake of oxygen. From the opposite perspective, we see that loss of the first O_2 from $Hb(O_2)_4$ facilitates loss of more O_2, and the fractional saturation curve for Hb dips below that of Mb. This phenomenon, called "positive cooperativity," originates in structural changes within the hemoglobin molecule as it binds successive molecules of oxygen.

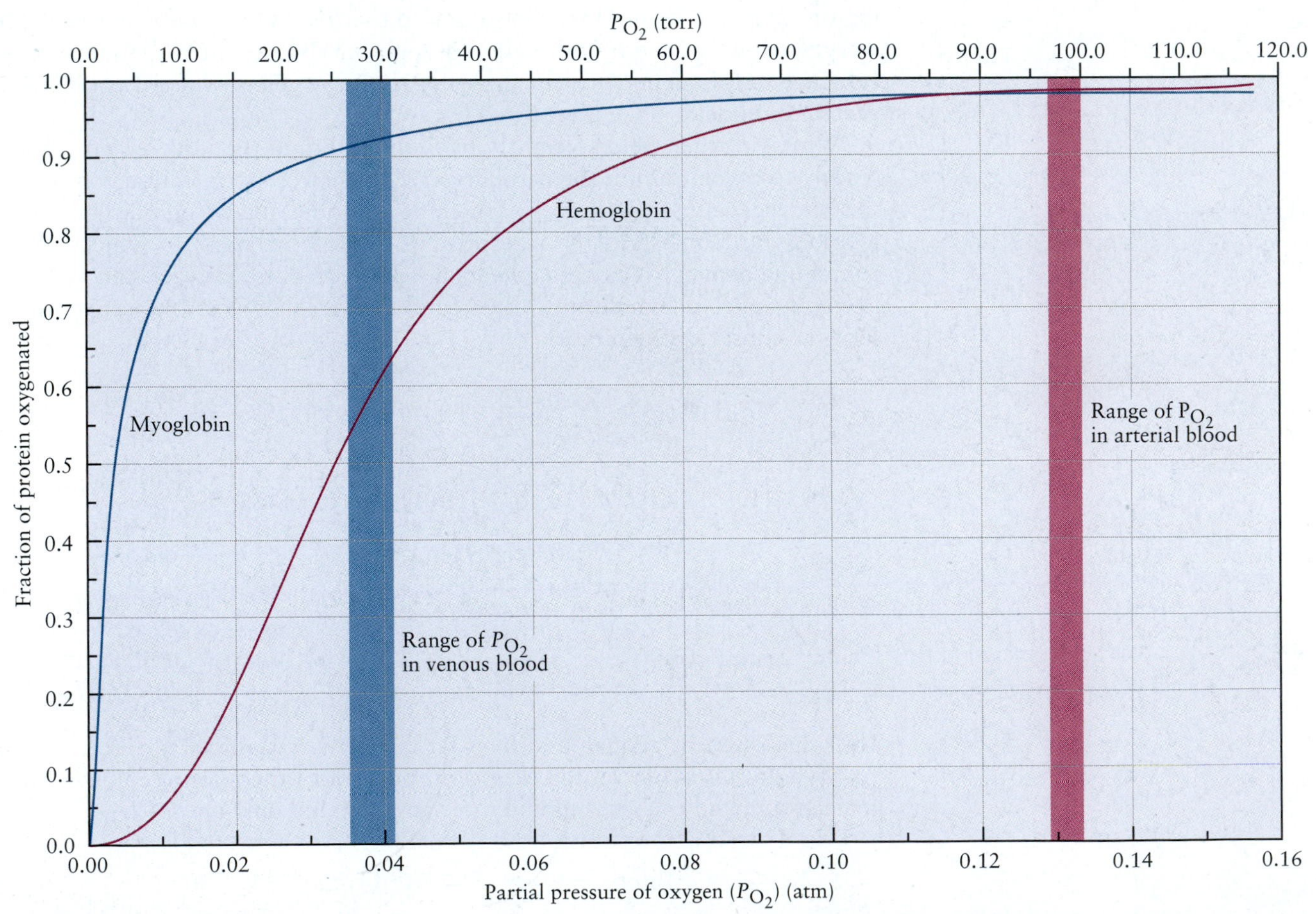

FIGURE 14.5 A plot of the fraction of the binding sites of hemoglobin (red) and myoglobin (blue) that are occupied as a function of the partial pressure of O_2.

14.5 Equilibrium Calculations for Gas-Phase and Heterogeneous Reactions

Equilibrium calculations, used throughout the science of chemistry in both basic and applied work, involve specific procedures. The present section presents these problem-solving techniques in the context of gas-phase and heterogeneous reactions, but they are applicable in *all* equilibrium calculations. We illustrate these calculation methods with several examples that fall into two broad classes: evaluating the equilibrium constant from reaction data, and calculating the amounts of products and reactants present at equilibrium when the equilibrium constant is known.

Evaluating Equilibrium Constants from Reaction Data

In Section 14.3 we showed how to evaluate K from calorimetric data on the pure reactants and products. Occasionally, these thermodynamic data may not be available for a specific reaction, or a quick estimate of the value of K may suffice. In these cases we can evaluate the equilibrium constant from measurements made directly on the reaction mixture. If we can measure the equilibrium partial pressures of all the reactants and products, we can calculate the equilibrium constant by writing the equilibrium expression and substituting the experimental values (in atmospheres) into it. In many cases it is not practical to measure directly the equilibrium partial pressure of each separate reactant and product. Nonetheless, the equilibrium constant can usually be derived from other available data, although the determination is less direct. We illustrate the method in the following two examples.

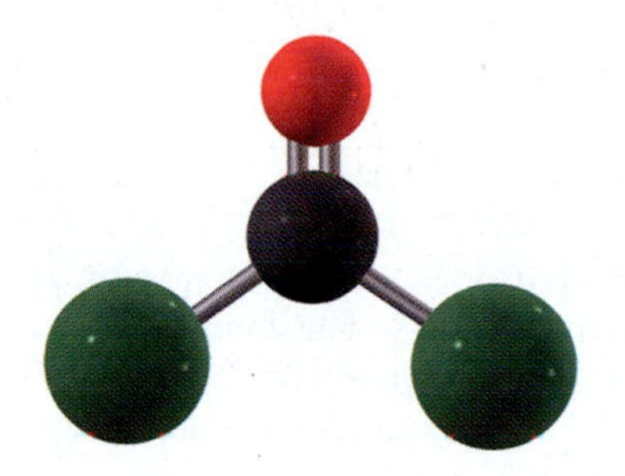

Phosgene, $COCl_2$, is a chemical intermediate used in making polyurethanes for foams and surface coatings.

EXAMPLE 14.8

Phosgene, $COCl_2$, forms from CO and Cl_2 according to the equilibrium

$$CO(g) + Cl_2(g) \rightleftharpoons COCl_2(g)$$

At 600°C, a gas mixture of CO and Cl_2 is prepared that has initial partial pressures (before reaction) of 0.60 atm for CO and 1.10 atm for Cl_2. After the reaction mixture has reached equilibrium, the partial pressure of $COCl_2(g)$ at this temperature is measured to be 0.10 atm. Calculate the equilibrium constant for this reaction. The reaction is carried out in a vessel of fixed volume.

SOLUTION

Only the equilibrium partial pressure of $COCl_2(g)$ and the *initial* partial pressures of the other two gases are given. To find the equilibrium constant, it is necessary to determine the equilibrium partial pressures of CO and Cl_2. To do this, we set up a simple table:

	$CO(g)$	+	$Cl_2(g)$	$\rightleftharpoons$	$COCl_2(g)$
Initial partial pressure (atm)	0.60		1.10		0
Change in partial pressure (atm)	?		?		+0.10
Equilibrium partial pressure (atm)	?		?		0.10

Note that the first two lines must add to give the third. We use the relationships built into the balanced chemical equation to fill in the blanks in the table. Because this is the only reaction taking place, every mole of $COCl_2$ produced consumes exactly 1 mol of CO and exactly 1 mol of Cl_2. According to the ideal gas equation, the partial pressures of gases are proportional to the number of moles of each gas present as long as the volume and temperature are held fixed. Therefore, the change in partial pressure of each gas must be proportional to the change in its number of moles as the mixture goes toward equilibrium. If the partial pressure of $COCl_2$ *increases* by 0.10 atm through reaction, the partial pressures of CO and Cl_2 must each *decrease* by 0.10 atm. Inserting these values into the table gives

	$CO(g)$	+	$Cl_2(g)$	$\rightleftarrows$	$COCl_2(g)$
Initial partial pressure (atm)	0.60		1.10		0
Change in partial pressure (atm)	−0.10		−0.10		+0.10
Equilibrium partial pressure (atm)	0.50		1.00		0.10

where the last line was obtained by adding the first two.

We now insert the equilibrium partial pressures into the equilibrium expression to calculate the equilibrium constant:

$$K = \frac{P_{COCl_2}}{(P_{CO})(P_{Cl_2})} = \frac{(0.10)}{(0.50)(1.00)} = 0.20$$

Related Problems: 21, 22, 23, 24

For a reaction in which some of the coefficients in the balanced equation are not equal to 1, deriving expressions for the changes in the partial pressures of the products and reactants requires care. Consider the combustion of ethane at constant volume:

$$2\ C_2H_6(g) + 7\ O_2(g) \rightleftarrows 4\ CO_2(g) + 6\ H_2O(g)$$

If the initial partial pressures are all 1.00 atm and there is a net reaction from left to right in this equation, the table is

	$2\ C_2H_6(g)$	+	$7\ O_2(g)$	$\rightleftarrows$	$4\ CO_2(g)$	+	$6\ H_2O(g)$
Initial partial pressure	1.00		1.00		1.00		1.00
Change in partial pressure	$-2y$		$-7y$		$+4y$		$+6y$
Equilibrium partial pressure	$1.00 - 2y$		$1.00 - 7y$		$1.00 + 4y$		$1.00 + 6y$

If the C_2H_6 partial pressure decreases by $2y$ atm, that of O_2 must decrease by $7y$, because the coefficients in the balanced equation are 2 and 7. The changes in partial pressures of products will have the opposite sign (positive instead of negative), because as reactants disappear, products appear. Their magnitudes are determined according to the coefficients in the balanced equation.

EXAMPLE 14.9

Graphite (a form of solid carbon) is added to a vessel that contains $CO_2(g)$ at a pressure of 0.824 atm at a certain high temperature. The pressure rises due to a reaction that produces $CO(g)$. The total pressure reaches an equilibrium value of 1.366 atm. **(a)** Write a balanced equation for the process. **(b)** Calculate the equilibrium constant.

SOLUTION

(a) The reaction can only be the oxidation of C by CO_2 during which the CO_2 is itself reduced to CO. The reaction and its equilibrium expression are written as

$$C(s) + CO_2(g) \rightleftarrows 2\ CO(g) \qquad \frac{(P_{CO})^2}{P_{CO_2}} = K$$

(b) To determine the equilibrium constant, we set up the standard table to describe reaction progress:

	$C(s)$	+	$CO_2(g)$	$\rightleftarrows$	$2\ CO(g)$
Initial partial pressure (atm)			0.824		0
Change in partial pressure (atm)			$-x$		$+2x$
Equilibrium partial pressure (atm)			$0.824 - x$		$2x$

The total pressure at equilibrium is

$$P_{tot} = 0.824 \text{ atm} - x + 2x = 0.824 + x = 1.366 \text{ atm}$$

Solving for x gives

$$x = 1.366 - 0.824 = 0.542 \text{ atm}$$

The equilibrium partial pressures of the two gases are

$$P_{CO} = 2x = 1.084 \text{ atm}$$

$$P_{CO_2} = 0.824 - 0.542 = 0.282 \text{ atm}$$

The equilibrium constant for the reaction is therefore

$$K = \frac{(1.084)^2}{0.282} = 4.17$$

Related Problems: 25, 26

Calculating Equilibrium Compositions When K Is Known

The law of mass action gives the *form* of the expression for the equilibrium constant for any chemical reaction. We can use these expressions to predict the outcomes of chemical reactions. How to approach the problem depends on the type of experimental data available.

Suppose you are asked to determine the partial pressures of reactants and products at equilibrium, given initial partial pressures of the reactants. This is illustrated by the following example.

EXAMPLE 14.10

Suppose $H_2(g)$ and $I_2(g)$ are sealed in a flask at $T = 400$ K with partial pressures $P_{H_2} = 1.320$ atm and $P_{I_2} = 1.140$ atm. At this temperature H_2 and I_2 do not react rapidly to form HI(g), although after a long enough time they would produce HI(g) at its equilibrium partial pressure. Suppose, instead, that the gases are heated in the sealed flask to 600 K, a temperature at which they quickly reach equilibrium:

$$H_2(g) + I_2(g) \rightleftarrows 2\ HI(g)$$

The equilibrium constant for the reaction is 92.6 at 600 K:

$$\frac{P_{HI}^2}{P_{H_2}P_{I_2}} = 92.6$$

(a) What are the equilibrium values of P_{H_2}, P_{I_2}, and P_{HI} at 600 K?

(b) What percentage of the I_2 originally present has reacted when equilibrium is reached?

SOLUTION

(a) Suppose H_2 and I_2 did *not* react at 600 K. From the ideal gas law at constant volume, their partial pressures would be

$$P^{\circ}_{H_2} = 1.320 \text{ atm} \times \left(\frac{600 \text{ K}}{400 \text{ K}}\right) = 1.980 \text{ atm}$$

$$P^{\circ}_{I_2} = 1.140 \text{ atm} \times \left(\frac{600 \text{ K}}{400 \text{ K}}\right) = 1.710 \text{ atm}$$

Of course, these gases *do* react, and the extent of the reaction can be calculated. To do this, set up a table as in Example 14.8:

	$H_2(g)$	+	$I_2(g)$	$\rightleftarrows$	$2\ HI(g)$
Initial partial pressure (atm)	1.980		1.710		0
Change in partial pressure (atm)	$-x$		$-x$		$+2x$
Equilibrium partial pressure (atm)	$1.980 - x$		$1.710 - x$		$2x$

If the partial pressure of H_2 decreases by x atm as the reaction proceeds, then the partial pressure of I_2 must also decrease by x atm because each mole of H_2 reacts with 1 mol of I_2. By similar reasoning, the partial pressure of HI increases by $2x$ atm: 2 mol of HI forms from each mole of H_2. Inserting the equilibrium partial pressures into the equilibrium expression results in the equation

$$\frac{(2x)^2}{(1.980 - x)(1.710 - x)} = 92.6$$

Multiplying and collecting terms gives

$$88.6x^2 - 341.694x + 313.525 = 0$$

Solving for x using the quadratic formula (see Appendix C) gives

$$x = \frac{-(-341.694) \pm \sqrt{(341.694)^2 - 4(88.6)(313.525)}}{2(88.6)}$$

$$= 1.5044 \text{ atm or } 2.3522 \text{ atm}$$

The second root is physically impossible because it leads to negative answers for the equilibrium partial pressures of the $H_2(g)$ and $I_2(g)$. Discarding it leaves

$$P_{HI} = 2 \times 1.5044 \text{ atm} = 3.0088 \text{ atm}$$

$$P_{H_2} = 1.980 \text{ atm} - 1.5044 \text{ atm} = 0.4756 \text{ atm}$$

$$P_{I_2} = 1.710 \text{ atm} - 1.5044 \text{ atm} = 0.2056 \text{ atm}$$

It is a good idea to check such results by inserting the calculated equilibrium partial pressures back into the equilibrium expression to make sure that the known value of K comes out.

$$\text{Check:} \quad \frac{(3.0088)^2}{(0.4756)(0.2056)} = 92.6$$

As the final step, round off each answer to the correct number of significant digits: $P_{HI} = 3.01$ atm, $P_{H_2} = 0.48$ atm, $P_{I_2} = 0.21$ atm. Rounding off sooner makes the K calculated in the check differ somewhat from 92.6.

(b) The fraction of I_2 that has *not* reacted is the ratio of the number of moles of I_2 present at the end to that present at the beginning. Because neither volume nor temperature changes during the reaction, this equals the ratio of the final partial pressure of I_2 (0.2056 atm) to its initial partial pressure (1.710 atm):

$$\text{percentage unreacted} = \frac{0.2056 \text{ atm}}{1.710 \text{ atm}} \times 100\% = 12\%$$

The percentage that *has* reacted is then 88%.

Related Problems: 27, 28, 29, 30, 31, 32

The concentrations of any products initially present must also be included in equilibrium calculations, as illustrated by the following example.

EXAMPLE 14.11

Hydrogen is made from natural gas (methane) for immediate consumption in industrial processes, such as ammonia production. The first step is called the "steam reforming of methane":

$$CH_4(g) + H_2O(g) \rightleftharpoons CO(g) + 3\,H_2(g)$$

The equilibrium constant for this reaction is 1.8×10^{-7} at 600 K. Gaseous CH_4, H_2O, and CO are introduced into an evacuated container at 600 K, and their initial partial pressures (before reaction) are 1.40 atm, 2.30 atm, and 1.60 atm, respectively. Determine the partial pressure of $H_2(g)$ that will result at equilibrium.

SOLUTION

The equilibrium expression is

$$\frac{(P_{CO})(P_{H_2})^3}{(P_{CH_4})(P_{H_2O})} = K$$

Set up the table of partial pressures:

	$CH_4(g)$	+	$H_2O(g)$	$\rightleftharpoons$	$CO(g)$	+	$3\,H_2(g)$
Initial partial pressure	1.40		2.30		1.60		0
Change in partial pressure	$-y$		$-y$		$+y$		$+3y$
Equilibrium partial pressure	$1.40 - y$		$2.30 - y$		$1.60 + y$		$3y$

Insert the equilibrium partial pressures into the equilibrium expression:

$$\frac{(1.60 + y)(3y)^3}{(1.40 - y)(2.30 - y)} = 1.8 \times 10^{-7}$$

If we expand this equation by multiplying through by the denominator, a polynomial equation of fourth order in y would result, for which the quadratic formula from the preceding problem would be useless. How can we solve the equation?

The equilibrium constant in this case is quite small, so the extent of reaction will also be small. This suggests that y will be a small number relative to the partial pressures of the gases present initially. Let's try the approximation that y can be ignored where it is added to a number that is close to one; that is, let's replace $1.60 + y$ with 1.60 in the preceding equation, and make the same approximation for the two terms in the denominator. When y *multiplies* something, as in the $(3y)^3$ term, of course we cannot set it equal to zero. The result of these steps is the approximate equation

$$\frac{(1.60)(3y)^3}{(1.40)(2.30)} = 1.8 \times 10^{-7}$$

$$y^3 = 1.34 \times 10^{-8}$$

The cube roots of both sides give

$$y = 2.38 \times 10^{-3}$$

This value is indeed small compared with 1.60, 1.40, and 2.30, so our approximation of neglecting y relative to these numbers is justified. Finally, at equilibrium we have

$$P_{H_2} = 3y = 7.1 \times 10^{-3} \text{ atm}$$

Related Problems: 33, 34, 35, 36

Suppose the data for a gas-phase equilibrium are given in terms of concentrations rather than partial pressures. In such cases we convert all concentrations to partial pressures before starting the calculations. Or, we can rewrite the equilibrium expression in terms of concentration variables. The concentration [A] of an ideal gas is related to its partial pressure P_A through

$$[A] = \frac{n_A}{V} = \frac{P_A}{RT}$$

which can be written

$$P_A = RT[A]$$

We can substitute such a relation for each species appearing in the equilibrium expression. It is best to put the factors of $P_{ref} = 1$ atm back into the equilibrium expression to examine the units of the resulting equations. For the general gas-phase reaction

$$aA(g) + bB(g) \rightleftharpoons cC(g) + dD(g)$$

the equilibrium expression is

$$\frac{(RT[C]/P_{ref})^c(RT[D]/P_{ref})^d}{(RT[A]/P_{ref})^a(RT[B]/P_{ref})^b} = K$$

Rearranging gives

$$\frac{[C]^c[D]^d}{[A]^a[B]^b} = K(RT/P_{ref})^{+a+b-c-d}$$

This expression relates concentrations of gas-phase species at equilibrium. If $a + b - c - d = 0$ (that is, if there is no change in the total number of moles of gases in the reaction mixture), the right side of the equilibrium expression reduces to K.

EXAMPLE 14.12

At elevated temperatures, PCl_5 dissociates extensively according to

$$PCl_5(g) \rightleftharpoons PCl_3(g) + Cl_2(g)$$

At 300°C, the equilibrium constant for this reaction is $K = 11.5$. The concentrations of PCl_3 and Cl_2 at equilibrium in a container at 300°C are both 0.0100 mol L^{-1}. Calculate $[PCl_5]$.

SOLUTION

Two moles of gases are produced for each mole of gas consumed, so RT/P_{ref} must be raised to the power $1 - 2 = -1$. Hence,

$$\frac{[PCl_3][Cl_2]}{[PCl_5]} = K\left(\frac{RT}{P_{ref}}\right)^{-1} = K\left(\frac{P_{ref}}{RT}\right)$$

$$= 11.5 \times \frac{1 \text{ atm}}{(0.08206 \text{ L atm mol}^{-1} \text{ K}^{-1})(573 \text{ K})} = 0.245 \frac{\text{mol}}{\text{L}}$$

Solving this equation for $[PCl_5]$ gives

$$[PCl_5] = \frac{[PCl_3][Cl_2]}{0.245 \text{ mol L}^{-1}} = \frac{(0.0100 \text{ mol L}^{-1})(0.0100 \text{ mol L}^{-1})}{0.245 \text{ mol L}^{-1}}$$

$$= 4.08 \times 10^{-4} \text{ mol L}^{-1}$$

Related Problems: 31, 32

14.6 The Direction of Change in Chemical Reactions: Empirical Description

The specific examples in Section 14.5 illustrate how the law of mass action gives information about the nature of the equilibrium state. The law of mass action also explains and predicts the direction in which a reaction will proceed spontaneously when reactants and products are initially mixed together with arbitrary partial pressures or compositions. This requires a new concept, the reaction quotient Q, which is related to the equilibrium constant. Through the *principle of Le Châtelier* (described below), the mass action law also explains how a reaction in equilibrium responds to an external perturbation.

The Reaction Quotient

The **reaction quotient** Q for a general gas-phase reaction is defined as

$$Q = \frac{(P_C)^c(P_D)^d}{(P_A)^a(P_B)^b}$$

where the partial pressures do *not* necessarily have their equilibrium values. The distinction between Q and K is crucial. The equilibrium constant K is determined by the partial pressures of reactants and products at equilibrium, and it is a constant, dependent only on the temperature. The reaction quotient Q depends on the actual instantaneous partial pressures, whatever they may be; thus, Q changes with time. As the reaction approaches equilibrium, Q approaches K. The initial partial pressures P°_A, P°_B, P°_C, and P°_D give an initial reaction quotient Q_0, whose magnitude relative to K determines the direction in which the reaction will proceed spontaneously toward equilibrium. If Q_0 is *less* than K, then Q must *increase* as time goes on. This requires an increase in the product partial pressures and a decrease in reactant partial pressures; in other words, the reaction proceeds from left to right. If Q_0 is *greater* than K, similar reasoning shows that the reaction will proceed from right to left, with Q *decreasing* with time until it becomes equal to K (Fig. 14.6).

EXAMPLE 14.13

The reaction between nitrogen and hydrogen to produce ammonia

$$N_2(g) + 3\ H_2(g) \rightleftharpoons 2\ NH_3(g)$$

is essential in making nitrogen-containing fertilizers. This reaction has an equilibrium constant equal to 1.9×10^{-4} at 400°C. Suppose that 0.10 mol of N_2, 0.040 mol of H_2, and 0.020 mol of NH_3 are sealed in a 1.00-L vessel at 400°C. In which direction will the reaction proceed?

SOLUTION

The initial pressures $P_i = n_iRT/V$ are readily calculated to be

$$P_{N_2} = 5.5 \text{ atm} \qquad P_{H_2} = 2.2 \text{ atm} \qquad P_{NH_3} = 1.1 \text{ atm}$$

The initial numerical value of Q is therefore

$$Q_0 = \frac{(P^\circ_{NH_3})^2}{P^\circ_{N_2}(P^\circ_{H_2})^3} = \frac{(1.1)^2}{(5.5)(2.2)^3} = 2.1 \times 10^{-2}$$

Because $Q_0 > K$, the reaction will proceed from right to left and ammonia will dissociate until equilibrium is reached.

Related Problems: 45, 46, 47, 48

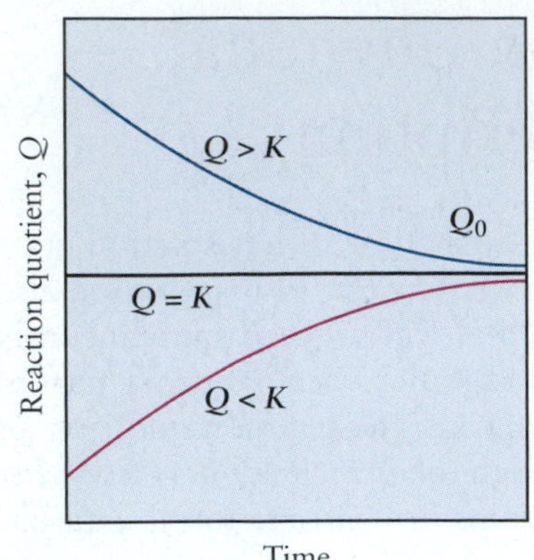

(a)

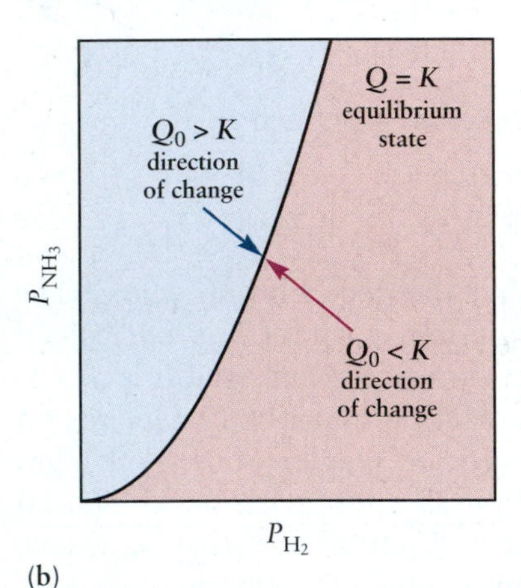

(b)

FIGURE 14.6 If nitrogen and hydrogen are mixed in 1:3 proportions together with some ammonia, they react according to the chemical equation

$$N_2(g) + 3\,H_2(g) \rightleftharpoons 2\,NH_3(g)$$

(a) If the initial reaction quotient Q_0 is less than K it increases with time; if it is greater than K, it decreases. (b) With three moles of H_2 available for each mole of N_2, a parabolic curve represents the partial pressures of ammonia and of hydrogen that coexist at equilibrium. From initial nonequilibrium conditions on either side, the partial pressures approach equilibrium along lines with slope $-2/3$, because three moles of H_2 are consumed to produce two moles of NH_3.

EXAMPLE 14.14

Solid ammonium chloride is in equilibrium with ammonia and hydrogen chloride gases:

$$NH_4Cl(s) \rightleftharpoons NH_3(g) + HCl(g)$$

The equilibrium constant at 275°C is 1.04×10^{-2}.

We place 0.980 g of solid NH_4Cl into a closed vessel with volume 1.000 L and heat to 275°C. (a) In what direction does the reaction proceed? (b) What is the partial pressure of each gas at equilibrium? (c) What is the mass of solid NH_4Cl at equilibrium?

SOLUTION

(a) We evaluate the reaction quotient

$$Q = P_{NH_3}P_{HCl}$$

Initially, $Q_0 = 0$ because neither gas is present. By comparison, $K = 1.04 \times 10^{-2}$. Because $Q < K$, the reaction will proceed spontaneously from left to right. Some of the solid NH_4Cl will decompose, and some gaseous NH_3 and HCl will appear in the vessel.

(b) We set up the standard table for the equilibrium calculation:

	$NH_4Cl(s)$	$\rightleftharpoons$	$NH_3(g)$	+	$HCl(g)$
Initial partial pressure (atm)			0		0
Change in partial pressure (atm)			$+x$		$+x$
Equilibrium partial pressure (atm)			$+x$		$+x$

Because NH_3 and HCl are formed in equimolar amounts, they will have the same partial pressure at equilibrium. The equilibrium expression is

$$P_{NH_3}P_{HCl} = K = 1.04 \times 10^{-2}$$

$$x^2 = 1.04 \times 10^{-2}$$

$$x = 0.102$$

At equilibrium, the partial pressures are

$$P_{NH_3} = P_{HCl} = 0.102 \text{ atm}$$

(c) The number of moles of NH_4Cl that decomposed is equal to the number of moles of each gas produced. We calculate this number, treating the gases as ideal

$$n_{NH_3} = n_{HCl} = \frac{(1.02 \times 10^{-1} \text{ atm})(1.000 \text{ L})}{(0.08206 \text{ L atm mol}^{-1}\text{ K}^{-1})(548.2 \text{ K})} = 2.268 \times 10^{-3} \text{ mol}$$

The mass of NH_4Cl consumed is $(2.268 \times 10^{-3} \text{ mol})(53.49 \text{ g mol}^{-1}) = 0.121$ g. The remaining mass is 0.859 g. The percentage decomposition of the original sample is 12.4%.

Related Problems: 49, 50

External Effects on K: Principle of Le Châtelier

Suppose a system at equilibrium is perturbed by some external stress such as a change in volume or temperature or a change in the partial pressure or concentration of one of the reactants or products. How will the system respond? The qualitative answer is embodied in a principle stated by Henri Le Châtelier in 1884:

> *A system in equilibrium that is subjected to a stress will react in a way that tends to counteract the stress.*

Le Châtelier's principle provides a way to predict qualitatively the direction of change of a system under an external perturbation. It relies heavily on Q as a predictive tool.

EFFECTS OF CHANGING THE CONCENTRATION OF A REACTANT OR PRODUCT As a simple example, consider what happens when a small quantity of a reactant is added to an equilibrium mixture. The addition of reactant lowers the reaction quotient Q below K and a net reaction takes place in the forward direction, partially converting reactants to products, until Q again equals K. The system partially counteracts the stress (the increase in the quantity of one of the reactants) and attains a new equilibrium state. If one of the *products* is added to an equilibrium mixture, Q temporarily becomes *greater* than K and a net *back* reaction occurs, partially counteracting the imposed stress by reducing the concentration of products (Fig. 14.7).

EXAMPLE 14.15

An equilibrium gas mixture of $H_2(g)$, $I_2(g)$, and $HI(g)$ at 600 K has

$$P_{H_2} = 0.4756 \text{ atm} \qquad P_{I_2} = 0.2056 \text{ atm} \qquad P_{HI} = 3.009 \text{ atm}$$

This is essentially the final equilibrium state of Example 14.10. Enough H_2 is added to increase its partial pressure to 2.000 atm at 600 K before any reaction takes place. The mixture then once again reaches equilibrium at 600 K. What are the final partial pressures of the three gases?

SOLUTION

Set up the usual table, in which "initial" now means the moment after the addition of the new H_2 but before it reacts further.

	$H_2(g)$	+	$I_2(g)$	$\rightleftarrows$	$2\ HI(g)$
Initial partial pressure (atm)	2.000		0.2056		3.009
Change in partial pressure (atm)	$-x$		$-x$		$+2x$
Equilibrium partial pressure (atm)	$2.000 - x$		$0.2056 - x$		$3.009 + 2x$

From Le Châtelier's principle it follows that net reaction to consume H_2 will occur after addition of H_2, and this fact has been used in assigning a negative sign to the change in the partial pressure of H_2 in the table.

Substitution of the equilibrium partial pressures into the equilibrium law gives

$$\frac{(3.009 + 2x)^2}{(2.000 - x)(0.2056 - x)} = 92.6$$

Expansion of this expression results in the quadratic equation

$$88.60x^2 - 216.275x + 29.023 = 0$$

which can be solved to give

$$x = 0.1425 \text{ or } 2.299$$

The second root would lead to negative partial pressures of H_2 and I_2 and is therefore physically impossible. Substitution of the first root into the expressions from the table gives

$$P_{H_2} = 2.000 - 0.1425 = 1.86 \text{ atm}$$

$$P_{I_2} = 0.2056 - 0.1425 = 0.063 \text{ atm}$$

$$P_{HI} = 3.009 + 2(0.1425) = 3.29 \text{ atm}$$

$$\text{Check:} \quad \frac{(3.29)^2}{(1.86)(0.063)} = 92.4$$

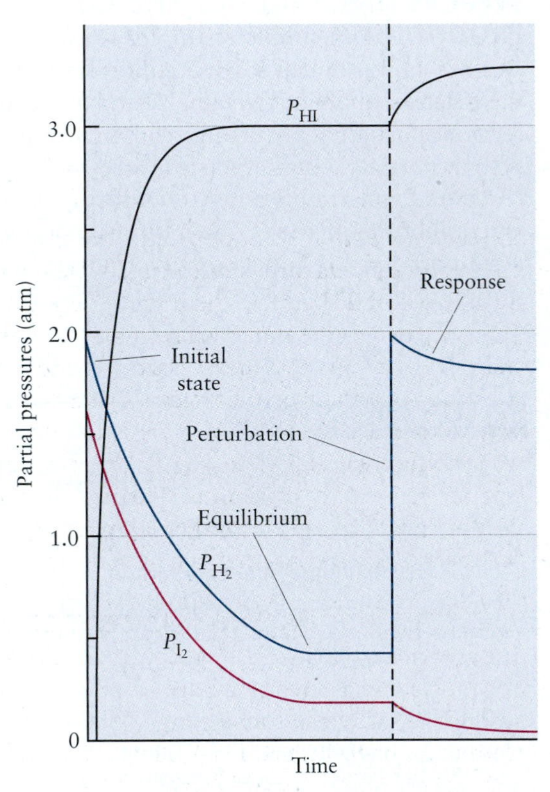

FIGURE 14.7 Partial pressures versus time for the equilibrium

$$H_2(g) + I_2(g) \rightleftharpoons 2\ HI(g)$$

The left part of the figure shows the attainment of equilibrium starting from the initial conditions of Example 14.10. Then the equilibrium state is abruptly perturbed by an increase in the partial pressure of H_2 to 2.000 atm. In accordance with Le Châtelier's principle, the system responds (Example 14.15) in such a way as to decrease the partial pressure of H_2—that is, to counteract the perturbation that moved it away from equilibrium in the first place.

When HI is made from the elements, iodine is a much more expensive reactant than hydrogen. It therefore makes sense to add hydrogen to the reaction mixture (as in Example 14.15) to ensure more complete reaction of the iodine. If one of the products is removed from an equilibrium mixture, the reaction will also occur in the forward direction to compensate partially by increasing the partial pressures of products. Most industrial operations are designed in such a way that products can be removed continuously to achieve high overall yields, even for reactions with small equilibrium constants.

EFFECTS OF CHANGING THE VOLUME Le Châtelier's principle also predicts the effect of a change in volume on gas-phase equilibrium. Decreasing the volume of a gaseous system increases its total pressure, and the system responds, if possible, to reduce the total pressure. For example, in the equilibrium

$$2\ P_2(g) \rightleftharpoons P_4(g)$$

the reaction shifts in the forward direction when the volume is decreased. This occurs because every two molecules of P_2 consumed produce only one molecule of P_4, thus reducing the total pressure and partially compensating for the external stress caused by the change in volume. In contrast, an *increase* in volume favors reactants over products in this system, and some P_4 dissociates to form P_2 (Fig. 14.8). If there is no difference in the total numbers of gas-phase molecules on the two sides of the equation, then a change in volume has no effect on the equilibrium.

This effect of changing the volume of an equilibrium reacting mixture can also be understood by using the reaction quotient. For the phosphorus equilibrium just described, the reaction quotient is

$$Q = \frac{P_{P_4}}{(P_{P_2})^2}$$

Initially, Q_0 equals K. Suppose the volume is then decreased by a factor of 2; because the temperature is unchanged, this initially increases each partial pressure by a factor of 2. Because there are two powers of the pressure in the denominator and only one in the numerator, this decreases Q by a factor of 2, making it lower than K. Reaction must then occur in the forward direction until Q again equals K.

When the volume of a system is decreased, its total pressure increases. Another way to increase the total pressure is to add an inert gas such as argon to the reaction mixture without changing the total volume. In this case the effect on the equilibrium is entirely different. Because the partial pressures of the reactant and product gases are unchanged by an inert gas, adding argon at constant volume has no effect on the position of the equilibrium.

EFFECTS OF CHANGING THE TEMPERATURE Chemical reactions are either **endothermic** (taking up heat from the surroundings) or **exothermic** (giving off heat). Raising the temperature of an equilibrium mixture by adding heat causes reactions to occur in such a way as to absorb some of the added heat. The equilibrium in an endothermic reaction shifts from left to right, while that in an exothermic reaction shifts from right to left, with "products" reacting to give "reactants."

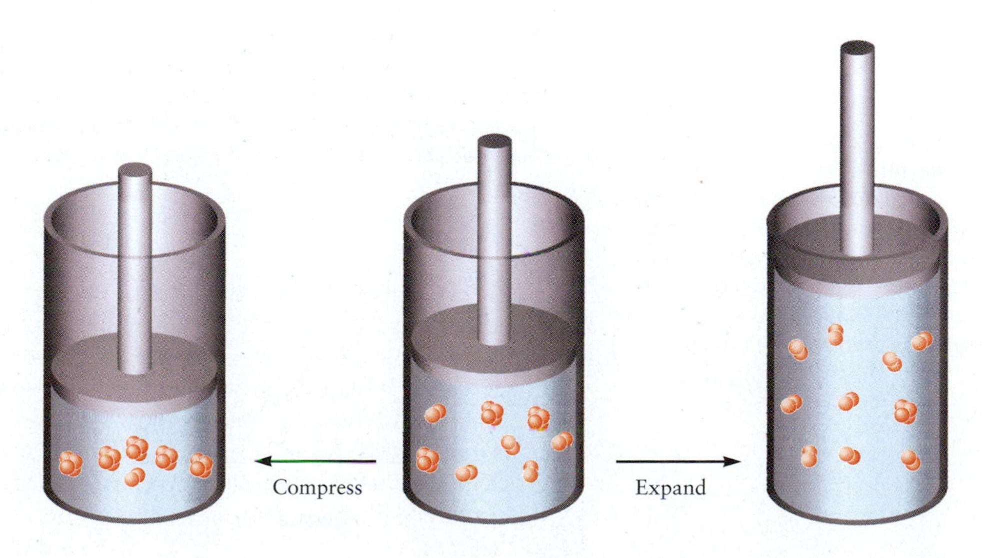

FIGURE 14.8 An equilibrium mixture of P_2 and P_4 (center) is compressed (left). Some P_2 molecules combine to give P_4 molecules, to reduce the total number of molecules and thus the total pressure. If the volume is increased (right), some P_4 molecules dissociate to pairs of P_2 molecules to increase the total number of molecules and the pressure exerted by those molecules.

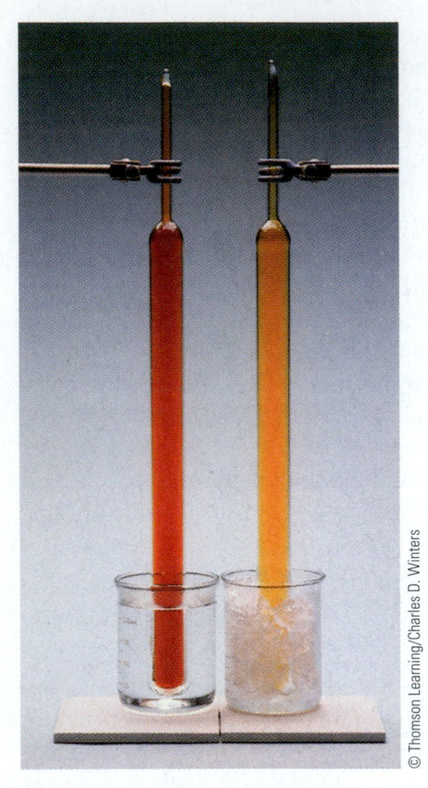

FIGURE 14.9 The equilibrium between N_2O_4 and NO_2 depends on temperature. The tube on the right, held in an ice bath at 0°C, contains mostly N_2O_4. Its color is pale because only NO_2 is colored. The deeper color in the tube on the left, which is held at 50°C, reflects the increased NO_2 present in equilibrium at the higher temperature. The tubes contain the same masses of substance, distributed in different ways between NO_2 and N_2O_4.

Equivalently, we can describe the shifts in terms of the effect of temperature on equilibrium constants. The equilibrium constant for an endothermic reaction increases with increasing temperature, while that for an exothermic reaction decreases with increasing temperature.

This effect is illustrated by the equilibrium between nitrogen dioxide (NO_2) and its dimer, dinitrogen tetraoxide (N_2O_4) (briefly considered in Example 9.6) expressed by the chemical equation

$$2\ NO_2(g) \rightleftharpoons N_2O_4(g)$$

Because NO_2 is a brown gas but N_2O_4 is colorless, the equilibrium between them can be studied by observing the color of a tube containing the two gases. At high temperatures, NO_2 predominates and a brown color results; as the temperature is lowered, the partial pressure of N_2O_4 increases and the color fades (Fig. 14.9).

The equilibrium expression for the N_2O_4–NO_2 equilibrium is

$$\frac{P_{N_2O_4}}{(P_{NO_2})^2} = K$$

K has the numerical value 8.8 at $T = 25°C$, provided the partial pressures of N_2O_4 and NO_2 are expressed in atmospheres. This reaction is exothermic ($\Delta H = -58.02\ kJ\ mol^{-1}$ at 298 K) because energy must be liberated when dimers are formed. Consequently, K decreases as the temperature T increases, so the amount of N_2O_4 present for a given partial pressure of NO_2 falls with increasing temperature as the dimer dissociates at elevated temperatures.

MAXIMIZING THE YIELD OF A REACTION As an application of Le Châtelier's principle, consider the reaction

$$N_2(g) + 3\ H_2(g) \rightleftharpoons 2\ NH_3(g)$$

which is the basis of the industrial synthesis of ammonia. Because this reaction is exothermic, the yield of ammonia is increased by working at as low a temperature as possible (Fig. 14.10a). At too low a temperature, the reaction is very slow, so a compromise temperature near 500°C is typically used. Because the number of moles of gas decreases as the reaction occurs, the yield of product is enhanced by decreasing the volume of the reaction vessel. Typically, total pressures of 150 to 300 atm are used (see Fig. 14.10b), although some plants work at up to 900 atm of pressure. Even at high pressures, the yield of ammonia is usually only 15% to 20% because the equilibrium constant is so small. To overcome this, ammonia plants operate in a cyclic process in which the gas mixture is cooled after ammonia is produced so that the ammonia liquefies (its boiling point is much higher than those of nitrogen and hydrogen) and is removed from the reaction vessel. Continuous removal of products helps drive the reaction to completion.

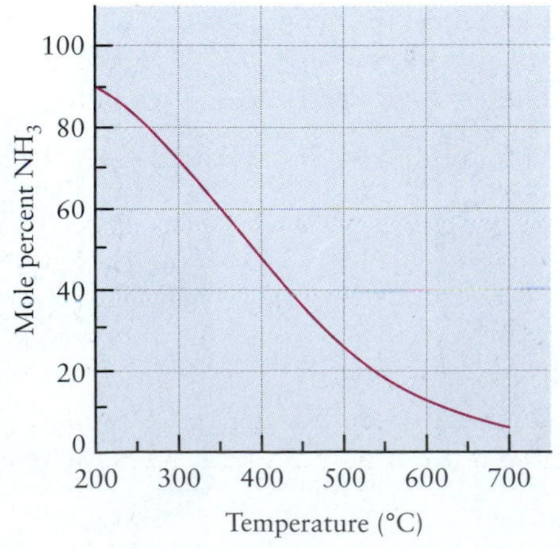

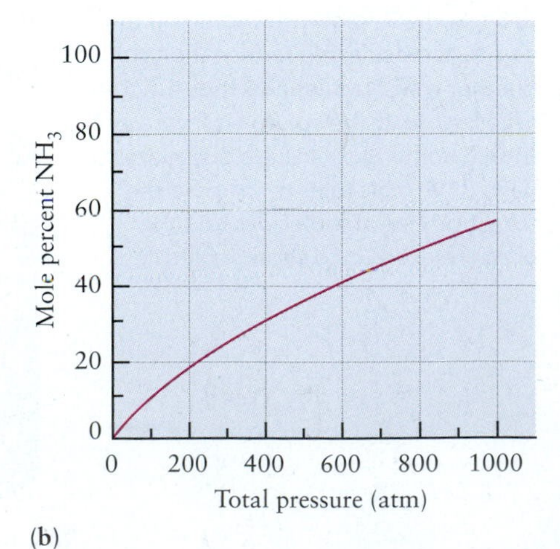

FIGURE 14.10 (a) The equilibrium mole percentage of ammonia in a 1:3 mixture of N_2 and H_2 varies with temperature; low temperatures favor high yields of NH_3. The data shown correspond to a fixed total pressure of 300 atm. (b) At a fixed temperature (here, 500°C), the yield of NH_3 increases with increasing total pressure.

14.7 The Direction of Change in Chemical Reactions: Thermodynamic Explanation

The specific examples in Section 14.5 demonstrate that when $K \gg 1$ the reaction has progressed far toward products, and when $K \ll 1$ the reaction has remained near reactants. The empirical discussion in Section 14.6 shows how the reaction quotient Q and the principle of Le Châtelier can predict the direction of spontaneous reaction and the response of an equilibrium state to an external perturbation. Here, we use the thermodynamic description of K from Section 14.3 to provide the thermodynamic basis for these results obtained empirically in Sections 14.5 and 14.6. We identify those thermodynamic factors that determine the magnitude of K. We also provide a thermodynamic criterion for predicting the direction in which a reaction proceeds from a given initial condition.

The Magnitude of the Equilibrium Constant

The expression connecting the standard Gibbs free energy change and the equilibrium constant can be rewritten as

$$\ln K = \frac{-\Delta G^\circ}{RT} = \frac{\Delta S^\circ}{R} - \frac{\Delta H^\circ}{RT}$$

so that

$$K = \exp\left[\frac{-\Delta G^\circ}{RT}\right] = \exp\left[\frac{\Delta S^\circ}{R}\right] \exp\left[\frac{-\Delta H^\circ}{RT}\right]$$

Here K is large (favoring the products) if ΔS° is positive and large and ΔH° is negative and large. In other words, an increase in the number of microstates ($\Delta S^\circ > 0$) and a decrease in enthalpy ($\Delta H^\circ < 0$) both favor a large K. Thus, the same factors that favor reaction spontaneity by making ΔG° negative also favor a large K if they can make ΔG° large in magnitude as well as negative in sign. If ΔH° and ΔS° have the *same* sign, the value of K will be a compromise between one effect that raises K and another that lowers it.

Free Energy Changes and the Reaction Quotient

The direction in which a spontaneous chemical reaction proceeds after it is initiated with a given initial concentration of products and reactants is the direction in which $\Delta G < 0$. If the initial condition is "to the left" of the equilibrium state, products will be formed at the expense of reactants; if the initial condition is "to the right" of the equilibrium state, products will be converted back to reactants. This criterion can be made quantitative and expressed in terms of the initial concentrations as follows.

Proceeding as in Section 14.3, we find ΔG for the general gas-phase reaction

$$a\text{A} + b\text{B} \longrightarrow c\text{C} + d\text{D}$$

to be

$$\Delta G = \Delta G^\circ + RT \ln \left[\frac{(P_\text{C}/P_\text{ref})^c(P_\text{D}/P_\text{ref})^d}{(P_\text{A}/P_\text{ref})^a(P_\text{B}/P_\text{ref})^b}\right]$$

At equilibrium, where $\Delta G = 0$, the combination of partial pressures appearing inside the brackets becomes the equilibrium constant, K. Away from equilibrium, this combination of partial pressures is the reaction quotient Q, introduced in Section 14.6:

$$\Delta G = \Delta G^\circ + RT \ln Q$$

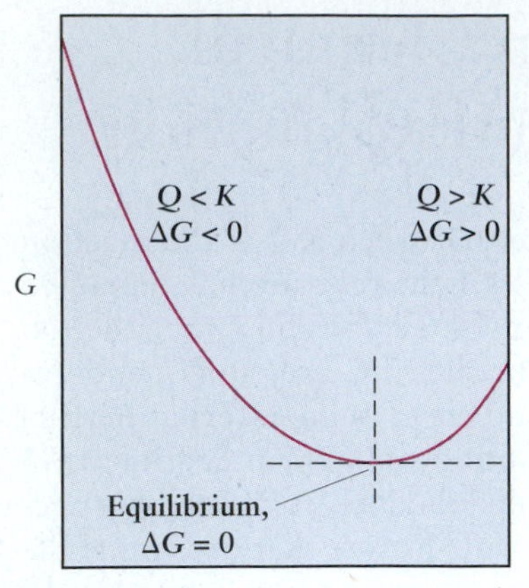

FIGURE 14.11 The free energy of a reaction system is plotted against its progress from pure reactants (left) to pure products (right). Equilibrium comes at the minimum of the curve. To the reactant side of equilibrium, $\Delta G < 0$ and $Q < K$. A reaction mixture with initial condition in this range will spontaneously move toward equilibrium by converting more reactants into products. To the product side of equilibrium, $\Delta G > 0$ and $Q > K$. A reaction mixture prepared in this range will spontaneously move toward equilibrium by converting products back into reactants.

The equilibrium constant can be substituted for $\Delta G°$ in this equation to obtain a very useful relation between ΔG, K, and Q as follows:

$$\Delta G = \Delta G° + RT \ln Q = -RT \ln K + RT \ln Q = RT \ln (Q/K) \qquad [14.10]$$

If the reaction quotient Q is *less* than K, $\Delta G < 0$ and the reaction will proceed spontaneously as written, from left to right. If $Q > K$, then $\Delta G > 0$ and the *reverse* reaction (right to left) will occur spontaneously until equilibrium is reached. These conditions are represented schematically in Figure 14.11. The second law of thermodynamics thus provides a very useful criterion for the direction of reaction in terms of the initial value of the reaction quotient.

There exists a deep relationship between Figure 14.6, which represents actual events occurring in the laboratory, and Figure 14.11, which represents the thermodynamic driving force (that is, the Gibbs free energy) governing these events.

Temperature Dependence of Equilibrium Constants

Le Châtelier's principle is a qualitative way of describing the *stability* of equilibrium states against sudden perturbations in concentration, pressure, and temperature. The responses of the system to all three effects can be described quantitatively by thermodynamics. Here we describe the effect of temperature, which is the most useful of these quantitative descriptions.

The temperature dependence of the equilibrium constant is determined by the equation

$$-RT \ln K = \Delta G° = \Delta H° - T\Delta S°$$

If $\Delta H°$ and $\Delta S°$ are independent of temperature, then all the temperature dependence of K lies in the factor of T and the equation can be used to relate the values of K at two different temperatures, as follows. At least over a limited temperature range, $\Delta H°$ and $\Delta S°$ do not vary much with temperature. To the extent that their temperature dependence may be neglected, it is evident that $\ln K$ is a linear function of $1/T$ as shown in Figure 14.12.

$$\ln K = -\frac{\Delta G°}{RT} = -\frac{\Delta H°}{RT} + \frac{\Delta S°}{R} \qquad [14.11]$$

A graph of $\ln K$ against $1/T$ is approximately a straight line with slope $-\Delta H°/R$ and intercept $\Delta S°/R$ (see Fig. 14.12). If the value of K is known for one temperature and $\Delta H°$ is also known, K can be calculated for other temperatures.

In addition to the graphical method, an equation can be obtained to connect the values of K at two different temperatures. Let K_1 and K_2 be the equilibrium constants for a reaction at temperatures T_1 and T_2, respectively. Then

$$\ln K_2 = -\frac{\Delta H°}{RT_2} + \frac{\Delta S°}{R}$$

$$\ln K_1 = -\frac{\Delta H°}{RT_1} + \frac{\Delta S°}{R}$$

Subtracting the second equation from the first gives

$$\ln\left(\frac{K_2}{K_1}\right) = -\frac{\Delta H°}{R}\left[\frac{1}{T_2} - \frac{1}{T_1}\right] \qquad [14.12]$$

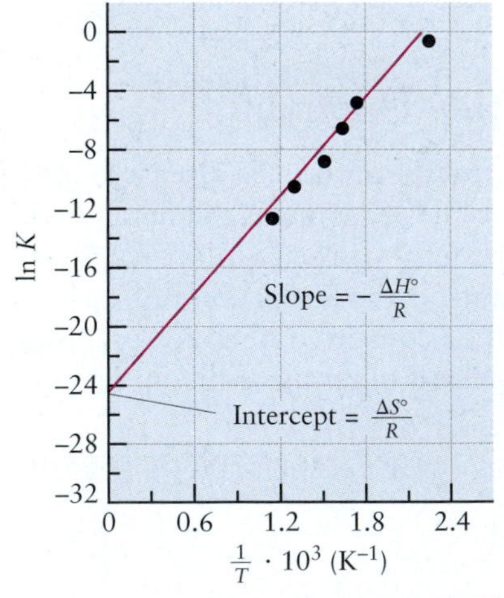

FIGURE 14.12 The temperature dependence of the equilibrium constant for the reaction

$$N_2(g) + 3\,H_2(g) \rightleftarrows 2\,NH_3(g)$$

Experimental data are shown by points.

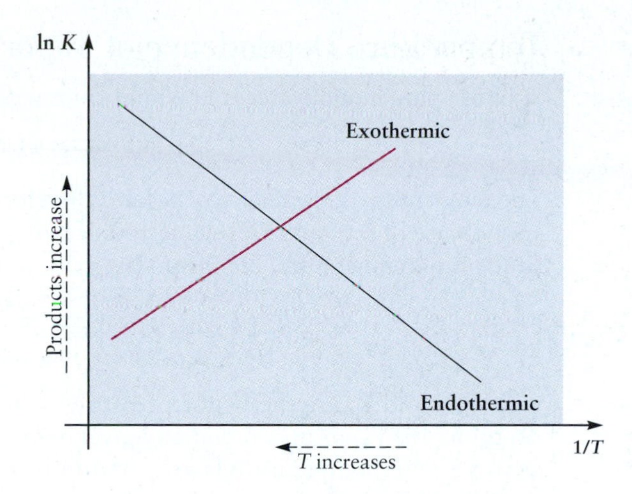

FIGURE 14.13 Sketch of ln K against $1/T$ for an exothermic reaction and for an endothermic reaction, as predicted by thermodynamics. Temperature increases to the left on this diagram. As T increases, K for the endothermic reaction increases and K for the exothermic reaction decreases in accordance with the principle of Le Châtelier.

This is known as the **van't Hoff equation** after the Dutch chemist Jacobus van't Hoff. Given $\Delta H°$ and K at one temperature, we can use the equation to calculate K at another temperature, within the approximation that $\Delta H°$ and $\Delta S°$ are independent of temperature. Alternatively, we can use it to determine $\Delta H°$ without a calorimeter if the values of K are known at several temperatures.

The effect of a temperature change on the equilibrium constant depends on the sign of $\Delta H°$. If $\Delta H°$ is negative (the reaction is exothermic, giving off energy as heat), then increasing the temperature *reduces* K. If $\Delta H°$ is positive (the reaction is endothermic, taking up energy as heat), then increasing the temperature *increases* K. These observations obtained from thermodynamics provide the quantitative basis for Le Châtelier's principle (Fig. 14.13).

EXAMPLE 14.16

Calculate K for the equilibrium of Example 14.4 at $T = 400$ K, assuming $\Delta H°$ to be approximately independent of temperature over the range from 298 to 400 K.

SOLUTION

The first step is to calculate $\Delta H°$ for the reaction. Appendix D provides data to calculate

$$\Delta H° = -155.52 \text{ kJ}$$

From the van't Hoff equation,

$$\ln\left(\frac{K_{400}}{K_{298}}\right) = -\frac{\Delta H°}{R}\left[\frac{1}{400 \text{ K}} - \frac{1}{298 \text{ K}}\right]$$

$$= \frac{155{,}520 \text{ J mol}^{-1}}{8.315 \text{ J K}^{-1} \text{ mol}^{-1}}\left[\frac{1}{400 \text{ K}} - \frac{1}{298 \text{ K}}\right] = -16.01$$

$$\frac{K_{400}}{K_{298}} = e^{-16.01} = 1.1 \times 10^{-7}$$

Taking K_{298} to be 1.8×10^{18} (from Example 14.4) gives

$$K_{400} = (1.8 \times 10^{18})(1.1 \times 10^{-7}) = 2.0 \times 10^{11}$$

Because the reaction is exothermic, an increase in temperature reduces the equilibrium constant.

An alternative way to do this calculation would be to determine both $\Delta H°$ and $\Delta S°$ and from them to calculate ΔG at 400 K.

Related Problems: 61, 62, 63, 64, 65, 66

Temperature Dependence of Vapor Pressure

Suppose pure liquid water is in equilibrium with its vapor at temperature T:

$$H_2O(\ell) \rightleftarrows H_2O(g) \qquad P_{H_2O(g)} = K$$

The temperature dependence of K (and therefore of the vapor pressure $P_{H_2O(g)}$) is a special case of the van't Hoff equation. If ΔH_{vap} and ΔS_{vap} are approximately independent of temperature, then from the van't Hoff equation,

$$\ln\left(\frac{K_2}{K_1}\right) = \ln\left(\frac{P_2}{P_1}\right) = -\frac{\Delta H_{vap}}{R}\left[\frac{1}{T_2} - \frac{1}{T_1}\right]$$

where P_2 and P_1 are the vapor pressures at temperatures T_2 and T_1. From this equation, the vapor pressure at any given temperature can be estimated if its value at some other temperature is known and if the enthalpy of vaporization is also known.

At the normal boiling point of a substance, T_b, the vapor pressure is 1 atm. If T_1 is taken to correspond to T_b and T_2 to some other temperature T, the van't Hoff equation is

$$\ln P = -\frac{\Delta H_{vap}}{R}\left[\frac{1}{T} - \frac{1}{T_b}\right] \qquad [14.13]$$

where P is the vapor pressure at temperature T, expressed in atmospheres.

EXAMPLE 14.17

The ΔH_{vap} for water is 40.66 kJ mol^{-1} at the normal boiling point, $T_b = 373$ K. Assuming ΔH_{vap} and ΔS_{vap} are approximately independent of temperature from 50°C to 100°C, estimate the vapor pressure of water at 50°C (323 K).

SOLUTION

$$\ln P_{323} = \frac{-40660 \text{ J mol}^{-1}}{8.315 \text{ J K}^{-1} \text{ mol}^{-1}}\left[\frac{1}{323 \text{ K}} - \frac{1}{373 \text{ K}}\right] = -2.03$$

$$P_{323} = 0.13 \text{ atm}$$

This differs slightly from the experimental value, 0.1217 atm, because ΔH_{vap} *does* change with temperature, an effect that was neglected in the approximate calculation.

Related Problems: 67, 68

14.8 Distribution of a Single Species between Immiscible Phases: Extraction and Separation Processes

An important type of heterogeneous equilibrium involves partitioning a solute species between two immiscible solvent phases. Such equilibria are used in many separation processes in chemical research and in industry.

FIGURE 14.14 Iodine is dissolved in water and poured on top of carbon tetrachloride in a separatory funnel (left). After the funnel is shaken (right), the iodine reaches a partition equilibrium between the upper (aqueous) phase and the lower (CCl_4) phase. The deeper color in the lower phase indicates that iodine dissolves preferentially in the denser CCl_4 phase.

Suppose two immiscible liquids, such as water and carbon tetrachloride, are put in a container. "Immiscible" means mutually insoluble; these liquids separate into two phases with the less dense liquid, in this case water, lying on top of the other liquid. A visible boundary, the meniscus, separates the two phases. If a solute such as iodine is added to the mixture and the vessel is shaken to distribute the iodine through the container (Fig. 14.14), the iodine is partitioned between the two phases at equilibrium with a characteristic concentration ratio, the **partition coefficient** K. This is the equilibrium constant for the process

$$I_2(aq) \rightleftharpoons I_2(CCl_4)$$

and can be written as

$$\frac{[I_2]_{CCl_4}}{[I_2]_{aq}} = K$$

in which $[I_2]_{CCl_4}$ and $[I_2]_{aq}$ are the concentrations (in moles per liter) of I_2 in the CCl_4 and aqueous phases, respectively. At 25°C, K has the value 85 for this equilibrium. The fact that K is greater than 1 shows that iodine is more soluble in CCl_4 than in water. If iodide ion (I^-) is dissolved in the water, then it can react with iodine to form the triiodide ion I_3^-:

$$I_2(aq) + I^-(aq) \longrightarrow I_3^-(aq)$$

This consumes $I_2(aq)$ and, by Le Châtelier's principle, causes more I_2 in the first equilibrium to move from the CCl_4 phase to the aqueous phase.

Extraction Processes

Extraction takes advantage of the partitioning of a solute between two immiscible solvents to remove that solute from one solvent into another. Suppose iodine is present as a contaminant in water that also contains other solutes that are insoluble in carbon tetrachloride. In such a case, most of the iodine could be removed by shaking the aqueous solution with CCl_4, allowing the two phases to separate, and then pouring off the water layer from the heavier layer of carbon tetrachloride. The greater the equilibrium constant for the partition of a solute from the original solvent into the extracting solvent, the more complete such a separation will be.

EXAMPLE 14.18

An aqueous solution has an iodine concentration of 2.00×10^{-3} M. Calculate the percentage of iodine remaining in the aqueous phase after extraction of 0.100 L of this aqueous solution with 0.050 L of CCl_4 at 25°C.

SOLUTION

The number of moles of I_2 present is

$$(2.00 \times 10^{-3} \text{ mol L}^{-1})(0.100\ L) = 2.00 \times 10^{-4} \text{ mol}$$

Suppose that y mol remains in the aqueous phase and $(2.00 \times 10^{-4} - y)$ mol passes into the CCl_4 phase. Then

$$\frac{[I_2]_{CCl_4}}{[I_2]_{aq}} = K = 85$$

$$= \frac{(2.00 \times 10^{-4} - y)/0.050}{y/0.100}$$

$$= \frac{2(2.00 \times 10^{-4} - y)}{y}$$

Note that the volumes of the two solvents used are unchanged because they are immiscible. Solving for y gives

$$y = 4.6 \times 10^{-6} \text{ mol}$$

The fraction remaining in the aqueous phase is $(4.6 \times 10^{-6})/(2.0 \times 10^{-4}) = 0.023$, or 2.3%. Additional extractions could be carried out to remove more of the I_2 from the aqueous phase.

Related Problems: 71, 72, 73, 74

One extraction process used industrially on a large scale is the purification of sodium hydroxide for use in the manufacture of rayon. The sodium hydroxide produced by electrolysis typically contains 1% sodium chloride and 0.1% sodium chlorate as impurities. If a concentrated aqueous solution of sodium hydroxide is extracted with liquid ammonia, the NaCl and $NaClO_3$ are partitioned into the ammonia phase in preference over the aqueous phase. The heavier aqueous phase is added to the top of an extraction vessel filled with ammonia, and equilibrium is reached as droplets of it settle through the ammonia phase to the bottom. This procedure reduces impurity concentrations in the sodium hydroxide solution to about 0.08% NaCl and 0.0002% $NaClO_3$.

Chromatographic Separations

Partition equilibria are the basis of an important class of separation techniques called **chromatography.** This word comes from the Greek root *chroma,* meaning "color," and was chosen because the original chromatographic separations involved colored substances. The technique can be applied to a variety of mixtures of substances.

Chromatography is a continuous extraction process in which solute species are exchanged between two phases. One, the mobile phase, moves with respect to the other, stationary phase. The partition ratio K of a solute A between the stationary and mobile phases is

$$\frac{[\mathrm{A}]_{\mathrm{stationary}}}{[\mathrm{A}]_{\mathrm{mobile}}} = K$$

Paper chromatography separates a line of ink, drawn across the bottom of the paper, into its component colors. As the water rises through the paper, the different components of the ink are attracted differently to the water and the paper and are separated.

As the mobile phase containing solute passes over the stationary phase, the solute molecules move between the two phases. True equilibrium is never fully established because the motion of the fluid phase continually brings fresh solvent into contact with the stationary phase. Nevertheless, the partition coefficient K provides a guide to the behavior of a particular solute. The greater K is, the more time the solute spends in the stationary phase and therefore the slower its progress through the separation system. Solutes with different values of K are separated by their different rates of travel.

Different types of chromatography use different mobile and stationary phases; Table 14.1 lists some of the most important. **Column chromatography** (Fig. 14.15) uses a tube packed with a porous material, often a silica gel on which water has been adsorbed. Water is therefore the stationary liquid phase in this case. Other solvents such as pyridine or benzene are used in the mobile phase; in some cases it is most efficient to use different solvents in succession to separate the components of a solute mixture. As the solute fractions reach the bottom of the column, they are separated for analysis or use. Column chromatography is important in industry because it is easily increased from laboratory to production scale.

Gas–liquid chromatography (Fig. 14.16) is one of the most important separation techniques for modern chemical research. The stationary phase is again a liquid

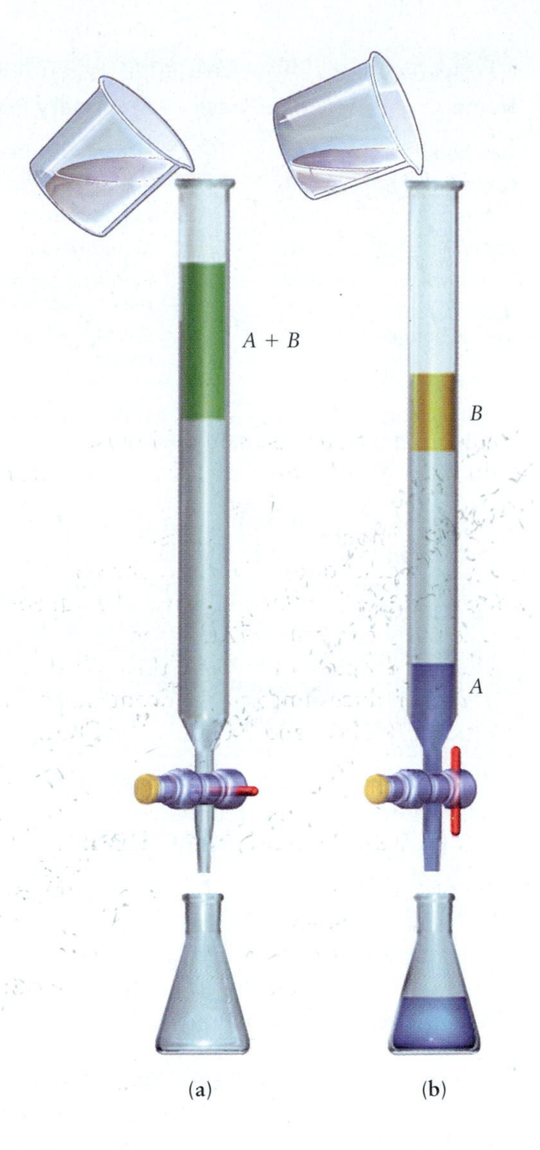

FIGURE 14.15 (a) In a column chromatograph, the top of the column is loaded with a mixture of solutes to be separated (green). (b) Upon addition of solvent, the different solutes travel at different rates, giving rise to bands. The separate fractions can be collected in different flasks for use or analysis.

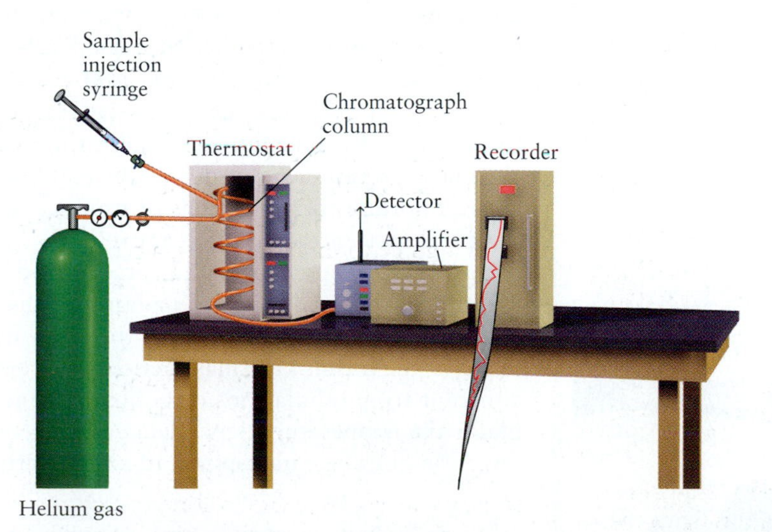

FIGURE 14.16 In a gas–liquid chromatograph, the sample is vaporized and passes through a column, carried in a stream of an inert gas such as helium or nitrogen. The residence time of any substance on the column depends on its partition coefficient from the vapor to the liquid in the column. A species leaving the column at a given time can be detected by a variety of techniques. The result is a gas chromatogram, with a peak corresponding to each substance in the mixture.

TABLE 14.1 Chromatographic Separation Techniques[†]

Name	Mobile Phase	Stationary Phase
Gas–liquid	Gas	Liquid adsorbed on a porous solid in a tube
Gas–solid	Gas	Porous solid in a tube
Column	Liquid	Liquid adsorbed on a porous solid in a tubular column
Paper	Liquid	Liquid held in the pores of a thick paper
Thin layer	Liquid	Liquid or solid; solid is held on glass plate and liquid may be adsorbed on it
Ion exchange	Liquid	Solid (finely divided ion-exchange resin) in a tubular column

[†]Adapted from D. A. Skoog and D. M. West, *Analytical Chemistry* (Saunders College Publishing, Philadelphia, 1980), Table 18-1.

adsorbed on a porous solid, but the mobile phase is now gaseous. The sample is vaporized and passes through the column, carried in a stream of an inert gas such as helium or nitrogen. The residence time in the column depends on the partition coefficient of the solute species, allowing an efficient separation of mixtures. The solute leaving the column at a given time can be detected by a variety of techniques that produce a **gas chromatogram** with a peak corresponding to each solute species in the mixture. Gas–liquid chromatography is widely used for separating the products of organic reactions. It can also be used to determine the purity of substances, because even very small amounts of impurities appear clearly as separate peaks in the chromatogram. The technique is important in the separation and identification of trace amounts of possibly toxic substances in environmental and biological samples. Amounts on the order of parts per trillion (10^{-12} g in a 1-g sample) can be detected and identified.

CHAPTER SUMMARY

Most chemical reactions do not go to completion. They arrive at the equilibrium state, after which there is no further net change in the amount of products or reactants. At the microscopic level, the reaction continues in both forward and reverse directions, at equal rates in these opposing directions. So, chemical equilibrium—which we measure at the macroscopic level—is characterized by a dynamical balance of events on the microscopic level. At the equilibrium state, concentrations of products and reactants always satisfy the mass action law, and so are related by the equilibrium constant. The equilibrium constant is a unique property of each chemical reaction. Knowledge of the equilibrium constant enables us to calculate the equilibrium concentrations of products and reactants that result from any set of initial concentrations for that reaction. Equilibrium calculations have enormous predictive power for interpreting and optimizing the outcome of chemical reactions. Thermodynamics explains all the empirical observations of chemical equilibrium and provides means for quantitative predictions. Thermodynamics explains the form of the mass action law, shows how to calculate the equilibrium constant from tabulations of Gibbs free energy for products and reactants, explains the temperature dependence of the equilibrium constant, and predicts the direction of change in response to any disturbance of the equilibrium state.

A stockpile of sulfur near chemical plants in Los Angeles.

CUMULATIVE EXERCISE

This cumulative exercise is divided into two parts. Readers who have not yet studied thermodynamics should stop after Part 1; those who have studied thermodynamics should continue with Part 2.

Cumulative Exercise Part 1

Sulfuric Acid

Sulfuric acid is produced in larger volume than any other chemical and has a tremendous number of applications, ranging from fertilizer manufacture to metal treatment and chemical synthesis.

The modern industrial production of sulfuric acid involves three steps, for which the balanced chemical equations are:

1. $S(s) + O_2(g) \rightleftharpoons SO_2(g)$

2. $SO_2(g) + \frac{1}{2}O_2(g) \rightleftharpoons SO_3(g)$

3. $SO_3(g) + H_2O(\ell) \rightleftharpoons H_2SO_4(\ell)$

(a) Write an equilibrium expression for each of these steps, with equilibrium constants K_1, K_2, and K_3.

(b) If these reactions could be carried out at 25°C, the equilibrium constants would be 3.9×10^{52}, 2.6×10^{12}, and 2.6×10^{14}. Write a balanced equation for the overall reaction and calculate its equilibrium constant at 25°C.

(c) Although the products of all three equilibria are strongly favored at 25°C [See the data given in part (b)], reactions 1 and 2 occur too slowly to be practical; they must be carried out at elevated temperatures. At 700°C, the partial pressures of SO_2, O_2, and SO_3 in an equilibrium mixture are measured to be 2.23 atm, 1.14 atm, and 6.26 atm, respectively. Calculate K_2 at 700°C.

(d) At 300°C, K_2 has the value 1.3×10^4. Suppose some SO_3 is introduced into an evacuated vessel at an initial partial pressure of 0.89 atm. Calculate the partial pressure of SO_2 that will be reached at equilibrium, assuming that only reaction 2 takes place under these conditions. (*Hint:* K_2 is large enough that you can assume the fraction of SO_3 dissociated is very small.)

(e) Some SO_2 is placed in a flask and heated with oxygen to 600°C, at which point $K_2 = 9.5$. At equilibrium, 62% of it has reacted to give SO_3. Calculate the partial pressure of oxygen at equilibrium in this reaction mixture.

(f) Equal numbers of moles of SO_2, O_2, and SO_3 are mixed and heated to 600°C, where their total pressure before reaction is 0.090 atm. Will reaction 2 occur from right to left or from left to right? Will the total pressure increase or decrease during the course of the reaction?

(g) Reactions 1 and 3 are both exothermic. State the effects on equilibria 1 and 3 of increasing the temperature and of decreasing the volume. (*Note:* A change in volume has little effect on liquids and solids taking part in a reaction.)

Answers

(a) $\frac{P_{SO_2}}{P_{O_2}} = K_1 \qquad \frac{P_{SO_3}}{P_{SO_2}P_{O_2}^{1/2}} = K_2 \qquad \frac{1}{P_{SO_3}} = K_3$

(b) $S(s) + \frac{3}{2}O_2(g) + H_2O(\ell) \rightleftharpoons H_2SO_4(\ell)$; $K = K_1K_2K_3 = 2.2 \times 10^{79}$

(c) K_2 is 2.63.

(d) The SO_2 partial pressure is 2.1×10^{-3} atm.

(e) The O_2 partial pressure is 0.029 atm.

(f) Left to right; pressure will decrease

(g) Increasing the temperature will shift both equilibria to the left. Decreasing the volume will not affect equilibrium 1 and will shift equilibrium 3 to the right.

Cumulative Exercise Part 2

We continue with examination of reactions 1, 2, and 3 in the production of sulfuric acid.

(h) Calculate $\Delta G°$ for each reaction at 25°C. (Standard state of sulfur is rhombic.)

(i) Write a balanced equation for the overall reaction, and calculate its value of $\Delta G°$ at 25°C.

(j) Part (h) shows all three reactions are spontaneous at 25°C. Nonetheless, reactions 1 and 2 occur too slowly at 25°C to be practical; they must be carried out at higher temperatures. Calculate $\Delta G°$ for 1, 2, and 3 at 700°C.

(k) Determine the highest temperature at which all three reactions are spontaneous.

Answers

(h) At 25°C: $\Delta G_1° = -300.19$ kJ mol^{-1}; $\Delta G_2° = -70.86$ kJ mol^{-1}; $\Delta G_3° = -81.64$ kJ mol^{-1}

(i) $S(s) + \frac{3}{2} O_2(g) + H_2O(\ell) \rightleftharpoons H_2SO_4(\ell)$; $\Delta G°_{net} = -452.89$ kJ mol^{-1}

(j) At 700°C: $\Delta G_1° = -307.81$ kJ mol^{-1}; $\Delta G_2° = -7.47$ kJ mol^{-1}; $\Delta G_3° = +32.64$ kJ mol^{-1}

(k) Because reaction 1 has $\Delta H° < 0$ and $\Delta S° > 0$, it is spontaneous at all temperatures. Because both reaction 2 and reaction 3 have $\Delta H° < 0$ and $\Delta S° < 0$, each is spontaneous at temperatures below the temperature T^* at which $\Delta G = 0$. With the higher ratio of $\Delta S°$ to $\Delta H°$, reaction 3 is the first to become nonspontaneous, at $T^* = 508°C$.

CHAPTER REVIEW

Fundamental Aspects of Chemical Equilibrium

- Approach to chemical equilibrium
  The equilibrium state for a particular reaction at a particular temperature is the same regardless of whether it is approached from reactants or from products.
- Characteristics of the equilibrium state
  No macroscopic evidence of change
  Reached through a spontaneous process
  Dynamic balance between forward and reverse processes
  The same regardless of direction of approach

The Law of Mass Action for Gas-Phase Reactions and Solution Reactions

- $\dfrac{P_C^c P_D^d}{P_A^a P_B^b} = K$
- $\dfrac{[C]^c[D]^d}{[A]^a[B]^b} = K$
- In these equations we describe reaction quantities as follows:
  The concentrations of gases are expressed in partial pressures in atm.
  The concentrations of solutes are expressed in moles per liter.
  The concentrations of pure liquids and pure solids are set equal to one, as is the concentration of the solvent in a dilute solution.
  Partial pressures and concentrations of products appear in the numerator and those of the reactants in the denominator; each is raised to a power equal to its stoichiometric coefficient in the balanced reaction.

Thermodynamic Description of the Equilibrium State

- Pressure dependence of the Gibbs Free Energy for ideal gas species
  $\Delta G = \Delta(H - TS) = \Delta H - T\,\Delta S = -T\,\Delta S$ for constant T
  $$\Delta S = nR \ln\left(\frac{V_2}{V_1}\right) = nR \ln\left(\frac{P_1}{P_2}\right) = -nR \ln\left(\frac{P_2}{P_1}\right)$$
  $\Delta G = nRT \ln P$ if P is expressed in atm.

- To obtain the equilibrium expression for ideal gas reactions:
  Take reactants from initial pressure to 1 atm.
  Run reaction at 1 atm.
  Take products from 1 atm to final pressure

$$\Delta G = \Delta G_1 + \Delta G_2 + \Delta G_3$$
$$-\Delta G^\circ = RT \ln K(T)$$

  Calculate K by calculating ΔG° from tables of ΔG_f° for reactants and products.

- The equilibrium expression for solution and heterogeneous reactions is developed in the same way.
  Represent each species by its activity.
  Replace the activity with its limiting form of partial pressure or concentration.
  For pure solid or liquid products or reactants in their reference states set the activity equal to 1.
  Calculate K by calculating ΔG° from tables of ΔG_f° for reactants and products, paying careful attention to reference state for each species.

The Law of Mass Action for Related and Simultaneous Reactions

- $K_{\text{reverse}} = -K_{\text{forward}}$
- K (equation multiplied by n) $= K^n$
- $K_3 = K_1K_2 \ldots$ when a series of reactions is added to give an overall reaction.
- $K_3 = K_1/K_2 \ldots$ if one reaction in a series is subtracted to give an overall reaction.

Equilibrium Calculations for Gas-Phase and Heterogeneous Reactions

- Evaluate equilibrium constants from reaction data.
  Substitute experimental results into equilibrium constant expression.
- Calculate equilibrium compositions when K is known.
  Procedures are illustrated in Example 14.1.

Empirical Description of the Direction of Change in Chemical Reactions

- The reaction quotient

$$Q = \frac{P_C^c P_D^d}{P_A^a P_D^d} \text{ is defined for any point along the reaction}$$

  $Q < K \Rightarrow$ reaction moves to the right
  $Q = K \Rightarrow$ equilibrium
  $Q > K \Rightarrow$ reaction moves to the left

- Le Châtelier's Principle—A system in equilibrium that is subjected to a stress will react in a way to counteract the stress. Several specific cases are:
  Stress: increase concentration or pressure of species A.
  Response: reaction will move in appropriate direction to decrease A.
  Stress: increase pressure.
  Response: reaction will move in the direction that gives fewer molecules in order to decrease the pressure.
  Stress: decrease volume. (Same as increase pressure.)
  Response: reaction moves in the direction that gives fewer molecules in order to decrease the pressure
  Stress: increase temperature.
  Response: reaction moves in appropriate direction to absorb head and decrease the temperature.
  Endothermic reactions move to the right as T increases.
  Exothermic reactions move to the left as T increases.

Thermodynamic Explanation of the Direction of Change in Chemical Reactions

- Temperature Dependence of Equilibrium Constants

$$\ln\left(\frac{K_2}{K_1}\right) = -\frac{\Delta H^\circ}{R}\left[\frac{1}{T_2} - \frac{1}{T_1}\right]$$

- Temperature Dependence of Vapor Pressure

$$\ln P = -\frac{\Delta H_{vap}}{R}\left[\frac{1}{T_2} - \frac{1}{T_1}\right]$$

Distribution of a Species between Immiscible Phases

- Methods for extracting a chemical species from preparative mixtures in order to analyze and purify
- Distribution between phases is a form of the mass action law, and is fully described by the methods of chemical equilibrium.

CONCEPTS & SKILLS

After studying this chapter and working the problems that follow, you should be able to:

1. Describe the nature of the equilibrium state in chemical reactions (Section 14.1).
2. Set up the equilibrium expression for homogeneous reactions in the gas phase and in solution, and for heterogeneous reactions (Section 14.2, Problems 1–12).
3. Relate the equilibrium constant of a reaction to its standard Gibbs free energy change (Section 14.3, Problems 13–16).
4. Combine equilibrium constants for individual reactions to obtain net equilibrium constants for combined reactions (Section 14.4, Problems 17–20).
5. Calculate equilibrium constants from experimental measurements of partial and total pressures (Section 14.5, Problems 21–26).
6. Calculate the equilibrium partial pressures of all species involved in a gas-phase chemical or gas-solid reaction from the initial pressure(s) of the reactants (Section 14.5, Problems 27–32).
7. Relate concentrations to partial pressures in equilibrium calculations (Section 14.5, Problems 37–38).
8. Determine the direction in which a chemical reaction will proceed spontaneously by calculating its reaction quotient (Section 14.6, Problems 45–46).
9. State Le Châtelier's principle and give several applications (Section 14.6, Problems 47–58).
10. Relate the change in the equilibrium constant of a reaction with temperature to its standard enthalpy change (Section 14.7, Problems 59–70).
11. Use the law of mass action to explain the distribution of a solute between two immiscible solvents (Section 14.8, Problems 71–74).
12. Outline the basis for separation of compounds by partition chromatography (Section 14.8).

KEY EQUATIONS

$$\frac{(P_C)^c(P_D)^d}{(P_A)^a(P_B)^b} = K \qquad \text{(Section 14.2)}$$

$$\frac{[C]^c[D]^d}{[A]^c[B]^b} = K \qquad \text{(Section 14.2)}$$

$$\Delta G = nRT \ln\left(\frac{P_2}{P_1}\right) \qquad \text{(Section 14.3)}$$

$$\Delta G = nRT \ln\left(\frac{P}{P_{ref}}\right) = nRT \ln P \qquad \text{(Section 14.3)}$$

$$-\Delta G° = RT \ln K(T) \qquad \text{(Section 14.3)}$$

$$\frac{a_C^c \cdot a_D^d}{a_A^a \cdot a_B^b} = K \qquad \text{(Section 14.3)}$$

$$\Delta G = \Delta G° + RT \ln Q = -RT \ln K + RT \ln Q$$

$$= RT \ln (Q/K) \qquad \text{(Section 14.7)}$$

$$\ln K = -\frac{\Delta G°}{RT} = -\frac{\Delta H°}{RT} + \frac{\Delta S°}{R} \qquad \text{(Section 14.7)}$$

$$\ln\left(\frac{K_2}{K_1}\right) = -\frac{\Delta H°}{R}\left[\frac{1}{T_2} - \frac{1}{T_1}\right] \qquad \text{(Section 14.7)}$$

$$\ln P = -\frac{\Delta H_{vap}}{R}\left[\frac{1}{T} - \frac{1}{T_b}\right] \qquad \text{(Section 14.7)}$$

PROBLEMS

Answers to problems whose numbers are boldface appear in Appendix G. Problems that are more challenging are indicated with asterisks.

The Empirical Law of Mass Action

1. Write equilibrium expressions for the following gas-phase reactions.
(a) $2\ H_2(g) + O_2(g) \rightleftharpoons 2\ H_2O(g)$
(b) $Xe(g) + 3\ F_2(g) \rightleftharpoons XeF_6(g)$
(c) $2\ C_6H_6(g) + 15\ O_2(g) \rightleftharpoons 12\ CO_2(g) + 6\ H_2O(g)$

2. Write equilibrium expressions for the following gas-phase reactions.
(a) $2\ Cl_2(g) + O_2(g) \rightleftharpoons 2\ Cl_2O(g)$
(b) $N_2(g) + O_2(g) + Br_2(g) \rightleftharpoons 2\ NOBr(g)$
(c) $C_3H_8(g) + 5\ O_2(g) \rightleftharpoons 3\ CO_2(g) + 4\ H_2O(g)$

3. At a moderately elevated temperature, phosphoryl chloride ($POCl_3$) can be produced in the vapor phase from the gaseous elements. Write a balanced chemical equation and an equilibrium expression for this system. Note that gaseous phosphorus consists of P_4 molecules at moderate temperatures.

4. If confined at high temperature, ammonia and oxygen quickly react and come to equilibrium with their products, water vapor and nitrogen oxide. Write a balanced chemical equation and an equilibrium expression for this system.

5. An important step in the industrial production of hydrogen is the reaction of carbon monoxide with water:

$$CO(g) + H_2O(g) \rightleftharpoons CO_2(g) + H_2(g)$$

(a) Use the law of mass action to write the equilibrium expression for this reaction.
(b) At 500°C, the equilibrium constant for this reaction is 3.9. Suppose that the equilibrium partial pressures of CO and H_2O are both 0.10 atm and that of CO_2 is 0.70 atm. Calculate the equilibrium partial pressure of $H_2(g)$.

6. Phosgene ($COCl_2$) is an important intermediate in the manufacture of certain plastics. It is produced by the reaction

$$CO(g) + Cl_2(g) \rightleftharpoons COCl_2(g)$$

(a) Use the law of mass action to write the equilibrium expression for this reaction.
(b) At 600°C, the equilibrium constant for this reaction is 0.20. Calculate the partial pressure of phosgene in equilibrium with a mixture of CO (at 0.0020 atm) and Cl_2 (at 0.00030 atm).

7. N_2O_4 is soluble in the solvent cyclohexane; however, dissolution does not prevent N_2O_4 from breaking down to give NO_2 according to the equation

$$N_2O_{4(cyclohexane)} \rightleftharpoons 2\ NO_{2(cyclohexane)}$$

An effort to compare this solution equilibrium with the similar equilibrium in the gas gave the following actual experimental data at 20°C:

$[N_2O_4]$ (mol L^{-1})	$[NO_2]$ (mol L^{-1})
0.190×10^{-3}	2.80×10^{-3}
0.686×10^{-3}	5.20×10^{-3}
1.54×10^{-3}	7.26×10^{-3}
2.55×10^{-3}	10.4×10^{-3}
3.75×10^{-3}	11.7×10^{-3}
7.86×10^{-3}	17.3×10^{-3}
11.9×10^{-3}	21.0×10^{-3}

(a) Graph the *square* of the concentration of NO_2 versus the concentration of N_2O_4.
(b) Compute the average equilibrium constant of this reaction.

8. NO_2 is soluble in carbon tetrachloride (CCl_4). As it dissolves, it dimerizes to give N_2O_4 according to the equation

$$2\ NO_{2(CCl_4)} \rightleftharpoons N_2O_{4(CCl_4)}$$

A study of this equilibrium gave the following experimental data at 20°C:

$[N_2O_4]$ (mol L^{-1})	$[NO_2]$ (mol L^{-1})
0.192×10^{-3}	2.68×10^{-3}
0.721×10^{-3}	4.96×10^{-3}
1.61×10^{-3}	7.39×10^{-3}
2.67×10^{-3}	10.2×10^{-3}
3.95×10^{-3}	11.0×10^{-3}
7.90×10^{-3}	16.6×10^{-3}
11.9×10^{-3}	21.4×10^{-3}

(a) Graph the concentration of N_2O_4 versus the *square* of the concentration of NO_2.
(b) Compute the average equilibrium constant of this reaction.

9. Using the law of mass action, write the equilibrium expression for each of the following reactions.
(a) $8\ H_2(g) + S_8(s) \rightleftharpoons 8\ H_2S(g)$
(b) $C(s) + H_2O(\ell) + Cl_2(g) \rightleftharpoons COCl_2(g) + H_2(g)$
(c) $CaCO_3(s) \rightleftharpoons CaO(s) + CO_2(g)$
(d) $3\ C_2H_2(g) \rightleftharpoons C_6H_6(\ell)$

10. Using the law of mass action, write the equilibrium expression for each of the following reactions.
(a) $3\ C_2H_2(g) + 3\ H_2(g) \rightleftharpoons C_6H_{12}(\ell)$
(b) $CO_2(g) + C(s) \rightleftharpoons 2\ CO(g)$
(c) $CF_4(g) + 2\ H_2O(\ell) \rightleftharpoons CO_2(g) + 4\ HF(g)$
(d) $K_2NiF_6(s) + TiF_4(s) \rightleftharpoons K_2TiF_6(s) + NiF_2(s) + F_2(g)$

11. Using the law of mass action, write the equilibrium expression for each of the following reactions.
(a) $Zn(s) + 2\ Ag^+(aq) \rightleftharpoons Zn^{2+}(aq) + 2\ Ag(s)$
(b) $VO_4^{3-}(aq) + H_2O(\ell) \rightleftharpoons VO_3(OH)^{2-}(aq) + OH^-(aq)$
(c) $2\ As(OH)_6^{3-}(aq) + 6\ CO_2(g) \rightleftharpoons As_2O_3(s) + 6\ HCO_3^-(aq) + 3\ H_2O(\ell)$

12. Using the law of mass action, write the equilibrium expression for each of the following reactions.
(a) $6\ I^-(aq) + 2\ MnO_4^-(aq) + 4\ H_2O(\ell) \rightleftharpoons 3\ I_2(aq) + 2\ MnO_2(s) + 8\ OH^-(aq)$
(b) $2\ Cu^{2+}(aq) + 4\ I^-(aq) \rightleftharpoons 2\ CuI(s) + I_2(aq)$
(c) $\frac{1}{2}\ O_2(g) + Sn^{2+}(aq) + 3\ H_2O(\ell) \rightleftharpoons SnO_2(s) + 2\ H_3O^+(aq)$

Thermodynamic Description of the Equilibrium State

13. Calculate $\Delta G°$ and the equilibrium constant K at 25°C for the reaction

$$2\ NH_3(g) + \tfrac{7}{2}\ O_2(g) \rightleftharpoons 2\ NO_2(g) + 3\ H_2O(g)$$

using data in Appendix D.

14. Write a reaction for the dehydrogenation of gaseous ethane (C_2H_6) to acetylene (C_2H_2). Calculate $\Delta G°$ and the equilibrium constant for this reaction at 25°C, using data from Appendix D.

15. Use the thermodynamic data from Appendix D to calculate the equilibrium constant at 25°C for the following reactions:
(a) $SO_2(g) + \frac{1}{2}\ O_2(g) \rightleftharpoons SO_3(g)$
(b) $3\ Fe_2O_3(s) \rightleftharpoons 2\ Fe_3O_4(s) + \frac{1}{2}\ O_2(g)$
(c) $CuCl_2(s) \rightleftharpoons Cu^{2+}(aq) + 2\ Cl^-(aq)$
Write the equilibrium expression for each reaction.

16. Use the thermodynamic data from Appendix D to calculate equilibrium constants at 25°C for the following reactions.
(a) $H_2(g) + N_2(g) + 2\ O_2(g) \rightleftharpoons 2\ HNO_2(g)$
(b) $Ca(OH)_2(s) \rightleftharpoons CaO(s) + H_2O(g)$
(c) $Zn^{2+}(aq) + 4\ NH_3(aq) \rightleftharpoons Zn(NH_3)_4^{2+}(aq)$
Write the equilibrium expression for each reaction.

The Law of Mass Action for Related and Simultaneous Equilibria

17. At a certain temperature, the value of the equilibrium constant for the reaction

$$CS_2(g) + 3\ O_2(g) \rightleftharpoons CO_2(g) + 2\ SO_2(g)$$

is K_1. How is K_1 related to the equilibrium constant K_2 for the related equilibrium

$$\tfrac{1}{3}\ CS_2(g) + O_2(g) \rightleftharpoons \tfrac{1}{3}\ CO_2(g) + \tfrac{2}{3}\ SO_2(g)$$

at the same temperature?

18. At 25°C, the equilibrium constant for the reaction

$$6\ ClO_3F(g) \rightleftharpoons 2\ ClF(g) + 4\ ClO(g) + 7\ O_2(g) + 2\ F_2(g)$$

is 32.6. Calculate the equilibrium constant at 25°C for the reaction

$$\tfrac{1}{3}\ ClF(g) + \tfrac{2}{3}\ ClO(g) + \tfrac{7}{6}\ O_2(g) + \tfrac{1}{3}\ F_2(g) \rightleftharpoons ClO_3F(g)$$

19. Suppose that K_1 and K_2 are the respective equilibrium constants for the two reactions

$$XeF_6(g) + H_2O(g) \rightleftharpoons XeOF_4(g) + 2\ HF(g)$$

$$XeO_4(g) + XeF_6(g) \rightleftharpoons XeOF_4(g) + XeO_3F_2(g)$$

Give the equilibrium constant for the reaction

$$XeO_4(g) + 2\ HF(g) \rightleftharpoons XeO_3F_2(g) + H_2O(g)$$

in terms of K_1 and K_2.

20. At 1330 K, germanium(II) oxide (GeO) and tungsten(VI) oxide (W_2O_6) are both gases. The following two equilibria are established simultaneously:

$$2\ GeO(g) + W_2O_6(g) \rightleftharpoons 2\ GeWO_4(g)$$

$$GeO(g) + W_2O_6(g) \rightleftharpoons GeW_2O_7(g)$$

The equilibrium constants for the two are respectively 7.0×10^3 and 38×10^3. Compute K for the reaction

$$GeO(g) + GeW_2O_7(g) \rightleftharpoons 2\ GeWO_4(g)$$

Equilibrium Calculations for Gas-Phase and Heterogeneous Reactions

21. At 454 K, $Al_2Cl_6(g)$ reacts to form $Al_3Cl_9(g)$ according to the equation

$$3\ Al_2Cl_6(g) \rightleftharpoons 2\ Al_3Cl_9(g)$$

In an experiment at this temperature, the equilibrium partial pressure of $Al_2Cl_6(g)$ is 1.00 atm and the equilibrium partial pressure of $Al_3Cl_9(g)$ is 1.02×10^{-2} atm. Compute the equilibrium constant of the preceding reaction at 454 K.

22. At 298 K, $F_3SSF(g)$ decomposes partially to $SF_2(g)$. At equilibrium, the partial pressure of $SF_2(g)$ is 1.1×10^{-4} atm and the partial pressure of F_3SSF is 0.0484 atm.
(a) Write a balanced equilibrium equation to represent this reaction.
(b) Compute the equilibrium constant corresponding to the equation you wrote.

23. The compound 1,3-di-*t*-butylcyclohexane exists in two forms that are known as the chair and boat conformations because their molecular structures resemble those objects. Equilibrium exists between the two forms, represented by the equation

$$\text{chair} \rightleftharpoons \text{boat}$$

At 580 K, 6.42% of the molecules are in the chair form. Calculate the equilibrium constant for the preceding reaction as written.

24. At 248°C and a total pressure of 1.000 atm, the fractional dissociation of $SbCl_5$ is 0.718 for the reaction

$$SbCl_5(g) \rightleftharpoons SbCl_3(g) + Cl_2(g)$$

This means that 718 of every 1000 molecules of $SbCl_5$ originally present have dissociated. Calculate the equilibrium constant.

25. Sulfuryl chloride (SO_2Cl_2) is a colorless liquid that boils at 69°C. Above this temperature, the vapors dissociate into sulfur dioxide and chlorine:

$$SO_2Cl_2(g) \rightleftharpoons SO_2(g) + Cl_2(g)$$

This reaction is slow at 100°C, but it is accelerated by the presence of some $FeCl_3$ (which does not affect the final position of the equilibrium). In an experiment, 3.174 g of $SO_2Cl_2(\ell)$ and a small amount of solid $FeCl_3$ are put into an evacuated 1.000-L flask, which is then sealed and heated to 100°C. The total pressure in the flask at that temperature is found to be 1.30 atm.
(a) Calculate the partial pressure of each of the three gases present.
(b) Calculate the equilibrium constant at this temperature.

26. A certain amount of $NOBr(g)$ is sealed in a flask, and the temperature is raised to 350 K. The following equilibrium is established:

$$NOBr(g) \rightleftharpoons NO(g) + \tfrac{1}{2}\,Br_2(g)$$

The total pressure in the flask when equilibrium is reached at this temperature is 0.675 atm, and the vapor density is 2.219 g L^{-1}.
(a) Calculate the partial pressure of each species.
(b) Calculate the equilibrium constant at this temperature.

27. The dehydrogenation of benzyl alcohol to make the flavoring agent benzaldehyde is an equilibrium process described by the equation

$$C_6H_5CH_2OH(g) \rightleftharpoons C_6H_5CHO(g) + H_2(g)$$

At 523 K, the value of its equilibrium constant is K = 0.558.
(a) Suppose 1.20 g of benzyl alcohol is placed in a 2.00-L vessel and heated to 523 K. What is the partial pressure of benzaldehyde when equilibrium is attained?
(b) What fraction of benzyl alcohol is dissociated into products at equilibrium?

28. Isopropyl alcohol can dissociate into acetone and hydrogen:

$$(CH_3)_2CHOH(g) \rightleftharpoons (CH_3)_2CO(g) + H_2(g)$$

At 179°C, the equilibrium constant for this dehydrogenation reaction is 0.444.
(a) If 10.00 g of isopropyl alcohol is placed in a 10.00-L vessel and heated to 179°C, what is the partial pressure of acetone when equilibrium is attained?
(b) What fraction of isopropyl alcohol is dissociated at equilibrium?

29. A weighed quantity of $PCl_5(s)$ is sealed in a 100.0-cm^3 glass bulb to which a pressure gauge is attached. The bulb is heated to 250°C, and the gauge shows that the pressure in the bulb rises to 0.895 atm. At this temperature, the solid PCl_5 is all vaporized and also partially dissociated into $Cl_2(g)$ and $PCl_3(g)$ according to the equation

$$PCl_5(g) \rightleftharpoons PCl_3(g) + Cl_2(g)$$

At 250°C, K = 2.15 for this reaction. Assume that the contents of the bulb are at equilibrium and calculate the partial pressure of the three different chemical species in the vessel.

30. Suppose 93.0 g of $HI(g)$ is placed in a glass vessel and heated to 1107 K. At this temperature, equilibrium is quickly established between $HI(g)$ and its decomposition products, $H_2(g)$ and $I_2(g)$:

$$2\,HI(g) \rightleftharpoons H_2(g) + I_2(g)$$

The equilibrium constant at 1107 K is 0.0259, and the total pressure at equilibrium is observed to equal 6.45 atm. Calculate the equilibrium partial pressures of $HI(g)$, $H_2(g)$, and $I_2(g)$.

31. The equilibrium constant at 350 K for the reaction

$$Br_2(g) + I_2(g) \rightleftharpoons 2\,IBr(g)$$

has a value of 322. Bromine at an initial partial pressure of 0.0500 atm is mixed with iodine at an initial partial pressure of 0.0400 atm and held at 350 K until equilibrium is reached. Calculate the equilibrium partial pressure of each of the gases.

32. The equilibrium constant for the reaction of fluorine and oxygen to form oxygen difluoride (OF_2) is 40.1 at 298 K:

$$F_2(g) + \tfrac{1}{2}\,O_2(g) \rightleftharpoons OF_2(g)$$

Suppose some OF_2 is introduced into an evacuated container at 298 K and allowed to dissociate until its partial pressure reaches an equilibrium value of 1.00 atm. Calculate the equilibrium partial pressures of F_2 and O_2 in the container.

33. At 25°C, the equilibrium constant for the reaction

$$N_2(g) + O_2(g) \rightleftharpoons 2\,NO(g)$$

is 4.2×10^{-31}. Suppose a container is filled with nitrogen (at an initial partial pressure of 0.41 atm), oxygen (at an initial partial pressure of 0.59 atm), and nitrogen oxide (at an initial partial pressure of 0.22 atm). Calculate the partial pressures of all three gases after equilibrium is reached at this temperature.

34. At 25°C, the equilibrium constant for the reaction

$$2\,NO_2(g) \rightleftharpoons 2\,NO(g) + O_2(g)$$

is 5.9×10^{-13}. Suppose a container is filled with nitrogen dioxide at an initial partial pressure of 0.89 atm. Calculate the partial pressures of all three gases after equilibrium is reached at this temperature.

* **35.** The equilibrium constant for the synthesis of ammonia

$$N_2(g) + 3\,H_2(g) \rightleftharpoons 2\,NH_3(g)$$

has the value $K = 6.78 \times 10^5$ at 25°C. Calculate the equilibrium partial pressures of $N_2(g)$, $H_2(g)$, and $NH_3(g)$ at 25°C if the total pressure is 1.00 atm and the H:N atom ratio in the system is 3:1. (*Hint:* Try the approximation that P_{N_2} and $P_{H_2} \ll P_{NH_3}$ and see if the resulting equations are simplified.)

* **36.** At 400°C, $K = 3.19 \times 10^{-4}$ for the reaction in problem 35. Repeat the calculation for P_{N_2}, P_{H_2}, and P_{NH_3}, assuming the same total pressure and composition. (*Hint:* Try the approximation that $P_{NH_3} \ll P_{N_2}$ and P_{H_2} and see if the resulting equations are simplified.)

37. Calculate the concentration of phosgene ($COCl_2$) that will be present at 600°C in equilibrium with carbon monoxide (at a concentration of 2.3×10^{-4} mol L^{-1}) and chlorine (at a concentration of 1.7×10^{-2} mol L^{-1}). (Use the data of problem 6.)

38. The reaction

$$SO_2Cl_2(g) \rightleftharpoons SO_2(g) + Cl_2(g)$$

has an equilibrium constant at 100°C of 2.40. Calculate the concentration of SO_2 that will be present at 100°C in equilibrium with SO_2Cl_2 (at a concentration of 3.6×10^{-4} mol L^{-1}) and chlorine (at a concentration of 6.9×10^{-3} mol L^{-1}).

39. At 298 K, the equilibrium constant for the reaction

$$Fe_2O_3(s) + 3\,H_2(g) \rightleftharpoons 2\,Fe(s) + 3\,H_2O(\ell)$$

is 4.0×10^{-6}, and that for

$$CO_2(g) + H_2(g) \rightleftharpoons CO(g) + H_2O(\ell)$$

is 3.2×10^{-4}. Suppose some solid Fe_2O_3, solid Fe, and liquid H_2O are brought into equilibrium with CO(*g*) and $CO_2(g)$ in a closed container at 298 K. Calculate the ratio of the partial pressure of CO(*g*) to that of $CO_2(g)$ at equilibrium.

40. A sample of ammonium carbamate placed in a glass vessel at 25°C undergoes the reaction

$$NH_4OCONH_2(s) \rightleftharpoons 2\,NH_3(g) + CO_2(g)$$

The total pressure of gases in equilibrium with the solid is found to be 0.115 atm.
(a) Calculate the partial pressures of NH_3 and CO_2.
(b) Calculate the equilibrium constant at 25°C.

41. The equilibrium constant for the reaction

$$NH_3(g) + HCl(g) \rightleftharpoons NH_4Cl(s)$$

at 340°C is $K = 4.0$.
(a) If the partial pressure of ammonia is $P_{NH_3} = 0.80$ atm and solid ammonium chloride is present, what is the equilibrium partial pressure of hydrogen chloride at 340°C?
(b) An excess of solid NH_4Cl is added to a container filled with ammonia at 340°C and a pressure of 1.50 atm. Calculate the pressures of $NH_3(g)$ and HCl(*g*) reached at equilibrium.

42. The equilibrium constant for the reaction

$$H_2(g) + I_2(s) \rightleftharpoons 2\,HI(g)$$

at 25°C is $K = 0.345$.
(a) If the partial pressure of hydrogen is $P_{H_2} = 1.00$ atm and solid iodine is present, what is the equilibrium partial pressure of hydrogen iodide, P_{HI}, at 25°C?
(b) An excess of solid I_2 is added to a container filled with hydrogen at 25°C and a pressure of 4.00 atm. Calculate the pressures of $H_2(g)$ and HI(*g*) reached at equilibrium.

43. Pure solid NH_4HSe is placed in an evacuated container at 24.8°C. Eventually, the pressure above the solid reaches the equilibrium pressure 0.0184 atm due to the reaction

$$NH_4HSe(s) \rightleftharpoons NH_3(g) + H_2Se(g)$$

(a) Calculate the equilibrium constant of this reaction at 24.8°C.
(b) In a different container, the partial pressure of $NH_3(g)$ in equilibrium with $NH_4HSe(s)$ at 24.8°C is 0.0252 atm. What is the partial pressure of $H_2Se(g)$?

44. The total pressure of the gases in equilibrium with solid sodium hydrogen carbonate at 110°C is 1.648 atm, corresponding to the reaction

$$2\,NaHCO_3(s) \rightleftharpoons Na_2CO_3(s) + H_2O(g) + CO_2(g)$$

($NaHCO_3$ is used in dry chemical fire extinguishers because the products of this decomposition reaction smother the fire.)
(a) Calculate the equilibrium constant at 110°C.
(b) What is the partial pressure of water vapor in equilibrium with $NaHCO_3(s)$ at 110°C if the partial pressure of $CO_2(g)$ is 0.800 atm?

The Direction of Change in Chemical Reactions: Empirical Description

45. Some Al_2Cl_6 (at a partial pressure of 0.473 atm) is placed in a closed container at 454 K with some Al_3Cl_9 (at a partial pressure of 1.02×10^{-2} atm). Enough argon is added to raise the total pressure to 1.00 atm.
(a) Calculate the initial reaction quotient for the reaction

$$3\,Al_2Cl_6(g) \rightleftharpoons 2\,Al_3Cl_9(g)$$

(b) As the gas mixture reaches equilibrium, will there be net production or consumption of Al_3Cl_9? (Use the data given in problem 21.)

46. Some SF_2 (at a partial pressure of 2.3×10^{-4} atm) is placed in a closed container at 298 K with some F_3SSF (at a partial pressure of 0.0484 atm). Enough argon is added to raise the total pressure to 1.000 atm.
(a) Calculate the initial reaction quotient for the decomposition of F_3SSF to SF_2.
(b) As the gas mixture reaches equilibrium, will there be net formation or dissociation of F_3SSF? (Use the data given in problem 22.)

47. The progress of the reaction

$$H_2(g) + Br_2(g) \rightleftharpoons 2\,HBr(g)$$

can be monitored visually by following changes in the color of the reaction mixture (Br_2 is reddish brown, and H_2 and

HBr are colorless). A gas mixture is prepared at 700 K, in which 0.40 atm is the initial partial pressure of both H_2 and Br_2 and 0.90 atm is the initial partial pressure of HBr. The color of this mixture then fades as the reaction progresses toward equilibrium. Give a condition that must be satisfied by the equilibrium constant K (for example, it must be greater than or smaller than a given number).

48. Recall from our discussion of the NO_2–N_2O_4 equilibrium that NO_2 has a brownish color. At elevated temperatures, NO_2 reacts with CO according to

$$NO_2(g) + CO(g) \rightleftharpoons NO(g) + CO_2(g)$$

The other three gases taking part in this reaction are colorless. When a gas mixture is prepared at 500 K, in which 3.4 atm is the initial partial pressure of both NO_2 and CO, and 1.4 atm is the partial pressure of both NO and CO_2, the brown color of the mixture is observed to fade as the reaction progresses toward equilibrium. Give a condition that must be satisfied by the equilibrium constant K (for example, it must be greater than or smaller than a given number).

49. The equilibrium constant for the "water gas" reaction

$$C(s) + H_2O(g) \rightleftharpoons CO(g) + H_2(g)$$

is $K = 2.6$ at a temperature of 1000 K. Calculate the reaction quotient Q for each of the following conditions, and state which direction the reaction shifts in coming to equilibrium.

(a) $P_{H_2O} = 0.600$ atm; $P_{CO} = 1.525$ atm; $P_{H_2} = 0.805$ atm
(b) $P_{H_2O} = 0.724$ atm; $P_{CO} = 1.714$ atm; $P_{H_2} = 1.383$ atm

50. The equilibrium constant for the reaction

$$H_2S(g) + I_2(g) \rightleftharpoons 2\ HI(g) + S(s)$$

at 110°C is equal to 0.0023. Calculate the reaction quotient Q for each of the following conditions and determine whether solid sulfur is consumed or produced as the reaction comes to equilibrium.

(a) $P_{I_2} = 0.461$ atm; $P_{H_2S} = 0.050$ atm; $P_{HI} = 0.0$ atm
(b) $P_{I_2} = 0.461$ atm; $P_{H_2S} = 0.050$ atm; $P_{HI} = 9.0$ atm

51. At $T = 1200°C$ the reaction

$$P_4(g) \rightleftharpoons 2\ P_2(g)$$

has an equilibrium constant $K = 0.612$.

(a) Suppose the initial partial pressure of P_4 is 5.00 atm and that of P_2 is 2.00 atm. Calculate the reaction quotient Q and state whether the reaction proceeds to the right or to the left as equilibrium is approached.
(b) Calculate the partial pressures at equilibrium.
(c) If the volume of the system is then increased, will there be net formation or net dissociation of P_4?

52. At $T = 100°C$ the reaction

$$SO_2Cl_2(g) \rightleftharpoons SO_2(g) + Cl_2(g)$$

has an equilibrium constant $K = 2.4$.

(a) Suppose the initial partial pressure of SO_2Cl_2 is 1.20 atm, and $P_{SO_2} = P_{Cl_2} = 0$. Calculate the reaction quotient Q and state whether the reaction proceeds to the right or to the left as equilibrium is approached.
(b) Calculate the partial pressures at equilibrium.
(c) If the volume of the system is then decreased, will there be net formation or net dissociation of SO_2Cl_2?

53. Explain the effect of each of the following stresses on the position of the following equilibrium:

$$3\ NO(g) \rightleftharpoons N_2O(g) + NO_2(g)$$

The reaction as written is exothermic.

(a) $N_2O(g)$ is added to the equilibrium mixture without change of volume or temperature.
(b) The volume of the equilibrium mixture is reduced at constant temperature.
(c) The equilibrium mixture is cooled.
(d) Gaseous argon (which does not react) is added to the equilibrium mixture while both the total gas pressure and the temperature are kept constant.
(e) Gaseous argon is added to the equilibrium mixture without changing the volume.

54. Explain the effect of each of the following stresses on the position of the equilibrium

$$SO_3(g) \rightleftharpoons SO_2(g) + \tfrac{1}{2}\ O_2(g)$$

The reaction as written is endothermic.

(a) $O_2(g)$ is added to the equilibrium mixture without changing volume or temperature.
(b) The mixture is compressed at constant temperature.
(c) The equilibrium mixture is cooled.
(d) An inert gas is pumped into the equilibrium mixture while the total gas pressure and the temperature are kept constant.
(e) An inert gas is added to the equilibrium mixture without changing the volume.

55. In a gas-phase reaction, it is observed that the equilibrium yield of products is increased by lowering the temperature and by reducing the volume.

(a) Is the reaction exothermic or endothermic?
(b) Is there a net increase or a net decrease in the number of gas molecules in the reaction?

56. The equilibrium constant of a gas-phase reaction increases as temperature is increased. When the nonreacting gas neon is admitted to a mixture of reacting gases (holding the temperature and the total pressure fixed and increasing the volume of the reaction vessel), the product yield is observed to decrease.

(a) Is the reaction exothermic or endothermic?
(b) Is there a net increase or a net decrease in the number of gas molecules in the reaction?

57. The most extensively used organic compound in the chemical industry is ethylene (C_2H_4). The two equations

$$C_2H_4(g) + Cl_2(g) \rightleftharpoons C_2H_4Cl_2(g)$$

$$C_2H_4Cl_2(g) \rightleftharpoons C_2H_3Cl(g) + HCl(g)$$

represent the way in which vinyl chloride (C_2H_3Cl) is synthesized for eventual use in polymeric plastics (polyvinyl chloride, PVC). The byproduct of the reaction, HCl, is now most cheaply made by this and similar reactions, rather than by direct combination of H_2 and Cl_2. Heat is given off in the first reaction and taken up in the second. Describe how you would design an industrial process to maximize the yield of vinyl chloride.

58. Methanol is made via the exothermic reaction

$$CO(g) + 2\ H_2(g) \longrightarrow CH_3OH(g)$$

Describe how you would control the temperature and pressure to maximize the yield of methanol.

The Direction of Change in Chemical Reactions: Thermodynamic Explanation

59. One way to manufacture ethanol is by the reaction

$$C_2H_4(g) + H_2O(g) \rightleftharpoons C_2H_5OH(g)$$

The ΔH_f° of $C_2H_4(g)$ is 52.3 kJ mol^{-1}; of $H_2O(g)$, −241.8 kJ mol^{-1}; and of $C_2H_5OH(g)$, −235.3 kJ mol^{-1}. Without doing detailed calculations, suggest the conditions of pressure and temperature that will maximize the yield of ethanol at equilibrium.

60. Dimethyl ether (CH_3OCH_3) is a good substitute for environmentally harmful propellants in aerosol spray cans. It is produced by the dehydration of methanol:

$$2\ CH_3OH(g) \rightleftharpoons CH_3OCH_3(g) + H_2O(g)$$

Describe reaction conditions that favor the equilibrium production of this valuable chemical. As a basis for your answer, compute ΔH° and ΔS° of the reaction from the data in Appendix D.

61. The equilibrium constant at 25°C for the reaction

$$2\ NO_2(g) \rightleftharpoons N_2O_4(g)$$

is 6.8. At 200°C the equilibrium constant is 1.21×10^{-3}. Calculate the enthalpy change (ΔH) for this reaction, assuming that ΔH and ΔS of the reaction are constant over the temperature range from 25°C to 200°C.

62. Stearic acid dimerizes when dissolved in hexane:

$$2\ C_{17}H_{35}COOH(hexane) \rightleftharpoons (C_{17}H_{35}COOH)_2(hexane)$$

The equilibrium constant for this reaction is 2900 at 28°C, but it drops to 40 at 48°C. Estimate ΔH° and ΔS° for the reaction.

63. The equilibrium constant for the reaction

$$\tfrac{1}{2}\ Cl_2(g) + \tfrac{1}{2}\ F_2(g) \rightleftharpoons ClF(g)$$

is measured to be 9.3×10^9 at 298 K and 3.3×10^7 at 398 K.

(a) Calculate ΔG° at 298 K for the reaction.

(b) Calculate ΔH° and ΔS°, assuming the enthalpy and entropy changes to be independent of temperature between 298 and 398 K.

64. Stearic acid also dimerizes when dissolved in carbon tetrachloride:

$$2\ C_{17}H_{35}COOH(CCl_4) \rightleftharpoons (C_{17}H_{35}COOH)_2(CCl_4)$$

The equilibrium constant for this reaction is 2780 at 22°C, but it drops to 93.1 at 42°C. Estimate ΔH° and ΔS° for the reaction.

65. For the synthesis of ammonia from its elements,

$$3\ H_2(g) + N_2(g) \rightleftharpoons 2\ NH_3(g)$$

the equilibrium constant $K = 5.9 \times 10^5$ at 298 K, and $\Delta H^\circ = -92.2$ kJ mol^{-1}. Calculate the equilibrium constant for the reaction at 600 K, assuming no change in ΔH° and ΔS° between 298 K and 600 K.

66. The cumulative exercise at the end of Chapter 14 explored reaction steps in the manufacture of sulfuric acid, including the oxidation of sulfur dioxide to sulfur trioxide:

$$SO_2(g) + \tfrac{1}{2}\ O_2(g) \rightleftharpoons SO_3(g)$$

At 25°C the equilibrium constant for this reaction is 2.6×10^{12}, but the reaction occurs very slowly. Calculate K for this reaction at 550°C, assuming ΔH° and ΔS° are independent of temperature in the range from 25°C to 550°C.

67. The vapor pressure of ammonia at −50°C is 0.4034 atm; at 0°C, it is 4.2380 atm.

(a) Calculate the molar enthalpy of vaporization (ΔH_{vap}) of ammonia.

(b) Calculate the normal boiling temperature of $NH_3(\ell)$.

68. The vapor pressure of butyl alcohol (C_4H_9OH) at 70.1°C is 0.1316 atm; at 100.8°C, it is 0.5263 atm.

(a) Calculate the molar enthalpy of vaporization (ΔH_{vap}) of butyl alcohol.

(b) Calculate the normal boiling point of butyl alcohol.

69. Although iodine is not very soluble in pure water, it dissolves readily in water that contains $I^-(aq)$ ion, thanks to the reaction

$$I_2(aq) + I^-(aq) \rightleftharpoons I_3^-(aq)$$

The equilibrium constant of this reaction was measured as a function of temperature with these results:

T:	3.8°C	15.3°C	25.0°C	35.0°C	50.2°C
K:	1160	841	689	533	409

(a) Plot ln K on the y axis as a function of $1/T$, the reciprocal of the absolute temperature.

(b) Estimate the ΔH° of this reaction.

70. Barium nitride vaporizes slightly at high temperature as it undergoes the dissociation

$$Ba_3N_2(s) \rightleftharpoons 3\ Ba(g) + N_2(g)$$

At 1000 K the equilibrium constant is 4.5×10^{-19}. At 1200 K the equilibrium constant is 6.2×10^{-12}.

(a) Estimate ΔH° for this reaction.

(b) The equation is rewritten as

$$2\ Ba_3N_2(s) \rightleftharpoons 6\ Ba(g) + 2\ N_2(g)$$

Now the equilibrium constant is 2.0×10^{-37} at 1000 K and 3.8×10^{-23} at 1200 K. Estimate ΔH° of *this* reaction.

Distribution of a Single Species between Immiscible Phases: Extraction and Separation Processes

71. An aqueous solution, initially 1.00×10^{-2} M in iodine (I_2), is shaken with an equal volume of an immiscible organic solvent, CCl_4. The iodine distributes itself between the aqueous and CCl_4 layers, and when equilibrium is reached at 27°C, the concentration of I_2 in the aqueous layer is 1.30×10^{-4} M. Calculate the partition coefficient K at 27°C for the reaction

$$I_2(aq) \rightleftharpoons I_2(CCl_4)$$

72. An aqueous solution, initially 2.50×10^{-2} M in iodine (I_2), is shaken with an equal volume of an immiscible organic solvent, CS_2. The iodine distributes itself between the aqueous and CS_2 layers, and when equilibrium is reached at 25°C, the concentration of I_2 in the aqueous layer is 4.16×10^{-5} M. Calculate the partition coefficient K at 25°C for the reaction

$$I_2(aq) \rightleftharpoons I_2(CS_2)$$

73. Benzoic acid (C_6H_5COOH) dissolves in water to the extent of 2.00 g L^{-1} at 15°C and in diethyl ether to the extent of 6.6×10^2 g L^{-1} at the same temperature.
(a) Calculate the equilibrium constants at 15°C for the two reactions

$$C_6H_5COOH(s) \rightleftharpoons C_6H_5COOH(aq)$$

and

$$C_6H_5COOH(s) \rightleftharpoons C_6H_5COOH(ether)$$

(b) From your answers to part (a), calculate the partition coefficient K for the reaction

$$C_6H_5COOH(aq) \rightleftharpoons C_6H_5COOH(ether)$$

74. Citric acid ($C_6H_8O_7$) dissolves in water to the extent of 1300 g L^{-1} at 15°C and in diethyl ether to the extent of 22 g L^{-1} at the same temperature.
(a) Calculate the equilibrium constants at 15°C for the two reactions

$$C_6H_8O_7(s) \rightleftharpoons C_6H_8O_7(aq)$$

and

$$C_6H_8O_7(s) \rightleftharpoons C_6H_8O_7(ether)$$

(b) From your answers to part (a), calculate the partition coefficient K for the reaction

$$C_6H_8O_7(aq) \rightleftharpoons C_6H_8O_7(ether)$$

ADDITIONAL PROBLEMS

75. At 298 K, unequal amounts of $BCl_3(g)$ and $BF_3(g)$ were mixed in a container. The gases reacted to form $BFCl_2(g)$ and $BClF_2(g)$. When equilibrium was finally reached, the four gases were present in these relative chemical amounts: BCl_3(90), BF_3(470), $BClF_2$(200), $BFCl_2$(45).
(a) Determine the equilibrium constants at 298 K of the two reactions

$$2\ BCl_3(g) + BF_3(g) \rightleftharpoons 3\ BFCl_2(g)$$

$$BCl_3(g) + 2\ BF_3(g) \rightleftharpoons 3\ BClF_2(g)$$

(b) Determine the equilibrium constant of the reaction

$$BCl_3(g) + BF_3(g) \rightleftharpoons BFCl_2 + BClF_2(g)$$

and explain why knowing this equilibrium constant really adds nothing to what you knew in part (a).

76. Methanol can be synthesized by means of the equilibrium reaction

$$CO(g) + 2\ H_2(g) \rightleftharpoons CH_3OH(g)$$

for which the equilibrium constant at 225°C is 6.08×10^{-3}. Assume that the ratio of the pressures of CO(g) and $H_2(g)$ is 1:2. What values should they have if the partial pressure of methanol is to be 0.500 atm?

77. At equilibrium at 425.6°C, a sample of *cis*-1-methyl-2-ethylcyclopropane is 73.6% converted into the *trans* form:

$$cis \rightleftharpoons trans$$

(a) Compute the equilibrium constant K for this reaction.
(b) Suppose that 0.525 mol of the *cis* compound is placed in a 15.00-L vessel and heated to 425.6°C. Compute the equilibrium partial pressure of the *trans* compound.

78. The equilibrium constant for the reaction

$$(CH_3)_3COH(g) \rightleftharpoons (CH_3)_2CCH_2(g) + H_2O(g)$$

is 2.42 at 450 K.
(a) A pure sample of the reactant, which is named "*t*-butanol," is confined in a container of fixed volume at a temperature of 450 K and at an original pressure of 0.100 atm. Calculate the fraction of this starting material that is converted to products at equilibrium.
(b) A second sample of the reactant is confined, this time at an original pressure of 5.00 atm. Again, calculate the fraction of the starting material that is converted to products at equilibrium.

79. At 627°C and 1 atm, SO_3 is partly dissociated into SO_2 and O_2:

$$SO_3(g) \rightleftharpoons SO_2(g) + \tfrac{1}{2}\ O_2(g)$$

The density of the equilibrium mixture is 0.925 g L^{-1}. What is the degree of dissociation of SO_3 under these circumstances?

80. Acetic acid in the vapor phase consists of both monomeric and dimeric forms in equilibrium:

$$2\ CH_3COOH(g) \rightleftharpoons (CH_3COOH)_2(g)$$

At 110°C the equilibrium constant for this reaction is 3.72.
(a) Calculate the partial pressure of the dimer when the total pressure is 0.725 atm at equilibrium.
(b) What percentage of the acetic acid is dimerized under these conditions?

* **81.** Repeat the calculation of problem 36 with a total pressure of 100 atm. (*Note:* Here it is necessary to solve the equation by successive approximations. See Appendix C.)

82. At 900 K the equilibrium constant for the reaction

$$\tfrac{1}{2}\ O_2(g) + SO_2(g) \rightleftharpoons SO_3(g)$$

has the value $K = 0.587$. What will be the equilibrium partial pressure of $O_2(g)$ if a sample of SO_3 weighing 0.800 g is heated to 900 K in a quartz-glass vessel whose volume is 100.0 cm^3? (*Note:* Here it is necessary to solve the equation by successive approximations. See Appendix C.)

* **83.** At 298 K chlorine is only slightly soluble in water. Thus, under a pressure of 1.00 atm of $Cl_2(g)$, 1.00 L of water at equilibrium dissolves just 0.091 mol of Cl_2.

$$Cl_2(g) \rightleftharpoons Cl_2(aq)$$

In such solutions the $Cl_2(aq)$ concentration is 0.061 M and the concentrations of $Cl^-(aq)$ and HOCl(aq) are both 0.030 M. These two additional species are formed by the equilibrium

$$Cl_2(aq) + H_2O(\ell) \rightleftharpoons H^+(aq) + Cl^-(aq) + HOCl(aq)$$

There are no other Cl-containing species. Compute the equilibrium constants K_1 and K_2 for the two reactions.

84. At 400°C the reaction

$$BaO_2(s) + 4\ HCl(g) \rightleftharpoons BaCl_2(s) + 2\ H_2O(g) + Cl_2(g)$$

has an equilibrium constant equal to K_1. How is K_1 related to the equilibrium constant K_2 of the reaction

$$2\ Cl_2(g) + 4\ H_2O(g) + 2\ BaCl_2(s) \rightleftharpoons 8\ HCl(g) + 2\ BaO_2(s)$$

at 400°C?

85. Ammonium hydrogen sulfide, a solid, decomposes to give $NH_3(g)$ and $H_2S(g)$. At 25°C, some $NH_4HS(s)$ is placed in an evacuated container. A portion of it decomposes, and the total pressure at equilibrium is 0.659 atm. Extra $NH_3(g)$ is then injected into the container, and when equilibrium is reestablished, the partial pressure of $NH_3(g)$ is 0.750 atm.
(a) Compute the equilibrium constant for the decomposition of ammonium hydrogen sulfide.
(b) Determine the final partial pressure of $H_2S(g)$ in the container.

86. The equilibrium constant for the reaction

$$KOH(s) + CO_2(g) \rightleftharpoons KHCO_3(s)$$

is 6×10^{15} at 25°C. Suppose 7.32 g of KOH and 9.41 g of $KHCO_3$ are placed in a closed evacuated container and allowed to reach equilibrium. Calculate the pressure of $CO_2(g)$ at equilibrium.

87. The equilibrium constant for the reduction of nickel(II) oxide to nickel at 754°C is 255.4, corresponding to the reaction

$$NiO(s) + CO(g) \rightleftharpoons Ni(s) + CO_2(g)$$

If the total pressure of the system at 754°C is 2.50 atm, calculate the partial pressures of CO(g) and $CO_2(g)$.

88. Both glucose (corn sugar) and fructose (fruit sugar) taste sweet, but fructose tastes sweeter. Each year in the United States, tons of corn syrup destined to sweeten food are treated to convert glucose as fully as possible to the sweeter fructose. The reaction is an equilibrium:

$$\text{glucose} \rightleftharpoons \text{fructose}$$

(a) A 0.2564 M solution of pure glucose is treated at 25°C with an enzyme (catalyst) that causes the preceding equilibrium to be reached quickly. The final concentration of fructose is 0.1175 M. In another experiment at the same temperature, a 0.2666 M solution of pure fructose is treated with the same enzyme and the final concentration of glucose is 0.1415 M. Compute an average equilibrium constant for the preceding reaction.
(b) At equilibrium under these conditions, what percentage of glucose is converted to fructose?

89. At 300°C the equilibrium constant for the reaction

$$PCl_5(g) \rightleftharpoons PCl_3(g) + Cl_2(g)$$

is $K = 11.5$.
(a) Calculate the reaction quotient Q if initially $P_{PCl_3} =$ 2.0 atm, $P_{Cl_2} = 6.0$ atm, and $P_{PCl_5} = 0.10$ atm. State whether the reaction proceeds to the right or to the left as equilibrium is approached.
(b) Calculate P_{PCl_3}, P_{Cl_2}, and P_{PCl_5} at equilibrium.
(c) If the volume of the system is then increased, will the amount of PCl_5 present increase or decrease?

90. Although the process of dissolution of helium gas in water is favored in terms of energy, helium is only very slightly soluble in water. What keeps this gas from dissolving in great quantities in water?

91. The hydrogenation of pyridine to piperidine

$$C_5H_5N(g) + 3\ H_2(g) \rightleftharpoons C_5H_{11}N(g)$$

is an equilibrium process whose equilibrium constant is given by the equation

$$\log_{10} K = -20.281 + \frac{10.560\ K}{T}$$

(a) Calculate the value of K at $T = 500$ K.
(b) If the partial pressure of hydrogen is 1.00 atm, what fraction of the nitrogen is in the form of pyridine molecules at $T = 500$ K?

92. The breaking of the O—O bond in peroxydisulfuryl difluoride (FO_2SOOSO_2F) gives FO_2SO:

$$(FO_2SO)_2 \rightleftharpoons 2\ FO_2SO$$

The compound on the left of this equation is a colorless liquid that boils at 67.1°C. Its vapor, when heated to about 100°C, turns brown as the product of the reaction forms. Suppose that, in a sample of the vapor, the intensity of the brown color doubles between 100°C and 110°C and that the total pressure increases only by the 2.7% predicted for an ideal gas. Estimate $\Delta H°$ for the preceding reaction.

93. Polychlorinated biphenyls (PCBs) are a major environmental problem. These oily substances have many uses, but they resist breakdown by bacterial action when spilled in the environment and, being fat-soluble, can accumulate to dangerous concentrations in the fatty tissues of fish and animals. One little-appreciated complication in controlling the problem is that there are *209* different PCBs, all now in the environment. They are generally similar, but their solubilities in fats differ considerably. The best measure of this is K_{ow}, the equilibrium constant for the partition of a PCB between the fat-like solvent octanol and water.

$$PCB(aq) \rightleftharpoons PCB(octanol)$$

An equimolar mixture of PCB-2 and PCB-11 in water is treated with an equal volume of octanol. Determine the ratio between the amounts of PCB-2 and PCB-11 in the water at equilibrium. At room temperature, K_{ow} is 3.98×10^4 for PCB-2 and 1.26×10^5 for PCB-11.

94. Refer to the data in problems 73 and 74. Suppose 2.00 g of a solid consisting of 50.0% benzoic acid and 50.0% citric acid by mass is added to 100.0 mL of water and 100.0 mL of diethyl ether and the whole assemblage is shaken. When the immiscible layers are separated and the solvents are removed by evaporation, two solids result. Calculate the percentage (by mass) of the major component in each solid.

95. At 25°C the partition coefficient for the equilibrium

$$I_2(aq) \rightleftharpoons I_2(CCl_4)$$

has the value $K = 85$. To 0.100 L of an aqueous solution, which is initially 2×10^{-3} M in I_2, we add 0.025 L of CCl_4. The mixture is shaken in a separatory funnel and allowed to separate into two phases, and the CCl_4 phase is withdrawn.

(a) Calculate the fraction of the I_2 remaining in the aqueous phase.

(b) Suppose the remaining aqueous phase is shaken with another 0.025 L of CCl_4 and again separated. What fraction of the I_2 from the original aqueous solution is *now* in the aqueous phase?

(c) Compare your answer with that of Example 9.18, in which the same total amount of CCl_4 (0.050 L) was used in a *single* extraction. For a given total amount of extracting solvent, which is the more efficient way to remove iodine from water?

96. From the data in Appendix D calculate $\Delta H°$, $\Delta G°$, and K, for the following reaction at 298 K:

$$6\ CH_4(g) + \tfrac{9}{2}\ O_2(g) \longrightarrow C_6H_6(\ell) + 9\ H_2O(\ell)$$

CUMULATIVE PROBLEMS

97. The reaction

$$P_4(g) \rightleftharpoons 2\ P_2(g)$$

is endothermic and begins to occur at moderate temperatures.

(a) In which direction do you expect deviations to occur from Boyle's law, $P \propto 1/V$ (constant T), for gaseous P_4?

(b) In which direction do you expect deviations to occur from Charles's law, $V \propto T$ (constant P), for gaseous P_4?

98. A 4.72-g mass of methanol, CH_3OH, is placed in an evacuated 1.00-L flask and heated to 250°C. It vaporizes and then reaches the following equilibrium:

$$CH_3OH(g) \rightleftharpoons CO(g) + 2\ H_2(g)$$

A tiny hole forms in the side of the container, allowing a small amount of gas to effuse out. Analysis of the escaping gases shows that the rate of escape of the hydrogen is 33 times the rate of escape of the methanol. Calculate the equilibrium constant for the preceding reaction at 250°C.

99. The triple bond in the N_2 molecule is very strong, but at high enough temperatures even it breaks down. At 5000 K, when the total pressure exerted by a sample of nitrogen is 1.00 atm, $N_2(g)$ is 0.65% dissociated at equilibrium:

$$N_2(g) \rightleftharpoons 2\ N(g)$$

At 6000 K with the same total pressure, the proportion of $N_2(g)$ dissociated at equilibrium rises to 11.6%. Use the van't Hoff equation to estimate the ΔH of this reaction.

100. At 25°C the equilibrium constant for the reaction

$$CaSO_4(s) + 2\ H_2O(g) \rightleftharpoons CaSO_4{\cdot}2H_2O(s)$$

is 1.6×10^3. Over what range of relative humidities do you expect $CaSO_4(s)$ to be converted to $CaSO_4{\cdot}2H_2O$? (*Note:* The relative humidity is the partial pressure of water vapor divided by its equilibrium vapor pressure and multiplied by 100%. Use Table 10.1.)

101. Calculate the equilibrium pressure (in atmospheres) of $O_2(g)$ over a sample of pure $NiO(s)$ in contact with pure $Ni(s)$ at 25°C. The $NiO(s)$ decomposes according to the equation

$$NiO(s) \longrightarrow Ni(s) + \tfrac{1}{2}\ O_2(g)$$

Use data from Appendix D.

102. (a) From the values in Appendix D, find the enthalpy change and the Gibbs free energy change when one mole of benzene C_6H_6 is vaporized at 25°C.

(b) Calculate the vapor pressure of benzene at 25°C.

(c) Assuming that the enthalpy and entropy of vaporization are constant, estimate the normal boiling point of benzene.

103. Snow and ice sublime spontaneously when the partial pressure of water vapor is below the equilibrium vapor pressure of ice. At 0°C the vapor pressure of ice is 0.0060 atm (the triple-point pressure of water). Taking the enthalpy of sublimation of ice to be 50.0 kJ mol^{-1}, calculate the partial pressure of water vapor below which ice will sublime spontaneously at −15°C.

104. The sublimation pressure of solid NbI_5 is the pressure of gaseous NbI_5 present in equilibrium with the solid. It is given by the empirical equation

$$\log P = -6762/T + 8.566$$

The vapor pressure of liquid NbI_5, on the other hand, is given by

$$\log P = -4653/T + 5.43$$

In these two equations, T is the absolute temperature in kelvins and P is the pressure in atmospheres.

(a) Determine the enthalpy and entropy of sublimation of $NbI_5(s)$.

(b) Determine the enthalpy and entropy of vaporization of $NbI_5(\ell)$.

(c) Calculate the normal boiling point of $NbI_5(\ell)$.

(d) Calculate the triple-point temperature and pressure of NbI_5. (*Hint:* At the triple point of a substance, the liquid and solid are in equilibrium and must have the same vapor pressure. If they did not, vapor would continually escape from the phase with the higher vapor pressure and collect in the phase with the lower vapor pressure.)

105. In an extraction process, a solute species is partitioned between two immiscible solvents. Suppose the two solvents are water and carbon tetrachloride. State which phase will have the higher concentration of each of the following solutes: (a) CH_3OH, (b) C_2Cl_6, (c) Br_2, (d) NaCl. Explain your reasoning.

106. The gaseous compounds allene and propyne are isomers with formula C_3H_4. Calculate the equilibrium constant and the standard enthalpy change at 25°C for the isomerization reaction

$$\text{allene}(g) \rightleftharpoons \text{propyne}(g)$$

from the following data, all of which apply to 298 K:

	ΔH_f° (kJ mol^{-1})	ΔG_f° (kJ mol^{-1})
Allene	192	202
Propyne	185	194

107. (a) Calculate the standard free-energy change and the equilibrium constant for the dimerization of NO_2 to N_2O_4 at 25°C (see Appendix D).
(b) Calculate ΔG for this reaction at 25°C when the pressures of NO_2 and N_2O_4 are each held at 0.010 atm. Which way will the reaction tend to proceed?

108. There are two isomeric hydrocarbons with formula C_4H_{10}, butane and isobutane, which we denote here B and I. The standard enthalpies of formation for the gaseous species are -124.7 kJ mol^{-1} for B, -131.3 kJ mol^{-1} for I; the standard free energies of formation are -15.9 kJ mol^{-1} for B, -18.0 kJ mol^{-1} for I.
(a) Which is the more stable under standard conditions, and which has the higher entropy?
(b) The reaction B $\rightleftharpoons$ I can occur in the presence of a catalyst. Calculate the equilibrium constant at 298 K for the conversion of B to I, and calculate the percentage of B in the equilibrium mixture.

109. At 3500 K the equilibrium constant for the reaction $CO_2(g) + H_2(g) \rightleftharpoons CO(g) + H_2O(g)$ is 8.28.
(a) What is $\Delta G°(3500)$ for this reaction?
(b) What is ΔG at 3500 K for transforming 1 mol CO_2 and 1 mol H_2, both held at 0.1 atm, to 1 mol CO and 1 mol H_2O, both held at 2 atm?
(c) In which direction would this last reaction run spontaneously?

110. At 1200 K in the presence of solid carbon, an equilibrium mixture of CO and CO_2 (called "producer gas") contains 98.3 mol percent CO and 1.69 mol percent of CO_2 when the total pressure is 1 atm. The reaction is

$$CO_2(g) + C(\textit{graphite}) \rightleftharpoons 2\ CO(g)$$

(a) Calculate P_{CO} and P_{CO_2}.
(b) Calculate the equilibrium constant.
(c) Calculate $\Delta G°$ for this reaction.

111. (a) Formulate the equilibrium expression for the endothermic reaction

$$AgCl{\cdot}NH_3(s) \rightleftharpoons AgCl(s) + NH_3(g)$$

(b) What is the effect on P_{NH_3} at equilibrium if additional $AgCl(s)$ is added?
(c) What is the effect on P_{NH_3} at equilibrium if additional $NH_3(g)$ is pumped into or out of the system, provided that neither of the two solid phases shown in the chemical equation is completely used up?
(d) What is the effect on P_{NH_3} of lowering the temperature?

112. Solid ammonium carbonate decomposes according to the equation

$$(NH_4)_2CO_3(s) \rightleftharpoons 2\ NH_3(g) + CO_2(g) + H_2O(g)$$

At a certain elevated temperature the total pressure of the gases NH_3, CO_2, and H_2O generated by the decomposition of, and at equilibrium with, pure solid ammonium carbonate is 0.400 atm. Calculate the equilibrium constant for the reaction considered. What would happen to P_{NH_3} and P_{CO_2} if P_{H_2O} were adjusted by external means to be 0.200 atm without changing the relative amounts of $NH_3(g)$ and $CO_2(g)$ and with $(NH_4)_2CO_3(s)$ still being present?

CHAPTER 15

Acid–Base Equilibria

Many naturally occurring dyes change color as the acidity of their surroundings changes. The compound cyanidin is blue in the basic sap of the cornflower and red in the acidic sap of the poppy. Such dyes can be used as indicators of the degree of acidity in a medium.

According to Chapter 11, an acid is a substance that upon dissolving in water increases the concentration of hydronium (H_3O^+) ions above the value found in pure water, and a base is a substance that increases the concentration of hydroxide (OH^-) ions above its value in pure water. Despite the careful language, it is commonplace to view acids and bases as substances that dissociate to give protons (which upon hydration become hydronium ions) and hydroxide ions, respectively. If the dissociation is complete, we can easily calculate the concentration of hydronium and hydroxide ions in the solution and then calculate the yield of acid–base neutralization reactions, and acid–base titrations, by the methods of stoichiometry in solution. But experience shows that many acid–base reactions do not go to completion. So, to predict the amount (or concentration) of their products, we have to use the ideas and methods of chemical equilibrium developed in Chapter 14. The present chapter deals with a fundamental aspect of acid–base chemistry key to its applications in the physical and biological sciences, in engineering, and in medicine: How far do such reactions proceed before reaching equilibrium?

Although many acid–base reactions occur in the gaseous and solid states and in nonaqueous solutions, this chapter focuses on acid–base reactions in aqueous solutions. Such solutions play important roles in everyday life. Vinegar, orange juice, and battery fluid are familiar acidic aqueous solutions. Basic solutions are produced when such common products as borax, baking soda, and antacids are dissolved in water. Acid–base chemistry will appear many times throughout the rest of this book. We consider the effects of acidity on the dissolution of solids in Chapter 16, on redox reactions in electrochemical cells in Chapter 17, and on the rates of reaction in Chapter 18. At the end of this chapter we relate acid strength to molecular structure in organic acids, and in Chapter 23 we point out the central roles of amino acids and nucleic acids in biochemistry. The reactions of acids and bases lie at the heart of nearly all branches of chemistry.

15.1 Classifications of Acids and Bases

In Section 11.3 we introduced acids and bases, as defined by Arrhenius. An **acid** is a substance that when dissolved in water increases the concentration of hydronium ion (H_3O^+) above the value it takes in pure water. A **base** increases the concentration of hydroxide ion (OH^-). In preparation for discussing acid–base equilibria we want to generalize these definitions to accommodate broader classes of compounds that are chemically similar to the familiar acids and bases. The first of these more general definitions was introduced by Brønsted and Lowry and the second by Lewis.

Brønsted–Lowry Acids and Bases

A broader definition of acids and bases, which will be useful in quantitative calculations in this chapter, was proposed independently by Johannes Brønsted and Thomas Lowry in 1923. A **Brønsted–Lowry acid** is defined as a substance that can *donate* a hydrogen ion, and a **Brønsted–Lowry base** is a substance that can *accept* a hydrogen ion. In a Brønsted–Lowry acid–base reaction, hydrogen ions are transferred from the acid to the base. When acetic acid is dissolved in water,

$$\underset{\text{Acid}_1}{CH_3COOH(aq)} + \underset{\text{Base}_2}{H_2O(\ell)} \rightleftharpoons \underset{\text{Acid}_2}{H_3O^+(aq)} + \underset{\text{Base}_1}{CH_3COO^-(aq)}$$

hydrogen ions are transferred from acetic acid to water. Throughout this chapter, the hydronium ion, $H_3O^+(aq)$, rather than $H^+(aq)$, will be used to represent the true nature of hydrogen ions in water (Fig. 15.1). Acids and bases occur as **conjugate acid–base pairs.** CH_3COOH and CH_3COO^- form such a pair, where CH_3COO^- is the conjugate base of CH_3COOH (equivalently, CH_3COOH is the conjugate acid of CH_3COO^-). In the same way, H_3O^+ and H_2O form a conjugate acid–base pair. The equilibrium that is established may be pictured as the competition between two bases for hydrogen ions. When ammonia is dissolved in water,

$$\underset{\text{Acid}_1}{H_2O(\ell)} + \underset{\text{Base}_2}{NH_3(aq)} \rightleftharpoons \underset{\text{Acid}_2}{NH_4^+(aq)} + \underset{\text{Base}_1}{OH^-(aq)}$$

the two bases NH_3 and OH^- compete for hydrogen ions.

FIGURE 15.1 The structure of the hydronium ion (H_3O^+).

The Brønsted–Lowry definition has the advantage of not being limited to aqueous solutions. An example with liquid ammonia as the solvent is

$$\underset{\text{Acid}_1}{HCl(\text{in } NH_3)} + \underset{\text{Base}_2}{NH_3(\ell)} \rightleftharpoons \underset{\text{Acid}_2}{NH_4^+(\text{in } NH_3)} + \underset{\text{Base}_1}{Cl^-(\text{in } NH_3)}$$

The NH_3 acts as a base even though hydroxide ion, OH^-, is not present.

Some molecules and ions function either as acids or bases depending on reaction conditions and are called **amphoteric.** The most common example is water itself. Water acts as an acid in donating a hydrogen ion to NH_3 (its conjugate base here is OH^-) and as a base in accepting a hydrogen ion from CH_3COOH (its conjugate acid here is H_3O^+). In the same way, the hydrogen carbonate ion can act as an acid

$$HCO_3^-(aq) + H_2O(\ell) \rightleftharpoons H_3O^+(aq) + CO_3^{2-}(aq)$$

or as a base:

$$HCO_3^-(aq) + H_2O(\ell) \rightleftharpoons H_2CO_3(aq) + OH^-(aq)$$

Lewis Acids and Bases

The Lewis bonding model with its electron pairs can be used to define a more general kind of acid–base behavior of which the Arrhenius and Brønsted–Lowry definitions are special cases. A **Lewis base** is any species that donates lone-pair electrons, and a **Lewis acid** is any species that accepts such electron pairs. The Arrhenius acids and bases considered so far fit this description (with the Lewis acid, H^+, acting as an acceptor toward various Lewis bases such as NH_3 and OH^-, the electron pair donors). Other reactions that do *not* involve hydrogen ions can still be considered Lewis acid–base reactions. An example is the reaction between electron-deficient BF_3 and electron-rich NH_3:

```
   :F:      H             :F: H
    |       |              |  |
 :F—B  +  :N—H   ⟶    :F—B—N—H
    |       |              |  |
   :F:      H             :F: H
```

Here ammonia, the Lewis base, donates lone-pair electrons to BF_3, the Lewis acid or electron acceptor. The bond that forms is called a **coordinate covalent bond,** in which both electrons are supplied by a lone pair on the Lewis base.

Octet-deficient compounds involving elements of Group III, such as boron and aluminum, are often strong Lewis acids because Group III atoms can achieve octet configurations by forming coordinate covalent bonds. Atoms and ions from Groups V through VII have the necessary lone pairs to act as Lewis bases. Compounds of main-group elements from the later periods can also act as Lewis acids through valence expansion. In such reactions the central atom accepts a share in additional lone pairs beyond the eight electrons needed to satisfy the octet rule. For example, $SnCl_4$ is a Lewis acid that accepts electrons from chloride ion lone pairs:

$$SnCl_4(\ell) + 2\ Cl^-(aq) \longrightarrow [SnCl_6]^{2-}(aq)$$

```
      :Cl:                          ⎡ Cl..Cl  ⎤2−
 :Cl:Sn:Cl:  +  2:Cl:⁻   ⟶        ⎢ Cl:Sn:Cl ⎥
      :Cl:                          ⎣ Cl˙˙Cl  ⎦
```

After the reaction, each tin atom is surrounded by 12 rather than 8 valence electrons.

The Lewis definition systematizes the chemistry of a great many binary oxides, which can be considered as anhydrides of acids or bases. An **acid anhydride** is obtained by removing water from an oxoacid (see Section 11.3) until only the oxide remains; thus, CO_2 is the anhydride of carbonic acid (H_2CO_3).

EXAMPLE 15.1

What is the acid anhydride of phosphoric acid (H_3PO_4)?

SOLUTION

If the formula unit H_3PO_4 (which contains an odd number of hydrogen atoms) is doubled, $H_6P_2O_8$ is obtained. Subtraction of 3 H_2O from this gives P_2O_5, which is the empirical formula of tetraphosphorus decaoxide (P_4O_{10}). This compound is the acid anhydride of phosphoric acid.

Related Problems: 9, 10

The Lewis definition provides a way to organize the chemistry of the main group oxides. The oxides of most of the nonmetals are **acid anhydrides,** which react with an excess of water to form acidic solutions. Examples are

$$N_2O_5(s) + H_2O(\ell) \longrightarrow 2\ HNO_3(aq) \longrightarrow 2\ H^+(aq) + 2\ NO_3^-(aq)$$

$$SO_3(g) + H_2O(\ell) \longrightarrow H_2SO_4(aq) \longrightarrow H^+(aq) + HSO_4^-(aq)$$

Silica (SiO_2) is the acid anhydride of the very weak silicic acid H_2SiO_3, a gelatinous material that is insoluble in water but readily dissolves in strongly basic aqueous solutions according to

$$H_2SiO_3(s) + 2\ OH^-(aq) \longrightarrow SiO_3^{2-}(aq) + 2\ H_2O(\ell)$$

Oxides of Group I and II metals are **base anhydrides,** obtained by removing water from the corresponding hydroxides. Calcium oxide, CaO, is the base anhydride of calcium hydroxide, $Ca(OH)_2$. The removal of water from $Ca(OH)_2$ is the reverse of the addition of water to the oxide:

$$CaO(s) + H_2O(\ell) \longrightarrow Ca(OH)_2(s)$$

$$Ca(OH)_2(s) \longrightarrow CaO(s) + H_2O(\ell)$$

The base anhydride of NaOH is Na_2O. Oxides of metals in the middle groups of the periodic table (III through V) lie on the border between ionic and covalent behavior and are frequently amphoteric. An example is aluminum oxide (Al_2O_3), which dissolves to only a limited extent in water but much more readily in either acids or bases:

Acting as a base: $Al_2O_3(s) + 6\ H_3O^+(aq) \longrightarrow 2\ Al^{3+}(aq) + 9\ H_2O(\ell)$

Acting as an acid: $Al_2O_3(s) + 2\ OH^-(aq) + 3\ H_2O(\ell) \longrightarrow 2\ Al(OH)_4^-(aq)$

Figure 15.2 summarizes the acid–base character of the main-group oxides in the first periods.

Increasing acidity →

I	II	III	IV	V	VI	VII
Li_2O	BeO	B_2O_3	CO_2	N_2O_5	(O_2)	OF_2
Na_2O	MgO	Al_2O_3	SiO_2	P_4O_{10}	SO_3	Cl_2O_7
K_2O	CaO	Ga_2O_3	GeO_2	As_2O_5	SeO_3	Br_2O_7
Rb_2O	SrO	In_2O_3	SnO_2	Sb_2O_5	TeO_3	I_2O_7
Cs_2O	BaO	Tl_2O_3	PbO_2	Bi_2O_5	PoO_3	At_2O_7

← Increasing basicity

Increasing basicity (↓) · Increasing acidity (↑)

FIGURE 15.2 Among the oxides of the main-group elements, acidity tends to increase from left to right and from bottom to top in the periodic table. Oxygen difluoride is only weakly acidic. Oxides shown in light blue are derived from metals, those in dark blue from metalloids, and those in dark red from nonmetals.

Although oxoacids and hydroxides are Arrhenius acids and bases (they release $H_3O^+(aq)$ or $OH^-(aq)$ into aqueous solution), acid and base anhydrides do not fall into this classification because they contain neither H^+ nor OH^-. Acid anhydrides are acids in the Lewis sense, (they accept electron pairs), and base anhydrides are bases in the Lewis sense, (their O^{2-} ions donate electron pairs). The reaction between an acid anhydride and a base anhydride is then a Lewis acid–base reaction. An example of such a reaction is

$$CaO(s) + CO_2(g) \longrightarrow CaCO_3(s)$$

Here, the Lewis base CaO donates an electron pair (one of the lone pairs of the oxygen atom) to the Lewis acid (CO_2) to form a coordinate covalent bond in the CO_3^{2-} ion. Similar Lewis acid–base reactions can be written for other acid–base anhydride pairs. Sulfur trioxide, for example, reacts with metal oxides to form sulfates:

$$MgO(s) + SO_3(g) \longrightarrow MgSO_4(s)$$

Note that these reactions are not redox reactions (oxidation numbers do not change). They are clearly not dissolution or precipitation reactions, nor are they acid–base reactions in the Arrhenius sense. But they are usefully classified as acid–base reactions in the Lewis sense.

Comparison of Arrhenius, Brønsted–Lowry, and Lewis Definitions

The neutralization reaction between HCl and NaOH

$$\underset{\text{Acid}}{HCl} + \underset{\text{Base}}{NaOH} \longrightarrow \underset{\text{Water}}{H_2O} + \underset{\text{Salt}}{NaCl}$$

introduced in Section 11.3 shows the progressive generality in these definitions. By the Arrhenius definition, HCl is the acid and NaOH is the base. By the Brønsted–Lowry definition, H_3O^+ is the acid and OH^- is the base. According to Lewis, H^+ is the acid and OH^- the base, because the proton accepts the lone pair donated by OH^- in the reaction

$$H^+(aq) + OH^-(aq) \longrightarrow H_2O(\ell)$$

15.2 Properties of Acids and Bases in Aqueous Solutions: The Brønsted–Lowry Scheme

Chapters 10 and 11 describe the special properties of liquid water. Because of its substantial dipole moment, water is especially effective as a solvent, stabilizing both polar and ionic solutes. Water is not only the solvent, but also participates in acid–base reactions as a reactant. Water plays an integral role in virtually all biochemical reactions essential to the survival of living organisms; these reactions involve acids, bases, and ionic species. In view of the wide-ranging importance of these reactions, we devote the remainder of this chapter to acid–base behavior and related ionic reactions in aqueous solution. The Brønsted–Lowry definition of acids and bases is especially well suited to describe these reactions.

TABLE 15.1 Temperature Dependence of K_w

T(°C)	K_w	pH of Water
0	0.114×10^{-14}	7.47
10	0.292×10^{-14}	7.27
20	0.681×10^{-14}	7.08
25	1.01×10^{-14}	7.00
30	1.47×10^{-14}	6.92
40	2.92×10^{-14}	6.77
50	5.47×10^{-14}	6.63
60	9.61×10^{-14}	6.51

Autoionization of Water

What happens when water acts as both acid and base in the same reaction? The resulting equilibrium is

$$\underset{\text{Acid}_1}{H_2O(\ell)} + \underset{\text{Base}_2}{H_2O(\ell)} \rightleftharpoons \underset{\text{Acid}_2}{H_3O^+(aq)} + \underset{\text{Base}_1}{OH^-(aq)}$$

or

$$2\,H_2O(\ell) \rightleftharpoons H_3O^+(aq) + OH^-(aq)$$

This reaction is responsible for the **autoionization** of water, which leads to small but measurable concentrations of hydronium and hydroxide ions at equilibrium. The equilibrium expression for this reaction is

$$[H_3O^+][OH^-] = K_w \qquad [15.1]$$

The equilibrium constant for this particular reaction has a special symbol: K_w, and a special name, the **ion product constant for water;** its value is 1.0×10^{-14} at 25°C. Because the liquid water appears in this equilibrium reaction equation as a pure substance, it is considered already to be in its reference state, and therefore contributes only the factor 1 to the mass action law equilibrium expression. The reasons for this are discussed more fully in Sections 14.2 and 14.3. The temperature dependence of K_w is given in Table 15.1; all problems in this chapter are assumed to refer to 25°C unless otherwise stated.

Pure water contains no ions other than H_3O^+ and OH^-, and to maintain overall electrical neutrality, an equal number of ions of each type must be present. Putting these facts into the equilibrium expression Equation 15.1 gives

$$[H_3O^+] = [OH^-] = y$$
$$y^2 = 1.0 \times 10^{-14}$$
$$y = 1.0 \times 10^{-7}$$

so that in pure water at 25°C the concentrations of both H_3O^+ and OH^- are 1.0×10^{-7} M.

Strong Acids and Bases

An aqueous acidic solution contains an excess of H_3O^+ over OH^- ions. A **strong acid** is one that ionizes almost completely in aqueous solution. When the strong acid HCl (hydrochloric acid) is put in water, the reaction

$$HCl(aq) + H_2O(\ell) \longrightarrow H_3O^+(aq) + Cl^-(aq)$$

occurs. A single rather than a double arrow indicates that the reaction is essentially complete. Another strong acid is perchloric acid ($HClO_4$). (See Table 11.1.) If 0.10 mol of either of these acids is dissolved in enough water to make 1.0 L of solution, 0.10 M concentration of $H_3O^+(aq)$ results. Because the acid–base properties of solutions are determined by their concentrations of $H_3O^+(aq)$, these two strong acids have the same effect in water despite differences we shall see shortly in their intrinsic abilities to donate hydrogen ions. Water is said to have a **leveling effect** on a certain group of acids (HCl, HBr, HI, H_2SO_4, HNO_3, and $HClO_4$) because they all behave as strong acids when water is the solvent. The reactions of these acids with water all lie so far to the right at equilibrium that the differences between the acids are negligible. The concentration of H_3O^+ in a 0.10 M solution of *any* strong acid that donates one hydrogen ion per molecule is simply 0.10 M. We use this result in Equation 15.1 to obtain the OH^- concentration

$$[OH^-] = \frac{K_w}{[H_3O^+]} = \frac{1.0 \times 10^{-14}}{0.10} = 1.0 \times 10^{-13}\text{ M}$$

In the same fashion, we define a **strong base** as one that reacts essentially completely to give $OH^-(aq)$ ion when put in water. The amide ion (NH_2^-) and the hydride ion (H^-) are both strong bases. For every mole per liter of either of these ions that is added to water, one mole per liter of $OH^-(aq)$ forms. Note that the other products in these reactions are gaseous NH_3 and H_2, respectively:

$$H_2O(\ell) + NH_2^-(aq) \longrightarrow NH_3(aq) + OH^-(aq)$$

$$H_2O(\ell) + H^-(aq) \longrightarrow H_2(aq) + OH^-(aq)$$

The important base sodium hydroxide, a solid, increases the OH^- concentration in water when it dissolves:

$$NaOH(s) \longrightarrow Na^+(aq) + OH^-(aq)$$

For every mole of NaOH that dissolves in water, one mole of $OH^-(aq)$ forms, so NaOH is a strong base. Strong bases are leveled in aqueous solution in the same way that strong acids are leveled. If 0.10 mol of NaOH or NH_2^- or H^- is put into enough water to make 1.0 L of solution, then in every case

$$[OH^-] = 0.10 \text{ M}$$

$$[H_3O^+] = \frac{1.0 \times 10^{-14}}{0.10} = 1.0 \times 10^{-13} \text{ M}$$

The OH^- contribution from the autoionization of water is negligible here, as was the contribution of H_3O^+ from autoionization in the 0.10 M HCl and $HClO_4$ solutions. When only a very small amount of strong acid or base is added to pure water (for example, 10^{-7} mol L^{-1}), we have to include the autoionization of water to describe the concentration of hydronium and hydroxide ions accurately.

The pH Function

In aqueous solution the concentration of hydronium ion can range from 10 M to 10^{-15} M. It is convenient to compress this enormous range by introducing a logarithmic acidity scale, called **pH** and defined by

$$pH = -\log_{10} [H_3O^+] \qquad [15.2]$$

Pure water at 25°C has $[H_3O^+] = 1.0 \times 10^{-7}$ M, so

$$pH = -\log_{10} (1.0 \times 10^{-7}) = -(-7.00) = 7.00$$

A 0.10 M solution of HCl has $[H_3O^+] = 0.10$ M, so

$$pH = -\log_{10} (0.10) = -\log_{10} (1.0 \times 10^{-1}) = -(-1.00) = 1.00$$

and at 25°C a 0.10 M solution of NaOH has

$$pH = -\log_{10}\left(\frac{1.0 \times 10^{-14}}{0.10}\right) = -\log_{10}(1.0 \times 10^{-13}) = -(-13.00) = 13.00$$

As these examples show, calculating the pH is especially easy when the concentration of H_3O^+ is exactly a power of 10, because the logarithm is then just the power to which 10 is raised. In other cases we need a calculator. When we know the pH, we calculate the concentration of H_3O^+ by raising 10 to the power (−pH).

Because the most commonly encountered H_3O^+ concentrations are less than 1 M, the negative sign is put in the definition of the pH function to give a positive

FIGURE 15.3 A simple pH meter with a digital readout.

value in most cases. A *high* pH means a *low* concentration of H_3O^+ and vice versa. At 25°C,

pH < 7	Acidic solution	[15.3a]
pH = 7	Neutral solution	[15.3b]
pH > 7	Basic solution	[15.3c]

At other temperatures the pH of water differs from 7.00 (see Table 15.1). A change of one pH unit implies that the concentrations of H_3O^+ and OH^- change by a factor of 10 (that is, one order of magnitude). The pH is most directly measured with a **pH meter** (Fig. 15.3). The mechanism by which pH meters operate is described in Chapter 17. Figure 15.4 shows the pH values for several common fluids.

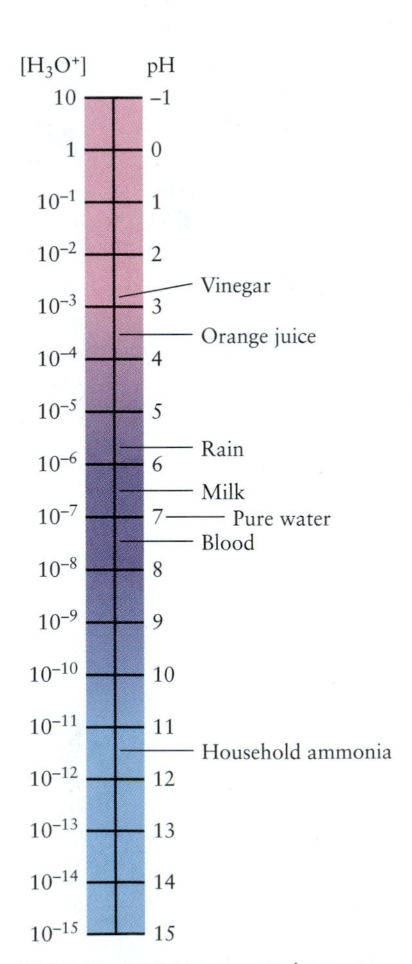

FIGURE 15.4 Many everyday materials are acidic or basic aqueous solutions with a wide range of pH values.

EXAMPLE 15.2

(a) A solution is prepared by dissolving 0.23 mol of NaH(*s*) in enough water to form 2.8 L of solution. Calculate its pH.

(b) The pH of some orange juice at 25°C is 2.85. Calculate $[H_3O^+]$ and $[OH^-]$.

SOLUTION

(a) Because H^- is a strong base, it reacts to give 0.23 mol of OH^-:

$$NaH(s) + H_2O(\ell) \longrightarrow Na^+(aq) + OH^-(aq) + H_2(g)$$

The concentration is

$$[OH^-] = \frac{0.23 \text{ mol}}{2.8 \text{ L}} = 8.2 \times 10^{-2} \text{ M}$$

$$[H_3O^+] = \frac{1.0 \times 10^{-14}}{8.2 \times 10^{-2}} = 1.22 \times 10^{-13} \text{ M}$$

The pH is then

$$pH = -\log_{10}(1.22 \times 10^{-13}) = 12.91$$

(b)

$$pH = 2.85 = -\log_{10}[H_3O^+]$$

$$[H_3O^+] = 10^{-2.85}$$

This can be evaluated by using a calculator to give

$$[H_3O^+] = 1.4 \times 10^{-3} \text{ M}$$

$$[OH^-] = \frac{1.0 \times 10^{-14}}{1.4 \times 10^{-3}} = 7.1 \times 10^{-12}$$

Related Problems: 13, 14, 15, 16

15.3 Acid and Base Strength

Hydrogen cyanide, HCN, is a highly toxic gas that dissolves in water to form equally toxic solutions of hydrocyanic acid.

Acids are classified as strong or weak, depending on the extent to which they are ionized in solution. In a **weak acid** the transfer of hydrogen ions to water does not proceed to completion. A weak acid such as acetic acid is thus also a **weak electrolyte;** its aqueous solutions do not conduct electricity as well as a strong acid of the same concentration because fewer ions are present. A weak acid shows smaller values for colligative properties than a strong acid (recall the effect of dissolved acetic acid on the freezing point of water in Fig. 11.13).

The Brønsted–Lowry theory helps to establish a quantitative scale for acid strength. The ionization of an acid (symbolized by "HA") in aqueous solution can be written as

$$\mathrm{HA}(aq) + \mathrm{H_2O}(\ell) \rightleftharpoons \mathrm{H_3O^+}(aq) + \mathrm{A^-}(aq)$$

where A^- is the conjugate base of HA. The equilibrium expression for this chemical reaction (see Section 14.2) is

$$\frac{[\mathrm{H_3O^+}][\mathrm{A^-}]}{[\mathrm{HA}]} = K_a \qquad [15.4]$$

where the subscript "a" stands for "acid."[1] For example, if the symbol A^- refers to the cyanide ion (CN^-), we write

$$\frac{[\mathrm{H_3O^+}][\mathrm{CN^-}]}{[\mathrm{HCN}]} = K_a$$

where K_a is the **acid ionization constant** for hydrogen cyanide in water and has a numerical value of 6.17×10^{-10} at 25°C. Table 15.2 gives values of K_a, and the useful quantity $\mathrm{p}K_a = -\log_{10} K_a$, for a number of important acids. The acid ionization constant is a quantitative measure of the strength of the acid in a given solvent (here, water). A strong acid has K_a greater than 1, so [HA] in the denominator is small and the acid ionizes almost completely. In a weak acid, K_a is smaller than 1 and the ionized species have low concentrations; the reaction with water proceeds to a limited extent before reaching equilibrium.

The strength of a base is inversely related to the strength of its conjugate acid; the weaker the acid, the stronger its conjugate base, and vice versa. To see this, note that the equation representing the ionization of a base such as ammonia in water can be written as

$$\underset{\text{Acid}_1}{\mathrm{H_2O}(\ell)} + \underset{\text{Base}_2}{\mathrm{NH_3}(aq)} \rightleftharpoons \underset{\text{Acid}_2}{\mathrm{NH_4^+}(aq)} + \underset{\text{Base}_1}{\mathrm{OH^-}(aq)}$$

which gives an equilibrium expression of the form

$$\frac{[\mathrm{NH_4^+}][\mathrm{OH^-}]}{[\mathrm{NH_3}]} = K_b$$

where the subscript "b" on K_b stands for "base." Because $[OH^-]$ and $[H_3O^+]$ are related through the water autoionization equilibrium expression

$$[\mathrm{OH^-}][\mathrm{H_3O^+}] = K_w$$

the K_b expression can be written as

$$K_b = \frac{[\mathrm{NH_4^+}]K_w}{[\mathrm{NH_3}][\mathrm{H_3O^+}]} = \frac{K_w}{K_a}$$

[1] $H_2O(\ell)$ is in its reference state, so it contributes the factor 1 (see Sections 14.2 and 14.3).

TABLE 15.2 Ionization Constants of Acids at 25°C

Acid	HA	A^-	K_a	pK_a
Hydroiodic	HI	I^-	$\sim 10^{11}$	~ -11
Hydrobromic	HBr	Br^-	$\sim 10^{9}$	~ -9
Perchloric	$HClO_4$	ClO_4^-	$\sim 10^{7}$	~ -7
Hydrochloric	HCl	Cl^-	$\sim 10^{7}$	~ -7
Chloric	$HClO_3$	ClO_3^-	$\sim 10^{3}$	~ -3
Sulfuric (1)	H_2SO_4	HSO_4^-	$\sim 10^{2}$	~ -2
Nitric	HNO_3	NO_3^-	~ 20	~ -1.3
Hydronium ion	H_3O^+	H_2O	1	0.0
Iodic	HIO_3	IO_3^-	1.6×10^{-1}	0.80
Oxalic (1)	$H_2C_2O_4$	$HC_2O_4^-$	5.9×10^{-2}	1.23
Sulfurous (1)	H_2SO_3	HSO_3^-	1.54×10^{-2}	1.81
Sulfuric (2)	HSO_4^-	SO_4^{2-}	1.2×10^{-2}	1.92
Chlorous	$HClO_2$	ClO_2^-	1.1×10^{-2}	1.96
Phosphoric (1)	H_3PO_4	$H_2PO_4^-$	7.52×10^{-3}	2.12
Arsenic (1)	H_3AsO_4	$H_2AsO_4^-$	5.0×10^{-3}	2.30
Chloroacetic	$CH_2ClCOOH$	CH_2ClCOO^-	1.4×10^{-3}	2.85
Hydrofluoric	HF	F^-	6.6×10^{-4}	3.18
Nitrous	HNO_2	NO_2^-	4.6×10^{-4}	3.34
Formic	$HCOOH$	$HCOO^-$	1.77×10^{-4}	3.75
Benzoic	C_6H_5COOH	$C_6H_5COO^-$	6.46×10^{-5}	4.19
Oxalic (2)	$HC_2O_4^-$	$C_2O_4^{2-}$	6.4×10^{-5}	4.19
Hydrazoic	HN_3	N_3^-	1.9×10^{-5}	4.72
Acetic	CH_3COOH	CH_3COO^-	1.76×10^{-5}	4.75
Propionic	CH_3CH_2COOH	$CH_3CH_2COO^-$	1.34×10^{-5}	4.87
Pyridinium ion	$HC_5H_5N^+$	C_5H_5N	5.6×10^{-6}	5.25
Carbonic (1)	H_2CO_3	HCO_3^-	4.3×10^{-7}	6.37
Sulfurous (2)	HSO_3^-	SO_3^{2-}	1.02×10^{-7}	6.91
Arsenic (2)	$H_2AsO_4^-$	$HAsO_4^{2-}$	9.3×10^{-8}	7.03
Hydrosulfuric	H_2S	HS^-	9.1×10^{-8}	7.04
Phosphoric (2)	$H_2PO_4^-$	HPO_4^{2-}	6.23×10^{-8}	7.21
Hypochlorous	$HClO$	ClO^-	3.0×10^{-8}	7.53
Hydrocyanic	HCN	CN^-	6.17×10^{-10}	9.21
Ammonium ion	NH_4^+	NH_3	5.6×10^{-10}	9.25
Carbonic (2)	HCO_3^-	CO_3^{2-}	4.8×10^{-11}	10.32
Arsenic (3)	$HAsO_4^{2-}$	AsO_4^{3-}	3.0×10^{-12}	11.53
Hydrogen peroxide	H_2O_2	HO_2^-	2.4×10^{-12}	11.62
Phosphoric (3)	HPO_4^{2-}	PO_4^{3-}	2.2×10^{-13}	12.67
Water	H_2O	OH^-	1.0×10^{-14}	14.00

(1) and (2) indicate the first and second ionization constants, respectively.

where K_a is the acid ionization constant for NH_4^+, the conjugate acid of the base NH_3. This general relationship between the K_b of a base and the K_a of its conjugate acid shows that K_b need not be tabulated separately from K_a, because the two are related through

$$K_w = K_a K_b \qquad [15.5]$$

It is also clear that if K_a is large (so that the acid is strong), then K_b is small (the conjugate base is weak). Figure 15.5 summarizes the strengths of conjugate acid–base pairs.

If two bases compete for hydrogen ions, the stronger base is favored in the equilibrium that is reached. The stronger acid donates hydrogen ions to the

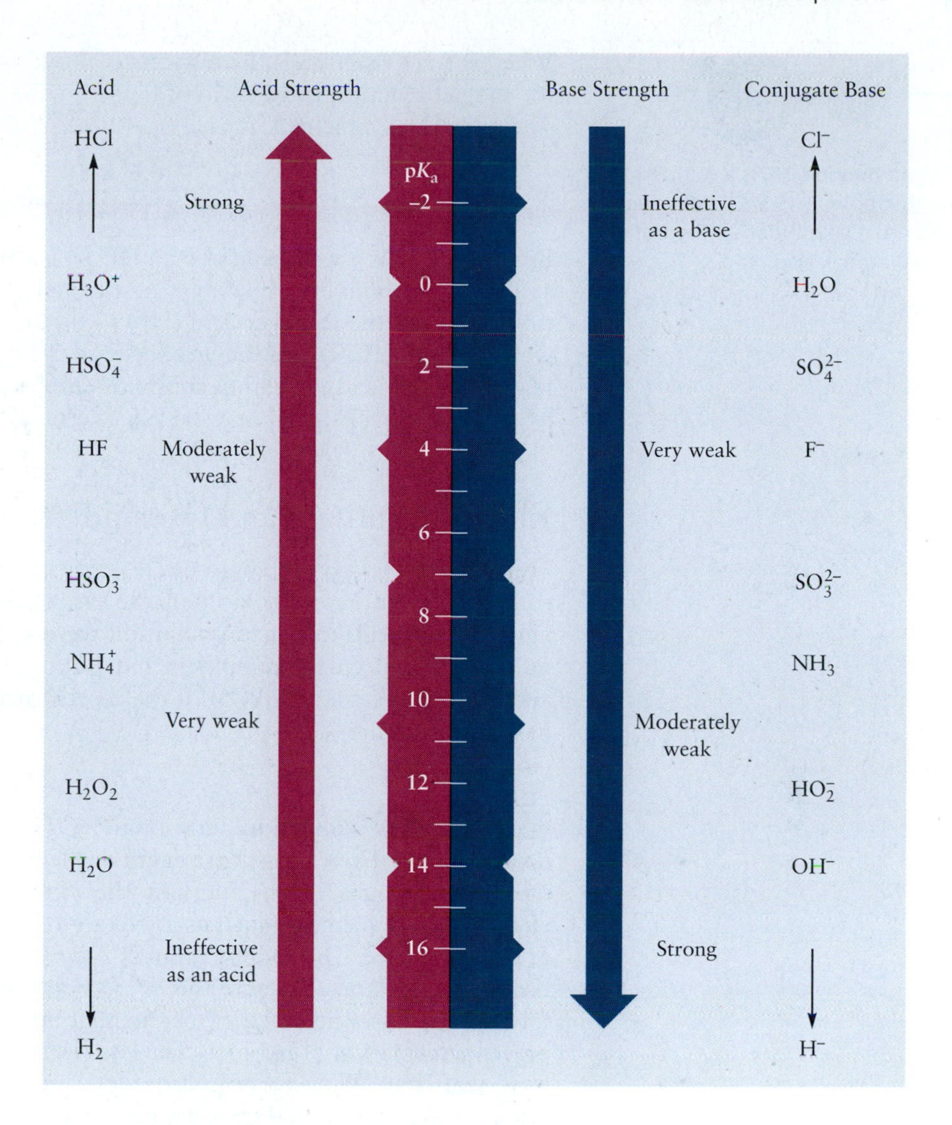

FIGURE 15.5 The relative strengths of some acids and their conjugate bases.

stronger base, producing a weaker acid and a weaker base. To see this, consider the equilibrium

$$\underset{\text{Acid}_1}{HF(aq)} + \underset{\text{Base}_2}{CN^-(aq)} \rightleftarrows \underset{\text{Acid}_2}{HCN(aq)} + \underset{\text{Base}_1}{F^-(aq)}$$

with equilibrium constant

$$\frac{[HCN][F^-]}{[HF][CN^-]} = K$$

The two bases F^- and CN^- compete for hydrogen ions. We can construct this net reaction by starting with one acid ionization reaction

$$HF(aq) + H_2O(\ell) \rightleftarrows H_3O^+(aq) + F^-(aq)$$

$$\frac{[H_3O^+][F^-]}{[HF]} = K_a = 6.6 \times 10^{-4}$$

and *subtracting* from it a second acid ionization reaction:

$$HCN(aq) + H_2O(\ell) \rightleftarrows H_3O^+(aq) + CN^-(aq)$$

$$\frac{[H_3O^+][CN^-]}{[HCN]} = K_a' = 6.17 \times 10^{-10}$$

Hydrogen fluoride, HF, is a colorless liquid that boils at 19.5°C. It dissolves in water to give solutions of hydrofluoric acid, a weak acid.

When the net reaction is the difference of two reactions, the equilibrium constant for the net reaction is the ratio of those for the separate reactions. (Section 14.4) The numerical value of K is

$$K = \frac{K_a}{K_a'} = \frac{6.6 \times 10^{-4}}{6.17 \times 10^{-10}} = 1.1 \times 10^6$$

Because HCN is a weaker acid than HF, K_a' is smaller than K_a, and K is larger than 1. The equilibrium described by K lies strongly to the right. The net result is donation of H^+ by the stronger acid (HF) to the stronger base (CN^-), to produce the weaker acid (HCN) and the weaker base (F^-). This example illustrates how the magnitudes of acid ionization constants can be used to predict the direction of net hydrogen ion transfer in reactions between acids and bases in aqueous solution.

Electronegativity and Oxoacid Strength

Trends in the relative strength of oxoacids are explained by the influence of electronegativity and bond polarity on the ease of donating a proton. The protons donated by **oxoacids** in aqueous solution were previously bonded to oxygen atoms on the acid molecule. Examples include sulfuric acid (H_2SO_4), nitric acid (HNO_3), and phosphoric acid (H_3PO_4). If the central atom is designated X, then oxoacids have the structure

$$-\text{X}-\text{O}-\text{H}$$

where X can be bonded to additional —OH groups, to oxygen atoms, or to hydrogen atoms. How does the strength of the oxoacid change as the electronegativity of X changes? Consider first the extreme case in which X is a highly electropositive element, such as an alkali metal. Of course, NaOH is not an acid at all, but a base. The sodium atom in Na—O—H gives up a full electron to make Na^+ and OH^- ions. Because the X—O bond here is almost completely ionic, the OH^- group has a net negative charge that holds the H^+ tightly to the oxygen and prevents formation of H^+ ions. The less electropositive alkaline-earth elements behave similarly. They form hydroxides, such as $Mg(OH)_2$, that are somewhat weaker bases than NaOH but in no way act as acids.

Now suppose the central atom X becomes more electronegative, reaching values between 2 and 3, as in the oxoacids of the elements B, C, P, As, S, Se, Br, and I. As X becomes more effective at withdrawing electron density from the oxygen atom, the X—O bond becomes more covalent. This leaves less negative charge on the oxygen atom, and consequently the oxoacid releases H^+ more readily (Fig. 15.6). Other things being equal, acid strength should increase with increasing electronegativity of the central atom. This trend is observed among the oxoacids listed in Table 15.2.

The strength of oxoacids with a given central element X increases with the number of lone oxygen atoms attached to the central atom. If the formula of these acids is written as $XO_n(OH)_m$, the corresponding acid strengths fall into distinct classes according to the value of n, the number of lone oxygen atoms (see

FIGURE 15.6 In part (a) the atom X is electropositive, so extra electron density (blue areas) accumulates on the OH group. The X—O bond then breaks easily, making the compound a base. In part (b) X is electronegative, so electron density is drawn from the H atom to the X—O bond. Now it is the O—H bond that breaks easily, and the compound is an acid.

(a) X—O—H → X + O—H

Bond breaks

(b) X—O—H → X—O + H

TABLE 15.3 Acid Ionization Constants for Oxoacids of the Nonmetals

$X(OH)_m$ Very Weak	K_a	$XO(OH)_m$ Weak	K_a	$XO_2(OH)_m$ Strong	K_a	$XO_3(OH)_m$ Very Strong	K_a
$Cl(OH)$	3×10^{-8}	$H_2PO(OH)$	8×10^{-2}	$SeO_2(OH)_2$	10^3	$ClO_3(OH)$	2×10^7
$Te(OH)_6$	2×10^{-8}	$IO(OH)_5$	2×10^{-2}	$ClO_2(OH)$	5×10^2		
$Br(OH)$	2×10^{-9}	$SO(OH)_2$	2×10^{-2}	$SO_2(OH)_2$	1×10^2		
$As(OH)_3$	6×10^{-10}	$ClO(OH)$	1×10^{-2}	$NO_2(OH)$	2×10^1		
$B(OH)_3$	6×10^{-10}	$HPO(OH)_2$	1×10^{-2}	$IO_2(OH)$	1.6×10^1		
$Ge(OH)_4$	4×10^{-10}	$PO(OH)_3$	8×10^{-3}				
$Si(OH)_4$	2×10^{-10}	$AsO(OH)_3$	5×10^{-3}				
$I(OH)$	4×10^{-11}	$SeO(OH)_2$	3×10^{-3}				
		$TeO(OH)_2$	3×10^{-3}				
		$NO(OH)$	5×10^{-4}				

Table 15.3). Each increase of 1 in n increases the acid ionization constant K_a by a factor of about 10^5. Another way to describe this effect is to focus on the stability of the conjugate base, $XO_{n+1}(OH)^-_{m-1}$, of the oxoacid. The greater the number of lone oxygen atoms attached to the central atom, the more easily the net negative charge can be spread out over the ion, and therefore the more stable the base. This leads to a larger K_a.

An unusual and interesting structural result can be obtained from Table 15.3. Figure 15.7a shows the simplest Lewis diagram for phosphorous acid (H_3PO_3) in which each atom achieves an octet configuration. Such a diagram could also be written $P(OH)_3$ and would be analogous to $As(OH)_3$, which has no lone oxygen atoms bonded to the central atom ($n = 0$). On the basis of this analogy, we would expect the value of K_a for $P(OH)_3$ to be on the order of 10^{-9} (a very weak acid). But in fact, H_3PO_3 is only a moderately weak acid ($K_a = 1 \times 10^{-2}$) and fits better into the class of acids with one lone oxygen atom bonded to the central atom. X-ray diffraction measurements support this structural conclusion inferred from chemical behavior. So, the structure of H_3PO_3 is best represented by Figure 15.7b and corresponds either to a Lewis diagram with more than eight electrons around the central phosphorus atom or to one with formal charges on the central phosphorus and lone oxygen atoms. The formula of this acid is written as $HPO(OH)_2$ in Table 15.3. Unlike phosphoric acid (H_3PO_4), which is a triprotic acid, H_3PO_3 is a diprotic acid. The third hydrogen atom, the one directly bonded to the phosphorus atom, does not ionize even in strongly basic aqueous solution.

H—O—P—O—H (with O—H above P)

(a)

H—O—P—O—H (with O⁻ above P⁺ and H below P)

(b)

FIGURE 15.7 (a) The simplest Lewis diagram that can be drawn for H_3PO_3 gives an incorrect structure. This acid would be triprotic, like H_3PO_4. (b) The observed structure of H_3PO_3 requires assigning formal charge to the P atom and the lone O atom. The hydrogen atom attached to the P is not released into acid solution, so the acid is diprotic.

Indicators

An **indicator** is a soluble dye that changes color noticeably over a fairly narrow range of pH. The typical indicator is a weak organic acid that has a different color from its conjugate base (Fig. 15.8). Litmus changes from red to blue as its acid form is converted to base. Good indicators have such intense colors that only a

FIGURE 15.8 Color differences in four indicators: bromophenol red, thymolphthalein, phenolphthalein, and bromocresol green. In each case the acidic form is on the left and the basic form is on the right.

few drops of a dilute indicator solution must be added to the solution being studied. The very low concentration of indicator molecules has almost no effect on the pH of solution. The color changes of the indicator reflect the effects of the *other* acids and bases present in the solution.

If the acid form of a given indicator is represented as HIn and the conjugate base form as In^-, their acid–base equilibrium is

$$HIn(aq) + H_2O(\ell) \rightleftarrows H_3O^+(aq) + In^-(aq) \qquad \frac{[H_3O^+][In^-]}{[HIn]} = K_a$$

where K_a is the acid ionization constant for the indicator. This expression can be rearranged to give

$$\frac{[H_3O^+]}{K_a} = \frac{[HIn]}{[In^-]} \qquad [15.6]$$

If the concentration of hydronium ion $[H_3O^+]$ is large relative to K_a, this ratio is large, and [HIn] is large compared with $[In^-]$. The solution has the color of the acid form of the indicator because most of the indicator molecules are in the acid form. Litmus, for example, has a K_a near 10^{-7}. If the pH is 5, then

$$\frac{[H_3O^+]}{K_a} = \frac{10^{-5}}{10^{-7}} = 100$$

Thus, approximately 100 times as many indicator molecules are in the acid form as in the base form, and the solution is red.

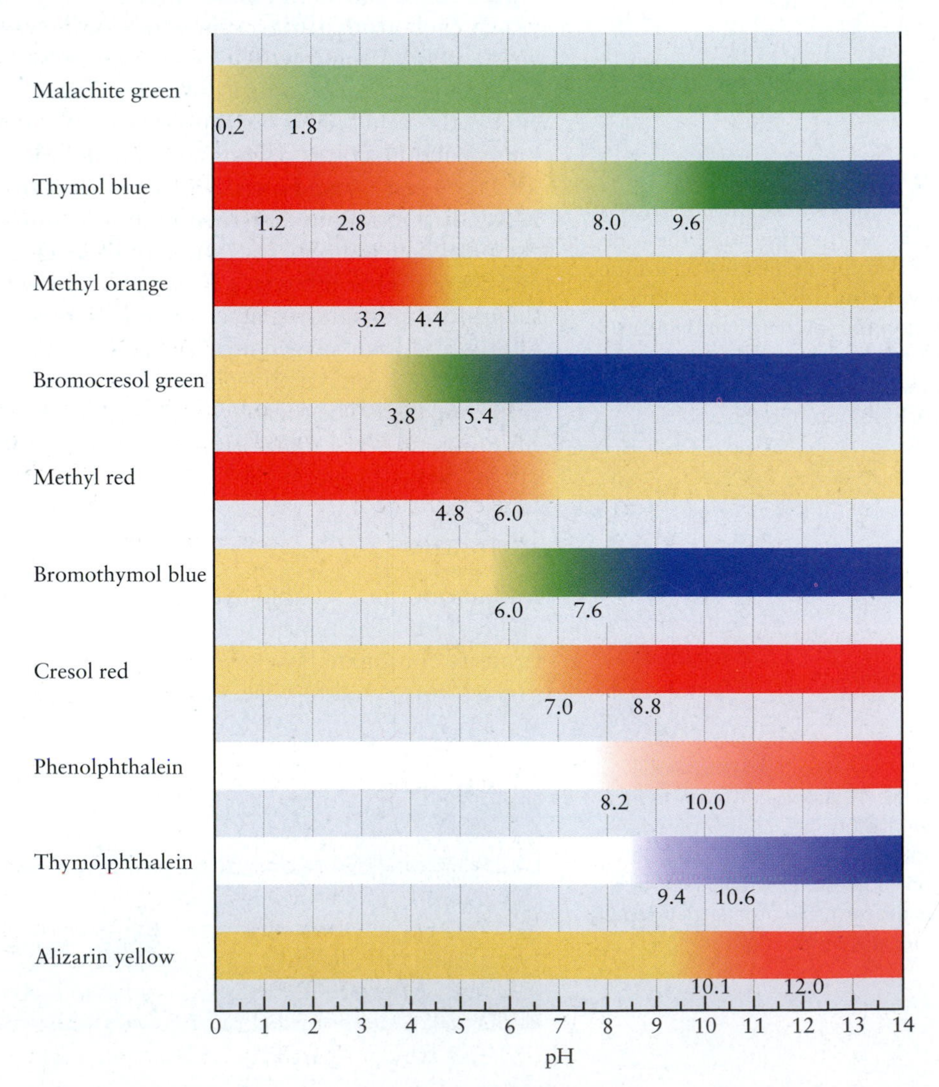

FIGURE 15.9 Indicators change their colors at very different pH values, so the best choice of indicator depends on the particular experimental conditions.

FIGURE 15.10 Red cabbage extract is a natural pH indicator. When the solution is highly acidic, the extract gives the solution a red color. As the solution becomes less and less acidic (more basic), the color changes from red to violet to yellow.

As the concentration of hydronium ion is reduced, more molecules of acid indicator ionize to give the base form. When $[H_3O^+]$ is near K_a, almost equal amounts of the two forms are present and the color is a mixture of the colors of the two indicator states (violet for litmus). A further decrease in $[H_3O^+]$ to a value much smaller than K_a then leads to a predominance of the base form, with the corresponding color being observed.

Different indicators have different values for K_a and thus show color changes at different pH values (Fig. 15.9). The weaker an indicator is as an acid, the higher the pH at which the color change takes place. Such color changes occur over a range of 1 to 2 pH units. Methyl red, for example, is red when the pH is below 4.8 and yellow above 6.0; shades of orange are seen at intermediate pH values. This limits the accuracy to which the pH can be determined through the use of indicators. Section 15.6 shows that this fact does not affect the analytical determination of acid or base concentrations through titration, provided that an appropriate indicator is used.

Many natural dyes found in fruits, vegetables, and flowers act as pH indicators by changing color with changes in acidity (Fig. 15.10). A particularly striking example is cyanidin, which is responsible both for the red color of poppies and the blue color of cornflowers. The sap of the poppy is sufficiently acidic to turn cyanidin red, but the sap of the cornflower is basic and makes the dye blue. (See the image on page 625.) Related natural dyes called anthocyanins contribute to the colors of raspberries, strawberries, and blackberries.

15.4 Equilibria Involving Weak Acids and Bases

The reactions of weak acids and bases with water do not go to completion. So, to calculate the pH of their solutions, we use K_a or K_b and the laws of chemical equilibrium. The calculations follow the pattern of Example 14.10 for gas equilibria. In that case the initial gas-phase pressures $P°$ are known, and we calculate the pressures of products resulting from the incomplete reaction. Here we know the initial concentration of acid or base, and calculate the concentrations of products resulting from its partial reaction with water.

Weak Acids

A weak acid has a K_a smaller than 1. Values of the pK_a start at zero for the strongest weak acid and range upward. (If the pK_a is greater than 14, the compound is ineffective as an acid in aqueous solution.) When a weak acid is dissolved in water, the original concentration is almost always known but partial reaction with water consumes some HA and generates A^- and H_3O^+:

$$HA(aq) + H_2O(\ell) \rightleftharpoons H_3O^+(aq) + A^-(aq)$$

To calculate the amounts of H_3O^+, A^-, and HA at equilibrium, we use the methods of Chapter 14, with partial pressures replaced by concentrations. A new feature here is that one of the products (H_3O^+) can also come from a second source, the autoionization of the solvent, water. In most of the applications we study, this second effect is small and can be neglected in the equations. Even so, it is a good idea to verify at the end of each calculation that the $[H_3O^+]$ from the acid ionization alone exceeds 10^{-7} M by at least one order of magnitude. Otherwise, we have to use the more complete method of analysis given in Section 15.8.

EXAMPLE 15.3

Acetic acid (CH_3COOH) has a K_a of 1.76×10^{-5} at 25°C. Suppose 1.000 mol is dissolved in enough water to give 1.000 L of solution. Calculate the pH and the fraction of acetic acid ionized at equilibrium.

SOLUTION

The initial concentration of acetic acid is 1.000 M. If y mol L^{-1} ionizes, then

	$CH_3COOH(aq)$ + $H_2O(\ell)$ $\rightleftharpoons$	$H_3O^+(aq)$ +	$CH_3COO^-(aq)$
Initial concentration (M)	1.000	≈0	0
Change in concentration (M)	$-y$	$+y$	$+y$
Equilibrium concentration (M)	$1.000 - y$	y	y

Note that we ignored the H_3O^+ initially present from the ionization of water because we expect it to be smaller than y for all but the weakest acids or the most dilute solutions. Then, the equilibrium expression states that

$$\frac{[H_3O^+][CH_3COO^-]}{[CH_3COOH]} = \frac{y^2}{1.000 - y} = K_a = 1.76 \times 10^{-5}$$

This equation could be solved using the quadratic formula, as we did in Chapter 14. A quicker way is based on the fact that acetic acid is a weak acid and therefore only a small fraction is ionized at equilibrium. Therefore it is reasonable to assume that y is small relative to 1.000 (that is, the final equilibrium concentration of CH_3COOH is close to its initial concentration). This gives

$$\frac{y^2}{1.000 - y} \approx \frac{y^2}{1.000} = 1.76 \times 10^{-5}$$

$$y = 4.20 \times 10^{-3}$$

so

$$[H_3O^+] = y = 4.20 \times 10^{-3} \text{ M}$$

Now we have to check whether the approximations are indeed valid. First, y is indeed much smaller than the original concentration of 1.000 M (by a factor of 200), so neglecting it in the denominator is justified. Second, the concentration of H_3O^+ from acetic acid (4.2×10^{-3} M) is large compared with 10^{-7} M, so neglecting the water ionization is also justified. Therefore,

$$\text{pH} = -\log_{10}(4.2 \times 10^{-3}) = 2.38$$

The fraction ionized is the ratio of the concentration of CH_3COO^- present at equilibrium to the concentration of CH_3COOH present in the first place:

$$\frac{y}{1.000} = y = 4.2 \times 10^{-3}$$

The percentage of the acetic acid that is ionized is 0.42%. Fewer than one in a hundred of the molecules of this typical weak acid dissociate in this solution.

Related Problems: 27, 28, 29, 30

Approximation methods of the type used in this example are also used in Example 14.6 and are discussed more extensively in Appendix C. As the solution of a weak acid becomes more dilute, a greater fraction of it ionizes, as shown by the following example.

EXAMPLE 15.4

Suppose that 0.00100 mol of acetic acid is used instead of the 1.000 mol in the preceding example. Calculate the pH and the percentage of acetic acid ionized.

SOLUTION

Once again we assume that the contribution of the ionization of water to $[H_3O^+]$ is negligible, and check the validity of the assumption at the end of the calculation. Hence,

$$[CH_3COOH] = 0.00100 - y$$

$$[CH_3COO^-] = [H_3O^+] = y$$

$$\frac{y^2}{0.00100 - y} = 1.76 \times 10^{-5}$$

In this case, y is *not* small relative to 0.00100 because a substantial fraction of the acetic acid molecules ionizes, so y cannot be neglected in the denominator. There are two alternatives for solving the problem in this case.

The first method is to solve the quadratic equation obtained by multiplying out the equilibrium expression:

$$y^2 + (1.76 \times 10^{-5})y - (1.76 \times 10^{-8}) = 0$$

Using of the quadratic formula gives

$$y = 1.24 \times 10^{-4}\ \text{M} = [H_3O^+] = [CH_3COO^-]$$

Because $y \gg 10^{-7}$ M, neglecting water ionization in setting up the problem is justified.

$$\text{pH} = -\log_{10}(1.24 \times 10^{-4}) = 3.91$$

$$\text{percentage ionized} = \frac{1.24 \times 10^{-4}\ \text{M}}{0.00100\ \text{M}} \times 100\% = 12.4\%$$

The other approach to solving the problem is to use successive approximations (see Appendix C). Begin by neglecting y in the denominator (relative to 0.00100), which gives

$$\frac{y^2}{0.00100} = 1.76 \times 10^{-5}$$

$$y = 1.33 \times 10^{-4}\ \text{M}$$

This is just what we do in Example 15.3. Now, continue by reinserting this *approximate* value of y into the *denominator* and then recalculate y:

$$\frac{y^2}{0.00100 - 0.00133} = 1.76 \times 10^{-5}$$

$$y^2 = 1.53 \times 10^{-8}$$

$$y = 1.33 \times 10^{-4}\ \text{M}$$

If we insert this new value of y into the denominator and iterate the calculation once again, we obtain 1.24×10^{-4} M, the same value produced by the quadratic formula. We terminate the iteration process when two successive results for y agree to the desired number of significant figures.

Related Problems: 31, 32

The method of successive approximations is often faster to apply than the quadratic formula. Keep in mind that the accuracy of a result is limited both by the accuracy of the input data (values of K_a and initial concentrations) and by the fact that solutions are not ideal. It is pointless to calculate equilibrium concentrations to any degree of accuracy higher than 1% to 3%.

Weak Bases

There is a perfect parallel between the behavior of weak bases and weak acids. The definition and description of weak acids, the existence of K_a, and the production of $H_3O^+(aq)$ ion when they are dissolved in water can be applied directly to weak bases, the existence of K_b, and the production of $OH^-(aq)$ ion. A **weak base** such as ammonia reacts only partially with water to produce $OH^-(aq)$:

$$H_2O(\ell) + NH_3(aq) \rightleftarrows NH_4^+(aq) + OH^-(aq)$$

$$\frac{[NH_4^+][OH^-]}{[NH_3]} = K_b = 1.8 \times 10^{-5}$$

The K_b of a weak base is smaller than 1, and the weaker the base, the smaller the K_b. If the K_b of a compound is smaller than 1×10^{-14}, that compound is ineffective as a base in aqueous solution.

The analogy to weak acids continues in the calculation of the aqueous equilibria of weak bases, as the following example shows.

EXAMPLE 15.5

Calculate the pH of a solution made by dissolving 0.0100 mol of NH_3 in enough water to give 1.000 L of solution at 25°C. The K_b for ammonia is 1.8×10^{-5}.

SOLUTION

Set up the table of the changes in concentrations that occur as the reaction goes to equilibrium, neglecting the small contribution to $[OH^-]$ from the autoionization of water:

	$H_2O(\ell)$ + $NH_3(aq)$	$\rightleftarrows NH_4^+(aq)$	+ $OH^-(aq)$
Initial concentration (M)	0.0100	0	≈0
Change in concentration (M)	$-y$	$+y$	$+y$
Equilibrium concentration (M)	$0.100 - y$	y	y

Substitution into the equilibrium expression gives

$$\frac{y^2}{0.0100 - y} = K_b = 1.8 \times 10^{-5}$$

which can be solved for y by either the quadratic formula or the method of successive approximations to obtain:

$$y = 4.15 \times 10^{-4}\ \text{M} = [OH^-]$$

The product of $[H_3O^+]$ and $[OH^-]$ is always K_w. Hence,

$$[H_3O^+] = \frac{K_w}{[OH^-]} = \frac{1.0 \times 10^{-14}}{4.15 \times 10^{-14}} = 2.4 \times 10^{-11}$$

$$pH = -\log_{10}(2.4 \times 10^{-11}) = 10.62$$

The pH is greater than 7, as expected for a solution of a base.

Related Problems: 35, 36

Hydrolysis

Most of the acids considered up to now have been uncharged species with the general formula HA. In the Brønsted–Lowry picture there is no reason why the acid should be an electrically neutral molecule. When NH_4Cl, a salt, dissolves in water, NH_4^+ ions are present. These ionize partially by transferring hydrogen ions to water, a straightforward Brønsted–Lowry acid–base reaction:

$$\underset{\text{Acid}_1}{NH_4^+(aq)} + \underset{\text{Base}_2}{H_2O(\ell)} \rightleftharpoons \underset{\text{Acid}_2}{H_3O^+(aq)} + \underset{\text{Base}_1}{NH_3(aq)}$$

$$\frac{[H_3O^+][NH_3]}{NH_4^+} = K_a = 5.6 \times 10^{-10}$$

NH_4^+ acts as an acid here, just as acetic acid did in Examples 15.3 and 15.4. Because K_a is much smaller than 1, only a small amount of H_3O^+ is generated. NH_4^+ is a weak acid, but it is nonetheless an acid, and a solution of ammonium chloride has a pH below 7 (Fig. 15.11).

(a)

(b)

(c)

FIGURE 15.11 Proof that ammonium chloride is an acid. (a) A solution of sodium hydroxide, with the pink color of phenolphthalein indicating its basic character. (b and c) As aqueous ammonium chloride is added, a neutralization reaction takes place and the solution turns colorless from the bottom up.

Hydrolysis is the label for the reaction of a substance with water. This term is applied especially to a reaction in which the pH changes from 7 upon dissolving a salt (in this case, NH_4Cl) in water. There is no need for any special description of hydrolysis. The hydrolysis that takes place when NH_4Cl dissolves in water is completely described as a Brønsted–Lowry reaction in which water acts as a base and NH_4^+ acts as an acid to give a pH below 7. In parallel fashion, dissolving a salt whose anion is a weak base produces a basic solution. This too is a case of hydrolysis; it is simply another Brønsted–Lowry acid–base reaction, with water acting now as an acid (a hydrogen ion donor).

Two salts that give basic solutions are sodium acetate and sodium fluoride. When these salts dissolve in water, they furnish acetate (CH_3COO^-) and fluoride (F^-) ions, respectively, both of which act as Brønsted–Lowry bases,

$$H_2O(\ell) + CH_3COO^-(aq) \rightleftharpoons CH_3COOH(aq) + OH^-(aq)$$

$$H_2O(\ell) + F^-(aq) \rightleftharpoons HF(aq) + OH^-(aq)$$

causing the OH^- concentration to increase and give a pH above 7.

EXAMPLE 15.6

Suppose 0.100 mol of $NaCH_3COO$ is dissolved in enough water to make 1.00 L of solution. What is the pH of the solution?

SOLUTION

The K_a for CH_3COOH, the conjugate acid of CH_3COO^-, is 1.76×10^{-5}. Therefore, K_b for the acetate ion is

$$\frac{[CH_3COOH][OH^-]}{[CH_3COO^-]} = K_b = \frac{K_w}{K_a} = 5.7 \times 10^{-10}$$

The equilibrium here is

	$H_2O(\ell)$ + $CH_3COO^-(aq)$ ⇌	$CH_3COOH(aq)$ +	$OH^-(aq)$
Initial concentration (M)	0.100	0	≈0
Change in concentration (M)	$-y$	$+y$	$+y$
Equilibrium concentration (M)	$0.100 - y$	y	y

Substitution into the equilibrium expression gives

$$\frac{y^2}{0.100 - y} = K_b = 5.7 \times 10^{-10}$$

$$y = 7.5 \times 10^{-6}\ \text{M} = [OH^-]$$

Note that y is small relative to 0.100 and fairly large relative to 10^{-7}.

$$[H_3O^+] = \frac{K_w}{[OH^-]} = \frac{1.0 \times 10^{-14}}{7.5 \times 10^{-6}} = 1.3 \times 10^{-9}\ \text{M}$$

$$\text{pH} = 8.89$$

Related Problems: 41, 42

Hydrolysis does not occur with all ions, only with those that are conjugate acids of weak bases or conjugate bases of weak acids. Chloride ion is the conjugate base of the strong acid HCl and consequently is ineffective as a base, unlike $F^-(aq)$ and $CH_3COO^-(aq)$. Its interaction with water would therefore scarcely change the $OH^-(aq)$ concentration. So, a solution of NaCl is neutral, whereas a solution of NaF is slightly basic.

15.5 Buffer Solutions

A **buffer solution** is any solution that maintains approximately constant pH upon small additions of acid or base. Typically, a buffer solution contains a weak acid and its conjugate weak base in approximately equal concentrations. Buffer solutions play important roles in controlling the solubility of ions in solution (see Chapter 16) and in maintaining the pH in biochemical and physiological processes. Many life processes are sensitive to pH and require regulation within a small range of H_3O^+ and OH^- concentrations. Organisms have built-in buffers to protect them against large changes in pH. Human blood has a pH near 7.4, which is maintained by a combination of carbonate, phosphate, and protein buffer systems. A blood pH below 7.0 or above 7.8 leads quickly to death.

Calculations of Buffer Action

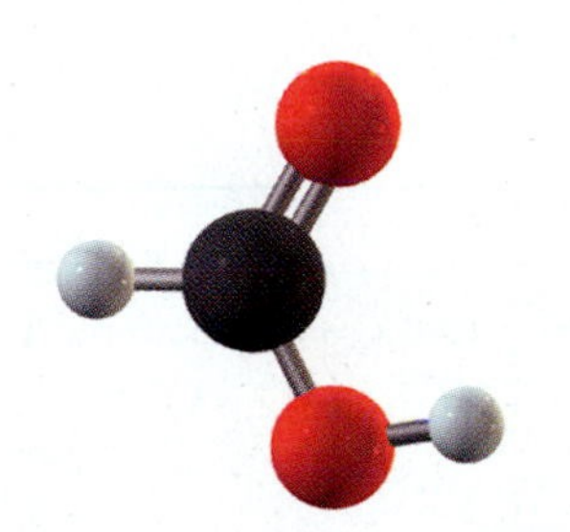

Formic acid, HCOOH, is the simplest carboxylic acid, with only a hydrogen atom attached to the —COOH.

Consider a typical weak acid, formic acid (HCOOH), and its conjugate base, formate ion ($HCOO^-$). The latter can be obtained by dissolving a salt such as sodium formate (NaHCOO) in water. The acid–base equilibrium established between these species is given by

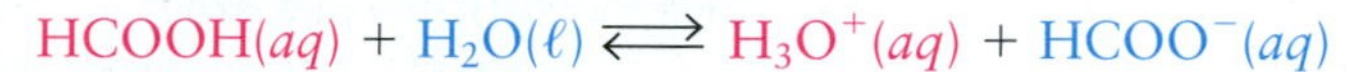

$$HCOOH(aq) + H_2O(\ell) \rightleftharpoons H_3O^+(aq) + HCOO^-(aq)$$

with an acid ionization constant

$$\frac{[H_3O^+][HCOO^-]}{[HCOOH]} = K_a = 1.77 \times 10^{-4}$$

Section 15.2 describes the pH of a solution containing only a weak acid (such as HCOOH) or only a weak base (such as $HCOO^-$). Suppose now that the weak acid and its conjugate base are *both* present initially. The resulting calculations resemble closely the calculation of Example 14.15, in which equilibrium was established from an initial mixture containing both reactants and products.

EXAMPLE 15.7

Suppose 1.00 mol of HCOOH and 0.500 mol of NaHCOO are added to water and diluted to 1.00 L. Calculate the pH of the solution.

SOLUTION

	$HCOOH(aq)$ + $H_2O(\ell)$ ⇌	$H_3O^+(aq)$ +	$HCOO^-(aq)$
Initial concentration (M)	1.00	≈0	0.500
Change in concentration (M)	$-y$	$+y$	$+y$
Equilibrium concentration (M)	$1.00 - y$	y	$0.500 + y$

The equilibrium expression is

$$\frac{y(0.500 + y)}{1.00 - y} = K_a = 1.77 \times 10^{-4}$$

Because y is likely to be small relative to 1.00 and to 0.500, we write the approximate expression

$$\frac{y(0.500)}{1.00} \approx 1.77 \times 10^{-4}$$

$$y \approx 3.54 \times 10^{-4} \text{ M} = [H_3O^+]$$

A glance verifies that y is indeed small relative to 1.00 and 0.500. Then

$$\text{pH} = -\log_{10}(3.54 \times 10^{-4}) = 3.45$$

Related Problems: 43, 44

To understand how buffer solutions work, let's write the equilibrium expression for the ionization of a weak acid HA in the form

$$[H_3O^+] = K_a \frac{[HA]}{[A^-]}$$

The concentration of hydronium ion depends on the ratio of the concentration of the weak acid to the concentration of its conjugate base. The key to effective buffer action is to keep these concentrations nearly equal and fairly large. Adding a small amount of base to an effective buffer takes away only a few percent of the HA molecules by converting them into A^- ions and adds only a few percent to the amount of A^- originally present. The ratio $[HA]/[A^-]$ decreases, but only very slightly. Added acid consumes a small fraction of the base A^- to generate a bit more HA. The ratio $[HA]/[A^-]$ now increases, but again the change is only slight. Because the concentration of H_3O^+ is tied directly to this ratio, it changes only slightly. The following example illustrates buffer action quantitatively.

EXAMPLE 15.8

Suppose 0.10 mol of a strong acid such as HCl is added to the solution in Example 15.7. Calculate the pH of the resulting solution.

SOLUTION

The strong acid HCl ionizes essentially completely in dilute aqueous solution. Initially assume that *all* the hydrogen ions from HCl are taken up by formate ions to produce formic acid, some of which ionizes back to formate ion and H_3O^+. This is simply a way of describing one possible route by which equilibrium is approached and does not define the sequence of reactions that actually occurs. The position of the final equilibrium does not depend on the route by which it is attained.

Because 0.10 mol of HCl reacts with an equal number of moles of $HCOO^-$, the concentrations of $HCOO^-$ and HCOOH *before* ionization are

$$[HCOO^-]_0 = 0.50 - 0.10 = 0.40 \text{ M}$$

$$[HCOOH]_0 = 1.00 + 0.10 = 1.10 \text{ M}$$

The table to calculate the concentrations at equilibrium is

	$HCOOH(aq)$ +	$H_2O(\ell) \rightleftarrows$	$H_3O^+(aq)$ +	$HCOO^-(aq)$
Initial concentration (M)	1.10		≈0	0.40
Change in concentration (M)	$-y$		$+y$	$+y$
Equilibrium concentration (M)	$1.10 - y$		y	$0.40 + y$

The equilibrium expression then becomes

$$\frac{y(0.40 + y)}{1.10 - y} = 1.77 \times 10^{-4}$$

Because y is again likely to be small relative to both 0.40 and 1.10,

$$y \approx \left(\frac{1.10}{0.40}\right)(1.77 \times 10^{-4}) = 4.9 \times 10^{-4}$$

$$\text{pH} = 3.31$$

Even though 0.10 mol of a strong acid was added, the pH changed only slightly, from 3.45 to 3.31. In contrast, the same amount of acid added to a liter of pure water would change the pH from 7 to 1.

Related Problems: 45, 46

Note that we solved this problem by first performing a stoichiometric (limiting reactant) calculation and then an equilibrium calculation. A similar strategy works if a strong base such as OH^- is added instead of a strong acid. The base reacts with formic acid to produce formate ions. Adding 0.10 mol of OH^- to the $HCOOH/HCOO^-$ buffer of Example 15.7 increases the pH only to 3.58. In the absence of the buffer system, the same base would raise the pH to 13.00.

In any buffer there is competition between the tendency of the acid to donate hydrogen ions to water (increasing the acidity) and the tendency of the base to accept hydrogen ions from water (increasing the basicity). The resulting pH depends on the magnitude of K_a. If K_a is large relative to 10^{-7}, the acid ionization will win out and acidity will increase, as in the $HCOOH/HCOO^-$ buffer. A basic buffer (with pH > 7) can be prepared by working with an acid–base pair with K_a smaller than 10^{-7}. In this case the net reaction produces OH^-, and we must use K_b to determine the equilibrium state. A typical example is the NH_4^+/NH_3 buffer prepared by mixing ammonium chloride with ammonia.

EXAMPLE 15.9

Calculate the pH of a solution made by adding 0.100 mol of NH_4Cl and 0.200 mol of NH_3 to water and diluting to 1.000 L. K_a for NH_4^+ is 5.6×10^{-10}.

SOLUTION

Because $K_a \ll 10^{-7}$ for NH_4^+ (equivalently, $K_b \gg 10^{-7}$ for NH_3), the net reaction is production of OH^- ions. Therefore, the equilibrium is written to show the net transfer of hydrogen ions from water to NH_3:

	$H_2O(\ell) + NH_3(aq)$	$\rightleftarrows NH_4^+(aq)$	$+ OH^-(aq)$
Initial concentration (M)	0.200	0.100	0
Change in concentration (M)	$-y$	$+y$	$+y$
Equilibrium concentration (M)	$0.200 - y$	$0.100 + y$	y

The equilibrium expression is

$$\frac{y(0.100 + y)}{0.200 - y} = K_b = \frac{K_w}{K_a} = 1.8 \times 10^{-5}$$

If y is small relative to the original concentrations of both NH_3 and NH_4^+, then

$$\frac{y(0.100)}{0.200} \approx 1.8 \times 10^{-5}$$

$$y \approx \left(\frac{0.200}{0.100}\right)(1.8 \times 10^{-5}) = 3.6 \times 10^{-5} \ll 0.100, 0.200$$

$$[H_3O^+] = \frac{1.0 \times 10^{-14}}{[OH^-]} = \frac{1.0 \times 10^{-14}}{3.6 \times 10^{-5}} = 2.8 \times 10^{-10} \text{ M}$$

$$\text{pH} = -\log_{10}(2.8 \times 10^{-10}) = 9.55$$

Designing Buffers

Control of pH is vital in synthetic and analytical chemistry, just as it is in living organisms. Procedures that work well at a pH of 5 may fail when the concentration of hydronium ion in the solution is raised tenfold to make the pH 4. Fortunately, it is possible to prepare buffer solutions that maintain the pH close to any desired value by the proper choice of a weak acid and the ratio of its concentration to that of its conjugate base. Let's see how to choose the best conjugate acid–base system and how to calculate the required acid–base ratio.

In Examples 15.7, 15.8, and 15.9, the equilibrium concentrations of acid and base in the buffer systems were close to the initial concentrations. When this is the case, the calculation of pH is simplified greatly, because for either an acidic *or* a basic buffer,

$$K_a = \frac{[H_3O^+][A^-]}{[HA]} \approx \frac{[H_3O^+][A^-]_0}{[HA]_0}$$

$$[H_3O^+] \approx \frac{[HA]_0}{[A^-]_0} K_a$$

and the pH is given by

$$\mathrm{pH} \approx \mathrm{p}K_a - \log_{10} \frac{[HA]_0}{[A^-]_0} \qquad [15.7]$$

We obtained Equation 15.7 by taking the logarithm and changing sign. It is easy to verify that this simple equation gives the correct result for the three preceding examples. However, it must be used with some care because it is only approximate. It is valid only when *both* $[H_3O^+]$ and $[OH^-]$ are small relative to $[HA]_0$ and $[A^-]_0$; this means the extent of ionization must be small.

We can use this expression relating pH to $\mathrm{p}K_a$ to design buffers with a specific value of pH. In an optimal buffer the acid and its conjugate base are purposely very nearly equal in concentration; if the difference in concentrations is too great, the buffer is less resistant to the effects of adding acid or base. To select a buffer system, we choose an acid with a $\mathrm{p}K_a$ as close as possible to the desired pH. Then we adjust the concentrations of acid and conjugate base to give exactly the desired pH.

EXAMPLE 15.10

Design a buffer system with pH 4.60.

SOLUTION

From Table 15.2, the $\mathrm{p}K_a$ for acetic acid is 4.75, so the CH_3COOH/CH_3COO^- buffer is a suitable one. The concentrations required to give the desired pH are related by

$$\mathrm{pH} = 4.60 = \mathrm{p}K_a - \log_{10} \frac{[CH_3COOH]_0}{[CH_3COO^-]_0}$$

$$\log_{10} \frac{[CH_3COOH]_0}{[CH_3CHOO^-]_0} = \mathrm{p}K_a - \mathrm{pH} = 4.75 - 4.60 = 0.15$$

$$\frac{[CH_3COOH]_0}{[CH_3COO^-]_0} = 10^{0.15} = 1.4$$

We can establish this ratio by dissolving 0.100 mol of sodium acetate and 0.140 mol of acetic acid in water and diluting to 1.00 L, or 0.200 mol of $NaCH_3COO$ and 0.280 mol of CH_3COOH in the same volume, and so on. As long as the ratio of the concentrations is 1.4, the solution will be buffered at approximately pH 4.60.

Related Problems: 47, 48, 49, 50

The preceding example shows that the *absolute* concentrations of acid and conjugate base in a buffer are much less important than is their ratio in determining the pH. Nonetheless, the absolute concentrations do affect the capacity of the solution to resist changes in pH when acid or base is added. The higher the concentrations of buffering species, the smaller the change in pH when a fixed amount of a

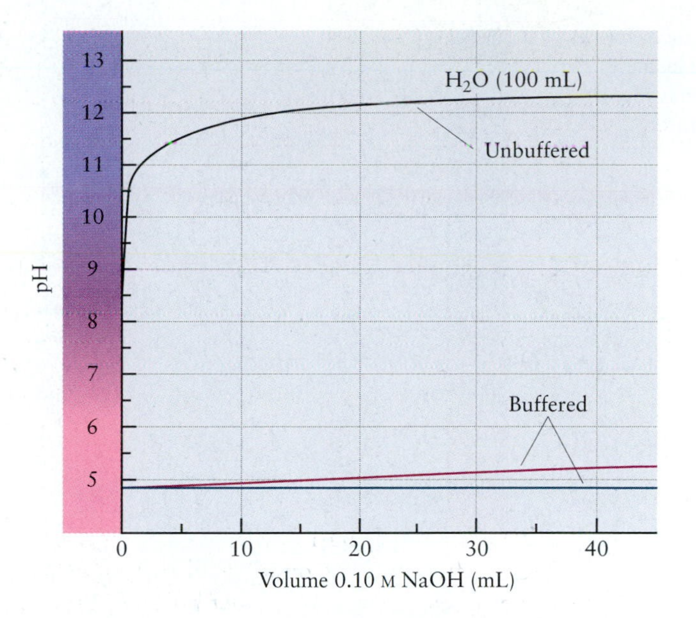

FIGURE 15.12 Addition of a given volume of base to buffered and unbuffered solutions causes a much greater change in the pH of the unbuffered solution. Of the two buffered solutions, the one with higher buffer concentration resists pH changes more effectively. The red line represents 100 mL of a buffer that is 0.1 M in both CH_3COOH and CH_3COO^-; the blue line represents the same volume of a buffer that is 1.0 M in both components.

strong acid or base is added. In Example 15.7 a change in buffer concentrations from 1.00 M and 0.500 M to 0.500 M and 0.250 M does not alter the original pH of 3.45 because the ratio of acid to base concentrations is unchanged. The pH after 0.100 mol of HCl is added does change to the value 3.15 rather than 3.31. The buffer at lower concentration is less resistant to pH change (Fig. 15.12). The buffering capacity of *any* buffer solution is exhausted if enough strong acid (or strong base) is added to use up the original amount of weak base (or weak acid) through chemical reaction.

15.6 Acid–Base Titration Curves

Section 11.3 describes an acid–base titration as the addition of carefully metered volumes of a basic solution of known concentration to an acidic solution of unknown concentration—or the addition of acid to base—to reach an *endpoint* at which the unknown solution has been completely neutralized. The endpoint is signaled by the color change of an indicator or by a sudden rise or fall in pH. The pH of the reaction mixture changes continuously over the course of an acid–base titration, but changes abruptly only near the end point. A graph of the pH versus the volume V of titrating solution added is a **titration curve.** Its shape depends on the value of K_a and on the concentrations of the acid and base reacting. The concepts and tools of acid–base equilibria predict the exact shapes of titration curves when these quantities are all known. The same concepts allow K_a and the concentration of the unknown solution to be calculated from an experimental titration curve. Here we examine three categories of titration and determine the titration curve for each: strong acid reacting with strong base, weak acid reacting with strong base, and strong acid reacting with weak base. Titrations of a weak acid with a weak base (and the reverse) are not useful for analytical purposes.

Titration of a Strong Acid with a Strong Base

The addition of a strong base to a strong acid (or the reverse) is the simplest type of titration. The chemical reaction is the neutralization:

$$H_3O^+(aq) + OH^-(aq) \longrightarrow 2\ H_2O(\ell)$$

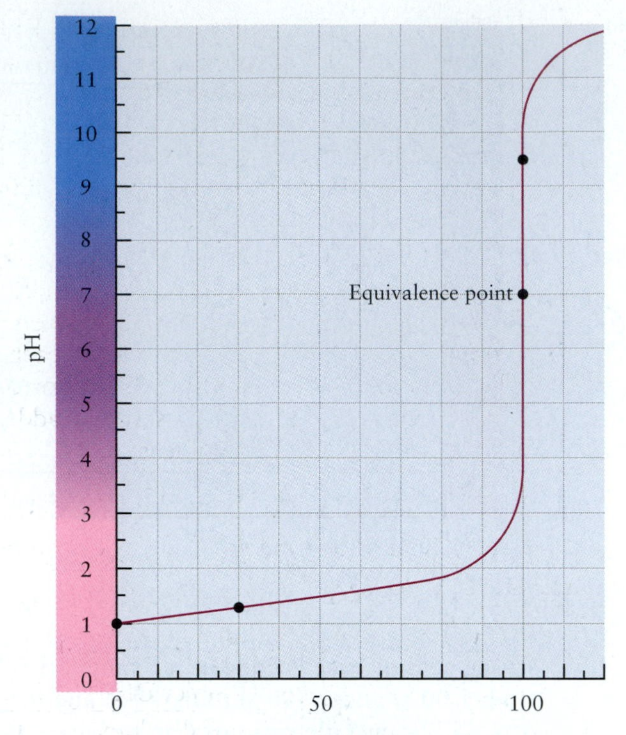

FIGURE 15.13 A titration curve for the titration of a strong acid by a strong base. The curve shown is for 100.0 mL of 0.1000 M HCl titrated with 0.1000 M NaOH.

Suppose a solution of 100.0 mL (0.1000 L) of 0.1000 M HCl is titrated with 0.1000 M NaOH. What does the titration curve look like? The curve can be measured experimentally, and the result is shown in Figure 15.13. We can also construct the curve theoretically by calculating the pH of the reaction mixture at many different points during the addition of the NaOH solution and plotting the results. The following example illustrates the theoretical procedure.

1. **V = 0 mL NaOH added**
   Initially, $[H_3O^+] = 0.1000$ M, so the pH is 1.000. The number of moles of H_3O^+ present initially is

$$n_{H_3O^+} = [H_3O^+](\text{volume}) = (0.1000 \text{ mol L}^{-1})(0.1000 \text{ L})$$
$$= 1.000 \times 10^{-2} \text{ mol}$$

2. **V = 30.00 mL NaOH added**
   30.00 mL of 0.1000 M NaOH solution contains

$$(0.1000 \text{ mol L}^{-1})(0.03000 \text{ L}) = 3.000 \times 10^{-3} \text{ mol OH}^-$$

   This reacts with (and neutralizes) an equal number of moles of the H_3O^+ ion present initially and reduces $n_{H_3O^+}$ to

$$n_{H_3O^+} = (1.000 \times 10^{-2} - 3.000 \times 10^{-3})\text{mol} = 7.00 \times 10^{-3} \text{ mol}$$

   In addition—and this is very important to remember—the volume of the titration mixture has increased from 100.0 to 130.0 mL (that is, from 0.1000 to 0.1300 L). The concentration of H_3O^+ at this point in the titration is

$$[H_3O^+] = \frac{n_{H_3O^+}}{V_{tot}} = \frac{7.00 \times 10^{-3} \text{ mol}}{0.1300 \text{ L}} = 0.0538 \text{ M}$$

$$\text{pH} = 1.27$$

3. **V = 100.00 mL NaOH added**
   This is called the **equivalence point,** that point in the titration at which the number of moles of base added equals the number of moles of acid originally present. The volume of base added up to the equivalence point is the **equivalent**

volume, V_e. At the equivalence point in the titration of a strong acid with a strong base, the concentrations of OH^- and H_3O^+ must be equal and the pH 7.0 due to the autoionization of water. At this point the solution is simply a nonhydrolyzing salt (in this case, NaCl) in water. The pH is 7 at the equivalence point only in the titration of a strong acid with a strong base (or vice versa). The pH at the equivalence point differs from 7 if the titration involves a weak acid or weak base.

4. **V = 100.05 mL NaOH added**
Beyond the equivalence point, OH^- is added to a neutral unbuffered solution. The OH^- concentration can be found from the number of moles of OH^- added after the equivalence point has been reached. The volume beyond the equivalence point at this stage is 0.05 mL, or 5×10^{-5} L. (This is the volume of approximately one drop of solution added from the buret.) The number of moles of OH^- in this volume is

$$(0.1000 \text{ mol L}^{-1})(5 \times 10^{-5} \text{ L}) = 5 \times 10^{-6} \text{ mol}$$

Meanwhile, the total volume of the titration mixture is

$$0.1000 \text{ L HCl solution} + 0.10005 \text{ L NaOH solution} = 0.20005 \text{ L solution}$$

Therefore, the concentration of OH^- is

$$[OH^-] = \frac{\text{moles } OH^-}{\text{total volume}} = \frac{5 \times 10^{-5} \text{ mol}}{0.2005 \text{ L}} = 2.5 \times 10^{-5} \text{ M}$$

$$[H_3O^+] = 4 \times 10^{-10} \text{ M}; \quad pH = 9.4$$

As the titration curve shows (see Fig. 15.13), the pH increases dramatically in the immediate vicinity of the equivalence point: $[H_3O^+]$ changes by four orders of magnitude between 99.98 mL and 100.02 mL NaOH! Any indicator whose color changes between pH = 5.0 and pH = 9.0 therefore signals the endpoint of the titration to an accuracy of ±0.02 mL in 100.0 mL, or ±0.02%. The titration **endpoint,** the experimentally measured volume at which the indicator changes color, is then almost identical to the equivalence point, the theoretical volume at which the chemical amount of added base equals that of acid present originally.

The titration of a strong base by a strong acid is entirely parallel. In this case the pH starts at a higher value and *drops* through a pH of 7 at the equivalence point. The roles of acid and base, and of H_3O^+ and OH^-, are reversed in the equations already given.

Titration of a Weak Acid with a Strong Base

We turn now to the titration of a weak acid with a strong base (the titration of a weak base with a strong acid is analogous). The equivalence point has the same meaning as for a strong acid titration. At the equivalence point, the number of moles of base added (in volume V_e) is equal to the number of moles of acid originally present (in volume V_0). So, once again

$$c_0 V_0 = c_t V_e$$

where c_0 is the original weak acid concentration and c_t is the OH^- concentration in the titrating solution.

The calculation of the titration curve differs from the strong acid–strong base case in that now equilibrium (reflected in the K_a of the weak acid) enters the picture. As an example, consider the titration of 100.0 mL of a 0.1000 M solution of acetic acid (CH_3COOH) with 0.1000 M NaOH. For this titration,

$$V_e = \frac{c_0}{c_t} V_0 = \left(\frac{0.1000 \text{ M}}{0.1000 \text{ M}}\right)(100.0 \text{ mL})$$
$$= 100.0 \text{ mL}$$

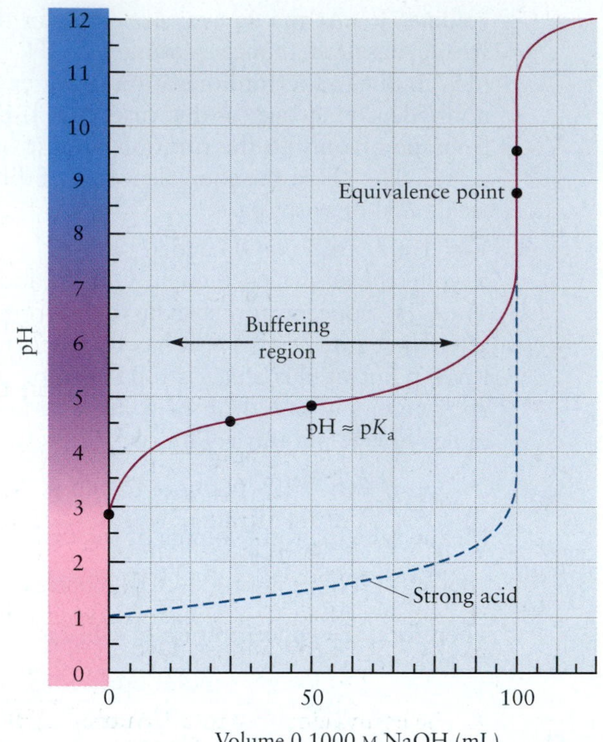

FIGURE 15.14 A titration curve for the titration of a weak acid by a strong base. The curve shown in red is for 100.0 mL of 0.1000 M CH_3COOH titrated with 0.1000 M NaOH. For comparison, the dashed blue line shows the titration curve for a strong acid of the same amount and concentration as presented in Figure 15.13.

There are four distinct ranges in the titration; each corresponds to a type of calculation that we have illustrated already. Figure 15.14 shows pH versus volume for this titration, and the following paragraphs outline the calculation of four typical points on the curve.

1. **$V = 0$ mL NaOH added**
   At the beginning of the titration—before any NaOH is added—the problem is simply the ionization of a weak acid, considered in Section 15.4. A calculation analogous to those of Examples 15.3 and 15.4 gives a pH of 2.88.
2. **$0 < V < V_e$**
   In this range the acid has been partially neutralized by added NaOH solution. Because OH^- is a stronger base than acetate ion, it reacts almost completely with the acid originally present:

$$CH_3COOH(aq) + OH^-(aq) \rightleftharpoons H_2O(\ell) + CH_3COO^-(aq)$$

$$K = \frac{1}{K_b} = \frac{K_a}{K_w} = 2 \times 10^9 \gg 1$$

As a specific example, suppose that 30.00 mL of 0.1000 M NaOH has been added. The 30.00 mL NaOH contains

$$(0.1000 \text{ mol L}^{-1})(0.03000 \text{ L}) = 3.000 \times 10^{-3} \text{ mol OH}^-$$

and the original solution contained

$$(0.1000 \text{ mol L}^{-1})(0.1000 \text{ L}) = 1.000 \times 10^{-2} \text{ mol CH}_3\text{COOH}$$

The neutralization reaction generates one CH_3COO^- ion for every OH^- ion added. Hence, 3.000×10^{-3} mol of CH_3COO^- ions is generated. The amount of acetic acid that remains unreacted is

$$1.000 \times 10^{-2} - 3.000 \times 10^{-3} = 7.00 \times 10^{-3} \text{ mol}$$

Because the total volume is now 130.0 mL (or 0.1300 L), the nominal concentrations after reaction are

$$[CH_3COOH] \approx \frac{7.00 \times 10^{-3} \text{ mol}}{0.1300 \text{ L}} = 5.38 \times 10^{-2} \text{ M}$$

$$[CH_3COO^-] \approx \frac{3.00 \times 10^{-3} \text{ mol}}{0.1300 \text{ L}} = 2.31 \times 10^{-2} \text{ M}$$

This is merely a buffer solution containing acetic acid at a concentration of 5.38×10^{-2} M and sodium acetate at a concentration of 2.31×10^{-2} M. Because the K_a for acetic acid is larger than 10^{-7}, hydronium ion (not hydroxide ion) predominates, and this is an acidic buffer. The pH can be found from the procedure used in Example 15.7 or, more approximately from Equation 15.7

$$\text{pH} \approx \text{p}K_a - \log_{10} \frac{[CH_3COOH]_0}{[CH_3COO^-]_0} = 4.75 - \log_{10} \frac{5.38 \times 10^{-2}}{2.31 \times 10^{-2}} = 4.38$$

This region of the titration shows clearly the buffering action of a mixture of a weak acid with its conjugate base. At the half-equivalence point $V = V_e/2$, $[CH_3COOH]_0 = [CH_3COO^-]_0$, which corresponds to an equimolar buffer; at this point pH $\approx$ pK_a. On either side of this point the pH rises relatively slowly as the NaOH solution is added.

3. $V = V_e$

   At the equivalence point, $c_t V_e$ mol of OH^- has been added to the same number of moles of CH_3COOH. An identical solution could have been prepared simply by adding $c_t V_e = 1.000 \times 10^{-2}$ mol of the base CH_3COO^- (in the form of $NaCH_3COO$) to 0.2000 L of water. The pH at the equivalence point corresponds to the hydrolysis of CH_3COO^-:

$$H_2O(\ell) + CH_3COO^-(aq) \rightleftharpoons CH_3COOH(aq) + OH^-(aq)$$

   as considered in Example 15.6. At the equivalence point the pH is 8.73.

   Note that in the titration of a weak acid by a strong base the equivalence point comes not at pH 7 but at a higher (more basic) value. By the same token, the equivalence point in the titration of a weak base by a strong acid occurs at a pH lower than 7.

4. $V > V_e$

   Beyond the equivalence point, $OH^-(aq)$ is added to a solution of the base CH_3COO^-. The $[OH^-]$ comes almost entirely from the hydroxide ion added beyond the equivalence point; very little comes from the reaction of the CH_3COO^- with water. Beyond V_e, the pH for the titration of a weak acid by a strong base is very close to that for a strong acid by a strong base.

The equivalent volume V_e is readily determined in the laboratory by using an indicator that changes color near pH 8.7, the pH at the equivalence point of the acetic acid titration. A suitable choice would be phenolphthalein, which changes from colorless to red over a pH range from 8.2 to 10.0. The slope of pH versus volume of strong base is less steep near the equivalence point for a weak acid than it is for a strong acid, so determination of the equivalent volume—and of the unknown weak acid concentration—is somewhat less accurate.

The following example shows how to determine concentrations and ionization constants for unknown acids and bases from titrations.

EXAMPLE 15.11

A volume of 50.00 mL of a weak acid of unknown concentration is titrated with a 0.1000 M solution of NaOH. The equivalence point is reached after 39.30 mL of NaOH solution has been added. At the half-equivalence point (19.65 mL) the pH is 4.85. Calculate the original concentration of the acid and its ionization constant K_a.

SOLUTION

The number of moles of acid originally present, $c_0 V_0$, is equal to the number of moles of base added at the equivalence point, $c_t V_e$, so

$$c_0 = \frac{V_e}{V_0} c_t = \frac{39.50 \text{ mL}}{50.00 \text{ mL}} \times 0.100 \text{ M} = .0786 \text{ M}$$

This is the original concentration of acid.

At the half-equivalence point,

$$\text{pH} = 4.85 \approx \text{p}K_a$$

$$K_a = 10^{-4.85} = 1.4 \times 10^{-5}$$

The unknown acid could be propionic acid, CH_3CH_2COOH (see Table 15.2).

Related Problems: 61, 62

15.7 Polyprotic Acids

So far, we have considered only **monoprotic acids,** whose molecules can donate only a single hydrogen ion to acceptor molecules. **Polyprotic acids** can donate two or more hydrogen ions to acceptors. Sulfuric acid is a familiar and important example. It reacts in two stages; first

$$H_2SO_4(aq) + H_2O(\ell) \longrightarrow H_3O^+(aq) + HSO_4^-(aq)$$

to give the hydrogen sulfate ion, and then

$$HSO_4^-(aq) + H_2O(\ell) \longrightarrow H_3O^+(aq) + SO_4^{2-}(aq)$$

The hydrogen sulfate ion is amphoteric, meaning that it is a base in the first reaction (with conjugate acid H_2SO_4) and an acid in the second (with conjugate base SO_4^{2-}). In its first ionization H_2SO_4 is a strong acid, but the product of that ionization (HSO_4^-) is a weak acid. So, the H_3O^+ produced in a solution of H_2SO_4 comes primarily from the first ionization, and the solution has a pH close to that of a monoprotic strong acid of the same concentration. But when this solution reacts with a strong base, its neutralizing power is twice that of a monoprotic acid of the same concentration, because each mole of sulfuric acid can react with and neutralize two moles of hydroxide ion.

Weak Polyprotic Acids

Weak polyprotic acids ionize in two or more stages. Examples are carbonic acid (H_2CO_3), formed from solvated CO_2 (carbonated water; Fig. 15.15), and phosphoric acid (H_3PO_4). Carbonic acid can give up one hydrogen ion to form HCO_3^- (hydrogen carbonate ion) or two hydrogen ions to form CO_3^{2-} (carbonate ion). Phosphoric acid ionizes in three stages, giving successively $H_2PO_4^-$, HPO_4^{2-}, and PO_4^{3-}.

Two simultaneous equilibria are involved in the ionization of a diprotic acid such as H_2CO_3:[2]

$$H_2CO_3(aq) + H_2O(\ell) \rightleftharpoons H_3O^+(aq) + HCO_3^-(aq)$$

$$\frac{[H_3O^+][HCO_3^-]}{[H_2CO_3]} = K_{a1} = 4.3 \times 10^{-7}$$

FIGURE 15.15 The indicator phenolphthalein is pink in basic solution *(left).* When dry ice (solid carbon dioxide) is placed in the bottom of the beaker *(right),* it dissolves to form carbonic acid. This neutralizes the base in solution, and causes the indicator to change to its colorless form characteristic of an acid solution.

[2]An accurate description of carbonic acid and solvated CO_2 is somewhat more complicated than indicated here. In fact, most of the dissolved CO_2 remains as $CO_2(aq)$, and only a small fraction actually reacts with water to give $H_2CO_3(aq)$. However, we will indicate by $[H_2CO_3]$ the total concentration of both of these species. Approximately 0.034 mol of CO_2 dissolves per liter of water at 25°C and atmospheric pressure.

and

$$HCO_3^-(aq) + H_2O(\ell) \rightleftharpoons H_3O^+(aq) + CO_3^{2-}(aq)$$

$$\frac{[H_3O^+][CO_3^{2-}]}{[HCO_3^-]} = K_{a2} = 4.8 \times 10^{-11}$$

We emphasize two important points at the outset:

1. The $[H_3O^+]$ in the two ionization equilibria are one and the same.
2. K_{a2} is invariably smaller than K_{a1} because the negative charge left behind by the loss of a hydrogen ion in the first ionization causes the second hydrogen ion to be more tightly bound.

Exact calculations of simultaneous equilibria can be complicated. They simplify considerably when the original acid concentration is not too small and the ionization constants K_{a1} and K_{a2} differ substantially in magnitude (by a factor of 100 or more). The latter condition is almost always satisfied. Under such conditions, the two equilibria can be treated sequentially, as in the following example.

EXAMPLE 15.12

Calculate the equilibrium concentrations of H_2CO_3, HCO_3^-, CO_3^{2-}, and H_3O^+ in a saturated aqueous solution of CO_2, in which the original concentration of H_2CO_3 is 0.034 M.

SOLUTION

The H_3O^+ arises both from the ionization of H_2CO_3 and from the subsequent ionization of HCO_3^-, but because $K_{a1} \gg K_{a2}$ it is reasonable to ignore the contribution of $[H_3O^+]$ from the second ionization (as well as from the autoionization of water). These approximations will be checked later in the calculation.

If y mol L^{-1} of H_2CO_3 ionizes, the following approximations apply:

	$H_2CO_3(aq)$ +	$H_2O(\ell) \rightleftharpoons$	$H_3O^+(aq)$ +	$HCO_3^-(aq)$
Initial concentration (M)	0.034		≈0	0
Change in concentration (M)	$-y$		$+y$	$+y$
Equilibrium concentration (M)	$0.034 - y$		y	y

where equating both $[HCO_3^-]$ and $[H_3O^+]$ to y involves the assumption that the subsequent ionization of HCO_3^- has only a small effect on its concentration. Then the first ionization equilibrium can be written as

$$\frac{[H_3O^+][HCO_3^-]}{[H_2CO_3]} = K_{a1}$$

$$\frac{y^2}{0.034 - y} = 4.3 \times 10^{-7}$$

Solving this equation for y gives

$$y = 1.2 \times 10^{-4} \text{ M} = [H_3O^+] = [HCO_3^-]$$

$$[H_2CO_3] = 0.034 - y = 0.034 \text{ M}$$

The second equilibrium can be written as

$$\frac{[H_3O^+][CO_3^{2-}]}{[HCO_3^-]} = K_{a2} = \frac{(1.2 \times 10^{-4})[CO_3^{2-}]}{1.2 \times 10^{-4}} = 4.8 \times 10^{-11}$$

$$[CO_3^{2-}] - 4.8 \times 10^{-11} \text{ M}$$

The concentration of the base produced in the second ionization, $[CO_3^{2-}]$, is numerically equal to K_{a2}.

Now we must check the validity of the assumptions. Because

$$[CO_3^{2-}] = 4.8 \times 10^{-11}\ \text{M} \ll 1.2 \times 10^{-4}\ \text{M} = [HCO_3^-]$$

we were justified in ignoring the effect of the second ionization on the concentrations of HCO_3^- and H_3O^+. The additional concentration of H_3O^+ furnished by HCO_3^- is only 4.8×10^{-11} M. Finally, $[H_3O^+]$ is much larger than 10^{-7} M, so neglecting the water autoionization was also justified.

Related Problems: 63, 64

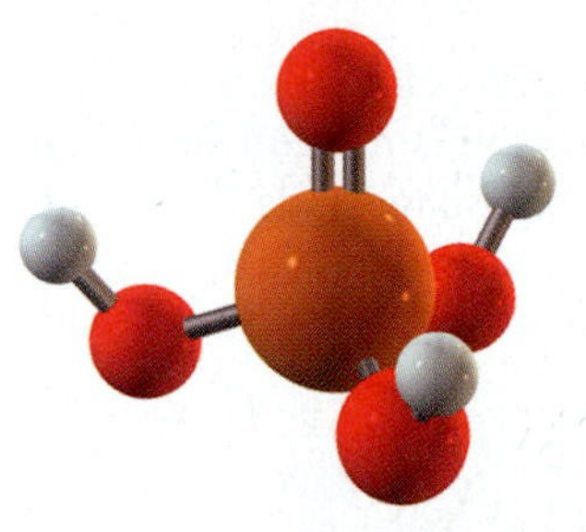

Phosphoric acid, H_3PO_4, is a triprotic acid that is used in the manufacture of phosphate fertilizers and in the food industry.

For a triprotic acid such as H_3PO_4, the concentration of the base (PO_4^{3-}) resulting from the third ionization could have been calculated in a similar manner.

An analogous procedure applies to the reactions of a base that can accept two or more hydrogen ions. In a solution of sodium carbonate (Na_2CO_3), the carbonate ion reacts with water to form first HCO_3^- and then H_2CO_3:

$$H_2O(\ell) + CO_3^{2-}(aq) \rightleftharpoons HCO_3^-(aq) + OH^-(aq)$$

$$\frac{[OH^-][HCO_3^-]}{[CO_3^{2-}]} = K_{b1} = \frac{K_w}{K_{a2}} = 2.1 \times 10^{-4}$$

$$H_2O(\ell) + HCO_3^-(aq) \rightleftharpoons H_2CO_3(aq) + OH^-(aq)$$

$$\frac{[OH^-][H_2CO_3]}{[HCO_3^-]} = K_{b2} = \frac{K_w}{K_{a1}} = 2.3 \times 10^{-8}$$

In this case $K_{b1} \gg K_{b2}$, so essentially all the OH^- arises from the first reaction.

The ensuing calculation of the concentrations of species present is just like that in Example 15.12.

Effect of pH on Solution Composition

Changing the pH of a solution shifts the positions of all acid–base equilibria, including those involving polyprotic acids. Acid–base equilibrium expressions and equilibrium constants are used to calculate the amount of the change. For example, the two equilibria that apply to solutions containing H_2CO_3, HCO_3^-, and CO_3^{2-} can be written as

$$\frac{[HCO_3^-]}{[H_2CO_3]} = \frac{K_{a1}}{[H_3O^+]} \qquad \frac{[CO_3^{2-}]}{[HCO_3^-]} = \frac{K_{a2}}{[H_3O^+]}$$

At a given pH, the right sides are known and the relative amounts of the three carbonate species can be calculated. This is illustrated by the following example.

EXAMPLE 15.13

Calculate the fraction of carbonate present as H_2CO_3, HCO_3^-, and CO_3^{2-} at pH 10.00.

SOLUTION

At this pH, $[H_3O^+] = 1.0 \times 10^{-10}$ M, and the values of K_{a1} and K_{a2} for the preceding equations are used to find

$$\frac{[HCO_3^-]}{[H_2CO_3]} = \frac{4.3 \times 10^{-7}}{1.0 \times 10^{-10}} = 4.3 \times 10^3$$

$$\frac{[CO_3^{2-}]}{[HCO_3^-]} = \frac{4.8 \times 10^{-11}}{1.0 \times 10^{-10}} = 0.48$$

It is most convenient to rewrite these ratios with the same species (say, HCO_3^-) in the denominator. The first equation becomes

$$\frac{[H_2CO_3]}{[HCO_3^-]} = \frac{1}{4.3 \times 10^3} = 2.3 \times 10^{-4}$$

The fraction of each species present is obtained by dividing the concentration of each species by the sum of the three concentrations. For H_2CO_3 this gives

$$\text{fraction } H_2CO_3 = \frac{[H_2CO_3]}{[H_2CO_3] + [HCO_3^-] + [CO_3^{2-}]}$$

This can be simplified by dividing numerator and denominator by $[HCO_3^-]$ and substituting the ratios that have already been calculated:

$$\text{fraction } H_2CO_3 = \frac{[H_2CO_3]/[HCO_3^-]}{([H_2CO_3]/[HCO_3^-]) + 1 + ([CO_3^{2-}]/[HCO_3^-])}$$

$$= \frac{2.3 \times 10^{-4}}{2.3 \times 10^{-4} + 1 + 0.48} = 1.6 \times 10^{-4}$$

In the same way the fractions of HCO_3^- and CO_3^{2-} present are calculated to be 0.68 and 0.32, respectively.

Related Problems: 67, 68

Repeating the calculation of Example 15.13 at a series of different pH values gives the graph shown in Figure 15.16. At high pH, CO_3^{2-} predominates, and at low pH, H_2CO_3 is the major species. At intermediate pH (near pH 8, the approximate pH of seawater), the hydrogen carbonate ion HCO_3^- is most prevalent. The variation in composition of sedimentary rocks from different locations can be traced back to the effect of pH on solution composition. Sediments containing carbonates were formed from alkaline (high-pH) lakes and oceans in which CO_2 was present mainly as CO_3^{2-} ion. Sediments deposited from waters with intermediate pH are hydrogen carbonates or mixtures between carbonates and hydrogen carbonates. An example of the latter is trona ($2Na_2CO_3 \cdot NaHCO_3 \cdot 2H_2O$), an ore from the western United States that is an important source of both carbonates of sodium. Acidic waters did not deposit carbonates, but instead released $CO_2(g)$ to the atmosphere.

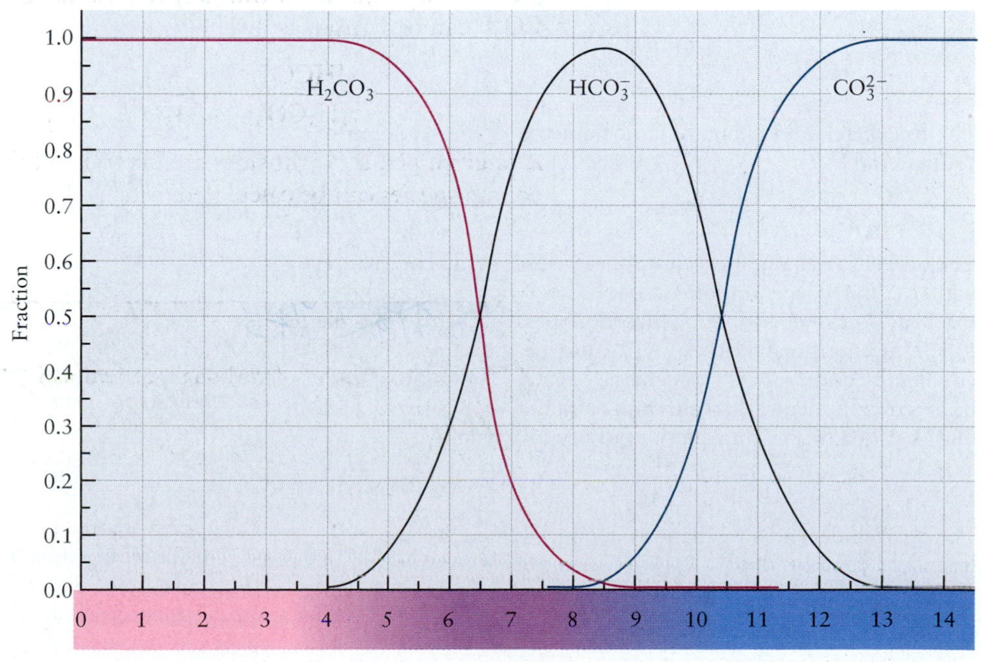

FIGURE 15.16 The equilibrium fractions of H_2CO_3, HCO_3^-, and CO_3^{2-} that are present in aqueous solution at different values of the pH.

A DEEPER LOOK

15.8 Exact Treatment of Acid–Base Equilibria

Sections 15.4 and 15.5 outline methods for calculating equilibria involving weak acids, bases, and buffer solutions. There we assume that the amount of hydronium ion (or hydroxide ion) resulting from the ionization of water can be neglected in comparison with that produced by the ionization of dissolved acids or bases. In this section, we replace that approximation by a treatment of acid–base equilibria that is exact, within the limits of the mass-action law. This approach leads to somewhat more complicated equations, but it serves several purposes. It has great practical importance in cases in which the previous approximations no longer hold, such as very weak acids or bases or very dilute solutions. It includes as special cases the various aspects of acid–base equilibrium considered earlier. Finally, it provides a foundation for treating amphoteric equilibrium later in this section.

Consider a general case in which the initial concentration of a weak acid HA is called c_a, and the initial concentration of its conjugate base (which for simplicity is assumed to come from the salt NaA) is c_b. In solution there will be five dissolved species: HA, A^-, Na^+, H_3O^+, and OH^-. It is necessary to write down and solve five independent equations that relate the equilibrium concentrations of these species to the initial concentrations c_a and c_b and to K_a, the acid ionization constant of HA. The first equation is simply

$$[Na^+] = c_b \qquad \text{(a)}$$

reflecting the fact that the $Na^+(aq)$ from the dissolved salt is a spectator ion that does not take part in the acid–base equilibrium. Next there are two equilibrium relations:

$$[H_3O^+][OH^-] = K_w \qquad \text{(b)}$$

$$\frac{[H_3O^+][A^-]}{[HA]} = K_a \qquad \text{(c)}$$

The fourth relation is one of stoichiometry, or conservation of "A-material":

$$c_a + c_b = [HA] + [A^-] \qquad \text{(d)}$$

The original A-material was introduced either as acid or as base, with a total concentration of $c_a + c_b$. When equilibrium is reached, some redistribution has doubtless occurred, but the *total* concentration, $[HA] + [A^-]$, must be the same. The fifth and final relation results from charge balance. The solution must be electrically neutral, so the total amount of positive charge must be equal to the total amount of negative charge:

$$[Na^+] + [H_3O^+] = [A^-] + [OH^-] \qquad \text{(e)}$$

These five independent equations completely determine the five unknown concentrations. To solve them, begin by substituting (a) into (e) and solving for $[A^-]$:

$$[A^-] = c_b + [H_3O^+] - [OH^-] \qquad \text{(e')}$$

Next, insert (e′) into (d) and solve for [HA]:

$$[HA] = c_a + c_b - [A^-] = c_a - [H_3O^+] + [OH^-] \qquad \text{(d')}$$

Next, substitute both (d′) and (e′) into (c) to find

$$\frac{[H_3O^+](c_b + [H_3O^+] - [OH^-])}{(c_a - [H_3O^+] + [OH^-])} = K_a \qquad \text{(c')}$$

There are two ways to proceed with the general equation (c′). First, we can set up the exact solution for $[H_3O^+]$, a procedure that is useful for very weak acids or bases or for very dilute solutions. Or, we can reduce (c′) in various limits to cases already considered.

For the exact solution, eliminate $[OH^-]$ in (c′) by using (b). This gives

$$\frac{[H_3O^+]\left(c_b + [H_3O^+] - \dfrac{K_w}{[H_3O^+]}\right)}{\left(c_a - [H_3O^+] + \dfrac{K_w}{[H_3O^+]}\right)} = K_a$$

The numerator and denominator are multiplied by $[H_3O^+]$ and the fraction is cleared by moving the denominator to the right side:

$$[H_3O^+](c_b[H_3O^+] + [H_3O^+]^2 - K_w) = K_a(c_a[H_3O^+] - [H_3O^+]^2 + K_w)$$

This can be rewritten as

$$[H_3O^+]^3 + (c_b + K_a)[H_3O^+]^2 - (K_w + c_aK_a)[H_3O^+] - K_aK_w = 0$$

This is a cubic equation for $[H_3O^+]$, which can be solved with a calculator. The concentrations of OH^-, A^-, and HA can then be found by successive substitutions into (b), (e′), and (d′).

Alternatively, the general equation (c′) can be examined in various limits. In an acidic buffer (as in Examples 15.7 and 15.8), if it can be assumed that $[H_3O^+] \gg [OH^-]$, then (c′) simplifies to

$$\frac{[H_3O^+](c_b + [H_3O^+])}{(c_a - [H_3O^+])} = K_a$$

which is exactly the equation used in those examples. If, in addition, $c_b = 0$, the weak-acid ionization limit of Examples 15.3 and 15.4 is reached. In a basic buffer, if it can be assumed that $[OH^-] \gg [H_3O^+]$, then (c′) simplifies to

$$\frac{(K_w/[OH^-])(c_b - [OH^-])}{(c_a + [OH^-])} = K_a$$

Here (b) was used to substitute for $[H_3O^+]$ where it multiplies the whole expression. This can be rewritten as

$$\frac{[OH^-](c_a + [OH^-])}{(c_b - [OH^-])} = \frac{K_w}{K_a} = K_b$$

which is exactly the equation used in Example 15.9. If no acid is present initially ($c_a = 0$), this expression reduces to the weak-base ionization limits of Examples 15.5 and 15.6. The general approach includes all of the previous calculations as special cases.

Unless conditions require the use of the exact solution, approximate equations are preferable because they are easier to apply and provide greater physical insight. If a calculation (ignoring water autoionization) of the ionization of a weak acid gives a concentration of H_3O^+ smaller than 10^{-6} M or if a calculation of base ionization gives a concentration of OH^- smaller than 10^{-6} M, then we have to use the more exact treatment. For buffer solutions, a pH near 7 does not necessarily mean that water ionization is important, unless the acid or base concentration becomes very small.

EXAMPLE 15.14

Calculate the pH of a 1.00×10^{-5} M solution of HCN(*aq*). The K_a of HCN(*aq*) is 6.17×10^{-10}.

SOLUTION

Suppose the autoionization of water is ignored and the method of Examples 15.3 and 15.4 is used. This gives $[H_3O^+] = 7.9 \times 10^{-8}$ M, which of course makes no sense, because it is *lower* than the concentration of hydronium ion in pure water. HCN is a very weak acid, but it is nonetheless an acid, not a base.

So, we have to use the exact cubic equation for $[H_3O^+]$, inserting into it the proper coefficients and taking $c_a = 1.00 \times 10^{-5}$ and $c_b = 0$. This gives

$$[H_3O^+]^3 + 6.17 \times 10^{-10}[H_3O^+]^2 - 1.617 \times 10^{-14}[H_3O^+] - 6.17 \times 10^{-24} = 0$$

Unfortunately, there is no method as simple as the quadratic formula to solve a cubic equation. The easiest way to solve this equation is to try a series of values for $[H_3O^+]$ on the left side, varying them to obtain a result as close as possible to 0 (see Appendix C). It is safe to assume that the final answer will be slightly larger than 1×10^{-7}, so the initial guesses should be of that magnitude. Carrying out the procedure gives

$$[H_3O^+] = 1.27 \times 10^{-7} \text{ M} \qquad \text{pH} = 6.90$$

Related Problems: 69, 70

Amphoteric Equilibria

A second situation in which an exact analysis of acid–base equilibrium is useful occurs when an amphoteric species is dissolved in water. The hydrogen carbonate ion (HCO_3^-) is amphoteric because it can act as an acid in the equilibrium

$$HCO_3^-(aq) + H_2O(\ell) \rightleftharpoons H_3O^+(aq) + CO_3^{2-}(aq)$$

$$\frac{[H_3O^+][CO_3^{2-}]}{[HCO_3^-]} = K_{a2} = 4.8 \times 10^{-11}$$

or as a base in the equilibrium

$$H_2O(\ell) + HCO_3^-(aq) \rightleftharpoons H_2CO_3(aq) + OH^-(aq)$$

$$\frac{[OH^-][H_2CO_3]}{[HCO_3^-]} = K_{b2} = 2.3 \times 10^{-8}$$

If sodium hydrogen carbonate ($NaHCO_3$) is dissolved in water, there is a competition between the tendency of HCO_3^- to accept hydrogen ions and to donate them. Because $K_{b2} > K_{a2}$, there should be more production of OH^- than of H_3O^+, so the solution should be basic.

In an exact treatment of this equilibrium, there are six unknown concentrations—those of Na^+, H_2CO_3, HCO_3^-, CO_3^{2-}, OH^-, and H_3O^+. Two equilibrium equations were already presented, and a third relates $[OH^-]$ and $[H_3O^+]$ to K_w. If $[HCO_3^-]_0$ is the original concentration of $NaHCO_3$, then from stoichiometry

$$[HCO_3^-]_0 = [HCO_3^-] + [H_2CO_3] + [CO_3^{2-}]$$

because the total amount of carbonate material is conserved. Any reduction in $[HCO_3^-]$ must be compensated by a corresponding increase in either $[H_2CO_3]$ or $[CO_3^{2-}]$. Next we use the principle of conservation of charge. The positively charged species present are Na^+ and H_3O^+, and the negatively charged species are HCO_3^-, CO_3^{2-}, and OH^-. Because there is overall charge neutrality,

$$[Na^+] + [H_3O^+] = [HCO_3^-] + 2\,[CO_3^{2-}] + [OH^-]$$

where the coefficient 2 for $[CO_3^{2-}]$ arises because each carbonate ion is doubly charged. In addition, the Na^+ concentration is unchanged, so

$$[Na^+] = [HCO_3^-]_0$$

In principle, these six equations can be solved simultaneously to calculate the exact $[H_3O^+]$ for an arbitrary initial concentration of HCO_3^-. The result is complex and gives little physical insight. Instead, we give only a simpler, approximate solution, which is sufficient in the cases considered here. Subtracting the carbonate balance equation from the charge balance equation gives

$$[H_3O^+] = [CO_3^{2-}] - [H_2CO_3] + [OH^-]$$

The three equilibrium expressions are used to rewrite this as

$$[H_3O^+] = K_{a2}\frac{[HCO_3^-]}{[H_3O^+]} - \frac{[H_3O^+][HCO_3^-]}{K_{a1}} + \frac{K_w}{[H_3O^+]}$$

where $[CO_3^{2-}]$ and $[H_2CO_3]$ have been eliminated in favor of $[HCO_3^-]$.

Multiplying by $K_{a1}[H_3O^+]$ gives

$$K_{a1}[H_3O^+]^2 + [HCO_3^-][H_3O^+]^2 = K_{a1}K_{a2}[HCO_3^-] + K_{a1}K_w$$

$$[H_3O^+]^2 = \frac{K_{a1}K_{a2}[HCO_3^-] + K_{a1}K_w}{K_{a1} + [HCO_3^-]}$$

This equation still contains two unknown quantities, $[H_3O^+]$ and $[HCO_3^-]$. Because both K_{a2} and K_{b2} are small, $[HCO_3^-]$ should be close to its original value, $[HCO_3^-]_0$. If $[HCO_3^-]$ is set equal to $[HCO_3^-]_0$, this becomes

$$[H_3O^+]^2 \approx \frac{K_{a1}K_{a2}[HCO_3^-]_0 + K_{a1}K_w}{K_{a1} + [HCO_3^-]_0}$$

which can be solved for $[H_3O^+]$. In many cases of interest, $[HCO_3^-]_0 \gg K_{a1}$, and $K_{a2}\,[HCO_3^-]_0 \gg K_w$. When this is so, the expression simplifies to

$$[H_3O^+]^2 \approx K_{a1}K_{a2}$$

$$[H_3O^+] \approx \sqrt{K_{a1}K_{a2}}$$

$$\text{pH} \approx \frac{1}{2}(\text{p}K_{a1} + \text{p}K_{a2})$$

so the pH of such a solution is the average of the pK_a values for the two ionizations.

EXAMPLE 15.15

What is the pH of a solution that is 0.100 M in $NaHCO_3$?

SOLUTION

First, the two assumptions are checked:

$$[HCO_3^-]_0 = 0.100 \gg 4.3 \times 10^{-7} = K_{a1}$$

$$[HCO_3^-]_0 K_{a2} = 4.8 \times 10^{-12} \gg 1.0 \times 10^{-14} = K_w$$

so both are satisfied. Therefore,

$$[H_3O^+] = \sqrt{K_{a1}K_{a2}} = 4.5 \times 10^{-9} \text{ M}$$

$$\text{pH} = 8.34$$

and the solution is basic, as expected.

Titration of a Polyprotic Acid

A polyprotic acid has more than one equivalence point. The first equivalence point occurs when the volume V_{e1} of base added is sufficient to remove one hydrogen ion from each acid molecule, V_{e2} is the volume sufficient to remove two hydrogen ions from each, and so forth. A diprotic acid shows two equivalence points, and a triprotic acid, three. The equivalent volumes are related to each other by

$$V_{e1} = \frac{1}{2} V_{e2} = \frac{1}{3} V_{e3}$$

Figure 15.17 shows a titration curve for triprotic phosphoric acid. The three equivalence points are at 100.0 mL, 200.0 mL, and 300.0 mL. Calculating the pH as a function of the volume of added base presents no new complications beyond those already considered. The initial pH is given by a calculation analogous to that of Example 15.12, and the pH in the flat regions between equivalence points is obtained by a buffer calculation like that for a diprotic acid. For example, the pH after addition of 50.0 mL of base is that of an equimolar $H_3PO_4/H_2PO_4^-$ buffer (subsequent ionization of $H_2PO_4^-$ can be ignored). Finally, the pH at the first equivalence point is that for a solution of NaH_2PO_4 and uses the amphoteric equilibria equations presented earlier in this section (PO_4^{3-} can be ignored in this case). The pH at the second equivalence point is an amphoteric equilibrium in which HPO_4^{2-} is in equilibrium with $H_2PO_4^-$ and with PO_4^{3-}.

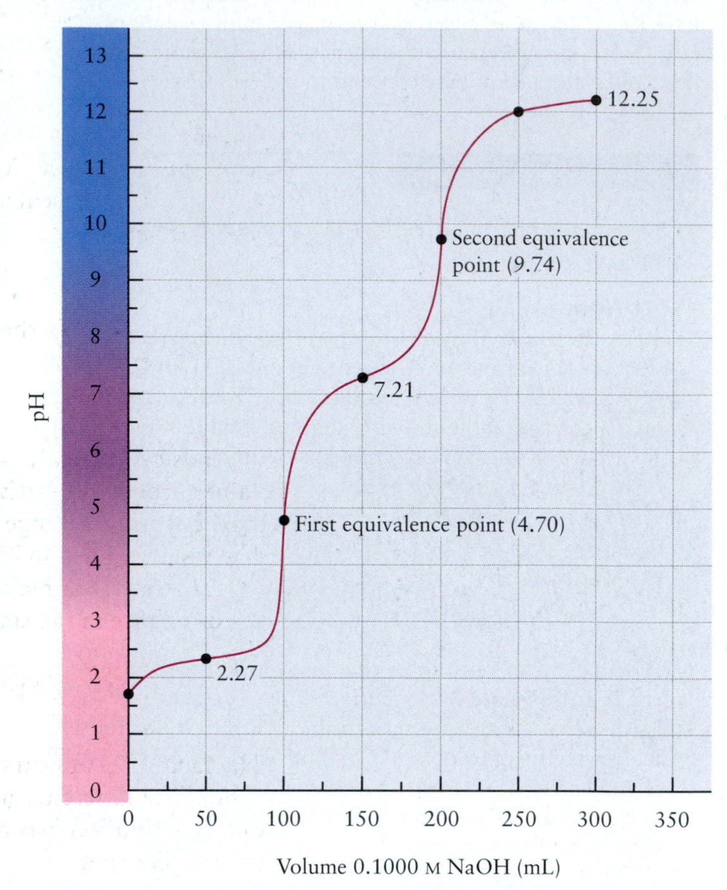

FIGURE 15.17 A titration curve for the titration of a polyprotic acid (phosphoric acid) by a strong base. The curve shown is for 100.0 mL of 0.1000 M H_3PO_4 titrated with 0.1000 M NaOH. No clear third equivalence point is seen at 300 mL because K_a for HPO_4^{2-} is not much greater than K_a for H_2O in aqueous solution.

15.9 Organic Acids and Bases: Structure and Reactivity

One of the goals of organic chemistry is to understand the relationship between *structure* and *function.* How does the molecular structure of an organic compound (including its attached functional groups) affect its chemical reactivity? As an example of the exploration of structure–function relationships we consider here the trends in acid strength in organic compounds. This extends the earlier discussion of acid strength in inorganic acids (see Section 15.3) and illustrates the usefulness of chemical concepts such as electronegativity and resonance in predicting reactivity. This section depends on the thermodynamic description in Section 14.3.

Table 15.2 lists a number of organic acids with pK_a values in aqueous solution between 0 (strong acid) and 14 (weak acid). Almost all of them are carboxylic acids with the characteristic —COOH functional group. Acid ionization constants can be also defined for many other functional groups besides carboxylic acids. They provide important information about reactivity, because many organic reactions involve the transfer of hydrogen ions or catalysis by acids or bases.

Aqueous solutions can be used to describe acidity only in compounds with values of the pK_a between about 14 and 0. Values much greater than 14 mean that there is negligible ionization in aqueous solution, and values much less than 0 mean that ionization is essentially complete. Because our goal is a scale of *relative* acidities, we can use information from nonaqueous solvents. In a sufficiently basic solvent, even a very weak acid will show some ionization; in a sufficiently acidic solvent, even a very strong acid will be only partially ionized. This allows an extension of the pK_a scale above 14 and below 0, provided we consistently think of it as only a relative scale. We thus generalize our earlier definition of K_a, which was limited to aqueous solvents, to describe acid ionization in a general solvent:

$$\text{HA} + \text{solvent} \rightleftharpoons \text{solvent-H}^+ + \text{A}^-$$

Our assumption is that the *difference* in pK_a values for two different acids in a given solvent

$$pK_a - pK_a' = -\log_{10}(K_a/K_a')$$

depends only weakly on the particular solvent chosen, even though the actual pK_a values are very sensitive to the solvent. If this is the case, we can extend the pK_a scale beyond the range that comes from measurements of pH in aqueous solution. The result is given in Table 15.4.

A thermodynamic analysis shows that this difference in pK_a values is related to the difference in the standard Gibbs free energy changes for the two reactions:

$$\Delta G° - \Delta G°' = -RT(\ln K_a - \ln K_a') = 2.3\ RT(pK_a - pK_a') \qquad [15.8]$$

The factor 2.3 appears when we relate the natural logarithm to the base 10 logarithm. The difference in standard Gibbs free energy changes can be separated into one contribution arising from standard enthalpy changes and one from absolute entropy changes:

$$\Delta G° - \Delta G°' = \Delta H° - \Delta H°' - T(\Delta S° - \Delta S°')$$

If we change the structure of an organic acid, what will be the effect on the pK_a? In general, modest structural changes such as the replacement of a functional group have only small effects on the entropy change of ionization, and most of the influence comes through changes in the enthalpy change of ionization. The quantitative magnitude of this effect is calculated in the following example.

EXAMPLE 15.16

How large a change in energy (or enthalpy) of reaction is needed for an acid ionization to lower the pK_a by one unit at 25°C? Assume that the entropy of ionization is unchanged.

SOLUTION

The entropy of ionization is unchanged, so $\Delta S° = \Delta S°'$, and $\Delta H° - \Delta H°' = \Delta G° - \Delta G°'$. If the change in pK_a is 1.00, then the change in Gibbs free energy must be

$$\Delta G° - \Delta G°' = -RT(\ln K_a - \ln K_a') = 2.3\ RT(pK_a - pK'_a) = 2.3\ RT(-1)$$

$$= 2.3(8.315\ \text{J mol}^{-1}\text{K}^{-1})(298\ \text{K})(-1) = -5700\ \text{J mol}^{-1} = -5.7\ \text{kJ mol}^{-1}$$

If we assume that this change in Gibbs free energy is entirely due to a change in enthalpy, then one unit of pK_a change corresponds to a lowering of the enthalpy of the products relative to the reactants by 5.7 kJ mol^{-1}.

Related Problems: 73, 74

TABLE 15.4 Acidities of Organic Compounds

Acid	pK_a	Conjugate Base
$CH_3{-}CH_3$	50	$CH_3{-}\ddot{C}H_2^-$
CH_4	49	$:CH_3^-$
$CH_2{=}CH_2$	44	$CH_2{=}\ddot{C}H^-$
C_6H_6	43	$C_6H_5^-$
$C_6H_5CH_3$	41	$C_6H_5\ddot{C}H_2^-$
$CH_3CH_2NH_3^+$	35	$CH_3CH_2NH_2$
$(C_6H_5)_3CH$	32	$(C_6H_5)_3\ddot{C}^-$
$HC{\equiv}CH$	25	$HC{\equiv}\ddot{C}^-$
CH_3COCH_3	20	$CH_3CO\ddot{C}H_2^-$
$(CH_3)_3C\ddot{O}H$	18	$(CH_3)_3C\ddot{O}:^-$
$C_2H_5\ddot{O}H$	16	$C_2H_5\ddot{O}:^-$
$H_2\ddot{O}$	15.7	$H\ddot{O}:^-$
$CH_3\ddot{O}H$	15	$CH_3\ddot{O}:^-$
$(NC)_2CH_2$	11.2	$(NC)_2\ddot{C}H^-$
CH_3NO_2	10.2	$:CH_2NO_2^-$
$C_6H_5\ddot{O}H$	10.0	$C_6H_5\ddot{O}:^-$
HCN	9.1	$:CN^-$
$(CH_3CO)_3CH$	5.9	$(CH_3CO)_3\ddot{C}^-$
CH_3CO_2H	4.8	$CH_3CO_2^-$
$C_6H_5CO_2H$	4.2	$C_6H_5CO_2^-$
HCO_2H	3.7	HCO_2^-
$CH_2(NO_2)_2$	3.6	$:CH(NO_2)_2^-$
$ClCH_2CO_2H$	2.9	$ClCH_2CO_2^-$
Cl_2CHCO_2H	1.3	$Cl_2CHCO_2^-$
Cl_3CCO_2H	0.7	$Cl_3CCO_2^-$
$CH_3CONH_3^+$	0.3	$CH_3CO\ddot{N}H_2$
HNO_3	−1.4	NO_3^-
H_3O^+	−1.7	H_2O
$CH_3\ddot{O}H_2^+$	−2.2	$CH_3\ddot{O}H$
$C_2H_5\ddot{O}H_2^+$	−2.4	$C_2H_5\ddot{O}H$
$(CH_3)_3C\ddot{O}H_2^+$	−3.8	$(CH_3)_3C\ddot{O}H$
$C_6H_5\ddot{O}H_2^+$	−6.7	$C_6H_5\ddot{O}H$
$(CH_3)_2C{=}\ddot{O}H^+$	−7.2	$(CH_3)_2C{=}\ddot{O}$
$R{-}C{\equiv}NH^+$	∼−10	$R{-}C{\equiv}N:$

The sign of the effect is clear from Example 15.16: If a change in structure of a molecule lowers its pK_a (making it a stronger acid), then the energy change for the acid ionization must be lowered. Another way to say this is that the effect of the structural change is to lower the energy of the conjugate base relative to that of the original acid.

An exploration of structure–function relations that lie behind the data in Table 15.4 must examine the effects of changes in molecular structure on the relative energetic stability of acid and conjugate base. We discuss three effects in turn: electronegativity, steric hindrance, and resonance.

Electronegativity

In Section 15.3 (see Fig. 15.6), we pointed out that in inorganic oxoacids with structures $-X-O-H$, the more electronegative the atom X, the more readily the $O^-{-}H^+$ bond breaks and the stronger the acid. A parallel effect of electronegativity explains the relative acid strengths of hydrocarbons, amines, and alcohols. Table 15.4 shows that ethane, C_2H_6, has a pK_a of 50, whereas that for ethylamine, $C_2H_5NH_2$, is 35 and that for ethanol, C_2H_5OH, is 16. These large changes occur because as the electronegativity increases from carbon to nitrogen to oxygen, the $-X^-$ conjugate base becomes more and more stable and thus the corresponding acid becomes stronger.

A similar situation occurs when electronegative atoms are substituted somewhat farther from the site where ionization takes place. Compare the acid strengths of acetic acid, CH_3COOH, with chloroacetic acid, $ClCH_2COOH$. The electronegative chlorine atom stabilizes the net negative charge of the conjugate base more strongly than does the hydrogen atom in acetic acid, reducing the pK_a from 4.8 to 2.9. Another way to describe this result is as an "inductive" effect involving bond dipoles. The negative charge on the chlorine atom in the Cl—C bond (and the corresponding positive charge on the carbon atom) interact favorably with the negative charge on the $-COO^-$ group of the conjugate base through a dipole-charge interaction (see Section 10.3), stabilizing the base and making the acid stronger:

$$Cl^{\delta-}-C^{\delta+}H_2-COO^-$$

As the substituted electronegative atom moves farther from the site of ionization, its effect on acid strength decreases. Thus, 4-chlorobutanoic acid, $ClCH_2CH_2CH_2COOH$, is a weaker acid ($pK_a = 4.5$) than 2-chlorobutanoic acid, $CH_3CH_2CHClCOOH$ ($pK_a = 2.9$).

Because most common functional groups in organic chemistry are more electronegative than hydrogen or carbon atoms (with the exception of metal atoms), their substitution tends to make organic acids stronger; the same is true of positively charged substituents. On the other hand, a negatively charged species such as $-COO^-$ interacts unfavorably with the additional negative change on the base and tends to reduce acidity.

EXAMPLE 15.17

Consider the dicarboxylic acid malonic acid, $HOOC-CH_2-COOH$. It ionizes in two stages with pK_a values pK_{a1} and pK_{a2}. How do you predict the magnitudes of these two will compare with the pK_a of acetic acid, CH_3COOH, which is 4.8?

SOLUTION

Malonic acid shows a substitution of a $-COOH$ functional group for one —H atom in acetic acid. Because this carboxylic acid functional group is more electronegative, it should make the compound more acidic, lowering its pK_a relative to acetic acid (measured value: $pK_{a1} = 2.8$). Once the first dissociation has taken place, however, the attached functional group in comparison with acetic acid is now $-COO^-$. Because this is negatively charged, it interacts unfavorably with the negative charge that results from the second acid ionization, raising the pK_a relative to that of acetic acid (measured value: $pK_{a2} = 5.7$).

Related Problem: 75

Steric Hindrance

As described in Section 11.2, ions in solution are stabilized through ion–dipole interactions with surrounding solvent molecules. This stabilization is reduced if the charged site is surrounded by bulky groups that prevent solvent molecules from coming close enough. This is an effect of **steric hindrance** on acidity. For example, compare the acidity of methanol, CH_3OH, with *tert*-butanol, $(CH_3)_3COH$. The latter compound is obtained by substituting $-CH_3$ groups for the three $-H$ atoms attached to the carbon atom in methanol. In the corresponding negatively charged conjugate bases CH_3O^- and $(CH_3)_3CO^-$, the former is more stable in solution because solvent molecules can approach the negatively charged site more closely. In the latter, the bulkier —CH_3 groups reduce the stability of the solvated base anion. The net effect is that methanol is a stronger acid in solution ($pK_a = 15$) than *tert*-butanol ($pK_a = 18$).

This observation depends on the presence of a liquid solvent around the acid and conjugate base. For the corresponding acid–base pair in the gas phase, *tert*-butanol is a stronger acid than methanol.

Resonance

A third and final contribution to relative acidity of organic compounds is resonance. Let's begin by noting that the acidity of carboxylic acids is considerably greater than that of alcohols. For example, the pK_a of acetic acid, CH_3COOH, is considerably lower than that of ethanol, CH_3CH_2OH (4.8 versus 16). Some of this may be due to the "inductive effect" of substituting an electronegative O atom for two H atoms. But, inductive effects are not typically so large (more than 11 units of pH change). An alternative explanation must be sought, and it can be found in the concept of resonance stabilization, which was introduced in Section 3.8 and discussed in the context of organic molecules in Section 7.8. A single Lewis structure can be drawn for the conjugate base of ethanol:

```
   H H -1
 H:C:C:O:
   H H
```

On the other hand, the conjugate base of acetic acid is represented by a resonance hybrid of two Lewis structures:

```
  H  :O.              H  :O:-1
H:C:C.        <-->  H:C:C.
  H  :O:-1            H  :O.
```

The existence of more than one resonance structure stabilizes the acetate ion and contributes to the greater acidity of acetic acid.

A second example of resonance stabilization affecting acidity arises in a comparison of phenols and alcohols. Phenol (also called carbolic acid) has a pK_a of 10, whereas the pK_a values for typical alcohols range from 16 to 18. The reason for this difference is the greater stability of the conjugate base (the phenoxide ion, $C_6H_5O^-$) due to the spreading out of the negative charge over the aromatic ring. Several different resonance structures contribute to the stability of the phenoxide ion.

```
(ring)—O: -1  <-->  (ring)—O: -1  <-->  -1 :(ring)=O:
```

Phenol, although not a strong acid, does react readily with sodium hydroxide to form the salt sodium phenoxide:

$$C_6H_5OH + NaOH \longrightarrow C_6H_5O^-Na^+ + H_2O$$

The corresponding reaction between NaOH and alcohols does not occur to a significant extent, although sodium ethoxide can be prepared by reaction of metallic sodium with anhydrous ethanol:

$$Na + C_2H_5OH \longrightarrow C_2H_5O^-Na^+ + \frac{1}{2}H_2$$

EXAMPLE 15.18

By using resonance Lewis structures, predict which will be the stronger acid: cyclopentane (C_5H_{10}) or cyclopentadiene (C_5H_6).

SOLUTION

We can write only one Lewis structure for the conjugate base of cyclopentane:

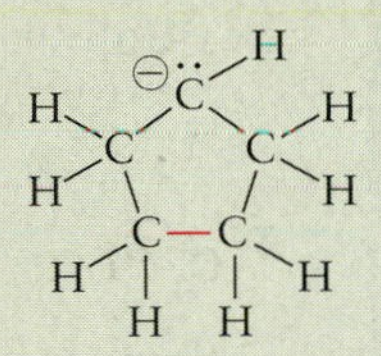

It is a very weak acid. In contrast, we can write five resonance structures for the conjugate base of cyclopentadiene:

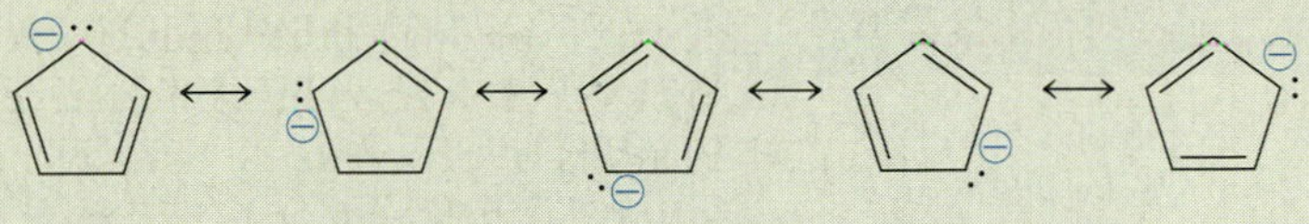

This is very stable, so cyclopentadiene will be a stronger acid (its pK_a is 16).

Related Problems: 77, 78

CHAPTER SUMMARY

Acid-base reactions in aqueous solutions are of central importance in a wide variety of systems that include industrial manufacturing processes, ordinary household substances, and the life-sustaining reactions in living systems. The common feature shared by all acids and bases is that they modify the properties of the water in which they are dissolved. An acid increases the concentration of the hydronium ion H_3O^+ above its value in pure water, which is controlled by the ion product constant of water. A base increases the concentration of the hydroxide ion OH^- above the value it takes in pure water, due to the ion product constant of water. The concentration of the hydronium ion is conveniently expressed by pH, and that of the hydroxide by pOH. These quantities are related as pH + pOH = 14 by the ion product constant of water. Some acids and bases dissociate completely in solution to produce concentrations of hydronium and hydroxide equivalent to the concentration of parent acid or base with which the solutions were prepared. These are called strong acids and bases, and reactions of their solutions are completely described by the laws of stoichiometry. Most acids and bases dissociate only partially, so we must use the principles of chemical equilibrium to describe the concentration of hydronium and hydroxide in their solutions. These are the weak acids and bases, and description of their reactions requires methods of stoichiometry combined with methods of equilibrium. Acid-base neutralization reactions produce water and salts as their products. These reactions can be followed quantitatively and precisely by titration, as a means to determine the amount of acid or base in an unknown sample. When the salt of a weak acid is dissolved in water, hydrolysis of the anion produces the undissociated acid, and the solution is basic. Similarly, hydrolysis of the salt of a weak base produces an acidic solution. In both cases the pH is calculated by equilibrium methods. An especially interesting situation arises when a weak acid is dissolved along with one of its salts. The

The effect of acid rain on a stand of trees in the Great Smoky Mountains of the United States.

combination of weak acid equilibrium and hydrolysis—called a buffer solution—keeps the pH of the solution constant as small amounts of acid are added. This stabilization mechanism is extremely important in biochemical and biological situations, where success of various reactions depends critically on keeping pH constant. The extent of all these acid-base reactions can be correlated with the structures of their molecules. The extent of reaction is governed by the equilibrium constant, which in turn depends on the Gibbs standard free energy of formation of the reactants and products, which depends on their molecular structure.

CUMULATIVE EXERCISE

Acid Precipitation

Acid precipitation is a major environmental problem throughout the industrialized world. One major source is the burning of fossil fuels containing sulfur (coal, oil, and natural gas). The sulfur dioxide released into the air dissolves in water or, more seriously, may be oxidized further to sulfur trioxide. The SO_3 dissolves in water to form sulfuric acid:

$$SO_3(g) + H_2O(\ell) \longrightarrow H_2SO_4(aq)$$

The net effect is to increase the acidity of the rain, which damages trees, kills fish in lakes, dissolves stone, and corrodes metal.

(a) A sample of rainwater is tested for acidity by using two indicators. Addition of methyl orange to half of the sample gives a yellow color, and addition of methyl red to the other half gives a red color. Estimate the pH of the sample.

(b) The pH in acid rain can range down to 3 or even lower in heavily polluted areas. Calculate the concentrations of H_3O^+ and OH^- in a raindrop at pH 3.30 at 25°C.

(c) When SO_2 dissolves in water to form sulfurous acid, $H_2SO_3(aq)$, that acid can donate a hydrogen ion to water. Write a balanced chemical equation for this reaction, and identify the stronger Brønsted–Lowry acid and base in the equation.

(d) Ignore the further ionization of HSO_3^-, and calculate the pH of a solution whose initial concentration of H_2SO_3 is 4.0×10^{-4} M. (*Hint:* Use the quadratic equation in this case.)

(e) Now suppose that all the dissolved SO_2 from part (d) has been oxidized further to SO_3, so that 4.0×10^{-4} mol of H_2SO_4 is dissolved per liter. Calculate the pH in this case. (*Hint:* Because the first ionization of H_2SO_4 is that of a strong acid, the concentration of H_3O^+ can be written as 4.0×10^{-4} plus the unknown amount of dissociation from $HSO_4^-(aq)$.)

(f) Lakes have a natural buffering capacity, especially in regions where limestone gives rise to dissolved calcium carbonate. Write an equation for the effect of a small amount of acid rain containing sulfuric acid if it falls into a lake containing carbonate (CO_3^{2-}) ions. Discuss how the lake will resist further pH changes. What happens if a large excess of acid rain is deposited?

(g) A sample of 1.00 L of rainwater known to contain only sulfurous (and not sulfuric) acid is titrated with 0.0100 M NaOH. The equivalence point of the H_2SO_3/HSO_3^- titration is reached after 31.6 mL has been added. Calculate the original concentration of sulfurous acid in the sample, again ignoring any effect of SO_3^{2-} on the equilibria.

(h) Calculate the pH at the half-equivalence point, after 15.8 mL has been added. (*Hint:* Use the quadratic equation.)

Answers

(a) 4.4 to 4.8

(b) $[H_3O^+] = 5.0 \times 10^{-4}$ M; $[OH^-] = 2.0 \times 10^{-11}$ M

(c) $H_2SO_3(aq) + H_2O(\ell) \rightleftharpoons HSO_3^-(aq) + H_3O^+(aq)$. The stronger acid is H_3O^+, and the stronger base is HSO_3^-.

(d) The pH is 3.41.

(e) The pH is 3.11.

(f) The H_3O^+ in the sulfuric acid solution reacts according to $H_3O^+(aq) + CO_3^{2-}(aq) \rightleftharpoons HCO_3^-(aq) + H_2O(\ell)$. The HSO_4^- in the sulfuric acid reacts according to $HSO_4^-(aq) + CO_3^{2-}(aq) \rightleftharpoons SO_4^{2-}(aq) + HCO_3^-(aq)$. This gives rise to a HCO_3^-/CO_3^{2-} buffer that can resist further changes in pH. An excess of acid rain overwhelms the buffer and leads to the formation of H_2CO_3.

(g) 3.16×10^{-4} M

(h) The pH is 3.81.

CHAPTER REVIEW

Classification of Acids and Bases

- Brønsted-Lowry concept
  - Acids are proton donors.
  - Bases are proton acceptors.
  - Acid–base reactions are proton transfer reactions.
- Lewis concept
  - Acids are electron pair acceptors.
  - Bases are electron pair donors.
  - Acid–base reactions are electron pair transfer reactions.

The Brønsted–Lowry Description of Acids and Bases in Aqueous Solutions

- Autoionization of water
  - $K_w = [H_3O^+][OH^-] = 10^{-14}$ for pure water at 25°C
  - $[H_3O^+] = [OH^-] = 10^{-7}$ M
- Strong acids and bases
  - Strong acids and bases dissociate essentially completely in solution.
- The pH function
  - $pH = -\log [H_3O^+]$
  - $pH < 7$ acidic solution
  - $pH > 7$ basic solution
  - $pH = 7$ neutral solution

Acid and Base Strength

- For the reaction $HA(aq) + H_2O(\ell) \rightleftharpoons H_3O^+(aq) + A^-(aq)$
  - $K_a = ([H_3O^+][A^-])/[HA]$
  - Strong acids have $K_a \gg 1$
- For the reaction $B(aq) + H_2O(\ell) \rightleftharpoons BH^+(aq) + OH^-(aq)$
  - $K_b = ([BH^+][OH^-])/[B]$
  - Strong bases have $K_b \gg 1$
- For a conjugate acid–base pair $K_a K_b = K_w$
- Stronger conjugate acids have weaker conjugate bases and *vice versa*.
- Strength of oxoacids: X—O—H increases as electronegativity of X increases
  - As X withdraws electron density from O, breaking the O—H bond to release H^+ becomes easier.
- Indicators respond to changes in pH because HIn and In^- have different colors. $K_a = ([H_3O^+][In^-])/[HIn]$

Equilibria Involving Weak Acids and Bases

- To describe weak acids
  let $x = [H_3O^+] = [A^-]$
  then $x^2/[HA] - x = K_a$
  If K_a is very small and initial concentration of HA ≥ 0.05 M, use the approximation $x^2/[HA] = K_a$.
  Otherwise, use the quadratic equation.
- Describe weak bases by the same procedure used for weak acids.
- Hydrolysis is the reaction of a salt of a weak base or a weak acid with water.
  The cation of a weak base acts as an acid to form the conjugate base and produce an acidic solution.
  The anion of a weak acid acts as a base to form the conjugate acid and produce a basic solution.
  For all conjugate acid–base pairs $K_b = K_w/K_a$.

Buffer Solutions

- A solution that contains a weak acid and its salt, or a weak base and its salt, in appreciable amounts resists changes in pH when either acid or base is added.
- The pH in a buffer solution is described by these approximate equations
  $[H_3O^+] \approx K_a[HA]_0/[A^-]_0$
  $pH \approx pK_a - \log_{10} [HA]_0/[A^-]_0$
- To describe the response of a buffer to added acid or base, follow the procedures demonstrated in Example 15.8

Acid–Base Titration Curves

- Titration curves for strong acids by strong bases and vice versa have four regions that can be calculated by using procedures of stoichiometry and definitions of solution concentration.
  1. Before any titrant is added. The concentration of the starting acid or base determines the pH.
  2. Before the equivalence point. Calculate the number of moles of acid or base remaining and divide by the total volume to calculate the concentration and determine the pH.
  3. At the equivalence point. pH = 7.
  4. After the equivalence point. Calculate the number of moles of excess acid or base and divide by the total volume to calculate the concentration and determine the pH.

  (**Note:** The choice of indicator is not critical in strong acid–strong base titrations since the slope of the transition at the equivalence point is very steep.)
- Titration curves for weak acids by strong bases and vice versa also have four regions. But each of these requires consideration of equilibrium in the solution and corresponds to a standard type of calculation illustrated in the chapter.
  1. Before any titrant is added. This is simply the ionization of a weak acid or based, as described in Examples 15.3 and 15.4.
  2. Before the equivalence point. This is a buffer solution, as described in Example 15.7.
  3. At the equivalence point. This is a hydrolysis problem. All of the initial acid or base has been converted to a salt, which hydrolyzes back to the acid or base, so the pH $\neq$ 7. See Example 15.6.
  4. After the equivalence point. The problem is very close to the titration of a strong acid or base. Calculate the number of moles of excess acid or base added and divide by the total volume to obtain the concentration and the pH.

  (**Note:** The choice of indicator is critical in weak acid–strong base and weak base–strong acid titrations because the slope of the transition is not so steep and the pH of the equivalence point can be very different from 7. pK_a (indicator) $\approx$ pH at the equivalence point.)

CONCEPTS & SKILLS

After studying this chapter and working the problems that follow, you should be able to:

1. Define an acid and a base in the Brønsted–Lowry and Lewis systems and give several examples of their reaction with a solvent (Section 15.1, Problems 1–12).
2. Define the pH function and convert between pH and $[H_3O^+]$ (Section 15.2, Problems 13–16).
3. State the relationship between the ionization constant for an acid and that for its conjugate base (Section 15.3, Problems 21–22).
4. Explain how indicators allow the pH of a solution to be estimated (Section 15.3, Problems 25–26).
5. Formulate the equilibrium expression for the ionization of a weak acid or base, and use it to determine the pH and fraction ionized (Section 15.4, Problems 27–36).
6. Explain the behavior of a buffer solution. Calculate its pH from the concentrations of its conjugate acid–base pair (Section 15.5, Problems 43–46).
7. Design a buffer system that will regulate the pH to a particular value (Section 15.5, Problems 47–50).
8. Calculate the pH at any stage in the titration of a strong acid or base by a strong base or acid (Section 15.6, Problems 51–52).
9. Calculate the pH at any stage in the titration of a weak acid or base by a strong base or acid (Section 15.6, Problems 53–62).
10. Calculate the concentrations of all the species present in a solution of a weak polyprotic acid (Section 15.7, Problems 63–68).
11. Outline the procedure for the exact treatment of acid–base equilibrium and use it to find the pH of a very dilute solution of a weak acid or base (Section 15.8, Problems 69–70).
12. Calculate the pH at selected points in the titration of a polyprotic acid (Section 15.8, Problems 71–72).
13. Use structure-function relations to predict effects of substitutions on relative acidities of organic acids (Section 15.9, Problems 73–80).

KEY EQUATIONS

$[H_3O^+][OH^-] = K_w$ (Section 15.2)

$pH = -\log_{10} [H_3O^+]$ (Section 15.2)

$pH < 7$ Acidic solution (Section 15.2)

$pH = 7$ Neutral solution (Section 15.2)

$pH > 7$ Basic solution (Section 15.2)

$\frac{[H_3O^+][A^-]}{[HA]} = K_a$ (Section 15.3)

$K_w = K_a K_b$ (Section 15.3)

$\frac{[H_3O^+]}{K_a} = \frac{[HIn]}{[In^-]}$ (Section 15.3)

$pH \approx pK_a - \log_{10}\frac{[HA]_0}{[A^-]_0}$ (Section 15.5)

$\Delta G° - \Delta G°' = -RT(\ln K_a - \ln K'_a) = 2.3\ RT(pK_a - pK'_a)$ (Section 15.9)

PROBLEMS

Answers to problems whose numbers are boldface appear in Appendix G. Problems that are more challenging are indicated with asterisks.

Classifications of Acids and Bases

1. Which of the following can act as Brønsted–Lowry acids? Give the formula of the conjugate Brønsted–Lowry base for each of them.
(a) Cl^- (b) HSO_4^- (c) NH_4^+
(d) NH_3 (e) H_2O

2. Which of the following can act as Brønsted–Lowry bases? Give the formula of the conjugate Brønsted–Lowry acid for each of them.
(a) F^- (b) SO_4^{2-} (c) O^{2-}
(d) OH^- (e) H_2O

3. Lemon juice contains citric acid ($C_6H_8O_7$). What species serves as a base when lemon juice is mixed with baking soda (sodium hydrogen carbonate) during the preparation of some lemon cookies?

4. A treatment recommended in case of accidental swallowing of ammonia-containing cleanser is to drink large amounts of diluted vinegar. Write an equation for the chemical reaction on which this procedure depends.

5. An important step in many industrial processes is the slaking of lime, in which water is added to calcium oxide to make calcium hydroxide.
(a) Write the balanced equation for this process.
(b) Can this be considered a Lewis acid–base reaction? If so, what is the Lewis acid and what is the Lewis base?

6. Silica (SiO_2) is an impurity that must be removed from a metal oxide or sulfide ore when the ore is being reduced to elemental metal. To do this, lime (CaO) is added. It reacts with the silica to form a slag of calcium silicate ($CaSiO_3$), which can be separated and removed from the ore.
(a) Write the balanced equation for this process.
(b) Can this be considered a Lewis acid–base reaction? If so, what is the Lewis acid and what is the Lewis base?

7. Chemists working with fluorine and its compounds sometimes find it helpful to think in terms of acid–base reactions in which the fluoride ion (F^-) is donated and accepted.
(a) Would the acid in this system be the fluoride donor or fluoride acceptor?
(b) Identify the acid and base in each of these reactions:

$$ClF_3O_2 + BF_3 \longrightarrow ClF_2O_2 \cdot BF_4$$

$$TiF_4 + 2\ KF \longrightarrow K_2[TiF_6]$$

8. Researchers working with glasses often think of acid–base reactions in terms of oxide donors and oxide acceptors. The oxide ion is O^{2-}.
(a) In this system, is the base the oxide donor or the oxide acceptor?
(b) Identify the acid and base in each of these reactions:

$$2\ CaO + SiO_2 \longrightarrow Ca_2SiO_4$$

$$Ca_2SiO_4 + SiO_2 \longrightarrow 2\ CaSiO_3$$

$$Ca_2SiO_4 + CaO \longrightarrow Ca_3SiO_5$$

9. Identify each of the following oxides as an acid or base anhydride. Write the chemical formula and give the name of the acid or base formed upon reaction with water.
(a) MgO (b) Cl_2O
(c) SO_3 (d) Cs_2O

10. Write the chemical formula and give the name of the anhydride corresponding to each of the following acids or bases, and identify it as an acid or base anhydride.
(a) H_3AsO_4 (b) H_2MoO_4
(c) RbOH (d) H_2SO_3

11. Tin(II) oxide is amphoteric. Write balanced chemical equations for its reactions with an aqueous solution of hydrochloric acid and with an aqueous solution of sodium hydroxide. (**Note:** The hydroxide complex ion of tin(II) is $[Sn(OH)_3]^-$.)

12. Zinc oxide is amphoteric. Write balanced chemical equations for its reactions with an aqueous solution of hydrochloric acid and with an aqueous solution of sodium hydroxide. (**Note:** The hydroxide complex ion of zinc is $[Zn(OH)_4]^{2-}$.)

Properties of Acids and Bases in Aqueous Solutions: The Brønsted–Lowry Scheme

13. The concentration of H_3O^+ in a sample of wine is 2.0×10^{-4} M. Calculate the pH of the wine.

14. The concentration of OH^- in a solution of household bleach is 3.6×10^{-2} M. Calculate the pH of the bleach.

15. The pH of normal human urine is in the range of 5.5 to 6.5. Compute the range of the H_3O^+ concentration and the range of the OH^- concentration in normal urine.

16. The pH of normal human blood is in the range of 7.35 to 7.45. Compute the range of the concentration of H_3O^+ and the range of the OH^- concentration in normal blood.

17. The pK_w of seawater at 25°C is 13.776. This differs from the usual pK_w of 14.00 at this temperature because dissolved salts make seawater a nonideal solution. If the pH in seawater is 8.00, what are the concentrations of H_3O^+ and OH^- in seawater at 25°C?

18. At body temperature (98.6°F = 37.0°C), K_w has the value 2.4×10^{-14}. If the pH of blood is 7.4 under these conditions, what are the concentrations of H_3O^+ and OH^-?

19. When placed in water, potassium starts to react instantly and continues to react with great vigor. On the basis of this information, select the better of the following two equations to represent the reaction.

$$2\ K(s) + 2\ H_2O(\ell) \longrightarrow 2\ KOH(aq) + H_2(g)$$

$$2\ K(s) + 2\ H_3O^+(aq) \longrightarrow 2\ K^+(aq) + H_2(g) + 2H_2O(\ell)$$

State the reason for your choice.

20. Molecules of *t*-butyl chloride, $(CH_3)_3CCl$, react very slowly when mixed with water at low pH to give *t*-butyl alcohol, $(CH_3)_3COH$. When the pH is raised, the reaction takes place rapidly. Write an equation or equations to explain these facts.

Acid and Base Strength

21. Ephedrine ($C_{10}H_{15}ON$) is a base that is used in nasal sprays as a decongestant.
(a) Write an equation for its equilibrium reaction with water.
(b) The K_b for ephedrine is 1.4×10^{-4}. Calculate the K_a for its conjugate acid.
(c) Is ephedrine a weaker or a stronger base than ammonia?

22. Niacin (C_5H_4NCOOH), one of the B vitamins, is an acid.
(a) Write an equation for its equilibrium reaction with water.
(b) The K_a for niacin is 1.5×10^{-5}. Calculate the K_b for its conjugate base.
(c) Is the conjugate base of niacin a stronger or a weaker base than pyridine, C_5H_5N?

23. Use the data in Table 15.2 to determine the equilibrium constant for the reaction.

$$HClO_2(aq) + NO_2^-(aq) \rightleftharpoons HNO_2(aq) + ClO_2^-(aq)$$

Identify the stronger Brønsted–Lowry acid and the stronger Brønsted–Lowry base.

24. Use the data in Table 15.2 to determine the equilibrium constant for the reaction

$$HPO_4^{2-} + HCO_3^- \rightleftharpoons PO_4^{3-} + H_2CO_3$$

Identify the stronger Brønsted–Lowry acid and the stronger Brønsted–Lowry base.

25. (a) Which is the stronger acid—the acidic form of the indicator bromocresol green or the acidic form of methyl orange?
(b) A solution is prepared in which bromocresol green is green and methyl orange is orange. Estimate the pH of this solution.

26. (a) Which is the stronger base—the basic form of the indicator cresol red or the basic form of thymolphthalein?
(b) A solution is prepared in which cresol red is red and thymolphthalein is colorless. Estimate the pH of this solution.

Equilibria Involving Weak Acids and Bases

27. Aspirin is acetylsalicylic acid, $HC_9H_7O_4$, which has a K_a of 3.0×10^{-4}. Calculate the pH of a solution made by dissolving 0.65 g of acetylsalicylic acid in water and diluting to 50.0 mL.

28. Vitamin C is ascorbic acid ($HC_6H_7O_6$), for which K_a is 8.0×10^{-5}. Calculate the pH of a solution made by dissolving a 500-mg tablet of pure vitamin C in water and diluting to 100 mL.

29. (a) Calculate the pH of a 0.20 M solution of benzoic acid at 25°C.
(b) How many moles of acetic acid must be dissolved per liter of water to obtain the same pH as that from part (a)?

30. (a) Calculate the pH of a 0.35 M solution of propionic acid at 25°C.
(b) How many moles of formic acid must be dissolved per liter of water to obtain the same pH as that from part (a)?

31. Iodic acid (HIO_3) is fairly strong for a weak acid, having a K_a equal to 0.16 at 25°C. Compute the pH of a 0.100 M solution of HIO_3.

32. At 25°C, the K_a of pentafluorobenzoic acid (C_6F_5COOH) is 0.033. Suppose 0.100 mol of pentafluorobenzoic acid is dissolved in 1.00 L of water. What is the pH of this solution?

33. Papaverine hydrochloride ($papH^+Cl^-$) is a drug used as a muscle relaxant. It is a weak acid. At 25°C, a 0.205 M solution of $papH^+Cl^-$ has a pH of 3.31. Compute the K_a of the $papH^+$ ion.

34. The unstable weak acid 2-germaacetic acid (GeH_3COOH) is derived structurally from acetic acid (CH_3COOH) by having a germanium atom replace one of the carbon atoms. At 25°C, a 0.050 M solution of 2-germaacetic acid has a pH of 2.42. Compute the K_1 of 2-germaacetic acid and compare it with that of acetic acid.

35. Morphine is a weak base for which K_b is 8×10^{-7}. Calculate the pH of a solution made by dissolving 0.0400 mol of morphine in water and diluting to 600.0 mL.

36. Methylamine is a weak base for which K_b is 4.4×10^{-4}. Calculate the pH of a solution made by dissolving 0.070 mol of methylamine in water and diluting to 800.0 mL.

37. The pH at 25°C of an aqueous solution of hydrofluoric acid, HF, is 2.13. Calculate the concentration of HF in this solution, in moles per liter.

38. The pH at 25°C of an aqueous solution of sodium cyanide (NaCN) is 11.50. Calculate the concentration of CN^- in this solution, in moles per liter.

39. You have 50.00 mL of a solution that is 0.100 M in acetic acid, and you neutralize it by adding 50.00 mL of a solution that is 0.100 M in sodium hydroxide. The pH of the resulting solution is not 7.00. Explain why. Is the pH of the solution greater than or less than 7?

40. A 75.00-mL portion of a solution that is 0.0460 M in $HClO_4$ is treated with 150.00 mL of 0.0230 M KOH(*aq*). Is the pH of the resulting mixture greater than, less than, or equal to 7.0? Explain.

41. Suppose a 0.100 M solution of each of the following substances is prepared. Rank the pH of the resulting solutions from lowest to highest: NH_4Br, NaOH, KI, $NaCH_3COO$, HCl.

42. Suppose a 0.100 M solution of each of the following substances is prepared. Rank the pH of the resulting solutions from lowest to highest: KF, NH_4I, HBr, NaCl, LiOH.

Buffer Solutions

43. "Tris" is short for tris(hydroxymethyl)aminomethane. This weak base is widely used in biochemical research for the preparation of buffers. It offers low toxicity and a pK_b (5.92 at 25°C) that is convenient for the control of pH in clinical applications. A buffer is prepared by mixing 0.050 mol of tris with 0.025 mol of HCl in a volume of 2.00 L. Compute the pH of the solution.

44. "Bis" is short for bis(hydroxymethyl)aminomethane. It is a weak base that is closely related to tris (see problem 43) and has similar properties and uses. Its pK_b is 8.8 at 25°C. A buffer is prepared by mixing 0.050 mol of bis with 0.025

mol of HCl in a volume of 2.00 L (the same proportions as in the preceding problem). Compute the pH of the solution.

45. (a) Calculate the pH in a solution prepared by dissolving 0.050 mol of acetic acid and 0.020 mol of sodium acetate in water and adjusting the volume to 500 mL.
(b) Suppose 0.010 mol of NaOH is added to the buffer from part (a). Calculate the pH of the solution that results.

46. Sulfanilic acid ($NH_2C_6H_4SO_3H$) is used in manufacturing dyes. It ionizes in water according to the equilibrium equation

$$NH_2C_6H_4SO_3H(aq) + H_2O(\ell) \rightleftharpoons NH_2C_6H_4SO_3^-(aq) + H_3O^+(aq) \quad K_a = 5.9 \times 10^{-4}$$

A buffer is prepared by dissolving 0.20 mol of sulfanilic acid and 0.13 mol of sodium sulfanilate ($NaNH_2C_6H_4SO_3$) in water and diluting to 1.00 L.
(a) Compute the pH of the solution.
(b) Suppose 0.040 mol of HCl is added to the buffer. Calculate the pH of the solution that results.

47. A physician wishes to prepare a buffer solution at pH = 3.82 that efficiently resists changes in pH yet contains only small concentrations of the buffering agents. Determine which one of the following weak acids, together with its sodium salt, would probably be best to use: *m*-chlorobenzoic acid, $K_a = 1.04 \times 10^{-4}$; *p*-chlorocinnamic acid, $K_a = 3.89 \times 10^{-5}$; 2,5-dihydroxybenzoic acid, $K_a = 1.08 \times 10^{-3}$; or acetoacetic acid, $K_a = 2.62 \times 10^{-4}$. Explain.

48. Suppose you were designing a buffer system for imitation blood and wanted the buffer to maintain the blood at the realistic pH of 7.40. All other things being equal, which buffer system would be preferable: H_2CO_3/HCO_3^- or $H_2PO_4^-/HPO_4^{2-}$? Explain.

49. You have at your disposal an ample quantity of a solution of 0.0500 M NaOH and 500 mL of a solution of 0.100 M formic acid (HCOOH). How much of the NaOH solution should be added to the acid solution to produce a buffer of pH 4.00?

50. You have at your disposal an ample quantity of a solution of 0.100 M HCl and 400 mL of a solution of 0.0800 M NaCN. How much of the HCl solution should be added to the NaCN solution to produce a buffer of pH 9.60?

Acid–Base Titration Curves

51. Suppose 100.0 mL of a 0.3750 M solution of the strong base $Ba(OH)_2$ is titrated with a 0.4540 M solution of the strong acid $HClO_4$. The neutralization reaction is

$$Ba(OH)_2(aq) + 2\,HClO_4(aq) \longrightarrow Ba(ClO_4)_2(aq) + 2\,H_2O(\ell)$$

Compute the pH of the titration solution before any acid is added, when the titration is 1.00 mL short of the equivalence point, when the titration is at the equivalence point, and when the titration is 1.00 mL past the equivalence point. (**Note:** Each mole of $Ba(OH)_2$ gives *two* moles of OH^- in solution.)

52. A sample containing 26.38 mL of 0.1439 M HBr is titrated with a solution of NaOH having a molarity of 0.1219 M. Compute the pH of the titration solution before any base is added, when the titration is 1.00 mL short of the equivalence point, when the titration is at the equivalence point, and when the titration is 1.00 mL past the equivalence point.

53. A sample containing 50.00 mL of 0.1000 M hydrazoic acid (HN_3) is being titrated with 0.1000 M sodium hydroxide. Compute the pH before any base is added, after the addition of 25.00 mL of the base, after the addition of 50.00 mL of the base, and after the addition of 51.00 mL of the base.

54. A sample of 50.00 mL of 0.1000 M aqueous solution of chloroacetic acid, $CH_2ClCOOH$ ($K_1 = 1.4 \times 10^{-3}$), is titrated with a 0.1000 M NaOH solution. Calculate the pH at the following stages in the titration, and plot the titration curve: 0, 5.00, 25.00, 49.00, 49.90, 50.00, 50.10, and 55.00 mL NaOH.

55. The base ionization constant of ethylamine ($C_2H_5NH_2$) in aqueous solution is $K_b = 6.41 \times 10^{-4}$ at 25°C. Calculate the pH for the titration of 40.00 mL of a 0.1000 M solution of ethylamine with 0.1000 M HCl at the following volumes of added HCl: 0, 5.00, 20.00, 39.90, 40.00, 40.10, and 50.00 mL.

56. Ammonia is a weak base with a K_b of 1.8×10^{-5}. A 140.0-mL sample of a 0.175 M solution of aqueous ammonia is titrated with a 0.106 M solution of the strong acid HCl. The reaction is

$$NH_3(aq) + HCl(aq) \longrightarrow NH_4^+(aq) + Cl^-(aq)$$

Compute the pH of the titration solution before any acid is added, when the titration is at the half-equivalence point, when the titration is at the equivalence point, and when the titration is 1.00 mL past the equivalence point.

57. Sodium benzoate, the sodium salt of benzoic acid, is used as a food preservative. A sample containing solid sodium benzoate mixed with sodium chloride is dissolved in 50.0 mL of 0.500 M HCl, giving an acidic solution (benzoic acid mixed with HCl). This mixture is then titrated with 0.393 M NaOH. After the addition of 46.50 mL of the NaOH solution, the pH is found to be 8.2. At this point, the addition of one more drop (0.02 mL) of NaOH raises the pH to 9.3. Calculate the mass of sodium benzoate (NaC_6H_5COO) in the original sample. (*Hint:* At the equivalence point, the *total* number of moles of acid [here HCl] equals the *total* number of moles of base [here, both NaOH and NaC_6H_5COO].)

58. An antacid tablet (such as Tums or Rolaids) weighs 1.3259 g. The only acid-neutralizing ingredient in this brand of antacid is $CaCO_3$. When placed in 12.07 mL of 1.070 M HCl, the tablet fizzes merrily as $CO_2(g)$ is given off. After all of the CO_2 has left the solution, an indicator is added, followed by 11.74 mL of 0.5310 M NaOH. The indicator shows that at this point the solution is definitely basic. Addition of 5.12 mL of 1.070 M HCl makes the solution acidic again. Then 3.17 mL of the 0.5310 M NaOH brings the titration exactly to an endpoint, as signaled by the indicator. Compute the percentage by mass of $CaCO_3$ in the tablet.

59. What is the mass of diethylamine, $(C_2H_5)_2NH$, in 100.0 mL of an aqueous solution if it requires 15.90 mL of 0.0750 M HCl to titrate it to the equivalence point? What is the pH at the equivalence point if $K_b = 3.09 \times 10^{-4}$? What would be a suitable indicator for the titration?

60. A chemist who works in the process laboratory of the Athabasca Alkali Company makes frequent analyses of ammonia recovered from the Solvay process for making sodium carbonate. What is the pH at the equivalence point if she titrates the aqueous ammonia solution (approximately 0.10 M) with a strong acid of comparable concentration? Select an indicator that would be suitable for the titration.

61. If 50.00 mL of a 0.200 M solution of the weak base *N*-ethylmorpholine ($C_6H_{13}NO$) is mixed with 8.00 mL of 1.00 M HCl and then diluted to a final volume of 100.0 mL with water, the result is a buffer with a pH of 7.0. Compute the K_b of *N*-ethylmorpholine.

62. The sodium salt of cacodylic acid, a weak acid, has the formula $NaO_2As(CH_3)_2 \cdot 3H_2O$. Its molar mass is 214.02 g mol^{-1}. A solution is prepared by mixing 21.40 g of this substance with enough water to make 1.000 L of solution. Then 50.00 mL of the sodium cacodylate solution is mixed with 29.55 mL of 0.100 M HCl and enough water to bring the volume to a total of 100.00 mL. The pH of the solution is 6.00. Determine the K_a of cacodylic acid.

Polyprotic Acids

63. Arsenic acid (H_3AsO_4) is a weak triprotic acid. Given the three acid ionization constants from Table 15.2 and an initial concentration of arsenic acid (before ionization) of 0.1000 M, calculate the equilibrium concentrations of H_3AsO_4, $H_2AsO_4^-$, $HAsO_4^{2-}$, AsO_4^{3-}, and H_3O^+.

64. Phthalic acid ($H_2C_8H_4O_4$, abbreviated H_2Ph) is a diprotic acid. Its ionization in water at 25°C takes place in two steps:

$$H_2Ph(aq) + H_2O(\ell) \rightleftharpoons H_3O^+(aq) + HPh^-(aq) \qquad K_{a1} = 1.26 \times 10^{-3}$$

$$HPh^-(aq) + H_2O(\ell) \rightleftharpoons H_3O^+(aq) + Ph^{2-}(aq) \qquad K_{a2} = 3.10 \times 10^{-6}$$

If 0.0100 mol of phthalic acid is dissolved per liter of water, calculate the equilibrium concentrations of H_2Ph, HPh^-, Ph^{2-}, and H_3O^+.

65. A solution as initially prepared contains 0.050 mol L^{-1} of phosphate ion (PO_4^{3-}) at 25°C. Given the three acid ionization constants from Table 15.2, calculate the equilibrium concentrations of PO_4^{3-}, HPO_4^{2-}, $H_2PO_4^-$, H_3PO_4, and OH^-.

66. Oxalic acid ionizes in two stages in aqueous solution:

$$H_2C_2O_4(aq) + H_2O(\ell) \rightleftharpoons H_3O^+(aq) + HC_2O_4^-(aq) \qquad K_{a1} = 5.9 \times 10^{-2}$$

$$HC_2O_4^-(aq) + H_2O(\ell) \rightleftharpoons H_3O^+(aq) + C_2O_4^{2-}(aq) \qquad K_{a2} = 6.4 \times 10^{-5}$$

Calculate the equilibrium concentrations of $C_2O_4^{2-}$, $HC_2O_4^-$, $H_2C_2O_4$, and OH^- in a 0.10 M solution of sodium oxalate ($Na_2C_2O_4$).

67. The pH of a normal raindrop is 5.60. Compute the concentrations of $H_2CO_3(aq)$, $HCO_3^-(aq)$, and $CO_3^{2-}(aq)$ in this raindrop if the total concentration of dissolved carbonates is 1.0×10^{-5} mol L^{-1}.

68. The pH of a drop of acid rain is 4.00. Compute the concentrations of $H_2CO_3(aq)$, $HCO_3^-(aq)$, and $CO_3^{2-}(aq)$ in the acid raindrop if the total concentration of dissolved carbonates is 3.6×10^{-5} mol L^{-1}.

A DEEPER LOOK . . . Exact Treatment of Acid–Base Equilibria

69. Thiamine hydrochloride (vitamin B_1 hydrochloride, $HC_{12}H_{17}ON_4SCl_2$) is a weak acid with $K_a = 3.4 \times 10^{-7}$. Suppose 3.0×10^{-5} g of thiamine hydrochloride is dissolved in 1.00 L of water. Calculate the pH of the resulting solution. (*Hint:* This is a sufficiently dilute solution that the autoionization of water cannot be neglected.)

70. A sample of vinegar contains 40.0 g of acetic acid (CH_3COOH) per liter of solution. Suppose 1.00 mL is removed and diluted to 1.00 L, and 1.00 mL of *that* solution is removed and diluted to 1.00 L. Calculate the pH of the resulting solution. (*Hint:* This is a sufficiently dilute solution that the autoionization of water cannot be neglected.)

71. At 25°C, 50.00 mL of a 0.1000 M solution of maleic acid, a diprotic acid whose ionization constants are $K_{a1} = 1.42 \times 10^{-2}$ and $K_{a2} = 8.57 \times 10^{-7}$, is titrated with a 0.1000 M NaOH solution. Calculate the pH at the following volumes of added base: 0, 5.00, 25.00, 50.00, 75.00, 99.90, 100.00, and 105.00 mL.

72. Quinine ($C_{20}H_{24}O_2N_2$) is a water-soluble base that ionizes in two stages, with $K_{b1} = 3.31 \times 10^{-6}$ and $K_{b2} = 1.35 \times 10^{-10}$, at 25°C. Calculate the pH during the titration of an aqueous solution of 1.622 g of quinine in 100.00 mL of water as a function of the volume of added 0.1000 M HCl solution at the following volumes: 0, 25.00, 50.00, 75.00, 99.90, 100.00, and 105.00 mL.

Organic Acids and Bases: Structure and Reactivity

73. Use data from Table 15.4 to estimate the stabilization (in kJ per mol) associated with substituting a phenyl ($-C_6H_5$) group for one of the hydrogen atoms in methane. Assume that the effect on the pK_a enters entirely through the greater energetic stability of the conjugate base.

74. Use data from Table 15.4 to estimate the stabilization (in kJ per mol) associated with substituting a nitro ($-NO_2$) group for one of the hydrogen atoms in methane. Assume that the effect on the pK_a enters entirely through the greater energetic stability of the conjugate base.

75. Propionic acid, CH_3CH_2COOH, has a pK_a of 4.9. Compare this with the diprotic succinic acid, $HOOCCH_2CH_2COOH$. Will the pK_{a1} and pK_{a2} of succinic acid be larger than or smaller than 4.9?

76. Predict the relative magnitudes of the pK_a's for a carboxylic acid, RCOOH, a ketone, $RCOCH_3$, and an amide, $RCONH_2$.

77. Which will be the stronger acid: benzene (C_6H_6) or cyclohexane (C_6H_{12})? Explain by using resonance Lewis structures.

78. Which will be the stronger acid: propene ($CH_2{=}CHCH_3$) or propane (C_3H_8)? Explain by using resonance Lewis structures.

79. For each of the following pairs of molecules, predict which is the stronger acid.
(a) CF_3COOH or CCl_3COOH.
(b) $CH_2FCH_2CH_2COOH$ or $CH_3CH_2CHFCOOH$
(c) [benzene ring]—COOH or [benzene ring with $C(CH_3)_3$ at both ortho positions]—COOH

80. For each of the following pairs of molecules, predict which is the stronger acid.
(a) CH_3CH_2COOH or $CH_3{-}C(C(CH_3)_3)_2{-}COOH$
(b) CI_3COOH or CCl_3COOH
(c) $CH_3CHClCH_2COOH$ or $CH_3CH_2CHClCOOH$

ADDITIONAL PROBLEMS

81. Although acetic acid is normally regarded as a weak acid, it is about 34% dissociated in a 10^{-4} M solution at 25°C. It is less than 1% dissociated in 1 M solution. Discuss this variation in degree of dissociation with dilution in terms of Le Châtelier's principle, and explain how it is consistent with the supposed constancy of equilibrium constants.

82. Suppose that a 0.10 M aqueous solution of a monoprotic acid HX has just 11 times the conductivity of a 0.0010 M aqueous solution of HX. What is the approximate dissociation constant of HX? (*Hint:* In thinking about this problem, consider what the ratio of the conductivities would be if HX were a strong acid and if HX were extremely weak, as limiting cases.)

83. The ionization constant of chloroacetic acid ($ClCH_2COOH$) in water is 1.528×10^{-3} at 0°C and 1.230×10^{-3} at 40°C. Calculate the enthalpy of ionization of the acid in water, assuming that ΔH and ΔS are constant over this range of temperature.

84. The autoionization constant of water (K_w) is 1.139×10^{-15} at 0°C and 9.614×10^{-14} at 60°C.
(a) Calculate the enthalpy of autoionization of water.

$$2\,H_2O(\ell) \rightleftharpoons H_3O^+(aq) + OH^-(aq)$$

(b) Calculate the entropy of autoionization of water.
(c) At what temperature will the pH of pure water be 7.00, from these data?

85. Calculate the concentrations of H_3O^+ and OH^- at 25°C in the following:
(a) Orange juice (pH 2.8)
(b) Tomato juice (pH 3.9)
(c) Milk (pH 4.1)
(d) Borax solution (pH 8.5)
(e) Household ammonia (pH 11.9)

86. Try to choose which of the following is the pH of a 6.44×10^{-10} M $Ca(OH)_2(aq)$ solution, without doing any written calculations.
(a) 4.81 (b) 5.11
(c) 7.00 (d) 8.89
(e) 9.19

87. $Cl_2(aq)$ reacts with $H_2O(\ell)$ as follows:

$$Cl_2(aq) + 2\,H_2O(\ell) \rightleftharpoons H_3O^+(aq) + Cl^-(aq) + HOCl(aq)$$

For an experiment to succeed, $Cl_2(aq)$ must be present, but the amount of $Cl^-(aq)$ in the solution must be minimized. For this purpose, should the pH of the solution be high, low, or neutral? Explain.

88. Use the data in Table 15.2 to determine the equilibrium constant for the reaction

$$H_2PO_4^-(aq) + 2\,CO_3^{2-}(aq) \rightleftharpoons PO_4^{3-}(aq) + 2\,HCO_3^-(aq)$$

89. The first acid ionization constant of the oxoacid H_3PO_2 is 8×10^{-2}. What molecular structure do you predict for H_3PO_2? Will this acid be monoprotic, diprotic, or triprotic in aqueous solution?

90. Oxoacids can be formed that involve several central atoms of the same chemical element. An example is $H_3P_3O_9$, which can be written $P_3O_6(OH)_3$. (Sodium salts of these polyphosphoric acids are used as "builders" in detergents to improve their cleaning power.) In such a case, we would expect acid strength to correlate approximately with the *ratio* of the number of lone oxygen atoms to the number of central atoms (this ratio is 6:3 for $H_3P_3O_9$, for example). Rank the following in order of increasing acid strength: H_3PO_4, $H_3P_3O_9$, $H_4P_2O_6$, $H_4P_2O_7$, $H_5P_3O_{10}$. Assume that no hydrogen atoms are directly bonded to phosphorus in these compounds.

91. Urea (NH_2CONH_2) is a component of urine. It is a very weak base, having an estimated pK_b of 13.8 at room temperature.
(a) Write the formula of the conjugate acid of urea.
(b) Compute the equilibrium concentration of urea in a solution that contains no urea but starts out containing 0.15 mol L^{-1} of the conjugate acid of urea.

92. Exactly 1.0 L of solution of acetic acid gives the same color with methyl red as 1.0 L of a solution of hydrochloric acid. Which solution will neutralize the greater amount of 0.10 M $NaOH(aq)$? Explain.

* 93. The K_a for acetic acid drops from 1.76×10^{-5} at 25°C to 1.63×10^{-5} at 50°C. Between the same two temperatures, K_w increases from 1.00×10^{-14} to 5.47×10^{-14}. At 50°C the density of a 0.10 M solution of acetic acid is 98.81% of its density at 25°C. Will the pH of a 0.10 M solution of acetic acid in water increase, decrease, or remain the same when it is heated from 25°C to 50°C? Explain.

94. Calculate the pH of a solution that is prepared by dissolving 0.23 mol of hydrofluoric acid (HF) and 0.57 mol of hypochlorous acid (HClO) in water and diluting to 3.60 L. Also, calculate the equilibrium concentrations of HF, F^-, HClO, and ClO^-. (*Hint:* The pH will be determined by the stronger acid of this pair.)

95. For each of the following compounds, indicate whether a 0.100 M aqueous solution is acidic (pH < 7), basic (pH > 7), or neutral (pH = 7): HCl, NH_4Cl, KNO_3, Na_3PO_4, $NaCH_3COO$.

*96. Calculate $[H_3O^+]$ in a solution that contains 0.100 mol of NH_4CN per liter.

$$NH_4^+(aq) + H_2O(\ell) \rightleftharpoons H_3O^+(aq) + NH_3(aq) \qquad K_a = 5.6 \times 10^{-10}$$

$$HCN(aq) + H_2O(\ell) \rightleftharpoons H_3O^+(aq) + CN^-(aq) \qquad K_a = 6.17 \times 10^{-10}$$

97. Discuss the justification for this statement: "Although one does not normally regard NH_4^+ as an acid, it is actually only slightly weaker as an acid than hydrocyanic acid, HCN, in aqueous solution."

*98. Imagine that you want to do physiological experiments at a pH of 6.0 and the organism with which you are working is sensitive to most available materials other than a certain weak acid, H_2Z, and its sodium salts. K_{a1} and K_{a2} for H_2Z are 3×10^{-1} M and 5×10^{-7} M. You have available 1.0 M aqueous H_2Z and 1.0 M NaOH. How much of the NaOH solution should be added to 1.0 L of the acid solution to give a buffer at pH = 6.0?

99. A buffer solution is prepared by mixing 1.00 L of 0.050 M pentafluorobenzoic acid (C_6F_5COOH) and 1.00 L of 0.060 M sodium pentafluorobenzoate (NaC_6F_5COO). The K_a of this weak acid is 0.033. Determine the pH of the buffer solution.

100. A chemist needs to prepare a buffer solution with pH = 10.00 and has both Na_2CO_3 and $NaHCO_3$ in pure crystalline form. What mass of each should be dissolved in 1.00 L of solution if the combined mass of the two salts is to be 10.0 g?

101. Which of these procedures would *not* make a pH = 4.75 buffer?
   (a) Mix 50.0 mL of 0.10 M acetic acid and 50.0 mL of 0.10 M sodium acetate.
   (b) Mix 50.0 mL of 0.20 M acetic acid and 50.0 mL of 0.10 M NaOH.
   (c) Start with 50.0 mL of 0.20 M acetic acid and add a solution of strong base until the pH equals 4.75.
   (d) Start with 50.0 mL of 0.20 M HCl and add a solution of strong base until the pH equals 4.75.
   (e) Start with 100.0 mL of 0.20 M sodium acetate and add 50.0 mL of 0.20 M HCl.

102. It takes 4.71 mL of 0.0410 M NaOH to titrate a 50.00-mL sample of flat (no CO_2) GG's Cola to a pH of 4.9. At this point the addition of one more drop (0.02 mL) of NaOH raises the pH to 6.1. The only acid in GG's Cola is phosphoric acid. Compute the concentration of phosphoric acid in this cola. Assume that the 4.71 mL of base removes only the first hydrogen from the H_3PO_4; that is, assume that the reaction is

$$H_3PO_4(aq) + OH^-(aq) \longrightarrow H_2O(\ell) + H_2PO_4^-(aq)$$

*103. Sodium carbonate exists in various crystalline forms with different amounts of water of crystallization, including Na_2CO_3, $Na_2CO_3{\cdot}10H_2O$, and others. The water of crystallization can be driven off by heating; the amount of water removed depends on the temperature and duration of heating.

A sample of $Na_2CO_3{\cdot}10H_2O$ had been heated inadvertently, and it was not known how much water had been removed. A 0.200-g sample of the solid that remained after the heating was dissolved in water, 30.0 mL of 0.100 M NaOH was added, and the CO_2 formed was removed. The solution was acidic; 6.4 mL of 0.200 M NaOH was needed to neutralize the excess acid. What fraction of the water had been driven from the $Na_2CO_3{\cdot}10H_2O$?

104. An aqueous solution of sodium carbonate, Na_2CO_3, is titrated with strong acid to a point at which two H^+ ions have reacted with each carbonate ion. (a) If 20.0 mL of the carbonate solution reacts with just 40.0 mL of 0.50 M acid, what is the molarity of the carbonate solution? (b) If the solution contains 5.0 percent by mass sodium carbonate, what is the density of the solution? (c) Suppose that you wanted to prepare a liter of an identical solution by starting with crystalline sodium carbonate decahydrate, $Na_2CO_3{\cdot}10H_2O$, rather than with solid Na_2CO_3 itself. How much of this substance would you need?

105. Three flasks, labeled *A*, *B*, and *C*, contained aqueous solutions of the same pH. It was known that one of the solutions was 1.0×10^{-3} M in nitric acid, one was 6×10^{-3} M in formic acid, and one was 4×10^{-2} M in the salt formed by the weak organic base aniline with hydrochloric acid ($C_6H_5NH_3Cl$). (Formic acid is monoprotic.) (a) Describe a procedure for identifying the solutions. (b) Compare qualitatively (on the basis of the preceding information) the strengths of nitric and formic acids with each other and with the acid strength of the anilinium ion, $C_6H_5NH_3^+$. (c) Show how the information given may be used to derive values for K_a for formic acid and K_b for aniline. Derive these values.

106. Novocain, the commonly used local anaesthetic, is a weak base with $K_b = 7 \times 10^{-6}$ M. (a) If you had a 0.0200 M solution of Novocain in water, what would be the approximate concentration of OH^- and the pH? (b) Suppose that you wanted to determine the concentration of Novocain in a solution that is about 0.020 M by titration with 0.020 M HCl. Calculate the expected pH at the equivalence point.

107. A 0.1000 M solution of a weak acid, HA, requires 50.00 mL of 0.1000 M NaOH to titrate it to its equivalence point. The pH of the solution is 4.50 when only 40.00 mL of the base has been added.
   (a) Calculate the ionization constant K_a of the acid.
   (b) Calculate the pH of the solution at the equivalence point.

108. The chief chemist of Victory Vinegar Works, Ltd., interviews two chemists for employment. He states, "Quality control requires that our high-grade vinegar contain 5.00 ± 0.01% acetic acid by mass. How would you analyze our product to ensure that it meets this specification?"

Anne Dalton says, "I would titrate a 50.00-mL sample of the vinegar with 1.000 M NaOH, using phenolphthalein to detect the equivalence point to within ±0.02 mL of base."

Charlie Cannizzarro says, "I would use a pH meter to determine the pH to ±0.01 pH units and interface it with a computer to print out the mass percentage of acetic acid."

Which candidate did the chief chemist hire? Why?

109. Phosphonocarboxylic acid

```
         :Ö  H  Ö:
          ‖  |  ‖   ..
  H—Ö—P—C—C—O—H
     ..   |  |      ..
       H—Ö: H
```

effectively inhibits the replication of the herpes virus. Structurally, it is a combination of phosphoric acid and acetic acid. It can donate three protons. The equilibrium constant values are $K_{a1} = 1.0 \times 10^{-2}$, $K_{a2} = 7.8 \times 10^{-6}$, and $K_{a3} = 2.0 \times 10^{-9}$. Enough phosphonocarboxylic acid is added to blood (pH 7.40) to make its total concentration 1.0×10^{-5} M. The pH of the blood does not change. Determine the concentrations of all four forms of the acid in this mixture.

110. Egg whites contain dissolved carbon dioxide and water, which react together to give carbonic acid (H_2CO_3). In the few days after an egg is laid, it loses carbon dioxide through its shell. Does the pH of the egg white increase or decrease during this period?

111. If you breathe too rapidly (hyperventilate), the concentration of dissolved CO_2 in your blood drops. What effect does this have on the pH of the blood?

112. A reference book states that a saturated aqueous solution of potassium hydrogen tartrate is a buffer with a pH of 3.56. Write two chemical equations that show the buffer action of this solution. (Tartaric acid is a diprotic acid with the formula $H_2C_4H_4O_6$. Potassium hydrogen tartrate is $KHC_4H_4O_6$.)

***113.** Glycine, the simplest amino acid, has both an acid group and a basic group in its structure ($H_2N{-}CH_2{-}COOH$). In aqueous solution it exists predominantly as a self-neutralized species called a zwitterion ($H_3\overset{+}{N}{-}CH_2{-}COO^-$). The zwitterion therefore behaves both as an acid and as a base, according to the equilibria at 25°C:

$$H_3\overset{+}{N}{-}CH_2{-}COO^-(aq) + H_2O(\ell) \rightleftharpoons H_2N{-}CH_2{-}COO^-(aq) + H_3O^+(aq) \qquad K_a = 1.7 \times 10^{-10}$$

$$H_3\overset{+}{N}{-}CH_2{-}COO^-(aq) + H_2O(\ell) \rightleftharpoons H_3\overset{+}{N}{-}CH_2{-}COOH(aq) + OH^-(aq) \qquad K_b = 2.2 \times 10^{-12}$$

Calculate the pH of a 0.10 M aqueous solution of glycine at 25°C. (*Hint:* You may need to take account of the autoionization of water.)

114. Use the effect of steric hindrance to predict whether a tertiary amine should be a stronger base than ammonia in aqueous solution. (*Hint:* Assume that the effect of solvation is greater for ions than for neutral species.)

***115.** Consider the two following nitrophenol structures:

OH / NO_2 — *p*-nitrophenol　　OH / NO_2 — *m*-nitrophenol

Predict which will be the stronger acid. (*Hint:* Consider possible resonance structures analogous to those given in the text for phenol.)

CUMULATIVE PROBLEMS

116. Use the data from Table 15.1, together with Le Châtelier's principle, to decide whether the autoionization of water is exothermic or endothermic.

117. Baking soda (sodium hydrogen carbonate, $NaHCO_3$) is used in baking because it reacts with acids in foods to form carbonic acid (H_2CO_3), which in turn decomposes to water and carbon dioxide. In a batter the carbon dioxide appears as gas bubbles that cause the bread or cake to rise.

(a) A rule of thumb in cooking is that $\frac{1}{2}$ teaspoon baking soda is neutralized by 1 cup of sour milk. The acid component of sour milk is lactic acid ($HC_3H_5O_3$). Write an equation for the neutralization reaction.

(b) If the density of baking soda is 2.16 g cm^{-3}, calculate the concentration of lactic acid in the sour milk, in moles per liter. Take 1 cup = 236.6 mL = 48 teaspoons.

(c) Calculate the volume of carbon dioxide that is produced at 1 atm pressure and 350°F (177°C) from the reaction of $\frac{1}{2}$ teaspoon of baking soda.

118. Boric acid, $B(OH)_3$, is an acid that acts differently from the usual Brønsted–Lowry acids. It reacts with water according to

$$B(OH)_3(aq) + 2\,H_2O(\ell) \rightleftharpoons B(OH)_4^-(aq) + H_3O^+(aq) \qquad K_a = 5.8 \times 10^{-10}$$

(a) Draw Lewis structures for $B(OH)_3$ and $B(OH)_4^-$. Can these be described as Lewis acids or Lewis bases?

(b) Calculate the pH of a 0.20 M solution of $B(OH)_3(aq)$.

119. At 40°C and 1.00 atm pressure, a gaseous monoprotic acid has a density of 1.05 g L^{-1}. After 1.85 g of this gas is dissolved in water and diluted to 450 mL, the pH is measured to be 5.01. Determine the K_a of this acid and use Table 15.2 to identify it.

120. At 25°C, the Henry's law constant for carbon dioxide dissolved in water is 1.8×10^3 atm. Calculate the pH of water saturated with $CO_2(g)$ at 25°C in Denver, where the barometric pressure is 0.833 atm.

CHAPTER 16

Solubility and Precipitation Equilibria

Sodium acetate crystals ($NaC_2H_3O_2$) form quickly in a supersaturated solution when a small speck of solute is added.

Dissolution and precipitation are chemical reactions by which solids pass into and out of solution. A brief introduction and several examples appear in Chapter 11. These reactions involve equilibria between dissolved species and species in the solid state, and so are described by the general principles of chemical equilibrium in Chapter 14. These reactions rank alongside acid–base reactions in practical importance. The dissolution and reprecipitation of solids permit chemists to isolate single products from reaction mixtures and to purify impure solid samples. Understanding the mechanisms of these reactions helps engineers prevent formation of deposits in water processing and distribution systems and helps doctors reduce the incidence of painful kidney stones. Dissolution and precipitation control the formation of mineral deposits and profoundly affect the ecologies of rivers, lakes, and oceans.

Our theme throughout this chapter is to manipulate equilibria to control the solubilities of ionic solids. In the first section we describe general features of the equilibria that govern dissolution and precipitation. In the remaining sections we explore quantitative aspects of these equilibria, including the effects of additional solutes, of acids and bases, and of ligands that can bind to metal ions to form complex ions.

16.1 The Nature of Solubility Equilibria

General Features of Solubility Equilibria

Solubility equilibria resemble the equilibria between volatile liquids (or solids) and their vapors in a closed container. In both cases, particles from a condensed phase tend to escape and spread through a larger, but limited, volume. In both cases, equilibrium is a dynamic compromise in which the rate of escape of particles from the condensed phase is equal to their rate of return. In a vaporization–condensation equilibrium, we assumed that the vapor above the condensed phase was an ideal gas. The analogous starting assumption for a dissolution–precipitation reaction is that the solution above the undissolved solid is an ideal solution. A solution in which sufficient solute has been dissolved to establish a dissolution–precipitation equilibrium between the solid substance and its dissolved form is called a **saturated solution.**

Le Châtelier's principle applies to these equilibria, as it does to all equilibria. One way to exert a stress on a solubility equilibrium is to change the amount of solvent. Adding solvent reduces the concentration of dissolved substance; more solid then tends to dissolve to restore the concentration of the dissolved substance to its equilibrium value. If an excess of solvent is added so that all of the solid dissolves, then obviously the solubility equilibrium ceases to exist and the solution is **unsaturated.** In a vaporization–condensation equilibrium, this corresponds to the complete evaporation of the condensed phase. Removing solvent from an already saturated solution forces additional solid to precipitate in order to maintain a constant concentration. A volatile solvent is often removed by simply letting a solution stand uncovered until the solvent evaporates. When conditions are right, the solid forms as crystals on the bottom and sides of the container (Fig. 16.1).

FIGURE 16.1 The beaker contains an aqueous solution of K_2PtCl_4, a substance used as a starting material in the synthesis of anticancer drugs. It is loosely covered to keep out dust and allowed to stand. As the water evaporates, the solution becomes saturated and deposits solid K_2PtCl_4 in the form of long, needle-like red crystals.

Controlled precipitation by manipulating solubility is a widely used technique for purifying reaction products in synthetic chemistry. Side reactions can generate significant amounts of impurities; other impurities enter with the starting materials or are put in deliberately to increase the reaction rate. Running a reaction may take only hours, but the *workup* (separation of crude product) and subsequent purification may require weeks. **Recrystallization,** one of the most powerful methods for purifying solids, relies on differences between the solubilities of the desired substance and its contaminants. An impure product is dissolved and reprecipitated, repeatedly if necessary, with careful control of the factors that influence solubility. Manipulating solubility requires an understanding of the equilibria that exist between an undissolved substance and its solution.

In recrystallization a solution begins to deposit a compound when it is brought to the point of saturation with respect to that compound. In dissolution the solvent attacks the solid and **solvates** it at the level of individual particles. In precipitation

FIGURE 16.2 Vastly different quantities of different compounds will dissolve in 1 L of water at 20°C. Clockwise from the front are borax, potassium permanganate, lead(II) chloride, sodium phosphate decahydrate, calcium oxide, and potassium dichromate.

the reverse occurs: Solute-to-solute attractions are reestablished as the solute leaves the solution. Often, solute-to-solvent attractions persist right through the process of precipitation, and solvent incorporates itself into the solid. When lithium sulfate (Li_2SO_4) precipitates from water, it brings with it into the solid one molecule of water per formula unit:

$$2\,Li^+(aq) + SO_4^{2-}(aq) + H_2O(\ell) \longrightarrow Li_2SO_4 \cdot H_2O(s)$$

Such loosely bound solvent is known as solvent of crystallization (Fig. 2.1). Dissolving and then reprecipitating a compound may thus furnish material that has a different chemical formula and a different mass. Consequently, recrystallization processes for purification of reaction products must be planned carefully.

Dissolution–precipitation reactions frequently come to equilibrium slowly. Days or even weeks of shaking a solid in contact with a solvent may be required before the solution becomes saturated. Moreover, solutions sometimes become **supersaturated,** a condition in which the concentration of dissolved solid *exceeds* its equilibrium value. The delay, then, is in forming, not dissolving, the solid. A supersaturated solution may persist for months or years and require extraordinary measures to be brought to equilibrium, although thermodynamics shows that equilibrium is possible all along. The generally sluggish approach to equilibrium in dissolution–precipitation reactions is quite the opposite of the rapid rates at which acid–base reactions reach equilibrium.

The Solubility of Ionic Solids

The **solubility** of a substance in a solvent is defined as the greatest amount (expressed either in grams or in moles) that will dissolve in equilibrium in a specified volume of solvent at a particular temperature. Although solvents other than water are used in many applications, aqueous solutions are the most important and are the exclusive concern here. Salts show a wide range of solubilities in water (Fig. 16.2). Silver perchlorate ($AgClO_4$) dissolves to the remarkable extent of about 5570 g (or almost 27 mol) per liter of water at 25°C, but at the same temperature only about 0.0018 g (or 1.3×10^{-5} mol) of silver chloride (AgCl)

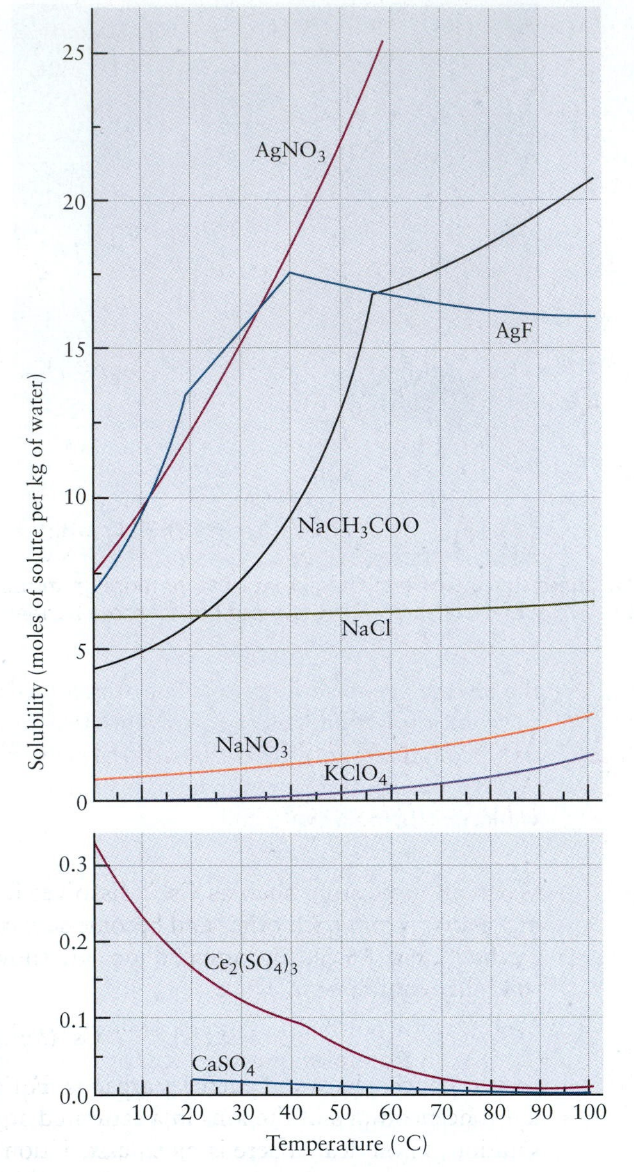

FIGURE 16.3 Most solubilities increase with increasing temperature, but some decrease. Note that the changes are not always smooth because different solid hydrates form over different temperature ranges.

FIGURE 16.4 A yellow precipitate of lead iodide forms when a drop of KI solution is added to a solution of $Pb(NO_3)_2$.

dissolves per liter of water. Many salts with even lower solubilities are known. Solubilities often depend strongly on temperature (Fig. 16.3). Most dissolution reactions for ionic solids are endothermic (heat is absorbed), so by Le Châtelier's principle the solubility increases with increasing temperature. Those dissolution reactions that are exothermic (such as for $CaSO_4$) show the opposite behavior.

Although all ionic compounds dissolve to some extent in water, those having solubilities (at 25°C) of less than 0.1 g L^{-1} are called *insoluble*. Those having solubilities of more than 10 g L^{-1} are *soluble*, and the intermediate cases (0.1 to 10 g L^{-1}) are said to be *slightly soluble*. Fortunately, it is not necessary to memorize long lists of solubility data. Table 16.1 lists some generalizations concerning groups of salts and gives enough factual data to support good predictions about precipitation or dissolution in thousands of situations of practical importance. Knowing the solubilities of ionic substances, even in these qualitative terms, provides a way to predict the courses of numerous reactions. For example, when a solution of KI is added to one of $Pb(NO_3)_2$, K^+ and NO_3^- ions are brought into contact, as are Pb^{2+} and I^- ions. From the table, KNO_3 is a soluble salt but PbI_2 is insoluble; therefore, a precipitate of PbI_2 will appear (Fig. 16.4).

TABLE 16.1 Solubilities of Ionic Compounds in Water

Anion	Soluble†	Slightly Soluble	Insoluble
NO_3^- (nitrate)	All	—	—
CH_3COO^- (acetate)	Most	—	$Be(CH_3COO)_2$
ClO_3^- (chlorate)	All	—	—
ClO_4^- (perchlorate)	Most	$KClO_4$	—
F^- (fluoride)	Group I, AgF, BeF_2	SrF_2, BaF_2, PbF_2	MgF_2, CaF_2
Cl^- (chloride)	Most	$PbCl_2$	AgCl, Hg_2Cl_2
Br^- (bromide)	Most	$PbBr_2$, $HgBr_2$	AgBr, Hg_2Br_2
I^- (iodide)	Most	—	AgI, Hg_2I_2, PbI_2, HgI_2
SO_4^{2-} (sulfate)	Most	$CaSO_4$, Ag_2SO_4, Hg_2SO_4	$SrSO_4$, $BaSO_4$, $PbSO_4$
S^{2-} (sulfide)	Groups I and II, $(NH_4)_2S$	—	Most
CO_3^{2-} (carbonate)	Group I, $(NH_4)_2CO_3$	—	Most
SO_3^{2-} (sulfite)	Group I, $(NH_4)_2SO_3$	—	Most
PO_4^{3-} (phosphate)	Group I, $(NH_4)_3PO_4$	—	Most
OH^- (hydroxide)	Group I, $Ba(OH)_2$	$Sr(OH)_2$, $Ca(OH)_2$	Most

†Soluble compounds are defined as those that dissolve to the extent of 10 or more grams per liter; slightly soluble compounds, 0.1 to 10 grams per liter; and insoluble compounds, less than 0.1 gram per liter at room temperature.

16.2 Ionic Equilibria between Solids and Solutions

When an ionic solid such as CsCl dissolves in water, it breaks up into ions that move apart from each other and become solvated by water molecules (*aquated*, or *hydrated;* Fig. 16.5). The aquated ions are shown in the chemical equation for the solubility equilibrium. Thus,

$$CsCl(s) \rightleftharpoons Cs^+(aq) + Cl^-(aq)$$

shows that the dissolved particles are ions. For a highly soluble salt (such as CsCl), the concentrations of the ions in a saturated aqueous solution are so large that the solution is nonideal. There is much association among the ions in solution, resulting in temporary pairs of oppositely charged ions and in larger clusters as well. In such cases the simple type of equilibrium expression developed in Sections 14.2 and 14.3 does not apply; therefore, we restrict our attention to sparingly soluble and "insoluble" salts, for which the concentrations in a saturated solution are 0.1 mol L^{-1} or less—low enough that the interactions among the solvated ions are relatively small.

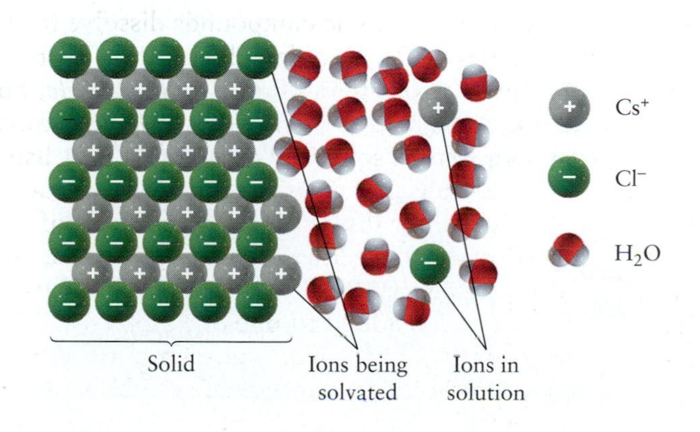

FIGURE 16.5 The dissolution of the ionic solid CsCl in water. Note the different solvations of positive and negative ions. Water molecules arrange themselves so the negative ends of their dipoles (the oxygen atoms) point toward the positively charged Cs^+ ions. The water molecules form hydrogen bonds to the Cl^- anions, so the O—H—Cl atoms line nearly in a straight line.

The sparingly soluble salt silver chloride, for example, establishes the following equilibrium when placed in water:

$$AgCl(s) \rightleftharpoons Ag^{+}(aq) + Cl^{-}(aq)$$

The equilibrium law for this reaction (written by following the rules for heterogeneous equilibria from Sections 14.2 and 14.3) is

$$[Ag^{+}][Cl^{-}] = K_{sp}$$

where the subscript "sp," standing for **"solubility product,"** distinguishes the K as referring to the dissolution of a slightly soluble ionic solid in water. At 25°C, K_{sp} has the numerical value 1.6×10^{-10} for silver chloride. In the AgCl solubility product expression, the concentrations of the two ions produced are raised to the first power because their coefficients are 1 in the chemical equation. Solid AgCl does not appear in the equilibrium expression; the amount of pure solid AgCl does not affect the equilibrium as long as *some* is present. If *no* solid is present, then the product of the two ion concentrations is no longer constrained by the solubility product expression.

Solubility and K_{sp}

The molar solubility of a salt in water is not the same as its solubility product constant, but a simple relation often exists between them. For example, let's define S

TABLE 16.2 Solubility Product Constants K_{sp} at 25°C

Salt	K_{sp}
Iodates	
$AgIO_3$	$[Ag^{+}][IO_3^{-}] = 3.1 \times 10^{-8}$
$CuIO_3$	$[Cu^{+}][IO_3^{-}] = 1.4 \times 10^{-7}$
$Pb(IO_3)_2$	$[Pb^{2+}][IO_3^{-}]^2 = 2.6 \times 10^{-13}$
Carbonates	
Ag_2CO_3	$[Ag^{+}]^2[CO_3^{2-}] = 6.2 \times 10^{-12}$
$BaCO_3$	$[Ba^{2+}][CO_3^{2-}] = 8.1 \times 10^{-9}$
$CaCO_3$	$[Ca^{2+}][CO_3^{2-}] = 8.7 \times 10^{-9}$
$PbCO_3$	$[Pb^{2+}][CO_3^{2-}] = 3.3 \times 10^{-14}$
$MgCO_3$	$[Mg^{2+}][CO_3^{2-}] = 4.0 \times 10^{-5}$
$SrCO_3$	$[Sr^{2+}][CO_3^{2-}] = 1.6 \times 10^{-9}$
Chromates	
Ag_2CrO_4	$[Ag^{+}]^2[CrO_4^{2-}] = 1.9 \times 10^{-12}$
$BaCrO_4$	$[Ba^{2+}][CrO_4^{2-}] = 2.1 \times 10^{-10}$
$PbCrO_4$	$[Pb^{2+}][CrO_4^{2-}] = 1.8 \times 10^{-14}$
Oxalates	
CuC_2O_4	$[Cu^{2+}][C_2O_4^{2-}] = 2.9 \times 10^{-8}$
FeC_2O_4	$[Fe^{2+}][C_2O_4^{2-}] = 2.1 \times 10^{-7}$
MgC_2O_4	$[Mg^{2+}][C_2O_4^{2-}] = 8.6 \times 10^{-5}$
PbC_2O_4	$[Pb^{2+}][C_2O_4^{2-}] = 2.7 \times 10^{-11}$
SrC_2O_4	$[Sr^{2+}][C_2O_4^{2-}] = 5.6 \times 10^{-8}$
Sulfates	
$BaSO_4$	$[Ba^{2+}][SO_4^{2-}] = 1.1 \times 10^{-10}$
$CaSO_4$	$[Ca^{2+}][SO_4^{2-}] = 2.4 \times 10^{-5}$
$PbSO_4$	$[Pb^{2+}][SO_4^{2-}] = 1.1 \times 10^{-8}$
Fluorides	
BaF_2	$[Ba^{2+}][F^{-}]^2 = 1.7 \times 10^{-6}$
CaF_2	$[Ca^{2+}][F^{-}]^2 = 3.9 \times 10^{-11}$
MgF_2	$[Mg^{2+}][F^{-}]^2 = 6.6 \times 10^{-9}$
PbF_2	$[Pb^{2+}][F^{-}]^2 = 3.6 \times 10^{-8}$
SrF_2	$[Sr^{2+}][F^{-}]^2 = 2.8 \times 10^{-9}$
Chlorides	
$AgCl$	$[Ag^{+}][Cl^{-}] = 1.6 \times 10^{-10}$
$CuCl$	$[Cu^{+}][Cl^{-}] = 1.0 \times 10^{-6}$
Hg_2Cl_2	$[Hg_2^{2+}][Cl^{-}]^2 = 2 \times 10^{-18}$
Bromides	
$AgBr$	$[Ag^{+}][Br^{-}] = 7.7 \times 10^{-13}$
$CuBr$	$[Cu^{+}][Br^{-}] = 4.2 \times 10^{-8}$
Hg_2Br_2	$[Hg_2^{2+}][Br^{-}]^2 = 1.3 \times 10^{-21}$
Iodides	
AgI	$[Ag^{+}][I^{-}] = 1.5 \times 10^{-16}$
CuI	$[Cu^{+}][I^{-}] = 5.1 \times 10^{-12}$
PbI_2	$[Pb^{2+}][I^{-}]^2 = 1.4 \times 10^{-8}$
Hg_2I_2	$[Hg_2^{2+}][I^{-}]^2 = 1.2 \times 10^{-28}$
Hydroxides	
$AgOH$	$[Ag^{+}][OH^{-}] = 1.5 \times 10^{-8}$
$Al(OH)_3$	$[Al^{3+}][OH^{-}]^3 = 3.7 \times 10^{-15}$
$Fe(OH)_3$	$[Fe^{3+}][OH^{-}]^3 = 1.1 \times 10^{-36}$
$Fe(OH)_2$	$[Fe^{2+}][OH^{-}]^2 = 1.6 \times 10^{-14}$
$Mg(OH)_2$	$[Mg^{2+}][OH^{-}]^2 = 1.2 \times 10^{-11}$
$Mn(OH)_2$	$[Mn^{2+}][OH^{-}]^2 = 2 \times 10^{-13}$
$Zn(OH)_2$	$[Zn^{2+}][OH^{-}]^2 = 4.5 \times 10^{-17}$

as the molar solubility of AgCl(s) in water at 25°C. Then, from stoichiometry, $[Ag^+] = [Cl^-] = S$ is the molarity of either ion at equilibrium. Hence,

$$[Ag^+][Cl^-] = S^2 = K_{sp} = 1.6 \times 10^{-10}$$

Taking the square roots of both sides of the equation gives

$$S = 1.26 \times 10^{-5} \text{ M}$$

which rounds off to 1.3×10^{-5} M. This is the molar solubility of AgCl in water. It is converted to a gram solubility by multiplying by the molar mass of AgCl:

$$(1.26 \times 10^{-5} \text{ mol L}^{-1})(143.3 \text{ g mol}^{-1}) = 1.8 \times 10^{-3} \text{ g L}^{-1}$$

Therefore, 1.8×10^{-3} g of AgCl dissolves per liter of water at 25°C.

Solubility product constants (like solubilities) can be sensitive to temperature. At 100°C the K_{sp} for silver chloride is 2.2×10^{-8}; hot water dissolves about 12 times as much silver chloride as does water at 25°C. Refer to Table 16.2 for the solubility product constants at 25°C of a number of important sparingly soluble salts.

EXAMPLE 16.1

The K_{sp} of calcium fluoride is 3.9×10^{-11}. Calculate the concentrations of calcium and fluoride ions in a saturated solution of CaF_2 at 25°C, and determine the solubility of CaF_2 in grams per liter.

SOLUTION

The solubility equilibrium is

$$CaF_2(s) \rightleftharpoons Ca^{2+}(aq) + 2\,F^-(aq)$$

and the expression for the solubility product is

$$[Ca^{2+}][F^-]^2 = K_{sp}$$

The concentration of fluoride ion is squared because it has a coefficient of 2 in the chemical equation.

If S mol of CaF_2 dissolves in 1 L, the equilibrium concentration of Ca^{2+} will be $[Ca^{2+}] = S$. The concentration of F^- will be $[F^-] = 2S$, because each mole of CaF_2 produces *two* moles of fluoride ions. Therefore,

$$[Ca^{2+}][F^-]^2 = S \times (2S)^2 = 4S^3 = K_{sp} = 3.9 \times 10^{-11}$$

Solving for S^3 gives

$$S^3 = \tfrac{1}{4}(3.9 \times 10^{-11})$$

Taking the cube roots of both sides of this equation gives

$$S = 2.1 \times 10^{-4}$$

The equilibrium concentrations are therefore

$$[Ca^{2+}] = S = 2.1 \times 10^{-4} \text{ M}$$

$$[F^-] = 2S = 4.3 \times 10^{-4} \text{ M}$$

Because the molar mass of CaF_2 is 78.1 g mol^{-1}, the gram solubility is

$$\text{gram solubility} = (2.1 \times 10^{-4} \text{ mol L}^{-1})(78.1 \text{ g mol}^{-1}) = 0.017 \text{ g L}^{-1}$$

Related Problems: 7, 8, 9, 10, 11, 12

It is possible to reverse the procedure just outlined, of course, and determine the value of K_{sp} from measured solubilities, as the following example illustrates.

EXAMPLE 16.2

Silver chromate (Ag_2CrO_4) is a red solid that dissolves in water to the extent of 0.029 g L^{-1} at 25°C. Estimate its K_{sp} and compare your estimate with the value in Table 16.2.

SOLUTION

The molar solubility is calculated from the gram solubility and the molar mass of silver chromate (331.73 g mol^{-1}):

$$\text{molar solubility} = \frac{0.029 \text{ g L}^{-1}}{331.73 \text{ g mol}^{-1}} = 8.74 \times 10^{-5} \text{ mol L}^{-1}$$

An extra significant digit is carried in this intermediate result to avoid round-off errors. Because each mole of Ag_2CrO_4 that dissolves gives *two* moles of silver ion, the concentration of $Ag^+(aq)$ is

$$[Ag^+] = 2 \times 8.74 \times 10^{-5} \text{ M} = 1.75 \times 10^{-4} \text{ M}$$

and that of CrO_4^{2-} is simply 8.74×10^{-5} M. The solubility product constant is then

$$K_{sp} = [Ag^+]^2[CrO_4^{2-}] = (1.75 \times 10^{-4})^2 \times (8.74 \times 10^{-5})$$
$$= 2.7 \times 10^{-12}$$

This estimate is about 42% greater than the tabulated value, 1.9×10^{-12}.

Related Problems: 13, 14, 15, 16

Computing K_{sp} from a solubility and solubility from K_{sp} is valid if the solution is ideal and if there are no side reactions that reduce the concentrations of the ions after they enter solution. If such reactions are present, they cause higher solubilities than are predicted from the K_{sp} expression. For example, the solubility computed for $PbSO_4(s)$ at 25°C from its K_{sp} is 0.032 g L^{-1}, whereas that measured experimentally is 0.0425 g L^{-1}. The difference is caused by the presence of species other than $Pb^{2+}(aq)$ in solution, such as $PbOH^+(aq)$. Further discussion of these side reactions is deferred to Section 16.6.

16.3 Precipitation and the Solubility Product

So far, we have considered only cases in which a single slightly soluble salt attains equilibrium with its component ions in water. The relative concentrations of the cations and anions in such solutions echo their relative numbers of moles in the original salt. Thus, when AgCl is dissolved, equal numbers of moles of $Ag^+(aq)$ and $Cl^-(aq)$ ions result, and when Ag_2SO_4 is dissolved, twice as many moles of $Ag^+(aq)$ ions as $SO_4^{2-}(aq)$ are produced. A solubility product relationship such as

$$[Ag^+][Cl^-] = K_{sp}$$

is more general than this, however, and continues in force even if the relative number of moles of the two ions in solution differ from those in the pure solid compound. Such a situation often results when two solutions are mixed to give a precipitate or when another salt is present that contains an ion common to the salt under consideration.

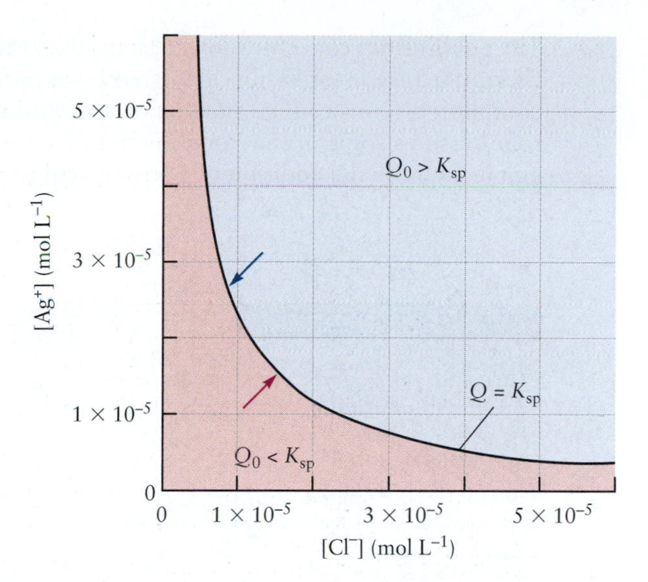

FIGURE 16.6 Some solid silver chloride is in contact with a solution containing $Ag^+(aq)$ and $Cl^-(aq)$ ions. If a solubility equilibrium exists, then the product Q of the concentrations of the ions $[Ag^+] \times [Cl^-]$ is a constant, K_{sp} (curved line). When Q exceeds K_{sp}, solid silver chloride tends to precipitate until equilibrium is attained. When Q is less than K_{sp}, additional solid tends to dissolve. If no solid is present, Q remains less than K_{sp}.

Precipitation from Solution

Suppose a solution is prepared by mixing one soluble salt, such as $AgNO_3$, with a solution of a second, such as NaCl. Will a precipitate of very slightly soluble silver chloride form? To reach an answer, the reaction quotient Q that was defined in connection with gaseous equilibria (see Section 14.6) is used. The initial reaction quotient Q_0, when the mixing of the solutions is complete but before any reaction occurs, is

$$Q_0 = [Ag^+]_0[Cl^-]_0$$

If $Q_0 < K_{sp}$, no solid silver chloride can appear. On the other hand, if $Q_0 > K_{sp}$, solid silver chloride precipitates until the reaction quotient Q reaches K_{sp} (Fig. 16.6).

EXAMPLE 16.3

An emulsion of silver chloride for photographic film is prepared by adding a soluble chloride salt to a solution of silver nitrate. Suppose 500 mL of a solution of $CaCl_2$ with a chloride ion concentration of 8.0×10^{-6} M is added to 300 mL of a 0.0040 M solution of $AgNO_3$. Will a precipitate of AgCl(*s*) form when equilibrium is reached?

SOLUTION

The "initial concentrations" to be used in calculating Q_0 are those *before* reaction but *after* dilution through mixing the two solutions. The initial concentration of $Ag^+(aq)$ after dilution from 300 mL to 800 mL of solution is

$$[Ag^+]_0 = 0.00400 \text{ M} \times \left(\frac{300 \text{ mL}}{800 \text{ mL}}\right) = 0.0015 \text{ M}$$

and that of $Cl^-(aq)$ is

$$[Cl^-]_0 = 8.0 \times 10^{-6} \text{ M} \times \left(\frac{500 \text{ mL}}{800 \text{ mL}}\right) = 5.0 \times 10^{-6} \text{ M}$$

The initial reaction quotient is

$$Q_0 = [Ag^+]_0[Cl^-]_0 = (0.0015)(5.0 \times 10^{-6}) = 7.5 \times 10^{-9}$$

Because $Q_0 > K_{sp}$, a precipitate of silver chloride appears at equilibrium, although there may be too little to detect visually. Another possible precipitate, calcium nitrate, is far too soluble to form in this experiment (see Table 16.1).

Related Problems: 17, 18, 19, 20

The equilibrium concentrations of ions after the mixing of two solutions to give a precipitate are most easily calculated by supposing that the reaction first goes to completion (consuming one type of ion) and that subsequent dissolution of the solid restores some of that ionic species to solution—just the approach used in Example 15.8 for the addition of a strong acid to a buffer solution.

EXAMPLE 16.4

Calculate the equilibrium concentrations of silver and chloride ions resulting from the precipitation reaction of Example 16.3.

SOLUTION

In this case the silver ion is clearly in excess; therefore, the chloride ion is the limiting reactant. If all of it were used up to make solid AgCl, the concentration of the remaining silver ion would be

$$[Ag^+] = 0.0015 - 5.0 \times 10^{-6} = 0.0015 \text{ M}$$

Set up the equilibrium calculation as

	$AgCl(s) \rightleftharpoons$ $Ag^+(aq)$	+ $Cl^-(aq)$
Initial concentration (M)	0.0015	0
Change in concentration (M)	$+y$	$+y$
Equilibrium concentration (M)	$0.0015 + y$	y

so that the equilibrium expression is

$$(0.0015 + y)y = K_{sp} = 1.6 \times 10^{-10}$$

This quadratic equation can be solved by use of the quadratic formula, provided the calculator carries ten significant figures. It can be solved more easily by making the approximation that y is much smaller than 0.0015; the equation therefore simplifies to

$$0.0015y \approx 1.6 \times 10^{-10}$$

$$y \approx 1.1 \times 10^{-7} \text{ M} = [Cl^-]$$

The assumption about the size of y was justified. The concentration of silver ion is

$$[Ag^+] = 0.0015 \text{ M}$$

Related Problems: 21, 22, 23, 24

The Common-Ion Effect

Suppose a small amount of NaCl(s) is added to a saturated solution of AgCl. What happens? Sodium chloride is quite soluble in water and dissolves to form $Na^+(aq)$ and $Cl^-(aq)$ ions, raising the concentration of chloride ion. The quantity $Q_0 = [Ag^+][Cl^-]$ then exceeds the K_{sp} of silver chloride, and silver chloride precipitates until the concentrations of $Ag^+(aq)$ and $Cl^-(aq)$ are sufficiently reduced that the solubility product expression once again is satisfied.

The same equilibrium may be approached from the other direction. The amount of AgCl(s) that can dissolve in a solution of sodium chloride is less than the amount that could dissolve in the same volume of pure water. Because $[Ag^+][Cl^-] = K_{sp}$, a graph of the equilibrium concentration of silver ion against chloride concentration has the form of a hyperbola (Fig. 16.7). The presence of excess $Cl^-(aq)$ reduces the concentration of $Ag^+(aq)$ permitted, and the solubility

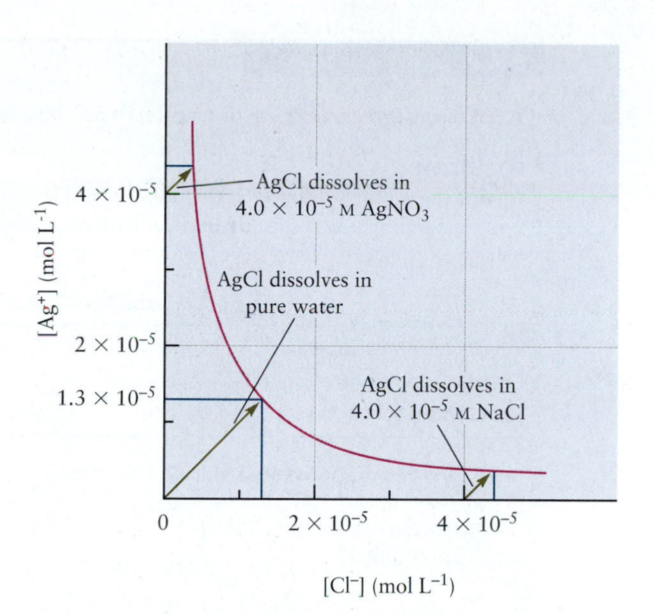

FIGURE 16.7 The presence of a dissolved common ion reduces the solubility of a salt in solution. As the AgCl dissolves, the concentrations of the ions follow the paths shown by the green arrows until they reach the red equilibrium curve. The molar solubilities are proportional to the lengths of the blue lines: 1.3×10^{-5} mol L^{-1} for AgCl in pure water, but only 0.37×10^{-5} mol L^{-1} in either 4.0×10^{-5} M $AgNO_3$ or 4.0×10^{-5} M NaCl.

of AgCl(*s*) is reduced. In the same way, the prior presence of $Ag^+(aq)$ in the solvent (for example, when an attempt is made to dissolve AgCl in water that already contains $AgNO_3$) reduces the amount of $Cl^-(aq)$ permitted at equilibrium and also reduces the solubility of AgCl. This is referred to as the **common-ion effect:** If the solution and the solid salt to be dissolved in it have an ion in common, then the solubility of the salt is depressed.

Let's examine the quantitative consequences of the common-ion effect. Suppose an excess of AgCl(*s*) is added to 1.00 L of a 0.100 M NaCl solution and the solubility is again determined. If *S* mol of AgCl dissolves per liter, the concentration of $Ag^+(aq)$ will be *S* mol L^{-1} and that of $Cl^-(aq)$ will be

$$[Cl^-] = 0.100 + S$$

because the chloride ion comes from two sources: the 0.100 M NaCl and the dissolution of AgCl. The expression for the solubility product is written as

$$[Ag^+][Cl^-] = S(0.100 + S) = K_{sp} = 1.6 \times 10^{-10}$$

The solubility of AgCl(*s*) in this solution must be smaller than it is in pure water, which is much smaller than 0.100. That is,

$$S < 1.3 \times 10^{-5} \ll 0.100$$

Thus, $(0.100 + S)$ can be approximated by 0.100 (as in Example 16.4), giving

$$(0.100)S \approx 1.6 \times 10^{-10}$$

$$S \approx 1.6 \times 10^{-9}$$

This is indeed much smaller than 0.100, so the approximation was a very good one. Therefore, at equilibrium,

$$[Ag^+] = S = 1.6 \times 10^{-9} \text{ M}$$

$$[Cl^-] = 0.100 \text{ M}$$

The gram solubility of AgCl in this example is

$$(1.6 \times 10^{-9} \text{ mol L}^{-1})(143.3 \text{ g mol}^{-1}) = 2.3 \times 10^{-7} \text{ g L}^{-1}$$

The solubility of AgCl in 0.100 M NaCl is lower than that in pure water by a factor of about 8000.

EXAMPLE 16.5

What is the gram solubility of $CaF_2(s)$ in a 0.100 M solution of NaF?

SOLUTION

Again, the molar solubility is denoted by S. The only source of the $Ca^{2+}(aq)$ in the solution at equilibrium is the dissolution of CaF_2, whereas the $F^-(aq)$ has two sources, the CaF_2 and the NaF. Hence,

	$CaF_2(s) \rightleftharpoons Ca^{2+}(aq)$	+	$2\ F^-(aq)$
Initial concentration (M)	0		0.100
Change in concentration (M)	$+S$		$+2S$
Equilibrium concentration (M)	S		$0.100 + 2S$

If $0.100 + 2S$ is approximated as 0.100, then

$$[Ca^{2+}][F^-]^2 = K_{sp}$$

$$S(0.100)^2 = 3.9 \times 10^{-11}$$

$$S = 3.9 \times 10^{-9}$$

Clearly,

$$2S = 7.8 \times 10^{-9} \ll 0.100$$

so the assumption was justified. The gram solubility of CaF_2 is

$$(3.9 \times 10^{-9}\ \text{mol L}^{-1})(78.1\ \text{g mol}^{-1}) = 3.0 \times 10^{-7}\ \text{g L}^{-1}$$

and the solubility in this case is reduced by a factor of 50,000.

Related Problems: 25, 26, 27, 28

16.4 The Effects of pH on Solubility

Some solids are only weakly soluble in water but dissolve readily in acidic solutions. Copper and nickel sulfides from ores, for example, can be brought into solution with strong acids, a fact that aids greatly in the separation and recovery of these valuable metals in their elemental forms. The effect of pH on solubility is shown dramatically in the damage done to buildings and monuments by acid precipitation (Fig. 16.8). Both marble and limestone are made up of small crystals of calcite ($CaCO_3$), which dissolves to only a limited extent in "natural" rain (with a pH of about 5.6) but dissolves much more extensively as the rainwater becomes more acidic. The reaction

$$CaCO_3(s) + H_3O^+(aq) \longrightarrow Ca^{2+}(aq) + HCO_3^-(aq) + H_2O(\ell)$$

causes this increase. This section examines the role of pH in solubility.

Solubility of Hydroxides

One direct effect of pH on solubility occurs with the metal hydroxides. The concentration of OH^- appears explicitly in the expression for the solubility product of such compounds. Thus, for the dissolution of $Zn(OH)_2(s)$,

$$Zn(OH)_2(s) \rightleftharpoons Zn^{2+}(aq) + 2\ OH^-(aq)$$

the solubility product expression is

$$[Zn^{2+}][OH^-]^2 = K_{sp} = 4.5 \times 10^{-17}$$

FIGURE 16.8 The calcium carbonate in marble and limestone is very slightly soluble in neutral water. Its solubility is much greater in acidic water. Objects carved of these materials dissolve relatively rapidly in areas where rain, snow, or fog is acidified from air pollution. Shown here is the damage to a marble statue of George Washington between 1935 (left) and 1994 (right).

As the solution is made more acidic, the concentration of hydroxide ion decreases, causing an increase in the concentration of $Zn^{2+}(aq)$ ion. Zinc hydroxide is thus more soluble in acidic solution than in pure water.

EXAMPLE 16.6

Compare the solubility of $Zn(OH)_2$ in pure water with that in a solution buffered at pH 6.00.

SOLUTION

In pure water the usual solubility product calculation applies:

$$[Zn^{2+}] = S \qquad [OH^-] = 2S$$

$$S(2S)^2 = 4S^3 = K_{sp} = 4.5 \times 10^{-17}$$

$$S = 2.2 \times 10^{-6}\ \text{M} = [Zn^{2+}]$$

so the solubility is 2.2×10^{-6} mol L^{-1}, or 2.2×10^{-4} g L^{-1}. Using

$$[OH^-] = 2S = 4.5 \times 10^{-6}\ \text{M}$$

the resulting solution is found to have pH 8.65.

In the second case it is assumed that the solution is buffered sufficiently that the pH remains 6.00 after dissolution of the zinc hydroxide. Then

$$[OH^-] = 1.0 \times 10^{-8}\ \text{M}$$

$$[Zn^{2+}] = \frac{K_{sp}}{[OH^-]^2} = \frac{4.5 \times 10^{-17}}{(1.0 \times 10^{-8})^2} = 0.45\ \text{M}$$

so that 0.45 mol L^{-1}, or 45 g L^{-1}, should dissolve in this case. When ionic concentrations are this high, the simple form of the solubility expression will likely break down, but the qualitative conclusion is still valid: $Zn(OH)_2$ is far more soluble at pH 6.00 than in pure water.

Related Problems: 31, 32

Solubility of Salts of Bases

Metal hydroxides can be described as salts of a strong base, the hydroxide ion. The solubility of salts in which the anion is a different weak or strong base is also affected by pH. For example, consider a solution of a slightly soluble fluoride, such as calcium fluoride. The solubility equilibrium is

$$CaF_2(s) \rightleftharpoons Ca^{2+}(aq) + 2\,F^-(aq) \qquad K_{sp} = 3.9 \times 10^{-11}$$

As the solution is made more acidic, some of the fluoride ion reacts with hydronium ion through

$$H_3O^+(aq) + F^-(aq) \rightleftharpoons HF(aq) + H_2O(\ell)$$

Because this is just the reverse of the acid ionization of HF, its equilibrium constant is the reciprocal of K_a for HF, or $1/(3.5 \times 10^{-4}) = 2.9 \times 10^3$. As acid is added, the concentration of fluoride ion is reduced, so the calcium ion concentration must increase to maintain the solubility product equilibrium for CaF_2. As a result, the solubility of fluoride salts increases in acidic solution. The same applies to other ionic substances in which the anion is a weak or a strong base. By contrast, the solubility of a salt such as AgCl is only very slightly affected by a decrease in pH. The reason is that HCl is a strong acid, so Cl^- is ineffective as a base. The reaction

$$Cl^-(aq) + H_3O^+(aq) \longrightarrow HCl(aq) + H_2O(\ell)$$

occurs to a negligible extent in acidic solution.

A DEEPER LOOK

16.5 Selective Precipitation of Ions

One way to analyze a mixture of ions in solution is to separate the mixture into its components by exploiting the differences in the solubilities of compounds containing the ions. To separate silver ions from lead ions, for example, a search is made for compounds of these elements that (1) have a common anion and (2) have widely different solubilities. The chlorides AgCl and $PbCl_2$ are two such compounds, for which the solubility equilibria are

$$AgCl(s) \rightleftharpoons Ag^+(aq) + Cl^-(aq) \qquad K_{sp} = 1.6 \times 10^{-10}$$

$$PbCl_2(s) \rightleftharpoons Pb^{2+}(aq) + 2\,Cl^-(aq) \qquad K_{sp} = 2.4 \times 10^{-4}$$

Lead chloride is far more soluble in water than is silver chloride. Consider a solution that is 0.10 M in both Ag^+ and Pb^{2+}. Is it possible to add enough Cl^- to precipitate almost all the Ag^+ ions but leave all the Pb^{2+} ions in solution? If so, a quantitative separation of the two species can be achieved.

For Pb^{2+} to remain in solution, its reaction quotient must remain smaller than K_{sp}: $Q = [Pb^{2+}][Cl^-]^2 < K_{sp}$. Inserting K_{sp} and the concentration of Pb^{2+} gives

$$[Cl^-]^2 < \frac{K_{sp}}{[Pb^{2+}]} = \frac{2.4 \times 10^{-4}}{0.10} = 2.4 \times 10^{-3}$$

The square root of this is

$$[Cl^-] < 4.9 \times 10^{-2}\ \text{M}$$

Thus, as long as the chloride ion concentration remains smaller than 0.049 M, no $PbCl_2$ should precipitate. To reduce the *silver* ion concentration in solution as far as possible (that is, to precipitate out as much silver chloride as possible), the chloride ion concentration should be kept as high as possible without exceeding 0.049 M. If exactly this concentration of $Cl^-(aq)$ is chosen, then at equilibrium,

$$[Ag^+] = \frac{K_{sp}}{[Cl^-]} = \frac{1.6 \times 10^{-10}}{0.049} = 3.3 \times 10^{-9}$$

At that concentration of Cl^-, the concentration of Ag^+ has been reduced to 3.3×10^{-9} from the original concentration of 0.10 M. In other words, only about three Ag^+ ions in 10^8 remain in solution, but all the Pb^{2+} ions are left in solution. A nearly perfect separation of the two ionic species has been achieved.

This calculation gave the optimal theoretical separation factor for the two ions. In practice it is necessary to keep the chloride concentration lower. If $[Cl^-]$ is ten times smaller, or 0.0049 M, about three Ag^+ ions in 10^7 remain in solution with the Pb^{2+}. This is ten times more Ag^+ than if $[Cl^-] = 0.049$ M, but the separation of Ag^+ from Pb^{2+} is still very good.

Figure 16.9 shows graphically how ions can be separated based on solubility. The relation between the concentration of Pb^{2+} and Cl^- ions in contact with solid $PbCl_2$

$$[Pb^{2+}] = \frac{K_{sp}}{[Cl^-]^2}$$

can be rewritten in a useful form by taking the common base-10 logarithms of both sides:

$$\log_{10}[Pb^{2+}] = -2\log_{10}[Cl^-] + \log_{10}K_{sp}$$

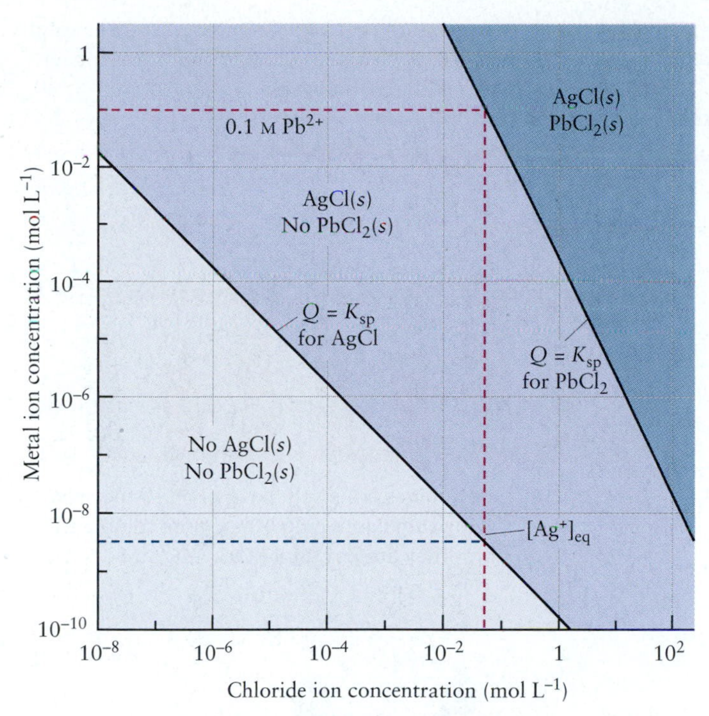

FIGURE 16.9 To separate a mixture of Ag^+ and Pb^{2+} ions, a chloride ion concentration is selected that gives a Pb^{2+} concentration below the equilibrium curve for $PbCl_2$ (so all Pb^{2+} remains in solution) but well above the equilibrium curve for Ag^+. As a result, nearly all the Ag^+ precipitates as AgCl. If $[Pb^{2+}]_0 = [Ag^+]_0 =$ 0.1 M, then the maximum $[Cl^-]$ is found by tracing the horizontal red line and then dropping down the vertical red line to find $[Cl^-]$ = 0.049 M. The concentration of Ag^+ still in solution is found by tracing the horizontal blue line from the intersection of the $[Cl^-]$ line of the AgCl equilibrium curve back to the vertical axis where $[Ag^+] = 3.3 \times 10^{-9}$ M.

That is, a graph of $\log_{10} [Pb^{2+}]$ against $\log_{10} [Cl^-]$ is a straight line with slope −2. The corresponding graph for the AgCl solubility equilibrium has slope −1.

$$\log_{10}[Ag^+] = -\log_{10}[Cl^-] + \log_{10}K_{sp}$$

If the concentration of Cl^- and the initial concentrations of the metal ions correspond to a point that lies between the two lines in Figure 16.9, then AgCl precipitates but $PbCl_2$ does not.

Metal Sulfides

Controlling the solubility of metal sulfides has important applications. According to Table 16.2, most metal sulfides are very slightly soluble in water; only a very small amount of a compound such as ZnS(*s*) will dissolve in water. Although it is tempting to write the resulting equilibrium as

$$ZnS(s) \rightleftharpoons Zn^{2+}(aq) + S^{2-}(aq) \qquad K_{sp} = ?$$

in analogy with the equilibria for other weakly soluble salts, this is misleading, because S^{2-}, like O^{2-}, is a very strong base (stronger than OH^-) and reacts almost quantitatively with water.

$$S^{2-}(aq) + H_2O(\ell) \longrightarrow HS^-(aq) + OH^-(aq)$$

TABLE 16.3 Equilibrium Constants for Metal Sulfide Dissolution at 25°C

Metal Sulfide	$K^\dagger$
CuS	5×10^{-37}
PbS	3×10^{-28}
CdS	7×10^{-28}
SnS	9×10^{-27}
ZnS	2×10^{-25}
FeS	5×10^{-19}
MnS	3×10^{-14}

$^\dagger K$ is the equilibrium constant for the reaction $MS(s) + H_2O(\ell) \rightleftharpoons M^{2+}(aq) + OH^-(aq) + HS^-(aq)$.

Recent research has shown that K_b for this reaction is on the order of 10^5; this means essentially no S^{2-} is present in aqueous solution. The net dissolution reaction is found by adding the two preceding equations,

$$ZnS(s) + H_2O(\ell) \rightleftharpoons Zn^{2+}(aq) + OH^-(aq) + HS^-(aq)$$

for which the equilibrium constant is

$$[Zn^{2+}][OH^-][HS^-] = K \approx 2 \times 10^{-25}$$

Table 16.3 gives values of the equilibrium constants for comparable reactions of other metal sulfides. As the pH decreases, the concentration of OH^- decreases. At the same time, the concentration of HS^- also decreases as the equilibrium

$$HS^-(aq) + H_3O^+(aq) \rightleftharpoons H_2S(aq) + H_2O(\ell)$$

shifts to the right upon addition of H_3O^+. If both $[OH^-]$ and $[HS^-]$ decrease, then $[Zn^{2+}]$ must increase in order to maintain a constant value for the product of the three concentrations. As a result, the solubility of ZnS(*s*) increases as the pH of the solution decreases. Other metal sulfides behave the same way, becoming more soluble in acidic solution.

A quantitative calculation of metal sulfide solubility requires treating several simultaneous equilibria, as the following example illustrates.

EXAMPLE 16.7

In a solution that is saturated with H_2S, $[H_2S]$ is fixed at 0.1 M. Calculate the molar solubility of FeS(*s*) in such a solution if it is buffered at pH 3.0.

SOLUTION

If the pH is 3.0, then

$$[OH^-] = 1 \times 10^{-11} \text{ M}$$

In addition, the acid ionization of H_2S must be considered:

$$H_2S(aq) + H_2O(\ell) \rightleftharpoons H_3O^+(aq) + HS^-(aq)$$

Substitution of $[H_2S] = 0.1$ M for a saturated solution and $[H_3O^+] = 1 \times 10^{-3}$ M (at pH 3.0) into the equilibrium expression for this reaction gives

$$\frac{[H_3O^+][HS^-]}{[H_2S]} = \frac{[1 \times 10^{-3}][HS^-]}{0.1} = K_a = 9.1 \times 10^{-8}$$

$$[HS^-] = 9 \times 10^{-6} \text{ M}$$

where K_a came from Table 15.2. For the reaction

$$FeS(s) + H_2O(\ell) \rightleftharpoons Fe^{2+}(aq) + HS^-(aq) + OH^-(aq)$$

the equilibrium constant from Table 16.3 is

$$[Fe^{2+}][HS^-][OH^-] = 5 \times 10^{-19}$$

Substituting the values of $[HS^-]$ and $[OH^-]$ and solving for $[Fe^{2+}]$ give

$$[Fe^{2+}](9 \times 10^{-6})(1 \times 10^{-11}) = 5 \times 10^{-19}$$

$$[Fe^{2+}] = 6 \times 10^{-3} \text{ M}$$

Hence, 6×10^{-3} mol of FeS dissolves per liter under these conditions.

Related Problems: 41, 42

By adjusting the pH through appropriate choice of buffers, as in Example 16.7, conditions can be selected so that metal ions of one element remain entirely in solution, whereas those of a second element in the mixture precipitate almost entirely as solid metal sulfide (Fig. 16.10). Such a procedure is important for separating metal ions in qualitative analysis.

FIGURE 16.10 These sulfides are insoluble at pH 1, so they can be separated out of a mixture containing other, more soluble sulfides. From left to right, they are PbS, Bi_2S_3, CuS, CdS, Sb_2S_3, SnS_2, As_2S_3, and HgS.

Hydrogen sulfide, H_2S, is a poisonous, foul-smelling gas. When dissolved in water, it gives a weak acid, hydrosulfuric acid.

16.6 Complex Ions and Solubility

Many transition-metal ions form **coordination complexes** in solution or in the solid state; these consist of a metal ion surrounded by a group of anions or neutral molecules called **ligands.** The interaction involves the sharing by the metal ion of a lone pair on each ligand molecule, giving a partially covalent bond with that ligand. Such complexes often have strikingly deep colors. When exposed to gaseous ammonia, greenish white crystals of copper sulfate ($CuSO_4$) give a deep blue crystalline solid with the chemical formula $[Cu(NH_3)_4]SO_4$ (Fig. 16.11). The anions in the solid are still sulfate ions (SO_4^{2-}), but the cations are now **complex ions,** or coordination complexes of the central Cu^{2+} ion with four ammonia molecules, $Cu(NH_3)_4^{2+}$. The ammonia molecules coordinate to the copper ion through their lone-pair electrons (Fig. 16.12), acting as Lewis bases toward the metal ion, the Lewis acid. When the solid is dissolved in water, the deep blue color remains. This is evidence that the complex persists in water, because when ordinary $CuSO_4$ (without ammonia ligands) is dissolved in water, a much paler blue color results (see Fig. 16.11b).

Here we explore the effects of the formation of complex ions on equilibria in aqueous solutions. The microscopic structure and bonding in these complexes is presented in Chapter 8.

(a)

(b)

FIGURE 16.11 (a) From left to right, crystals of $Cu(NH_3)_4SO_4$, $CuSO_4 \cdot 5H_2O$, and $CuSO_4$. (b) Aqueous solutions of copper sulfate containing (*left*) and not containing (*right*) ammonia.

(© Thomson Learning/Leon Lewandowski)

Complex-Ion Equilibria

When silver ions are dissolved in an aqueous ammonia solution, doubly coordinated silver–ammonia complexes, shown in Figure 16.13, form in two stepwise reactions:

$$Ag^+(aq) + NH_3(aq) \rightleftarrows Ag(NH_3)^+(aq)$$

$$\frac{[Ag(NH_3)^+]}{[Ag^+][NH_3]} = K_1 = 2.1 \times 10^3$$

$$Ag(NH_3)^+(aq) + NH_3(aq) \rightleftarrows Ag(NH_3)_2^+(aq)$$

$$\frac{[Ag(NH_3)_2^+]}{[Ag(NH_3)^+][NH_3]} = K_2 = 8 \times 10^3$$

If these two chemical equations are added (and their corresponding equilibrium laws are multiplied), the result is

$$Ag^+(aq) + 2\,NH_3(aq) \rightleftarrows Ag(NH_3)_2^+(aq)$$

$$\frac{[Ag(NH_3)_2^+]}{[Ag^+][NH_3]^2} = K_f = K_1K_2 = 1.7 \times 10^7$$

where K_f is the **formation constant** of the full complex ion $Ag(NH_3)_2^+$. Table 16.4 lists formation constants for a representative selection of complex ions. The larger the formation constant K_f, the more stable the corresponding complex ion (for ions with the same number of ligands).

Because K_1 and K_2 of the $Ag(NH_3)_2^+$ complex ion are both large, a silver salt dissolved in water that contains a high concentration of ammonia will be primarily in the form of the complex ion $[Ag(NH_3)_2]^+$ at equilibrium.

TABLE 16.4 Formation Constants of Coordination Complexes in Aqueous Solution

	K_f	K_1	K_2	K_3	K_4	K_5	K_6
Ammines							
$Ag(NH_3)_2^+$	1.7×10^7	2.1×10^3	8.2×10^3				
$Co(NH_3)_6^{2+}$	2.5×10^4	1.0×10^2	32	8.5	4.4	1.1	0.18
$Cu(NH_3)_4^{2+}$	1.1×10^{12}	1.0×10^4	2×10^3	5×10^2	90		
$Ni(NH_3)_6^{2+}$	8×10^{-7}	5×10^2	1.3×10^2	40	12	3.3	0.8
$Zn(NH_3)_4^{2+}$	5×10^8	1.5×10^2	1.8×10^2	2×10^2	90		
Chlorides							
$AgCl_2^-$	1.8×10^5	1.7×10^3	1.0×10^2				
$FeCl_4^-$	0.14	28	4.5	0.1	1.1×10^2		
$HgCl_4^{2-}$	1.2×10^{15}	5.5×10^6	3×10^6	7	10		
$PbCl_4^{2-}$	25	40	1.5	0.8	0.5		
$SnCl_4^{2-}$	30	32	5.4	0.6	0.3		
Hydroxides							
$Co(OH)_3^-$	3×10^{10}	4×10^4	1	8×10^5			
$Cu(OH)_4^{2-}$	3×10^{18}	1×10^7	5×10^6	2×10^3	30		
$Ni(OH)_3^-$	2×10^{11}	9×10^4	4×10^3	6×10^2			
$Pb(OH)_3^-$	4×10^{14}	7×10^7	1.1×10^3	5×10^3			
$Zn(OH)_4^{2-}$	5×10^{14}	2.5×10^4	8×10^6	70	33		

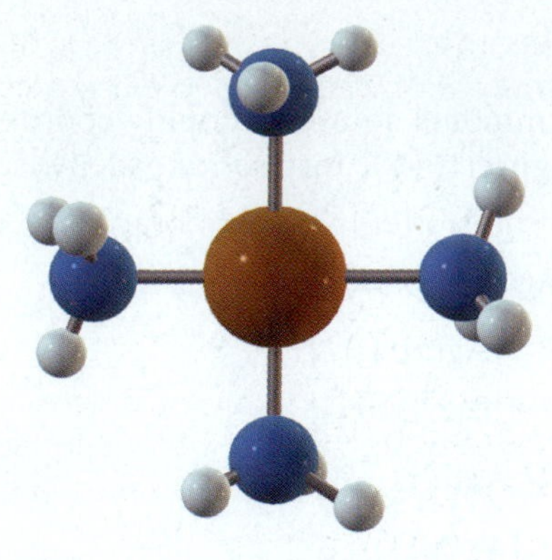

FIGURE 16.12 The structure of $Cu(NH_3)_4^{2+}$.

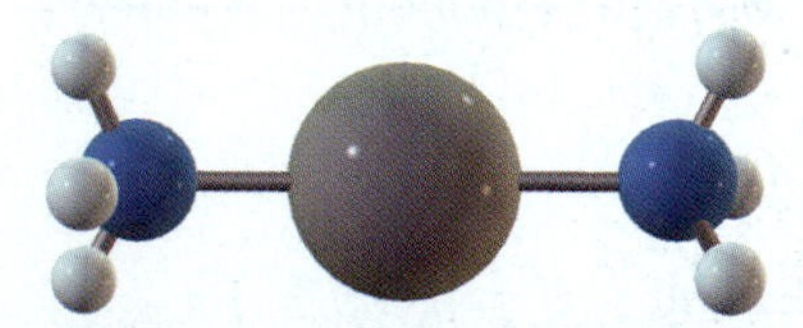

FIGURE 16.13 The structure of $Ag(NH_3)_2^+$.

EXAMPLE 16.8

Suppose 0.100 mol of $AgNO_3$ is dissolved in 1.00 L of a 1.00 M in solution of NH_3. Calculate the concentrations of the Ag^+ and $Ag(NH_3)^+$ ions present at equilibrium.

SOLUTION

Suppose that most of the Ag^+ is present as $Ag(NH_3)_2^+$ (this will be checked later). Then

$$[Ag(NH_3)_2^+]_0 = 0.100 \text{ M}$$

$$[NH_3]_0 = 1.00 \text{ M} - (2 \times 0.100) \text{ M} = 0.80 \text{ M}$$

after each silver ion has become complexed with two ammonia molecules. The two stages of the dissociation of the $Ag(NH_3)_2^+$ ion are the reverse reaction of the complexation, so their equilibrium constants are the reciprocals of K_2 and K_1, respectively:

$$Ag(NH_3)_2^+(aq) \rightleftharpoons Ag(NH_3)^+(aq) + NH_3(aq)$$

$$\frac{[Ag(NH_3)^+][NH_3]}{[Ag(NH_3)_2^+]} = \frac{1}{K_2} = \frac{1}{8.2 \times 10^3}$$

$$Ag(NH_3)^+(aq) \rightleftharpoons Ag^+(aq) + NH_3(aq)$$

$$\frac{[Ag^+][NH_3]}{[Ag(NH_3)^+]} = \frac{1}{K_1} = \frac{1}{2.1 \times 10^3}$$

If y mol L^{-1} of $Ag(NH_3)_2^+$ dissociates at equilibrium according to the first equation,

	$Ag(NH_3)_2^+(aq)$ ⇌	$Ag(NH_3)^+(aq)$ +	$NH_3(aq)$
Initial concentration (M)	0.100	0	0.80
Change in concentration (M)	$-y$	$+y$	$+y$
Equilibrium concentration (M)	$0.100 - y$	y	$0.80 + y$

then the first equilibrium expression becomes

$$\frac{y(0.80 + y)}{0.10 - y} = \frac{1}{K_2} = \frac{1}{8.2 \times 10^3}$$

$$y = 1.5 \times 10^{-5} \text{ M} = [Ag(NH_3)^+]$$

We can then calculate the concentration of free Ag^+ ions from the equilibrium law for the second step of the dissociation of the complex ion:

$$\frac{[Ag^+][NH_3]}{[Ag(NH_3)^+]} = \frac{1}{K_1}$$

$$\frac{[Ag^+](0.80)}{1.5 \times 10^{-5}} = \frac{1}{2.1 \times 10^3}$$

$$[Ag^+] = 9 \times 10^{-9} \text{ M}$$

It is clear that the original assumption was correct, and most of the silver present is tied up in the $Ag(NH_3)_2^+$ complex.

Related Problems: 43, 44

There is a close similarity between the working of Example 16.8 and an acid–base calculation. The first step (the assumption that the reaction goes to completion and is followed by a small amount of back dissociation) is analogous to the procedure for dealing with the addition of a small amount of a strong acid to a solution of a weak base. The subsequent calculation of the successive dissociation steps resembles the calculation of polyprotic acid equilibria in Example 15.12. The only difference is that in complex-ion equilibria it is conventional to work with formation constants, which are the inverse of the dissociation constants used in acid–base equilibria.

FIGURE 16.14 An illustration of the effect of complex ion formation on solubility. Each test tube contains 2.0 g AgBr, but the one on the left also contains dissolved thiosulfate ion, which forms a complex ion with Ag^+. Almost none of the white solid AgBr has dissolved in pure water, but all of it has dissolved in the solution containing thiosulfate.

The formation of coordination complexes can have a large effect on the solubility of a compound in water. Silver bromide is only very weakly soluble in water,

$$AgBr(s) \rightleftharpoons Ag^+(aq) + Br^-(aq) \qquad K_{sp} = 7.7 \times 10^{-13}$$

but addition of thiosulfate ion ($S_2O_3^{2-}$) to the solution allows the complex ion $Ag(S_2O_3)_2^{3-}$ to form:

$$AgBr(s) + 2\,S_2O_3^{2-}(aq) \rightleftharpoons Ag(S_2O_3)_2^{3-}(aq) + Br^-(aq)$$

This greatly increases the solubility of the silver bromide (Fig. 16.14). The formation of this complex ion is an important step in the development of photographic images; thiosulfate ion is a component of the fixer that brings silver bromide into solution from the unexposed portion of the film.

EXAMPLE 16.9

Calculate the solubility of AgBr in a 1.00 M aqueous solution of ammonia.

SOLUTION

We tentatively assume that almost all the silver that dissolves is complexed as $Ag(NH_3)_2^+$ (this will be checked later). The overall reaction is then

$$AgBr(s) + 2\,NH_3(aq) \rightleftharpoons Ag(NH_3)_2^+(aq) + Br^-(aq)$$

Note that this is the sum of the two reactions

$$AgBr(s) \rightleftharpoons Ag^+(aq) + Br^-(aq) \qquad K_{sp} = 7.7 \times 10^{-13}$$

$$Ag^+(aq) + 2\,NH_3(aq) \rightleftharpoons Ag(NH_3)_2^+(aq) \qquad K_f = 1.7 \times 10^7$$

so its equilibrium constant is the product $K_{sp}K_f = 1.3 \times 10^{-5}$.

If S mol L^{-1} of AgBr dissolves, then

$$S = [Br^-] \approx [Ag(NH_3)_2^+]$$

$$[NH_3] = 1.00 - 2S$$

because 2 mol of NH_3 is used up for each mole of complex formed. The equilibrium expression is

$$\frac{S^2}{(1.00 - 2S)^2} = K_{sp}K_f = 1.3 \times 10^{-5}$$

$$S = 3.6 \times 10^{-3}\ \text{M} = [Ag(NH_3)_2^+] = [Br^-]$$

To check the original assumption, calculate the concentration of free silver ion:

$$[Ag^+] = \frac{K_{sp}}{[Br^-]} = \frac{K_{sp}}{S} = 2.1 \times 10^{-10} \ll [Ag(NH_3)_2^+]$$

verifying that almost all the silver is complexed. The solubility is therefore 3.6×10^{-3} mol L^{-1}, significantly greater than the solubility in pure water:

$$\text{solubility in pure water} = \sqrt{K_{sp}} = 8.8 \times 10^{-7}\ \text{mol L}^{-1}$$

Related Problems: 47, 48

The thiosulfate ion (*top*) is related to the sulfate ion, SO_4^{2-} (*bottom*), by the replacement of one oxygen atom with one sulfur atom. It is prepared, however, by the reaction of elemental sulfur with the sulfite ion (SO_3^{2-}).

Another interesting effect of complex ions on solubilities is illustrated by the addition of iodide ion to a solution containing mercury(II) ion. After a moderate amount of iodide ion has been added, an orange precipitate forms (Fig. 16.15) through the reaction

$$Hg^{2+}(aq) + 2\,I^-(aq) \rightleftharpoons HgI_2(s)$$

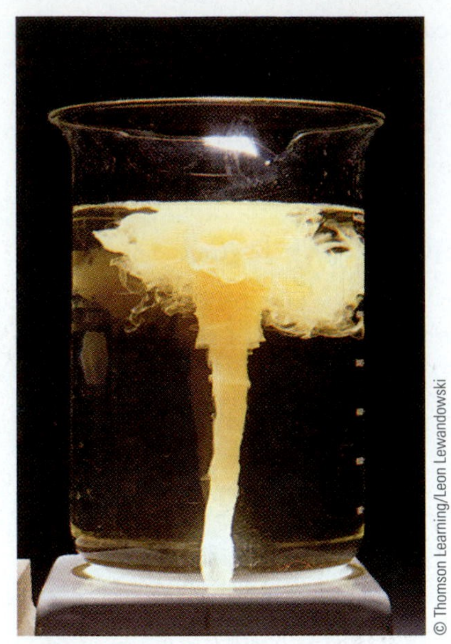

FIGURE 16.15 The "orange tornado" is a striking demonstration of the effects of complex ions on solubility. A solution is prepared with an excess of $I^-(aq)$ over $Hg^{2+}(aq)$ so that the Hg^{2+} is complexed as HgI_3^- and HgI_4^{2-}. A magnetic stirrer is used to create a vortex in the solution. Addition of a solution containing Hg^{2+} down the center of the vortex then causes the orange solid HgI_2 to form in a layer at the edges of the vortex, giving the tornado effect.

With further addition of iodide ion, however, the orange solid redissolves because complex ions form:

$$HgI_2(s) + I^-(aq) \rightleftharpoons HgI_3^-(aq)$$

$$HgI_3^-(aq) + I^-(aq) \rightleftharpoons HgI_4^{2-}(aq)$$

In the same way, silver chloride will dissolve in a concentrated solution of sodium chloride by forming soluble $AgCl_2^-$ complex ions. Complex ion formation affects solubility in the opposite direction from the common-ion effect of Section 16.3.

Acidity and Amphoterism of Complex Ions

When dissolved in water, many metal ions increase the acidity of the solution. The iron(III) ion is an example: Each dissolved Fe^{3+} ion is strongly solvated by six water molecules, leading to a complex ion $Fe(H_2O)_6^{3+}$. This complex ion can act as a Brønsted–Lowry acid, donating hydrogen ions to the solvent, water:

$$\underset{\text{Acid}_1}{Fe(H_2O)_6^{3+}(aq)} + \underset{\text{Base}_2}{H_2O(\ell)} \rightleftharpoons \underset{\text{Acid}_2}{H_3O^+(aq)} + \underset{\text{Base}_1}{Fe(H_2O)_5OH^{2+}(aq)}$$

$$\frac{[H_3O^+][Fe(H_2O)_5OH^{2+}]}{[Fe(H_2O)_6^{3+}]} = K_a = 7.7 \times 10^{-3}$$

Metal ion hydrolysis fits into the general scheme of the Brønsted–Lowry acid–base reaction.

EXAMPLE 16.10

Calculate the pH of a solution that is 0.100 M in $Fe(NO_3)_3$.

SOLUTION

The iron(III) is present as $Fe(H_2O)_6^{3+}$, which reacts as a weak acid:

$$Fe(H_2O)_6^{3+}(aq) + H_2O(\ell) \rightleftharpoons Fe(H_2O)_5OH^{2+}(aq) + H_3O^+(aq)$$

with K_a equal to 7.7×10^{-3}. If y mol L^{-1} of $[Fe(H_2O)_6]^{3+}$ reacts, then (neglecting the ionization of water itself)

$$[H_3O^+] = [Fe(H_2O)_5OH^{2+}] = y$$

$$[Fe(H_2O)_6^{3+}] = 0.100 - y$$

The equilibrium expression has the form

$$\frac{y^2}{0.100 - y} = 7.7 \times 10^{-3}$$

$$y = 2.4 \times 10^{-2}\ \text{M} = [H_3O^+]$$

so the pH is 1.62. Solutions of iron(III) salts are strongly acidic.

Related Problems: 51, 52

TABLE 16.5 pH of 0.1 M Aqueous Metal Nitrate Solutions at 25°C

Metal Nitrate	PH
$Fe(NO_3)_3$	1.6
$Pb(NO_3)_2$	3.6
$Cu(NO_3)_2$	4.0
$Zn(NO_3)_2$	5.3
$Ca(NO_3)_2$	6.7
$NaNO_3$	7.0

Another acceptable way to write the reaction that makes iron(III) solutions acidic is

$$Fe^{3+}(aq) + 2\ H_2O(\ell) \rightleftharpoons H_3O^+(aq) + FeOH^{2+}(aq)$$

in which the specific mention of the six waters of hydration is now omitted. The $FeOH^{2+}$ complex ions are brown, but Fe^{3+} ions are almost colorless. This reaction occurs to a sufficient extent to make a solution of $Fe(NO_3)_3$ in water pale brown. When strong acid is added, the equilibrium is driven back to the left and the color fades. Table 16.5 gives values of the pH for 0.1 M solutions of several metal ions. Those that form strong complexes with hydroxide ion have low pH

FIGURE 16.16 The acidity of metal ions is illustrated by the vigorous reaction of anhydrous $AlCl_3$ with water to generate hydrated aluminum oxides, $HCl(aq)$, and heat. The HCl turns the indicator to its red acid form.

(Fig. 16.16), whereas those that do not form such complexes give neutral solutions (pH 7).

Different cations behave differently as water ligands are replaced by hydroxide ions in an increasingly basic solution. A particularly interesting example is Zn^{2+}. It forms a series of hydroxo complex ions:

$$Zn^{2+}(aq) + OH^-(aq) \rightleftharpoons ZnOH^+(aq)$$

$$ZnOH^+(aq) + OH^-(aq) \rightleftharpoons Zn(OH)_2(s)$$

$$Zn(OH)_2(s) + OH^-(aq) \rightleftharpoons Zn(OH)_3^-(aq)$$

$$Zn(OH)_3^-(aq) + OH^-(aq) \rightleftharpoons Zn(OH)_4^{2-}(aq)$$

In the Brønsted–Lowry theory, a polyprotic acid, $Zn(H_2O)_4^{2+}(aq)$, donates hydrogen ions in succession to make all the product ions. The second product, $Zn(OH)_2$, is amphoteric; it can react as either acid or base. It is only slightly soluble in pure water (its K_{sp} is only 1.9×10^{-17}). If enough acid is added to solid $Zn(OH)_2$, the OH^- ligands are removed, forming the soluble Zn^{2+} ion; if enough base is added, OH^- ligands attach to form the soluble $Zn(OH)_4^{2+}$ (zincate) ion. Thus, $Zn(OH)_2$ is soluble in strongly acidic *or* strongly basic solutions but is only slightly soluble at intermediate pH values (Fig. 16.17). This amphoterism can be used to separate Zn^{2+} from other cations that do not share the property. For example, Mg^{2+} adds a maximum of two OH^- ions to form $Mg(OH)_2$, a sparingly soluble hydroxide. Further addition of OH^- does not lead to the formation of new complex ions. If a mixture of Mg^{2+} and Zn^{2+} ions is made sufficiently basic, the Mg^{2+} precipitates as $Mg(OH)_2$ but the zinc remains in solution as $Zn(OH)_4^{2-}$, allowing the two to be separated. In the same way, aluminum is separated from iron industrially by dissolving solid $Al(OH)_3$ in strong base as $Al(OH)_4^-(aq)$ while $Fe(OH)_3$ remains as a precipitate.

FIGURE 16.17 Zinc hydroxide is insoluble in water (*center*) but dissolves readily in acid (*left*) and base (*right*). The indicator used is bromocresol red, which turns from red to yellow in acidic solution.

CHAPTER SUMMARY

Dissolution–precipitation reactions involve equilibrium between a substance in its solid form and molecules or ions of that same substance dissolved in solution. This equilibrium is described by the mass action law, so knowledge of the equilibrium constant permits manipulation of the concentrations in solution through the principles described in Chapter 14. Numerous separation and purification procedures are based on such manipulations. The significance of these dissolution–precipitation equilibria in practical applications is comparable to that of acid–base reactions. Solutions of slightly soluble solids are said to be saturated when the dissolved concentration is at equilibrium with the solid. In accordance with the principle of Le Châtelier, if the solution goes to a state of supersaturation, it will return to equilibrium by precipitating solid out of solution. Similarly, according to Le Châtelier, solubility can be controlled by increasing temperature (for endothermic reactions) or decreasing temperature (for exothermic reactions). Solubility is greatly influenced by the presence of other species to the solution, through the way they shift the equilibrium concentration of the dissolved solid species. This general principle explains the common ion effect and the influence of pH on solubility.

CUMULATIVE EXERCISE

Carbonate Minerals

The carbonates are among the most abundant and important minerals in the earth's crust. When these minerals come into contact with fresh water or seawater, solubility equilibria are established that greatly affect the chemistry of the natural waters. Calcium carbonate ($CaCO_3$), the most important natural carbonate, makes up limestone and other forms of rock such as marble. Other carbonate minerals include dolomite, $CaMg(CO_3)_2$, and magnesite, $MgCO_3$. These compounds are sufficiently soluble that their solutions are nonideal, so calculations based on solubility product expressions are only approximate.

(a) The rare mineral nesquehonite contains $MgCO_3$ together with water of hydration. A sample containing 21.7 g of nesquehonite is acidified and heated, and the volume of $CO_2(g)$ produced is measured to be 3.51 L at 0°C and $P = 1$ atm. Assuming all the carbonate has reacted to form CO_2, give the chemical formula for nesquehonite.

(b) Write a chemical equation and a solubility product expression for the dissolution of dolomite in water.

(c) In a sufficiently basic solution, the carbonate ion does not react significantly with water to form hydrogen carbonate ion. Calculate the solubility (in grams per liter) of limestone (calcium carbonate) in a 0.10 M solution of sodium hydroxide. Use the K_{sp} from Table 16.2.

(d) In a strongly basic 0.10 M solution of Na_2CO_3, the concentration of CO_3^{2-} is 0.10 M. What is the gram solubility of limestone in this solution? Compare your answer with that for part (c).

(e) In a mountain lake having a pH of 8.1, the total concentration of carbonate species, $[CO_3^{2-}] + [HCO_3^-]$, is measured to be 9.6×10^{-4} M, whereas the concentration of Ca^{2+} is 3.8×10^{-4} M. Calculate the concentration of CO_3^{2-} in this lake, using $K_a = 4.8 \times 10^{-11}$ for the acid ionization of HCO_3^- to CO_3^{2-}. Is the lake unsaturated, saturated, or supersaturated with respect to $CaCO_3$?

(f) Will acid rainfall into the lake increase or decrease the solubility of limestone rocks in the lake's bed?

(g) Seawater contains a high concentration of Cl^- ions, which form weak complexes with calcium, such as the ion pair $CaCl^+$. Does the presence of such complexes increase or decrease the equilibrium solubility of $CaCO_3$ in seawater?

Answers

(a) $MgCO_3 \cdot 3H_2O$

(b) $CaMg(CO_3)_2(s) \rightleftharpoons Ca^{2+}(aq) + Mg^{2+}(aq) + 2\ CO_3^{2-}(aq)$

$$[Ca^{2+}][Mg^{2+}][CO_3^{2-}]^2 = K_{sp}$$

(c) 9.3×10^{-3} g L^{-1}

(d) 8.7×10^{-6} g L^{-1}, smaller than in part (c) because of the common-ion effect

(e) 5.8×10^{-6} M. $Q = 2.2 \times 10^{-9} < K_{sp} = 8.7 \times 10^{-9}$, so the lake is slightly less than saturated

(f) Increase

(g) Increase

Carbonate minerals. Calcite (*left*) and aragonite (*middle*) are both $CaCO_3$, and smithsonite (*right*) is $ZnCO_3$.

(Left, © Thomson Learning/Charles D. Winters; middle, Ken Lucas/Visuals Unlimited; right, copyright Tom McHugh/Photo Researchers, Inc.)

CHAPTER REVIEW

The Nature of Solubility Equilibria

- A substance in its solid state may be in contact with a solution in which this same substance is dissolved. Molecules of this substance pass back and forth between the solid state and the dissolved state.
- When there is equilibrium between the solid substance and its dissolved form, the solution is *saturated* and cannot accommodate more solute. A nonequilibrium state can exist in which there is temporarily more dissolved solute than the equilibrium concentration. This state is called supersaturated.
- The solubility of a substance is the greatest amount that will dissolve in a specified volume of solvent at a specified temperature. Solubility is expressed in $g\ L^{-1}$ or $mol\ L^{-1}$.
- Solubility is temperature dependent. By Le Châtelier's principle it can either increase or decrease with increasing temperature depending on whether the dissolution reaction is endothermic or exothermic.

Ionic Equilibria between Solids and Solutions

- Slightly soluble ionic solids dissociate in solution, so the solid is in equilibrium with its dissolved cations and anions.
- $AB(s) \rightleftharpoons A^+(aq) + B^-(aq)$
  $K_{sp} = [A^+]\,[B^-] = S^2$ if S is the molar solubility.
- $AB_2(s) \rightleftharpoons A^+(aq) + 2\,B^-(aq)$
  $K_{sp} = [A^+]\,[B^-]^2 = 4S^3$, and so on, for different stoichiometric ratios.

Precipitation and the Solubility Product

- $Q = [A^+]\,[B^-]$ at any point in the reaction. If $Q > K_{sp}$ a precipitate will form.
- The common ion effect—no matter what the sources of the ions, the product of their concentrations cannot exceed K_{sp} under equilibrium conditions.

Special Chemical Effects in Solubility

- Solubility of metal hydroxides increases at lower pH because the acid present ties up hydroxide ions in solution and so shifts the dissolution equilibrium to the right.
- Solubility of salts in which the anion is a weak base increases at lower pH because the hydronium ion in solution consumes the anion and so shifts the dissolution equilibrium to the right.
- Solubility of metal ions can be increased by forming coordination complexes with ligands.
- Metal ions undergo hydrolysis reactions in water by forming hydrated complex ions that increase the acidity of the solution.

CONCEPTS & SKILLS

After studying this chapter and working the problems that follow, you should be able to:

1. Discuss the dynamical processes that lead to solubility equilibria (Section 16.1).
2. Relate the solubilities of sparingly soluble salts in water to their solubility product constants (Section 16.2, Problems 7–16).
3. Use the reaction quotient to predict whether a precipitate will form when two solutions are mixed, and then calculate the equilibrium concentrations that result (Section 16.3, Problems 17–24).

4. Calculate the solubility of a sparingly soluble salt in a solution that contains a given concentration of a common ion (Section 16.3, Problems 25–30).
5. Determine the dependence on pH of the solubility of a salt of a weak base (Section 16.4, Problems 33–34).
6. Specify the optimal conditions for the separation of two elements on the basis of the differing solubilities of their ionic compounds (Section 16.5, Problems 35–42).
7. Calculate the concentrations of molecular and ionic species in equilibrium with complex ions (Section 16.6, Problems 43–46).
8. Determine the effect of complex-ion formation on the solubility of sparingly soluble salts involving a common cation (Section 16.6, Problems 47–48).
9. Calculate the pH of aqueous solutions containing metal cations (Section 16.6, Problems 49–54).

PROBLEMS

Answers to problems whose numbers are boldface appear in Appendix G. Problems that are more challenging are indicated with asterisks.

The Nature of Solubility Equilibria

1. Gypsum has the formula $CaSO_4 \cdot 2H_2O$. Plaster of Paris has the chemical formula $CaSO_4 \cdot \frac{1}{2}\ H_2O$. In making wall plaster, water is added to plaster of Paris and the mixture then hardens into solid gypsum. How much water (in liters, at a density of 1.00 kg L^{-1}) should be added to 25.0 kg of plaster of Paris to turn it into gypsum, assuming no loss from evaporation?

2. A 1.00-g sample of magnesium sulfate is dissolved in water, and the water is then evaporated away until the residue is bone dry. If the temperature of the water is kept between 48°C and 69°C, the solid that remains weighs 1.898 g. If the experiment is repeated with the temperature held between 69°C and 100°C, however, the solid has a mass of 1.150 g. Determine how many waters of crystallization per $MgSO_4$ there are in each of these two solids.

3. The following graph shows the solubility of KBr in water in units of grams of KBr per 100 g of H_2O. If 80 g of KBr is added to 100 g of water at 10°C and the mixture is heated slowly, at what temperature will the last KBr dissolve?

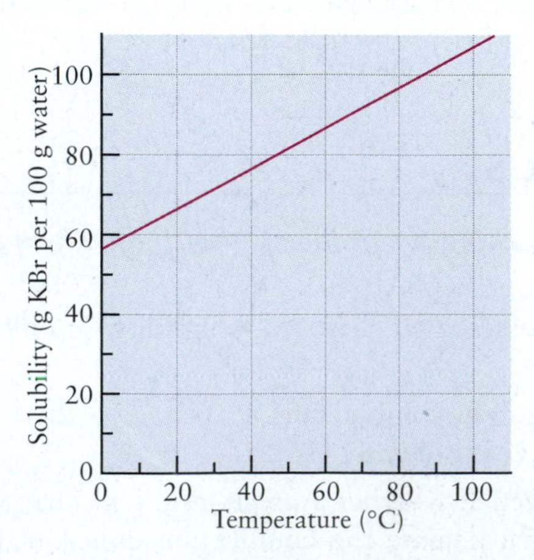

4. Figure 16.3 shows the solubility of $AgNO_3$ in water in units of moles of $AgNO_3$ per kilogram of H_2O. If 255 g of $AgNO_3$ is added to 100 g of water at 95°C and cooled slowly, at what temperature will the solution become saturated?

Ionic Equilibria between Solids and Solutions

5. Iron(III) sulfate, $Fe_2(SO_4)_3$, is a yellow compound that is used as a coagulant in water treatment. Write a balanced chemical equation and a solubility product expression for its dissolution in water.

6. Lead antimonate, $Pb_3(SbO_4)_2$, is used as an orange pigment in oil-based paints and in glazes. Write a balanced chemical equation and a solubility product expression for its dissolution in water.

7. Thallium(I) iodate ($TlIO_3$) is only slightly soluble in water. Its K_{sp} at 25°C is 3.07×10^{-6}. Estimate the solubility of thallium(I) iodate in water in units of grams per 100.0 mL of water.

8. Thallium thiocyanate (TlSCN) is only slightly soluble in water. Its K_{sp} at 25°C is 1.82×10^{-4}. Estimate the solubility of thallium thiocyanate in units of grams per 100.0 mL of water.

9. Potassium perchlorate, $KClO_4$, has a K_{sp} at 25°C of 1.07×10^{-2}. Compute its solubility in grams per liter of solution.

10. Ammonium hexachloroplatinate(IV), $(NH_4)_2(PtCl_6)$, is one of the few sparingly soluble ammonium salts. Its K_{sp} at 20°C is 5.6×10^{-6}. Compute its solubility in grams per liter of solution.

11. The solubility product constant of mercury(I) iodide is 1.2×10^{-28} at 25°C. Estimate the concentration of Hg_2^{2+} and I^- in equilibrium with solid Hg_2I_2.

12. The solubility product constant of Hg_2Cl_2 is 2×10^{-18} at 25°C. Estimate the concentration of Hg_2^{2+} and Cl^- in equilibrium with solid Hg_2Cl_2 at 25°C.

13. The solubility of silver chromate (Ag_2CrO_4) in 500 mL of water at 25°C is 0.0129 g. Calculate its solubility product constant.

14. At 25°C, 400 mL of water can dissolve 0.00896 g of lead iodate, $Pb(IO_3)_2$. Calculate K_{sp} for lead iodate.

15. At 100°C, water dissolves 1.8×10^{-2} g of AgCl per liter. Compute the K_{sp} of AgCl at this temperature.

16. A mass of 0.017 g of silver dichromate ($Ag_2Cr_2O_7$) will dissolve in 300 mL of water at 25°C. Calculate the solubility product constant K_{sp} of silver dichromate.

Precipitation and the Solubility Product

17. A solution of barium chromate ($BaCrO_4$) is prepared by dissolving 6.3×10^{-3} g of this yellow solid in 1.00 L of hot water. Will solid barium chromate precipitate upon cooling to 25°C, according to the solubility product expression? Explain.

18. A solution is prepared by dissolving 0.090 g of PbI_2 in 1.00 L of hot water and cooling the solution to 25°C. Will solid precipitate result from this process, according to the solubility product expression? Explain.

19. A solution is prepared by mixing 250.0 mL of 2.0×10^{-3} M $Ce(NO_3)_3$ and 150.0 mL of 10×10^{-2} M KIO_3 at 25°C. Determine whether $Ce(IO_3)_3(s)$ ($K_{sp} = 1.9 \times 10^{-10}$) tends to precipitate from this mixture.

20. Suppose 100.0 mL of a 0.0010 M $CaCl_2$ solution is added to 50.0 mL of a 6.0×10^{-5} M NaF solution at 25°C. Determine whether $CaF_2(s)$ ($K_{sp} = 3.9 \times 10^{-11}$) tends to precipitate from this mixture.

21. Suppose 50.0 mL of a 0.0500 M solution of $Pb(NO_3)_2$ is mixed with 40.0 mL of a 0.200 M solution of $NaIO_3$ at 25°C. Calculate the $[Pb^{2+}]$ and $[IO_3^-]$ when the mixture comes to equilibrium. At this temperature, K_{sp} for $Pb(IO_3)_2$ is 2.6×10^{-13}.

22. Silver iodide (AgI) is used in place of silver chloride for the fastest photographic film because it is more sensitive to light and can therefore form an image in a very short exposure time. A silver iodide emulsion is prepared by adding 6.60 L of 0.10 M NaI solution to 1.50 L of 0.080 M $AgNO_3$ solution at 25°C. Calculate the concentration of silver ion remaining in solution when the mixture comes to equilibrium and its chemical amount relative to the amount present initially.

23. When 50.0 mL of 0.100 M $AgNO_3$ and 30.0 mL of 0.0600 M Na_2CrO_4 are mixed, a precipitate of silver chromate (Ag_2CrO_4) is formed. The solubility product K_{sp} of silver chromate in water at 25°C is 1.9×10^{-12}. Calculate the $[Ag^+]$ and $[CrO_4^{2-}]$ remaining in solution at equilibrium.

24. When 40.0 mL of 0.0800 M $Sr(NO_3)_2$ and 80.0 mL of 0.0500 M KF are mixed, a precipitate of strontium fluoride (SrF_2) is formed. The solubility product K_{sp} of strontium fluoride in water at 25°C is 2.8×10^{-9}. Calculate the $[Sr^{2+}]$ and $[F^-]$ remaining in solution at equilibrium.

25. Calculate the solubility (in mol L^{-1}) of $CaF_2(s)$ at 25°C in a 0.040 M aqueous solution of NaF.

26. Calculate the mass of AgCl that can dissolve in 100 mL of 0.150 M NaCl solution.

27. The solubility product of nickel(II) hydroxide, $Ni(OH)_2$, at 25°C is $K_{sp} = 1.6 \times 10^{-16}$.

(a) Calculate the molar solubility of $Ni(OH)_2$ in pure water at 25°C.

(b) Calculate the molar solubility of $Ni(OH)_2$ in 0.100 M NaOH.

28. Silver arsenate (Ag_3AsO_4) is a slightly soluble salt having a solubility product of $K_{sp} = 1.0 \times 10^{-22}$ at 25°C for the equilibrium

$$Ag_3AsO_4(s) \rightleftharpoons 3\,Ag^+(aq) + AsO_4^{3-}(aq)$$

(a) Calculate the molar solubility of silver arsenate in pure water at 25°C.

(b) Calculate the molar solubility of silver arsenate in 0.10 M $AgNO_3$.

29. A saturated solution of $Mg(OH)_2$ at 25°C is prepared by equilibrating solid $Mg(OH)_2$ with water. Concentrated NaOH is then added until the solubility of $Mg(OH)_2$ is 0.0010 times that in H_2O alone. (Ignore the change in volume resulting from the addition of NaOH.) The solubility product K_{sp} of $Mg(OH)_2$ is 1.2×10^{-11} at 25°C. Calculate the concentration of hydroxide ion in the solution after the addition of the NaOH.

30. A saturated solution of BaF_2 at 25°C is prepared by equilibrating solid BaF_2 with water. Powdered NaF is then dissolved in the solution until the solubility of BaF_2 is 1.0% of that in H_2O alone. The solubility product K_{sp} of BaF_2 is 1.7×10^{-6} at 25°C. Calculate the concentration of fluoride ion in the solution after addition of the powdered NaF.

The Effects of pH on Solubility

31. Compare the molar solubility of AgOH in pure water with that in a solution buffered at pH 7.00. Note the difference between the two: When AgOH is dissolved in pure water, the pH does not remain at 7.

32. Compare the molar solubility of $Mg(OH)_2$ in pure water with that in a solution buffered at pH 9.00.

33. For each of the following ionic compounds, state whether the solubility will increase, decrease, or remain unchanged as a solution at pH 7 is made acidic.

(a) PbI_2 (b) AgOH (c) $Ca_3(PO_4)_2$

34. For each of the following ionic compounds, state whether the solubility will increase, decrease, or remain unchanged as a solution at pH 7 is made acidic.

(a) $SrCO_3$ (b) Hg_2Br_2 (c) MnS

A DEEPER LOOK . . . Selective Precipitation of Ions

35. An aqueous solution at 25°C is 0.10 M in both Mg^{2+} and Pb^{2+} ions. We wish to separate the two kinds of metal ions by taking advantage of the different solubilities of their oxalates, MgC_2O_4 and PbC_2O_4.

(a) What is the highest possible oxalate ion concentration that allows only one solid oxalate salt to be present at equilibrium? Which ion is present in the solid—Mg^{2+} or Pb^{2+}?

(b) What fraction of the less soluble ion still remains in solution under the conditions of part (a)?

36. An aqueous solution at 25°C is 0.10 M in Ba^{2+} and 0.50 M in Ca^{2+} ions. We wish to separate the two by taking advantage of the different solubilities of their fluorides, BaF_2 and CaF_2.

(a) What is the highest possible fluoride ion concentration that allows only one solid fluoride salt to be present at equilibrium? Which ion is present in the solid—Ba^{2+} or Ca^{2+}?

(b) What fraction of the less soluble ion still remains in solution under the conditions of part (a)?

37. The cations in an aqueous solution that contains 0.100 M $Hg_2(NO_3)_2$ and 0.0500 M $Pb(NO_3)_2$ are to be separated by taking advantage of the difference in the solubilities of their iodides. $K_{sp}(PbI_2) = 1.4 \times 10^{-8}$ and $K_{sp}(Hg_2I_2) = 1.2 \times 10^{-28}$. What should be the concentration of iodide ion for the best separation? In the "best" separation, one of the cations should remain entirely in solution and the other should precipitate as fully as possible.

38. The cations in an aqueous solution that contains 0.150 M $Ba(NO_3)_2$ and 0.0800 M $Ca(NO_3)_2$ are to be separated by taking advantage of the difference in the solubilities of their sulfates. $K_{sp}(BaSO_4) = 1.1 \times 10^{-10}$ and $K_{sp}(CaSO_4) = 2.4 \times 10^{-5}$. What should be the concentration of sulfate ion for the best separation?

39. Calculate the $[Zn^{2+}]$ in a solution that is in equilibrium with $ZnS(s)$ and in which $[H_3O^+] = 1.0 \times 10^{-5}$ M and $[H_2S] = 0.10$ M.

40. Calculate the $[Cd^{2+}]$ in a solution that is in equilibrium with $CdS(s)$ and in which $[H_3O^+] = 1.0 \times 10^{-3}$ M and $[H_2S] = 0.10$ M.

41. What is the highest pH at which 0.10 M Fe^{2+} will remain entirely in a solution that is saturated with H_2S at a concentration of $[H_2S] = 0.10$ M? At this pH, what would be the concentration of Pb^{2+} in equilibrium with solid PbS in this solution?

42. What is the highest pH at which 0.050 M Mn^{2+} will remain entirely in a solution that is saturated with H_2S at a concentration of $[H_2S] = 0.10$ M? At this pH, what would be the concentration of Cd^{2+} in equilibrium with solid CdS in this solution?

Complex Ions and Solubility

43. Suppose 0.10 mol of $Cu(NO_3)_2$ and 1.50 mol of NH_3 are dissolved in water and diluted to a total volume of 1.00 L. Calculate the concentrations of $Cu(NH_3)_4^{2+}$ and of Cu^{2+} at equilibrium.

44. The formation constant of the $TlCl_4^-$ complex ion is 1×10^{18}. Suppose 0.15 mol of $Tl(NO_3)_3$ is dissolved in 1.00 L of a 0.50 M solution of NaCl. Calculate the concentration at equilibrium of $TlCl_4^-$ and of Tl^{3+}.

45. The organic compound "18-crown-6" binds alkali metals in aqueous solution by wrapping around and enfolding the ion. It presents a niche that nicely accommodates the K^+ ion but is too small for the Rb^+ ion and too large for the Na^+ ion. The values of the equilibrium constants show this:

$$Na^+(aq) + \text{18-crown-6}(aq) \rightleftharpoons \text{Na-crown}^+(aq) \qquad K = 6.6$$

$$K^+(aq) + \text{18-crown-6}(aq) \rightleftharpoons \text{K-crown}^+(aq) \qquad K = 111.6$$

$$Rb^+(aq) + \text{18-crown-6}(aq) \rightleftharpoons \text{Rb-crown}^+(aq) \qquad K = 36$$

An aqueous solution is initially 0.0080 M in 18-crown-6(aq) and also 0.0080 M in $K^+(aq)$. Compute the equilibrium concentration of free K^+. ("Free" means not tied up with the 18-crown-6.) Compute the concentration of free Na^+ if the solution contains 0.0080 M $Na^+(aq)$ instead of $K^+(aq)$.

46. The organic compound 18-crown-6 (see preceding problem) also binds strongly with the alkali metal ions in methanol.

$$K^+ + \text{18-crown-6} \rightleftharpoons [\text{complex}]^+$$

In methanol solution the equilibrium constant is 1.41×10^6. A similar reaction with Cs^+ has an equilibrium constant of only 2.75×10^4. A solution is made (in methanol) containing 0.020 mol L^{-1} each of K^+ and Cs^+. It also contains 0.30 mol L^{-1} of 18-crown-6. Compute the equilibrium concentrations of both the uncomplexed K^+ and the uncomplexed Cs^+.

47. Will silver chloride dissolve to a significantly greater extent in a 1.00 M NaCl solution than in pure water due to the possible formation of $AgCl_2^-$ ions? Use data from Tables 16.2 and 16.4 to provide a quantitative answer to this question. What will happen in a 0.100 M NaCl solution?

48. Calculate how many grams of silver chloride will dissolve in 1.0 L of a 1.0 M NH_3 solution through formation of the complex ion $Ag(NH_3)_2^+$.

49. The pH of a 0.2 M solution of $CuSO_4$ is 4.0. Write chemical equations to explain why a solution of this salt is neither basic [from the reaction of $SO_4^{2-}(aq)$ with water] nor neutral, but acidic.

50. Will a 0.05 M solution of $FeCl_3$ be acidic, basic, or neutral? Explain your answer by writing chemical equations to describe any reactions taking place.

51. The acid ionization constant for $Co(H_2O)_6^{2+}(aq)$ is 3×10^{-10}. Calculate the pH of a 0.10 M solution of $Co(NO_3)_2$.

52. The acid ionization constant for $Fe(H_2O)_6^{2+}(aq)$ is 3×10^{-6}. Calculate the pH of a 0.10 M solution of $Fe(NO_3)_2$, and compare it with the pH of the corresponding iron(III) nitrate solution from Example 16.10.

53. A 0.15 M aqueous solution of the chloride salt of the complex ion $Pt(NH_3)_4^{2+}$ is found to be weakly acidic with a pH of 4.92. This is initially puzzling because the Cl^- ion in water is not acidic and NH_3 in water is *basic*, not acidic. Finally, it is suggested that the $Pt(NH_3)_4^{2+}$ ion as a group donates hydrogen ions. Compute the K_a of this acid, assuming that just one hydrogen ion is donated.

54. The pH of a 0.10 M solution of $Ni(NO_3)_2$ is 5.0. Calculate the acid ionization constant of $Ni(H_2O)_6^{2+}(aq)$.

* **55.** K_{sp} for $Pb(OH)_2$ is 4.2×10^{-15}, and K_f for $Pb(OH)_3^-$ is 4×10^{14}. Suppose a solution whose initial concentration of $Pb^{2+}(aq)$ is 1.00 M is brought to pH 13.0 by addition of solid NaOH. Will solid $Pb(OH)_2$ precipitate, or will the lead be dissolved as $Pb(OH)_3^-(aq)$? What will be $[Pb^{2+}]$ and $[Pb(OH)_3^-]$ at equilibrium? Repeat the calculation for an initial Pb^{2+} concentration of 0.050 M. (*Hint:* One way to solve this problem is to *assume* that $Pb(OH)_2(s)$ is present and calculate $[Pb^{2+}]$ and $[Pb(OH)_3^-]$ that would be in equilibrium with the solid. If the sum of these is less than the original $[Pb^{2+}]$, the remainder can be assumed to have precipitated. If not, there is a contradiction and we must assume that *no* $Pb(OH)_2(s)$ is present. In this case we can calculate $[Pb^{2+}]$ and $[Pb(OH)_3^-]$ directly from K_f.)

* **56.** K_{sp} from $Zn(OH)_2$ is 4.5×10^{-17}, and K_f for $Zn(OH)_4^{2-}$ is 5×10^{14}. Suppose a solution whose initial concentration of $Zn^{2+}(aq)$ is 0.010 M is brought to pH 14.0 by addition of solid NaOH. Will solid $Zn(OH)_2$ precipitate, or will the zinc be dissolved as $Zn(OH)_4^{2-}(aq)$? What will be $[Zn^{2+}]$ and $[Zn(OH)_4^{2-}]$ at equilibrium? Repeat the calculation at pH 13 for an initial Zn^{2+} concentration of 0.10 M. See the hint in problem 55.

ADDITIONAL PROBLEMS

57. Write a chemical equation for the dissolution of mercury(I) chloride in water, and give its solubility product expression.

* 58. Magnesium ammonium phosphate has the formula $MgNH_4PO_4 \cdot 6H_2O$. It is only slightly soluble in water (its K_{sp} is 2.3×10^{-13}). Write a chemical equation and the corresponding equilibrium law for the dissolution of this compound in water.

59. Soluble barium compounds are poisonous, but barium sulfate is routinely ingested as a suspended solid in a "barium cocktail" to improve the contrast in x-ray images. Calculate the concentration of dissolved barium per liter of water in equilibrium with solid barium sulfate.

60. A saturated aqueous solution of silver perchlorate ($AgClO_4$) contains 84.8% by mass $AgClO_4$, but a saturated solution of $AgClO_4$ in 60% aqueous perchloric acid contains only 5.63% by mass $AgClO_4$. Explain this large difference using chemical equations.

61. Suppose 140 mL of 0.0010 M $Sr(NO_3)_2$ is mixed with enough 0.0050 M NaF to make 1.00 L of solution. Will $SrF_2(s)$ ($K_{sp} = 2.8 \times 10^{-9}$) precipitate at equilibrium? Explain.

62. The concentration of calcium ion in a town's supply of drinking water is 0.0020 M. (This water is referred to as hard water because it contains such a large concentration of Ca^{2+}.) Suppose the water is to be fluoridated by the addition of NaF for the purpose of reducing tooth decay. What is the maximum concentration of fluoride ion that can be achieved in the water before precipitation of CaF_2 begins? Will the water supply attain the level of fluoride ion recommended by the U.S. Public Health Service, about 5×10^{-5} M (1 mg fluorine per liter)?

63. Suppose that 150 mL of 0.200 M K_2CO_3 and 100 mL of 0.400 M $Ca(NO_3)_2$ are mixed together. Assume that the volumes are additive, that $CaCO_3$ is completely insoluble, and that all other substances that might be formed are soluble. Calculate the mass of $CaCO_3$ precipitated, and calculate the concentrations in the final solution of the four ions that were present initially.

64. The solubility of $CaCO_3$ in water is about 7 mg L^{-1}. Show how one can calculate the solubility product of $BaCO_3$ from this information and from the fact that when sodium carbonate solution is added slowly to a solution containing equimolar concentrations of Ca^{2+} and Ba^{2+}, no $CaCO_3$ is formed until about 90% of the Ba^{2+} has been precipitated as $BaCO_3$.

65. It is sometimes asserted that carbonates are soluble in strong acids because a gas is formed that escapes (CO_2). Suppose that CO_2 were extremely soluble in water (as, for example, ammonia is) and therefore it did not leave the site of the reaction, but that otherwise, its chemistry was unchanged. Would calcium carbonate be soluble in strong acids? Explain.

66. The solubility products of $Fe(OH)_3$ and $Ni(OH)_2$ are about 10^{-36} and 6×10^{-18}, respectively. Find the approximate pH range suitable for the separation of Fe^{3+} and Ni^{2+} by precipitation of $Fe(OH)_3$ from a solution initially 0.01 M in each ion, as follows: (a) Calculate the lowest pH at which all but 0.1% of the Fe^{3+} will be precipitated as $Fe(OH)_3$; (b) calculate the highest pH possible without precipitation of $Ni(OH)_2$.

67. The two solids $CuBr(s)$ and $AgBr(s)$ are only very slightly soluble in water: $K_{sp}(CuBr) = 4.2 \times 10^{-8}$ and $K_{sp}(AgBr) = 7.7 \times 10^{-13}$. Some $CuBr(s)$ and $AgBr(s)$ are both mixed into a quantity of water that is then stirred until it is saturated with respect to both solutes. Next, a small amount of KBr is added and dissolves completely. Compute the ratio of $[Cu^+]$ to $[Ag^+]$ after the system reestablishes equilibrium.

* 68. The two salts $BaCl_2$ and Ag_2SO_4 are both far more soluble in water than either $BaSO_4$ ($K_{sp} = 1.1 \times 10^{-10}$) or AgCl ($K_{sp} = 1.6 \times 10^{-10}$) at 25°C. Suppose 50.0 mL of 0.040 M $BaCl_2(aq)$ is added to 50.0 mL of 0.020 M $Ag_2SO_4(aq)$. Calculate the concentrations of $SO_4^{2-}(aq)$, $Cl^-(aq)$, $Ba^{2+}(aq)$, and $Ag^+(aq)$ that remain in solution at equilibrium.

69. The Mohr method is a technique for determining the amount of chloride ion in an unknown sample. It is based on the difference in solubility between silver chloride (AgCl; $K_{sp} = 1.6 \times 10^{-10}$) and silver chromate ($Ag_2CrO_4$; $K_{sp} = 1.9 \times 10^{-12}$). In using this method, one adds a small amount of chromate ion to a solution with unknown chloride concentration. By measuring the volume of $AgNO_3$ added before the appearance of the red silver chromate, one can determine the amount of Cl^- originally present. Suppose we have a solution that is 0.100 M in Cl^- and 0.00250 M in CrO_4^{2-}. If we add 0.100 M $AgNO_3$ solution drop by drop, will AgCl or Ag_2CrO_4 precipitate first? When $Ag_2CrO_4(s)$ first appears, what fraction of the Cl^- originally present remains in solution?

70. Oxide ion, like sulfide ion, is a strong base. Write an equation for the dissolution of CaO in water and give its equilibrium constant expression. Write the corresponding equation for the dissolution of CaO in an aqueous solution of a strong acid, and relate its equilibrium constant to the previous one.

71. Water that has been saturated with magnesia (MgO) at 25°C has a pH of 10.16. Write a balanced chemical equation for the equilibrium between $MgO(s)$ and the ions it furnishes in aqueous solution, and calculate the equilibrium constant at 25°C. What is the solubility, in moles per liter, of MgO in water?

72. To 1.00 L of a 0.100 M $AgNO_3$ solution is added an excess of sodium chloride. Then 1.00 L of 0.500 M $NH_3(aq)$ is added. Finally, sufficient nitric acid is added until the pH of the resulting solution is 1.0. Write balanced equations for the reactions that take place (if any) at each of three steps in this process.

73. Only about 0.16 mg of $AgBr(s)$ will dissolve in 1.0 L of water (this volume of solid is smaller than the head of a pin). In a solution of ammonia that contains 0.10 mol ammonia per liter of water, there are about 555 water molecules for every molecule of ammonia. However, more than 400 times as much AgBr (68 mg) will dissolve in this solution as in plain water. Explain how such a tiny change in the composition of the solution can have such a large effect on the solubility of AgBr.

* 74. (a) Calculate the solubility of calcium oxalate (CaC_2O_4) in 1.0 M oxalic acid ($H_2C_2O_4$) at 25°C, using the two acid ionization constants for oxalic acid from Table 15.2 and the solubility product $K_{sp} = 2.6 \times 10^{-9}$ for CaC_2O_4.
(b) Calculate the solubility of calcium oxalate in pure water at 25°C.
(c) Account for the difference between the results of (a) and (b).

* 75. When 6 M HCl is added to solid CdS, some of the solid dissolves to give the complex ion $CdCl_4^{2-}(aq)$.
(a) Write a balanced equation for the reaction that occurs.
(b) Use data from Tables 15.2 and 16.3 and the formation constant of $CdCl_4^{2-}$ ($K_f = 8 \times 10^2$) to calculate the equilibrium constant for the reaction of part (a).
(c) What is the molar solubility of CdS per liter of 6 M HCl?

* 76. Using data from Table 16.4, calculate the concentrations of $Hg^{2+}(aq)$, $HgCl^+(aq)$, and $HgCl_2(aq)$ that result when 1.00 L of a 0.100 M $Hg(NO_3)_2$ solution is mixed with an equal volume of a 0.100 M $HgCl_2$ solution. (*Hint:* Use the analogy with amphoteric equilibria discussed in Section 10.8.)

* 77. Calculate the concentration of $Cu^{2+}(aq)$ in a solution that contains 0.020 mol of $CuCl_2$ and 0.100 mol of NaCN in 1.0 L.

$$Cu^{2+}(aq) + 4\,CN^-(aq) \rightleftharpoons Cu(CN)_4^{2-}(aq) \qquad K = 2.0 \times 10^{30}$$

(*Hint:* Do not overlook the reaction of CN^- with water to give HCN.)

78. An aqueous solution of $K_2[Pt(OH)_6]$ has a pH greater than 7. Explain this fact by writing an equation showing the $Pt(OH)_6^{2-}$ ion acting as a Brønsted–Lowry base and accepting a hydrogen ion from water.

79. In Example 16.10 we included only the first acid dissociation K_{a1} of a 0.100 M aqueous solution of $Fe(H_2O)_6^{3+}$. Subsequent dissociation can also occur, with $K_{a2} = 2.0 \times 10^{-5}$, to give $Fe(H_2O)_4(OH)_2^+$.
(a) Calculate the concentration of $Fe(H_2O)_4(OH)_2^+$ at equilibrium. Does the pH change significantly when this second dissociation is taken into account?
(b) We can describe the same reaction as the dissociation of a complex ion $Fe(OH)_2^+$ to Fe^{3+} and two OH^- ions. Calculate K_f, the formation constant for $Fe(OH)_2^+$.

CUMULATIVE PROBLEMS

80. The volume of a certain saturated solution is greater than the sum of the volumes of the water and salt from which it is made. Predict the effect of increased pressure on the solubility of this salt.

81. Codeine has the molecular formula $C_{18}H_{21}NO_3$. It is soluble in water to the extent of 1.00 g per 120 mL of water at room temperature and 1.00 g per 60 mL of water at 80°C. Compute the molal solubility (in mol kg^{-1}) at both temperatures, taking the density of water to be fixed at 1.00 g cm^{-3}. Is the dissolution of codeine in water endothermic or exothermic?

82. Suppose 1.44 L of a saturated solution of strontium carbonate ($SrCO_3$) in boiling water at 100°C is prepared. The solution is then strongly acidified and shaken to drive off all the gaseous CO_2 that forms. The volume of this gas (at a temperature of 100°C and a partial pressure of 0.972 atm) is measured to be 0.20 L (200 mL).
(a) Calculate the molar solubility of $SrCO_3$ in water at 100°C.
(b) Estimate the solubility product constant K_{sp} of $SrCO_3$ at this temperature.
(c) Explain why the actual K_{sp} at this temperature may be lower than you predicted in part (b).

83. A buffer is prepared by adding 50.0 mL of 0.15 M $HNO_3(aq)$ to 100.0 mL of 0.12 M $NaHCOO(aq)$ (sodium formate). Calculate the solubility of $CaF_2(s)$ in this solution.

CHAPTER 17

Electrochemistry

Corrosion (rusted iron shown here) has enormous adverse economic consequences. Its prevention requires an understanding of the fundamental underlying chemical processes.

Electrochemistry connects the chemistry of oxidation–reduction reactions to the physics of charge flow. It is the branch of chemistry concerned with the principles and methods for interconverting chemical and electrical energy. Spontaneous chemical reactions can be used as a source of electrical energy; conversely, electrical energy can be used to drive nonspontaneous chemical reactions. The principles of electrochemistry are the foundations for numerous practical applications such as the storage of energy in batteries and the efficient conversion of energy from readily available sources, (e.g., solar or chemical) to other forms useful in technology.

Electrochemistry occurs through redox reactions, and its description requires balanced equations for redox reactions. Both topics are discussed in Section 11.4, which you should review now.

Electrochemical reactions are carried out in devices called **electrochemical cells,** which couple the redox reactions to external electrical circuits. We start our discussion of electrochemistry by using redox reactions in aqueous solution to explain how electrochemical cells operate and relating the current produced in the cell to the stoichiometry of the reaction. Then we apply thermodynamic principles to describe the interconversion of energy in the operation of cells. Because cells operate at constant temperature and pressure, the Gibbs free energy is the most important thermodynamic state function in this description. We relate the change in Gibbs free energy of the redox reaction in the cell to changes in the electrical potential in the external circuit, which in turn measure the capacity of the external circuit to perform electrical work. With this background established, we show how thermodynamics guides the search for new electrochemical technologies. We illustrate practical applications of electrochemical cells in batteries, fuel cells, and the spontaneous corrosion of metals in the environment. We end the chapter by describing practical applications of electrochemical cells in metallurgy and in the electrolysis of water.

17.1 Electrochemical Cells

Electrochemical cells couple oxidation–reduction reactions to external electrical circuits. Electrochemical cells have the fascinating feature that the oxidizing reactant can be physically separate from the reducing reactant as long as they are connected through the external circuit. This is the key feature that allows us to intervene in the conversion between chemical energy and electrical energy. If the reaction is spontaneous, the cell utilizes the Gibbs free energy available from the reaction to convert chemical energy into electrical energy. The electrical energy is provided to the external circuit, where it can perform useful work. This type of cell is called **galvanic. Electrolytic cells** operate in the opposite sense; they take electrical energy from the external circuit and use it to drive reactions that are not spontaneous. In both cases the direction and magnitude of current flow in the external circuit are governed by the Gibbs free energy change of the oxidation–reduction reaction in the cell.

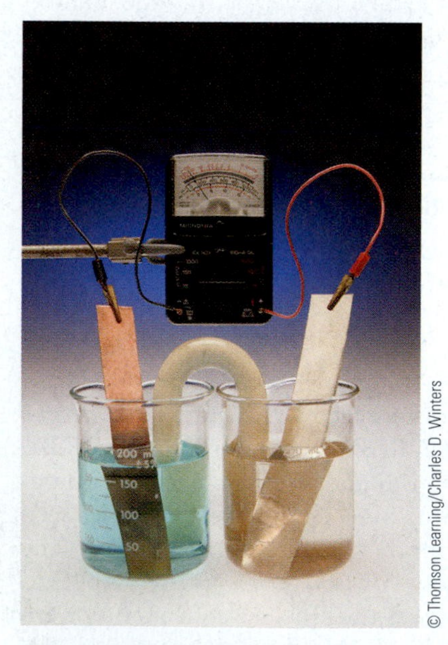

FIGURE 17.1 A metallic copper anode reacts to give a blue solution containing copper(II) ions as silver ions plate out on a silver cathode in a galvanic cell.

Let's examine a specific case where an electrochemical cell presents an alternative means to carry out a familiar oxidation–reduction reaction. The reaction of copper metal with aqueous silver ions shown in Figure 11.7 is clearly spontaneous. This same reaction can be carried out quite differently without ever bringing the two reactants into *direct* contact with each other if a galvanic cell (a battery) is constructed from them. A copper strip is partially immersed in a solution of $Cu(NO_3)_2$ and a silver strip in a solution of $AgNO_3$, as Figures 17.1 and 17.2 illustrate. The two solutions are connected by a **salt bridge,** which is an inverted U-shaped tube containing a solution of a salt such as $NaNO_3$. The ends of the bridge are stuffed with porous plugs that prevent the two solutions from mixing but allow ions to pass through. The two metal strips are connected to an **ammeter,** an instrument that measures the direction and magnitude of electric current through it.

As copper is oxidized on the left side, Cu^{2+} ions enter the solution. The electrons released in the reaction pass through the external circuit from left to right, as shown by the deflection of the ammeter needle. The electrons enter the silver strip, and at the metal–solution interface, they are picked up by Ag^+ ions, which plate out as atoms on the surface of the silver. This process would lead to an increase of positive charge in the left beaker and a decrease in the right one were it not for the salt bridge; the bridge permits a net flow of positive ions through

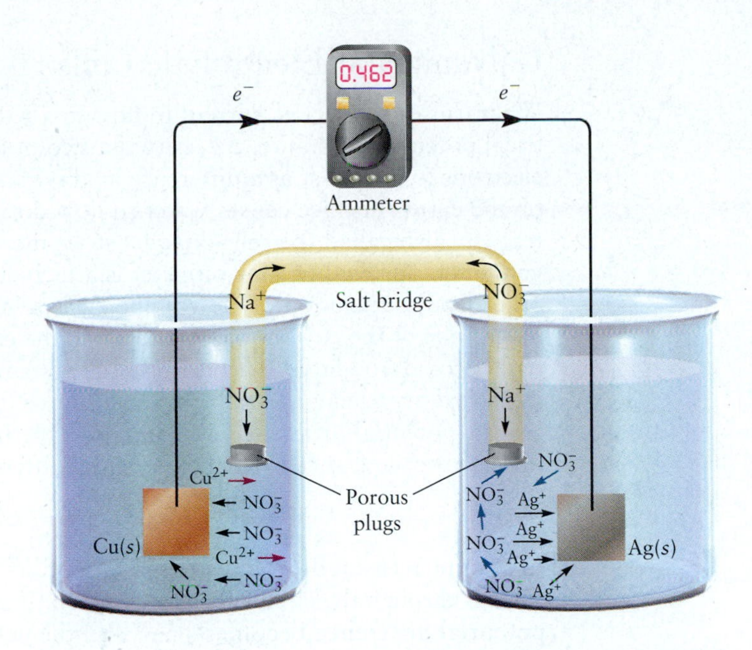

FIGURE 17.2 In the galvanic cell of Figure 17.1, charged particles move when the circuit is completed. Electrons flow from the copper to the silver electrode through the wire. In solution, anions migrate toward the copper electrode and cations move toward the silver electrode. Sodium and nitrate ions migrate through the salt bridge to maintain electrical neutrality.

it into the right beaker and of negative ions into the left beaker, preserving charge neutrality in each.

As discussed in Section 11.4, this oxidation–reduction reaction is composed of two separate half-reactions. The oxidation half-reaction in the left-hand beaker is

$$Cu(s) \longrightarrow Cu^{2+}(aq) + 2e^-$$

and the reduction half-reaction in the right beaker is

$$Ag^+(aq) + e^- \longrightarrow Ag(s)$$

Following Michael Faraday, chemists call the site at which oxidation occurs in an electrochemical cell the **anode** and the site at which reduction occurs the **cathode.** In the galvanic cell just discussed, copper is the anode (because it is oxidized) and silver is the cathode (because Ag^+ is reduced). Electrons flow in the external circuit from anode to cathode. In solution both positive and negative ions are free to move. In an electrochemical cell, negative ions (anions) move toward the anode and positive ions (cations) move toward the cathode. We will adopt a symbolic representation for galvanic cells in which the anode is shown on the left and the cathode on the right so that electrons flow through the external circuit from left to right. Schematically the copper–silver galvanic cell is represented as

$$Cu|Cu^{2+}||Ag^+|Ag$$

with the anode on the left and the cathode on the right and with the metal–solution interface represented by $|$ and the salt bridge by $||$.

The net chemical reaction in this simple galvanic cell is the same one that takes place when a copper strip is placed in an aqueous solution of silver nitrate (see Fig. 11.7), but there is an essential difference in the process. Because the reaction components are separated into two compartments while electrical continuity is preserved through the external circuit, electrons cannot go directly from copper atoms to silver ions. Electrons are forced to travel through the external circuit (wire) before they ultimately accomplish the same net effect as in Figure 11.7. The current of electrons through the wire can be used for a variety of purposes. A light bulb in the electric circuit would glow from the current passing through it, and the electrochemical cell would have converted chemical energy to electrical energy and ultimately to heat and radiant energy. Or, with a small electric motor in the circuit, the energy change of the chemical reaction can be used to perform mechanical work.

Galvanic and Electrolytic Cells

What causes an electric current to flow in a galvanic cell? There must be an electrical **potential difference,** $\Delta\mathscr{E}$, between two points in the external circuit to cause electrons to flow, just as a difference in gravitational potential between two points on the earth's surface causes water to flow downhill. This electrical potential difference, also called the **cell voltage,** can be measured with an instrument called a **voltmeter;** one lead of the voltmeter is attached to the cathode of the cell and the other lead to the anode. The voltage measured in a galvanic cell depends on the magnitude of the current passing through the cell, and the voltage falls if the current becomes too large. The intrinsic cell voltage (the value at zero current) can be measured by placing a variable voltage source in the external circuit in such a way that its potential difference $\Delta\mathscr{E}_{ext}$ *opposes* the intrinsic potential difference $\Delta\mathscr{E}$ of the electrochemical cell. The net potential difference is then

$$\Delta\mathscr{E}_{net} = \Delta\mathscr{E} - \Delta\mathscr{E}_{ext}$$

$\Delta\mathscr{E}$ can be measured by adjusting $\Delta\mathscr{E}_{ext}$ until $\Delta\mathscr{E}_{net}$ becomes 0, at which point the current through the circuit falls to 0 as well. If $\Delta\mathscr{E}_{ext}$ is held just below $\Delta\mathscr{E}$, the net potential difference becomes small and the cell operation is close to reversible, with only a small current and a slow rate of reaction at the electrodes.

If the opposing external voltage is increased *above* the intrinsic potential difference of the cell, the electrons reverse direction and move toward the copper electrode. Copper ions in solution accept electrons and deposit as copper metal, and silver metal dissolves and furnishes additional Ag^+ ions. The net result is to drive the reverse of the spontaneous reaction, namely,

$$2\ Ag(s) + Cu^{2+}(aq) \longrightarrow 2\ Ag^+ + Cu(s)$$

A galvanic (or voltaic) cell is an electrochemical cell that operates spontaneously. Such a cell converts chemical energy to electrical energy, which can be used to perform work. A cell in which an opposing external potential causes the reaction to occur in the direction opposite the spontaneous direction is called an electrolytic cell; such a cell uses electrical energy provided by the external circuit to carry out chemical reactions that would otherwise not occur. When a galvanic cell is changed into an electrolytic cell by adding an external potential source that reverses the direction of electron flow, the sites of the anode and the cathode are reversed. In the electrolytic cell, oxidation takes place at the silver electrode, which therefore becomes the anode, and the copper electrode becomes the cathode.

EXAMPLE 17.1

The final step in the production of magnesium from seawater is the electrolysis of molten magnesium chloride, in which the overall reaction is

$$Mg^{2+} + 2\ Cl^- \longrightarrow Mg(\ell) + Cl_2(g)$$

Write equations for the half-reactions occurring at the anode and at the cathode, and indicate the direction in which electrons flow through the external circuit.

SOLUTION

The anode is the site at which oxidation takes place, that is, where electrons are given up. The anode half-reaction must be

$$2\ Cl^- \longrightarrow Cl_2(g) + 2\ e^-$$

At the cathode, reduction takes place and the electrons are taken up. The cathode half-reaction is

$$Mg^{2+} + 2\,e^- \longrightarrow Mg(\ell)$$

Electrons move from the anode, where chlorine is liberated, through the external circuit to the cathode, where molten magnesium is produced.

Related Problems: 1, 2

Faraday's Laws

The dual aspect of the electrochemical cell—galvanic or electrolytic—was recognized shortly after the cell's discovery in 1800 by Alessandro Volta. Volta constructed a "battery of cells" consisting of a number of plates of silver and zinc that were separated from one another by porous strips of paper saturated with a salt solution. By 1807, Sir Humphry Davy had prepared elemental sodium and potassium by using a battery to electrolyze their respective hydroxides. But, the underlying scientific basis of the electrochemical cell was not understood. Michael Faraday's research showed a direct quantitative relationship between the amounts of substances that react at the cathode and the anode and the total electric charge that passes through the cell. This observation is the substance of **Faraday's laws,** which we state as follows:

1. The mass of a given substance that is produced or consumed at an electrode is proportional to the quantity of electric charge passed through the cell.
2. Equivalent masses[1] of different substances are produced or consumed at an electrode by the passage of a given quantity of electric charge through the cell.

These laws, which summarize the stoichiometry of electrochemical processes, were discovered by Michael Faraday in 1833, more than half a century before the electron was discovered and the atomic basis of electricity was understood.

The charge e on a single electron (expressed in coulombs) has been very accurately determined to be

$$e = 1.60217646 \times 10^{-19}\ \text{C}$$

so the quantity of charge represented by one mole of electrons is

$$Q = (6.0221420 \times 10^{23}\ \text{mol}^{-1})(1.60217646 \times 10^{-19}\ \text{C}) = 96{,}485.34\ \text{C mol}^{-1}$$

This quantity of charge is called the **Faraday constant** (symbol $\mathcal{F}$):

$$\mathcal{F} = 96{,}485.34\ \text{C mol}^{-1}$$

Electric current is the amount of charge flowing through a circuit per unit time. If Q is the magnitude of the charge in coulombs and t is the time in seconds that it takes to pass a point in the circuit, then the current I is

$$I = \frac{Q}{t} \qquad [17.1]$$

[1]The equivalent mass of an element in a redox reaction, or of a compound containing that element, is its molar mass divided by the number of moles of electrons transferred per mole of substance in the corresponding half-reaction.

where the units for I are amperes (A) or coulombs per second. A current of I amperes flowing for t seconds causes It coulombs of charge to pass through the circuit. The number of moles of electrons in that amount of charge is

$$\frac{It}{96{,}485 \text{ C mol}^{-1}} = \text{moles of electrons}$$

From the number of moles of electrons that pass through a circuit, the number of moles (and therefore the number of grams) of substances reacting at the electrodes in the electrochemical cell can be calculated. Suppose a zinc–silver galvanic cell is constructed in which the anode half-reaction is

$$\text{Zn}(s) \longrightarrow \text{Zn}^{2+}(aq) + 2\,e^-$$

and the cathode half-reaction is

$$\text{Ag}^+(aq) + e^- \longrightarrow \text{Ag}(s)$$

Each mole of electrons that passes through the cell arises from the oxidation of $\frac{1}{2}$ mol of Zn(s) (because each Zn atom gives up two electrons) and reduces 1 mol of silver ions. From the molar masses of silver and zinc, we calculate that 65.38/2 = 32.69 g of zinc is dissolved at the anode and 107.87 g of silver is deposited at the cathode. The same relationships hold if the cell is operated as an electrolytic cell, but in that case silver is dissolved and zinc is deposited.

EXAMPLE 17.2

An electrolytic cell is constructed in which the silver ions in silver chloride are reduced to silver at the cathode and copper is oxidized to $\text{Cu}^{2+}(aq)$ at the anode. A current of 0.500 A is passed through the cell for 101 minutes. Calculate the mass of copper metal dissolved and the mass of silver metal deposited.

SOLUTION

$$t = (101 \text{ min})(60 \text{ s min}^{-1}) = 6.06 \times 10^3 \text{ s}$$

The number of moles of electrons passed through the circuit during this time is

$$\frac{(0.500 \text{ C s}^{-1})(6.06 \times 10^3 \text{ s})}{96{,}485 \text{ C mol}^{-1}} = 3.14 \times 10^{-2} \text{ mol } e^-$$

The half-cell reactions are

$$\text{AgCl}(s) + e^- \longrightarrow \text{Ag}(s) + \text{Cl}^-(aq) \quad \text{(cathode)}$$

$$\text{Cu}(s) \longrightarrow \text{Cu}^{2+}(aq) + 2\,e^- \quad \text{(anode)}$$

so the masses of silver deposited and copper dissolved are

$$(3.14 \times 10^{-2} \text{ mol } e^-) \times \left(\frac{1 \text{ mol Ag}}{1 \text{ mol } e^-}\right) \times (107.78 \text{ g mol}^{-1}) = 3.39 \text{ g Ag deposited}$$

$$(3.14 \times 10^{-2} \text{ mol } e^-) \times \left(\frac{1 \text{ mol Cu}}{2 \text{ mol } e^-}\right) \times (63.55 \text{ g mol}^{-1}) = 0.998 \text{ g Cu dissolved}$$

Related Problems: 3, 4, 5, 6

17.2 The Gibbs Free Energy and Cell Voltage

Chapters 12 and 13 discussed the pressure–volume work associated with expanding and compressing gases. In electrochemistry a different kind of work, **electrical work,** is fundamental. If an amount of charge, Q, moves through a potential difference $\Delta\mathscr{E}$, the electrical work is

$$w_{\text{elec}} = -Q\,\Delta\mathscr{E} \qquad [17.2]$$

The minus sign appears in this equation because the same thermodynamic convention is followed as in Chapters 12 and 13—that work done by the system (here, an electrochemical cell) has a negative sign. Because work is measured in joules and charge in coulombs, $\Delta\mathscr{E}$ has units of joules per coulomb. One volt (abbreviated V) is 1 joule per coulomb. Because the total charge Q is the current I multiplied by the time t in seconds during which the current is flowing, this equation for the electrical work can also be written

$$w_{elec} = -It\,\Delta\mathscr{E}$$

The potential difference $\Delta\mathscr{E}$ is positive for a galvanic cell, so w_{elec} is negative in this case and net electrical work is performed by the galvanic cell. In an electrolytic cell, in contrast, $\Delta\mathscr{E}$ is negative and w_{elec} is positive, corresponding to net electrical work done on the system by an external source such as an electric generator.

EXAMPLE 17.3

A 6.00-V battery delivers a steady current of 1.25 A for a period of 1.50 hours. Calculate the total charge Q, in coulombs, that passes through the circuit and the electrical work done *by* the battery.

SOLUTION

The total charge is

$$Q = It = (1.25\text{ C s}^{-1})(1.50\text{ hr})(3600\text{ s hr}^{-1}) = 6750\text{ C}$$

The electrical work is

$$w_{elec} = -Q\,\Delta\mathscr{E} = -(6750\text{ C})(6.00\text{ J C}^{-1}) = -4.05 \times 10^4\text{ J}$$

This is the work done *on* the battery, so the work done *by* the battery is the negative of this, or +40.5 kJ.

Thermodynamics demonstrates a fundamental relationship between the change in free energy, ΔG, of a spontaneous chemical reaction at constant temperature and pressure, and the maximum electrical work that such a reaction is capable of producing:

$$-w_{elec,max} = |\Delta G| \qquad \text{(at constant } T \text{ and } P) \qquad [17.3]$$

To show this, recall the definition of the Gibbs free energy function G:

$$G = H - TS = U + PV - TS$$

For processes at constant pressure P and constant temperature T (the usual case in electrochemical cells),

$$\Delta G = \Delta U + P\,\Delta V - T\,\Delta S$$

From the first law of thermodynamics,

$$\Delta U = q + w$$

or

$$\Delta U = q + w_{elec} - P\,\Delta V$$

because there are now two kinds of work, electrical work w_{elec} and pressure–volume work $-P_{ext}\,\Delta V = -P\,\Delta V$. Combining this with the equation for the change in free energy gives

$$\Delta G = q + w_{elec} - P\,\Delta V + P\,\Delta V - T\,\Delta S = q + w_{elec} - T\,\Delta S$$

If the condition of reversibility is imposed upon the galvanic cell, then

$$q = q_{rev} = T\,\Delta S$$

and the free energy change is

$$\Delta G = w_{\text{elec,rev}}$$

In the same cell operated irreversibly (i.e., with a large current permitted to flow), less electrical work is accomplished. The maximum electrical work is done by the galvanic cell when it is operated reversibly.

If n mol of electrons (or $n\mathcal{F}$ coulombs of charge) passes through the external circuit of the galvanic cell when it is operated reversibly, and if $\Delta\mathcal{E}$ is the reversible cell voltage, then

$$\Delta G = w_{\text{elec}} = -Q\,\Delta\mathcal{E} = -n\mathcal{F}\,\Delta\mathcal{E} \qquad \text{(reversible)}$$

Electrical work is produced by an electrochemical cell only if $\Delta G < 0$, or when $\Delta\mathcal{E} > 0$, which amounts to the same thing. This relationship provides a direct way to determine free energy changes for chemical reactions from measurements of cell voltage.

Standard States and Cell Voltages

Recall from Chapters 12 and 13 that the *standard state* of a substance means a pressure of 1 atm and a specified temperature. In addition, the standard state of a solute is that for which its concentration in ideal solution is 1 M.[2] The standard free energy change $\Delta G°$ for a reaction in which all reactants and products are in their standard states can be calculated from a table of standard free energies of formation $\Delta G_f°$ of the substances taking part in the reaction (see Appendix D). For reactions that can be carried out in electrochemical cells, the standard free energy change $\Delta G°$ is related to a **standard cell voltage** $\Delta\mathcal{E}°$ by

$$\Delta G° = -n\mathcal{F}\,\Delta\mathcal{E}° \qquad [17.4]$$

Here $\Delta\mathcal{E}°$ is the cell voltage (potential difference) of a galvanic cell in which reactants and products are in their standard states (gases at 1 atm pressure, solutes at 1 M concentration, metals in their pure stable states, and a specified temperature). This standard cell voltage is an intrinsic electrical property of the cell, which can be calculated from the standard free energy change $\Delta G°$ by means of Equation 17.4. Conversely, $\Delta G°$ (and hence equilibrium constants for cell reactions) can be determined by measuring the standard cell voltage $\Delta\mathcal{E}°$ of a reaction in which n mol of electrons passes through the external circuit.

EXAMPLE 17.4

A $Zn^{2+}|Zn$ half-cell is connected to a $Cu^{2+}|Cu$ half-cell to make a galvanic cell, in which $[Zn^{2+}] = [Cu^{2+}] = 1.00$ M. The cell voltage at 25°C is measured to be $\Delta\mathcal{E}° = 1.10$ V, and Cu is observed to plate out as the reaction proceeds. Calculate $\Delta G°$ for the chemical reaction that takes place in the cell, for 1.00 mol of zinc dissolved.

SOLUTION

The reaction is

$$Zn(s) + Cu^{2+}(aq) \longrightarrow Zn^{2+}(aq) + Cu(s)$$

[2]Recall that the standard state of a solute has unit activity, which can differ significantly from 1 M concentration for a real electrolyte solution. In what follows, we will nevertheless refer to 1 M concentration as the standard state.

because Cu is a product. For the reaction as written, in which 1 mol of Zn(*s*) and 1 mol of $Cu^{2+}(aq)$ react, 2 mol of electrons passes through the external circuit, so $n = 2$. Therefore,

$$\Delta G° = -n\mathcal{F}\,\Delta\mathcal{E}° = -(2.00\text{ mol})(96{,}485\text{ C mol}^{-1})(1.10\text{ V})$$
$$= -2.12 \times 10^5\text{ J} = -212\text{ kJ}$$

Related Problems: 11, 12

Half-Cell Voltages

We could tabulate all the conceivable galvanic cells and their standard voltages, but the list would be very long. To avoid this, the half-cell reduction potentials $\mathcal{E}°$ are tabulated; they can be combined to obtain the standard cell voltage $\Delta\mathcal{E}°$ for any complete cell.

To see this, return to the standard cell just considered, made up of $Zn^{2+}|Zn$ and $Cu^{2+}|Cu$ half-cells. Each half-cell reaction is written as a reduction:

$$Zn^{2+}(aq) + 2\,e^- \longrightarrow Zn(s)$$
$$Cu^{2+}(aq) + 2\,e^- \longrightarrow Cu(s)$$

Each half-reaction has a certain tendency to occur. If **reduction potentials** $\mathcal{E}°(Zn^{2+}|Zn)$ and $\mathcal{E}°(Cu^{2+}|Cu)$ are associated with the two half-cells, the magnitude and sign of each of these potentials are related to the tendency of the reaction to occur as written. The more positive the half-cell reaction's potential, the greater its tendency to take place as a reduction. Of course, a single half-cell reduction reaction cannot take place in isolation because a source of electrons is required. This can be achieved only by combining two half-cells. When two half-cells are combined, the reaction that has the more positive (or algebraically greater) reduction potential occurs as a reduction, and the one with the less positive potential is forced to run in reverse (as an oxidation) to supply electrons. The net cell potential (voltage) is then the *difference* between the individual half-cell reduction potentials. In the preceding example, copper is observed to plate out on the cathode, so its potential must be more positive than that of zinc:

$$\Delta\mathcal{E}° = \mathcal{E}°(Cu^{2+}|Cu) - \mathcal{E}°(Zn^{2+}|Zn) = 1.10\text{ V}$$

When two half-cells are combined to make a galvanic cell, reduction occurs in the half-cell with the algebraically greater potential (making it the cathode) and oxidation occurs in the other half-cell (making it the anode). We therefore write

$$\Delta\mathcal{E}° = \mathcal{E}°(\text{cathode}) - \mathcal{E}°(\text{anode}) \qquad [17.5]$$

for a galvanic cell. Cell voltages and half-cell reduction potentials are *intensive* quantities, independent of the amount of matter in a cell or the amount reacting. Half-cell reduction potentials are therefore *not* multiplied by numbers of moles reacting (as Gibbs free energies would be) to obtain the overall cell potential.

In thermodynamics, only energy differences are measurable; absolute energies are not. Therefore, energies (or enthalpies or free energies) are defined relative to a reference state for which these quantities are arbitrarily set at 0 by international agreement. The same reasoning applies to half-cells: Because only differences are measured, we are free to define a reference reduction potential for a particular half-cell and measure other half-cell reduction potentials relative to it. The convention used is to define $\mathcal{E}°$ for the half-cell reduction of $H_2(g)$ to $H_3O^+(aq)$ to be 0 at all temperatures, when the gas pressure at the electrode is 1 atm and the $H_3O^+(aq)$ concentration in solution is 1 M (Fig. 17.3).

$$2\,H_3O^+(aq) + 2\,e^- \longrightarrow H_2(g) + 2\,H_2O(\ell) \qquad \mathcal{E}° = 0\text{ V (by definition)}$$

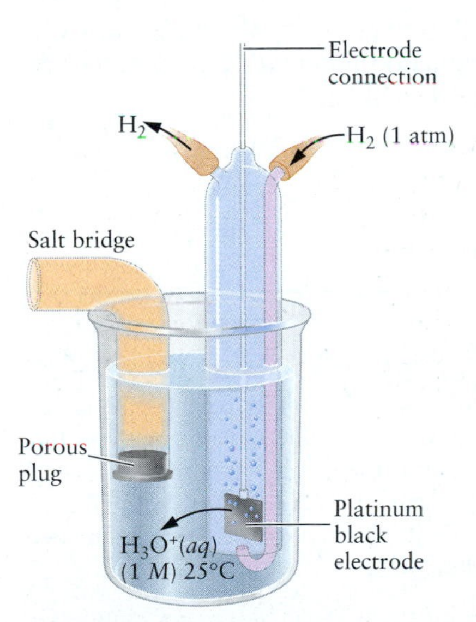

FIGURE 17.3 Hydrogen is a gas at room conditions, and electrodes cannot be constructed from it directly. In the hydrogen half-cell shown here, a piece of platinum covered with a fine coating of platinum black is dipped into the solution, and a stream of hydrogen is passed over the surface. The platinum itself does not react but provides a support surface for the reaction of the H_2 and water to give H_3O^+, or the reverse. It also provides the necessary electrical connection to carry away or supply electrons.

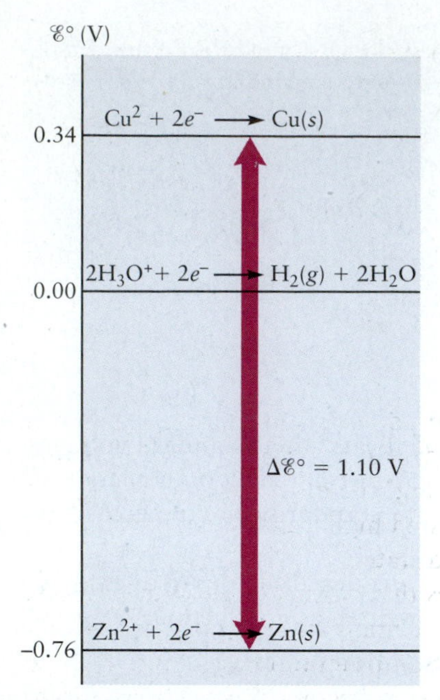

FIGURE 17.4 Sometimes standard reduction potentials are represented to scale. In this diagram the distance between a pair of standard half-cell potentials is proportional to the $\Delta ℰ°$ generated by the electrochemical cell that combines the two. The half-reaction with the higher reduction potential forces the other half-reaction to occur in the reverse direction.

All other half-cell potentials are then determined from this reference potential by combining their standard half-cells with the standard $H_3O^+|H_2$ half-cell in a galvanic cell and measuring the cell voltage. The magnitude and sign of that cell voltage give the standard potential of the half-cell in question, its reaction being written as an oxidation or a reduction according to whether the electrode behaves as the anode or the cathode of the galvanic cell.

For example, if a $Cu^{2+}(1\ \text{M})|Cu$ half-cell is connected to the standard $H_3O^+(1\ \text{M})|H_2$ half-cell, copper is observed to plate out; therefore the copper half-cell is the cathode and the hydrogen half-cell is the anode. The observed cell voltage is 0.34 V; thus,

$$\begin{aligned}\Delta ℰ° &= ℰ°(\text{cathode}) - ℰ°(\text{anode}) \\ &= ℰ°(Cu^{2+}|Cu) - ℰ°(H_3O^+|H_2) = ℰ°(Cu^{2+}|Cu) - 0 \\ &= 0.34\ \text{V}\end{aligned}$$

Therefore, the standard $Cu^{2+}|Cu$ half-cell potential is 0.34 V on a scale in which the standard $H_3O^+|H_2$ half-cell potential is 0:

$$Cu^{2+}(1\ \text{M}) + 2\,e^- \longrightarrow Cu(s) \qquad ℰ° = 0.34\ \text{V}$$

When a $Zn^{2+}(1\ \text{M})|Zn$ half-cell is connected to the standard hydrogen half-cell, zinc dissolves; its half-cell is the anode because oxidation occurs in it. The measured cell voltage is 0.76 V, so

$$\Delta ℰ° = 0.76\ \text{V} = ℰ°(\text{cathode}) - ℰ°(\text{anode}) = 0 - ℰ°(Zn^{2+}|Zn)$$

The $Zn^{2+}(1\ \text{M})|Zn$ half-cell reduction potential is therefore -0.76 V:

$$Zn^{2+}(1\ \text{M}) + 2\,e^- \longrightarrow Zn(s) \qquad ℰ° = -0.76\ \text{V}$$

If the zinc and copper half-cells are combined, the copper half-cell will be the cathode because it has the more positive half-cell reduction potential. The galvanic cell voltage under standard-state conditions (Fig. 17.4) will be

$$\Delta ℰ° = ℰ°(\text{cathode}) - ℰ°(\text{anode}) = 0.34\ \text{V} - (-0.76\ \text{V}) = 1.10\ \text{V}$$

in agreement with the measured value.

EXAMPLE 17.5

An aqueous solution of potassium permanganate ($KMnO_4$) appears deep purple. In aqueous acidic solution, the permanganate ion can be reduced to the pale-pink manganese(II) ion (Mn^{2+}). Under standard conditions, the reduction potential of an $MnO_4^-|Mn^{2+}$ half-cell is $ℰ° = 1.49$ V. Suppose this half-cell is combined with a $Zn^{2+}|Zn$ half-cell in a galvanic cell, with $[Zn^{2+}] = [MnO_4^-] = [Mn^{2+}] = [H_3O^+] = 1$ M. **(a)** Write equations for the reactions at the anode and the cathode. **(b)** Write a balanced equation for the overall cell reaction. **(c)** Calculate the standard cell potential difference, $\Delta ℰ°$.

SOLUTION

(a) Because $ℰ°(MnO_4^-|Mn^{2+}) = 1.49$ V is more positive than $ℰ°(Zn^{2+}|Zn) = -0.76$ V, permanganate ions will be reduced at the cathode. The balanced half-cell reaction requires the presence of H_3O^+ ions and water, giving

$$MnO_4^-(aq) + 8\,H_3O^+(aq) + 5\,e^- \longrightarrow Mn^{2+}(aq) + 12\,H_2O(\ell)$$

The half-equation for the oxidation of Zn at the anode is

$$Zn(s) \longrightarrow Zn^{2+}(aq) + 2\,e^-$$

(b) In the overall reaction, the number of electrons taken up at the cathode must equal the number released at the anode, so the first equation must be multiplied by 2 and the second by 5. Adding the two gives the overall reaction:

$$2\,MnO_4^-(aq) + 16\,H_3O^+(aq) + 5\,Zn(s) \longrightarrow 2\,Mn^{2+}(aq) + 24\,H_2O(\ell) + 5\,Zn^{2+}(aq)$$

(c) The galvanic cell potential is the difference between the standard reduction potential for permanganate (at the cathode) and that for zinc (at the anode):

$$\Delta\mathscr{E}° = \mathscr{E}°(MnO_4^-|Mn^{2+}) - \mathscr{E}°(Zn^{2+}|Zn) = 1.49 - (-0.76) = 2.25 \text{ V}$$

Note that the half-cell potentials are not multiplied by their coefficients (2 and 5) before subtraction. Half-cell potentials are *intensive* properties of a galvanic cell and are therefore independent of the amount of the reacting species.

Related Problems: 13, 14

Appendix E summarizes the standard reduction potentials for a large number of half-reactions. The table lists the reactions in order of decreasing reduction potentials—that is, with the most positive at the top and the most negative at the bottom. In any galvanic cell, the half-cell that is listed higher in the table will act as the cathode (if both half-cells are in the standard state).

A strong **oxidizing agent** is a chemical species that is itself easily reduced. Such species are marked by large positive reduction potentials and appear at the top left in Appendix E. Fluorine has the largest reduction potential listed, and fluorine molecules are extremely eager to accept electrons to become fluoride ions. Other strong oxidizing agents include hydrogen peroxide (H_2O_2) and solutions of permanganate ion (MnO_4^-). A strong **reducing agent,** on the other hand, is easily oxidized, so its corresponding reduction potential is large and negative. Such reducing agents appear at the lower right in Appendix E. The alkali and alkaline earth metals are especially good reducing agents.

Oxygen itself is a good oxidizing agent in acidic solution at pH 0 because it has a fairly high reduction potential:

$$O_2(g) + 4\,H_3O^+(aq) + 4\,e^- \longrightarrow 6\,H_2O(\ell) \qquad \mathscr{E}° = 1.229 \text{ V}$$

Ozone (O_3) is still stronger because its high free energy relative to oxygen provides an additional driving force for oxidation reactions, as shown by its high half-cell reduction potential in acidic aqueous solution:

$$O_3(g) + 2\,H_3O^+(aq) + 2\,e^- \longrightarrow O_2 + 3\,H_2O(\ell) \qquad \mathscr{E}° = 2.07 \text{ V}$$

Because of its great oxidizing power, ozone is used commercially as a bleach for wood pulp and as a disinfectant and sterilizing agent for water, where it oxidizes algae and organic impurities but leaves no undesirable residue. In basic aqueous solution, the half-reactions of oxygen-containing species involve hydroxide ions instead of hydronium ions, and the reduction potentials are correspondingly changed. At pH 14 (standard basic conditions),

$$O_2(g) + 2\,H_2O(\ell) + 4\,e^- \longrightarrow 4\,OH^-(aq) \qquad \mathscr{E}° = 0.401 \text{ V}$$

$$O_3(g) + H_2O(\ell) + 2\,e^- \longrightarrow O_2 + 2\,OH^-(aq) \qquad \mathscr{E}° = 1.24 \text{ V}$$

Both oxygen and ozone are less effective oxidizing agents in basic solutions than in acidic solutions.

Adding and Subtracting Half-Cell Reactions

When two half-cells are combined to form a galvanic cell, their half-cell potentials are *not* multiplied by the coefficients that appear in the overall balanced chemical equation. This shows that potentials cannot be combined as thermodynamic state functions are, because potentials are intensive rather than extensive properties. The same observation applies when two half-cells are combined to obtain *another half-cell.* For example, suppose we wish to know the standard half-cell potential corresponding to the reaction

$$Cu^{2+}(aq) + e^- \longrightarrow Cu^+(aq)$$

when the standard half-cell potentials are known for the reactions

$$Cu^{2+}(aq) + 2\,e^- \longrightarrow Cu(s) \qquad \mathscr{E}_1^\circ = \mathscr{E}^\circ(Cu^{2+}|Cu) = 0.340 \text{ V}$$

$$Cu^{+}(aq) + e^- \longrightarrow Cu(s) \qquad \mathscr{E}_2^\circ = \mathscr{E}^\circ(Cu^{+}|Cu) = 0.522 \text{ V}$$

Could we not simply subtract the second half-reaction from the first and, correspondingly, the second half-cell potential from the first half-cell potential, to obtain

$$Cu^{2+}(aq) + e^- \longrightarrow Cu^{+}(aq) \qquad \mathscr{E}_3^\circ = \mathscr{E}^\circ(Cu^{2+}|Cu^{+}) = -0.182 \text{ V?}$$

The answer is "No." Instead, the *free energy change* for each half-reaction must be calculated from

$$\Delta G_{\text{hc}}^\circ = -n_{\text{hc}}\mathscr{F}\mathscr{E}^\circ$$

and combined before calculating $\mathscr{E}^\circ$ for their resultant. Thus, for the free energy change for the difference of the two half-reactions,

$$\Delta G_3^\circ = \Delta G_1^\circ - \Delta G_2^\circ = -n_1\mathscr{F}\mathscr{E}_1^\circ + n_2\mathscr{F}\mathscr{E}_2^\circ$$

Setting this equation equal to $-n_3\mathscr{F}\mathscr{E}_3^\circ$ gives

$$-n_3\mathscr{F}\mathscr{E}_3^\circ = -n_1\mathscr{F}\mathscr{E}_1^\circ + n_2\mathscr{F}\mathscr{E}_2^\circ$$

$$\mathscr{E}_3^\circ = \frac{n_1\mathscr{E}_1^\circ - n_2\mathscr{E}_2^\circ}{n_3}$$

The correct standard half-cell potential for the half-cell reaction

$$Cu^{2+}(aq) + e^- \longrightarrow Cu^{+}(aq)$$

is

$$\mathscr{E}^\circ(Cu^{2+}|Cu^{+}) = \frac{(2 \text{ mol})(0.340 \text{ V}) - (1 \text{ mol})(0.522 \text{ V})}{1 \text{ mol}} = 0.158 \text{ V}$$

This same procedure using free energies can be used to calculate overall cell voltages from half-cell potentials. Here, the number of moles of electrons n_1 released at the anode is equal to the number n_2 taken up at the cathode and is the number n_3 implied by the overall chemical equation. The electrons then cancel out, and the result reduces to the simple form found before:

$$\Delta\mathscr{E}^\circ = \mathscr{E}^\circ(\text{cathode}) - \mathscr{E}^\circ(\text{anode})$$

Reduction Potential Diagrams and Disproportionation

We can summarize the half-reactions of copper in a **reduction potential diagram** of the form

$$Cu^{2+} \xrightarrow{0.158 \text{ V}} Cu^{+} \xrightarrow{0.522 \text{ V}} Cu$$

$$\underbrace{\qquad\qquad\qquad}_{0.340 \text{ V}}$$

The line connecting each pair of species stands for the entire half-reaction, written as a reduction and balanced by the addition of electrons (and, where necessary, water and H_3O^+ or OH^-). The number over each line is the corresponding reduction potential in volts.

Recall from Example 11.10 that *disproportionation* is the process in which a single substance is both oxidized and reduced. Reduction potential diagrams enable us to determine which species are stable with respect to disproportionation.

A species can disproportionate if and only if a reduction potential that lies immediately to its right is larger than one that appears immediately to its left. To demonstrate this, note that a disproportionation reaction is the sum of a reduction half-reaction and an oxidation half-reaction. The driving force is the *difference* between the two reduction potentials, and if this difference is positive the disproportionation occurs spontaneously. In the case of Cu^+, the disproportionation reaction is

$$2\ Cu^+ \longrightarrow Cu^{2+} + Cu(s) \qquad \Delta\mathscr{E}^\circ = 0.522 - 0.158 = 0.364\ V$$

Because $\Delta\mathscr{E}^\circ > 0$, $\Delta G^\circ < 0$ and the reaction occurs spontaneously, although the rate may be slow.

EXAMPLE 17.6

Hydrogen peroxide, H_2O_2, is a possible product of the reduction of oxygen in acidic solution:

$$O_2 + 2\ H_3O^+ + 2\ e^- \longrightarrow H_2O_2 + 2\ H_2O \qquad \mathscr{E}_3^\circ = ?$$

It can then be further reduced to water:

$$H_2O_2 + 2\ H_3O^+ + 2\ e^- \longrightarrow 4\ H_2O \qquad \mathscr{E}_2^\circ = 1.77\ V$$

(a) Use the half-cell potential just given for the reduction of H_2O_2, together with that given earlier,

$$O_2 + 4\ H_3O^+ + 4\ e^- \longrightarrow 6\ H_2O \qquad \mathscr{E}_1^\circ = 1.229\ V$$

to calculate the standard half-cell potential for the reduction of O_2 to H_2O_2 in acidic solution.

(b) Write a reduction potential diagram for O_2, H_2O_2, and H_2O.

(c) Is H_2O_2 stable with respect to disproportionation in acidic solution?

SOLUTION

(a) The desired half-cell reaction is obtained by *subtracting* the reaction with potential $\mathscr{E}_2^\circ$ from that with potential $\mathscr{E}_1^\circ$. The half-cell reduction potentials are not subtracted, however, but rather combined as described earlier in this section. Taking $n_1 = 4$, $n_2 = 2$, and $n_3 = 2$ gives

$$\mathscr{E}_3^\circ = \frac{n_1\mathscr{E}_1^\circ - n_2\mathscr{E}_2^\circ}{n_3}$$

$$= \frac{(4\ \text{mol})(1.229\ V) - (2\ \text{mol})(1.77\ V)}{2\ \text{mol}} = 0.69\ V$$

(b) The reduction potential diagram is obtained by omitting the electrons, water, and H_3O^+ from the corresponding half-equations:

$$O_2 \xrightarrow{0.69\ V} H_2O_2 \xrightarrow{1.77\ V} H_2O$$

$$O_2 \xrightarrow{1.229\ V} H_2O \ \text{(bracket spanning } O_2 \text{ to } H_2O)$$

(c) H_2O_2 is thermodynamically unstable to disproportionation in acidic solution because the half-cell potential to its right (1.77 V) is higher than that to its left (0.69 V). The disproportionation of H_2O_2 is also spontaneous in neutral solution, but it is slow enough that aqueous solutions of hydrogen peroxide can be stored for a long time without deteriorating, as long as they are kept out of the light.

Related Problems: 23, 24, 25, 26

17.3 Concentration Effects and the Nernst Equation

In real-world applications, concentrations and pressures are rarely conveniently fixed at their standard state values. It is thus necessary to understand how concentration and pressure affect the cell voltage by applying the thermodynamic principles of Chapter 14 to electrochemical cells. In Chapter 14, we showed that the free energy change is related to the reaction quotient Q through

$$\Delta G = \Delta G^\circ + RT \ln Q$$

Combining this equation with

$$\Delta G = -n\mathcal{F}\,\Delta\mathcal{E}$$

and

$$\Delta G^\circ = -n\mathcal{F}\,\Delta\mathcal{E}^\circ$$

gives

$$-n\mathcal{F}\,\Delta\mathcal{E} = -n\mathcal{F}\,\Delta\mathcal{E}^\circ + RT \ln Q$$

and

$$\Delta\mathcal{E} = \Delta\mathcal{E}^\circ - \frac{RT}{n\mathcal{F}} \ln Q \qquad [17.6]$$

which is known as the **Nernst equation.**

The Nernst equation can be rewritten in terms of common (base-10) logarithms by using the fact that

$$\ln Q \approx 2.303 \log_{10} Q$$

At 25°C (298.15 K), the combination of constants $2.303\ RT/\mathcal{F}$ becomes

$$2.3\frac{RT}{\mathcal{F}} = (2.3)\left[\frac{(8.315\ \text{J K}^{-1}\ \text{mol}^{-1})(298.15\ \text{K})}{96{,}485\ \text{C mol}^{-1}}\right]$$

$$= 0.0592\ \text{J C}^{-1} = 0.0592\ \text{V}$$

because 1 joule per coulomb is 1 volt. The Nernst equation then becomes

$$\Delta\mathcal{E} = \Delta\mathcal{E}^\circ - \frac{0.0592\ \text{V}}{n}\log_{10} Q \quad (\text{at } 25^\circ\text{C}) \qquad [17.7]$$

which is its most familiar form. Here n is the number of moles of electrons transferred in the overall chemical reaction as written, and V signifies the units of the constant 0.0592. In a galvanic cell made from zinc, aluminum, and their ions,

$$3\ Zn^{2+}(aq) + 6\ e^- \longrightarrow 3\ Zn(s)$$

$$2\ Al(s) \longrightarrow 2\ Al^{3+}(aq) + 6\ e^-$$

$$3\ Zn^{2+}(aq) + 2\ Al(s) \longrightarrow 2\ Al^{3+}(aq) + 3\ Zn(s)$$

there is a net transfer of $n = 6$ mol of electrons through the external circuit for the equation as written.

The Nernst equation applies to half-cells in exactly the same way as it does to complete electrochemical cells. For any half-cell potential at 25°C,

$$\mathcal{E} = \mathcal{E}^\circ - \frac{0.0592\ \text{V}}{n_{hc}}\log_{10} Q_{hc}$$

where n_{hc} is the number of electrons appearing in the half-reaction and Q_{hc} is the reaction quotient for the half-cell reaction written as a reduction. The electrons themselves do not appear in Q_{hc}. In the $Zn^{2+}|Zn$ case, the half-reaction written as a reduction is

$$Zn^{2+}(aq) + 2\,e^- \longrightarrow Zn(s)$$

so $n_{hc} = 2$ and $Q_{hc} = 1/[Zn^{2+}]$. If two half-cells in which reactants and products are not in their standard states are combined in a galvanic cell, the one with the more positive value of $\mathscr{E}$ will be the cathode, where reduction takes place, and the cell voltage will be

$$\Delta\mathscr{E} = \mathscr{E}(\text{cathode}) - \mathscr{E}(\text{anode})$$

EXAMPLE 17.7

Suppose the $Zn|Zn^{2+}||MnO_4^-|Mn^{2+}$ cell from Example 17.5 is operated at pH 2.00 with $[MnO_4^-] = 0.12$ M, $[Mn^{2+}] = 0.0010$ M, and $[Zn^{2+}] = 0.015$ M. Calculate the cell voltage $\Delta\mathscr{E}$ at 25°C.

SOLUTION

Recall that the overall equation for this cell is

$$2\,MnO_4^-(aq) + 5\,Zn(s) + 16\,H_3O^+(aq) \longrightarrow 2\,Mn^{2+}(aq) + 5\,Zn^{2+}(aq) + 24\,H_2O(\ell)$$

From Example 17.5, for every 5 mol of Zn oxidized (or 2 mol of MnO_4^- reduced), 10 mol of electrons passes through the external circuit, so $n = 10$. A pH of 2.00 corresponds to a hydronium ion concentration of 0.010 M. Putting this and the other concentrations into the expression for the reaction quotient and substituting both Q and the value of $\Delta\mathscr{E}°$ (from Example 17.5) into the Nernst equation give

$$\Delta\mathscr{E} = 2.25\text{ V} - \frac{0.0592\text{ V}}{10}\log_{10}\frac{[Mn^{2+}]^2[Zn^{2+}]^5}{[MnO_4^-]^2[H_3O^+]^{16}}$$

$$= 2.25\text{ V} - \frac{0.0592\text{ V}}{10}\log_{10}\frac{(0.0010)^2(0.015)^5}{(0.12)^2(0.010)^{16}}$$

$$= 2.25\text{ V} - \frac{0.0592\text{ V}}{10}\log_{10}(5.3 \times 10^{18}) = 2.14\text{ V}$$

Related Problems: 27, 28

Measuring Equilibrium Constants

Electrochemistry provides a convenient and accurate way to measure equilibrium constants for many solution-phase reactions. For an overall cell reaction,

$$\Delta G° = -n\mathscr{F}\,\Delta\mathscr{E}°$$

In addition, $\Delta G°$ is related to the equilibrium constant K through

$$\Delta G° = -RT\ln K$$

so

$$RT\ln K = n\mathscr{F}\,\Delta\mathscr{E}°$$

$$\ln K = \frac{n\mathscr{F}}{RT}\Delta\mathscr{E}°$$

$$\log_{10} K = \frac{n}{0.0592\text{ V}}\Delta\mathscr{E}° \quad (\text{at } 25°\text{C}) \qquad [17.8]$$

The same result can be obtained in a slightly different way. Return to the Nernst equation, which reads (at 25°C)

$$\Delta\mathscr{E} = \Delta\mathscr{E}° - \frac{0.0592\ \text{V}}{n} \log_{10} Q$$

Suppose all species are present under standard-state conditions; then $Q = 1$ initially and the cell voltage $\Delta\mathscr{E}$ is the standard cell voltage $\Delta\mathscr{E}°$. As the reaction takes place, Q increases and $\Delta\mathscr{E}$ decreases. As reactants are used up and products are formed, the cell voltage approaches 0. Eventually, equilibrium is reached and $\Delta\mathscr{E}$ becomes 0. At that point,

$$\Delta\mathscr{E}° = \frac{0.0592\ \text{V}}{n} \log_{10} K$$

which is the same as the relationship just obtained. This relation lets us calculate equilibrium constants from standard cell voltages, as the following example illustrates.

EXAMPLE 17.8

Calculate the equilibrium constant for the redox reaction

$$2\ MnO_4^-(aq) + 5\ Zn(s) + 16\ H_3O^+(aq) \longrightarrow 2\ Mn^{2+}(aq) + 5\ Zn^{2+}(aq) + 24\ H_2O(\ell)$$

at 25°C using the cell voltage calculated in Example 17.5.

SOLUTION

From Example 17.5, $\Delta\mathscr{E}° = 2.25$ V, and from the analysis there and in Example 17.7, $n = 10$ for the preceding reaction. Therefore,

$$\log_{10} K = \frac{n}{0.0592\ \text{V}} \Delta\mathscr{E}° = \frac{10}{0.0592\ \text{V}}(2.25\ \text{V}) = 380$$

$$K = 10^{380}$$

This overwhelmingly large equilibrium constant reflects the strength of permanganate ion as an oxidizing agent and of zinc as a reducing agent. It means that for all practical purposes no MnO_4^- ions are present at equilibrium.

Related Problems: 35, 36

The foregoing example illustrates how equilibrium constants for overall cell reactions can be determined electrochemically. Although the example dealt with redox equilibrium, related procedures can be used to measure the solubility product constants of sparingly soluble ionic compounds or the ionization constants of weak acids and bases. Suppose that the solubility product constant of AgCl is to be determined by means of an electrochemical cell. One half-cell contains solid AgCl and Ag metal in equilibrium with a known concentration of Cl^- (aq) (established with 0.00100 M NaCl, for example) so that an unknown but definite concentration of $Ag^+(aq)$ is present. A silver electrode is used so that the half-cell reaction involved is either the reduction of $Ag^+(aq)$ or the oxidation of Ag. This is, in effect, an $Ag^+|Ag$ half-cell whose potential is to be determined. The second half-cell can be any whose potential is accurately known, and its choice is a matter of convenience. In the following example, the second half-cell is a standard $H_3O^+|H_2$ half-cell.

EXAMPLE 17.9

A galvanic cell is constructed using a standard hydrogen half-cell (with platinum electrode) and a half-cell containing silver and silver chloride:

$$\mathrm{Pt}|\mathrm{H_2(1\ atm)}|\mathrm{H_3O^+(1\ M)}||\mathrm{Cl^-(1.00 \times 10^{-3}\ M)} + \mathrm{Ag^+(?\ M)}|\mathrm{AgCl}|\mathrm{Ag}$$

The $H_2|H_3O^+$ half-cell is observed to be the anode, and the measured cell voltage is $\Delta\mathscr{E} = 0.397$ V. Calculate the silver ion concentration in the cell and the K_{sp} of AgCl at 25°C.

SOLUTION

The half-cell reactions are

$$\mathrm{H_2}(g) + 2\,\mathrm{H_2O}(\ell) \longrightarrow 2\,\mathrm{H_3O^+}(aq) + 2\,e^- \qquad \text{(anode)}$$

$$2\,\mathrm{Ag^+}(aq) + 2\,e^- \longrightarrow 2\,\mathrm{Ag}(s) \qquad \text{(cathode)}$$

$$\Delta\mathscr{E}° = \mathscr{E}°(\text{cathode}) - \mathscr{E}°(\text{anode}) = 0.800 - 0.000\ \text{V} = 0.800\ \text{V}$$

Note that $n = 2$ for the overall cell reaction, and the reaction quotient simplifies to

$$Q = \frac{[\mathrm{H_3O^+}]^2}{[\mathrm{Ag^+}]^2\,P_{\mathrm{H_2}}} = \frac{1}{[\mathrm{Ag^+}]^2}$$

because $[H_3O^+] = 1$ M and $P_{H_2} = 1$ atm. The Nernst equation is

$$\Delta\mathscr{E} = \Delta\mathscr{E}° - \frac{0.0592\ \text{V}}{n}\log_{10} Q$$

$$\log_{10} Q = \frac{n}{0.0592\ \text{V}}(\Delta\mathscr{E}° - \Delta\mathscr{E}) = \frac{2}{0.0592\ \text{V}}(0.800\ \text{V} - 0.397\ \text{V}) = 13.6$$

$$Q = 10^{13.6} = 4 \times 10^{13} = 1/[\mathrm{Ag^+}]^2$$

This can be solved for the silver ion concentration $[Ag^+]$ to give

$$[\mathrm{Ag^+}] = 1.6 \times 10^{-7}\ \text{M}$$

so that

$$[\mathrm{Ag^+}][\mathrm{Cl^-}] = (1.6 \times 10^{-7})(1.00 \times 10^{-3})$$

$$K_{sp} = 1.6 \times 10^{-10}$$

Related Problems: 41, 42

pH Meters

The voltage of a galvanic cell is sensitive to the pH if one of its electrodes is a $Pt|H_2$ electrode that dips into a solution of variable pH. A simple cell to measure pH is

$$\mathrm{Pt}|\mathrm{H_2(1\ atm)}|\mathrm{H_3O^+(variable)}\ ||\ \mathrm{H_3O^+(1\ M)}|\mathrm{H_2(1\ atm)}|\mathrm{Pt}$$

If the half-cell reactions are written as

$$\mathrm{H_2(1\ atm)} + 2\,\mathrm{H_2O}(\ell) \longrightarrow 2\,\mathrm{H_3O^+(var)} + 2\,e^- \qquad \text{(anode)}$$

$$2\,\mathrm{H_3O^+(1\ M)} + 2\,e^- \longrightarrow \mathrm{H_2(1\ atm)} + 2\,\mathrm{H_2O}(\ell) \qquad \text{(cathode)}$$

then $n = 2$ and $Q = [H_3O^+(\text{variable})]^2$ because the other concentrations and gas pressures are 1 M and 1 atm, respectively. From the Nernst equation,

$$\Delta\mathscr{E} = \Delta\mathscr{E}° - \frac{0.0592\ \text{V}}{n}\log_{10} Q$$

and because $\Delta\mathscr{E}^\circ = 0$,

$$\Delta\mathscr{E} = -\frac{0.0592\text{ V}}{2}\log_{10}[H_3O^+]^2$$

$$= -0.0592\text{ V}\log_{10}[H_3O^+] = (0.0592\text{ V})\text{ pH}$$

The measured cell voltage is proportional to the pH.

We have just described a simple **pH meter.** Because it is inconvenient to have to bubble hydrogen gas through the unknown and reference half-cells, a more portable and miniaturized pair of electrodes is used to replace the hydrogen half-cells. In a typical commercial pH meter, two electrodes are dipped into the solution of unknown pH. One of these, the **glass electrode,** usually consists of an AgCl-coated silver electrode in contact with an HCl solution of known (e.g., 1.0 M) concentration in a thin-walled glass bulb. A pH-dependent potential develops across this thin glass membrane when the glass electrode is immersed in a solution of different, unknown $[H_3O^+]$. The second half-cell is often a **saturated calomel electrode,** consisting of a platinum wire in electrical contact with a paste of liquid mercury, calomel ($Hg_2Cl_2(s)$), and a saturated solution of KCl. The overall cell (Fig. 17.5) can be represented as

$$\text{Ag}|\text{AgCl}|\text{Cl}^- + \text{H}_3\text{O}^+(1.0\text{ M})|\text{glass}|\text{H}_3\text{O}^+(\text{var})||\text{Cl}^-(\text{sat})|\text{Hg}_2\text{Cl}_2(s)|\text{Hg}|\text{Pt}$$

whose half-reactions are

$$2\text{Ag}(s) + 2\text{Cl}^-(1.0\text{ M}) \longrightarrow 2\text{AgCl}(s) + 2e^- \qquad \text{(anode)}$$

$$\text{H}_3\text{O}^+(1.0\text{ M}) \longrightarrow \text{H}_3\text{O}^+(\text{var})$$

$$\text{Hg}_2\text{Cl}_2(s) + 2\,e^- \longrightarrow 2\text{Hg}(\ell) + 2\text{Cl}^-(\text{sat}) \qquad \text{(cathode)}$$

The first and third half-reactions have half-cell potentials $\mathscr{E}(\text{AgCl}|\text{Cl}^+|\text{Ag})$ and $\mathscr{E}(\text{Hg}_2\text{Cl}_2|\text{Cl}^+|\text{Hg})$ that can be combined and called $\Delta\mathscr{E}_{\text{ref}}$ because they make a constant contribution to the cell voltage. The second reaction is the source of a variable potential in the cell, corresponding to the free energy of dilution of H_3O^+ from a concentration of 1.0 M to an unknown and variable concentration, and its potential exists across the thin glass membrane of the glass electrode. The Nernst equation for the cell can therefore be written as

$$\Delta\mathscr{E} = \Delta\mathscr{E}(\text{ref}) - \frac{0.0592\text{ V}}{1}\log_{10}\frac{[\text{H}_3\text{O}^+(\text{var})]}{1.00}$$

$$= \Delta\mathscr{E}(\text{ref}) + (0.0592\text{V})\text{ pH}$$

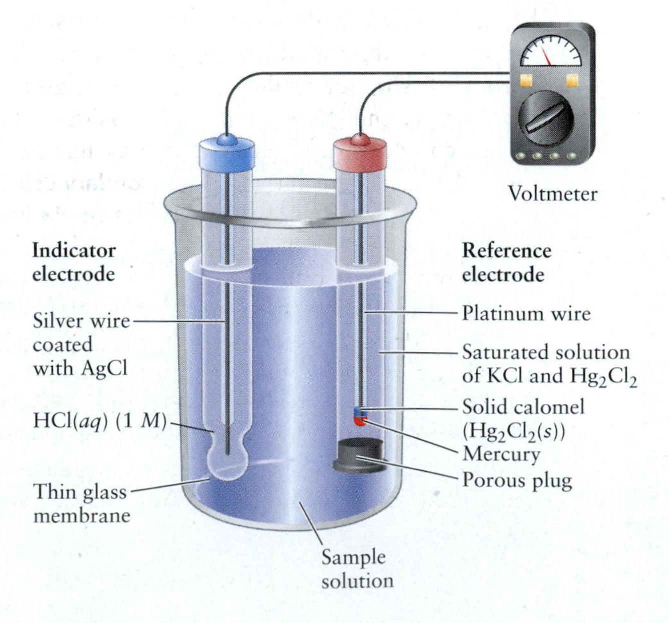

FIGURE 17.5 A pH meter consists of a glass electrode (left) and a calomel electrode (right), both of which dip into a solution of unknown hydronium ion concentration.

and the pH of the unknown solution is

$$\text{pH} = \frac{\Delta\mathscr{E} - \Delta\mathscr{E}(\text{ref})}{0.0592\text{V}}$$

The glass electrode has a number of advantages. It responds only to changes in $[H_3O^+]$ and does so over a wide range of pH. It is unaffected by strong oxidizing agents that would make a hydrogen electrode unreliable. Highly colored solutions that would render acid–base indicators useless do not interfere with the glass electrode. Finally, the glass electrode can be miniaturized to permit insertion into individual living cells and therefore finds wide use in biology.

Other types of electrodes have been designed to measure the concentrations of ions other than H_3O^+. The simplest example of such an **ion-selective electrode** is a metal wire, which can be used to detect the concentration of the corresponding metal ion in solution. Silver and copper wires can be used reproducibly in this way to determine the concentrations of Ag^+ and Cu^{2+}, respectively. Still other electrodes have been developed to detect specific ions. For example, glasses of chemically modified composition are used to construct membrane electrodes to determine potassium and sodium ions or halogen ions.

17.4 Batteries and Fuel Cells

The electrochemistry of galvanic cells leads to a variety of applications in industry and in everyday life. This section focuses on two of the most important: batteries to store energy and fuel cells to convert chemical energy to electrical energy.

Batteries

The origin of the **battery** is lost in history, but it is believed that Persian artisans must have used some form of battery to gold-plate jewelry as long ago as the second century B.C. The modern development of the battery began with Alessandro Volta in 1800. Volta constructed a stack of alternating zinc and silver disks, between pairs of which were inserted disks of paper saturated with a salt solution. This device generated a potential difference across its ends that could cause electrical shock if the stack contained enough disks. In Volta's invention, a large number of galvanic cells were arranged in series (cathode to anode) so that their individual voltages added together. Such an arrangement is, strictly speaking, a **battery of cells,** but modern usage makes no distinction between a single cell and a voltaic pile, and the word *battery* has come to mean the cell itself. Cells that are discarded when their electrical energy has been spent are called **primary** cells, and those that can be recharged are **secondary** cells.

Batteries vary in size and chemistry. Shown here are an automobile lead-storage battery, rechargeable nickel–cadmium cells, alkaline cells, and zinc–carbon dry cells.

The most familiar primary cell is the **Leclanché cell** (also called a *zinc–carbon dry cell*) used for flashlights, portable radios, and a host of other purposes. Each year, more than 5 billion such dry cells are used worldwide, and estimates place the quantity of zinc consumed for this purpose at more than 30 metric tons per day. The "dry cell" is not really dry at all. Rather, its electrolyte is a moist powder containing ammonium chloride and zinc chloride. Figure 17.6 is a cutaway illustration of a dry cell, which consists of a zinc shell for an anode (negative pole) and an axial graphite rod for a cathode (positive pole), with the rod surrounded by a densely packed layer of graphite and manganese dioxide. Each of these components performs an interesting and essential function. At the zinc anode, oxidation takes place:

$$Zn(s) \longrightarrow Zn^{2+}(aq) + 2\,e^- \text{ (anode)}$$

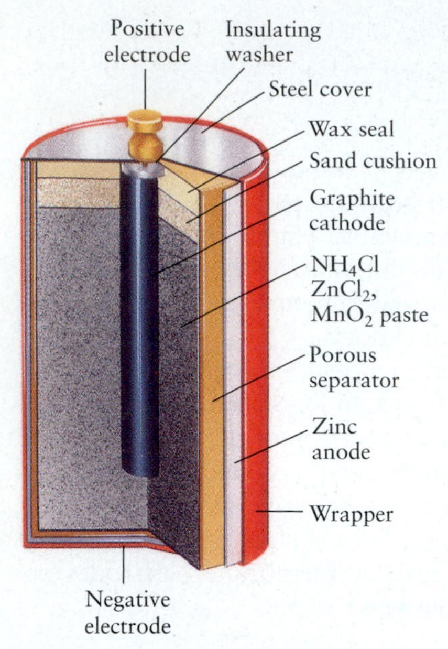

FIGURE 17.6 In a Leclanché dry cell, electrons are released to an external circuit at the anode and enter the cell again at the cathode, where the reduction of MnO_2 occurs.

The moist salt mixture permits ions to transport through the cell a net charge equal to that carried by electrons in the external circuit, just as a salt bridge does in the cells we have considered until now. Manganese dioxide is the ultimate electron acceptor and is reduced to Mn_2O_3 by electrons that pass to it from the graphite rod through the graphite particles:

$$2\,MnO_2(s) + 2\,NH_4^+(aq) + 2\,e^- \longrightarrow Mn_2O_3(s) + 2\,NH_3(aq) + H_2O(\ell) \quad \text{(cathode)}$$

The ingenious procedure of mixing powdered graphite with powdered MnO_2 greatly increases the effective surface area of the cathode, reduces the internal resistance of the cell, and enables currents of several amperes to flow. The overall cell reaction is

$$Zn(s) + 2\,MnO_2(s) + 2\,NH_4^+(aq) \longrightarrow Zn^{2+}(aq) + Mn_2O_3(s) + 2\,NH_3(aq) + H_2O(\ell)$$

The cell components are hermetically sealed in a steel shell that is in contact with the zinc and acts as the negative terminal of the battery. A fresh zinc–carbon dry cell generates a potential difference of 1.5 V.

The Leclanché cell has the disadvantage that its concentrations change with time, and thus the voltage of the battery falls as it is used. In an **alkaline dry cell,** the ammonium chloride is replaced by potassium hydroxide, and the half-cell reactions become

$$Zn(s) + 2\,OH^-(aq) \longrightarrow Zn(OH)_2(s) + 2\,e^- \quad \text{(anode)}$$

$$2\,MnO_2(s) + H_2O(\ell) + 2\,e^- \longrightarrow Mn_2O_3(s) + 2\,OH^-(aq) \quad \text{(cathode)}$$

The overall reaction is then

$$Zn(s) + 2\,MnO_2(s) + H_2O(\ell) \longrightarrow Zn(OH)_2(s) + Mn_2O_3(s)$$

Because dissolved species do not appear in the overall reaction, concentrations do not change significantly and a steadier voltage results.

A third primary dry cell is the **zinc–mercuric oxide cell** depicted in Figure 17.7. It is commonly given the shape of a small button and is used in automatic cameras, hearing aids, digital calculators, and quartz–electric watches. This battery has an anode that is a mixture of mercury and zinc and a steel cathode in contact with solid mercury(II) oxide (HgO). The electrolyte is a 45% KOH solution that saturates an absorbent material. The anode half-reaction is the same as that in an alkaline dry cell,

$$Zn(s) + 2\,OH^-(aq) \longrightarrow Zn(OH)_2(s) + 2\,e^- \quad \text{(anode)}$$

but the cathode half-reaction is now

$$HgO(s) + H_2O(\ell) + 2\,e^- \longrightarrow Hg(\ell) + 2\,OH^-(aq) \quad \text{(cathode)}$$

The overall reaction is

$$Zn(s) + HgO(s) + H_2O(\ell) \longrightarrow Zn(OH)_2(s) + Hg(\ell)$$

This cell has a very stable output of 1.34 V, a fact that makes it especially valuable for use in communication equipment and scientific instruments.

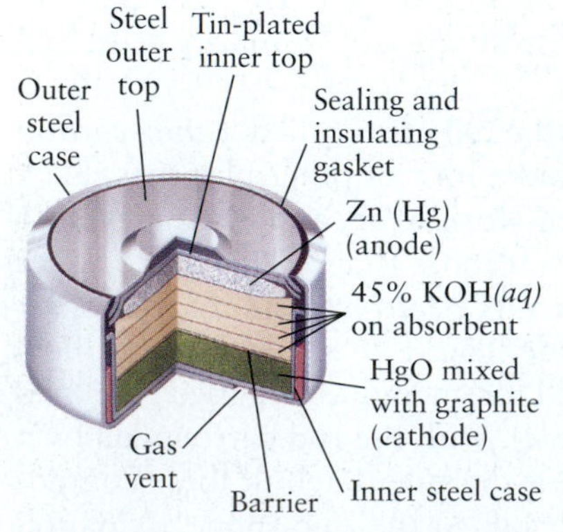

FIGURE 17.7 A zinc–mercuric oxide dry cell, used in electric watches and cameras.

Rechargeable Batteries

In some batteries the electrodes can be regenerated after depletion by imposing an external potential across them that reverses the direction of current flow through the cell. These are called **secondary batteries,** and the process of reconstituting them to their original state is called "recharging." To recharge a run-down secondary battery, the voltage of the external source must be larger than that of the battery in its original state and, of course, opposite in polarity.

© Thomson Learning/Charles D. Winters

FIGURE 17.8 Rechargeable nickel–cadmium batteries.

The **nickel–cadmium cell** (or *nicad battery;* Fig. 17.8) is used in handheld electronic calculators and other cordless electric implements such as portable shavers. Its half-cell reactions during discharge are

$$Cd(s) + 2\ OH^-(aq) \rightarrow Cd(OH)_2(s) + 2\ e^- \qquad \text{(anode)}$$

$$2\ NiO(OH)(s) + 2\ H_2O(\ell) + 2\ e^- \rightarrow 2\ Ni(OH)_2(s) + 2\ OH^-(aq) \qquad \text{(cathode)}$$

$$Cd(s) + 2\ NiO(OH)(s) + 2\ H_2O(\ell) \longrightarrow Cd(OH)_2(s) + 2\ Ni(OH)_2(s)$$

This battery gives a fairly constant voltage of 1.4 V. When it is connected to an external voltage source, the preceding reactions are reversed as the battery is recharged.

One technically important secondary battery is the **lead–acid storage battery,** used in automobiles. A 12-V lead storage battery consists of six 2.0-V cells (Fig. 17.9) connected in series (cathode to anode) by an internal lead linkage and housed in a hard rubber or plastic case. In each cell the anode consists of metallic lead in porous form to maximize its contact area with the electrolyte. The cathode is of similar design, but its lead has been converted to lead dioxide. A sulfuric acid solution (37% by mass) serves as the electrolyte.

When the external circuit is completed, electrons are released from the anode to the external circuit and the resulting Pb^{2+} ions precipitate on the electrode as insoluble lead sulfate. At the cathode, electrons from the external circuit reduce PbO_2 to water and Pb^{2+} ions, which also precipitate as $PbSO_4$ on that electrode. The half-cell reactions are

$$Pb(s) + SO_4^{2-}(aq) \longrightarrow PbSO_4(s) + 2\ e^- \qquad \text{(anode)}$$

$$PbO_2(s) + SO_4^{2-}(aq) + 4\ H_3O^+(aq) + 2\ e^- \longrightarrow PbSO_4(s) + 6\ H_2O(\ell) \qquad \text{(cathode)}$$

$$Pb(s) + PbO_2(s) + 2\ SO_4^{2-}(aq) + 4\ H_3O^+(aq) \longrightarrow 2\ PbSO_4(s) + 6\ H_2O(\ell)$$

The anode and cathode are both largely converted to $PbSO_4(s)$ when the storage battery is fully discharged, and because sulfuric acid is a reactant, its concentration falls. Measuring the density of the electrolyte provides a quick way to estimate the state of charge of the battery.

When a voltage in excess of 12 V is applied across the terminals of the battery in the opposite direction, the half-cell reactions are reversed. This part of the cycle restores the battery to its initial state, ready to be used in another work-producing discharge half-cycle. Lead–acid storage batteries endure many thousands of cycles of discharge and charge before they ultimately fail because of the flaking off of $PbSO_4$ from the electrodes or the development of internal short circuits. In automobiles they are usually not designed to undergo complete discharge–recharge cycles. Instead, a generator converts some of the kinetic energy of the vehicle to electrical energy for continuous or intermittent charging. About 1.8×10^7 J can be obtained in the discharge of an average automobile battery, and currents as large as 100 A are drawn for the short time needed to start the engine.

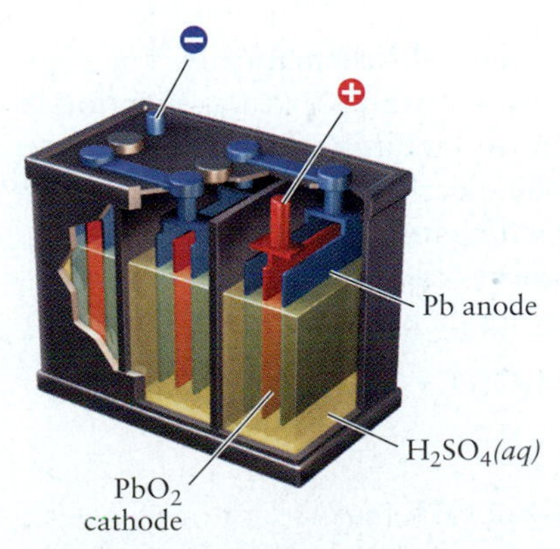

FIGURE 17.9 In a lead–acid storage battery, anodes made of Pb alternate with cathodes of Pb coated with PbO_2. The electrolyte is sulfuric acid.

A drawback of the lead–acid storage battery is its low energy density, the amount of energy obtainable per kilogram of battery mass. This is not important when a battery is used to start a gasoline-powered automobile, but it precludes the battery's use in a vehicle driven by an electric motor. This difficulty of a low energy-to-mass ratio, which limits the range of vehicle operation before recharge is necessary, has spurred electrochemists to develop secondary batteries that have much higher energy densities.

One promising line of approach has been the development of rechargeable batteries that use an alkali metal (lithium or sodium) as the anode and sulfur as the electron acceptor. Sulfur is a nonconductor of electricity, so graphite is used as the cathode that conducts electrons to it. The elements must be in their liquid states, so these batteries are high-temperature cells (sulfur melts at 112°C, lithium at

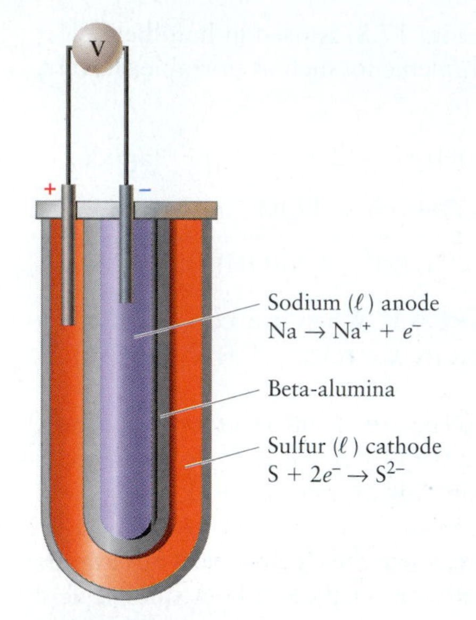

FIGURE 17.10 In a sodium–sulfur battery, Na^+ ions migrate through beta-alumina to the cathode to equalize the charge as electrons flow spontaneously from the anode to the cathode through the external circuit.

186°C, and sodium at 98°C). The sodium–sulfur cell (Fig. 17.10), for example, has an optimal operating temperature of 250°C. The half-cell reactions are

$$2\ \text{Na} \longrightarrow 2\ \text{Na}^+ + 2\ e^- \quad \text{(anode)}$$

$$\text{S} + 2\ e^- \longrightarrow \text{S}^{2-} \quad \text{(cathode)}$$

$$2\ \text{Na} + \text{S} \longrightarrow 2\ \text{Na}^+ + \text{S}^{2-}$$

This is actually an oversimplification of the cathodic process because sulfide ion forms polysulfides with sulfur:

$$\text{S}^{2-} + n\text{S} \longrightarrow \text{S}_{n+1}^{2-}$$

but the fundamental principles of the cell operation are the same.

What makes the sodium–sulfur cell possible is a remarkable property of a compound called beta-alumina, which has the composition $NaAl_{11}O_{17}$. Beta-alumina allows sodium ions to migrate through its structure very easily, but it blocks the passage of polysulfide ions. Therefore, it can function as a semipermeable medium like the membranes used in osmosis (see Section 11.5). Such an ion-conducting solid electrolyte is essential to prevent direct chemical reaction between sulfur and sodium. The lithium–sulfur battery operates on similar principles, and other solid electrolytes such as calcium fluoride, which permits ionic transport of fluoride ion, may find use in cells based on those elements.

The considerable promise of the alkali metal–sulfur cell lies in its high energy density, which makes it possible to construct lightweight batteries capable of generating large currents. Their use in electric cars is an attractive possibility. As the cost of petroleum-based fuels increases and their supply becomes less reliable, cells such as the alkali metal–sulfur battery may come into practical applications.

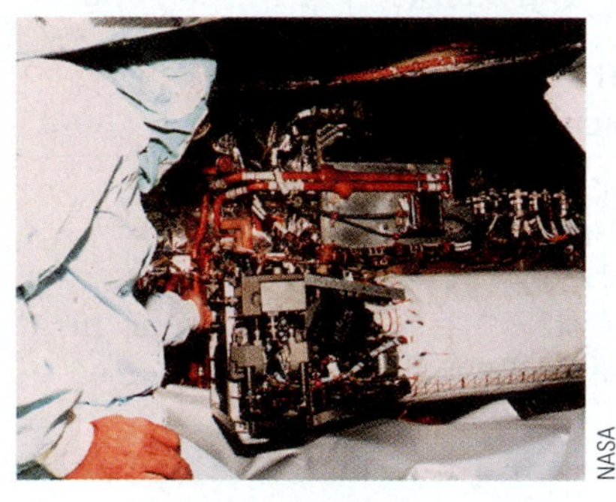

FIGURE 17.11 A hydrogen–oxygen fuel cell used on U.S. space missions.

Fuel Cells

A battery is a closed system that delivers electrical energy by electrochemical reactions. Once the chemicals originally present are consumed, the battery must be either recharged or discarded. In contrast, a **fuel cell** is designed for continuous operation, with reactants (fuel) being supplied and products removed continuously. It is an energy converter, transforming chemical energy into electrical energy. Fuel cells based on the reaction

$$2\ \text{H}_2(g) + \text{O}_2(g) \longrightarrow 2\ \text{H}_2\text{O}(\ell)$$

were used on the Gemini and Apollo space vehicles to help meet the electrical requirements of the missions (Fig. 17.11). After the product was purified by ion exchange, the crew consumed it as drinking water.

Figure 17.12 represents the hydrogen–oxygen fuel cell schematically. The electrodes can be any nonreactive conductor (graphite, for example); their function is to conduct electrons into and out of the cell and to facilitate electron exchange between the gases and the ions in solution. The electrolyte transports charge through the cell, and the ions dissolved in it participate in the half-reactions at each electrode. Acidic solutions present corrosion problems, so an alkaline solution is preferable (1 M NaOH, for instance). The anode half-reaction is then

$$\text{H}_2(g) + 2\ \text{OH}^-(aq) \longrightarrow 2\ \text{H}_2\text{O}(\ell) + 2\ e^-$$

and the cathode reaction is

$$\tfrac{1}{2}\ \text{O}_2(g) + \text{H}_2\text{O}(\ell) + 2\ e^- \longrightarrow 2\ \text{OH}^-(aq)$$

The standard reduction potentials at 25°C (1 M concentrations, 1 atm pressure) are

$$\mathscr{E}°\ (\text{H}_2|\text{H}_2\text{O}) = -0.828\ \text{V} \quad \text{and} \quad \mathscr{E}°\ (\text{O}_2|\text{OH}^-) = 0.401\ \text{V}$$

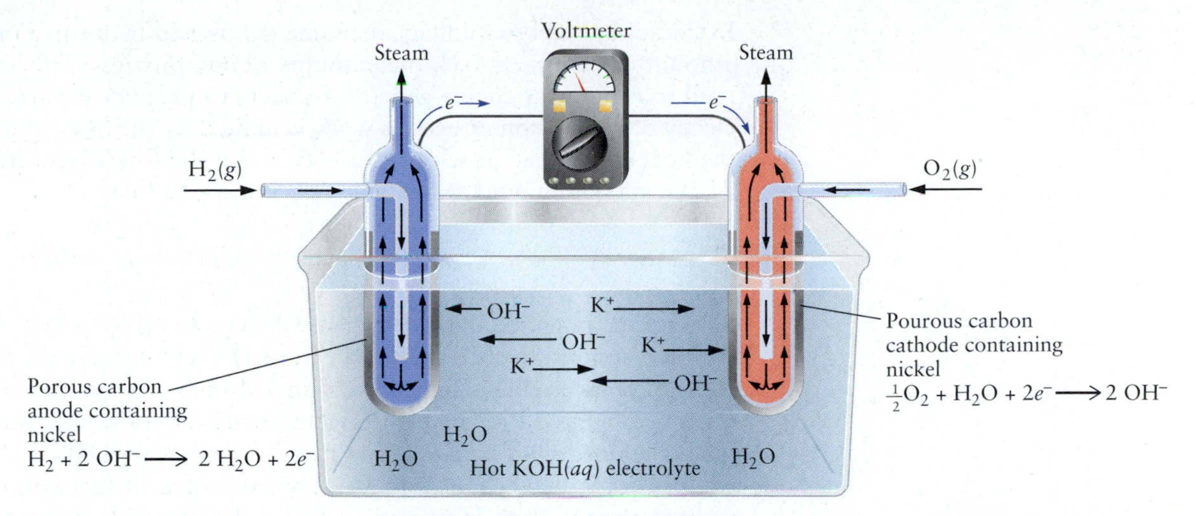

FIGURE 17.12 In a hydrogen–oxygen fuel cell, the two gases are fed in separately and are oxidized or reduced on the electrodes. A hot solution of potassium hydroxide between the electrodes completes the circuit, and the steam produced in the reaction evaporates from the cell continuously.

The overall cell reaction is the production of water,

$$H_2(g) + \tfrac{1}{2}\,O_2(g) \longrightarrow H_2O(\ell)$$

and the cell voltage is

$$\Delta\mathscr{E}° = \mathscr{E}°\,(\text{cathode}) - \mathscr{E}°\,(\text{anode}) = 0.401 - (-0.828) = 1.229\text{ V}$$

The overall cell voltage does not depend on pH because the OH^- has dropped out; the same voltage would be obtained in an acidic fuel cell.

Another practical fuel cell accomplishes the overall reaction

$$CO(g) + \tfrac{1}{2}\,O_2(g) \longrightarrow CO_2(g)$$

for which the half-reactions are

$$CO(g) + 3\,H_2O(\ell) \longrightarrow CO_2(g) + 2\,H_3O^+(aq) + 2\,e^- \qquad \text{(anode)}$$

$$\tfrac{1}{2}\,O_2(g) + 2\,H_3O^+(aq) + 2\,e^- \longrightarrow 3\,H_2O(\ell) \qquad \text{(cathode)}$$

The electrolyte is usually concentrated phosphoric acid, and the operating temperature is between 100°C and 200°C. Platinum is the electrode material of choice because it facilitates the electron transfer reactions.

Natural gas (largely CH_4) and even fuel oil can be "burned" electrochemically to provide electrical energy by either of two approaches. They can be converted to CO and H_2 or to CO_2 and H_2 before use in a fuel cell by reaction with steam,

$$CH_4(g) + H_2O(g) \longrightarrow CO(g) + 3\,H_2(g)$$

$$CO(g) + H_2O(g) \longrightarrow CO_2(g) + H_2(g)$$

at temperatures of about 500°C. Alternatively, they can be used directly in a fuel cell at higher temperatures (up to 750°C) with a molten alkali metal carbonate as the electrolyte. Either type of fuel cell is attractive as an electrochemical energy converter in regions where hydrocarbon fuels are readily available but large-scale power plants (fossil or nuclear) are remote.

The theoretical advantage of electrochemical fuel cells over more traditional fuel technology can be seen from a thermodynamic analysis. If a chemical reaction such as the oxidation of a fuel can be carried out electrochemically, the maximum (reversible) work obtainable is equal to the free energy change ΔG (see Section 17.2):

$$+w_{\text{max}}(\text{fuel cell}) = |\Delta G|$$

In traditional fuel technology, the same fuel would be burned in air, producing an amount of heat $q_P = \Delta H$, the enthalpy of combustion. The heat would then be used to run a heat engine-generator system to produce electrical power. The efficiency of conversion of heat to work is limited by the laws of thermodynamics. If the heat is supplied at temperature T_h and if the lower operating temperature is T_l, the maximum work obtainable (see Section 13.4) is

$$-w_{max}(\text{heat engine}) = \epsilon|q_P| = \frac{T_h - T_l}{T_h}|\Delta H|$$

Because the magnitude of ΔH is generally comparable to that of ΔG for fuel oxidation reactions, the fuel cell will be more efficient because the factor $(T_h - T_l)/T_h$ for a thermal engine is much less than 1. In practice, fuel cells and heat engines must be operated irreversibly (to increase the rate of energy production), and the work obtained with both is less than w_{max}. The overall efficiency of practical heat engines rarely exceeds 30% to 35%, whereas that of fuel cells can be in the 60% to 70% range. This advantage of fuel cell technology is partially offset by the greater expense of constructing and maintaining fuel cells, however.

17.5 Corrosion and Its Prevention

The **corrosion** of metals is one of the most significant problems faced by advanced industrial societies (Fig. 17.13). It has been estimated that in the United States alone, the annual cost of corrosion amounts to tens of *billions* of dollars. Effects of corrosion are both visible (the formation of rust on exposed iron surfaces) and invisible (the cracking and resulting loss of strength of metal beneath the surface). The mechanism of corrosion must be understood before processes can be developed for its prevention.

Although corrosion is a serious problem for many metals, we will focus on the spontaneous electrochemical reactions of iron. Corrosion can be pictured as a "short-circuited" galvanic cell, in which some regions of the metal surface act as cathodes and others as anodes, and the electric "circuit" is completed by electron flow through the iron itself. These electrochemical cells form in parts of the metal where there are impurities or in regions that are subject to stress. The anode reaction is

$$Fe(s) \longrightarrow Fe^{2+}(aq) + 2\,e^-$$

Various cathode reactions are possible. In the absence of oxygen (for example, at the bottom of a lake), the corrosion reactions are

$$\begin{array}{ll} Fe(s) \longrightarrow Fe^{2+}(aq) + 2\,e^- & \text{(anode)} \\ 2\,H_2O(\ell) + 2\,e^- \longrightarrow 2\,OH^-(aq) + H_2(g) & \text{(cathode)} \\ \hline Fe(s) + 2\,H_2O(\ell) \longrightarrow Fe^{2+}(aq) + 2\,OH^-(aq) + H_2(g) & \end{array}$$

These reactions are generally slow and do not cause serious amounts of corrosion. Far more extensive corrosion takes place when the iron is in contact with both oxygen and water. In this case the cathode reaction is

$$\tfrac{1}{2}\,O_2(g) + 2\,H_3O^+(aq) + 2\,e^- \longrightarrow 3\,H_2O(\ell)$$

The Fe^{2+} ions formed simultaneously at the anode migrate to the cathode, where they are further oxidized by O_2 to the +3 oxidation state to form rust ($Fe_2O_3{\cdot}xH_2O$), a hydrated form of iron(III) oxide:

$$2\,Fe^{2+}(aq) + \tfrac{1}{2}\,O_2(g) + (6 + x)H_2O(\ell) \longrightarrow Fe_2O_3{\cdot}xH_2O + 4\,H_3O^+(aq)$$

The hydronium ions produced in this reaction allow the corrosion cycle to continue.

FIGURE 17.13 Rust.

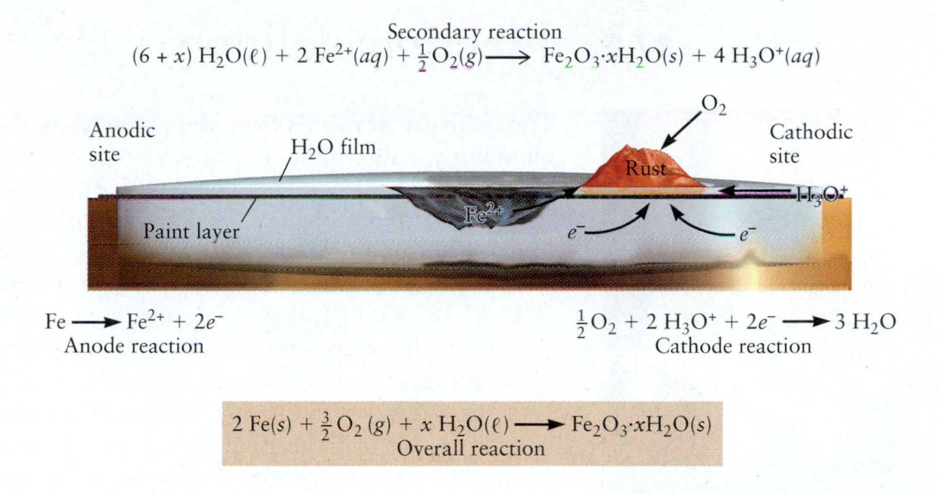

FIGURE 17.14 The corrosion of iron. Note that pitting occurs in the anodic region, and rust appears in the cathodic region.

When a portion of the paint that protects a piece of iron or steel is chipped off (Fig. 17.14), the exposed area acts as the cathode because it is open to the atmosphere (air and water) and is therefore rich in oxygen, whereas oxygen-poor areas *under* the paint act as anodes. Rust forms on the cathode (the visible, exposed region), and pitting (loss of metal through oxidation of iron and flow of metal ions to the cathode) occurs at the anode. This pitting can lead to loss of structural strength in girders and other supports. The most serious harm done by corrosion is not the visible rusting but the damage done beneath the painted surface.

A number of factors speed corrosion. Dissolved salt provides an electrolyte that improves the flow of charge through solution; a well-known example is the rapid rusting of cars in areas where salt is spread on icy roads. Higher acidity also increases corrosion, as seen by the role of H_3O^+ as a reactant in the reduction process at the cathode. Acidity is enhanced by the presence of dissolved CO_2 (which produces H_3O^+ and HCO_3^- ions) and by air pollution from oxides of sulfur, which leads to the formation of dissolved sulfuric acid in acid precipitation.

Corrosion of iron can be inhibited in a number of ways. Coatings of paint or plastics obviously protect the metal, but they can crack or suffer other damage, thereby localizing and accentuating the process. An important method of protecting metals arises from the phenomenon of **passivation,** in which a thin metal oxide layer forms on the surface and prevents further electrochemical reactions. Some metals become passivated spontaneously upon exposure to air; aluminum, for example, reacts with oxygen to form a thin protective layer of Al_2O_3. Special paints designed to prevent rusting contain potassium dichromate ($K_2Cr_2O_7$) and lead oxide (Pb_3O_4), which cause the superficial oxidation and passivation of iron. Stainless steel is an alloy of iron with chromium in which the chromium leads to passivation and prevents rusting.

A different way of preventing iron corrosion is to use a **sacrificial anode.** A comparison of the standard reduction potentials of iron and magnesium

$$Fe^{2+} + 2\ e^- \longrightarrow Fe(s) \qquad \mathscr{E}° = -0.41\ \text{V}$$

$$Mg^{2+} + 2\ e^- \longrightarrow Mg(s) \qquad \mathscr{E}° = -2.39\ \text{V}$$

shows that Mg^{2+} is much harder to reduce than Fe^{2+} or, conversely, that $Mg(s)$ is more easily oxidized than $Fe(s)$. A piece of magnesium in electrical contact with iron is oxidized in preference to the iron, and the iron is therefore protected. The magnesium is the sacrificial anode, and once it is consumed by oxidation it must be replaced. This method is used to protect ship hulls, bridges, and iron water pipes from corrosion. Magnesium plates are attached at regular intervals along a piece of buried pipe, and it is far easier to replace them periodically than to replace the entire pipe.

17.6 Electrometallurgy

© Thomson Learning/Charles D. Winters

FIGURE 17.15 A specimen of native copper.

The recovery of metals from their sources in the earth is the science of **extractive metallurgy,** a discipline that draws on chemistry, physics, and engineering for its methods. As a science it is a comparatively recent subject, but its beginnings, which were evidently in the Near East about 6000 years ago, marked the emergence of humanity from the Stone Age. The earliest known metals were undoubtedly gold, silver, and copper because they could be found in their native (elemental) states (Fig. 17.15). Gold and silver were valued for their ornamental uses, but they are too soft to have been made into tools. Iron was also found in elemental form—although rarely—in meteorites.

Most metals in nature are combined with other elements such as oxygen and sulfur in ores, and chemical processes are required to free them. As Table 17.1 shows, the free energies of formation of most metal oxides are negative, indicating that the reverse reactions, which would yield the free metal and oxygen, have positive free energy changes. Scientists can carry out one of these reverse reactions to obtain the free metal only by coupling it with a second, spontaneous chemical reaction. The greater the cost in free energy, the more difficult the production of the free metal. Thus, silver and gold (at the bottom of Table 17.1) exist in nature as elements, and mercury can be released from its oxide or sulfide ore (cinnabar) merely by moderate heating (see Fig. 1.6). Extracting pure copper, zinc, and iron requires more stringent conditions; the ores of these metals are reduced in chemical reactions at high temperatures, collectively called **pyrometallurgy.** These reactions are carried out in huge furnaces, in which a fuel such as coke (coal from which volatile components have been expelled) serves both as the reducing agent and as the source of heat to maintain the required high temperatures. The process, called **smelting,** involves both chemical change and melting. The combustion of carbon

$$C(s) + O_2(g) \longrightarrow CO_2(g) \qquad \Delta G^\circ = -394 \text{ kJ mol}^{-1}$$

provides the driving force for the overall reaction.

TABLE 17.1 Metal Oxides Arranged According to Ease of Reduction

Metal Oxide	Metal	$n^\dagger$	$\Delta G_f^\circ/n^\dagger$ (kJ mol^{-1})	A Method of Production of the Metal
MgO	Mg	2	−285	Electrolysis of $MgCl_2$
Al_2O_3	Al	6	−264	Electrolysis
TiO_2	Ti	4	−222	Reaction with Mg
Na_2O	Na	2	−188	Electrolysis of NaCl
Cr_2O_3	Cr	6	−176	Electrolysis, reduction by Al
ZnO	Zn	2	−159	Smelting of ZnS
SnO_2	Sn	4	−130	Smelting
Fe_2O_3	Fe	6	−124	Smelting
NiO	Ni	2	−106	Smelting of nickel sulfides
PbO	Pb	2	−94	Smelting of PbS
CuO	Cu	2	−65	Smelting of $CuFeS_2$
HgO	Hg	2	−29	Moderate heating of HgS
Ag_2O	Ag	2	−6	Found in elemental form
Au_2O_3	Au	6	>0	Found in elemental form

†The standard free energies of formation of the metal oxides in kJ mol^{-1} are adjusted for fair comparison by dividing them by *n,* the total decrease in oxidation state required to reduce the metal atoms contained in the oxide to oxidation states of 0. Thus, the reduction of Cr_2O_3 involves a change in the oxidation state of two chromium atoms from +3 to 0, so $n = 2 \times 3 = 6$.

Even smelting is not sufficient to recover the metals at the top of Table 17.1, which have the most negative free energies of formation for their oxides (and for their sulfides as well). These metals have particularly high values of free energy relative to their compounds found in readily available ores. **Electrometallurgy,** or electrolytic production, provides the best ways of recovering such elements from their ores. Electrochemical cells are also used to purify the metals produced by the techniques of pyrometallurgy.

Aluminum

Aluminum is the third most abundant element in the earth's crust (after oxygen and silicon), accounting for 8.2% of the total mass. It occurs most commonly in association with silicon in the aluminosilicates of feldspars and micas and in clays, the products of weathering of these rocks. The most important ore for aluminum production is bauxite, a hydrated aluminum oxide that contains 50% to 60% Al_2O_3; 1% to 20% Fe_2O_3; 1% to 10% silica; minor concentrations of titanium, zirconium, vanadium, and other transition-metal oxides; and the balance (20% to 30%) water. Bauxite is purified via the **Bayer process,** which takes advantage of the fact that the amphoteric oxide alumina is soluble in strong bases but iron(III) oxide is not. Crude bauxite is dissolved in sodium hydroxide

$$Al_2O_3(s) + 2\ OH^-(aq) + 3\ H_2O(\ell) \longrightarrow 2\ Al(OH)_4^-(aq)$$

and separated from hydrated iron oxide and other insoluble impurities by filtration. Pure hydrated aluminum oxide precipitates when the solution is cooled to supersaturation and seeded with crystals of the product:

$$2\ Al(OH)_4^-(aq) \longrightarrow Al_2O_3 \cdot 3H_2O(s) + 2\ OH^-(aq)$$

The water of hydration is removed by calcining at high temperature (1200°C).

Compared with copper, iron, gold, and lead, which were known in antiquity, aluminum is a relative newcomer. Sir Humphry Davy obtained it as an alloy of iron and proved its metallic nature in 1809. It was first prepared in relatively pure form in 1825 by H. C. Oersted through reduction of aluminum chloride with an amalgam of potassium dissolved in mercury,

$$AlCl_3(s) + 3\ K(Hg)_x(\ell) \longrightarrow 3\ KCl(s) + Al(Hg)_{3x}(\ell)$$

after which the mercury was removed by distillation. Aluminum remained largely a laboratory curiosity until 1886, when Charles Hall in the United States (then a 21-year-old graduate of Oberlin College) and Paul Héroult (a Frenchman of the same age) independently invented an efficient process for its production. In the 1990s the worldwide production of aluminum by the **Hall–Héroult process** was approximately 1.5×10^7 metric tons per year.

The Hall–Héroult process involves the cathodic deposition of aluminum, from molten cryolite (Na_3AlF_6) containing dissolved Al_2O_3, in electrolysis cells (Fig. 17.16). Each cell consists of a rectangular steel box some 6 m long, 2 m wide, and 1 m high, which serves as the cathode, and massive graphite anodes that extend through the roof of the cell into the molten cryolite bath. Enormous currents (50,000 to 100,000 A) are passed through the cell, and as many as 100 such cells may be connected in series.

Molten cryolite, which is completely dissociated into Na^+ and AlF_6^{3-} ions, is an excellent solvent for aluminum oxide, giving rise to an equilibrium distribution of ions such as Al^{3+}, AlF^{2+}, AlF_2^+, . . . , AlF_6^{3-}, and O^{2-} in the electrolyte. Cryolite melts at 1000°C, but its melting point is lowered by dissolved aluminum oxide, so the operating temperature of the cell is about 950°C. Compared with the melting point of pure Al_2O_3 (2050°C), this is a low temperature, and it is the reason the Hall–Héroult process has succeeded. Molten aluminum is somewhat denser than the melt at 950°C and therefore collects at the bottom of the cell, from which it is

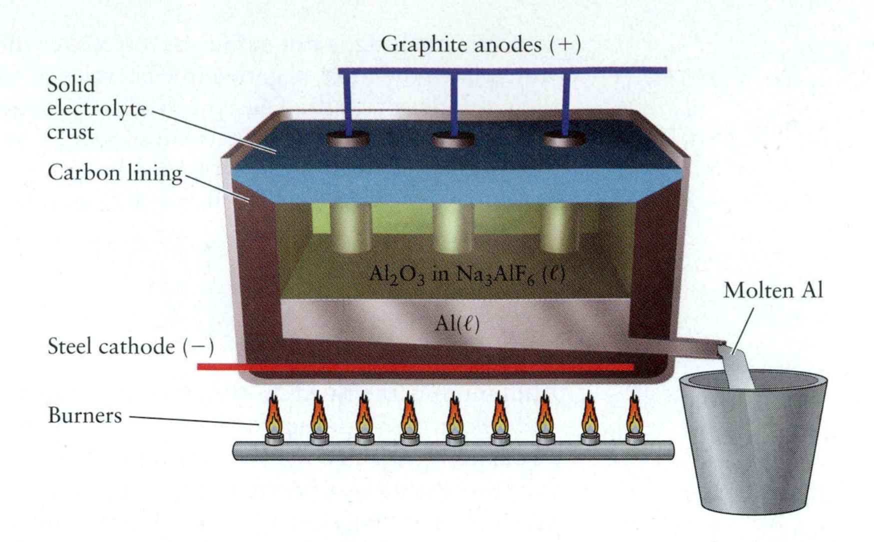

FIGURE 17.16 An electrolytic cell used in the Hall–Héroult process for production of aluminum.

tapped periodically. Oxygen is the primary anode product, but it reacts with the graphite electrode to produce carbon dioxide. The overall cell reaction is

$$2\ Al_2O_3 + 3\ C \longrightarrow 4\ Al + 3\ CO_2$$

Aluminum and its alloys have a tremendous variety of applications. Many of these make use of aluminum's low density (Table 17.2), an advantage over iron or steel when weight savings are desirable—such as in the transportation industry, which uses aluminum in vehicles from automobiles to satellites. Aluminum's high electrical conductivity and low density make it useful for electrical transmission lines. For structural and building applications, its resistance to corrosion is an important feature, as is the fact that it becomes stronger at subzero temperatures. (Steel and iron sometimes become brittle under these circumstances.) Household products that contain aluminum include foil, soft drink cans, and cooking utensils.

Magnesium

Like aluminum, magnesium is an abundant element on the surface of the earth, but it is not easy to prepare in elemental form. Although ores such as dolomite ($CaMg(CO_3)_2$) and carnallite ($KCl \cdot MgCl_2 \cdot 6H_2O$) exist, the major commercial source of magnesium and its compounds is seawater. Magnesium forms the second most abundant positive ion in the sea, and scientists separate Mg^{2+} from the other cations in seawater (Na^+, Ca^{2+}, and K^+, in particular) by taking advantage of the fact that magnesium hydroxide is the least soluble hydroxide of the group. Economical recovery of magnesium requires a low-cost base to treat large volumes of seawater and efficient methods for separating the $Mg(OH)_2(s)$ that precipitates from the solution. One base that is used in this way is calcined dolomite, prepared by heating dolomite to high temperatures to drive off carbon dioxide:

$$CaMg(CO_3)_2(s) \longrightarrow CaO \cdot MgO(s) + 2\ CO_2(g)$$

The greater solubility of calcium hydroxide ($K_{sp} = 5.5 \times 10^{-6}$) relative to magnesium hydroxide ($K_{sp} = 1.2 \times 10^{-11}$) leads to the reaction

$$CaO \cdot MgO(s) + Mg^{2+}(aq) + 2\ H_2O(\ell) \longrightarrow 2\ Mg(OH)_2(s) + Ca^{2+}(aq)$$

The magnesium hydroxide that is produced in this process includes not only the magnesium from the seawater but also that from the dolomite.

An interesting alternative to dolomite as a base for magnesium production is a process used off the coast of Texas (Fig. 17.17). Oyster shells (composed largely of

TABLE 17.2 Densities of Selected Metals

Metal	Density (g cm^{-3}) at Room Conditions
Li	0.534
Na	0.971
Mg	1.738
Al	2.702
Ti	4.54
Zn	7.133
Fe	7.874
Ni	8.902
Cu	8.96
Ag	10.500
Pb	11.35
U	18.95
Au	19.32
Pt	21.45

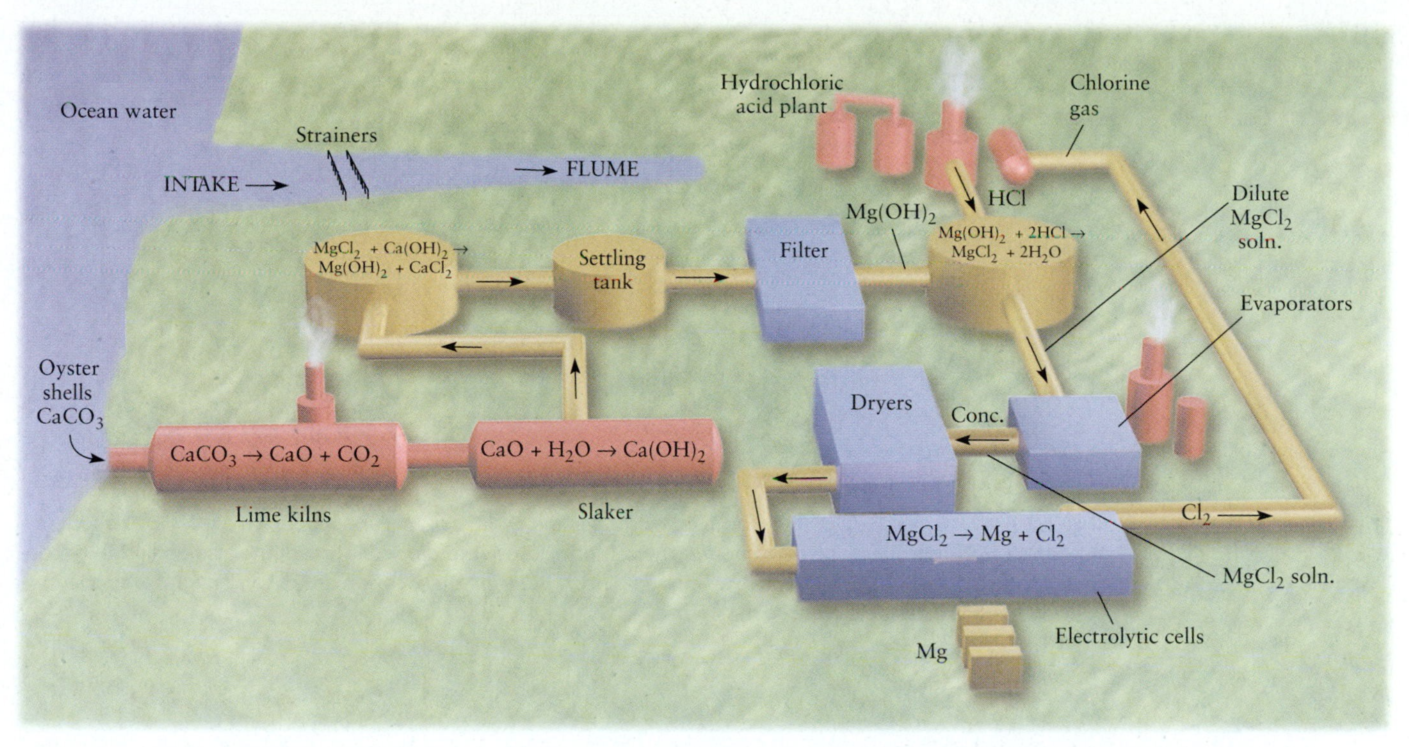

FIGURE 17.17 The production of magnesium hydroxide starts with the addition of lime (CaO) to seawater. Reaction of the magnesium hydroxide with hydrochloric acid produces magnesium chloride, which, after drying, is electrolyzed to give magnesium.

$CaCO_3$) are calcined to give lime (CaO), which is added to the seawater to yield magnesium hydroxide. The $Mg(OH)_2$ slurry (a suspension in water) is washed and filtered in huge nylon filters.

After purification, $Mg(OH)_2$ can be reacted with carbon dioxide to give magnesium carbonate, used for coating sodium chloride in table salt to prevent caking and for antacid remedies. Another alternative is to add hydrochloric acid to the magnesium hydroxide to neutralize it and yield hydrated magnesium chloride:

$$Mg(OH)_2(s) + 2\,HCl(aq) \longrightarrow MgCl_2(aq) + 2\,H_2O(\ell)$$

After the water is evaporated, the solid magnesium chloride is melted (m.p. 708°C) in a large steel electrolysis cell that holds as much as 10 tons of the molten salt. The steel in the cell acts as the cathode during electrolysis, with graphite anodes suspended from the top. The cell reaction is

$$MgCl_2(\ell) \longrightarrow Mg(\ell) + Cl_2(g)$$

The molten magnesium liberated at the cathode floats to the surface and is dipped out periodically, while the chlorine released at the anodes is collected and reacted with steam at high temperatures to produce hydrochloric acid. This is recycled for further reaction with magnesium hydroxide.

Until 1918, elemental magnesium was used mainly in fireworks and flashbulbs, which took advantage of its great reactivity with the oxygen in air and the bright light given off in that reaction (see Fig. 11.6). Since then, many further uses for the metal and its alloys have been developed. Magnesium is even less dense than aluminum and is used in alloys with aluminum to lower its density and improve its resistance to corrosion under basic conditions. As discussed in Section 17.5 magnesium is used as a sacrificial anode to prevent the oxidation of another metal with which it is in contact. It is also used as a reducing agent to produce other metals such as titanium, uranium, and beryllium from their compounds.

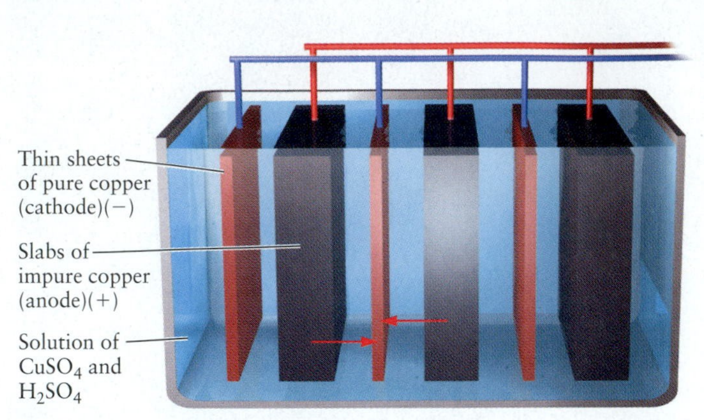

FIGURE 17.18 In the electrolytic refining of copper, many slabs of impure copper, which serve as anodes, alternate with thin sheets of pure copper (the cathodes). Both are dipped into a dilute acidic solution of copper. As the copper is oxidized from the impure anodes, it enters the solution as Cu^{2+} and migrates to the cathodes, where it plates out in purer form.

Electrorefining and Electroplating

Metals that have been produced by pyrometallurgical methods, such as copper, silver, nickel, and tin, are too impure for many purposes, and **electrorefining** is used to purify them further. Crude metallic copper is cast into slabs, which are used as anodes in electrolysis cells that contain a solution of $CuSO_4$ in aqueous H_2SO_4. Thin sheets of pure copper serve as cathodes, and the copper that dissolves at the anodes is deposited in purer form on the cathodes (Fig. 17.18). Impurities that are more easily oxidized than copper, such as nickel, dissolve along with the copper but remain in solution; elements that are less easily oxidized, such as silver and gold, do not dissolve but fall away from the anode as a metallic slime. Periodically, the anode slime and the solution are removed and further processed for recovery of the elements they contain.

A related process is **electroplating,** in which electrolysis is used to plate out a thin layer of a metal on another material, often a second metal. In chrome plating, the piece of metal to be plated is placed in a hot bath of sulfuric acid and chromic acid (H_2CrO_4) and is made the cathode in an electrolytic cell. As current passes through the cell, chromium is reduced from the +6 oxidation state in chromic acid to elemental chromium and plates out on the cathode. A decorative chromium layer can be as thin as 2.5×10^{-5} cm (corresponding to 2 g of Cr per square meter of surface). Thicker layers ranging up to 10^{-2} cm are found in hard chromium plate, prized for its resistance to wear and used in automobile trim. Steel can be plated with cadmium to improve its resistance to corrosion in marine environments. Gold and silver are used both for decorative plating and (because they are good conductors of electricity) on electronic devices.

EXAMPLE 17.10

Suppose a layer of chromium 3.0×10^{-3} cm thick is to be plated onto an automobile bumper with a surface area of 2.0×10^{3} cm^2. If a current of 250 A is used, how long must current be passed through the cell to achieve the desired thickness? The density of chromium is 7.2 g cm^{-3}.

SOLUTION

The volume of the Cr is the product of the thickness of the layer and the surface area:

$$\text{volume} = (3.0 \times 10^{-3}\text{ cm})(2.0 \times 10^{3}\text{ cm}^2) = 6.0\text{ cm}^3$$

The mass of chromium is the product of this volume and the density:

$$\text{mass Cr} = (6.0\text{ cm}^3)(7.2\text{ g cm}^{-3}) = 43.2\text{ g}$$

From this, the number of moles of Cr that must be reduced is

$$\frac{43.2\text{ g}}{52.00\text{ g mol}^{-1}} = 0.831\text{ mol Cr}$$

Because Cr is being reduced from oxidation state +6 in H_2CrO_4 to 0 in the elemental form, six electrons are required for each atom of Cr deposited. The number of moles of electrons is then

$$0.831\text{ mol Cr} \times \left(\frac{6\text{ mol } e^-}{1\text{ mol Cr}}\right) = 4.98\text{ mol } e^-$$

$$\text{total charge} = (4.98\text{ mol})(96{,}485\text{ C mol}^{-1}) = 4.81 \times 10^5\text{ C}$$

The required electrolysis time is the total charge divided by the current (in amperes):

$$\text{time} = \frac{4.81 \times 10^5\text{ C}}{250\text{ C s}^{-1}} = 1.9 \times 10^3\text{ s} = 32\text{ min}$$

Related Problems: 61, 62

A DEEPER LOOK

17.7 Electrolysis of Water and Aqueous Solutions

In Section 17.6 we discussed applications of electrolysis in the extraction and purification of metals from their ore sources. Here we examine the electrolysis of water and aqueous solutions. Consider first the electrolysis of water between inert electrodes such as platinum, for which the half-cell reactions are

$$2\,H_3O^+(aq) + 2\,e^- \longrightarrow H_2(g) + 2\,H_2O(\ell) \qquad \text{(cathode)}$$

$$3\,H_2O(\ell) \longrightarrow \tfrac{1}{2}\,O_2(g) + 2\,H_3O^+(aq) + 2\,e^- \qquad \text{(anode)}$$

$$H_2O(\ell) \longrightarrow H_2(g) + \tfrac{1}{2}\,O_2(g)$$

A practical problem immediately arises. The concentration of $H_3O^+(aq)$ and $OH^-(aq)$ ions in pure water at 25°C is only 1.0×10^{-7} M, so the *rate* of electrolysis will be exceedingly small. This practical consideration is put aside for the moment because it does not alter the *thermodynamic* analysis.

The potential $\mathscr{E}°$ for the cathode reaction is by definition 0 V, but because the $H_3O^+(aq)$ concentration in pure water is not 1 M but 1.0×10^{-7} M, $\mathscr{E}$ differs from $\mathscr{E}°$ and equals

$$\mathscr{E}(\text{cathode}) = \mathscr{E}°(\text{cathode}) - \frac{0.0592\text{ V}}{n_{hc}} \log Q_{hc}$$

$$= 0.00 - \frac{0.0592}{2} \log \frac{P_{H_2}}{[H_3O^+]^2}$$

If $H_2(g)$ is produced at atmospheric pressure, this simplifies to

$$\mathscr{E}(\text{cathode}) = 0.00 - \frac{0.0592\text{ V}}{2} \log \frac{1}{(10^{-7})^2}$$

$$= -0.414\text{ V}$$

The anode half-reaction written as a reduction (*i.e.*, in reverse) is

$$\tfrac{1}{2}\,O_2(g) + 2\,H_3O^+(aq) + 2\,e^- \longrightarrow 3\,H_2O(\ell)$$

A table of standard reduction potentials (see Appendix E) gives $\mathscr{E}° = 1.229$ V. In the present case, the $H_3O^+(aq)$ concentration is 1.0×10^{-7} M rather than 1 M, so

$$\mathscr{E}(\text{anode}) = \mathscr{E}° - \frac{0.0592\text{ V}}{2} \log \frac{1}{(P_{O_2})^{1/2}[H_3O^+]^2}$$

$$= 1.229\text{ V} - \frac{0.0592\text{ V}}{2} \log \frac{1}{(10^{-7})^2}$$

$$= 0.815\text{ V}$$

if $P_{O_2} = 1$ atm. The overall cell voltage is

$$\Delta\mathscr{E} = \mathscr{E}(\text{cathode}) - \mathscr{E}(\text{anode}) = -0.414 - 0.815\text{ V} = -1.229\text{ V}$$

A negative $\Delta\mathscr{E}$ means that the process does not occur spontaneously; it can be made to occur only by applying an external voltage sufficient to overcome the intrinsic negative voltage of the cell. In the electrolysis of water, this minimum external voltage, which is called the **decomposition potential** of water, is 1.229 V. When this potential difference has been applied, an overall cell reaction will occur, given by the sum of the reduction and oxidation half-reactions:

$$H_2O(\ell) \longrightarrow \tfrac{1}{2}\,O_2(g) + H_2(g)$$

Hydrogen will bubble off at the cathode and oxygen at the anode.

Suppose an external voltage is now used to electrolyze an electrolyte solution instead of pure water. The resultant products depend on the concentrations of the ions present and their half-cell potentials. In a 0.10 M NaCl solution, we could conceive of the following processes taking place:

Cathode:

$$Na^+(0.1\ \text{M}) + e^- \longrightarrow Na(s)$$

$$2\,H_3O^+(10^{-7}\ \text{M}) + 2\,e^- \longrightarrow H_2(g) + 2\,H_2O(\ell)$$

Anode:

$$Cl^-(0.1\ \text{M}) \longrightarrow \tfrac{1}{2}\,Cl_2(g) + e^-$$

$$3\,H_2O(\ell) \longrightarrow \tfrac{1}{2}\,O_2(g) + 2\,H_3O^+\ (10^{-7}\ \text{M}) + 2\,e^-$$

In each of these pairs of possible processes, which one will actually occur?

For the first half-reaction the reduction potential is

$$\mathscr{E}(Na^+|Na) = \mathscr{E}°(Na^+|Na) - \frac{0.0592\ \text{V}}{1}\log\frac{1}{[Na^+]}$$

$$= -2.71 - 0.06 = -2.77\ \text{V}$$

Because this result is more negative than the half-cell voltage $\mathscr{E}(H_3O^+(10^{-7}\ \text{M})|\,H_2) = -0.414$ V for pure water, the reduction of $Na^+(aq)$ is impossible, and $H_2(g)$ is the cathode product.

For the third half-reaction, the reduction potential is

$$\mathscr{E}(Cl_2|Cl^-) = \mathscr{E}°(Cl_2|Cl^-) - \frac{0.0592\ \text{V}}{1}\log\frac{[Cl^-]}{P_{Cl_2}^{1/2}}$$

$$= 1.36 + 0.06 = 1.42\ \text{V}$$

Because this is more positive than the half-cell voltage $\mathscr{E}(O_2,H_3O^+(10^{-7}\ \text{M})|\,H_2O) = 0.815$ V for pure water, Cl_2 has a *greater* tendency to be reduced than O_2; therefore, Cl^- has a *lesser* tendency to be oxidized than H_2O, and the anode product is $O_2(g)$. If we try to increase the external potential above 1.229 V, all that will happen is that water will be electrolyzed at a greater rate to produce hydrogen and oxygen. Sodium and chlorine will not appear as long as sufficient water is present.

Suppose now that 0.10 M NaI is substituted for the 0.10 M NaCl solution. Sodium ions still will not be reduced; however, $\mathscr{E}(I_2|\,I^-)$ is

$$\mathscr{E}(I_2|I^-) = \mathscr{E}°(I_2|I^-) - \frac{0.0592\ \text{V}}{1}\log\,[I^-]$$

$$= 0.535 + 0.059 = 0.594\ \text{V}$$

The half-cell potential for the reduction of iodine is *less* positive than the reduction potential of $O_2(g)$ in water at pH = 7(0.815 V), so the oxidation of 0.10 M I^- occurs in preference to the oxidation of water. The anode reaction is therefore

$$I^-(0.10\ \text{M}) \longrightarrow \tfrac{1}{2}\,I_2(s) + e^-$$

and the overall cell reaction is

$$H_3O^+(10^{-7}\ \text{M}) + I^-(0.10\ \text{M}) \longrightarrow \tfrac{1}{2}\,H_2(g) + \tfrac{1}{2}\,I_2(s) + H_2O(\ell)$$

The intrinsic cell voltage is

$$\Delta\mathscr{E} = \mathscr{E}(\text{cathode}) - \mathscr{E}(\text{anode}) = -0.414 - 0.594\ \text{V} = -1.008\ \text{V}$$

When the applied voltage exceeds 1.008 V, $H_2(g)$ and $I_2(s)$ begin to form. Of course, this causes $[I^-]$ to decrease and $\mathscr{E}(I_2|\,I^-)$ to increase. When the iodide ion concentration reaches about 2×10^{-5} M, $\mathscr{E}(I_2|\,I^-)$ will have increased to 0.815 V and the external voltage required to maintain electrolysis will have increased to 1.229 V. At this point, water will start to be electrolyzed and oxygen will be produced at the anode.

Our results for the electrolysis of *neutral* aqueous solutions are summarized as follows:

1. A species can be reduced only if its reduction potential is algebraically greater than −0.414 V.
2. A species can be oxidized only if its reduction potential is algebraically smaller than 0.815 V.

In solutions with pH different from 7, these results must be modified, as shown by the following example.

EXAMPLE 17.11

An aqueous 0.10 M solution of $NiCl_2$ is electrolyzed under 1 atm pressure. Determine the products formed at the anode and the cathode and the decomposition potential, if the pH is **(a)** 7.0; **(b)** 0.0.

SOLUTION

The reduction of Ni^{2+} at the cathode has the half-cell potential

$$Ni^{2+}(aq) + 2e^- \longrightarrow Ni(s)$$

$$\mathscr{E}(Ni^{2+}|Ni) = \mathscr{E}°(Ni^{2+}|Ni) - \frac{0.0592\ \text{V}}{2}\log\frac{1}{[Ni^{2+}]}$$

$$= -0.23 - 0.03 = -0.26\ \text{V}$$

and the oxidation of the Cl^- at the anode has the *reduction* potential

$$\tfrac{1}{2}\,Cl_2(g) + e^- \longrightarrow Cl^-(aq)$$

$$\mathscr{E}(Cl_2|Cl^-) = \mathscr{E}°(Cl_2|Cl^-) - \frac{0.0592\ \text{V}}{2}\log\frac{[Cl^-]}{P_{Cl_2}^{1/2}}$$

$$= 1.36\ \text{V} - (0.0592\ \text{V})\log 0.2 = 1.40\ \text{V}$$

(a) In neutral solution the cathode half-reaction is

$$Ni^{2+}(aq) + 2\,e^- \longrightarrow Ni(s)$$

because $\mathscr{E}(Ni^{2+}|\,Ni) = -0.26\ \text{V} > -0.414$ V. The anode half-reaction is

$$3\,H_2O(\ell) \longrightarrow 2\,H_3O^+(10^{-7}\ \text{M}) + \tfrac{1}{2}\,O_2(g) + 2\,e^-$$

because $\mathscr{E}(Cl_2|\,Cl^-) = 1.40\ \text{V} > 0.815$ V (the reduction potential for this half-reaction). The cell voltage is

$$\Delta\mathscr{E} = \mathscr{E}(\text{cathode}) - \mathscr{E}(\text{anode}) = -0.26 - 0.815\ \text{V} = -1.08\ \text{V}$$

so the decomposition potential is 1.08 V.

(b) In 1.0 M acid solution (pH = 0.0), the anode half-reaction is still

$$3\ H_2O(\ell) \longrightarrow 2\ H_3O^+(1\ \text{M}) + \tfrac{1}{2}\ O_2(g) + 2\ e^-$$

The reduction potential for this reaction is now $\mathscr{E}_{hc} = \mathscr{E}^\circ_{hc} =$ 1.229 V, which is still less than 1.40 V, the reduction potential for the competing reaction involving chlorine.

The cathode half-reaction now becomes

$$2\ H_3O^+(1\ \text{M}) + 2\ e^- \longrightarrow H_2(g) + 2\ H_2O(\ell)$$

because $\mathscr{E}(H_3O^+|H_2) = 0.0\ V > -0.26\ V$ for $\mathscr{E}(Ni^{2+}|Ni)$. We now have

$$\Delta\mathscr{E} = \mathscr{E}(\text{cathode}) - \mathscr{E}(\text{anode}) = 0.00 - 1.229 = -1.229\ V$$

so the decomposition potential of the solution is now 1.229 V, just as it is for pure water.

Related Problems: 63, 64

CHAPTER SUMMARY

Electrochemical reactions are an important class of oxidation–reduction (redox) reactions that interconvert chemical and electrical energy. The free energy released in a spontaneous chemical reaction can be used to generate electricity, or electrical energy can be provided from an external source to drive chemical reactions that are not normally spontaneous. The key to this flexibility is the separation of the oxidation and reduction parts of the reaction with the electrons being transferred through an external circuit. Electrochemistry is enormously important in many existing technologies such as energy conversion and storage, and large-scale chemical syntheses of commodity chemicals like chlorine. It is also an enabling science for alternative energy sources, such as fuel cells for large scale deployment in transportation and small dedicated power plants, and for the efficient capture and storage of solar energy in batteries and as chemical energy in hydrogen produced by photoelectrochemical water splitting. The cell potential $\Delta\mathscr{E}$ is simply related to the Gibbs free energy for an electrochemical reaction, providing a quantitative measure of the driving force for the reaction. The cell potential goes to zero for reactions at equilibrium, so the standard cell potential $\Delta\mathscr{E}^\circ$ is a direct measure of the equilibrium constant in electrochemical reactions. Because $\mathscr{E}$ is a state function, the equilibrium constant for a particular reaction is the same whether carried out electrochemically or otherwise, so electrochemistry is a powerful way to measure equilibrium constants for reactions that can be difficult to measure in other environments.

CUMULATIVE EXERCISE

A pile of manganese metal.

Manganese

Manganese is the 12th most abundant element on the earth's surface. Its most important ore source is pyrolusite (MnO_2). The preparation and uses of manganese and its compounds (which range up to +7 in oxidation state) are intimately bound up with electrochemistry.

(a) Elemental manganese in a state of high purity can be prepared by electrolyzing aqueous solutions of Mn^{2+}. At which electrode (anode or cathode) does the Mn appear? Electrolysis is also used to make MnO_2 in high purity from Mn^{2+} solutions. At which electrode does the MnO_2 appear?

(b) The Winkler method is an analytical procedure for determining the amount of oxygen dissolved in water. In the first step, $Mn(OH)_2(s)$ is oxidized by gaseous oxygen to $Mn(OH)_3(s)$ in basic aqueous solution. Write the oxidation and reduction half-equations for this step, and write the balanced overall equation. Then use Appendix E to calculate the standard voltage that would be measured if this reaction were carried out in an electrochemical cell.

(c) Calculate the equilibrium constant at 25°C for the reaction in part (b).

(d) In the second step of the Winkler method, the $Mn(OH)_3$ is acidified to give Mn^{3+} and iodide ion is added. Will Mn^{3+} spontaneously oxidize I^-? Write a balanced equation for its reaction with I^-, and use data from Appendix E to calculate its equilibrium constant. Titration of the I_2 produced completes the use of the Winkler method.

(e) Manganese(IV) is an even stronger oxidizing agent than manganese(III). It oxidizes zinc to Zn^{2+} in the dry cell. Such a battery has a cell voltage of 1.5 V. Calculate the electrical work done by this battery in 1.00 hour if it produces a steady current of 0.70 A.

(f) Calculate the mass of zinc reacting in the process described in part (e).

(g) The reduction potential of permanganate ion (+7 oxidation state) in acidic aqueous solution is given by

$$MnO_4^-(aq) + 8\,H_3O^+(aq) + 5\,e^- \longrightarrow Mn^{2+}(aq) + 12\,H_2O(\ell) \quad \mathscr{E}° = 1.491V$$

whereas that of the analogous fifth-period species, pertechnetate ion, is

$$TcO_4^-(aq) + 8\,H_3O^+(aq) + 5\,e^- \longrightarrow Tc^{2+}(aq) + 12\,H_2O(\ell) \quad \mathscr{E}° = 0.500V$$

Which is the stronger oxidizing agent, permanganate ion or pertechnetate ion?

(h) A galvanic cell is made from two half-cells. In the first, a platinum electrode is immersed in a solution at pH 2.00 that is 0.100 M in both MnO_4^- and Mn^{2+}. In the second, a zinc electrode is immersed in a 0.0100 M solution of $Zn(NO_3)_2$. Calculate the cell voltage that will be measured.

Answers

(a) Mn appears at the cathode and MnO_2 at the anode.

(b)

$$Mn(OH)_2(s) + OH^-(aq) \longrightarrow Mn(OH)_3(s) + e^- \quad \text{(oxidation)}$$

$$O_2(g) + 2\,H_2O(\ell) + 4\,e^- \longrightarrow 4\,OH^-(aq) \quad \text{(reduction)}$$

$$4\,Mn(OH)_2(s) + O_2(g) + 2\,H_2O(\ell) \longrightarrow 4\,Mn(OH)_3(s)$$

$$\Delta\mathscr{E}° = 0.401 - (-0.40) = 0.80\text{ V}$$

(c) $K = 1 \times 10^{54}$

(d) Mn^{3+} will spontaneously oxidize I^-.

$$2\,Mn^{3+}(aq) + 2\,I^-(aq) \longrightarrow 2\,Mn^{2+}(aq) + I_2(s) \qquad K = 9 \times 10^{32}$$

(e) 3.8×10^3 J

(f) 0.85 g Zn is oxidized.

(g) Permanganate ion

(h) 2.12 V

CHAPTER REVIEW

- Electrochemical reactions are oxidation–reduction reactions that interconvert chemical and electrical energy.
- The free energy released in spontaneous chemical reactions conducted in a galvanic cell is converted to electrical energy, which is transferred through an external circuit to be stored or converted into work.

- Electrical energy provided through an external circuit to an electrolytic cell drives reactions that are not spontaneous, converting electrical energy into chemical potential energy.
- Electrodes in electrochemical cells are immersed in electrolytes in separate containers. Wires connect the electrodes to form an external circuit that includes meters to measure the cell potential (voltage) and current. A salt bridge transports ions between the two containers to maintain electrical neutrality.
- The electrode at which oxidation occurs is called the anode and the electrode at which reduction occurs is called the cathode.
- The number of moles of a substance that is oxidized or reduced in an electrochemical reaction is proportional to the number of moles of electrons passed through the cell. One mole of electrons is 96,485 C, and it is called the Faraday constant, $\mathcal{F}$.
- $\Delta\mathcal{E}$, the electrical potential difference or the cell potential, is the change in potential energy per unit of charge as electrons move between regions with different potentials.
  - $\Delta\mathcal{E}$ is the driving force for electrochemical reactions.
  - Because the charge of the electron is negative, spontaneous processes are those for which $\Delta\mathcal{E}$ increases.
  - $\Delta\mathcal{E}$ is measured in joules per coulomb (J C^{-1}) or volts (V) in SI units.
- The electrical work done by a galvanic cell on the surroundings is $w_{\text{elec}} = -Q\Delta\mathcal{E}$.
- The reversible work done per mol of electrons is the change in the Gibbs free energy, $\Delta G = w_{\text{elec}} - n\mathcal{F}\,\Delta\mathcal{E}$. The standard cell potential is related to the standard free energy change by $\Delta G° = w_{\text{elec}} = n\mathcal{F}\,\Delta\mathcal{E}°$.
- Half-cell reactions provide a convenient way to balance electrochemical reactions and to calculate cell potentials.
  - Half-cell reactions have the form $X^{n+}(aq) + ne^- \longrightarrow X(aq)$ for the reduction reaction and $Y(aq) \longrightarrow Y^{n+}(aq) + ne^-$ for the oxidation reaction.
  - The overall reaction is written as the sum of a reduction half-cell reaction and an oxidation half-cell reaction.
  - A set of standard reduction potentials has been established by measuring the cell potentials of a number of half-cell reactions against the half-cell reduction of H_3O^+ to form hydrogen gas and water. This reaction has been arbitrarily assigned a standard reduction potential $\mathcal{E}° = 0$.
  - $\Delta\mathcal{E}° = \mathcal{E}°(\text{cathode}) - \mathcal{E}°(\text{anode})$.
- Cell potentials can be calculated for conditions other than standard state conditions using the Nernst equation.

  $\Delta\mathcal{E} = \Delta\mathcal{E}° = \dfrac{RT}{n\mathcal{F}} \ln Q$ where Q is the reaction quotient discussed in Section 14.6.

  At 25°C the Nernst equation can be written as $\Delta\mathcal{E} = \Delta\mathcal{E}° - \dfrac{0.0592 \text{ V}}{n} \log_{10} Q$ which is more convenient for quick calculations and estimates.
- Measuring equilibrium constants is one of the most important applications of electrochemistry. Since $\Delta\mathcal{E}° = 0$ at equilibrium (no thermodynamic driving force for change) and $Q = K$, the Nernst equation can be rearranged to give $\Delta\mathcal{E}° = \dfrac{0.0592 \text{ V}}{n} \log_{10} K$ allowing equilibrium constants to be determined simply by measuring standard cell potentials.

CONCEPTS & SKILLS

After studying this chapter and working the problems that follow, you should be able to:

1. Define the terms *anode* and *cathode* and give the convention that is used to represent a galvanic cell (Section 17.1, problems 1–2).
2. Use Faraday's laws to calculate the quantities of substances produced or consumed at the electrodes of electrochemical cells in relation to the total charge passing through the circuit (Section 17.1, problems 3–10).
3. Explain the relationship between the free energy change in a galvanic cell and the amount of electrical work (Section 17.2, problems 11–12).
4. Combine half-cell reactions and their standard reduction potentials to obtain the overall reactions and standard voltages of electrochemical cells (Section 17.2, problems 13–16).
5. Combine half-reactions and their standard reduction potentials to form other half-reactions and their standard reduction potentials (Section 17.2, problems 23–24).
6. Use reduction potential diagrams to determine strengths of oxidizing and reducing agents and stability toward disproportionation (Section 17.2, problems 25–26).
7. Apply the Nernst equation to calculate the voltage of a cell in which reactants and products are not in their standard states and to calculate the value of an equilibrium constant from the voltage of an electrochemical cell (Section 17.3, problems 27–38).
8. Describe the principles that underlie the use of electrochemical cells as pH meters (Section 17.3, problems 39–40).
9. Discuss the electrochemistry of a primary battery and contrast it with that of a secondary battery and a fuel cell (Section 17.4, problems 43–50).
10. Discuss the electrochemical corrosion of metals and describe measures that may be used to minimize it (Section 17.5, problems 51–54).
11. Describe the Hall–Héroult process for the production of aluminum and the methods used to recover magnesium from seawater (Section 17.6, problems 57–58).
12. Use Faraday's laws to relate current to metal deposited in electrorefining and electroplating operations (Section 17.6, problems 61–62).
13. Predict the products liberated at the anode and cathode of an electrolysis cell with a given aqueous electrolyte composition (Section 17.7, problems 63–64).

KEY EQUATIONS

$$I = \frac{Q}{t}$$ (Section 17.1)

$$w_{\text{elec}} = -Q\Delta\mathscr{E}$$ (Section 17.2)

$$-w_{\text{elec,max}} = |\Delta G| \qquad \text{(at constant } T \text{ and } P\text{)}$$ (Section 17.2)

$$\Delta G^\circ = -n\mathscr{F}\,\Delta\mathscr{E}^\circ$$ (Section 17.2)

$$\Delta\mathscr{E}^\circ = \mathscr{E}^\circ(\text{cathode}) - \mathscr{E}^\circ(\text{anode})$$ (Section 17.2)

$$\Delta\mathscr{E} = \Delta\mathscr{E}^\circ - \frac{RT}{n\mathscr{F}} \ln Q \qquad \text{(Section 17.3)}$$

$$\Delta\mathscr{E} = \Delta\mathscr{E}^\circ - \frac{0.0592 \text{ V}}{n} \log_{10} Q \quad \text{(at 25°C)} \qquad \text{(Section 17.3)}$$

$$\log_{10} K = \frac{n}{0.0592 \text{ V}} \Delta\mathscr{E}^\circ \quad \text{(at 25°C)} \qquad \text{(Section 17.3)}$$

PROBLEMS

Answers to problems whose numbers are boldface appear in Appendix G. Problems that are more challenging are indicated with asterisks.

Electrochemical Cells

1. Diagram the following galvanic cell, indicating the direction of flow of electrons in the external circuit and the motion of ions in the salt bridge.

$$Pt(s)|Cr^{2+}(aq), Cr^{3+}(aq)||Cu^{2+}(aq)|Cu(s)$$

Write a balanced equation for the overall reaction in this cell.

2. Diagram the following galvanic cell, indicating the direction of flow of electrons in the external circuit and the motion of ions in the salt bridge.

$$Ni(s)|Ni^{2+}(aq)||HCl(aq)|H_2(g)|Pt(s)$$

Write a balanced equation for the overall reaction in this cell.

3. A quantity of electricity equal to 6.95×10^4 C passes through an electrolytic cell that contains a solution of $Sn^{4+}(aq)$ ions. Compute the maximum chemical amount, in moles, of $Sn(s)$ that can be deposited at the cathode.

4. A quantity of electricity equal to 9.263×10^4 C passes through a galvanic cell that has an $Ni(s)$ anode. Compute the maximum chemical amount, in moles, of $Ni^{2+}(aq)$ that can be released into solution.

5. A galvanic cell is constructed that has a zinc anode immersed in a $Zn(NO_3)_2$ solution and a platinum cathode immersed in an NaCl solution equilibrated with $Cl_2(g)$ at 1 atm and 25°C. A salt bridge connects the two half-cells.
(a) Write a balanced equation for the cell reaction.
(b) A steady current of 0.800 A is observed to flow for a period of 25.0 minutes. How much charge passes through the circuit during this time? How many moles of electrons is this charge equivalent to?
(c) Calculate the change in mass of the zinc electrode.
(d) Calculate the volume of gaseous chlorine generated or consumed as a result of the reaction.

6. A galvanic cell consists of a cadmium cathode immersed in a $CdSO_4$ solution and a zinc anode immersed in a $ZnSO_4$ solution. A salt bridge connects the two half-cells.
(a) Write a balanced equation for the cell reaction.
(b) A current of 1.45 A is observed to flow for a period of 2.60 hours. How much charge passes through the circuit during this time? How many moles of electrons is this charge equivalent to?
(c) Calculate the change in mass of the zinc electrode.
(d) Calculate the change in mass of the cadmium electrode.

7. An acidic solution containing copper ions is electrolyzed, producing gaseous oxygen (from water) at the anode and copper at the cathode. For every 16.0 g of oxygen generated, 63.5 g of copper plates out. What is the oxidation state of the copper in the solution?

8. Michael Faraday reported that passing electricity through one solution liberated 1 mass of hydrogen at the cathode and 8 masses of oxygen at the anode. The same quantity of electricity liberated 36 masses of chlorine at the anode and 58 masses of tin at the cathode from a second solution. What were the oxidation states of hydrogen, oxygen, chlorine, and tin in these solutions?

9. Liquid potassium chloride, $KCl(\ell)$, is decomposed in an electrolytic cell to form potassium and chlorine. Liquid KCl consists of K^+ and Cl^- ions.
(a) Write balanced equations for the half-cell reactions at the anode and at the cathode and for the overall cell reaction.
(b) If a current of 2.00 A is passed through the cell for a period of 5.00 hours, calculate the mass of metal deposited and of gas liberated.

10. In the Hall–Héroult process for the electrolytic production of aluminum, Al^{3+} ions from Al_2O_3 dissolved in molten cryolite (Na_3AlF_6) are reduced to $Al(\ell)$ while carbon (graphite) is oxidized to CO_2 by reaction with oxide ions.
(a) Write balanced equations for the half-reactions at the anode and at the cathode and for the overall cell reaction.
(b) If a current of 50,000 A is passed through the cell for a period of 24 hours, what mass of aluminum will be recovered?

The Gibbs Free Energy and Cell Voltage

11. A $Ni|Ni^{2+}||Ag^+|Ag$ galvanic cell is constructed in which the standard cell voltage is 1.03 V. Calculate the free energy change at 25°C when 1.00 g of silver plates out, if all concentrations remain at their standard value of 1 M throughout the process. What is the maximum electrical work done by the cell on its surroundings during this experiment?

12. A $Zn|Zn^{2+}||Co^{2+}|Co$ galvanic cell is constructed in which the standard cell voltage is 0.48 V. Calculate the free energy change at 25°C per gram of zinc lost at the anode, if all concentrations remain at their standard value of 1 M throughout the process. What is the maximum electrical work done by the cell on its surroundings during this experiment?

13. A galvanic cell is constructed in which a $Br_2|Br^+$ half-cell is connected to a $Co^{2+}|Co$ half-cell.
    (a) By referring to Appendix E, write balanced chemical equations for the half-reactions at the anode and the cathode and for the overall cell reaction.
    (b) Calculate the cell voltage, assuming that all reactants and products are in their standard states.

14. A galvanic cell is constructed in which a $Pt|Fe^{2+}, Fe^{3+}$ half-cell is connected to a $Cd^{2+}|Cd$ half-cell.
    (a) By referring to Appendix E, write balanced chemical equations for the half-reactions at the anode and the cathode and for the overall cell reaction.
    (b) Calculate the cell voltage, assuming that all reactants and products are in their standard states.

15. In a galvanic cell, one half-cell consists of a zinc strip dipped into a 1.00 M solution of $Zn(NO_3)_2$. In the second half-cell, solid indium adsorbed on graphite is in contact with a 1.00 M solution of $In(NO_3)_3$. Indium is observed to plate out as the galvanic cell operates, and the initial cell voltage is measured to be 0.425 V at 25°C.
    (a) Write balanced equations for the half-reactions at the anode and the cathode.
    (b) Calculate the standard reduction potential of an $In^{3+}|In$ half-cell. Consult Appendix E for the reduction potential of the $Zn^{2+}|Zn$ electrode.

16. In a galvanic cell, one half-cell consists of gaseous chlorine bubbled over a platinum electrode at a pressure of 1.00 atm into a 1.00 M solution of NaCl. The second half-cell has a strip of solid gallium immersed in a 1.00 M $Ga(NO_3)_3$ solution. The initial cell voltage is measured to be 1.918 V at 25°C, and as the cell operates, the concentration of chloride ion is observed to increase.
    (a) Write balanced equations for the half-reactions at the anode and the cathode.
    (b) Calculate the standard reduction potential of a $Ga^{3+}|Ga$ half-cell. Consult Appendix E for the reduction potential of the $Cl_2|Cl^-$ electrode.

17. Would you expect powdered solid aluminum to act as an oxidizing agent or as a reducing agent?

18. Would you expect potassium perchlorate, $KClO_4(aq)$, in a concentrated acidic solution to act as an oxidizing agent or as a reducing agent?

19. Bromine is sometimes used in place of chlorine as a disinfectant in swimming pools. If the effectiveness of a chemical as a disinfectant depends solely on its strength as an oxidizing agent, do you expect bromine to be better or worse than chlorine as a disinfectant, at a given concentration?

20. Many bleaches, including chlorine and its oxides, oxidize dye compounds in cloth. Predict which of the following will be the strongest bleach at a given concentration and pH 0: $NaClO_3(aq)$, $NaClO(aq)$, $Cl_2(aq)$. How does the strongest chlorine-containing bleach compare in strength with ozone, $O_3(g)$?

21. Suppose you have the following reagents available at pH 0, atmospheric pressure, and 1 M concentration:

$$Co(s),\ Ag^+(aq),\ Cl^-(aq),\ Cr(s),\ BrO_3^-(aq),\ I_2(s)$$

    (a) Which is the strongest oxidizing agent?
    (b) Which is the strongest reducing agent?
    (c) Which reagent will reduce $Pb^{2+}(aq)$ while leaving $Cd^{2+}(aq)$ unreacted?

22. Suppose you have the following reagents available at pH 0, atmospheric pressure, and 1 M concentration:

$$Sc(s),\ Hg_2^{2+}(aq),\ Cr_2O_7^{2-}(aq),\ H_2O_2(aq),\ Sn^{2+}(aq),\ Ni(s)$$

    (a) Which is the strongest oxidizing agent?
    (b) Which is the strongest reducing agent?
    (c) Which reagent will oxidize Fe(s) while leaving Cu(s) unreacted?

23. (a) Use the data from Appendix E to calculate the half-cell potential $\mathscr{E}°$ for the half-reaction

$$Mn^{3+}(aq) + 3\,e^- \longrightarrow Mn(s)$$

    (b) Consider the disproportionation reaction

$$3\,Mn^{2+}(aq) \rightleftharpoons Mn(s) + 2\,Mn^{3+}(aq)$$

    Will Mn^{2+} disproportionate in aqueous solution?

24. The following standard reduction potentials have been measured in aqueous solution at 25°C:

$$Tl^{3+} + e^- \longrightarrow Tl^{2+} \qquad \mathscr{E}° = -0.37\ V$$

$$Tl^{3+} + 2\,e^- \longrightarrow Tl^+ \qquad \mathscr{E}° = 1.25\ V$$

    (a) Calculate the half-cell potential for the half-reaction

$$Tl^{2+} + e^- \longrightarrow Tl^+$$

    (b) Consider the disproportionation reaction

$$2\,Tl^{2+}(aq) \rightleftharpoons Tl^{3+}(aq) + Tl^+(aq)$$

    Will Tl^{2+} disproportionate in aqueous solution?

25. The following reduction potentials are measured at pH 0:

$$BrO_3^- + 6\,H_3O^+ + 5\,e^- \longrightarrow \tfrac{1}{2}\,Br_2(\ell) + 9\,H_2O \qquad \mathscr{E}° = 1.52\ V$$

$$Br_2(\ell) + 2\,e^- \longrightarrow 2\,Br^- \qquad \mathscr{E}° = 1.065\ V$$

    (a) Will bromine disproportionate spontaneously in acidic solution?
    (b) Which is the stronger reducing agent at pH 0: $Br_2(\ell)$ or Br^-?

26. The following reduction potentials are measured at pH 14:

$$ClO^- + H_2O + 2\,e^- \longrightarrow Cl^- + 2\,OH^- \qquad \mathscr{E}° = 0.90\ V$$

$$ClO_2^- + H_2O + 2\,e^- \longrightarrow ClO^- + 2\,OH^- \qquad \mathscr{E}° = 0.59\ V$$

    (a) Will ClO^- disproportionate spontaneously in basic solution?
    (b) Which is the stronger reducing agent at pH 14: ClO^- or Cl^-?

Concentration Effects and the Nernst Equation

27. A galvanic cell is constructed that carries out the reaction

$$Pb^{2+}(aq) + 2\,Cr^{2+}(aq) \longrightarrow Pb(s) + 2\,Cr^{3+}(aq)$$

If the initial concentration of $Pb^{2+}(aq)$ is 0.15 M, that of $Cr^{2+}(aq)$ is 0.20 M, and that of $Cr^{3+}(aq)$ is 0.0030 M, calculate the initial voltage generated by the cell at 25°C.

28. A galvanic cell is constructed that carries out the reaction

$$2\,Ag(s) + Cl_2(g) \longrightarrow 2\,Ag^+(aq) + 2\,Cl^-(aq)$$

If the partial pressure of $Cl_2(g)$ is 1.00 atm, the initial concentration of $Ag^+(aq)$ is 0.25 M, and that of $Cl^-(aq)$ is 0.016 M, calculate the initial voltage generated by the cell at 25°C.

29. Calculate the reduction potential for a $Pt|Cr^{3+}, Cr^{2+}$ half-cell in which $[Cr^{3+}]$ is 0.15 M and $[Cr^{2+}]$ is 0.0019 M.

30. Calculate the reduction potential for an $I_2(s)|I^-$ half-cell in which $[I^-]$ is 1.5×10^{-6} M.

31. An $I_2(s)|I^-(1.00\text{ M})$ half-cell is connected to an $H_3O^+|H_2$ (1 atm) half-cell in which the concentration of the hydronium ion is unknown. The measured cell voltage is 0.841 V, and the $I_2|I^-$ half-cell is the cathode. What is the pH in the $H_3O^+|H_2$ half-cell?

32. A $Cu^{2+}(1.00\text{ M})|Cu$ half-cell is connected to a $Br_2(\ell)|Br^-$ half-cell in which the concentration of bromide ion is unknown. The measured cell voltage is 0.963 V, and the $Cu^{2+}|Cu$ half-cell is the anode. What is the bromide ion concentration in the $Br_2(\ell)|Br^-$ half-cell?

33. The following reaction occurs in an electrochemical cell:

$$3\,HClO_2(aq) + 2\,Cr^{3+}(aq) + 12\,H_2O(\ell) \longrightarrow 3\,HClO(aq) + Cr_2O_7^{2-}(aq) + 8\,H_3O^+(aq)$$

(a) Calculate $\Delta\mathscr{E}°$ for this cell.
(b) At pH 0, with $[Cr_2O_7^{2-}] = 0.80$ M, $[HClO_2] = 0.15$ M, and $[HClO] = 0.20$ M, the cell voltage is found to be 0.15 V. Calculate the concentration of $Cr^{3+}(aq)$ in the cell.

34. A galvanic cell is constructed in which the overall reaction is

$$Cr_2O_7^{2-}(aq) + 14\,H_3O^+(aq) + 6\,I^-(aq) \longrightarrow 2\,Cr^{3+}(aq) + 3\,I_2(s) + 21\,H_2O(\ell)$$

(a) Calculate $\Delta\mathscr{E}°$ for this cell.
(b) At pH 0, with $[Cr_2O_7^{2-}] = 1.5$ M and $[I^-] = 0.40$ M, the cell voltage is found to equal 0.87 V. Calculate the concentration of $Cr^{3+}(aq)$ in the cell.

35. By using the half-cell potentials in Appendix E, calculate the equilibrium constant at 25°C for the reaction in problem 33. Dichromate ion ($Cr_2O_7^{2-}$) is orange, and Cr^{3+} is light green in aqueous solution. If 2.00 L of 1.00 M $HClO_2$ solution is added to 2.00 L of 0.50 M $Cr(NO_3)_3$ solution, what color will the resulting solution have?

36. By using the half-cell potentials in Appendix E, calculate the equilibrium constant at 25°C for the reaction

$$6\,Hg^{2+}(aq) + 2\,Au(s) \rightleftharpoons 3\,Hg_2^{2+}(aq) + 2\,Au^{3+}(aq)$$

If 1.00 L of a 1.00 M $Au(NO_3)_3$ solution is added to 1.00 L of a 1.00 M $Hg_2(NO_3)_2$ solution, calculate the concentrations of Hg^{2+}, Hg_2^{2+}, and Au^{3+} at equilibrium.

37. The following standard reduction potentials have been determined for the aqueous chemistry of indium:

$$In^{3+}(aq) + 2\,e^- \longrightarrow In^+(aq) \qquad \mathscr{E}° = -0.40\text{ V}$$
$$In^+(aq) + e^- \longrightarrow In(s) \qquad \mathscr{E}° = -0.21\text{ V}$$

Calculate the equilibrium constant (K) for the disproportionation of $In^+(aq)$ at 25°C.

$$3\,In^+(aq) \rightleftharpoons 2\,In(s) + In^{3+}(aq)$$

38. Use data from Appendix E to compute the equilibrium constant for the reaction

$$Hg^{2+}(aq) + Hg(\ell) \rightleftharpoons Hg_2^{2+}(aq)$$

39. A galvanic cell consists of a $Pt|H_3O^+(1.00\text{ M})|H_2(g)$ cathode connected to a $Pt|H_3O^+(aq)|H_2(g)$ anode in which the concentration of H_3O^+ is unknown but is kept constant by the action of a buffer consisting of a weak acid, HA(0.10 M), mixed with its conjugate base, A^-(0.10 M). The measured cell voltage is $\Delta\mathscr{E} = 0.150$ V at 25°C, with a hydrogen pressure of 1.00 atm at both electrodes. Calculate the pH in the buffer solution, and from it determine the K_a of the weak acid.

40. In a galvanic cell, the cathode consists of a $Ag^+(1.00\text{ M})|Ag$ half-cell. The anode is a platinum wire, with hydrogen bubbling over it at 1.00-atm pressure, that is immersed in a buffer solution containing benzoic acid and sodium benzoate. The concentration of benzoic acid (C_6H_5COOH) is 0.10 M, and that of benzoate ion ($C_6H_5COO^-$) is 0.050 M. The overall cell reaction is then

$$Ag^+(aq) + \tfrac{1}{2}\,H_2(g) + H_2O(\ell) \longrightarrow Ag(s) + H_3O^+(aq)$$

and the measured cell voltage is 1.030 V. Calculate the pH in the buffer solution and determine the K_a of benzoic acid.

41. A galvanic cell is constructed in which the overall reaction is

$$Br_2(\ell) + H_2(g) + 2\,H_2O(\ell) \longrightarrow 2\,Br^-(aq) + 2\,H_3O^+(aq)$$

(a) Calculate $\Delta\mathscr{E}°$ for this cell.
(b) Silver ions are added until AgBr precipitates at the cathode and $[Ag^+]$ reaches 0.060 M. The cell voltage is then measured to be 1.710 V at pH = 0 and $P_{H_2} = 1.0$ atm. Calculate $[Br^-]$ under these conditions.
(c) Calculate the solubility product constant K_{sp} for AgBr.

42. A galvanic cell is constructed in which the overall reaction is

$$Pb(s) + 2\,H_3O^+(aq) \longrightarrow Pb^{2+}(aq) + H_2(g) + 2\,H_2O(\ell)$$

(a) Calculate $\Delta\mathscr{E}°$ for this cell.
(b) Chloride ions are added until $PbCl_2$ precipitates at the anode and $[Cl^-]$ reaches 0.15 M. The cell voltage is then measured to be 0.22 V at pH = 0 and $P_{H_2} = 1.0$ atm. Calculate $[Pb^{2+}]$ under these conditions.
(c) Calculate the solubility product constant K_{sp} of $PbCl_2$.

Batteries and Fuel Cells

43. Calculate the voltage $\Delta\mathscr{E}°$ of a lead–acid cell if all reactants and products are in their standard states. What will be the voltage if six such cells are connected in series?

44. Calculate the standard voltage of the zinc–mercuric oxide cell shown in Figure 17.7. (*Hint:* The easiest way to proceed is to calculate $\Delta G°$ for the corresponding overall reaction, and then find $\Delta\mathscr{E}°$ from it.) Take $\Delta_f^\circ(Zn(OH)_2(s)) = -553.5$ kJ mol^{-1}.

45. (a) What quantity of charge (in coulombs) is a fully charged 12-V lead–acid storage battery theoretically capable of furnishing if the spongy lead available for reaction at the anodes weighs 10 kg and there is excess PbO_2?

(b) What is the theoretical maximum amount of work (in joules) that can be obtained from this battery?

46. (a) What quantity of charge (in coulombs) is a fully charged 1.34-V zinc–mercuric oxide watch battery theoretically capable of furnishing if the mass of HgO in the battery is 0.50 g?

(b) What is the theoretical maximum amount of work (in joules) that can be obtained from this battery?

47. The concentration of the electrolyte, sulfuric acid, in a lead–acid storage battery diminishes as the battery is discharged. Is a discharged battery recharged by replacing the dilute H_2SO_4 with fresh, concentrated H_2SO_4? Explain.

48. One cold winter morning the temperature is well below 0°F. In trying to start your car, you run the battery down completely. Several hours later, you return to replace your fouled spark plugs and find that the liquid in the battery has now frozen even though the air temperature is actually a bit higher than it was in the morning. Explain how this can happen.

49. Consider the fuel cell that accomplishes the overall reaction

$$H_2(g) + \tfrac{1}{2} O_2(g) \longrightarrow H_2O(\ell)$$

If the fuel cell operates with 60% efficiency, calculate the amount of electrical work generated per gram of water produced. The gas pressures are constant at 1 atm, and the temperature is 25°C.

50. Consider the fuel cell that accomplishes the overall reaction

$$CO(g) + \tfrac{1}{2} O_2(g) \longrightarrow CO_2(g)$$

Calculate the maximum electrical work that could be obtained from the conversion of 1.00 mol of $CO(g)$ to $CO_2(g)$ in such a fuel cell operated with 100% efficiency at 25°C and with the pressure of each gas equal to 1 atm.

Corrosion and Its Prevention

51. Two half-reactions proposed for the corrosion of iron in the absence of oxygen are

$$Fe(s) \longrightarrow Fe^{2+}(aq) + 2\,e^-$$

$$2\,H_2O(\ell) + 2\,e^- \longrightarrow 2\,OH^-(aq) + H_2(g)$$

Calculate the standard voltage generated by a galvanic cell running this pair of half-reactions. Is the overall reaction spontaneous under standard conditions? As the pH falls from 14, will the reaction become spontaneous?

52. In the presence of oxygen, the cathode half-reaction written in the preceding problem is replaced by

$$\tfrac{1}{2} O_2(g) + 2\,H_3O^+(aq) + 2\,e^- \longrightarrow 3\,H_2O(\ell)$$

but the anode half-reaction is unchanged. Calculate the standard cell voltage for *this* pair of reactions operating as a galvanic cell. Is the overall reaction spontaneous under standard conditions? As the water becomes more acidic, does the driving force for the rusting of iron increase or decrease?

53. Could sodium be used as a sacrificial anode to protect the iron hull of a ship?

54. If it is shown that titanium can be used as a sacrificial anode to protect iron, what conclusion can be drawn about the standard reduction potential of its half-reaction?

$$Ti^{3+}(aq) + 3\,e^- \longrightarrow Ti(s)$$

Electrometallurgy

55. In the Downs process, molten sodium chloride is electrolyzed to produce sodium. A valuable byproduct is chlorine. Write equations representing the processes taking place at the anode and at the cathode in the Downs process.

56. The first element to be prepared by electrolysis was potassium. In 1807, Humphry Davy, then 29 years old, passed an electric current through molten potassium hydroxide (KOH), obtaining liquid potassium at one electrode and water and oxygen at the other. Write equations to represent the processes taking place at the anode and at the cathode.

57. A current of 55,000 A is passed through a series of 100 Hall–Héroult cells for 24 hours. Calculate the maximum theoretical mass of aluminum that can be recovered.

58. A current of 75,000 A is passed through an electrolysis cell containing molten $MgCl_2$ for 7.0 days. Calculate the maximum theoretical mass of magnesium that can be recovered.

59. An important use for magnesium is to make titanium. In the Kroll process, magnesium reduces titanium(IV) chloride to elemental titanium in a sealed vessel at 800°C. Write a balanced chemical equation for this reaction. What mass of magnesium is needed, in theory, to produce 100 kg of titanium from titanium(IV) chloride?

60. Calcium is used to reduce vanadium(V) oxide to elemental vanadium in a sealed steel vessel. Vanadium is used in vanadium steel alloys for jet engines, high-quality knives, and tools. Write a balanced chemical equation for this process. What mass of calcium is needed, in theory, to produce 20.0 kg of vanadium from vanadium(V) oxide?

61. Galvanized steel consists of steel with a thin coating of zinc to reduce corrosion. The zinc can be deposited electrolytically, by making the steel object the cathode and a block of zinc the anode in an electrochemical cell containing a dissolved zinc salt. Suppose a steel garbage can is to be galvanized and requires that a total mass of 7.32 g of zinc be coated to the required thickness. How long should a current of 8.50 A be passed through the cell to achieve this?

62. In the electroplating of a silver spoon, the spoon acts as the cathode and a piece of pure silver as the anode. Both dip into a solution of silver cyanide (AgCN). Suppose that a current of 1.5 A is passed through such a cell for 22 minutes and that the spoon has a surface area of 16 cm^2. Calculate the average thickness of the silver layer deposited on the spoon, taking the density of silver to be 10.5 g cm^{-3}.

A DEEPER LOOK . . . Electrolysis of Water and Aqueous Solutions

63. An electrolytic cell consists of a pair of inert metallic electrodes in a solution buffered to pH = 5.0 and containing nickel sulfate ($NiSO_4$) at a concentration of 1.00 M. A current of 2.00 A is passed through the cell for 10.0 hours.

(a) What product is formed at the cathode?

(b) What is the mass of this product?

(c) If the pH is changed to pH = 1.0, what product will form at the cathode?

64. A 0.100 M neutral aqueous $CaCl_2$ solution is electrolyzed using platinum electrodes. A current of 1.50 A passes through the solution for 50.0 hours.
   (a) Write the half-reactions occurring at the anode and at the cathode.
   (b) What is the decomposition potential?
   (c) Calculate the mass, in grams, of the product formed at the cathode.

ADDITIONAL PROBLEMS

65. The drain cleaner Drano consists of aluminum turnings mixed with sodium hydroxide. When it is added to water, the sodium hydroxide dissolves and releases heat. The aluminum reacts with water to generate bubbles of hydrogen and aqueous ions. Write a balanced net ionic equation for this reaction.

66. Sulfur-containing compounds in the air tarnish silver, giving black Ag_2S. A practical method of cleaning tarnished silverware is to place the tarnished item in electrical contact with a piece of zinc and dip both into water containing a small amount of salt. Write balanced half-equations to represent what takes place.

67. A current passed through inert electrodes immersed in an aqueous solution of sodium chloride produces chlorate ion, $ClO_3^-(aq)$, at the anode and gaseous hydrogen at the cathode. Given this fact, write a balanced equation for the chemical reaction if gaseous hydrogen and aqueous sodium chlorate are mixed and allowed to react spontaneously until they reach equilibrium.

68. A galvanic cell is constructed by linking a $Co^{2+}|Co(s)$ half-cell to an $Ag^+|Ag(s)$ half-cell through a salt bridge and then connecting the cobalt and silver electrodes through an external circuit. When the circuit is closed, the cell voltage is measured to be 1.08 V and silver is seen to plate out while cobalt dissolves.
   (a) Write the half-reactions that occur at the anode and at the cathode and the balanced overall cell reaction.
   (b) The cobalt electrode is weighed after 150 minutes of operation and is found to have decreased in mass by 0.36 g. By what amount has the silver electrode increased in mass?
   (c) What is the average current drawn from the cell during this period?

69. The galvanic cell $Zn(s)|Zn^{2+}(aq)||Ni^{2+}(aq)|Ni(s)$ is constructed using a completely immersed zinc electrode that weighs 32.68 g and a nickel electrode immersed in 575 mL of 1.00 M $Ni^{2+}(aq)$ solution. A steady current of 0.0715 A is drawn from the cell as the electrons move from the zinc electrode to the nickel electrode.
   (a) Which reactant is the limiting reactant in this cell?
   (b) How long does it take for the cell to be completely discharged?
   (c) How much mass has the nickel electrode gained when the cell is completely discharged?
   (d) What is the concentration of the $Ni^{2+}(aq)$ when the cell is completely discharged?

70. A newly discovered bacterium can reduce selenate ion, $SeO_4^{2-}(aq)$, to elemental selenium, Se(*s*), in reservoirs. This is significant because the soluble selenate ion is potentially toxic, but elemental selenium is insoluble and harmless. Assume that water is oxidized to oxygen as the selenate ion is reduced. Compute the mass of oxygen produced if all the selenate in a 10^{12}-L reservoir contaminated with 100 mg L^{-1} of selenate ion is reduced to selenium.

71. Thomas Edison invented an electric meter that was nothing more than a simple coulometer, a device to measure the amount of electricity passing through a circuit. In this meter, a small, fixed fraction of the total current supplied to a household was passed through an electrolytic cell, plating out zinc at the cathode. Each month the cathode could then be removed and weighed to determine the amount of electricity used. If 0.25% of a household's electricity passed through such a coulometer and the cathode increased in mass by 1.83 g in a month, how many coulombs of electricity were used during that month?

72. The chief chemist of the Brite-Metal Electroplating Co. is required to certify that the rinse solutions that are discharged from the company's tin-plating process into the municipal sewer system contain no more than 10 ppm (parts per million) by mass of Sn^{2+}. The chemist devises the following analytical procedure to determine the concentration. At regular intervals, a 100-mL (100-g) sample is withdrawn from the waste stream and acidified to pH = 1.0. A starch solution and 10 mL of 0.10 M potassium iodide are added, and a 25.0-mA current is passed through the solution between platinum electrodes. Iodine appears as a product of electrolysis at the anode when the oxidation of Sn^{2+} to Sn^{4+} is practically complete and signals its presence with the deep blue color of a complex formed with starch. What is the maximum duration of electrolysis to the appearance of the blue color that ensures that the concentration of Sn^{2+} does not exceed 10 ppm?

73. Estimate the cost of the electrical energy needed to produce 1.5×10^{10} kg (a year's supply for the world) of aluminum from $Al_2O_3(s)$ if electrical energy costs 10 cents per kilowatt-hour (1 kWh = 3.6 MJ = 3.6×10^6 J) and if the cell voltage is 5 V.

74. Titanium can be produced by electrolytic reduction from an anhydrous molten salt electrolyte that contains titanium(IV) chloride and a spectator salt that furnishes ions to make the electrolyte conduct electricity. The standard enthalpy of formation of $TiCl_4(\ell)$ is -750 kJ mol^{-1}, and the standard entropies of $TiCl_4(\ell)$, Ti(*s*), and $Cl_2(g)$ are 253, 30, and 223 J K^{-1} mol^{-1}, respectively. What minimum applied voltage will be necessary at 100°C?

75. A half-cell has a graphite electrode immersed in an acidic solution (pH 0) of Mn^{2+} (concentration 1.00 M) in contact with solid MnO_2. A second half-cell has an acidic solution (pH 0) of H_2O_2 (concentration 1.00 M) in contact with a platinum electrode past which gaseous oxygen at a pressure of 1.00 atm is bubbled. The two half-cells are connected to form a galvanic cell.
   (a) By referring to Appendix E, write balanced chemical equations for the half-reactions at the anode and the cathode and for the overall cell reaction.
   (b) Calculate the cell voltage.

76. By considering these half-reactions and their standard reduction potentials,

$$Pt^{2} + 2\,e^- \longrightarrow Pt \qquad \mathscr{E}° = 1.2\ V$$

$$NO_3^- + 4\,H_3O^+ + 3\,e^- \longrightarrow NO + 6\,H_2O \qquad \mathscr{E}° = 0.96\ V$$

$$PtCl_4^{2-} + 2\,e^- \longrightarrow Pt + 4\,Cl^- \qquad \mathscr{E}° = 0.73\ V$$

account for the fact that platinum will dissolve in a mixture of hydrochloric acid and nitric acid (*aqua regia*) but will not dissolve in either acid alone.

77. (a) One method to reduce the concentration of unwanted $Fe^{3+}(aq)$ in a solution of $Fe^{2+}(aq)$ is to drop a piece of metallic iron into the storage container. Write the reaction that removes the Fe^{3+}, and compute its standard cell potential.
(b) By referring to problem 23, suggest a way to remove unwanted $Mn^{3+}(aq)$ from solutions of $Mn^{2+}(aq)$.

78. (a) Based only on the standard reduction potentials for the $Cu^{2+}|Cu^{+}$ and the $I_2(s)|I^-$ half-reactions, would you expect $Cu^{2+}(aq)$ to be reduced to $Cu^{+}(aq)$ by $I^-(aq)$?
(b) The formation of solid CuI plays a role in the interaction between $Cu^{2+}(aq)$ and $I^-(aq)$.

$$Cu^{2+}(aq) + I^-(aq) + e^- \rightleftharpoons CuI(s) \qquad \mathscr{E}° = 0.86 \text{ V}$$

Taking into account this added information, do you expect Cu^{2+} to be reduced by iodide ion?

79. In some old European churches, the stained-glass windows have so darkened from corrosion and age that hardly any light comes through. Microprobe analysis showed that tiny cracks and defects on the glass surface were enriched in insoluble Mn(III) and Mn(IV) compounds. From Appendix E, suggest a reducing agent and conditions that might successfully convert these compounds to soluble Mn(II) without simultaneously reducing Fe(III) (which gives the glass its colors) to Fe(II). Take MnO_2 as representative of the insoluble Mn(III) and Mn(IV) compounds.

80. (a) Calculate the half-cell potential for the reaction

$$O_2(g) + 4\,H_3O^+(aq) + 4\,e^- \longrightarrow 6\,H_2O(\ell)$$

at pH 7 with the oxygen pressure at 1 atm.
(b) Explain why aeration of solutions of I^- leads to their decomposition. Write a balanced equation for the redox reaction that occurs.
(c) Will the same problem arise with solutions containing Br^- or Cl^-? Explain.
(d) Will decomposition be favored or opposed by increasing acidity?

81. An engineer needs to prepare a galvanic cell that uses the reaction

$$2\,Ag^+(aq) + Zn(s) \longrightarrow Zn^{2+}(aq) + 2\,Ag(s)$$

and generates an initial voltage of 1.50 V. She has 0.010 M $AgNO_3(aq)$ and 0.100 M $Zn(NO_3)_2(aq)$ solutions, as well as electrodes of metallic copper and silver, wires, containers, water, and a KNO_3 salt bridge. Sketch the cell. Clearly indicate the concentrations of all solutions.

82. Consider a galvanic cell for which the anode reaction is

$$Pb(s) \longrightarrow Pb^{2+}(1.0 \times 10^{-2}\text{ M}) + 2\,e^-$$

and the cathode reaction is

$$VO^{2+}(0.10\text{ M}) + 2\,H_3O^+(0.10\text{ M}) + e^- \longrightarrow V^{3+}(1.0 \times 10^{-5}\text{ M}) + 3\,H_2O(\ell)$$

The measured cell voltage is 0.640 V.
(a) Calculate $\mathscr{E}°$ for the $VO^{2+}|V^{3+}$ half-reaction, using $\mathscr{E}°(Pb^{2+}|Pb)$ from Appendix E.
(b) Calculate the equilibrium constant (K) at 25°C for the reaction

$$Pb(s) + 2\,VO^{2+}(aq) + 4\,H_3O^+(aq) \rightleftharpoons Pb^{2+}(aq) + 2\,V^{3+}(aq) + 6\,H_2O(\ell)$$

83. Suppose we construct a pressure cell in which the gas pressures differ in the two half-cells. Suppose such a cell consists of a $Cl_2(0.010\text{ atm})|Cl^-(1\text{ M})$ half-cell connected to a $Cl_2(0.50\text{ atm})|Cl^-(1\text{ M})$ half-cell. Determine which half-cell will be the anode, write the overall equation for the reaction, and calculate the cell voltage.

84. A student decides to measure the solubility of lead sulfate in water and sets up the electrochemical cell

$$Pb|PbSO_4|SO_4^{2-}(aq, 0.0500\text{ M})||Cl^-(aq, 1.00\text{ M})|AgCl|Ag$$

At 25°C the student finds the cell voltage to be 0.546 V, and from Appendix E the student finds

$$AgCl(s) + e \rightleftharpoons Ag(s) + Cl^-(aq) \qquad \mathscr{E}° = 0.222 \text{ V}$$

What does he find for the K_{sp} of $PbSO_4$?

85. A wire is fastened across the terminals of the Leclanché cell in Figure 17.6. Indicate the direction of electron flow in the wire.

86. Overcharging a lead–acid storage battery can generate hydrogen. Write a balanced equation to represent the reaction taking place.

87. An ambitious chemist discovers an alloy electrode that is capable of catalytically converting ethanol reversibly to carbon dioxide at 25°C according to the half-reaction

$$C_2H_5OH(\ell) + 15\,H_2O(\ell) \longrightarrow 2\,CO_2(g) + 12\,H_3O^+(aq) + 12\,e^-$$

Believing that this discovery is financially important, the chemist patents its composition and designs a fuel cell that may be represented as

$$\text{Alloy}|C_2H_5OH(\ell)|CO_2(g) + H_3O^+(1\text{ M})||H_3O^+(1\text{ M})|O_2|Ni$$

(a) Write the half-reaction occurring at the cathode.
(b) Using data from Appendix D, calculate $\Delta\mathscr{E}°$ for the cell at 25°C.
(c) What is the $\mathscr{E}°$ value for the ethanol half-cell?

88. Iron or steel is often covered by a thin layer of a second metal to prevent rusting: Tin cans consist of steel covered with tin, and galvanized iron is made by coating iron with a layer of zinc. If the protective layer is broken, however, iron will rust more readily in a tin can than in galvanized iron. Explain this observation by comparing the half-cell potentials of iron, tin, and zinc.

89. An electrolysis cell contains a solution of 0.10 M $NiSO_4$. The anode and cathode are both strips of Pt foil. Another electrolysis cell contains the same solution, but the electrodes are strips of Ni foil. In each case a current of 0.10 A flows through the cell for 10 hours.
(a) Write a balanced equation for the chemical reaction that occurs at the anode in each cell.
(b) Calculate the mass, in grams, of the product formed at the anode in each cell. (The product may be a gas, a solid, or an ionic species in solution.)

90. A potential difference of 2.0 V is impressed across a pair of inert electrodes (e.g., platinum) that are immersed in a 0.050 M aqueous KBr solution. What are the products that form at the anode and the cathode?

91. An aqueous solution is simultaneously 0.10 M in $SnCl_2$ and in $CoCl_2$.
(a) If the solution is electrolyzed, which metal will appear first?
(b) At what decomposition potential will that metal first appear?
(c) As the electrolysis proceeds, the concentration of the metal being reduced will drop and the voltage will change. How complete a separation of the metals using electrolysis is theoretically possible? In other words, at the point where the second metal begins to form, what fraction of the first metal is left in solution?

92. A 55.5-kg slab of crude copper from a smelter has a copper content of 98.3%. Estimate the time required to purify it electrochemically if it is used as the anode in a cell that has acidic copper(II) sulfate as its electrolyte and a current of 2.00×10^3 A is passed through the cell.

93. Sheet iron can be galvanized by passing a direct current through a cell containing a solution of zinc sulfate between a graphite anode and the iron sheet. Zinc plates out on the iron. The process can be made continuous if the iron sheet is a coil that unwinds as it passes through the electrolysis cell and coils up again after it emerges from a rinse bath. Calculate the cost of the electricity required to deposit a 0.250-mm-thick layer of zinc on both sides of an iron sheet that is 1.00 m wide and 100 m long, if a current of 25 A at a voltage of 3.5 V is used and the energy efficiency of the process is 90%. The cost of electricity is 10 cents per kilowatt–hour (1 kWh = 3.6 MJ). Consult Appendix F for data on zinc.

CUMULATIVE PROBLEMS

94. A 1.0 M solution of NaOH is electrolyzed, generating $O_2(g)$ at the anode. A current of 0.15 A is passed through the cell for 75 minutes. Calculate the volume of (wet) oxygen generated in this period if the temperature is held at 25°C and the total pressure is 0.985 atm. (*Hint:* Use the vapor pressure of water at this temperature from Table 10.1.)

95. Use standard entropies from Appendix D to predict whether the standard voltage of the $Cu|Cu^{2+}||Ag^{+}|Ag$ cell (diagrammed in Figure 17.2) will increase or decrease if the temperature is raised above 25°C.

96. About 50,000 kJ of electrical energy is required to produce 1.0 kg of Al from its $Al(OH)_3$ ore. The major energy cost in recycling aluminum cans is the melting of the aluminum. The enthalpy of fusion of Al(*s*) is 10.7 kJ mol^{-1}. Compare the energy cost for making new aluminum with that for recycling.

97. (a) Use the following half-reactions and their reduction potentials to calculate the K_{sp} of AgBr:

$$Ag^{+}(aq) + e^{-} \longrightarrow Ag(s) \qquad \mathscr{E}° = 0.7996 \text{ V}$$

$$AgBr(s) + e^{-} \longrightarrow Ag(s) + Br^{-}(aq) \qquad \mathscr{E}° = 0.0713 \text{ V}$$

(b) Estimate the solubility of AgBr in 0.10 M NaBr(*aq*).

98. Amounts of iodine dissolved in aqueous solution, $I_2(aq)$, can be determined by titration with thiosulfate ion ($S_2O_3^{2-}$). The thiosulfate ion is oxidized to $S_4O_6^{2-}$ while the iodine is reduced to iodide ion. Starch is used as an indicator because it has a strong blue color in the presence of dissolved iodine.
(a) Write a balanced equation for this reaction.
(b) If 56.40 mL of 0.100 M $S_2O_3^{-}$ solution is used to reach the endpoint of a titration of an unknown amount of iodine, calculate the number of moles of iodine originally present.
(c) Combine the appropriate half-cell potentials from Appendix E with thermodynamic data from Appendix D for the equilibrium

$$I_2(s) \rightleftharpoons I_2(aq)$$

to calculate the equilibrium constant at 25°C for the reaction in part (a).

UNIT V

Rates of Chemical and Physical Processes

Zinc reacts readily with aqueous hydrochloric acid to give hydrogen gas and aqueous zinc chloride.

Thermodynamics explains *why* chemical reactions occur; minimizing the Gibbs free energy is the driving force toward chemical equilibrium. Thermodynamics provides deep insight into the nature of chemical equilibrium, but it gives no answer to the crucial question of how rapidly that equilibrium is achieved.

Chemical kinetics explains *how* reactions occur by studying their rates and mechanisms. Chemical kinetics explains how the speeds of different chemical reactions vary from explosive rapidity to glacial sluggishness and how slow reactions can be accelerated by materials called catalysts. Chemical kinetics has enormous practical importance because it provides the basis for optimizing conditions to carry out chemical reactions at reasonable speed, under proper control.

The central goal in chemical kinetics is to find the relationship between the rate of a reaction and the amount of reactants present. Once this connection is established, the influence of external conditions—principally the temperature—can be explored.

Nuclear chemistry represents a particularly simple limiting form of kinetics in which unstable nuclei decay with a constant probability during any time interval. Its richness arises from the multiplicity of decay paths that are possible, which arise from the mass-energy relationships that determine nuclear stability.

Molecules absorb and emit light at specific wavelengths determined by the separations between molecular energy levels. Through quantum mechanics, molecular spectroscopy connects these specific wavelengths to models of molecular structure, and so determines the structure of molecules from their interaction with light. Light can induce chemical reactions through pathways not available by thermal excitation. Both spectroscopy and photochemistry rely on tools of chemical kinetics to describe rates of light-induced processes.

UNIT CHAPTERS

UNIT GOALS

- To relate the rate of a chemical reaction to the instantaneous concentration(s) of reactants by developing the rate law and the rate constant for the reaction
- To describe the influence of temperature on the reaction rate by identifying the activation energy for the reaction
- To explain the mechanism of a complex reaction by identifying the separate elementary reaction steps through which it proceeds
- To explain the role of catalysts in manipulating reaction rates
- To develop an elementary description of the rates of nuclear reactions, emphasizing the half-life of radioactive species
- To survey the applications and consequences of nuclear reactions in medicine, biology, energy production, and the environment
- To develop methods of molecular spectroscopy for structure determination
- To relate the initiation of photochemical reactions to the wavelength of light, and survey consequences of photochemical reactions in the atmosphere

CHAPTER 18

Chemical Kinetics

Powdered chalk (mostly calcium carbonate $CaCO_3$) reacts rapidly with dilute hydrochloric acid because it has a large total surface area. A stick of chalk has a much smaller surface area, so it reacts much more slowly.

Why do some chemical reactions proceed with lightning speed when others require days, months, or even years to produce detectable amounts of products? How do catalysts increase the rates of chemical reactions? Why do small changes in temperature often have such large effects on the cooking rate of food? How does a study of the rate of a chemical reaction inform us about the way in which molecules combine to form products? All of these questions involve studies of reaction rates, which is the subject of chemical kinetics.

Chemical kinetics is a complex subject, and at present it is not understood nearly as well as chemical thermodynamics. For many reactions the equilibrium constants are known accurately, but the rates and detailed reaction pathways remain poorly understood. This is particularly true of reactions in which many species participate in the overall process connecting reactants to products. One good example is the reaction

$$5\ Fe^{2+}(aq) + MnO_4^-(aq) + 8\ H_3O^+(aq) \longrightarrow 5\ Fe^{3+}(aq) + Mn^{2+}(aq) + 12\ H_2O(\ell)$$

We can measure the equilibrium constant for this reaction easily from the voltage of a galvanic cell, and then calculate from it the equilibrium concentrations that will result from arbitrary initial conditions. It is considerably harder to determine the exact pathway by which the reaction goes from reactants to products. This path certainly does *not* involve the simultaneous collision of five Fe^{2+} ions and one MnO_4^- ion with eight H_3O^+ ions, because such a collision would be exceedingly rare. Instead, the path proceeds through a series of elementary steps involving two or at most three ions, such as

$$Fe^{2+}(aq) + MnO_4^-(aq) \longrightarrow Fe^{3+}(aq) + MnO_4^{2-}(aq)$$

$$MnO_4^{2-}(aq) + H_3O^+(aq) \longrightarrow HMnO_4^-(aq) + H_2O(\ell)$$

$$HMnO_4^-(aq) + Fe^{2+}(aq) \longrightarrow HMnO_4^{2-}(aq) + Fe^{3+}(aq)$$

Other postulated steps take the process to its final products. Some of these steps are slow, and others fast; taken together they constitute the **reaction mechanism.**

The primary goal of chemical kinetics is to deduce the mechanism of a reaction from experimental studies of its rate. For this, we have to measure how the rate depends on the concentrations of the reacting species. This chapter lays out the methods and concepts for measuring and interpreting reaction rates and for identifying the mechanism.

18.1 Rates of Chemical Reactions

The speed of a reaction depends on many factors. Concentrations of reacting species certainly play a major role in speeding up or slowing down a particular reaction (Fig. 18.1). As we will see in Section 18.5, many reaction rates are extremely sensitive to temperature. This means that careful control of temperature is critical

FIGURE 18.1 The rate of reaction of zinc with aqueous sulfuric acid depends on the concentration of the acid. The dilute solution reacts slowly (left), and the more concentrated solution reacts rapidly (right).

FIGURE 18.2 Steel wool burning in oxygen.

for quantitative measurements in chemical kinetics. Finally, the rate often depends crucially on the physical forms of the reactants. An iron nail oxidizes only very slowly in dry air to iron oxide, but steel wool burns spectacularly in oxygen (Fig. 18.2). Because the quantitative study of heterogeneous reactions—those involving two or more phases, such as a solid and a gas—is difficult, we begin with homogeneous reactions, which take place entirely within the gas phase or solution. In Section 18.7 we turn briefly to some important aspects of heterogeneous reactions.

Measuring Reaction Rates

A kinetics experiment measures the rate of change of the concentration of a substance participating in a chemical reaction. How can we experimentally monitor a changing concentration? If the reaction is slow enough, we can let it run for a measured time and then abruptly "quench" (effectively stop) it by rapidly cooling the reaction mixture sufficiently. At that low temperature the composition of the reaction mixture remains constant, so we have time to analyze the mixture for some particular reactant or product. This procedure is not useful for rapid reactions, especially those involving gas mixtures, because they are difficult to cool quickly. An alternative is to probe the concentrations by the absorption of light. Chapter 20 shows that different molecules absorb at different wavelengths. If a wavelength is absorbed by only one particular reactant or product, measuring the amount of light absorbed by the reaction mixture at that wavelength determines the concentration of the absorbing species. A series of such measurements at different times reveals the rate of change of the concentration. Often, a flash of light can also be used to initiate a very fast reaction, whose rate is then tracked by measuring absorption at a particular wavelength.

The average rate of a reaction is analogous to the average speed of a car. If the average position of a car is recorded at two different times, then

$$\text{average speed} = \frac{\text{distance traveled}}{\text{time elapsed}} = \frac{\text{change in location}}{\text{change in time}}$$

In the same way, the **average reaction rate** is obtained by dividing the change in concentration of a reactant or product by the time interval over which that change occurs:

$$\text{average reaction rate} = \frac{\text{change in concentration}}{\text{change in time}}$$

If concentration is measured in mol L^{-1} and time in seconds, then the rate of a reaction has units of mol $L^{-1}\ s^{-1}$.

Consider a specific example. In the gas-phase reaction

$$NO_2(g) + CO(g) \longrightarrow NO(g) + CO_2(g)$$

NO_2 and CO are consumed as NO and CO_2 are produced. If a probe can measure the NO concentration, the average rate of reaction can be estimated from the ratio of the change in NO concentration $\Delta[NO]$ to the time interval Δt:

$$\text{average rate} = \frac{\Delta[NO]}{\Delta t} = \frac{[NO]_f - [NO]_i}{t_f - t_i}$$

This estimate depends on the time interval Δt that is selected, because the rate at which NO is produced changes with time. From the data in Figure 18.3, the average rate of reaction during the first 50 s is

$$\text{average rate} = \frac{\Delta[NO]}{\Delta t} = \frac{(0.0160 - 0)\text{ mol L}^{-1}}{(50 - 0)\text{ s}} = 3.2 \times 10^{-4}\text{ mol L}^{-1}\text{ s}^{-1}$$

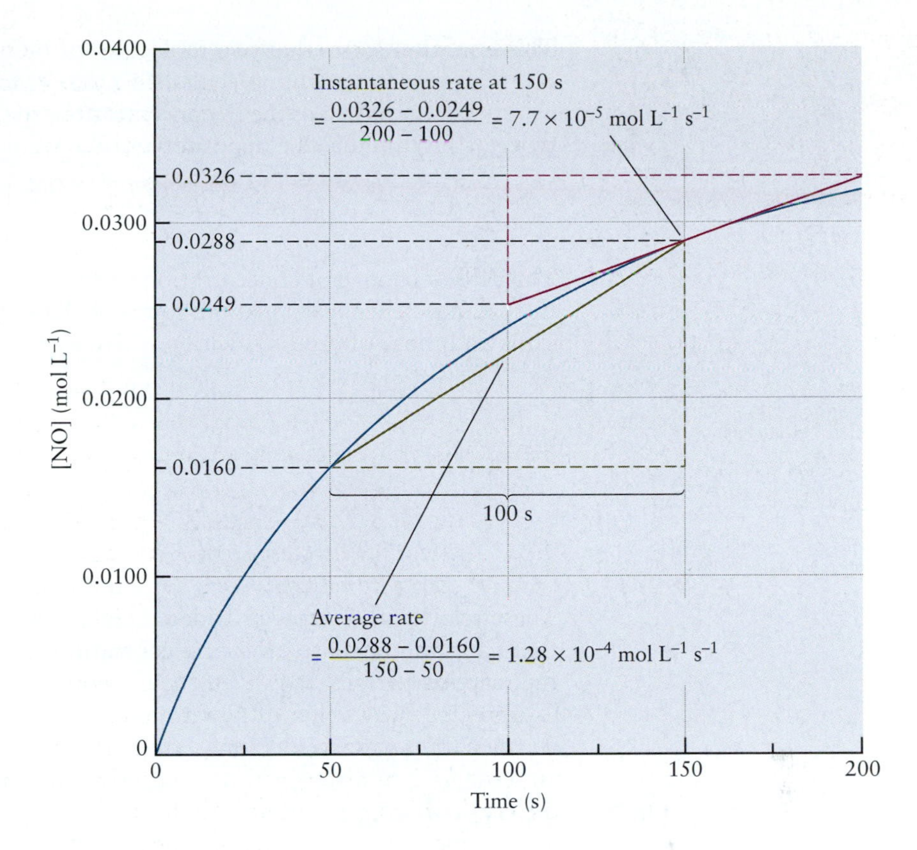

FIGURE 18.3 A graph of the concentration of NO against time in the reaction $NO_2 + CO \longrightarrow NO + CO_2$. The average rate during the time interval from 50 to 150 seconds is obtained by dividing the change in NO concentration by the duration of the interval (green box). Note that the average rate (green line) underestimates the true production rate over the time interval. The instantaneous rate 150 seconds after the start of the reaction is found by calculating the slope of the line tangent to the curve at that point (red box).

During the second 50 s, the average rate is 1.6×10^{-4} mol L^{-1} s^{-1}, and during the third 50 s it is 9.6×10^{-5} mol L^{-1} s^{-1}. Clearly, this reaction slows as it progresses, and its average rate indeed depends on the time interval chosen. Figure 18.3 shows a graphical method for determining average rates. The average rate is the slope of the straight line connecting the concentrations at the initial and final points of a time interval.

The **instantaneous rate** of a reaction is obtained by considering smaller and smaller time increments Δt (with correspondingly smaller values of Δ[NO]). As Δt approaches 0, the rate becomes the slope of the line tangent to the curve at time t (see Fig. 18.3). This slope is written as the derivative of [NO] with respect to time:

$$\text{instantaneous rate} = \text{limit}_{\Delta t \to 0} \frac{[\text{NO}]_{t+\Delta t} - [\text{NO}]_t}{\Delta t} = \frac{d[\text{NO}]}{dt}$$

Throughout the rest of this book, we refer to the instantaneous rate simply as the *rate*. The instantaneous rate of a reaction at the moment that it begins (at $t = 0$) is the **initial rate** of that reaction.

The rate of this sample reaction could just as well have been measured by monitoring changes in the concentration of CO_2, NO_2, or CO instead of NO. Because every molecule of NO produced is accompanied by one molecule of CO_2, the rate of increase of CO_2 concentration is the same as that of NO. The concentrations of the two reactants, NO_2 and CO, *decrease* at the same rate that the concentrations of the products increase, because the coefficients in the balanced equation are also both equal to 1. This is summarized as

$$\text{rate} = -\frac{d[\text{NO}_2]}{dt} = -\frac{d[\text{CO}]}{dt} = \frac{d[\text{NO}]}{dt} = \frac{d[\text{CO}_2]}{dt}$$

Another gas-phase reaction is

$$2\ \text{NO}_2(g) + \text{F}_2(g) \longrightarrow 2\ \text{NO}_2\text{F}(g)$$

This equation states that two molecules of NO_2 disappear and two molecules of NO_2F appear for each molecule of F_2 that reacts. Thus, the NO_2 concentration changes twice as fast as the F_2 concentration; the NO_2F concentration also changes twice as fast and has the opposite sign. We write the rate in this case as

$$\text{rate} = -\frac{1}{2}\frac{d[NO_2]}{dt} = -\frac{d[F_2]}{dt} = \frac{1}{2}\frac{d[NO_2F]}{dt}$$

The rate of change of concentration of each species is divided by its coefficient in the balanced chemical equation. Rates of change of reactants appear with negative signs and those of products with positive signs. For the general reaction

$$aA + bB \longrightarrow cC + dD$$

the rate is

$$\text{rate} = -\frac{1}{a}\frac{d[A]}{dt} = -\frac{1}{b}\frac{d[B]}{dt} = \frac{1}{c}\frac{d[C]}{dt} = \frac{1}{d}\frac{d[D]}{dt} \quad [18.1]$$

These relations hold true provided there are no transient intermediate species or, if there are intermediates, their concentrations are independent of time for most of the reaction period.

18.2 Rate Laws

In discussing chemical equilibrium we stressed that both forward and reverse reactions can occur; once products are formed, they can react back to give the original reactants. The net rate is the difference:

$$\text{net rate} = \text{forward rate} - \text{reverse rate}$$

Strictly speaking, measurements of concentration give the net rate rather than simply the forward rate. Near the beginning of a reaction that starts from pure reactants the concentrations of reactants are far higher than those of products, and the reverse rate can be neglected. In addition, many reactions go to "completion" ($K \gg 1$). This means they have a measurable rate only in the forward direction, or else the experiment can be arranged so that the products are removed as they are formed. This section focuses on forward rates exclusively.

Order of a Reaction

The forward rate of a chemical reaction depends on the concentrations of the reactants. As an example, consider the decomposition of gaseous dinitrogen pentaoxide (N_2O_5). This compound is a white solid that is stable below 0°C but decomposes when vaporized:

$$N_2O_5(g) \longrightarrow 2\ NO_2(g) + \tfrac{1}{2}\ O_2(g)$$

The rate of the reaction depends on the concentration of $N_2O_5(g)$. Figure 18.4 shows the graph of rate versus concentration to be a straight line that can be extrapolated to pass through the origin. So, the rate can be written

$$\text{rate} = k[N_2O_5]$$

This relation between the rate of a reaction and concentration is called a **empirical rate expression** or **rate law,** and the proportionality constant k is called the **rate constant** for the reaction. Like an equilibrium constant, a rate constant is independent of concentration but depends on temperature, as we describe in Section 18.5.

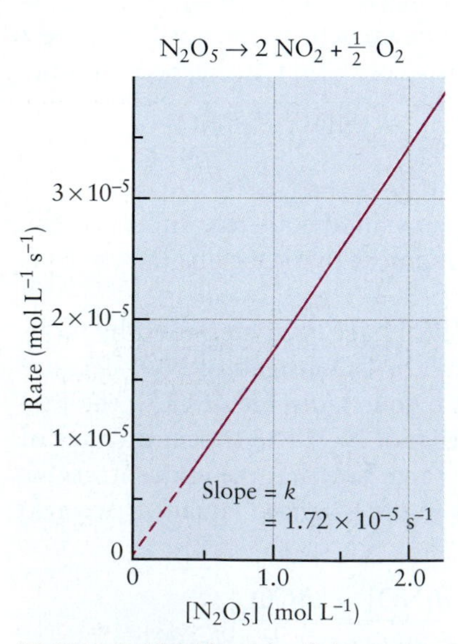

FIGURE 18.4 The rate of decomposition of $N_2O_5(g)$ at 25°C is proportional to its concentration. The slope of this line is equal to the rate constant k for the reaction.

For many (but not all) reactions with a single reactant, the rate is proportional to the concentration of that reactant raised to a power. That is, the rate expression for

$$aA \longrightarrow \text{products}$$

frequently has the form

$$\text{rate} = k[A]^n$$

It is important to note that the power n in the rate expression has no direct relation to the coefficient a in the balanced chemical equation. This number has to be determined experimentally for each rate law. For the decomposition of ethane at high temperatures and low pressures,

$$C_2H_6(g) \longrightarrow 2\ CH_3(g)$$

the rate expression has the form

$$\text{rate} = k[C_2H_6]^2$$

Therefore, $n = 2$ even though the coefficient in the chemical equation is 1.

The power to which the concentration is raised is called the **order** of the reaction with respect to that reactant. Thus, the decomposition of N_2O_5 is **first order,** whereas that of C_2H_6 is **second order.** Some processes are **zeroth order** over a range of concentrations. Because $[A]^0 = 1$, such reactions have rates that are independent of concentration:

$$\text{rate} = k \qquad \text{(for zeroth-order kinetics)}$$

The order of a reaction does not have to be an integer; fractional powers are sometimes found. At 450 K, the decomposition of acetaldehyde (CH_3CHO) is described by the rate expression

$$\text{rate} = k[CH_3CHO]^{3/2}$$

These examples demonstrate that reaction order is an experimentally determined property that cannot be predicted from the form of the chemical equation.

The following example illustrates how the order of a reaction can be deduced from experimental data.

EXAMPLE 18.1

At elevated temperatures, HI reacts according to the chemical equation

$$2\ HI(g) \longrightarrow H_2(g) + I_2(g)$$

At 443°C, the rate of the reaction increases with concentration of HI as follows:

[HI] (mol L^{-1})	0.0050	0.010	0.020
Rate (mol L^{-1} s^{-1})	7.5×10^{-4}	3.0×10^{-3}	1.2×10^{-2}

(a) Determine the order of the reaction and write the rate expression.

(b) Calculate the rate constant, and give its units.

(c) Calculate the reaction rate for a 0.0020 M concentration of HI.

SOLUTION

(a) The rate expressions at two different concentrations $[HI]_1$ and $[HI]_2$ are

$$\text{rate}_1 = k([HI]_1)^n$$

$$\text{rate}_2 = k([HI]_2)^n$$

After dividing the second equation by the first, the rate constant k drops out, leaving the reaction order n as the only unknown quantity.

$$\frac{\text{rate}_2}{\text{rate}_1} = \left(\frac{[\text{HI}]_2}{[\text{HI}]_1}\right)^n$$

We can now substitute any two sets of data into this equation and solve for n. Taking the first two sets with $[\text{HI}]_1 = 0.0050$ M and $[\text{HI}]_2 = 0.010$ M gives

$$\frac{3.0 \times 10^{-3}}{7.5 \times 10^{-4}} = \left(\frac{0.010}{0.0050}\right)^n$$

which simplifies to

$$4 = (2)^n$$

By inspection, $n = 2$, so the reaction is second order in HI. When the solution of the equation is less obvious, we can take the logarithms of both sides, giving (in this case)

$$\log_{10} 4 = n \log_{10} 2$$

$$n = \frac{\log_{10} 4}{\log_{10} 2} = \frac{0.602}{0.301} = 2$$

The rate expression has the form

$$\text{rate} = k[\text{HI}]^2$$

(b) The rate constant k is calculated by inserting any of the sets of data into the rate expression. Taking the first set gives

$$7.5 \times 10^{-4} \text{ mol L}^{-1} \text{ s}^{-1} = k(0.0050 \text{ mol L}^{-1})^2$$

Solving for k gives

$$k = 30 \text{ L mol}^{-1} \text{ s}^{-1}$$

(c) Finally, the rate is calculated for $[\text{HI}] = 0.0020$ M:

$$\text{rate} = k[\text{HI}]^2 = (30 \text{ L mol}^{-1} \text{ s}^{-1})(0.0020 \text{ mol L}^{-1})^2$$
$$= 1.2 \times 10^{-4} \text{ mol L}^{-1} \text{ s}^{-1}$$

So far, each example reaction rate has depended only on a single concentration. In reality, many rates depend on the concentrations of two or more different chemical species, and the rate expression is written in a form such as

$$\text{rate} = -\frac{1}{a}\frac{d[\text{A}]}{dt} = k[\text{A}]^m[\text{B}]^n$$

Again the exponents m and n do not derive from the coefficients in the balanced equation for the reaction; they must be determined experimentally and are usually integers or half-integers.

The exponents $m, n, \ldots$ give the order of the reaction, just as in the simpler case where only one concentration appeared in the rate expression. The preceding reaction is said to be mth order in A, meaning that a change in the concentration of A by a certain factor leads to a change in the rate by that factor raised to the mth power. The reaction is nth order in B, and the **overall reaction order** is $m + n$. For the reaction

$$H_2PO_2^-(aq) + OH^-(aq) \longrightarrow HPO_3^{2-}(aq) + H_2(g)$$

the experimentally determined rate expression is

$$\text{rate} = k[H_2PO_2^-][OH^-]^2$$

so the reaction is said to be first order in $H_2PO_2^-(aq)$ and second order in $OH^-(aq)$, with an overall reaction order of 3. The units of k depend on the reaction order. If all concentrations are expressed in mol L^{-1} and if $p = m + n + \cdots$ is the overall reaction order, then k has units of $\text{mol}^{-(p-1)} \text{ L}^{p-1} \text{ s}^{-1}$.

EXAMPLE 18.2

Use the preceding rate expression to determine the effect of the following changes on the rate of decomposition of $H_2PO_2^-(aq)$:

(a) Tripling the concentration of $H_2PO_2^-(aq)$ at constant pH

(b) Changing the pH from 13 to 14 at a constant concentration of $H_2PO_2^-(aq)$

SOLUTION

(a) Because the reaction is first order in $H_2PO_2^-(aq)$, tripling this concentration will triple the reaction rate.

(b) A change in pH from 13 to 14 corresponds to an increase in the $OH^-(aq)$ concentration by a factor of 10. Because the reaction is second order in $OH^-(aq)$ (that is, this term is squared in the rate expression), this will increase the reaction rate by a factor of 10^2, or 100.

Related Problems: 5, 6

Rate expressions that depend on more than one concentration are more difficult to obtain experimentally than those that depend just on one. One method is to find the instantaneous initial rates of reaction for several values of one of the concentrations, holding the other initial concentrations fixed from one run to the next. The experiment can then be repeated, changing one of the other concentrations. The following example illustrates this procedure.

EXAMPLE 18.3

The reaction of $NO(g)$ with $O_2(g)$ gives $NO_2(g)$:

$$2\ NO(g) + O_2(g) \longrightarrow 2\ NO_2(g)$$

From the dependence of the initial rate ($-\frac{1}{2}\ d[NO]/dt$) on the initial concentrations of NO and O_2, determine the rate expression and the value of the rate constant.

[NO] (mol L^{-1})	[O_2] (mol L^{-1})	Initial Rate (mol L^{-1} s^{-1})
1.0×10^{-4}	1.0×10^{-4}	2.8×10^{-6}
1.0×10^{-4}	3.0×10^{-4}	8.4×10^{-6}
2.0×10^{-4}	3.0×10^{-4}	3.4×10^{-5}

SOLUTION

When $[O_2]$ is multiplied by 3 (with [NO] constant), the rate is also multiplied by 3 (from 2.8×10^{-6} to 8.4×10^{-6}), so the reaction is first order in O_2. When [NO] is multiplied by 2 (with $[O_2]$ constant), the rate is multiplied by

$$\frac{3.4 \times 10^{-5}}{8.4 \times 10^{-6}} \approx 4 = 2^2$$

so the reaction is second order in NO. Thus, the form of the rate expression is

$$\text{rate} = k[O_2][NO]^2$$

To evaluate k, we insert any set of data into the equation. From the first set,

$$2.8 \times 10^{-6}\ \text{mol L}^{-1}\ \text{s}^{-1} = k(1.0 \times 10^{-4}\ \text{mol L}^{-1})(1.0 \times 10^{-4}\ \text{mol L}^{-1})^2$$

$$k = 2.8 \times 10^{6}\ \text{L}^2\ \text{mol}^{-2}\ \text{s}^{-1}$$

Related Problems: 7, 8

Integrated Rate Laws

Measuring an initial rate involves determining small changes in concentration $\Delta[A]$ that occur during a short time interval Δt. Sometimes it can be difficult to obtain sufficiently precise experimental data for these small changes. An alternative is to use an **integrated rate law,** which expresses the concentration of a species directly as a function of the time. For any simple rate expression, a corresponding integrated rate law can be obtained.

FIRST-ORDER REACTIONS

Consider again the reaction

$$N_2O_5(g) \longrightarrow 2\ NO_2(g) + \tfrac{1}{2}\ O_2(g)$$

whose rate law has been determined experimentally to be

$$\text{rate} = -\frac{d[N_2O_5]}{dt} = k[N_2O_5]$$

This is a first-order reaction. If we let $[N_2O_5] = c$, a function of time, we have

$$\frac{dc}{dt} = -kc$$

We seek a function whose slope at every time is proportional to the value of the function itself. This function can be found through the use of calculus. Separating the variables (with concentration c on the left and time t on the right) gives

$$\frac{1}{c}\,dc = -k\,dt$$

Integrating from an initial concentration c_0 at time $t = 0$ to a concentration c at time t (see Appendix C, Section C.5) gives

$$\int_{c_0}^{c} \frac{1}{c}\,dc = -k\int_0^t dt$$

$$\ln c - \ln c_0 = -kt$$

$$\ln (c/c_0) = -kt$$

$$c = c_0\,e^{-kt} \qquad [18.2]$$

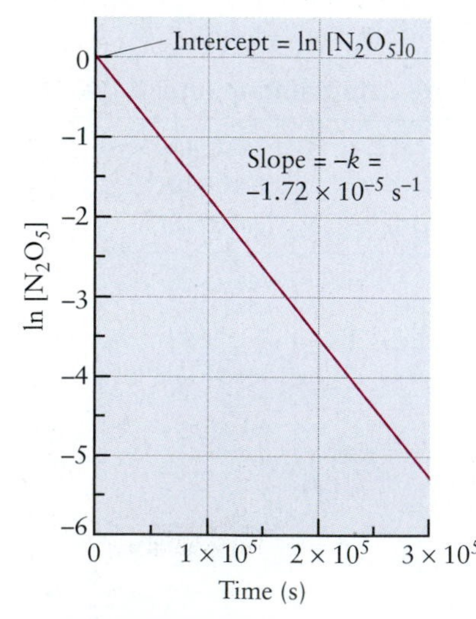

FIGURE 18.5 In a first-order reaction such as the decomposition of N_2O_5, a graph of the natural logarithm of the concentration against time is a straight line, the negative of whose slope gives the rate constant for the reaction.

The concentration falls off exponentially with time. For a first-order reaction, a plot of ln c against t is a straight line with slope $-k$ (Fig. 18.5).

A useful concept in discussions of first-order reactions is the **half-life** $t_{1/2}$—the time it takes for the original concentration c_0 to be reduced to half its value, $c_0/2$. Setting $c = c_0/2$ gives

$$\ln\left(\frac{c}{c_0}\right) = \ln\left(\frac{c_0/2}{c_0}\right) = -\ln 2 = -kt_{1/2}$$

$$t_{1/2} = \frac{\ln 2}{k} = \frac{0.6931}{k} \qquad [18.3]$$

If k has units of s^{-1}, (ln 2)/k is the half-life in seconds. During each half-life, the concentration of A falls to half its value again (Fig. 18.6).

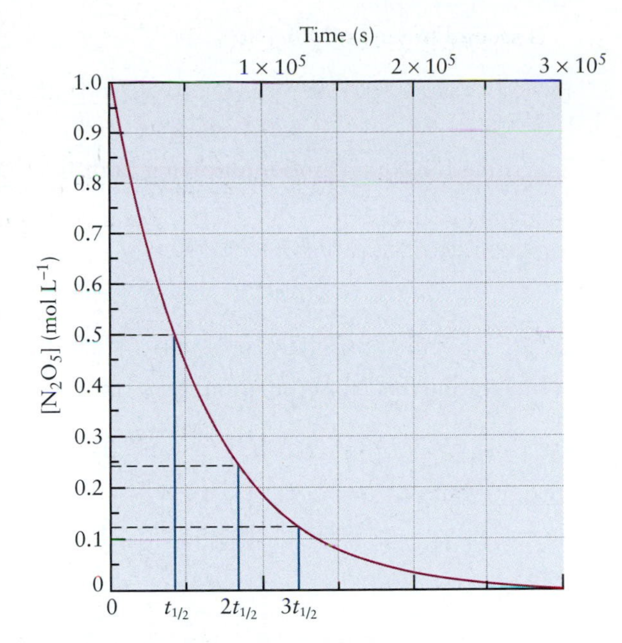

FIGURE 18.6 The same data as in Figure 18.5 are graphed in a concentration-versus-time picture. The half-life $t_{1/2}$ is the time it takes for the concentration to be reduced to half its initial value. In two half-lives, the concentration falls to one quarter of its initial value.

EXAMPLE 18.4

(a) What is the rate constant k for the first-order decomposition of $N_2O_5(g)$ at 25°C if the half-life of $N_2O_5(g)$ at that temperature is 4.03×10^4 s?

(b) What percentage of the N_2O_5 molecules will *not* have reacted after one day?

SOLUTION

(a)
$$t_{1/2} = \frac{\ln 2}{k} = 4.03 \times 10^4 \text{ s}$$

Solving for the rate constant k gives

$$k = \frac{\ln 2}{t_{1/2}} = \frac{0.6931}{4.03 \times 10^4 \text{ s}} = 1.72 \times 10^{-5} \text{ s}^{-1}$$

(b) From the integrated rate law for a first-order reaction,

$$\frac{c}{c_0} = e^{-kt}$$

Putting in the value for k and setting t to 1 day $= 8.64 \times 10^4$ s gives

$$\frac{c}{c_0} = \exp\left[-(1.72 \times 10^{-5} \text{ s}^{-1})(8.64 \times 10^4 \text{ s})\right]$$

$$= e^{-1.49} = 0.226$$

Therefore, 22.6% of the molecules will not yet have reacted after one day.

Related Problems: 11, 12

SECOND-ORDER REACTIONS

Integrated rate laws can be obtained for reactions of other orders. The observed rate of the reaction

$$2\ NO_2(g) \longrightarrow 2\ NO(g) + O_2(g)$$

is second order in $[NO_2]$:

$$\text{rate} = -\frac{1}{2}\frac{d[NO_2]}{dt} = k[NO_2]^2$$

Writing $[NO_2] = c$ and multiplying both sides of the equation by -1 gives

$$\frac{dc}{dt} = -2kc^2$$

$$\frac{1}{c^2}\,dc = -2k\,dt$$

Integrating this from the initial concentration c_0 at time 0 to c at time t gives

$$\int_{c_0}^{c}\frac{1}{c^2}\,dc = -2k\int_0^t dt$$

$$-\frac{1}{c} + \frac{1}{c_0} = -2kt$$

$$\frac{1}{c} = \frac{1}{c_0} + 2kt \qquad [18.4]$$

For such a second-order reaction, a plot of $1/c$ against t is linear (Fig. 18.7). The factor 2 multiplying kt in this expression arises from the stoichiometric coefficient 2 for NO_2 in the balanced equation for the specific example reaction. For other second-order reactions with different stoichiometric coefficients for the reactant (see the thermal decomposition of ethane described on page 755), we must modify the integrated rate law accordingly.

The concept of half-life has little use for second-order reactions. Setting $[NO_2]$ equal to $[NO_2]_0/2$ in the preceding equation and solving for t gives

$$\frac{2}{[NO_2]} = 2kt_{1/2} + \frac{1}{[NO_2]_0}$$

$$t_{1/2} = \frac{1}{2k[NO_2]_0}$$

For second-order reactions, the half-life is not a constant; it depends on the initial concentration.

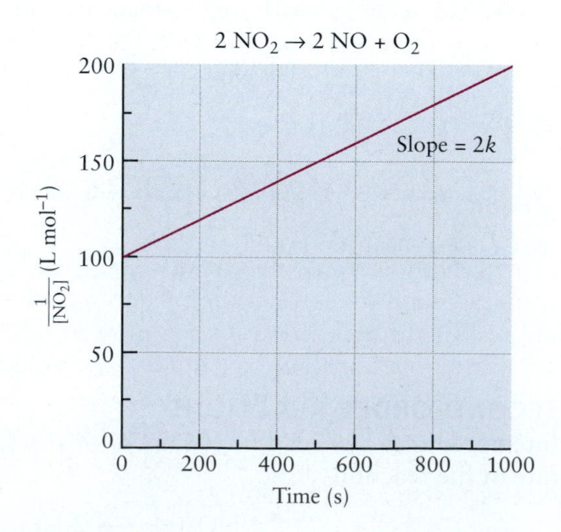

FIGURE 18.7 For a second-order reaction such as $2\ NO_2 \longrightarrow 2\ NO + O_2$, a graph of the reciprocal of the concentration against time is a straight line with slope $2k$.

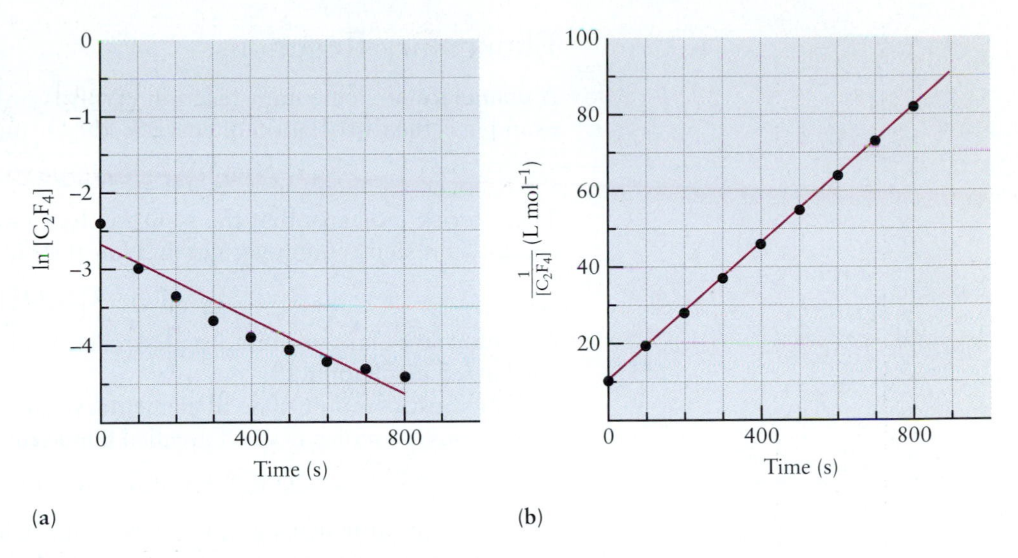

FIGURE 18.8 For the reaction in Example 18.5, (a) plotting the logarithm of the concentration of C_2F_4 against time tests for first-order kinetics and (b) plotting the reciprocal of the concentration of C_2F_4 against time tests for second-order kinetics. It is clear that the assumption of first-order kinetics does not fit the data as well; no straight line will pass through the data.

EXAMPLE 18.5

The dimerization of tetrafluoroethylene (C_2F_4) to octafluorocyclobutane (C_4F_8) is second order in the reactant C_2F_4, and at 450 K its rate constant is $k = 0.0448$ L mol^{-1} s^{-1}. If the initial concentration of C_2F_4 is 0.100 mol L^{-1}, what will its concentration be after 205 s?

SOLUTION

For this second-order reaction,

$$\frac{1}{c} - \frac{1}{c_0} = 2kt$$

Solving for the concentration c after a time t = 205 s gives

$$\frac{1}{c} = (2)(0.0448 \text{ L mol}^{-1} \text{ s}^{-1})(205 \text{ s}) + \frac{1}{0.100 \text{ mol L}^{-1}} = 28.4 \text{ L mol}^{-1}$$

$$c = 3.53 \times 10^{-2} \text{ mol L}^{-1}$$

Related Problems: 15, 16

In a real experimental study the reaction order is usually not known. In such a case we try several plots to find the one that best fits the data (Fig. 18.8).

18.3 Reaction Mechanisms

Many reactions do not occur in a single step, but rather proceed through a sequence of steps to arrive at the products. Each step is called an **elementary reaction** and occurs through the collisions of atoms, ions, or molecules. The rate expression for an overall reaction cannot be derived from the stoichiometry of the balanced equation, and must be determined experimentally. But the rate of an elementary reaction *is* directly proportional to the product of the concentrations of the reacting species, each raised to a power equal to its coefficient in the balanced elementary equation.

Elementary Reactions

A **unimolecular** elementary reaction involves only a single reactant molecule. An example is the dissociation of energized N_2O_5 molecules in the gas phase:

$$N_2O_5^*(g) \longrightarrow NO_2(g) + NO_3(g)$$

The asterisk indicates that the N_2O_5 molecules have far more than ground state energy. This step is unimolecular and has the rate expression

$$\text{rate} = k[N_2O_5^*]$$

An important class of unimolecular reactions is the decay of radioactive nuclei, considered in Chapter 19.

The most common type of elementary reaction involves the collision of two atoms, ions, or molecules and is called **bimolecular.** An example is the reaction

$$NO(g) + O_3(g) \longrightarrow NO_2(g) + O_2(g)$$

The frequency at which a given NO molecule collides with ozone molecules is proportional to the concentration of ozone: If there are twice as many ozone molecules per unit volume, each NO molecule will undergo twice as many collisions as it moves through space, and ozone will react twice as rapidly. The rate of collisions of *all* the NO molecules in the container is proportional to the concentration of NO as well, so the rate law of a bimolecular reaction like this one has the form

$$\text{rate} = k[NO][O_3]$$

A **termolecular** reaction step involves the simultaneous collision of three molecules, which is a much less likely event. An example is the recombination of iodine atoms in the gas phase to form iodine molecules. So much energy is released in forming the I—I bond that the molecule would simply fly apart as soon as it was formed if the event were a binary collision. A third atom or molecule is necessary to take away some of the excess energy. If iodine recombination takes place in the presence of a sufficiently high concentration of an inert gas such as argon, termolecular reactions

$$I(g) + I(g) + Ar(g) \longrightarrow I_2(g) + Ar(g)$$

occur in which the argon atom leaves with more kinetic energy than it had initially. The rate law for this termolecular reaction is

$$\text{rate} = k[I]^2[Ar]$$

Elementary reactions involving collisions of four or more molecules are not observed, and even termolecular collisions are rare if other pathways are possible.

Elementary reactions in liquid solvents involve encounters of solute species with one another. If the solution is ideal, the rates of these processes are proportional to the product of the concentrations of the solute species involved. Solvent molecules are always present and may affect the reaction, even though they do not appear in the rate expression because the solvent concentration cannot be varied appreciably. A reaction such as the recombination of iodine atoms occurs readily in a liquid. It appears to be second order with rate law

$$\text{rate} = k[I]^2$$

only because the third body involved is a solvent molecule. In the same way, a reaction between a solvent molecule and a solute molecule appears to be unimolecular, and only the concentration of solute molecules enters the rate expression for that step.

Reaction Mechanisms

A **reaction mechanism** is a detailed sequence of elementary reactions, with their rates, that are combined to yield the overall reaction. It is often possible to write several reaction mechanisms, each of which is consistent with a given overall reaction. One of the goals of chemical kinetics is to use the observed rate of a reaction to choose among various conceivable reaction mechanisms.

The gas-phase reaction of nitrogen dioxide with carbon monoxide provides a good example of a reaction mechanism. The generally accepted mechanism at low temperatures has two steps, both bimolecular:

$$NO_2(g) + NO_2(g) \longrightarrow NO_3(g) + NO(g) \qquad \text{(slow)}$$

$$NO_3(g) + CO(g) \longrightarrow NO_2(g) + CO_2(g) \qquad \text{(fast)}$$

For any reaction mechanism, combining the steps must give the overall reaction. When each elementary step occurs the same number of times in the course of the reaction, the chemical equations can simply be added. (If one step occurs twice as often as the others, it must be multiplied by 2 before the elementary reactions are added.) In this case, we add the two chemical equations to give

$$2\,NO_2(g) + NO_3(g) + CO(g) \longrightarrow NO_3(g) + NO(g) + NO_2(g) + CO_2(g)$$

Canceling out the NO_3 and one molecule of NO_2 from each side leads to

$$NO_2(g) + CO(g) \longrightarrow NO(g) + CO_2(g)$$

A **reaction intermediate** (here, NO_3) is a chemical species that is formed and consumed in the reaction but does not appear in the overall balanced chemical equation. One of the major challenges in chemical kinetics is to identify intermediates, which are often so short-lived that they are difficult to detect directly.

EXAMPLE 18.6

Consider the following reaction mechanism:

$$Cl_2(g) \longrightarrow 2\,Cl(g)$$

$$Cl(g) + CHCl_3(g) \longrightarrow HCl(g) + CCl_3(g)$$

$$CCl_3(g) + Cl(g) \longrightarrow CCl_4(g)$$

(a) What is the molecularity of each elementary step?

(b) Write the overall equation for the reaction.

(c) Identify the reaction intermediate(s).

SOLUTION

(a) The first step is unimolecular, and the other two are bimolecular.

(b) Adding the three steps gives

$$Cl_2(g) + 2\,Cl(g) + CHCl_3(g) + CCl_3(g) \longrightarrow 2\,Cl(g) + HCl(g) + CCl_3(g) + CCl_4(g)$$

The two species that appear in equal amounts on both sides cancel out to leave

$$Cl_2(g) + CHCl_3(g) \longrightarrow HCl(g) + CCl_4(g)$$

(c) The two reaction intermediates are Cl and CCl_3.

Related Problems: 21, 22

Kinetics and Chemical Equilibrium

There is a direct connection between the rates for the elementary steps in a chemical reaction mechanism and the overall equilibrium constant K. To see this connection, consider the reaction

$$2\,NO(g) + 2\,H_2(g) \rightleftarrows N_2(g) + 2\,H_2O(g)$$

This reaction is believed to occur in a three-step process involving N_2O_2 and N_2O as intermediates:

$$NO(g) + NO(g) \underset{k_{-1}}{\overset{k_1}{\rightleftharpoons}} N_2O_2(g)$$

$$N_2O_2(g) + H_2(g) \underset{k_{-2}}{\overset{k_2}{\rightleftharpoons}} N_2O(g) + H_2O(g)$$

$$N_2O(g) + H_2(g) \underset{k_{-3}}{\overset{k_3}{\rightleftharpoons}} N_2(g) + H_2O(g)$$

These elementary reactions are shown as equilibria, so the reverse reactions (from products to reactants) are included here as well; k_1, k_2, and k_3 are the rate constants for the forward elementary steps, and k_{-1}, k_{-2}, and k_{-3} are the rate constants for the corresponding reverse reactions.

We now invoke the principle of **detailed balance,** which states that at equilibrium the rate of *each* elementary process is balanced by (equal to) the rate of its reverse process. For the preceding mechanism we conclude that

$$k_1[NO]_{eq}^2 = k_{-1}[N_2O_2]_{eq}$$

$$k_2[N_2O_2]_{eq}[H_2]_{eq} = k_{-2}[N_2O]_{eq}[H_2O]_{eq}$$

$$k_3[N_2O]_{eq}[H_2]_{eq} = k_{-3}[N_2]_{eq}[H_2O]_{eq}$$

The equilibrium constants[1] K_1, K_2, and K_3 for the elementary reactions are equal to the ratio of the forward and reverse reaction rate constants:

$$K_1 = \frac{[N_2O_2]_{eq}}{[NO]_{eq}^2} = \frac{k_1}{k_{-1}}$$

$$K_2 = \frac{[N_2O]_{eq}[H_2O]_{eq}}{[N_2O_2]_{eq}[H_2]} = \frac{k_2}{k_{-2}}$$

$$K_3 = \frac{[N_2]_{eq}[H_2O]_{eq}}{[N_2O]_{eq}[H_2]_{eq}} = \frac{k_3}{k_{-3}}$$

The steps of the mechanism are now added together to obtain the overall reaction. Recall from Section 14.4 that when reactions are added, their equilibrium constants are multiplied. Therefore, the overall equilibrium constant K is

$$K = K_1K_2K_3 = \frac{k_1k_2k_3}{k_{-1}k_{-2}k_{-3}} = \frac{[N_2O_2]_{eq}[N_2O]_{eq}[H_2O]_{eq}[N_2]_{eq}[H_2O]_{eq}}{[NO]_{eq}^2[N_2O_2]_{eq}[H_2]_{eq}[N_2O]_{eq}[H_2]_{eq}}$$

$$= \frac{[H_2O]_{eq}^2[N_2]_{eq}}{[NO]_{eq}^2[H_2]_{eq}^2}$$

The concentrations of the intermediates N_2O_2 and N_2O cancel out, giving the usual expression of the mass-action law.

This result can be generalized to any reaction mechanism. The product of the forward rate constants for the elementary reactions divided by the product of the reverse rate constants is always equal to the equilibrium constant of the overall reaction. If there are several possible mechanisms for a given reaction (which might involve intermediates other than N_2O_2 and N_2O), their forward and reverse rate constants will all be consistent in this way with the equilibrium constant of the overall reaction.

[1]Thermodynamic equilibrium constants are dimensionless because they are expressed in terms of activities rather than partial pressure or concentration. The convention in chemical kinetics is to use concentrations rather than activities, even for gaseous species. Therefore, the equilibrium constants K_1, K_2, and K_3 introduced here are the empirical equilibrium constants K_c described briefly in Section 14.2. These constants are not dimensionless and must be multiplied by the concentration of the reference state, $c_{ref} = RT/P_{ref}$, raised to the appropriate power to be made equal to the thermodynamic equilibrium constant. Nevertheless, to maintain consistency with the conventions of chemical kinetics, such constants as K_1, K_2, and K_3 are referred to as equilibrium constants in this section and are written without the subscript c.

18.4 Reaction Mechanisms and Rate

In many reaction mechanisms, one step is significantly slower than all the others; this step is called the **rate-determining step.** Because an overall reaction can occur only as fast as its slowest step, that step is crucial in determining the rate of the reaction. This is analogous to the flow of automobile traffic on a highway which has a slowdown at some point. The rate at which cars can complete a trip down the full length of the highway (in cars per minute) is approximately equal to the rate at which they pass through the bottleneck.

If the rate-determining step is the first one, the analysis is particularly simple. An example is the reaction

$$2\,NO_2(g) + F_2(g) \longrightarrow 2\,NO_2F(g)$$

for which the experimental rate law is

$$\text{rate} = k_{obs}[NO_2][F_2]$$

A possible mechanism for the reaction is

$$NO_2(g) + F_2(g) \xrightarrow{k_1} NO_2F(g) + F(g) \qquad \text{(slow)}$$

$$NO_2(g) + F(g) \xrightarrow{k_2} NO_2F(g) \qquad \text{(fast)}$$

The first step is slow and determines the rate, $k_1[NO_2][F_2]$, in agreement with the observed rate expression. The subsequent fast step does not affect the reaction rate because fluorine atoms react with NO_2 almost as soon as they are produced.

Mechanisms in which the rate-determining step occurs after one or more fast steps are often signaled by a reaction order greater than 2, by a nonintegral reaction order, or by an inverse concentration dependence on one of the species taking part in the reaction. An example is the reaction

$$2\,NO(g) + O_2(g) \longrightarrow 2\,NO_2(g)$$

for which the experimental rate law is

$$\text{rate} = k_{obs}[NO]^2[O_2]$$

One possible mechanism would be a single-step termolecular reaction of two NO molecules with one O_2 molecule. This would be consistent with the form of the rate expression, but termolecular collisions are rare, and if there is an alternative pathway it is usually followed.

One such alternative is the two-step mechanism

$$NO(g) + NO(g) \underset{}{\overset{k_1}{\rightleftharpoons}} N_2O_2(g) \qquad \text{(fast equilibrium)}$$

$$N_2O_2(g) + O_2(g) \overset{k_2}{\rightleftharpoons} 2\,NO_2(g) \qquad \text{(slow)}$$

Because the slow step determines the overall rate, we can write

$$\text{rate} = k_2[N_2O_2][O_2]$$

The concentration of a reactive intermediate such as N_2O_2 cannot be varied at will. Because the N_2O_2 reacts only slowly with O_2, the reverse reaction (to 2 NO) is possible and must be taken into account. In fact, it is reasonable to assume that all of the elementary reactions that occur *before* the rate-determining step are in equilibrium, with the forward and reverse reactions occurring at the same rate. In this case, we have

$$\frac{[N_2O_2]}{[NO]^2} = \frac{k_1}{k_{-1}} = K_1$$

$$[N_2O_2] = K_1[NO]^2$$

$$\text{rate} = k_2K_1[NO]^2[O_2]$$

This result is consistent with the observed reaction order, with $k_2k_1 = k_{obs}$.

EXAMPLE 18.7

In basic aqueous solution the reaction

$$I^- + OCl^- \longrightarrow Cl^- + OI^-$$

follows a rate law that is consistent with the following mechanism:

$$OCl^-(aq) + H_2O(\ell) \underset{k_{-1}}{\overset{k_1}{\rightleftarrows}} HOCl(aq) + OH^-(aq) \qquad \text{(fast equilibrium)}$$

$$I^-(aq) + HOCl(aq) \xrightarrow{k_2} HOI(aq) + Cl^-(aq) \qquad \text{(slow)}$$

$$OH^-(aq) + HOI(aq) \xrightarrow{k_3} H_2O(\ell) + OI^-(aq) \qquad \text{(fast)}$$

What rate law is predicted by this mechanism?

SOLUTION

The rate is determined by the slowest elementary step, the second one:

$$\text{rate} = k_2[I^-][HOCl]$$

But, the HOCl is in equilibrium with OCl^- and OH^- due to the first step:

$$\frac{[HOCl][OH^-]}{[OCl^-]} = K_1 = \frac{k_1}{k_{-1}}$$

Solving this for [HOCl] and inserting it into the previous expression gives the prediction

$$\text{rate} = k_2K_1\frac{[I^-][OCl^-]}{[OH^-]}$$

which is, in fact, the experimentally observed rate law.

Related Problems: 25, 26, 27, 28, 29, 30

The rate law of the foregoing example depends on the inverse of the concentration of OH^- ion. Such a form is often a clue that a rapid equilibrium occurs in the first steps of a reaction, preceding the rate-determining step. Fractional orders of reaction provide a similar clue, as in the reaction of H_2 with Br_2 to form HBr,

$$H_2 + Br_2 \longrightarrow 2\ HBr$$

for which the initial reaction rate (before very much HBr builds up) is

$$\text{rate} = k_{obs}[H_2][Br_2]^{1/2}$$

How can such a fractional power appear? One reaction mechanism that predicts this rate law is

$$Br_2 + M \underset{k_{-1}}{\overset{k_1}{\rightleftarrows}} Br + Br + M \qquad \text{(fast equilibrium)}$$

$$Br + H_2 \xrightarrow{k_2} HBr + H \qquad \text{(slow)}$$

$$H + Br_2 \xrightarrow{k_3} HBr + Br \qquad \text{(fast)}$$

Here M stands for a second molecule that does not react but that supplies the energy to break up the bromine molecules. For such a mechanism the reaction rate is determined by the slow step:

$$\text{rate} = k_2[Br][H_2]$$

However, [Br] is fixed by the establishment of equilibrium in the first reaction,

$$\frac{[Br]^2}{[Br_2]} = K_1 = \frac{k_1}{k_{-1}}$$

so

$$[\text{Br}] = K_1^{1/2}[\text{Br}_2]^{1/2}$$

The rate expression predicted by this mechanism is thus

$$\text{rate} = k_2 K_1^{1/2}[\text{H}_2][\text{Br}_2]^{1/2}$$

This is in accord with the observed fractional power in the rate law. On the other hand, the simple bimolecular mechanism

$$\text{H}_2 + \text{Br}_2 \xrightarrow{k_1} 2\ \text{HBr} \qquad \text{(slow)}$$

predicts a rate law:

$$\text{rate} = k_1[\text{H}_2][\text{Br}_2]$$

This disagrees with the observed rate law, so it can be ruled out as the major contributor to the measured rate.

This discussion shows that deducing a rate law from a proposed mechanism is relatively straightforward, but doing the reverse is much harder. In fact, several competing mechanisms often give rise to the same rate law, and only some independent type of measurement can distinguish between them. A proposed reaction mechanism cannot be proven to be correct if its predictions agree with an experimental rate law, but it can be proven wrong if its predictions disagree with the experimental results.

A classic example is the reaction

$$\text{H}_2 + \text{I}_2 \longrightarrow 2\ \text{HI}$$

for which the observed rate law is

$$\text{rate} = k_{\text{obs}}[\text{H}_2][\text{I}_2]$$

(Contrast this with the rate law already given for the analogous reaction of H_2 with Br_2.) This is one of the earliest and most extensively studied reactions in chemical kinetics, and until 1967 it was widely believed to occur as a one-step elementary reaction. At that time J. H. Sullivan investigated the effect of illuminating the reacting sample with light, which splits some of the I_2 molecules into iodine atoms. If the mechanism we proposed for H_2 and Br_2 is correct here as well, the effect of the light on the reaction should be small because it leads only to a small decrease in the I_2 concentration.

Instead, Sullivan observed a dramatic *increase* in the rate of reaction under illumination, which could be explained only by the participation of iodine *atoms* in the reaction mechanism. One such mechanism is

$$\text{I}_2 + \text{M} \underset{k_{-1}}{\overset{k_1}{\rightleftharpoons}} \text{I} + \text{I} + \text{M} \qquad \text{(fast equilibrium)}$$

$$\text{H}_2 + \text{I} + \text{I} \xrightarrow{k_2} 2\ \text{HI} \qquad \text{(slow)}$$

for which the rate law is

$$\text{rate} = k_2[\text{H}_2][\text{I}]^2 = k_2 K_1[\text{H}_2][\text{I}_2]$$

This mechanism gives the same rate law that is observed experimentally, and it is consistent with the effect of light on the reaction. The other reaction mechanism also appears to contribute significantly to the overall rate.

This example illustrates the hazards of trying to determine reaction mechanisms from rate laws: several mechanisms can fit any given empirical rate law, and it is always possible that a new piece of information suggesting a different mechanism will be found. The problem is that, under ordinary conditions, reaction intermediates cannot be isolated and studied like the reactants and products. This situation is changing with the development of experimental techniques that allow the direct study of the transient intermediates that form in small concentration during the course of a chemical reaction.

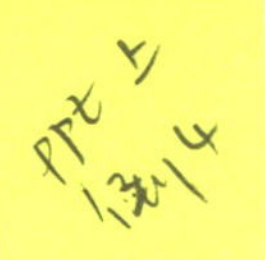

The Steady-State Approximation

In some reaction mechanisms there is no single step that is much slower than the others, so the methods discussed so far cannot predict the rate law. In such cases we use the **steady-state approximation,** which states that the concentrations of reactive intermediates remain nearly constant through most of the reaction.

To illustrate this approximation, let's examine the mechanism proposed by F. A. Lindemann for the dissociation of molecules in the gas phase. A molecule such as N_2O_5 undergoes collisions with neighboring molecules M, where M can stand for another N_2O_5 molecule or for an inert gas such as argon. Through such collisions the N_2O_5 molecule can become excited (or activated) to a state indicated by $N_2O_5^*$:

$$N_2O_5(g) + M(g) \underset{k_{-1}}{\overset{k_1}{\rightleftarrows}} N_2O_5^*(g) + M(g)$$

The reverse process, with rate constant k_{-1}, is also indicated because the activated molecule can be deactivated by collisions with other molecules. The second step is the unimolecular decomposition of $N_2O_5^*$:

$$N_2O_5^*(g) \xrightarrow{k_2} NO_3(g) + NO_2(g)$$

Subsequent reaction steps to form O_2 and NO_2 from NO_3 occur rapidly and do not affect the measured rate:

$$NO_3(g) + NO_2(g) \xrightarrow{k_3} NO(g) + NO_2(g) + O_2(g) \qquad \text{(fast)}$$

$$NO_3(g) + NO(g) \xrightarrow{k_4} 2\ NO_2(g) \qquad \text{(fast)}$$

The $N_2O_5^*$ is a reactive intermediate; it is produced at a rate $k_1[N_2O_5][M]$ from collisions of N_2O_5 molecules with other molecules and is lost at a rate $k_{-1}[N_2O_5^*][M]$ due to deactivation and at a rate $k_{-2}[N_2O_5^*][M]$ due to dissociation. The net rate of change of $[N_2O_5^*]$ is then

$$\frac{d[N_2O_5^*]}{dt} = k_1[N_2O_5][M] - k_{-1}[N_2O_5^*][M] - k_2[N_2O_5^*]$$

At the beginning of the reaction, $[N_2O_5^*] = 0$, but this concentration builds up after a short time to a small value. The steady-state approximation consists of the assumption that after this short time the rates of production and loss of $N_2O_5^*$ become equal, and

$$\frac{d[N_2O_5^*]}{dt} = 0$$

The steady-state concentration of $[N_2O_5^*]$ persists practically unchanged throughout most of the course of the reaction.

Setting the net rate of change of the $N_2O_5^*$ concentration to 0 gives

$$\frac{d[N_2O_5^*]}{dt} = 0 = k_1[N_2O_5][M] - k_{-1}[N_2O_5^*][M] - k_2[N_2O_5^*]$$

Solving for $[N_2O_5^*]$ gives

$$[N_2O_5^*](k_2 + k_{-1}[M]) = k_1[N_2O_5][M]$$

$$[N_2O_5^*] = \frac{k_1[N_2O_5][M]}{k_2 + k_{-1}[M]}$$

The rate of the overall reaction $N_2O_5 \longrightarrow 2\ NO_2 + \frac{1}{2}O_2$ is

$$\text{rate} = \frac{1}{2}\frac{d[NO_2]}{dt} = k_2[N_2O_5^*] = \frac{k_1k_2[N_2O_5][M]}{k_2 + k_{-1}[M]}$$

This expression has two limiting cases:

1. *Low pressure* When [M] is small enough, $k_2 \gg k_{-1}[M]$ and we can use the approximation

$$\text{rate} = k_1[N_2O_5][M] \qquad \text{(second order)}$$

This same result would be found by assuming the first step to be rate-determining.

2. *High pressure* When [M] is large enough, $k_{-1}[M] \gg k_2$ and we can use the approximation

$$\text{rate} = \frac{k_1}{k_{-1}} k_2[N_2O_5] \qquad \text{(first order)}$$

This same result would be found by assuming the second step to be rate-determining.

The steady-state approximation is a more general approach than those considered earlier and can be used when no single rate-determining step exists.

Chain Reactions

A **chain reaction** proceeds through a series of elementary steps, some of which are repeated many times. Chain reactions have three stages: (1) **initiation,** in which two or more reactive intermediates are generated; (2) **propagation,** in which products are formed but reactive intermediates are continuously regenerated; and (3) **termination,** in which two intermediates combine to give a stable product.

An example of a chain reaction is the reaction of methane with fluorine to give CH_3F and HF:

$$CH_4(g) + F_2(g) \longrightarrow CH_3F(g) + HF(g)$$

Although in principle this reaction could occur through a one-step bimolecular process, that route turns out to be too slow to contribute significantly under normal reaction conditions. Instead, the mechanism involves a chain reaction of the following type:

$$CH_4(g) + F_2(g) \longrightarrow CH_3(g) + HF(g) + F(g) \qquad \text{(initiation)}$$

$$CH_3(g) + F_2(g) \longrightarrow CH_3F(g) + F(g) \qquad \text{(propagation)}$$

$$CH_4(g) + F(g) \longrightarrow CH_3(g) + HF(g) \qquad \text{(propagation)}$$

$$CH_3(g) + F(g) + M(g) \longrightarrow CH_3F(g) + M(g) \qquad \text{(termination)}$$

In the initiation step, two reactive intermediates (CH_3 and F) are produced. During the propagation steps, these intermediates are not used up while reactants (CH_4 and F_2) are being converted to products (CH_3F and HF). The propagation steps can be repeated again and again, until eventually two reactive intermediates come together in a termination step. As we see in Chapter 23, chain reactions are important in building up long-chain molecules called polymers.

The chain reaction just considered proceeds at a constant rate, because each propagation step both uses up and produces a reactive intermediate. The concentrations of the reactive intermediates remain approximately constant and are determined by the rates of chain initiation and termination. Another type of chain reaction is possible in which the number of reactive intermediates increases during one or more propagation steps. This is called a **branching chain reaction.** An example is the reaction of oxygen with hydrogen. The mechanism is complex and can be initiated in various ways, leading to the formation of several reactive intermediates such as O, H, and OH. Some propagation steps are of the type already seen for CH_4 and F_2, such as

$$OH(g) + H_2(g) \longrightarrow H_2O(g) + H(g)$$

in which one reactive intermediate (OH) is used up and one (H) is produced. Other propagation steps are branching:

$$H(g) + O_2(g) \longrightarrow OH(g) + O(g)$$

$$O(g) + H_2(g) \longrightarrow OH(g) + H(g)$$

In these steps, each reactive intermediate used up causes the generation of two others. This leads to rapid growth in the number of reactive species, speeding the rate further and possibly causing an explosion. Branching chain reactions are critical in the fission of uranium (see Chapter 19).

18.5 Effect of Temperature on Reaction Rates

The first four sections of this chapter describe the experimental determination of rate laws and their relation to assumed mechanisms for chemical reactions. Now we have to find out what determines the actual magnitudes of rate constants (either for elementary reactions or for overall rates of multistep reactions), and how temperature affects reaction rates. To consider these matters, it is necessary to connect molecular collision rates to the rates of chemical reactions. We limit the discussion to gas-phase reactions, for which the kinetic theory of Chapter 9 is applicable.

Gas-Phase Reaction Rate Constants

In Section 9.8, we applied the kinetic theory of gases to estimate the frequency of collisions between a particular molecule and other molecules in a gas. In Example 9.14, we calculated this frequency to be $4.1 \times 10^9 \text{ s}^{-1}$ under room conditions for a typical small molecule such as oxygen. If every collision led to reaction, the reaction would be practically complete in about 10^{-9} s. Some reactions do proceed at rates almost this high. An example is the bimolecular reaction between two CH_3 radicals to give ethane, C_2H_6,

$$2\ CH_3 \longrightarrow C_2H_6$$

for which the observed rate constant is $1 \times 10^{10} \text{ L mol}^{-1} \text{ s}^{-1}$. For initial pressure of CH_3 near 1 atm at 25°C, the concentration initially is about 0.04 M. The second-order integrated rate law from Section 18.2 predicts that the concentration would drop to 0.02 M after a period of 10^{-9} s. But, reaction rates that are much lower—by factors of 10^{12} or more—are common. The naive idea that "to collide is to react" clearly must be modified if we are to understand these lower rates.

We find a clue in the observed temperature dependence of reaction rate constants. The rates of many reactions increase extremely rapidly as temperature increases; typically a 10°C rise in temperature may double the rate. In 1889 Svante Arrhenius suggested that rate constants vary exponentially with inverse temperature,

$$k = Ae^{-E_a/RT} \qquad [18.5]$$

where E_a is a constant with dimensions of energy and A is a constant with the same dimensions as k. Taking the natural logarithm of this equation gives

$$\ln k = \ln A - \frac{E_a}{RT} \qquad [18.6]$$

So a plot of $\ln k$ against $1/T$ should be a straight line with slope $-E_a/R$ and intercept $\ln A$. Many rate constants do show just this kind of temperature dependence (Fig. 18.9).

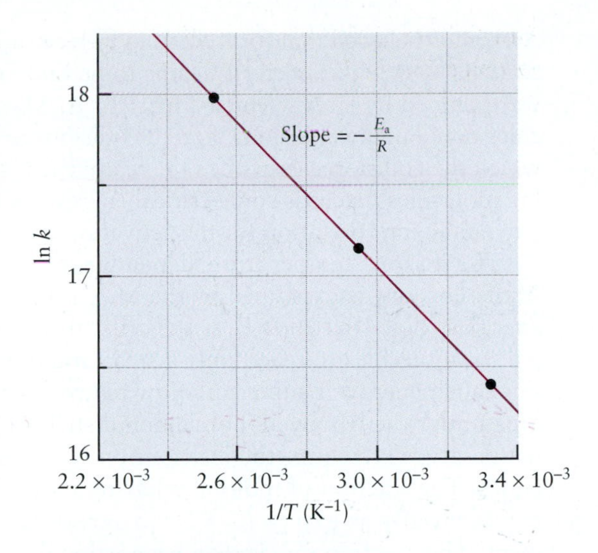

FIGURE 18.9 An Arrhenius plot of ln k against $1/T$ for the reaction of benzene vapor with oxygen atoms. An extrapolation to $1/T = 0$ gives the constant ln A from the intercept of this line.

EXAMPLE 18.8

The decomposition of hydroxylamine (NH_2OH) in the presence of oxygen follows the rate law

$$-\frac{d[NH_2OH]}{dt} = k_{obs}[NH_2OH][O_2]$$

where k_{obs} is 0.237×10^{-4} L mol^{-1} s^{-1} at 0°C and 2.64×10^{-4} L mol^{-1} s^{-1} at 25°C. Calculate E_a and the factor A for this reaction.

SOLUTION

Let us write the Arrhenius equation at two different temperatures T_1 and T_2:

$$\ln k_1 = \ln A - \frac{E_a}{RT_1} \quad \text{and} \quad \ln k_2 = \ln A - \frac{E_a}{RT_2}$$

If the first equation is subtracted from the second, the term ln A cancels out, leaving

$$\ln k_2 - \ln k_1 = \ln \frac{k_2}{k_1} = -\frac{E_a}{R}\left(\frac{1}{T_2} - \frac{1}{T_1}\right)$$

which can be solved for E_a. In the present case, $T_1 = 273$ K and $T_2 = 298$ K; therefore,

$$\ln \frac{2.64 \times 10^{-4}}{0.237 \times 10^{-4}} = \frac{-E_a}{8.315 \text{ J K}^{-1} \text{ mol}^{-1}}\left(\frac{1}{298 \text{ K}} - \frac{1}{273 \text{ K}}\right)$$

$$2.410 = \frac{E_a}{8.315 \text{ J K}^{-1} \text{ mol}^{-1}}(3.07 \times 10^{-4} \text{ K}^{-1})$$

$$E_a = 6.52 \times 10^4 \text{ J mol}^{-1} = 65.2 \text{ kJ mol}^{-1}$$

Now that E_a is known, the constant A can be calculated by using data at either temperature. At 273 K,

$$\ln A = \ln k_1 + \frac{E_a}{RT}$$

$$= \ln(0.237 \times 10^{-4}) + \frac{6.52 \times 10^4 \text{ J mol}^{-1}}{(8.315 \text{ J K}^{-1} \text{ mol}^{-1})(273 \text{ K})}$$

$$= -10.65 + 28.73 = 18.08$$

$$A = e^{18.08} = 7.1 \times 10^7 \text{ L mol}^{-1} \text{ s}^{-}$$

We could determine E_a and A more accurately from measurements at a series of temperatures and a least-squares fit to a plot such as that in Figure 18.9.

Related Problems: 35, 36

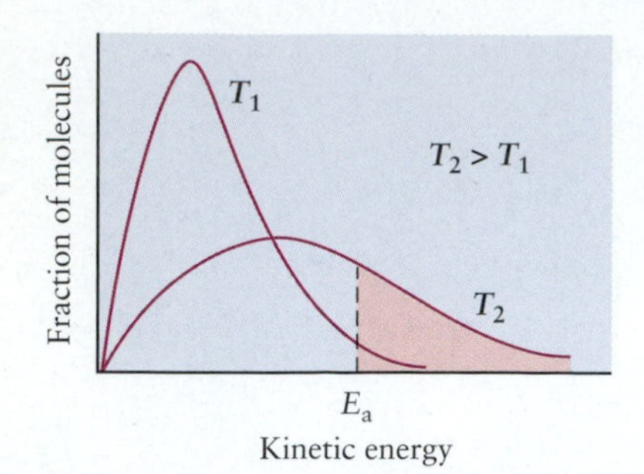

FIGURE 18.10 The Maxwell–Boltzmann distribution of molecular kinetic energies, showing the effect of temperature on the fraction of molecules having large enough kinetic energies to react. This figure shows only translational energy; internal vibrational and rotational energy also promote reactions.

Arrhenius believed that for molecules to react upon collision they must become "activated," so the parameter E_a came to be known as the **activation energy.** His ideas were refined by later scientists. In 1915 A. Marcelin pointed out that, while molecules make many collisions, not all collisions are reactive. Only those collisions for which the collision energy (i.e., the relative translational kinetic energy of the colliding molecules) exceeds some critical energy result in reaction. Thus, Marcelin gave a dynamic interpretation for the activation energy inferred from reaction rates.

The strong temperature dependence of rate constants, described by the Arrhenius law, is explained by the Maxwell–Boltzmann distribution of molecular energies (Fig. 18.10). If E_a is the critical relative collision energy required for a pair of molecules to react, only a small fraction of the molecules will have at least this much energy at sufficiently low temperature. This fraction corresponds to the area under the Maxwell–Boltzmann distribution curve between E_a and ∞. As the temperature increases, the distribution function spreads out to include higher energies. The fraction of molecules having more than the critical energy E_a increases exponentially as $\exp(-E_a/RT)$, in agreement with Arrhenius's law and experiment. The reaction rate is then proportional to $\exp(-E_a/RT)$. So, both the strong temperature dependence and the order of magnitude of the experimental rate constants are explained by the kinetic theory of gases.

The Reaction Coordinate and the Activated Complex

Why should there be a critical collision energy E_a for reaction to occur between two molecules? To understand this, let's consider the physical analogy of marbles rolling on a hilly surface. As a marble rolls up a hill, its potential energy increases and its kinetic energy decreases; it slows down as it climbs the hill. If it can reach the top of the hill, it will fall down the other side, whereupon its kinetic energy will increase and its potential energy will decrease. Not every marble will make it over the hill. If its initial speed, and therefore kinetic energy, is too small, a marble will roll only part way up and then fall back down. Only those marbles with initial kinetic energy higher than some critical threshold will pass over the hill.

We can use this physical model to describe molecular collisions and reactions. As two reactant molecules, atoms, or ions approach each other along a **reaction path,** their potential energy increases as the bonds within them distort. At some maximum potential energy the collision partners become connected in an unstable entity called the **activated complex** or **transition state.** The activated complex is the cross-over stage where the smooth ascent in potential energy as the reactants come together becomes a smooth descent as the product molecules separate. As in the case of the marbles, not all pairs of colliding bodies react. Only those pairs with sufficient kinetic energy can stretch bonds and rearrange atoms enough to become the transition state through which reactants become products. If the barrier to reach the transition state is too high, almost all colliding pairs of reactant molecules separate from each other without reacting. The height of the barrier is close to the measured activation energy for the reaction.

Figure 18.11 shows a graph of the potential energy versus position along the reaction path for the reaction

$$NO_2(g) + CO(g) \longrightarrow NO(g) + CO_2(g)$$

Two activation energies are shown in Figure 18.11: $E_{a,f}$ is the activation energy for the forward reaction, and $E_{a,r}$ is that for the reverse reaction, in which NO takes an oxygen atom from CO_2 to form NO_2 and CO. The difference between the two is ΔU, the change in internal energy of the chemical reaction:

$$\Delta U = E_{a,f} - E_{a,r}$$

Whereas ΔU is a thermodynamic quantity that can be obtained from calorimetric measurements, $E_{a,f}$ and $E_{a,r}$ must be found from the temperature dependence of the rate constants for the forward and reverse reactions. In this reaction the forward and reverse activation energies are 132 and 358 kJ mol^{-1}, respectively, and ΔU from thermodynamics is -226 kJ mol^{-1}.

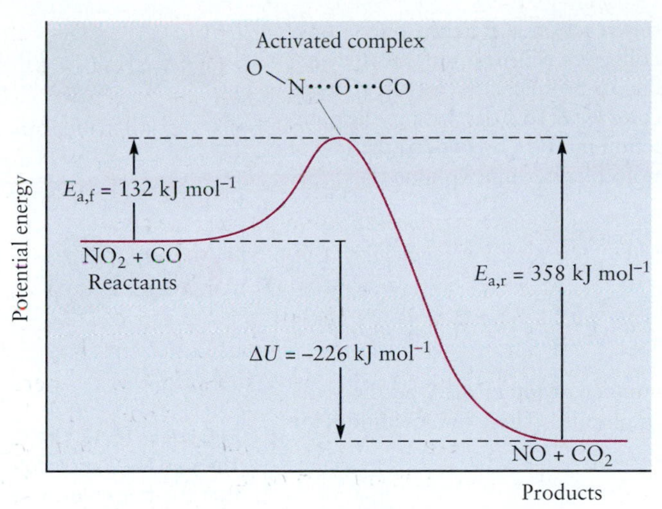

FIGURE 18.11 The energy profile along the reaction coordinate for the reaction $NO_2 + CO \longrightarrow NO + CO_2$. This direct reaction dominates the kinetics at high temperatures (above about 500 K).

The activation energy for an elementary reaction is always positive—although in some cases it can be quite small—because there is always some energy barrier to surmount. Rates of elementary reactions therefore increase with increasing temperature. This is not necessarily true for rates of overall reactions consisting of more than one elementary reaction. These sometimes have "negative activation energies," which means that the overall reaction rate is slower at higher temperature. How can this be? Let's examine a specific example: the reaction of NO with oxygen

$$2\,NO(g) + O_2(g) \longrightarrow 2\,NO_2(g)$$

has the observed rate law

$$\text{rate} = k_{obs}[NO]^2[O_2]$$

where k_{obs} *decreases* with increasing temperature. In Section 18.4 we accounted for this rate expression with a two-step mechanism. The first step is a rapid equilibrium (with equilibrium constant K_1) between two NO molecules and their dimer, N_2O_2. The second step is the slow reaction (with rate constant k_2) of N_2O_2 with O_2 to form products. The overall rate constant is therefore the product of k_2 and K_1. Whereas k_2 is the rate constant for an elementary reaction, and so increases with increasing temperature, K_1 is an *equilibrium* constant and may decrease as temperature increases. Provided the reaction is sufficiently exothermic (as it is in this case), K_1 will decrease so rapidly with increasing temperature that the product k_2K_1 will decrease as well. This combination of effects explains the observation of "negative activation energies" in some overall chemical reactions.

A DEEPER LOOK

18.6 Reaction Dynamics

In this section we first study the dynamics of reactive collisions in the gas phase and connect the predicted results to experimentally measured rate constants. Then we compare these results with the corresponding dynamics for reactions in solution. The starting point is Equation 9.35 from Section 9.8, which gives the rate of collisions of a single molecule with other molecules of the same type:

$$Z_1 = \sqrt{2}\,\pi\, d^2 \bar{u}\,\frac{N}{V} = 4\,d^2\sqrt{\frac{\pi RT}{\mathcal{M}}}\,\frac{N}{V}$$

Here d is the molecular diameter, $\bar{u}$ the average speed, $\mathcal{M}$ the molar mass, and N/V the number density of molecules in the gas. If there are N molecules in the volume, the *total* number of collisions per unit time is $\frac{1}{2}N \times Z_1$. (The factor $\frac{1}{2}$ appears because the collision of two A molecules counts as only one collision—not as one collision for the first A molecule with the second, plus another for the second with the first.) The rate of collisions *per unit volume* is then this result divided by V, or

$$\text{rate of collisions per unit volume} = Z_{AA} = 2\,d^2\sqrt{\frac{\pi RT}{\mathcal{M}}}\left(\frac{N}{V}\right)^2$$

The next step is to relate Z_{AA} to the second-order rate constant for the reaction

$$A + A \longrightarrow \text{products}$$

If the activation energy for this reaction is E_a, then only a fraction $\exp(-E_a/RT)$ of these collisions will have sufficient energy to overcome the barrier leading to products. Each such *effective* collision of a pair of A molecules leads to a decrease in the number of A molecules in the reaction mixture by two, so the rate of change of the number of A molecules per unit volume is

$$\frac{d(N/V)}{dt} = -2Z_{AA}e^{-E_a/RT}$$

$$= -2 \times 2\, d^2\sqrt{\frac{\pi RT}{\mathcal{M}}}e^{-E_a/RT}\left(\frac{N}{V}\right)^2$$

Rate constants involve the number of *moles* of A per unit volume, [A], not the number of molecules. These two quantities are related by

$$N_A[A] = \frac{N}{V}$$

where N_A is Avogadro's number. Substituting this equation and bringing a factor of $-\frac{1}{2}$ back to the left side (as in our usual definition of reaction rate), we find

$$\text{rate} = -\frac{1}{2}\frac{d[A]}{dt} = 2\, d^2\, N_A\sqrt{\frac{\pi RT}{\mathcal{M}}}e^{-E_a/RT}[A]^2$$

The rate constant predicted by simple collision dynamics is thus

$$k = 2\, d^2\, N_A\sqrt{\frac{\pi RT}{\mathcal{M}}}\, e^{-E_a/RT} \qquad [18.7]$$

How well does this simple theory agree with experiment? By fitting data for gas-phase elementary reaction rates to the Arrhenius form, we can obtain the activation energy and the factor A. The value of A can be compared with the theory, once we estimate the molecular diameter. For the elementary reaction

$$2\, NOCl(g) \longrightarrow 2\, NO(g) + Cl_2(g)$$

the measured rate constant is 0.16 times the calculated rate constant. This indicates that not all collisions lead to reaction, even if the molecules have sufficient relative kinetic energy. What other effect have we overlooked? The relative orientations of the colliding molecules certainly should play a role in determining whether a particular collision results in a reaction. In order to split off a Cl_2 molecule as a reaction product, it seems obvious that the two NOCl reactant molecules must approach each other in orientations that bring the chlorine atoms close together (Fig. 18.12). So, the calculated collision frequency must be multiplied by a **steric factor** P (0.16 for this reaction) to account for the fact that only a fraction of the collisions occur with the proper orientation to lead to reaction.

We can extend the collision theory to calculate the rate constant for bimolecular reactions of two species, A and B. Comparing observed and predicted rate constants gives the values of P shown in Table 18.1. As the colliding molecules become larger and more complex, P becomes smaller because a smaller fraction of collisions is effective in causing reaction. The steric factor is an empirical correction that has to be identified by comparing results of the simple theory with experimental data. It can be predicted in more advanced theories but only for especially simple reactions.

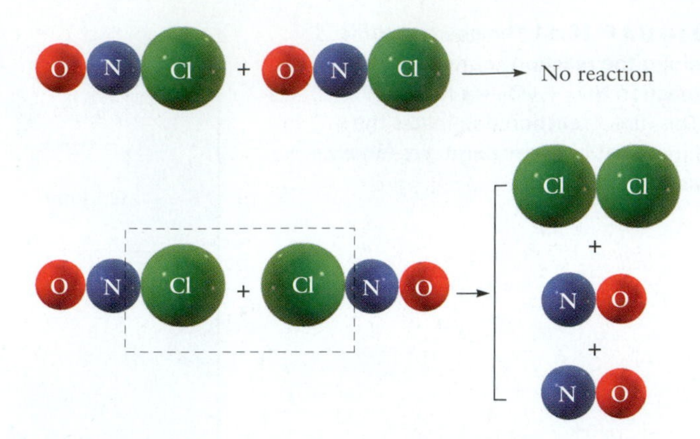

FIGURE 18.12 The steric effect on the probability of a reaction. The two NOCl molecules must approach each other in such a way that the two chlorine atoms are close together, if the encounter is to produce $Cl_2(g)$ and $NO(g)$.

TABLE 18.1 Steric Factors for Gas-Phase Reactions

Reaction	Steric Factor P
$2\, NOCl \longrightarrow 2\, NO + Cl_2$	0.16
$2\, NO_2 \longrightarrow 2\, NO + O_2$	5.0×10^{-2}
$2\, ClO \longrightarrow Cl_2 + O_2$	2.5×10^{-3}
$H_2 + C_2H_4 \longrightarrow C_2H_6$	1.7×10^{-6}

Adapted from P. W. Atkins, *Physical Chemistry,* 5th ed. New York: W. H. Freeman, 1994, p. C30.

Molecular Beams

The kinetics experiments described so far are all carried out by changing reactant concentrations and the temperature. These are macroscopic properties of a reacting mixture, and allow the colliding molecules to have a broad range of energy. So, the results represent averages over range of collision energies. An alternative way to study kinetics and obtain molecular information directly is to use the *crossed molecular beam technique*. In this device two beams of reactant molecules intersect in a chamber under high vacuum, and the reaction products are identified by some appropriate detector (Fig. 18.13).

This apparatus allows the reactant molecules to be prepared in highly selective conditions. A velocity selector between the beam source and the collision region (see Fig. 9.13) will pass only those molecules whose velocities fall within a small range. This permits much finer tuning of reaction energies than can be achieved simply by changing the temperature in a macroscopic kinetics experiment. Electric and magnetic fields or laser light can be applied to select reactant according to how fast they rotate and how much they vibrate. The velocity of products can be measured, their angular distribution relative to the directions of the initial beams can be determined, and their vibration and rotation energies can be determined. All this information allows a much more detailed examination of the way in which molecules collide and react. For example, if the activated complex rotates many times before it finally breaks up into products, the angular distribution of products should be uniform. But if it breaks up before it has a chance to rotate, the angular distribution of products

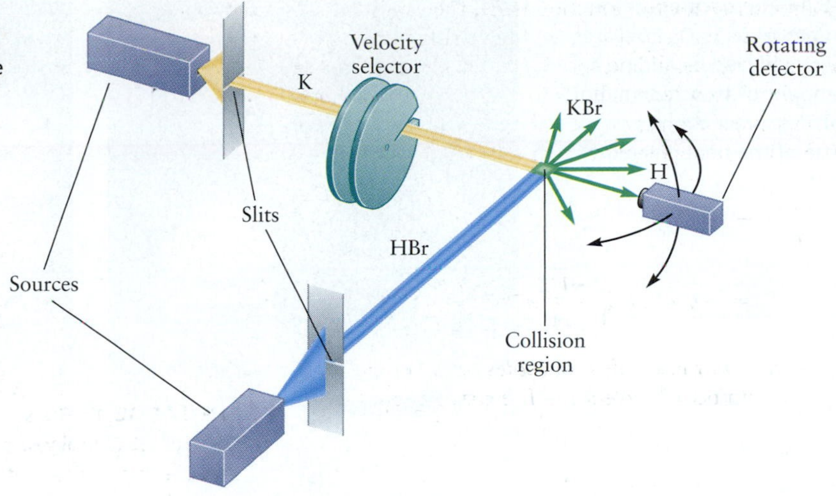

FIGURE 18.13 Diagram of a crossed molecular beam experiment. The reaction being studied is K + HBr ⟶ KBr + H. The curved arrows on the right side of the figure represent rotation of the detector in the vertical and horizontal planes.

should be very nonuniform and should depend on the directions of the original beams. This type of measurement gives information on the lifetime of an activated complex.

Molecular beams are limited to reactions that are carried out in vacuum, where well-defined beams of reactant molecules can be prepared. This limits their application to gas-phase reactions and to reactions of gaseous molecules with solid surfaces. Molecular beam methods cannot be used to study kinetics in liquid solvents. The detailed information they provide for gas–gas and gas–surface reactions allows precise testing of models and theories for the dynamics of these classes of reactions.

Reaction Kinetics in Liquids

In a gas, atoms or molecules move in straight lines between occasional collisions (see Fig. 9.22). In a liquid the concept of collision has no meaning because molecules interact continuously with not one, but many neighbors. Their trajectories can nonetheless be envisioned as rattling motions of a "test" molecule in a temporary "cage" formed by its neighbors, superimposed on its random and erratic displacement by diffusion. In a time interval t, molecules undergo various net displacements Δr. The average of the *square* of their displacements during the interval is found experimentally to be proportional to the time interval:

$$\overline{(\Delta r)^2} = 6\,Dt$$

This is the same law given for gases in Section 9.8, and it applies also to solids (see Section 21.5). The magnitude of the diffusion constant D varies widely from one phase to another. Typical values of D are on the order of 10^{-9} m^2 s^{-1} for liquids (four orders of magnitude smaller than for a typical gas at atmospheric pressure). So in a time interval of 1 s, the mean-square displacement through diffusion is of order 10^{-8} m^2 and the root-mean-square displacement is of order 10^{-4} m (0.1 mm). For larger particles or in more viscous fluids the diffusion coefficient can be much smaller and the resulting displacement smaller as well.

The motion of molecules in a liquid has a significant effect on the kinetics of chemical reactions in solution. Molecules must diffuse together before they can react, so their diffusion constants affect the rate of reaction. If the intrinsic reaction rate of two molecules that come into contact is fast enough (that is, if almost every encounter leads to reaction), then diffusion is the rate-limiting step. Such **diffusion-controlled reactions** have a maximum bimolecular rate constant on the order of 10^{10} L mol^{-1} s^{-1} in aqueous solution for the reaction of two neutral species. If the two species have opposite charges, the reaction rate can be even higher. One of the fastest known reactions in aqueous solution is the neutralization of hydronium ion (H_3O^+) by hydroxide ion (OH^-):

$$H_3O^+(aq) + OH^-(aq) \longrightarrow 2\,H_2O(\ell)$$

for which the diffusion-controlled rate constant is greater than 10^{11} L mol^{-1} s^{-1}.

18.7 Kinetics of Catalysis

A **catalyst** is a substance that takes part in a chemical reaction and speeds up the rate but undergoes no permanent chemical change itself. Catalysts therefore do not appear in the overall balanced chemical equation. But their presence very much affects the rate law, modifying and speeding existing pathways or, more commonly, providing completely new pathways by which a reaction can occur (Fig. 18.14). Catalysts have significant effects on reaction rates even when they are present in very small amounts. Industrial chemistry devotes great effort to finding catalysts to accelerate particular desired reactions without increasing the generation of undesired products.

(a)

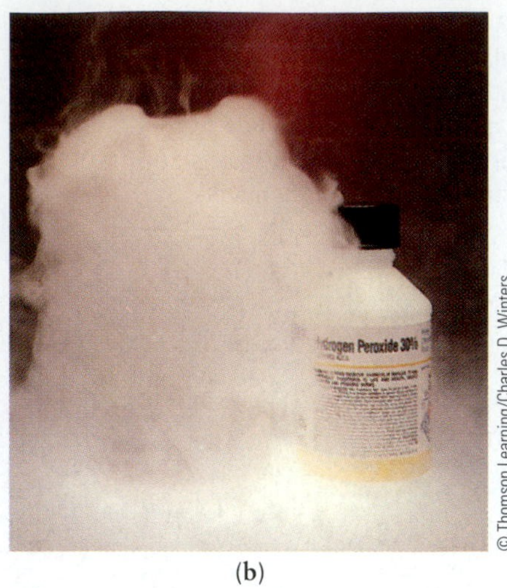

(b)

FIGURE 18.14 The decomposition of hydrogen peroxide, H_2O_2, to water and oxygen is catalyzed by adding a very small amount of transition metal oxide (a). (b) The water evolves as steam because of the heat given off in the reaction.

Catalysis are classified into two types: homogeneous and heterogeneous. In **homogeneous catalysis** the catalyst is present in the same phase as the reactants, as when a gas-phase catalyst speeds up a gas-phase reaction, or a species dissolved in solution speeds up a reaction in solution. Chlorofluorocarbons and oxides of nitrogen are homogeneous catalysts responsible for the destruction of ozone in the stratosphere. These reactions are examined in more detail in Section 20.5. A second example is the catalysis of the oxidation–reduction reaction

$$Tl^+(aq) + 2\ Ce^{4+}(aq) \longrightarrow Tl^{3+}(aq) + 2\ Ce^{3+}(aq)$$

by silver ions in solution. The direct reaction of Tl^+ with a single Ce^{4+} ion to give Tl^{2+} as an intermediate is slow. The reaction can be speeded up by adding Ag^+ ions, which take part in a reaction mechanism of the form

$$Ag^+ + Ce^{4+} \underset{k_{-1}}{\overset{k_1}{\rightleftharpoons}} Ag^{2+} + Ce^{3+} \qquad \text{(fast)}$$

$$Tl^+ + Ag^{2+} \xrightarrow{k_2} Tl^{2+} + Ag^+ \qquad \text{(slow)}$$

$$Tl^{2+} + Ce^{4+} \xrightarrow{k_3} Tl^{3+} + Ce^{3+} \qquad \text{(fast)}$$

The Ag^+ ions are not permanently transformed by this reaction because those used up in the first step are regenerated in the second; they play the role of catalyst in significantly speeding the rate of the overall reaction.

In **heterogeneous catalysis** the catalyst is present as a phase distinct from the reaction mixture. The most important case is the catalytic action of certain solid surfaces on gas-phase and solution-phase reactions. A critical step in the production of sulfuric acid relies on a solid oxide of vanadium (V_2O_5) as catalyst. Many other solid catalysts are used in industrial processes. One of the best studied is the addition of hydrogen to ethylene to form ethane:

$$C_2H_4(g) + H_2(g) \longrightarrow C_2H_6(g)$$

The process occurs extremely slowly in the gas phase but is catalyzed by a platinum surface (Fig. 18.15).

Another example is the solid catalyst used to reduce the emission of pollutants such as unburned hydrocarbons, carbon monoxide, and nitrogen oxides in the exhaust streams of automobile engines (Fig. 18.16). A **catalytic converter** is designed to simultaneously *oxidize* hydrocarbons and CO through the reactions

$$CO, C_xH_y, O_2 \xrightarrow{\text{Catalyst}} CO_2, H_2O$$

and *reduce* nitrogen oxides through the reactions

$$NO, NO_2 \xrightarrow{\text{Catalyst}} N_2, O_2$$

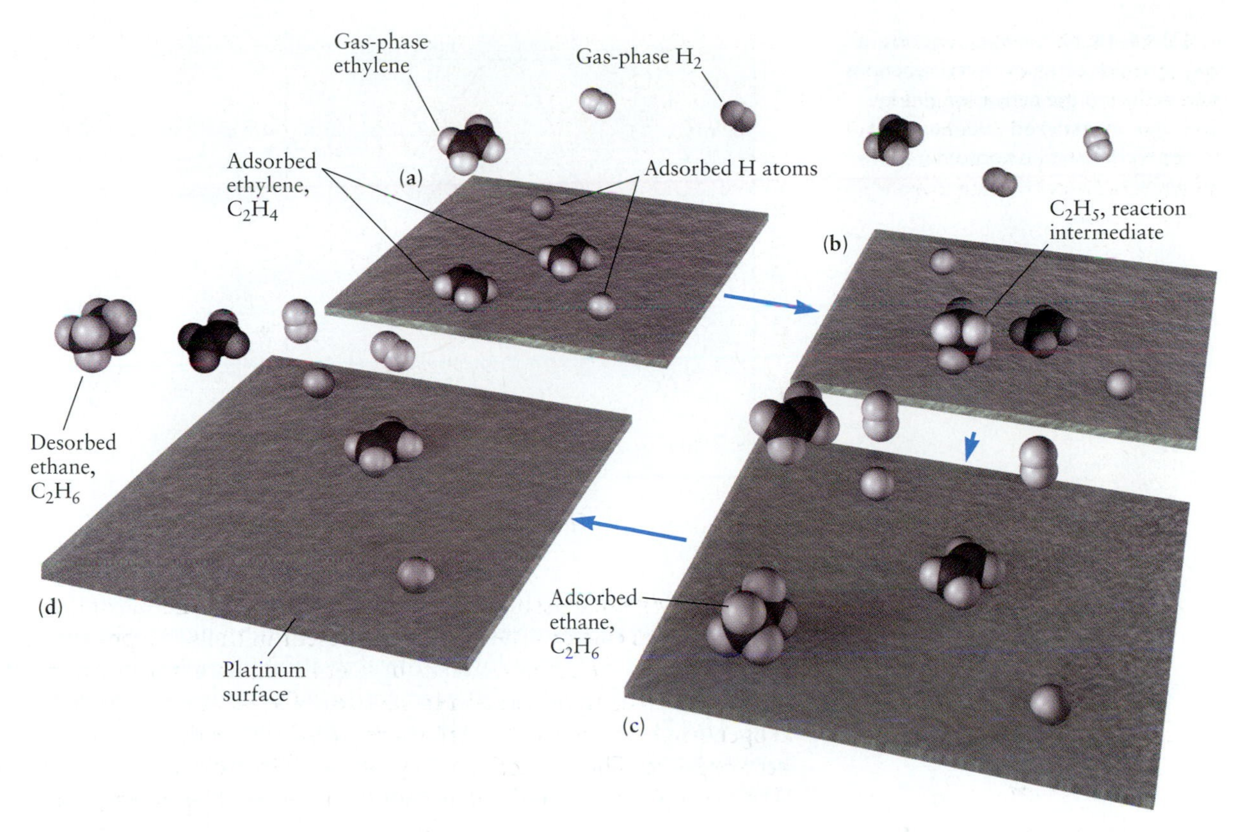

FIGURE 18.15 Platinum catalyzes the reaction $H_2 + C_2H_4$ by providing a surface that promotes the dissociation of H_2 to H atoms, which can then add to the C_2H_4 stepwise to give ethane, C_2H_6.

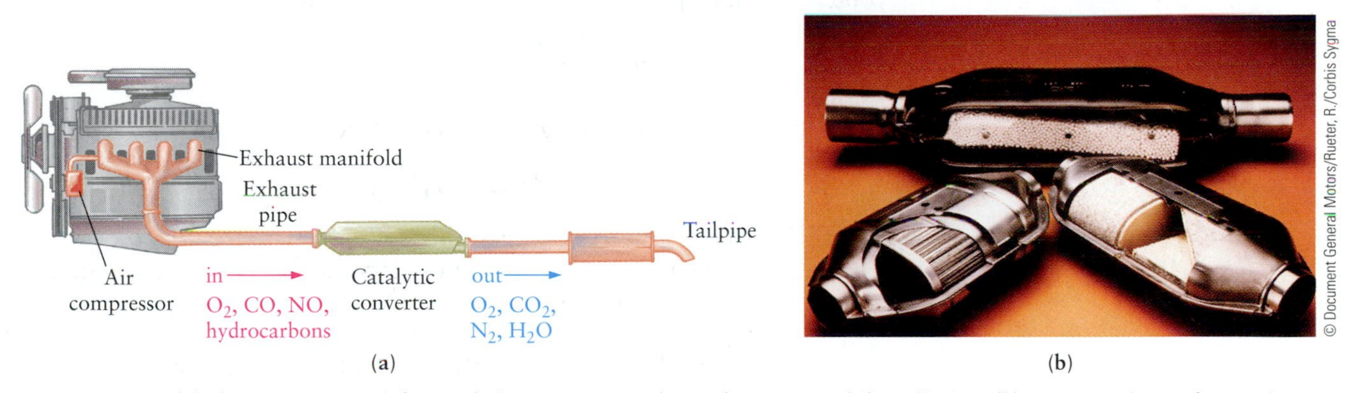

FIGURE 18.16 (a) The arrangement of a catalytic converter used to reduce automobile pollution. (b) Cutaway views of several catalytic converters showing different structures for organizing metal catalysts, platinum, palladium, and rhodium on different substrates and supports. A steel-alloy heating element raises the temperature to 400°C in seconds, activating the catalysts and reducing the pollution emitted in the first minutes after the car is started.

Clearly, the best catalyst for the reduction reactions may not be the best for the oxidation reactions, so two catalysts are combined. The noble metals, although expensive, are particularly useful. Typically, platinum and rhodium are deposited on a fine honeycomb mesh of alumina (Al_2O_3) to give a large surface area that increases the contact time of the exhaust gas with the catalysts. The platinum serves primarily as an oxidation catalyst and the rhodium as a reduction catalyst. Catalytic converters can be poisoned with certain metals that block their active sites and reduce their effectiveness. Because lead is one of the most serious such poisons, automobiles with catalytic converters must use unleaded fuel.

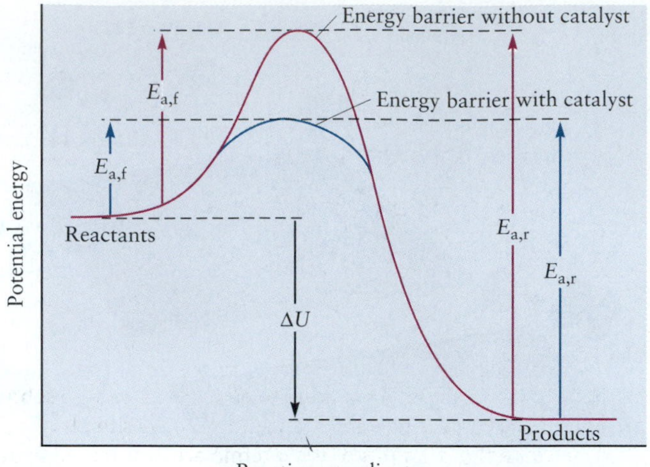

FIGURE 18.17 The most important way in which catalysts speed reactions is by reducing the activation energy. Both the uncatalyzed (blue) and catalyzed (red) reaction coordinates are shown.

A catalyst speeds up the rate of a reaction by increasing the Arrhenius factor A. More often a catalyst lowers the activation energy E_a by providing a new reaction pathway through a different activated complex with lower potential energy (Fig. 18.17). A corollary of this fact is that the same catalyst will speed both the forward *and* reverse reactions, because it lowers both the forward and reverse activation energies equally. It is important to remember that a catalyst does not change the thermodynamics of the overall reaction. Because the free energy is a function of state, ΔG is independent of the path followed and therefore the equilibrium constant is not changed by a catalyst. Reaction products not favored by thermodynamics still do not form. The role of catalysts is to speed up the rate of production of products that *are* allowed by thermodynamics.

An **inhibitor** plays an opposite role to a catalyst. It slows the rate of a reaction, often by increasing the activation energy. Like catalysts, inhibitors are important in the chemical industries because they reduce the rates of undesirable side reactions, allowing desired products to form in greater yield.

Enzyme Catalysis

Many chemical reactions in living systems are catalyzed by enzymes. An **enzyme** is a large protein molecule (typically of molar mass 20,000 g mol^{-1} or more) with a structure capable of carrying out a specific reaction or series of reactions. One or more reactant molecules (called **substrates**) bind to an enzyme at its **active sites.** These are regions on the surface of the enzyme where the local structures and chemical properties will selectively bind a specific substrate so particular chemical transformations of it can be carried out (Fig. 18.18). Many enzymes are quite specific in their active sites. The enzyme urease catalyzes the hydrolysis of urea, $(NH_2)_2CO$,

$$H_3O^+(aq) + (NH_2)_2CO(aq) + H_2O(\ell) \xrightarrow{\text{Urease}} 2\ NH_4^+(aq) + HCO_3^-(aq)$$

but will not bind most other kinds of molecules, even those of similar structure. In some cases a second species *does* bind to the enzyme and acts as an inhibitor, preventing the enzyme from its usual role as catalyst.

We summarize the kinetics of enzyme catalysis schematically by the reaction mechanism

$$E + S \underset{k_{-1}}{\overset{k_1}{\rightleftarrows}} ES$$

$$ES \xrightarrow{k_2} E + P$$

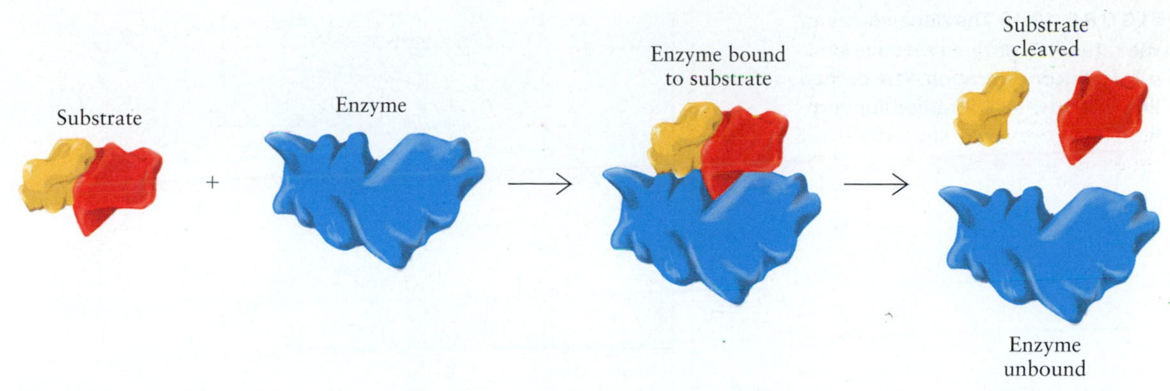

FIGURE 18.18 The binding of a substrate to an enzyme and the subsequent reaction of the substrate. As this schematic figure suggests, the size and shape of the active site play a role in determining which substrates bind. Equally important are the strengths of the intermolecular forces between nearby groups on the enzyme and substrate.

Here E stands for the free enzyme, S for the substrate, ES for the complex formed when the substrate binds to the active site, and P for the product of the chemical transformation. We can obtain the rate of formation of P through this mechanism from the steady-state approximation for the concentration of the enzyme-substrate complex, [ES]. As Section 18.4 shows, this approximation involves setting the rate of change of [ES] to 0:

$$\frac{d[\mathrm{ES}]}{dt} = 0 = k_1[\mathrm{E}][\mathrm{S}] - k_{-1}[\mathrm{ES}] - k_2[\mathrm{ES}]$$

This equation could be solved to find [ES] in terms of [E], but that is not quite what we want. Here [E] is the concentration of free enzyme which does not have substrate bound to it, and [ES] is the concentration of bound enzyme. Their sum is $[\mathrm{E}]_0$, the total amount of enzyme present:

$$[\mathrm{E}]_0 = [\mathrm{E}] + [\mathrm{ES}]$$

Because $[\mathrm{E}]_0$ is the quantity accessible to experiment, we replace [E] in the preceding steady-state equation with $[\mathrm{E}]_0 - [\mathrm{ES}]$. This gives

$$\frac{d[\mathrm{ES}]}{dt} = 0 = k_1[\mathrm{E}]_0[\mathrm{S}] - k_1[\mathrm{ES}][\mathrm{S}] - k_{-1}[\mathrm{ES}] - k_2[\mathrm{ES}]$$

which can be solved for [ES] to give

$$[\mathrm{ES}] = \frac{k_1[\mathrm{E}]_0[\mathrm{S}]}{k_1[\mathrm{S}] + (k_{-1} + k_2)}$$

If K_m is defined by

$$K_\mathrm{m} = \frac{k_{-1} + k_2}{k_1}$$

then

$$[\mathrm{ES}] = \frac{[\mathrm{E}]_0[\mathrm{S}]}{[\mathrm{S}] + K_\mathrm{m}}$$

The rate of formation of product is

$$\frac{d[\mathrm{P}]}{dt} = k_2[\mathrm{ES}] = \frac{k_2[\mathrm{E}]_0[\mathrm{S}]}{[\mathrm{S}] + K_\mathrm{m}} \qquad [18.8]$$

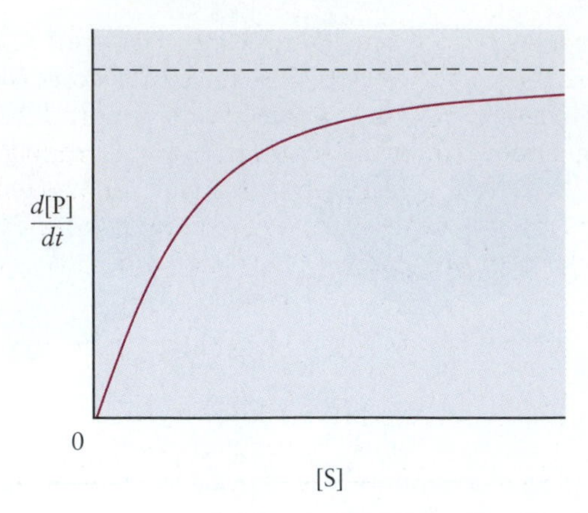

FIGURE 18.19 The dependence of the rate of an enzyme reaction on substrate concentration. The dashed line gives the rate attained for very high [S].

This is the **Michaelis–Menten equation,** which describes the kinetics of many enzyme-catalyzed reactions. Figure 18.19 shows a typical experimental curve for the dependence of the rate of such a reaction on substrate concentration [S]. For low substrate concentrations the rate is linear in [S], but at high concentrations it levels off. This behavior is easily understood on physical grounds. When [S] is small, the more substrate is added, the greater is the rate of formation of product. As the amount of substrate becomes large, all the active sites of the enzyme molecules are bound to substrate, so adding more S does not affect the rate of the reaction. Saturation occurs, and the rate levels off.

By taking the inverse of each side of the Michaelis–Menten equation, we obtain the very useful alternative form

$$\frac{1}{d[\mathrm{P}]/dt} = \frac{1}{k_2[\mathrm{E}]_0} + \frac{K_\mathrm{m}}{k_2[\mathrm{E}]_0[\mathrm{S}]}$$

Using this form of the Michaelis–Menten equation, we see that plotting the inverse of the reaction rate against the inverse of the substrate concentration should produce a straight line whose slope and intercept are $k_2[\mathrm{E}]_0$ and K_m, respectively.

CHAPTER SUMMARY

Chemical kinetics is the study of the rates of chemical reactions and the factors that control those rates. The rates of chemical reactions depend on the concentrations of all of the reactants and products as well as the temperature. The concentration dependence of reaction rates is given by empirical rate laws in which the rate depends on the concentration of each reactant raised to some power that is not, in general, related to the stoichiometric coefficients in the balanced equation for the reaction. Rate laws may be determined by measuring the initial rate of a reaction as a function of initial concentration or by inspecting the concentration-time profile over the course of the reaction. Concentrations decrease exponentially with time for first order reactions and the half-life is a characteristic time scale for the reaction. Most chemical reactions occur as a series of elementary reactions, which constitute the reaction mechanism. The steady state approximation provides a simple framework to analyze the kinetics of multistep reactions. This approximation, which is widely used in many areas of chemistry and biochemistry, shows how a reaction may display either first or second order kinetics, depending on starting concentrations. It also accounts for the rate-limiting step, which determines the overall rate of the reaction. Spectroscopic detection or chemical trapping of reaction intermediates is a powerful way to establish a reaction

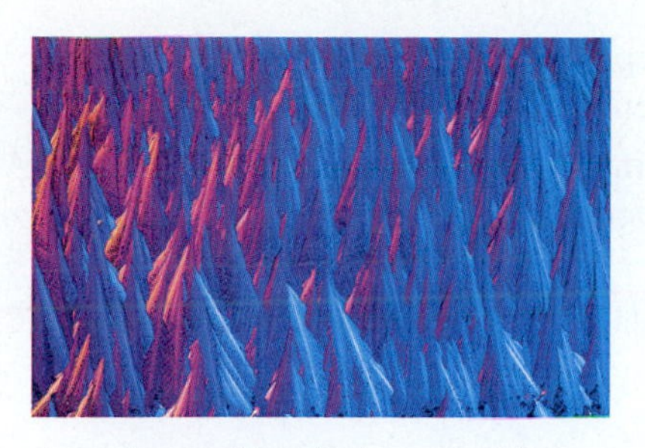

Crystals of sodium hydrogen sulfite, under polarized light.

(© Stefan Eberhard/Fran Heyl Associates)

mechanism. The rates of all elementary reactions increase with increasing temperature. As temperatures rise, more molecules have enough energy to overcome the activation barrier and proceed to form products. Chemical kinetics is a fascinating subject covering time scales that range from femtoseconds (primary photochemical processes) to eons (geochemical and astrochemical processes). The fundamental principles developed in this chapter can be applied to understand the rates and mechanism of chemical reactions in these very different areas of science.

CUMULATIVE EXERCISE

Sulfite and Sulfate Kinetics

Sulfur dioxide dissolves in water droplets (fog, clouds, and rain) in the atmosphere and reacts according to the equation

$$SO_2(aq) + 2\,H_2O(\ell) \longrightarrow HSO_3^-(aq) + H_3O^+(aq)$$

$HSO_3^-(aq)$ (hydrogen sulfite ion) is then slowly oxidized by oxygen that is also dissolved in the droplets:

$$2\,HSO_3^-(aq) + O_2(aq) + 2\,H_2O(\ell) \longrightarrow 2\,SO_4^{2-}(aq) + 2\,H_3O^+(aq)$$

Although the second reaction has been studied for many years, only recently was it shown to proceed by the steps

$$2\,HSO_3^-(aq) + O_2(aq) \longrightarrow S_2O_7^{2-}(aq) + H_2O(\ell) \qquad \text{(fast)}$$

$$S_2O_7^{2-}(aq) + 3\,H_2O(\ell) \longrightarrow 2\,SO_4^{2-}(aq) + 2\,H_3O^+(aq) \qquad \text{(slow)}$$

The reactive intermediate, $S_2O_7^{2-}(aq)$, which had not been detected in earlier studies of this reaction, is well known in other reactions. It is the *disulfate* ion.

In an experiment at 25°C, a solution was mixed with the realistic initial concentrations of 0.270 M $HSO_3^-(aq)$ and 0.0135 M $O_2(aq)$. The initial pH was 3.90. The following table tells what happened in the solution, beginning at the moment of mixing.

Time(s)	$[HSO_3^-]$ (M)	$[O_2]$ (M)	$[S_2O_7^{2-}]$ (M)	$[HSO_4^-] + [SO_4^{2-}]$ (M)
0.000	0.270	0.0135	0.000	0.000
0.010	0.243	0.000	13.5×10^{-3}	0.000
10.0	0.243	0.000	11.8×10^{-3}	03.40×10^{-3}
45.0	0.243	0.000	7.42×10^{-3}	12.2×10^{-3}
90.0	0.243	0.000	4.08×10^{-3}	18.8×10^{-3}
150.0	0.243	0.000	1.84×10^{-3}	23.3×10^{-3}
450.0	0.243	0.000	0.034×10^{-3}	26.9×10^{-3}
600.0	0.243	0.000	0.005×10^{-3}	27.0×10^{-3}

(a) Determine the average rate of increase of the total of the concentrations of the sulfate plus hydrogen sulfate ions during the first 10 s of the experiment.

(b) Determine the average rate of disappearance of hydrogen sulfite ion during the first 0.010 s of the experiment.

(c) Explain why the hydrogen sulfite ion stops disappearing after 0.010 s.

(d) Plot the concentration of disulfate ion versus time on graph paper, and use the graph to estimate the instantaneous rate of disappearance of disulfate ion 90.0 s after the reaction starts.

(e) Determine the order with respect to the disulfate ion of the second step of the conversion, and the rate constant of that step.

(f) Determine the half-life of the second step of the conversion process.

(g) At 15°C the rate constant of the second step of the conversion is only 62% of its value at 25°C. Compute the activation energy of the second step.

(h) The first step of the conversion occurs much faster when 1.0×10^{-6} M $Fe^{2+}(aq)$ ion is added (but the rate of the second step is unaffected). What role does Fe^{2+} play?

(i) Write a balanced equation for the overall reaction that gives sulfuric acid from SO_2 dissolved in water droplets in the air.

Answers

(a) 3.40×10^{-4} mol L^{-1} s^{-1}

(b) 2.7 mol L^{-1} s^{-1}

(c) All of the oxygen is consumed, so the first step of the process is over.

(d) 5.43×10^{-5} mol L^{-1} s^{-1}

(e) First order; $k = 0.0133$ s^{-1}

(f) 52 s

(g) 34 kJ mol^{-1}

(h) Fe^{2+} acts as a catalyst.

(i) $2\ SO_2(aq) + O_2(aq) + 6\ H_2O(\ell) \longrightarrow SO_4^{2-}(aq) + 4\ H_3O^+(aq)$

CHAPTER REVIEW

- The average rate of a chemical reaction is the change in concentration of one of the reactants or products over a finite interval of time. The instantaneous rate is the slope of a plot of concentration versus time at a particular point in time. The rate of the reaction is given by the rate for a particular species divided by its stoichiometric coefficient.
- The rates of all chemical reactions depend on the concentrations of reactants. This concentration dependence is expressed in the empirical rate law in which the rate is proportional to the concentration of reactants raised to some power. The power is not necessarily related to the stoichiometric coefficient in the balanced reaction.
- The order of the reaction with respect to each species is the power to which the concentration of that species is raised in the empirical rate law. The overall order of the reaction is the sum of the individual reaction orders.
- Integrated rate laws show how the concentration of a particular species changes over time, and they can be useful in deducing empirical rate laws and rate constants when it is difficult to measure initial rates.
- $c(t) = c_0 e^{-kt}$ for a first order reaction.
  A plot of ln $[c(t)]$ versus t is a straight line with slope $-k$.
  The half-life, $t_{1/2}$, is the time it takes for the concentration to decrease to half its initial value.
  $t_{1/2}$ is independent of concentration for first order reactions and it provides a convenient way to characterize the rates of first order reactions.
- $\frac{1}{c} = \frac{1}{c_0} + 2kt$ for second order reactions written in the form $2A \longrightarrow$ products.
  A plot of $1/c$ versus t is a straight line with slope $2k$.
  The half life in second order reactions is not independent of time and is not a very useful measure of the reaction rate.
- A reaction mechanism is the detailed sequence of elementary reactions that lead to the overall chemical reaction.
  An elementary reaction occurs exactly as its balanced equation implies.

Dissociation of a reactant in the absence of any other reactants is a unimolecular elementary reaction.
Bimolecular elementary reactions are the most common in chemistry; they result from single bimolecular collisions between two reactants.
Termolecular reactions resulting from simultaneous collisions of three reactants are rare.

- The principle of detailed balance connects kinetics and chemical equilibrium. The rates of forward and reverse reactions for elementary reactions must be the same at equilibrium so $\frac{k_f}{k_r} = K$.
- Writing empirical rate laws for reactions whose mechanisms have been established is straightforward; inferring mechanisms from empirical rate laws is not. Direct detection or trapping of reactive intermediates provides the strongest evidence in support of a proposed mechanism.
- One elementary reaction is often slower than all the others in a multi-step mechanism. The rate of the overall reaction is limited by this rate-determining step. Elementary reactions that precede the rate-limiting step are often at equilibrium.
- The steady-state approximation provides a simple framework for analyzing the kinetics of many chemical reactions.
  The mechanism is a sequence of two reactions; the first step is reversible and the second step is irreversible.
  At low pressures or concentrations the first step is rate limiting and second order kinetics are observed.
  At high pressures or concentrations equilibrium is established in the first step and the second step is rate limiting. First order kinetics are observed.
- Chain reactions proceed through a series of elementary reactions, one or more of which is repeated many times.
  Reactive intermediates are generated at initiation.
  Products are formed by propagation (with intermediate concentrations at steady state).
  The reaction terminates when two reactive intermediates combine to form a stable product.
- The rates of all elementary reactions increase with increasing temperatures.
  Collision rates increase with increasing temperature.
  The number of effective collisions increases dramatically because the fraction of molecules (calculated using the Boltzmann distribution) with energy greater than the activation energy is greater.
  The empirical Arrhenius equation, $k = Ae^{-E_A/RT}$, accounts for the increase in the number of effective collisions in the exponential term, whereas the pre-exponential factor, A, is assumed to be temperature independent.
- The activated complex or transition state is the configuration of the system with the highest potential energy. The energy difference between the reactants and the transition state is the activation energy for the reaction. Only those molecules with enough energy to surmount the barrier will react.
- Reaction dynamics provide a microscopic picture of chemical reactions that connect to the macroscopic kinetics.
  Rates of bimolecular collisions are calculated using the kinetic theory of gases to arrive at the Arrhenius pre-exponential factor and the activation energy is measured experimentally.
  Observed rates can be factors of 10^{-6} smaller than those predicted from simple reaction dynamics suggesting that a steric factor, which accounts for the relative orientations of the reactants, is extremely important in determining reaction rates.

- Catalysts increase the rates of chemical reactions by providing alternate paths that have lower activation barriers.
  The transition state in the catalyzed reaction is generally different than in the uncatalyzed reaction.
  Homogeneous catalysts are present in the same phase as the reaction being catalyzed whereas heterogeneous catalysts are present in a different phase.
- Enzymes are biological catalysts that bind substrates with exquisite selectivity, position reactants at optimal locations and stabilize transition states, all of which leads to lower activation barriers than those in the uncatalyzed reaction.

CONCEPTS & SKILLS

After studying this chapter and working the problems that follow, you should be able to:

1. Describe experimental methods for measuring average and instantaneous rates (Section 18.1, Problems 1–2).
2. Deduce rate laws and reaction orders from experimental measurements of the dependence of reaction rates on concentrations (Section 18.2, Problems 5–8).
3. Use the integrated rate laws for first- and second-order reactions to calculate the concentrations remaining after a certain elapsed time (Section 18.2, Problems 9–18).
4. Describe the relationship between the equilibrium constant for a reaction and the corresponding forward and reverse rate constants (Section 18.3, Problems 23–24).
5. Deduce the rate law from a mechanism characterized by a single rate-determining step (Section 18.4, Problems 25–30).
6. Use the steady-state approximation to deduce rate laws when no single rate-determining step exists (Section 18.4, Problems 31–34).
7. Calculate Arrhenius factors and activation energies from measurements of the temperature dependence of rate constants (Section 18.5, Problems 35–40).
8. Discuss the connection between activation energy and the energy distribution of molecules, and relate the forward and reverse activation energies to each other through thermodynamics (Section 18.5, Problems 41–42).
9. Outline the quantitative calculation of rate constants, using the collision theory of gases (Section 18.6, Problems 43–44).
10. Describe several types of catalysts and their effects on chemical reactions (Section 18.7).
11. Relate the rate of an enzyme-catalyzed reaction to the concentrations of substrate and enzyme in the reaction mixture (Section 18.7, Problems 45–46).

KEY EQUATIONS

$$\text{rate} = -\frac{1}{a}\frac{d[\text{A}]}{dt} = -\frac{1}{b}\frac{d[\text{B}]}{dt} = \frac{1}{c}\frac{d[\text{C}]}{dt} = \frac{1}{d}\frac{d[\text{D}]}{dt} \qquad \textbf{(Section 18.1)}$$

$$c = c_0\, e^{-kt} \qquad \textbf{(Section 18.2)}$$

$$t_{1/2} = \frac{\ln 2}{k} = \frac{0.6931}{k} \qquad \textbf{(Section 18.2)}$$

$$\frac{1}{c} = \frac{1}{c_0} + 2kt \qquad \text{(Section 18.2)}$$

$$k = Ae^{-E_a/RT} \qquad \text{(Section 18.5)}$$

$$\ln k = \ln A - \frac{E_a}{RT} \qquad \text{(Section 18.5)}$$

$$k = 2\,d^2\,N_A\sqrt{\frac{\pi RT}{\mathcal{M}}}\,e^{-E_a/RT} \qquad \text{(Section 18.6)}$$

$$\frac{d[P]}{dt} = k_2[ES] = \frac{k_2[E]_0[S]}{[S] + K_m} \qquad \text{(Section 18.7)}$$

PROBLEMS

Answers to problems whose numbers are boldface appear in Appendix G. Problems that are more challenging are indicated with asterisks.

Rates of Chemical Reactions

1. Use Figure 18.3 to estimate graphically the instantaneous rate of production of NO at $t = 200$ s.

2. Use Figure 18.3 to estimate graphically the instantaneous rate of production of NO at $t = 100$ s.

3. Give three related expressions for the rate of the reaction

$$N_2(g) + 3\,H_2(g) \longrightarrow 2\,NH_3(g)$$

assuming that the concentrations of any intermediates are constant and that the volume of the reaction vessel does not change.

4. Give four related expressions for the rate of the reaction

$$2\,H_2CO(g) + O_2(g) \longrightarrow 2\,CO(g) + 2\,H_2O(g)$$

assuming that the concentrations of any intermediates are constant and that the volume of the reaction vessel does not change.

Rate Laws

5. Nitrogen oxide reacts with hydrogen at elevated temperatures according to the following chemical equation:

$$2\,NO(g) + 2\,H_2(g) \longrightarrow N_2(g) + 2\,H_2O(g)$$

It is observed that, when the concentration of H_2 is cut in half, the rate of the reaction is also cut in half. When the concentration of NO is multiplied by 10, the rate of the reaction increases by a factor of 100.

(a) Write the rate expression for this reaction, and give the units of the rate constant k.

(b) If [NO] were multiplied by 3 and $[H_2]$ by 2, what change in the rate would be observed?

6. In the presence of vanadium oxide, $SO_2(g)$ reacts with an excess of oxygen to give $SO_3(g)$:

$$SO_2(g) + \tfrac{1}{2}\,O_2(g) \xrightarrow{V_2O_5} SO_3(g)$$

This reaction is an important step in the manufacture of sulfuric acid. It is observed that tripling the SO_2 concentration increases the rate by a factor of 3, but tripling the SO_3 concentration decreases the rate by a factor of $1.7 \approx \sqrt{3}$. The rate is insensitive to the O_2 concentration as long as an excess of oxygen is present.

(a) Write the rate expression for this reaction, and give the units of the rate constant k.

(b) If $[SO_2]$ is multiplied by 2 and $[SO_3]$ by 4 but all other conditions are unchanged, what change in the rate will be observed?

7. In a study of the reaction of pyridine (C_5H_5N) with methyl iodide (CH_3I) in a benzene solution, the following set of initial reaction rates was measured at 25°C for different initial concentrations of the two reactants:

$[C_5H_5N]$ (mol L^{-1})	$[CH_3I]$ (mol L^{-1})	Rate (mol L^{-1} s^{-1})
1.00×10^{-4}	1.00×10^{-4}	7.5×10^{-7}
2.00×10^{-4}	2.00×10^{-4}	3.0×10^{-6}
2.00×10^{-4}	4.00×10^{-4}	6.0×10^{-6}

(a) Write the rate expression for this reaction.

(b) Calculate the rate constant k, and give its units.

(c) Predict the initial reaction rate for a solution in which $[C_5H_5N]$ is 5.0×10^{-5} M and $[CH_3I]$ is 2.0×10^{-5} M.

8. The rate for the oxidation of iron(II) by cerium(IV)

$$Ce^{4+}(aq) + Fe^{2+}(aq) \longrightarrow Ce^{3+}(aq) + Fe^{3+}(aq)$$

is measured at several different initial concentrations of the two reactants:

$[Ce^{4+}]$ (mol L^{-1})	$[Fe^{2+}]$ (mol L^{-1})	Rate (mol L^{-1} s^{-1})
1.1×10^{-5}	1.8×10^{-5}	2.0×10^{-7}
1.1×10^{-5}	2.8×10^{-5}	3.1×10^{-7}
3.4×10^{-5}	2.8×10^{-5}	9.5×10^{-7}

(a) Write the rate expression for this reaction.

(b) Calculate the rate constant k, and give its units.

(c) Predict the initial reaction rate for a solution in which $[Ce^{4+}]$ is 2.6×10^{-5} M and $[Fe^{2+}]$ is 1.3×10^{-5} M.

9. The reaction $SO_2Cl_2(g) \longrightarrow SO_2(g) + Cl_2(g)$ is first order, with a rate constant of $2.2 \times 10^{-5}\ s^{-1}$ at 320°C. The partial pressure of $SO_2Cl_2(g)$ in a sealed vessel at 320°C is 1.0 atm. How long will it take for the partial pressure of $SO_2Cl_2(g)$ to fall to 0.50 atm?

10. The reaction $FClO_2(g) \longrightarrow FClO(g) + O(g)$ is first order with a rate constant of $6.76 \times 10^{-4}\ s^{-1}$ at 322°C.
(a) Calculate the half-life of the reaction at 322°C.
(b) If the initial partial pressure of $FClO_2$ in a container at 322°C is 0.040 atm, how long will it take to fall to 0.010 atm?

11. The decomposition of benzene diazonium chloride

$$C_6H_5N_2Cl \longrightarrow C_6H_5Cl + N_2$$

follows first-order kinetics with a rate constant of $4.3 \times 10^{-5}\ s^{-1}$ at 20°C. If the initial partial pressure of $C_6H_5N_2Cl$ is 0.0088 atm, calculate its partial pressure after 10.0 hours.

12. At 600 K, the rate constant for the first-order decomposition of nitroethane

$$CH_3CH_2NO_2(g) \longrightarrow C_2H_4(g) + HNO_2(g)$$

is $1.9 \times 10^{-4}\ s^{-1}$. A sample of $CH_3CH_2NO_2$ is heated to 600 K, at which point its initial partial pressure is measured to be 0.078 atm. Calculate its partial pressure after 3.0 hours.

13. Chloroethane decomposes at elevated temperatures according to the reaction

$$C_2H_5Cl(g) \longrightarrow C_2H_4(g) + HCl(g)$$

This reaction obeys first-order kinetics. After 340 s at 800 K, a measurement shows that the concentration of C_2H_5Cl has decreased from 0.0098 mol L^{-1} to 0.0016 mol L^{-1}. Calculate the rate constant k at 800 K.

14. The isomerization reaction

$$CH_3NC \longrightarrow CH_3CN$$

obeys the first-order rate law

$$\text{rate} = -k[CH_3NC]$$

in the presence of an excess of argon. Measurements at 500 K reveal that in 520 s the concentration of CH_3NC decreases to 71% of its original value. Calculate the rate constant k of the reaction at 500 K.

15. At 25°C in CCl_4 solvent, the reaction

$$I + I \longrightarrow I_2$$

is second order in the concentration of the iodine atoms. The rate constant k has been measured as 8.2×10^9 L $mol^{-1}\ s^{-1}$. Suppose the initial concentration of I atoms is 1.00×10^{-4} M. Calculate their concentration after 2.0×10^{-6} s.

16. HO_2 is a highly reactive chemical species that plays a role in atmospheric chemistry. The rate of the gas-phase reaction

$$HO_2(g) + HO_2(g) \longrightarrow H_2O_2(g) + O_2(g)$$

is second order in $[HO_2]$, with a rate constant at 25°C of 1.4×10^9 L $mol^{-1}\ s^{-1}$. Suppose some HO_2 with an initial concentration of 2.0×10^{-8} M could be confined at 25°C. Calculate the concentration that would remain after 1.0 s, assuming no other reactions take place.

17. The rate for the reaction

$$OH^-(aq) + NH_4^+(aq) \longrightarrow H_2O(\ell) + NH_3(aq)$$

is first order in both OH^- and NH_4^+ concentrations, and the rate constant k at 20°C is 3.4×10^{10} L $mol^{-1}\ s^{-1}$. Suppose 1.00 L of a 0.0010 M NaOH solution is rapidly mixed with the same volume of 0.0010 M NH_4Cl solution. Calculate the time (in seconds) required for the OH^- concentration to decrease to a value of 1.0×10^{-5} M.

18. The rate for the reaction

$$OH^-(aq) + HCN(aq) \longrightarrow H_2O(\ell) + CN^-(aq)$$

is first order in both OH^- and HCN concentrations, and the rate constant k at 25°C is 3.7×10^9 L $mol^{-1}\ s^{-1}$. Suppose 0.500 L of a 0.0020 M NaOH solution is rapidly mixed with the same volume of a 0.0020 M HCN solution. Calculate the time (in seconds) required for the OH^- concentration to decrease to a value of 1.0×10^{-4} M.

Reaction Mechanisms

19. Identify each of the following elementary reactions as unimolecular, bimolecular, or termolecular, and write the rate expression.
(a) $HCO + O_2 \longrightarrow HO_2 + CO$
(b) $CH_3 + O_2 + N_2 \longrightarrow CH_3O_2 + N_2$
(c) $HO_2NO_2 \longrightarrow HO_2 + NO_2$

20. Identify each of the following elementary reactions as unimolecular, bimolecular, or termolecular, and write the rate expression.
(a) $BrONO_2 \longrightarrow BrO + NO_2$
(b) $HO + NO_2 + Ar \longrightarrow HNO_3 + Ar$
(c) $O + H_2S \longrightarrow HO + HS$

21. Consider the following reaction mechanism:

$$H_2O_2 \longrightarrow H_2O + O$$
$$O + CF_2Cl_2 \longrightarrow ClO + CF_2Cl$$
$$ClO + O_3 \longrightarrow Cl + 2\,O_2$$
$$Cl + CF_2Cl \longrightarrow CF_2Cl_2$$

(a) What is the molecularity of each elementary step?
(b) Write the overall equation for the reaction.
(c) Identify the reaction intermediate(s).

22. Consider the following reaction mechanism:

$$NO_2Cl \longrightarrow NO_2 + Cl$$
$$Cl + H_2O \longrightarrow HCl + OH$$
$$OH + NO_2 + N_2 \longrightarrow HNO_3 + N_2$$

(a) What is the molecularity of each elementary step?
(b) Write the overall equation for the reaction.
(c) Identify the reaction intermediate(s).

23. The rate constant of the elementary reaction

$$BrO(g) + NO(g) \longrightarrow Br(g) + NO_2(g)$$

is 1.3×10^{10} L $mol^{-1}\ s^{-1}$ at 25°C, and its equilibrium constant is 5.0×10^{10} at this temperature. Calculate the rate constant at 25°C of the elementary reaction

$$Br(g) + NO_2(g) \longrightarrow BrO(g) + NO(g)$$

24. The compound $IrH_3(CO)(P(C_6H_5)_3)_2$ exists in two forms: the meridional ("mer") and facial ("fac"). At 25°C in a nonaqueous solvent, the reaction mer $\longrightarrow$ fac has a rate constant of 2.33 s^{-1}, and the reaction fac $\longrightarrow$ mer has a rate constant of 2.10 s^{-1}. What is the equilibrium constant of the mer-to-fac reaction at 25°C?

Reaction Mechanisms and Rate

25. Write the overall reaction and rate laws that correspond to the following reaction mechanisms. Be sure to eliminate intermediates from the answers.

(a) $A + B \underset{k_{-1}}{\overset{k_1}{\rightleftharpoons}} C + D$ (fast equilibrium)

$C + E \xrightarrow{k_2} F$ (slow)

(b) $A \underset{k_{-1}}{\overset{k_1}{\rightleftharpoons}} B + C$ (fast equilibrium)

$C + D \underset{k_{-2}}{\overset{k_2}{\rightleftharpoons}} E$ (fast equilibrium)

$E \xrightarrow{k_3} F$ (slow)

26. Write the overall reaction and the rate laws that correspond to the following reaction mechanisms. Be sure to eliminate intermediates from the answers.

(a) $2\,A + B \underset{k_{-1}}{\overset{k_1}{\rightleftharpoons}} D$ (fast equilibrium)

$D + B \xrightarrow{k_2} E + F$ (slow)

$F \xrightarrow{k_3} G$ (fast)

(b) $A + B \underset{k_{-1}}{\overset{k_1}{\rightleftharpoons}} C$ (fast equilibrium)

$C + D \underset{k_{-2}}{\overset{k_2}{\rightleftharpoons}} F$ (fast equilibrium)

$F \xrightarrow{k_3} G$ (slow)

27. HCl reacts with propene (CH_3CHCH_2) in the gas phase according to the overall reaction

$$HCl + CH_3CHCH_2 \longrightarrow CH_3CHClCH_3$$

The experimental rate expression is

$$\text{rate} = k[HCl]^3[CH_3CHCH_2]$$

Which, if any, of the following mechanisms are consistent with the observed rate expression?

(a) $HCl + HCl \rightleftharpoons H + HCl_2$ (fast equilibrium)
$H + CH_3CHCH_2 \longrightarrow CH_3CHCH_3$ (slow)
$HCl_2 + CH_3CHCH_3 \longrightarrow CH_3CHClCH_3 + HCl$ (fast)

(b) $HCl + HCl \rightleftharpoons H_2Cl_2$ (fast equilibrium)
$HCl + CH_3CHCH_2 \rightleftharpoons CH_3CHClCH_3^*$ (fast equilibrium)
$CH_3CHClCH_3^* + H_2Cl_2 \longrightarrow CH_3CHClCH_3 + 2\,HCl$ (slow)

(c) $HCl + CH_3CHCH_2 \rightleftharpoons H + CH_3CHClCH_2$ (fast equilibrium)
$H + HCl \rightleftharpoons H_2Cl$ (fast equilibrium)
$H_2Cl + CH_3CHClCH_2 \longrightarrow HCl + CH_3CHClCH_3$ (slow)

28. Chlorine reacts with hydrogen sulfide in aqueous solution

$$Cl_2(aq) + H_2S(aq) \longrightarrow S(s) + 2\,H^+(aq) + 2\,Cl^-(aq)$$

in a second-order reaction that follows the rate expression

$$\text{rate} = k[Cl_2][H_2S]$$

Which, if any, of the following mechanisms are consistent with the observed rate expression?

(a) $Cl_2 + H_2S \longrightarrow H^+ + Cl^- + Cl^+ + HS^-$ (slow)
$Cl^+ + HS^- \longrightarrow H^+ + Cl^- + S$ (fast)

(b) $H_2S \rightleftharpoons HS^- + H^+$ (fast equilibrium)
$HS^- + Cl_2 \longrightarrow 2\,Cl^- + S + H^+$ (slow)

(c) $H_2S \rightleftharpoons HS^- + H^+$ (fast equilibrium)
$H^+ + Cl_2 \rightleftharpoons H^+ + Cl^- + Cl^+$ (fast equilibrium)
$Cl^+ + HS^- \longrightarrow H^+ + Cl^- + S$ (slow)

29. Nitryl chloride is a reactive gas with a normal boiling point of −16°C. Its decomposition to nitrogen dioxide and chlorine is described by the equation

$$2\,NO_2Cl \longrightarrow 2\,NO_2 + Cl_2$$

The rate expression for this reaction has the form

$$\text{rate} = k[NO_2Cl]$$

Which, if any, of the following mechanisms are consistent with the observed rate expression?

(a) $NO_2Cl \longrightarrow NO_2 + Cl$ (slow)
$Cl + NO_2Cl \longrightarrow NO_2 + Cl_2$ (fast)

(b) $2\,NO_2Cl \rightleftharpoons N_2O_4 + Cl_2$ (fast equilibrium)
$N_2O_4 \longrightarrow 2\,NO_2$ (slow)

(c) $2\,NO_2Cl \rightleftharpoons ClO_2 + N_2O + ClO$ (fast equilibrium)
$N_2O + ClO_2 \rightleftharpoons NO_2 + NOCl$ (fast equilibrium)
$NOCl + ClO \longrightarrow NO_2 + Cl_2$ (slow)

30. Ozone in the upper atmosphere is decomposed by nitrogen oxide through the reaction

$$O_3 + NO \longrightarrow O_2 + NO_2$$

The experimental rate expression for this reaction is

$$\text{rate} = k[O_3][NO]$$

Which, if any, of the following mechanisms are consistent with the observed rate expression?

(a) $O_3 + NO \longrightarrow O + NO_3$ (slow)
$O + O_3 \longrightarrow 2\,O_2$ (fast)
$NO_3 + NO \longrightarrow 2\,NO_2$ (fast)

(b) $O_3 + NO \longrightarrow O_2\,NO_2$ (slow)

(c) $NO + NO \rightleftharpoons N_2O_2$ (fast equilibrium)
$N_2O_2 + O_3 \longrightarrow NO_2 + 2\,O_2$ (slow)

31. Consider the mechanism of problem 25(a). Suppose *no* assumptions are made about the relative rates of the steps. By making a steady-state approximation for the concentration of the intermediate (C), express the rate of production of the product (F) in terms of the concentrations of A, B, D, and E. In what limit does this reduce to the result of problem 25(a)?

32. Consider the mechanism of problem 25(b). Suppose *no* assumptions are made about the relative rates of the steps. By making a steady-state approximation for the concentrations of the intermediates (C and E), express the rate of production of the product (F) in terms of the concentrations of A, B, and D. In what limit does this reduce to the result of problem 25(b)?

33. The mechanism for the decomposition of NO_2Cl is

$$NO_2Cl \underset{k_{-1}}{\overset{k_1}{\rightleftharpoons}} NO_2 + Cl$$

$$NO_2Cl + Cl \xrightarrow{k_2} NO_2 + CL_2$$

By making a steady-state approximation for [Cl], express the rate of appearance of Cl_2 in terms of the concentrations of NO_2Cl and NO_2.

34. A key step in the formation of sulfuric acid from dissolved SO_2 in acid precipitation is the oxidation of hydrogen sulfite ion by hydrogen peroxide:

$$HSO_3^-(aq) + H_2O_2(aq) \longrightarrow HSO_4^-(aq) + H_2O(\ell)$$

The mechanism involves peroxymonosulfurous acid, SO_2OOH^-:

$$HSO_3^-(aq) + H_2O_2(aq) \underset{k_{-1}}{\overset{k_1}{\rightleftharpoons}} SO_2OOH^-(aq) + H_2O(\ell)$$

$$SO_2OOH^-(aq) + H_3O^+(aq) \xrightarrow{k_2} HSO_4^-(aq) + H_3O^+(aq)$$

By making a steady-state approximation for the reactive intermediate concentration, $[SO_2OOH^-(aq)]$, express the rate of formation of $HSO_4^-(aq)$ in terms of the concentrations of $HSO_3^-(aq)$, $H_2O_2(aq)$, and $H_3O^+(aq)$.

Effect of Temperature on Reaction Rates

35. The rate of the elementary reaction

$$Ar + O_2 \longrightarrow Ar + O + O$$

has been studied as a function of temperature between 5000 and 18,000 K. The following data were obtained for the rate constant k:

Temperature (K)	k (L mol^{-1} s^{-1})
5,000	5.49×10^6
10,000	9.86×10^8
15,000	5.09×10^9
18,000	8.60×10^9

(a) Calculate the activation energy of this reaction.
(b) Calculate the factor A in the Arrhenius equation for the temperature dependence of the rate constant.

36. The gas-phase reaction

$$H + D_2 \longrightarrow HD + D$$

is the exchange of isotopes of hydrogen of atomic mass 1 (H) and 2 (D, deuterium). The following data were obtained for the rate constant k of this reaction:

Temperature (K)	k (L mol^{-1} s^{-1})
299	1.56×10^4
327	3.77×10^4
346	7.6×10^4
440	10^6
549	1.07×10^6
745	8.7×10^7

(a) Calculate the activation energy of this reaction.
(b) Calculate the factor A in the Arrhenius equation for the temperature dependence of the rate constant.

37. The rate constant of the elementary reaction

$$BH_4^-(aq) + NH_4^+(aq) \longrightarrow BH_3NH_3(aq) + H_2(g)$$

is $k = 1.94 \times 10^{-4}$ L mol^{-1} s^{-1} at 30.0°C, and the reaction has an activation energy of 161 kJ mol^{-1}.
(a) Compute the rate constant of the reaction at a temperature of 40.0°C.
(b) After equal concentrations of $BH_4^-(aq)$ and $NH_4^+(aq)$ are mixed at 30.0°C, 1.00×10^4 s is required for half of them to be consumed. How long will it take to consume half of the reactants if an identical experiment is performed at 40.0°C?

38. Dinitrogen tetraoxide (N_2O_4) decomposes spontaneously at room temperature in the gas phase:

$$N_2O_4(g) \longrightarrow 2\ NO_2(g)$$

The rate law governing the disappearance of N_2O_4 with time is

$$-\frac{d[N_2O_4]}{dt} = k[N_2O_4]$$

At 30°C, $k = 5.1 \times 10^6$ s^{-1} and the activation energy for the reaction is 54.0 kJ mol^{-1}.
(a) Calculate the time (in seconds) required for the partial pressure of $N_2O_4(g)$ to decrease from 0.10 atm to 0.010 atm at 30°C.
(b) Repeat the calculation of part (a) at 300°C.

39. The activation energy for the isomerization reaction of CH_3NC in Problem 14 is 161 kJ mol^{-1}, and the reaction rate constant at 600 K is 0.41 s^{-1}.
(a) Calculate the Arrhenius factor A for this reaction.
(b) Calculate the rate constant for this reaction at 1000 K.

40. Cyclopropane isomerizes to propylene according to a first-order reaction:

$$\text{cyclopropane} \longrightarrow \text{propylene}$$

The activation energy is $E_a = 272$ kJ mol^{-1}. At 500°C, the reaction rate constant is 6.1×10^{-4} s^{-1}.
(a) Calculate the Arrhenius factor A for this reaction.
(b) Calculate the rate constant for this reaction at 25°C.

41. The activation energy of the gas-phase reaction

$$OH(g) + HCl(g) \longrightarrow H_2O(g) + Cl(g)$$

is 3.5 kJ mol^{-1}, and the change in the internal energy in the reaction is $\Delta U = -66.8$ kJ mol^{-1}. Calculate the activation energy of the reaction

$$H_2O(g) + Cl(g) \longrightarrow OH(g) + HCl(g)$$

42. The compound HOCl is known, but the related compound HClO, with a different order for the atoms in the molecule, is not known. Calculations suggest that the activation energy for the conversion HOCl $\longrightarrow$ HClO is 311 kJ mol^{-1} and that for the conversion HClO $\longrightarrow$ HOCl is 31 kJ mol^{-1}. Estimate ΔU for the reaction HOCl $\longrightarrow$ HClO.

Reaction Dynamics

43. Use collision theory to estimate the preexponential factor in the rate constant for the elementary reaction

$$NOCl + NOCl \longrightarrow 2\ NO + Cl_2$$

at 25°C. Take the average diameter of an NOCl molecule to be 3.0×10^{-10} m and use the steric factor P from Table 18.1.

44. Use collision theory to estimate the preexponential factor in the rate constant for the elementary reaction

$$NO_2 + NO_2 \longrightarrow 2\ NO + O_2$$

at 500 K. Take the average diameter of an NO_2 molecule to be 2.6×10^{-10} m and use the steric factor P from Table 18.1.

Kinetics of Catalysis

45. Certain bacteria use the enzyme penicillinase to decompose penicillin and render it inactive. The Michaelis–Menten constants for this enzyme and substrate are $K_m = 5 \times 10^{-5}$ mol L^{-1} and $k_2 = 2 \times 10^3\ s^{-1}$.
(a) What is the maximum rate of decomposition of penicillin if the enzyme concentration is 6×10^{-7} M?
(b) At what substrate concentration will the rate of decomposition be half that calculated in part (a)?

46. The conversion of dissolved carbon dioxide in blood to HCO_3^- and H_3O^+ is catalyzed by the enzyme carbonic anhydrase. The Michaelis–Menten constants for this enzyme and substrate are $K_m = 8 \times 10^{-5}$ mol L^{-1} and $k_2 = 6 \times 10^5\ s^{-1}$.
(a) What is the maximum rate of reaction of carbon dioxide if the enzyme concentration is 5×10^{-6} M?
(b) At what CO_2 concentration will the rate of decomposition be 30% of that calculated in part (a)?

ADDITIONAL PROBLEMS

47. Hemoglobin molecules in blood bind oxygen and carry it to cells, where it takes part in metabolism. The binding of oxygen

$$\text{hemoglobin}(aq) + O_2(aq) \longrightarrow \text{hemoglobin}\cdot O_2(aq)$$

is first order in hemoglobin and first order in dissolved oxygen, with a rate constant of 4×10^7 L $mol^{-1}\ s^{-1}$. Calculate the initial rate at which oxygen will be bound to hemoglobin if the concentration of hemoglobin is 2×10^{-9} M and that of oxygen is 5×10^{-5} M.

* 48. Suppose 1.00 L of 9.95×10^{-3} M $S_2O_3^{2-}$ is mixed with 1.00 L of 2.52×10^{-3} M H_2O_2 at a pH of 7.0 and a temperature of 25°C. These species react by two competing pathways, represented by the balanced equations

$$S_2O_3^{2-} + 4\ H_2O_2 \longrightarrow 2\ SO_4^{2-} + H_2O + 2\ H_3O^+$$

$$2\ S_2O_3^{2-} + H_2O_2 + 2\ H_3O^+ \longrightarrow S_4O_6^{2-} + 4\ H_2O$$

At the instant of mixing, the thiosulfate ion ($S_2O_3^{2-}$) is observed to be disappearing at the rate of 7.9×10^{-7} mol $L^{-1}\ s^{-1}$. At the same moment, the H_2O_2 is disappearing at the rate of 8.8×10^{-7} mol $L^{-1}\ s^{-1}$.
(a) Compute the percentage of the $S_2O_3^{2-}$ that is, at that moment, reacting according to the first equation.
(b) It is observed that the hydronium ion concentration drops. Use the data and answer from part (a) to compute how many milliliters per minute of 0.100 M H_3O^+ must be added to keep the pH equal to 7.0.

49. 8.23×10^{-3} mol of InCl(s) is placed in 1.00 L of 0.010 M HCl(aq) at 75°C. The InCl(s) dissolves quite quickly, and then the following reaction occurs:

$$3\ In^+(aq) \longrightarrow 2\ In(s) + In^{3+}(aq)$$

As this disproportionation proceeds, the solution is analyzed at intervals to determine the concentration of $In^+(aq)$ that remains.

Time (s)	$[In^+]$ (mol L^{-1})
0	8.23×10^{-3}
240	6.41×10^{-3}
480	5.00×10^{-3}
720	3.89×10^{-3}
1000	3.03×10^{-3}
1200	3.03×10^{-3}
10,000	3.03×10^{-3}

(a) Plot ln $[In^+]$ versus time, and determine the apparent rate constant for this first-order reaction.
(b) Determine the half-life of this reaction.
(c) Determine the equilibrium constant K for the reaction under the experimental conditions.

* 50. A compound called di-*t*-butyl peroxide [abbreviation DTBP, formula $(CH_3)_3COOC(CH_3)_3$] decomposes to give acetone [$(CH_3)_2CO$] and ethane (C_2H_6):

$$(CH_3)_3COOC(CH_3)_3(g) \longrightarrow 2\ (CH_3)_2CO(g) + C_2H_6(g)$$

The *total* pressure of the reaction mixture changes with time, as shown by the following data at 147.2°C:

Time (min)	P_{tot} (atm)	Time (min)	P_{tot} (atm)
0	0.2362	26	0.3322
2	0.2466	30	0.3449
6	0.2613	34	0.3570
10	0.2770	38	0.3687
14	0.2911	40	0.3749
18	0.3051	42	0.3801
20	0.3122	46	0.3909
22	0.3188		

(a) Calculate the partial pressure of DTBP at each time from these data. Assume that at time 0, DTBP is the only gas present.
(b) Are the data better described by a first-order or a second-order rate expression with respect to DTBP concentration?

51. The reaction of OH^- with HCN in aqueous solution at 25°C has a forward rate constant k_f of 3.7×10^9 L $mol^{-1}\ s^{-1}$. Using this information and the measured acid ionization constant of HCN (see Table 15.2), calculate the rate constant k_r in the first-order rate law rate $= k_r[CN^-]$ for the transfer of hydrogen ions to CN^- from surrounding water molecules:

$$H_2O(\ell) + CN^-(aq) \longrightarrow OH^-(aq) + HCN(aq)$$

52. Carbon dioxide reacts with ammonia to give ammonium carbamate, a solid. The reverse reaction also occurs:

$$CO_2(g) + 2\ NH_3(g) \rightleftharpoons NH_4OCONH_2(s)$$

The forward reaction is first order in $CO_2(g)$ and second order in $NH_3(g)$. Its rate constant is 0.238 $atm^{-2}\ s^{-1}$ at

0.0°C (expressed in terms of partial pressures rather than concentrations). The reaction in the reverse direction is zero order, and its rate constant, at the same temperature, is 1.60×10^{-7} atm s^{-1}. Experimental studies show that, at all stages in the progress of this reaction, the net rate is equal to the forward rate minus the reverse rate. Compute the equilibrium constant of this reaction at 0.0°C.

53. For the reactions

$$I + I + M \longrightarrow I_2 + M$$

$$Br + Br + M \longrightarrow Br_2 + M$$

the rate laws are

$$-\frac{d[I]}{dt} = k_I[I]^2[M]$$

$$-\frac{d[Br]}{dt} = k_{Br}[Br]^2[M]$$

The ratio k_I/k_{Br} at 500°C is 3.0 when M is an Ar molecule. Initially, $[I]_0 = 2[Br]_0$, while [M] is the same for both reactions and is much greater than $[I]_0$. Calculate the ratio of the time required for [I] to decrease to half its initial value to the same time for [Br] at 500°C.

54. In some reactions there is a competition between kinetic control and thermodynamic control over product yields. Suppose compound A can undergo two elementary reactions to stable products:

$$A \underset{k_{-1}}{\overset{k_1}{\rightleftharpoons}} B \quad \text{or} \quad A \underset{k_{-2}}{\overset{k_2}{\rightleftharpoons}} C$$

For simplicity we assume first-order kinetics for both forward and reverse reactions. We take the numerical values $k_1 = 1 \times 10^8$ s^{-1}, $k_{-1} = 1 \times 10^2$ s^{-1}, $k_2 = 1 \times 10^9$ s^{-1}, and $k_{-2} = 1 \times 10^4$ s^{-1}.

(a) Calculate the equilibrium constant for the equilibrium

$$B \rightleftharpoons C$$

From this value, give the ratio of the concentration of B to that of C at equilibrium. This is an example of thermodynamic control.

(b) In the case of kinetic control, the products are isolated (or undergo additional reaction) before the back reactions can take place. Suppose the back reactions in the preceding example (k_{-1} and k_{-2}) can be ignored. Calculate the concentration ratio of B to C reached in this case.

55. Compare and contrast the mechanisms for the two gas-phase reactions

$$H_2 + Br_2 \longrightarrow 2\,HBr$$

$$H_2 + I_2 \longrightarrow 2\,HI$$

56. In Section 18.4 the steady-state approximation was used to derive a rate expression for the decomposition of $N_2O_5(g)$:

$$\text{rate} = \frac{k_1 k_2[M][N_2O_5]}{k_2 + k_{-1}[M]} = k_{eff}[N_2O_5]$$

At 300 K, with an excess of nitrogen present, the following values of k_{eff} as a function of total pressure were found:

P (atm)	k_{eff}(s^{-1})	*P* (atm)	k_{eff} (s^{-1})
9.21	0.265	0.625	0.116
5.13	0.247	0.579	0.108
3.16	0.248	0.526	0.104
3.03	0.223	0.439	0.092
		0.395	0.086

Use the data to estimate the value of k_{eff} at very high total pressure and the value of k_1 in L mol^{-1} s^{-1}.

57. The decomposition of ozone by light can be described by the mechanism

$$O_3 + \text{light} \xrightarrow{k_1} O_2 + O$$

$$O + O_2 + M \xrightarrow{k_2} O_3 + M$$

$$O + O_3 \xrightarrow{k_3} 2\,O_2$$

with the overall reaction being

$$2\,O_3 + \text{light} \longrightarrow 3\,O_2$$

The rate constant k_1 depends on the light intensity and the type of light source used. By making a steady-state approximation for the concentration of oxygen atoms, express the rate of formation of O_2 in terms of the O_2, O_3, and M concentrations and the elementary rate constants. Show that only the ratio k_3/k_2, and not the individual values of k_2 and k_3, affects the rate.

* 58. In Section 18.4 we considered the following mechanism for the reaction of Br_2 with H_2:

$$Br_2 + M \underset{k_{-1}}{\overset{k_1}{\rightleftharpoons}} Br + Br + M$$

$$Br + H_2 \xrightarrow{k_2} HBr + H$$

$$H + Br_2 \xrightarrow{k_3} HBr + Br$$

Although this is adequate for calculating the *initial* rate of reaction, before product HBr builds up, there is an additional process that can participate as the reaction continues:

$$HBr + H \xrightarrow{k_4} H_2 + Br$$

(a) Write an expression for the rate of change of [H].

(b) Write an expression for the rate of change of [Br].

(c) As hydrogen and bromine atoms are both short-lived species, we can make the steady-state approximation and set the rates from parts (a) and (b) to 0. Express the steady-state concentrations [H] and [Br] in terms of concentrations of H_2, Br_2, HBr, and M. [*Hint:* Try adding the rate for part (a) to that for part (b).]

(d) Express the rate of production of HBr in terms of concentrations of H_2, Br_2, HBr, and M.

59. The following observations have been made about a certain reacting system: (i) When A, B, and C are mixed at about equal concentrations in neutral solution, two different products are formed, D and E, with the amount of D about 10 times as great as the amount of E. (ii) If everything is

done as in (i) except that a trace of acid is added to the reaction mixture, the same products are formed, except that now the amount of D produced is much smaller than (about 1% of) the amount of E. The acid is not consumed in the reaction. The following mechanism has been proposed to account for some of these observations and others about the order of the reactions:

(1) $A + B \underset{k_{-1}}{\overset{k_2}{\rightleftharpoons}} F$ (rapid equilibrium)

(2) $C + F \xrightarrow{k_2} D$ (negligible reverse rate)

(3) $C + F \xrightarrow{k_3} E$ (negligible reverse rate)

(a) Explain what this proposed scheme of reactions implies about the dependence (if any) of the rate of formation of D on the concentrations of A, of B, and of C. What about the dependence (if any) of the rate of formation of E on these same concentrations? (b) What can you say about the relative magnitudes of k_2 and k_3? (c) What explanation can you give for observation (ii) in view of your answer to (b)?

60. Iron(II) ion is oxidized by chlorine in aqueous solution, the overall equation being

$$2\,Fe^{2+} + Cl_2 \longrightarrow 2\,Fe^{3+} + 2\,Cl^-$$

It is found experimentally that the rate of the overall reaction is decreased when either the iron(III) ion or the chloride-ion concentration is increased. Which of the following possible mechanisms is consistent with the experimental observations?

(a) (1) $Fe^{2+} + Cl_2 \underset{k_{-1}}{\overset{k_1}{\rightleftharpoons}} Fe^{3+} + Cl^- + Cl$ (rapid equilibrium)

(2) $Fe^{2+} + Cl \xrightarrow{k_2} Fe^{3+} + Cl^-$ (negligible reverse rate)

(b) (3) $Fe^{2+} + Cl_2 \underset{k_{-3}}{\overset{k_3}{\rightleftharpoons}} Fe(IV) + 2\,Cl^-$ (rapid equilibrium)

(4) $Fe(IV) + Fe^{2+} \xrightarrow{k_4} 2\,Fe^{3+}$ (negligible reverse rate)

where Fe(IV) is Fe in the (+IV) oxidation state.

61. Manfred Eigen, a German physical chemist working during the 1970s and 1980s, earned a Nobel Prize for developing the "temperature-jump" method for studying kinetics of very rapid reactions in solution, such as proton transfer. Eigen and his co-workers found that the specific rate of proton transfer from a water molecule to an ammonia molecule in a dilute aqueous solution is $k = 2 \times 10^5\ s^{-1}$. The equilibrium constant K_b for the reaction of ammonia with water is 1.8×10^{-5} M. What, if anything, can be deduced from this information about the rate of transfer of a proton from NH_4^+ to a hydroxide ion? Write equations for any reactions you mention, making it clear to which reaction(s) any quoted constant(s) apply.

62. Consider the reaction

$$A + B \rightleftharpoons C + D$$

with all reactants and products gaseous (for simplicity) and an equilibrium constant K. (a) Assume that the elementary steps in the reaction are those indicated by the stoichiometric equation (in each direction), with specific rate constants for the forward reaction and the reverse reaction, respectively, k_f and k_r. Derive the relation between k_f, k_r, and K. Comment on the general validity of the assumptions made about the relation of elementary steps and the stoichiometric equation and also on the general validity of K. (b) Assume that the reaction as written is exothermic. Explain what this implies about the change of K with temperature. Explain also what it implies about the relation of the activation energies of the forward and reverse reactions and how this relation is consistent with your statement about the variation of K with temperature.

63. The gas-phase decomposition of acetaldehyde can be represented by the overall chemical equation

$$CH_3CHO \longrightarrow CH_4 + CO$$

It is thought to occur through the sequence of reactions

$$CH_3CHO \longrightarrow CH_3 + CHO$$

$$CH_3 + CH_3CHO \longrightarrow CH_4 + CH_2CHO$$

$$CH_2CHO \longrightarrow CO + CH_3$$

$$CH_3 + CH_3 \longrightarrow CH_3CH_3$$

Show that this reaction mechanism corresponds to a chain reaction, and identify the initiation, propagation, and termination steps.

64. Lanthanum(III) phosphate crystallizes as a hemihydrate, $LaPO_4 \cdot \frac{1}{2}\,H_2O$. When it is heated, it loses water to give anhydrous lanthanum(III) phosphate:

$$2(LaPO_4 \cdot \tfrac{1}{2}\,H_2O(s)) \longrightarrow 2\,LaPO_4(s) + H_2O(g)$$

This reaction is first order in the chemical amount of $LaPO_4 \cdot \frac{1}{2}\,H_2O$. The rate constant varies with temperature as follows:

Temperature (°C)	k (s^{-1})
205	2.3×10^{-4}
219	3.69×10^{-4}
246	7.75×10^{-4}
260	12.3×10^{-4}

Compute the activation energy of this reaction.

65. The water in a pressure cooker boils at a temperature greater than 100°C because it is under pressure. At this higher temperature, the chemical reactions associated with the cooking of food take place at a greater rate.

(a) Some food cooks fully in 5 min in a pressure cooker at 112°C and in 10 minutes in an open pot at 100°C. Calculate the average activation energy for the reactions associated with the cooking of this food.

(b) How long will the same food take to cook in an open pot of boiling water in Denver, where the average atmospheric pressure is 0.818 atm and the boiling point of water is 94.4°C?

66. (a) A certain first-order reaction has an activation energy of 53 kJ mol^{-1}. It is run twice, first at 298 K and then at 308 K (10°C higher). All other conditions are identical. Show that, in the second run, the reaction occurs at double its rate in the first run.
(b) The same reaction is run twice more at 398 K and 408 K. Show that the reaction goes 1.5 times as fast at 408 K as it does at 398 K.

* 67. The gas-phase reaction between hydrogen and iodine

$$H_2(g) + I_2(g) \underset{k_r}{\overset{k_f}{\rightleftharpoons}} 2\,HI(g)$$

proceeds with a forward rate constant at 1000 K of $k_f = 240$ L mol^{-1} s^{-1} and an activation energy of 165 kJ mol^{-1}. By using this information and data from Appendix D, calculate the activation energy for the reverse reaction and the value of k_r at 1000 K. Assume that ΔH and ΔS for the reaction are independent of temperature between 298 and 1000 K.

68. The following reaction mechanism has been proposed for a chemical reaction:

$$A_2 \underset{k_{-1}}{\overset{k_1}{\rightleftharpoons}} A + A \quad \text{(fast equilibrium)}$$

$$A + B \underset{k_{-2}}{\overset{k_2}{\rightleftharpoons}} AB \quad \text{(fast equilibrium)}$$

$$AB + CD \xrightarrow{k_3} AC + BD \quad \text{(slow)}$$

(a) Write a balanced equation for the overall reaction.
(b) Write the rate expression that corresponds to the preceding mechanism. Express the rate in terms of concentrations of reactants only (A_2, B, CD).
(c) Suppose that the first two steps in the preceding mechanism are endothermic and the third one is exothermic. Will an increase in temperature increase the reaction rate constant, decrease it, or cause no change? Explain.

69. How would you describe the role of the CF_2Cl_2 in the reaction mechanism of Problem 21?

70. In Section 18.7 we wrote a mechanism in which silver ions catalyze the reaction of Tl^+ with Ce^{4+}. Determine the rate law for this mechanism by making a steady-state approximation for the concentration of the reactive intermediate Ag^{2+}.

71. The rates of enzyme catalysis can be lowered by the presence of inhibitor molecules I, which bind to the active site of the enzyme. This adds the following additional step to the reaction mechanism considered in Section 18.7:

$$E + I \underset{k_{-3}}{\overset{k_3}{\rightleftharpoons}} EI \quad \text{(fast equilibrium)}$$

Determine the effect of the presence of inhibitor at total concentration $[I]_0 = [I] + [EI]$ on the rate expression for formation of products derived at the end of this chapter.

72. The enzyme lysozyme kills certain bacteria by attacking a sugar called *N*-acetylglucosamine (NAG) in their cell walls. At an enzyme concentration of 2×10^{-6} M, the maximum rate for substrate (NAG) reaction, found at high substrate concentration, is 1×10^{-6} mol L^{-1} s^{-1}. The rate is reduced by a factor of 2 when the substrate concentration is reduced to 6×10^{-6} M. Determine the Michaelis–Menten constants K_m and k_2 for lysozyme.

CUMULATIVE PROBLEMS

73. The rate of the gas-phase reaction

$$H_2 + I_2 \longrightarrow 2\,HI$$

is given by

$$\text{rate} = -\frac{d[I_2]}{dt} = k[H_2][I_2]$$

with $k = 0.0242$ L mol^{-1} s^{-1} at 400°C. If the initial concentration of H_2 is 0.081 mol L^{-1} and that of I_2 is 0.036 mol L^{-1}, calculate the initial rate at which heat is absorbed or emitted during the reaction. Assume that the enthalpy change at 400°C is the same as that at 25°C.

74. The rate of the reaction

$$2\,ClO_2(aq) + 2\,OH^-(aq) \longrightarrow ClO_3^-(aq) + ClO_2^-(aq) + H_2O(\ell)$$

is given by

$$\text{rate} = k[ClO_2]^2[OH^-]$$

with $k = 230$ L^2 mol^{-2} s^{-1} at 25°C. A solution is prepared that has initial concentrations $[ClO_2] = 0.020$ M, $[HCN] = 0.095$ M, and $[CN^-] = 0.17$ M. Calculate the initial rate of the reaction.

75. A gas mixture was prepared at 500 K with total pressure 3.26 atm and a mole fraction of 0.00057 of NO and 0.00026 of O_3. The elementary reaction

$$NO + O_3 \longrightarrow NO_2 + O_2$$

has a second-order rate constant of 7.6×10^7 L mol^{-1} s^{-1} at this temperature. Calculate the initial rate of the reaction under these conditions.

76. The activation energy for the reaction

$$2\,NO_2 \longrightarrow 2\,NO + O_2$$

is $E_a = 111$ kJ mol^{-1}. Calculate the root-mean-square velocity of an NO_2 molecule at 400 K and compare it to the velocity of an NO_2 molecule with kinetic energy E_a/N_A.

CHAPTER 19

Nuclear Chemistry

A fuel element being removed at the High Flux Isotope Reactor at Oak Ridge National Laboratory. The fuel element and the reactor head are submerged in water in this design. The blue glow is due to Cerenkov radiation, emitted by energetic charged particles traveling through the water. This 100-megawatt reactor is used to conduct research on synthetic heavy elements, and its principal product is the isotope ^{235}Cf.

In the preceding chapters, we discuss chemical reactions essentially as rearrangements of atoms from one set of "bonding partners" to another set. We turn now to a different kind of reaction in which the elemental identity of the atomic nucleus changes in the course of the reaction. This is a fascinating subject that arose early in the 20th century from experimental studies of the physical structure of the atom (see Section 1.4).

A fundamentally different new feature distinguishes this field from the material in earlier chapters. The separate laws of conservation of matter and conservation of energy must be generalized. Nuclear reactions occur at energies sufficiently high that energy and matter are interconverted. This is the origin of the enormous amounts of energy that can be released during nuclear reactions. Mass and energy are related through Einstein's relation $E = mc^2$, and their *sum* is conserved during nuclear reactions.

19.1 Mass–Energy Relationships in Nuclei

Recall from Section 1.4 that almost all the mass of an atom is concentrated in a very small volume in the nucleus. The small size of the nucleus (which occupies less than one *trillionth* of the space in the atom) and the strong forces between the protons and neutrons that make it up largely isolate its behavior from the outside world of electrons and other nuclei. This greatly simplifies our analysis of nuclear chemistry, allowing us to examine single nuclei without concern for the atoms, ions, or molecules in which they may be found.

As defined in Section 1.4, a nuclide is characterized by the number of protons, Z, and the number of neutrons, N, it contains. The atomic number Z determines the charge $+Ze$ on the nucleus and therefore decides the identity of the element; the sum $Z + N = A$ is the mass number of the nuclide and is the integer closest to the relative atomic mass of the nuclide. Nuclides are designated by the symbol ${}^{A}_{Z}X$ where X is the chemical symbol for the element.

The absolute masses of nuclides and of elementary particles such as the proton, neutron, and electron are far too small to express in kilograms or grams without the constant use of 10 raised to some large negative exponent. A unit of mass that is better sized for measuring on the submicroscopic scale is needed. An **atomic mass unit** (abbreviated u) is defined as exactly $\frac{1}{12}$ the mass of a single atom of ^{12}C. Because 1 mol of ^{12}C atoms weighs exactly 12 g, one atom weighs $12/N_A$ grams, and

$$1\text{ u} = \frac{1}{12}\left(\frac{12\text{ g mol}^{-1}}{6.0221420 \times 10^{23}\text{ mol}^{-1}}\right)$$

$$= 1.6605387 \times 10^{-24}\text{ g} = 1.6605387 \times 10^{-27}\text{ kg}$$

The conversion factor between atomic mass units and grams is numerically equal to the inverse of Avogadro's number N_A, and the mass of a single atom in atomic mass units is numerically equal to the mass of one mole of atoms in grams. Thus, one atom of ^{1}H has a mass of 1.007825 u because 1 mol of ^{1}H has a mass of 1.007825 g. The *dalton* is a mass unit that is equivalent to the atomic mass unit and is used frequently in biochemistry.

Table 19.1 lists the masses of elementary particles and selected neutral atoms, obtained from mass spectrometry (see Fig. 1.12). Note that each atomic mass includes the contribution from the surrounding Z electrons as well as that from the nucleus.

The Einstein Mass–Energy Relationship

Some nuclides are stable and can exist indefinitely, but others are unstable (radioactive) and decay spontaneously to form other nuclides in processes to be considered in Section 19.2. Figure 19.1 shows the distribution of the number of neutrons N $(= A - Z)$ against atomic number Z (protons in the nucleus) for the known nuclides of the elements. For the light elements, $N/Z \approx 1$ for stable nuclides, meaning nearly equal numbers of protons and neutrons are found in the nucleus. For the heavier elements, $N/Z > 1$ and there are progressively more

TABLE 19.1 Masses of Selected Elementary Particles and Atoms

Elementary Particle	Symbol	Mass (u)	Mass (kg)
Electron, beta particle	${}^{0}_{-1}e^{-}$	0.000548579911	$9.1093819 \times 10^{-31}$
Positron	${}^{0}_{1}e^{+}$	0.000548579911	$9.1093819 \times 10^{-31}$
Proton	${}^{1}_{1}p^{+}$	1.0072764669	$1.6726216 \times 10^{-27}$
Neutron	${}^{1}_{0}n$	1.0086649158	$1.6749272 \times 10^{-27}$

Atom	Mass (u)	Atom	Mass (u)
${}^{1}_{1}H$	1.007825032	${}^{23}_{11}Na$	22.9897697
${}^{2}_{1}H$	2.014101778	${}^{24}_{12}Mg$	23.9850419
${}^{3}_{1}H$	3.016049268	${}^{30}_{14}Si$	29.97377022
${}^{3}_{2}He$	3.016029310	${}^{30}_{15}P$	29.9783138
${}^{4}_{2}He$	4.002603250	${}^{32}_{16}S$	31.9720707
${}^{7}_{3}Li$	7.0160040	${}^{35}_{17}Cl$	34.96885271
${}^{8}_{4}Be$	8.00530509	${}^{40}_{20}Ca$	39.9625912
${}^{9}_{4}Be$	9.0121821	${}^{49}_{22}Ti$	48.947871
${}^{10}_{4}Be$	10.0135337	${}^{81}_{35}Br$	80.916291
${}^{8}_{5}B$	8.024607	${}^{87}_{37}Rb$	86.909183
${}^{10}_{5}B$	10.0129370	${}^{87}_{38}Sr$	86.908879
${}^{11}_{5}B$	11.0093055	${}^{127}_{153}I$	126.904468
${}^{11}_{6}C$	11.011433	${}^{226}_{88}Ra$	226.025403
${}^{12}_{6}C$	12 exactly	${}^{228}_{88}Ra$	228.031064
${}^{13}_{6}C$	13.003354838	${}^{228}_{89}Ac$	228.031015
${}^{14}_{6}C$	14.003241988	${}^{232}_{90}Th$	232.038050
${}^{14}_{7}N$	14.003074005	${}^{234}_{90}Th$	234.043595
${}^{16}_{8}O$	15.994914622	${}^{231}_{91}Pa$	231.035879
${}^{17}_{8}O$	16.9991315	${}^{231}_{92}U$	231.036289
${}^{18}_{8}O$	17.999160	${}^{234}_{92}U$	234.040945
${}^{19}_{9}F$	18.9984032	${}^{235}_{92}U$	235.043923
${}^{21}_{11}Na$	20.99764	${}^{238}_{92}U$	238.050783

neutrons than protons in the nucleus. One of the goals of nuclear chemistry is to understand the differing stabilities of the isotopes. Nuclear mass provides the key to this understanding.

Note from Table 19.1 that the neutron has a slightly greater mass than the proton. The neutron is stable inside a nucleus, but in free space it is unstable, decaying into a proton and an electron with a half-life of about 12 minutes:

$$ {}^{1}_{0}n \longrightarrow {}^{1}_{1}p^{+} + {}^{0}_{-1}e^{-} $$

Here ${}^{1}_{0}n$ and ${}^{1}_{1}p^{+}$ represent the neutron and the proton, respectively. We have written this as a **balanced nuclear equation.** To check such an equation for balance, first verify that the total mass numbers (the superscripts) are equal on the two sides. Then verify that the nuclear charges are balanced by checking the sum of the subscripts.

Mass is not conserved in the reaction just written. By subtracting the mass of the reactant from the masses of the products (using Table 19.1), the change in mass is calculated to be

$$ \Delta m = -1.3947 \times 10^{-30} \text{ kg} = -8.398690 \times 10^{-4} \text{ u} $$

What has happened to the "lost" mass? It has been converted to energy—specifically, kinetic energy carried off by the electron and proton. According to Einstein's special theory of relativity, a change in mass always accompanies a change in energy:

$$ E = mc^2 $$

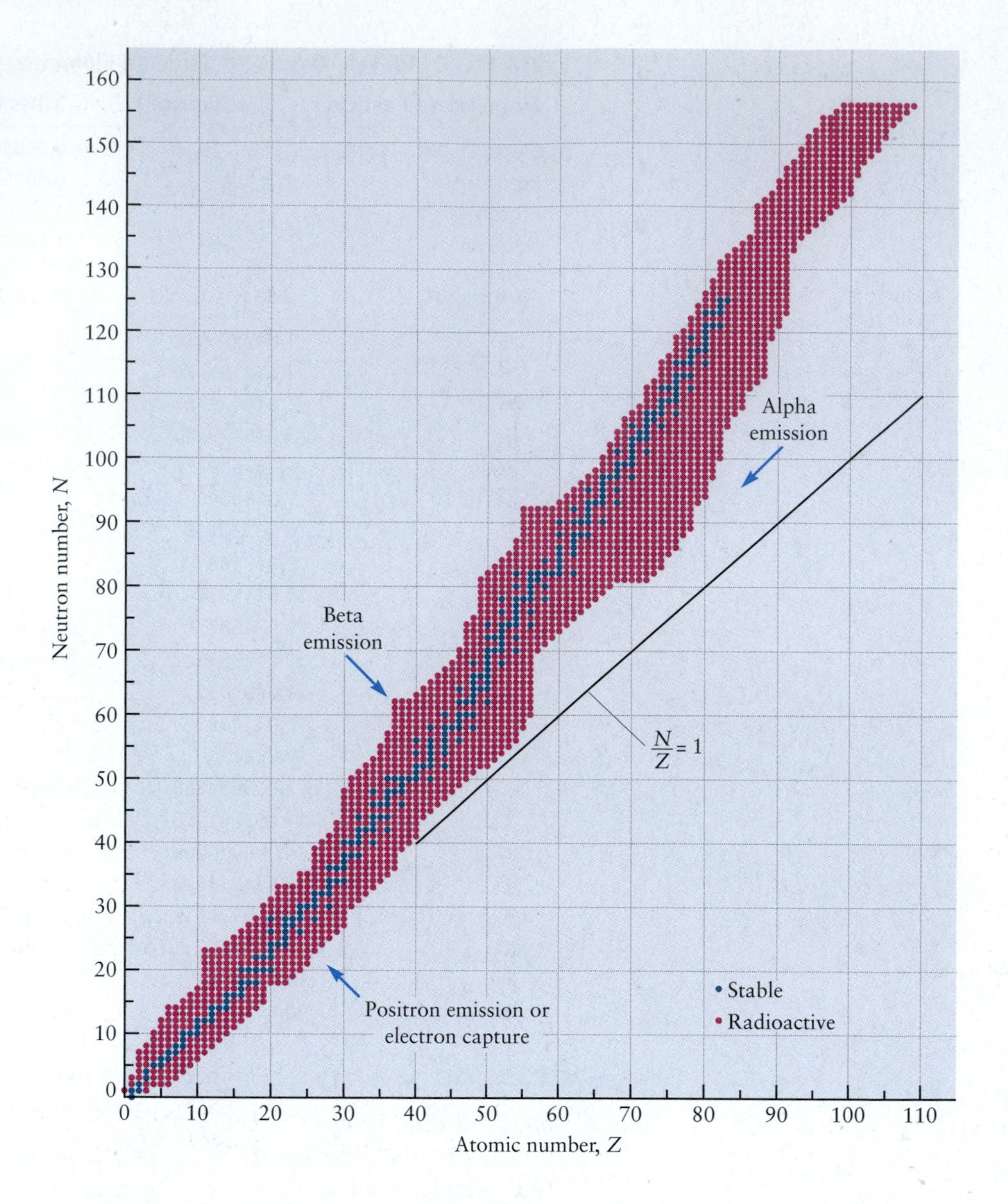

FIGURE 19.1 A plot of N versus Z for nuclides (represented by dots). Arrows show the directions of decay of unstable nuclides via alpha emission, beta emission, positron emission, or electron capture.

or, better,

$$\Delta E = c^2 \, \Delta m \qquad [19.1]$$

where c is the speed of light in a vacuum and Δm is the change in mass (mass of products minus mass of reactants). To express ΔE in joules, c must have units of meters per second, and Δm, units of kilograms. In the decay of one neutron, for example,

$$\Delta E = c^2 \, \Delta m = (2.9979 \times 10^8 \text{ m s}^{-1})^2(-1.3947 \times 10^{-30} \text{ kg}) = -1.2535 \times 10^{-13} \text{ J}$$

In ordinary chemical reactions, the change in mass is negligibly small. In the combustion of 1 mol of carbon to CO_2, the energy change is about 400 kJ, corresponding to a change in mass of only 4×10^{-9} g.

Changes in energy in nuclear reactions are nearly always expressed in more convenient energy units than the joule—namely, the **electron volt** (eV) and the **million electron volt** (MeV). The electron volt is defined as the energy given to an electron when it is accelerated through a potential difference of exactly 1 V:

$$\Delta E \text{ (joules)} = Q\text{(coulombs)} \times \Delta\mathscr{E}\text{(volts)}$$

$$1 \text{ eV} = 1.60217646 \times 10^{-19} \text{ C} \times 1 \text{ V} = 1.60217646 \times 10^{-19} \text{ J}$$

The million electron volt is 1 million times larger:

$$1 \text{ MeV} = 1.60217646 \times 10^{-13} \text{ J}$$

A change in mass of 1 u = $1.6605387 \times 10^{-27}$ kg corresponds to

$$\Delta E = c^2 \, \Delta m = (2.9979246 \times 10^8 \text{ m s}^{-1})^2 (1.6605387 \times 10^{-27} \text{ kg})$$
$$= 1.4924178 \times 10^{-10} \text{ J}$$

Converting to MeV,

$$\frac{1.4924178 \times 10^{-10} \text{ J}}{1.60217646 \times 10^{-13} \text{ J MeV}^{-1}} = 931.494 \text{ MeV}$$

We say that the **energy equivalent** of 1 u is 931.494 MeV. In the decay of a neutron,

$$\Delta E = (-8.39869 \times 10^{-4} \text{ u})(931.494 \text{ MeV u}^{-1}) = 0.782 \text{ MeV}$$

Binding Energies of Nuclei

The **binding energy** E_B of a nucleus is defined as the negative of the energy change ΔE that would occur if that nucleus were formed from its component protons and neutrons. For the ${}^4_2\text{He}$ nucleus, for example,

$$2\,{}^1_1p^+ + 2\,{}^1_0n \longrightarrow {}^4_2\text{He}^{2+} \qquad \Delta E = ?$$

Binding energies of nuclei are calculated with Einstein's relationship and accurate data from a mass spectrometer. Most measurements determine the mass of the atom with most or all of its electrons, not the mass of the bare nucleus (recall that it is the ${}^{12}\text{C}$ *atom,* not the ${}^{12}\text{C}$ nucleus, that is defined to have atomic mass 12). Therefore, we take the nuclear binding energy E_B of the helium nucleus to be the negative of the energy change ΔE for the formation of an *atom* of helium from hydrogen *atoms* and neutrons.

$$2\,{}^1_1\text{H} + 2\,{}^1_0n \longrightarrow {}^4_2\text{He}$$

The correction for the differences in binding energies of electrons (which of course are present both in the hydrogen atoms and in the atom being formed) is a very small one that need not concern us here.

EXAMPLE 19.1

Calculate the binding energy of ${}^4\text{He}$ from the data in Table 19.1, and express it both in joules and in million electron volts (MeV).

SOLUTION

The mass change is

$$\Delta m = m[{}^4_2\text{He}] - 2m[{}^1_1\text{H}] - 2m[{}^1_0n]$$
$$= 4.00260325 - 2(1.00782503) - 2(1.00866492) = -0.03037665 \text{ u}$$

Einstein's relation then gives

$$\Delta E = (-0.03037665 \text{ u})(1.6605387 \times 10^{-27} \text{ kg u}^{-1})(2.9979246 \times 10^8 \text{ m s}^{-1})^2$$
$$= -4.533465 \times 10^{-12} \text{ J}$$
$$E_B = 4.533465 \times 10^{-12} \text{ J}$$

If 1 *mol* of ${}^4_2\text{He}$ atoms were formed in this way, the energy change would be greater by a factor of Avogadro's number N_A, giving $\Delta E = -2.73 \times 10^{12} \text{ J mol}^{-1}$. This is an enormous quantity, seven orders of magnitude greater than typical energy changes in chemical reactions.

The energy change, in MeV, accompanying the formation of a $^{4}_{2}He$ atom is

$$\Delta E = (-0.03037665 \text{ u})(931.494 \text{ MeV u}^{-1}) = -28.2957 \text{ MeV}$$

Related Problems: 3, 4

There are four nucleons in the $^{4}_{2}He$ nucleus, so the binding energy *per nucleon* is

$$\frac{E_B}{4} = \frac{28.2957 \text{ MeV}}{4} = 7.07392 \text{ MeV}$$

The binding energy per nucleon is a direct measure of the stability of the nucleus. It varies with atomic number among the stable elements, increasing to a maximum of about 8.8 MeV for iron and nickel, as Figure 19.2 illustrates.

19.2 Nuclear Decay Processes

Why are some nuclei **radioactive,** decaying spontaneously, when others are stable? Thermodynamics gives the criterion $\Delta G < 0$ for a process to be spontaneous at constant T and P. The energy change ΔE in a nuclear decay process is so great that

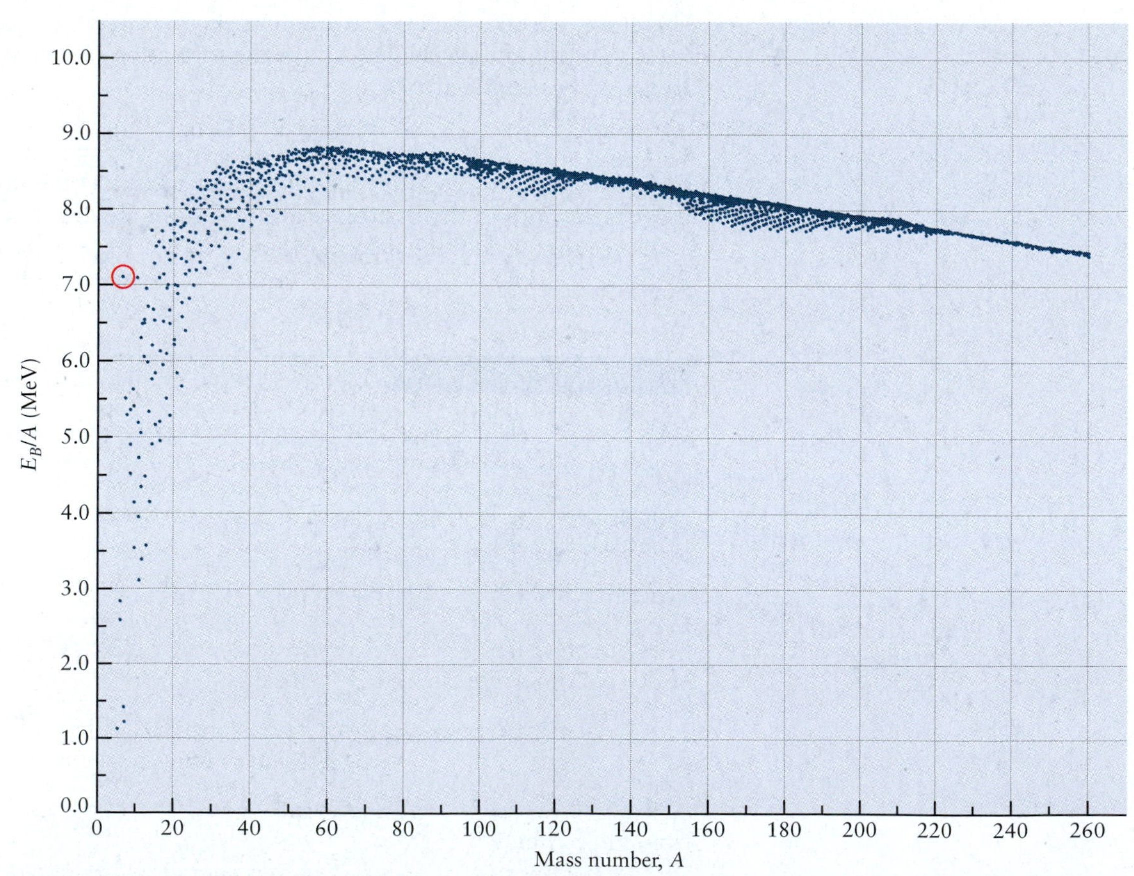

FIGURE 19.2 The variation of the binding energy per nucleon with mass number. A total of 2212 nuclides are shown, of which 274 are stable. The greatest nuclear stability is seen in the vicinity of ^{56}Fe. Note that ^{4}He (red circle) is unusually stable for its mass number.

the free energy change is essentially equal to it, and the thermodynamic criterion for spontaneous nuclear reaction consequently simplifies to

$$\Delta E < 0 \qquad \text{or} \qquad \Delta m < 0$$

The equivalence of these criteria follows from Einstein's mass–energy relationship. Spontaneous transformations of one nucleus into others can occur only if the combined mass of products is less than the mass of the original nuclide.

Before we address the details of nuclear decay, it is necessary to mention **antiparticles.** Each subatomic particle is thought to have an antiparticle of the same mass but opposite charge. Thus, the antiparticle of an electron is a **positron** (e^+), and the antiparticle of a proton is a negatively charged particle called an **antiproton.** Antiparticles have only transient existences because matter and antimatter annihilate one another when they come together, emitting radiation carrying an equivalent amount of energy. In the case of an electron and a positron, two oppositely directed **gamma rays** (γ-rays; high-energy photons), called the **annihilation radiation,** are emitted. Another important particle–antiparticle pair is the neutrino (symbol ν) and antineutrino ($\tilde{\nu}$), both of which are uncharged and almost massless. (They differ only in parity, a quantum-mechanical symmetry upon reflection.) Neutrinos interact so weakly with matter that elaborate detectors are required to record the extremely rare events induced by them in selected nuclides.

Beta Decay

If an unstable nuclide contains fewer protons than do stable isotopes of the same mass number, it is called "proton deficient." Such a nucleus can decay by transforming one of its neutrons into a proton and emitting a high-energy electron ${}_{-1}^{0}e^-$, also called a **beta particle,** and an antineutrino ($\tilde{\nu}$). The superscript 0 indicates a mass number of 0 because an electron contains neither protons nor neutrons and has a much smaller mass than a nucleus. The subscript -1 indicates the negative charge on the particle. It is not a true "atomic number," but writing the symbol in this way is helpful in balancing nuclear equations. The nuclide that results has the same mass number A, but its atomic number Z is increased by 1 because a neutron has been transformed into a proton. Examples of **beta decay** of unstable nuclei are

$$ {}_{6}^{14}\text{C} \longrightarrow {}_{7}^{14}\text{N} + {}_{-1}^{0}e^- + \tilde{\nu} \qquad {}_{11}^{24}\text{Na} \longrightarrow {}_{12}^{24}\text{Mg} + {}_{-1}^{0}e^- + \tilde{\nu}$$

The criterion $\Delta m < 0$ for spontaneous beta decay requires that

$$\underset{\text{(Parent nucleus)}}{m[{}^{A}\text{Z}]} > \underset{\text{(Daughter nucleus)}}{m[{}^{A}(\text{Z}+1)]} + m[{}_{-1}^{0}e^-]$$

because the mass of the antineutrino ($\tilde{\nu}$) is almost zero. This inequality refers to the masses of the parent and daughter *nuclei* and the emitted electron. It must be reformulated in terms of neutral *atom* masses because the mass spectrometer yields masses of atoms rather than nuclei. To do this, add the mass of Z electrons to both sides, giving a total of $Z + 1$ electrons on the right side (because the emitted ${}_{-1}^{0}e^-$ is also counted). The inequality becomes

$$\underset{\text{(Parent atom)}}{m[{}^{A}\text{Z}]} > \underset{\text{(Daughter atom)}}{m[{}^{A}(\text{Z}+1)]}$$

A direct comparison of atomic masses (Table 19.1) lets us determine whether beta emission can take place.

The condition that energy be released follows from the Einstein mass–energy relation:

$$\Delta E = c^2\{m[{}^{A}(\text{Z}+1)] - m[{}^{A}\text{Z}]\} < 0$$

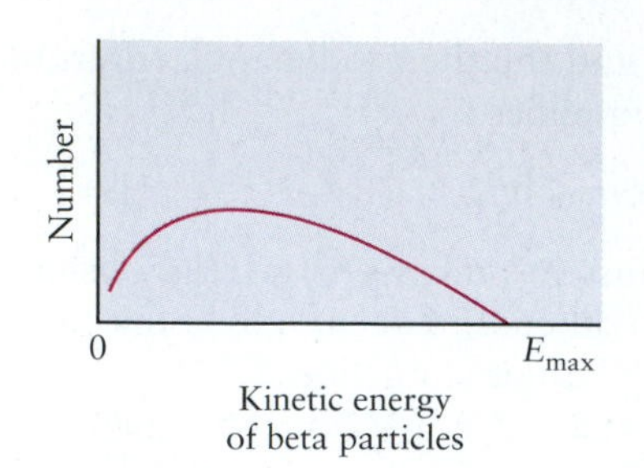

FIGURE 19.3 Emitted beta particles have a distribution of kinetic energies up to a cutoff value of E_{max}.

The liberated energy is carried off as kinetic energy by the beta particle (electron) and the antineutrino, because the nucleus produced is heavy enough that its recoil energy is small and can be neglected. The kinetic energy of the electron can fall anywhere in a continuous range from 0 up to $-\Delta E$ (Fig. 19.3), with the antineutrino carrying off the balance of the energy.

EXAMPLE 19.2

Calculate the maximum kinetic energy of the electron in the decay

$$^{14}_{6}\text{C} \longrightarrow {}^{14}_{7}\text{N} + {}^{0}_{-1}e^{-} + \tilde{\nu}$$

SOLUTION

Table 19.1 gives the masses of the relevant atoms. From them we calculate

$$\Delta m = m[^{14}_{7}\text{N}] - m[^{14}_{6}\text{C}] = 14.0030740 - 14.0032420 = -0.0001680 \text{ u}$$

$$\Delta E = (-1.68 \times 10^{-4} \text{ u})(931.5 \text{ MeV u}^{-1}) = -0.156 \text{ MeV}$$

The maximum kinetic energy of the electron is 0.156 MeV.

Related Problem: 8

Positron Emission

When a nucleus has too many protons for stability relative to the number of neutrons it contains, it may decay by emitting a positron. In this event a proton is converted to a neutron, and a high-energy positron (symbolized $^{0}_{1}e^{+}$) and a neutrino (ν) are emitted. As in beta decay, the mass number of the nuclide remains the same, but now the atomic number Z *decreases* by 1. Examples of **positron emission** are

$$^{11}_{6}\text{C} \longrightarrow {}^{11}_{5}\text{B} + {}^{0}_{1}e^{+} + \nu \qquad {}^{19}_{10}\text{Ne} \longrightarrow {}^{19}_{9}\text{F} + {}^{0}_{1}e^{+} + \nu$$

The positron has the same mass as an electron but has a positive charge. The mass criterion $\Delta m < 0$ is different from that for beta decay. For the bare nuclei the criterion is

$$\underset{\text{(Parent nucleus)}}{m[^{A}\text{Z}]} > \underset{\text{(Daughter nucleus)}}{m[^{A}(\text{Z}-1)]} + m[^{0}_{1}e^{+}]$$

Adding the mass of Z electrons to both sides gives the mass of the neutral atom ^{A}Z on the left, but only $Z - 1$ electrons are needed on the right side to form a neutral atom $^{A}(Z - 1)$. This leaves additional electron and positron masses on the right. These masses are the same, so the criterion for spontaneous positron emission is

$$\underset{\text{(Parent atom)}}{m[^{A}\text{Z}]} > \underset{\text{(Daughter atom)}}{m[^{A}(\text{Z}-1)]} + 2m[^{0}_{1}e^{+}]$$

The energy change in positron emission is

$$\Delta E = c^{2}\{m[^{A}(\text{Z}-1)] + 2m[^{0}_{1}e^{+}] - m[^{A}\text{Z}]\} < 0$$

The kinetic energy ($-\Delta E$) is distributed between the positron and the neutrino.

EXAMPLE 19.3

Calculate the maximum kinetic energy of the positron emitted in the decay

$$^{11}_{6}\text{C} \longrightarrow {}^{11}_{5}\text{B} + {}^{0}_{1}e^{+} + \nu$$

SOLUTION

The change in mass is

$$\Delta m = m[{}^{11}_{5}\text{B}] + 2m[{}^{0}_{1}e^{+}] - m[{}^{11}_{6}\text{C}] = 11.0093055 + 2(0.00054858) - 11.011433$$
$$= -0.00130 \text{ u}$$

The energy change is then

$$\Delta E = (-0.00130 \text{ u})(931.5 \text{ MeV u}^{-1}) = -0.960 \text{ MeV}$$

The maximum kinetic energy of the positron is +0.960 MeV.

Related Problem: 7

Electron Capture

If $m[{}^{A}Z]$ exceeds $m[{}^{A}(Z-1)]$ (both referring to neutral atoms) but by less than 1.0972×10^{-3} u (equivalent to $2m[{}^{0}_{1}e^{+}]$), positron emission cannot occur. An alternative process that *is* allowed is **electron capture,** in which an electron outside the nucleus of the parent atom is captured by the nucleus and a proton is converted to a neutron. As in positron emission, the mass number is unchanged and the atomic number decreases by 1. In this case, however, the only particle emitted is a neutrino. An example is

$${}^{231}_{92}\text{U} + {}^{0}_{-1}e^{-} \longrightarrow {}^{231}_{91}\text{Pa} + \nu$$

The mass criterion for electron capture is simply that

$$\underset{\text{(Parent atom)}}{m[{}^{A}Z]} > \underset{\text{(Daughter atom)}}{m[{}^{A}(Z-1)]}$$

so the energy change is

$$\Delta E = c^{2}\{m[{}^{A}(Z-1)] - m[{}^{A}Z]\} < 0$$

Electron capture is common for heavier neutron-deficient nuclei, for which the mass change with atomic number is too small for positron emission to be possible. In the preceding example,

$$\Delta m = m[{}^{231}_{91}\text{Pa}] - m[{}^{231}_{92}\text{U}] = 231.035879 - 231.03689 = 0.00041 \text{ u}$$

Because $\Delta m < 0$, this process can and does occur by electron capture, but because 0.00041 u is less than $2m[{}^{0}_{1}e^{+}]$, the analogous process could not occur through positron emission.

Alpha Decay

The three forms of decay discussed so far all lead to changes in the atomic number Z but not in the mass number A. Another process, **alpha decay,** involves the emission of an alpha particle (${}^{4}_{2}\text{He}^{2+}$ ion) and a decrease in mass number A by 4 (the atomic number Z and the neutron number N each decrease by 2). An example is

$${}^{238}_{92}\text{U} \longrightarrow {}^{234}_{90}\text{Th} + {}^{4}_{2}\text{He}$$

Such alpha decay occurs chiefly for elements in the unstable region beyond bismuth ($Z = 83$) in the periodic table.

The criterion for alpha decay is once again $\Delta m < 0$. In the preceding example,

$$\Delta m = \underset{\text{(Atom)}}{m[{}^{234}_{90}\text{Th}]} + \underset{\text{(Atom)}}{m[{}^{4}_{2}\text{He}]} - \underset{\text{(Atom)}}{m[{}^{238}_{92}\text{U}]}$$
$$= 234.043595 + 4.002603 - 238.050783 \text{ u} = -0.004585 \text{ u} < 0$$
$$\Delta E = (-0.004585 \text{ u})(931.5 \text{ MeV u}^{-1}) = -4.27 \text{ MeV}$$

Most of the energy is carried away by the lighter helium atom, with a small fraction appearing as recoil energy of the heavy thorium atom.

Other Modes of Decay

The arrows in Figure 19.1 show the directions of the types of decay discussed so far. Nuclei whose values of N/Z are too great move toward the line of stability by beta emission, whereas those whose values are too small undergo positron emission or electron capture. Nuclei that are simply too massive can move to lower values of both N and Z by alpha emission. The range of stable nuclides is limited by the interplay of attractive nuclear forces and repulsive Coulomb forces between protons in the nucleus. Scientists believe that it is unlikely that additional stable elements will be found, although long-lived radioactive nuclides may well exist at still higher Z.

When nuclei are very proton-deficient or very neutron-deficient, an excess particle may "boil off," that is, be ejected directly from the nucleus. These decay modes are called **neutron emission** and **proton emission,** respectively, and move nuclides down or to the left in Figure 19.1. Finally, certain unstable nuclei undergo spontaneous **fission,** in which they split into two nuclei of roughly equal size. Nuclear fission will be discussed in more detail in Section 19.5.

Detecting and Measuring Radioactivity

Many methods have been developed to detect, identify, and quantitatively measure the products of nuclear reactions. Some are quite simple, but others require complex electronic instrumentation. Perhaps the simplest radiation detector is the **photographic emulsion,** first used by Henri Becquerel, the discoverer of radioactivity. In 1896 Becquerel reported his observation that potassium uranyl sulfate ($K_2UO_2(SO_4)_2 \cdot 2H_2O$) could expose a photographic plate even in the dark. Such detectors are still used today in the film badges that are worn to monitor exposure to penetrating radiation. The degree of darkening of the film is proportional to the quantity of radiation received.

Rutherford and his students used a screen coated with zinc sulfide to detect the arrival of alpha particles by the pinpoint scintillations of light they produce. That simple device has been developed into the modern **scintillation counter.** Instead of a ZnS screen, the modern scintillation counter uses a crystal of sodium iodide, in which a small fraction of the Na^+ ions have been replaced by thallium (Tl^+) ions. The crystal emits a pulse of light when it absorbs a beta particle or a gamma ray, and a photomultiplier tube detects and counts the light pulses.

The **Geiger counter** (Fig. 19.4) consists of a cylindrical tube, usually of glass, coated internally with metal to provide a negative electrode and with a wire down the center for a positive electrode. The tube is filled to a total pressure of about 0.1 atm with a mixture of 90% argon and 10% ethyl alcohol vapor, and a potential difference of about 1000 V is applied across the electrodes. When a high-energy electron (beta particle) enters the tube, it produces positive ions and electrons. The light electrons are quickly accelerated toward the positively

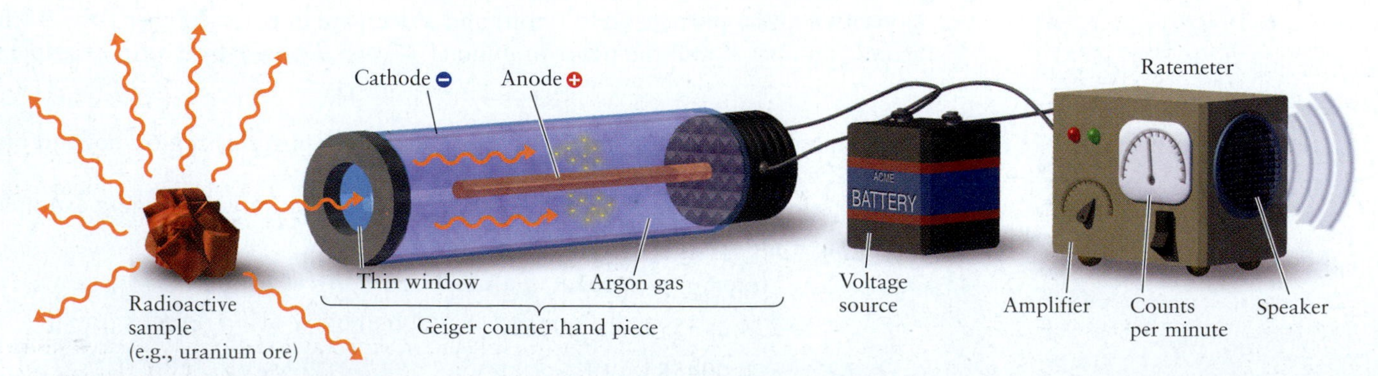

FIGURE 19.4 In a Geiger tube, radiation ionizes gas in the tube, freeing electrons that are accelerated to the anode wire in a cascade. Their arrival creates an electrical pulse, which is detected by a ratemeter. The ratemeter displays the accumulated pulses as the number of ionization events per minute.

charged wire. As they advance, they encounter and ionize other neutral atoms. An avalanche of electrons builds up, and a large electron current flows into the central wire. This causes a drop in the potential difference, which is recorded, and the multiplicative electron discharge is quenched by the alcohol molecules. In this way single beta particles produce electrical pulses that can be amplified and counted. Portable Geiger counters are widely used in uranium prospecting and to measure radiation in workplaces.

19.3 Kinetics of Radioactive Decay

The decay of any given unstable nucleus is a random event and is independent of the number of surrounding nuclei that have decayed. When the number of nuclei is large, we can be confident that during any given period a definite fraction of the original number of nuclei will have undergone a transformation into another nuclear species. In other words, the rate of decay of a collection of nuclei is proportional to the number of nuclei present, showing that nuclear decay follows a first-order rate equation of the type discussed in Chapter 18. All the results developed in that chapter apply to the present situation; for example, the integrated rate law has the form

$$N = N_i \, e^{-kt}$$

where N_i is the number of nuclei originally present at $t = 0$. The decay constant k is related to a half-life $t_{1/2}$ through

$$t_{1/2} = \frac{\ln 2}{k} = \frac{0.6931}{k}$$

just as in the first-order gas-phase chemical kinetics of Section 18.2. The half-life is the time required for the nuclei in a sample to decay to one-half their initial number, and it can range from less than 10^{-21} s to more than 10^{24} years for unstable nuclei. Table 19.2 lists the half-lives and decay modes of some unstable nuclides.

There is one important practical difference between chemical kinetics and nuclear kinetics. In chemical kinetics the concentration of a reactant or product is monitored over time, and the rate of a reaction is then found from the rate of change of that concentration. In nuclear kinetics the rate of occurrence of decay events, $-dN/dt$, is measured directly with a Geiger counter or other radiation detector. This decay rate—the average disintegration rate in numbers of nuclei per unit time—is called the **activity *A*.**

$$A = -\frac{dN}{dt} = kN \qquad [19.2]$$

Because the activity is proportional to the number of nuclei N, it also decays exponentially with time:

$$A = A_i \, e^{-kt} \qquad [19.3]$$

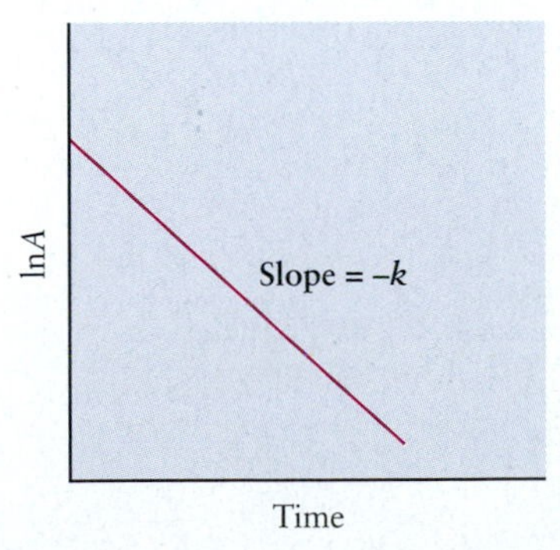

FIGURE 19.5 A graph of the logarithm of the activity of a radioactive nuclide against time is a straight line with slope $-k = -(\ln 2)/t_{1/2}$. The decay rate can also be measured as the ratio of the activity to the number of atoms of the radioisotope, if the latter is known.

A plot of ln A against time t is linear with slope $-k = -(\ln 2)/t_{1/2}$, as Figure 19.5 shows. The activity A is reduced to half its initial value in a time $t_{1/2}$. Once A and k are known, the number of nuclei N at that time can be calculated from

$$N = \frac{A}{k} = \frac{At_{1/2}}{\ln 2} = \frac{At_{1/2}}{0.6931}$$

The S.I. unit of activity is the becquerel (Bq), defined as 1 radioactive disintegration per second. An older and much larger unit of activity is the curie (abbreviated Ci), which is defined as 3.7×10^{10} disintegrations per second. The activity of 1 g of radium is 1 Ci.

TABLE 19.2 Decay Characteristics of Some Radioactive Nuclei

Nuclide	$t_{1/2}$	Decay Mode†	Daughter
$^{3}_{1}$H (tritium)	12.26 years	e^-	$^{3}_{2}$He
$^{8}_{4}$Be	~1 × 10^{-16} s	α	$^{4}_{2}$He
$^{14}_{6}$C	5730 years	e^-	$^{14}_{7}$N
$^{22}_{11}$Na	2.601 years	e^+	$^{22}_{10}$Ne
$^{24}_{11}$Na	15.02 hours	e^-	$^{24}_{12}$Mg
$^{32}_{15}$P	14.28 days	e^-	$^{32}_{16}$S
$^{35}_{16}$S	87.2 days	e^-	$^{35}_{17}$Cl
$^{36}_{17}$Cl	3.01 × 10^{5} years	e^-	$^{36}_{18}$Ar
$^{40}_{19}$K	1.28 × 10^{9} years	e^- (89.3%)	$^{40}_{20}$Ca
		E.C. (10.7%)	$^{40}_{18}$Ar
$^{59}_{26}$Fe	44.6 days	e^-	$^{59}_{27}$Co
$^{60}_{27}$Co	5.27 years	e^-	$^{60}_{28}$Ni
$^{90}_{38}$Sr	29 years	e^-	$^{90}_{39}$Y
$^{109}_{48}$Cd	453 days	E.C.	$^{109}_{47}$Ag
$^{125}_{53}$I	59.7 days	E.C.	$^{125}_{52}$Te
$^{131}_{53}$I	8.041 days	e^-	$^{131}_{54}$Xe
$^{127}_{54}$Xe	36.41 days	E.C.	$^{127}_{53}$I
$^{137}_{57}$La	~6 × 10^{4} years	E.C.	$^{137}_{56}$Ba
$^{222}_{86}$Rn	3.824 days	α	$^{218}_{84}$Po
$^{226}_{88}$Ra	1600 years	α	$^{222}_{86}$Rn
$^{232}_{90}$Th	1.40 × 10^{10} years	α	$^{228}_{88}$Ra
$^{235}_{92}$U	7.04 × 10^{8} years	α	$^{231}_{90}$Th
$^{238}_{92}$U	4.468 × 10^{9} years	α	$^{234}_{90}$Th
$^{239}_{93}$Np	2.350 days	e^-	$^{239}_{94}$Pu
$^{239}_{94}$Pu	2.411 × 10^{4} years	α	$^{235}_{92}$U

†E.C. stands for electron capture; e^+ for positron emission; e^- for beta emission; α, for alpha emission.

EXAMPLE 19.4

Tritium (^{3}H) decays by beta emission to ^{3}He with a half-life of 12.26 years. A sample of a tritiated compound has an initial activity of 0.833 Bq. Calculate the number N_i of tritium nuclei in the sample initially, the decay constant k, and the activity after 2.50 years.

SOLUTION

Convert the half-life to seconds:

$$t_{1/2} = (12.26 \text{ yr})(60 \times 60 \times 24 \times 365 \text{ s yr}^{-1}) = 3.866 \times 10^{8} \text{ s}$$

The number of nuclei originally present was

$$N_i = \frac{A_i t_{1/2}}{\ln 2} = \frac{(0.833 \text{ s}^{-1})(3.866 \times 10^{8} \text{ s})}{0.6931} = 4.65 \times 10^{8}\ {}^{3}\text{H nuclei}$$

The decay constant k is calculated directly from the half-life:

$$k = \frac{\ln 2}{t_{1/2}} = \frac{0.6931}{3.866 \times 10^{8} \text{ s}} = 1.793 \times 10^{-9} \text{ s}^{-1}$$

To find the activity after 2.50 years, convert this time to seconds (7.884×10^{7} s) and use

$$A = A_i\, e^{-kt} = (0.833 \text{ Bq}) \exp[-(1.793 \times 10^{-9} \text{ s}^{-1})(7.884 \times 10^{7} \text{ s})] = 0.723 \text{ Bq}$$

Related Problems: 21, 22

Radioactive Dating

The decay of radioactive nuclides with known half-lives enables geochemists to measure the ages of rocks from their isotopic compositions. Suppose that a uranium-bearing mineral was deposited some 2 billion (2.00×10^9) years ago and has remained geologically unaltered to the present time. The ^{238}U in the mineral decays with a half-life of 4.51×10^9 years to form a series of short-lived intermediates, ending in the stable lead isotope ^{206}Pb (Fig. 19.6). The fraction of uranium remaining after 2.00×10^9 years should be

$$\frac{N}{N_i} = e^{-kt} = e^{-0.6931\, t/t_{1/2}} = \exp\left(\frac{-0.6931 \times 2.00 \times 10^9 \text{ yr}}{4.51 \times 10^9 \text{ yr}}\right) = 0.735$$

and the number of ^{206}Pb atoms should be approximately

$$(1 - 0.735)N_i = 0.265\ N_i(^{238}\text{U})$$

The ratio of abundances

$$\frac{N(^{206}\text{Pb})}{N(^{238}\text{U})} = \frac{0.265}{0.735} = 0.361$$

therefore is determined by the time elapsed since the deposit was originally formed—in this case, 2.00×10^9 years. Of course, to calculate the age of the mineral, we work backward from the measured $N(^{206}\text{Pb})/N(^{238}\text{U})$ ratio.

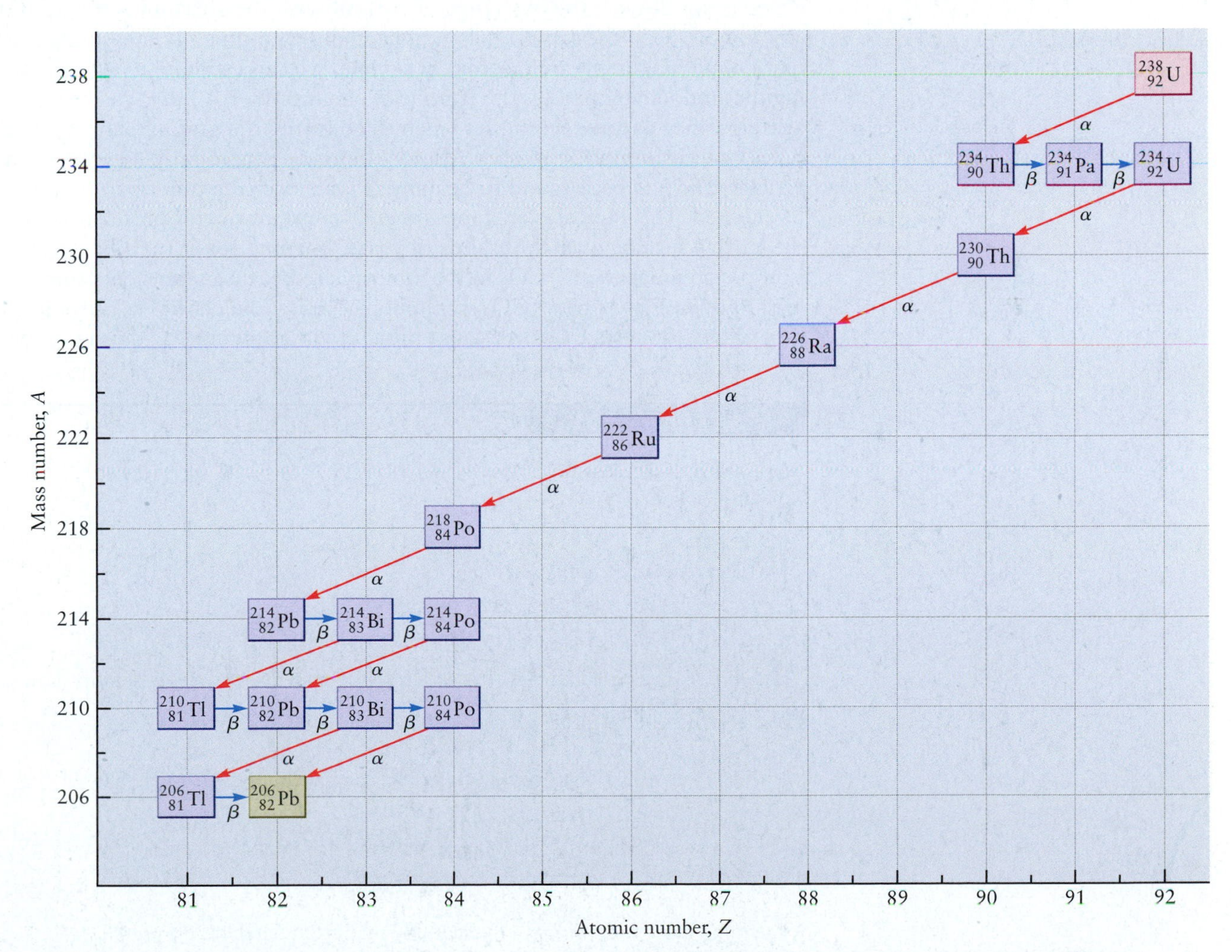

FIGURE 19.6 The radioactive nuclide ^{238}U decays via a series of alpha and beta emissions to the stable nuclide ^{206}Pb.

To use the method, it is necessary to be certain that the stable nuclide generated (^{206}Pb in this case) arises only from the parent species (^{238}U here) and that neither lead nor uranium has left or entered the rock over the course of geologic time. If possible, it is desirable to measure the ratios of several different isotopically paired species in the same rock sample. For example, ^{87}Rb decays to ^{87}Sr with a half-life of 4.9×10^{10} years, and one mode of ^{40}K decay (with a half-life of 1.28×10^{9} years) is to ^{40}Ar. Each pair should ideally yield the same age. Detailed analysis of a large number of samples suggests that the oldest surface rocks on earth are about 3.8 billion years old. An estimate of 4.5 billion years for the age of the earth and solar system comes from indirect evidence involving an isotopic analysis of meteorites, which are believed to have formed at the same time.

A somewhat different type of dating uses measurements of ^{14}C decay, which covers the range of human history and prehistory back to about 30,000 years ago. This unstable species (with a half-life of 5730 years) is produced continuously in the atmosphere. Cosmic rays of very high energy cause nuclear reactions that produce neutrons. These can collide with $^{14}_{7}N$ nuclei to produce $^{14}_{6}C$ by the reaction

$$^{14}_{7}N + ^{1}_{0}n \longrightarrow ^{14}_{6}C + ^{1}_{1}H$$

The resulting ^{14}C enters the carbon reservoir on the Earth's surface, mixing with stable ^{12}C as dissolved $H^{14}CO_3^-$ in the oceans, as $^{14}CO_2$ in the atmosphere, and in the tissues of plants and animals. This mixing, which is believed to have occurred at a fairly constant rate over the past 50,000 years, means that the ^{14}C in a living organism has a specific activity of close to 15.3 disintegrations per minute per gram of total carbon—that is, 0.255 Bq g^{-1}. When a plant or animal dies (for example, when a tree is cut down), the exchange of carbon with the surroundings stops and the amount of ^{14}C in the sample falls exponentially with time. By measuring the ^{14}C activity remaining in an archaeological sample, we can estimate its age. This ^{14}C dating method, developed by the American chemist W. F. Libby, has been calibrated against other dating techniques (such as counting the annual rings of bristlecone pines or examining the written records that may accompany a carbon-containing artifact) and has been found to be quite reliable over the time span for which it can be checked. This indicates an approximately constant rate of production of ^{14}C near the earth's surface over thousands of years. Burning fossil fuels has decreased the isotopic abundance of $^{14}CO_2$ in the atmosphere because plants (the source of fossil fuels) preferentially use $^{12}CO_2$ in photosynthesis. The changing isotopic composition of atmospheric CO_2 will cause difficulty in applying ^{14}C dating in the future.

EXAMPLE 19.5

A wooden implement has a specific activity of ^{14}C of 0.195 Bq g^{-1}. Estimate the age of the implement.

SOLUTION

The decay constant for ^{14}C is

$$k = \frac{0.6931}{5730 \text{ yr}} = 1.21 \times 10^{-4} \text{ yr}^{-1}$$

The initial specific activity was 0.255 Bq g^{-1}, and the measured activity now (after t years) is 0.195 Bq g^{-1}, so

$$A = A_i e^{-kt}$$

$$0.195 \text{ Bq g}^{-1} = 0.255 \text{ Bq g}^{-1} e^{-(1.21 \times 10^{-4})t}$$

$$\ln\left(\frac{0.195}{0.255}\right) = -(1.21 \times 10^{-4} \text{ yr}^{-1})t$$

$$t = 2200 \text{ yr}$$

The implement comes from a tree cut down approximately 2200 years ago.

Related Problems: 25, 26

19.4 Radiation in Biology and Medicine

Radiation has both harmful and beneficial effects for living organisms. All forms of radiation cause damage in direct proportion to the amount of energy they deposit in cells and tissues. The damage takes the form of chemical changes in cellular molecules, which alter their functions and lead either to uncontrolled multiplication and growth of cells or to their death. Alpha particles lose their kinetic energy over very short distances in matter (typically 10 cm in air or 0.05 cm in water or tissues), producing intense ionization in their wakes until they accept electrons and are neutralized to harmless helium atoms. Radium, for example, is an alpha emitter that substitutes for calcium in bone tissue and destroys its capacity to produce both red and white blood cells. Beta particles, gamma rays, and x-rays have greater penetrating power than alpha particles and so present a radiation hazard even when their source is well outside an organism.

The amount of damage produced in tissue by any of these kinds of radiation is proportional to the number of particles or photons and to their energy. A given activity of tritium causes less damage than the same activity of ^{14}C, because the beta particles from tritium have a maximum kinetic energy of 0.0179 MeV, whereas those from ^{14}C have an energy of 0.156 MeV. What is important is the amount of ionization produced or the quantity of energy deposited by radiation. For this purpose, several units are used. The *rad* (*r*adiation *a*bsorbed *d*ose) is defined as the amount of radiation that deposits 10^{-2} J of energy per kilogram of tissue. The damage produced in human tissue depends on still other factors, such as the nature of the tissue, the kind of radiation, the total radiation dose, and the dose rate. To take all these into account, the *rem* (*r*oentgen *e*quivalent in *m*an) has been defined to measure the effective dosages of radiation received by humans. A physical dose of 1 rad of beta or gamma radiation translates into a human dose of 1 rem. Alpha radiation is more toxic; a physical dose of 1 rad of alpha radiation equals about 10 rems. The SI unit for absorbed radiation, analogous to the rad, is the gray (Gy); 1 Gy = 1 joule per kg of material. So, 1 Gy = 100 rad. The SI unit for effective dosage of radiation is the sievert (Sv). The sievert is defined in the same way as the rem, except the delivered dose is expressed in Gy instead of rad; consequently, 1 Sv = 100 rem.

Exposure to radiation is unavoidable. The average person in the United States receives about 100 millirems (mrem) or 1 millisievert (mSv) annually from natural sources that include cosmic radiation and radioactive nuclides such as ^{40}K and ^{222}Rn. Another 50 to 100 mrem or 0.5 to 1.0 mSv (variable) comes from human activities (including dental and medical x-ray examinations and airplane flights, which increase exposure to cosmic rays higher in the atmosphere). The safe level of exposure to radiation is a controversial issue among biologists; one group maintains that the effects of radiation are cumulative, another that a threshold dose is necessary for pathological change. The problem is made even more complex by the necessity to distinguish between tissue damage in an exposed individual and genetic damage, which may not become apparent for several generations. It is much easier to define the radiation level that, with high probability, will cause death in an exposed person. The LD_{50} level in human beings (the level carrying a 50% probability that death will result within 30 days after a single exposure) is 500 rad, or 5 Gy.

EXAMPLE 19.6

The beta decay of ^{40}K that is a natural part of the body makes all human beings slightly radioactive. An adult weighing 70.0 kg contains about 170 g of potassium. The relative natural abundance of ^{40}K is 0.0118%, its half-life is 1.28×10^9 years, and its beta particles have an average kinetic energy of 0.55 MeV.

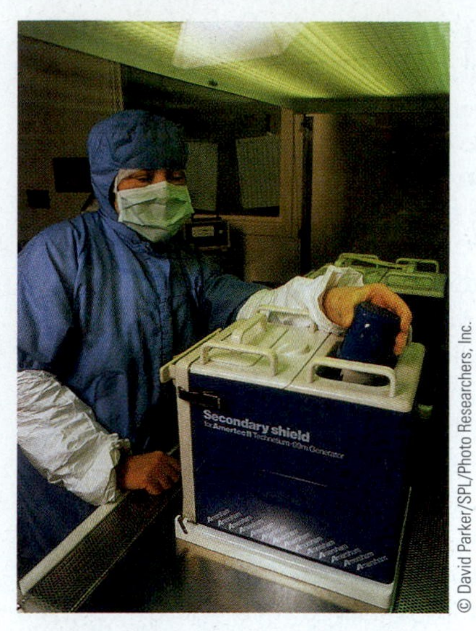

© David Parker/SPL/Photo Researchers, Inc.

FIGURE 19.7 A medical physicist unloading ^{99m}Tc from an Amertec II generator in preparation for gamma scintigraphy (gamma camera scanning). ^{99m}Tc is an excited state of ^{99}Tc that decays to the nuclear ground state by emission of gamma rays. This isotope is used in nuclear medicine to study the heart. The ^{99m}Tc is taken up by heart tissue; a gamma-ray detector then provides an image of the heart.

(a) Calculate the total activity of ^{40}K in this person.

(b) Determine (in Gy per year) the annual radiation absorbed dose arising from this internal ^{40}K.

SOLUTION

(a) First calculate the decay constant of ^{40}K in s^{-1}:

$$k = \frac{0.693}{t_{1/2}} = \frac{0.693}{(1.28 \times 10^9 \text{ yr})(365 \times 24 \times 60 \times 60 \text{ s yr}^{-1})}$$

$$= 1.72 \times 10^{-17} \text{ s}^{-1}$$

$$\text{number of } ^{40}\text{K atoms} = \frac{170 \text{ g}}{40.0 \text{ g mol}^{-1}}(1.18 \times 10^{-4})(6.02 \times 10^{23} \text{ mol}^{-1})$$

$$= 3.02 \times 10^{20}$$

$$A = -\frac{dN}{dt} = kN = (1.72 \times 10^{-17} \text{ s}^{-1})(3.02 \times 10^{20}) = 5.19 \times 10^3 \text{ s}^{-1}$$

(b) Each disintegration of ^{40}K emits an average of 0.55 MeV of energy, and we assume that all of this energy is deposited within the body. From part (a), 5.19×10^3 disintegrations occur per second, and we know how many seconds are in a year. The total energy deposited per year is then

$$5.19 \times 10^3 \text{ s}^{-1} \times (60 \times 60 \times 24 \times 365 \text{ s yr}^{-1}) \times 0.55 \text{ MeV} = 9.0 \times 10^{10} \text{ MeV yr}^{-1}$$

Next, because a Gy is 1 J per kilogram of tissue, we express this answer in joules per year:

$$(9.0 \times 10^{10} \text{ MeV yr}^{-1})(1.602 \times 10^{-13} \text{ J MeV}^{-1}) = 0.0144 \text{ J yr}^{-1}$$

Each kilogram of body tissue receives 1/70.0 of this amount of energy per year, because the person weighs 70.0 kg. The dose is thus 21×10^{-5} J kg^{-1} yr^{-1}, which is equivalent to 0.21 Gy yr^{-1} or 21 mrad yr^{-1}. This is about a fifth of the annual background dosage received by a person.

Related Problems: 35, 36

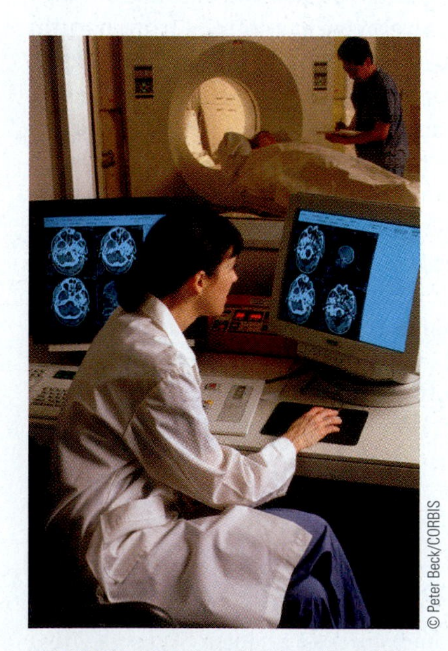

© Peter Beck/CORBIS

FIGURE 19.8 A patient undergoing a PET brain scan, as seen from the radiographer's control room.

Although radiation can do great harm, it confers great benefits in medical applications (Fig. 19.7). The diagnostic importance of x-ray imaging hardly needs mention. Both x-rays and gamma rays are used selectively in cancer therapy to destroy malignant cells. The beta-emitting ^{131}I nuclide finds use in the treatment of cancer of the thyroid because iodine is taken up preferentially by the thyroid gland. Heart pacemakers use the decay of tiny amounts of radioactive ^{238}Pu, converted to electrical energy.

Positron emission tomography (PET) is an important diagnostic technique using radiation (Fig.19.8). It employs radioisotopes such as ^{11}C (half-life 20.3 min) or ^{15}O (half-life 124 s) that emit positrons when they decay. These are incorporated (quickly, because of their short half-lives) into substances such as glucose, which are injected into the patient. By following the pattern of positron emission from the body, researchers can study blood flow and glucose metabolism in healthy and diseased individuals. Computer-reconstructed pictures of positron emissions from the brain are particularly useful, because the locations of glucose metabolism appear to differ between healthy persons and patients with ailments such as manic depression (bipolar disorder) and schizophrenia.

Less direct benefits come from other applications. An example is the study of the mechanism of photosynthesis, in which carbon dioxide and water are combined to form glucose in the green leaves of plants.

$$6\ CO_2(g) + 6\ H_2O(\ell) \longrightarrow C_6H_{12}O_6(s) + 6\ O_2(g)$$

Exposure of plants to CO_2 containing a higher-than-normal proportion of ^{14}C as a tracer enables scientists to follow the mechanism of the reaction. The plants are analyzed at intervals to find what compounds contain ^{14}C in their molecules and thereby to identify the intermediates in photosynthesis. Radioactive tracers are also widely used in medical diagnosis. The radioimmunoassay technique, invented by Nobel laureate Rosalind Yalow, determines the levels of drugs and hormones in body fluids.

19.5 Nuclear Fission

The nuclear reactions considered so far have been spontaneous, first-order decays of unstable nuclides. By the 1920s, physicists and chemists were using particle accelerators to bombard samples with high-energy particles to *induce* nuclear reactions. One of the first results of this program was the identification in 1932 by the English physicist James Chadwick (a student of Rutherford) of the neutron as a product of the reaction of alpha particles with light nuclides such as ^{9}Be:

$$^{4}_{2}\text{He} + {}^{9}_{4}\text{Be} \longrightarrow {}^{1}_{0}n + {}^{12}_{6}\text{C}$$

Shortly after Chadwick's discovery, a group of physicists in Rome, led by Enrico Fermi, began to study the interaction of neutrons with the nuclei of various elements. The experiments produced a number of radioactive species, and it was evident that the absorption of a neutron increased the $N:Z$ ratio in target nuclei above the stability line (see Fig. 19.1). One of the targets used was uranium, the heaviest naturally occurring element. Several radioactive products resulted, none of which had chemical properties characteristic of the elements between $Z = 86$ (radon) and $Z = 92$ (uranium). It appeared to the Italian scientists in 1934 that several new transuranic elements ($Z > 92$) had been synthesized, and an active period of investigation followed.

In 1938 in Berlin, Otto Hahn and Fritz Strassmann sought to characterize the supposed transuranic elements. To their bewilderment, they found instead that barium ($Z = 56$) appeared among the products of the bombardment of uranium by neutrons. Hahn informed his former colleague Lise Meitner, and she conjectured that the products of the bombardment of uranium by neutrons were not transuranic elements but fragments of uranium atoms resulting from a process she termed **fission**. The implication of this phenomenon—the possible release of enormous amounts of energy—was immediately evident. When the outbreak of World War II appeared imminent in the summer of 1939, Albert Einstein wrote to President Franklin Roosevelt to inform him of the possible military uses of fission and of his concern that Germany might develop a nuclear explosive. As a result, in 1942 President Roosevelt authorized the Manhattan District Project, an intense, coordinated effort by a large number of physicists, chemists, and engineers to make a fission bomb of unprecedented destructive power.

The operation of the first atomic bomb hinged on the fission of uranium in a chain reaction induced by absorption of neutrons. The two most abundant isotopes of uranium are ^{235}U and ^{238}U, whose natural relative abundances are 0.720% and 99.275%, respectively. Both undergo fission upon absorbing a neutron, the latter only with "fast" neutrons and the former with both "fast" and "slow" neutrons. In the early days of neutron research, it was not known that the absorption of neutrons by nuclei depends strongly upon neutron velocity. By accident, Fermi and his colleagues discovered that experiments conducted on a wooden table led to a much higher yield of radioactive products than those performed on a marble-topped table. Fermi then repeated the irradiation experiments with a block of paraffin wax interposed between the radium–beryllium neutron source and the target sample, with the startling result that the induced level of radioactivity was greatly enhanced. Within hours Fermi had found the explanation: The high-energy neutrons emitted from the radium–beryllium source were reduced to thermal energy by collision with the low-mass nuclei of the paraffin molecules. Because of their lower energies, their probability of reaction with ^{235}U was greater and a higher yield was achieved. Hydrogen nuclei and the nuclei of other light elements such as ^{12}C (in graphite) are very effective in reducing the energies of high-velocity neutrons and are called **moderators.**

The fission of ^{235}U follows many different patterns, and some 34 elements have been identified among the fission products. In any single fission event two particular nuclides are produced together with two or three secondary neutrons; collectively, they carry away about 200 MeV of kinetic energy. Usually, the daughter nuclei have different Z and A numbers, so the fission process is asymmetric. Three of the many pathways are

$$ {}^{1}_{0}n + {}^{235}_{92}U \longrightarrow \begin{cases} {}^{72}_{30}Zn + {}^{162}_{62}Sm + 2\,{}^{1}_{0}n \\ {}^{80}_{38}Sr + {}^{153}_{54}Xe + 3\,{}^{1}_{0}n \\ {}^{94}_{36}Kr + {}^{139}_{56}Ba + 3\,{}^{1}_{0}n \end{cases} $$

Figure 19.9 shows the distribution of the nuclides produced. The emission of more than one neutron per neutron absorbed in the fission process means that this is a branching chain reaction in which the number of neutrons grows exponentially with time (Fig. 19.10). Permitted to proceed unchecked, this reaction would quickly lead to the release of enormous quantities of energy. Neutrons can be lost by various processes, so it is not obvious that a self-propagating reaction will occur. On December 2, 1942, Fermi and his associates demonstrated that a self-sustaining neutron chain reaction occurred in a uranium "pile" with a graphite moderator. They limited its power output to $\frac{1}{2}$ J s^{-1} by inserting cadmium control rods to absorb neutrons, thereby balancing the rates of neutron generation and loss.

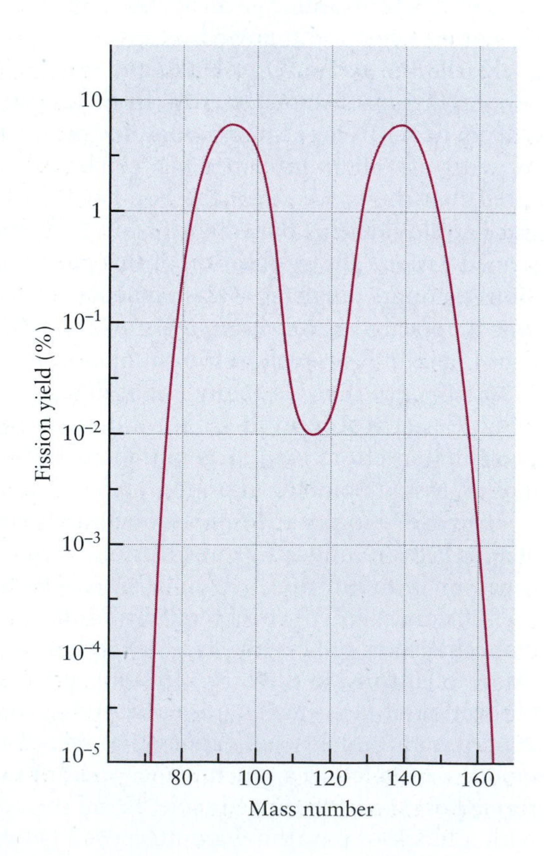

FIGURE 19.9 The distribution of nuclides produced in the fission of ^{235}U has two peaks. Nuclei having mass numbers in the vicinity of $A = 95$ and $A = 139$ are formed with the highest yield; those with $A \approx 117$ are produced with lower probability.

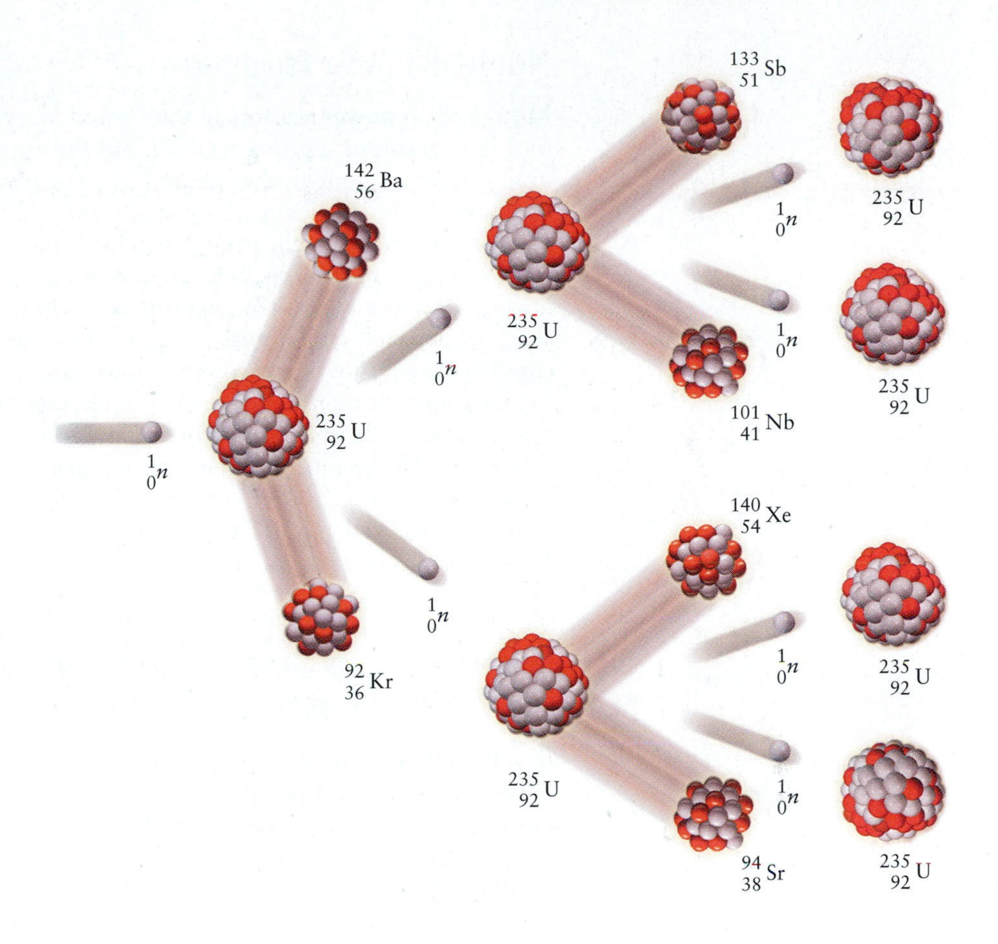

FIGURE 19.10 In a self-propagating nuclear chain reaction, the number of neutrons grows exponentially during fission.

Fermi's work made two developments possible: (1) the exploitation of nuclear fission for the controlled generation of energy in nuclear reactors and (2) the production of ^{239}Pu, a slow- and fast-neutron fissionable isotope of plutonium, as an alternative to ^{235}U for the construction of atomic bombs. For the sudden release of energy required in an explosive, it was necessary to obtain ^{235}U or ^{239}Pu in a state free of neutron-absorbing impurities. Both alternatives were pursued simultaneously. The first required enriching ^{235}U from its relative abundance of 0.72% in natural uranium. This was accomplished through gaseous diffusion (see Section 9.8). For the second alternative, ^{239}Pu was recovered from the partially spent uranium fuel of large nuclear reactors by means of redox reactions, precipitation, and solvent extraction.

Although Fermi and his associates were the first scientists to demonstrate a self-sustaining nuclear chain reaction, a natural uranium fission reactor "went critical" about 1.8 billion years ago in a place now called Oklo, in the Gabon Republic of equatorial Africa. In 1972 French scientists discovered that the ^{235}U content of ore from a site in the open-pit mine at Oklo was only 0.7171%; the normal content of ore from other areas of the mine was 0.7207%. Although this deviation was not large, it was significant, and an investigation revealed that other elements were also present in the ore, in the exact proportions expected after nuclear fission. This discovery established that a self-sustaining nuclear reaction had occurred at Oklo. The geological age of the ore body was found to be about 1.8×10^9 years, and the original ^{235}U concentration is calculated to have been about 3%. From the size of the active ore mass and the depletion of ^{235}U, it is estimated that the reactor generated about 15,000 megawatt-years (5×10^{17} J) of energy over about 100,000 years.

Nuclear Power Reactors

Most **nuclear power reactors** in the United States (Fig. 19.11) use rods of U_3O_8 as fuel. The uranium is primarily ^{238}U, but the amount of ^{235}U is enriched above natural abundance to a level of about 3%. The moderator used to slow the neutrons (to increase the efficiency of the fission) is ordinary water in most cases, so these reactors are called "light-water" reactors. The controlled release of energy by nuclear fission in power reactors demands a delicate balance between neutron generation and neutron loss. As mentioned earlier, this is accomplished by means of steel control rods containing ^{112}Cd or ^{10}B, isotopes that have a very large neutron-capture probability. These rods are automatically inserted into or withdrawn from the fissioning system in response to a change in the neutron flux. As the nuclear reaction proceeds, the moderator (water) is heated and transfers its heat to a steam generator. The steam then goes to turbines that generate electricity (Fig. 19.12).

The power reactors discussed so far rely on the fission of ^{235}U, an isotope in extremely limited supply. An alternative is to convert the much more abundant ^{238}U to fissionable plutonium (^{239}Pu) by neutron bombardment:

$$^{238}_{92}\text{U} + {}^{1}_{0}n \longrightarrow {}^{239}_{93}\text{Np} + {}^{0}_{-1}e^{-} \longrightarrow {}^{239}_{94}\text{Pu} + 2\,{}^{0}_{-1}e^{-}$$

Fissionable ^{233}U can also be made from thorium:

$$^{232}_{90}\text{Th} + {}^{1}_{0}n \longrightarrow {}^{233}_{91}\text{Pa} + {}^{0}_{-1}e^{-} \longrightarrow {}^{233}_{92}\text{U} + 2\,{}^{0}_{-1}e^{-}$$

In a **breeder reactor,** in addition to heat being generated by fission, neutrons are absorbed in a blanket of uranium or thorium. This causes the preceding reactions to occur and generates additional fuel for the reactor to use. An advanced technology for breeder reactors uses liquid sodium instead of water as the coolant, allowing the use of faster neutrons than in water-cooled reactors. The faster neutrons cause more complete consumption of radioactive fuels, increasing efficiency and greatly reducing radioactive waste.

The risks associated with the operation of nuclear reactors are small but not negligible, as the failure of the Three Mile Island reactor in the United States in 1979 and the disaster at Chernobyl in the former Soviet Union in 1987 demonstrated. If a reactor has to be shut down quickly, there is danger of a meltdown, in which the heat from the continuing fission processes melts the uranium fuel. Coolant must be circulated until heat from the decay of short-lived isotopes has

FIGURE 19.11 A nuclear power plant. The large structure on the left is a cooling tower; the containment building is the smaller building on the right with the domed top.

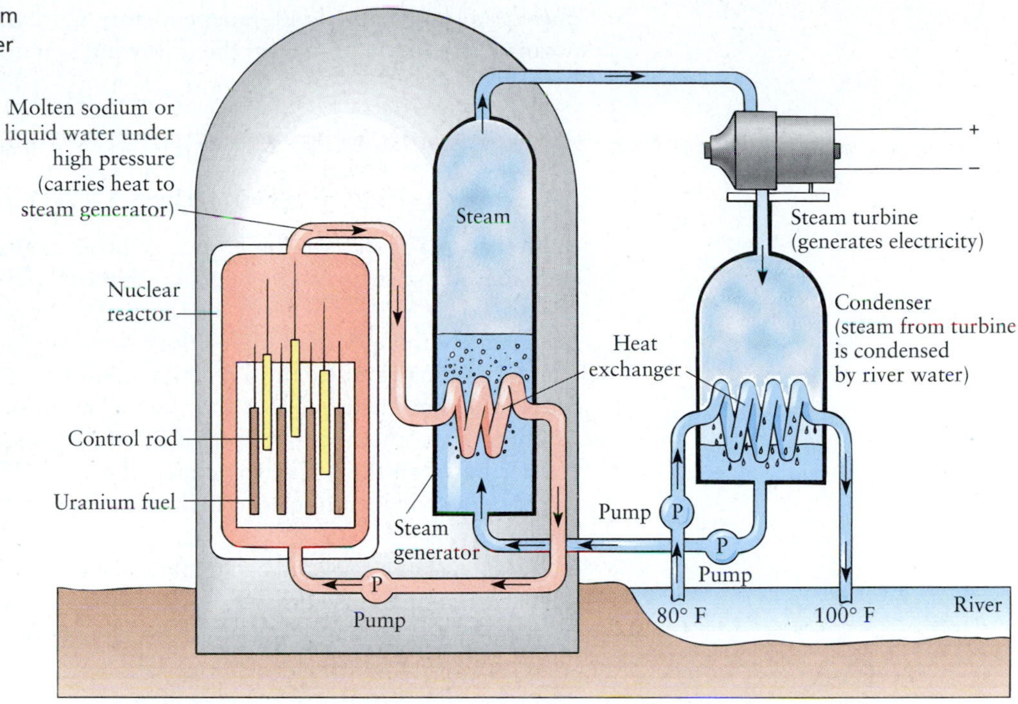

FIGURE 19.12 A schematic diagram of a pressurized-water nuclear power reactor.

been dissipated. The Three Mile Island accident resulted in a partial meltdown because some water coolant pumps were inoperative and others were shut down too soon, causing damage to the core and a slight release of radioactivity into the environment. The Chernobyl disaster was caused by a failure of the water-cooling system and a meltdown. The rapid and uncontrolled nuclear reaction that took place set the graphite moderator on fire and caused the reactor building to rupture, spreading radioactive nuclides with an activity estimated at 2×10^{20} Bq into the atmosphere. A major part of the problem was that the reactor at Chernobyl, unlike those in the United States, was not in a massive containment building.

The safe disposal of the radioactive wastes from nuclear reactors is an important and controversial matter. A variety of proposals have been made, including the burial of radioactive waste in deep mines on either a recoverable or a permanent basis, burial at sea, and launching the waste into outer space. The first alternative is the only one that appears credible. The essential requirement is that the disposal site(s) be stable with respect to possible earthquakes or invasion by underground water. Spent nuclear fuel can be encased in blocks of borosilicate glass, packed in metal containers, and buried in stable rock formations. For a nuclide such as ^{239}Pu, whose half-life is 24,000 years, a storage site that is stable over 240,000 years is needed before the activity drops to 0.1% of its original value. Some shorter lived isotopes are more hazardous over short periods, but their threat diminishes more quickly.

19.6 Nuclear Fusion and Nucleosynthesis

Nuclear fusion is the union of two light nuclides to form a heavier nuclide with the release of energy. Fusion processes are often called **thermonuclear reactions** because they require that the colliding particles possess very high kinetic energies, corresponding to temperatures of millions of degrees, before they are initiated. They are the processes that occur in the sun and other stars. In 1939 the German physicists Hans Bethe (and, independently, Carl von Weizsäcker) proposed that in normal stars (main sequence) the following reactions take place:

$$^{1}_{1}H + ^{1}_{1}H \longrightarrow ^{2}_{1}H + ^{0}_{1}e^{+} + \nu$$

$$^{2}_{1}H + ^{1}_{1}H \longrightarrow ^{3}_{2}He + \gamma$$

$$^{3}_{2}He + ^{3}_{2}He \longrightarrow ^{4}_{2}He + 2\,^{1}_{1}H$$

In the first reaction, two high-velocity protons fuse to form a deuteron, with the emission of a positron and a neutrino that carry away (as kinetic energy) the additional 0.415 MeV of energy released. In the second reaction, a high-energy deuteron combines with a high-velocity proton to form a helium nucleus of mass 3 and a gamma ray. The third reaction completes the cycle with the formation of a normal helium nucleus ($^{4}_{2}He$) and the regeneration of two protons. Each of these reactions is exothermic, but up to 1.25 MeV is required to overcome the repulsive barrier between the positively charged nuclei. The overall result of the cycle is to convert hydrogen nuclei to helium nuclei, and the process is called **hydrogen burning.**

As such a star ages and accumulates helium, it begins to contract under the influence of its immense gravity. As it contracts, its helium core heats up; when it reaches a temperature of about 10^8 K, a stage of **helium burning** begins. The first reaction that occurs is

$$2\,^{4}_{2}He \rightleftharpoons ^{8}_{4}Be$$

This reaction is written as an equilibrium because the ^{8}Be quickly reverts to helium nuclei with a half-life of only 2×10^{-16} s. Even with this short half-life, the ^{8}Be nuclei are believed to occasionally react with alpha particles to form stable ^{12}C:

$$^{8}_{4}Be + ^{4}_{2}He \longrightarrow ^{12}_{6}C$$

The overall effect of the helium-burning phase of a star's life is to convert three helium nuclei to a carbon nucleus, just as helium was formed from four hydrogen nuclei in the hydrogen-burning phase. The density of the core of a star that is burning helium is on the order of 10^5 g cm^{-3}.

This process of **nucleosynthesis** continues beyond the formation of ^{12}C to produce ^{13}N, ^{13}C, ^{14}N, ^{15}O, ^{15}N, and ^{16}O. The stars in this stage are classified as red giants. Similar cycles occur until the temperature of a star core is about 4×10^9 K, its density is about 3×10^6 g cm^{-3}, and the nuclei are ^{56}Fe, ^{59}Co, and ^{60}Ni. These are the nuclei that have the maximum binding energy per nucleon (see Fig. 19.2). It is thought that the synthesis of still heavier nuclei occurs in the immense explosions of supernovae.

Heavy elements can also be produced in particle accelerators, which accelerate ions to high speeds, causing collisions that generate the new elements. Technetium, for example, is not found in nature but was first produced in 1937 when high-energy deuterons were directed at a molybdenum source:

$$^{96}_{42}Mo + ^{2}_{1}H \longrightarrow ^{97}_{43}Tc + ^{1}_{0}n$$

The first **transuranic element** was produced in 1940. Neptunium ($Z = 93$) results from the capture of a neutron by ^{238}U, followed by beta decay. Subsequent work by the American chemist Glenn Seaborg and others led to the production of plutonium ($Z = 94$) and heavier elements. In recent years, nuclides with Z as high as 116 have been made, but in tiny quantities. These nuclides have very short half-lives.

Major efforts are now under way to achieve controlled nuclear fusion as a source of energy. One approach is based on the reaction between deuterium and tritium atoms,

$$^{2}_{1}H + ^{3}_{1}H \longrightarrow ^{4}_{2}He + ^{1}_{0}n$$

(a)

(b)

FIGURE 19.13 (a) The target chamber of Nova, a 16-foot-diameter aluminum sphere inside of which ten powerful laser beams converge. (b) The target itself is a tiny capsule (1 mm in diameter) of deuterium mixed with tritium. A laser pulse one billionth of a second in duration heats the target hotter than the sun's core, raising its pressure to more than 100 million atm, to initiate nuclear fusion. (a, © Phototake; b, Courtesy of the University of California, Lawrence Livermore National Laboratory, and the U.S. Department of Energy)

with a predicted energy release of 17.6 MeV. The neutrons produced in this reaction can be used to make the needed tritium by means of the reaction

$$^{6}_{3}\text{Li} + ^{1}_{0}n \longrightarrow ^{3}_{1}\text{H} + ^{4}_{2}\text{He}$$

in a lithium "blanket" that surrounds the region of the reaction. Temperatures on the order of 10^8 K are required to initiate these reactions, by which point atoms are stripped of all their electrons and the system is a plasma (ionized gas) of nuclei, electrons, and neutrons. There are two major problems: heating the plasma to this high temperature and confining it. In a laser fusion reactor, pellets of deuterium–tritium fuel are raised to the thermonuclear ignition temperature by bursts of light fired from lasers (Fig. 19.13). Contact between the plasma and the wall of the confining vessel would immediately result in prohibitive energy losses and extinguish the fusion reaction. A plasma can be contained in an intense toroidal (doughnut-shaped) magnetic field. Research reactors using this confinement method have progressed and are reaching the point where they produce more energy than they consume. Formidable problems still must be solved, including the development of materials for containment vessels that can withstand the corrosive effects of the intense x-ray and neutron radiation that is present.

The rewards of a workable nuclear fusion process would be great. Fusion produces neither the long-lived radioactive nuclides that accompany nuclear fission (although tritium requires care in handling) nor the environmental pollutants released by the burning of fossil fuels. Although deuterium is present in only 1/6000 of the abundance of ordinary hydrogen, its separation from the latter by the electrolysis of water is readily accomplished, and the oceans contain a virtually unlimited quantity of deuterium.

CHAPTER SUMMARY

The identities of the elements are not preserved in nuclear reactions—elements decay into lighter daughter elements in fission reactions and heavier elements are synthesized from lighter elements in fusion reactions. Mass changes in nuclear reactions are relatively small, but the accompanying energy changes are enormous; they are related by Einstein's famous formula $E = mc^2$. The isotopes of the lighter elements ($Z < 40$ or so) are stable when the ratio of the number of neutrons to the number of protons (N/Z) is approximately equal to 1. Isotopes with $N/Z < 1$ will decay via positron emission or electron capture to increase the number of protons in the nucleus, whereas those with $N/Z > 1$ will decay via beta emission to decrease the number of protons in the nucleus. A fourth decay channel, alpha particle emission, becomes important for heavier nuclei. Radioactive decay follows first order kinetics and the half-life $t_{1/2}$ is a convenient measure of the timescale of the reaction. Half-lives range from 10^{-16} s to 10^{10} years, which is an incredibly wide range of timescales. The half-life of ^{238}U has been used to calibrate geological timescales and ^{14}C dating is a well-established method for dating human artifacts in anthropology. X-rays, gamma rays, electrons, positrons, and alpha particles are all used in medical diagnostics and therapy with increasing efficacy and fewer side effects. Nuclear fusion, which led to the formation of the heavier elements from hydrogen in the process called nucleosynthesis, continues to hold promise as a source of clean power for the future.

CUMULATIVE EXERCISE

Radon

Radioactive ^{222}Rn and ^{220}Rn form constantly from the decay of uranium and thorium in rocks and soil and, being gaseous, seep out of the ground. The radon isotopes decay fairly quickly, but their products, which are also radioactive, are then in the air and attach themselves to dust particles. Thus, airborne radioactivity can accumulate to worrisome levels in poorly ventilated basements in ground that is rich in uranium and thorium.

(a) Describe the composition of an atom of ^{222}Rn and compare it with that of an atom of ^{220}Rn.

(b) Although ^{222}Rn is a decay product of ^{238}U, ^{220}Rn comes from ^{232}Th. How many alpha particles are emitted in the formation of these radon isotopes from their uranium or thorium starting points? (*Hint:* Alpha decay changes the mass number *A*, but other decay processes do not.)

(c) Can alpha decay alone explain the formation of these radon isotopes from ^{238}U and ^{232}Th? If not, state what other types of decay must occur.

(d) Can $^{222}_{86}$Rn and $^{220}_{86}$Rn decay by alpha particle emission? Write balanced nuclear equations for these two decay processes, and calculate the changes in mass that would result. The masses of ^{222}Rn and ^{220}Rn atoms are 222.01757 and 220.01140 u, respectively; those of ^{218}Po and ^{216}Po are 218.0089 and 216.00192 u, respectively.

(e) Calculate the energy change in the alpha decay of one ^{220}Rn nucleus, in million electron volts and in joules.

(f) The half-life of ^{222}Rn is 3.82 days. Calculate the initial activity of 2.00×10^{-8} g of ^{222}Rn, in disintegrations per second.

(g) What will be the activity of the ^{222}Rn from part (f) after 14 days?

(h) The half-life of ^{220}Rn is 54 s. Are the health risks of exposure to a given amount of radon for a given short length of time greater or smaller for ^{220}Rn than for ^{222}Rn?

Radon most commonly enters houses through the foundation or basement walls.

Answers

(a) An atom of ^{222}Rn has 86 electrons outside the nucleus. Inside the nucleus are 86 protons and $222 - 86 = 136$ neutrons. An atom of ^{220}Rn has the same number of electrons and protons, but only 134 neutrons in its nucleus.

(b) Four alpha particles are produced to make ^{222}Rn from ^{238}U; three are produced to make ^{220}Rn from ^{232}Th.

(c) If $^{238}_{92}U$ were to lose four alpha particles, $^{222}_{84}Po$ would result instead of $^{222}_{84}Rn$. Two $^{0}_{-1}e^{-}$ beta particles must be ejected from the nucleus along the way to raise the atomic number to $Z = 86$. The same is true of the production of ^{220}Rn from ^{232}Th.

(d) $^{222}_{86}Rn \longrightarrow {}^{218}_{84}Po + {}^{4}_{2}He$; $\Delta m = -0.00617$ u < 0; allowed

$^{220}_{86}Rn \longrightarrow {}^{216}_{84}Po + {}^{4}_{2}He$; $\Delta m = -0.0069$ u < 0; allowed

(e) $\Delta E = -6.4$ MeV $= -1.03 \times 10^{-12}$ J

(f) $A = 1.14 \times 10^{8}\ s^{-1}$

(g) $A = 9.0 \times 10^{6}\ s^{-1}$

(h) Greater

CHAPTER REVIEW

- One atomic mass unit (u) is 1.6605×10^{-27} kg.
- The energy change ΔE that results from a mass change in a nuclear reaction (Δm) is given by the Einstein mass-energy relationship $\Delta E = c^2\Delta m$.
- $\Delta E = 931.5$ MeV for $\Delta u = 1$ so the energy equivalent of 1 u is 931.5 MeV.
- Nuclear reactions are spontaneous when $\Delta E < 0$ because ΔE is so large compared with pressure-volume work or any entropy changes associated with nuclear reactions, making $\Delta G = \Delta E$.
- Electron–positron and proton–antiproton are particle–antiparticle pairs in which the members of each pair have the same masses but opposite charges. The neutrino and antineutrino are neutral particles that form another important particle–antiparticle pair.
- Nuclear decay processes

  The equations for nuclear reactions are balanced using the same methods developed for chemical reactions. Charge, mass, and atomic number are conserved. The electron is assigned an effective atomic number $Z = -1$ for this purpose.

  Beta decay: unstable nuclides with fewer protons than stable nuclides of the same mass number convert neutrons into protons by emitting an electron (beta particle).

  Positron emission: unstable nuclides with more protons than stable nuclides of the same mass number convert protons into neutrons by emitting a positron.

  Electron capture: electron capture is more important than positron emission for converting protons into neutrons when the mass change is less than twice the mass of the positron.

  Alpha decay: very heavy elements can decay by emitting a He nucleus (alpha particle), decreasing the atomic number by 2 and the mass number by 4.
- Radiation detectors

  Photographic film

  Scintillators: particles that emit light when excited by radiation

  Geiger counters: gas tubes designed to produce large current pulses from avalanches of ionized electrons

  Array detectors: scintillators coupled to very large arrays of semiconductor detectors that have largely replaced film in medical diagnostics

- Kinetics of radioactive decay: first order decay with half-life $t_{1/2} = \ln 2/k = 0.693/k$.
- Radioactive dating: living organisms have $^{14}C/^{12}C$ ratios characteristic of the atmosphere of the time. When the organism dies, ^{14}C decays with $t_{1/2} = 5730$ years, enabling dating by counting half-lives.
- Radiation dosimetry in biology and medicine
  rad (radiation adsorbed dose): 10^{-2} J per kg of tissue (S.I. unit gray (Gy) = 100 rad = 1 J per kg)
  rem (roentgen equivalent in man): accounts for relative biological damage (S.I. unit sievert (Sv) = 100 rem by analogy to gray)
  1 rad (beta) $\longrightarrow$ 1 rem
  1 rad (alpha) $\longrightarrow$ 10 rem
- Nuclear chain reactions
  ^{235}U decay produces more than one neutron per disintegration, which can lead to a chain reaction. Rates of fission reactions are controlled using moderators to reduce the energy of the neutrons and enhance the rate and control rods to limit the neutron flux and reduce the rate.
  Fusion of hydrogen isotopes in normal stars produces He, which reacts further to produce all of the heavier elements in a series of fusion reactions called nucleosynthesis.

CONCEPTS & SKILLS

After studying this chapter and working the problems that follow, you should be able to:

1. Calculate the binding energy of a nucleus (Section 19.1, Problems 3–6).
2. Write balanced nuclear equations for beta decay, positron emission, electron capture, and alpha decay processes and calculate the maximum kinetic energies of particles emitted (Section 19.2, Problems 7–18).
3. Describe several methods that are used to detect the products of radioactive decay (Section 19.2).
4. Solve problems involving the half-life or decay constant of a radioactive sample and its activity (Section 19.3, Problems 19–24).
5. Apply the kinetics of nuclear decay to the dating of rocks or artifacts (Section 19.3, Problems 25–30).
6. Discuss the interactions of radiation with various kinds of matter and the measurement of radiation dosage (Section 19.4, Problems 33–36).
7. Describe the processes of nuclear fission and fusion, and calculate the amounts of energy released when they occur (Sections 19.5 and 19.6, Problems 37–47).
8. Explain the benefits and risks associated with the use of nuclear reactions for power generation (Sections 19.5 and 19.6).

KEY EQUATIONS

$$\Delta E = c^2 \Delta m$$ **(Section 19.1)**

$$A = -\frac{dN}{dt} = kN$$ **(Section 19.3)**

$$A = A_i e^{-kt}$$ **(Section 19.3)**

PROBLEMS

Answers to problems whose numbers are boldface appear in Appendix G. Problems that are more challenging are indicated with asterisks.

Mass–Energy Relationships in Nuclei

1. Complete and balance the following equations for nuclear reactions that are thought to take place in stars:
(a) $2\ ^{12}_{6}C \longrightarrow ? + ^{1}_{0}n$
(b) $? + ^{1}_{1}H \longrightarrow ^{12}_{6}C + ^{4}_{2}He$
(c) $2\ ^{3}_{2}He \longrightarrow ? + 2\ ^{1}_{1}H$

2. Complete and balance the following equations for nuclear reactions that are used in particle accelerators to make elements beyond uranium:
(a) $^{4}_{2}He + ^{253}_{99}Es \longrightarrow ? + 2^{1}_{0}n$
(b) $^{249}_{98}Cf + ? \longrightarrow ^{257}_{103}Lr + 2\ ^{1}_{0}n$
(c) $^{238}_{92}U + ^{12}_{6}C \longrightarrow ^{244}_{98}Cf + ?$

3. Calculate the total binding energy, in both kJ per mole and MeV per atom, and the binding energy per nucleon of the following nuclides, using the data from Table 19.1.
(a) $^{40}_{20}Ca$ (b) $^{87}_{37}Rb$ (c) $^{238}_{92}U$

4. Calculate the total binding energy, in both kilojoules per mole and MeV per atom, and the binding energy per nucleon of the following nuclides, using the data from Table 19.1.
(a) $^{10}_{4}Be$ (b) $^{35}_{17}Cl$ (c) $^{49}_{22}Ti$

5. Use the data from Table 19.1 to predict which is more stable: four protons, four neutrons, and four electrons organized as two ^{4}He atoms or as one ^{8}Be atom. What is the mass difference?

6. Use the data from Table 19.1 to predict which is more stable: 16 protons, 16 neutrons, and 16 electrons organized as two ^{16}O atoms or as one ^{32}S atom. What is the mass difference?

Nuclear Decay Processes

7. The nuclide $^{8}_{5}B$ decays by positron emission to $^{8}_{4}Be$. What is the energy released (in MeV)?

8. The nuclide $^{10}_{4}Be$ undergoes spontaneous radioactive decay to $^{10}_{5}B$ with emission of a beta particle. Calculate the maximum kinetic energy of the emitted beta particle.

9. Write balanced equations that represent the following nuclear reactions.
(a) Beta emission by $^{39}_{17}Cl$
(b) Positron emission by $^{22}_{11}Na$
(c) Alpha emission by $^{224}_{88}Ra$
(d) Electron capture by $^{82}_{38}Sr$

10. Write balanced equations that represent the following nuclear reactions.
(a) Alpha emission by $^{155}_{70}Yb$
(b) Positron emission by $^{26}_{14}Si$
(c) Electron capture by $^{65}_{30}Zn$
(d) Beta emission by $^{100}_{41}Nb$

11. The stable isotopes of neon are ^{20}Ne, ^{21}Ne, and ^{22}Ne. Predict the nuclides formed when ^{19}Ne and ^{23}Ne decay.

12. The two stable isotopes of carbon are ^{12}C and ^{13}C. Predict the nuclides formed when ^{11}C and ^{14}C decay. Is alpha emission by ^{14}C possible?

13. The free neutron is an unstable particle that decays into a proton. What other particle is formed in neutron decay, and what is the maximum kinetic energy (in MeV) that it can possess?

14. The radionuclide $^{210}_{84}Po$ decays by alpha emission to a daughter nuclide. The atomic mass of $^{210}_{84}Po$ is 209.9829 u, and that of its daughter is 205.9745 u.
(a) Identify the daughter, and write the nuclear equation for the radioactive decay process.
(b) Calculate the total energy released per disintegration (in MeV).
(c) Calculate the kinetic energy of the emitted alpha particle.

15. The natural abundance of ^{30}Si is 3.1%. Upon irradiation with neutrons, this isotope is converted to ^{31}Si, which decays to the stable isotope ^{31}P. This provides a way of introducing trace amounts of phosphorus into silicon in a much more uniform fashion than is possible by ordinary mixing of silicon and phosphorus and gives semiconductor devices the capability of handling much higher levels of power. Write balanced nuclear equations for the two steps in the preparation of ^{31}P from ^{30}Si.

16. The most convenient way to prepare the element polonium is to expose bismuth (which is 100% ^{209}Bi) to neutrons. Write balanced nuclear equations for the two steps in the preparation of polonium.

17. One convenient source of neutrons is the reaction of an alpha particle from an emitter such as polonium (^{210}Po) with an atom of beryllium (^{9}Be). Write nuclear equations for the reactions that occur.

18. Three atoms of element 111 were produced in 1994 by bombarding ^{209}Bi with ^{64}Ni.
(a) Write a balanced equation for this nuclear reaction. What other species is produced?
(b) Write a balanced equation for the alpha decay process of this nuclide of element 111.

Kinetics of Radioactive Decay

19. How many radioactive disintegrations occur per minute in a 0.0010-g sample of ^{209}Po that has been freshly separated from its decay products? The half-life of ^{209}Po is 103 years.

20. How many alpha particles are emitted per minute by a 0.0010-g sample of ^{238}U that has been freshly separated from its decay products? Assume that each decay emits one alpha particle. The half-life of ^{238}U is 4.47×10^{9} years.

21. The nuclide ^{19}O, prepared by neutron irradiation of ^{19}F, has a half-life of 29 s.
(a) How many ^{19}O atoms are in a freshly prepared sample if its decay rate is $2.5 \times 10^{4}\ s^{-1}$?
(b) After 2.00 min, how many ^{19}O atoms remain?

22. The nuclide ^{35}S decays by beta emission with a half-life of 87.1 days.
(a) How many grams of ^{35}S are in a sample that has a decay rate from that nuclide of $3.70 \times 10^{2}\ s^{-1}$?
(b) After 365 days, how many grams of ^{35}S remain?

23. Astatine is the rarest naturally occurring element, with ^{219}At appearing as the product of a very minor side branch in the decay of ^{235}U (itself not a very abundant isotope). It is estimated that the mass of all the naturally occurring ^{219}At in the upper kilometer of the earth's surface has a steady-state value of only 44 mg. Calculate the total activity (in disintegrations per second) caused by all the naturally occurring astatine in this part of the earth. The half-life of ^{219}At is 54 s, and its atomic mass is 219.01 u.

24. Technetium has not been found in nature. It can be obtained readily as a product of uranium fission in nuclear power plants, however, and is now produced in quantities of many kilograms per year. One medical use relies on the tendency of ^{99m}Tc (an excited nuclear state of ^{99}Tc) to concentrate in abnormal heart tissue. Calculate the total activity (in disintegrations per second) caused by the decay of 1.0 μg of ^{99m}Tc, which has a half-life of 6.0 hours.

25. The specific activity of ^{14}C in the biosphere is 0.255 Bq g^{-1}. What is the age of a piece of papyrus from an Egyptian tomb if its beta counting rate is 0.153 Bq g^{-1}? The half-life of ^{14}C is 5730 years.

26. The specific activity of an article found in the Lascaux Caves in France is 0.0375 Bq g^{-1}. Calculate the age of the article.

27. Over geological time, an atom of ^{238}U decays to a stable ^{206}Pb atom in a series of eight alpha emissions, each of which leads to the formation of one helium atom. A geochemist analyzes a rock and finds that it contains 9.0×10^{-5} cm^3 of helium (at 0°C and atmospheric pressure) per gram and 2.0×10^{-7} g of ^{238}U per gram. Estimate the age of the mineral, given that $t_{1/2}$ of ^{238}U is 4.47×10^9 years.

28. The isotope ^{232}Th decays to ^{208}Pb by the emission of six alpha particles, with a half-life of 1.39×10^{10} years. Analysis of 1.00 kg of ocean sediment shows it to contain 7.4 mg of ^{232}Th and 4.9×10^{-3} cm^3 of gaseous helium at 0°C and atmospheric pressure. Estimate the age of the sediment, assuming no loss or gain of thorium or helium from the sediment since its formation and assuming that the helium arose entirely from the decay of thorium.

29. The half-lives of ^{235}U and ^{238}U are 7.04×10^8 years and 4.47×10^9 years, respectively, and the present abundance ratio is $^{238}U/^{235}U = 137.7$. It is thought that their abundance ratio was 1 at some time *before* our earth and solar system were formed about 4.5×10^9 years ago. Estimate how long ago the supernova occurred that supposedly produced all the uranium isotopes in equal abundance, including the two longest lived isotopes, ^{238}U and ^{235}U.

30. Using the result of problem 29 and the accepted age of the earth, 4.5×10^9 yr, calculate the $^{238}U/^{235}U$ ratio at the time the earth was formed.

Radiation in Biology and Medicine

31. Write balanced equations for the decays of ^{11}C and ^{15}O, both of which are used in positron emission tomography to scan the uptake of glucose in the body.

32. Write balanced equations for the decays of ^{13}N and ^{18}F, two other radioisotopes that are used in positron emission tomography. What is the ultimate fate of the positrons?

33. The positrons emitted by ^{11}C have a maximum kinetic energy of 0.99 MeV, and those emitted by ^{15}O have a maximum kinetic energy of 1.72 MeV. Calculate the ratio of the number of millisieverts of radiation exposure caused by ingesting a given fixed chemical amount (equal numbers of atoms) of each of these radioisotopes.

34. Compare the relative health risks of contact with a given amount of ^{226}Ra, which has a half-life of 1622 years and emits 4.78-MeV alpha particles, with contact with the same chemical amount of ^{14}C, which has a half-life of 5730 years and emits beta particles with energies of up to 0.155 MeV.

35. The nuclide ^{131}I undergoes beta decay with a half-life of 8.041 days. Large quantities of this nuclide were released into the environment in the Chernobyl accident. A victim of radiation poisoning has absorbed 5.0×10^{-6} g (5.0 μg) of ^{131}I.

(a) Compute the activity, in becquerels, of the ^{131}I in this person, taking the atomic mass of the nuclide to equal 131 g mol^{-1}.

(b) Compute the radiation absorbed dose, in milligrays, caused by this nuclide during the first *second* after its ingestion. Assume that beta particles emitted by ^{131}I have an average kinetic energy of 0.40 MeV, that all of this energy is deposited within the victim's body, and that the victim weighs 60 kg.

(c) Is this dose likely to be lethal? Remember that the activity of the ^{131}I diminishes as it decays.

36. The nuclide ^{239}Pu undergoes alpha decay with a half-life of 2.411×10^4 years. An atomic energy worker breathes in 5.0×10^{-6} g (5.0 μg) of ^{239}Pu, which lodges permanently in a lung.

(a) Compute the activity, in becquerels, of the ^{239}Pu ingested, taking the atomic mass of the nuclide to be 239 g mol^{-1}.

(b) Determine the radiation absorbed dose, in milligrays, during the first *year* after its ingestion. Assume that alpha particles emitted by ^{239}Pu have an average kinetic energy of 5.24 MeV, that all of this energy is deposited within the worker's body, and that the worker weighs 60 kg.

(c) Is this dose likely to be lethal?

Nuclear Fission

37. Strontium-90 is one of the most hazardous products of atomic weapons testing because of its long half-life ($t_{1/2}$ = 28.1 years) and its tendency to accumulate in bone.

(a) Write nuclear equations for the decay of ^{90}Sr via the successive emission of two beta particles.

(b) The atomic mass of ^{90}Sr is 89.9073 u and that of ^{90}Zr is 89.9043 u. Calculate the energy released per ^{90}Sr atom, in MeV, in decaying to ^{90}Zr.

(c) What will be the initial activity of 1.00 g of ^{90}Sr released into the environment, in disintegrations per second?

(d) What activity will the material from part (c) show after 100 years?

38. Plutonium-239 is the fissionable isotope produced in breeder reactors; it is also produced in ordinary nuclear plants and in weapons tests. It is an extremely poisonous substance with a half-life of 24,100 years.

(a) Write an equation for the decay of ^{239}Pu via alpha emission.

(b) The atomic mass of ^{239}Pu is 239.05216 u and that of ^{235}U is 235.04393 u. Calculate the energy released per ^{239}Pu atom, in MeV, in decaying via alpha emission.
(c) What will be the initial activity, in disintegrations per second, of 1.00 g of ^{239}Pu buried in a disposal site for radioactive wastes?
(d) What activity will the material from part (c) show after 100,000 years?

39. The three naturally occurring isotopes of uranium are ^{234}U (half-life 2.5×10^5 years), ^{235}U (half-life 7.0×10^8 years), and ^{238}U (half-life 4.5×10^9 years). As time passes, will the average atomic mass of the uranium in a sample taken from nature increase, decrease, or remain constant?

40. Natural lithium consists of 7.42% ^{6}Li and 92.58% ^{7}Li. Much of the tritium ($^3_1$H) used in experiments with fusion reactions is made by the capture of neutrons by ^{6}Li atoms.
(a) Write a balanced nuclear equation for the process. What is the other particle produced?
(b) After ^{6}Li is removed from natural lithium, the remainder is sold for other uses. Is the molar mass of the left-over lithium greater or smaller than that of natural lithium?

41. Calculate the amount of energy released, in kilojoules per *gram* of uranium, in the fission reaction

$$^{235}_{92}\text{U} + ^{1}_{0}n \longrightarrow ^{94}_{36}\text{Kr} + ^{130}_{56}\text{Ba} + 3\ ^{1}_{0}n$$

Use the atomic masses in Table 19.1. The atomic mass of ^{94}Kr is 93.919 u and that of ^{139}Ba is 138.909 u.

Nuclear Fusion and Nucleosynthesis

42. Calculate the amount of energy released, in kilojoules per *gram* of deuterium (^{2}H), for the fusion reaction

$$^{2}_{1}\text{H} + ^{2}_{1}\text{H} \longrightarrow \quad + ^{4}_{2}\text{He}$$

Use the atomic masses in Table 19.1. Compare your answer with that from the preceding problem.

ADDITIONAL PROBLEMS

43. When an electron and a positron meet, they are replaced by two gamma rays, called the "annihilation radiation." Calculate the energies of these radiations, assuming that the kinetic energies of the incoming particles are 0.

44. The nuclide $^{231}_{92}$U converts spontaneously to $^{231}_{91}$Pa.
(a) Write two balanced nuclear equations for this conversion, one if it proceeds by electron capture and the other if it proceeds by positron emission.
(b) Using the nuclidic masses in Table 19.1, calculate the change in mass for each process. Explain why electron capture can occur spontaneously in this case but positron emission cannot.

45. The radioactive nuclide $^{64}_{29}$Cu decays by beta emission to $^{64}_{30}$Zn or by positron emission to $^{64}_{28}$Ni. The maximum kinetic energy of the beta particles is 0.58 MeV, and that of the positrons is 0.65 MeV. The mass of the neutral $^{64}_{29}$Cu atom is 63.92976 u.
(a) Calculate the mass, in atomic mass units, of the neutral $^{64}_{30}$Zn atom.
(b) Calculate the mass, in atomic mass units, of the neutral $^{64}_{28}$Ni atom.

46. A puzzling observation that led to the discovery of isotopes was the fact that lead obtained from uranium-containing ores had an atomic mass lower by two full atomic mass units than lead obtained from thorium-containing ores. Explain this result, using the fact that decay of radioactive uranium and thorium to stable lead occurs via alpha and beta emission.

47. By 1913, the elements radium, actinium, thorium, and uranium had all been discovered, but element 91, between thorium and uranium in the periodic table, was not yet known. The approach used by Meitner and Hahn was to look for the parent that decays to form actinium. Alpha and beta emission are the most important decay pathways among the heavy radioactive elements. What elements would decay to actinium by each of these two pathways? If radium salts show no sign of actinium, what does this suggest about the parent of actinium? What is the origin of the name of element 91, discovered by Meitner and Hahn in 1918?

48. Working in Rutherford's laboratory in 1932, Cockcroft and Walton bombarded a lithium target with 700-keV protons and found that the following reaction occurred:

$$^{7}_{3}\text{Li} + ^{1}_{1}\text{H} \longrightarrow \quad + 2\ ^{4}_{2}\text{He}$$

Each of the alpha particles was found to have a kinetic energy of 8.5 MeV. This research provided the first experimental test of Einstein's $\Delta E = c^2 \Delta m$ relationship. Discuss. Using the atomic masses from Table 19.1, calculate the value of c needed to account for this result.

49. (a) Calculate the binding energy per nucleon in $^{30}_{15}$P.
(b) The radioactive decay of the $^{30}_{15}$P occurs through positron emission. Calculate the maximum kinetic energy carried off by the positron.
(c) The half-life for this decay is 150 s. Calculate the rate constant k and the fraction remaining after 450 s.

50. Selenium-82 undergoes *double* beta decay:

$$^{82}_{34}\text{Se} \longrightarrow ^{82}_{36}\text{Kr} + 2\ ^{0}_{-1}e^{-} + 2\bar{\nu}$$

This low-probability process occurs with a half-life of 3.5×10^{27} s, one of the longest half-lives ever measured. Estimate the activity in an 82.0-g (1.00 mol) sample of this isotope. How many ^{82}Se nuclei decay in a day?

51. Gallium citrate, which contains the radioactive nuclide ^{67}Ga, is used in medicine as a tumor-seeking agent. Gallium-67 decays with a half-life of 77.9 hours. How much time is required for it to decay to 5.0% of its initial activity?

52. The nuclide ^{241}Am is used in smoke detectors. As it decays (with a half-life of 458 years), the emitted alpha particles ionize the air. When combustion products enter the detector, the number of ions changes and with it the conductivity of the air, setting off an alarm. If the activity of ^{241}Am in the detector is 3×10^4 Bq, calculate the mass of ^{241}Am present.

53. The half-life of ^{14}C is $t_{1/2}$ = 5730 years, and 1.00 g of modern wood charcoal has an activity of 0.255 Bq.
(a) Calculate the number of ^{14}C atoms per gram of carbon in modern wood charcoal.
(b) Calculate the fraction of carbon atoms in the biosphere that are ^{14}C.

54. Carbon-14 is produced in the upper atmosphere by the reaction

$$^{14}_{7}N + ^{1}_{0}n \longrightarrow ^{14}_{6}C + ^{1}_{1}H$$

where the neutrons come from nuclear processes induced by cosmic rays. It is estimated that the steady-state ^{14}C activity in the biosphere is 1.1×10^{19} Bq.
(a) Estimate the total mass of carbon in the biosphere, using the data in problem 53.
(b) The earth's crust has an average carbon content of 250 parts per million by mass, and the total crustal mass is 2.9×10^{25} g. Estimate the fraction of the carbon in the earth's crust that is part of the biosphere. Speculate on the whereabouts of the rest of the carbon in the earth's crust.

55. Analysis of a rock sample shows that it contains 0.42 mg of ^{40}Ar for every 1.00 mg of ^{40}K. Assuming that all the argon resulted from decay of the potassium and that neither element has left or entered the rock since its formation, estimate the age of the rock. (*Hint:* Use data from Table 19.2.) Note that not all the ^{40}K decays to ^{40}Ar.

* 56. Cobalt-60 and iodine-131 are used in treatments for some types of cancer. Cobalt-60 decays with a half-life of 5.27 years, emitting beta particles with a maximum energy of 0.32 MeV. Iodine-131 decays with a half-life of 8.04 days, emitting beta particles with a maximum energy of 0.60 MeV.
(a) Suppose a fixed small number of moles of each of these isotopes were to be ingested and remain in the body indefinitely. What is the *ratio* of the number of millisieverts of total lifetime radiation exposure that would be caused by the two radioisotopes?
(b) Now suppose that the contact with each of these isotopes is for a fixed short period, such as 1 hour. What is the ratio of millisieverts of radiation exposure for the two in this case?

57. Boron is used in control rods in nuclear power reactors because it is a good neutron absorber. When the isotope ^{10}B captures a neutron, an alpha particle (helium nucleus) is emitted. What other atom is formed? Write a balanced equation.

* 58. The average energy released in the fission of a ^{235}U nucleus is about 200 MeV. Suppose the conversion of this energy to electrical energy is 40% efficient. What mass of ^{235}U is converted to its fission products in a year's operation of a 1000-megawatt nuclear power station? Recall that 1 W is $1\ J\ s^{-1}$.

59. The energy released by a bomb is sometimes expressed in tons of TNT (trinitrotoluene). When one ton of TNT explodes, 4×10^{9} J of energy is released. The fission of 1 mol of uranium releases approximately 2×10^{13} J of energy. Calculate the energy released by the fission of 1.2 kg of uranium in a small atomic bomb. Express your answer in tons of TNT.

60. The solar system abundances of the elements Li, Be, and B are four to seven orders of magnitude lower than those of the elements that immediately follow them: C, N, and O. Explain.

* 61. The sun's distance from earth is approximately 1.50×10^{8} km, and the earth's radius is 6371 km. The earth receives radiant energy from hydrogen burning in the sun at a rate of $0.135\ J\ s^{-1}\ cm^{-2}$. Using the data of Table 19.1, calculate the mass of hydrogen converted per second in the sun.

CUMULATIVE PROBLEMS

62. In 1951 wood from two sequoia trees was dated by the ^{14}C method. In one tree, clean borings located between the growth rings associated with the years A.D. 1057 and 1087 (that is, wood known to have grown 880 ± 15 year prior to the date of measurement) had a ^{14}C activity about 0.892 of that of wood growing in 1951. A sample from a second tree had an activity about 0.838 of that of new wood, and its age was established as 1377 ± 4 year by tree-ring counting.
(a) What ages does carbon dating associate with the wood samples?
(b) What values of $t_{1/2}$ can be deduced if the tree-ring dates given are used as the starting point?
(c) Discuss assumptions underlying the calculations in (a) and (b), and indicate in what direction failures of these assumptions might affect the calculations.

63. A typical electrical generating plant has a capacity of 500 megawatt (MW; 1 MW = $10^{6}\ J\ s^{-1}$) and an overall efficiency of about 25%. (a) The combustion of 1 kg of bituminous coal releases about 3.2×10^{4} kJ and leaves an ash residue of 100 g. What weight of coal must be used to operate a 500-MW generating plant for 1 year, and what weight of ash must be disposed of? (b) Enriched fuel for nuclear reactors contains about 4% ^{235}U, fission of which gives 1.9×10^{10} kJ per mole ^{235}U. What weight of ^{235}U is needed to operate a 500-MW power plant, assumed to have 25% efficiency, for 1 year, and what weight of fuel must be reprocessed to remove radioactive wastes? (c) The radiation from the sun striking the earth's surface on a sunny day corresponds to a power of $1.5\ kW\ m^{-2}$. How large must the collection surface be for a 500-MW solar-generating plant? (Assume that there are 6 hours of bright sun each day and that storage facilities continue to produce power at other times. The efficiency for solar-power generation would be about 25%.)

64. Examine the ratio of atomic mass to atomic number for the elements with *even* atomic number through calcium. This ratio is approximately the ratio of the average mass number to the atomic number.
(a) Which two elements stand out as different in this set of ten?
(b) What would be the "expected" atomic mass of argon, based on the correlation considered here?
(c) Show how the anomaly in the ordering of natural atomic masses of argon and potassium can be accounted for by the formation of "extra" ^{40}Ar via decay of ^{40}K atoms.

65. Hydrazine, $N_2H_4(\ell)$, reacts with oxygen in a rocket engine to form nitrogen and water vapor:

$$N_2H_4(\ell) + O_2(g) \longrightarrow N_2(g) + 2\,H_2O(g)$$

(a) Calculate $\Delta H°$ for this highly exothermic reaction at 25°C, using data from Appendix D.
(b) Calculate $\Delta U°$ of this reaction at 25°C.
(c) Calculate the total change in mass, in grams, during the reaction of 1.00 mol of hydrazine.

66. The long-lived isotope of radium, ^{226}Ra, decays by alpha particle emission to its daughter radon, ^{222}Rn, with a half-life of 1622 years. The energy of the alpha particle is 4.79 MeV. Suppose 1.00 g of ^{226}Ra, freed of all its radioactive progeny, were placed in a calorimeter that contained 10.0 g of water, initially at 25°C. Neglecting the heat capacity of the calorimeter and heat loss to the surroundings, calculate the temperature the water would reach after 1.00 hour. Take the specific heat of water to be 4.18 J K^{-1} g^{-1}.

67. The radioactive nuclide $^{232}_{90}Th$ has a half-life of 1.39×10^{10} years. It decays by a series of consecutive steps, the first two of which involve $^{228}_{88}Ra$ (half-life 6.7 years) and $^{228}_{89}Ac$ (half-life 6.13 hours).

(a) Write balanced equations for the first two steps in the decay of ^{232}Th, indicating all decay products. Calculate the total kinetic energy carried off by the decay products.
(b) After a short initial time, the rate of formation of ^{228}Ra becomes equal to its rate of decay. Express the number of ^{228}Ra nuclei in terms of the number of ^{232}Th nuclei, using the steady-state approximation from Section 18.4.

68. Zirconium is used in the fuel rods of most nuclear power plants. The following half-cell reduction potential applies to aqueous acidic solution:

$$ZrO_2(s) + 4\,H_3O^+(aq) + 4e^- \longrightarrow Zr(s) + 6H_2O\,(\ell) \qquad \mathscr{E}° = -1.43\text{ V}$$

(a) Predict whether zirconium can reduce water to hydrogen. Write a balanced equation for the overall reaction.
(b) Calculate $\Delta\mathscr{E}°$ and K for the reaction in part (a).
(c) Can your answer to part (b) explain the release of hydrogen in the Three Mile Island accident and the much greater release of hydrogen (which subsequently exploded) at Chernobyl?

CHAPTER 20

Interaction of Molecules with Light

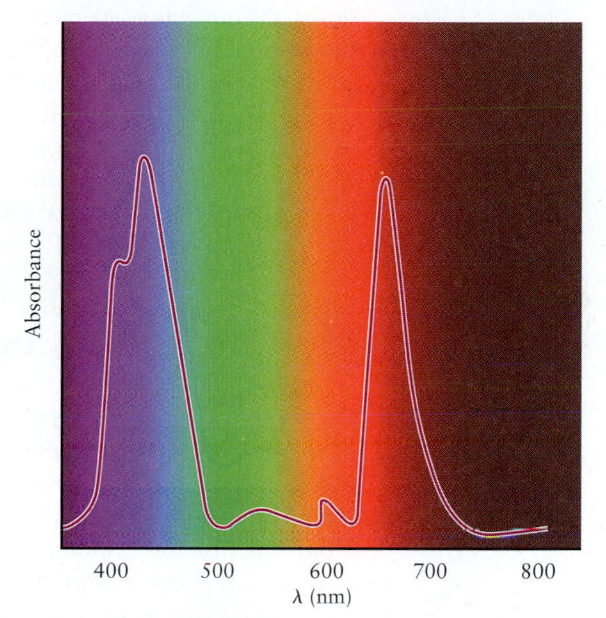

Optical absorption spectrum of chlorophyll *a* superimposed on the solar spectrum, showing how efficiently this key photosynthetic pigment absorbs a significant fraction of the available light. The color of green leaves is due to chlorophyll's relatively weak abosorption of green light.

Molecules have definite three-dimensional structure, described by their bond lengths and bond angles and determined by the covalent bonds that hold the molecules together. Lewis's electron dot diagrams rationalize the formation of particular compounds, and VSEPR theory rationalizes their structures (see Chapter 3). The quantum theory fully explains the covalent bond, and in its quantitative form predicts the structures and energy levels of the molecules (see Chapter 6). Despite these theoretical advances, there is still a need for experimental determination of molecular structure. Molecular structure is determined by x-ray diffraction in solid-state samples (see Chapter 21), by molecular spectroscopy (the absorption and emission of light by molecules), and by magnetic resonance spectroscopy (the absorption and emission of long-wavelength electromagnetic radiation by molecules in the presence of a magnetic field). In this chapter we survey the spectroscopy techniques that depend on interaction of molecules with light, and we summarize the molecular properties they determine:

1. Bond length, from rotational spectroscopy
2. Bond force constant, from vibrational spectroscopy

3. Pathways of energy flow through molecules, from electronic excitation
4. Identification of bonding groups and their interactions, from magnetic resonance spectroscopy

Light also plays crucial roles in changing molecules by transferring the energy that causes bonds to break or rearrange and by inducing chemical reactions via photochemical pathways very different from ordinary thermochemical pathways. We provide a brief introduction to photochemistry by examining photochemical events that are modifying the properties of the atmosphere of the earth. Light from the sun is the ultimate source for almost all the energy used on the earth. The atmosphere plays a crucial role in transmitting some radiation from the sun to the surface of the earth and absorbing other portions or reradiating them into space. Living species have transformed the composition of the atmosphere over the last 200 years, and industrialization has accelerated the rate of change. These changes raise questions about the sustainability of life on the earth, should present trends continue, and demonstrate the need for enlightened management of energy resources and consumption.

We conclude this chapter with a brief introduction to **photosynthesis,** the natural process through which light energy from the sun is harvested by plants. Energy stored in plants is converted to other forms by combustion, which is one contributor to the changes induced in the atmosphere through industrialization.

20.1 General Aspects of Molecular Spectroscopy

In Chapter 5 we discussed atomic spectroscopy and showed how the frequencies of the light absorbed or emitted by an atom are related to the energy differences between the atom's quantum states. In particular, the spectrum of the hydrogen atom can be interpreted in terms of solutions of the Schrödinger equation for that atom. Because molecules include several nuclei as well as electrons, their internal motions are more complicated than those of atoms, and their energy-level diagrams and spectra consequently exhibit new features not seen in those for atoms. Analysis of these features gives useful information about molecular structure, bond lengths, and bond energies (Table 20.1). Along with diffraction, molecular spectroscopy is a main source of experimental information about chemical bonding and the structure and shape of molecules.

Molecular Energy Levels

The Born–Oppenheimer approximation (see Section 6.1) treats the motion of electrons and nuclei on very different time scales. The electrons, being much less massive, move much more rapidly than the nuclei in a molecule. In the Born–Oppenheimer approximation, we consider the nuclei to be frozen at particular locations and calculate the electronic energy levels and wave functions (molecular orbitals) for the rapidly moving electrons (see Chapter 6). We find that the allowed

TABLE 20.1 Spectroscopic Experiments

Spectral Region	Frequency (s^{-1})	Energy Levels Involved	Information Obtained
Radio waves	10^7–10^9	Nuclear spin states	Electronic structure near the nucleus
Microwave, far infrared	10^9–10^{12}	Rotational	Bond lengths and bond angles
Near infrared	10^{12}–10^{14}	Vibrational	Stiffness of bonds
Visible, ultraviolet	10^{14}–10^{17}	Valence electrons	Electron configuration
X-ray	10^{17}–10^{19}	Core electrons	Core electron energies

energy levels for the electrons depend explicitly on the positions of the frozen nuclei. Then we turn the problem around. In a particular electronic state, we consider the bonding energy of the rapidly moving electrons plus the nuclear–nuclear repulsion to be the potential energy function that governs the motions (vibrations and rotations) of the more sluggish nuclei. Therefore, each quantized energy state for the electrons will have an associated set of quantized vibrational and rotational energy states for the nuclei. Figure 20.1 schematically illustrates the Born–Oppenheimer picture for the energy levels of a typical diatomic molecule. Two electronic states with numerous rotational and vibrational states are shown. The separation between electronic states is much greater than that for rotational and vibrational states. Transitions between electronic states involve absorption or emission of electromagnetic radiation in the visible and ultraviolet regions of the electromagnetic spectrum. Transitions between rotational states involve microwave radiation, and infrared radiation accompanies transitions between vibration states. Rotation of the molecule as a rigid body leads to molecular transitions in the microwave region, and vibrational motions of one nucleus relative to another lead to molecular transitions in the infrared region (see Fig. 4.3).

To an excellent level of approximation, the total energy of a molecule can be viewed as the sum of the energies associated with each of these separate motions:

$$E_{tot} = E_{trans} + E_{rot} + E_{vib} + E_{el}$$

When the total energy of a molecule changes during absorption or emission of radiation, the transition usually involves changes in more than one kind of energy. For example, a line in the visible region of the emission spectrum may connect two energy levels that differ in rotational, vibrational, and electronic energy (see the blue arrow in Fig. 20.1).

A polyatomic molecule would have an even more complicated set of energy levels, each characterized by several vibrational quantum numbers and by up to three rotational quantum numbers.

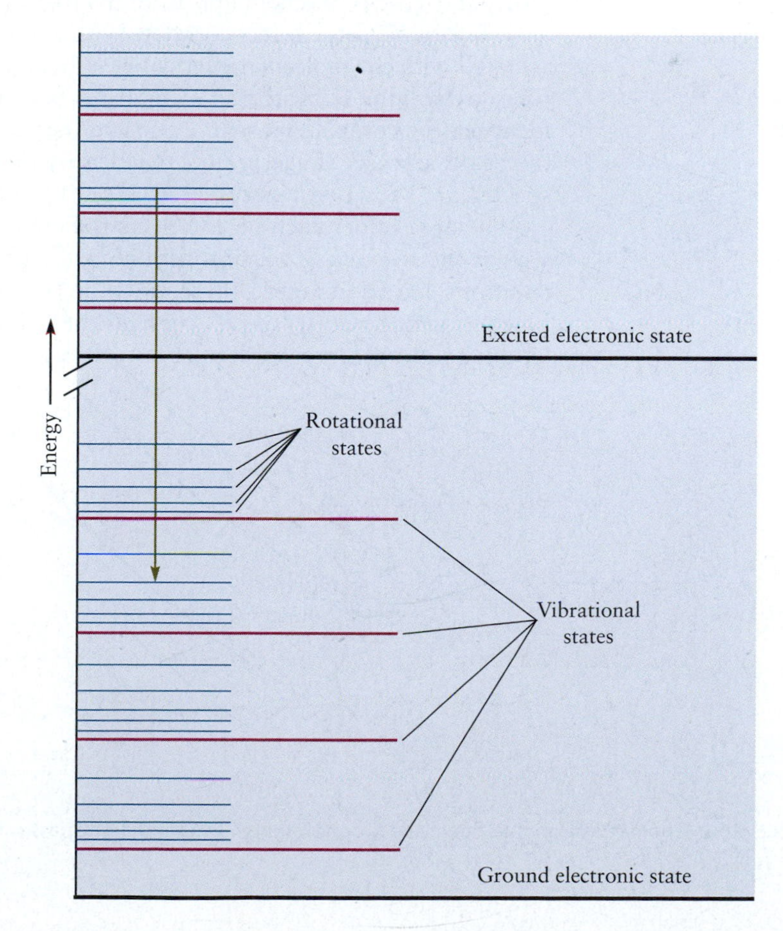

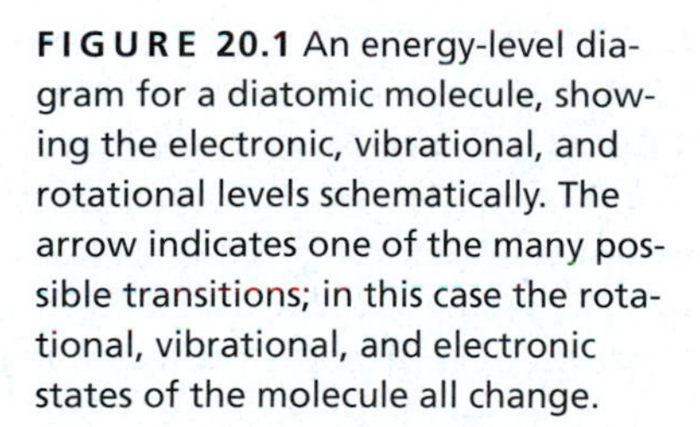
FIGURE 20.1 An energy-level diagram for a diatomic molecule, showing the electronic, vibrational, and rotational levels schematically. The arrow indicates one of the many possible transitions; in this case the rotational, vibrational, and electronic states of the molecule all change.

Experimental Methods in Molecular Spectroscopy

Molecular absorption spectra are recorded by *spectrophotometers* (Fig. 20.2), which differ from the spectrographs illustrated with atomic spectra in Figure 4.9. Light from the source is directed to a prism or grating to select a specific wavelength λ, which then is passed through the sample confined in a cell. The intensity of light transmitted through the sample cell, I_S, is measured. To remove spurious effects due to absorption or scattering of light at the cell walls, the incoming beam is actually split into two parts, one of which is passed through a reference cell of the same size and shape as the sample cell. The spectrophotometer can be calibrated to record a graph of either the **transmittance** $T = I_S/I_R$ or the **absorbance** $A = \ln [I_R/I_S]$ versus λ as wavelength is scanned over the range of interest. The magnitude of the signal is related to properties of the sample as follows. The transmittance T decreases (and the absorbance A increases) as cell length ℓ increases:

$$-\ln \frac{I_S}{I_R} = a\ell = A = -\ln T$$

The parameter a is called the **absorption,** or **extinction, coefficient.** If the sample is a solution, then T decreases and A increases as the concentration c (in mol L^{-1}) of the light-absorbing solute increases, as described by the Beer–Lambert law:

$$-\ln \frac{I_S}{I_R} = c\epsilon\ell = A \qquad [20.1]$$

The parameter ϵ, called the **molar extinction coefficient,** is a property of the light-absorbing solute in the solution; it measures the extent to which that species can absorb light at a particular wavelength.

Peaks in the graph at particular values of λ correspond to transitions between molecular energy levels E_i and E_f that satisfy the relation $\Delta E = E_f - E_i = h\nu = hc/\lambda$. Positions of the peaks correlate with some feature of molecular structure associated with the molecular energy levels E_i and E_f. In principle, these can be identified by solving Schrödinger's equation for the molecule. In practice, they are identified by comparison with extensive tables of spectral data already compiled; the peaks serve as "fingerprints" for identifying structural features. Representative spectra are shown in Figures 20.10, 20.14, and 20.22, which are discussed later.

The area under each peak reflects the concentration of molecules present, as well as the *strength* of the absorption governed by ϵ. This fact is illustrated by the solutions shown in Figure 14.1a and c, in which merely 2% concentration of the blue tetrachloro Co(II) species changes the color of the solution from pink to violet due to its very large molecular extinction coefficient.

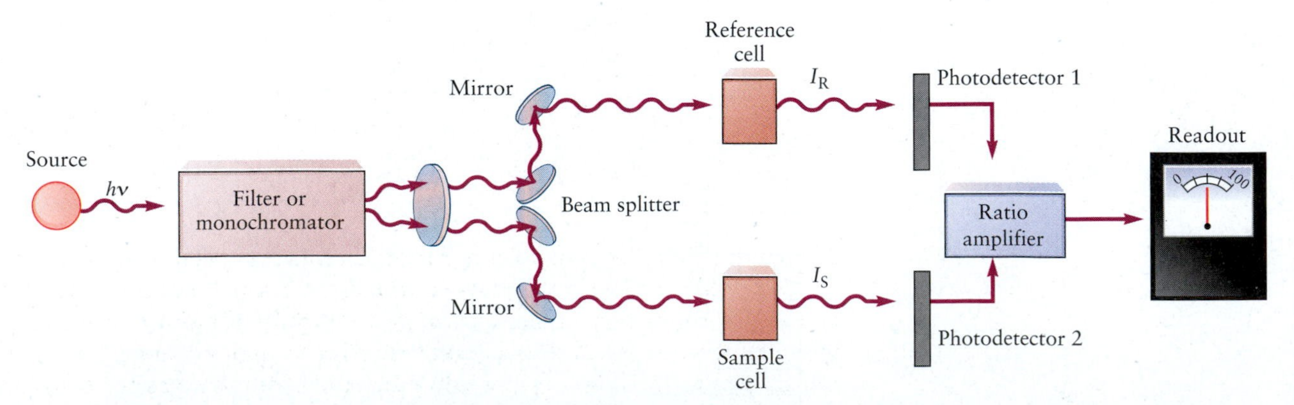

FIGURE 20.2 Schematic of double-beam spectrophotometer. Incoming light passes through a reference cell identical to the sample cell.

The magnitude of ϵ is determined by the strength of the interaction between the absorbing molecule and the light wave and the difference in populations of the initial and final states involved in the transition. Detailed discussion of the interaction of light with molecules requires quantum concepts beyond the scope of this book. The essential physical effect is that the changing dipole moment of the molecule couples with the oscillating electric field of the light wave (see Fig. 4.2) and functions as a "molecular antenna" to absorb energy from the electric field. For energy to be absorbed at a particular wavelength, the number of molecules in the initial state with energy E_i must be greater than the number in the final state with energy E_f.

Thermal Occupation of Molecular Energy Levels

The intensity of an absorption line depends on the strength of the transition and on the population ratio of the initial and final states. The probability that the energy level E_i is occupied is $N(E_i)$, the number of molecules in that level, divided by N_A, the total number of molecules in the system. This is given by the **Boltzmann distribution**

$$P(E_i) = N(E_i)/N_A \propto g(E_i) \exp(-E_i/k_B T) \quad [20.2]$$

where k_B is the Boltzmann constant, equal to 1.38×10^{-23} J K^{-1} and $g(E_i)$ is the degeneracy of E_i, that is the number of distinct quantum states that correspond to E_i. (See Section 9.6 for an introduction to the Boltzmann distribution and its application to vibrational energy levels of diatomic molecules.) If energies per *mole* are used, k_B is replaced by the gas constant R. This exponential falloff means that at thermal equilibrium very few molecules have energies that are large compared with k_BT.

The population ratio for two energy levels E_i and E_f is

$$P(E_i)/P(E_f) = g(E_i)/g(E_f) \exp[-(E_i - E_f)/k_B T] \quad [20.3]$$

At room temperature, k_BT has the value 4×10^{-21} J (equivalent to 2.5 kJ mol^{-1}). This is large compared with the typical spacing of rotational levels, so at room temperature many rotational states are occupied in a collection of molecules. This value is also a little smaller than typical vibrational level spacings, so most small molecules at room temperature are in their ground vibrational states, but a measurable fraction occupy excited states. Finally, k_BT is small compared with typical electronic energy spacings, so only the ground electronic state is occupied at room temperature. When molecules are heated, the occupation of more highly excited levels increases.

20.2 Vibrations and Rotations of Molecules: Infrared and Microwave Spectroscopy

Three types of nuclear motion occur in gas-phase molecules: overall translational motion of a molecule through its container, rotational motion in which the molecule turns about one or more axes, and vibrational motion in which the nuclei move relative to each other as bond lengths or angles change. All three motions are subject to the laws of quantum mechanics, but in a gas the translational energy states are so close in energy (they correspond to the particle-in-a-box states of Section 4.6) that quantum effects are not apparent.

In this section we consider the interaction between nuclear motion in molecules and infrared and microwave photons.

Rotations of Molecules

We consider the molecule to be a rigid body, having no distortions in its dimensions while moving. Rotational energies are described in terms of the molecular **moments of inertia** for rotation of the molecule about its center of gravity. (See Appendix B for background on rotational motion.) Figure 20.3 shows the rotational motion for a diatomic molecule, which depends on a single moment of inertia I, defined by

$$I = \mu R_e^2 \qquad [20.4]$$

where $\mu = m_1 m_2/(m_1 + m_2)$ is the **reduced mass** of the molecule with m_1 and m_2 being the masses of the two atoms and R_e the bond length. Quantum mechanics demonstrates that rotational motion is quantized and only certain discrete rotational energy levels are permitted. In a linear molecule, the rotational energy can take on only the values

$$E_{\text{rot},J} = \frac{h^2}{8\pi^2 I} J(J+1) = hBJ(J+1) \qquad J = 0, 1, 2, \ldots \qquad [20.5]$$

where J is the rotational quantum number and $B = h/(8\pi^2 I)$. Typical rotational energy differences range from 0.001 kJ mol^{-1} to 1 kJ mol^{-1}. In addition to the energy, the angular momentum of the rotating molecule is also quantized, and the rotational wave function is identical to the angular portion of the hydrogen atom wave functions. For $J = 0$, it resembles an s orbital, for $J = 1$ it resembles p_x, p_y, and p_z orbitals. The degeneracy of the rotational energy levels is $g(E_J) = 2J + 1$.

Radiation in the far (long-wavelength) infrared and microwave regions of the electromagnetic spectrum excites rotational states of molecules. In order to absorb radiation, the molecule must have a permanent dipole moment. Moreover, for a heteronuclear diatomic molecule, absorption of light is possible only between

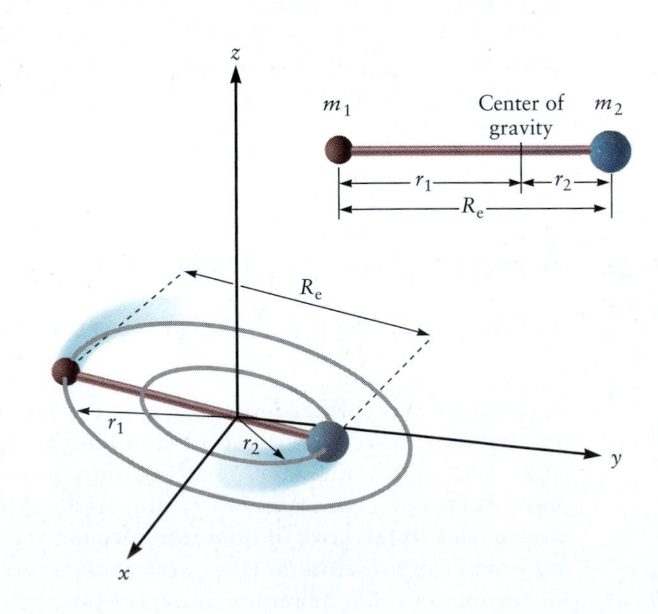

FIGURE 20.3 A diatomic molecule rotates about its center of gravity. The center of gravity is located by the distances r_1 and r_2, determined from the condition $m_1 r_1 = m_2 r_2$.

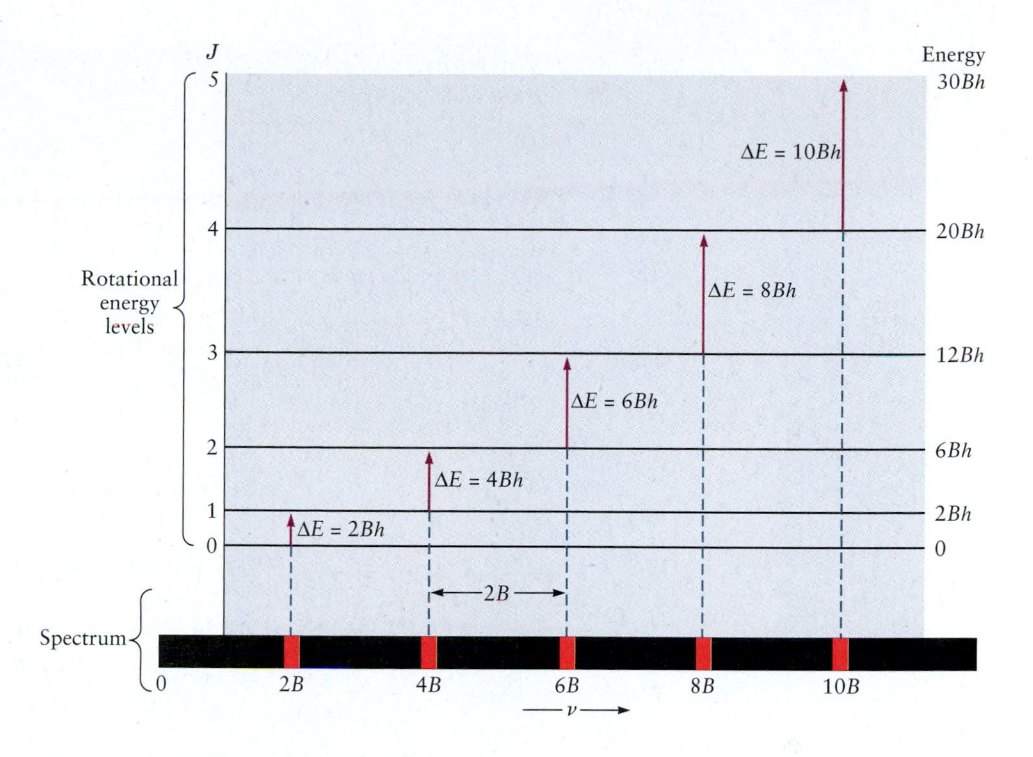

FIGURE 20.4 The energy levels and allowed transitions for rotational motion of a rigid diatomic molecule. The restriction that $\Delta J = +1$ leads to a series of uniformly spaced spectral lines separated in frequency by $2B = 2h/(8\pi^2 I)$. The measured frequency for any one of these lines enables determination of I, as illustrated in Example 20.1.

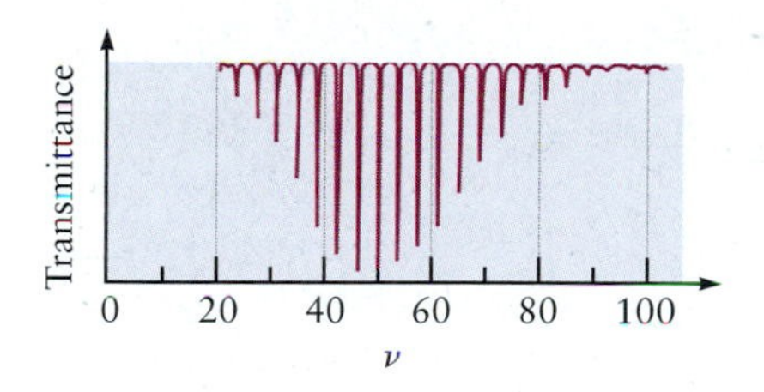

FIGURE 20.5 Schematic representation of an experimental rotational spectrum. Because many rotational states are occupied at room temperature (see Problems 20.9 and 20.10), the measured spectrum includes many of the allowed transitions identified in Figure 20.4.

states that differ by 1 in rotational quantum number ($\Delta J = +1$). The allowed absorption frequencies are

$$\nu = \frac{\Delta E}{h} = \frac{h}{8\pi^2 I}[J_f(J_f+1) - J_i(J_i+1)]$$

$$= \frac{h}{8\pi^2 I}[(J_i+1)(J_i+2) - J_i(J_i+1)] \qquad [20.6]$$

$$h/8\pi^2 I\,(2J+2) = 2B(J+1)$$

because the quantum number of the final rotational state J_f must be 1 greater than that of the initial rotational state J_i. Consequently, the rotational absorption spectrum of such a molecule shows a series of equally spaced lines with a frequency separation of $2B$. Such a spectrum is shown schematically in relation to the allowed energy transitions in Figure 20.4. Because numerous rotational states are populated at room temperature, an actual spectrum has the appearance shown in Figure 20.5.

From gas-phase absorption spectra in these regions, the moment of inertia for the diatomic molecule can be calculated and (because the atomic masses are known) the bond length can be determined.

EXAMPLE 20.1

The microwave absorption spectrum of gaseous NaH (isotope: $^{23}Na^{1}H$) was determined experimentally. The photon wavelength required to excite the molecules rotationally from the state $J = 0$ to $J = 1$ was measured to be 1.02×10^{-3} m. Calculate the bond length of the NaH molecule. Use isotope atomic masses from Table 19.1.

SOLUTION

The reduced mass is

$$\mu = \frac{m_{Na}m_H}{m_{Na} + m_H} = \frac{(22.9898\ u)(1.0078\ u)}{22.9898 + 1.0078\ u}$$

$$= 0.9655\ u = 1.603 \times 10^{-27}\ kg$$

The energy change is

$$\Delta E = h\nu = \frac{hc}{\lambda}$$

which in this case is

$$\Delta E_{rot} = \frac{(6.626 \times 10^{-34}\ J\ s)(2.998 \times 10^{8}\ m\ s^{-1})}{1.02 \times 10^{-3}\ m} = 1.95 \times 10^{-22}\ J$$

$$= \frac{h^2}{8\pi^2 I}[(1)(2) - (0)(1)] = \frac{h^2}{4\pi^2 I}$$

Solving for the moment of inertia I gives

$$I = \frac{h^2}{4\pi^2 \Delta E_{rot}} = \frac{(6.626 \times 10^{-34}\ J\ s)^2}{4\pi^2(1.95 \times 10^{-22}\ J)} = 5.70 \times 10^{-47}\ kg\ m^2$$

The moment of inertia I is related to the bond length R_e by $I = \mu R_e^2$, so

$$R_e^2 = \frac{I}{\mu} = \frac{5.70 \times 10^{-47}\ kg\ m^2}{1.603 \times 10^{-27}\ kg} = 3.56 \times 10^{-20}\ m^2$$

$$R_e = 1.89 \times 10^{-10}\ m = 1.89\ \text{Å}$$

Related Problems: 5, 6, 7, 8

Polyatomic molecules have up to three different moments of inertia, corresponding to rotations about three axes (Fig. 20.6). The rotational spectra for nonlinear polyatomic molecules are more complex than the example just illustrated, but their interpretation is carried out in the same way and has enabled chemists to determine with high accuracy the molecular geometries for many small polyatomic molecules.

Vibrations of Molecules

Now, we give up the restriction that the molecule moves as a rigid body, and we examine its distortions. The information obtained allows us to estimate the "stiffness" of the bond, which is a measure of the bond strength. Figure 20.7 shows the potential energy curve for a diatomic molecule (see also Figs. 3.9 and 6.7). The minimum of the curve corresponds to the equilibrium bond length. The steep rise to the left indicates that the molecule has been "squeezed" to a distance shorter than the equilibrium bond length and repulsive forces are appearing between the two nuclei. The rise to the right indicates that potential energy must be overcome to stretch the bond away from its equilibrium value. The flat region to the extreme right side indicates the amount of energy that must be supplied to dissociate the molecule into free atoms. Consider first the vibrational motion of the diatomic molecule about its equilibrium bond length. If the bond is stretched, the two atoms experience a restoring force that tends to bring them back to their original separation R_e, just as a restoring force acts on the ends of a spring that has been

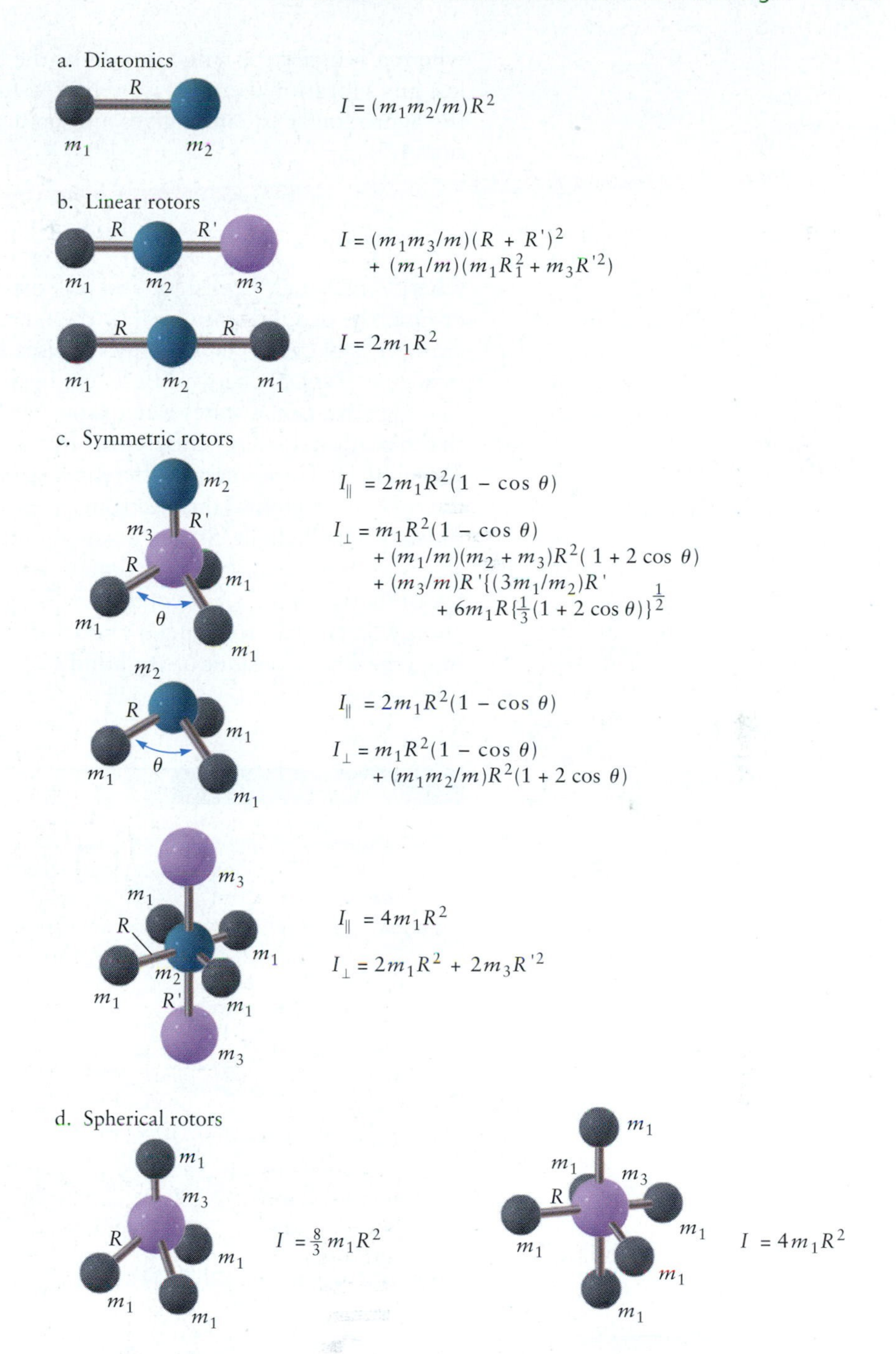

FIGURE 20.6 Moments of inertia for selected classes of molecular structures. In each case, m represents the sum of all the atomic masses in the molecule. (a) Diatomic molecules have a single moment of inertia, uniquely related to the bond length. (b) Linear triatomic molecules also have a single moment of inertia that can be determined from rotational spectra. Because I depends on two bond lengths, additional information is required to determine both. This is accomplished by comparing rotational spectra of isotopically substituted molecules. (c) Symmetric rotors have three moments of inertia, of which two are equal. The moment about the vertical z-axis through the molecule is called $I_\parallel$ and the two moments about the x and y axes are called $I_\perp$. The two moments of inertia can be determined from rotational spectroscopy, but additional information (for example from x-ray diffraction) is needed to determine all the bond lengths. (d) Spherical rotors have all three moments of inertia equal. The symmetry of polyatomic molecules guides interpretation of their rotational spectra.
(Adapted from P.W. Atkins, Physical Chemistry (Third Edition), Freeman, 1986.)

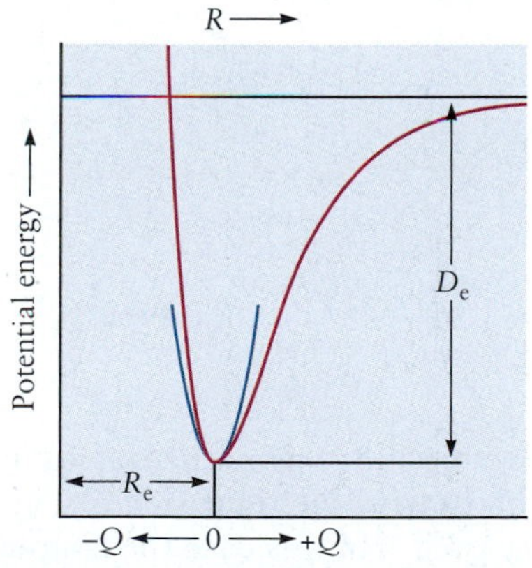

FIGURE 20.7 The potential energy for a diatomic molecule has its minimum at the equilibrium bond length R_e. The displacement coordinate $Q = R - R_e$ represents stretching the bond ($Q > 0$) or compressing the bond ($Q < 0$). For small Q, the potential energy is approximated by the harmonic oscillator equation $V(Q) = (1/2)kQ^2$. The bond dissociation energy D_e is defined from the bottom of the potential well at R_e.

stretched. For a small change in bond length $R - R_e$, the force is proportional to that change:

$$F = -k(R - R_e)$$

where the **force constant** k determines the stiffness of the bond. In SI units, k is measured in newtons per meter (Nm^{-1}). The potential energy corresponding to this restoring force is shown as the blue parabolic curve in Figure 20.7.

According to classical mechanics, when a stretched bond is released, the atoms will oscillate back and forth about their average separation, much as two balls connected by a spring do. The oscillation frequency ν is given by

$$\nu = \frac{1}{2\pi}\sqrt{\frac{k}{\mu}}$$

where μ is again the reduced mass of the harmonic oscillator. In classical mechanics, any vibrational energy is permitted for the oscillator. In quantum mechanics, the Schrödinger equation gives a discrete set of vibrational energy levels (see Section 4.7):

$$E_{\text{vib},n} = (n + \tfrac{1}{2})h\nu \qquad n = 0, 1, 2, \ldots \qquad [20.7]$$

where h is Planck's constant and n is the vibrational quantum number. Note that even in the ground state ($n = 0$) the vibrational energy is not 0. The residual energy $\frac{1}{2}h\nu$ is the **zero-point energy;** it arises from the requirements of the uncertainty principle. Typical energy differences between the ground ($n = 0$) and first excited ($n = 1$) vibrational states range from 2 kJ mol^{-1} to 40 kJ mol^{-1}; they are greater than rotational energy differences but less than electronic energy differences (see Table 20.1). These energy differences correspond to the absorption of radiation in the infrared region of the electromagnetic spectrum, with wavelengths longer than those of visible light. Strong absorption is observed only for transitions between states that differ by 1 in vibrational quantum number ($\Delta n = 1$). Because the separation between adjacent energy levels is independent of quantum number, these energy levels lead to a single vibrational frequency, which can be used to determine the force constant of the bond.

EXAMPLE 20.2

The infrared spectrum of gaseous NaH (isotope: $^{23}Na^{1}H$) was determined via absorption spectroscopy. The photon wavelength needed to excite the molecule vibrationally from the state $n = 0$ to $n = 1$ was 8.53×10^{-6} m. Calculate the vibrational force constant of the NaH molecule. Use isotope atomic masses from Table 19.1.

SOLUTION

The vibrational frequency is

$$\nu = \frac{c}{\lambda} = \frac{2.998 \times 10^{8} \text{ m s}^{-1}}{8.53 \times 10^{-6}} = 3.515 \times 10^{13} \text{ s}^{-1} = \frac{1}{2\pi}\sqrt{\frac{k}{\mu}}$$

Solving for the constant k (using the reduced mass from Example 20.1) gives

$$k = \mu(2\pi\nu)^2 = (1.603 \times 10^{-27} \text{ kg})(2\pi)^2(3.515 \times 10^{13} \text{ s}^{-1})^2$$

$$= 78.2 \text{ kg s}^{-2} = 78.2 \text{ J m}^{-2}$$

Related Problems: 11, 12, 13, 14

FIGURE 20.8 The Morse potential is defined by $V(R - R_e) = D_e\{1 - \exp[-a(R - R_e)]\}^2$, where $a = (k/2D_e)^{1/2}$. The potential is plotted here in dimensionless variables. Unlike the harmonic oscillator, the Morse potential has a finite number of bound states.

At higher vibrational energies, the motions of the molecule range rather far from the equilibrium value, and they sample the region toward the right side of Figure 20.7, where the motion is said to be "anharmonic." In this case the simple parabolic approximation breaks down, and it is necessary to find the energy levels for the anharmonic potential. Although beyond the scope of this book, this can be carried out for the specific model of the anharmonic potential called the Morse potential (Fig. 20.8). The energy levels are no longer uniformly spaced, and the experimental spectrum shows a progression of frequencies corresponding to these smaller energy gaps. The experimental spectrum can be fit to the Morse potential, and the dissociation energy of the diatomic molecule can be determined.

In a polyatomic molecule, several types of vibrational motion are possible (Fig. 20.9). Each has a different frequency, and each gives rise to a series of allowed quantum vibrational states. Infrared absorption spectra provide useful information about vibrational frequencies and force constants in molecules. For

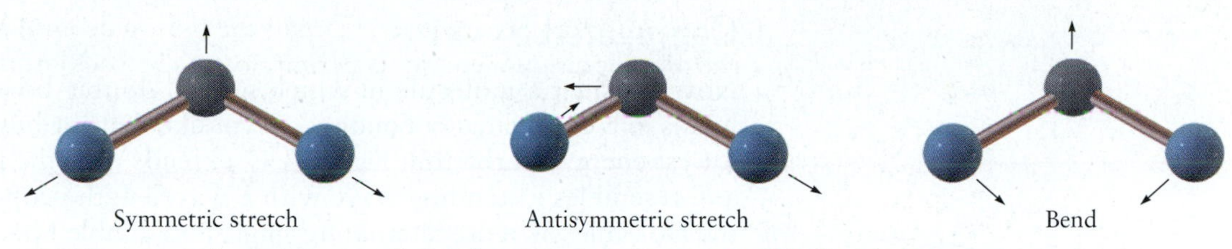

FIGURE 20.9 The three types of vibrational motion that are possible for a bent triatomic molecule. Arrows show the displacement of each atom during each type of vibration.

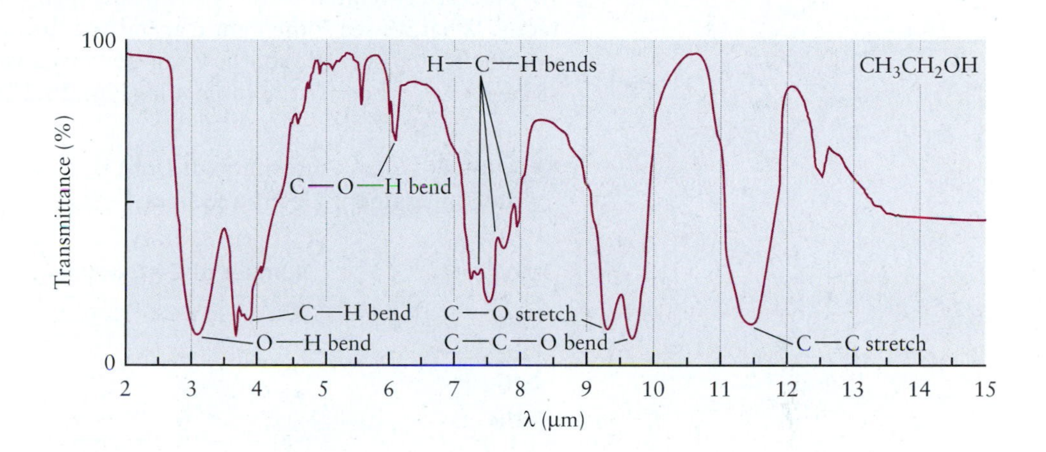

FIGURE 20.10 The infrared absorption spectrum of ethanol, CH_3CH_2OH. Each peak is labeled with the vibrational motion from which it arises.

larger molecules, vibrational spectra are useful for identifying compounds, because each group (such as a C=O bond or a C—OH bond) has its own characteristic frequency in the spectrum, largely independent of the mass, geometry, and other features of the molecule of which it is a part. The spectrum thus provides information about the functional groups in a given molecule (Fig. 20.10).

20.3 Excited Electronic States: Electronic Spectroscopy of Molecules

Excited states of atoms result from the promotion of one or more electrons into orbitals of higher energy. Excited electronic states in molecules arise similarly and can be described in the framework of molecular orbital theory.

As an example, consider the ethylene molecule (C_2H_4). The bonding in ethylene can be described by the combine localized and de-localized method (see Sections 6.5 and 7.3). The two carbon atoms in the molecular "backbone" can be described through sp^2 hybridization. The two remaining carbon $2p_z$ orbitals combined to form a π (bonding) molecular orbital and a π^* (antibonding) molecular orbital for the de-localized electrons. The energy level diagram for the two π orbitals is shown in Figure 20.11. The lowest energy state of ethylene has two electrons in the π orbital. An excited state results when one electron is moved from the π orbital to the π^* orbital. This $\pi \rightarrow \pi^*$ transition in ethylene is caused by absorption of light at the wavelength 162 nm, which lies in the ultraviolet region of the spectrum. An ethylene molecule in this excited state has properties quite different from one in the ground state. The electron in the π^* antibonding orbital cancels the effect of the electron in the π bonding orbital, so the net bond order is reduced from 2 to 1 (the σ bond). Consequently, the molecule in the excited state has a lower bond energy and a longer C—C bond length than in the ground state.

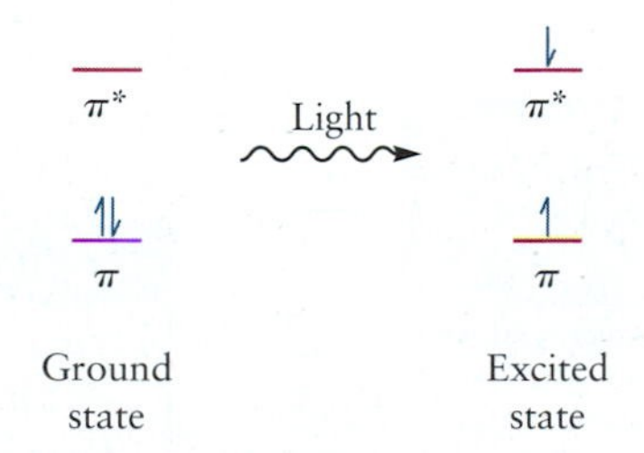

FIGURE 20.11 Occupancy of the π molecular orbitals in the ground state and first excited state of ethylene, C_2H_4. Ethylene in the excited state is produced by irradiation of ground-state ethylene with ultraviolet light in the appropriate frequency range.

Conjugated Systems

Now consider a molecule in which several double bonds alternate with single bonds in a conjugated π-bonding system like that for butadiene. Observe that the lowest energy π orbital in Figure 7.17 extends over the full conjugated π system and resembles a standing wave with a wavelength proportional to the length of the molecule. As more alternating single and double bonds are added, the characteristic wavelengths become longer and the frequencies and energies become lower (Table 20.2). When the chain becomes sufficiently long, the first absorption shifts into the visible region of the spectrum, and the substance takes on color. The color we perceive is related to the absorption spectrum of the material, but only indirectly. What we see is the light transmitted through or reflected from the material, not the light absorbed. So, the color perceived is **complementary** to the color most strongly absorbed by the molecule (Fig. 20.12a and b). Examples are the dyes

TABLE 20.2 Absorption of Light by Molecules with Conjugated π Electron Systems

Molecule	Number of C=C Bonds	Wavelength of Maximum Absorption (nm)
C_2H_4	1	162
C_4H_6	2	217
C_6H_8	3	251
C_8H_{10}	4	304

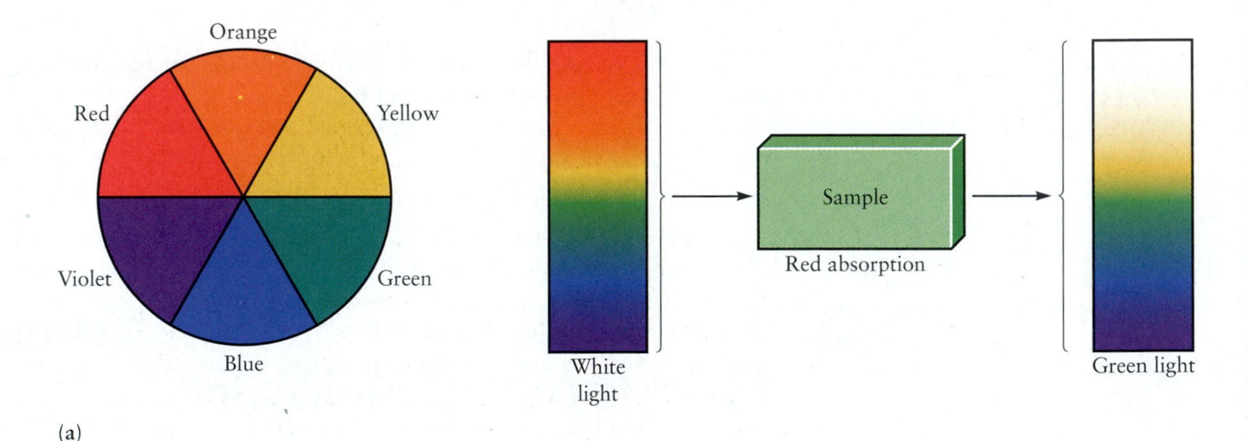

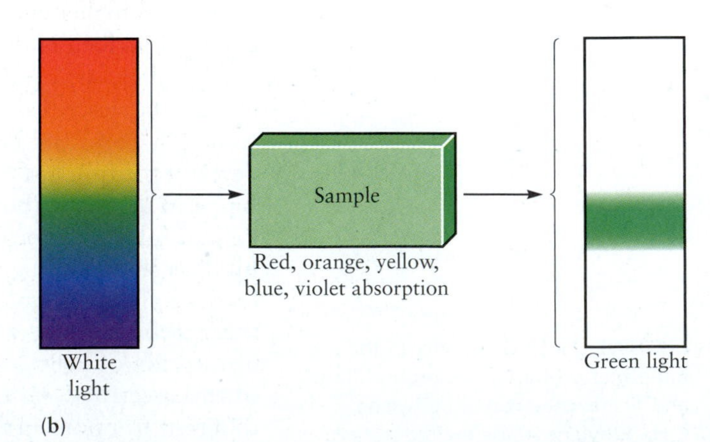

FIGURE 20.12 Two ways in which the color green can be produced. (a) If the sample strongly absorbs red light but transmits yellow and blue light, the sample appears green. (b) If all visible light *except* green is absorbed by the sample, the sample also appears green. The color wheel in (a) shows complementary colors opposite each other.

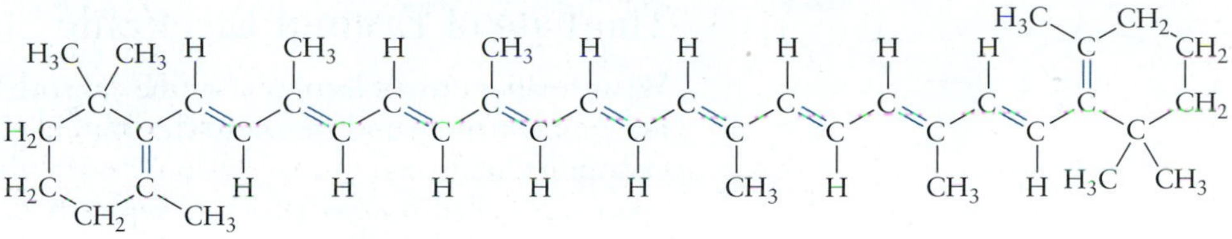

FIGURE 20.13 The molecular structure of the dye beta-carotene. Eleven double bonds alternate with single bonds in this conjugated structure.

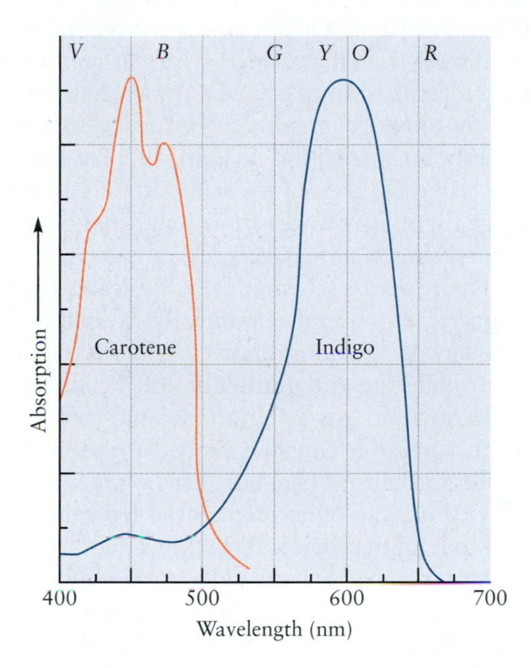

FIGURE 20.14 The absorption spectra of the dyes beta-carotene and indigo differ in the visible region and hence have different colors. The letters stand for the colors of the light at particular wavelengths (violet, blue, green, yellow, orange, and red).

indigo and beta-carotene (Fig. 20.13). Figure 20.14 plots the strength of the absorption of these two dyes at different wavelengths. Indigo absorbs in the yellow-orange region of the spectrum, so it appears violet-blue (the colors complementary to yellow and orange). Beta-carotene absorbs at shorter wavelengths in the blue and violet regions and appears orange or yellow. Indigo is used to dye blue jeans; beta-carotene is responsible for the orange color of carrots, the colors of some processed foods, and the yellow and orange in certain bird feathers.

EXAMPLE 20.3

Suppose you set out to design a new green dye. Over what range of wavelengths would you want your trial compound to absorb light?

SOLUTION

A good green dye must transmit green light and absorb other colors of light. This can be achieved by a molecule that absorbs in the violet-blue *and* orange-red regions of the spectrum (see Fig. 20.12b). Thus, you would want a dye with strong absorptions in both these regions.

The naturally occurring substance chlorophyll, which is responsible for the green colors of grass and leaves, absorbs light over just these wavelength ranges, converting solar energy to chemical energy for the growth of the plant. It is used commercially as a green dye.

Related Problems: 19, 20

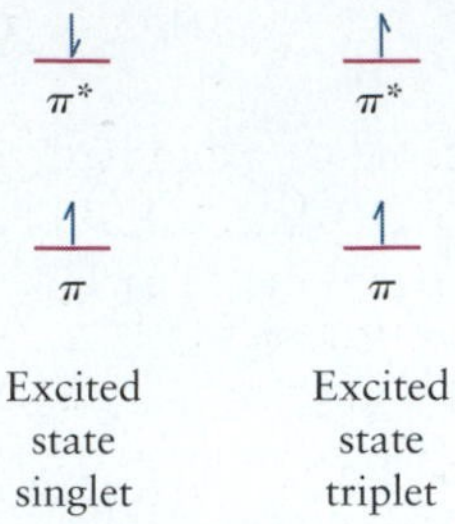

π^* π^*

π π

Excited state singlet | Excited state triplet

FIGURE 20.15 The first excited singlet and triplet states have the same orbital occupancy but different spin states. The triplet is generally lower in energy.

The Fate of Excited Electronic States

What happens to molecules after they absorb radiation? Some reemit a photon almost immediately and return to the ground electronic state, although they usually end up in rotational and vibrational states different from the original ones. This process is called **fluorescence** and typically occurs within about 10^{-9} s. A second type of light emission called **phosphorescence** occurs much more slowly, over a period of seconds (or in some cases hours). To understand phosphorescence, we have to consider the spins of the electrons in an excited state. The absorption of light (shown in Fig. 20.11) does not change the electron spin and therefore gives an excited state (called a *singlet*) with one spin up and one down. Another possible excited electronic state (called a *triplet*) has both spins up (Fig. 20.15) and, by Hund's rule, has lower energy than the singlet. Although the triplet state cannot be reached directly by absorption of light, there is a small probability that an excited molecule will cross over to the triplet state as it evolves in time or undergoes collisions. Once there, it remains for a long time because emission from a triplet state to a ground singlet state is very slow.

Figure 20.16 schematically illustrates these two pathways for loss of energy by radiation from an electronically excited molecule, along with one nonradiative pathway. In the nonradiative pathway, the molecule may cross over to the ground electronic state and gradually lose its energy as heat to the surroundings as it cascades down to lower vibrational and rotational levels.

It is possible that the energy of the excited molecule is high enough to break bonds and cause a chemical reaction. Molecules in excited states have enhanced reactivity and can often fragment or rearrange to give new molecules. **Photochemistry,** the study of the chemical reactions that follow the excitation of molecules to higher electronic states through absorption of photons, is an active area of research.

Some compounds are so photochemically sensitive that they must be stored in the dark, because they react rapidly to form products when exposed to light. An example is anhydrous hydrogen peroxide, which reacts explosively in light to give water and oxygen:

$$H_2O_2 \longrightarrow 2\ OH \longrightarrow H_2O + \tfrac{1}{2}O_2$$

When the energy of a photon of visible light is added to the molecule, the rather weak O—O bond in H_2O_2 breaks apart. This process is called **photodissociation.**

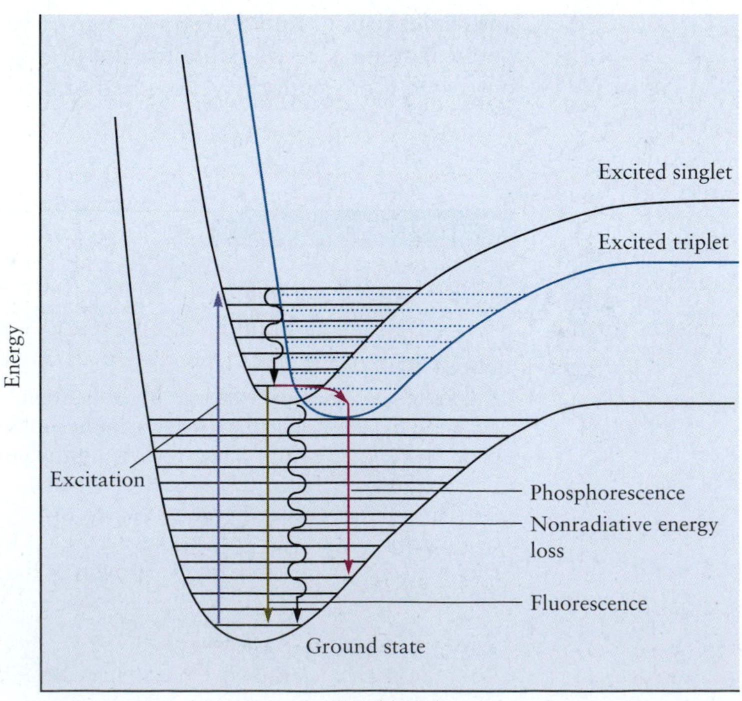

FIGURE 20.16 A schematic picture of the loss of energy by a molecule in an excited electronic state. After the molecule absorbs a photon (purple arrow), it cascades down through a series of levels to the bottom of the excited singlet state (short wavy black arrow). Then three outcomes are possible. The molecule can return to the ground state by fluorescence (green arrow), it can cross over to the triplet state and return from there to the ground state by phosphorescence (red arrow), or it can cross over to the ground state and cascade down without radiating (long wavy black arrow).

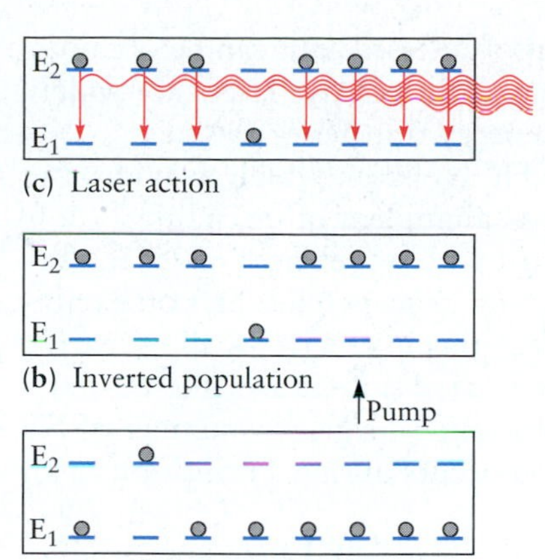

FIGURE 20.17 Operation of a laser device requires population inversion and stimulated emission.
(Adapted from P.W. Atkins, Physical Chemistry (Third Edition), Freeman, 1986.)

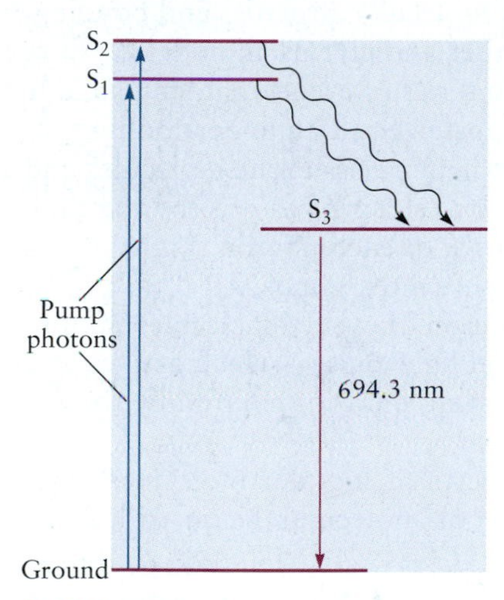

FIGURE 20.18 The ruby laser operates between the ground state and an excited state of Cr^{3+} ions.

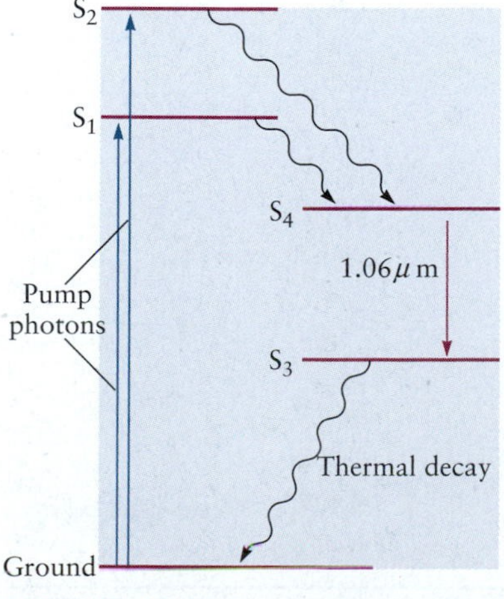

FIGURE 20.19 The neodymium-YAG laser operates between two excited states of the neodymium ions.

Lasers

Through a process called **spontaneous emission,** a molecule or atom in an excited state E_2 can undergo a transition to a lower state E_1 and emit light of frequency $\nu_{12} = (E_2 - E_1)/h$. This radiation will be randomly distributed in phase and in direction; this is "ordinary light." A different process, called **stimulated emission,** can operate between these same two states to emit light of the same frequency ν_{12} but all with a single phase and traveling in a single direction. This is "laser light" whose special properties make it uniquely suited for entertainment functions such as laser light shows and reading information from compact discs (CDs) and digital video discs (DVDs), as well as for a wide range of technical and scientific applications. "**Laser**" is an acronym for **l**ight **a**mplification by **s**timulated **e**mission of **r**adiation.

Suppose we have a material medium containing molecules that have the states E_2 and E_1. Depending on the temperature, molecules will be distributed between these states according to the Boltzmann distribution. Now imagine that we bring a beam of light of frequency ν_{12} into this medium. Whenever the beam hits a molecule in state E_2, the beam will stimulate the excited molecule to emit a new photon of frequency ν_{12}; this increases the intensity of light at frequency ν_{12} in the medium. If we can maintain a so-called "population inversion" in which N_2, the number of molecules in the excited state, is greater than the number N_1 in the lower state and if we can pass this light beam back and forth through the medium many times, then the beam at frequency ν_{12} will be amplified in intensity and will have the special properties of laser light. These are the conditions a practical device must meet to function as a laser.

In Figure 20.17 we sketch the operation of a laser device. The population inversion is achieved by "pumping" in extra energy by thermal excitation, chemical reaction, or optical excitation. The laser medium is contained between two reflecting mirrors that form the "optical cavity" of the laser. When the first few photons are emitted spontaneously from the excited states, they immediately stimulate the emission of more photons. The resulting wave is reflected between the mirrors many times, stimulating further emission on each passage. This growing beam is spatially coherent, meaning that all the waves have the same phase. All the waves are traveling in the same direction, because any wave that might hit one of the mirrors at an angle is reflected out of the pathway and is not amplified. If one of the cavity mirrors is less reflecting than the other, some of the laser light escapes at that end of the cavity and is ready for applications. This phenomenon was first demonstrated by the American physicist Charles Townes at Columbia University using microwave radiation, and the device was called a *maser* (**m**icrowave **a**mplification by **s**timulated **e**mission of **r**adiation).

The first demonstration of laser operation with visible light was achieved in 1960 by the American physicist Theodore Maiman at Hughes Aircraft Research Laboratories in Malibu, California, using the process sketched in Figure 20.18. The laser medium was a rod of ruby (aluminum oxide doped with Cr^{3+} ions), and the laser action occurred between an excited triplet state of the Cr^{3+} ions and their ground state. The population inversion was achieved by optical pumping with a flash lamp to raise ions to the excited singlet state, which then populated the excited triplet state by a radiationless transition. The ruby laser emits visible red light with wavelength 694.3 nm. The chief disadvantage of the ruby laser is that it directly involves the ground state, relative to which it is difficult to maintain a population inversion.

In Figure 20.19 we sketch the operation of the neodymium-YAG laser. The laser medium is a crystal of yttrium-aluminum-garnet doped with neodymium ions. This laser has the advantage over the ruby laser that the laser action occurs between two excited states, and the population inversion is consequently easier to maintain. The Nd-YAG laser is widely used in science and technology. One major application is to pump the so-called "dye lasers" in which the medium is intensely colored dye molecules (usually with conjugated double bonds) dissolved in

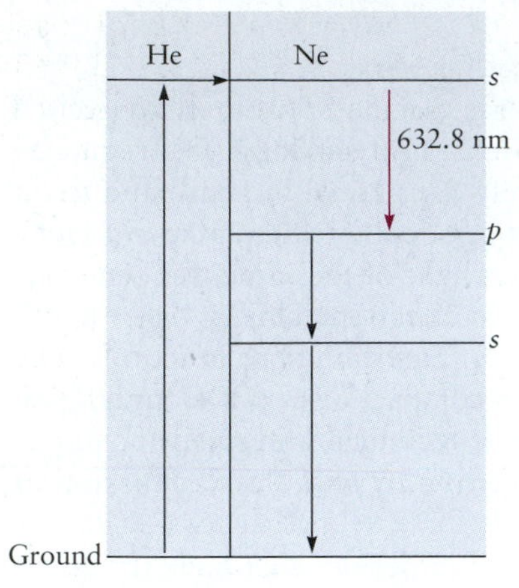

FIGURE 20.20 The He-Ne laser emits from an excited state of the Ne atom, which is produced by collisions with He atoms excited in an electrical discharge.

methanol. Dye lasers are "tunable," which means that laser light can be obtained over essentially the entire visible wavelength range. Pumped dye lasers are widely used in research on molecular spectroscopy and energy transfer processes.

In Figure 20.20 we sketch operation of the He-Ne laser, which provides the red laser light widely used in applications such as alignment of mechanical tools and intruder detection systems. The light is emitted with wavelength 632.8 nm between two states in excited Ne atoms. The pumping process relies first on excitation of He atoms by collisions with energetic electrons in an electrical discharge. This excitation level happens to coincide with an excited state of Ne, and the excitation is transferred from He to Ne during a collision. This gives a number of Ne atoms in highly excited states, and the laser action occurs during a transition to an unpopulated state above the ground state.

Lasers for CDs and DVDs are based on emission from solid state semiconductor materials; their operation is described in Chapter 22.

Newer Methods in Spectroscopy

All the spectroscopy methods described up to this point are carried out in the manner shown in Figure 20.2. They identify transitions between energy levels of molecules, from which we extract information about molecular structure and bonding. Modern molecular spectroscopy aims to go further and give us an understanding of the dynamics of excited molecules—a detailed series of events that may include nuclear motion, light emission, and electron transfer. Such an understanding requires **time-resolved spectroscopy,** in which short light pulses excite molecules and the emitted photons are identified not only by wavelength but also by the time delay from the exciting pulse until the appearance of each photon. Light pulses with durations of picoseconds (10^{-12} s) and even femtoseconds (10^{-15} s) can be generated by modern lasers. The very rapid evolution of excited molecules is monitored by measuring their absorption of a second light pulse—which arrives at a variable time after the initial pulse—or by counting photons emitted by the excited molecules as a function of time (Fig. 20.21).

A promising approach to simplifying the spectra of complex molecules relies upon cooling the molecules in a special type of molecular beam called the

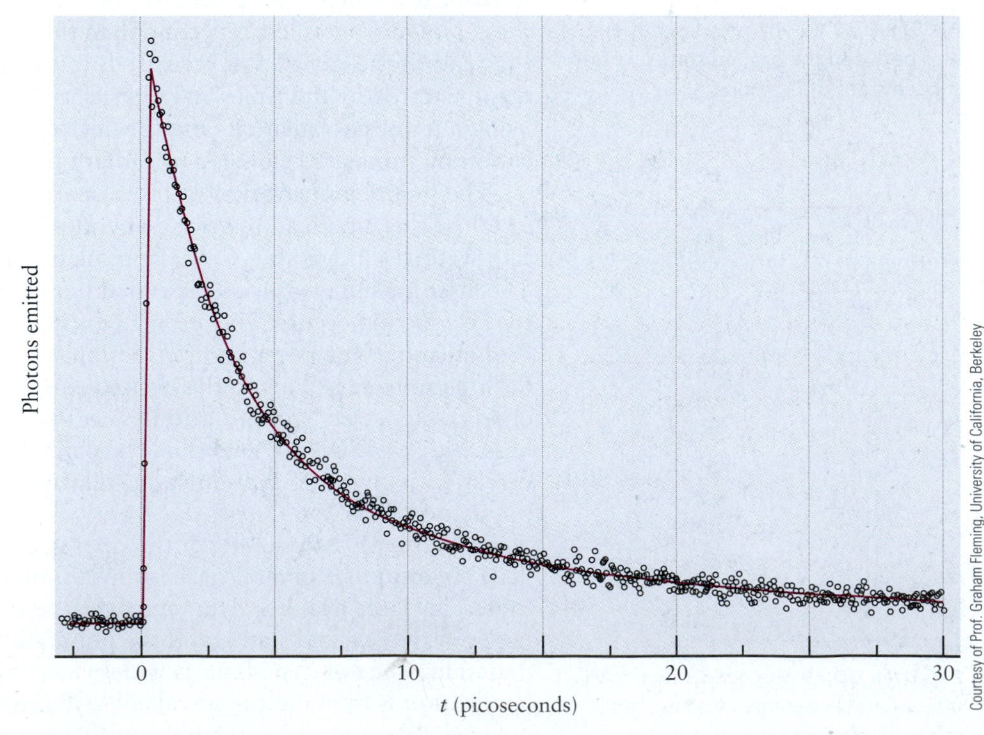

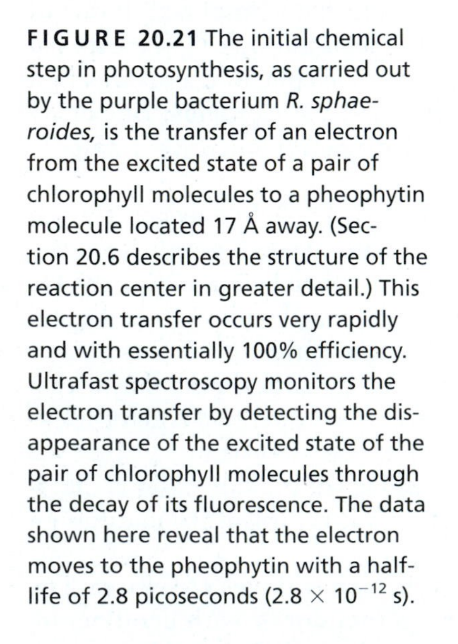

FIGURE 20.21 The initial chemical step in photosynthesis, as carried out by the purple bacterium *R. sphaeroides,* is the transfer of an electron from the excited state of a pair of chlorophyll molecules to a pheophytin molecule located 17 Å away. (Section 20.6 describes the structure of the reaction center in greater detail.) This electron transfer occurs very rapidly and with essentially 100% efficiency. Ultrafast spectroscopy monitors the electron transfer by detecting the disappearance of the excited state of the pair of chlorophyll molecules through the decay of its fluorescence. The data shown here reveal that the electron moves to the pheophytin with a half-life of 2.8 picoseconds (2.8×10^{-12} s).

supersonic jet. A typical, moderate-size molecule in solution has a rather broad and featureless spectrum in the visible and ultraviolet regions (Fig. 20.22, red curve). The dissolved molecule has numerous vibrational and rotational levels. Interactions between the dissolved molecule and solvent molecules at different distances shift these vibrational and rotational energies, giving a blurred overall picture of the spectrum. (The same blurring of rotational levels within a particular vibrational state causes the broad bands in Fig. 20.10.) When a molecule is vaporized into the gas state, the effects of solvent are removed and additional structure appears in its spectrum (see Fig. 20.22, blue curve). Even so, molecular collisions and the numerous thermally occupied vibrational and rotational energy levels still give bands that are too broad to be interpreted quantitatively. To simplify the spectrum even further, it would help to cool the molecule to low temperatures so that very few levels are occupied. But then the gas would then condense to a liquid or solid and defeat the purpose of isolating the molecules. The supersonic jet molecular beam apparatus solves the problem by cooling and isolating the molecules. In a supersonic jet, the molecules to be studied—along with a carrier gas such as helium—are forced through a nozzle into a vacuum at supersonic speeds. The gas expands adiabatically, which cools the molecules (see Section 12.6). Collisions redistribute molecules into lower energy levels, until downstream in the jet the effective temperatures are quite low. As the beam expands into the vacuum, collisions cease, and the cooled molecules fly along in isolation. Under these conditions much more detailed structure appears in the spectra (see Fig. 20.22, inset).

Surprising as it may seem, another way to simplify spectra is to examine one molecule at a time. Chemists have studied the spectra of molecules present in low concentrations in a low-temperature solid. Working with thin samples and

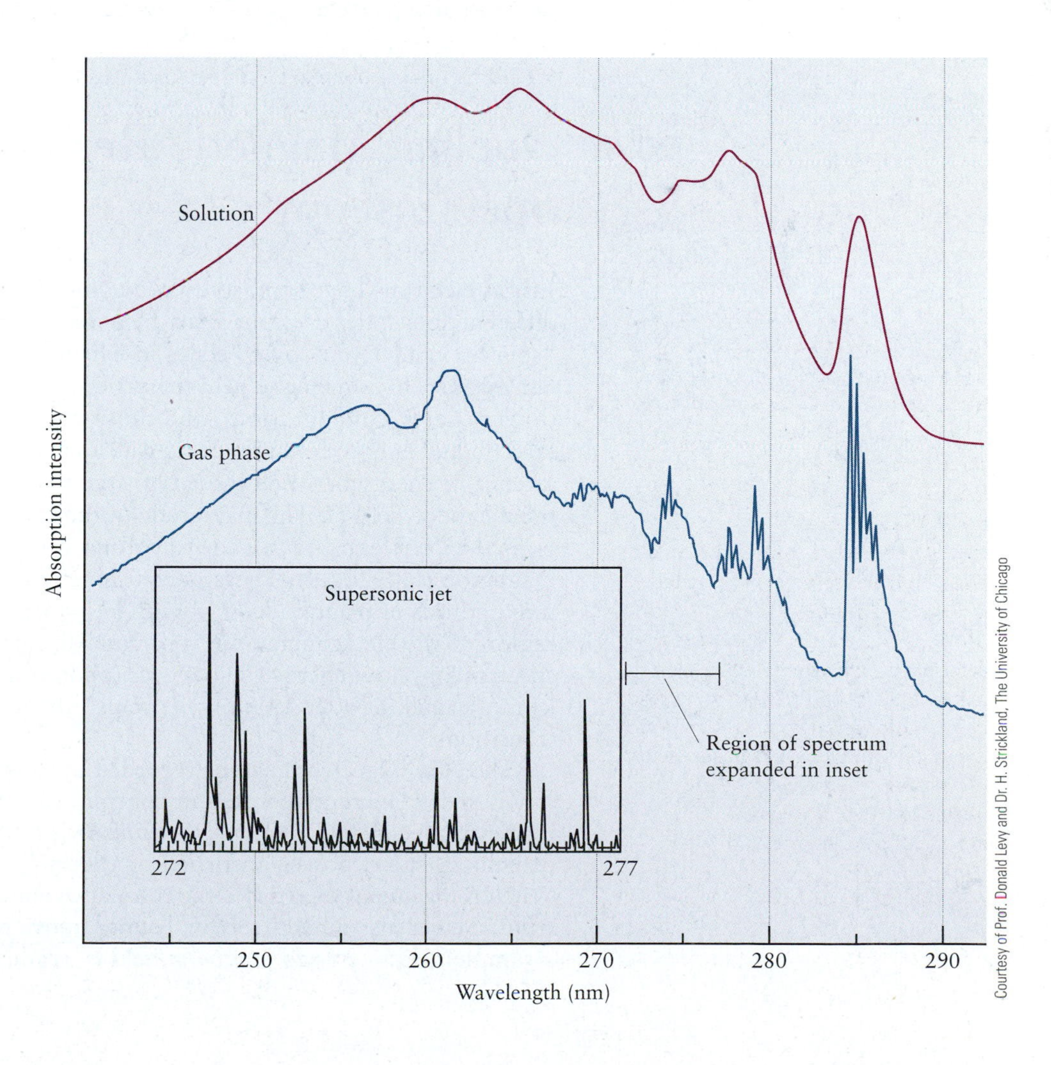

FIGURE 20.22 A portion of the absorption spectrum of indole, C_8H_7N, in the ultraviolet region. In heptane solution (red curve) the spectrum has broad and featureless bands, whereas in the gas phase (blue curve) additional structure appears. Cooling indole molecules in a supersonic jet reveals a large number of sharp absorption lines (see inset, which covers a very small range of wavelengths).

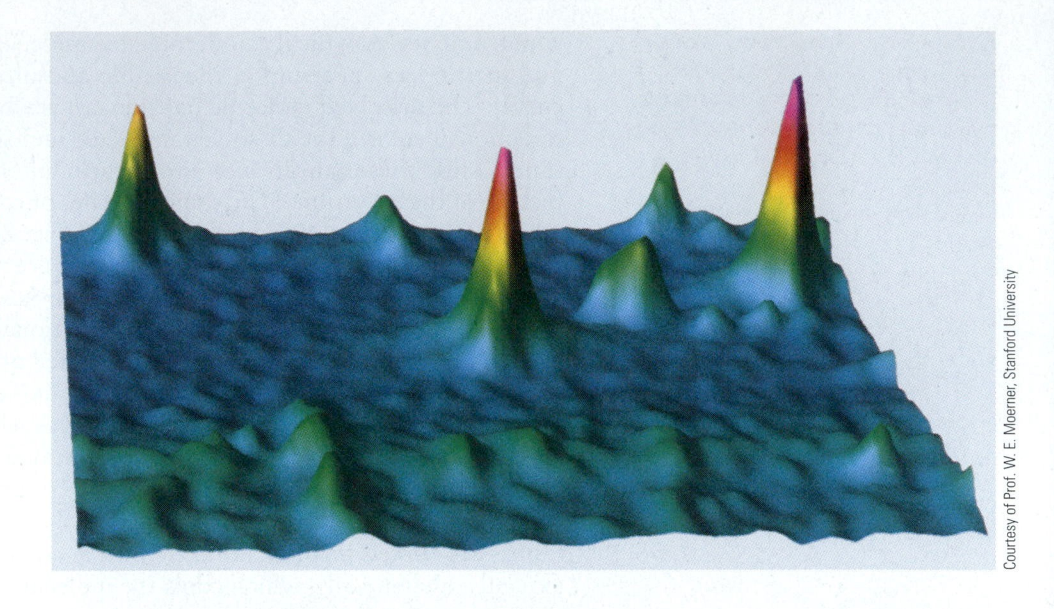

Courtesy of Prof. W. E. Moerner, Stanford University

FIGURE 20.23 Images of the fluorescence of single molecules. The height of each peak gives the intensity of light emitted by that molecule. The axis into the page is spatial position (scanned over a distance of 4×1^{-5} m), and the horizontal axis scans a narrow range of wavelengths for the exciting light.

focused laser beams, they can probe a region of the sample with a relatively small number of those molecules. Using extraordinarily sensitive detectors, they measure signals from one molecule at a time (Fig. 20.23). Chemists can even see how a signal changes with time; as the molecules near the probed molecule move, the frequencies of transitions in the probed molecule shift as well. Thus, chemists not only can "see" single atoms with scanning tunneling microscopes (see Fig. 1.12), they can also probe the spectra of single molecules.

20.4 Nuclear Magnetic Resonance Spectroscopy

Just as electrons have spin, so do some nuclei have spins, and the energies of their different spin states are split apart by a magnetic field. Hydrogen (^{1}H) nuclei have "spin-up" and "spin-down" states in which the proton behaves as if it were a tiny bar magnet. In a magnetic field the protons with spin up align parallel to the field (lower energy configuration), and those with spin down align antiparallel to the field (higher energy configuration). The analogy with bar magnets suggests the difference in energy between these two states will be determined by the strength of the magnetic field H. The magnitude of the energy splitting is $\Delta E = g_N \beta_N H$ where g_N is the "nuclear g-factor" for protons and β_N is a constant called the nuclear magneton (Fig. 20.24). These nuclei can change spin state by absorbing or emitting photons of magnitude $h\nu = g_N \beta_N H$, which fall in the FM radio frequency (rf) region of the electromagnetic spectrum. In nuclear magnetic resonance (NMR) spectroscopy, low-energy radio waves (photons carrying energies between 0.00002 kJ mol^{-1} and 0.00020 kJ mol^{-1}) "tickle" the nuclei in a molecule, and cause spin transitions.

The NMR spectrum can be recorded in various ways. The earliest commercial NMR spectrometers operated in continuous-wave mode, in which the sample is irradiated at constant frequency ν while the magnetic field is swept through a range of values. The rf power absorbed by the sample is recorded at each value of H. When the value of H satisfies the resonance condition, a peak appears in the spectrum. Newer instruments rely on **Fourier transform (FT NMR) spectroscopy,** in which a sample held in a fixed magnetic field is irradiated with a short, intense burst of

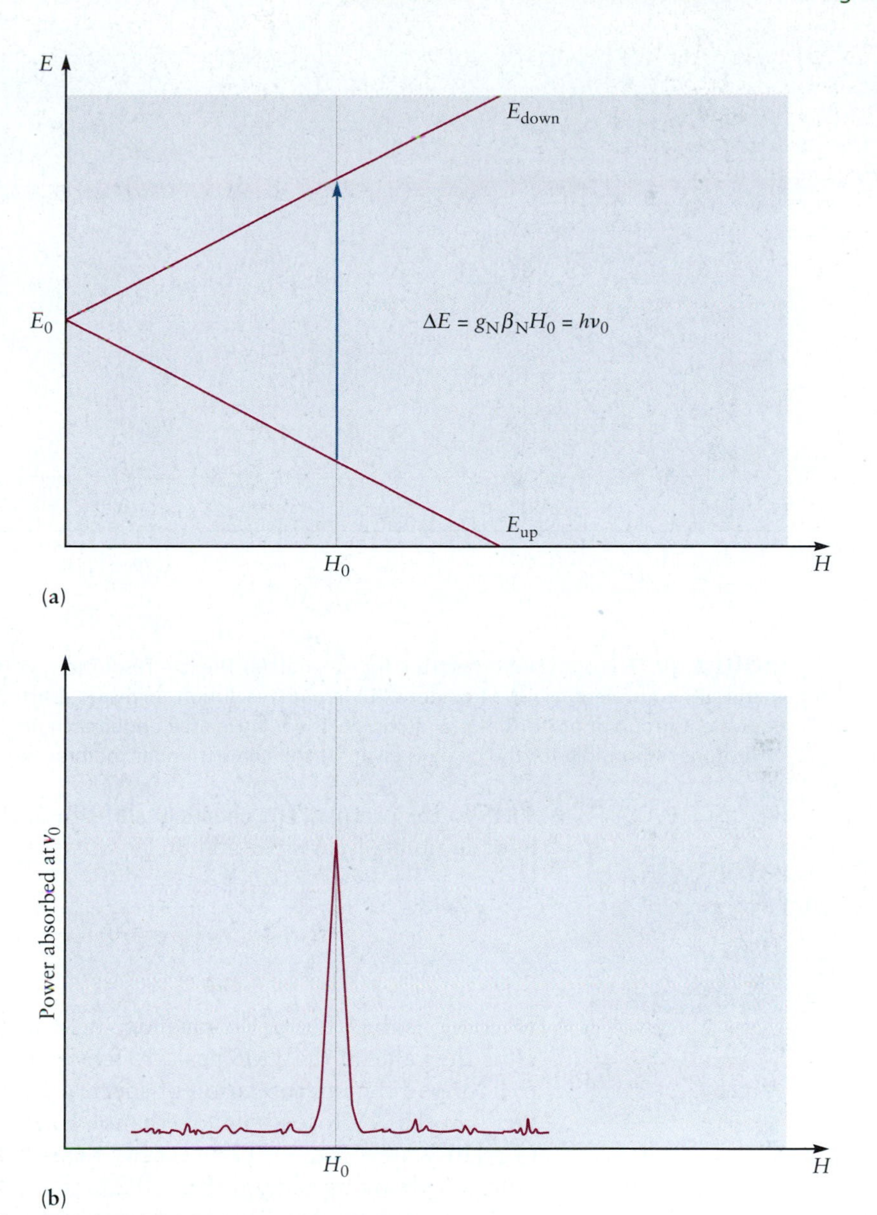

FIGURE 20.24 (a) Energy-level diagram for proton NMR measurements. The magnitude of the energy gap between the spin states depends on the strength of the applied magnetic field. (b) Energy is absorbed when the magnetic field and the rf radiation satisfy the resonance condition.

rf power spanning a frequency range broad enough to stimulate many transitions. The specific frequencies of all the many transitions are measured simultaneously and stored in computer memory. A complete spectrum can be acquired within a few seconds. The measurement can be repeated many times and the results averaged in computer memory to reduce the effects of random background noise. FT NMR spectrometers detect numerous nuclei besides ^{1}H (^{13}C holds special interest in organic chemistry and biochemistry), but our discussion is limited to ^{1}H.

Nuclear spins are sensitive to the chemical environment of a nucleus. Electrons moving near the nucleus establish an internal magnetic field that modifies the local effective field felt by each proton to a value different from that of the externally applied field. The resulting **chemical shift** causes protons within different structural units of the molecule to show NMR peaks at different values of magnetic field. All protons in chemically equivalent environments will contribute to a single absorption peak in the spectrum. The relative area under each absorption peak is proportional to the number of protons within each equivalent group. In order to standardize procedures, chemical shift values are recorded relative to the selected reference compound tetramethylsilane (TMS) by adding a very small amount of

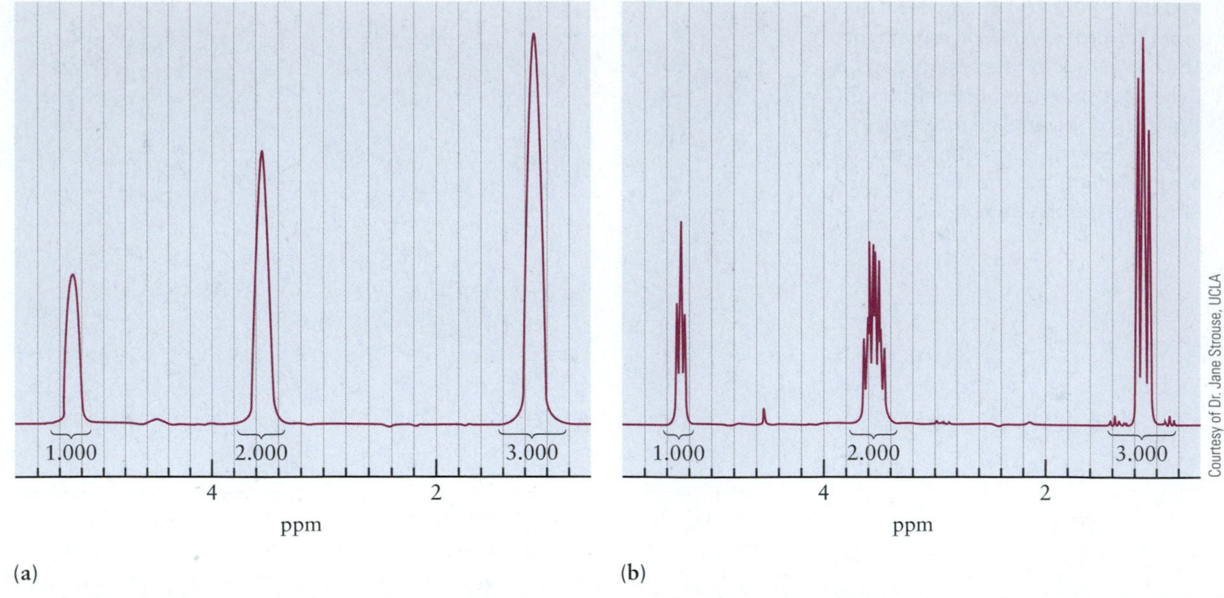

FIGURE 20.25 Proton NMR spectrum for ethanol. (a) The low-resolution spectrum shows a single broad peak for each chemically equivalent group of protons. (b) In high resolution, spin-spin splitting separates the peak for each chemically equivalent group of protons into a multiplet. The relative area under each peak, which is proportional to the number of protons responsible for the peak, is given by the number under the brackets below each peak.

TMS to the sample. The chemical shift for each proton in the spectrum is then defined in units of parts per million (ppm) as

$$\delta = \frac{H_s - H_r}{H_r} \times 10^6 \qquad [20.8]$$

where H_s is the value of the magnetic field at the peak in the sample spectrum and H_r is the value at the TMS peak. Extensive tables of chemical shift values relative to TMS aid the interpretation of spectra. For example, these considerations predict that the NMR spectrum of ethanol, CH_3-CH_2-OH should have three peaks of relative area 3:2:1. This is exactly the result obtained when the spectrum is acquired with low resolution (Fig. 20.25a).

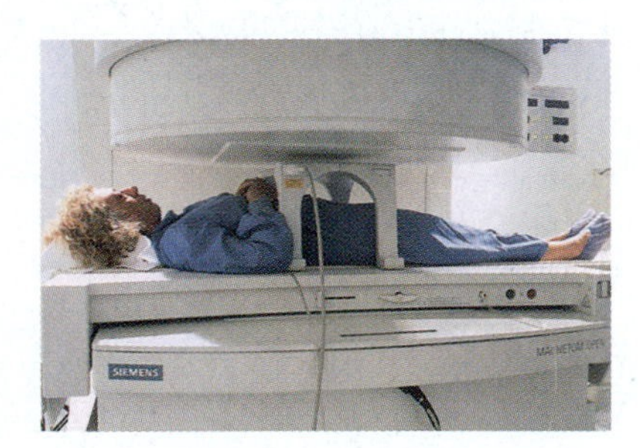

FIGURE 20.26 A magnetic resonance imaging (MRI) machine. The patient lies underneath the magnet opening.
(© Mauro Fermariello/Science Photo Library/Photo Researchers, Inc.)

Nuclear spins are also sensitive to other nearby nuclear spins. Because each spin behaves as if it were a small bar magnet, each influences the local value of magnetic field felt by others nearby. The result is **spin–spin splitting,** which breaks the broad peak of each chemically equivalent group of protons into a multiplet of narrower, sharper peaks. So long as the chemical shift between chemically nonequivalent groups of protons is large, the multiplet patterns can be interpreted by qualitative rules that we do not describe here. For example, when applied to ethanol, these rules predict splitting of the CH_3 protons into a triplet and splitting of the CH_2 protons into a quartet, as seen in high-resolution experimental spectra (see Fig. 20.25b). Through chemical shifts and spin-spin splitting, nuclear magnetic resonance spectroscopy offers a way to identify the bonding groups in a molecule and interpret molecular structure.

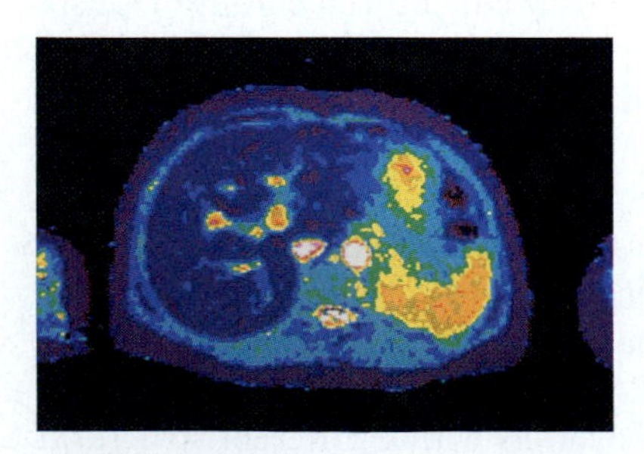

FIGURE 20.27 An MRI scan of the abdomen of a healthy person. The liver is the large blue mass (top and left); the spleen is at the lower right (yellow). Stomach contents are at the top right (red and yellow). The purple outer layer is fat.
(© Alexander Tsiaris/Science Source/Photo Researchers, Inc.)

An adaptation of NMR spectroscopy used in medical diagnosis is called **magnetic resonance imaging (MRI).** It relies on emission by the protons in the water contained in the organs of the body. The patient is placed under the opening of a large magnet (Fig. 20.26), and a radio transmitter raises the proton spins in the relevant part of the body to their high-energy state. The rf photons subsequently emitted are then detected by a radio receiver coil. The amplitude of the signal indicates the excess or deficiency of water present. Thus, MRI can identify tumors by the excess water in their cells. The time delay until emission occurs is related to the type of tissue being examined. Figure 20.27 shows an MRI scan of the abdomen.

20.5 Introduction to Atmospheric Photochemistry

In photosynthesis, green plants and photosynthetic bacteria harvest energy from the sun through photochemical reactions. The ability of light to cause chemical reactions is also apparent in the earth's atmosphere. In fact, photosynthesis and atmospheric chemistry are intimately connected. Reactions in the atmosphere determine the intensities and wavelengths of light that reach the earth's surface to be harvested by living species. At the same time, oxygen, the gas produced by green-plant photosynthesis, has transformed the atmosphere; before photosynthesis began, there was almost no free oxygen at the earth's surface.

Although the chemical composition summarized in Table 9.1 describes the average makeup of the portion of the atmosphere closest to the earth's surface, it does not do justice to the variation in chemical properties with altitude, to the dramatic role of local fluctuations in trace gases, or to the dynamics underlying observed average concentrations. The atmosphere is a complex chemical system that is far from equilibrium. Its properties are determined by an intricate interplay of thermodynamic and kinetic factors. It is a multilayered structure, bathed in radiation from the sun and interacting at the bottom with the oceans and land masses. At least four layers are identifiable, each with a characteristic variation of temperature (Fig. 20.28). In the outer two layers (the **thermosphere** and the **mesosphere**), atmospheric density is low and the intense radiation from the sun causes extensive ionization of the particles that are present. The third layer, the **stratosphere,** is the region from 12 km to 50 km (approximately) above the earth's surface. The **troposphere** is the lowest region, extending 12 km out from the earth's surface. In the troposphere, warmer air lies beneath cooler air. This is a dynamically unstable situation because warm air is less dense and tends to rise, so convection takes place, mixing the gases in the troposphere and determining the

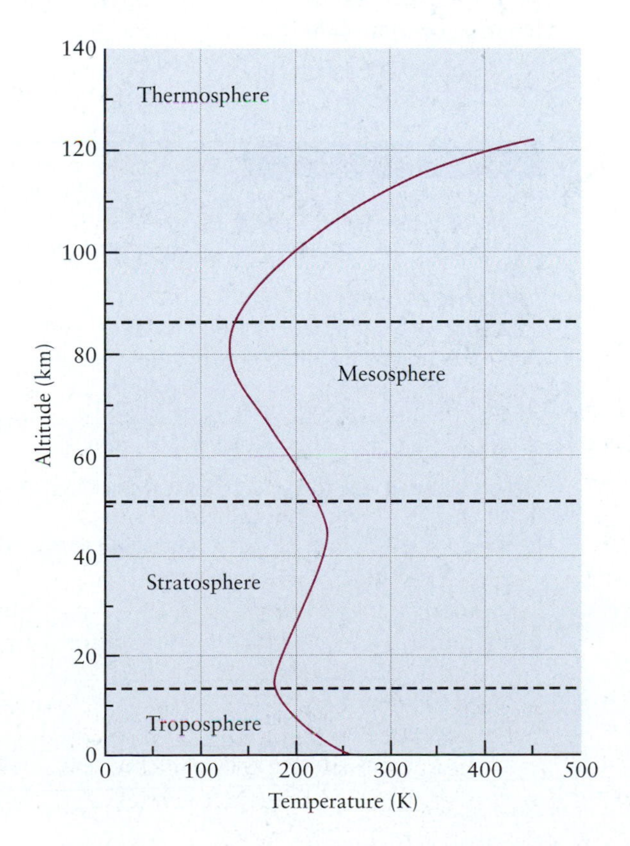

FIGURE 20.28 The variation of the temperature of the atmosphere with altitude, showing the layered structure of the atmosphere.

weather. In the stratosphere the temperature increases with altitude and there is little vertical mixing from convection. Mixing across the borders between the layers is also slow, so most chemical processes in each layer can be described separately.

Research in atmospheric chemistry dates back to the 18th century. Cavendish, Priestley, Lavoisier, and Ramsay were the first scientists to study the composition of the atmosphere. In recent years, atmospheric chemistry has developed in two different but related directions. First, the sensitivity of chemical analysis has greatly improved, and analyses for substances at concentrations below the part-per-billion (ppb) level are now carried out routinely. Airplanes and satellites enable scientists to map the global distributions of trace substances. Second, advances in gas-phase chemical kinetics have led to a better quantitative understanding of the ways in which substances in the atmosphere react with one another and with light. Much of the impetus for these studies of atmospheric chemistry comes from concern about the effect of air pollution on life.

Stratospheric Chemistry

As Figure 4.6 shows, the sun emits light over a broad range of wavelengths, with the highest intensity at about 500 nm, in the visible region of the spectrum. The intensities at wavelengths down to 100 nm in the ultraviolet region are quite substantial, and because the energy $h\nu$ carried by a photon is inversely proportional to the wavelength $\lambda (h\nu = hc/\lambda)$, the ultraviolet photons carry much more energy than do photons of visible light. If these photons could penetrate to the earth's surface in substantial numbers, they could greatly damage living organisms. The outer portions of the atmosphere (especially the thermosphere) play a crucial role in preventing this penetration through the photodissociation of oxygen molecules:

$$O_2 + h\nu \longrightarrow 2O$$

where $h\nu$ symbolizes a photon. This reaction reduces the number of high-energy photons reaching the lower parts of the atmosphere, especially those with wavelengths less than 200 nm.

EXAMPLE 20.4

The bond dissociation energy of O_2 is 496 kJ mol^{-1}. Calculate the maximum wavelength of light that can photodissociate an oxygen molecule.

SOLUTION

Because 496 kJ dissociates 1.00 mol of O_2 molecules, the energy to dissociate one molecule is found by dividing by Avogadro's number:

$$\frac{496 \times 10^3 \text{ J mol}^{-1}}{6.022 \times 10^{23} \text{ mol}^{-1}} = 8.24 \times 10^{-19} \text{ J}$$

A photon carrying this energy has a wavelength λ, given by

$$8.24 \times 10^{-19} \text{ J} = h\nu = \frac{hc}{\lambda}$$

so

$$\lambda = \frac{hc}{8.24 \times 10^{-19} \text{ J}} = \frac{(6.626 \times 10^{-34} \text{ J s})(2.998 \times 10^{8} \text{ m s}^{-1})}{8.24 \times 10^{-19} \text{ J}}$$

$$= 2.41 \times 10^{-7} \text{ m} = 241 \text{ nm}$$

Any photons with wavelengths *shorter* than this are energetic enough to dissociate oxygen molecules. Those with wavelengths shorter than 200 nm are the most efficient in causing photodissociation.

Related Problems: 29, 30

Rather few photons with wavelengths below 200 nm can penetrate to the stratosphere, but those that do establish a small concentration of oxygen atoms in that layer. These atoms can collide with the much more prevalent oxygen molecules to form highly excited molecules of ozone, symbolized by O_3^*:

$$O + O_2 \rightleftharpoons O_3^*$$

The excited-state O_3^* can dissociate in a unimolecular reaction back to O and O_2, as indicated by the reverse arrow in this equilibrium. Alternatively, if another atom or molecule collides with it soon enough, some of its excess energy can be transferred to that atom or molecule. This process is represented as

$$O_3^* + M \longrightarrow O_3 + M$$

where M stands for the atom or molecule with which the O_3^* collides. The most likely candidates for M are oxygen and nitrogen molecules, because they are the most abundant species in the atmosphere. The net effect of these two reactions is to produce a small concentration of ozone in the stratosphere.

Under laboratory conditions, ozone is an unstable compound. Its conversion to oxygen is thermodynamically spontaneous

$$O_3(g) \longrightarrow \tfrac{3}{2} O_2(g) \qquad \Delta G° = -163 \text{ kJ}$$

but the rate is quite slow in the absence of light. In the stratosphere, ozone photodissociates readily to O_2 and O in a reaction that requires about 106 kJ per mole of ozone, much less than that required for the dissociation of O_2:

$$O_3 + h\nu \longrightarrow O_2 + O$$

This process occurs most efficiently for wavelengths between 200 and 350 nm. The energy of light in this range of wavelengths is too small to be absorbed by molecular oxygen but large enough to damage organisms at the earth's surface. The ozone layer shields the earth's surface from 200-nm to 350-nm ultraviolet radiation coming from the sun. The balance between formation and photodissociation leads to a steady-state concentration of more than 10^{15} molecules of ozone per liter in the stratosphere.

It is important to know whether molecules being released in the lower atmosphere can reach the stratosphere and affect the amount of ozone in it. Certain types of air pollution give rise to **radicals** that catalyze ozone depletion. A radical is a chemical species that contains an odd (unpaired) electron, and it is usually formed by the rupture of a covalent bond to form a pair of neutral species. One pressing concern involves chlorofluorocarbons (CFCs)—compounds of chlorine, fluorine, and carbon used as refrigerants and as propellants in some aerosol sprays. CFCs are nonreactive at sea level but can photodissociate in the stratosphere:

$$CCl_2F_2 + h\nu \longrightarrow CClF_2 + Cl$$

The atomic chlorine released thereby soon reacts with O atoms or O_3 to give ClO:

$$Cl + O \longrightarrow ClO$$

$$Cl + O_3 \longrightarrow ClO + O_2$$

The ClO radical is the immediate culprit in the destruction of stratospheric ozone. Local increases in ClO concentration are directly correlated with decreases in O_3 concentration. ClO catalyzes the destruction of ozone, probably by the mechanism

$$\begin{aligned} 2\,\text{ClO} + \text{M} &\longrightarrow \text{ClOOCl} + \text{M} \\ \text{ClOOCl} + h\nu &\longrightarrow \text{ClOO} + \text{Cl} \\ \text{ClOO} + \text{M} &\longrightarrow \text{Cl} + \text{O}_2 + \text{M} \\ 2 \times (\text{Cl} + \text{O}_3 &\longrightarrow \text{ClO} + \text{O}_2) \\ \hline \text{net reaction: } 2\,\text{O}_3 &\longrightarrow 3\,\text{O}_2 \end{aligned}$$

where M stands for N_2 and O_2 molecules. This catalytic cycle is ordinarily disrupted rather quickly as Cl reacts with other stratospheric species to form less reactive "reservoir molecules" such as HCl and $ClONO_2$. But, the lack of mixing in the stratosphere keeps the reservoir molecules around for long periods.

It is believed that atomic Cl escapes its reservoir molecules through heterogeneous reactions on the polar stratospheric clouds that form during the intense Antarctic winter. The liberated Cl gives rise to the annual "ozone holes" above the Antarctic. During these episodes, more than 70% of the total column ozone is depleted before the values rise again (Fig. 20.29). Researchers have demonstrated that the extent of depletion above the Antarctic increases each year (Fig. 20.30), and measurements have shown smaller but still serious depletions over other parts of the globe. International agreement (the Montreal Protocol) banned the production of cholorfluorocarbons by industrialized nations by the end of the 20th century. Developing countries have until 2010 to comply with the regulation.

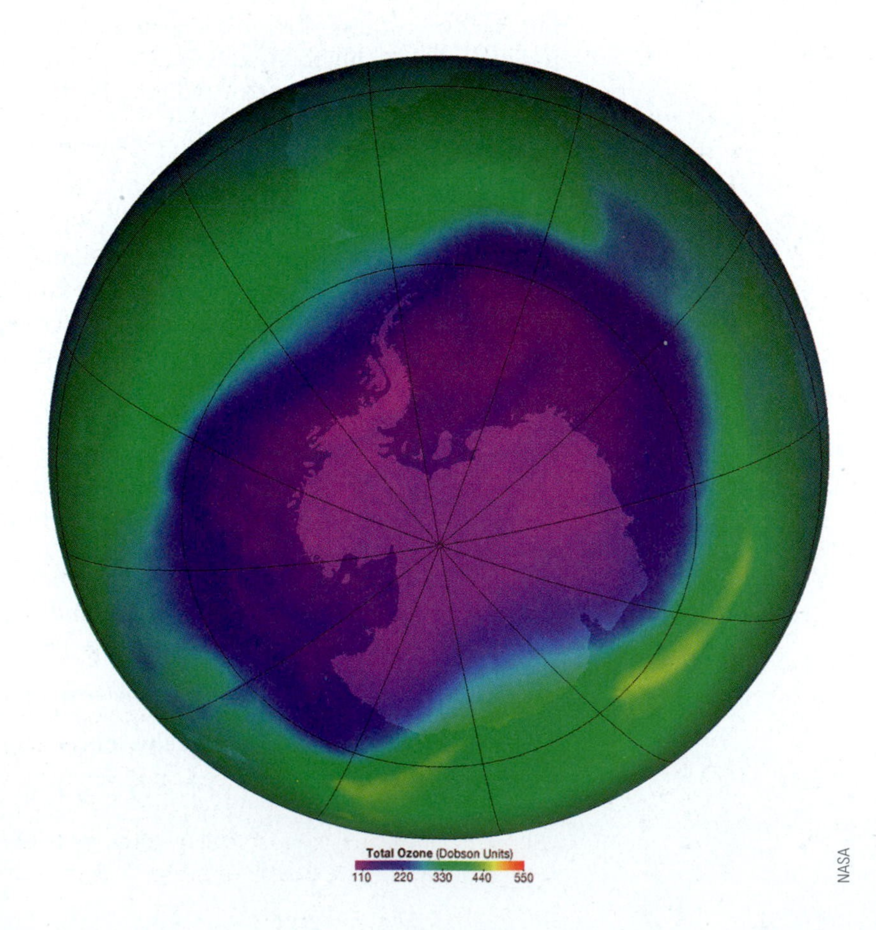

FIGURE 20.29 This false color image shows total stratospheric ozone amounts over the southern hemisphere for September 24, 2006, as recorded by the Ozone Monitoring Instrument (OMI) mounted on the Aura spacecraft. The dramatic depletion of the ozone layer over Antarctica is revealed with the help of the false color scale at the bottom of the figure. Ozone amounts are commonly expressed in Dobson units; 300 Dobson units is a typical global average over the course of a year. The size of the Antarctic ozone hole was near a record high and the levels of ozone near a record low on this date.

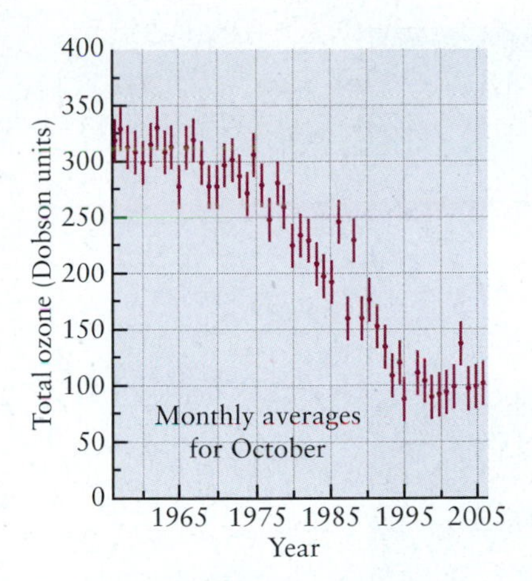

FIGURE 20.30 Worsening ozone depletion over Halley Bay, Antarctica. These measurements were all taken in the Antarctic spring (October) when depletion is at its worst.

Tropospheric Chemistry

The troposphere is the part of the atmosphere in contact with the earth's surface. It is therefore most directly and immediately influenced by human activities, especially by the gases or small particles put into the air by automobiles, power plants, and factories. Some pollutants have long lifetimes and are spread fairly evenly over the earth's surface; others attain large concentrations only around particular cities or industrial areas.

The oxides of nitrogen are major air pollutants. Their persistence in the atmosphere demonstrates the importance of kinetics, as opposed to thermodynamics, in the chemistry of the atmosphere. All of the oxides of nitrogen are thermodynamically unstable with respect to the elements at 25°C, as shown by their positive free energies of formation. They form whenever air is heated to high enough temperatures, either in an industrial process or in the engine of a car. They accumulate to build up concentrations much higher than the equilibrium values because they decompose slowly. The interconversion among NO, NO_2, and N_2O_4 is rapid and strongly temperature-dependent, so they are generally grouped together as "NO_x" in pollution reports. Photochemical smog is formed by the action of light on nitrogen dioxide, followed by reaction to produce ozone:

$$NO_2 + h\nu \longrightarrow NO + O$$

$$O + O_2 + M \longrightarrow O_3 + M \qquad (M = N_2 \text{ or } O_2)$$

It may seem paradoxical that concern about ozone in the stratosphere involves its potential *depletion,* whereas concern about the troposphere involves the *production* of ozone. Although ozone is beneficial in preventing radiation from penetrating to the earth's surface, it is quite harmful in direct contact with organisms because of its strong oxidizing power. Levels of 10 ppm to 15 ppm are sufficient to kill small mammals, and a concentration as low as 3 ppm is enough to trigger an "ozone alert." In addition, ozone reacts with incompletely oxidized organic compounds from gasoline and with nitrogen oxides in the air to produce harmful irritants such as methyl nitrate (CH_3NO_3).

The oxides of sulfur create global pollution problems because they have longer lifetimes in the atmosphere than the oxides of nitrogen. Some of the SO_2 and SO_3 in the air originates from biological processes and from volcanoes, but much comes from the oxidation of sulfur in petroleum and in coal burned for fuel. If the sulfur is not removed from the fuel or the exhaust gas, SO_2 enters the atmosphere as a stable but reactive pollutant. Further oxidation by radicals leads to sulfur trioxide:

$$SO_2 + OH \longrightarrow SO_2OH$$

$$SO_2OH + O_2 \longrightarrow SO_3 + OOH$$

$$OOH \longrightarrow O + OH$$

The atomic oxygen thus produced can react with molecular oxygen to form ozone. **Acid rain** results from the reaction of NO_2 and SO_3 with hydroxyl radical and water vapor in the air to form nitric acid (HNO_3) and sulfuric acid (H_2SO_4). The acids dissolve in water and return to the earth in the rain.

The Greenhouse Effect

As we have seen, most of the short-wavelength photons from the sun are absorbed by the outer atmosphere and do not reach the earth's surface. The radiation that does reach the earth is in steady state balance with re-radiation into space, which helps to maintain a livable temperature. Certain gases in the troposphere play a crucial role in this balance because they absorb infrared radiation emitted by the warm surface of the earth rather than letting it pass out to space (Fig. 20.31). In

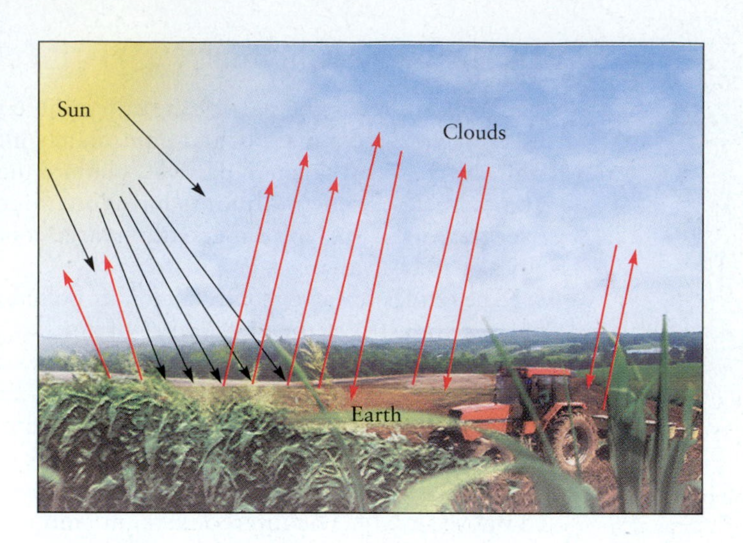

FIGURE 20.31 Visible light from the sun strikes the earth's surface and heats it. Much energy is radiated back into space in the longer wavelength infrared region of the spectrum. Molecules in the atmosphere and in clouds can block some of this loss, keeping the lower atmosphere warmer than it would be without these molecules.

winter, the temperature does not fall as low on cloudy nights as on clear nights because the water vapor in the clouds provides a thermal blanket, absorbing outgoing radiation with wavelengths near 2×10^{-5} m and reradiating it to the earth's surface.

Two other gases that absorb infrared radiation to a significant extent are carbon dioxide and methane. Both are uniformly distributed in fairly low concentrations through the troposphere, but their concentrations have increased steadily over the 200 to 300 years since the beginning of the industrial revolution (Fig. 20.32). The tremendous increase in the burning of fossil fuels for energy is a major contributor to the increase of CO_2 (Fig. 20.33). It is estimated that by the year 2050 the concentration of CO_2 in the atmosphere will be double the premodern value. Chlorofluorocarbons also absorb infrared radiation, especially over some key wavelength ranges left open by CO_2 and water vapor.

Such changes are viewed with alarm because of a phenomenon called the **greenhouse effect.** This term refers to a global increase in average surface temperature of the earth's surface that occurs when heat given off at the earth's surface is prevented from escaping into space by gases that absorb and reradiate infrared radiation. An increase in average surface temperature of 2°C to 5°C over the next century would have major consequences, including melting some polar ice—and a consequent rise in the level of the oceans—and the conversion of arable land to desert. To prevent these undesirable changes, it will be necessary in the next decades to develop new energy sources not based on burning fossil fuels.

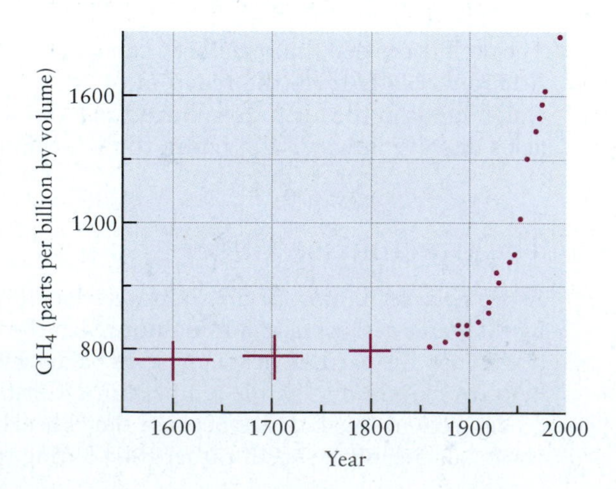

FIGURE 20.32 The concentration of methane in the lower atmosphere has doubled since 1600. Note the dramatic increase in the last 100 years.

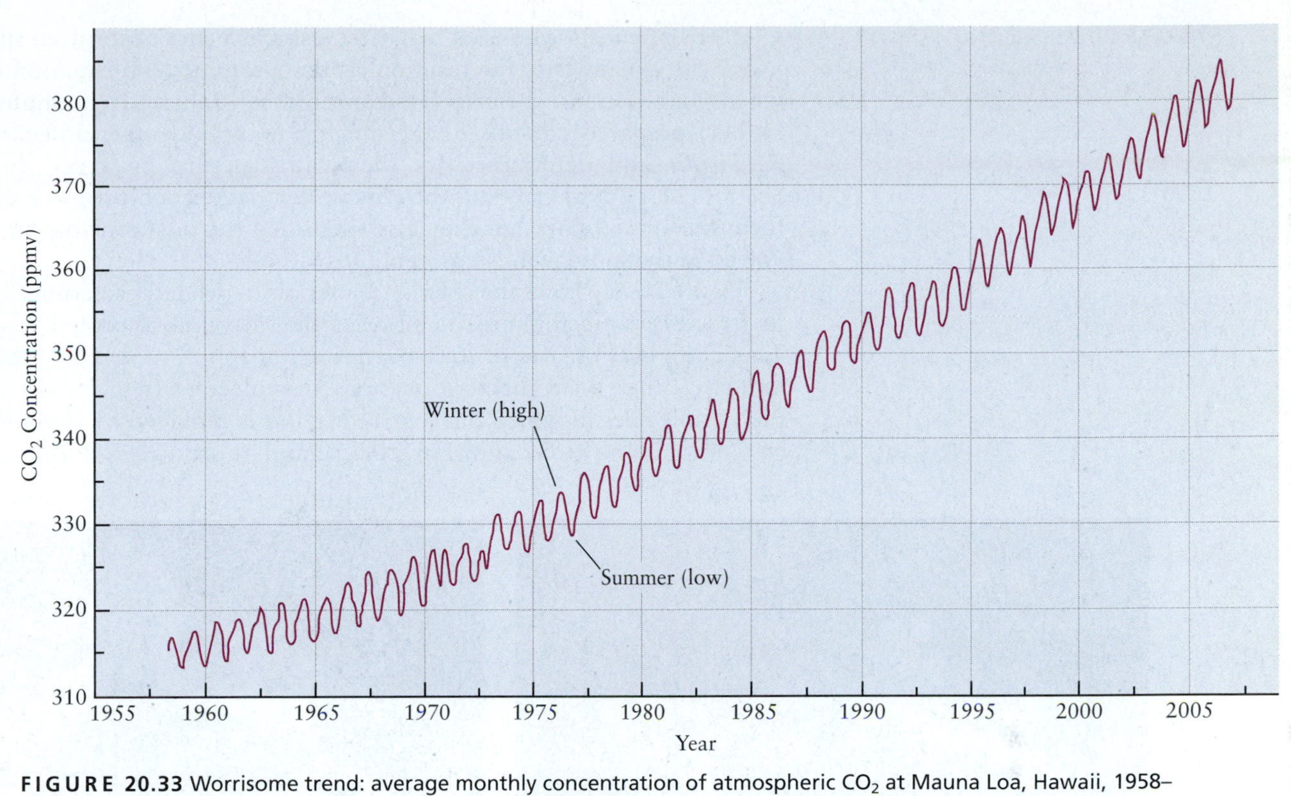

FIGURE 20.33 Worrisome trend: average monthly concentration of atmospheric CO_2 at Mauna Loa, Hawaii, 1958–2007. The concentration varies seasonally, caused by changes in carbon dioxide absorption by plants. (Data from www.cmdl.noaa.gov/ccgg/trends/co2_data_ml0.php, accessed on January 20, 2007.)

20.6 Photosynthesis

Even the most efficient manufactured solar collectors (see Section 22.8) fall far short of nature in their ability to convert solar energy from the sun into other useful forms of energy. Living species harvest light and store its energy by carrying out chemical reactions with positive free energy changes. For this purpose they use compounds dominated by carbon, in which metal ions play a critical role. This process of **photosynthesis** has transformed our planet and permitted the evolution of human life. Before we consider green plants (which provide most of the photosynthetic energy storage today), let us examine the reaction mechanisms in existent photosynthetic bacteria that are thought to have evolved earlier, and are considerably simpler, than plants. These bacteria have recently yielded some important secrets in experimental investigations.

The purple photosynthetic bacteria include the species *Rhodobacter sphaeroides* and *Rhodopseudomonas viridis*. Although these two differ in the details of their chemical makeup, the broad features of their mechanisms appear to be closely related. Their purple color reflects their absorption of light predominantly in the red region of the spectrum, with wavelengths up to 925 nm. *R. sphaeroides* and *R. viridis* use a variety of pigment molecules, including derivatives of chlorophyll (Fig. 8.22b), as "antennas" to absorb photons that reach their surface. The antenna molecules store the light energy through promotion of electrons to excited singlet states (see Fig. 20.11), then transfer this energy rapidly to other molecules nearby. In such a transfer the initially excited molecule returns to its ground state while a neighboring molecule reaches an excited singlet state. Within about 10^{-10} s (100 picoseconds), the energy migrates to the **reaction center** where it is trapped and is ready to be used in chemical reactions.

The initial trapping site in the reaction center consists of a **special pair** of bacteriochlorophyll molecules. The existence of this special pair was postulated in the

1960s by scientists who used magnetic resonance and absorption spectroscopy to probe the dynamics of the reaction center. Dramatic confirmation came in 1984 when three German chemists (Hartmut Michel, Johann Deisenhofer, and Robert Huber) prepared crystals of the membrane-bound reaction-center protein in *R. viridis* and determined its chemical structure by x-ray diffraction (see Section 21.1). Not only did they locate the bacteriochlorophyll molecules that form the special pair, but they also identified the positions of the molecules involved in the subsequent steps in photosynthesis.

Figure 20.34 shows the reaction center of the related bacterium *R. sphaeroides;* its structure is much more evident when the surrounding protein is stripped away. This complex consists of four bacteriochlorophyll molecules (the special pair and two others), two bacteriopheophytin molecules (which are bacteriochlorophyll molecules in which the central Mg ion is replaced by two hydrogen ions), two ubiquinone molecules (Fig. 20.35), and an iron(II) ion. Interestingly, these

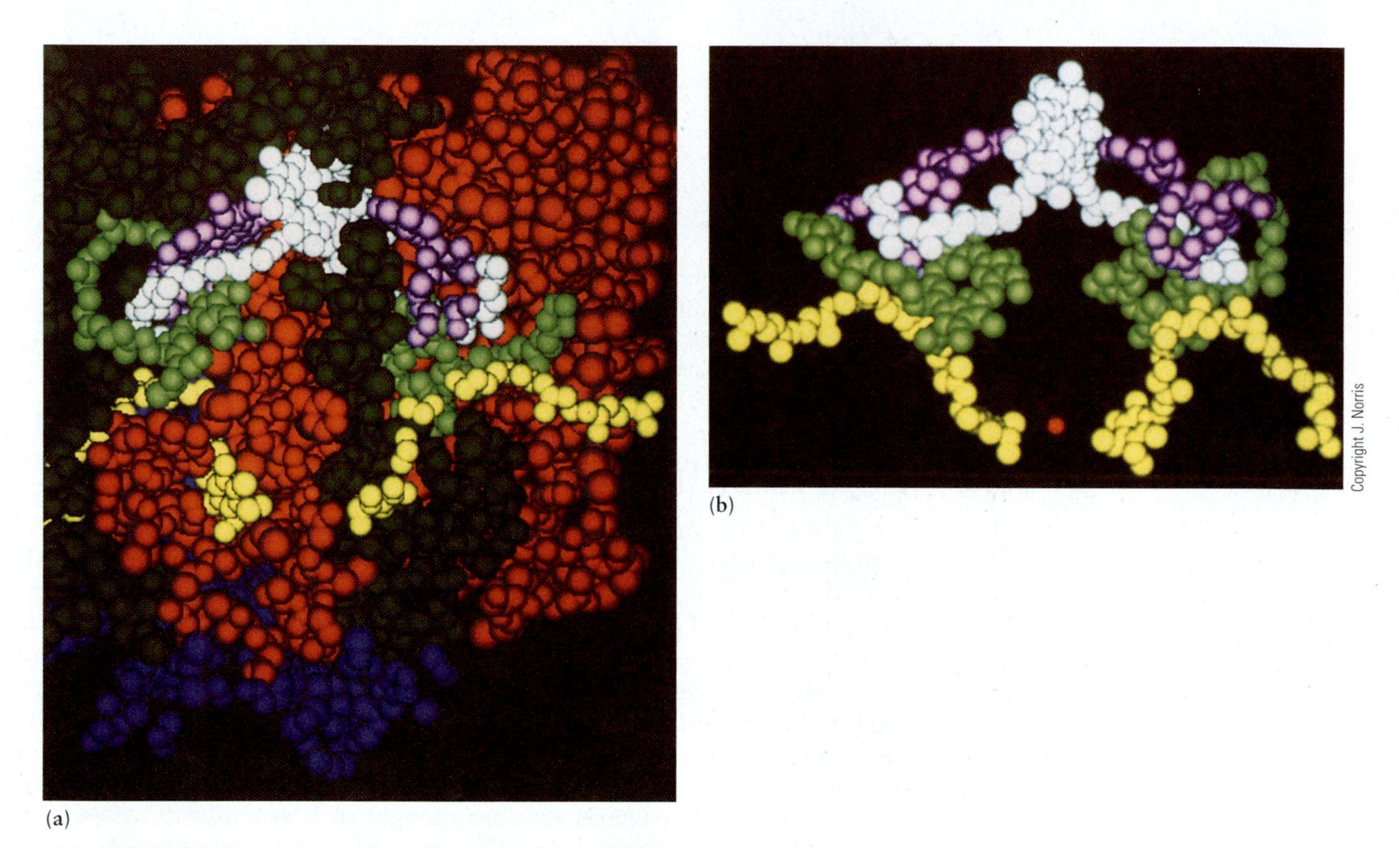

FIGURE 20.34 Computer-generated diagrams of the reaction center of the purple bacterium *R. sphaeroides* based on x-ray diffraction data. (a) The reaction center has been sliced through the middle to show the donor-acceptor molecules surrounded by protein (dark green, red, and blue portions). (b) The protein has been removed entirely. The four bacteriochlorophyll molecules are shown in white (for the special pair) and purple (for the other two); bacteriopheophytin molecules are light green; ubiquinone molecules are yellow. An iron(II) ion is shown between the ubiquinone molecules. The A branch is on the right, and the B branch is on the left.

FIGURE 20.35 The structures of (a) ubiquinone (UQ) and (b) its reduced form, UQH_2. Note the tail that contains a C_5H_8 group repeated ten times.

(a) (b)

molecules are arranged in an almost perfectly symmetric fashion, with two branches (called A and B). But, the surrounding protein breaks the symmetry of the branches and causes the energy flow to pass almost entirely through the A branch (on the right in Fig. 20.34). Any energy passing through the B branch is too little to detect.

Time-resolved spectroscopy (see Fig. 20.21) has shown that within about 2.8 picoseconds (2.8×10^{-12} s) an electron is transferred from the special pair to the bacteriopheophytin molecule in the A branch, creating an anion on that site and leaving a cation on the special pair. Within 200 picoseconds, the electron reaches the ubiquinone molecule in the A branch and then is transferred across to the other ubiquinone molecule, in the B branch. The special pair picks up an electron from the iron atom in the heme group (see Fig. 8.22c) on a cytochrome protein outside the membrane and is reduced back to its original state. It then absorbs a second photon and transfers a second electron to the same B-branch ubiquinone molecule. This doubly reduced ubiquinone is less tightly bound to the reaction-center protein and moves away, being replaced by an unreduced ubiquinone molecule. The reduced ubiquinone gives up its electrons and hydrogen ions to another protein, which releases H^+ ions outside the cell membrane. The resulting hydrogen ion (proton) gradient across the cell wall stores energy (Recall from the Nernst equation in Chapter 17 that a voltage difference will exist across a concentration gradient.) and can carry out chemical reactions needed by the cell, ultimately transferring hydrogen ions back into the cell to close the cycle.

The chemical reactions at the reaction center can be summarized by the following six steps:

1. $(BChl)_2 + UQ \xrightarrow[\text{light}]{} (BChl)_2^+ + UQ^-$
2. $(BChl)_2^+ + Cyt \longrightarrow (BChl)_2 + Cyt^+$
3. $(BChl)_2 + UQ^- \xrightarrow[\text{light}]{} (BChl)_2^+ + UQ^{2-}$
4. $(BChl)_2^+ + Cyt \longrightarrow (BChl)_2 + Cyt^+$
5. $UQ^{2-} + 2\,H_{in}^+ \longrightarrow UQH_2$
6. $UQH_2 + 2\,H_{in}^+ + 2\,Cyt^+ \longrightarrow UQ + 4\,H_{out}^+ + 2\,Cyt$

Overall: $4\,H_{in}^+ \longrightarrow 4\,H_{out}^+$

Here $(BChl)_2$ stands for the special pair of bacteriochlorophyll molecules, UQ for ubiquinone, and Cyt for the cytochrome protein. Steps 1 and 3 involve excitation of bacteriochlorophyll and transfer of a pair of electrons to a ubiquinone molecule. Steps 2 and 4 restore the special pair to its initial state. Steps 5 and 6 transfer hydrogen ions outside the membrane wall and restore the cytochrome to its reduced form. The net reaction is the light-driven movement of hydrogen ions from inside the cell to outside the cell.

Purple bacteria have only a single type of reaction center, but green plants have two types, referred to as photosystem I (PS I) and photosystem II (PS II). These absorb light at somewhat different wavelengths, allowing plants to use light energy from the sun more efficiently than purple bacteria do. Moreover, the two photosystems are linked to one another, so the excitation of PS II is followed by the excitation of PS I with a photon of longer wavelength. The cumulative effect of the two photons absorbed in series is the storage of a larger amount of energy than is available to purple bacteria. Whereas the bacteria use H_2S and other reduced compounds as sources of hydrogen, green plants are able to split abundant water molecules. In the process they give off oxygen, which is vital to the existence and survival of animal species on the planet.

Chemists are studying the structure and kinetics of the photosynthetic reaction center both to understand the fundamentals of this important natural process and to design new materials that mimic nature's ability to harvest light energy at such high efficiency. Artificial photosynthesis may lead to carbon-based materials that will replace the silicon collectors in solar cells in the 21st century. This will help reduce human dependence on stored fossil fuels as energy sources in the future.

CHAPTER SUMMARY

Spectroscopy—the study of the interaction between matter and radiation—has provided us with most of the data on which our understanding of the nature of molecules and reactions has been built. Each of the different kinds of motions of a molecule is quantized, and each one can be selectively excited by the absorption of electromagnetic radiation of the appropriate frequency. The energies of the motions increase in the order: nuclear spin flips, molecular rotations, molecular vibrations, electronic excitations. They are excited by radiofrequency, microwave, infrared, and visible and ultraviolet radiation, respectively. We obtain bond lengths and angles, bond strengths, and the energies of molecular orbitals from the spectra of molecules. Molecular spectroscopy is increasingly used in time-resolved and imaging modes, allowing us to follow molecular dynamics on time scales as short as 10^{-14} seconds and on length scales as short as 100 nm. Atmospheric chemistry is a vibrant science that is attacking problems of truly global scope—ozone destruction, air pollution, and global warming. Understanding the fundamentals of molecular spectroscopy and photochemistry is vital for us to make progress in solving these important problems.

CUMULATIVE EXERCISE

Bromine

Elemental bromine is a brownish red liquid that was first isolated in 1826. It is currently produced by oxidation of bromide ion in natural brines with elemental chlorine.

(a) What is the ground-state configuration of the valence electrons of bromine molecules (Br_2)? Is bromine paramagnetic or diamagnetic?

(b) What is the electron configuration of the Br_2^+ molecular ion? Is its bond stronger or weaker than that in Br_2? What is its bond order?

(c) Bromine compounds have been known and used for centuries. The deep purple color that symbolized imperial power in ancient Rome originated with the compound dibromoindigo, which was extracted in tiny quantities from purple snails (about 8000 snails per gram of compound). What color and maximum wavelength of *absorbed* light would give a deep purple (violet) color?

(d) What excited electronic state is responsible for the brownish red color of bromine? Refer to Figures 6.16 and 20.11.

(e) The two naturally occurring isotopes of bromine are ^{79}Br and ^{81}Br, with masses of 78.918 and 80.916 u, respectively. The wavelength of the $J = 0$ to $J = 1$ rotational transition in $^{79}Br^{81}Br$ is measured to be 6.18 cm. Use this information to calculate the bond length in the Br_2 molecule, and compare the result with that listed in Table 3.3.

(f) The wavelength of the vibrational transition in the $^{79}Br^{81}Br$ molecule is 3.09×10^{-5} m. Calculate the force constant for the bond in this molecule.

(g) The action of light on bromine compounds released into the air (such as by leaded gasoline) causes the formation of the BrO radical. Give the bond order of this species by comparing it with the related radical OF.

(h) There is concern that synthetic bromine-containing compounds, in addition to chlorofluorocarbons, are helping to destroy ozone in the stratosphere. The BrO [see part (g)] can take part with ClO in the following catalytic cycle:

$$Cl + O_3 \longrightarrow ClO + O_2$$

$$Br + O_3 \longrightarrow BrO + O_2$$

$$ClO + BrO \longrightarrow Cl + Br + O_2$$

Write the overall equation for this cycle.

Answers

(a) $(\sigma_{g4s})^2(\sigma^*_{u4s})^2(\sigma_{g4p_z})^2(\pi_{u4p})^4(\pi^*_{g4p})^4$; diamagnetic

(b) $(\sigma_{g4s})^2(\sigma^*_{u4s})^2(\sigma_{g4p_z})^2(\pi_{u4p})^4(\pi^*_{g4p})^3$; stronger; bond order is $\frac{3}{2}$ versus 1

(c) Yellow light, near 530 nm (see Figures 20.12 and 4.3).

(d) The lowest energy excited state, which arises from the excitation of an electron from the filled π^*_{g4p} orbital to the unfilled σ^*_{u4pz} orbital

(e) 2.28 Å (from Table 3.3: 2.286 Å)

(f) 247 J Nm^{-1}

(g) $\frac{3}{2}$ order

(h) Overall: $2\ O_3 \longrightarrow 3\ O_2$

CHAPTER REVIEW

Regions of the Electromagnetic Spectrum and Associated Molecular Transitions

- Radiofrequency—nuclear spin states, NMR spectroscopy
- Microwave—molecular rotations, bond lengths and bond angles
- Infrared—molecular vibrations, force constants
- Visible, ultraviolet—valence electron excitations

Experimental Methods in Molecular Spectroscopy

- A spectrophotometer measures the sample transmittance, $T = I_S/I_R$, which is the ratio of the light transmitted through the sample and that transmitted through a reference.
- The absorbance, $A = -\ln T$, measures the strength of the interaction between the molecule and the radiation.
- For most samples the absorption strength is proportional to the concentration, as expressed by the Beer–Lambert law: $A = \epsilon c \ell$, where A is the absorbance (dimensionless), ϵ is the molar extinction coefficient (L mol^{-1} cm^{-1}), c is the concentration in mols L^{-1}, and ℓ is the path length in cm.

Thermal Population of Molecular Energy Levels According to the Boltzmann Distribution

- Nuclear spin states are equally populated.
- Many excited rotational states are populated.
- Only the ground vibrational and electronic states are significantly populated at room temperature.

Rotational Spectroscopy

- Energy levels are given by

$$E_{\text{rot},J} = \frac{h^2}{8\pi^2 I} J(J+1) = hBJ(J+1) \qquad J = 0, 1, 2, \ldots$$

- Spectra consist of a series of lines equally spaced by $2B$, the rotational constant, which depends on the molecule's moment of inertia $I = \mu R_e^2$.
- Because many levels are populated at room temperature the intensities of the lines form an envelope, the shape of which can be used to measure temperature.
- Only molecules with permanent dipole moments absorb microwave radiation to excite rotations.

Vibrational Spectroscopy

- The harmonic oscillator model and the Morse potential are used to describe the vibrational motions of molecules.
- The dipole moment of the molecule must change during a vibration for that vibrational motion to be excited by the absorption of infrared radiation.

Excited Electronic States and Electronic Spectroscopy

- Electrons may be excited from occupied π molecular orbitals to unoccupied π^* molecular orbitals by absorbing visible or ultraviolet radiation.
- The particle-in-a-box model provides a good qualitative understanding of the orbital energies and absorption wavelengths of conjugated π systems, those that contain alternating single and double carbon-carbon bonds.

The Fate of Excited Electronic States

- Excited electronic states may return to the ground state of the same total spin by emitting radiation called fluorescence—it is generally intense and the excited state is very short-lived (nanoseconds).
- Excited electronic states may return to states with different total spin by emitting radiation called phosphorescence—it is generally very weak and the excited state lives for a rather long time (seconds).

Laser—Light Amplification by Stimulated Emission of Radiation

- The relative population of two electronic energy levels must be inverted from its equilibrium value to create an excess of excited states.
- Spontaneous emission initiates a chain reaction in which a series of stimulated emissions amplifies the number of photons enormously.

Time-resolved Spectroscopy

- Pulsed lasers with pulse durations as short as a few femtoseconds (10^{-15} s) provide detailed snapshots of fundamental physical and chemical processes such as the relaxation of excited electronic states and the formation and breaking of chemical bonds.

NMR Spectroscopy

- The energy of nuclear spin states is split in an externally applied magnetic field. The absorption of radiofrequency radiation can cause transitions between these two states.
- The chemical shift is a measure of the chemical environment of the proton.
- Spin-spin splitting identifies the number of equivalent protons attached to a carbon atom and is useful in identifying functional groups.

Stratospheric Chemistry

- Ozone is formed and destroyed in a series of stratospheric reactions. Its steady state concentration is described by the Chapman cycle.
- A class of refrigerants and propellants called chlorofluorocarbons (CFCs) produce chlorine radicals that catalytically destroy ozone.
- The ozone hole over Antarctica results from the storage of chlorine-containing molecules on the surfaces of polar stratospheric clouds during the Antarctic winter followed by sudden photolysis and release in the spring.

Tropospheric Chemistry

- Oxides of nitrogen (NO_x) and sulfur (SO_x), produced by the combustion of fossil fuels in vehicles and power plants are pollutants and also precursors to acid rain.

The Greenhouse Effect

- Energy provided by ultraviolet and visible radiation from the sun warms the Earth, which re-radiates in the infrared region of the electromagnetic spectrum.
- CO_2 absorbs infrared radiation very efficiently, causing the planet to be much warmer that if there were no CO_2 in the atmosphere.
- Atmospheric CO_2 concentrations and global temperatures are strongly correlated, as established by ice core data.
- The rise in CO_2 concentrations since the Industrial Revolution is thought to be largely due to human activities, and there is considerable concern over the possibility of significant warming in the coming decades.

CONCEPTS & SKILLS

After studying this chapter and working the problems that follow, you should be able to:

1. Apply the Beer–Lambert law to absorption spectra (Section 20.1, Problems 1–4).
2. Discuss the Boltzmann distribution of population among possible molecular quantum states (Sections 20.1 and 20.2, Problems 9, 10, 15, 16).
3. Relate the moments of inertia, bond lengths, and vibrational force constants of diatomic molecules to their rotational and vibrational spectra (Section 20.2, Problems 5–8, 11–14).
4. Describe the preparation of excited electronic states by absorption of radiation and the subsequent flow of energy (Section 20.3, Problems 17–24).
5. Describe the operation of laser devices (Section 20.3).
6. Discuss several new techniques in molecular spectroscopy and the purposes they serve (Section 20.3).
7. Outline the principles of magnetic resonance spectroscopy (Section 20.4, Problems 25–26).
8. Indicate processes in the atmosphere that are beneficial and those that are potentially damaging to the ecosystem (Section 20.5, Problems 27–30).
9. Explain the mechanism by which photosynthetic bacteria and green plants convert light energy to chemical energy. (Section 20.6, Problems 31–32).

KEY EQUATIONS

$$-\ln \frac{I_S}{I_R} = c\epsilon\ell = A \qquad \text{(Section 20.1)}$$

$$P(E_i) = N(E_i)/N_0 \propto g(E_i)\exp(-E_i/k_BT) \qquad \text{(Section 20.1)}$$

$$P(E_i)/P(E_f) = g(E_i)/g(E_f)\exp[-(E_i - E_f)/k_BT] \qquad \text{(Section 20.1)}$$

$$I = \mu R_e^2 \qquad \text{(Section 20.2)}$$

$$E_{\text{rot},J} = \frac{h^2}{8\pi^2 I}J(J+1) = hBJ(J+1) \qquad J = 0, 1, 2, \ldots \qquad \text{(Section 20.2)}$$

$$\nu = \frac{\Delta E}{h} = \frac{h}{8\pi^2 I}[J_f(J_f + 1) - J_i(J_i + 1)]$$

$$= \frac{h}{8\pi^2 I}[(J_i + 1)(J_i + 2) - J_i(J_i + 1)]$$

$$= \frac{h}{4\pi^2 I}(J_i + 1) = 2B(J_i + 1) \qquad J_i = 0, 1, 2, \ldots \qquad \text{(Section 20.2)}$$

$$E_{\text{vib},n} = h\nu\left(n + \tfrac{1}{2}\right) \qquad n = 0, 1, 2, \ldots \qquad \text{(Section 20.2)}$$

$$\delta = \frac{H_s - H_r}{H_r} \times 10^6 \qquad \text{(Section 20.4)}$$

PROBLEMS

Answers to problems whose numbers are boldface appear in Appendix G. Problems that are more challenging are indicated with asterisks.

General Aspects of Molecular Spectroscopy

1. The percentage transmittance of light at 250 nm through a certain aqueous solution is 20.0% at 25°C. The experimental cell length is 1.0 cm, and the concentration of the solution is 5×10^{-4} mol L^{-1}. Calculate the absorbance. Calculate the molar absorption coefficient.

2. Beer's law is used to measure the concentration of species in solutions, once a "standardization curve" has been prepared for that species. In one such experiment, percent transmission was measured for a series of solutions with known concentrations and the results were as follows:

Concentration μg mL^{-1}	1.0	2.0	3.0	4.0	5.0
Transmission, percent	66.8	44.7	29.2	19.9	13.3

Plot these results to obtain the standardization curve. An unknown concentration of the same species, measured in the same transmission cell, transmitted 35% of the incoming light. Calculate the concentration of the unknown solution.

3. Beer's law can be used to determine the concentration of two substances A and B in solution, provided they do not react or interact, so they absorb radiation independently. The following data were obtained for A and B in three different solutions:

	[A] mol L^{-1}	[B] mol L^{-1}	% Transmittance at λ = 400 nm	% Transmittance at λ = 500 nm
Solution 1	0.0010	0	10.0	60.0
Solution 2	0	0.0050	80.0	20.0
Solution 3	?	?	40.0	50.0

Calculate the concentrations of A and B in Solution 3.

4. The absorption of ultraviolet light by proteins at wavelength 280 nm is caused mostly by the amino acids tyrosine and tryptophan along the protein molecular chains. The molecular absorption coefficients for these two amino acids are:

$$\epsilon_{tryp}^{280} = 5690 \text{ L cm}^{-1} \text{ mol}^{-1}$$

$$\epsilon_{tyro}^{280} = 1280 \text{ L cm}^{-1} \text{ mol}^{-1}$$

Experiments are carried out on a protein with molecular weight 26,000, which contains two units of tryptophan and six units of tyrosine along the chain. The absorption is measured in a cell 1 cm long, and the protein concentration is 1.0 mg mL^{-1}. Calculate the absorbance and the percent transmission.

Vibrations and Rotations of Molecules: Infrared and Microwave Spectroscopy

5. Use data from Tables 19.1 and 3.3 to predict the energy spacing between the ground state and the first excited rotational state of the $^{14}N^{16}O$ molecule.

6. Use data from Tables 19.1 and 3.3 to predict the energy spacing between the ground state and the first excited rotational state of the $^{1}H^{19}F$ molecule.

7. The first three absorption lines in the pure rotational spectrum of gaseous $^{12}C^{16}O$ are found to have the frequencies 1.15×10^{11}, 2.30×10^{11}, and 3.46×10^{11} s^{-1}. Calculate:
(a) The moment of inertia I of CO (in kg m^2)
(b) The energies of the $J = 1$, $J = 2$, and $J = 3$ rotational levels of CO, measured from the $J = 0$ state (in joules)
(c) The C—O bond length (in angstroms)

8. Four consecutive absorption lines in the pure rotational spectrum of gaseous $^{1}H^{35}Cl$ are found to have the frequencies 2.50×10^{12}, 3.12×10^{12}, 3.74×10^{12}, and 4.37×10^{12} s^{-1}. Calculate:
(a) The moment of inertia I of HCl (in kg m^2)
(b) The energies of the $J = 1$, $J = 2$, and $J = 3$ rotational levels of HCl, measured from the $J = 0$ state (in joules)
(c) The H—Cl bond length (in angstroms)
(d) The initial and final J states for the observed absorption lines

9. In Example 20.1, we determined that the moment of inertia of the NaH molecule is 5.70×10^{-47} kg m^2.
(a) Calculate the relative population of the $J = 5$ level and the ground state at 25°C.
(b) Calculate the relative population of the $J = 15$ level and the ground state at 25°C.
(c) Calculate the relative population of the $J = 25$ level and the ground state at 25°C.

10. The linear molecule N_2O has been examined by rotational spectroscopy. The N—N bond length was found to be 1.126 Å, and the N—O bond length was found to be 1.191 Å. From these results, we can calculate the moment of inertia to be 66.8×10^{-47} kg m^2 as shown in Fig. 20.6.
(a) Calculate the relative population of the $J = 5$ level and the ground state at 25°C.
(b) Calculate the relative population of the $J = 15$ level and the ground state at 25°C.
(c) Calculate the relative population of the $J = 25$ level and the ground state at 25°C.

11. The Li_2 molecule (^{7}Li isotope) shows a very weak infrared line in its vibrational spectrum at a wavelength of 2.85×10^{-5} m. Calculate the force constant for the Li_2 molecule.

12. The Na_2 molecule (^{23}Na isotope) shows a very weak infrared line in its vibrational spectrum at a wavelength of 6.28×10^{-5} m. Calculate the force constant for the Na_2 molecule, and compare your result with that of problem 11. Give a reason for any difference.

13. The "signature" infrared absorption that indicates the presence of a C—H stretching motion in a molecule occurs at wavelengths near 3.4×10^{-6} m. Use this information to estimate the force constant of the C—H stretch. Take the reduced mass in this motion to be approximately equal to the mass of the hydrogen atom (a good approximation when the H atom is attached to a heavy group).

14. Repeat the calculation of the preceding problem for the N—H stretch, where absorption occurs near 2.9×10^{-6} m. Which bond is stiffer: N—H or C—H?

15. Estimate the ratio of the number of molecules in the first excited vibrational state of the molecule N_2 to the number in the ground state, at a temperature of 450 K. The vibrational frequency of N_2 is $7.07 \times 10^{13}\ s^{-1}$.

16. The vibrational frequency of the ICl molecule is $1.15 \times 10^{13}\ s^{-1}$. For every million ($1.00 \times 10^6$) molecules in the ground vibrational state, how many will be in the first excited vibrational state at a temperature of 300 K?

Excited Electronic States: Electronic Spectroscopy of Molecules

17. Suppose that an ethylene molecule gains an additional electron to give the $C_2H_4^-$ ion. Will the bond order of the carbon–carbon bond increase or decrease? Explain.

18. Suppose that an ethylene molecule is ionized by a photon to give the $C_2H_4^+$ ion. Will the bond order of the carbon–carbon bond increase or decrease? Explain.

19. The color of the dye "indanthrene brilliant orange" is evident from its name. In what wavelength range would you expect to find the maximum in the absorption spectrum of this molecule? Refer to the color spectrum in Figure 4.3.

20. In what wavelength range would you expect to find the maximum in the absorption spectrum of the dye "crystal violet"?

21. The structure of the molecule cyclohexene is shown below:

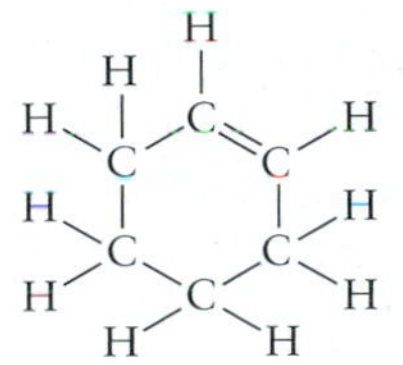

Does the absorption of ultraviolet light by cyclohexene occur at shorter wavelengths than in benzene? Explain.

22. The naphthalene molecule has a structure that corresponds to two benzene molecules fused together:

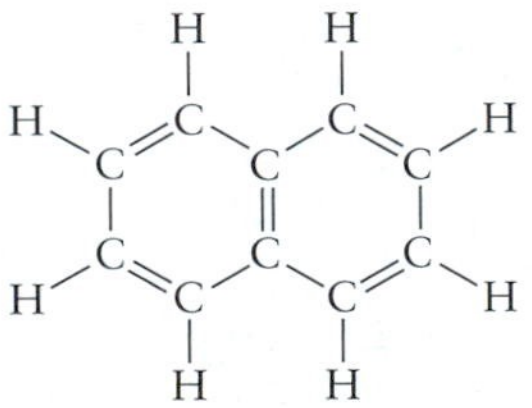

The π-electrons in this molecule are delocalized over the entire molecule. The wavelength of maximum absorption in the UV-visible part of the spectrum in benzene is 255 nm. Is the corresponding wavelength shorter or longer than 255 nm for naphthalene?

23. Use data from Table 3.3 to give an upper bound on the wavelengths of light that are capable of dissociating a molecule of ClF.

24. Use data from Table 3.3 to give an upper bound on the wavelengths of light that are capable of dissociating a molecule of ICl.

Nuclear Magnetic Resonance Spectroscopy

25. Give the number of peaks and the relative peak areas that should be observed in the low-resolution proton magnetic resonance spectra of the following molecules:

$$CH_3CH_2CH_2CH_3,\ CH_3OCH_3,\ CH_3NHCH_3.$$

26. The organic compound 1,4-dimethylbenzene (also known as *p*-xylene) has the formula $(CH_3)_2C_6H_4$. Its structure has two CH_3 (methyl) groups substituted at opposite positions on the benzene (C_6H_6) ring. Predict the number of peaks in the low-resolution proton NMR spectrum of this compound and the relative areas of the peaks.

Introduction to Atmospheric Photochemistry

27. The bond dissociation energy of a typical C—F bond in a chlorofluorocarbon is approximately 440 kJ mol^{-1}. Calculate the maximum wavelength of light that can photodissociate a molecule of CCl_2F_2, breaking such a C—F bond.

28. The bond dissociation energy of a typical C—Cl bond in a chlorofluorocarbon is approximately 330 kJ mol^{-1}. Calculate the maximum wavelength of light that can photodissociate a molecule of CCl_2F_2, breaking such a C—Cl bond.

29. Draw a Lewis diagram(s) for the ozone molecule (O_3). Determine the steric number and hybridization of the central oxygen atom, and identify the molecular geometry. Describe the nature of the π bonds and give the bond order of the O—O bonds in ozone.

30. The compounds carbon dioxide (CO_2) and sulfur dioxide (SO_2) are formed by the burning of coal. Their apparently similar formulas mask underlying differences in molecular structure. Determine the shapes of these two types of molecules, identify the hybridization at the central atom of each, and compare the natures of their π bonds.

Photosynthesis

31. One way in which photosynthetic bacteria store chemical energy is through the conversion of a compound called adenosine diphosphate (ADP), together with hydrogen phosphate ion, to adenosine triphosphate (ATP):

$$ADP^{3-} + HPO_4^{2-} + H_3O^+ \longrightarrow ATP^{4-} + 2\ H_2O$$

$$\Delta G = +34.5\ kJ\ (pH\ 7)$$

Suppose some chlorophyll molecules absorb 1.00 mol of photons of blue light with wavelength 430 nm. If *all* this energy could be used to convert ADP to ATP at room conditions and pH 7, how many molecules of ATP would be produced per photon absorbed? (The actual number is smaller because the conversion is not 100% efficient.)

32. Repeat the calculation of the preceding problem for red light with wavelength 700 nm.

ADDITIONAL PROBLEMS

33. What are the moments of inertia of $^1H^{19}F$ and $^1H^{81}Br$, expressed in kg m^2? Compute the spacings $\nu = \Delta E/h$ of the rotational states, in s^{-1}, between $J = 0$ and 1 and between $J = 1$ and 2. Explain, in one sentence, why the large change in mass from 19 to 81 causes only a small change in rotational energy differences.

34. The average bond length of a molecule can change slightly with vibrational state. In $^{23}Na^{35}Cl$, the frequency of light absorbed in a change from the $J = 1$ to the $J = 2$ rotational state in the ground vibrational state ($n = 0$) was measured to be $\nu = 2.60511 \times 10^{10}\ s^{-1}$, and that for a change from $J = 1$ to $J = 2$ in the first excited vibrational state ($n = 1$) was $\nu = 2.58576 \times 10^{10}\ s^{-1}$. Calculate the average bond lengths of NaCl in these two vibrational states, taking the relative atomic mass of ^{23}Na to be 22.9898 and that of ^{35}Cl to be 34.9689.

35. The vibrational frequencies of $^{23}Na^{1}H$, $^{23}Na^{35}Cl$, and $^{23}Na^{127}I$ are $3.51 \times 10^{13}\ s^{-1}$, $1.10 \times 10^{13}\ s^{-1}$, and $0.773 \times 10^{13}\ s^{-1}$, respectively. Their bond lengths are 1.89 Å, 2.36 Å, and 2.71 Å. What are their reduced masses? What are their force constants? If NaH and NaD have the same force constant, what is the vibrational frequency of NaD? D is ^{2}H.

36. Recall that nuclear spin states in nuclear magnetic resonance are typically separated by energies of 2×10^{-5} kJ mol^{-1} to 2×10^{-4} kJ mol^{-1}. What are the ratios of occupation probability between a pair of such levels at thermal equilibrium and a temperature of 25°C?

37. The vibrational temperature of a molecule prepared in a supersonic jet can be estimated from the observed populations of its vibrational levels, assuming a Boltzmann distribution. The vibrational frequency of HgBr is $5.58 \times 10^{12}\ s^{-1}$, and the ratio of the number of molecules in the $n = 1$ state to the number in the $n = 0$ state is 0.127. Estimate the vibrational temperature under these conditions.

38. An electron in the π orbital of ethylene (C_2H_4) is excited by a photon to the π^* orbital. Do you expect the equilibrium bond length in the excited ethylene molecule to be greater or less than that in ground-state ethylene? Will the vibrational frequency in the excited state be higher or lower than in the ground state? Explain your reasoning.

39. One isomer of retinal is converted to a second isomer by the absorption of a photon:

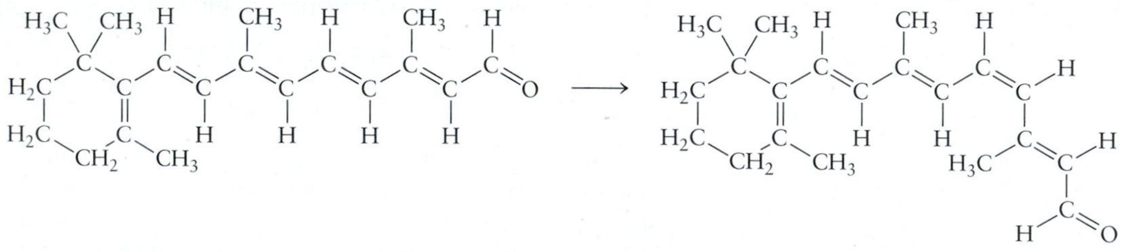

This process is a key step in the chemistry of vision. Although free retinal (in the form shown to the left of the arrow) has an absorption maximum at 376 nm, in the ultraviolet region of the spectrum this absorption shifts into the visible range when the retinal is bound in a protein, as it is in the eye.
   (a) How many of the C=C bonds are *cis* and how many are *trans* in each of the preceding structures? (When assigning labels, consider the relative positions of the two largest groups attached at each double bond.) Describe the motion that takes place upon absorption of a photon.
   (b) If the ring and the —CHO group in retinal were replaced by $-CH_3$ groups, would the absorption maximum in the molecule shift to longer or shorter wavelengths?

* 40. The ground-state electron configuration of the H_2^+ molecular ion is $(\sigma_{g1s})^1$.
   (a) A molecule of H_2^+ absorbs a photon and is excited to the σ_{u1s}^* molecular orbital. Predict what happens to the molecule.
   (b) Another molecule of H_2^+ absorbs even more energy in an interaction with a photon and is excited to the σ_{g2s} molecular orbital. Predict what happens to this molecule.

* 41. (a) Draw a Lewis diagram for formaldehyde (H_2CO), and decide the hybridization of the central carbon atom.
   (b) Formulate the molecular orbitals for the molecule.
   (c) A strong absorption is observed in the ultraviolet region of the spectrum and is attributed to a $\pi \longrightarrow \pi^*$ transition. Another, weaker transition is observed at lower frequencies. What electronic excitation causes the weaker transition?

42. Write balanced chemical equations that describe the formation of nitric acid and sulfuric acid in rain, starting with the sulfur in coal and the oxygen, nitrogen, and water vapor in the atmosphere.

43. Compare and contrast the roles of ozone (O_3) and nitrogen dioxide (NO_2) in the stratosphere and in the troposphere.

44. Describe the greenhouse effect and its mechanism of operation. Give three examples of energy sources that contribute to increased CO_2 in the atmosphere and three that do not.

45. Draw a schematic diagram of the steps in bacterial photosynthesis, numbering them in sequence and showing the approximate spatial relations of the involved molecules.

46. Do you expect the energy of the special pair of bacteriochlorophyll molecules to be higher or lower than the energy of an isolated bacteriochlorophyll? (*Hint:* Think about the analogy between the mixing of atomic orbitals to make molecular orbitals and the mixing of molecular orbitals on two nearby molecules.)

CUMULATIVE PROBLEMS

47. At thermal equilibrium, is the rate as a molecule is excited from $n = 0$ to the $n = 1$ level greater than or less than the rate for the reverse process? What is the ratio of the rate constants? (*Hint:* Think of the analogy with the chemical equilibrium between two species.)

* **48.** It is important to know the dissociation constant of an indicator in order to use it properly in acid–base titrations. Spectrophotometry can be used to measure the concentration of these intensely colored species in acidic versus basic solutions, and from these data the equilibrium between the acidic and basic forms can be calculated. In one such study on the indicator *m*-nitrophenol, a 6.36×10^{-4} M solution was examined by spectrophotometry at 390 nm and 25°C in the following experiments. In highly acidic solution, where essentially all the indicator was in the form HIn, the absorbance was 0.142. In highly basic solution, where essentially all of the indicator was in the form In^-, the absorbance was 0.943. In a further series of experiments, the pH was adjusted using a buffer solution of ionic strength I, and absorbance was measured at each pH value. The following results were obtained:

pH	I	A
8.321	0.10	0.527
8.302	0.08	0.518
8.280	0.06	0.505
8.251	0.04	0.493
8.207	0.02	0.470

Calculate pK_a for the indicator at each value of ionic strength.

49. The hydroxyl radical has been referred to as the "chief clean-up agent in the troposphere." Its concentration is approximately zero at night and becomes as high as 1×10^7 molecules per cm^3 in highly polluted air.

(a) Calculate the maximum mole fraction and partial pressure of OH in polluted air at 25°C and atmospheric pressure.

(b) Write an equation for the reaction of HO with NO_2 in the atmosphere. How does the oxidation state of nitrogen change in this reaction? What is the ultimate fate of the product of this reaction?

50. In unpolluted air at 300 K, the hydroxyl radical OH reacts with CO with a bimolecular rate constant of 1.6×10^{11} L mol^{-1} s^{-1} and with CH_4 with a rate constant of 3.8×10^9 L mol^{-1} s^{-1}. Take the partial pressure of CO in air to be constant at 1.0×10^{-7} atm and that of CH_4 to be 1.7×10^{-6} atm, and assume that these are the primary mechanisms by which OH is consumed in the atmosphere. Calculate the half-life of OH under these conditions.

UNIT VI

Materials

Courtesy of Imago Scientific Instruments Corporation

Cross-sectional view of a three-dimensional map of dopant atoms (light blue spheres) implanted into a typical silicon transistor structure. Red dots represent the silicon atoms (only 2% are shown for clarity) and the gray spheres represent a native silicon dioxide layer located at the interface between the crystalline silicon substrate and layer of deposited polycrystalline silicon.

Throughout history, the discovery of new materials from which to fashion the structures, machines, and devices of everyday life has set off great change in human affairs. Modern science and engineering—with chemistry in the central role—provide routes for modifying properties of materials to meet specific applications and for synthesizing and processing new materials designed from the beginning to have specific properties. The mechanical, thermal, electrical, and optical properties of a material depend on the extended nano-structural arrangement of chemical bonds within it. This arrangement can be created, modified, and tailored through chemical reactions. The contemporary disciplines of materials chemistry, solid state chemistry, and materials science and engineering—among the most active branches of chemistry today—all rely on the "properties↔structure↔reactions" correlation to make the leap from chemical bonding in isolated molecules (Unit II) to engineering applications.

UNIT CHAPTERS

UNIT GOALS

- To study the relationship between crystal symmetry and atomic-level structure as revealed by x-ray diffraction experiments
- To survey chemical bonding in classes of solids and correlate bonding with the properties of solids
- To explore three essential classes of materials—ceramics, optical and electronic materials, and polymers
- To illustrate the role of modern chemistry in measuring properties and identifying applications for both natural and manufactured materials

CHAPTER 21

Structure and Bonding in Solids

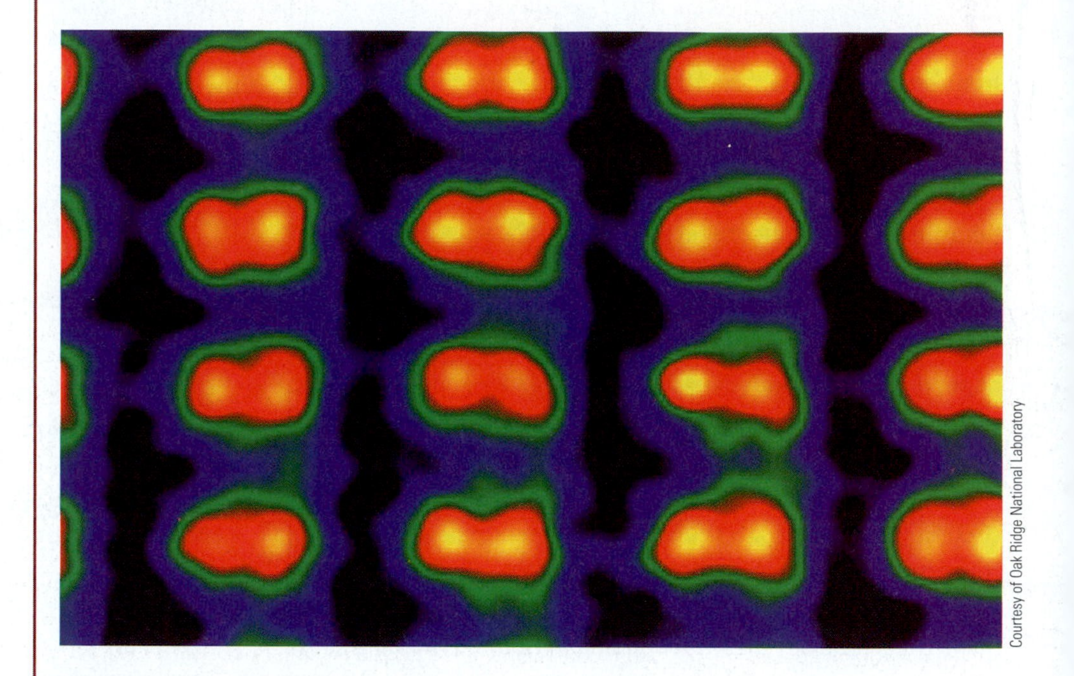

Courtesy of Oak Ridge National Laboratory

Pairs of silicon atoms separated by only 0.78 Å are clearly resolved in this ultrahigh-resolution electron microscope image.

In this chapter we begin a radical departure from the emphasis on the single molecule that permeates the previous twenty chapters. Unit II describes the chemical bonding and structure of single, isolated molecules. Unit III explains the macroscopic properties of gases and liquids through intermolecular forces that originate in the structures of individual molecules. While the behavior of many molecules contributes to these properties, the mechanism of their contribution is essentially that a few molecules come close enough to a "target" molecule to experience the intermolecular forces it sets up. Units IV and V deal with the equilibrium and rate aspects of chemical reactions as consequences of collisions between individual molecules.

The behavior of solids is a different story altogether. Chapter 10 identifies *rigidity* as the unique characteristic of solids, in dramatic contrast to the fluidity of gases and liquids. A rigid material retains its shape when an external mechanical force, called stress, is applied. A rigid material shows structural strength by not flowing under stress. All classes of solids behave this way because—with one exception—they are not collections of molecules held together by intermolecular

FIGURE 21.1 Microcrystals of realgar, As_4S_4.

forces. Rather, solids are extended arrays of strong chemical bonds between atoms, almost as if they were "super-molecules." These arrays extend over macroscopic distances, and it is the *collective behavior* of this set of chemical bonds that imparts strength and rigidity to solids.

Solids whose structures are highly ordered and symmetrical over macroscopic distances are called **crystals.** Although the symmetry and beauty of crystals have always excited curiosity and wonder, the science of **crystallography** began only in the latter part of the 18th century. In those closing days of the Age of Enlightenment—while Lavoisier led the modern approach to chemistry—another brilliant French thinker established the fundamental laws of crystallography. René-Just Haüy was struck by the observation that when he accidentally dropped a crystal of calcite (a form of calcium carbonate), it fractured into smaller crystals with the same interfacial angles between their planar surfaces as in the original crystal. The statement that constant interfacial angles are observed when crystals are cleaved is now known as **Haüy's law.** Haüy concluded that the outward symmetry of crystals (Fig. 21.1) implies a highly regular internal structure and the existence of a smallest crystal unit. His inferences were correct. What distinguishes the crystalline state from the gaseous and liquid states is the nearly perfect positional order of the atoms, ions, or molecules in crystals. This crystalline order has been confirmed experimentally by x-ray diffraction and explained theoretically by the quantum theory of solids.

We begin this chapter with a look at the microscopic structure of a perfect crystal, and establish the methods for determining structure and the language for describing it. We then examine the types of chemical bonding in solids, identifying the forces that hold together different kinds of solids. The perfect crystal is the idealized model for investigations in solid state science, just as the ideal gas is the starting point for studies of fluid behavior. We use it as the point of reference for describing less ordered condensed phases of matter—defective crystals and amorphous solids—in terms of their deviations from perfect order and the consequent changes in properties. We end the chapter with a brief introduction to diffusion in solids, the mechanism by which a free atom migrates through an extended solid state structure. Diffusion has great influence on the rate and equilibrium of chemical reactions in the solid state and on tailoring the properties of the solids through carefully controlled incorporation of impurities.

21.1 Crystal Symmetry and the Unit Cell

The unifying aspect of crystal structure is the repetition, over long distances, of the same basic structural features in the arrangement of the atoms. The most fundamental way to characterize and classify these structures is based on the numbers and kinds of their **symmetry elements.** When the result of rotation, reflection, or inversion of an object can be exactly superimposed on the original object—that is, matched point for point to the original object—the structure is said to contain the corresponding symmetry element. Examples include an axis of rotation, a plane of reflection (mirror plane), or a central point (inversion center), as shown in Figure 21.2. These symmetry operations can be applied to geometrical shapes, to physical objects, and to molecular structures.

Consider a cube as an example. Suppose the center of the cube is placed at the origin of its coordinate system and the symmetry operations that transform it into identity with itself are counted (Fig. 21.3). The x, y, and z coordinate axes are 4-fold axes of rotational symmetry, denoted by C_4, because a cube that is rotated through a multiple of 90° (= 360°/4) about any one of these axes is indistinguishable from the original cube. Similarly, a cube has four 3-fold axes of rotational

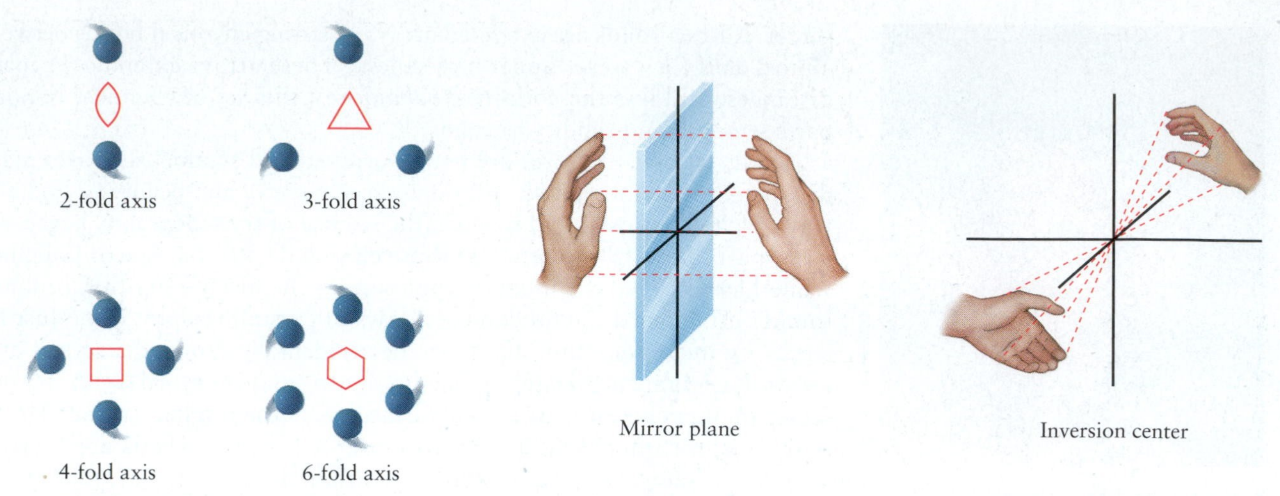

FIGURE 21.2 Three types of symmetry operation: rotation about an *n*-fold axis, reflection in a plane, and inversion through a point.

symmetry, designated C_3, that are the body diagonals of the cube, connecting opposite vertices. In addition, a cube has six 2-fold rotational axes of symmetry, defined by the six axes that pass through the centers of edges and through the coordinate origin. Next, the cube has nine mirror planes of symmetry (designated by the symbol m), which reflect any point in one half of the cube into an equivalent point in the other half. Finally, a cube has a center of inversion (reflection through a point, designated i).

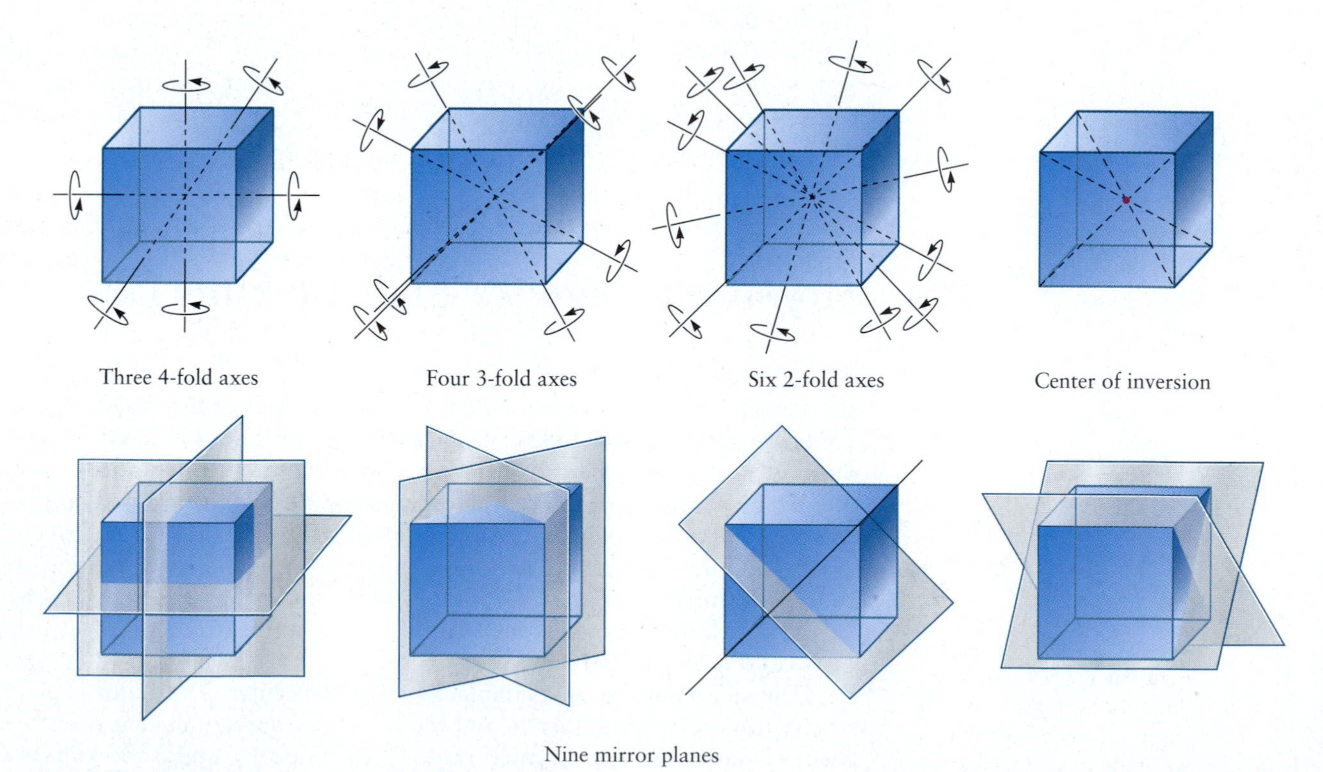

FIGURE 21.3 Symmetry operations acting on a cube. The various rotations of the cube are suggested by the curved arrows around the rotational axes.

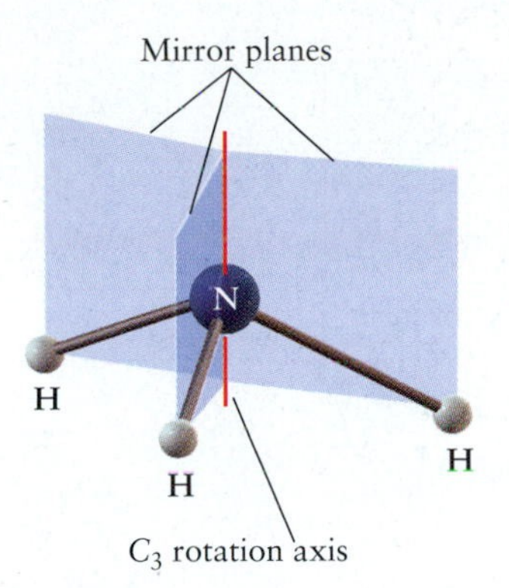

FIGURE 21.4 An ammonia molecule has a 3-fold axis of rotation and three mirror planes of symmetry. Each of the N—H bonds lies in a mirror plane.

EXAMPLE 21.1

Identify the symmetry elements of the ammonia molecule (NH_3).

SOLUTION

If the ammonia molecule is drawn as a pyramid with the nitrogen atom at the top (Fig. 21.4), then the only axis of rotational symmetry is a 3-fold axis passing downward through the N atom. Three mirror planes intersect at this 3-fold axis.

Related Problems: 3, 4

Unit Cells in Crystals

The symmetry operations just described can be applied to crystals as well as to individual molecules or shapes. Identical sites within a crystal recur regularly because of long-range order in the organization of the atoms. The three-dimensional array made up of all the points within a crystal that have the same environment in the same orientation is a **crystal lattice.** Such a lattice is an abstraction "lifted away" from a real crystal, embodying the scheme of repetition at work in that crystal. The lattice of highest possible symmetry is that of the **cubic system.** This lattice is obtained by filling space with a series of identical cubes, which are the **unit cells** in the system. A single unit cell contains all structural information about its crystal, because in principle the crystal could be constructed by making a great many copies of a single original unit cell and stacking them in a three-dimensional array. Unit cells fill space.

Other crystal systems besides cubic can be defined by their own unique unit cells. Constraints of symmetry permit only seven types of three-dimensional lattices. Each type has a unit cell with the shape of a parallelepiped (Fig. 21.5), whose size and shape are fully described by three edge lengths (a, b, and c) and the three angles between those edges (α, β, and γ). These lengths and angles are the **cell constants.** The symmetry that defines each of the seven crystal systems imposes conditions on the shape of the unit cell summarized in relations among the cell constants (Table 21.1, Fig. 21.6). The unit cell chosen is the smallest unit that has all the symmetry elements of the crystal lattice; there is no benefit in using large cells once all the symmetry elements have been included. A unit cell of the minimum size is **primitive** and shares each of the eight **lattice points** at its corners with seven other unit cells, giving one lattice point per unit cell.

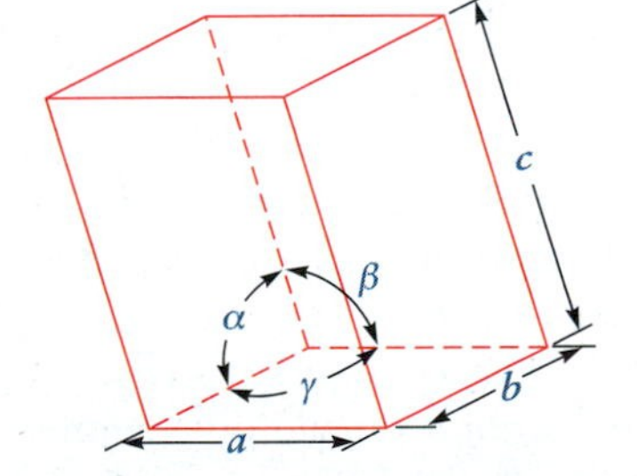

FIGURE 21.5 Unit cells always have three pairs of mutually parallel faces. Only six pieces of information are required to construct a scale model of a unit cell: the three cell edges (a, b, and c) and the three angles between the edges (α, β, and γ). By convention, γ is the angle between edges a and b, α the angle between b and c, and β the angle between a and c.

TABLE 21.1 The Seven Crystal Systems

Crystal System	Minimum Essential Symmetry	Conditions on Unit-Cell Edges and Angles
Hexagonal	One 6-fold rotation	$a = b$; $\alpha = \beta = 90°$, $\gamma = 120°$
Cubic	Four independent 3-fold rotations†	$a = b = c$; $\alpha = \beta = \gamma = 90°$
Tetragonal	One 4-fold rotation	$a = b$; $\alpha = \beta = \gamma = 90°$
Trigonal	One 3-fold rotation	$a = b = c$; $\alpha = \beta = \gamma \neq 90°$
Orthorhombic	Three mutually perpendicular 2-fold rotations	$\alpha = \beta = \gamma = 90°$
Monoclinic	One 2-fold rotation	$\alpha = \beta = 90°$
Triclinic	No symmetry required	None

†Each of these axes makes 70.53° angles with the other three.

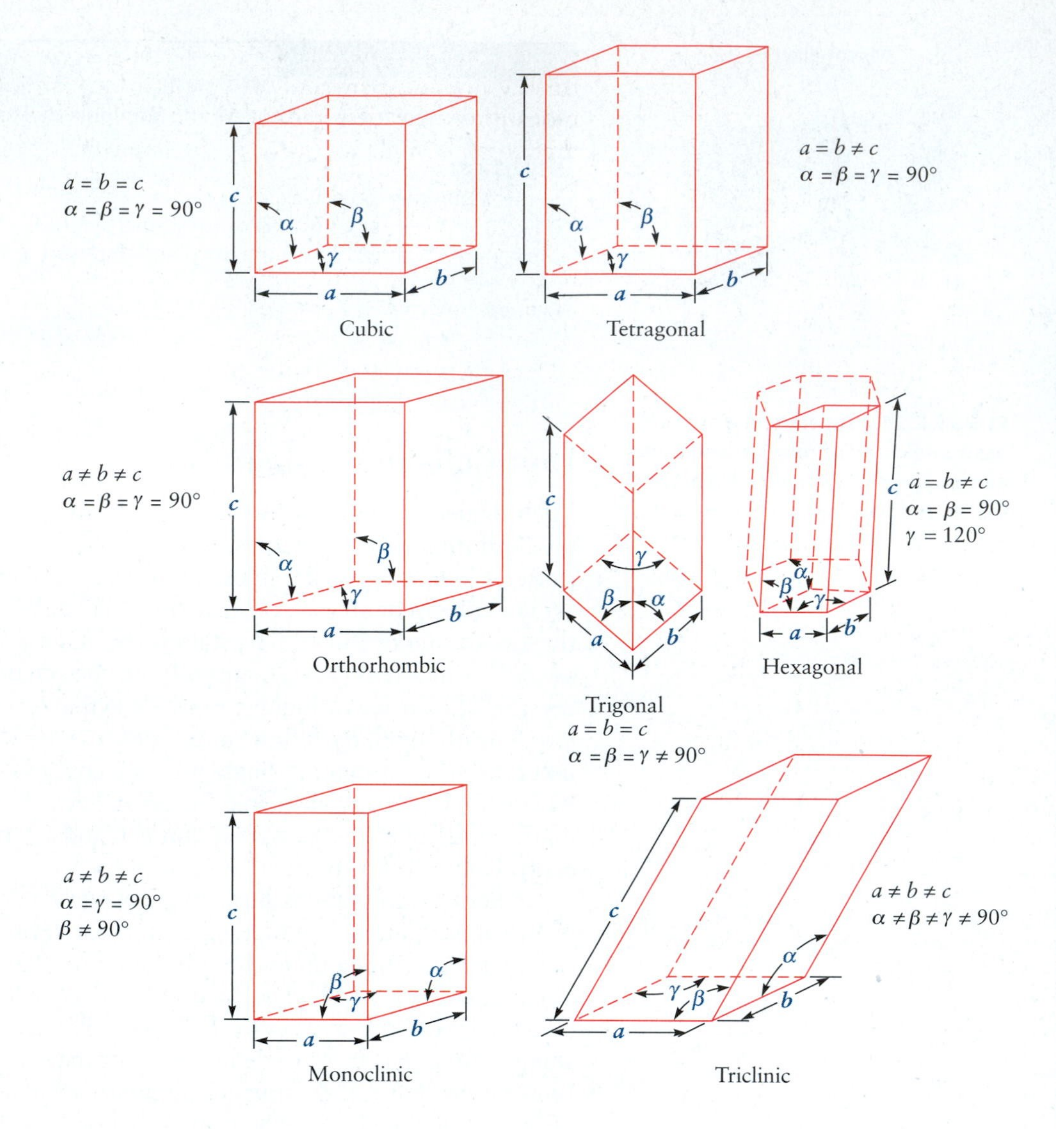

FIGURE 21.6 Shapes of the unit cells in the seven crystal systems. The more symmetric crystal systems have more symmetric cells.

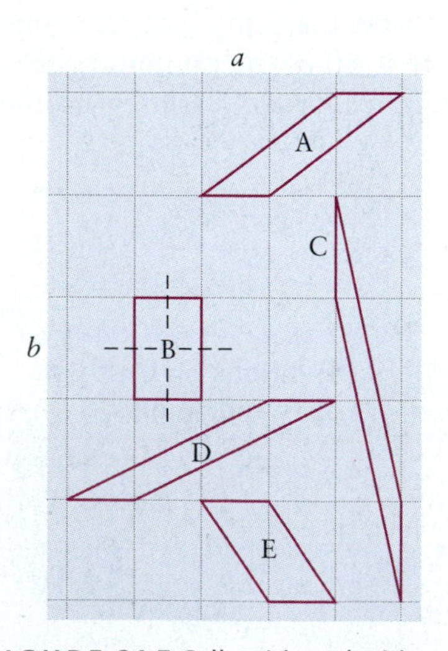

FIGURE 21.7 Cells with and without mirror planes in a two-dimensional lattice. Each cell has the same area, but only cell B possesses mirror planes in addition to the 2-fold rotation axes required by the lattice.

Other possible unit cells with the same volume (an infinite number, in fact) could be constructed, and each could generate the macroscopic crystal by repeated elementary translations, but only those shown in Figure 21.6 possess the symmetry elements of their crystal systems. Figure 21.7 illustrates a few of the infinite number of cells that can be constructed for a two-dimensional rectangular lattice. Only the rectangular cell B in the figure has three 2-fold rotation axes and two mirror planes. Although the other cells all have the same area, each of them has only one 2-fold axis and no mirror planes; they are therefore not acceptable unit cells.

Sometimes the smallest, or primitive, unit cell does not have the full symmetry of the crystal lattice. If so, a larger *nonprimitive* unit cell that does have the characteristic symmetry is deliberately chosen (Fig. 21.8). Only three types of nonprimitive cells are commonly used in the description of crystals: **body-centered, face-centered,** and **side-centered.** They are shown in Figure 21.9.

Scattering of X-Rays by Crystals

In the 19th century, crystallographers could classify crystals into the seven crystal systems only on the basis of their external symmetries. They could not measure the dimensions of unit cells or the positions of atoms within them. Several

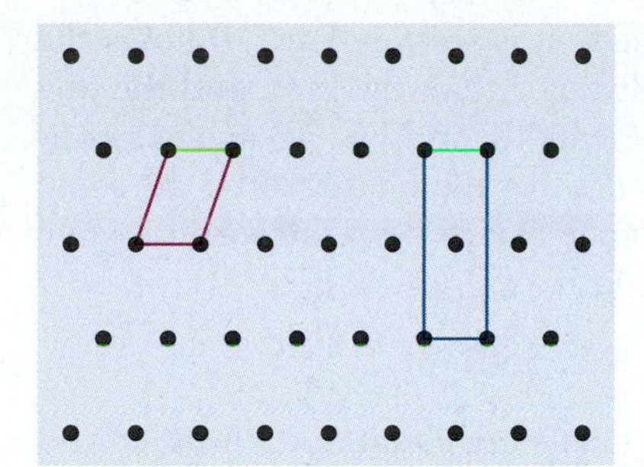

FIGURE 21.8 In this two-dimensional lattice, every lattice point is at the intersection of a horizontal mirror line and a vertical mirror line. It is possible to draw a primitive unit cell (red), but the larger, centered unit cell (blue) is preferred because it also has two mirror lines, the full symmetry of the lattice. A similar argument applies to the choice of unit cells on three-dimensional lattices.

developments by German physicists changed this situation at the turn of the century. Wilhelm Roentgen's discovery of x-rays in 1895 provided a tool of enormous power for determining the structures of crystals. Max von Laue suggested that crystals might serve as three-dimensional gratings for the diffraction of electromagnetic radiation with a wavelength comparable to the distance between planes of atoms. Friedrich and Knipping demonstrated experimentally in 1912 that this was indeed the case, and von Laue was awarded the Nobel Prize in physics in 1914 for his theory of the diffraction of x-rays by crystals. At about the same time, W. H. Bragg and W. L. Bragg (father and son) at Cambridge University in England also demonstrated the diffraction of x-rays by crystals and shared the Nobel Prize in physics the following year. (W. L. Bragg was 22 years old and still a student at Cambridge when he discovered the diffraction law.) The formulation of the diffraction law proposed by the Braggs is equivalent to von Laue's suggestion and somewhat simpler to visualize. So we will follow an approach similar to theirs.

When electromagnetic radiation passes through matter, it interacts with the electrons in atoms, and some of it is scattered as spherical waves going out from the atoms in the solid. Suppose that x-radiation strikes two neighboring scattering centers. The expanding spheres of scattered waves soon encounter each other and interfere. In some directions, the waves are in phase and reinforce each other, or interfere *constructively* (Fig. 21.10a); in others they are out of phase and cancel each other out, or interfere *destructively* (see Fig. 21.10b). Constructive interference occurs when the paths traversed by two waves differ in length by a whole number of wavelengths. The amplitudes of waves that interfere constructively add to one another, and the intensity of the scattered radiation in that direction is proportional to the square of the total amplitude.

Figure 21.11 illustrates the constructive interference of x-rays scattered by the electrons in atoms in equally spaced planes separated by the distance d. A parallel bundle of coherent x-rays of a single known wavelength is allowed to fall on the surface of a crystal, making an angle θ with a set of parallel planes of atoms in the crystal. The scattering angle 2θ is then varied by rotating the crystal about an axis perpendicular to the plane of the figure. Line AD in Figure 21.11 represents a wave front of waves that are in phase as they approach the crystal. The wave that is

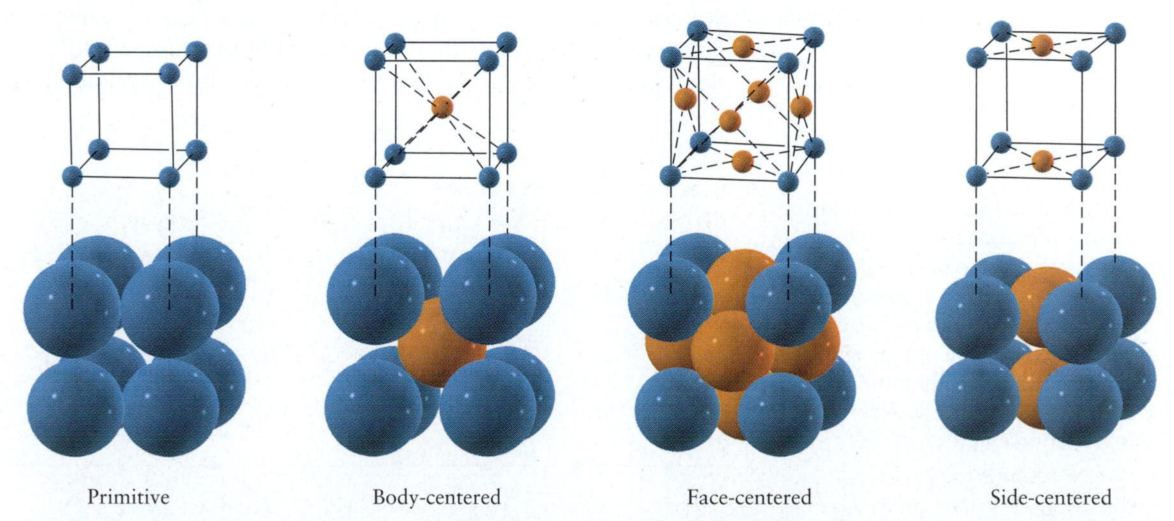

FIGURE 21.9 Centered lattices, like all lattices, have lattice points at the eight corners of the unit cell. A body-centered lattice has an additional lattice point at the center of the cell, a face-centered lattice has additional points at the centers of the six faces, and a side-centered lattice has points at the centers of two parallel sides of the unit cell. (**Note:** The colored dots in the lattice diagrams represent lattice points, not atoms.)

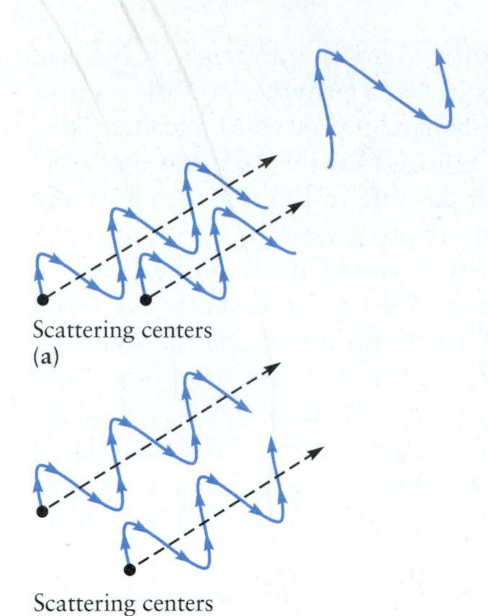

FIGURE 21.10 A beam of x-rays (not shown) is striking two scattering centers, which emit scattered radiation. The difference in the lengths of the paths followed by the scattered waves determines whether they interfere (a) constructively or (b) destructively. This path difference depends on both the distance between the centers and the direction in which the scattered waves are moving.

scattered at B follows the path ABC, and the one that is scattered at F follows the path DFH. The second wave travels a greater distance than the first, and the difference in path length is the sum of the two segments EF and FG. To achieve constructive interference in the scattered waves (that is, for the phases to be the same along the wave front CH), this additional distance traveled by the second wave must be an integral multiple of the x-ray wavelength λ:

$$\text{EF} + \text{FG} = n\lambda \qquad n = 1, 2, 3, \ldots$$

From trigonometry, the lengths of these two segments are equal to each other and to $d \sin \theta$, where d is the interplanar spacing. Therefore, constructive interference occurs only when

$$n\lambda = 2d \sin \theta \qquad n = 1, 2, 3, \ldots \qquad [21.1]$$

It is easy to verify that for angles that meet this condition, the waves scattered from the third and subsequent planes are also in phase with the waves scattered from the first two planes.

The preceding condition on allowed wavelengths is called the **Bragg law,** and the corresponding angles are called **Bragg angles** for that particular set of parallel planes of atoms. It appears as though the beam of x-rays has been reflected symmetrically from those crystal planes, and we often speak colloquially of the "Bragg reflection" of x-rays. The x-rays have not been reflected, however, but have undergone constructive interference, more commonly called diffraction. The case $n = 1$ is called first-order Bragg diffraction, $n = 2$ is second-order, and so forth.

We now possess a tool of immense value for determining the interplanar spacings of crystals. If a crystal is turned through different directions, other parallel sets of planes with different separations are brought into the Bragg condition. The symmetry of the resulting diffraction pattern identifies the crystal system, and the Bragg angles determine the cell constants. Moreover, the *intensities* of the diffracted beams permit the locations of the atoms in the unit cell to be determined.

Analogous scattering techniques use beams of neutrons. In that case the scattering interaction is between the magnetic moments of the incident neutrons and nuclei in the solid, but the principles are the same as for x-ray diffraction. Of course, it is the wave character of neutrons (in particular, their de Broglie wavelength; see Chapter 4) that is responsible for neutron diffraction. Recall that the de Broglie wavelength and neutron momentum are related by $\lambda = h/p$.

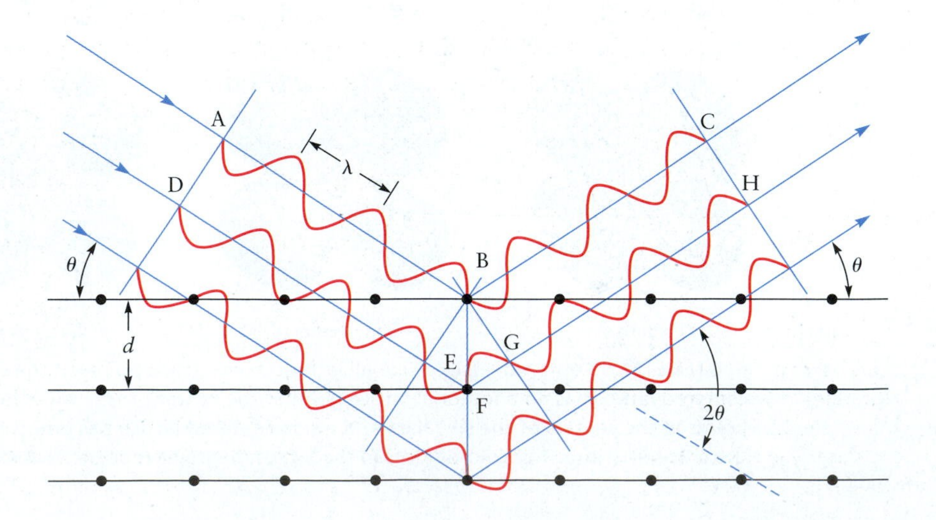

FIGURE 21.11 Constructive interference of x-rays scattered by atoms in lattice planes. Three beams of x-rays, scattered by atoms in three successive layers of a simple cubic crystal, are shown. Note that the phases of the waves are the same along the line CH, indicating constructive interference at this scattering angle 2θ.

EXAMPLE 21.2

A diffraction pattern of aluminum is obtained by using x-rays with wavelength λ = 0.709 Å. The second-order Bragg diffraction from the parallel faces of the cubic unit cells is observed at the angle $2\theta = 20.2°$. Calculate the lattice parameter a.

SOLUTION

From the Bragg condition for $n = 2$,

$$2\lambda = 2d \sin \theta$$

the spacing between planes, which is the lattice parameter, is

$$d = \frac{\lambda}{\sin \theta} = \frac{0.709 \text{ Å}}{\sin (10.1°)} = 4.04 \text{ Å} = a$$

Related Problems: 5, 6

21.2 Crystal Structure

The crystal lattice is an abstract construction whose points of intersection describe the underlying symmetry of a crystal. To flesh out the description of a particular solid state structure, we must identify some structural elements that are "pinned" to the lattice points. These structural elements can be atoms, ions, or even groups of atoms as we see in this and the next chapter. We begin with some illustrative simple cases. Some of the chemical elements crystallize in particularly simple solid structures, in which a single atom is situated at each point of the lattice.

Polonium is the only element known to crystallize in the **simple cubic lattice,** with its atoms at the intersections of three sets of equally spaced planes that meet at right angles. Each unit cell contains one Po atom, separated from each of its six nearest neighbors by 3.35 Å.

The alkali metals crystallize in the **body-centered cubic (bcc) structure** at atmospheric pressure (Fig. 21.12). A unit cell of this structure contains two lattice points, one at the center of the cube and the other at any one of the eight corners. A single alkali-metal atom is associated with each lattice point. An alternative way to visualize this is to realize that each of the eight atoms that lie at the corners of a bcc unit cell is shared by the eight unit cells that meet at those corners. The contribution of the atoms to one unit cell is therefore $8 \times \frac{1}{8} = 1$ atom, to which is added the atom that lies wholly within that cell at its center.

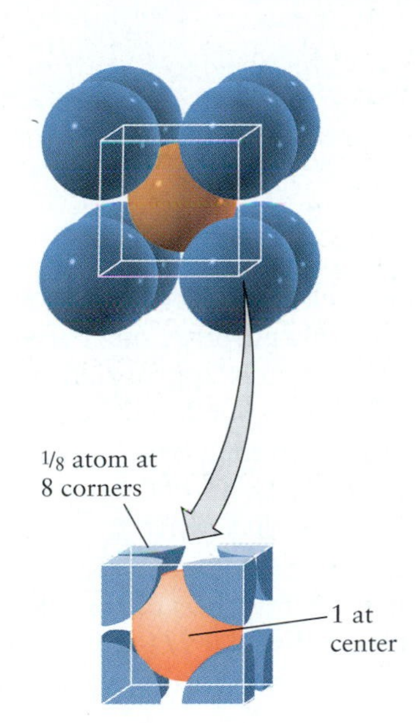

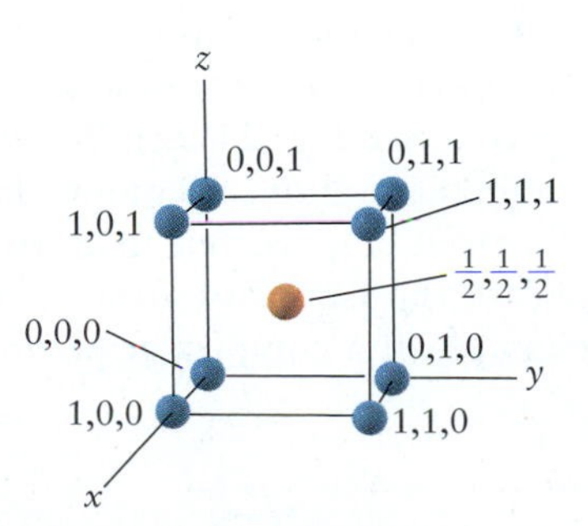

FIGURE 21.12 The bcc structure. An atom is located at the center of each cubic cell (orange) as well as at each corner of the cube (blue). The atoms are reduced slightly in size to make positions clear.

The metals aluminum, nickel, copper, and silver, among others, crystallize in the **face-centered cubic (fcc) structure** shown in Figure 21.13. This unit cell contains four lattice points, with a single atom associated with each point. No atom lies wholly within the unit cell; there are atoms at the centers of its six faces, each of which is shared with another cell (contributing $6 \times \frac{1}{2} = 3$ atoms), and an atom at each corner of the cell (contributing $8 \times \frac{1}{8} = 1$ atom), for a total of four atoms per unit cell.

The volume of a unit cell is given by the formula

$$V_c = abc\sqrt{1 - \cos^2\alpha - \cos^2\beta - \cos^2\gamma + 2\cos\alpha\cos\beta\cos\gamma} \quad [21.2]$$

When the angles are all 90° (so that their cosines are 0), this formula reduces to the simple result $V = abc$ for the volume of a rectangular box. If the mass of the unit cell contents is known, the theoretical cell density can be computed. This density must come close to the measured density of the crystal, a quantity that can be

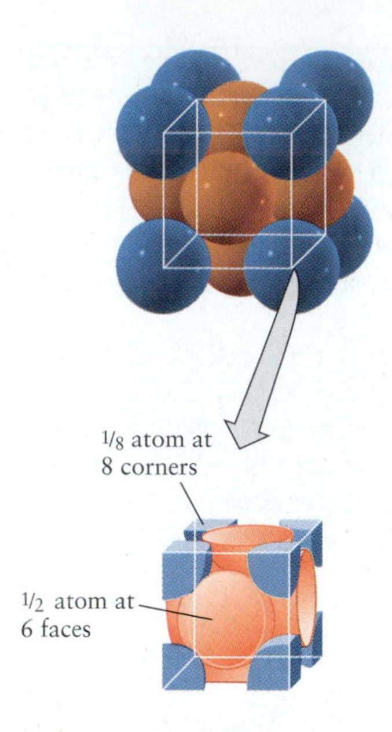

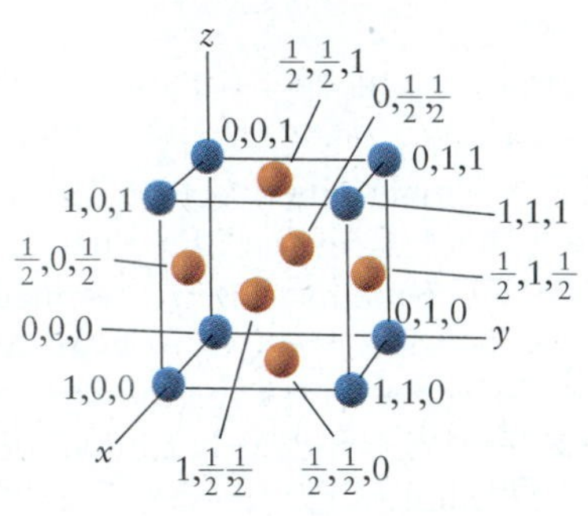

FIGURE 21.13 The fcc structure. Atoms are located at the centers of the faces (orange) as well as at the corners of the cube (blue). The atoms are reduced slightly in size to make positions clear.

determined by entirely independent experiments. For an *element* whose crystal contains n_c atoms per unit cell, the calculated cell density is

$$\text{density} = \rho = \frac{\text{mass}}{\text{volume}} = \frac{n_c \dfrac{\mathcal{M}}{N_A}}{V_c} = \frac{n_c\mathcal{M}}{N_A V_c}$$

where we have used the fact that the molar mass $\mathcal{M}$ of the element divided by Avogadro's number N_A is the mass of a single atom. This equation can also be used to calculate Avogadro's number from the measured density and cell constants, as the following example illustrates.

EXAMPLE 21.3

Sodium has a density of $\rho = 0.9700 \text{ g cm}^{-3}$ at 20°C, and its lattice parameter is $a =$ 4.2856 Å. What is the value of Avogadro's number, given that the molar mass of sodium is $22.9898 \text{ g mol}^{-1}$?

SOLUTION

Sodium has two atoms per unit cell because its structure is bcc, and the volume of the unit cell is a^3. Solve the foregoing equation for Avogadro's number:

$$N_A = \frac{n_c\mathcal{M}}{\rho a^3} = \frac{(2)(22.9898 \text{ g mol}^{-1})}{(0.9700 \text{ g cm}^{-3})(4.2856 \times 10^{-8} \text{ cm})^3} = 6.022 \times 10^{23} \text{ mol}^{-1}$$

Related Problems: 15, 16

It is useful to define the locations of atoms in the unit cell with a set of three numbers measured in units of the lattice parameter(s). For this purpose, one corner of the unit cell is taken to be at the origin of the coordinate axes appropriate to the crystal system, and an atom at that lattice point has the coordinates (0, 0, 0). Equivalent atoms at the seven remaining corners of the cell then have the coordinates (1, 0, 0), (0, 1, 0), (0, 0, 1), (1, 0, 1), (1, 1, 0), (0, 1, 1), and (1, 1, 1). These atom positions are generated from an atom at (0, 0, 0) by successive translations through unit distances along the three axes. An atom in a body-centered site has coordinates $(\frac{1}{2}, \frac{1}{2}, \frac{1}{2})$, and the prescription for locating it is to proceed from the coordinate origin at (0, 0, 0) a distance $a/2$ along a, then a distance $b/2$ along b, and finally $c/2$ along c. In the same way, an atom in a face-centered site has coordinates such as $(\frac{1}{2}, \frac{1}{2}, 0)$, $(\frac{1}{2}, 0, \frac{1}{2})$, or $(0, \frac{1}{2}, \frac{1}{2})$.

So far, we have considered only cubic metals in which one atom corresponds to each lattice point. More complicated structures also occur, even for the elements, in which atoms occupy positions that are not lattice points. Diamond has an fcc structure with eight atoms (not four) per unit cell (See Fig. 21.22). Boron has a tetragonal structure with a very complex unit cell that contains 50 atoms. In molecular crystals the number of atoms per unit cell can be still greater; in a protein with a molecular mass of 10^5, there may be tens of thousands of atoms per unit cell. The protein shown in Figure 23.13 is an example of a complex structure revealed by modern x-ray crystallography.

Atomic Packing in Crystals

As we begin to add structural features like atoms onto crystal lattice sites, we have to pay attention to the size of the features. How efficiently can we pack atoms onto a lattice? We have already discussed two measures of the "size" of an atom or molecule. In Section 9.7, the van der Waals parameter b was related to the volume excluded per mole of molecules, so b/N_A is one measure of molecular size. In

Section 5.5, we defined an approximate radius of an atom as the distance at which the electron density had fallen off to a particular value, or as the radius of a sphere containing a certain fraction of the total electron density. A third related measure of atomic size is based on the interatomic separations in a crystal. The radius of a noble-gas or metallic atom can be approximated as half the distance between the center of an atom and the center of its nearest neighbor in the crystal. We picture crystal structures as resulting from packing spheres in which nearest neighbors are in contact.

The nearest neighbor separation in a simple cubic crystal is equal to the lattice parameter a, so the atomic radius in that case is $a/2$. For the bcc lattice, the central atom in the unit cell "touches" each of the eight atoms at the corners of the cube, but those at the corners do not touch one another, as Figure 21.12 shows. The nearest neighbor distance is calculated from the Cartesian coordinates of the atom at the origin (0, 0, 0) and that at the cell center ($a/2$, $a/2$, $a/2$). By the Pythagorean theorem, the distance between these points is $\sqrt{(a/2)^2 + (a/2)^2 + (a/2)^2} = a\sqrt{3}/2$, so the atomic radius is $a\sqrt{3}/4$. Figure 21.13 shows that in an fcc crystal the atom at the center of a face [such as at (0, $a/2$, $a/2$)] touches each of the neighboring corner atoms [such as at (0, 0, 0)], so the nearest-neighbor distance is $\sqrt{(a/2)^2 + (a/2)^2} = a\sqrt{2}/2$ and the atomic radius is $a\sqrt{2}/4$. Table 21.2 summarizes the results for cubic lattices.

What is the most dense crystal packing that can be achieved? To answer this question, construct a crystal by first putting down a plane of atoms with the highest possible density, shown in Figure 21.14a. Each sphere is in contact with six other spheres in the plane. Then put down a second close-packed plane on top of the first one (see Fig. 21.14b) in such a way that each sphere in the second plane is in contact with three spheres in the plane below it; that is, each sphere in the second plane forms a tetrahedron with three spheres beneath. When the third plane is laid down, there are two possibilities. In Figure 21.14c, the atoms in the third plane lie on sites not directly over those in the first layer, whereas in Figure 21.14d the third-plane atoms are directly over the first-plane atoms.

TABLE 21.2 Structural Properties of Cubic Lattices

	Simple Cubic	Body-Centered Cubic	Face-Centered Cubic
Lattice points per cell	1	2	4
Number of nearest neighbors	6	8	12
Nearest-neighbor distance	A	$a\sqrt{3}/2 = 0.866a$	$a\sqrt{2}/2 = 0.707a$
Atomic radius	$a/2$	$a\sqrt{3}/4 = 0.433a$	$a\sqrt{2}/4 = 0.354a$
Packing fraction	$\frac{\pi}{6} = 0.524$	$\frac{\sqrt{3}\pi}{8} = 0.680$	$\frac{\sqrt{2}\pi}{6} = 0.740$

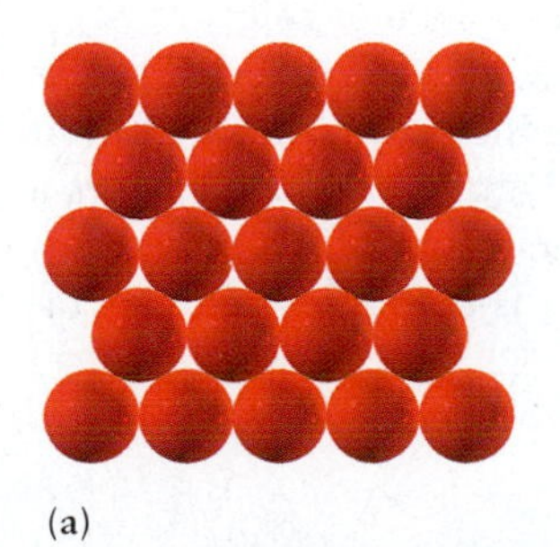
(a)

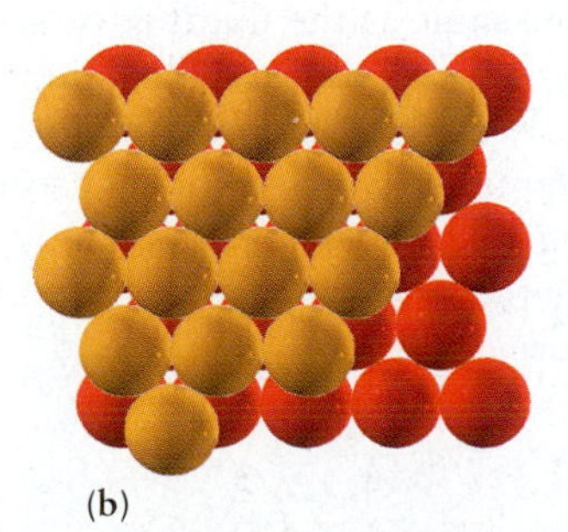
(b)

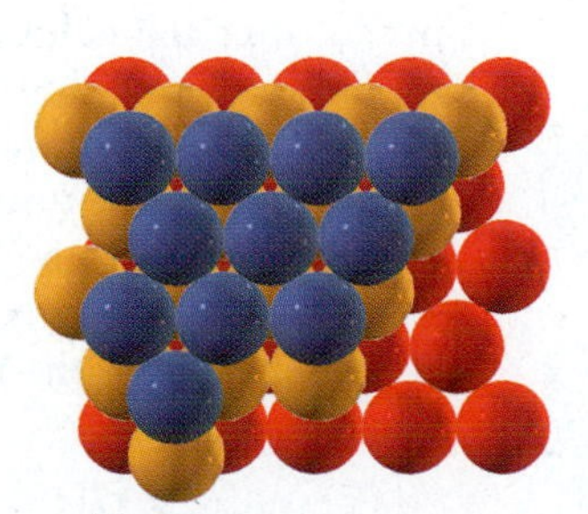
(c)

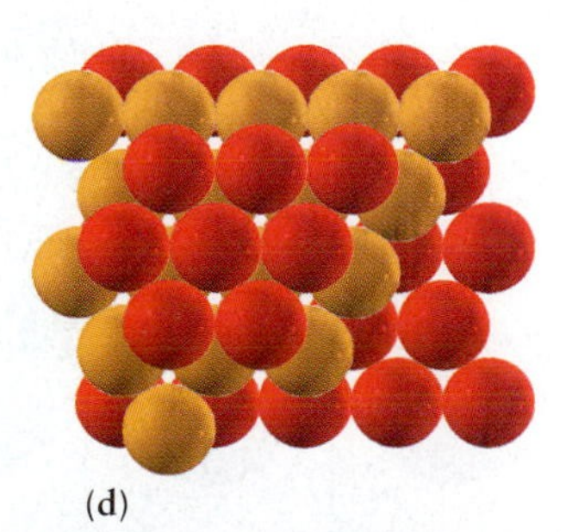
(d)

FIGURE 21.14 Close-packing of spheres. (a) One layer, with each atom surrounded by six nearest neighbors. (b) Two layers, with the atoms of the second layer centered on top of indentations in the layer below. (c) The third layer can be placed on sites that are not directly above the sites in the first layer (note that the red spheres show through). If this pattern is repeated as *abcabc . . .* it gives cubic close-packing. (d) Alternatively, the third layer can be placed directly over the sites in the first layer (note that the white triangular spaces show through). The repeated pattern *ababab . . .* gives hexagonal close-packing.

Clearly, there are two choices for placing each plane, and an infinite number of crystal structures can be generated that have the same atomic packing density. The two simplest such structures correspond to the periodic layer sequences *abcabcab* . . . and *abababa*. . . . The first of these is the fcc structure already discussed, and the second is a close-packed structure in the hexagonal crystal system termed **hexagonal close-packed (hcp).** In each of these simple structures, atoms occupy 74.0% of the unit cell volume, as the following example shows. (Atoms that crystallize in the bcc structure occupy only 68.0% of the crystal volume, and the packing fraction for a simple cubic array is only 52.4%.)

EXAMPLE 21.4

Calculate (**a**) the atomic radius of an aluminum atom and (**b**) the fraction of the volume of aluminum that is occupied by its atoms.

SOLUTION

(**a**) Aluminum crystallizes in the fcc crystal system, and its unit cell therefore contains four atoms. Because a face-centered atom touches each of the atoms at the corners of its face, the atomic radius r_1 (see the triangle in Fig. 21.15) can be expressed by

$$4r_1 = a\sqrt{2}$$

Using the value $a = 4.04$ Å derived from x-ray diffraction (see Example 21.2) and solving for r_1 gives

$$r_1 = 1.43 \text{ Å}$$

(**b**) The fraction of the volume of an aluminum single crystral that is occupied by its atoms is

$$f = \frac{4\left[\frac{4}{3}\pi r_1^3\right]}{a^3} = \frac{4\frac{4}{3}\pi\left[\frac{a\sqrt{2}}{4}\right]^3}{a^3} = 0.740$$

Related Problem: 23

Interstitial Sites

The ways in which the empty volume is distributed in a crystal are both interesting and important. For the close-packed fcc structure, two types of **interstitial sites,** upon which the free volume in the unit cell is centered, are identifiable. An **octahedral site** is surrounded at equal distances by six nearest neighbor atoms. Figure 21.15 shows that such sites lie at the midpoints of the edges of the fcc unit cell. A cell has 12 edges, each of which is shared by four unit cells, so the edges contribute three octahedral interstitial sites per cell. In addition, the site at the center of the unit cell is also octahedral, so the total number of octahedral sites per fcc unit cell is four, the same as the number of atoms in the unit cell.

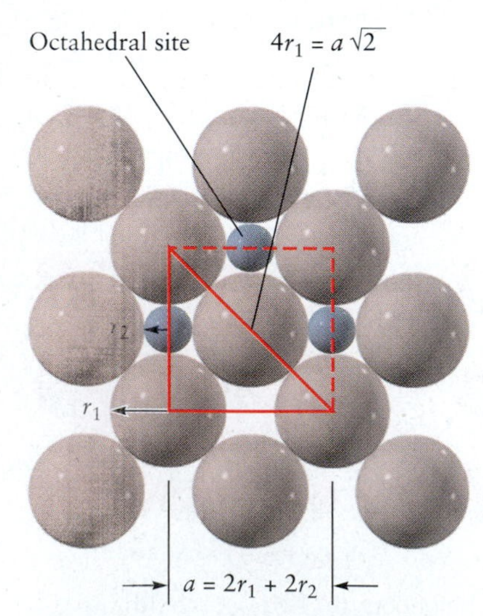

FIGURE 21.15 Octahedral sites in an fcc lattice. The geometric procedure for relating the site radius r_2 to the atom radius r_1 is shown.

With a bit of simple geometry, we can calculate the size of an octahedral site in an fcc structure or, more precisely, the radius r_2 of a smaller atom that would fit in the site without overlapping its neighboring atoms. Figure 21.15 represents a cell face in which the length of the diagonal is $4r_1$ and the length of the cell edge is $2r_1 + 2r_2$, where r_1 is the radius of the host atoms and r_2 is the radius of the octahedral site. From the figure,

$$r_1 = a\frac{\sqrt{2}}{4}$$

$$a = 2r_2 + 2r_1 = 2r_2 + 2a\frac{\sqrt{2}}{4}$$

$$r_2 = \frac{a}{2} - a\frac{\sqrt{2}}{4} = 0.146a$$

and the ratio of the octahedral-site radius to the host-atom radius is

$$\frac{r_2}{r_1} = \frac{0.146a}{a\sqrt{2}/4} = 0.414$$

The second type of interstitial site, known as a **tetrahedral site,** lies at the center of the space defined by four touching spheres. In an fcc cell, a tetrahedral site occurs in the volume between a corner atom and the three face-centered atoms nearest to it. Geometrical reasoning like that used for the octahedral site gives the following ratio of the radius of a tetrahedral site to that of a host atom:

$$\frac{r_2}{r_1} = 0.225$$

The fcc unit cell contains eight tetrahedral sites, twice the number of atoms in the cell.

Interstitial sites are important when a crystal contains atoms of several kinds with considerably different radii. We will return to this shortly when we consider the structures of ionic crystals.

21.3 Cohesion in Solids

In addition to symmetry, the nature of the bonding forces between atoms provides a useful way to classify solids. This classification does indeed lead to an understanding of the remarkable differences in the chemical and physical properties of different materials. We now consider crystals held together through ionic, metallic, or covalent bonding interactions, and the one class of solids held together by intermolecular forces.

Ionic Crystals

Compounds formed by atoms with significantly different electronegativities are largely ionic, and to a first approximation the ions can be treated as hard, charged spheres that occupy positions on the crystal lattice (see the ionic radii in Appendix F). All the elements of Groups I and II of the periodic table react with Group VI and VII elements to form ionic compounds, the great majority of which crystallize in the cubic system. The alkali-metal halides (except for the cesium halides), the ammonium halides, and the oxides and sulfides of the alkaline-earth metals all crystallize in the **rock-salt,** or **sodium chloride, structure** shown in Figure 21.16. It may be viewed as an fcc lattice of anions whose octahedral sites are all occupied by cations or, equivalently, as an fcc lattice of cations whose octahedral sites are

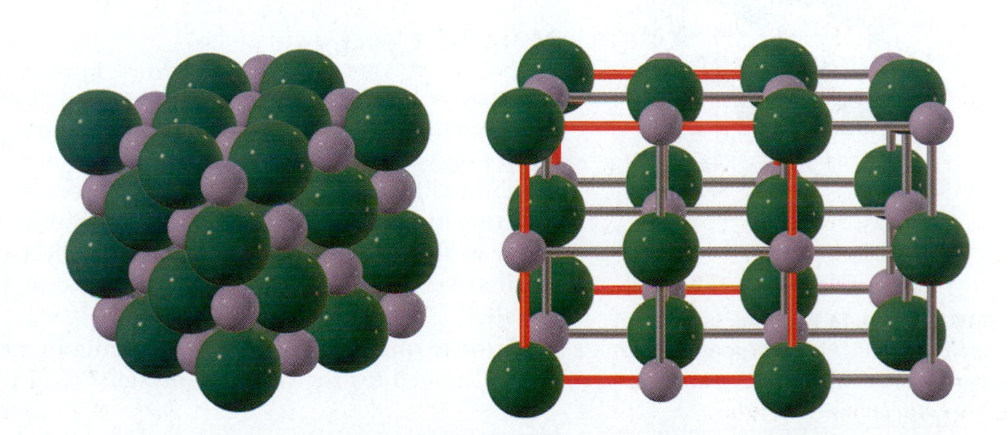

FIGURE 21.16 The sodium chloride, or rock-salt, structure. On the left, the sizes of the Na^+ ions (purplish pink) and the Cl^- ions (green) are drawn to scale. On the right, the ions are reduced in size to allow a unit cell (shown by red lines) to be outlined clearly.

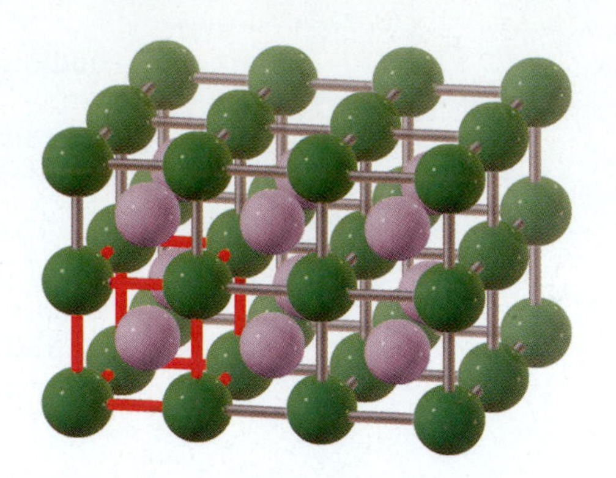

FIGURE 21.17 The structure of cesium chloride. On the left, the sizes of the Cs^+ ions (purplish pink) and the Cl^- ions (green) are drawn to scale. On the right, they are reduced in size to allow a unit cell (shown by red lines) to be outlined clearly. Note that the lattice in this structure is simple cubic, with one Cs^+ ion and one Cl^- ion per unit cell.

all occupied by anions. Either way, each ion is surrounded by six equidistant ions of the opposite charge. The rock-salt structure is a stable crystal structure when the cation-anion radius ratio lies between 0.414 and 0.732, if cations and anions are assumed to behave as incompressible charged spheres.

When the hard-sphere cation-anion radius ratio exceeds 0.732, as it does for the cesium halides, a different crystal structure called the **cesium chloride structure,** is more stable. It may be viewed as two interpenetrating simple cubic lattices, one of anions and the other of cations, as shown in Figure 21.17. When the cation–anion radius ratio is less than 0.414, the **zinc blende,** or **sphalerite, structure** (named after the structure of ZnS) results. This crystal consists of an fcc lattice of S^{2-} ions, with Zn^{2+} ions occupying half of the available tetrahedral sites in alternation, as Figure 21.18a illustrates. **Fluorite** (CaF_2) has yet another structure; the unit cell is based on an fcc lattice of Ca^{2+} ions. The F^- ions occupy all eight of the tetrahedral sites, so the unit cell contains four Ca^{2+} and eight F^- ions (see Fig. 21.18b). The radius ratios (0.414 and 0.732) at which crossovers from one type of crystal to another occur are not accidental numbers. Recall from Section 21.2 that 0.414 is the ratio of the octahedral-site radius to the host-atom radius for an fcc lattice; only when this size ratio is exceeded does the ion inserted into that site come into contact with ions of the opposite sign in the rock-salt structure. The number 0.732 comes from a corresponding calculation of the radius ratio of the interstitial site at the center of a simple cubic unit cell (see problem 25). It is important to realize that the radius-ratio criterion for the stability limits of the structures of binary ionic compounds assumes that the ions are incompressible and that the wave functions do not overlap. The criterion fails when these approximations are not met.

The strength and range of the electrostatic attractions make ionic crystals hard, high-melting, brittle solids that are electrical insulators. Melting an ionic crystal, however, disrupts the lattice and sets the ions free to move, so ionic liquids are good electrical conductors.

S^{2-} Zn^{2+}

Sphalerite (ZnS)

Ca^{2+} F^-

Fluorite (CaF_2)

FIGURE 21.18 Two ionic lattices in the fcc system. A single (nonprimitive) cubic unit cell of each is shown.

Metallic Crystals

The type of bonding found in metals is quite different from that in other crystals. As we compare the various main group and transition metals in the periodic table we see only small differences in electronegativity. So, there is little tendency for ionic bonding in metals. The electronic configurations of metal atoms, even in the transition metals, do not have nearly-filled subshells, so there is little tendency to form covalent bonds by sharing electrons to achieve a stable octet. The familiar classical models of chemical bonding (see Chapter 3) do not extend to metals.

Prior to quantum mechanics, bonding in metals was described by the Drude model, named for the German physicist Paul Drude. The solid was viewed as a fixed array of positively charged metal ions, each localized at a site of the crystal lattice. These fixed ions were surrounded by a sea of mobile electrons, one contributed by each of the atoms in the solid. The number density of electrons was equal to the number density of positively charged ions, so the metal was electrically neutral. The sea of delocalized electrons would interact with the stable ions to give a strong cohesive force keeping the metal bound together. The Drude model accounts for the malleability (deformation in response to mechanical force, like hammering) and ductility (ease of drawing into a fine wire) of metals. As some ions move to new positions in response to these mechanical disturbances, delocalized electrons can rapidly adjust to maintain metallic bonding in the deformed or drawn solid. The ease with which metals conduct electricity is explained because these delocalized electrons can respond to any applied electric fields.

The delocalized sea of electrons picture of metal bonding survives in the quantum theory of solids, which is an extension of the molecular orbital description of molecular bonding (see Chapter 6). The valence electrons in a metal are delocalized in huge molecular orbitals that extend over the entire crystal and provide the "glue" that holds together the positively charged ion cores of the metal atoms. To understand the origin of these molecular orbitals, suppose just two sodium atoms are brought together, each in its electronic ground state with the configuration $1s^2 2s^2 2p^6 3s^1$. As the atoms approach each other, the wave functions of their 3*s* electrons combine to form two molecular orbitals—one in which their phases are symmetric (σ_{3s}) and another in which they are antisymmetric (σ^*_{3s}). Solving the Schrödinger equation yields two energy states, one above and the other below the energy of the atomic 3*s* levels, analogous to the formation of a hydrogen molecule from two hydrogen atoms described in Chapter 6. If both valence electrons are put into the level of lower energy with spins opposed, the result is a Na_2 molecule. If a third sodium atom is added, the 3*s* atomic levels of the atoms split into three sublevels (Fig. 21.19). Two electrons occupy the lowest level with their spins opposed, and the third electron occupies the middle level. The three energy levels and the three electrons belong collectively to the three sodium atoms. A fourth sodium atom could be added so that there would be four closely spaced energy sublevels, and this process could be carried on without limit.

The foregoing is not a mere "thought experiment." Sodium vapor contains about 17% Na_2 molecules at its normal boiling point. Larger sodium clusters have been produced in molecular beam experiments, and mass spectrometry shows that such clusters can contain any desired number of atoms. For each added Na atom, another energy sublevel is added. Because the sublevels are so very closely spaced in a solid (with, say, 10^{23} atoms), the collection of sublevels can be regarded as an **energy band.** Figure 21.19 depicts the formation of bands of sublevels that broaden (become delocalized) as the spacing between the nuclear centers decreases, and the 3*s* electrons go into these bands. The electrons that belong to the 1*s*, 2*s*, and 2*p* atomic levels of sodium are only very slightly broadened at the equilibrium internuclear separation of the crystal, so they retain their distinct, localized character as the core levels of the ions at the lattice sites. Chapter 22 explores the electrical properties of metals, which arise from this band structure.

Most metals have crystal structures of high symmetry and crystallize in bcc, fcc, or hcp lattices (Fig. 21.20). Relatively few metals (Ga, In, Sn, Sb, Bi, and Hg) have more complex crystal structures. Many metals undergo phase transitions to other structures when the temperature or pressure is changed. Both liquid and crystal phases can be metals; in fact, the conductivity usually drops by only a small amount when a metal melts. The electron sea provides very strong binding in most metals, as shown by their high boiling points. Metals have a very large range of melting points. Gallium melts at 29.78°C (Fig. 21.21), and mercury stays liquid at temperatures that freeze water. Many transition metals require temperatures in excess of 1000°C to melt, and tungsten, the highest melting elemental metal, melts at 3410°C (see Section 8.1).

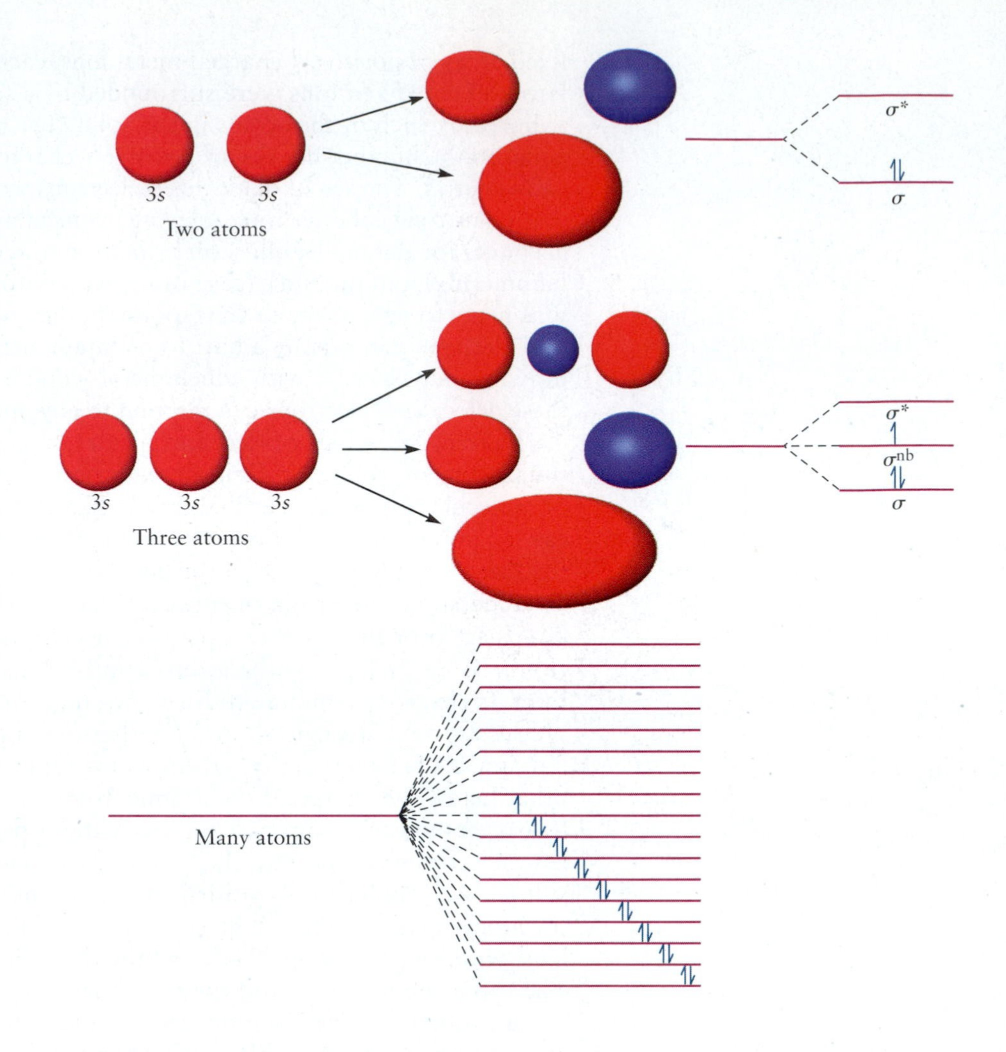

FIGURE 21.19 As sodium atoms are brought together, the molecular orbitals formed from their 3*s* atomic orbitals spread out into a band of levels, half occupied by electrons.

Covalent Crystals

We turn finally to a class of crystalline solids whose atoms are linked by covalent bonds rather than by the electrostatic attractions of ions or the valence electron "glue" in a metal. The archetype of the covalent crystal is diamond, which belongs to the cubic system. The ground-state electron configuration of a carbon atom is $1s^22s^22p^2$, and as shown in Section 6.5, its bonding is described by four hybrid sp^3 orbitals directed to the four corners of a regular tetrahedron. Each of the equivalent hybrid orbitals contains one electron that can spin-pair with the electron in one of the sp^3 orbitals of another carbon atom. Each carbon atom can thus link covalently to four others to yield the space-filling network shown in Figure 21.22. Covalent crystals are also called "network crystals," for obvious reasons. In a sense, every atom in a covalent crystal is part of one giant molecule that is the crystal itself. These crystals have very high melting points because of the strong attractions between covalently bound atoms. They are hard and brittle. Chapter 22 describes the electrical properties of covalent crystals.

Molecular Crystals

Molecular crystals include the noble gases; oxygen; nitrogen; the halogens; compounds such as carbon dioxide; metal halides of low ionicity such as Al_2Cl_6, $FeCl_3$, and $BiCl_3$; and the vast majority of organic compounds. All these molecules are held in their lattice sites by the intermolecular forces discussed in Sections 9.7 and 10.2. The trade-off between attractive and repulsive forces among even small mol-

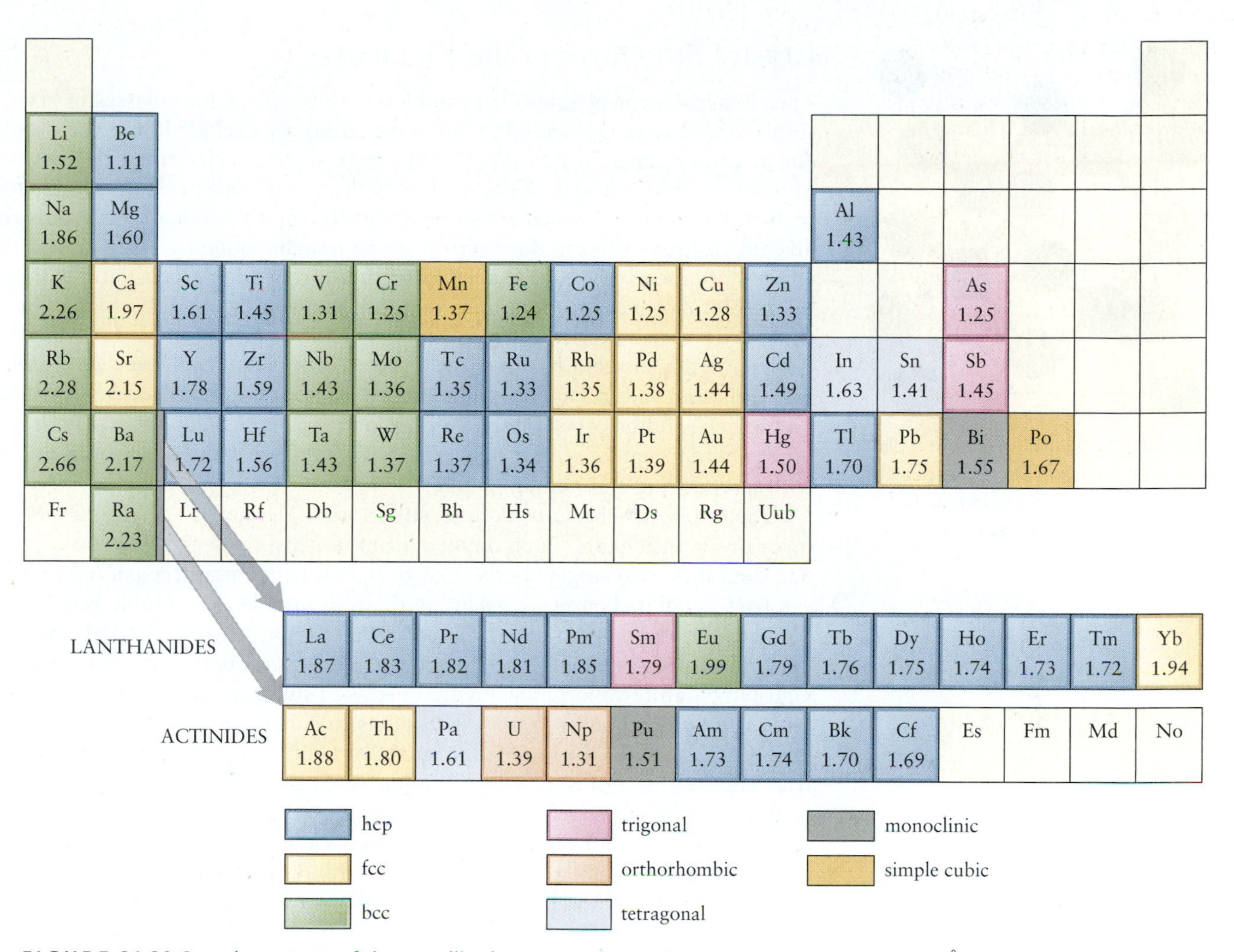

FIGURE 21.20 Crystal structures of the metallic elements at 25°C and 1 atm pressure. Atomic radii (Å) are calculated as one half the closest atom–atom distance in each structure; in most cases this is the same radius as calculated using the hard sphere contact model of Example 21.4. There are no known crystal structures for those elements for which atomic radii are not listed.

ecules in a molecular crystal is complex because so many atoms are involved. A useful simplification is to picture a molecule as a set of fused spheres centered at each nucleus. The radius of each sphere is the van der Waals radius of the element involved. In molecular crystals, such shapes pack together so that no molecules overlap but empty space is minimized. Figure 21.23 depicts such a "space-filling model" of cyanuric triazide (C_3N_{12}), showing how nature solves the problem of efficiently packing many copies of the rather complicated molecular shape of C_3N_{12} in a single layer. In the three-dimensional molecular crystal of C_3N_{12}, many such layers stack up with a slight offset that minimizes unfilled space between layers.

FIGURE 21.21 Solid gallium has a low melting point, low enough to melt from the heat of the body.

Van der Waals forces are much weaker than the forces that operate in ionic, metallic, and covalent crystals. Consequently, molecular crystals typically have low melting points and are soft and easily deformed. Although at atmospheric pressure the noble-gas elements crystallize in the highly symmetric fcc lattice shown in Figure 21.13, molecules (especially those with complex geometries) more often form crystals of low symmetry in the monoclinic or triclinic systems. Molecular crystals are of great scientific value. If proteins and other macromolecules are obtained in the crystalline state, their structures can be determined by x-ray diffraction. Knowing the three-dimensional structures of biological molecules is the starting point for understanding their functions.

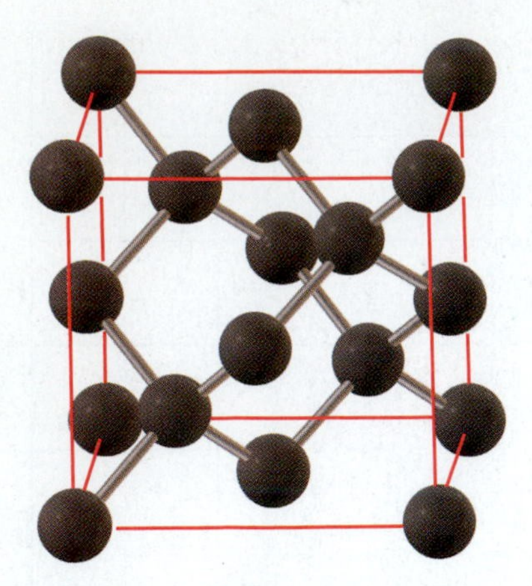

FIGURE 21.22 The structure of diamond. Each carbon atom has four nearest neighbors surrounding it at the corners of a tetrahedron.

Crystal Structures of the Elements

The chemical elements provide examples of three of the four classes of crystalline solids described in this section. Only ionic solids are excluded, because a single element cannot have the two types of atoms of different electronegativities needed to form an ionic material. We have already discussed some of the structures formed by metallic elements, which are sufficiently electropositive that their atoms readily give up electrons to form the electron sea of metallic bonding. The nonmetallic elements are more complex in their structures, reflecting a competition between intermolecular and intramolecular bonding and producing molecular or covalent solids with varied properties.

Each halogen atom has seven valence electrons and can react with one other halogen atom to form a diatomic molecule. Once this single bond forms, there is no further bonding capacity; the halogen diatomic molecules interact with one another only through relatively weak van der Waals forces and form molecular solids with low melting and boiling points.

The Group VI elements oxygen, sulfur, and selenium display dissimilar structures in the solid state. Each oxygen atom (with six valence electrons) can form one double or two single bonds. Except in ozone, its high-free-energy form, oxygen uses up all its bonding capacity with an intramolecular double bond, forming a molecular liquid and a molecular solid that are only weakly bound. In contrast, diatomic sulfur molecules (S=S) are relatively rare, being encountered only in high-temperature vapors. The favored forms of sulfur involve the bonding of every atom to two other sulfur atoms. This leads to either rings or chains, and both are observed. The stable form of sulfur at room temperature consists of S_8 molecules, with eight sulfur atoms arranged in a puckered ring (Fig. 21.24). The weak interactions between S_8 molecules make elemental sulfur a rather soft molecular solid. Above 160°C the rings in molten sulfur break open and relink to form long, tangled chains, producing a highly viscous liquid. An unstable ring form of selenium (Se_8) is known, but the thermodynamically stable form of this element is a gray

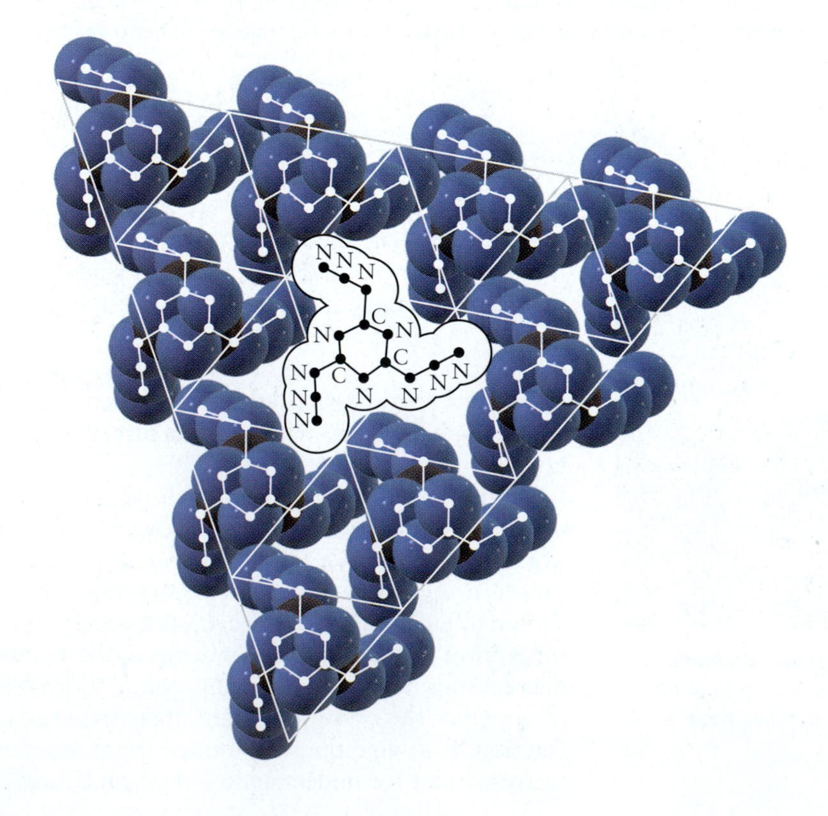

FIGURE 21.23 The van der Waals radii of the carbon and nitrogen atoms superimposed on an outline of the molecular structure of cyanuric triazide, C_3N_{12}, to show the volume of space from which each molecule excludes the others. Van der Waals forces in the molecular crystal hold the molecules in contact in a pattern that minimizes empty space. The thin white lines emphasize the 3-fold symmetry of the pattern.

FIGURE 21.24 The structure of the S_8 sulfur molecule. The orthorhombic unit cell of rhombic sulfur, the most stable form of elemental sulfur at room temperature, is large and contains 16 of these S_8 molecules for a total of 128 atoms of sulfur.

crystal of metallic appearance that consists of very long spiral chains with weak interchain interaction. Crystalline tellurium has a similar structure. The Group VI elements thus show a trend (moving down the periodic table) from the formation of multiple bonds toward the chains and rings characteristic of atoms that each form two single bonds.

A similar trend is evident in Group V. Only nitrogen forms diatomic molecules with triple bonds, in which all the bonding capacity is used between pairs of atoms. Elemental phosphorus exists in three forms, in all of which each phosphorus atom forms three single bonds rather than one triple bond. White phosphorus (Fig. 21.25a) consists of tetrahedral P_4 molecules, which interact with each other through weak van der Waals forces. Black phosphorus and red phosphorus (see Figs. 21.25b, c) are higher melting network solids in which the three bonds formed by each atom connect it directly or indirectly with all the other atoms in the sample. Unstable solid forms of arsenic and antimony that consist of As_4 or Sb_4 tetrahedra like those in white phosphorus can be prepared by rapid cooling of the vapor. The stable forms of these elements have structures related to that of black phosphorus.

The elements considered so far lie on the border between covalent and molecular solids. Other elements, those of intermediate electronegativity, exist as solids on the border between metallic and covalent; these are called **metalloids.** Antimony has a metallic luster, for example, but is a rather poor conductor of electricity and heat. Silicon and germanium are **semiconductors,** with electrical conductivities far lower than those of metals but still significantly higher than those of true insulators such as diamond. Section 22.7 examines the special properties of these materials more closely.

Some elements of intermediate electronegativity exist in two crystalline forms with very different properties. White tin has a tetragonal crystal structure and is a metallic conductor. Below 13°C it crumbles slowly to form a powder of gray tin (with the diamond structure) that is a poor conductor. Its formation at low temperature is known as the "tin disease" and can be prevented by the addition of

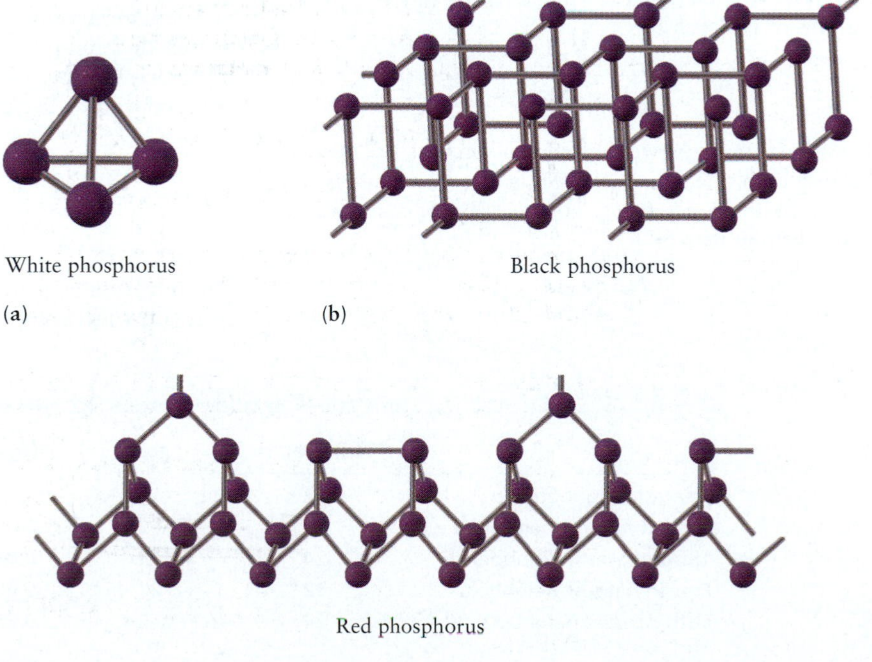

FIGURE 21.25 Structures of elemental phosphorus.

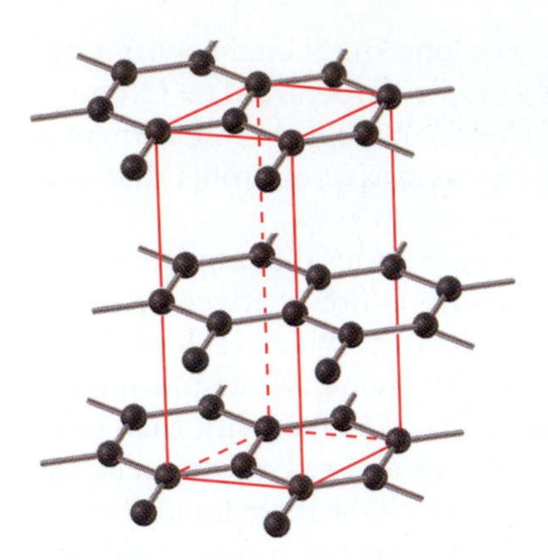

FIGURE 21.26 The structure of graphite.

small amounts of bismuth or antimony. The thermodynamically stable form of carbon at room conditions is not the insulator diamond, but graphite. Graphite consists of sheets of fused hexagonal rings with only rather weak interactions between layers (Fig. 21.26). Each carbon atom shows sp^2 hybridization, with its remaining p orbital (perpendicular to the graphite layers) taking part in extended π-bonding interactions over the whole plane. Graphite can be pictured as a series of interlocked benzene rings, with π-electron delocalization contributing significantly to its stability. The delocalized electrons give graphite a significant value for conductivity in the planes of fused hexagons approaching that of the metallic elements. The conductivity and relative chemical inertness of graphite make it useful for electrodes in electrochemistry.

A DEEPER LOOK

21.4 Lattice Energies of Crystals

The **lattice energy** of a crystal is the energy required to separate the crystal into its component atoms, molecules, or ions at 0 K. In this section we examine the calculation and measurement of lattice energies for molecular and ionic crystals.

Lattice Energy of a Molecular Crystal

The lattice energy of a molecular crystal can be estimated by using the simple Lennard-Jones potential of Section 9.7:

$$V_{\text{LJ}}(R) = 4\varepsilon\left[\left(\frac{\sigma}{R}\right)^{12} - \left(\frac{\sigma}{R}\right)^{6}\right]$$

Table 9.4 lists the values of e and σ for various atoms and molecules. To obtain the total potential energy for 1 mol, sum over all pairs of atoms or molecules:

$$V_{\text{tot}} = \tfrac{1}{2}\sum_{i=1}^{N_A}\sum_{j=1}^{N_A} V_{\text{LJ}}(R_{ij})$$

where R_{ij} is the distance between atom i and atom j. The factor $\frac{1}{2}$ arises because each interaction between a pair of atoms should be counted only once, not twice. For a crystal of macroscopic size, this can be rewritten as

$$V_{\text{tot}} = \frac{N_A}{2}\sum_{j=1}^{N_A} V_{\text{LJ}}(R_{ij})$$

where i is taken to be some atom in the middle of the crystal. Taking the nearest neighbor distance to be R_0, we define a ratio of distances $p_{ij} = R_{ij}/R_0$ and rewrite V_{tot} for the Lennard-Jones potential as

$$V_{\text{tot}} = \frac{N_A}{2}(4\varepsilon)\left[\sum_j\left(\frac{\sigma}{p_{ij}R_0}\right)^{12} - \sum_j\left(\frac{\sigma}{p_{ij}R_0}\right)^{6}\right]$$

$$= 2\varepsilon N_A\left[\left(\frac{\sigma}{R_0}\right)^{12}\sum_j (p_{ij})^{-12} - \left(\frac{\sigma}{R_0}\right)^{6}\sum_j (p_{ij})^{-6}\right]$$

The two summations are dimensionless properties of the lattice structure, and accurate values can be obtained by summing over the first few sets of nearest neighbors (Table 21.3). The resulting total energy for the fcc lattice is

$$V_{\text{tot}} = 2\varepsilon N_A\left[12.132\left(\frac{\sigma}{R_0}\right)^{12} - 14.454\left(\frac{\sigma}{R_0}\right)^{6}\right]$$

The equilibrium atomic spacing at $T = 0$ K should be close to the one that gives a minimum in V_{tot}, which can be calculated by differentiating the preceding expression with respect to R_0 and setting the derivative to 0. The result is

$$R_0 = 1.09\sigma$$

and the value of V_{tot} at this value of R_0 is

$$V_{\text{tot}} = -8.61\varepsilon N_A$$

The corresponding potential energy when the atoms or molecules are completely separated from one another is zero. The

TABLE 21.3 Lattice Sums for Molecular Crystals (fcc Structure)

	Number, n	p_{ij}	$n(p_{ij})^{-12}$	$n(p_{ij})^{-6}$
Nearest neighbors	12	1	12	12
Second nearest neighbors	6	$\sqrt{2}$	0.0938	0.750
Third nearest neighbors	24	$\sqrt{3}$	0.0329	0.889
Fourth nearest neighbors	12	2	0.0029	0.188
Fifth nearest neighbors	24	$\sqrt{5}$	0.0015	0.192
	⋮	⋮	⋮	⋮
Total			12.132	14.454

TABLE 21.4 Properties of Noble-Gas Crystals†

	R_0 (Å)		Lattice Energy (kJ mol^{-1})	
	Predicted	Observed	Predicted	Observed
Ne	3.00	3.13	1.83	1.88
Ar	3.71	3.76	7.72	7.74
Kr	3.92	4.01	11.50	11.20
Xe	4.47	4.35	15.20	16.00

†All data are extrapolated to 0 K and zero pressure.

lattice energy is the difference between these quantities and is a positive number:

$$\text{lattice energy} = -V_{\text{tot}} = 8.61\varepsilon N_{\text{A}}$$

This overestimates the true lattice energy because of the quantum effect of zero-point energy (see Sections 4.6 and 4.7). When a quantum correction is applied, the binding energy is reduced by 28%, 10%, 6%, and 4% for Ne, Ar, Kr, and Xe, respectively. Table 21.4 shows the resulting predictions for crystal lattice energies and nearest neighbor distances. The agreement with experiment is quite reasonable, considering the approximations inherent in the use of a Lennard–Jones potential derived entirely from gas-phase data. For helium the amplitude of zero-point motion is so great that if a crystal did form, it would immediately melt. Consequently, helium remains liquid down to absolute zero at atmospheric pressure.

Lattice Energy of an Ionic Crystal

In Section 3.6, we calculated the potential energy of a gaseous diatomic ionic molecule relative to the separated ions by means of Coulomb's law:

$$V = \frac{q_1 q_2}{4\pi\epsilon_0 R_0}$$

where R_0 is the equilibrium internuclear separation. Coulomb's law can also be used to calculate the lattice energies of ionic compounds in the crystalline state.

For simplicity, consider a hypothetical one-dimensional crystal (Fig. 21.27), in which ions of charge $+e$ and $-e$ alternate with an internuclear separation of R_0. One ion, selected to occupy an arbitrary origin, will interact attractively with all ions of opposite sign to make the following contribution to the crystal energy:

$$V_{\text{attraction}} = -\frac{e^2}{4\pi\epsilon_0 R_0}\left[2(1) + 2(\tfrac{1}{3}) + 2(\tfrac{1}{5}) + \ldots\right]$$

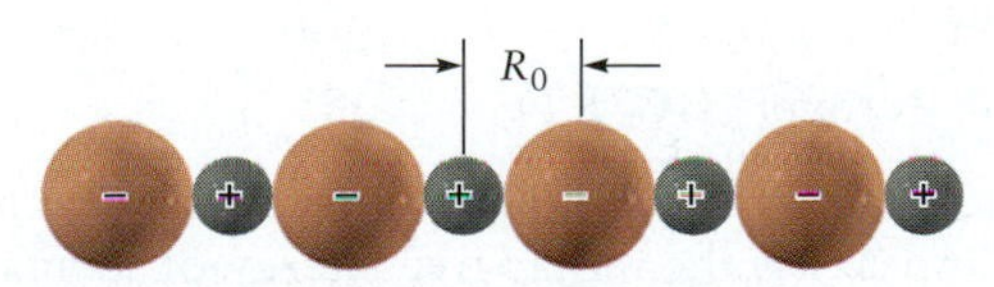

FIGURE 21.27 Lattice energy for a one-dimensional ionic crystal.

Here the factors of 2 come from the fact that there are *two* ions of opposite sign at a distance R_0 from a given ion, two at a distance $3R_0$, two at $5R_0$, and so forth. The negative sign occurs because the ions that occupy odd-numbered sites have a charge opposite that of the ion at the origin, and their interaction with the ion at the origin is attractive. The ion at the origin also interacts repulsively with all ions of the same sign to make the following contribution to the crystal energy:

$$V_{\text{repulsion}} = +\frac{e^2}{4\pi\epsilon_0 R_0}\left[2(\tfrac{1}{2}) + 2(\tfrac{1}{4}) + 2(\tfrac{1}{6}) + \ldots\right]$$

The net interaction of N_{A} such ions of each sign with one another is:

$$V_{\text{net}} = -\frac{N_{\text{A}}e^2}{4\pi\epsilon_0 R_0}\left[2 - \frac{2}{2} + \frac{2}{3} - \frac{2}{4} + \frac{2}{5} - \frac{2}{6} + \ldots\right]$$

We must be very careful with factors of 2. To obtain the potential energy for the interaction of N_{A} positive ions with N_{A} negative ions, it is necessary to multiply the total potential energy of a given ion due to all others by $2N_{\text{A}}$ and then divide by 2 to avoid counting the interaction of a given pair of ions twice. This gives the preceding result.

If such a calculation is carried out for a real three-dimensional crystal, the result is a series (such as that just given in brackets) whose value sums to a dimensionless number that depends upon the crystal structure. That number is called the **Madelung constant,** M, and its value is independent of the unit-cell dimensions. Table 21.5 lists the values of the Madelung constant for several crystal structures. The lattice energy is again the opposite of the total potential energy. Expressed in terms of the Madelung constant, it is

$$\text{lattice energy} = \frac{N_{\text{A}}e^2}{4\pi\epsilon_0 R_0}M \qquad [21.3]$$

TABLE 21.5 Madelung Constants

Lattice	M
Rock salt	1.7476
CsCl	1.7627
Zinc blende	1.6381
Fluorite	2.5194

EXAMPLE 21.5

Calculate the electrostatic part of the lattice energy of sodium chloride, given that the internuclear separation between Na^+ and Cl^- ions is 2.82 Å.

SOLUTION

Using the Madelung constant of 1.7476 for this structure gives

$$\frac{(6.02 \times 10^{23}\ \text{mol}^{-1})(1.602 \times 10^{-19}\ \text{C})^2(1.7476)}{(4\pi)(8.854 \times 10^{-12}\ \text{C}^2\ \text{J}^{-1}\ \text{m}^{-1})(2.82 \times 10^{-10}\ \text{m})} =$$

$$8.61 \times 10^5\ \text{J mol}^{-1} = 861\ \text{kJ mol}^{-1}$$

Related Problems: 35, 36

Ionic lattice energies are measured experimentally by means of a thermodynamic cycle developed by Max Born and Fritz Haber. The **Born–Haber cycle** is an application of Hess's law (the first law of thermodynamics). It is illustrated by a determination of the lattice energy of sodium chloride, which is ΔE for the reaction

$$\text{NaCl}(s) \longrightarrow \text{Na}^+(g) + \text{Cl}^-(g) \qquad \Delta E = ?$$

This reaction can be represented as a series of steps, each with a measurable energy or enthalpy change. In the first step, the ionic solid is converted to the elements in their standard states:

$$\text{NaCl}(s) \longrightarrow \text{Na}(s) + \tfrac{1}{2}\,\text{Cl}_2(g)$$

$$\Delta E_1 \approx \Delta H = -\Delta H_f^\circ(\text{NaCl}(s)) = +411.2\ \text{kJ}$$

In the second step, the elements are transformed into gas-phase atoms:

$$\text{Na}(s) \longrightarrow \text{Na}(g) \qquad \Delta E \approx \Delta H = \Delta H_f^\circ(\text{Na}(g)) = +107.3\ \text{kJ}$$

$$\tfrac{1}{2}\,\text{Cl}_2(g) \longrightarrow \text{Cl}(g) \qquad \Delta E \approx \Delta H = \Delta H_f^\circ(\text{Cl}(g)) = +121.7\ \text{kJ}$$

$$\text{Na}(s) + \tfrac{1}{2}\,\text{Cl}_2(g) \longrightarrow \text{Na}(g) + \text{Cl}(g) \qquad \Delta E_2 = +229.0\ \text{kJ}$$

Finally, in the third step, electrons are transferred from the sodium atoms to the chlorine atoms to give ions:

$$\text{Na}(g) \longrightarrow \text{Na}^+(g) + e^- \qquad \Delta E = IE_1(\text{Na}) = 496\ \text{kJ}$$

$$\text{Cl}(g) + e^- \longrightarrow \text{Cl}^-(g) \qquad \Delta E = -EA(\text{Cl}) = -349\ \text{kJ}$$

$$\text{Na}(g) + \text{Cl}(g) \longrightarrow \text{Na}^+(g) + \text{Cl}^-(g) \qquad \Delta E_3 = +147\ \text{kJ}$$

Here $EA(\text{Cl})$ is the electron affinity of Cl, and $IE_1(\text{Na})$ is the first ionization energy of Na. The total energy change is

$$\Delta E = \Delta E_1 + \Delta E_2 + \Delta E_3 = 411 + 229 + 147 = +787\ \text{kJ}$$

The small differences between ΔE and ΔH were neglected in this calculation. If their difference is taken into account (using $\Delta H = \Delta E + RT\Delta n_g$, where Δn_g is the change in the number of moles of gas molecules in each step of the reaction), then ΔE_2 is decreased by $\frac{3}{2}RT$ and ΔE_1 by $\frac{1}{2}RT$, giving a net decrease of $2RT$ and changing ΔE to $+782$ kJ mol^{-1}. Comparing this experimental lattice energy with the energy calculated in Example 21.5 shows that the latter is approximately 10% greater, presumably because short-range repulsive interactions and zero-point energy were not taken into account in the lattice energy calculation.

21.5 Defects and Amorphous Solids

Although real crystals display beautiful symmetries to the eye, they are not perfect. As a practical matter, it is impossible to rid a crystal of all impurities or to ensure that it contains perfect periodic ordering. So, we describe real crystals as "perfect crystals with defects," and define means to characterize these defects. If so many defects are present that crystalline order is destroyed, we describe the material as an **amorphous solid.**

Point Defects

Point defects in a pure crystalline substance include **vacancies,** in which atoms are missing from lattice sites, and **interstitials,** in which atoms are inserted in sites different from their normal sites. In real crystals, a small fraction of the normal atom sites remain unoccupied. Such vacancies are called **Schottky defects,** and their concentration depends on temperature:

$$N = N_s \exp(-\Delta G/RT)$$

where N is the number of lattice vacancies per unit volume, N_s is the number of atom sites per unit volume, and ΔG is the molar free energy of formation of

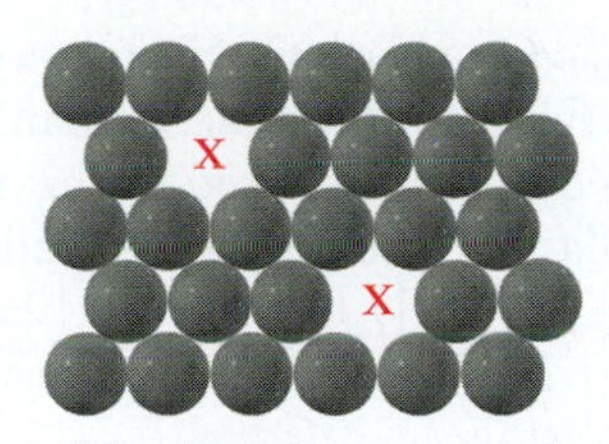

Schottky defects in a metal or noble gas crystal

(a)

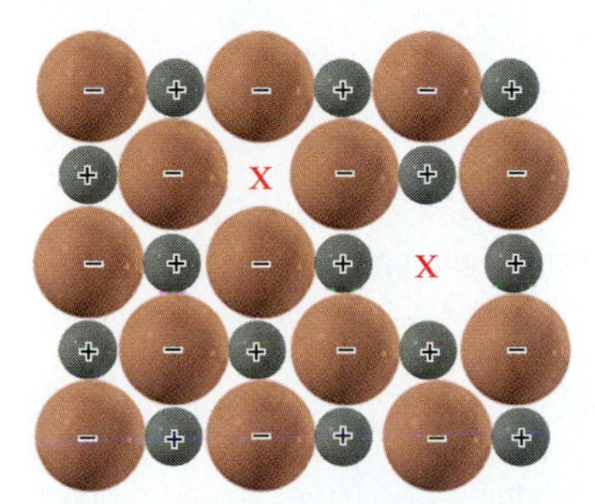

Schottky defects in an ionic crystal

(b)

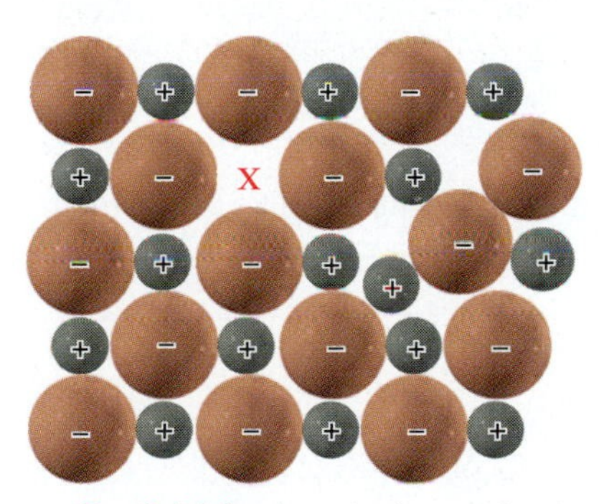

Frenkel defect in an ionic crystal

(c)

FIGURE 21.28 Point imperfections in a lattice. The red Xs denote vacancies.

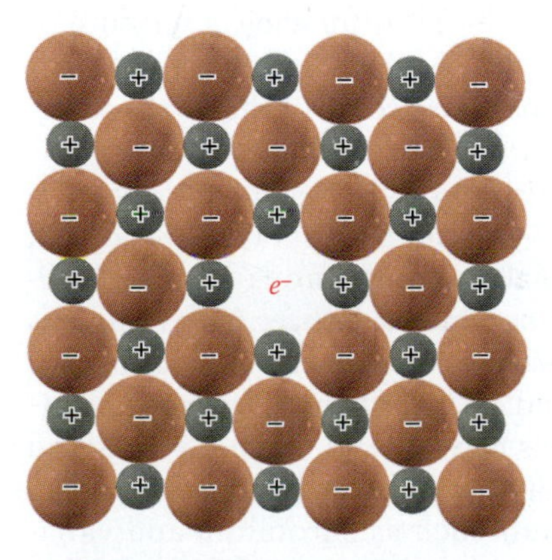

FIGURE 21.29 An F-center in a crystal.

vacancies. Figure 21.28a illustrates Schottky defects in the crystal structure of a metal or noble gas. Schottky defects also occur in ionic crystals but with the restriction that the imperfect crystals remain electrically neutral. Thus, in sodium chloride, for every missing Na^+ ion there must also be a missing Cl^- ion (see Fig. 21.28b).

In certain kinds of crystals, atoms or ions are displaced from their regular lattice sites to interstitial sites, and the crystal defect consists of the lattice vacancy plus the interstitial atom or ion. Figure 21.28c illustrates this type of lattice imperfection, known as a **Frenkel defect.** The silver halides (AgCl, AgBr, AgI) are examples of crystals in which Frenkel disorder is extreme. The crystal structures of these compounds are established primarily by the anion lattice, and the silver ions occupy highly disordered, almost random, sites. The rate of diffusion of silver ions in these solids is exceptionally high, as studies using radioactive isotopes of silver have shown. Both Frenkel and Schottky defects in crystals are mobile, jumping from one lattice site to a neighboring site with frequencies that depend on the temperature and the strengths of the atomic forces. Diffusion in crystalline solids is due largely to the presence and mobility of point defects; it is a thermally activated process, just like the rates of chemical reactions considered in Chapter 18. The coefficient of self-diffusion has the form

$$D = D_0 \exp[-E_a/RT] \qquad [21.4]$$

where E_a is the activation energy. The rates of diffusive motion in crystalline solids vary enormously from one substance to another. In a crystal of a low-melting metal such as sodium, an average atom undergoes about 10^8 diffusive jumps per second at 50°C, whereas in a metal such as tungsten that melts at 3410°C, an average atom jumps to another lattice site less than once per year at 1000°C!

If an alkali halide crystal such as NaCl is irradiated with x-rays, ultraviolet radiation, or high-energy electrons, some Cl^- ions may lose an electron:

$$Cl^- + h\nu \longrightarrow Cl + e^-$$

The resulting Cl atom, being uncharged and much smaller than a Cl^- ion, is no longer strongly bound in the crystal and can diffuse to the surface and escape. The electron can migrate through the crystal quite freely until it encounters an anion vacancy and is trapped in the Coulomb field of the surrounding cations (Fig. 21.29). This crystal defect is called an **F-center** (from the German word *Farbenzentrum,* meaning **color center**). It is the simplest of a family of electronic crystal defects. As the name suggests, it imparts a color to ionic crystals (Fig. 21.30).

Nonstoichiometric Compounds

As Chapter 1 emphasizes, the law of definite proportions was one of the principal pieces of evidence that led to the acceptance of Dalton's atomic theory. It is now recognized that a great many solid-state binary compounds do *not* have fixed and unvarying compositions but exist over a range of compositions in a single phase. Thus, FeO (wüstite) has the composition range $Fe_{0.85}O_{1.00}$ to $Fe_{0.95}O_{1.00}$ and is never found with its nominal 1:1 composition. The compounds NiO and Cu_2S also deviate considerably from their nominal stoichiometries.

The explanation depends on the existence of more than one oxidation state for the metal. In wüstite, iron can exist in either the +2 or the +3 oxidation state. Suppose a solid were to begin at the hypothetical composition of $Fe_{1.00}O_{1.00}$, with iron entirely in the +2 oxidation state. For every two Fe^{3+} ions introduced, three Fe^{2+} ions must be removed to maintain overall charge neutrality. The total number of moles of iron is then less than that in the ideal FeO stoichiometry. The departure from the nominal stoichiometry can be far more extreme than that found

FIGURE 21.30 Pure calcium fluoride, CaF2, is white but the natural sample of calcium fluoride (fluorite) shown here is a rich purple because F-centers are present. These are lattice sites where the F^- anion is replaced by an electron only.

in wüstite. The composition of "TiO" ranges from $Ti_{0.75}O$ to $Ti_{1.45}O$. Nickel oxide varies only from $Ni_{0.97}O$ to NiO in composition, but the variation is accompanied by a dramatic change in properties. When the compound is prepared in the 1:1 composition, it is pale green and is an electrical insulator. When it is prepared in an excess of oxygen, it is black and conducts electricity fairly well. In the black material, a small fraction of Ni^{2+} ions are replaced by Ni^{3+} ions, and compensating vacancies occur at some nickel atom sites in the crystal.

EXAMPLE 21.6

The composition of a sample of wüstite is $Fe_{0.930}O_{1.00}$. What percentage of the iron is in the form of iron(III)?

SOLUTION

For every 1.00 mol of oxygen atoms in this sample, there is 0.930 mol of iron atoms. Suppose y mol of the iron is in the +3 oxidation state and $0.930 - y$ is in the +2 oxidation state. Then the total positive charge from the iron (in moles of electron charge) is

$$+3y + 2(0.930 - y)$$

This positive charge must exactly balance the 2 mol of negative charge carried by the mole of oxygen atoms (recall that each oxygen atom has oxidation number −2). We conclude that

$$3y + 2(0.930 - y) = +2$$

Solving this equation for y gives

$$y = 0.140$$

The percentage of iron in the form of Fe^{3+} is then the ratio of this number to the total number of moles of iron, 0.930, multiplied by 100%:

$$\% \text{ iron in form of } Fe^{3+} = \frac{0.140}{0.930} \times 100\% = 15.1\%$$

Related Problems: 41, 42

Alloys

The nonstoichiometric compounds just described are ionic materials with compositional disorder. A related type of disorder is exhibited by an **alloy,** a mixture of elements that displays metallic properties.

There are two types of alloys. In a **substitutional alloy,** some of the metal atoms in a crystal lattice are replaced by other atoms (usually of comparable size). Examples are brass, in which approximately one third of the atoms in a copper crystal are replaced by zinc atoms, and pewter, an alloy of tin that contains 7% copper, 6% bismuth, and 2% antimony. In an **interstitial alloy,** atoms of one or more additional elements enter the interstitial sites of the host metal lattice. An example is steel, in which carbon atoms occupy interstitial sites of an iron crystal, making the material stronger and harder than pure iron. Mild steel contains less than 0.2% C and is used for nails, whereas high-carbon steels can contain up to 1.5% C and are used in specialty applications such as tools and springs. *Alloy steels* are both substitutional and interstitial; atoms from metals such as chromium and vanadium substitute for iron atoms, with carbon remaining in interstitial sites. Alloy steels have a variety of specialized purposes, ranging from cutlery to bicycle frames.

© Stockbyte Silver/Getty Images

FIGURE 21.31 Optical fibers (extremely thin glass fibers of specialized composition) carry information in the form of light waves.

Amorphous Solids and Glasses

The arrangements of atoms, ions, or molecules in crystalline solids exhibit high degrees of spatial order. Now let us briefly consider solids that lack this characteristic. **Amorphous solids,** commonly called **glasses,** resemble crystalline solids in many respects. They may have chemical compositions, mechanical properties such as hardness and elasticity, and electrical and magnetic properties that are similar to those of crystals. Like crystals, glasses may have molecular, ionic, covalent, or metallic bonding. On an atomic scale, however, amorphous solids lack the regular periodic structure of crystals. They are states of matter in which so many defects are present that crystalline order is destroyed.

Some substances have a strong tendency to solidify as glasses. The best example is the material used in common window panes, with the approximate chemical formula $Na_2O \cdot CaO \cdot (SiO_2)_6$. This is a partly ionic, partly covalent material with Na^+ and Ca^{2+} ions distributed through a covalently bonded Si—O network. Glass-forming ability is not restricted to a few special materials, however. If a substance can be liquefied, it can almost certainly be prepared in an amorphous state. Even metals, which are known primarily in the crystalline state, have been made into amorphous solids. The trick is to bypass crystallization by cooling molten material very fast. One technique involves shooting a jet of liquid metal at a rapidly rotating cold cylinder, which produces a continuous ribbon of amorphous metal at a rate up to 2 km per minute.

On the molecular level, a strong tendency to form glasses is associated with the presence of long or irregularly shaped molecules that can easily become tangled and disordered. Even slowly cooling a liquid assembly of such molecules may not afford enough time for them to organize into a crystalline lattice before solidification. Instead of a sharp liquid-to-crystal transition, such glass-formers transform continuously, over a range of temperature, into amorphous solids. They lend themselves to fabrication into articles of every conceivable shape, because the flow properties of the work piece can be managed by controlling its temperature. This plasticity is the reason that glass has played an indispensable role in science, industry, and the arts.

One of the most exciting new uses of a glass is the transmission of voice messages, television images, and data as light pulses. Tens of thousands of audio messages can be transmitted simultaneously through glass fibers no greater in diameter than a human hair. This is done by encoding the audio signal into electronic impulses that modulate light from a laser source. The light then passes down the glass fiber as though it were a tube. Chemical control of the glass composition reduces light loss and permits messages to travel many kilometers without amplification (Fig. 21.31).

CHAPTER SUMMARY

This chapter describes the distinctive properties of solids as the consequences of the collective behavior of an extended array of chemical bonds. The chemical bonds involved are already familiar in the context of isolated molecules. What is new here is the extended array of these bonds in the solid state. We can determine the array experimentally by x-ray diffraction. To describe the array we introduce the concepts of local symmetry and repetitive, long-range order; together, these define the unit cell and the crystal lattice. By placing structural elements at the points of the lattice, we complete the description of the structure of a perfect crystal. The number of atoms per unit cell depends on packing density of the atoms in the lattice. The properties of different classes of solids depend on the type of bonding involved: ionic, covalent, metallic, or molecular. The perfect crystal is the fundamental starting point for investigations in the solid state. Practical materials can be described and categorized by the nature and extent of their deviations from perfect crystal structure.

CUMULATIVE EXERCISE

Phosphorus

Solid elemental phosphorus appears in a rich variety of forms, with crystals in all seven crystal systems reported under various conditions of temperature, pressure, and sample preparation.

(a) The thermodynamically stable form of phosphorus under room conditions is black phosphorus. Its unit cell is orthorhombic with edges of lengths 3.314, 4.376, and 10.48 Å. Calculate the volume of one unit cell, and determine the number of phosphorus atoms per unit cell, if the density of this form of phosphorus is 2.69 g cm^{-3}.

(b) The form of phosphorus that is easiest to prepare from the liquid or gaseous state is white phosphorus, which consists of P_4 molecules in a cubic lattice. When x-rays of wavelength 2.29 Å are scattered from the parallel faces of its unit cells, the first-order Bragg diffraction is observed at an angle 2θ of 7.10°. Calculate the length of the unit-cell edge for white phosphorus. At what angle will third-order Bragg diffraction be seen?

(c) Amorphous red phosphorus has been reported to convert to monoclinic, triclinic, tetragonal, and cubic red forms with different heat treatments. Identify the changes in the shape of the unit cell as a cubic lattice is converted first to tetragonal, then monoclinic, then triclinic.

(d) A monoclinic form of red phosphorus has been studied that has cell edge lengths 9.21, 9.15, and 22.60 Å, with an angle β of 106.1°. Each unit cell contains 84 atoms of phosphorus. Estimate the density of this form of phosphorus.

(e) Phosphorus forms many compounds with other elements. Describe the nature of the bonding in the solids white elemental phosphorus (P_4), black elemental phosphorus, sodium phosphate (Na_3PO_4), and phosphorus trichloride (PCl_3).

Answers

(a) Volume is 152.0 Å^3; eight atoms per unit cell

(b) 18.5 Å; angle $2\theta = 21.4°$

(c) Cubic to tetragonal: One cell edge is stretched or shrunk. Tetragonal to monoclinic: A second cell edge is stretched or shrunk, and the angles between two adjacent faces (and their opposite faces) are changed from 90°. Monoclinic to triclinic: The remaining two angles between faces are deformed from 90°.

(d) 2.36 g cm^{-3}

(e) P_4(white) and PCl_3 are molecular solids; P(black) is covalent; Na_3PO_4 is ionic.

Two forms of elemental phosphorus: white and red.

CHAPTER REVIEW

- Solids are distinguished from liquids and gases by their rigidity. Solids retain their shape and exhibit structural strength when external forces are applied. These properties originate in the strong, directional chemical bonds between the atoms in solids.
- Solids whose structures are highly ordered and symmetrical over macroscopic distances are called crystals. Concepts and methods have been developed to define and measure the structures of crystals.

Crystal structure is defined in terms of the crystal lattice, a mathematical abstraction that represents the ordered and repetitive nature of the structure. In effect the lattice is a set of coordinates for locating each atom in the structure of a network. The lattice embodies all the symmetry in the structure. Lattice points are identified by the fundamental symmetry operations of rotation, reflection, and inversion.

The unit cell is the smallest region of a crystal lattice that contains all the structural information about the crystal. So, the crystal lattice is visualized as a stack of multiple identical unit cells. Each unit cell has characteristic lengths and angles. There are seven types of crystal structures, each defined by the properties of its unit cell: hexagonal, cubic, tetragonal, trigonal, orthorhombic, monoclinic, and triclinic.

The distance d between planes in a crystal can be measured experimentally by x-ray diffraction. It is related to the wavelength λ of the radiation and the diffraction angle θ by Bragg's law $n\lambda = 2d \sin \theta$, where n is the order of the diffraction.

Once the crystal lattice has been identified, description of the structure is completed by specifying the structural elements that are located at lattice points. The resulting structures are named by characteristics of the unit cell. Three examples are displayed by the elemental metals.

Simple cubic (one atom per unit cell)	Po
Body-centered cubic (two atoms per unit cell)	alkali metals
Face-centered cubic (four atoms per unit cell)	Al, Ni, Cu, Ag

- Different types of chemical bonding appear in solids and are responsible for the differences in mechanical and structural properties of different types of solids.
- Because electrostatic forces are strong and operate over large distances, ionic crystals are hard, brittle solids that have high melting points and are poor conductors of heat and electricity They crystallize in structures determined primarily by atomic packing density.
  zinc blende structure for cation-anion radius ratio smaller than 0.414
  rock-salt structure for cation-anion radius ratio between 0.414 and 0.732
  cesium chloride structure for cation-anion radius ratio greater than 0.732
- Bonding in metallic crystals is explained as a sea of delocalized electrons around positively charged ions located at the lattice sites. The number density of electrons is equal to the number density of positive ions, so the metal is electrically neutral. The bonds are quite strong, evidenced by the high boiling points of metals. Metals are malleable and ductile because the highly mobile electrons can rapidly adjust when lattice ions are pushed to new locations by external mechanical forces. Metals are good conductors of heat and electricity because the delocalized electrons respond easily to applied external fields.
- Covalent crystals are held together by strong, highly directional bonds usually described by the valence bond hybrid orbital method. Each atom is part of a large extended single molecule that is the crystal itself. Because of the nature of their bonds, covalent crystals have very high melting points and are hard and brittle.
- Molecular crystals are held together by van der Waals forces, the same as the intermolecular forces in gases and liquids. Because these are much weaker than ionic, metallic, and covalent bonds, the molecular crystals are usually soft, easily deformed, and have low melting points.

- Real-world crystals do not have perfect symmetry and order. It is useful to characterize practical materials by the ways in which they deviate from the structure of the perfect crystal models described earlier.

  Point defects where atoms are missing from lattice sites are called vacancies, or Schottky defects. Their number density depends on the temperature and on the Gibbs free energy of formation of defects.

  Point defects where atoms are located between lattice sites are called interstitials. If a lattice atom is displaced to an interstitial site, the combination of the defect and the interstitial is called a Frenkel defect.

- Diffusion in solids occurs as Schottky or Frenkel defects hop from one lattice site to another by a thermally activated process akin to chemical reactions.
- Amorphous solids, or glasses, may have chemical compositions and mechanical properties similar to crystalline materials, and they may have ionic, metallic, covalent, or molecular bonding. But at the microscopic level they lack crystalline order. This arises from kinetic effects during solidification that reduce the mobility of atoms or molecules and prevent them from achieving ordered structures. Long or irregularly shaped molecules that are easily entangled lead to glass formation.

CONCEPTS & SKILLS

After studying this chapter and working the problems that follow, you should be able to:

1. Identify the symmetry elements of different crystal systems (Section 21.1, Problems 1–4).
2. Explain how x-rays and neutrons are diffracted by crystals, and use information from such experiments to calculate lattice spacings (Section 21.1, Problems 5–10).
3. Describe the packing of atoms in simple crystal lattices (Section 21.2, Problems 11–24).
4. Compare the natures of the forces that hold atoms or molecules in their lattice sites in ionic, metallic, covalent, and molecular crystals (Section 21.3, Problems 27–32).
5. Calculate lattice energies of molecular and ionic crystals (Section 21.4, Problems 35–38).
6. Describe the kinds of equilibrium defects that are present in crystalline solids and the properties of amorphous solids (Section 21.5).
7. Determine the oxidation states present in nonstoichiometric solids (Section 21.5, Problems 41–42).

KEY EQUATIONS

$$n\lambda = 2d \sin\theta \qquad n = 1, 2, 3, \ldots \qquad \textbf{(Section 21.1)}$$

$$V_c = abc\sqrt{1 - \cos^2\alpha - \cos^2\beta - \cos^2\gamma + 2\cos\alpha\cos\beta\cos\gamma} \qquad \textbf{(Section 21.2)}$$

$$\text{lattice energy} = \frac{N_A e^2}{4\pi\epsilon_0 R_0} M \qquad \textbf{(Section 21.4)}$$

$$D = D_0 \exp[-E_a/RT] \qquad \textbf{(Section 21.5)}$$

PROBLEMS

Answers to problems whose numbers are boldface appear in Appendix G. Problems that are more challenging are indicated with asterisks.

Crystal Symmetry and the Unit Cell

1. Which of the following has 3-fold rotational symmetry? Explain.
(a) An isosceles triangle
(b) An equilateral triangle
(c) A tetrahedron
(d) A cube

2. Which of the following has 4-fold rotational symmetry? Explain.
(a) A cereal box (exclusive of the writing on the sides)
(b) A stop sign (not counting the writing)
(c) A tetrahedron
(d) A cube

3. Identify the symmetry elements of the CCl_2F_2 molecule (see Fig. 12.15).

4. Identify the symmetry elements of the PF_5 molecule (see Fig. 3.20a).

5. The second-order Bragg diffraction of x-rays with λ = 1.660 Å from a set of parallel planes in copper occurs at an angle 2θ = 54.70°. Calculate the distance between the scattering planes in the crystal.

6. The second-order Bragg diffraction of x-rays with λ = 1.237 Å from a set of parallel planes in aluminum occurs at an angle 2θ = 35.58°. Calculate the distance between the scattering planes in the crystal.

7. The distance between members of a set of equally spaced planes of atoms in crystalline lead is 4.950 Å. If x-rays with λ = 1.936 Å are diffracted by this set of parallel planes, calculate the angle 2θ at which fourth-order Bragg diffraction will be observed.

8. The distance between members of a set of equally spaced planes of atoms in crystalline sodium is 4.28 Å. If x-rays with λ = 1.539 Å are diffracted by this set of parallel planes, calculate the angle 2θ at which second-order Bragg diffraction will be observed.

9. The members of a series of equally spaced parallel planes of ions in crystalline LiCl are separated by 2.570 Å. Calculate all the angles 2θ at which diffracted beams of various orders may be seen, if the x-ray wavelength used is 2.167 Å.

10. The members of a series of equally spaced parallel planes in crystalline vitamin B_{12} are separated by 16.02 Å. Calculate all the angles 2θ at which diffracted beams of various orders may be seen, if the x-ray wavelength used is 2.294 Å.

Crystal Structure

11. A crucial protein at the photosynthetic reaction center of the purple bacterium *Rhodopseudomonas viridis* (see Section 20.6) has been separated from the organism, crystallized, and studied by x-ray diffraction. This substance crystallizes with a primitive unit cell in the tetragonal system. The cell dimensions are $a = b$ = 223.5 Å and c = 113.6 Å.
(a) Determine the volume, in cubic angstroms, of this cell.
(b) One of the crystals in this experiment was box-shaped, with dimensions 1 × 1 × 3 mm. Compute the number of unit cells in this crystal.

12. Compute the volume (in cubic angstroms) of the unit cell of potassium hexacyanoferrate(III) ($K_3Fe(CN)_6$), a substance that crystallizes in the monoclinic system with a = 8.40 Å, b = 10.44 Å, and c = 7.04 Å and with β = 107.5°.

13. The compound $Pb_4In_3B_{17}S_{18}$ crystallizes in the monoclinic system with a unit cell having a = 21.021 Å, b = 4.014 Å, c = 18.898 Å, and the only non-90° angle equal to 97.07°. There are two molecules in every unit cell. Compute the density of this substance.

14. Strontium chloride hexahydrate ($SrCl_2 \cdot 6H_2O$) crystallizes in the trigonal system in a unit cell with a = 8.9649 Å and α = 100.576°. The unit cell contains three formula units. Compute the density of this substance.

15. At room temperature, the edge length of the cubic unit cell in elemental silicon is 5.431 Å, and the density of silicon at the same temperature is 2.328 g cm^{-3}. Each cubic unit cell contains eight silicon atoms. Using only these facts, perform the following operations.
(a) Calculate the volume (in cubic centimeters) of one unit cell.
(b) Calculate the mass (in grams) of silicon present in a unit cell.
(c) Calculate the mass (in grams) of an atom of silicon.
(d) The mass of an atom of silicon is 28.0855 u. Estimate Avogadro's number to four significant figures.

16. One form of crystalline iron has a bcc lattice with an iron atom at every lattice point. Its density at 25°C is 7.86 g cm^{-3}. The length of the edge of the cubic unit cell is 2.87 Å. Use these facts to estimate Avogadro's number.

17. Sodium sulfate (Na_2SO_4) crystallizes in the orthorhombic system in a unit cell with a = 5.863 Å, b = 12.304 Å, and c = 9.821 Å. The density of these crystals is 2.663 g cm^{-3}. Determine how many Na_2SO_4 formula units are present in the unit cell.

18. The density of turquoise, $CuAl_6(PO_4)_4(OH)_8(H_2O)_4$, is 2.927 g cm^{-3}. This gemstone crystallizes in the triclinic system with cell constants a = 7.424 Å, b = 7.629 Å, c = 9.910 Å, α = 68.61°, β = 69.71°, and γ = 65.08°. Calculate the volume of the unit cell, and determine how many copper atoms are present in each unit cell of turquoise.

19. An oxide of rhenium has a structure with a Re atom at each corner of the cubic unit cell and an O atom at the center of each edge of the cell. What is the chemical formula of this compound?

20. The mineral perovskite has a calcium atom at each corner of the unit cell, a titanium atom at the center of the unit cell, and an oxygen atom at the center of each face. What is the chemical formula of this compound?

* **21.** Iron has a body-centered cubic structure with a density of 7.86 g cm^{-3}.
(a) Calculate the nearest neighbor distance in crystalline iron.
(b) What is the lattice parameter for the cubic unit cell of iron?
(c) What is the atomic radius of iron?

22. The structure of aluminum is fcc and its density is $\rho = 2.70 \text{ g cm}^{-3}$.
(a) How many Al atoms belong to a unit cell?
(b) Calculate a, the lattice parameter, and d, the nearest neighbor distance.

23. Sodium has the body-centered cubic structure, and its lattice parameter is 4.28 Å.
(a) How many Na atoms does a unit cell contain?
(b) What fraction of the volume of the unit cell is occupied by Na atoms, if they are represented by spheres in contact with one another?

24. Nickel has a fcc structure with a density of 8.90 g cm^{-3}.
(a) Calculate the nearest neighbor distance in crystalline nickel.
(b) What is the atomic radius of nickel?
(c) What is the radius of the largest atom that could fit into the interstices of a nickel lattice, approximating the atoms as spheres?

25. Calculate the ratio of the maximum radius of an interstitial atom at the center of a simple cubic unit cell to the radius of the host atom.

26. Calculate the ratio of the maximum radius of an interstitial atom at the center of each face of a bcc unit cell to the radius of the host atom.

Cohesion in Solids

27. Classify each of the following solids as molecular, ionic, metallic, or covalent.
(a) $BaCl_2$ (b) SiC
(c) CO (d) Co

28. Classify each of the following solids as molecular, ionic, metallic, or covalent.
(a) Rb (b) C_5H_{12}
(c) B (d) Na_2HPO_4

29. The melting point of cobalt is 1495°C, and that of barium chloride is 963°C. Rank the four substances in problem 27 from lowest to highest in melting point.

30. The boiling point of pentane (C_5H_{12}) is slightly less than the melting point of rubidium. Rank the four substances in problem 28 from lowest to highest in melting point.

31. Explain the relationship between the number of bonds that can be formed by a typical atom in a crystal and the possibility of forming linear, two-dimensional, and three-dimensional network structures.

32. Although large crystals of sugar (rock candy) and large crystals of salt (rock salt) have different geometric shapes, they look much the same to the untrained observer. What physical tests other than taste might be performed to distinguish between these two crystalline substances?

33. By examining Figure 21.17, determine the number of nearest neighbors, second nearest neighbors, and third nearest neighbors of a Cs^+ ion in crystalline CsCl. The nearest neighbors of the Cs^+ ion are Cl^- ions, and the second nearest neighbors are Cs^+ ions.

34. Repeat the determinations of the preceding problem for the NaCl crystal, referring to Figure 21.16.

A DEEPER LOOK *Lattice Energies of Crystals*

35. Calculate the energy needed to dissociate 1.00 mol of crystalline RbCl into its gaseous ions if the Madelung constant for its structure is 1.7476 and the radii of Rb^+ and Cl^- are 1.48 Å and 1.81 Å, respectively. Assume that the repulsive energy reduces the lattice energy by 10% from the pure Coulomb energy.

36. Repeat the calculation of problem 35 for CsCl, taking the Madelung constant from Table 21.5 and taking the radii of Cs^+ and Cl^- to be 1.67 Å and 1.81 Å.

37. (a) Use the Born–Haber cycle, with data from Appendices D and F, to calculate the lattice energy of LiF.
(b) Compare the result of part (a) with the Coulomb energy calculated by using an Li—F separation of 2.014 Å in the LiF crystal, which has the rock-salt structure.

38. Repeat the calculations of problem 37 for crystalline KBr, which has the rock-salt structure with a K—Br separation of 3.298 Å.

Defects and Amorphous Solids

39. Will the presence of Frenkel defects change the measured density of a crystal?

40. What effect will the (unavoidable) presence of Schottky defects have on the determination of Avogadro's number via the method described in problems 15 and 16?

41. Iron(II) oxide is nonstoichiometric. A particular sample was found to contain 76.55% iron and 23.45% oxygen by mass.
(a) Calculate the empirical formula of the compound (four significant figures).
(b) What percentage of the iron in this sample is in the +3 oxidation state?

42. A sample of nickel oxide contains 78.23% Ni by mass.
(a) What is the empirical formula of the nickel oxide to four significant figures?
(b) What fraction of the nickel in this sample is in the +3 oxidation state?

ADDITIONAL PROBLEMS

43. Some water waves with a wavelength of 3.0 m are diffracted by an array of evenly spaced posts in the water. If the rows of posts are separated by 5.0 m, calculate the angle 2θ at which the first-order "Bragg diffraction" of these water waves will be seen.

44. A crystal scatters x-rays of wavelength $\lambda = 1.54$ Å at an angle 2θ of 32.15°. Calculate the wavelength of the x-rays in another experiment if this same diffracted beam from the same crystal is observed at an angle 2θ of 34.46°.

45. The number of beams diffracted by a single crystal depends on the wavelength λ of the x-rays used and on the volume associated with one lattice point in the crystal—that is, on the volume V_p of a primitive unit cell. An approximate formula is

$$\text{number of diffracted beams} = \frac{4}{3}\pi\left(\frac{2}{\lambda}\right)^3 V_p$$

(a) Compute the volume of the conventional unit cell of crystalline sodium chloride. This cell is cubic and has an edge length of 5.6402 Å.
(b) The NaCl unit cell contains four lattice points. Compute the volume of a primitive unit cell for NaCl.
(c) Use the formula given in this problem to estimate the number of diffracted rays that will be observed if NaCl is irradiated with x-rays of wavelength 2.2896 Å.
(d) Use the formula to estimate the number of diffracted rays that will be observed if NaCl is irradiated with x-rays having the shorter wavelength 0.7093 Å.

46. If the wavelength λ of the x-rays is too large relative to the spacing of planes in the crystal, no Bragg diffraction will be seen because $\sin\theta$ would be larger than 1 in the Bragg equation, even for $n = 1$. Calculate the longest wavelength of x-rays that can give Bragg diffraction from a set of planes separated by 4.20 Å.

47. The crystal structure of diamond is fcc, and the atom coordinates in the unit cell are $(0, 0, 0)$, $(\frac{1}{2}, \frac{1}{2}, 0)$, $(\frac{1}{2}, 0, \frac{1}{2})$, $(0, \frac{1}{2}, \mathrm{v})$, $(\frac{1}{4}, \frac{1}{4}, \frac{1}{4})$, $(\frac{3}{4}, \frac{1}{4}, \frac{3}{4})$, $(\frac{3}{4}, \frac{3}{4}, \frac{1}{4})$, and $(\frac{1}{4}, \frac{3}{4}, \frac{3}{4})$. The lattice parameter is $a = 3.57$ Å. What is the C—C bond distance in diamond?

48. Polonium is the only element known to crystallize in the simple cubic lattice.
(a) What is the distance between nearest neighbor polonium atoms if the first-order diffraction of x-rays with $\lambda = 1.785$ Å from the parallel faces of its unit cells appears at an angle of $2\theta = 30.96°$ from these planes?
(b) What is the density of polonium in this crystal (in g cm^{-3})?

49. At room temperature, monoclinic sulfur has the unit-cell dimensions $a = 11.04$ Å, $b = 10.98$ Å, $c = 10.92$ Å, and $\beta = 96.73°$. Each cell contains 48 atoms of sulfur.
(a) Explain why it is not necessary to give the values of the angles α and γ in this cell.
(b) Compute the density of monoclinic sulfur (in g cm^{-3}).

50. A compound contains three elements: sodium, oxygen, and chlorine. It crystallizes in a cubic lattice. The oxygen atoms are at the corners of the unit cells, the chlorine atoms are at the centers of the unit cells, and the sodium atoms are at the centers of the faces of the unit cells. What is the formula of the compound?

***51.** Show that the radius of the largest sphere that can be placed in a tetrahedral interstitial site in an fcc lattice is $0.225r_1$, where r_1 is the radius of the atoms making up the lattice. (*Hint:* Consider a cube with the centers of four spheres placed at alternate corners, and visualize the tetrahedral site at the center of the cube. What is the relationship between r_1 and the length of a diagonal of a face? The length of a body diagonal?)

52. What is the closest packing arrangement possible for a set of thin circular discs lying in a plane? What fraction of the area of the plane is occupied by the discs? Show how the same reasoning can be applied to the packing of infinitely long, straight cylindrical fibers.

53. Name two elements that form molecular crystals, two that form metallic crystals, and two that form covalent crystals. What generalizations can you make about the portions of the periodic table where each type is found?

54. The nearest-neighbor distance in crystalline LiCl (rock-salt structure) is 2.570 Å; the bond length in a gaseous LiCl molecule is significantly shorter, 2.027 Å. Explain.

55. (a) Using the data of Table 9.4, estimate the lattice energy and intermolecular separation of nitrogen in its solid state, assuming an fcc structure for the solid lattice.
(b) The density of cubic nitrogen is 1.026 g cm^{-3}. Calculate the lattice parameter a and the nearest neighbor distance. Compare your answer with that from part (a).

***56.** Solid CuI_2 is unstable relative to CuI at room temperature, but $CuBr_2$, $CuCl_2$, and CuF_2 are all stable relative to the copper(I) halides. Explain by considering the steps in the Born–Haber cycle for these compounds.

57. A crystal of sodium chloride has a density of 2.165 g cm^{-3} in the absence of defects. Suppose a crystal of NaCl is grown in which 0.15% of the sodium ions and 0.15% of the chloride ions are missing. What is the density in this case?

58. The activation energy for the diffusion of sodium atoms in the crystalline state is 42.22 kJ mol^{-1}, and $D_0 = 0.145$ cm^2 s^{-1}.
(a) Calculate the diffusion constant $D = D_0 \exp(-E_a/RT)$ of sodium in the solid at its melting point (97.8°C).
(b) What is the root-mean-square displacement of an average sodium atom from an arbitrary origin after the lapse of 1.0 hour at $t = 97.8$°C? (*Hint:* Use Equation 9.37 in Chapter 9.)

59. A compound of titanium and oxygen contains 28.31% oxygen by mass.
(a) If the compound's empirical formula is Ti_xO, calculate x to four significant figures.
(b) The nonstoichiometric compounds Ti_xO can be described as having a Ti^{2+}—O^{2-} lattice in which certain Ti^{2+} ions are missing or are replaced by Ti^{3+} ions. Calculate the fraction of Ti^{2+} sites in the nonstoichiometric compound that are vacant and the fraction that are occupied by Ti^{3+} ions.

60. Classify the bonding in the following amorphous solids as molecular, ionic, metallic, or covalent.
(a) Amorphous silicon, used in photocells to collect light energy from the sun
(b) Polyvinyl chloride, a plastic of long-chain molecules composed of $-CH_2CHCl-$ repeating units, used in pipes and siding
(c) Soda-lime-silica glass, used in windows
(d) Copper-zirconium glass, an alloy of the two elements with approximate formula Cu_3Zr_2, used for its high strength and good conductivity

CUMULATIVE PROBLEMS

61. Sodium hydride (NaH) crystallizes in the rock-salt structure, with four formula units of NaH per cubic unit cell. A beam of monoenergetic neutrons, selected to have a velocity of 2.639×10^3 m s^{-1}, is scattered in second order through an angle of $2\theta = 36.26°$ by the parallel faces of the unit cells of a sodium hydride crystal.
   (a) Calculate the wavelength of the neutrons.
   (b) Calculate the edge length of the cubic unit cell.
   (c) Calculate the distance from the center of an Na^+ ion to the center of a neighboring H^- ion.
   (d) If the radius of a Na^+ ion is 0.98 Å, what is the radius of an H^- ion, assuming the two ions are in contact?

62. Chromium(III) oxide has a structure in which chromium ions occupy two thirds of the octahedral interstitial sites in a hexagonal close-packed lattice of oxygen ions. What is the *d*-electron configuration on the chromium ion?

63. A useful rule of thumb is that in crystalline compounds every nonhydrogen atom occupies 18 Å^3, and the volume occupied by hydrogen atoms can be neglected. Using this rule, estimate the density of ice (in g cm^{-3}). Explain why the answer is so different from the observed density of ice.

64. Estimate, for the F-centers in CaF_2, the wavelength of maximum absorption in the visible region of the spectrum that will give rise to the color shown in Figure 21.30.

CHAPTER 22

Inorganic Materials

Azurite is a basic copper carbonate with chemical formula $Cu_3(CO_3)_2(OH)_2$.

Having laid the conceptual foundation for relating properties of solids to chemical bonding in Chapter 21, we turn now to applications of these concepts to three important classes of materials: ceramics, electronic materials, and optical materials. All of these are synthetic materials fashioned from inorganic, nonmetallic substances by chemical methods of synthesis and processing.

Ceramics are one of the oldest classes of materials prepared by humankind. New discoveries in ceramics are occurring at a startling rate, and major new technologic advances will certainly come from these discoveries. Ceramics have value both as structural materials—the role emphasized in this chapter—and for their wide range of electronic and optical properties.

High-speed computing, fast communication, and rapid display of information were major technologic achievements in the second half of the 20th century. These developments will grow even more rapidly in the 21st century, when every home will be digitally connected to the Internet and every cell phone will display images requiring broadband transmission. These developments were made possible by the "microelectronics revolution," beginning with the first integrated circuits fabricated on microchips by Jack Kilby at Texas Instruments and Robert Noyce at Fairchild in 1958 to 1959.[1] The speed of computers increased dramatically through advances in transistor design; advances in the solid-state laser enabled high-speed communication via fiber optics. All these advances in device design relied critically on equally dramatic accomplishments in the growth and processing of materials, which in every case involved making and breaking chemical bonds in solid-state materials. Our goals in this chapter are to introduce the optical and electronic properties of materials and to show how they depend on chemical structure.

Electronic properties describe the movement of charged particles in a material in response to an applied electric field. If the charges are free to move throughout the material, the process is *electrical conduction*, measured by the *electrical conductivity* of the material. Differences in the magnitude of the conductivity distinguish metals, semiconductors, and insulators. If the charges can move only limited distances and are then halted by opposing binding forces, separation of positive and negative charges leads to *electric polarization* of the material, measured by its *dielectric constant*. Conduction involves dissipation of energy as heat, whereas polarization involves storage of potential energy in the material.

Optical properties describe the response of a material to electromagnetic radiation, particularly visible light. The list of optical properties is long, including reflection of light from a surface, refraction (bending the direction) of light as it passes from one medium into another, absorption, and transmission. We limit the discussion here to the generation and detection of light in solid materials, as extensions of the molecular processes of emission and excitation already described in Chapter 20. The absorption of light creates the bright colors of inorganic pigments and the conversion of solar energy into electrical energy in solar cells.

In this survey of mechanical, electrical, and optical properties, keep two questions in mind: (1) How does a material respond on an atomic level to applied mechanical stress and to electrical and optical fields? (2) If the chemical structure is modified, how does this change influence the response of the material to these forces?

22.1 Minerals: Naturally Occurring Inorganic Materials

This chapter begins with a survey of the naturally occurring inorganic, nonmetallic minerals that are the starting materials for synthesis and processing of inorganic materials.

Silicates

Silicon and oxygen make up most of the earth's crust, with oxygen accounting for 47% and silicon for 28% of its mass. The silicon–oxygen bond is strong and partially ionic. It forms the basis for a class of minerals called **silicates,** which make up the bulk of the rocks, clays, sand, and soils in the earth's crust. From time immemorial, silicates have provided the ingredients for building materials such as bricks, cement, concrete, and glass (which are considered later in this chapter).

The structure-building properties of silicates (Table 22.1) originate in the tetrahedral orthosilicate anion (SiO_4^{4-}), in which the negative charge of the silicate ion

[1]Kilby was awarded the Nobel Prize in Physics in 2000, but Noyce, who had died, did not share the award. The Nobel Prize is not awarded posthumously.

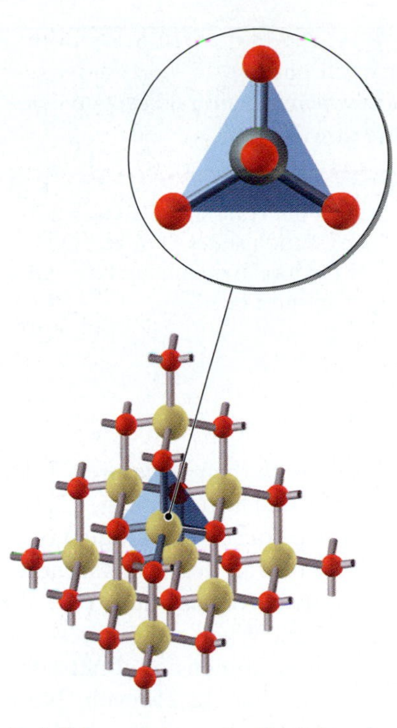
Each SiO_2 unit shares O—Si—O bonds with other SiO_2 units arranged in a lattice of tetrahedra. Si atoms are shown in yellow and O atoms in red.

is balanced by the compensating charge of one or more cations. The simplest silicates consist of individual SiO_4^{4-} anions (Fig. 22.1a), with cations arranged around them on a regular crystalline lattice. Such silicates are properly called **orthosilicates.** Examples are forsterite (Mg_2SiO_4) and fayalite (Fe_2SiO_4), which are the extreme members of a class of minerals called **olivines,** $[Mg,Fe]_2SiO_4$. There is a continuous range of proportions of magnesium and iron in the olivines.

Other silicate structures form when two or more SiO_4^{4-} tetrahedra link and share oxygen vertices. The simplest such minerals are the **disilicates,** such as thortveitite, $Sc_2(Si_2O_7)$, in which two tetrahedra are linked (see Fig. 22.1b). Additional linkages of tetrahedra create the ring, chain, sheet, and network structures shown in Figure 22.1 and listed in Table 22.1. In each of these, the fundamental tetrahedron is readily identified, but the Si:O ratio is no longer 4 because oxygen (O) atoms are shared at the linkages.

TABLE 22.1 Silicate Structures

Structure	Figure Number	Corners Shared at Each Si atom	Repeat Unit	Si:O Ratio	Example
Tetrahedra	22.1a	0	SiO_4^{4-}	1:4	Olivines
Pairs of tetrahedra	22.1b	1	$Si_2O_7^{6-}$	$1:3\frac{1}{2}$	Thortveitite
Closed rings	22.1c	2	SiO_3^{2-}	1:3	Beryl
Infinite single chains	22.1d	2	SiO_3^{2-}	1:3	Pyroxenes
Infinite double chains	22.1e	$2\frac{1}{2}$	$Si_4O_{11}^{6-}$	$1:2\frac{3}{4}$	Amphiboles
Infinite sheets	22.1f	3	$Si_2O_5^{2-}$	$1:2\frac{1}{2}$	Talc
Infinite network	22.1g	4	SiO_2	1:2	Quartz

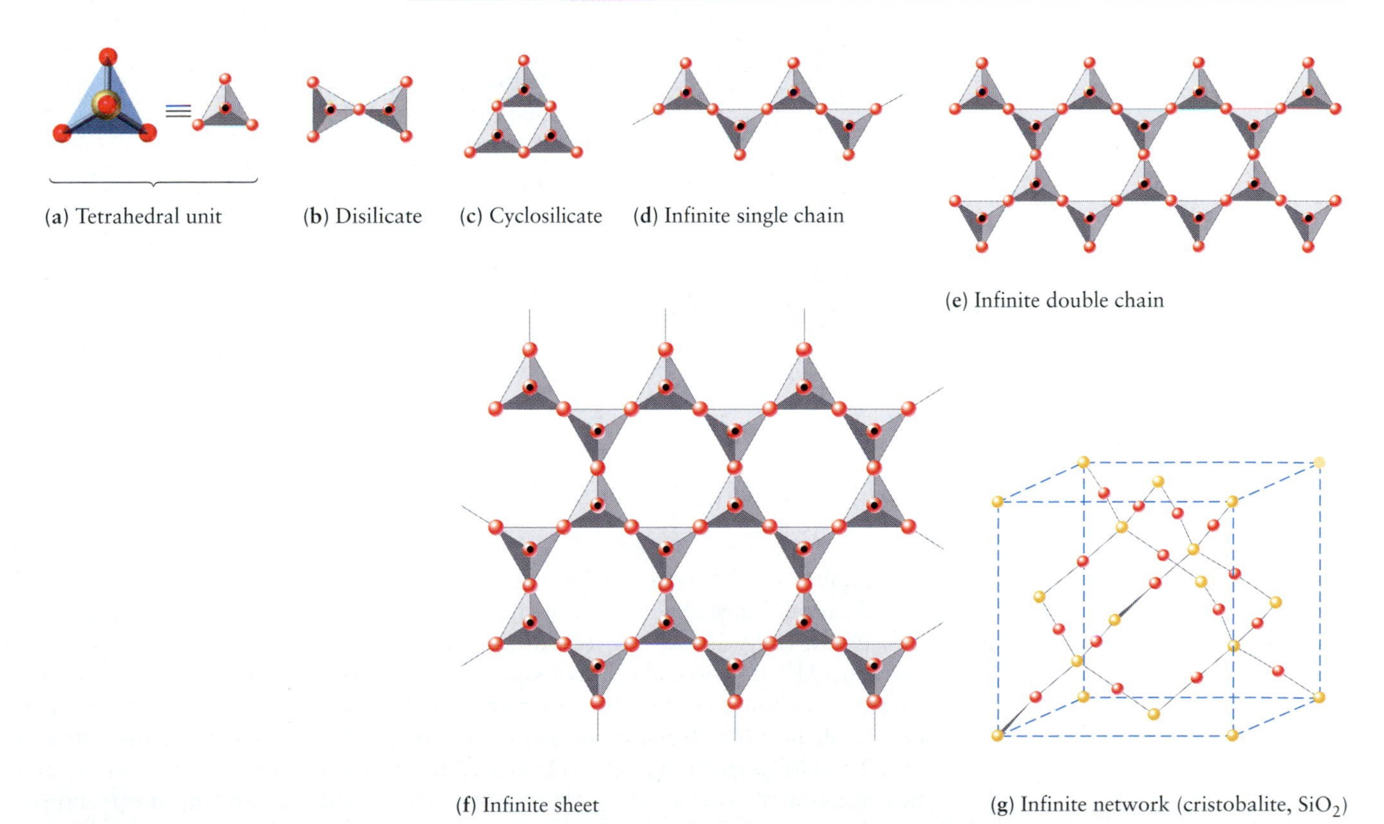

FIGURE 22.1 Classes of silicate structures. (a) Symbol used for the SiO_4^{4-} tetrahedron. This top view of the symbol shows a red circle to represent the fourth oxygen atom at the top of the tetrahedron. The black dot represents the silicon atom at the center of the tetrahedron. Bear in mind that all of these structures are actually three dimensional. Planar projections are used here for convenience of representation. (b) Disilicate. (c) Cyclosilicate. (d) Infinite single chain. (e) Infinite double chain. (f) Infinite sheet. (g) Infinite network (cristobalite, SiO_2).

FIGURE 22.2 The mineral quartz is one form of silica, SiO_2.

FIGURE 22.3 The fibrous structure of asbestos is apparent in this sample.

EXAMPLE 22.1

By referring to Table 22.1, predict the structural class in which the mineral Egyptian blue ($CaCuSi_4O_{10}$) belongs. Give the oxidation state of each of its atoms.

SOLUTION

Because the silicon/oxygen (Si:O) ratio is 4:10, or $1:2\frac{1}{2}$, this mineral should have an infinite sheet structure with the repeating unit $Si_2O_5^{2-}$. The oxidation states of Si and O are +4 and −2, as usual, and that of calcium (Ca) is +2. For the total oxidation number per formula unit to sum to 0, the oxidation state of copper (Cu) must be +2.

Related Problems: 3, 4

The physical properties of the silicates correlate closely with their structures. Talc, $Mg_3(Si_4O_{10})(OH)_2$, is an example of an infinite layered structure (see Fig. 22.1f). In talc, all of the bonding interactions among the atoms occur in a single layer. Layers of talc sheets are attracted to one another only by van der Waals interactions, which (being weak) permit one layer to slip easily across another. This accounts for the slippery feel of talc (called talcum powder). When all four vertices of each tetrahedron are linked to other tetrahedra, three-dimensional network structures such as cristobalite (see Fig. 22.1g) or quartz (Fig. 22.2) result. Note that the quartz network carries no charge; consequently, there are no cations in its structure. Three-dimensional network silicates such as quartz are much stiffer and harder than the linear and layered silicates, and they resist deformation well.

Asbestos is a generic term for a group of naturally occurring, hydrated silicates that can be processed mechanically into long fibers (Fig. 22.3). Some of these silicates, such as tremolite, $Ca_2Mg_5(Si_4O_{11})_2(OH)_2$, show the infinite double-chain structure of Figure 22.1e. Another kind of asbestos mineral is chrysotile, $Mg_3(Si_2O_5)(OH)_4$. As the formula indicates, this mineral has a sheet structure (see Fig. 22.1f), but the sheets are rolled into long tubes. Asbestos minerals are fibrous because the bonds along the strandlike tubes are stronger than those that hold different tubes together. Asbestos is an excellent thermal insulator that does not burn, resists acids, and is strong. For many years, it was used in cement for pipes and ducts and woven into fabric to make fire-resistant roofing paper and floor tiles. Its use has decreased significantly in recent years because inhalation of its small fibers can cause the lung disease asbestosis. The risk comes with breathing asbestos dust that is raised during mining and manufacturing processes or that is released in buildings in which asbestos-containing materials are fraying, crumbling, or being removed.

Aluminosilicates

An important class of minerals called **aluminosilicates** results from the replacement of some of the Si atoms in silicates with aluminum (Al) atoms. Aluminum is the third most abundant element in the earth's crust (8% by mass), where it occurs largely in the form of aluminosilicates. Aluminum in minerals can be a simple cation (Al^{3+}), or it can replace silicon in tetrahedral coordination. When it replaces silicon, it contributes only three electrons to the bonding framework in place of the four electrons of Si atoms. The additional required electron is supplied by the ionization of a metal atom such as sodium (Na) or potassium (K); the resulting alkali-metal ions occupy nearby sites in the aluminosilicate structure.

The most abundant and important of the aluminosilicate minerals in the earth's surface are the **feldspars,** which result from the substitution of aluminum for silicon in three-dimensional silicate networks such as quartz. The Al ions must be accompanied by other cations such as sodium, potassium, or calcium to maintain overall charge neutrality. Albite is a feldspar with the chemical formula $NaAlSi_3O_8$.

FIGURE 22.4 Naturally occurring muscovite mica. The mechanical properties of crystals of mica are quite anisotropic. Thin sheets can be peeled off a crystal of mica by hand, but the sheets resist stresses in other directions more strongly. Transparent, thin sheets of mica, sometimes called isinglass, have been used for heat-resistant windows in stoves or in place of window glass.

In the high-temperature form of this mineral, the Al and Si atoms are distributed at random (in 1 : 3 proportion) over the tetrahedral sites available to them. At lower temperatures, other crystal structures become thermodynamically stable, with partial ordering of the Al and Si sites.

If one of the four Si atoms in the structural unit of talc, $Mg_3(Si_4O_{10})(OH)_2$, is replaced by an Al atom and a K atom is furnished to supply the fourth electron needed for bonding in the tetrahedral silicate framework, the result is the composition $KMg_3(AlSi_3O_{10})(OH)_2$, which belongs to the family of **micas** (Fig. 22.4). Mica is harder than talc, and its layers slide less readily over one another, although the crystals still cleave easily into sheets. The cations occupy sites between the infinite sheets, and the van der Waals bonding that holds adjacent sheets together in talc is augmented by an ionic contribution. The further replacement of the three Mg^{2+} ions in $KMg_3(AlSi_3O_{10})(OH)_2$ with two Al^{3+} ions gives the mineral muscovite, $KAl_2(AlSi_3O_{10})(OH)_2$. Writing its formula in this way indicates that there are Al atoms in two kinds of sites in the structure: One Al atom per formula unit occupies a tetrahedral site, substituting for one Si atom, and the other two Al atoms are between the two adjacent layers. The formulas that mineralogists and crystallographers use convey more information than the usual empirical chemical formula of a compound.

Clay Minerals

Clays are minerals produced by the weathering action of water and heat on primary minerals. Their compositions can vary widely as a result of the replacement of one element with another. Invariably, they are microcrystalline or powdered in form and are usually hydrated. Often, they are used as supports for catalysts, as fillers in paint, and as ion-exchange vehicles. The clays that readily absorb water and swell are used as lubricants and bore-hole sealers in the drilling of oil wells.

The derivation of clays from talcs and micas provides a direct way to understand the structures of the clays. The infinite-sheet mica pyrophyllite, $Al_2(Si_4O_{10})(OH)_2$, serves as an example. If one of six Al^{3+} ions in the pyrophyllite structure is replaced by one Mg^{2+} ion and one Na^+ ion (which together carry the same charge), a type of clay called montmorillonite, $MgNaAl_5(Si_4O_{10})_3(OH)_6$, results. This clay readily absorbs water, which infiltrates between the infinite sheets and hydrates the Mg^{2+} and Na^+ ions there, causing the montmorillonite to swell (Fig. 22.5).

A different clay derives from the layered mineral talc, $Mg_3(Si_4O_{10})(OH)_2$. If iron(II) and aluminum replace magnesium and silicon in varying proportions and water molecules are allowed to take up positions between the layers, the swelling clay vermiculite results. When heated, vermiculite pops like popcorn, as the steam generated by the vaporization of water between the layers puffs the flakes up into a light, fluffy material with air inclusions. Because of its porous structure, vermiculite is used for thermal insulation or as an additive to loosen soils.

Zeolites

Zeolites are a class of three-dimensional aluminosilicates. Like the feldspars, they carry a negative charge on the aluminosilicate framework that is compensated by neighboring alkali-metal or alkaline-earth cations. Zeolites differ from feldspars in having much more open structures that consist of polyhedral cavities connected by tunnels (Fig. 22.6). Many zeolites are found in nature, but they can also be synthesized under conditions controlled to favor cavities of uniform size and shape. Most zeolites accommodate water molecules in their cavities, where they provide a mobile phase for the migration of the charge-compensating cations. This enables zeolites to serve as ion-exchange materials (in which one kind of positive ion can be readily exchanged for another) and is the key to their ability to soften water. Water "hardness" arises from soluble calcium and magnesium salts such as

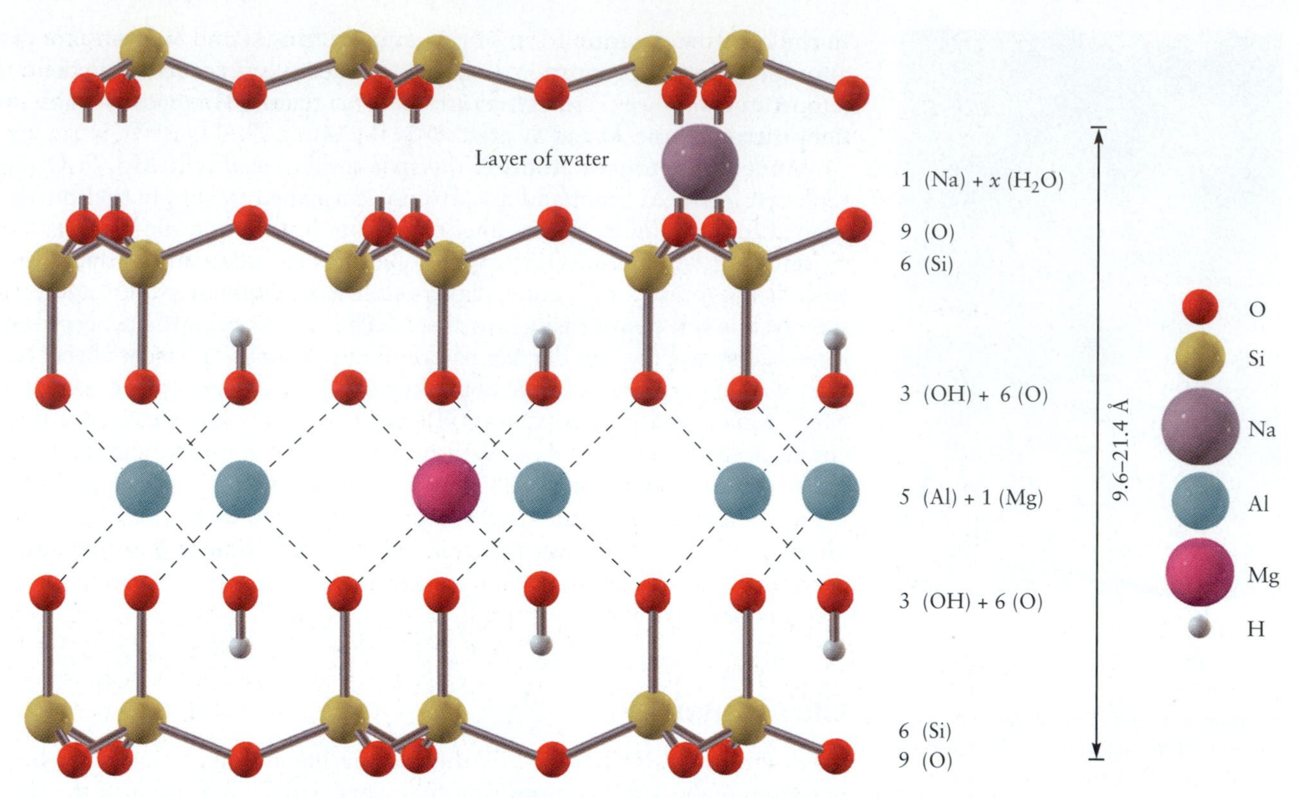

FIGURE 22.5 Structure of the clay mineral montmorillonite. Insertion of variable amounts of water causes the distance between layers to swell from 9.6 Å to more than 20 Å. When 1 Al^{3+} ion is replaced by an Mg^{2+} ion, an additional ion such as Na^{+} is introduced into the water layers to maintain overall charge neutrality.

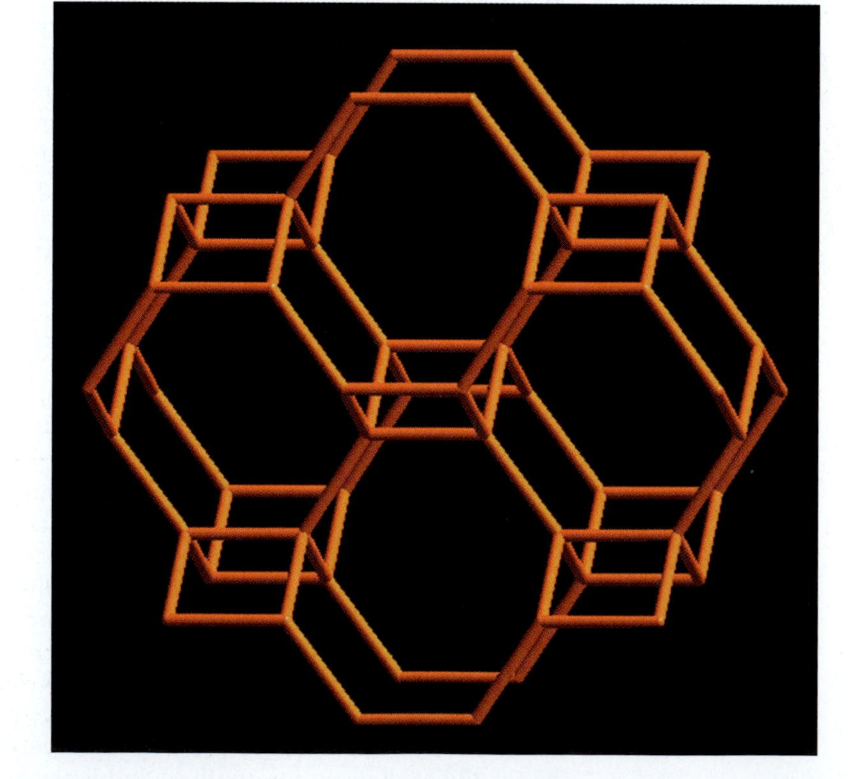

FIGURE 22.6 Structure of the synthetic zeolite Li-ABW [$Li_4(Si_4Al_4O_{16})\cdot 4H_2O$].
(C. Baerlocher, L.B. McCusker: Atlas of Zeolite Structure Types. http://www.iza-sc.ethz.ch/IZA-SC/Atlas/AtlasHome.html)

$Ca(HCO_3)_2$ and $Mg(HCO_3)_2$. Such salts are converted to insoluble carbonates (boiler scale) when the water is heated and form objectionable precipitates (bathtub ring) with soaps. When hard water is passed through a column packed with a zeolite that has Na ions in its structure, the Ca and Mg ions exchange with the Na ions and are removed from the water phase:

$$2\ NaZ(s) + Ca^{2+}(aq) \rightleftharpoons CaZ_2(s) + 2\ Na^{+}(aq)$$

When the ion-exchange capacity of the zeolite is exhausted, this reaction can be reversed by passing a concentrated solution of sodium chloride through the zeolite to regenerate it in the sodium form.

A second use of zeolites derives from the ease with which they adsorb small molecules. Their spongelike affinity for water makes them useful as drying agents; they are put between the panes of double-pane glass windows to prevent moisture from condensing on the inner surfaces. The pore size of zeolites can be selected to allow molecules that are smaller than a certain size to pass through but hold back larger molecules. Such zeolites serve as "molecular sieves"; they have been used to capture nitrogen molecules in a gas stream while permitting oxygen molecules to pass through.

Perhaps the most exciting use of zeolites is as catalysts. Molecules of varying sizes and shapes have different rates of diffusion through a zeolite; this feature enables chemists to enhance the rates and yields of desired reactions and suppress unwanted reactions. The most extensive applications of zeolites currently are in the catalytic cracking of crude oil, a process that involves breaking down long-chain hydrocarbons and re-forming them into branched-chain molecules of lower molecular mass for use in high-octane unleaded gasoline. A relatively new process uses "shape-selective" zeolite catalysts to convert methanol (CH_3OH) to high-quality gasoline. Plants have been built to make gasoline by this process, using methanol derived from coal or from natural gas.

22.2 Properties of Ceramics

The term ***ceramics*** covers synthetic materials that have as their essential components inorganic, nonmetallic materials. This broad definition includes cement, concrete, and glass, in addition to the more traditional fired clay products such as bricks, roof tiles, pottery, and porcelain. The use of ceramics predates recorded history; the emergence of civilization from a primitive state is chronicled in fragments of pottery. No one knows when small vessels were first shaped by human hands from moist clay and left to harden in the heat of the sun. Such containers held nuts, grains, and berries well, but they lost their shape and slumped into formless mud when water was poured into them. Then someone discovered (perhaps by accident) that if clay was placed in the glowing embers of a fire, it became as hard as rock and withstood water well. Molded figures (found in what is now the Czech Republic) that were made 24,000 years ago are the earliest fired ceramic objects discovered so far, and fired clay vessels from the Near East date from 8000 B.C. With the action of fire on clay, the art and science of ceramics began.

Ceramics offer stiffness, hardness, resistance to wear, and resistance to corrosion (particularly by oxygen and water), even at high temperature. They are less dense than most metals, which makes them desirable metal substitutes when weight is a factor. Most are good electrical insulators at ordinary temperatures, a property that is exploited in electronics and power transmission. Ceramics retain their strength well at high temperatures. Several important structural metals soften or melt at temperatures 1000°C below the melting points of their chemical compounds in ceramics. Aluminum, for example, melts at 660°C, whereas aluminum oxide (Al_2O_3), an important compound in many ceramics, does not melt until a temperature of 2051°C is reached.

Against these advantages must be listed some serious disadvantages. Ceramics are generally brittle and low in tensile strength. They tend to have high thermal expansion but low thermal conductivity, making them subject to **thermal shock,** in which sudden local temperature change causes cracking or shattering. Metals and plastics dent or deform under stress, but ceramics cannot absorb stress in this way: instead, they break. A major drawback of ceramics as structural materials is their tendency to fail unpredictably and catastrophically in use. Moreover, some ceramics lose mechanical strength as they age, an insidious and serious problem.

Composition and Structure of Ceramics

Ceramics use a variety of chemical compounds, and useful ceramic bodies are nearly always mixtures of several compounds. **Silicate ceramics,** which include the commonplace pots, dishes, and bricks, are made from aluminosilicate clay minerals. All contain the tetrahedral SiO_4 grouping discussed in Section 22.1. In **oxide ceramics,** silicon is a minor or nonexistent component. Instead, a number of metals combine with oxygen to give compounds such as alumina (Al_2O_3), magnesia (MgO), or yttria (Y_2O_3). **Nonoxide ceramics** contain compounds that are free of oxygen as principal components. Some important compounds in nonoxide ceramics are silicon nitride (Si_3N_4), silicon carbide (SiC), and boron carbide (approximate composition B_4C).

One important property of ceramics is their porosity. Porous ceramics have small openings into which fluids (typically air or water) can infiltrate. Fully dense ceramics have no channels of this sort. Two ceramic pieces can have the same chemical composition but quite different densities if the first is porous and the second is not.

A **ceramic phase** is any portion of the whole body that is physically homogeneous and bounded by a surface that separates it from other parts. Distinct phases are visible at a glance in coarse-grained ceramic pieces; in a fine-grained piece, phases can be seen with a microscope. When examined on a still finer scale, most ceramics, like metals, are microcrystalline, consisting of small crystalline grains cemented together (Fig. 22.7). The **microstructure** of such objects includes the sizes and shapes of the grains, the sizes and distribution of voids (openings between grains) and cracks, the identity and distribution of impurity grains, and the presence of stresses within the structure. Microstructural variations have enormous importance in ceramics because slight changes at this level strongly influence the properties of individual ceramic pieces. This is less true for plastic and metallic objects.

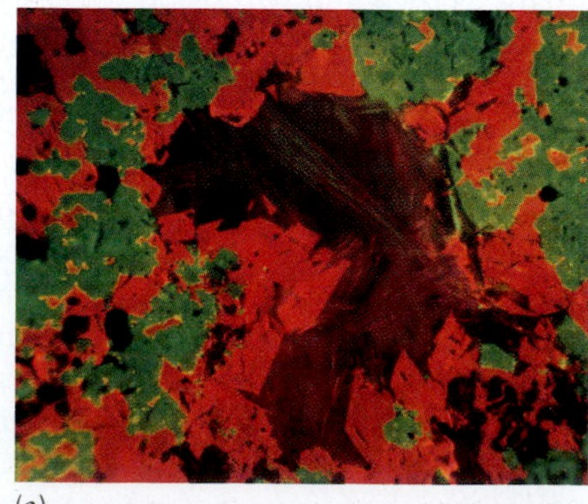

(a)

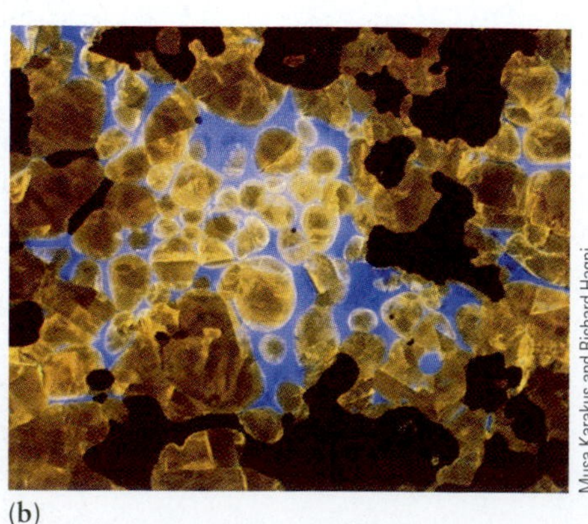

(b)

Musa Karakus and Richard Hagni

FIGURE 22.7 Microstructures of aluminosilicate ceramics, viewed by the different colors of light emitted after bombardment by electrons. (a) Forsterite (red), spinel (green), and periclase (dark brown) grains. (b) Periclase (blue) and oldhamite (yellow) grains.

FIGURE 22.8 A part freshly fabricated from silicon nitride for use in a gas turbine engine. Other parts of the engine are shown in the background.

The microstructure of a ceramic body depends markedly on the details of its fabrication. The techniques of forming and firing a ceramic piece are as important as its chemical composition in determining ultimate behavior because they confer a unique microstructure. This fact calls attention to the biggest problem with ceramics as structural materials: inconsistent quality. Ceramic engineers can produce parts that are stronger than steel, but not reliably so because of the difficulties of monitoring and controlling microstructure. Gas turbine engines fabricated of silicon nitride (Fig. 22.8), for example, run well at 1370°C, which is hot enough to soften or melt most metals. The higher operating temperature increases engine efficiency, and the ceramic turbines weigh less, which further boosts fuel economy. Despite these advantages, there is no commercial ceramic gas turbine. Acceptable ceramic turbines have to be built from selected, pretested components. The testing costs and rejection rates are so high that economical mass production has been impossible so far.

Making Ceramics

The manufacture of most ceramics involves four steps: (1) the preparation of the raw material; (2) the forming of the desired shape, often achieved by mixing a powder with water or other binder and molding the resulting plastic mass; (3) the drying and firing of the piece, also called its **densification,** because pores (voids) in the dried ceramic fill in; and (4) the finishing of the piece by sawing, grooving, grinding, or polishing.

The raw materials for traditional ceramics are natural clays that come from the earth as powders or thick pastes and become plastic enough after adjustment of their water content to be formed freehand or on a potter's wheel. Special ceramics (both oxide and nonoxide) require chemically pure raw materials that are produced synthetically. Close control of the purity of the starting materials for these ceramics is essential to produce finished pieces with the desired properties. In addition to being formed by hand or in open molds, ceramic pieces are shaped by the squeezing (compacting) of the dry or semidry powders in a strong, closed mold of the desired shape, at either ordinary or elevated temperatures (hot pressing).

Firing a ceramic causes **sintering** to occur. In sintering, the fine particles of the ceramic start to merge together by diffusion at high temperatures. The density of the material increases as the voids between grains are partially filled. Sintering occurs below the melting point of the material and shrinks the ceramic body. In addition to the merging of the grains of the ceramic, firing causes partial melting, chemical reactions among different phases, reactions with gases in the atmosphere of the firing chamber, and recrystallization of compounds with an accompanying growth in crystal size. All of these changes influence the microstructure of a piece and must be understood and controlled. Firing accelerates physical and chemical changes, of course, but thermodynamic equilibrium in a fired ceramic piece is rarely reached. Kinetic factors—including the rate of heating, the length of time at which each temperature is held, and the rate of cooling—influence microstructure. As a result, the use of microwave radiation (as in microwave ovens) rather than kilns to fire ceramics is under development in ceramic factories, because it promises more exact control of the heating rate, and thereby more reliable quality.

22.3 Silicate Ceramics

The silicate ceramics include materials that vary widely in composition, structure, and use. They range from simple earthenware bricks and pottery to cement, fine porcelain, and glass. Their structural strength is based on the same linking of silicate ion tetrahedra that gives structure to silicate minerals in nature.

Pottery and Structural Clay Products

Aluminosilicate clays are products of the weathering of primary minerals. When water is added to such clays in moderate amount, a thick paste results that is easily molded into different shapes. Clays expand as water invades the space between adjacent aluminosilicate sheets of the mineral, but they release most of this water to a dry atmosphere and shrink. A small fraction of the water or hydroxide ions remains rather tightly bound by ion–dipole forces to cations between the aluminosilicate sheets and is lost only when the clay is heated to a high temperature. The firing of aluminosilicate clays simultaneously causes irreversible chemical changes to occur. The clay kaolinite ($Al_2Si_2O_5(OH)_4$) undergoes the following reaction:

$$\underset{\text{(kaolinite)}}{3\ Al_2Si_2O_5(OH)_4} \longrightarrow \underset{\text{(mullite)}}{Al_6Si_2O_{13}(s)} + \underset{\text{(silica)}}{4\ SiO_2(s)} + \underset{\text{(water)}}{6\ H_2O(g)}$$

The fired ceramic body is a mixture of two phases: mullite and silica. Mullite, a rare mineral in nature, takes the form of needlelike crystals that interpenetrate and confer strength on the ceramic. When the temperature is above 1470°C, the silica phase forms as minute grains of cristobalite, one of the several crystalline forms of SiO_2.

If chemically pure kaolinite is fired, the finished ceramic object is white. Such purified clay minerals are the raw material for fine china. As they occur in nature, clays contain impurities, such as transition-metal oxides, that affect the color of both the unfired clay and the fired ceramic object if they are not removed. The colors of the metal oxides arise from their absorption of light at visible wavelengths, as explained by crystal field theory (see Section 8.5). Common colors for ceramics are yellow or greenish yellow, brown, and red. Bricks are red when the clay used to make them has high iron content.

Before a clay is fired in a kiln, it must first be freed of moisture by slowly heating to about 500°C. If a clay body dried at room temperature were to be placed directly into a hot kiln, it would literally explode from the sudden, uncontrolled expulsion of water. A fired ceramic shrinks somewhat as it cools, causing cracks to form. These imperfections limit the strength of the fired object and are undesirable. The occurrence of imperfections can be reduced by coating the surface of a partially fired clay object with a **glaze,** a thin layer that minimizes crack formation in the underlying ceramic by holding it in a state of tension as it cools. Glazes, as their name implies, are glasses that have no sharp melting temperature, but rather harden and develop resistance to shear stresses increasingly as the temperature of the high-fired clay object is gradually reduced. Glazes generally are aluminosilicates that have high aluminum content to raise their viscosity, and thereby reduce the tendency to run off the surface during firing. They also provide the means of coloring the surfaces of fired clays and imparting decorative designs to them. Transition-metal oxides (particularly those of titanium [Ti], vanadium [V], chromium [Cr], manganese [Mn], iron [Fe], cobalt [Co], nickel [Ni], and Cu) are responsible for the colors. The oxidation state of the transition metal in the glaze is critical in determining the color produced and is controlled by regulating the composition of the atmosphere in the kiln; atmospheres rich in oxygen give high oxidation states, and those poor or lacking in oxygen give low oxidation states.

Glass

Glassmaking probably originated in the Near East about 3500 years ago. It is one of the oldest domestic arts, but its beginnings, like those of metallurgy, are obscure. Both required high-temperature, charcoal-fueled ovens and vessels made of materials that did not easily melt to initiate and contain the necessary chemical reactions. In the early period of glassmaking, desired shapes were fabricated by sculpting them from solid chunks of glass. At a later date, molten glass was poured in successive layers over a core of sand. A great advance was the invention of

© Charles D. Winters/Photo Researchers, Inc.

FIGURE 22.9 Handcrafted glassware is trimmed after being blown into the desired shape.

© David Guizak/Phototake

FIGURE 22.10 The strains associated with internal stresses in rapidly cooled glass can be made visible by viewing the glass in polarized light. They appear as colored regions.

glassblowing, which probably occurred in the first century B.C. A long iron tube was dipped into molten glass and a rough ball of viscous material was caused to accumulate on its end by rotating the tube. Blowing into the iron tube forced the soft glass to take the form of a hollow ball (Fig. 22.9) that could be further shaped into a vessel and severed from the blowing tube with a blade. Artisans in the Roman Empire developed glassblowing to a high degree, but with the decline of that civilization, the skill of glassmaking in Europe deteriorated until the Venetians redeveloped the lost techniques a thousand years later.

Glasses are amorphous solids of widely varying composition (see Section 21.5 for a discussion of the physical properties of glass). In this chapter, the term *glass* is used in the restricted and familiar sense to refer to materials formed from silica, usually in combination with metal oxides. The absence of long-range order in glasses has the consequence that they are **isotropic**—that is, their physical properties are the same in all directions. This has advantages in technology, including that glasses expand uniformly in all directions with an increase in temperature. The mechanical strength of glass is intrinsically high, exceeding the tensile strength of steel, provided the surface is free of scratches and other imperfections. Flaws in the surface provide sites where fractures can start when the glass is stressed. When a glass object of intricate shape and nonuniform thickness is suddenly cooled, internal stresses are locked in; these stresses may be relieved catastrophically when the object is heated or struck (Fig. 22.10). Slowly heating a strained object to a temperature somewhat below its softening point and holding it there for a while before allowing it to cool slowly is called **annealing;** it gives short-range diffusion of atoms a chance to occur and to eliminate internal stresses.

The softening and annealing temperatures of a glass and other properties, such as density, depend on its chemical composition (Table 22.2). Silica itself (SiO_2) forms a glass if it is heated above its melting point and then cooled rapidly to avoid

TABLE 22.2 Composition and Properties of Various Glasses

Silica Glass	Soda-lime Glass	Borosilicate Glass	Aluminosilicate Glass	Leaded Glass
Composition				
SiO_2, 99.9% H_2O, 0.1%	SiO_2, 73% Na_2O, 17% CaO, 5% MgO, 4% Al_2O_3, 1%	SiO_2, 81% B_2O_3, 13% Na_2O, 4% Al_2O_3, 2% B_2O_3, 5%	SiO_2, 63% Al_2O_3, 17% CaO, 8% MgO, 7% Al_2O_3, 2%	SiO_2, 56% PbO, 29% K_2O, 9% Na_2O, 4%
Coefficient of Linear Thermal Expansion ($°C^{-1} \times 10^7$)[†]				
5.5	93	33	42	89
Softening Point (°C)				
1580	695	820	915	630
Annealing Point (°C)				
1050	510	565	715	435
Density ($g\ cm^{-3}$)				
2.20	2.47	2.23	2.52	3.05
Refractive Index[‡] **at λ = 589 nm**				
1.459	1.512	1.474	1.530	1.560

[†]The coefficient of linear thermal expansion is defined as the fractional increase in length of a body when its temperature is increased by 1°C.
[‡]The refractive index is a vital property of glass for optical applications. It is defined by $n = \sin \theta_i / \sin \theta_r$, where θ_i is the angle of incidence of a ray of light on the surface of the glass and θ_r is the angle of refraction of the ray of light in the glass.

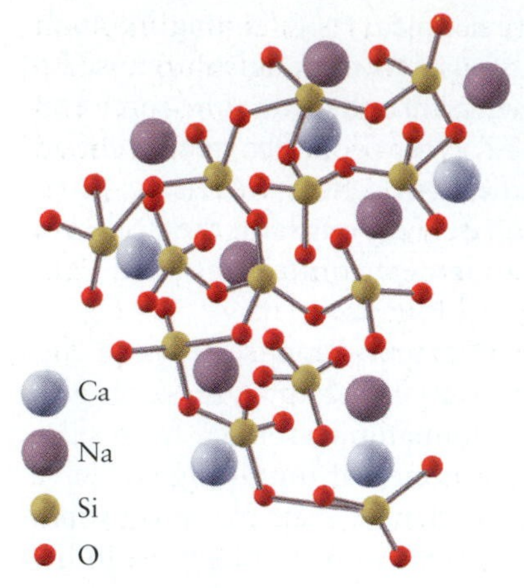

FIGURE 22.11 Structure of a soda-lime glass. Note the tetrahedral coordination of oxygen atoms around each silicon atom.

crystallization. The resulting vitreous (glassy) silica has limited use because the high temperatures required to shape it make it quite expensive. Sodium silicate glasses are formed in the high-temperature reaction of silica sand with anhydrous sodium carbonate (soda ash, Na_2CO_3):

$$Na_2CO_3(s) + n\ SiO_2(s) \longrightarrow Na_2O\cdot(SiO_2)_n(s) + CO_2(g)$$

The melting point of the nonvolatile product is about 900°C, and the glassy state results if cooling through that temperature is rapid. The product, called "water glass," is water-soluble; thus, it is unsuitable for making vessels. Its aqueous solutions, however, are used in some detergents and as adhesives for sealing cardboard boxes.

An insoluble glass with useful structural properties results if lime (CaO) is added to the sodium carbonate–silica starting materials. **Soda-lime glass** is the resulting product, with the approximate composition $Na_2O\cdot CaO\cdot(SiO_2)_6$. Soda-lime glass is easy to melt and shape and is used in applications ranging from bottles to window glass. It accounts for more than 90% of all the glass manufactured today. The structure of this ionic glass is shown schematically in Figure 22.11. It is a three-dimensional network of the type discussed in Section 22.1, but with random coordination of the silicate tetrahedra. The network is a giant "polyanion," with Na^+ and Ca^{2+} ions distributed in the void spaces to compensate for the negative charge on the network.

Replacing lime and some of the silica in a glass by other oxides (Al_2O_3, B_2O_3, K_2O, or PbO) modifies its properties noticeably. For example, the thermal conductivity of ordinary (soda-lime) glass is quite low, and its coefficient of thermal expansion is high. This means that internal stresses are created when its surface is subjected locally to extreme heat or cold, and it may shatter. The coefficient of thermal expansion is appreciably lower in certain borosilicate glasses, in which many of the silicon sites are occupied by boron. Pyrex, the most familiar of these glasses, has a coefficient of linear expansion about one-third that of ordinary soda-lime glass and is the preferred material for laboratory glassware and household ovenware. Vycor has an even smaller coefficient of expansion (approximating that of fused silica) and is made by chemical treatment of a borosilicate glass to leach out its sodium. This leaves a porous structure that is densified by increasing the temperature and shrinking the glass to its final volume.

Cements

Hydraulic cement was first developed by the ancient Romans, who found that a mixture of lime (CaO) and dry volcanic ash reacts slowly with water, even at low temperatures, to form a durable solid. They used this knowledge to build the Pantheon in Rome, a circular building whose concrete dome, spanning 143 feet without internal support, still stands nearly 2000 years after its construction! The knowledge of cement making was lost for centuries after the fall of the Roman Empire. It was rediscovered in 1824 by an English bricklayer, Joseph Aspdin, who patented a process for calcining a mixture of limestone and clay. He called the product **Portland cement** because, when mixed with water, it hardened to a material that resembled a kind of limestone found on the Isle of Portland. Portland cement is now manufactured in every major country, and annual worldwide production is currently about 800 million metric tons, exceeding the production of all other materials. Portland cement opened up a new age in the methods of constructing highways and buildings: Rock could be crushed and then molded in cement, rather than shaped with cutting tools.

Portland cement is a finely ground, powdered mixture of compounds produced by the high-temperature reaction of lime, silica, alumina, and iron oxide. The lime (CaO) may come from limestone or chalk deposits, and the silica (SiO_2) and alumina (Al_2O_3) are often obtained in clays or slags. The blast furnaces of steel mills are a common source of slag, which is a byproduct of the smelting of iron ore.

TABLE 22.3 Composition of Portland Cement

Oxide	Percentage by Mass
Lime (CaO)	61–69
Silica (SiO_2)	18–24
Alumina (Al_2O_3)	4–8
Iron(III) oxide (Fe_2O_3)	1–8
Minor oxides (MgO, Na_2O, K_2O, SO_3)	2–4

The composition of slag varies, but it can be represented as a calcium aluminum silicate of approximate formula $CaO{\cdot}Al_2O_3{\cdot}(SiO_2)_2$. Molten slag solidifies into "blast furnace clinkers" on quenching in water. This material is crushed and ground to a fine powder, blended with lime in the correct proportion, and burned again in a horizontal rotary kiln at temperatures up to 1500°C to produce "cement clinker." A final stage of grinding and the addition of about 5% gypsum ($CaSO_4{\cdot}2\ H_2O$) to lengthen the setting time completes the process of manufacture. Table 22.3 presents the composition of a typical Portland cement. Table 22.3 gives percentages of the separate oxides; these simple materials, which are the "elements" of cement making, combine in the cement in more complex compounds such as tricalcium silicate, $(CaO)_3{\cdot}SiO_2$, and tricalcium aluminate, $(CaO)_3{\cdot}Al_2O_3$.

Cement *sets* when the semiliquid slurry first formed by the addition of water to the powder becomes a solid of low strength. Subsequently, it gains strength in a slower *hardening* process. Setting and hardening involve a complex group of exothermic reactions in which several hydrated compounds form. Portland cement is a **hydraulic cement,** because it hardens not by loss of admixed water, but by chemical reactions that incorporate water into the final body. The main reaction during setting is the hydration of the tricalcium aluminate, which can be approximated by the following equation:

$$(CaO)_3{\cdot}Al_2O_3(s) + 3\ (CaSO_4{\cdot}2H_2O)(s) + 26\ H_2O(\ell) \longrightarrow (CaO)_3{\cdot}Al_2O_3{\cdot}(CaSO_4)_3{\cdot}32H_2O(s)$$

The product forms after 5 or 6 hours as a microscopic forest of long crystalline needles that lock together to solidify the cement. Later, the calcium silicates react with water to harden the cement. For example:

$$6\ (CaO)_3{\cdot}SiO_2(s) + 18\ H_2O(\ell) \longrightarrow (CaO)_5{\cdot}(SiO_2)_6{\cdot}5H_2O(s) + 13\ Ca(OH)_2(s)$$

The hydrated calcium silicates develop as strong tendrils that coat and enclose unreacted grains of cement, each other, and other particles that may be present, binding them in a robust network. Most of the strength of cement comes from these entangled networks, which, in turn, depend ultimately for strength on chains of O—Si—O—Si silicate bonds. Hardening is slower than setting; it may take as long as 1 year for the final strength of a cement to be attained.

Portland cement is rarely used alone. Generally, it is combined with sand, water, and lime to make **mortar,** which is applied with a trowel to bond bricks or stone together in an assembled structure. When Portland cement is mixed with sand and aggregate (crushed stone or pebbles) in the proportions of 1:3.75:5 by volume, the mixture is called **concrete.** Concrete is outstanding in its resistance to compressive forces and is therefore the primary material in use for the foundations of buildings and the construction of dams, in which the compressive loads are enormous. The stiffness (resistance to bending) of concrete is high, but its fracture toughness (resistance to impact) is substantially lower and its tensile strength is relatively poor. For this reason, concrete is usually reinforced with steel rods when it is used in structural elements such as beams that are subject to transverse or tensile stresses.

As excess water evaporates from cement during hardening, pores form that typically comprise 25% to 30% of the volume of the solid. This porosity weakens concrete, and recent research has shown that the fracture strength is related inversely to the size of the largest pores in the cement. A new material, called "macro defect–free" (MDF) cement, has been developed in which the size of the pores is reduced from about a millimeter to a few micrometers by the addition of water-soluble polymers that make a doughlike "liquid" cement that is moldable with the use of far less water. Unset MDF cement is mechanically kneaded and extruded into the desired shape. The final result possesses substantially increased bending resistance and fracture toughness. MDF cement can even be molded into springs and shaped on a conventional lathe. When it is reinforced with organic fibers, its toughness is further increased. The development of MDF cement is a good illustration of the way in which chemistry and engineering collaborate to furnish new materials.

22.4 Nonsilicate Ceramics

Many useful ceramics exist that are *not* based on the Si—O bond and the SiO_4 tetrahedron. They have important uses in electronics, optics, and the chemical industry. Some of these materials are oxides, but others contain neither silicon nor oxygen.

Oxide Ceramics

Oxide ceramics are materials that contain oxygen in combination with any of a number of metals. These materials are named by adding an *-ia* ending to the stem of the name of the metallic element. Thus, if the main chemical component of an oxide ceramic is Be_2O_3, it is a *beryllia* ceramic; if the main component is Y_2O_3, it is an *yttria* ceramic; and if it is MgO, it is a *magnesia* ceramic. As Table 22.4 shows, the melting points of these and other oxides are substantially higher than the melting points of the elements themselves. Such high temperatures are hard to achieve and maintain, and the molten oxides corrode most container materials. Oxide ceramic bodies are therefore not shaped by melting the appropriate oxide and pouring it into a mold. Instead, these ceramics are fabricated by sintering, like the silicate ceramics.

Alumina (Al_2O_3) is the most important nonsilicate ceramic material. It melts at a temperature of 2051°C and retains strength even at temperatures of 1500°C to 1700°C. Alumina has a large electrical resistivity and withstands both thermal shock and corrosion well. These properties make it a good material for spark plug insulators, and most spark plugs now use a ceramic that is 94% alumina.

High-density alumina is fabricated in such a way that open pores between the grains are nearly completely eliminated; the grains are small, with an average diameter as low as 1.5 μm. Unlike most others, this ceramic has good mechanical strength against impact, which has led to its use in armor plating. The ceramic absorbs the energy of an impacting projectile by breaking; thus, penetration does not occur. High-density alumina is also used in high-speed cutting tools for machining metals. The temperature resistance of the ceramic allows much faster cutting speeds, and a ceramic cutting edge has no tendency to weld to the metallic work piece, as metallic tools do. These properties make alumina cutting tools superior to metallic tools, as long as they do not break too easily. High-density alumina is also used in artificial joints (Fig. 22.12).

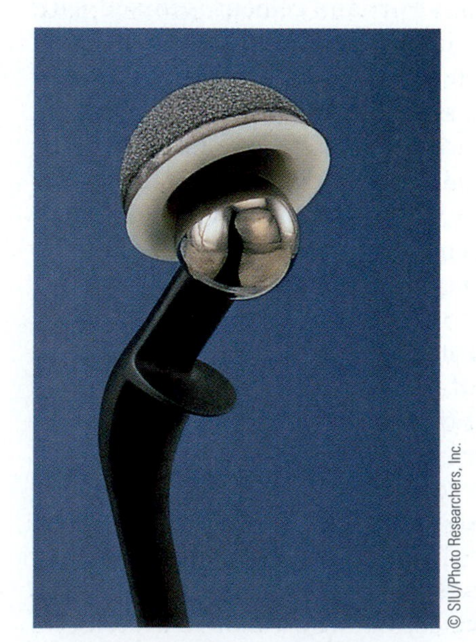

FIGURE 22.12 Socket in this artificial hip is made of high-density alumina.

If Al_2O_3 doped with a small percentage of MgO is fired in a vacuum or a hydrogen atmosphere (instead of air) at a temperature of 1800°C to 1900°C, even very small pores, which scatter light and make the material white, are removed. The resulting ceramic is translucent. This material is used to contain the sodium in high-intensity sodium discharge lamps. With envelopes of high-density alumina, these lamps can be operated at temperatures of 1500°C to give a whiter and more intense light. Old-style sodium-vapor lamps with glass envelopes were limited to a temperature of 600°C because the sodium vapor reacted with the glass at higher

TABLE 22.4 Melting Points of Some Metals and Their Oxides

Metal	Melting Point (°C)	Oxide	Melting Point (°C)
Be	1287	BeO	2570
Mg	651	MgO	2800
Al	660	Al_2O_3	2051
Si	1410	SiO_2	1723
Ca	865	CaO	2572
Y	1852	Y_2O_3	2690

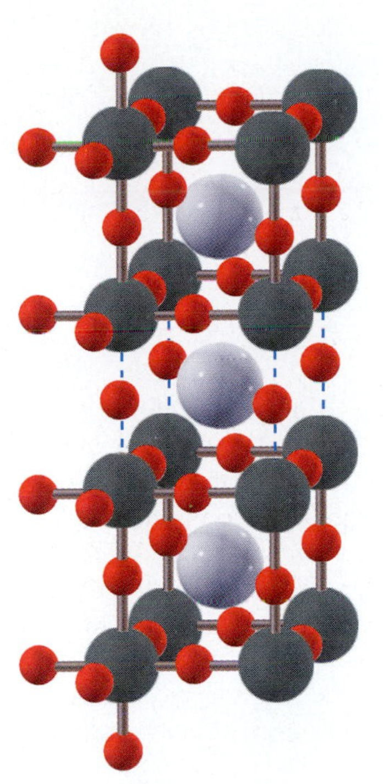

FIGURE 22.13 Structure of perovskite ($CaTiO_3$). A stack of three unit cells is shown, with some additional O atoms (red) from neighboring cells. Each Ti atom (gray) is surrounded by six O atoms; each Ca atom (white) has eight O atoms as nearest neighbors.

temperatures. At low temperatures, the light from a sodium-vapor lamp has an undesirable yellow color.

Magnesia (MgO) is mainly used as a **refractory**—a ceramic material that withstands a temperature of more than 1500°C without melting (MgO melts at 2800°C). A major use of magnesia is as insulation in electrical heating devices, because it combines high thermal conductivity with excellent electrical resistance. Magnesia is prepared from magnesite ores, which consist of $MgCO_3$ and a variety of impurities. When purified magnesite is heated to 800°C to 900°C, carbon dioxide is driven off to form MgO(*s*) in fine grains:

$$MgCO_3(s) \longrightarrow MgO(s) + CO_2(g)$$

After cooling, fine-grained MgO reacts vigorously with water to form magnesium hydroxide:

$$MgO(s) + H_2O(\ell) \longrightarrow Mg(OH)_2(s)$$

Heating fine-grained MgO(*s*) to 1700°C causes the MgO grains to sinter, giving a "dead-burned magnesia" that consists of large crystals and does not react with water. The $\Delta G°$ of the reaction between MgO and H_2O does not change when magnesia is dead burned. The altered microstructure (larger crystals vs small) makes the reaction with water exceedingly slow, however.

Superconducting Ceramics

The oxide ceramics discussed so far in this chapter all consisted of single chemical compounds, except for minor additives. A natural idea for new ceramics is to make materials that contain two (or more) oxides in equal or nearly equal molar amounts. Thus, if $BaCO_3$ and TiO_2 are mixed and heated to high temperature, they react to give the ceramic barium titanate:

$$BaCO_3(s) + TiO_2(s) \longrightarrow BaTiO_3(s) + CO_2(g)$$

Barium titanate, which has many novel properties, is a **mixed oxide ceramic.** It has the same structure as the mineral **perovskite,** $CaTiO_3$ (Fig. 22.13), except, of course, that Ba replaces Ca. Perovskites typically have two metal atoms for every three O atoms, giving them the general formula ABO_3, where A stands for a metal atom at the center of the unit cube and B stands for an atom of a different metal at the cube corners.

FIGURE 22.14 Levitation of a small magnet above a disk of superconducting material. A superconducting substance cannot be penetrated by an external magnetic field. At room temperature, the magnet rests on the ceramic disk. When the disk is cooled with liquid nitrogen, it becomes superconducting and excludes the magnet's field, forcing the magnet into the air.
(© DOE/Science Source/Photo Researchers, Inc.)

Research interest in perovskite ceramic compositions has grown explosively after the discovery that some of them become **superconducting** at relatively high temperatures. A superconducting material offers no resistance whatsoever to the flow of an electric current. The phenomenon was discovered by the Dutch physicist Heike Kamerlingh-Onnes in 1911, when he cooled mercury below its superconducting transition temperature of 4 K. Such low temperatures are difficult to achieve and maintain, but if practical superconductors could be made to work at higher temperatures (or even at room temperature!), then power transmission, electronics, transportation, medicine, and many other aspects of human life would be transformed. More than 60 years of research with metallic systems culminated in 1973 with the discovery of a niobium–tin alloy with a world-record superconducting transition temperature of 23.3 K. Progress toward higher temperature superconductors then stalled until 1986, when K. Alex Müller and J. Georg Bednorz, who had had the inspired notion to look for higher transition temperatures among perovskite ceramics, found a Ba—La—Cu—O perovskite phase having a transition temperature of 35 K. This result motivated other scientists, who soon discovered another rare-earth–containing perovskite ceramic that became a superconductor at 90 K. This result was particularly exciting because 90 K exceeds the boiling point of liquid nitrogen (77 K), a relatively cheap refrigerant (Fig. 22.14). This 1-2-3 compound (so called because its formula, $YBa_2Cu_3O_{(9-x)}$,

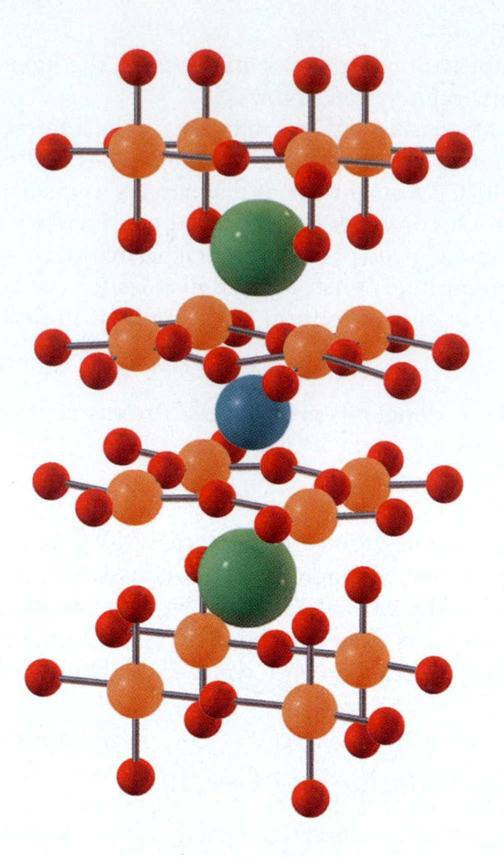

FIGURE 22.15 The structure of $YBa_2Cu_3O_{9-x}$ is a layered perovskite. Each Cu atom (orange) is bonded to O atoms (red). The layers differ because one-third of them contain Y atoms (blue-gray), whereas two-thirds contain Ba atoms (green). This structure is a variation of the perovskite structure in Figure 22.13. In a hypothetical "$BaCuO_3$" structure, every third Ba has been replaced by a Y, and the O atoms in the layer containing the Y have been removed. Note that the Cu—O layers above and below the Y atom are puckered.

has one Y, two Ba, and three Cu atoms per formula unit) is not an ideal perovskite because it has fewer than nine O atoms in combination with its six metal atoms. The deficiency makes x in the formula somewhat greater than 2, depending on the exact method of preparation. The structure of this nonstoichiometric solid is shown in Figure 22.15. More recently, the maximum superconducting transition temperature has increased to 125 K in another class of ceramics that does not contain rare-earth elements.

A crucial concern in the application of superconducting ceramics is to devise ways to fabricate the new materials in desired shapes such as wires. This will be quite a challenge because these superconductors are ceramics and have the brittleness and fragility typical of ceramic materials.

Nonoxide Ceramics

In nonoxide ceramics, nitrogen (N) or carbon (C) takes the place of oxygen in combination with silicon or boron. Specific substances are boron nitride (BN), boron carbide (B_4C), the silicon borides (SiB_4 and SiB_6), silicon nitride (Si_3N_4), and silicon carbide (SiC). All of these compounds possess strong, short covalent bonds. They are hard and strong, but brittle. Table 22.5 lists the enthalpies of the chemical bonds in these compounds.

Much research has aimed at making gas-turbine and other engines from ceramics. Of the oxide ceramics, only alumina and zirconia (ZrO_2) are strong enough, but both resist thermal shock too poorly for this application. Attention has therefore turned to the nonoxide **silicon nitride** (Si_3N_4). In this **network solid** (Fig. 22.16), every Si atom bonds to four N atoms that surround it at the corners of a tetrahedron; these tetrahedra link into a three-dimensional network by sharing corners. The Si—N bond is covalent and strong (the bond enthalpy is 439 kJ mol^{-1}). The similarity to the joining of SiO_4 units in silicate minerals (see Section 22.1) is clear.

TABLE 22.5 Bond Enthalpies in Nonoxide Ceramics

Bond	Bond Enthalpy (kJ mol^{-1})
B—N	389
B—C	448
Si—N	439
Si—C	435
B—Si	289
C—C	350

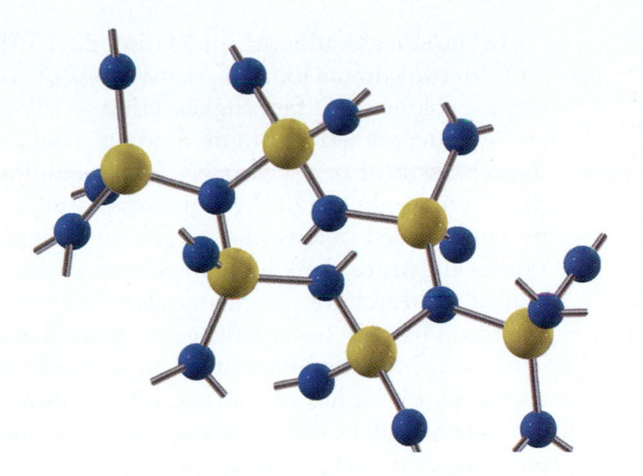

FIGURE 22.16 Structure of silicon nitride (Si_3N_4). Each Si atom is bonded to four N atoms, and each N atom is bonded to three Si atoms. The result is a strong network.

At first, silicon nitride appears chemically unpromising as a high-temperature structural material. It is unstable in contact with water because the following reaction has a negative $\Delta G°$:

$$Si_3N_4(s) + 6\ H_2O(\ell) \longrightarrow 3\ SiO_2(s) + 4\ NH_3(g)$$

In fact, this reaction causes finely ground Si_3N_4 powder to give off an odor of ammonia in moist air at room temperature. Silicon nitride is also thermodynamically unstable in air, reacting spontaneously with oxygen:

$$Si_3N_4(s) + 3\ O_2(g) \longrightarrow 3\ SiO_3(s) + 2\ N_2(g)$$

The reaction has a $\Delta G°$ of -1927 kJ mol^{-1}. In practice, neither reaction occurs at a perceptible rate when Si_3N_4 is in bulk form. Initial contact of oxygen or water with $Si_3N_4(s)$ forms a surface film of $SiO_2(s)$ that protects the bulk of the $Si_3N_4(s)$ from further attack. When *strongly* heated in air (to about 1900°C), Si_3N_4 does decompose, violently, but until that temperature is reached, it resists attack.

Fully dense silicon nitride parts are stronger than metallic alloys at high temperatures. Ball bearings made of dense silicon nitride work well without lubrication at temperatures up to 700°C; for example, they last longer than steel ball bearings. Because the strength of silicon nitride increases with the density attained in the production process, the trick is to form a dense piece of silicon nitride in the desired shape. One method for making useful shapes of silicon nitride is "reaction bonding." Powdered silicon is compacted in molds, removed, and then fired under an atmosphere of nitrogen at 1250°C to 1450°C. The following reaction forms the ceramic:

$$3\ Si(s) + 2\ N_2(g) \longrightarrow Si_3N_4(s)$$

The parts neither swell nor shrink significantly during the chemical conversion from Si to Si_3N_4, making it possible to fabricate complex shapes reliably. Unfortunately, reaction-bonded Si_3N_4 is still somewhat porous and is not strong enough for many applications. In the "hot-pressed" forming process, Si_3N_4 powder is prepared in the form of exceedingly small particles by reaction of silicon tetrachloride with ammonia:

$$3\ SiCl_4(g) + 4\ NH_3(g) \longrightarrow Si_3N_4(s) + 12\ HCl(g)$$

The solid Si_3N_4 forms as a smoke. It is captured as a powder, mixed with a carefully controlled amount of MgO additive, placed in an enclosed mold, and sintered at 1850°C under a pressure of 230 atm. The resulting ceramic shrinks to nearly full density (no pores). Because the material does not flow well (to fill a complex mold completely), only simple shapes are possible. Hot-pressed silicon nitride is impressively tough and can be machined only with great difficulty and with diamond tools.

In the silicate minerals of Section 22.1, AlO_4^{5-} units routinely substitute for SiO_4^{4-} tetrahedra as long as positive ions of some type are present to balance the electric charge. This fact suggests that in silicon nitride some Si^{4+} ions, which lie at the centers of tetrahedra of N atoms, could be replaced by Al^{3+} if a compensating replacement of O^{2-} for N^{3-} were simultaneously made. Experiments show that ceramic alloying of this type works well, giving many new ceramics with great potential called **sialons** (named for the four elements Si—Al—O—N). These ceramics illustrate the way a structural theme in a naturally occurring material guided the search for new materials.

Boron has one fewer valence electron than carbon, and nitrogen has one more valence electron. **Boron nitride** (BN) is therefore isoelectronic with C_2, and it is not surprising that it has two structural modifications that resemble the structures of graphite and diamond. In hexagonal boron nitride, the B and N atoms take alternate places in an extended "chicken-wire" sheet in which the B—N distance is 1.45 Å. The sheets stack in such a way that each B atom has a N atom directly above it and directly below it, and vice versa. The cubic form of boron nitride has the diamond structure, is comparable in hardness to diamond, and resists oxidation better. Boron nitride is often prepared by **chemical vapor deposition,** a method used in fabricating several other ceramics as well. In this method, a controlled chemical reaction of gases on a contoured, heated surface gives a solid product of the desired shape. If a cup made of BN is needed, a cup-shaped mold is heated to a temperature exceeding 1000°C and a mixture of $BCl_3(g)$ and $NH_3(g)$ is passed over its surface. The reaction

$$BCl_3(g) + NH_3(g) \longrightarrow BN(s) + 3\,HCl(g)$$

deposits a cup-shaped layer of BN(*s*). Boron nitride cups and tubes are used to contain and evaporate molten metals.

Silicon carbide (SiC) is diamond in which half of the C atoms are replaced by Si atoms. Also known by its trade name of Carborundum, silicon carbide was developed originally as an abrasive, but it is now used primarily as a refractory and as an additive in steel manufacture. It is formed and densified by methods similar to those used with silicon nitride. Silicon carbide is often produced in the form of small plates or whiskers to reinforce other ceramics. Fired silicon carbide whiskers are quite small (0.5 μm in diameter and 50 μm long) but are strong. They are mixed with a second ceramic material before that material is formed. Firing then gives a **composite ceramic.** Such composites are stronger and tougher than unreinforced bodies of the same primary material. Whiskers serve to reinforce the main material by stopping cracks; they either deflect advancing cracks or soak up their energy, and a widening crack must dislodge them to proceed (Fig. 22.17). Recently, SiC has been used in high power, high temperature semiconductor devices (Section 22.7).

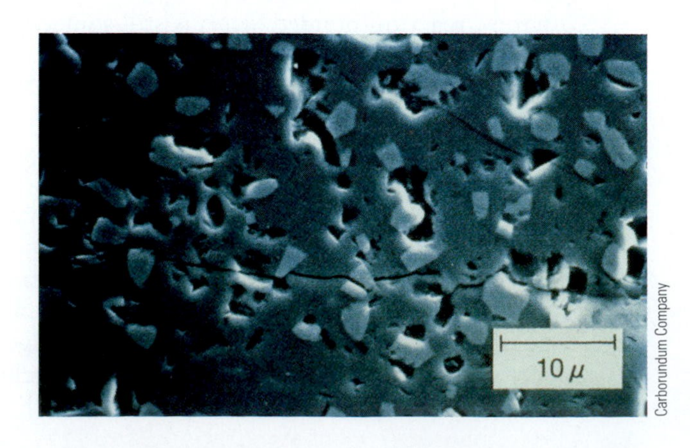

FIGURE 22.17 Microstructure of a composite ceramic. The light-colored areas are TiB_2, the gray areas are SiC, and the dark areas are voids. The TiB_2 acts to toughen the SiC matrix. Note that the crack shown passing through from left to right is forced to deflect around the TiB_2 particles.

Carborundum Company

22.5 Electrical Conduction in Materials

Electronic properties describe the movement of charged particles in a material in response to an applied electric field. If the charges are free to move throughout the material, the process is *electrical conduction*, measured by the *electrical conductivity* of the material. Differences in the magnitude of the conductivity distinguish metals, semiconductors, and insulators. If the charges can move only limited distances and are then halted by opposing binding forces, separation of positive and negative charges leads to *electric polarization* of the material, measured by its *dielectric constant*. Conduction involves dissipation of energy as heat, whereas polarization involves storage of potential energy in the material.

An electric field applied to a material that has free charged particles causes these particles to flow through the material and into the external circuit. How do we define and measure the conductivity of a material? On what macroscopic properties does it depend? How does it relate to the detailed chemical structure of the material?

Measurement of Conductivity

The electrical conductivity of a material is measured by placing a cylindrical sample of cross-sectional area, A, and length, ℓ, in a simple electrical circuit as a resistor in series with a power supply and ammeter; a voltmeter measures the actual voltage across the sample (Fig. 22.18). If the voltage, V, is varied and the resulting current, I, is measured at each voltage, a plot of I versus V is a straight line (Fig. 22.19):

$$I = GV \qquad [22.1]$$

The resulting slope is called the **conductance** and is denoted by G. If V is measured in volts and I in amperes (A), then G has units of siemens (1 siemens = 1 A/V). If the value of G is constant, the material follows Ohm's law

$$V = IR$$

and G is the reciprocal of the **resistance** R, which is measured in ohms (Ω; 1 Ω = 1 V/A). So far, the measurement depends on the size and shape of the material sample, in addition to its composition. To remove these geometric effects, we note from experiment that R increases as ℓ increases, and it decreases as A increases.

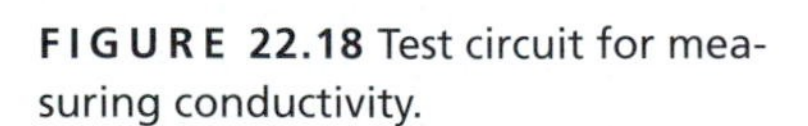
FIGURE 22.18 Test circuit for measuring conductivity.

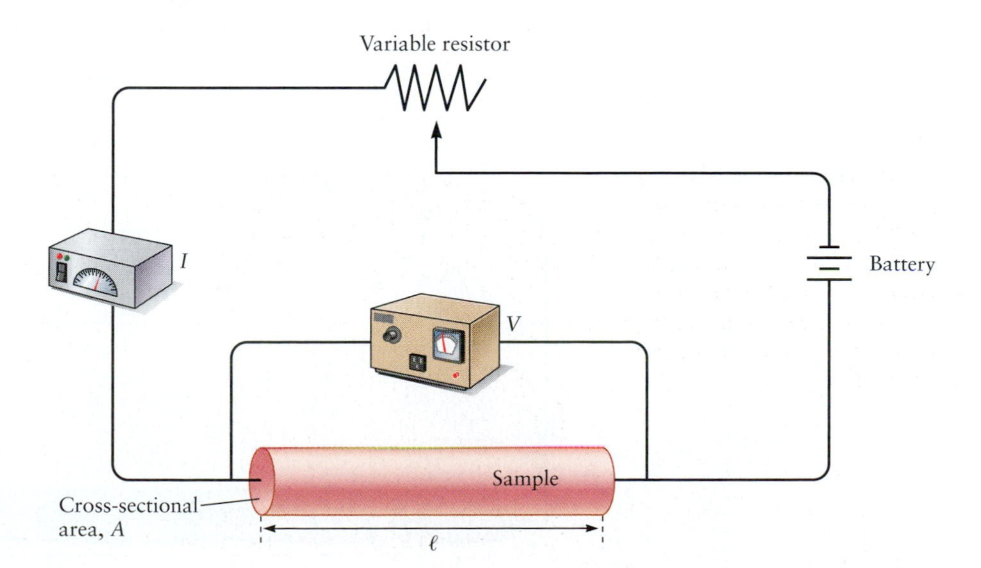

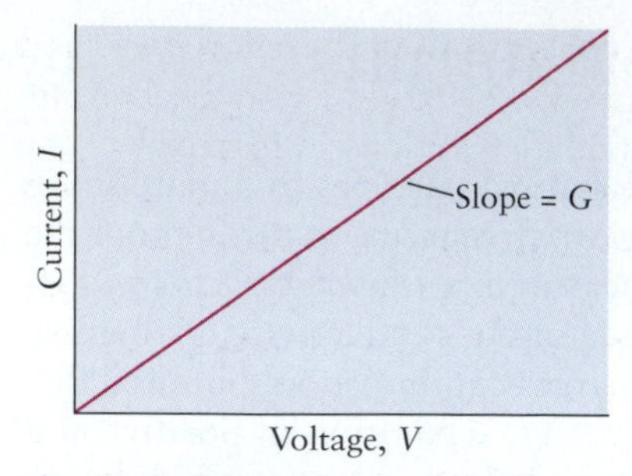

FIGURE 22.19 Plot of current against voltage is a straight line whose slope is the conductance G.

Therefore, we define the material property **resistivity,** denoted by ρ, as the proportionality constant that summarizes these two effects:

$$R = \frac{\ell}{A}\rho \qquad [22.2]$$

From this definition, ρ has dimensions of Ω m. Finally, we define the material property **conductivity,** σ, as the reciprocal of the resistivity:

$$\sigma = \frac{1}{\rho} = \frac{\ell}{RA} \qquad [22.3]$$

from which σ has units of $(\Omega \text{ m})^{-1}$ or S m^{-1}. We now rewrite Ohm's law in a form independent of sample geometry. First, define the current density $J(\text{A m}^{-2})$ through the sample as I/A and define the electric field $E(\text{V m}^{-1})$ through the sample as V/ℓ. Inserting these definitions and the definition of σ from Equation 22.3 into Equation 22.1 gives:

$$J = \sigma E \qquad [22.4]$$

According to Equation 22.4, the current density flowing through a material sample is proportional to the electric field applied to the sample, and the proportionality constant is the conductivity of the material of which the sample is made. This is the equation we use to relate the conductivity of a material to its microstructural properties. Table 22.6 lists conductivities for several common metals at room temperature.[2]

TABLE 22.6 Electrical Conductivity of Selected Metals at Room Temperature

Metal	Conductivity $[(\Omega \text{ m})^{-1}]$
Silver	6.8×10^7
Copper	6.0×10^7
Gold	4.3×10^7
Aluminum	3.8×10^7
Iron	1.0×10^7
Platinum	0.94×10^7
Stainless steel	0.2×10^7

Microscopic Origins of the Conductivity

Insight into the conductivity is provided by measuring the electrical conductivity of aqueous ionic solutions (Fig. 22.20; this topic is referred to in Chapters 11 and 15). The conductivity of pure water, multiply distilled to remove all impurities, is about 0.043×10^{-6} $(\Omega \text{ cm})^{-1}$. Exposed to the air, pure water dissolves CO_2, which forms carbonic acid, H_2CO_3; dissociation produces H_3O^+ and HCO_3^-, which increase the conductivity to about 1×10^{-6} $(\Omega \text{ cm})^{-1}$. As ionic solutes are added to water, the conductivity increases rapidly; a 1.0-M solution in NaOH has conductivity of about 0.180 $(\Omega \text{ cm})^{-1}$ at 25°C. The conductivity depends strongly on both concentration and ionic species. The concentration dependence is summarized by the **molar**

[2]Because the meter is inconveniently large for measuring the dimensions of most samples in material studies, you will frequently see ρ expressed in units of Ω cm and σ in units of $(\Omega \text{ cm})^{-1}$. Be alert to the actual units used. In base International System of Units (SI), resistivity has units expressed as kg m^3 s^{-3} A^{-2}.

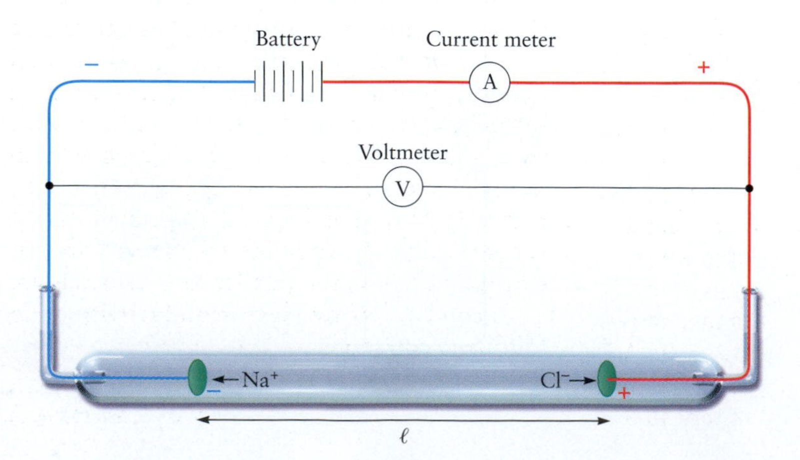

FIGURE 22.20 Apparatus for measuring the conductivity of an aqueous solution of NaCl.

conductivity, defined by $\Lambda_m = \sigma/c$, where c is the concentration of the ion measured in moles per liter. To gain fundamental understanding of the conduction mechanism, we relate Λ_m to the details of the ionic motion in dilute solutions in response to the electric field. This leads to the concept of *mobility* of the ion, denoted by μ, which shows how its surroundings influence the response of an ion to an applied field.

As the electric field is turned on, each ion is accelerated by the field and experiences a force whose magnitude and direction are given by the vector relation

$$\vec{F} = ze\vec{E}$$

where z is the charge on the ion in units of the elementary charge, e, and the expression is valid for either cations or anions (the force points in opposite directions in the two cases). As the ion is accelerated by the field, however, its forward motion is retarded by friction from the surrounding solvent molecules. This retarding force $\vec{F}'$ opposes the direction of motion of the ion and is proportional to its velocity, $\vec{v}$:

$$\vec{F}' = -f\vec{v}$$

where f is the frictional drag coefficient. In due course, these forces balance and the ion achieves its steady drift velocity $\vec{v}_d$. This force balance condition provides the following relation:

$$\vec{v}_d = ze\frac{\vec{E}}{f}$$

This equation shows that the electric field increases the drift velocity, whereas the drag coefficient reduces it. We define the proportionality constant between the *drift speed* (the magnitude of the drift velocity) and the field to be the **mobility,** μ, of the ion (a positive number), giving the following relation:

$$|\vec{v}_d| = \mu|\vec{E}| \qquad [22.5]$$

Mobility has physical units of $m^2\ V^{-1}\ s^{-1}$. For ions in solution, the mobility is given by

$$\mu = \frac{|z|e}{f}$$

In more advanced work, it is possible to estimate the drag coefficient, f, and provide a theoretical prediction of the mobility. Experimentally, the mobility for ions is obtained from the molar conductivity through the following equation, which we do not justify:

$$\Lambda_m = |z|\mu\mathcal{F} \qquad [22.6]$$

where $\mathcal{F}$ is the Faraday constant, 96,485.34 C mol^{-1} (the Faraday constant is introduced in Chapter 17).

Mobilities are shown in Table 22.7 for several ions in aqueous solution. One interesting feature is that the smaller cations have lower mobilities than the larger cations; for example, μ for Na^+ is smaller than that for K^+, and μ for Mg^{2+} is smaller than that for Ca^{2+}. One might expect that the smaller ions would have larger mobility, because their smaller size should encounter less frictional drag when moving through the solvent. However, the greater charge density on the smaller cations attracts a larger solvation shell than occurs on the larger cations (see Section 11.2 and Figure 11.4). The entity moving through the solution in response to the electric field is not the "bare" cation but the cation "dressed" with its solvation shell. The larger solvation shell on the smaller ions causes more frictional drag and lower mobility.

It can be shown that the conductivity of an ion is related to its mobility by the following equation:

$$\sigma_{\text{ion}} = |z|en_{\text{ion}}\mu_{\text{ion}} \qquad [22.7]$$

TABLE 22.7 Mobilities of Selected Ions in Aqueous Solution at 25°C

Ion	Mobility ($cm^2\ V^{-1}\ s^{-1}$)
Li^+	4.01×10^{-4}
Na^+	5.19×10^{-4}
K^+	7.62×10^{-4}
Mg^{2+}	5.50×10^{-4}
Ca^{2+}	6.17×10^{-4}
Ba^{2+}	6.60×10^{-4}
Cl^-	7.91×10^{-4}
Br^-	8.10×10^{-4}
NO_3^-	7.40×10^{-4}
ClO_4^-	6.98×10^{-4}
CH_3COO^-	4.24×10^{-4}

where n_{ion} is the number density of ions present, expressed in number per cubic meter. The conductivity depends on two fundamental microscopic variables: the number density of carriers and the carrier mobility. Both positive and negative ions can respond to the electric field in the solution, so the total conductivity must include a contribution from each:

$$\sigma_{\text{tot}} = |z_{\text{cation}}| e n_{\text{cation}} \mu_{\text{cation}} + |z_{\text{anion}}| e n_{\text{anion}} \mu_{\text{anion}} \qquad [22.8]$$

Equations 22.7 and 22.8 are valid for all electrical conduction processes: motion of positive and negative ions in solution and in ionic solids, and motion of electrons in solids. These equations are used in subsequent sections to discuss conduction in a variety of materials. These equations should remind you that the conductivity in any material depends on two separate microscopic parameters: the number of charge carriers present and their mobilities.

EXAMPLE 22.2

The molar conductivity of Na^+ ions in aqueous solution at 25°C has been determined to be $5.01 \times 10^{-2}\ (\Omega\ \text{cm})^{-1}\ \text{mol}^{-1}\ \text{L}$. Assume an electric field of $1.0 \times 10^2\ \text{V cm}^{-1}$ is applied to the solution. Calculate the mobility and the drift velocity of Na^+ ions.

SOLUTION

From Equation 22.6, the mobility and molar conductivity are related by

$$\mu = \frac{\Lambda_m}{|z|\mathcal{F}} = \frac{5.01 \times 10^{-2}\ \Omega^{-1}\ \text{cm}^{-1}\ \text{mol}^{-1}\ \text{L}}{(1)(96{,}485\ \text{C mol}^{-1})} = 5.19 \times 10^{-4}\ \text{cm}^2\ \text{V}^{-1}\ \text{s}^{-1}$$

We used Ohm's law $V = IR$ or $V = \text{C s}^{-1}\ \Omega$ to simplify the units, and we took $1\ \text{L} = 10^3\ \text{cm}^3$.

From Equation 22.5, the drift velocity is given by

$$v_d = \mu E = (5.19 \times 10^{-4}\ \text{cm}^2\ \text{V}^{-1}\ \text{s}^{-1})(1.0 \times 10^2\ \text{V cm}^{-1}) = 5.19 \times 10^{-2}\ \text{cm s}^{-1}$$

Related Problems: 21, 22, 23, 24, 25, 26

Illustration: Conductivity in Metals

Before the quantum theory of solids (see description in Chapter 21), microscopic descriptions of metals were based on the Drude model, named for the German physicist Paul Drude. The solid was viewed as a fixed array of positively charged metal ions, each localized to a site on the solid lattice. These fixed ions were surrounded by a sea of mobile electrons, one contributed by each of the atoms in the solid. The number density of the electrons, n_{el}, is then equal to the number density of atoms in the solid. As the electrons move through the ions in response to an applied electric field, they can be scattered away from their straight-line motions by collisions with the fixed ions; this influences the mobility of the electrons. As temperature increases, the electrons move more rapidly and the number of their collisions with the ions increases; therefore, the mobility of the electrons decreases as temperature increases. Equation 22.7 applied to the electrons in the Drude model gives

$$\sigma_{\text{el}} = e n_{\text{el}} \mu_{\text{el}} \qquad [22.9]$$

which predicts that the electrical conductivity of a metal will decrease as temperature increases, because the electron mobility decreases with temperature whereas the electron number density is independent of temperature. This simple model prediction agrees with the experimental fact that the resistivity of metals increases as temperature increases.

22.6 Band Theory of Conduction

The characteristic property of metals is their good ability to conduct electricity and heat. Both phenomena are due to the ease with which valence electrons move; electrical conduction is a result of the flow of electrons from regions of high potential energy to those of low potential energy, and heat conduction is a result of the flow of electrons from high-temperature regions (where their kinetic energies are high) to low-temperature regions (where their kinetic energies are low). Why are electrons so mobile in a metal but so tightly bound to atoms in an insulating solid, such as diamond or sodium chloride?

Figure 21.19 shows how the quantum energy levels of Na atoms in a crystal are spread out into a continuous band of states, which is half occupied by electrons. Now, suppose a small electric potential difference is applied across a sodium crystal. The spin-paired electrons that lie deep in the band cannot be accelerated by a weak electric field because occupied levels exist just above them. They have no place to go (recall the Pauli principle, which states that an energy level contains at most two electrons). At the top of the "sea" of occupied levels, however, there is an uppermost electron-occupied or half-occupied level called the **Fermi level.** Electrons that lie near that level have the highest kinetic energy of all the valence electrons in the crystal and can be accelerated by the electric field to occupy the levels above. They are free to migrate in response to the electric field so that they conduct an electric current. These same electrons at the Fermi level are responsible for the high thermal conductivities of metals. They are also the electrons freed by the photoelectric effect when a photon gives them sufficient kinetic energy to escape from the metal (see Section 4.4).

A natural question arises: Why are the alkaline-earth elements metals, given the argument just presented? A metal such as magnesium contains two 3*s* electrons, and one might expect the band derived from the broadened 3*s* level to be completely filled. The answer to this question is that the energies of the 3*s* orbitals and 3*p* orbitals for magnesium are not greatly different. When the internuclear separation becomes small enough, the 3*p* band overlaps the 3*s* band, and many unoccupied sublevels are then available above the highest filled level.

Band Picture of Bonding in Silicon

An analogous procedure to that shown in Figure 21.19 can be used to construct a band picture for silicon. In this case, the $4N_A$ valence atomic orbitals (both 3*s* and 3*p*) from 1 mol (N_A atoms) silicon split into *two* bands in the silicon crystal, each containing $2N_A$ closely spaced levels (Fig. 22.21). The lower band is called the **valence band** and the upper one the **conduction band.** Between the top of the valence band and the bottom of the conduction band is an energy region that is forbidden to electrons. The magnitude of the separation in energy between the valence band and the lowest level of the conduction band is called the **band gap,** E_g, which for pure silicon is 1.94×10^{-19} J. This is the amount of energy that an electron must gain to be excited from the top of the valence band to the bottom of the conduction band. For 1 mol of electrons to be excited in this way, the energy is larger by a factor of Avogadro's number N_A, giving 117 kJ mol^{-1}. (Another unit used for band gaps is the *electron volt,* defined in Section 3.2 as 1.60218×10^{-19} J. In this unit, the band gap in Si is 1.21 eV.)

Each Si atom in the crystal contributes four valence electrons to the bands of orbitals in Figure 22.21, for a total of $4N_A$ per mole. This is a sufficient number to place two electrons in each level of the valence band (with opposing spins) and leave the conduction band empty. There are no low-lying energy levels for those

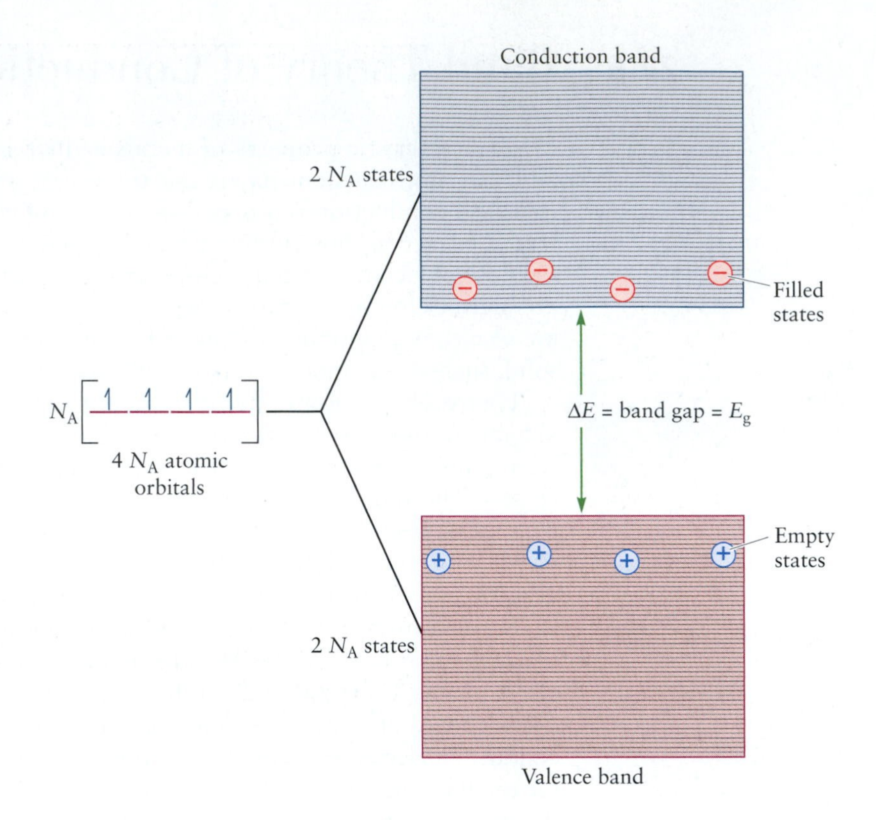

FIGURE 22.21 The valence orbitals of the silicon atom combine in crystalline silicon to give two bands of closely spaced levels. The valence band is almost completely filled, and the conduction band is almost empty.

electrons at the top of the valence band to enter if given a small increment in their energy. The band gap means that an electron at the top of the filled valence band must acquire an energy of at least 1.94×10^{-19} J, equivalent to 117 kJ mol^{-1}, to jump to the lowest empty level of the conduction band. This is a large amount of energy. If it were to be supplied by a thermal source, the temperature of the source would have to be on the order of

$$T = \frac{\Delta E}{R} = \frac{117{,}000 \text{ J mol}^{-1}}{8.315 \text{ J K}^{-1} \text{ mol}^{-1}} = 14{,}000 \text{ K}$$

which is far above the temperature at which a crystalline sample could exist. At room temperature, only a few electrons per mole in the extreme tail of the Boltzmann distribution have enough energy to jump the gap; therefore, the conduction band in pure silicon is sparsely populated with electrons. The result is that silicon is not a good conductor of electricity; good electrical conductivity requires a net motion of many electrons under the impetus of a small electric potential difference. Silicon has an electrical conductivity that is 11 orders of magnitude smaller than that of copper at room temperature. It is called a **semiconductor,** because its electrical conductivity, although smaller than that of a metal, is far greater than that of an **insulator** such as diamond, which has a larger band gap. The conductivity of a semiconductor is increased by increasing the temperature, which excites more electrons into levels in the conduction band. Another way to increase the conductivity of a semiconductor is to irradiate it with a beam of electromagnetic radiation with a frequency high enough to excite electrons from the valence band to the conduction band. This process resembles the photoelectric effect described in Chapter 4, with the difference that now the electrons are not removed from the material but only moved into the conduction band, ready to conduct a current if a potential difference is imposed.

EXAMPLE 22.3

Calculate the longest wavelength of light that can excite electrons from the valence to the conduction band in silicon. In what region of the spectrum does this wavelength fall?

SOLUTION

The energy carried by a photon is $h\nu = hc/\lambda$, where h is Planck's constant, ν the photon frequency, c the speed of light, and λ the photon wavelength. For a photon to just excite an electron across the band gap, this energy must be equal to that band gap energy, $E_g = 1.94 \times 10^{-19}$ J, where

$$\frac{hc}{\lambda} = E_g$$

Solving for the wavelength gives

$$\lambda = \frac{hc}{E_g} = \frac{(6.626 \times 10^{-34}\ \text{J s}) \times (2.998 \times 10^{8}\ \text{m s}^{-1})}{1.94 \times 10^{-19}\ \text{J}}$$

$$= 1.02 \times 10^{-6}\ \text{m} = 1020\ \text{nm}$$

This wavelength falls in the infrared region of the spectrum. Photons with shorter wavelengths (for example, visible light) carry more than enough energy to excite electrons to the conduction band in silicon.

Related Problems: 27, 28

22.7 Semiconductors

Silicon in very high purity is said to display its **intrinsic properties.** When certain other elements are added to pure silicon in a process called **doping,** it acquires interesting electronic properties. For example, if atoms of a Group V element such as arsenic or antimony are diffused into silicon, they substitute for Si atoms in the network. Such atoms have five valence electrons, so each introduces one more electron into the silicon crystal than is needed for bonding. The extra electrons occupy energy levels just below the lowest level of the conduction band. Little energy is required to promote electrons from such a **donor impurity** level into the conduction band, and the electrical conductivity of the silicon crystal is increased without the necessity of increasing the temperature. Silicon doped with atoms of a Group V element is called an ***n*-type semiconductor** to indicate that the charge carrier is negative.

However, if a Group III element such as gallium is used as a dopant, there is one fewer electron in the valence band per dopant atom because Group III elements have only three valence electrons, not four. This situation corresponds to creating one **hole** in the valence band, with an effective charge of +1, for each Group III atom added. If a voltage difference is impressed across a crystal that is doped in this way, it causes the positively charged holes to move toward the negative source of potential. Equivalently, a valence band electron next to the (positive) hole will move in the opposite direction (that is, toward the positive source of potential). Whether we think of holes or electrons as the mobile charge carrier in the valence band, the result is the same; we are simply using different words to describe the same physical phenomenon (Fig. 22.22). Silicon that has been doped with a Group III element is called a ***p*-type semiconductor** to indicate that the carrier has an effective positive charge.

A different class of semiconductors is based not on silicon, but on equimolar compounds of Group III with Group V elements. Gallium arsenide, for example, is isoelectronic to the Group IV semiconductor germanium. When GaAs is doped with the Group VI element tellurium, an *n*-type semiconductor is produced; doping

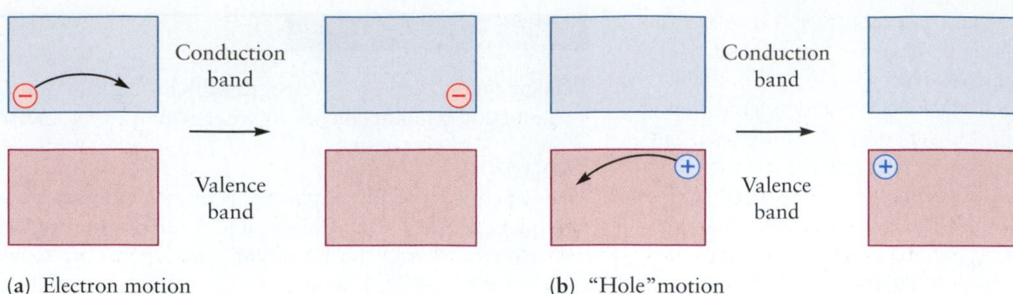

FIGURE 22.22 (a) In an *n*-type semiconductor, a small number of electrons occupy levels in the conduction band. When an electric field is imposed, each electron moves (black arrows) into one of the numerous nearby vacant energy levels in the conduction band. (b) In a *p*-type semiconductor, a small number of levels in the valence band are unoccupied. Conduction occurs as electrons in the numerous occupied levels of the valence band jump into the sparsely distributed unoccupied levels. The arrow shows the motion of an electron to occupy a previously empty level. The process can also be described as the motion of positively charged holes in the opposite direction.

with zinc, which has one *fewer* valence electron than gallium, gives a *p*-type semiconductor. Other III–V combinations have different band gaps and are useful in particular applications. Indium antimonide (InSb), for example, has a small enough band gap that absorption of infrared radiation causes electrons to be excited from the valence to the conduction band, and an electric current then flows when a small potential difference is applied. This compound is therefore used as a detector of infrared radiation. Still other compounds formed between the zinc group (zinc, cadmium, and mercury) and Group VI elements such as sulfur also have the same average number of valence electrons per atom as silicon and make useful semiconductors.

p–n Junctions and Device Performance

Of what value is it to have a semiconductor that can conduct an electric current by the flow of electrons if it is *n*-type or by the flow of holes if it is *p*-type? Many electronic functions can be fulfilled by semiconductors that possess these properties, but the simplest is **rectification**—the conversion of alternating current into

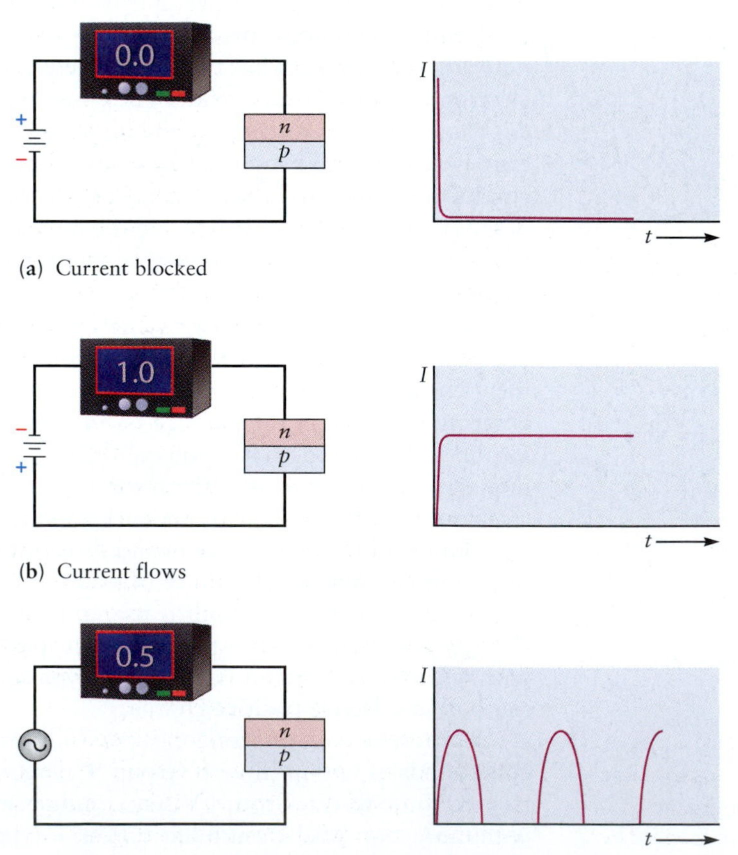

FIGURE 22.23 Current rectification by a *p–n* transistor.

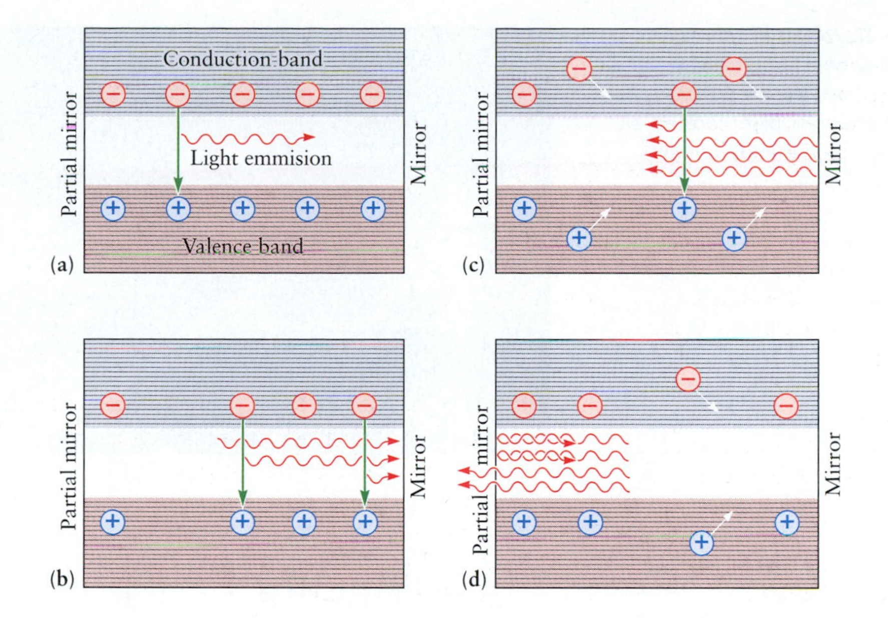

FIGURE 22.24 (a) In this solid-state laser, photons emitted as electrons and holes recombine to stimulate the emission of additional photons. (b) Reflection by a mirror on the right side sends coherent waves back through the laser medium. (c) Further amplification occurs by stimulated emission. (d) Some of the waves pass through a partially reflecting mirror on the left side.

direct current. Suppose thin crystals of *n*- and *p*-type silicon are placed in contact with one another and connected to a battery. In Figure 22.23a, the positive pole of the battery is connected to the *n*-type silicon, and the *p*-type silicon is connected to the negative pole. Only a small transient current can flow through the circuit in this case, because when the electrons in the conduction band of the *n*-type silicon have flowed out to the positive pole of the battery, there is none to take their place and current ceases to flow. However, if the negative pole of the battery is connected to the *n*-type silicon and the positive pole to the *p*-type silicon (see Fig. 22.23b), a steady current flows because electrons and holes move in opposite directions and recombine at the *n–p* junction. In effect, electrons flow toward the *n–p* junction in the *n*-type material, holes flow toward the *n–p* junction in the *p*-type material, and the junction is a "sink" where electrons fill the holes in the valence band and neutralize one another. If, instead of a galvanic cell, an alternating current source was connected to the *n–p* rectifier, current would flow in one direction only, creating pulsed direct current (see Fig. 22.23c).

Gallium arsenide and other semiconductors also provide materials for making solid-state lasers, which have applications ranging from reading compact discs to performing delicate eye surgery. When an electric current is passed through a material containing *n–p* junctions, electrons from the *n* regions and holes from the *p* regions flow toward the junctions, where they recombine and emit light. The light moves through the material, stimulating additional recombinations at other junctions and the emission of additional photons (Fig. 22.24). Critical to the operation of a laser is that these photons are coherent (in phase with one another), so the corresponding electric fields add constructively to create a plane wave. The light is reflected by a mirror at one end of the material and sets up a standing wave inside the semiconductor. At the other end, a partially reflecting mirror allows an intense beam of coherent light with a fixed wavelength to leave.

FIGURE 22.25 An integrated circuit, like the microprocessor in a laptop computer, can contain hundreds of millions of transistors in a chip whose area is about 1 cm^2. The integrated circuit shown in this figure is connected to a printed circuit board by rows of wires on all four sides.

Semiconductors perform a wide range of electronic functions that formerly required the use of vacuum tubes. Vacuum tubes occupy much more space, generate large amounts of heat, and require considerably more energy to operate than **transistors,** their semiconductor counterparts. More important, semiconductors can be built into integrated circuits (Fig. 22.25) and made to store information and process it at great speeds.

Solar cells based on silicon or gallium arsenide provide a way to convert the radiant energy of the sun directly into electrical work by a technology that is virtually nonpolluting (Fig. 22.26). The high capital costs of solar cells make them uncompetitive with conventional fossil fuel sources of energy at this time, but as reserves of fossil fuels dwindle, solar energy will become an important option.

FIGURE 22.26 Silicon solar collectors are used on a large scale to harvest energy from the sun in this photovoltaic power plant located on one of the Tremiti Islands in Italy. (© Tommaso Guicciardini/Science Photo Library/Photo Researchers, Inc.)

22.8 Pigments and Phosphors: Optical Displays

The band gap of an insulator or semiconductor has a significant effect on its color. Pure diamond has a large band gap, so even blue light does not have enough energy to excite electrons from the valence band to the conduction band. As a result, light passes through diamonds without being absorbed and the diamonds are colorless. Cadmium sulfide (CdS; Fig. 22.27) has a band gap of 4.2×10^{-19} J, which corresponds to a wavelength of 470 nm in the visible region of the spectrum. Cadmium sulfide, therefore, absorbs violet and blue light but strongly transmits yellow, giving it a deep yellow color. Cadmium sulfide is the pigment called cadmium yellow. Cinnabar (HgS, Fig. 22.28) has a smaller band gap of 3.2×10^{-19} J and absorbs all light except red. It has a deep red color and is the pigment vermilion. Semiconductors with band gaps of less than 2.8×10^{-19} J absorb all wavelengths of visible light and appear black. These include silicon (see Example 22.2), germanium, and gallium arsenide.

Doping silicon brings donor levels close enough to the conduction band or acceptor levels close enough to the valence band that thermal excitation can cause electrons to move into a conducting state. The corresponding doping of *insulators* or wide band-gap semiconductors can bring donor or acceptor states into positions where *visible light* can be absorbed or emitted. This changes the colors and optical properties of the materials. Nitrogen doped in diamond gives a donor impurity level in the band gap. Transitions to this level can absorb some blue light,

FIGURE 22.27 Mixed crystals of two semiconductors with different band gaps, CdS (yellow) and CdSe (black), show a range of colors, illustrating a decrease in the band gap energy as the composition of the mixture becomes richer in Se. © Kurt Nassau/Phototake

FIGURE 22.28 Crystalline cinnabar, HgS.

giving the diamond an undesirable yellowish color. In contrast, boron doped into diamond gives an acceptor level that absorbs red light most strongly and gives the highly prized and rare "blue diamond."

Phosphors are wide band-gap materials with dopants selected to create new levels such that particular colors of light are emitted. Electrons in these materials are excited by light of other wavelengths or by electrons hitting their surfaces, and light is then emitted as they return to lower energy states. A fluorescent lamp, for example, is a mercury-vapor lamp in which the inside of the tube has been coated with phosphors. The phosphors absorb the violet and ultraviolet light emitted by mercury vapor and emit at lower energies and longer wavelengths, giving a nearly white light that is more desirable than the bluish light that comes from a mercury-vapor lamp without the phosphors.

Phosphors are also used in television screens. The picture is formed by scanning a beam of electrons (from an electron gun) over the screen. The electrons strike the phosphors coating the screen, exciting their electrons and causing them to emit light. In a black-and-white television tube, the phosphors are a mixture of silver doped into ZnS, which gives blue light, and silver doped into $Zn_xCd_{1-x}S$, which gives yellow light. The combination of the two provides a reasonable approximation of white. A color television uses three different electron guns, with three corresponding types of phosphor on the screen. Silver doped in ZnS gives blue, manganese doped in Zn_2SiO_4 is used for green, and europium doped in YVO_4 gives red light. Masks are used to ensure that each electron beam encounters only the phosphors corresponding to the desired color.

CHAPTER SUMMARY

The properties of solid materials are determined by their microscopic structure, which in turn depends on the nature of the chemical bonds created during synthesis and processing. Mechanical and structural properties originate in strong ionic and covalent bonding. Electrical conductivity measures the movement of charged particles throughout the material. Metals, semiconductors, and insulators are distinguished by differences in conductivity values, as explained by the differences in their band gaps. In effect, this difference measures the extent to which some valence electrons from the atoms comprising the solid are delocalized while the remainder are involved in formation of localized bonds. Optical properties measure the response of the solid to visible light. The magnitude of the band gap determines what wavelength of light is absorbed, and therefore the color of the material.

CHAPTER REVIEW

- The mechanical properties of solid materials are determined by their internal structure, which in turn depends on the nature of the chemical bonds created during synthesis and processing.

    Silicate ceramics are well suited for structural applications because of their strength, which originates in the partially ionic, strong silicon–oxygen bonds in the tetrahedral orthosilicate anion. This structural unit appears in naturally occurring minerals and clays, which are fashioned into ceramic pieces through sintering and densification processes.

    Nonsilicate ceramics derive comparable properties from other inorganic structural units.

Oxide ceramics are made from oxides of numerous metals including beryllium, aluminum, calcium, and yttrium; these metal-oxide bonds are essentially ionic.
Nonoxide ceramics are based on the nitrogen and carbon compounds of silicon and boron. These compounds have short, strong, highly directional covalent bonds, so the materials have great structural strength but are brittle.

- The electrical properties measure the movement of charged particles in a material in response to an applied electric field. If the particles are free to move throughout the material, electrical conduction is the result, and the magnitude is measured by the electrical conductivity of the material.

Electrical conductivity depends on two separate microscopic parameters of a material: the number density of charge carriers present and the mobility of the carriers.
The band theory of solids explains the three broad classes of electronic conductivity seen in nature in terms of the number density of charge carriers available in classes of solids.
Metals have high conductivity values because the number density of free, mobile electrons is quite high—at least one per atom in the solid is in the conduction band. The "electron sea" is delocalized throughout the solid, and the free electrons respond easily to applied electric fields.
Insulating materials, such as the ceramics, have very low electrical conductivity because they have essentially no free electrons to carry current. There are no electrons in the conduction band, and the band gap is too large for electrons to be promoted from the valence band to the conduction band.
Semiconductors have conductivity values intermediate between metals and insulators because their bandgaps are small enough that electrons can be promoted from the valence band to the conduction band with modest thermal or optical excitation.
Doping a semiconductor by adding an electron donor impurity creates new states very near the bottom of the conduction band, so electrons can be promoted from the donor into the conduction band without increasing the temperature. These promoted electrons are now free to respond to an applied electric field. This doping process leads to an *n*-type semiconductor, in which the charge carrier is an electron.
Doping a semiconductor by adding an electron acceptor impurity creates new states very near the top of the valence band. An electron can move to this impurity state from the valence band, leaving a positively charged hole in the valence band. The hole can move in response to an applied electric field. This doping process leads to a *p*-type semiconductor, in which the charge carrier is a hole.
When a *p*-type semiconductor is placed in contact with an *n*-type semiconductor and the resulting *p–n* junction is placed in an electrical circuit, the junction can either pass or block DC current, or rectify AC current depending on details of the circuit. The device can emit light due to recombination of holes and electrons at the junction.

- Optical properties describe the response of a material to light in the visible range of the electromagnetic spectrum.

The bandgap of an insulator or semiconductor determines the wavelength of light absorbed by the material. Because the remaining portions of white light are transmitted through the material, color of the material is complementary to the wavelength of the light absorbed.
Many insulators are not colored because their bandgaps are so large that no visible light is absorbed. Adding dopants can introduce donor or acceptor states that enable absorption of visible light, so the doped materials are colored.

Phosphors are wide bandgap materials doped to emit light at specific wavelengths upon excitation by electron impact or incident light at other wavelengths. Phosphors are used to generate visible light in fluorescent fixtures, and to display signals on video monitors.

CONCEPTS & SKILLS

After studying this chapter and working the problems that follow, you should be able to:

1. Show how the fundamental silicate tetrahedral unit (SiO_4^{4-}) links to other silicate tetrahedra to form rings, chains, double chains, sheets, and space-filling crystalline networks (Section 22.1, Problems 1–4).
2. Describe the chemical compositions and structures of aluminosilicates, clays, and zeolites (Section 22.1, Problems 5–6).
3. Describe the structure of ceramic materials and the ways in which they are formed (Section 22.2).
4. Outline the properties of pottery, glass, and cement and the chemical reactions that give them structural strength (Section 22.3, Problems 7–14).
5. List several important oxide and mixed oxide ceramics and give some of their uses (Section 22.4, Problems 15–16).
6. Discuss the special properties of nonoxide ceramics and the kinetic and thermodynamic factors that make them useful (Section 22.4, Problems 17–20).
7. Explain how the conductivity of a material is measured and relate it to the number and mobility of charge carriers (Section 22.5, Problems 21–26).
8. Use the band model to describe the conductivity of metals, semiconductors, and insulators (Section 22.6, Problems 27–30).
9. Describe the mechanism of action of intrinsic, n-type, and p-type semiconductors (Section 22.7, Problems 31–34).
10. Describe how silicon-based solar energy collectors operate (Section 22.7).
11. Relate the band gap of a semiconductor or phosphor to the frequencies of electromagnetic radiation absorbed or emitted when electrons make transitions between the valence and conduction bands (Section 22.8, Problems 35–36).

KEY EQUATIONS

$$I = GV$$ (Section 22.5)

$$R = \frac{\ell}{A}\rho$$ (Section 22.5)

$$\sigma = \frac{1}{\rho} = \frac{\ell}{RA}$$ (Section 22.5)

$$J = \sigma E$$ (Section 22.5)

$$|\vec{v}_d| = \mu|\vec{E}|$$ (Section 22.5)

$$\Lambda_m = |z|\mu\mathcal{F}$$ (Section 22.5)

$$\sigma_{\text{ion}} = |z|en_{\text{ion}}\,\mu_{\text{ion}}$$ (Section 22.5)

$$\sigma_{\text{tot}} = |z_{\text{cation}}|en_{\text{cation}}\,\mu_{\text{cation}} + |z_{\text{anion}}|en_{\text{anion}}\,\mu_{\text{anion}}$$ (Section 22.5)

$$\sigma_{\text{el}} = en_{\text{el}}\,\mu_{\text{el}}$$ (Section 22.5)

PROBLEMS

Answers to problems whose numbers are boldface appear in Appendix G. Problems that are more challenging are indicated with asterisks.

Minerals: Naturally Occurring Inorganic Materials

1. Draw a Lewis electron-dot diagram for the disilicate ion ($Si_2O_7^{6-}$). What changes in this structure would be necessary to produce the structure of the pyrophosphate ion ($P_2O_7^{4-}$) and the pyrosulfate ion ($S_2O_7^{2-}$)? What is the analogous compound of chlorine?

2. Draw a Lewis electron-dot diagram for the cyclosilicate ion ($Si_6O_{18}^{12-}$), which forms part of the structures of beryl and emerald.

3. Using Table 22.1, predict the structure of each of the following silicate minerals (network, sheets, double chains, and so forth). Give the oxidation state of each atom.
(a) Andradite, $Ca_3Fe_2(SiO_4)_3$
(b) Vlasovite, $Na_2ZrSi_4O_{10}$
(c) Hardystonite, $Ca_2ZnSi_2O_7$
(d) Chrysotile, $Mg_3Si_2O_5(OH)_4$

4. Using Table 22.1, predict the structure of each of the following silicate minerals (network, sheets, double chains, and so forth). Give the oxidation state of each atom.
(a) Tremolite, $Ca_2Mg_5(Si_4O_{11})_2(OH)_2$
(b) Gillespite, $BaFeSi_4O_{10}$
(c) Uvarovite, $Ca_3Cr_2(SiO_4)_3$
(d) Barysilate, $MnPb_8(Si_2O_7)_3$

5. Using Table 22.1, predict the structure of each of the following aluminosilicate minerals (network, sheets, double chains, and so forth). In each case, the Al atoms grouped with the Si and O in the formula substitute for Si in tetrahedral sites. Give the oxidation state of each atom.
(a) Keatite, $Li(AlSi_2O_6)$
(b) Muscovite, $KAl_2(AlSi_3O_{10})(OH)_2$
(c) Cordierite, $Al_3Mg_2(AlSi_5O_{18})$

6. Using Table 22.1, predict the structure of each of the following aluminosilicate minerals (network, sheets, double chains, and so forth). In each case, the Al atoms grouped with the Si and O in the formula substitute for Si in tetrahedral sites. Give the oxidation state of each atom.
(a) Amesite, $Mg_2Al(AlSiO_5)(OH)_4$
(b) Phlogopite, $KMg_3(AlSi_3O_{10})(OH)_2$
(c) Thomsonite, $NaCa_2(Al_5Si_5O_{20}) \cdot 6\ H_2O$

Silicate Ceramics

7. A ceramic that has been much used by artisans and craftsmen for the carving of small figurines is based on the mineral steatite (commonly known as soapstone). Steatite is a hydrated magnesium silicate that has the composition $Mg_3Si_4O_{10}(OH)_2$. It is remarkably soft—a fingernail can scratch it. When heated in a furnace to about 1000°C, chemical reaction transforms it into a hard, two-phase composite of magnesium silicate ($MgSiO_3$) and quartz in much the same way that clay minerals are converted into mullite ($Al_6Si_2O_{13}$) and cristobalite (SiO_2) on firing. Write a balanced chemical equation for this reaction.

8. A clay mineral that is frequently used together with or in place of kaolinite is pyrophyllite ($Al_2Si_4O_{10}(OH)_2$). Write a balanced chemical equation for the production of mullite and cristobalite on the firing of pyrophyllite.

9. Calculate the volume of carbon dioxide produced at standard temperature and pressure when a sheet of ordinary glass of mass 2.50 kg is made from its starting materials—sodium carbonate, calcium carbonate, and silica. Take the composition of the glass to be $Na_2O \cdot CaO \cdot (SiO_2)_6$.

10. Calculate the volume of steam produced when a 4.0-kg brick made from pure kaolinite is completely dehydrated at 600°C and a pressure of 1.00 atm.

11. A sample of soda-lime glass for tableware is analyzed and found to contain the following percentages by mass of oxides: SiO_2, 72.4%; Na_2O, 18.1%; CaO, 8.1%; Al_2O_3, 1.0%; MgO, 0.2%; BaO, 0.2%. (The elements are not actually present as binary oxides, but this is the way compositions are usually given.) Calculate the chemical amounts of Si, Na, Ca, Al, Mg, and Ba atoms per mole of O atoms in this sample.

12. A sample of Portland cement is analyzed and found to contain the following percentages by mass of oxides: CaO, 64.3%; SiO_2, 21.2%; Al_2O_3, 5.9%; Fe_2O_3, 2.9%; MgO, 2.5%; SO_3, 1.8%; Na_2O, 1.4%. Calculate the chemical amounts of Ca, Si, Al, Fe, Mg, S, and Na atoms per mole of O atoms in this sample.

13. The most important contributor to the strength of hardened Portland cement is tricalcium silicate, $(CaO)_3 \cdot SiO_2$, for which the measured standard enthalpy of formation is -2929.2 kJ mol^{-1}. Calculate the standard enthalpy change for the production of 1.00 mol tricalcium silicate from quartz and lime.

14. One of the simplest of the heat-generating reactions that occur when water is added to cement is the production of calcium hydroxide (slaked lime) from lime. Write a balanced chemical equation for this reaction, and use data from Appendix D to calculate the amount of heat generated by the reaction of 1.00 kg lime with water at room conditions.

Nonsilicate Ceramics

15. Calculate the average oxidation number of the copper in $YBa_2Cu_3O_{9-x}$ if $x = 2$. Assume that the rare-earth element yttrium is in its usual +3 oxidation state.

16. The mixed oxide ceramic $Tl_2Ca_2Ba_2Cu_3O_{10+x}$ has zero electrical resistance at 125 K. Calculate the average oxidation number of the copper in this compound if $x = 0.50$ and thallium is in the +3 oxidation state.

17. Silicon carbide (SiC) is made by the high-temperature reaction of silica sand (quartz) with coke; the byproduct is carbon monoxide.
(a) Write a balanced chemical equation for this reaction.
(b) Calculate the standard enthalpy change per mole of SiC produced.
(c) Predict (qualitatively) the following physical properties of silicon carbide: conductivity, melting point, and hardness.

18. Boron nitride (BN) is made by the reaction of boron trichloride with ammonia.
 (a) Write a balanced chemical equation for this reaction.
 (b) Calculate the standard enthalpy change per mole of BN produced, given that the standard molar enthalpy of formation of BN(*s*) is (254.4 kJ mol^{-1}.
 (c) Predict (qualitatively) the following physical properties of boron nitride: conductivity, melting point, and hardness.

19. The standard free energy of formation of cubic silicon carbide (SiC) is (62.8 kJ mol^{-1}. Determine the standard free energy change when 1.00 mol SiC reacts with oxygen to form SiO_2 (*s*, quartz) and $CO_2(g)$. Is silicon carbide thermodynamically stable in the air at room conditions?

20. The standard free energy of formation of boron carbide (B_4C) is (71 kJ mol^{-1}. Determine the standard free energy change when 1.00 mol B_4C reacts with oxygen to form $B_2O_3(s)$ and $CO_2(g)$. Is boron carbide thermodynamically stable in the air at room conditions?

Electrical Conduction in Materials

21. A cylindrical sample of solid germanium has length 55.0 mm and diameter 5.0 mm. In a test circuit, 0.150 A of current flowed through this sample when the voltage applied between its ends was 17.5 V. What is the electrical conductivity of this sample?

22. A gold wire 4.0 mm in diameter and 1.5 m in length is to be used in a test circuit. (a) Calculate the resistance of the wire. (b) Calculate the current density in the wire when the voltage applied between its ends is 0.070 V. (c) Calculate the electric field in the wire.

23. The mobilities for Na^+ and Cl^- in aqueous solution are given in Table 22.7. Calculate the conductivity of a 0.10-M solution of NaCl in water at 25°C.

24. Explain why the ionic mobility of CH_3COO^- is smaller than that for Cl^-.

25. The electrical conductivity for copper is given in Table 22.6. The electron mobility in copper at room temperature is 3.0×10^{-3} m^2 V^{-1} s^{-1}. Using the Drude model for metallic conductivity, calculate the number of free electrons per Cu atom. The density of copper is 8.9 g cm^{-3}.

26. A variety of useful metallic alloys can be prepared by dissolving Ni in Cu. The room temperature resistivity of pure copper is 1.6×10^{-8} Ω m. As nickel is dissolved in copper up to 50% mass, the resistivity increases in a nearly linear fashion to the value 47.0×10^{-8} Ω m. Explain this increase qualitatively.

Band Theory of Conduction

27. Electrons in a semiconductor can be excited from the valence band to the conduction band through the absorption of photons with energies exceeding the band gap. At room temperature, indium phosphide (InP) is a semiconductor that absorbs light only at wavelengths less than 920 nm. Calculate the band gap in InP.

28. Both GaAs and CdS are semiconductors that are being studied for possible use in solar cells to generate electric current from sunlight. Their band gaps are 2.29×10^{-19} J and 3.88×10^{-19} J, respectively, at room temperature. Calculate the longest wavelength of light that is capable of exciting electrons across the band gap in each of these substances. In which region of the electromagnetic spectrum do these wavelengths fall? Use this result to explain why CdS-based sensors are used in some cameras to estimate the proper exposure conditions.

29. The number of electrons excited to the conduction band per cubic centimeter in a semiconductor can be estimated from the following equation:

$$n_e = (4.8 \times 10^{15}\ \mathrm{cm^{-3}\ K^{-3/2}})\ T^{3/2}\ e^{-E_g/(2RT)}$$

where T is the temperature in kelvins and E_g the band gap in joules *per mole*. The band gap of diamond at 300 K is 8.7×10^{-19} J. How many electrons are thermally excited to the conduction band at this temperature in a 1.00-cm^3 diamond crystal?

30. The band gap of pure crystalline germanium is 1.1×10^{-19} J at 300 K. How many electrons are excited from the valence band to the conduction band in a 1.00-cm^3 crystal of germanium at 300 K? Use the equation given in the preceding problem.

Semiconductors

31. Describe the nature of electrical conduction in (a) silicon doped with phosphorus and (b) indium antimonide doped with zinc.

32. Describe the nature of electrical conduction in (a) germanium doped with indium and (b) cadmium sulfide doped with arsenic.

33. In a light-emitting diode (LED), which is used in displays on electronic equipment, watches, and clocks, a voltage is imposed across an *n–p* semiconductor junction. The electrons on the *n* side combine with the holes on the *p* side and emit light at the frequency of the band gap. This process can also be described as the emission of light as electrons fall from levels in the conduction band to empty levels in the valence band. It is the reverse of the production of electric current by illumination of a semiconductor.

 Many LEDs are made from semiconductors that have the general composition $GaAs_{1-x}P_x$. When x is varied between 0 and 1, the band gap changes and, with it, the color of light emitted by the diode. When $x = 0.4$, the band gap is 2.9×10^{-19} J. Determine the wavelength and color of the light emitted by this LED.

34. When the LED described in Problem 33 has the composition $GaAs_{0.14}P_{0.86}$ (i.e., $x = 0.86$), the band gap has increased to 3.4×10^{-19} J. Determine the wavelength and color of the light emitted by this LED.

Pigments and Phosphors: Optical Displays

35. The pigment zinc white (ZnO) turns bright yellow when heated, but the white color returns when the sample is cooled. Does the band gap increase or decrease when the sample is heated?

36. Mercury(II) sulfide (HgS) exists in two different crystalline forms. In cinnabar, the band gap is 3.2×10^{-19} J; in metacinnabar, it is 2.6×10^{-19} J. In some old paintings with improperly formulated paints, the pigment vermilion (cinnabar) has transformed to metacinnabar on exposure to light. Describe the color change that results.

ADDITIONAL PROBLEMS

37. Predict the structure of each of the following silicate minerals (network, sheets, double chains, and so forth). Give the oxidation state of each atom.
(a) Apophyllite, $KCa_4(Si_8O_{20})F \cdot 8\ H_2O$
(b) Rhodonite, $CaMn_4(Si_5O_{15})$
(c) Margarite, $CaAl_2(Al_2Si_2O_{10})(OH)_2$

38. Using Table 22.1, predict the kind of structure formed by manganpyrosmalite, a silicate mineral with chemical formula $Mn_{12}FeMg_3(Si_{12}O_{30})(OH)_{10}Cl_{10}$. Give the oxidation state of each atom in this formula unit.

39. A reference book lists the chemical formula of one form of vermiculite as

$$[(Mg_{2.36}Fe_{0.48}Al_{0.16})(Si_{2.72}Al_{1.28})O_{10}(OH)_2][Mg_{0.32}(H_2O)_{4.32}]$$

Determine the oxidation state of the iron in this mineral.

40. The most common feldspars are those that contain potassium, sodium, and calcium cations. They are called, respectively, orthoclase ($KAlSi_3O_8$), albite ($NaAlSi_3O_8$), and anorthite ($CaAl_2Si_2O_8$). The solid solubility of orthoclase in albite is limited, and its solubility in anorthite is almost negligible. Albite and anorthite, however, are completely miscible at high temperatures and show complete solid solution. Offer an explanation for these observations, based on the tabulated radii of the K^+, Na^+, and Ca^{2+} ions from Appendix F.

41. The clay mineral kaolinite ($Al_2Si_2O_5(OH)_4$) is formed by the weathering action of water containing dissolved carbon dioxide on the feldspar mineral anorthite ($CaAl_2Si_2O_8$). Write a balanced chemical equation for the reaction that occurs. The CO_2 forms H_2CO_3 as it dissolves. As the pH is lowered, will the weathering occur to a greater or a lesser extent?

42. Certain kinds of zeolite have the general formula $M_2O{\cdot}Al_2O_3{\cdot}ySiO_2{\cdot}wH_2O$, where M is an alkali metal such as sodium or potassium, y is 2 or more, and w is any integer. Compute the mass percentage of aluminum in a zeolite that has M = K, $y = 4$, and $w = 6$.

43. (a) Use data from Tables 15.2 and 16.2 to calculate the solubility of $CaCO_3$ in water at pH 7.
(b) Will the solubility increase or decrease if the pH is lowered and the water becomes more acidic?
(c) Calculate the maximum amount of limestone (primarily calcium carbonate) that could dissolve per year in a river at pH 7 with an average flow rate of $1.0 \times 10^6\ m^3/h$.

* 44. Silica (SiO_2) exists in several forms, including quartz (molar volume 22.69 $cm^3\ mol^{-1}$) and cristobalite (molar volume 25.74 $cm^3\ mol^{-1}$).
(a) Use data from Appendix D to calculate $\Delta H°$, $\Delta S°$, and ΔG at 25°C.
(b) Which form is thermodynamically stable at 25°C?
(c) Which form is stable at very high temperatures, provided that melting does not occur first?

45. Talc, $Mg_3Si_4O_{10}(OH)_2$, reacts with forsterite (Mg_2SiO_4) to form enstatite ($MgSiO_3$) and water vapor.
(a) Write a balanced chemical equation for this reaction.
(b) If the water pressure is equal to the total pressure, will formation of products be favored or disfavored with increasing total pressure?
(c) The entropy change for this reaction is positive. Will the slope of the coexistence curve (pressure plotted against temperature) be positive or negative?

46. In what ways does soda-lime glass resemble and in what ways does it differ from a pot made from the firing of kaolinite? Include the following aspects in your discussion: composition, structure, physical properties, and method of preparation.

47. Iron oxides are red when the average oxidation state of iron is high and black when it is low. To impart each of these colors to a pot made from clay that contains iron oxides, would you use an air-rich or a smoky atmosphere in the kiln? Explain.

48. Refractories can be classified as acidic or basic, depending on the properties of the oxides in question. A basic refractory must not be used in contact with acid, and an acidic refractory must not be used in contact with a base. Classify magnesia and silica as acidic or basic refractories.

49. Dolomite bricks are used in the linings of furnaces in the cement and steel industries. Pure dolomite contains 45.7% $MgCO_3$ and 54.3% $CaCO_3$ by mass. Determine the empirical formula of dolomite.

50. Beryllia (BeO) ceramics have some use but show only poor resistance to strong acids and bases. Write likely chemical equations for the reaction of BeO with a strong acid and with a strong base.

51. Silicon nitride resists all acids except hydrofluoric, with which it reacts to give silicon tetrafluoride and ammonia. Write a balanced chemical equation for this reaction.

52. Compare oxide ceramics such as alumina (Al_2O_3) and magnesia (MgO), which have significant ionic character, with covalently bonded nonoxide ceramics such as silicon carbide (SiC) and boron carbide (B_4C; see Problems 19 and 20) with respect to thermodynamic stability at ordinary conditions.

53. Compare the hybridization of Si atoms in Si(s) with that of C atoms in graphite (see Fig. 21.26). If silicon were to adopt the graphite structure, would its electrical conductivity be high or low?

54. Describe how the band gap varies from a metal to a semiconductor to an insulator.

55. Suppose some people are sitting in a row at a movie theater, with a single empty seat on the left end of the row. Every 5 minutes, a person moves into a seat on his or her left if it is empty. In what direction and with what speed does the empty seat "move" along the row? Comment on the connection with hole motion in p-type semiconductors.

56. A sample of silicon doped with antimony is an n-type semiconductor. Suppose a small amount of gallium is added to such a semiconductor. Describe how the conduction properties of the solid will vary with the amount of gallium added.

CHAPTER 23

Polymeric Materials and Soft Condensed Matter

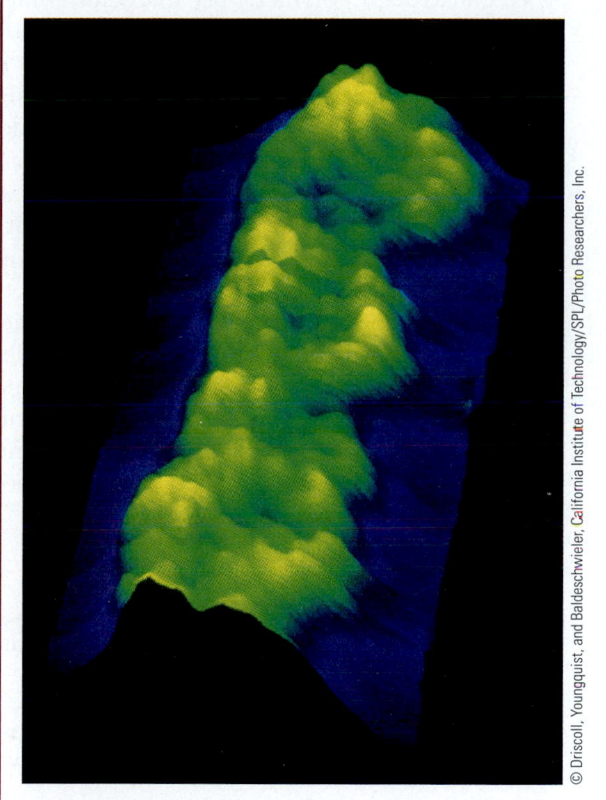

A false-color scanning tunneling micrograph (STM) of a DNA double-helix molecule adsorbed on a graphite substrate.

The organic compounds discussed in Chapter 7 were relatively small molecules, ranging from four or five atoms (such as methane or formaldehyde) to long-chain hydrocarbons up to 30 carbon atoms with relative molecular masses of several hundred. In addition to these smaller molecules, carbon atoms string together in stable chains of essentially unlimited length. Such chains provide the backbones of truly huge molecules that may contain hundreds of thousands or even millions of atoms. Such compounds, called **polymers,** are formed by linking numerous separate small **monomer units** in strands and webs.

Although many polymers are based on the ability of carbon to form stable long-chain molecules with various functional groups attached, carbon is not unique in this ability. Recall from Chapter 22 the chains, sheets, and networks found in natural silicates, in which the elements silicon and oxygen join together to form extended structures. This chapter focuses on organic polymers, whose chemical and physical properties depend on the bonding and functional group chemistry discussed in Section 7.6. We examine both synthetic polymers, which are built largely from the hydrocarbon raw materials discussed in Section 7.1, and naturally occurring biopolymers such as starch, proteins, and nucleic acids, which are built from products of biological synthesis.

23.1 Polymerization Reactions for Synthetic Polymers

To construct a polymer, very many monomers must add to a growing polymer molecule, and the reaction must not falter after the first few molecules have reacted. This is achieved by having the polymer molecule retain highly reactive functional groups at all times during its synthesis. The two major types of polymer growth are addition polymerization and condensation polymerization.

In **addition polymerization,** monomers react to form a polymer chain without net loss of atoms. The most common type of addition polymerization involves the free-radical chain reaction of molecules that have C=C bonds. As in the chain reactions considered in Section 18.4, the overall process consists of three steps: initiation, propagation (repeated many times to build up a long chain), and termination. As an example, consider the polymerization of vinyl chloride (chloroethene, $CH_2{=}CHCl$) to polyvinyl chloride (Fig. 23.1). This process can be initiated by a small concentration of molecules that have bonds weak enough to be broken by the action of light or heat, giving radicals. An example of such an **initiator** is a peroxide, which can be represented as R—O—O—R′, where R and R′ represent alkyl groups. The weak O—O bonds break

$$R{-}\ddot{\underset{\cdot\cdot}{O}}{-}\ddot{\underset{\cdot\cdot}{O}}{-}R' \longrightarrow R{-}\ddot{\underset{\cdot\cdot}{O}}\cdot + \cdot\ddot{\underset{\cdot\cdot}{O}}{-}R' \quad \text{(initiation)}$$

to give radicals, whose oxygen valence shells are incomplete. The radicals remedy this by reacting readily with vinyl chloride, accepting electrons from the C=C bonds to reestablish a closed-shell electron configuration on the oxygen atoms:

$$R{-}\ddot{\underset{\cdot\cdot}{O}}\cdot + CH_2{=}CHCl \longrightarrow R{-}\ddot{\underset{\cdot\cdot}{O}}{-}CH_2{-}\overset{\displaystyle H}{\underset{\displaystyle Cl}{C}}\cdot \quad \text{(propagation)}$$

(a)

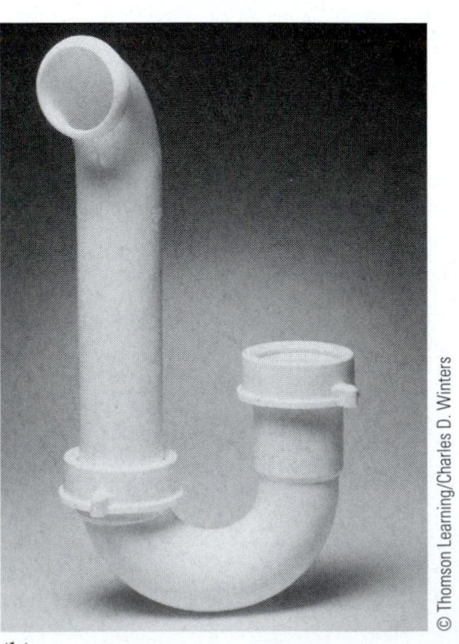

(b)

FIGURE 23.1 (a) This chemical plant in Texas produces several billion kilograms of polyvinyl chloride each year. The spherical tanks store gaseous raw material. (b) A pipefitting of polyvinyl chloride.

One of the two π electrons in the vinyl chloride double bond has been used to form a single bond with the $R{-}O\cdot$ radical. The other remains on the second carbon atom, leaving it as a seven–valence-electron atom that will react with another vinyl chloride molecule:

$$R{-}O{-}CH_2{-}\overset{H}{\underset{Cl}{C}}\cdot + CH_2{=}\overset{H}{\underset{Cl}{C}} \longrightarrow R{-}O{-}CH_2{-}\overset{H}{\underset{Cl}{C}}{-}CH_2{-}\overset{H}{\underset{Cl}{C}}\cdot \quad \text{(propagation)}$$

At each stage, the end group of the lengthening chain is one electron short of a valence octet and remains quite reactive. The reaction can continue, building up long-chain molecules of high molecular mass. The vinyl chloride monomers always attach to the growing chain with their CH_2 group because the odd electron is more stable on a CHCl end group. This gives the polymer a regular alternation of $-CH_2-$ and $-CHCl-$ groups. Its chemical formula is $(-CH_2CHCl-)_n$.

Termination occurs when the radical end groups on two different chains encounter each other and the two chains couple to give a longer chain:

$$R{-}O{-}(CH_2{-}CHCl)_m{-}CH_2{-}\overset{H}{\underset{Cl}{C}}\cdot + \cdot\overset{H}{\underset{Cl}{C}}{-}CH_2{-}(CHCl{-}CH_2)_n{-}O{-}R' \longrightarrow$$

$$R{-}O{-}(CH_2{-}CHCl)_m{-}CH_2{-}\overset{H}{\underset{Cl}{C}}{-}\overset{H}{\underset{Cl}{C}}{-}CH_2{-}(CHCl{-}CH_2)_n{-}O{-}R'$$

Alternatively, a hydrogen atom may transfer from one end group to the other:

$$R{-}O{-}(CH_2{-}CHCl)_m{-}CH_2{-}\overset{(H)}{\underset{H}{C}}{-}\overset{H}{\underset{Cl}{C}}\cdot + \cdot\overset{H}{\underset{Cl}{C}}{-}CH_2{-}(CHCl{-}CH_2)_n{-}O{-}R' \longrightarrow$$

$$R{-}O{-}(CH_2{-}CHCl)_m{-}CH{=}CHCl + CH_2Cl{-}CH_2{-}(CHCl{-}CH_2)_n{-}O{-}R' \quad \text{(termination)}$$

The latter termination step leaves a double bond on one chain end and a $-CH_2Cl$ group on the other. When the polymer molecules are long, the exact natures of the end groups have little effect on the physical and chemical properties of the material. A different type of hydrogen transfer step often has a much greater effect on the properties of the resulting polymer. Suppose that a hydrogen atom transfers not from the monomer unit on the *end* of a second chain but from a monomer unit in the middle of that chain (Fig. 23.2). Then the first chain stops growing, but

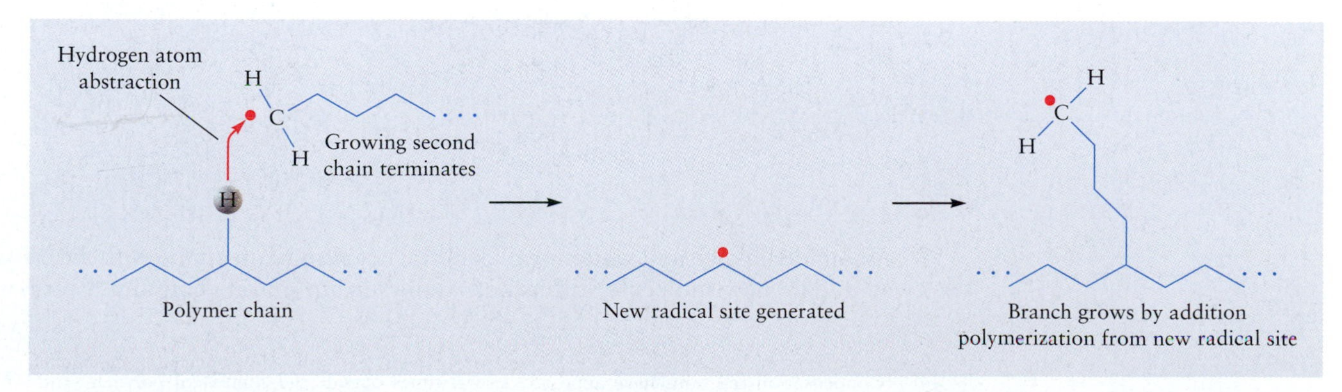

FIGURE 23.2 Chain branching can occur when a hydrogen atom is transferred (abstracted) from the middle of one chain to the free radical end of a second chain, thus terminating the growth of the second chain. The newly generated free radical in the middle of the polymer chain provides a site for the growth of a branched chain via continued monomer addition polymerization.

the radical site moves to the middle of the second chain, and growth resumes from that point, forming a *branched* polymeric chain with very different properties.

Addition polymerization can be initiated by ions as well as by free radicals. An example is the polymerization of acrylonitrile:

$$n\ \mathrm{CH_2{=}CH(C{\equiv}N)} \longrightarrow \mathrm{\left[CH_2CH(C{\equiv}N)\right]_n}$$

A suitable initiator for this process is butyl lithium, $(CH_3CH_2CH_2CH_2)^-Li^+$. The butyl anion (abbreviated Bu^-) reacts with the end carbon atom in a molecule of acrylonitrile to give a new anion:

$$\mathrm{Bu^{\ominus}Li^{\oplus} + CH_2{=}CH(C{\equiv}N) \longrightarrow Bu{-}CH_2{-}CH(C{\equiv}N)^{\ominus}Li^{\oplus}} \quad \text{(initiation)}$$

The new anion then reacts with an additional molecule of acrylonitrile:

$$\mathrm{Bu{-}CH_2{-}CH(C{\equiv}N)^{\ominus}Li^{\oplus} + CH_2{=}CH(C{\equiv}N) \longrightarrow Bu{-}CH_2{-}CH(C{\equiv}N){-}CH_2{-}CH(C{\equiv}N)^{\ominus}Li^{\oplus}} \quad \text{(propagation)}$$

The process continues, building up a long-chain polymer.

Ionic polymerization differs from free-radical polymerization because the negatively charged end groups repel one another, ruling out termination by the coupling of two chains. The ionic group at the end of the growing polymer is stable at each stage. Once the supply of monomer has been used up, the polymer can exist indefinitely with its ionic end group, in contrast with the free-radical case, in which some reaction must take place to terminate the process. Ion-initiated polymers are called "living" polymers because, when additional monomer is added (even months later), they resume growth and increase in molecular mass. Termination can be achieved by adding water to replace the Li^+ with a hydrogen ion:

$$\mathrm{{-}\!\left(CH_2{-}CH(C{\equiv}N)\right)_n{-}CH_2{-}CH(C{\equiv}N)^{\ominus}Li^{\oplus} + H_2O \longrightarrow {-}\!\left(CH_2{-}CH(C{\equiv}N)\right)_n{-}CH_2{-}CH_2(C{\equiv}N) + Li^{\oplus} + OH^{\ominus}} \quad \text{(termination)}$$

A second important mechanism of polymerization is **condensation polymerization,** in which a small molecule (frequently water) is split off as each monomer unit is attached to the growing polymer.[1] An example is the polymerization of 6-aminohexanoic acid. The first two molecules react upon heating according to

$$\mathrm{HO{-}C({=}O){-}(CH_2)_5{-}NH{-}H + HO{-}C({=}O){-}(CH_2)_5{-}NH_2 \longrightarrow HO{-}C({=}O){-}(CH_2)_5{-}N(H){-}C({=}O){-}(CH_2)_5{-}NH_2 + H_2O}$$

An amide linkage and water form from the reaction of an amine with a carboxylic acid. The new molecule still has an amine group on one end and a carboxylic

[1]Condensation reactions have appeared several times outside the context of polymer synthesis. For example, two molecules of H_2SO_4 condense to form disulfuric acid ($H_2S_2O_7$), and a carboxylic acid condenses with an alcohol to form an ester (see Section 7.6).

acid group on the other; it can therefore react with two more molecules of 6-aminohexanoic acid. The process repeats to build up a long-chain molecule. For each monomer unit added, one molecule of water is split off. The final polymer in this case is called nylon 6 and is used in fiber-belted radial tires and in carpets.

Both addition and condensation polymerization can be carried out with mixtures of two or more types of monomers present in the reaction mixture. The result is a **random copolymer** that incorporates both types of monomers in an irregular sequence along the chain. For example, a 1:6 molar ratio of styrene to butadiene monomers is used to make styrene–butadiene rubber (SBR) for automobile tires, and a 2:1 ratio gives a copolymer that is an ingredient in latex paints.

Cross-Linking: Nonlinear Synthetic Polymers

If every monomer forming a polymer has only two reactive sites, then only chains and rings can be made. The 6-aminohexanoic acid used in making nylon 6 has one amine group and one carboxylic acid group per molecule. When both functional groups react, one link is forged in the polymer chain, but that link cannot react further. If some or all of the monomers in a polymer have three or more reactive sites, however, then cross-linking to form sheets or networks is possible.

One important example of cross-linking involves phenol–formaldehyde copolymers (Fig. 23.3). When these two compounds are mixed (with the phenol in excess in the presence of an acid catalyst), straight-chain polymers form. The first step is the addition of formaldehyde to phenol to give methylolphenol:

$$C_6H_5\text{–OH} + H_2C{=}O \longrightarrow HOCH_2\text{–}C_6H_4\text{–OH}$$

Molecules of methylolphenol then undergo condensation reactions (releasing water) to form a linear polymer called novalac:

$$2n\ HOCH_2\text{–}C_6H_4\text{–OH} \longrightarrow \left[\text{–}C_6H_3(OH)\text{–}CH_2\text{–}C_6H_3(OH)\text{–}CH_2\text{–}\right]_n + 2n\ H_2O$$

FIGURE 23.3 When a mixture of phenol (C_6H_5OH) and formaldehyde (CH_2O) dissolved in acetic acid is treated with concentrated hydrochloric acid, a phenol–formaldehyde polymer grows.

If, on the other hand, the reaction is carried out with an excess of formaldehyde, dimethylolphenols and trimethylolphenols form:

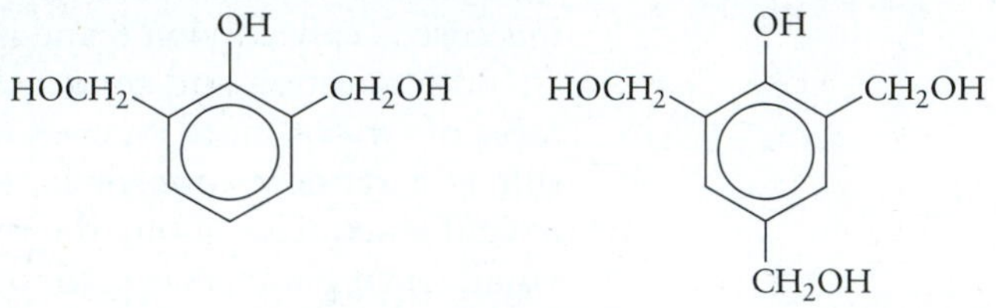

OH
$HOCH_2$ CH_2OH $HOCH_2$ CH_2OH
CH_2OH

Each of these monomers has more than two reactive sites and can react with up to three others to form a cross-linked polymer that is much stronger and more impact-resistant than the linear polymer. The very first synthetic plastic, Bakelite, was made in 1907 from cross-linked phenol and formaldehyde. Modern phenol–formaldehyde polymers are used as adhesives for plywood; more than a billion kilograms are produced per year in the United States.

Cross-linking is often desirable because it leads to a stronger material. Sometimes cross-linking agents are added deliberately to form additional bonds between polymer chains. Polybutadiene contains double bonds that can be linked upon addition of appropriate oxidizing agents. One especially important kind of cross-linking occurs through sulfur chains in rubber, as we will see.

23.2 Applications for Synthetic Polymers

The three largest uses for polymers are in fibers, plastics, and elastomers (rubbers). We can distinguish between these three types of materials on the basis of their physical properties, especially their resistance to stretching. A typical fiber strongly resists stretching and can be elongated by less than 10% before breaking. Plastics are intermediate in their resistance to stretching and elongate 20% to 100% before breaking. Finally, elastomers stretch readily, with elongations of 100% to 1000% (that is, some types of rubber can be stretched by a factor of 10 without breaking). A fourth important class is the more recently developed electrically conducting polymers, which combine the optical and electronic properties of inorganic semiconductors with the processibility of conventional polymers. This section examines the major kinds of synthetic polymers and their uses.

Fibers

Many important fibers, including cotton and wool, are naturally occurring polymers. The first commercially successful synthetic polymers were made not by polymerization reactions but through the chemical regeneration of the natural polymer cellulose, a condensation polymer of the sugar glucose that is made by plants:

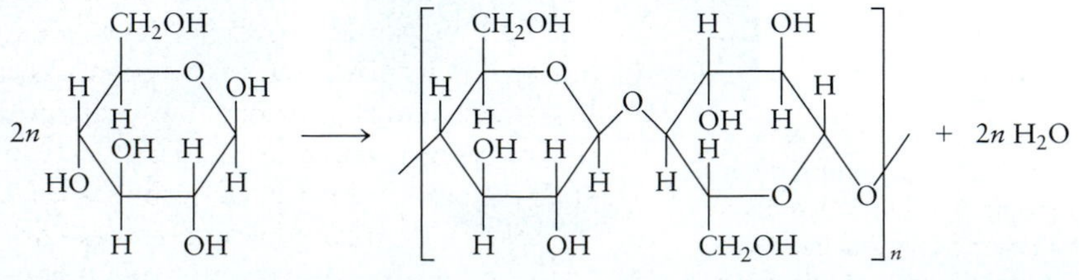

$2n$ (glucose) $\longrightarrow$ [cellulose repeat unit]$_n$ $+ \ 2n\, H_2O$

In the viscose rayon process, still used today, cellulose is digested in a concentrated solution of NaOH to convert the $-OH$ groups to $-O^- \, Na^+$ ionic groups. Reac-

FIGURE 23.4 Filter paper (cellulose) will dissolve in a concentrated ammonia solution containing $[Cu(NH_3)_4]^{2+}$ ions. When the solution is extruded into aqueous sulfuric acid, a dark blue thread of rayon (regenerated cellulose) precipitates.

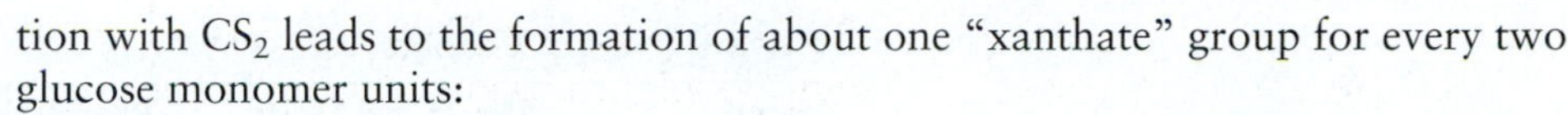

tion with CS_2 leads to the formation of about one "xanthate" group for every two glucose monomer units:

$$-\overset{|}{\underset{|}{C}}-O^{\ominus}Na^{\oplus} + CS_2 \longrightarrow -\overset{|}{\underset{|}{C}}-O-\overset{\overset{\displaystyle S}{\|}}{C}-S^{\ominus}Na^{\oplus}$$

Xanthate

Such substitutions reduce the hydrogen-bond forces holding polymer chains together. In the ripening step, some of these xanthate groups are removed with regeneration of CS_2, and others migrate to the $-CH_2OH$ groups from the ring $-OH$ groups. Afterward, sulfuric acid is added to neutralize the NaOH and to remove the remaining xanthate groups. At the same time, the viscose rayon is spun out to form fibers (Fig. 23.4) while new hydrogen bonds form.

Rayon is a "semisynthetic" fiber because it is prepared from a natural polymeric starting material. The first truly synthetic polymeric fiber was nylon, developed in the 1930s by the American chemist Wallace Carothers at DuPont Company. He knew of the condensation of an amine with a carboxylic acid to form an amide linkage (see Section 7.6) and noted that, if each molecule had *two* amine or carboxylic acid functional groups, long-chain polymers could form. The specific starting materials upon which Carothers settled, after numerous attempts, were adipic acid and hexamethylenediamine:

$$HO-\overset{\overset{\displaystyle O}{\|}}{C}-(CH_2)_4-\overset{\overset{\displaystyle O}{\|}}{C}-OH \qquad H_2N-(CH_2)_6-NH_2$$

Adipic acid   Hexamethylenediamine

The two react with loss of water, according to the equation

$$HO-\overset{\overset{\displaystyle O}{\|}}{C}-(CH_2)_4-\overset{\overset{\displaystyle O}{\|}}{C}-\boxed{OH + H}-\underset{\underset{\displaystyle H}{|}}{N}-(CH_2)_6-NH_2 \longrightarrow$$

$$HO-\overset{\overset{\displaystyle O}{\|}}{C}-(CH_2)_4-\overset{\overset{\displaystyle O}{\|}}{C}-\underset{\underset{\displaystyle H}{|}}{N}-(CH_2)_6-NH_2 + H_2O$$

The resulting molecule has a carboxylic acid group on one end (which can react with another molecule of hexamethylenediamine) and an amine group on the other end (which can react with another molecule of adipic acid). The process can continue indefinitely, leading to a polymer with the formula

$$\left[-\overset{\overset{\displaystyle O}{\|}}{C}-(CH_2)_4-\overset{\overset{\displaystyle O}{\|}}{C}-\underset{\underset{\displaystyle H}{|}}{N}-(CH_2)_6-\underset{\underset{\displaystyle H}{|}}{N}-\right]_n$$

FIGURE 23.5 Hexamethylenediamine is dissolved in water (lower layer), and adipyl chloride, a derivative of adipic acid, is dissolved in hexane (upper layer). At the interface between the layers, nylon forms and is drawn out onto the stirring bar.

called nylon 66 (Fig. 23.5). The nylon is extruded as a thread or spun as a fiber from the melt. The combination of well-aligned polymer molecules and $N-H\cdots O$ hydrogen bonds between chains makes nylon one of the strongest materials known. The designation "66" indicates that this nylon has six carbon atoms on the starting carboxylic acid and six on the diamine. Other nylons can be made with different numbers of carbon atoms.

Just as a carboxylic acid reacts with an amine to give an amide, it also reacts with an alcohol to give an ester. This suggests the possible reaction of a dicarboxylic acid and a glycol (dialcohol) to form a polymer. The polymer produced most extensively in this way is polyethylene terephthalate, which is built up from

TABLE 23.1 Fibers

Name	Structural Units	Properties	Sample Uses
Rayon	Regenerated cellulose	Absorbent, soft, easy to dye, poor wash and wear	Dresses, suits, coats, curtains, blankets
Acetate	Acetylated cellulose	Fast drying, supple, shrink-resistant	Dresses, shirts, draperies, upholstery
Nylon	Polyamide	Strong, lustrous, easy to wash, smooth, resilient	Carpeting, upholstery, tents, sails, hosiery, stretch fabrics, rope
Dacron	Polyester	Strong, easy to dye, shrink-resistant	Permanent-press fabrics, rope, sails, thread
Acrylic (Orlon)	$+CH_2-CH(C{\equiv}N)+_n$	Warm, lightweight, resilient, quick-drying	Carpeting, sweaters, baby clothes, socks

Adapted from P. J. Chenier, *Survey of Industrial Chemistry.* New York: John Wiley & Sons, 1986, Table 18.4.

terephthalic acid (a benzene ring with $-COOH$ groups on both ends) and ethylene glycol. The first two molecules react according to

$$HO-C(=O)-C_6H_4-C(=O)-OH + HO-CH_2-CH_2-OH \longrightarrow$$

Terephthalic acid Ethylene glycol

$$HO-C(=O)-C_6H_4-C(=O)-O-CH_2-CH_2-OH + H_2O$$

Further reaction then builds up the polymer, which is called polyester and sold under trade names such as Dacron. The planar benzene rings in this polymer make it stiffer than nylon, which has no aromatic groups in its backbone, and help make polyester fabrics crush-resistant. The same polymer formed in a thin sheet rather than a fiber becomes Mylar, a very strong film used for audio and video tapes.

Table 23.1 summarizes the structures, properties, and uses of some important fibers.

Plastics

Plastics are loosely defined as polymeric materials that can be molded or extruded into desired shapes and that harden upon cooling or solvent evaporation. Rather than being spun into threads in which their molecules are aligned, as in fibers, plastics are cast into three-dimensional forms or spread into films for packaging applications. Although celluloid articles were fabricated by plastic processing by the late 1800s, the first important synthetic plastic was Bakelite, the phenol–formaldehyde resin whose cross-linking was discussed earlier in this section. Table 23.2 lists some of the most important plastics and their properties.

Ethylene ($CH_2{=}CH_2$) is the simplest monomer that will polymerize. Through free-radical–initiated addition polymerization at high pressures (1000 atm to 3000 atm) and temperatures (300°C to 500°C), it forms polyethylene:

$$n\,CH_2{=}CH_2 \longrightarrow +CH_2OCH_2+_n$$

The polyethylene formed in this way is not the perfect linear chain implied by this simple equation. Free radicals frequently abstract hydrogen from the middles of chains in this synthesis, so the polyethylene is heavily branched with hydrocarbon side chains of varying length. It is called **low-density polyethylene** (LDPE) because the difficulty of packing the irregular side chains gives it a lower density (<0.94 g

TABLE 23.2 Plastics

Name	Structural Units	Properties	Sample Uses
Polyethylene	$\text{-}(CH_2-CH_2)_n\text{-}$	High density: hard, strong, stiff	Molded containers, lids, toys, pipe
		Low density: soft, flexible, clear	Packaging, trash bags, squeeze bottles
Polypropylene	$\text{-}(CH_2-CH(CH_3))_n\text{-}$	Stiffer, harder than high-density polyethylene, higher melting point	Containers, lids, carpeting, luggage, rope
Polyvinyl chloride	$\text{-}(CH_2-CH(Cl))_n\text{-}$	Nonflammable, resistant to chemicals	Water pipes, roofing, credit cards, records
Polystyrene	$\text{-}(CH_2-CH(C_6H_5))_n\text{-}$	Brittle, flammable, not resistant to chemicals, easy to process and dye	Furniture, toys, refrigerator linings, insulation
Phenolics	Phenol–formaldehyde copolymer	Resistant to heat, water, chemicals	Plywood adhesive, Fiberglass binder, circuit boards

Adapted from P. J. Chenier, *Survey of Industrial Chemistry.* New York: John Wiley & Sons, 1986, pp. 252–264.

cm^{-3}) than that of perfectly linear polyethylene. This irregularity also makes it relatively soft, so its primary uses are in coatings, plastic packaging, trash bags, and squeeze bottles in which softness is an advantage, not a drawback.

A major breakthrough occurred in 1954, when the German chemist Karl Ziegler showed that ethylene could also be polymerized with a catalyst consisting of $TiCl_4$ and an organoaluminum compound [for example, $Al(C_2H_5)_3$]. The addition of ethylene takes place at each stage within the coordination sphere of the titanium atom, so monomers can add only at the end of the growing chain. The result is linear polyethylene, also called **high-density polyethylene** (HDPE) because of its density (0.96 g cm^{-3}). Because its linear chains are regular, HDPE contains large crystalline regions, which make it much harder than LDPE and thus suitable for molding into plastic bowls, lids, and toys.

A third kind of polyethylene introduced in the late 1970s is called **linear low-density polyethylene** (LLDPE). It is made by the same metal-catalyzed reactions as HDPE, but it is a deliberate copolymer with other 1-alkenes such as 1-butene. It has some side groups (which reduce the crystallinity and density), but they have a controlled short length instead of the irregular, long side branches in LDPE. LLDPE is stronger and more rigid than LDPE; it is also less expensive because lower pressures and temperatures are used in its manufacture.

If one of the hydrogen atoms of the ethylene monomer unit is replaced with a different type of atom or functional group, the plastics that form upon polymerization have different properties. Substitution of a methyl group (that is, the use of propylene as monomer) leads to polypropylene:

$$\left[CH_2-\underset{\displaystyle CH_3}{\underset{|}{CH}} \right]_n$$

This reaction cannot be carried out successfully by free-radical polymerization. It was first achieved in 1953–1954 by Ziegler and the Italian chemist Giulio Natta, who used the Ziegler catalyst later employed in making HDPE. In polypropylene, the methyl groups attached to the carbon backbone can be arranged in different conformations (Fig. 23.6). In the **isotactic form,** all the methyl groups are arranged on the same side, whereas in the **syndiotactic form** they alternate in a regular fashion. The **atactic form** shows a random positioning of methyl groups. Natta showed that the Ziegler catalyst led to isotactic polypropylene, and he developed another catalyst, using VCl_4, that gave the syndiotactic form. Polypropylene plastic is

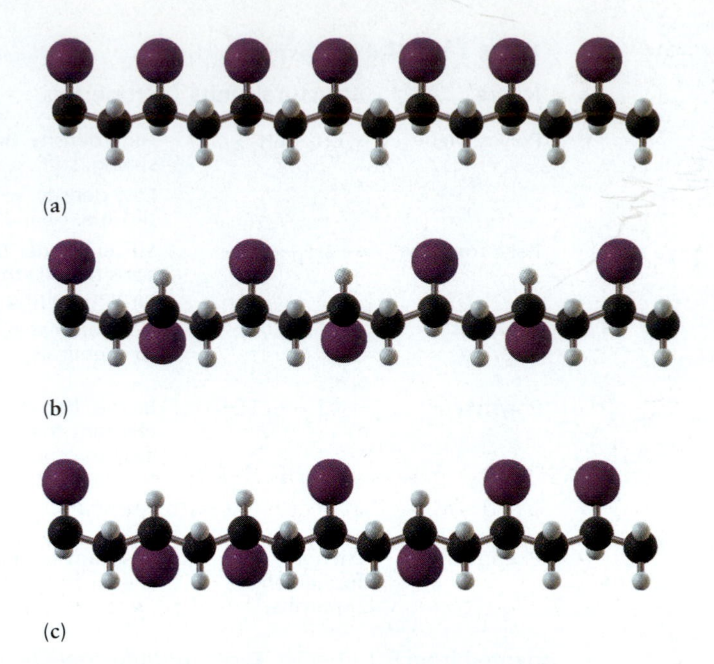

FIGURE 23.6 The structures of (a) isotactic, (b) syndiotactic, and (c) atactic polypropylene. In these structures the purplish-pink spheres represent $-CH_3$ (methyl) side groups on the long chain.

stiffer and harder than HDPE and has a higher melting point, so it is particularly useful in applications requiring high temperatures (such as the sterilization of medical instruments).

In polystyrene, a benzene ring replaces one hydrogen atom of each ethylene monomer unit. Because such a ring is bulky, atactic polystyrene does not crystallize to any significant extent. The most familiar application of this polymer is in the polystyrene foam used in disposable containers for food and drinks and as insulation. A volatile liquid or a compound that dissociates to gaseous products on heating is added to the molten polystyrene. It forms bubbles that remain as the polymer is cooled and molded. The gas-filled pockets in the final product make it a good thermal insulator.

Synthetic polymers with other elements beyond carbon and hydrogen offer many additional possibilities for making plastics. Polyvinylchloride was already discussed in Section 23.1. Another well-known plastic is the solid perfluorocarbon called polytetrafluoroethylene (Teflon), formulated as $(-CF_2-CF_2-)_n$, where n is a large number. Chemically, this compound is nearly completely inert, resisting attack from boiling sulfuric acid, molten potassium hydroxide, gaseous fluorine, and other aggressive chemicals. Physically, it has excellent heat stability (a working temperature up to 260°C), is a very good electrical insulator, and has a low coefficient of friction that makes it useful for bearing surfaces in machines as well as coating frying pans ("non-stick" Teflon). In Teflon the carbon atoms lie in a long chain that is encased by tightly bound fluorine atoms (Fig. 23.7). Even reactants with a strong innate ability to disrupt C—C bonds (such as fluorine itself) fail to attack Teflon at observable rates because there is no way to get past the surrounding fluorine atoms and their tightly held electrons.

FIGURE 23.7 A model of the molecular structure of Teflon shows that the fluorine atoms shield the carbon chain very effectively. Note that the chain must twist to accommodate the bulk of the fluorine atoms, completing a full spiral every 26 C atoms along the chain.

Rubber

An **elastomer** is a polymer that can be deformed to a great extent and still recover its original form when the deforming stress is removed. The term *rubber* was introduced by Joseph Priestley, who observed that such materials can be used to rub out pencil marks. Natural rubber is a polymer of isoprene (2-methylbutadiene). The isoprene molecule contains two double bonds of which polymerization removes only one; natural rubber is therefore unsaturated, containing one double bond per isoprene unit. In polymeric isoprene, the geometry at each double bond

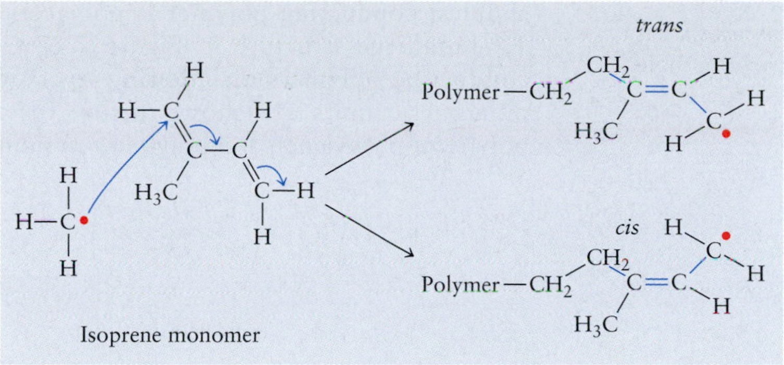

FIGURE 23.8 In the polymerization of isoprene, a *cis* or *trans* configuration can form at each double bond in the polymer. The blue arrows show the redistribution of the electrons upon bond formation.

can be either *cis* or *trans* (Fig. 23.8). Natural rubber is all-*cis* polyisoprene. The all-*trans* form also occurs in nature in the sap of certain trees and is called gutta-percha. This material is used to cover golf balls because it is particularly tough. Isoprene can be polymerized by free-radical addition polymerization, but the resulting polymer contains a mixture of *cis* and *trans* double bonds and is useless as an elastomer.

Even pure natural rubber has limited utility because it melts, is soft, and does not fully spring back to its original form after being stretched. In 1839 the American inventor Charles Goodyear discovered that if sulfur is added to rubber and the mixture is heated, the rubber hardens, becomes more resilient, and does not melt. This process is referred to as **vulcanization** and involves the formation of sulfur bridges between the methyl side groups on different chains. Small amounts of sulfur ($<5\%$) yield an elastic material in which sulfur links between chains remain after stretching and enable the rubber to regain its original form when the external force is removed. Large amounts of sulfur give the very hard, nonelastic material ebonite.

Research on synthetic substitutes for natural rubber began in the United States and Europe before World War II. Attention focused on copolymers of butadiene with styrene (now called SBR rubber) and with acrylonitrile (NBR rubber). The Japanese occupation of the rubber-producing countries of Southeast Asia sharply curtailed the supply of natural rubber to the Allied nations, and rapid steps were taken to increase production of synthetic rubber. The initial production goal was 40,000 tons per year of SBR. By 1945, U.S. production had reached an incredible total of more than 600,000 tons per year. During those few years, many advances were made in production techniques, quantitative analysis, and basic understanding of rubber elasticity. Styrene–butadiene rubber production continued after the war, and in 1950 SBR exceeded natural rubber in overall production volume for the first time. More recently, several factors have favored natural rubber: the increasing cost of the hydrocarbon feed stock for synthetic rubber, gains in productivity of natural rubber, and the growing preference for belted radial tires, which use more natural rubber.

The development of the Ziegler–Natta catalysts has affected rubber production as well. First, it facilitated the synthesis of all-*cis* polyisoprene and the demonstration that its properties were nearly identical to those of natural rubber. (A small amount of "synthetic natural rubber" is produced today.) Second, a new kind of synthetic rubber was developed: all-*cis* polybutadiene. It now ranks second in production after styrene–butadiene rubber.

Electrically Conducting Polymers

Electrically conducting polymers, sometimes called *synthetic metals,* have a backbone that is a π-conjugated system, with alternating double and single bonds. This system is formed by overlap of carbon $2p_z$-orbitals, as in Figure 7.17. The polymers are named after the monomer units on which their structures are based. The simplest conducting polymer is polyacetylene, which is a continuation of the 1,3-butadiene structure in Figure 7.17 to much longer chain lengths. Other conducting polymers include ring structures in the conjugated backbone. The monomeric units are shown below for *trans*-polyacetylene, polythiophene, poly(*para*-phenylene), and poly(*para*-pyridine).

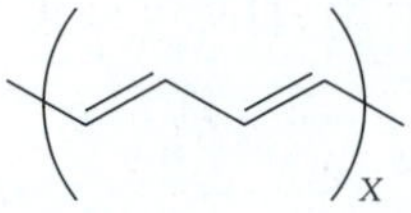

trans-Polyacetylene 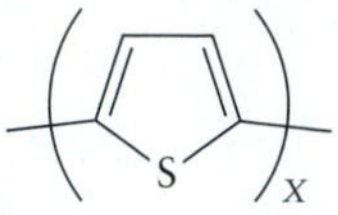 Polythiophene

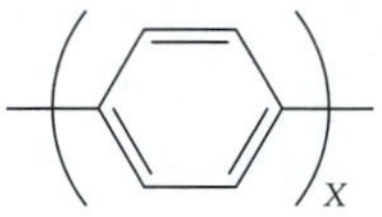

Poly(*para*-phenylene) 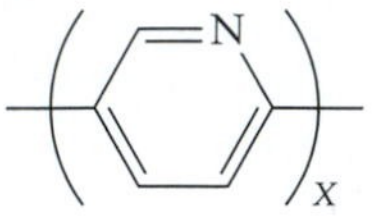 Poly(*para*-pyridine)

As the chain length increases, the energy levels shown in Figure 7.17 for 1,3-butadiene increase in number and coalesce into bands. Thus, the conjugated electronic structure for the individual linear polymer molecule is described by bands, which previously we have seen only for extended three-dimensional solids (see Figs. 21.20 and 22.21). The ground state for the polymer chain is that of an insulator, with an energy gap between occupied and empty levels.

The pure polymers are made conductive by doping; the conductivity increases as the doping level increases. Room temperature conductivity for polyacetylene doped with iodine has reached values of 5×10^4 S cm^{-1}, which is about one-tenth the value for copper. (See Section 23.1 for the definition and dimensions of conductivity.) Doping of conductive polymers does not involve substitutional replacement of lattice atoms as in the inorganic semiconductors (see Section 22.7). Rather, doping proceeds by partial oxidation or reduction of the polymer. Electron-donating dopants like Na, K, and Li produce *n*-type material (partly reduced), whereas electron acceptors like I_2, PF_6, and BF_4 produce *p*-type material (partly oxidized). The dopant ions appear interstitially between the polymer chains and promote conductivity by exchanging charges with the conjugated polymer backbones. A wide variety of interesting structural arrangements of polymer chains and dopants can be produced. The details of the conduction process depend strongly on structure and on the degree of ordering of the polymer chains. Research in this area is a fascinating interplay between concepts of solid state physics and synthetic organic chemistry.

Applications of conductive polymers rely on their combination of electrical and optical properties (comparable to metallic conductors and inorganic semiconductors) with the mechanical flexibility and the chemical processibility of organic polymers. Applications have already appeared in packaging materials for items that are sensitive to electrostatic discharges, in flexible materials for shielding against electromagnetic interference (previously achieved only with rigid metal enclosures), and in rechargeable batteries. Applications are envisioned for electrochemical drug delivery in medicine. Very recent applications include light-emitting diodes, transistors, and memory cells. This field holds rich opportunity for cross-disciplinary developments in chemistry, physics, materials science and engineering, and electrical engineering.

23.3 Liquid Crystals

Liquid crystals constitute an interesting state of matter with properties intermediate between those of true liquids and those of crystals. Unlike glasses, liquid-crystal states are thermodynamically stable. Many organic materials do not show a single solid-to-liquid transition but rather a cascade of transitions involving new intermediate phases. In recent years, liquid crystals have been used in a variety of practical applications, ranging from temperature sensors to displays on calculators and other electronic devices.

The Structure of Liquid Crystals

Substances that form liquid crystals are usually characterized by molecules with elongated, rod-like shapes. An example is terephthal-bis-(4-*n*-butylaniline), called TBBA, whose molecular structure can be represented as

$$H_9C_4 - \bigcirc - N{=}\overset{\displaystyle H}{\overset{|}{C}} - \bigcirc - \overset{\displaystyle H}{\overset{|}{C}}{=}N - \bigcirc - C_4H_9$$

with hydrocarbon groups at the ends separated by a relatively rigid backbone of benzene rings and N=C bonds. Such rod-like molecules tend to line up even in the liquid phase, as Figure 23.9a shows. Ordering in this phase persists only over small distances, however, and on average a given molecule is equally likely to take any orientation.

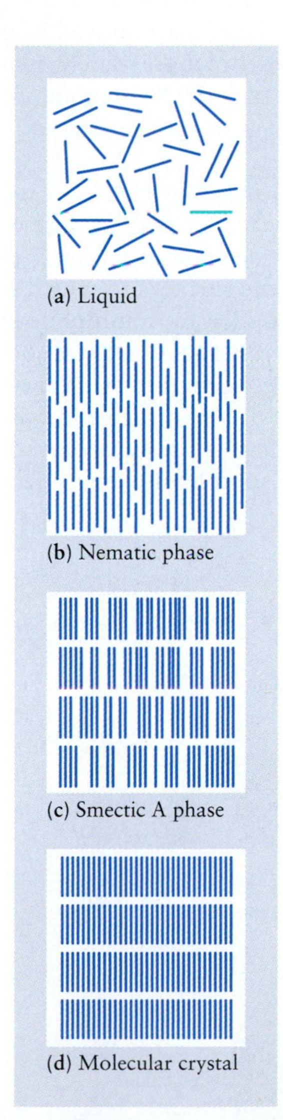

FIGURE 23.9 Different states of structural order for rod-shaped molecules. The figure is only schematic; in a real sample, the lining up of the molecules would not be so nearly perfect.

The simplest type of liquid-crystal phase is the **nematic phase** (see Fig. 23.9b); TBBA undergoes a transition from liquid to nematic at 237°C. In a nematic liquid crystal, the molecules display a preferred orientation in a particular direction, but their centers are distributed at random, as they would be in an ordinary liquid. Although liquid-crystal phases are characterized by a net orientation of molecules over large distances, not all the molecules point in exactly the same direction. There are fluctuations in the orientation of each molecule, and only *on average* do the molecules have a greater probability of pointing in a particular direction.

Some liquid crystals form one or more **smectic phases.** These display a variety of microscopic structures that are indicated by the letters A, B, C, and so forth. Figure 23.9c shows one of them, the smectic A structure; the molecules continue to display net orientational ordering, but now, unlike in the nematic phase, the centers of the molecules also tend to lie in layers. Within each layer, however, these centers are distributed at random as in an ordinary liquid. TBBA enters the smectic A phase at 200°C, before undergoing transitions to two other more ordered smectic phases at lower temperatures.

At low enough temperatures (below 113°C for TBBA), a liquid crystal freezes into a crystalline solid (see Fig. 23.9d) in which the molecules' orientations are ordered and their centers lie on a regular three-dimensional lattice. The progression of structures in Figure 23.9 illustrates the meaning of the term *liquid crystal.* Liquid crystals are solid-like in showing orientational ordering but liquid-like in the random distribution of the centers of their molecules.

A third type of liquid crystal is called **cholesteric.** The name stems from the fact that many of these liquid crystals involve derivatives of the cholesterol molecule. The structure of a cholesteric liquid crystal is shown schematically in Figure 23.10. In each plane the molecules show a nematic type of ordering, but the orientation of the molecules changes by a regular amount from plane to plane, leading to a helical structure. The distance between planes with the same orientations is referred to as the **pitch** P, which can be quite large (on the order of hundreds of nanometers or longer). A cholesteric liquid crystal will strongly diffract light with wavelengths λ comparable to the pitch. As the temperature changes, the pitch changes as well; the color of the diffracted light can therefore be used as a simple temperature sensor.

The particular orientation taken by a liquid crystal is very sensitive to both the nature of the surfaces with which it is in contact and small electric or magnetic fields. This sensitivity is the basis for the use of nematic liquid crystals in electronic display devices such as digital watches and calculators (Fig. 23.11), as well as in large-screen liquid-crystal displays.

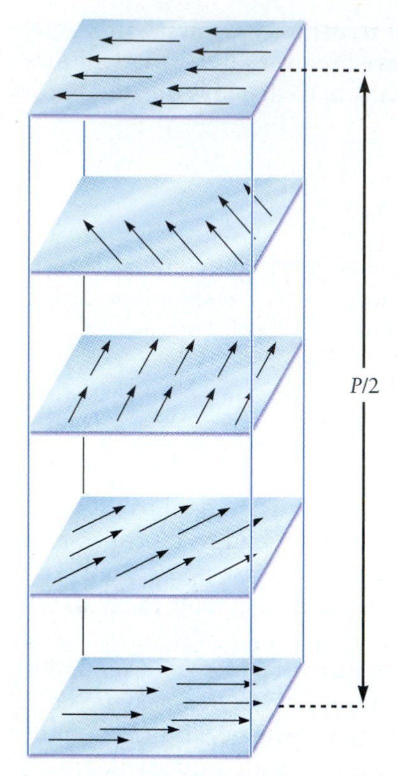

FIGURE 23.10 Several cuts through a cholesteric liquid crystal, showing how the molecular orientation changes with position. The pitch *P* is the distance over which the orientation repeats itself; here, one half of this distance is shown.

Micelles and Membranes

The liquid crystals we have discussed so far have all been single-component systems, but an interesting second type can be formed from two-component mixtures. One component of the mixture is frequently water, and a typical second component is sodium stearate, which has the chemical formula $CH_3(CH_2)_{16}COO^-Na^+$. Preparation of this soap from animal fat was described in Section 7.6. It is a salt analogous to sodium acetate, and its special properties arise from the different natures of the two ends of the molecules. The long hydrocarbon tail is **hydrophobic** ("water fearing") because hydrocarbons do not dissolve in water and avoid contact with it. The ionic carboxylate group ($-COO^-$), on the other hand, is **hydrophilic** and dissolves readily in water both because of its ionic nature and because it can participate in hydrogen bonds. Such molecules are called **amphiphiles.**

If a very small amount of an amphiphile is dissolved in water, it will separate into solvated individual molecules; as soon as a critical concentration is exceeded, however, the molecules organize into **micelles** containing 40 to 100 molecules (Fig. 23.12a). These are small, nearly spherical clusters of molecules whose hydrocarbon tails are in the nonpolar interior and whose ionic groups are exposed to the water. This organization requires a decrease in entropy but leads to a significant lowering of the energy because the hydrophobic chains are removed from direct contact with water. If, on the other hand, a hydrocarbon solvent is used, **reverse micelles** can form, in which the hydrocarbon tails of the long-chain ions make contact with the solvent and small amounts of water are collected in the polar interior of the micelle (see Fig. 23.12b).

Micelle formation is critical to the action of soaps and detergents. Grease and fat are oily substances that are more soluble in hydrocarbons than in pure water. The function of a soap such as sodium stearate in the cleaning of fabrics is to detach the grease and associated materials (dirt) from the surfaces to which it has adhered and to form a suspension of oil drops surrounded by amphiphile molecules, which can then be rinsed off. The main disadvantage of natural soaps is that their salts with ions such as Ca^{2+} and Mg^{2+} (present in dirt or hard water) are not soluble and precipitate, leaving a scum or residue in the objects being washed. To prevent this, a number of analogs to natural soaps have been developed whose calcium and magnesium salts are more soluble in water. Such amphiphilic synthetic agents are called **detergents.**

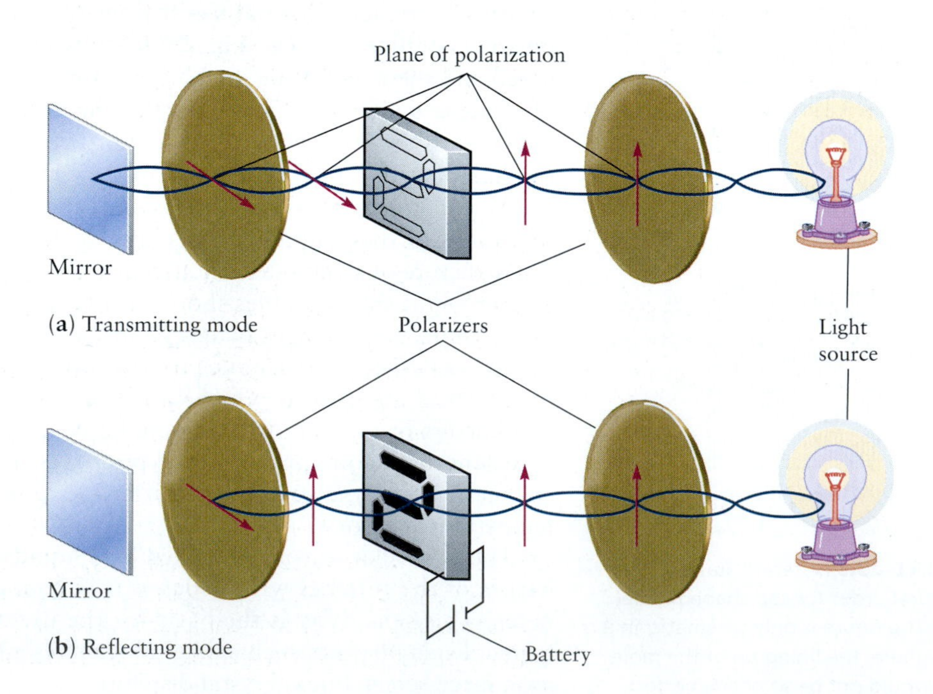

FIGURE 23.11 The mode of operation of a liquid-crystal display device. (a) The light has a polarization that permits it to pass through the second polarizing filter and strike the mirror, giving a bright display. (b) Imposition of a potential difference across some portion of the display causes the liquid-crystal molecules to rotate, creating a different polarization of light. Because the "rotated" light is blocked by the second filter, it does not reach the mirror, and that part of the display appears black.

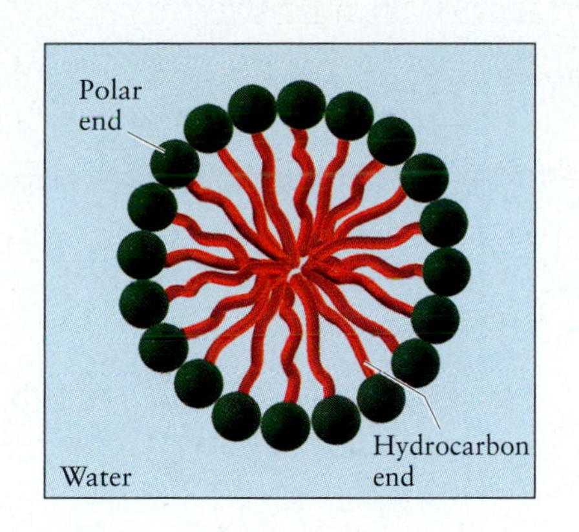

(a)

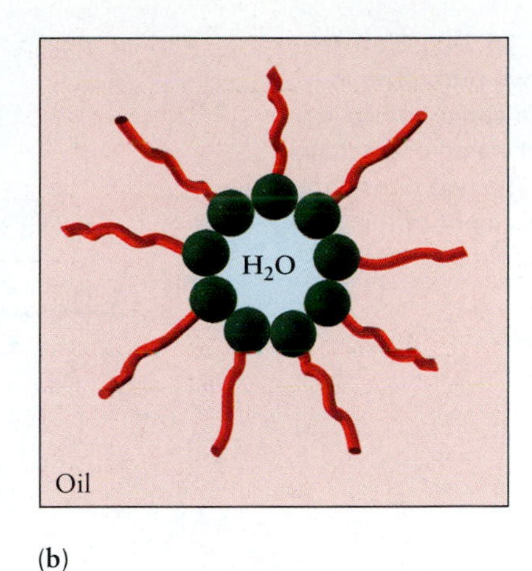

(b)

FIGURE 23.12 The structure of a micelle (a) and a reverse micelle (b).

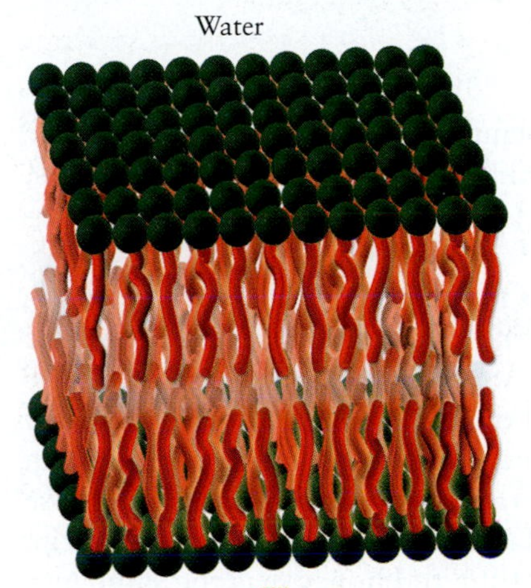

FIGURE 23.13 A bilayer membrane.

Micelles are not the only structures that can form when molecules with hydrophilic and hydrophobic sections are dissolved in water. At higher concentrations, flat **bilayer membranes** form (Fig. 23.13) and can stack into layered, or **lamellar,** phases that resemble smectic liquid crystals in their macroscopic properties. A bilayer membrane consists of two planar layers of molecules, with the hydrophilic portions in contact with water and with the hydrophobic portions of one layer in contact with the corresponding hydrophobic portions of a second layer. Such membranes can be made artificially with detergent solutions and serve as models for biological membranes that enclose living cells. Biological membranes contain embedded proteins that control the passage of ions and molecules through the cell wall. In this way they affect the response of the cell to nerve signals and hormones.

23.4 Natural Polymers

All the products of human ingenuity in the design of polymers pale beside the products of nature. Plants and animals employ a tremendous variety of long-chain molecules with different functions: some for structural strength, others to act as catalysts, and still others to provide instructions for the synthesis of vital components of the cell. In this section we discuss these three important classes of natural polymers: polysaccharides, proteins, and nucleic acids.

Carbohydrates and Polysaccharides

Carbohydrates form a class of compounds of carbon with hydrogen and oxygen. The name comes from the chemical formulas of these compounds, which can be written $C_n(H_2O)_m$, suggesting a "hydrate" of carbon. Simple **sugars,** or **monosaccharides,** are carbohydrates with the chemical formula $C_nH_{2n}O_n$. Sugars with three, four, five, and six carbon atoms are called trioses, tetroses, pentoses, and hexoses, respectively.

Glucose is a hexose sugar that exists in several forms in solution (Fig. 23.14). There is a rapid equilibrium between a straight-chain form (a six-carbon molecule with five —OH groups and one aldehyde —CHO group) and a cyclic form, in which the ring is composed of five carbon atoms and one oxygen, with four —OH side groups and one $-CH_2OH$ side group. In the straight-chain form, four of the carbon atoms (those numbered 2 through 5) are chiral centers, with four different groups bonded to them. As discussed in Section 7.2 (see Fig. 7.9), each such carbon atom can exist in two configurations, each labeled L- for *levo* or D- for *dextro* (Latin for *left* and *right,* respectively). These configurations give rise to $2^4 = 16$

FIGURE 23.14 D-Glucose exists in two ring forms in solution (a and c), which interconvert via an open-chain form (b). The two rings differ in the placement of the —OH and —H groups on carbon atom 1.

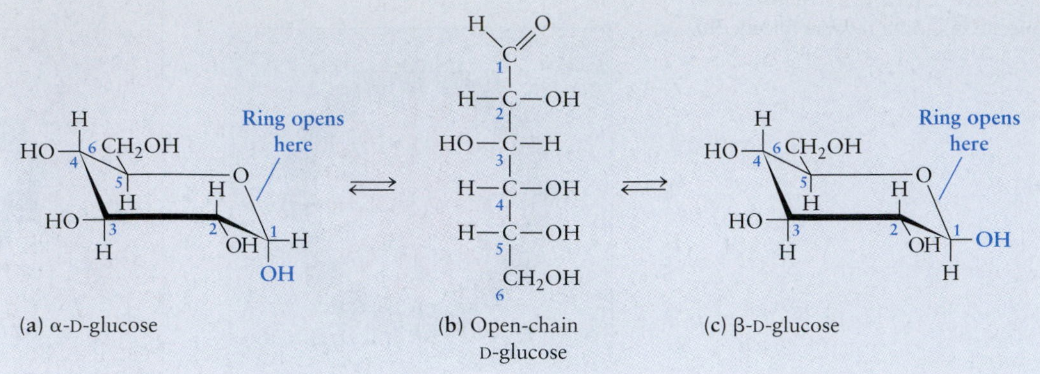

(a) α-D-glucose (b) Open-chain D-glucose (c) β-D-glucose

distinct hexose sugars. The glucose formed in plant photosynthesis always has the chirality shown in Figure 23.14b. Of the 15 other straight-chain hexose sugars, the only ones found in nature are D-galactose (in the milk sugar lactose) and D-mannose (a plant sugar).

Figure 23.14 shows that glucose actually has two different ring forms, depending on whether the —OH group created from the aldehyde by the closing of the ring lies above or below the plane of the ring. Another way to see this is to note that closing the ring creates a fifth chiral carbon atom. The two ring forms of D-glucose are called α-D-glucose (Fig. 23.14a) and β-D-glucose (see Fig. 23.14c). In aqueous solution, these two forms interconvert rapidly via the open-chain glucose form and cannot be separated. They can be isolated separately in crystalline form, however. D-fructose, a common sugar found in fruit and honey, has the same molecular formula as D-glucose but is a member of a class of hexose sugars that are ketones rather than aldehydes. In their straight-chain forms, these sugars have the C=O bond at carbon atom 2 rather than carbon atom 1 (Fig. 23.15).

Many plant cells do not stop the synthesis process with simple sugars such as glucose, but rather continue by linking sugars together to form more complex carbohydrates. **Disaccharides** are composed of two simple sugars linked together by a condensation reaction with the elimination of water. Examples shown in Figure 23.16 are the milk sugar lactose and the plant sugar sucrose (ordinary table sugar, extracted from sugarcane and sugar beets). Further linkages of sugar units lead to polymers called **polysaccharides.** The position of the oxygen atom linking the monomer units has a fundamental effect on the properties and functions of the polymers that result. Starch (Fig. 23.17a) is a polymer of α-D-glucose and is metabolized by humans and animals. Cellulose (see Fig. 23.17b), a polymer of β-D-glucose, cannot be digested except by certain bacteria that live in the digestive tracts of goats, cows, and other ruminants and in some insects, such as termites. It forms the structural fiber of trees and plants and is present in linen, cotton, and paper. It is the most abundant organic compound on earth.

FIGURE 23.15 In aqueous solutions of the sugar D-fructose, an equilibrium exists among a five-atom ring, an open chain, and a six-atom ring. In addition to the β isomers shown here, both ring forms have α isomers, in which the $-CH_2OH$ and —OH on carbon 2 are exchanged.

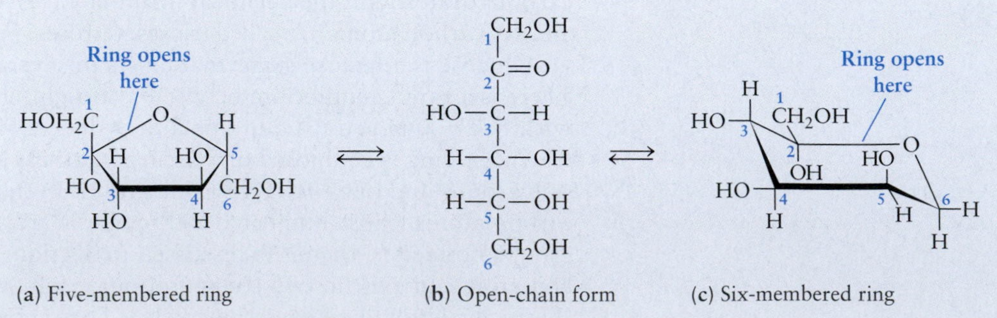

(a) Five-membered ring (b) Open-chain form (c) Six-membered ring

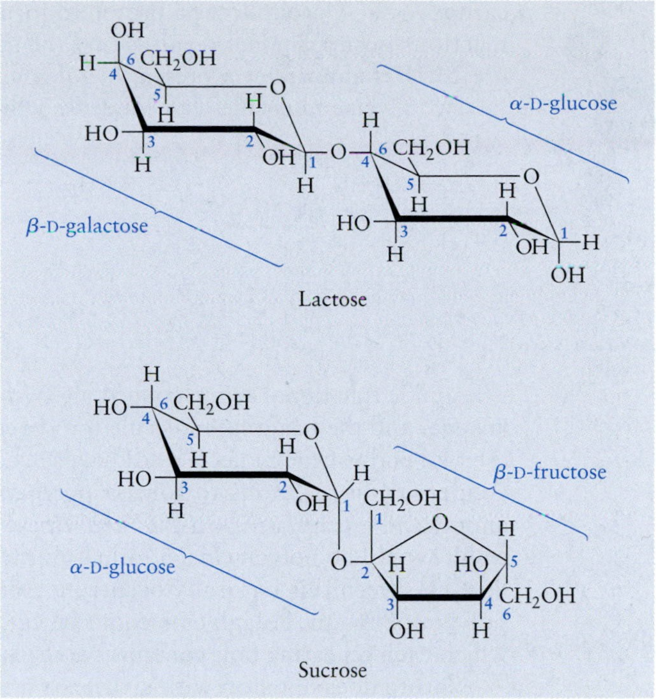

FIGURE 23.16 Two disaccharides. Their derivations from monosaccharide building blocks are shown.

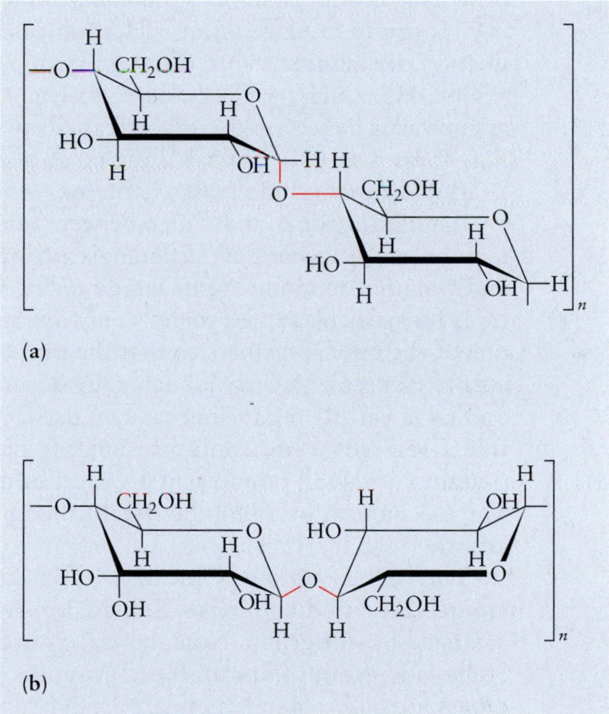

Lactose

Sucrose

FIGURE 23.17 Both starch (a) and cellulose (b) are polymers of glucose. In starch, all the cyclic glucose units are α-D-glucose. In cellulose, all the monomer units are β-D-glucose.

Amino Acids and Proteins

The monomeric building blocks of the biopolymers, called proteins, are the α-amino acids. The simplest amino acid is glycine, which has the molecular structure shown in Figure 23.18. An amino acid, as indicated by the name, must contain an amine group ($-NH_2$) and a carboxylic acid group ($-COOH$). In α-amino acids, the two groups are bonded to the same carbon atom. In acidic aqueous solution, the amine group is protonated to form $-NH_3^+$; in basic solution, the

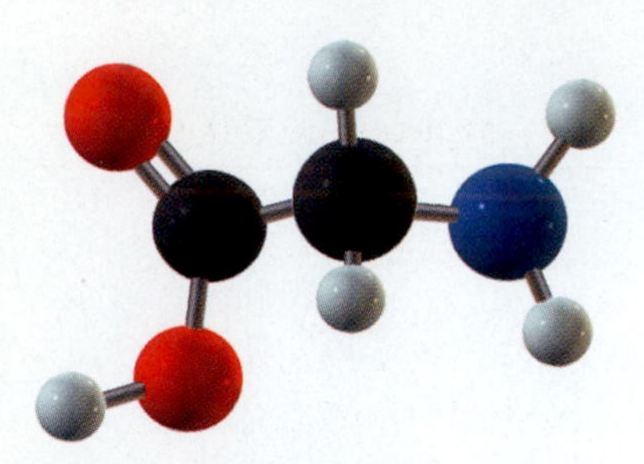

FIGURE 23.18 The structure of glycine. On the left side is the carboxylic acid group (—COOH), and on the right is the amine group ($-NH_2$).

carboxylic acid group loses a proton to form $-COO^-$. At intermediate pH, both reactions occur. The net result is that the simple amino acid form shown in Figure 23.18 is almost never present in aqueous solution (see Problem 27).

Two glycine molecules can condense with loss of water to form an amide:

$$HO(O=)C-CH_2-N(H)-H + HO-C(=O)-CH_2-NH_2 \longrightarrow$$

$$HO(O=)C-CH_2-N(H)-C(=O)-CH_2-NH_2 + H_2O$$

The amide functional group connecting two amino acids is referred to as a **peptide linkage,** and the resulting molecule is a *dipeptide*—in this case, diglycine. Because the two ends of the molecule still have carboxylic acid and amine groups, further condensation reactions to form a **polypeptide,** a polymer comprised of many amino acid groups, are possible. If glycine were the only amino acid available, the result would be polyglycine, a rather uninteresting protein. There is a close similarity between this naturally occurring condensation polymer and the synthetic polyamide nylon. Polyglycine could be called "nylon 2," a simple polyamide in which each repeating unit contains two carbon atoms.

Nature does not stop with glycine as a monomer unit. Instead, any of 20 different α-amino acids are found in most natural polypeptides. In each of these, one of the hydrogen atoms on the central carbon atom of glycine is replaced by another side group. Alanine is the next simplest α-amino acid after glycine; it has a $-CH_3$ group in place of an —H atom. This substitution has a profound consequence. In alanine, four different groups are attached to a central carbon: —COOH, $-NH_2$, $-CH_3$, and —H. There are two ways in which four different groups can be arranged in a tetrahedral structure about a central atom (see Fig. 7.9). The two optical isomers of alanine are designated by the prefixes L- and D- for *levo* and *dextro* (Latin for "left" and "right," respectively).

If a mixture of L- and D-alanine were caused to polymerize, nearly all the polymer molecules would have different structures because their sequences of D-alanine and L-alanine monomer units would differ. To create polymers with definite structures for particular roles, there is only one recourse: to build all polypeptides from one of the optical isomers so that the properties will be reproducible from molecule to molecule. Nearly all naturally occurring α-amino acids are of the L form, and most earthly organisms have no use for D-α-amino acids in making polypeptides. Terrestrial life could presumably have begun equally well using mainly D-amino acids (all biomolecules would be mirror images of their present forms). The mechanism by which the established preference was initially selected is not known.

The —H group of glycine and $-CH_3$ group of alanine give just the first two amino acid building blocks. Table 23.3 shows all 20 important α-amino acids, arranged by side group. Note the variety in their chemical and physical properties. Some side groups contain basic groups; others are acidic. Some are compact; others are bulky. Some can take part in hydrogen bonds; others can complex readily with metal ions to form coordination complexes.

This variety in properties of the α-amino acids leads to even more variety in the polymers derived from them, called **proteins.** The term *protein* is usually applied to polymers with more than about 50 amino acid groups; large proteins may contain many thousand such groups. Given the fact that any one of 20 α-amino acids may appear at each point in the chain, the number of possible sequences of amino acids in even small proteins is staggering. Moreover, the amino acid sequence describes only one aspect of the molecular structure of a protein. It contains no information about the three-dimensional conformation adopted by the protein. The carbonyl group and the amine group in each amino acid along the

TABLE 23.3 α-Amino Acid Side Groups

	Symbol	Structure of Side Group
Hydrogen "Side Group"		
Glycine	Gly	$-H$
Alkyl Side Groups		
Alanine	Ala	$-CH_3$
Valine	Val	$-CH(CH_3)-CH_3$
Leucine	Leu	$-CH_2-CH(CH_3)-CH_3$
Isoleucine	Ile	$-CH(CH_3)-CH_2-CH_3$
Proline	Pro (structure of entire amino acid)	five-membered ring: $H_2C-CH_2-CH_2-CH(COOH)-HN$ (ring closed between HN and H_2C)
Aromatic Side Groups		
Phenylalanine	Phe	$-CH_2-C_6H_5$ (benzene ring)
Tyrosine	Tyr	$CH_2-C_6H_4-OH$
Tryptophan	Trp	$-CH_2-C=CH-NH-$ (indole ring fused to benzene)
Alcohol-Containing Side Groups		
Serine	Ser	$-CH_2OH$
Threonine	Thr	$-CH(OH)-CH_3$
Basic Side Groups		
Lysine	Lys	$-CH_2CH_2CH_2CH_2NH_2$
Arginine	Arg	$-CH_2CH_2CH_2NH-C(=NH)NH_2$
Histidine	His	$-CH_2-C=CH-N=CH-HN-$ (imidazole ring)
Acidic Side Groups		
Aspartic acid	Asp	$-CH_2COOH$
Glutamic acid	Glu	$-CH_2CH_2COOH$
Amide-Containing Side Groups		
Asparagine	Asn	$-CH_2C(=O)-NH_2$
Glutamine	Gln	$-CH_2CH_2C(=O)-NH_2$
Sulfur-Containing Side Groups		
Cysteine	Cys	$-CH_2-SH$
Methionine	Met	$-CH_2CH_2-S-CH_3$

protein chain are potential sites for hydrogen bonds, which may also involve functional groups on the amino acid side chains. Also, the cysteine side groups ($-CH_2-SH$) can react with one another, with loss of hydrogen, to form $-CH_2-S-S-CH_2-$ disulfide bridges between different cysteine groups in a single chain or between neighboring chains (the same kind of cross-linking by sulfur occurs in the vulcanization of rubber). As a result of these strong intrachain interactions, the molecules of a given protein have a rather well-defined conformation even in solution, as compared with the much more varied range of conformations available to a simple alkane chain (see Fig. 7.3). The three-dimensional structures of many proteins have been determined by x-ray diffraction.

There are two primary categories of proteins: fibrous and globular. **Fibrous proteins** are usually structural materials and consist of polymer chains linked in sheets or twisted in long fibers. Silk is a fibrous protein in which the monomer units are primarily glycine and alanine, with smaller amounts of serine and tyrosine. The protein chains are cross-linked by hydrogen bonds to form sheet-like structures (Fig. 23.19) that are arranged so that the nonhydrogen side groups all lie on one side of the sheet; the sheets then stack in layers. The relatively weak forces between sheets give silk its characteristic smooth feel. The amino acids in wool and hair have side chains that are larger, bulkier, and less regularly distributed than those in silk; therefore, sheet structures do not form. Instead, the protein molecules twist into a right-handed coil called an **α-helix** (Fig. 23.20). In this

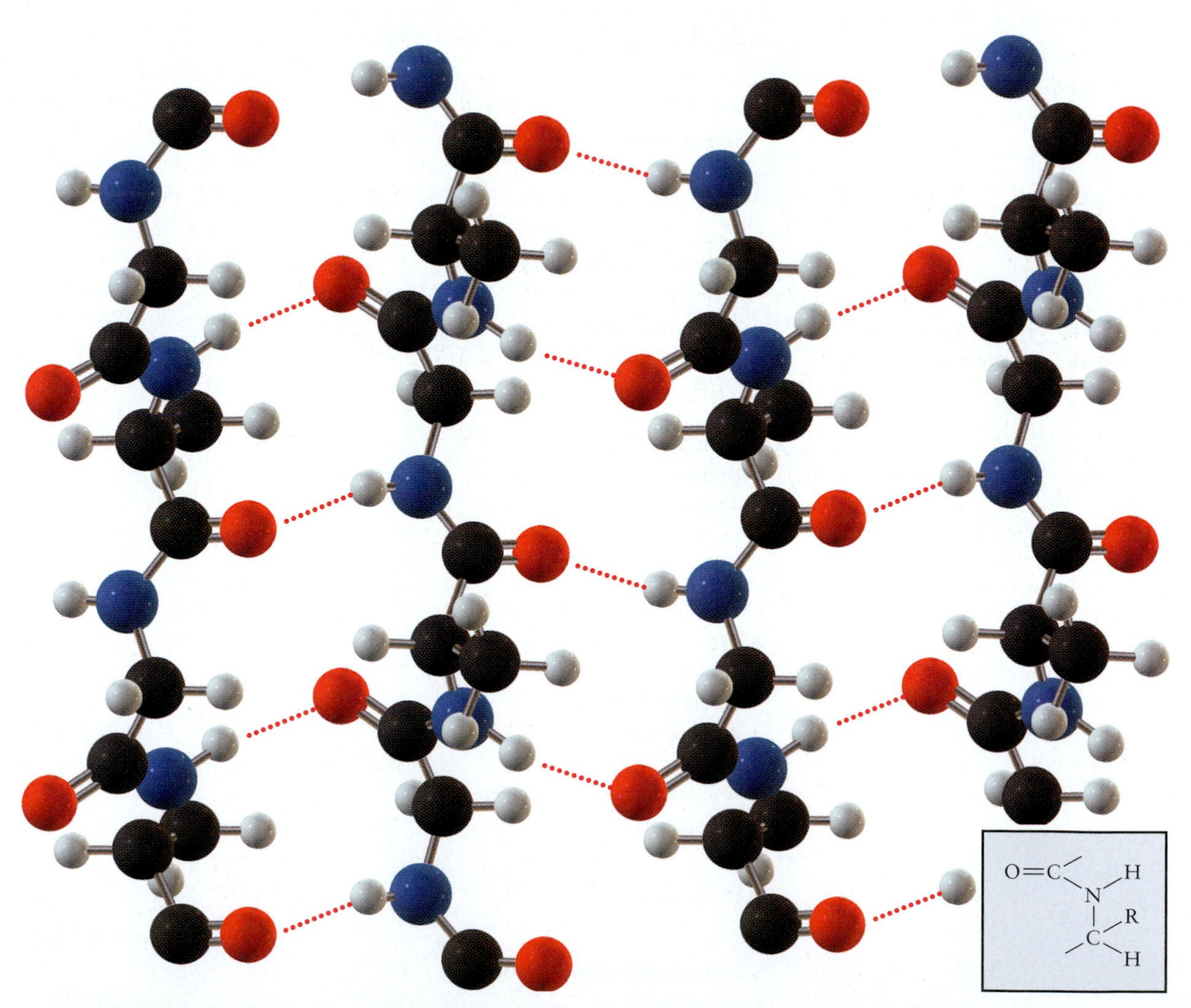

FIGURE 23.19 The structure of silk. The amino acid side groups (shown as R) must be small to fit into a sheet-like structure. Glycine (R=H) and alanine ($R=CH_3$) predominate. The repeating unit is shown as an inset.

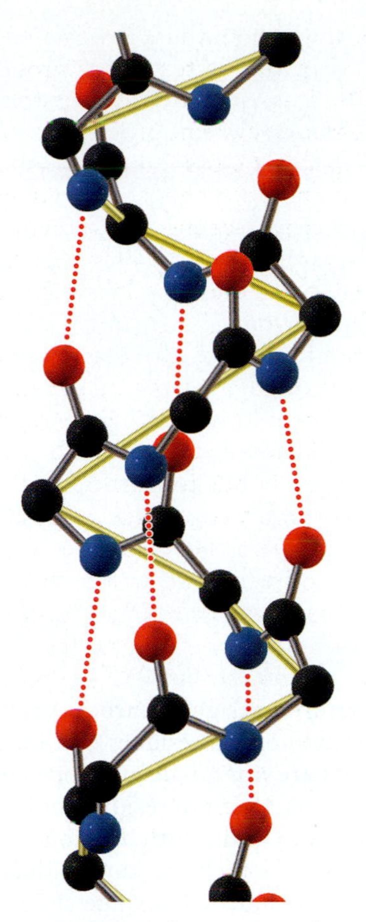

FIGURE 23.20 The structure of an α-helix for a stretch of linked glycine monomer units. The superimposed yellow line highlights the helical structure, which is maintained by hydrogen bonds (red dotted lines). The hydrogen atoms themselves are omitted for clarity.

structure, each carbonyl group is hydrogen-bonded to the amine group of the fourth amino acid farther along the chain; the bulky side groups jut out from the helix and do not interfere with one another.

The second type of protein is the **globular protein.** Globular proteins include the carriers of oxygen in the blood (hemoglobin) and in cells (myoglobin). They have irregular folded structures (Fig. 23.21) and typically consist of 100 to 1000 amino acid groups in one or more chains. Globular proteins frequently have parts of their structures in α-helices and sheets, with other portions in more disordered forms. Hydrocarbon side groups tend to cluster in regions that exclude water, whereas charged and polar side groups tend to remain in close contact with water. The sequences of amino acid units for many such proteins have been worked out by cleaving them into smaller pieces and analyzing the structure of the fragments. It took Frederick Sanger 10 years to complete the first such determination of sequence for the 51 amino acids in bovine insulin, an accomplishment that earned him the Nobel Prize in chemistry in 1958. Now, automated procedures enable scientists to rapidly determine amino acid sequences in much longer protein molecules.

Enzymes constitute a very important class of globular proteins. They catalyze particular reactions in the cell, such as the synthesis and breakdown of proteins, the transport of substances across cell walls, and the recognition and resistance of foreign bodies. Enzymes act by lowering the activation barrier for a reaction, and they must be selective so as to act only on a restricted group of substrates.

Let's examine the enzyme carboxypeptidase A, whose structure has been determined by x-ray diffraction. It removes amino acids one at a time from the carboxylic acid end of a polypeptide. Figure 23.22 shows the structure of the active site (with a peptide chain in place, ready to be cleaved). A special feature of this enzyme is the role played by the zinc ion, which is coordinated to two histidine residues in the enzyme and to a carboxylate group on a nearby glutamic acid residue. The zinc ion helps remove electrons from the carbonyl group of the peptide linkage, making it more positive and thereby more susceptible to attack by water or by the carboxylate group of a second glutamic acid residue. The side chain on the outer amino acid of the peptide being cleaved is positioned in a hydrophobic cavity, which favors large aromatic or branched side chains (such as that in tyrosine) over smaller hydrophilic side chains (such as that in aspartic acid). Carboxypeptidase A is thus selective about the sites at which it cleaves peptide chains.

FIGURE 23.21 A computer-generated model of the structure of myoglobin. The central heme group (red) is shown in greater detail in Figure 8.22; two histidine groups (green) extend toward the central iron atom. Much of the protein (blue tube) is coiled in α-helices.

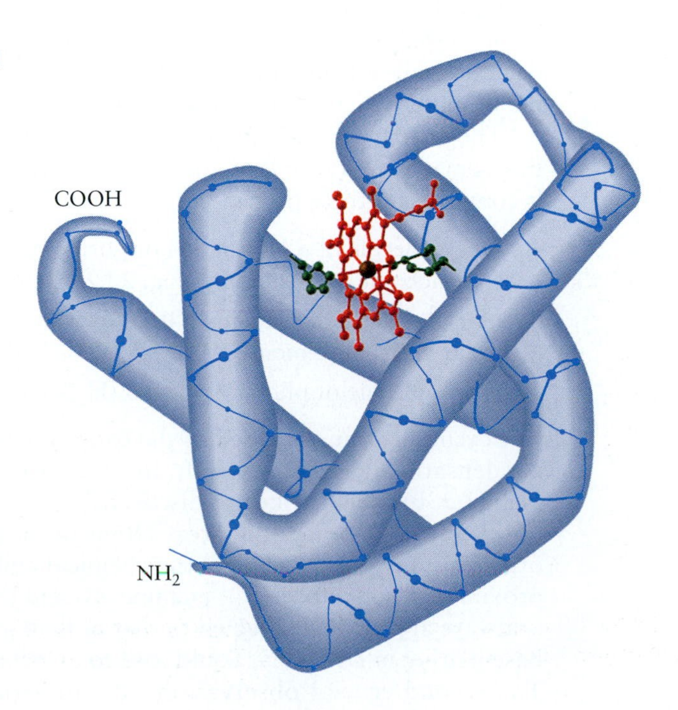

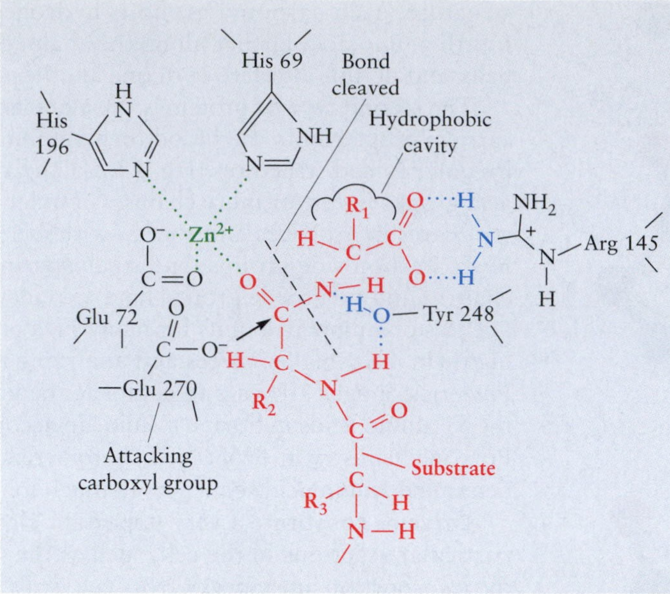

FIGURE 23.22 The active site of carboxypeptidase A. Shown in red is a substrate polypeptide that is being cleaved by the enzyme. Green is used to show the role of Zn^{2+} as a complexing ion, and blue is used to show the hydrogen bonds that maintain the geometry.

The molecular "engineering" that lies behind nature's design of carboxypeptidase A and other enzymes is truly remarkable. The amino acid residues that form the active site and determine its catalytic properties are *not* adjacent to one another in the protein chain. As indicated by the numbers after the residues in Figure 23.22, the two glutamic acid residues are the 72nd and 270th amino acids along the chain. The enzyme adopts a conformation in which the key residues, distant from one another in terms of chain position, are nonetheless quite close in three-dimensional space, allowing the enzyme to carry out its specialized function.

Nucleotides and Nucleic Acids

We have seen that proteins are copolymers made up typically of 20 types of monomer units. Simply mixing the amino acids and letting them dehydrate to form polymer chains at random would never lead to the particular structures needed by living cells. How does the cell preserve information about the amino acid sequences that make up its proteins, and how does it transmit this information to daughter cells through the reproductive process? These questions lie in the field of molecular genetics, an area in which chemistry plays the central role.

The primary genetic material is deoxyribonucleic acid (DNA). This biopolymer is made up of four types of monomer units called **nucleotides.** Each nucleotide is composed of three parts:

1. One molecule of a pyrimidine or purine base. The four bases are thymine, cytosine, adenine, and guanine (Fig. 23.23a).
2. One molecule of the sugar D-deoxyribose ($C_5H_{10}O_4$). D-Ribose is a pentose sugar with a five-membered ring.
3. One molecule of phosphoric acid (H_3PO_4).

The cyclic sugar molecule links the base to the phosphate group, undergoing two condensation reactions, with loss of water, to form the nucleotide (see Fig. 23.23b). The first key to discovering the structure of DNA was the following observation: Although the proportions of the four bases in DNA from different organisms are quite variable, the chemical amount of cytosine (C) is always approximately equal to that of guanine (G) and the chemical amount of adenine (A) is always approximately equal to that of thymine (T). This suggested some type of base pairing in DNA that could lead to association of C with G and of A with T. The second crucial observation was an x-ray diffraction study by Rosalind

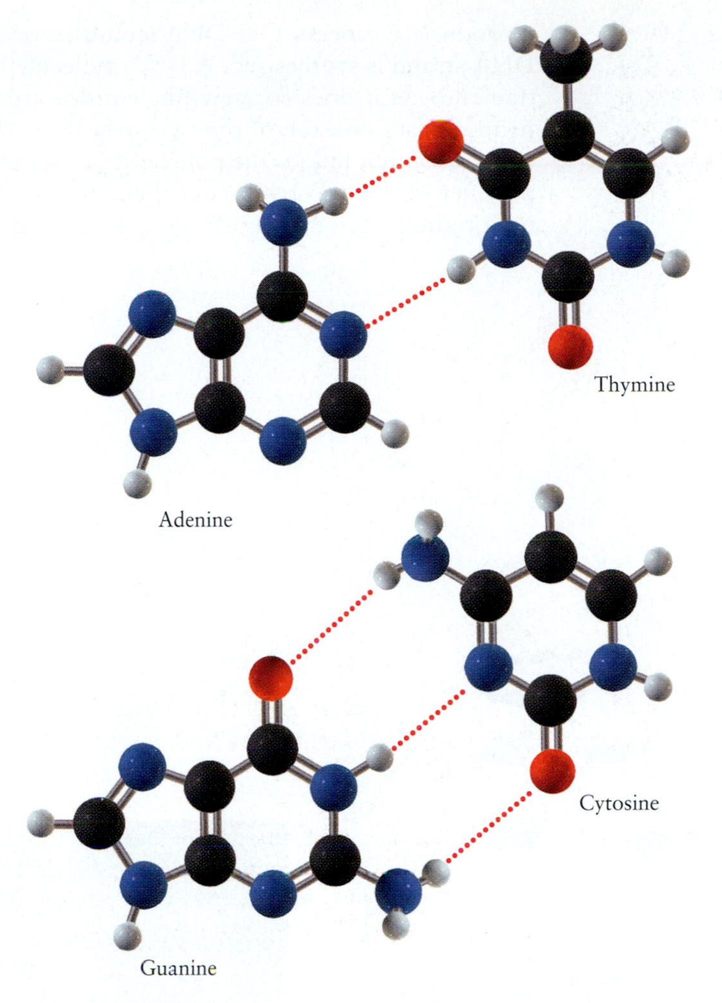

FIGURE 23.23 (a) The structures of the purine and pyrimidine bases. Hydrogen-bonding between pairs of bases is indicated by red dots. (b) The structure of the nucleotide adenosine monophosphate (AMP).

Franklin and Maurice Wilkins that suggested the presence of helical structures of more than one chain in DNA.

James Watson and Francis Crick put together these two pieces of information in their famous 1953 proposal of a double-helix structure for DNA. They concluded that DNA consists of two interacting helical strands of nucleic acid polymer (Fig. 23.24), with each cytosine on one strand linked through hydrogen bonds to a guanine on the other and each adenine to a thymine. This accounted for the observed molar ratios of the bases, and it also provided a model for the replication of the molecule, which is crucial for passing on information during the reproductive process. One DNA strand serves as a template upon which a second DNA strand is synthesized. A DNA molecule reproduces by starting to unwind at one end. As it does so, new nucleotides are guided into position opposite the proper bases on each of the two strands. If the nucleotide does not fit the template, it cannot link to the polymeric strand under construction. The result of the polymer synthesis is two double-helix molecules, each containing one strand from the original and one new strand that is identical to the original in every respect.

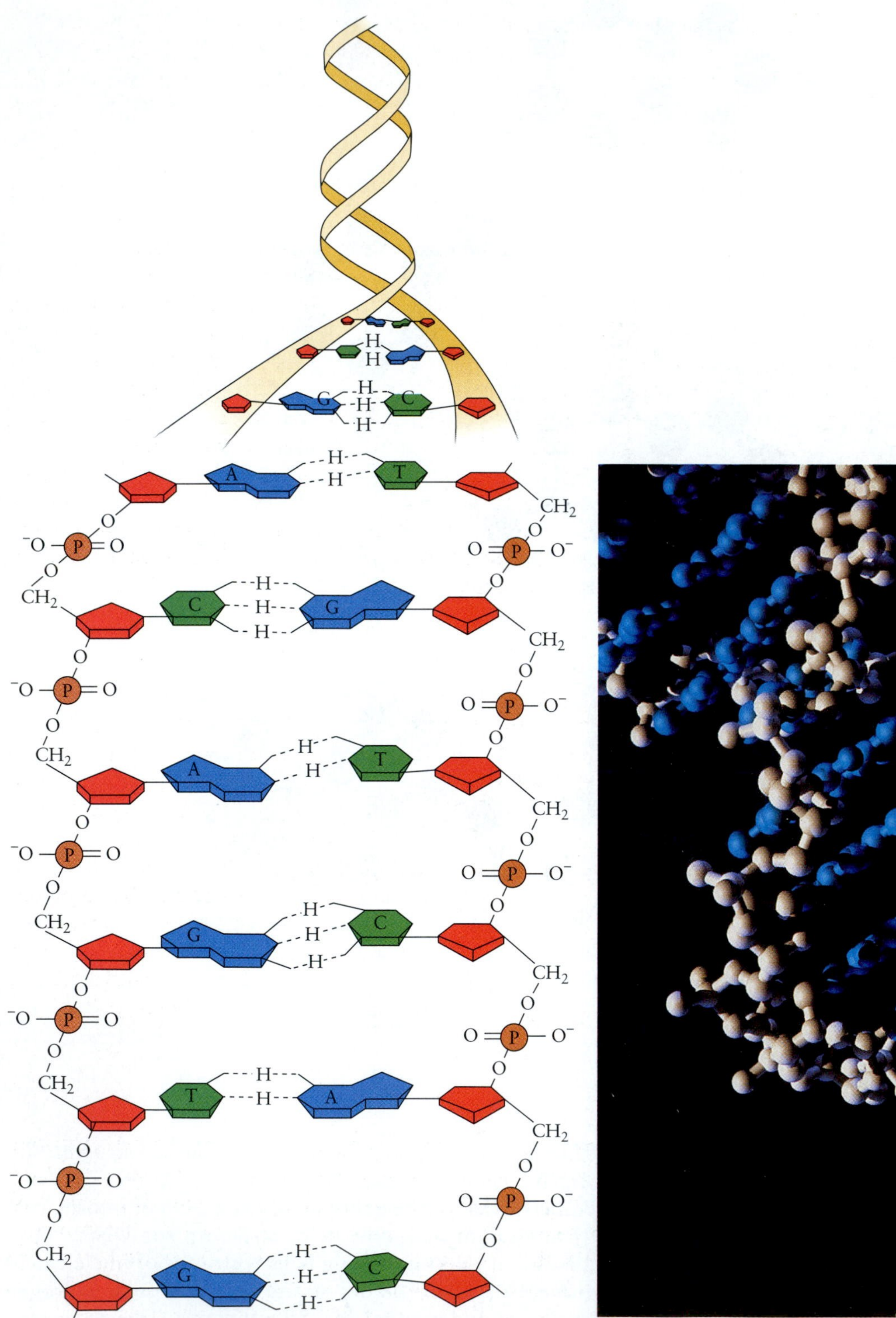

FIGURE 23.24 The double-helix structure of DNA.

Information is encoded in DNA in the sequence of the base pairs. Subsequent research has broken this genetic code and established the connection between the base sequence in a segment of DNA and the amino acid sequence of the protein synthesized according to the directions in that segment. The code in a nucleic acid is read as consecutive, nonoverlapping triplets of bases, with each triplet standing for a particular amino acid. Thus, a nucleic acid strand consisting of pure cytosine gives a polypeptide of pure proline, meaning that the triplet CCC codes for proline. The nucleic acid strand AGAGAGAG . . . is read as the alternating triplets AGA and GAG and gives a polypeptide consisting of alternating arginine (coded by AGA) and glutamic acid (coded by GAG) monomer units. There are 64 (4^3) possible triplets, so typically more than one code exists for a particular amino acid. Some triplets serve as signals to terminate a polypeptide chain. Remarkably, the genetic code appears to be universal, independent of the species of plant or animal, a finding that suggests a common origin for all terrestrial life.

A single change in a base pair in DNA causes a change in one amino acid of the protein that is coded for by that DNA. Such a change may seem small, but it may have dramatic (even fatal) effects for the organism in question. For example, the triplets GAA and GAG both code for glutamic acid (an acidic residue), whereas the triplets GTA and GTG code for valine (a nonpolar residue). A single change in the central A to T in the DNA thus changes an amino acid in a protein produced by the organism. This might seem like a small effect, but it can change the structure and mode of action of the protein. This particular change is responsible for the presence of hemoglobin S (instead of hemoglobin A) in the blood of people who have sickle cell anemia. In two of the four hemoglobin S chains a glutamic acid amino acid is changed to a valine, and the result is a decrease in the solubility of S relative to A by a factor of 25. This leads to polymerization of the hemoglobin to form insoluble structures that bend red blood cells into sickle shapes, a change that can lead to early death. One of the challenges of modern genetic engineering is to use chemistry to modify coding molecules in living species to eliminate fatal or disease-causing mutants.

CHAPTER SUMMARY

Mankind has long used naturally occurring polymers, obtained from plants and animals, as useful materials. These are familiar in cotton, wool, and silk fabrics, and in structural applications of wood and rubber. Inspired by these natural products, chemists have invented synthetic routes to these and similar materials, often with modified or improved properties. Curiosity about life processes in living systems has inspired research on natural polymers, such as polysaccharides, proteins, enzymes, and DNA, to discern the relation between their structure and function. The structures of these solid synthetic and natural polymeric materials range from purely amorphous to highly crystalline. The route to controlling their molecular and crystalline structures lies through understanding their local bonding.

CHAPTER REVIEW

- Polymer formation requires that many monomers must be attached to a growing polymer molecule. This requires that highly reactive functional groups must be available at each growth step. This is achieved by two main mechanisms.

    Addition polymerization requires monomers to join the polymer without net loss of atoms. This usually involves free radical reaction of molecules that have C=C double bonds, and proceeds through three steps: initiation, propagation, and termination.

Condensation polymerization requires that a small molecule such as water is split off as each monomer is added to the polymer.

Both addition and condensation polymerization can occur with mixtures of monomers to produce a random copolymer of the two monomers.

If monomers have three or more reactive sites, cross-linking reactions to form sheets or networks is possible. Cross-linking is often brought about deliberately to obtain stronger materials.

- Synthetic polymers are selected for four major classes of applications based on their physical properties

  Fibers for use in fabrics are spun into threads in which their molecules are aligned. They must resist stretching, and usually break after only 10% elongation. The first purely synthetic fiber was nylon, developed by condensation polymerization.

  Plastics are molded or extruded into desired shapes that harden upon cooling or solvent evaporation. They typically elongate 20% to 100% before breaking. Many of these are formed by addition polymerization of ethylene or its derivatives. Examples include polyethylene, polystyrene, and Teflon. By using proper catalysts and manipulating the size of side group substituents, it is possible to overcome the tendency of addition polymerization to form highly branched chains. These tools make it possible to control the degree of crystallinity in polymer materials.

  Rubbers, or elastomers, stretch readily to elongate by a factor of 10 before breaking. Natural rubber, obtained from the sap of certain trees, can be hardened and toughened by addition of sulfur in the vulcanization process. Synthetic rubber is produced by addition copolymerization of butadiene and styrene.

  Electrically conducting polymers combine the optical and electronic properties of inorganic semiconductors with the processing ease of conventional polymers. Their structures are continuations of the 1,3-butadiene structure to greater lengths, and the electronic structure for the individual molecule is described by bands. These polymers are made electrically conducting by doping.

- Liquid crystals have properties intermediate between those of true liquids and crystals. Unlike glasses, they are thermodynamically stable.

  Based on microscopic structural details liquid crystals form three separate ordered phases: nematic, smectic, and cholesteric.

  Orientation of liquid crystals depends sensitively on small electric and magnetic fields. This is the basis of liquid crystal displays for digital information.

- Naturally occurring polymers are synthesized by plants and animals to support a variety of life processes.

  Polysaccharides are grown by linkage of simple sugar units (carbohydrates). One example is starch, which is readily metabolized by humans and animals. Another is cellulose, which forms the structural fibers of plants and trees and appears in linen and cotton fabrics and in paper.

  The biological polymers called proteins are built up from the amino acid units, ranging in number from 50 to several thousand per polymer. Since 20 amino acids are available, a very large number of sequences can be realized. Fibrous proteins consist of polymers linked in sheets or twisted in long fibers; they are structural materials. Globular proteins include hemoglobin and enzymes; they generally control chemical reactions in living systems.

  The primary genetic material DNA is a biopolymer made up of four types of monomer units called nucleotides. DNA serves as a template during reproduction to enable a cell to preserve information about the amino acid sequences that make up its proteins, and transfer this information to daughter cells through the genetic code.

CONCEPTS & SKILLS

After studying this chapter and working the problems that follow, you should be able to:

1. Contrast the methods of addition and condensation polymerization (Section 23.1, Problems 1–6).
2. Give several examples of fibers, plastics, and rubbers and describe how they are made and used (Section 23.2, Problems 7–10).
3. Explain what a liquid crystal is, and state how nematic and smectic phases differ from ordinary liquids and crystalline solids (Section 23.3, Problems 11-12).
4. Describe the formation of ordered structures such as micelles and membranes in surfactant solutions (Section 23.3 Problems 13-14).
5. Describe the formation of polysaccharides from sugars, proteins from amino acids, and DNA from nucleotides and the roles of these biopolymers in living cells (Section 23.4, Problems 15–22).

PROBLEMS

Answers to problems whose numbers are boldface appear in Appendix G. Problems that are more challenging are indicated with asterisks.

Polymerization Reactions for Synthetic Polymers

1. Write a balanced chemical equation to represent the addition polymerization of 1,1-dichloroethylene. The product of this reaction is Saran, used as a plastic wrap.

2. Write a balanced chemical equation to represent the addition polymerization of tetrafluoroethylene. The product of this reaction is Teflon.

3. A polymer produced by addition polymerization consists of $(-CH_2-O-)$ groups joined in a long chain. What was the starting monomer?

4. The polymer polymethyl methacrylate is used to make Plexiglas. It has the formula

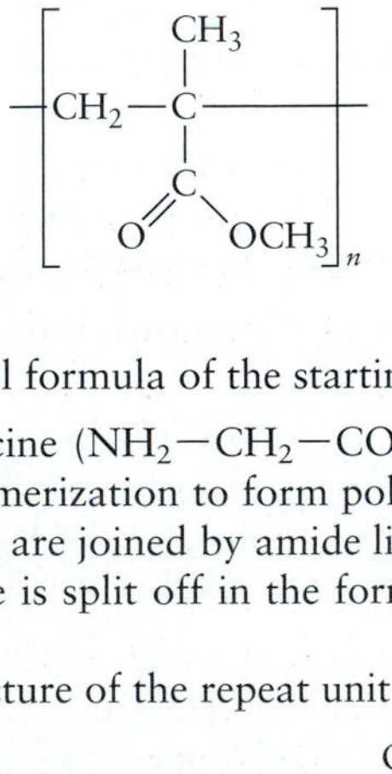

$$\left[-CH_2-\overset{\displaystyle CH_3}{\underset{\displaystyle \overset{}{C}(=O)OCH_3}{\overset{|}{\underset{|}{C}}}}- \right]_n$$

Draw the structural formula of the starting monomer.

5. The monomer glycine (NH_2-CH_2-COOH) can undergo condensation polymerization to form polyglycine, in which the structural units are joined by amide linkages.

(a) What molecule is split off in the formation of polyglycine?

(b) Draw the structure of the repeat unit in polyglycine.

6. The polymer forms $-(NH-CH(CH_3)-\overset{\displaystyle O}{\overset{\|}{C}})_n-$ upon condensation polymerization with loss of water. Draw the structure of the starting monomer.

Applications for Synthetic Polymers

7. Determine the mass of adipic acid and the mass of hexamethylenediamine needed to make 1.00×10^3 kg of nylon 66 fiber.

8. Determine the mass of terephthalic acid and the mass of ethylene glycol needed to make 10.0 kg of polyester fiber.

9. In a recent year, 4.37 billion kilograms of low-density polyethylene was produced in the United States. What volume of gaseous ethylene at 0°C and 1.00 atm would give this amount?

10. In a recent year, 2.84 billion kilograms of polystyrene was produced in the United States. Polystyrene is the addition polymer formed from the styrene monomer, $C_6H_5CH{=}CH_2$. How many styrene monomer units were incorporated in that 2.84 billion kilograms of polymer?

Liquid Crystals

11. Compare the natures and extents of order in the smectic liquid-crystal and isotropic liquid phases of a substance. Which has the higher entropy? Which has the higher enthalpy?

12. Nematic liquid crystals form when a liquid of long rod-like molecules is cooled. What *additional* types of intermolecular interactions would you expect to favor the formation of a *smectic* phase?

13. Consider a ternary (three-component) system of amphiphile, hydrocarbon, and water. What structure do you expect to form if small amounts of the first two components are mixed with a large amount of water?

14. In addition to spherical micelles and lamellar phases, a binary amphiphile-water mixture can also form extended cylindrical rolls, with the hydrophilic groups pointing out and the hydrophobic chains on the interior. Over what composition range are such cylinders most likely to be found, relative to spherical micelles and planar layers?

Natural Polymers

15. By referring to Figure 23.16a, draw the structure of the ring form of β-D-galactose. How many asymmetric carbon atoms (chiral centers) are there in the molecule?

16. By referring to Figure 23.23b, draw the structure of the ring form of D-ribose. How many asymmetric carbon atoms (chiral centers) are there in the molecule?

17. How many tripeptides can be synthesized using just three different species of α-amino acids?

18. How many different polypeptides, each containing ten amino acids, can be made from the amino acids listed in Table 23.3? How many different polypeptides, each containing 100 amino acids, can be made?

19. Draw the structure of the pentapeptide alanine—leucine—phenylalanine—glycine—isoleucine. Assume that the free $-NH_2$ group is at the alanine end of the peptide chain. Would this compound be more likely to dissolve in water or in octane? Explain.

20. Draw the structure of the pentapeptide aspartic acid—serine—lysine—glutamic acid—tyrosine. Assume that the free $-NH_2$ group is at the aspartic acid end of the peptide chain. Would this compound be more likely to dissolve in water or in octane? Explain.

21. Suppose a long-chain polypeptide is constructed entirely from phenylalanine monomer units. What is its empirical formula? How many amino acids does it contain if its molar mass is 17,500 g mol^{-1}?

22. A typical bacterial DNA has a molar mass of 4×10^9 g mol^{-1}. Approximately how many nucleotides does it contain?

ADDITIONAL PROBLEMS

23. In the addition polymerization of acrylonitrile, a very small amount of butyl lithium causes a reaction that can consume hundreds of pounds of the monomer; however, the butyl lithium is called an initiator, not a catalyst. Explain why.

24. Based on the facts that the free-radical polymerization of ethylene is spontaneous and that polymer molecules are less disorganized than the starting monomers, decide whether the polymerization reaction is exothermic or endothermic. Explain.

25. According to a trade journal, approximately 950 million lb of ethylene dichloride was exported from the United States in a recent year. The article states that "between 500 million and 550 million pounds of PVC could have been made from that ethylene dichloride." Compute the range of percentage yields of PVC from ethylene dichloride that is implied by these figures.

26. The complete hydrogenation of natural rubber (the addition of H_2 to all double bonds) gives a product that is indistinguishable from the product of the complete hydrogenation of gutta-percha. Explain how this strengthens the conclusion that these two substances are isomers of each other.

27. A reducing solution breaks S—S bonds in proteins, whereas an oxidizing solution allows them to re-form. Discuss how such solutions might be used to carry out the curling of hair.

28. L-Sucrose tastes sweet, but it is not metabolized. It has been suggested as a potential nonnutritive sweetener. Draw the molecular structure of L-sucrose, using Figure 23.16b as a starting point.

29. Polypeptides are synthesized from a 50:50 mixture of L-alanine and D-alanine. How many different isomeric molecules containing 22 monomer units are possible?

30. An osmotic pressure measurement taken on a solution containing hemoglobin shows that the molar mass of that protein is approximately 65,000 g mol^{-1}. A chemical analysis shows it to contain 0.344% of iron by mass. How many iron atoms does each hemoglobin molecule contain?

* 31. At very low pH, alanine is a diprotic acid that can be represented as $H_3N^+-CH(CH_3)-COOH$. The pK_a of the carboxyl group is 2.3, and the pK_a of the $-NH_3^+$ group is 9.7.
 (a) At pH 7, what fraction of the amino acid molecules dissolved in an aqueous solution will have the form $H_3N^+-CH(CH_3)-COO^-$?
 (b) What fraction of the molecules at this pH will have the form $H_2N-CH(CH_3)-COOH$?

32. The sequence of bases in one strand of DNA reads ACTTGACCG. Write the sequence of bases in the complementary strand.

33. Nucleic acids are diesters of phosphoric acid. Esters are not usually acidic. Why are nucleic acids acidic, or is this name inappropriate?

34. The codons in the genetic code are sequences of three bases. Explain why sequences of only two bases could not be used to code for the 20 different amino acids commonly found in proteins.

35. The average distance between base pairs measured parallel to the axis of a double-helical DNA molecule is 3.4 angstroms. The average molecular weight of a pair of nucleotides is about 650 g mol^{-1}. What is the approximate length in millimeters of a single DNA molecule of molecular weight 2.8×10^9 g mol^{-1} (a value typical for the DNA of some bacteria)? About how many base pairs does this DNA contain?

Appendices

A crystal of elemental bismuth.

APPENDIX A

Scientific Notation and Experimental Error

A.1 Scientific Notation

Very large and very small numbers are common in chemistry. Repeatedly writing such numbers in the ordinary way (for example, the important number 602,214,200,000,000,000,000,000) would be tedious and would engender errors. **Scientific notation** offers a better way. A number in scientific notation is expressed as a number from 1 to 10 multiplied by 10 raised to some power. Any number can be represented in this way, as the following examples show.

$$643.8 = 6.438 \times 10^2$$

$$-19{,}000{,}000 = -1.9 \times 10^7$$

$$0.0236 = 2.36 \times 10^{-2}$$

$$602{,}214{,}200{,}000{,}000{,}000{,}000{,}000 = 6.022142 \times 10^{23}$$

A simple rule of thumb is that the power to which 10 is raised is n if the decimal point is moved n places to the left and is $-n$ if the decimal is moved n places to the right.

When two or more numbers written in scientific notation are to be added or subtracted, they should first be expressed as multiples of the *same* power of 10:

$$\begin{array}{rcr} 6.431 \times 10^4 & \longrightarrow & 6.431 \times 10^4 \\ +2.1 \times 10^2 & \longrightarrow & +0.021 \times 10^4 \\ \underline{+3.67 \times 10^3} & \longrightarrow & \underline{+0.367 \times 10^4} \\ ? & & 6.819 \times 10^4 \end{array}$$

When two numbers in scientific notation are multiplied, the coefficients are multiplied and then the powers of 10 are multiplied (by adding the exponents):

$$1.38 \times 10^{-16} \times 8.80 \times 10^3 = (1.38 \times 8.80) \times 10^{(-16+3)}$$

$$= 12.1 \times 10^{-13} = 1.21 \times 10^{-12}$$

We divide one number by a second by dividing the coefficients and then multiplying by 10 raised to the first exponent minus the second (exponents are subtracted):

$$\frac{6.63 \times 10^{-27}}{2.34 \times 10^{-16}} = \frac{6.63}{2.34} \times \frac{10^{-27}}{10^{-16}}$$

$$= 2.83 \times 10^{[-27-(-16)]} = 2.83 \times 10^{-11}$$

Any calculator or computer equipped to perform scientific and engineering calculations can accept and display numbers in scientific notation. It cannot determine whether the input has an error, however, or whether the answer makes sense. That is your responsibility! Develop the habit of mentally estimating the order of magnitude of the answer as a rough check on your calculator's result.

A.2 Experimental Error

Chemistry is an experimental science in which every quantitative measurement is subject to some degree of error. We can seek to reduce error by carrying out additional measurements or by changing our experimental apparatus, but we can never eliminate error altogether. It is important, therefore, to be able to assess the results of an experiment quantitatively to establish the limits of the experiment's validity. Errors are of two types: random (lack of precision) and systematic (lack of accuracy).

Precision and Random Errors

Precision refers to the degree of agreement in a collection of experimental results and is estimated by repeating the measurement under conditions as nearly identical as possible. If the conditions are truly identical, then differences among the trials are due to random error. As a specific example, consider some actual results of an early, important experiment by American physicist Robert Millikan in 1909, to measure the charge e on the electron. The experiment (discussed in greater detail in Chapter 1) involved a study of the motion of charged oil drops suspended in air in an electric field. Millikan made hundreds of measurements on many different oil drops, but we shall consider only a set of results for e found for one particular drop (Table A.1). The values he found ranged from 4.894 to 4.941 $\times$ 10^{-10} esu. What do we choose to report as the best estimate for e? The proper procedure is to first examine the data to see whether any of the results are especially far from the rest (a value above 5×10^{-10} esu would fall into this category). Such values are likely to result from some mistake in carrying out or reporting that particular measurement and therefore are excluded from further consideration (although there have been cases in science where just such exceptional results have led to significant breakthroughs). In Millikan's data, no such points should be excluded. To obtain our best estimate for e, we calculate the **mean,** or **average value,** by adding up the values found and dividing by the number of measurements. We can write the average value of any property after a series of N measurements $x_1, x_2, \ldots, x_N$ as

$$\overline{x} = \frac{1}{N}(x_1 + x_2 + \cdots + x_N) = \frac{1}{N}\sum_{i=1}^{N} x_i$$

where a capital Greek sigma (Σ) is introduced to indicate a summation of x_i over values of i from 1 to N. In the present case, this gives an average for e of 4.917 $\times$ 10^{-10} esu.

This average by itself does not convey any estimate of uncertainty. If all of the measurements had given results between 4.91×10^{-10} and 4.92×10^{-10} esu, the uncertainty would be less than if the results had ranged from 4×10^{-10} to 6×10^{-10} esu. Furthermore, an average of 100 measurements should have less uncertainty than an average of 5. How are these ideas made quantitative? A statistical

TABLE A.1

Measurement Number
e (10^{-10} esu)

1	2	3	4	5	6	7	8	9	10	11	12	13
4.915	4.920	4.937	4.923	4.931	4.936	4.941	4.902	4.927	4.900	4.904	4.897	4.894

From R. A. Millikan, *Phys. Rev.* 32:349, 1911. [1 esu = 3.3356×10^{-10} C]

measure of the spread of data, called the **standard deviation** σ, is useful in this regard. It is given by the formula

$$\sigma = \sqrt{\frac{(x_1 - \bar{x})^2 + (x_2 - \bar{x})^2 + \cdots + (x_N - \bar{x})^2}{N - 1}}$$

$$= \sqrt{\frac{1}{N-1} \sum_{i=1}^{N} (x_i - \bar{x})^2}$$

The standard deviation is found by adding up the squares of the deviations of the individual data points from the average value $\bar{x}$, dividing by $N - 1$, and taking the square root. Table A.2 shows how σ is used quantitatively. A **confidence limit** is defined as

$$\text{confidence limit} = \pm\frac{t\sigma}{\sqrt{N}}$$

The table gives the factor t for various numbers of measurements, N, and for various levels of confidence.

For Millikan's data, $N = 13$ and $\sigma = 0.017 \times 10^{-10}$. For 95% confidence with 13 measurements, the table shows $t = 2.18$ and the confidence limit is

$$\text{confidence limit} = \pm\frac{(2.18)(0.017 \times 10^{-10})}{\sqrt{13}} = \pm 0.010 \times 10^{-10} \text{ esu}$$

Thus, a 95% probability exists that the *true* average (obtained by repeating the experiment under the same conditions an infinite number of times) will lie within $\pm 0.010 \times 10^{-10}$ esu of the average 4.917×10^{-10} esu. Within this 95% confidence level, our best estimate for e is written as

$$(4.917 \pm 0.010) \times 10^{-10} \text{ esu}$$

For other confidence levels and other numbers of measurements, the factor t and therefore the confidence limit change.

TABLE A.2

N	Factor *t* for Confidence Interval of			
(Number of Observations)	80%	90%	95%	99%
2	3.08	6.31	12.7	63.7
3	1.89	2.92	4.30	9.92
4	1.64	2.35	3.18	5.84
5	1.53	2.13	2.78	4.60
6	1.48	2.02	2.57	4.03
7	1.44	1.94	2.45	3.71
8	1.42	1.90	2.36	3.50
9	1.40	1.86	2.31	3.36
10	1.38	1.83	2.26	3.25
11	1.37	1.81	2.23	3.17
12	1.36	1.80	2.20	3.11
13	1.36	1.78	2.18	3.06
14	1.35	1.77	2.16	3.01
15	1.34	1.76	2.14	2.98
∞	1.29	1.64	1.96	2.58

Accuracy and Systematic Errors

The charge e on the electron has been measured by several different techniques since Millikan's day. The current best estimate for e is

$$e = (4.80320775 \pm 0.0000015) \times 10^{-10} \text{ esu}$$
$$= (1.60217646 \pm 0.00000049) \times 10^{-19} \text{ C}$$

This value lies outside the range of uncertainty we estimated from Millikan's original data. In fact, it lies well below the smallest of the 13 measurements of e. Why?

To understand this discrepancy, we need to remember that there is a second source of error in any experiment: *systematic* error that causes a shift in the measured values from the true value and reduces the **accuracy** of the result. By making more measurements, we can reduce the uncertainty due to *random* errors and improve the *precision* of our result; however, if systematic errors are present, the average value will continue to deviate from the true value. Such systematic errors may result from a miscalibration of the experimental apparatus or from a fundamental inadequacy in the technique for measuring a property. In the case of Millikan's experiment, the then-accepted value for the viscosity of air (used in calculating the charge e) was subsequently found to be wrong. This caused his results to be systematically too high.

Error thus arises from two sources. Lack of precision (random errors) can be estimated by a statistical analysis of a series of measurements. Lack of accuracy (systematic errors) is much more problematic. If a systematic error is known to be present, we should do our best to correct for it before reporting the result. (For example, if our apparatus has not been calibrated correctly, it should be recalibrated.) The problem is that systematic errors of which we have no knowledge may be present. In this case the experiment should be repeated with different apparatus to eliminate the systematic error caused by a particular piece of equipment; better still, a different and independent way to measure the property might be devised. Only after enough independent experimental data are available can we be convinced of the accuracy of a result—that is, how closely it approximates the true result.

A.3 Significant Figures

The number of **significant figures** is the number of digits used to express a measured or calculated quantity, excluding zeros that may precede the first nonzero digit. Suppose the mass of a sample of sodium chloride is measured to be 8.241 g and the uncertainty is estimated to be ±0.001 g. The mass is said to be given to four significant figures because we are confident of the first three digits (8, 2, 4) and the uncertainty appears in the fourth (1), which nevertheless is still significant. Writing additional digits beyond the 1 would not be justified, however, unless the accuracy of the weighing could be improved. When we record a volume as 22.4 L, we imply that the uncertainty in the measurement is in the last digit written ($V = 22.4 \pm 0.3$ L, for example). A volume written as 22.43 L, on the other hand, implies that the uncertainty is far less and appears only in the fourth significant figure.

In the same way, writing 20.000 m is quite different from writing 20.0 m. The second measurement (with three significant figures) could easily be made with a common meterstick. The first (with five significant figures) would require a more precise method. We should avoid reporting results such as "700 m," however, because the two trailing zeros may or may not be significant. The uncertainty in the measurement could be of order ±1 m or ±10 m or perhaps ±100 m; it is impossible to tell which without further information. To avoid this ambiguity, we can write such measurements using the scientific notation described in Section A.1. The measurement "700 m" translates into any of the following:

7.00×10^2 m	Three significant figures
7.0×10^2 m	Two significant figures
7×10^2 m	One significant figure

Frequently, it is necessary to combine several different experimental measurements to obtain a final result. Some operations involve addition or subtraction, and others entail multiplication or division. These operations affect the number of significant figures that should be retained in the calculated result. Suppose, for example, that a weighed sample of 8.241 g of sodium chloride is dissolved in 160.1 g of water. What will be the mass of the solution that results? It is tempting to simply write $160.1 + 8.241 = 168.341$ g, but this is *not* correct. In saying that the mass of water is 160.1 g, we imply that there is some uncertainty about the number of tenths of a gram measured. This uncertainty must also apply to the sum of the masses, so the last two digits in the sum are not significant and should be **rounded off,** leaving 168.3 as the final result.

Following addition or subtraction, round off the result to the leftmost decimal place that contained an uncertain digit in the original numbers.

Rounding off is a straightforward operation. It consists of first discarding the digits that are not significant and then adjusting the last remaining digit. If the first discarded digit is less than 5, the remaining digits are left as they are (for example, 168.341 is rounded down to 168.3 because the first discarded digit, 4, is less than 5). If the first discarded digit is greater than 5, or if it is equal to 5 and is followed by one or more nonzero digits, then the last digit is increased by 1 (for example, 168.364 and 168.3503 both become 168.4 when rounded off to four digits). Finally, if the first digit discarded is 5 and all subsequent digits are zeros, the last digit remaining is rounded to the nearest even digit (for example, both 168.35 and 168.45 would be rounded to 168.4). This last rule is chosen so that, on the average, as many numbers are rounded up as down. Other conventions are sometimes used.

In multiplication or division it is not the number of decimal places that matters (as in addition or subtraction) but the number of significant figures in the least precisely known quantity. Suppose, for example, the measured volume of a sample is 4.34 cm^3 and its mass is 8.241 g. The density, found by dividing the mass by the volume on a calculator, for example, is

$$\frac{8.241 \text{ g}}{4.34 \text{ cm}^3} = 1.89884\ldots \text{ g cm}^{-3}$$

How many significant figures should we report? Because the volume is the less precisely known quantity (three significant figures as opposed to four for the mass), it controls the precision that may properly be reported in the answer. Only three significant figures are justified, so the result is rounded to 1.90 g cm^{-3}.

The number of significant figures in the result of a multiplication or division is the smallest of the numbers of significant figures used as input.

It is best to carry out the arithmetical operations and *then* round the final answer to the correct number of significant figures, rather than round off the input data first. The difference is usually small, but this recommendation is nevertheless worth following. For example, the correct way to add the three distances 15 m, 6.6 m, and 12.6 m is

$$\begin{array}{r} 15 \quad \text{m} \\ +\ 6.6 \text{ m} \\ +12.6 \text{ m} \\ \hline 34.2 \text{ m} \end{array} \longrightarrow 34 \text{ m} \qquad \text{rather than} \qquad \begin{array}{r} 15 \text{ m} \longrightarrow 15 \text{ m} \\ 6.6 \text{ m} \longrightarrow 7 \text{ m} \\ 12.6 \text{ m} \longrightarrow 13 \text{ m} \\ \hline 35 \text{ m} \end{array}$$

For the same reason, we frequently carry extra digits through the intermediate steps of a worked example and round off only for the final answer. If a calculation is done entirely on a scientific calculator or a computer, several extra digits are usually carried along automatically. Before the final answer is reported, however, it is important to round off to the proper number of significant figures.

Sometimes pure constants appear in expressions. In this case the accuracy of the result is determined by the accuracy of the other factors. The uncertainty in the volume of a sphere, $\frac{4}{3}\pi r^3$, depends only on the uncertainty in the radius r; 4 and 3 are pure constants (4.000 . . . and 3.000 . . . , respectively), and π can be given to as many significant figures (3.14159265 . . .) as are warranted by the radius.

PROBLEMS

Answers to problems whose numbers are boldface appear in Appendix G.

Scientific Notation

1. Express the following in scientific notation.
(a) 0.0000582
(b) 402
(c) 7.93
(d) −6593.00
(e) 0.002530
(f) 1.47

2. Express the following in scientific notation.
(a) 4579
(b) −0.05020
(c) 2134.560
(d) 3.825
(e) 0.0000450
(f) 9.814

3. Convert the following from scientific notation to decimal form.
(a) 5.37×10^{-4}
(b) 9.390×10^{6}
(c) -2.47×10^{-3}
(d) 6.020×10^{-3}
(e) 2×10^{4}

4. Convert the following from scientific notation to decimal form.
(a) 3.333×10^{-3}
(b) -1.20×10^{7}
(c) 2.79×10^{-5}
(d) 3×10^{1}
(e) 6.700×10^{-2}

5. A certain chemical plant produces 7.46×10^{8} kg of polyethylene in one year. Express this amount in decimal form.

6. A microorganism contains 0.0000046 g of vanadium. Express this amount in scientific notation.

Experimental Error

7. A group of students took turns using a laboratory balance to weigh the water contained in a beaker. The results they reported were 111.42 g, 111.67 g, 111.21 g, 135.64 g, 111.02 g, 111.29 g, and 111.42 g.
(a) Should any of the data be excluded before the average is calculated?
(b) From the remaining measurements, calculate the average value of the mass of the water in the beaker.
(c) Calculate the standard deviation σ and, from it, the 95% confidence limit.

8. By measuring the sides of a small box, a group of students made the following estimates for its volume: 544 cm^3, 590 cm^3, 523 cm^3, 560 cm^3, 519 cm^3, 570 cm^3, and 578 cm^3.
(a) Should any of the data be excluded before the average is calculated?
(b) Calculate the average value of the volume of the box from the remaining measurements.
(c) Calculate the standard deviation σ and, from it, the 90% confidence limit.

9. Of the measurements in problems 7 and 8, which is more precise?

10. A more accurate determination of the mass in problem 7 (using a better balance) gives the value 104.67 g, and an accurate determination of the volume in problem 8 gives the value 553 cm^3. Which of the two measurements in problems 7 and 8 is more accurate, in the sense of having the smaller systematic error relative to the actual value?

Significant Figures

11. State the number of significant figures in each of the following measurements.
(a) 13.604 L
(b) −0.00345°C
(c) 340 lb
(d) 3.40×10^{2} miles
(e) 6.248×10^{-27} J

12. State the number of significant figures in each of the following measurements.
(a) -0.0025 in
(b) 7000 g
(c) 143.7902 s
(d) 2.670×10^{7} Pa
(e) 2.05×10^{-19} J

13. Round off each of the measurements in problem 11 to two significant figures.

14. Round off each of the measurements in problem 12 to two significant figures.

15. Round off the measured number 2,997,215.548 to nine significant digits.

16. Round off the measured number in problem 15 to eight, seven, six, five, four, three, two, and one significant digits.

17. Express the results of the following additions and subtractions to the proper number of significant figures. All of the numbers are measured quantities.
(a) $67.314 + 8.63 - 243.198 =$
(b) $4.31 + 64 + 7.19 =$
(c) $3.1256 \times 10^{15} - 4.631 \times 10^{13} =$
(d) $2.41 \times 10^{-26} - 7.83 \times 10^{-25} =$

18. Express the results of the following additions and subtractions to the proper number of significant figures. All of the numbers are measured quantities.
(a) $245.876 + 4.65 + 0.3678 =$
(b) $798.36 - 1005.7 + 129.652 =$
(c) $7.98 \times 10^{17} + 6.472 \times 10^{19} =$
(d) $(4.32 \times 10^{-15}) + (6.257 \times 10^{-14}) - (2.136 \times 10^{-13}) =$

19. Express the results of the following multiplications and divisions to the proper number of significant figures. All of the numbers are measured quantities.
(a) $\dfrac{-72.415}{8.62} =$
(b) $52.814 \times 0.00279 =$
(c) $(7.023 \times 10^{14}) \times (4.62 \times 10^{-27}) =$
(d) $\dfrac{4.3 \times 10^{-12}}{9.632 \times 10^{-26}} =$

20. Express the results of the following multiplications and divisions to the proper number of significant figures. All of the numbers are measured quantities.
(a) $129.587 \times 32.33 =$
(b) $\dfrac{4.7791}{3.21 \times 5.793} =$
(c) $\dfrac{10566.9}{3.584 \times 10^{29}} =$
(d) $(5.247 \times 10^{13}) \times (1.3 \times 10^{-17}) =$

21. Compute the area of a triangle (according to the formula $A = \frac{1}{2}ba$) if its base and altitude are measured to equal 42.07 cm and 16.0 cm, respectively. Justify the number of significant figures in the answer.

22. An inch is defined as exactly 2.54 cm. The length of a table is measured as 505.16 cm. Compute the length of the table in inches. Justify the number of significant figures in the answer.

APPENDIX B

SI Units, Unit Conversions, and Physics for General Chemistry

This appendix reviews essential concepts of physics, as well as systems of units of measure, essential for work in general chemistry.

B.1 SI Units and Unit Conversions

Scientific work requires measurement of quantities or properties observed in the laboratory. Results are expressed not as pure numbers but rather as **dimensions** that reflect the nature of the property under study. For example, mass, time, length, area, volume, energy, and temperature are fundamentally distinct quantities, each of which is characterized by its unique dimension. The magnitude of each dimensioned quantity can be expressed in various **units** (for example, feet or centimeters for length). Several systems of units are available for use, and facility at conversion among them is essential for scientific work. Over the course of history, different countries evolved different sets of units to express length, mass, and many other physical dimensions. Gradually, these diverse sets of units are being replaced by international standards that facilitate comparison of measurements made in different localities and that help avoid complications and confusion. The unified system of units recommended by international agreement is called SI, which stands for "Système International d'Unités," or the International System of Units. In this section we outline the use of SI units and discuss interconversions with other systems of units.

The SI uses seven **base units,** which are listed in Table B.1. All other units can be written as combinations of the base units. In writing the units for a measurement, we abbreviate them (see Table B.1), and we use exponential notation to denote the power to which a unit is raised; a minus sign appears in the exponent

TABLE B.1 Base Units in the International System of Units

Quantity	Unit	Symbol
Length	meter	m
Mass	kilogram	kg
Time	second	s
Temperature	kelvin	K
Number of moles (of substance)	mole	mol
Electric current	ampere	A
Luminous intensity	candela	cd

TABLE B.2 Derived Units in SI

Quantity	Unit	Symbol	Definition
Energy	joule	J	$kg\ m^2\ s^{-2}$
Force	newton	N	$kg\ m\ s^{-2} = J\ m^{-1}$
Power	watt	W	$kg\ m^2\ s^{-3} = J\ s^{-1}$
Pressure	pascal	Pa	$kg\ m^{-1}\ s^{-2} = N\ m^{-2}$
Electric charge	coulomb	C	A s
Electric potential difference	volt	V	$kg\ m^2\ s^{-3}\ A^{-1} = J\ C^{-1}$

when the unit is in the denominator. For example, velocity is a quantity with dimensions of length divided by time, so in SI it is expressed in meters per second, or $m\ s^{-1}$. Some derived units that are used frequently have special names. Energy, for example, is the product of mass and the square of the velocity. Therefore, it is measured in units of kilogram square meters per square seconds ($kg\ m^2\ s^{-2}$), and $1\ kg\ m^2\ s^{-2}$ is called a *joule*. Other derived units, such as the pascal for measuring pressure, appear in Table B.2. Although these names provide a useful shorthand, it is important to remember their meanings in terms of the base units.

Because scientists work on scales ranging from the microscopic to the astronomical, there is a tremendous range in the magnitudes of measured quantities. Consequently, a set of **prefixes** has been incorporated into the International System of Units to simplify the description of small and large quantities (Table B.3). The prefixes specify various powers of 10 times the base and derived units. Some of them are quite familiar in everyday use: the kilometer, for example, is 10^3 m. Others may sound less familiar—for instance, the femtosecond ($1\ fs = 10^{-15}$ s) or the gigapascal ($1\ GPa = 10^9$ Pa).

In addition to base and derived SI units, several units that are not officially sanctioned are used in this book. The first is the liter (abbreviated L), a very convenient size for volume measurements in chemistry; a liter is $10^{-3}\ m^3$, or 1 cubic decimeter (dm^3):

$$1\ L = 1\ dm^3 = 10^{-3}\ m^3 = 10^3\ cm^3$$

Second, we use the *angstrom* (abbreviated Å) as a unit of length for atoms and molecules:

$$1\ Å = 10^{-10}\ m = 100\ pm = 0.1\ nm$$

This unit is used simply because most atomic sizes and chemical bond lengths fall in the range of one to several angstroms, and the use of either picometers or nanometers is slightly awkward. Next, we use the *atmosphere* (abbreviated atm) as a unit of pressure. It is not a simple power of 10 times the SI unit of the pascal, but rather is defined as follows:

$$1\ atm = 101{,}325\ Pa = 0.101325\ MPa$$

TABLE B.3 Prefixes in SI

Fraction	Prefix	Symbol	Factor	Prefix	Symbol
10^{-1}	*deci-*	d	10	*deca-*	da
10^{-2}	*centi-*	c	10^2	*hecto-*	h
10^{-3}	*milli-*	m	10^3	*kilo-*	k
10^{-6}	*micro-*	μ	10^6	*mega-*	M
10^{-9}	*nano-*	n	10^9	*giga-*	G
10^{-12}	*pico-*	p	10^{12}	*tera-*	T
10^{-15}	*femto-*	f			
10^{-18}	*atto-*	a			

This unit is used because most chemistry procedures are carried out at pressures near atmospheric pressure, for which the pascal is too small a unit to be convenient. In addition, expressions for equilibrium constants (see Chapter 14) are simplified when pressures are expressed in atmospheres.

The non-SI temperature units require special mention. The two most important temperature scales in the United States are the Fahrenheit scale and the Celsius scale, which employ the Fahrenheit degree (°F) and the Celsius degree (°C), respectively. The *size* of the Celsius degree is the same as that of the SI temperature unit, the kelvin, but the two scales are shifted relative to each other by 273.15°C:

$$T_K = \frac{1\ \mathrm{K}}{1°\mathrm{C}}(t°\mathrm{C}) + 273.15\ \mathrm{K}$$

The size of the degree Fahrenheit is $\frac{5}{9}$ the size of the Celsius degree. The two scales are related by

$$t°\mathrm{C} = \frac{5°\mathrm{C}}{9°\mathrm{F}}(t°\mathrm{F} - 32°\mathrm{F}) \qquad \text{or} \qquad t°\mathrm{F} = \frac{9°\mathrm{F}}{5°\mathrm{C}}(t°\mathrm{C}) + 32°\mathrm{F}$$

The advantage of a unified system of units is that if all the quantities in a calculation are expressed in SI units, the final result must come out in SI units. Nevertheless, it is important to become familiar with the ways in which units are interconverted because units other than SI base units often appear in calculations. The **unit conversion method** provides a systematic approach to this problem.

As a simple example, suppose the mass of a sample is measured to be 64.3 g. If this is to be used in a formula involving SI units, it should be converted to kilograms (the SI base unit of mass). To do this, we use the fact that 1 kg = 1000 g and write

$$\frac{64.3\ \mathrm{g}}{1000\ \mathrm{g\ kg^{-1}}} = 0.0643\ \mathrm{kg}$$

Note that this is, in effect, division by 1; because 1000 g = 1 kg, 1000 g $\mathrm{kg^{-1}}$ = 1, and we cancel units between numerator and denominator to obtain the final result. This unit conversion could also be written as

$$64.3\ \mathrm{g}\left(\frac{1\ \mathrm{kg}}{1000\ \mathrm{g}}\right) = 0.0643\ \mathrm{kg}$$

In this book we use the more compact first version of the unit conversion. Instead of *dividing* by 1000 g $\mathrm{kg^{-1}}$, we can equally well *multiply* by 1 = 10^{-3} kg $\mathrm{g^{-1}}$:

$$(64.3\ \mathrm{g})(10^{-3}\ \mathrm{kg\ g^{-1}}) = 0.0643\ \mathrm{kg}$$

Other unit conversions may involve more than just powers of 10, but they are equally easy to carry out. For example, to express 16.4 inches in meters, we use the fact that 1 inch = 0.0254 m (or 1 = 0.0254 m $\mathrm{inch^{-1}}$), so

$$(16.4\ \mathrm{inches})(0.0254\ \mathrm{m\ inch^{-1}}) = 0.417\ \mathrm{m}$$

More complicated combinations are possible. For example, to convert from liter-atmospheres to joules (the SI unit of energy), two separate unit conversions are used:

$$\begin{aligned}(1\ \mathrm{L\ atm})(10^{-3}\ \mathrm{m^3\ L^{-1}})(101{,}325\ \mathrm{Pa\ atm^{-1}}) &= 101.325\ \mathrm{kg\ m^2\ s^{-2}}\\ &= 101.325\ \mathrm{Pa\ m^3} = 101.325\ \mathrm{J}\end{aligned}$$

When doing chemical calculations, it is very important to write out units explicitly and cancel units in intermediate steps to obtain the correct units for the final result. This practice is a way of checking to make sure that units have not been incorrectly mixed without unit conversions or that an incorrect formula has not been used. If a result that is supposed to be a temperature comes out with units of $\mathrm{m^3\ s^{-2}}$, then a mistake has been made!

B.2 Background in Physics

Although physics and chemistry are distinct sciences with distinct objectives and methods of investigation, concepts from one can aid investigations in the other. Physical reasoning aids chemical understanding in cases where applied forces move chemical systems to new positions—or change their sizes and shapes—and change their energy. These energy changes can have chemical consequences, as shown in two specific examples. First, the total energy content of a system increases when the system is compressed by externally applied pressure. Second, when significant forces exist between molecules, the mutual energy of a pair of molecules changes as the molecules are pushed closer together. Intermolecular forces profoundly influence the organization of matter into solid, liquid, and gaseous states, as well as the effectiveness of molecular collisions in causing chemical reactions. Specific illustrations appear at many places in this book. This section reviews general aspects of motion, forces, and energy as background for these specific applications. For simplicity, we limit the discussion to a **point mass,** which is an object characterized only by its total mass m. We do not inquire into the internal structure of the object. In various contexts, we represent planets, projectiles, molecules, atoms, nuclei, or electrons by this model to predict their response to applied forces.

Describing the Motion of an Object

Our goal here is to find a precise way to describe how an object changes its location and how that change occurs at various rates. We need several definitions. First, we define the position of the object precisely by stating its **displacement** x from a selected reference point. For example, an automobile could be located 1.5 miles north of the intersection of Wilshire and Westwood boulevards in Los Angeles. The electron in a hydrogen atom could be located 5×10^{-11} m from the nucleus. Displacement has dimensions of length (L). The rate of change of location is given by the **average velocity** v, defined over the time interval t_1 to t_2 as $v = [x_2 - x_1]/[t_2 - t_1]$. Both the displacement and the average velocity are defined relative to the selected reference point and therefore possess direction as well as magnitude. The **average speed** s gives the magnitude of the average velocity but not its direction. For example, an automobile can have average velocity of 35 mph southbound from Wilshire and Westwood, while its average speed is 35 mph. Both velocity and speed have dimension of length per time ($\mathrm{L\ t^{-1}}$). The **momentum** p of a body is defined as $p = mv$. The momentum indicates the ability of a moving body to exert impact on another body upon collision. For example, a slow-moving automobile "packs a bigger wallop" than a fast-moving bicycle. Momentum has dimensions of $\mathrm{M\ L\ t^{-1}}$. The rate at which the velocity changes is given by the **acceleration** a, defined as $a = [v_2 - v_1]/[t_2 - t_1]$. Acceleration has direction as well as magnitude and has dimensions of $\mathrm{L\ t^{-2}}$.

Forces Change the Motion of an Object

Force is defined as the agent that changes the motion of an object. This restatement of our daily experience that pushing or pulling an object causes it to move is the basis of Sir Isaac Newton's first law of motion:

> *Every object persists in a state of rest or of uniform motion in a straight line unless compelled to change that state by forces impressed upon it.*

We determine the properties of a force experimentally in the laboratory by measuring the consequences of applying the force. Suppose we have arbitrarily selected a standard test object of known mass m_1. We apply a force F_1 and, by making the distance and time measurements described earlier, determine the acceleration $a_{1,1}$ imparted to the object by the force. Experience shows that a different force F_2 imparts different acceleration $a_{1,2}$ to the test object. Forces can be ranked by the magnitude of the acceleration they impart to the standard test object. Now suppose we choose a second test object of mass m_2. Applying the original force F_1 to this object produces acceleration $a_{2,1}$, which is different from the acceleration $a_{1,1}$ that it gave to the first test object. The results of such experiments are summarized in Newton's second law of motion:

> *The acceleration imparted to an object by an applied force is proportional to the magnitude of the force, parallel to the direction of the force, and inversely proportional to the mass of the object.*

In mathematical form this statement becomes

$$F = ma \quad \text{or} \quad a = F/m$$

This equation demonstrates that force must have dimensions of M L t^{-2}. In SI units, force is expressed in newtons (N).

One familiar example is the force caused by gravity at the surface of the earth. Measurements show this force produces a downward acceleration (toward the center of the earth) of constant magnitude 9.80665 m s^{-2}, conventionally denoted by g, the gravitational constant. The gravitational force exerted on a body of mass m is

$$F = mg$$

Another familiar example is the restoring force exerted on an object by a spring. In chemistry this is a useful model of the binding forces that keep atoms together in a chemical bond or near their "home" positions in a solid crystal. Imagine an object of mass m located on a smooth tabletop. The object is connected to one end of a coiled metallic spring; the other end of the spring is anchored to a post in the tabletop. At rest, the object is located at position x_0. Now suppose the object is pulled away from the post in a straight line to a new position x beyond x_0 so the spring is stretched. Measurements show that at the stretched position x the magnitude of the force exerted on the object by the spring is directly proportional to the displacement from the rest position:

$$|F| = K(x - x_0)$$

In the preceding equation, K is a constant whose magnitude must be determined empirically in each case studied. Clearly, this force is directed back toward the rest position, and when the object is released, the recoiling spring accelerates the object back toward the rest position. Because the displacement is directed *away* from the rest position and the restoring force is directed *toward* the rest position, we insert a minus sign in the equation:

$$F(x) = -K(x - x_0)$$

When studying the dependence of force on position, we must pay careful attention to how displacement is defined in each particular problem. The force and the acceleration at any point are parallel to each other, but the displacement (defined relative to some convenient origin of coordinates in each particular case) may point in a different direction. In applications concerned with the height of an object above the earth, the gravitational force is usually written as

$$F(y) = -mg$$

to emphasize that the vertical displacement y is positive and pointing away from the surface of the earth, while the gravitational force is clearly directed toward the earth.

Once a force has been determined, the motion of the object under that force can be predicted from Newton's second law. If a constant force (that is, a force with constant acceleration) is applied to an object initially at rest for a period of time starting at t_1 and ending at t_2, the velocity at t_2 is given by

$$v(t_2) = a(t_2 - t_1)$$

and the position at t_2 is given by

$$x(t_2) = x(t_1) + \tfrac{1}{2}a(t_2 - t_1)^2$$

If the force depends on position, predicting the motion is slightly more complicated and requires the use of calculus. We merely quote the results for the linear restoring force. Under this force, the object oscillates about the rest position x_0 with a period τ given by

$$\frac{1}{\tau} = \frac{1}{2\pi}\left(\frac{K}{m}\right)^{1/2}$$

The frequency ν of the oscillation (the number of cycles per unit time) is given by $\nu = \tau^{-1}$. If we represent the motion in terms of the angular frequency $\omega = 2\pi\nu$, which has dimensions radians s^{-1}, the position of the oscillating object is given by

$$x(t) - x_0 = A\cos(\omega t + \delta)$$

where the amplitude A and the phase factor δ are determined by the initial position and velocity of the object.

Forms of Energy

The concept of **energy** originated in the science of mechanics and was first defined as the capacity to perform work, that is, to move an object from one position to another. It is now understood that energy appears in many different forms, each of which can cause particular kinds of physical and chemical changes. From daily experience, we recognize the kinetic energy due to the speed of an onrushing automobile. To stop the automobile, its *kinetic energy* must be overcome by work performed by its brakes. Otherwise, the automobile will crash into other objects and expend its kinetic energy by deforming these objects, as well as itself. The *potential energy* of a mass of snow on a ski slope becomes the kinetic energy of an avalanche. The *electrical energy* stored in a battery can move objects by driving motors, or warm objects through electric heaters. The chemical energy stored in gasoline can move objects by powering an internal combustion engine. The *thermal energy* of hot steam can move objects by driving a steam engine.

Each of these forms of energy plays a role in chemistry, and each is described at the appropriate point in this textbook. Here we concentrate on the nature of potential and kinetic energy and their interconversion. The understanding we gain here is essential background for understanding these other forms of energy.

The **kinetic energy** of a moving object is defined by

$$\mathcal{T} = \frac{1}{2}mv^2$$

where m is the mass of the object and v is its speed. A stationary object has no kinetic energy. In the SI systems of units (see Appendix B.1), energy is expressed in joules (J). Thus, 0.5 joule is the kinetic energy of an object with mass 1 kg moving at a speed of 1 m s^{-1}. Experience shows that applying a force to a moving object changes the kinetic energy of the object. If the force is applied in the same direction as the velocity of the object, the kinetic energy is increased; if the force is opposed to the velocity, the kinetic energy is decreased.

Potential energy is the energy stored in an object due to its location relative to a specified reference position. Potential energy therefore depends explicitly on the position x of the object and is expressed as a mathematical function $V(x)$. The change in potential energy when an object is moved from position x_1 to position x_2 by a constant force is equal to the work done in moving the object:

$$\text{change in } V(x) = \text{force} \times \text{displacement} = \text{force} \times (x_2 - x_1)$$

To lift an object of mass m from the surface of the earth to the height h, some agency must perform work in the amount mgh against the downward-directed force of gravity, which has constant acceleration denoted by g. The change in potential energy is mgh. Potential energy is expressed in joules in the SI system of units.

Conservation of Energy

The science of mechanics deals with idealized motions of objects in which friction does not occur. The motions of an object interconvert its potential and kinetic energy subject to the restriction that their sum always remains constant. For example, consider a soccer ball rolling down the side of a steep gully at the edge of the playing field, and assume there is no friction between the ball and the surface on which it rolls. The ball rolls down one side, goes across the bottom, climbs partway up the opposite side, then stops and reverses direction. This pattern is repeated many times as the ball continues to oscillate back and forth across the bottom of the gully. On each downward leg of its journey, the ball loses potential energy and gains kinetic energy, but their sum remains constant. On each upward leg, the ball loses kinetic energy and gains potential energy, but their sum remains constant.

The description in the previous paragraph is clearly an idealization, because eventually the ball comes to rest at the bottom of the gully. In reality, the ball loses some of its kinetic energy through friction with the surface of the gully on each upward leg and on each downward leg of its journey. Both the ball and the gully surface become slightly warmer as a result. The energy lost from the purely mechanical motion is added to the *internal energy* of the ball and the gully surface. The total amount of energy has not changed; we have neither created nor destroyed energy in this process. Rather, we have identified a new mode of energy storage (called internal energy) to interpret the new effects beyond those explained by basic mechanics. Internal energy, heat, and friction are discussed in detail in Chapter 12 as part of the science of thermodynamics.

Similar arguments have extended the law of conservation of energy from the idealized motions of mechanics to include a broad variety of phenomena in which several different forms of energy are involved. This law is one of the securest building blocks in scientific reasoning. It provides the starting point for interpreting and relating a great variety of superficially different processes. In chemistry this law provides the foundation for studying complex processes in which kinetic, potential, electrical, chemical, and thermal energy are interconverted without net loss or gain.

Representing Energy Conservation by Potential Energy Curves

Consider again the soccer ball rolling down the walls of the gully. This process is shown schematically in Figure B.1, where the curve $V(x)$ suggests a "cross-sectional" sketch of the gully. The curve $V(x)$ actually represents the potential energy of the ball relative to its value at the bottom of the gully. The x coordinate locates the distance of the ball from the bottom of the gully to a position along its side. Suppose the ball is held in place at the position x_1. It possesses only potential energy, which we represent by the value E_1. If the ball is released, it falls down the slope and passes across the bottom, where its potential energy is zero and its kinetic energy is E_1. It then climbs the opposite side until it rises to position $-x_1$, where its kinetic energy is zero. The ball promptly reverses direction and retraces its path back to position x_1, where its total energy E_1 is all potential energy.

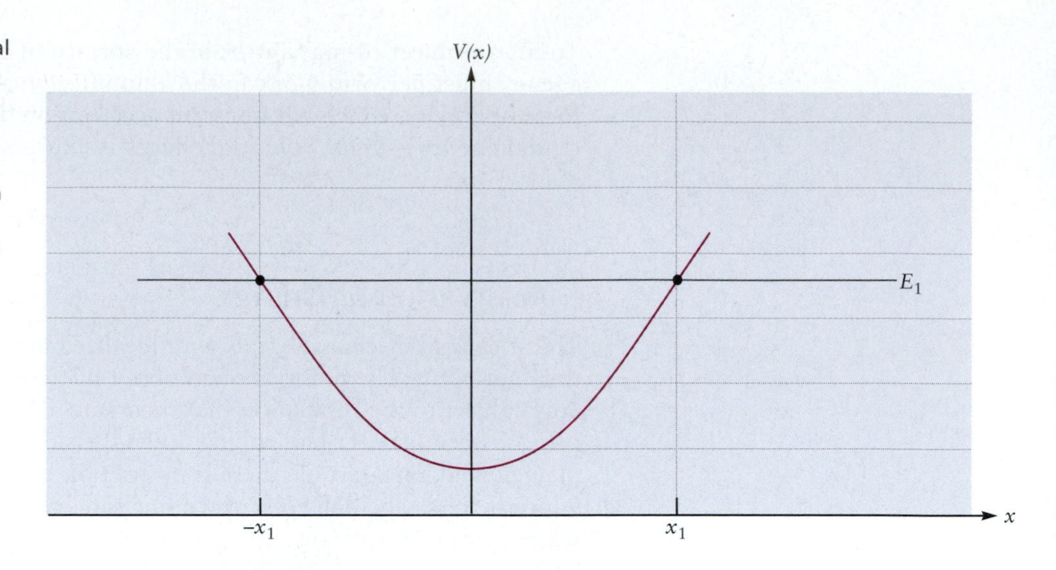

FIGURE B.1 Sketch of the potential energy of a soccer ball rolling down the sides of a gully. The ball is released at position x_1 with total energy E_1. In idealized motion, it continues to oscillate back and forth across the bottom of the gully. The net force on the ball is always in the direction in which the potential energy has negative slope. In real systems there is sufficient friction between the ball and the gully to bring the ball to rest at the bottom of the gully.

Potential Energy Curves, Force, and Stability

One very important application of the potential energy function in this textbook is to provide a way to qualitatively predict the motion of an object without solving Newton's second law. The method determines the direction of the force applied to an object from knowledge of the potential energy curve as shown in Figure B.1. In preparation for this method, let's examine the potential energy curves for some familiar forces.

The value of the potential energy at a general point x is always stated relative to its value at a specially selected reference point x_0. The value $V(x_0)$ is a constant and is usually assigned the value zero, and the point x_0 is frequently selected as the origin of coordinates for specifying the location of the object.

In the presence of a particular force F, the potential energy of the object at x is defined as the work required to move the object to x from the reference position x_0. Since the work done by a constant force in moving an object is defined as (force) × (displacement), the definition of potential energy in mathematical terms is

$$V(x) - V(x_0) = (F)(x - x_0)$$

Note carefully that the displacement referred to in the definition of potential energy starts at x_0 and ends at x. For example, the potential energy of an object of mass m at height h above the surface of the earth is the work done by a force $+mg$ opposing gravity in lifting the object from h_0 to h:

$$\begin{aligned} V(h) - V(h_0) &= mg(h - h_0) \\ &= mgh - mgh_0 \\ &= mgh \end{aligned}$$

where the value $V(h_0)$ has been set to zero. For a variable force, the definition of potential energy becomes slightly more complicated and requires explicit use of calculus. The potential energy for the linear restoring force is

$$V(x) = \tfrac{1}{2}K(x)^2$$

where the value $V(x_0)$ has been set to 0 and x_0 has been selected as the origin of coordinates.

It is instructive to plot these two potential energy functions as graphs (see Appendix C). The gravitational potential energy function is a straight line through the origin with slope mg, while the potential energy function for the linear restoring force is a parabola. In each case the potential energy of the object increases as the distance from the reference point x_0 increases.

Knowledge of the potential energy curve as a function of position lets us predict the net force on the object at each position. From either side, the force is directed toward the bottom; the force is always in the direction in which the slope of the potential energy curve is negative. From the definition of potential energy stated earlier, it can be shown that

$$\text{force} = -(\text{slope of PE})$$

The net force drives the object toward the position where the potential energy is a minimum and its slope is zero. For both the gravitational force and the linear restoring force, the object experiences negative forces that attract it back toward the center of force. In general, we identify **attractive forces** as those that correspond to positive slope of the potential energy curve; **repulsive forces** operate in regions where the slope of the potential energy curve is negative.

The minimum of the potential energy curve shown in Figure B.1 is called a point of **stable equilibrium** because the net force (that is, the slope of the potential energy curve) at that point is zero. As the ball tries to climb the wall on either side of the minimum, the restoring force always drives it back toward this position of stable equilibrium. This qualitative interpretation predicts that the ball will oscillate about the equilibrium position, as predicated by the exact solution to Newton's second law quoted earlier.

We use similar potential energy diagrams to represent the interaction between a pair of objects, such as Earth–moon, Earth–Mars, electron–nucleus, electron–electron, nucleus–nucleus, atom–atom, or molecule–molecule. We construct such diagrams at several points in this textbook and use them to interpret the relative motions of the pair of objects. These methods are extremely important in describing the formation of chemical bonds, the states of matter, and the role of molecular collisions in chemical reactions.

Electrical Forces

The concepts summarized so far in this section also are used to describe the mutual interactions and the motions of electrically charged particles. The only new features are to identify the force that represents electrical interactions and to obtain the corresponding potential energy function. Positive and negative charges and the electrical forces between them were first quantified by Charles Coulomb late in the 18th century. In his honor the unit of charge in SI units is the **coulomb** (C). Electrical charge is fundamentally quantized in units of the charge carried by a single electron e, which is equal to 1.60218×10^{-19} C. The coulomb is thus an inconveniently large unit for chemical reasoning. Nonetheless, for consistency and for quantitative accuracy, physical equations involving charge use SI units.

Suppose a charge of magnitude q_1 is held at the origin of coordinates and another charge of magnitude q_2 is brought near it, at the distance r from the origin. The magnitude of the acceleration imparted to the charge q_2 by the charge q_1 fixed at the origin can be determined as described earlier for uncharged objects. From such measurements, Coulomb determined that the magnitude of the force was directly proportional to the magnitudes of the two charges and inversely proportional to the distance between them:

$$F \propto \frac{q_1 q_2}{r^2}$$

The radial displacement variable r has its origin at the same location as charge q_1, and its value increases in the outward direction. If the charges have the same sign, the acceleration pushes them apart in the same direction as r. Consequently, the force between charges is **repulsive** and defined to be **positive.** If the charges have opposite signs, the acceleration pulls them together in the direction opposite to the displacement variable r. In this case the force between charges is **attractive** and defined to be **negative.** It is instructive to sketch the results graphically in these two cases. The quantitative form of Coulomb's law with force expressed in newtons (N) and charge in coulombs (C) is

$$F = \frac{q_1 q_2}{4\pi \epsilon_0 r^2}$$

where ε_0, called the permittivity of the vacuum, is a constant with value 8.854×10^{-12} C^2 J^{-1} m^{-1}. In this and related equations, the symbol q for each charge represents the magnitude and carries implicitly the sign of the charge.

The **Coulomb potential energy** corresponding to this force is

$$V(r) = \frac{q_1 q_2}{4\pi \epsilon_0 r}$$

Note that the force varies as r^{-2} while the potential energy varies as r^{-1}. Note also that if the charges have the same sign, the potential energy is positive and repulsive; if the charges have opposite sign, the potential energy is negative and attractive. If we know the potential energy curve between two charged particles, we can predict the direction of their relative motion.

The Coulomb potential energy function holds great importance in chemistry for examining the structure of atoms and molecules. In 1912, Ernest Rutherford proposed that an atom of atomic number Z comprises a dense, central nucleus of positive charge with magnitude Ze surrounded by a total of Z individual electrons moving around the nucleus. Thus, each individual electron has a potential energy of interaction with the nucleus given by

$$V(r) = -\frac{Ze^2}{4\pi \epsilon_0 r}$$

which is clearly negative and attractive. Each electron has a potential energy of interaction with every other electron in the atom given by

$$V(r) = \frac{e^2}{4\pi \epsilon_0 r}$$

which is clearly positive and repulsive. Rutherford's model of the atom, firmly based on experimental results, was completely at odds with the physical theories of the day. Attempts to reconcile these results with theory led to the development of the new theory called quantum mechanics.

Circular Motion and Angular Momentum

An object executing uniform circular motion about a point (for example, a ball on the end of a rope being swung in circular motion) is described by position, velocity, speed, momentum, and force, just as defined earlier for objects in linear motion. It is most convenient to describe this motion in polar coordinates r and θ, which represent, respectively, the distance of the object from the center and its angular displacement from the x-axis in ordinary cartesian coordinates. Because r is constant for circular motion, the motion variables depend directly on θ. The angular velocity during the time interval from t_1 to t_2 is given by

$$\omega = \frac{\theta_2 - \theta_1}{t_2 - t_1}$$

and the angular acceleration is given by

$$\alpha = \frac{\omega_2 - \omega_1}{t_2 - t_1}$$

In linear motion, we are concerned with the momentum $p = mv$ of an object as it heads toward a particular point; the linear momentum measures the impact that the object can transfer in a collision as it arrives at the point. To extend this concept to circular motion, we define the **angular momentum** of an object as it revolves around a point as $L = mvr$. This is in effect the *moment* of the linear momentum over the distance r, and it is a measure of the torque felt by the object as it executes angular motion. The angular momentum of an electron around a nucleus is a crucial feature of atomic structure, which is discussed in Chapter 5.

PROBLEMS

Answers to problems whose numbers are boldface appear in Appendix G.

SI Units and Unit Conversions

1. Rewrite the following in scientific notation, using only the base units of Table B.1, without prefixes.
(a) 65.2 nanograms
(b) 88 picoseconds
(c) 5.4 terawatts
(d) 17 kilovolts

2. Rewrite the following in scientific notation, using only the base units of Table B.1, without prefixes.
(a) 66 μK
(b) 15.9 MJ
(c) 0.13 mg
(d) 62 GPa

3. Express the following temperatures in degrees Celsius.
(a) 9001°F
(b) 98.6°F (the normal body temperature of human beings)
(c) 20°F above the boiling point of water at 1 atm pressure
(d) −40°F

4. Express the following temperatures in degrees Fahrenheit.
(a) 5000°C
(b) 40.0°C
(c) 212°C
(d) −40°C

5. Express the temperatures given in problem 3 in kelvins.

6. Express the temperatures given in problem 4 in kelvins.

7. Express the following in SI units, either base or derived.
(a) 55.0 miles per hour (1 mile = 1609.344 m)
(b) 1.15 g cm^{-3}
(c) 1.6×10^{-19} C Å
(d) 0.15 mol L^{-1}
(e) 5.7×10^3 L atm day^{-1}

8. Express the following in SI units, either base or derived.
(a) 67.3 atm
(b) 1.0×10^4 V cm^{-1}
(c) 7.4 Å year^{-1}
(d) 22.4 L mol^{-1}
(e) 14.7 lb inch^{-2} (1 inch = 2.54 cm; 1 lb = 453.59 g)

9. The kilowatt-hour (kWh) is a common unit in measurements of the consumption of electricity. What is the conversion factor between the kilowatt-hour and the joule? Express 15.3 kWh in joules.

10. A car's rate of fuel consumption is often measured in miles per gallon (mpg). Determine the conversion factor between miles per gallon and the SI unit of meters per cubic decimeter (1 gallon = 3.785 dm^3, and 1 mile = 1609.344 m). Express 30.0 mpg in SI units.

11. A certain V-8 engine has a displacement of 404 in^3. Express this volume in cubic centimeters (cm^3) and in liters.

12. Light travels in a vacuum at a speed of 3.00×10^8 m s^{-1}.
(a) Convert this speed to miles per second.
(b) Express this speed in furlongs per fortnight, a little-used unit of speed. (A furlong, a distance used in horse racing, is 660 ft; a fortnight is exactly 2 weeks.)

The Concept of Energy: Forms, Measurements, and Conservation

13. After being spiked, a volleyball travels with speed near 100 miles per hour. Calculate the kinetic energy of the volleyball. The mass of a volleyball is 0.270 kg.

14. The fastball of a famous pitcher in the National League has been clocked in excess of 95 mph. Calculate the work done by the pitcher in accelerating the ball to that speed. The mass of a baseball is 0.145 kg.

15. A tennis ball weighs approximately 2 ounces on a postage scale. A student practices his serve against the wall of the chemistry building, and the ball achieves the speed of 98 miles per hour. Calculate the kinetic energy of the ball after the serve. How much work is done on the chemistry building in one collision?

APPENDIX C

Mathematics for General Chemistry

Mathematics is an essential tool in chemistry. This appendix reviews some of the most important mathematical techniques for general chemistry.

C.1 Using Graphs

In many situations in science, we are interested in how one quantity (measured or predicted) depends on another quantity. The position of a moving car depends on the time, for example, or the pressure of a gas depends on the volume of the gas at a given temperature. A very useful way to show such a relation is through a **graph,** in which one quantity is plotted against another.

The usual convention in drawing graphs is to use the horizontal axis for the variable over which we have control and use the vertical axis for the measured or calculated quantity. After a series of measurements, the points on the graph frequently lie along a recognizable curve, and that curve can be drawn through the points. Because any experimental measurement involves some degree of uncertainty, there will be some scatter in the points, so there is no purpose in drawing a curve that passes precisely through every measured point. If there is a relation between the quantities measured, however, the points will display a systematic trend and a curve can be drawn that represents that trend.

The curves plotted on graphs can have many shapes. The simplest and most important shape is a straight line. A straight line is a graph of a relation such as

$$y = 4x + 7$$

or, more generally,

$$y = mx + b$$

where the variable y is plotted along the vertical axis and the variable x along the horizontal axis (Fig. C.1). The quantity m is the **slope** of the line that is plotted, and b is the **intercept**—the point at which the line crosses the y axis. This can be demonstrated by setting x equal to 0 and noting that y is then equal to b. The slope is a measure of the steepness of the line; the greater the value of m, the steeper the line. If the line goes up and to the right, the slope is positive; if it goes down and to the right, the slope is negative.

The slope of a line can be determined from the coordinates of two points on the line. Suppose, for example, that when $x = 3$, $y = 5$, and when $x = 4$, $y = 7$. These two points can be written in shorthand notation as (3, 5) and (4, 7). The

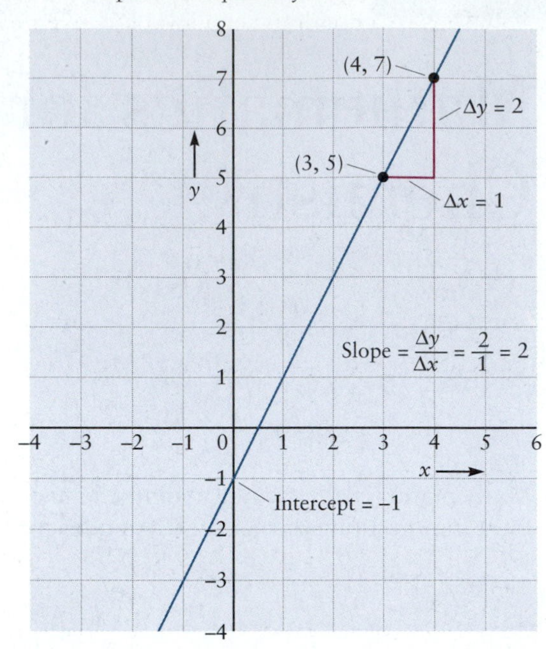

FIGURE C.1 A straight-line, or linear, relationship between two experimental quantities is a very desirable result because it is easy to graph and easy to represent mathematically. The equation of this straight line ($y = 2x - 1$) fits the general form $y = mx + b$. The line's y intercept is -1 and its slope is 2.

slope of the line is then defined as the "rise over the run"—the change in the y coordinate divided by the change in the x coordinate:

$$\text{slope} = m = \frac{\Delta y}{\Delta x} = \frac{7 - 5}{4 - 3} = \frac{2}{1} = 2$$

The symbol Δ (capital Greek delta) indicates the change in a quantity—the final value minus the initial value. In chemistry, if the two quantities being graphed have dimensions, the slope has dimensions as well. If a graph of distance traveled (in meters) against time (in seconds) is a straight line, its slope has dimensions of meters per second (m s^{-1}).

C.2 Solution of Algebraic Equations

In chemistry it is often necessary to solve an algebraic equation for an unknown quantity, such as a concentration or a partial pressure in an equilibrium-constant expression. Let us represent the unknown quantity with the symbol x. If the equation is linear, the method of solution is straightforward:

$$5x + 9 = 0$$

$$5x = -9$$

$$x = -\frac{9}{5}$$

or, more generally, if $ax + b = 0$, then $x = -b/a$.

Nonlinear equations are of many kinds. One of the most common in chemistry is the quadratic equation, which can always be rearranged into the form

$$ax^2 + bx + c = 0$$

where each of the constants (a, b, and c) may be positive, negative, or 0. The two solutions to a quadratic equation are given by the quadratic formula:

$$x = \frac{-b \pm \sqrt{b^2 - 4ac}}{2a}$$

As an example, suppose that the equation

$$x = 3 + \frac{7}{x}$$

arises in a chemistry problem. Multiplying by x and rearranging the terms gives

$$x^2 - 3x - 7 = 0$$

Inserting $a = 1$, $b = -3$, and $c = -7$ into the quadratic formula gives

$$x = \frac{-(-3) \pm \sqrt{(-3)^2 - 4(1)(-7)}}{2} = \frac{3 \pm \sqrt{9 + 28}}{2} = \frac{3 \pm \sqrt{37}}{2}$$

The two roots of the equation are

$$x = 4.5414 \qquad \text{and} \qquad x = -1.5414$$

In a chemistry problem, the choice of the proper root can frequently be made on physical grounds. If x corresponds to a concentration, for example, the negative root is unphysical and can be discarded.

The solution of cubic or higher order algebraic equations (or more complicated equations involving sines, cosines, logarithms, or exponentials) becomes more difficult, and approximate or numerical methods must be used. As an illustration, consider the equation

$$x^2\left(\frac{2.00 + x}{3.00 - x}\right) = 1.00 \times 10^{-6}$$

If x is assumed to be small relative to both 2.00 and 3.00, we obtain the simpler approximate equation

$$x^2\left(\frac{2.00}{3.00}\right) \approx 1.00 \times 10^{-6}$$

which leads to the roots $x = \pm 0.00122$. We immediately confirm that the solutions obtained in this way *are* small compared with 2.00 (and 3.00) and that our approximation was a good one. When a quantity (x in this case) is added to or subtracted from a larger quantity in a complicated equation, it is usually worthwhile, in solving the equation, to simply neglect the occurrence of the small quantity. Note that this tactic works only for addition and subtraction, never for multiplication or division.

Suppose now that the equation is changed to

$$x^2\left(\frac{2.00 + x}{3.00 - x}\right) = 1.00 \times 10^{-2}$$

In this case the equation that comes from neglecting x compared with 2.00 and 3.00 is

$$x^2\left(\frac{2.00}{3.00}\right) \approx 1.00 \times 10^{-2}$$

which is solved by $x \approx \pm 0.122$. The number 0.122 is smaller than 2.00 and 3.00, but not so small that it can be ignored. In this case more precise results can be obtained by **iteration.** Let us simply add and subtract the approximate positive root $x \approx +0.122$ as specified inside the parentheses and solve again for x:

$$x^2\left(\frac{2.00 + 0.122}{3.00 - 0.122}\right) = 1.00 \times 10^{-2}$$

$$x = 0.116$$

This new value can again be inserted into the original equation and the process repeated:

$$x^2\left(\frac{2.00 + 0.116}{3.00 - 0.116}\right) = 1.00 \times 10^{-2}$$

$$x = 0.117$$

Once the successive values of x agree to within the desired accuracy, the iteration can be stopped. Another root of the equation is obtained in a similar fashion if the iterative procedure starts with $x = -0.122$.

In some cases iteration fails. Suppose, for example, we have the equation

$$x^2\left(\frac{2.00 + x}{3.00 - x}\right) = 10.0$$

There is no particular reason to believe that x should be small compared with 2.00 or 3.00, but if we nevertheless assume that it is, we find $x = \pm\sqrt{15} = \pm 3.873$. Putting $x = +3.873$ back in for an iterative cycle leads to the equation

$$x^2 = -1.49$$

which has no real solutions. Starting with $x = -3.873$ succeeds no better; it gives

$$x^2 = -36.7$$

One way to overcome these difficulties is to solve the original equation graphically. We plot the left side of the equation against x and see at which values it becomes equal to 10 (the right side). We might initially calculate

x	$x^2\left(\frac{2.00 + x}{3.00 - x}\right)$
0	0
1	1.5
2	16

We observe that for $x = 1$ the left side is less than 10, whereas for $x = 2$ it is greater than 10. Somewhere in between there must be an x for which the left side is *equal to* 10. We can pinpoint it by selecting values of x between 1 and 2; if the left side is less than 10, x should be increased, and if it is greater than 10, x should be decreased.

x	$x^2\left(\frac{2.00 + x}{3.00 - x}\right)$
1.5	5.25
Increase to 1.8	10.26
Decrease to 1.75	9.19
Increase to 1.79	10.04

Thus, 1.79 is quite close to a solution of the equation. Improved values are easily obtained by further adjustments of this type.

C.3 Powers and Logarithms

Raising a number to a power and the inverse operation, taking a logarithm, are important in many chemical problems. Although the ready availability of electronic calculators makes the mechanical execution of these operations quite routine, it remains important to understand what is involved in such "special functions."

The mathematical expression 10^4 implies multiplying 10 by itself 4 times to give 10,000. Ten, or any other number, when raised to the power 0 always gives 1:

$$10^0 = 1$$

Negative powers of 10 give numbers less than 1 and are equivalent to raising 10 to the corresponding *positive* power and then taking the reciprocal:

$$10^{-3} = 1/10^3 = 0.001$$

We can extend the idea of raising to a power to include powers that are not whole numbers. For example, raising to the power $\frac{1}{2}$ (or 0.5) is the same as taking the square root:

$$10^{0.5} = \sqrt{10} = 3.1623\ldots$$

Scientific calculators have a 10^x (or INV LOG) key that can be used for calculating powers of 10 in cases where the power is not a whole number.

Numbers other than 10 can be raised to powers, as well; these numbers are referred to as **bases.** Many calculators have a y^x key that lets any positive number y be raised to any power x. One of the most important bases in scientific problems is the transcendental number called e (2.7182818 . . .). The e^x (or INV LN) key on a calculator is used to raise e to any power x. The quantity e^x also denoted as $\exp(x)$, is called the **exponential** of x. A key property of powers is that a base raised to the sum of two powers is equivalent to the product of the base raised separately to these powers. Thus, we can write

$$10^{21+6} = 10^{21} \times 10^6 = 10^{27}$$

The same type of relationship holds for any base, including e.

Logarithms also occur frequently in chemistry problems. The logarithm of a number is the exponent to which some base has to be raised to obtain the number. The base is almost always either 10 or the transcendental number e. Thus,

$$B^a = n \qquad \text{and} \qquad \log_B n = a$$

where a is the logarithm, B is the base, and n is the number.

Common logarithms are base-10 logarithms; that is, they are powers to which 10 has to be raised in order to give the number. For example, $10^3 = 1000$, so $\log_{10} 1000 = 3$. We shall frequently omit the 10 when showing common logarithms and write this equation as $\log 1000 = 3$. Only the logarithms of 1, 10, 100, 1000, and so on are whole numbers; the logarithms of other numbers are decimal fractions. The decimal point in a logarithm divides it into two parts. To the left of the decimal point is the *characteristic;* to the right is the *mantissa.* Thus, the logarithm in the equation

$$\log (7.310 \times 10^3) = 3.8639$$

has a characteristic of 3 and a mantissa of 0.8639. As may be verified with a calculator, the base-10 logarithm of the much larger (but closely related) number 7.310×10^{23} is 23.8639. As this case illustrates, the characteristic is determined solely by the location of the decimal point in the number and not by the number's precision, so it is *not* included when counting significant figures. The mantissa should be written with as many significant figures as the original number.

A logarithm is truly an exponent and as such follows the same rules of multiplication and division as other exponents. In multiplication and division we have

$$\log (n \times m) = \log n + \log m$$

$$\log \left(\frac{n}{m}\right) = \log n - \log m$$

Furthermore,

$$\log n^m = m \log n$$

so the logarithm of $3^5 = 243$ is

$$\log 3^5 = 5 \log 3 = 5 \times 0.47712 = 2.3856$$

There is no such thing as the logarithm of a negative number, because there is no power to which 10 (or any other base) can be raised to give a negative number.

A frequently used base for logarithms is the number e ($e = 2.7182818. . . .$). The logarithm to the base e is called the **natural logarithm** and is indicated by $\log_e$ or ln. Base-e logarithms are related to base-10 logarithms by the formula

$$\ln n = 2.3025851 \log n$$

As already stated, calculations of logarithms and powers are inverse operations. Thus, if we want to find the number for which 3.8639 is the common logarithm, we simply calculate

$$10^{3.8639} = 7.310 \times 10^3$$

If we need the number for which the natural logarithm is 2.108, we calculate

$$e^{2.108} = 8.23$$

As before, the number of significant digits in the answer should correspond to the number of digits in the *mantissa* of the logarithm.

C.4 Slopes of Curves and Derivatives

Very frequently in science, one measured quantity depends on a second one. If the property y depends on a second property x, we can write $y = f(x)$, where f is a function that expresses the dependence of y on x. Often, we are interested in the effect of a small change Δx on the dependent variable y. If x changes to $x + \Delta x$, then y will change to $y + \Delta y$. How is Δy related to Δx? Suppose we have the simple linear relation between y and x

$$y = mx + b$$

where m and b are constants. If we substitute $y + \Delta y$ and $x + \Delta x$, we find

$$y + \Delta y = m(x + \Delta x) + b$$

Subtracting the first equation from the second leaves

$$\Delta y = m\Delta x$$

or

$$\frac{\Delta y}{\Delta x} = m$$

The change in y is proportional to the change in x, with a proportionality constant equal to the slope of the line in the graph of y against x.

Suppose now we have a slightly more complicated relationship such as

$$y = ax^2$$

where a is a constant. If we substitute $y = y + \Delta y$ and $x = x + \Delta x$ here, we find

$$y + \Delta y = a(x + \Delta x)^2$$
$$= ax^2 + ax\,\Delta x + (\Delta x)^2$$

Subtracting as before leaves

$$\Delta y = 2ax\,\Delta x + (\Delta x)^2$$

Here we have a more complicated, nonlinear relationship between Δy and Δx. If Δx is small enough, however, the term $(\Delta x)^2$ will be small relative to the term proportional to Δx, and we may write

$$\Delta y \approx 2ax\,\Delta x \qquad (\Delta x \text{ small})$$

$$\frac{\Delta y}{\Delta x} \approx 2ax \qquad (\Delta x \text{ small})$$

How do we indicate this graphically? The graph of y against x is no longer a straight line, so we need to generalize the concept of slope. We define a **tangent line** at the point x_0 as the line that touches the graph of $f(x)$ at $x = x_0$ without crossing it.[1] If Δy and Δx are small (Fig. C.2), the slope of the tangent line is

$$\frac{\Delta y}{\Delta x} = 2ax_0 \quad \text{at} \quad x = x_0$$

for the example just discussed. Clearly, the slope is not constant but changes with x_0. If we define the slope at each point on the curve by the slope of the corresponding tangent line, we obtain a *new* function that gives the slope of the curve $f(x)$ at each point x. We call this new function the **derivative** of $f(x)$ and represent it with df/dx. We have already calculated the derivatives of two functions:

$$f(x) = mx + b \Rightarrow \frac{df}{dx} = m$$

$$f(x) = ax^2 \Rightarrow \frac{df}{dx} = ax$$

[1]This is an intuitive, rather than a rigorously mathematical, definition of a tangent line.

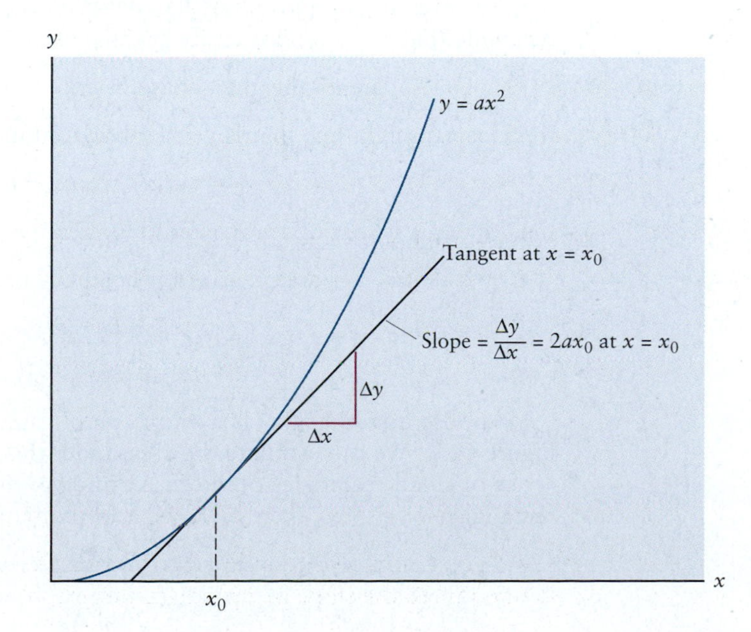

FIGURE C.2 The slope of a curve at a point.

TABLE C.1 Derivatives of Simple Functions

Function $f(x)$	Derivative $\frac{df}{dx}$
$mx + b$	m
ax^2	$2ax$
$\frac{a}{x} = ax^{-1}$	$-\frac{a}{x^2} = -ax^{-2}$
ax^n	nax^{n-1}
e^{ax}	ae^{ax}
$\ln ax$	$\frac{1}{x}$
$\sin ax$	$a \cos ax$
$\cos ax$	$-a \sin ax$

The constants a, b, m, and n may be positive or negative.

Table C.1 shows derivatives of several other functions that are important in basic chemistry.

The derivative gives the slope of the tangent line to $f(x)$ at each point x and can be used to approximate the response to a small perturbation Δx in the independent variable x:

$$\Delta y \approx \frac{df}{dx} \Delta x$$

As an example, consider 1.00 mol of an ideal gas at 0°C that obeys the law

$$P \text{ (atm)} = \frac{22.414 \text{ L atm}}{V(\text{L})}$$

Suppose V is 20.00 L. If the volume V is changed by a small amount ΔV at fixed T and number of moles, what will be the corresponding change ΔP in the pressure? We write

$$P = f(V) = \frac{22.414}{V} = \frac{a}{V}$$

$$\frac{dP}{dV} = \frac{df}{dV} = -\frac{a}{V^2} = -\frac{22.414}{V^2} \qquad \text{(from Table C.1)}$$

$$\Delta P \approx \frac{dP}{dV} \Delta V = -\frac{22.414}{V^2} (\Delta V)$$

If $V = 20.00$ L, then

$$\Delta P \approx -\frac{22.414 \text{ L atm}}{(200.00 \text{ L})^2} \Delta V = -(0.0560 \text{ atm L}^{-1}) \Delta V$$

C.5 Areas under Curves and Integrals

Another mathematical operation that arises frequently in science is the calculation of the area under a curve. Some areas are those of simple geometric shapes and are easy to calculate. If the function $f(x)$ is a constant,

$$f(x) = a$$

then the area under a graph of $f(x)$ between the two points x_1 and x_2 is that of a rectangle (Fig. C.3a) and is easily calculated as

$$\text{area} = \text{height} \times \text{base} = a(x_2 - x_1) = [ax_2] - [ax_1]$$

If $f(x)$ is a straight line that is neither horizontal nor vertical,

$$f(x) = mx + b$$

then the area is that of a trapezoid (Fig. C.3b) and is equal to

$$\begin{aligned}\text{area} &= \text{average height} \times \text{base}\\ &= \tfrac{1}{2}(mx_1 + b + mx_2 + b)(x_2 - x_1)\\ &= [\tfrac{1}{2} mx_2^2 + bx_2] - [\tfrac{1}{2} mx_1^2 + bx_1]\end{aligned}$$

Suppose now that $f(x)$ is a more complicated function, such as that shown in Figure C.3c. We can estimate the area under this graph by approximating it with a series of small rectangles of width Δx and varying heights. If the height of the ith rectangle is $y_i = f(x_i)$, then we have, approximately,

$$\begin{aligned}\text{area} &\approx f(x_1)\,\Delta x + f(x_2)\,\Delta x + \ldots\\ &= \sum_i f(x_i)\,\Delta x\end{aligned}$$

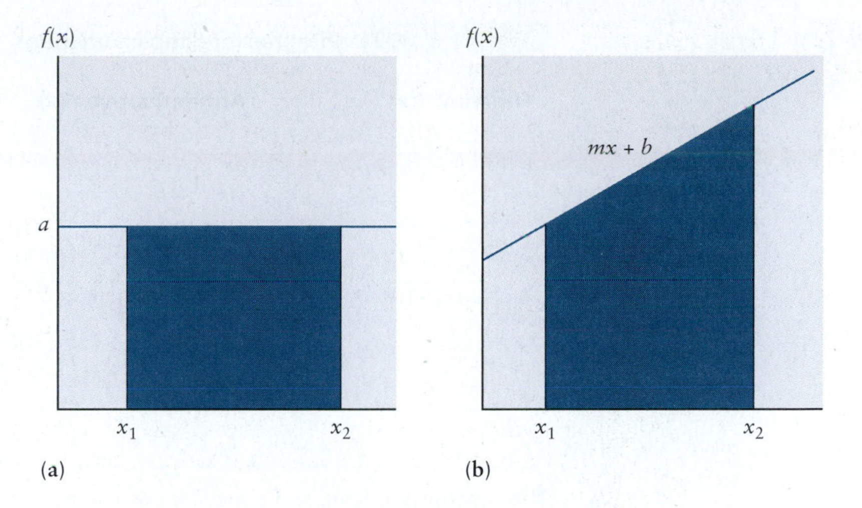

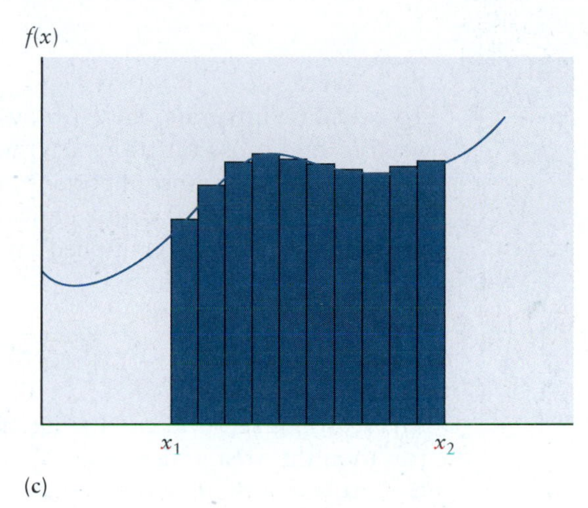

FIGURE C.3 The integral as the area under a curve.

As the widths of the rectangles Δx become small, this becomes a better approximation. We *define* the area under a curve $f(x)$ between two points as the limiting value of this sum as Δx approaches 0, called the **definite integral:**

$$\text{area} = \int_{x_1}^{x_2} f(x)\, dx$$

We have already worked out two examples of such integrals:

$$f(x) = a \Rightarrow \int_{x_1}^{x_2} a\, dx = [ax_2] - [ax_1]$$

$$f(x) = ax + b \Rightarrow \int_{x_1}^{x_2} (ax + b)\, dx = [\tfrac{1}{2}ax_2^2 + bx_2] - [\tfrac{1}{2}ax_1^2 + bx_1]$$

These integrals have an interesting form, which can be generalized to other cases as well: some function F evaluated at the upper limit (x_2) minus that function F evaluated at the lower limit (x_1).

$$\int_{x_1}^{x_2} f(x)\, dx = F(x_2) - F(x_1)$$

F is called the **antiderivative** of f because, for reasons we shall not go into here, it is obtained by the inverse of a derivative operation. In other words, if F is the antiderivative of f, then f is the derivative of F:

$$\frac{dF}{dx} = f(x)$$

TABLE C.2 Integrals of Simple Functions

Function $f(x)$	Antiderivative $F(x)$	Integral $\int_{x_1}^{x_2} f(x)\,dx$
a	ax	$ax_2 - ax_1$
$mx + b$	$\frac{1}{2}mx^2 + bx$	$[\frac{1}{2}mx_2^2 + bx_2] - [\frac{1}{2}mx_1^2 - bx_1]$
x^n $(n \neq -1)$	$\frac{1}{n+1}x^{n+1}$	$\frac{1}{n+1}(x_2^{n+1} - x_1^{n+1})$
$\frac{1}{x} = x^{-1}$ $(x > 0)$	$\ln x$	$(\ln x_2 - \ln x_1) = \ln \frac{x_2}{x_1}$
$\frac{1}{x^2} = x^{-2}$	$-\frac{1}{x}$	$-\left(\frac{1}{x_2} - \frac{1}{x_1}\right)$
e^{ax}	$\frac{e^{ax}}{a}$	$\frac{1}{a}(e^{ax_2} - e^{ax_1})$

The constants a, b, m, and n may be positive or negative.

To calculate integrals, therefore, we invert the results of Table C.1, because we need to find those functions $F(x)$ whose derivatives are equal to $f(x)$. Table C.2 lists several of the most important integrals for basic chemistry.

Several additional mathematical properties of integrals are important at this point. If a function is multiplied by a constant c, then the integral is multiplied by the same constant:

$$\int_{x_1}^{x_2} cf(x)\,dx = c\int_{x_1}^{x_2} f(x)\,dx$$

The reason is self-evident: if a function is multiplied everywhere by a constant factor, then the area under its graph must be increased by the same factor. Second, the integral of the sum of two functions is the sum of the separate integrals:

$$\int_{x_1}^{x_2} [f(x) + g(x)]\,dx = \int_{x_1}^{x_2} f(x)\,dx + \int_{x_1}^{x_2} g(x)\,dx$$

Finally, if the upper and lower limits of integration are reversed, the sign of the integral changes as well. This is easily seen from the antiderivative form:

$$\int_{x_2}^{x_1} f(x)\,dx = F(x_1) - F(x_2) = -[F(x_2) - F(x_1)] = -\int_{x_1}^{x_2} f(x)\,dx$$

C.6 Probability

Many applications in chemistry require us to interpret—and even predict—the results of measurements where we have only limited information about the system and the process involved. In such cases the best we can do is identify the possible outcomes of the experiment and assign a *probability* to each of them. Two examples illustrate the issues we face. In discussions of atomic structure, we would like to know the position of an electron relative to the nucleus. The principles of quantum mechanics tell us we can never know the exact location or trajectory of an electron; the most information we can have is the probability of finding an electron at each point in space around the nucleus. In discussing the behavior of a macroscopic amount of helium gas confined at a particular volume, pressure, and temperature we would like to know the speed with which an atom is moving in the container. We do not have experimental means to "tag" a particular atom, track its motions in the container, and measure its speed. The best we can do is estimate the probability that some typical atom is moving with each possible speed.

Everyone is familiar with the common sense concept of probability as a way to assess the likelihood of a desirable outcome in a game of chance. The purpose of this section is to give a brief introduction to probability in a form suited for scientific work.

Random Variable

The first step in setting up a probability model of a statistical experiment is to define the **random variable X,** the measurable quantity whose values *fluctuate,* or change, as we carry out many repetitions of the experiment. While defining the random variable, we also identify the **outcomes** of the statistical experiment, the possible values that X can take as we carry out many repetitions. If the experiment consists of flipping a coin, then X is simply the label on the side of the coin facing up, and the only possible outcomes are H for "heads" and T for "tails." If the experiment consists of rolling a die, X is the number of dots on the side facing up, with integral values from 1 to 6. If the experiment is rolling a pair of dice, X is the sum of the number of dots on the two sides facing up, with integral values from $1 + 1 = 2$ to $6 + 6 = 12$. If the experiment is to find the position of an electron in an atom, X is then r, the distance from the nucleus to the electron, and it can range from 0 to infinity. If the experiment is to find the speed of an atom in a sample of gas, X is then u, the speed of the atom, which can range from 0 to some large value.

Probability and Probability Functions

It is convenient to set up a graphical representation of the probability model. We represent the random variable X along the horizontal axis of an ordinary Cartesian graph. The possible outcomes of measuring X are shown as points along the horizontal axis. We see from the earlier examples that these can be either discrete or continuous values, depending on the nature of X. Along the vertical axis we want to plot $P(X)$, the probability of observing the random variable to have the value X, for each of the possible outcomes.

The **probability function $P(X)$** is generated by making multiple measurements of the random variable X and recording the results in a histogram. The performance of a chemistry class with 50 students on a quiz with maximum score 20 is a good example of a statistical experiment. Here the random variable X is the score a student achieves, and the possible outcomes are the integers from 0 to 20. The set of graded papers constitute 50 repetitions of the statistical experiment, and the results are summarized in Figure C.4. We define the probability that a particular score was achieved on the exam to be the fraction of the papers with that score:

$$P(X_i) = \frac{n_i(X_i)}{N}$$

where $n_i(X_i)$ is the number of papers with score X_i and N is the total number of papers. Figure C.5 shows the probability function generated from the data in Figure C.4. The function has a maximum at $X = 14$, which is labeled the **most probable value X_{mp}.** This is the value that appears most often in the experimental data. Probability is a pure number; it does not have physical dimensions. Probability cannot be determined from a single measurement; many repetitions of the statistical experiment are required for the definition of probability.

The probability function has the interesting property that its sum over all the possible outcomes is 1:

$$P_0 + P_1 + \ldots + P_{19} + P_{20} = \frac{1}{50}(n_0 + n_1 + \ldots + n_{19} + n_{20}) = \frac{50}{50} = 1$$

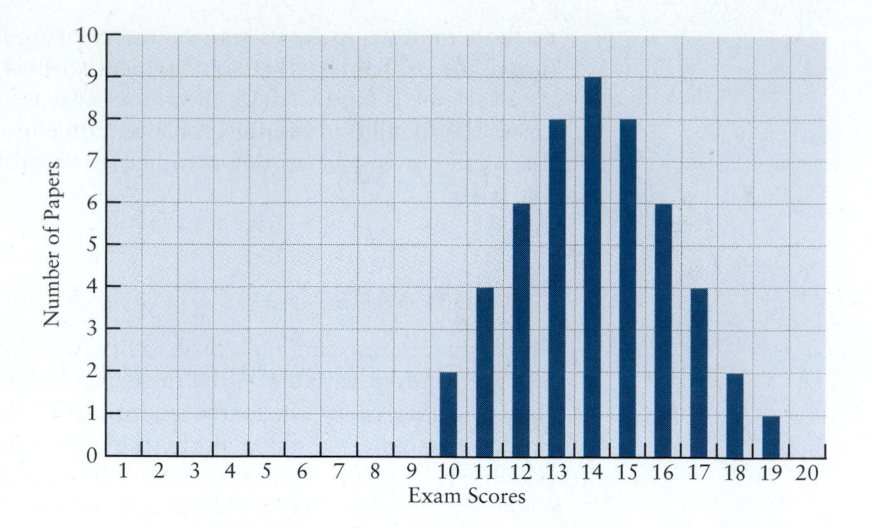

FIGURE C.4 Distribution of exam scores in a class of 50 students.

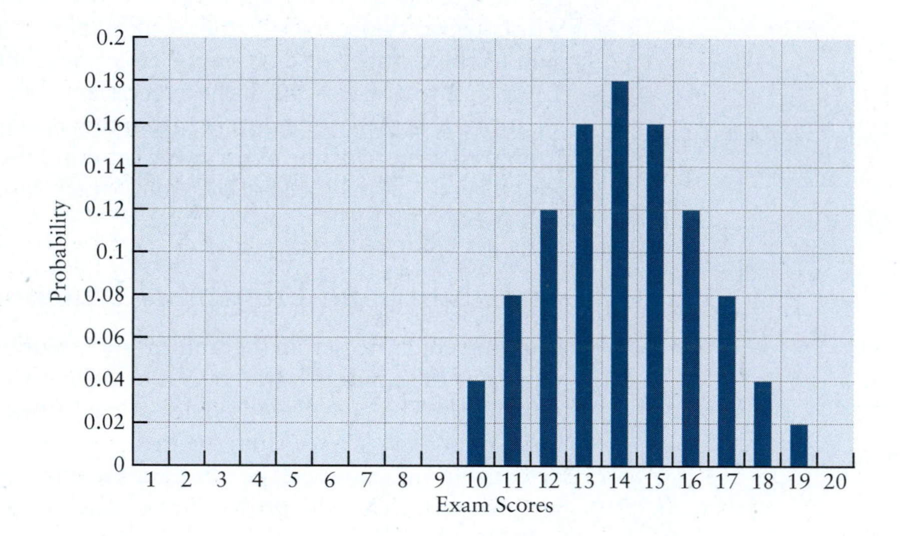

FIGURE C.5 Probability function for the exam scores in Figure C.4.

We represent this long sum in the following compact way where the uppercase Greek sigma means "add up all the terms," and the running index *i* indicates the term for each possible outcome from 0 to 20:

$$\sum_{i=0}^{20} P_i = \frac{1}{50}\sum_{i=0}^{20} n_i = \frac{50}{50} = 1$$

A probability function with this property is said to be **normalized.** Normalization signifies that we have accounted for all the possible outcomes: the probability that *some* score is achieved is 1. We should always check that a probability function is normalized before using it in further calculations.

Average Values of the Random Variable

The probability function is used in numerous statistical analyses of the experiment, only one of which we describe here. We calculate $\overline{X}$, the *mean,* or average, value of the random variable, by multiplying each possible outcome by its probability and summing over all the possible outcomes. In our compact form this operation for the probability function in Figure C.5 is

$$\overline{X} = \sum_{i=0}^{20} X_i P(X_i) = \frac{1}{50}\sum_{i=0}^{20} X_i n_i = 14.7$$

The mean value is one way to convey the essence of the probability function in a single number: Is the overall class performance on the exam good, fair, or poor? Another way is to state the most probable value X_{mp}. In this example the mean value is slightly larger than the most probable value because the distribution slightly favors the high end of the range, courtesy of the student who scored 19 of 20! The mean and most probable values coincide only when the distribution is perfectly symmetrical about the maximum. The complete probability function gives much more insight into the results of the experiment than the mean or most probable values standing alone.

Probability Density: Continuous Random Variables and Probability Functions

When the random variable is continuous, as in the case of the position of the electron and speed of the helium atom referred to earlier, we must consider every point of the X axis—not just the integers—as a possible outcome. Therefore, we cannot define the probability of observing a specific value, which we call X_0. Rather, we define the probability that X falls within a narrow, infinitesimal interval between X_0 and $X_0 + dX$. When the interval dX is sufficiently small, the probability will be proportional to the width of the interval, and we define the probability as follows:

$$[\text{Probability of } X \text{ falling in the interval between } X_0 \text{ and } X_0 + dX] = \mathcal{P}(X_0)dX$$

Because this analysis is required at each point X, we drop the subscript 0, and discuss the properties of $\mathcal{P}(X)$ in general. The function $\mathcal{P}(X)$, called the **probability density function** for X, is plotted as a continuous function above the X-axis. It is defined for all points in an interval (a, b) whose end points depend on the nature of X. In some cases the end points can include $+\infty, -\infty$, or both. Note that $\mathcal{P}(X)$ has physical dimensions of X^{-1}, since the product $\mathcal{P}(X)dX$ must be dimensionless.

The probability density plays the same role for continuous variables as does $P(X)$ for discrete random variables. The normalization condition becomes

$$\int_a^b \mathcal{P}(X)dX = 1$$

and the mean value of X is calculated as

$$\overline{X} = \int_a^b X \cdot \mathcal{P}(X)dX$$

The symmetry of the probability density function determines whether the most probable and mean values of X coincide. We use a special symbol here to distinguish probability density from probability. In the main portion of the textbook we do not always make that distinction in symbols. But, it is always clear from context whether we are discussing a probability function or a probability density function.

Experimental Tests of Theoretical Probability

Theoretical probability identifies the possible outcomes of a statistical experiment, and uses theoretical arguments to predict the probability of each. Many applications in chemistry take this form. In atomic and molecular structure problems, the general principles of quantum mechanics predict the probability functions. In other cases the theoretical predictions are based on assumptions about the chemical or physical behavior of a system. In all cases, the validity of these predictions must be tested by comparison with laboratory measurements of the behavior of the same random variable. A full determination of experimental probability, and the mean values that come from it, must be obtained and compared with the theoretical predictions. A theoretical prediction of probability can never be tested or interpreted with a single measurement. A large number of repeated measurements is necessary to reveal the true statistical behavior.

PROBLEMS

Answers to problems whose numbers are boldface appear in Appendix G.

Using Graphs

1. A particular plot of distance traveled against time elapsed is found to be a straight line. After an elapsed time of 1.5 hours, the distance traveled was 75 miles, and after an elapsed time of 3.0 hours, the distance traveled was 150 miles. Calculate the slope of the plot of distance against time, and give its units.

2. The pressure of a gas in a rigid container is measured at several different temperatures, and it is found that a plot of pressure against temperature is a straight line. At 20.0°C the pressure is 4.30 atm, and at 100.0°C the pressure is 5.47 atm. Calculate the slope of the plot of pressure against temperature and give its units.

3. Rewrite each of the following linear equations in the form $y = mx + b$, and give the slope and intercept of the corresponding plot. Then draw the graph.
(a) $y = 4x - 7$
(b) $7x - 2y = 5$
(c) $3y + 6x - 4 = 0$

4. Rewrite each of the following linear equations in the form $y = mx + b$, and give the slope and intercept of the corresponding plot. Then draw the graph.
(a) $y = -2x - 8$
(b) $-3x + 4y = 7$
(c) $7y - 16x + 53 = 0$

5. Graph the relation

$$y = 2x^3 - 3x^2 + 6x - 5$$

from $y = -2$ to $y = +2$. Is the plotted curve a straight line?

6. Graph the relation

$$y = \frac{8 - 10x - 3x^2}{2 - 3x}$$

from $x = -3$ to $x = +3$. Is the plotted curve a straight line?

Solution of Algebraic Equations

7. Solve the following linear equations for x.
(a) $7x + 5 = 0$
(b) $-4x + 3 = 0$
(c) $-3x = -2$

8. Solve the following linear equations for x.
(a) $6 - 8x = 0$
(b) $-2x - 5 = 0$
(c) $4x = -8$

9. Solve the following quadratic equations for x.
(a) $4x^2 + 7x - 5 = 0$
(b) $2x^2 = -3 - 6x$
(c) $2x + \frac{3}{x} = 6$

10. Solve the following quadratic equations for x.
(a) $6x^2 + 15x + 2 = 0$
(b) $4x = 5x^2 - 3$
(c) $\frac{1}{2 - x} + 3x = 4$

11. Solve each of the following equations for x using the approximation of small x, iteration, or graphical solution, as appropriate.
(a) $x(2.00 + x)^2 = 2.6 \times 10^{-6}$
(b) $x(3.00 - 7x)(2.00 + 2x) = 0.230$
(c) $2x^3 + 3x^2 + 12x = -16$

12. Solve each of the following equations for x using the approximation of small x, iteration, or graphical solution, as appropriate.
(a) $x(2.00 + x)(3.00 - x)(5.00 + 2x) = 1.58 \times 10^{-15}$
(b) $x\frac{(3.00 + x)(1.00 - x)}{2.00 - x} = 0.122$
(c) $12x^3 - 4x^2 + 35x = 10$

Powers and Logarithms

13. Calculate each of the following expressions, giving your answers the proper numbers of significant figures.
(a) $\log(3.56 \times 10^4)$
(b) $e^{-15.69}$
(c) $10^{8.41}$
(d) $\ln(6.893 \times 10^{-22})$

14. Calculate each of the following expressions, giving your answers the proper numbers of significant figures.
(a) $10^{-16.528}$
(b) $\ln(4.30 \times 10^{13})$
(c) $e^{14.21}$
(d) $\log(4.983 \times 10^{-11})$

15. What number has a common logarithm of 0.4793?

16. What number has a natural logarithm of -15.824?

17. Determine the common logarithm of 3.00×10^{121}. It is quite likely that your calculator will *not* give a correct answer. Explain why.

18. Compute the value of $10^{-107.8}$. It is quite likely that your calculator will *not* give the correct answer. Explain why.

19. The common logarithm of 5.64 is 0.751. Without using a calculator, determine the common logarithm of 5.64×10^7 and of 5.64×10^{-3}.

20. The common logarithm of 2.68 is 0.428. Without using a calculator, determine the common logarithm of 2.68×10^{192} and of 2.68×10^{-289}.

21. Use the graphical method and a calculator to solve the equation

$$\log \ln x = -x$$

for x. Give x to four significant figures.

22. Use a calculator to find a number that is equal to the reciprocal of its own natural logarithm. Report the answer to four significant figures.

Slopes of Curves and Derivatives

23. Calculate the derivatives of the following functions.
(a) $y = 4x^2 + 4$
(b) $y = \sin 3x + 4 \cos 2x$
(c) $y = 3x + 2$
(d) $y = \ln 7x$

24. Calculate the derivatives of the following functions.
(a) $y = 6x^{19}$
(b) $y = 7x^2 + 6x + 2$
(c) $y = e^{-6x}$
(d) $y = \cos 2x + \frac{7}{x}$

Areas under Curves and Integrals

25. Calculate the following integrals.
(a) $\int_2^4 (3x + 1)dx$
(b) $\int_0^5 x^6\, dx$
(c) $\int_1^4 e^{-2x}\, dx$

26. Calculate the following integrals.
(a) $\int_{-1}^3 4\, dx$
(b) $\int_2^{100} \frac{1}{x}\, dx$
(c) $\int_2^4 \frac{5}{x^2}\, dx$

APPENDIX D

Standard Chemical Thermodynamic Properties

This table lists standard enthalpies of formation ΔH_f°, standard third-law entropies S°, standard free energies of formation ΔG_f°, and molar heat capacities at constant pressure, C_P, for a variety of substances, all at 25°C (298.15 K) and 1 atm. The table proceeds from the left side to the right side of the periodic table. Binary compounds are listed under the element that occurs to the left in the periodic table, except that binary oxides and hydrides are listed with the other element. Thus, KCl is listed with potassium and its compounds, but ClO_2 is listed with chlorine and its compounds.

Note that the *solution-phase* entropies are not absolute entropies but are measured relative to the arbitrary standard $S^\circ(H^+(aq)) = 0$. Consequently, some of them (as well as some of the heat capacities) are negative.

Most of the thermodynamic data in these tables were taken from the *NBS Tables of Chemical Thermodynamic Properties* (1982) and changed, where necessary, from a standard pressure of 0.1 MPa to 1 atm. The data for organic compounds $C_nH_m (n > 2)$ were taken from the *Handbook of Chemistry and Physics* (1981).

Substance	ΔH_f° (25°C) kJ mol^{-1}	S° (25°C) J K^{-1} mol^{-1}	ΔG_f° (25°C) kJ mol^{-1}	C_P (25°C) J K^{-1} mol^{-1}
$H(g)$	217.96	114.60	203.26	20.78
$H_2(g)$	0	130.57	0	28.82
$H^+(aq)$	0	0	0	0
$H_3O^+(aq)$	−285.83	69.91	−237.18	75.29
$Li(s)$	0	29.12	0	24.77
$Li(g)$	159.37	138.66	126.69	20.79
$Li^+(aq)$	−278.49	13.4	−293.31	68.6
$LiH(s)$	−90.54	20.01	−68.37	27.87
$Li_2O(s)$	−597.94	37.57	−561.20	54.10
$LiF(s)$	−615.97	35.65	−587.73	41.59
$LiCl(s)$	−408.61	59.33	−384.39	47.99
$LiBr(s)$	−351.21	74.27	−342.00	—
$LiI(s)$	−270.41	86.78	−270.29	51.04
$Na(s)$	0	51.21	0	28.24
$Na(g)$	107.32	153.60	76.79	20.79
$Na^+(aq)$	−240.12	59.0	−261.90	46.4
$Na_2O(s)$	−414.22	75.06	−375.48	69.12
$NaOH(s)$	−425.61	64.46	−379.53	59.54
$NaF(s)$	−573.65	51.46	−543.51	48.86
$NaCl(s)$	−411.15	72.13	−384.15	50.50

continued

Substance	ΔH_f° (25°C) kJ mol^{-1}	S° (25°C) J K^{-1} mol^{-1}	ΔG_f° (25°C) kJ mol^{-1}	C_P (25°C) J K^{-1} mol^{-1}
$NaBr(s)$	−361.06	86.82	−348.98	51.38
$NaI(s)$	−287.78	98.53	−286.06	52.09
$NaNO_3(s)$	−467.85	116.52	−367.07	92.88
$Na_2S(s)$	−364.8	83.7	−349.8	—
$Na_2SO_4(s)$	−1387.08	149.58	−1270.23	128.20
$NaHSO_4(s)$	−1125.5	113.0	−992.9	—
$Na_2CO_3(s)$	−1130.68	134.98	−1044.49	112.30
$NaHCO_3(s)$	−950.81	101.7	−851.1	87.61
$K(s)$	0	64.18	0	29.58
$K(g)$	89.24	160.23	60.62	20.79
$K^+(aq)$	−252.38	102.5	−283.27	21.8
$KO_2(s)$	−284.93	116.7	−239.4	77.53
$K_2O_2(s)$	−494.1	102.1	−425.1	—
$KOH(s)$	−424.76	78.9	−379.11	64.9
$KF(s)$	−567.27	66.57	−537.77	49.04
$KCl(s)$	−436.75	82.59	−409.16	51.30
$KClO_3(s)$	−397.73	143.1	−296.25	100.25
$KBr(s)$	−393.80	95.90	−380.66	52.30
$KI(s)$	−327.90	106.32	−324.89	52.93
$KMnO_4(s)$	−837.2	171.71	−737.7	117.57
$K_2CrO_4(s)$	−1403.7	200.12	−1295.8	145.98
$K_2Cr_2O_7(s)$	−2061.5	291.2	−1881.9	219.24
$Rb(s)$	0	76.78	0	31.06
$Rb(g)$	80.88	169.98	53.09	20.79
$Rb^+(aq)$	−251.17	121.50	−283.98	—
$RbCl(s)$	−435.35	95.90	−407.82	52.38
$RbBr(s)$	−394.59	109.96	−381.79	52.84
$RbI(s)$	−333.80	118.41	−328.86	53.18
$Cs(s)$	0	85.23	0	32.17
$Cs(g)$	76.06	175.49	49.15	70.79
$Cs^+(aq)$	−258.28	133.05	−292.02	−10.5
$CsF(s)$	−553.5	92.80	−525.5	51.09
$CsCl(s)$	−443.04	101.17	−414.55	52.47
$CsBr(s)$	−405.81	113.05	−391.41	52.93
$CsI(s)$	−346.60	123.05	−340.58	52.80
II				
$Be(s)$	0	9.50	0	16.44
$Be(g)$	324.3	136.16	286.6	20.79
$BeO(s)$	−609.6	14.14	−580.3	25.52
$Mg(s)$	0	32.68	0	24.89
$Mg(g)$	147.70	148.54	113.13	20.79
$Mg^{2+}(aq)$	−466.85	−138.1	−454.8	—
$MgO(s)$	−601.70	26.94	−569.45	37.15
$MgCl_2(s)$	−641.32	89.62	−591.82	71.38
$MgSO_4(s)$	−1284.9	91.6	−1170.7	96.48
$Ca(s)$	0	41.42	0	25.31
$Ca(g)$	178.2	154.77	144.33	20.79
$Ca^{2+}(aq)$	−542.83	−53.1	−553.58	—
$CaH_2(s)$	−186.2	42	−147.2	—
$CaO(s)$	−635.09	39.75	−604.05	42.80
$CaS(s)$	−482.4	56.5	−477.4	47.40
$Ca(OH)_2(s)$	−986.09	83.39	−898.56	87.49
$CaF_2(s)$	−1219.6	68.87	−1167.3	67.03
$CaCl_2(s)$	−795.8	104.6	−748.1	72.59
$CaBr_2(s)$	−682.8	130	−663.6	—
$CaI_2(s)$	−533.5	142	−528.9	—
$Ca(NO_3)_2(s)$	−938.39	193.3	−743.20	149.37
$CaC_2(s)$	−59.8	69.96	−64.9	62.72

continued

Substance	ΔH_f° (25°C) kJ mol^{-1}	S° (25°C) J K^{-1} mol^{-1}	ΔG_f° (25°C) kJ mol^{-1}	C_P (25°C) J K^{-1} mol^{-1}
$CaCO_3$(*s*, calcite)	−1206.92	92.9	−1128.84	81.88
$CaCO_3$(*s*, aragonite)	−1207.13	88.7	−1127.80	81.25
$CaSO_4$(*s*)	−1434.11	106.9	−1321.86	99.66
$CaSiO_3$(*s*)	−1634.94	81.92	−1549.66	85.27
$CaMg(CO_3)_2$ (*s*, dolomite)	−2326.3	155.18	−2163.4	157.53
Sr(*s*)	0	52.3	0	26.4
Sr(*g*)	164.4	164.51	130.9	20.79
Sr^{2+}(*aq*)	−545.80	−32.6	−559.48	—
$SrCl_2$(*s*)	−828.9	114.85	−781.1	75.60
$SrCO_3$(*s*)	−1220.0	97.1	−1140.1	81.42
Ba(*s*)	0	62.8	0	28.07
Ba(*g*)	180	170.24	146	20.79
Ba^{2+}(*aq*)	−537.64	9.6	−560.77	—
$BaCl_2$(*s*)	−858.6	123.68	−810.4	75.14
$BaCO_3$(*s*)	−1216.3	112.1	−1137.6	85.35
$BaSO_4$(*s*)	−1473.2	132.2	−1362.3	101.75
Sc(*s*)	0	34.64	0	25.52
Sc(*g*)	377.8	174.68	336.06	22.09
Sc^{3+}(*aq*)	−614.2	−255	−586.6	—
Ti(*s*)	0	30.63	0	25.02
Ti(*g*)	469.9	180.19	425.1	24.43
TiO_2(*s*, rutile)	−944.7	50.33	−889.5	55.02
$TiCl_4$(*ℓ*)	−804.2	252.3	−737.2	145.18
$TiCl_4$(*g*)	−763.2	354.8	−726.8	95.4
Cr(*s*)	0	23.77	0	23.35
Cr(*g*)	396.6	174.39	351.8	20.79
Cr_2O_3(*s*)	−1139.7	81.2	−1058.1	118.74
CrO_4^{2-}(*aq*)	−881.15	50.21	−727.75	—
$Cr_2O_7^{2-}$(*aq*)	−1490.3	261.9	−1301.1	—
W(*s*)	0	32.64	0	24.27
W(*g*)	849.4	173.84	807.1	21.31
WO_2(*s*)	−589.69	50.54	−533.92	56.11
WO_3(*s*)	−842.87	75.90	−764.08	73.76
Mn(*s*)	0	32.01	0	26.32
Mn(*g*)	280.7	238.5	173.59	20.79
Mn^{2+}(*aq*)	−220.75	−73.6	−228.1	50
MnO(*s*)	−385.22	59.71	−362.92	45.44
MnO_2(*s*)	−520.03	53.05	−465.17	54.14
MnO_4^-(*s*)	−541.4	191.2	−447.2	−82.0
Fe(*s*)	0	27.28	0	25.10
Fe(*g*)	416.3	180.38	370.7	25.68
Fe^{2+}(*aq*)	−89.1	−137.7	−78.9	—
Fe^{3+}(*aq*)	−48.5	−315.9	−4.7	—
$Fe_{0.947}O$(*s*, wüstite)	−266.27	57.49	−245.12	48.12
Fe_2O_3(*s*, hematite)	−824.2	87.40	−742.2	103.85
Fe_3O_4(*s*, magnetite)	−1118.4	146.4	−1015.5	143.43
$Fe(OH)_3$(*s*)	−823.0	106.7	−696.5	—
FeS(*s*)	−100.0	60.29	−100.4	50.54
$FeCO_3$(*s*)	−740.57	93.1	−666.72	82.13
$Fe(CN)_6^{3-}$(*aq*)	561.9	270.3	729.4	—
$Fe(CN)_6^{4-}$(*aq*)	455.6	95.0	695.1	—

continued

Substance	ΔH_f° (25°C) kJ mol^{-1}	S° (25°C) J K^{-1} mol^{-1}	ΔG_f° (25°C) kJ mol^{-1}	C_P (25°C) J K^{-1} mol^{-1}
Co(*s*)	0	30.04	0	24.81
Co(*g*)	424.7	179.41	380.3	23.02
Co^{2+}(*aq*)	−58.2	−113	−54.4	—
Co^{3+}(*aq*)	92	−305	134	—
CoO(*s*)	−237.94	52.97	−214.22	55.23
$CoCl_2$(*s*)	−312.5	109.16	−269.8	78.49
Ni(*s*)	0	29.87	0	26.07
Ni(*g*)	429.7	182.08	384.5	25.36
Ni^{2+}(*aq*)	−54.0	−128.9	−45.6	—
NiO(*s*)	−239.7	37.99	−211.7	44.31
Pt(*s*)	0	41.63	0	25.86
Pt(*g*)	565.3	192.30	520.5	25.53
$PtCl_6^{2-}$(*aq*)	−668.2	219.7	−482.7	—
Cu(*s*)	0	33.15	0	24.44
Cu(*g*)	338.32	166.27	298.61	20.79
Cu^{+}(*aq*)	71.67	40.6	49.98	—
Cu^{2+}(*aq*)	64.77	−99.6	65.49	—
CuO(*s*)	−157.3	42.63	−129.7	42.30
Cu_2O(*s*)	−168.6	93.14	−146.0	63.64
CuCl(*s*)	−137.2	86.2	−119.88	48.5
$CuCl_2$(*s*)	−220.1	108.07	−175.7	71.88
$CuSO_4$(*s*)	−771.36	109	−661.9	100.0
$Cu(NH_3)_4^{2+}$(*aq*)	−348.5	273.6	−111.07	—
Ag(*s*)	0	42.55	0	25.35
Ag(*g*)	284.55	172.89	245.68	20.79
Ag^{+}(*aq*)	105.58	72.68	77.11	21.8
AgCl(*s*)	−127.07	96.2	−109.81	50.79
$AgNO_3$(*s*)	−124.39	140.92	−33.48	93.05
$Ag(NH_3)_2^{+}$(*aq*)	−111.29	245.2	−17.12	—
Au(*s*)	0	47.40	0	25.42
Au(*g*)	366.1	180.39	326.3	20.79
Zn(*s*)	0	41.63	0	25.40
Zn(*g*)	130.73	160.87	95.18	20.79
Zn^{2+}(*aq*)	−153.89	−112.1	−147.06	46
ZnO(*s*)	−348.28	43.64	−318.32	40.25
ZnS(*s*, sphalerite)	−205.98	57.7	−201.29	46.0
$ZnCl_2$(*s*)	−415.05	111.46	−369.43	71.34
$ZnSO_4$(*s*)	−982.8	110.5	−871.5	99.2
$Zn(NH_3)_4^{2+}$(*aq*)	−533.5	301	−301.9	—
Hg(*ℓ*)	0	76.02	0	27.98
Hg(*g*)	61.32	174.85	31.85	20.79
HgO(*s*)	−90.83	70.29	−58.56	44.06
$HgCl_2$(*s*)	−224.3	146.0	−178.6	—
Hg_2Cl_2(*s*)	−265.22	192.5	−210.78	—
B(*s*)	0	5.86	0	11.09
B(*g*)	562.7	153.34	518.8	20.80
B_2H_6(*g*)	35.6	232.00	86.6	56.90
B_5H_9(*g*)	73.2	275.81	174.9	96.78
B_2O_3(*s*)	−1272.77	53.97	−1193.70	62.93
H_3BO_3(*s*)	−1094.33	88.83	−969.02	81.38
BF_3(*g*)	−1137.00	254.01	−1120.35	50.46
BF_4^{-}(*aq*)	−1574.9	180	−1486.9	—
BCl_3(*g*)	−403.76	289.99	−388.74	62.72
BBr_3(*g*)	−205.64	324.13	−232.47	67.78
Al(*s*)	0	28.33	0	24.35

continued

Substance	ΔH_f° (25°C) kJ mol^{-1}	S° (25°C) J K^{-1} mol^{-1}	ΔG_f° (25°C) kJ mol^{-1}	C_P (25°C) J K^{-1} mol^{-1}
$Al(g)$	326.4	164.43	285.7	21.38
$Al^{3+}(aq)$	−531	−321.7	−485	—
$Al_2O_3(s)$	−1675.7	50.92	−1582.3	79.04
$AlCl_3(s)$	−704.2	110.67	−628.8	91.84
$Ga(s)$	0	40.88	0	25.86
$Ga(g)$	277.0	168.95	238.9	25.36
$Tl(s)$	0	64.18	0	26.32
$Tl(g)$	182.21	180.85	147.44	20.79
IV				
$C(s$, graphite)	0	5.74	0	8.53
$C(s$, diamond)	1.895	2.377	2.900	6.11
$C(g)$	716.682	157.99	671.29	20.84
$CH_4(g)$	−74.81	186.15	−50.75	35.31
$C_2H_2(g)$	226.73	200.83	209.20	43.93
$C_2H_4(g)$	52.26	219.45	68.12	43.56
$C_2H_6(g)$	−84.68	229.49	−32.89	52.63
$C_3H_8(g)$	−103.85	269.91	−23.49	73.0
n-$C_4H_{10}(g)$	−124.73	310.03	−15.71	97.5
$C_4H_{10}(g$, isobutane)	−131.60	294.64	−17.97	96.8
n-$C_5H_{12}(g)$	−146.44	348.40	−8.20	120
$C_6H_6(g)$	82.93	269.2	129.66	81.6
$C_6H_6(\ell)$	49.03	172.8	124.50	136
$CO(g)$	−110.52	197.56	−137.15	29.14
$CO_2(g)$	−393.51	213.63	−394.36	37.11
$CO_2(aq)$	−413.80	117.6	−385.98	—
$CS_2(\ell)$	89.70	151.34	65.27	75.7
$CS_2(g)$	117.36	237.73	67.15	45.40
$H_2CO_3(aq)$	−699.65	187.4	−623.08	—
$HCO_3^-(aq)$	−691.99	91.2	−586.77	—
$CO_3^{2-}(aq)$	−677.14	−56.9	−527.81	—
$HCOOH(\ell)$	−424.72	128.95	−361.42	99.04
$HCOOH(aq)$	−425.43	163	−372.3	—
$COOH^-(aq)$	−425.55	92	−351.0	−87.9
$CH_2O(g)$	−108.57	218.66	−102.55	35.40
$CH_3OH(\ell)$	−238.66	126.8	−166.35	81.6
$CH_3OH(g)$	−200.66	239.70	−162.01	43.89
$CH_3OH(aq)$	−245.93	133.1	−175.31	—
$H_2C_2O_4(s)$	−827.2	—	—	117
$HC_2O_4^-(aq)$	−818.4	149.4	−698.34	—
$C_2O_4^{2-}(aq)$	−825.1	45.6	−673.9	—
$CH_3COOH(\ell)$	−484.5	159.8	−390.0	124.3
$CH_3COOH(g)$	−432.25	282.4	−374.1	66.5
$CH_3COOH(aq)$	−485.76	178.7	−396.46	—
$CH_3COO^-(aq)$	−486.01	86.6	−369.31	−6.3
$CH_3CHO(\ell)$	−192.30	160.2	−128.12	—
$C_2H_5OH(\ell)$	−277.69	160.7	−174.89	111.46
$C_2H_5OH(g)$	−235.10	282.59	−168.57	65.44
$C_2H_5OH(aq)$	−288.3	148.5	−181.64	—
$CH_3OCH_3(g)$	−184.05	266.27	−112.67	64.39
$CF_4(g)$	−925	261.50	−879	61.09
$CCl_4(\ell)$	−135.44	216.40	−65.28	131.75
$CCl_4(g)$	−102.9	309.74	−60.62	83.30
$CHCl_3(g)$	−103.14	295.60	−70.37	65.69
$COCl_2(g)$	−218.8	283.53	−204.6	57.66
$CH_2Cl_2(g)$	−92.47	270.12	−65.90	50.96
$CH_3Cl(g)$	−80.83	234.47	−57.40	40.75
$CBr_4(s)$	18.8	212.5	47.7	144.3
$CH_3I(\ell)$	−15.5	163.2	13.4	126

continued

Substance	ΔH_f° (25°C) kJ mol^{-1}	S° (25°C) J K^{-1} mol^{-1}	ΔG_f° (25°C) kJ mol^{-1}	C_P (25°C) J K^{-1} mol^{-1}
$HCN(g)$	135.1	201.67	124.7	35.86
$HCN(aq)$	107.1	124.7	119.7	—
$CN^-(aq)$	150.6	94.1	172.4	—
$CH_3NH_2(g)$	−22.97	243.30	32.09	53.1
$CO(NH_2)_2(s)$	−333.51	104.49	−197.44	93.14
$Si(s)$	0	18.83	0	20.00
$Si(g)$	455.6	167.86	411.3	22.25
$SiC(s)$	−65.3	16.61	−62.8	26.86
$SiO_2(s$, quartz)	−910.94	41.84	−856.67	44.43
$SiO_2(s$, cristobalite)	−909.48	42.68	−855.43	44.18
$Ge(s)$	0	31.09	0	23.35
$Ge(g)$	376.6	335.9	167.79	30.73
$Sn(s$, white)	0	51.55	0	26.99
$Sn(s$, gray)	−2.09	44.14	0.13	25.77
$Sn(g)$	302.1	168.38	267.3	21.26
$SnO(s)$	−285.8	56.5	−256.9	44.31
$SnO_2(s)$	−580.7	52.3	−519.6	52.59
$Sn(OH)_2(s)$	−561.1	155	−491.7	—
$Pb(s)$	0	64.81	0	26.44
$Pb(g)$	195.0	161.9	175.26	20.79
$Pb^{2+}(aq)$	−1.7	10.5	−24.43	—
$PbO(s$, yellow)	−217.32	68.70	−187.91	45.77
$PbO(s$, red)	−218.99	66.5	−188.95	45.81
$PbO_2(s)$	−277.4	68.6	−217.36	64.64
$PbS(s)$	−100.4	91.2	−98.7	49.50
$PbI_2(s)$	−175.48	174.85	−173.64	77.36
$PbSO_4(s)$	−919.94	148.57	−813.21	103.21
$N_2(g)$	0	191.50	0	29.12
$N(g)$	472.70	153.19	455.58	20.79
$NH_3(g)$	−46.11	192.34	−16.48	35.06
$NH_3(aq)$	−80.29	111.3	−26.50	—
$NH_4^+(aq)$	−132.51	113.4	−79.31	79.9
$N_2H_4(\ell)$	50.63	121.21	149.24	98.87
$N_2H_4(aq)$	34.31	138	128.1	—
$NO(g)$	90.25	210.65	86.55	29.84
$NO_2(g)$	33.18	239.95	51.29	37.20
$NO_2^-(aq)$	−104.6	123.0	−32.2	−97.5
$NO_3^-(aq)$	−205.0	146.4	−108.74	−86.6
$N_2O(g)$	82.05	219.74	104.18	38.45
$N_2O_4(g)$	9.16	304.18	97.82	77.28
$N_2O_5(s)$	−43.1	178.2	113.8	143.1
$HNO_2(g)$	−79.5	254.0	−46.0	45.6
$HNO_3(\ell)$	−174.10	155.49	−80.76	109.87
$NH_4NO_3(s)$	−365.56	151.08	−184.02	139.3
$NH_4Cl(s)$	−314.43	94.6	−202.97	84.1
$(NH_4)_2SO_4(s)$	−1180.85	220.1	−901.90	187.49
$P(s$, white)	0	41.09	0	23.84
$P(s$, red)	−17.6	22.80	−12.1	21.21
$P(g)$	314.64	163.08	278.28	20.79
$P_2(g)$	144.3	218.02	103.7	32.05
$P_4(g)$	58.91	279.87	24.47	67.15
$PH_3(g)$	5.4	210.12	13.4	37.11
$H_3PO_4(s)$	−1279.0	110.50	−1119.2	106.06
$H_3PO_4(aq)$	−1288.34	158.2	−1142.54	—
$H_2PO_4^-(aq)$	−1296.29	90.4	−1130.28	—
$HPO_4^{2-}(aq)$	−1292.14	−33.5	−1089.15	—
$PO_4^{3-}(aq)$	−1277.4	−222	−1018.7	—

continued

	Substance	ΔH°_f (25°C) kJ mol^{-1}	S° (25°C) J K^{-1} mol^{-1}	ΔG°_f (25°C) kJ mol^{-1}	C_P (25°C) J K^{-1} mol^{-1}
	$PCl_3(g)$	−287.0	311.67	−267.8	71.84
	$PCl_5(g)$	−374.9	364.47	−305.0	112.80
	As(*s*, gray)	0	35.1	0	24.64
	$As(g)$	302.5	174.10	261.0	20.79
	$As_2(g)$	222.2	239.3	171.9	35.00
	$As_4(g)$	143.9	314	92.4	—
	$AsH_3(g)$	66.44	222.67	68.91	38.07
	$As_4O_6(s)$	−1313.94	214.2	−1152.53	191.29
	$Sb(s)$	0	45.69	0	25.33
	$Sb(g)$	262.3	180.16	222.1	20.79
	$Bi(s)$	0	56.74	0	25.52
	$Bi(g)$	207.1	186.90	168.2	20.79
VI	$O_2(g)$	0	205.03	0	29.36
	$O(g)$	249.17	160.95	231.76	21.91
	$O_3(g)$	142.7	238.82	163.2	39.20
	$OH^-(aq)$	−229.99	−10.75	−157.24	−148.5
	$H_2O(\ell)$	−285.83	69.91	−237.18	75.29
	$H_2O(g)$	−241.82	188.72	−228.59	35.58
	$H_2O_2(\ell)$	−187.78	109.6	−120.42	89.1
	$H_2O_2(aq)$	−191.17	143.9	−134.03	—
	S(*s*, rhombic)	0	31.80	0	22.64
	S(*s*, monoclinic)	0.30	32.6	0.096	—
	$S(g)$	278.80	167.71	238.28	23.67
	$S_8(g)$	102.30	430.87	49.66	156.44
	$H_2S(g)$	−20.63	205.68	−33.56	34.23
	$H_2S(aq)$	−39.7	121	−27.83	—
	$HS^-(aq)$	−17.6	62.8	12.08	—
	$SO(g)$	6.26	221.84	−19.87	30.17
	$SO_2(g)$	−296.83	248.11	−300.19	39.87
	$SO_3(g)$	−395.72	256.65	−371.08	50.67
	$H_2SO_3(aq)$	−608.81	232.2	−537.81	—
	$HSO_3^-(aq)$	−626.22	139.7	−527.73	—
	$SO_3^{2-}(aq)$	−635.5	−29	−486.5	—
	$H_2SO_4(\ell)$	−813.99	156.90	−690.10	138.91
	$HSO_4^-(aq)$	−887.34	131.8	−755.91	−84
	$SO_4^{2-}(aq)$	−909.27	20.1	−744.53	−293
	$SF_6(g)$	−1209	291.71	−1105.4	97.28
	Se(*s*, black)	0	42.44	0	25.36
	$Se(g)$	227.07	176.61	187.06	20.82
VII	$F_2(g)$	0	202.67	0	31.30
	$F(g)$	78.99	158.64	61.94	22.74
	$F^-(aq)$	−332.63	−13.8	−278.79	−106.7
	$HF(g)$	−271.1	173.67	−273.2	29.13
	$HF(aq)$	−320.08	88.7	−296.82	—
	$XeF_4(s)$	−261.5	—	—	—
	$Cl_2(g)$	0	222.96	0	33.91
	$Cl(g)$	121.68	165.09	105.71	21.84
	$Cl^-(aq)$	−167.16	56.5	−131.23	−136.4
	$HCl(g)$	−92.31	186.80	−95.30	29.12
	$ClO^-(aq)$	−107.1	42	−36.8	—
	$ClO_2(g)$	102.5	256.73	120.5	41.97
	$ClO_2^-(aq)$	−66.5	101.3	17.2	—
	$ClO_3^-(aq)$	−103.97	162.3	−7.95	—
	$ClO_4^-(aq)$	−129.33	182.0	−8.52	—
	$Cl_2O(g)$	80.3	266.10	97.9	45.40
	$HClO(aq)$	−120.9	142	−79.9	—

continued

	Substance	ΔH_f° (25°C) kJ mol^{-1}	S° (25°C) J K^{-1} mol^{-1}	ΔG_f° (25°C) kJ mol^{-1}	C_P (25°C) J K^{-1} mol^{-1}
	$ClF_3(g)$	−163.2	281.50	−123.0	63.85
	$Br_2(\ell)$	0	152.23	0	75.69
	$Br_2(g)$	30.91	245.35	3.14	36.02
	$Br_2(aq)$	−2.59	130.5	3.93	—
	$Br(g)$	111.88	174.91	82.41	20.79
	$Br^-(aq)$	−121.55	82.4	−103.96	−141.8
	$HBr(g)$	−36.40	198.59	−53.43	29.14
	$BrO_3^-(aq)$	−67.07	161.71	18.60	—
	$I_2(s)$	0	116.14	0	54.44
	$I_2(g)$	62.44	260.58	19.36	36.90
	$I_2(aq)$	22.6	137.2	16.40	—
	$I(g)$	106.84	180.68	70.28	20.79
	$I^-(aq)$	−55.19	111.3	−51.57	−142.3
	$I_3^-(aq)$	−51.5	239.3	−51.4	—
	$HI(g)$	26.48	206.48	1.72	29.16
	$ICl(g)$	17.78	247.44	−5.44	35.56
	$IBr(g)$	40.84	258.66	3.71	36.44
VIII	$He(g)$	0	126.04	0	20.79
	$Ne(g)$	0	146.22	0	20.79
	$Ar(g)$	0	154.73	0	20.79
	$Kr(g)$	0	163.97	0	20.79
	$Xe(g)$	0	169.57	0	20.79

APPENDIX E

Standard Reduction Potentials at 25°C

Half-Reaction	$\mathscr{E}°$ (volts)
$F_2(g) + 2\ e^- \longrightarrow 2\ F^-$	2.87
$H_2O_2 + 2\ H_3O^+ + 2\ e^- \longrightarrow 4\ H_2O$	1.776
$PbO_2(s) + SO_4^{2-} + 4\ H_3O^+ + 2\ e^- \longrightarrow PbSO_4(s) + 6\ H_2O$	1.685
$Au^+ + e^- \longrightarrow Au(s)$	1.68
$MnO_4^- + 4\ H_3O^+ + 3\ e^- \longrightarrow MnO_2(s) + 6\ H_2O$	1.679
$HClO_2 + 2\ H_3O^+ + 2\ e^- \longrightarrow HClO + 3\ H_2O$	1.64
$HClO + H_3O^+ + e^- \longrightarrow \frac{1}{2}\ Cl_2(g) + 2\ H_2O$	1.63
$Ce^{4+} + e^- \longrightarrow Ce^{3+}$ (1 M HNO_3 solution)	1.61
$2\ NO(g) + 2\ H_3O^+ + 2\ e^- \longrightarrow N_2O(g) + 3\ H_2O$	1.59
$BrO_3^- + 6\ H_3O^+ + 5\ e^- \longrightarrow \frac{1}{2}\ Br_2(\ell) + 9\ H_2O$	1.52
$Mn^{3+} + e^- \longrightarrow Mn^{2+}$	1.51
$MnO_4^- + 8\ H_3O^+ + 5\ e^- \longrightarrow Mn^{2+} + 12\ H_2O$	1.491
$ClO_3^- + 6\ H_3O^+ + 5\ e^- \longrightarrow \frac{1}{2}\ Cl_2(g) + 9\ H_2O$	1.47
$PbO_2(s) + 4\ H_3O^+ + 2\ e^- \longrightarrow Pb^{2+} + 6\ H_2O$	1.46
$Au^{3+} + 3\ e^- \longrightarrow Au(s)$	1.42
$Cl_2(g) + 2\ e^- \longrightarrow 2\ Cl^-$	1.3583
$Cr_2O_7^{2-} + 14\ H_3O^+ + 6\ e^- \longrightarrow 2\ Cr^{3+} + 21\ H_2O$	1.33
$O_3(g) + H_2O + 2\ e^- \longrightarrow O_2 + 2\ OH^-$	1.24
$O_2(g) + 4\ H_3O^+ + 4\ e^- \longrightarrow 6\ H_2O$	1.229
$MnO_2(s) + 4\ H_3O^+ + 2\ e^- \longrightarrow Mn^{2+} + 6\ H_2O$	1.208
$ClO_4^- + 2\ H_3O^+ + 2\ e^- \longrightarrow ClO_3 + 3\ H_2O$	1.19
$Br_2(\ell) + 2\ e^- \longrightarrow 2\ Br^-$	1.065
$NO_3^- + 4\ H_3O^+ + 3\ e^- \longrightarrow NO(g) + 6\ H_2O$	0.96
$2\ Hg^{2+} + 2\ e^- \longrightarrow Hg_2^{2+}$	0.905
$Ag^+ + e^- \longrightarrow Ag(s)$	0.7996
$Hg_2^{2+} + 2\ e^- \longrightarrow 2\ Hg(\ell)$	0.7961
$Fe^{3+} + e^- \longrightarrow Fe^{2+}$	0.770
$O_2(g) + 2\ H_3O^+ + 2\ e^- \longrightarrow H_2O_2 + 2\ H_2O$	0.682
$BrO_3^- + 3\ H_2O + 6\ e^- \longrightarrow Br^- + 6\ OH^-$	0.61
$MnO_4^- + 2\ H_2O + 3\ e^- \longrightarrow MnO_2(s) + 4\ OH^-$	0.588
$I_2(s) + 2\ e^- \longrightarrow 2\ I^-$	0.535
$Cu^+ + e^- \longrightarrow Cu(s)$	0.522
$O_2(g) + 2\ H_2O + 4\ e^- \longrightarrow 4\ OH^-$	0.401
$Cu^{2+} + 2\ e^- \longrightarrow Cu(s)$	0.3402
$PbO_2(s) + H_2O + 2\ e^- \longrightarrow PbO(s) + 2\ OH^-$	0.28
$Hg_2Cl_2(s) + 2\ e^- \longrightarrow 2\ Hg(\ell) + 2\ Cl^-$	0.2682
$AgCl(s) + e^- \longrightarrow Ag(s) + Cl^-$	0.2223
$SO_4^{2-} + 4\ H_3O^+ + 2\ e^- \longrightarrow H_2SO_3 + 5\ H_2O$	0.20

continued

Half-Reaction	$\mathscr{E}°$ (volts)
$Cu^{2+} + e^- \longrightarrow Cu^+$	0.158
$S_4O_6^{2-} + 2\ e^- \longrightarrow 2\ S_2O_3^{2-}$	0.0895
$NO_3^- + H_2O + 2\ e^- \longrightarrow NO_2^- + 2\ OH^-$	0.01
$2\ H_3O^+ + 2\ e^- \longrightarrow H_2(g) + 2\ H_2O(\ell)$	0.000 exactly
$Pb^{2+} + 2\ e^- \longrightarrow Pb(s)$	−0.1263
$Sn^{2+} + 2\ e^- \longrightarrow Sn(s)$	−0.1364
$Ni^{2+} + 2\ e^- \longrightarrow Ni(s)$	−0.23
$Co^{2+} + 2\ e^- \longrightarrow Co(s)$	−0.28
$PbSO_4(s) + 2\ e^- \longrightarrow Pb(s) + SO_4^{2-}$	−0.356
$Mn(OH)_3(s) + e^- \longrightarrow Mn(OH)_2(s) + OH^-$	−0.40
$Cd^{2+} + 2\ e^- \longrightarrow Cd(s)$	−0.4026
$Fe^{2+} + 2\ e^- \longrightarrow Fe(s)$	−0.409
$Cr^{3+} + e^- \longrightarrow Cr^{2+}$	−0.424
$Fe(OH)_3(s) + e^- \longrightarrow Fe(OH)_2(s) + OH^-$	−0.56
$PbO(s) + H_2O + 2\ e^- \longrightarrow Pb(s) + 2\ OH^-$	−0.576
$2\ SO_3^{2-} + 3\ H_2O + 4\ e^- \longrightarrow S_2O_3^{2-} + 6\ OH^-$	−0.58
$Ni(OH)_2(s) + 2\ e^- \longrightarrow Ni(s) + 2\ OH^-$	−0.66
$Co(OH)_2(s) + 2\ e^- \longrightarrow Co(s) + 2\ OH^-$	−0.73
$Cr^{3+} + 3\ e^- \longrightarrow Cr(s)$	−0.74
$Zn^{2+} + 2\ e^- \longrightarrow Zn(s)$	−0.7628
$2\ H_2O + 2\ e^- \longrightarrow H_2(g) + 2\ OH^-$	−0.8277
$Cr^{2+} + 2\ e^- \longrightarrow Cr(s)$	−0.905
$SO_4^{2-} + H_2O + 2\ e^- \longrightarrow SO_3^{2-} + 2\ OH^-$	−0.92
$Mn^{2+} + 2\ e^- \longrightarrow Mn(s)$	−1.029
$Mn(OH)_2(s) + 2\ e^- \longrightarrow Mn(s) + 2\ OH^-$	−1.47
$Al^{3+} + 3\ e^- \longrightarrow Al(s)$	−1.706
$Sc^{3+} + 3\ e^- \longrightarrow Sc(s)$	−2.08
$Ce^{3+} + 3\ e^- \longrightarrow Ce(s)$	−2.335
$La^{3+} + 3\ e^- \longrightarrow La(s)$	−2.37
$Mg^{2+} + 2\ e^- \longrightarrow Mg(s)$	−2.375
$Mg(OH)_2(s) + 2\ e^- \longrightarrow Mg(s) + 2\ OH^-$	−2.69
$Na^+ + e^- \longrightarrow Na(s)$	−2.7109
$Ca^{2+} + 2\ e^- \longrightarrow Ca(s)$	−2.76
$Ba^{2+} + 2\ e^- \longrightarrow Ba(s)$	−2.90
$K^+ + e^- \longrightarrow K(s)$	−2.925
$Li^+ + e^- \longrightarrow Li(s)$	−3.045

All voltages are standard reduction potentials (relative to the standard hydrogen electrode) at 25°C and 1 atm pressure. All species are in aqueous solution unless otherwise indicated.

APPENDIX F

Physical Properties of the Elements

Hydrogen and the Alkali Metals (Group I Elements)

	Hydrogen	Lithium	Sodium	Potassium	Rubidium	Cesium	Francium
Atomic number	1	3	11	19	37	55	87
Atomic mass	1.00794	6.941	22.98976928	39.0983	85.4678	132.9054519	(223.0197)
Melting point (°C)	−259.14	180.54	97.81	63.65	38.89	28.40	25
Boiling point (°C)	−252.87	1347	903.8	774	688	678.4	677
Density at 25°C (g cm^{-3})	0.070 (−253°C)	0.534	0.971	0.862	1.532	1.878	
Color	Colorless	Silver	Silver	Silver	Silver	Silver	
Ground-state electron configuration	$1s^1$	$[He]2s^1$	$[Ne]3s^1$	$[Ar]4s^1$	$[Kr]5s^1$	$[Xe]6s^1$	$[Rn]7s^1$
Ionization energy†	1312.0	520.2	495.8	418.8	403.0	375.7	≈400
Electron affinity†	72.770	59.63	52.867	48.384	46.884	45.505	est. 44
Electronegativity	2.20	0.98	0.93	0.82	0.82	0.79	0.70
Ionic radius (Å)	1.46(H^-)	0.68	0.98	1.33	1.48	1.67	≈1.8
Atomic radius (Å)	0.37	1.52	1.86	2.27	2.47	2.65	≈2.7
Enthalpy of fusion†	0.1172	3.000	2.602	2.335	2.351	2.09	
Enthalpy of vaporization†	0.4522	147.1	97.42	89.6	76.9	67.8	
Bond enthalpy of M_2†	436	102.8	72.6	54.8	51.0	44.8	
Standard reduction potential (volts)	0 H^+/H_2	−3.045 Li^+/Li	−2.7109 Na^+/Na	−2.924 K^+/K	−2.925 Rb^+/Rb	−2.923 Cs^+/Cs	≈2.9 Fr^+/Fr

†In kilojoules per mole.

The Alkaline-Earth Metals (Group II Elements)

	Beryllium	Magnesium	Calcium	Strontium	Barium	Radium
Atomic number	4	12	20	38	56	88
Atomic mass	9.012182	24.3050	40.078	87.62	137.327	(226.0254)
Melting point (°C)	1283	648.8	839	769	725	700
Boiling point (°C)	2484	1105	1484	1384	1640	
Density at 25°C (g cm^{-3})	1.848	1.738	1.55	2.54	3.51	5
Color	Gray	Silver	Silver	Silver	Silver-yellow	Silver
Ground-state electron configuration	[He]$2s^2$	[Ne]$3s^2$	[Ar]$4s^2$	[Kr]$5s^2$	[Xe]$6s^2$	[Rn]$7s^2$
Ionization energy†	899.4	737.7	589.8	549.5	502.9	509.3
Electron affinity†	<0	<0	2.0	4.6	13.95	>0
Electronegativity	1.57	1.31	1.00	0.95	0.89	0.90
Ionic radius (Å)	0.31	0.66	0.99	1.13	1.35	1.43
Atomic radius (Å)	1.13	1.60	1.97	2.15	2.17	2.23
Enthalpy of fusion†	11.6	8.95	8.95	9.62	7.66	7.15
Enthalpy of vaporization†	297.6	127.6	154.7	154.4	150.9	136.7
Bond enthalpy of M_2†	9.46					
Standard reduction potential (volts)	−1.70	−2.375	−2.76	−2.89	−2.90	−2.916
	Be^{2+}/Be	Mg^{2+}/Mg	Ca^{2+}/Ca	Sr^{2+}/Sr	Ba^{2+}/Ba	Ra^{2+}/Ra

Group III Elements

	Boron	Aluminum	Gallium	Indium	Thallium
Atomic number	5	13	31	49	81
Atomic mass	10.811	26.9815386	69.723	114.818	204.3833
Melting point (°C)	2300	660.37	29.78	156.61	303.5
Boiling point (°C)	3658	2467	2403	2080	1457
Density at 25°C (g cm^{-3})	2.34	2.702	5.904	7.30	11.85
Color	Yellow	Silver	Silver	Silver	Blue-white
Ground-state electron configuration	[He]$2s^22p^1$	[Ne]$3s^23p^1$	[Ar]$3d^{10}4s^24p^1$	[Kr]$4d^{10}5s^25p^1$	[Xe]$4f^{14}5d^{10}6s^26p^1$
Ionization energy†	800.6	577.6	578.8	558.3	589.3
Electron affinity†	26.7	42.6	29	29	≈20
Electronegativity	2.04	1.61	1.81	1.78	1.83
Ionic radius (Å)	0.23 (+3)	0.51 (+3)	0.62 (+3)	0.81 (+3)	0.95 (+3)
Atomic radius (Å)	0.88	1.43	1.22	1.63	1.70
Enthalpy of fusion†	22.6	10.75	5.59	3.26	4.08
Enthalpy of vaporization†	508	291	272	243	182
Bond enthalpy of M_2†	295	167	116	106	≈63
Standard reduction potential (volts)	−0.890	−1.706	−0.560	−0.338	0.719
	$B(OH)_3/B$	Al^{3+}/Al	Ga^{3+}/Ga	In^{3+}/In	Tl^{3+}/Tl

†In kilojoules per mole.

Group IV Elements

	Carbon	Silicon	Germanium	Tin	Lead
Atomic number	6	14	32	50	82
Atomic mass	12.0107	28.0855	72.64	118.710	207.2
Melting point (°C)	3550	1410	937.4	231.9681	327.502
Boiling point (°C)	4827	2355	2830	2270	1740
Density at 25°C (g cm^{-3})	2.25 (gr) 3.51 (dia)	2.33	5.323	5.75 (gray) 7.31 (white)	11.35
Color	Black (gr) Colorless (dia)	Gray	Gray-white	Silver	Blue-white
Ground-state electron configuration	$[He]2s^22p^2$	$[Ne]3s^23p^2$	$[Ar]3d^{10}4s^24p^2$	$[Kr]4d^{10}5s^25p^2$	$[Xe]4f^{14}5d^{10}6s^26p^2$
Ionization energy†	1086.4	786.4	762.2	708.6	715.5
Electron affinity†	121.85	133.6	≈120	≈120	35.1
Electronegativity	2.55	1.90	2.01	1.88	2.10
Ionic radius (Å)	0.15 (+4) 2.60 (−4)	0.42 (+4) 2.71 (−4)	0.53 (+4) 0.73 (+2) 2.72 (−4)	0.71 (+4) 0.93 (+2)	0.84 (+4) 1.20 (+2)
Atomic radius (Å)	0.77	1.17	1.22	1.40	1.75
Enthalpy of fusion†	105.0	50.2	34.7	6.99	4.774
Enthalpy of vaporization†	718.9	359	328	302	195.6
Bond enthalpy of M_2†	178	317	280	192	61
Standard reduction potential (volts)			−0.13 $H_2GeO_3,H^+/Ge$	−0.1364 Sn^{2+}/Sn	−0.1263 Pb^{2+}/Pb

Group V Elements

	Nitrogen	Phosphorus	Arsenic	Antimony	Bismuth
Atomic number	7	15	33	51	83
Atomic mass	14.00674	30.973762	74.92160	121.760	208.98040
Melting point (°C)	−209.86	44.1	817 (28 atm.)	630.74	271.3
Boiling point (°C)	−195.8	280	613 (subl.)	1750	1560
Density at 25°C (g cm^{-3})	0.808 (−196°C)	1.82 (white) 2.20 (red) 2.69 (black)	5.727	6.691	9.747
Color	Colorless		Gray	Blue-white	White
Ground-state electron configuration	$[He]2s^22p^3$	$[Ne]3s^23p^3$	$[Ar]3d^{10}4s^24p^3$	$[Kr]4d^{10}5s^25p^3$	$[Xe]4f^{14}5d^{10}6s^26p^3$
Ionization energy†	1402.3	1011.7	947	833.7	703.3
Electron affinity†	−7	72.03	≈80	103	91.3
Electronegativity	3.04	2.19	2.18	2.05	2.02
Ionic radius (Å)	1.71 (−3)	0.44 (+3) 2.12 (−3)	0.46 (+5) 0.58 (+3) 2.22 (−3)	0.62 (+5) 0.76 (+3) 2.45 (−3)	0.96 (+3)
Atomic radius (Å)	0.70	1.10	1.21	1.41	1.55
Enthalpy of fusion†	0.720	6.587	27.72	20.91	10.88
Enthalpy of vaporization†	5.608	59.03	334	262.5	184.6
Bond enthalpy of M_2†	945	485	383	289	194
Standard reduction potential (volts)	0.96 $NO_3^-,H^+/NO$	−0.276 H_3PO_4/H_3PO_3	0.234 $As_2O_3,H^+/As$	0.1445 $Sb_2O_3,H^+/Sb$	−0.46 $Bi_2O_3,OH^-/Bi$

†In kilojoules per mole.

The Chalcogens (Group VI Elements)

	Oxygen	Sulfur	Selenium	Tellurium	Polonium
Atomic number	8	16	34	52	84
Atomic mass	15.9994	32.065	78.96	127.60	(208.9824)
Melting point (°C)	−218.4	119.0 (mon.) 112.8 (rhom.)	217	449.5	254
Boiling point (°C)	−182.962	444.674	684.9	989.8	962
Density at 25°C (g cm^{-3})	1.14 (−183°C)	1.957 (mon.) 2.07 (rhom.)	4.79	6.24	9.32
Color	Pale blue (ℓ)	Yellow	Gray	Silver	Silver-gray
Ground-state electron configuration	[He]$2s^2 2p^4$	[Ne]$3s^2 3p^4$	[Ar]$3d^{10} 4s^2 4p^4$	[Kr]$4d^{10} 5s^2 5p^4$	[Xe]$4f^{14} 5d^{10} 6s^2 6p^4$
Ionization energy†	1313.9	999.6	940.9	869.3	812
Electron affinity†	140.97676	200.4116	194.967	190.15	≈180
Electronegativity	3.44	2.58	2.55	2.10	2.00
Ionic radius (Å)	1.40 (−2)	0.29 (+6) 1.84 (−2)	0.42 (+6) 1.98 (−2)	0.56 (+6) 2.21 (−2)	0.67 (+6) 2.30 (−2)
Atomic radius (Å)	0.66	1.04	1.17	1.43	1.67
Enthalpy of fusion†	0.4187	1.411	5.443	17.50	10
Enthalpy of vaporization†	6.819	238	207	195	90
Bond enthalpy of M_2†	498	429	308	225	
Standard reduction potential (volts)	1.229 $O_2, H^+/H_2O$	−0.508 S/S^{2-}	−0.78 Se/Se^{2-}	−0.92 Te/Te^{2-}	≈−1.4 Po/Po^{2-}

The Halogens (Group VII Elements)

	Fluorine	Chlorine	Bromine	Iodine	Astatine
Atomic number	9	17	35	53	85
Atomic mass	18.9984032	35.453	79.904	126.90447	(209.9871)
Melting point (°C)	−219.62	−100.98	−7.25	113.5	302
Boiling point (°C)	−188.14	−34.6	58.78	184.35	337
Density at 25°C (g cm^{-3})	1.108 (−189°C)	1.367 (−34.6°C)	3.119	4.93	
Color	Yellow	Yellow-green	Deep red	Violet-black	
Ground-state electron configuration	[He]$2s^2 2p^5$	[Ne]$3s^2 3p^5$	[Ar]$3d^{10} 4s^2 4p^5$	[Kr]$4d^{10} 5s^2 5p^5$	[Xe]$4f^{14} 5d^{10} 6s^2 6p^5$
Ionization energy†	1681.0	1251.1	1139.9	1008.4	≈930
Electron affinity†	328.0	349.0	324.7	295.2	≈270
Electronegativity	3.98	3.16	2.96	2.66	2.20
Ionic radius (Å)	1.33	1.81	1.96	2.20	≈2.27
Atomic radius (Å)	0.64	0.99	1.14	1.33	1.40
Enthalpy of fusion†	0.511	6.410	10.55	15.78	23.9
Enthalpy of vaporization†	6.531	20.347	29.56	41.950	
Bond enthalpy of M_2†	158	243	193	151	110
Standard reduction potential (volts)	2.87 F_2/F^-	1.358 Cl_2/Cl^-	1.065 Br_2/Br^-	0.535 I_2/I^-	≈0.2 At_2/At^-

†In kilojoules per mole.

The Noble Gases (Group VIII Elements)

	Helium	Neon	Argon	Krypton	Xenon	Radon
Atomic number	2	10	18	36	54	86
Atomic mass	4.002602	20.1797	39.948	83.798	131.293	(222.0176)
Melting point (°C)	−272.2 (26 atm)	−248.67	−189.2	−156.6	−111.9	−71
Boiling point (°C)	−268.934	−246.048	−185.7	−152.30	−107.1	−61.8
Density at 25°C (g cm^{-3})	0.147 (−270.8°C)	1.207 (−246.1°C)	1.40 (−186°C)	2.155 (−152.9°C)	3.52 (−109°C)	4.4 (−52°C)
Color	Colorless	Colorless	Colorless	Colorless	Colorless	Colorless
Ground-state electron configuration	$1s^2$	$[He]2s^22p^6$	$[Ne]3s^23p^6$	$[Ar]3d^{10}4s^24p^6$	$[Kr]4d^{10}5s^25p^6$	$[Xe]4f^{14}5d^{10}6s^26p^6$
Ionization energy†	2372.3	2080.6	1520.5	1350.7	1170.4	1037.0
Electron affinity†	<0	<0	<0	<0	<0	<0
Atomic radius (Å)	0.32	0.69	0.97	1.10	1.30	1.45
Enthalpy of fusion†	0.02093	0.3345	1.176	1.637	2.299	2.9
Enthalpy of vaporization†	0.1005	1.741	6.288	9.187	12.643	18.4

†In kilojoules per mole.

The Transition Elements

	Scandium	Yttrium	Lutetium	Titanium	Zirconium	Hafnium
Atomic number	21	39	71	22	40	72
Atomic mass	44.955912	88.90585	174.967	47.867	91.224	178.49
Melting point (°C)	1541	1522	1656	1660	1852	2227
Boiling point (°C)	2831	3338	3315	3287	4504	4602
Density at 25°C (g cm^{-3})	2.989	4.469	9.840	4.54	6.506	13.31
Color	Silver	Silver	Silver	Silver	Gray-white	Silver
Ground-state electron configuration	$[Ar]3d^14s^2$	$[Kr]4d^15s^2$	$[Xe]4f^{14}5d^16s^2$	$[Ar]3d^24s^2$	$[Kr]4d^25s^2$	$[Xe]4f^{14}5d^26s^2$
Ionization energy†	631	616	523.5	658	660	654
Electron affinity†	18.1	29.6	≈50	7.6	41.1	≈0
Electronegativity	1.36	1.22	1.27	1.54	1.33	1.30
Ionic radius (Å)	0.81	0.93	0.848 (+3)	0.68	0.80	0.78
Atomic radius (Å)	1.61	1.78	1.72	1.45	1.59	1.56
Enthalpy of fusion†	11.4	11.4	19.2	18.62	20.9	25.5
Enthalpy of vaporization†	328	425	247	426	590	571
Standard reduction potential (volts)	−2.08 Sc^{3+}/Sc	−2.37 Y^{3+}/Y	−2.30 Lu^{3+}/Lu	−0.86 $TiO_2,H^+/Ti$	−1.43 $ZrO_2,H^+/Zr$	−1.57 $HfO_2,H^+/Hf$

†In kilojoules per mole.

The Transition Elements (cont.)

	Vanadium	Niobium	Tantalum	Chromium	Molybdenum	Tungsten
Atomic number	23	41	73	24	42	74
Atomic mass	50.9415	92.90638	180.94788	51.9961	95.94	183.84
Melting point (°C)	1890	2468	2996	1857	2617	3410
Boiling point (°C)	3380	4742	5425	2672	4612	5660
Density at 25°C (g cm^{-3})	6.11	8.57	16.654	7.18	10.22	19.3
Color	Silver-white	Gray-white	Steel gray	Silver	Silver	Steel gray
Ground-state electron configuration	$[Ar]3d^34s^2$	$[Kr]4d^45s^1$	$[Xe]4f^{14}5d^36s^2$	$[Ar]3d^54s^1$	$[Kr]4d^55s^1$	$[Xe]4f^{14}5d^46s^2$
Ionization energy†	650	664	761	652.8	684.9	770
Electron affinity†	50.7	86.2	31.1	64.3	72.0	78.6
Electronegativity	1.63	1.60	1.50	1.66	2.16	2.36
Ionic radius (Å)	0.59 (+5) 0.63 (+4) 0.74 (+3) 0.88 (+2)	0.69 (+5) 0.74 (+4)	0.68 (+5)	0.63 (+3) 0.89 (+2)	0.62 (+6) 0.70 (+4)	0.62 (+6) 0.70 (+4)
Atomic radius (Å)	1.31	1.43	1.43	1.25	1.36	1.37
Enthalpy of fusion†	21.1	26.4		20.9	27.8	35.4
Enthalpy of vaporization†	512	722	781	394.7	589.2	819.3
Standard reduction potential (volts)	−1.2 V^{2+}/V	−0.62 $Nb_2O_5,H^+/Nb$	−0.71 $Ta_2O_5,H^+/Ta$	−0.74 Cr^{3+}/Cr	0.0 $H_2MoO_4,H^+/Mo$	−0.09 $WO_3,H^+/W$

	Manganese	Technetium	Rhenium	Iron	Ruthenium	Osmium
Atomic number	25	43	75	26	44	76
Atomic mass	54.938045	(97.9064)	186.207	55.845	101.07	190.23
Melting point (°C)	1244	2172	3180	1535	2310	3045
Boiling point (°C)	1962	4877	5627	2750	3900	5027
Density at 25°C (g cm^{-3})	7.21	11.50	21.02	7.874	12.41	22.57
Color	Gray-white	Silver-gray	Silver	Gray	White	Blue-white
Ground-state electron configuration	$[Ar]3d^54s^2$	$[Kr]4d^55s^2$	$[Xe]4f^{14}5d^56s^2$	$[Ar]3d^64s^2$	$[Kr]4d^75s^1$	$[Xe]4f^{14}5d^66s^2$
Ionization energy†	717.4	702	760	759.3	711	840
Electron affinity†	<0	≈ 53	≈ 14	15.7	≈ 100	≈ 106
Electronegativity	1.55	1.90	1.90	1.90	2.2	2.20
Ionic radius (Å)	0.80 (+2)		0.56 (+7) 0.27 (+4)	0.60 (+3) 0.72 (+2)	0.67 (+4)	0.69 (+6) 0.88 (+4)
Atomic radius (Å)	1.37	1.35	1.34	1.24	1.32	1.34
Enthalpy of fusion†	14.6	23.8	33.1	15.19	26.0	31.8
Enthalpy of vaporization†	279	585	778	414	649	678
Standard reduction potential (volts)	−0.183 Mn^{3+}/Mn	0.738 $TcO_4^-,H^+/TcO_2$	0.3 Re^{3+}/Re	−0.036 Fe^{3+}/Fe	0.49 Ru^{4+}/Ru^{3+}	0.85 $OsO_4,H^+/Os$

† In kilojoules per mole.

The Transition Elements (cont.)

	Cobalt	Rhodium	Iridium	Nickel	Palladium	Platinum
Atomic number	27	45	77	28	46	78
Atomic mass	58.933195	102.90550	192.217	58.6934	106.42	195.084
Melting point (°C)	1459	1966	2410	1453	1552	1772
Boiling point (°C)	2870	3727	4130	2732	3140	3827
Density at 25°C (g cm^{-3})	8.9	12.41	22.42	8.902	12.02	21.45
Color	Steel gray	Silver	Silver	Silver	Steel white	Silver
Ground-state electron configuration	$[Ar]3d^7 4s^2$	$[Kr]4d^8 5s^1$	$[Xe]4f^{14}5d^7 6s^2$	$[Ar]3d^8 4s^2$	$[Kr]4d^{10}$	$[Xe]4f^{14}5d^9 6s^1$
Ionization energy†	758	720	880	736.7	805	868
Electron affinity†	63.8	110	151	111.5	51.8	205.1
Electronegativity	1.88	2.28	2.20	1.91	2.20	2.28
Ionic radius (Å)	0.63 (+3)	0.68 (+3)	0.68 (+4)	0.69 (+2)	0.65 (+4)	0.65 (+4)
	0.72 (+2)				0.80 (+2)	0.80 (+2)
Atomic radius (Å)	1.25	1.34	1.36	1.25	1.38	1.37
Enthalpy of fusion†	16.2	21.5	26.4	17.6	17.6	19.7
Enthalpy of vaporization†	373	557	669	428	353	564
Standard reduction potential (volts)	−0.28	1.43	0.1	−0.23	0.83	1.2
	Co^{2+}/Co	Rh^{4+}/Rh^{3+}	$Ir_2O_3/Ir,OH^-$	Ni^{2+}/Ni	Pd^{2+}/Pd	Pt^{2+}/Pt

	Copper	Silver	Gold	Zinc	Cadmium	Mercury
Atomic number	29	47	79	30	48	80
Atomic mass	63.546	107.8682	196.966569	65.409	112.411	200.59
Melting point (°C)	1083.4	961.93	1064.43	419.58	320.9	−38.87
Boiling point (°C)	2567	2212	2807	907	765	356.58
Density at 25°C (g cm^{-3})	8.96	10.50	19.32	7.133	8.65	13.546
Color	Red	Silver	Yellow	Blue-white	Blue-white	Silver
Ground-state electron configuration	$[Ar]3d^{10}4s^1$	$[Kr]4d^{10}5s^1$	$[Xe]4f^{14}5d^{10}6s^1$	$[Ar]3d^{10}4s^2$	$[Kr]4d^{10}5s^2$	$[Xe]4f^{14}5d^{10}6s^2$
Ionization energy†	745.4	731.0	890.1	906.4	867.7	1007.0
Electron affinity†	118.5	125.6	222.749	<0	<0	<0
Electronegativity	1.90	1.93	2.54	1.65	1.69	2.00
Ionic radius (Å)	0.72 (+2)	0.89 (+2)	0.85 (+2)	0.74 (+2)	0.97 (+2)	1.10 (+2)
	0.96 (+1)	1.26 (+1)	1.37 (+1)		1.14 (+1)	1.27 (+1)
Atomic radius (Å)	1.28	1.44	1.44	1.34	1.49	1.50
Enthalpy of fusion†	13.3	11.95	12.36	7.39	6.11	2.300
Enthalpy of vaporization†	304	285	365	131	112	59.1
Standard reduction potential (volts)	0.340	0.800	1.42	−0.763	−0.403	0.796
	Cu^{2+}/Cu	Ag^+/Ag	Au^{3+}/Au	Zn^{2+}/Zn	Cd^{2+}/Cd	Hg_2^{2+}/Hg

† In kilojoules per mole.

The Lanthanide Elements

	Lanthanum	Cerium	Praseodymium	Neodymium	Promethium	Samarium	Europium
Atomic number	57	58	59	60	61	62	63
Atomic mass	138.90547	140.116	140.90765	144.242	(144.9127)	150.36	151.964
Melting point (°C)	921	798	931	1010	≈1080	1072	822
Boiling point (°C)	3457	3257	3212	3127	≈2400	1778	1597
Density at 25°C (g cm^{-3})	6.145	6.657	6.773	6.80	7.22	7.520	5.243
Color	Silver	Gray	Silver	Silver		Silver	Silver
Ground-state electron configuration	[Xe]$5d^1 6s^2$	[Xe]$4f^1 5d^1 6s^2$	[Xe]$4f^3 6s^2$	[Xe]$4f^4 6s^2$	[Xe]$4f^5 6s^2$	[Xe]$4f^6 6s^2$	[Xe]$4f^7 6s^2$
Ionization energy†	538.1	528	523	530	536	543	547
Electron affinity†	50			est. 50			
Electronegativity	1.10	1.12	1.13	1.14		1.17	
Ionic radius (Å)	1.15	0.92 (+4) 1.034 (+3)	0.90 (+4) 1.013 (+3)	0.995 (+3)	0.979 (+3)	0.964 (+3)	0.950 (+3) 1.09 (+2)
Atomic radius (Å)	1.87	1.82	1.82	1.81	1.81	1.80	2.00
Enthalpy of fusion†	5.40	5.18	6.18	7.13	12.6	8.91	(10.5)
Enthalpy of vaporization†	419	389	329	324		207	172
Standard reduction potential (volts)	−2.37 La^{3+}/La	−2.335 Ce^{3+}/Ce	−2.35 Pr^{3+}/Pr	−2.32 Nd^{3+}/Nd	−2.29 Pm^{3+}/Pm	−2.30 Sm^{3+}/Sm	−1.99 Eu^{3+}/Eu

	Gadolinium	Terbium	Dysprosium	Holmium	Erbium	Thulium	Ytterbium
Atomic number	64	65	66	67	68	69	70
Atomic mass	157.25	158.92535	162.500	164.93032	167.259	168.93421	173.04
Melting point (°C)	1311	1360	1409	1470	1522	1545	824
Boiling point (°C)	3233	3041	2335	2720	2510	1727	1193
Density at 25°C (g cm^{-3})	7.900	8.229	8.550	8.795	9.066	9.321	6.965
Color	Silver	Silver-gray	Silver	Silver	Silver	Silver	Silver
Ground-state electron configuration	[Xe]$4f^7 5d^1 6s^2$	[Xe]$4f^9 6s^2$	[Xe]$4f^{10} 6s^2$	[Xe]$4f^{11} 6s^2$	[Xe]$4f^{12} 6s^2$	[Xe]$4f^{13} 6s^2$	[Xe]$4f^{14} 6s^2$
Ionization energy†	592	564	572	581	589	596.7	603.4
Electron affinity†				est. 50			
Electronegativity	1.20		1.22	1.23	1.24	1.25	
Ionic radius (Å)	0.938 (+3)	0.84 (+4) 0.923 (+3)	0.908 (+3)	0.894 (+3)	0.881 (+3)	0.869 (+3)	0.858 (+3) 0.93 (+2)
Atomic radius (Å)	1.79	1.76	1.75	1.74	1.73	1.72	1.94
Enthalpy of fusion†	15.5	16.3	17.2	17.2	17.2	18.2	9.2
Enthalpy of vaporization†	301	293	165	285	280	240	165
Standard reduction potential (volts)	−2.28 Gd^{3+}/Gd	−2.31 Tb^{3+}/Tb	−2.29 Dy^{3+}/Dy	−2.33 Ho^{3+}/Ho	−2.32 Er^{3+}/Er	−2.32 Tm^{3+}/Tm	−2.22 Yb^{3+}/Yb

†In kilojoules per mole.

The Actinide Elements

	Actinium	Thorium	Protactinium	Uranium	Neptunium	Plutonium	Americium
Atomic number	89	90	91	92	93	94	95
Atomic mass	(227.0277)	232.0381	231.0359	238.0289	(237.0482)	(244.0642)	(243.0614)
Melting point (°C)	1050	1750	1600	1132.3	640	624	994
Boiling point (°C)	3200	4790		3818	2732	3232	2607
Density at 25°C (g cm^{-3})	10.07	11.72	15.37	18.95	20.25	19.84	13.67
Color	Silver	Silver	Silver	Silver	Silver	Silver	Silver
Ground-state electron configuration	$[Rn]6d^17s^2$	$[Rn]6d^27s^2$	$[Rn]5f^26d^17s^2$	$[Rn]5f^36d^17s^2$	$[Rn]5f^46d^17s^2$	$[Rn]5f^67s^2$	$[Rn]5f^77s^2$
Ionization energy†	499	587	568	587	597	585	578
Electronegativity	1.1	1.3	1.5	1.38	1.36	1.28	1.3
Ionic radius (Å)	1.11 (+3)	0.99 (+4)	0.89 (+5) 0.96 (+4) 1.05 (+3)	0.80 (+6) 0.93 (+4) 1.03 (+3)	0.71 (+7) 0.92 (+4) 1.01 (+3)	0.90 (+4) 1.00 (+3)	0.89 (+4) 0.99 (+3)
Atomic radius (Å)	1.88	1.80	1.61	1.38	1.30	1.51	1.84
Enthalpy of fusion†	14.2	18.8	16.7	12.9	9.46	3.93	14.4
Enthalpy of vaporization†	293	575	481	536	337	348	238
Standard reduction potential (volts)	−2.6 Ac^{3+}/Ac	−1.90 Th^{4+}/Th	−1.0 $PaO_2^+,H^+/Pa$	−1.8 U^{3+}/U	−1.9 Np^{3+}/U	−2.03 Pu^{3+}/Pu	−2.32 Am^{3+}/Am

	Curium	Berkelium	Californium	Einsteinium	Fermium	Mendelevium	Nobelium
Atomic number	96	97	98	99	100	101	102
Atomic mass	(247.0703)	(247.0703)	(251.0796)	(252.0830)	(257.0951)	(258.0984)	(259.1011)
Melting point (°C)	1340						
Boiling point (°C)							
Density at 25°C (g cm^{-3})	13.51	14					
Color	Silver	Silver	Silver	Silver			
Ground-state electron configuration	$[Rn]5f^76d^17s^2$	$[Rn]5f^97s^2$	$[Rn]5f^{10}7s^2$	$[Rn]5f^{11}7s^2$	$[Rn]5f^{12}7s^2$	$[Rn]5f^{13}7s^2$	$[Rn]5f^{14}7s^2$
Ionization energy†	581	601	608	619	627	635	642
Electronegativity	1.3	1.3	1.3	1.3	1.3	1.3	1.3
Ionic radius (Å)	0.88 (+4) 1.01 (+3) 1.19 (+2)	0.87 (+4) 1.00 (+3) 1.18 (+2)	0.86 (+4) 0.99 (+3) 1.17 (+2)	0.85 (+4) 0.98 (+3) 1.16 (+2)	0.84 (+4) 0.97 (+3) 1.15 (+2)	0.84 (+4) 0.96 (+3) 1.14 (+2)	0.83 (+4) 0.95 (+3) 1.13 (+2)
Standard reduction potential (volts)	−2.06 Cm^{3+}/Cm	−1.05 Bk^{3+}/Bk	−1.93 Cf^{3+}/Cf	−2.0 Es^{3+}/Es	−1.96 Fm^{3+}/Fm	−1.7 Md^{3+}/Md	−1.2 No^{3+}/No

†In kilojoules per mole.

The Transactinide Elements†

	Lawrencium	Rutherfordium	Dubnium	Seaborgium	Bohrium	Hassium	Meitnerium
Atomic number	103	104	105	106	107	108	109
Atomic mass	(262)	(261)	(262)	(263)	(262)	(265)	(266)
Melting point (°C)	1600						
Ground-state electron configuration	$[Rn]5f^{14}7s^27p^1$	$[Rn]5f^{14}6d^27s^2$	$[Rn]5f^{14}6d^37s^2$	$[Rn]5f^{14}6d^47s^2$	$[Rn]5f^{14}6d^57s^2$	$[Rn]5f^{14}6d^67s^2$	$[Rn]5f^{14}6d^77s^2$
Ionization energy		490	640	730	660	750	840

†All missing data are unknown.

APPENDIX G

Answers to Odd-Numbered Problems

CHAPTER 1

1. Mercury is an element; water and sodium chloride are compounds. The other materials are mixtures: seawater and air are homogeneous; table salt and wood are heterogeneous. Mayonnaise appears homogeneous to the naked eye, but under magnification shows itself as water droplets suspended in oil.

3. Substances

5. 16.9 g

7. (a) 2.005 and 1.504 g
(b) 2.005/1.504 = 1.333 = 4/3; SiN (or a multiple)

9. 2, 3, 4, 5

11. (a) HO (or any multiple, such as H_2O_2)
(b) All would give H_2 and O_2 in 1 : 1 ratio.

13. 2.0 L N_2O, 3.0 L O_2

15. 28.086

17. 11.01

19. (a) 145/94 = 1.54 (b) 94 electrons

21. 95 protons, 146 neutrons, 95 electrons

CHAPTER 2

1. 2.107298×10^{-22} g

3. (a) 283.89 (b) 115.36 (c) 164.09 (d) 158.03 (e) 132.13

5. The total count of 2.52×10^9 atoms of gold has a mass of only 8.3×10^{-13} g, which is far too small to detect with a balance.

7. 1.041 mol

9. 2540 cm^3 = 2.54 L

11. 7.03×10^{23} atoms

13. Pt: 47.06%; F: 36.67%; Cl: 8.553%; O: 7.720%

15. N_4H_6, H_2O, LiH, $C_{12}H_{26}$

17. 0.225%

19. $Zn_3P_2O_8$

21. Fe_3Si_7

23. BaN, Ba_3N_2

25. (a) 0.923 g C, 0.077 g H
(b) No
(c) 92.3% C, 7.7% H
(d) CH

27. C_4F_8

29. (a) 62.1
(b) 6
(c) 56
(d) Si (atomic mass 28.1), N (atomic mass 14.0)
(e) Si_2H_6

31. (a) $3\ H_2 + N_2 \longrightarrow 2\ NH_3$
(b) $2\ K + O_2 \longrightarrow K_2O_2$
(c) $PbO_2 + Pb + 2\ H_2SO_4 \longrightarrow 2\ PbSO_4 + 2\ H_2O$
(d) $2\ BF_3 + 3\ H_2O \longrightarrow B_2O_3 + 6\ HF$
(e) $2\ KClO_3 \longrightarrow 2\ KCl + 3\ O_2$
(f) $CH_3COOH + 2\ O_2 \longrightarrow 2\ CO_2 + 2\ H_2O$
(g) $2\ K_2O_2 + 2\ H_2O \longrightarrow 4\ KOH + O_2$
(h) $3\ PCl_5 + 5\ AsF_3 \longrightarrow 3\ PF_5 + 5\ AsCl_3$

33. (a) 12.06 g (b) 1.258 g (c) 4.692 g

35. 7.83 g $K_2Zn_3[Fe(CN)_6]_2$

37. 0.134 g SiO_2

39. 1.18×10^3 g

41. 418 g KCl; 199 g Cl_2

43. (a) 58.8 (b) Probably nickel (Ni)

45. 42.49% NaCl, 57.51% KCl

47. 14.7 g NH_4Cl, 5.3 g NH_3

49. 303.0 g Fe; 83.93%

CHAPTER 3

1. Melting point 1250°C (obs. 1541°C), boiling point 2386°C (obs. 2831°C), density 3.02 g cm^{-3} (obs. 2.99)

3. SbH_3, HBr, SnH_4, H_2Se

5. (a) $F(r) = 7.1999 \times 10^{-9}$ N
(b) $V(r) = 8.984741$ eV

7. (a) -8.63994×10^{-8} N
(b) -17.96933 eV
(c) 1.951188×10^7 m s^{-1}

9. (a) Sr (b) Rn (c) Xe (d) Sr

11. Using data for Be from Table 3.l, calculate $\log(I_n)$ for $n = 1, 2, 3$, and 4. The graph of $\log(I_n)$ versus n shows a dramatic increase between $n = 2$ and $n = 3$, suggesting two easily removed electrons outside a stable helium-like inner shell containing two electrons.

13. (a) Cs (b) F (c) K (d) At

15. K < Si < S < O < F

17.

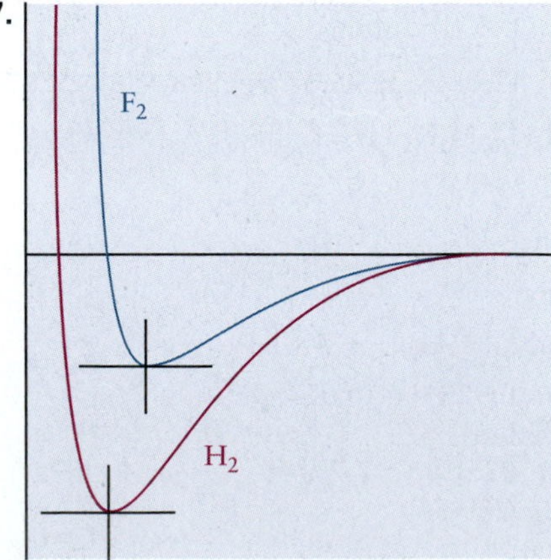

19.

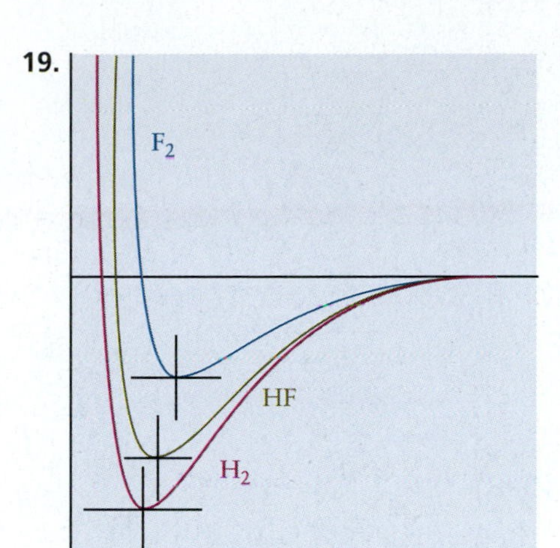

21. (a) $86\ e^-$ (8 valence, 78 core)
(b) $37\ e^-$ (1 valence, 36 core)
(c) $36\ e^-$ (8 valence, 28 core)
(d) $52\ e^-$ (6 valence, 46 core)

23. $K(g) + Cl(g) \longrightarrow K^+(g) + Cl^-(g) \qquad \Delta E = 70\ \text{kJ mol}^{-1}$
$K(g) + Cl(g) \longrightarrow K^-(g) + Cl^+(g) \qquad \Delta E = 1203\ \text{kJ mol}^{-1}$

25. 450 kJ mol^{-1}

27. The As—H bond length will lie between 1.42 and 1.71 Å (observed: 1.52 Å). SbH_3 will have the weakest bond.

29. Bond lengths are often shorter than the sum of atomic radii in polar molecules because of the electrostatic attraction between the oppositely charged ends of the dipole. Such shortening is slight in HI, indicating that it is not very polar.

31. Most polar: N—P > C—N > N—O > N—N :least polar

33. (a) CI_4 (b) OF_2 (c) SiH_4

35. ClO, 16%; KI, 74%; TlCl, 38%; InCl, 33%

37. HF, 40%; HCl, 19%; HBr, 14%; HI, 8%; CsF, 87%

39. (a) (b)

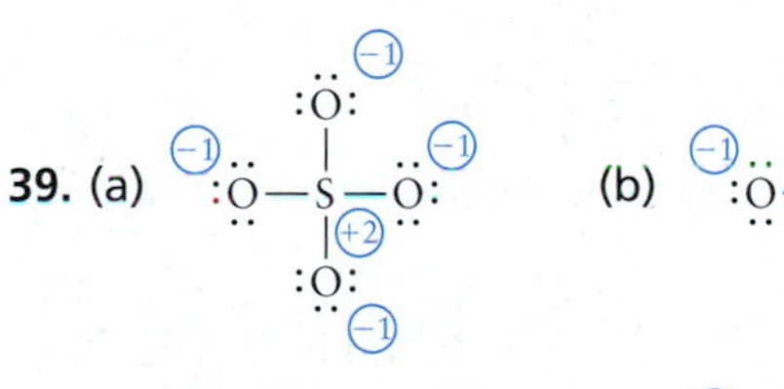

(c) :F—Sb—F: with :F: below (formal charges 0, 0, 0, 0)
(d) :S—C≡N: (formal charges −1, 0, 0)

41. The first is favored. H—N=O (0, 0, 0)   H—O=N (0, +1, −1)

43. (a) Group IV, CO_2 (b) Group VII, Cl_2O_7
(c) Group V, NO_2^- (d) Group VI, HSO_4^-

45. (a) H—As—H with H below (b) H—O—Cl:
(c) $[:Kr:F:]^+$ (Kr +1) (d) :O:P:O: with :Cl: above and below (formal charges +1, −1, −1)

47. H—N—C—N—H with :O: double-bonded to C, and H below each N

Bond lengths: N—H 1.01×10^{-10} m, N—C 1.47×10^{-10} m, C=O 1.20×10^{-10} m

49.

```
:S̈—S̈—S̈:
 |      |
:S̈:    :S̈:
 |      |
:S̈—S̈—S̈:
```

51. (a)

```
     H  :F̈:
     |   |(−1)
H—N—B—F̈:
  (+1)|   |
     H  :F̈:
```

(b)

```
    H     O:              H      O:(−1)
    |    //               |     /
H—C—C          ↔   H—C—C
    |    \                |     \\
    H     O:(−1)          H      O:
```

(c)

```
           O:                      O:(−1)
          //                      /
H—Ö—C             ↔   H—Ö—C
          \                       \\
           O:(−1)                  O:
```

53. $\overset{(-1)}{:\ddot{O}}—\ddot{N}=\ddot{O} \leftrightarrow \ddot{O}=\ddot{N}—\overset{(-1)}{\ddot{O}:}$

Between 1.18×10^{-10} m and 1.43×10^{-10} m.

55.

```
    H                    H (+1)       (−1)         H  (−1)     (+1)
    |                    |                         |
H—C—N̈=C=Ö   ↔   H—C—N≡C—Ö:   ↔   H—C—N̈—C≡O:
    |                    |                         |
    H                    H                         H
```

57. (a)

```
     :F̈:
:F̈    |   F̈:
   \  |  /
     P
    / \
 :F̈:  :F̈:
```

(b)

```
 :F̈     F̈:
    \  /
:F̈——S̈——F̈:
```

(c)

```
:O:
 ||   F̈:
:Xe /
 ||  \
:O:   F̈:
```

59. (a) SN = 4, tetrahedral (b) SN = 3, trigonal planar
(c) SN = 6, octahedral (d) SN = 4, pyramidal
(e) SN = 5, distorted T-shape

61. (a) SN = 6, square planar (b) SN = 4, bent, angle < 109.5°
(c) SN = 4, pyramidal, angle < 109.5° (d) SN = 2, linear

63. (a) SO_3 (b) NF_3 (c) NO_2^- (d) CO_3^{2-}

65. Only (d) and (e) are polar.

67. No, because VSEPR theory predicts a steric number of 3 and a bent molecule in both cases.

69. (a) Linear (b) The N end

71. $\overset{+2\,-1}{SrBr_2}$ $\overset{+2\,-2\,+1}{Zn(OH)_4^{2-}}$ $\overset{-4\,+1}{SiH_4}$ $\overset{+2\,+4\,-2}{CaSiO_3}$ $\overset{+6\,-2}{Cr_2O_7^{2-}}$ $\overset{+2\;\;+5\,-2\;\;-1}{Ca_5(PO_4)_3F}$ $\overset{+1\,-\frac{1}{2}}{KO_2}$ $\overset{+1\,-1}{CsH}$

73. (a) CsCl, cesium chloride $Cs\cdot + :\ddot{Cl}\cdot \rightarrow (Cs^+)(:\ddot{Cl}:^-)$

(b) $CaAt_2$, calcium astatide $\dot{Ca}\cdot + 2:\ddot{At}\cdot \rightarrow (Ca^{2+})(:\ddot{At}:^-)_2$

(c) Al_2S_3, aluminum sulfide $2\cdot\dot{Al}\cdot + 3\cdot\ddot{S}\cdot \rightarrow (Al^{3+})_2(:\ddot{S}:^{2-})_3$

(d) K_2Te, potassium telluride $2\ K\cdot + \cdot\ddot{Te}\cdot \rightarrow (K^+)_2(:\ddot{Te}:^{2-})$

75. (a) Aluminum oxide (b) Rubidium selenide
(c) Ammonium sulfide (d) Calcium nitrate
(e) Cesium sulfate (f) Potassium hydrogen carbonate

77. (a) AgCN (b) $Ca(OCl)_2$ (c) K_2CrO_4 (d) Ga_2O_3 (e) KO_2
(f) $Ba(HCO_3)_2$

79. Na_3PO_4, sodium phosphate

81. (a) SiO_2 (b) $(NH_4)_2CO_3$ (c) PbO_2 (d) P_2O_5 (e) CaI_2
(f) $Fe(NO_3)_3$

83. (a) Copper(I) sulfide and copper(II) sulfide (b) Sodium sulfate
(c) Tetraarsenic hexaoxide (d) Zirconium(IV) chloride
(e) Dichlorine heptaoxide (f) Gallium(I) oxide

CHAPTER 4

1. 0.66 m s^{-1}

3. 3.04 m

5. (a) $5.00 \times 10^5 \text{ s}^{-1}$ (b) 4.4 min

7. $\lambda = 1.313$ m; time = 0.0873 s

9. 2.7 K

11. Red

13. 550 nm, green

15. (a) 3.371×10^{-19} J (b) $203.0 \text{ kJ mol}^{-1}$ (c) $4.926 \times 10^{-3} \text{ mol s}^{-1}$

17. 5.895×10^{-7} m

19. $r_3 = 0.0952$ nm, $E_3 = -6.06 \times 10^{-18}$ J, energy per mole = $3.65 \times 10^3 \text{ kJ mol}^{-1}$, $\nu = 1.14 \times 10^{16} \text{ s}^{-1}$, $\lambda = 2.63 \times 10^{-8}$ m = 26.3 nm

21. 72.90 nm, ultraviolet

23. Blue

25. Part of the yellow light, together with green and blue light, will eject electrons from cesium. No visible light will eject electrons from selenium; ultraviolet light is required.

27. (a) 7.4×10^{-20} J (b) $4.0 \times 10^5 \text{ m s}^{-1}$

29. (a) 100 cm, 33 cm (b) two nodes

31. (a) 7.27×10^{-7} m (b) 3.96×10^{-10} m (c) 2.2×10^{-34} m

35. (a) $5.8 \times 10^4 \text{ m s}^{-1}$ (b) 7.9 m s^{-1}

37. $E_1 = 3.36 \times 10^{-18}$ J; $E_2 = 1.34 \times 10^{-17}$ J; $E_3 = 3.02 \times 10^{-17}$ J; $\lambda = 1.97 \times 10^{-8}$ m

39. (a) $\tilde{\Psi}_{21} = \tilde{\Psi}_{12}$ because the x- and y-axes are equivalent in a square box
(b) Exchanging x and y corresponds to a 90-degree rotation
(c) Exchanging labels cannot change the energy of the particle, a physically observable quantity

43. 1.6 Hz

45. 5.72×10^{-12} J = 0.357 eV

CHAPTER 5

1. Only (b) is allowed.

3. (a) $4p$ (b) $2s$ (c) $6f$

5. (a) 2 radial, 1 angular (b) 1 radial, 0 angular (c) 2 radial, 3 angular

7. $R^2_{pz} \propto \cos^2 q = 0$ for $q = p/2$; d_{xz} nodal planes are the y-z and x-y planes; $d_{x^2-y^2}$ nodal planes are two planes containing the z-axis at 45° from the x- and y-axes.

9. 3.17 Å, 2.64 Å

11. $\epsilon_{2s} = -0.396$ Ry $= -520 \text{ kJ mol}^{-1}$

13. −1 Ry exactly, −0.396 Ry, −0.376 Ry

15. (a) $1s^22s^22p^2$ (b) $[Ar]3d^{10}4s^24p^4$ (c) $[Ar]3d^64s^2$

17. Be^+: $1s^22s^1$; C^-: $1s^22s^22p^3$; Ne^{2+}: $1s^22s^22p^4$; Mg^+: $[Ne]3s^1$; P^{2+}: $[Ne]3s^23p^1$; Cl^-: $[Ne]3s^23p^6$; As^+: $[Ar]3d^{10}4s^24p^2$; I^-: $[Kr]4d^{10}5s^25p^6$. All except Cl^- and I^- are paramagnetic.

19. (a) In (b) S^{2-} (c) Mn^{4+}

21. 117

23. First "noble gases" at $Z = 1, 5, 9$

25. 9.52 eV or 1.52×10^{-18} J

29. (a) $Z_{eff,1s} = 7.12$ (b) $Z_{eff,2s} = 0.79$ (c) $Z_{eff,2p} = 0.47$

31. (a) K (b) Cs (c) Kr (d) K (e) Cl^-

33. (a) S^{2-} (b) Ti^{2+} (c) Mn^{2+} (d) Sr^{2+}

35. (a) For definition see Section 3.1. Helium has two electrons in the 1s orbital, which has the smallest radius. Moreover, each electron only partially screens the other from the nucleus, which has charge +2. Therefore, removal of the first electron requires considerable energy.
(b) Li, because Li^+ is essentially like He and explanation in (a) applies.
(c) 50.4 nm

37. (a) Start with Ca^{2+} and Ar, which are isoelectronic; that is, they have the same number of electrons. Ca^{2+} has the higher nuclear charge and is therefore smaller. Mg is above Ca in Group 2, so its 2+ ion is smaller than that of Ca. Br^- is larger than Cl^- in Group 7, which is, in turn, larger than Ar. Therefore, $Mg^{2+} < Ca^{2+} < Ar < Br^-$.
(b) Na has one s electron, well shielded from the nucleus; Ne and Na+ are isoelectronic, but Na+ has a net positive charge; O has an unfilled shell, but no special stability relative to Ne. Therefore, Na < O < Ne < Na+.
(c) Al is metallic, and thus electropositive; electronegativity sequence of others follows their relative horizontal positions in the periodic table. Therefore, Al < H < O < F.

39. 318.4 nm, near ultraviolet

CHAPTER 6

1. 0, 1, 2, 3, 2, 3 (6 σ orbitals in order of increasing energy)

5. The $1\sigma_g$ MO

7. Multiplying one of the two H atom orbitals by −1 is required for constructive interference; combination with two + signs leads to destruction interference.

9. H_2

11. He_2^+

15. Smaller bond energy; larger bond length

17. (a) F_2: $(\sigma_{g2s})^2(\sigma^*_{u2s})^2(\sigma_{g2p})^2(\pi_{u2p})^4(\pi^*_{g2p})^4$; F_2^+: $(\sigma_{g2s})^2(\sigma^*_{u2s})^2(\sigma_{g2p})^2(\pi_{u2p})^4(\pi^*_{g2p})^3$
(b) F_2: 1, F_2^+: $\frac{3}{2}$
(c) F_2^+ should be paramagnetic
(d) $F_2 < F_2^+$

19. $(\sigma_{g3s})^2(\sigma^*_{u3s})^2(\sigma_{g3pz})^2(\pi_{u3p})^4(\pi^*_{g3p})^2$; bond order = 2; paramagnetic

21. (a) F, 1 (b) N, $2\frac{1}{2}$ (c) O, $1\frac{1}{2}$

23. (a) is diamagnetic; (b) and (c) are paramagnetic

25. Bond order $2\frac{1}{2}$; paramagnetic

27. The outermost electron in CF is in a π^*_{2p} molecular orbital. Removing it gives a stronger bond.

29. $(\sigma_{1s})^2(\sigma^*_{1s})^2$; bond order = 0. It should be unstable.

31. Smaller because an antibonding electron is lost

33. Little change in dissociation energy or bond lengths because a nonbonding electron is lost

35. (a) IE = 15.6 eV, from $\sigma_{g_{2pz}}$, which is highest occupied level
(b) IE = 16.7 eV, from $\pi_{u_{2pxy}}$, which is second highest occupied level

39. 11.88 eV: the 4p_x or the 4p_y orbital; 15.2 eV: the σ orbital

41. Li_2: $\psi_\sigma^{bond}(1, 2; R_{AB}) = c_1[2s^A(1)2s^B(2) + 2s^A(2)2s^B(1)]$
C_2: $\psi_\pi^{bond}(1, 2, 3, 4; R_{AB} = c_1 R_{AB}[2p_y^A(1)\ 2p_y^B(2)][2p_x^A(3)\ 2p_x^B(4)] + c_1 R_A[2p_y^A(2)\ 2p_y^B(1)]$
$[2p_x^A(4)\ 2p_x^B(3)]$

43. Because the ground state of Be has no unpaired electrons, the simple VB model predicts that a bond will not form between two Be atoms. The same result is predicted by the LCAO approach. Two of the valence electrons in Be_2 occupy a bonding molecular orbital, but the other two must occupy an antibonding molecular orbital, leading to bond order zero.

45. $\psi_\sigma^{\text{bond}}(1, 2; R_{\text{BH}}) = C_1[1s^{\text{H}}(1)2p_z^{\text{B}}(2)] + C_2[1s^{\text{H}}(2)2p_z^{\text{B}}(1)]$
The simple VB model predicts a linear BO molecule. This is incorrect. The correct prediction is BH_3.

47. N has one unpaired electron in each of its $2p$ orbitals. Each of these can overlap with H $1s$ to form a σ bond whose wavefunction has the form

$\psi_\sigma = C_1[2p_a^{\text{N}}(1)1s^{\text{H}}(2)] + C_2[2p_a^{\text{N}}(2)1s^{\text{H}}(1)]$ where a is x, y, or z.

Because the $2p$ orbitals are mutally perpendicular, the simple VB model predicts a trigonal pyramid with angles of 90 degrees.

49. sp^3 hybridization of central N^-, bent molecular ion

51. (a) sp^3 on C, tetrahedral
(b) sp on C, linear
(c) sp^3 on O, bent
(d) sp^3 on C, pyramidal
(e) sp on Be, linear

53. sp^2 hybrid orbitals. ClO_3^+ is trigonal planar, ClO_2^+ is bent.

55. sp^3 hybrid orbitals, tetrahedral

59. Sixteen electrons, so molecule is linear; sp hybridization of central N gives two σ bonds, with $2p_z$ orbitals on outer nitrogen atoms (four electrons). Lone pairs on both $2s$ orbitals on outer nitrogen atoms (four atoms). π system as in Figure 16.21: $(\pi)^4(\pi^{\text{nb}})^4$ with eight electrons. Total bond order = 4; bond order 2 per N—N bond. N_3 and N_3^+ should be bound. N_3 and N_3^+ are paramagnetic.

61. 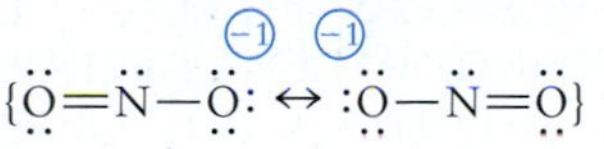 $\{\ddot{\text{O}}=\ddot{\text{N}}-\ddot{\text{O}}:^{(-1)} \leftrightarrow :^{(-1)}\ddot{\text{O}}-\ddot{\text{N}}=\ddot{\text{O}}\}$

$SN = 3$, sp^2 hybridization, bent molecule. The $2p_z$ orbitals perpendicular to the plane of the molecule can be combined into a π molecular orbital containing one pair of electrons. This orbital adds bond order 1/2 to each N—O bond, for a total bond order of 3/2 per bond.

CHAPTER 7

1. Yes, if the amount of knocking is less than that of iso-octane. Examples are the BTX compounds.

3. Ethane; $2\ C_2H_6(g) + 7\ O_2(g) \longrightarrow 4\ CO_2(g) + 6\ H_2O(\ell)$

5. (a) $C_{10}H_{22} \longrightarrow C_5H_{10} + C_5H_{12}$
(b) Two: 1-pentene and 2-pentene

7.

(a) 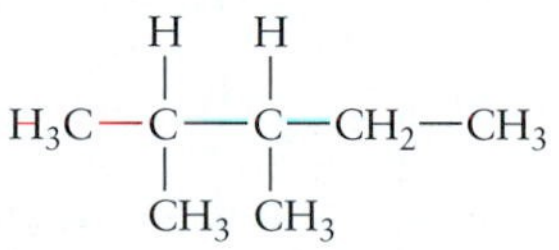
```
          H    H
          |    |
    H3C—C——C—CH2—CH3
          |    |
        CH3  CH3
```

(b)
```
    H3C          CH2—CH3
        \       /
         C=C
        /       \
      H          CH2—CH3
```

(c)
```
      H    CH3
        \  /
   H     C     H
     \  / \  /
      C———C
     /      \
    H        H
```

(d)
```
          CH3  H
          |    |
    H3C—C——C—CH3
          |    |
        CH3  H
```

(e)
```
    H          CH2—CH2—CH3
      \       /
       C=C
      /       \
   H3C         CH2—CH2—CH3
```

(f)
```
    H          H
      \       /
       C=C      CH3
      /     \   /
    H         C
            /   \
          H      CH2—CH2—CH3
```

(g)
```
        CH3          CH2—CH3
         |            |
    H3C—C—CH2—C—CH2—CH2—CH3
         |            |
         H            H
```

(h)
```
                      CH2—CH3
                       |
    H3C—C≡C—C—CH2—CH2—CH3
                       |
                       H
```

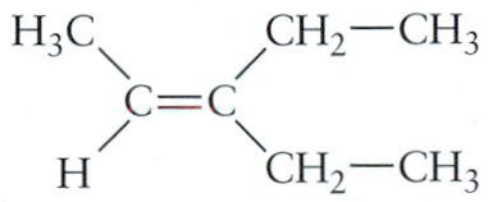

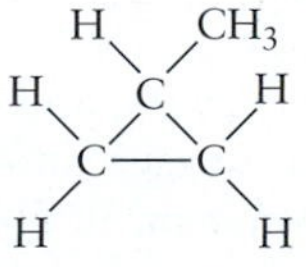

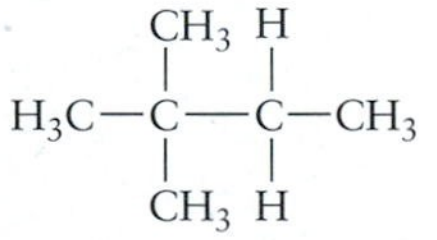

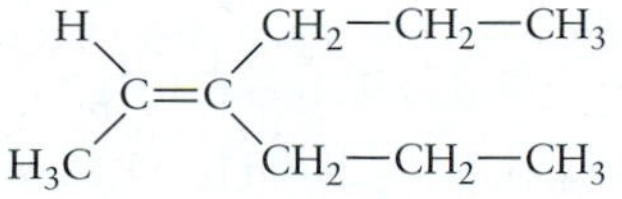

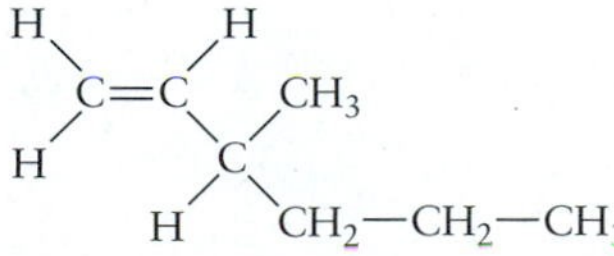

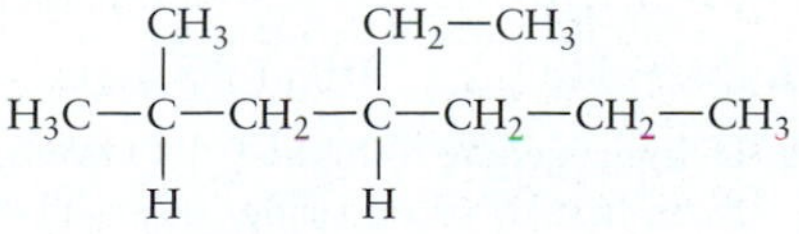

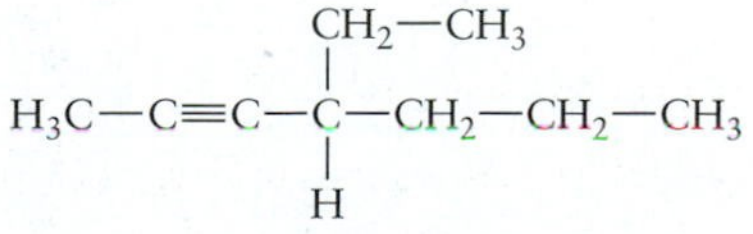

9.

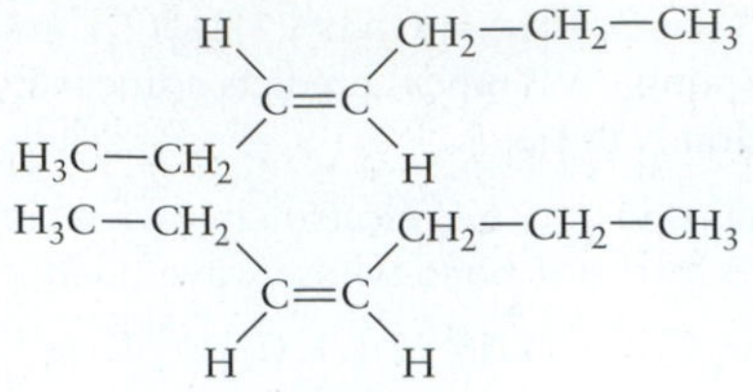

$$\begin{array}{c} H \quad\quad CH_2{-}CH_2{-}CH_3 \\ C{=}C \\ H_3C{-}CH_2 \quad\quad H \end{array}$$

$$\begin{array}{c} H_3C{-}CH_2 \quad\quad CH_2{-}CH_2{-}CH_3 \\ C{=}C \\ H \quad\quad H \end{array}$$

11. (a) 1,2-Hexadiene
(b) 1,3,5-Hexatriene
(c) 2-Methyl-1-hexene
(d) 3-Hexyne

13. (a) sp^2, sp, sp^2, sp^3, sp^3, sp^3
(b) All sp^2
(c) sp^2, sp^2; all others sp^3
(d) sp^3, sp^3, sp, sp, sp^3, sp^3

15. 30 double bonds; on the bonds shared by two hexagonal faces

17. 11.2%; 4.49×10^9 kg

19. (a) $CH_3CH_2CH_2CH_2OH + CH_3COOH \longrightarrow CH_3COOCH_2CH_2CH_2CH_3 + H_2O$
(b) $NH_4CH_3COO \longrightarrow CH_3CONH_2 + H_2O$
(c) $CH_3CH_2CH_2OH \longrightarrow CH_3CH_2CHO$ (propionaldehyde) $+ H_2$
(d) $CH_3(CH_2)_5CH_3 + 11\ O_2 \longrightarrow 7\ CO_2 + 8\ H_2O$

21. (a) $CH_2{=}CH_2 + Br_2 \longrightarrow CH_2BrCH_2Br$; $CH_2BrCH_2Br \longrightarrow CH_2{=}CHBr + HBr$
(b) $CH_3CH_2CH{=}CH_2 + H_2O \longrightarrow CH_3CH_2CH(OH)CH_3$ (using H_2SO_4)
(c) $CH_3CH{=}CH_2 + H_2O \longrightarrow CH_3CH(OH)CH_3$ (using H_2SO_4);
$CH_3CH(OH)CH_3 \longrightarrow CH_3COCH_3 + H_2$ (copper or zinc oxide catalyst)

23. 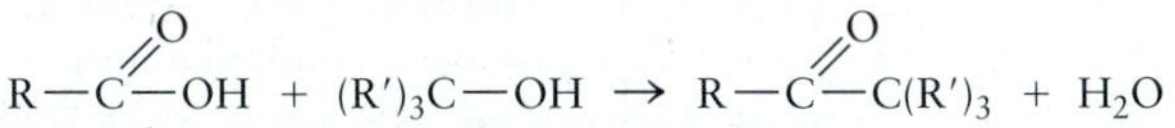 $R{-}\overset{O}{\overset{\|}{C}}{-}OH + (R')_3C{-}OH \rightarrow R{-}\overset{O}{\overset{\|}{C}}{-}C(R')_3 + H_2O$

25. 79.9 L

27. $-CH_3$ carbon sp^3, other carbon sp^2. A π orbital with two electrons bonds the second carbon atom with the oxygen atom. The three groups around the second carbon form an approximately trigonal planar structure, with bond angles near 120 degrees. The geometry about the first carbon atom is approximately tetrahedral, with angles near 109.5 degrees.

29. The Lewis diagrams for HCOOH and $HCOO^-$ are

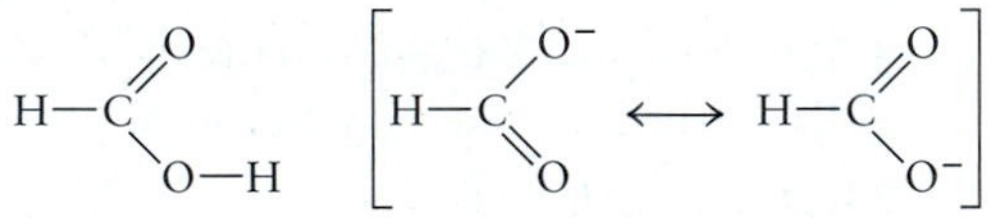

$$H{-}C({=}O){-}O{-}H \qquad \left[H{-}C(O^-){=}O \longleftrightarrow H{-}C({=}O){-}O^- \right]$$

One resonance form is given for HCOOH, but two are given for the formate anion $HCOO^-$. In formic acid, one oxygen atom is doubly bonded to the carbon atom, and the other is singly bonded. In the anion, there is some double-bond character in both C—O bonds. The carbon atom in HCOOH is sp^2 hybridized (*SN* 3), and the OH oxygen atom is sp^3 hybridized (*SN* 4). The immediate surroundings of the carbon atom have trigonal planar geometry, and the C—O—H group is bent. In the $HCOO^-$, the carbon atom and both oxygen atoms are sp^2 hybridized (*SN* 3), possessing a three-center four-electron *p* system. In HCOOH, *p* overlap occurs between orbitals on the carbon atom and only one oxygen atom. Both the C— to —O bond lengths in the formate ion should lie somewhere between the value for the single bond (1.36 Å) and the value for the double bond (1.23 Å).

31. (a) Alcohol: $CH_3CH(OH)CH_3$, isopropyl alcohol; carboxylic acid:
$CH_3C(CH_3)(COCH_3)(CH_2)_3CH(CH_3)\ CH_2CH{=}CHC(CH_3){=}CHCOOH$
(b) 3,7,10-Trimethyl-2,4-dodecadiene

33. (a) $C_9H_8O_4$ (b) 1.80×10^{-3} mol

35. Dehydrogenate to make C=O bond from C—OH bond on first ring; move C=C bond from first to second ring; insert C=O group on third ring; remove hydrocarbon side chain on fourth ring together with hydrogen atom and replace it with an —OH group and a $-COCH_2OH$ group.

CHAPTER 8

1. (a) PtF_4 (b) PtF_6

3. $V_{10}O_{28}^{6-} + 16\ H_3O^+ \longrightarrow 10\ VO_2^+ + 24\ H_2O + 5$ state, V_2O_5

5. $2\ TiO_2 + H_2 \longrightarrow Ti_2O_3 + H_2O$ Titanium(III) oxide

7. Monodentate, at the N-atom lone pair

9. +2, +2, +2, 0

11. (a) $Na_2[Zn(OH)_4]$ (b) $[CoCl_2(en)_2]NO_3$
(c) $[Pt(H_2O)_3Br]Cl$ (d) $[Pt(NH_3)_4(NO_2)_2]Br_2$

13. (a) Ammonium diamminetetraisothiocyanatochromate(III)
(b) Pentacarbonyltechnetium(I) iodide
(c) Potassium pentacyanomanganate(IV)
(d) Tetra-ammineaquachlorocobalt(III) bromide

15. $[Cu(NH_3)_2Cl_2] < KNO_3 < Na_2[PtCl_6] < [Co(NH_3)_6]Cl_3$

21. (a) Strong: 1, weak: 5 (b) Strong or weak: 0
(c) Strong or weak: 3 (d) Strong: 2, weak: 4
(e) Strong: 0, weak: 4

23. $[Fe(CN)_6]^{3-}$: 1 unpaired, CFSE = $-2\ \Delta_o$; $[Fe(H_2O)_6]^{3+}$: 5 unpaired, CFSE = 0

25. d^3 gives half-filled and d^8 filled and half-filled shells for metal ions in an octahedral field. d^5 will be half filled and stable for high spin (small Δ_o), and d^6 will be a filled subshell and stable for low spin (large Δ_o).

27. This ion does not absorb any significant amount of light in the visible range.

29. $\lambda \approx 480$ nm; $\Delta_o \approx 250$ kJ mol^{-1}

31. -500 kJ mol^{-1}

33. (a) Orange–yellow
(b) Approximately 600 nm (actual: 575 nm)
(c) Decrease, because CN^- is a strong-field ligand and will increase Δ_o.

35. (a) Pale, because F^- is an even weaker field ligand than H_2O, so it should be high-spin d^5. This is a half-filled shell and the complex will absorb only weakly.
(b) Colorless, because Hg(II) is a d^{10} filled-subshell species

CHAPTER 9

1. $NH_4HS(s) \longrightarrow NH_3(g) + H_2S(g)$

3. $NH_4Br(s) + NaOH(aq) \longrightarrow NH_3(g) + H_2O(\ell) + NaBr(aq)$

5. 10.3 m

7. 1.40×10^4 ft

9. 1697.5 atm, 1.7200×10^3 bar

11. 0.857 atm

13. 8.00 L

15. 14.3 gill

17. 134 L

19. 35.2 psi

21. (a) 19.8 atm (b) 23.0 atm

23. The mass of a gas in a given volume changes proportionately to the absolute temperature. This statement will be true only at -19.8°C.

25. (a) $2\ Na(s) + 2\ HCl(g) \longrightarrow 2\ NaCl(s) + H_2(g)$ (b) 4.23 L

27. 3.0×10^6 L

29. 24.2 L

31. (a) 932 L H_2S (b) 1.33 kg, 466 L SO_2

33. $X_{SO_3} = 0.135$; $P_{SO_3} = 0.128$ atm

35. $X_{N_2} = 0.027$; $P_{N_2} = 1.6 \times 10^{-4}$ atm

37. (a) $X_{CO} = 0.444$ (b) $X_{CO} = 0.33$

39. (a) 5.8×10^{17} (b) 520 L

41. (a) 1.93×10^3 m s^{-1} = 1.93 km s^{-1} (b) 226 m s^{-1}

43. 6100 m s^{-1} (6000 K), 790 m s^{-1} (100 K)

45. Greater. As T increases, the Maxwell–Boltzmann distribution shifts to higher speeds.

47. $N_{high}/N_{low} = 0.91$

49. 5.3×10^{-4}

51. 2.42×10^{-9}

53. $P_h = 0.91$ atm

55. 162 atm = 2.38×10^3 psi

57. (a) 27.8 atm (b) 24.6 atm; attractive forces dominate.

59. 6.1×10^{-9} m

61. 92.6 g mol^{-1}

63. 1830 stages

65. 7.4×10^{-7} atm; 20 m^2 s^{-1}

CHAPTER 10

1. Gas

3. (a) Condensed (b) 10.3 cm^3 mol^{-1}

5. Condensed

7. The liquid water has vaporized to steam.

9. Harder, because forces binding the particles in NaCl are stronger and resist deformation better.

11. In all three phases, the diffusion constant should decrease as its density is increased. At higher densities, molecules are closer to each other. In gases, they will collide more often and travel shorter distances between collisions. In liquids and solids, there will be less space for molecules to move around each other.

13. Both involve the interaction of an ion with a dipole. In the first case, the dipole is permanent (preexisting), whereas in the second, it is induced by the approach of the ion. Induced dipole forces are weaker than ion–dipole forces. Examples: Na^+ with HCl (ion–dipole); Na^+ with Cl_2 (induced dipole).

15. (a) *Ion–ion,* dispersion (b) *Dipole–dipole,* dispersion
(c) *Dispersion* (d) *Dispersion*

17. Bromide ion

19. (a) 2.0×10^{-10} m; 2.5×10^{-10} m
(b) KCl has a longer bond yet lower potential energy (greater bond strength).

21. Heavier gases have stronger attractive forces, favoring the liquid and solid states.

23. Ne < NO < NH_3 < RbCl; nonpolar < polar < hydrogen-bonded < ionic

25. An eight-membered ring of alternating H and O atoms is reasonable. The ring is probably not planar.

27. The two have comparable molar masses, but the hydrogen-bonding in hydrazine should give it a higher boiling point.

29. 6.7×10^{25}

31. 6.16 L; several times smaller than the volume of 1 mol at standard temperature and pressure (22.4 L mol^{-1})

33. 7.02×10^{13} atoms cm^{-3}

35. 0.9345 g L^{-1}

37. 2.92 g $CaCO_3$

39. 0.69 atm; 31% lies below

41. Iridium. Its higher melting and boiling points indicate that the intermolecular forces in iridium are stronger than those in sodium.

43. No phase change will occur.

47. (a) Liquid (b) Gas (c) Solid (d) Gas

49. (a) Above. If gas and solid coexist at $-84.0°C$, their coexistence must extend upward in temperature to the triple point.
(b) The solid will sublime at some temperature below $-84.0°C$.

51. The meniscus between gas and liquid phases will disappear at the critical temperature, 126.19 K.

CHAPTER 11

1. (a) 5.53×10^{-3} M (b) 5.5×10^{-3} molal (c) 3.79 L

3. Molarity = 12.39 M in HCl; mole fraction of HCl = 0.2324; molality = 16.81 molal in HCl

5. 8.9665 molal

7. 0.00643 g H_2O

9. (a) 1.33 g mL^{-1} (b) 164 mL

11. 1.06 M in NaOH

13. (a) $Ag^+(aq) + Cl^-(aq) \longrightarrow AgCl(s)$
(b) $K_2CO_3(s) + 2\,H^+(aq) \longrightarrow 2\,K^+(aq) + CO_2(g) + H_2O(\ell)$
(c) $2\,Cs(s) + 2\,H_2O(\ell) \longrightarrow 2\,Cs^+(aq) + 2\,OH^-(aq) + H_2(g)$
(d) $2\,MnO_4^-(aq) + 16\,H^+(aq) + 10\,Cl^-(aq) \longrightarrow 5\,Cl_2(g) + 2\,Mn^{2+}(aq) + 8\,H_2O(\ell)$

15. 16.8 mL of 7.91 M HNO_3

17. 6.74×10^3 L $CO_2(g)$

19. (a) $Ca(OH)_2(aq) + 2\,HF(aq) \longrightarrow CaF_2(aq) + 2\,H_2O(\ell)$
Hydrofluoric acid, calcium hydroxide, calcium fluoride
(b) $2\,RbOH(aq) + H_2SO_4(aq) \longrightarrow Rb_2SO_4(aq) + 2\,H_2O(\ell)$
Sulfuric acid, rubidium hydroxide, rubidium sulfate
(c) $Zn(OH)_2(s) + 2\,HNO_3(aq) \longrightarrow Zn(NO_3)_2(aq) + 2\,H_2O(\ell)$
Nitric acid, zinc hydroxide, zinc nitrate
(d) $KOH(aq) + CH_3COOH(aq) \longrightarrow KCH_3COO(aq) + H_2O(\ell)$
Acetic acid, potassium hydroxide, potassium acetate

21. Sodium sulfide

23. (a) $PF_3 + 3\,H_2O \longrightarrow H_3PO_3 + 3\,HF$
(b) $[H_3PO_3] = 0.0882$ M; $[HF] = 0.265$ M

25. 0.04841 M HNO_3

27. (a) $2\,\overset{+3}{P}F_2I(\ell) + 2\,\overset{0}{Hg}(\ell) \longrightarrow \overset{+2}{P}_2F_4(g) + \overset{+1}{Hg}_2I_2(s)$
(b) $2\,K\overset{+5}{Cl}\overset{-2}{O}_3(s) \longrightarrow 2\,K\overset{-1}{Cl}(s) + 3\,\overset{0}{O}_2(g)$
(c) $4\,\overset{-3}{N}H_3(g) + 5\,\overset{0}{O}_2(g) \longrightarrow 4\,\overset{+2}{N}\overset{-2}{O}(g) + 6\,H_2\overset{-2}{O}(g)$
(d) $2\,\overset{0}{As}(s) + 6\,NaO\overset{+1}{H}(\ell) \longrightarrow 2Na_3\overset{+3}{As}O_3(s) + 3\overset{0}{H}_2(g)$

29. $2\,\overset{0}{Au}(s) + 6\,H_2\overset{+6}{Se}O_4 \longrightarrow \overset{+3}{Au}_2(\overset{+6}{Se}O_4)_3(aq) + 3\,H_2\overset{+4}{Se}O_3(aq) + 3\,H_2O_6(\ell)$
Au is oxidized
Se is reduced (half of it)

31. (a) $2\,VO_2^+(aq) + SO_2(g) \longrightarrow 2\,VO^{2+}(aq) + SO_4^{2-}(aq)$
(b) $Br_2(\ell) + SO_2(g) + 6\,H_2O(\ell) \longrightarrow 2\,Br^-(aq) + SO_4^{2-}(aq) + 4\,H_3O^+(aq)$
(c) $Cr_2O_7^{2-}(aq) + 3\,Np^{4+}(aq) + 2\,H_3O^+(aq) \longrightarrow$
$2\,Cr^{3+}(aq) + 3\,NpO_2^{2+}(aq) + 3\,H_2O(\ell)$
(d) $5\,HCOOH(aq) + 2\,MnO_4^-(aq) + 6\,H_3O^+(aq) \longrightarrow$
$5\,CO_2(g) + 2\,Mn^{2+}(aq) + 14\,H_2O(\ell)$
(e) $3\,Hg_2HPO_4(s) + 2\,Au(s) + 8\,Cl^-(aq) + 3\,H_3O^+(aq) \longrightarrow$
$6\,Hg(\ell) + 3\,H_2PO_4^-(aq) + 2\,AuCl_4^-(aq) + 3\,H_2O(\ell)$

33. (a) $2\ Cr(OH)_3(s) + 3\ Br_2(aq) + 10\ OH^-(aq) \longrightarrow$
$2\ CrO_4^{2-}(aq) + 6\ Br^-(aq) + 8\ H_2O(\ell)$
(b) $ZrO(OH)_2(s) + 2\ SO_3^{2-}(aq) \longrightarrow Zr(s) + 2\ SO_4^{2-}(aq) + H_2O(\ell)$
(c) $7\ HPbO_2^-(aq) + 2\ Re(s) \longrightarrow 7\ Pb(s) + 2\ ReO_4^-(aq) + H_2O(\ell) + 5\ OH^-(aq)$
(d) $4\ HXeO_4^-(aq) + 8\ OH^-(aq) \longrightarrow 3\ XeO_6^{4-}(aq) + Xe(g) + 6\ H_2O(\ell)$
(e) $N_2H_4(aq) + 2\ CO_3^{2-}(aq) \longrightarrow N_2(g) + 2\ CO(g) + 4\ OH^-(aq)$

35. (a) $Fe^{2+}(aq) \longrightarrow Fe^{3+}(aq) + e^-$ oxidation
$H_2O_2(aq) + 2\ H_3O^+(aq) + 2\ e^- \longrightarrow 4\ H_2O(\ell)$ reduction
(b) $5\ H_2O(\ell) + SO_2(aq) \longrightarrow HSO_4^-(aq) + 3\ H_3O^+(aq) + 2\ e^-$ oxidation
$Mn_4^-(aq) + 8\ H_3O^+(aq) + 5\ e^- \longrightarrow Mn^{2+}(aq) + 12\ H_2O(\ell)$ reduction
(c) $ClO_2^-(aq) \longrightarrow ClO_2(g) + e^-$ oxidation
$ClO_2^-(aq) + 4\ H_3O^+(aq) + 4\ e^- \longrightarrow Cl^-(aq) + 6\ H_2O(\ell)$ reduction

37. $3\ HNO_2(aq) \longrightarrow NO_3^-(aq) + 2\ NO(g) + H_3O^+(aq)$

39. 7.175×10^{-3} M $Fe^{2+}(aq)$

41. 0.2985 atm

43. 3.34 K kg mol^{-1}

45. 340 g mol^{-1}

47. 1.7×10^2 g mol^{-1}

49. −2.8°C. As the solution becomes more concentrated, its freezing point decreases further.

51. 2.70 particles (complete dissociation of Na_2SO_4 gives 3 particles)

53. 1.88×10^4 g mol^{-1}

55. 7.46×10^3 g mol^{-1}

57. (a) 0.17 mol CO_2
(b) Because the partial pressure of CO_2 in the atmosphere is much less than 1 atm, the excess CO_2 will bubble out from the solution and escape when the cap is removed.

59. 4.13×10^2 atm

61. 0.774

63. (a) 0.491 (b) 0.250 atm (c) 0.575

CHAPTER 12

1. -2.16×10^4 L atm $= -2.19 \times 10^6$ J

3. 86.6 m

5. 24.8, 28.3, 29.6, 31.0, and 32.2 J K^{-1} mol^{-1}, respectively; extrapolating gives about 33.5 J K^{-1} mol^{-1} for Fr

7. 26.1, 25.4, 25.0, 24.3, 25.4, and 27.6 J K^{-1} mol^{-1}

9. (a) w is zero, q is positive, and ΔU is positive.
(b) w is zero, q is negative, and ΔU is negative.
(c) $(w_1 + w_2)$ is zero. $(\Delta U_1 + \Delta U_2) = (q_1 + q_2)$. The latter two sums could be any of three possibilities: both positive, both negative, or both zero.

11. 0.468 J K^{-1} g^{-1}

13. $q_1 = Mc_{s_1}\Delta T_1 = -q_2 = -Mc_{s_2}\Delta T_2$ $\quad \dfrac{C_{s_1}}{C_{s_2}} = \dfrac{\Delta T_2}{\Delta T_1}$

15. 314 J g^{-1} (modern value is 333 J g^{-1})

17. $w = -323$ J; $\Delta U = -393$ J; $q = -70$ J

19. (a) 38.3 L
(b) $w = -1.94 \times 10^3$ J; $q = 0$; $\Delta U = -1.94 \times 10^3$ J
(c) 272 K

21. $\Delta UE = +11.2$ kJ; $q = 0$; $w = +11.2$ kJ

23. (a) −6.68 kJ (b) +7.49 kJ (c) +0.594 kJ

25. 41.3 kJ

27. +513 J

29. 10.4°C

31. −623.5 kJ

33. A pound of diamonds. (This is recommended as a source of heat only as a last resort, however!)

35. −555.93 kJ

37. (a) −878.26 kJ (b) -1.35×10^7 kJ of heat absorbed

39. (a) −81.4 kJ (b) 55.1°C

41. $\Delta H_f^\circ = -152.3$ kJ mol^{-1}

43. (a) $C_{10}H_8(s) + 12\ O_2(g) \longrightarrow 10\ CO_2(g) + 4\ H_2O(\ell)$
(b) −5157 kJ (c) −5162 kJ (d) +84 kJ mol^{-1}

45. −264 kJ mol^{-1}

47. -1.58×10^3 kJ

49. Each side of the equation has three B—Br and three B—Cl bonds.

51. $w = -6.87$ kJ; $q = +6.87$ kJ; $\Delta U = \Delta H = 0$

53. $T_f = 144$ K; $w = \Delta U = -3.89$ kJ; $\Delta H = -6.48$ kJ

CHAPTER 13

1. (a) The system is all the matter participating in the reaction $NH_4NO_3(s) \longrightarrow NH_4^+(aq) + NO_3^-(aq)$. This includes solid ammonium nitrate, the water in which it dissolves, and the aquated ions that are the products of the dissolution process. The surroundings include the flask or beaker in which the system is held, the air above the system, and other neighboring materials. The dissolution of ammonium nitrate is spontaneous after any physical separation (such as a glass wall or a space of air) between the water and the ammonium nitrate has been removed.
(b) The system is all the matter participating in the reaction $H_2(g) + O_2(g) \longrightarrow$ products. The surroundings are the walls of the bomb and other portions of its environment that might deliver heat or work or absorb heat or work. The reaction of hydrogen with oxygen is spontaneous. Once hydrogen and oxygen are mixed in a closed bomb, no constraint exists to prevent their reaction. It is found experimentally that this system gives products quite slowly at room temperature (no immediate explosion). It explodes instantly at higher temperatures.
(c) The system is the rubber band. The surroundings consist of the weight (visualized as attached to the lower end of the rubber band), a hanger at the top of the rubber band, and the air in contact with the rubber band. The change is spontaneous once a constraint such as a stand or support underneath the weight is removed.
(d) The system is the gas contained in the chamber. The surroundings are the walls of the chamber and the moveable piston head. The process is spontaneous if the force exerted by the weight on the piston exceeds the force exerted by the collisions of the molecules of the gas on the bottom of the piston. (The forces due to the mass of the piston itself and friction between the piston and the walls within which it moves are neglected.)
(e) The system is the drinking glass in the process glass ⟶ fragments. The surroundings are the floor, the air, and the other materials in the room. The change is spontaneous. It occurs when the constraint, which is whatever portion of the surroundings holds the glass above the floor, is removed.

3. (a) $6 \times 6 = 36$ (b) 1 in 36

5. The tendency for entropy to increase

7. $10^{-3.62 \times 10^{23}}$ (that is, 1 part in $10^{+3.62 \times 10^{23}}$)

9. (a) $\Delta S > 0$ (b) $\Delta S > 0$ (c) $\Delta S < 0$

11. (a) 0.333 (b) $q = -1000$ J (c) $w = -500$ J

13. 9.61 J K^{-1} mol^{-1}

15. 29 kJ mol^{-1}

17. $\Delta U = \Delta H = 0$; $w = -1.22 \times 10^4$ J; $q = +1.22 \times 10^4$ J; $\Delta S = +30.5$ J K^{-1}

19. $\Delta S_{sys} = +30.2\ J\ K^{-1}$; $\Delta S_{surr} = -30.2\ J\ K^{-1}$; $\Delta S_{univ} = 0$

21. $\Delta S_{Fe} = -8.24\ J\ K^{-1}$; $\Delta S_{H_2O} = +9.49\ J\ K^{-1}$; $\Delta S_{tot} = +1.25\ J\ K^{-1}$

23. (a) $-116.58\ J\ K^{-1}$ (b) Lower (more negative)

25. DS° = −162.54, −181.12, −186.14, −184.72, −191.08 $J\ K^{-1}$. The entropy changes in these reactions become increasingly negative with increasing atomic mass, except that the rubidium reaction is out of line.

27. ΔS_{surr} must be positive and greater in magnitude than 44.7 $J\ K^{-1}$.

29. $\Delta S_{sys} > 0$ because the gas produced (oxygen) has many possible microstates.

31. (a) +740 J (b) 2.65 kJ (c) No (d) 196 K

33. $q = +38.7$ kJ; $w = -2.92$ kJ; $\Delta U = +35.8$ kJ; $\Delta S_{sys} = +110\ J\ K^{-1}$; $\Delta G = 0$

35. Overall reaction: $2\ Fe_2O_3(s) + 3\ C(s) \longrightarrow 4\ Fe(s) + 3\ CO_2(g)$ $\Delta G = +840\ J + 3(-400)\ J = -360\ J < 0$

37. (a) $0 < T < 3000$ K
(b) $0 < K < 1050$ K
(c) Spontaneous at all temperatures

39. $WO_3(s) + 3\ H_2(g) \longrightarrow W(s) + 3\ H_2O(g)$
$\Delta H° = 117.41$ kJ; $\Delta S° = 131.19\ J\ K^{-1}$; $\Delta G < 0$ for $T > \Delta H°/\Delta S° = 895$ K

CHAPTER 14

1. (a) $\dfrac{P^2_{H_2O}}{P^2_{H_2}P_{O_2}} = K$ (b) $\dfrac{P_{XeF_6}}{P_{Xe}P^3_{F_2}} = K$ (c) $\dfrac{P^{12}_{CO_2}P^6_{H_2O}}{P^2_{C_6H_6}P^{15}_{O_2}} = K$

3. $P_4(g) + 6\ Cl_2(g) + 2\ O_2(g) \rightleftharpoons 4\ POCl_3(g)$ $\dfrac{P^4_{POCl_3}}{P_{P_4}P^6_{Cl_2}P^2_{O_2}} = K$

5. (a) $\dfrac{P_{CO_2}P_{H_2}}{P_{CO}P_{H_2O}} = K$ (b) 0.056 atm

7. (a) The graph is a straight line passing through the origin.
(b) The experimental K's range from 3.42×10^{-2} to 4.24×10^{-2}, with a mean of 3.84×10^{-2}.

9. (a) $\dfrac{(P_{H_2S})^8}{(P_{H_2})^8} = K$

(b) $\dfrac{(P_{COCl_2})(P_{H_2})}{(P_{Cl_2})} = K$

(c) $P_{CO_2} = K$

(d) $\dfrac{1}{(P_{C_2H_2})^3} = K$

11. (a) $\dfrac{[Zn^{2+}]}{[Ag^+]^2} = K$

(b) $\dfrac{[VO_3(OH)^{2-}][OH^-]}{[VO_4^{3-}]} = K$

(c) $\dfrac{[HCO_3^-]^6}{[As(OH)_6^{3-}]^2P^6_{CO_2}} = K$

13. $\Delta G° = -550.23$ kJ; $K = 2.5 \times 10^{96}$

15. (a) 2.6×10^{12} $\dfrac{P_{SO_3}}{(P_{SO_2})(P_{O_2})^{1/2}} = K$
(b) 5.4×10^{-35} $(P_{O_2})^{1/2} = K$
(c) 5.3×10^3 $[Cu^{2+}][Cl^-]^2 = K$

17. $K_1 = (K_2)^3$

19. K_2/K_1

21. 1.04×10^{-4}

23. 14.6

25. (a) $P_{SO_2} = P_{Cl_2} = 0.58$ atm; $P_{SO_2Cl_2} = 0.14$ atm (b) 2.4

27. (a) 0.180 atm (b) 0.756

29. $P_{PCl_5} = 0.078$ atm; $P_{PCl_3} = P_{Cl_2} = 0.409$ atm

31. $P_{Br_2} = 0.0116$ atm; $P_{I_2} = 0.0016$ atm; $P_{IBr} = 0.0768$ atm

33. $P_{N_2} = 0.52$ atm; $P_{O_2} = 0.70$ atm; $P_{NO} = 3.9 \times 10^{-16}$ atm

35. $P_{N_2} = 0.0148$ atm; $P_{H_2} = 0.0445$ atm; $P_{NH_3} = 0.941$ atm

37. 5.6×10^{-5} mol L^{-1}

39. 2.0×10^{-2}

41. (a) 0.31 atm (b) 1.65 atm, 0.15 atm

43. (a) 8.46×10^{-5} (b) 0.00336 atm

45. (a) 9.83×10^{-4} (b) Net consumption

47. $K > 5.1$

49. (a) $Q = 2.05$; reaction shifts to right.
(b) $Q = 3.27$; reaction shifts to left.

51. (a) 0.800, left
(b) $P_{P_4} = 5.12$ atm, $P_{P_2} = 1.77$ atm
(c) Dissociation

53. (a) Shifts left
(b) Shifts right
(c) Shifts right
(d) The volume must have been increased to keep the total pressure constant; shifts left
(e) No effect

55. (a) Exothermic (b) Decrease

57. Run the first step at low temperature and high pressure, and the second step at high temperature and low pressure.

59. Low temperature and high pressure

61. -58 kJ

63. (a) -56.9 kJ
(b) $\Delta H° = -55.6$ kJ, $\Delta S° = +4.2$ J K^{-1}

65. 4.3×10^{-3}

67. (a) 23.8 kJ mol^{-1}
(b) $T_b = 240$ K

69. (a)

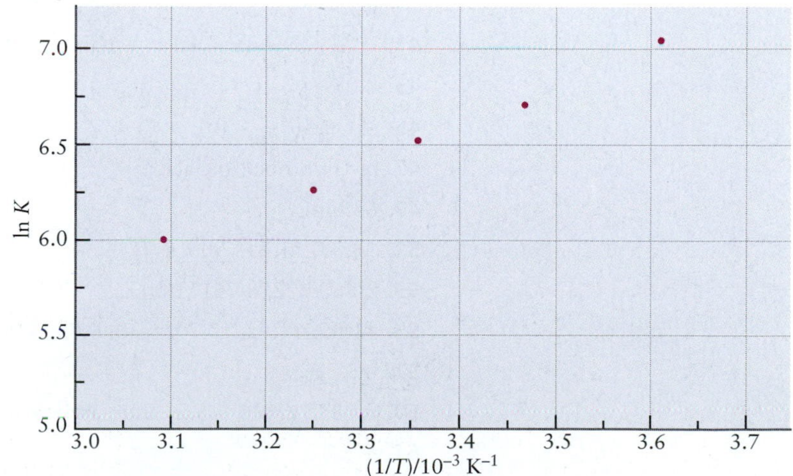

(b) -16.8×10^3 J mol^{-1}

71. 76

73. (a) $K_1 = 1.64 \times 10^{-2}$; $K_2 = 5.4$ (b) 330

CHAPTER 15

1. (a) Cl^- cannot act as a Brønsted–Lowry acid.
(b) SO_4^{2-}
(c) NH_3
(d) NH_2^-
(e) OH^-

3. $HCO_3^-(aq)$ serves as the base.

5. (a) $CaO(s) + H_2O(\ell) \longrightarrow Ca(OH)_2(s)$
(b) The CaO acts as a Lewis base, donating a pair of electrons (located on the oxide ion) to one of the hydrogen ions (the Lewis acid) of the water molecule.

7. (a) Fluoride acceptor (b) Acids: BF_3, TiF_4; bases: ClF_3O_2, KF

9. (a) Base, $Mg(OH)_2$ (b) Acid, HOCl
(c) Acid, H_2SO_4 (d) Base, CsOH

11. $SnO(s) + 2\ HCl(aq) \longrightarrow Sn^{2+}(aq) + 2\ Cl^-(aq) + H_2O(\ell)$;
$SnO(s) + NaOH(aq) + H_2O(\ell) \longrightarrow Sn(OH)_3^-(aq) + Na^+(aq)$

13. 3.70

15. 3×10^{-7} M $< [H_3O^+] < 3 \times 10^{-6}$ M; 3×10^{-9} M $< [OH^-] < 3 \times 10^{-8}$ M

17. $[H_3O^+] = 1.0 \times 10^{-8}$ M; $[OH^-] = 1.7 \times 10^{-6}$ M

19. The first reaction is more likely to be correct. In the second case, the reactant H_3O^+ would be present at very low concentration (10^{-7} M) and would give neither a fast nor a vigorous reaction.

21. (a) $C_{10}H_{15}ON(aq) + H_2O(\ell) \rightleftharpoons C_{10}H_{15}ONH^+(aq) + OH^-(aq)$
(b) 7.1×10^{-11}
(c) Stronger

23. $K = 24$; $HClO_2$ is the stronger acid and NO_2^- the stronger base.

25. (a) Methyl orange (b) 3.8 to 4.4

27. 2.35

29. (a) 2.45 (b) 0.72 mol

31. 1.16

33. 1.2×10^{-6}

35. 10.4

37. 0.083 M

39. The reaction gives a base of moderate strength, the acetate ion, in solution, so the pH > 7.

41. HCl < NH_4Br < KI < $NaCH_3COO$ < NaOH

43. 8.08

45. (a) 4.36 (b) 4.63

47. *m*-Chlorobenzoic acid

49. 639 mL

51. 13.88, 11.24, 7.00, 2.77

53. 2.86, 4.72, 8.71, 11.00

55. 11.89, 11.52, 10.79, 8.20, 6.05, 3.90, 1.95

57. 0.97 g

59. 0.0872 g, pH = 6.23 if no approximations are made, bromothymol blue

61. 4×10^{-7}

63. $[H_3AsO_4] = 8.0 \times 10^{-2}$ M; $[H_2AsO_4^-] = [H_3O^+] = 2.0 \times 10^{-2}$ M;
$[HAsO_4^{2-}] = 9.3 \times 10^{-8}$ M; $[AsO_4^{3-}] = 1.4 \times 10^{-17}$ M

65. $[PO_4^{3-}] = 0.020$ M; $[HPO_4^{2-}] = [OH^-] = 0.030$ M; $[H_2PO_4^-] = 1.61 \times 10^{-7}$ M;
$[H_3PO_4] = 7.1 \times 10^{-18}$ M

67. $[H_2CO_3] = 8.5 \times 10^{-6}$ M; $[HCO_3^-] = 1.5 \times 10^{-6}$ M; $[CO_3^{2-}] = 2.8 \times 10^{-11}$ M

69. 6.86

71. 1.51, 1.61, 2.07, 4.01, 6.07, 8.77, 9.29, 11.51

73. 46 kJ mol^{-1}

75. pK_{a1} should be smaller than 4.9; pK_{a2} should be larger.

77. Benzene

79. (a) CF_3COOH (b) $CH_3CH_2CHFCOOH$ (c) C_6H_5COOH

CHAPTER 16

1. 4.65 L

3. About 48°C

5. $Fe_2(SO_4)_3(s) \rightleftharpoons 2\ Fe^{3+}(aq) + 3\ SO_4^{2-}(aq)$ $[Fe^+]^2[SO_4^{2-}]^3 = K_{sp}$

7. 0.0665 g per 100 mL water

9. 14.3 g L^{-1}

11. $[I^-] = 6.2 \times 10^{-10}$ M, $[Hg_2^{2+}] = 3.1 \times 10^{-10}$ M

13. 1.9×10^{-12}

15. 1.6×10^{-8}

17. Yes. The initial reaction quotient is $6.2 \times 10^{-10} > K_{sp}$.

19. No

21. $[Pb^{2+}] = 2.3 \times 10^{-10}$ M; $[IO_3^-] = 0.033$ M

23. $[Ag^+] = 1.8 \times 10^{-2}$ M; $[CrO_4^{2-}] = 6.2 \times 10^{-9}$ M

25. 2.4×10^{-8} mol L^{-1}

27. (a) 3.4×10^{-6} M (b) 1.6×10^{-14} M

29. 9.1×10^{-3} M

31. In pure water: 1.2×10^{-4} M; at pH 7: 0.15 M

33. (a) Unchanged (b) Increase (c) Increase

35. (a) 8.6×10^{-4} M, Pb^{2+} in solid (b) 3.1×10^{-7}

37. $[I^-] = 5.3 \times 10^{-4}$ M

39. 2×10^{-13} M

41. pH = 2.4; $[Pb^{2+}] = 6 \times 10^{-11}$ M

43. $[Cu(NH_3)_4^{2+}] = 0.10$ M; $[Cu^{2+}] = 7 \times 10^{-14}$ M

45. $[K^+] = 0.0051$ M and $[Na^+] = 0.0076$ M

47. More will dissolve in 1 M NaCl (3×10^{-5} versus 1.3×10^{-5} mol L^{-1}). Less will dissolve in 0.100 M NaCl (3×10^{-6} versus 1.3×10^{-5} mol L^{-1}).

49. $Cu^{2+}(aq) + 2\ H_2O(\ell) \rightleftharpoons CuOH^+(aq) + H_3O^+(aq)$ or
$Cu(H_2O)_4^{2+}(aq) + H_2O(\ell) \rightleftharpoons Cu(H_2O)_3OH^+(aq) + H_3O^+(aq)$

51. 5.3

53. $K_a = 9.6 \times 10^{-10}$

55. In the first case, solid $Pb(OH)_2$ will precipitate; $[Pb^{2+}] = 4.2 \times 10^{-13}$ M, $[Pb(OH)_3^-] =$ 0.17 M. In the second case, solid $Pb(OH)_2$ will not precipitate; $[Pb^{2+}] = 1 \times 10^{-13}$ M, $[Pb(OH)_3^-] = 0.050$ M.

CHAPTER 17

3. 0.180 mol

5. (a) $Zn(s) + Cl_2(g) \longrightarrow Zn^{2+}(aq) + 2\ Cl^-(aq)$
(b) 1.20×10^3 C; 1.24×10^{-2} mol e^-
(c) Decreases by 0.407 g (d) 0.152 L Cl_2 consumed

7. +2

9. (a) Anode: $Cl^- \longrightarrow \frac{1}{2}Cl_2(g) + e^-$; cathode: $K^+ + e^- \longrightarrow K(\ell)$
total: $Cl^- + K^+ \longrightarrow K(\ell) + \frac{1}{2}Cl_2(g)$
(b) Mass K = 14.6 g; mass Cl_2 = 13.2 g

11. $\Delta G° = -921$ J; $w_{max} = +921$ J

13. (a) Anode: $Co \longrightarrow Co^{2+} + 2\,e^-$; cathode: $Br_2 + 2\,e^- \longrightarrow 2\,Br^-$;
total: $Co + Br_2 \longrightarrow Co^{2+} + 2\,Br^-$
(b) 1.34 V

15. (a) Anode: $Zn \longrightarrow Zn^{2+} + 2\,e^-$; cathode: $In^{3+} + 3\,e^- \longrightarrow In$
(b) −0.338 V

17. Reducing agent

19. Br_2 less effective than Cl_2

21. (a) BrO_3^- (b) Cr (c) Co

23. (a) $\mathscr{E}° = -0.183$ V (b) It will not disproportionate.

25. (a) No (b) Br^-

27. 0.37 V

29. −0.31 V

31. 5.17

33. (a) 0.31 V (b) 1×10^{-8} M

35. $K = 3 \times 10^{31}$, orange

37. 3×10^6

39. pH = 2.53; $K_a = 0.0029$

41. (a) 1.065 V (b) 1.3×10^{-11} M (c) 7.6×10^{-13}

43. 2.041 V; 12.25 V

45. (a) 9.3×10^6 C (b) 1.1×10^8 J

47. No, because $H_2SO_4(aq)$ is not the only substance whose amount changes during discharge. The accumulated $PbSO_4$ must also be removed and replaced by Pb and PbO_2.

49. 7900 J g^{-1}

51. $\Delta\mathscr{E}° = -0.419$ V, not spontaneous under standard conditions (pH = 14). If $[OH^-]$ is small enough, the equilibrium will shift to the right and the reaction will become spontaneous.

53. According to its reduction potential, yes. In practice, however, the sodium would react instantly and explosively with the water and would therefore be useless for this purpose.

55. $2\,Cl^- \longrightarrow Cl_2(g) + 2\,e^-$ anode
$Na^+ + e^- \longrightarrow Na(\ell)$ cathode

57. 4.4×10^4 kg

59. $2\,Mg + TiCl_4 \longrightarrow Ti + 2\,MgCl_2$; 102 kg

61. 42.4 min

63. (a) Ni
(b) 21.9 g (provided the electrolyte volume is very large)
(c) H_2

CHAPTER 18

1. 5.3×10^{-5} mol L^{-1} s^{-1}

3. $\text{rate} = -\dfrac{\Delta[N_2]}{\Delta t} = -\dfrac{1}{3}\dfrac{\Delta[H_2]}{\Delta t} = \dfrac{1}{2}\dfrac{\Delta[NH_3]}{\Delta t}$

5. (a) Rate = $k[NO]^2[H_2]$; k has the units L^2 mol^{-2} s^{-1}
(b) An increase by a factor of 18

7. (a) Rate = $k[C_5H_5N][CH_3I]$ (b) $k = 75$ L mol^{-1} s^{-1}
(c) 7.5×10^{-8} mol L^{-1} s^{-1}

9. 3.2×10^4 s

11. 0.0019 atm

13. 5.3×10^{-3} s^{-1}

15. 2.34×10^{-5} M

17. 2.9×10^{-6} s

19. (a) Bimolecular, rate = k[HCO][O_2]
(b) Termolecular, rate = k[CH_3][O_2][N_2]
(c) Unimolecular, rate = k[HO_2NO_2]

21. (a) The first step is unimolecular; the others are bimolecular.
(b) $H_2O_2 + O_3 \longrightarrow H_2O + 2\ O_2$
(c) O, ClO, CF_2Cl, Cl

23. 0.26 L mol^{-1} s^{-1}

25. (a) $A + B + E \longrightarrow D + F$ $\quad$ $\text{rate} = \dfrac{k_1k_2}{k_{-1}}\dfrac{[A][B][E]}{[D]}$

(b) $A + D \longrightarrow B + F$ $\quad$ $\text{rate} = \dfrac{k_1k_2k_3}{k_{-1}k_{-2}}\dfrac{[A][D]}{[B]}$

27. Only mechanism (b)

29. Only mechanism (a)

31. $\text{rate} = \dfrac{k_2k_1[A][B][E]}{k_2[E] + k_{-1}[D]}$ $\quad$ When $k_2[E] \ll k_{-1}[D]$

33. $\dfrac{d[Cl_2]}{dt} = \dfrac{k_1k_2[NO_2Cl]^2}{k_{-1}[NO_2] + k_2[NO_2Cl]}$

35. (a) 4.25×10^5 J $\quad$ (b) 1.54×10^{11} L mol^{-1} s^{-1}

37. (a) 1.49×10^{-3} L mol^{-1} s^{-1} $\quad$ (b) 1.30×10^3 s

39. (a) 4.3×10^{13} s^{-1} $\quad$ (b) 1.7×10^5 s^{-1}

41. 70.3 kJ mol^{-1}

43. 6.0×10^9 L mol^{-1} s^{-1}

45. (a) 1.2×10^{-3} mol L^{-1} s^{-1} $\quad$ (b) 5×10^{-5} M

CHAPTER 19

1. (a) $2\ {}^{12}_{6}C \longrightarrow {}^{23}_{12}Mg + {}^{1}_{0}n$
(b) ${}^{15}_{7}N + {}^{1}_{1}H \longrightarrow {}^{12}_{6}C + {}^{4}_{2}He$
(c) $2\ {}^{3}_{2}He \longrightarrow {}^{4}_{2}He + 2\ {}^{1}_{1}H$

3. (a) 3.300×10^{10} kJ mol^{-1} = 342.1 MeV total; 8.551 MeV per nucleon
(b) 7.312×10^{10} kJ mol^{-1} = 757.9 MeV total; 8.711 MeV per nucleon
(c) 1.738×10^{11} kJ mol^{-1} = 1801.7 MeV total; 7.570 MeV per nucleon

5. The two ^{4}He atoms are more stable, with a mass lower than that of one ^{8}Be atom by 9.86×10^{-5} u.

7. 16.96 MeV

9. (a) ${}^{39}_{17}Cl \longrightarrow {}^{39}_{18}Ar + {}^{0}_{-1}e^- + \tilde{\nu}$
(b) ${}^{22}_{11}Na \longrightarrow {}^{22}_{10}Ne + {}^{0}_{-1}e^+ + \nu$
(c) ${}^{224}_{88}Ra \longrightarrow {}^{220}_{86}Rn + {}^{4}_{2}He$
(d) ${}^{82}_{38}Sr + {}^{0}_{-1}e^- \longrightarrow {}^{82}_{37}Rb + \nu$

11. ^{19}Ne decays to ^{19}F by positron emission; ^{23}Ne decays to ^{23}Na by beta decay.

13. An electron, 0.78 MeV

15. ${}^{30}_{14}Si + {}^{1}_{0}n \longrightarrow {}^{31}_{14}Si \longrightarrow {}^{31}_{15}P + {}^{0}_{-1}e^- + \tilde{\nu}$

17. ${}^{210}_{84}Po \longrightarrow {}^{206}_{82}Pb + {}^{4}_{2}He$ $\quad$ ${}^{4}_{2}He + {}^{9}_{4}Be \longrightarrow {}^{12}_{6}C + {}^{1}_{0}n$

19. 3.7×10^{10} min^{-1}

21. (a) 1.0×10^6 $\quad$ (b) 5.9×10^4

23. 1.6×10^{18} disintegrations s^{-1}

25. 4200 years

27. 3.0×10^9 years

29. 6.0×10^9 years

31. ${}^{11}_{6}C \longrightarrow {}^{11}_{5}B + {}^{0}_{1}e^+ + n$; ${}^{15}_{8}O \longrightarrow {}^{15}_{7}N + {}^{0}_{1}e^+ + n$

33. Exposure from ${}^{15}O$ is greater by a factor of $1.72/0.99 = 1.74$.

35. (a) 2.3×10^{10} Bq
(b) 0.024 mGy
(c) Yes. A dose of 5 Gy has a 50% chance of being lethal; this dose is greater than 8.5 Gy in the first 8 days.

37. (a) ${}^{90}_{38}Sr \longrightarrow {}^{90}_{40}Zr + 2\,{}^{0}_{-1}e^- + 2\tilde{\nu}$
(b) 2.8 MeV
(c) 5.23×10^{12} disintegrations s^{-1}
(d) 4.44×10^{11} disintegrations s^{-1}

39. Increase (assuming no light isotopes of U are products of the decay)

41. 7.59×10^7 kJ g^{-1}

CHAPTER 20

1. (a) $A = 0.699$ (b) $\epsilon = 1.4 \times 10^3$ L mol^{-1} cm^{-1}

3. $c_A = 0.000368$ mol L^{-1}
$c_B = 0.00157$ mol L^{-1}

5. 6.736×10^{-23} J $= 40.56$ J mol^{-1}

7. (a) 1.45×10^{-46} kg m^2
(b) $E_1 = 7.65 \times 10^{-23}$ J; $E_2 = 2.30 \times 10^{-22}$ J; $E_3 = 4.59 \times 10^{-22}$ J
(c) 1.13 Å

9. (a) 5.12 (b) 6.80×10^{-2} (c) 3.23×10^{-6}

11. 25.4 kg s^{-2}

13. 5.1×10^2 kg s^{-2}

15. 5.3×10^{-4}

17. Decrease, because the electron will enter a π^* antibonding orbital

19. In the blue range, around 450 nm

21. Shorter; the π bonding is localized in cyclohexene.

23. 474 nm

25. 2, 3:2; 1; 2, 6:1

27. 2.72×10^{-7} m

29. $[\ddot{O}{=}\ddot{O}^{(+1)}{-}\ddot{O}^{(-1)}{:} \longleftrightarrow {:}\ddot{O}^{(-1)}{-}\ddot{O}^{(+1)}{=}\ddot{O}]$
$SN = 3$, sp^2 hybridization, bent molecule
Two electrons in a π orbital formed from the three $2p_z$ orbitals perpendicular to the molecular plane, total bond order of $\frac{3}{2}$ for each O—O bond.

31. 8 molecules

CHAPTER 21

1. (b), (c), and (d)

3. Two mirror planes, one 2-fold rotation axis

5. 3.613 Å

7. 102.9°

9. 49.87° and 115.0°

11. (a) 5.675×10^6 Å^3 (b) 5×10^{14}

13. 4.059 g cm^{-3}

15. (a) 1.602×10^{-22} cm^3 (b) 3.729×10^{-22} g
(c) 4.662×10^{-23} g (d) 6.025×10^{23} (in error by 0.05%)

17. 8

19. ReO_3

21. (a) 2.48 Å (b) 2.87 Å (c) 1.24 Å

23. (a) 2 (b) 0.680

25. 0.732

27. (a) Ionic (b) Covalent (c) Molecular (d) Metallic

29. $CO < BaCl_2 < Co < SiC$

33. 8, 6, 12

35. 664 kJ mol^{-1}

37. (a) 1041 kJ mol^{-1} (b) 1205 kJ mol^{-1}

39. Not by a significant amount because each vacancy is accompanied by an interstitial. In large numbers, such defects could cause a small bulging of the crystal and a decrease in its density.

41. (a) $Fe_{0.9352}O$ (b) 13.86% of the iron

CHAPTER 22

1. The structures of $P_2O_7^{4-}$ and $S_2O_7^{2-}$ have the same number of electrons and are found by writing "P" or "S" in place of "Si" and adjusting the formal charges. The chlorine compound is Cl_2O_7, dichlorine heptaoxide.

3. (a) Tetrahedra; Ca: +2, Fe: +3, Si: +4, O: −2
(b) Infinite sheets; Na: +1, Zr: +2, Si: +4, O: −2
(c) Pairs of tetrahedra; Ca: +2, Zn: +2, Si: +4, O: −2
(d) Infinite sheets; Mg: +2, Si: +4, O: −2, H: +1

5. (a) Infinite network; Li: +1, Al: +3, Si: +4, O: −2
(b) Infinite sheets; K: +1, Al: +3, Si: +4, O: −2, H: +1
(c) Closed rings or infinite single chains; Al: +3, Mg: +2, Si: +4, O: −2

7. $Mg_3Si_4O_{10}(OH)_2(s) \longrightarrow 3\ MgSiO_3(s) + SiO_2(s) + H_2O(g)$

9. 234 L

11. 0.418 mol Si, 0.203 mol Na, 0.050 mol Ca, 0.0068 mol Al, 0.002 mol Mg, 0.0005 mol Ba

13. −113.0 kJ

15. $\frac{7}{3}$

17. (a) $SiO_2(s) + 3\ C(s) \longrightarrow SiC(s) + 2\ CO(g)$
(b) +624.6 kJ mol^{-1}
(c) Low conductivity, high melting point, very hard

19. −1188.2 kJ, unstable

21. 24 $(\Omega\ m)^{-1}$ or 24 S m^{-1}

23. 1.26 C V^{-1} s^{-1} m^{-1} or 1.26 $(\Omega\ m)^{-1}$ or 1.26 S m^{-1}

25. 1.48 electrons per Cu atom

27. 2.16×10^{-19} J

29. 6.1×10^{-27} electrons; that is, no electrons

31. (a) Electron movement (*n*-type)
(b) Hole movement (*p*-type)

33. 680 nm, red

35. Decrease

CHAPTER 23

1. $n\ CCl_2{=}CH_2 \longrightarrow -(CCl_2-CH_2-)_n$

3. Formaldehyde, $CH_2{=}O$

5. (a) H_2O (b) $-(NH-CH_2-CO-)$

7. 646 kg adipic acid and 513 kg hexamethylenediamine

9. 3.49×10^{12} L = 3.49×10^{9} m^3 = 3.49 km^3

11. The isotropic phase has higher entropy and higher enthalpy.

13. A micelle (containing hydrocarbon in the interior and water on the outside)

15. 5 chiral centers

17. 27

19. In octane, because the side groups are all hydrocarbon

21. C_9H_9NO; 119 units

APPENDIX A

1. (a) 5.82×10^{-5} (b) 1.402×10^{3} (c) 7.93
(d) -6.59300×10^{3} (e) 2.530×10^{-3} (f) 1.47

3. (a) 0.000537 (b) 9,390,000 (c) −0.00247
(d) 0.00620 (e) 20,000

5. 746,000,000 kg

7. (a) The value is 135.6 (b) 111.34 g (c) 0.22 g, 0.23 g

9. That of the mass in problem 7

11. (a) 5 (b) 3 (c) Either two or three (d) 3 (e) 4

13. (a) 14 L (b) −0.0034°C (c) 340 lb = 3.4×10^{2} lb
(d) 3.4×10^{2} miles (e) 6.2×10^{-27} J

15. 2,997,215.55

17. (a) −167.25 (b) 76 (c) 3.1693×10^{15} (d) -7.59×10^{-25}

19. (a) −8.40 (b) 0.147 (c) 3.24×10^{-12} (d) 4.5×10^{13}

21. A = 337 cm^2

APPENDIX B

1. (a) 6.52×10^{-11} kg (b) 8.8×100^{-11} s (c) 5.4×10^{12} kg m^2 s^{-3}
(d) 1.7×10^{4} kg m^2 s^{-3} A^{-1}

3. (a) 4983°C (b) 37.0°C (c) 111°C (d) −40°C

5. (a) 5256 K (b) 310.2 K (c) 384 K (d) 233 K

7. (a) 24.6 m s^{-1} (b) 1.51×103 kg m^{-3} (c) 1.6×10^{-29} A s m
(d) 1.5×10^{2} mol m^{-3} (e) 6.7 kg m^2 s^{-3} = 6.7 W

9. 1 kW-hr = 3.6×10^{6} J; 15.3 kW-hr = 5.51×10^{7} J

11. 6620 cm^3 or 6.62 L

13. 3×10^{2} J

15. (a) 55 J
(b) O (No work was done because no displacement was achieved.)

APPENDIX C

1. 50 miles per hour

3. (a) Slope = 4, intercept = −7
(b) Slope = $\frac{7}{2}$ = 3.5, intercept = $-\frac{5}{2}$ = −2.5
(c) Slope = −2, intercept = $\frac{4}{3}$ = 1.3333 . . .

5. The graph is not a straight line.

7. (a) $-\frac{5}{7}$ = −0.7142857 . . . (b) $\frac{3}{4}$ = 0.75 (c) $\frac{2}{3}$ = 0.6666 . . .

9. (a) 0.5447 and −2.295 (b) −0.6340 and −2.366 (c) 2.366 and 0.6340

11. (a) 6.5×10^{-7} (also two complex roots)
(b) 4.07×10^{-2} (also 0.399 and −1.011)
(c) −1.3732 (also two complex roots)

13. (a) 4.551 (b) 1.53×10^{-7} (c) 2.6×10^{8} (d) -48.7264

15. 3.015

17. 121.477

19. 7.751 and -2.249

21. $x = 1.086$

23. (a) $8x$ (b) $3 \cos 3x - 8 \sin 2x$ (c) 3 (d) $\frac{1}{x}$

25. (a) 20 (b) $\frac{78125}{7}$ (c) 0.0675

Index/Glossary

Page numbers followed by "f" denote figures; those followed by "t" denote tables.

P

Q

T

Locations of Some Important Tables of Data